FORMULAS/EQUATIONS

Distance Formula

The distance from (x_1, y_1) to (x_2, y_2) is $\sqrt{(x_2 - x_1)^2 + \ldots}$

Midpoint Formula

The midpoint of the line segment with endpoints $(x_1, y_1) \ldots$

Standard Equation of a Circle

The standard equation of a circle of radius r with center at (h, k) is
$$(x - h)^2 + (y - k)^2 = r^2$$

Slope Formula

The slope m of the line containing the points (x_1, y_1) and (x_2, y_2) is
$$\text{slope } (m) = \frac{\text{change in } y}{\text{change in } x} = \frac{y_2 - y_1}{x_2 - x_1} \quad (x_1 \neq x_2)$$
m is undefined if $x_1 = x_2$

Slope-Intercept Equation of a Line

The equation of a line with slope m and y-intercept $(0, b)$ is $y = mx + b$

Point-Slope Equation of a Line

The equation of a line with slope m containing the point (x_1, y_1) is $y - y_1 = m(x - x_1)$

Quadratic Formula

The solutions of the equation $ax^2 + bx + c = 0$, $a \neq 0$, are $x = \dfrac{-b \pm \sqrt{b^2 - 4ac}}{2a}$

If $b^2 - 4ac > 0$, there are two distinct real solutions.

If $b^2 - 4ac = 0$, there is a repeated real solution.

If $b^2 - 4ac < 0$, there are two complex solutions (complex conjugates).

GEOMETRY FORMULAS

Circle

$r = $ Radius, $A = $ Area, $C = $ Circumference
$A = \pi r^2 \quad C = 2\pi r$

Triangle

$b = $ Base, $h = $ Height (Altitude), $A = $ area
$A = \frac{1}{2}bh$

Rectangle

$l = $ Length, $w = $ Width, $A = $ area, $P = $ perimeter
$A = lw \quad P = 2l + 2w$

Rectangular Box

$l = $ Length, $w = $ Width, $h = $ Height, $V = $ Volume, $S = $ Surface area
$V = lwh \quad S = 2lw + 2lh + 2wh$

Sphere

$r = $ Radius, $V = $ Volume, $S = $ Surface area
$V = \frac{4}{3}\pi r^3 \quad S = 4\pi r^2$

Right Circular Cylinder

$r = $ Radius, $h = $ Height, $V = $ Volume, $S = $ Surface area
$V = \pi r^2 h \quad S = 2\pi r^2 + 2\pi rh$

CONVERSION TABLE

1 centimeter ≈ 0.394 inch
1 meter ≈ 39.370 inches
≈ 3.281 feet
1 kilometer ≈ 0.621 mile
1 liter ≈ 0.264 gallon
1 newton ≈ 0.225 pound

1 joule ≈ 0.738 foot-pound
1 gram ≈ 0.035 ounce
1 kilogram ≈ 2.205 pounds
1 inch ≈ 2.540 centimeters
1 foot ≈ 30.480 centimeters
≈ 0.305 meter

1 mile ≈ 1.609 kilometers
1 gallon ≈ 3.785 liters
1 pound ≈ 4.448 newtons
1 foot-lb ≈ 1.356 Joules
1 ounce ≈ 28.350 grams
1 pound ≈ 0.454 kilogram

FUNCTIONS

Constant Function	$f(x) = b$
Linear Function	$f(x) = mx + b$, where m is the slope and b is the y-intercept
Quadratic Function	$f(x) = ax^2 + bx + c, a \neq 0$ or $f(x) = a(x - h)^2 + k$ parabola vertex (h, k)
Polynomial Function	$f(x) = a_n x^n + a_{n-1} x^{n-1} + \cdots + a_1 x + a_0$
Rational Function	$R(x) = \dfrac{n(x)}{d(x)} = \dfrac{a_n x^n + a_{n-1} x^{n-1} + \cdots + a_1 x + a_0}{b_m x^m + a_{m-1} x^{m-1} + \cdots + b_1 x + b_0}$
Exponential Function	$f(x) = b^x, b > 0, b \neq 1$
Logarithmic Function	$f(x) = \log_b x, b > 0, b \neq 1$

GRAPHS OF COMMON FUNCTIONS

Constant Function

Identity Function

Absolute Value Function

Square Root Function

Square Function

Cube Function

Exponential Function

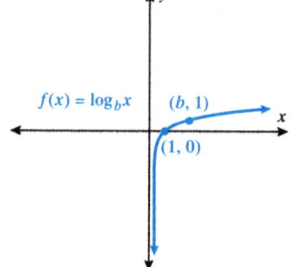

Logarithmic Function

TRANSFORMATIONS

In each case, c represents a positive real number.

Function		Draw the graph of f and:
Vertical translations	$\begin{cases} y = f(x) + c \\ y = f(x) - c \end{cases}$	Shift f upward c units. Shift f downward c units.
Horizontal translations	$\begin{cases} y = f(x - c) \\ y = f(x + c) \end{cases}$	Shift f to the right c units. Shift f to the left c units.
Reflections	$\begin{cases} y = -f(x) \\ y = f(-x) \end{cases}$	Reflect f about the x-axis. Reflect f about the y-axis.

HERON'S FORMULA FOR AREA

If the semiperimeter, s, of a triangle is

$$s = \frac{a + b + c}{2}$$

then the area of that triangle is

$$A = \sqrt{s(s - a)(s - b)(s - c)}$$

WileyPLUS

WileyPLUS is a research-based online environment for effective teaching and learning.

WileyPLUS builds students' confidence because it takes the guesswork out of studying by providing students with a clear roadmap:

- **what to do**
- **how to do it**
- **if they did it right**

It offers interactive resources along with a complete digital textbook that help students learn more. With *WileyPLUS*, students take more initiative so you'll have greater impact on their achievement in the classroom and beyond.

For more information, visit www.wileyplus.com

WileyPLUS

ALL THE HELP, RESOURCES, AND PERSONAL SUPPORT YOU AND YOUR STUDENTS NEED!
www.wileyplus.com/resources

1st DAY OF CLASS ... AND BEYOND!

2-Minute Tutorials and all of the resources you and your students need to get started

WileyPLUS

Student Partner Program

Student support from an experienced student user

Wiley Faculty Network

Collaborate with your colleagues, find a mentor, attend virtual and live events, and view resources
www.WhereFacultyConnect.com

WileyPLUS

Quick Start

Pre-loaded, ready-to-use assignments and presentations created by subject matter experts

Technical Support 24/7 FAQs, online chat, and phone support
www.wileyplus.com/support

© Courtney Keating/ iStockphoto

Your *WileyPLUS* Account Manager, providing personal training and support

Algebra and Trigonometry

Third Edition

Algebra and Trigonometry

Third Edition

CYNTHIA Y. YOUNG | *Professor of Mathematics*
UNIVERSITY OF CENTRAL FLORIDA

ANNOTATED INSTRUCTOR'S EDITION

WILEY

PUBLISHER	Laurie Rosatone
ACQUISITIONS EDITOR	Joanna Dingle
PROJECT EDITOR	Jennifer Brady
ASSOCIATE CONTENT EDITOR	Beth Pearson
EDITORIAL ASSISTANT	Elizabeth Baird
SENIOR PRODUCTION EDITOR	Kerry Weinstein/Ken Santor
DESIGNER	Madelyn Lesure
OPERATIONS MANAGER	Melissa Edwards
ILLUSTRATION EDITOR	Sandra Rigby; Electronic illustrations provided by Techsetters, Inc.
SENIOR PHOTO EDITOR	Jennifer MacMillan
COVER DESIGN	Madelyn Lesure
COVER PHOTO	Front Cover: Combined image: Ty Milford/Masterfile and Leslie Banks/iStockphoto
	Back Cover: Rustem Gurler/iStockphoto

"Modeling Our World" image: © David Woodfall/Getty Images

This book was set in 10/12 Times by MPS Limited, and printed and bound by Quad/Graphics, Inc. The cover was printed by Quad/Graphics, Inc.

This book is printed on acid free paper. ∞

Founded in 1807, John Wiley & Sons, Inc. has been a valued source of knowledge and understanding for more than 200 years, helping people around the world meet their needs and fulfill their aspirations. Our company is built on a foundation of principles that include responsibility to the communities we serve and where we live and work. In 2008, we launched a Corporate Citizenship Initiative, a global effort to address the environmental, social, economic, and ethical challenges we face in our business. Among the issues we are addressing are carbon impact, paper specifications and procurement, ethical conduct within our business and among our vendors, and community and charitable support. For more information, please visit our website: www.wiley.com/go/citizenship.

Evaluation copies are provided to qualified academics and professionals for review purposes only, for use in their courses during the next academic year. These copies are licensed and may not be sold or transferred to a third party. Upon completion of the review period, please return the evaluation copy to Wiley. Return instructions and a free of charge return mailing label are available at www.wiley.com/go/returnlabel. If you have chosen to adopt this textbook for use in your course, please accept this book as your complimentary desk copy. Outside of the United States, please contact your local sales representative.

ISBN: 978-0-470-64803-2
BRV ISBN: 978-1-118-12930-2
AIE ISBN: 978-1-118-13491-7

Printed in the United States of America

10 9 8 7 6 5 4 3 2 1

The Wiley Faculty Network
The Place Where Faculty Connect

The Wiley Faculty Network (WFN) is a global community of faculty connected by a passion for teaching and a drive to learn, share, and collaborate. Whether you're seeking guidance, training, and resources or simply looking to re-energize your course, you'll find what you need with the WFN. The WFN also partners with institutions to provide customized professional development opportunities. Connect with the Wiley Faculty Network to collaborate with your colleagues, find a Mentor, attend virtual and live events, and view a wealth of resources all designed to help you grow as an educator.

Attend

Discover innovative ideas and gain knowledge you can use.
Learn from instructors around the world, as well as recognized leaders across disciplines. Join thousands of faculty just like you who participate in virtual and live events each semester. You'll connect with fresh ideas, best practices, and practical tools for a wide range of timely topics.

View

Explore your resources and development opportunities.
See all that is available to you when you connect with the Wiley Faculty Network. From Learning Modules and archived Guest Lectures to faculty-development and peer-reviewed resources, there is a wealth of materials at your fingertips.

Collaborate

Connect with colleagues—your greatest resource.
Tap into your greatest resource—your peers. Exchange ideas and teaching tools, while broadening your perspective. Whether you choose to blog, join interest groups, or connect with a Mentor—you've come to the right place!

Wiley Faculty Network

Embrace the art of teaching – Great things happen where faculty connect!

Web: www.WhereFacultyConnect.com
EMAIL: FacultyNetwork@wiley.com
PHONE: 1-866-4FACULTY (1-866-432-2858)

For Christopher and Caroline

About the Author

Cynthia Y. Young is a native of Tampa, Florida. She currently is a Professor of Mathematics at the University of Central Florida (UCF) and the author of *College Algebra*, *Trigonometry*, *Algebra and Trigonometry*, and *Precalculus*. She holds a B.A. degree in Secondary Mathematics Education from the University of North Carolina (Chapel Hill), an M.S. degree in Mathematical Sciences from UCF, and both an M.S. in Electrical Engineering and a Ph.D in Applied Mathematics from the University of Washington. She has taught high school in North Carolina and Florida, developmental mathematics at Shoreline Community College in Washington, and undergraduate and graduate students at UCF. Dr. Young's two main research interests are laser propagation through random media and improving student learning in STEM. She has authored or co-authored over 60 books and articles and been involved in over $2.5M in external funding. Her atmospheric propagation research was recognized by the Office of Naval Research Young Investigator award, and in 2007 she was selected as a Fellow of the International Society for Optical Engineers. She is currently the co-director of UCF's EXCEL program whose goal is to improve the retention of STEM majors.

Although Dr. Young excels in research, she considers teaching her true calling. She has been the recipient of the UCF Excellence in Undergraduate Teaching Award, the UCF Scholarship of Teaching and Learning Award, and a two-time recipient of the UCF Teaching Incentive Program. Dr. Young is committed to improving student learning in mathematics and has shared her techniques and experiences with colleagues around the country through talks at colleges, universities, and conferences.

Dr. Young and her husband, Dr. Christopher Parkinson, enjoy spending time outdoors and competing in Field Trials with their Labrador Retrievers. *Laird's Cynful Wisdom* (call name "*Wiley*") is titled in Canada and currently pursuing her U.S. title. *Laird's Cynful Ellegance* (call name "*Ellie*") was a finalist in the Canadian National in 2009 and is retired (relaxing at home).

Dr. Young is pictured here with Ellie's 2011 litter of puppies!

Bonnie Farris

Preface

As a mathematics professor I would hear my students say, "I understand you in class, but when I get home I am lost." When I would probe further, students would continue with "I can't read the book." As a mathematician I always found mathematics textbooks quite easy to read—and then it dawned on me: don't look at this book through a mathematician's eyes; look at it through the eyes of students who might not view mathematics the same way that I do. What I found was that the books were not at all like my class. Students understood me in class, but when they got home they couldn't understand the book. It was then that the folks at Wiley lured me into writing. My goal was to write a book that is seamless with how we teach and is an ally (not an adversary) to student learning. I wanted to give students a book they could read without sacrificing the rigor needed for conceptual understanding. The following quote comes from a reviewer of this third edition when asked about the rigor of the book:

> *I would say that this text comes across as a little less rigorous than other texts, but I think that stems from how easy it is to read and how clear the author is. When one actually looks closely at the material, the level of rigor is high.*

Distinguishing Features

Four key features distinguish this book from others, and they came directly from my classroom.

PARALLEL WORDS AND MATH

Have you ever looked at your students' notes? I found that my students were only scribbling down the mathematics that I would write—never the words that I would say in class. I started passing out handouts that had two columns: one column for math and one column for words. Each Example would have one or the other; either the words were there and students had to fill in the math, or the math was there and students had to fill in the words. If you look at the Examples in this book, you will see that the words (your voice) are on the left and the mathematics is on the right. In most math books, when the author illustrates an Example, the mathematics is usually down the center of the page, and if the students don't know what mathematical operation was performed, they will look to the right for some brief statement of help. That's not how we teach; we don't write out an Example on the board and then say, "Class, guess what I just did!" Instead we lead our students, telling them what step is coming and then performing that mathematical step *together*—and reading naturally from left to right. Student reviewers have said that the Examples in this book are easy to read; that's because *your* voice is right there with them, working through problems *together*.

EXAMPLE 1 **Graphing a Quadratic Function Given in Standard Form**

Graph the quadratic function $f(x) = (x - 3)^2 - 1$.

Solution:

STEP 1 The parabola opens up. $a = 1$, so $a > 0$

STEP 2 Determine the vertex. $(h, k) = (3, -1)$

STEP 3 Find the y-intercept. $f(0) = (-3)^2 - 1 = 8$

SKILLS AND CONCEPTS (LEARNING OBJECTIVES AND EXERCISES)

In my experience as a mathematics teacher/instructor/professor, I find skills to be on the micro level and concepts on the macro level of understanding mathematics. I believe that too often skills are emphasized at the expense of conceptual understanding. I have purposely separated *learning objectives*

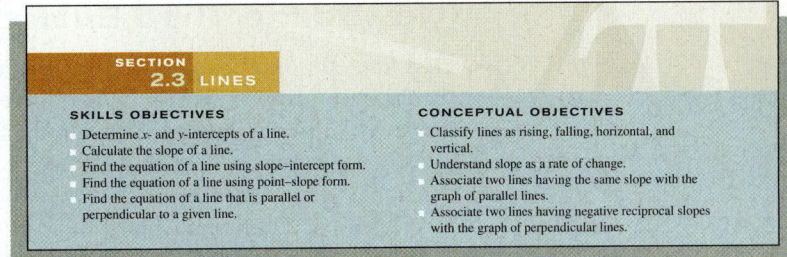

at the beginning of every section into two categories: *skills objectives*—what students should be able to do; and *conceptual objectives*—what students should understand. At the beginning of every class I discuss the learning objectives for the day—both skills and concepts. These are reinforced with both skills exercises and conceptual exercises.

CATCH THE MISTAKE

Have you ever made a mistake (or had a student bring you his or her homework with a mistake) and you go over it and over it and can't find the mistake? It's often easier to simply take out a new sheet of paper and solve it from scratch again than it is to actually find the mistake. Finding the mistake demonstrates a higher level of understanding. I include a few *Catch the Mistake* exercises in each section that demonstrate a common mistake that I have seen in my experience. I use these in class (either as a whole or often in groups), which leads to student discussion and offers an opportunity for formative assessment in real time.

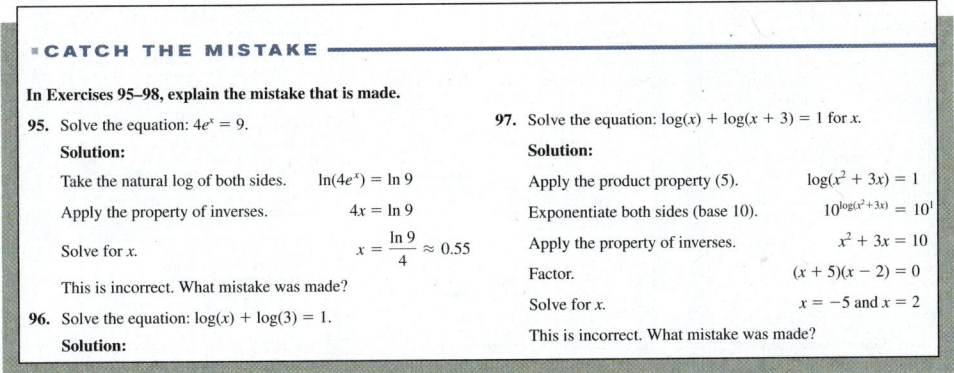

LECTURE VIDEOS BY THE AUTHOR

To ensure consistency in the students' learning experiences, I authored the videos myself. Throughout the book wherever a student sees the video icon, that indicates a video. These videos provide a mini lecture in that the chapter openers and chapter summaries are more like class discussion and selected Examples. Your Turns throughout the book also have an accompanying video of me working out that exact problem.

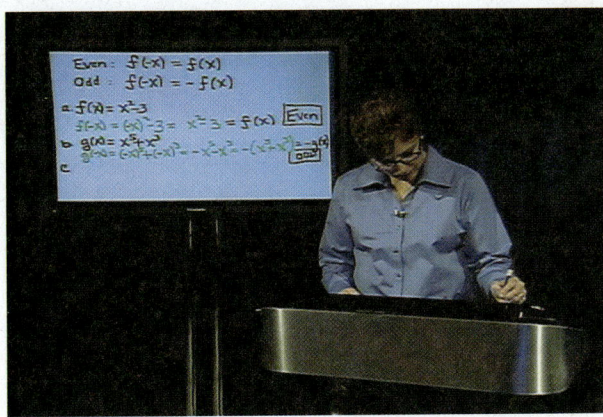

New to the Third Edition

The first edition was *my* book, the second edition was *our* book, and this third edition is *our even better* book. I've incorporated some specific line-by-line suggestions from reviewers throughout the exposition, added some new Examples, and added over 200 new Exercises. The three main global upgrades to the third edition are a new Chapter Map with Learning Objectives, End-of-chapter Inquiry-Based Learning Projects, and additional Applications Exercises in areas such as Business, Economics, Life Sciences, Health Sciences, and Medicine. A section (2.5*) on Linear Regression was added, as well as some technology exercises on Quadratic, Exponential, and Logarithmic Regression.

LEARNING OBJECTIVES

LEARNING OBJECTIVES

- Find the domain and range of a function.
- Sketch the graphs of common functions.
- Sketch graphs of general functions employing translations of common functions.
- Perform composition of functions.
- Find the inverse of a function.
- Model applications with functions using variation.

INQUIRY-BASED LEARNING PROJECTS

APPLICATIONS TO BUSINESS, ECONOMICS, HEALTH SCIENCES, AND MEDICINE

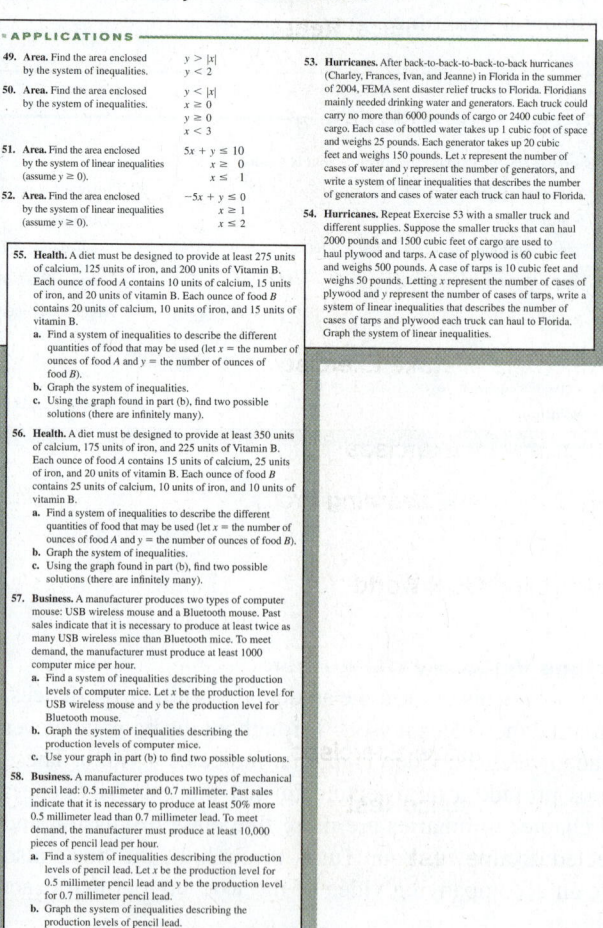

FEATURE	BENEFIT TO STUDENT
Chapter Opening Vignette	Piques the student's interest with a real-world application of material presented in the chapter. Later in the chapter, the same concept from the vignette is reinforced.
Chapter Overview, Flowchart, and Learning Objectives	Students see the big picture of how topics relate and overarching learning objectives are presented.
Skills and Conceptual Objectives	Skills objectives represent what students should be able to do. Conceptual objectives emphasize a higher level global perspective of concepts.
Clear, Concise, and Inviting Writing Style, Tone, and Layout	Students are able to *read* this book, which reduces math anxiety and promotes student success.
Parallel Words and Math	Increases students' ability to read and understand examples with a seamless representation of their instructor's class (instructor's voice and what they would write on the board).
Common Mistakes	Addresses a different learning style: teaching by counter-example. Demonstrates common mistakes so that students understand why a step is incorrect and reinforces the correct mathematics.
Color for Pedagogical Reasons	Particularly helpful for visual learners when they see a function written in red and then its corresponding graph in red or a function written in blue and then its corresponding graph in blue.
Study Tips	Reinforces specific notes that you would want to emphasize in class.
Author Videos	Gives students a mini class of several examples worked by the author.
Your Turn	Engages students during class, builds student confidence, and assists instructor in real-time assessment.
Catch the Mistake Exercises	Encourages students to assume the role of teacher—demonstrating a higher mastery level.
Conceptual Exercises	Teaches students to think more globally about a topic.
Inquiry-Based Learning Project	Lets students *discover* a mathematical identify, formula, etc. that is derived in the book.
Modeling OUR World	Engages students in a modeling project of a timely subject: global climate change.
Chapter Review	Key ideas and formulas are presented section by section in a chart. Improves study skills.
Chapter Review Exercises	Improves study skills.
Chapter Practice Test	Offers self-assessment and improves study skills.
Cumulative Test	Improves retention.

Instructor Supplements

INSTRUCTOR'S SOLUTIONS MANUAL (ISBN: 978-1-118-13759-8)

- Contains worked-out solutions to all exercises in the text.

INSTRUCTOR'S MANUAL

Authored by Cynthia Young, the manual provides practical advice on teaching with the text, including:

- sample lesson plans and homework assignments
- suggestions for the effective utilization of additional resources and supplements
- sample syllabi
- Cynthia Young's Top 10 Teaching Tips & Tricks
- online component featuring the author presenting these Tips & Tricks

ANNOTATED INSTRUCTOR'S EDITION (ISBN: 978-1-118-13491-7)

- Displays answers to all exercise questions, which can be found in the back of the book.
- Provides additional classroom examples within the standard difficulty range of the in-text exercises, as well as challenge problems to assess your students' mastery of the material.

POWERPOINT SLIDES

- For each section of the book, a corresponding set of lecture notes and worked-out examples are presented as PowerPoint slides, available on the Book Companion Site (www.wiley.com/college/young) and *WileyPLUS*.

TEST BANK

- Contains approximately 900 questions and answers from every section of the text.

COMPUTERIZED TEST BANK

Electonically enhanced version of the Test Bank that

- contains approximately 900 algorithmically-generated questions.
- allows instructors to freely edit, randomize, and create questions.
- allows instructors to create and print different versions of a quiz or exam.
- recognizes symbolic notation.

BOOK COMPANION WEBSITE (WWW.WILEY.COM/COLLEGE/YOUNG)

- Contains all instructor supplements listed plus a selection of personal response system questions.

WileyPLUS

- Features a full-service, digital learning environment, including additional resources for instructors, such as assignable homework exercises, tutorials, gradebook, and integrated links between the online version of the text and supplements.

Student Supplements

STUDENT SOLUTIONS MANUAL (ISBN: 978-1-118-13758-1)

- Includes worked-out solutions for all odd problems in the text.

BOOK COMPANION WEBSITE (WWW.WILEY.COM/COLLEGE/YOUNG)

- Provides additional resources for students to enhance the learning experience.

What Do Students Receive with *WileyPLUS*?

A RESEARCH-BASED DESIGN

WileyPLUS provides an online environment that integrates relevant resources, including the entire digital textbook, in an easy-to-navigate framework that helps students study more effectively.

- *WileyPLUS* adds structure by organizing textbook content into smaller, more manageable "chunks."
- Related media, examples, and sample practice items reinforce the learning objectives.
- Innovative features such as visual progress tracking, and self-evaluation tools improve time management and strengthen areas of weakness.

ONE-ON-ONE ENGAGEMENT

With *WileyPLUS*, students receive 24/7 access to resources that promote positive learning outcomes. Students engage with related examples (in various media) and sample practice items, including:

- Self-Study Quizzes
- Video Quizzes
- Proficiency Exams
- Guided Online (GO) Tutorial Problems
- Concept Questions
- Lecture Videos by Cynthia Young, including chapter introductions, chapter summaries, and selected video examples.

MEASURABLE OUTCOMES

Throughout each study session, students can assess their progress and gain immediate feedback. *WileyPLUS* provides precise reporting of strengths and weaknesses, as well as individualized quizzes, so that students are confident they are spending their time on the right things. With *WileyPLUS*, students always know the exact outcome of their efforts.

What Do Instructors Receive with *WileyPLUS*?

WileyPLUS provides reliable, customizable resources that reinforce course goals inside and outside of the classroom, as well as visibility into individual student progress. Pre-created materials and activities help instructors optimize their time.

CUSTOMIZABLE COURSE PLAN

WileyPLUS comes with a pre-created Course Plan designed by a subject matter expert uniquely for this course.

PRE-CREATED ACTIVITY TYPES INCLUDE:

- Questions
- Readings and Resources
- Print Tests

COURSE MATERIALS AND ASSESSMENT CONTENT

- Lecture Notes PowerPoint Slides
- Instructor's Manual
- Question Assignments (all end-of-chapter problems coded algorithmically with hints, links to text, whiteboard/show work feature, and instructor controlled problem solving help)

GRADEBOOK

WileyPLUS provides instant access to reports on trends in class performance, student use of course materials, and progress toward learning objectives, helping inform decisions and drive classroom discussions.

Acknowledgments

I want to express my sincerest gratitude to the entire Wiley team. I've said this before, and I will say it again: Wiley is the right partner for me. There is a reason that my dog is named Wiley—she's smart, competitive, a team player, and most of all, a joy to be around. There are several people within Wiley to whom I feel the need to express my appreciation: first and foremost to Laurie Rosatone who convinced Wiley Higher Ed to invest in a young assistant professor's vision for a series and who has been unwavering in her commitment to student learning. To my editor Joanna Dingle whose judgment I trust in both editorial and preschool decisions; thank you for surpassing my greatest expectations for an editor. To the rest of the ladies on the math editorial team (Jen Brady, Beth Pearson, and Liz Baird), you are all first class! This revision was planned and executed exceptionally well thanks to you three. To the math marketing specialists Jonathan Cottrell and Jen Wreyford, thank you for helping reps tell my story: you both are outstanding at your jobs. To Kerry Weinstein, thank you for your attention to detail. To the art and illustration folks (Jennifer MacMillan, Sandra Rigby, and Dennis Ormond), thank you for bringing to life all of the sketches and figures. And finally, I'd like to thank all of the Wiley reps: thank you for your commitment to my series and your tremendous efforts to get professors to adopt this book for their students.

I would also like to thank all of the contributors who helped us make this *our even better book*. I'd first like to thank Mark McKibben. He is known as the author of the solutions manuals that accompany this series, but he is much more than that. Mark, thank you for making this series a priority, for being so responsive, and most of all for being my "go-to" person to think through ideas. I'd also like to especially thank Jodi B.A. McKibben who is a statistician and teamed with Mark to develop the new regression material. I'd like to thank Steve Davis who was the inspiration for the Inquiry-Based Learning Projects and a huge thanks to Lyn Riverstone who developed all of the IBLPs. Special thanks to Laura Watkins for finding applications that are real and timely and to Ricki Alexander for updating all of the Technology Tips. I'd also like to thank Becky Schantz for her environmental problems (I now use AusPens because of Becky).

I'd also like to thank the following reviewers whose input helped make this book even better.

Aaron Anderson, *Hillsborough Community College*
Bernadette Antkoviak, *Harrisburg Area Community College*
Jan Archibald, *Ventura College*
Shari Beck, *Navarro College*
Patricia K. Bezona, *Valdosta State University*
Connie Buller, *Metropolitan Community College*
James Carolan, *Wharton County Junior College*
Diane Cook, *Collegiate High School at Northwest Florida State College*
Doris C. Cowan, *Polk State College*
Jean Davis, *Texas State University*

Nerissa Felder, *Polk State College*
Sunshine Gibbons, *Southeast Missouri State University*
Mehran Hassanpour, *South Texas College*
Tom Hayes, *Montana State University – Bozeman*
Celeste Hernandez, *Richland College*
Carolyn Horseman, *Polk State College*
Dianne Marquart, *Valdosta State University*
Maria Luisa Mendez, *Laredo Community College*
Lily Rai, *South Texas College*
Leela Rakesh, *Central Michigan University*
Denise Reid, *Valdosta State University*
Linda Tansil, *Southeast Missouri State University*

And a special thanks to our student reviewer Luis Suarez del Rio.

Table of Contents

Getty Images/Blend Images/Getty Images

Garry Wade/Taxi/Getty Images

Radius Images/©Corbis

Vetta/Getty Images

John Giustina/Superstock

Focus on Sport/Getty Images

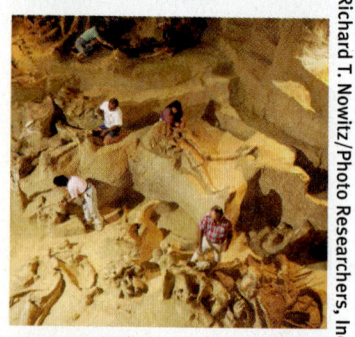

Richard T. Nowitz/Photo Researchers, Inc.

© Kathy Yturralde/Alamy

© Jason Brindel Commercial/Alamy

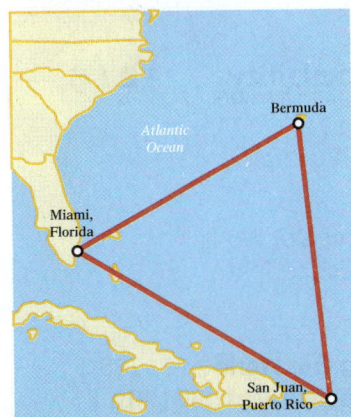

Q. Sakamaki/Redux Pictures

Q. Sakamaki/Redux Pictures

Paul J.Richards/AFP/Getty Images

Paul Souders/The Image Bank/Getty Images

A BASIC STRATEGY FOR BLACKJACK

Dealer's Up Card

HIT STAND DOUBLE DOWN SPLIT

11 Analytic Geometry and Systems of Nonlinear Equations and Inequalities 1108

12 Sequences, Series, and Probability 1216

A Note from the Author to the Student

I wrote this text with careful attention to ways in which to make your learning experience more successful. If you take full advantage of the unique features and elements of this textbook, I believe your experience will be fulfilling and enjoyable. Let's walk through some of the special book features that will help you in your study of algebra and trigonometry.

Prerequisites and Review (Chapter 0)

A comprehensive review of prerequisite knowledge (intermediate algebra topics) in Chapter 0 provides a brush up on knowledge and skills necessary for success in the course.

Clear, Concise, and Inviting Writing

Special attention has been made to present an engaging, clear, precise narrative in a layout that is easy to use and designed to reduce any math anxiety you may have.

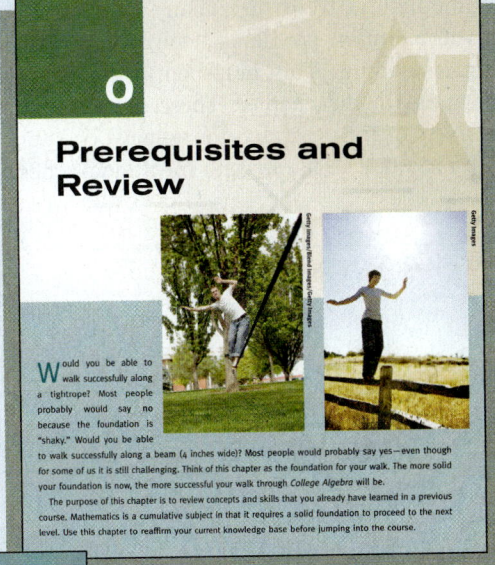

Chapter Introduction, Flow Chart, Section Headings, and Objectives

An opening vignette, flow chart, list of chapter sections, and chapter learning objectives give you an overview of the chapter.

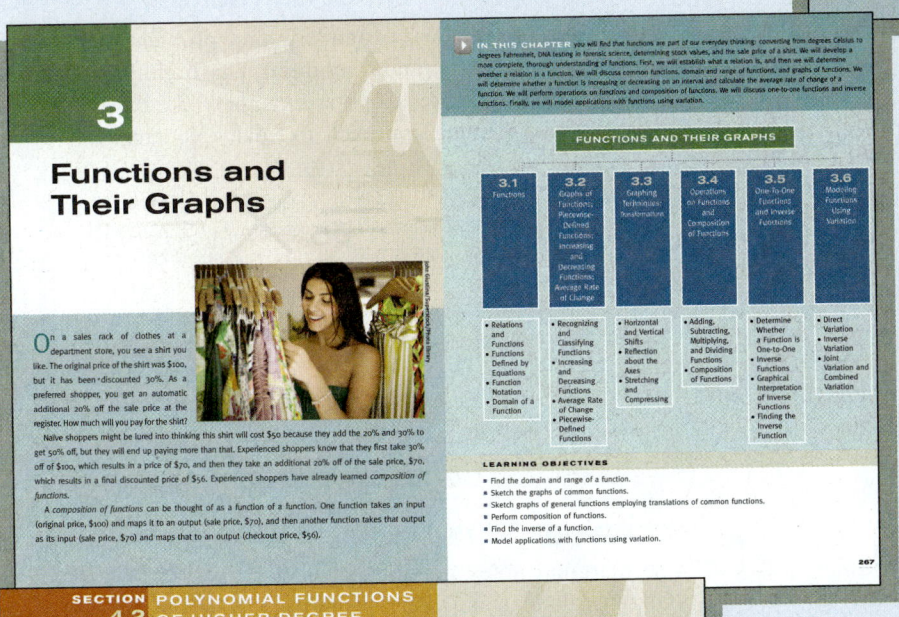

Skills and Conceptual Objectives

For every section, objectives are further divided by skills *and* concepts so you can see the difference between solving problems and truly understanding concepts.

Examples

Examples pose a specific problem using concepts already presented and then work through the solution. These serve to enhance your understanding of the subject matter.

Your Turn

Immediately following many examples, you are given a similar problem to reinforce and check your understanding. This helps build confidence as you progress in the chapter. These are ideal for in-class activity or for preparing for homework later. Answers are provided in the margin for a quick check of your work.

COMMON MISTAKE

A common misunderstanding is to interpret the notation $f(x + 1)$ as a sum: $f(x + 1) \neq f(x) + f(1)$.

✪ CORRECT

Write the original function.

$f(x) = x^2 - 3x$

Replace the argument x with a placeholder.

$f(\square) = (\square)^2 - 3(\square)$

Substitute $x + 1$ for the argument.

$f(x + 1) = (x + 1)^2 - 3(x + 1)$

Eliminate the parentheses.

$f(x + 1) = x^2 + 2x + 1 - 3x - 3$

Combine like terms.

$f(x + 1) = x^2 - x - 2$

✗ INCORRECT

The **ERROR** is in interpreting the notation as a sum.

$f(x + 1) \neq f(x) + f(1)$

$\neq x^2 - 3x - 2$

EXAMPLE 9   **Evaluating the Difference Quotient**

For the function $f(x) = x^2 - x$, find $\dfrac{f(x + h) - f(x)}{h}$, $h \neq 0$.

Solution:

Use placeholder notation for the function $f(x) = x^2 - x$.　$f(\square) = (\square)^2 - (\square)$

Calculate $f(x + h)$.　$f(x + h) = (x + h)^2 - (x + h)$

Write the difference quotient.　$\dfrac{f(x + h) - f(x)}{h}$

Let $f(x + h) = (x + h)^2 - (x + h)$ and $f(x) = x^2 - x$.

$$\dfrac{f(x + h) - f(x)}{h} = \dfrac{\overbrace{[(x + h)^2 - (x + h)]}^{f(x+h)} - \overbrace{[x^2 - x]}^{f(x)}}{h} \quad h \neq 0$$

Eliminate the parentheses inside the first set of brackets.　$= \dfrac{[x^2 + 2xh + h^2 - x - h] - [x^2 - x]}{h}$

Eliminate the brackets in the numerator.　$= \dfrac{x^2 + 2xh + h^2 - x - h - x^2 + x}{h}$

Combine like terms.　$= \dfrac{2xh + h^2 - h}{h}$

Factor the numerator.　$= \dfrac{h(2x + h - 1)}{h}$

Divide out the common factor, h.　$= \boxed{2x + h - 1} \quad h \neq 0$

■ YOUR TURN   Evaluate the difference quotient for $f(x) = x^2 - 1$.

Common Mistake/ Correct vs. Incorrect

In addition to standard examples, some problems are worked out both correctly and incorrectly to highlight common errors students make. Counter examples like these are often an effective learning approach for many students.

Parallel Words and Math

This text reverses the common textbook presentation of examples by placing the explanation in words *on the left* and the mathematics in parallel *on the right*. This makes it easier for students to read through examples as the material flows more naturally from left to right and as commonly presented in class.

WORDS	MATH
Write the interest formula for compounding continuously.	$A = Pe^{rt}$
Let $A = 2P$ (investment doubles).	$2P = Pe^{rt}$
Divide both sides of the equation by P.	$2 = e^{rt}$
Take the natural log of both sides of the equation.	$\ln 2 = \ln e^{rt}$
Simplify the right side by applying the property $\ln e^x = x$.	$\ln 2 = rt$
Divide both sides by r.	$t = \dfrac{\ln 2}{r}$
Approximate $\ln 2 \approx 0.7$.	$t \approx \dfrac{0.7}{r}$

Study Tips and Caution Notes

These marginal reminders call out important hints or warnings to be aware of related to the topic or problem.

Technology Tips

These marginal notes provide problem solving instructions and visual examples using graphing calculators.

Technology Tip

A graphing utility can be used to evaluate $P(2)$. Enter $P(x) = 4x^5 - 3x^4 + 2x^3 - 7x^2 + 9x - 5$ as Y_1.

```
Plot1 Plot2 Plot3
\Y1■4X^5-3X^4+2X
^3-7X²+9X-5
\Y2=
```

To evaluate $P(2)$, press VARS ▶

Y-VARS 1:Function 1:Y_1

(2) ENTER

```
Y1(2)
              81
■
```

Study Tip

The largest number of zeros a polynomial can have is equal to the degree of the polynomial.

▼ **CAUTION**

$f \circ g \neq f \cdot g$

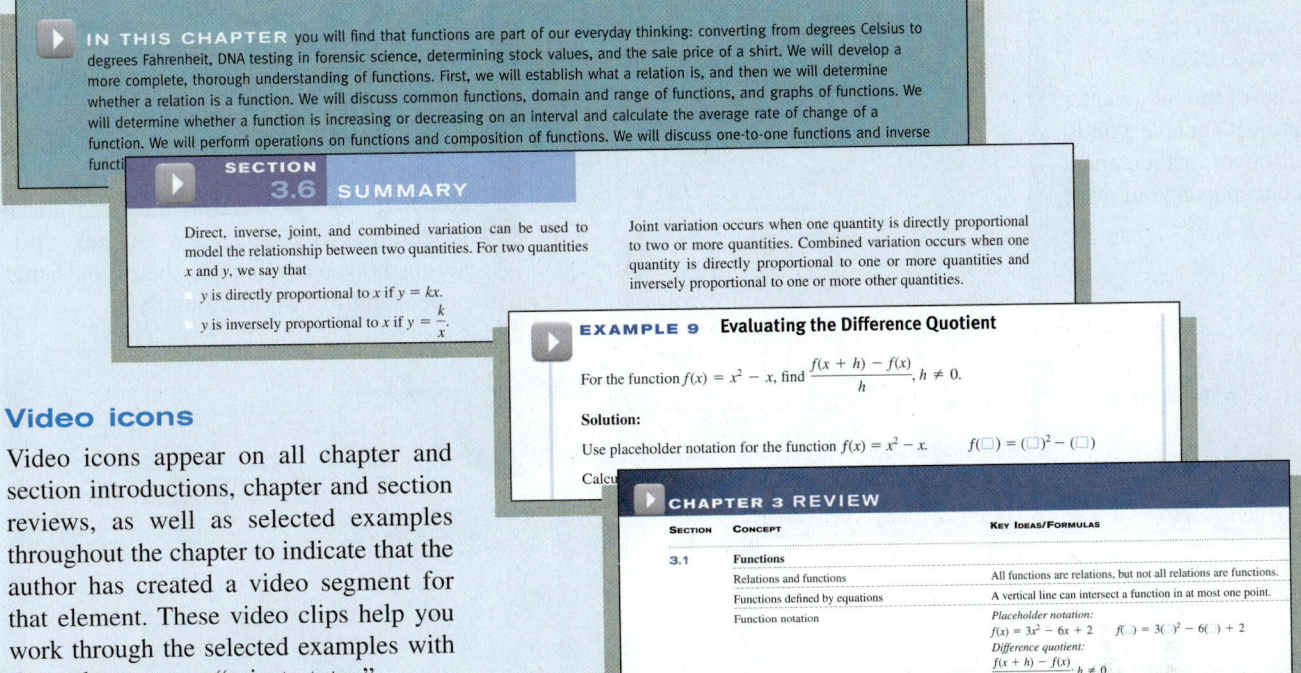

IN THIS CHAPTER you will find that functions are part of our everyday thinking: converting from degrees Celsius to degrees Fahrenheit, DNA testing in forensic science, determining stock values, and the sale price of a shirt. We will develop a more complete, thorough understanding of functions. First, we will establish what a relation is, and then we will determine whether a relation is a function. We will discuss common functions, domain and range of functions, and graphs of functions. We will determine whether a function is increasing or decreasing on an interval and calculate the average rate of change of a function. We will perform operations on functions and composition of functions. We will discuss one-to-one functions and inverse functi

SECTION 3.6 SUMMARY

Direct, inverse, joint, and combined variation can be used to model the relationship between two quantities. For two quantities x and y, we say that

- y is directly proportional to x if $y = kx$.
- y is inversely proportional to x if $y = \dfrac{k}{x}$.

Joint variation occurs when one quantity is directly proportional to two or more quantities. Combined variation occurs when one quantity is directly proportional to one or more quantities and inversely proportional to one or more other quantities.

EXAMPLE 9   Evaluating the Difference Quotient

For the function $f(x) = x^2 - x$, find $\dfrac{f(x + h) - f(x)}{h}, h \neq 0$.

Solution:

Use placeholder notation for the function $f(x) = x^2 - x$. $f(\square) = (\square)^2 - (\square)$

Calcu

CHAPTER 3 REVIEW

SECTION	CONCEPT	KEY IDEAS/FORMULAS
3.1	Functions	
	Relations and functions	All functions are relations, but not all relations are functions.
	Functions defined by equations	A vertical line can intersect a function in at most one point.
	Function notation	Placeholder notation: $f(x) = 3x^2 - 6x + 2 \quad f(\ ) = 3(\ )^2 - 6(\ ) + 2$ Difference quotient: $\dfrac{f(x + h) - f(x)}{h}; h \neq 0$

Video icons

Video icons appear on all chapter and section introductions, chapter and section reviews, as well as selected examples throughout the chapter to indicate that the author has created a video segment for that element. These video clips help you work through the selected examples with the author as your "private tutor."

SECTION 3.5 EXERCISES

SKILLS

In Exercises 1–16, determine whether the given relation is a function. If it is a function, determine whether it is a one-to-one function.

1. Domain → Range   2. Domain → Range

MONTH
October
January
April

APPLICATIONS

65. Temperature. The equation used to convert from degrees Celsius to degrees Fahrenheit is $f(x) = \frac{9}{5}x + 32$. Determine the inverse function $f^{-1}(x)$. What does the inverse function represent?

66. Temperature. The equation used to convert from degrees Fahrenheit to degrees Celsius is $C(x) = \frac{5}{9}(x - 32)$. Determine the inverse function $C^{-1}(x)$. What does the inverse function represent?

67. Budget. The Richmond rowing club is planning to enter the Head of the Charles figure out how much per boat for the first boat. Find the cost f number of boats t

Security, write a function $E(x)$ that expresses the student's take-home pay each week. Find the inverse function $E^{-1}(x)$.

70. Salary. A grocery store pays you $8 per hour for the first 40 hours per week and time and a half for overtime. Write a piecewise-defined function that represents your weekly earnings $E(x)$ as a function of the number of hours worked x. Find the inverse function $E^{-1}(x)$. What does the inverse function tell you?

CATCH THE MISTAKE

In Exercises 75–78, explain the mistake that is made.

75. Is $x = y^2$ a one-to-one function?

Solution:

Yes, this graph represents a one-to-one function because it passes the horizontal line test.

This is incorrect. Wha

76. A linear one-to-one function is graphed below. Draw its inverse.

Solution:

Note that the points $(3, 3)$ and $(0, -4)$ lie on the graph of the function.

CONCEPTUAL

In Exercises 91–94, determine if each statement is true or false.

91. The graph of a polynomial function might not have any y-intercepts.

92. The graph of a polynomial function might not have any x-intercepts.

93. The domain of all polyn

94. The range of all polynomial functions is $(-\infty, \infty)$.

95. What is the maximum number of zeros that a polynomial of degree n can have?

96. What is the maximum number of turning points a graph of an

CHALLENGE

91. For the functions $f(x) = x + a$ and $g(x) = \dfrac{1}{x - a}$, find $g \circ f$ and state its domain.

92. For the functions $f(x) = ax^2 + bx + c$ and $g(x) = \dfrac{1}{x - c}$, find $g \circ f$ and state its domain.

93. For the functions $f(x) = \sqrt{x + a}$ and $g(x) = x^2 - a$ find $g \circ f$ and state its domain.

94. For the functions $f(x) = \dfrac{1}{x^a}$ and $g(x) = \dfrac{1}{x^b}$, find $g \circ f$ and state its domain. Assume $a > 1$ and $b > 1$.

TECHNOLOGY

95. Using a graphing utility, plot $y_1 = \sqrt{x + 7}$ and $y_2 = \sqrt{9 - x}$. Plot $y_3 = y_1 + y_2$. What is the domain of y_3?

96. Using a graphing utility, plot $y_1 = \sqrt{x + 5}$, $y_2 = \dfrac{1}{\sqrt{3 - x}}$, and $y_3 = \dfrac{y_1}{y_2}$. What is the domain of y_3?

97. Using a graphing utility, plot $y_1 = \sqrt{x^2 - 3x - 4}$, $y_2 = \dfrac{1}{x^2 - 14}$, and $y_3 = \dfrac{1}{y_1^2 - 14}$. If y_1 represents a function f and y_2 represents a function g, then y_3 represents the composite function $g \circ f$. The graph of y_3 is only defined for the domain of $g \circ f$. State the domain of $g \circ f$.

98. Using a graphing utility, plot

Six Different Types of Exercises

Every text section ends with **Skills, Applications, Catch the Mistake, Conceptual, Challenge, and Technology** exercises. The exercises gradually increase in difficulty and vary in skill and conceptual emphasis. Catch the Mistake exercises increase the depth of understanding and reinforce what you have learned. Conceptual and Challenge exercises specifically focus on assessing conceptual understanding. Technology exercises enhance your understanding and ability using scientific and graphing calculators.

Inquiry-Based Learning Projects

These end of chapter projects enable you to discover mathematical concepts on your own!

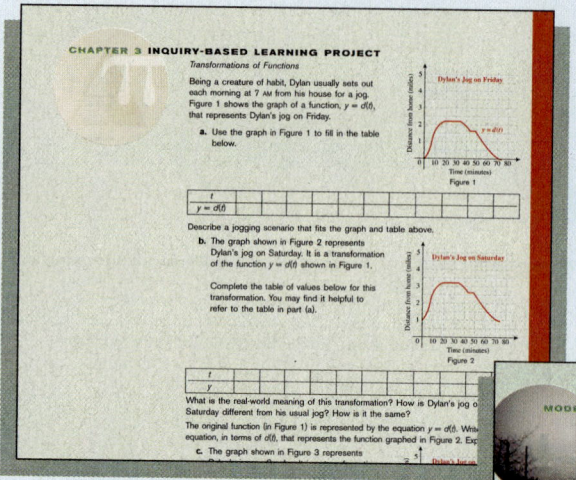

Modeling Our World

These unique end-of-chapter exercises provide a fun and interesting way to take what you have learned and model a real world problem. By using climate change as the continuous theme, these exercises can help you to develop more advanced modeling skills with each chapter while seeing how modeling can help you better understand the world around you.

Chapter Review, Review Exercises, Practice Test, Cumulative Test

At the end of every chapter, a summary review chart organizes the key learning concepts in an easy to use one or two-page layout. This feature includes key ideas and formulas, as well as indicating relevant pages and review exercises so that you can quickly summarize a chapter and study smarter. Review Exercises, arranged by section heading, are provided for extra study and practice. A Practice Test, without section headings, offers even more self-practice before moving on. A new Cumulative Test feature offers study questions based on all previous chapters' content, thus helping you build upon previously learned concepts.

Algebra and Trigonometry

Third Edition

0

Prerequisites and Review

Getty Images/Blend Images/Getty Images

Garry Wade/Taxi/Getty Images

Would you be able to walk successfully along a tightrope? Most people probably would say no because the foundation is "shaky." Would you be able to walk successfully along a beam (4 inches wide)? Most people would probably say yes—even though for some of us it is still challenging. Think of this chapter as the foundation for your walk. The more solid your foundation is now, the more successful your walk through *College Algebra* will be.

The purpose of this chapter is to review concepts and skills that you already have learned in a previous course. Mathematics is a cumulative subject in that it requires a solid foundation to proceed to the next level. Use this chapter to reaffirm your current knowledge base before jumping into the course.

IN THIS CHAPTER real numbers, integer exponents, and scientific notation will be discussed, followed by rational exponents and radicals. Simplification of radicals and rationalization of denominators will be reviewed. Basic operations such as addition, subtraction, and multiplication of polynomials will be discussed followed by a review of how to factor polynomials. Rational expressions will be discussed and a brief overview of solving simple algebraic equations will be given. After reviewing all of these aspects of real numbers, this chapter will conclude with a review of complex numbers.

PREREQUISITES AND REVIEW

0.1 Real Numbers	**0.2** Integer Exponents and Scientific Notation	**0.3** Polynomials: Basic Operations	**0.4** Factoring Polynomials	**0.5** Rational Expressions	**0.6** Rational Exponents and Radicals	**0.7** Complex Numbers
• The Set of Real Numbers • Approximations: Rounding and Truncation • Order of Operations • Properties of Real Numbers	• Integer Exponents • Scientific Notation	• Adding and Subtracting Polynomials • Multiplying Polynomials • Special Products	• Greatest Common Factor • Factoring Formulas: Special Polynomial Forms • Factoring a Trinomial as a Product of Two Binomials • Factoring by Grouping • A Strategy for Factoring Polynomials	• Rational Expressions and Domain Restrictions • Simplifying Rational Expressions • Multiplying and Dividing Rational Expressions • Adding and Subtracting Rational Expressions • Complex Rational Expressions	• Square Roots • Other (nth) Roots • Rational Exponents	• The Imaginary Unit, i • Adding and Subtracting Complex Numbers • Multiplying Complex Numbers • Dividing Complex Numbers • Raising Complex Numbers to Integer Powers

LEARNING OBJECTIVES

- Understand that rational and irrational numbers together constitute the real numbers.
- Apply properties of exponents.
- Perform operations on polynomials.
- Factor polynomials.
- Simplify expressions that contain rational exponents.
- Simplify radicals.
- Write complex numbers in standard form.

SKILLS OBJECTIVES

- Classify real numbers as rational or irrational.
- Round or truncate real numbers.
- Simplify expressions using correct order of operations.
- Evaluate algebraic expressions.
- Apply properties of real numbers.

CONCEPTUAL OBJECTIVES

- Understand that rational and irrational numbers are mutually exclusive and complementary subsets of real numbers.
- Learn the order of operations for real numbers.

The Set of Real Numbers

A **set** is a group or collection of objects that are called **members** or **elements** of the set. If *every* member of set B is also a member of set A, then we say B is a **subset** of A and denote it as $B \subset A$.

For example, the starting lineup on a baseball team is a subset of the entire team. The set of **natural numbers**, $\{1, 2, 3, 4, \ldots\}$, is a subset of the set of **whole numbers**, $\{0, 1, 2, 3, 4, \ldots\}$, which is a subset of the set of **integers**, $\{\ldots, -4, -3, -2, -1, 0, 1, 2, 3, \ldots\}$, which is a subset of the set of *rational numbers*, which is a subset of the set of *real numbers*. The three dots, called an **ellipsis**, indicate that the pattern continues indefinitely.

If a set has no elements, it is called the **empty set**, or **null set**, and is denoted by the symbol $\varnothing$. The **set of real numbers** consists of two main subsets: *rational* and *irrational* numbers.

DEFINITION   **Rational Number**

A **rational number** is a number that can be expressed as a quotient (ratio) of two integers, $\dfrac{a}{b}$, where the integer a is called the **numerator** and the integer b is called the **denominator** and where $b \neq 0$.

Rational numbers include all integers or all fractions that are ratios of integers. Note that any integer can be written as a ratio whose denominator is equal to 1. In decimal form, the rational numbers are those that terminate or are nonterminating with a repeated decimal pattern, which is represented with an overbar. Those decimals that do not repeat and do not terminate are **irrational numbers**. The numbers

$$5, \quad -17, \quad \frac{1}{3}, \quad \sqrt{2}, \quad \pi, \quad 1.37, \quad 0, \quad -\frac{19}{17}, \quad 3.66\overline{6}, \quad 3.2179\ldots$$

are examples of **real** numbers, where $5, -17, \frac{1}{3}, 1.37, 0, -\frac{19}{17}$, and $3.66\overline{6}$ are rational numbers, and $\sqrt{2}, \pi$, and $3.2179\ldots$ are irrational numbers. It is important to note that the ellipsis following the last decimal digit denotes continuing in an irregular fashion, whereas the absence of such dots to the right of the last decimal digit implies the decimal expansion terminates.

RATIONAL NUMBER (FRACTION)	CALCULATOR DISPLAY	DECIMAL REPRESENTATION	DESCRIPTION
$\frac{7}{2}$	3.5	3.5	Terminates
$\frac{15}{12}$	1.25	1.25	Terminates
$\frac{2}{3}$	0.666666666	$0.\overline{6}$	Repeats
$\frac{1}{11}$	0.09090909	$0.\overline{09}$	Repeats

Notice that the overbar covers the entire repeating pattern. The following figure and table illustrate the subset relationship and examples of different types of real numbers.

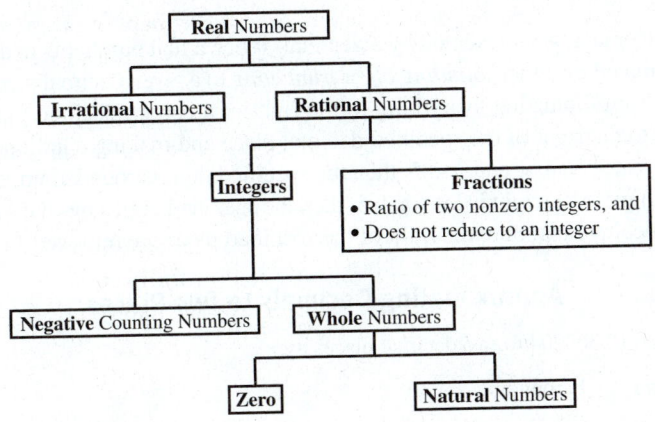

SYMBOL	NAME	DESCRIPTION	EXAMPLES
$\mathbb{N}$	Natural numbers	Counting numbers	$1, 2, 3, 4, 5, \ldots$
$\mathbb{W}$	Whole numbers	Natural numbers and zero	$0, 1, 2, 3, 4, 5, \ldots$
$\mathbb{Z}$	Integers	Whole numbers and negative natural numbers	$\ldots, -5, -4, -3, -2, -1, 0, 1, 2, 3, 4, 5, \ldots$
$\mathbb{Q}$	Rational numbers	Ratios of integers: $\frac{a}{b}$ $(b \neq 0)$ • Decimal representation terminates, or • Decimal representation repeats	$-17, -\frac{19}{7}, 0, \frac{1}{3}, 1.37, 3.66\overline{6}, 5$
$\mathbb{I}$	Irrational numbers	Numbers whose decimal representation does *not* terminate or repeat	$\sqrt{2}, 1.2179\ldots, \pi$
$\mathbb{R}$	Real numbers	Rational and irrational numbers	$\pi, 5, -\frac{2}{3}, 17.25, \sqrt{7}$

Since the set of real numbers can be formed by combining the set of rational numbers and the set of irrational numbers, then every real number is either rational or irrational. The set of rational numbers and the set of irrational numbers are both mutually exclusive (no shared elements) and complementary sets. The **real number line** is a graph used to represent the set of all real numbers.

EXAMPLE 1   Classifying Real Numbers

Classify the following real numbers as rational or irrational:

$$-3, \quad 0, \quad \tfrac{1}{4}, \quad \sqrt{3}, \quad \pi, \quad 7.51, \quad \tfrac{1}{3}, \quad -\tfrac{8}{5}, \quad 6.66666$$

Solution:

Rational: $-3,\quad 0,\quad \tfrac{1}{4},\quad 7.51,\quad \tfrac{1}{3},\quad -\tfrac{8}{5},\quad -6.66\overline{66}$ Irrational: $\sqrt{3}, \pi$

■ **Answer:**
Rational: $-\tfrac{7}{3}, 5.999\overline{9}, 12, 0, -5.27$
Irrational: $\sqrt{5}, 2.010010001\ldots$

■ **YOUR TURN** Classify the following real numbers as rational or irrational:

$$-\tfrac{7}{3}, 5.999\overline{9}, 12, 0, -5.27, \sqrt{5}, 2.010010001\ldots$$

Approximations: Rounding and Truncation

Every real number can be represented by a decimal. When a real number is in decimal form, it can be approximated by either *rounding off* or *truncating* to a given decimal place. **Truncation** is "cutting off" or eliminating everything to the right of a certain decimal place. **Rounding** means looking to the right of the specified decimal place and making a judgment. If the digit to the right is greater than or equal to 5, then the specified digit is rounded up, or increased by one unit. If the digit to the right is less than 5, then the specified digit stays the same. In both of these cases all decimal places to the right of the specified place are removed.

EXAMPLE 2   Approximating Decimals to Two Places

Approximate 17.368204 to two decimal places by

a. truncation   **b.** rounding

Solution:

a. To truncate, eliminate all digits to the right of the 6. $\boxed{17.36}$

b. To round, look to the right of the 6.
Because "8" is greater than 5, round up (add 1 to the 6). $\boxed{17.37}$

■ **Answer: a.** Truncation: 23.02
b. Rounding: 23.02

■ **YOUR TURN** Approximate 23.02492 to two decimal places by

a. truncation   **b.** rounding

EXAMPLE 3   Approximating Decimals to Four Places

Approximate 7.293516 to four decimal places by

a. truncation   **b.** rounding

Solution:

The "5" is in the fourth decimal place.

a. To truncate, eliminate all digits to the right of 5. $\boxed{7.2935}$

b. To round, look to the right of the 5.
Because "1" is less than 5, the 5 remains the same. $\boxed{7.2935}$

■ **Answer: a.** Truncation: -2.3818
b. Rounding: -2.3819

■ **YOUR TURN** Approximate -2.381865 to four decimal places by

a. truncation   **b.** rounding

It is important to note that *rounding and truncation sometimes yield the same approximation* (Example 3), *but not always* (Example 2).

Order of Operations

Addition, subtraction, multiplication, and division are called arithmetic operations. The results of these operations are called the sum, difference, product, and quotient, respectively. These four operations are summarized in the following table.

OPERATION	NOTATION	RESULT
Addition	$a + b$	Sum
Subtraction	$a - b$	Difference
Multiplication	$a \cdot b$ or ab or $(a)(b)$	Product
Division	$\dfrac{a}{b}$ or a/b $(b \neq 0)$	Quotient (Ratio)

Since algebra involves *variables* such as x, the traditional multiplication sign $\times$ is not used. Three alternatives are shown in the preceding table. Similarly, the arithmetic sign for division $\div$ is often represented by vertical or slanted fractions.

The symbol $=$ is called the **equal sign**, and is pronounced "equals" or "is," and it implies that the expression on one side of the equal sign is equivalent to (has the same value as) the expression on the other side of the equal sign.

WORDS	MATH
The sum of seven and eleven equals eighteen:	$7 + 11 = 18$
Three times five is fifteen:	$3 \cdot 5 = 15$
Four times six equals twenty-four:	$4(6) = 24$
Eight divided by two is four:	$\dfrac{8}{2} = 4$
Three subtracted from five is two:	$5 - 3 = 2$

When evaluating expressions involving real numbers, it is important to remember the correct *order of operations*. For example, how do we simplify the expression $3 + 2 \cdot 5$? Do we multiply first and then add, or add first and then multiply? In mathematics, conventional order implies multiplication first, and then addition: $3 + 2 \cdot 5 = 3 + 10 = 13$. Parentheses imply grouping of terms, and the necessary operations should always be performed inside them first. If there are nested parentheses, always start with the innermost parentheses and work your way out. Within parentheses follow the conventional order of operations. Exponents are an important part of order of operations and will be discussed in Section 0.2.

ORDER OF OPERATIONS

1. Start with the innermost parentheses (grouping symbols) and work outward.
2. Perform all indicated multiplications and divisions, working from left to right.
3. Perform all additions and subtractions, working from left to right.

EXAMPLE 4   Simplifying Expressions Using the Correct Order of Operations

Simplify the expressions.

a. $4 + 3 \cdot 2 - 7 \cdot 5 + 6$ **b.** $\dfrac{7 - 6}{2 \cdot 3 + 8}$

Solution (a):

Perform multiplication first.

$$4 + \underbrace{3 \cdot 2}_{6} - \underbrace{7 \cdot 5}_{35} + 6$$

Then perform the indicated additions and subtractions. $= 4 + 6 - 35 + 6 = \boxed{-19}$

Solution (b):

The numerator and the denominator are similar to expressions in parentheses. Simplify these separately first, following the correct order of operations.

Perform multiplication in the denominator first.

$$\frac{7 - 6}{\underbrace{2 \cdot 3}_{6} + 8}$$

Then perform subtraction in the numerator and addition in the denominator.

$$= \frac{7 - 6}{6 + 8} = \boxed{\frac{1}{14}}$$

▪ **YOUR TURN** Simplify the expressions.

a. $-7 + 4 \cdot 5 - 2 \cdot 6 + 9$ **b.** $\dfrac{9 - 6}{2 \cdot 5 + 6}$

Parentheses ( ) and brackets [ ] are the typical notations for grouping and are often used interchangeably. When nesting (groups within groups), use parentheses on the innermost and then brackets on the outermost.

EXAMPLE 5   Simplifying Expressions That Involve Grouping Signs Using the Correct Order of Operations

Simplify the expression $3[5 \cdot (4 - 2) - 2 \cdot 7]$.

Solution:

Simplify the inner parentheses. $3[5 \cdot (4 - 2) - 2 \cdot 7] = 3[5 \cdot 2 - 2 \cdot 7]$

Inside the brackets, perform the multiplication
$5 \cdot 2 = 10$ and $2 \cdot 7 = 14$. $= 3[10 - 14]$

Inside the brackets, perform the subtraction. $= 3[-4]$

Multiply. $= \boxed{-12}$

▪ **YOUR TURN** Simplify the expression $2[-3 \cdot (13 - 5) + 4 \cdot 3]$.

Algebraic Expressions

Everything discussed until now has involved real numbers (explicitly). In algebra, however, numbers are often represented by letters (such as x and y), which are called **variables**. A **constant** is a fixed (known) number such as 5. A **coefficient** is the constant that is multiplied by a variable. Quantities within the *algebraic expression* that are separated by addition or subtraction are referred to as **terms**.

DEFINITION   **Algebraic Expression**

An **algebraic expression** is the combination of variables and constants using basic operations such as addition, subtraction, multiplication, and division. Each term is separated by addition or subtraction.

Algebraic Expression	Variable Term	Constant Term	Coefficient
$5x + 3$	$5x$	3	5

When we know the value of the variables, we can **evaluate an algebraic expression** using the **substitution principle**:

Algebraic expression: $5x + 3$
Value of the variable: $x = 2$
Substitute $x = 2$: $5(2) + 3 = 10 + 3 = 13$

EXAMPLE 6   **Evaluating Algebraic Expressions**

Evaluate the algebraic expression $7x + 2$ for $x = 3$.

Solution:

Start with the algebraic expression.	$7x + 2$
Substitute $x = 3$.	$7(3) + 2$
Perform the multiplication.	$= 21 + 2$
Perform the addition.	$= \boxed{23}$

■ **YOUR TURN** Evaluate the algebraic expression $6y + 4$ for $y = 2$.

Classroom Example 0.1.6
Evaluate the expressions.
a. $5(2x - 3)$ for $x = -2$
b.* $\dfrac{6y - 2(y - 1)}{-1 - y(3 - y)}$ for $y = 1$

Answer:
a. -35 **b.** -2

■ **Answer:** 16

In Example 6, the value for the variable was specified in order for us to evaluate the algebraic expression. What if the value of the variable is not specified; can we simplify an expression like $3(2x - 5y)$? In this case, we cannot subtract $5y$ from $2x$. Instead, we rely on the basic *properties of real numbers*, or the *basic rules of algebra*.

Properties of Real Numbers

You probably already know many properties of real numbers. For example, if you add up four numbers, it does not matter in which order you add them. If you multiply five numbers, it does not matter what order you multiply them. If you add 0 to a real number or multiply a real number by 1, the result yields the original real number. **Basic properties of real numbers** are summarized in the following table. Because these properties are true for variables and algebraic expressions, these properties are often called the **basic rules of algebra**.

PROPERTIES OF REAL NUMBERS (BASIC RULES OF ALGEBRA)

NAME	DESCRIPTION	MATH (LET a, b, AND c EACH BE ANY REAL NUMBER)	EXAMPLE
Commutative property of addition	Two real numbers can be added in any order.	$a + b = b + a$	$3x + 5 = 5 + 3x$
Commutative property of multiplication	Two real numbers can be multiplied in any order.	$ab = ba$	$y \cdot 3 = 3y$
Associative property of addition	When three real numbers are added, it does not matter which two numbers are added first.	$(a + b) + c = a + (b + c)$	$(x + 5) + 7 = x + (5 + 7)$
Associative property of multiplication	When three real numbers are multiplied, it does not matter which two numbers are multiplied first.	$(ab)c = a(bc)$	$(-3x)y = -3(xy)$
Distributive property	Multiplication is distributed over *all* the terms of the sums or differences within the parentheses.	$a(b + c) = ab + ac$ $a(b - c) = ab - ac$	$5(x + 2) = 5x + 10$ $5(x - 2) = 5x - 10$
Additive identity property	Adding zero to any real number yields the same real number.	$a + 0 = a$ $0 + a = a$	$7y + 0 = 7y$
Multiplicative identity property	Multiplying any real number by 1 yields the same real number.	$a \cdot 1 = a$ $1 \cdot a = a$	$(8x)(1) = 8x$
Additive inverse property	The sum of a real number and its additive inverse (opposite) is zero.	$a + (-a) = 0$	$4x + (-4x) = 0$
Multiplicative inverse property	The product of a nonzero real number and its multiplicative inverse (reciprocal) is 1.	$a \cdot \dfrac{1}{a} = 1 \qquad a \neq 0$	$(x + 2) \cdot \left(\dfrac{1}{x + 2} \right) = 1$ $x \neq -2$

The properties in the previous table govern addition and multiplication. Subtraction can be defined in terms of addition of the *additive inverse*, and division can be defined in terms of multiplication by *the multiplicative inverse (reciprocal)*.

SUBTRACTION AND DIVISION

Let a and b be real numbers.

	MATH	TYPE OF INVERSE	WORDS
Subtraction	$a - b = a + (-b)$	$-b$ is the **additive inverse** or **opposite** of b	Subtracting a real number is equal to adding its opposite.
Division	$a \div b = a \cdot \dfrac{1}{b}$ $b \neq 0$	$\dfrac{1}{b}$ is the **multiplicative inverse** or **reciprocal** of b	Dividing by a real number is equal to multiplying by its reciprocal.

EXAMPLE 7 Using the Distributive Property

Use the distributive property to eliminate the parentheses.

a. $3(x + 5)$ **b.** $2(y - 6)$

Solution (a):

Use the distributive property.

Perform the multiplication.

$$3(x + 5) = 3(x) + 3(5)$$
$$= \boxed{3x + 15}$$

Solution (b):

Use the distributive property.

Perform the multiplication.

$$2(y - 6) = 2(y) - 2(6)$$
$$= \boxed{2y - 12}$$

■ **YOUR TURN** Use the distributive property to eliminate the parentheses.

a. $2(x + 3)$ **b.** $5(y - 3)$

> **Classroom Example 0.1.7**
> Use the distributive property to eliminate the parentheses.
> **a.** $-5(2 - z)$
> **b.** $3[2(1 - y) - 1]$
>
> **Answer:**
> **a.** $5z - 10$ **b.** $3 - 6y$

■ **Answer: a.** $2x + 6$ **b.** $5y - 15$

You also probably know the rules that apply when multiplying a negative real number. For example, "a negative times a negative is a positive."

PROPERTIES OF NEGATIVES

DESCRIPTION	MATH (LET a AND b BE POSITIVE REAL NUMBERS)	EXAMPLE
A *negative* quantity *times* a *positive* quantity is a *negative* quantity.	$(-a)(b) = -ab$	$(-8)(3) = -24$
A *negative* quantity *divided* by a *positive* quantity is a *negative* quantity. or A *positive* quantity *divided* by a *negative* quantity is a *negative* quantity.	$\dfrac{-a}{b} = -\dfrac{a}{b}$ or $\dfrac{a}{-b} = -\dfrac{a}{b}$	$\dfrac{-16}{4} = -4$ or $\dfrac{15}{-3} = -5$
A *negative* quantity times a *negative* quantity is a *positive* quantity.	$(-a)(-b) = ab$	$(-2x)(-5) = 10x$
A *negative* quantity *divided* by a *negative* quantity is a *positive* quantity.	$\dfrac{-a}{-b} = \dfrac{a}{b}$	$\dfrac{-12}{-3} = 4$
The opposite of a negative quantity is a positive quantity (subtracting a negative quantity is equivalent to adding a positive quantity).	$-(-a) = a$	$-(-9) = 9$
A negative sign preceding an expression is distributed throughout the expression.	$-(a + b) = -a - b$ $-(a - b) = -a + b$	$-3(x + 5) = -3x - 15$ $-3(x - 5) = -3x + 15$

EXAMPLE 8 Using Properties of Negatives

Eliminate the parentheses and perform the operations.

a. $-5 + 7 - (-2)$ **b.** $-(-3)(-4)(-6)$

Solution:

a. Distribute the negative.

$$-5 + 7 \underbrace{- (-2)}_{+2}$$

Combine the three quantities.

$$= -5 + 7 + 2$$
$$= \boxed{4}$$

b. Group the terms.

Perform the multiplication inside the [ ].

Multiply.

$$[-(-3)][(-4)(-6)]$$
$$= [3][24]$$
$$= \boxed{72}$$

Technology Tip

a.

b. Here are the calculator keystrokes for $-(-3)(-4)(-6)$.

We use properties of negatives to define the *absolute value* of any real number. The **absolute value** of a real number a, denoted $|a|$, is its magnitude. On a number line this is the distance from the origin, 0, to the point. For example, algebraically, the absolute value of 5 is 5, that is, $|5| = 5$; and the absolute value of -5 is 5, or $|-5| = 5$. Graphically, the distance on the real number line from 0 to either -5 or 5 is 5.

Notice that the absolute value does not change a positive real number, but changes a negative real number to a positive number. A negative number becomes a positive number if it is multiplied by -1.

IF a IS A ...	$|a|$	EXAMPLE
Positive real number	$|a| = a$	$|5| = 5$
Negative real number	$|a| = -a$	$|-5| = -(-5) = 5$
Zero	$|a| = a$	$|0| = 0$

EXAMPLE 9   Finding the Absolute Value of a Real Number

Evaluate the expressions.

a. $|-3 + 7|$

b. $|2 - 8|$

Solution:

a. $|-3 + 7| = |4|$
$\qquad\qquad = \boxed{4}$

b. $|2 - 8| = |-6|$
$\qquad\qquad = \boxed{6}$

Properties of the absolute value will be discussed in Section 1.7.

EXAMPLE 10   Using Properties of Negatives and the Distributive Property

Eliminate the parentheses $-(2x - 3y)$.

COMMON MISTAKE

A common mistake is applying a negative only to the first term.

★ CORRECT

$-(2x - 3y)$

$= -(2x) - (-3y)$

$= \boxed{-2x + 3y}$

✗ INCORRECT

Error: $-2x - 3y$

The negative $(-)$ was not distributed through the second term.

▪ **YOUR TURN** Eliminate the parentheses.

a. $-2(x + 5y)$ b. $-(3 - 2b)$

What is the product of any real number and zero? The answer is zero. This property also leads to the zero product property, which is the basis for factoring (one of the methods used to solve quadratic equations, which will be discussed in Section 1.3).

PROPERTIES OF ZERO

DESCRIPTION	MATH (LET *a* BE A REAL NUMBER)	EXAMPLE
A real number multiplied by zero is zero.	$a \cdot 0 = 0$	$0 \cdot x = 0$
Zero divided by a nonzero real number is zero.	$\dfrac{0}{a} = 0 \qquad a \neq 0$	$\dfrac{0}{3 - x} = 0 \qquad x \neq 3$
A real number divided by zero is undefined.	$\dfrac{a}{0}$ is undefined	$\dfrac{x + 2}{0}$ is undefined

ZERO PRODUCT PROPERTY

DESCRIPTION	MATH	EXAMPLE
If the product of two real numbers is zero, then one of those numbers has to be zero.	If $ab = 0$, then $a = 0$ or $b = 0$	If $x(x + 2) = 0$, then $x = 0$ or $x + 2 = 0$ therefore $x = 0$ or $x = -2$

Note: If a and b are *both* equal to zero, then the product is still zero.

Fractions always seem to intimidate students. In fact, many instructors teach students to eliminate fractions in algebraic equations. It is important to realize that you can never divide by zero. Therefore, in the following table of fractional properties it is assumed that no denominators are zero.

FRACTIONAL PROPERTIES

DESCRIPTION	MATH	ZERO CONDITION	EXAMPLE
Equivalent fractions	$\dfrac{a}{b} = \dfrac{c}{d}$ if and only if $ad = bc$	$b \neq 0$ and $d \neq 0$	$\dfrac{y}{2} = \dfrac{6y}{12}$ since $12y = 12y$
Multiplying two fractions	$\dfrac{a}{b} \cdot \dfrac{c}{d} = \dfrac{ac}{bd}$	$b \neq 0$ and $d \neq 0$	$\dfrac{3}{5} \cdot \dfrac{x}{7} = \dfrac{3x}{35}$
Adding fractions that have the same denominator	$\dfrac{a}{b} + \dfrac{c}{b} = \dfrac{a + c}{b}$	$b \neq 0$	$\dfrac{x}{3} + \dfrac{2}{3} = \dfrac{x + 2}{3}$
Subtracting fractions that have the same denominator	$\dfrac{a}{b} - \dfrac{c}{b} = \dfrac{a - c}{b}$	$b \neq 0$	$\dfrac{7}{3} - \dfrac{5}{3} = \dfrac{7 - 5}{3} = \dfrac{2}{3}$
Adding fractions with different denominators using a common denominator	$\dfrac{a}{b} + \dfrac{c}{d} = \dfrac{ad}{bd} + \dfrac{cb}{bd} = \dfrac{ad + bc}{bd}$	$b \neq 0$ and $d \neq 0$	$\dfrac{1}{2} + \dfrac{5}{3} = \dfrac{(1)(3) + (5)(2)}{(2)(3)} = \dfrac{13}{6}$
Subtracting fractions with different denominators using a common denominator	$\dfrac{a}{b} - \dfrac{c}{d} = \dfrac{ad}{bd} - \dfrac{cb}{bd} = \dfrac{ad - bc}{bd}$	$b \neq 0$ and $d \neq 0$	$\dfrac{1}{3} - \dfrac{1}{4} = \dfrac{(1)(4) - (1)(3)}{(3)(4)} = \dfrac{1}{12}$
Dividing by a fraction is equivalent to multiplying by its reciprocal	$\dfrac{a}{b} \div \dfrac{c}{d} = \dfrac{a}{b} \cdot \dfrac{d}{c}$	$b \neq 0, c \neq 0,$ and $d \neq 0$	$\dfrac{x}{3} \div \dfrac{2}{7} = \dfrac{x}{3} \cdot \dfrac{7}{2} = \dfrac{7x}{6}$

Technology Tip

To change the decimal number to a fraction, press: MATH 1 ▶ Frac
ENTER

a.

```
2/3-1/4
        .4166666667
Ans▶Frac
             5/12
■
```

b.

```
2/3/4
        .1666666667
Ans▶Frac
             1/6
■
```

■ **Answer:**

a. $\frac{11}{10}$ **b.** $\frac{2}{3}$ **c.** $\frac{10 - 3x}{15}$

The **least common multiple** of two or more integers is the smallest integer that is evenly divisible by each of the integers. For example, the least common multiple (LCM) of 3 and 4 is 12. The LCM of 8 and 6 is 24. The reason the LCM of 8 and 6 is not 48 is that 8 and 6 have a common factor of 2. When adding and subtracting fractions, a common denominator can be found by multiplying the denominators. When there are common factors in the denominators, the LCM is the *least* common denominator (LCD) of the original denominators.

EXAMPLE 11   Performing Operations with Fractions

Perform the indicated operations involving fractions and simplify.

a. $\frac{2}{3} - \frac{1}{4}$ **b.** $\frac{2}{3} \div 4$ **c.** $\frac{x}{2} + \frac{3}{5}$

Solution (a):

Determine the LCD. $\qquad\qquad\qquad\qquad 3 \cdot 4 = 12$

Rewrite fractions applying the LCD. $\qquad \dfrac{2}{3} - \dfrac{1}{4} = \dfrac{2 \cdot 4}{3 \cdot 4} - \dfrac{1 \cdot 3}{4 \cdot 3}$

$$= \dfrac{2(4) - 1(3)}{3(4)}$$

Eliminate the parentheses. $\qquad\qquad\qquad = \dfrac{8 - 3}{12}$

Combine terms in the numerator. $\qquad\qquad = \boxed{\dfrac{5}{12}}$

Solution (b):

Rewrite 4 with an understood 1 in the denominator. $\qquad = \dfrac{2}{3} \div \dfrac{4}{1}$

Dividing by a fraction is equivalent to multiplying by its reciprocal. $\qquad = \dfrac{2}{3} \cdot \dfrac{1}{4}$

Multiply numerators and denominators, respectively. $\qquad = \dfrac{2}{12}$

Reduce the fraction to simplest form. $\qquad = \boxed{\dfrac{1}{6}}$

Solution (c):

Determine the LCD. $\qquad\qquad\qquad\qquad 2 \cdot 5 = 10$

Rewrite fractions in terms of the LCD. $\qquad \dfrac{x}{2} + \dfrac{3}{5} = \dfrac{5x + 3(2)}{(2)(5)}$

Simplify the numerator. $\qquad\qquad\qquad = \boxed{\dfrac{5x + 6}{10}}$

■ **YOUR TURN** Perform the indicated operations involving fractions.

a. $\frac{3}{5} + \frac{1}{2}$ **b.** $\frac{1}{5} \div \frac{3}{10}$ **c.** $\frac{2}{3} - \frac{x}{5}$

SECTION 0.1 SUMMARY

In this section, real numbers were defined as the set of all rational and irrational numbers. Decimals are approximated by either truncating or rounding.

- *Truncating*: Eliminate all values after a particular digit.
- *Rounding*: Look to the right of a particular digit. If the number is 5 or greater, increase the digit by 1; otherwise, leave it as is and eliminate all digits to the right.

The *order* in which we perform operations is

1. parentheses (grouping); work from inside outward.
2. multiplication/division; work from left to right.
3. addition/subtraction; work from left to right.

The *properties of real numbers* are employed as the basic rules of algebra when dealing with algebraic expressions.

- Commutative property of addition: $a + b = b + a$
- Commutative property of multiplication: $ab = ba$
- Associative property of addition:
  $(a + b) + c = a + (b + c)$
- Associative property of multiplication: $(ab)c = a(bc)$
- Distributive property:
  $a(b + c) = ab + ac$ or $a(b - c) = ab - ac$
- Additive identity: $a + 0 = a$
- Multiplicative identity: $a \cdot 1 = a$
- Additive inverse (opposite): $a + (-a) = 0$
- Multiplicative inverse (reciprocal): $a \cdot \dfrac{1}{a} = 1$ $a \neq 0$

Subtraction and division can be defined in terms of addition and multiplication.

- *Subtraction*: $a - b = a + (-b)$ (add the opposite)
- *Division*: $a \div b = a \cdot \dfrac{1}{b}$, where $b \neq 0$

  (multiply by the reciprocal)

Properties of negatives were reviewed. If a and b are positive real numbers, then:

- $(-a)(b) = -ab$
- $(-a)(-b) = ab$
- $-(-a) = a$
- $-(a + b) = -a - b$ and $-(a - b) = -a + b$
- $\dfrac{-a}{b} = -\dfrac{a}{b}$
- $\dfrac{-a}{-b} = \dfrac{a}{b}$

Absolute value of real numbers: $|a| = a$ if a is nonnegative, and $|a| = -a$ if a is negative.

Properties of zero were reviewed.

- $a \cdot 0 = 0$ and $\dfrac{0}{a} = 0$ $a \neq 0$
- $\dfrac{a}{0}$ is undefined
- Zero product property: If $ab = 0$, then $a = 0$ or $b = 0$

Properties of fractions were also reviewed.

- $\dfrac{a}{b} \pm \dfrac{c}{d} = \dfrac{ad \pm bc}{bd}$ $b \neq 0$ and $d \neq 0$
- $\dfrac{a}{b} \div \dfrac{c}{d} = \dfrac{a}{b} \cdot \dfrac{d}{c}$ $b \neq 0, c \neq 0$, and $d \neq 0$

SECTION 0.1 EXERCISES

▪ SKILLS

In Exercises 1–8, classify the following real numbers as rational or irrational.

1. $\dfrac{11}{3}$

2. $\dfrac{22}{3}$

3. $2.07172737\ldots$

4. π

5. $2.7766\overline{776677}$

6. $5.22222\overline{2}$

7. $\sqrt{5}$

8. $\sqrt{17}$

In Exercises 9–16, approximate the real number to three decimal places by (a) rounding and (b) truncation.

9. 7.3471

10. 9.2549

11. 2.9949

12. 6.9951

13. 0.234492

14. 1.327491

15. 5.238473

16. 2.118465

In Exercises 17–40, perform the indicated operations in the correct order.

17. $5 + 2 \cdot 3 - 7$

18. $2 + 5 \cdot 4 + 3 \cdot 6$

19. $2 \cdot (5 + 7 \cdot 4 - 20)$

20. $-3 \cdot (2 + 7) + 8 \cdot (7 - 2 \cdot 1)$

21. $2 - 3[4(2 \cdot 3 + 5)]$

22. $4 \cdot 6(5 - 9)$

23. $8 - (-2) + 7$

24. $-10 - (-9)$

25. $-3 - (-6)$

26. $-5 + 2 - (-3)$

27. $x - (-y) - z$

28. $-a + b - (-c)$

29. $-(3x + y)$

30. $-(4a - 2b)$

31. $\dfrac{-3}{(5)(-1)}$

32. $-\dfrac{12}{(-3)(-4)}$

33. $-4 - 6[(5 - 8)(4)]$

34. $\dfrac{-14}{5 - (-2)}$

35. $-(6x - 4y) - (3x + 5y)$

36. $\dfrac{-4x}{6 - (-2)}$

37. $-(3 - 4x) - (4x + 7)$

38. $2 - 3[(4x - 5) - 3x - 7]$

39. $\dfrac{-4(5) - 5}{-5}$

40. $-6(2x + 3y) - [3x - (2 - 5y)]$

In Exercises 41–56, write as a single fraction and simplify.

41. $\dfrac{1}{3} + \dfrac{5}{4}$

42. $\dfrac{1}{2} - \dfrac{1}{5}$

43. $\dfrac{5}{6} - \dfrac{1}{3}$

44. $\dfrac{7}{3} - \dfrac{1}{6}$

45. $\dfrac{3}{2} + \dfrac{5}{12}$

46. $\dfrac{1}{3} + \dfrac{5}{9}$

47. $\dfrac{1}{9} - \dfrac{2}{27}$

48. $\dfrac{3}{7} - \dfrac{(-4)}{3} - \dfrac{5}{6}$

49. $\dfrac{x}{5} + \dfrac{2x}{15}$

50. $\dfrac{y}{3} - \dfrac{y}{6}$

51. $\dfrac{x}{3} - \dfrac{2x}{7}$

52. $\dfrac{y}{10} - \dfrac{y}{15}$

53. $\dfrac{4y}{15} - \dfrac{(-3y)}{4}$

54. $\dfrac{6x}{12} - \dfrac{7x}{20}$

55. $\dfrac{3}{40} + \dfrac{7}{24}$

56. $\dfrac{-3}{10} - \left(\dfrac{-7}{12}\right)$

In Exercises 57–68, perform the indicated operation and simplify, if possible.

57. $\dfrac{2}{7} \cdot \dfrac{14}{3}$

58. $\dfrac{2}{3} \cdot \dfrac{9}{10}$

59. $\dfrac{2}{7} \div \dfrac{10}{3}$

60. $\dfrac{4}{5} \div \dfrac{7}{10}$

61. $\dfrac{4b}{9} \div \dfrac{a}{27} \quad a \neq 0$

62. $\dfrac{3a}{7} \div \dfrac{b}{21} \quad b \neq 0$

63. $\dfrac{3x}{10} \div \dfrac{6x}{15} \quad x \neq 0$

64. $4\dfrac{1}{5} \div 7\dfrac{1}{20}$

65. $\dfrac{3x}{4} \div \dfrac{9}{16y} \quad y \neq 0$

66. $\dfrac{14m}{2} \cdot \dfrac{4}{7}$

67. $\dfrac{6x}{7} \div \dfrac{3y}{28} \quad y \neq 0$

68. $2\dfrac{1}{3} \cdot 7\dfrac{5}{6}$

In Exercises 69–72, evaluate the algebraic expression for the specified values.

69. $\dfrac{-c}{2d}$ for $c = -4, d = 3$

70. $2l + 2w$ for $l = 5, w = 10$

71. $\dfrac{m_1 \cdot m_2}{r^2}$ for $m_1 = 3, m_2 = 4, r = 10$

72. $\dfrac{x - \mu}{\sigma}$ for $x = 100, \mu = 70, \sigma = 15$

▪ APPLICATIONS

On December 16, 2007, the United States debt was estimated at $9,176,366,494,947, and at that time the estimated population was 303,818,361 citizens.

73. **U.S. National Debt.** Round the debt to the nearest million.

74. **U.S. Population.** Round the number of citizens to the nearest thousand.

75. **U.S. Debt.** If the debt is distributed evenly to all citizens, what is the national debt per citizen? Round your answer to the nearest dollar.

76. **U.S. Debt.** If the debt is distributed evenly to all citizens, what is the national debt per citizen? Round your answer to the nearest cent.

■ CATCH THE MISTAKE

In Exercises 77–80, explain the mistake that is made.

77. Round 13.2749 to two decimal places.

Solution:

The 9, to the right of the 4, causes
the 4 to round to 5.   13.275

The 5, to the right of the 7, causes
the 7 to be rounded to 8.   13.28

This is incorrect. What mistake was made?

78. Simplify the expression $\frac{2}{3} + \frac{1}{9}$.

Solution:

Add the numerators and denominators. $\dfrac{2+1}{3+9} = \dfrac{3}{12}$

Reduce. $= \dfrac{1}{4}$

This is incorrect. What mistake was made?

79. Simplify the expression $3(x+5) - 2(4+y)$.

Solution:

Eliminate parentheses. $3x + 15 - 8 + y$
Simplify. $3x + 7 + y$

This is incorrect. What mistake was made?

80. Simplify the expression $-3(x+2) - (1-y)$.

Solution:

Eliminate parentheses. $-3x - 6 - 1 - y$
Simplify. $-3x - 7 - y$

This is incorrect. What mistake was made?

■ CONCEPTUAL

In Exercises 81–84, determine whether each of the following statements is true or false.

81. Student athletes are a subset of the students in the honors program.

82. The students who are members of fraternities or sororities are a subset of the entire student population.

83. Every integer is a rational number.

84. A real number can be both rational and irrational.

85. What restrictions are there on x for the following to be true:

$$\frac{3}{x} \div \frac{5}{x} = \frac{3}{5}$$

86. What restrictions are there on x for the following to be true:

$$\frac{x}{2} \div \frac{x}{6} = 3$$

■ CHALLENGE

In Exercises 87 and 88, simplify the expressions.

87. $-2[3(x-2y) + 7] + [3(2-5x) + 10] - 7[-2(x-3) + 5]$

88. $-2\{-5(y-x) - 2[3(2x-5) + 7(2) - 4] + 3\} + 7$

■ TECHNOLOGY

89. Use your calculator to evaluate $\sqrt{1260}$. Does the answer appear to be a rational or an irrational number? Why?

90. Use your calculator to evaluate $\sqrt{\dfrac{144}{25}}$. Does the answer appear to be a rational or an irrational number? Why?

91. Use your calculator to evaluate $\sqrt{4489}$. Does the answer appear to be a rational or an irrational number? Why?

92. Use your calculator to evaluate $\sqrt{\dfrac{882}{49}}$. Does the answer appear to be a rational or an irrational number? Why?

SKILLS OBJECTIVES

- Evaluate expressions involving integer exponents.
- Apply properties of exponents.
- Use scientific notation.

CONCEPTUAL OBJECTIVES

- Visualize negative exponents as reciprocals.
- Understand that scientific notation is an effective way to represent very large or very small real numbers.

Integer Exponents

Exponents represent repeated multiplication. For example, $2 \cdot 2 \cdot 2 \cdot 2 \cdot 2 = 2^5$. The 2 that is repeatedly multiplied is called the *base*, and the small number 5 above and to the right of the 2 is called the *exponent*.

Study Tip

a^n: "a raised to the nth power"
a^2: "a squared"
a^3: "a cubed"

> **DEFINITION**   **Natural-Number Exponent**
>
> Let a be a real number and n be a natural number (positive integer); then a^n is defined as
>
> $$a^n = \underbrace{a \cdot a \cdot a \cdots a}_{n \text{ factors}} \qquad (a \text{ appears as a factor } n \text{ times})$$
>
> where n is the **exponent**, or **power**, and a is the **base**.

Technology Tip

```
4^3
              64
8^1
               8
```

```
5^4
             625
(1/2)^5
          .03125
Ans▶Frac
            1/32
```

■ Answer: **a.** 216   **b.** $\frac{1}{81}$

Classroom Example 0.2.1
Evaluate the expressions.
a. 2^4 **b.** $\left(\frac{1}{5}\right)^3$ **c.** $\left(\frac{1}{3}\right)^2 + 3^2$
Answer:
a. 16   **b.** $\frac{1}{125}$ **c.** $\frac{82}{9}$

EXAMPLE 1   Evaluating Expressions Involving Natural-Number Exponents

Evaluate the expressions.

a. 4^3 **b.** 8^1 **c.** 5^4 **d.** $\left(\frac{1}{2}\right)^5$

Solution:

a. $4^3 = 4 \cdot 4 \cdot 4 = \boxed{64}$ **b.** $8^1 = \boxed{8}$

c. $5^4 = 5 \cdot 5 \cdot 5 \cdot 5 = \boxed{625}$ **d.** $\left(\frac{1}{2}\right)^5 = \frac{1}{2} \cdot \frac{1}{2} \cdot \frac{1}{2} \cdot \frac{1}{2} \cdot \frac{1}{2} = \boxed{\frac{1}{32}}$

■ **YOUR TURN** Evaluate the expressions.

a. 6^3 **b.** $\left(\frac{1}{3}\right)^4$

We now include exponents in our order of operations:

1. Parentheses
2. Exponents
3. Multiplication/Division
4. Addition/Subtraction

EXAMPLE 2   Evaluating Expressions Involving Natural-Number Exponents

Evaluate the expressions.

a. $(-3)^4$ **b.** -3^4 **c.** $(-2)^3 \cdot 5^2$

Solution:

a. $(-3)^4 = (-3)(-3)(-3)(-3) = \boxed{81}$

b. $-3^4 = -\underbrace{(3 \cdot 3 \cdot 3 \cdot 3)}_{81} = \boxed{-81}$

c. $(-2)^3 \cdot 5^2 = \underbrace{(-2)(-2)(-2)}_{-8} \cdot \underbrace{5 \cdot 5}_{25} = \boxed{-200}$

> **Classroom Example 0.2.2**
> Evaluate the expressions.
> **a.** $-2^4 \cdot 3^2$
> **b.*** $-(-2^3 \cdot 5)\left(-\frac{1}{2} \cdot \frac{1}{5}\right)^2$
>
> **Answer:**
> **a.** -144 **b.** $\frac{2}{5}$

■ **YOUR TURN** Evaluate the expression $-4^3 \cdot 2^3$.

■ **Answer:** -512

So far, we have discussed only exponents that are natural numbers (positive integers). When the exponent is a negative integer, we use the following property.

Study Tip

A negative exponent implies a reciprocal.

NEGATIVE-INTEGER EXPONENT PROPERTY

Let a be any nonzero real number and n be a natural number (positive integer); then

$$a^{-n} = \frac{1}{a^n} \qquad a \neq 0$$

In other words, a base raised to a negative-integer exponent is equivalent to the reciprocal of the base raised to the opposite (positive) integer exponent.

EXAMPLE 3   Evaluating Expressions Involving Negative-Integer Exponents

Evaluate the expressions.

a. 2^{-4} **b.** $\dfrac{1}{3^{-3}}$ **c.** $4^{-3} \cdot \dfrac{1}{2^{-4}}$ **d.** $-2^3 \cdot \dfrac{1}{(-6)^{-2}}$

Solution:

a. $2^{-4} = \dfrac{1}{2^4} = \boxed{\dfrac{1}{16}}$

b. $\dfrac{1}{3^{-3}} = \dfrac{1}{\left(\dfrac{1}{3^3}\right)} = 1 \div \left(\dfrac{1}{3^3}\right) = 1 \cdot \dfrac{3^3}{1} = 3^3 = \boxed{27}$

c. $4^{-3} \cdot \dfrac{1}{2^{-4}} = \dfrac{1}{4^3} \cdot 2^4 = \dfrac{16}{64} = \boxed{\dfrac{1}{4}}$

d. $-2^3 \cdot \dfrac{1}{(-6)^{-2}} = \underbrace{-2^3}_{-8} \cdot \underbrace{(-6)^2}_{36} = (-8)(36) = \boxed{-288}$

> **Classroom Example 0.2.3***
> Evaluate the expressions.
> **a.** $-(-3)^2(-1)^5\, 6^{-2}$
> **b.** $\dfrac{(-1)^{-3}\, 2^{-3}}{(-3)^{-2}}$
>
> **Answer:**
> **a.** $\frac{1}{4}$ **b.** $-\frac{9}{8}$

Technology Tip

■ **YOUR TURN** Evaluate the expressions.

a. $-\dfrac{1}{5^{-2}}$ **b.** $\dfrac{1}{3^{-2}} \cdot 6^{-2}$

■ **Answer: a.** -25 **b.** $\frac{1}{4}$

Now we can evaluate expressions involving positive and negative exponents. How do we evaluate an expression with a zero exponent? We define any nonzero real number raised to the zero power as 1.

ZERO-EXPONENT PROPERTY

Let a be any nonzero real number; then

$$a^0 = 1 \qquad a. \neq 0$$

Classroom Example 0.2.4
Evaluate the expressions.
a. $-\pi^0 - (2)^0$
b.* $-(-1 - 2^0)(3 - 5)^{-1}$
c.* $\dfrac{-4^{-2}(-3)^0}{2^{-2}}$

Answer:
a. -2 **b.** -1 **c.** $-\frac{1}{4}$

EXAMPLE 4   Evaluating Expressions Involving Zero Exponents

Evaluate the expressions.

a. 5^0 **b.** $\dfrac{1}{2^0}$ **c.** $(-3)^0$ **d.** -4^0

Solution:

a. $5^0 = \boxed{1}$ **b.** $\dfrac{1}{2^0} = \dfrac{1}{1} = \boxed{1}$ **c.** $(-3)^0 = \boxed{1}$ **d.** $-\dfrac{4^0}{1} = \boxed{-1}$

We now can evaluate expressions involving integer (positive, negative, or zero) exponents. What about when expressions involving integer exponents are multiplied, divided, or raised to a power?

WORDS	**MATH**
When expressions with the *same base* are *multiplied*, the exponents are *added*.	$2^3 \cdot 2^4 = \underbrace{2 \cdot 2 \cdot 2}_{2^3} \cdot \underbrace{2 \cdot 2 \cdot 2 \cdot 2}_{2^4} = 2^{3+4} = 2^7$
When expressions with the *same base* are *divided*, the exponents are *subtracted*.	$\dfrac{2^5}{2^3} = \dfrac{2 \cdot 2 \cdot 2 \cdot 2 \cdot 2}{2 \cdot 2 \cdot 2} = \dfrac{2 \cdot 2}{1} = 2^2$ or $2^{5-3} = 2^2$
When an expression involving an exponent is *raised to a power*, the exponents are *multiplied*.	$(2^3)^2 = (8)^2 = 64$ or $(2^3)^2 = 2^{3 \cdot 2} = 2^6 = 64$

The following table summarizes the *properties of integer exponents*.

PROPERTIES OF INTEGER EXPONENTS

NAME	DESCRIPTION	MATH (LET a AND b BE NONZERO REAL NUMBERS AND m AND n BE INTEGERS)	EXAMPLE
Product property	When multiplying exponentials with the same base, **add** exponents.	$a^m \cdot a^n = a^{m+n}$	$x^2 \cdot x^5 = x^{2+5} = x^7$
Quotient property	When dividing exponentials with the same base, **subtract** the exponents (numerator − denominator).	$\dfrac{a^m}{a^n} = a^{m-n}$	$\dfrac{x^5}{x^3} = x^{5-3} = x^2 \qquad x \neq 0$
Power property	When raising an exponential to a power, **multiply** exponents.	$(a^m)^n = a^{mn}$	$(x^2)^4 = x^{2 \cdot 4} = x^8$
Product to a power property	A product raised to a power is equal to the product of each factor raised to the power.	$(ab)^n = a^n b^n$	$(2x)^3 = 2^3 \cdot x^3 = 8x^3$
Quotient to a power property	A quotient raised to a power is equal to the quotient of the factors raised to the power.	$\left(\dfrac{a}{b}\right)^n = \dfrac{a^n}{b^n}$	$\left(\dfrac{x}{y}\right)^4 = \dfrac{x^4}{y^4} \qquad y \neq 0$

Common Errors Made Using Properties of Exponents

INCORRECT	CORRECT	ERROR
$x^4 \cdot x^3 = x^{12}$	$x^4 \cdot x^3 = x^7$	Exponents should be added (not multiplied).
$\dfrac{x^{18}}{x^6} = x^3$	$\dfrac{x^{18}}{x^6} = x^{12}; x \neq 0$	Exponents should be subtracted (not divided).
$(x^2)^3 = x^8$	$(x^2)^3 = x^6$	Exponents should be multiplied (not raised to a power).
$(2x)^3 = 2x^3$	$(2x)^3 = 8x^3$	Both factors (the 2 and the x) should be cubed.
$2^3 \cdot 2^4 = 4^7$	$2^3 \cdot 2^4 = 2^7$	The original common base should be retained.
$2^3 \cdot 3^5 = 6^8$	$2^3 \cdot 3^5$	The properties of integer exponents require the *same* base.

We will now use properties of integer exponents to *simplify exponential expressions.*

An exponential expression is **simplified** when:

- All parentheses (groupings) have been eliminated.
- A base appears only once.
- No powers are raised to other powers.
- All exponents are positive.

EXAMPLE 5   Simplifying Exponential Expressions

Simplify the expressions (assume all variables are nonzero).

a. $(-2x^2y^3)(5x^3y)$ **b.** $(2x^2yz^3)^3$ **c.** $\dfrac{25x^3y^6}{-5x^5y^4}$

Solution (a):

Parentheses imply multiplication.

Group the same bases together.

$(-2x^2y^3)(5x^3y) = (-2)(5)x^2x^3y^3y$

Apply the product property.

$= (-2)(5)\underbrace{x^2x^3}_{x^{2+3}}\underbrace{y^3y}_{y^{3+1}}$

Multiply the constants.

$= \boxed{-10x^5y^4}$

Solution (b):

Apply the product to a power property.

$(2x^2yz^3)^3 = (2)^3(x^2)^3(y)^3(z^3)^3$

Apply the power property.

$= 8x^{2\cdot3}y^{1\cdot3}z^{3\cdot3}$

Simplify.

$= \boxed{8x^6y^3z^9}$

Solution (c):

Group the same bases together.

$\dfrac{25x^3y^6}{-5x^5y^4} = \left(\dfrac{25}{-5}\right)\left(\dfrac{x^3}{x^5}\right)\left(\dfrac{y^6}{y^4}\right)$

Apply the quotient property.

$= (-5)x^{3-5}y^{6-4}$

$= -5x^{-2}y^2$

Apply the negative exponent property.

$= \boxed{\dfrac{-5y^2}{x^2}}$

■ **YOUR TURN** Simplify the expressions (assume all variables are nonzero).

a. $(-3x^3y^2)(4xy^3)$ **b.** $(-3xy^3z^2)^3$ **c.** $\dfrac{-16x^4y^3}{4xy^7}$

Classroom Example 0.2.5*

Simplify the expressions.

a. $(-x^2y)(2xy^2z)^3(yz^3)^0$

b. $\dfrac{(-xy^3z^2w^2)^3}{-x(yz)^2w^3}$

c. $\dfrac{(5x)(2y)^2}{-(-2y)^2x^2}$

Answer:

a. $-8x^5y^7z^3$

b. $x^2y^7z^4w^3$

c. $-\dfrac{5}{x}$

Study Tip

It is customary not to leave negative exponents. Instead we use the negative exponent property to write exponential expressions with only positive exponents.

■ **Answer:**

a. $-12x^4y^5$ **b.** $-27x^3y^9z^6$ **c.** $\dfrac{-4x^3}{y^4}$

EXAMPLE 6 Simplifying Exponential Expressions

Write each expression so that all exponents are positive (assume all variables are nonzero).

a. $(3x^2z^{-4})^{-3}$ **b.** $\dfrac{(x^2y^{-3})^2}{(x^{-1}y^4)^{-3}}$ **c.** $\dfrac{(-2xy^2)^3}{-(6xz^3)^2}$

Solution (a):

Apply the product to a power property.	$(3x^2z^{-4})^{-3} = (3)^{-3}(x^2)^{-3}(z^{-4})^{-3}$
Apply the power property.	$= 3^{-3}x^{-6}z^{12}$
Apply the negative-integer exponent property.	$= \dfrac{z^{12}}{3^3x^6}$
Evaluate 3^3.	$= \boxed{\dfrac{z^{12}}{27x^6}}$

Solution (b):

Apply the product to a power property.	$\dfrac{(x^2y^{-3})^2}{(x^{-1}y^4)^{-3}} = \dfrac{x^4y^{-6}}{x^3y^{-12}}$
Apply the quotient property.	$= x^{4-3}y^{-6-(-12)}$
Simplify.	$= \boxed{xy^6}$

Solution (c):

Apply the product to a power property on both the numerator and denominator.	$\dfrac{(-2xy^2)^3}{-(6xz^3)^2} = \dfrac{(-2)^3(x)^3(y^2)^3}{-(6)^2(x)^2(z^3)^2}$
Apply the power property.	$= \dfrac{-8x^3y^6}{-36x^2z^6}$
Group constant terms and x terms.	$= \left(\dfrac{-8}{-36}\right)\left(\dfrac{x^3}{x^2}\right)\left(\dfrac{y^6}{z^6}\right)$
Apply the quotient property.	$= \left(\dfrac{8}{36}\right)(x^{3-2})\left(\dfrac{y^6}{z^6}\right)$
Simplify.	$= \boxed{\dfrac{2xy^6}{9z^6}}$

■ **YOUR TURN** Simplify the exponential expression and express it in terms of positive exponents

$$\dfrac{(tv^2)^{-3}}{(2t^4v^3)^{-1}}.$$

Scientific Notation

You are already familiar with base 10 raised to positive-integer powers. However, it can be inconvenient to write all the zeros out, so we give certain powers of 10 particular names: thousand, million, billion, trillion, and so on. For example, we say there are 300 million U.S. citizens as opposed to writing out 300,000,000 citizens. Or we say that the national debt is $14 trillion as opposed to writing out $14,000,000,000,000. The following table contains scientific notation for positive exponents and examples of some common prefixes and abbreviations. One of the fundamental applications of scientific notation is *measurement*.

EXPONENTIAL FORM	REAL NUMBER	NUMBER OF ZEROS FOLLOWING THE 1	PREFIX	ABBREVIATION	EXAMPLE
10^1	10	1			
10^2	100	2			
10^3	1000 (one thousand)	3	kilo-	k	The relay-for-life team ran a total of 80 km (kilometers).
10^4	10,000	4			
10^5	100,000	5			
10^6	1,000,000 (one million)	6	mega-	M	Modern high-powered diesel–electric railroad locomotives typically have a peak power output of 3 to 5 MW (megawatts).
10^7	10,000,000	7			
10^8	100,000,000	8			
10^9	1,000,000,000 (one billion)	9	giga-	G	A flash drive typically has 1 to 4 GB (gigabytes) of storage.
10^{10}	10,000,000,000	10			
10^{11}	100,000,000,000	11			
10^{12}	1,000,000,000,000 (one trillion)	12	tera-	T	Laser systems offer higher frequencies on the order of THz (terahertz).

Notice that 10^8 is a 1 followed by 8 zeros; alternatively, you can start with 1.0 and move the decimal point 8 places to the right (insert zeros). The same type of table can be made for negative-integer powers with base 10. To find the real number associated with exponential form, start with 1.0 and move the decimal a certain number of places to the left (fill in missing decimal places with zeros).

EXPONENTIAL FORM	REAL NUMBER	NUMBER OF PLACES DECIMAL (1.0) MOVES TO THE LEFT	PREFIX	ABBREVIATION	EXAMPLE
10^{-1}	0.1	1			
10^{-2}	0.01	2			
10^{-3}	0.001 (one thousandth)	3	milli-	m	Excedrin Extra Strength tablets each have 250 mg (milligrams) of acetaminophen.
10^{-4}	0.0001	4			
10^{-5}	0.00001	5			
10^{-6}	0.000001 (one millionth)	6	micro-	μ	A typical laser has a wavelength of 1.55 μm (micrometers*).
10^{-7}	0.0000001	7			
10^{-8}	0.00000001	8			
10^{-9}	0.000000001 (one billionth)	9	nano-	n	PSA levels less than 4 ng/ml (nanogram per milliliter of blood) represent low risk for prostate cancer.
10^{-10}	0.0000000001	10			
10^{-11}	0.00000000001	11			
10^{-12}	0.000000000001 (one trillionth)	12	pico-	p	A single yeast cell weighs 44 pg (picograms).

*In optics a micrometer is called a micron.

Classroom Example 0.2.7
Express in scientific notation.
a. 0.0000109   **b.** 4,000,000,000
c. 0.0010005   **d.** $0.06\overline{87}$

Answer:
a. 1.09×10^{-5} **b.** 4.0×10^{9}
c. 1.0005×10^{-3} **d.** $6.\overline{87} \times 10^{-2}$

SCIENTIFIC NOTATION

A positive real number can be written in **scientific notation** with the form $c \times 10^{n}$, where $1 \le c < 10$ and n is an integer.

Note that c is a real number between 1 and 10. Therefore, 22.5×10^{3} is not in scientific notation, but we can convert it to scientific notation: 2.25×10^{4}.

For example, there are approximately 50 trillion cells in the human body. We write 50 trillion as 50 followed by 12 zeros 50,000,000,000,000. An efficient way of writing such a large number is using **scientific notation**. Notice that 50,000,000,000,000 is **5** followed by **13** zeros, or in scientific notation, $\mathbf{5 \times 10^{13}}$. Very small numbers can also be written using scientific notation. For example, in laser communications a pulse width is 2 femtoseconds, or 0.000000000000002 second. Notice that if we start with **2.0** and move the decimal point **15** places to the left (adding zeros in between), the result is 0.000000000000002, or in scientific notation, $\mathbf{2 \times 10^{-15}}$.

EXAMPLE 7   Expressing a Positive Real Number in Scientific Notation

Express the numbers in scientific notation.

a. 3,856,000,000,000,000   **b.** 0.00000275

Solution:

a. Rewrite the number with the implied decimal point.   3,856,000,000,000,000.

Move the decimal point to the left 15 places. $= \boxed{3.856 \times 10^{15}}$

b. Move the decimal point to the right 6 places. $0.00000275 = \boxed{2.75 \times 10^{-6}}$

■ **YOUR TURN** Express the numbers in scientific notation.

a. 4,520,000,000   **b.** 0.00000043

EXAMPLE 8   Converting from Scientific Notation to Decimals

Write each number as a decimal.

a. 2.869×10^{5} **b.** 1.03×10^{-3}

Solution:

a. Move the decimal point 5 places to the right (add zeros in between).

b. Move the decimal point 3 places to the left (add zeros in between).

Classroom Example 0.2.8
Write as a decimal.
a. 1.123×10^{6} **b.** -2.0001×10^{-3}

Answer:
a. 1,123,000   **b.** $-.0020001$

286,900. or $\boxed{286,900}$

$\boxed{0.00103}$

■ **YOUR TURN** Write each number as a decimal.

a. 8.1×10^{4} **b.** 3.7×10^{-8}

SECTION
0.2 SUMMARY

In this section we discussed properties of exponents.

Integer Exponents

The following table summarizes integer exponents. Let a be any real number and n be a natural number.

NAME	DESCRIPTION	MATH
Natural-number exponent	Multiply n factors of a.	$a^n = \underbrace{a \cdot a \cdot a \cdots a}_{n \text{ factors}}$
Negative-integer exponent property	A negative exponent implies a reciprocal.	$a^{-n} = \dfrac{1}{a^n} \qquad a \neq 0$
Zero-exponent property	Any nonzero real number raised to the zero power is equal to one.	$a^0 = 1 \qquad a \neq 0$

Properties of Integer Exponents

The following table summarizes properties of integer exponents. Let a and b be nonzero real numbers and m and n be integers.

NAME	DESCRIPTION	MATH
Product property	When multiplying exponentials with the same base, **add** exponents.	$a^m \cdot a^n = a^{m+n}$
Quotient property	When dividing exponentials with the same base, **subtract** the exponents (numerator − denominator).	$\dfrac{a^m}{a^n} = a^{m-n}$
Power property	When raising an exponential to a power, **multiply** exponents.	$(a^m)^n = a^{mn}$
Product to a power property	A product raised to a power is equal to the product of each factor raised to the power.	$(ab)^n = a^n b^n$
Quotient to a power property	A quotient raised to a power is equal to the quotient of the factors raised to the power.	$\left(\dfrac{a}{b}\right)^n = \dfrac{a^n}{b^n}$

Scientific Notation

Scientific notation is a convenient way of using exponents to represent either very small or very large numbers. Real numbers greater than 1 correspond to positive exponents in scientific notation, whereas real numbers greater than 0 but less than 1 correspond to negative exponents in scientific notation. Scientific notation offers the convenience of multiplying and dividing real numbers by applying properties of exponents.

REAL NUMBER (DECIMAL FORM)	PROCESS	SCIENTIFIC NOTATION
2,357,000,000	Move the implied decimal point to the *left* 9 places	2.357×10^9
0.00000465	Move the decimal point to the *right* 6 places	4.65×10^{-6}

SECTION
0.2 EXERCISES

SKILLS

In Exercises 1–20, evaluate each expression.

1. 4^4
2. 5^3
3. $(-3)^5$
4. $(-4)^2$
5. -5^2
6. -7^2
7. $-2^2 \cdot 4$
8. $-3^2 \cdot 5$
9. 9^0
10. $-8x^0$
11. 10^{-1}
12. a^{-1}
13. 8^{-2}
14. 3^{-4}
15. $-6 \cdot 5^2$
16. $-2 \cdot 4^2$
17. $8 \cdot 2^{-3} \cdot 5$
18. $5 \cdot 2^{-4} \cdot 32$
19. $-6 \cdot 3^{-2} \cdot 81$
20. $6 \cdot 4^2 \cdot 4^{-4}$

In Exercises 21–50, simplify and write the resulting expression with only positive exponents.

21. $x^2 \cdot x^3$

22. $y^3 \cdot y^5$

23. $x^2 x^{-3}$

24. $y^3 \cdot y^{-7}$

25. $(x^2)^3$

26. $(y^3)^2$

27. $(4a)^3$

28. $(4x^2)^3$

29. $(-2t)^3$

30. $(-3b)^4$

31. $(5xy^2)^2(3x^3y)$

32. $(4x^2y)(2xy^3)^2$

33. $\dfrac{x^5 y^3}{x^7 y}$

34. $\dfrac{y^5 x^2}{y^{-2} x^{-5}}$

35. $\dfrac{(2xy)^2}{(-2xy)^3}$

36. $\dfrac{(-3x^3y)}{-4(x^2y^3)^3}$

37. $\left(\dfrac{b}{2}\right)^{-4}$

38. $\left(\dfrac{c}{3}\right)^{-2}$

39. $(9a^{-2}b^3)^{-2}$

40. $(-9x^{-3}y^2)^{-4}$

41. $\dfrac{a^{-2}b^3}{a^4b^5}$

42. $\dfrac{x^{-3}y^2}{y^{-4}x^5}$

43. $\dfrac{(x^3y^{-1})^2}{(xy^2)^{-2}}$

44. $\dfrac{(x^3y^{-2})^2}{(x^4y^3)^{-3}}$

45. $\dfrac{3(x^2y)^3}{12(x^{-2}y)^4}$

46. $\dfrac{(-4x^{-2})^2 y^3 z}{(2x^3)^{-2}(y^{-1}z)^4}$

47. $\dfrac{(x^{-4}y^5)^{-2}}{[-2(x^3)^2 y^{-4}]^5}$

48. $-2x^2(-2x^3)^5$

49. $\left[\dfrac{a^2(-xy^4)^3}{x^4(-a^3y^2)^2}\right]^3$

50. $\left[\dfrac{b^{-3}(-x^3y^2)^4}{y^2(-b^2x^5)^3}\right]^5$

51. Write $2^8 \cdot 16^3 \cdot (64)$ as a power of 2 : $2^?$

52. Write $3^9 \cdot 81^5 \cdot (9)$ as a power of 3 : $3^?$

In Exercises 53–60, express the given number in scientific notation.

53. 27,600,000

54. 144,000,000,000

55. 93,000,000

56. 1,234,500,000

57. 0.0000000567

58. 0.00000828

59. 0.000000123

60. 0.000000005

In Exercises 61–66, write the number as a decimal.

61. 4.7×10^7

62. 3.9×10^5

63. 2.3×10^4

64. 7.8×10^{-3}

65. 4.1×10^{-5}

66. 9.2×10^{-8}

▪ APPLICATIONS

In Exercises 67 and 68, refer to the following:

It is estimated that there are currently 5.0×10^9 cell phones being used worldwide. Assume that the average cell phone measures 5 inches in length and there are 5280 feet in a mile.

67. Cell Phones Spanning the Earth.

   a. If all of the cell phones currently in use were to be lined up next to each other tip to tip, how many *feet* would the line of cell phones span? Write the answer in scientific notation.

   b. The circumference of the Earth (measured at the equator) is approximately 25,000 miles. If the cell phones in part (a) were to be wrapped around the Earth at the equator, would they circle the Earth completely? If so, approximately how many times?

68. Cell Phones Reaching the Moon.

   a. If all of the cell phones currently in use were to be lined up next to each other tip to tip, how many *miles* would the line of cell phones span? Write the answer in scientific notation.

   b. The Moon traces an elliptical path around the Earth, with the average distance between them being approximately 239,000 miles. Would the line of cell phones in part (a) reach the Moon?

69. Astronomy. The distance from Earth to Mars on a particular day can be 200 million miles. Express this distance in scientific notation.

70. Astronomy. The distance from Mars to the Sun on a particular day can be 142 million miles. Express this distance in scientific notation.

71. Lasers. The wavelength of a typical laser used for communication systems is 1.55 microns (or 1.55×10^{-6} meters). Express the wavelength in decimal representation in terms of meters.

72. Lasers. A ruby-red laser has a wavelength of 694 nanometers (or 6.93×10^{-7} meters). Express the wavelength in decimal representation in terms of meters.

▪ CATCH THE MISTAKE

In Exercises 73–76, explain the mistake that is made.

73. Simplify $(-2y^3)(3x^2y^2)$.

| Group like factors together. | $(-2)(3)x^2y^3y^2$ |
| Use the product property. | $-6x^2y^6$ |

This is incorrect. What mistake was made?

74. Simplify $(2xy^2)^3$.

| Eliminate the parentheses. | $(2xy^2)^3 = 2x^3y^6$ |

This is incorrect. What mistake was made?

75. Simplify $(-2xy^3)^2(5x^2y)^2$.

Apply the product to a power property.	$= (-2)^2x^2(y^3)^2(5)^2(x^2)^2y^2$
Apply the power rule.	$= 4x^2y^9 25x^4y^2$
Group like factors.	$= (4)(25)x^2x^4y^9y^2$
Apply the product property.	$= 100x^6y^{11}$

This is incorrect. What mistake was made?

76. Simplify $\dfrac{-4x^{16}y^9}{8x^2y^3}$.

| Group like factors. | $= \left(\dfrac{-4}{8}\right)\left(\dfrac{x^{16}}{x^2}\right)\left(\dfrac{y^9}{y^3}\right)$ |
| Use the quotient property. | $= -\dfrac{1}{2}x^8y^3$ |

This is incorrect. What mistake was made?

▪ CONCEPTUAL

In Exercises 77–80, determine whether each of the following statements is true or false.

77. $-2^n = (-2)^n$, if n is an integer.

78. Any nonzero real number raised to the zero power is one.

79. $\dfrac{x^{n+1}}{x^n} = x$ for $x =$ any real number.

80. $x^{-1} + x^{-2} = x^{-3}$

81. Simplify $\left((a^m)^n\right)^k$.

82. Simplify $\left((a^{-m})^{-n}\right)^{-k}$.

In Exercises 83–86, evaluate the expression for the given value.

83. $-a^2 + 2ab$ for $a = -2$, $b = 3$

84. $2a^3 - 7a^2$ for $a = 4$

85. $-16t^2 + 100t$ for $t = 3$

86. $\dfrac{a^3 - 27}{a - 4}$ for $a = -2$

▪ CHALLENGE

87. The Earth's population is approximately 6.6×10^9 people, and there are approximately 1.5×10^8 square kilometers of land on the surface of the Earth. If one square kilometer is approximately 247 acres, how many acres per person are there on Earth? Round to the nearest tenth of an acre.

89. Evaluate: $\dfrac{(4 \times 10^{-23})(3 \times 10^{12})}{(6 \times 10^{-10})}$. Express your answer in both scientific and decimal notation.

88. The population of the United States is approximately 3.0×10^8 people, and there are approximately 3.79×10^6 square miles of land in the United States. If one square mile is approximately 640 acres, how many acres per person are there in the United States? Round to the nearest tenth of an acre.

90. Evaluate: $\dfrac{(2 \times 10^{-17})(5 \times 10^{13})}{(1 \times 10^{-6})}$. Express your answer in both scientific and decimal notation.

▪ TECHNOLOGY

Scientific calculators have an EXP button that is used for scientific notation. For example, 2.5×10^3 can be input into the calculator by pressing 2.5 EXP 3.

91. Repeat Exercise 87 and confirm your answer with a calculator.

92. Repeat Exercise 88 and confirm your answer with a calculator.

In Exercises 93 and 94, use a graphing utility or scientific calculator to evaluate the expression. Express your answer in scientific notation.

93. $\dfrac{(7.35 \times 10^{-26})(2.19 \times 10^{19})}{(3.15 \times 10^{-21})}$

94. $\dfrac{(1.6849 \times 10^{32})}{(8.12 \times 10^{16})(3.32 \times 10^{-9})}$

SKILLS OBJECTIVES

- Add and subtract polynomials.
- Multiply polynomials.
- Recognize special products.

CONCEPTUAL OBJECTIVES

- Recognize like terms.
- Learn formulas for special products.

Adding and Subtracting Polynomials

Polynomials in Standard Form

The expressions

$$3x^2 - 7x - 1 \qquad 4y^3 - y \qquad 5z$$

are all examples of *polynomials* in one variable. A monomial in one variable, ax^k, is the product of a constant and a variable raised to a nonnegative-integer power. The constant a is called the **coefficient** of the monomial, and k is called the **degree** of the monomial. A **polynomial** is the sum of monomials. The monomials that are part of a polynomial are called **terms**.

> **DEFINITION**   **Polynomial**
>
> A **polynomial in x** is an algebraic expression of the form
>
> $$a_n x^n + a_{n-1} x^{n-1} + a_{n-2} x^{n-2} + \cdots + a_2 x^2 + a_1 x + a_0$$
>
> where $a_0, a_1, a_2, \ldots, a_n$ are real numbers, with $a_n \neq 0$, and n is a nonnegative integer. The polynomial is of **degree** n, a_n is the **leading coefficient**, and a_0 is the **constant term**.

Polynomials with one, two, and three terms are called **monomials**, **binomials**, and **trinomials**, respectively. Polynomials are typically written in **standard form** in order of decreasing degrees, and the **degree** of the polynomial is determined by the highest degree (exponent) of any single term.

POLYNOMIAL	STANDARD FORM	SPECIAL NAME	DEGREE	DESCRIPTION
$4x^3 - 5x^7 + 2x - 6$	$-5x^7 + 4x^3 + 2x - 6$	Polynomial	7	A *seventh*-degree polynomial in x
$5 + 2y^3 - 4y$	$2y^3 - 4y + 5$	Trinomial	3	A *third*-degree polynomial in y
$7z^2 + 2$	$7z^2 + 2$	Binomial	2	A *second*-degree polynomial in z
$-17x^5$	$-17x^5$	Monomial	5	A *fifth*-degree monomial in x

EXAMPLE 1 **Writing Polynomials in Standard Form**

Write the polynomials in standard form and state their degree, leading coefficient, and constant term.

a. $4x - 9x^5 + 2$ **b.** $3 - x^2$
c. $3x^2 - 8 + 14x^3 - 20x^8 + x$ **d.** $-7x^3 + 25x$

Solution:

	Standard Form	Degree	Leading Coefficient	Constant Term
a.	$-9x^5 + 4x + 2$	5	-9	2
b.	$-x^2 + 3$	2	-1	3
c.	$-20x^8 + 14x^3 + 3x^2 + x - 8$	8	-20	-8
d.	$-7x^3 + 25x$	3	-7	0

■ **YOUR TURN** Write the polynomial in standard form and state its degree, leading coefficient, and constant term.

$$17x^2 - 4x^3 + 5 - x$$

Classroom Example 0.3.1
Write in standard form and state the degree.
a. $\sqrt{2}x - 2x^3 + 4$
b. $-1 - 3x + \pi x^7 - x^4$
c.* $-(2x^2)^3 - x^8 + 3$

Answer:
a. $-2x^3 + \sqrt{2}x + 4$ degree 3
b. $\pi x^7 - x^4 - 3x - 1$ degree 7
c. $-x^8 - 8x^6 + 3$ degree 8

■ **Answer:** $-4x^3 + 17x^2 - x + 5$
Degree: 3
Leading coefficient: -4
Constant term: 5

Adding and Subtracting Polynomials

Polynomials are added and subtracted by combining *like terms*. **Like terms** are terms having the same variables and exponents. Like terms can be combined by adding their coefficients.

WORDS	MATH
Identify like terms.	$\underline{3x^2} + 2x + \underline{4x^2} + 5$
Add coefficients of like terms.	$7x^2 + 2x + 5$

Note: The $2x$ and 5 could not be combined because they are *not* like terms.

EXAMPLE 2 **Adding Polynomials**

Find the sum and simplify $(5x^2 - 2x + 3) + (3x^3 - 4x^2 + 7)$.

Solution:

Eliminate parentheses.	$5x^2 - 2x + 3 + 3x^3 - 4x^2 + 7$
Identify like terms.	$\underline{5x^2} - 2x + \underset{=}{3} + 3x^3 - \underline{4x^2} + \underset{=}{7}$
Combine like terms.	$x^2 - 2x + 10 + 3x^3$
Write in standard form.	$\boxed{3x^3 + x^2 - 2x + 10}$

■ **YOUR TURN** Find the sum and simplify:

$$(3x^2 + 5x - 2x^5) + (6x^3 - x^2 + 11)$$

Classroom Example 0.3.2
Compute:
a. $(3x - x^2 - 1) + (2 - 2x + 3x^2)$
b. $(5x^3 + 2) + (3x^2 - 2)$

Answer:
a. $2x^2 + x + 1$ **b.** $5x^3 + 3x^2$

■ **Answer:**
$-2x^5 + 6x^3 + 2x^2 + 5x + 11$

Study Tip

When subtracting polynomials, it is important to distribute the negative through *all* of the terms in the *second* polynomial.

■ **Answer:** $x^3 - 7x^2 - 4x + 3$

EXAMPLE 3   Subtracting Polynomials

Find the difference and simplify $(3x^3 - 2x + 1) - (x^2 + 5x - 9)$.

COMMON MISTAKE

Distributing the negative to only the first term in the second polynomial.

★ **CORRECT**

Eliminate the parentheses.

$3x^3 - 2x + 1 - x^2 - 5x + 9$

Identify like terms.

$3x^3 - \underline{2x} + \underline{\underline{1}} - x^2 - \underline{5x} + \underline{\underline{9}}$

Combine like terms.

$\boxed{3x^3 - x^2 - 7x + 10}$

✕ **INCORRECT**

ERROR:

$3x^3 - 2x + 1 - x^2 + 5x - 9$

Don't forget to distribute the negative through the entire second polynomial.

■ **YOUR TURN** Find the difference and simplify:

$$(-7x^2 - x + 5) - (2 - x^3 + 3x)$$

Multiplying Polynomials

The product of two monomials is found by using the properties of exponents (Section 0.2). For example,

$$(-5x^3)(9x^2) = (-5)(9)x^{3+2} = -45x^5$$

To multiply a monomial and a polynomial we use the distributive property (Section 0.1).

EXAMPLE 4   Multiplying a Monomial and a Polynomial

Find the product and simplify $5x^2(3x^5 - x^3 + 7x - 4)$.

Solution:

Use the distributive property.

$5x^2(3x^5 - x^3 + 7x - 4)$

$= 5x^2(3x^5) - 5x^2(x^3) + 5x^2(7x) - 5x^2(4)$

Multiply each set of monomials.

$= \boxed{15x^7 - 5x^5 + 35x^3 - 20x^2}$

■ **Answer:** $12x^4 - 6x^3 + 3x^2$

■ **YOUR TURN** Find the product and simplify $3x^2(4x^2 - 2x + 1)$.

How do we multiply two polynomials if neither one is a monomial? For example, how do we find the product of a binomial and a trinomial such as $(2x - 5)(x^2 - 2x + 3)$? Notice that the binomial is a combination of two monomials. Therefore, we treat each monomial, $2x$ and -5, separately and then combine our results. In other words, use the distributive property repeatedly.

WORDS	**MATH**
Apply the distributive property.	$(2x - 5)(x^2 - 2x + 3) = 2x(x^2 - 2x + 3) - 5(x^2 - 2x + 3)$
Apply the distributive property.	$= (2x)(x^2) + (2x)(-2x) + (2x)(3) - 5(x^2) - 5(-2x) - 5(3)$
Multiply the monomials.	$= 2x^3 - 4x^2 + 6x - 5x^2 + 10x - 15$
Combine like terms.	$= 2x^3 - 9x^2 + 16x - 15$

EXAMPLE 5 **Multiplying Two Polynomials**

Multiply and simplify $(2x^2 - 3x + 1)(x^2 - 5x + 7)$.

Solution:

Multiply each term of the
first trinomial by the
entire second trinomial. $= 2x^2(x^2 - 5x + 7) - 3x(x^2 - 5x + 7) + 1(x^2 - 5x + 7)$

Identify like terms. $= 2x^4 - \underline{10x^3} + \underline{14x^2} - 3x^3 + \underline{15x^2} - \underline{21x} + x^2 - \underline{5x} + 7$

Combine like terms. $= \boxed{2x^4 - 13x^3 + 30x^2 - 26x + 7}$

■ **YOUR TURN** Multiply and simplify $(-x^3 + 2x - 4)(3x^2 - x + 5)$.

Special Products

The method outlined for multiplying polynomials works for *all* products of polynomials. For
the special case when both polynomials are binomials, the **FOIL method** can also be used.

WORDS	**MATH**
Apply the distributive property.	$(5x - 1)(2x + 3) = 5x(2x + 3) - 1(2x + 3)$
Apply the distributive property.	$= 5x(2x) + 5x(3) - 1(2x) - 1(3)$
Multiply each set of monomials.	$= 10x^2 + 15x - 2x - 3$
Combine like terms.	$= 10x^2 + 13x - 3$

The FOIL method finds the
products of the **F**irst terms,
Outer terms, **I**nner terms, and
Last terms.

$(5x - 1)(2x + 3) = 10x^2 + 15x - 2x - 3$

 EXAMPLE 6 **Multiplying Binomials Using the FOIL Method**

Multiply $(3x + 1)(2x - 5)$ using the FOIL method.

Solution:

Multiply the **first** terms. $(3x)(2x) = 6x^2$

Multiply the **outer** terms. $(3x)(-5) = -15x$

Multiply the **inner** terms. $(1)(2x) = 2x$

Multiply the **last** terms. $(1)(-5) = -5$

Add the first, outer, inner, and last terms,
and identify the like terms. $(3x + 1)(2x - 5) = 6x^2 - \underline{15x} + \underline{2x} - 5$

Combine like terms. $= \boxed{6x^2 - 13x - 5}$

■ **YOUR TURN** Multiply $(2x - 3)(5x - 2)$.

Some products of binomials occur frequently in algebra and are given special names. Example 7 illustrates the *difference of two squares* and *perfect squares*.

EXAMPLE 7   Multiplying Binomials Resulting in Special Products

Find the following: **a.** $(x - 5)(x + 5)$ **b.** $(x + 5)^2$ **c.** $(x - 5)^2$

Solution:

a. $(x - 5)(x + 5) = \overset{\text{First}}{x^2} + \underset{\text{Outer}}{5x} - \overset{\text{Inner}}{5x} - \underset{\text{Last}}{5^2} = \overset{\text{Difference of two squares}}{x^2 - 5^2} = \boxed{x^2 - 25}$

b. $(x + 5)^2 = (x + 5)(x + 5) = \overset{\text{First}}{x^2} + \underset{\text{Outer}}{5x} + \overset{\text{Inner}}{5x} + \underset{\text{Last}}{5^2} = x^2 + 2(5x) + 5^2 = \boxed{x^2 + 10x + 25}$

c. $(x - 5)^2 = (x - 5)(x - 5) = \overset{\text{First}}{x^2} - \underset{\text{Outer}}{5x} - \overset{\text{Inner}}{5x} + \underset{\text{Last}}{5^2} = x^2 - 2(5x) + 5^2 = \boxed{x^2 - 10x + 25}$

Let a and b be any real number, variable, or algebraic expression in the following special products.

DIFFERENCE OF TWO SQUARES

$$(a + b)(a - b) = a^2 - b^2$$

PERFECT SQUARES

Square of a binomial sum: $(a + b)^2 = (a + b)(a + b) = a^2 + 2ab + b^2$

Square of a binomial difference: $(a - b)^2 = (a - b)(a - b) = a^2 - 2ab + b^2$

EXAMPLE 8   Finding the Square of a Binomial Sum

Find $(x + 3)^2$.

COMMON MISTAKE

Forgetting the middle term, which is *twice* the product of the two terms in the binomial.

✪ CORRECT

$(x + 3)^2 = (x + 3)(x + 3)$
$\qquad = x^2 + 3x + 3x + 9$
$\qquad = x^2 + 6x + 9$

✗ INCORRECT

ERROR:

$(x + 3)^2 \neq x^2 + 9$

Don't forget the middle term, which is twice the product of the two terms in the binomial.

EXAMPLE 9 **Using Special Product Formulas**

Find the following:

a. $(2x - 1)^2$ **b.** $(3 + 2y)^2$ **c.** $(4x + 3)(4x - 3)$

Solution (a):

Write the square of a binomial
difference formula.
$$(a - b)^2 = a^2 - 2ab + b^2$$

Let $a = 2x$ and $b = 1$.
$$(2x - 1)^2 = (2x)^2 - 2(2x)(1) + 1^2$$

Simplify.
$$= \boxed{4x^2 - 4x + 1}$$

Solution (b):

Write the square of a binomial
sum formula.
$$(a + b)^2 = a^2 + 2ab + b^2$$

Let $a = 3$ and $b = 2y$.
$$(3 + 2y)^2 = (3)^2 + 2(3)(2y) + (2y)^2$$

Simplify.
$$= 9 + 12y + 4y^2$$

Write in standard form.
$$= \boxed{4y^2 + 12y + 9}$$

Solution (c):

Write the difference of two
squares formula.
$$(a + b)(a - b) = a^2 - b^2$$

Let $a = 4x$ and $b = 3$.
$$(4x + 3)(4x - 3) = (4x)^2 - 3^2$$

Simplify.
$$= \boxed{16x^2 - 9}$$

■ **YOUR TURN** Find the following:

a. $(3x + 1)^2$ **b.** $(1 - 3y)^2$ **c.** $(3x + 2)(3x - 2)$

■ **Answer: a.** $9x^2 + 6x + 1$
b. $9y^2 - 6y + 1$
c. $9x^2 - 4$

EXAMPLE 10 **Cubing a Binomial**

Find the following:

a. $(x + 2)^3$ **b.** $(x - 2)^3$

Solution (a):

Write the cube as a product of
three binomials.
$$(x + 2)^3 = (x + 2)\underbrace{(x + 2)(x + 2)}_{(x + 2)^2}$$

Apply the perfect square formula.
$$= (x + 2)(x^2 + 4x + 4)$$

Apply the distributive property.
$$= x(x^2 + 4x + 4) + 2(x^2 + 4x + 4)$$

Apply the distributive property.
$$= x^3 + 4x^2 + 4x + 2x^2 + 8x + 8$$

Combine like terms.
$$= \boxed{x^3 + 6x^2 + 12x + 8}$$

Solution (b):

Write the cube as a product of
three binomials.
$$(x - 2)^3 = (x - 2)\underbrace{(x - 2)(x - 2)}_{(x - 2)^2}$$

Apply the perfect square formula.
$$= (x - 2)(x^2 - 4x + 4)$$

Apply the distributive property.
$$= x(x^2 - 4x + 4) - 2(x^2 - 4x + 4)$$

Apply the distributive property.
$$= x^3 - 4x^2 + 4x - 2x^2 + 8x - 8$$

Combine like terms.
$$= \boxed{x^3 - 6x^2 + 12x - 8}$$

▼ **CAUTION**

$(x + 2)^3 \neq x^3 + 8$
$(x - 2)^3 \neq x^3 - 8$

PERFECT CUBES

Cube of a binomial sum:	$(a + b)^3 = a^3 + 3a^2b + 3ab^2 + b^3$
Cube of a binomial difference:	$(a - b)^3 = a^3 - 3a^2b + 3ab^2 - b^3$

EXAMPLE 11   **Applying the Special Product Formulas**

Find the following:

a. $(2x + 1)^3$ **b.** $(2x - 5)^3$

Solution (a):

Write the cube of a binomial
sum formula. $(a + b)^3 = a^3 + 3a^2b + 3ab^2 + b^3$

Let $a = 2x$ and $b = 1$. $(2x + 1)^3 = (2x)^3 + 3(2x)^2(1) + 3(2x)(1)^2 + 1^3$

Simplify. $= \boxed{8x^3 + 12x^2 + 6x + 1}$

Solution (b):

Write the cube of a binomial
difference formula. $(a - b)^3 = a^3 - 3a^2b + 3ab^2 - b^3$

Let $a = 2x$ and $b = 5$. $(2x - 5)^3 = (2x)^3 - 3(2x)^2(5) + 3(2x)(5)^2 - 5^3$

Simplify. $= \boxed{8x^3 - 60x^2 + 150x - 125}$

■ **Answer:**
$27x^3 - 108x^2 + 144x - 64$

■ **YOUR TURN** Find $(3x - 4)^3$.

EXAMPLE 12   **Applying the Special Product Formulas
for Binomials in Two Variables**

Find $(2x - 3y)^2$.

Solution:

Write the square of a binomial
difference formula. $(a - b)^2 = (a - b)(a - b) = a^2 - 2ab + b^2$

Let $a = 2x$ and $b = 3y$. $(2x - 3y)^2 = (2x)^2 - 2(2x)(3y) + (3y)^2$

Simplify. $= \boxed{4x^2 - 12xy + 9y^2}$

■ **Answer:** $9x^2 - 12xy + 4y^2$

■ **YOUR TURN** Find $(3x - 2y)^2$.

SECTION
0.3 SUMMARY

In this section, polynomials were defined. Polynomials with one, two, and three terms are called monomials, binomials, and trinomials, respectively. Polynomials are added and subtracted by combining like terms. Polynomials are multiplied by distributing the monomials in the first polynomial throughout the second polynomial. In the special case of the product of two binomials, the FOIL method can also be used. The following are special products of binomials.

Difference of Two Squares

$$(a + b)(a - b) = a^2 - b^2$$

Perfect Squares

Square of a binomial sum.

$$(a + b)^2 = (a + b)(a + b) = a^2 + 2ab + b^2$$

Square of a binomial difference.

$$(a - b)^2 = (a - b)(a - b) = a^2 - 2ab + b^2$$

Perfect Cubes

Cube of a binomial sum.

$$(a + b)^3 = a^3 + 3a^2b + 3ab^2 + b^3$$

Cube of a binomial difference.

$$(a - b)^3 = a^3 - 3a^2b + 3ab^2 - b^3$$

SECTION
0.3 EXERCISES

▪ SKILLS

In Exercises 1–8, write the polynomial in standard form and state the degree of the polynomial.

1. $5x^2 - 2x^3 + 16 - 7x^4$ **2.** $7x^3 - 9x^2 + 5x - 4$ **3.** $4x + 3 - 6x^3$ **4.** $5x^5 - 7x^3 + 8x^4 - x^2 + 10$

5. 15 **6.** -14 **7.** $y - 2$ **8.** $x - 5$

In Exercises 9–24, add or subtract the polynomials, gather like terms, and write the simplified expression in standard form.

9. $(2x^2 - x + 7) + (-3x^2 + 6x - 2)$ **10.** $(3x^2 + 5x + 2) + (2x^2 - 4x - 9)$

11. $(-7x^2 - 5x - 8) - (-4x - 9x^2 + 10)$ **12.** $(8x^3 - 7x^2 - 10) - (7x^3 + 8x^2 - 9x)$

13. $(2x^4 - 7x^2 + 8) - (3x^2 - 2x^4 + 9)$ **14.** $(4x^2 - 9x - 2) - (5 - 3x - 5x^2)$

15. $(7z^2 - 2) - (5z^2 - 2z + 1)$ **16.** $(25y^3 - 7y^2 + 9y) - (14y^2 - 7y + 2)$

17. $(3y^3 - 7y^2 + 8y - 4) - (14y^3 - 8y + 9y^2)$ **18.** $(2x^2 + 3xy) - (x^2 + 8xy - 7y^2)$

19. $(6x - 2y) - 2(5x - 7y)$ **20.** $3a - [2a^2 - (5a - 4a^2 + 3)]$

21. $(2x^2 - 2) - (x + 1) - (x^2 - 5)$ **22.** $(3x^3 + 1) - (3x^2 - 1) - (5x - 3)$

23. $(4t - t^2 - t^3) - (3t^2 - 2t + 2t^3) + (3t^3 - 1)$ **24.** $(-z^3 - 2z^2) + (z^2 - 7z + 1) - (4z^3 + 3z^2 - 3z + 2)$

In Exercises 25–64, multiply the polynomials and write the expressions in standard form.

25. $5xy^2(7xy)$ **26.** $6z(4z^3)$ **27.** $2x^3(1 - x + x^2)$ **28.** $-4z^2(2 + z - z^2)$

29. $-2x^2(5 + x - 5x^2)$ **30.** $-\frac{1}{2}z(2z + 4z^2 - 10)$ **31.** $(x^2 + x - 2)2x^3$ **32.** $(x^2 - x + 2)3x^3$

33. $2ab^2(a^2 + 2ab - 3b^2)$ **34.** $bc^3d^2(b^2c + cd^3 - b^2d^4)$ **35.** $(2x + 1)(3x - 4)$ **36.** $(3z - 1)(4z + 7)$

37. $(x + 2)(x - 2)$ **38.** $(y - 5)(y + 5)$ **39.** $(2x + 3)(2x - 3)$ **40.** $(5y + 1)(5y - 1)$

41. $(2x - 1)(1 - 2x)$ **42.** $(4b - 5y)(4b + 5y)$ **43.** $(2x^2 - 3)(2x^2 + 3)$ **44.** $(4xy - 9)(4xy + 9)$

45. $(7y - 2y^2)(y - y^2 + 1)$ **46.** $(4 - t^2)(6t + 1 - t^2)$ **47.** $(x + 1)(x^2 - 2x + 3)$ **48.** $(x + 3)(x^2 - 3x + 9)$

49. $(t - 2)^2$ **50.** $(t - 3)^2$ **51.** $(z + 2)^2$ **52.** $(z + 3)^2$

53. $[(x + y) - 3]^2$ **54.** $(2x^2 + 3y)^2$ **55.** $(5x - 2)^2$ **56.** $(x + 1)(x^2 + x + 1)$

57. $y(3y + 4)(2y - 1)$ **58.** $p^2(p + 1)(p - 2)$ **59.** $(x^2 + 1)(x^2 - 1)$ **60.** $(t - 5)^2(t + 5)^2$

61. $(b - 3a)(a + 2b)(b + 3a)$ **62.** $(x - 2y)(x^2 + 2xy + 4y^2)$ **63.** $(x + y - z)(2x - 3y + 5z)$ **64.** $(5b^2 - 2b + 1)(3b - b^2 + 2)$

▪ APPLICATIONS

In Exercises 65–68, profit is equal to revenue minus cost: $P = R - C$.

65. Profit. Donna decides to sell fabric cord covers on eBay for $20 a piece. The material for each cord cover costs $9, and it costs her $100 a month to advertise on eBay. Let x be the number of cord covers sold. Write a polynomial representing her monthly profit.

66. Profit. Calculators are sold for $25 each. Advertising costs are $75 per month. Let x be the number of calculators sold. Write a polynomial representing the monthly profit earned by selling x calculators.

67. Profit. If the revenue associated with selling x units of a product is $R = -x^2 + 100x$, and the cost associated with producing x units of the product is $C = -100x + 7500$, find the polynomial that represents the profit of making and selling x units.

68. Profit. A business sells a certain quantity x of items. The revenue generated by selling x items is given by the equation $R = -\frac{1}{2}x^2 + 50x$. The costs are given by $C = 8000 - 150x$. Find a polynomial representing the net profit of this business when x items are sold.

69. Volume of a Box. A rectangular sheet of cardboard is to be used in the construction of a box by cutting out squares of side length x from each corner and turning up the sides. Suppose the dimensions of the original rectangle are 15 inches by 8 inches. Determine a polynomial in x that would give the volume of the box.

70. Volume of a Box. Suppose a box is to be constructed from a square piece of material of side length x by cutting out a 2-inch square from each corner and turning up the sides. Express the volume of the box as a polynomial in the variable x.

71. Geometry. Suppose a running track is constructed of a rectangular portion that measures $2x$ feet wide by $2x + 5$ feet long. Each end of the rectangular portion consists of a semicircle whose diameter is $2x$. Write a polynomial that determines the

a. perimeter of the track in terms of the variable x.
b. area of the track in terms of x.

72. Geometry. A right circular cylinder whose radius is r and whose height is $2r$ is surmounted by a hemisphere of radius r.

a. Find a polynomial in the variable r that represents the volume of the "silo" shape.
b. Find a polynomial in r that represents the total surface area of the "silo."

73. Engineering. The force of an electrical field is given by the equation $F = k\frac{q_1 q_2}{r^2}$. Suppose $q_1 = x$, $q_2 = 3x$, and $r = 10x$. Find a polynomial representing the force of the electrical field in terms of the variable x.

74. Engineering. If a football (or other projectile) is thrown upward, its height above the ground is given by the equation $s = 16t^2 + v_0 t + s_0$, where v_0 and s_0 are the initial velocity and initial height of the football, respectively, and t is the time in seconds. Suppose the football is thrown from the top of a building that is 192 feet tall, with an initial speed of 96 feet per second.

a. Write the polynomial that gives the height of the football in terms of the variable t (time).
b. What is the height of the football after 2 seconds have elapsed? Will the football hit the ground after 2 seconds?

▪ CATCH THE MISTAKE

In Exercises 75 and 76, explain the mistake that is made.

75. Subtract and simplify $(2x^2 - 5) - (3x - x^2 + 1)$.

Solution:

Eliminate the parentheses. $2x^2 - 5 - 3x - x^2 + 1$

Collect like terms. $x^2 - 3x - 4$

This is incorrect. What mistake was made?

76. Simplify $(2 + x)^2$.

Solution:

Write the square of the binomial as the sum of the squares. $(2 + x)^2 = 2^2 + x^2$

Simplify. $= x^2 + 4$

This is incorrect. What mistake was made?

▪ CONCEPTUAL

In Exercises 77–80, determine whether each of the following statements is true or false.

77. All binomials are polynomials.

78. The product of two monomials is a binomial.

79. $(x + y)^3 = x^3 + y^3$

80. $(x - y)^2 = x^2 + y^2$

In Exercises 81 and 82, let m and n be real numbers and $m > n$.

81. What degree is the *product* of a polynomial of degree n and a polynomial of degree m?

82. What degree is the *sum* of a polynomial of degree n and a polynomial of degree m?

▪ CHALLENGE

In Exercises 83–86, perform the indicated operations and simplify.

83. $(7x - 4y^2)^2 (7x + 4y^2)^2$

84. $(3x - 5y^2)^2 (3x + 5y^2)^2$

85. $(x - a)(x^2 + ax + a^2)$

86. $(x + a)(x^2 - ax + a^2)$

▪ TECHNOLOGY

87. Use a graphing utility to plot the graphs of the three expressions $(2x + 3)(x - 4)$, $2x^2 + 5x - 12$, and $2x^2 - 5x - 12$. Which two graphs agree with each other?

88. Use a graphing utility to plot the graphs of the three expressions $(x + 5)^2$, $x^2 + 25$, and $x^2 + 10x + 25$. Which two graphs agree with each other?

SECTION

0.4 FACTORING POLYNOMIALS

SKILLS OBJECTIVES

- Factor out the greatest common factor.
- Factor the difference of two squares.
- Factor perfect squares.
- Factor the sum or difference of two cubes.
- Factor a trinomial as a product of binomials.
- Factor by grouping.

CONCEPTUAL OBJECTIVES

- Understand that factoring has its basis in the distributive property.
- Identify prime (irreducible) polynomials.
- Develop a general strategy for factoring polynomials.

In Section 0.3 we discussed multiplying polynomials. In this section we examine the reverse of that process, which is called *factoring*. Consider the following product:

$$(x + 3)(x + 1) = x^2 + 4x + 3$$

To *factor* the resulting polynomial, you reverse the process to undo the multiplication:

$$x^2 + 4x + 3 = (x + 3)(x + 1)$$

The polynomials $(x + 3)$ and $(x + 1)$ are called **factors** of the polynomial $x^2 + 4x + 3$. The process of writing a polynomial as a product is called **factoring**. In Chapter 1 we will solve quadratic equations by factoring.

In this section we will restrict our discussion to factoring polynomials with integer coefficients, which is called **factoring over the integers**. If a polynomial cannot be factored using integer coefficients, then it is **prime** or irreducible over the integers. When a polynomial is written as a product of prime polynomials, then the polynomial is said to be **factored completely**.

Greatest Common Factor

The simplest type of factoring of polynomials occurs when there is a factor common to every term of the polynomial. This **common factor** is a monomial that can be "factored out" by applying the distributive property in reverse:

$$ab + ac = a(b + c)$$

For example, $4x^2 - 6x$ can be written as $2x(x) - 2x(3)$. Notice that $2x$ is a common factor to both terms, so the distributive property tells us we can factor this polynomial to yield $2x(x - 3)$. Although 2 is a common factor and x is a common factor, the monomial $2x$ is called the *greatest common factor*.

> **GREATEST COMMON FACTOR**
>
> The monomial ax^k is called the **greatest common factor (GCF)** of a polynomial in x with integer coefficients if *both* of the following are true:
>
> - a is the *greatest* integer factor common to all of the polynomial coefficients.
> - k is the *smallest* exponent on x found in all of the terms of the polynomial.

POLYNOMIAL	GCF	WRITE EACH TERM AS A PRODUCT OF GCF AND REMAINING FACTOR	FACTORED FORM
$7x + 21$	7	$7(x) + 7(3)$	$7(x + 3)$
$3x^2 + 12x$	$3x$	$3x(x) + 3x(4)$	$3x(x + 4)$
$4x^3 + 2x + 6$	2	$2(2x^3) + 2x + 2(3)$	$2(2x^3 + x + 3)$
$6x^4 - 9x^3 + 12x^2$	$3x^2$	$3x^2(2x^2) - 3x^2(3x) + 3x^2(4)$	$3x^2(2x^2 - 3x + 4)$
$-5x^4 + 25x^3 - 20x^2$	$-5x^2$	$-5x^2(x^2) - 5x^2(-5x) - 5x^2(4)$	$-5x^2(x^2 - 5x + 4)$

Classroom Example 0.4.1
Factor:
a. $-27x^3 - 9x^5$
b. $x^4 - 3x^3 - 9x^2 - 9x$
c.* $4a^2bx^2 - 16ab^2x$, where a, b are real numbers

Answer:
a. $-9x^3(3 + x^2)$
b. $x(x^3 - 3x^2 - 9x - 9)$
c. $4abx(ax - 4b)$

EXAMPLE 1 **Factoring Polynomials by Extracting the Greatest Common Factor**

Factor:

a. $6x^5 - 18x^4$ **b.** $6x^5 - 10x^4 - 8x^3 + 12x^2$

Solution (a):

Identify the greatest common factor. $6x^4$

Write each term as a product with the GCF as a factor. $6x^5 - 18x^4 = 6x^4(x) - 6x^4(3)$

Factor out the GCF. $= \boxed{6x^4(x - 3)}$

Solution (b):

Identify the greatest common factor. $\qquad\qquad\qquad\qquad 2x^2$

Write each term as a product with the GCF as a factor.

$$6x^5 - 10x^4 - 8x^3 + 12x^2 = 2x^2(3x^3) - 2x^2(5x^2) - 2x^2(4x) + 2x^2(6)$$

Factor out the GCF. $\qquad\qquad\qquad = \boxed{2x^2(3x^3 - 5x^2 - 4x + 6)}$

■ **YOUR TURN** Factor:

    a. $12x^3 - 4x$ **b.** $3x^5 - 9x^4 + 12x^3 - 6x^2$

■ **Answer: a.** $4x(3x^2 - 1)$
         b. $3x^2(x^3 - 3x^2 + 4x - 2)$

Factoring Formulas: Special Polynomial Forms

The first step in factoring polynomials is to look for a common factor. If there is no common factor, then we look for special polynomial forms that we learned were special products in Section 0.3 and reverse the process.

Difference of two squares	$a^2 - b^2 = (a + b)(a - b)$
Perfect squares	$a^2 + 2ab + b^2 = (a + b)^2$
	$a^2 - 2ab + b^2 = (a - b)^2$
Sum of two cubes	$a^3 + b^3 = (a + b)(a^2 - ab + b^2)$
Difference of two cubes	$a^3 - b^3 = (a - b)(a^2 + ab + b^2)$

EXAMPLE 2 **Factoring the Difference of Two Squares**

Factor:

a. $x^2 - 9$ **b.** $4x^2 - 25$ **c.** $x^4 - 16$

Solution (a):

Rewrite as the difference of two squares. $\qquad\qquad x^2 - 9 = x^2 - 3^2$

Let $a = x$ and $b = 3$ in $a^2 - b^2 = (a + b)(a - b)$. $\qquad = \boxed{(x + 3)(x - 3)}$

Solution (b):

Rewrite as the difference of two squares. $\qquad\qquad 4x^2 - 25 = (2x)^2 - 5^2$

Let $a = 2x$ and $b = 5$ in $a^2 - b^2 = (a + b)(a - b)$. $\qquad = \boxed{(2x + 5)(2x - 5)}$

Solution (c):

Rewrite as the difference of two squares. $\qquad\qquad x^4 - 16 = (x^2)^2 - 4^2$

Let $a = x^2$ and $b = 4$ in $a^2 - b^2 = (a + b)(a - b)$. $\qquad = (x^2 + 4)(x^2 - 4)$

Note that $x^2 - 4$ is also a difference of two squares (part a of the following Your Turn). $\qquad = \boxed{(x + 2)(x - 2)(x^2 + 4)}$

■ **YOUR TURN** Factor:

    a. $x^2 - 4$ **b.** $9x^2 - 16$ **c.** $x^4 - 81$

Classroom Example 0.4.2
Factor:
a. $49x^2 - 16$
b. $625 - 16x^4$

Answer:
a. $(7x - 4)(7x + 4)$
b. $(5 - 2x)(5 + 2x)(25 + 4x^2)$

■ **Answer: a.** $(x + 2)(x - 2)$
         b. $(3x + 4)(3x - 4)$
         c. $(x^2 + 9)(x - 3)(x + 3)$

A trinomial is a perfect square if it has the form $a^2 \pm 2ab + b^2$. Notice that:

- The first term and third term are perfect squares.
- The middle term is twice the product of the bases of these two perfect squares.
- The sign of the middle term determines the sign of the factored form:

$$a^2 \pm 2ab + b^2 = (a \pm b)^2$$

Classroom Example 0.4.3
Factor:
a. $9x^2 - 30x + 25$
b.* $a^2x^2 - 4abx + 4b^2$,
   where $a, b \neq 0$

Answer:
a. $(3x - 5)^2$
b. $(ax - 2b)^2$

EXAMPLE 3   **Factoring Trinomials That Are Perfect Squares**

Factor:

a. $x^2 + 6x + 9$ **b.** $x^2 - 10x + 25$ **c.** $9x^2 - 12x + 4$

Solution (a):

Rewrite the trinomial so that
the first and third terms
are perfect squares. $x^2 + 6x + 9 = x^2 + 6x + 3^2$

Notice that if we let $a = x$ and $b = 3$ in
$a^2 + 2ab + b^2 = (a + b)^2$, then the
middle term $6x$ is $2ab$. $x^2 + 6x + 9 = x^2 + 2(3x) + 3^2 = \boxed{(x + 3)^2}$

Solution (b):

Rewrite the trinomial so that
the first and third terms
are perfect squares. $x^2 - 10x + 25 = x^2 - 10x + 5^2$

Notice that if we let $a = x$ and $b = 5$ in
$a^2 - 2ab + b^2 = (a - b)^2$, then the
middle term $-10x$ is $-2ab$. $x^2 - 10x + 25 = x^2 - 2(5x) + 5^2 = \boxed{(x - 5)^2}$

Solution (c):

Rewrite the trinomial so that
the first and third terms
are perfect squares. $9x^2 - 12x + 4 = (3x)^2 - 12x + 2^2$

Notice that if we let $a = 3x$ and $b = 2$ in
$a^2 - 2ab + b^2 = (a - b)^2$, then the
middle term $-12x$ is $-2ab$. $9x^2 - 12x + 4 = (3x)^2 - 2(3x)(2) + 2^2 = \boxed{(3x - 2)^2}$

- **Answer: a.** $(x + 4)^2$
   b. $(x - 2)^2$
   c. $(5x - 2)^2$

■ **YOUR TURN** Factor:

a. $x^2 + 8x + 16$ **b.** $x^2 - 4x + 4$ **c.** $25x^2 - 20x + 4$

Classroom Example 0.4.4
Factor:
a. $343 + x^3$
b.* $8a^3 + 125x^3$, where $a \neq 0$

Answer:
a. $(7 + x)(49 - 7x + x^2)$
b. $(2a + 5x)(4a^2 - 10ax + 25x^2)$

EXAMPLE 4   **Factoring the Sum of Two Cubes**

Factor $x^3 + 27$.

Solution:

Rewrite as the sum of two cubes. $x^3 + 27 = x^3 + 3^3$

Write the sum of two cubes formula. $a^3 + b^3 = (a + b)(a^2 - ab + b^2)$

Let $a = x$ and $b = 3$. $x^3 + 27 = x^3 + 3^3 = \boxed{(x + 3)(x^2 - 3x + 9)}$

EXAMPLE 5 **Factoring the Difference of Two Cubes**

Factor $x^3 - 125$.

Solution:

Rewrite as the difference of two cubes. $\qquad x^3 - 125 = x^3 - 5^3$

Write the difference of two cubes formula. $\qquad a^3 - b^3 = (a - b)(a^2 + ab + b^2)$

Let $a = x$ and $b = 5$. $\qquad x^3 - 125 = x^3 - 5^3 = \boxed{(x - 5)(x^2 + 5x + 25)}$

▪ **YOUR TURN** Factor:

$\quad$ **a.** $x^3 + 8$ $\quad$ **b.** $x^3 - 64$

▪ **Answer: a.** $(x + 2)(x^2 - 2x + 4)$
$\qquad\qquad$ **b.** $(x - 4)(x^2 + 4x + 16)$

Factoring a Trinomial as a Product of Two Binomials

The first step in factoring is to look for a common factor. If there is no common factor, look to see whether the polynomial is a special form for which we know the factoring formula. If it is not of such a special form and if it is a trinomial, then we proceed with a general factoring strategy.

We know that $(x + 3)(x + 2) = x^2 + 5x + 6$, so we say the **factors** of $x^2 + 5x + 6$ are **$(x + 3)$** and **$(x + 2)$**. In factored form we have $x^2 + 5x + 6 = (x + 3)(x + 2)$. Recall the FOIL method from Section 0.3. The product of the last terms (3 and 2) is 6, and the sum of the products of the inner terms ($3x$) and the outer terms ($2x$) is $5x$. Let's pretend for a minute that we didn't know this factored form but had to work with the general form:

$$x^2 + 5x + 6 = (x + a)(x + b)$$

The goal is to find a and b. We start by multiplying the two binomials on the right.

$$x^2 + 5x + 6 = (x + a)(x + b) = x^2 + ax + bx + ab = x^2 + (a + b)x + ab$$

Compare the expression we started with on the left with the expression on the far right $x^2 + \mathbf{5x} + \mathbf{6} = x^2 + \mathbf{(a + b)x} + \mathbf{ab}$. We see that $\mathbf{ab = 6}$ and $\mathbf{(a + b) = 5}$. *Start* with the possible combinations of a and b whose product is 6, and *then* look among those for the combination whose sum is 5.

$ab = 6$	a, b:	1, 6	$-1, -6$	2, 3	$-2, -3$
$a + b$		7	-7	**5**	-5

All of the possible a, b combinations in the first row have a product equal to 6, but only one of those has a sum equal to 5. Therefore the factored form is

$$x^2 + 5x + 6 = (x + a)(x + b) = (x + 2)(x + 3)$$

EXAMPLE 6 Factoring a Trinomial

Factor $x^2 + 10x + 9$.

Solution:

Write the trinomial as a product of two binomials in general form.

$$x^2 + 10x + 9 = (x + \square)(x + \square)$$

Write all of the integers whose product is 9.

Integers whose product is 9	1, 9	−1, −9	3, 3	−3, −3

Determine the sum of the integers.

Integers whose product is 9	**1, 9**	−1, −9	3, 3	−3, −3
Sum	**10**	−10	6	−6

Select 1, 9 because the product is 9 (last term of the trinomial) and the sum is 10 (middle term coefficient of the trinomial).

$$x^2 + 10x + 9 = (x + 9)(x + 1)$$

Check: $(x + 9)(x + 1) = x^2 + 1x + 9x + 9 = x^2 + 10x + 9$ ✓

■ **YOUR TURN** Factor $x^2 + 9x + 20$.

■ **Answer:** $(x + 4)(x + 5)$

In Example 6, all terms in the trinomial are positive. When the constant term is negative, then (regardless of whether the middle term is positive or negative) the factors will be opposite in sign, as illustrated in Example 7.

EXAMPLE 7 Factoring a Trinomial

Factor $x^2 − 3x − 28$.

Solution:

Write the trinomial as a product of two binomials in general form.

$$x^2 − 3x − 28 = (x + \square)(x − \square)$$

Write all of the integers whose product is −28.

Integers whose product is −28	1, −28	−1, 28	2, −14	−2, 14	4, −7	−4, 7

Determine the sum of the integers.

Integers whose product is −28	1, −28	−1, 28	2, −14	−2, 14	**4, −7**	−4, 7
Sum	−27	27	−12	12	**−3**	3

Select 4, −7 because the product is −28 (last term of the trinomial) and the sum is −3 (middle term coefficient of the trinomial).

$$x^2 − 3x − 28 = (x + 4)(x − 7)$$

Check: $(x + 4)(x − 7) = x^2 − 7x + 4x − 28 = x^2 − 3x − 28$ ✓

■ **YOUR TURN** Factor $x^2 + 3x − 18$.

■ **Answer:** $(x + 6)(x − 3)$

When the leading coefficient of the trinomial is not equal to 1, then we consider all possible factors using the following procedure, which is based on the FOIL method in reverse.

FACTORING A TRINOMIAL WHOSE LEADING COEFFICIENT IS NOT 1

$$ax^2 + bx + c = (\boxed{}x + \boxed{})(\boxed{}x + \boxed{})$$

Factors of a

Factors of c

Step 1: Find two **F**irst terms whose product is the first term of the trinomial.

Step 2: Find two **L**ast terms whose product is the last term of the trinomial.

Step 3: Consider all possible combinations found in Steps 1 and 2 until the sum of the **O**uter and **I**nner products are equal to the middle term of the trinomial.

EXAMPLE 8   Factoring a Trinomial Whose Leading Coefficient Is Not 1

Factor $5x^2 + 9x - 2$.

Solution:

STEP 1 Start with the <u>first</u> term. Note that $5x \cdot x = 5x^2$. $(5x \pm \square)(x \pm \square)$

STEP 2 The product of the <u>last</u> terms should yield -2. $-1, 2$ or $1, -2$

STEP 3 Consider all possible factors based on Steps 1 and 2.
$$(5x - 1)(x + 2)$$
$$(5x + 1)(x - 2)$$
$$(5x + 2)(x - 1)$$
$$(5x - 2)(x + 1)$$

Since the <u>outer</u> and <u>inner</u> products must sum to $9x$, the factored form must be: $\boxed{5x^2 + 9x - 2 = (5x - 1)(x + 2)}$

Check: $(5x - 1)(x + 2) = 5x^2 + 10x - 1x - 2 = 5x^2 + 9x - 2$ ✓

■ **YOUR TURN** Factor $2t^2 + t - 3$.

■ **Answer:** $(2t + 3)(t - 1)$

EXAMPLE 9   Factoring a Trinomial Whose Leading Coefficient Is Not 1

Factor $15x^2 - x - 6$.

Solution:

STEP 1 Start with the <u>first</u> term. $(5x \pm \square)(3x \pm \square)$ or $(15x \pm \square)(x \pm \square)$

STEP 2 The product of the <u>last</u> terms should yield -6. $-1, 6$ or $1, -6$ or $2, -3$ or $-2, 3$

STEP 3 Consider all possible factors based on Steps 1 and 2.

$(5x - 1)(3x + 6)$ $(15x - 1)(x + 6)$
$(5x + 6)(3x - 1)$ $(15x + 6)(x - 1)$
$(5x + 1)(3x - 6)$ $(15x + 1)(x - 6)$
$(5x - 6)(3x + 1)$ $(15x - 6)(x + 1)$
$(5x + 2)(3x - 3)$ $(15x + 2)(x - 3)$
$(5x - 3)(3x + 2)$ $(15x - 3)(x + 2)$
$(5x - 2)(3x + 3)$ $(15x - 2)(x + 3)$
$(5x + 3)(3x - 2)$ $(15x + 3)(x - 2)$

Since the <u>outer</u> and <u>inner</u> products must sum to $-x$, the factored form must be:

$$15x^2 - x - 6 = (5x + 3)(3x - 2)$$

Check: $(5x + 3)(3x - 2) = 15x^2 - 10x + 9x - 6 = 15x^2 - x - 6$ ✓

■ **Answer:** $(3x - 4)(2x + 3)$

■ **YOUR TURN** Factor $6x^2 + x - 12$.

EXAMPLE 10 **Identifying Prime (Irreducible) Polynomials**

Factor $x^2 + x - 8$.

Solution:

Write the trinomial as a product of two binomials in general form.

$$x^2 + x - 8 = (x + \square)(x - \square)$$

Write all of the integers whose product is -8.

Integers whose product is -8	1, -8	$-1, 8$	4, -2	$-4, 2$

Determine the sum of the integers.

Integers whose product is -8	1, -8	$-1, 8$	4, -2	$-4, 2$
Sum	-7	7	2	-2

The middle term of the trinomial is $-x$, so we look for the sum of the integers that equals -1. Since no sum exists for the given combinations, we say that this polynomial is *prime* (irreducible) over the integers.

Factoring by Grouping

Much of our attention in this section has been on factoring trinomials. For polynomials with more than three terms we first look for a common factor to all terms. If there is no common factor to all terms of the polynomial, we look for a group of terms that have a common factor. This strategy is called **factoring by grouping**.

EXAMPLE 11 **Factoring a Polynomial by Grouping**

Factor $x^3 - x^2 + 2x - 2$.

Solution:

Group the terms that have a common factor. $= (x^3 - x^2) + (2x - 2)$

Factor out the common factor in each pair of parentheses. $= x^2(x - 1) + 2(x - 1)$

Use the distributive property. $= (x^2 + 2)(x - 1)$

EXAMPLE 12 Factoring a Polynomial by Grouping

Factor $2x^2 + 2x - x - 1$.

Solution:

Group the terms that have a common factor.	$= (2x^2 + 2x) + (-x - 1)$
Factor out the common factor in each pair of parentheses.	$= 2x(x + 1) - 1(x + 1)$
Use the distributive property.	$= \boxed{(2x - 1)(x + 1)}$

■ **YOUR TURN** Factor $x^3 + x^2 - 3x - 3$.

■ **Answer:** $(x + 1)(x^2 - 3)$

A Strategy for Factoring Polynomials

The first step in factoring a polynomial is to look for the greatest common factor. When specifically factoring trinomials, look for special known forms: a perfect square or a difference of two squares. A general approach to factoring a trinomial uses the FOIL method in reverse. Finally, we look for factoring by grouping. The following strategy for factoring polynomials is based on the techniques discussed in this section.

STRATEGY FOR FACTORING POLYNOMIALS

1. Factor out the greatest common factor (monomial).
2. Identify any special polynomial forms and apply factoring formulas.
3. Factor a trinomial into a product of two binomials: $(ax + b)(cx + d)$.
4. Factor by grouping.

Study Tip

When factoring, always start by factoring out the GCF.

EXAMPLE 13 Factoring Polynomials

Factor:

a. $3x^2 - 6x + 3$ **b.** $-4x^3 + 2x^2 + 6x$ **c.** $15x^2 + 7x - 2$ **d.** $x^3 - x + 2x^2 - 2$

Solution (a):

Factor out the greatest common factor.	$3x^2 - 6x + 3 = 3(x^2 - 2x + 1)$
The trinomial is a perfect square.	$= \boxed{3(x - 1)^2}$

Solution (b):

Factor out the greatest common factor.	$-4x^3 + 2x^2 + 6x = -2x(2x^2 - x - 3)$
Use the FOIL method in reverse to factor the trinomial.	$= \boxed{-2x(2x - 3)(x + 1)}$

Solution (c):

There is no common factor.	$15x^2 + 7x - 2$
Use the FOIL method in reverse to factor the trinomial.	$= \boxed{(3x + 2)(5x - 1)}$

Solution (d):

Factor by grouping.	$x^3 - x + 2x^2 - 2 = (x^3 - x) + (2x^2 - 2)$
	$= x(x^2 - 1) + 2(x^2 - 1)$
	$= (x + 2)(x^2 - 1)$
Factor the difference of two squares.	$= \boxed{(x + 2)(x - 1)(x + 1)}$

SECTION
0.4 SUMMARY

In this section, we discussed factoring polynomials, which is the reverse process of multiplying polynomials. Four main techniques were discussed.

Greatest Common Factor: ax^k

- a is the greatest common factor for all coefficients of the polynomial.
- k is the smallest exponent found on all of the terms in the polynomial.

Factoring Formulas: Special Polynomial Forms

Difference of two squares: $a^2 - b^2 = (a + b)(a - b)$
Perfect squares: $a^2 + 2ab + b^2 = (a + b)^2$
   $a^2 - 2ab + b^2 = (a - b)^2$
Sum of two cubes: $a^3 + b^3 = (a + b)(a^2 - ab + b^2)$
Difference of two cubes: $a^3 - b^3 = (a - b)(a^2 + ab + b^2)$

Factoring a Trinomial as a Product of Two Binomials

$$x^2 + bx + c = (x + ?)(x + ?)$$

1. Find all possible combinations of factors whose product is c.
2. Of the combinations in Step 1, look for the sum of factors that equals b.

$$ax^2 + bx + c = (?x + ?)(?x + ?)$$

1. Find all possible combinations of the first terms whose product is ax^2.
2. Find all possible combinations of the last terms whose product is c.
3. Consider all possible factors based on Steps 1 and 2.

Factoring by Grouping

- Group terms that have a common factor.
- Use the distributive property.

SECTION
0.4 EXERCISES

SKILLS

In Exercises 1–12, factor each expression. Start by finding the greatest common factor (GCF).

1. $5x + 25$
2. $x^2 + 2x$
3. $4t^2 - 2$
4. $16z^2 - 20z$
5. $2x^3 - 50x$
6. $4x^2y - 8xy^2 + 16x^2y^2$
7. $3x^3 - 9x^2 + 12x$
8. $14x^4 - 7x^2 + 21x$
9. $x^3 - 3x^2 - 40x$
10. $-9y^2 + 45y$
11. $4x^2y^3 + 6xy$
12. $3z^3 - 6z^2 + 18z$

In Exercises 13–20, factor the difference of two squares.

13. $x^2 - 9$
14. $x^2 - 25$
15. $4x^2 - 9$
16. $1 - x^4$
17. $2x^2 - 98$
18. $144 - 81y^2$
19. $225x^2 - 169y^2$
20. $121y^2 - 49x^2$

In Exercises 21–32, factor the perfect squares.

21. $x^2 + 8x + 16$
22. $y^2 - 10y + 25$
23. $x^4 - 4x^2 + 4$
24. $1 - 6y + 9y^2$
25. $4x^2 + 12xy + 9y^2$
26. $x^2 - 6xy + 9y^2$
27. $9 - 6x + x^2$
28. $25x^2 - 20xy + 4y^2$
29. $x^4 + 2x^2 + 1$
30. $x^6 - 6x^3 + 9$
31. $p^2 + 2pq + q^2$
32. $p^2 - 2pq + q^2$

In Exercises 33–42, factor the sum or difference of two cubes.

33. $t^3 + 27$
34. $z^3 + 64$
35. $y^3 - 64$
36. $x^3 - 1$
37. $8 - x^3$
38. $27 - y^3$
39. $y^3 + 125$
40. $64x - x^4$
41. $27 + x^3$
42. $216x^3 - y^3$

In Exercises 43–52, factor each trinomial into a product of two binomials.

43. $x^2 - 6x + 5$

44. $t^2 - 5t - 6$

45. $y^2 - 2y - 3$

46. $y^2 - 3y - 10$

47. $2y^2 - 5y - 3$

48. $2z^2 - 4z - 6$

49. $3t^2 + 7t + 2$

50. $4x^2 - 2x - 12$

51. $-6t^2 + t + 2$

52. $-6x^2 - 17x + 10$

In Exercises 53–60, factor by grouping.

53. $x^3 - 3x^2 + 2x - 6$

54. $x^5 + 5x^3 - 3x^2 - 15$

55. $a^4 + 2a^3 - 8a - 16$

56. $x^4 - 3x^3 - x + 3$

57. $3xy - 5rx - 10rs + 6sy$

58. $6x^2 - 10x + 3x - 5$

59. $20x^2 + 8xy - 5xy - 2y^2$

60. $9x^5 - a^2x^3 - 9x^2 + a^2$

In Exercises 61–92, factor each of the polynomials completely, if possible. If the polynomial cannot be factored, state that it is prime.

61. $x^2 - 4y^2$

62. $a^2 + 5a + 6$

63. $3a^2 + a - 14$

64. $ax + b + bx + a$

65. $x^2 + 16$

66. $x^2 + 49$

67. $4z^2 + 25$

68. $\frac{1}{16} - b^4$

69. $6x^2 + 10x + 4$

70. $x^2 + 7x + 5$

71. $6x^2 + 13xy - 5y^2$

72. $15x + 15xy$

73. $36s^2 - 9t^2$

74. $3x^3 - 108x$

75. $a^2b^2 - 25c^2$

76. $2x^3 + 54$

77. $4x^2 - 3x - 10$

78. $10x - 25 - x^2$

79. $3x^3 - 5x^2 - 2x$

80. $2y^3 + 3y^2 - 2y$

81. $x^3 - 9x$

82. $w^3 - 25w$

83. $xy - x - y + 1$

84. $a + b + ab + b^2$

85. $x^4 + 5x^2 + 6$

86. $x^6 - 7x^3 - 8$

87. $x^2 - 2x - 24$

88. $25x^2 + 30x + 9$

89. $x^4 + 125x$

90. $x^4 - 1$

91. $x^4 - 81$

92. $10x^2 - 31x + 15$

■ APPLICATIONS

93. **Geometry.** A rectangle has a length of $2x + 4$ and a width of x. Express the perimeter of the rectangle as a factored polynomial in x.

94. **Geometry.** The volume of a box is given by the expression $x^3 + 7x^2 + 12x$. Express the volume as a factored polynomial in the variable x.

95. **Business.** The profit of a business is given by the expression $P = 2x^2 - 15x + 4x - 30$. Express the profit as a factored polynomial in the variable x.

96. **Business.** The break-even point for a company is given by solving the equation $3x^2 + 9x - 4x - 12 = 0$. Factor the polynomial on the left side of the equation.

97. **Engineering.** The height of a projectile is given by the equation $s = -16t^2 - 78t + 10$. Factor the expression on the right side of the equal sign.

98. **Engineering.** The electrical field at a point P between two charges is given by $k = \dfrac{10x - x^2}{100}$. Factor the numerator of this expression.

■ CATCH THE MISTAKE

In Exercises 99 and 100, explain the mistake that is made.

99. Factor $x^3 - x^2 - 9x + 9$.

Solution:

Group terms with common factors.	$(x^3 - x^2) + (-9x + 9)$
Factor out common factors.	$x^2(x - 1) - 9(x - 1)$
Distributive property.	$(x - 1)(x^2 - 9)$
Factor $x^2 - 9$.	$(x - 1)(x - 3)^2$

This is incorrect. What mistake was made?

100. Factor $4x^2 + 12x - 40$.

Solution:

Factor the trinomial into a product of binomials.	$(2x - 4)(2x + 10)$
Factor out a 2.	$= 2(x - 2)(x + 5)$

This is incorrect. What mistake was made?

▪ CONCEPTUAL

In Exercises 101–104, determine whether each of the following statements is true or false.

101. All trinomials can be factored into a product of two binomials.

102. All polynomials can be factored into prime factors with respect to the integers.

103. $x^2 - y^2 = (x - y)(x + y)$

104. $x^2 + y^2 = (x + y)^2$

▪ CHALLENGE

105. Factor $a^{2n} - b^{2n}$ completely, assuming a, b, and n are positive integers.

106. Find all the values of c such that the trinomial $x^2 + cx - 14$ can be factored.

▪ TECHNOLOGY

107. Use a graphing utility to plot the graphs of the three expressions $8x^3 + 1$, $(2x + 1)(4x^2 - 2x + 1)$, and $(2x - 1)(4x^2 + 2x + 1)$. Which two graphs agree with each other?

108. Use a graphing utility to plot the graphs of the three expressions $27x^3 - 1$, $(3x - 1)^3$, and $(3x - 1)(9x^2 + 3x + 1)$. Which two graphs agree with each other?

SECTION

0.5 RATIONAL EXPRESSIONS

SKILLS OBJECTIVES

▪ Find the domain of an algebraic expression.
▪ Reduce a rational expression to lowest terms.
▪ Multiply and divide rational expressions.
▪ Add and subtract rational expressions.
▪ Simplify complex rational expressions.

CONCEPTUAL OBJECTIVES

▪ Understand why rational expressions have domain restrictions.
▪ Understand the least common denominator method for rational expressions.

Rational Expressions and Domain Restrictions

Recall that a rational number is the ratio of two integers with the denominator not equal to zero. Similarly, the ratio, or quotient, of two polynomials is a **rational expression**.

$$\text{Rational Numbers:} \qquad \frac{3}{7} \qquad \frac{5}{9} \qquad \frac{9}{11}$$

$$\text{Rational Expressions:} \qquad \frac{3x + 2}{x - 5} \qquad \frac{5x^2}{x^2 + 1} \qquad \frac{9}{3x - 2}$$

As with rational numbers, the denominators of rational expressions are never equal to zero. In the first and third rational expressions, there are values of the variable that would correspond to a denominator equal to zero; these values are not permitted:

$$\frac{3x + 2}{x - 5} \qquad x \neq 5 \qquad\qquad \frac{9}{3x - 2} \qquad x \neq \frac{2}{3}$$

In the second rational expression, $\dfrac{5x^2}{x^2 + 1}$, there are no real numbers that will correspond to a zero denominator.

The set of real numbers for which an algebraic expression is *defined* is called the **domain**. Since a rational expression is not defined if its denominator is zero, we must eliminate from the domain those values of the variable that would result in a zero denominator.

To find the domain of an algebraic expression we ask the question, "What can x (the variable) be?" For rational expressions the answer in general is "any values except those that make the denominator equal to zero."

EXAMPLE 1   Finding the Domain of an Algebraic Expression

Find the domain of the expressions.

a. $2x^2 - 5x + 3$ **b.** $\dfrac{2x + 1}{x - 4}$ **c.** $\dfrac{x}{x^2 + 1}$ **d.** $\dfrac{3x + 1}{x}$

Solution:

ALGEBRAIC EXPRESSION	TYPE	DOMAIN	NOTE
a. $2x^2 - 5x + 3$	Polynomial	All real numbers	The domain of all polynomials is the set of all real numbers.
b. $\dfrac{2x + 1}{x - 4}$	Rational expression	All real numbers except $x = 4$	When $x = 4$, the rational expression is undefined.
c. $\dfrac{x}{x^2 + 1}$	Rational expression	All real numbers	There are no real numbers that will result in the denominator being equal to zero.
d. $\dfrac{3x + 1}{x}$	Rational expression	All real numbers except $x = 0$	When $x = 0$, the rational expression is undefined.

■ **YOUR TURN** Find the domain of the expressions.

a. $\dfrac{3x - 1}{x + 1}$ **b.** $\dfrac{5x - 1}{x}$ **c.** $3x^2 + 2x - 7$ **d.** $\dfrac{2x + 5}{x^2 + 4}$

Classroom Example 0.5.1
Find the domain of the expressions.

a. $\dfrac{x(3x - 1)}{(x + 1)^2}$

b. $\dfrac{x^2 + 8}{x^2 + 1} + 3x$

c.* $\dfrac{x}{x^2 + 4} + \dfrac{2x^{-1}}{x - 3}$

Answer:
a. $x \neq -1$
b. all real numbers
c. $x \neq 0, 3$

■ **Answer: a.** $x \neq -1$
   b. $x \neq 0$
   c. all real numbers
   d. all real numbers

In this text, it will be assumed that the domain is the set of all real numbers except the real numbers shown to be excluded.

Classroom Example 0.5.2
What real numbers must be excluded from the domain of these expressions?

a. $\dfrac{(x-1)(x^2+2)}{1-2x} + x$

b.* $\dfrac{3 + \dfrac{1}{x} + \dfrac{3}{x^2}}{x^2 - 4}$

Answer:
a. $\dfrac{1}{2}$ b. $\pm 2, 0$

EXAMPLE 2   Excluding Values from the Domain of Rational Expressions

Determine what real numbers must be excluded from the domain of the following rational expressions.

a. $\dfrac{7x+5}{x^2-4}$ b. $\dfrac{3x+2}{x^2-5x}$

Solution (a):

Factor the denominator.

$$\frac{7x+5}{x^2-4} = \frac{7x+5}{(x+2)(x-2)}$$

Determine the values of x that will make the denominator equal to zero.

$\boxed{x = -2}$ and $\boxed{x = 2}$ must be excluded from the domain.

Solution (b):

Factor the denominator.

$$\frac{3x+2}{x^2-5x} = \frac{3x+2}{x(x-5)}$$

Determine the values of x that will make the denominator equal to zero.

$\boxed{x = 0}$ and $\boxed{x = 5}$ must be excluded from the domain.

In this section, we will simplify rational expressions and perform operations on rational expressions such as multiplication, division, addition, and subtraction. The resulting expressions may not have *explicit* domain restrictions, but it is important to note that there are *implicit* domain restrictions, because the domain restrictions on the original rational expression still apply.

Technology Tip

Compare the graphs of
$Y_1 = \dfrac{x^2 - x - 2}{2x + 2}$ and

$Y_2 = \dfrac{x-2}{2}, x \neq -1.$

To show the graph and table of Y_1, unhighlight ☰ for Y_2 and press GRAPH .

Note that $x = -1$ is not defined.

Simplifying Rational Expressions

Recall that a fraction is *reduced* when it is written with no common factors.

$$\frac{16}{12} = \frac{4 \cdot 4}{4 \cdot 3} = \left(\frac{4}{4}\right) \cdot \left(\frac{4}{3}\right) = (1) \cdot \left(\frac{4}{3}\right) = \frac{4}{3} \quad \text{or} \quad \frac{16}{12} = \frac{\cancel{4} \cdot 4}{\cancel{4} \cdot 3} = \frac{4}{3}$$

Similarly, rational expressions are **reduced to lowest terms**, or **simplified**, if the numerator and denominator have no common factors other than ± 1. As with real numbers, the ability to write fractions in reduced form is dependent upon your ability to factor.

REDUCING A RATIONAL EXPRESSION TO LOWEST TERMS (SIMPLIFYING)

1. Factor the numerator and denominator completely.
2. State any domain restrictions.
3. Cancel (divide out) the common factors in the numerator and denominator.

EXAMPLE 3   **Reducing a Rational Expression to Lowest Terms**

Simplify $\dfrac{x^2 - x - 2}{2x + 2}$ and state any domain restrictions.

Solution:

Factor the numerator and denominator.

$$\frac{x^2 - x - 2}{2x + 2} = \frac{(x - 2)(x + 1)}{2(x + 1)}$$

State any domain restrictions.

$$x \neq -1$$

Cancel (divide out) the common factor, $x + 1$.

$$= \frac{(x - 2)\cancel{(x + 1)}}{2\cancel{(x + 1)}}$$

The rational expression is now in lowest terms (simplified).

$$= \boxed{\frac{x - 2}{2} \qquad x \neq -1}$$

■ **YOUR TURN** Simplify $\dfrac{x^2 + x - 2}{2x - 2}$ and state any domain restrictions.

The following table summarizes two **common mistakes** made with rational expressions.

CORRECT	INCORRECT	COMMENT
$\dfrac{x + 5}{y + 5}$ is already simplified.	**Error:** $\dfrac{x + \cancel{5}}{y + \cancel{5}} = \dfrac{x}{y}$	*Factors* can be divided out (canceled). Terms or parts of terms cannot be divided out. Remember to factor the numerator and denominator first, and then divide out common factors.
$\dfrac{x}{x^2 + x}$ $= \dfrac{\cancel{x}}{\cancel{x}(x + 1)} \qquad x \neq 0, x \neq -1$ $= \boxed{\dfrac{1}{x + 1} \qquad x \neq 0, x \neq -1}$	**Error:** $\dfrac{x}{x^2 + x} = \dfrac{1}{x + 1} \qquad x \neq -1$ *Note:* Missing $x \neq 0$.	Determine the domain restrictions *before* dividing out common factors.

EXAMPLE 4   **Simplifying Rational Expressions**

Reduce $\dfrac{x^2 - x - 6}{x^2 + x - 2}$ to lowest terms and state any domain restrictions.

Solution:

Factor the numerator and denominator.

$$\frac{x^2 - x - 6}{x^2 + x - 2} = \frac{(x - 3)(x + 2)}{(x - 1)(x + 2)}$$

State domain restrictions.

$$x \neq -2, x \neq 1$$

Divide out the common factor, $x + 2$.

$$= \frac{(x - 3)\cancel{(x + 2)}}{(x - 1)\cancel{(x + 2)}}$$

Simplify.

$$= \boxed{\frac{x - 3}{x - 1} \qquad x \neq -2, x \neq 1}$$

■ **YOUR TURN** Reduce $\dfrac{x^2 + x - 6}{x^2 + 2x - 3}$ to lowest terms and state any domain restrictions.

Classroom Example 0.5.3
Simplify and state any domain restrictions.

a. $\dfrac{5x^2 + 3x - 2}{2x^2 + 4x + 2}$

b. $\dfrac{15x^2 - 51x + 18}{-5x^2 + 17x - 6}$

Answer:

a. $\dfrac{5x - 2}{2(x + 1)}, x \neq -1$

b. $-3, x \neq 3, \frac{2}{5}$

■ **Answer:** $\dfrac{x + 2}{2} \qquad x \neq 1$

Study Tip

Factors can be divided out (canceled). Factor the numerator and denominator first, and then divide out common factors.

Study Tip

Determine domain restrictions of a rational expression *before* dividing out (canceling) common factors.

Classroom Example 0.5.4
Simplify and state any domain restrictions.

$\dfrac{14 - 29x - 15x^2}{5x^2 - 7x + 2}$

Answer:

$\dfrac{3x + 7}{1 - x}, x \neq 1, \frac{2}{5}$

■ **Answer:** $\dfrac{x - 2}{x - 1} \qquad x \neq -3, x \neq 1$

EXAMPLE 5 Simplifying Rational Expressions

Reduce $\dfrac{x^2 - 4}{2 - x}$ to lowest terms and state any domain restrictions.

Solution:

Factor the numerator and denominator.

$$\frac{x^2 - 4}{2 - x} = \frac{(x - 2)(x + 2)}{(2 - x)}$$

State domain restrictions.

$$x \neq 2$$

Factor out a negative in the denominator.

$$= \frac{(x - 2)(x + 2)}{-(x - 2)}$$

Cancel (divide out) the common factor, $x - 2$.

$$= \frac{\cancel{(x - 2)}(x + 2)}{-\cancel{(x - 2)}}$$

Simplify.

$$= \boxed{-(x + 2) \qquad x \neq 2}$$

■ **YOUR TURN** Reduce $\dfrac{x^2 - 25}{5 - x}$ to lowest terms and state any domain restrictions.

Classroom Example 0.5.5
Simplify and state any domain restrictions.

a. $\dfrac{4x^3 - 4x^2 + x}{4x^2 - 1}$

b.* $\dfrac{a^2x^2 - b^2}{2ab - 2a^2x}$, where $a, b \neq 0$

Answer:

a. $\dfrac{2x^2 - x}{2x + 1}, x \neq \pm\frac{1}{2}$

b. $-\dfrac{ax + b}{2a}, x \neq \dfrac{b}{a}$

■ **Answer:** $-(x + 5) \qquad x \neq 5$

Multiplying and Dividing Rational Expressions

The same rules that apply to multiplying and dividing rational numbers also apply to rational expressions.

PROPERTY	RESTRICTION	DESCRIPTION
$\dfrac{a}{b} \cdot \dfrac{c}{d} = \dfrac{ac}{bd}$	$b \neq 0, d \neq 0$	Multiply numerators and denominators, respectively.
$\dfrac{a}{b} \div \dfrac{c}{d} = \dfrac{a}{b} \cdot \dfrac{d}{c}$	$b \neq 0, d \neq 0, c \neq 0$	Dividing is equivalent to multiplying by a reciprocal.

Multiplying Rational Expressions

1. Factor all numerators and denominators completely.
2. State any domain restrictions.
3. Divide the numerators and denominators by any common factors.
4. Multiply the remaining numerators and denominators, respectively.

Dividing Rational Expressions

1. Factor all numerators and denominators completely.
2. State any domain restrictions.
3. Rewrite division as multiplication by a reciprocal.
4. State any additional domain restrictions.
5. Divide the numerators and denominators by any common factors.
6. Multiply the remaining numerators and denominators, respectively.

EXAMPLE 6 Multiplying Rational Expressions

Multiply and simplify $\dfrac{3x + 1}{4x^2 + 4x} \cdot \dfrac{x^3 + 3x^2 + 2x}{9x + 3}$.

Solution:

Factor the numerators and denominators.

$$= \frac{(3x + 1)}{4x(x + 1)} \cdot \frac{x(x + 1)(x + 2)}{3(3x + 1)}$$

State any domain restrictions.

$$x \neq 0, x \neq -1, x \neq -\frac{1}{3}$$

Divide the numerators and denominators by common factors.

$$= \frac{(3x + 1)}{4x(x + 1)} \cdot \frac{x(x + 1)(x + 2)}{3(3x + 1)}$$

Simplify.

$$= \boxed{\frac{x + 2}{12} \qquad x \neq 0, x \neq -1, x \neq -\frac{1}{3}}$$

■ **YOUR TURN** Multiply and simplify $\dfrac{2x + 1}{3x^2 - 3x} \cdot \dfrac{x^3 + 2x^2 - 3x}{8x + 4}$.

EXAMPLE 7 Dividing Rational Expressions

Divide and simplify $\dfrac{x^2 - 4}{x} \div \dfrac{3x^3 - 12x}{5x^3}$.

Solution:

Factor numerators and denominators.

$$\frac{(x - 2)(x + 2)}{x} \div \frac{3x(x - 2)(x + 2)}{5x^3}$$

State any domain restrictions.

$$x \neq 0$$

Write the quotient as a product.

$$= \frac{(x - 2)(x + 2)}{x} \cdot \frac{5x^3}{3x(x - 2)(x + 2)}$$

State any additional domain restrictions.

$$x \neq -2, x \neq 2$$

Divide out the common factors.

$$= \frac{(x - 2)(x + 2)}{x} \cdot \frac{5x^3}{3x(x - 2)(x + 2)}$$

Simplify.

$$= \boxed{\frac{5x}{3} \qquad x \neq -2, x \neq 0, x \neq 2}$$

■ **YOUR TURN** Divide and simplify $\dfrac{x^2 - 9}{x} \div \dfrac{2x^3 - 18x}{7x^4}$.

Classroom Example 0.5.6
Multiply and simplify.

a. $\dfrac{4 + x^2}{4 + x} \cdot \dfrac{x^2 - 16}{8 + 2x^2}$

b. $\dfrac{(x^2 + 1)(2x + 1)}{x^2} \cdot \dfrac{2x^5 + x^4}{x^4 - 1}$

Answer:

a. $\dfrac{x - 4}{2}, x \neq -4$

b. $\dfrac{x^2(2x + 1)^2}{x^2 - 1}, x \neq 0, \pm 1$

■ **Answer:**

$\dfrac{x + 3}{12} \qquad x \neq 0, x \neq 1, x \neq -\dfrac{1}{2}$

Classroom Example 0.5.7
Divide and simplify.

a. $\dfrac{(x^2 + 1)(2x + 1)}{x^2(x^2 - 1)} \div \dfrac{2x^5 + x^4}{x^4 - 1}$

b.* $\dfrac{x^2 - 5}{16x} \div \left(\dfrac{x^2}{x^4 - 25} \right)^{-1}$

Answer:

a. $\dfrac{(x^2 + 1)^2}{x^6}$

$x \neq -\dfrac{1}{2}, x \neq 0, x \neq \pm 1$

b. $\dfrac{x}{16(x^2 + 5)}$

$x \neq \pm\sqrt{5}, x \neq 0$

■ **Answer:**

$\dfrac{7x^2}{2} \qquad x \neq -3, x \neq 0, x \neq 3$

Adding and Subtracting Rational Expressions

The same rules that apply to adding and subtracting rational numbers also apply to rational expressions.

PROPERTY	RESTRICTION	DESCRIPTION
$\dfrac{a}{b} \pm \dfrac{c}{b} = \dfrac{a \pm c}{b}$	$b \neq 0$	Adding or subtracting rational expressions when the denominators are the same
$\dfrac{a}{b} \pm \dfrac{c}{d} = \dfrac{ad \pm bc}{bd}$	$b \neq 0$ and $d \neq 0$	Adding or subtracting rational expressions when the denominators are different

EXAMPLE 8   **Adding and Subtracting Rational Expressions: Equal Denominators**

Perform the indicated operation and simplify.

a. $\dfrac{x+7}{(x+2)^2} + \dfrac{3x+1}{(x+2)^2}$ b. $\dfrac{6x+7}{2x-1} - \dfrac{2x+9}{2x-1}$

Solution (a):

Write as a single expression. $= \dfrac{x+7+3x+1}{(x+2)^2}$

State any domain restrictions. $x \neq -2$

Combine like terms in the numerator. $= \dfrac{4x+8}{(x+2)^2}$

Factor out the common factor in the numerator. $= \dfrac{4(x+2)}{(x+2)^2}$

Cancel (divide out) the common factor, $x + 2$. $= \boxed{\dfrac{4}{x+2} \qquad x \neq -2}$

Solution (b):

Write as a single expression. Use parentheses around
the second numerator to ensure that the negative will $= \dfrac{6x+7-(2x+9)}{2x-1}$
be distributed throughout all terms.

State any domain restrictions. $x \neq \dfrac{1}{2}$

Eliminate parentheses. Distribute the negative. $= \dfrac{6x+7-2x-9}{2x-1}$

Combine like terms in the numerator. $= \dfrac{4x-2}{2x-1}$

Factor out the common factor in the numerator. $= \dfrac{2(2x-1)}{2x-1}$

Divide out (cancel) the common factor, $2x - 1$. $= \boxed{2 \qquad x \neq \dfrac{1}{2}}$

EXAMPLE 9   **Adding and Subtracting Rational Expressions: No Common Factors in Denominators**

Perform the indicated operation and simplify.

a. $\dfrac{3-x}{2x+1} + \dfrac{x}{x-1}$ b. $\dfrac{1}{x^2} - \dfrac{2}{x+1}$

Solution (a):

The common denominator is the product of the denominators.

$$\dfrac{3-x}{2x+1} + \dfrac{x}{x-1} = \dfrac{(3-x)(x-1)}{(2x+1)(x-1)} + \dfrac{x(2x+1)}{(x-1)(2x+1)}$$

$$= \dfrac{(3-x)(x-1) + x(2x+1)}{(2x+1)(x-1)}$$

Eliminate parentheses in
the numerator. $= \dfrac{3x-3-x^2+x+2x^2+x}{(2x+1)(x-1)}$

Combine like terms in
the numerator. $= \boxed{\dfrac{x^2+5x-3}{(2x+1)(x-1)} \qquad x \neq -\dfrac{1}{2}, x \neq 1}$

Solution (b):

The common denominator is the product of the denominators.

$$\frac{1}{x^2} - \frac{2}{x+1} = \frac{(1)(x+1) - 2(x^2)}{x^2(x+1)}$$

Eliminate parentheses in the numerator.

$$= \frac{x+1-2x^2}{x^2(x+1)} = \frac{-2x^2+x+1}{x^2(x+1)} = \frac{-(2x^2-x-1)}{x^2(x+1)}$$

Write the numerator in factored form to ensure no further simplification is possible.

$$= \boxed{\frac{-(2x+1)(x-1)}{x^2(x+1)} \qquad x \ne -1, x \ne 0}$$

Study Tip

When adding or subtracting rational expressions whose denominators have no common factors, the least common denominator is the product of the two denominators.

■ **YOUR TURN** Perform the indicated operation and simplify.

a. $\dfrac{2x-1}{x+3} + \dfrac{x}{2x+1}$ **b.** $\dfrac{2}{x^2+1} - \dfrac{1}{x}$

■ **Answer:**

a. $\dfrac{5x^2+3x-1}{(x+3)(2x+1)}$
$x \ne -3, x \ne -\frac{1}{2}$

b. $\dfrac{-(x-1)^2}{x(x^2+1)} \qquad x \ne 0$

When combining two or more fractions through addition or subtraction, recall that the **least common multiple**, or **least common denominator (LCD)**, is the smallest real number that all of the denominators divide into evenly (that is, the smallest of which all are factors). For example,

$$\frac{2}{3} + \frac{1}{6} - \frac{4}{9}$$

To find the LCD of these three fractions, factor the denominators into prime factors:

$$
\begin{aligned}
3 &= \mathbf{3} \\
6 &= \mathbf{3 \cdot 2} \\
9 &= \mathbf{3} \quad \cdot \mathbf{3} \\
\hline
&\mathbf{3 \cdot 2 \cdot 3 = 18}
\end{aligned}
\qquad
\frac{2}{3} + \frac{1}{6} - \frac{4}{9} = \frac{12+3-8}{18} = \frac{7}{18}
$$

Rational expressions follow this same procedure, only now variables are also considered:

$$\frac{1}{2x} + \frac{1}{x^3} = \frac{1}{2x} + \frac{1}{x \cdot x \cdot x} \qquad \text{LCD} = 2x^3$$

$$\frac{2}{x+1} - \frac{x}{2x+1} + \frac{3-x}{(x+1)^2} = \frac{2}{(x+1)} - \frac{x}{(2x+1)} + \frac{3-x}{(x+1)(x+1)} \qquad \text{LCD} = (x+1)^2(2x+1)$$

The following box summarizes the LCD method for adding and subtracting rational expressions whose denominators have common factors.

THE LCD METHOD FOR ADDING AND SUBTRACTING RATIONAL EXPRESSIONS

1. Factor each of the denominators completely.
2. The LCD is the product of each of these distinct factors raised to the highest power to which that factor appears in any of the denominators.
3. Write each rational expression using the LCD for each denominator.
4. Add or subtract the resulting numerators.
5. Factor the resulting numerator to check for common factors.

EXAMPLE 10 **Subtracting Rational Expressions: Common Factors in Denominators (LCD)**

Perform the indicated operation and write in simplified form.

$$\frac{5x}{2x - 6} - \frac{7x - 2}{x^2 - x - 6}$$

Solution:

Factor the denominators.

$$= \frac{5x}{2(x - 3)} - \frac{7x - 2}{(x - 3)(x + 2)}$$

Identify the LCD.

$$\text{LCD} = 2(x - 3)(x + 2)$$

Write each expression using the LCD as the denominator.

$$= \frac{5x(x + 2)}{2(x - 3)(x + 2)} - \frac{2(7x - 2)}{2(x - 3)(x + 2)}$$

Combine into one expression. Distribute the negative through the entire second numerator.

$$= \frac{5x^2 + 10x - 14x + 4}{2(x - 3)(x + 2)}$$

Simplify.

$$= \boxed{\frac{5x^2 - 4x + 4}{2(x - 3)(x + 2)} \qquad x \neq -2, x \neq 3}$$

▪ **YOUR TURN** Perform the indicated operation and write in simplified form.

$$\frac{2x}{3x - 6} - \frac{5x + 1}{x^2 + 2x - 8}$$

Classroom Example 0.5.10
Perform the indicated operation and simplify.

a. $\dfrac{1 - x}{x^2 + 1} - \dfrac{-3}{2x}$

b.* $\dfrac{a}{x + a} - \dfrac{a}{x - a}$,
where $a \neq 0$

Answer:

a. $\dfrac{x^2 + 2x + 3}{2x(x^2 + 1)} \quad x \neq 0$

b. $\dfrac{-2a^2}{x^2 - a^2} \quad x \neq \pm a$

▪ **Answer:**
$\dfrac{2x^2 - 7x - 3}{3(x - 2)(x + 4)} \quad x \neq -4, x \neq 2$

Complex Rational Expressions

A rational expression that contains another rational expression in either its numerator or denominator is called a **complex rational expression**. The following are examples of complex rational expressions.

$$\frac{\dfrac{1}{x} - 5}{2 + x} \qquad\qquad \frac{2 - x}{4 + \dfrac{3}{x - 1}} \qquad\qquad \frac{\dfrac{3}{x} - 7}{\dfrac{6}{2x - 5} - 1}$$

TWO METHODS FOR SIMPLIFYING COMPLEX RATIONAL EXPRESSIONS

Procedure 1: Write a sum or difference of rational expressions that appear in either the numerator or denominator as a single rational expression. Once the complex rational expression contains a single rational expression in the numerator and one in the denominator, then rewrite the division as multiplication by the reciprocal.

OR

Procedure 2: Find the LCD of all rational expressions contained in both the numerator and denominator. Multiply the numerator and denominator by this LCD and simplify.

EXAMPLE 11   Simplifying a Complex Rational Expression

Write the rational expression in simplified form.

$$\dfrac{\dfrac{2}{x} + 1}{1 + \dfrac{1}{x + 1}}$$

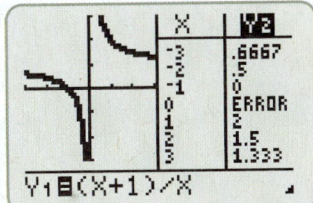
Solution:

State the domain restrictions. $\boxed{x \neq 0}$, $\boxed{x \neq -1}$, and $\boxed{x \neq -2}$

Procedure 1:

Add the expressions in both the
numerator and denominator.

$$= \dfrac{\dfrac{2}{x} + \dfrac{x}{x}}{\dfrac{x + 1}{x + 1} + \dfrac{1}{x + 1}} = \dfrac{\dfrac{2 + x}{x}}{\dfrac{(x + 1) + 1}{x + 1}}$$

Simplify.

$$= \dfrac{\dfrac{2 + x}{x}}{\dfrac{x + 2}{x + 1}}$$

Express the quotient as a product.

$$= \dfrac{2 + x}{x} \cdot \dfrac{x + 1}{x + 2}$$

Divide out the common factors.

$$= \dfrac{\cancel{2 + x}}{x} \cdot \dfrac{x + 1}{\cancel{x + 2}}$$

Write in simplified form.

$$= \boxed{\dfrac{x + 1}{x} \qquad x \neq -2, x \neq -1, x \neq 0}$$

Procedure 2:

Find the LCD of the numerator
and denominator.

$$\dfrac{\dfrac{2}{x} + 1}{1 + \dfrac{1}{x + 1}}$$

Identify the LCDs.

Numerator LCD: x
Denominator LCD: $x + 1$
Combined LCD: $x(x + 1)$

Multiply both numerator and
denominator by their combined LCD.

$$= \dfrac{\dfrac{2}{x} + 1}{1 + \dfrac{1}{x + 1}} \cdot \dfrac{x(x + 1)}{x(x + 1)}$$

Multiply the numerators and
denominators, respectively, applying
the distributive property.

$$= \dfrac{\dfrac{2}{x} \cdot x(x + 1) + 1x(x + 1)}{1 \cdot x(x + 1) + \dfrac{1}{x + 1} \cdot x(x + 1)}$$

Divide out common factors.

$$= \dfrac{\dfrac{2}{\cancel{x}} \cdot \cancel{x}(x + 1) + 1x(x + 1)}{x(x + 1) + \dfrac{1}{\cancel{x + 1}} \cdot x(\cancel{x + 1})}$$

Simplify.

$$= \dfrac{2(x + 1) + x(x + 1)}{x(x + 1) + x}$$

Apply the distributive property.

$$= \frac{2x + 2 + x^2 + x}{x^2 + x + x}$$

Combine like terms.

$$= \frac{x^2 + 3x + 2}{x^2 + 2x}$$

Factor the numerator and denominator.

$$= \frac{(x+2)(x+1)}{x(x+2)}$$

Divide out the common factor.

$$= \frac{\cancel{(x+2)}(x+1)}{x\cancel{(x+2)}}$$

Write in simplified form.

$$= \boxed{\frac{x+1}{x} \qquad x \neq -2, x \neq -1, x \neq 0}$$

EXAMPLE 12   Simplifying a Complex Rational Expression

Write the rational expression in simplified form.

$$\frac{\dfrac{1}{x^2 - 9} + 3}{1 - \dfrac{x}{2x+6}} \qquad x \neq -6, -3, 3$$

Solution: Using Procedure 1

Factor the respective denominators.

$$\frac{\dfrac{1}{(x-3)(x+3)} + 3}{1 - \dfrac{x}{2(x+3)}}$$

Identify the LCDs.

Numerator LCD: $(x-3)(x+3)$
Denominator LCD: $2(x+3)$
Combined LCD: $2(x-3)(x+3)$

Multiply both the numerator and the denominator by the combined LCD.

$$= \frac{\dfrac{1}{(x-3)(x+3)} + 3}{1 - \dfrac{x}{2(x+3)}} \cdot \frac{2(x+3)(x-3)}{2(x+3)(x-3)}$$

Multiply the numerators and denominators, respectively, applying the distributive property.

$$= \frac{\dfrac{2(x+3)(x-3)}{(x-3)(x+3)} + 3 \cdot 2(x+3)(x-3)}{2(x+3)(x-3) - \dfrac{x \cdot 2(x+3)(x-3)}{2(x+3)}}$$

Simplify.

$$= \frac{2 + 6(x+3)(x-3)}{2(x+3)(x-3) - x(x-3)}$$

Eliminate the parentheses.

$$= \frac{2 + 6x^2 - 54}{2x^2 - 18 - x^2 + 3x}$$

Combine like terms.

$$= \frac{6x^2 - 52}{x^2 + 3x - 18}$$

Factor the numerator and denominator to make sure there are no common factors.

$$= \boxed{\frac{2(3x^2 - 26)}{(x+6)(x-3)} \qquad x \neq -6, x \neq -3, x \neq 3}$$

In this section, rational expressions were defined as quotients of polynomials. The domain of any polynomial is the set of all real numbers. Since rational expressions are ratios of polynomials, the domain of rational expressions is the set of all real numbers *except those values that make the denominator equal to zero.* In this section, rational expressions were simplified (written with no common factors), multiplied, divided, added, and subtracted.

OPERATION	EXAMPLE	NOTE
Multiplying rational expressions	$\dfrac{2}{x+1} \cdot \dfrac{3x}{x-1} = \dfrac{6x}{x^2-1}$ $\quad x \neq \pm 1$	State domain restrictions.
Dividing rational expressions	$\dfrac{2}{x+1} \div \dfrac{3x}{x-1}$ $\quad x \neq \pm 1$ $ = \dfrac{2}{x+1} \cdot \dfrac{x-1}{3x}$ $\quad x \neq 0$ $ = \dfrac{2(x-1)}{3x(x+1)}$ $\quad x \neq 0, x \neq \pm 1$	When dividing rational expressions, remember to check for additional domain restrictions once the division is rewritten as multiplication by a reciprocal.
Adding/subtracting rational expressions with no common factors	$\dfrac{2}{x+1} + \dfrac{3x}{x-1}$ $\quad x \neq \pm 1$ $\text{LCD} = (x+1)(x-1)$ $ = \dfrac{2(x-1) + 3x(x+1)}{(x+1)(x-1)}$ $ = \dfrac{2x-2+3x^2+3x}{(x+1)(x-1)}$ $ = \dfrac{3x^2+5x-2}{(x+1)(x-1)}$ $ = \dfrac{(3x-1)(x+2)}{(x+1)(x-1)}$ $\quad x \neq \pm 1$	The least common denominator (LCD) is the product of the two denominators.
Adding/subtracting rational expressions with common factors	$\dfrac{3}{x(x+1)} - \dfrac{2}{x(x+2)}$ $\quad x \neq -2, -1, 0$ $\text{LCD} = x(x+1)(x+2)$ $ = \dfrac{3(x+2) - 2(x+1)}{x(x+1)(x+2)}$ $ = \dfrac{3x+6-2x-2}{x(x+1)(x+2)}$ $ = \dfrac{x+4}{x(x+1)(x+2)}$ $\quad x \neq -2, -1, 0$	The LCD is the product of each of these distinct factors raised to the highest power that appears in any of the denominators.

Complex rational expressions are simplified in one of two ways:

1. Combine the sum or difference of rational expressions in a numerator or denominator as a single rational expression. The result is a rational expression in the numerator and a rational expression in the denominator. Then write the division as multiplication by the reciprocal.

2. Multiply the numerator and denominator by the overall LCD (LCD for all rational expressions that appear). The result is a single rational expression. Then simplify, if possible.

SKILLS

In Exercises 1–10, state any real numbers that must be excluded from the domain of each rational expression.

1. $\dfrac{3}{x}$ **2.** $\dfrac{5}{x}$ **3.** $\dfrac{3}{x-1}$ **4.** $\dfrac{6}{y-1}$ **5.** $\dfrac{5x-1}{x+1}$ **6.** $\dfrac{2x}{3-x}$

7. $\dfrac{2p^2}{p^2-1}$ **8.** $\dfrac{3t}{t^2-9}$ **9.** $\dfrac{3p-1}{p^2+1}$ **10.** $\dfrac{2t-2}{t^2+4}$

In Exercises 11–30, reduce the rational expression to lowest terms and state any real numbers that must be excluded from the domain.

11. $\dfrac{(x+3)(x-9)}{2(x+3)(x+9)}$ **12.** $\dfrac{4y(y-8)(y+7)}{8y(y+7)(y+8)}$ **13.** $\dfrac{(x-3)(x+1)}{2(x+1)}$ **14.** $\dfrac{(2x+1)(x-3)}{3(x-3)}$

15. $\dfrac{2(3y+1)(2y-1)}{3(2y-1)(3y)}$ **16.** $\dfrac{7(2y+1)(3y-1)}{5(3y-1)(2y)}$ **17.** $\dfrac{(5y-1)(y+1)}{25y-5}$ **18.** $\dfrac{(2t-1)(t+2)}{4t+8}$

19. $\dfrac{(3x+7)(x-4)}{4x-16}$ **20.** $\dfrac{(t-7)(2t+5)}{3t-21}$ **21.** $\dfrac{x^2-4}{x-2}$ **22.** $\dfrac{t^3-t}{t-1}$

23. $\dfrac{x+7}{x+7}$ **24.** $\dfrac{2y+9}{2y+9}$ **25.** $\dfrac{x^2+9}{2x+9}$ **26.** $\dfrac{x^2+4}{2x+4}$

27. $\dfrac{x^2+5x+6}{x^2-3x-10}$ **28.** $\dfrac{x^2+19x+60}{x^2+8x+16}$ **29.** $\dfrac{6x^2-x-1}{2x^2+9x-5}$ **30.** $\dfrac{15x^2-x-2}{5x^2+13x-6}$

In Exercises 31–48, multiply the rational expressions and simplify. State any real numbers that must be excluded from the domain.

31. $\dfrac{x-2}{x+1} \cdot \dfrac{3x+5}{x-2}$ **32.** $\dfrac{4x+5}{x-2} \cdot \dfrac{3x+4}{4x+5}$ **33.** $\dfrac{5x+6}{x} \cdot \dfrac{2x}{5x-6}$

34. $\dfrac{4(x-2)(x+5)}{8x} \cdot \dfrac{16x}{(x-5)(x+5)}$ **35.** $\dfrac{2x-2}{3x} \cdot \dfrac{x^2+x}{x^2-1}$ **36.** $\dfrac{5x-5}{10x} \cdot \dfrac{x^2+x}{x^2-1}$

37. $\dfrac{3x^2-12}{x} \cdot \dfrac{x^2+5x}{x^2+3x-10}$ **38.** $\dfrac{4x^2-32x}{x} \cdot \dfrac{x^2+3x}{x^2-5x-24}$ **39.** $\dfrac{t+2}{3t-9} \cdot \dfrac{t^2-6t+9}{t^2+4t+4}$

40. $\dfrac{y+3}{3y+9} \cdot \dfrac{y^2-10y+25}{y^2+3y-40}$ **41.** $\dfrac{t^2+4}{t-3} \cdot \dfrac{3t}{t+2}$ **42.** $\dfrac{7a^2+21a}{14(a^2-9)} \cdot \dfrac{a+3}{7}$

43. $\dfrac{y^2-4}{y-3} \cdot \dfrac{3y}{y+2}$ **44.** $\dfrac{t^2+t-6}{t^2-4} \cdot \dfrac{8t}{2t^2}$ **45.** $\dfrac{3x^2-15x}{2x^3-50x} \cdot \dfrac{2x^2-7x-15}{3x^2+15x}$

46. $\dfrac{5t-1}{4t} \cdot \dfrac{4t^2+3t}{16t^2-9}$ **47.** $\dfrac{6x^2-11x-35}{8x^2-22x-21} \cdot \dfrac{4x^2-49}{9x^2-25}$ **48.** $\dfrac{3x^2-2x}{12x^3-8x^2} \cdot \dfrac{x^2-7x-18}{2x^2-162}$

In Exercises 49–66, divide the rational expressions and simplify. State any real numbers that must be excluded from the domain.

49. $\dfrac{3}{x} \div \dfrac{12}{x^2}$

50. $\dfrac{5}{x^2} \div \dfrac{10}{x^3}$

51. $\dfrac{6}{x-2} \div \dfrac{12}{(x-2)(x+2)}$

52. $\dfrac{5(x+6)}{10(x-6)} \div \dfrac{20(x+6)}{8}$

53. $\dfrac{1}{x-1} \div \dfrac{5}{x^2-1}$

54. $\dfrac{5}{3x-4} \div \dfrac{10}{9x^2-16}$

55. $\dfrac{2-p}{p^2-1} \div \dfrac{2p-4}{p+1}$

56. $\dfrac{4-x}{x^2-16} \div \dfrac{12-3x}{x-4}$

57. $\dfrac{36-n^2}{n^2-9} \div \dfrac{n+6}{n+3}$

58. $\dfrac{49-y^2}{y^2-25} \div \dfrac{7+y}{2y+10}$

59. $\dfrac{3t^3-6t^2-9t}{5t-10} \div \dfrac{6+6t}{4t-8}$

60. $\dfrac{x^3+8x^2+12x}{5x^2-10x} \div \dfrac{4x+8}{x^2-4}$

61. $\dfrac{w^2-w}{w} \div \dfrac{w^3-w}{5w^3}$

62. $\dfrac{y^2-3y}{2y} \div \dfrac{y^3-3y^2}{8y}$

63. $\dfrac{x^2+4x-21}{x^2+3x-10} \div \dfrac{x^2-2x-63}{x^2+x-20}$

64. $\dfrac{2y^2-5y-3}{2y^2-9y-5} \div \dfrac{3y-9}{y^2-5y}$

65. $\dfrac{20x^2-3x-2}{25x^2-4} \div \dfrac{12x^2+23x+5}{3x^2+5x}$

66. $\dfrac{x^2-6x-27}{2x^2+13x-7} \div \dfrac{2x^2-15x-27}{2x^2+9x-5}$

In Exercises 67–82, add or subtract the rational expression and simplify. State any real numbers that must be excluded from the domain.

67. $\dfrac{3}{x} - \dfrac{2}{5x}$

68. $\dfrac{5}{7x} - \dfrac{3}{x}$

69. $\dfrac{3}{p-2} + \dfrac{5p}{p+1}$

70. $\dfrac{4}{9+x} - \dfrac{5x}{x-2}$

71. $\dfrac{2x+1}{5x-1} - \dfrac{3-2x}{1-5x}$

72. $\dfrac{7}{2x-1} - \dfrac{5}{1-2x}$

73. $\dfrac{3y^2}{y+1} + \dfrac{1-2y}{y-1}$

74. $\dfrac{3}{1-x} + \dfrac{4}{x-1}$

75. $\dfrac{3x}{x^2-4} + \dfrac{3+x}{x+2}$

76. $\dfrac{x-1}{4-x^2} - \dfrac{x+1}{2+x}$

77. $\dfrac{x-1}{x-2} + \dfrac{x-6}{x^2-4}$

78. $\dfrac{2}{y-3} + \dfrac{7}{y+2}$

79. $\dfrac{5a}{a^2-b^2} - \dfrac{7}{b-a}$

80. $\dfrac{1}{y} + \dfrac{4}{y^2-4} - \dfrac{2}{y^2-2y}$

81. $7 + \dfrac{1}{x-3}$

82. $\dfrac{3}{5y+6} - \dfrac{4}{y-2} + \dfrac{y^2-y}{5y^2-4y-12}$

In Exercises 83–90, simplify the complex rational expressions. State any real numbers that must be excluded from the domain.

83. $\dfrac{\dfrac{1}{x}-1}{1-\dfrac{2}{x}}$

84. $\dfrac{\dfrac{3}{y}-5}{4-\dfrac{2}{y}}$

85. $\dfrac{3+\dfrac{1}{x}}{9-\dfrac{1}{x^2}}$

86. $\dfrac{\dfrac{1}{x}+\dfrac{2}{x^2}}{\dfrac{9}{x}-\dfrac{5}{x^2}}$

87. $\dfrac{\dfrac{1}{x-1}+1}{1-\dfrac{1}{x+1}}$

88. $\dfrac{\dfrac{7}{y+7}}{\dfrac{1}{y+7}-\dfrac{1}{y}}$

89. $\dfrac{\dfrac{1}{x-1}+1}{\dfrac{1}{x+1}+1}$

90. $\dfrac{\dfrac{3}{x+1}-\dfrac{3}{x-1}}{\dfrac{5}{x^2-1}}$

■ **APPLICATIONS**

91. **Finance.** The amount of payment made on a loan is given by the formula $A = \dfrac{pi}{1-1/(1+i)^n}$, where p is the principal (amount borrowed), and $i = \dfrac{r}{n}$, where r is the interest rate expressed as a decimal and n is the number of payments per year. Suppose $n = 5$. Simplify the formula as much as possible.

92. **Finance.** Use the formula $A = \dfrac{pi}{1-\dfrac{1}{(1+i)^{nt}}}$ to calculate the amount your monthly payment will be on a loan of $150,000 at an interest rate of 6.5% for 30 years ($nt = 360$).

93. **Circuits.** If two resistors are connected in parallel, the combined resistance is given by the formula $R = \dfrac{1}{1/R_1 + 1/R_2}$, where R_1 and R_2 are the individual resistances. Simplify the formula.

94. **Optics.** The focal length of a lens can be calculated by applying the formula $f = \dfrac{1}{1/p + 1/q}$, where p is the distance that the object is from the lens and q is the distance that the image is from the lens. Simplify the formula.

▪CATCH THE MISTAKE

In Exercises 95 and 96, explain the mistake that is made.

95. Simplify $\dfrac{x^2 + 2x + 1}{x + 1}$.

Solution:

Factor the numerator.

$$\frac{x^2 + 2x + 1}{x + 1} = \frac{(x + 1)(x + 1)}{(x + 1)}$$

Cancel the common factor, $x + 1$.

$$= \frac{(x + 1)(\cancel{x + 1})}{(\cancel{x + 1})}$$

Write in simplified form.

$$= x + 1$$

This is incorrect. What mistake was made?

96. Simplify $\dfrac{x + 1}{x^2 + 2x + 1}$.

Solution:

Cancel the common 1s.

$$\frac{x \cancel{+ 1}}{x^2 + 2x \cancel{+ 1}}$$

Factor the denominator.

$$= \frac{x}{x(x + 2)}$$

Cancel the common x.

$$= \frac{\cancel{x}}{\cancel{x}(x + 2)}$$

Write in simplified form.

$$= \frac{1}{x + 2} \qquad x \neq -2, x \neq 0$$

This is incorrect. What mistake was made?

▪CONCEPTUAL

In Exercises 97–100, determine whether each of the statements is true or false.

97. $\dfrac{x^2 - 81}{x - 9} = x + 9$

98. $\dfrac{x - 9}{x^2 - 81} = \dfrac{1}{x + 9} \qquad x \neq -9, 9$

99. When adding or subtracting rational expressions, the LCD is always the product of all the denominators.

100. $\dfrac{x - c}{c - x} = -1$ for all values of x.

▪CHALLENGE

101. Perform the operation and simplify (remember to state domain restrictions).

$$\frac{x + a}{x + b} \div \frac{x + c}{x + d}$$

102. Write the numerator as the product of two binomials. Divide out any common factors of the numerator and denominator.

$$\frac{a^{2n} - b^{2n}}{a^n - b^n}$$

▪TECHNOLOGY

103. Utilizing a graphing technology, plot the expression $y = \dfrac{x + 7}{x + 7}$. Zoom in near $x = -7$. Does this agree with what you found in Exercise 23?

104. Utilizing a graphing technology, plot the expression $y = \dfrac{x^2 - 4}{x - 2}$. Zoom in near $x = 2$. Does this agree with what you found in Exercise 21?

In Exercises 105 and 106, for each given expression:
(a) simplify the expression, (b) use a graphing utility to plot the expression and the answer in (a) in the same viewing window, and (c) determine the domain restriction(s) where the graphs will agree with each other.

105. $\dfrac{1 + \dfrac{1}{x - 2}}{1 - \dfrac{1}{x + 2}}$

106. $\dfrac{1 - \dfrac{2}{x + 3}}{1 + \dfrac{1}{x + 4}}$

SKILLS OBJECTIVES

- Simplify square roots.
- Simplify radicals.
- Add and subtract radicals.
- Rationalize denominators containing radicals.
- Apply properties of rational exponents.

CONCEPTUAL OBJECTIVES

- Understand that radicals are equivalent to rational exponents.
- Understand that a radical implies one number (the principal root), not two ($\pm$ the principal root).

In Section 0.2, we discussed integer exponents and their properties. For example, $4^2 = 16$ and $x^2 \cdot x^3 = x^5$. In this section we expand our discussion of exponents to include any rational numbers. For example, $16^{1/2} = ?$ and $(x^{1/2})^{3/4} = ?$. We will first start with a more familiar notation (*roots*) and discuss operations on *radicals*, and then *rational exponents* will be discussed.

Square Roots

> **DEFINITION** **Principal Square Root**
>
> Let a be any nonnegative real number; then the nonnegative real number b is called the **principal square root of a**, denoted $b = \sqrt{a}$, if $b^2 = a$. The symbol $\sqrt{}$ is called a **radical sign**, and a is called the **radicand**.

It is important to note that the principal square root b is nonnegative. "The principal square root of 16 is 4" implies $4^2 = 16$. Although it is also true that $(-4)^2 = 16$, the principal square root is defined to be nonnegative.

It is also important to note that negative real numbers do not have real square roots. For example, $\sqrt{-9}$ is not a real number because there are no real numbers that when squared yield -9. Since principal square roots are defined to be nonnegative, this means they must be zero or positive. The square root of zero is equal to zero: $\sqrt{0} = 0$. All other nonnegative principal square roots are positive.

Study Tip

Incorrect: $\sqrt{25} = \pm 5$
Correct: $\sqrt{25} = 5$
The principal square root, $\sqrt{}$, is defined as a nonnegative real number.

EXAMPLE 1 Evaluating Square Roots

Evaluate the square roots, if possible.

a. $\sqrt{169}$ **b.** $\sqrt{\dfrac{4}{9}}$ **c.** $\sqrt{-36}$

Solution:

a. What positive real number squared results in 169?

$\sqrt{169} = \boxed{13}$

Check: $13^2 = 169$

b. What positive real number squared results in $\frac{4}{9}$?

$\sqrt{\dfrac{4}{9}} = \boxed{\dfrac{2}{3}}$

Check: $\left(\frac{2}{3}\right)^2 = \frac{4}{9}$

c. What positive real number squared results in -36?

$\boxed{\text{No real number}}$

Classroom Example 0.6.1
Evaluate these, if possible.

a. $\sqrt{\dfrac{25}{49}}$ **b.** $\sqrt{-4}$

Answer:

a. $\frac{5}{7}$ **b.** not a real number

SQUARE ROOTS OF PERFECT SQUARES

Let a be any real number; then:

$$\sqrt{a^2} = |a|$$

EXAMPLE 2 Finding Square Roots of Perfect Squares

Evaluate the following:

a. $\sqrt{6^2}$ **b.** $\sqrt{(-7)^2}$ **c.** $\sqrt{x^2}$

Solution:

a. $\sqrt{6^2} = \sqrt{36} = \boxed{6}$ **b.** $\sqrt{(-7)^2} = \sqrt{49} = \boxed{7}$ **c.** $\sqrt{x^2} = \boxed{|x|}$

Classroom Example 0.6.2*
Evaluate these, if possible.

a. $\sqrt{(-\pi)^2}$ **b.** $\sqrt{-3^2}$

Answer:
a. π **b.** not a real number

Simplifying Square Roots

So far only square roots of perfect squares have been discussed. Now we consider how to simplify square roots such as $\sqrt{12}$. We rely on the following properties.

PROPERTIES OF SQUARE ROOTS

Let a and b be nonnegative real numbers, then:

Property	Description	Example
$\sqrt{a \cdot b} = \sqrt{a} \cdot \sqrt{b}$	The square root of a product is the product of the square roots.	$\sqrt{20} = \sqrt{4} \cdot \sqrt{5} = 2\sqrt{5}$
$\sqrt{\dfrac{a}{b}} = \dfrac{\sqrt{a}}{\sqrt{b}}$ $b \neq 0$	The square root of a quotient is the quotient of the square roots.	$\sqrt{\dfrac{40}{49}} = \dfrac{\sqrt{40}}{\sqrt{49}} = \dfrac{\sqrt{4} \cdot \sqrt{10}}{7} = \dfrac{2\sqrt{10}}{7}$

Classroom Example 0.6.3
Simplify:

a. $\sqrt{56x^5}$ **b.** $\sqrt{63x^3}$
c.* $\sqrt{12x} \cdot \sqrt{12x^5}$

d.* $\dfrac{\sqrt{243x^7}}{\sqrt{3x^3}}$

Answer:
a. $2x^2\sqrt{14x}$ **b.** $3|x|\sqrt{7x}$
c. $12x^3$ **d.** $9x^2$

EXAMPLE 3 Simplifying Square Roots

Simplify:

a. $\sqrt{48x^2}$ **b.** $\sqrt{28x^3}$ **c.** $\sqrt{12x} \cdot \sqrt{6x}$ **d.** $\dfrac{\sqrt{45x^3}}{\sqrt{5x}}$

Solution:

a. $\sqrt{48x^2} = \sqrt{48} \cdot \sqrt{x^2} = \sqrt{16 \cdot 3} \cdot \sqrt{x^2} = \underbrace{\sqrt{16}}_{4} \cdot \sqrt{3} \cdot \underbrace{\sqrt{x^2}}_{|x|} = \boxed{4|x|\sqrt{3}}$

b. $\sqrt{28x^3} = \sqrt{28} \cdot \sqrt{x^3} = \sqrt{4 \cdot 7} \cdot \sqrt{x^2 \cdot x} = \underbrace{\sqrt{4}}_{2} \cdot \sqrt{7} \cdot \underbrace{\sqrt{x^2}}_{|x|} \cdot \sqrt{x}$

$\qquad = 2|x|\sqrt{7}\,\sqrt{x} = 2|x|\sqrt{7x} = \boxed{2x\sqrt{7x}}$ since $x \geq 0$

c. $\sqrt{12x} \cdot \sqrt{6x} = \sqrt{72x^2} = \sqrt{36 \cdot 2 \cdot x^2} = \underbrace{\sqrt{36}}_{6} \cdot \sqrt{2} \cdot \underbrace{\sqrt{x^2}}_{|x|} = 6|x|\sqrt{2}$

$\qquad\qquad\qquad\qquad\qquad\qquad\qquad = \boxed{6x\sqrt{2}}$ since $x \geq 0$

d. $\dfrac{\sqrt{45x^3}}{\sqrt{5x}} = \sqrt{\dfrac{45x^3}{5x}} = \sqrt{9x^2} = \underbrace{\sqrt{9}}_{3} \cdot \underbrace{\sqrt{x^2}}_{|x|} = 3|x| = \boxed{3x}$ since $x > 0$

Note: $x \neq 0, \quad x > 0$

Study Tip

If you have a single odd power under the square root, like $\sqrt{x^3}$, the variable is forced to be nonnegative because a negative cubed results in a negative inside the radical; therefore, the absolute value is not necessary.

Answer:
a. $2x\sqrt{15x}$ **b.** $x\sqrt{5}$

■ **YOUR TURN** Simplify: **a.** $\sqrt{60x^3}$ **b.** $\dfrac{\sqrt{125x^5}}{\sqrt{25x^3}}$

Other (nth) Roots

We now expand our discussion from square roots to other *n*th roots.

DEFINITION   **Principal *n*th Root**

Let a be a real number and n be a positive integer. Then the real number b is called the **principal *n*th root of a**, denoted $b = \sqrt[n]{a}$, if $b^n = a$. If n is even, then a and b are nonnegative real numbers. The positive integer n is called the **index**. The square root corresponds to $n = 2$, and the **cube root** corresponds to $n = 3$.

n	a	b	EXAMPLE
Even	Positive	Positive	$\sqrt[4]{16} = 2$ because $2^4 = 16$
Even	Negative	Not a real number	$\sqrt[4]{-16}$ is not a real number
Odd	Positive	Positive	$\sqrt[3]{27} = 3$ because $3^3 = 27$
Odd	Negative	Negative	$\sqrt[3]{-125} = -5$ because $(-5)^3 = -125$

A radical sign, $\sqrt{}$, combined with a radicand is called a **radical**.

PROPERTIES OF RADICALS

Let a and b be real numbers, then

PROPERTY	DESCRIPTION	EXAMPLE
$\sqrt[n]{ab} = \sqrt[n]{a} \cdot \sqrt[n]{b}$ if $\sqrt[n]{a}$ and $\sqrt[n]{b}$ both exist	The *n*th root of a product is the product of the *n*th roots.	$\sqrt[3]{16} = \sqrt[3]{8} \cdot \sqrt[3]{2} = 2\sqrt[3]{2}$
$\sqrt[n]{\dfrac{a}{b}} = \dfrac{\sqrt[n]{a}}{\sqrt[n]{b}} \quad b \neq 0$ if $\sqrt[n]{a}$ and $\sqrt[n]{b}$ both exist	The *n*th root of a quotient is the quotient of the *n*th roots.	$\sqrt[4]{\dfrac{81}{16}} = \dfrac{\sqrt[4]{81}}{\sqrt[4]{16}} = \dfrac{3}{2}$
$\sqrt[n]{a^m} = \left(\sqrt[n]{a}\right)^m$	The *n*th root of a power is the power of the *n*th root.	$\sqrt[3]{8^2} = \left(\sqrt[3]{8}\right)^2 = (2)^2 = 4$
$\sqrt[n]{a^n} = a \qquad n$ is odd	When *n* is odd, the *n*th root of a raised to the *n*th power is a.	$\sqrt[3]{x^3} = x$
$\sqrt[n]{a^n} = |a| \qquad n$ is even	When *n* is even, the *n*th root of a raised to the *n*th power is the absolute value of a.	$\sqrt[4]{x^4} = |x|$

▶ **EXAMPLE 4   Simplifying Radicals**

Simplify:

a. $\sqrt[3]{-24x^5}$ b. $\sqrt[4]{32x^5}$

Solution:

a. $\sqrt[3]{-24x^5} = \sqrt[3]{(-8)(3)x^3 x^2} = \underbrace{\sqrt[3]{-8}}_{-2} \cdot \sqrt[3]{3} \cdot \underbrace{\sqrt[3]{x^3}}_{x} \cdot \sqrt[3]{x^2} = \boxed{-2x\sqrt[3]{3x^2}}$

b. $\sqrt[4]{32x^5} = \sqrt[4]{16 \cdot 2 \cdot x^4 \cdot x} = \underbrace{\sqrt[4]{16}}_{2} \cdot \sqrt[4]{2} \cdot \underbrace{\sqrt[4]{x^4}}_{|x|} \cdot \sqrt[4]{x} = 2|x|\sqrt[4]{2x} = \boxed{2x\sqrt[4]{2x}}$ since $x \geq 0$

Classroom Example 0.6.4
Simplify:

a. $\sqrt[5]{243x^9}$ b.* $\sqrt[4]{400x^{11}}$

Answer:

a. $3x\sqrt[5]{x^4}$ b. $2x^2\sqrt[4]{25x^3}$

Combining Like Radicals

We have already discussed properties for multiplying and dividing radicals. Now we focus on combining (adding or subtracting) radicals. Radicals with the same index and radicand are called **like radicals**. Only like radicals can be added or subtracted.

Classroom Example 0.6.5
Combine the radicals, if possible.

a. $5\sqrt{2} - \sqrt{2} + 2\sqrt{2}$
b. $2\sqrt{3} - 3\sqrt{2}$
c. $\sqrt[4]{x} - \sqrt[3]{x} + \sqrt{x}$

Answer:

a. $6\sqrt{2}$
b. not possible
c. not possible

EXAMPLE 5   Combining Like Radicals

Combine the radicals if possible.

a. $4\sqrt{3} - 6\sqrt{3} + 7\sqrt{3}$ b. $2\sqrt{5} - 3\sqrt{7} + 6\sqrt{3}$

c. $3\sqrt{5} + \sqrt{20} - 2\sqrt{45}$ d. $\sqrt[4]{10} - 2\sqrt[3]{10} + 3\sqrt{10}$

Solution (a):

Use the distributive property. $4\sqrt{3} - 6\sqrt{3} + 7\sqrt{3} = (4 - 6 + 7)\sqrt{3}$

Eliminate the parentheses. $= \boxed{5\sqrt{3}}$

Solution (b):

None of these radicals are alike.
The expression is in simplified form. $\boxed{2\sqrt{5} - 3\sqrt{7} + 6\sqrt{3}}$

Solution (c):

Write the radicands as
products with a factor of 5. $3\sqrt{5} + \sqrt{20} - 2\sqrt{45} = 3\sqrt{5} + \sqrt{4 \cdot 5} - 2\sqrt{9 \cdot 5}$

The square root of a product is the
product of square roots. $= 3\sqrt{5} + \sqrt{4} \cdot \sqrt{5} - 2\sqrt{9} \cdot \sqrt{5}$

Simplify the square roots of perfect squares. $= 3\sqrt{5} + 2\sqrt{5} - \underbrace{2(3)}_{6}\sqrt{5}$

All three radicals are now like radicals. $= 3\sqrt{5} + 2\sqrt{5} - 6\sqrt{5}$

Use the distributive property. $= (3 + 2 - 6)\sqrt{5}$

Simplify. $= \boxed{-\sqrt{5}}$

Solution (d):

None of these radicals are alike because
they have different indices. The expression
is in simplified form. $\boxed{\sqrt[4]{10} - 2\sqrt[3]{10} + 3\sqrt{10}}$

■ **Answer:**

a. $7\sqrt[3]{7}$ b. $4\sqrt{6}$

■ **YOUR TURN** Combine the radicals.

a. $4\sqrt[3]{7} - 6\sqrt[3]{7} + 9\sqrt[3]{7}$ b. $5\sqrt{24} - 2\sqrt{54}$

Rationalizing Denominators

When radicals appear in a quotient, it is customary to write the quotient with no radicals in the denominator. This process is called **rationalizing the denominator** and involves multiplying by an expression that will eliminate the radical in the denominator.

For example, the expression $\dfrac{1}{\sqrt{3}}$ contains a single radical in the denominator. In a case like this, multiply the numerator and denominator by an appropriate radical expression, so that the resulting denominator will be radical free:

$$\frac{1}{\sqrt{3}} \cdot \underset{1}{\frac{(\sqrt{3})}{(\sqrt{3})}} = \frac{\sqrt{3}}{\sqrt{3} \cdot \sqrt{3}} = \frac{\sqrt{3}}{3}$$

If the denominator contains a sum of the form $a + \sqrt{b}$, multiply both the numerator and the denominator by the **conjugate** of the denominator, $a - \sqrt{b}$, which uses the difference of two squares to eliminate the radical term. Similarly, if the denominator contains a difference of the form $a - \sqrt{b}$, multiply both the numerator and the denominator by the conjugate of the denominator, $a + \sqrt{b}$. For example, to rationalize $\dfrac{1}{3 - \sqrt{5}}$, take the conjugate of the denominator, which is $3 + \sqrt{5}$:

$$\frac{1}{(3 - \sqrt{5})} \cdot \frac{(3 + \sqrt{5})}{(3 + \sqrt{5})} = \frac{3 + \sqrt{5}}{3^2 + \underbrace{3\sqrt{5} - 3\sqrt{5}}_{\text{like terms}} - (\sqrt{5})^2} = \frac{3 + \sqrt{5}}{9 - 5} = \frac{3 + \sqrt{5}}{4}$$

In general we apply the difference of two squares:

$$(\sqrt{a} + \sqrt{b})(\sqrt{a} - \sqrt{b}) = (\sqrt{a})^2 - (\sqrt{b})^2 = a - b$$

Notice that the product does not contain a radical. Therefore, to simplify the expression

$$\frac{1}{(\sqrt{a} + \sqrt{b})}$$

multiply the numerator and denominator by $(\sqrt{a} - \sqrt{b})$:

$$\frac{1}{(\sqrt{a} + \sqrt{b})} \cdot \frac{(\sqrt{a} - \sqrt{b})}{(\sqrt{a} - \sqrt{b})}$$

The denominator now contains no radicals:

$$\frac{(\sqrt{a} - \sqrt{b})}{(a - b)}$$

EXAMPLE 6   Rationalizing Denominators

Rationalize the denominators and simplify.

a. $\dfrac{2}{3\sqrt{10}}$ **b.** $\dfrac{5}{3 - \sqrt{2}}$ **c.** $\dfrac{\sqrt{5}}{\sqrt{2} - \sqrt{7}}$

Solution (a):

Multiply the numerator and denominator by $\sqrt{10}$.

$$= \frac{2}{3\sqrt{10}} \cdot \frac{\sqrt{10}}{\sqrt{10}}$$

Simplify.

$$= \frac{2\sqrt{10}}{3(\sqrt{10})^2} = \frac{2\sqrt{10}}{3(10)} = \frac{2\sqrt{10}}{30}$$

Divide out the common 2 in the numerator and denominator.

$$= \boxed{\frac{\sqrt{10}}{15}}$$

Classroom Example 0.6.6
Rationalize and simplify.

a. $\dfrac{-1}{2\sqrt{5}}$

b.* $\dfrac{-1}{a - \sqrt{a}}$, where $a > 0^-$

Answer:

a. $\dfrac{-\sqrt{5}}{10}$ b. $\dfrac{-(a + \sqrt{a})}{a(a - 1)}$

■ **Answer:** $-\dfrac{7(1 + \sqrt{3})}{2}$

Solution (b):

Multiply the numerator and denominator by the conjugate, $3 + \sqrt{2}$.

$$= \dfrac{5}{(3 - \sqrt{2})} \cdot \dfrac{(3 + \sqrt{2})}{(3 + \sqrt{2})}$$

$$= \dfrac{5(3 + \sqrt{2})}{(3 - \sqrt{2})(3 + \sqrt{2})}$$

The denominator now contains no radicals.

$$= \dfrac{15 + 5\sqrt{2}}{9 - 2}$$

Simplify.

$$= \boxed{\dfrac{15 + 5\sqrt{2}}{7}}$$

Solution (c):

Multiply the numerator and denominator by the conjugate, $\sqrt{2} + \sqrt{7}$.

$$= \dfrac{\sqrt{5}}{(\sqrt{2} - \sqrt{7})} \cdot \dfrac{(\sqrt{2} + \sqrt{7})}{(\sqrt{2} + \sqrt{7})}$$

Multiply the numerators and denominators, respectively.

$$= \dfrac{\sqrt{5}(\sqrt{2} + \sqrt{7})}{(\sqrt{2} - \sqrt{7})(\sqrt{2} + \sqrt{7})}$$

The denominator now contains no radicals.

$$= \dfrac{\sqrt{10} + \sqrt{35}}{2 - 7}$$

Simplify.

$$= \boxed{-\dfrac{\sqrt{10} + \sqrt{35}}{5}}$$

■ **YOUR TURN** Write the expression $\dfrac{7}{1 - \sqrt{3}}$ in simplified form.

SIMPLIFIED FORM OF A RADICAL EXPRESSION

A radical expression is in **simplified form** if
■ No factor in the radicand is raised to a power greater than or equal to the index.
■ The power of the radicand does not share a common factor with the index.
■ The denominator does not contain a radical.
■ The radical does not contain a fraction.

Classroom Example 0.6.7*
Simplify:

a. $\sqrt[4]{\dfrac{625x^2 y^{-5}}{81x^{-6} y^2}}$ b. $\sqrt[3]{\dfrac{-2x^3 y}{125xy^4}}$

Answer:

a. $\dfrac{5x^2}{3y} \cdot \sqrt[4]{\dfrac{1}{y^3}} = \dfrac{5x^2}{3y\sqrt[4]{y^3}} \dfrac{5x^2 \sqrt[4]{y}}{3y^2}$

b. $\dfrac{-1}{5y} \sqrt[3]{2x^2}$

EXAMPLE 7   Expressing a Radical Expression in Simplified Form

Express the radical expression in simplified form: $\sqrt[3]{\dfrac{16x^5}{81y^7}}$ $x \geq 0, y > 0$

Solution:

Rewrite the expression so that the radical does not contain a fraction.

$$\sqrt[3]{\dfrac{16x^5}{81y^7}} = \dfrac{\sqrt[3]{16x^5}}{\sqrt[3]{81y^7}}$$

Let $16 = 2^4$ and $81 = 3^4$.

$$= \dfrac{\sqrt[3]{2^4 \cdot x^5}}{\sqrt[3]{3^4 \cdot y^7}}$$

Factors in both radicands are raised to powers greater than the index (3). Rewrite the expression so that each power in the radicand is less than the index.

$$= \dfrac{2x\sqrt[3]{2x^2}}{3y^2\sqrt[3]{3y}}$$

The denominator contains a radical. In order to eliminate the radical in the denominator, we multiply the numerator and denominator by $\sqrt[3]{9y^2}$.

$$= \frac{2x\sqrt[3]{2x^2}}{3y^2\sqrt[3]{3y}} \cdot \frac{\sqrt[3]{9y^2}}{\sqrt[3]{9y^2}}$$

$$= \frac{2x\sqrt[3]{18x^2y^2}}{3y^2\sqrt[3]{27y^3}}$$

$$= \frac{2x\sqrt[3]{18x^2y^2}}{9y^3}$$

The radical expression now satisfies the conditions for simplified form.

$$\boxed{= \frac{2x\sqrt[3]{18x^2y^2}}{9y^3}}$$

Rational Exponents

We now use radicals to define rational exponents.

RATIONAL EXPONENTS: $\dfrac{1}{n}$

Let a be any real number and n be a positive integer, then

$$a^{1/n} = \sqrt[n]{a}$$

where $\dfrac{1}{n}$ is the **rational exponent** of a.

- When n is even and a is negative, then $a^{1/n}$ and $\sqrt[n]{a}$ are not real numbers.
- Furthermore, if m is a positive integer with m and n having no common factors, then

$$a^{m/n} = \left(a^{1/n}\right)^m = \left(a^m\right)^{1/n} = \sqrt[n]{a^m}$$

Note: Any of the four notations can be used.

Classroom Example 0.6.8
Simplify:
a. $125^{4/3}$
b. $(-125)^{2/3}$
c. $-(25x^3)^{3/2}$

Answer: .
a. 625   **b.** 25   **c.** $-125x^4\sqrt{x}$

EXAMPLE 8   Simplifying Expressions with Rational Exponents

Simplify:

a. $16^{3/2}$ **b.** $(-8)^{2/3}$

Solution:

a. $16^{3/2} = \left(16^{1/2}\right)^3 = \left(\sqrt{16}\right)^3 = 4^3 = 64$

b. $(-8)^{2/3} = \left[(-8)^{1/3}\right]^2 = (-2)^2 = 4$

■ YOUR TURN Simplify $27^{2/3}$.

Technology Tip

```
16^(3/2)
              64
(-8)^(2/3)
               4
■
```

■ **Answer:** 9

The properties of exponents that hold for integers also hold for rational numbers:

$$a^{-1/n} = \frac{1}{a^{1/n}} \quad \text{and} \quad a^{-m/n} = \frac{1}{a^{m/n}} \qquad a \neq 0$$

EXAMPLE 9   Simplifying Expressions with Negative Rational Exponents

Simplify $\dfrac{(9x)^{-1/2}}{4x^{-3/2}}$ $x > 0$.

Solution:

Negative exponents correspond to positive exponents in the reciprocal.

$$\frac{(9x)^{-1/2}}{4x^{-3/2}} = \frac{x^{3/2}}{4 \cdot (9x)^{1/2}}$$

Eliminate the parentheses.

$$= \frac{x^{3/2}}{4 \cdot 9^{1/2}x^{1/2}}$$

Apply the quotient property on x.

$$= \frac{x^{3/2-1/2}}{4 \cdot \underset{3}{9^{1/2}}}$$

$$= \frac{x^1}{4 \cdot 3}$$

Simplify.

$$= \boxed{\frac{x}{12}}$$

■ Answer: $18x^2$

■ YOUR TURN Simplify $\dfrac{9x^{3/2}}{(4x)^{-1/2}}$ $x > 0$.

EXAMPLE 10   Simplifying Algebraic Expressions with Rational Exponents

Simplify $\dfrac{(-8x^2y)^{1/3}}{(9xy^4)^{1/2}}$ $x > 0, y > 0$.

Solution:

$$\frac{(-8x^2y)^{1/3}}{(9xy^4)^{1/2}} = \frac{(-8)^{1/3}(x^2)^{1/3}y^{1/3}}{9^{1/2}x^{1/2}(y^4)^{1/2}} = \frac{(-8)^{1/3}x^{2/3}y^{1/3}}{9^{1/2}x^{1/2}y^2} = \left(\frac{-2}{3}\right)x^{2/3-1/2}y^{1/3-2} = -\frac{2}{3}x^{1/6}y^{-5/3}$$

Write in terms of positive exponents.

$$= \boxed{-\frac{2x^{1/6}}{3y^{5/3}}}$$

■ Answer: $\dfrac{4x^{5/6}}{3y^{1/2}}$

■ YOUR TURN Simplify $\dfrac{(16x^3y)^{1/2}}{(27x^2y^3)^{1/3}}$ and write your answer with only positive exponents.

EXAMPLE 11   Factoring Expressions with Rational Exponents

Factor completely $x^{8/3} - 5x^{5/3} - 6x^{2/3}$.

Solution:

Factor out the greatest common factor $x^{2/3}$.

$$\underset{x^{2/3}x^{6/3}}{x^{8/3}} - \underset{x^{2/3}x^{3/3}}{5x^{5/3}} - 6x^{2/3} = x^{2/3}(x^2 - 5x - 6)$$

Factor the trinomial.

$$= \boxed{x^{2/3}(x - 6)(x + 1)}$$

■ Answer: $x^{1/3}(x - 2)(x + 1)$

■ YOUR TURN Factor completely $x^{7/3} - x^{4/3} - 2x^{1/3}$.

SECTION
0.6 SUMMARY

In this section, we defined radicals as "$b = \sqrt[n]{a}$ means $a = b^n$" for a and b positive real numbers when n is a positive even integer, and a and b any real numbers when n is a positive odd integer.

Properties of Radicals

PROPERTY	EXAMPLE
$\sqrt[n]{ab} = \sqrt[n]{a} \cdot \sqrt[n]{b}$	$\sqrt[3]{16} = \sqrt[3]{8} \cdot \sqrt[3]{2} = 2\sqrt[3]{2}$
$\sqrt[n]{\dfrac{a}{b}} = \dfrac{\sqrt[n]{a}}{\sqrt[n]{b}}$ $b \neq 0$	$\sqrt[4]{\dfrac{81}{16}} = \dfrac{\sqrt[4]{81}}{\sqrt[4]{16}} = \dfrac{3}{2}$
$\sqrt[n]{a^m} = \left(\sqrt[n]{a}\right)^m$	$\sqrt[3]{8^2} = \left(\sqrt[3]{8}\right)^2 = (2)^2 = 4$
$\sqrt[n]{a^n} = a$ n is odd	$\sqrt[3]{x^5} = \sqrt[3]{x^3} \cdot \sqrt[3]{x^2} = x\sqrt[3]{x^2}$
$\sqrt[n]{a^n} = |a|$ n is even	$\sqrt[4]{x^6} = \sqrt[4]{x^4} \cdot \sqrt[4]{x^2} = |x|\sqrt[4]{x^2}$

Radicals can be combined only if they are like radicals (same radicand and index). Quotients with radicals in the denominator are usually rewritten with no radicals in the denominator.

Rational exponents were defined in terms of radicals: $a^{1/n} = \sqrt[n]{a}$. The properties for integer exponents we learned in Section 0.2 also hold true for rational exponents:

$$a^{m/n} = \left(a^{1/n}\right)^m = \left(\sqrt[n]{a}\right)^m \quad \text{and} \quad a^{m/n} = \left(a^m\right)^{1/n} = \sqrt[n]{a^m}$$

Negative rational exponents: $a^{-m/n} = \dfrac{1}{a^{m/n}}$, for m and n positive integers with no common factors, $a \neq 0$.

SECTION
0.6 EXERCISES

▪ SKILLS

In Exercises 1–24, evaluate each expression or state that it is not a real number.

1. $\sqrt{100}$ **2.** $\sqrt{121}$ **3.** $-\sqrt{144}$ **4.** $\sqrt{-169}$ **5.** $\sqrt[3]{-216}$ **6.** $\sqrt[3]{-125}$

7. $\sqrt[3]{343}$ **8.** $-\sqrt[3]{-27}$ **9.** $\sqrt[8]{1}$ **10.** $\sqrt[7]{-1}$ **11.** $\sqrt[3]{0}$ **12.** $\sqrt[5]{0}$

13. $\sqrt{-16}$ **14.** $\sqrt[5]{-1}$ **15.** $(-27)^{1/3}$ **16.** $(-64)^{1/3}$ **17.** $8^{2/3}$ **18.** $(-64)^{2/3}$

19. $(-32)^{1/5}$ **20.** $(-243)^{1/3}$ **21.** $(-1)^{1/3}$ **22.** $1^{5/2}$ **23.** $9^{3/2}$ **24.** $(27)^{2/3}$

In Exercises 25–40, simplify (if possible) the radical expressions.

25. $\sqrt{2} - 5\sqrt{2}$ **26.** $3\sqrt{5} - 7\sqrt{5}$ **27.** $3\sqrt{5} - 2\sqrt{5} + 7\sqrt{5}$ **28.** $6\sqrt{7} + 7\sqrt{7} - 10\sqrt{7}$

29. $\sqrt{12} \cdot \sqrt{2}$ **30.** $2\sqrt{5} \cdot 3\sqrt{40}$ **31.** $\sqrt[3]{12} \cdot \sqrt[3]{4}$ **32.** $\sqrt[4]{8} \cdot \sqrt[4]{4}$

33. $\sqrt{3}\sqrt{7}$ **34.** $\sqrt{5}\sqrt{2}$ **35.** $8\sqrt{25x^2}$ **36.** $16\sqrt{36y^4}$

37. $\sqrt{4x^2y}$ **38.** $\sqrt{16x^3y}$ **39.** $\sqrt[3]{-81x^6y^8}$ **40.** $\sqrt[5]{-32x^{10}y^8}$

In Exercises 41–56, rationalize the denominators.

41. $\sqrt{\dfrac{1}{3}}$ **42.** $\sqrt{\dfrac{2}{5}}$ **43.** $\dfrac{2}{3\sqrt{11}}$ **44.** $\dfrac{5}{3\sqrt{2}}$

45. $\dfrac{3}{1 - \sqrt{5}}$ **46.** $\dfrac{2}{1 + \sqrt{3}}$ **47.** $\dfrac{1 + \sqrt{2}}{1 - \sqrt{2}}$ **48.** $\dfrac{3 - \sqrt{5}}{3 + \sqrt{5}}$

49. $\dfrac{3}{\sqrt{2} - \sqrt{3}}$

50. $\dfrac{5}{\sqrt{2} + \sqrt{5}}$

51. $\dfrac{4}{3\sqrt{2} + 2\sqrt{3}}$

52. $\dfrac{7}{2\sqrt{3} + 3\sqrt{2}}$

53. $\dfrac{4 + \sqrt{5}}{3 + 2\sqrt{5}}$

54. $\dfrac{6}{3\sqrt{2} + 4}$

55. $\dfrac{\sqrt{7} + 3}{\sqrt{2} - \sqrt{5}}$

56. $\dfrac{\sqrt{y}}{\sqrt{x} - \sqrt{y}}$

In Exercises 57–64, simplify by applying the properties of rational exponents. Express your answers in terms of positive exponents.

57. $\left(x^{1/2} y^{2/3}\right)^{6}$

58. $\left(y^{2/3} y^{1/4}\right)^{12}$

59. $\dfrac{\left(x^{1/3} y^{1/2}\right)^{-3}}{\left(x^{-1/2} y^{1/4}\right)^{2}}$

60. $\dfrac{\left(x^{-2/3} y^{-3/4}\right)^{-2}}{\left(x^{1/3} y^{1/4}\right)^{4}}$

61. $\dfrac{x^{1/2} y^{1/5}}{x^{-2/3} y^{-9/5}}$

62. $\dfrac{\left(y^{-3/4} x^{-2/3}\right)^{12}}{\left(y^{1/4} x^{7/3}\right)^{24}}$

63. $\dfrac{\left(2x^{2/3}\right)^{3}}{\left(4x^{-1/3}\right)^{2}}$

64. $\dfrac{\left(2x^{-2/3}\right)^{3}}{\left(4x^{-4/3}\right)^{2}}$

In Exercises 65–68, factor each expression completely.

65. $x^{7/3} - x^{4/3} - 2x^{1/3}$

66. $8x^{1/4} + 4x^{5/4}$

67. $7x^{3/7} - 14x^{6/7} + 21x^{10/7}$

68. $7x^{-1/3} + 70x$

▪APPLICATIONS

69. **Gravity.** If a penny is dropped off a building, the time it takes (seconds) to fall d feet is given by $\sqrt{\dfrac{d}{16}}$. If a penny is dropped off a 1280-foot-tall building, how long will it take until it hits the ground? Round to the nearest second.

70. **Gravity.** If a ball is dropped off a building, the time it takes (seconds) to fall d meters is approximately given by $\sqrt{\dfrac{d}{5}}$.

    If a ball is dropped off a 600-meter-tall building, how long will it take until it hits the ground? Round to the nearest second.

71. **Kepler's Law.** The square of the period p (in years) of a planet's orbit around the Sun is equal to the cube of the planet's maximum distance from the Sun, d (in astronomical units or AU). This relationship can be expressed mathematically as $p^2 = d^3$. If this formula is solved for d, the resulting equation is $d = p^{2/3}$. If Saturn has an orbital period of 29.46 Earth years, calculate Saturn's maximum distance from the Sun to the nearest hundredth of an AU.

72. **Period of a Pendulum.** The period (in seconds) of a pendulum of length L (in meters) is given by $P = 2 \cdot \pi \cdot \left(\dfrac{L}{9.8}\right)^{1/2}$. If a certain pendulum has a length of 19.6 meters, determine the period P of this pendulum to the nearest tenth of a second.

▪CATCH THE MISTAKE

In Exercises 73 and 74, explain the mistake that is made.

73. Simplify $\left(4x^{1/2} y^{1/4}\right)^{2}$

    Solution:

    Use properties of exponents. $4\left(x^{1/2}\right)^{2} \left(y^{1/4}\right)^{2}$

    Simplify. $4xy^{1/2}$

    This is incorrect. What mistake was made?

74. Simplify $\dfrac{2}{5 - \sqrt{11}}$.

    Solution:

    Multiply numerator and denominator by $5 - \sqrt{11}$. $\dfrac{2}{5 - \sqrt{11}} \cdot \dfrac{\left(5 - \sqrt{11}\right)}{\left(5 - \sqrt{11}\right)}$

    Multiply numerators and denominators. $\dfrac{2\left(5 - \sqrt{11}\right)}{25 - 11}$

    Simplify. $\dfrac{2\left(5 - \sqrt{11}\right)}{14} = \dfrac{5 - \sqrt{11}}{7}$

    This is incorrect. What mistake was made?

■ **CONCEPTUAL**

In Exercises 75–78, determine whether each statement is true or false.

75. $\sqrt{121} = \pm 11$

76. $\sqrt{x^2} = x$, where x is any real number.

77. $\sqrt{a^2 + b^2} = \sqrt{a} + \sqrt{b}$

78. $\sqrt{-4} = -2$

In Exercises 79 and 80, $a, m, n,$ and k are any positive real numbers.

79. Simplify $((a^m)^n)^k$.

80. Simplify $(a^{-k})^{-1/k}$.

In Exercises 81 and 82, evaluate each algebraic expression for the specified values.

81. $\dfrac{\sqrt{b^2 - 4ac}}{2a}$ for $a = 1, b = 7, c = 12$

82. $\sqrt{b^2 - 4ac}$ for $a = 1, b = 7, c = 12$

■ **CHALLENGE**

83. Rationalize the denominator and simplify: $\dfrac{1}{(\sqrt{a} + \sqrt{b})^2}$.

84. Rationalize the denominator and simplify: $\dfrac{\sqrt{a+b} - \sqrt{a}}{\sqrt{a+b} + \sqrt{a}}$.

■ **TECHNOLOGY**

85. Use a calculator to approximate $\sqrt{11}$ to three decimal places.

86. Use a calculator to approximate $\sqrt[3]{7}$ to three decimal places.

87. Given $\dfrac{4}{5\sqrt{2} + 4\sqrt{3}}$

  a. Rationalize the denominator.
  b. Use a graphing utility to evaluate the expression and the answer.
  c. Do they agree?

88. Given $\dfrac{2}{4\sqrt{5} - 3\sqrt{6}}$.

  a. Rationalize the denominator.
  b. Use a graphing utility to evaluate the expression and the answer.
  c. Do they agree?

SECTION
0.7 COMPLEX NUMBERS

SKILLS OBJECTIVES

- Write radicals with negative radicands as imaginary numbers.
- Add and subtract complex numbers.
- Multiply complex numbers.
- Divide complex numbers.
- Raise complex numbers to powers.

CONCEPTUAL OBJECTIVES

- Understand that real numbers and imaginary numbers are subsets of complex numbers.
- Understand how to eliminate imaginary numbers in denominators.

The Imaginary Unit, *i*

In Section 1.3, we will be studying equations whose solutions sometimes involve the square roots of negative numbers. In Section 0.6, when asked to evaluate the square root of a negative number, like $\sqrt{-16}$, we said "it is not a real number," because there is no real number such that $x^2 = -16$. To include such roots in the number system, mathematicians created a new expanded set of numbers, called the *complex numbers*. The foundation of this new set of numbers is the imaginary unit *i*.

> **DEFINITION** | **The Imaginary Unit *i***
>
> The **imaginary unit** is denoted by the letter *i* and is defined as
>
> $$i = \sqrt{-1}$$
>
> where $i^2 = -1$.

Recall that for positive real numbers *a* and *b* we defined the principal square root as

$$b = \sqrt{a} \quad \text{which means} \quad b^2 = a$$

Similarly, we define the *principal square root of a negative number* as $\sqrt{-a} = i\sqrt{a}$, since $\left(i\sqrt{a}\right)^2 = i^2 a = -a$.

> If $-a$ is a negative real number, then the **principal square root** of $-a$ is
>
> $$\sqrt{-a} = i\sqrt{a}$$
>
> where *i* is the imaginary unit and $i^2 = -1$.

We write $i\sqrt{a}$ instead of $\sqrt{a}\,i$ to avoid any confusion as to what is included in the radical.

EXAMPLE 1 Using Imaginary Numbers to Simplify Radicals

Simplify using imaginary numbers.

a. $\sqrt{-9}$ **b.** $\sqrt{-8}$

Solution:

a. $\sqrt{-9} = i\sqrt{9} = \boxed{3i}$ **b.** $\sqrt{-8} = i\sqrt{8} = i \cdot 2\sqrt{2} = \boxed{2i\sqrt{2}}$

■ **YOUR TURN** Simplify $\sqrt{-144}$.

> **DEFINITION** | **Complex Number**
>
> A **complex number** in standard form is defined as
>
> $$a + bi$$
>
> where *a* and *b* are real numbers and *i* is the imaginary unit. We denote *a* as the **real part** of the complex number and *b* as the **imaginary part** of the complex number.

A complex number written as $a + bi$ is said to be in **standard form**. If $a = 0$ and $b \neq 0$, then the resulting complex number bi is called a **pure imaginary number**. If $b = 0$, then $a + bi$ is a real number. The set of all real numbers and the set of all imaginary numbers are both subsets of the set of complex numbers.

The following are examples of complex numbers.

$$17 \qquad 2 - 3i \qquad -5 + i \qquad 3 - i\sqrt{11} \qquad -9i$$

DEFINITION   **Equality of Complex Numbers**

The complex numbers $a + bi$ and $c + di$ are **equal** if and only if $a = c$ and $b = d$. In other words, two complex numbers are equal if and only if both real parts are equal *and* both imaginary parts are equal.

Classroom Example 0.7.2
a. Simplify:
$(-1 - 3i) + (-2 + i)$
b.* Find x and y such that
$(2 - 3i) - (x + iy) = i$

Answer:
a. $-3 - 2i$
b. $x = 2$ and $y = -4$

Adding and Subtracting Complex Numbers

Complex numbers in the standard form $a + bi$ are treated in much the same way as binomials of the form $a + bx$. We can add, subtract, and multiply complex numbers the same way we performed these operations on binomials. When adding or subtracting complex numbers, combine real parts with real parts and combine imaginary parts with imaginary parts.

EXAMPLE 2   Adding and Subtracting Complex Numbers

Perform the indicated operation and simplify.

a. $(3 - 2i) + (-1 + i)$ **b.** $(2 - i) - (3 - 4i)$

Solution (a):

Eliminate the parentheses. $(3 - 2i) + (-1 + i) = 3 - 2i - 1 + i$

Group real and imaginary numbers, respectively. $= (3 - 1) + (-2i + i)$

Simplify. $= \boxed{2 - i}$

Solution (b):

Eliminate the parentheses (distribute the negative). $(2 - i) - (3 - 4i) = 2 - i - 3 + 4i$

Group real and imaginary numbers, respectively. $= (2 - 3) + (-i + 4i)$

Simplify. $= \boxed{-1 + 3i}$

▪ **YOUR TURN** Perform the indicated operation and simplify: $(4 + i) - (3 - 5i)$.

Technology Tip

Be sure to put the graphing calculator in $\boxed{a + bi}$ mode.

a. $(3 - 2i) + (-1 + i)$
b. $(2 - i) - (3 - 4i)$

```
(3-2i)+(-1+i)
               2-i
(2-i)-(3-4i)
              -1+3i
■
```

▪ **Answer:** $1 + 6i$

Study Tip

When multiplying complex numbers, remember that $i^2 = -1$.

Multiplying Complex Numbers

When multiplying complex numbers, you apply all the same methods as you did when multiplying binomials. It is important to remember that $i^2 = -1$.

WORDS	MATH
Multiply the complex numbers.	$(5 - i)(3 - 4i)$
Multiply using the distributive property.	$= 5(3) + 5(-4i) - i(3) - (i)(-4i)$
Eliminate the parentheses.	$= 15 - 20i - 3i + 4i^2$
Let $i^2 = -1$.	$= 15 - 20i - 3i + 4(-1)$
Simplify.	$= 15 - 20i - 3i - 4$
Combine real parts and imaginary parts, respectively.	$= 11 - 23i$

Technology Tip

Be sure to put the graphing calculator in $\boxed{a + bi}$ mode.

a. $(3 - i)(2 + i)$
b. $i(-3 + i)$

```
(3-i)(2+i)
              7+i
i(-3+i)
              -1-3i
■
```

EXAMPLE 3   **Multiplying Complex Numbers**

Multiply the complex numbers and express the result in standard form, $a \pm bi$.

a. $(3 - i)(2 + i)$ **b.** $i(-3 + i)$

Solution (a):

Use the distributive property.	$(3 - i)(2 + i) = 3(2) + 3(i) - i(2) - i(i)$
Eliminate the parentheses.	$= 6 + 3i - 2i - i^2$
Substitute $i^2 = -1$.	$= 6 + 3i - 2i - (-1)$
Group like terms.	$= (6 + 1) + (3i - 2i)$
Simplify.	$= \boxed{7 + i}$

Solution (b):

Use the distributive property.	$i(-3 + i) = -3i + i^2$
Substitute $i^2 = -1$.	$= -3i - 1$
Write in standard form.	$= \boxed{-1 - 3i}$

■ **Answer:** $2 + 11i$

■ **YOUR TURN** Multiply the complex numbers and express the result in standard form, $a \pm bi$: $(4 - 3i)(-1 + 2i)$.

Classroom Example 0.7.3
Simplify:
a. $-i(-1 + 3i)$
b.* $(a - 2bi)(2a + bi)$, where $a, b \neq 0$

Answer:
a. $3 + i$
b. $(2a^2 + 2b^2) - 3abi$

Dividing Complex Numbers

Recall the special product that produces a difference of two squares, $(a + b)(a - b) = a^2 - b^2$. This special product has only first and last terms because the outer and inner terms cancel each other out. Similarly, if we multiply complex numbers in the same manner, the result is a real number because the imaginary terms cancel each other out.

COMPLEX CONJUGATE

The product of a complex number, $z = a + bi$, and its **complex conjugate**, $\bar{z} = a - bi$, is a real number.

$$z\bar{z} = (a + bi)(a - bi) = a^2 - b^2i^2 = a^2 - b^2(-1) = a^2 + b^2$$

In order to write a quotient of complex numbers in standard form, $a + bi$, multiply the numerator and the denominator by the complex conjugate of the denominator. It is important to note that if i is present in the denominator, then the complex number is *not* in standard form.

EXAMPLE 4   Dividing Complex Numbers

Write the quotient in standard form: $\dfrac{2 - i}{1 + 3i}$.

Solution:

Multiply numerator and denominator by the complex conjugate of the denominator, $1 - 3i$.	$\left(\dfrac{2 - i}{1 + 3i}\right)\left(\dfrac{1 - 3i}{1 - 3i}\right)$
Multiply the numerators and denominators, respectively.	$= \dfrac{(2 - i)(1 - 3i)}{(1 + 3i)(1 - 3i)}$
Use the FOIL method (or distributive property).	$= \dfrac{2 - 6i - i + 3i^2}{1 - 3i + 3i - 9i^2}$
Combine imaginary parts.	$= \dfrac{2 - 7i + 3i^2}{1 - 9i^2}$
Substitute $i^2 = -1$.	$= \dfrac{2 - 7i - 3}{1 - 9(-1)}$
Simplify the numerator and denominator.	$= \dfrac{-1 - 7i}{10}$
Write in standard form. Recall that $\dfrac{a + b}{c} = \dfrac{a}{c} + \dfrac{b}{c}$.	$= \boxed{-\dfrac{1}{10} - \dfrac{7}{10}i}$

■ **YOUR TURN**   Write the quotient in standard form: $\dfrac{3 + 2i}{4 - i}$.

Technology Tip

Be sure to put the graphing calculator in $\boxed{a + bi}$ mode.

$$\frac{2 - i}{1 + 3i}$$

To change the answer to the fraction form, press $\boxed{\text{MATH}}$, highlight $\boxed{\text{1: Frac}}$, press $\boxed{\text{ENTER}}$ and $\boxed{\text{ENTER}}$.

```
(2-i)/(1+3i)
              -.1-.7i
Ans▶Frac
           -1/10-7/10i
■
```

■ **Answer:** $\dfrac{10}{17} + \dfrac{11}{17}i$

Raising Complex Numbers to Integer Powers

Note that i raised to the fourth power is 1. In simplifying imaginary numbers, we factor out i raised to the largest multiple of 4.

$$i = \sqrt{-1}$$
$$i^2 = -1$$
$$i^3 = i^2 \cdot i = (-1)i = -i$$
$$i^4 = i^2 \cdot i^2 = (-1)(-1) = 1$$
$$\overline{i^5 = i^4 \cdot i = (1)(i) = i}$$
$$i^6 = i^4 \cdot i^2 = (1)(-1) = -1$$
$$i^7 = i^4 \cdot i^3 = (1)(-i) = -i$$
$$i^8 = (i^4)^2 = 1$$

Classroom Example 0.7.5
Simplify:
a. i^{35}
b.* i^{2-4n}, where n is a positive integer

Answer:
a. $-i$ **b.** -1

Classroom Example 0.7.4
Write these quotients in standard form.

a. $\dfrac{-i}{1 - 5i}$

b.* $\dfrac{a + 2bi}{b + 2ai}$, where $a, b \neq 0$

Answer:

a. $\dfrac{5 - i}{26}$

b. $\dfrac{5ab}{b^2 + 4a^2} - \dfrac{2(a^2 - b^2)}{b^2 + 4a^2}i$

Technology Tip

```
i.^7
        -3E -13-i.
```

This is due to rounding off error in the programming. Since -3×10^{-13} can be approximated as 0, $i^7 = -i$.

```
i.^13
        -3E -13+i.
i.^100
              1
```

■ **Answer:** $-i$

Technology Tip

Be sure to put the graphing calculator in $\boxed{a + bi}$ mode.

$$(2 - i)^3$$

```
(2-i.)^3
        2-11i.
```

■ **Answer:** $2 + 11i$

EXAMPLE 5 Raising the Imaginary Unit to Integer Powers

Simplify:

a. i^7 **b.** i^{13} **c.** i^{100}

Solution:

a. $i^7 = i^4 \cdot i^3 = (1)(-i) = \boxed{-i}$

b. $i^{13} = i^{12} \cdot i = (i^4)^3 \cdot i = 1^3 \cdot i = \boxed{i}$

c. $i^{100} = (i^4)^{25} = 1^{25} = \boxed{1}$

■ **YOUR TURN** Simplify i^{27}.

EXAMPLE 6 Raising a Complex Number to an Integer Power

Write $(2 - i)^3$ in standard form.

Solution:

Recall the formula for cubing a binomial. $(a - b)^3 = a^3 - 3a^2 b + 3ab^2 - b^3$

Let $a = 2$ and $b = i$. $(2 - i)^3 = 2^3 - 3(2)^2(i) + 3(2)(i)^2 - i^3$

Let $i^2 = -1$ and $i^3 = -i$. $= 2^3 - 3(2)^2(i) + (3)(2)(-1) - (-i)$

Eliminate parentheses. $= 8 - 6 - 12i + i$

Combine the real parts and imaginary parts, respectively. $= \boxed{2 - 11i}$

■ **YOUR TURN** Write $(2 + i)^3$ in standard form.

SECTION 0.7 SUMMARY

The Imaginary Unit i
- $i = \sqrt{-1}$
- $i^2 = -1$

Complex Numbers
- *Standard Form*: $a + bi$, where a is the real part and b is the imaginary part.
- The set of real numbers and the set of pure imaginary numbers are subsets of the set of complex numbers.

Adding and Subtracting Complex Numbers
- $(a + bi) + (c + di) = (a + c) + (b + d)i$
- $(a + bi) - (c + di) = (a - c) + (b - d)i$
- To add or subtract complex numbers, add or subtract the real parts and imaginary parts, respectively.

Multiplying Complex Numbers
- $(a + bi)(c + di) = (ac - bd) + (ad + bc)i$
- Apply the same methods as for multiplying binomials. It is important to remember that $i^2 = -1$.

Dividing Complex Numbers
- Complex conjugate of $a + bi$ is $a - bi$.
- In order to write a quotient of complex numbers in standard form, multiply the numerator and the denominator by the complex conjugate of the denominator:

$$\frac{a + bi}{c + di} \cdot \frac{(c - di)}{(c - di)}$$

SECTION
0.7 EXERCISES

SKILLS

In Exercises 1–12, write each expression as a complex number in standard form. Some expressions simplify to either a real number or a pure imaginary number.

1. $\sqrt{-16}$
2. $\sqrt{-100}$
3. $\sqrt{-20}$
4. $\sqrt{-24}$

5. $\sqrt[3]{-64}$
6. $\sqrt[3]{-27}$
7. $\sqrt{-64}$
8. $\sqrt{-27}$

9. $3 - \sqrt{-100}$
10. $4 - \sqrt{-121}$
11. $-10 - \sqrt{-144}$
12. $7 - \sqrt[3]{-125}$

In Exercises 13–40, perform the indicated operation, simplify, and express in standard form.

13. $(3 - 7i) + (-1 - 2i)$
14. $(1 + i) + (9 - 3i)$
15. $(3 - 4i) + (7 - 10i)$
16. $(5 + 7i) + (-10 - 2i)$

17. $(4 - 5i) - (2 - 3i)$
18. $(-2 + i) - (1 - i)$
19. $(-3 + i) - (-2 - i)$
20. $(4 + 7i) - (5 + 3i)$

21. $3(4 - 2i)$
22. $4(7 - 6i)$
23. $12(8 - 5i)$
24. $-3(16 + 4i)$

25. $-3(16 - 9i)$
26. $5(-6i + 3)$
27. $-6(17 - 5i)$
28. $-12(8 + 3i)$

29. $(1 - i)(3 + 2i)$
30. $(-3 + 2i)(1 - 3i)$
31. $(5 - 7i)(-3 + 4i)$
32. $(16 - 5i)(-2 - i)$

33. $(7 - 5i)(6 + 9i)$
34. $(-3 - 2i)(7 - 4i)$
35. $(12 - 18i)(-2 + i)$
36. $(-4 + 3i)(-4 - 3i)$

37. $\left(\frac{1}{2} + 2i\right)\left(\frac{4}{9} - 3i\right)$
38. $\left(-\frac{3}{4} + \frac{9}{16}i\right)\left(\frac{2}{3} + \frac{4}{9}i\right)$
39. $(-i + 17)(2 + 3i)$
40. $(-3i - 2)(-2 - 3i)$

In Exercises 41–48, for each complex number z, write the complex conjugate $\bar{z}$ and find $z\bar{z}$.

41. $z = 4 + 7i$
42. $z = 2 + 5i$
43. $z = 2 - 3i$
44. $z = 5 - 3i$

45. $z = 6 + 4i$
46. $z = -2 + 7i$
47. $z = -2 - 6i$
48. $z = -3 - 9i$

In Exercises 49–64, write each quotient in standard form.

49. $\dfrac{2}{i}$
50. $\dfrac{3}{i}$
51. $\dfrac{1}{3 - i}$
52. $\dfrac{2}{7 - i}$
53. $\dfrac{1}{3 + 2i}$
54. $\dfrac{1}{4 - 3i}$
55. $\dfrac{2}{7 + 2i}$
56. $\dfrac{8}{1 + 6i}$

57. $\dfrac{1 - i}{1 + i}$
58. $\dfrac{3 - i}{3 + i}$
59. $\dfrac{2 + 3i}{3 - 5i}$
60. $\dfrac{2 + i}{3 - i}$
61. $\dfrac{4 - 5i}{7 + 2i}$
62. $\dfrac{7 + 4i}{9 - 3i}$
63. $\dfrac{8 + 3i}{9 - 2i}$
64. $\dfrac{10 - i}{12 + 5i}$

In Exercises 65–76, simplify.

65. i^{15}
66. i^{99}
67. i^{40}
68. i^{18}
69. $(5 - 2i)^2$
70. $(3 - 5i)^2$

71. $(2 + 3i)^2$
72. $(4 - 9i)^2$
73. $(3 + i)^3$
74. $(2 + i)^3$
75. $(1 - i)^3$
76. $(4 - 3i)^3$

APPLICATION

Electrical impedance is the ratio of voltage to current in ac circuits. Let Z represent the total impedance of an electrical circuit. If there are two resistors in a circuit, let $Z_1 = 3 - 6i$ ohms and $Z_2 = 5 + 4i$ ohms.

77. **Electrical Circuits in Series.** When the resistors in the circuit are placed in series, the total impedance is the sum of the two impedances $Z = Z_1 + Z_2$. Find the total impedance of the electrical circuit in series.

78. **Electrical Circuits in Parallel.** When the resistors in the circuit are placed in parallel, the total impedance is given by $\dfrac{1}{Z} = \dfrac{1}{Z_1} + \dfrac{1}{Z_2}$. Find the total impedance of the electrical circuit in parallel.

■ CATCH THE MISTAKE

In Exercises 79 and 80, explain the mistake that is made.

79. Write the quotient in standard form: $\dfrac{2}{4-i}$.

Solution:

Multiply the numerator and the denominator by $4-i$.

$$\dfrac{2}{4-i} \cdot \dfrac{4-i}{4-i}$$

Multiply the numerator using the distributive property and the denominator using the FOIL method.

$$\dfrac{8-2i}{16-1}$$

Simplify.

$$\dfrac{8-2i}{15}$$

Write in standard form.

$$\dfrac{8}{15} - \dfrac{2}{15}i$$

This is incorrect. What mistake was made?

80. Write the product in standard form: $(2-3i)(5+4i)$.

Solution:

Use the FOIL method to multiply the complex numbers.

$$10 - 7i - 12i^2$$

Simplify.

$$-2 - 7i$$

This is incorrect. What mistake was made?

■ CONCEPTUAL

In Exercises 81–84, determine whether each statement is true or false.

81. The product is a real number: $(a+bi)(a-bi)$.

82. Imaginary numbers are a subset of the complex numbers.

83. Real numbers are a subset of the complex numbers.

84. There is no complex number that equals its conjugate.

■ CHALLENGE

85. Factor completely over the complex numbers: $x^4 + 2x^2 + 1$.

86. Factor completely over the complex numbers: $x^4 + 18x^2 + 81$.

■ TECHNOLOGY

In Exercises 87–90, use a graphing utility to simplify the expression. Write your answer in standard form.

87. $(1+2i)^5$

88. $(3-i)^6$

89. $\dfrac{1}{(2-i)^3}$

90. $\dfrac{1}{(4+3i)^2}$

Understanding Rules through Patterns and Examples

In this chapter, you reviewed many definitions, rules, and properties from your previous algebra classes. It can be a lot to remember, and sometimes it's easy to misremember something. *Observing patterns* can help you see that rules are not arbitrary; rather they make sense in the context of other mathematics you already know. And by *looking at examples*, you can test whether you've correctly remembered a rule. You will explore both of these strategies here.

1. In this part, you will observe patterns to try to discover some of the rules for exponents given in this chapter.

   a. For instance, to discover *a rule for zero as an exponent*, first complete the next four steps.

   $$a^5 = a \cdot a \cdot a \cdot a \cdot a$$
   $$a^4 = \underline{\hspace{2cm}}$$
   $$a^3 = \underline{\hspace{2cm}}$$
   $$a^2 = \underline{\hspace{2cm}}$$
   $$a^1 = \underline{\hspace{2cm}}$$

   As the power decreases by one at each step, what pattern do you notice? Extend this pattern to find the next step.

   Now complete this rule:

   | Let a be any *nonzero* whole number. |
   | Then $a^0 =$ |

   Notice in the statement of the above rule, the base a is required to be nonzero. To see why, consider these two patterns:

   $$4^0 = 1 \qquad\qquad 0^4 = 0$$
   $$3^0 = 1 \qquad\qquad 0^3 = 0$$
   $$2^0 = 1 \qquad\qquad 0^2 = 0$$
   $$1^0 = 1 \qquad\qquad 0^1 = 0$$

   According to this pattern,   According to this pattern,
     0^0 should be _____. 0^0 should be _____.

   Since these two patterns are not consistent, we say that 0^0 **is undefined**.

   b. Look again at the pattern in part (a) that shows why a^0 ought to be defined as 1, for nonzero a. Write the next several steps after a^0, following the same pattern.

   To be consistent with the pattern, how should we define a^{-n}?

   | Let a be any *nonzero* whole number. |
   | Then $a^{-n} =$ |

   Observing patterns can be helpful for making sense of rules, but an equally useful tool is looking at examples. Trying several examples can help you better understand given rules, or decide whether an equation is in fact a rule, as you will see in the next several parts.

2. A property of radicals in your text is stated as follows: "$\sqrt[n]{a^n} = a$ when n is odd, and $\sqrt[n]{a^n} = |a|$ when n is even." Investigate several examples with various values of a and n, to try to discover the reasons for the two different rules for odd and even n. For instance, a positive value of a and an even value for n are given in the following chart.

a	n	a^n	$\sqrt[n]{a^n}$
3	2		

Based on your examples in this chart, explain why the property is $\sqrt[n]{a^n} = a$ when n is odd, but $\sqrt[n]{a^n} = |a|$ when n is even.

3. In this chapter you practiced using the distributive property of multiplication over addition: $a(b + c) = ab + ac$. Suppose a fellow student wonders, "Does it work the same way if there is a multiplication in the parentheses?" Investigate whether the equation $a(b \times c) = ab \times ac$ is a property of whole-number multiplication, and then explain to this student what you did to decide.

4. Another student says, "$\dfrac{a + b}{b} = a$ is a rule, because the b's cancel out." Do you agree with the student? Explain your answer by looking at some examples.

SECTION	CONCEPT	KEY IDEAS/FORMULAS
0.1	**Real numbers**	
	The set of real numbers	*Rational:* $\dfrac{a}{b}$, where a and b are integers or a decimal that terminates or repeats. *Irrational:* Nonrepeating/nonterminating decimal.

	Approximations: Rounding and truncation	*Rounding:* Examine the digit to the right of the last desired digit.
		   Digit < 5: Keep last desired digit as is.
		   Digit ≥ 5: Round the last desired digit up 1.
		Truncating: Eliminate all digits to the right of the desired digit.

	Order of operations	1. Parentheses 2. Multiplication and Division 3. Addition and Subtraction
	Properties of real numbers	■ $a(b + c) = ab + ac$ ■ If $xy = 0$, then $x = 0$ or $y = 0$ ■ $\dfrac{a}{b} \pm \dfrac{c}{d} = \dfrac{ad \pm bc}{bd}$ $b \neq 0$ and $d \neq 0$ ■ $\dfrac{a}{b} \div \dfrac{c}{d} = \dfrac{a}{b} \cdot \dfrac{d}{c}$ $b \neq 0,\ c \neq 0,$ and $d \neq 0$
0.2	**Integer exponents and scientific notation**	$a^n = \underbrace{a \cdot a \cdot a \cdots a}_{n \text{ factors}}$
	Integer exponents	$a^m \cdot a^n = a^{m+n}$ $\dfrac{a^m}{a^n} = a^{m-n}$ $a^0 = 1$ $(a^m)^n = a^{mn}$ $a^{-n} = \dfrac{1}{a^n} = \left(\dfrac{1}{a}\right)^n$ $a \neq 0$
	Scientific notation	$c \times 10^n$ where c is a positive real number $1 \leq c < 10$ and n is an integer.
0.3	**Polynomials: Basic operations**	
	Adding and subtracting polynomials	Combine like terms.
	Multiplying polynomials	Distributive property

	Special products	$(x + a)(x + b) = x^2 + (a + b)x + ab$
		Perfect Squares
		   $(a + b)^2 = (a + b)(a + b) = a^2 + 2ab + b^2$
		   $(a - b)^2 = (a - b)(a - b) = a^2 - 2ab + b^2$
		Difference of Two Squares
		   $(a + b)(a - b) = a^2 - b^2$
		Perfect Cubes
		   $(a + b)^3 = a^3 + 3a^2b + 3ab^2 + b^3$
		   $(a - b)^3 = a^3 - 3a^2b + 3ab^2 - b^3$

SECTION	CONCEPT	KEY IDEAS/FORMULAS
0.4	**Factoring polynomials**	
	Greatest common factor	Factor out using distributive property: ax^k
	Factoring formulas: Special polynomial forms	*Difference of Two Squares* $\quad a^2 - b^2 = (a + b)(a - b)$ *Perfect Squares* $\quad a^2 + 2ab + b^2 = (a + b)^2$ $\quad a^2 - 2ab + b^2 = (a - b)^2$ *Sum of Two Cubes* $\quad a^3 + b^3 = (a + b)(a^2 - ab + b^2)$ *Difference of Two Cubes* $\quad a^3 - b^3 = (a - b)(a^2 + ab + b^2)$
	Factoring a trinomial as a product of two binomials	■ $x^2 + bx + c = (x + ?)(x + ?)$ ■ $ax^2 + bx + c = (?x + ?)(?x + ?)$
	Factoring by grouping	Group terms with common factors.
	A strategy for factoring polynomials	1. Factor out any common factors. 2. Recognize any special products. 3. Use the foil method in reverse for trinomials. 4. Look for factoring by grouping.
0.5	**Rational expressions**	
	Rational expressions and domain restrictions	Note domain restrictions when denominator is equal to zero.
	Simplifying rational expressions	
	Multiplying and dividing rational expressions	■ Use properties of rational numbers. ■ State additional domain restrictions once division is rewritten as multiplication of a reciprocal.
	Adding and subtracting rational expressions	Least common denominator (LCD)
	Complex rational expressions	Two strategies: 1. Write sum/difference in numerator/ denominator as a rational expression. 2. Multiply by the LCD of the numerator and denominator.
0.6	**Rational exponents and radicals**	
	Square roots	$\sqrt{25} = 5$
	Other (*n*th) roots	$b = \sqrt[n]{a}$ means $a = b^n$ for a and b positive real numbers and n a positive even integer, or for a and b any real numbers and n a positive odd integer. $\sqrt[n]{ab} = \sqrt[n]{a} \cdot \sqrt[n]{b} \qquad \sqrt[n]{\dfrac{a}{b}} = \dfrac{\sqrt[n]{a}}{\sqrt[n]{b}} \qquad b \neq 0$ $\sqrt[n]{a^m} = \left(\sqrt[n]{a}\right)^m$ $\sqrt[n]{a^n} = a \qquad n$ is odd $\sqrt[n]{a^n} = |a| \qquad n$ is even

SECTION	CONCEPT	KEY IDEAS/FORMULAS
	Rational exponents	$a^{1/n} = \sqrt[n]{a}$
		$a^{m/n} = (a^{1/n})^m = (\sqrt[n]{a})^m$
		$a^{-m/n} = \dfrac{1}{a^{m/n}}$ for m and n positive integers with
		no common factors, $a \neq 0$.
0.7	**Complex numbers**	
	The Imaginary Unit, i	$i = \sqrt{-1}$
	Adding and subtracting complex numbers	Complex Numbers: $a + bi$ where a and b are real numbers. Combine real parts with real parts and imaginary parts with imaginary parts.
	Multiplying complex numbers	Use the FOIL method and $i^2 = -1$ to simplify.
	Dividing complex numbers	If $a + bi$ is in the denominator, then multiply the numerator and the denominator by $a - bi$. The result is a real number in the denominator.
	Raising complex numbers to integer powers	$i = \sqrt{-1} \quad i^2 = -1 \quad i^3 = -i \quad i^4 = 1$

0.1 Real Numbers

Approximate to two decimal places by (a) rounding and (b) truncating.

1. 5.21597

2. 7.3623

Simplify.

3. $7 - 2 \cdot 5 + 4 \cdot 3 - 5$

4. $-2(5 + 3) + 7(3 - 2 \cdot 5)$

5. $-\dfrac{16}{(-2)(-4)}$

6. $-3(x - y) + 4(3x - 2y)$

Perform the indicated operation and simplify.

7. $\dfrac{x}{4} - \dfrac{x}{3}$

8. $\dfrac{y}{3} + \dfrac{y}{5} - \dfrac{y}{6}$

9. $\dfrac{12}{7} \cdot \dfrac{21}{4}$

10. $\dfrac{a^2}{b^3} \div \dfrac{2a}{b^2}$

0.2 Integer Exponents and Scientific Notation

Simplify using properties of exponents.

11. $(-2z)^3$

12. $(-4z^2)^3$

13. $\dfrac{(3x^3 y^2)^2}{2(x^2 y)^4}$

14. $\dfrac{(2x^2 y^3)^2}{(4xy)^3}$

15. Express 0.00000215 in scientific notation.

16. Express 7.2×10^9 as a real number.

0.3 Polynomials: Basic Operations

Perform the indicated operation and write the results in standard form.

17. $(14z^2 + 2) + (3z - 4)$

18. $(27y^2 - 6y + 2) - (y^2 + 3y - 7)$

19. $(36x^2 - 4x - 5) - (6x - 9x^2 + 10)$

20. $[2x - (4x^2 - 7x)] - [3x - (2x^2 + 5x - 4)]$

21. $5xy^2(3x - 4y)$

22. $-2st^2(-t + s - 2st)$

23. $(x - 7)(x + 9)$

24. $(2x + 1)(3x - 2)$

25. $(2x - 3)^2$

26. $(5x - 7)(5x + 7)$

27. $(x^2 + 1)^2$

28. $(1 - x^2)^2$

0.4 Factoring Polynomials

Factor out the common factor.

29. $14x^2 y^2 - 10xy^3$

30. $30x^4 - 20x^3 + 10x^2$

Factor the trinomial into a product of two binomials.

31. $2x^2 + 9x - 5$

32. $6x^2 - 19x - 7$

33. $16x^2 - 25$

34. $9x^2 - 30x + 25$

Factor the sum or difference of two cubes.

35. $x^3 + 125$

36. $1 - 8x^3$

Factor into a product of three polynomials.

37. $2x^3 + 4x^2 - 30x$

38. $6x^3 - 5x^2 + x$

Factor into a product of two binomials by grouping.

39. $x^3 + x^2 - 2x - 2$

40. $2x^3 - x^2 + 6x - 3$

0.5 Rational Expressions

State the domain restrictions on each of the rational expressions.

41. $\dfrac{4x^2 - 3}{x^2 - 9}$

42. $\dfrac{1}{x^2 + 1}$

Simplify.

43. $\dfrac{x^2 - 4}{x - 2}$

44. $\dfrac{x - 5}{x - 5}$

45. $\dfrac{t^2 + t - 6}{t^2 - t - 2}$

46. $\dfrac{z^3 - z}{z^2 + z}$

Perform the indicated operation and simplify.

47. $\dfrac{x^2 + 3x - 10}{x^2 + 2x - 3} \cdot \dfrac{x^2 + x - 2}{x^2 + x - 6}$

48. $\dfrac{x^2 - x - 2}{x^3 + 3x^2} \div \dfrac{x + 1}{x^2 + 2x}$

49. $\dfrac{1}{x + 1} - \dfrac{1}{x + 3}$

50. $\dfrac{1}{x} - \dfrac{1}{x + 1} + \dfrac{1}{x + 2}$

Simplify.

51. $\dfrac{2 + \dfrac{1}{x-3}}{\dfrac{1}{5x-15} + 4}$

52. $\dfrac{\dfrac{1}{x} + \dfrac{2}{x^2}}{3 - \dfrac{1}{x^2}}$

0.6 Rational Exponents and Radicals

Simplify.

53. $\sqrt{20}$

54. $\sqrt{80}$

55. $\sqrt[3]{-125x^5y^4}$

56. $\sqrt[4]{32x^4y^5}$

57. $3\sqrt{20} + 5\sqrt{80}$

58. $4\sqrt{27x} - 8\sqrt{12x}$

59. $(2 + \sqrt{5})(1 - \sqrt{5})$

60. $(3 + \sqrt{x})(4 - \sqrt{x})$

61. $\dfrac{1}{2 - \sqrt{3}}$

62. $\dfrac{1}{3 - \sqrt{x}}$

63. $\dfrac{(3x^{2/3})^2}{(4x^{1/3})^2}$

64. $\dfrac{(4x^{3/4})^2}{(2x^{-1/3})^2}$

65. $\dfrac{5^{1/2}}{5^{1/3}}$

66. $(x^{-2/3}y^{1/4})^{12}$

0.7 Complex Numbers

Simplify.

67. $\sqrt{-169}$

68. $\sqrt{-32}$

69. i^{19}

70. i^9

Perform the indicated operation, simplify, and express in standard form.

71. $(3 - 2i) + (5 - 4i)$

72. $(-4 + 7i) + (-2 - 3i)$

73. $(12 - i) - (-2 - 5i)$

74. $(9 + 8i) - (4 - 2i)$

75. $(2 + 2i)(3 - 3i)$

76. $(1 + 6i)(1 + 5i)$

77. $(4 + 7i)^2$

78. $(7 - i)^2$

Express the quotient in standard form.

79. $\dfrac{1}{2 - i}$

80. $\dfrac{1}{3 + i}$

81. $\dfrac{7 + 2i}{4 + 5i}$

82. $\dfrac{6 - 5i}{3 - 2i}$

83. $\dfrac{10}{3i}$

84. $\dfrac{7}{2i}$

Technology

Section 0.1

85. Use your calculator to evaluate $\sqrt{272.25}$. Does the answer appear to be a rational or an irrational number? Why?

86. Use your calculator to evaluate $\sqrt{\dfrac{1053}{81}}$. Does the answer appear to be a rational or an irrational number? Why?

Section 0.2

Use a graphing utility to evaluate the expression. Express your answer in scientific notation.

87. $\dfrac{(8.2 \times 10^{11})(1.167 \times 10^{-35})}{(4.92 \times 10^{-18})}$

88. $\dfrac{(1.4805 \times 10^{21})}{(5.64 \times 10^{26})(1.68 \times 10^{-9})}$

Section 0.3

89. Use a graphing utility to plot the graphs of the three expressions $(2x + 3)^3$, $8x^3 + 27$, and $8x^3 + 36x^2 + 54x + 27$. Which two graphs agree with each other?

90. Use a graphing utility to plot the graphs of the three expressions $(x - 3)^2$, $8x^2 + 9$, and $x^2 - 6x + 9$. Which two graphs agree with each other?

Section 0.4

91. Use a graphing utility to plot the graphs of the three expressions $x^2 - 3x + 18$, $(x + 6)(x - 3)$, and $(x - 6)(x + 3)$. Which two graphs agree with each other?

92. Use a graphing utility to plot the graphs of the three expressions $x^2 - 8x + 16$, $(x + 4)^2$, and $(x - 4)^2$. Which two graphs agree with each other?

Section 0.5

For each given expression: (a) simplify the expression, (b) use a graphing utility to plot the expression and the answer in (a) in the same viewing window, and (c) determine the domain restriction(s) where the graphs will agree with each other.

93. $\dfrac{1 - (4/x)}{1 - (4/x^2)}$

94. $\dfrac{1 - (3/x)}{1 + (9/x^2)}$

Section 0.6

95. Given $\dfrac{6}{\sqrt{5} - \sqrt{2}}$

    a. Rationalize the denominator.
    b. Use a graphing utility to evaluate the expression and the answer.
    c. Do they agree?

96. Given $\dfrac{11}{2\sqrt{6} + \sqrt{13}}$

    a. Rationalize the denominator.
    b. Use a graphing utility to evaluate the expression and the answer.
    c. Do they agree?

Section 0.7

In Exercises 97 and 98, use a graphing utility to simplify the expression. Write your answer in standard form.

97. $(3 + 5i)^5$

98. $\dfrac{1}{(1 + 3i)^4}$

99. Apply a graphing utility to simplify the expression and write your answer in standard form.

$$\dfrac{1}{(6 + 2i)^4}$$

Simplify.

1. $\sqrt{16}$

2. $\sqrt[3]{54x^6}$

3. $-3(2 + 5^2) + 2(3 - 7) - (3^2 - 1)$

4. $\sqrt[5]{-32}$

5. $\sqrt{-12x^2}$

6. i^{17}

7. $\dfrac{(x^2 y^{-3} z^{-1})^{-2}}{(x^{-1} y^2 z^3)^{1/2}}$

8. $3\sqrt{x} - 4\sqrt{x} + 5\sqrt{x}$

9. $3\sqrt{18} - 4\sqrt{32}$

10. $\left(5\sqrt{6} - 2\sqrt{2}\right)\left(\sqrt{6} + 3\sqrt{2}\right)$

Perform the indicated operation and simplify.

11. $(3y^2 - 5y + 7) - (y^2 + 7y - 13)$

12. $(2x - 3)(5x + 7)$

Factor.

13. $x^2 - 16$

14. $3x^2 + 15x + 18$

15. $4x^2 + 12xy + 9y^2$

16. $x^4 - 2x^2 + 1$

17. $2x^2 - x - 1$

18. $6y^2 - y - 1$

19. $2t^3 - t^2 - 3t$

20. $2x^3 - 5x^2 - 3x$

21. $x^2 - 3yx + 4yx - 12y^2$

22. $x^4 + 5x^2 - 3x^2 - 15$

23. $81 + 3x^3$

24. $27x - x^4$

Perform the indicated operations and simplify.

25. $\dfrac{2}{x} + \dfrac{3}{x - 1}$

26. $\dfrac{5x}{x^2 - 7x + 10} - \dfrac{4}{x^2 - 25}$

27. $\dfrac{x - 1}{x^2 - 1} \cdot \dfrac{x^2 + x + 1}{x^3 - 1}$

28. $\dfrac{4x^2 - 9}{x^2 - 11x - 60} \cdot \dfrac{x^2 - 16}{2x + 3}$

29. $\dfrac{x - 3}{2x - 5} \div \dfrac{x^2 - 9}{5 - 2x}$

30. $\dfrac{1 - t}{3t + 1} \div \dfrac{t^2 - 2t + 1}{7t + 21t^2}$

Write the resulting expression in standard form.

31. $(1 - 3i)(7 - 5i)$

32. $\dfrac{2 - 11i}{4 + i}$

33. Rationalize the denominator: $\dfrac{7 - 2\sqrt{3}}{4 - 5\sqrt{3}}$.

34. Represent 0.0000155 in scientific notation.

35. Simplify $\dfrac{\dfrac{1}{x} - \dfrac{2}{x + 1}}{x - 1}$ and state any domain restrictions.

36. For the given expression:

$$\dfrac{1 + \dfrac{5}{x}}{1 - \dfrac{25}{x^2}}$$

   a. Simplify the expression.
   b. Use a graphing utility to plot the expression and the answer in (a) in the same viewing window.
   c. Determine the domain restriction(s) where the graphs will agree with each other.

37. Apply a graphing utility to evaluate the expression. Round your answer to three decimal places.

$$\dfrac{\sqrt{5}}{\sqrt{13} - \sqrt{7}}$$

1

Equations and Inequalities

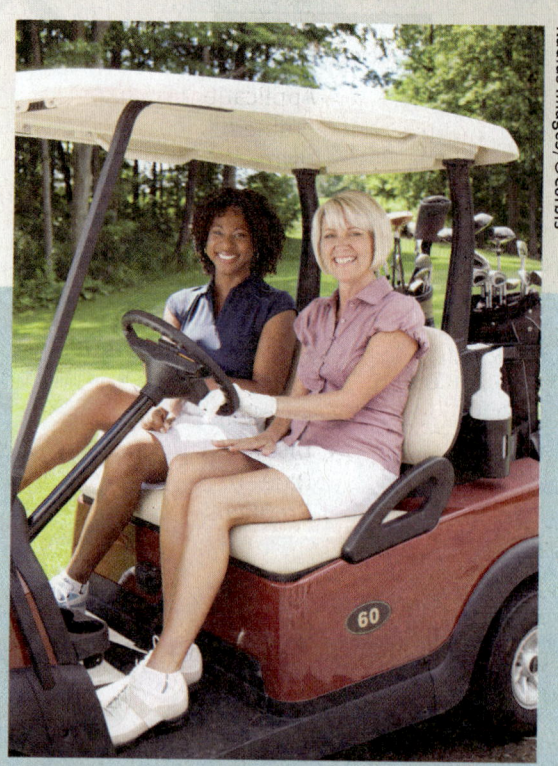

Radius Images/©Corbis

Golf courses usually charge both greens fees (cost of playing the course) and cart fees (cost of renting a golf cart). Two friends who enjoy playing golf decide to investigate becoming members at a golf course.

The course they enjoy playing the most charges $40 for greens fees and $15 for cart rental (per person), so it currently costs each of them $55 every time they play. The membership offered at that course costs $160 per month with no greens fees, but there is still the per person cart rental fee.

How many times a month would they have to play golf in order for the membership option to be the better deal?* This is just one example of how the real world can be modeled with equations and inequalities.

*See Section 1.5, Exercises 109 and 110.

IN THIS CHAPTER you will solve linear and quadratic equations. You will then solve more complicated equations (polynomial, rational, radical, and absolute value) by first transforming them into linear or quadratic equations. Then you will solve linear, quadratic, polynomial, rational, and absolute value inequalities. Throughout this chapter you will solve applications of equations and inequalities.

EQUATIONS AND INEQUALITIES

1.1 Linear Equations	1.2 Applications Involving Linear Equations	1.3 Quadratic Equations	1.4 Other Types of Equations	1.5 Linear Inequalities	1.6 Polynomial and Rational Inequalities	1.7 Absolute Value Equations and Inequalities
• Solving Linear Equations in One Variable • Solving Rational Equations That Are Reducible to Linear Equations	• Solving Application Problems Using Mathematical Models • Geometry Problems • Interest Problems • Mixture Problems • Distance–Rate–Time Problems	• Factoring • Square Root Method • Completing the Square • Quadratic Formula	• Radical Equations • Equations Quadratic in Form: u-Substitution • Factorable Equations	• Graphing Inequalities and Interval Notation • Solving Linear Inequalities	• Polynomial Inequalities • Rational Inequalities	• Equations Involving Absolute Value • Inequalities Involving Absolute Value

LEARNING OBJECTIVES

- Solve linear equations.
- Solve application problems involving linear equations.
- Solve quadratic equations.
- Solve rational, polynomial, and radical equations.
- Solve linear inequalities.
- Solve polynomial and rational inequalities.
- Solve absolute value equations and inequalities.

SKILLS OBJECTIVES

- Solve linear equations in one variable.
- Solve rational equations that are reducible to linear equations.

CONCEPTUAL OBJECTIVE

- Eliminate values that result in a denominator being equal to zero.

Solving Linear Equations in One Variable

An **algebraic expression** (see Chapter 0) consists of one or more terms that are combined through basic operations such as addition, subtraction, multiplication, or division; for example:

$$3x + 2 \qquad 5 - 2y \qquad x + y$$

An **equation** is a statement that says two expressions are equal. For example, the following are all equations in one variable, x:

$$x + 7 = 11 \qquad x^2 = 9 \qquad 7 - 3x = 2 - 3x \qquad 4x + 7 = x + 2 + 3x + 5$$

To **solve** an equation means to find all the values of x that make the equation true. These values are called **solutions**, or **roots**, of the equation. The first of these statements shown above, $x + 7 = 11$, is true when $x = 4$ and false for any other values of x. We say that $x = 4$ is the solution to the equation. Sometimes an equation can have more than one solution, as in $x^2 = 9$. In this case, there are actually two values of x that make this equation true, $x = -3$ and $x = 3$. We say the **solution set** of this equation is $\{-3, 3\}$. In the third equation, $7 - 3x = 2 - 3x$, no values of x make the statement true. Therefore, we say this equation has **no solution**. And the fourth equation, $4x + 7 = x + 2 + 3x + 5$, is true for any values of x. An equation that is true for any value of the variable x is called an **identity**. In this case, we say the solution set is the **set of all real numbers**.

Two equations that have the same solution set are called **equivalent equations**. For example,

$$3x + 7 = 13 \qquad 3x = 6 \qquad x = 2$$

are all equivalent equations because each of them has the solution set $\{2\}$. Note that $x^2 = 4$ is not equivalent to these three equations because it has the solution set $\{-2, 2\}$.

When solving equations it helps to find a simpler equivalent equation in which the variable is isolated (alone). The following table summarizes the procedures for generating equivalent equations.

Generating Equivalent Equations

ORIGINAL EQUATION	DESCRIPTION	EQUIVALENT EQUATION
$3(x - 6) = 6x - x$	■ Eliminate parentheses. ■ Combine like terms on one or both sides of an equation.	$3x - 18 = 5x$
$7x + 8 = 29$	Add (or subtract) the same quantity to (from) *both* sides of an equation. $7x + 8 - \mathbf{8} = 29 - \mathbf{8}$	$7x = 21$
$5x = 15$	Multiply (or divide) both sides of an equation by the same nonzero quantity: $\dfrac{5x}{5} = \dfrac{15}{5}$.	$x = 3$
$-7 = x$	Interchange the two sides of the equation.	$x = -7$

You probably already know how to solve simple linear equations. Solving a linear equation in one variable is done by finding an equivalent equation. In generating an equivalent equation, remember that whatever operation is performed on one side of an equation must also be performed on the other side of the equation.

Technology Tip

Use a graphing utility to display graphs of $y_1 = 3x + 4$ and $y_2 = 16$.

The x-coordinate of the point of intersection is the solution to the equation $3x + 4 = 16$.

EXAMPLE 1   Solving a Linear Equation

Solve the equation $3x + 4 = 16$.

Solution:

Subtract 4 from both sides of the equation.

$$3x + 4 = 16$$
$$\underline{-4 \quad -4}$$
$$3x \quad = 12$$

Divide both sides by 3.

$$\frac{3x}{3} = \frac{12}{3}$$

The solution is $x = 4$.

$$\boxed{x = 4}$$

The solution set is $\{4\}$.

■ **YOUR TURN** Solve the equation $2x + 3 = 9$.

■ **Answer:** The solution is $x = 3$. The solution set is $\{3\}$.

Example 1 illustrates solving linear equations in one variable. What is a linear equation in one variable?

Classroom Example 1.1.1
Solve the equation $3x - 3 = 15$.
Answer: $x = 6$

DEFINITION   Linear Equation

A **linear equation in one variable**, x, can be written in the form

$$ax + b = 0$$

where a and b are real numbers and $a \neq 0$.

What makes this equation linear is that x is raised to the first power. We can also classify a linear equation as a **first-degree** equation.

Equation	Degree	General Name
$x - 7 = 0$	First	Linear
$x^2 - 6x - 9 = 0$	Second	Quadratic
$x^3 + 3x^2 - 8 = 0$	Third	Cubic

Technology Tip

Use a graphing utility to display graphs of $y_1 = 5x - (7x - 4) - 2$ and $y_2 = 5 - (3x + 2)$.

The x-coordinate of the point of intersection is the solution to the equation $5x - (7x - 4) - 2 = 5 - (3x + 2)$.

■ **Answer:** The solution is $x = 2$. The solution set is $\{2\}$.

Study Tip

Prime Factors

$$
\begin{array}{r}
2 = 2 \\
6 = 2 \cdot 3 \\
5 = \cdot 5 \\
\hline
\text{LCD} = 2 \cdot 3 \cdot 5 = 30
\end{array}
$$

Technology Tip

Use a graphing utility to display graphs of $y_1 = \frac{1}{2}p - 5$ and $y_2 = \frac{3}{4}x$.

The x-coordinate of the point of intersection is the solution.

■ **Answer:** The solution is $m = -18$. The solution set is $\{-18\}$.

EXAMPLE 2   Solving a Linear Equation

Solve the equation $5x - (7x - 4) - 2 = 5 - (3x + 2)$.

Solution:

Eliminate the parentheses.

Don't forget to distribute the negative sign through *both* terms inside the parentheses.

$$5x - (7x - 4) - 2 = 5 - (3x + 2)$$

$$5x - 7x + 4 - 2 = 5 - 3x - 2$$

Combine x terms on the left, constants on the right. Add $3x$ to both sides.

$$
\begin{array}{r}
-2x + 2 = 3 - 3x \\
+3x + 3x \\
\hline
x + 2 = 3
\end{array}
$$

Subtract 2 from both sides.

$$
\begin{array}{r}
-2 - 2 \\
\hline
x = 1
\end{array}
$$

Check to verify that $x = 1$ is a solution to the original equation.

$$5 \cdot 1 - (7 \cdot 1 - 4) - 2 = 5 - (3 \cdot 1 + 2)$$
$$5 - (7 - 4) - 2 = 5 - (3 + 2)$$
$$5 - (3) - 2 = 5 - (5)$$
$$0 = 0$$

Since the solution $x = 1$ makes the equation true, the solution set is $\{1\}$.

■ **YOUR TURN** Solve the equation $4(x - 1) - 2 = x - 3(x - 2)$.

Classroom Example 1.1.2
Solve the equation
$5(x - 1) + 20 = x - 7(x + 1)$.
Answer: $x = -2$

To solve a linear equation involving fractions, find the least common denominator (LCD) of all terms and multiply both sides of the equation by the LCD. We will first review how to find the LCD.

To add the fractions $\frac{1}{2} + \frac{1}{6} + \frac{2}{5}$, we must first find a common denominator. Some people are taught to find the lowest number that 2, 6, and 5 all divide evenly into. Others prefer a more systematic approach in terms of prime factors.

EXAMPLE 3   Solving a Linear Equation Involving Fractions

Solve the equation $\frac{1}{2}p - 5 = \frac{3}{4}p$.

Solution:

Write the equation.

$$\frac{1}{2}p - 5 = \frac{3}{4}p$$

Multiply each term in the equation by the LCD, 4.

$$(4)\frac{1}{2}p - (4)5 = (4)\frac{3}{4}p$$

The result is a linear equation with no fractions.

$$2p - 20 = 3p$$

Subtract $2p$ from both sides.

$$
\begin{array}{r}
-2p -2p \\
\hline
-20 = p
\end{array}
$$

$$\boxed{p = -20}$$

Since $p = -20$ satisfies the original equation, the solution set is $\{-20\}$.

■ **YOUR TURN** Solve the equation $\frac{1}{4}m = \frac{1}{12}m - 3$.

Solving a Linear Equation in One Variable

STEP	DESCRIPTION	EXAMPLE
1	Simplify the algebraic expressions on both sides of the equation.	$-3(x-2) + 5 = 7(x-4) - 1$ $-3x + 6 + 5 = 7x - 28 - 1$ $-3x + 11 = 7x - 29$
2	Gather all variable terms on one side of the equation and all constant terms on the other side.	$\begin{aligned} -3x + 11 &= 7x - 29 \\ \underline{+3x \qquad\quad +3x} \\ 11 &= 10x - 29 \\ \underline{+29 \qquad\quad +29} \\ 40 &= 10x \end{aligned}$
3	Isolate the variable.	$10x = 40$ $\boxed{x = 4}$

Solving Rational Equations That Are Reducible to Linear Equations

A **rational equation** is an equation that contains one or more rational expressions (Chapter 0). Some rational equations can be transformed into linear equations that you can then solve, but as you will see momentarily, you must be certain that the solution to the linear equation also satisfies the original rational equation.

EXAMPLE 4 **Solving a Rational Equation That Can Be Reduced to a Linear Equation**

Solve the equation $\dfrac{2}{3x} + \dfrac{1}{2} = \dfrac{4}{x} + \dfrac{4}{3}$.

Solution:

State the excluded values (those that make any denominator equal 0).

$$\frac{2}{3x} + \frac{1}{2} = \frac{4}{x} + \frac{4}{3} \qquad x \neq 0$$

Eliminate fractions by multiplying *each term* by the LCD, $6x$.

$$6x\left(\frac{2}{3x}\right) + 6x\left(\frac{1}{2}\right) = 6x\left(\frac{4}{x}\right) + 6x\left(\frac{4}{3}\right)$$

Simplify both sides.

$$4 + 3x = 24 + 8x$$

Subtract 4.

$$\underline{-4 \qquad\qquad -4}$$

> **Classroom Example 1.1.4**
>
> Solve the equation $\dfrac{2}{a} - 2 = \dfrac{12}{7a}$.
>
> **Answer:** $a = \frac{1}{7}$ (Note: $a = 0$ must be excluded.)

Subtract $8x$.

$$3x = 20 + 8x$$
$$\underline{-8x \qquad\qquad -8x}$$
$$-5x = 20$$

Divide by -5.

$$\boxed{x = -4}$$

Since $x = -4$ satisfies the original equation, the solution set is $\{-4\}$.

■ **YOUR TURN** Solve the equation $\dfrac{3}{y} + 2 = \dfrac{7}{2y}$.

■ **Answer:** The solution is $y = \frac{1}{4}$. The solution set is $\left\{\frac{1}{4}\right\}$.

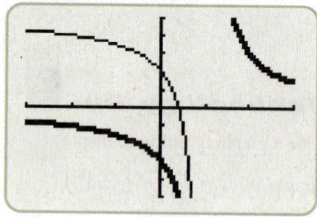
Extraneous solutions are solutions that satisfy a transformed equation but do not satisfy the original equation. It is important to first state any values of the variable that must be eliminated based on the original rational equation. Once the rational equation is transformed to a linear equation and solved, remove any excluded values of the variable.

EXAMPLE 5 Solving Rational Equations That Can Be Reduced to Linear Equations

Solve the equation $\dfrac{3x}{x-1} + 2 = \dfrac{3}{x-1}$.

Solution:

State the excluded values (those that make any denominator equal 0).

$$\frac{3x}{x-1} + 2 = \frac{3}{x-1} \quad \boxed{x \neq 1}$$

Eliminate the fractions by multiplying each term by the LCD, $x - 1$.

$$\frac{3x}{x-1} \cdot (x-1) + 2 \cdot (x-1) = \frac{3}{x-1} \cdot (x-1)$$

Simplify.

$$\frac{3x}{x-1} \cdot (x-1) + 2 \cdot (x-1) = \frac{3}{x-1} \cdot (x-1)$$

$$3x + 2(x-1) = 3$$

Distribute the 2. $\qquad 3x + 2x - 2 = 3$

Combine x terms on the left. $\qquad 5x - 2 = 3$

Add 2 to both sides. $\qquad 5x = 5$

Divide both sides by 5. $\qquad x = 1$

It may seem that $x = 1$ is the solution. However, the original equation had the restriction $x \neq 1$. Therefore, $x = 1$ is an extraneous solution and must be eliminated as a possible solution.

Thus, the equation $\dfrac{3x}{x-1} + 2 = \dfrac{3}{x-1}$ has $\boxed{\text{no solution}}$.

■ **YOUR TURN** Solve the equation $\dfrac{2x}{x-2} - 3 = \dfrac{4}{x-2}$.

We have reviewed finding the least common denominator (LCD) for real numbers. Now let us consider finding the LCD for rational equations that have different denominators. We multiply the denominators in order to get a common denominator.

$$\text{Rational expression:} \quad \frac{1}{x} + \frac{2}{x-1} \qquad \text{LCD: } x(x-1)$$

In order to find a *least* common denominator, it is useful to first factor the denominators to identify common multiples.

$$\text{Rational equation:} \quad \frac{1}{3x-3} + \frac{1}{2x-2} = \frac{1}{x^2 - x}$$

$$\text{Factor the denominators:} \quad \frac{1}{3(x-1)} + \frac{1}{2(x-1)} = \frac{1}{x(x-1)}$$

$$\text{LCD:} \quad 6x(x-1)$$

EXAMPLE 6   Solving Rational Equations

Solve the equation $\dfrac{1}{3x+18}-\dfrac{1}{2x+12}=\dfrac{1}{x^2+6x}$.

Solution:

Factor the denominators.

$$\frac{1}{3(x+6)}-\frac{1}{2(x+6)}=\frac{1}{x(x+6)}$$

State the excluded values.

$$\boxed{x\neq 0,-6}$$

Multiply the equation by the LCD, $6x(x+6)$.

$$6x(x+6)\cdot\frac{1}{3(x+6)}-6x(x+6)\cdot\frac{1}{2(x+6)}=6x(x+6)\cdot\frac{1}{x(x+6)}$$

Divide out the common factors.

$$6x\cancel{(x+6)}\cdot\frac{1}{3\cancel{(x+6)}}-6x\cancel{(x+6)}\cdot\frac{1}{2\cancel{(x+6)}}=6x\cancel{(x+6)}\cdot\frac{1}{x\cancel{(x+6)}}$$

Simplify.

$$2x-3x=6$$

Solve the linear equation.

$$x=-6$$

Since one of the excluded values is $x\neq -6$, we say that $x=-6$ is an extraneous solution. Therefore, this rational equation has $\boxed{\text{no solution}}$.

■ **YOUR TURN**   Solve the equation $\dfrac{2}{x}+\dfrac{1}{x+1}=-\dfrac{1}{x(x+1)}$.

■ **Answer:** no solution

EXAMPLE 7   Solving Rational Equations

Solve the equation $\dfrac{2}{x-3}=\dfrac{-3}{2-x}$.

Solution:

What values make *either* denominator equal to zero? The values $x=2$ and $x=3$ must be excluded from possible solutions to the equation.

$$\frac{2}{x-3}=\frac{-3}{2-x}\qquad\boxed{x\neq 2,x\neq 3}$$

Multiply the equation by the LCD, $(x-3)(2-x)$.

$$\frac{2}{x-3}(x-3)(2-x)=\frac{-3}{2-x}(x-3)(2-x)$$

Divide out the common factors.

$$2(2-x)=-3(x-3)$$

Eliminate the parentheses.

$$4-2x=-3x+9$$

Collect x terms on the left, constants on the right.

$$\boxed{x=5}$$

Since $x=5$ satisfies the original equation, the solution set is $\{5\}$.

■ **YOUR TURN**   Solve the equation $\dfrac{-4}{x+8}=\dfrac{3}{x-6}$.

Technology Tip

Use a graphing utility to display graphs of $y_1=\dfrac{2}{x-3}$ and $y_2=\dfrac{-3}{2-x}$.

The x-coordinate of the point of intersection is the solution to the equation $\dfrac{2}{x-3}=\dfrac{-3}{2-x}$.

■ **Answer:** The solution is $x=0$. The solution set is $\{0\}$.

EXAMPLE 8   Automotive Service

A car dealership charges for parts and an hourly rate for labor. If parts cost $273, labor is $53 per hour, and the total bill is $458.50, how many hours did the dealership spend working on your car?

Solution:

Let x equal the number of hours the dealership worked on your car.

Write the cost equation.

$$\overbrace{53x}^{\text{labor}} + \overbrace{273}^{\text{parts}} = \overbrace{458.50}^{\text{total cost}}$$

Subtract 273 from both sides of the equation.

$$53x = 185.50$$

Divide both sides of the equation by 53.

$$\boxed{x = 3.5}$$

$$\boxed{\text{The dealership charged for 3.5 hours of labor.}}$$

EXAMPLE 9   Grades

Dante currently has the following three test scores: 82, 79, and 90. If the score on the final exam is worth two test scores and his goal is to earn an 85 for his class average, what score on the final exam does Dante need to achieve his course goal?

Solution:

Let x equal final exam grade.

Write the equation that determines the course grade.

$$\dfrac{\overbrace{82 + 79 + 90}^{\text{scores 1, 2, and 3}} + \overbrace{2x}^{\substack{\text{final is worth} \\ \text{two test scores}}}}{\underbrace{5}_{\text{total of five test scores}}} = \underbrace{85}_{\text{average}}$$

Simplify the numerator.

$$\dfrac{251 + 2x}{5} = 85$$

Multiply the equation by 5 (or cross multiply).

$$251 + 2x = 425$$

Solve the linear equation.

$$\boxed{x = 87}$$

$$\boxed{\text{Dante needs to score at least an 87 on the final exam.}}$$

SECTION 1.1 SUMMARY

Linear equations, $ax + b = 0$, are solved by:

1. Simplifying the algebraic expressions on both sides of the equation.
2. Gathering all variable terms on one side of the equation and all constant terms on the other side.
3. Isolating the variable.

Rational equations are solved by:

1. Determining any excluded values (denominator equals 0).
2. Multiplying the equation by the LCD.
3. Solving the resulting equation.
4. Eliminating any extraneous solutions.

SECTION 1.1 EXERCISES

■ SKILLS

In Exercises 1–36, solve for the indicated variable.

1. $5x = 35$

2. $4t = 32$

3. $-3 + n = 12$

4. $4 = -5 + y$

5. $24 = -3x$

6. $-50 = -5t$

7. $\frac{1}{5}n = 3$

8. $6 = \frac{1}{3}p$

9. $3x - 5 = 7$

10. $4p + 5 = 9$

11. $9m - 7 = 11$

12. $2x + 4 = 5$

13. $5t + 11 = 18$

14. $7x + 4 = 21 + 24x$

15. $3x - 5 = 25 + 6x$

16. $5x + 10 = 25 + 2x$

17. $20n - 30 = 20 - 5n$

18. $14c + 15 = 43 + 7c$

19. $4(x - 3) = 2(x + 6)$

20. $5(2y - 1) = 2(4y - 3)$

21. $-3(4t - 5) = 5(6 - 2t)$

22. $2(3n + 4) = -(n + 2)$

23. $2(x - 1) + 3 = x - 3(x + 1)$

24. $4(y + 6) - 8 = 2y - 4(y + 2)$

25. $5p + 6(p + 7) = 3(p + 2)$

26. $3(z + 5) - 5 = 4z + 7(z - 2)$

27. $7x - (2x + 3) = x - 2$

28. $3x - (4x + 2) = x - 5$

29. $2 - (4x + 1) = 3 - (2x - 1)$

30. $5 - (2x - 3) = 7 - (3x + 5)$

31. $2a - 9(a + 6) = 6(a + 3) - 4a$

32. $25 - [2 + 5y - 3(y + 2)] = -3(2y - 5) - [5(y - 1) - 3y + 3]$

33. $32 - [4 + 6x - 5(x + 4)] = 4(3x + 4) - [6(3x - 4) + 7 - 4x]$

34. $12 - [3 + 4m - 6(3m - 2)] = -7(2m - 8) - 3[(m - 2) + 3m - 5]$

35. $20 - 4[c - 3 - 6(2c + 3)] = 5(3c - 2) - [2(7c - 8) - 4c + 7]$

36. $46 - [7 - 8y + 9(6y - 2)] = -7(4y - 7) - 2[6(2y - 3) - 4 + 6y]$

Exercises 37–48 involve fractions. Clear the fractions by first multiplying by the least common denominator, and then solve the resulting linear equation.

37. $\frac{1}{5}m = \frac{1}{60}m + 1$

38. $\frac{1}{12}z = \frac{1}{24}z + 3$

39. $\frac{x}{7} = \frac{2x}{63} + 4$

40. $\frac{a}{11} = \frac{a}{22} + 9$

41. $\frac{1}{3}p = 3 - \frac{1}{24}p$

42. $\frac{3x}{5} - x = \frac{x}{10} - \frac{5}{2}$

43. $\frac{5y}{3} - 2y = \frac{2y}{84} + \frac{5}{7}$

44. $2m - \frac{5m}{8} = \frac{3m}{72} + \frac{4}{3}$

45. $p + \frac{p}{4} = \frac{5}{2}$

46. $\frac{c}{4} - 2c = \frac{5}{4} - \frac{c}{2}$

47. $\frac{x - 3}{3} - \frac{x - 4}{2} = 1 - \frac{x - 6}{6}$

48. $1 - \frac{x - 5}{3} = \frac{x + 2}{5} - \frac{6x - 1}{15}$

In Exercises 49–70, specify any values that must be excluded from the solution set and then solve the equation.

49. $\frac{4}{y} - 5 = \frac{5}{2y}$

50. $\frac{4}{x} + 10 = \frac{2}{3x}$

51. $7 - \frac{1}{6x} = \frac{10}{3x}$

52. $\frac{7}{6t} = 2 + \frac{5}{3t}$

53. $\frac{2}{a} - 4 = \frac{4}{3a}$

54. $\frac{4}{x} - 2 = \frac{5}{2x}$

55. $\frac{x}{x - 2} + 5 = \frac{2}{x - 2}$

56. $\frac{n}{n - 5} + 2 = \frac{n}{n - 5}$

57. $\frac{2p}{p - 1} = 3 + \frac{2}{p - 1}$

58. $\frac{4t}{t + 2} = 3 - \frac{8}{t + 2}$

59. $\frac{3x}{x + 2} - 4 = \frac{2}{x + 2}$

60. $\frac{5y}{2y - 1} - 3 = \frac{12}{2y - 1}$

61. $\frac{1}{n} + \frac{1}{n + 1} = \frac{-1}{n(n + 1)}$

62. $\frac{1}{x} + \frac{1}{x - 1} = \frac{1}{x(x - 1)}$

63. $\frac{3}{a} - \frac{2}{a + 3} = \frac{9}{a(a + 3)}$

64. $\frac{1}{c - 2} + \frac{1}{c} = \frac{2}{c(c - 2)}$

65. $\frac{n - 5}{6n - 6} = \frac{1}{9} - \frac{n - 3}{4n - 4}$

66. $\frac{5}{m} + \frac{3}{m - 2} = \frac{6}{m(m - 2)}$

67. $\frac{2}{5x + 1} = \frac{1}{2x - 1}$

68. $\frac{3}{4n - 1} = \frac{2}{2n - 5}$

69. $\frac{t - 1}{1 - t} = \frac{3}{2}$

70. $\frac{2 - x}{x - 2} = \frac{3}{4}$

▪APPLICATIONS

71. Temperature. To calculate temperature in degrees Fahrenheit we use the formula $F = \frac{9}{5}C + 32$, where F is degrees Fahrenheit and C is degrees Celsius. Find the formula to convert from Fahrenheit to Celsius.

72. Geometry. The perimeter P of a rectangle is related to the length L and width W of the rectangle through the equation $P = 2L + 2W$. Determine the width in terms of the perimeter and length.

73. Costs: Cellular Phone. Your cell phone plan charges $15 a month plus 12 cents per minute. If your monthly bill is $25.08, how many minutes did you use?

74. Costs: Rental Car. Becky rented a car on her Ft. Lauderdale vacation. The car was $25 a day plus 10 cents per mile. She kept the car for 5 days and her bill was $185. How many miles did she drive the car?

75. Costs: Internet. When traveling in London, Charlotte decided to check her e-mail at an Internet café. There was a flat charge of $2 plus a charge of 10 cents a minute. How many minutes was she logged on if her bill was $3.70?

76. Sales: Income. For a summer job, Dwayne decides to sell magazine subscriptions. He will be paid $20 a day plus $1 for each subscription he sells. If he works for 25 days and makes $645, how many subscriptions did he sell?

77. Business. The operating costs for a local business are a fixed amount of $15,000 and $2500 per day.

  a. Find C that represents operating costs for the company which depends on the number of days open, x.

  b. If the business accrues $5,515,000 in annual operating costs, how many days did the business operate during the year?

78. Business. Negotiated contracts for a technical support provider produce monthly revenue of $5000 and $0.75 per minute per phone call.

  a. Find R that represents the revenue for the technical support provider which depends on the number of minutes of phone calls x.

  b. In one month the provider received $98,750 in revenue. How many minutes of technical support were provided?

In Exercises 79 and 80 refer to the following:

Medications are often packaged in liquid form (known as a suspension) so that a precise dose of a drug is delivered within a volume of inert liquid; for example, 250 milligrams amoxicillin in 5 milliliters of a liquid suspension. If a patient is prescribed a dose of a drug, medical personnel must compute the volume of the liquid with a known concentration to administer. The formula

$$a = \frac{d}{c}$$

defines the relationship between the dose of the drug prescribed d, the concentration of the liquid suspension c, and the amount of the liquid administered a.

79. Medicine. A physician has ordered a 600-milligram dose of amoxicillin. The pharmacy has a suspension of amoxicillin with a concentration of 125 milligrams per 5 milliliters. How much liquid suspension must be administered to the patient?

80. Medicine. A physician has ordered a 600-milligram dose of carbamazepine. The pharmacy has a suspension of carbamazepine with a concentration of 100 milligrams per 5 milliliters. How much liquid suspension must be administered to the patient?

81. Speed of Light. The frequency f of an optical signal in hertz (Hz) is related to the wavelength λ in meters (m) of a laser through the equation $f = \frac{c}{\lambda}$, where c is the speed of light in a vacuum and is typically taken to be $c = 3.0 \times 10^8$ meters per second (m/s). What values must be eliminated from the wavelengths?

82. Optics. For an object placed near a lens, an image forms on the other side of the lens at a distinct position determined by the distance from the lens to the object. The position of the image is found using the thin lens equation:

$$\frac{1}{f} = \frac{1}{d_o} + \frac{1}{d_i}$$

where d_o is the distance from the object to the lens, d_i is the distance from the lens to the image, and f is the *focal length* of the lens. Solve for the object distance d_o in terms of the focal length and image distance.

▪CATCH THE MISTAKE

In Exercises 83–86, explain the mistake that is made.

83. Solve the equation $4x + 3 = 6x - 7$.

  Solution:

  Subtract $4x$ and add 7 to the equation. $3 = 6x$

  Divide by 3. $x = 2$

  This is incorrect. What mistake was made?

84. Solve the equation $3(x + 1) + 2 = x - 3(x - 1)$.

  Solution:

$$3x + 3 + 2 = x - 3x - 3$$
$$3x + 5 = -2x - 3$$
$$5x = -8$$
$$x = -\frac{8}{5}$$

  This is incorrect. What mistake was made?

85. Solve the equation $\dfrac{4}{p} - 3 = \dfrac{2}{5p}$.

Solution:

Cross multiply.

$$(p - 3)2 = 4(5p)$$
$$2p - 6 = 20p$$
$$-6 = 18p$$
$$p = -\dfrac{6}{18}$$
$$p = -\dfrac{1}{3}$$

This is incorrect. What mistake was made?

86. Solve the equation $\dfrac{1}{x} + \dfrac{1}{x-1} = \dfrac{1}{x(x-1)}$.

Solution:

Multiply by the LCD, $x(x - 1)$.

$$\dfrac{x(x-1)}{x} + \dfrac{x(x-1)}{x-1} = \dfrac{x(x-1)}{x(x-1)}$$

Simplify.

$$(x - 1) + x = 1$$
$$x - 1 + x = 1$$
$$2x = 2$$
$$x = 1$$

This is incorrect. What mistake was made?

▪ CONCEPTUAL

In Exercises 87–90, determine whether each of the statements is true or false.

87. The solution to the equation $x = \dfrac{1}{1/x}$ is the set of all real numbers.

88. The solution to the equation

$$\dfrac{1}{(x-1)(x+2)} = \dfrac{1}{x^2 + x - 2}$$

is the set of all real numbers.

89. $x = -1$ is a solution to the equation $\dfrac{x^2 - 1}{x - 1} = x + 1$.

90. $x = 1$ is a solution to the equation $\dfrac{x^2 - 1}{x - 1} = x + 1$.

91. Solve for x, given that a, b, and c are real numbers and $a \neq 0$:

$$ax + b = c$$

92. Solve for x, given that a, b, and c are real numbers and $c \neq 0$:

$$\dfrac{a}{x} - \dfrac{b}{x} = c$$

▪ CHALLENGE

93. Solve the equation for x: $\dfrac{b+c}{x+a} = \dfrac{b-c}{x-a}$. Are there any restrictions given that $a \neq 0$, $x \neq 0$?

94. Solve the equation for y: $\dfrac{1}{y-a} + \dfrac{1}{y+a} = \dfrac{2}{y-1}$.

Does y have any restrictions?

95. Solve for x: $\dfrac{1 - 1/x}{1 + 1/x} = 1$.

96. Solve for t: $\dfrac{t + 1/t}{1/t - 1} = 1$.

97. Solve the equation for x in terms of y:

$$y = \dfrac{a}{1 + b/x + c}$$

98. Find the number a for which $y = 2$ is a solution of the equation $y - a = y + 5 - 3ay$.

▪ TECHNOLOGY

In Exercises 99–106, graph the function represented by each side of the equation in the same viewing rectangle and solve for x.

99. $3(x + 2) - 5x = 3x - 4$

100. $-5(x - 1) - 7 = 10 - 9x$

101. $2x + 6 = 4x - 2x + 8 - 2$

102. $10 - 20x = 10x - 30x + 20 - 10$

103. $\dfrac{x(x-1)}{x^2} = 1$

104. $\dfrac{2x(x+3)}{x^2} = 2$

105. $0.035x + 0.029(8706 - x) = 285.03$

106. $\dfrac{1}{0.75x} - \dfrac{0.45}{x} = \dfrac{1}{9}$

SKILLS OBJECTIVES

- Solve application problems involving common formulas.
- Solve geometry problems.
- Solve interest problems.
- Solve mixture problems.
- Solve distance–rate–time problems.

CONCEPTUAL OBJECTIVE

- Understand the mathematical modeling process.

Solving Application Problems Using Mathematical Models

In this section, we will use algebra to solve problems that occur in our day-to-day lives. You typically will read the problem in words, develop a mathematical model (equation) for the problem, solve the equation, and write the answer in words.

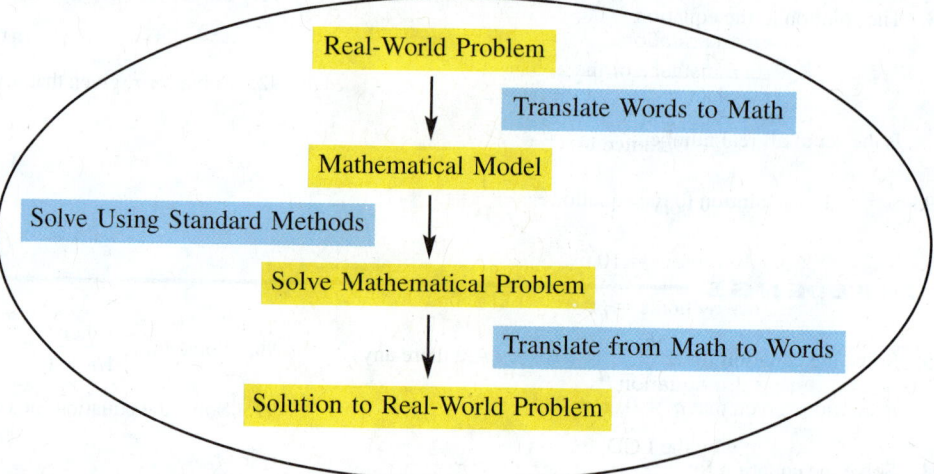

You will have to come up with a unique formula to solve each kind of word problem, but there is a universal *procedure* for approaching all word problems.

PROCEDURE FOR SOLVING WORD PROBLEMS

Step 1: Identify the question. Read the problem *one* time and note what you are asked to find.

Step 2: Make notes. Read until you can note something (an amount, a picture, anything). Continue reading and making notes until you have read the problem a second* time.

Step 3: Assign a variable to whatever is being asked for (if there are two choices, then let it be the smaller of the two).

Step 4: Set up an equation.

Step 5: Solve the equation.

Step 6: Check the solution. Run the solution past the "common sense department" using estimation.

*Step 2 often requires multiple readings of the problem.

EXAMPLE 1   How Long Was the Trip?

During a camping trip in North Bay, Ontario, a couple went one-third of the way by boat, 10 miles by foot, and one-sixth of the way by horse. How long was the trip?

Solution:

STEP 1 **Identify the question.**
How many miles was the trip?

STEP 2 **Make notes.**

Read	**Write**
... one-third of the way by boat	BOAT: $\frac{1}{3}$ of the trip
... 10 miles by foot	FOOT: 10 miles
... one-sixth of the way by horse	HORSE: $\frac{1}{6}$ of the trip

STEP 3 **Assign a variable.**
Distance of total trip in miles = x

STEP 4 **Set up an equation.**
The total distance of the trip is the sum of all the distances by boat, foot, and horse.

Distance by boat + Distance by foot + Distance by horse = Total distance of trip

Distance by boat = $\frac{1}{3}x$

Distance by foot = 10 miles

Distance by horse = $\frac{1}{6}x$

$$\overbrace{\frac{1}{3}x}^{\text{boat}} + \overbrace{10}^{\text{foot}} + \overbrace{\frac{1}{6}x}^{\text{horse}} = \overbrace{x}^{\text{total}}$$

STEP 5 **Solve the equation.**

$$\frac{1}{3}x + 10 + \frac{1}{6}x = x$$

Multiply by the LCD, 6. $2x + 60 + x = 6x$

Collect x terms on the right. $60 = 3x$

Divide by 3. $20 = x$

The trip was 20 miles. $x = 20$

STEP 6 **Check the solution.**
Estimate: The boating distance, $\frac{1}{3}$ of 20 miles, is approximately 7 miles; the riding distance on horse, $\frac{1}{6}$ of 20 miles, is approximately 3 miles. Adding these two distances to the 10 miles by foot gives a trip distance of 20 miles.

■ **YOUR TURN** A family arrives at the Walt Disney World parking lot. To get from their car in the parking lot to the gate at the Magic Kingdom they walk $\frac{1}{4}$ mile, take a tram for $\frac{1}{3}$ of their total distance, and take a monorail for $\frac{1}{2}$ of their total distance. How far is it from their car to the gate of Magic Kingdom?

■ **Answer:** The distance from their car to the gate is 1.5 miles.

EXAMPLE 2   Find the Numbers

Find three consecutive even integers so that the sum of the three numbers is 2 more than twice the third.

Solution:

STEP 1   Identify the question.

What are the three consecutive even integers?

STEP 2   Make notes.

Examples of three consecutive even integers are 14, 16, 18 or $-8, -6, -4$ or 2, 4, 6.

STEP 3   Assign a variable.

Let n represent the first even integer. The next consecutive even integer is $n + 2$ and the next consecutive even integer after that is $n + 4$.

$$n = \text{1st integer}$$
$$n + 2 = \text{2nd consecutive even integer}$$
$$n + 4 = \text{3rd consecutive even integer}$$

STEP 4   Set up an equation.

Read	Write
... sum of the three numbers	$n + (n + 2) + (n + 4)$
... is	$=$
... two more than	$+2$
... twice the third	$2(n + 4)$

$$\underbrace{n + (n + 2) + (n + 4)}_{\text{sum of the three numbers}} \underbrace{=}_{\text{is}} \underbrace{2}_{\text{2 more than}} + \underbrace{2(n + 4)}_{\text{twice the third}}$$

STEP 5   Solve the equation. $n + (n + 2) + (n + 4) = 2 + 2(n + 4)$

Eliminate the parentheses. $n + n + 2 + n + 4 = 2 + 2n + 8$

Simplify both sides. $3n + 6 = 2n + 10$

Collect n terms on the left and constants on the right. $n = 4$

The three consecutive even integers are 4, 6, and 8.

STEP 6   Check the solution.

Substitute the solution into the problem to see whether it makes sense. The sum of the three integers $(4 + 6 + 8)$ is 18. Twice the third is 16. Since 2 more than twice the third is 18, the solution checks.

Answer: The three consecutive odd integers are 11, 13, and 15.

YOUR TURN Find three consecutive odd integers so that the sum of the three integers is 5 less than 4 times the first.

Geometry Problems

Some problems require geometric formulas in order to be solved. The following geometric formulas may be useful.

Geometric Formulas

Rectangle	Perimeter	Area
w [rectangle] l	$P = 2l + 2w$	$A = l \cdot w$

Circle	Circumference	Area
[circle with radius r]	$C = 2\pi r$	$A = \pi r^2$

Triangle	Perimeter	Area
[triangle with sides a, c, height h, base b]	$P = a + b + c$	$A = \frac{1}{2}bh$

EXAMPLE 3　Geometry

A rectangle 24 meters long has the same area as a square with 12-meter sides. What are the dimensions of the rectangle?

Solution:

STEP 1 Identify the question.

What are the dimensions (length and width) of the rectangle?

STEP 2 Make notes.

Read	**Write/Draw**
A rectangle 24 meters long	w [rectangle] $l = 24$
	area of rectangle $= l \cdot w = 24w$
... a square with 12-meter sides	12 m [square] 12 m
	area of square $= 12 \cdot 12 = 144$

STEP 3 Assign a variable.

Let w = width of the rectangle.

STEP 4 Set up an equation.

The area of the rectangle is
equal to the area of the square.　　　rectangle area = square area

Substitute in known quantities.　　　$24w = 144$

STEP 5 Solve the equation.

Divide by 24.　　　$w = \dfrac{144}{24} = 6$

The rectangle is 24 meters long and 6 meters wide.

STEP 6 Check the solution.

A 24 meter by 6 meter rectangle has an area of 144 square meters.

■ **YOUR TURN**　A rectangle 3 inches wide has the same area as a square with 9-inch sides. What are the dimensions of the rectangle?

Classroom Example 1.2.3*
Consider two circles, a smaller one and a larger one. If the larger one has a radius that is 8 feet longer than the smaller one and the ratio of the circumferences is 3:1, then what are the radii of the two circles?

Answer: The radius of the smaller is 4 feet, and the radius of the larger is 12 feet.

■ **Answer:** The rectangle is 27 inches long and 3 inches wide.

Interest Problems

In our personal or business financial planning, a particular concern we have is interest. **Interest** is money paid for the use of money; it is the cost of borrowing money. The total amount borrowed is called the **principal**. The principal can be the price of our new car; we pay the bank interest for loaning us money. The principal can also be the amount we keep in a CD or money market account; the bank uses this money and pays us interest. Typically interest rate, expressed as a percentage, is the amount charged for the use of the principal for a given time, usually in years.

Simple interest is interest that is paid only on principal during a period of time. Later we will discuss *compound interest*, which is interest paid on both principal and the interest accrued over a period of time.

DEFINITION   **Simple Interest**

If a principal of P dollars is borrowed for a period of t years at an annual interest rate r (expressed in decimal form), the interest I charged is

$$I = Prt$$

This is the formula for **simple interest**.

Classroom Example 1.2.4
An ambitious 14-year-old has saved $1800 from chores and odd jobs around the neighborhood. If he puts this money into a CD that pays a simple interest rate of 4% a year, how much money will he have in his CD at the end of 18 months?

Answer: $1908

EXAMPLE 4   Simple Interest

Through a summer job Morgan is able to save $2500. If she puts that money into a 6-month certificate of deposit (CD) that pays a simple interest rate of 3% a year, how much money will she have in her CD at the end of the 6 months?

Solution:

STEP 1   **Identify the question.**
How much money does Morgan have after 6 months?

STEP 2   **Make notes.**
The principal is $2500.

The annual interest rate is 3%, which in decimal form is 0.03.

The time the money spends accruing interest is 6 months, or $\frac{1}{2}$ of a year.

STEP 3   **Assign a variable.**
Label the known quantities. $P = 2500$, $r = 0.03$, and $t = 0.5$

STEP 4   **Set up an equation.**
Write the simple interest formula. $I = Prt$

STEP 5   **Solve the equation.** $I = Prt$

$$I = (2500)(0.03)(0.5) = 37.5$$

The interest paid on the CD is $37.50. Adding this to the principal gives a total of

$$\$2500 + \$37.50 = \boxed{\$2537.50}$$

STEP 6   **Check the solution.**
This answer agrees with our intuition. Had we made a mistake, say, of moving one decimal place to the right, then the interest would have been $375, which is much larger than we would expect on a principal of only $2500.

EXAMPLE 5 **Multiple Investments**

Theresa earns a full athletic scholarship for college, and her parents have given her the $20,000 they had saved to pay for her college tuition. She decides to invest that money with an overall goal of earning 11% interest. She wants to put some of the money in a low-risk investment that has been earning 8% a year and the rest of the money in a medium-risk investment that typically earns 12% a year. How much money should she put in each investment to reach her goal?

Solution:

STEP 1 **Identify the question.**

How much money is invested in each (the 8% and the 12%) account?

STEP 2 **Make notes.**

Read	**Write/Draw**
Theresa has $20,000 to invest.	
If part is invested at 8% and the rest at 12%, how much should be invested at each rate to yield 11% on the total amount invested?	

STEP 3 **Assign a variable.**

If we let x represent the amount Theresa puts into the 8% investment, how much of the $20,000 is left for her to put in the 12% investment?

Amount in the 8% investment: x

Amount in the 12% investment: $20{,}000 - x$

STEP 4 **Set up an equation.**

Simple interest formula: $I = Prt$

INVESTMENT	PRINCIPAL	RATE	TIME (YR)	INTEREST
8% Account	x	0.08	1	$0.08x$
12% Account	$20{,}000 - x$	0.12	1	$0.12(20{,}000 - x)$
Total	20,000	0.11	1	$0.11(20{,}000)$

Adding the interest earned in the 8% investment to the interest earned in the 12% investment should earn an average of 11% on the total investment.

$$0.08x + 0.12(20{,}000 - x) = 0.11(20{,}000)$$

STEP 5 **Solve the equation.**

Eliminate the parentheses. $0.08x + 2400 - 0.12x = 2200$

Collect x terms on the left,
constants on the right. $-0.04x = -200$

Divide by -0.04. $x = 5000$

Calculate the amount at 12%. $20{,}000 - 5000 = 15{,}000$

Theresa should invest $5000 at 8% and $15,000 at 12% to reach her goal.

Classroom Example 1.2.5
Jason wants to invest $65,000 to earn a total of $5240 in one year. He will put some of his money in a lower risk investment that earns 7% a year, and the rest of the money in an investment of slightly higher risk that earns 10% per year. How much money should he put in each investment to reach his goal?

Answer: $42,000 in the 7% investment and $23,000 in the 10% investment.

STEP 6 Check the solution.

If money is invested at 8% and 12% with a goal of averaging 11%, our intuition tells us that more should be invested at 12% than 8%, which is what we found. The exact check is as follows:

$$0.08(5000) + 0.12(15,000) = 0.11(20,000)$$

$$400 + 1800 = 2200$$

$$2200 = 2200$$

■ **Answer:** $10,000 is invested at 18% and $14,000 is invested at 12%.

■ **YOUR TURN** You win $24,000 and you decide to invest the money in two different investments: one paying 18% and the other paying 12%. A year later you have $27,480 total. How much did you originally invest in each account?

Mixture Problems

Mixtures are something we come across every day. Different candies that sell for different prices may make up a movie snack. New blends of coffees are developed by coffee connoisseurs. Chemists mix different concentrations of acids in their labs. Whenever two or more distinct ingredients are combined the result is a **mixture**.

Our choice at a gas station is typically 87, 89, and 93 octane. The octane number is the number that represents the percentage of iso-octane in fuel; 89 octane is significantly overpriced. Therefore, if your car requires 89 octane, it would be more cost effective to mix 87 and 93 octane.

EXAMPLE 6   Mixture Problem

The manual for your new car suggests using gasoline that is 89 octane. In order to save money, you decide to use some 87 octane and some 93 octane in combination with the 89 octane currently in your tank in order to have an approximate 89 octane mixture. Assuming you have 1 gallon of 89 octane remaining in your tank (your tank capacity is 16 gallons), how many gallons of 87 and 93 octane should be used to fill up your tank to achieve a mixture of 89 octane?

Classroom Example 1.2.6
A mechanic is working on the coolant system of a vehicle with a capacity of 11.0 liters. Currently the system is filled with coolant that is 45% ethylene glycol. How much fluid must be drained and replaced with 100% ethylene glycol so that the system will be filled with coolant that is 60% ethylene glycol?

Answer: 3 liters

Solution:

STEP 1 Identify the question.

How many gallons of 87 octane and how many gallons of 93 octane should be used?

STEP 2 Make notes.

Read

Assuming you have one gallon of 89 octane remaining in your tank (your tank capacity is 16 gallons), how many gallons of 87 and 93 octane should you add?

Write/Draw

89 octane   +   87 octane   +   93 octane   =   89 octane
[1 gallon]   [? gallons]   [? gallons]   [16 gallons]

STEP 3 Assign a variable.

x = gallons of 87 octane gasoline added at the pump

$15 - x$ = gallons of 93 octane gasoline added at the pump

1 = gallons of 89 octane gasoline already in the tank

STEP 4 Set up an equation.

$$0.89(1) + 0.87x + 0.93(15 - x) = 0.89(16)$$

STEP 5 Solve the equation.

$$0.89(1) + 0.87x + 0.93(15 - x) = 0.89(16)$$

Eliminate the parentheses.

$$0.89 + 0.87x + 13.95 - 0.93x = 14.24$$

Collect x terms on the left side.

$$-0.06x + 14.84 = 14.24$$

Subtract 14.84 from both sides
of the equation.

$$-0.06x = -0.6$$

Divide both sides by -0.06.

$$x = 10$$

Calculate the amount of 93 octane.

$$15 - 10 = 5$$

Add 10 gallons of 87 octane and 5 gallons of 93 octane.

STEP 6 Check the solution.

Estimate: Our intuition tells us that if the desired mixture is 89 octane, then we should add approximately one part 93 octane and two parts 87 octane. The solution we found, 10 gallons of 87 octane and 5 gallons of 93 octane, agrees with this.

■ YOUR TURN For a certain experiment, a student requires 100 milliliters of a solution that is 11% HCl (hydrochloric acid). The storeroom has only solutions that are 5% HCl and 15% HCl. How many milliliters of each available solution should be mixed to get 100 milliliters of 11% HCl?

■ **Answer:** 40 milliliters of 5% HCl and 60 milliliters of 15% HCl

Distance–Rate–Time Problems

The next example deals with distance, rate, and time. On a road trip, you see a sign that says your destination is 90 miles away, and your speedometer reads 60 miles per hour. Dividing 90 miles by 60 miles per hour tells you that your arrival will be in 1.5 hours. Here is how you know.

If the rate, or speed, is assumed to be constant, then the equation that relates distance (d), rate (r), and time (t) is given by $d = r \cdot t$. In the above driving example,

$$d = 90 \text{ miles} \qquad r = 60 \frac{\text{miles}}{\text{hour}}$$

Substituting these into
$d = r \cdot t$, we arrive at

$$90 \text{ miles} = \left[60 \frac{\text{miles}}{\text{hour}} \right] \cdot t$$

Solving for t, we get

$$t = \frac{90 \text{ miles}}{60 \frac{\text{miles}}{\text{hour}}} = 1.5 \text{ hours}$$

Classroom Example 1.2.7
You and your roommate decided to take a road trip to the beach one weekend. You drove all the way to the beach at an average speed of 60 miles per hour. Your roommate drove all the way back (on the same route, but with no traffic) at an average rate of 75 miles per hour. If the total round trip drive took a total of 9 hours, how many miles was the trip to the beach?

Answer: 300 miles, taking 5 hours to get there and 4 hours to return.

EXAMPLE 7   Distance–Rate–Time

It takes 8 hours to fly from Orlando to London and 9.5 hours to return. If an airplane averages 550 miles per hour in still air, what is the average rate of the wind blowing in the direction from Orlando to London? Assume the speed of the wind (jet stream) is constant and the same for both legs of the trip. Round your answer to the nearest miles per hour.

Solution:

STEP 1   **Identify the question.**
At what rate in mph is the wind blowing?

STEP 2   **Make notes.**

Read	**Write/Draw**
It takes 8 hours to fly from Orlando to London and 9.5 hours to return.	
If the airplane averages 550 miles per hour in still air...	

STEP 3   **Assign a variable.** w = wind speed

STEP 4   **Set up an equation.**
The formula relating distance, rate, and time is $d = r \cdot t$. The distance d of each flight is the same. On the Orlando to London flight the time is 8 hours due to an increased speed from a tailwind. On the London to Orlando flight the time is 9.5 hours, and the speed is decreased due to the headwind. Let w represent the wind speed.

Orlando to London: $d = (550 + w)8$

London to Orlando: $d = (550 - w)9.5$

These distances are the same, so set them equal to each other:

$$(550 + w)8 = (550 - w)9.5$$

STEP 5   **Solve the equation.**
Eliminate the parentheses. $4400 + 8w = 5225 - 9.5w$

Collect w terms on the left, constants on the right. $17.5w = 825$

Divide by 17.5. $w = 47.1429 \approx 47$

The wind is blowing approximately $\boxed{47 \text{ miles per hour}}$ in the direction from Orlando to London.

STEP 6   **Check the solution.**
Estimate: Going from Orlando to London, the tailwind is approximately 50 miles per hour, which added to the plane's 550 miles per hour speed yields a ground speed of 600 miles per hour. The Orlando to London route took 8 hours. The distance of that flight is (600 mph)(8 hr), which is 4800 miles. The return trip experienced a headwind of approximately 50 miles per hour, so subtracting the 50 from 550 gives an average speed of 500 miles per hour. That route took 9.5 hours, so the distance of the London to Orlando flight was (500 mph)(9.5 hr), which is 4750 miles. Note that the estimates of 4800 and 4750 miles are close.

■ **Answer:** The wind is blowing 30 mph.

■ **YOUR TURN**   A Cessna 150 averages 150 miles per hour in still air. With a tailwind it is able to make a trip in $2\frac{1}{3}$ hours. Because of the headwind, it is only able to make the return trip in $3\frac{1}{2}$ hours. What is the average wind speed?

EXAMPLE 8   Work

Connie can clean her house in 2 hours. If Alvaro helps her, they can clean the house in 1 hour and 15 minutes together. How long would it take Alvaro to clean the house by himself?

Solution:

STEP 1   Identify the question.

How long would it take Alvaro to clean the house by himself?

STEP 2   Make notes.

Connie can clean her house in 2 hours, so Connie can clean $\frac{1}{2}$ of the house per hour. Together Connie and Alvaro can clean the house in 1 hour and 15 minutes, or $\frac{5}{4}$ of an hour. Therefore together they can clean $\dfrac{1}{5/4} = \dfrac{4}{5}$ of the house per hour.

STEP 3   Assign a variable.

Let x = number of hours it takes Alvaro to clean the house by himself. So Alvaro can clean $\dfrac{1}{x}$ of the house per hour.

STEP 4   Set up an equation.

	AMOUNT OF TIME TO DO ONE JOB	AMOUNT OF JOB DONE PER UNIT OF TIME
Connie	2	$\dfrac{1}{2}$
Alvaro	x	$\dfrac{1}{x}$
Together	$\dfrac{5}{4}$	$\dfrac{4}{5}$

$$
\underbrace{\frac{1}{2}}_{\substack{\text{Amount of house Connie can}\\ \text{clean per hour}}} + \underbrace{\frac{1}{x}}_{\substack{\text{Amount of house Alvaro can}\\ \text{clean per hour}}} = \underbrace{\frac{4}{5}}_{\substack{\text{Amount of house they can clean per hour}\\ \text{if they work together}}}
$$

STEP 5   Solve the equation.

Multiply $\frac{1}{2} + \frac{1}{x} = \frac{4}{5}$ by the LCD, $10x$. $5x + 10 = 8x$

Solve for x. $x = \dfrac{10}{3} = 3\dfrac{1}{3}$

It takes Alvaro 3 hours and 20 minutes to clean the house by himself.

STEP 6   Check the solution.

Connie cleans the house in 2 hours. If Alvaro could clean it in 2 hours, then together it would take them 1 hour. Since together it takes them 1 hour and 15 minutes, we expect that it takes Alvaro more than 2 hours by himself.

SECTION 1.2   SUMMARY

In the real world many kinds of application problems can be solved through modeling with linear equations. The following six-step procedure will help you develop the model. Some problems require development of a mathematical model, while others rely on common formulas.

1. Identify the quantity you are to determine.
2. Make notes on any clues that will help you set up an equation.
3. Assign a variable.
4. Set up the equation.
5. Solve the equation.
6. Check the solution against your intuition.

APPLICATIONS

1. **Discount Price.** Donna redeems a 10% off coupon at her local nursery. After buying azaleas, bougainvillea, and bags of potting soil, her checkout price before tax is $217.95. How much would she have paid without the coupon?

2. **Discount Price.** The original price of a pair of binoculars is $74. The sale price is $51.80. How much was the markdown?

3. **Cost: Fair Share.** Jeff, Tom, and Chelsea order a large pizza. They decide to split the cost according to how much they will eat. Tom pays $5.16, Chelsea eats $\frac{1}{8}$ of the pizza, and Jeff eats $\frac{1}{2}$ of the pizza. How much did the pizza cost?

4. **Event Planning.** A couple decide to analyze their monthly spending habits. The monthly bills are 50% of their take-home pay, and they invest 20% of their take-home pay. They spend $560 on groceries, and 23% goes to miscellaneous. How much is their take-home pay per month?

5. **Discounts.** A builder of tract homes reduced the price of a model by 15%. If the new price is $125,000, what was its original price? How much can be saved by purchasing the model?

6. **Markups.** A college bookstore marks up the price it pays the publisher for a book by 25%. If the selling price of a book is $79, how much did the bookstore pay for the book?

7. **Puzzle.** Angela is on her way from home in Jersey City into New York City for dinner. She walks 1 mile to the train station, takes the train $\frac{3}{4}$ of the way, and takes a taxi $\frac{1}{6}$ of the way to the restaurant. How far does Angela live from the restaurant?

8. **Puzzle.** An employee at Kennedy Space Center (KSC) lives in Daytona Beach and works in the vehicle assembly building (VAB). She carpools to work with a colleague. She drives 7 miles from her house to the park-and-ride. Then she rides with her colleague from the park-and-ride in Daytona Beach to the KSC headquarters building, and then takes the KSC shuttle from the headquarters building to the VAB. The drive from the park-and-ride to the headquarters building is $\frac{5}{6}$ of her total trip, and the shuttle ride is $\frac{1}{20}$ of her total trip. How many miles does she travel from her house to the VAB on days when her colleague drives?

9. **Puzzle.** A typical college student spends $\frac{1}{3}$ of her waking time in class, $\frac{1}{5}$ of her waking time eating, $\frac{1}{10}$ of her waking time working out, 3 hours studying, and $2\frac{1}{2}$ hours doing other things. How many hours of sleep does the typical college student get?

10. **Diet.** A particular 1550-calories-per-day diet suggests eating breakfast, lunch, dinner, and two snacks. Dinner is twice the calories of breakfast. Lunch is 100 calories more than breakfast. The two snacks are 100 and 150 calories. How many calories are each meal?

11. **Budget.** A company has a total of $20,000 allocated for monthly costs. Fixed costs are $15,000 per month and variable costs are $18.50 per unit. How many units can be manufactured in a month?

12. **Budget.** A woman decides to start a small business making monogrammed cocktail napkins. She can set aside $1870 for monthly costs. Fixed costs are $1329.50 per month and variable costs are $3.70 per set of napkins. How many sets of napkins can she afford to make per month?

13. **Numbers.** Find a number such that 10 less than $\frac{2}{3}$ the number is $\frac{1}{4}$ the number.

14. **Numbers.** Find a positive number such that 10 times the number is 16 more than twice the number.

15. **Numbers.** Find two consecutive even integers such that 4 times the smaller number is 2 more than 3 times the larger number.

16. **Numbers.** Find three consecutive integers such that the sum of the three is equal to 2 times the sum of the first two integers.

17. **Geometry.** Find the perimeter of a triangle if one side is 11 inches, another side is $\frac{1}{5}$ the perimeter, and the third side is $\frac{1}{4}$ the perimeter.

18. **Geometry.** Find the dimensions of a rectangle whose length is a foot longer than twice its width and whose perimeter is 20 feet.

19. **Geometry.** An NFL playing field is a rectangle. The length of the field (excluding the end zones) is 40 more yards than twice the width. The perimeter of the playing field is 260 yards. What are the dimensions of the field in yards?

20. **Geometry.** The length of a rectangle is 2 more than 3 times the width, and the perimeter is 28 inches. What are the dimensions of the rectangle?

21. **Geometry.** Consider two circles, a smaller one and a larger one. If the larger one has a radius that is 3 feet larger than that of the smaller circle and the ratio of the circumferences is 2:1, what are the radii of the two circles?

22. **Geometry.** The perimeter of a semicircle is doubled when the radius is increased by 1. Find the radius of the semicircle.

23. **Home Improvement.** A man wants to remove a tall pine tree from his yard. Before he goes to Home Depot, he needs to know how tall an extension ladder he needs to purchase. He measures the shadow of the tree to be 225 feet long. At the same time he measures the shadow of a 4-foot stick to be 3 feet. Approximately how tall is the pine tree?

24. **Home Improvement.** The same man in Exercise 23 realizes he also wants to remove a dead oak tree. Later in the day he measures the shadow of the oak tree to be 880 feet long, and the 4-foot stick now has a shadow of 10 feet. Approximately how tall is the oak tree?

25. **Biology: Alligators.** It is common to see alligators in ponds, lakes, and rivers in Florida. The ratio of head size (back of the head to the end of the snout) to the full body length of an alligator is typically constant. If a $3\frac{1}{2}$-foot alligator has a head length of 6 inches, how long would you expect an alligator to be whose head length is 9 inches?

26. **Biology: Snakes.** In the African rainforest there is a snake called a Gaboon viper. The fang size of this snake is proportional to the length of the snake. A 3-foot snake typically has 2-inch fangs. If a herpetologist finds Gaboon viper fangs that are 2.6-inches long, how long a snake would she expect to find?

27. **Investing.** Ashley has $120,000 to invest and decides to put some in a CD that earns 4% interest per year and the rest in a low-risk stock that earns 7%. How much did she invest in each to earn $7800 interest in the first year?

28. **Investing.** You inherit $13,000 and you decide to invest the money in two different investments: one paying 10% and the other paying 14%. A year later your investments are worth $14,580. How much did you originally invest in each account?

29. **Investing.** Wendy was awarded a volleyball scholarship to the University of Michigan, so on graduation her parents gave her the $14,000 they had saved for her college tuition. She opted to invest some money in a privately held company that pays 10% per year and evenly split the remaining money between a money market account yielding 2% and a high-risk stock that yielded 40%. At the end of the first year she had $16,610 total. How much did she invest in each of the three?

30. **Interest.** A high school student was able to save $5000 by working a part-time job every summer. He invested half the money in a money market account and half the money in a stock that paid three times as much interest as the money market account. After a year he earned $150 in interest. What were the interest rates of the money market account and the stock?

31. **Budget: Home Improvement.** When landscaping their yard, a couple budgeted $4200. The irrigation system costs $2400 and the sod costs $1500. The rest they will spend on trees and shrubs. Trees each cost $32 and shrubs each cost $4. They plant a total of 33 trees and shrubs. How many of each did they plant in their yard?

32. **Budget: Shopping.** At the deli Jennifer bought spicy turkey and provolone cheese. The turkey costs $6.32 per pound and the cheese costs $4.27 per pound. In total, she bought 3.2 pounds and the price was $17.56. How many pounds of each did she buy?

33. **Chemistry.** For a certain experiment, a student requires 100 milliliters of a solution that is 8% HCl (hydrochloric acid). The storeroom has only solutions that are 5% HCl and 15% HCl. How many milliliters of each available solution should be mixed to get 100 milliliters of 8% HCl?

34. **Chemistry.** How many gallons of pure alcohol must be mixed with 5 gallons of a solution that is 20% alcohol to make a solution that is 50% alcohol?

35. **Automobiles.** A mechanic has tested the amount of antifreeze in your radiator. He says it is only 40% antifreeze and the remainder is water. How many gallons must be drained from your 5 gallon radiator and replaced with pure antifreeze to make the mixture in your radiator 80% antifreeze?

36. **Costs: Overhead.** A professor is awarded two research grants, each having different overhead rates. The research project conducted on campus has a rate of 42.5% overhead, and the project conducted in the field, off campus, has a rate of 26% overhead. If she was awarded $1,170,000 total for the two projects with an average overhead rate of 39%, how much was the research project on campus and how much was the research project off campus?

37. **Theater.** On the way to the movies a family picks up a custom-made bag of candies. The parents like caramels ($1.50/lb) and the children like gummy bears ($2.00/lb). They bought a 1.25-pound bag of combined candies that cost $2.50. How much of each candy did they buy?

38. **Coffee.** Joy is an instructional assistant in one of the college labs. She is on a very tight budget. She loves Jamaican Blue Mountain coffee, but it costs $12 a pound. She decides to blend this with regular coffee beans that cost $4.20 a pound. If she spends $14.25 on 2 pounds of coffee, how many pounds of each did she purchase?

39. **Communications.** The speed of light is approximately 3.0×10^8 meters per second (670,616,629 mph). The distance from Earth to Mars varies because of the orbits of the planets around the Sun. On average, Mars is 100 million miles from Earth. If we use laser communication systems, what will be the delay between Houston and NASA astronauts on Mars?

40. **Speed of Sound.** The speed of sound is approximately 760 miles per hour in air. If a gun is fired $\frac{1}{2}$ mile away, how long will it take the sound to reach you?

41. **Business.** During the month of February 2011, the average price of gasoline rose 4.7% in the United States. If the average price of gasoline at the end of February 2011 was $3.21 per gallon, what was the price of gasoline at the beginning of February?

42. **Business.** During the Christmas shopping season of 2010, the average price of a flat screen television fell by 40%. A shopper purchased a 42-inch flat screen television for $299 in late November 2010. How much would the shopper have paid, to the nearest dollar, for the same television if it was purchased in September 2010?

43. **Medicine.** A patient requires an IV of 0.9% saline solution, also known as normal saline solution. How much distilled water, to the nearest milliliter, must be added to 100 milliliters of a 3% saline solution to produce normal saline?

44. **Medicine.** A patient requires an IV of D5W, a 5% solution of Dextrose (sugar) in water. To the nearest milliliter, how much D20W, a 20% solution of Dextrose in water, must be added to 100 milliliters of distilled water to produce a D5W solution?

45. Boating. A motorboat can maintain a constant speed of 16 miles per hour relative to the water. The boat makes a trip upstream to a marina in 20 minutes. The return trip takes 15 minutes. What is the speed of the current?

46. Aviation. A Cessna 175 can average 130 miles per hour. If a trip takes 2 hours one way and the return takes 1 hour and 15 minutes, find the wind speed, assuming it is constant.

47. Exercise. A jogger and a walker cover the same distance. The jogger finishes in 40 minutes. The walker takes an hour. How fast is each exerciser moving if the jogger runs 2 miles per hour faster than the walker?

48. Travel. A high school student in Seattle, Washington, attended the University of Central Florida. On the way to UCF he took a southern route. After graduation he returned to Seattle via a northern trip. On both trips he had the same average speed. If the southern trek took 45 hours and the northern trek took 50 hours, and the northern trek was 300 miles longer, how long was each trip?

49. Distance–Rate–Time. College roommates leave for their first class in the same building. One walks at 2 miles per hour and the other rides his bike at a slow 6 miles per hour pace. How long will it take each to get to class if the walker takes 12 minutes longer to get to class and they travel on the same path?

50. Distance–Rate–Time. A long-distance delivery service sends out a truck with a package at 7 A.M. At 7:30 A.M., the manager realizes there was another package going to the same location. He sends out a car to catch the truck. If the truck travels at an average speed of 50 miles per hour and the car travels at 70 miles per hour, how long will it take the car to catch the truck?

51. Work. Christopher can paint the interior of his house in 15 hours. If he hires Cynthia to help him, they can do the same job together in 9 hours. If he lets Cynthia work alone, how long will it take her to paint the interior of his house?

52. Work. Jay and Morgan work in the summer for a landscaper. It takes Jay 3 hours to complete the company's largest yard alone. If Morgan helps him, it takes only 1 hour. How much time would it take Morgan alone?

53. Work. Tracey and Robin deliver Coke products to local convenience stores. Tracey can complete the deliveries in 4 hours alone. Robin can do it in 6 hours alone. If they decide to work together on a Saturday, how long will it take?

54. Work. Joshua can deliver his newspapers in 30 minutes. It takes Amber 20 minutes to do the same route. How long would it take them to deliver the newspapers if they worked together?

55. Music. A major chord in music is composed of notes whose frequencies are in the ratio 4:5:6. If the first note of a chord has a frequency of 264 hertz (middle C on the piano), find the frequencies of the other two notes. *Hint:* Set up two proportions using 4:5 and 4:6.

56. Music. A minor chord in music is composed of notes whose frequencies are in the ratio 10:12:15. If the first note of a minor chord is A, with a frequency of 220 hertz, what are the frequencies of the other two notes?

57. Grades. Danielle's test scores are 86, 80, 84, and 90. The final exam will count as $\frac{2}{3}$ of the final grade. What score does Danielle need on the final in order to earn a B, which requires an average score of 80? What score does she need to earn an A, which requires an average of 90?

58. Grades. Sam's final exam will count as two tests. His test scores are 80, 83, 71, 61, and 95. What score does Sam need on the final in order to have an average score of 80?

59. Sports. In Super Bowl XXXVII, the Tampa Bay Buccaneers scored a total of 48 points. All of their points came from field goals and touchdowns. Field goals are worth 3 points and each touchdown was worth 7 points (Martin Gramatica was successful in every extra point attempt). They scored a total of 8 times. How many field goals and touchdowns were scored?

60. Sports. A tight end can run the 100-yard dash in 12 seconds. A defensive back can do it in 10 seconds. The tight end catches a pass at his own 20 yard line with the defensive back at the 15 yard line. If no other players are nearby, at what yard line will the defensive back catch up to the tight end?

61. Recreation. How do two children of different weights balance on a seesaw? The heavier child sits closer to the center and the lighter child sits farther away. When the product of the weight of the child and the distance from the center is equal on both sides, the seesaw should be horizontal to the ground. Suppose Max weighs 42 pounds and Maria weighs 60 pounds. If Max sits 5 feet from the center, how far should Maria sit from the center in order to balance the seesaw horizontal to the ground?

62. Recreation. Refer to Exercise 61. Suppose Martin, who weighs 33 pounds, sits on the side of the seesaw with Max. If their average distance to the center is 4 feet, how far should Maria sit from the center in order to balance the seesaw horizontal to the ground?

63. Recreation. If a seesaw has an adjustable bench, then the board can slide along the fulcrum. Maria and Max in Exercise 61 decide to sit on the very edge of the board on each side. Where should the fulcrum be placed along the board in order to balance the seesaw horizontally to the ground? Give the answer in terms of the distance from each child's end.

64. Recreation. Add Martin (Exercise 62) to Max's side of the seesaw and recalculate Exercise 63.

In Exercises 65–68, refer to this lens law. (See Exercise 82 in Section 1.1.)

The position of the image is found using the thin lens equation:

$$\frac{1}{f} = \frac{1}{d_o} + \frac{1}{d_i},$$

where d_o is the distance from the object to the lens, d_i is the distance from the lens to the image, and f is the focal length of the lens.

65. Optics. If the focal length of a lens is 3 centimeters and the image distance is 5 centimeters from the lens, what is the distance from the object to the lens?

66. Optics. If the focal length of the lens is 8 centimeters and the image distance is 2 centimeters from the lens, what is the distance from the object to the lens?

67. Optics. The focal length of a lens is 2 centimeters. If the image distance from the lens is half the distance from the object to the lens, find the object distance.

68. Optics. The focal length of a lens is 8 centimeters. If the image distance from the lens is half the distance from the object to the lens, find the object distance.

■ **CONCEPTUAL**

In Exercises 69–76, solve each formula for the specified variable.

69. $P = 2l + 2w$ for w

70. $P = 2l + 2w$ for l

71. $A = \frac{1}{2} bh$ for h

72. $C = 2\pi r$ for r

73. $A = lw$ for w

74. $d = rt$ for t

75. $V = lwh$ for h

76. $V = \pi r^2 h$ for h

■ **CHALLENGE**

77. Tricia and Janine are roommates and leave Houston on Interstate 10 at the same time to visit their families for a long weekend. Tricia travels west and Janine travels east. If Tricia's average speed is 12 miles per hour faster than Janine's, find the speed of each if they are 320 miles apart in 2 hours and 30 minutes.

78. Rick and Mike are roommates and leave Gainesville on Interstate 75 at the same time to visit their girlfriends for a long weekend. Rick travels north and Mike travels south. If Mike's average speed is 8 miles per hour faster than Rick's, find the speed of each if they are 210 miles apart in 1 hour and 30 minutes.

■ **TECHNOLOGY**

79. Suppose you bought a house for $132,500 and sold it 3 years later for $168,190. Plot these points using a graphing utility. Assuming a linear relationship, how much could you have sold the house for had you waited 2 additional years?

80. Suppose you bought a house for $132,500 and sold it 3 years later for $168,190. Plot these points using a graphing utility. Assuming a linear relationship, how much could you have sold the house for had you sold it 1 year after buying it?

81. A golf club membership has two options. Option A is a $300 monthly fee plus $15 cart fee every time you play. Option B has a $150 monthly fee and a $42 fee every time you play. Write a mathematical model for monthly costs for each plan and graph both in the same viewing rectangle using a graphing utility. Explain when Option A is the better deal and when Option B is the better deal.

82. A phone provider offers two calling plans. Plan A has a $30 monthly charge and a $0.10 per minute charge on every call. Plan B has a $50 monthly charge and a $0.03 per minute charge on every call. Explain when Plan A is the better deal and when Plan B is the better deal.

SKILLS OBJECTIVES

- Solve quadratic equations by factoring.
- Use the square root method to solve quadratic equations.
- Solve quadratic equations by completing the square.
- Use the quadratic formula to solve quadratic equations.

CONCEPTUAL OBJECTIVES

- Choose appropriate methods for solving quadratic equations.
- Interpret different types of solution sets (real, imaginary, complex conjugates, repeated roots).
- Derive the Quadratic Formula.

Factoring

In a linear equation, the variable is raised only to the first power in any term where it occurs. In a *quadratic equation*, the variable is raised to the second power in at least one term. Examples of *quadratic equations*, also called second-degree equations, are:

$$x^2 + 3 = 7 \qquad 5x^2 + 4x - 7 = 0 \qquad x^2 - 3 = 0$$

DEFINITION **Quadratic Equation**

A **quadratic equation** in x is an equation that can be written in the **standard form**

$$ax^2 + bx + c = 0$$

where a, b, and c are real numbers and $a \neq 0$.

There are several methods for solving quadratic equations: *factoring*, the *square root method*, *completing the square*, and the *Quadratic Formula*.

FACTORING METHOD

The **factoring method** applies the **zero product property**:

WORDS

If a product is zero, then at least one of its factors has to be zero.

MATH

If $B \cdot C = 0$, then $B = 0$ or $C = 0$ or both.

Consider $(x - 3)(x + 2) = 0$. The zero product property says that $x - 3 = 0$ or $x + 2 = 0$, which leads to $x = -2$ or $x = 3$. The solution set is $\{-2, 3\}$.

When a quadratic equation is written in the standard form $ax^2 + bx + c = 0$ it may be possible to factor the left side of the equation as a product of two first-degree polynomials. We use the zero product property and set each linear factor equal to zero. We solve the resulting two linear equations to obtain the solutions of the quadratic equation.

EXAMPLE 1   Solving a Quadratic Equation by Factoring

Solve the equation $x^2 - 6x - 16 = 0$.

Solution:

The quadratic equation is already in
standard form.

$$x^2 - 6x - 16 = 0$$

Factor the left side into a product of two
linear factors.

$$(x - 8)(x + 2) = 0$$

If a product equals zero, one of its factors
has to be equal to zero.

$$x - 8 = 0 \quad \text{or} \quad x + 2 = 0$$

Solve both linear equations.

$$\boxed{x = 8 \quad \text{or} \quad x = -2}$$

The solution set is $\boxed{\{-2, 8\}}$.

■ **YOUR TURN**   Solve the quadratic equation $x^2 + x - 20 = 0$ by factoring.

Classroom Example 1.3.1
Solve the equation
$x^2 - 12x + 35 = 0$.

Answer: $x = 5$ or $x = 7$

■ **Answer:** The solution is $x = -5, 4$.
The solution set is $\{-5, 4\}$.

▼ **CAUTION**

Don't forget to put the quadratic
equation in standard form first.

EXAMPLE 2   Solving a Quadratic Equation by Factoring

Solve the equation $x^2 - 6x + 5 = -4$.

COMMON MISTAKE

A common mistake is to forget to put the equation in standard form first and then use
the zero product property incorrectly.

⭐ **CORRECT**

Write the original equation.

$$x^2 - 6x + 5 = -4$$

Write the equation in standard form by
adding 4 to both sides.

$$x^2 - 6x + 9 = 0$$

Factor the left side.

$$(x - 3)(x - 3) = 0$$

Use the zero product property and set
each factor equal to zero.

$$x - 3 = 0 \quad \text{or} \quad x - 3 = 0$$

Solve each linear equation.

$$\boxed{x = 3}$$

❌ **INCORRECT**

Factor the left side.

$$(x - 5)(x - 1) = -4$$

The **error** occurs here.

$$x - 5 = -4 \quad \text{or} \quad x - 1 = -4$$

▶**Don't forget to put the quadratic
equation in standard form first.**

Note: The equation has one solution, or root, which is 3. The solution set is $\{3\}$. Since the
linear factors were the same, or repeated, we say that 3 is a **double root**, or **repeated root**.

■ **YOUR TURN**   Solve the quadratic equation $9p^2 = 24p - 16$ by factoring.

Technology Tip

Use a graphing utility to display
graphs of $y_1 = x^2 - 6x + 5$ and
$y_2 = -4$.

The point of intersection is the
solution to this equation.

Classroom Example 1.3.2*
Solve the equation
$3t^2 - 10t - 6 = 2$.

Answer: $t = -\frac{2}{3}$ or $t = 4$

■ **Answer:** The solution is $p = \frac{4}{3}$,
which is a double root. The
solution set is $\left\{\frac{4}{3}\right\}$.

Classroom Example 1.3.3
Solve the equation
$5Y^2 = -12Y$.

Answer: $Y = -\frac{12}{5}$ or $Y = 0$

Technology Tip

Use a graphing utility to display graphs of $y_1 = 2x^2$ and $y_2 = 3x$.

```
Plot1 Plot2 Plot3
\Y1■2X²
\Y2■3X
\Y3=■
```

The points of intersection are the solutions to this equation.

EXAMPLE 3   Solving a Quadratic Equation by Factoring

Solve the equation $2x^2 = 3x$.

COMMON MISTAKE

The common mistake here is dividing both sides by x, which is not allowed because x might be zero.

⭐ **CORRECT**

Write the equation in standard form by subtracting $3x$.

$$2x^2 - 3x = 0$$

Factor the left side.

$$x(2x - 3) = 0$$

Use the zero product property and set each factor equal to zero.

$$x = 0 \quad \text{or} \quad 2x - 3 = 0$$

Solve each linear equation.

$$\boxed{x = 0 \quad \text{or} \quad x = \frac{3}{2}}$$

The solution set is $\boxed{\left\{0, \frac{3}{2}\right\}}$.

❌ **INCORRECT**

Write the original equation.

$$2x^2 = 3x$$

The **error** occurs here when both sides are divided by x.

$$2x = 3$$

In Example 3, the root $x = 0$ is lost when the original quadratic equation is divided by x. Remember to put the equation in standard form first and then factor.

Square Root Method

The square root of 16, $\sqrt{16}$, is 4, *not* ± 4. In the review (Chapter 0) the **principal square root** was discussed. The solutions to $x^2 = 16$, however, are $x = -4$ and $x = 4$. Let us now investigate quadratic equations that do not have a first-degree term. They have the form

$$ax^2 + c = 0 \quad a \neq 0$$

The method we use to solve such equations uses the square root property.

SQUARE ROOT PROPERTY

WORDS

If an expression squared is equal to a constant, then that expression is equal to the positive or negative square root of the constant.

MATH

If $x^2 = P$, then $x = \pm\sqrt{P}$.

Note: The variable squared must be isolated first (coefficient equal to 1).

EXAMPLE 4 **Using the Square Root Property**

Solve the equation $3x^2 - 27 = 0$.

Solution:

Add 27 to both sides. $3x^2 = 27$

Divide both sides by 3. $x^2 = 9$

Apply the square root property. $x = \pm\sqrt{9} = \pm 3$

The solution set is $\boxed{\{-3, 3\}}$.

Classroom Example 1.3.4
Solve the equation
$2v^2 - 32 = 0$.
Answer: $v = 4$ or $v = -4$

If we alter Example 4 by changing subtraction to addition, we see in Example 5 that we get imaginary roots (as opposed to real roots), which we discussed in Chapter 0.

EXAMPLE 5 **Using the Square Root Property**

Solve the equation $3x^2 + 27 = 0$.

Solution:

Subtract 27 from both sides. $3x^2 = -27$

Divide by 3. $x^2 = -9$

Apply the square root property. $x = \pm\sqrt{-9}$

Simplify. $x = \pm i\sqrt{9} = \pm 3i$

The solution set is $\boxed{\{-3i, 3i\}}$.

Classroom Example 1.3.5
Solve the equation
$5a^2 + 10 = 0$.
Answer:
$a = i\sqrt{2}$ or $a = -i\sqrt{2}$

■ **YOUR TURN** Solve the equations $y^2 - 147 = 0$ and $v^2 + 64 = 0$.

■ **Answer:** The solution is
$y = \pm 7\sqrt{3}$. The solution set is
$\left\{-7\sqrt{3}, 7\sqrt{3}\right\}$. The solution is
$v = \pm 8i$. The solution set is
$\{-8i, 8i\}$.

EXAMPLE 6 **Using the Square Root Property**

Solve the equation $(x - 2)^2 = 16$.

Solution:

Approach 1:
If an expression squared is 16, then
the expression equals $\pm\sqrt{16}$. $(x - 2) = \pm\sqrt{16}$

Separate into two equations. $x - 2 = \sqrt{16}$ or $x - 2 = -\sqrt{16}$

                                                                         $x - 2 = 4$ $x - 2 = -4$

                                                                                  $x = 6$ $x = -2$

The solution set is $\boxed{\{-2, 6\}}$.

Approach 2:
It is acceptable notation to keep the $(x - 2) = \pm\sqrt{16}$
equations together. $x - 2 = \pm 4$

                                                                             $x = 2 \pm 4$

                                                                         $\boxed{x = -2, 6}$

Classroom Example 1.3.6
Solve the equation
$(x - 3)^2 = 25$.
Answer: $x = -2$ or $x = 8$

Completing the Square

Factoring and the square root method are two efficient, quick procedures for solving many quadratic equations. However, some equations, such as $x^2 - 10x - 3 = 0$, cannot be solved directly by these methods. A more general procedure to solve this kind of equation is called **completing the square**. The idea behind completing the square is to transform any standard quadratic equation $ax^2 + bx + c = 0$ into the form $(x + A)^2 = B$, where A and B are constants and the left side, $(x + A)^2$, has the form of a **perfect square**. This last equation can then be solved by the square root method. How do we transform the first equation into the second equation?

Note that the above-mentioned example, $x^2 - 10x - 3 = 0$, cannot be factored into expressions in which all numbers are integers (or even rational numbers). We can, however, transform this quadratic equation into a form that contains a perfect square.

WORDS	MATH
Write the original equation.	$x^2 - 10x - 3 = 0$
Add 3 to both sides.	$x^2 - 10x = 3$
Add 25 to both sides.*	$x^2 - 10x + \mathbf{25} = 3 + \mathbf{25}$
The left side can be written as a perfect square.	$(x - 5)^2 = 28$
Apply the square root method.	$x - 5 = \pm\sqrt{28}$
Add 5 to both sides.	$x = 5 \pm 2\sqrt{7}$

*Why did we add 25 to both sides? Recall that $(x - c)^2 = x^2 - 2xc + c^2$. In this case $c = 5$ in order for $-2xc = -10x$. Therefore the desired perfect square $(x - 5)^2$ results in $x^2 - 10x + 25$. Applying this product we see that $+25$ is needed. A systematic approach is to take the coefficient of the first-degree term $x^2 - 10x - 3 = 0$, which is -10. Take half of (-10), which is (-5), and then square it $(-5)^2 = 25$.

SOLVING A QUADRATIC EQUATION BY COMPLETING THE SQUARE

WORDS	MATH
Express the quadratic equation in the following form.	$x^2 + bx = c$
Divide b by 2 and square the result, then add the square to both sides.	$x^2 + bx + \left(\dfrac{b}{2}\right)^2 = c + \left(\dfrac{b}{2}\right)^2$
Write the left side of the equation as a perfect square.	$\left(x + \dfrac{b}{2}\right)^2 = c + \left(\dfrac{b}{2}\right)^2$
Solve using the square root method.	

EXAMPLE 7 Completing the Square

Solve the quadratic equation $x^2 + 8x - 3 = 0$ by completing the square.

Solution:

Add 3 to both sides.

$$x^2 + 8x = 3$$

Add $\left(\frac{1}{2} \cdot 8\right)^2 = 4^2$ to both sides.

$$x^2 + 8x + \mathbf{4^2} = 3 + \mathbf{4^2}$$

Write the left side as a perfect square and simplify the right side.

$$(x + 4)^2 = 19$$

Apply the square root method to solve.

$$x + 4 = \pm\sqrt{19}$$

Subtract 4 from both sides.

$$\boxed{x = -4 \pm \sqrt{19}}$$

The solution set is $\boxed{\left\{ -4 - \sqrt{19}, \, -4 + \sqrt{19} \right\}}$.

Technology Tip

Graph $y_1 = x^2 + 8x - 3$.

The x-intercepts are the solutions to this equation.

In Example 7, the leading coefficient (the coefficient of the x^2 term) is 1. When the leading coefficient is not 1, start by first dividing the equation by that leading coefficient.

EXAMPLE 8 Completing the Square When the Leading Coefficient Is Not Equal to 1

Solve the equation $3x^2 - 12x + 13 = 0$ by completing the square.

Solution:

Divide by the leading coefficient, 3.

$$x^2 - 4x + \frac{13}{3} = 0$$

Collect variables to one side of the equation and constants to the other side.

$$x^2 - 4x = -\frac{13}{3}$$

Add $\left(-\frac{4}{2}\right)^2 = 4$ to both sides.

$$x^2 - 4x + \mathbf{4} = -\frac{13}{3} + \mathbf{4}$$

Write the left side of the equation as a perfect square and simplify the right side.

$$(x - 2)^2 = -\frac{1}{3}$$

Solve using the square root method.

$$x - 2 = \pm\sqrt{-\frac{1}{3}}$$

Simplify.

$$x = 2 \pm i\sqrt{\frac{1}{3}}$$

Rationalize the denominator (Chapter 0).

$$x = 2 \pm \frac{i}{\sqrt{3}} \cdot \frac{\sqrt{3}}{\sqrt{3}}$$

Simplify.

$$\boxed{x = 2 - \frac{i\sqrt{3}}{3}, \quad x = 2 + \frac{i\sqrt{3}}{3}}$$

The solution set is $\left\{ 2 - \frac{i\sqrt{3}}{3}, \, 2 + \frac{i\sqrt{3}}{3} \right\}$.

Technology Tip

Graph $y_1 = 3x^2 - 12x + 13$.

The graph does not cross the x-axis, so there is no real solution to this equation.

Study Tip

When the leading coefficient is not 1, start by first dividing the equation by that leading coefficient.

■ **Answer:** The solution is

$$x = 1 \pm \frac{i\sqrt{2}}{2}.$$ The solution set is

$$\left\{ 1 - \frac{i\sqrt{2}}{2}, \, 1 + \frac{i\sqrt{2}}{2} \right\}.$$

■ **YOUR TURN** Solve the equation $2x^2 - 4x + 3 = 0$ by completing the square.

Quadratic Formula

Let us now consider the most general quadratic equation:

$$ax^2 + bx + c = 0 \quad a \neq 0$$

We can solve this equation by completing the square.

WORDS	MATH
Divide the equation by the leading coefficient a.	$x^2 + \dfrac{b}{a}x + \dfrac{c}{a} = 0$
Subtract $\dfrac{c}{a}$ from both sides.	$x^2 + \dfrac{b}{a}x = -\dfrac{c}{a}$
Square half of $\dfrac{b}{a}$ and add the result $\left(\dfrac{b}{2a}\right)^2$ to both sides.	$x^2 + \dfrac{b}{a}x + \left(\dfrac{b}{2a}\right)^2 = \left(\dfrac{b}{2a}\right)^2 - \dfrac{c}{a}$
Write the left side of the equation as a perfect square and the right side as a single fraction.	$\left(x + \dfrac{b}{2a}\right)^2 = \dfrac{b^2 - 4ac}{4a^2}$
Solve using the square root method.	$x + \dfrac{b}{2a} = \pm\sqrt{\dfrac{b^2 - 4ac}{4a^2}}$
Subtract $\dfrac{b}{2a}$ from both sides and simplify the radical.	$x = -\dfrac{b}{2a} \pm \dfrac{\sqrt{b^2 - 4ac}}{2a}$
Write as a single fraction.	$x = \dfrac{-b \pm \sqrt{b^2 - 4ac}}{2a}$

We have derived the **Quadratic Formula**.

Study Tip

$x = \dfrac{-b \pm \sqrt{b^2 - 4ac}}{2a}$

Read as "negative b plus or minus the square root of the quantity b squared minus $4ac$ all over $2a$."

QUADRATIC FORMULA

If $ax^2 + bx + c = 0$, $a \neq 0$, then the solution is

$$x = \dfrac{-b \pm \sqrt{b^2 - 4ac}}{2a}$$

Note: The quadratic equation must be in standard form ($ax^2 + bx + c = 0$) in order to identify the parameters:

$\quad a$—coefficient of x^2 $\qquad b$—coefficient of x $\qquad c$—constant

Study Tip

The Quadratic Formula works for *any* quadratic equation.

We read this formula as *negative b plus or minus the square root of the quantity b squared minus $4ac$ all over $2a$.* It is important to note that negative b could be positive (if b is negative). For this reason, an alternate form is "opposite b. . ." The Quadratic Formula should be memorized and used when simpler methods (factoring and the square root method) cannot be used. The Quadratic Formula works for *any* quadratic equation.

EXAMPLE 9 **Using the Quadratic Formula and Finding Two Distinct Real Roots**

Use the Quadratic Formula to solve the quadratic equation $x^2 - 4x - 1 = 0$.

Solution:

For this problem $a = 1$, $b = -4$, and $c = -1$.

Write the Quadratic Formula.

$$x = \frac{-b \pm \sqrt{b^2 - 4ac}}{2a}$$

Use parentheses to avoid losing a minus sign.

$$x = \frac{-(\square) \pm \sqrt{(\square)^2 - 4(\square)(\square)}}{2(\square)}$$

Substitute values for a, b, and c into the parentheses.

$$x = \frac{-(-4) \pm \sqrt{(-4)^2 - 4(1)(-1)}}{2(1)}$$

Simplify. $\quad x = \dfrac{4 \pm \sqrt{16 + 4}}{2} = \dfrac{4 \pm \sqrt{20}}{2} = \dfrac{4 \pm 2\sqrt{5}}{2} = \dfrac{4}{2} \pm \dfrac{2\sqrt{5}}{2} = \boxed{2 \pm \sqrt{5}}$

The solution set $\boxed{\left\{2 - \sqrt{5}, 2 + \sqrt{5}\right\}}$ contains two distinct real numbers.

■ **YOUR TURN** Use the Quadratic Formula to solve the quadratic equation $x^2 + 6x - 2 = 0$.

Study Tip

Using parentheses as place holders helps avoid $\pm$ errors.

$$x = \frac{-b \pm \sqrt{b^2 - 4ac}}{2a}$$

$$x = \frac{-(\square) \pm \sqrt{(\square)^2 - 4(\square)(\square)}}{2(\square)}$$

Classroom Example 1.3.9
Use the Quadratic Formula to solve the quadratic equation $4n^2 + 6n - 9 = 0$.

Answer: $n = \dfrac{-3 \pm 3\sqrt{5}}{4}$

■ **Answer:** The solution is $x = -3 \pm \sqrt{11}$. The solution set is $\left\{-3 - \sqrt{11}, -3 + \sqrt{11}\right\}$.

EXAMPLE 10 **Using the Quadratic Formula and Finding Two Complex Roots**

Use the Quadratic Formula to solve the quadratic equation $x^2 + 8 = 4x$.

Solution:

Write this equation in standard form $x^2 - 4x + 8 = 0$ in order to identify $a = 1$, $b = -4$, and $c = 8$.

Write the Quadratic Formula.

$$x = \frac{-b \pm \sqrt{b^2 - 4ac}}{2a}$$

Use parentheses to avoid overlooking a minus sign.

$$x = \frac{-(\square) \pm \sqrt{(\square)^2 - 4(\square)(\square)}}{2(\square)}$$

Substitute the values for a, b, and c into the parentheses.

$$x = \frac{-(-4) \pm \sqrt{(-4)^2 - 4(1)(8)}}{2(1)}$$

Simplify. $\quad x = \dfrac{4 \pm \sqrt{16 - 32}}{2} = \dfrac{4 \pm \sqrt{-16}}{2} = \dfrac{4 \pm 4i}{2} = \dfrac{4}{2} \pm \dfrac{4i}{2} = \boxed{2 \pm 2i}$

The solution set $\boxed{\{2 - 2i, 2 + 2i\}}$ contains two complex numbers. Note that they are complex conjugates of each other.

■ **YOUR TURN** Use the Quadratic Formula to solve the quadratic equation $x^2 + 2 = 2x$.

Classroom Example 1.3.10
Use the Quadratic Formula to solve $5x^2 - 10x + 6 = 0$.

Answer: $x = 1 \pm \dfrac{i\sqrt{5}}{5}$

Technology Tip

Set the graphing utility in complex number mode. Keystrokes:

| MODE | NORMAL | $a + bi$ |

■ **Answer:** The solution set is $\{1 - i, 1 + i\}$.

**EXAMPLE 11   Using the Quadratic Formula and Finding
One Repeated Real Root**

Use the Quadratic Formula to solve the quadratic equation $4x^2 - 4x + 1 = 0$.

Solution:

Identify a, b, and c. $a = 4$, $b = -4$, $c = 1$

Write the Quadratic Formula. $x = \dfrac{-b \pm \sqrt{b^2 - 4ac}}{2a}$

Use parentheses to avoid losing a $x = \dfrac{-(\square) \pm \sqrt{(\square)^2 - 4(\square)(\square)}}{2(\square)}$
minus sign.

Substitute values $a = 4$, $b = -4$, $x = \dfrac{-(-4) \pm \sqrt{(-4)^2 - 4(4)(1)}}{2(4)}$
and $c = 1$.

Simplify. $x = \dfrac{4 \pm \sqrt{16 - 16}}{8} = \dfrac{4 \pm 0}{8} = \dfrac{1}{2}$

The solution set is a repeated real root $\left\{\dfrac{1}{2}\right\}$.

Note: This quadratic equation also could have been solved by factoring: $(2x - 1)^2 = 0$.

■**Answer:** $\left\{\frac{1}{3}\right\}$

■**YOUR TURN** Use the Quadratic Formula to solve the quadratic equation
$9x^2 - 6x + 1 = 0$.

TYPES OF SOLUTIONS

The term inside the radical, $b^2 - 4ac$, is called the **discriminant**. The discriminant
gives important information about the corresponding solutions or roots of
$ax^2 + bx + c = 0$, where a, b, and c are real numbers.

$b^2 - 4ac$	SOLUTIONS (ROOTS)
Positive	Two distinct real roots
0	One real root (a double or repeated root)
Negative	Two complex roots (complex conjugates)

In Example 9, the discriminant is positive and the solution has two distinct real roots. In
Example 10, the discriminant is negative and the solution has two complex (conjugate)
roots. In Example 11, the discriminant is zero and the solution has one repeated real root.

Applications Involving Quadratic Equations

In Section 1.2, we developed a procedure for solving word problems involving linear
equations. The procedure is the same for applications involving quadratic equations. The
only difference is that the mathematical equations will be quadratic, as opposed to linear.

EXAMPLE 12   Stock Value

From 1999 to 2001 the price of Abercrombie & Fitch's (ANF) stock was approximately given by $P = 0.2t^2 - 5.6t + 50.2$, where P is the price of stock in dollars, t is in months, and $t = 1$ corresponds to January 1999. When was the value of the stock worth \$30?

ANF Quarterly   5/26/04

Solution:

STEP 1   Identify the question.

When is the price of the stock equal to \$30?

STEP 2   Make notes.

Stock price:

$$P = 0.2t^2 - 5.6t + 50.2$$
$$P = 30$$

STEP 3   Set up an equation.

$$0.2t^2 - 5.6t + 50.2 = 30$$

STEP 4   Solve the equation.

Subtract 30 from both sides.

$$0.2t^2 - 5.6t + 20.2 = 0$$

Solve for t using the Quadratic Formula.

$$t = \frac{-(-5.6) \pm \sqrt{(-5.6)^2 - 4(0.2)(20.2)}}{2(0.2)}$$

Simplify.

$$t \approx \frac{5.6 \pm 3.9}{0.4} \approx 4.25, 23.75$$

Rounding these two numbers, we find that $t \approx 4$ and $t \approx 24$. Since $t = 1$ corresponds to January 1999, these two solutions correspond to $\boxed{\text{April 1999 and December 2000}}$.

STEP 5   Check the solution.

Look at the figure. The horizontal axis represents the year (2000 corresponds to January 2000), and the vertical axis represents the stock price. Estimating when the stock price is approximately \$30, we find April 1999 and December 2000.

EXAMPLE 13   Pythagorean Theorem

Hitachi makes a 60-inch HDTV that has a 60-inch diagonal. If the width of the screen is approximately 52-inches, what is the approximate height of the screen?

Solution:

STEP 1   Identify the question.

What is the approximate height of the HDTV screen?

STEP 2   Make notes.

52 inches
?
60 inches

STEP 3   Set up an equation.

Recall the Pythagorean theorem. $a^2 + b^2 = c^2$

Substitute in the known values. $h^2 + 52^2 = 60^2$

STEP 4   Solve the equation.

Simplify the constants. $h^2 + 2704 = 3600$

Subtract 2704 from both sides. $h^2 = 896$

Solve using the square root method. $h = \pm\sqrt{896} \approx \pm 30$

Distance is positive, so the negative value is eliminated.

The height is approximately $\boxed{30 \text{ inches}}$.

STEP 5   Check the solution.

$$30^2 + 52^2 \stackrel{?}{=} 60^2$$
$$900 + 2704 \stackrel{?}{=} 3600$$
$$3604 \approx 3600$$

SECTION
1.3 SUMMARY

The four methods for solving quadratic equations

$$ax^2 + bx + c = 0 \qquad a \neq 0$$

are factoring, the square root method, completing the square, and the Quadratic Formula. Factoring and the square root method are the quickest and easiest but cannot always be used. The quadratic formula and completing the square work for all quadratic equations and can yield three types of solutions: two distinct real roots, one real root (repeated), or two complex roots (conjugates of each other).

Quadratic Formula: $\quad x = \dfrac{-b \pm \sqrt{b^2 - 4ac}}{2a}$

SECTION
1.3 EXERCISES

▪ SKILLS

In Exercises 1–22, solve by factoring.

1. $x^2 - 5x + 6 = 0$
2. $v^2 + 7v + 6 = 0$
3. $p^2 - 8p + 15 = 0$
4. $u^2 - 2u - 24 = 0$

5. $x^2 = 12 - x$
6. $11x = 2x^2 + 12$
7. $16x^2 + 8x = -1$
8. $3x^2 + 10x - 8 = 0$

9. $9y^2 + 1 = 6y$
10. $4x = 4x^2 + 1$
11. $8y^2 = 16y$
12. $3A^2 = -12A$

13. $9p^2 = 12p - 4$
14. $4u^2 = 20u - 25$
15. $x^2 - 9 = 0$
16. $16v^2 - 25 = 0$

17. $x(x + 4) = 12$
18. $3t^2 - 48 = 0$
19. $2p^2 - 50 = 0$
20. $5y^2 - 45 = 0$

21. $3x^2 = 12$
22. $7v^2 = 28$

In Exercises 23–34, solve using the square root method.

23. $p^2 - 8 = 0$
24. $y^2 - 72 = 0$
25. $x^2 + 9 = 0$
26. $v^2 + 16 = 0$

27. $(x - 3)^2 = 36$
28. $(x - 1)^2 = 25$
29. $(2x + 3)^2 = -4$
30. $(4x - 1)^2 = -16$

31. $(5x - 2)^2 = 27$
32. $(3x + 8)^2 = 12$
33. $(1 - x)^2 = 9$
34. $(1 - x)^2 = -9$

In Exercises 35–44, what number should be added to complete the square of each expression?

35. $x^2 + 6x$
36. $x^2 - 8x$
37. $x^2 - 12x$
38. $x^2 + 20x$
39. $x^2 - \frac{1}{2}x$

40. $x^2 - \frac{1}{3}x$
41. $x^2 + \frac{2}{5}x$
42. $x^2 + \frac{4}{5}x$
43. $x^2 - 2.4x$
44. $x^2 + 1.6x$

In Exercises 45–56, solve by completing the square.

45. $x^2 + 2x = 3$
46. $y^2 + 8y - 2 = 0$
47. $t^2 - 6t = -5$
48. $x^2 + 10x = -21$

49. $y^2 - 4y + 3 = 0$
50. $x^2 - 7x + 12 = 0$
51. $2p^2 + 8p = -3$
52. $2x^2 - 4x + 3 = 0$

53. $2x^2 - 7x + 3 = 0$
54. $3x^2 - 5x - 10 = 0$
55. $\dfrac{x^2}{2} - 2x = \dfrac{1}{4}$
56. $\dfrac{t^2}{3} + \dfrac{2t}{3} + \dfrac{5}{6} = 0$

In Exercises 57–68, solve using the Quadratic Formula.

57. $t^2 + 3t - 1 = 0$
58. $t^2 + 2t = 1$
59. $s^2 + s + 1 = 0$
60. $2s^2 + 5s = -2$

61. $3x^2 - 3x - 4 = 0$
62. $4x^2 - 2x = 7$
63. $x^2 - 2x + 17 = 0$
64. $4m^2 + 7m + 8 = 0$

65. $5x^2 + 7x = 3$
66. $3x^2 + 5x = -11$
67. $\frac{1}{4}x^2 + \frac{2}{3}x - \frac{1}{2} = 0$
68. $\frac{1}{4}x^2 - \frac{2}{3}x - \frac{1}{3} = 0$

In Exercises 69–74, determine whether the discriminant is positive, negative, or zero, and indicate the number and type of root to expect. Do not solve.

69. $x^2 - 22x + 121 = 0$

70. $x^2 - 28x + 196 = 0$

71. $2y^2 - 30y + 68 = 0$

72. $-3y^2 + 27y + 66 = 0$

73. $9x^2 - 7x + 8 = 0$

74. $-3x^2 + 5x - 7 = 0$

In Exercises 75–94, solve using any method.

75. $v^2 - 8v = 20$

76. $v^2 - 8v = -20$

77. $t^2 + 5t - 6 = 0$

78. $t^2 + 5t + 6 = 0$

79. $(x + 3)^2 = 16$

80. $(x + 3)^2 = -16$

81. $(p - 2)^2 = 4p$

82. $(u + 5)^2 = 16u$

83. $8w^2 + 2w + 21 = 0$

84. $8w^2 + 2w - 21 = 0$

85. $3p^2 - 9p + 1 = 0$

86. $3p^2 - 9p - 1 = 0$

87. $\dfrac{2}{3}t^2 + \dfrac{4}{3}t = \dfrac{1}{5}$

88. $\dfrac{1}{2}x^2 + \dfrac{2}{3}x = \dfrac{2}{5}$

89. $x + \dfrac{12}{x} = 7$

90. $x - \dfrac{10}{x} = -3$

91. $\dfrac{4(x - 2)}{x - 3} + \dfrac{3}{x} = \dfrac{-3}{x(x - 3)}$

92. $\dfrac{5}{y + 4} = 4 + \dfrac{3}{y - 2}$

93. $x^2 - 0.1x = 0.12$

94. $y^2 - 0.5y = -0.06$

▪ APPLICATIONS

95. Stock Value. From June 2003 until April 2004 JetBlue airlines stock (JBLU) was approximately worth $P = -4t^2 + 80t - 360$, where P denotes the price of the stock in dollars and t corresponds to months, with $t = 1$ corresponding to January 2003. During what months was the stock equal to $24?

96. Stock Value. From November 2003 until March 2004, Wal-Mart stock (WMT) was approximately worth $P = 2t^2 - 12t + 70$, where P denotes the price of the stock in dollars and t corresponds to months, with $t = 1$ corresponding to November 2003. During what months was the stock equal to $60?

In Exercises 97 and 98 refer to the following:

Research indicates that monthly profit for Widgets R Us is modeled by the function

$$P = -100 + (0.2q - 3)q$$

where P is profit measured in millions of dollars and q is the quantity of widgets produced measured in thousands.

97. Business. Find the break-even point for a month to the nearest unit.

98. Business. Find the production level that produces a monthly profit of $40 million.

In Exercises 99 and 100 refer to the following:

In response to economic conditions, a local business explores the effect of a price increase on weekly profit. The function

$$P = -5(x + 3)(x - 24)$$

models the effect that a price increase of x dollars on a bottle of wine will have on the profit P measured in dollars.

99. Business/Economics. What is the smallest price increase that will produce a weekly profit of $460?

100. Business/Economics. What is the smallest price increase that will produce a weekly profit of $630?

In Exercises 101 and 102 refer to the following:

An epidemiological study of the spread of the flu in a small city finds that the total number P of people who contracted the flu t days into an outbreak is modeled by the function

$$P = -t^2 + 13t + 130 \qquad 1 \le t \le 6$$

101. Health/Medicine. After approximately how many days will 160 people have contracted the flu?

102. Health/Medicine. After approximately how many days will 172 people have contracted the flu?

103. Environment: Reduce Your Margins, Save a Tree.
Let's define the *usable area* of an 8.5-inch by 11-inch piece of paper as the rectangular space between the margins of that piece of paper. Assume the default margins in a word processor in a college's computer lab are set up to be 1.25 inches wide (top and bottom) and 1 inch wide (left and right). Answer the following questions using this information.

a. Determine the amount of usable space, in square inches, on one side of an 8.5-inch by 11-inch piece of paper with the default margins of 1.25-inch and 1-inch.

b. The Green Falcons, a campus environmental club, has convinced their college's computer lab to reduce the default margins in their word-processing software by x inches. Create and simplify the quadratic expression that represents the new usable area, in square inches, of one side of an 8.5-inch by 11-inch piece of paper if the default margins at the computer lab are each reduced by x inches.

c. Subtract the usable space in part (a) from the expression in part (b). Explain what this difference represents.

d. If 10 pages are printed using the new margins and as a result the computer lab saved one whole sheet of paper, then by how much did the computer lab reduce the margins? Round to the nearest tenth of an inch.

104. Environment: Reduce Your Margins, Save a Tree. Repeat Exercise 103 assuming the computer lab's default margins are 1 inch all the way around (left, right, top, and bottom). If 15 pages are printed using the new margins and as a result the computer lab saved one whole sheet of paper, then by how much did the computer lab reduce the margins? Round to the nearest tenth of an inch.

105. Television. A standard 32-inch television has a 32-inch diagonal and a 25-inch width. What is the height of the 32-inch television?

106. Television. A 42-inch LCD television has a 42-inch diagonal and a 20-inch height. What is the width of the 42-inch LCD television?

107. Numbers. Find two consecutive numbers such that their sum is 35 and their product is 306.

108. Numbers. Find two consecutive odd integers such that their sum is 24 and their product is 143.

109. Geometry. The area of a rectangle is 135 square feet. The width is 6 feet less than the length. Find the dimensions of the rectangle.

110. Geometry. A rectangle has an area of 31.5 square meters. If the length is 2 more than twice the width, find the dimensions of the rectangle.

111. Geometry. A triangle has a height that is 2 more than 3 times the base and an area of 60 square units. Find the base and height.

112. Geometry. A square's side is increased by 3 yards, which corresponds to an increase in the area by 69 square yards. How many yards is the side of the initial square?

113. Falling Objects. If a person drops a water balloon off the rooftop of a 100-foot building, the height of the water balloon is given by the equation $h = -16t^2 + 100$, where t is in seconds. When will the water balloon hit the ground?

114. Falling Objects. If the person in Exercise 113 throws the water balloon downward with a speed of 5 feet per second, the height of the water balloon is given by the equation $h = -16t^2 - 5t + 100$, where t is in seconds. When will the water balloon hit the ground?

115. Gardening. A square garden has an area of 900 square feet. If a sprinkler (with a circular pattern) is placed in the center of the garden, what is the minimum radius of spray the sprinkler would need in order to water all of the garden?

116. Sports. A baseball diamond is a square. The distance from base to base is 90 feet. What is the distance from home plate to second base?

117. Volume. A flat square piece of cardboard is used to construct an open box. Cutting a 1-foot by 1-foot square off of each corner and folding up the edges will yield an open box (assuming these edges are taped together). If the desired volume of the box is 9 cubic feet, what are the dimensions of the original square piece of cardboard?

118. Volume. A rectangular piece of cardboard whose length is twice its width is used to construct an open box. Cutting a 1-foot by 1-foot square off of each corner and folding up the edges will yield an open box. If the desired volume is 12 cubic feet, what are the dimensions of the original rectangular piece of cardboard?

119. Gardening. A landscaper has planted a rectangular garden that measures 8 feet by 5 feet. He has ordered 1 cubic yard (27 cubic feet) of stones for a border along the outside of the garden. If the border needs to be 4 inches deep and he wants to use all of the stones, how wide should the border be?

120. Gardening. A gardener has planted a semicircular rose garden with a radius of 6 feet, and 2 cubic yards of mulch (1 cubic yard = 27 cubic feet) are being delivered. Assuming she uses all of the mulch, how deep will the layer of mulch be?

121. Work. Lindsay and Kimmie, working together, can balance the financials for the Kappa Kappa Gamma sorority in 6 days. Lindsay by herself can complete the job in 5 days less than Kimmie. How long will it take Lindsay to complete the job by herself?

122. Work. When Jack cleans the house, it takes him 4 hours. When Ryan cleans the house, it takes him 6 hours. How long would it take both of them if they worked together?

■ **CATCH THE MISTAKE**

In Exercises 123–126, explain the mistake that is made.

123. $t^2 - 5t - 6 = 0$

$(t - 3)(t - 2) = 0$

$t = 2, 3$

124. $(2y - 3)^2 = 25$

$2y - 3 = 5$

$2y = 8$

$y = \frac{5}{4}$

$y = 4$

125. $16a^2 + 9 = 0$

$16a^2 = -9$

$a^2 = -\frac{9}{16}$

$a = \pm\sqrt{\frac{9}{16}}$

$a = \pm\frac{3}{4}$

126. $2x^2 - 4x = 3$

$2(x^2 - 2x) = 3$

$2(x^2 - 2x + 1) = 3 + 1$

$2(x - 1)^2 = 4$

$(x - 1)^2 = 2$

$x - 1 = \pm\sqrt{2}$

$x = 1 \pm\sqrt{2}$

■ **CONCEPTUAL**

In Exercises 127–130, determine whether the following statements are true or false.

127. The equation $(3x + 1)^2 = 16$ has the same solution set as the equation $3x + 1 = 4$.

128. The quadratic equation $ax^2 + bx + c = 0$ can be solved by the square root method only if $b = 0$.

129. All quadratic equations can be solved exactly.

130. The Quadratic Formula can be used to solve any quadratic equation.

131. Write a quadratic equation in general form that has $x = a$ as a repeated real root.

132. Write a quadratic equation in general form that has $x = bi$ as a root.

133. Write a quadratic equation in general form that has the solution set $\{2, 5\}$.

134. Write a quadratic equation in general form that has the solution set $\{-3, 0\}$.

In Exercises 135–138, solve for the indicated variable in terms of other variables.

135. Solve $s = \frac{1}{2}gt^2$ for t.

136. Solve $A = P(1 + r)^2$ for r.

137. Solve $a^2 + b^2 = c^2$ for c.

138. Solve $P = EI - RI^2$ for I.

139. Solve the equation by factoring: $x^4 - 4x^2 = 0$.

140. Solve the equation by factoring: $3x - 6x^2 = 0$.

141. Solve the equation using factoring by grouping:
$x^3 + x^2 - 4x - 4 = 0$.

142. Solve the equation using factoring by grouping:
$x^3 + 2x^2 - x - 2 = 0$.

■ **CHALLENGE**

143. Show that the sum of the roots of a quadratic equation is equal to $-\frac{b}{a}$.

144. Show that the product of the roots of a quadratic equation is equal to $\frac{c}{a}$.

145. Write a quadratic equation in general form whose solution set is $\left\{3 + \sqrt{5}, 3 - \sqrt{5}\right\}$.

146. Write a quadratic equation in general form whose solution set is $\{2 - i, 2 + i\}$.

147. Aviation. An airplane takes 1 hour longer to go a distance of 600 miles flying against a headwind than on the return trip with a tailwind. If the speed of the wind is 50 miles per hour, find the speed of the plane in still air.

148. Boating. A speedboat takes 1 hour longer to go 24 miles up a river than to return. If the boat cruises at 10 miles per hour in still water, what is the rate of the current?

149. Find a quadratic equation whose two distinct real roots are the negatives of the two distinct real roots of the equation $ax^2 + bx + c = 0$.

150. Find a quadratic equation whose two distinct real roots are the reciprocals of the two distinct real roots of the equation $ax^2 + bx + c = 0$.

151. A small jet and a 757 leave Atlanta at 1 P.M. The small jet is traveling due west. The 757 is traveling due south. The speed of the 757 is 100 miles per hour faster than the small jet. At 3 P.M. the planes are 1000 miles apart. Find the average speed of each plane.

152. Two boats leave Key West at noon. The smaller boat is traveling due west. The larger boat is traveling due south. The speed of the larger boat is 10 miles per hour faster than the speed of the smaller boat. At 3 P.M. the boats are 150 miles apart. Find the average speed of each boat.

■ TECHNOLOGY

153. Solve the equation $x^2 - x = 2$ by first writing it in standard form and then factoring. Now plot both sides of the equation in the same viewing screen ($y_1 = x^2 - x$ and $y_2 = 2$). At what x-values do these two graphs intersect? Do those points agree with the solution set you found?

154. Solve the equation $x^2 - 2x = -2$ by first writing it in standard form and then using the quadratic formula. Now plot both sides of the equation in the same viewing screen ($y_1 = x^2 - 2x$ and $y_2 = -2$). Do these graphs intersect? Does this agree with the solution set you found?

155. a. Solve the equation $x^2 - 2x = b$, $b = 8$ by first writing it in standard form. Now plot both sides of the equation in the same viewing screen ($y_1 = x^2 - 2x$ and $y_2 = b$). At what x values do these two graphs intersect? Do those points agree with the solution set you found?
   b. Repeat part (a) for $b = -3, -1, 0$, and 5.

156. a. Solve the equation $x^2 + 2x = b$, $b = 8$ by first writing it in standard form. Now plot both sides of the equation in the same viewing screen ($y_1 = x^2 + 2x$ and $y_2 = b$). At what x values do these two graphs intersect? Do those points agree with the solution set you found?
   b. Repeat part (a) for $b = -3, -1, 0$, and 5.

SECTION
1.4   OTHER TYPES OF EQUATIONS

SKILLS OBJECTIVES

■ Solve radical equations.
■ Solve equations that are quadratic in form.
■ Solve equations that are factorable.

CONCEPTUAL OBJECTIVES

■ Transform a difficult equation into a simpler linear or quadratic equation.
■ Recognize the need to check solutions when the transformation process may produce extraneous solutions.
■ Realize that not all polynomial equations are factorable.

Radical Equations

Radical equations are equations in which the variable is inside a radical (that is, under a square root, cube root, or higher root). Examples of radical equations follow.

$$\sqrt{x - 3} = 2 \qquad \sqrt{2x + 3} = x \qquad \sqrt{x + 2} + \sqrt{7x + 2} = 6$$

Until now your experience has been with linear and quadratic equations. Often you can transform a radical equation into a simple linear or quadratic equation. Sometimes the transformation process yields **extraneous solutions**, or apparent solutions that may solve the transformed problem but are not solutions of the original radical equation. Therefore, it is very important to check your answers.

EXAMPLE 1 Solving an Equation Involving a Radical

Solve the equation $\sqrt{x - 3} = 2$.

Solution:

Square both sides of the equation. $\left(\sqrt{x - 3}\right)^2 = 2^2$

Simplify. $x - 3 = 4$

Solve the resulting linear equation. $\boxed{x = 7}$

The solution set is $\{7\}$.

Check: $\sqrt{7 - 3} = \sqrt{4} = 2$

■ **YOUR TURN** Solve the equation $\sqrt{3p + 4} = 5$.

■ **Answer:** $p = 7$ or $\{7\}$

When both sides of an equation are squared, extraneous solutions can arise. For example, take the equation

$$x = 2$$

If we square both sides of this equation, then the resulting equation, $x^2 = 4$, has two solutions: $x = -2$ and $x = 2$. Notice that the value $x = -2$ is not in the solution set of the original equation $x = 2$. Therefore, we say that $x = -2$ is an extraneous solution.

In solving a radical equation we square both sides of the equation and then solve the resulting equation. The solutions to the resulting equation can sometimes be extraneous in that they do not satisfy the *original* radical equation.

Study Tip

Extraneous solutions are common when we deal with radical equations, so remember to check your answers.

Technology Tip

Use a graphing utility to display graphs of $y_1 = \sqrt{2x + 3}$ and $y_2 = x$.

The *x*-coordinate of the point of intersection is the solution to the equation $\sqrt{2x + 3} = x$.

EXAMPLE 2 Solving an Equation Involving a Radical

Solve the equation $\sqrt{2x + 3} = x$.

Solution:

Square both sides of the equation. $\left(\sqrt{2x + 3}\right)^2 = x^2$

Simplify. $2x + 3 = x^2$

Write the quadratic equation in standard form. $x^2 - 2x - 3 = 0$

Factor. $(x - 3)(x + 1) = 0$

Use the zero product property. $x = 3$ or $x = -1$

Check these values to see whether they *both* make the original equation statement true.

$x = 3$: $\sqrt{2(3) + 3} = 3 \Rightarrow \sqrt{6 + 3} = 3 \Rightarrow \sqrt{9} = 3 \Rightarrow 3 = 3$ ✓

$x = -1$: $\sqrt{2(-1) + 3} = -1 \Rightarrow \sqrt{-2 + 3} = -1 \Rightarrow \sqrt{1} = -1 \Rightarrow 1 \neq -1$ X

The solution is $\boxed{x = 3}$. The solution set is $\{3\}$.

■ **YOUR TURN** Solve the equation $\sqrt{12 + t} = t$.

■ **YOUR TURN** Solve the equation $\sqrt{2x + 6} = x + 3$.

■ **Answer:** $t = 4$ or $\{4\}$

■ **Answer:** $x = -1$ and $x = -3$ or $\{-3, -1\}$

What happened in Example 2? When we transformed the radical equation into a quadratic equation, we created an **extraneous solution**, $x = -1$, a solution that appears to solve the original equation but does not. When solving radical equations, answers must be checked to avoid including extraneous solutions in the solution set.

EXAMPLE 3   Solving an Equation That Involves a Radical

Solve the equation $4x - 2\sqrt{x + 3} = -10$.

Solution:

Subtract $4x$ from both sides. $\qquad\qquad\qquad\qquad -2\sqrt{x + 3} = -10 - 4x$

Divide both sides by -2. $\qquad\qquad\qquad\qquad\qquad \sqrt{x + 3} = 2x + 5$

Square both sides. $\qquad\qquad\qquad\qquad\qquad x + 3 = \underbrace{(2x + 5)(2x + 5)}_{(2x + 5)^2}$

Eliminate the parentheses. $\qquad\qquad\qquad\qquad x + 3 = 4x^2 + 20x + 25$

Rewrite the quadratic equation in standard form. $\qquad 4x^2 + 19x + 22 = 0$

Factor. $\qquad\qquad\qquad\qquad\qquad\qquad\qquad (4x + 11)(x + 2) = 0$

Solve. $\qquad\qquad\qquad\qquad\qquad\qquad\qquad x = -\dfrac{11}{4}$ and $x = -2$

The apparent solutions are $-\frac{11}{4}$ and -2. Note that $-\frac{11}{4}$ does not satisfy the original equation; therefore it is extraneous. The solution is $\boxed{x = -2}$. The solution set is $\{-2\}$.

■ **YOUR TURN** Solve the equation $2x - 4\sqrt{x + 2} = -6$.

COMMON MISTAKE

⭐ CORRECT	❌ INCORRECT
Square the expression.	Square the expression.
$(3 + \sqrt{x + 2})^2$	$(3 + \sqrt{x + 2})^2$
Write the square as a product of two factors.	The **error** occurs here when only individual terms are squared.
$(3 + \sqrt{x + 2})(3 + \sqrt{x + 2})$	$\neq 9 + (x + 2)$
Use the FOIL method.	
$9 + 6\sqrt{x + 2} + (x + 2)$	

In Examples 1 through 3 each equation only contained one radical each. The next example contains two radicals. Our technique will be to isolate one radical on one side of the equation with the other radical on the other side of the equation.

EXAMPLE 4 Solving an Equation with More Than One Radical

Solve the equation $\sqrt{x + 2} + \sqrt{7x + 2} = 6$.

Solution:

Subtract $\sqrt{x + 2}$ from both sides.
$$\sqrt{7x + 2} = 6 - \sqrt{x + 2}$$

Square both sides.
$$\left(\sqrt{7x + 2}\right)^2 = \left(6 - \sqrt{x + 2}\right)^2$$

Simplify.
$$7x + 2 = \left(6 - \sqrt{x + 2}\right)\left(6 - \sqrt{x + 2}\right)$$

Multiply the expressions on the right side of the equation.
$$7x + 2 = 36 - 12\sqrt{x + 2} + (x + 2)$$

Isolate the term with the radical on the left side.
$$12\sqrt{x + 2} = 36 + x + 2 - 7x - 2$$

Combine like terms on the right side.
$$12\sqrt{x + 2} = 36 - 6x$$

Divide by 6.
$$2\sqrt{x + 2} = 6 - x$$

Square both sides.
$$4(x + 2) = (6 - x)^2$$

Simplify.
$$4x + 8 = 36 - 12x + x^2$$

Rewrite the quadratic equation in standard form.
$$x^2 - 16x + 28 = 0$$

Factor.
$$(x - 14)(x - 2) = 0$$

Solve.
$$x = 14 \quad \text{and} \quad x = 2$$

The apparent solutions are 2 and 14. Note that $x = 14$ does not satisfy the original equation; therefore it is extraneous. The solution is $\boxed{x = 2}$. The solution set is {2}.

▪ **YOUR TURN** Solve the equation $\sqrt{x - 4} = 5 - \sqrt{x + 1}$.

Technology Tip

Use a graphing utility to display graphs of
$$y_1 = \sqrt{x + 2} + \sqrt{7x + 2}$$
and $y_2 = 6$.

The x-coordinate of the point of intersection is the solution to the equation
$$\sqrt{x + 2} + \sqrt{7x + 2} = 6.$$

Study Tip

Remember to check both solutions.

▪ **Answer:** $x = 8$ or {8}

PROCEDURE FOR SOLVING RADICAL EQUATIONS

Step 1: Isolate the term with a radical on one side.
Step 2: Raise both (*entire*) sides of the equation to the power that will eliminate this radical, and simplify the equation.
Step 3: If a radical remains, repeat Steps 1 and 2.
Step 4: Solve the resulting linear or quadratic equation.
Step 5: Check the solutions and eliminate any extraneous solutions.

Note: If there is more than one radical in the equation, it does not matter which radical is isolated first.

Equations Quadratic in Form: *u*-Substitution

Equations that are higher order or that have fractional powers often can be transformed into a quadratic equation by introducing a *u*-substitution. When this is the case, we say that equations are **quadratic in form**. In the following table, the two original equations are quadratic in form because they can be transformed into a quadratic equation given the correct substitution.

ORIGINAL EQUATION	SUBSTITUTION	NEW EQUATION
$x^4 - 3x^2 - 4 = 0$	$u = x^2$	$u^2 - 3u - 4 = 0$
$t^{2/3} + 2t^{1/3} + 1 = 0$	$u = t^{1/3}$	$u^2 + 2u + 1 = 0$
$\dfrac{2}{y} - \dfrac{1}{\sqrt{y}} + 1 = 0$	$u = y^{-1/2}$	$2u^2 - u + 1 = 0$

For example, the equation $x^4 - 3x^2 - 4 = 0$ is a fourth-degree equation in x. How did we know that $u = x^2$ would transform the original equation into a quadratic equation? If we rewrite the original equation as $(x^2)^2 - 3(x^2) - 4 = 0$, the expression in parentheses is the *u*-substitution.

Let us introduce the substitution $u = x^2$. Note that squaring both sides implies $u^2 = x^4$. We then replace x^2 in the original equation with u, and x^4 in the original equation with u^2, which leads to a quadratic equation in u: $u^2 - 3u - 4 = 0$.

WORDS	**MATH**
Solve for x.	$x^4 - 3x^2 - 4 = 0$
Introduce *u*-substitution.	$u = x^2$ [Note that $u^2 = x^4$.]
Write the quadratic equation in u.	$u^2 - 3u - 4 = 0$
Factor.	$(u - 4)(u + 1) = 0$
Solve for u.	$u = 4$ or $u = -1$
Transform back to x, $u = x^2$.	$x^2 = 4$ or $x^2 = -1$
Solve for x.	$\boxed{x = \pm 2 \quad \text{or} \quad x = \pm i}$

The solution set is $\{\pm 2, \pm i\}$.

It is important to correctly determine the appropriate substitution in order to arrive at an equation quadratic in form. For example, $t^{2/3} + 2t^{1/3} + 1 = 0$ is an original equation given in the above table. If we rewrite this equation as $(t^{1/3})^2 + 2(t^{1/3}) + 1 = 0$, then it becomes apparent that the correct substitution is $u = t^{1/3}$, which transforms the equation in t into a quadratic equation in u: $u^2 + 2u + 1 = 0$.

PROCEDURE FOR SOLVING EQUATIONS QUADRATIC IN FORM

Step 1: Identify the substitution.
Step 2: Transform the equation into a quadratic equation.
Step 3: Solve the quadratic equation.
Step 4: Apply the substitution to rewrite the solution in terms of the original variable.
Step 5: Solve the resulting equation.
Step 6: Check the solutions in the original equation.

EXAMPLE 5 **Solving an Equation Quadratic in Form with Negative Exponents**

Find the solutions to the equation $x^{-2} - x^{-1} - 12 = 0$.

Solution:

Rewrite the original equation.	$(x^{-1})^2 - (x^{-1}) - 12 = 0$
Determine the u-substitution.	$u = x^{-1}$ [Note that $u^2 = x^{-2}$.]
The original equation in x corresponds to a quadratic equation in u.	$u^2 - u - 12 = 0$
Factor.	$(u - 4)(u + 3) = 0$
Solve for u.	$u = 4$ or $u = -3$

The most common mistake is forgetting to transform back to x.

Transform back to x. Let $u = x^{-1}$.	$x^{-1} = 4$ or $x^{-1} = -3$
Write x^{-1} as $\dfrac{1}{x}$.	$\dfrac{1}{x} = 4$ or $\dfrac{1}{x} = -3$
Solve for x.	$\boxed{x = \dfrac{1}{4}}$ or $\boxed{x = -\dfrac{1}{3}}$

The solution set is $\left\{-\frac{1}{3}, \frac{1}{4}\right\}$.

■ **YOUR TURN** Find the solutions to the equation $x^{-2} - x^{-1} - 6 = 0$.

■ **Answer:** The solution is $x = -\frac{1}{2}$ or $x = \frac{1}{3}$. The solution set is $\left\{-\frac{1}{2}, \frac{1}{3}\right\}$.

 EXAMPLE 6 **Solving an Equation Quadratic in Form with Fractional Exponents**

Find the solutions to the equation $x^{2/3} - 3x^{1/3} - 10 = 0$.

Solution:

Rewrite the original equation.	$(x^{1/3})^2 - 3x^{1/3} - 10 = 0$
Identify the substitution as $u = x^{1/3}$.	$u^2 - 3u - 10 = 0$
Factor.	$(u - 5)(u + 2) = 0$
Solve for u.	$u = 5$ or $u = -2$
Let $u = x^{1/3}$ again.	$x^{1/3} = 5$ $x^{1/3} = -2$
Cube both sides of the equations.	$(x^{1/3})^3 = (5)^3$ $(x^{1/3})^3 = (-2)^3$
Simplify.	$\boxed{x = 125}$ $\boxed{x = -8}$

The solution set is $\boxed{\{-8, 125\}}$, which a check will confirm.

■ **YOUR TURN** Find the solution to the equation $2t - 5t^{1/2} - 3 = 0$.

■ **Answer:** $t = 9$ or $\{9\}$.

Factorable Equations

Technology Tip

Use a graphing utility to graph $y_1 = x^{7/3} - 3x^{4/3} - 4x^{1/3}$.

The x-intercepts are the solutions to this equation.

Some equations (both polynomial and with rational exponents) that are factorable can be solved using the zero product property.

EXAMPLE 7   **Solving an Equation with Rational Exponents by Factoring**

Solve the equation $x^{7/3} - 3x^{4/3} - 4x^{1/3} = 0$.

Solution:

Factor the left side of the equation.

$$x^{1/3}(x^2 - 3x - 4) = 0$$

Factor the quadratic expression.

$$x^{1/3}(x - 4)(x + 1) = 0$$

Apply the zero product property.

$$x^{1/3} = 0 \quad \text{or} \quad x - 4 = 0 \quad \text{or} \quad x + 1 = 0$$

Solve for x.

$$\boxed{x = 0} \quad \text{or} \quad \boxed{x = 4} \quad \text{or} \quad \boxed{x = -1}$$

The solution set is $\{-1, 0, 4\}$.

> **Classroom Example 1.4.7***
> Solve the equation
> $2x^{8/3} + x^{5/3} - x^{2/3} = 0$.
> **Answer:** $x = 0, \frac{1}{2}, -1$

> **Classroom Example 1.4.8***
> Solve the equation
> $6x^3 - 12x^2 - 2x + 4 = 0$.
>
> **Answer:** $x = 2, \pm\dfrac{\sqrt{3}}{3}$

EXAMPLE 8   **Solving a Polynomial Equation Using Factoring by Grouping**

Solve the equation $x^3 + 2x^2 - x - 2 = 0$.

Solution:

Factor by grouping (Chapter 0).

$$(x^3 - x) + (2x^2 - 2) = 0$$

Identify the common factors.

$$x(x^2 - 1) + 2(x^2 - 1) = 0$$

Factor.

$$(x + 2)(x^2 - 1) = 0$$

Factor the quadratic expression.

$$(x + 2)(x - 1)(x + 1) = 0$$

Apply the zero product property.

$$x + 2 = 0 \quad \text{or} \quad x - 1 = 0 \quad \text{or} \quad x + 1 = 0$$

Solve for x.

$$\boxed{x = -2} \quad \text{or} \quad \boxed{x = 1} \quad \text{or} \quad \boxed{x = -1}$$

The solution set is $\{-2, -1, 1\}$.

■ **Answer:** $x = -1$ or $x = \pm 2$
or $\{-2, -1, 2\}$

■ **YOUR TURN** Solve the equation $x^3 + x^2 - 4x - 4 = 0$.

SECTION
1.4 **SUMMARY**

Radical equations, equations quadratic in form, and factorable equations can often be solved by transforming them into simpler linear or quadratic equations.

■ **Radical Equations:** Isolate the term containing a radical and raise it to the appropriate power that will eliminate the radical. If there is more than one radical, it does not matter which radical is isolated first. Raising radical equations to powers may cause extraneous solutions, so check each solution.

■ **Equations Quadratic in Form:** Identify the u-substitution that transforms the equation into a quadratic equation. Solve the quadratic equation and then remember to transform back to the original variable.

■ **Factorable Equations:** Look for a factor common to all terms or factor by grouping.

▪ SKILLS

In Exercises 1–40, solve the radical equation for the given variable.

1. $\sqrt{t-5}=2$

2. $\sqrt{2t-7}=3$

3. $(4p-7)^{1/2}=5$

4. $11=(21-p)^{1/2}$

5. $\sqrt{u+1}=-4$

6. $-\sqrt{3-2u}=9$

7. $\sqrt[3]{5x+2}=3$

8. $\sqrt[3]{1-x}=-2$

9. $(4y+1)^{1/3}=-1$

10. $(5x-1)^{1/3}=4$

11. $\sqrt{12+x}=x$

12. $x=\sqrt{56-x}$

13. $y=5\sqrt{y}$

14. $\sqrt{y}=\dfrac{y}{4}$

15. $s=3\sqrt{s-2}$

16. $-2s=\sqrt{3-s}$

17. $\sqrt{2x+6}=x+3$

18. $\sqrt{8-2x}=2x-2$

19. $\sqrt{1-3x}=x+1$

20. $\sqrt{2-x}=x-2$

21. $3x-6\sqrt{x-1}=3$

22. $5x-10\sqrt{x+2}=-10$

23. $3x-6\sqrt{x+2}=3$

24. $2x-4\sqrt{x+1}=4$

25. $3\sqrt{x+4}-2x=9$

26. $2\sqrt{x+1}-3x=-5$

27. $\sqrt{x^2-4}=x-1$

28. $\sqrt{25-x^2}=x+1$

29. $\sqrt{x^2-2x-5}=x+1$

30. $\sqrt{2x^2-8x+1}=x-3$

31. $\sqrt{3x+1}-\sqrt{6x-5}=1$

32. $\sqrt{2-x}+\sqrt{6-5x}=6$

33. $\sqrt{x+12}+\sqrt{8-x}=6$

34. $\sqrt{5-x}+\sqrt{3x+1}=4$

35. $\sqrt{2x-1}-\sqrt{x-1}=1$

36. $\sqrt{8-x}=2+\sqrt{2x+3}$

37. $\sqrt{3x-5}=7-\sqrt{x+2}$

38. $\sqrt{x+5}=1+\sqrt{x-2}$

39. $\sqrt{2+\sqrt{x}}=\sqrt{x}$

40. $\sqrt{2-\sqrt{x}}=\sqrt{x}$

In Exercises 41–70, solve the equations by introducing a substitution that transforms these equations to quadratic form.

41. $x^{2/3}+2x^{1/3}=0$

42. $x^{1/2}-2x^{1/4}=0$

43. $x^4-3x^2+2=0$

44. $x^4-8x^2+16=0$

45. $2x^4+7x^2+6=0$

46. $x^8-17x^4+16=0$

47. $(2x+1)^2+5(2x+1)+4=0$

48. $(x-3)^2+6(x-3)+8=0$

49. $4(t-1)^2-9(t-1)=-2$

50. $2(1-y)^2+5(1-y)-12=0$

51. $x^{-8}-17x^{-4}+16=0$

52. $2u^{-2}+5u^{-1}-12=0$

53. $3y^{-2}+y^{-1}-4=0$

54. $5a^{-2}+11a^{-1}+2=0$

55. $z^{2/5}-2z^{1/5}+1=0$

56. $2x^{1/2}+x^{1/4}-1=0$

57. $(x+3)^{5/3}=32$

58. $(x+2)^{4/3}=16$

59. $(x+1)^{2/3}=4$

60. $(x-7)^{4/3}=81$

61. $6t^{-2/3}-t^{-1/3}-1=0$

62. $t^{-2/3}-t^{-1/3}-6=0$

63. $3=\dfrac{1}{(x+1)^2}+\dfrac{2}{(x+1)}$

64. $\dfrac{1}{(x+1)^2}+\dfrac{4}{(x+1)}+4=0$

65. $\left(\dfrac{1}{2x-1}\right)^2+\left(\dfrac{1}{2x-1}\right)-12=0$

66. $\dfrac{5}{(2x+1)^2}-\dfrac{3}{(2x+1)}=2$

67. $u^{4/3}-5u^{2/3}=-4$

68. $u^{4/3}+5u^{2/3}=-4$

69. $t=\sqrt[4]{t^2+6}$

70. $u=\sqrt[4]{-2u^2-1}$

In Exercises 71–86, solve by factoring.

71. $x^3-x^2-12x=0$

72. $2y^3-11y^2+12y=0$

73. $4p^3-9p=0$

74. $25x^3=4x$

75. $u^5-16u=0$

76. $t^5-81t=0$

77. $x^3-5x^2-9x+45=0$

78. $2p^3-3p^2-8p+12=0$

79. $y(y-5)^3-14(y-5)^2=0$

80. $v(v+3)^3-40(v+3)^2=0$

81. $x^{9/4}-2x^{5/4}-3x^{1/4}=0$

82. $u^{7/3}+u^{4/3}-20u^{1/3}=0$

83. $t^{5/3}-25t^{-1/3}=0$

84. $4x^{9/5}-9x^{-1/5}=0$

85. $y^{3/2}-5y^{1/2}+6y^{-1/2}=0$

86. $4p^{5/3}-5p^{2/3}-6p^{-1/3}=0$

▪APPLICATIONS

In Exercises 87 and 88 refer to the following:

An analysis of sales indicates that demand for a product during a calendar year is modeled by

$$d = 3\sqrt{t + 1} - 0.75t$$

where d is demand in millions of units and t is the month of the year where $t = 0$ represents January.

87. Economics. During which month(s) is demand 3 million units?

88. Economics. During which month(s) is demand 4 million units?

In Exercises 89 and 90 refer to the following:

Body Surface Area (BSA) is used in physiology and medicine for many clinical purposes. BSA can be modeled by the function

$$BSA = \sqrt{\frac{wh}{3600}}$$

where w is weight in kilograms and h is height in centimeters.

89. Health. The BSA of a 72 kilogram female is 1.8. Find the height of the female to the nearest centimeter.

90. Health. The BSA of a 177 centimeter tall male is 2.1. Find the weight of the male to the nearest kilogram.

91. Insurance: Health. Cost for health insurance with a private policy is given by $C = \sqrt{10 + a}$, where C is the cost per day and a is the insured's age in years. Health insurance for a 6-year-old, $a = 6$, is $4 a day (or $1460 per year). At what age would someone be paying $9 a day (or $3285 per year)?

92. Insurance: Life. Cost for life insurance is given by $C = \sqrt{5a + 1}$, where C is the cost per day and a is the insured's age in years. Life insurance for a newborn, $a = 0$, is $1 a day (or $365 per year). At what age would someone be paying $20 a day (or $7300 per year)?

93. Stock Value. The stock price of MGI Pharmaceutical (MOGN) from March 2004 to June 2004 can be approximately modeled by the equation $P = 5\sqrt{t^2 + 1} + 50$, where P is the price of the stock in dollars and t is the month with $t = 0$ corresponding to March 2004. Assuming this trend continues, when would the stock be worth $85?

MOGN Daily ▬   5/28/04

Apr   May

94. Grades. The average combined math and verbal SAT score of incoming freshmen at a university is given by the equation $S = 1000 + 10\sqrt{2t}$, where t is in years and $t = 0$ corresponds to 1990. What year will the incoming class have an average SAT score of 1230?

95. Speed of Sound. A man buys a house with an old well but does not know how deep the well is. To get an estimate he decides to drop a rock in the opening of the well and time how long it takes until he hears the splash. The total elapsed time T given by $T = t_1 + t_2$, is the sum of the time it takes for the rock to reach the water, t_1, and the time it takes for the sound of the splash to travel to the top of the well, t_2. The time (seconds) that it takes for the rock to reach the water is given by $t_1 = \dfrac{\sqrt{d}}{4}$, where d is the depth of the well in feet. Since the speed of sound is 1100 ft/s, the time (seconds) it takes for the sound to reach the top of the well is $t_2 = \dfrac{d}{1100}$. If the splash is heard after 3 seconds, how deep is the well?

96. Speed of Sound. If the owner of the house in Exercise 91 forgot to account for the speed of sound, what would he have calculated the depth of the well to be?

97. Physics: Pendulum. The period (T) of a pendulum is related to the length (L) of the pendulum and acceleration due to gravity (g) by the formula $T = 2\pi\sqrt{\dfrac{L}{g}}$. If gravity is 9.8 m/s^2 and the period is 1 second, find the approximate length of the pendulum. Round to the nearest centimeter. *Note:* 100 cm = 1 m.

98. Physics: Pendulum. The period (T) of a pendulum is related to the length (L) of the pendulum and acceleration due to gravity (g) by the formula $T = 2\pi\sqrt{\dfrac{L}{g}}$. If gravity is 32 ft/s^2 and the period is 1 second, find the approximate length of the pendulum. Round to the nearest inch. *Note:* 12 in. = 1 ft.

In Exercises 99 and 100, refer to the following:

Einstein's special theory of relativity states that time is relative: Time speeds up or slows down, depending on how fast one object is moving with respect to another. For example, a space probe traveling at a velocity v near the speed of light c will have "clocked" a time t hours, but for a stationary observer on Earth that corresponds to a time t_0. The formula governing this relativity is given by

$$t = t_0\sqrt{1 - \frac{v^2}{c^2}}$$

99. Physics: Special Theory of Relativity. If the time elapsed on a space probe mission is 18 years but the time elapsed on Earth during that mission is 30 years, how fast is the space probe traveling? Give your answer relative to the speed of light.

100. Physics: Special Theory of Relativity. If the time elapsed on a space probe mission is 5 years but the time elapsed on Earth during that mission is 30 years, how fast is the space probe traveling? Give your answer relative to the speed of light.

■**CATCH THE MISTAKE**

In Exercises 101–104, explain the mistake that is made.

101. Solve the equation $\sqrt{3t + 1} = -4$.

Solution:

$$3t + 1 = 16$$
$$3t = 15$$
$$t = 5$$

This is incorrect. What mistake was made?

102. Solve the equation $x = \sqrt{x + 2}$.

Solution:

$$x^2 = x + 2$$
$$x^2 - x - 2 = 0$$
$$(x - 2)(x + 1) = 0$$
$$x = -1, x = 2$$

This is incorrect. What mistake was made?

103. Solve the equation $x^{2/3} - x^{1/3} - 20 = 0$.

Solution:

$$u = x^{1/3}$$
$$u^2 - u - 20 = 0$$
$$(u - 5)(u + 4) = 0$$
$$x = 5, x = -4$$

This is incorrect. What mistake was made?

104. Solve the equation $x^4 - 2x^2 = 3$.

Solution:

$$x^4 - 2x^2 - 3 = 0$$
$$u = x^2$$
$$u^2 - 2u - 3 = 0$$
$$(u - 3)(u + 1) = 0$$
$$u = -1, u = 3$$
$$u = x^2$$
$$x^2 = -1, x^2 = 3$$
$$x = \pm 1, x = \pm 3$$

This is incorrect. What mistake was made?

■**CONCEPTUAL**

In Exercises 105–108, determine whether each statement is true or false.

105. The equation $(2x - 1)^6 + 4(2x - 1)^3 + 3 = 0$ is quadratic in form.

106. The equation $t^{25} + 2t^5 + 1 = 0$ is quadratic in form.

107. If two solutions are found and one does not check, then the other does not check.

108. Squaring both sides of $\sqrt{x + 2} + \sqrt{x} = \sqrt{x + 5}$ leads to $x + 2 + x = x + 5$.

■**CHALLENGE**

109. Solve $\sqrt{x^2} = x$.

110. Solve $\sqrt{x^2} = -x$.

111. Solve the equation $3x^2 + 2x = \sqrt{3x^2 + 2x}$ *without squaring* both sides.

112. Solve the equation $3x^{7/12} - x^{5/6} - 2x^{1/3} = 0$.

113. Solve the equation $\sqrt{x + 6} + \sqrt{11 + x} = 5\sqrt{3 + x}$.

114. Solve the equation $\sqrt[4]{2x\sqrt[3]{x\sqrt{x}}} = 2$.

■**TECHNOLOGY**

115. Solve the equation $\sqrt{x - 3} = 4 - \sqrt{x + 2}$. Plot both sides of the equation in the same viewing screen, $y_1 = \sqrt{x - 3}$ and $y_2 = 4 - \sqrt{x + 2}$, and zoom in on the x-coordinate of the point of intersection. Does the graph agree with your solution?

116. Solve the equation $2\sqrt{x + 1} = 1 + \sqrt{3 - x}$. Plot both sides of the equation in the same viewing screen, $y_1 = 2\sqrt{x + 1}$ and $y_2 = 1 + \sqrt{3 - x}$, and zoom in on the x-coordinate of the points of intersection. Does the graph agree with your solution?

117. Solve the equation $-4 = \sqrt{x + 3}$. Plot both sides of the equation in the same viewing screen, $y_1 = -4$ and $y_2 = \sqrt{x + 3}$. Does the graph agree or disagree with your solution?

118. Solve the equation $x^{1/4} = -4x^{1/2} + 21$. Plot both sides of the equation in the same viewing screen, $y_1 = x^{1/4}$ and $y_2 = -4x^{1/2} + 21$. Does the point(s) of intersection agree with your solution?

119. Solve the equation $x^{1/2} = -4x^{1/4} + 21$. Plot both sides of the equation in the same viewing screen, $y_1 = x^{1/2}$ and $y_2 = -4x^{1/4} + 21$. Does the point(s) of intersection agree with your solution?

120. Solve the equation $x^{-1} = 3x^{-2} - 10$. Plot both sides of the equation in the same viewing screen, $y_1 = x^{-1}$ and $y_2 = 3x^{-2} - 10$. Does the point(s) of intersection agree with your solution?

121. Solve the equation $x^{-2} = 3x^{-1} - 10$. Plot both sides of the equation in the same viewing screen, $y_1 = x^{-2}$ and $y_2 = 3x^{-1} - 10$. Does the point(s) of intersection agree with your solution?

SKILLS OBJECTIVES

- Use interval notation.
- Solve linear inequalities.
- Solve application problems involving linear inequalities.

CONCEPTUAL OBJECTIVES

- Apply intersection and union concepts.
- Compare and contrast equations and inequalities.
- Understand that linear inequalities may have one solution, no solution, or an interval solution.

Graphing Inequalities and Interval Notation

An example of a linear equation is $3x - 2 = 7$. On the other hand, $3x - 2 \leq 7$ is an example of a **linear inequality**. One difference between a linear equation and a linear inequality is that the equation has at most only one solution, or value of x, that makes the statement true, whereas the inequality can have a range or continuum of numbers that make the statement true. For example, the inequality $x \leq 4$ denotes all real numbers x that are less than or equal to 4. Four inequality symbols are used.

Symbol	In Words
$<$	Less than
$>$	Greater than
$\leq$	Less than or equal to
$\geq$	Greater than or equal to

We call $<$ and $>$ **strict inequalities**. For any two real numbers a and b, one of three things *must* be true:

$$a < b \quad \text{or} \quad a = b \quad \text{or} \quad a > b$$

This property is called the **trichotomy property** of real numbers.

If x is less than 5 ($x < 5$) and x is greater than or equal to -2 ($x \geq -2$), then we can represent this as a **double** (or **combined**) **inequality**, $-2 \leq x < 5$, which means that x is greater than or equal to -2 and less than 5.

We will express solutions to inequalities in four ways: an inequality, a solution set, an interval, and a graph. The following are ways of expressing all real numbers greater than or equal to a and less than b.

Inequality Notation	Solution Set	Interval Notation	Graph/Number Line
$a \leq x < b$	$\{x \mid a \leq x < b\}$	$[a, b)$	or

In this example, a is referred to as the **left endpoint** and b is referred to as the **right endpoint**. If an inequality is a strict inequality ($<$ or $>$), then the graph and interval notation use *parentheses*. If it includes an endpoint ($\geq$ or $\leq$), then the graph and interval notation use *brackets*. Number lines are drawn with either closed/open circles or brackets/parentheses. In this text the brackets/parentheses notation will be used. Intervals are classified as follows:

Open $(,)$ Closed $[,]$ Half open $(,]$ or $[,)$

LET *x* BE A REAL NUMBER. *x* IS...	INEQUALITY	SET NOTATION	INTERVAL	GRAPH
greater than *a* and less than *b*	$a < x < b$	$\{x \mid a < x < b\}$	(a, b)	⟵(a———)b⟶
greater than or equal to *a* and less than *b*	$a \le x < b$	$\{x \mid a \le x < b\}$	$[a, b)$	[a———)b
greater than *a* and less than or equal to *b*	$a < x \le b$	$\{x \mid a < x \le b\}$	$(a, b]$	(a———]b
greater than or equal to *a* and less than or equal to *b*	$a \le x \le b$	$\{x \mid a \le x \le b\}$	$[a, b]$	[a———]b
less than *a*	$x < a$	$\{x \mid x < a\}$	$(-\infty, a)$	⟵———)a
less than or equal to *a*	$x \le a$	$\{x \mid x \le a\}$	$(-\infty, a]$	⟵———]a
greater than *b*	$x > b$	$\{x \mid x > b\}$	(b, ∞)	(b——⟶
greater than or equal to *b*	$x \ge b$	$\{x \mid x \ge b\}$	$[b, \infty)$	[b——⟶
all real numbers	$\mathbb{R}$	$\mathbb{R}$	$(-\infty, \infty)$	⟵———⟶

1. *Infinity* (∞) is not a number. It is a symbol that means continuing indefinitely to the right on the number line. Similarly, *negative infinity* ($-\infty$) means continuing indefinitely to the left on the number line. Since both are unbounded, we use a parenthesis, never a bracket.

2. In interval notation, the lower number is always written to the left.

   Write the inequality in interval notation: $-1 \le x < 3$.

⭐ **CORRECT** $[-1, 3)$ ❌ **INCORRECT** $(3, -1]$

EXAMPLE 1   **Expressing Inequalities Using Interval Notation and a Graph**

Express the following as an inequality, an interval, and a graph.

a. *x* is greater than -3.
b. *x* is less than or equal to 5.
c. *x* is greater than or equal to -1 and less than 4.
d. *x* is greater than or equal to 0 and less than or equal to 4.

Solution:

Inequality	Interval	Graph
a. $x > -3$	$(-3, \infty)$	
b. $x \le 5$	$(-\infty, 5]$	
c. $-1 \le x < 4$	$[-1, 4)$	
d. $0 \le x \le 4$	$[0, 4]$	

Since the solutions to inequalities are sets of real numbers, it is useful to discuss two operations on sets called **intersection** and **union**.

DEFINITION   **Union and Intersection**

The **union** of sets A and B, denoted $A \cup B$, is the set formed by combining all the elements in A with all the elements in B.

$$A \cup B = \{x \,|\, x \text{ is in } A \textbf{ or } B \textbf{ or } \text{both}\}$$

The **intersection** of sets A and B, denoted $A \cap B$, is the set formed by the elements that are in both A and B.

$$A \cap B = \{x \,|\, x \text{ is in } A \textbf{ and } B\}$$

The notation "$x \,|\, x$ is in" is read "all x such that x is in." The vertical line represents "such that."

As an example of intersection and union, consider the following sets of people:

$$A = \{\text{Austin, Brittany, Jonathan}\} \qquad B = \{\text{Anthony, Brittany, Elise}\}$$

Intersection: $A \cap B = \{\text{Brittany}\}$

Union: $A \cup B = \{\text{Anthony, Austin, Brittany, Elise, Jonathan}\}$

EXAMPLE 2   **Determining Unions and Intersections: Intervals and Graphs**

If $A = [-3, 2]$ and $B = (1, 7)$, determine $A \cup B$ and $A \cap B$. Write these sets in interval notation, and graph.

Solution:

Set	Interval notation	Graph
A	$[-3, 2]$	
B	$(1, 7)$	
$A \cup B$	$[-3, 7)$	
$A \cap B$	$(1, 2]$	

■ Answer:

$$C \cup D = [-3, 5]$$

$$C \cap D = (0, 3)$$

■ YOUR TURN   If $C = [-3, 3)$ and $D = (0, 5]$, find $C \cup D$ and $C \cap D$. Express the intersection and union in interval notation, and graph.

Solving Linear Inequalities

As mentioned at the beginning of this section, if we were to solve the equation $3x - 2 = 7$, we would add 2 to both sides, divide by 3, and find that $x = 3$ is the solution, the *only* value that makes the equation true. If we were to solve the linear inequality $3x - 2 \leq 7$, we would follow the same procedure: add 2 to both sides, divide by 3, and find that $x \leq 3$, which is an *interval* or *range* of numbers that make the inequality true.

In solving linear inequalities we follow the same procedures that we used in solving linear equations with one general exception: *if you multiply or divide an inequality by a negative number, then you must change the direction of the inequality sign.* For example, if $-2x < -10$, then the solution set includes real numbers such as $x = 6$ and $x = 7$. Note that real numbers such as $x = -6$ and $x = -7$ are not included in the solution set. Therefore, when this inequality is divided by -2, the inequality sign must also be reversed: $x > 5$. If $a < b$, then $ac < bc$ if $c > 0$ and $ac > bc$ if $c < 0$.

The most common mistake that occurs when solving an inequality is forgetting to change the direction of, or reverse, the inequality symbol when the inequality is multiplied or divided by a negative number.

INEQUALITY PROPERTIES

Procedures That Do Not Change the Inequality Sign

1. Simplifying by eliminating parentheses and collecting like terms.
$$3(x - 6) < 6x - x$$
$$3x - 18 < 5x$$

2. Adding or subtracting the same quantity on both sides.
$$7x + 8 \geq 29$$
$$7x \geq 21$$

3. Multiplying or dividing by the same *positive* real number.
$$5x \leq 15$$
$$x \leq 3$$

Procedures That Change (Reverse) the Inequality Sign

1. Interchanging the two sides of the inequality.
$$x \leq 4 \text{ is equivalent to } 4 \geq x$$

2. Multiplying or dividing by the same *negative* real number.
$$-5x \leq 15 \text{ is equivalent to } x \geq -3$$

Technology Tip

Use a graphing utility to display graphs of $y_1 = 5 - 3x$ and $y_2 = 23$.

Ploti Plot2 Plot3
\Y₁◼5-3X
\Y₂◼23
\Y₃=◼

The solutions are the *x*-values such that the graph of $y_1 = 5 - 3x$ is below that of $y_2 = 23$.

Intersection
X=-6   Y=23

 EXAMPLE 3   Solving a Linear Inequality

Solve and graph the inequality $5 - 3x < 23$.

Solution:

Write the original inequality.
$$5 - 3x < 23$$

Subtract 5 from both sides.
$$-3x < 18$$

Divide both sides by -3 and reverse the inequality sign.
$$\frac{-3x}{-3} > \frac{18}{-3}$$

Simplify.
$$x > -6$$

Solution set: $\boxed{\{x \mid x > -6\}}$ Interval notation: $\boxed{(-6, \infty)}$ Graph:

$$\xleftarrow{\quad\quad} \underset{-6}{\bullet} \longrightarrow$$

> **Classroom Example 1.5.3**
> Solve the linear inequality
> $5 - 3x \geq 17$.
>
> **Answer:** $x \leq -4$ or $(-\infty, -4]$

■ **YOUR TURN** Solve the inequality $5 \leq 3 - 2x$. Express the solution in set and interval notation, and graph.

■ **Answer:** Solution set: $\{x \mid x \leq -1\}$
Interval notation: $(-\infty, -1]$
Graph: $\xleftarrow{\quad} \underset{-1}{\quad} \longrightarrow$

▼ CAUTION

Cross multiplication should not be used in solving inequalities.

EXAMPLE 4   Solving Linear Inequalities with Fractions

Solve the inequality $\dfrac{5x}{3} \le \dfrac{4+3x}{2}$.

COMMON MISTAKE

A common mistake is using cross multiplication to solve inequalities. The reason cross multiplication should not be used is because the expression by which you are multiplying might be negative for some values of x, and that would require the direction of the inequality sign to be reversed.

★ CORRECT

Eliminate the fractions by multiplying by the LCD, 6.

$$6\left(\frac{5x}{3}\right) \le 6\left(\frac{4+3x}{2}\right)$$

Simplify.

$$10x \le 3(4+3x)$$

Eliminate the parentheses.

$$10x \le 12 + 9x$$

Subtract $9x$ from both sides.

$$\boxed{x \le 12}$$

✗ INCORRECT

Cross multiply. $3(4+3x) \le 2(5x)$

The error is in cross multiplying.

Although it is not possible to "check" inequalities since the solutions are often intervals, it is possible to confirm that some points that lie in your solution do satisfy the inequality. It is important to remember that cross multiplication cannot be used in solving inequalities.

Technology Tip

Use a graphing utility to display graphs of $y_1 = -2$, $y_2 = 3x + 4$, and $y_3 = 16$.

The solutions are the x-values such that the graph of $y_2 = 3x + 4$ is between the graphs of $y_1 = -2$ and $y_3 = 16$ and overlaps that of $y_3 = 16$.

EXAMPLE 5   Solving a Double Linear Inequality

Solve the inequality $-2 < 3x + 4 \le 16$.

Solution:

This double inequality can be written as two inequalities. $-2 < \overbrace{3x + 4 \le 16}$

Both inequalities must be satisfied. $-2 < 3x + 4$ and $3x + 4 \le 16$

Subtract 4 from both sides of each inequality. $-6 < 3x$ and $3x \le 12$

Divide each inequality by 3. $-2 < x$ and $x \le 4$

Combining these two inequalities gives us $-2 < x \le 4$ in inequality notation; in interval notation we have $(-2, \infty) \cap (-\infty, 4]$ or $(-2, 4]$.

Notice that the steps we took in solving these inequalities individually were identical. This leads us to a **shortcut method** in which we solve them together:

Write the combined inequality. $-2 < 3x + 4 \le 16$

Subtract 4 from each part. $-6 < 3x \le 12$

Divide each part by 3. $-2 < x \le 4$

Interval notation: $\boxed{(-2, 4]}$

For the remainder of this section we will use the shortcut method for solving inequalities.

EXAMPLE 6   Solving a Double Linear Inequality

Solve the inequality $1 \le \dfrac{-2 - 3x}{7} < 4$. Express the solution set in interval notation, and graph.

Solution:

Write the original double inequality.

$$1 \le \frac{-2 - 3x}{7} < 4$$

Multiply each part by 7.

$$7 \le -2 - 3x < 28$$

Add 2 to each part.

$$9 \le -3x < 30$$

Divide each part by -3 and reverse the inequality signs.

$$-3 \ge x > -10$$

Write in standard form.

$$-10 < x \le -3$$

Interval notation: $\boxed{(-10, -3]}$ Graph:

$$\xleftarrow{\hspace{2cm}}\overset{(]}{\underset{\underset{-10 \quad -3}{}}{\rule{3cm}{0.4pt}}}\xrightarrow{\hspace{2cm}}$$

EXAMPLE 7   Solving a Double Linear Inequality

Solve the inequality $x - 1 \le 4x - 4 \le x + 8$. Express the solution in interval notation.

Solution:

Subtract x from all three parts.

$$-1 \le 3x - 4 \le 8$$

Add 4 to all three parts.

$$3 \le 3x \le 12$$

Divide all three parts by 3.

$$1 \le x \le 4$$

Express the solution in interval notation.

$$[1, 4]$$

▪ **YOUR TURN**   Solve the inequality $2x + 1 < 4x + 2 < 2x + 5$. Express the solution in interval notation.

Applications Involving Linear Inequalities

EXAMPLE 8   Temperature Ranges

New York City on average has a yearly temperature range of 23 degrees Fahrenheit to 86 degrees Fahrenheit. What is the range in degrees Celsius given that the conversion relation is $F = 32 + \frac{9}{5}C$?

Solution:

The temperature ranges from 23°F to 86°F.

$$23 \le F \le 86$$

Replace F using the Celsius conversion.

$$23 \le 32 + \frac{9}{5}C \le 86$$

Subtract 32 from all three parts.

$$-9 \le \frac{9}{5}C \le 54$$

Multiply all three parts by $\frac{5}{9}$.

$$-5 \le C \le 30$$

New York City has an average yearly temperature range of $\boxed{-5°\text{C to } 30°\text{C}}$.

Classroom Example 1.5.6

Classroom Example 1.5.6
Solve the linear inequality
$$-4 < \frac{x + 7}{3} < 20.$$
Answer:
$-19 < x < 53$ or $(-19, 53)$

▪ **Answer:** $\left(-\frac{1}{2}, \frac{3}{2}\right)$

Technology Tip

Use a graphing utility to display graphs of $y_1 = 23$, $y_2 = 32 + \frac{9}{5}x$, and $y_3 = 86$.

The solutions are the x-values such that the graph of $y_2 = 32 + \frac{9}{5}x$ is between the graphs of $y_1 = 23$ and $y_3 = 86$ and overlaps both graphs.

EXAMPLE 9 Comparative Shopping

Two car rental companies have advertised weekly specials on full-size cars. Hertz is advertising an $80 rental fee plus an additional $0.10 per mile. Thrifty is advertising $60 and $0.20 per mile. How many miles must you drive for the rental car from Hertz to be the better deal?

Solution:

Let x = number of miles driven during the week.

Write the cost for the Hertz rental.	$80 + 0.1x$
Write the cost for the Thrifty rental.	$60 + 0.2x$
Write the inequality if Hertz is less than Thrifty.	$80 + 0.1x < 60 + 0.2x$
Subtract $0.1x$ from both sides.	$80 < 60 + 0.1x$
Subtract 60 from both sides.	$20 < 0.1x$
Divide both sides by 0.1.	$200 < x$

You must drive more than 200 miles for Hertz to be the better deal.

> **Classroom Example 1.5.9**
> Executive Car Service charges a $25 pick-up fee with a charge of $0.55 for each mile driven to the drop-off location. Star Car Service charges a $20 pick-up fee with a $0.60 for each mile to the drop-off location. How many miles must a customer travel for Executive to be the better deal?
>
> **Answer:** more than 100 miles

SECTION 1.5 SUMMARY

The solutions of linear inequalities are solution sets that can be expressed four ways:

1. Inequality notation $a < x \leq b$
2. Set notation $\{x \mid a < x \leq b\}$
3. Interval notation $(a, b]$
4. Graph (number line)

Linear inequalities are solved using the same procedures as linear equations with one exception: **When you multiply or divide by a negative number, you must reverse the inequality sign.**

Note: Cross multiplication cannot be used with inequalities.

SECTION 1.5 EXERCISES

■ SKILLS

In Exercises 1–16, rewrite in interval notation and graph.

1. $x \geq 3$
2. $x < -2$
3. $x \leq -5$
4. $x > -7$
5. $-2 \leq x < 3$
6. $-4 \leq x \leq -1$
7. $-3 < x \leq 5$
8. $0 < x < 6$
9. $0 \leq x \leq 0$
10. $-7 \leq x \leq -7$
11. $x \leq 6$ and $x \geq 4$
12. $x > -3$ and $x \leq 2$
13. $x \leq -6$ and $x \geq -8$
14. $x < 8$ and $x < 2$
15. $x > 4$ and $x \leq -2$
16. $x \geq -5$ and $x < -6$

In Exercises 17–24, rewrite in set notation.

17. $[0, 2)$
18. $(0, 3]$
19. $(-7, -2)$
20. $[-3, 2]$
21. $(-\infty, 6]$
22. $(5, \infty)$
23. $(-\infty, \infty)$
24. $[4, 4]$

In Exercises 25–32, write in inequality and interval notation.

25.

26.

27.

28.

29.

30.

31.

32.

In Exercises 33–50, graph the indicated set and write as a single interval, if possible.

33. $(-5, 2] \cup (-1, 3)$

34. $(2, 7) \cup [-5, 3)$

35. $[-6, 4) \cup [-2, 5)$

36. $[-3, 1) \cup [-6, 0)$

37. $(-\infty, 1] \cap [-1, \infty)$

38. $(-\infty, -5) \cap (-\infty, 7]$

39. $(-\infty, 4) \cap [1, \infty)$

40. $(-3, \infty) \cap [-5, \infty)$

41. $[-5, 2) \cap [-1, 3]$

42. $[-4, 5) \cap [-2, 7)$

43. $(-\infty, 4) \cup (4, \infty)$

44. $(-\infty, -3] \cup [-3, \infty)$

45. $(-\infty, -3] \cup [3, \infty)$

46. $(-2, 2) \cap [-3, 1]$

47. $(-\infty, \infty) \cap (-3, 2]$

48. $(-\infty, \infty) \cup (-4, 7)$

49. $(-6, -2) \cap [1, 4)$

50. $(-\infty, -2) \cap (-1, \infty)$

In Exercises 51–58, write in interval notation.

51.

52.

53.

54.

55.

56.

57.

58.

In Exercises 59–90, solve and express the solution in interval notation.

59. $x - 3 < 7$

60. $x + 4 > 9$

61. $3x - 2 \le 4$

62. $3x + 7 \ge -8$

63. $-5p \ge 10$

64. $-4u < 12$

65. $3 - 2x \le 7$

66. $4 - 3x > -17$

67. $-1.8x + 2.5 > 3.4$

68. $2.7x - 1.3 < 6.8$

69. $3(t + 1) > 2t$

70. $2(y + 5) \le 3(y - 4)$

71. $7 - 2(1 - x) > 5 + 3(x - 2)$

72. $4 - 3(2 + x) < 5$

73. $\dfrac{x + 2}{3} - 2 \ge \dfrac{x}{2}$

74. $\dfrac{y - 3}{5} - 2 \le \dfrac{y}{4}$

75. $\dfrac{t - 5}{3} \le -4$

76. $\dfrac{2p + 1}{5} > -3$

77. $\dfrac{2}{3}y - \dfrac{1}{2}(5 - y) < \dfrac{5y}{3} - (2 + y)$

78. $\dfrac{s}{2} - \dfrac{(s - 3)}{3} > \dfrac{s}{4} - \dfrac{1}{12}$

79. $-2 < x + 3 < 5$

80. $1 < x + 6 < 12$

81. $-8 \le 4 + 2x < 8$

82. $0 < 2 + x \le 5$

83. $-3 < 1 - x \le 9$

84. $3 \le -2 - 5x \le 13$

85. $0 < 2 - \frac{1}{3}y < 4$

86. $3 < \frac{1}{2}A - 3 < 7$

87. $\dfrac{1}{2} \le \dfrac{1 + y}{3} \le \dfrac{3}{4}$

88. $-1 < \dfrac{2 - z}{4} \le \dfrac{1}{5}$

89. $-0.7 \le 0.4x + 1.1 \le 1.3$

90. $7.1 > 4.7 - 1.2x > 1.1$

▪ APPLICATIONS

91. Weight. A healthy weight range for a woman is given by the following formula:

- 110 pounds for the first 5 feet (tall)
- 2–6 pounds per inch for every inch above 5 feet

Write an inequality representing a healthy weight, w, for a 5 foot 9 inch woman.

92. Weight. NASA has more stringent weight allowances for its astronauts. Write an inequality representing allowable weight for a female 5 foot 9 inch mission specialist given 105 pounds for the first 5 feet, and 1–5 pounds per inch for every additional inch.

93. Profit. A seamstress decides to open a dress shop. Her fixed costs are $4000 per month, and it costs her $20 to make each dress. If the price of each dress is $100, how many dresses does she have to sell per month to make a profit?

94. Profit. Labrador retrievers that compete in field trials typically cost $2000 at birth. Professional trainers charge $400 to $1000 per month to train the dogs. If the dog is a champion by age 2, it sells for $30,000. What is the range of profit for a champion at age 2?

In Exercises 95 and 96 refer to the following:

The annual revenue for a small company is modeled by

$$R = 5000 + 1.75x$$

where x is hundreds of units sold and R is revenue in thousands of dollars.

95. Business. Find the number of units (to the nearest 100) that must be sold to generate at least $10 million in revenue.

96. Business. Find the number of units (to the nearest 100) that must be sold to generate at least $7.5 million in revenue.

In Exercises 97 and 98 refer to the following:

The Target or Training Heart Rate (THR) is a range of heart rate (measured in beats per minute) that enables a person's heart and lungs to benefit the most from an aerobic workout. THR can be modeled by the formula

$$THR = (HR_{max} - HR_{rest}) \times I + HR_{rest}$$

where HR_{max} is the maximum heart rate that is deemed safe for the individual, HR_{rest} is the resting heart rate, and I is the intensity of the workout that is reported as a percentage.

97. Health. A female with a resting heart rate of 65 beats per minute has a maximum safe heart rate of 170 beats per minute. If her target heart rate is between 100 and 140 beats per minute, what percent intensities of workout can she consider?

98. Health. A male with a resting heart rate of 75 beats per minute has a maximum safe heart rate of 175 beats per minute. If his target heart rate is between 110 and 150 beats per minute, what percent intensities of workout can he consider?

99. Cost: Cell Phones. A cell phone company charges $50 for an 800-minute monthly plan, plus an additional $0.22 per minute for every minute over 800. If a customer's bill ranged from a low of $67.16 to a high of $96.86 over a 6-month period, what were the most minutes used in a single month? What were the least?

100. Cost: Internet. An Internet provider charges $30 per month for 1000 minutes of DSL service plus $0.08 for each additional minute. In a one-year period the customer's bill ranged from $36.40 to $47.20. What were the most and least minutes used?

101. Grades. In your general biology class, your first three test scores are 67, 77, and 84. What is the lowest score you can get on the fourth test to earn at least a B for the course? Assume that each test is of equal weight and the minimum score required to earn a B is an 80.

102. Grades. In your Economics I class there are four tests and a final exam, all of which count equally. Your four test grades are 96, 87, 79, and 89. What grade on your final exam is needed to earn between 80 and 90 for the course?

103. Markups. Typical markup on new cars is 15–30%. If the sticker price is $27,999, write an inequality that gives the range of the invoice price (what the dealer paid the manufacturer for the car).

104. Markups. Repeat Exercise 103 with a sticker price of $42,599.

105. Lasers. A circular laser beam with a radius r_T is transmitted from one tower to another tower. If the received beam radius r_R fluctuates 10% from the transmitted beam radius due to atmospheric turbulence, write an inequality representing the received beam radius.

106. Electronics: Communications. Communication systems are often evaluated based on their signal-to-noise ratio (SNR), which is the ratio of the average power of received signal, S, to average power of noise, N, in the system. If the SNR is required to be at least 2 at all times, write an inequality representing the received signal power if the noise can fluctuate 10%.

107. Real Estate. The Aguileras are listing their house with a real estate agent. They are trying to determine a listing price, L, for the house. Their realtor advises them that most buyers traditionally offer a buying price, B, that is 85–95% of the listing price. Write an inequality that relates the buying price to the listing price.

108. Humidity. The National Oceanic and Atmospheric Administration (NOAA) has stations on buoys in the oceans to measure atmosphere and ocean characteristics such as temperature, humidity, and wind. The humidity sensors have an error of 5%. Write an inequality relating the measured humidity h_m, and the true humidity h_t.

109. Recreation: Golf. Two friends enjoy playing golf. Their favorite course charges $40 for greens fees (to play the course) and a $15 cart rental (per person), so it currently costs each of them $55 every time they play. The membership offered at that course is $160 per month. The membership allows them to play as much as they want (no greens fees), but does still charge a cart rental fee of $10 every time they play. What is the least number of times they should play a month in order for the membership to be the better deal?

110. Recreation: Golf. The same friends in Exercise 109 have a second favorite course. That golf course charges $30 for greens fees (to play the course) and a $10 cart rental (per person), so it currently costs each of them $40 every time they play. The membership offered at that course is $125 per month. The membership allows them to play as much as they want (no greens fees), but does still charge a cart rental fee of $10. What is the least number of times they should play a month in order for the membership to be the better deal?

The following table is the 2007 Federal Tax Rate Schedule for people filing as *single:*

TAX BRACKET #	IF TAXABLE INCOME IS OVER—	BUT NOT OVER—	THE TAX IS:
I	$0	$7,825	10% of the amount over $0
II	$7,825	$31,850	$782.50 plus 15% of the amount over 7,825
III	$31,850	$77,100	$4,386.25 plus 25% of the amount over 31,850
IV	$77,100	$160,850	$15,698.75 plus 28% of the amount over 77,100
V	$160,850	$349,700	$39,148.75 plus 33% of the amount over 160,850
VI	$349,700	No limit	$101,469.25 plus 35% of the amount over 349,700

111. Federal Income Tax. What is the range of federal income taxes a person in tax bracket III will pay the IRS?

112. Federal Income Tax. What is the range of federal income taxes a person in tax bracket IV will pay the IRS?

▪CATCH THE MISTAKE

In Exercises 113–116, explain the mistake that is made.

113. Rewrite in interval notation.

$-1 \leq x < 4$

$(-1, 4]$

This is incorrect. What mistake was made?

114. Graph the indicated set and write as a single interval if possible.

$[-2, 4) \cap (3, 6]$

$[-2, 6]$

This is incorrect. What mistake was made?

115. Solve the inequality $2 - 3p \leq -4$ and express the solution in interval notation.

Solution:
$$2 - 3p \leq -4$$
$$-3p \leq -6$$
$$p \leq 2$$
$$(-\infty, 2]$$

This is incorrect. What mistake was made?

116. Solve the inequality $3 - 2x \leq 7$ and express the solution in interval notation.

Solution:
$$3 - 2x \leq 7$$
$$-2x \leq 4$$
$$x \geq -2$$
$$(-\infty, -2]$$

This is incorrect. What mistake was made?

▪CONCEPTUAL

In Exercises 117 and 118, determine whether each statement is true or false.

117. If $x < a$, then $a > x$.   **118.** If $-x \geq a$, then $x \geq -a$.

In Exercises 119–122, select any of the statements $a - d$ that could be true.

   a. $m > 0$ and $n > 0$ **b.** $m < 0$ and $n < 0$

   c. $m > 0$ and $n < 0$ **d.** $m < 0$ and $n > 0$

119. $mn > 0$ **120.** $mn < 0$

121. $\dfrac{m}{n} > 0$ **122.** $\dfrac{m}{n} < 0$

In Exercises 123 and 124, select any of the statements $a - c$ that could be true.

   a. $n = 0$ **b.** $n > 0$ **c.** $n < 0$

123. $m + n < m - n$ **124.** $m + n \geq m - n$

▪CHALLENGE

125. Solve the inequality $x \leq -x$ mentally (without doing any algebraic manipulation).

126. Solve the inequality $x > -x$ mentally (without doing any algebraic manipulation).

127. Solve the inequality $ax + b < ax - c$, where $0 < b < c$.

128. Solve the inequality $-ax + b < -ax + c$, where $0 < b < c$.

▪TECHNOLOGY

129. a. Solve the inequality $2.7x + 3.1 < 9.4x - 2.5$.
   b. Graph each side of the inequality in the same viewing screen. Find the range of x-values when the graph of the left side lies *below* the graph of the right side.
   c. Do (a) and (b) agree?

130. a. Solve the inequality $-0.5x + 2.7 > 4.1x - 3.6$.
   b. Graph each side of the inequality in the same viewing screen. Find the range of x-values when the graph of the left side lies *above* the graph of the right side.
   c. Do (a) and (b) agree?

131. a. Solve the inequality $x - 3 < 2x - 1 < x + 4$.
   b. Graph all three expressions of the inequality in the same viewing screen. Find the range of x-values when the graph of the middle expression lies above the graph of the left side and below the graph of the right side.
   c. Do (a) and (b) agree?

132. a. Solve the inequality $x - 2 < 3x + 4 \leq 2x + 6$.
   b. Graph all three expressions of the inequality in the same viewing screen. Find the range of x-values when the graph of the middle expression lies above the graph of the left side and on top of and below the graph of the right side.
   c. Do (a) and (b) agree?

133. a. Solve the inequality $x + 3 < x + 5$.
   b. Graph each side of the inequality in the same viewing screen. Find the range of x-values when the graph of the left side lies *below* the graph of the right side.
   c. Do (a) and (b) agree?

134. a. Solve the inequality $\frac{1}{2}x - 3 > -\frac{2}{3}x + 1$.
   b. Graph each side of the inequality in the same viewing screen. Find the range of x-values when the graph of the left side lies *above* the graph of the right side.
   c. Do (a) and (b) agree?

Polynomial Inequalities

In this section we will focus primarily on quadratic inequalities, but the procedures outlined are also valid for higher degree polynomial inequalities. An example of a quadratic inequality is $x^2 + x - 2 < 0$. This statement is true when the value of the polynomial on the left side is negative. For any value of x, a polynomial has either a positive, negative, or zero value. A polynomial must pass through zero before its value changes from positive to negative or from negative to positive. **Zeros** of a polynomial are the values of x that make the polynomial equal to zero. These zeros divide the real number line into **test intervals** where the value of the polynomial is either positive or negative. For example, if we set the above polynomial equal to zero and solve:

$$x^2 + x - 2 = 0$$
$$(x + 2)(x - 1) = 0$$
$$x = -2 \quad \text{or} \quad x = 1$$

we find that $x = -2$ and $x = 1$ are the zeros. These zeros divide the real number line into three test intervals: $(-\infty, -2)$, $(-2, 1)$, and $(1, \infty)$.

Since the polynomial is equal to zero at $x = -2$ and $x = 1$, the value of the polynomial in each of these three intervals is either positive or negative. We select one real number that lies in each of the three intervals and test to see whether the value of the polynomial at each point is either positive or negative. In this example, we select the real numbers: $x = -3$, $x = 0$, and $x = 2$. At this point, there are two ways we can determine whether the value of the polynomial is positive or negative on the interval. One approach is to substitute each of the test points into the polynomial $x^2 + x - 2$.

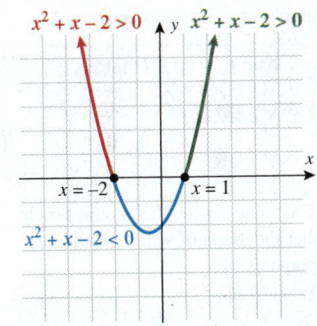

$x = -3$	$(-3)^2 + (-3) - 2 = 9 - 3 - 2 = 4$	Positive
$x = 0$	$(0)^2 + (0) - 2 = 0 - 0 - 2 = -2$	Negative
$x = 2$	$(2)^2 + (2) - 2 = 4 + 2 - 2 = 4$	Positive

The second approach is to simply determine the sign of the result as opposed to actually calculating the exact number. This alternate approach is often used when the expressions or test points get more complicated to evaluate. The polynomial is written as the product $(x + 2)(x - 1)$; therefore, we simply look for the sign in each set of parentheses.

$$(x + 2)(x - 1)$$

$x = -3$: $\quad (-3 + 2)(-3 - 1) = (-1)(-4) \rightarrow (-)(-) = (+)$

$x = 0$: $\quad (0 + 2)(0 - 1) = (2)(-1) \rightarrow (+)(-) = (-)$

$x = 2$: $\quad (2 + 2)(2 - 1) = (4)(1) \rightarrow (+)(+) = (+)$

In this second approach we find the same result: $(-\infty, -2)$ and $(1, \infty)$ correspond to a positive value of the polynomial, and $(-2, 1)$ corresponds to a negative value of the polynomial.

In this example, the statement $x^2 + x - 2 < 0$ is true when the value of the polynomial (in factored form), $(x + 2)(x - 1)$, is negative. In the interval $(-2, 1)$, the value of the polynomial is negative. Thus, the solution to the inequality $x^2 + x - 2 < 0$ is $(-2, 1)$. To check the solution, select any number in the interval and substitute it into the original inequality to make sure it makes the statement true. The value $x = -1$ lies in the interval $(-2, 1)$. Upon substituting into the original inequality, we find that $x = -1$ satisfies the inequality $(-1)^2 + (-1) - 2 = -2 < 0$.

Study Tip

If the original polynomial is < 0, then the interval(s) that yield(s) *negative* products should be selected. If the original polynomial is > 0, then the interval(s) that yield(s) *positive* products should be selected.

PROCEDURE FOR SOLVING POLYNOMIAL INEQUALITIES

Step 1: Write the inequality in *standard form*.
Step 2: Identify zeros.
Step 3: Draw the number line with zeros labeled.
Step 4: Determine the sign of the polynomial in each interval.
Step 5: Identify which interval(s) make the inequality true.
Step 6: Write the solution in interval notation.

Note: Be careful in Step 5. If the original polynomial is < 0, then the interval(s) that correspond(s) to the value of the polynomial being negative should be selected. If the original polynomial is > 0, then the interval(s) that correspond(s) to the value of the polynomial being positive should be selected.

Technology Tip

Use a graphing utility to display graphs of $y_1 = x^2 - x$ and $y_2 = 12$.

The solutions are the *x*-values such that the graph of y_1 lies above the graph of y_2.

■ **Answer:** $[-1, 6]$

▶ **EXAMPLE 1   Solving a Quadratic Inequality**

Solve the inequality $x^2 - x > 12$.

Solution:

STEP 1   Write the inequality in standard form. $x^2 - x - 12 > 0$

Factor the left side. $(x + 3)(x - 4) > 0$

STEP 2   Identify the zeros. $(x + 3)(x - 4) = 0$

$x = -3 \quad \text{or} \quad x = 4$

STEP 3   Draw the number line with the zeros labeled.

$-3 \qquad 4$

STEP 4   Determine the sign of $(x + 3)(x - 4)$ in each interval.

$(-)(-) = (+) \qquad (+)(-) = (-) \qquad (+)(+) = (+)$

$-4 \quad -3 \qquad 0 \qquad 4 \quad 5$

STEP 5   Intervals in which the value of the polynomial is *positive* make this inequality true.

$(-\infty, -3) \quad \text{or} \quad (4, \infty)$

STEP 6   Write the solution in interval notation.

$(-\infty, -3) \cup (4, \infty)$

Classroom Example 1.6.1
Solve the inequality
$6t^2 - 7 \le t$.

Answer: $\left[-1, \frac{7}{6}\right]$

■ **YOUR TURN**   Solve the inequality $x^2 - 5x \le 6$ and express the solution in interval notation.

The inequality in Example 1, $x^2 - x > 12$, is a strict inequality, so we use parentheses when we express the solution in interval notation $(-\infty, -3) \cup (4, \infty)$. It is important to note that if we change the inequality sign from $>$ to $\geq$, then the zeros $x = -3$ and $x = 4$ also make the inequality true. Therefore the solution to $x^2 - x \geq 12$ is $(-\infty, -3] \cup [4, \infty)$.

EXAMPLE 2 Solving a Quadratic Inequality

Solve the inequality $x^2 \leq 4$.

COMMON MISTAKE

Do not take the square root of both sides. You must write the inequality in standard form and factor.

⭐ **CORRECT**

Step 1: Write the inequality in standard form.

$$x^2 - 4 \leq 0$$

Factor.

$$(x - 2)(x + 2) \leq 0$$

Step 2: Identify the zeros.

$$(x - 2)(x + 2) = 0$$

$$x = 2 \text{ and } x = -2$$

Step 3: Draw the number line with the zeros labeled.

Step 4: Determine the sign of $(x - 2)(x + 2)$, in each interval.

Step 5: Intervals in which the value of the polynomial is *negative* make the inequality true.

$$(-2, 2)$$

The endpoints, $x = -2$ and $x = 2$, satisfy the inequality, so they are included in the solution.

Step 6: Write the solution in interval notation.

$$\boxed{[-2, 2]}$$

When solving quadratic inequalities, you must first write the inequality in standard form and then factor to identify zeros.

❌ **INCORRECT**

ERROR:
Take the square root of both sides.

$$x \leq \pm 2$$

▾ **CAUTION**

The square root method cannot be used for quadratic inequalities.

Classroom Example 1.6.2
Solve the inequality
$x^2 - 1 > 15$.

Answer: $(-\infty, -4) \cup (4, \infty)$

Not all inequalities have a solution. For example, $x^2 < 0$ has no real solution. Any real number squared is always nonnegative, so there are no real values that when squared will yield a negative number. The zero is $x = 0$, which divides the real number line into two intervals: $(-\infty, 0)$ and $(0, \infty)$. Both of these intervals, however, correspond to the value of x^2 being positive, so there are no intervals that satisfy the inequality. We say that this inequality has no real solution.

Classroom Example 1.6.3
Solve $(u + 2)^2 > -4$.

Answer: $(-\infty, \infty)$

EXAMPLE 3 **Solving a Quadratic Inequality**

Solve the inequality $x^2 + 2x \geq -3$.

Solution:

STEP 1 Write the inequality in standard form. $\qquad x^2 + 2x + 3 \geq 0$

STEP 2 Identify the zeros. $\qquad\qquad\qquad\qquad x^2 + 2x + 3 = 0$

Apply the quadratic formula. $\qquad\qquad x = \dfrac{-2 \pm \sqrt{2^2 - 4(1)(3)}}{2(1)}$

Simplify. $\qquad x = \dfrac{-2 \pm \sqrt{-8}}{2} = \dfrac{-2 \pm 2i\sqrt{2}}{2} = -1 \pm i\sqrt{2}$

Since there are no real zeros, the quadratic expression $x^2 + 2x + 3$ never equals zero; hence its value is either always positive or always negative. If we select any value for x, say, $x = 0$, we find that $(0)^2 + 2(0) + 3 \geq 0$. Therefore the quadratic expression is always positive, so the solution is the set of all real numbers, $\boxed{(-\infty, \infty)}$.

EXAMPLE 4 **Solving a Quadratic Inequality**

Solve the inequality $x^2 > -5x$.

Classroom Example 1.6.4
Solve the inequality $u^2 < 7u$.

Answer: $(0, 7)$

COMMON MISTAKE

A common mistake is to divide by x. Never divide by a variable, because the value of the variable might be zero. Always start by writing the inequality in standard form and then factor to determine the zeros.

⭐ **CORRECT**

Step 1: Write the inequality in standard form.

$$x^2 + 5x > 0$$

Factor.

$$x(x + 5) > 0$$

Step 2: Identify the zeros.

$$x = 0, x = -5$$

Step 3: Draw the number line with the zeros labeled.

Step 4: Determine the sign of $x(x + 5)$ in each interval.

$(-)(-) = (+)\quad (-)(+) = (-)\quad (+)(+) = (+)$

Step 5: Intervals in which the value of the polynomial is *positive* satisfy the inequality.

$(-\infty, -5)$ and $(0, \infty)$

Step 6: Express the solution in interval notation.

$$\boxed{(-\infty, -5) \cup (0, \infty)}$$

❌ **INCORRECT**

Write the original inequality.

$$x^2 > -5x$$

ERROR:
Divide both sides by x.

$$x > -5$$

Dividing by x is the mistake. If x is negative, the inequality sign must be reversed. What if x is zero?

EXAMPLE 5 **Solving a Quadratic Inequality**

Solve the inequality $x^2 + 2x < 1$.

Solution:

Write the inequality in standard form.

$$x^2 + 2x - 1 < 0$$

Identify the zeros.

$$x^2 + 2x - 1 = 0$$

Apply the quadratic formula.

$$x = \frac{-2 \pm \sqrt{2^2 - 4(1)(-1)}}{2(1)}$$

Simplify.

$$x = \frac{-2 \pm \sqrt{8}}{2} = \frac{-2 \pm 2\sqrt{2}}{2} = -1 \pm \sqrt{2}$$

Draw the number line with the intervals labeled.

Note: $-1 - \sqrt{2} \approx -2.41$
$-1 + \sqrt{2} \approx 0.41$

Test each interval.

$(-\infty, -1 - \sqrt{2})$	$x = -3:$	$(-3)^2 + 2(-3) - 1 = 2 > 0$
$(-1 - \sqrt{2}, -1 + \sqrt{2})$	$x = 0:$	$(0)^2 + 2(0) - 1 = -1 < 0$
$(-1 + \sqrt{2}, \infty)$	$x = 1:$	$(1)^2 + 2(1) - 1 = 2 > 0$

Intervals in which the value of the polynomial is *negative* make this inequality true.

$$\boxed{\left(-1 - \sqrt{2}, -1 + \sqrt{2}\right)}$$

■ **YOUR TURN** Solve the inequality $x^2 - 2x \geq 1$.

Classroom Example 1.6.5*

Solve the inequality $\dfrac{5t^2}{t + 3} \geq 8t$.

Answer: $(-\infty, -8] \cup (-3, 0]$

Technology Tip

Using a graphing utility, graph $y_1 = x^2 + 2x$ and $y_2 = 1$.

Plot1 Plot2 Plot3
\Y₁◻X²+2X
\Y₂◻1

The solutions are the *x*-values such that the graph of y_1 lies below the graph of y_2.

Intersection
X=.41421356 Y=1

Note that:

-1-√(2)
      -2.414213562
-1+√(2)
      .4142135624

■ **Answer:**
$\left(-\infty, 1 - \sqrt{2}\right] \cup \left[1 + \sqrt{2}, \infty\right)$

EXAMPLE 6 **Solving a Polynomial Inequality**

Solve the inequality $x^3 - 3x^2 \geq 10x$.

Solution:

Write the inequality in standard form.

$$x^3 - 3x^2 - 10x \geq 0$$

Factor.

$$x(x - 5)(x + 2) \geq 0$$

Identify the zeros.

$$x = 0, x = 5, x = -2$$

Draw the number line with the zeros (intervals) labeled.

Test each interval.

Intervals in which the value of the polynomial is *positive* make this inequality true.

$$\boxed{[-2, 0] \cup [5, \infty)}$$

Classroom Example 1.6.6*
Solve $5x^2 - 2x < 3x^3$.

Answer: $\left(0, \frac{2}{3}\right) \cup (1, \infty)$

■ **Answer:** $(-\infty, -2) \cup (0, 3)$

■ **YOUR TURN** Solve the inequality $x^3 - x^2 - 6x < 0$.

Rational Inequalities

Rational expressions have numerators and denominators. Recalling the properties of negative real numbers (Chapter 0), we see that the following possible combinations correspond to either positive or negative rational expressions.

$$\frac{(+)}{(+)} = (+) \qquad \frac{(-)}{(+)} = (-) \qquad \frac{(-)}{(-)} = (+) \qquad \frac{(+)}{(-)} = (-)$$

A rational expression can change signs if either the numerator or denominator changes signs. In order to go from positive to negative or vice versa, you must pass through zero. Therefore, to solve rational inequalities such as $\frac{x-3}{x^2-4} \geq 0$ we use a similar procedure to the one used for solving polynomial inequalities, with one exception. You must eliminate values for x that make the denominator equal to zero. In this example, we must eliminate $x = -2$ and $x = 2$ because these values make the denominator equal to zero. Rational inequalities have implied domains. In this example, $x \neq \pm2$ is a domain restriction and these values ($x = -2$ and $x = 2$) must be eliminated from a possible solution.

We will proceed with a similar procedure involving zeros and test intervals that was outlined for polynomial inequalities. However, in rational inequalities once expressions are combined into a single fraction, any values that make *either* the numerator *or* the denominator equal to zero divide the number line into intervals.

Technology Tip

Use a graphing utility to display graph of $y_1 = \frac{x-3}{x^2-4}$.

```
Plot1 Plot2 Plot3
\Y1◻(X-3)/(X²-4)
```

The solutions are the x-values such that the graph of y_1 lies on top and above the x-axis, excluding $x = \pm2$.

Classroom Example 1.6.7
Solve the inequality
$\frac{s+1}{64-s^2} > 0$.
Answer: $(-\infty, -8) \cup (-1, 8)$

■ **Answer:** $[-2, 1]$

EXAMPLE 7 **Solving a Rational Inequality**

Solve the inequality $\frac{x-3}{x^2-4} \geq 0$.

Solution:

Factor the denominator.
$$\frac{(x-3)}{(x-2)(x+2)} \geq 0$$

State the domain restrictions on the variable.
$$x \neq 2, x \neq -2$$

Identify the zeros of numerator and denominator.
$$x = -2, x = 2, x = 3$$

Draw the number line and divide into intervals.

Test the intervals.

$$\frac{(x-3)}{(x-2)(x+2)}$$

$$\frac{(-)}{(-)(-)} = (-) \qquad \frac{(-)}{(-)(+)} = (+) \qquad \frac{(-)}{(+)(+)} = (-) \qquad \frac{(+)}{(+)(+)} = (+)$$

$$-3 \qquad -2 \qquad 0 \qquad 2 \qquad 2.5 \qquad 3 \qquad 4$$

Intervals in which the value of the rational expression is *positive* satisfy this inequality. $(-2, 2)$ and $(3, \infty)$

Since this inequality is greater than or equal to, we include $x = 3$ in our solution because it satisfies the inequality. However, $x = -2$ and $x = 2$ are not included in the solution because they make the denominator equal to zero.

The solution is $\boxed{(-2, 2) \cup [3, \infty)}$.

■ **YOUR TURN** Solve the inequality $\frac{x+2}{x-1} \leq 0$.

EXAMPLE 8 **Solving a Rational Inequality**

Solve the inequality $\dfrac{x-5}{x^2+9} < 0$.

Solution:

Identify the zero(s) of the numerator. $x = 5$

Note that the denominator is never
equal to zero when x is any
real number.

Draw the number line and divide into intervals.

Test the intervals.

The denominator is always positive.

Intervals in which the value of the rational
expression is *negative* satisfy the inequality. $(-\infty, 5)$

Notice that $x = 5$ is not included in the solution because of the strict inequality.

▪ YOUR TURN Solve the inequality $\dfrac{x+4}{x^2+25} \geq 0$.

■ **Answer:** $[-4, \infty)$

EXAMPLE 9 **Solving a Rational Inequality**

Solve the inequality $\dfrac{x}{x+2} \leq 3$.

▾ CAUTION

Rational inequalities should not be
solved using cross multiplication.

COMMON MISTAKE

Do not cross multiply. The LCD or expression by which you are multiplying might be
negative for some values of x, and that would require the direction of the inequality
sign to be reversed.

✪ CORRECT

Subtract 3 from both sides.

$$\frac{x}{x+2} - 3 \leq 0$$

Write as a single rational expression.

$$\frac{x-3(x+2)}{x+2} \leq 0$$

Eliminate the parentheses.

$$\frac{x-3x-6}{x+2} \leq 0$$

Simplify the numerator.

$$\frac{-2x-6}{x+2} \leq 0$$

Factor the numerator.

$$\frac{-2(x+3)}{x+2} \leq 0$$

✘ INCORRECT

ERROR:
Do not cross multiply.

$$x \leq 3(x+2)$$

Identify the zeros of the numerator and the denominator.

$$x = -3 \quad \text{and} \quad x = -2$$

Draw the number line and test the intervals.

$$\frac{-2(x + 3)}{x + 2} \le 0$$

$$\frac{(-)(-)}{(-)} = (-) \quad \frac{(-)(+)}{(-)} = (+) \quad \frac{(-)(+)}{(+)} = (-)$$

$$\xleftarrow{\hspace{1cm}} \underset{-3}{\bullet} \hspace{1cm} \underset{-2}{\circ} \xrightarrow{\hspace{1cm}}$$

Intervals in which the value of the rational expression is *negative* satisfy the inequality, $(-\infty, -3]$ and $(-2, \infty)$. Note that $x = -2$ is not included in the solution because it makes the denominator zero, and $x = -3$ is included because it satisfies the inequality.

The solution is:

$$\boxed{(-\infty, -3] \cup (-2, \infty)}$$

Applications

EXAMPLE 10 Stock Prices

From November 2003 until March 2004 Wal-Mart stock (WMT) was worth approximately $P = 2t^2 - 12t + 70$, where P denotes the price of the stock in dollars and t corresponds to months. November 2003 is represented by $t = 1$, December 2003 by $t = 2$, January 2004 by $t = 3$, and so on. During what months was the stock value at least \$54?

Solution:

Set the price greater than or equal to 54.	$2t^2 - 12t + 70 \ge 54 \qquad 1 \le t \le 5$
Write in standard form.	$2t^2 - 12t + 16 \ge 0$
Divide by 2.	$(t^2 - 6t + 8) \ge 0$
Factor.	$(t - 4)(t - 2) \ge 0$
Identify the zeros.	$t = 4 \quad \text{and} \quad t = 2$
Test the intervals.	$(t - 4)(t - 2) \ge 0$

$$(-)(-) = (+) \quad (-)(+) = (-) \quad (+)(+) = (+)$$

$$\xleftarrow{\hspace{1cm}} \underset{1}{\hspace{0.3cm}} \underset{2}{\bullet} \hspace{0.5cm} \underset{3}{\hspace{0.3cm}} \underset{4}{\bullet} \hspace{0.5cm} \underset{5}{\hspace{0.3cm}} \xrightarrow{\hspace{1cm}}$$

Positive intervals satisfy the inequality. $[1, 2]$ and $[4, 5]$

The Wal-Mart stock price was at least \$54 during November 2003, December 2003, February 2004, and March 2004.

SECTION 1.6 SUMMARY

The following procedure can be used for solving polynomial and rational inequalities.

1. Write in standard form—zero on one side.
2. Determine the zeros; if it is a rational function, note the domain restrictions.

■ Polynomial Inequality
  – Factor if possible.
  – Otherwise, use the quadratic formula.

■ Rational Inequality
  – Write as a single fraction.
  – Determine values that make the numerator or denominator equal to zero. Always exclude values that make the denominator $= 0$.

3. Draw the number line labeling the intervals.
4. Test the intervals to determine whether they are positive or negative.
5. Select the intervals according to the sign of the inequality.
6. Write the solution in interval notation.

SECTION 1.6 EXERCISES

■ SKILLS

In Exercises 1–28, solve the polynomial inequality and express the solution set in interval notation.

1. $x^2 - 3x - 10 \geq 0$
2. $x^2 + 2x - 3 < 0$
3. $u^2 - 5u - 6 \leq 0$
4. $u^2 - 6u - 40 > 0$

5. $p^2 + 4p < -3$
6. $p^2 - 2p \geq 15$
7. $2t^2 - 3 \leq t$
8. $3t^2 \geq -5t + 2$

9. $5v - 1 > 6v^2$
10. $12t^2 < 37t + 10$
11. $2s^2 - 5s \geq 3$
12. $8s + 12 \leq -s^2$

13. $y^2 + 2y \geq 4$
14. $y^2 + 3y \leq 1$
15. $x^2 - 4x < 6$
16. $x^2 - 2x > 5$

17. $u^2 \geq 3u$
18. $u^2 \leq -4u$
19. $-2x \leq -x^2$
20. $-3x \leq x^2$

21. $x^2 > 9$
22. $x^2 \geq 16$
23. $t^2 < 81$
24. $t^2 \leq 49$

25. $z^2 > -16$
26. $z^2 \geq -2$
27. $y^2 < -4$
28. $y^2 \leq -25$

In Exercises 29–58, solve the rational inequality and graph the solution on the real number line.

29. $-\dfrac{3}{x} \leq 0$
30. $\dfrac{3}{x} \leq 0$
31. $\dfrac{y}{y + 3} > 0$
32. $\dfrac{y}{2 - y} \leq 0$

33. $\dfrac{t + 3}{t - 4} \geq 0$
34. $\dfrac{2t - 5}{t - 6} < 0$
35. $\dfrac{s + 1}{4 - s^2} \geq 0$
36. $\dfrac{s + 5}{4 - s^2} \leq 0$

37. $\dfrac{x - 3}{x^2 - 25} \geq 0$
38. $\dfrac{1 - x}{x^2 - 9} \leq 0$
39. $\dfrac{2u^2 + u}{3} < 1$
40. $\dfrac{u^2 - 3u}{3} \geq 6$

41. $\dfrac{3t^2}{t + 2} \geq 5t$
42. $\dfrac{-2t - t^2}{4 - t} \geq t$
43. $\dfrac{3p - 2p^2}{4 - p^2} < \dfrac{3 + p}{2 - p}$
44. $-\dfrac{7p}{p^2 - 100} \leq \dfrac{p + 2}{p + 10}$

45. $\dfrac{x^2}{5 + x^2} < 0$
46. $\dfrac{x^2}{5 + x^2} \leq 0$
47. $\dfrac{x^2 + 10}{x^2 + 16} > 0$
48. $-\dfrac{x^2 + 2}{x^2 + 4} < 0$

49. $\dfrac{v^2 - 9}{v - 3} \geq 0$
50. $\dfrac{v^2 - 1}{v + 1} \leq 0$
51. $\dfrac{2}{t - 3} + \dfrac{1}{t + 3} \geq 0$
52. $\dfrac{1}{t - 2} + \dfrac{1}{t + 2} \leq 0$

53. $\dfrac{3}{x + 4} - \dfrac{1}{x - 2} \leq 0$
54. $\dfrac{2}{x - 5} - \dfrac{1}{x - 1} \geq 0$
55. $\dfrac{1}{p + 4} + \dfrac{1}{p - 4} > \dfrac{p^2 - 48}{p^2 - 16}$
56. $\dfrac{1}{p - 3} - \dfrac{1}{p + 3} \leq 2$

57. $\dfrac{1}{p - 2} - \dfrac{1}{p + 2} \geq \dfrac{3}{p^2 - 4}$
58. $\dfrac{2}{2p - 3} - \dfrac{1}{p + 1} \leq \dfrac{1}{2p^2 - p - 3}$

▪ APPLICATIONS

59. Profit. A Web-based embroidery company makes monogrammed napkins. The profit associated with producing x orders of napkins is governed by the equation

$$P = -x^2 + 130x - 3000$$

Determine the range of orders the company should accept in order to make a profit.

60. Profit. Repeat Exercise 59 using $P = x^2 - 130x + 3600$.

61. Car Value. The term "upside down" on car payments refers to owing more than a car is worth. Assume you buy a new car and finance 100% over 5 years. The difference between the value of the car and what is owed on the car is governed by the expression $\dfrac{t}{t-3}$ where t is age (in years) of the car. Determine the time period when the car is worth more than you owe $\left(\dfrac{t}{t-3} > 0\right)$.

When do you owe more than it's worth $\left(\dfrac{t}{t-3} < 0\right)$?

62. Car Value. Repeat Exercise 61 using the expression $-\dfrac{2-t}{4-t}$.

63. Bullet Speed. A .22-caliber gun fires a bullet at a speed of 1200 feet per second. If a .22-caliber gun is fired straight upward into the sky, the height of the bullet in feet is given by the equation $h = -16t^2 + 1200t$, where t is the time in seconds with $t = 0$ corresponding to the instant the gun is fired. How long is the bullet in the air?

64. Bullet Speed. A .38-caliber gun fires a bullet at a speed of 600 feet per second. If a .38-caliber gun is fired straight upward into the sky, the height of the bullet in feet is given by the equation $h = -16t^2 + 600t$. How many seconds is the bullet in the air?

65. Geometry. A rectangular area is fenced in with 100 feet of fence. If the minimum area enclosed is to be 600 square feet, what is the range of feet allowed for the length of the rectangle?

66. Stock Value. From June 2003 until April 2004, JetBlue airlines stock (JBLU) was approximately worth $P = -4t^2 + 80t - 360$, where P denotes the price of the stock in dollars and t corresponds to months, with $t = 1$ corresponding to January 2003. During what months was the stock value at least $36?

In Exercises 67 and 68 refer to the following:

In response to economic conditions, a local business explores the effect of a price increase on weekly profit. The function

$$P = -5(x + 3)(x - 24)$$

models the effect that a price increase of x dollars on a bottle of wine will have on the profit P measured in dollars.

67. Economics. What price increase will lead to a weekly profit of less than $460?

68. Economics. What price increases will lead to a weekly profit of more than $550?

69. Real Estate. A woman is selling a piece of land that she advertises as 400 acres (±7 acres) for $1.36 million. If you pay that price, what is the range of dollars per acre you have paid? Round to the nearest dollar.

70. Real Estate. A woman is selling a piece of land that she advertises as 1000 acres (±10 acres) for $1 million. If you pay that price, what is the range of dollars per acre you have paid? Round to the nearest dollar.

▪ CATCH THE MISTAKE

In Exercises 71–74, explain the mistake that is made.

71. Solve the inequality $3x < x^2$.

Solution:

Divide by x.	$3 < x$
Write the solution in interval notation.	$(3, \infty)$

This is incorrect. What mistake was made?

72. Solve the inequality $u^2 < 25$.

Solution:

Take the square root of both sides.	$u < -5$
Write the solution in interval notation.	$(-\infty, -5)$

This is incorrect. What mistake was made?

73. Solve the inequality $\dfrac{x^2 - 4}{x + 2} > 0$.

Solution:

Factor the numerator and denominator.

$$\dfrac{(x - 2)(x + 2)}{(x + 2)} < 0$$

Cancel the $(x + 2)$ common factor. $x - 2 > 0$
Solve. $x > 2$

This is incorrect. What mistake was made?

74. Solve the inequality $\dfrac{x + 4}{x} < -\dfrac{1}{3}$.

Solution:

Cross multiply.	$3(x + 4) < -1(x)$
Eliminate the parentheses.	$3x + 12 < -x$
Combine like terms.	$4x < -12$
Divide both sides by 4.	$x < -3$

This is incorrect. What mistake was made?

■ **CONCEPTUAL**

In Exercises 75 and 76, determine whether each statement is true or false. Assume that a is a positive real number.

75. If $x < a^2$, then the solution is $(-\infty, a)$.

76. If $x \geq a^2$, then the solution is $[a, \infty)$.

77. Assume the quadratic inequality $ax^2 + bx + c < 0$ is true. If $b^2 - 4ac < 0$, then describe the solution.

78. Assume the quadratic inequality $ax^2 + bx + c > 0$ is true. If $b^2 - 4ac < 0$, then describe the solution.

■ **CHALLENGE**

In Exercises 79–82, solve for x given that a and b are both positive real numbers.

79. $-x^2 \leq a^2$

80. $\dfrac{x^2 - b^2}{x + b} < 0$

81. $\dfrac{x^2 + a^2}{x^2 + b^2} \geq 0$

82. $\dfrac{a}{x^2} < -b$

■ **TECHNOLOGY**

In Exercises 83–90, plot the left side and the right side of each inequality in the same screen and use the zoom feature to determine the range of values for which the inequality is true.

83. $1.4x^2 - 7.2x + 5.3 > -8.6x + 3.7$

84. $17x^2 + 50x - 19 < 9x^2 + 2$

85. $11x^2 < 8x + 16$

86. $0.1x + 7.3 > 0.3x^2 - 4.1$

87. $x < x^2 - 3x < 6 - 2x$

88. $x^2 + 3x - 5 \geq -x^2 + 2x + 10$

89. $\dfrac{2p}{5 - p} > 1$

90. $\dfrac{3p}{4 - p} < 1$

SKILLS OBJECTIVES

- Solve absolute value equations.
- Solve absolute value inequalities.

CONCEPTUAL OBJECTIVE

- Understand absolute value in terms of distance on the number line.

Equations Involving Absolute Value

The **absolute value** of a real number can be interpreted algebraically and graphically. Algebraically, the absolute value of 5 is 5, or in mathematical notation, $|5| = 5$; and the absolute value of -5 is 5 or $|-5| = 5$. Graphically, the absolute value of a real number is the distance on the real number line between the real number and the origin; thus the distance from 0 to either -5 or 5 is 5.

DEFINITION   **Absolute Value**

The **absolute value** of a real number a, denoted by the symbol $|a|$, is defined by

$$|a| = \begin{cases} a, & \text{if } a \geq 0 \\ -a, & \text{if } a < 0 \end{cases}$$

The absolute value of a real number is never negative. When $a = -5$, this definition says $|-5| = -(-5) = 5$.

PROPERTIES OF ABSOLUTE VALUE

For all real numbers a and b,

1. $|a| \geq 0$ **2.** $|-a| = |a|$ **3.** $|ab| = |a|\,|b|$ **4.** $\left|\dfrac{a}{b}\right| = \dfrac{|a|}{|b|}$ $b \neq 0$

Absolute value can be used to define the distance between two points on the real number line.

Classroom Example 1.7.1
Find the distance between -5 and -27 on the real number line.

Answer: 22

DISTANCE BETWEEN TWO POINTS ON THE REAL NUMBER LINE

If a and b are real numbers, the **distance between a and b** is the absolute value of their difference given by $|a - b|$ or $|b - a|$.

EXAMPLE 1 **Finding the Distance Between Two Points on the Number Line**

Find the distance between -4 and 3 on the real number line.

Solution:

The distance between -4 and 3 is given by the absolute value of the difference.

$$|-4 - 3| = |-7| = 7$$

Note that if we reverse the numbers the result is the same.

$$|3 - (-4)| = |7| = 7$$

We check this by counting the units between -4 and 3 on the number line.

When absolute value is involved in algebraic equations, we interpret the definition of absolute value as follows.

DEFINITION **Absolute Value Equation**

If $|x| = a$, then $x = -a$ or $x = a$, where $a \geq 0$.

In words, "if the absolute value of a number is a, then that number equals $-a$ or a." For example, the equation $|x| = 7$ is true if $x = -7$ or $x = 7$. We say the equation $|x| = 7$ has the solution set $\{-7, 7\}$. *Note:* $|x| = -3$ does not have a solution because there is no value of x such that its absolute value is -3.

 EXAMPLE 2 **Solving an Absolute Value Equation**

Solve the equation $|x - 3| = 8$ algebraically and graphically.

Solution:

Using the absolute value equation definition, we see that if the absolute value of an expression is 8, then that expression is either -8 or 8. Rewrite as two equations:

$$x - 3 = -8 \qquad \text{or} \qquad x - 3 = 8$$
$$x = -5 \qquad\qquad\qquad x = 11$$

The solution set is $\boxed{\{-5, 11\}}$.

Graph: The absolute value equation $|x - 3| = 8$ is interpreted as "what numbers are eight units away from 3 on the number line?" We find that eight units to the right of 3 is 11 and eight units to the left of 3 is -5.

■ **YOUR TURN** Solve the equation $|x + 5| = 7$.

Technology Tip

Use a graphing utility to display graphs of $y_1 = |x - 3|$ and $y_2 = 8$.

The x-coordinates of the points of intersection are the solutions to $|x - 3| = 8$.

■ **Answer:** $x = 2$ or $x = -12$. The solution set is $\{-12, 2\}$.

Study Tip

Rewrite an absolute value equation as two equations.

Classroom Example 1.7.3
Solve $|-1 + 2x| = 1$.

Answer: $x = 1, 0$

■ **Answer:** $x = -3$ or $x = 2$. The solution set is $\{-3, 2\}$.

EXAMPLE 3   Solving an Absolute Value Equation

Solve the equation $|1 - 3x| = 7$.

Solution:

If the absolute value of an expression is 7, then that expression is -7 or 7.

$$1 - 3x = -7 \quad \text{or} \quad 1 - 3x = 7$$
$$-3x = -8 \qquad\qquad -3x = 6$$
$$x = \frac{8}{3} \qquad\qquad\quad x = -2$$

The solution set is $\boxed{\left\{-2, \frac{8}{3}\right\}}$.

■ **YOUR TURN** Solve the equation $|1 + 2x| = 5$.

▶

EXAMPLE 4   Solving an Absolute Value Equation

Solve the equation $2 - 3|x - 1| = -4|x - 1| + 7$.

Solution:

Isolate the absolute value expressions to one side.

Add $4|x - 1|$ to both sides. $2 + |x - 1| = 7$

Subtract 2 from both sides. $|x - 1| = 5$

If the absolute value of an expression $x - 1 = -5 \quad \text{or} \quad x - 1 = 5$
is equal to 5, then the expression is $x = -4 \qquad\qquad x = 6$
equal to either -5 or 5.

The solution set is $\boxed{\{-4, 6\}}$.

Classroom Example 1.7.4
Solve $-1 + 2|2x + 1| = -5|2x + 1| + 6$.

Answer: $x = 0, -1$

■ **Answer:** $x = -4$ or $x = 12$. The solution set is $\{-4, 12\}$.

■ **YOUR TURN** Solve the equation $3 - 2|x - 4| = -3|x - 4| + 11$.

Classroom Example 1.7.5
Solve the equation $|3x + 1| = -5$.

Answer: no solution

EXAMPLE 5   Finding That an Absolute Value Equation Has No Solution

Solve the equation $|1 - 3x| = -7$.

Solution:

The absolute value of an expression is never negative. Therefore no values of x make this equation true.

$$\boxed{\text{No solution}}$$

EXAMPLE 6 **Solving a Quadratic Absolute Value Equation**

Solve the equation $|5 - x^2| = 1$.

Solution:

If the absolute value of an expression is 1, that expression is either -1 or 1, which leads to two equations.

$$5 - x^2 = -1 \quad \text{or} \quad 5 - x^2 = 1$$
$$-x^2 = -6 \qquad\qquad -x^2 = -4$$
$$x^2 = 6 \qquad\qquad x^2 = 4$$
$$x = \pm\sqrt{6} \qquad\qquad x = \pm\sqrt{4} = \pm 2$$

The solution set is $\boxed{\left\{\pm 2, \pm\sqrt{6}\right\}}$.

■ **YOUR TURN** Solve the equation $|7 - x^2| = 2$.

> **Classroom Example 1.7.6***
> Solve the equation
> $|x^2 + 7| = 32$.
>
> **Answer:** $x = \pm 5$ with no
> solution for $x^2 + 7 = -32$.

■ **Answer:** $x = \pm\sqrt{5}$ or $x = \pm 3$.
The solution set is $\left\{\pm\sqrt{5}, \pm 3\right\}$.

Inequalities Involving Absolute Value

To solve the inequality $|x| < 3$, look for all real numbers that make this statement true. Some numbers that make it true are $-2, -\frac{3}{2}, -1, 0, \frac{1}{5}, 1$, and 2. Some numbers that make it false are $-7, -5, -3.5, -3, 3$, and 4. If we interpret this inequality as distance, we ask *what numbers are less than three units from the origin?* We can represent the solution in the following ways.

Inequality notation: $-3 < x < 3$

Interval notation: $(-3, 3)$

Graph:

Similarly, to solve the inequality $|x| \geq 3$, look for all real numbers that make the statement true. If we interpret this inequality as a distance, we ask *what numbers are at least three units from the origin?* We can represent the solution in the following three ways.

Inequality notation: $x \leq -3$ or $x \geq 3$

Interval notation: $(-\infty, -3] \cup [3, \infty)$

Graph:

This discussion leads us to the following equivalence relations.

PROPERTIES OF ABSOLUTE VALUE INEQUALITIES

1. $|x| < a$ is equivalent to $-a < x < a$

2. $|x| \leq a$ is equivalent to $-a \leq x \leq a$

3. $|x| > a$ is equivalent to $x < -a$ or $x > a$

4. $|x| \geq a$ is equivalent to $x \leq -a$ or $x \geq a$

Note: $a > 0$.

It is important to realize that in the above four properties the variable x can be any algebraic expression.

Technology Tip

Use a graphing utility to display graphs of $y_1 = |3x - 2|$ and $y_2 = 7$.

The values of x where the graph of y_1 lies on top and below the graph of y_2 are the solutions to this inequality.

■ **Answer:** Inequality notation: $-6 < x < 5$. Interval notation: $(-6, 5)$.

Classroom Example 1.7.7
Solve the inequality $|3 - x| < 2$.

Answer: $(1, 5)$

Study Tip

Less than inequalities can be written as a single statement.

Greater than inequalities must be written as two statements.

Classroom Example 1.7.8
Solve $|3 - x| \geq 2$.

Answer: $(-\infty, 1] \cup [5, \infty)$

■ **Answer:** Inequality notation: $x \leq 2$ or $x \geq 3$. Interval notation: $(-\infty, 2] \cup [3, \infty)$.

EXAMPLE 7 **Solving an Inequality Involving an Absolute Value**

Solve the inequality $|3x - 2| \leq 7$.

Solution:

We apply property (2) and squeeze the absolute value expression between -7 and 7. $-7 \leq 3x - 2 \leq 7$

Add 2 to all three parts. $-5 \leq 3x \leq 9$

Divide all three parts by 3. $-\dfrac{5}{3} \leq x \leq 3$

The solution in interval notation is $\boxed{\left[-\dfrac{5}{3}, 3\right]}$.

Graph:

■ **YOUR TURN** Solve the inequality $|2x + 1| < 11$.

It is often helpful to note that for absolute value inequalities,

- *less than* inequalities can be written as a single statement (see Example 7).
- *greater than* inequalities must be written as two statements (see Example 8).

EXAMPLE 8 **Solving an Inequality Involving an Absolute Value**

Solve the inequality $|1 - 2x| > 5$.

Solution:

Apply property (3). $1 - 2x < -5$ or $1 - 2x > 5$

Subtract 1 from all expressions. $-2x < -6$ $-2x > 4$

Divide by -2 and reverse the inequality sign. $x > 3$ $x < -2$

Express the solution in interval notation. $\boxed{(-\infty, -2) \cup (3, \infty)}$

Graph:

■ **YOUR TURN** Solve the inequality $|5 - 2x| \geq 1$.

Notice that if we change the problem in Example 8 to $|1 - 2x| > -5$, the answer is all real numbers because the absolute value of any expression is greater than or equal to zero. Similarly, $|1 - 2x| < -5$ would have no solution because the absolute value of an expression can never be negative.

EXAMPLE 9 **Solving an Inequality Involving an Absolute Value**

Solve the inequality $2 - |3x| < 1$.

Solution:

Subtract 2 from both sides.

$$-|3x| < -1$$

Multiply by (-1) and reverse the inequality sign.

$$|3x| > 1$$

Apply property (3).

$$3x < -1 \quad \text{or} \quad 3x > 1$$

Divide both inequalities by 3.

$$x < -\frac{1}{3} \qquad x > \frac{1}{3}$$

Express in interval notation.

$$\left(-\infty, -\frac{1}{3}\right) \cup \left(\frac{1}{3}, \infty\right)$$

Graph.

$$-\frac{1}{3} \quad 0 \quad \frac{1}{3}$$

Classroom Example 1.7.9
Solve the inequality
$$\left|\frac{2x + 5}{3}\right| \geq 7.$$
Answer: $(-\infty, -13] \cup [8, \infty)$

Classroom Example 1.7.9*
Solve $-1 - |2x| \geq -3$.

Answer: $[-1, 1]$

 SECTION
1.7 **SUMMARY**

Absolute value equations and absolute value inequalities are solved by writing the equations or inequalities in terms of two equations or inequalities. *Note: $A > 0$.*

Equations

$$|x| = A \quad \text{is equivalent to} \quad x = -A \text{ or } x = A$$

Inequalities

$$|x| < A \quad \text{is equivalent to} \quad -A < x < A$$

$$|x| > A \quad \text{is equivalent to} \quad x < -A \text{ or } x > A$$

SECTION
1.7 **EXERCISES**

SKILLS

In Exercises 1–38, solve the equation.

1. $|x| = 3$

2. $|x| = 2$

3. $|x| = -4$

4. $|x| = -2$

5. $|t + 3| = 2$

6. $|t - 3| = 2$

7. $|p - 7| = 3$

8. $|p + 7| = 3$

9. $|4 - y| = 1$

10. $|2 - y| = 11$

11. $|3x| = 9$

12. $|5x| = 50$

13. $|2x + 7| = 9$

14. $|2x - 5| = 7$

15. $|3t - 9| = 3$

16. $|4t + 2| = 2$

17. $|7 - 2x| = 9$

18. $|6 - 3y| = 12$

19. $|1 - 3y| = 1$

20. $|5 - x| = 2$

21. $|4.7 - 2.1x| = 3.3$

22. $|5.2x + 3.7| = 2.4$

23. $\left|\frac{2}{3}x - \frac{4}{7}\right| = \frac{5}{3}$

24. $\left|\frac{1}{2}x + \frac{3}{4}\right| = \frac{1}{16}$

25. $|x - 5| + 4 = 12$

26. $|x + 3| - 9 = 2$

27. $3|x - 2| + 1 = 19$

28. $2|1 - x| - 4 = 2$

29. $5 = 7 - |2 - x|$

30. $-1 = 3 - |x - 3|$

31. $2|p + 3| - 15 = 5$

32. $8 - 3|p - 4| = 2$

33. $5|y - 2| - 10 = 4|y - 2| - 3$ **34.** $3 - |y + 9| = 11 - 3|y + 9|$ **35.** $|4 - x^2| = 1$ **36.** $|7 - x^2| = 3$

37. $|x^2 + 1| = 5$ **38.** $|x^2 - 1| = 5$

In Exercises 39–70, solve the inequality and express the solution in interval notation.

39. $|x| < 7$ **40.** $|y| < 9$ **41.** $|y| \geq 5$ **42.** $|x| \geq 2$

43. $|x + 3| < 7$ **44.** $|x + 2| \leq 4$ **45.** $|x - 4| > 2$ **46.** $|x - 1| < 3$

47. $|4 - x| \leq 1$ **48.** $|1 - y| < 3$ **49.** $|2x| > -3$ **50.** $|2x| < -3$

51. $|2t + 3| < 5$ **52.** $|3t - 5| > 1$ **53.** $|7 - 2y| \geq 3$ **54.** $|6 - 5y| \leq 1$

55. $|4 - 3x| \geq 0$ **56.** $|4 - 3x| \geq 1$ **57.** $2|4x| - 9 \geq 3$ **58.** $5|x - 1| + 2 \leq 7$

59. $2|x + 1| - 3 \leq 7$ **60.** $3|x - 1| - 5 > 4$ **61.** $3 - 2|x + 4| < 5$ **62.** $7 - 3|x + 2| \geq -14$

63. $9 - |2x| < 3$ **64.** $4 - |x + 1| > 1$ **65.** $|1 - 2x| < \dfrac{1}{2}$ **66.** $\left|\dfrac{2 - 3x}{5}\right| \geq \dfrac{2}{5}$

67. $|2.6x + 5.4| < 1.8$ **68.** $|3.7 - 5.5x| > 4.3$ **69.** $|x^2 - 1| \leq 8$ **70.** $|x^2 + 4| \geq 29$

In Exercises 71–76, write an inequality that fits the description.

71. Any real numbers less than seven units from 2.

72. Any real numbers more than three units from -2.

73. Any real numbers at least $\frac{1}{2}$ unit from $\frac{3}{2}$.

74. Any real number no more than $\frac{5}{3}$ units from $\frac{11}{3}$.

75. Any real numbers no more than two units from a.

76. Any real number at least a units from -3.

■ **APPLICATIONS**

77. Temperature. If the average temperature in Hawaii is 83°F ($\pm 15°$), write an absolute value inequality representing the temperature in Hawaii.

78. Temperature. If the average temperature of a human is 97.8°F (± 1.2), write an absolute value inequality describing normal human body temperature.

79. Sports. Two women tee off the green of a par-3 hole on a golf course. They are playing "closest to the pin." If the first woman tees off and lands exactly 4 feet from the hole, write an inequality that describes where the second woman must land in order to win the hole. What equation would suggest a tie? Let d = the distance from where the second woman lands to the tee.

80. Electronics. A band-pass filter in electronics allows certain frequencies within a range (or band) to pass through to the receiver and eliminates all other frequencies. Write an absolute value inequality that allows any frequency f within 15 Hertz of the carrier frequency f_c to pass.

In Exercises 81 and 82 refer to the following:

A company is reviewing revenue for the prior sales year. The model for projected revenue and the model for actual revenue are

$$R_{projected} = 200 + 5x$$

$$R_{actual} = 210 + 4.8x$$

where x represents the number of units sold and R represents the revenue in thousands of dollars. Since the two revenue models are not identical, an error in projected revenue occurred. This error is represented by

$$E = |R_{projected} - R_{actual}|$$

81. Business. For what number of units sold was the error in projected revenue less than $5000?

82. Business. For what number of units sold was the error in projected revenue less than $3000?

▪ CATCH THE MISTAKE

In Exercises 83–86, explain the mistake that is made.

83. Solve the absolute value equation $|x - 3| = 7$.

Solution:

Eliminate the absolute value symbols. $x - 3 = 7$

Add 3 to both sides. $x = 10$

Check. $|10 - 3| = 7$

This is incorrect. What mistake was made?

84. Solve the inequality $|x - 3| < 7$.

Solution:

Eliminate the absolute value symbols. $x - 3 < -7$ or $x - 3 > 7$

Add 3 to both sides. $x < -4$ $x > 10$

The solution is $(-\infty, -4) \cup (10, \infty)$.

This is incorrect. What mistake was made?

85. Solve the inequality $|5 - 2x| \le 1$.

Solution:

Eliminate the absolute value symbols. $-1 \le 5 - 2x \le 1$

Subtract 5. $-6 \le -2x \le -4$

Divide by -2. $3 \le x \le 2$

Write the solution in interval notation. $(-\infty, 2] \cup [3, \infty)$

This is incorrect. What mistake was made?

86. Solve the equation $|5 - 2x| = -1$.

Solution: $5 - 2x = -1$ or $5 - 2x = 1$

$-2x = -6$ $-2x = -4$

$x = 3$ $x = 2$

The solution is $\{2, 3\}$.

This is incorrect. What mistake was made?

▪ CONCEPTUAL

In Exercises 87–90, determine whether each statement is true or false.

87. $-|m| \le m \le |m|$

88. $|n^2| = n^2$

89. $|m + n| = |m| + |n|$ is true only when m and n are both nonnegative.

90. For what values of x does the absolute value equation $|x - 7| = x - 7$ hold?

In Exercises 91–96, assuming a and b are real positive numbers, solve the equation or inequality and express the solution in interval notation.

91. $|x - a| < b$

92. $|a - x| > b$

93. $|x| \ge -a$

94. $|x| \le -b$

95. $|x - a| = b$

96. $|x - a| = -b$

▪ CHALLENGE

97. For what values of x does the absolute value equation $|x + 1| = 4 + |x - 2|$ hold?

98. Solve the inequality $|3x^2 - 7x + 2| > 8$.

▪ TECHNOLOGY

99. Graph $y_1 = |x - 7|$ and $y_2 = x - 7$ in the same screen. Do the x-values where these two graphs coincide agree with your result in Exercise 90?

100. Graph $y_1 = |x + 1|$ and $y_2 = |x - 2| + 4$ in the same screen. Do the x-values where these two graphs coincide agree with your result in Exercise 97?

101. Graph $y_1 = |3x^2 - 7x + 2|$ and $y_2 = 8$ in the same screen. Do the x-values where y_1 lies above y_2 agree with your result in Exercise 98?

102. Solve the inequality $|2.7x^2 - 7.9x + 5| \le |5.3x^2 - 9.2|$ by graphing both sides of the inequality and identify which x-values make this statement true.

103. Solve the inequality $\left| \dfrac{x}{x + 1} \right| < 1$ by graphing both sides of the inequality, and identify which x-values make this statement true.

104. Solve the inequality $\left| \dfrac{x}{x + 1} \right| < 2$ by graphing both sides of the inequality, and identify which x-values make this statement true.

Equivalent Equations and Extraneous Solutions

A general strategy for solving all the various types of equations you encountered in this chapter can be summarized as follows: From a given equation, perform algebraic operations on both sides in order to generate equivalent equations. Remember, *equivalent equations* have the same solution set.

1. Consider first a linear equation: $3x - 1 = 5$.

   a. Use a graphing utility to show $y_1 = 3x - 1$ and $y_2 = 5$ and determine the point of intersection. Make a sketch and label it.

   b. How does the graph in part (a) relate to the solution set of the equation $3x - 1 = 5$?

   c. To solve the equation $3x - 1 = 5$ algebraically, the first step is to add 1 to both sides of the equation, as follows:

   $$\begin{aligned} 3x - 1 &= 5 \\ +\ 1 \quad\ &+1 \\ \hline 3x \quad\ &=\ \ 6 \end{aligned}$$

   Use a graphing utility to show $y_1 = 3x$ and $y_2 = 6$, and determine the point of intersection. Make a sketch and label it. How does this graph relate to the equation $3x = 6$?

   d. The final algebraic step to solve the equation is to divide both sides of the equation by 3.

   $$\frac{3x}{3} = \frac{6}{3}$$
   $$x = 2$$

   Use a graphing utility to show $y_1 = x$ and $y_2 = 2$, and determine the point of intersection. Make a sketch and label it. How does this graph relate to the equation $x = 2$?

   e. The algebraic steps to solve the equation $3x - 1 = 5$ produce two equations: $3x = 6$ and $x = 2$. How do the graphs you sketched above represent the fact that these equations are equivalent to the original?

2. Next consider the equation $x - 2 = \sqrt{4 - x}$

   a. Use a graphing utility to show $y_1 = x - 2$ and $y_2 = \sqrt{4 - x}$ and determine any points of intersection. What do you learn about the solution set of the equation $x - 2 = \sqrt{4 - x}$? Make a sketch to explain.

   b. The algebraic steps to solve this equation are as follows:

$$x - 2 = \sqrt{4 - x}$$

$$(x - 2)^2 = (\sqrt{4 - x})^2 \qquad \text{Square both sides.}$$

$$x^2 - 4x + 4 = 4 - x \qquad \text{Simplify.}$$

$$x^2 - 3x = 0 \qquad \text{Write the quadratic equation in standard form.}$$

$$x(x - 3) = 0 \qquad \text{Factor.}$$

$$x = 0 \quad \text{or} \quad x = 3 \qquad \text{Use the zero product property.}$$

Use a graphing utility to show the first step above: $y_1 = (x - 2)^2$ and $y_2 = (\sqrt{4 - x})^2$. What do you learn about the solution set of $(x - 2)^2 = (\sqrt{4 - x})^2$?

   c. Discuss whether $x - 2 = \sqrt{4 - x}$ and $(x - 2)^2 = (\sqrt{4 - x})^2$ are equivalent equations.

   d. The algebraic process of squaring both sides introduced an *extraneous solution*. What do you think that means?

   e. Why is it important to always check the solutions you obtain when solving equations?

3. Consider $x^4 - x^2 = 0$

   a. A fellow student suggests dividing both sides of the equation by x^2. What will be the resulting equation?

   b. Is the equation you wrote in part (a) equivalent to the original equation? How can you use a graphing utility to illustrate this?

   c. Show the algebraic steps you should take to solve $x^4 - x^2 = 0$.

MODELING OUR WORLD

Used in fields of study ranging from engineering to economics to sociology, a mathematical model is a tool that uses mathematical language to describe a system. There are many types of models, which help us not only better understand the world as it is, but by projecting different scenarios based on available data, allow us glimpses into possible futures.

Current changes in the environment, which will affect all of our futures, have brought about a fierce debate. Some scientists believe that human activities have played a large part in bringing about global warming, which impacts not only day-to-day temperatures, but species extinction, loss of glacial ice, and the quality of the air we breathe. Others feel that current changes in the climate are simply part of a natural cycle and are not a cause for concern. The "Modeling Our World" feature at the end of every chapter allows you to use modeling to explore the topic of global climate change and become an informed participant in this debate.

1. Write an absolute value equation that models the increase in the Earth's near-surface air temperature from 1905 to 2005 in degrees Celsius (°C). Let t represent the increase in temperature.

2. Use the temperature scale conversion $F = \frac{9}{5}C + 32$ equation to write an absolute value equation that models the increase in the Earth's near-surface air temperature from 1905 to 2005 in degrees Fahrenheit (°F). Let t represent the increase in temperature. The following chart illustrates different global warming projections for the next 100 years.

Global Temperatures

The Earth's near-surface global air temperature increased approximately 0.74 ± 0.18°C over 1905–2005.*

*Climate Change 2007: The Physical Science Basis. Contribution of Working Group 1 to the Fourth Assessment Report of the Intergovernmental Panel on Climate Change. Intergovernmental Panel on Climate Change (2007-02-05).

Global Warming Projections

3. Let t represent the increase in temperature in degrees Celsius (let $t = 0$ correspond to the year 2000) and write an approximate absolute value inequality such that the NCAR PCM projection is the lowest possible temperature anomaly and the CCSR/NIES projection is the highest possible temperature anomaly.

4. Repeat Problem 3 for temperature in degrees Fahrenheit.

172

SECTION	CONCEPT	KEY IDEAS/FORMULAS
1.1	**Linear equations**	$ax + b = 0$
	Solving linear equations in one variable	Isolate variable on one side and constants on the other side.
	Solving rational equations that are reducible to linear equations	Any values that make the denominator equal to 0 must be eliminated as possible solutions.
1.2	**Applications involving linear equations**	
	Solving application problems using mathematical models	Five-step procedure: Step 1: Identify the question. Step 2: Make notes. Step 3: Set up an equation. Step 4: Solve the equation. Step 5: Check the solution.
	Geometry problems	Formulas for rectangles, triangles, and circles
	Interest problems	Simple interest: $I = Prt$
	Mixture problems	Whenever two *distinct* quantities are mixed, the result is a mixture.
	Distance–rate–time problems	$d = r \cdot t$
1.3	**Quadratic equations**	$ax^2 + bx + c = 0 \qquad a \neq 0$
	Factoring	If $(x - h)(x - k) = 0$, then $x = h$ or $x = k$.
	Square root method	If $x^2 = P$, then $x = \pm\sqrt{P}$.
	Completing the square	Find half of b; square that quantity; add the result to both sides.
	Quadratic Formula	$x = \dfrac{-b \pm \sqrt{b^2 - 4ac}}{2a}$
1.4	**Other types of equations**	
	Radical equations	Check solutions to avoid extraneous solutions.
	Equations quadratic in form: *u*-substitution	Use a *u*-substitution to write the equation in quadratic form.
	Factorable equations	Extract common factor or factor by grouping.
1.5	**Linear inequalities**	Solutions are a range of real numbers.
	Graphing inequalities and interval notation	■ $a < x < b$ is equivalent to (a, b). ■ $x \leq a$ is equivalent to $(-\infty, a]$. ■ $x > a$ is equivalent to (a, ∞).
	Solving linear inequalities	If an inequality is multiplied or divided by a *negative* number, the inequality sign must be reversed.
1.6	**Polynomial and rational inequalities**	
	Polynomial inequalities	Zeros are values that make the polynomial equal to 0.
	Rational inequalities	The number line is divided into intervals. The endpoints of these intervals are values that make either the numerator or denominator equal to 0. Always exclude values that make the denominator $= 0$.
1.7	**Absolute value equations and inequalities**	$|b - a|$ is the distance between points a and b on the number line.
	Equations involving absolute value	If $|x| = a$, then $x = -a$ or $x = a$.
	Inequalities involving absolute value	■ $|x| \leq a$ is equivalent to $-a \leq x \leq a$. ■ $|x| > a$ is equivalent to $x < -a$ or $x > a$.

1.1 Linear Equations

Solve for the variable.

1. $7x - 4 = 12$

2. $13d + 12 = 7d + 6$

3. $20p + 14 = 6 - 5p$

4. $4(x - 7) - 4 = 4$

5. $3(x + 7) - 2 = 4(x - 2)$

6. $7c + 3(c - 5) = 2(c + 3) - 14$

7. $14 - [-3(y - 4) + 9] = [4(2y + 3) - 6] + 4$

8. $[6 - 4x + 2(x - 7)] - 52 = 3(2x - 4) + 6[3(2x - 3) + 6]$

9. $\dfrac{12}{b} - 3 = \dfrac{6}{b} + 4$

10. $\dfrac{g}{3} + g = \dfrac{7}{9}$

11. $\dfrac{13x}{7} - x = \dfrac{x}{4} - \dfrac{3}{14}$

12. $5b + \dfrac{b}{6} = \dfrac{b}{3} - \dfrac{29}{6}$

Specify any values that must be excluded from the solution set and then solve.

13. $\dfrac{1}{x} - 4 = \dfrac{3}{x} - 5$

14. $\dfrac{4}{x + 1} - \dfrac{8}{x - 1} = 3$

15. $\dfrac{2}{t + 4} - \dfrac{7}{t} = \dfrac{6}{t(t + 4)}$

16. $\dfrac{3}{2x - 7} = \dfrac{-2}{3x + 1}$

17. $\dfrac{3}{2x} - \dfrac{6}{x} = 9$

18. $\dfrac{3 - (5/m)}{2 + (5/m)} = 1$

19. $7x - (2 - 4x) = 3[-6 + (4 - 2x + 7)] + 12$

20. $\dfrac{x}{5} - \dfrac{x - 3}{15} = -6$

Solve for the specified variable.

21. Solve for x in terms of y:

$$3x - 2[(y + 4)3 - 7] = y - 2x + 6(x - 3)$$

22. If $y = \dfrac{x + 3}{1 + 2x}$, find $\dfrac{y + 2}{1 - 2y}$ in terms of x.

1.2 Applications Involving Linear Equations

23. Transportation. Maria is on her way from her home near Orlando to the Sundome in Tampa for a rock concert. She drives 16 miles to the Orlando park-n-ride, takes a bus $\frac{3}{4}$ of the way to a bus station in Tampa, and then takes a cab $\frac{1}{12}$ of the way to the Sundome. How far does Maria live from the Sundome?

24. Diet. A particular 2000 calorie per day diet suggests eating breakfast, lunch, dinner, and four snacks. Each snack is $\frac{1}{4}$ the calories of lunch. Lunch has 100 calories less than dinner. Dinner has 1.5 times as many calories as breakfast. How many calories are in each meal and snack?

25. Numbers. Find a number such that 12 more than $\frac{1}{4}$ the number is $\frac{1}{3}$ the number.

26. Numbers. Find four consecutive odd integers such that the sum of the four numbers is equal to three more than three times the fourth integer.

27. Geometry. The length of a rectangle is one more than two times the width, and the perimeter is 20 inches. What are the dimensions of the rectangle?

28. Geometry. Find the perimeter of a triangle if one side is 10 inches, another side is $\frac{1}{3}$ of the perimeter, and the third side is $\frac{1}{6}$ of the perimeter.

29. Investments. You win $25,000 and you decide to invest the money in two different investments: one paying 20% and the other paying 8%. A year later you have $27,600 total. How much did you originally invest in each account?

30. Investments. A college student on summer vacation was able to make $5000 by working a full-time job every summer. He invested half the money in a mutual fund and half the money in a stock that yielded four times as much interest as the mutual fund. After a year he earned $250 in interest. What were the interest rates of the mutual fund and the stock?

31. Chemistry. For an experiment, a student requires 150 milliliters of a solution that is 8% NaCl (sodium chloride). The storeroom has only solutions that are 10% NaCl and 5% NaCl. How many milliliters of each available solution should be mixed to get 150 milliliters of 8% NaCl?

32. Chemistry. A mixture containing 8% salt is to be mixed with 4 ounces of a mixture that is 20% salt, in order to obtain a solution that is 12% salt. How much of the first solution must be used?

33. Grades. Going into the College Algebra final exam, which will count as two tests, Danny has test scores of 95, 82, 90, and 77. If his final exam is higher than his lowest test score, then it will count for the final exam and replace the lowest test score. What score does Danny need on the final in order to have an average score of at least 90?

34. Car Value. A car salesperson reduced the price of a model car by 20%. If the new price is $25,000, what was its original price? How much can be saved by purchasing the model?

1.3 Quadratic Equations

Solve by factoring.

35. $b^2 = 4b + 21$

36. $x(x - 3) = 54$

37. $x^2 = 8x$

38. $6y^2 - 7y - 5 = 0$

Solve by the square root method.

39. $q^2 - 169 = 0$

40. $c^2 + 36 = 0$

41. $(2x - 4)^2 = -64$

42. $(d + 7)^2 - 4 = 0$

Solve by completing the square.

43. $x^2 - 4x - 12 = 0$

44. $2x^2 - 5x - 7 = 0$

45. $\dfrac{x^2}{2} = 4 + \dfrac{x}{2}$

46. $8m = m^2 + 15$

Solve by the Quadratic Formula.

47. $3t^2 - 4t = 7$

48. $4x^2 + 5x + 7 = 0$

49. $8f^2 - \tfrac{1}{3}f = \tfrac{7}{6}$

50. $x^2 = -6x + 6$

Solve by any method.

51. $5q^2 - 3q - 3 = 0$

52. $(x - 7)^2 = -12$

53. $2x^2 - 3x - 5 = 0$

54. $(g - 2)(g + 5) = -7$

55. $7x^2 = -19x + 6$

56. $7 = 2b^2 + 1$

Solve for the indicated variable.

57. $S = \pi r^2 h$ for r

58. $V = \dfrac{\pi r^3 h}{3}$ for r

59. $h = vt - 16t^2$ for v

60. $A = 2\pi r^2 + 2\pi rh$ for h

61. Geometry. Find the base and height of a triangle with an area of 2 square feet if its base is 3 feet longer than its height.

62. Falling Objects. A man is standing on top of a building 500 feet tall. If he drops a penny off the roof, the height of the penny is given by $h = -16t^2 + 500$, where t is in seconds. Determine how many seconds it takes until the penny hits the ground.

1.4 Other Types of Equations

Solve the radical equation for the given variable.

63. $\sqrt[3]{2x - 4} = 2$

64. $\sqrt{x - 2} = -4$

65. $(2x - 7)^{1/5} = 3$

66. $x = \sqrt{7x - 10}$

67. $x - 4 = \sqrt{x^2 + 5x + 6}$

68. $\sqrt{2x - 7} = \sqrt{x + 3}$

69. $\sqrt{x + 3} = 2 - \sqrt{3x + 2}$

70. $4 + \sqrt{x - 3} = \sqrt{x - 5}$

71. $x - 2 = \sqrt{49 - x^2}$

72. $\sqrt{2x - 5} - \sqrt{x + 2} = 3$

73. $-x = \sqrt{3 - x}$

74. $\sqrt{15 + 2\sqrt{x - 4}} + \sqrt{x} = 5$

Solve the equation by introducing a substitution that transforms the equation to quadratic form.

75. $-28 = (3x - 2)^2 - 11(3x - 2)$

76. $x^4 - 6x^2 + 9 = 0$

77. $\left(\dfrac{x}{1 - x}\right)^2 = 15 - 2\left(\dfrac{x}{1 - x}\right)$

78. $3(x - 4)^4 - 11(x - 4)^2 - 20 = 0$

79. $y^{-2} - 5y^{-1} + 4 = 0$

80. $p^{-2} + 4p^{-1} = 12$

81. $3x^{1/3} + 2x^{2/3} = 5$

82. $2x^{2/3} - 3x^{1/3} - 5 = 0$

83. $x^{-2/3} + 3x^{-1/3} + 2 = 0$

84. $y^{-1/2} - 2y^{-1/4} + 1 = 0$

85. $x^4 + 5x^2 = 36$

86. $3 - 4x^{-1/2} + x^{-1} = 0$

Solve the equation by factoring.

87. $x^3 + 4x^2 - 32x = 0$

88. $9t^3 - 25t = 0$

89. $p^3 - 3p^2 - 4p + 12 = 0$

90. $4x^3 - 9x^2 + 4x - 9 = 0$

91. $p(2p - 5)^2 - 3(2p - 5) = 0$

92. $2(t^2 - 9)^3 - 20(t^2 - 9)^2 = 0$

93. $y - 81y^{-1} = 0$

94. $9x^{3/2} - 37x^{1/2} + 4x^{-1/2} = 0$

1.5 Linear Inequalities

Rewrite using interval notation.

95. $x \le -4$

96. $-1 < x \le 7$

97. $2 \le x \le 6$

98. $x > -1$

Rewrite using inequality notation.

99. $(-6, \infty)$

100. $(-\infty, 0]$

101. $[-3, 7]$

102. $(-5, 2]$

Express each interval using inequality and interval notation.

103.

104.

Graph the indicated set and write as a single interval, if possible.

105. $(4, 6] \cup [5, \infty)$

106. $(-\infty, -3) \cup [-7, 2]$

107. $(3, 12] \cap [8, \infty)$

108. $(-\infty, -2) \cap [-2, 9)$

Solve and graph.

109. $2x < 5 - x$

110. $6x + 4 \leq 2$

111. $4(x - 1) > 2x - 7$

112. $\dfrac{x + 3}{3} \geq 6$

113. $6 < 2 + x \leq 11$

114. $-6 \leq 1 - 4(x + 2) \leq 16$

115. $\dfrac{2}{3} \leq \dfrac{1 + x}{6} \leq \dfrac{3}{4}$

116. $\dfrac{x}{3} + \dfrac{x + 4}{9} > \dfrac{x}{6} - \dfrac{1}{3}$

Applications

117. Grades. In your algebra class your first four exam grades are 72, 65, 69, and 70. What is the lowest score you can get on the fifth exam to earn a C for the course? Assume that each exam is equal in weight and a C is any score greater than or equal to 70.

118. Profit. A tailor decided to open a men's custom suit business. His fixed costs are $8500 per month, and it costs him $50 for the materials to make each suit. If the price he charges per suit is $300, how many suits does he have to tailor per month to make a profit?

1.6 Polynomial and Rational Inequalities

Solve the polynomial inequality and express the solution set using interval notation.

119. $x^2 \leq 36$

120. $6x^2 - 7x < 20$

121. $4x \leq x^2$

122. $-x^2 \geq 9x + 14$

123. $-x^2 < -7x$

124. $x^2 < -4$

125. $4x^2 - 12 > 13x$

126. $3x \leq x^2 + 2$

Solve the rational inequality and express the solution set using interval notation.

127. $\dfrac{x}{x - 3} < 0$

128. $\dfrac{x - 1}{x - 4} > 0$

129. $\dfrac{x^2 - 3x}{3} \geq 18$

130. $\dfrac{x^2 - 49}{x - 7} \geq 0$

131. $\dfrac{3}{x - 2} - \dfrac{1}{x - 4} \leq 0$

132. $\dfrac{4}{x - 1} \leq \dfrac{2}{x + 3}$

133. $\dfrac{x^2 + 9}{x - 3} \geq 0$

134. $x < \dfrac{5x + 6}{x}$

1.7 Absolute Value Equations and Inequalities

Solve the equation.

135. $|x - 3| = -4$

136. $|2 + x| = 5$

137. $|3x - 4| = 1.1$

138. $|x^2 - 6| = 3$

Solve the inequality and express the solution using interval notation.

139. $|x| < 4$

140. $|x - 3| < 6$

141. $|x + 4| > 7$

142. $|-7 + y| \leq 4$

143. $|2x| > 6$

144. $\left| \dfrac{4 + 2x}{3} \right| \geq \dfrac{1}{7}$

145. $|2 + 5x| \geq 0$

146. $|1 - 2x| \leq 4$

Applications

147. Temperature. If the average temperature in Phoenix is 85°F ($\pm 10°$) write an inequality representing the average temperature T in Phoenix.

148. Blood Alcohol Level. If a person registers a 0.08 blood alcohol level, he will be issued a DUI ticket in the state of Florida. If the test is accurate within 0.007, write a linear inequality representing an actual blood alcohol level that will not be issued a ticket.

Technology Exercises

Section 1.1

Graph the function represented by each side of the question in the same viewing rectangle, and solve for x.

149. $0.031x + 0.017(4000 - x) = 103.14$

150. $\dfrac{1}{0.16x} - \dfrac{0.2}{x} = \dfrac{1}{4}$

Section 1.3

151. a. Solve the equation $x^2 + 4x = b$, $b = 5$ by first writing in standard form and then factoring. Now plot both sides of the equation in the same viewing screen ($y_1 = x^2 + 4x$ and $y_2 = b$). At what x-values do these two graphs intersect? Do those points agree with the solution set you found?

b. Repeat part (a) for $b = -5, 0, 7$, and 12.

152. a. Solve the equation $x^2 - 4x = b$, $b = 5$ by first writing in standard form and then factoring. Now plot both sides of the equation in the same viewing screen ($y_1 = x^2 - 4x$ and $y_2 = b$). At what x-values do these two graphs intersect? Do those points agree with the solution set you found?

b. Repeat part (a) for $b = -5, 0, 7$, and 12.

Section 1.4

153. Solve the equation $2x^{1/4} = -x^{1/2} + 6$. Round your answer to two decimal places. Plot both sides of the equation in the same viewing screen, $y_1 = 2x^{1/4}$ and $y_2 = -x^{1/2} + 6$. Does the point(s) of intersection agree with your solution?

154. Solve the equation $2x^{-1/2} = x^{-1/4} + 6$. Plot both sides of the equation in the same viewing screen, $y_1 = 2x^{-1/2}$ and $y_2 = x^{-1/4} + 6$. Does the point(s) of intersection agree with your solution?

Section 1.5

155. **a.** Solve the inequality $-0.61x + 7.62 > 0.24x - 5.47$. Express the solution set using interval notation.
  b. Graph each side of the inequality in the same viewing screen. Find the range of x-values when the graph of the left side lies above the graph of the right side.
  c. Do parts (a) and (b) agree?

156. **a.** Solve the inequality $-\frac{1}{2}x + 7 < \frac{3}{4}x - 5$. Express the solution set using interval notation.
  b. Graph each side of the inequality in the same viewing screen. Find the range of x-values when the graph of the left side lies below the graph of the right side.
  c. Do parts (a) and (b) agree?

Section 1.6

Plot the left side and the right side of each inequality in the same screen, and use the zoom feature to determine the range of values for which the inequality is true.

157. $0.2x^2 - 2 > 0.05x + 3.25$

158. $12x^2 - 7x - 10 < 2x^2 + 2x - 1$

159. $\dfrac{3p}{7 - 2p} > 1$

160. $\dfrac{7p}{15 - 2p} < 1$

Section 1.7

161. Solve the inequality $|1.6x^2 - 4.5| < 3.2$ by graphing both sides of the inequality, and identify which x-values make this statement true. Express the solution using interval notation and round to two decimal places.

162. Solve the inequality $|0.8x^2 - 5.4x| > 4.5$ by graphing both sides of the inequality, and identify which x-values make this statement true. Express the solution using interval notation and round to two decimal places.

PRACTICE TEST

Solve the equation.

1. $4p - 7 = 6p - 1$

2. $-2(z - 1) + 3 = -3z + 3(z - 1)$

3. $3t = t^2 - 28$

4. $8x^2 - 13x = 6$

5. $6x^2 - 13x = 8$

6. $\dfrac{3}{x - 1} = \dfrac{5}{x + 2}$

7. $\dfrac{5}{y - 3} + 1 = \dfrac{30}{y^2 - 9}$

8. $x^4 - 5x^2 - 36 = 0$

9. $\sqrt{2x + 1} + x = 7$

10. $2x^{2/3} + 3x^{1/3} - 2 = 0$

11. $\sqrt{3y - 2} = 3 - \sqrt{3y + 1}$

12. $x(3x - 5)^3 - 2(3x - 5)^2 = 0$

13. $x^{7/3} - 8x^{4/3} + 12x^{1/3} = 0$

Solve for the specified variable.

14. $F = \dfrac{9}{5}C + 32$ for C

15. $P = 2L + 2W$ for L

Solve the inequality and express the solution in interval notation.

16. $7 - 5x > -18$

17. $3x + 19 \geq 5(x - 3)$

18. $-1 \leq 3x + 5 < 26$

19. $\dfrac{2}{5} < \dfrac{x + 8}{4} \leq \dfrac{1}{2}$

20. $3x \geq 2x^2$

21. $3p^2 \geq p + 4$

22. $|5 - 2x| > 1$

23. $\dfrac{x - 3}{2x + 1} \leq 0$

24. $\dfrac{x + 4}{x^2 - 9} \geq 0$

25. **Puzzle.** A piling supporting a bridge sits so that $\frac{1}{4}$ of the piling is in the sand, 150 feet is in the water, and $\frac{3}{5}$ of the piling is in the air. What is the total height of the piling?

26. **Real Estate.** As a realtor you earn 7% of the sale price. The owners of a house you have listed at $150,000 will entertain offers within 10% of the list price. Write an inequality that models the commission you could make on this sale.

27. **Costs: Cell Phones.** A cell phone company charges $49 for a 600-minute monthly plan, plus an additional $0.17 per minute for every minute over 600. If a customer's bill ranged from a low of $53.59 to a high of $69.74 over a 6-month period, write an inequality expressing the number of monthly minutes used over the 6-month period.

28. **Television.** Television and film formats are classified as ratios of width to height. Traditional televisions have a 4:3 ratio (1.33:1), and movies are typically made in widescreen format with a 21:9 ratio (2.35:1). If you own a traditional 25-inch television (20 inch × 15 inch screen) and you play a widescreen DVD on it, there will be black bars above and below the image. What are the dimensions of the movie and of the black bars?

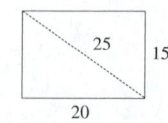

Andy Washnik

29. Solve the equation $\dfrac{1}{0.75x} - \dfrac{0.45}{x} = \dfrac{1}{9}$. Graph the function represented by each side in the same viewing rectangle and solve for x.

30. Solve the inequality $0.3 + |2.4x^2 - 1.5| \leq 6.3$ by graphing both sides of the inequality, and identify which x-values make this statement true. Express the solution using interval notation.

Simplify.

1. $5 \cdot (7 - 3 \cdot 4 + 2)$

Simplify and express in terms of positive exponents.

2. $(4x^{-3}b^4)^{-3}$

3. $\dfrac{(x^2 y^{-2})^3}{(x^2 y)^{-3}}$

Perform the operations and simplify.

4. $(-x^4 + 2x^3) + (x^3 - 5x - 6) - (5x^4 + 4x^3 - 6x + 8)$

5. $x^2 (x + 5)(x - 3)$

Factor completely.

6. $3x^3 - 3x^2 - 60x$

7. $2a^3 + 2000$

Perform the operations and simplify.

8. $\dfrac{3 - x}{x^2 - 1} \div \dfrac{5x - 15}{x + 1}$

9. $\dfrac{6x}{x - 2} - \dfrac{5x}{x + 2}$

Solve for x.

10. $x^3 - x^2 - 30x = 0$

11. $\frac{2}{7}x = \frac{1}{8}x + 9$

12. Perform the operation and express in standard form: $\dfrac{45}{6 - 3i}$.

Solve for x.

13. $\dfrac{6x}{5} - \dfrac{8x}{3} = 4 - \dfrac{7x}{15}$

14. $\dfrac{x - 6}{6 - x} = \dfrac{3}{2}$

15. Tim can paint the interior of a condo in 9 hours. If Chelsea is hired to help him, they can do a similar condo in 5 hours. Working alone, how long will it take Chelsea to paint a similar condo?

16. Solve using the square root method: $y^2 + 36 = 0$.

17. Solve by completing the square: $x^2 + 12x + 40 = 0$.

18. Solve using the Quadratic Formula: $x^2 + x + 9 = 0$.

19. Solve and check: $\sqrt{4 - x} = x - 4$.

20. Solve using substitution: $3x^{-2} + 8x^{-1} + 4 = 0$.

Solve and express the solution in interval notation.

21. $0 < 4 - x \le 7$

22. $4x^2 < 9x - 11$

23. $\dfrac{x + 2}{9 - x^2} \ge 0$

24. $\left| \dfrac{4 - 5x}{7} \right| \ge \dfrac{3}{14}$

25. Solve for x: $\left| \frac{1}{5}x + \frac{2}{3} \right| = \frac{7}{15}$.

26. Solve the equation $x^6 + \frac{37}{8}x^3 = 27$. Plot both sides of the equation in the same viewing screen, $y_1 = x^6 + \frac{37}{8}x^3$ and $y_2 = 27$. Does the point(s) of intersection agree with your solution?

27. Solve the inequality $\left| \dfrac{3x}{x - 2} \right| < 1$ by graphing both sides of the inequality, and identify which x-values make this statement true.

Graphs

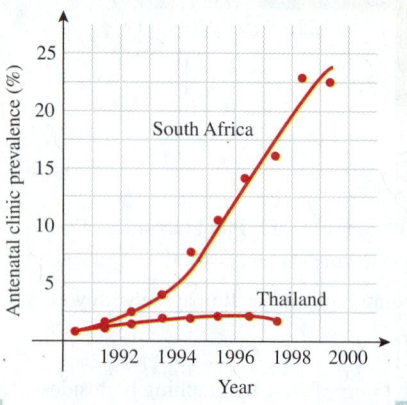

Antenatal clinic prevalence (%)

South Africa

Thailand

1992 1994 1996 1998 2000
Year

HIV infection rates

COKE 5–Minute   4:53 PM

55.25
55.00
54.75
54.50

10 11 12 1 2 3

Stock prices fluctuating throughout the day

°F

200

100

50 100 °C

The conversion between degrees Fahrenheit and degrees Celsius is a linear relationship. Notice that 0°C corresponds to 32°F.

Emperor penguins walking in a line, Weddell Sea, Antarctica

Vetta/Getty Images

G raphs are used in many ways. There is only one temperature that yields the same number in degrees Celsius and degrees Fahrenheit. Do you know what it is?* The penguins are a clue.

*See Section 2.3, Exercise 107.

IN THIS CHAPTER you will review the Cartesian plane. You will calculate the distance between two points and find the midpoint of a line segment joining two points. You will then apply point-plotting techniques to sketch graphs of equations. Special attention is given to two types of equations: lines and circles.

GRAPHS

2.1 Basic Tools: Cartesian Plane, Distance, and Midpoint	**2.2** Graphing Equations: Point-Plotting, Intercepts, and Symmetry	**2.3** Lines	**2.4** Circles	**2.5*** Linear Regression: Best Fit
• Cartesian Plane • Distance Between Two Points • Midpoint of a Line Segment Joining Two Points	• Point-Plotting • Intercepts • Symmetry • Using Intercepts and Symmetry as Graphing Aids	• Graphing a Line • Equations of Lines • Parallel and Perpendicular Lines	• Standard Equation of a Circle • Transforming Equations of Circles to the Standard Form by Completing the Square	• Scatterplots • Identifying Patterns • Linear Regression

LEARNING OBJECTIVES

- Calculate the distance between two points and the midpoint of a line segment joining two points.
- Sketch the graph of an equation using intercepts and symmetry as graphing aids.
- Find the equation of a line.
- Graph circles.
- Find the line of best fit for a given set of data.*

---

*Optional Technology Required Section.

SKILLS OBJECTIVES

- Plot points in the Cartesian plane.
- Calculate the distance between two points.
- Find the midpoint of a line segment joining two points.

CONCEPTUAL OBJECTIVE

- Expand the concept of a one-dimensional number line to a two-dimensional plane.

Cartesian Plane

HIV infection rates, stock prices, and temperature conversions are all examples of relationships between two quantities that can be expressed in a two-dimensional graph. Because it is two dimensional, such a graph lies in a **plane**.

Two perpendicular real number lines, known as the **axes** in the plane, intersect at a point we call the **origin**. Typically, the horizontal axis is called the **x-axis**, and the vertical axis is denoted as the **y-axis**. The axes divide the plane into four **quadrants**, numbered by Roman numerals and ordered counterclockwise.

Points in the plane are represented by **ordered pairs**, denoted (x, y). The first number of the ordered pair indicates the position in the horizontal direction and is often called the **x-coordinate** or **abscissa**. The second number indicates the position in the vertical direction and is often called the **y-coordinate** or **ordinate**. The **origin** is denoted $(0, 0)$.

Examples of other coordinates are given on the graph to the right.

The point $(2, 4)$ lies in quadrant I. To **plot** this point, start at the origin $(0, 0)$ and move to the right two units and up four units.

All points in quadrant I have positive coordinates, and all points in quadrant III have negative coordinates. Quadrant II has negative x-coordinates and positive y-coordinates; quadrant IV has positive x-coordinates and negative y-coordinates.

This representation is called the **rectangular coordinate system** or **Cartesian coordinate system**, named after the French mathematician René Descartes.

EXAMPLE 1 Plotting Points in a Cartesian Plane

a. Plot and label the points $(-1, -4)$, $(2, 2)$, $(-2, 3)$, $(2, -3)$, $(0, 5)$, and $(-3, 0)$ in the Cartesian plane.

b. List the points and corresponding quadrant or axis in a table.

Solution:

a.

b.

POINT	QUADRANT
$(2, 2)$	I
$(-2, 3)$	II
$(-1, -4)$	III
$(2, -3)$	IV
$(0, 5)$	y-axis
$(-3, 0)$	x-axis

Distance Between Two Points

Suppose you want to find the distance between any two points in the plane. In the previous graph, to find the distance between the points $(2, -3)$ and $(2, 2)$, count the units between the two points. The distance is 5. What if the two points do not lie along a horizontal or vertical line? Example 2 uses the Pythagorean theorem to help find the distance between any two points.

EXAMPLE 2   **Finding the Distance Between Two Points**

Find the distance between the points $(-2, -1)$ and $(1, 3)$.

Solution:

STEP 1   Plot and label the two points in the Cartesian plane and draw a line segment indicating the distance d between the two points.

STEP 2   Form a right triangle by connecting the points to a third point, $(1, -1)$.

STEP 3   Calculate the length of the horizontal segment. $3 = |1 - (-2)|$

Calculate the length of the vertical segment. $4 = |3 - (-1)|$

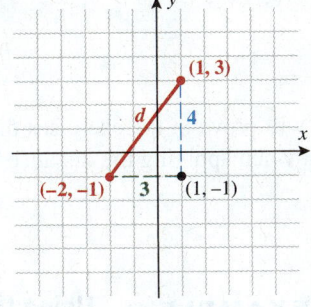

STEP 4   Use the Pythagorean theorem to calculate the distance d.

$$d^2 = 3^2 + 4^2$$
$$d^2 = 25$$
$$\boxed{d = 5}$$

WORDS

For any two points, (x_1, y_1) and (x_2, y_2):

MATH

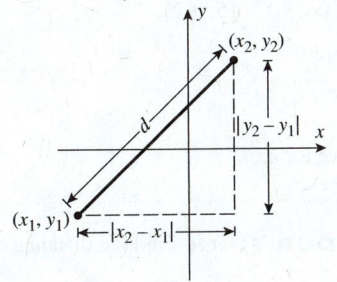

The distance along the horizontal segment is the absolute value of the difference between the x-values.

$|x_2 - x_1|$

The distance along the vertical segment is the absolute value of the difference between the y-values.

$|y_2 - y_1|$

Use the Pythagorean theorem to calculate the distance d.

$d^2 = |x_2 - x_1|^2 + |y_2 - y_1|^2$

$|a|^2 = a^2$ for all real numbers a.

$d^2 = (x_2 - x_1)^2 + (y_2 - y_1)^2$

Use the square root property.

$d = \pm\sqrt{(x_2 - x_1)^2 + (y_2 - y_1)^2}$

Distance can be only positive.

$d = \sqrt{(x_2 - x_1)^2 + (y_2 - y_1)^2}$

Study Tip

It does not matter which point is taken to be the first point or the second point.

DEFINITION   **Distance Formula**

The **distance** d between two points $P_1 = (x_1, y_1)$ and $P_2 = (x_2, y_2)$ is given by

$$d = \sqrt{(x_2 - x_1)^2 + (y_2 - y_1)^2}$$

The distance between two points is the square root of the sum of the square of the distance between the x-coordinates and the square of the distance between the y-coordinates.

You will prove in the exercises that it does not matter which point you take to be the first point when applying the distance formula.

 EXAMPLE 3   Using the Distance Formula to Find the Distance Between Two Points

Classroom Example 2.1.3*

Let a and b be positive real numbers.

a. Find the distance between the points $(a, -7)$ and $(a, -8)$.

b. Find the distance between the points $(a, 2b)$ and $(a, -2b)$.

Answer:

a. 1   **b.** $4b$

■ **Answer:** $d = \sqrt{58}$

Find the distance between $(-3, 7)$ and $(5, -2)$.

Solution:

Write the distance formula.

$d = \sqrt{[x_2 - x_1]^2 + [y_2 - y_1]^2}$

Substitute $(x_1, y_1) = (-3, 7)$ and $(x_2, y_2) = (5, -2)$.

$d = \sqrt{[5 - (-3)]^2 + [-2 - 7]^2}$

Simplify.

$d = \sqrt{[5 + 3]^2 + [-2 - 7]^2}$

$d = \sqrt{8^2 + (-9)^2} = \sqrt{64 + 81} = \sqrt{145}$

Solve for d.

$d = \sqrt{145}$

■ **YOUR TURN** Find the distance between $(4, -5)$ and $(-3, -2)$.

Midpoint of a Line Segment Joining Two Points

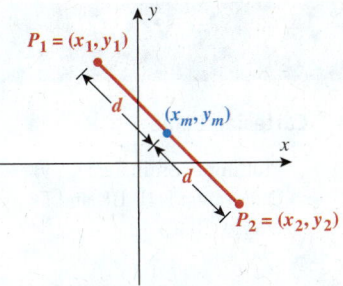

The **midpoint**, (x_m, y_m), of a line segment joining two points (x_1, y_1) and (x_2, y_2) is defined as the point that lies on the segment which has the same distance d from both points. In other words, the midpoint of a segment lies halfway between the given endpoints. The coordinates of the midpoint are found by averaging the x-coordinates and averaging the y-coordinates.

DEFINITION **Midpoint Formula**

The **midpoint**, (x_m, y_m), of the line segment with endpoints (x_1, y_1) and (x_2, y_2) is given by

$$(x_m, y_m) = \left(\frac{x_1 + x_2}{2}, \frac{y_1 + y_2}{2} \right)$$

The midpoint can be found by averaging the x-coordinates and averaging the y-coordinates.

Classroom Example 2.1.4*
Show that every point on the x-axis is equidistant to both $Q_1(a, 2b)$ and $Q_2(a, -2b)$.

Answer: A point P on the x-axis is of the form $(x, 0)$ for some real number x.
$d(P, Q_1) = \sqrt{(x - a)^2 + (0 - 2b)^2}$
$d(P, Q_2) = \sqrt{(x - a)^2 + (0 - (-2b))^2}$
Since $(0 - 2b)^2 = (0 - (-2b))^2$, these distances are the same.

EXAMPLE 4 **Finding the Midpoint of a Line Segment**

Find the midpoint of the line segment joining the points $(2, 6)$ and $(-4, -2)$.

Solution:

Write the midpoint formula.

$$(x_m, y_m) = \left(\frac{x_1 + x_2}{2}, \frac{y_1 + y_2}{2} \right)$$

Substitute $(x_1, y_1) = (2, 6)$ and $(x_2, y_2) = (-4, -2)$.

$$(x_m, y_m) = \left(\frac{2 + (-4)}{2}, \frac{6 + (-2)}{2} \right)$$

Simplify.

$$(x_m, y_m) = (-1, 2)$$

One way to verify your answer is to plot the given points and the midpoint to make sure your answer looks reasonable.

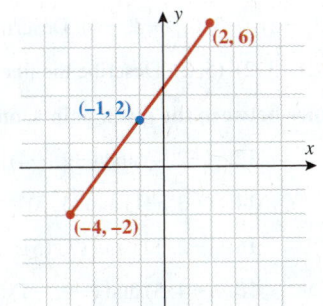

Classroom Example 2.1.4
Find the midpoint of the line segment joining the points $(-10, -17)$ and $(-6, -7)$.

Answer: $(-8, -12)$

Technology Tip

Show a screen display of how to enter $\dfrac{2 + (-4)}{2}$ and $\dfrac{6 + (-2)}{2}$.

Scientific calculators:

Or

```
(2+-4)/2
                -1
(6+-2)/2
                 2
```

Answer: Midpoint = $(4, 2)$

■ YOUR TURN Find the midpoint of the line segment joining the points $(3, -4)$ and $(5, 8)$.

SECTION 2.1 SUMMARY

Cartesian Plane

- Plotting coordinates: (x, y)
- Quadrants: I, II, III, and IV
- Origin: $(0, 0)$

```
              y-axis
   (−, +)  |  (+, +)
     II    |   I
           | Origin
  ---------+--------→ x-axis
     III   |  IV
   (−, −)  |  (+, −)
```

Distance Between Two Points
$$d = \sqrt{(x_2 - x_1)^2 + (y_2 - y_1)^2}$$

Midpoint of Line Segment Joining Two Points
$$\text{Midpoint} = (x_m, y_m) = \left(\frac{x_1 + x_2}{2}, \frac{y_1 + y_2}{2} \right)$$

SECTION 2.1 EXERCISES

■ **SKILLS**

In Exercises 1–6, give the coordinates for each point labeled.

1. Point A
2. Point B
3. Point C
4. Point D
5. Point E
6. Point F

In Exercises 7 and 8, plot each point in the Cartesian plane and indicate in which quadrant or on which axis the point lies.

7. $A: (-2, 3)$ $B: (1, 4)$ $C: (-3, -3)$ $D: (5, -1)$ $E: (0, -2)$ $F: (4, 0)$

8. $A: (-1, 2)$ $B: (1, 3)$ $C: (-4, -1)$ $D: (3, -2)$ $E: (0, 5)$ $F: (-3, 0)$

9. Plot the points $(-3, 1), (-3, 4), (-3, -2), (-3, 0), (-3, -4)$. Describe the line containing points of the form $(-3, y)$.

10. Plot the points $(-1, 2), (-3, 2), (0, 2), (3, 2), (5, 2)$. Describe the line containing points of the form $(x, 2)$.

In Exercises 11–32, calculate the distance between the given points, and find the midpoint of the segment joining them.

11. $(1, 3)$ and $(5, 3)$
12. $(-2, 4)$ and $(-2, -4)$
13. $(-1, 4)$ and $(3, 0)$
14. $(-3, -1)$ and $(1, 3)$
15. $(-10, 8)$ and $(-7, -1)$
16. $(-2, 12)$ and $(7, 15)$
17. $(-3, -1)$ and $(-7, 2)$
18. $(-4, 5)$ and $(-9, -7)$
19. $(-6, -4)$ and $(-2, -8)$
20. $(0, -7)$ and $(-4, -5)$
21. $\left(-\frac{1}{2}, \frac{1}{3}\right)$ and $\left(\frac{7}{2}, \frac{10}{3}\right)$
22. $\left(\frac{1}{5}, \frac{7}{3}\right)$ and $\left(\frac{9}{5}, -\frac{2}{3}\right)$
23. $\left(-\frac{2}{3}, -\frac{1}{5}\right)$ and $\left(\frac{1}{4}, \frac{1}{3}\right)$
24. $\left(\frac{7}{5}, \frac{1}{9}\right)$ and $\left(\frac{1}{2}, -\frac{7}{3}\right)$
25. $(-1.5, 3.2)$ and $(2.1, 4.7)$
26. $(-1.2, -2.5)$ and $(3.7, 4.6)$
27. $(-14.2, 15.1)$ and $(16.3, -17.5)$
28. $(1.1, 2.2)$ and $(3.3, 4.4)$
29. $(\sqrt{3}, 5\sqrt{2})$ and $(\sqrt{3}, \sqrt{2})$
30. $(3\sqrt{5}, -3\sqrt{3})$ and $(-\sqrt{5}, -\sqrt{3})$
31. $(1, \sqrt{3})$ and $(-\sqrt{2}, -2)$
32. $(2\sqrt{5}, 4)$ and $(1, 2\sqrt{3})$

In Exercises 33 and 34, calculate (to two decimal places) the perimeter of the triangle with the following vertices:

33. Points A, B, and C

34. Points C, D, and E

In Exercises 35–38, determine whether the triangle with the given vertices is a right triangle, an isosceles triangle, neither, or both. (Recall that a right triangle satisfies the Pythagorean theorem and an isosceles triangle has at least two sides of equal length.)

35. $(0, -3)$, $(3, -3)$, and $(3, 5)$

36. $(0, 2)$, $(-2, -2)$, and $(2, -2)$

37. $(1, 1)$, $(3, -1)$, and $(-2, -4)$

38. $(-3, 3)$, $(3, 3)$, and $(-3, -3)$

■ APPLICATIONS

39. Cell Phones. A cellular phone company currently has three towers: one in Tampa, one in Orlando, and one in Gainesville to serve the central Florida region. If Orlando is 80 miles east of Tampa and Gainesville is 100 miles north of Tampa, what is the distance from Orlando to Gainesville?

40. Cell Phones. The same cellular phone company in Exercise 39 has decided to add additional towers at each "halfway" between cities. How many miles from Tampa is each "halfway" tower?

41. Travel. A retired couple who live in Columbia, South Carolina, decide to take their motor home and visit two children who live in Atlanta and in Savannah, Georgia. Savannah is 160 miles south of Columbia, and Atlanta is 215 miles west of Columbia. How far apart do the children live from each other?

42. Sports. In the 1984 Orange Bowl, Doug Flutie, the 5 foot 9 inch quarterback for Boston College, shocked the world as he threw a "hail Mary" pass that was caught in the end zone with no time left on the clock, defeating the Miami Hurricanes 47–45. Although the record books have it listed as a 48 yard pass, what was the actual distance the ball was thrown? The following illustration depicts the path of the ball.

43. NASCAR Revenue. Action Performance Inc., the leading seller of NASCAR merchandise, recorded $260 million in revenue in 2002 and $400 million in revenue in 2004. Calculate the midpoint to estimate the revenue Action Performance Inc. recorded in 2003. Assume the horizontal axis represents the year and the vertical axis represents the revenue in millions.

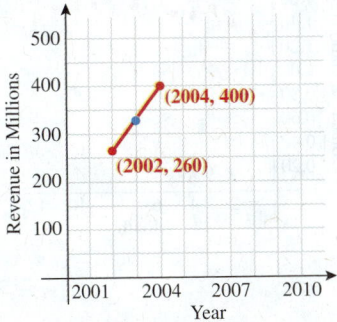

44. **Ticket Price.** In 1993, the average Miami Dolphins ticket price was $28, and in 2001 the average price was $56. Find the midpoint of the segment joining these two points to estimate the ticket price in 1997.

In Exercises 45 and 46, refer to the following:

It is often useful to display data in visual form by plotting the data as a set of points. This provides a graphical display between the two variables. The following table contains data on the average monthly price of gasoline.

U.S. All Grades Conventional Retail Gasoline Prices, 1994–2010 (Dollars per Gallon)

YEAR	JAN	FEB	MAR	APR	MAY	JUN	JUL	AUG	SEP	OCT	NOV	DEC
1994											1.175	1.112
1995	1.107	1.099	1.099	1.143	1.213	1.226	1.189	1.161	1.148	1.122	1.098	1.105
1996	1.123	1.121	1.169	1.259	1.302	1.282	1.254	1.238	1.238	1.243	1.273	1.273
1997	1.270	1.263	1.237	1.228	1.229	1.227	1.206	1.250	1.254	1.222	1.198	1.159
1998	1.115	1.082	1.055	1.064	1.088	1.086	1.078	1.049	1.033	1.045	1.020	0.964
1999	0.957	0.940	1.000	1.137	1.143	1.134	1.177	1.237	1.279	1.271	1.280	1.302
2000	1.319	1.409	1.538	1.476	1.496	1.645	1.568	1.480	1.562	1.546	1.533	1.458
2001	1.467	1.471	1.423	1.557	1.689	1.586	1.381	1.422	1.539	1.312	1.177	1.111
2002	1.134	1.129	1.259	1.402	1.394	1.380	1.402	1.398	1.403	1.466	1.424	1.389
2003	1.464	1.622	1.675	1.557	1.477	1.489	1.519	1.625	1.654	1.551	1.512	1.488
2004	1.595	1.654	1.728	1.794	1.981	1.950	1.902	1.880	1.880	1.993	1.973	1.843
2005	1.852	1.927	2.102	2.251	2.155	2.162	2.287	2.489	2.907	2.736	2.265	2.216
2006	2.343	2.293	2.454	2.762	2.873	2.849	2.964	2.952	2.548	2.258	2.254	2.328
2007	2.237	2.276	2.546	2.831	3.157	3.067	2.989	2.821	2.858	2.838	3.110	3.032
2008	3.068	3.064	3.263	3.468	3.783	4.038	4.051	3.789	3.760	3.065	2.153	1.721
2009	1.821	1.942	1.987	2.071	2.289	2.645	2.530	2.613	2.530	2.549	2.665	2.620
2010	2.730	2.657	2.793	2.867	2.847	2.733	2.728	2.733	2.727	2.816	2.866	3.004

Source: http://www.eia.doe.gov/dnav/pet/hist/LeafHandler.ashx?n=PET&s=EMM_EPM0U_PTE_NUS_DPG&f=M

The following graph displays the data for the year 2000.

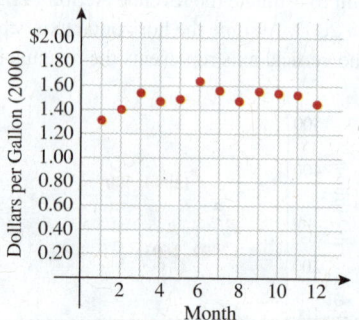

45. **Economics.** Create a graph displaying the price of gasoline for the year 2008.

46. **Economics.** Create a graph displaying the price of gasoline for the year 2009.

▪ CATCH THE MISTAKE

In Exercises 47–50, explain the mistake that is made.

47. Calculate the distance between $(2, 7)$ and $(9, 10)$.

Solution:

Write the distance formula. $d = \sqrt{(x_2 - x_1)^2 + (y_2 - y_1)^2}$

Substitute $(2, 7)$ and
$(9, 10)$. $\qquad\qquad d = \sqrt{(7 - 2)^2 + (10 - 9)^2}$

Simplify. $\qquad\qquad d = \sqrt{(5)^2 + (1)^2} = \sqrt{26}$

This is incorrect. What mistake was made?

48. Calculate the distance between $(-2, 1)$ and $(3, -7)$.

Solution:

Write the distance formula. $d = \sqrt{(x_2 - x_1)^2 + (y_2 - y_1)^2}$

Substitute $(-2, 1)$ and
$(3, -7)$. $\qquad\qquad d = \sqrt{(3 - 2)^2 + (-7 - 1)^2}$

Simplify. $\qquad\qquad d = \sqrt{(1)^2 + (-8)^2} = \sqrt{65}$

This is incorrect. What mistake was made?

49. Compute the midpoint of the segment with endpoints $(-3, 4)$ and $(7, 9)$.

Solution:

Write the midpoint
formula. $\qquad (x_m, y_m) = \left(\dfrac{x_1 + x_2}{2}, \dfrac{y_1 + y_2}{2} \right)$

Substitute $(-3, 4)$
and $(7, 9)$. $\qquad (x_m, y_m) = \left(\dfrac{-3 + 4}{2}, \dfrac{7 + 9}{2} \right)$

Simplify. $\qquad (x_m, y_m) = \left(\dfrac{1}{2}, \dfrac{16}{2} \right) = \left(\dfrac{1}{2}, 4 \right)$

This is incorrect. What mistake was made?

50. Compute the midpoint of the segment with endpoints $(-1, -2)$ and $(-3, -4)$.

Solution:

Write the midpoint
formula. $\qquad (x_m, y_m) = \left(\dfrac{x_1 - x_2}{2}, \dfrac{y_1 - y_2}{2} \right)$

Substitute $(-1, -2)$
and $(-3, -4)$. $\qquad (x_m, y_m) = \left(\dfrac{-1 - (-3)}{2}, \dfrac{-2 - (-4)}{2} \right)$

Simplify. $\qquad (x_m, y_m) = (1, 1)$

This is incorrect. What mistake was made?

▪ CONCEPTUAL

In Exercises 51–54, determine whether each statement is true or false.

51. The distance from the origin to the point (a, b) is

$d = \sqrt{a^2 + b^2}$.

52. The midpoint of the line segment joining the origin and the

point (a, a) is $\left(\dfrac{a}{2}, \dfrac{a}{2} \right)$.

53. The midpoint of any segment joining two points in quadrant I also lies in quadrant I.

54. The midpoint of any segment joining a point in quadrant I to a point in quadrant III also lies in either quadrant I or III.

55. Calculate the length and the midpoint of the line segment joining the points (a, b) and (b, a).

56. Calculate the length and the midpoint of the line segment joining the points (a, b) and $(-a, -b)$.

▪ CHALLENGE

57. Assume that two points (x_1, y_1) and (x_2, y_2) are connected by a segment. Prove that the distance from the midpoint of the segment to either of the two points is the same.

58. Prove that the diagonals of a parallelogram in the figure intersect at their midpoints.

59. Assume that two points (a, b) and (c, d) are the endpoints of a line segment. Calculate the distance between the two points. Prove that it does not matter which point is labeled as the "first" point in the distance formula.

60. Show that the points $(-1, -1)$, $(0, 0)$, and $(2, 2)$ are collinear (lie on the same line) by showing that the sum of the distance from $(-1, -1)$ to $(0, 0)$ and the distance from $(0, 0)$ to $(2, 2)$ is equal to the distance from $(-1, -1)$ to $(2, 2)$.

▪ TECHNOLOGY

In Exercises 61–64, calculate the distance between the two points. Use a graphing utility to graph the segment joining the two points and find the midpoint of the segment.

61. $(-2.3, 4.1)$ and $(3.7, 6.2)$

62. $(-4.9, -3.2)$ and $(5.2, 3.4)$

63. $(1.1, 2.2)$ and $(3.3, 4.4)$

64. $(-1.3, 7.2)$ and $(2.3, -4.5)$

SECTION 2.2 GRAPHING EQUATIONS: POINT-PLOTTING, INTERCEPTS, AND SYMMETRY

SKILLS OBJECTIVES

- ▪ Sketch graphs of equations by plotting points.
- ▪ Find intercepts for graphs of equations.
- ▪ Conduct a test for symmetry about the x-axis, y-axis, and origin.
- ▪ Use intercepts and symmetry as graphing aids.

CONCEPTUAL OBJECTIVE

- ▪ Relate symmetry graphically and algebraically.

In this section, you will learn how to graph equations by plotting points. However, when we discuss graphing principles in Chapter 3, you will see that other techniques can be more efficient.

Point-Plotting

Most equations in two variables, such as $y = x^2$, have an infinite number of ordered pairs as solutions. For example, $(0, 0)$ is a solution to $y = x^2$ because when $x = 0$ and $y = 0$, the equation is true. Two other solutions are $(-1, 1)$ and $(1, 1)$.

The **graph of an equation** in two variables, x and y, consists of all the points in the xy-plane whose coordinates (x, y) satisfy the equation. A procedure for plotting the graphs of equations is outlined below and is illustrated with the example $y = x^2$.

WORDS

Step 1: In a table, list several pairs of coordinates that make the equation true.

MATH

x	$y = x^2$	(x, y)
0	0	$(0, 0)$
-1	1	$(-1, 1)$
1	1	$(1, 1)$
-2	4	$(-2, 4)$
2	4	$(2, 4)$

Step 2: Plot these points on a graph and connect the points with a smooth curve. Use arrows to indicate that the graph continues.

In graphing an equation, first select arbitrary values for x and then use the equation to find the corresponding value of y, or vice versa.

EXAMPLE 1   Graphing an Equation of a Line by Plotting Points

Graph the equation $y = 2x - 1$.

Solution:

STEP 1   In a table, list several pairs of coordinates that make the equation true.

x	$y = 2x - 1$	(x, y)
0	-1	$(0, -1)$
-1	-3	$(-1, -3)$
1	1	$(1, 1)$
-2	-5	$(-2, -5)$
2	3	$(2, 3)$

STEP 2   Plot these points on a graph and connect the points, resulting in a line.

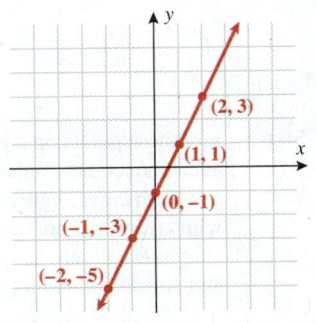

Classroom Example 2.2.1

Graph $y = -\frac{3}{2}x - \frac{1}{2}$.

Answer:

■ **Answer:**

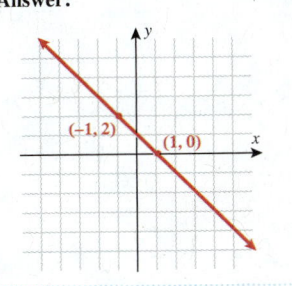

■ **YOUR TURN** The graph of the equation $y = -x + 1$ is a line. Graph the line.

Technology Tip

Graph of $y_1 = x^2 - 5$ is shown in a [−5, 5] by [−8, 8] viewing rectangle.

EXAMPLE 2 **Graphing an Equation by Plotting Points**

Graph the equation $y = x^2 - 5$.

Solution:

STEP 1 In a table, list several pairs of coordinates that make the equation true.

x	$y = x^2 - 5$	(x, y)
0	−5	(0, −5)
−1	−4	(−1, −4)
1	−4	(1, −4)
−2	−1	(−2, −1)
2	−1	(2, −1)
−3	4	(−3, 4)
3	4	(3, 4)

STEP 2 Plot these points on a graph and connect the points with a smooth curve, indicating with arrows that the curve continues.

This graph is called a *parabola* and will be discussed in further detail in Chapter 8.

YOUR TURN Graph the equation $y = x^2 - 1$.

■ **Answer:**

EXAMPLE 3 **Graphing an Equation by Plotting Points**

Graph the equation $y = x^3$.

Solution:

STEP 1 In a table, list several pairs of coordinates that satisfy the equation.

x	$y = x^3$	(x, y)
0	0	(0, 0)
−1	−1	(−1, −1)
1	1	(1, 1)
−2	−8	(−2, −8)
2	8	(2, 8)

STEP 2 Plot these points on a graph and connect the points with a smooth curve, indicating with arrows that the curve continues in both the positive and negative directions.

Intercepts

When point-plotting graphs of equations, which points should be selected? Points where a graph crosses (or touches) either the *x*-axis or *y*-axis are called *intercepts*, and identifying these points helps define the graph unmistakably.

An **x-intercept** of a graph is a point where the graph intersects the *x*-axis. Specifically, an *x*-intercept is the *x*-coordinate of such a point. For example, if a graph intersects the *x*-axis at the point (3, 0), then we say that 3 is the *x*-intercept. Since the value for *y* along the *x*-axis is zero, all points corresponding to *x*-intercepts have the form $(a, 0)$.

A **y-intercept** of a graph is a point where the graph intersects the *y*-axis. Specifically, a *y*-intercept is the *y*-coordinate of such a point. For example, if a graph intersects the *y*-axis at the point (0, 2), then we say that 2 is the *y*-intercept. Since the value for *x* along the *y*-axis is zero, all points corresponding to *y*-intercepts have the form $(0, b)$.

It is important to note that graphs of equations do not have to have intercepts, and if they do have intercepts, they can have one or more of each type.

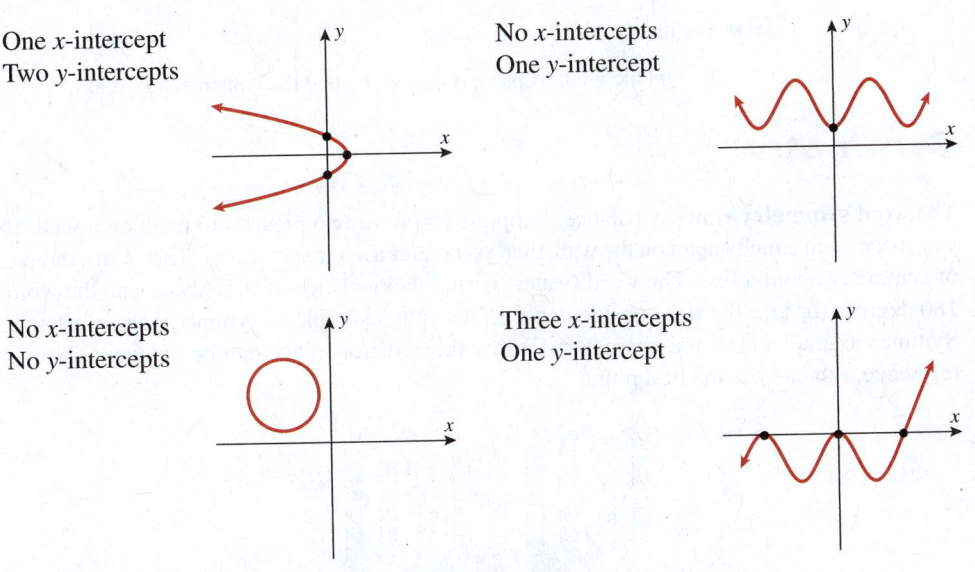

One *x*-intercept
Two *y*-intercepts

No *x*-intercepts
One *y*-intercept

No *x*-intercepts
No *y*-intercepts

Three *x*-intercepts
One *y*-intercept

Classroom Example 2.2.3*
a. Graph $x = (y + 1)^3 - 1$.
b. Graph $x = 1 - (y + 1)^3$.
c. Discuss the relationship between (a) and (b).

Answer:
a.

b.

c. They are reflections of each other over the *y*-axis.

Note: The origin (0, 0) corresponds to both an *x*-intercept and a *y*-intercept.

The graph given to the right has two *y*-intercepts and one *x*-intercept.

- The *x*-intercept is -1, which corresponds to the point $(-1, 0)$.
- The *y*-intercepts are -1 and 1, which correspond to the points $(0, -1)$ and $(0, 1)$, respectively.

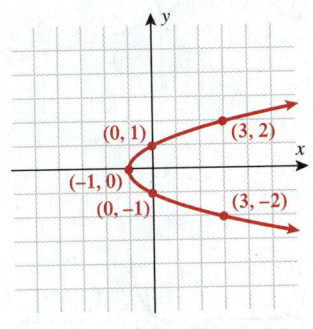

Algebraically, how do we find intercepts from an equation? The graph in the margin corresponds to the equation $x = y^2 - 1$. The *x*-intercepts are located on the *x*-axis, which corresponds to $y = 0$. If we let $y = 0$ in the equation $x = y^2 - 1$ and solve for *x*, the result is $x = -1$. This corresponds to the *x*-intercept we identified above. Similarly, the *y*-intercepts are located on the *y*-axis, which corresponds to $x = 0$. If we let $x = 0$ in the equation $x = y^2 - 1$ and solve for *y*, the result is $y = \pm 1$. These correspond to the *y*-intercepts we identified above.

Classroom Example 2.2.4

Find the x- and y-intercepts of the following graphs, if they exist.

a. $y = -|x - 2|$
b. $y = -3|x + 3| - 6$
c. $x = -2 - y^2$
d. $x = 1 + 2y^2$

Answer:

a. $(0, -2), (2, 0)$

b. $(0, -15)$, no x-intercept

c. $(-2, 0)$, no y-intercept

d. $(1, 0)$, no y-intercept

EXAMPLE 4 Finding Intercepts from an Equation

Given the equation $y = x^2 + 1$, find the indicated intercepts of its graph, if any.

a. x-intercept(s)   **b.** y-intercept(s)

Solution (a):

Let $y = 0$.

Solve for x.

$$0 = x^2 + 1$$
$$x^2 = -1 \quad \text{no real solution}$$

There are no x-intercepts.

Solution (b):

Let $x = 0$.

Solve for y.

$$y = 0^2 + 1$$
$$y = 1$$

The y-intercept is located at the point $(0, 1)$.

■ **YOUR TURN** For the equation $y = x^2 - 4$,

a. find the x-intercept(s), if any.   **b.** find the y-intercept(s), if any.

■ **Answer:**

a. x-intercepts: -2 and 2
b. y-intercept: -4

Symmetry

The word **symmetry** conveys balance. Suppose you have two pictures to hang on a wall. If you space them equally apart on the wall, then you prefer a symmetric décor. This is an example of symmetry about a line. The word (water) written below is identical if you rotate the word 180 degrees (or turn the page upside down). This is an example of symmetry about a point. Symmetric graphs have the characteristic that their mirror image can be obtained about a reference, typically a line or a point.

In Example 2, the points $(-2, -1)$ and $(2, -1)$ both lie on the graph of $y = x^2 - 5$, as do the points $(-1, -4)$ and $(1, -4)$. Notice that the graph on the right side of the y-axis is a mirror image of the part of the graph to the left of the y-axis. This graph illustrates *symmetry* with respect to the *y-axis* (the line $x = 0$).

In the graph of the equation $x = y^2 - 1$ in the margin, the points $(0, 1)$ and $(0, -1)$ both lie on the graph, as do the points $(3, 2)$ and $(3, -2)$. Notice that the part of the graph above the x-axis is a mirror image of the part of the graph below the x-axis. This graph illustrates *symmetry* with respect to the *x-axis* (the line $y = 0$).

In Example 3, the points $(-1, -1)$ and $(1, 1)$ both lie on the graph. Notice that rotating this graph 180 degrees (or turning your page upside down) results in an identical graph. This is an example of *symmetry* with respect to the *origin* $(0, 0)$.

Symmetry aids in graphing by giving information "for free." For example, if a graph is symmetric about the y-axis, then once the graph to the right of the y-axis is found, the left side of the graph is the mirror image of that. If a graph is symmetric about the origin, then once the graph is known in quadrant I, the graph in quadrant III is found by rotating the known graph 180 degrees.

It would be beneficial to know whether a graph of an equation is symmetric about a line or point before the graph of the equation is sketched. Although a graph can be symmetric about any line or point, we will discuss only symmetry about the *x*-axis, *y*-axis, and origin. These types of symmetry and the algebraic procedures for testing for symmetry are outlined below.

Types and Tests for Symmetry

TYPE OF SYMMETRY	GRAPH	IF THE POINT (*a*, *b*) IS ON THE GRAPH, THEN THE POINT . . .	ALGEBRAIC TEST FOR SYMMETRY
Symmetric with respect to the **x-axis**	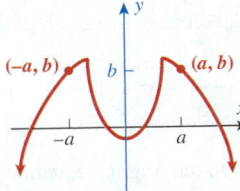	(*a*, −*b*) is on the graph.	Replacing *y* with −*y* leaves the equation unchanged.
Symmetric with respect to the **y-axis**	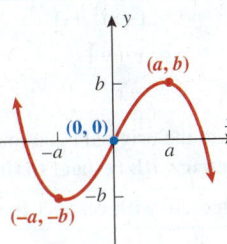	(−*a*, *b*) is on the graph.	Replacing *x* with −*x* leaves the equation unchanged.
Symmetric with respect to the **origin**		(−*a*, −*b*) is on the graph.	Replacing *x* with −*x* and *y* with −*y* leaves the equation unchanged.

EXAMPLE 5   Testing for Symmetry with Respect to the Axes

Test the equation $y^2 = x^3$ for symmetry with respect to the axes.

Solution:

Test for symmetry with respect to the x-axis.

Replace *y* with −*y*. $(-y)^2 = x^3$

Simplify. $y^2 = x^3$

The resulting equation is the same as the original equation, $y^2 = x^3$.

> Therefore $y^2 = x^3$ is **symmetric with respect to the x-axis**.

Test for symmetry with respect to the y-axis.

Replace *x* with −*x*. $y^2 = (-x)^3$

Simplify. $y^2 = -x^3$

The resulting equation, $y^2 = -x^3$, is not the same as the original equation, $y^2 = x^3$.

> Therefore $y^2 = x^3$ is **not** symmetric with respect to the y-axis.

When testing for symmetry about the *x*-axis, *y*-axis, and origin, there are *five* possibilities:

- No symmetry
- Symmetry with respect to the *x*-axis
- Symmetry with respect to the *y*-axis
- Symmetry with respect to the origin
- Symmetry with respect to the *x*-axis, *y*-axis, and origin

Technology Tip

Graph of $y_1 = x^2 + 1$ is shown.

Graph of $y_1 = x^3 + 1$ is shown.

Classroom Example 2.2.6*

Let *a* and *b* be positive real numbers. Test for symmetry in the following graphs.
a. $x = -a|y| + b$
b. $y^3|x| - |y| = x^2$

Answer:
a. *x*-axis   b. *y*-axis

- **Answer:** The graph of the equation is symmetric with respect to the *x*-axis.

EXAMPLE 6   Testing for Symmetry

Determine what type of symmetry (if any) the graphs of the equations exhibit.

a. $y = x^2 + 1$ **b.** $y = x^3 + 1$

Solution (a):

Replace *x* with $-x$. $y = (-x)^2 + 1$

Simplify. $y = x^2 + 1$

The resulting equation is equivalent to the original equation, so the graph of the equation $y = x^2 + 1$ is **symmetric with respect to the *y*-axis**.

Replace *y* with $-y$. $(-y) = x^2 + 1$

Simplify. $y = -x^2 - 1$

The resulting equation $y = -x^2 - 1$ is not equivalent to the original equation $y = x^2 + 1$, so the graph of the equation $y = x^2 + 1$ is **not symmetric with respect to the *x*-axis**.

Replace *x* with $-x$ and *y* with $-y$. $(-y) = (-x)^2 + 1$

Simplify. $-y = x^2 + 1$

$y = -x^2 - 1$

The resulting equation $y = -x^2 - 1$ is not equivalent to the original equation $y = x^2 + 1$, so the graph of the equation $y = x^2 + 1$ is **not symmetric with respect to the origin**.

The graph of the equation $y = x^2 + 1$ is **symmetric with respect to the *y*-axis**.

Solution (b):

Replace *x* with $-x$. $y = (-x)^3 + 1$

Simplify. $y = -x^3 + 1$

The resulting equation $y = -x^3 + 1$ is not equivalent to the original equation $y = x^3 + 1$. Therefore, the graph of the equation $y = x^3 + 1$ is **not symmetric with respect to the *y*-axis**.

Replace *y* with $-y$. $(-y) = x^3 + 1$

Simplify. $y = -x^3 - 1$

The resulting equation $y = -x^3 - 1$ is not equivalent to the original equation $y = x^3 + 1$. Therefore, the graph of the equation $y = x^3 + 1$ is **not symmetric with respect to the *x*-axis**.

Replace *x* with $-x$ and *y* with $-y$. $(-y) = (-x)^3 + 1$

Simplify. $-y = -x^3 + 1$

$y = x^3 - 1$

The resulting equation $y = x^3 - 1$ is not equivalent to the original equation $y = x^3 + 1$. Therefore, the graph of the equation $y = x^3 + 1$ is **not symmetric with respect to the origin**.

The graph of the equation $y = x^3 + 1$ exhibits **no symmetry**.

- **YOUR TURN** Determine the symmetry (if any) for $x = y^2 - 1$.

Using Intercepts and Symmetry as Graphing Aids

How can we use intercepts and symmetry to assist us in graphing? Intercepts are a good starting point—though not the only one. For symmetry, look back at Example 2, $y = x^2 - 5$. We selected seven x-coordinates and solved the equation to find the corresponding y-coordinates. If we had known that this graph was symmetric with respect to the y-axis, then we would have had to find the solutions to only the positive x-coordinates, since we get the negative x-coordinates for free. For example, we found the point $(1, -4)$ to be a solution to the equation. The rules of symmetry tell us that $(-1, -4)$ is also on the graph.

EXAMPLE 7   Using Intercepts and Symmetry as Graphing Aids

For the equation $x^2 + y^2 = 25$, use intercepts and symmetry to help you graph the equation using the point-plotting technique.

Solution:

STEP 1 Find the intercepts.

For the x-intercepts, let $y = 0$. $x^2 + 0^2 = 25$

Solve for x. $x = \pm 5$

The two x-intercepts correspond to the points $(-5, 0)$ and $(5, 0)$.

For the y-intercepts, let $x = 0$. $0^2 + y^2 = 25$

Solve for y. $y = \pm 5$

The two y-intercepts correspond to the points $(0, -5)$ and $(0, 5)$.

STEP 2 Identify the points on the graph corresponding to the intercepts.

STEP 3 Test for symmetry with respect to the y-axis, x-axis, and origin.

Test for symmetry with respect to the **y-axis**.

Replace x with $-x$. $(-x)^2 + y^2 = 25$

Simplify. $x^2 + y^2 = 25$

The resulting equation is equivalent to the original, so the graph of $x^2 + y^2 = 25$ is **symmetric with respect to the y-axis**.

Classroom Example 2.2.7
Use symmetry and intercepts with the point-plotting technique to graph $|x| = |y|$.

Answer:

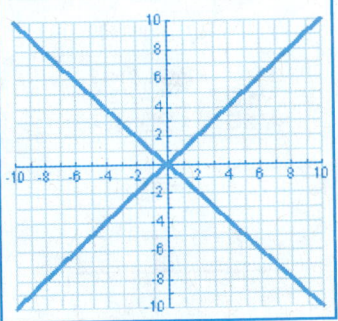

Test for symmetry with respect to the **x-axis**.

Replace y with $-y$. $\qquad\qquad\qquad\qquad\qquad\qquad\qquad\qquad\qquad$ $x^2 + (-y)^2 = 25$

Simplify. $\qquad\qquad\qquad\qquad\qquad\qquad\qquad\qquad\qquad\qquad\qquad$ $x^2 + y^2 = 25$

The resulting equation is equivalent to the original, so the graph of $x^2 + y^2 = 25$ is **symmetric with respect to the x-axis**.

Test for symmetry with respect to the **origin**.

Replace x with $-x$ and y with $-y$. $\qquad\qquad\qquad\qquad\qquad$ $(-x)^2 + (-y)^2 = 25$

Simplify. $\qquad\qquad\qquad\qquad\qquad\qquad\qquad\qquad\qquad\qquad\qquad$ $x^2 + y^2 = 25$

The resulting equation is equivalent to the original, so the graph of $x^2 + y^2 = 25$ is **symmetric with respect to the origin**.

Since the graph is symmetric with respect to the **y-axis**, **x-axis**, and **origin**, we need to determine solutions to the equation on only the positive x- and y-axes and in quadrant I because of the following symmetries:

- **Symmetry with respect to the y-axis gives the solutions in quadrant II.**
- **Symmetry with respect to the origin gives the solutions in quadrant III.**
- **Symmetry with respect to the x-axis yields solutions in quadrant IV.**

Solutions to $x^2 + y^2 = 25$.

Quadrant I: $(3, 4)$, $(4, 3)$

Additional points due to symmetry:

Quadrant II: $(-3, 4)$, $(-4, 3)$

Quadrant III: $(-3, -4)$, $(-4, -3)$,

Quadrant IV: $(3, -4)$, $(4, -3)$

Technology Tip

To enter the graph of $x^2 + y^2 = 25$, solve for y first. The graphs of $y_1 = \sqrt{25 - x^2}$ and $y_2 = -\sqrt{25 - x^2}$ are shown.

Connecting the points with a smooth curve yields a **circle**. We will discuss circles in more detail in Section 2.4.

SECTION 2.2 SUMMARY

Sketching graphs of equations can be accomplished using a point-plotting technique. Intercepts are defined as points where a graph intersects an axis or the origin.

Symmetry about the x-axis, y-axis, and origin is defined both algebraically and graphically. **Intercepts** and **symmetry** provide much of the information useful for sketching graphs of equations.

Intercepts

INTERCEPTS	POINT WHERE THE GRAPH INTERSECTS THE...	HOW TO FIND INTERCEPTS	POINT ON GRAPH
x-intercept	x-axis	Let $y = 0$ and solve for x.	$(a, 0)$
y-intercept	y-axis	Let $x = 0$ and solve for y.	$(0, b)$

Types and Tests for Symmetry

TYPE OF SYMMETRY	GRAPH	IF THE POINT (a, b) IS ON THE GRAPH, THEN THE POINT . . .	ALGEBRAIC TEST FOR SYMMETRY
Symmetric with respect to the *x*-axis		$(a, -b)$ is on the graph.	Replacing y with $-y$ leaves the equation unchanged.
Symmetric with respect to the *y*-axis		$(-a, b)$ is on the graph.	Replacing x with $-x$ leaves the equation unchanged.
Symmetric with respect to the **origin**		$(-a, -b)$ is on the graph.	Replacing x with $-x$ and y with $-y$ leaves the equation unchanged.

SECTION
2.2 EXERCISES

■ **SKILLS**

In Exercises 1–8, determine whether each point lies on the graph of the equation.

1. $y = 3x - 5$ **a.** $(1, 2)$ **b.** $(-2, -11)$ 2. $y = -2x + 7$ **a.** $(-1, 9)$ **b.** $(2, -4)$

3. $y = \frac{2}{5}x - 4$ **a.** $(5, -2)$ **b.** $(-5, 6)$ 4. $y = -\frac{3}{4}x + 1$ **a.** $(8, 5)$ **b.** $(-4, 4)$

5. $y = x^2 - 2x + 1$ **a.** $(-1, 4)$ **b.** $(0, -1)$ 6. $y = x^3 - 1$ **a.** $(-1, 0)$ **b.** $(-2, -9)$

7. $y = \sqrt{x + 2}$ **a.** $(7, 3)$ **b.** $(-6, 4)$ 8. $y = 2 + |3 - x|$ **a.** $(9, -4)$ **b.** $(-2, 7)$

In Exercises 9–14, complete the table and use the table to sketch a graph of the equation.

9.

x	$y = 2 + x$	(x, y)
-2		
0		
1		

10.

x	$y = 3x - 1$	(x, y)
-1		
0		
2		

11.

x	$y = x^2 - x$	(x, y)
-1		
0		
$\frac{1}{2}$		
1		
2		

12.

x	$y = 1 - 2x - x^2$	(x, y)
-3		
-2		
-1		
0		
1		

13.

x	$y = \sqrt{x - 1}$	(x, y)
1		
2		
5		
10		

14.

x	$y = -\sqrt{x + 2}$	(x, y)
-2		
-1		
2		
7		

In Exercises 15–22, graph the equation by plotting points.

15. $y = -3x + 2$

16. $y = 4 - x$

17. $y = x^2 - x - 2$

18. $y = x^2 - 2x + 1$

19. $x = y^2 - 1$

20. $x = |y + 1| + 2$

21. $y = \frac{1}{2}x - \frac{3}{2}$

22. $y = 0.5|x - 1|$

In Exercises 23–32, find the x-intercept(s) and y-intercepts(s) (if any) of the graphs of the given equations.

23. $2x - y = 6$

24. $4x + 2y = 10$

25. $y = x^2 - 9$

26. $y = 4x^2 - 1$

27. $y = \sqrt{x - 4}$

28. $y = \sqrt[3]{x - 8}$

29. $y = \dfrac{1}{x^2 + 4}$

30. $y = \dfrac{x^2 - x - 12}{x}$

31. $4x^2 + y^2 = 16$

32. $x^2 - y^2 = 9$

In Exercises 33–38, match the graph with the corresponding symmetry.

  a. No symmetry

  b. Symmetry with respect to the x-axis

  c. Symmetry with respect to the y-axis

  d. Symmetry with respect to the origin

  e. Symmetry with respect to the x-axis, y-axis, and origin

33.

34.

35.

36.

37.

38.

In Exercises 39–44, a point that lies on a graph is given along with that graph's symmetry. State the other known points that must also lie on the graph.

Point on a Graph	The Graph Is Symmetric about the	Point on a Graph	The Graph Is Symmetric about the
39. $(-1, 3)$	x-axis	**40.** $(-2, 4)$	y-axis
41. $(7, -10)$	origin	**42.** $(-1, -1)$	origin
43. $(3, -2)$	x-axis, y-axis, and origin	**44.** $(-1, 7)$	x-axis, y-axis, and origin

In Exercises 45–58, test algebraically to determine whether the equation's graph is symmetric with respect to the x-axis, y-axis, or origin.

45. $x = y^2 + 4$ **46.** $x = 2y^2 + 3$ **47.** $y = x^3 + x$ **48.** $y = x^5 + 1$

49. $x = |y|$ **50.** $x = |y| - 2$ **51.** $x^2 - y^2 = 100$ **52.** $x^2 + 2y^2 = 30$

53. $y = x^{2/3}$ **54.** $x = y^{2/3}$ **55.** $x^2 + y^3 = 1$ **56.** $y = \sqrt{1 + x^2}$

57. $y = \dfrac{2}{x}$ **58.** $xy = 1$

In Exercises 59–72, plot the graph of the given equation.

59. $y = x$ **60.** $y = -\dfrac{1}{2}x + 3$ **61.** $y = x^2 - 1$ **62.** $y = 9 - 4x^2$

63. $y = \dfrac{x^3}{2}$ **64.** $x = y^2 + 1$ **65.** $y = \dfrac{1}{x}$ **66.** $xy = -1$

67. $y = |x|$ **68.** $|x| = |y|$ **69.** $x^2 + y^2 = 16$ **70.** $\dfrac{x^2}{4} + \dfrac{y^2}{9} = 1$

71. $x^2 - y^2 = 16$ **72.** $x^2 - \dfrac{y^2}{25} = 1$

■ APPLICATIONS

73. Sprinkler. A sprinkler will water a grassy area in the shape of $x^2 + y^2 = 9$. Apply symmetry to draw the watered area, assuming the sprinkler is located at the origin.

74. Sprinkler. A sprinkler will water a grassy area in the shape of $x^2 + \dfrac{y^2}{9} = 1$. Apply symmetry to draw the watered area, assuming the sprinkler is located at the origin.

75. Electronic Signals: Radio Waves. The received power of an electromagnetic signal is a fraction of the power transmitted. The relationship is given by

$$P_{\text{received}} = P_{\text{transmitted}} \cdot \dfrac{1}{R^2}$$

where R is the distance that the signal has traveled in meters. Plot the percentage of transmitted power that is received for $R = 100$ m, 1 km, and 10,000 km.

76. Electronic Signals: Laser Beams. The wavelength λ and the frequency f of a signal are related by the equation

$$f = \frac{c}{\lambda}$$

where c is the speed of light in a vacuum, $c = 3.0 \times 10^8$ meters per second. For the values, $\lambda = 0.001$, $\lambda = 1$, and $\lambda = 100$ mm, plot the points corresponding to frequency, f. What do you notice about the relationship between frequency and wavelength? Note that the frequency will have units Hz = 1/seconds.

77. Profit. The profit associated with making a particular product is given by the equation

$$y = -x^2 + 6x - 8$$

where y represents the profit in millions of dollars and x represents the number of thousands of units sold. ($x = 1$ corresponds to 1000 units and $y = 1$ corresponds to \$1M.) Graph this equation and determine how many units must be sold to break even (profit = 0). Determine the range of units sold that correspond to making a profit.

78. Profit. The profit associated with making a particular product is given by the equation

$$y = -x^2 + 4x - 3$$

where y represents the profit in millions of dollars and x represents the number of thousands of units sold. ($x = 1$ corresponds to 1000 units and $y = 1$ corresponds to \$1M.) Graph this equation and determine how many units must be sold to break even (profit = 0). Determine the range of units sold that correspond to making a profit.

79. Economics. The demand for an electronic device is modeled by

$$p = 2.95 - \sqrt{0.01x - 0.01}$$

where x is thousands of units demanded per day and p is the price (in dollars) per unit.

a. Find the domain of the demand equation. Interpret your result.

b. Plot the demand equation.

80. Economics. The demand for a new electronic game is modeled by

$$p = 39.95 - \sqrt{0.01x - 0.4}$$

where x is thousands of units demanded per day and p is the price (in dollars) per unit.

a. Find the domain of the demand equation. Interpret your result.

b. Plot the demand equation.

■ **CATCH THE MISTAKE**

In Exercises 81–84, explain the mistake that is made.

81. Graph the equation $y = x^2 + 1$.

Solution:

x	$y = x^2 + 1$	(x, y)
0	1	$(0, 1)$
1	2	$(1, 2)$

This is incorrect. What mistake was made?

82. Test $y = -x^2$ for symmetry with respect to the y-axis.

Solution:

Replace x with $-x$. $y = -(-x)^2$

Simplify. $y = x^2$

The resulting equation is not equivalent to the original equation; $y = -x^2$ is not symmetric with respect to the y-axis.

This is incorrect. What mistake was made?

83. Test $x = |y|$ for symmetry with respect to the y-axis.

Solution:

Replace y with $-y$. $x = |-y|$

Simplify. $x = |y|$

The resulting equation is equivalent to the original equation; $x = |y|$ is symmetric with respect to the y-axis.

This is incorrect. What mistake was made?

84. Use symmetry to help you graph $x^2 = y - 1$.

Solution:

Replace x with $-x$. $\qquad (-x)^2 = y - 1$

Simplify. $\qquad\qquad\qquad x^2 = y - 1$

$x^2 = y - 1$ is symmetric with respect to the x-axis.

Determine points that lie on the graph in quadrant I.

y	$x^2 = y - 1$	(x, y)
1	0	(0, 1)
2	1	(1, 2)
5	2	(2, 5)

Symmetry with respect to the x-axis implies that $(0, -1)$, $(1, -2)$, and $(2, -5)$ are also points that lie on the graph.

This is incorrect. What mistake was made?

■ CONCEPTUAL

In Exercises 85–88, determine whether each statement is true or false.

85. If the point (a, b) lies on a graph that is symmetric about the x-axis, then the point $(-a, b)$ also must lie on the graph.

86. If the point (a, b) lies on a graph that is symmetric about the y-axis, then the point $(-a, b)$ also must lie on the graph.

87. If the point $(a, -b)$ lies on a graph that is symmetric about the x-axis, y-axis, and origin, then the points (a, b), $(-a, -b)$, and $(-a, b)$ must also lie on the graph.

88. Two points are all that is needed to plot the graph of an equation.

■ CHALLENGE

89. Determine whether the graph of $y = \dfrac{ax^2 + b}{cx^3}$ has any symmetry, where a, b, and c are real numbers.

90. Find the intercepts of $y = (x - a)^2 - b^2$, where a and b are real numbers.

■ TECHNOLOGY

In Exercises 91–96, graph the equation using a graphing utility and state whether there is any symmetry.

91. $y = 16.7x^4 - 3.3x^2 + 7.1$

92. $y = 0.4x^5 + 8.2x^3 - 1.3x$

93. $2.3x^2 = 5.5\,|y|$

94. $3.2x^2 - 5.1y^2 = 1.3$

95. $1.2x^2 + 4.7y^2 = 19.4$

96. $2.1y^2 = 0.8\,|x + 1|$

SKILLS OBJECTIVES

- Determine *x*- and *y*-intercepts of a line.
- Calculate the slope of a line.
- Find the equation of a line using slope–intercept form.
- Find the equation of a line using point–slope form.
- Find the equation of a line that is parallel or perpendicular to a given line.

CONCEPTUAL OBJECTIVES

- Classify lines as rising, falling, horizontal, and vertical.
- Understand slope as a rate of change.
- Associate two lines having the same slope with the graph of parallel lines.
- Associate two lines having negative reciprocal slopes with the graph of perpendicular lines.

Graphing a Line

What is the shortest path between two points? The answer is a *straight line.* In this section, we will discuss characteristics of lines such as slope and intercepts. We will also discuss types of lines such as horizontal, vertical, falling, and rising, and recognize relationships between lines such as perpendicular and parallel. At the end of this section, you should be able to find the equation of a line when given two specific pieces of information about the line.

First-degree equations, such as

$$y = -2x + 4, \qquad 3x + y = 6, \qquad y = 2, \qquad \text{and} \qquad x = -3,$$

have graphs that are straight lines. The first two equations given represent inclined or "slant" lines, whereas $y = 2$ represents a horizontal line and $x = -3$ represents a vertical line. One way of writing an equation of a straight line is called *general form.*

EQUATION OF A STRAIGHT LINE: GENERAL* FORM

If *A*, *B*, and *C* are constants and *x* and *y* are variables, then the equation

$$Ax + By = C$$

is in **general form**, and its graph is a straight line.

Note: *A* or *B* (but not both) can be zero.

The equation $2x - y = -2$ is a first-degree equation, so its graph is a straight line. To graph this line, list two solutions in a table, plot those points, and use a straight edge to draw the **line**.

x	*y*	*(x, y)*
−2	−2	**(−2, −2)**
1	4	**(1, 4)**

---

*Some books refer to this as standard form.

Intercepts

The point where a line crosses, or intersects, the x-axis is called the **x-intercept**. The point where a line crosses, or intersects, the y-axis is called the **y-intercept**. By inspecting the graph of the previous line, we see that the **x-intercept** is **$(-1, 0)$** and the **y-intercept** is **$(0, 2)$**.

Intercepts were discussed in Section 2.2. There, we found that the graphs of some equations could have no intercepts or one or multiple intercepts. Slant lines, however, have exactly one x-intercept and exactly one y-intercept.

DETERMINING INTERCEPTS

	Coordinates	Axis Crossed	Algebraic Method
x-intercept:	$(a, 0)$	x-axis	Set $y = 0$ and solve for x.
y-intercept:	$(0, b)$	y-axis	Set $x = 0$ and solve for y.

Horizontal lines and vertical lines, however, each have only one intercept.

TYPE OF LINE	EQUATION	X-INTERCEPT	Y-INTERCEPT	GRAPH
Horizontal	$y = b$	None	b	
Vertical	$x = a$	a	None	

Note: The special cases of $x = 0$ (y-axis) and $y = 0$ (x-axis) have infinitely many y-intercepts and x-intercepts, respectively.

EXAMPLE 1 Determining *x*- and *y*-Intercepts

Determine the *x*- and *y*-intercepts (if they exist) for the lines given by the following equations.

a. $2x + 4y = 10$ **b.** $x = -2$

Solution (a): $2x + 4y = 10$

To find the *x*-intercept, set $y = 0$. $2x + 4(0) = 10$

Solve for *x*. $2x = 10$

$x = 5$

The *x*-intercept corresponds to the point $\boxed{(5, 0)}$.

To find the *y*-intercept, set $x = 0$. $2(0) + 4y = 10$

Solve for *y*. $4y = 10$

$y = \dfrac{5}{2}$

The *y*-intercept corresponds to the point $\boxed{\left(0, \frac{5}{2}\right)}$.

Classroom Example 2.3.1
Determine the *x*- and
y-intercepts, if they exist.
a. $x - 2y = 1$
b. $3 - x = 0$
c. $2y - 3 = x + 6$
Answer: a. $\left(0, -\frac{1}{2}\right), (1, 0)$
b. $(3, 0)$, no *y*-intercept
c. $\left(0, \frac{9}{2}\right), (-9, 0)$

Solution (b): $x = -2$

This vertical line consists of all points $(-2, y)$.

The graph shows that the *x*-intercept is -2.

We also find that the line never crosses the *y*-axis, so the *y*-intercept does not exist.

■ **Answer: a.** *x*-intercept: $\left(\frac{2}{3}, 0\right)$;
y-intercept: $(0, -2)$
b. *x*-intercept does not exist;
y-intercept: $(0, 5)$

■ **YOUR TURN** Determine the *x*- and *y*-intercepts (if they exist) for the lines given by the following equations.

     a. $3x - y = 2$ **b.** $y = 5$

Slope

If the graph of $2x - y = -2$ represented an incline that you were about to walk on, would you classify that incline as steep? In the language of mathematics, we use the word **slope** as a measure of steepness. Slope is the ratio of the change in *y* over the change in *x*. An easy way to remember this is *rise over run*.

SLOPE OF A LINE

A nonvertical line passing through two points (x_1, y_1) and (x_2, y_2) has slope *m*, given by the formula

$$m = \frac{y_2 - y_1}{x_2 - x_1}, \text{ where } x_1 \neq x_2 \text{ or}$$

$$m = \frac{\text{rise}}{\text{run}} = \frac{\text{vertical change}}{\text{horizontal change}}$$

Study Tip

To get the correct sign (±) for the slope, remember to start with the same point for both *x* and *y*.

Note: Always start with the same point for both the *x*-coordinates and the *y*-coordinates.

Let's find the slope of our graph $2x - y = -2$. We'll let $(x_1, y_1) = (-2, -2)$ and $(x_2, y_2) = (1, 4)$ in the slope formula:

$$m = \frac{y_2 - y_1}{x_2 - x_1} = \frac{[4 - (-2)]}{[1 - (-2)]} = \frac{6}{3} = 2$$

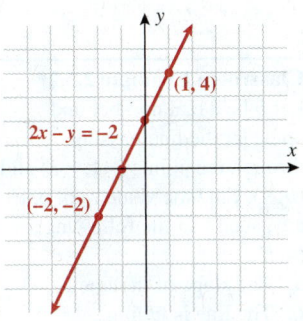

Notice that if we had chosen the two intercepts $(x_1, y_1) = (0, 2)$ and $(x_2, y_2) = (-1, 0)$ instead, we still would have found the slope to be $m = 2$.

COMMON MISTAKE

The most common mistake in calculating slope is writing the coordinates in the wrong order, which results in the slope being opposite in sign.

Find the slope of the line containing the two points $(1, 2)$ and $(3, 4)$.

▼ **CAUTION**

Interchanging the coordinates can result in a sign error in the slope.

⭐ CORRECT

Label the points.

$(x_1, y_1) = (1, 2)$ $(x_2, y_2) = (3, 4)$

Write the slope formula.

$$m = \frac{y_2 - y_1}{x_2 - x_1}$$

Substitute the coordinates.

$$m = \frac{4 - 2}{3 - 1}$$

Simplify. $m = \dfrac{2}{2} = \boxed{1}$

❌ INCORRECT

The **ERROR** is interchanging the coordinates of the first and second points.

$$m = \frac{4 - 2}{1 - 3}$$

The calculated slope is **INCORRECT** by a negative sign.

$$m = \frac{2}{-2} = -1$$

When interpreting slope, always read the graph from *left to right*. Since we have determined the slope to be 2, or $\frac{2}{1}$, we can interpret this as rising two units and running (to the right) one unit. If we start at the point $(-2, -2)$ and move two units up and one unit to the right, we end up at the x-intercept, $(-1, 0)$. Again, moving two units up and one unit to the right puts us at the y-intercept, $(0, 2)$. Another rise of two and run of one takes us to the point $(1, 4)$. See the figure on the right.

Lines fall into one of four categories: rising, falling, horizontal, or vertical.

Line	Slope
Rising	**Positive ($m > 0$)**
Falling	**Negative ($m < 0$)**
Horizontal	**Zero ($m = 0$), hence $y = b$**
Vertical	**Undefined, hence $x = a$**

The slope of a horizontal line is 0 because the y-coordinates of any two points are the same. The change in y in the slope formula's numerator is 0, hence $m = 0$. The slope of a vertical line is undefined because the x-coordinates of any two points are the same. The change in x in the slope formula's denominator is zero; hence m is undefined.

EXAMPLE 2 **Graph, Classify the Line, and Determine the Slope**

Sketch a line through each pair of points, classify the line as rising, falling, vertical, or horizontal, and determine its slope.

a. $(-1, -3)$ and $(1, 1)$ **b.** $(-3, 3)$ and $(3, 1)$
c. $(-1, -2)$ and $(3, -2)$ **d.** $(1, -4)$ and $(1, 3)$

Solution (a): $(-1, -3)$ and $(1, 1)$

This line is **rising**, so its slope is positive.

$$m = \frac{1 - (-3)}{1 - (-1)} = \frac{4}{2} = \frac{2}{1} = 2$$

Solution (b): $(-3, 3)$ and $(3, 1)$

This line is **falling**, so its slope is negative.

$$m = \frac{3 - 1}{-3 - 3} = -\frac{2}{6} = -\frac{1}{3}$$

Solution (c): $(-1, -2)$ and $(3, -2)$

This is a **horizontal** line, so its slope is zero.

$$m = \frac{-2 - (-2)}{3 - (-1)} = \frac{0}{4} = 0$$

Solution (d): $(1, -4)$ and $(1, 3)$

This is a **vertical** line, so its slope is undefined.

$$m = \frac{3 - (-4)}{1 - 1} = \frac{7}{0}, \text{ which is undefined.}$$

■ **YOUR TURN** For each pair of points classify the line that passes through them as rising, falling, vertical, or horizontal, and determine its slope. Do not graph.

a. $(2, 0)$ and $(1, 5)$ **b.** $(-2, -3)$ and $(2, 5)$
c. $(-3, -1)$ and $(-3, 4)$ **d.** $(-1, 2)$ and $(3, 2)$

Equations of Lines

Slope–Intercept Form

As mentioned earlier, the general form for an equation of a line is $Ax + By = C$. A more standard way to write an equation of a line is in slope–intercept form, because it identifies the slope and the y-intercept.

> **EQUATION OF A STRAIGHT LINE: SLOPE–INTERCEPT FORM**
>
> The **slope–intercept form** for the equation of a nonvertical line is
>
> $$y = mx + b$$
>
> Its graph has slope m and y-intercept b.

For example, $2x - y = -3$ is in general form. To write this equation in **slope–intercept form**, we isolate the y variable:

$$y = 2x + 3$$

The **slope** of this line is **2** and the **y-intercept** is **3**.

EXAMPLE 3   Using Slope–Intercept Form to Graph an Equation of a Line

Write $2x - 3y = 15$ in slope–intercept form and graph it.

Solution:

STEP 1 *Write in slope–intercept form.*

Subtract $2x$ from both sides. $-3y = -2x + 15$

Divide both sides by -3.

$$y = \frac{2}{3}x - 5$$

STEP 2 *Graph.*

Identify the slope and y-intercept.   Slope: $m = \dfrac{2}{3}$ y-intercept: $b = -5$

Plot the point corresponding to the y-intercept $(0, -5)$.

From the point $(0, -5)$, rise two units and run (to the right) three units, which corresponds to the point $(3, -3)$.

Draw the line passing through the two points.

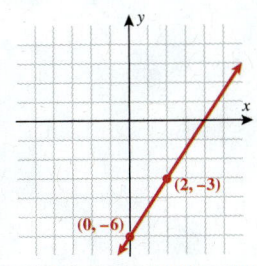
■ **YOUR TURN** Write $3x - 2y = 12$ in slope–intercept form and graph it.

Instead of starting with equations of lines and characterizing them, let us now start with particular features of a line and derive its governing equation. Suppose that you are given the y-intercept and the slope of a line. Using the slope–intercept form of an equation of a line, $y = mx + b$, you could find its equation.

EXAMPLE 4   Using Slope–Intercept Form to Find the Equation of a Line

Find the equation of a line that has slope $\frac{2}{3}$ and y-intercept $(0, 1)$.

Solution:

Write the slope–intercept form of an equation of a line. $y = mx + b$

Label the slope. $m = \dfrac{2}{3}$

Label the y-intercept. $b = 1$

The equation of the line in slope–intercept form is $\boxed{y = \frac{2}{3}x + 1}$.

■ **Answer:** $y = -\frac{3}{2}x + 2$

■ **YOUR TURN** Find the equation of the line that has slope $-\frac{3}{2}$ and y-intercept $(0, 2)$.

Point–Slope Form

Now, suppose that the two pieces of information you are given about an equation are its slope and one point that lies on its graph. You still have enough information to write an equation of the line. Recall the formula for slope:

$$m = \frac{y_2 - y_1}{x_2 - x_1}, \qquad \text{where } x_2 \neq x_1$$

We are given the slope m, and we know a particular point that lies on the line (x_1, y_1). We refer to all other points that lie on the line as (x, y). Substituting these values into the slope formula gives us

$$m = \frac{y - y_1}{x - x_1}$$

Cross multiplying yields

$$y - y_1 = m(x - x_1)$$

This is called the *point–slope form* of an equation of a line.

EQUATION OF A STRAIGHT LINE: POINT–SLOPE FORM

The **point–slope form** for the equation of a line is

$$y - y_1 = m(x - x_1)$$

Its graph passes through the point (x_1, y_1), and its slope is m.

Note: This formula does not hold for vertical lines since their slope is undefined.

EXAMPLE 5 **Using Point–Slope Form to Find the Equation of a Line**

Find the equation of the line that has slope $-\frac{1}{2}$ and passes through the point $(-1, 2)$.

Solution:

Write the point–slope form of an equation of a line.
$$y - y_1 = m(x - x_1)$$

Substitute the values $m = -\frac{1}{2}$ and $(x_1, y_1) = (-1, 2)$.
$$y - 2 = -\frac{1}{2}(x - (-1))$$

Distribute.
$$y - 2 = -\frac{1}{2}x - \frac{1}{2}$$

Isolate y.
$$y = -\frac{1}{2}x + \frac{3}{2}$$

We can also express the equation in general form $\boxed{x + 2y = 3}$.

■ **YOUR TURN** Derive the equation of the line that has slope $\frac{1}{4}$ and passes through the point $\left(1, -\frac{1}{2}\right)$. Give the answer in general form.

Suppose the slope of a line is not given at all. Instead, two points that lie on the line are given. If we know two points that lie on the line, then we can calculate the slope. Then, using the slope and *either* of the two points, the equation of the line can be derived.

EXAMPLE 6 **Finding the Equation of a Line Given Two Points**

Find the equation of the line that passes through the points $(-2, -1)$ and $(3, 2)$.

Solution:

Write the equation of a line.
$$y = mx + b$$

Calculate the slope.
$$m = \frac{y_2 - y_1}{x_2 - x_1}$$

Substitute $(x_1, y_1) = (-2, -1)$ and $(x_2, y_2) = (3, 2)$.
$$m = \frac{2 - (-1)}{3 - (-2)} = \frac{3}{5}$$

Substitute $\frac{3}{5}$ for the slope.
$$y = \frac{3}{5}x + b$$

Let $(x, y) = (3, 2)$. (Either point satisfies the equation.)
$$2 = \frac{3}{5}(3) + b$$

Solve for b.
$$b = \frac{1}{5}$$

Write the equation in slope–intercept form.
$$y = \frac{3}{5}x + \frac{1}{5}$$

Write the equation in general form.
$$\boxed{-3x + 5y = 1}$$

■ **YOUR TURN** Find the equation of the line that passes through the points $(-1, 3)$ and $(2, -4)$.

■ **Answer:** $y = \frac{1}{4}x - \frac{3}{4}$

Study Tip
When two points that lie on a line are given, first calculate the slope of the line, then use either point and the slope–intercept form (shown in Example 6) or the point–slope form:
$$m = \frac{3}{5}, (3, 2)$$
$$y - y_1 = m(x - x_1)$$
$$y - 2 = \frac{3}{5}(x - 3)$$
$$5y - 10 = 3(x - 3)$$
$$5y - 10 = 3x - 9$$
$$\boxed{-3x + 5y = 1}$$

■ **Answer:** $y = -\frac{7}{3}x + \frac{2}{3}$ or $7x + 3y = 2$

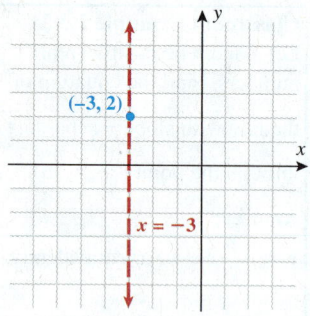

EXAMPLE 7   Finding the Equation of a Horizontal or Vertical Line

Find the equation of each of the following lines given their slope and a point that the line passes through:

a. Slope: undefined; $(-3, 2)$

b. Slope: $m = 0$; $(-4, 5)$

Solution (a):

A vertical line has undefined slope. $x = a$

The x-coordinate of the point the line passes through $(-3, 2)$ is -3. $x = -3$

Graph of the line $x = -3$ and the point $(-3, 2)$ indicated.

Solution (b):

A horizontal line has slope $m = 0$. $y = b$

The y-coordinate of the point the line passes through $(-4, 5)$ is 5. $y = 5$

Graph of the line $y = 5$ with the point $(-4, 5)$ indicated.

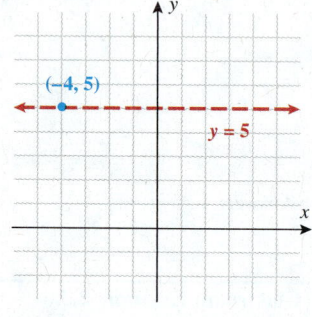

■ **Answer:**

a. $y = 7$

b. $x = 5$

■ **YOUR TURN** Find the equation of each of the following lines given their slope and a point that the line passes through:

a. Slope: $m = 0$; $(3, 7)$

b. Slope: undefined; $(5, -2)$

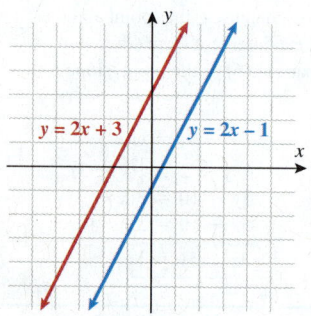

Parallel and Perpendicular Lines

Two distinct nonintersecting lines in a plane are *parallel*. How can we tell whether the two lines in the graph on the left are parallel? Parallel lines must have the same steepness. In other words, parallel lines must have the same slope. The two lines shown on the left are parallel because they have the same slope, 2.

> **DEFINITION**   **Parallel Lines**
>
> Two distinct lines in a plane are **parallel** if and only if their slopes are equal.

In other words, if two lines in a plane are parallel, then their slopes are equal, and if the slopes of two lines in a plane are equal, then the lines are parallel.

WORDS	MATH
Lines L_1 and L_2 are parallel.	$L_1 \| L_2$
Two parallel lines have the same slope.	$m_1 = m_2$

EXAMPLE 8   Determining Whether Two Lines Are Parallel

Determine whether the lines $-x + 3y = -3$ and $y = \frac{1}{3}x - 6$ are parallel.

Solution:

Write the first line in slope–intercept form. $\qquad\qquad -x + 3y = -3$

Add x to both sides. $\qquad\qquad\qquad\qquad\qquad\qquad 3y = x - 3$

Divide by 3. $\qquad\qquad\qquad\qquad\qquad\qquad\qquad y = \dfrac{1}{3}x - 1$

Compare the two lines. $\qquad\qquad\qquad y = \dfrac{1}{3}x - 1 \quad$ and $\quad y = \dfrac{1}{3}x - 6$

Both lines have the same slope, $\frac{1}{3}$. These are distinct lines because they have different y-intercepts. Thus, the $\boxed{\text{two lines are parallel}}$.

Classroom Example 2.3.8
Are the lines $2y - 3 = 4x$ and $6x - 3y = -1$ parallel?

Answer: yes

▶ EXAMPLE 9   Finding an Equation of a Parallel Line

Find the equation of the line that passes through the point $(1, 1)$ and is parallel to the line $y = 3x + 1$.

Solution:

Write the slope–intercept equation of a line. $\qquad\qquad y = mx + b$

Parallel lines have equal slope. $\qquad\qquad\qquad\qquad m = 3$

Substitute the slope into the equation of the line. $\qquad y = 3x + b$

Since the line passes through $(1, 1)$, this point must
satisfy the equation. $\qquad\qquad\qquad\qquad\qquad 1 = 3(1) + b$

Solve for b. $\qquad\qquad\qquad\qquad\qquad\qquad\qquad b = -2$

The equation of the line is $\boxed{y = 3x - 2}$.

■ **YOUR TURN** Find the equation of the line parallel to $y = 2x - 1$ that passes through the point $(-1, 3)$.

Classroom Example 2.3.9*
a. Determine the value(s) of b, if any, such that the lines $y + 2 = (b + 6)x$ and $y - b^2x = 3$ are parallel.
b. Determine the value(s) of b, if any, for which the point $(b^2, 2b)$ lies on the line that passes through $\left(0, \frac{1}{2}\right)$ and which is parallel to the line $2y + 5x = 10$.

Answer:
a. $b = -2, 3$
b. $b = -1, \frac{1}{5}$

■ **Answer:** $y = 2x + 5$

Two *perpendicular* lines form a right angle at their point of intersection. Notice the slopes of the two perpendicular lines in the figure to the right. They are $-\frac{1}{2}$ and 2, negative reciprocals of each other. It turns out that almost all perpendicular lines share this property. Horizontal ($m = 0$) and vertical (m undefined) lines do not share this property.

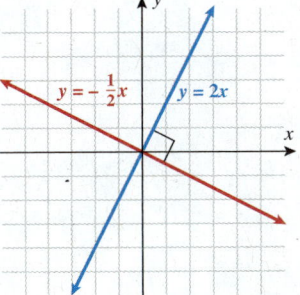

DEFINITION   Perpendicular Lines

Except for the special case of a vertical and a horizontal line, two lines in a plane are **perpendicular** if and only if their slopes are negative reciprocals of each other.

In other words, if two lines in a plane are perpendicular, their slopes are negative reciprocals, provided their slopes are defined. Similarly, if the slopes of two lines in a plane are negative reciprocals, then the lines are perpendicular.

WORDS	MATH
Lines L_1 and L_2 are perpendicular.	$L_1 \perp L_2$
Two perpendicular lines have negative reciprocal slopes.	$m_1 = -\dfrac{1}{m_2} \quad m_1 \neq 0,\, m_2 \neq 0$

Study Tip

If a line has slope equal to 3, then a line perpendicular to it has slope $-\frac{1}{3}$.

■ **Answer:** $y = 2x - 7$

EXAMPLE 10   **Finding an Equation of a Line That Is Perpendicular to Another Line**

Find the equation of the line that passes through the point $(3, 0)$ and is perpendicular to the line $y = 3x + 1$.

Solution:

Identify the slope of the given line $y = 3x + 1$. $m_1 = 3$

The slope of a line perpendicular to the given line is the negative reciprocal of the slope of the given line. $m_2 = -\dfrac{1}{m_1} = -\dfrac{1}{3}$

Write the equation of the line we are looking for in slope–intercept form. $y = m_2x + b$

Substitute $m_2 = -\dfrac{1}{3}$ into $y = m_2x + b$. $y = -\dfrac{1}{3}x + b$

Since the desired line passes through $(3, 0)$, this point must satisfy the equation. $0 = -\dfrac{1}{3}(3) + b$

                                                                                          $0 = -1 + b$

Solve for b. $b = 1$

The equation of the line is $\boxed{y = -\dfrac{1}{3}x + 1}$.

■ **YOUR TURN**   Find the equation of the line that passes through the point $(1, -5)$ and is perpendicular to the line $y = -\dfrac{1}{2}x + 4$.

Applications Involving Linear Equations

Slope is the ratio of the change in y over the change in x. In applications, slope can often be interpreted as the **rate of change**, as illustrated in the next example.

EXAMPLE 11   **Slope as a Rate of Change**

The average age that a person first marries has been increasing over the last several decades. In 1970 the average age was 20, in 1990 it was 25 years old, and in 2010 it is expected that the average age will be 30 years old at the time of a person's first marriage. Find the slope of the line passing through these points. Describe what that slope represents.

Solution:

If we let x represent the year and y represent the age, then two points* that lie on the line are $(1970, 20)$ and $(2010, 30)$.

Write the slope formula. $m = \dfrac{\text{change in } y}{\text{change in } x}$

Substitute the points into the slope formula. $m = \dfrac{30 - 20}{2010 - 1970} = \dfrac{10}{40} = \dfrac{1}{4}$

The slope is $\dfrac{1}{4}$ and can be interpreted as the rate of change of the average age when a person is first married. $\boxed{\text{Every 4 years the average age at the first marriage is 1 year older.}}$

---

*In Example 11 we chose to use the points $(1970, 20)$ and $(2010, 30)$. We could have also used the point $(1990, 25)$ with either of the other points.

EXAMPLE 12   Service Charges

Suppose that your two neighbors both use the same electrician. One neighbor had a 2-hour job that cost her $100, and another neighbor had a 3-hour job that cost him $130. Assuming that a linear equation governs the service charge of this electrician, what will your cost be for a 5-hour job?

Solution:

STEP 1 Identify the question.

Determine the linear equation for this electrician's service charge and calculate the charge for a 5-hour job.

STEP 2 Make notes.

A 2-hour job costs $100 and a 3-hour job costs $130.

STEP 3 Set up an equation.

Let x equal the number of hours and y equal the service charge in dollars.

Linear equation: $y = mx + b$

Two points that must satisfy this equation are (2, 100) and (3, 130).

STEP 4 Solve the equation.

Calculate the rate of change (slope). $\qquad m = \dfrac{130 - 100}{3 - 2} = \dfrac{30}{1} = 30$

Substitute the slope into the linear equation. $\qquad y = 30x + b$

Either point must satisfy the equation. Use (2, 100).

$$100 = 30(2) + b$$
$$100 = 60 + b$$
$$b = 40$$

The service charge y is given by $y = 30x + 40$.

Substitute $x = 5$ into this equation for a 5-hour job. $\qquad y = 30(5) + 40 = 190$

The 5-hour job will cost $\boxed{\$190}$.

STEP 5 Check the solution.

The service charge $y = 30x + 40$ can be interpreted as a fixed cost of $40 for coming to your home and a $30 per hour fee for the job. A 5-hour job would cost $190. Additionally, a 5-hour job should cost less than the sum of a 2-hour job and a 3-hour job ($100 + $130 = $230), since the $40 fee is charged only once.

....

■ **YOUR TURN** You decide to hire a tutor that some of your friends recommended. The tutor comes to your home, so she charges a flat fee per session and then an hourly rate. One friend prefers 2-hour sessions, and the charge is $60 per session. Another friend has 5-hour sessions that cost $105 per session. How much should you be charged for a 3-hour session?

In this section we discussed graphs and equations of lines. Lines are often expressed in two forms:

- General Form: $Ax + By = C$
- Slope–Intercept Form: $y = mx + b$

All lines (except horizontal and vertical) have exactly one x-intercept and exactly one y-intercept. The slope of a line is a measure of steepness.

- Slope of a line passing through (x_1, y_1) and (x_2, y_2):

$$m = \frac{y_2 - y_1}{x_2 - x_1} = \frac{\text{rise}}{\text{run}}$$

- Horizontal lines: $m = 0$
- Vertical lines: m is undefined

We found equations of lines, given either two points or the slope and a point. We found the point–slope form, $y - y_1 = m(x - x_1)$, useful when the slope and a point are given. We also discussed both parallel (nonintersecting) and perpendicular (forming a right angle) lines. Parallel lines have the same slope. Perpendicular lines have negative reciprocal slopes, provided their slopes are defined.

SECTION
2.3 **EXERCISES**

■ **SKILLS**

In Exercises 1–10, find the slope of the line that passes through the given points.

1. $(1, 3)$ and $(2, 6)$

2. $(2, 1)$ and $(4, 9)$

3. $(-2, 5)$ and $(2, -3)$

4. $(-1, -4)$ and $(4, 6)$

5. $(-7, 9)$ and $(3, -10)$

6. $(11, -3)$ and $(-2, 6)$

7. $(0.2, -1.7)$ and $(3.1, 5.2)$

8. $(-2.4, 1.7)$ and $(-5.6, -2.3)$

9. $\left(\frac{2}{3}, -\frac{1}{4}\right)$ and $\left(\frac{5}{6}, -\frac{3}{4}\right)$

10. $\left(\frac{1}{2}, \frac{3}{5}\right)$ and $\left(-\frac{3}{4}, \frac{7}{5}\right)$

For each graph in Exercises 11–16, identify (by inspection) the x- and y-intercepts and slope if they exist, and classify the line as rising, falling, horizontal, or vertical.

11.

12.

13.

14.

15.

16.

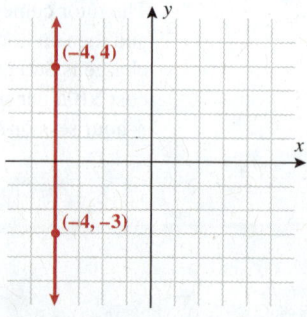

In Exercises 17–30, find the *x*- and *y*-intercepts if they exist and graph the corresponding line.

17. $y = 2x - 3$

18. $y = -3x + 2$

19. $y = -\frac{1}{2}x + 2$

20. $y = \frac{1}{3}x - 1$

21. $2x - 3y = 4$

22. $-x + y = -1$

23. $\frac{1}{2}x + \frac{1}{2}y = -1$

24. $\frac{1}{3}x - \frac{1}{4}y = \frac{1}{12}$

25. $x = -1$

26. $y = -3$

27. $y = 1.5$

28. $x = -7.5$

29. $x = -\frac{7}{2}$

30. $y = \frac{5}{3}$

In Exercises 31–42, write the equation in slope–intercept form. Identify the slope and the *y*-intercept.

31. $2x - 5y = 10$

32. $3x - 4y = 12$

33. $x + 3y = 6$

34. $x + 2y = 8$

35. $4x - y = 3$

36. $x - y = 5$

37. $12 = 6x + 3y$

38. $4 = 2x - 8y$

39. $0.2x - 0.3y = 0.6$

40. $0.4x + 0.1y = 0.3$

41. $\frac{1}{2}x + \frac{2}{3}y = 4$

42. $\frac{1}{4}x + \frac{2}{5}y = 2$

In Exercises 43–50, write the equation of the line, given the slope and intercept.

43. Slope: $m = 2$
y-intercept: $(0, 3)$

44. Slope: $m = -2$
y-intercept: $(0, 1)$

45. Slope: $m = -\frac{1}{3}$
y-intercept: $(0, 0)$

46. Slope: $m = \frac{1}{2}$
y-intercept: $(0, -3)$

47. Slope: $m = 0$
y-intercept: $(0, 2)$

48. Slope: $m = 0$
y-intercept: $(0, -1.5)$

49. Slope: undefined
x-intercept: $\left(\frac{3}{2}, 0\right)$

50. Slope: undefined
x-intercept: $(-3.5, 0)$

In Exercises 51–60, write an equation of the line in slope–intercept form, if possible, given the slope and a point that lies on the line.

51. Slope: $m = 5$
$(-1, -3)$

52. Slope: $m = 2$
$(1, -1)$

53. Slope: $m = -3$
$(-2, 2)$

54. Slope: $m = -1$
$(3, -4)$

55. Slope: $m = \frac{3}{4}$
$(1, -1)$

56. Slope: $m = -\frac{1}{7}$
$(-5, 3)$

57. Slope: $m = 0$
$(-2, 4)$

58. Slope: $m = 0$
$(3, -3)$

59. Slope: undefined
$(-1, 4)$

60. Slope: undefined
$(4, -1)$

In Exercises 61–80, write the equation of the line that passes through the given points. Express the equation in slope–intercept form or in the form $x = a$ or $y = b$.

61. $(-2, -1)$ and $(3, 2)$

62. $(-4, -3)$ and $(5, 1)$

63. $(-3, -1)$ and $(-2, -6)$

64. $(-5, -8)$ and $(7, -2)$

65. $(20, -37)$ and $(-10, -42)$

66. $(-8, 12)$ and $(-20, -12)$

67. $(-1, 4)$ and $(2, -5)$

68. $(-2, 3)$ and $(2, -3)$

69. $\left(\frac{1}{2}, \frac{3}{4}\right)$ and $\left(\frac{3}{2}, \frac{9}{4}\right)$

70. $\left(-\frac{2}{3}, -\frac{1}{2}\right)$ and $\left(\frac{7}{3}, \frac{1}{2}\right)$

71. $(3, 5)$ and $(3, -7)$

72. $(-5, -2)$ and $(-5, 4)$

73. $(3, 7)$ and $(9, 7)$

74. $(-2, -1)$ and $(3, -1)$

75. $(0, 6)$ and $(-5, 0)$

76. $(0, -3)$ and $(0, 2)$

77. $(-6, 8)$ and $(-6, -2)$

78. $(-9, 0)$ and $(-9, 2)$

79. $\left(\frac{2}{5}, -\frac{3}{4}\right)$ and $\left(\frac{2}{5}, \frac{1}{2}\right)$

80. $\left(\frac{1}{3}, \frac{2}{5}\right)$ and $\left(\frac{1}{3}, \frac{1}{2}\right)$

In Exercises 81–86, write the equation corresponding to each line. Express the equation in slope–intercept form.

81.

82.

83.

84.

85.

86.

In Exercises 87–100, find the equation of the line that passes through the given point and also satisfies the additional piece of information. Express your answer in slope–intercept form, if possible.

87. $(-3, 1)$; parallel to the line $y = 2x - 1$

88. $(1, 3)$; parallel to the line $y = -x + 2$

89. $(0, 0)$; perpendicular to the line $2x + 3y = 12$

90. $(0, 6)$; perpendicular to the line $x - y = 7$

91. $(3, 5)$; parallel to the x-axis

92. $(3, 5)$; parallel to the y-axis

93. $(-1, 2)$; perpendicular to the y-axis

94. $(-1, 2)$; perpendicular to the x-axis

95. $(-2, -7)$; parallel to the line $\frac{1}{2}x - \frac{1}{3}y = 5$

96. $(1, 4)$; perpendicular to the line $-\frac{2}{3}x + \frac{3}{2}y = -2$

97. $\left(-\frac{2}{3}, \frac{2}{3}\right)$; perpendicular to the line $8x + 10y = -45$

98. $\left(\frac{6}{5}, 3\right)$; perpendicular to the line $6x + 14y = 7$

99. $\left(\frac{7}{2}, 4\right)$; parallel to the line $-15x + 35y = 7$

100. $\left(-\frac{1}{4}, -\frac{13}{9}\right)$; parallel to the line $10x + 45y = -9$

▪ APPLICATIONS

101. Budget: Home Improvement. The cost of having your bathroom remodeled is the combination of material costs and labor costs. The materials (tile, grout, toilet, fixtures, etc.) cost is $1200, and the labor cost is $25 per hour. Write an equation that models the total cost C of having your bathroom remodeled as a function of hours h. How much will the job cost if the worker estimates 32 hours?

102. Budget: Rental Car. The cost of a one-day car rental is the sum of the rental fee, $50, plus $0.39 per mile. Write an equation that models the total cost associated with the car rental.

103. Budget: Monthly Driving Costs. The monthly costs associated with driving a new Honda Accord are the monthly loan payment plus $25 every time you fill up with gasoline. If you fill up 5 times in a month, your total monthly cost is $500. How much is your loan payment?

104. Budget: Monthly Driving Costs. The monthly costs associated with driving a Ford Explorer are the monthly loan payment plus the cost of filling up your tank with gasoline. If you fill up 3 times in a month, your total monthly cost is $520. If you fill up 5 times in a month, your total monthly cost is $600. How much is your monthly loan, and how much does it cost every time you fill up with gasoline?

105. Business. The operating costs for a local business are a fixed amount of $1300 plus $3.50 per unit sold, while revenue is $7.25 per unit sold. How many units does the business have to sell in order to break even?

106. Business. The operating costs for a local business are a fixed amount of $12,000 plus $13.50 per unit sold, while revenue is $27.25 per unit sold. How many units does the business have to sell in order to break even?

107. Weather: Temperature. The National Oceanic and Atmospheric Administration (NOAA) has an online conversion chart that relates degrees Fahrenheit, °F, to degrees Celsius, °C. 77°F is equivalent to 25°C, and 68°F is equivalent to 20°C. Assuming the relationship is linear, write the equation relating degrees Celsius C to degrees Fahrenheit F. What temperature is the same in both degrees Celsius and degrees Fahrenheit?

108. Weather: Temperature. According to NOAA, a "standard day" is 15°C at sea level, and every 500 feet elevation above sea level corresponds to a 1°C temperature drop. Assuming the relationship between temperature and elevation is linear, write an equation that models this relationship. What is the expected temperature at 2500 feet on a "standard day"?

109. Life Sciences: Height. The average height of a man has increased over the last century. What is the rate of change in inches per year of the average height of men?

110. Life Sciences: Height. The average height of a woman has increased over the last century. What is the rate of change in inches per year of the average height of women?

111. Life Sciences: Weight. The average weight of a baby born in 1900 was 6 pounds 4 ounces. In 2000, the average weight of a newborn was 6 pounds 10 ounces. What is the rate of change of birth weight in ounces per year? What do we expect babies to weigh at birth in 2040?

112. Sports. The fastest a man could run a mile in 1906 was 4 minutes and 30 seconds. In 1957, Don Bowden became the first American to break the 4-minute mile. Calculate the rate of change in mile speed per year.

113. Monthly Phone Costs. Mike's home phone plan charges a flat monthly fee plus a charge of $0.05 per minute for long-distance calls. The total monthly charge is represented by $y = 0.05x + 35$, $x \geq 0$, where y is the total monthly charge and x is the number of long-distance minutes used. Interpret the meaning of the y-intercept.

114. Cost: Automobile. The value of a Daewoo car is given by $y = 11,100 - 1850x$, $x \geq 0$, where y is the value of the car and x is the age of the car in years. Find the x-intercept and y-intercept and interpret the meaning of each.

115. Weather: Rainfall. The average rainfall in Norfolk, Virginia, for July was 5.2 inches in 2003. The average July rainfall for Norfolk was 3.8 inches in 2007. What is the rate of change of rainfall in inches per year? If this trend continues, what is the expected average rainfall in 2010?

116. Weather: Temperature. The average temperature for Boston in January 2005 was 43°F. In 2007 the average January temperature was 44.5°F. What is the rate of change of the temperature per year? If this trend continues, what is the expected average temperature in January 2010?

117. Environment. In 2000, Americans used approximately 380 billion plastic bags. In 2005, approximately 392 billion were used. What is the rate of change of plastic bags used per year? How many plastic bags will be expected to be used in 2010?

118. Finance: Debt. According to the Federal Reserve, Americans individually owed $744 in revolving credit in 2004. In 2006, they owed approximately $788. What is the rate of change of the amount of revolving credit owed per year? How much should Americans be expected to owe in 2008?

119. Business. A website that supplies Asian specialty foods to restaurants advertises a 64 ounce bottle of Hoisin Sauce for $16.00. Shipping cost for one bottle is $15.93. The shipping cost for two bottles is $19.18. The cost for five bottles, including shipping, is $111.83. Answer the following questions based on this scenario. Round to the nearest cent, when necessary.

a. Write the three ordered pairs where x represents the number of bottles purchased and y represents the total cost (including shipping) for one, two, or five bottles purchased.

b. Calculate the slope between the origin and the ordered pair that represents the purchase of one bottle of Hoisin Sauce. Explain what this amount means in terms of the sauce purchase.

c. Calculate the slope between the origin and the ordered pair that represents the purchase of two bottles of Hoisin (including shipping). Explain what this amount means in terms of the sauce purchase.

d. Calculate the slope between the origin and the ordered pair that represents the purchase of five bottles of Hoisin (including shipping). Explain what this amount means in terms of the sauce purchase.

120. Business. A website that supplies Asian specialty foods to restaurants advertises an 8 ounce bottle of Plum Sauce for $4.00, but shipping for one bottle is $14.27. The shipping cost for two bottles is $14.77. The cost for five bottles, including shipping, is $35.93. Answer the following questions based on this scenario. Round to the nearest cent, when necessary.

a. Write the three ordered pairs where x represents the number of bottles purchased and y represents the total cost, including shipping for one, two, or five bottles purchased.

b. Calculate the slope between the origin and the ordered pair that represents the purchase of one bottle of Plum Sauce. Explain what this amount means in terms of the sauce purchase.

c. Calculate the slope between the origin and the ordered pair that represents the purchase of two bottles of Plum Sauce (including shipping). Explain what this amount means in terms of the sauce purchase.

d. Calculate the slope between the origin and the ordered pair that represents the purchase of five bottles of Plum Sauce (including shipping). Explain what this amount means in terms of the sauce purchase.

▪ CATCH THE MISTAKE

In Exercises 121–124, explain the mistake that is made.

121. Find the x- and y-intercepts of the line with equation $2x - 3y = 6$.

Solution:

x-intercept: set $x = 0$ and solve for y.　　$-3y = 6$
$$y = -2$$

The x-intercept is $(0, -2)$.

y-intercept: set $y = 0$ and solve for x.　　$2x = 6$
$$x = 3$$

The y-intercept is $(3, 0)$.

This is incorrect. What mistake was made?

122. Find the slope of the line that passes through the points $(-2, 3)$ and $(4, 1)$.

Solution:

Write the slope formula.　　$m = \dfrac{y_2 - y_1}{x_2 - x_1}$

Substitute $(-2, 3)$ and $(4, 1)$.　　$m = \dfrac{1 - 3}{-2 - 4} = \dfrac{-2}{-6} = \dfrac{1}{3}$

This is incorrect. What mistake was made?

123. Find the slope of the line that passes through the points $(-3, 4)$ and $(-3, 7)$.

Solution:

Write the slope formula.　　$m = \dfrac{y_2 - y_1}{x_2 - x_1}$

Substitute $(-3, 4)$ and $(-3, 7)$.　　$m = \dfrac{-3 - (-3)}{4 - 7} = 0$

This is incorrect. What mistake was made?

124. Given the slope, classify the line as rising, falling, horizontal, or vertical.

a. $m = 0$

b. m undefined

c. $m = 2$

d. $m = -1$

Solution:

a. vertical line

b. horizontal line

c. rising

d. falling

These are incorrect. What mistakes were made?

▪ CONCEPTUAL

In Exercises 125–128, determine whether each statement is true or false.

125. A line can have at most one x-intercept.

126. A line must have at least one y-intercept.

127. If the slopes of two lines are $-\frac{1}{5}$ and 5, then the lines are parallel.

128. If the slopes of two lines are -1 and 1, then the lines are perpendicular.

129. If a line has slope equal to zero, describe a line that is perpendicular to it.

130. If a line has no slope, describe a line that is parallel to it.

▪ CHALLENGE

131. Find an equation of a line that passes through the point $(-B, A + 1)$ and is parallel to the line $Ax + By = C$. Assume that B is not equal to zero.

132. Find an equation of a line that passes through the point $(B, A - 1)$ and is parallel to the line $Ax + By = C$. Assume that B is not equal to zero.

133. Find an equation of a line that passes through the point $(-A, B - 1)$ and is perpendicular to the line $Ax + By = C$. Assume that A and B are both nonzero.

134. Find an equation of a line that passes through the point $(A, B + 1)$ and is perpendicular to the line $Ax + By = C$. Assume that A and B are both nonzero.

135. Show that two lines with equal slopes and different y-intercepts have no point in common. *Hint:* Let $y_1 = mx + b_1$ and $y_2 = mx + b_2$ with $b_1 \neq b_2$. What equation must be true for there to be a point of intersection? Show that this leads to a contradiction.

136. Let $y_1 = m_1x + b_1$ and $y_2 = m_2x + b_2$ be two nonparallel lines ($m_1 \neq m_2$). What is the x-coordinate of the point where they intersect?

■ **TECHNOLOGY**

For Exercises 137–142, determine whether the lines are parallel, perpendicular, or neither, and then graph both lines in the same viewing screen using a graphing utility to confirm your answer.

137. $y_1 = 17x + 22$

$y_2 = -\frac{1}{17}x - 13$

138. $y_1 = 0.35x + 2.7$

$y_2 = 0.35x - 1.2$

139. $y_1 = 0.25x + 3.3$

$y_2 = -4x + 2$

140. $y_1 = \frac{1}{2}x + 5$

$y_2 = 2x - 3$

141. $y_1 = 0.16x + 2.7$

$y_2 = 6.25x - 1.4$

142. $y_1 = -3.75x + 8.2$

$y_2 = \frac{4}{15}x + \frac{5}{6}$

SECTION

2.4 CIRCLES

SKILLS OBJECTIVES

■ Identify the center and radius of a circle from the standard equation.
■ Graph a circle.
■ Transform equations of circles to the standard form by completing the square.

CONCEPTUAL OBJECTIVE

■ Understand algebraic and graphical representations of circles.

Standard Equation of a Circle

Most people understand the shape of a circle. The goal in this section is to develop the equation of a circle.

DEFINITION   Circle

A **circle** is the set of all points in a plane that are a fixed distance from a point, the **center**. The center, C, is typically denoted by (h, k), and the fixed distance, or **radius**, is denoted by r.

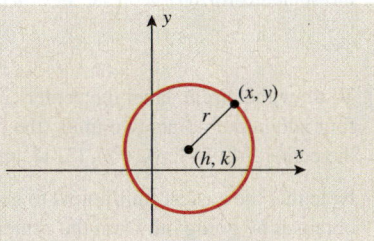

What is the equation of a circle? We'll use the distance formula from Section 2.1.

Distance formula: $d = \sqrt{(x_2 - x_1)^2 + (y_2 - y_1)^2}$

The distance between the center (h, k) and any point (x, y) on the circle is the radius r. Substitute these values $d = r$, $(x_1, y_1) = (h, k)$, and $(x_2, y_2) = (x, y)$ into the distance formula $r = \sqrt{(x - h)^2 + (y - k)^2}$ and square both sides: $(x - h)^2 + (y - k)^2 = r^2$. All circles can be written in *standard form*, which makes it easy to identify the center and radius.

EQUATION OF A CIRCLE

The standard form of the equation of a **circle** with **radius** r and **center** (h, k) is

$$(x - h)^2 + (y - k)^2 = r^2$$

For the special case of a circle with center at the origin $(0, 0)$, the equation simplifies to $x^2 + y^2 = r^2$.

UNIT CIRCLE

A circle with radius 1 and center $(0, 0)$ is called the **unit circle**:

$$x^2 + y^2 = 1$$

The unit circle plays an important role in the study of trigonometry. Note that if $x^2 + y^2 = 0$, the radius is 0, so the "circle" is just a point.

Technology Tip

To enter the graph of $(x - 2)^2 + (y + 1)^2 = 4$, solve for y first. The graphs of

$y_1 = \sqrt{4 - (x - 2)^2} - 1$ and

$y_2 = -\sqrt{4 - (x - 2)^2} - 1$ are shown.

■ **Answer:** Center: $(-1, -2)$
Radius: 3

EXAMPLE 1   Finding the Center and Radius of a Circle

Identify the center and radius of the given circle and graph.

$$(x - 2)^2 + (y + 1)^2 = 4$$

Solution:

Rewrite this equation in standard form. $[x - 2]^2 + [y - (-1)]^2 = 2^2$

Identify h, k, and r by comparing this equation with the standard form of a circle: $(x - h)^2 + (y - k)^2 = r^2$. $h = 2$, $k = -1$, and $r = 2$

$\boxed{\text{Center } (2, -1) \text{ and } r = 2}$

To draw the circle, label the center $(2, -1)$. Label four additional points two units (the radius) away from the center: $(4, -1)$, $(0, -1)$, $(2, 1)$, and $(2, -3)$.

Note that the easiest four points to get are those obtained by going out from the center both horizontally and vertically. Connect those four points with a smooth curve.

■ **YOUR TURN** Identify the center and radius of the given circle and graph.

$$(x + 1)^2 + (y + 2)^2 = 9$$

EXAMPLE 2   Graphing a Circle: Fractions and Radicals

Identify the center and radius of the given circle and sketch its graph.

$$\left(x - \frac{1}{2}\right)^2 + \left(y + \frac{1}{3}\right)^2 = 20$$

Solution:

If $r^2 = 20$, then $r = \sqrt{20} = 2\sqrt{5}$.
Write the equation in standard form.

$$\left(x - \frac{1}{2}\right)^2 + \left[y - \left(-\frac{1}{3}\right)\right]^2 = \left(2\sqrt{5}\right)^2$$

Identify the center and radius.

Center $\left(\dfrac{1}{2}, -\dfrac{1}{3}\right)$ and $r = 2\sqrt{5}$

To graph the circle, we'll use decimal
approximations of the fractions and
radicals: $(0.5, -0.3)$ for the center and
4.5 for the radius. Four points on the
circle that are 4.5 units from the center
are $(-4, -0.3)$, $(5, -0.3)$, $(0.5, 4.2)$,
and $(0.5, -4.8)$. Connect them with
a smooth curve.

Technology Tip

To enter the graph of
$\left(x - \frac{1}{2}\right)^2 + \left(y + \frac{1}{3}\right)^2 = 20$,
solve for y first. The graphs of
$y_1 = \sqrt{20 - \left(x - \frac{1}{2}\right)^2} - \frac{1}{3}$ and
$y_2 = -\sqrt{20 - \left(x - \frac{1}{2}\right)^2} - \frac{1}{3}$ are shown.

EXAMPLE 3   Determining the Equation of a Circle
Given the Center and Radius

Find the equation of a circle with radius 5 and center $(-2, 3)$. Graph the circle.

Solution:

Substitute $(h, k) = (-2, 3)$ and $r = 5$ into the
standard equation of a circle.

$$[x - (-2)]^2 + (y - 3)^2 = 5^2$$

Simplify.

$$(x + 2)^2 + (y - 3)^2 = 25$$

To graph the circle, plot the center $(-2, 3)$
and four points 5 units away from the center:
$(-7, 3)$, $(3, 3)$, $(-2, -2)$, and $(-2, 8)$. Connect
them with a smooth curve.

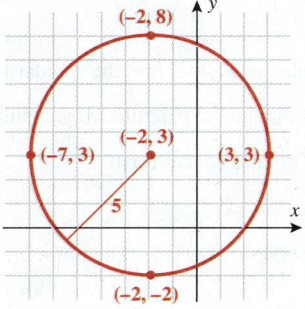

■ **Answer:** $x^2 + (y - 1)^2 = 9$

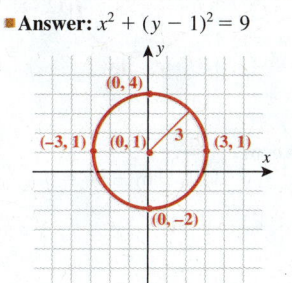

■ **YOUR TURN** Find the equation of a circle with radius 3 and center $(0, 1)$ and graph.

Let's change the look of the equation given in Example 1.

In Example 1 the equation of the circle was given as: $(x - 2)^2 + (y + 1)^2 = 4$

Eliminate the parentheses. $x^2 - 4x + 4 + y^2 + 2y + 1 = 4$

Group like terms and subtract 4 from both sides. $x^2 + y^2 - 4x + 2y + 1 = 0$

We have written the *general form* of the equation of the circle in Example 1.

> The **general form** of the **equation of a circle** is $x^2 + y^2 + ax + by + c = 0$

Suppose you are given a point that lies on a circle and the center of the circle. Can you find the equation of the circle?

EXAMPLE 4 Finding the Equation of a Circle Given Its Center and One Point

The point $(10, -4)$ lies on a circle centered at $(7, -8)$. Find the equation of the circle in general form.

Solution:

This circle is centered at $(7, -8)$, so its standard equation is $(x - 7)^2 + (y + 8)^2 = r^2$.

All that remains is to find the radius.

Approach 1:
Since the point $(10, -4)$ lies on the circle, it must satisfy the equation of the circle.

Substitute $(x, y) = (10, -4)$. $(10 - 7)^2 + (-4 + 8)^2 = r^2$

Simplify. $3^2 + 4^2 = r^2$

The distance from $(10, -4)$ to $(7, -8)$ is five units. $r = 5$

Approach 2:
Find the distance between $(10, -4)$ and $(7, -8)$.

$$r = d = \sqrt{(10 - 7)^2 + (-4 - (-8))^2}$$
$$= \sqrt{3^2 + 4^2}$$
$$= \sqrt{25}$$
$$= 5$$

Substitute $r = 5$ into the standard equation. $(x - 7)^2 + (y + 8)^2 = 5^2$

Eliminate the parentheses and simplify. $x^2 - 14x + 49 + y^2 + 16y + 64 = 25$

Write in **general form**. $\boxed{x^2 + y^2 - 14x + 16y + 88 = 0}$

YOUR TURN The point $(1, 11)$ lies on a circle centered at $(-5, 3)$. Find the equation of the circle in general form.

Transforming Equations of Circles to the Standard Form by Completing the Square

If the equation of a circle is given in general form, it must be rewritten in standard form in order to identify its center and radius. To transform equations of circles from general to standard form, complete the square (Section 1.3) on both the *x*- and *y*-variables.

EXAMPLE 5 **Finding the Center and Radius of a Circle by Completing the Square**

Find the center and radius of the circle with the equation:

$$x^2 - 8x + y^2 + 20y + 107 = 0$$

Solution:

Our goal is to transform this equation into standard form

$$(x - h)^2 + (y - k)^2 = r^2$$

Group x and y terms, respectively, on the left side of the equation; move constants to the right side.

$$(x^2 - 8x) + (y^2 + 20y) = -107$$

Complete the square on both the x and y expressions.

$$(x^2 - 8x + \square) + (y^2 + 20y + \square) = -107$$

Add $\left(-\frac{8}{2}\right)^2 = \mathbf{16}$ and $\left(\frac{20}{2}\right)^2 = \mathbf{100}$ to both sides.

$$\left(x^2 - 8x + \underset{16}{\mathbf{4^2}}\right) + \left(y^2 + 20y + \underset{100}{\mathbf{10^2}}\right) = -107 + \mathbf{16} + \mathbf{100}$$

Factor the perfect squares on the left side and simplify the right side.

$$(x - 4)^2 + (y + 10)^2 = 9$$

Write in standard form.

$$(x - 4)^2 + [y - (-10)]^2 = 3^2$$

> The center is $(4, -10)$ and the radius is 3.

■ **YOUR TURN** Find the center and radius of the circle with the equation:

$$x^2 + y^2 + 4x - 6y - 12 = 0$$

COMMON MISTAKE

A common mistake is forgetting to add *both* constants to the right side of the equation. Identify the center and radius of the circle with the equation:

$$x^2 + y^2 + 16x + 8y + 44 = 0$$

★ CORRECT

$$x^2 + y^2 + 16x + 8y + 44 = 0$$

$$\left(x^2 + 16x\right) + \left(y^2 + 8y\right) = -44$$

$$\left(x^2 + 16x + \square\right) + \left(y^2 + 8y + \square\right) = -44$$

$$\left(x^2 + 16x + \mathbf{64}\right) + \left(y^2 + 8y + \mathbf{16}\right) = -44 + \mathbf{64} + \mathbf{16}$$

$$(x + 8)^2 + (y + 4)^2 = 36$$

Center: $(-8, -4)$ Radius: 6

✗ INCORRECT

$$\left(x^2 + 16x + \mathbf{64}\right) + \left(y^2 + 8y + \mathbf{16}\right) = -44$$

ERROR: Don't forget to add 16 and 64 to the right.

Technology Tip

To graph $x^2 - 8x + y^2 + 20y + 107 = 0$ without transforming it into a standard form, solve for y using the Quadratic Formula:

$$y = \frac{-20 \pm \sqrt{20^2 - 4(1)(x^2 - 8x + 107)}}{2}$$

Next, set the window to $[-5, 30]$ by $[-30, 5]$ and use $\boxed{ZSquare}$ under $\boxed{ZOOM}$ to adjust the window variable to make the circle look circular. The graphs of

$$y_1 = \frac{-20 + \sqrt{20^2 - 4(x^2 - 8x + 107)}}{2}$$

and

$$y_2 = \frac{-20 - \sqrt{20^2 - 4(x^2 - 8x + 107)}}{2}$$

are shown.

■ **Answer:** Center: $(-2, 3)$ Radius: 5

▼ CAUTION

Don't forget to add *both* constants to each side of the equation when completing the square for x and y.

> **Classroom Example 2.4.5**
> Find the center and radius of the circle whose equation is $2x^2 + 32x + 2y^2 - 12 = -128$
>
> **Answer:**
> Center: $(-8, 0)$ Radius: $\sqrt{6}$

SECTION
2.4 SUMMARY

The equation of a circle is given by
- Standard form: $(x - h)^2 + (y - k)^2 = r^2$.
  - Center: (h, k)
  - Radius: r

- General form: $x^2 + y^2 + ax + by + c = 0$.
  - Complete the square to transform the equation to standard form.

SECTION
2.4 EXERCISES

• **SKILLS**

In Exercises 1–20, write the equation of the circle in standard form.

1. Center $(1, 2)$
$r = 3$

2. Center $(3, 4)$
$r = 5$

3. Center $(-3, -4)$
$r = 10$

4. Center $(-1, -2)$
$r = 4$

5. Center $(5, 7)$
$r = 9$

6. Center $(2, 8)$
$r = 6$

7. Center $(-11, 12)$
$r = 13$

8. Center $(6, -7)$
$r = 8$

9. Center $(0, 0)$
$r = 2$

10. Center $(0, 0)$
$r = 3$

11. Center $(0, 2)$
$r = 3$

12. Center $(3, 0)$
$r = 2$

13. Center $(0, 0)$
$r = \sqrt{2}$

14. Center $(-1, 2)$
$r = \sqrt{7}$

15. Center $(5, -3)$
$r = 2\sqrt{3}$

16. Center $(-4, -1)$
$r = 3\sqrt{5}$

17. Center $\left(\frac{2}{3}, -\frac{3}{5}\right)$
$r = \frac{1}{4}$

18. Center $\left(-\frac{1}{3}, -\frac{2}{7}\right)$
$r = \frac{2}{5}$

19. Center $(1.3, 2.7)$
$r = 3.2$

20. Center $(-3.1, 4.2)$
$r = 5.5$

In Exercises 21–32, find the center and radius of the circle with the given equations.

21. $(x - 1)^2 + (y - 3)^2 = 25$

22. $(x + 1)^2 + (y + 3)^2 = 11$

23. $(x - 2)^2 + (y + 5)^2 = 49$

24. $(x + 3)^2 + (y - 7)^2 = 81$

25. $(x - 4)^2 + (y - 9)^2 = 20$

26. $(x + 1)^2 + (y + 2)^2 = 8$

27. $\left(x - \frac{2}{5}\right)^2 + \left(y - \frac{1}{7}\right)^2 = \frac{4}{9}$

28. $\left(x - \frac{1}{2}\right)^2 + \left(y - \frac{1}{3}\right)^2 = \frac{9}{25}$

29. $(x - 1.5)^2 + (y + 2.7)^2 = 1.69$

30. $(x + 3.1)^2 + (y - 7.4)^2 = 56.25$

31. $x^2 + y^2 - 50 = 0$

32. $x^2 + y^2 - 8 = 0$

In Exercises 33–50, state the center and radius of each circle.

33. $x^2 + y^2 + 4x + 6y - 3 = 0$

34. $x^2 + y^2 + 2x + 10y + 17 = 0$

35. $x^2 + y^2 + 6x + 8y - 75 = 0$

36. $x^2 + y^2 + 2x + 4y - 9 = 0$

37. $x^2 + y^2 - 10x - 14y - 7 = 0$

38. $x^2 + y^2 - 4x - 16y + 32 = 0$

39. $x^2 + y^2 - 2y - 15 = 0$

40. $x^2 + y^2 + 2x - 8 = 0$

41. $x^2 + y^2 - 2x - 6y + 1 = 0$

42. $x^2 + y^2 - 8x - 6y + 21 = 0$

43. $x^2 + y^2 - 10x + 6y + 22 = 0$

44. $x^2 + y^2 + 8x + 2y - 28 = 0$

45. $x^2 + y^2 - 6x - 4y + 1 = 0$

46. $x^2 + y^2 - 2x - 10y + 2 = 0$

47. $x^2 + y^2 - x + y + \dfrac{1}{4} = 0$

48. $x^2 + y^2 - \dfrac{x}{2} - \dfrac{3y}{2} + \dfrac{3}{8} = 0$

49. $x^2 + y^2 - 2.6x - 5.4y - 1.26 = 0$

50. $x^2 + y^2 - 6.2x - 8.4y - 3 = 0$

In Exercises 51–56, find the equation of each circle.

51. Centered at $(-1, -2)$ and passing through the point $(1, 0)$.

52. Centered at $(4, 9)$ and passing through the point $(2, 5)$.

53. Centered at $(-2, 3)$ and passing through the point $(3, 7)$.

54. Centered at $(1, 1)$ and passing through the point $(-8, -5)$.

55. Centered at $(-2, -5)$ and passing through the point $(1, -9)$.

56. Centered at $(-3, -4)$ and passing through the point $(-1, -8)$.

▪ APPLICATIONS

57. Cell Phones. If a cellular phone tower has a reception radius of 100 miles and you live 95 miles north and 33 miles east of the tower, can you use your cell phone while at home?

58. Cell Phones. Repeat Exercise 57, assuming you live 45 miles south and 87 miles west of the tower.

59. Construction/Home Improvement. A couple and their dog moved into a new house that does not have a fenced-in backyard. The backyard is square with dimensions 100 feet by 100 feet. If they put a stake in the center of the backyard with a long leash, write the equation of the circle that will map out the dog's outer perimeter.

100 ft

100 ft

60. Construction/Home Improvement. Repeat Exercise 59 except that the couple put in a pool and a garden and want to restrict the dog to quadrant I. What coordinates represent the center of the circle? What is the radius?

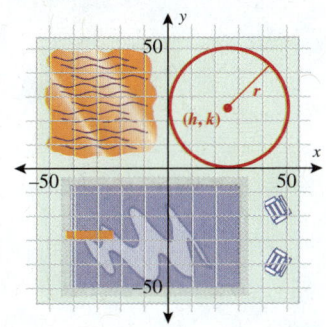

61. Design. A university designs its campus with a master plan of two concentric circles. All of the academic buildings are within the inner circle (so that students can get between classes in less than 10 minutes), and the outer circle contains all the dormitories, the Greek park, cafeterias, the gymnasium, and intramural fields. Assuming the center of campus is the origin, write an equation for the inner circle if the diameter is 3000 feet.

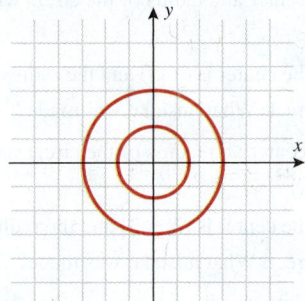

62. Design. Repeat Exercise 61 for the outer circle with a diameter of 6000 feet.

63. Cell Phones. A cellular phone tower has a reception radius of 200 miles. Assuming the tower is located at the origin, write the equation of the circle that represents the reception area.

64. Environment. In a state park, a fire has spread in the form of a circle. If the radius is 2 miles, write an equation for the circle.

For Exercises 65 and 66, refer to the following:

A cell phone provider is expanding its coverage and needs to place four cell phone towers to provide complete coverage of a 100-square-mile area formed by a 10 mile by 10 mile square. This area can be represented by a region on the Cartesian coordinate system; see figure.

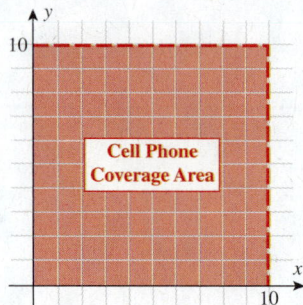

The placement of the four towers is very important in that the cell phone provider needs to provide coverage of the entire 100-square-mile area. The cell phone towers being installed can process signals from cell phones within a 3.5-mile radius.

65. **Engineering.** One plan under consideration is to place the four towers in locations that correspond to the points (2.5, 2.5), (2.5, 7.5), (7.5, 2.5), and (7.5, 7.5) on the graph.

 a. Write an equation that describes the perimeter of the cell phone coverage for each of the four towers.

 b. Draw the coverage provided by each of these towers. Will this placement of towers provide the needed coverage?

66. **Engineering.** One plan under consideration is to place the four towers in locations that correspond to the points (3, 3), (3, 7), (7, 3), and (7, 7) on the graph.

 a. Write an equation that describes the perimeter of the cell phone coverage for each of the four towers.

 b. Draw the coverage provided by each of these towers. Will this placement of towers provide the needed coverage?

▪ CATCH THE MISTAKE

In Exercises 67–70, explain the mistake that is made.

67. Identify the center and radius of the circle with equation $(x - 4)^2 + (y + 3)^2 = 25$.

 Solution: The center is (4, 3) and the radius is 5.

 This is incorrect. What mistake was made?

68. Identify the center and radius of the circle with equation $(x - 2)^2 + (y + 3)^2 = 2$.

 Solution: The center is $(2, -3)$ and the radius is 2.

 This is incorrect. What mistake was made?

69. Graph the solution to the equation $(x - 1)^2 + (y + 2)^2 = -16$.

 Solution: The center is $(1, -2)$ and the radius is 4.

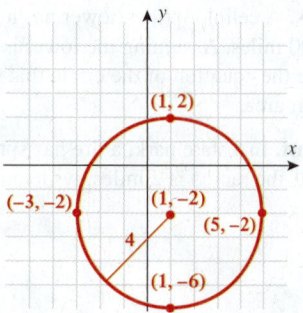

 This is incorrect. What mistake was made?

70. Find the center and radius of the circle with the equation $x^2 + y^2 - 6x + 4y - 3 = 0$.

 Solution:

 Group like terms. $(x^2 - 6x) + (y^2 + 4y) = 3$

 Complete the $(x^2 - 6x + 9) + (y^2 + 4y + 4) = 12$
 square. $(x - 3)^2 + (y + 2)^2 = (2\sqrt{3})^2$

 The center is $(3, -2)$ and the radius is $2\sqrt{3}$.

 This is incorrect. What mistake was made?

▪ CONCEPTUAL

In Exercises 71–74, determine whether each statement is true or false.

71. The equation whose graph is depicted has infinitely many solutions.

72. The equation $(x - 7)^2 + (y + 15)^2 = -64$ has no solution.

73. The equation $(x - 2)^2 + (y + 5)^2 = -20$ has no solution.

74. The equation $(x - 1)^2 + (y + 3)^2 = 0$ has only one solution.

75. Describe the graph (if it exists) of:

$$x^2 + y^2 + 10x - 6y + 34 = 0$$

76. Describe the graph (if it exists) of:

$$x^2 + y^2 - 4x + 6y + 49 = 0$$

77. Find the equation of a circle that has a diameter with endpoints $(5, 2)$ and $(1, -6)$.

78. Find the equation of a circle that has a diameter with endpoints $(3, 0)$ and $(-1, -4)$.

▪ CHALLENGE

79. For the equation $x^2 + y^2 + ax + by + c = 0$, specify conditions on a, b, and c so that the graph is a single point.

80. For the equation $x^2 + y^2 + ax + by + c = 0$, specify conditions on a, b, and c so that there is no corresponding graph.

81. Determine the center and radius of the circle given by the equation $x^2 + y^2 - 2ax = 100 - a^2$.

82. Determine the center and radius of the circle given by the equation $x^2 + y^2 + 2by = 49 - b^2$.

▪ TECHNOLOGY

In Exercises 83–86, use a graphing utility to graph each equation. Does this agree with the answer you gave in the Conceptual section?

83. $(x - 2)^2 + (y + 5)^2 = -20$ (See Exercise 73 for comparison.)

84. $(x - 1)^2 + (y + 3)^2 = 0$ (See Exercise 74 for comparison.)

85. $x^2 + y^2 + 10x - 6y + 34 = 0$ (See Exercise 75 for comparison.)

86. $x^2 + y^2 - 4x + 6y + 49 = 0$ (See Exercise 76 for comparison.)

In Exercises 87–88, (a) with the equation of the circle in standard form, state the center and radius, and graph; (b) use the quadratic formula to solve for y; and (c) use a graphing utility to graph each equation found in (b). Does the graph in (a) agree with the graphs in (c)?

87. $x^2 + y^2 - 11x + 3y - 7.19 = 0$

88. $x^2 + y^2 + 1.2x - 3.2y + 2.11 = 0$

SKILLS OBJECTIVES

- Draw a scatterplot.
- Use linear regression to determine the line of best fit associated with some data.
- Use the line of best fit to predict values of one variable from the values of another.

CONCEPTUAL OBJECTIVES

- Recognize positive or negative association.
- Recognize linear or nonlinear association.
- Understand what "best fit" means.

Scatterplots

An important aspect of applied research across disciplines is to discover and understand relationships between variables, and often how to use such a relationship to predict values of one variable in terms of another. You have likely encountered such issues while watching TV, reading a magazine or newspaper, or simply talking with friends. Some *questions* include

- Is age predictive of texting speed?
- Is the level of pollution in a country related to the prevalence of asthma in that country?
- Do the ratings of car reliability necessarily increase with the price of the car?

In this section we focus on situations involving relationships between two variables x and y, so that the experimental data gathered consists of ordered pairs $(x_1, y_1), \ldots, (x_n, y_n)$.

A first step in understanding a data set of the form $\{(x_1, y_1), \ldots, (x_n, y_n)\}$ is to create a pictorial representation of it. Identifying the first coordinates of these ordered pairs as values of an **independent variable** (or **predictor variable**) x and the second coordinates as the values of a **dependent variable** (or **response variable**) y, we simply plot them all on a single xy-plane. The resulting picture is called a **scatterplot**.

EXAMPLE 1 **Drawing a Scatterplot of Olympic Decathlon Data**

The 2004 Men's Olympic Decathlon consisted of the following 10 events: 100 meter, long jump, shot put, high jump, 400 meter, 110 meter hurdles, discus, pole vault, javelin throw, and 1500 meter. Actual scores are converted to a point system where points are assigned to each of these events based on performance. Events are equally weighted when converting to points. These points are then summed to obtain total scores, and, in turn, medals are assigned based on these total scores.

It would be interesting to know if certain events are more predictive of the total scores than are others. If someone does exceedingly well in the javelin throw, for example, is that person more likely to do well across all events and therefore obtain a large total points score?

Data from the Men's 2004 Olympic Decathlon are presented on the next page and were retrieved from the following Web source: **http://rss.acs.unt.edu/Rdoc/library/FactoMineR/html/decathlon.html**.

Let's consider the paired data set $\{(x, y)\}$ where x = score on the 400 m and y = total score.

---

*Optional Technology Required Section.

One scatterplot using the 400 m and total points information from this data set is shown in the following graph.

Several natural questions arise: How are different pairs of these data related? Are there any discernible patterns present, and if so, how strong are they? Is there a single curve that can be used to describe the general trend present in the data? We shall answer these questions one by one in this section.

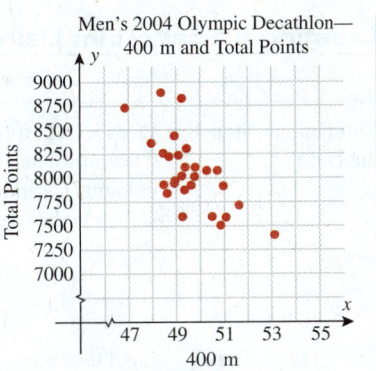

Men's 2004 Olympic Decathlon— 400 m and Total Points

Olympians	X100m	Long Jump	Shot-put	High Jump	X400m	X110m Hurdle	Discus	Pole Vault	Javelin	X1500m	Rank	Total Score
Sebrle	10.85	7.84	16.36	2.12	48.36	14.05	48.72	5.00	70.52	280.01	1	8893
Clay	10.44	7.96	15.23	2.06	49.19	14.13	50.11	4.90	69.71	282.00	2	8820
Karpov	10.50	7.81	15.93	2.09	46.81	13.97	51.65	4.60	55.54	278.11	3	8725
Macey	10.89	7.47	15.73	2.15	48.97	14.56	48.34	4.40	58.46	265.42	4	8414
Warners	10.62	7.74	14.48	1.97	47.97	14.01	43.73	4.90	55.39	278.05	5	8343
Zsivoczky	10.91	7.14	15.31	2.12	49.40	14.95	45.62	4.70	63.45	269.54	6	8287
Hernu	10.97	7.19	14.65	2.03	48.73	14.25	44.72	4.80	57.76	264.35	7	8237
Nool	10.80	7.53	14.26	1.88	48.81	14.80	42.05	5.40	61.33	276.33	8	8235
Bernard	10.69	7.48	14.80	2.12	49.13	14.17	44.75	4.40	55.27	276.31	9	8225
Schwarzl	10.98	7.49	14.01	1.94	49.76	14.25	42.43	5.10	56.32	273.56	10	8102
Pogorelov	10.95	7.31	15.10	2.06	50.79	14.21	44.60	5.00	53.45	287.63	11	8084
Schoenbeck	10.90	7.30	14.77	1.88	50.30	14.34	44.41	5.00	60.89	278.82	12	8077
Barras	11.14	6.99	14.91	1.94	49.41	14.37	44.83	4.60	64.55	267.09	13	8067
Smith	10.85	6.81	15.24	1.91	49.27	14.01	49.02	4.20	61.52	272.74	14	8023
Averyanov	10.55	7.34	14.44	1.94	49.72	14.39	39.88	4.80	54.51	271.02	15	8021
Ojaniemi	10.68	7.50	14.97	1.94	49.12	15.01	40.35	4.60	59.26	275.71	16	8006
Smirnov	10.89	7.07	13.88	1.94	49.11	14.77	42.47	4.70	60.88	263.31	17	7993
Qi	11.06	7.34	13.55	1.97	49.65	14.78	45.13	4.50	60.79	272.63	18	7934
Drews	10.87	7.38	13.07	1.88	48.51	14.01	40.11	5.00	51.53	274.21	19	7926
Parkhomenko	11.14	6.61	15.69	2.03	51.04	14.88	41.90	4.80	65.82	277.94	20	7918
Terek	10.92	6.94	15.15	1.94	49.56	15.12	45.62	5.30	50.62	290.36	21	7893
Gomez	11.08	7.26	14.57	1.85	48.61	14.41	40.95	4.40	60.71	269.70	22	7865
Turi	11.08	6.91	13.62	2.03	51.67	14.26	39.83	4.80	59.34	290.01	23	7708
Lorenzo	11.10	7.03	13.22	1.85	49.34	15.38	40.22	4.50	58.36	263.08	24	7592
Karlivans	11.33	7.26	13.30	1.97	50.54	14.98	43.34	4.50	52.92	278.67	25	7583
Korkizoglou	10.86	7.07	14.81	1.94	51.16	14.96	46.07	4.70	53.05	317.00	26	7573
Uldal	11.23	6.99	13.53	1.85	50.95	15.09	43.01	4.50	60.00	281.70	27	7495
Casarsa	11.36	6.68	14.92	1.94	53.20	15.39	48.66	4.40	58.62	296.12	28	7404

Creating a scatterplot by hand can be tedious, especially for large data sets. You can also very easily lose precision and detail. Using technology to create a scatterplot is very appropriate and quite easy. Below are the procedures for how you would create the scatterplot shown in Example 1 using the TI-83+ (or TI-84) and *Excel* 2007.

Creating a Scatterplot Using the TI-83+ (or TI-84)

	INSTRUCTION	SCREENSHOT

Entering the Data

Step 1 Press **STAT**, followed by **1:Edit**. . . . Clear any data already present in columns L1 and L2 so that the screen looks like the one to the right.

Step 2 Input the values of the *x*-variable (first entries in the ordered pairs) in column L1, pressing **ENTER** after each entry. Then, right arrow over to column L2 and input the values of the *y*-variable. The screen (starting from the beginning of the data set) should look like the one to the right when you are done.

Plotting the Data

Step 3 Press **Y=** and then select **Plot1** in the top row of the screen. If either **Plot2** or **Plot3** is darkened, move the cursor onto it and press **ENTER** to undarken it. The screen should look like the one to the right when you are done.

Step 4 Press **2nd**, followed by **Y=** (for **StatPlot**). Select 1: and modify the entries to make the screen look like the one to the right.

Step 5 Press **2nd**, followed by **Y=** and make certain that both **Plot2** and **Plot3** are OFF. The screen should look like the one to the right.

Step 6 Make certain the ranges for the *x* and *y* values are appropriate for the given data set. Here, we use the window shown to the right.

Step 7 Press **GRAPH** and you should get the scatterplot shown to the right.

Creating a Scatterplot Using *Excel* 2007

	INSTRUCTION	SCREENSHOT

Entering the Data — **Step 1** Open a new *Excel* spreadsheet. Input the values of the *x*-variable (first entries in the ordered pairs) in column A, starting with cell 1A. Then, input the values of the *y*-variable in column B, starting with cell 1B. The screen should look like the one to the right when you are done.

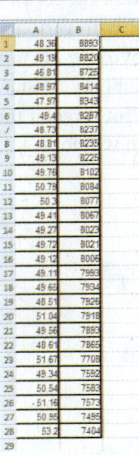

Plotting the Data — **Step 2** Highlight the data. The screen should look like the one to the right when you are done.

Step 3 Go to the **Insert** tab and select the icon labeled **Scatter**. A window list of five possible choices pops up.

Step 2

Step 3

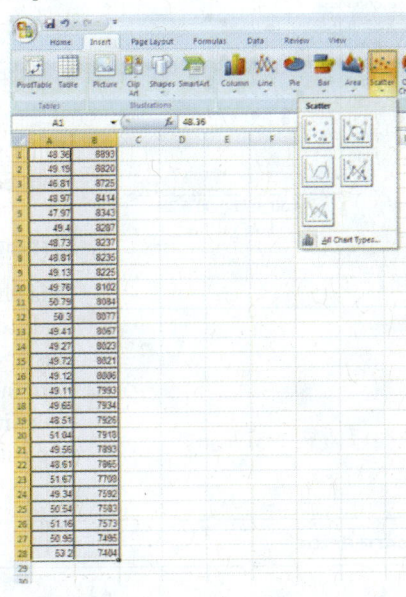

Step 4 Select the leftmost choice in the top row. Press **ENTER**. The scatterplot shown to the right should appear on the screen.

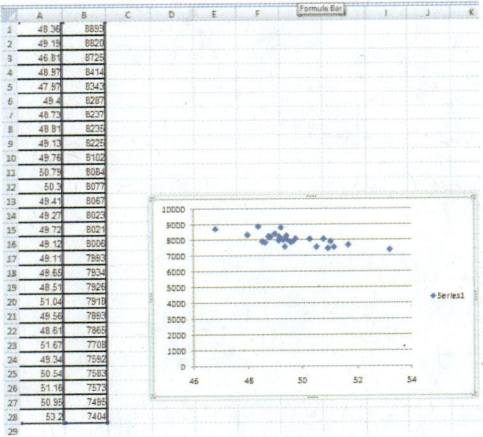

INSTRUCTION	SCREENSHOT
Step 5 You can alter the format of the scatterplot with various bells and whistles by *right-clicking* anywhere near the data points and then selecting **Format Plot Area** at the bottom of the pop-up window.	

■ **Answer:**

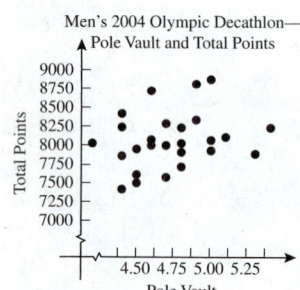

Men's 2004 Olympic Decathlon—Pole Vault and Total Points

■ **YOUR TURN** Using the data in Example 1, identify x = *score on the pole vault* and z = *total score*. Use technology to create a scatterplot for the data set consisting of the ordered pairs (x, z).

Identifying Patterns

While scatterplots are comprised merely of clusters of ordered pairs, patterns of various types can emerge that can provide insight into how the variables x and y are related.

Direction of Association

This characteristic is analogous to the concept of slope of a line. If the cluster of points tends to rise from left to right, we say that x and y are **positively associated**, whereas if the cluster of points falls from left to right, we say that x and y are **negatively associated**. Certainly, the more closely packed together the points are to an identifiable curve, the easier it is to make such a determination. Some examples of scatterplots of varying degrees of positive and negative association are shown in the following table.

SCATTERPLOT	DIRECTION OF ASSOCIATION	VERBAL DESCRIPTION
	Positive Association	A shape that increases from left to right is very discernible in each case.
		The association is a bit loose, but you can still tell the data points tend to rise from left to right.

SCATTERPLOT	DIRECTION OF ASSOCIATION	VERBAL DESCRIPTION
	No Association	Haphazard scattering of points suggests neither positive nor negative association.
	Negative Association	The association is a bit loose, but you can still tell the data points tend to fall from left to right.
	Negative Association	A shape that decreases from left to right is very discernible in each case.

Linearity

Depending on the phenomena being studied and the actual sample being used, the data points comprising a scatterplot can conform very closely to an actual curve. If the curve is a line, we say that the relationship between x and y is **linear**; otherwise, we say the relationship is **nonlinear**. Some illustrative examples follow.

SCATTERPLOT	LINEARITY	VERBAL DESCRIPTION
	Linear	Perfect linear relationship; positive association
	Linear	Perfect linear relationship; negative association

SCATTERPLOT	LINEARITY	VERBAL DESCRIPTION
	Linear	Fairly tight linear relationship; positive association
	Linear	Fairly tight linear relationship; negative association
	Linear	Rather loose, but still somewhat discernible, linear relationship; positive association
	Linear	Rather loose, but still somewhat discernible, linear relationship; negative association
	Nonlinear	Perfect nonlinear relationship
	Nonlinear	Fairly tight nonlinear relationship
	No specifically identifiable relationship	No discernible linear relationship or degree of association

EXAMPLE 2   Describing Patterns in a Data Set

Describe the patterns present in the paired data set (x, y) considered in Example 1, where x = *points on the 400 m* and y = *total score*. Is this intuitive?

Solution:

We can surmise that the variable *score on the 400 m* has a relatively strong linear, negative association with the variable *total score*. This means that as the 400 m score decreases, the total score tends to increase. A negative relationship makes sense here in that 400 m scores reflect how much time it took to complete this race. So, lower scores (less time) reflect better performance and therefore more total points.

■ **YOUR TURN**   Using the data in Example 1, identify x = *score on the pole vault* and z = *total score*. Comment on the degree of association and linearity of the scatterplot consisting of the ordered pairs (x, z). Is this intuitive?

■ **Answer:** The variable *score the pole vault* has a rather weak linear, positive association with the variable *total score*. This means that as the pole vault score increases, the total score tends to increase. In this case, a positive relationship makes sense in that pole vault scores reflect the height achieved. So, higher scores (greater height) reflect better performance and therefore more total points.

Strength of Linear Relationship

The variability in the data can render it difficult to determine if there is a linear relationship between two variables. As such, it is useful to have a way of measuring how tightly a paired data set conforms to a linear shape. This measure is called the **correlation coefficient**, r, and is defined by the following formula:

DEFINITION

For a paired data set $\{(x_1, y_1), \ldots, (x_n, y_n)\}$, the *correlation coefficient*, r, is defined by

$$r = \frac{n\sum xy - (\sum x)(\sum y)}{\sqrt{n\sum x^2 - (\sum x)^2} \cdot \sqrt{n\sum y^2 - (\sum y)^2}}$$

The symbol $\sum z$ is a shorthand way of writing $z_1 + \cdots + z_n$. So, for instance, $\sum x^2 = x_1^2 + \cdots + x_n^2$.

This is tedious to calculate by hand but is easily computed using technology. Below are the procedures for how you would compute the correlation coefficient for the data set introduced in Example 1 using the TI-83+ (or TI-84) and *Excel* 2007.

Computing a Correlation Coefficient Using the TI-83+ (or TI-84)

	INSTRUCTION	SCREENSHOT
Step 1	**Enter the data** following the procedure outlined earlier in this section. The screen should look like the one to the right.	L1: 48.36, 49.19, 46.81, 48.97, 47.97, 49.4, 48.73 — L2: 8893, 8820, 8725, 8414, 8343, 8287, 8237 — L2(1)=8893

INSTRUCTION		SCREENSHOT

Step 2	**Set up what will display!** In order for the desired output to display once we execute the commands to follow, we must tell the calculator to do so. As such, do the following:
	
	i. Press **2nd**, followed by 0 to get **CATALOG**.
	ii. Scroll down until you get to **DiagnosticOn**. Press **ENTER**. Then, this command will appear on the home screen. Press **ENTER** again. The resulting screen should look like the one to the right.
		`DiagnosticOn`
		`            Done`
		`■`

Step 3	Press **STAT**, followed by **CALC**, and then by **4:LinReg**(ax+b). The resulting screen should look like the one to the right. Press **ENTER**.	`EDIT CALC TESTS`
		`1:1-Var Stats`
		`2:2-Var Stats`
		`3:Med-Med`
		`4▮LinReg(ax+b)`
		`5:QuadReg`
		`6:CubicReg`
		`7↓QuartReg`

Step 4	Press **ENTER** again. After a brief moment, your screen should look like the one to the right. The value we want is in the bottom row of the screen, about $r = -0.7045$.	`LinReg`
		` y=ax+b`
		` a=-206.9238808`
		` b=18317.02944`
		` r²=.4963043621`
		` r=-.7044887239`

Note: The other information provided will be pertinent once we define the *best fit line* in the next subsection.

Computing a Correlation Coefficient Using *Excel* 2007

INSTRUCTION	SCREENSHOT
Step 1 **Enter the data** following the procedure outlined earlier in this section. The screen should look like the one to the right.	Step 1 Step 2
Step 2 Select the **Formulas** tab at the top of the screen and then choose the **More Functions** within the *Function Library* grouping (on the left). The pull-down menu should be as shown to the right.	

Step 3   From here, select **Statistical**, and then from this list, scroll down and choose **CORREL**. A pop-up window should appear, as shown to the right.

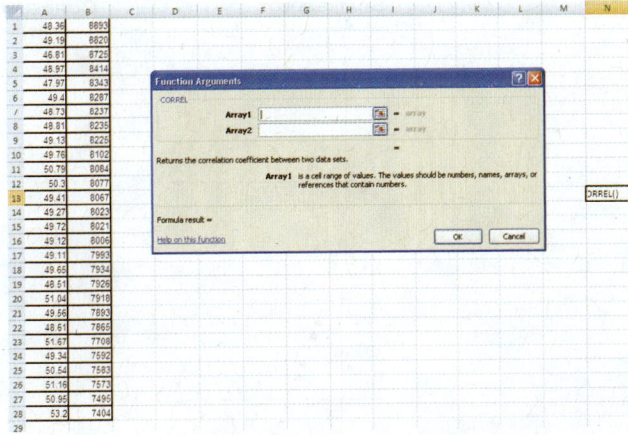

Step 4   Enter **A1:A28** in Array 1 and **B1:B28** in Array 2, as shown to the right. Press **OK**. You will notice that the correlation coefficient appears directly beneath Array 2. In this case, r is about -0.7045.

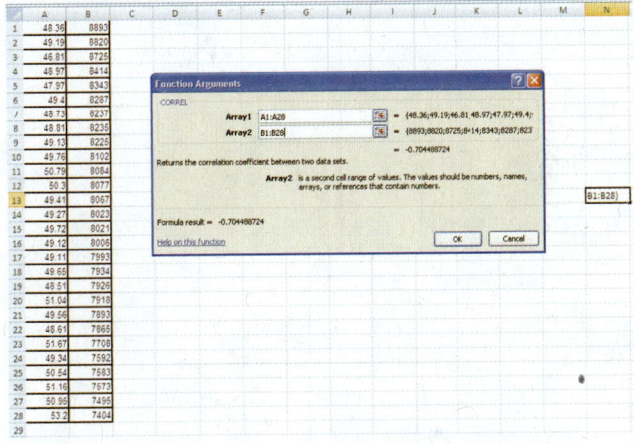

The square of the correlation coefficient is interpreted as a signed percentage of the variability among the y-values that is actually explained by the linear relationship, where the sign corresponds to the direction of the association. For instance, an r-value of $+1$ means that 100% of the variability among the y-values is explained by a line with positive slope; in such a case, all of the points in the data set actually lie on a single line. An r-value of -1 means the same thing, but the line has a negative slope. As the r-values get closer to zero, the more dispersed the points become from a line describing the pattern, so that an r-value very close to zero suggests no linear relationship whatsoever is discernible. The following sample of scatterplots with the associated correlation coefficients should provide you with a feel for the strength of linearity suggested by various values of r.

SCATTERPLOT	CORRELATION COEFFICIENT r	STRENGTH OF LINEARITY
	$r = 1.0$	Perfect positive linear relationship
	$r = -1.0$	Perfect negative linear relationship
	$r = 0.80$	Reasonably strong, though not perfect, positive linear relationship
	$r = -0.45$	Pretty weak, barely discernible, negative linear relationship
	$r = 0.10$	Essentially no discernible linear relationship whatsoever

EXAMPLE 3   Calculating the Correlation Coefficient Associated with a Data Set

Use technology to calculate the correlation coefficient r for the paired data set (x, y) considered in Example 1, where x = *points on the 400 m* and y = *total score*. Interpret the strength of the linear relationship.

Solution:

We see that using either form of technology yields $r = -0.7045$. This suggests that the data follow a relatively strong negative (i.e., negative slope) linear pattern.

■ **YOUR TURN** Using the data in Example 1, identify x = *score on the pole vault* and z = *total score*, and calculate the correlation coefficient for the data set consisting of the ordered pairs (x, z) using technology. Interpret the strength of the linear relationship.

■ **Answer:** The correlation coefficient is approximately $r = 0.28$. This suggests that while the data follow a positive (i.e., positive slope) pattern, the degree to which an actual line describes the trend in the data is rather weak.

Linear Regression

Determining the "Best Fit" Line

Assuming that a data set follows a reasonably strong linear pattern, it is natural to ask which *single straight line* best describes this pattern. Having such a line would enable us to not only describe the relationship between the two variables x and y precisely, but it would also enable us to predict values of y from values of x not present among the points of the data set.

Consider the paired data set (x, y) from Example 1, where $x = $ *points on the 400 m* and $y = $ *total score*. You learned in Section 2.3 that between any two points there is a unique line whose equation can be determined. Three such lines passing through various pairs of points in the data set are illustrated below.

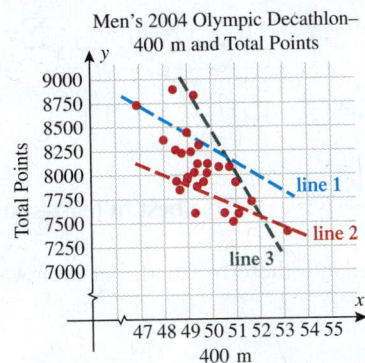

Men's 2004 Olympic Decathlon–
400 m and Total Points

The unavoidable shortcoming of all of these lines, however, is that not all of the data points lie on a single one of them. Each has a negative slope, which *is* characteristic of the data set, and each of the lines is close to some of the data points, but not close to others. In fact, we could draw infinitely many such lines and make a similar assessment. But which one *best fits* the data?

The answer to this question depends on how you define "best." Reasonably, for the line that *best fits* the data, the error incurred in using it to describe *all* of the points in the data set should be as small as possible. The conventional approach is to define this error by summing the n distances d_i between the y-coordinates of the data points and the corresponding y-value on the line $y = Mx + B$ (that is, the y-value of the point on the line corresponding to the same x-value). These distances are, in effect, the error in making the approximation. This is illustrated below:

Using the distance formula, we find that

$$d_i = \sqrt{(x_i - x_i)^2 + (y_i - (Mx_i + B))^2} = |y_i - (Mx_i + B)|.$$

Note that $d_i = 0$ precisely when the data point (x_i, y_i) lies directly on the line $y = Mx + B$, and that the closer d_i is to 0, the closer the point (x_i, y_i) is to the line $y = Mx + B$. As such, the goal is to determine the values of the slope M and y-intercept B for which the sum $d_1 + \cdots + d_n$ is as small as possible. Then, the resulting straight line $y = Mx + B$ best fits the data set $\{(x_1, y_1), \ldots, (x_n, y_n)\}$.

This is fine, in theory, but it turns out to be inconvenient to work with a sum of absolute value expressions. It is actually much more convenient to work with the *squared distances* d_i^2. The values of M and B that minimize $d_1 + \cdots + d_n$ are precisely the same as those that minimize $d_1^2 + \cdots + d_n^2$. Using calculus, it can be shown that the formulas for M and B are as follows:

$$M = \frac{n\sum xy - (\sum x)(\sum y)}{n\sum x^2 - (\sum x)^2}, \qquad B = \frac{\sum y}{n} - M\frac{\sum x}{n}$$

The resulting line $y = Mx + B$ is called the **best fit least-squares regression line** for the data set $\{(x_1, y_1), \ldots, (x_n, y_n)\}$.

Again, it is tedious to compute these by hand, but their values are actually produced easily using technology.

EXAMPLE 4   Finding the Line of Best Fit by Linear Regression

Find the line of best fit (best fit least-squares regression line) for the paired data set (x, y) from Example 1, where $x = $ *points on the 400 m* and $y = $ *total score,* using (a) the TI-83+ (or TI-84) and (b) *Excel 2007.*

Solution (a):

Determining the Best Fit Least-Squares Regression Line Using the TI-83+ (or TI-84).

INSTRUCTION	SCREENSHOT
STEP 1 **Enter the data** following the procedure outlined earlier in this section. The screen should look like the one to the right.	L1 L2 L3 2 48.36 8893 49.19 8820 46.81 8725 48.97 8414 47.97 8343 49.4 8287 48.73 8237 L2(1)=8893
STEP 2 Press **STAT**, followed by **CALC**, and then by **4:LinReg(ax+b)**. The resulting screen should look like the one to the right.	LinReg(ax+b)

INSTRUCTION	SCREENSHOT

STEP 3 For this example, the data is stored in lists L1 and L2. And, since we will want to graph our best fit line on the scatterplot, it will need to be stored as a function of *x*, say as Y1. In order to do this, proceed as follows:

Directly next to **LinReg(ax+b)** on the home screen, we need to type the following: **L1, L2, Y1**.

Use the following key strokes: **2nd, 1, ⟐, 2nd, 2, ⟐ VARS, Y-VARS, 1:FUNCTION, Y1** The resulting screen should look like the one to the right.

◀ax+b) L₁,L₂,Y₁

STEP 4 Press **ENTER**. The equation of the best fit least-squares line with the slope (labeled as *a*) and the *y*-intercept (labeled as *b*) appears on the screen as shown to the right.

So, the equation of the best fit least-squares regression line is approximately
$y = -206.9x + 18317.03$.

LinReg
y=ax+b
a=-206.9238808
b=18317.02944
r²=.4963043621
r=-.7044887239

STEP 5 In order to obtain a graph of the scatterplot *with* the best fit line from Step 4 superimposed on it, press **ZOOM**, then **9:ZoomStat**. The resulting screen should look like the one to the right.

Step 1

Solution (b):

Determining the Best Fit Least-Squares Regression Line Using *Excel* 2007.

INSTRUCTION	SCREENSHOT

STEP 1 **Enter the data** following the procedure outlined earlier in this section. The screen should look like the one to the right.

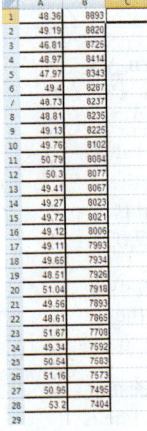

	A	B	C
1	48.36	8893	
2	49.19	8820	
3	46.81	8725	
4	48.97	8414	
5	47.97	8343	
6	49.4	8287	
7	48.73	8237	
8	48.81	8235	
9	49.13	8225	
10	49.76	8102	
11	50.79	8084	
12	50.3	8077	
13	49.41	8067	
14	49.27	8023	
15	49.72	8021	
16	49.12	8006	
17	49.11	7993	
18	49.65	7934	
19	48.51	7926	
20	51.04	7918	
21	49.56	7893	
22	48.61	7865	
23	51.67	7708	
24	49.34	7692	
25	50.54	7683	
26	51.16	7673	
27	50.95	7495	
28	53.2	7404	
29			

INSTRUCTION	SCREENSHOT

STEP 2 Select the **Formulas** tab at the top of the screen and then choose the **More Functions** within the *Function Library* grouping (on the left). The pull-down menu should be as shown to the right.

STEP 3 From here, select **Statistical**, and then from this list, scroll down and choose **LINEST**. A pop-up window should appear, as shown to the right.

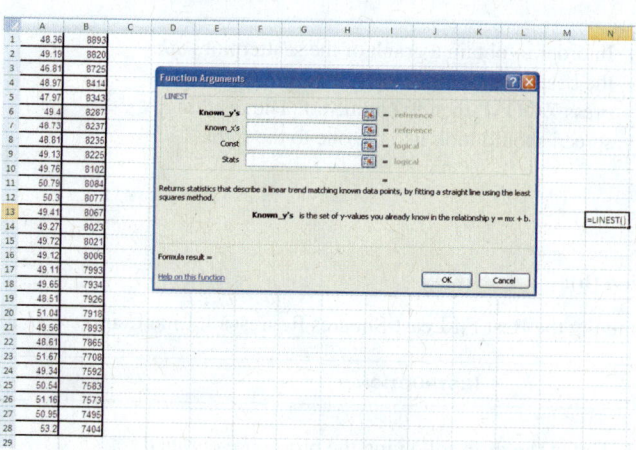

STEP 4 Enter **B1:B28** in **Known_y's** and **A1:A28** in **Known_x's**, as shown to the right. You will notice that a set of two values occurs directly beneath the entry boxes—the output is about $\{-206.92, 18{,}317.03\}$.

The first value is the slope M, and the second value is the y-intercept B of the best fit line.

So, the equation of the best fit least-squares regression line is approximately
$$y = -206.92x + 18{,}317.03.$$

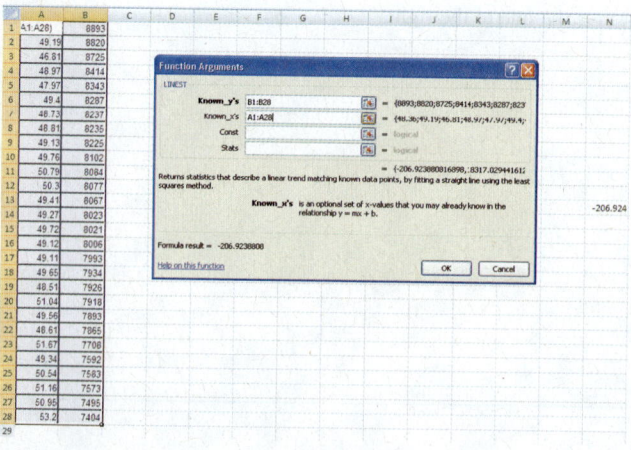

INSTRUCTION	SCREENSHOT

STEP 5 In order to obtain a graph of the scatterplot *with* the best fit line from Step 4 superimposed on it, construct the scatterplot as before, right-click on the scatterplot near the data points, choose **Add Trendline**, and press **ENTER**.

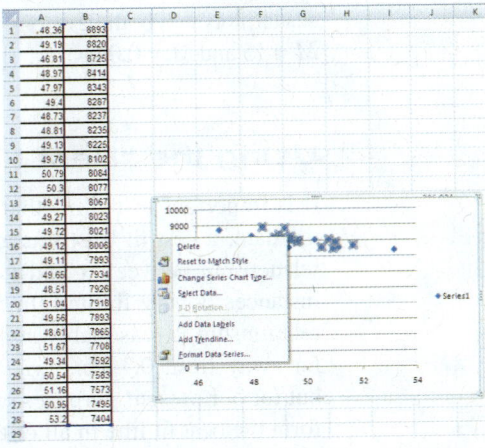

STEP 6 The best fit line will appear on the scatterplot, along with a pop-up window allowing you to change the format of the line/curve that is displayed.

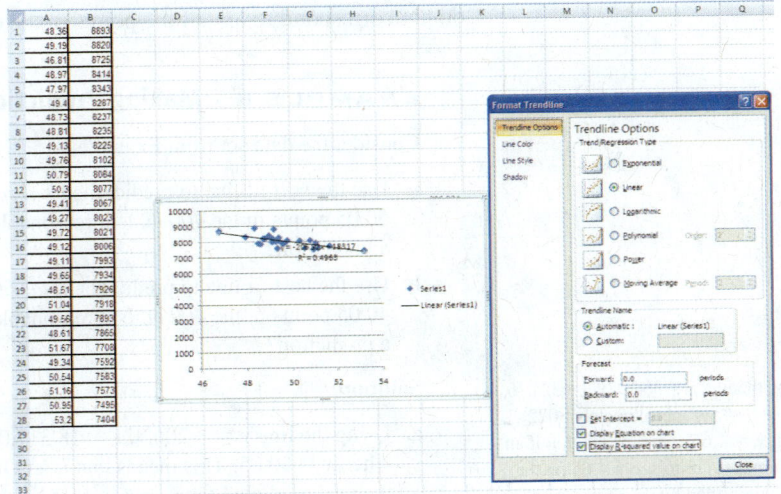

■ **YOUR TURN**   Using the data in Example 1, identify *x = score on the pole vault* and *z = total score*, and determine the best fit least-squares regression line for the data set consisting of the ordered pairs (*x, z*). Superimpose the graph of this line on the scatterplot.

■ **Answer:**

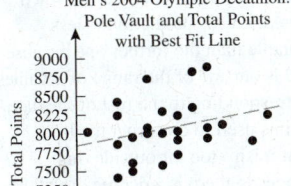

Men's 2004 Olympic Decathlon: Pole Vault and Total Points with Best Fit Line

Here, the best fit line displays a positive relationship, and its equation is $z = 364.97x + 6324.46$.

It is important to realize that the correlation coefficient is NOT equal, or even related to, the actual slope of the best fit line. In fact, two distinct perfectly linear, positively associated scatterplots will both have $r = 1$, even though the actual lines that fit the data might have slope $M = 15$ and $M = 0.04$.

Using the "Best Fit" Line for Prediction

It is important to realize that for any scatterplot, no matter how haphazardly dispersed the data, a best fit least-squares regression line can be created. This is true even when the relationship between x and y is *non*linear. However, the utility of such a line in these instances is very limited. In fact, a best fit line should only be created when the linear relationship is reasonably strong, which means the correlation coefficient is "reasonably far away from 0." This criterion can be made more precise using statistical methods, but for our present purposes, we shall make the blanket assumption that it makes sense to form the best fit line in all of the scenarios we present.

Once we have the best fit line in hand, we can use it to predict y-values for values of x that do not correspond to any of the points in the data set. For instance, consider the following:

EXAMPLE 5 ·**Making Predictions Using the Line of Best Fit**

Consider the data set from Example 1.

a. Use the best fit line to predict the *total score* given that an Olympian scored 50.05 points in the *400 m*. Is it reasonable to use the best fit line to make such a prediction?

b. Use the best fit line to predict the *total score* given that an Olympian scored 40.05 points in the *400 m*. Is it reasonable to use the best fit line to make such a prediction?

Solution:

a. Using the line $y = -206.92x + 18{,}317.03$, we see that y is approximately 7961 when $x = 50.05$. This means that if an Olympian were to score 50.05 on the *400 m*, then his predicted *total score* would be approximately 7961. Using the best fit line to predict the total score in this case is reasonable because the value 50.05 is well within the range of x-values already present in the data set.

b. Using the line $y = -206.92x + 18{,}317.03$, we see that y is approximately 10,030 when $x = 40.05$. This means that if an Olympian were to score 40.05 on the *400 m*, then his predicted *total score* would be approximately 10,030. This prediction is questionable because the x-value at which you are using the best fit line to predict y is sufficiently far away from the rest of the data points that were used to construct the line. As such, there is no reason to believe that the line is valid for such x-values.

■ **Answer:**

Using the line $z = 364.97x + 6324.46$, we see that z is approximately 8022 when $x = 4.65$. This means that if an Olympian were to score 4.65 on the *pole vault*, then his predicted *total score* would be approximately 8022. Using the best fit line to predict the total score in this case is reasonable because the value 4.65 is well within the range of x-values already present in the data set.

Next, using the same line, we see that z is approximately 8478 when $x = 5.9$. This prediction is less reliable than the former one because 5.9 is outside of the range of x-values corresponding to the rest of the data points used to construct the line. But it isn't too far outside this range, so there is a degree of validity to the prediction.

■ **YOUR TURN** Using the data in Example 1, identify $x = $ *score on the pole vault* and $z = $ *total score*, and use the best fit line to predict the *total score* given that an Olympian scored 4.65 points on the *pole vault* and then, given that an Olympian scored 5.9 points on the *pole vault*. Comment on the validity of these predictions.

EXAMPLE 6   The Power of TV Advertisement

Video Board Tests, Inc., an advertising testing agency, collected data based on 4000 adult participants of a survey. The participants (who were regular product users) were asked to recall a commercial that they had viewed for a given product category in the previous week. The goal was to examine the relationship between retained impressions of commercials and the corresponding TV advertising budget for a given product. The data were published in the *Wall Street Journal* in March 1984.

The following is an adaptation of the original data set (**TV Ad Yields was obtained from the following Web source: http://lib.stat.cmu.edu/DASL/Datafiles/tvadsdat.html**), but contains three modifications made for illustrative purposes. Specifically, ATT/BELL, FORD, and MCDONALD'S have been replaced with DIALTONE USA, CARZ, and HAPPY BURGERS, respectively. These changes have been highlighted in green.

COMPANY	TV ADVERTISING BUDGET, 1983 ($ MILLIONS)	MILLIONS RETAINED IMPRESSIONS PER WEEK
MILLER_LITE	50.1	32.1
PEPSI	74.1	99.6
STROH'S	19.3	11.7
FEDERAL_EXPRESS	22.9	21.9
BURGER_KING	82.4	60.8
COCA-COLA	40.1	78.6
HAPPY_BURGERS	165.0	10.0
MCI	26.9	50.7
DIET_COLA	20.4	21.4
CARZ	165.0	50.0
LEVI'S	27.0	40.8
BUD_LITE	45.6	10.4
DIALTONE_USA	70.0	88.9
CALVIN_KLEIN	5.0	12.0
WENDY'S	49.7	29.2
POLAROID	26.9	38.0
SHASTA	5.7	10.0
MEOW_MIX	7.6	12.3
OSCAR_MEYER	9.2	23.4
CREST	32.4	71.1
KIBBLES_'N_BITS	6.1	4.4

First, a scatterplot for this data set (formed using PASW Statistics 18) is shown below. The approximate regression line is $y = 0.18x + 29.04$ and $r = 0.28$.

It appears that there *might* be a relationship between the budget and the retained impressions. Both CARZ and HAPPY_BURGERS are pretty far away from the bulk of the data and might be skewing an otherwise tighter relationship between x and y; such points are potential *outliers*. What would happen to the regression line if we removed each of these points, one at a time? Would the new line be dramatically different from the original one, or might there be very little change?

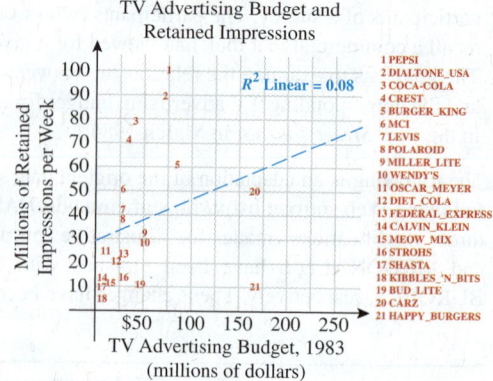

Let's start by removing CARZ. The resulting regression line is $y = 0.21x + 27.96$. We observe only a small change in both the slope and intercept. Next, let's put CARZ back into the data set and remove HAPPY_BURGERS instead. In this case, the best fit regression line is $y = 0.40x + 22.61$. This time, we have a slightly larger change in the y-intercept, but more importantly, the slope is more than double the slope of the regression line from the original data set. The resulting best fit regression line is displayed to the right.

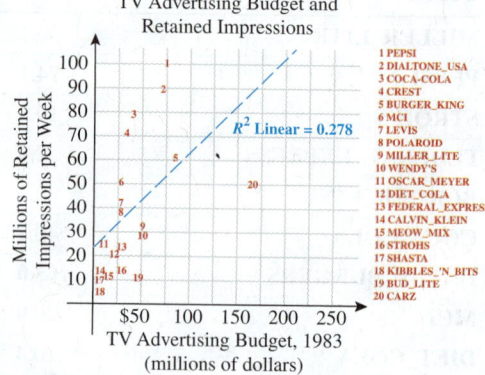

Clearly, HAPPY_BURGERS was a very influential data point since removing it dramatically changed the slope between the budget and retained impressions, thereby considerably changing the mathematical description of the relationship between these two variables. But why did this happen?

There was a large distance between it and both the x- and y-directions from the bulk of the data set. When a potential outlier is distant in only one of the two directions (x or y), as was the case with CARZ, it is far less likely to be influential.

SECTION 2.5* SUMMARY

Two variables x and y can be related in different ways. A paired data set $\{(x_1, y_1), \ldots, (x_n, y_n)\}$ obtained experimentally can be illustrated using a *scatterplot*. Patterns concerning the direction of association and linearity can be used to describe the relationship between x and y, and the strength of the linear relationship can be measured using the correlation coefficient r. If r is sufficiently far from 0, a *best fit least-squares regression line* can be formed to precisely describe the linear relationship and used for reasonable prediction purposes.

SECTION
2.5* EXERCISES

▪ SKILLS

In Exercises 1–4, for each of the following scatterplots, identify the pattern as

a. having a positive association, negative association, or no identifiable association.

b. being linear or nonlinear.

1.

2.

3.

4.

In Exercises 5–8, match the following scatterplots with the following correlation coefficients.

a. $r = -0.90$ **c.** $r = -0.68$

b. $r = 0.80$ **d.** $r = 0.20$

5.

6.

7.

8.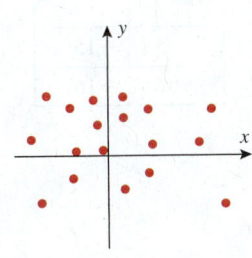

For each of the following data sets,

a. create a scatterplot.

b. guess the value of the correlation coefficient r.

c. use technology to determine the equation of the best fit line and to calculate r.

d. give a verbal description of the relationship between x and y.

9.

x	y
-3	14
-1	8
0	5
1	2
3	-4
5	-10

10.

x	y
-8	-16
-6	-12
-4	-8
-2	-4
1	2
3	6

11.

x	y
-10	1
-6	0
0	-2
8	-10
14	-11
20	-16

12.

x	y
-1	-17
$-1/2$	-11
$-1/4$	-5
$-1/10$	0
0	1
1/10	1
1/5	8
1	12

13.

x	y
-3	-6
-2	3
-1	1
0	1
1	5
2	-1
4	1

14.

x	y
-6	-1
-3	4
-1	3
1	0
2	-6
5	-4
8	1

In Exercises 15–18, for each of the data sets,

a. use technology to create a scatterplot, to determine the best fit line, and to compute *r*.

b. indicate whether or not the best fit line can be used for predictive purposes for the following *x*-values. For those for which it can be used, give the predicted value of *y*:

   i. $x = 0$ **iii.** $x = 12$

   ii. $x = -6$ **iv.** $x = -15$

c. Using the best fit line, at what *x*-value would you expect *y* to be equal to 2?

15.

x	y
−5	−8
−3	0
−2	0
2	1.5
5	4
7	2
10	8

16.

x	y
5	0
5	3
6	3
7	6
7	9
8	9
8	15
9	9
9	15
10	15
10	18

17.

x	y
−20	15
−18	10
−14	3
−14	8
−13	3
−8	0
−8	−3
−5	−6
1	−11
1	−15

18.

x	y
−15	4
−15	12
−15	16
−13	3
−10	4
−10	8
−10	12
−5	4
−2	3
−2	−2
2	3
2	6
4	−1
4	0
4	4
7	−2
7	3
10	−4
10	−2
10	3

For Exercises 19–22,

a. use technology to create a scatterplot, to determine the best fit line, and to compute *r* for the *entire* data set.

b. repeat (a), but with the data set obtained by removing the *starred* (***) data points.

c. compare the *r*-values from (a) and (b), as well as the slopes of the best fit lines. Comment on any differences, whether they are substantive, and why this seems reasonable.

19.

x	y
−3	14
−1	8
0	5
1	2
3	−4
*** 5	−10

20.

x	y
−10	1
−6	0
0	−2
8	−10
14	−11
*** 20	−16

21.

x	y
−3	14
−1	8
0	5
1	2
*** 3	−4
5	−15
*** 6	−16

22.

x	y
0	0
1	2
2	4
3	6
4	8
6	12
*** 7	25

■ CONCEPTUAL

23. Consider the data set from Exercise 17.

a. Reverse the roles of x and y so that now y is the *explanatory* variable and x is the *response* variable. Create a scatterplot for the ordered pairs of the form (y, x) using this data set.

b. Compute r. How does it compare to the r-value from Exercise 17? Why does this make sense?

c. The best fit line for the scatterplot in (a) will be of the form $x = my + b$. Determine this line.

d. Using the line from (c), find the predicted x-value for the following y-values, if appropriate. If it is not appropriate, tell why.

    i. $y = 23$ **ii.** $y = 2$ **iii.** $y = -16$

24. Consider the data set from Exercise 16. Redo the parts in Exercise 23.

25. Consider the following data set.

x	y
3	0
3	1
3	-1
3	-2
3	4
3	15
3	-6
3	8
3	10

Guess the values of r and the best fit line. Then, check your answers using technology. What happens? Can you reason why this is the case?

26. Consider the following data set.

x	y
-5	-2
-4	-2
-1	-2
0	-2
1	-2
3	-2
8	-2
17	-2

Guess the values of r and the best fit line. Then, check your answers using technology. What happens? Can you reason why this is the case?

■ CATCH THE MISTAKE

27. The following screenshot was taken when using the TI-83+ to determine the equation of the best fit line for paired data (x, y):

Using the regression line, we observe that there is a strong positive linear association between x and y, and that for every unit increase in x, the y-value increases by about 1.257 units.

28. The following scatterplot was produced using the TI-83+ for paired data (x, y).

The equation of the best fit line was reported to be $y = -3.207x + 0.971$ with $r^2 = 0.9827$. Thus, the correlation coefficient is given by $r = 0.9913$, which indicates a strong linear association between x and y.

▪APPLICATIONS

For Exercises 29 and 30, refer to the data set in Example 1.

29. a. Examine the relationship between each of the decathlon events and the total points by computing the correlation coefficient in each case.

  b. Using the information from part (a), which event has the strongest relationship to the total points?

  c. What is the equation of the best fit line that describes the relationship between the event from part (b) and the total points?

  d. Using the best fit line, if you had a score of 40 for this event, what would the predicted total points score be?

30. a. Using the information from part (a), which event has the second strongest relationship to the total points?

  b. What is the equation of the best fit line that describes the relationship between the event in part (b) and the total points?

  c. Is it reasonable to expect the best fit line from part (c) to produce accurate predictions of total points using this event?

  d. Using the best fit line, if you had a score of 40 for this event, what would the total points score be?

For Exercises 31 and 32, refer to the following scenario: *Texting Speed.*

According to the CTIA—The Wireless Association, as of December 2010, 187.7 billion messages were sent per month or 2.1 trillion messages in that year.[1] According to a 2010 Pew Internet survey, 72% of all teens—or 88% of teen cell phone users—are text-messagers. Teens make and receive far fewer phone calls than text messages on their cell phones.

A number of competitions regarding texting speed have taken place worldwide. According to the Guinness World Records, "The fastest completion of a prescribed 160-character text message is 34.65 seconds and was achieved by Frode Ness (Norway) at the Norwegian SMS championships held at the Oslo City shopping centre in Oslo, Østlandet, Norway, on 13 November 2010."[2]

The data set regarding texting speed on the next page was provided by AP Central. (**http://apcentral.collegeboard.com/apc/public/courses/teachers_corner/195435.html**)

In the data given, the *A total score* is the amount of time (in seconds) it took to text the following message, "Statistics students are above average." The *B total score* is the amount of time (in seconds) to type, "Meet me at my car after school today." The *Total both scores* is the sum of the *A total score* and *B total score*.

What influences texting speed in this group? Let's consider thumb length.

31. What is the relationship between the variables *left thumb length* and *total both scores*?

  a. Create a scatterplot to show the relationship between *left thumb length* and *total both scores*.

  b. What is the correlation coefficient between *left thumb length* and *total both scores*?

  c. Describe the strength of the relationship between *left thumb length* and *total both scores*.

  d. What is the equation of the best fit line that describes the relationship between *left thumb length* and *total both scores*?

  e. Could you use the best fit line to produce accurate predictions of *total both scores* using *left thumb length*?

32. Repeat Exercise 31 for *right thumb length* and *total both scores*.

---

[1] http://www.cita.org/advocacy/research/index.cfm/aid/10323
[2] http://www.guinnessworldrecords.com/Search/Details/Fastest-text-message/57979.htm

Gender	Texting Style	Left Thumb Length	Right Thumb Length	A Total Score	B Total Score	Total Both Scores	A Minus B	Avg Thumb	Diff Thumb
Male	Char	6.5	6.5	35	25	60	10	6.5	0
Female	Char	5	5	61	57	118	4	5	0
Female	Word	6	6	24	20	44	4	6	0
Male	Word	7	7	43	60	103	−17	7	0
Female	Word	6	6	14	19	33	−5	6	0
Male	Word	7	6	15	18	33	−3	6.5	−1
Female	Word	6	6	13	14	27	−1	6	0
Female	Word	6	6	22	10	22	12	6	0
Male	Word	6.5	6	13	15	28	−2	6.25	−0.5
Female	Word	5.5	5.5	16	16	32	0	5.5	0
Male	Char	6	5	85	78	163	7	5.5	−1
Male	Char	6	6	126	120	246	6	6	0
Male	Word	7.5	6.5	67	69	136	−2	7	−1
Female	Char	5.5	5.5	11	7	18	4	5.5	0
Female	Word	5.5	5.7	14	17	31	−3	5.6	0.2
Female	Word	6	6	17	14	31	3	6	0
Female	Word	5	5	20	15	35	5	5	0
Male	Word	6.5	6.5	15	13	28	2	6.5	0
Male	Word	7	7	30	31	61	−1	7	0
Male	Word	6	6.1	120	117	237	3	6.05	0.1
Male	Word	6	6	74	25	99	49	6	0
Male	Word	6.3	6	23	21	44	2	6.15	−0.3
Male	Char	6	5.9	45	50	95	−5	5.95	−0.1
Male	Word	6	6.1	86	100	186	−14	6.05	0.1
Male	Char	6	6	25	23	48	2	6	0
Male	Char	6.3	6.3	81	57	138	24	6.3	0
Female	Char	5.5	5.5	88	66	154	22	5.5	0
Female	Word	7	6.8	10	9	19	1	6.9	−0.2
Male	Char	6.5	7	21	18	39	3	6.75	0.5
Female	Char	5.4	5.2	72	48	121	24	5.3	−0.2
Female	Char	8	8	36	23	59	13	8	0
Male	Char	7	6.5	46	45	91	1	6.75	−0.5
Female	Char	6	6.8	48	39	87	9	6.4	0.8
Female	Char	7.1	7.1	84	57	141	27	7.1	0
Female	Word	5.9	5.5	25	23	48	2	5.7	−0.4
Female	Char	7.6	7.2	32	45	77	−13	7.4	−0.4
Male	Word	6.9	7	23	28	51	−5	6.95	0.1
Female	Char	7.7	7.5	18	15	33	3	7.6	−0.2
Male	Char	8.6	(cast)	22	20	42	2	N/A	N/A
Male	Char	7.3	7.1	54	50	104	4	7.2	−0.2

For Exercises 33 and 34, refer to the following data set:
Herd Immunity.

According to the U.S. Department of Health and Human Services, herd immunity is defined as "a concept of protecting a community against certain diseases by having a high percentage of the community's population immunized. Even if a few members of the community are unable to be immunized, the entire community will be indirectly protected because the disease has little opportunity for an outbreak. However, with a low percentage of population immunity, the disease would have great opportunity for an outbreak."[3]

Suppose a study is conducted in the year 2016 looking at the outbreak of *Haemophilus influenzae type b* in the winter of 2015 across 22 nursing homes. We might look at the percentage of residents in each of the nursing homes that were immunized and the percentage of residents who were infected with this type of influenza.

The fictional data set is as follows.

NURSING HOME	% RESIDENTS IMMUNIZED	% RESIDENTS WITH INFLUENZA
1	70	11
2	68	9
3	80	8
4	10	34
5	12	30
6	18	31
7	27	22
8	64	18
9	73	6
10	9	31
11	35	19
12	56	16
13	57	22
14	83	10
15	74	13
16	64	15
17	16	28
18	23	25
19	29	24
20	33	20
21	82	28
22	67	9

33. What is the relationship between the variables *% residents immunized* and *% residents with influenza*?

    a. Create a scatterplot to illustrate the relationship between *% residents immunized* and *% residents with influenza*.

    b. What is the correlation coefficient between *% residents immunized* and *% residents with influenza*?

    c. Describe the strength of the relationship between *% residents immunized* and *% residents with influenza*.

    d. What is the equation of the best fit line that describes the relationship between *% residents immunized* and *% residents with influenza*?

    e. Could you use the best fit line to produce accurate predictions of *% residents with influenza* using *% residents immunized*?

34. What is the impact of the outlier(s) on this data set?

    a. Identify the outlier in this data set. What is the nursing home number for this outlier?

    b. Remove the outlier and re-create the scatterplot to show the relationship between *% residents immunized* and *% residents with influenza*.

    c. What is the revised correlation coefficient between *% residents immunized* and *% residents with influenza*?

    d. By removing the outlier is the strength of the relationship between *% residents immunized* and *% residents with influenza* increased or decreased?

    e. What is the revised equation of the best fit line that describes the relationship between *% residents immunized* and *% residents with influenza*?

For Exercises 35–38, refer to the following data set:
Amusement Park Rides.

According to the International Association of Amusement Parks and Attractions (IAAPA), "There are more than 400 amusement parks and traditional attractions in the United States alone. In 2008, amusement parks in the United States entertained 300 million visitors who safely enjoyed more than 1.7 billion rides."[4] Despite the popularity of amusement parks, the wait times, especially for the most popular rides, are not so highly regarded. There are different approaches and tactics that people take to get the most rides in their visit to the park. Now, there are even apps for the iPhone and Android to track waiting times at various amusement parks.

One might ask, "Are the wait times worth it? Are the rides with the longest wait times, the most enjoyable?"

Consider the following fictional data.

[3] http://www.hhs.gov/nvpo/glossary1.htm

[4] http://www.iaapa.org/pressroom/Amusement_Park_Industry_Statistics.asp

Ride ID	Ride Name	Avg Wait Time	Avg Enjoyment Rating	Park	Park Location
1	Xoom	45	58	1	Florida
2	Accentuator	35	40	1	Florida
3	Wobbler	15	15	1	Florida
4	Arctic_Attack	75	75	1	Florida
5	Gusher	60	70	1	Florida
6	Alley_Cats	5	60	1	Florida
7	Moon_Swing	10	15	1	Florida
8	Speedster	70	50	1	Florida
9	Hailstorm	80	90	1	Florida
10	DragonFire	70	88	1	Florida
1	Xoom	50	10	2	California
2	Accentuator	35	40	2	California
3	Wobbler	20	75	2	California
4	Arctic_Attack	70	60	2	California
5	Gusher	70	80	2	California
6	Alley_Cats	10	18	2	California
7	Moon_Swing	15	80	2	California
8	Speedster	80	35	2	California
9	Hailstorm	95	40	2	California
10	DragonFire	55	60	2	California

The data shows 10 popular rides in two sister parks located in Florida and California. For each ride in each park, average wait times (in minutes) in the summer of 2010 and the average rating of ride enjoyment (on a scale of 1–100) are provided.

35. What is the relationship between the variables *average wait times* and *average rating of enjoyment*?

    a. Create a scatterplot to show the relationship between *average wait times* and *average rating of enjoyment*.

    b. What is the correlation coefficient between *average wait times* and *average rating of enjoyment*?

    c. Describe the strength of the relationship between *average wait times* and *average rating of enjoyment*.

    d. What is the equation of the best fit line that describes the relationship between *average wait times* and *average rating of enjoyment*?

    e. Could you use the best fit line to produce accurate predictions of *average wait times* using *average rating of enjoyment*?

36. Examine the relationship between *average wait times* and *average rating of enjoyment* for *Park 1* in *Florida* by repeating Exercise 35 for only *Park 1*.

37. Examine the relationship between *average wait times* and *average rating of enjoyment* for *Park 2* in *California* repeating Exercise 35 for only *Park 2*.

38. Compare the relationship between *average wait times* and *average rating of enjoyment* for *Park 1 in Florida* versus *Park 2 in California*.

■ **CHALLENGE**

For Exercises 39–42, refer to the following:

Exploring other types of best-fit curves

When describing the patterns that emerge in paired data sets, there are many more possibilities other than best fit *lines*. Indeed, once you have drawn a scatterplot and are ready to identify the curve that *best fits* the data, there is a substantive collection of other curves that might more accurately describe the data. The following are listed among those in **STATS/CALC** on the TI-83+, along with some comments:

Name of Regression Curve	Form of the Curve	Comments
5: QuadReg	$y = ax^2 + bx + c$	The data set must have at least 3 points to be able to select this option.
6: CubicReg	$y = ax^3 + bx^2 + cx + d$	The data set must have at least 4 points to be able to select this option.
7: QuartReg	$y = ax^4 + bx^3 + cx^2 + dx + e$	The data set must have at least 5 points to be able to select this option.
9: LnReg	$y = a + b \ln x$	The data set must have at least 2 points to be able to select this option, and x cannot take on negative values.
0: ExpReg	$y = a * b^x$	The data set must have at least 2 points to be able to select this option, and y cannot take on the value of 0.
A: PwrReg	$y = a * x^b$	The data set must have at least 2 points to be able to select this option.

For each of the following data sets,

a. Create a scatterplot.

b. Use **LinReg(ax+b)** to determine the best fit *line* and *r*. Does the line seem to accurately describe the pattern in the data?

c. For each of the different choices listed in the above chart, find the equation of the best fit curve and its associated r^2 value. Of all of the curves, which seems to provide the best fit?

Note: The r^2-value reported in each case is NOT the *linear* correlation coefficient reported when running LinReg(ax+B). Rather, the value will typically change depending on the curve. The reason why is that each time, the r^2-value is measuring how accurate the fit is between the data and *that type of curve*. A value of r^2 close to 1 still corresponds to a good fit with whichever curve you are fitting to the data.

41.

x	y
1	0.2
1.5	0.93
2	1.46
3	2.25
10	4.51
15	5.50

42.

x	y
1	32.3
2	8.12
3	−16.89
5	−45.2
6	0.89
8	62.1

39.

x	y
1	16.2
2	21
3	23.7
4	24.8
5	23.9
6	20.7
7	15.8
8	9.1
9	0.3

40.

x	y
0.5	1.20
1.0	0.760
1.5	0.412
2.1	0.196
2.9	0.131
3.3	0.071

The following table shows U.S. population estimates for the years 1991 to 2002, along with the number of tons of municipal waste generated (in 100 million tons) and the percentage that was recycled in the United States during those years.

YEAR	U.S. POPULATION ESTIMATE (100 MILLION PEOPLE)	MUNICIPAL WASTE GENERATED (100 MILLION TONS)	PERCENTAGE RECYCLED
1991	2.46	2.69	8%
1992	2.49	2.93	11.5%
1993	2.52	2.80	14%
1994	2.55	2.91	17%
1995	2.58	3.06	19%
1996	2.60	3.22	23%
1997	2.63	3.26	27%
1998	2.65	3.27	28%
1999	2.68	3.40	30%
2000	2.73	3.74	31.5%
2001	2.80	3.82	33%
2002	2.86	4.09	32%

Adapted from: A Yearly Snapshot of U.S. (Municipal) Waste and Recycling (Data Source: BIOCYCLE/ Table & Conversion: ZWA http://www.zerowasteamerica.org/statistics.htm)

1. For parts (a)–(f), consider the columns for U.S. Population Estimate, x, and Municipal Waste Generated, y.

   a. Write the equation (in slope–intercept form) of the line that passes through the points (2.60, 3.22) and (2.80, 3.82).

   b. How can you interpret the slope of the line in part (a)?

   c. Now choose two other data points and write the equation of the line that passes through your chosen points. How can you interpret the slope of this line?

   d. How does the slope of this line compare to the line in part (a)?

   e. Sketch a graph of the two lines. What do you notice about their y-intercepts?

   f. Finally, plot all the other ordered points (Population, Waste Generated).

The graph of all the data points is called a **scatterplot**. Since the data fall in approximately a straight line, each of the lines you graphed above is an approximation of the data. Section 2.5* presented methods for finding the line that *best* fits a set of data points, called the least-squares regression line.

2. For parts (a)–(c) below, consider the columns U.S. Population Estimate, x, and Percentage Recycled, y.

   a. Graph the scatterplot for this data.

   b. Sketch the graph of a line that contains two data points. Choose a line you think fits the data well.

   c. Write the equation of the line.

   d. How can you interpret the slope of this line?

The Intergovernmental Panel on Climate Change (IPCC) claims that carbon dioxide (CO_2) production from industrial activity (such as fossil fuel burning and other human activities) has increased the CO_2 concentrations in the atmosphere. Because it is a greenhouse gas, elevated CO_2 levels will increase global mean (average) temperature. In this section, we will examine the increasing rate of carbon emissions on Earth.

In 1955, there were (globally) 2 billion tons of carbon emitted per year. In 2005, the carbon emissions more than tripled to reach approximately 7 billion tons of carbon emitted per year. Currently, we are on the path to doubling our current carbon emissions in the next 50 years.

Two Princeton professors* (Stephen Pacala and Rob Socolow) introduced the Climate Carbon Wedge concept. A "wedge" is a strategy to reduce carbon emissions over a 50-year time period by 1.0 GtC/yr (gigatons of carbon per year).

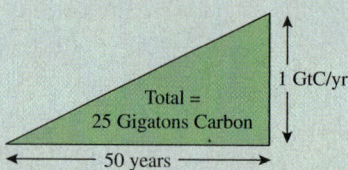

1. Draw the Cartesian plane. Label the vertical axis C, where C represents the number of gigatons (billions of tons) of carbon emitted, and label the horizontal axis t, where t is the number of years. Let $t = 0$ correspond to 2005.

2. Find the equations of the flat path and the seven lines corresponding to the seven wedges.

   a. Flat path (no increase) over 50 years (2005 to 2055)
   b. Increase of 1 GtC over 50 years (2005 to 2055)
   c. Increase of 2 GtC over 50 years (2005 to 2055)
   d. Increase of 3 GtC over 50 years (2005 to 2055)
   e. Increase of 4 GtC over 50 years (2005 to 2055)
   f. Increase of 5 GtC over 50 years (2005 to 2055)
   g. Increase of 6 GtC over 50 years (2005 to 2055)
   h. Increase of 7 GtC over 50 years (2005 to 2055) [projected path]

3. For each of the seven wedges and the flat path, determine how many **total** gigatons of carbon will be reduced over a 50-year period. In other words, how many gigatons of carbon would the world have to reduce in each of the eight cases?

   a. Flat path
   b. Increase of 1 GtC over 50 years
   c. Increase of 2 GtC over 50 years
   d. Increase of 3 GtC over 50 years
   e. Increase of 4 GtC over 50 years
   f. Increase of 5 GtC over 50 years
   g. Increase of 6 GtC over 50 years
   h. Increase of 7 GtC over 50 years (projected path)

4. Research the "climate carbon wedge" concept and discuss the types of changes (transportation efficiency, transportation conservation, building efficiency, efficiency in electricity production, alternate energies, etc.) the world would have to make that would correspond to each of the seven wedges and the flat path.

   a. Flat path
   b. Wedge 1
   c. Wedge 2
   d. Wedge 3
   e. Wedge 4
   f. Wedge 5
   g. Wedge 6

---

*S. Pacala and R. Socolow, "Stabilization Wedges: Solving the Climate Problem for the Next 50 Years with Current Technologies," *Science*, Vol. 305 (2004).

Section	Concept	Key Ideas/Formulas
2.1	**Basic tools: Cartesian plane, distance, and midpoint**	Two points in the xy-plane: (x_1, y_1) and (x_2, y_2)
	Cartesian plane	x-axis, y-axis, origin, and quadrants
	Distance between two points	$d = \sqrt{(x_2 - x_1)^2 + (y_2 - y_1)^2}$
	Midpoint of a line segment joining two points	$(x_m, y_m) = \left(\dfrac{x_1 + x_2}{2}, \dfrac{y_1 + y_2}{2} \right)$
2.2	**Graphing equations: Point-plotting, intercepts, and symmetry**	
	Point-plotting	List a table with several coordinates that are solutions to the equation; plot and connect.
	Intercepts	x-intercept: let $y = 0$ y-intercept: let $x = 0$
	Symmetry	The graph of an equation can be symmetric about the x-axis, y-axis, or origin.
	Using intercepts and symmetry as graphing aids	If (a, b) is on the graph of the equation, then $(-a, b)$ is on the graph if symmetric about the y-axis, $(a, -b)$ is on the graph if symmetric about the x-axis, and $(-a, -b)$ is on the graph if symmetric about the origin.
2.3	**Lines**	General form: $Ax + By = C$
	Graphing a line	Vertical: $x = a$ Slant: $Ax + By = C$ Horizontal: $y = b$ where $A \neq 0$ and $B \neq 0$ **Intercepts** x-intercept $(a, 0)$ y-intercept $(0, b)$ **Slope** $m = \dfrac{y_2 - y_1}{x_2 - x_1}$, where $x_1 \neq x_2$ $\dfrac{\text{"rise"}}{\text{"run"}}$
	Equations of lines	**Slope–intercept form:** $y = mx + b$ m is the slope and b is the y-intercept. **Point–slope form:** $y - y_1 = m(x - x_1)$
	Parallel and perpendicular lines	$L_1 \parallel L_2$ if and only if $m_1 = m_2$ (slopes are equal). $L_1 \perp L_2$ if and only if $m_1 = -\dfrac{1}{m_2} \begin{cases} m_1 \neq 0 \\ m_2 \neq 0 \end{cases}$ (slopes are negative reciprocals).
2.4	**Circles**	
	Standard equation of a circle	$(x - h)^2 + (y - k)^2 = r^2$ $C: (h, k)$
	Transforming equations of circles to the standard form by completing the square	General form: $x^2 + y^2 + ax + by + c = 0$
2.5*	**Linear regression: Best fit**	Fitting data with a line
	Scatterplots	Creating a scatterplot: ■ Using Microsoft Excel ■ Using a graphing calculator
	Identifying patterns	Association Linearity ■ Positive ■ Negative ■ Linear ■ Nonlinear Correlation coefficient, r
	Linear regression	■ Determine the "best fit" line ■ Making predictions

2.1 Basic Tools: Cartesian Plane, Distance, and Midpoint

Plot each point and indicate which quadrant the point lies in.

1. $(-4, 2)$

2. $(4, 7)$

3. $(-1, -6)$

4. $(2, -1)$

Calculate the distance between the two points.

5. $(-2, 0)$ and $(4, 3)$

6. $(1, 4)$ and $(4, 4)$

7. $(-4, -6)$ and $(2, 7)$

8. $\left(\frac{1}{4}, \frac{1}{12}\right)$ and $\left(\frac{1}{3}, -\frac{7}{3}\right)$

Calculate the midpoint of the segment joining the two points.

9. $(2, 4)$ and $(3, 8)$

10. $(-2, 6)$ and $(5, 7)$

11. $(2.3, 3.4)$ and $(5.4, 7.2)$

12. $(-a, 2)$ and $(a, 4)$

Applications

13. **Sports.** A quarterback drops back to pass. At the point $(-5, -20)$ he throws the ball to his wide receiver located at $(10, 30)$. Find the distance the ball has traveled. Assume the width of the football field is $[-15, 15]$ and the length is $[-50, 50]$. Units of measure are yards.

14. **Sports.** Suppose that in the above exercise a defender was midway between the quarterback and the receiver. At what point was the defender located when the ball was thrown over his head?

2.2 Graphing Equations: Point-Plotting, Intercepts, and Symmetry

Find the x-intercept(s) and y-intercept(s) if any.

15. $x^2 + 4y^2 = 4$

16. $y = x^2 - x + 2$

17. $y = \sqrt{x^2 - 9}$

18. $y = \dfrac{x^2 - x - 12}{x - 12}$

Use algebraic tests to determine symmetry with respect to the x-axis, y-axis, or origin.

19. $x^2 + y^3 = 4$

20. $y = x^2 - 2$

21. $xy = 4$

22. $y^2 = 5 + x$

Use symmetry as a graphing aid and point-plot the given equations.

23. $y = x^2 - 3$

24. $y = |x| - 4$

25. $y = \sqrt[3]{x}$

26. $x = y^2 - 2$

27. $y = x\sqrt{9 - x^2}$

28. $x^2 + y^2 = 36$

Applications

29. **Sports.** A track around a high school football field is in the shape of the graph $8x^2 + y^2 = 8$. Graph using symmetry and by plotting points.

30. **Transportation.** A "bypass" around a town follows the graph $y = x^3 + 2$, where the origin is the center of town. Graph the equation.

2.3 Lines

Express the equation for each line in slope–intercept form. Identify the slope and y-intercept of each line.

31. $6x + 2y = 12$

32. $3x + 4y = 9$

33. $-\frac{1}{2}x - \frac{1}{3}y = \frac{1}{6}$

34. $-\frac{2}{3}x - \frac{1}{4}y = \frac{1}{8}$

Find the x- and y-intercepts and the slope of each line if they exist and graph.

35. $y = 4x - 5$

36. $y = -\frac{3}{4}x - 3$

37. $x + y = 4$

38. $x = -4$

39. $y = 2$

40. $-\frac{1}{2}x - \frac{1}{2}y = 3$

Write the equation of the line, given the slope and the intercepts.

41. Slope: $m = 4$
y-intercept: $(0, -3)$

42. Slope: $m = 0$
y-intercept: $(0, 4)$

43. Slope: m is undefined
x-intercept: $(-3, 0)$

44. Slope: $m = \frac{-2}{3}$
y-intercept: $\left(0, \frac{3}{4}\right)$

Write an equation of the line, given the slope and a point that lies on the line.

45. $m = -2$ $(-3, 4)$

46. $m = \frac{3}{4}$ $(2, 16)$

47. $m = 0$ $(-4, 6)$

48. m is undefined $(2, -5)$

Write the equation of the line that passes through the given points. Express the equation in slope–intercept form or in the form of $x = a$ or $y = b$.

49. $(-4, -2)$ and $(2, 3)$

50. $(-1, 4)$ and $(-2, 5)$

51. $\left(-\frac{3}{4}, \frac{1}{2}\right)$ and $\left(-\frac{7}{4}, \frac{5}{2}\right)$

52. $(3, -2)$ and $(-9, 2)$

Find the equation of the line that passes through the given point and also satisfies the additional piece of information.

53. $(-2, -1)$ parallel to the line $2x - 3y = 6$

54. $(5, 6)$ perpendicular to the line $5x - 3y = 0$

55. $\left(-\frac{3}{4}, \frac{5}{2}\right)$ perpendicular to the line $\frac{2}{3}x - \frac{1}{2}y = 12$

56. $(a + 2, b - 1)$ parallel to the line $Ax + By = C$

Applications

57. Grades. For a GRE prep class, a student must take a pretest and then a posttest after the completion of the course. Two students' results are shown below. Give a linear equation to represent the given data.

Pretest	Posttest
1020	1324
950	1240

58. Budget: Car Repair. The cost of having the air conditioner in your car repaired is the combination of material costs and labor costs. The materials (tubing, coolant, etc.) are \$250, and the labor costs \$38 per hour. Write an equation that models the total cost C of having your air conditioner repaired as a function of hours t. Graph this equation with t as the horizontal axis and C representing the vertical axis. How much will the job cost if the mechanic works 1.5 hours?

2.4 Circles

Write the equation of the circle in standard form.

59. center $(-2, 3)$
$r = 6$

60. center $(-6, -8)$
$r = 3\sqrt{6}$

61. center $\left(\frac{3}{4}, \frac{5}{2}\right)$
$r = \frac{2}{5}$

62. center $(1.2, -2.4)$
$r = 3.6$

Find the center and the radius of the circle given by the equation.

63. $(x + 2)^2 + (y + 3)^2 = 81$

64. $(x - 4)^2 + (y + 2)^2 = 32$

65. $\left(x + \frac{3}{4}\right)^2 + \left(y - \frac{1}{2}\right)^2 = \frac{16}{36}$

66. $x^2 + y^2 + 4x - 2y = 0$

67. $x^2 + y^2 + 2y - 4x + 11 = 0$

68. $3x^2 + 3y^2 - 6x - 7 = 0$

69. $9x^2 + 9y^2 - 6x + 12y - 76 = 0$

70. $x^2 + y^2 + 3.2x - 6.6y - 2.4 = 0$

71. Find the equation of a circle centered at $(2, 7)$ and passing through $(3, 6)$.

72. Find the equation of a circle that has the diameter with endpoints $(-2, -1)$ and $(5, 5)$.

Technology Exercises

Section 2.1

Determine whether the triangle with the given vertices is a right triangle, isosceles triangle, neither, or both.

73. $(-10, -5)$, $(20, -45)$, $(10, 10)$

74. $(4.2, 8.4)$, $(-4.2, 2.1)$, $(6.3, -10.5)$

Section 2.2

Graph the equation using a graphing utility and state whether there is any symmetry.

75. $y^2 = |x^2 - 4|$

76. $0.8x^2 - 1.5y^2 = 4.8$

Section 2.3

Determine whether the lines are parallel, perpendicular, or neither, then graph both lines in the same viewing screen using a graphing utility to confirm your answer.

77. $y_1 = 0.875x + 1.5$

$y_2 = -\frac{8}{7}x - \frac{9}{14}$

78. $y_1 = -0.45x - 2.1$

$y_2 = \frac{5}{6} - \frac{9}{20}x$

Section 2.4

79. Use the Quadratic Formula to solve for y, and use a graphing utility to graph each equation. Do the graphs agree with the graph in Exercise 69?

$$9x^2 + 9y^2 - 6x + 12y - 76 = 0$$

80. Use the Quadratic Formula to solve for y, and use a graphing utility to graph each equation. Do the graphs agree with the graph in Exercise 70?

$$x^2 + y^2 + 3.2x - 6.6y - 2.4 = 0$$

1. Find the distance between the points $(-7, -3)$ and $(2, -2)$.

2. Find the midpoint between $(-3, 5)$ and $(5, -1)$.

3. Determine the length and the midpoint of a segment that joins the points $(-2, 4)$ and $(3, 6)$.

4. **Research Triangle.** The Research Triangle in North Carolina was established as a collaborative research center among Duke University (Durham), North Carolina State University (Raleigh), and the University of North Carolina (Chapel Hill).

Durham is 10 miles north and 8 miles east of Chapel Hill, and Raleigh is 28 miles east and 15 miles south of Chapel Hill. What is the perimeter of the research triangle? Round your answer to the nearest mile.

5. Determine the two values for y so that the point $(3, y)$ is 5 units away from the point $(6, 5)$.

6. If the point $(3, -4)$ is on a graph that is symmetric with respect to the y-axis, what point must also be on the graph?

7. Determine whether the graph of the equation $x - y^2 = 5$ has any symmetry (x-axis, y-axis, and origin).

8. Find the x-intercept(s) and the y-intercept(s), if any: $4x^2 - 9y^2 = 36$.

Graph the following equations.

9. $2x^2 + y^2 = 8$

10. $y = \dfrac{4}{x^2 + 1}$

11. Find the x-intercept and the y-intercept of the line $x - 3y = 6$.

12. Express the line in slope–intercept form: $4x - 6y = 12$.

13. Express the line in slope–intercept form: $\frac{2}{3}x - \frac{1}{4}y = 2$.

Find the equation of the line that is characterized by the given information. Graph the line.

14. Slope $= 4$; y-intercept $(0, 3)$

15. Passes through the points $(-3, 2)$ and $(4, 9)$

16. Parallel to the line $y = 4x + 3$ and passes through the point $(1, 7)$

17. Perpendicular to the line $2x - 4y = 5$ and passes through the point $(1, 1)$

18. x-intercept $(3, 0)$; y-intercept $(0, 6)$

For Exercises 19 and 20, write the equation of the line that corresponds to the graph.

19.

20.

21. Write the equation of a circle that has center $(6, -7)$ and radius $r = 8$.

22. Determine the center and radius of the circle $x^2 + y^2 - 10x + 6y + 22 = 0$.

23. Find the equation of the circle that is centered at $(4, 9)$ and passes through the point $(2, 5)$.

24. **Solar System.** Earth is approximately 93 million miles from the Sun. Approximating Earth's orbit around the Sun as circular, write an equation governing Earth's path around the Sun. Locate the Sun at the origin.

25. Determine whether the triangle with the given vertices is a right triangle, isosceles triangle, neither, or both.

$$(-8.4, 16.8), (0, 37.8), (12.6, 8.4)$$

26. Graph the given equation using a graphing utility and state whether there is any symmetry.

$$0.25y^2 + 0.04x^2 = 1$$

1. Simplify $\dfrac{7-2}{7+3}$.

2. Simplify and express in terms of positive exponents: $\dfrac{\left(5x^{3/4}\right)^4}{25x^{-1/4}}$.

3. Perform the operation and simplify: $(x-4)^2(x+4)^2$.

4. Factor completely $8x^3 - 27y^3$.

5. Perform the operations and simplify: $\dfrac{1/x - 1/5}{1/x + 1/5}$.

6. Solve for x: $x^3 - 5x^2 - 4x + 20 = 0$.

7. Perform the operations and write in standard form:
$$\sqrt{-36}(5 - 2i).$$

Solve for x.

8. $15 - [5 + 3x - 4(2x - 6)]$
$= 4(6x - 7) - [5(3x - 7) - 6x + 10]$

9. $\dfrac{5}{4x + 1} = \dfrac{3}{4x - 1}$

10. Ashley inherited $17,000. She invested some money in a CD that earns 5% and the rest in a stock that earns 8%. How much was invested in each account, if the interest for the first year is $1075?

11. Solve by factoring: $5x^2 = 45$.

12. Solve by completing the square: $3x^2 + 6x = 7$.

13. Use the discriminant to determine the number and type of roots: $5x^2 + 2x + 7 = 0$.

14. Solve for r: $p^2 + q^2 = r^2$.

15. Solve and check: $\sqrt{x^2 + 3x - 10} = x - 2$.

16. Solve using substitution: $\dfrac{1}{(x + 2)^2} - \dfrac{5}{x + 2} + 4 = 0$.

Solve and express the solution in interval notation:

17. $6 < \frac{1}{4}x + 6 < 9$

18. $x^2 - x \geq 20$

19. $|2 - x| < 4$

20. Solve for x: $|5 - 4x| = 23$.

21. Use algebraic tests to determine whether the graph of the equation $y = 4x$ is symmetric with respect to the x-axis, y-axis, or origin.

22. Write an equation of a line in slope–intercept form with slope $m = \frac{4}{5}$ that passes through the point $(5, 1)$.

23. Write an equation of a line that is perpendicular to the x-axis and passes through the point $(5, 3)$.

24. Write an equation of a line in slope–intercept form that passes through the two points $\left(\frac{1}{7}, \frac{5}{3}\right)$ and $\left(-\frac{6}{7}, -\frac{2}{3}\right)$.

25. Find the center and radius of the circle:
$(x + 5)^2 + (y + 3)^2 = 30$.

26. Calculate the distance between the two points $\left(-\sqrt{11}, 5\right)$ and $\left(2, \sqrt{7}\right)$, and find the midpoint of the segment joining the two points. Round your answers to one decimal place.

27. Determine whether the lines $y_1 = 0.32x + 1.5$ and $y_2 = -\frac{5}{16}x + \frac{1}{4}$ are parallel, perpendicular, or neither, then graph both lines in the same viewing screen using a graphing utility to confirm your answer.

3

Functions and Their Graphs

On a sales rack of clothes at a department store, you see a shirt you like. The original price of the shirt was $100, but it has been discounted 30%. As a preferred shopper, you get an automatic additional 20% off the sale price at the register. How much will you pay for the shirt?

Naïve shoppers might be lured into thinking this shirt will cost $50 because they add the 20% and 30% to get 50% off, but they will end up paying more than that. Experienced shoppers know that they first take 30% off of $100, which results in a price of $70, and then they take an additional 20% off of the sale price, $70, which results in a final discounted price of $56. Experienced shoppers have already learned *composition of functions*.

A *composition of functions* can be thought of as a function of a function. One function takes an input (original price, $100) and maps it to an output (sale price, $70), and then another function takes that output as its input (sale price, $70) and maps that to an output (checkout price, $56).

IN THIS CHAPTER you will find that functions are part of our everyday thinking: converting from degrees Celsius to degrees Fahrenheit, DNA testing in forensic science, determining stock values, and the sale price of a shirt. We will develop a more complete, thorough understanding of functions. First, we will establish what a relation is, and then we will determine whether a relation is a function. We will discuss common functions, domain and range of functions, and graphs of functions. We will determine whether a function is increasing or decreasing on an interval and calculate the average rate of change of a function. We will perform operations on functions and composition of functions. We will discuss one-to-one functions and inverse functions. Finally, we will model applications with functions using variation.

FUNCTIONS AND THEIR GRAPHS

3.1
Functions

- Relations and Functions
- Functions Defined by Equations
- Function Notation
- Domain of a Function

3.2
Graphs of Functions; Piecewise-Defined Functions; Increasing and Decreasing Functions; Average Rate of Change

- Recognizing and Classifying Functions
- Increasing and Decreasing Functions
- Average Rate of Change
- Piecewise-Defined Functions

3.3
Graphing Techniques: Transformations

- Horizontal and Vertical Shifts
- Reflection about the Axes
- Stretching and Compressing

3.4
Operations on Functions and Composition of Functions

- Adding, Subtracting, Multiplying, and Dividing Functions
- Composition of Functions

3.5
One-to-One Functions and Inverse Functions

- Determine Whether a Function Is One-to-One
- Inverse Functions
- Graphical Interpretation of Inverse Functions
- Finding the Inverse Function

3.6
Modeling Functions Using Variation

- Direct Variation
- Inverse Variation
- Joint Variation and Combined Variation

LEARNING OBJECTIVES

- Find the domain and range of a function.
- Sketch the graphs of common functions.
- Sketch graphs of general functions employing translations of common functions.
- Perform composition of functions.
- Find the inverse of a function.
- Model applications with functions using variation.

SKILLS OBJECTIVES

- Determine whether a relation is a function.
- Determine whether an equation represents a function.
- Use function notation.
- Find the value of a function.
- Determine the domain and range of a function.

CONCEPTUAL OBJECTIVES

- Think of function notation as a placeholder or mapping.
- Understand that all functions are relations but not all relations are functions.

Relations and Functions

What do the following pairs have in common?

- Every person has a blood type.
- Temperature is some specific value at a particular time of day.
- Every working household phone in the United States has a 10-digit phone number.
- First-class postage rates correspond to the weight of a letter.
- Certain times of the day are start times of sporting events at a university.

They all describe a particular correspondence between two groups. A **relation** is a correspondence between two sets. The first set is called the **domain**, and the corresponding second set is called the **range**. Members of these sets are called **elements**.

DEFINITION **Relation**

A **relation** is a correspondence between two sets where each element in the first set, called the **domain**, corresponds to *at least* one element in the second set, called the **range**.

A relation is a set of ordered pairs. The domain is the set of all the first components of the ordered pairs, and the range is the set of all the second components of the ordered pairs.

PERSON	BLOOD TYPE	ORDERED PAIR
Michael	A	(Michael, A)
Tania	A	(Tania, A)
Dylan	AB	(Dylan, AB)
Trevor	O	(Trevor, O)
Megan	O	(Megan, O)

WORDS

The domain is the set of all the first components.

The range is the set of all the second components.

MATH

{Michael, Tania, Dylan, Trevor, Megan}

{A, AB, O}

A relation in which each element in the domain corresponds to exactly one element in the range is a **function**.

DEFINITION   Function

A **function** is a correspondence between two sets where each element in the first set, called the **domain**, corresponds to *exactly* one element in the second set, called the **range**.

Note that the definition of a function is more restrictive than the definition of a relation. For a relation, each input corresponds to *at least* one output, whereas, for a function, each input corresponds to *exactly* one output. The blood-type example given is both a relation and a function.

Also note that the range (set of values to which the elements of the domain correspond) is a subset of the set of all blood types. However, although all functions are relations, not all relations are functions.

For example, at a university, four primary sports typically overlap in the late fall: football, volleyball, soccer, and basketball. On a given Saturday, the following table indicates the start times for the competitions.

TIME OF DAY	COMPETITION
1:00 P.M.	Football
2:00 P.M.	Volleyball
7:00 P.M.	Soccer
7:00 P.M.	Basketball

WORDS	**MATH**
The 1:00 start time corresponds to exactly one event, Football.	(1:00 P.M., Football)
The 2:00 start time corresponds to exactly one event, Volleyball.	(2:00 P.M., Volleyball)
The 7:00 start time corresponds to two events, Soccer and Basketball.	(7:00 P.M., Soccer) (7:00 P.M., Basketball)

Because an element in the domain, 7:00 P.M., corresponds to more than one element in the range, Soccer and Basketball, this is not a function. It is, however, a relation.

EXAMPLE 1   Determining Whether a Relation Is a Function

Determine whether the following relations are functions.

a. $\{(-3, 4), (2, 4), (3, 5), (6, 4)\}$
b. $\{(-3, 4), (2, 4), (3, 5), (2, 2)\}$
c. Domain = Set of all items for sale in a grocery store; Range = Price

Solution:

a. No x-value is repeated. Therefore, each x-value corresponds to exactly one y-value.
   │ This relation is a function. │

b. The value $x = 2$ corresponds to *both* $y = 2$ and $y = 4$. │ This relation is not a function. │

c. Each item in the grocery store corresponds to exactly one price. │ This relation is a function. │

■ **YOUR TURN** Determine whether the following relations are functions.

a. $\{(1, 2), (3, 2), (5, 6), (7, 6)\}$
b. $\{(1, 2), (1, 3), (5, 6), (7, 8)\}$
c. $\{(11:00 \text{ A.M.}, 83°F), (2:00 \text{ P.M.}, 89°F), (6:00 \text{ P.M.}, 85°F)\}$

Study Tip

All functions are relations but not all relations are functions.

Classroom Example 3.1.1
Determine whether these relations are functions.
a. $\{(1, -1), (-1, 1), (0, -1), (0, 1)\}$
b. $\{(-1, -2), (-2, -3), (-3, -4), (-4, -5)\}$
c. Domain = Set of high school seniors; Range = GPA

Answer:
a. no   **b.** yes   **c.** yes

■ **Answer:** **a.** function
            b. not a function
            c. function

All of the examples we have discussed thus far are **discrete** sets in that they represent a countable set of distinct pairs of (x, y). A function can also be defined algebraically by an equation.

Functions Defined by Equations

Let's start with the equation $y = x^2 - 3x$, where x can be any real number. This equation assigns to each x-value exactly one corresponding y-value.

x	$y = x^2 - 3x$	y
1	$y = (1)^2 - 3(1)$	-2
5	$y = (5)^2 - 3(5)$	10
$-\frac{2}{3}$	$y = \left(-\frac{2}{3}\right)^2 - 3\left(-\frac{2}{3}\right)$	$\frac{22}{9}$
1.2	$y = (1.2)^2 - 3(1.2)$	-2.16

Since the variable y *depends* on what value of x is selected, we denote y as the **dependent variable**. The variable x can be any number in the domain; therefore, we denote x as the **independent variable**.

Although functions are defined by equations, it is important to recognize that *not all equations are functions.* The requirement for an equation to define a function is that each element in the domain corresponds to exactly one element in the range. Throughout the ensuing discussion, we assume x to be the independent variable and y to be the dependent variable.

▼ **CAUTION**

Not all equations are functions.

Equations that represent functions of x: $y = x^2$ $y = |x|$ $y = x^3$

Equations that do not represent functions of x: $x = y^2$ $x^2 + y^2 = 1$ $x = |y|$

Study Tip

We say that $x = y^2$ is not a function of x. However, if we reverse the independent and dependent variables, then $x = y^2$ is a function of y.

In the "equations that represent functions of x," every x-value corresponds to exactly one y-value. Some ordered pairs that correspond to these functions are

$$y = x^2: \quad (-1, 1)\,(0, 0)\,(1, 1)$$
$$y = |x|: \quad (-1, 1)\,(0, 0)\,(1, 1)$$
$$y = x^3: \quad (-1, -1)\,(0, 0)\,(1, 1)$$

The fact that $x = -1$ and $x = 1$ both correspond to $y = 1$ in the first two examples does not violate the definition of a function.

In the "equations that do not represent functions of x," some x-values correspond to *more than one* y-value. Some ordered pairs that correspond to these equations are

RELATION	SOLVE RELATION FOR Y	POINTS THAT LIE ON THE GRAPH	
$x = y^2$	$y = \pm\sqrt{x}$	$(1, -1)\,(0, 0)\,(1, 1)$	$x = 1$ maps to **both** $y = -1$ and $y = 1$
$x^2 + y^2 = 1$	$y = \pm\sqrt{1 - x^2}$	$(0, -1)\,(0, 1)\,(-1, 0)\,(1, 0)$	$x = 0$ maps to **both** $y = -1$ and $y = 1$
$x = |y|$	$y = \pm x$	$(1, -1)\,(0, 0)\,(1, 1)$	$x = 1$ maps to **both** $y = -1$ and $y = 1$

Let's look at the graphs of the three **functions of *x***:

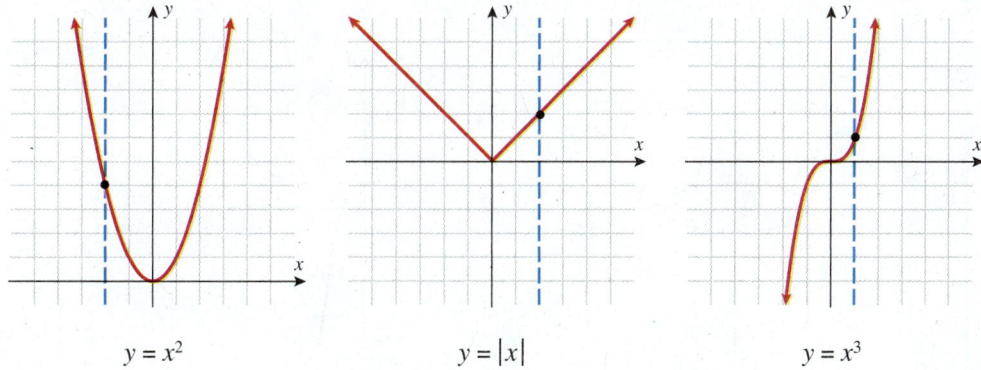

$$y = x^2 \qquad\qquad y = |x| \qquad\qquad y = x^3$$

Let's take any value for *x*, say *x* = *a*. The graph of *x* = *a* corresponds to a vertical line. A function of *x* maps each *x*-value to exactly one *y*-value; therefore, there should be at most one point of intersection with any vertical line. We see in the three graphs of the functions above that if a vertical line is drawn at any value of *x* on any of the three graphs, the vertical line only intersects the graph in one place. Look at the graphs of the three equations that do **not** represent **functions of *x***.

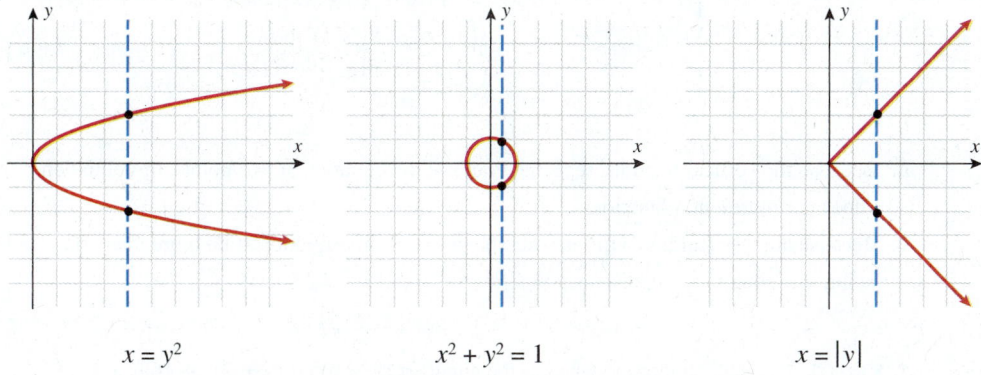

$$x = y^2 \qquad\qquad x^2 + y^2 = 1 \qquad\qquad x = |y|$$

A vertical line can be drawn on any of the three graphs such that the vertical line will intersect each of these graphs at two points. Thus, there are two *y*-values that correspond to some *x*-value in the domain, which is why these equations do not define *y* as a function of *x*.

DEFINITION **Vertical Line Test**

Given the graph of an equation, if any vertical line that can be drawn intersects the graph at no more than one point, the equation defines *y* as a function of *x*. This test is called the **vertical line test**.

Study Tip

If any *x*-value corresponds to more than one *y*-value, then *y* is **not** a function of *x*.

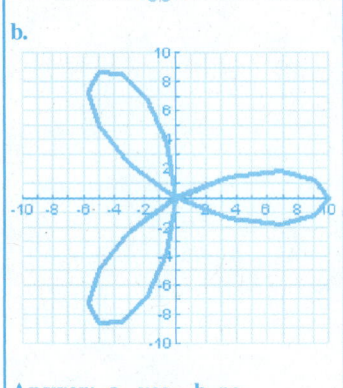
EXAMPLE 2   Using the Vertical Line Test

Use the vertical line test to determine whether the graphs of equations define functions of x.

a.   **b.**

Solution:

Apply the vertical line test.

a.   **b.**

          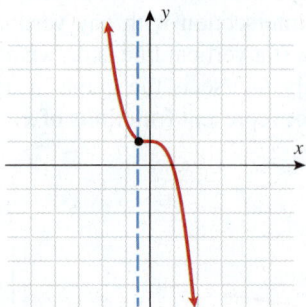

a. Because the vertical line intersects the graph of the equation at two points, this equation
   does not represent a function .

b. Because any vertical line will intersect the graph of this equation at no more than one
   point, this equation   represents a function .

■ **YOUR TURN** Determine whether the equation $(x - 3)^2 + (y + 2)^2 = 16$ is a
                 function of x.

To recap, a function can be expressed one of four ways: verbally, numerically, algebraically, and graphically. This is sometimes called the Rule of 4.

Expressing a Function

VERBALLY	**NUMERICALLY**	**ALGEBRAICALLY**	**GRAPHICALLY**
Every real number has a corresponding absolute value.	$\{(-3, 3), (-1, 1), (0, 0), (1, 1), (5, 5)\}$	$y = \lvert x \rvert$	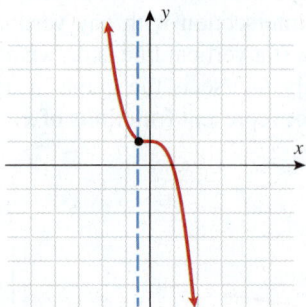

Function Notation

We know that the equation $y = 2x + 5$ defines y as a function of x because its graph is a nonvertical line and thus passes the vertical line test. We can select x-values (input) and determine unique corresponding y-values (output). The output is found by taking 2 times the input and then adding 5. If we give the function a name, say, "f", then we can use **function notation**:

$$f(x) = 2x + 5$$

The symbol $f(x)$ is read "f evaluated at x" or "f of x" and represents the y-value that corresponds to a particular x-value. In other words, $y = f(x)$.

INPUT	FUNCTION	OUTPUT	EQUATION
x	f	$f(x)$	$f(x) = 2x + 5$
Independent variable	Mapping	Dependent variable	Mathematical rule

It is important to note that f is the function name, whereas $f(x)$ is the value of the function. In other words, the function f maps some value x in the domain to some value $f(x)$ in the range.

x	$f(x) = 2x + 5$	$f(x)$
0	$f(0) = 2(0) + 5$	$f(0) = 5$
1	$f(1) = 2(1) + 5$	$f(1) = 7$
2	$f(2) = 2(2) + 5$	$f(2) = 9$

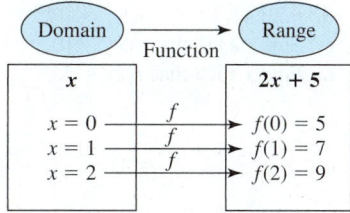

The independent variable is also referred to as the **argument** of a function. To evaluate functions, it is often useful to think of the independent variable or argument as a placeholder. For example, $f(x) = x^2 - 3x$ can be thought of as

$$f(\square) = (\square)^2 - 3(\square)$$

In other words, "f of the argument is equal to the argument squared minus 3 times the argument." Any expression can be substituted for the argument:

$$f(1) = (1)^2 - 3(1)$$
$$f(x + 1) = (x + 1)^2 - 3(x + 1)$$
$$f(-x) = (-x)^2 - 3(-x)$$

It is important to note:

- $f(x)$ does *not* mean f times x.
- The most common function names are f and F since the word *function* begins with an "f". Other common function names are g and G, but any letter can be used.
- The letter most commonly used for the independent variable is x. The letter t is also common because in real-world applications it represents time, but any letter can be used.
- Although we can think of y and $f(x)$ as interchangeable, the function notation is useful when we want to consider two or more functions of the same independent variable.

EXAMPLE 3 **Evaluating Functions by Substitution**

Given the function $f(x) = 2x^3 - 3x^2 + 6$, find $f(-1)$.

Solution:

Consider the independent variable x to be a placeholder.

$$f(\square) = 2(\square)^3 - 3(\square)^2 + 6$$

To find $f(-1)$, substitute $x = -1$ into the function.

$$f(-1) = 2(-1)^3 - 3(-1)^2 + 6$$

Evaluate the right side.

$$f(-1) = -2 - 3 + 6$$

Simplify.

$$\boxed{f(-1) = 1}$$

EXAMPLE 4 **Finding Function Values from the Graph of a Function**

The graph of f is given on the right.

a. Find $f(0)$.
b. Find $f(1)$.
c. Find $f(2)$.
d. Find $4f(3)$.
e. Find x such that $f(x) = 10$.
f. Find x such that $f(x) = 2$.

Solution (a): The value $x = 0$ corresponds to the value $y = 5$. $\boxed{f(0) = 5}$

Solution (b): The value $x = 1$ corresponds to the value $y = 2$. $\boxed{f(1) = 2}$

Solution (c): The value $x = 2$ corresponds to the value $y = 1$. $\boxed{f(2) = 1}$

Solution (d): The value $x = 3$ corresponds to the value $y = 2$. $4f(3) = 4 \cdot 2 = \boxed{8}$

Solution (e): The value $y = 10$ corresponds to the value $\boxed{x = 5}$.

Solution (f): The value $y = 2$ corresponds to the values $\boxed{x = 1}$ and $\boxed{x = 3}$.

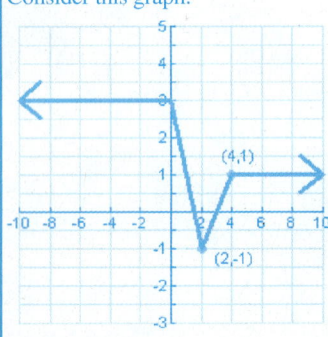
■ **YOUR TURN** For the following graph of a function, find:

a. $f(-1)$ **b.** $f(0)$ **c.** $3f(2)$

d. the value of x that corresponds to $f(x) = 0$

EXAMPLE 5 **Evaluating Functions with Variable Arguments (Inputs)**

For the given function $f(x) = x^2 - 3x$, evaluate $f(x + 1)$ and simplify if possible.

COMMON MISTAKE

A common misunderstanding is to interpret the notation $f(x + 1)$ as a sum: $f(x + 1) \neq f(x) + f(1)$.

⭐ CORRECT

Write the original function.

$$f(x) = x^2 - 3x$$

Replace the argument x with a placeholder.

$$f(\square) = (\square)^2 - 3(\square)$$

Substitute $x + 1$ for the argument.

$$f(x + 1) = (x + 1)^2 - 3(x + 1)$$

Eliminate the parentheses.

$$f(x + 1) = x^2 + 2x + 1 - 3x - 3$$

Combine like terms.

$$\boxed{f(x + 1) = x^2 - x - 2}$$

❌ INCORRECT

The **ERROR** is in interpreting the notation as a sum.

$$f(x + 1) \neq f(x) + f(1)$$

$$\neq x^2 - 3x - 2$$

■ **YOUR TURN** For the given function $g(x) = x^2 - 2x + 3$, evaluate $g(x - 1)$.

Classroom Example 3.1.5
Let $f(x) = 1 - (x - 3)^2$.
Compute:
a. $f(x + 3)$ **b.** $f(3 - x)$
c. $f(2x + 1)$

Answer:
a. $1 - x^2$ **b.** $1 - x^2$
c. $-4x^2 + 8x - 3$

▼ **CAUTION**

$f(x + 1) \neq f(x) + f(1)$

■ **Answer:** $g(x - 1) = x^2 - 4x + 6$

Classroom Example 3.1.6
Let $f(x) = 2 - x^2$. Compute:
a. $f(x + 1)$ **b.** $f(x) + f(1)$

Answer:
a. $-x^2 - 2x + 1$
b. $3 - x^2$

EXAMPLE 6 **Evaluating Functions: Sums**

For the given function $H(x) = x^2 + 2x$, evaluate:

a. $H(x + 1)$ **b.** $H(x) + H(1)$

Solution (a):

Write the function H in placeholder notation. $H(\square) = (\square)^2 + 2(\square)$

Substitute $x + 1$ for the argument of H. $H(x + 1) = (x + 1)^2 + 2(x + 1)$

Eliminate the parentheses on the right side. $H(x + 1) = x^2 + 2x + 1 + 2x + 2$

Combine like terms on the right side. $\boxed{H(x + 1) = x^2 + 4x + 3}$

Solution (b):

Write $H(x)$. $H(x) = x^2 + 2x$

Evaluate H at $x = 1$. $H(1) = (1)^2 + 2(1) = 3$

Evaluate the sum $H(x) + H(1)$. $H(x) + H(1) = x^2 + 2x + 3$

$\boxed{H(x) + H(1) = x^2 + 2x + 3}$

Note: Comparing the results of part (a) and part (b), we see that

$$H(x + 1) \neq H(x) + H(1).$$

Technology Tip

Use a graphing utility to display graphs of
$y_1 = H(x + 1) = (x + 1)^2 + 2(x + 1)$
and $y_2 = H(x) + H(1) = x^2 + 2x + 3$.

```
Plot1  Plot2  Plot3
\Y1▉(X+1)²+2(X+1
)
\Y2▉X²+2X+3
```

The graphs are not the same.

Technology Tip

Use a graphing utility to display graphs of $y_1 = G(-x) = (-x)^2 - (-x)$ and $y_2 = -G(x) = -(x^2 - x)$.

The graphs are not the same.

Classroom Example 3.1.7*
Let $f(x) = -3 - x - 2x^2$.
Compute $-f(-x)$.

Answer: $3 - x + 2x^2$

Classroom Example 3.1.8
Let $f(x) = 1 - (x - 1)^2$.
Compute:
a. $\dfrac{f(2)}{f(3)}$ b. $f\left(\dfrac{2}{3}\right)$

Answer: a. 0 **b.** $\dfrac{8}{9}$

▼ **CAUTION**

$$f\left(\frac{a}{b}\right) \neq \frac{f(a)}{f(b)}$$

■ **Answer: a.** $G(t - 2) = 3t - 10$
b. $G(t) - G(2) = 3t - 6$
c. $\dfrac{G(1)}{G(3)} = -\dfrac{1}{5}$
d. $G\left(\dfrac{1}{3}\right) = -3$

EXAMPLE 7   Evaluating Functions: Negatives

For the given function $G(t) = t^2 - t$, evaluate:

a. $G(-t)$ **b.** $-G(t)$

Solution (a):

Write the function G in placeholder notation.	$G(\square) = (\square)^2 - (\square)$
Substitute $-t$ for the argument of G.	$G(-t) = (-t)^2 - (-t)$
Eliminate the parentheses on the right side.	$\boxed{G(-t) = t^2 + t}$

Solution (b):

Write $G(t)$.	$G(t) = t^2 - t$
Multiply by -1.	$-G(t) = -(t^2 - t)$
Eliminate the parentheses on the right side.	$\boxed{-G(t) = -t^2 + t}$

Note: Comparing the results of part (a) and part (b), we see that $\mathbf{G(-t) \neq -G(t)}$.

EXAMPLE 8   Evaluating Functions: Quotients

For the given function $F(x) = 3x + 5$, evaluate:

a. $F\left(\dfrac{1}{2}\right)$ **b.** $\dfrac{F(1)}{F(2)}$

Solution (a):

Write F in placeholder notation.	$F(\square) = 3(\square) + 5$
Replace the argument with $\frac{1}{2}$.	$F\left(\dfrac{1}{2}\right) = 3\left(\dfrac{1}{2}\right) + 5$
Simplify the right side.	$\boxed{F\left(\dfrac{1}{2}\right) = \dfrac{13}{2}}$

Solution (b):

Evaluate $F(1)$.	$F(1) = 3(1) + 5 = 8$
Evaluate $F(2)$.	$F(2) = 3(2) + 5 = 11$
Divide $F(1)$ by $F(2)$.	$\boxed{\dfrac{F(1)}{F(2)} = \dfrac{8}{11}}$

Note: Comparing the results of part (a) and part (b), we see that $\mathbf{F\left(\dfrac{1}{2}\right) \neq \dfrac{F(1)}{F(2)}}$.

■ **YOUR TURN**   Given the function $G(t) = 3t - 4$, evaluate:

a. $G(t - 2)$ **b.** $G(t) - G(2)$ **c.** $\dfrac{G(1)}{G(3)}$ **d.** $G\left(\dfrac{1}{3}\right)$

Examples 6, 7, and 8 illustrate the following:

$$f(a + b) \neq f(a) + f(b) \qquad f(-t) \neq -f(t) \qquad f\left(\frac{a}{b}\right) \neq \frac{f(a)}{f(b)}$$

Now that we have shown that $f(x + h) \neq f(x) + f(h)$, we turn our attention to one of the fundamental expressions in calculus: the **difference quotient**.

$$\frac{f(x + h) - f(x)}{h} \qquad h \neq 0$$

Example 9 illustrates the difference quotient, which will be discussed in detail in Section 3.2. For now, we will concentrate on the algebra involved when finding the difference quotient. In Section 3.2, the application of the difference quotient will be the emphasis.

EXAMPLE 9 **Evaluating the Difference Quotient**

For the function $f(x) = x^2 - x$, find $\dfrac{f(x + h) - f(x)}{h}$, $h \neq 0$.

Solution:

Use placeholder notation for the function $f(x) = x^2 - x$. $\qquad f(\square) = (\square)^2 - (\square)$

Calculate $f(x + h)$. $\qquad f(x + h) = (x + h)^2 - (x + h)$

Write the difference quotient. $\qquad \dfrac{f(x + h) - f(x)}{h}$

Let $f(x + h) = (x + h)^2 - (x + h)$ and $f(x) = x^2 - x$.

$$\frac{f(x + h) - f(x)}{h} = \frac{\overbrace{\left[(x + h)^2 - (x + h)\right]}^{f(x+h)} - \overbrace{\left[x^2 - x\right]}^{f(x)}}{h} \qquad h \neq 0$$

Eliminate the parentheses inside the first set of brackets.

$$= \frac{[x^2 + 2xh + h^2 - x - h] - [x^2 - x]}{h}$$

Eliminate the brackets in the numerator.

$$= \frac{x^2 + 2xh + h^2 - x - h - x^2 + x}{h}$$

Combine like terms.

$$= \frac{2xh + h^2 - h}{h}$$

Factor the numerator.

$$= \frac{h(2x + h - 1)}{h}$$

Divide out the common factor, h.

$$= \boxed{2x + h - 1} \qquad h \neq 0$$

■ **YOUR TURN** Evaluate the difference quotient for $f(x) = x^2 - 1$.

■ **Answer:** $2x + h$

Domain of a Function

Sometimes the domain of a function is stated *explicitly*. For example,

$$f(x) = |x| \qquad \underbrace{x < 0}_{\text{domain}}$$

Here, the **explicit domain** is the set of all negative real numbers, $(-\infty, 0)$. Every negative real number in the domain is mapped to a positive real number in the range through the absolute value function.

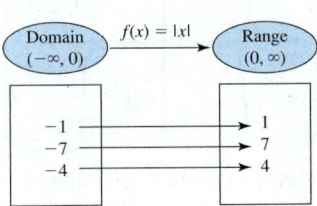

If the expression that defines the function is given but the domain is not stated explicitly, then the domain is implied. The **implicit domain** is the largest set of real numbers for which the function is defined and the output value $f(x)$ is a real number. For example,

$$f(x) = \sqrt{x}$$

does not have the domain explicitly stated. There is, however, an implicit domain. Note that if the argument is negative, that is, if $x < 0$, then the result is an imaginary number. In order for the output of the function, $f(x)$, to be a real number, we must restrict the domain to nonnegative numbers, that is, if $x \geq 0$.

FUNCTION	IMPLICIT DOMAIN
$f(x) = \sqrt{x}$	$[0, \infty)$

In general, we ask the question, "what can x be?" The implicit domain of a function excludes values that cause a function to be undefined or have outputs that are not real numbers.

EXPRESSION THAT DEFINES THE FUNCTION	EXCLUDED x-VALUES	EXAMPLE	IMPLICIT DOMAIN
Polynomial	None	$f(x) = x^3 - 4x^2$	All real numbers
Rational	x-values that make the denominator equal to 0	$g(x) = \dfrac{2}{x^2 - 9}$	$x \neq \pm 3$ or $(-\infty, -3) \cup (-3, 3) \cup (3, \infty)$
Radical	x-values that result in a square (even) root of a negative number	$h(x) = \sqrt{x - 5}$	$x \geq 5$ or $[5, \infty)$

Technology Tip

To visualize the domain of each function, ask the question: What are the excluded x-values in the graph?

a. Graph of $F(x) = \dfrac{3}{x^2 - 25}$ is shown.

The excluded x-values are -5 and 5.

▶ **EXAMPLE 10   Determining the Domain of a Function**

State the domain of the given functions.

a. $F(x) = \dfrac{3}{x^2 - 25}$ **b.** $H(x) = \sqrt[4]{9 - 2x}$ **c.** $G(x) = \sqrt[3]{x - 1}$

Solution (a):

Write the original equation.	$F(x) = \dfrac{3}{x^2 - 25}$
Determine any restrictions on the values of x.	$x^2 - 25 \neq 0$
Solve the restriction equation.	$x^2 \neq 25$ or $x \neq \pm\sqrt{25} = \pm 5$
State the domain restrictions.	$x \neq \pm 5$
Write the domain in interval notation.	$\boxed{(-\infty, -5) \cup (-5, 5) \cup (5, \infty)}$

Solution (b):

Write the original equation.	$H(x) = \sqrt[4]{9 - 2x}$
Determine any restrictions on the values of x.	$9 - 2x \geq 0$
Solve the restriction inequality.	$9 \geq 2x$
State the domain restrictions.	$x \leq \dfrac{9}{2}$
Write the domain in interval notation.	$\boxed{\left(-\infty, \dfrac{9}{2}\right]}$

Solution (c):

Write the original equation.	$G(x) = \sqrt[3]{x - 1}$
Determine any restrictions on the values of x.	no restrictions
State the domain.	$\mathbb{R}$
Write the domain in interval notation.	$\boxed{(-\infty, \infty)}$

▪ **YOUR TURN** State the domain of the given functions.

    a. $f(x) = \sqrt{x - 3}$ **b.** $g(x) = \dfrac{1}{x^2 - 4}$

Classroom Example 3.1.10
Find the domain of these functions.

a. $f(x) = \dfrac{x - 1}{2x^2 - 32}$

b. $f(x) = \sqrt[6]{-9 - 3x}$

c.* $f(x) = \sqrt[3]{\dfrac{x - 1}{x}}$

d.* $f(x) = \dfrac{\sqrt[3]{-x}}{x^2 + 1}$

Answer:
a. $(-\infty, -4) \cup (-4, 4) \cup (4, \infty)$
b. $(-\infty, -3]$
c. $(-\infty, 0) \cup (0, \infty)$
d. $(-\infty, \infty)$

▪ **Answer: a.** $x \geq 3$ or $[3, \infty)$
  b. $x \neq \pm 2$ or
  $(-\infty, -2) \cup (-2, 2) \cup (2, \infty)$

Applications

Functions that are used in applications often have restrictions on the domains due to physical constraints. For example, the volume of a cube is given by the function $V(x) = x^3$, where x is the length of a side. The function $f(x) = x^3$ has no restrictions on x, and therefore the domain is the set of all real numbers. However, the volume of any cube has the restriction that the length of a side can never be negative or zero.

EXAMPLE 11　Price of Gasoline

Following the capture of Saddam Hussein in Iraq in 2003, gas prices in the United States escalated and then finally returned to their precapture prices. Over a 6-month period, the average price of a gallon of 87 octane gasoline was given by the function $C(x) = -0.05x^2 + 0.3x + 1.7$, where C is the cost function and x represents the number of months after the capture.

a. Determine the domain of the cost function.
b. What was the average price of gas per gallon 3 months after the capture?

Solution (a):

Since the cost function $C(x) = -0.05x^2 + 0.3x + 1.7$ modeled the price of gas only for 6 months after the capture, the domain is $0 \leq x \leq 6$ or $\boxed{[0, 6]}$.

Solution (b):

Write the cost function.	$C(x) = -0.05x^2 + 0.3x + 1.7 \quad 0 \leq x \leq 6$
Find the value of the function when $x = 3$.	$C(3) = -0.05(3)^2 + 0.3(3) + 1.7$
Simplify.	$C(3) = 2.15$

The average price per gallon 3 months after the capture was $\boxed{\$2.15}$.

EXAMPLE 12 The Dimensions of a Pool

Express the volume of a 30 ft × 10 ft rectangular swimming pool as a function of its depth.

Solution:

The volume of any rectangular box is $V = lwh$, where V is the volume, l is the length, w is the width, and h is the height. In this example, the length is 30 ft, the width is 10 ft, and the height represents the depth d of the pool.

Write the volume as a function of depth d. $\qquad$ $V(d) = (30)(10)d$

Simplify. $\qquad$ $V(d) = 300d$

Determine any restrictions on the domain. $\qquad$ $d > 0$

SECTION 3.1 SUMMARY

Relations and Functions (Let x represent the independent variable and y the dependent variable.)

TYPE	MAPPING/CORRESPONDENCE	EQUATION	GRAPH
Relation	Every x-value in the domain maps to **at least one** y-value in the range.	$x = y^2$	
Function	Every x-value in the domain maps to **exactly one** y-value in the range.	$y = x^2$	Passes vertical line test

All functions are relations, but not all relations are functions. Functions can be represented by equations. In the following table, each column illustrates an alternative notation.

INPUT	CORRESPONDENCE	OUTPUT	EQUATION
x	Function	y	$y = 2x + 5$
Independent variable	Mapping	Dependent variable	Mathematical rule
Argument	f	$f(x)$	$f(x) = 2x + 5$

The **domain** is the set of all inputs (x-values), and the **range** is the set of all corresponding outputs (y-values). Placeholder notation is useful when evaluating functions.

$$f(x) = 3x^2 + 2x$$

$$f(\square) = 3(\square)^2 + 2(\square)$$

Explicit domain is stated, whereas **implicit domain** is found by *excluding* x-values that:

- make the function undefined (denominator = 0).
- result in a nonreal output (even roots of negative real numbers).

SECTION

3.1 EXERCISES

▪ SKILLS

In Exercises 1–24, determine whether each relation is a function. Assume that the coordinate pair (x, y) represents the independent variable x and the dependent variable y.

1.

2.

3.

4.

5.

6.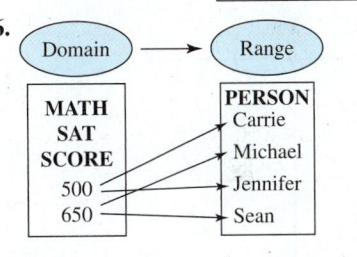

7. $\{(0, -3), (0, 3), (-3, 0), (3, 0)\}$

8. $\{(2, -2), (2, 2), (5, -5), (5, 5)\}$

9. $\{(0, 0), (9, -3), (4, -2), (4, 2), (9, 3)\}$

10. $\{(0, 0), (-1, -1), (-2, -8), (1, 1), (2, 8)\}$

11. $\{(0, 1), (1, 0), (2, 1), (-2, 1), (5, 4), (-3, 4)\}$

12. $\{(0, 1), (1, 1), (2, 1), (3, 1)\}$

13. $x^2 + y^2 = 9$ **14.** $x = |y|$ **15.** $x = y^2$ **16.** $y = x^3$ **17.** $y = |x - 1|$ **18.** $y = 3$

19.

20.

21.

22.

23.

24.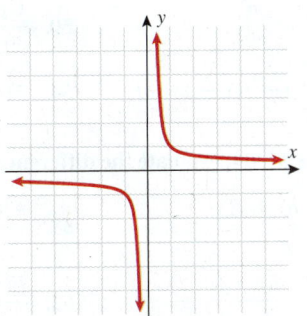

In Exercises 25–32, use the given graphs to evaluate the functions.

25. $y = f(x)$

a. $f(2)$ **b.** $f(0)$ **c.** $f(-2)$

26. $y = g(x)$

a. $g(-3)$ **b.** $g(0)$ **c.** $g(5)$

27. $y = p(x)$

a. $p(-1)$ **b.** $p(0)$ **c.** $p(1)$

28. $y = r(x)$

a. $r(-4)$ **b.** $r(-1)$ **c.** $r(3)$

29. $y = C(x)$

a. $C(2)$ **b.** $C(0)$ **c.** $C(-2)$

30. $y = q(x)$

a. $q(-4)$ **b.** $q(0)$ **c.** $q(2)$

31. $y = S(x)$

a. $S(-3)$ **b.** $S(0)$ **c.** $S(2)$

32. $y = T(x)$

a. $T(-5)$ **b.** $T(-2)$ **c.** $T(4)$

33. Find x if $f(x) = 3$ in Exercise 25.

34. Find x if $g(x) = -2$ in Exercise 26.

35. Find x if $p(x) = 5$ in Exercise 27.

36. Find x if $C(x) = -7$ in Exercise 29.

37. Find x if $C(x) = -5$ in Exercise 29.

38. Find x if $q(x) = -2$ in Exercise 30.

39. Find x if $S(x) = 1$ in Exercise 31.

40. Find x if $T(x) = 4$ in Exercise 32.

In Exercises 41–56, evaluate the given quantities applying the following four functions.

$$f(x) = 2x - 3 \qquad F(t) = 4 - t^2 \qquad g(t) = 5 + t \qquad G(x) = x^2 + 2x - 7$$

41. $f(-2)$

42. $G(-3)$

43. $g(1)$

44. $F(-1)$

45. $f(-2) + g(1)$

46. $G(-3) - F(-1)$

47. $3f(-2) - 2g(1)$

48. $2F(-1) - 2G(-3)$

49. $\dfrac{f(-2)}{g(1)}$

50. $\dfrac{G(-3)}{F(-1)}$

51. $\dfrac{f(0) - f(-2)}{g(1)}$

52. $\dfrac{G(0) - G(-3)}{F(-1)}$

53. $f(x + 1) - f(x - 1)$

54. $F(t + 1) - F(t - 1)$

55. $g(x + a) - f(x + a)$

56. $G(x + b) + F(b)$

In Exercises 57–64, evaluate the difference quotients using the same f, F, G, and g given for Exercises 41–56.

57. $\dfrac{f(x + h) - f(x)}{h}$

58. $\dfrac{F(t + h) - F(t)}{h}$

59. $\dfrac{g(t + h) - g(t)}{h}$

60. $\dfrac{G(x + h) - G(x)}{h}$

61. $\dfrac{f(-2 + h) - f(-2)}{h}$

62. $\dfrac{F(-1 + h) - F(-1)}{h}$

63. $\dfrac{g(1 + h) - g(1)}{h}$

64. $\dfrac{G(-3 + h) - G(-3)}{h}$

In Exercises 65–96, find the domain of the given function. Express the domain in interval notation.

65. $f(x) = 2x - 5$ **66.** $f(x) = -2x - 5$ **67.** $g(t) = t^2 + 3t$ **68.** $h(x) = 3x^4 - 1$

69. $P(x) = \dfrac{x + 5}{x - 5}$ **70.** $Q(t) = \dfrac{2 - t^2}{t + 3}$ **71.** $T(x) = \dfrac{2}{x^2 - 4}$ **72.** $R(x) = \dfrac{1}{x^2 - 1}$

73. $F(x) = \dfrac{1}{x^2 + 1}$ **74.** $G(t) = \dfrac{2}{t^2 + 4}$ **75.** $q(x) = \sqrt{7 - x}$ **76.** $k(t) = \sqrt{t - 7}$

77. $f(x) = \sqrt{2x + 5}$ **78.** $g(x) = \sqrt{5 - 2x}$ **79.** $G(t) = \sqrt{t^2 - 4}$ **80.** $F(x) = \sqrt{x^2 - 25}$

81. $F(x) = \dfrac{1}{\sqrt{x - 3}}$ **82.** $G(x) = \dfrac{2}{\sqrt{5 - x}}$ **83.** $f(x) = \sqrt[3]{1 - 2x}$ **84.** $g(x) = \sqrt[5]{7 - 5x}$

85. $P(x) = \dfrac{1}{\sqrt[5]{x + 4}}$ **86.** $Q(x) = \dfrac{x}{\sqrt[3]{x^2 - 9}}$ **87.** $R(x) = \dfrac{x + 1}{\sqrt[4]{3 - 2x}}$ **88.** $p(x) = \dfrac{x^2}{\sqrt{25 - x^2}}$

89. $H(t) = \dfrac{t}{\sqrt{t^2 - t - 6}}$ **90.** $f(t) = \dfrac{t - 3}{\sqrt[4]{t^2 + 9}}$ **91.** $f(x) = (x^2 - 16)^{1/2}$ **92.** $g(x) = (2x - 5)^{1/3}$

93. $r(x) = x^2(3 - 2x)^{-1/2}$ **94.** $p(x) = (x - 1)^2(x^2 - 9)^{-3/5}$ **95.** $f(x) = \frac{2}{5}x - \frac{2}{4}$ **96.** $g(x) = \frac{2}{3}x^2 - \frac{1}{6}x - \frac{3}{4}$

97. Let $g(x) = x^2 - 2x - 5$ and find the values of x that correspond to $g(x) = 3$.

98. Let $g(x) = \frac{5}{6}x - \frac{3}{4}$ and find the value of x that corresponds to $g(x) = \frac{2}{3}$.

99. Let $f(x) = 2x(x - 5)^3 - 12(x - 5)^2$ and find the values of x that correspond to $f(x) = 0$.

100. Let $f(x) = 3x(x + 3)^2 - 6(x + 3)^3$ and find the values of x that correspond to $f(x) = 0$.

▪ APPLICATIONS

101. Budget: Event Planning. The cost associated with a catered wedding reception is $45 per person for a reception for more than 75 people. Write the cost of the reception in terms of the number of guests and state any domain restrictions.

102. Budget: Long-Distance Calling. The cost of a local home phone plan is $35 for basic service and $.10 per minute for any domestic long-distance calls. Write the cost of monthly phone service in terms of the number of monthly long-distance minutes and state any domain restrictions.

103. Temperature. The average temperature in Tampa, Florida, in the springtime is given by the function $T(x) = -0.7x^2 + 16.8x - 10.8$, where T is the temperature in degrees Fahrenheit and x is the time of day in military time and is restricted to $6 \le x \le 18$ (sunrise to sunset). What is the temperature at 6 A.M.? What is the temperature at noon?

104. Falling Objects: Firecrackers. A firecracker is launched straight up, and its height is a function of time, $h(t) = -16t^2 + 128t$, where h is the height in feet and t is the time in seconds with $t = 0$ corresponding to the instant it launches. What is the height 4 seconds after launch? What is the domain of this function?

105. Collectibles. The price of a signed Alex Rodriguez baseball card is a function of how many are for sale. When Rodriguez was traded from the Texas Rangers to the New York Yankees in 2004, the going rate for a signed baseball card on eBay was $P(x) = 10 + \sqrt{400{,}000 - 100x}$, where x represents the number of signed cards for sale. What was the value of the card when there were 10 signed cards for sale? What was the value of the card when there were 100 signed cards for sale?

106. Collectibles. In Exercise 105, what was the lowest price on eBay, and how many cards were available then? What was the highest price on eBay, and how many cards were available then?

107. Volume. An open box is constructed from a square 10-inch piece of cardboard by cutting squares of length x inches out of each corner and folding the sides up. Express the volume of the box as a function of x, and state the domain.

108. Volume. A cylindrical water basin will be built to harvest rainwater. The basin is limited in that the largest radius it can have is 10 feet. Write a function representing the volume of water V as a function of height h. How many additional gallons of water will be collected if you increase the height by 2 feet? *Hint:* 1 cubic foot = 7.48 gallons.

For Exercises 109–110, refer to the following:

The weekly exchange rate of the U.S. dollar to the Japanese yen is shown in the graph as varying over an 8-week period. Assume the exchange rate $E(t)$ is a function of time (week); let $E(1)$ be the exchange rate during Week 1.

109. Economics. Approximate the exchange rates of the U.S. dollar to the nearest yen during Weeks 4, 7, and 8.

110. Economics. Find the increase or decrease in the number of Japanese yen to the U.S. dollar exchange rate, to the nearest yen, from (a) Week 2 to Week 3 and (b) Week 6 to Week 7.

For Exercises 111–112, refer to the following:

An epidemiological study of the spread of malaria in a rural area finds that the total number P of people who contracted malaria t days into an outbreak is modeled by the function

$$P(t) = -\frac{1}{4}t^2 + 7t + 180 \qquad 1 \leq t \leq 14$$

111. Medicine/Health. How many people have contracted malaria 14 days into the outbreak?

112. Medicine/Health. How many people have contracted malaria 6 days into the outbreak?

113. Environment: Tossing the Envelopes. The average American adult receives 24 pieces of mail per week, usually of some combination of ads and envelopes with windows. Suppose each of these adults throws away a dozen envelopes per week.

   a. The width of the window of an envelope is 3.375 inches less than its length x. Create the function $A(x)$ that represents the area of the window in square inches. Simplify, if possible.

   b. Evaluate $A(4.5)$ and explain what this value represents.

   c. Assume the dimensions of the envelope are 8 inches by 4 inches. Evaluate $A(8.5)$. Is this possible for this particular envelope? Explain.

114. Environment: Tossing the Envelopes. Each month, Jack receives his bank statement in a 9.5 inch by 6 inch envelope. Each month, he throws away the envelope after removing the statement.

   a. The width of the window of the envelope is 2.875 inches less than its length x. Create the function $A(x)$ that represents the area of the window in square inches. Simplify, if possible.

   b. Evaluate $A(5.25)$ and explain what this value represents.

   c. Evaluate $A(10)$. Is this possible for this particular envelope? Explain.

Refer to the table below for Exercises 115 and 116. It illustrates the average federal funds rate for the month of January (2000 to 2008).

YEAR	FED. RATE
2000	5.45
2001	5.98
2002	1.73
2003	1.24
2004	1.00
2005	2.25
2006	4.50
2007	5.25
2008	3.50

115. Finance. Is the relation whose domain is the year and whose range is the average federal funds rate for the month of January a function? Explain.

116. Finance. Write five ordered pairs whose domain is the set of even years from 2000 to 2008 and whose range is the set of corresponding average federal funds rate for the month of January.

For Exercises 117 and 118, use the following figure:

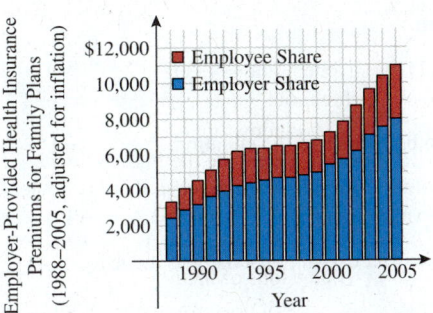

Source: Kaiser Family Foundation Health Research and Education Trust. *Note:* The following years were interpolated: 1989–1992; 1994–1995; 1997–1998.

117. Health Care Costs. Fill in the following table. Round dollars to the nearest $1000.

YEAR	TOTAL HEALTH CARE COST FOR FAMILY PLANS
1989	
1993	
1997	
2001	
2005	

Write the five ordered pairs resulting from the table.

118. Health Care Costs. Using the table found in Exercise 117, let the years correspond to the domain and the total costs correspond to the range. Is this relation a function? Explain.

For Exercises 119 and 120, use the following information:

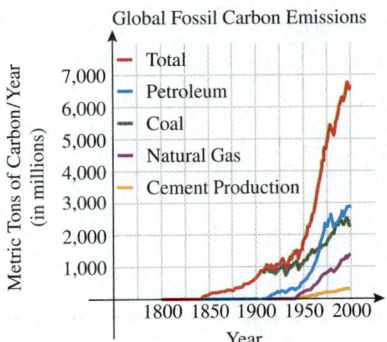

Source: http://www.naftc.wvu.edu

Let the functions f, F, g, G, and H represent the number of tons of carbon emitted per year as a function of year corresponding to cement production, natural gas, coal, petroleum, and the total amount, respectively. Let t represent the year, with $t = 0$ corresponding to 1900.

119. Environment: Global Climate Change. Estimate (to the nearest thousand) the value of
a. $F(50)$ **b.** $g(50)$ **c.** $H(50)$

120. Environment: Global Climate Change. Explain what the sum $F(100) + g(100) + G(100)$ represents.

▪ CATCH THE MISTAKE

In Exercises 121–126, explain the mistake that is made.

121. Determine whether the relationship is a function.

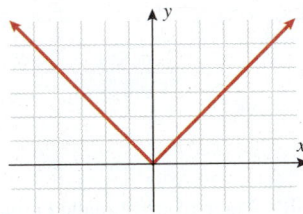

Solution:

Apply the horizontal line test.

Because the horizontal line intersects the graph in two places, this is not a function.

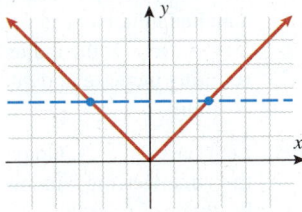

This is incorrect. What mistake was made?

122. Given the function $H(x) = 3x - 2$, evaluate the quantity $H(3) - H(-1)$.

Solution: $H(3) - H(-1) = H(3) + H(1) = 7 + 1 = 8$

This is incorrect. What mistake was made?

123. Given the function $f(x) = x^2 - x$, evaluate the quantity $f(x + 1)$.

Solution: $f(x + 1) = f(x) + f(1) = x^2 - x + 0$
$f(x + 1) = x^2 - x$

This is incorrect. What mistake was made?

124. Determine the domain of the function $g(t) = \sqrt{3 - t}$ and express it in interval notation.

Solution:

What can t be? Any nonnegative real number.
$3 - t > 0$
$3 > t$ or $t < 3$
Domain: $(-\infty, 3)$

This is incorrect. What mistake was made?

125. Given the function $G(x) = x^2$, evaluate

$$\frac{G(-1 + h) - G(-1)}{h}.$$

Solution:

$$\frac{G(-1 + h) - G(-1)}{h} = \frac{G(-1) + G(h) - G(-1)}{h}$$

$$= \frac{G(h)}{h} = \frac{h^2}{h} = h$$

This is incorrect. What mistake was made?

126. Given the functions $f(x) = |x - A| - 1$ and $f(1) = -1$, find A.

Solution:

Since $f(1) = -1$, the point $(-1, 1)$ must satisfy the function. $\qquad -1 = |-1 - A| - 1$

Add 1 to both sides of the equation. $\qquad |-1 - A| = 0$

The absolute value of zero is zero, so there is no need for the absolute value signs: $-1 - A = 0 \Rightarrow A = -1$.

This is incorrect. What mistake was made?

■ CONCEPTUAL

In Exercises 127–130, determine whether each statement is true or false.

127. If a vertical line does not intersect the graph of an equation, then that equation does not represent a function.

128. If a horizontal line intersects a graph of an equation more than once, the equation does not represent a function.

129. If $f(-a) = f(a)$, then f does not represent a function.

130. If $f(-a) = f(a)$, then f may or may not represent a function.

131. If $f(x) = Ax^2 - 3x$ and $f(1) = -1$, find A.

132. If $g(x) = \dfrac{1}{b - x}$ and $g(3)$ is undefined, find b.

■ CHALLENGE

133. If $F(x) = \dfrac{C - x}{D - x}$, $F(-2)$ is undefined, and $F(-1) = 4$, find C and D.

134. Construct a function that is undefined at $x = 5$ and whose graph passes through the point $(1, -1)$.

In Exercises 135 and 136, find the domain of each function, where a is any positive real number.

135. $f(x) = \dfrac{-100}{x^2 - a^2}$

136. $f(x) = -5\sqrt{x^2 - a^2}$

■ TECHNOLOGY

137. Using a graphing utility, graph the temperature function in Exercise 103. What time of day is it the warmest? What is the temperature? Looking at this function, explain why this model for Tampa, Florida, is valid only from sunrise to sunset (6 to 18).

138. Using a graphing utility, graph the height of the firecracker in Exercise 104. How long after liftoff is the firecracker airborne? What is the maximum height that the firecracker attains? Explain why this height model is valid only for the first 8 seconds.

139. Using a graphing utility, graph the price function in Exercise 105. What are the lowest and highest prices of the cards? Does this agree with what you found in Exercise 106?

140. The makers of malted milk balls are considering increasing the size of the spherical treats. The thin chocolate coating on a malted milk ball can be approximated by the surface area, $S(r) = 4\pi r^2$. If the radius is increased 3 mm, what is the resulting increase in required chocolate for the thin outer coating?

141. Let $f(x) = x^2 + 1$. Graph $y_1 = f(x)$ and $y_2 = f(x - 2)$ in the same viewing window. Describe how the graph of y_2 can be obtained from the graph of y_1.

142. Let $f(x) = 4 - x^2$. Graph $y_1 = f(x)$ and $y_2 = f(x + 2)$ in the same viewing window. Describe how the graph of y_2 can be obtained from the graph of y_1.

SKILLS OBJECTIVES

- Classify functions as even, odd, or neither.
- Determine whether functions are increasing, decreasing, or constant.
- Calculate the average rate of change of a function.
- Evaluate the difference quotient for a function.
- Graph piecewise-defined functions.

CONCEPTUAL OBJECTIVES

- Identify common functions.
- Develop and graph piecewise-defined functions.
  - Identify and graph points of discontinuity.
  - State the domain and range.
- Understand that even functions have graphs that are symmetric about the y-axis.
- Understand that odd functions have graphs that are symmetric about the origin.

Recognizing and Classifying Functions

Common Functions

Point-plotting techniques were introduced in Section 2.2, and we noted there that we would explore some more efficient ways of graphing functions in Chapter 3. The nine main functions you will read about in this section will constitute a "library" of functions that you should commit to memory. We will draw on this library of functions in the next section when graphing transformations are discussed. Several of these functions have been shown previously in this chapter, but now we will classify them specifically by name and identify properties that each function exhibits.

In Section 2.3, we discussed equations and graphs of lines. All lines (with the exception of vertical lines) pass the vertical line test, and hence are classified as functions. Instead of the traditional notation of a line, $y = mx + b$, we use function notation and classify a function whose graph is a *line* as a *linear* function.

LINEAR FUNCTION

$$f(x) = mx + b \qquad m \text{ and } b \text{ are real numbers.}$$

The domain of a linear function $f(x) = mx + b$ is the set of all real numbers $\mathbb{R}$. The graph of this function has slope m and y-intercept b.

LINEAR FUNCTION: $f(x) = mx + b$	SLOPE: m	y-INTERCEPT: b
$f(x) = 2x - 7$	$m = 2$	$b = -7$
$f(x) = -x + 3$	$m = -1$	$b = 3$
$f(x) = x$	$m = 1$	$b = 0$
$f(x) = 5$	$m = 0$	$b = 5$

One special case of the linear function is the *constant function* ($m = 0$).

CONSTANT FUNCTION

$$f(x) = b \qquad \qquad b \text{ is any real number.}$$

The graph of a constant function $f(x) = b$ is a horizontal line. The y-intercept corresponds to the point $(0, b)$. The domain of a constant function is the set of all real numbers $\mathbb{R}$. The range, however, is a single value b. In other words, all x-values correspond to a single y-value.

Points that lie on the graph of a constant function $f(x) = b$ are

$(-5, b)$

$(-1, b)$

$(0, b)$

$(2, b)$

$(4, b)$

. . .

(x, b)

Domain: $(-\infty, \infty)$ Range: $[b, b]$ or $\{b\}$

Another specific example of a linear function is the function having a slope of one ($m = 1$) and a y-intercept of zero ($b = 0$). This special case is called the *identity function*.

IDENTITY FUNCTION

$$f(x) = x$$

The graph of the identity function has the following properties: It passes through the origin, and every point that lies on the line has equal x- and y-coordinates. Both the domain and the range of the identity function are the set of all real numbers $\mathbb{R}$.

A function that squares the input is called the *square function*.

Identity Function
Domain: $(-\infty, \infty)$ Range: $(-\infty, \infty)$

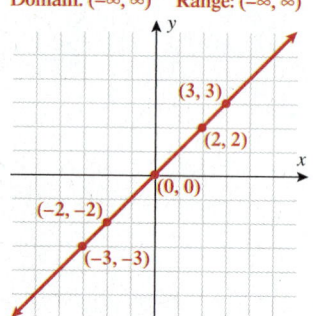

SQUARE FUNCTION

$$f(x) = x^2$$

The graph of the square function is called a parabola and will be discussed in further detail in Chapters 4 and 8. The domain of the square function is the set of all real numbers $\mathbb{R}$. Because squaring a real number always yields a positive number or zero, the range of the square function is the set of all nonnegative numbers. Note that the intercept is the origin and the square function is symmetric about the y-axis. This graph is contained in quadrants I and II.

Square Function
Domain: $(-\infty, \infty)$ Range: $[0, \infty)$

A function that cubes the input is called the *cube function*.

CUBE FUNCTION

$$f(x) = x^3$$

Cube Function
Domain: $(-\infty, \infty)$ Range: $(-\infty, \infty)$

The domain of the cube function is the set of all real numbers $\mathbb{R}$. Because cubing a negative number yields a negative number, cubing a positive number yields a positive number, and cubing 0 yields 0, the range of the cube function is also the set of all real numbers $\mathbb{R}$. Note that the only intercept is the origin and the cube function is symmetric about the origin. This graph extends only into quadrants I and III.

The next two functions are counterparts of the previous two functions: square root and cube root. When a function takes the square root of the input or the cube root of the input, the function is called the *square root function* or the *cube root function*, respectively.

SQUARE ROOT FUNCTION

$$f(x) = \sqrt{x} \quad \text{or} \quad f(x) = x^{1/2}$$

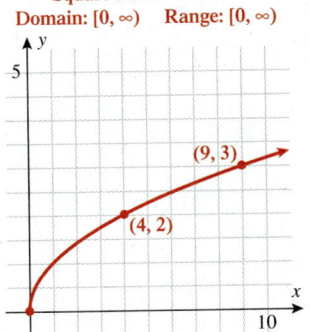

Square Root Function
Domain: $[0, \infty)$ Range: $[0, \infty)$

In Section 3.1, we found the domain to be $[0, \infty)$. The output of the function will be all real numbers greater than or equal to zero. Therefore, the range of the square root function is $[0, \infty)$. The graph of this function will be contained in quadrant I.

CUBE ROOT FUNCTION

$$f(x) = \sqrt[3]{x} \quad \text{or} \quad f(x) = x^{1/3}$$

Cube Root Function
Domain: $(-\infty, \infty)$ Range: $(-\infty, \infty)$

In Section 3.1, we stated the domain of the cube root function to be $(-\infty, \infty)$. We see by the graph that the range is also $(-\infty, \infty)$. This graph is contained in quadrants I and III and passes through the origin. This function is symmetric about the origin.

In Section 1.7, you read about absolute value equations and inequalities. Now we shift our focus to the graph of the *absolute value function*.

ABSOLUTE VALUE FUNCTION

$$f(x) = |x|$$

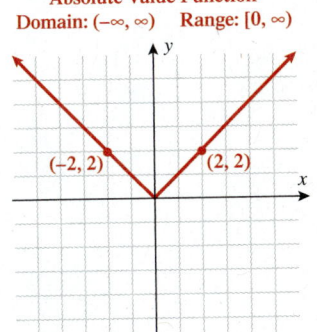

Absolute Value Function
Domain: $(-\infty, \infty)$ Range: $[0, \infty)$

Some points that are on the graph of the absolute value function are $(-1, 1)$, $(0, 0)$, and $(1, 1)$. The domain of the absolute value function is the set of all real numbers $\mathbb{R}$, yet the range is the set of nonnegative real numbers. The graph of this function is symmetric with respect to the y-axis and is contained in quadrants I and II.

A function whose output is the reciprocal of its input is called the *reciprocal function*.

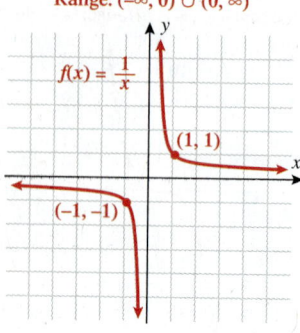

Reciprocal Function
Domain: $(-\infty, 0) \cup (0, \infty)$
Range: $(-\infty, 0) \cup (0, \infty)$

$f(x) = \frac{1}{x}$

$(1, 1)$

$(-1, -1)$

RECIPROCAL FUNCTION

$$f(x) = \frac{1}{x} \qquad x \neq 0$$

The only restriction on the domain of the reciprocal function is that $x \neq 0$. Therefore, we say the domain is the set of all real numbers excluding zero. The graph of the reciprocal function illustrates that its range is also the set of all real numbers except zero. Note that the reciprocal function is symmetric with respect to the origin and is contained in quadrants I and III.

Even and Odd Functions

Of the nine functions discussed above, several have similar properties of symmetry. The constant function, square function, and absolute value function are all symmetric with respect to the *y*-axis. The identity function, cube function, cube root function, and reciprocal function are all symmetric with respect to the origin. The term **even** is used to describe functions that are symmetric with respect to the *y*-axis, or vertical axis, and the term **odd** is used to describe functions that are symmetric with respect to the origin. Recall from Section 2.2 that symmetry can be determined both graphically and algebraically. The box below summarizes the graphic and algebraic characteristics of even and odd functions.

EVEN AND ODD FUNCTIONS

Function	Symmetric with Respect to	On Replacing x with $-x$
Even	y-axis or vertical axis	$f(-x) = f(x)$
Odd	origin	$f(-x) = -f(x)$

The algebraic method for determining symmetry with respect to the *y*-axis, or vertical axis, is to substitute $-x$ for x. If the result is an equivalent equation, the function is symmetric with respect to the *y*-axis. Some examples of even functions are $f(x) = b$, $f(x) = x^2$, $f(x) = x^4$; and $f(x) = |x|$. In any of these equations, if $-x$ is substituted for x, the result is the same; that is, $f(-x) = f(x)$. Also note that, with the exception of the absolute value function, these examples are all even-degree polynomial equations. All constant functions are degree zero and are even functions.

The algebraic method for determining symmetry with respect to the origin is to substitute $-x$ for x. If the result is the negative of the original function, that is, if $f(-x) = -f(x)$, then the function is symmetric with respect to the origin and, hence, classified as an odd function. Examples of odd functions are $f(x) = x$, $f(x) = x^3$, $f(x) = x^5$, and $f(x) = x^{1/3}$. In any of these functions, if $-x$ is substituted for x, the result is the negative of the original function. Note that with the exception of the cube root function, these equations are odd-degree polynomials.

Be careful, though, because functions that are combinations of even- and odd-degree polynomials can turn out to be neither even nor odd, as we will see in Example 1.

EXAMPLE 1 **Determining Whether a Function Is Even, Odd, or Neither**

Determine whether the functions are even, odd, or neither.

a. $f(x) = x^2 - 3$ **b.** $g(x) = x^5 + x^3$ **c.** $h(x) = x^2 - x$

Solution (a):

Original function.	$f(x) = x^2 - 3$
Replace x with $-x$.	$f(-x) = (-x)^2 - 3$
Simplify.	$f(-x) = x^2 - 3 = f(x)$

Because $f(-x) = f(x)$, we say that $\boxed{f(x) \text{ is an } even \text{ function}}$.

Solution (b):

Original function.	$g(x) = x^5 + x^3$
Replace x with $-x$.	$g(-x) = (-x)^5 + (-x)^3$
Simplify.	$g(-x) = -x^5 - x^3 = -(x^5 + x^3) = -g(x)$

Because $g(-x) = -g(x)$, we say that $\boxed{g(x) \text{ is an } odd \text{ function}}$.

Solution (c):

Original function.	$h(x) = x^2 - x$
Replace x with $-x$.	$h(-x) = (-x)^2 - (-x)$
Simplify.	$h(-x) = x^2 + x$

$h(-x)$ is neither $-h(x)$ nor $h(x)$; therefore the function $h(x)$ is $\boxed{\text{neither even nor odd}}$.

In parts (a), (b), and (c), we classified these functions as either even, odd, or neither, using the algebraic test. Look back at them now and reflect on whether these classifications agree with your intuition. In part (a), we combined two functions: the square function and the constant function. Both of these functions are even, and adding even functions yields another even function. In part (b), we combined two odd functions: the fifth-power function and the cube function. Both of these functions are odd, and adding two odd functions yields another odd function. In part (c), we combined two functions: the square function and the identity function. The square function is even, and the identity function is odd. In this part, combining an even function with an odd function yields a function that is neither even nor odd and, hence, has no symmetry with respect to the vertical axis or the origin.

▪ **YOUR TURN** Classify the functions as even, odd, or neither.

a. $f(x) = |x| + 4$ **b.** $f(x) = x^3 - 1$

Classroom Example 3.2.1
Determine whether these functions are even, odd, or neither.

a. $f(x) = 3x^3 + x^5$
b. $f(x) = 3x^3 - x^5$
c.* $f(x) = -(3x^3 + x^5)$
d. $f(x) = 3x^3 + x^5 + 1$
e.* $f(x) = (3x^3 + x^5)^2$

Answer:
a. odd **b.** odd **c.** odd
d. neither **e.** even

Technology Tip

a. Graph $y_1 = f(x) = x^2 - 3$.

Even; symmetric with respect to the y-axis.

b. Graph $y_1 = g(x) = x^5 + x^3$.

Odd; symmetric with respect to origin.

c. Graph $y_1 = h(x) = x^2 - x$.

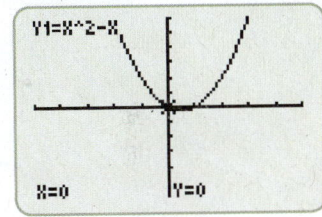

No symmetry with respect to y-axis or origin.

■ **Answer: a.** even **b.** neither

Increasing and Decreasing Functions

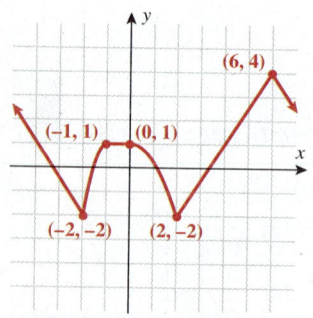

Look at the figure in the margin to the left. Graphs are read from **left to right**. If we start at the left side of the graph and trace the red curve with our pen, we see that the function values (values in the vertical direction) are decreasing until arriving at the point $(-2, -2)$. Then, the function values increase until arriving at the point $(-1, 1)$. The values then remain constant ($y = 1$) between the points $(-1, 1)$ and $(0, 1)$. Proceeding beyond the point $(0, 1)$, the function values decrease again until the point $(2, -2)$. Beyond the point $(2, -2)$, the function values increase again until the point $(6, 4)$. Finally, the function values decrease and continue to do so.

When specifying a function as increasing, decreasing, or constant, the **intervals are classified according to the x-coordinate**. For instance, in this graph, we say the function is increasing when x is between $x = -2$ and $x = -1$ and again when x is between $x = 2$ and $x = 6$. The graph is classified as decreasing when x is less than -2 and again when x is between 0 and 2 and again when x is greater than 6. The graph is classified as constant when x is between -1 and 0. In interval notation, this is summarized as

Decreasing	Increasing	Constant
$(-\infty, -2) \cup (0, 2) \cup (6, \infty)$	$(-2, -1) \cup (2, 6)$	$(-1, 0)$

An algebraic test for determining whether a function is increasing, decreasing, or constant is to compare the value $f(x)$ of the function for particular points in the intervals.

INCREASING, DECREASING, AND CONSTANT FUNCTIONS

1. A function f is **increasing** on an open interval I if for any x_1 and x_2 in I, where $x_1 < x_2$, then $f(x_1) < f(x_2)$.
2. A function f is **decreasing** on an open interval I if for any x_1 and x_2 in I, where $x_1 < x_2$, then $f(x_1) > f(x_2)$.
3. A function f is **constant** on an open interval I if for any x_1 and x_2 in I, then $f(x_1) = f(x_2)$.

In addition to classifying a function as increasing, decreasing, or constant, we can also determine the domain and range of a function by inspecting its graph from left to right:

- The domain is the set of all x-values (from left to right) where the function is defined.
- The range is the set of all y-values (from bottom to top) that the graph of the function corresponds to.
- A solid dot on the left or right end of a graph indicates that the graph terminates there and the point is included in the graph.
- An open dot indicates that the graph terminates there and the point is not included in the graph.
- Unless a dot is present, it is assumed that a graph continues indefinitely in the same direction. (An arrow is used in some books to indicate direction.)

EXAMPLE 2 **Finding Intervals When a Function Is Increasing or Decreasing**

Given the graph of a function:

a. State the domain and range of the function.

b. Find the intervals when the function is increasing, decreasing, or constant.

Solution (a):

Domain: $[-5, \infty)$ Range: $[0, \infty)$

Solution (b):

Reading the graph from **left to right**, we see that the graph

- decreases from the point $(-5, 7)$ to the point $(-2, 4)$.

- is constant from the point $(-2, 4)$ to the point $(0, 4)$.

- decreases from the point $(0, 4)$ to the point $(2, 0)$.

- increases from the point $(2, 0)$ on.

The intervals of increasing and decreasing correspond to the **x-coordinates**.

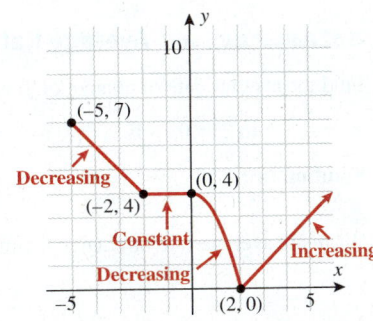

We say that this function is

- increasing on the interval $(2, \infty)$.

- decreasing on the interval $(-5, -2) \cup (0, 2)$.

- constant on the interval $(-2, 0)$.

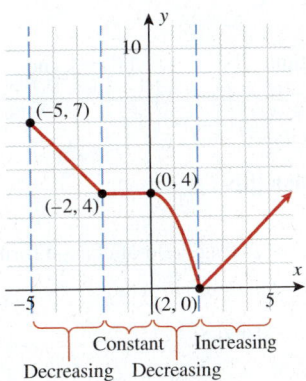

Note: The intervals of increasing or decreasing are defined on *open* intervals. This should not be confused with the domain. For example, the point $x = -5$ is included in the domain of the function but not in the interval where the function is classified as decreasing.

Classroom Example 3.2.2
Consider the following graph of a function.

a. Find the domain and range.
b. Find the interval(s) where f is ↑, ↓, or constant.

Answer:
a. Domain: $(-\infty, 4]$
   Range: $[-3, 2)$
b. ↑: $(0, 2)$
   ↓: $(-1, 0) \cup (2, 4)$
   Constant: $(-\infty, -1)$

Average Rate of Change

How do we know *how much* a function is increasing or decreasing? For example, is the price of a stock slightly increasing or is it doubling every week? One way we determine how much a function is increasing or decreasing is by calculating its *average rate of change*.

Let (x_1, y_1) and (x_2, y_2) be two points that lie on the graph of a function f. Draw the line that passes through these two points (x_1, y_1) and (x_2, y_2). This line is called a **secant line**.

Note that the slope of the secant line is given by $m = \dfrac{y_2 - y_1}{x_2 - x_1}$, and recall that the

slope of a line is the rate of change of that line. The **slope of the secant line** is used to represent the *average rate of change* of the function.

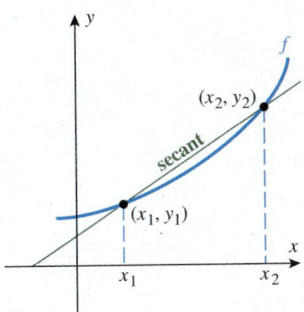

AVERAGE RATE OF CHANGE

Let $(x_1, f(x_1))$ and $(x_2, f(x_2))$ be two distinct points, $(x_1 \neq x_2)$, on the graph of the function f. The **average rate of change** of f between x_1 and x_2 is given by

Average rate of change $= \dfrac{f(x_2) - f(x_1)}{x_2 - x_1}$

EXAMPLE 3 Average Rate of Change

Find the average rate of change of $f(x) = x^4$ from:

a. $x = -1$ to $x = 0$ **b.** $x = 0$ to $x = 1$ **c.** $x = 1$ to $x = 2$

Solution (a):

Write the average rate of change formula.

$$\dfrac{f(x_2) - f(x_1)}{x_2 - x_1}$$

Let $x_1 = -1$ and $x_2 = 0$.

$$= \dfrac{f(0) - f(-1)}{0 - (-1)}$$

Substitute $f(-1) = (-1)^4 = 1$ and $f(0) = 0^4 = 0$.

$$= \dfrac{0 - 1}{0 - (-1)}$$

Simplify.

$$= \boxed{-1}$$

Solution (b):

Write the average rate of change formula.

$$\dfrac{f(x_2) - f(x_1)}{x_2 - x_1}$$

Let $x_1 = 0$ and $x_2 = 1$.

$$= \dfrac{f(1) - f(0)}{1 - 0}$$

Substitute $f(0) = 0^4 = 0$ and $f(1) = (1)^4 = 1$.

$$= \dfrac{1 - 0}{1 - 0}$$

Simplify.

$$= \boxed{1}$$

Solution (c):

Write the average rate of change formula.

$$\dfrac{f(x_2) - f(x_1)}{x_2 - x_1}$$

Let $x_1 = 1$ and $x_2 = 2$.

$$= \dfrac{f(2) - f(1)}{2 - 1}$$

Substitute $f(1) = 1^4 = 1$ and $f(2) = (2)^4 = 16$.

$$= \dfrac{16 - 1}{2 - 1}$$

Simplify.

$$= \boxed{15}$$

Graphical Interpretation: Slope of the Secant Line

a. Average rate of change of f from $x = -1$ to $x = 0$:
Decreasing at a rate of 1

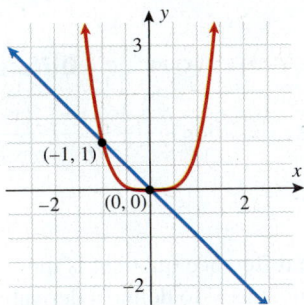

b. Average rate of change of f from $x = 0$ to $x = 1$:
Increasing at a rate of 1

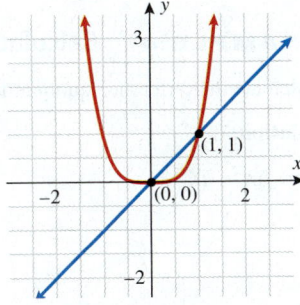

c. Average rate of change of f from $x = 1$ to $x = 2$:
Increasing at a rate of 15

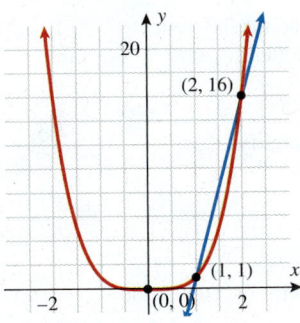

■ **YOUR TURN** Find the average rate of change of $f(x) = x^2$ from:

　　　a. $x = -2$ to $x = 0$　　**b.** $x = 0$ to $x = 2$

■ **Answer: a.** -2　　**b.** 2

The average rate of change can also be written in terms of the difference quotient.

WORDS

Let the difference between
x_1 and x_2 be h.

Solve for x_2.

Substitute $x_2 - x_1 = h$ into
the denominator and

$x_2 = x_1 + h$ into the numerator
of the average rate of change.

Let $x_1 = x$.

MATH

$$x_2 - x_1 = h$$

$$x_2 = x_1 + h$$

$$\text{Average rate of change} = \frac{f(x_2) - f(x_1)}{x_2 - x_1}$$

$$= \frac{f(x_1 + h) - f(x_1)}{h}$$

$$\boxed{= \frac{f(x + h) - f(x)}{h}}$$

When written in this form, the average rate of change is called the **difference quotient**.

> **DEFINITION**   **Difference Quotient**
>
> The expression $\dfrac{f(x + h) - f(x)}{h}$, where $h \neq 0$, is called the **difference quotient**.

The difference quotient is more meaningful when h is small. In calculus the difference quotient is used to define a derivative.

Study Tip

Use brackets or parentheses around $f(x)$ to avoid forgetting to distribute the negative sign:

$$\frac{f(x + h) - [f(x)]}{h}$$

EXAMPLE 4   Calculating the Difference Quotient

Calculate the difference quotient for the function $f(x) = 2x^2 + 1$.

Solution:

Find $f(x + h)$.

$$f(x + h) = 2(x + h)^2 + 1$$
$$= 2(x^2 + 2xh + h^2) + 1$$
$$= 2x^2 + 4xh + 2h^2 + 1$$

Find the difference quotient.

$$\frac{f(x + h) - f(x)}{h} = \frac{\overbrace{2x^2 + 4xh + 2h^2 + 1}^{f(x+h)} - \overbrace{(2x^2 + 1)}^{f(x)}}{h}$$

Simplify.

$$\frac{f(x + h) - f(x)}{h} = \frac{2x^2 + 4xh + 2h^2 + 1 - 2x^2 - 1}{h}$$

$$\frac{f(x + h) - f(x)}{h} = \frac{4xh + 2h^2}{h}$$

Factor the numerator.

$$\frac{f(x + h) - f(x)}{h} = \frac{h(4x + 2h)}{h}$$

Cancel (divide out) the common h.

$$\frac{f(x + h) - f(x)}{h} = \boxed{4x + 2h} \qquad h \neq 0$$

Classroom Example 3.2.4
Calculate the difference quotient for $f(x) = -2 - 3x^2$.

Answer: $-6x - 3h$

■ **Answer:**

$$\frac{f(x + h) - f(x)}{h} = -2x - h$$

■ **YOUR TURN** Calculate the difference quotient for the function $f(x) = -x^2 + 2$.

Piecewise-Defined Functions

Most of the functions that we have seen in this text are functions defined by polynomials. Sometimes the need arises to define functions in terms of *pieces*. For example, most plumbers charge a flat fee for a house call and then an additional hourly rate for the job. For instance, if a particular plumber charges $100 to drive out to your house and work for 1 hour and then an additional $25 an hour for every additional hour he or she works on your job, we would define this function in pieces. If we let h be the number of hours worked, then the charge is defined as

$$\text{Plumbing charge} = \begin{cases} 100 & h \leq 1 \\ 100 + 25(h - 1) & h > 1 \end{cases}$$

If we were to graph this function, we would see that there is 1 hour that is constant and after that the function continually increases.

Another piecewise-defined function is the absolute value function. The absolute value function can be thought of as two pieces: the line $y = -x$ (when x is negative) and the line $y = x$ (when x is nonnegative). We start by graphing these two lines on the same graph.

The absolute value function behaves like the line $y = -x$ when x is negative (erase the blue graph in quadrant IV) and like the line $y = x$ when x is positive (erase the red graph in quadrant III).

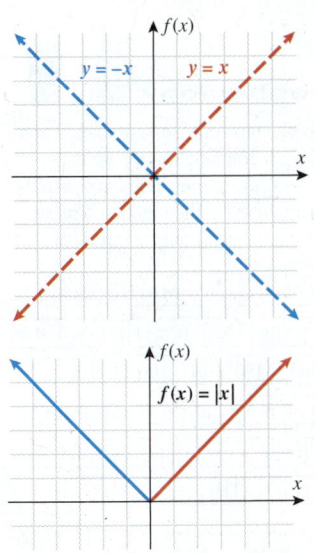

Absolute value function

$$f(x) = |x| = \begin{cases} -x & x < 0 \\ x & x \ge 0 \end{cases}$$

The next example is a piecewise-defined function given in terms of functions in our "library of functions." Because the function is defined in terms of pieces of other functions, we draw the graph of each individual function, and then for each function, darken the piece corresponding to its part of the domain. This is like the procedure above for the absolute value function.

EXAMPLE 5   Graphing Piecewise-Defined Functions

Graph the piecewise-defined function, and state the domain, range, and intervals when the function is increasing, decreasing, or constant.

$$G(x) = \begin{cases} x^2 & x < -1 \\ 1 & -1 \le x \le 1 \\ x & x > 1 \end{cases}$$

Solution:

Graph each of the functions on the same plane.

Square function:
$f(x) = x^2$

Constant function:
$f(x) = 1$

Identity function:
$f(x) = x$

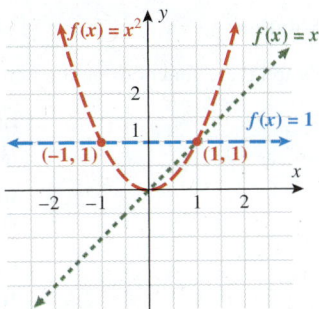

The points to focus on in particular are the x-values where the pieces change over—that is, $x = -1$ and $x = 1$.

Let's now investigate each piece. When $x < -1$, this function is defined by the square function, $f(x) = x^2$, so darken that particular function to the left of $x = -1$. When $-1 \le x \le 1$, the function is defined by the constant function, $f(x) = 1$, so darken that particular function between the x values of -1 and 1. When $x > 1$, the function is defined by the identity function, $f(x) = x$, so darken that function to the right of $x = 1$. Erase everything that is not darkened, and the resulting graph of the piecewise-defined function is given on the right.

Technology Tip

To plot a piecewise-defined function using a graphing utility, use the TEST menu operations to define the inequalities in the function. Press:

Y= ((1 − X, T, θ, n)) ((
X, T, θ, n 2nd MATH 5 < 0
) + (X, T, θ, n)) (X, T, θ, n
2nd MATH 4 ≥ 0)) ((
X, T, θ, n 2nd MATH 5 < 2
) + (((−) 1)) (X, T, θ, n
2nd MATH 3 > 2)

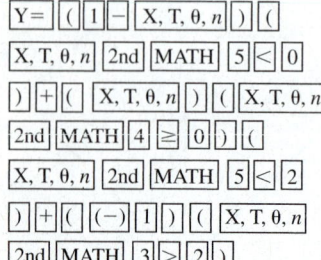

To avoid connecting graphs of the pieces, press MODE and Dot. Set the viewing rectangle as [−3, 4] by [−2, 5]; then press GRAPH.

Be sure to include the open circle and closed circle at the appropriate endpoints of each piece in the function.

The table of values supports the graph, except at $x = 2$. The function is not defined at $x = 2$.

This function is defined for all real values of x, so the domain of this function is the set of all real numbers. The values that this function yields in the vertical direction are all real numbers greater than or equal to 1. Hence, the range of this function is $[1, \infty)$. The intervals of increasing, decreasing, and constant are as follows:

$$\text{Decreasing: } (-\infty, -1)$$

$$\text{Constant: } (-1, 1)$$

$$\text{Increasing: } (1, \infty)$$

The term **continuous** implies that there are no holes or jumps and that the graph can be drawn without picking up your pencil. A function that does have holes or jumps and cannot be drawn in one motion without picking up your pencil is classified as **discontinuous**, and the points where the holes or jumps occur are called *points of discontinuity*.

The previous example illustrates a *continuous* piecewise-defined function. At the $x = -1$ junction, the square function and constant function both pass through the point $(-1, 1)$. At the $x = 1$ junction, the constant function and the identity function both pass through the point $(1, 1)$. Since the graph of this piecewise-defined function has no holes or jumps, we classify it as a continuous function.

The next example illustrates a *discontinuous* piecewise-defined function.

EXAMPLE 6 Graphing a Discontinuous Piecewise-Defined Function

Graph the piecewise-defined function, and state the intervals where the function is increasing, decreasing, or constant, along with the domain and range.

$$f(x) = \begin{cases} 1 - x & x < 0 \\ x & 0 \le x < 2 \\ -1 & x > 2 \end{cases}$$

Solution:

Graph these functions on the same plane.

Linear function:
$f(x) = 1 - x$

Identity function:
$f(x) = x$

Constant function:
$f(x) = -1$

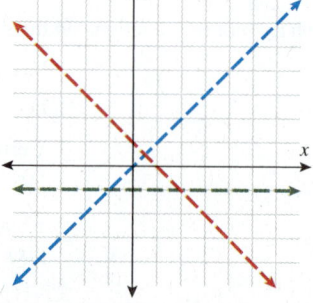

Darken the piecewise-defined function on the graph. For all values less than zero $(x < 0)$ the function is defined by the **linear function**. Note the use of an open circle, indicating up to but not including $x = 0$. For values $0 \le x < 2$, the function is defined by the **identity function**. The circle is filled in at the left endpoint, $x = 0$. An open circle is used at $x = 2$. For all values greater than 2, $x > 2$, the function is defined by the **constant function**. Because this interval does not include the point $x = 2$, an open circle is used.

At what intervals is the function increasing, decreasing, or constant? Remember that the intervals correspond to the *x*-values.

Decreasing: $(-\infty, 0)$ Increasing: $(0, 2)$ Constant: $(2, \infty)$

The function is defined for all values of *x* except $x = 2$.

Domain: $(-\infty, 2) \cup (2, \infty)$

The output of this function (vertical direction) takes on the *y*-values $y \geq 0$ and the additional single value $y = -1$.

Range: $[-1, -1] \cup [0, \infty)$ or $\{-1\} \cup [0, \infty)$

We mentioned earlier that a discontinuous function has a graph that exhibits holes or jumps. In this example, the point $x = 0$ corresponds to a jump, because you would have to pick up your pencil to continue drawing the graph. The point $x = 2$ corresponds to both a hole and a jump. The hole indicates that the function is not defined at that point, and there is still a jump because the identity function and the constant function do not meet at the same *y*-value at $x = 2$.

■ **YOUR TURN** Graph the piecewise-defined function, and state the intervals where the function is increasing, decreasing, or constant, along with the domain and range.

$$f(x) = \begin{cases} -x & x \leq -1 \\ 2 & -1 < x < 1 \\ x & x > 1 \end{cases}$$

■ **Answer:** Increasing: $(1, \infty)$
Decreasing: $(-\infty, -1)$
Constant: $(-1, 1)$
Domain:
$(-\infty, 1) \cup (1, \infty)$
Range: $[1, \infty)$

Piecewise-defined functions whose "pieces" are constants are called **step functions**. The reason for this name is that the graph of a step function looks like steps of a staircase. A common step function used in engineering is the **Heaviside step function** (also called the **unit step function**):

$$H(t) = \begin{cases} 0 & t < 0 \\ 1 & t \geq 0 \end{cases}$$

This function is used in signal processing to represent a signal that turns on at some time and stays on indefinitely.

A common step function used in business applications is the *greatest integer function*.

GREATEST INTEGER FUNCTION

$f(x) = [[x]] =$ greatest integer less than or equal to *x*.

x	1.0	1.3	1.5	1.7	1.9	2.0
$f(x) = [[x]]$	1	1	1	1	1	2

**SECTION
3.2 SUMMARY**

NAME	FUNCTION	DOMAIN	RANGE	GRAPH	EVEN/ODD
Linear	$f(x) = mx + b, m \neq 0$	$(-\infty, \infty)$	$(-\infty, \infty)$		Neither (unless $y = x$)
Constant	$f(x) = c$	$(-\infty, \infty)$	$[c, c]$ or $\{c\}$		Even
Identity	$f(x) = x$	$(-\infty, \infty)$	$(-\infty, \infty)$		Odd
Square	$f(x) = x^2$	$(-\infty, \infty)$	$[0, \infty)$		Even
Cube	$f(x) = x^3$	$(-\infty, \infty)$	$(-\infty, \infty)$		Odd
Square Root	$f(x) = \sqrt{x}$	$[0, \infty)$	$[0, \infty)$		Neither
Cube Root	$f(x) = \sqrt[3]{x}$	$(-\infty, \infty)$	$(-\infty, \infty)$		Odd
Absolute Value	$f(x) = |x|$	$(-\infty, \infty)$	$[0, \infty)$		Even
Reciprocal	$f(x) = \dfrac{1}{x}$	$(-\infty, 0) \cup (0, \infty)$	$(-\infty, 0) \cup (0, \infty)$		Odd

Domain and Range of a Function

■ Implied domain: Exclude any values that lead to the function being undefined (dividing by zero) or imaginary outputs (square root of a negative real number).

■ Inspect the graph to determine the set of all inputs (domain) and the set of all outputs (range).

Finding Intervals Where a Function Is Increasing, Decreasing, or Constant

- Increasing: Graph of function rises from left to right.
- Decreasing: Graph of function falls from left to right.
- Constant: Graph of function does not change height from left to right.

Average Rate of Change $\quad \dfrac{f(x_2) - f(x_1)}{x_2 - x_1} \qquad x_1 \neq x_2$

Difference Quotient $\quad \dfrac{f(x + h) - f(x)}{h} \qquad h \neq 0$

Piecewise-Defined Functions

- Continuous: You can draw the graph of a function without picking up the pencil.
- Discontinuous: Graph has holes and/or jumps.

SECTION

3.2 EXERCISES

■**SKILLS**

In Exercises 1–24, determine whether the function is even, odd, or neither.

1. $G(x) = x + 4$
2. $h(x) = 3 - x$
3. $f(x) = 3x^2 + 1$
4. $F(x) = x^4 + 2x^2$
5. $g(t) = 5t^3 - 3t$
6. $f(x) = 3x^5 + 4x^3$
7. $h(x) = x^2 + 2x$
8. $G(x) = 2x^4 + 3x^3$
9. $h(x) = x^{1/3} - x$
10. $g(x) = x^{-1} + x$
11. $f(x) = |x| + 5$
12. $f(x) = |x| + x^2$
13. $f(x) = |x|$
14. $f(x) = |x^3|$
15. $G(t) = |t - 3|$
16. $g(t) = |t + 2|$
17. $G(t) = \sqrt{t - 3}$
18. $f(x) = \sqrt{2 - x}$
19. $g(x) = \sqrt{x^2 + x}$
20. $f(x) = \sqrt{x^2 + 2}$
21. $h(x) = \dfrac{1}{x} + 3$
22. $h(x) = \dfrac{1}{x} - 2x$
23.
24.

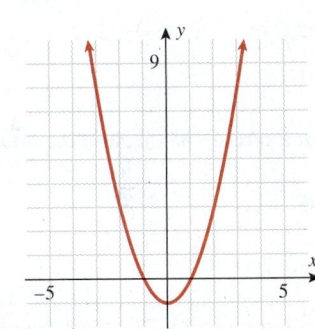

In Exercises 25–36, state the (a) domain, (b) range, and (c) x-interval(s) where the function is increasing, decreasing, or constant. Find the values of (d) $f(0)$, (e) $f(-2)$, and (f) $f(2)$.

25.
26.
27.
28.

29.

30.

31.

32.

33.

34.

35.

36.

In Exercises 37–44, find the difference quotient $\dfrac{f(x + h) - f(x)}{h}$ **for each function.**

37. $f(x) = x^2 - x$

38. $f(x) = x^2 + 2x$

39. $f(x) = 3x + x^2$

40. $f(x) = 5x - x^2$

41. $f(x) = x^2 - 3x + 2$

42. $f(x) = x^2 - 2x + 5$

43. $f(x) = -3x^2 + 5x - 4$

44. $f(x) = -4x^2 + 2x - 3$

In Exercises 45–52, find the average rate of change of the function from $x = 1$ to $x = 3$.

45. $f(x) = x^3$

46. $f(x) = \dfrac{1}{x}$

47. $f(x) = |x|$

48. $f(x) = 2x$

49. $f(x) = 1 - 2x$

50. $f(x) = 9 - x^2$

51. $f(x) = |5 - 2x|$

52. $f(x) = \sqrt{x^2 - 1}$

In Exercises 53–78, graph the piecewise-defined functions. State the domain and range in interval notation. Determine the intervals where the function is increasing, decreasing, or constant.

53. $f(x) = \begin{cases} x & x < 2 \\ 2 & x \geq 2 \end{cases}$

54. $f(x) = \begin{cases} -x & x < -1 \\ -1 & x \geq -1 \end{cases}$

55. $f(x) = \begin{cases} 1 & x < -1 \\ x^2 & x \geq -1 \end{cases}$

56. $f(x) = \begin{cases} x^2 & x < 2 \\ 4 & x \geq 2 \end{cases}$

57. $f(x) = \begin{cases} x & x < 0 \\ x^2 & x \geq 0 \end{cases}$

58. $f(x) = \begin{cases} -x & x \leq 0 \\ x^2 & x > 0 \end{cases}$

59. $f(x) = \begin{cases} -x + 2 & x < 1 \\ x^2 & x \geq 1 \end{cases}$

60. $f(x) = \begin{cases} 2 + x & x \leq -1 \\ x^2 & x > -1 \end{cases}$

61. $f(x) = \begin{cases} 5 - 2x & x < 2 \\ 3x - 2 & x > 2 \end{cases}$

62. $f(x) = \begin{cases} 3 - \dfrac{1}{2}x & x < -2 \\ 4 + \dfrac{3}{2}x & x > -2 \end{cases}$

63. $G(x) = \begin{cases} -1 & x < -1 \\ x & -1 \leq x \leq 3 \\ 3 & x > 3 \end{cases}$

64. $G(x) = \begin{cases} -1 & x < -1 \\ x & -1 < x < 3 \\ 3 & x > 3 \end{cases}$

65. $G(t) = \begin{cases} 1 & t < 1 \\ t^2 & 1 \le t \le 2 \\ 4 & t > 2 \end{cases}$

66. $G(t) = \begin{cases} 1 & t < 1 \\ t^2 & 1 < t < 2 \\ 4 & t > 2 \end{cases}$

67. $f(x) = \begin{cases} -x - 1 & x < -2 \\ x + 1 & -2 < x < 1 \\ -x + 1 & x \ge 1 \end{cases}$

68. $f(x) = \begin{cases} -x - 1 & x \le -2 \\ x + 1 & -2 < x < 1 \\ -x + 1 & x > 1 \end{cases}$

69. $G(x) = \begin{cases} 0 & x < 0 \\ \sqrt{x} & x \ge 0 \end{cases}$

70. $G(x) = \begin{cases} 1 & x < 1 \\ \sqrt[3]{x} & x > 1 \end{cases}$

71. $G(x) = \begin{cases} 0 & x = 0 \\ \dfrac{1}{x} & x \ne 0 \end{cases}$

72. $G(x) = \begin{cases} 0 & x = 0 \\ -\dfrac{1}{x} & x \ne 0 \end{cases}$

73. $G(x) = \begin{cases} -\sqrt[3]{x} & x \le -1 \\ x & -1 < x < 1 \\ -\sqrt{x} & x > 1 \end{cases}$

74. $G(x) = \begin{cases} -\sqrt[3]{x} & x < -1 \\ x & -1 \le x < 1 \\ \sqrt{x} & x > 1 \end{cases}$

75. $f(x) = \begin{cases} x + 3 & x \le -2 \\ |x| & -2 < x < 2 \\ x^2 & x \ge 2 \end{cases}$

76. $f(x) = \begin{cases} |x| & x < -1 \\ 1 & -1 < x < 1 \\ |x| & x > 1 \end{cases}$

77. $f(x) = \begin{cases} x & x \le -1 \\ x^3 & -1 < x < 1 \\ x^2 & x > 1 \end{cases}$

78. $f(x) = \begin{cases} x^2 & x \le -1 \\ x^3 & -1 < x < 1 \\ x & x \ge 1 \end{cases}$

▪ APPLICATIONS

For Exercises 79 and 80, refer to the following:

A manufacturer determines that his *profit* and *cost* functions over one year are represented by the following graphs.

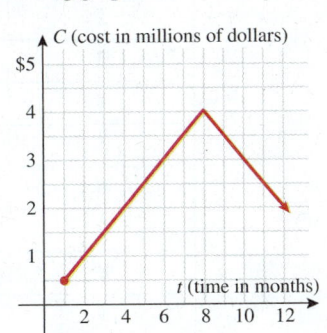

79. Business. Find the intervals on which profit is increasing, decreasing, and constant.

80. Business. Find the intervals on which cost is increasing, decreasing, and constant.

81. Budget: Costs. The Kappa Kappa Gamma sorority decides to order custom-made T-shirts for its *Kappa Krush* mixer with the Sigma Alpha Epsilon fraternity. If the sorority orders 50 or fewer T-shirts, the cost is $10 per shirt. If it orders more than 50 but less than or equal to 100, the cost is $9 per shirt. If it orders more than 100, the cost is $8 per shirt. Find the cost function $C(x)$ as a function of the number of T-shirts x ordered.

82. Budget: Costs. The marching band at a university is ordering some additional uniforms to replace existing uniforms that are worn out. If the band orders 50 or fewer, the cost is $176.12 per uniform. If it orders more than 50 but less than or equal to 100, the cost is $159.73 per uniform. Find the cost function $C(x)$ as a function of the number of new uniforms x ordered.

83. Budget: Costs. The Richmond rowing club is planning to enter the *Head of the Charles* race in Boston and is trying to figure out how much money to raise. The entry fee is $250 per boat for the first 10 boats and $175 for each additional boat. Find the cost function $C(x)$ as a function of the number of boats x the club enters.

84. Phone Cost: Long-Distance Calling. A phone company charges $.39 per minute for the first 10 minutes of an international long-distance phone call and $.12 per minute every minute after that. Find the cost function $C(x)$ as a function of the length of the phone call x in minutes.

85. Event Planning. A young couple are planning their wedding reception at a yacht club. The yacht club charges a flat rate of $1000 to reserve the dining room for a private party. The cost of food is $35 per person for the first 100 people and $25 per person for every additional person beyond the first 100. Write the cost function $C(x)$ as a function of the number of people x attending the reception.

86. Home Improvement. An irrigation company gives you an estimate for an eight-zone sprinkler system. The parts are $1400, and the labor is $25 per hour. Write a function $C(x)$ that determines the cost of a new sprinkler system if you choose this irrigation company.

87. Sales. A famous author negotiates with her publisher the monies she will receive for her next suspense novel. She will receive $50,000 up front and a 15% royalty rate on the first 100,000 books sold, and 20% on any books sold beyond that. If the book sells for $20 and royalties are based on the selling price, write a royalties function $R(x)$ as a function of total number x of books sold.

88. Sales. Rework Exercise 87 if the author receives $35,000 up front, 15% for the first 100,000 books sold, and 25% on any books sold beyond that.

89. Profit. Some artists are trying to decide whether they will make a profit if they set up a Web-based business to market and sell stained glass that they make. The costs associated with this business are $100 per month for the website and $700 per month for the studio they rent. The materials cost $35 for each work in stained glass, and the artists charge $100 for each unit they sell. Write the monthly profit as a function of the number of stained-glass units they sell.

90. Profit. Philip decides to host a shrimp boil at his house as a fundraiser for his daughter's AAU basketball team. He orders gulf shrimp to be flown in from New Orleans. The shrimp costs $5 per pound. The shipping costs $30. If he charges $10 per person, write a function $F(x)$ that represents either his loss or profit as a function of the number of people x that attend. Assume that each person will eat 1 pound of shrimp.

91. Postage Rates. The following table corresponds to first-class postage rates for the U.S. Postal Service. Write a piecewise-defined function in terms of the greatest integer function that models this cost of mailing flat envelopes first class.

WEIGHT LESS THAN (OUNCES)	FIRST-CLASS RATE (FLAT ENVELOPES)
1	$0.80
2	$0.97
3	$1.14
4	$1.31
5	$1.48
6	$1.65
7	$1.82
8	$1.99
9	$2.16
10	$2.33
11	$2.50
12	$2.67
13	$2.84

92. Postage Rates. The following table corresponds to first-class postage rates for the U.S. Postal Service. Write a piecewise-defined function in terms of the greatest integer function that models this cost of mailing parcels first class.

WEIGHT LESS THAN (OUNCES)	FIRST-CLASS RATE (PARCELS)
1	$1.13
2	$1.30
3	$1.47
4	$1.64
5	$1.81
6	$1.98
7	$2.15
8	$2.32
9	$2.49
10	$2.66
11	$2.83
12	$3.00
13	$3.17

A square wave is a waveform used in electronic circuit testing and signal processing. A square wave alternates regularly and instantaneously between two levels.

sciencephotos/Alamy

93. Electronics: Signals. Write a step function $f(t)$ that represents the following square wave.

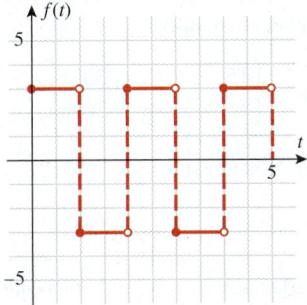

94. Electronics: Signals. Write a step function $f(x)$ that represents the following square wave, where x represents frequency in Hz.

For Exercises 95 and 96, refer to the following table:

Global Carbon Emissions from Fossil Fuel Burning

Year	Millions of Tons of Carbon
1900	500
1925	1000
1950	1500
1975	5000
2000	7000

95. Climate Change: Global Warming. What is the average rate of change in global carbon emissions from fossil fuel burning from

   a. 1900 to 1950?
   b. 1950 to 2000?

96. Climate Change: Global Warming. What is the average rate of change in global carbon emissions from fossil fuel burning from

   a. 1950 to 1975?
   b. 1975 to 2000?

For Exercises 97 and 98, use the following information:

The height (in feet) of a falling object with an initial velocity of 48 feet per second launched straight upward from the ground is given by $h(t) = -16t^2 + 48t$, where t is time (in seconds).

97. Falling Objects. What is the average rate of change of the height as a function of time from $t = 1$ to $t = 2$?

98. Falling Objects. What is the average rate of change of the height as a function of time from $t = 1$ to $t = 3$?

For Exercises 99 and 100, refer to the following:

An analysis of sales indicates that demand for a product during a calendar year (no leap year) is modeled by

$$d(t) = 3\sqrt{t^2 + 1} - 2.75t$$

where d is demand in thousands of units and t is the day of the year and $t = 1$ represents January 1.

99. Economics. Find the average rate of change of the demand of the product over the first quarter.

100. Economics. Find the average rate of change of the demand of the product over the fourth quarter.

▪ CATCH THE MISTAKE

In Exercises 101–104, explain the mistake that is made.

101. Graph the piecewise-defined function. State the domain and range.

$$f(x) = \begin{cases} -x & x < 0 \\ x & x > 0 \end{cases}$$

Solution:

Draw the graphs of $f(x) = -x$ and $f(x) = x$.

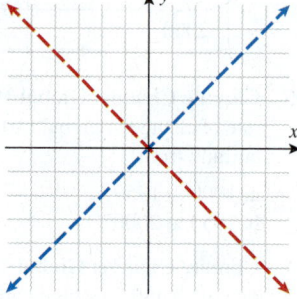

Darken the function $f(x) = -x$ when $x < 0$ and the function $f(x) = x$ when $x > 0$. This gives us the familiar absolute value graph.

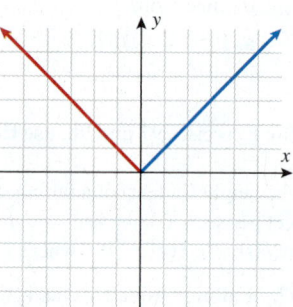

Domain: $(-\infty, \infty)$ or $\mathbb{R}$
Range: $[0, \infty)$

This is incorrect. What mistake was made?

102. Graph the piecewise-defined function. State the domain and range.

$$f(x) = \begin{cases} -x & x \le 1 \\ x & x > 1 \end{cases}$$

Solution:

Draw the graphs of $f(x) = -x$ and $f(x) = x$.

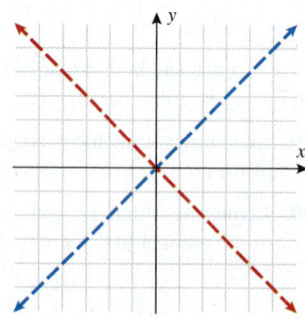

Darken the function $f(x) = -x$ when $x < 1$ and the function $f(x) = x$ when $x > 1$.

The resulting graph is as shown.

Domain: $(-\infty, \infty)$ or $\mathbb{R}$
Range: $(-1, \infty)$

This is incorrect. What mistake was made?

103. The cost of airport Internet access is \$15 for the first 30 minutes and \$1 per minute for each additional minute. Write a function describing the cost of the service as a function of minutes used online.

Solution: $C(x) = \begin{cases} 15 & x \le 30 \\ 15 + x & x > 30 \end{cases}$

This is incorrect. What mistake was made?

104. Most money market accounts pay a higher interest with a higher principal. If the credit union is offering 2% on accounts with less than or equal to \$10,000 and 4% on the additional money over \$10,000, write the interest function $I(x)$ that represents the interest earned on an account as a function of dollars in the account.

Solution: $I(x) = \begin{cases} 0.02x & x \le 10{,}000 \\ 0.02(10{,}000) + 0.04x & x > 10{,}000 \end{cases}$

This is incorrect. What mistake was made?

■ CONCEPTUAL

In Exercises 105–108, determine whether each statement is true or false.

105. The identity function is a special case of the linear function.

106. The constant function is a special case of the linear function.

107. If an odd function has an interval where the function is increasing, then it also has to have an interval where the function is decreasing.

108. If an even function has an interval where the function is increasing, then it also has to have an interval where the function is decreasing.

■ CHALLENGE

In Exercises 109 and 110, for a and b real numbers, can the function given ever be a continuous function? If so, specify the value for a and b that would make it so.

109. $f(x) = \begin{cases} ax & x \le 2 \\ bx^2 & x > 2 \end{cases}$

110. $f(x) = \begin{cases} -\dfrac{1}{x} & x < a \\ \dfrac{1}{x} & x \ge a \end{cases}$

■ TECHNOLOGY

111. In trigonometry you will learn about the sine function, $\sin x$. Plot the function $f(x) = \sin x$, using a graphing utility. It should look like the graph on the right. Is the sine function even, odd, or neither?

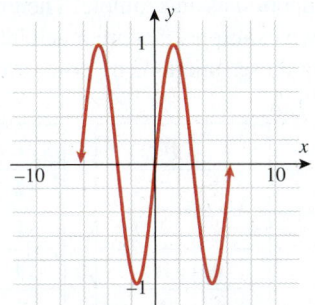

112. In trigonometry you will learn about the cosine function, $\cos x$. Plot the function $f(x) = \cos x$, using a graphing utility. It should look like the graph on the right. Is the cosine function even, odd, or neither?

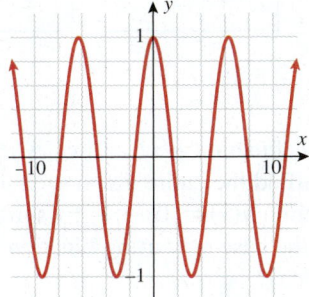

113. In trigonometry you will learn about the tangent function, $\tan x$. Plot the function $f(x) = \tan x$, using a graphing utility. If you restrict the values of x so that $-\dfrac{\pi}{2} < x < \dfrac{\pi}{2}$, the graph should resemble the graph below. Is the tangent function even, odd, or neither?

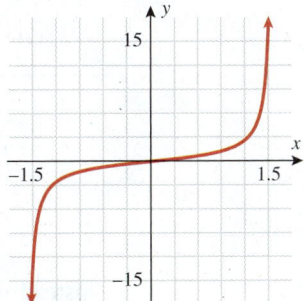

114. Plot the function $f(x) = \dfrac{\sin x}{\cos x}$. What function is this?

115. Graph the function $f(x) = [[3x]]$ using a graphing utility. State the domain and range.

116. Graph the function $f(x) = \left[\left[\frac{1}{3}x\right]\right]$ using a graphing utility. State the domain and range.

SKILLS OBJECTIVES

- Sketch the graph of a function using horizontal and vertical shifting of common functions.
- Sketch the graph of a function by reflecting a common function about the *x*-axis or *y*-axis.
- Sketch the graph of a function by stretching or compressing a common function.
- Sketch the graph of a function using a sequence of transformations.

CONCEPTUAL OBJECTIVES

- Identify the common functions by their graphs.
- Apply multiple transformations of common functions to obtain graphs of functions.
- Understand that domain and range also are transformed.

Horizontal and Vertical Shifts

The focus of the previous section was to learn the graphs that correspond to particular functions such as identity, square, cube, square root, cube root, absolute value, and reciprocal. Therefore, at this point, you should be able to recognize and generate the graphs of $y = x$, $y = x^2$, $y = x^3$, $y = \sqrt{x}$, $y = \sqrt[3]{x}$, $y = |x|$, and $y = \dfrac{1}{x}$. In this section, we will discuss how to sketch the graphs of functions that are very simple modifications of these functions. For instance, a common function may be shifted (horizontally or vertically), reflected, or stretched (or compressed). Collectively, these techniques are called **transformations**.

Let's take the absolute value function as an example. The graph of $f(x) = |x|$ was given in the last section. Now look at two examples that are much like this function: $g(x) = |x| + 2$ and $h(x) = |x - 1|$. Graphing these functions by point-plotting yields

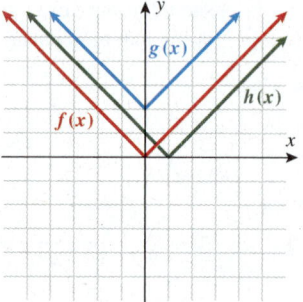

x	$f(x)$
-2	2
-1	1
0	0
1	1
2	2

x	$g(x)$
-2	4
-1	3
0	2
1	3
2	4

x	$h(x)$
-2	3
-1	2
0	1
1	0
2	1

Instead of point-plotting the function $g(x) = |x| + 2$, we could have started with the function $f(x) = |x|$ and shifted the entire graph *up* 2 units. Similarly, we could have generated the graph of the function $h(x) = |x - 1|$ by shifting the function $f(x) = |x|$ to the *right* 1 unit. In both cases, the base or starting function is $f(x) = |x|$. Why did we go up for $g(x)$ and to the right for $h(x)$?

Note that we could rewrite the functions $g(x)$ and $h(x)$ in terms of $f(x)$:

$$g(x) = |x| + 2 = f(x) + 2$$

$$h(x) = |x - 1| = f(x - 1)$$

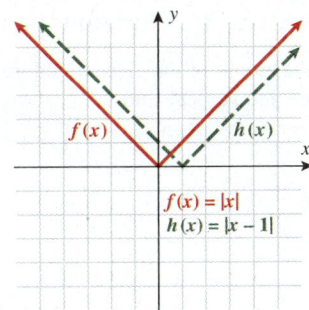

In the case of $g(x)$, the shift $(+2)$ occurs "outside" the function—that is, outside the parentheses showing the argument. Therefore, the output for $g(x)$ is two more than the typical output for $f(x)$. Because the output corresponds to the vertical axis, this results in a shift *upward* of two units. In general, shifts that occur *outside* the function correspond to a *vertical* shift corresponding to the sign of the shift. For instance, had the function been $G(x) = |x| - 2$, this graph would have started with the graph of the function $f(x)$ and shifted down two units.

In the case of $h(x)$, the shift occurs "inside" the function—that is, inside the parentheses showing the argument. Note that the point $(0, 0)$ that lies on the graph of $f(x)$ was shifted to the point $(1, 0)$ on the graph of the function $h(x)$. The y-value remained the same, but the x-value shifted to the right one unit. Similarly, the points $(-1, 1)$ and $(1, 1)$ were shifted to the points $(0, 1)$ and $(2, 1)$, respectively. In general, shifts that occur *inside* the function correspond to a *horizontal* shift opposite the sign. In this case, the graph of the function $h(x) = |x - 1|$ shifted the graph of the function $f(x)$ to the right one unit. If, instead, we had the function $H(x) = |x + 1|$, this graph would have started with the graph of the function $f(x)$ and shifted to the left one unit.

VERTICAL SHIFTS

Assuming that c is a positive constant,

To Graph	Shift the Graph of $f(x)$
$f(x) + c$	c units upward
$f(x) - c$	c units downward

Adding or subtracting a constant **outside** the function corresponds to a **vertical** shift that goes **with the sign**.

HORIZONTAL SHIFTS

Assuming that c is a positive constant,

To Graph	Shift the Graph of $f(x)$
$f(x + c)$	c units to the left
$f(x - c)$	c units to the right

Adding or subtracting a constant **inside** the function corresponds to a **horizontal** shift that goes **opposite the sign**.

Technology Tip

a. Graphs of $y_1 = x^2$ and $y_2 = g(x) = x^2 - 1$ are shown.

b. Graphs of $y_1 = x^2$ and $y_2 = H(x) = (x + 1)^2$ are shown.

■ **Answer:**

a.

b.

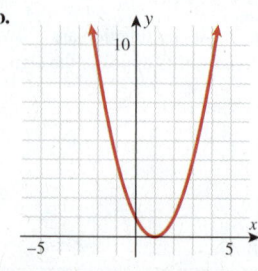

EXAMPLE 1   Horizontal and Vertical Shifts

Sketch the graphs of the given functions using horizontal and vertical shifts:

a. $g(x) = x^2 - 1$ **b.** $H(x) = (x + 1)^2$

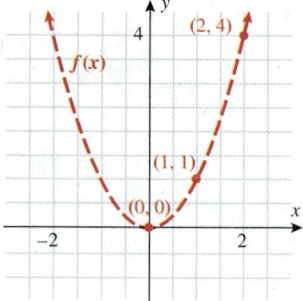

Solution:

In both cases, the function to start with is $f(x) = x^2$.

a. $g(x) = x^2 - 1$ can be rewritten as
$g(x) = f(x) - 1$.
  1. The shift (one unit) occurs *outside* of the function. Therefore, we expect a vertical shift that goes with the sign.
  2. Since the sign is *negative*, this corresponds to a *downward* shift.
  3. Shifting the graph of the function $f(x) = x^2$ down one unit yields the graph of $g(x) = x^2 - 1$.

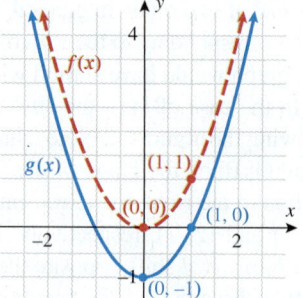

b. $H(x) = (x + 1)^2$ can be rewritten as
$H(x) = f(x + 1)$.
  1. The shift (one unit) occurs *inside* of the function. Therefore, we expect a horizontal shift that goes *opposite* the sign.
  2. Since the sign is *positive*, this corresponds to a shift to the *left*.
  3. Shifting the graph of the function $f(x) = x^2$ to the left one unit yields the graph of $H(x) = (x + 1)^2$.

■ **YOUR TURN** Sketch the graphs of the given functions using horizontal and vertical shifts.

a. $g(x) = x^2 + 1$ **b.** $H(x) = (x - 1)^2$

It is important to note that the domain and range of the resulting function can be thought of as also being shifted. Shifts in the domain correspond to horizontal shifts, and shifts in the range correspond to vertical shifts.

EXAMPLE 2 **Horizontal and Vertical Shifts and Changes in the Domain and Range**

Graph the functions using translations and state the domain and range of each function.

a. $g(x) = \sqrt{x + 1}$ **b.** $G(x) = \sqrt{x} - 2$

Solution:

In both cases the function to start with is $f(x) = \sqrt{x}$.

Domain: $[0, \infty)$

Range: $[0, \infty)$

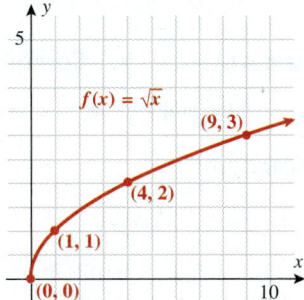

a. $g(x) = \sqrt{x + 1}$ can be rewritten as $g(x) = f(x + 1)$.

1. The shift (one unit) is *inside* the function, which corresponds to a *horizontal* shift *opposite the sign*.

2. Shifting the graph of $f(x) = \sqrt{x}$ to the *left* one unit yields the graph of $g(x) = \sqrt{x + 1}$. Notice that the point $(0, 0)$, which lies on the graph of $f(x)$, gets shifted to the point $(-1, 0)$ on the graph of $g(x)$.

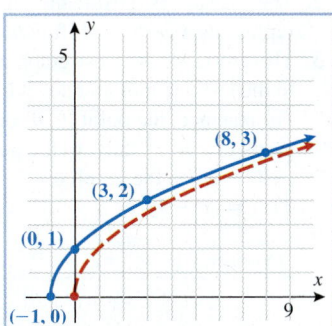

Although the original function $f(x) = \sqrt{x}$ had an implicit restriction on the domain: $[0, \infty)$, the function $g(x) = \sqrt{x + 1}$ has the implicit restriction that $x \geq -1$. We see that the output or range of $g(x)$ is the same as the output of the original function $f(x)$.

Domain: $[-1, \infty)$ **Range:** $[0, \infty)$

b. $G(x) = \sqrt{x} - 2$ can be rewritten as $G(x) = f(x) - 2$.

1. The shift (two units) is *outside* the function, which corresponds to a *vertical* shift *with the sign*.

2. The graph of $G(x) = \sqrt{x} - 2$ is found by shifting $f(x) = \sqrt{x}$ down two units. Note that the point $(0, 0)$, which lies on the graph of $f(x)$, gets shifted to the point $(0, -2)$ on the graph of $G(x)$.

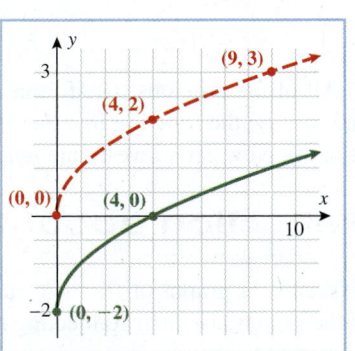

The original function $f(x) = \sqrt{x}$ has an implicit restriction on the domain: $[0, \infty)$. The function $G(x) = \sqrt{x} - 2$ also has the implicit restriction that $x \geq 0$. The output or range of $G(x)$ is always two units less than the output of the original function $f(x)$.

Domain: $[0, \infty)$ **Range:** $[-2, \infty)$

■ **YOUR TURN** Sketch the graph of the functions using shifts and state the domain and range.

a. $G(x) = \sqrt{x} - 2$ **b.** $h(x) = |x| + 1$

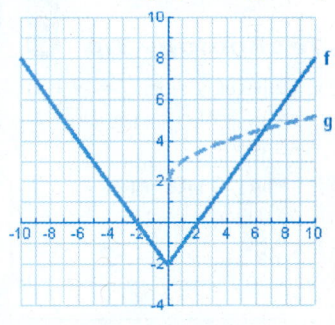
■ **Answer:**
a. $G(x) = \sqrt{x} - 2$

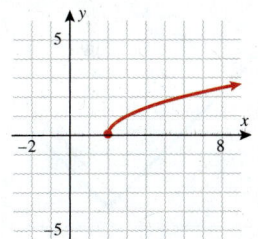

Domain: $[2, \infty)$ Range: $[0, \infty)$

b. $h(x) = |x| + 1$

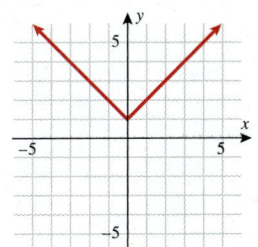

Domain: $(-\infty, \infty)$ Range: $[1, \infty)$

The previous examples have involved graphing functions by shifting a known function either in the horizontal or vertical direction. Let us now look at combinations of horizontal and vertical shifts.

EXAMPLE 3 Combining Horizontal and Vertical Shifts

Sketch the graph of the function $F(x) = (x + 1)^2 - 2$. State the domain and range of F.

Solution:

The base function is $y = x^2$.

1. The shift (one unit) is *inside* the function, so it represents a *horizontal* shift *opposite the sign.*
2. The -2 shift is *outside* the function, which represents a *vertical* shift *with the sign.*
3. Therefore, we shift the graph of $y = x^2$ to the left one unit and down two units. For instance, the point **(0, 0)** on the graph of $y = x^2$ shifts to the point $(-1, -2)$ on the graph of $F(x) = (x + 1)^2 - 2$.

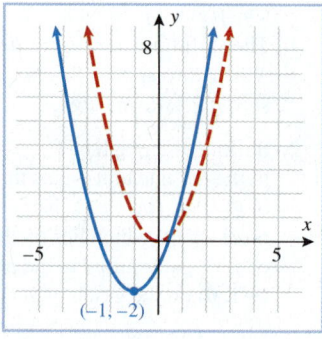

Domain: $(-\infty, \infty)$	Range: $[-2, \infty)$

■ **YOUR TURN** Sketch the graph of the function $f(x) = |x - 2| + 1$. State the domain and range of f.

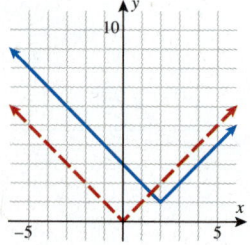
All of the previous transformation examples involve starting with a common function and shifting the function in either the horizontal or the vertical direction (or a combination of both). Now, let's investigate *reflections* of functions about the *x*-axis or *y*-axis.

Reflection about the Axes

To sketch the graphs of $f(x) = x^2$ and $g(x) = -x^2$ start by first listing points that are on each of the graphs and then connecting the points with smooth curves.

x	$f(x)$
-2	4
-1	1
0	0
1	1
2	4

x	$g(x)$
-2	-4
-1	-1
0	0
1	-1
2	-4

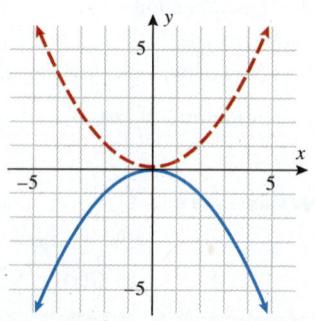

Note that if the graph of $f(x) = x^2$ is reflected about the x-axis, the result is the graph of $g(x) = -x^2$. Also note that the function $g(x)$ can be written as the negative of the function $f(x)$; that is, $g(x) = -f(x)$. In general, **reflection about the x-axis** is produced by multiplying a function by -1.

Let's now investigate reflection about the y-axis. To sketch the graphs of $f(x) = \sqrt{x}$ and $g(x) = \sqrt{-x}$ start by listing points that are on each of the graphs and then connecting the points with smooth curves.

x	$f(x)$
0	0
1	1
4	2
9	3

x	$g(x)$
-9	3
-4	2
-1	1
0	0

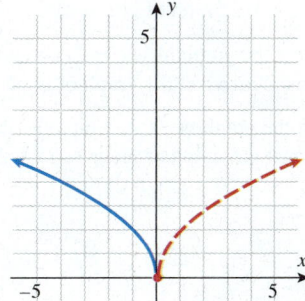

Note that if the graph of $f(x) = \sqrt{x}$ is reflected about the y-axis, the result is the graph of $g(x) = \sqrt{-x}$. Also note that the function $g(x)$ can be written as $g(x) = f(-x)$. In general, **reflection about the y-axis** is produced by replacing x with $-x$ in the function. Notice that the domain of f is $[0, \infty)$, whereas the domain of g is $(-\infty, 0]$.

REFLECTION ABOUT THE AXES

The graph of $-f(x)$ is obtained by reflecting the graph of $f(x)$ about the **x-axis**.
The graph of $f(-x)$ is obtained by reflecting the graph of $f(x)$ about the **y-axis**.

EXAMPLE 4 Sketching the Graph of a Function Using Both Shifts and Reflections

Sketch the graph of the function $G(x) = -\sqrt{x + 1}$.

Solution:

Start with the square root function.

$f(x) = \sqrt{x}$

Shift the graph of $f(x)$ to the left one unit to arrive at the graph of $f(x + 1)$.

$f(x + 1) = \sqrt{x + 1}$

Reflect the graph of $f(x + 1)$ about the x-axis to arrive at the graph of $-f(x + 1)$.

$-f(x + 1) = -\sqrt{x + 1}$

a. Graphs of $y_1 = \sqrt{x}$, $y_2 = \sqrt{x+2}$,
$y_3 = \sqrt{-x+2}$, and
$y_4 = f(x) = \sqrt{2-x} + 1$ are
shown.

■ **Answer:**
Domain: $[1, \infty)$
Range: $(-\infty, 2]$

Classroom Example 3.3.5*
Graph:
a. $f(x) = -|x+2|$
b. $g(x) = -(x-1)^3 + 2$
c. $h(x) = 2 - (1-x)^3$

Answer: The graphs are given
by:

EXAMPLE 5 **Sketching the Graph of a Function Using Both Shifts and Reflections**

Sketch the graph of the function $f(x) = \sqrt{2-x} + 1$.

Solution:

Start with the square root function. $\qquad g(x) = \sqrt{x}$

Shift the graph of $g(x)$ to the left two units to
arrive at the graph of $g(x+2)$. $\qquad g(x+2) = \sqrt{x+2}$

Reflect the graph of $g(x+2)$ about the
y-axis to arrive at the graph of $g(-x+2)$. $\qquad g(-x+2) = \sqrt{-x+2}$

Shift the graph $g(-x+2)$ up one unit to
arrive at the graph of $g(-x+2) + 1$. $\qquad g(-x+2) + 1 = \sqrt{2-x} + 1$

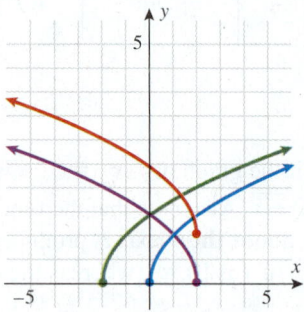

■ **YOUR TURN** Use shifts and reflections to sketch the graph of the function
$f(x) = -\sqrt{x-1} + 2$. State the domain and range of $f(x)$.

Look back at the order in which transformations were performed in Example 5: horizontal
shift, reflection, and then vertical shift. Let us consider an alternate order of transformations.

WORDS	**MATH**
Start with the square root function.	$g(x) = \sqrt{x}$
Shift the graph of $g(x)$ up one unit to arrive at the graph of $g(x) + 1$.	$g(x) + 1 = \sqrt{x} + 1$
Reflect the graph of $g(x) + 1$ about the y-axis to arrive at the graph of $g(-x) + 1$.	$g(-x) + 1 = \sqrt{-x} + 1$
Replace x with $x - 2$, which corresponds to a shift of the graph of $g(-x) + 1$ to the right two units to arrive at the graph of $g[-(x-2)] + 1$.	$g(-x+2) + 1 = \sqrt{2-x} + 1$

In the last step we replaced x with $x - 2$,
which required us to think ahead knowing
the desired result was $2 - x$ inside the radical.
To avoid any possible confusion, follow this
order of transformations:

1. Horizontal shifts: $f(x \pm c)$

2. Reflection: $f(-x)$ and/or $-f(x)$

3. Vertical shifts: $f(x) \pm c$

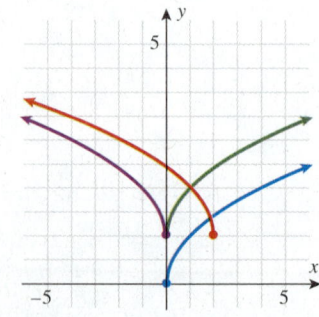

Stretching and Compressing

Horizontal shifts, vertical shifts, and reflections change only the position of the graph in the Cartesian plane, leaving the basic shape of the graph unchanged. These transformations (shifts and reflections) are called **rigid transformations** because they alter only the *position*. **Nonrigid transformations**, on the other hand, distort the *shape* of the original graph. We now consider *stretching* and *compressing* of graphs in both the vertical and the horizontal direction.

A vertical stretch or compression of a graph occurs when the function is multiplied by a positive constant. For example, the graphs of the functions $f(x) = x^2$, $g(x) = 2f(x) = 2x^2$, and $h(x) = \frac{1}{2}f(x) = \frac{1}{2}x^2$ are illustrated below. Depending on if the constant is larger than 1 or smaller than 1 will determine whether it corresponds to a stretch (expansion) or compression (contraction) in the vertical direction.

x	$f(x)$
-2	4
-1	1
0	0
1	1
2	4

x	$g(x)$
-2	8
-1	2
0	0
1	2
2	8

x	$h(x)$
-2	2
-1	$\frac{1}{2}$
0	0
1	$\frac{1}{2}$
2	2

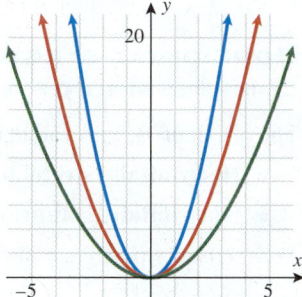

Note that when the function $f(x) = x^2$ is multiplied by 2, so that $g(x) = 2f(x) = 2x^2$, the result is a graph stretched in the vertical direction. When the function $f(x) = x^2$ is multiplied by $\frac{1}{2}$, so that $h(x) = \frac{1}{2}f(x) = \frac{1}{2}x^2$, the result is a graph that is compressed in the vertical direction.

VERTICAL STRETCHING AND VERTICAL COMPRESSING OF GRAPHS

The graph of $cf(x)$ is found by:

- **Vertically stretching** the graph of $f(x)$ if $c > 1$
- **Vertically compressing** the graph of $f(x)$ if $0 < c < 1$

Note: c is any positive real number.

EXAMPLE 6   Vertically Stretching and Compressing Graphs

Graph the function $h(x) = \frac{1}{4}x^3$.

Solution:

1. Start with the cube function. $f(x) = x^3$

2. Vertical compression is expected because $\frac{1}{4}$ is less than 1. $h(x) = \dfrac{1}{4}x^3$

Classroom Example 3.3.6
Graph:

a. $f(x) = \frac{1}{2}|x|$
b. $g(x) = \frac{1}{2}|x - 1|$
c. $h(x) = 2|x|$
d. $j(x) = 2|x - 1|$

Answer: The graphs are given by:

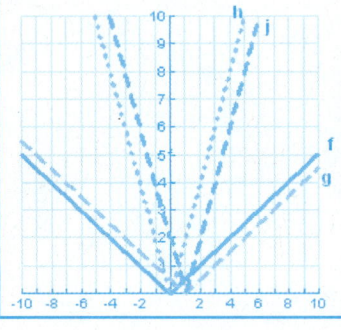

3. Determine a few points that lie on the graph of h.

$$(0, 0) \quad (2, 2) \quad (-2, -2)$$

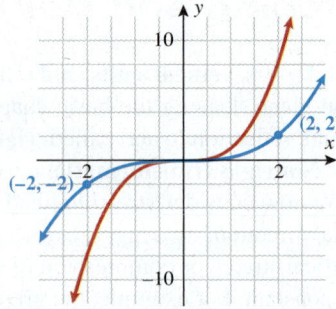

Conversely, if the argument x of a function f is multiplied by a positive real number c, then the result is a *horizontal* stretch of the graph of f if $0 < c < 1$. If $c > 1$, then the result is a *horizontal* compression of the graph of f.

HORIZONTAL STRETCHING AND HORIZONTAL COMPRESSING OF GRAPHS

The graph of $f(cx)$ is found by:

- **Horizontally stretching** the graph of $f(x)$ if $0 < c < 1$
- **Horizontally compressing** the graph of $f(x)$ if $c > 1$

Note: c is any positive real number.

EXAMPLE 7   Vertically Stretching and Horizontally Compressing Graphs

Given the graph of $f(x)$, graph:

a. $2f(x)$ **b.** $f(2x)$

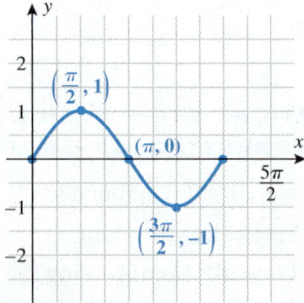

Classroom Example 3.3.7
Graph:

a. $f = \frac{1}{2}\sqrt{x}$
b. $g = \sqrt{\frac{1}{2}x}$
c. $h = \sqrt{3x}$

Answer: The graphs are given by:

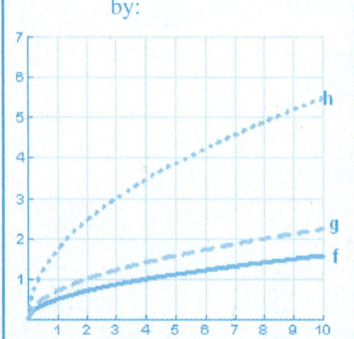

Solution (a):

Since the function is multiplied (on the outside) by 2, the result is that each **y-value** of $f(x)$ is *multiplied* **by 2**, which corresponds to vertical stretching.

Solution (b):

Since the argument of the function is multiplied (on the inside) by 2, the result is that each **x-value** of $f(x)$ is **divided by 2**, which corresponds to horizontal compression.

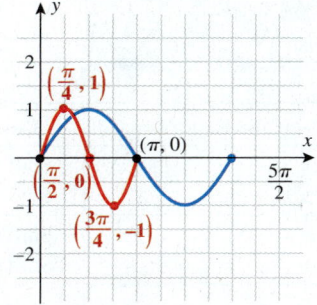

■ **Answer:** *Stretching* of the graph $f(x) = x^3$.

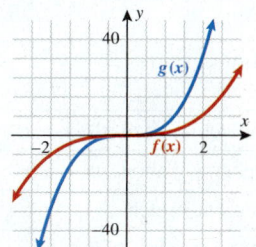

■ **YOUR TURN** Graph the function $g(x) = 4x^3$.

EXAMPLE 8 **Sketching the Graph of a Function Using Multiple Transformations**

Sketch the graph of the function $H(x) = -2(x - 3)^2$.

Solution:

Start with the square function. $f(x) = x^2$

Shift the graph of $f(x)$ to the right three units to arrive at the graph of $f(x - 3)$. $f(x - 3) = (x - 3)^2$

Vertically stretch the graph of $f(x - 3)$ by a factor of 2 to arrive at the graph of $2f(x - 3)$. $2f(x - 3) = 2(x - 3)^2$

Reflect the graph $2f(x - 3)$ about the x-axis to arrive at the graph of $-2f(x - 3)$. $-2f(x - 3) = -2(x - 3)^2$

Technology Tip

Graphs of $y_1 = x^2$, $y_2 = (x - 3)^2$, $y_3 = 2(x - 3)^2$, and $y_4 = H(x) = -2(x - 3)^2$ are shown.

Classroom Example 3.3.8*

Graph:

a. $f = -2\sqrt{3x}$

b. $g = \frac{1}{2}\sqrt{-x} - 1$

c. $h = -\frac{1}{2}\sqrt{\frac{1}{2}x} + 1$

Answer: The graphs are given by:

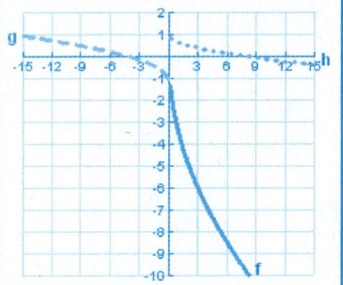

In Example 8 we followed the same "inside out" approach with the functions to determine the order for the transformations: horizontal shift, vertical stretch, and reflection.

SECTION 3.3 SUMMARY

TRANSFORMATION	TO GRAPH THE FUNCTION...	DRAW THE GRAPH OF *f* AND THEN...	DESCRIPTION
Horizontal shifts	$f(x + c)$ $f(x - c)$	Shift the graph of f to the left c units. Shift the graph of f to the right c units.	Replace x by $x + c$. Replace x by $x - c$.
Vertical shifts	$f(x) + c$ $f(x) - c$	Shift the graph of f up c units. Shift the graph of f down c units.	Add c to $f(x)$. Subtract c from $f(x)$.
Reflection about the x-axis	$-f(x)$	Reflect the graph of f about the x-axis.	Multiply $f(x)$ by -1.
Reflection about the y-axis	$f(-x)$	Reflect the graph of f about the y-axis.	Replace x by $-x$.
Vertical stretch	$cf(x)$, where $c > 1$	Vertically stretch the graph of f.	Multiply $f(x)$ by c.
Vertical compression	$cf(x)$, where $0 < c < 1$	Vertically compress the graph of f.	Multiply $f(x)$ by c.
Horizontal stretch	$f(cx)$, where $0 < c < 1$	Horizontally stretch the graph of f.	Replace x by cx.
Horizontal compression	$f(cx)$, where $c > 1$	Horizontally compress the graph of f.	Replace x by cx.

SECTION 3.3 EXERCISES

▪ SKILLS

In Exercises 1–12, match the function to the graph.

1. $f(x) = x^2 + 1$

2. $f(x) = (x - 1)^2$

3. $f(x) = -(1 - x)^2$

4. $f(x) = -x^2 - 1$

5. $f(x) = -(x + 1)^2$

6. $f(x) = -(1 - x)^2 + 1$

7. $f(x) = \sqrt{x - 1} + 1$

8. $f(x) = -\sqrt{x} - 1$

9. $f(x) = \sqrt{1 - x} - 1$

10. $f(x) = \sqrt{-x} + 1$

11. $f(x) = -\sqrt{-x} + 1$

12. $f(x) = -\sqrt{1 - x} - 1$

a.

b.

c.

d.

e.

f.

g.

h.

i.

j.

k.

l.

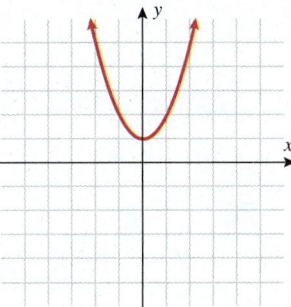

In Exercises 13–18, write the function whose graph is the graph of $y = |x|$, but is transformed accordingly.

13. Shifted up three units

14. Shifted to the left four units

15. Reflected about the y-axis

16. Reflected about the x-axis

17. Vertically stretched by a factor of 3

18. Vertically compressed by a factor of 3

In Exercises 19–24, write the function whose graph is the graph of $y = x^3$, but is transformed accordingly.

19. Shifted down four units

20. Shifted to the right three units

21. Shifted up three units and to the left one unit

22. Reflected about the x-axis

23. Reflected about the y-axis

24. Reflected about both the x-axis and the y-axis

In Exercises 25–48, use the given graph to sketch the graph of the indicated functions.

25.

a. $y = f(x - 2)$
b. $y = f(x) - 2$

26.

a. $y = f(x + 2)$
b. $y = f(x) + 2$

27.

a. $y = f(x) - 3$
b. $y = f(x - 3)$

28.

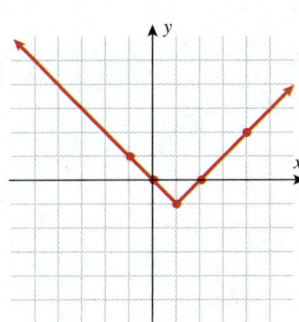

a. $y = f(x) + 3$
b. $y = f(x + 3)$

29.

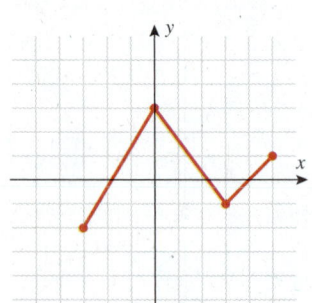

a. $y = -f(x)$
b. $y = f(-x)$

30.

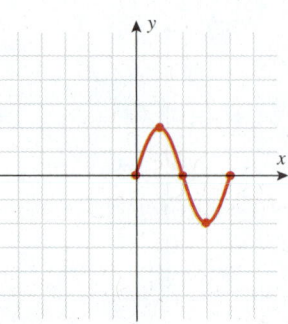

a. $y = -f(x)$
b. $y = f(-x)$

31.

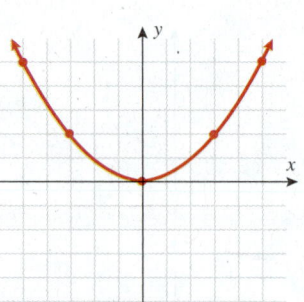

a. $y = 2f(x)$
b. $y = f(2x)$

32.

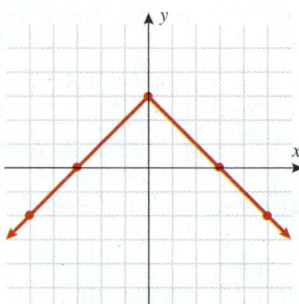

a. $y = 2f(x)$
b. $y = f(2x)$

33. $y = f(x - 2) - 3$

34. $y = f(x + 1) - 2$

35. $y = -f(x - 1) + 2$

36. $y = -2f(x) + 1$

37. $y = -\frac{1}{2}g(x)$

38. $y = \frac{1}{4}g(-x)$

39. $y = -g(2x)$

40. $y = g\left(\frac{1}{2}x\right)$

41. $y = \frac{1}{2}F(x - 1) + 2$

42. $y = \frac{1}{2}F(-x)$

43. $y = -F(1 - x)$

44. $y = -F(x - 2) - 1$

45. $y = 2G(x + 1) - 4$

46. $y = 2G(-x) + 1$

47. $y = -2G(x - 1) + 3$

48. $y = -G(x - 2) - 1$

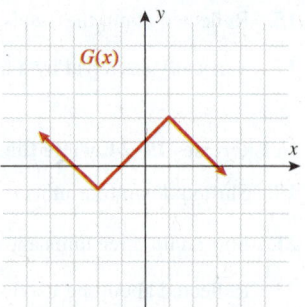

In Exercises 49–74, graph the function using transformations.

49. $y = x^2 - 2$

50. $y = x^2 + 3$

51. $y = (x + 1)^2$

52. $y = (x - 2)^2$

53. $y = (x - 3)^2 + 2$

54. $y = (x + 2)^2 + 1$

55. $y = -(1 - x)^2$

56. $y = -(x + 2)^2$

57. $y = |-x|$

58. $y = -|x|$

59. $y = -|x + 2| - 1$

60. $y = |1 - x| + 2$

61. $y = 2x^2 + 1$

62. $y = 2|x| + 1$

63. $y = -\sqrt{x - 2}$

64. $y = \sqrt{2 - x}$

65. $y = -\sqrt{2 + x} - 1$

66. $y = \sqrt{2 - x} + 3$

67. $y = \sqrt[3]{x - 1} + 2$

68. $y = \sqrt[3]{x + 2} - 1$

69. $y = \dfrac{1}{x + 3} + 2$

70. $y = \dfrac{1}{3 - x}$

71. $y = 2 - \dfrac{1}{x + 2}$

72. $y = 2 - \dfrac{1}{1 - x}$

73. $y = 5\sqrt{-x}$

74. $y = -\frac{1}{5}\sqrt{x}$

In Exercises 75–80, transform the function into the form $f(x) = c(x - h)^2 + k$, where c, k, and h are constants, by completing the square. Use graph-shifting techniques to graph the function.

75. $y = x^2 - 6x + 11$ **76.** $f(x) = x^2 + 2x - 2$ **77.** $f(x) = -x^2 - 2x$ **78.** $f(x) = -x^2 + 6x - 7$

79. $f(x) = 2x^2 - 8x + 3$ **80.** $f(x) = 3x^2 - 6x + 5$

■ APPLICATIONS

81. Salary. A manager hires an employee at a rate of $10 per hour. Write the function that describes the current salary of the employee as a function of the number of hours worked per week, x. After a year, the manager decides to award the employee a raise equivalent to paying him for an additional 5 hours per week. Write a function that describes the salary of the employee after the raise.

82. Profit. The profit associated with St. Augustine sod in Florida is typically $P(x) = -x^2 + 14,000x - 48,700,000$, where x is the number of pallets sold per year in a normal year. In rainy years Sod King gives away 10 free pallets per year. Write the function that describes the profit of x pallets of sod in rainy years.

83. Taxes. Every year in the United States each working American typically pays in taxes a percentage of his or her earnings (minus the standard deduction). Karen's 2011 taxes were calculated based on the formula $T(x) = 0.22(x - 6500)$. That year the standard deduction was $6500 and her tax bracket paid 22% in taxes. Write the function that will determine her 2012 taxes, assuming she receives the raise that places her in the 33% bracket.

84. Medication. The amount of medication that an infant requires is typically a function of the baby's weight. The number of milliliters of an antiseizure medication A is given by $A(x) = \sqrt{x} + 2$, where x is the weight of the infant in ounces. In emergencies there is often not enough time to weigh the infant, so nurses have to estimate the baby's weight. What is the function that represents the actual amount of medication the infant is given if his weight is overestimated by 3 ounces?

For Exercises 85 and 86, refer to the following:

Body Surface Area (BSA) is used in physiology and medicine for many clinical purposes. BSA can be modeled by the function

$$BSA = \sqrt{\frac{wh}{3600}}$$

where w is weight in kilograms and h is height in centimeters. Since BSA depends on weight and height, it is often thought of as a function of both weight and height. However, for an individual adult height is generally considered constant; thus BSA can be thought of as a function of weight alone.

85. Health/Medicine. (a) If an adult female is 162 centimeters tall, find her BSA as a function of weight. (b) If she loses 3 kilograms, find a function that represents her new BSA.

86. Health/Medicine. (a) If an adult male is 180 centimeters tall, find his BSA as a function of weight. (b) If he gains 5 kilograms, find a function that represents his new BSA.

■ CATCH THE MISTAKE

In Exercises 87–90, explain the mistake that is made.

87. Describe a procedure for graphing the function $f(x) = \sqrt{x - 3} + 2$.

Solution:
a. Start with the function $f(x) = \sqrt{x}$.
b. Shift the function to the left three units.
c. Shift the function up two units.

This is incorrect. What mistake was made?

88. Describe a procedure for graphing the function $f(x) = -\sqrt{x + 2} - 3$.

Solution:
a. Start with the function $f(x) = \sqrt{x}$.
b. Shift the function to the left two units.
c. Reflect the function about the y-axis.
d. Shift the function down three units.

This is incorrect. What mistake was made?

89. Describe a procedure for graphing the function $f(x) = |3 - x| + 1$.

Solution:
a. Start with the function $f(x) = |x|$.
b. Reflect the function about the y-axis.
c. Shift the function to the left three units.
d. Shift the function up one unit.

This is incorrect. What mistake was made?

90. Describe a procedure for graphing the function $f(x) = -2x^2 + 1$.

Solution:
a. Start with the function $f(x) = x^2$.
b. Reflect the function about the y-axis.
c. Shift the function up one unit.
d. Expand in the vertical direction by a factor of 2.

This is incorrect. What mistake was made?

■ **CONCEPTUAL**

In Exercises 91–94, determine whether each statement is true or false.

91. The graph of $y = |-x|$ is the same as the graph of $y = |x|$.

92. The graph of $y = \sqrt{-x}$ is the same as the graph of $y = \sqrt{x}$.

93. If the graph of an odd function is reflected about the x-axis and then the y-axis, the result is the graph of the original odd function.

94. If the graph of $y = \frac{1}{x}$ is reflected about the x-axis, it produces the same graph as if it had been reflected about the y-axis.

■ **CHALLENGE**

95. The point (a, b) lies on the graph of the function $y = f(x)$. What point is guaranteed to lie on the graph of $f(x - 3) + 2$?

96. The point (a, b) lies on the graph of the function $y = f(x)$. What point is guaranteed to lie on the graph of $-f(-x) + 1$?

■ **TECHNOLOGY**

97. Use a graphing utility to graph:

a. $y = x^2 - 2$ and $y = |x^2 - 2|$

b. $y = x^3 - 1$ and $y = |x^3 + 1|$

What is the relationship between $f(x)$ and $|f(x)|$?

98. Use a graphing utility to graph:

a. $y = x^2 - 2$ and $y = |x|^2 - 2$

b. $y = x^3 + 1$ and $y = |x|^3 + 1$

What is the relationship between $f(x)$ and $f(|x|)$?

99. Use a graphing utility to graph:

a. $y = \sqrt{x}$ and $y = \sqrt{0.1x}$

b. $y = \sqrt{x}$ and $y = \sqrt{10x}$

What is the relationship between $f(x)$ and $f(ax)$, assuming that a is positive?

100. Use a graphing utility to graph:

a. $y = \sqrt{x}$ and $y = 0.1\sqrt{x}$

b. $y = \sqrt{x}$ and $y = 10\sqrt{x}$

What is the relationship between $f(x)$ and $af(x)$, assuming that a is positive?

101. Use a graphing utility to graph $y = f(x) = [[0.5x]] + 1$. Use transforms to describe the relationship between $f(x)$ and $y = [[x]]$.

102. Use a graphing utility to graph $y = g(x) = 0.5[[x]] + 1$. Use transforms to describe the relationship between $g(x)$ and $y = [[x]]$.

SKILLS OBJECTIVES

- Add, subtract, multiply, and divide functions.
- Evaluate composite functions.
- Determine the domain of functions resulting from operations on and composition of functions.

CONCEPTUAL OBJECTIVES

- Understand domain restrictions when dividing functions.
- Realize that the domain of a composition of functions excludes values that are not in the domain of the inside function.

Two different functions can be combined using mathematical operations such as addition, subtraction, multiplication, and division. Also, there is an operation on functions called composition, which can be thought of as a function of a function. When we combine functions, we do so algebraically. Special attention must be paid to the domain and range of the combined functions.

Adding, Subtracting, Multiplying, and Dividing Functions

Consider the two functions $f(x) = x^2 + 2x - 3$ and $g(x) = x + 1$. The domain of both of these functions is the set of all real numbers. Therefore, we can add, subtract, or multiply these functions for any real number x.

$$\text{Addition: } f(x) + g(x) = x^2 + 2x - 3 + x + 1 = x^2 + 3x - 2$$

The result is in fact a new function, which we denote:

$$(f + g)(x) = x^2 + 3x - 2 \qquad \text{This is the \textbf{sum function}.}$$

$$\text{Subtraction: } f(x) - g(x) = x^2 + 2x - 3 - (x + 1) = x^2 + x - 4$$

The result is in fact a new function, which we denote:

$$(f - g)(x) = x^2 + x - 4 \qquad \text{This is the \textbf{difference function}.}$$

$$\text{Multiplication: } f(x) \cdot g(x) = (x^2 + 2x - 3)(x + 1) = x^3 + 3x^2 - x - 3$$

The result is in fact a new function, which we denote:

$$(f \cdot g)(x) = x^3 + 3x^2 - x - 3 \qquad \text{This is the \textbf{product function}.}$$

Although both f and g are defined for all real numbers x, we must restrict x so that $x \neq -1$ to form the quotient $\dfrac{f}{g}$.

$$\text{Division: } \frac{f(x)}{g(x)} = \frac{x^2 + 2x - 3}{x + 1}, \quad x \neq -1$$

The result is in fact a new function, which we denote:

$$\left(\frac{f}{g}\right)(x) = \frac{x^2 + 2x - 3}{x + 1}, \quad x \neq -1 \qquad \text{This is called the \textbf{quotient function}.}$$

Two functions can be added, subtracted, and multiplied. The resulting function domain is therefore the intersection of the domains of the two functions. However, for division, any value of x (input) that makes the denominator equal to zero must be eliminated from the domain.

The previous examples involved polynomials. The domain of any polynomial is the set of all real numbers. Adding, subtracting, and multiplying polynomials result in other polynomials, which have domains of all real numbers. Let's now investigate operations applied to functions that have a restricted domain.

The domain of the sum function, difference function, or product function is the *intersection* of the individual domains of the two functions. The quotient function has a similar domain in that it is the intersection of the two domains. However, any values that make the denominator zero must also be eliminated.

Function	Notation	Domain
Sum	$(f + g)(x) = f(x) + g(x)$	{domain of f} $\cap$ {domain of g}
Difference	$(f - g)(x) = f(x) - g(x)$	{domain of f} $\cap$ {domain of g}
Product	$(f \cdot g)(x) = f(x) \cdot g(x)$	{domain of f} $\cap$ {domain of g}
Quotient	$\left(\dfrac{f}{g}\right)(x) = \dfrac{f(x)}{g(x)}$	{domain of f} $\cap$ {domain of g} $\cap$ {$g(x) \neq 0$}

We can think of this in the following way: Any number that is in the domain of *both* the functions is in the domain of the combined function. The exception to this is the quotient function, which also eliminates values that make the denominator equal to zero.

Classroom Example 3.4.1

Let $f(x) = 1 + 2\sqrt[3]{x}$ and $g(x) = -4\sqrt[3]{x}$. Compute $f + g, f - g, fg,$ and $\dfrac{f}{g}$. State the domains.

Answer:

$f + g = 1 - 2\sqrt[3]{x}$
Domain: $(-\infty, \infty)$,

$f - g = 1 + 6\sqrt[3]{x}$
Domain: $(-\infty, \infty)$,

$fg = -4\sqrt[3]{x} - 8\sqrt[3]{x^2}$
Domain: $(-\infty, \infty)$,

$\dfrac{f}{g} = \dfrac{1 + 2\sqrt[3]{x}}{-4\sqrt[3]{x}}$
Domain: $x \neq 0$.

EXAMPLE 1 Operations on Functions: Determining Domains of New Functions

For the functions $f(x) = \sqrt{x - 1}$ and $g(x) = \sqrt{4 - x}$, determine the sum function, difference function, product function, and quotient function. State the domain of these four new functions.

Solution:

Sum function: $f(x) + g(x) = \sqrt{x - 1} + \sqrt{4 - x}$

Difference function: $f(x) - g(x) = \sqrt{x - 1} - \sqrt{4 - x}$

Product function: $f(x) \cdot g(x) = \sqrt{x - 1} \cdot \sqrt{4 - x}$
$$= \sqrt{(x - 1)(4 - x)} = \sqrt{-x^2 + 5x - 4}$$

Quotient function: $\dfrac{f(x)}{g(x)} = \dfrac{\sqrt{x - 1}}{\sqrt{4 - x}} = \sqrt{\dfrac{x - 1}{4 - x}}$

The domain of the square root function is determined by setting the argument under the radical greater than or equal to zero.

$$\text{Domain of } f(x): \ [1, \infty)$$

$$\text{Domain of } g(x): \ (-\infty, 4]$$

The domain of the sum, difference, and product functions is

$$[1, \infty) \cap (-\infty, 4] = [1, 4]$$

The quotient function has the additional constraint that the denominator cannot be zero. This implies that $x \neq 4$, so the domain of the quotient function is $[1, 4)$.

Answer:
$(f + g)(x) = \sqrt{x + 3} + \sqrt{1 - x}$
Domain: $[-3, 1]$

■ **YOUR TURN** Given the function $f(x) = \sqrt{x + 3}$ and $g(x) = \sqrt{1 - x}$, find $(f + g)(x)$ and state its domain.

EXAMPLE 2   Quotient Function and Domain Restrictions

Given the functions $F(x) = \sqrt{x}$ and $G(x) = |x - 3|$, find the quotient function, $\left(\dfrac{F}{G}\right)(x)$, and state its domain.

Solution:

The quotient function is written as

$$\left(\frac{F}{G}\right)(x) = \frac{F(x)}{G(x)} = \frac{\sqrt{x}}{|x - 3|}$$

Classroom Example 3.4.2

Let $f(x) = 3\sqrt{x}$ and

$g(x) = 3|x| - 6$. Form $\dfrac{f}{g}$ and

state the domain.

Answer: $\dfrac{f}{g} = \dfrac{\sqrt{x}}{|x| - 2}$

Domain: $[0, 2) \cup (2, \infty)$

Domain of $F(x)$: $[0, \infty)$ Domain of $G(x)$: $(-\infty, \infty)$

The real numbers that are in both the domain for $F(x)$ and the domain for $G(x)$ are represented by the intersection $[0, \infty) \cap (-\infty, \infty) = [0, \infty)$. Also, the denominator of the quotient function is equal to zero when $x = 3$, so we must eliminate this value from the domain.

$$\boxed{\text{Domain of } \left(\frac{F}{G}\right)(x): \ [0, 3) \cup (3, \infty)}$$

■ YOUR TURN For the functions given in Example 2, determine the quotient

function $\left(\dfrac{G}{F}\right)(x)$, and state its domain.

The graphs of $y_1 = F(x) = \sqrt{x}$,

$y_2 = G(x) = |x - 3|$, and

$y_3 = \dfrac{F(x)}{G(x)} = \dfrac{\sqrt{x}}{|x - 3|}$ are shown.

■ Answer:

$\left(\dfrac{G}{F}\right)(x) = \dfrac{G(x)}{F(x)} = \dfrac{|x - 3|}{\sqrt{x}}$

Domain: $(0, \infty)$

Composition of Functions

Recall that a function maps every element in the domain to exactly one corresponding element in the range as shown in the figure on the right.

Suppose there is a sales rack of clothes in a department store. Let x correspond to the original price of each item on the rack. These clothes have recently been marked down 20%. Therefore, the function $g(x) = 0.80x$ represents the current sale price of each item. You have been invited to a special sale that lets you take 10% off the current sale price and an additional $5 off every item at checkout. The function $f(g(x)) = 0.90g(x) - 5$ determines the checkout price. Note that the output of the function g is the input of the function f as shown in the figure below.

Domain   Range

Domain of f Range of f

Domain of g Range of g

Original price   Sale price 20%   Additional 10% off
                       off original price   sale price and $5 off
                                                at checkout

This is an example of a **composition of functions**, when the output of one function is the input of another function. It is commonly referred to as a function of a function.

An algebraic example of this is the function $y = \sqrt{x^2 - 2}$. Suppose we let $g(x) = x^2 - 2$ and $f(x) = \sqrt{x}$. Recall that the independent variable in function notation is a placeholder. Since $f(\square) = \sqrt{(\square)}$, then $f(g(x)) = \sqrt{(g(x))}$. Substituting the expression for $g(x)$, we find $f(g(x)) = \sqrt{x^2 - 2}$. The function $y = \sqrt{x^2 - 2}$ is said to be a composite function, $y = f(g(x))$.

Note that the domain of $g(x)$ is the set of all real numbers, and the domain of $f(x)$ is the set of all nonnegative numbers. The domain of a composite function is the set of all x such that $g(x)$ is in the domain of f. For instance, in the composite function $y = f(g(x))$, we know that the allowable inputs into f are all numbers greater than or equal to zero. Therefore, we restrict the outputs of $g(x) \geq 0$ and find the corresponding x-values. Those x-values are the only allowable inputs and constitute the domain of the composite function $y = f(g(x))$.

The symbol that represents composition of functions is a small open circle; thus $(f \circ g)(x) = f(g(x))$ and is read aloud as "f of g." It is important not to confuse this with the multiplication sign: $(f \cdot g)(x) = f(x)g(x)$.

▼ **CAUTION**

$f \circ g \neq f \cdot g$

COMPOSITION OF FUNCTIONS

Given two functions f and g, there are two **composite functions** that can be formed.

NOTATION	WORDS	DEFINITION	DOMAIN
$f \circ g$	f composed with g	$f(g(x))$	The set of all real numbers x in the domain of g such that $g(x)$ is also in the domain of f.
$g \circ f$	g composed with f	$g(f(x))$	The set of all real numbers x in the domain of f such that $f(x)$ is also in the domain of g.

Study Tip

Order is important:

$(f \circ g)(x) = f(g(x))$
$(g \circ f)(x) = g(f(x))$

It is important to realize that there are two "filters" that allow certain values of x into the domain. The first filter is $g(x)$. If x is not in the domain of $g(x)$, it cannot be in the domain of $(f \circ g)(x) = f(g(x))$. Of those values for x that are in the domain of $g(x)$, only some pass through, because we restrict the output of $g(x)$ to values that are allowable as input into f. This adds an additional filter.

Study Tip

The domain of $f \circ g$ is always a subset of the domain of g, and the range of $f \circ g$ is always a subset of the range of f.

The domain of $f \circ g$ is always a subset of the domain of g, and the range of $f \circ g$ is always a subset of the range of f.

$(f \circ g)(x) = f(g(x))$

▶ **EXAMPLE 3 Finding a Composite Function**

Given the functions $f(x) = x^2 + 1$ and $g(x) = x - 3$, find $(f \circ g)(x)$.

Solution:

Write $f(x)$ using placeholder notation. $f(\square) = (\square)^2 + 1$

Express the composite function $f \circ g$. $f(g(x)) = (g(x))^2 + 1$

Substitute $g(x) = x - 3$ into f. $f(g(x)) = (x - 3)^2 + 1$

Eliminate the parentheses on the right side. $f(g(x)) = x^2 - 6x + 10$

$$(f \circ g)(x) = f(g(x)) = x^2 - 6x + 10$$

■ **Answer:** $g \circ f = g(f(x)) = x^2 - 2$

■ **YOUR TURN** Given the functions in Example 3, find $(g \circ f)(x)$.

EXAMPLE 4   Determining the Domain of a Composite Function

Given the functions $f(x) = \dfrac{1}{x-1}$ and $g(x) = \dfrac{1}{x}$, determine $f \circ g$, and state its domain.

Solution:

Write $f(x)$ using placeholder notation. $f(\square) = \dfrac{1}{(\square) - 1}$

Express the composite function $f \circ g$. $f(g(x)) = \dfrac{1}{g(x) - 1}$

Substitute $g(x) = \dfrac{1}{x}$ into f. $f(g(x)) = \dfrac{1}{\dfrac{1}{x} - 1}$

Multiply the right side by $\dfrac{x}{x}$. $f(g(x)) = \dfrac{1}{\dfrac{1}{x} - 1} \cdot \dfrac{x}{x} = \dfrac{x}{1 - x}$

$$(f \circ g) = f(g(x)) = \dfrac{x}{1 - x}$$

What is the domain of $(f \circ g)(x) = f(g(x))$? By inspecting the final result of $f(g(x))$, we see that the denominator is zero when $x = 1$. Therefore, $x \neq 1$. Are there any other values for x that are not allowed? The function $g(x)$ has the domain $x \neq 0$; therefore we must also exclude zero.
The domain of $(f \circ g)(x) = f(g(x))$ excludes $x = 0$ and $x = 1$ or, in interval notation,

$$(-\infty, 0) \cup (0, 1) \cup (1, \infty)$$

■ **YOUR TURN**   For the functions f and g given in Example 4, determine the composite function $g \circ f$ and state its domain.

The domain of the composite function cannot always be determined by examining the final form of $f \circ g$.

EXAMPLE 5   Determining the Domain of a Composite Function (Without Finding the Composite Function)

Let $f(x) = \dfrac{1}{x - 2}$ and $g(x) = \sqrt{x + 3}$. Find the domain of $f(g(x))$. Do not find the composite function.

Solution:

Find the domain of g. $[-3, \infty)$

Find the range of g. $[0, \infty)$

In $f(g(x))$, the output of g becomes the input for f. Since the domain of f is the set of all real numbers except 2, we eliminate any values of x in the domain of g that correspond to $g(x) = 2$.

Let $g(x) = 2$. $\sqrt{x + 3} = 2$

Square both sides. $x + 3 = 4$

Solve for x. $x = 1$

Eliminate $x = 1$ from the domain of g, $[-3, \infty)$.

State the domain of $f(g(x))$. $\boxed{[-3, 1) \cup (1, \infty)}$

Technology Tip

The graphs of $y_1 = f(x) = \dfrac{1}{x-1}$,

$y_2 = g(x) = \dfrac{1}{x}$, and $y_3 = (f \circ g)(x)$

$= \dfrac{1}{1/x - 1} = \dfrac{x}{1 - x}$ are shown.

■ **Answer:** $g(f(x)) = x - 1$. Domain of $g \circ f$ is $x \neq 1$, or in interval notation, $(-\infty, 1) \cup (1, \infty)$.

▼ **CAUTION**

The domain of the composite function cannot always be determined by examining the final form of $f \circ g$.

Classroom Example 3.4.4*
State the domains for $f \circ g$ and $g \circ f$ in Classroom Example 3.4.3.

Answer:
Domain $f \circ g = [-2, 2]$
Domain $g \circ f = [0, \infty)$

Classroom Example 3.4.5*
Let $f(x) = \sqrt{1 + 2x}$ and $g(x) = \frac{1}{2}|x| - \frac{1}{4}$. What is the domain of $f \circ g$?

Answer: $\mathbb{R}$

Classroom Example 3.4.6
Let $f(x) = 2\sqrt{x}$ and
$g(x) = -x^2 + 4$. Calculate, if
possible:
a. $(f \circ g)(-1)$
b. $(g \circ f)(-1)$
c. $(f \circ g)(2)$
d. $g(f(-3))$

Answer:
a. $2\sqrt{3}$
b. not possible
c. 0
d. not possible

▶ **EXAMPLE 6**   **Evaluating a Composite Function**

Given the functions $f(x) = x^2 - 7$ and $g(x) = 5 - x^2$, evaluate:

a. $f(g(1))$ **b.** $f(g(-2))$ **c.** $g(f(3))$ **d.** $g(f(-4))$

Solution:

One way of evaluating these composite functions is to calculate the two individual composites in terms of x: $f(g(x))$ and $g(f(x))$. Once those functions are known, the values can be substituted for x and evaluated.

Another way of proceeding is as follows:

a. Write the desired quantity. $f(g(1))$
    Find the value of the inner function g. $g(1) = 5 - 1^2 = 4$
    Substitute $g(1) = 4$ into f. $f(g(1)) = f(4)$
    Evaluate $f(4)$. $f(4) = 4^2 - 7 = 9$

$$\boxed{f(g(1)) = 9}$$

b. Write the desired quantity. $f(g(-2))$
    Find the value of the inner function g. $g(-2) = 5 - (-2)^2 = 1$
    Substitute $g(-2) = 1$ into f. $f(g(-2)) = f(1)$
    Evaluate $f(1)$. $f(1) = 1^2 - 7 = -6$

$$\boxed{f(g(-2)) = -6}$$

c. Write the desired quantity. $g(f(3))$
    Find the value of the inner function f. $f(3) = 3^2 - 7 = 2$
    Substitute $f(3) = 2$ into g. $g(f(3)) = g(2)$
    Evaluate $g(2)$. $g(2) = 5 - 2^2 = 1$

$$\boxed{g(f(3)) = 1}$$

d. Write the desired quantity. $g(f(-4))$
    Find the value of the inner function f. $f(-4) = (-4)^2 - 7 = 9$
    Substitute $f(-4) = 9$ into g. $g(f(-4)) = g(9)$
    Evaluate $g(9)$. $g(9) = 5 - 9^2 = -76$

$$\boxed{g(f(-4)) = -76}$$

■ **Answer:** $f(g(1)) = 5$ and
          $g(f(1)) = -7$

■ **YOUR TURN** Given the functions $f(x) = x^3 - 3$ and $g(x) = 1 + x^3$, evaluate $f(g(1))$ and $g(f(1))$.

Application Problems

Recall the example at the beginning of this chapter regarding the clothes that are on sale. Often, real-world applications are modeled with composite functions. In the clothes example, x is the original price of each item. The first function maps its input (original price) to an output (sale price). The second function maps its input (sale price) to an output (checkout price). Example 7 is another real-world application of composite functions.

Three temperature scales are commonly used:

■ The degree Celsius (°C) scale
  ● This scale was devised by dividing the range between the freezing (0°C) and boiling (100°C) points of pure water at sea level into 100 equal parts. This scale is used in science and is one of the standards of the "metric" (SI) system of measurements.

■ The Kelvin (K) temperature scale
  ● This scale shifts the Celsius scale down so that the zero point is equal to absolute zero (about $-273.15°C$), a hypothetical temperature at which there is a complete absence of heat energy.
  ● Temperatures on this scale are called **kelvins**, *not* degrees kelvin, and kelvin is not capitalized. The symbol for the kelvin is K.
■ The degree Fahrenheit (°F) scale
  ● This scale evolved over time and is still widely used mainly in the United States, although Celsius is the preferred "metric" scale.
  ● With respect to pure water at sea level, the **degrees Fahrenheit** are gauged by the spread from 32°F (freezing) to 212°F (boiling).

The equations that relate these temperature scales are

$$F = \frac{9}{5}C + 32 \qquad C = K - 273.15$$

EXAMPLE 7   Applications Involving Composite Functions

Determine degrees Fahrenheit as a function of kelvins.

Solution:

Degrees Fahrenheit is a function of degrees Celsius.

$$F = \frac{9}{5}C + 32$$

Now substitute $C = K - 273.15$ into the equation for F.

$$F = \frac{9}{5}(K - 273.15) + 32$$

Simplify.

$$F = \frac{9}{5}K - 491.67 + 32$$

$$\boxed{F = \frac{9}{5}K - 459.67}$$

SECTION 3.4 SUMMARY

Operations on Functions

Function	Notation
Sum	$(f + g)(x) = f(x) + g(x)$
Difference	$(f - g)(x) = f(x) - g(x)$
Product	$(f \cdot g)(x) = f(x) \cdot g(x)$
Quotient	$\left(\dfrac{f}{g}\right)(x) = \dfrac{f(x)}{g(x)} \qquad g(x) \neq 0$

The domain of the sum, difference, and product functions is the intersection of the domains, or common domain shared by both f and g. The domain of the quotient function is also the intersection of the domain shared by both f and g with an additional restriction that $g(x) \neq 0$.

Composition of Functions

$$(f \circ g)(x) = f(g(x))$$

The domain restrictions cannot always be determined simply by inspecting the final form of $f(g(x))$. Rather, the domain of the composite function is a subset of the domain of $g(x)$. Values of x must be eliminated if their corresponding values of $g(x)$ are not in the domain of f.

SECTION
3.4 EXERCISES

SKILLS

In Exercises 1–10, given the functions f and g, find $f + g$, $f - g$, $f \cdot g$, and $\dfrac{f}{g}$, and state the domain of each.

1. $f(x) = 2x + 1$

$g(x) = 1 - x$

2. $f(x) = 3x + 2$

$g(x) = 2x - 4$

3. $f(x) = 2x^2 - x$

$g(x) = x^2 - 4$

4. $f(x) = 3x + 2$

$g(x) = x^2 - 25$

5. $f(x) = \dfrac{1}{x}$

$g(x) = x$

6. $f(x) = \dfrac{2x + 3}{x - 4}$

$g(x) = \dfrac{x - 4}{3x + 2}$

7. $f(x) = \sqrt{x}$

$g(x) = 2\sqrt{x}$

8. $f(x) = \sqrt{x - 1}$

$g(x) = 2x^2$

9. $f(x) = \sqrt{4 - x}$

$g(x) = \sqrt{x + 3}$

10. $f(x) = \sqrt{1 - 2x}$

$g(x) = \dfrac{1}{x}$

In Exercises 11–20, for the given functions f and g, find the composite functions $f \circ g$ and $g \circ f$, and state their domains.

11. $f(x) = 2x + 1$

$g(x) = x^2 - 3$

12. $f(x) = x^2 - 1$

$g(x) = 2 - x$

13. $f(x) = \dfrac{1}{x - 1}$

$g(x) = x + 2$

14. $f(x) = \dfrac{2}{x - 3}$

$g(x) = 2 + x$

15. $f(x) = |x|$

$g(x) = \dfrac{1}{x - 1}$

16. $f(x) = |x - 1|$

$g(x) = \dfrac{1}{x}$

17. $f(x) = \sqrt{x - 1}$

$g(x) = x + 5$

18. $f(x) = \sqrt{2 - x}$

$g(x) = x^2 + 2$

19. $f(x) = x^3 + 4$

$g(x) = (x - 4)^{1/3}$

20. $f(x) = \sqrt[3]{x^2 - 1}$

$g(x) = x^{2/3} + 1$

In Exercises 21–38, evaluate the functions for the specified values, if possible.

$$f(x) = x^2 + 10 \qquad g(x) = \sqrt{x - 1}$$

21. $(f + g)(2)$

22. $(f + g)(10)$

23. $(f - g)(2)$

24. $(f - g)(5)$

25. $(f \cdot g)(4)$

26. $(f \cdot g)(5)$

27. $\left(\dfrac{f}{g}\right)(10)$

28. $\left(\dfrac{f}{g}\right)(2)$

29. $f(g(2))$

30. $f(g(1))$

31. $g(f(-3))$

32. $g(f(4))$

33. $f(g(0))$

34. $g(f(0))$

35. $f(g(-3))$

36. $g(f(\sqrt{7}))$

37. $(f \circ g)(4)$

38. $(g \circ f)(-3)$

In Exercises 39–50, evaluate $f(g(1))$ and $g(f(2))$, if possible.

39. $f(x) = \dfrac{1}{x}$, $\quad g(x) = 2x + 1$

40. $f(x) = x^2 + 1$, $\quad g(x) = \dfrac{1}{2 - x}$

41. $f(x) = \sqrt{1 - x}$, $\quad g(x) = x^2 + 2$

42. $f(x) = \sqrt{3 - x}$, $\quad g(x) = x^2 + 1$

43. $f(x) = \dfrac{1}{|x - 1|}$, $\quad g(x) = x + 3$

44. $f(x) = \dfrac{1}{x}$, $\quad g(x) = |2x - 3|$

45. $f(x) = \sqrt{x - 1}$, $\quad g(x) = x^2 + 5$

46. $f(x) = \sqrt[3]{x - 3}$, $\quad g(x) = \dfrac{1}{x - 3}$

47. $f(x) = \dfrac{1}{x^2 - 3}$, $\quad g(x) = \sqrt{x - 3}$

48. $f(x) = \dfrac{x}{2 - x}$, $\quad g(x) = 4 - x^2$

49. $f(x) = (x - 1)^{1/3}$, $\quad g(x) = x^2 + 2x + 1$

50. $f(x) = (1 - x^2)^{1/2}$, $\quad g(x) = (x - 3)^{1/3}$

In Exercises 51–60, show that $f(g(x)) = x$ and $g(f(x)) = x$.

51. $f(x) = 2x + 1$, $g(x) = \dfrac{x - 1}{2}$

52. $f(x) = \dfrac{x - 2}{3}$, $g(x) = 3x + 2$

53. $f(x) = \sqrt{x - 1}$, $g(x) = x^2 + 1$ for $x \geq 1$

54. $f(x) = 2 - x^2$, $g(x) = \sqrt{2 - x}$ for $x \leq 2$

55. $f(x) = \dfrac{1}{x}$, $g(x) = \dfrac{1}{x}$ for $x \neq 0$

56. $f(x) = (5 - x)^{1/3}$, $g(x) = 5 - x^3$

57. $f(x) = 4x^2 - 9$, $g(x) = \dfrac{\sqrt{x + 9}}{2}$ for $x \geq 0$

58. $f(x) = \sqrt[3]{8x - 1}$, $g(x) = \dfrac{x^3 + 1}{8}$

59. $f(x) = \dfrac{1}{x - 1}$, $g(x) = \dfrac{x + 1}{x}$ for $x \neq 0$, $x \neq 1$

60. $f(x) = \sqrt{25 - x^2}$, $g(x) = \sqrt{25 - x^2}$ for $0 \leq x \leq 5$

In Exercises 61–66, write the function as a composite of two functions f and g. (More than one answer is correct.)

61. $f(g(x)) = 2(3x - 1)^2 + 5(3x - 1)$

62. $f(g(x)) = \dfrac{1}{1 + x^2}$

63. $f(g(x)) = \dfrac{2}{|x - 3|}$

64. $f(g(x)) = \sqrt{1 - x^2}$

65. $f(g(x)) = \dfrac{3}{\sqrt{x + 1} - 2}$

66. $f(g(x)) = \dfrac{\sqrt{x}}{3\sqrt{x} + 2}$

■ **APPLICATIONS**

Exercises 67 and 68 depend on the relationship between degrees Fahrenheit, degrees Celsius, and kelvins:

$$F = \frac{9}{5}C + 32 \qquad C = K - 273.15$$

67. **Temperature.** Write a composite function that converts kelvins into degrees Fahrenheit.

68. **Temperature.** Convert the following degrees Fahrenheit to kelvins: 32°F and 212°F.

69. **Dog Run.** Suppose that you want to build a *square* fenced-in area for your dog. Fencing is purchased in linear feet.
  a. Write a composite function that determines the area of your dog pen as a function of how many linear feet are purchased.
  b. If you purchase 100 linear feet, what is the area of your dog pen?
  c. If you purchase 200 linear feet, what is the area of your dog pen?

70. **Dog Run.** Suppose that you want to build a *circular* fenced-in area for your dog. Fencing is purchased in linear feet.
  a. Write a composite function that determines the area of your dog pen as a function of how many linear feet are purchased.
  b. If you purchase 100 linear feet, what is the area of your dog pen?
  c. If you purchase 200 linear feet, what is the area of your dog pen?

71. **Market Price.** Typical supply and demand relationships state that as the number of units for sale increases, the market price decreases. Assume that the market price p and the number of units for sale x are related by the demand equation:

$$p = 3000 - \frac{1}{2}x$$

Assume that the cost $C(x)$ of producing x items is governed by the equation

$$C(x) = 2000 + 10x$$

and the revenue $R(x)$ generated by selling x units is governed by

$$R(x) = 100x$$

  a. Write the cost as a function of price p.
  b. Write the revenue as a function of price p.
  c. Write the profit as a function of price p.

72. **Market Price.** Typical supply and demand relationships state that as the number of units for sale increases, the market price decreases. Assume that the market price p and the number of units for sale x are related by the demand equation:

$$p = 10{,}000 - \frac{1}{4}x$$

Assume that the cost $C(x)$ of producing x items is governed by the equation

$$C(x) = 30{,}000 + 5x$$

and the revenue $R(x)$ generated by selling x units is governed by

$$R(x) = 1000x$$

  a. Write the cost as a function of price p.
  b. Write the revenue as a function of price p.
  c. Write the profit as a function of price p.

In Exercises 73 and 74, refer to the following:

The cost of manufacturing a product is a function of the number of hours t the assembly line is running per day. The number of products manufactured n is a function of the number of hours t the assembly line is operating and is given by the function $n(t)$. The cost of manufacturing the product C measured in thousands of dollars is a function of the quantity manufactured, that is, the function $C(n)$.

73. **Business.** If the quantity of a product manufactured during a day is given by
$$n(t) = 50t - t^2$$
and the cost of manufacturing the product is given by
$$C(n) = 10n + 1375$$
   a. Find a function that gives the cost of manufacturing the product in terms of the number of hours t the assembly line was functioning, $C(n(t))$.
   b. Find the cost of production on a day when the assembly line was running for 16 hours. Interpret your answer.

74. **Business.** If the quantity of a product manufactured during a day is given by
$$n(t) = 100t - 4t^2$$
and the cost of manufacturing the product is given by
$$C(n) = 8n + 2375$$
   a. Find a function that gives the cost of manufacturing the product in terms of the number of hours t the assembly line was functioning, $C(n(t))$.
   b. Find the cost of production on a day when the assembly line was running for 24 hours. Interpret your answer.

In Exercises 75 and 76, refer to the following:

Surveys performed immediately following an accidental oil spill at sea indicate the oil moved outward from the source of the spill in a nearly circular pattern. The radius of the oil spill r measured in miles is a function of time t measured in days from the start of the spill, while the area of the oil spill is a function of radius, that is, the function $A(r)$.

75. **Environment: Oil Spill.** If the radius of the oil spill is given by
$$r(t) = 10t - 0.2t^2$$
and the area of the oil spill is given by
$$A(r) = \pi r^2$$
   a. Find a function that gives the area of the oil spill in terms of the number of days since the start of the spill, $A(r(t))$.
   b. Find the area of the oil spill to the nearest square mile 7 days after the start of the spill.

76. **Environment: Oil Spill.** If the radius of the oil spill is given by
$$r(t) = 8t - 0.1t^2$$
and the area of the oil spill is given by
$$A(r) = \pi r^2$$
   a. Find a function that gives the area of the oil spill in terms of the number of days since the start of the spill, $A(r(t))$.
   b. Find the area of the oil spill to the nearest square mile 5 days after the start of the spill.

77. **Environment: Oil Spill.** An oil spill makes a circular pattern around a ship such that the radius in feet grows as a function of time in hours $r(t) = 150\sqrt{t}$. Find the area of the spill as a function of time.

78. **Pool Volume.** A 20 foot by 10 foot rectangular pool has been built. If 50 cubic feet of water is pumped into the pool per hour, write the water-level height (feet) as a function of time (hours).

79. **Fireworks.** A family is watching a fireworks display. If the family is 2 miles from where the fireworks are being launched and the fireworks travel vertically, what is the distance between the family and the fireworks as a function of height above ground?

80. **Real Estate.** A couple are about to put their house up for sale. They bought the house for $172,000 a few years ago, and when they list it with a realtor they will pay a 6% commission. Write a function that represents the amount of money they will make on their home as a function of the asking price p.

■ **CATCH THE MISTAKE**

In Exercises 81–86, for the functions $f(x) = x + 2$ and $g(x) = x^2 - 4$, find the indicated function and state its domain. Explain the mistake that is made in each problem.

81. $\dfrac{g}{f}$

   Solution:
   $$\frac{g(x)}{f(x)} = \frac{x^2 - 4}{x + 2}$$
   $$= \frac{(x - 2)(x + 2)}{x + 2}$$
   $$= x - 2$$

   Domain: $(-\infty, \infty)$

   This is incorrect. What mistake was made?

82. $\dfrac{f}{g}$

   Solution:
   $$\frac{f(x)}{g(x)} = \frac{x + 2}{x^2 - 4}$$
   $$= \frac{x + 2}{(x - 2)(x + 2)} = \frac{1}{x - 2}$$
   $$= \frac{1}{x - 2}$$

   Domain: $(-\infty, 2) \cup (2, \infty)$

   This is incorrect. What mistake was made?

83. $f \circ g$

Solution: $f \circ g = f(x)g(x)$

$$= (x + 2)(x^2 - 4)$$

$$= x^3 + 2x^2 - 4x - 8$$

Domain: $(-\infty, \infty)$

This is incorrect. What mistake was made?

84. Given the function $f(x) = x^2 + 7$ and $g(x) = \sqrt{x - 3}$, find $f \circ g$, and state the domain.

Solution: $f \circ g = f(g(x)) = \left(\sqrt{x - 3}\right)^2 + 7$

$$= f(g(x)) = x - 3 + 7$$

$$= x - 4$$

Domain: $(-\infty, \infty)$

This is incorrect. What mistake was made?

85. $(f + g)(2) = (x + 2 + x^2 - 4)(2)$

$$= (x^2 + x - 2)(2)$$

$$= 2x^2 + 2x - 4$$

Domain: $(-\infty, \infty)$

This is incorrect. What mistake was made?

86. $f(x) - g(x) = x + 2 - x^2 - 4$

$$= -x^2 + x - 2$$

Domain: $(-\infty, \infty)$

This is incorrect. What mistake was made?

■ **CONCEPTUAL**

In Exercises 87–90, determine whether each statement is true or false.

87. When adding, subtracting, multiplying, or dividing two functions, the domain of the resulting function is the union of the domains of the individual functions.

88. For any functions f and g, $f(g(x)) = g(f(x))$ for all values of x that are in the domain of both f and g.

89. For any functions f and g, $(f \circ g)(x)$ exists for all values of x that are in the domain of $g(x)$, provided the range of g is a subset of the domain of f.

90. The domain of a composite function can be found by inspection, without knowledge of the domain of the individual functions.

■ **CHALLENGE**

91. For the functions $f(x) = x + a$ and $g(x) = \dfrac{1}{x - a}$, find $g \circ f$ and state its domain.

92. For the functions $f(x) = ax^2 + bx + c$ and $g(x) = \dfrac{1}{x - c}$, find $g \circ f$ and state its domain.

93. For the functions $f(x) = \sqrt{x + a}$ and $g(x) = x^2 - a$ find $g \circ f$ and state its domain.

94. For the functions $f(x) = \dfrac{1}{x^a}$ and $g(x) = \dfrac{1}{x^b}$, find $g \circ f$ and state its domain. Assume $a > 1$ and $b > 1$.

■ **TECHNOLOGY**

95. Using a graphing utility, plot $y_1 = \sqrt{x + 7}$ and $y_2 = \sqrt{9 - x}$. Plot $y_3 = y_1 + y_2$. What is the domain of y_3?

96. Using a graphing utility, plot $y_1 = \sqrt[3]{x + 5}$, $y_2 = \dfrac{1}{\sqrt{3 - x}}$, and $y_3 = \dfrac{y_1}{y_2}$. What is the domain of y_3?

97. Using a graphing utility, plot $y_1 = \sqrt{x^2 - 3x - 4}$, $y_2 = \dfrac{1}{x^2 - 14}$, and $y_3 = \dfrac{1}{y_1^2 - 14}$. If y_1 represents a function f and y_2 represents a function g, then y_3 represents the composite function $g \circ f$. The graph of y_3 is only defined for the domain of $g \circ f$. State the domain of $g \circ f$.

98. Using a graphing utility, plot $y_1 = \sqrt{1 - x}$, $y_2 = x^2 + 2$, and $y_3 = y_1^2 + 2$. If y_1 represents a function f and y_2 represents a function g, then y_3 represents the composite function $g \circ f$. The graph of y_3 is only defined for the domain of $g \circ f$. State the domain of $g \circ f$.

Every human being has a blood type, and every human being has a DNA sequence. These are examples of functions, where a person is the input and the output is blood type or DNA sequence. These relationships are classified as functions because each person can have one and only one blood type or DNA strand. The difference between these functions is that many people have the same blood type, but DNA is unique to each individual. Can we map backwards? For instance, if you know the blood type, do you know specifically which person it came from? No, but, if you know the DNA sequence, you know exactly to which person it corresponds. When a function has a one-to-one correspondence, like the DNA example, then mapping backwards is possible. The map back is called the *inverse function*.

Determine Whether a Function Is One-to-One

In Section 3.1, we defined a function as a relationship that maps an input (contained in the domain) to exactly one output (found in the range). Algebraically, each value for x can correspond to only a single value for y. Recall the square, identity, absolute value, and reciprocal functions from our library of functions in Section 3.3.

All of the graphs of these functions satisfy the vertical line test. Although the square function and the absolute value function map each value of x to exactly one value for y, these two functions map two values of x to the same value for y. For example, $(-1, 1)$ and $(1, 1)$ lie on both graphs. The identity and reciprocal functions, on the other hand, map each x to a single value for y, and no two x-values map to the same y-value. These two functions are examples of *one-to-one functions*.

DEFINITION   **One-to-One Function**

A function $f(x)$ is **one-to-one** if no two elements in the domain correspond to the same element in the range; that is,

$$\text{if } x_1 \neq x_2, \text{ then } f(x_1) \neq f(x_2).$$

In other words, it is one-to-one if no two inputs map to the same output.

EXAMPLE 1 **Determining Whether a Function Defined as a Set of Points Is a One-to-One Function**

For each of the three relations, determine whether the relation is a function. If it is a function, determine whether it is a one-to-one function.

$$f = \{(0, 0), (1, 1), (1, -1)\}$$

$$g = \{(-1, 1), (0, 0), (1, 1)\}$$

$$h = \{(-1, -1), (0, 0), (1, 1)\}$$

Solution:

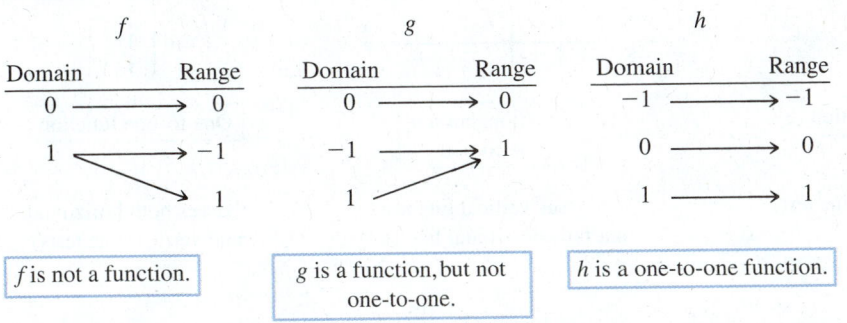

f is not a function.

g is a function, but not one-to-one.

h is a one-to-one function.

Just as there is a graphical test for functions, the vertical line test, there is a graphical test for one-to-one functions, the *horizontal line test*. Note that a horizontal line can be drawn on the square and absolute value functions so that it intersects the graph of each function at two points. The identity and reciprocal functions, however, will intersect a horizontal line in at most only one point. This leads us to the horizontal line test for one-to-one functions.

DEFINITION **Horizontal Line Test**

If every horizontal line intersects the graph of a function in at most one point, then the function is classified as a one-to-one function.

EXAMPLE 2 **Using the Horizontal Line Test to Determine Whether a Function Is One-to-One**

For each of the three relations, determine whether the relation is a function. If it is a function, determine whether it is a one-to-one function. Assume that x is the independent variable and y is the dependent variable.

$$x = y^2 \qquad y = x^2 \qquad y = x^3$$

Solution:

$x = y^2$ $y = x^2$ $y = x^3$

| Not a function | Function, but not one-to-one | One-to-one function |

(fails vertical line test)   (passes vertical line test but fails horizontal line test)   (passes both horizontal and vertical line tests)

■ **Answer:**
a. yes   **b.** no

■ **YOUR TURN** Determine whether each of the functions is a one-to-one function.

a. $f(x) = x + 2$ **b.** $f(x) = x^2 + 1$

Another way of writing the definition of a one-to-one function is:

If $f(x_1) = f(x_2)$, then $x_1 = x_2$.

In the Your Turn following Example 2, we found (using the horizontal line test) that $f(x) = x + 2$ is a one-to-one function, but that $f(x) = x^2 + 1$ is not a one-to-one function. We can also use this alternative definition to determine algebraically whether a function is one-to-one.

WORDS	**MATH**
State the function.	$f(x) = x + 2$
Let there be two real numbers, x_1 and x_2, such that $f(x_1) = f(x_2)$.	$x_1 + 2 = x_2 + 2$
Subtract 2 from both sides of the equation.	$x_1 = x_2$

$f(x) = x + 2$ is a one-to-one function.

WORDS	**MATH**
State the function.	$f(x) = x^2 + 1$
Let there be two real numbers, x_1 and x_2, such that $f(x_1) = f(x_2)$.	$x_1^2 + 1 = x_2^2 + 1$
Subtract 1 from both sides of the equation.	$x_1^2 = x_2^2$
Solve for x_1.	$x_1 = \pm x_2$

$f(x) = x^2 + 2$ is *not* a one-to-one function.

EXAMPLE 3 Determining Algebraically Whether a Function Is One-to-One

Determine algebraically whether the following functions are one-to-one:

a. $f(x) = 5x^3 - 2$ **b.** $f(x) = |x + 1|$

Solution (a):

Find $f(x_1)$ and $f(x_2)$. $f(x_1) = 5x_1^3 - 2$ and $f(x_2) = 5x_2^3 - 2$

Let $f(x_1) = f(x_2)$. $5x_1^3 - 2 = 5x_2^3 - 2$

Add 2 to both sides of the equation. $5x_1^3 = 5x_2^3$

Divide both sides of the equation by 5. $x_1^3 = x_2^3$

Take the cube root of both sides of the equation. $(x_1^3)^{1/3} = (x_2^3)^{1/3}$

Simplify. $x_1 = x_2$

> $f(x) = 5x^3 - 2$ is a one-to-one function.

Solution (b):

Find $f(x_1)$ and $f(x_2)$. $f(x_1) = |x_1 + 1|$ and $f(x_2) = |x_2 + 1|$

Let $f(x_1) = f(x_2)$. $|x_1 + 1| = |x_2 + 1|$

Solve the absolute value equation. $(x_1 + 1) = (x_2 + 1)$ or $(x_1 + 1) = -(x_2 + 1)$

                                       $x_1 = x_2$ or $x_1 = -x_2 - 2$

> $f(x) = |x + 1|$ is **not** a one-to-one function.

Classroom Example 3.5.3
Determine whether these are one-to-one algebraically.

a. $y = -3\sqrt{2x + 1}$
b. $y = |3x - 1| + 1$
c.* $y = (x - 2)^3 - 2$

Answer:
a. yes
b. no
c. yes

Inverse Functions

If a function is one-to-one, then the function maps each x to exactly one y, and no two x-values map to the same y-value. This implies that there is a one-to-one correspondence between the inputs (domain) and outputs (range) of a one-to-one function $f(x)$. In the special case of a one-to-one function, it would be possible to map from the output (range of f) back to the input (domain of f), and this mapping would also be a function. The function that maps the output back to the input of a function f is called the **inverse function** and is denoted $f^{-1}(x)$.

A one-to-one function f maps every x in the domain to a unique and distinct corresponding y in the range. Therefore, the inverse function f^{-1} maps every y back to a unique and distinct x.

The function notations $f(x) = y$ and $f^{-1}(y) = x$ indicate that if the point (x, y) satisfies the function, then the point (y, x) satisfies the inverse function.

For example, let the function $h(x) = \{(-1, 0), (1, 2), (3, 4)\}$.

$h = \{(-1, 0), (1, 2), (3, 4)\}$

Domain	Range
-1	0
1	2
3	4
Range	Domain

h is a one-to-one function

$h^{-1} = \{(0, -1), (2, 1), (4, 3)\}$

The inverse function undoes whatever the function does. For example, if $f(x) = 5x$, then the function f maps any value x in the domain to a value $5x$ in the range. If we want to map backwards or undo the $5x$, we develop a function called the inverse function that takes $5x$ as input and maps back to x as output. The inverse function is $f^{-1}(x) = \frac{1}{5}x$. Note that if we input $5x$ into the inverse function, the output is x: $f^{-1}(5x) = \frac{1}{5}(5x) = x$.

▼ **CAUTION**

$$f^{-1} \neq \frac{1}{f}$$

DEFINITION   **Inverse Function**

If f and g denote two one-to-one functions such that

$$f(g(x)) = x \text{ for every } x \text{ in the domain of } g$$
and
$$g(f(x)) = x \text{ for every } x \text{ in the domain of } f,$$

then g is the **inverse** of the function f. The function g is denoted by f^{-1} (read "f-inverse").

Note: f^{-1} is used to denote the inverse of f. The -1 is not used as an exponent and, therefore, does not represent the reciprocal of f: $\dfrac{1}{f}$.

Two properties hold true relating one-to-one functions to their inverses: (1) the range of the function is the domain of the inverse, and the range of the inverse is the domain of the function, and (2) the composite function that results with a function and its inverse (and vice versa) is the identity function x.

> Domain of f = range of f^{-1} and range of f = domain of f^{-1}
>
> $$f^{-1}(f(x)) = x \quad \text{and} \quad f(f^{-1}(x)) = x$$

Classroom Example 3.5.4
Verify that
$f^{-1}(x) = (x - 1)^3 + 2$ is the inverse of
$f(x) = \sqrt[3]{x - 2} + 1$.

Answer:
Show that $f(f^{-1}(x)) = x$ and $f^{-1}(f(x)) = x$ for all real numbers x.

EXAMPLE 4   Verifying Inverse Functions

Verify that $f^{-1}(x) = \frac{1}{2}x - 2$ is the inverse of $f(x) = 2x + 4$.

Solution:

Show that $\boldsymbol{f^{-1}(f(x)) = x}$ and $\boldsymbol{f(f^{-1}(x)) = x}$.

Write f^{-1} using placeholder notation.

$$f^{-1}(\square) = \frac{1}{2}(\square) - 2$$

Substitute $f(x) = 2x + 4$ into f^{-1}.

$$f^{-1}(f(x)) = \frac{1}{2}(2x + 4) - 2$$

Simplify.

$$f^{-1}(f(x)) = x + 2 - 2 = x$$
$$f^{-1}(f(x)) = x$$

Write f using placeholder notation.

$$f(\square) = 2(\square) + 4$$

Substitute $f^{-1}(x) = \frac{1}{2}x - 2$ into f.

$$f(f^{-1}(x)) = 2\left(\frac{1}{2}x - 2\right) + 4$$

Simplify.

$$f(f^{-1}(x)) = x - 4 + 4 = x$$
$$f(f^{-1}(x)) = x$$

Note the relationship between the domain and range of f and f^{-1}.

	DOMAIN	RANGE
$f(x) = 2x + 4$	$(-\infty, \infty)$	$(-\infty, \infty)$
$f^{-1}(x) = \frac{1}{2}x - 2$	$(-\infty, \infty)$	$(-\infty, \infty)$

EXAMPLE 5   Verifying Inverse Functions with Domain Restrictions

Verify that $f^{-1}(x) = x^2$, for $x \geq 0$, is the inverse of $f(x) = \sqrt{x}$.

Solution:

Show that $f^{-1}(f(x)) = x$ and $f(f^{-1}(x)) = x$.

Write f^{-1} using placeholder notation.

Substitute $f(x) = \sqrt{x}$ into f^{-1}.

$$f^{-1}(\square) = (\square)^2$$
$$f^{-1}(f(x)) = (\sqrt{x})^2 = x$$
$$f^{-1}(f(x)) = x \text{ for } x \geq 0$$

Write f using placeholder notation.

Substitute $f^{-1}(x) = x^2$, $x \geq 0$ into f.

$$f(\square) = \sqrt{(\square)}$$
$$f(f^{-1}(x)) = \sqrt{x^2} = x, \ x \geq 0$$
$$f(f^{-1}(x)) = x \text{ for } x \geq 0$$

	DOMAIN	RANGE
$f(x) = \sqrt{x}$	$[0, \infty)$	$[0, \infty)$
$f^{-1}(x) = x^2, x \geq 0$	$[0, \infty)$	$[0, \infty)$

Graphical Interpretation of Inverse Functions

In Example 4, we showed that $f^{-1}(x) = \frac{1}{2}x - 2$ is the inverse of $f(x) = 2x + 4$. Let's now investigate the graphs that correspond to the function f and its inverse f^{-1}.

x	y
-3	-2
-2	0
-1	2
0	4

$f^{-1}(x)$

x	y
-2	-3
0	-2
2	-1
4	0

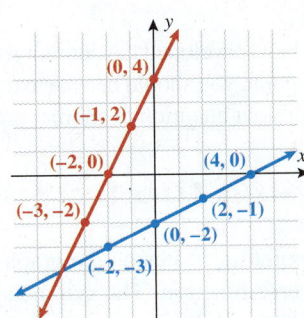

Note that the point $(-3, -2)$ lies on the function and the point $(-2, -3)$ lies on the inverse. In fact, every point (a, b) that lies on the function corresponds to a point (b, a) that lies on the inverse.

Draw the line $y = x$ on the graph. In general, the point (b, a) on the inverse $f^{-1}(x)$ is the reflection (about $y = x$) of the point (a, b) on the function $f(x)$.

In general, if the point (a, b) is on the graph of a function, then the point (b, a) is on the graph of its inverse.

Study Tip

If the point (a, b) is on the function, then the point (b, a) is on the inverse. Notice the interchanging of the x- and y-coordinates.

EXAMPLE 6 **Graphing the Inverse Function**

Given the graph of the function $f(x)$, plot the graph of its inverse $f^{-1}(x)$.

Solution:

Because the points $(-3, -2)$, $(-2, 0)$, $(0, 2)$, and $(2, 4)$ lie on the graph of f, then the points $(-2, -3)$, $(0, -2)$, $(2, 0)$, and $(4, 2)$ lie on the graph of f^{-1}.

■ **Answer:**

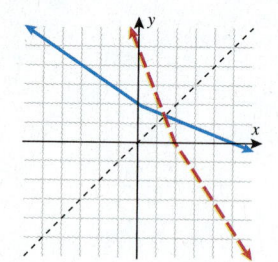

■ **YOUR TURN** Given the graph of a function f, plot the inverse function.

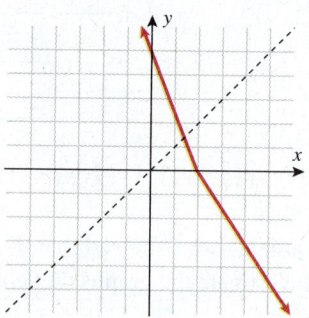

We have developed the definition of an inverse function and described properties of inverses. At this point, you should be able to determine whether two functions are inverses of one another. Let's turn our attention to another problem: How do you find the inverse of a function?

Finding the Inverse Function

If the point (a, b) lies on the graph of a function, then the point (b, a) lies on the graph of the inverse function. The symmetry about the line $y = x$ tells us that the roles of x and y interchange. Therefore, if we start with every point (x, y) that lies on the graph of a function, then every point (y, x) lies on the graph of its inverse. Algebraically, this corresponds to interchanging x and y. Finding the inverse of a finite set of ordered pairs is easy: simply interchange the x- and y-coordinates. Earlier, we found that if $h(x) = \{(-1, 0), (1, 2), (3, 4)\}$, then $h^{-1}(x) = \{(0, -1), (2, 1), (4, 3)\}$. But how do we find the inverse of a function defined by an equation?

Recall the mapping relationship if f is a one-to-one function. This relationship implies that $f(x) = y$ and $f^{-1}(y) = x$. Let's use these two identities to find the inverse. Now consider the function defined by $f(x) = 3x - 1$. To find f^{-1}, we let $f(x) = y$, which yields $y = 3x - 1$. Solve for the variable x: $x = \frac{1}{3}y + \frac{1}{3}$.

Recall that $f^{-1}(y) = x$, so we have found the inverse to be $f^{-1}(y) = \frac{1}{3}y + \frac{1}{3}$. It is customary to write the independent variable as x, so we write the inverse as $f^{-1}(x) = \frac{1}{3}x + \frac{1}{3}$. Now that we have found the inverse, let's confirm that the properties $f^{-1}(f(x)) = x$ and $f(f^{-1}(x)) = x$ hold.

$$f(f^{-1}(x)) = 3\left(\frac{1}{3}x + \frac{1}{3}\right) - 1 = x + 1 - 1 = x$$

$$f^{-1}(f(x)) = \frac{1}{3}(3x - 1) + \frac{1}{3} = x - \frac{1}{3} + \frac{1}{3} = x$$

FINDING THE INVERSE OF A FUNCTION

Let f be a one-to-one function. Then the following procedure can be used to find the inverse function f^{-1} if the inverse exists.

STEP	PROCEDURE	EXAMPLE
1	Let $y = f(x)$.	$f(x) = -3x + 5$ $y = -3x + 5$
2	Solve the resulting equation for x in terms of y (if possible).	$3x = -y + 5$ $x = -\frac{1}{3}y + \frac{5}{3}$
3	Let $x = f^{-1}(y)$.	$f^{-1}(y) = -\frac{1}{3}y + \frac{5}{3}$
4	Let $y = x$ (interchange x and y).	$f^{-1}(x) = -\frac{1}{3}x + \frac{5}{3}$

The same result is found if we first interchange x and y and then solve for y in terms of x.

STEP	PROCEDURE	EXAMPLE
1	Let $y = f(x)$.	$f(x) = -3x + 5$ $y = -3x + 5$
2	Interchange x and y.	$x = -3y + 5$
3	Solve for y in terms of x.	$3y = -x + 5$ $y = -\frac{1}{3}x + \frac{5}{3}$
4	Let $y = f^{-1}(x)$	$f^{-1}(x) = -\frac{1}{3}x + \frac{5}{3}$

Note the following:

- Verify first that a function is one-to-one prior to finding an inverse (if it is not one-to-one, then the inverse does not exist).
- State the domain restrictions on the inverse function. The domain of f is the range of f^{-1} and vice versa.
- To verify that you have found the inverse, show that $f(f^{-1}(x)) = x$ for all x in the domain of f^{-1} and $f^{-1}(f(x)) = x$ for all x in the domain of f.

Given the graph of $f(x)$ below, plot the graph of the inverse $f^{-1}(x)$.

a.

b.

(Assume there is a closed hole at $(0, 2)$ and an open hole at $(0, 4)$.)

Answer:

a.

b.

(There is a closed hole at $(2, 0)$ and an open hole at $(4, 0)$.)

Technology Tip

Using a graphing utility, plot

$y_1 = f(x) = \sqrt{x + 2}$,
$y_2 = f^{-1}(x) = x^2 - 2$, and
$y_3 = x$.

Note that the function $f(x)$ and its inverse $f^{-1}(x)$ are symmetric about the line $y = x$.

Study Tip

Had we ignored the domain and range in Example 7, we would have found the inverse function to be the square function $f(x) = x^2 - 2$, which is not a one-to-one function. It is only when we restrict the domain of the square function that we get a one-to-one function.

Classroom Example 3.5.7
Determine the inverse of the following functions and state the domain and range.

a. $f(x) = \sqrt{1 - 3x}$

b.* $g(x) = -\sqrt{2 + x} + 1$

Answer:

a. $f^{-1}(x) = \dfrac{1 - x^2}{3}$.

  Range: $\left(-\infty, \frac{1}{3}\right]$
  Domain: $[0, \infty)$

b. $g^{-1}(x) = (1 - x)^2 - 2$
  Range: $[-2, \infty)$
  Domain: $(-\infty, 1]$

■ **Answers:**

a. $f^{-1}(x) = \dfrac{x + 3}{7}$, Domain:

  $(-\infty, \infty)$, Range: $(-\infty, \infty)$

b. $g^{-1}(x) = x^2 + 1$, Domain:
  $[0, \infty)$, Range: $[1, \infty)$

EXAMPLE 7   The Inverse of a Square Root Function

Find the inverse of the function $f(x) = \sqrt{x + 2}$. State the domain and range of both f and f^{-1}.

Solution:

$f(x)$ is a one-to-one function because it passes the horizontal line test.

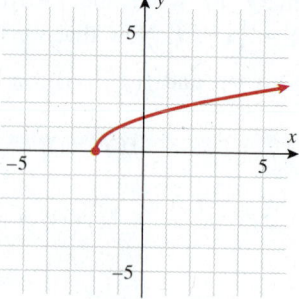

STEP 1	Let $y = f(x)$.	$y = \sqrt{x + 2}$
STEP 2	Interchange x and y.	$x = \sqrt{y + 2}$
STEP 3	Solve for y.	
	Square both sides of the equation.	$x^2 = y + 2$
	Subtract 2 from both sides.	$x^2 - 2 = y$ or $y = x^2 - 2$
STEP 4	Let $y = f^{-1}(x)$.	$f^{-1}(x) = x^2 - 2$

Note any domain restrictions. (State the domain and range of both f and f^{-1}.)

f: Domain: $[-2, \infty)$ Range: $[0, \infty)$

f^{-1}:   Domain: $[0, \infty)$ Range: $[-2, \infty)$

The inverse of $f(x) = \sqrt{x + 2}$ is $\boxed{f^{-1}(x) = x^2 - 2 \text{ for } x \geq 0}$.

Check.

$f^{-1}(f(x)) = x$ for all x in the domain of f.

$f^{-1}(f(x)) = \left(\sqrt{x + 2}\right)^2 - 2$
$= x + 2 - 2$ for $x \geq -2$
$= x$

$f(f^{-1}(x)) = x$ for all x in the domain of f^{-1}.

$f(f^{-1}(x)) = \sqrt{(x^2 - 2) + 2}$
$= \sqrt{x^2}$ for $x \geq 0$
$= x$

Note that the function $f(x) = \sqrt{x + 2}$ and its inverse $f^{-1}(x) = x^2 - 2$ for $x \geq 0$ are symmetric about the line $y = x$.

■ **YOUR TURN** Find the inverse of the given function. State the domain and range of the inverse function.

a. $f(x) = 7x - 3$ **b.** $g(x) = \sqrt{x - 1}$

EXAMPLE 8 A Function That Does Not Have an Inverse Function

Find the inverse of the function $f(x) = |x|$ if it exists.

Solution:

The function $f(x) = |x|$ fails the horizontal line test and therefore is not a one-to-one function. Because f is not a one-to-one function, its inverse function does not exist.

EXAMPLE 9 Finding the Inverse Function

The function $f(x) = \dfrac{2}{x + 3}, x \neq -3$, is a one-to-one function. Find its inverse.

Solution:

STEP 1 Let $y = f(x)$.

$$y = \frac{2}{x + 3}$$

STEP 2 Interchange x and y.

$$x = \frac{2}{y + 3}$$

STEP 3 Solve for y.
Multiply the equation by $(y + 3)$.

$$x(y + 3) = 2$$

Eliminate the parentheses.

$$xy + 3x = 2$$

Subtract $3x$ from both sides.

$$xy = -3x + 2$$

Divide the equation by x.

$$y = \frac{-3x + 2}{x} = -3 + \frac{2}{x}$$

STEP 4 Let $y = f^{-1}(x)$.

$$f^{-1}(x) = -3 + \frac{2}{x}$$

Note any domain restrictions on $f^{-1}(x)$.

$$x \neq 0$$

The inverse of the function $f(x) = \dfrac{2}{x + 3}, x \neq -3$, is $\boxed{f^{-1}(x) = -3 + \dfrac{2}{x}, x \neq 0}$.

Check.

$$f^{-1}(f(x)) = -3 + \frac{2}{\left(\dfrac{2}{x + 3}\right)} = -3 + (x + 3) = x, x \neq -3$$

$$f(f^{-1}(x)) = \frac{2}{\left(-3 + \dfrac{2}{x}\right) + 3} = \frac{2}{\left(\dfrac{2}{x}\right)} = x, x \neq 0$$

■ YOUR TURN The function $f(x) = \dfrac{4}{x - 1}, x \neq 1$, is a one-to-one function. Find its inverse.

The graphs of $y_1 = f(x) = \dfrac{2}{x + 3}$, $x \neq -3$, and $y_2 = f^{-1}(x) = -3 + \dfrac{2}{x}$, $x \neq 0$, are shown.

Note that the function $f(x)$ and its inverse $f^{-1}(x)$ are symmetric about the line $y = x$.

Study Tip

The range of the function is equal to the domain of its inverse function.

■ Answer: $f^{-1}(x) = 1 + \dfrac{4}{x}, x \neq 0$

Note in Example 9 that the domain of f is $(-\infty, -3) \cup (-3, \infty)$ and the domain of f^{-1} is $(-\infty, 0) \cup (0, \infty)$. Therefore, we know that the range of f is $(-\infty, 0) \cup (0, \infty)$, and the range of f^{-1} is $(-\infty, -3) \cup (-3, \infty)$.

EXAMPLE 10 Finding the Inverse of a Piecewise-Defined Function

The function $f(x) = \begin{cases} 3x & x < 0 \\ x^2 & x \geq 0 \end{cases}$, is a one-to-one function. Find its inverse.

Solution:

From the graph of f we can make a table with corresponding domain and range values.

DOMAIN OF f	RANGE OF f
$(-\infty, 0)$	$(-\infty, 0)$
$[0, \infty)$	$[0, \infty)$

From this information we can also list domain and range values for f^{-1}.

DOMAIN OF f / RANGE OF f^{-1}	RANGE OF f / DOMAIN OF f^{-1}
$(-\infty, 0)$	$(-\infty, 0)$
$[0, \infty)$	$[0, \infty)$

$f(x) = 3x$ on $(-\infty, 0)$; find $f^{-1}(x)$ on $(-\infty, 0)$.

STEP 1 Let $y = f(x)$. $y = 3x$

STEP 2 Solve for x in terms of y. $x = 3y$

STEP 3 Solve for y. $y = \frac{1}{3}x$

STEP 4 Let $y = f^{-1}(x)$. $f^{-1}(x) = \frac{1}{3}x$ on $(-\infty, 0)$

$f(x) = x^2$ on $[0, \infty)$; find $f^{-1}(x)$ on $[0, \infty)$.

STEP 1 Let $y = f(x)$. $y = x^2$

STEP 2 Solve for x in terms of y. $x = y^2$

STEP 3 Solve for y. $y = \pm\sqrt{x}$

STEP 4 Let $y = f^{-1}(x)$. $f^{-1}(x) = \pm\sqrt{x}$

STEP 5 The range of f^{-1} is $[0, \infty)$ $f^{-1}(x) = \sqrt{x}$

Combining the two pieces yields a piecewise-defined inverse function.

$$f^{-1}(x) = \begin{cases} \dfrac{1}{3}x & x < 0 \\ \sqrt{x} & x \geq 0 \end{cases}$$

SECTION 3.5 SUMMARY

One-to-One Functions

Each input in the domain corresponds to exactly one output in the range, and no two inputs map to the same output. There are three ways to test a function to determine whether it is a one-to-one function.

1. Discrete points: For the set of all points (a, b) verify that no y-values are repeated.
2. Algebraic equations: Let $f(x_1) = f(x_2)$; if it can be shown that $x_1 = x_2$, then the function is one-to-one.
3. Graphs: Use the horizontal line test; if any horizontal line intersects the graph of the function in more than one point, then the function is not one-to-one.

Properties of Inverse Functions

1. If f is a one-to-one function, then f^{-1} exists.
2. Domain and range
   - Domain of f = range of f^{-1}
   - Domain of f^{-1} = range of f
3. Composition of inverse functions
   - $f^{-1}(f(x)) = x$ for all x in the domain of f.
   - $f(f^{-1}(x)) = x$ for all x in the domain of f^{-1}.
4. The graphs of f and f^{-1} are symmetric with respect to the line $y = x$.

Procedure for Finding the Inverse of a Function

1. Let $y = f(x)$.
2. Interchange x and y.
3. Solve for y.
4. Let $y = f^{-1}(x)$.

SECTION 3.5 EXERCISES

■ SKILLS

In Exercises 1–16, determine whether the given relation is a function. If it is a function, determine whether it is a one-to-one function.

1.

2.

3.

4.

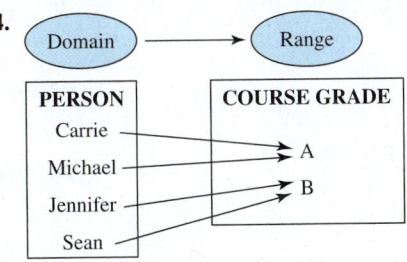

5. $\{(0, 1), (1, 2), (2, 3), (3, 4)\}$

6. $\{(0, -2), (2, 0), (5, 3), (-5, -7)\}$

7. $\{(0, 0), (9, -3), (4, -2), (4, 2), (9, 3)\}$

8. $\{(0, 1), (1, 1), (2, 1), (3, 1)\}$

9. $\{(0, 1), (1, 0), (2, 1), (-2, 1), (5, 4), (-3, 4)\}$

10. $\{(0, 0), (-1, -1), (-2, -8), (1, 1), (2, 8)\}$

11.

12.

13.

14.

15.

16.

In Exercises 17–24, determine algebraically and graphically whether the function is one-to-one.

17. $f(x) = |x - 3|$

18. $f(x) = (x - 2)^2 + 1$

19. $f(x) = \dfrac{1}{x - 1}$

20. $f(x) = \sqrt[3]{x}$

21. $f(x) = x^2 - 4$

22. $f(x) = \sqrt{x + 1}$

23. $f(x) = x^3 - 1$

24. $f(x) = \dfrac{1}{x + 2}$

In Exercises 25–34, verify that the function $f^{-1}(x)$ is the inverse of $f(x)$ by showing that $f(f^{-1}(x)) = x$ and $f^{-1}(f(x)) = x$. Graph $f(x)$ and $f^{-1}(x)$ on the same axes to show the symmetry about the line $y = x$.

25. $f(x) = 2x + 1;\; f^{-1}(x) = \dfrac{x - 1}{2}$

26. $f(x) = \dfrac{x - 2}{3};\; f^{-1}(x) = 3x + 2$

27. $f(x) = \sqrt{x - 1},\, x \geq 1;\; f^{-1}(x) = x^2 + 1,\, x \geq 0$

28. $f(x) = 2 - x^2,\, x \geq 0;\; f^{-1}(x) = \sqrt{2 - x},\, x \leq 2$

29. $f(x) = \dfrac{1}{x};\; f^{-1}(x) = \dfrac{1}{x},\, x \neq 0$

30. $f(x) = (5 - x)^{1/3};\; f^{-1}(x) = 5 - x^3$

31. $f(x) = \dfrac{1}{2x + 6},\, x \neq -3;\; f^{-1}(x) = \dfrac{1}{2x} - 3,\, x \neq 0$

32. $f(x) = \dfrac{3}{4 - x},\, x \neq 4;\; f^{-1}(x) = 4 - \dfrac{3}{x},\, x \neq 0$

33. $f(x) = \dfrac{x + 3}{x + 4},\, x \neq -4;\; f^{-1}(x) = \dfrac{3 - 4x}{x - 1},\, x \neq 1$

34. $f(x) = \dfrac{x - 5}{3 - x},\, x \neq 3;\; f^{-1}(x) = \dfrac{3x + 5}{x + 1},\, x \neq -1$

In Exercises 35–42, graph the inverse of the one-to-one function that is given.

35.

36.

37.

38.

39.

40.

41.

42.

In Exercises 43–60, the function f is one-to-one. Find its inverse, and check your answer. State the domain and range of both f and f^{-1}.

43. $f(x) = x - 1$

44. $f(x) = 7x$

45. $f(x) = -3x + 2$

46. $f(x) = 2x + 3$

47. $f(x) = x^3 + 1$

48. $f(x) = x^3 - 1$

49. $f(x) = \sqrt{x - 3}$

50. $f(x) = \sqrt{3 - x}$

51. $f(x) = x^2 - 1, x \geq 0$

52. $f(x) = 2x^2 + 1, x \geq 0$

53. $f(x) = (x + 2)^2 - 3, x \geq -2$

54. $f(x) = (x - 3)^2 - 2, x \geq 3$

55. $f(x) = \dfrac{2}{x}$

56. $f(x) = -\dfrac{3}{x}$

57. $f(x) = \dfrac{2}{3 - x}$

58. $f(x) = \dfrac{7}{x + 2}$

59. $f(x) = \dfrac{7x + 1}{5 - x}$

60. $f(x) = \dfrac{2x + 5}{7 + x}$

In Exercises 61–64, graph the piecewise-defined function to determine whether it is a one-to-one function. If it is a one-to-one function, find its inverse.

61. $G(x) = \begin{cases} 0 & x < 0 \\ \sqrt{x} & x \geq 0 \end{cases}$

62. $G(x) = \begin{cases} \dfrac{1}{x} & x < 0 \\ \sqrt{x} & x \geq 0 \end{cases}$

63. $f(x) = \begin{cases} x & x \leq -1 \\ x^3 & -1 < x < 1 \\ x & x \geq 1 \end{cases}$

64. $f(x) = \begin{cases} x + 3 & x \leq -2 \\ |x| & -2 < x < 2 \\ x^2 & x \geq 2 \end{cases}$

▪ APPLICATIONS

65. Temperature. The equation used to convert from degrees Celsius to degrees Fahrenheit is $f(x) = \frac{9}{5}x + 32$. Determine the inverse function $f^{-1}(x)$. What does the inverse function represent?

66. Temperature. The equation used to convert from degrees Fahrenheit to degrees Celsius is $C(x) = \frac{5}{9}(x - 32)$. Determine the inverse function $C^{-1}(x)$. What does the inverse function represent?

67. Budget. The Richmond rowing club is planning to enter the Head of the Charles race in Boston and is trying to figure out how much money to raise. The entry fee is $250 per boat for the first 10 boats and $175 for each additional boat. Find the cost function $C(x)$ as a function of the number of boats the club enters x. Find the inverse function that will yield how many boats the club can enter as a function of how much money it will raise.

68. Long-Distance Calling Plans. A phone company charges $.39 per minute for the first 10 minutes of a long-distance phone call and $.12 per minute every minute after that. Find the cost function $C(x)$ as a function of the length of the phone call in minutes x. Suppose you buy a "prepaid" phone card that is planned for a single call. Find the inverse function that determines how many minutes you can talk as a function of how much you prepaid.

69. Salary. A student works at Target making $10 per hour and the weekly number of hours worked per week x varies. If Target withholds 25% of his earnings for taxes and Social Security, write a function $E(x)$ that expresses the student's take-home pay each week. Find the inverse function $E^{-1}(x)$. What does the inverse function tell you?

70. Salary. A grocery store pays you $8 per hour for the first 40 hours per week and time and a half for overtime. Write a piecewise-defined function that represents your weekly earnings $E(x)$ as a function of the number of hours worked x. Find the inverse function $E^{-1}(x)$. What does the inverse function tell you?

In Exercises 71–74, refer to the following:

By analyzing available empirical data it was determined that during an illness a patient's body temperature fluctuated during one 24-hour period according to the function

$$T(t) = 0.0003(t - 24)^3 + 101.70$$

where T represents that patient's temperature in degrees Fahrenheit and t represents the time of day in hours measured from 12:00 A.M. (midnight).

71. Health/Medicine. Find the domain and range of the function $T(t)$.

72. Health/Medicine. Find time as a function of temperature, that is, the inverse function $t(T)$.

73. Health/Medicine. Find the domain and range of the function $t(T)$ found in Exercise 72.

74. Health/Medicine. At what time, to the nearest hour, was the patient's temperature 99.5°F?

▪ CATCH THE MISTAKE

In Exercises 75–78, explain the mistake that is made.

75. Is $x = y^2$ a one-to-one function?

Solution:

Yes, this graph represents a one-to-one function because it passes the horizontal line test.

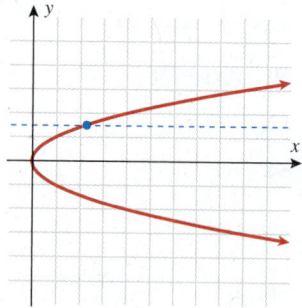

This is incorrect. What mistake was made?

76. A linear one-to-one function is graphed below. Draw its inverse.

Solution:

Note that the points **(3, 3)** and **(0, −4)** lie on the graph of the function.

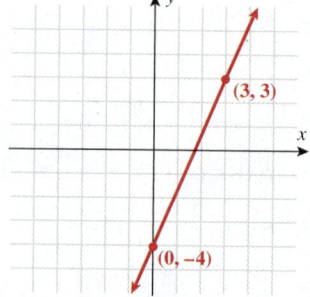

By symmetry, the points **(−3, −3)** and **(0, 4)** lie on the graph of the inverse.

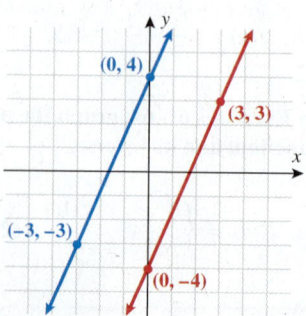

This is incorrect. What mistake was made?

77. Given the function $f(x) = x^2$, find the inverse function $f^{-1}(x)$.

Solution:

Step 1: Let $y = f(x)$. $\quad\quad\quad y = x^2$

Step 2: Solve for x. $\quad\quad\quad x = \sqrt{y}$

Step 3: Interchange x and y. $\quad y = \sqrt{x}$

Step 4: Let $y = f^{-1}(x)$. $\quad\quad f^{-1}(x) = \sqrt{x}$

Check: $f(f^{-1}(x)) = (\sqrt{x})^2 = x$ and $f^{-1}(f(x)) = \sqrt{x^2} = x$.

The inverse of $f(x) = x^2$ is $f^{-1}(x) = \sqrt{x}$.

This is incorrect. What mistake was made?

78. Given the function $f(x) = \sqrt{x - 2}$, find the inverse function $f^{-1}(x)$, and state the domain restrictions on $f^{-1}(x)$.

Solution:

Step 1: Let $y = f(x)$. $\quad\quad\quad\quad y = \sqrt{x - 2}$

Step 2: Interchange x and y. $\quad\quad x = \sqrt{y - 2}$

Step 3: Solve for y. $\quad\quad\quad\quad\quad y = x^2 + 2$

Step 4: Let $f^{-1}(x) = y$. $\quad\quad\quad f^{-1}(x) = x^2 + 2$

Step 5: Domain restrictions: $f(x) = \sqrt{x - 2}$ has the domain restriction that $x \geq 2$.

The inverse of $f(x) = \sqrt{x - 2}$ is $f^{-1}(x) = x^2 + 2$.

The domain of $f^{-1}(x)$ is $x \geq 2$.

This is incorrect. What mistake was made?

■ **CONCEPTUAL**

In Exercises 79–82, determine whether each statement is true or false.

79. Every even function is a one-to-one function.

80. Every odd function is a one-to-one function.

81. It is not possible that $f = f^{-1}$.

82. A function f has an inverse. If the function lies in quadrant II, then its inverse lies in quadrant IV.

83. If $(0, b)$ is the y-intercept of a one-to-one function f, what is the x-intercept of the inverse f^{-1}?

84. If $(a, 0)$ is the x-intercept of a one-to-one function f, what is the y-intercept of the inverse f^{-1}?

■ **CHALLENGE**

85. The unit circle is not a function. If we restrict ourselves to the semicircle that lies in quadrants I and II, the graph represents a function, but it is not a one-to-one function. If we further restrict ourselves to the quarter circle lying in quadrant I, the graph does represent a one-to-one function. Determine the equations of both the one-to-one function and its inverse. State the domain and range of both.

86. Find the inverse of $f(x) = \dfrac{c}{x}, c \neq 0$.

87. Under what conditions is the linear function $f(x) = mx + b$ a one-to-one function?

88. Assuming that the conditions found in Exercise 87 are met, determine the inverse of the linear function.

■ **TECHNOLOGY**

In Exercises 89–92, graph the following functions and determine whether they are one-to-one.

89. $f(x) = |4 - x^2|$

90. $f(x) = \dfrac{3}{x^3 + 2}$

91. $f(x) = x^{1/3} - x^5$

92. $f(x) = \dfrac{1}{x^{1/2}}$

In Exercises 93–96, graph the functions f and g and the line $y = x$ in the same screen. Do the two functions appear to be inverses of each other?

93. $f(x) = \sqrt{3x - 5};\quad g(x) = \dfrac{x^2}{3} + \dfrac{5}{3}$

94. $f(x) = \sqrt{4 - 3x};\quad g(x) = \dfrac{4}{3} - \dfrac{x^2}{3}, x \geq 0$

95. $f(x) = (x - 7)^{1/3} + 2;\quad g(x) = x^3 - 6x^2 + 12x - 1$

96. $f(x) = \sqrt[3]{x + 3} - 2;\quad g(x) = x^3 + 6x^2 + 12x + 6$

SKILLS OBJECTIVES

- Develop mathematical models using direct variation.
- Develop mathematical models using inverse variation.
- Develop mathematical models using combined variation.
- Develop mathematical models using joint variation.

CONCEPTUAL OBJECTIVES

- Understand the difference between direct variation and inverse variation.
- Understand the difference between combined variation and joint variation.

In this section we discuss mathematical models for different applications. Two quantities in the real world often *vary* with respect to one another. Sometimes, they vary *directly*. For example, the more money we make, the more total dollars of federal income tax we expect to pay. Sometimes, quantities vary *inversely*. For example, when interest rates on mortgages decrease, we expect the number of homes purchased to increase because a buyer can afford "more house" with the same mortgage payment when rates are lower. In this section we discuss quantities varying *directly*, *inversely*, and *jointly*.

Direct Variation

When one quantity is a constant multiple of another quantity, we say that the quantities are *directly proportional* to one another.

DIRECT VARIATION

Let x and y represent two quantities. The following are equivalent statements:

- $y = kx$, where k is a nonzero constant.
- y **varies directly** with x.
- y is **directly proportional** to x.

The constant k is called the **constant of variation** or the **constant of proportionality**.

In 2005, the national average cost of residential electricity was 9.53 ¢/kWh (cents per kilowatt-hour). For example, if a residence used 3400 kWh, then the bill would be $324.02, and if a residence used 2500 kWh, then the bill would be $238.25.

EXAMPLE 1   Finding the Constant of Variation

In the United States, the cost of electricity is directly proportional to the number of kilowatt · hours (kWh) used. If a household in Tennessee on average used 3098 kWh per month and had an average monthly electric bill of $179.99, find a mathematical model that gives the cost of electricity in Tennessee in terms of the number of kilowatt · hours used.

Solution:

Write the direct variation model. $y = kx$

Label the variables and constant. x = number of kWh
                                                                    y = cost (dollars)
                                                                    k = cost per kWh

Substitute the given data x = 3098 kWh
and y = $179.99 into $y = kx$. $179.99 = 3098k$

Solve for k. $k = \dfrac{179.99}{3098} = 0.05810$

                                                                    $y = 0.0581x$

In Tennessee the cost of electricity is $\boxed{5.81 \text{ ¢/kWh}}$.

....

■ **YOUR TURN** Find a mathematical model that describes the cost of electricity in California if the cost is directly proportional to the number of kWh used and a residence that consumes 4000 kWh is billed $480.

■ **Answer:** $y = 0.12x$; the cost of electricity in California is 12 ¢/kWh.

Not all variation we see in nature is direct variation. Isometric growth, where the various parts of an organism grow in direct proportion to each other, is rare in living organisms. If organisms grew isometrically, young children would look just like adults, only smaller. In contrast, most organisms grow nonisometrically; the various parts of organisms do not increase in size in a one-to-one ratio. The relative proportions of a human body change dramatically as the human grows. Children have proportionately larger heads and shorter legs than adults. *Allometric growth* is the pattern of growth whereby different parts of the body grow at different rates with respect to each other. Some human body characteristics vary directly, and others can be mathematically modeled by *direct variation with powers*.

DIRECT VARIATION WITH POWERS

Let x and y represent two quantities. The following are equivalent statements:

- $y = kx^n$, where k is a nonzero constant.
- y **varies directly with the *n*th power** of x.
- y **is directly proportional to the *n*th power** of x.

One example of direct variation with powers is height and weight of humans. Weight (in pounds) is directly proportional to the cube of height (feet).

$$W = kH^3$$

EXAMPLE 2 Direct Variation with Powers

The following is a personal ad:

Single professional male (6 ft/194 lbs) seeks single professional female for long-term relationship. Must be athletic, smart, like the movies and dogs, and have height and weight similarly proportioned to mine.

Find a mathematical equation that describes the height and weight of the male who wrote the ad. How much would a 5′6″ woman weigh who has the same proportionality as the male?

Solution:

Write the direct variation (cube) model for height versus weight.

$$W = kH^3$$

Substitute the given data $W = 194$ and $H = 6$ into $W = kH^3$.

$$194 = k(6)^3$$

Solve for k.

$$k = \frac{194}{216} = 0.898148 \approx 0.90$$

$$W = 0.9H^3$$

Let $H = 5.5$ ft.

$$W = 0.9(5.5)^3 = 149.73$$

A woman 5′6″ tall with the same height and weight proportionality as the male would weigh $\boxed{150 \text{ lb}}$.

■ **YOUR TURN** A brother and sister both have weight (pounds) that varies as the cube of height (feet) and they share the same proportionality constant. The sister is 6 feet tall and weighs 170 pounds. Her brother is 6 feet 4 inches. How much does he weigh?

Inverse Variation

Two fundamental topics covered in economics are supply and demand. Supply is the quantity that producers are willing to sell at a given price. For example, an artist may be willing to paint and sell 5 portraits if each sells for $50, but that same artist may be willing to sell 100 portraits if each sells for $10,000. Demand is the quantity of a good that consumers are not only willing to purchase but also have the capacity to buy at a given price. For example, consumers may purchase 1 billion Big Macs from McDonald's every year, but perhaps only 1 million filet mignons are sold at Outback. There may be 1 billion people who want to buy the filet mignon but don't have the financial means to do so. Economists study the equilibrium between supply and demand.

Demand can be modeled with an *inverse variation* of price: when the price increases, demand decreases, and vice versa.

INVERSE VARIATION

Let x and y represent two quantities. The following are equivalent statements:

■ $y = \dfrac{k}{x}$, where k is a nonzero constant.

■ y **varies inversely** with x.

■ y is **inversely proportional** to x.

The constant k is called the **constant of variation** or the **constant of proportionality**.

EXAMPLE 3   Inverse Variation

The number of potential buyers of a house decreases as the price of the house increases (see graph on the right). If the number of potential buyers of a house in a particular city is inversely proportional to the price of the house, find a mathematical equation that describes the demand for houses as it relates to price. How many potential buyers will there be for a $2 million house?

Solution:

Write the inverse variation model.	$y = \dfrac{k}{x}$
Label the variables and constant.	x = price of house in thousands of dollars y = number of buyers
Select *any* point that lies on the curve.	(200, 500)
Substitute the given data $x = 200$ and $y = 500$ into $y = \dfrac{k}{x}$.	$500 = \dfrac{k}{200}$
Solve for k.	$k = 200 \cdot 500 = 100{,}000$
	$y = \dfrac{100{,}000}{x}$
Let $x = 2000$.	$y = \dfrac{100{,}000}{2000} = 50$

There are only 50 potential buyers for a $2 million house in this city.

■ **Answer:** 5000

■ **YOUR TURN**   In New York City, the number of potential buyers in the housing market is inversely proportional to the price of a house. If there are 12,500 potential buyers for a $2 million condominium, how many potential buyers are there for a $5 million condominium?

Two quantities can vary inversely with the *n*th power of *x*.

If x and y are related by the equation $y = \dfrac{k}{x^n}$, then we say that y varies **inversely** with the *n*th **power of x**, or y is inversely **proportional to the *n*th power of x**.

Joint Variation and Combined Variation

We now discuss combinations of variations. When one quantity is proportional to the product of two or more other quantities, the variation is called **joint variation**. When direct variation and inverse variation occur at the same time, the variation is called **combined variation**.

An example of a **joint variation** is simple interest (Section 1.2), which is defined as

$$I = Prt$$

where

- I is the interest in dollars
- P is the principal (initial) dollars
- r is the interest rate (expressed in decimal form)
- t is time in years

The interest earned is proportional to the product of three quantities (principal, interest rate, and time). Note that if the interest rate increases, then the interest earned also increases. Similarly, if either the initial investment (principal) or the time the money is invested increases, then the interest earned also increases.

An example of **combined variation** is the combined gas law in chemistry,

$$P = k\frac{T}{V}$$

where

- P is pressure
- T is temperature (kelvins)
- V is volume
- k is a gas constant

This relation states that the pressure of a gas is directly proportional to the temperature and inversely proportional to the volume containing the gas. For example, as the temperature increases, the pressure increases, but when the volume decreases, pressure increases.

As an example, the gas in the headspace of a soda bottle has a fixed volume. Therefore, as temperature increases, the pressure increases. Compare the different pressures of opening a twist-off cap on a bottle of soda that is cold versus one that is hot. The hot one feels as though it "releases more pressure."

EXAMPLE 4   Combined Variation

The gas in the headspace of a soda bottle has a volume of 9.0 ml, pressure of 2 atm (atmospheres), and a temperature of 298 K (standard room temperature of 77°F). If the soda bottle is stored in a refrigerator, the temperature drops to approximately 279 K (42°F). What is the pressure of the gas in the headspace once the bottle is chilled?

Solution:

Write the combined gas law. $P = k\dfrac{T}{V}$

Let $P = 2$ atm, $T = 298$ K, and $V = 9.0$ ml. $2 = k\dfrac{298}{9}$

Solve for k. $k = \dfrac{18}{298}$

Let $k = \dfrac{18}{298}, T = 279,$ and $V = 9.0$ in $P = k\dfrac{T}{V}.$ $P = \dfrac{18}{298}\cdot\dfrac{279}{9} \approx 1.87$

Since we used the same physical units for both the chilled and room-temperature soda bottles, the pressure is in atmospheres. $P = 1.87$ atm

SECTION

3.6 SUMMARY

Direct, inverse, joint, and combined variation can be used to model the relationship between two quantities. For two quantities x and y, we say that

- y is directly proportional to x if $y = kx$.

- y is inversely proportional to x if $y = \dfrac{k}{x}$.

Joint variation occurs when one quantity is directly proportional to two or more quantities. Combined variation occurs when one quantity is directly proportional to one or more quantities and inversely proportional to one or more other quantities.

SECTION

3.6 EXERCISES

▪ SKILLS

In Exercises 1–16, write an equation that describes each variation. Use k as the constant of variation.

1. y varies directly with x.

2. s varies directly with t.

3. V varies directly with x^3.

4. A varies directly with x^2.

5. z varies directly with m.

6. h varies directly with $\sqrt{t}$.

7. f varies inversely with λ.

8. P varies inversely with r^2.

9. F varies directly with w and inversely with L.

10. V varies directly with T and inversely with P.

11. v varies directly with both g and t.

12. S varies directly with both t and d.

13. R varies inversely with both P and T.

14. y varies inversely with both x and z.

15. y is directly proportional to the square root of x.

16. y is inversely proportional to the cube of t.

In Exercises 17–36, write an equation that describes each variation.

17. d is directly proportional to t. $d = r$ when $t = 1$.

18. F is directly proportional to m. $F = a$ when $m = 1$.

19. V is directly proportional to both l and w. $V = 6h$ when $w = 3$ and $l = 2$.

20. A is directly proportional to both b and h. $A = 10$ when $b = 5$ and $h = 4$.

21. A varies directly with the square of r. $A = 9\pi$ when $r = 3$.

22. V varies directly with the cube of r. $V = 36\pi$ when $r = 3$.

23. V varies directly with both h and r^2. $V = 1$ when $r = 2$ and $h = \dfrac{4}{\pi}$.

24. W is directly proportional to both R and the square of I. $W = 4$ when $R = 100$ and $I = 0.25$.

25. V varies inversely with P. $V = 1000$ when $P = 400$.

26. I varies inversely with the square of d. $I = 42$ when $d = 16$.

27. F varies inversely with both λ and L. $F = 20\pi/m^2$ when $\lambda = 1$ μm and $L = 100$ kilometers.

28. y varies inversely with both x and z. $y = 32$ when $x = 4$ and $z = 0.05$.

29. t varies inversely with s. $t = 2.4$ when $s = 8$.

30. W varies inversely with the square of d. $W = 180$ when $d = 0.2$.

31. R varies inversely with the square of I. $R = 0.4$ when $I = 3.5$.

32. y varies inversely with both x and the square root of z. $y = 12$ when $x = 0.2$ and $z = 4$.

33. R varies directly with L and inversely with A. $R = 0.5$ when $L = 20$ and $A = 0.4$.

34. F varies directly with m and inversely with d. $F = 32$ when $m = 20$ and $d = 8$.

35. F varies directly with both m_1 and m_2 and inversely with the square of d. $F = 20$ when $m_1 = 8$, $m_2 = 16$, and $d = 0.4$.

36. w varies directly with the square root of g and inversely with the square of t. $w = 20$ when $g = 16$ and $t = 0.5$.

▪ APPLICATIONS

37. **Wages.** Jason and Valerie both work at Panera Bread and have the following paycheck information for a certain week. Find an equation that shows their wages W varying directly with the number of hours worked H.

EMPLOYEE	HOURS WORKED	WAGES
Jason	23	$172.50
Valerie	32	$240.00

38. **Sales Tax.** The sales tax in Orange and Seminole counties in Florida differs by only 0.5%. A new resident knows this but doesn't know which of the counties has the higher tax. The resident lives near the border of the counties and is in the market for a new plasma television and wants to purchase it in the county with the lower tax. If the tax on a pair of $40 sneakers is $2.60 in Orange County and the tax on a $12 T-shirt is $0.84 in Seminole County, write two equations: one for each county that describes the tax T, which is directly proportional to the purchase price P.

For Exercises 39 and 40, refer to the following:

The ratio of the speed of an object to the speed of sound determines the Mach number. Aircraft traveling at a subsonic speed (less than the speed of sound) have a Mach number less than 1. In other words, the speed of an aircraft is directly proportional to its Mach number. Aircraft traveling at a supersonic speed (greater than speed of sound) have a Mach number greater than 1. The speed of sound at sea level is approximately 760 miles per hour.

39. **Military.** The U.S. Navy Blue Angels fly F-18 Hornets that are capable of Mach 1.7. How fast can F-18 Hornets fly at sea level?

40. **Military.** The U.S. Air Force's newest fighter aircraft is the F-35, which is capable of Mach 1.9. How fast can an F-35 fly at sea level?

Exercises 41 and 42 are examples of the golden ratio, or phi, a proportionality constant that appears in nature. The numerical approximate value of phi is 1.618. From www.goldenratio.net.

41. **Human Anatomy.** The length of your forearm F (wrist to elbow) is directly proportional to the length of your hand H (length from wrist to tip of middle finger). Write the equation that describes this relationship if the length of your forearm is 11 inches and the length of your hand is 6.8 inches.

Kim Steele/Getty Images, Inc.

42. **Human Anatomy.** Each section of your index finger, from the tip to the base of the wrist, is larger than the preceding one by about the golden (Fibonacci) ratio. Find an equation that represents the ratio of each section of your finger related to the previous one if one section is eight units long and the next section is five units long.

For Exercises 43 and 44, refer to the following:

Hooke's law in physics states that if a spring at rest (equilibrium position) has a weight attached to it, then the distance the spring stretches is directly proportional to the force (weight), according to the formula:

$$F = kx$$

where F is the force in Newtons (N), x is the distance stretched in meters (m), and k is the spring constant (N/m).

43. Physics. A force of 30 N will stretch the spring 10 centimeters. How far will a force of 72 N stretch the spring?

44. Physics. A force of 30 N will stretch the spring 10 centimeters. How much force is required to stretch the spring 18 centimeters?

45. Business. A cell phone company develops a pay-as-you-go cell phone plan in which the monthly cost varies directly as the number of minutes used. If the company charges $17.70 in a month when 236 minutes are used, what should the company charge for a month in which 500 minutes are used?

46. Economics. Demand for a product varies inversely with the price per unit of the product. Demand for the product is 10,000 units when the price is $5.75 per unit. Find the demand for the product (to the nearest hundred units) when the price is $6.50.

47. Sales. Levi's makes jeans in a variety of price ranges for juniors. The Flare 519 jeans sell for about $20, whereas the 646 Vintage Flare jeans sell for $300. The demand for Levi's jeans is inversely proportional to the price. If 300,000 pairs of the 519 jeans were bought, approximately how many of the Vintage Flare jeans were bought?

48. Sales. Levi's makes jeans in a variety of price ranges for men. The Silver Tab Baggy jeans sell for about $30, whereas the Offender jeans sell for about $160. The demand for Levi's jeans is inversely proportional to the price. If 400,000 pairs of the Silver Tab Baggy jeans were bought, approximately how many of the Offender jeans were bought?

For Exercises 49 and 50, refer to the following:

In physics, the inverse square law states that any physical quantity or strength is inversely proportional to the square of the distance from the source of that physical quantity. In particular, the intensity of light radiating from a point source is inversely proportional to the square of the distance from the source. Below is a table of average distances from the Sun:

PLANET	DISTANCE TO THE SUN
Mercury	58,000 km
Earth	150,000 km
Mars	228,000 km

49. Solar Radiation. The solar radiation on the Earth is approximately 1400 watts per square meter (w/m^2). How much solar radiation is there on Mars? Round to the nearest hundred watts per square meter.

50. Solar Radiation. The solar radiation on the Earth is approximately 1400 watts per square meter. How much solar radiation is there on Mercury? Round to the nearest hundred watts per square meter.

51. Investments. Marilyn receives a $25,000 bonus from her company and decides to put the money toward a new car that she will need in two years. Simple interest is directly proportional to the principal and the time invested. She compares two different banks' rates on money market accounts. If she goes with Bank of America, she will earn $750 in interest, but if she goes with the Navy Federal Credit Union, she will earn $1500. What is the interest rate on money market accounts at both banks?

52. Investments. Connie and Alvaro sell their house and buy a fixer-upper house. They made $130,000 on the sale of their previous home. They know it will take 6 months before the general contractor will start their renovation, and they want to take advantage of a 6-month CD that pays simple interest. What is the rate of the 6-month CD if they will make $3250 in interest?

53. Chemistry. A gas contained in a 4 milliliter container at a temperature of 300 K has a pressure of 1 atmosphere. If the temperature decreases to 275 K, what is the resulting pressure?

54. Chemistry. A gas contained in a 4 milliliter container at a temperature of 300 K has a pressure of 1 atmosphere. If the container changes to a volume of 3 milliliters, what is the resulting pressure?

▪ CATCH THE MISTAKE

In Exercises 55 and 56, explain the mistake that is made.

55. y varies directly with t and indirectly with x. When $x = 4$ and $t = 2$, then $y = 1$. Find an equation that describes this variation.

Solution:

Write the variation equation. $y = ktx$

Let $x = 4$, $t = 2$, and $y = 1$. $1 = k(2)(4)$

Solve for k. $k = \dfrac{1}{8}$

Substitute $k = \frac{1}{8}$ into $y = ktx$. $y = \dfrac{1}{8}tx$

This is incorrect. What mistake was made?

56. y varies directly with t and the square of x. When $x = 4$ and $t = 1$, then $y = 8$. Find an equation that describes this variation.

Solution:

Write the variation equation. $y = kt\sqrt{x}$

Let $x = 4$, $t = 1$, and $y = 8$. $8 = k(1)\sqrt{4}$

Solve for k. $k = 4$

Substitute $k = 4$ into $y = kt\sqrt{x}$. $y = 4t\sqrt{x}$

This is incorrect. What mistake was made?

▪ CONCEPTUAL

In Exercises 57 and 58, determine whether each statement is true or false.

57. The area of a triangle is directly proportional to both the base and the height of the triangle (joint variation).

58. Average speed is directly proportional to both distance and time (joint variation).

In Exercises 59 and 60, match the variation with the graph.

59. Inverse variation

60. Direct variation

a.

b.

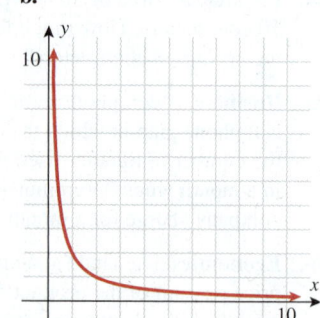

▪ CHALLENGE

Exercises 61 and 62 involve the theory governing laser propagation through the Earth's atmosphere.

The three parameters that help classify the strength of optical turbulence are:

- C_n^2, index of refraction structure parameter
- k, wave number of the laser, which is inversely proportional to the wavelength λ of the laser:

$$k = \frac{2\pi}{\lambda}$$

- L, propagation distance

The variance of the irradiance of a laser σ^2 is directly proportional to C_n^2, $k^{7/6}$, and $L^{11/16}$.

61. When $C_n^2 = 1.0 \times 10^{-13}$ m$^{-2/3}$, $L = 2$ km, and $\lambda = 1.55$ μm, the variance of irradiance for a plane wave σ_{pl}^2 is 7.1. Find the equation that describes this variation.

62. When $C_n^2 = 1.0 \times 10^{-13}$ m$^{-2/3}$, $L = 2$ km, and $\lambda = 1.55$ μm, the variance of irradiance for a spherical wave σ_{sp}^2 is 2.3. Find the equation that describes this variation.

■ **TECHNOLOGY**

For Exercises 63–66, refer to the following:

Data from 1995 to 2006 for oil prices in dollars per barrel, the U.S. Dow Jones Utilities Stock Index, New Privately Owned Housing, and 5-year Treasury Constant Maturity Rate are given in the table. (Data are from Forecast Center's Historical Economic and Market Home Page at www.neatideas.com/djutil.htm.)

Use the calculator STAT EDIT commands to enter the table with L_1 as the oil price, L_2 as the utilities stock index, L_3 as number of housing units, and L_4 as the 5-year maturity rate.

JANUARY OF EACH YEAR	OIL PRICE, $ PER BARREL	U.S. DOW JONES UTILITIES STOCK INDEX	NEW, PRIVATELY OWNED HOUSING UNITS	5-YEAR TREASURY CONSTANT MATURITY RATE
1995	17.99	193.12	1407	7.76
1996	18.88	230.85	1467	5.36
1997	25.17	232.53	1355	6.33
1998	16.71	263.29	1525	5.42
1999	12.47	302.80	1748	4.60
2000	27.18	315.14	1636	6.58
2001	29.58	372.32	1600	4.86
2002	19.67	285.71	1698	4.34
2003	32.94	207.75	1853	3.05
2004	32.27	271.94	1911	3.12
2005	46.84	343.46	2137	3.71
2006	65.51	413.84	2265	4.35

63. An increase in oil price in dollars per barrel will drive the U.S. Dow Jones Utilities Stock Index to soar.

a. Use the calculator commands STAT, linReg $(ax + b)$, and STATPLOT to model the data using the least squares regression. Find the equation of the least-squares regression line using x as the oil price in dollars per barrel.

b. If the U.S. Dow Jones Utilities Stock Index varies directly as the oil price in dollars per barrel, then use the calculator commands STAT, PwrReg, and STATPLOT to model the data using the power function. Find the variation constant and equation of variation using x as the oil price in dollars per barrel.

c. Use the equations you found in (a) and (b) to predict the stock index when the oil price hits $72.70 per barrel in September 2006. Which answer is closer to the actual stock index of 417? Round all answers to the nearest whole number.

64. An increase in oil price in dollars per barrel will affect the interest rates across the board—in particular, the 5-year Treasury constant maturity rate.

a. Use the calculator commands STAT, linReg $(ax + b)$, and STATPLOT to model the data using the least-squares regression. Find the equation of the least-squares regression line using x as the oil price in dollars per barrel.

b. If the 5-year Treasury constant maturity rate varies inversely as the oil price in dollars per barrel, then use the calculator commands STAT, PwrReg, and STATPLOT to model the data using the power function. Find the variation constant and equation of variation using x as the oil price in dollars per barrel.

c. Use the equations you found in (a) and (b) to predict the maturity rate when the oil price hits $72.70 per barrel in September 2006. Which answer is closer to the actual maturity rate at 5.02%? Round all answers to two decimal places.

65. An increase in interest rates—in particular, the 5-year Treasury constant maturity rate—will affect the number of new, privately owned housing units.

    a. Use the calculator commands $\boxed{\text{STAT}}$, $\boxed{\text{linReg}}$ $(ax + b)$, and $\boxed{\text{STATPLOT}}$ to model the data using the least-squares regression. Find the equation of the least-squares regression line using x as the 5-year rate.

    b. If the number of new privately owned housing units varies inversely as the 5-year Treasury constant maturity rate, then use the calculator commands $\boxed{\text{STAT}}$, $\boxed{\text{PwrReg}}$, and $\boxed{\text{STATPLOT}}$ to model the data using the power function. Find the variation constant and equation of variation using x as the 5-year rate.

    c. Use the equations you found in (a) and (b) to predict the number of housing units when the maturity rate is 5.02% in September 2006. Which answer is closer to the actual number of new, privately owned housing units of 1861? Round all answers to the nearest unit.

66. An increase in the number of new, privately owned housing units will affect the U.S. Dow Jones Utilities Stock Index.

    a. Use the calculator commands $\boxed{\text{STAT}}$, $\boxed{\text{linReg}}$ $(ax + b)$, and $\boxed{\text{STATPLOT}}$ to model the data using the least-squares regression. Find the equation of the least-squares regression line using x as the number of housing units.

    b. If the U.S. Dow Jones Utilities Stock Index varies directly as the number of new, privately owned housing units, then use the calculator commands $\boxed{\text{STAT}}$, $\boxed{\text{PwrReg}}$, and $\boxed{\text{STATPLOT}}$ to model the data using the power function. Find the variation constant and equation of variation using x as the number of housing units.

    c. Use the equations you found in (a) and (b) to predict the utilities stock index if there are 1861 new, privately owned housing units in September 2006. Which answer is closer to the actual stock index of 417? Round all answers to the nearest whole number.

For Exercises 67 and 68, refer to the following:

Data for retail gasoline price in dollars per gallon for the period March 2000 to March 2008 are given in the following table. (Data are from Energy Information Administration, Official Energy Statistics from the U.S. government at http://tonto.eia.doe.gov/oog/info/gdu/gaspump.html.) Use the calculator $\boxed{\text{STAT}}$ $\boxed{\text{EDIT}}$ command to enter the table below with L_1 as the year ($x = 1$ for year 2000) and L_2 as the gasoline price in dollars per gallon.

MARCH OF EACH YEAR	2000	2001	2002	2003	2004	2005	2006	2007	2008
RETAIL GASOLINE PRICE $ PER GALLON	1.517	1.409	1.249	1.693	1.736	2.079	2.425	2.563	3.244

67. a. Use the calculator commands $\boxed{\text{STAT}}$ $\boxed{\text{LinReg}}$ to model the data using the least-squares regression. Find the equation of the least-squares regression line using x as the year ($x = 1$ for year 2000) and y as the gasoline price in dollars per gallon. Round all answers to three decimal places.

    b. Use the equation to predict the gasoline price in March 2006. Round all answers to three decimal places. Is the answer close to the actual price?

    c. Use the equation to predict the gasoline price in March 2009. Round all answers to three decimal places.

68. a. Use the calculator commands $\boxed{\text{STAT}}$ $\boxed{\text{PwrReg}}$ to model the data using the power function. Find the variation constant and equation of variation using x as the year ($x = 1$ for year 2000) and y as the gasoline price in dollars per gallon. Round all answers to three decimal places.

    b. Use the equation to predict the gasoline price in March 2006. Round all answers to three decimal places. Is the answer close to the actual price?

    c. Use the equation to predict the gasoline price in March 2009. Round all answers to three decimal places.

Transformations of Functions

Being a creature of habit, Dylan usually sets out each morning at 7 AM from his house for a jog. Figure 1 shows the graph of a function, $y = d(t)$, that represents Dylan's jog on Friday.

a. Use the graph in Figure 1 to fill in the table below.

Figure 1

t										
$y = d(t)$										

Describe a jogging scenario that fits the graph and table above.

b. The graph shown in Figure 2 represents Dylan's jog on Saturday. It is a transformation of the function $y = d(t)$ shown in Figure 1.

Complete the table of values below for this transformation. You may find it helpful to refer to the table in part (a).

Figure 2

t										
y										

What is the real-world meaning of this transformation? How is Dylan's jog on Saturday different from his usual jog? How is it the same?

The original function (in Figure 1) is represented by the equation $y = d(t)$. Write an equation, in terms of $d(t)$, that represents the function graphed in Figure 2. Explain.

c. The graph shown in Figure 3 represents Dylan's jog on Sunday. It is a transformation of the function $y = d(t)$ shown in Figure 1.

Complete the table of values below for this transformation.

Figure 3

t										
y										

What is the real-world meaning of this transformation? How is Dylan's jog on Sunday different from his usual jog?

The original function (in Figure 1) is represented by the equation $y = d(t)$. Use function notation to represent the function graphed in Figure 3. Explain.

d. Suppose Dylan's jog on Monday can be represented by the equation $y = \frac{1}{2}d(t)$.

Complete the table of values below and sketch a graph at the right for this transformation.

t										
y										

What is the real-world meaning of this transformation? How does Dylan's jog on Monday differ from his usual jog? How is it the same?

e. Suppose Dylan has a goal of cutting his usual jogging time in half, while covering the same distance. Represent this scenario as a transformation of $y = d(t)$ shown in Figure 1. Complete the table, sketch a graph, and write an equation in function notation. Explain why your equation makes sense. Finally, discuss whether you think Dylan's goal is realistic.

$y = $ _____

t										
y										

The U.S. National Oceanic and Atmospheric Association (NOAA) monitors temperature and carbon emissions at its observatory in Mauna Loa, Hawaii. NOAA's goal is to help foster an informed society that uses a comprehensive understanding of the role of the oceans, coasts, and atmosphere in the global ecosystem to make the best social and economic decisions. The data presented in this chapter is from the Mauna Loa Observatory, where historical atmospheric measurements have been recorded for the last 50 years. You will develop linear models based on this data to predict temperature and carbon emissions in the future.

The following table summarizes average yearly temperature in degrees Fahrenheit °F and carbon dioxide emissions in parts per million (ppm) for Mauna Loa, Hawaii.

Year	1960	1965	1970	1975	1980	1985	1990	1995	2000	2005
Temperature (°F)	44.45	43.29	43.61	43.35	46.66	45.71	45.53	47.53	45.86	46.23
CO_2 Emissions (ppm)	316.9	320.0	325.7	331.1	338.7	345.9	354.2	360.6	369.4	379.7

1. Plot the temperature data with time on the horizontal axis and temperature on the vertical axis. Let $t = 0$ correspond to 1960.

2. Find a *linear function* that models the temperature in Mauna Loa.
   a. Use data from 1965 and 1995.
   b. Use data from 1960 and 1990.
   c. Use linear regression and all data given.

3. Predict what the temperature will be in Mauna Loa in 2020.
   a. Apply the line found in Exercise 2(a).
   b. Apply the line found in Exercise 2(b).
   c. Apply the line found in Exercise 2(c).

4. Predict what the temperature will be in Mauna Loa in 2100.
   a. Apply the line found in Exercise 2(a).
   b. Apply the line found in Exercise 2(b).
   c. Apply the line found in Exercise 2(c).

5. Do you think your models support the claim of "global warming"? Explain.

6. Plot the carbon dioxide emissions data with time on the horizontal axis and carbon dioxide levels on the vertical axis. Let $t = 0$ correspond to 1960.

7. Find a *linear function* that models the CO_2 emissions (ppm) in Mauna Loa.
   a. Use data from 1965 and 1995.
   b. Use data from 1960 and 1990.
   c. Use linear regression and all data given.

8. Predict the expected CO_2 levels in Mauna Loa in 2020.
   a. Apply the line found in Exercise 7(a).
   b. Apply the line found in Exercise 7(b).
   c. Apply the line found in Exercise 7(c).

9. Predict the expected CO_2 levels in Mauna Loa in 2100.
   a. Apply the line found in Exercise 7(a).
   b. Apply the line found in Exercise 7(b).
   c. Apply the line found in Exercise 7(c).

10. Do you think your models support the claim of the "greenhouse effect"? Explain.

SECTION	CONCEPT	KEY IDEAS/FORMULAS
3.1	**Functions**	
	Relations and functions	All functions are relations, but not all relations are functions.
	Functions defined by equations	A vertical line can intersect a function in at most one point.
	Function notation	*Placeholder notation:* $f(x) = 3x^2 - 6x + 2$ $\quad f(\square) = 3(\square)^2 - 6(\square) + 2$ *Difference quotient:* $\dfrac{f(x + h) - f(x)}{h}; h \neq 0$
	Domain of a function	Are there any restrictions on x?
3.2	**Graphs of functions; piecewise-defined functions; increasing and decreasing functions; average rate of change**	
	Recognizing and classifying functions	*Common functions* $f(x) = mx + b, f(x) = x, f(x) = x^2,$ $f(x) = x^3, f(x) = \sqrt{x}, f(x) = \sqrt[3]{x},$ $f(x) = |x|, f(x) = \dfrac{1}{x}$ *Even and odd functions* Even: Symmetry about y-axis: $f(-x) = f(x)$ Odd: Symmetry about origin: $f(-x) = -f(x)$
	Increasing and decreasing functions	• Increasing: rises (left to right) • Decreasing: falls (left to right)
	Average rate of change	$\dfrac{f(x_2) - f(x_1)}{x_2 - x_1} \qquad x_1 \neq x_2$
	Piecewise-defined functions	Points of discontinuity
3.3	**Graphing techniques: Transformations**	Shift the graph of $f(x)$.
	Horizontal and vertical shifts	$f(x + c)$ $\quad$ c units to the left, $c > 0$ $f(x - c)$ $\quad$ c units to the right, $c > 0$ $f(x) + c$ $\quad$ c units upward, $c > 0$ $f(x) - c$ $\quad$ c units downward, $c > 0$
	Reflection about the axes	$-f(x)$ $\quad$ Reflection about the x-axis $f(-x)$ $\quad$ Reflection about the y-axis
	Stretching and compressing	$cf(x)$ if $c > 1$; stretch vertically $cf(x)$ if $0 < c < 1$; compress vertically $f(cx)$ if $c > 1$; compress horizontally $f(cx)$ if $0 < c < 1$; stretch horizontally

Section	Concept	Key Ideas/Formulas
3.4	**Operations on functions and composition of functions**	
	Adding, subtracting, multiplying, and dividing functions	$(f + g)(x) = f(x) + g(x)$ $(f - g)(x) = f(x) - g(x)$ $(f \cdot g)(x) = f(x) \cdot g(x)$ Domain of the resulting function is the intersection of the individual domains. $\left(\dfrac{f}{g}\right)(x) = \dfrac{f(x)}{g(x)}, \ g(x) \neq 0$ Domain of the quotient is the intersection of the domains of f and g, and any points when $g(x) = 0$ must be eliminated.
	Composition of functions	$(f \circ g)(x) = f(g(x))$ The domain of the composite function is a subset of the domain of $g(x)$. Values for x must be eliminated if their corresponding values $g(x)$ are not in the domain of f.
3.5	**One-to-one functions and inverse functions**	
	Determine whether a function is one-to-one	• No two x-values map to the same y-value. If $f(x_1) = f(x_2)$, then $x_1 = x_2$. • A horizontal line may intersect a one-to-one function in at most one point.
	Inverse functions	• Only one-to-one functions have inverses. • $f^{-1}(f(x)) = x$ and $f(f^{-1}(x)) = x$. • Domain of f = range of f^{-1}. Range of f = domain of f^{-1}.
	Graphical interpretation of inverse functions	• The graph of a function and its inverse are symmetric about the line $y = x$. • If the point (a, b) lies on the graph of a function, then the point (b, a) lies on the graph of its inverse.
	Finding the inverse function	1. Let $y = f(x)$. 2. Interchange x and y. 3. Solve for y. 4. Let $y = f^{-1}(x)$.
3.6	**Modeling functions using variation**	
	Direct variation	$y = kx$
	Inverse variation	$y = \dfrac{k}{x}$
	Joint variation and combined variation	Joint: One quantity is directly proportional to the product of two or more other quantities. Combined: Direct variation and inverse variation occur at the same time.

3.1 Functions

Determine whether each relation is a function.

1.

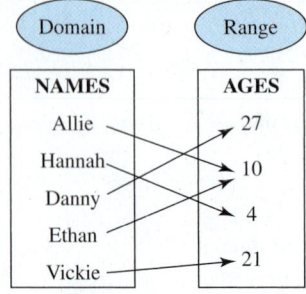

Domain	Range
NAMES	AGES

Allie, Hannah, Danny, Ethan, Vickie → 27, 10, 4, 21

2. {(1, 2), (3, 4), (2, 4), (3, 7)}

3. {(−2, 3), (1, −3), (0, 4), (2, 6)}

4. {(4, 7), (2, 6), (3, 8), (1, 7)}

5. $x^2 + y^2 = 36$

6. $x = 4$

7. $y = |x + 2|$

8. $y = \sqrt{x}$

9.

10.

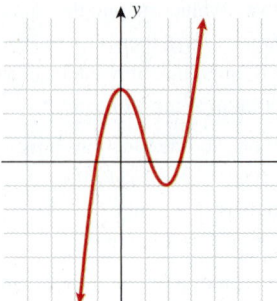

Use the graphs of the functions to find:

11.

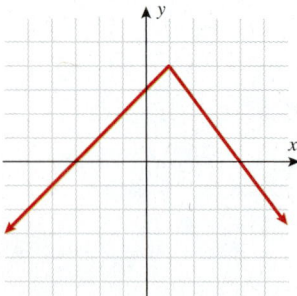

a. $f(-1)$ **b.** $f(1)$

c. x, where $f(x) = 0$

12.

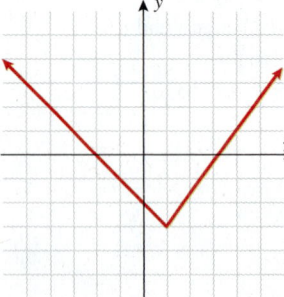

a. $f(-4)$ **b.** $f(0)$

c. x, where $f(x) = 0$

13.

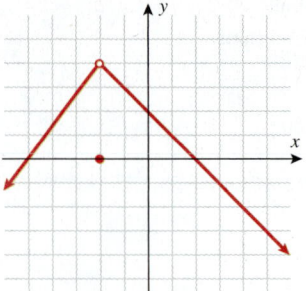

a. $f(-2)$ **b.** $f(4)$

c. x, where $f(x) = 0$

14.

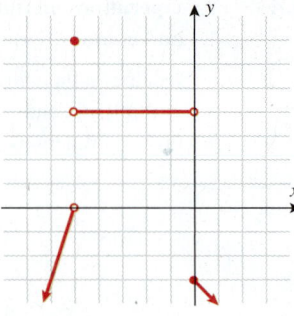

a. $f(-5)$ **b.** $f(0)$

c. x, where $f(x) = 0$

Evaluate the given quantities using the following three functions.

$$f(x) = 4x - 7 \qquad F(t) = t^2 + 4t - 3 \qquad g(x) = |x^2 + 2x + 4|$$

15. $f(3)$

16. $F(4)$

17. $f(-7) \cdot g(3)$

18. $\dfrac{F(0)}{g(0)}$

19. $\dfrac{f(2) - F(2)}{g(0)}$

20. $f(3 + h)$

21. $\dfrac{f(3 + h) - f(3)}{h}$

22. $\dfrac{F(t + h) - F(t)}{h}$

Find the domain of the given function. Express the domain in interval notation.

23. $f(x) = -3x - 4$

24. $g(x) = x^2 - 2x + 6$

25. $h(x) = \dfrac{1}{x + 4}$

26. $F(x) = \dfrac{7}{x^2 + 3}$

27. $G(x) = \sqrt{x - 4}$

28. $H(x) = \dfrac{1}{\sqrt{2x - 6}}$

Challenge

29. If $f(x) = \dfrac{D}{x^2 - 16}$, $f(4)$ and $f(-4)$ are undefined, and $f(5) = 2$, find D.

30. Construct a function that is undefined at $x = -3$ and $x = 2$ such that the point $(0, -4)$ lies on the graph of the function.

3.2 Graphs of Functions

Determine whether the function is even, odd, or neither.

31. $f(x) = 2x - 7$

32. $g(x) = 7x^5 + 4x^3 - 2x$

33. $h(x) = x^3 - 7x$

34. $f(x) = x^4 + 3x^2$

35. $f(x) = x^{1/4} + x$

36. $f(x) = \sqrt{x + 4}$

37. $f(x) = \dfrac{1}{x^3} + 3x$

38. $f(x) = \dfrac{1}{x^2} + 3x^4 + |x|$

Use the graph of the functions to find:

a. Domain
b. Range
c. Intervals on which the function is increasing, decreasing, or constant.

39.   **40.**

41. Find the average rate of change of $f(x) = 4 - x^2$ from $x = 0$ to $x = 2$.

42. Find the average rate of change of $f(x) = |2x - 1|$ from $x = 1$ to $x = 5$.

Graph the piecewise-defined function. State the domain and range in interval notation.

43. $F(x) = \begin{cases} x^2 & x < 0 \\ 2 & x \geq 0 \end{cases}$

44. $f(x) = \begin{cases} -2x - 3 & x \leq 0 \\ 4 & 0 < x \leq 1 \\ x^2 + 4 & x > 1 \end{cases}$

45. $f(x) = \begin{cases} x^2 & x \leq 0 \\ -\sqrt{x} & 0 < x \leq 1 \\ |x + 2| & x > 1 \end{cases}$

46. $F(x) = \begin{cases} x^2 & x < 0 \\ x^3 & 0 < x < 1 \\ -|x| - 1 & x \geq 1 \end{cases}$

Applications

47. Tutoring Costs. A tutoring company charges $25.00 for the first hour of tutoring and $10.50 for every 30-minute period after that. Find the cost function $C(x)$ as a function of the length of the tutoring session. Let x = number of 30-minute periods.

48. Salary. An employee who makes $30.00 per hour also earns time and a half for overtime (any hours worked above the normal 40-hour work week). Write a function $E(x)$ that describes her weekly earnings as a function of the number of hours worked x.

3.3 Graphing Techniques: Transformations

Graph the following functions using graphing aids.

49. $y = -(x - 2)^2 + 4$ **50.** $y = |-x + 5| - 7$

51. $y = \sqrt[3]{x - 3} + 2$ **52.** $y = \dfrac{1}{x - 2} - 4$

53. $y = -\dfrac{1}{2}x^3$ **54.** $y = 2x^2 + 3$

Use the given graph to graph the following:

55.   **56.**

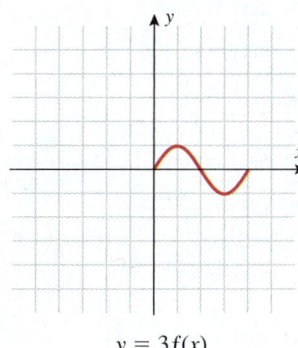

$y = f(x - 2)$ $y = 3f(x)$

57.   **58.**

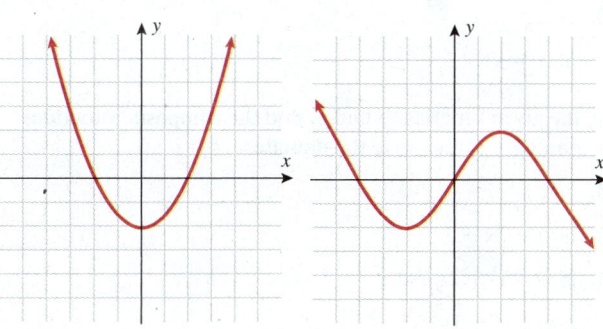

$y = -2f(x)$ $y = f(x) + 3$

Write the function whose graph is the graph of $y = \sqrt{x}$, but is transformed accordingly, and state the domain of the resulting function.

59. Shifted to the left three units

60. Shifted down four units

61. Shifted to the right two units and up three units

62. Reflected about the y-axis

63. Stretched by a factor of 5 and shifted down six units

64. Compressed by a factor of 2 and shifted up three units

Transform the function into the form $f(x) = c(x - h)^2 + k$ by completing the square and graph the resulting function using transformations.

65. $y = x^2 + 4x - 8$ **66.** $y = 2x^2 + 6x - 5$

3.4 Operations on Functions and Composition of Functions

Given the functions g and h, find $g + h, g - h, g \cdot h$, and $\dfrac{g}{h}$, and state the domain.

67. $g(x) = -3x - 4$

$h(x) = x - 3$

68. $g(x) = 2x + 3$

$h(x) = x^2 + 6$

69. $g(x) = \dfrac{1}{x^2}$

$h(x) = \sqrt{x}$

70. $g(x) = \dfrac{x + 3}{2x - 4}$

$h(x) = \dfrac{3x - 1}{x - 2}$

71. $g(x) = \sqrt{x - 4}$

$h(x) = \sqrt{2x + 1}$

72. $g(x) = x^2 - 4$

$h(x) = x + 2$

For the given functions f and g, find the composite functions $f \circ g$ and $g \circ f$, and state the domains.

73. $f(x) = 3x - 4$

$g(x) = 2x + 1$

74. $f(x) = x^3 + 2x - 1$

$g(x) = x + 3$

75. $f(x) = \dfrac{2}{x + 3}$

$g(x) = \dfrac{1}{4 - x}$

76. $f(x) = \sqrt{2x^2 - 5}$

$g(x) = \sqrt{x + 6}$

77. $f(x) = \sqrt{x - 5}$

$g(x) = x^2 - 4$

78. $f(x) = \dfrac{1}{\sqrt{x}}$

$g(x) = \dfrac{1}{x^2 - 4}$

Evaluate $f(g(3))$ and $g(f(-1))$, if possible.

79. $f(x) = 4x^2 - 3x + 2$

$g(x) = 6x - 3$

80. $f(x) = \sqrt{4 - x}$

$g(x) = x^2 + 5$

81. $f(x) = \dfrac{x}{|2x - 3|}$

$g(x) = |5x + 2|$

82. $f(x) = \dfrac{1}{x - 1}$

$g(x) = x^2 - 1$

83. $f(x) = x^2 - x + 10$

$g(x) = \sqrt[3]{x - 4}$

84. $f(x) = \dfrac{4}{x^2 - 2}$

$g(x) = \dfrac{1}{x^2 - 9}$

Write the function as a composite $f(g(x))$ of two functions f and g.

85. $h(x) = 3(x - 2)^2 + 4(x - 2) + 7$

86. $h(x) = \dfrac{\sqrt[3]{x}}{1 - \sqrt[3]{x}}$

87. $h(x) = \dfrac{1}{\sqrt{x^2 + 7}}$

88. $h(x) = \sqrt{|3x + 4|}$

Applications

89. Rain. A rain drop hitting a lake makes a circular ripple. If the radius, in inches, grows as a function of time, in minutes, $r(t) = 25\sqrt{t + 2}$, find the area of the ripple as a function of time.

90. Geometry. Let the area of a rectangle be given by $42 = l \cdot w$, and let the perimeter be $36 = 2 \cdot l + 2 \cdot w$. Express the perimeter in terms of w.

3.5 One-to-One Functions and Inverse Functions

Determine whether the given function is a one-to-one function.

91.

92.

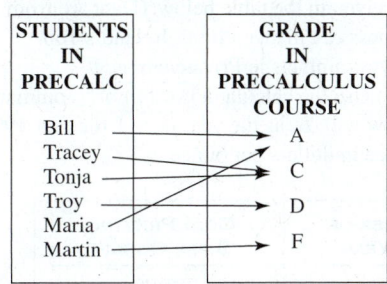

93. $\{(2, 3), (-1, 2), (3, 3), (-3, -4), (-2, 1)\}$

94. $\{(-3, 9), (5, 25), (2, 4), (3, 9)\}$

95. $\{(-2, 0), (4, 5), (3, 7)\}$

96. $\{(-8, -6), (-4, 2), (0, 3), (2, -8), (7, 4)\}$

97. $y = \sqrt{x}$ **98.** $y = x^2$ **99.** $f(x) = x^3$ **100.** $f(x) = \dfrac{1}{x^2}$

Verify that the function $f^{-1}(x)$ is the inverse of $f(x)$ by showing that $f(f^{-1}(x)) = x$. Graph $f(x)$ and $f^{-1}(x)$ on the same graph and show the symmetry about the line $y = x$.

101. $f(x) = 3x + 4;\ f^{-1}(x) = \dfrac{x - 4}{3}$

102. $f(x) = \dfrac{1}{4x - 7};\ f^{-1}(x) = \dfrac{1 + 7x}{4x}$

103. $f(x) = \sqrt{x + 4};\ f^{-1}(x) = x^2 - 4 \quad x \ge 0$

104. $f(x) = \dfrac{x + 2}{x - 7};\ f^{-1}(x) = \dfrac{7x + 2}{x - 1}$

The function f is one-to-one. Find its inverse and check your answer. State the domain and range of both f and f^{-1}.

105. $f(x) = 2x + 1$ **106.** $f(x) = x^5 + 2$

107. $f(x) = \sqrt{x + 4}$ **108.** $f(x) = (x + 4)^2 + 3 \quad x \ge -4$

109. $f(x) = \dfrac{x + 6}{x + 3}$ **110.** $f(x) = 2\sqrt[3]{x - 5} - 8$

Applications

111. Salary. A pharmaceutical salesperson makes \$22,000 base salary a year plus 8% of the total products sold. Write a function $S(x)$ that represents her yearly salary as a function of the total dollars worth of products sold x. Find $S^{-1}(x)$. What does this inverse function tell you?

112. Volume. Express the volume V of a rectangular box that has a square base of length s and is 3 feet high as a function of the square length. Find V^{-1}. If a certain volume is desired, what does the inverse tell you?

3.6 Modeling Functions Using Variation

Write an equation that describes each variation.

113. C is directly proportional to r. $C = 2\pi$ when $r = 1$.

114. V is directly proportional to both l and w. $V = 12h$ when $w = 6$ and $l = 2$.

115. A varies directly with the square of r. $A = 25\pi$ when $r = 5$.

116. F varies inversely with both λ and L. $F = 20\pi$ when $\lambda = 10\ \mu m$ and $L = 10$ km.

Applications

117. Wages. Cole and Dickson both work at the same museum and have the following paycheck information for a certain week. Find an equation that shows their wages (W) varying directly with the number of hours (H) worked.

EMPLOYEE	HOURS WORKED	WAGE
Cole	27	\$229.50
Dickson	30	\$255.00

118. Sales Tax. The sales tax in two neighboring counties differs by 1%. A new resident knows the difference but doesn't know which county has the higher tax rate. The resident lives near the border of the two counties and wants to buy a new car. If the tax on a \$50.00 jacket is \$3.50 in County A and the tax on a \$20.00 calculator is \$1.60 in County B, write two equations (one for each county) that describe the tax (T), which is directly proportional to the purchase price (P).

Technology Exercises

Section 3.1

119. Use a graphing utility to graph the function and find the domain. Express the domain in interval notation.

$$f(x) = \dfrac{1}{\sqrt{x^2 - 2x - 3}}$$

120. Use a graphing utility to graph the function and find the domain. Express the domain in interval notation.

$$f(x) = \dfrac{x^2 - 4x - 5}{x^2 - 9}$$

Section 3.2

121. Use a graphing utility to graph the function. State the (a) domain, (b) range, and (c) x intervals where the function is increasing, decreasing, and constant.

$$f(x) = \begin{cases} 1 - x & x < -1 \\ [[x]] & -1 \le x < 2 \\ x + 1 & x > 2 \end{cases}$$

122. Use a graphing utility to graph the function. State the (a) domain, (b) range, and (c) x intervals where the function is increasing, decreasing, and constant.

$$f(x) = \begin{cases} |x^2 - 1| & -2 < x < 2 \\ \sqrt{x - 2} + 4 & x > 2 \end{cases}$$

Section 3.3

123. Use a graphing utility to graph $f(x) = x^2 - x - 6$ and $g(x) = x^2 - 5x$. Use transforms to describe the relationship between $f(x)$ and $g(x)$?

124. Use a graphing utility to graph $f(x) = 2x^2 - 3x - 5$ and $g(x) = -2x^2 - x + 6$. Use transforms to describe the relationship between $f(x)$ and $g(x)$?

Section 3.4

125. Using a graphing utility, plot $y_1 = \sqrt{2x + 3}$, $y_2 = \sqrt{4 - x}$, and $y_3 = \dfrac{y_1}{y_2}$. What is the domain of y_3?

126. Using a graphing utility, plot $y_1 = \sqrt{x^2 - 4}$, $y_2 = x^2 - 5$, and $y_3 = y_1^2 - 5$. If y_1 represents a function f and y_2 represents a function g, then y_3 represents the composite function $g \circ f$. The graph of y_3 is only defined for the domain of $g \circ f$. State the domain of $g \circ f$.

Section 3.5

127. Use a graphing utility to graph the function and determine whether it is one-to-one.

$$f(x) = \frac{6}{\sqrt[5]{x^3 - 1}}$$

128. Use a graphing utility to graph the functions f and g and the line $y = x$ in the same screen. Are the two functions inverses of each other?

$$f(x) = \sqrt[4]{x - 3} + 1, \, g(x) = x^4 - 4x^3 + 6x^2 - 4x + 3$$

Section 3.6

From December 1999 to December 2007, data for gold price in dollars per ounce are given in the table below. (Data are from Finfacts Ireland Business & Finance Portal, Ireland's Top Business website at www.finfacts.ie/Private/curency/goldmarketprice.htm.) Use the calculator $\boxed{\text{STAT}}$ $\boxed{\text{EDIT}}$ commands to enter the table below with L_1 as the year ($x = 1$ for year 1999) and L_2 as the gold price in dollars per ounce.

DECEMBER OF EACH YEAR	GOLD PRICE IN $ PER OUNCE
1999	287.90
2000	272.15
2001	278.70
2002	346.70
2003	414.80
2004	438.10
2005	517.20
2006	636.30
2007	833.20

129. a. Use the calculator commands $\boxed{\text{STAT}}$ $\boxed{\text{LinReg}}$ to model the data using the least-squares regression. Find the equation of the least-squares regression line using x as the year ($x = 1$ for year 1999) and y as the gold price in dollars per ounce. Round all answers to two decimal places.

b. Use the equation to predict the gold price in December 2005. Round all answers to two decimal places. Is the answer close to the actual price?

c. Use the equation to predict the gold price in December 2008. Round all answers to two decimal places.

130. a. Use the calculator commands $\boxed{\text{STAT}}$ $\boxed{\text{PwrReg}}$ to model the data using the power function. Find the variation constant and equation of variation using x as the year ($x = 1$ for year 1999) and y as the gold price in dollars per ounce. Round all answers to two decimal places.

b. Use the equation to predict the gold price in December 2005. Round all answers to two decimal places. Is the answer close to the actual price?

c. Use the equation to predict the gold price in December 2008. Round all answers to two decimal places.

REVIEW EXERCISES

Assuming that x represents the independent variable and y represents the dependent variable, classify the relationships as:

a. not a function
b. a function, but not one-to-one
c. a one-to-one function

1. $f(x) = |2x + 3|$ 2. $x = y^2 + 2$ 3. $y = \sqrt[3]{x + 1}$

Use $f(x) = \sqrt{x - 2}$ and $g(x) = x^2 + 11$, and determine the desired quantity or expression. In the case of an expression, state the domain.

4. $f(11) - 2g(-1)$

5. $\left(\dfrac{f}{g}\right)(x)$

6. $\left(\dfrac{g}{f}\right)(x)$

7. $g(f(x))$

8. $(f + g)(6)$

9. $f(g(\sqrt{7}))$

Determine whether the function is odd, even, or neither.

10. $f(x) = |x| - x^2$ 11. $f(x) = 9x^3 + 5x - 3$ 12. $f(x) = \dfrac{2}{x}$

Graph the functions. State the domain and range of each function.

13. $f(x) = -\sqrt{x - 3} + 2$

14. $f(x) = -2(x - 1)^2$

15. $f(x) = \begin{cases} -x & x < -1 \\ 1 & -1 < x < 2 \\ x^2 & x \geq 2 \end{cases}$

Use the graphs of the function to find:

16.

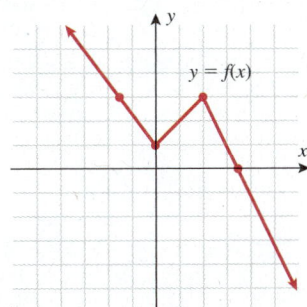

a. $f(3)$ b. $f(0)$ c. $f(-4)$
d. x, where $f(x) = 3$ e. x, where $f(x) = 0$

17.

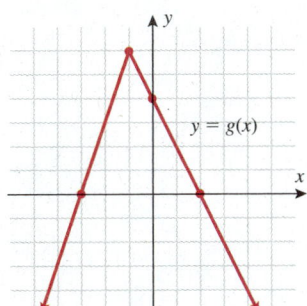

a. $g(3)$ b. $g(0)$ c. $g(-4)$
d. x, where $g(x) = 0$

18.

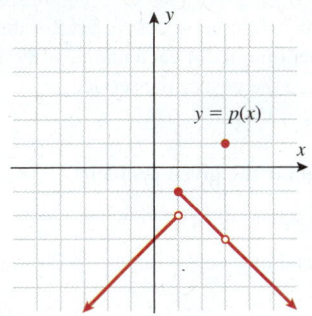

a. $p(0)$ b. x, where $p(x) = 0$
c. $p(1)$ d. $p(3)$

Find $\dfrac{f(x + h) - f(x)}{h}$ for:

19. $f(x) = 3x^2 - 4x + 1$ 20. $f(x) = 5 - 7x$

Find the average rate of change of the given functions.

21. $f(x) = 64 - 16x^2$ for $x = 0$ to $x = 2$

22. $f(x) = \sqrt{x - 1}$ for $x = 2$ to $x = 10$

Given the function f, find the inverse if it exists. State the domain and range of both f and f^{-1}.

23. $f(x) = \sqrt{x - 5}$

24. $f(x) = x^2 + 5$

25. $f(x) = \dfrac{2x + 1}{5 - x}$

26. $f(x) = \begin{cases} -x & x \leq 0 \\ -x^2 & x > 0 \end{cases}$

27. What domain restriction can be made so that $f(x) = x^2$ has an inverse?

28. If the point $(-2, 5)$ lies on the graph of a function, what point lies on the graph of its inverse function?

29. **Discount.** Suppose a suit has been marked down 40% off the original price. An advertisement in the newspaper has an "additional 30% off the sale price" coupon. Write a function that determines the "checkout" price of the suit.

30. **Temperature.** Degrees Fahrenheit (°F), degrees Celsius (°C), and kelvins (K) are related by the two equations: $F = \frac{9}{5}C + 32$ and $K = C + 273.15$. Write a function whose input is kelvins and output is degrees Fahrenheit.

31. **Circles.** If a quarter circle is drawn by tracing the unit circle in quadrant III, what does the inverse of that function look like? Where is it located?

32. **Sprinkler.** A sprinkler head malfunctions at midfield in an NFL football field. The puddle of water forms a circular pattern around the sprinkler head with a radius in yards that grows as a function of time, in hours: $r(t) = 10\sqrt{t}$. When will the puddle reach the sidelines? (A football field is 30 yards from sideline to sideline.)

33. Internet. The cost of airport Internet access is $15 for the first 30 minutes and $1 per minute for each minute after that. Write a function describing the cost of the service as a function of minutes used.

Use variation to find a model for the given problem.

34. y varies directly with the square of x. $y = 8$ when $x = 5$.

35. F varies directly with m and inversely with p. $F = 20$ when $m = 2$ and $p = 3$.

36. Use a graphing utility to graph the function. State the (a) domain, (b) range, and (c) x intervals where the function is increasing, decreasing, and constant.

$$f(x) = \begin{cases} 5 & -4 \le x < -2 \\ |3 + 2x - x^2| & -2 < x \le 4 \end{cases}$$

37. Use a graphing utility to graph the function and determine whether it is one-to-one.

$$y = x^3 - 12x^2 + 48x - 65$$

1. Simplify $\dfrac{2}{3 - \sqrt{5}}$.

2. Factor completely: $10x^2 - 29x - 21$.

3. Simplify and state the domain: $\dfrac{x^3 - 4x}{x + 2}$.

4. Solve for x: $\dfrac{1}{6}x = -\dfrac{1}{5}x + 11$.

5. Perform the operation, simplify, and express in standard form: $(8 - 9i)(8 + 9i)$.

6. Solve for x, and give any excluded values: $\dfrac{5}{x} - 10 = \dfrac{10}{3x}$.

7. The original price of a hiking stick is $59.50. The sale price is $35.70. Find the percent of the markdown.

8. Solve by factoring: $x(6x + 1) = 12$.

9. Solve by completing the square: $\dfrac{x^2}{2} - x = \dfrac{1}{5}$.

10. Solve and check: $\sqrt[3]{x + 2} = -3$.

11. Solve using substitution: $x^4 - x^2 - 12 = 0$.

Solve and express the solution in interval notation.

12. $-7 < 3 - 2x \le 5$

13. $\dfrac{x}{x - 5} < 0$

14. $|2.7 - 3.2x| \le 1.3$

15. Calculate the distance and midpoint between the segment joining the points $(-2.7, -1.4)$ and $(5.2, 6.3)$.

16. Find the slope of the line passing through the points $(0.3, -1.4)$ and $(2.7, 4.3)$.

17. Write an equation of a line that passes through the points $(1.2, -3)$ and $(-0.2, -3)$.

18. Transform the equation into standard form by completing the square, and state the center and radius of the circle: $x^2 + y^2 + 12x - 18y - 4 = 0$.

19. Find the equation of a circle with center $(-2, -1)$ and passing through the point $(-4, 3)$.

20. If a cellular phone tower has a reception radius of 100 miles and you live 85 miles north and 23 miles east of the tower, can you use your cell phone at home? Explain.

21. Use interval notation to express the domain of the function $g(x) = \dfrac{1}{x - 1}$.

22. Find the average rate of change for $f(x) = 5x^2$, from $x = 2$ to $x = 4$.

23. Evaluate $g(f(-1))$ for $f(x) = |6 - x|$ and $g(x) = x^2 - 3$.

24. Find the inverse of the function $f(x) = x^2 + 3$ for $x \ge 0$.

25. Write an equation that describes the variation: r is inversely proportional to t. $r = 45$ when $t = 3$.

26. Use a graphing utility to graph the function. State the (a) domain, (b) range, and (c) x intervals where the function is increasing, decreasing, and constant.

$$f(x) = \begin{cases} 1 - |x| & -1 \le x < 1 \\ 1 - |x - 2| & 1 < x \le 3 \end{cases}$$

27. Use a graphing utility to graph the function $f(x) = x^2 - 3x$ and $g(x) = x^2 + x - 2$ in the same screen. Find the function h such that $g \circ h = f$.

4

Polynomial and Rational Functions

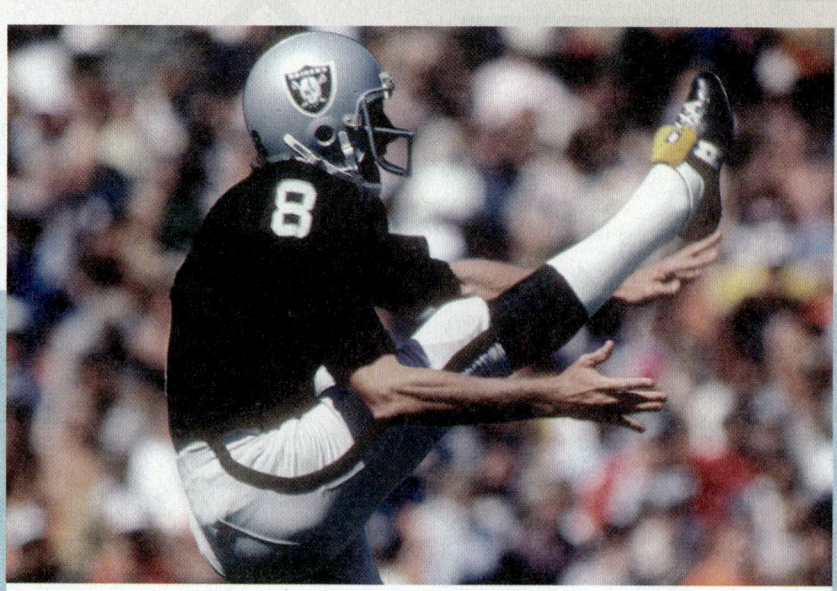

Focus on Sport/Getty Images

Ray Guy

Indoor football stadiums are designed so that football punters will not hit the roof with the football. One of the greatest NFL punters of all time was Ray Guy, who played 14 seasons from 1973 to 1986. "In the 1976 Pro Bowl, one of his punts hit the giant TV screen hanging from the rafters in the Louisiana Superdome. Not only did Guy punt high and far—the term 'hang time' came into the NFL lexicon during his tenure—once he even had an opponent take a ball he punted and test it for helium!" (www.prokicker.com; Ray Guy Fact Sheet)

The path that punts typically follow is called a *parabola* and is classified as a *quadratic function*. The distance of a punt is measured in the horizontal direction. The yard line where the punt is kicked from and the yard line where the punt either hits the field or is caught are the zeros of the quadratic function. Zeros are the points where the function value is equal to zero.

IN THIS CHAPTER we will start by discussing quadratic functions (polynomial functions of degree 2) whose graphs are parabolas. We will find the vertex, which is the maximum or minimum point on the graph. Then we will expand our discussion to higher degree polynomial functions. We will discuss techniques to find zeros of polynomial functions and strategies for graphing polynomial functions. Last we will discuss rational functions, which are ratios of polynomial functions.

POLYNOMIAL AND RATIONAL FUNCTIONS

4.1 Quadratic Functions	4.2 Polynomial Functions of Higher Degree	4.3 Dividing Polynomials: Long Division and Synthetic Division	4.4 The Real Zeros of a Polynomial Function	4.5 Complex Zeros: The Fundamental Theorem of Algebra	4.6 Rational Functions
• Graphs of Quadratic Functions: Parabolas • Finding the Equation of a Parabola	• Identifying Polynomial Functions • Graphing Polynomial Functions Using Transformations of Power Functions • Real Zeros of a Polynomial Function • Graphing General Polynomial Functions	• Long Division of Polynomials • Synthetic Division of Polynomials	• The Remainder Theorem and the Factor Theorem • The Rational Zero Theorem and Descartes' Rule of Signs • Factoring Polynomials • The Intermediate Value Theorem • Graphing Polynomial Functions	• Complex Zeros • Factoring Polynomials	• Domain of Rational Functions • Vertical, Horizontal, and Slant Asymptotes • Graphing Rational Functions

LEARNING OBJECTIVES

- Find the vertex (maximum or minimum) of the graph of a quadratic function.
- Graph polynomial functions.
- Divide polynomials using long division and synthetic division.
- Develop strategies for searching for zeros of a polynomial function.
- Understand that complex zeros come in conjugate pairs.
- Graph rational functions.

Graphs of Quadratic Functions: Parabolas

In Chapter 3, we studied functions in general. In this chapter we will learn about a special group of functions called *polynomial functions*. Polynomial functions are simple functions; often, more complicated functions are approximated by polynomial functions. Polynomial functions model many real-world applications such as the stock market, football punts, business costs, revenues and profits, and the flight path of NASA's "vomit comet." Let's start by defining a polynomial function.

DEFINITION   **Polynomial Function**

Let n be a nonnegative integer, and let $a_n, a_{n-1}, \ldots, a_2, a_1, a_0$ be real numbers with $a_n \neq 0$. The function

$$f(x) = a_n x^n + a_{n-1} x^{n-1} + \cdots + a_2 x^2 + a_1 x + a_0$$

is called a **polynomial function of x with degree n**. The coefficient a_n is called the **leading coefficient**, and a_0 is the constant.

Polynomials of particular degrees have special names. In Chapter 3, the library of functions included the constant function $f(x) = b$, which is a horizontal line; the linear function $f(x) = mx + b$, which is a line with slope m and y-intercept $(0, b)$; the square function $f(x) = x^2$; and the cube function $f(x) = x^3$. These are all special cases of a polynomial function.

Here are more examples of polynomial functions of particular degree together with their names:

Polynomial	Degree	Special Name
$f(x) = 3$	0	Constant function
$f(x) = -2x + 1$	1	Linear function
$f(x) = 7x^2 - 5x + 19$	2	Quadratic function
$f(x) = 4x^3 + 2x - 7$	3	Cubic function

The leading coefficients of these functions are 3, -2, 7, and 4, respectively. The constants of these functions are 3, 1, 19, and -7, respectively. In Section 2.3, we discussed graphs of linear functions, which are first-degree polynomial functions. In this section we will discuss graphs of quadratic functions, which are second-degree polynomial functions.

In Section 3.2, the library of functions that we compiled included the square function $f(x) = x^2$, whose graph is a parabola. See the graph on the right.

In Section 3.3, we graphed functions using transformation techniques such as $F(x) = (x + 1)^2 - 2$, which can be graphed by starting with the square function $y = x^2$ and shifting one unit to the left and down two units. See the graph on the right.

Note that if we eliminate the parentheses in $F(x) = (x + 1)^2 - 2$ to get

$$F(x) = x^2 + 2x + 1 - 2$$
$$= x^2 + 2x - 1$$

the result is a function defined by a second-degree polynomial (a polynomial with x^2 as the highest degree term), which is also called a *quadratic function*.

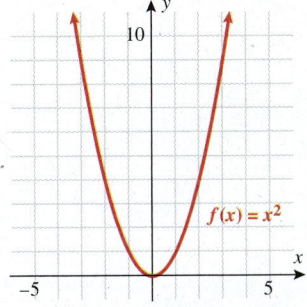

DEFINITION   Quadratic Function

Let a, b, and c be real numbers with $a \neq 0$. The function

$$f(x) = ax^2 + bx + c$$

is called a **quadratic function**.

The graph of any quadratic function is a **parabola**. If the leading coefficient a is *positive*, then the parabola opens *up*. If the leading coefficient a is *negative*, then the parabola opens *down*. The **vertex** (or turning point) is the minimum point, or low point, on the graph if the parabola opens up, whereas it is the maximum point, or high point, on the graph if the parabola opens down. The vertical line that intersects the parabola at the vertex is called the **axis of symmetry**.

The axis of symmetry is the line $x = h$, and the vertex is located at the point (h, k), as shown in the following two figures.

At the start of this section, the function $F(x) = x^2 + 2x - 1$, which can also be written as $F(x) = (x + 1)^2 - 2$, was shown through graph-shifting techniques to have a graph of a parabola. Looking at the graph on top of the margin, we see that the parabola opens up ($a = 1 > 0$), has a vertex at the point $(-1, -2)$, and has an axis of symmetry of $x = -1$.

Graphing Quadratic Functions Given in Standard Form

In general, writing a quadratic function in the form

$$f(x) = a(x - h)^2 + k$$

allows the vertex (h, k) and the axis of symmetry $x = h$ to be determined by inspection. This form is a convenient way to express a quadratic function in order to quickly determine its corresponding graph. Hence, this form is called *standard form*.

QUADRATIC FUNCTION: STANDARD FORM

The quadratic function

$$f(x) = a(x - h)^2 + k$$

is in **standard form**. The graph of f is a parabola whose vertex is the point (h, k). The parabola is symmetric with respect to the line $x = h$. If $a > 0$, the parabola opens up. If $a < 0$, the parabola opens down.

Recall that graphing linear functions requires finding two points on the line, or a point and the slope of the line. However, for quadratic functions simply knowing two points that lie on its graph is no longer sufficient. Below is a general step-by-step procedure for graphing quadratic functions given in standard form.

GRAPHING QUADRATIC FUNCTIONS

To graph $f(x) = a(x - h)^2 + k$

Step 1: Determine whether the parabola opens up or down.

$$a > 0 \quad \text{up}$$
$$a < 0 \quad \text{down}$$

Step 2: Determine the vertex (h, k).
Step 3: Find the y-intercept (by setting $x = 0$).
Step 4: Find any x-intercepts (by setting $f(x) = 0$ and solving for x).
Step 5: Plot the vertex and intercepts and connect them with a smooth curve.

Note that Step 4 says to "find any x-intercepts." Parabolas opening up or down will always have a y-intercept. However, they can have **one**, **two**, or **no** x-intercepts. The figures above illustrate this for parabolas opening up, and the same can be said about parabolas opening down.

EXAMPLE 1   Graphing a Quadratic Function Given in Standard Form

Graph the quadratic function $f(x) = (x - 3)^2 - 1$.

Solution:

STEP 1 The parabola opens up.

$a = 1$, so $a > 0$

STEP 2 Determine the vertex.

$(h, k) = (3, -1)$

STEP 3 Find the y-intercept.

$f(0) = (-3)^2 - 1 = 8$
$(0, 8)$ corresponds to the y-intercept

STEP 4 Find any x-intercepts.

$f(x) = (x - 3)^2 - 1 = 0$
$(x - 3)^2 = 1$

Use square-root method.

$x - 3 = \pm 1$

Solve.

$x = 2 \text{ or } x = 4$
$(2, 0)$ and $(4, 0)$ correspond to the x-intercepts

STEP 5 Plot the vertex and intercepts $(3, -1), (0, 8), (2, 0), (4, 0)$.

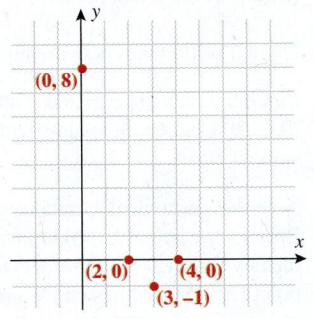

Connect the points with a smooth curve opening up.

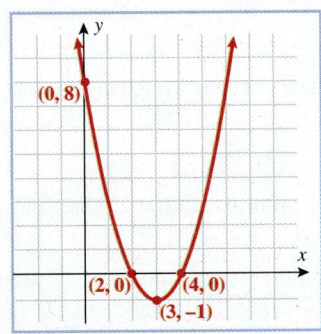

The graph in Example 1 could also have been found by shifting the square function to the right three units and down one unit.

■ **YOUR TURN** Graph the quadratic function $f(x) = (x - 1)^2 - 4$.

Classroom Example 4.1.1
Graph $f(x) = (x + 2)^2 - 2$.

Answer:

Study Tip

A quadratic function given in standard form can be graphed using the transformation techniques shown in Section 3.3 for the square function.

■ **Answer:**

EXAMPLE 2 Graphing a Quadratic Function Given in Standard Form with a Negative Leading Coefficient

Graph the quadratic function $f(x) = -2(x-1)^2 - 3$.

Solution:

STEP 1 The parabola opens down. $a = -2$, so $a < 0$

STEP 2 Determine the vertex. $(h, k) = (1, -3)$

STEP 3 Find the y-intercept. $f(0) = -2(-1)^2 - 3 = -2 - 3 = -5$
$(0, -5)$ corresponds to the y-intercept

STEP 4 Find any x-intercepts.

$$f(x) = -2(x-1)^2 - 3 = 0$$
$$-2(x-1)^2 = 3$$
$$(x-1)^2 = -\frac{3}{2}$$

The square root of a negative number is not a real number.

$$x - 1 = \pm\sqrt{-\frac{3}{2}}$$

No real solutions. There are no x-intercepts.

STEP 5 Plot the vertex $(1, -3)$ and y-intercept $(0, -5)$. Connect the points with a smooth curve.

Note that the axis of symmetry is $x = 1$. Because the point $(0, -5)$ lies on the parabola, then by symmetry with respect to $x = 1$, the point $(2, -5)$ also lies on the graph.

▪ YOUR TURN Graph the quadratic function $f(x) = -3(x+1)^2 - 2$.

When graphing quadratic functions (parabolas), have *at least 3 points* labeled on the graph.

- When there are x-intercepts (Example 1), label the vertex, y-intercept, and x-intercepts.
- When there are no x-intercepts (Example 2), label the vertex, y-intercept, and another point.

Graphing Quadratic Functions in General Form

A quadratic function is often written in one of two forms:

$$\text{Standard form: } f(x) = a(x-h)^2 + k$$
$$\text{General form: } f(x) = ax^2 + bx + c$$

When the quadratic function is expressed in standard form, the graph is easily obtained by identifying the vertex (h, k) and the intercepts and drawing a smooth curve that opens either up or down, depending on the sign of a.

Typically, quadratic functions are expressed in general form and a graph is the ultimate goal, so we must first express the quadratic function in standard form. One technique for transforming a quadratic function from general form to standard form was introduced in Section 1.3 and is called *completing the square*.

EXAMPLE 3 **Graphing a Quadratic Function Given in General Form**

Graph the quadratic function $f(x) = x^2 - 6x + 4$.

Solution:

Express the quadratic function in standard form by completing the square.

Write the original function.	$f(x) = x^2 - 6x + 4$
Group the variable terms together.	$= (x^2 - 6x) + 4$

Complete the square.

Half of -6 is -3; -3 squared is 9.

Add and subtract 9 within the parentheses.	$= (x^2 - 6x + 9 - 9) + 4$
Write the -9 outside the parentheses.	$= (x^2 - 6x + 9) - 9 + 4$
Write the expression inside the parentheses as a perfect square and simplify.	$= (x - 3)^2 - 5$

Now that the quadratic function is written in standard form, $f(x) = (x - 3)^2 - 5$, we follow our step-by-step procedure for graphing a quadratic function in standard form.

STEP 1 The parabola opens up. $\quad a = 1$, so $a > 0$

STEP 2 Determine the vertex. $\quad (h, k) = (3, -5)$

STEP 3 Find the y-intercept. $\quad f(0) = (0)^2 - 6(0) + 4 = 4$
$(0, 4)$ corresponds to the y-intercept

STEP 4 Find any x-intercepts.
$$f(x) = 0$$
$$f(x) = (x - 3)^2 - 5 = 0$$
$$(x - 3)^2 = 5$$
$$x - 3 = \pm\sqrt{5}$$
$$x = 3 \pm \sqrt{5}$$

$(3 + \sqrt{5}, 0)$ and $(3 - \sqrt{5}, 0)$ correspond to the x-intercepts.

STEP 5 Plot the vertex and intercepts $(3, -5)$, $(0, 4)$, $(3 + \sqrt{5}, 0)$, and $(3 - \sqrt{5}, 0)$.

Connect the points with a smooth parabolic curve.

Note: $3 + \sqrt{5} \approx 5.24$ and
$3 - \sqrt{5} \approx 0.76$.

YOUR TURN Graph the quadratic function $f(x) = x^2 - 8x + 14$.

Technology Tip

Use a graphing utility to graph the function $f(x) = x^2 - 6x + 4$ as y_1.

Classroom Example 4.1.3
Graph $f(x) = x^2 + 2x + 4$.

Answer:

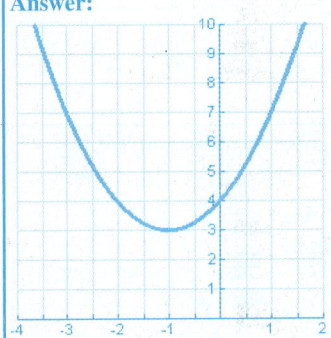

Study Tip

Although either form (standard or general) can be used to find the intercepts, it is often more convenient to use the general form when finding the y-intercept and the standard form when finding the x-intercept.

■ **Answer:**

When the leading coefficient of a quadratic function is not equal to 1, the leading coefficient must be factored out before completing the square.

EXAMPLE 4   Graphing a Quadratic Function Given in General Form with a Negative Leading Coefficient

Graph the quadratic function $f(x) = -3x^2 + 6x + 2$.

Solution:

Express the function in standard form by completing the square.

Write the original function.	$f(x) = -3x^2 + 6x + 2$
Group the variable terms together.	$= (-3x^2 + 6x) + 2$
Factor out -3 in order to make the coefficient of x^2 equal to 1 inside the parentheses.	$= -3(x^2 - 2x) + 2$
Add and subtract 1 inside the parentheses to create a perfect square.	$= -3(x^2 - 2x + 1 - 1) + 2$
Regroup the terms.	$= -3(x^2 - 2x + 1) - 3(-1) + 2$
Write the expression inside the parentheses as a perfect square and simplify.	$= -3(x - 1)^2 + 5$

Now that the quadratic function is written in standard form, $f(x) = -3(x - 1)^2 + 5$, we follow our step-by-step procedure for graphing a quadratic function in standard form.

STEP 1 The parabola opens down. $a = -3$, therefore $a < 0$

STEP 2 Determine the vertex. $(h, k) = (1, 5)$

STEP 3 Find the y-intercept using the general form.
$f(0) = -3(0)^2 + 6(0) + 2 = 2$
$(0, 2)$ corresponds to the y-intercept

STEP 4 Find any x-intercepts using the standard form.

$$f(x) = -3(x - 1)^2 + 5 = 0$$
$$-3(x - 1)^2 = -5$$
$$(x - 1)^2 = \frac{5}{3}$$
$$x - 1 = \pm\sqrt{\frac{5}{3}}$$
$$x = 1 \pm \sqrt{\frac{5}{3}}$$
$$= 1 \pm \frac{\sqrt{15}}{3}$$

The x-intercepts are $\left(1 + \dfrac{\sqrt{15}}{3}, 0\right)$ and $\left(1 - \dfrac{\sqrt{15}}{3}, 0\right)$.

STEP 5 Plot the vertex and intercepts

$(1, 5), (0, 2), \left(1 + \dfrac{\sqrt{15}}{3}, 0\right)$, and

$\left(1 - \dfrac{\sqrt{15}}{3}, 0\right)$.

Connect the points with a smooth curve.

Note: $1 + \dfrac{\sqrt{15}}{3} \approx 2.3$ and

$1 - \dfrac{\sqrt{15}}{3} \approx -0.3$.

■ **Answer:**

■ **YOUR TURN** Graph the quadratic function $f(x) = -2x^2 - 4x + 1$.

In Examples 3 and 4, the quadratic functions were given in general form, and they were transformed into standard form by completing the square. It can be shown (by completing the square) that the vertex of a quadratic function in general form, $f(x) = ax^2 + bx + c$, is located at $x = -\dfrac{b}{2a}$.

Another approach to sketching the graphs of quadratic functions is to first find the vertex and then find additional points through point-plotting.

VERTEX OF A PARABOLA

The graph of a quadratic function $f(x) = ax^2 + bx + c$ is a parabola with the **vertex** located at the point

$$\left(-\frac{b}{2a}, f\left(-\frac{b}{2a}\right)\right) = \left(-\frac{b}{2a}, \frac{4ac - b^2}{4a}\right)$$

GRAPHING A QUADRATIC FUNCTION IN GENERAL FORM

Step 1: Find the vertex.
Step 2: Determine whether the parabola opens up or down.
  ▪ If $a > 0$, the parabola opens up.
  ▪ If $a < 0$, the parabola opens down.
Step 3: Find additional points near the vertex.
Step 4: Sketch the graph with a parabolic curve.

EXAMPLE 5 Graphing a Quadratic Function Given in General Form

Sketch the graph of $f(x) = -2x^2 + 4x + 5$.

Solution: Let $a = -2$, $b = 4$, and $c = 5$.

STEP 1 Find the vertex.

$$x = -\frac{b}{2a} = -\frac{4}{2(-2)} = 1$$
$$f(1) = -2(1)^2 + 4(1) + 5 = 7$$
$$\text{Vertex} = (1, 7)$$

STEP 2 The parabola opens down. $a = -2$

STEP 3 Find additional points near the vertex.

x	−1	0	1	2	3
$f(x)$	$f(-1) = -1$	$f(0) = 5$	$f(1) = 7$	$f(2) = 5$	$f(3) = -1$

STEP 4 Label the vertex and additional points, then sketch the graph.

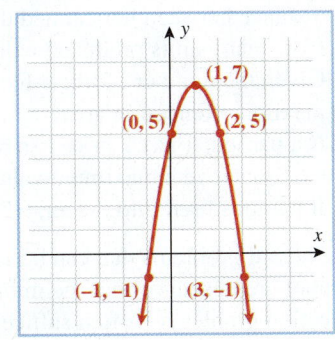

▪ **YOUR TURN** Sketch the graph of $f(x) = 3x^2 - 6x + 4$.

Classroom Example 4.1.5*
a. Graph $f(x) = x^2 - 4x$.
b. Let $b > 0$.
  (i) What is the vertex of $g(x) = x^2 - bx$?
  (ii) By how much would you need to vertically translate $g(x)$ such that the vertex lies on the x-axis (i.e., is tangent to it)?

Answer:
a.

b. (i) $\left(\dfrac{b}{2}, -\dfrac{b^2}{4}\right)$

  (ii) Translate up $\dfrac{b^2}{4}$ units. The new equation would be
  $$g(x) = x^2 - bx + \frac{b^2}{4}.$$

▪ **Answer:**

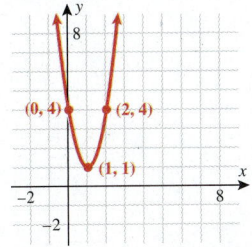

Finding the Equation of a Parabola

It is important to understand that the equation $y = x^2$ is equivalent to the quadratic function $f(x) = x^2$. Both have the same parabolic graph. Thus far, we have been given the equation and then asked to find characteristics (vertex and intercepts) in order to graph. We now turn our attention to the problem of determining the equation, or function, given certain characteristics.

Classroom Example 4.1.6
Find the equation of a parabola where vertex is $(-1, 4)$ and passes through $(2, -14)$.

Answer:
$f(x) = -2(x + 1)^2 + 4$

Classroom Example 4.1.6*
Suppose a parabola has
$y = ax^2 + bx + c$
x-intercepts $(1, 0)$ and $(5, 0)$, and $a = -2$. Find the equation of this parabola.

Answer:
$f(x) = -2x^2 + 12x - 10$

EXAMPLE 6 Finding the Equation of a Parabola Given the Vertex and Another Point

Find the equation of a parabola whose graph has a vertex at $(3, 4)$ and that passes through the point $(2, 3)$. Express the quadratic function in both standard and general forms.

Solution:

Write the standard form of the equation of a parabola.

$$f(x) = a(x - h)^2 + k$$

Substitute the coordinates of the vertex $(h, k) = (3, 4)$.

$$f(x) = a(x - 3)^2 + 4$$

Use the point $(2, 3)$ to find a.

The point $(2, 3)$ implies $f(2) = 3$.

$$f(2) = a(2 - 3)^2 + 4 = 3$$

Solve for a.

$$a(2 - 3)^2 + 4 = 3$$
$$a(-1)^2 + 4 = 3$$
$$a + 4 = 3$$
$$a = -1$$

Write both forms of the equation of this parabola.

Standard form: $\boxed{f(x) = -(x - 3)^2 + 4}$ General form: $\boxed{f(x) = -x^2 + 6x - 5}$

■ **Answer:** Standard form:
$f(x) = (x + 3)^2 - 5$

■ **YOUR TURN** Find the standard form of the equation of a parabola whose graph has a vertex at $(-3, -5)$ and that passes through the point $(-2, -4)$.

As we have seen in Example 6, once the vertex is known, the leading coefficient a can be found from any point that lies on the parabola.

Application Problems That Involve Quadratic Functions

- What is the minimum distance a driver has to maintain in order to be at a safe distance between her car and the car in front as a function of speed?
- How many units produced will yield a maximum profit for a company?
- Given the number of linear feet of fence, what rectangular dimensions will yield a maximum fenced-in area?
- If a particular stock price has been shown to follow a quadratic trend, when will the stock achieve a maximum value?
- If a gun is fired into the air, where will the bullet land?

These are all problems that can be solved using quadratic functions. Because the vertex of a parabola represents either the minimum or maximum value of the quadratic function, in application problems it often suffices simply to find the vertex.

In Example 6, the function $F(x) = -x^2 + 6x - 5$, which can also be written as $F(x) = -(x - 3)^2 + 4$, was shown to "open down" and has a vertex at the point $(3, 4)$. Suppose this function represents profit, where x is the number (millions) of units made. The vertex would then represent the *maximum* profit. Instead of rewriting the function in standard form through completing the square, we use the vertex formula $x = -\dfrac{b}{2a}$.

WORDS	**MATH**
Quadratic function.	$F(x) = -x^2 + 6x - 5$
Coefficients.	$a = -1, b = 6, c = -5$
Find the x-coordinate of the vertex.	$x = -\dfrac{b}{2a} = -\dfrac{6}{2(-1)} = 3$
Find the value of the function at $x = 3$.	$f(3) = -(3)^2 + 6(3) - 5 = 4$

Therefore, the vertex is located at the point $\boxed{(3, 4)}$.

Technology Tip

Use a graphing utility to graph the cost function
$C(x) = 2000 - 15x + 0.05x^2$ as y_1.

EXAMPLE 7 Finding the Minimum Cost of Manufacturing a Motorcycle

A company that produces motorcycles has a daily production cost of

$$C(x) = 2000 - 15x + 0.05x^2$$

where C is the cost in dollars to manufacture a motorcycle and x is the number of motorcycles produced. How many motorcycles can be produced each day in order to minimize the cost of each motorcycle? What is the corresponding minimum cost?

Solution:

The graph of the quadratic function is a parabola.

Rewrite the quadratic function in general form.	$C(x) = 0.05x^2 - 15x + 2000$
The parabola opens up, because a is positive.	$a = 0.05 > 0$

Because the parabola opens up, the vertex of the parabola is a *minimum*.

Find the x-coordinate of the vertex. $\qquad x = -\dfrac{b}{2a} = -\dfrac{(-15)}{2(0.05)} = 150$

The company keeps costs to a minimum when 150 motorcycles are produced each day.

The minimum cost is \$875 per motorcycle. $\qquad C(150) = 875$

■ **Answer:** 250,000 bottles

■ **YOUR TURN** The revenue associated with selling vitamins is

$$R(x) = 500x - 0.001x^2$$

where R is the revenue in dollars and x is the number of bottles of vitamins sold. Determine how many bottles of vitamins should be sold to maximize the revenue.

Technology Tip

Use a graphing utility to graph the area function $A(x) = -2x^2 + 100x$.

```
WINDOW
 Xmin=-20
 Xmax=60
 Xscl=10
 Ymin=-500
 Ymax=1400
 Yscl=100
 Xres=1
```

```
Plot1 Plot2 Plot3
\Y1目-2X²+100X
\Y2=■
```

The maximum occurs when $x = 25$. The maximum area is $y = 1250$ square feet.

```
Y1=-2X²+100X
```

```
X=25        Y=1250
```

A table of values supports the solution.

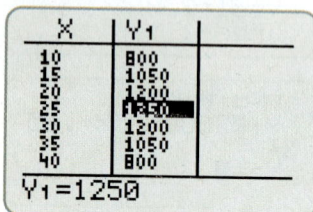

```
  X  | Y1
 10  | 800
 15  | 1050
 20  | 1200
 25  | 1250
 30  | 1200
 35  | 1050
 40  | 800
Y1=1250
```

■ **Answer:** 50 feet by 50 feet

EXAMPLE 8   Finding the Dimensions That Yield a Maximum Area

You have just bought a puppy and want to fence in an area in the backyard for her. You buy 100 linear feet of fence from Home Depot and have decided to make a rectangular fenced-in area using the back of your house as one side. Determine the dimensions of the rectangular pen that will maximize the area in which your puppy may roam. What is the maximum area of the rectangular pen?

Solution:

STEP 1 **Identify the question.**

Find the dimensions of the rectangular pen.

STEP 2 **Draw a picture.**

STEP 3 **Set up a function.**

If we let x represent the length of one side of the rectangle, then the opposite side is also of length x. Because there are 100 feet of fence, the remaining fence left for the side opposite the house is $100 - 2x$.

The area of a rectangle is equal to length times width:

$$A(x) = x(100 - 2x)$$

STEP 4 **Find the maximum value of the function.**

$$A(x) = x(100 - 2x) = -2x^2 + 100x$$

Find the maximum of the parabola that corresponds to the quadratic function for area $A(x) = -2x^2 + 100x$.

$a = -2$ and $b = 100$; therefore, the maximum occurs when

$$x = -\frac{b}{2a} = -\frac{100}{2(-2)} = 25$$

Replacing x with 25 in our original diagram:

The dimensions of the rectangle are

$\boxed{25 \text{ feet by } 50 \text{ feet}}$.

The maximum area $A(25) = 1250$ is

$\boxed{1250 \text{ square feet}}$.

STEP 5 **Check the solution.**

Two sides are 25 feet and one side is 50 feet, and together they account for all 100 feet of fence.

■ **YOUR TURN**   Suppose you have 200 linear feet of fence to enclose a rectangular garden. Determine the dimensions of the rectangle that will yield the greatest area.

EXAMPLE 9 **Path of a Punted Football**

The path of a particular punt follows the quadratic function: $h(x) = -\frac{1}{8}(x - 5)^2 + 50$, where $h(x)$ is the height of the ball in yards and x corresponds to the horizontal distance in yards. Assume $x = 0$ corresponds to midfield (the 50 yard line). For example, $x = -20$ corresponds to the punter's own 30 yard line, whereas $x = 20$ corresponds to the other team's 30 yard line.

a. Find the maximum height the ball achieves.
b. Find the horizontal distance the ball covers. Assume the height is zero when the ball is kicked and when the ball is caught.

Solution (a):

Identify the vertex since it is given in standard form. $(h, k) = (5, 50)$

The maximum height of the punt occurs at the other team's 45 yard line, and the height the ball achieves is $\boxed{50 \text{ yards } (150 \text{ feet})}$.

Solution (b):

The height when the ball is kicked
or caught is zero.

$$h(x) = -\frac{1}{8}(x - 5)^2 + 50 = 0$$

Solve for x.

$$\frac{1}{8}(x - 5)^2 = 50$$
$$(x - 5)^2 = 400$$
$$(x - 5) = \pm\sqrt{400}$$
$$x = 5 \pm 20$$
$$x = -15 \text{ and } x = 25$$

The horizontal distance is the distance between these two points: $|25 - (-15)| = \boxed{40 \text{ yards}}$.

SECTION
4.1 SUMMARY

All quadratic functions $f(x) = ax^2 + bx + c$ or $f(x) = a(x - h)^2 + k$ have graphs that are parabolas:

■ If $a > 0$, the parabola opens up.
■ If $a < 0$, the parabola opens down.
■ The vertex is at the point

$$(h, k) = \left(-\frac{b}{2a}, f\left(-\frac{b}{2a}\right)\right) = \left(-\frac{b}{2a}, \frac{4ac - b^2}{4a}\right)$$

■ When the quadratic function is given in general form, completing the square can be used to rewrite the function in standard form.
■ At least three points are needed to graph a quadratic function.
  ● vertex
  ● y-intercept
  ● x-intercept(s) or other point(s)

SECTION
4.1 EXERCISES

▪SKILLS

In Exercises 1–4, match the quadratic function with its graph.

1. $f(x) = 3(x + 2)^2 - 5$ **2.** $f(x) = 2(x - 1)^2 + 3$ **3.** $f(x) = -\frac{1}{2}(x + 3)^2 + 2$ **4.** $f(x) = -\frac{1}{3}(x - 2)^2 + 3$

a.   b.   c.   d.

In Exercises 5–8, match the quadratic function with its graph.

5. $f(x) = 3x^2 + 5x - 2$ **6.** $f(x) = 3x^2 - x - 2$ **7.** $f(x) = -x^2 + 2x - 1$ **8.** $f(x) = -2x^2 - x + 3$

a.   b.   c.   d.

In Exercises 9–22, graph the quadratic function, which is given in standard form.

9. $f(x) = (x + 1)^2 - 2$ **10.** $f(x) = (x + 2)^2 - 1$ **11.** $f(x) = (x - 2)^2 - 3$ **12.** $f(x) = (x - 4)^2 + 2$

13. $f(x) = -(x - 3)^2 + 9$ **14.** $f(x) = -(x - 5)^2 - 4$ **15.** $f(x) = -(x + 1)^2 - 3$ **16.** $f(x) = -(x - 2)^2 + 6$

17. $f(x) = 2(x - 2)^2 + 2$ **18.** $f(x) = -3(x + 2)^2 - 15$ **19.** $f(x) = \left(x - \frac{1}{3}\right)^2 + \frac{1}{9}$ **20.** $f(x) = \left(x + \frac{1}{4}\right)^2 - \frac{1}{2}$

21. $f(x) = -0.5(x - 0.25)^2 + 0.75$ **22.** $f(x) = -0.2(x + 0.6)^2 + 0.8$

In Exercises 23–32, rewrite the quadratic function in standard form by completing the square.

23. $f(x) = x^2 + 6x - 3$ **24.** $f(x) = x^2 + 8x + 2$ **25.** $f(x) = -x^2 - 10x + 3$ **26.** $f(x) = -x^2 - 12x + 6$

27. $f(x) = 2x^2 + 8x - 2$ **28.** $f(x) = 3x^2 - 9x + 11$ **29.** $f(x) = -4x^2 + 16x - 7$ **30.** $f(x) = -5x^2 + 100x - 36$

31. $f(x) = \frac{1}{2}x^2 - 4x + 3$ **32.** $f(x) = -\frac{1}{3}x^2 + 6x + 4$

In Exercises 33–40, graph the quadratic function.

33. $f(x) = x^2 + 6x - 7$ **34.** $f(x) = x^2 - 3x + 10$ **35.** $f(x) = x^2 + 2x - 15$ **36.** $f(x) = -x^2 + 3x + 4$

37. $f(x) = -2x^2 - 12x - 16$ **38.** $f(x) = -3x^2 + 12x - 12$ **39.** $f(x) = \frac{1}{2}x^2 - \frac{1}{2}$ **40.** $f(x) = -\frac{1}{3}x^2 + \frac{4}{3}$

In Exercises 41–48, find the vertex of the parabola associated with each quadratic function.

41. $f(x) = 33x^2 - 2x + 15$ **42.** $f(x) = 17x^2 + 4x - 3$ **43.** $f(x) = \frac{1}{2}x^2 - 7x + 5$ **44.** $f(x) = -\frac{1}{3}x^2 + \frac{2}{5}x + 4$

45. $f(x) = -0.002x^2 - 0.3x + 1.7$ **46.** $f(x) = 0.05x^2 + 2.5x - 1.5$ **47.** $f(x) = -\frac{2}{5}x^2 + \frac{3}{7}x + 2$ **48.** $f(x) = -\frac{1}{7}x^2 - \frac{2}{3}x + \frac{1}{9}$

In Exercises 49–58, find the quadratic function that has the given vertex and goes through the given point.

49. vertex: $(-1, 4)$ point: $(0, 2)$ **50.** vertex: $(2, -3)$ point: $(0, 1)$ **51.** vertex: $(2, 5)$ point: $(3, 0)$

52. vertex: $(1, 3)$ point: $(-2, 0)$ **53.** vertex: $(-1, -3)$ point: $(-4, 2)$ **54.** vertex: $(0, -2)$ point: $(3, 10)$

55. vertex: $\left(\frac{1}{2}, -\frac{3}{4}\right)$ point: $\left(\frac{3}{4}, 0\right)$ **56.** vertex: $\left(-\frac{5}{6}, \frac{2}{3}\right)$ point: $(0, 0)$ **57.** vertex: $(2.5, -3.5)$ point: $(4.5, 1.5)$

58. vertex: $(1.8, 2.7)$ point: $(-2.2, -2.1)$

▪ APPLICATIONS

59. Business. The annual profit for a company that manufactures cell phone accessories can be modeled by the function

$$P(x) = -0.0001x^2 + 70x + 12{,}500$$

where x is the number of units sold and P is the total profit in dollars.

a. What sales level maximizes the company's annual profit?

b. Find the maximum annual profit for the company.

60. Business. A manufacturer of office supplies has daily production costs of

$$C(x) = 0.5x^2 - 20x + 1600$$

where x is the number of units produced measured in thousands and C is cost in hundreds of dollars.

a. What production level will minimize the manufacturer's daily production costs?

b. Find the minimum daily production costs for the manufacturer.

For Exercises 61 and 62, refer to the following:

An adult male's weight, in kilograms, can be modeled by the function

$$W(t) = -\frac{2}{3}t^2 + \frac{13}{5}t + \frac{433}{5}; \quad 1 \le t \le 18$$

where t measures months ($t = 1$ is January 2010, $t = 2$ is February 2010, etc.) and W is the male's weight.

61. Health/Medicine. During which months was the male losing weight and gaining weight?

62. Health/Medicine. Find the maximum weight to the nearest kilogram of the adult male during the 18 months.

Exercises 63 and 64 concern the path of a punted football. Refer to the diagram in Example 9.

63. Sports. The path of a particular punt follows the quadratic function

$$h(x) = -\frac{8}{125}(x + 5)^2 + 40$$

where $h(x)$ is the height of the ball in yards and x corresponds to the horizontal distance in yards. Assume $x = 0$ corresponds to midfield (the 50 yard line). For example, $x = -20$ corresponds to the punter's own 30 yard line, whereas $x = 20$ corresponds to the other team's 30 yard line.

a. Find the maximum height the ball achieves.

b. Find the horizontal distance the ball covers. Assume the height is zero when the ball is kicked and when the ball is caught.

64. Sports. The path of a particular punt follows the quadratic function

$$h(x) = -\frac{5}{40}(x - 30)^2 + 50$$

where $h(x)$ is the height of the ball in yards and x corresponds to the horizontal distance in yards. Assume $x = 0$ corresponds to midfield (the 50 yard line). For example, $x = -20$ corresponds to the punter's own 30 yard line, whereas $x = 20$ corresponds to the other team's 30 yard line.

a. Find the maximum height the ball achieves.

b. Find the horizontal distance the ball covers. Assume the height is zero when the ball is kicked and when the ball is caught.

65. Ranching. A rancher has 10,000 linear feet of fencing and wants to enclose a rectangular field and then divide it into two equal pastures with an internal fence parallel to one of the rectangular sides. What is the maximum area of each pasture? Round to the nearest square foot.

66. Ranching. A rancher has 30,000 linear feet of fencing and wants to enclose a rectangular field and then divide it into four equal pastures with three internal fences parallel to one of the rectangular sides. What is the maximum area of each pasture?

67. Gravity. A person standing near the edge of a cliff 100 feet above a lake throws a rock upward with an initial speed of 32 feet per second. The height of the rock above the lake at the bottom of the cliff is a function of time and is described by

$$h(t) = -16t^2 + 32t + 100$$

a. How many seconds will it take until the rock reaches its maximum height? What is that height?

b. At what time will the rock hit the water?

68. Gravity. A person holds a pistol straight upward and fires. The initial velocity of most bullets is around 1200 feet per second. The height of the bullet is a function of time and is described by

$$h(t) = -16t^2 + 1200t$$

How long, after the gun is fired, does the person have to get out of the way of the bullet falling from the sky?

69. Zero Gravity. As part of their training, astronauts ride the "vomit comet," NASA's reduced gravity KC 135A aircraft that performs parabolic flights to simulate weightlessness. The plane starts at an altitude of 20,000 feet and makes a steep climb at 52° with the horizon for 20–25 seconds and then dives at that same angle back down, repeatedly. The equation governing the altitude of the flight is

$$A(x) = -0.0003x^2 + 9.3x - 46,075$$

where $A(x)$ is altitude and x is horizontal distance in feet.

a. What is the maximum altitude the plane attains?

b. Over what horizontal distance is the entire maneuver performed? (Assume the starting and ending altitude is 20,000 feet.)

NASA's "Vomit Comet"

70. Sports. A soccer ball is kicked from the ground at a 45° angle with an initial velocity of 40 feet per second. The height of the soccer ball above the ground is given by $H(x) = -0.0128x^2 + x$, where x is the horizontal distance the ball travels.

a. What is the maximum height the ball reaches?

b. What is the horizontal distance the ball travels?

71. Profit. A small company in Virginia Beach manufactures handcrafted surfboards. The profit of selling x boards is given by

$$P(x) = 20,000 + 80x - 0.4x^2$$

a. How many boards should be made to maximize the profit?

b. What is the maximum profit?

72. Environment: Fuel Economy. Gas mileage (miles per gallon, mpg) can be approximated by a quadratic function of speed. For a particular automobile, assume the vertex occurs when the speed is 50 miles per hour (mph); the mpg will be 30.

a. Write a quadratic function that models this relationship, assuming 70 mph corresponds to 25 mpg.

b. What gas mileage would you expect for this car driving 90 mph?

For Exercises 73 and 74, use the following information:

One function of particular interest in economics is the **profit function**. We denote this function by $P(x)$. It is defined to be the difference between revenue $R(x)$ and cost $C(x)$ so that

$$P(x) = R(x) - C(x)$$

The total revenue received from the sale of x goods at price p is given by

$$R(x) = px$$

The total cost function relates the cost of production to the level of output x. This includes both fixed costs C_f and variable costs C_v (costs per unit produced). The total cost in producing x goods is given by

$$C(x) = C_f + C_v x$$

Thus, the profit function is

$$P(x) = px - C_f - C_v x$$

Assume fixed costs are $1000, variable costs per unit are $20, and the demand function is

$$p = 100 - x$$

73. Profit. How many units should the company produce to break even?

74. Profit. What is the maximum profit?

75. Cell Phones. The number of cell phones in the United States can be approximated by a quadratic function. In 1996 there were approximately 16 million cell phones, and in 2005 there were approximately 100 million. Let t be the number of years since 1996. The number of cell phones in 1996 is represented by $(0, 16)$, and the number in 2005 is $(9, 100)$. Let $(0, 16)$ be the vertex.

a. Find a quadratic function that represents the number of cell phones.

b. Based on this model, how many cell phones were used in 2010?

76. Underage Smoking. The number of underage cigarette smokers (ages 10–17) has declined in the United States. The peak percent was in 1998 at 49%. In 2006 this had dropped to 36%. Let t be time in years after 1998 ($t = 0$ corresponds to 1998).

a. Find a quadratic function that models the percent of underage smokers as a function of time. Let $(0, 49)$ be the vertex.

b. Based on this model, what was the percent of underage smokers in 2010?

77. Drug Concentration. The concentration of a drug in the bloodstream, measured in parts per million, can be modeled with a quadratic function. In 50 minutes the concentration is 93.75 parts per million. The maximum concentration of the drug in the bloodstream occurs in 225 minutes and is 400 parts per million.

a. Find a quadratic function that models the concentration of the drug as a function of time in minutes.

b. In how many minutes will the drug be eliminated from the bloodstream?

78. Revenue. Jeff operates a mobile car washing business. When he charged $20 a car, he washed 70 cars a month. He raised the price to $25 a car and his business dropped to 50 cars a month.

a. Find a linear function that represents the demand equation (the price per car as a function of the number of cars washed).

b. Find the revenue function $R(x) = xp$.

c. How many cars should he wash to maximize the revenue?

d. What price should he charge to maximize revenue?

▪CATCH THE MISTAKE

In Exercises 79–82, explain the mistake that is made. There may be a single mistake, or there may be more than one mistake.

79. Plot the quadratic function $f(x) = (x + 3)^2 - 1$.

Solution:

Step 1: The parabola opens up because $a = 1 > 0$.

Step 2: The vertex is $(3, -1)$.

Step 3: The y-intercept is $(0, 8)$.

Step 4: The x-intercepts are $(2, 0)$ and $(4, 0)$.

Step 5: Plot the vertex and intercepts, and connect the points with a smooth curve.

This is incorrect. What mistake(s) was made?

80. Determine the vertex of the quadratic function $f(x) = 2x^2 - 6x - 18$.

Solution:

Step 1: The vertex is given by $(h, k) = \left(-\dfrac{b}{2a}, f\left(-\dfrac{b}{2a}\right)\right)$.

In this case, $a = 2$ and $b = 6$.

Step 2: The x-coordinate of the vertex is

$$x = -\frac{6}{2(2)} = -\frac{6}{4} = -\frac{3}{2}$$

Step 3: The y-coordinate of the vertex is

$$f\left(-\frac{3}{2}\right) = -2\left(-\frac{3}{2}\right)^2 + 6\left(-\frac{3}{2}\right) - 18$$

$$= -2\left(\frac{9}{4}\right) - \frac{18}{2} - 18$$

$$= -\frac{9}{2} - 9 - 18$$

$$= -\frac{63}{2}$$

This is incorrect. What mistake(s) was made?

81. Rewrite the following quadratic function in standard form:

$$f(x) = -x^2 + 2x + 3$$

Solution:

Step 1: Group the variables together. $(-x^2 + 2x) + 3$

Step 2: Factor out a negative. $-(x^2 + 2x) + 3$

Step 3: Add and subtract 1 inside the parentheses. $-(x^2 + 2x + 1 - 1) + 3$

Step 4: Factor out the -1. $-(x^2 + 2x + 1) + 1 + 3$

Step 5: Simplify. $-(x + 1)^2 + 4$

This is incorrect. What mistake(s) was made?

82. Find the quadratic function whose vertex is $(2, -3)$ and whose graph passes through the point $(9, 0)$.

Solution:

Step 1: Write the quadratic function in standard form. $f(x) = a(x - h)^2 + k$

Step 2: Substitute $(h, k) = (2, -3)$. $f(x) = a(x - 2)^2 - 3$

Step 3: Substitute the point $(9, 0)$ and solve for a.
$$f(0) = a(0 - 2)^2 - 3 = 9$$
$$4a - 3 = 9$$
$$4a = 12$$
$$a = 3$$

The quadratic function sought is: $f(x) = 3(x - 2)^2 - 3$.

This is incorrect. What mistake(s) was made?

▪CONCEPTUAL

In Exercises 83–86, determine whether each statement is true or false.

83. A quadratic function must have a y-intercept.

84. A quadratic function must have an x-intercept.

85. A quadratic function may have more than one y-intercept.

86. A quadratic function may have more than one x-intercept.

■ CHALLENGE

87. For the general quadratic equation, $f(x) = ax^2 + bx + c$, show that the vertex is $(h, k) = \left(-\dfrac{b}{2a}, f\left(-\dfrac{b}{2a}\right)\right)$.

88. Given the quadratic function $f(x) = a(x - h)^2 + k$, determine the x- and y-intercepts in terms of a, h, and k.

89. A rancher has 1000 feet of fence to enclose a pasture.
  a. Determine the maximum area if a rectangular fence is used.
  b. Determine the maximum area if a circular fence is used.

90. A 600-room hotel in Orlando is filled to capacity every night when the rate is \$90 per night. For every \$5 increase in the rate, 10 fewer rooms are filled. How much should the hotel charge to produce the maximum income? What is the maximum income?

■ TECHNOLOGY

91. On a graphing calculator, plot the quadratic function $f(x) = -0.002x^2 + 5.7x - 23$.
  a. Identify the vertex of this parabola.
  b. Identify the y-intercept.
  c. Identify the x-intercepts (if any).
  d. What is the axis of symmetry?

92. Determine the quadratic function whose vertex is $(-0.5, 1.7)$ and whose graph passes through the point $(0, 4)$.
  a. Write the quadratic function in general form.
  b. Plot this quadratic function with a graphing calculator.
  c. Zoom in on the vertex and y-intercept. Do they agree with the given values?

In Exercises 93 and 94: (a) use the calculator commands STAT QuadReg to model the data using a quadratic function; (b) write the quadratic function in standard form and identify the vertex; (c) plot this quadratic function with a graphing calculator and use the TRACE key to highlight the given points. Do they agree with the given values?

93.

x	-2	2	5
y	-29.28	21.92	18.32

94.

x	-9	-2	4
y	-2.72	-16.18	6.62

For Exercises 95 and 96, refer to the following discussion of quadratic regression:

The "least-squares" criterion used to create a *quadratic regression* curve $y = ax^2 + bx + c$ that fits a set of n data points (x_1, y_1), $(x_2, y_2), \ldots, (x_n, y_n)$ is that the sum of the squares of the vertical distances from the points to the curve be minimum. This means that we need to determine values of a, b, and c for which

$$\sum_{i=1}^{n} (y_i - (ax_i^2 + bx_i + c))^2$$ is as small as possible. Calculus can be used to determine formulas for a, b, and c that do the job, but computing them by hand is tedious and unnecessary because the TI-83+ has a built-in program called *QuadReg* that does this. In fact, this was introduced in Section 2.5*, Problems 39–42. The following are application problems that involve experimental data for which the best fit curve is a parabola.

Projectile Motion
It is well known that the trajectory of an object thrown with initial velocity v_0 from an initial height s_0 is described by the quadratic function $s(t) = -16t^2 + v_0 t + s_0$. If we have data points obtained in such a context, we can apply the procedure outlined in Section 2.5* with *QuadReg* in place of *LinReg(ax+b)* to find a best fit *parabola* of the form $y = ax^2 + bx + c$.

Each year during Halloween season, it is tradition to hold the Pumpkin Launching Contest where students literally hurl their pumpkins in the hope of throwing them the farthest horizontal distance. Ben claims that it is better to use a steeper trajectory since it will have more air time, while Rick believes in throwing the pumpkin with all of his might, but with less inclination. The following data points that describe the pumpkin's horizontal x and vertical y distances (measured in feet) are collected during the flights of their pumpkins:

BEN'S DATA		RICK'S DATA	
x	y	x	y
0	4	0	4
1	14.2	1	8.5
2	28.4	2	10.6
3	30.1	3	13.3
4	35.9	4	16.2
5	37.8	5	17.3
6	41.1	6	19.3
7	38.2	7	19.5

95. a. Form a scatterplot for Ben's data.
  b. Determine the equation of the best fit parabola and report the value of the associated correlation coefficient.
  c. Use the best fit curve from (b) to answer the following:
    i. What is the initial height of the pumpkin's trajectory and with what initial velocity was it thrown?
    ii. What is the maximum height of Ben's pumpkin's trajectory?
    iii. How much horizontal distance has the pumpkin traveled by the time it lands?

96. Repeat Exercise 95 for Rick's data.

SKILLS OBJECTIVES

- Identify a polynomial function and determine its degree.
- Graph polynomial functions using transformations.
- Identify real zeros of a polynomial function and their multiplicities.
- Determine the end behavior of a polynomial function.
- Graph polynomial functions.
  - *x*-intercepts
  - multiplicity (touch/cross) of each zero
  - end behavior

CONCEPTUAL OBJECTIVES

- Understand that real zeros of polynomial functions correspond to *x*-intercepts.
- Understand the intermediate value theorem and how it assists in graphing polynomial functions.
- Realize that end behavior is a result of the leading term dominating.
- Understand that zeros correspond to factors of the polynomial.

Identifying Polynomial Functions

Polynomial functions model many real-world applications. Often, the input is time, and the output of the function can be many things. For example, the number of active-duty military personnel in the United States, the number of new cases in the spread of an epidemic, and the stock price as a function of time t can all be modeled with polynomial functions.

DEFINITION   **Polynomial Function**

Let n be a nonnegative integer and let $a_n, a_{n-1}, \ldots, a_2, a_1, a_0$ be real numbers with $a_n \neq 0$. The function

$$f(x) = a_n x^n + a_{n-1}x^{n-1} + \cdots + a_2 x^2 + a_1 x + a_0$$

is called a **polynomial function of x with degree n**. The coefficient a_n is called the leading coefficient.

Note: If n is a nonnegative integer, then $n - 1, n - 2, \ldots, 2, 1, 0$ are also nonnegative integers.

Classroom Example 4.2.1
Determine whether these are polynomials. If yes, state its degree.
a. $f(x) = x(3 - 2x)^{-2}$
b.* $g(x) = (\sqrt{x} + 1)(\sqrt{x} - 1) + 3$
c.* $h(x) = (\sqrt[3]{2x} + 1)^3$
d.* $j(x) = (x - \sqrt{2})^2(1 - 3x)^3$

Answer:
a. no   **b.** yes, degree 1
c. no   **d.** yes, degree 5

EXAMPLE 1   **Identifying Polynomials and Their Degree**

For each of the functions given, determine whether the function is a polynomial function. If it is a polynomial function, then state the degree of the polynomial. If it is not a polynomial function, justify your answer.

a. $f(x) = 3 - 2x^5$

b. $F(x) = \sqrt{x} + 1$

c. $g(x) = 2$

d. $h(x) = 3x^2 - 2x + 5$

e. $H(x) = 4x^5(2x - 3)^2$

f. $G(x) = 2x^4 - 5x^3 - 4x^{-2}$

Solution:

a. $f(x)$ is a polynomial function of degree 5.

b. $F(x)$ is not a polynomial function. The variable x is raised to the power of $\frac{1}{2}$, which is not an integer.

c. $g(x)$ is a polynomial function of degree zero, also known as a constant function. Note that $g(x) = 2$ can also be written as $g(x) = 2x^0$ (assuming $x \neq 0$).

d. $h(x)$ is a polynomial function of degree 2. A polynomial function of degree 2 is called a quadratic function.

e. $H(x)$ is a polynomial function of degree 7. *Note:* $4x^5(4x^2 - 12x + 9) = 16x^7 - 48x^6 + 36x^5$.

f. $G(x)$ is not a polynomial function. $-4x^{-2}$ has an exponent that is negative.

....

■ **YOUR TURN** For each of the functions given, determine whether the function is a polynomial function. If it is a polynomial function, then state the degree of the polynomial. If it is not a polynomial function, justify your answer.

a. $f(x) = \dfrac{1}{x} + 2$ b. $g(x) = 3x^8(x - 2)^2(x + 1)^3$

■ **Answer:**
a. $f(x)$ is not a polynomial because x is raised to the power of -1, which is a negative integer.
b. $g(x)$ is a polynomial of degree 13.

Whenever we have discussed a particular polynomial, we have graphed it too. The graph of a constant function (degree 0) is a horizontal line. The graph of a general linear function (degree 1) is a slant line. The graph of a quadratic function (degree 2) is a parabola. These functions are summarized in the table below.

POLYNOMIAL	DEGREE	SPECIAL NAME	GRAPH
$f(x) = c$	0	Constant function	Horizontal line
$f(x) = mx + b$	1	Linear function	Line • Slope $= m$ • y-intercept: $(0, b)$
$f(x) = ax^2 + bx + c$	2	Quadratic function	Parabola • Opens up if $a > 0$. • Opens down if $a < 0$.

How do we graph polynomial functions that are of degree 3 or higher, and why do we care? Polynomial functions model real-world applications, as mentioned earlier. One example is the percentage of fat in our bodies as we age. We can model the weight of a baby after it comes home from the hospital as a function of time. When a baby comes home from the hospital, it usually experiences weight loss. Then typically there is an increase in the percent of body fat when the baby is nursing. When infants start to walk, the increase in exercise is associated with a drop in the percentage of fat. Growth spurts in children are examples of the percent of body fat increasing and decreasing. Later in life, our metabolism slows down, and typically the percent of body fat increases. We will model this with a polynomial function. Other examples are stock prices, the federal funds rate, and yo-yo dieting as functions of time.

Polynomial functions are considered simple functions. Graphs of all polynomial functions are both *continuous* and *smooth*. A **continuous** graph is one you can draw completely without picking up your pencil (the graph has no jumps or holes). A **smooth** graph has no sharp corners. The following graphs illustrate what it means to be smooth (no sharp corners or cusps) and continuous (no holes or jumps).

The graph is *not continuous.*

The graph is *not continuous.*

The graph is *continuous* but *not smooth.*

The graph is *continuous* and *smooth.*

All polynomial functions have graphs that are both continuous and smooth.

Graphing Polynomial Functions Using Transformations of Power Functions

Recall from Chapter 3 that graphs of functions can be drawn by hand using graphing aids such as intercepts and symmetry. The graphs of polynomial functions can be graphed using these same aids. Let's start with the simplest types of polynomial functions called **power functions**. Power functions are monomial functions of the form $f(x) = x^n$, where n is an integer greater than zero.

> **DEFINITION** **Power Function**
>
> Let n be a positive integer and the coefficient $a \neq 0$ be a real number. The function
>
> $$f(x) = ax^n$$
>
> is called a **power function of degree n**.

Power functions with *even* powers look similar to the square function.

Power functions with *odd* powers (other than $n = 1$) look similar to the cube function.

All even power functions have similar characteristics to a quadratic function (parabola), and all odd ($n > 1$) power functions have similar characteristics to a cubic function. For example, all even functions are symmetric with respect to the y-axis, whereas all odd functions are symmetric with respect to the origin. The following table summarizes their characteristics.

CHARACTERISTICS OF POWER FUNCTIONS: $f(x) = x^n$

	n EVEN	n ODD
Symmetry	y-axis	Origin
Domain	$(-\infty, \infty)$	$(-\infty, \infty)$
Range	$[0, \infty)$	$(-\infty, \infty)$
Some key points that lie on the graph	$(-1, 1), (0, 0),$ and $(1, 1)$	$(-1, -1), (0, 0),$ and $(1, 1)$
Increasing	$(0, \infty)$	$(-\infty, \infty)$
Decreasing	$(-\infty, 0)$	N/A

We now have the tools to graph polynomial functions that are transformations of power functions. We will use the power functions summarized above, combined with our graphing techniques such as horizontal and vertical shifting and reflection (Section 3.3).

EXAMPLE 2 Graphing Transformations of Power Functions

Graph the function $f(x) = (x - 1)^3$.

Solution:

STEP 1 Start with the graph of $y = x^3$.

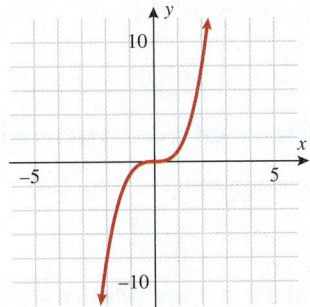

STEP 2 Shift $y = x^3$ to the right one unit to yield the graph of $f(x) = (x - 1)^3$.

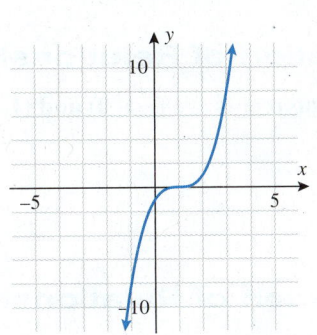

■ **YOUR TURN** Graph the function $f(x) = 1 - x^4$.

Classroom Example 4.2.2
Graph:
a. $f(x) = -(x + 2)^4$
b. $g(x) = 2 - x^5$
c. $h(x) = (x - 3)^5 + 1$
Answer: The graphs are given by:

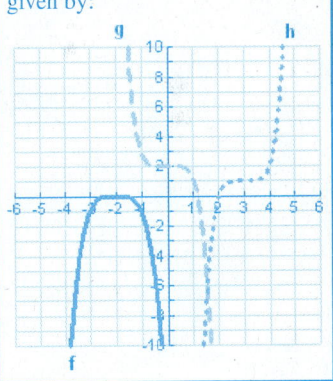

f

■ **Answer:** $f(x) = 1 - x^4$

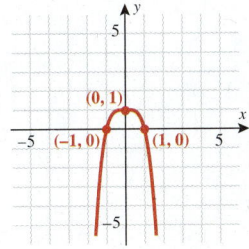

Real Zeros of a Polynomial Function

How do we graph general polynomial functions of degree greater than or equal to 3 if they cannot be written as transformations of power functions? We start by identifying the x-intercepts of the polynomial function. Recall that we determine the x-intercepts by setting the function equal to *zero* and solving for x. Therefore, an alternative name for an x-intercept of a function is a *zero* of the function. In our experience, to set a quadratic function equal to zero, the first step is to factor the quadratic expression into linear factors and then set each factor equal to zero. Therefore, there are four equivalent relationships that are summarized in the following box.

REAL ZEROS OF POLYNOMIAL FUNCTIONS

If $f(x)$ is a polynomial function and a is a *real* number, then the following statements are equivalent.

Relationship 1: $x = a$ is a **solution**, or **root**, of the equation $f(x) = 0$.
Relationship 2: $(a, 0)$ is an **x-intercept** of the graph of $f(x)$.
Relationship 3: $x = a$ is a **zero** of the function $f(x)$.
Relationship 4: $(x - a)$ is a **factor** of $f(x)$.

Let's use a simple polynomial function to illustrate these four relationships. We'll focus on the quadratic function $f(x) = x^2 - 1$. The graph of this function is a parabola that opens up and has as its vertex the point $(0, -1)$.

Illustration of Relationship 1

Set the function equal to zero, $f(x) = 0$. $\qquad\qquad x^2 - 1 = 0$
Factor. $\qquad\qquad\qquad\qquad\qquad\qquad\qquad (x - 1)(x + 1) = 0$
Solve. $\qquad\qquad\qquad\qquad\qquad\qquad\qquad\qquad x = 1 \text{ or } x = -1$

$x = -1$ **and** $x = 1$ **are solutions, or roots, of the equation** $x^2 - 1 = 0$.

Illustration of Relationship 2

The **x-intercepts** are $(-1, 0)$ **and** $(1, 0)$.

Illustration of Relationship 3

The value of the function at $x = -1$ and $x = 1$ is 0.

$f(-1) = (-1)^2 - 1 = \mathbf{0}$
$f(1) = (1)^2 - 1 = \mathbf{0}$

Illustration of Relationship 4

$$f(x) = x^2 - 1$$
$$= (x - 1)(x + 1)$$

$x = 1$ **is a zero of the polynomial, so** $(x - 1)$ **is a factor.**
$x = -1$ **is a zero of the polynomial, so** $(x + 1)$ **is a factor.**

We have a good reason for wanting to know the x-intercepts, or zeros. When the value of a continuous function transitions from negative to positive and vice versa, it must pass through zero.

DEFINITION | **Intermediate Value Theorem**

Let a and b be real numbers such that $a < b$ and let f be a polynomial function. If $f(a)$ and $f(b)$ have opposite signs, then there is at least one zero between a and b.

The intermediate value theorem will be used later in this chapter to assist us in finding real zeros of a polynomial function. For now, it tells us that in order to change signs, the polynomial function must pass through the x-axis. In other words, once we know the zeros, then we know that between two consecutive zeros the polynomial is either entirely above the x-axis or entirely below the x-axis. This enables us to break the x-axis down into intervals that we can test, which will assist us in graphing polynomial functions. Keep in mind, though, that the existence of a zero does not imply that the function will change signs—as you will see in the subsection on graphing general polynomial functions.

Technology Tip

Graph $y_1 = x^3 + x^2 - 2x$.

The zeros of the function $-2, 0,$ and 1 correspond to the x-intercepts $(-2, 0), (0, 0),$ and $(1, 0)$.

The table supports the real zeros shown by the graph.

 EXAMPLE 3   Identifying the Real Zeros of a Polynomial Function

Find the zeros of the polynomial function $f(x) = x^3 + x^2 - 2x$.

Solution:

Set the function equal to zero. $x^3 + x^2 - 2x = 0$

Factor out an x common to all three terms. $x(x^2 + x - 2) = 0$

Factor the quadratic expression
inside the parentheses. $x(x + 2)(x - 1) = 0$

Apply the zero product property. $x = 0$ or $(x + 2) = 0$ or $(x - 1) = 0$

Solve. $x = -2, x = 0,$ and $x = 1$

The zeros are $\boxed{-2, 0, \text{ and } 1}$.

■ YOUR TURN Find the zeros of the polynomial function $f(x) = x^3 - 7x^2 + 12x$.

■ **Answer:** The zeros are 0, 3, and 4.

Recall that when we were factoring a quadratic equation, if the factor was raised to a power greater than 1, the corresponding root, or zero, was repeated. For example, the quadratic equation $x^2 - 2x + 1 = 0$ when factored is written as $(x - 1)^2 = 0$. The solution, or root, in this case is $x = 1$, and we say that it is a **repeated** root. Similarly, when determining zeros of higher order polynomial functions, if a factor is repeated, we say that the zero is a repeated, or **multiple**, zero of the function. The number of times that a zero repeats is called its *multiplicity*.

Classroom Example 4.2.3
Find the zeros of these polynomials.
a. $f(x) = 2x^3 - 3x^2 - 5x$
b. $g(x) = -18x^3 + 3x^2 + 3x$

Answer: a. $0, -1, \frac{5}{2}$
       b. $0, -\frac{1}{3}, \frac{1}{2}$

DEFINITION	**Multiplicity of a Zero**

If $(x - a)^n$ is a factor of a polynomial f, then a is called a **zero of multiplicity n** of f.

EXAMPLE 4 **Finding the Multiplicities of Zeros of a Polynomial Function**

Find the zeros, and state their multiplicities, of the polynomial function
$g(x) = (x - 1)^2\left(x + \frac{3}{5}\right)^7(x + 5)$.

Solution:

> 1 is a zero of multiplicity 2.
>
> $-\frac{3}{5}$ is a zero of multiplicity 7.
>
> -5 is a zero of multiplicity 1.

Note: Adding the multiplicities yields the degree of the polynomial. The polynomial $g(x)$ is of degree 10, since $2 + 7 + 1 = 10$.

■ **YOUR TURN** For the polynomial $h(x)$, determine the zeros and state their multiplicities.

$$h(x) = x^2(x - 2)^3\left(x + \frac{1}{2}\right)^5$$

EXAMPLE 5 **Finding a Polynomial from Its Zeros**

Find a polynomial of degree 7 whose zeros are:

$$-2 \text{ (multiplicity 2)} \quad 0 \text{ (multiplicity 4)} \quad 1 \text{ (multiplicity 1)}$$

Solution:

If $x = a$ is a zero, then $(x - a)$ is a factor.	$f(x) = (x + 2)^2(x - 0)^4(x - 1)^1$
Simplify.	$= x^4(x + 2)^2(x - 1)$
Square the binomial.	$= x^4(x^2 + 4x + 4)(x - 1)$
Multiply the two polynomials.	$= x^4(x^3 + 3x^2 - 4)$
Distribute x^4.	$= \boxed{x^7 + 3x^6 - 4x^4}$

Graphing General Polynomial Functions

Let's develop a strategy for sketching an approximate graph of any polynomial function. First, we determine the x- and y-intercepts. Then we use the x-intercepts, or zeros, to divide the domain into intervals where the polynomial is positive or negative so that we can find points in those intervals to assist in sketching a smooth and continuous graph. *Note:* It is not always possible to find x-intercepts. Sometimes there are no x-intercepts.

EXAMPLE 6 **Using a Strategy for Sketching the Graph of a Polynomial Function**

Sketch the graph of $f(x) = (x + 2)(x - 1)^2$.

Solution:

STEP 1 Find the y-intercept.
(Let $x = 0$.)

$f(0) = (2)(-1)^2 = 2$
(0, 2) is the y-intercept

STEP 2 Find any x-intercepts.
(Set $f(x) = 0$.)

$f(x) = (x + 2)(x - 1)^2 = 0$
$x = -2$ or $x = 1$
(−2, 0) and **(1, 0)** are the x-intercepts

STEP 3 Plot the intercepts.

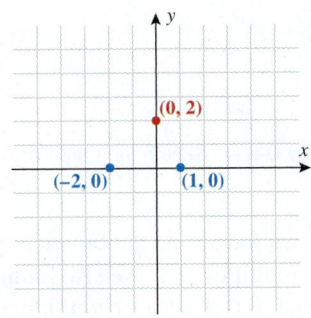

STEP 4 Divide the x-axis into intervals:* $(-\infty, -2)$, $(-2, 1)$, and $(1, \infty)$

*The x-intercepts $(-2, 0)$ and $(1, 0)$ divide the x-axis into three intervals similar to those created by zeros when we studied inequalities in Section 1.5.

STEP 5 Select a number in each interval and test each interval. The function $f(x)$ either *crosses* the x-axis at an x-intercept or *touches* the x-axis at an x-intercept. Therefore, we need to check each of these intervals to determine whether the function is positive (above the x-axis) or negative (below the x-axis). We do so by selecting numbers in the intervals and determining the value of the function at the corresponding points.

Interval	$(-\infty, -2)$	$(-2, 1)$	$(1, \infty)$
Number Selected in Interval	-3	-1	2
Value of Function	$f(-3) = -16$	$f(-1) = 4$	$f(2) = 4$
Point on Graph	$(-3, -16)$	$(-1, 4)$	$(2, 4)$
Interval Relation to x-Axis	Below x-axis	Above x-axis	Above x-axis

Technology Tip

The graph of $f(x) = (x + 2)(x - 1)^2$ is shown.

Note: The graph crosses the x-axis at the point $x = -2$ and touches the x-axis at the point $x = 1$.

A table of values supports the graph.

Study Tip

We do not know for sure that the points $(-1, 4)$ and $(1, 0)$ are turning points. A graphing utility can confirm this, and later in calculus you will learn how to find relative maximum points and relative minimum points.

■ **Answer:**

From the table, we find three additional points on the graph: $(-3, -16)$, $(-1, 4)$, and $(2, 4)$. The point $(-2, 0)$ is an intercept where the function *crosses* the x-axis, because it is below the x-axis to the left of -2 and above the x-axis to the right of -2. The point $(1, 0)$ is an intercept where the function *touches* the x-axis, because it is above the x-axis on both sides of $x = 1$. Connecting these points with a smooth curve yields the graph.

STEP 6 Sketch a plot of the function.

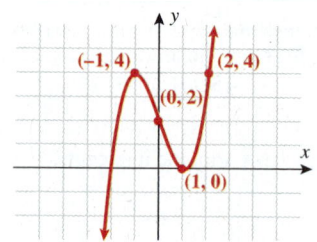

■ **YOUR TURN** Sketch the graph of $f(x) = x^2(x + 3)^2$.

In Example 6, we found that the function crosses the x-axis at the point $(-2, 0)$. Note that -2 is a zero of multiplicity 1. We also found that the function touches the x-axis at the point $(1, 0)$. Note that 1 is a zero of multiplicity 2. In general, zeros with even multiplicity correspond to intercepts where the function touches the x-axis and zeros with odd multiplicity correspond to intercepts where the function crosses the x-axis.

Study Tip

In general, zeros with *even* multiplicity correspond to intercepts where the function *touches* the x-axis and zeros with *odd* multiplicity correspond to intercepts where the function *crosses* the x-axis.

MULTIPLICITY OF A ZERO AND RELATION TO THE GRAPH OF A POLYNOMIAL

If a is a zero of $f(x)$, then:

MULTIPLICITY OF a	$f(x)$ ON EITHER SIDE OF $x = a$	GRAPH OF FUNCTION AT THE INTERCEPT
Even	Does not change sign	Touches the x-axis (turns around) at point $(a, 0)$
Odd	Changes sign	Crosses the x-axis at point $(a, 0)$

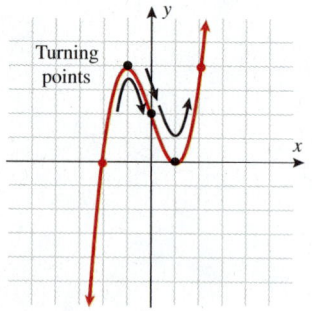

Also in Example 6, we know that somewhere in the interval $(-2, 1)$ the function must reach a maximum and then turn back toward the x-axis, because both points $(-2, 0)$ and $(1, 0)$ correspond to x-intercepts. When we sketch the graph, it "appears" that the point $(-1, 4)$ is a *turning point*. The point $(1, 0)$ also corresponds to a turning point. In general, if f is a polynomial of degree n, then the graph of f has at most $n - 1$ turning points.

The point $(-1, 4)$, which we call a turning point, is also a "high point" on the graph in the vicinity of the point $(-1, 4)$. Also note that the point $(1, 0)$, which we call a turning point, is a "low point" on the graph in the vicinity of the point $(1, 0)$. We call a "high point" on a graph a **local (relative) maximum** and a "low point" on a graph a **local (relative) minimum**. For quadratic functions we can find the maximum or minimum point by finding the vertex. However, for higher degree polynomial functions, we rely on graphing utilities to locate such points. Later in calculus, techniques will be developed for finding such points exactly. For now, we use the ⃞zoom⃞ and ⃞trace⃞ features to locate such points on a graph, and we can use the ⃞table⃞ feature of a graphing utility to approximate relative minima or maxima.

Study Tip

If f is a polynomial of degree n, then the graph of f has at most $n - 1$ turning points.

Let us take the polynomial $f(x) = x^3 - 2x^2 - 5x + 6$. Using methods discussed thus far, we can find that the x-intercepts of its graph are $(-2, 0)$, $(1, 0)$, and $(3, 0)$ and the y-intercept is the point $(0, 6)$. We can also find additional points that lie on the graph such as $(-1, 8)$ and $(2, -4)$. Plotting these points we might "think" that the points $(-1, 8)$ and $(2, -4)$ might be turning points, but a graphing utility reveals an approximate relative maximum at the point $(-0.7863, 8.2088207)$ and an approximate relative minimum at the point $(2.1196331, -4.060673)$.

Intercepts and turning points assist us in sketching graphs of polynomial functions. Another piece of information that will assist us in graphing polynomial functions is knowledge of the *end behavior*. All polynomials eventually rise or fall without bound as x gets large in both the positive $(x \to \infty)$ and negative $(x \to -\infty)$ directions. The highest degree monomial within the polynomial dominates the *end behavior*. In other words, the highest power term is eventually going to overwhelm the other terms as x grows without bound.

Technology Tip

Use TI to graph the function
$f(x) = x^3 - 2x^2 - 5x + 6$ as Y_1.

END BEHAVIOR

As x gets large in the positive $(x \to \infty)$ and negative $(x \to -\infty)$ directions, the graph of the polynomial

$$f(x) = a_n x^n + a_{n-1} x^{n-1} + \cdots + a_2 x^2 + a_1 x + a_0$$

has the same behavior as the power function

$$y = a_n x^n$$

Power functions behave much like a quadratic function (parabola) for even-degree polynomial functions and much like a cubic function for odd-degree polynomial functions. There are four possibilities because the leading coefficient can be positive or negative with either an odd or even power.

Let $y = a_n x^n$; then

n	Even	Even	Odd	Odd
a_n	Positive	Negative	Negative	Positive
$x \to -\infty$ (Left)	The graph of the function *rises*	The graph of the function *falls*	The graph of the function *rises*	The graph of the function *falls*
$x \to \infty$ (Right)	The graph of the function *rises*	The graph of the function *falls*	The graph of the function *falls*	The graph of the function *rises*
Graph	$a_n > 0$	$a_n < 0$	$a_n < 0$	$a_n > 0$

Classroom Example 4.2.7
Graph:

a. $f(x) = x(x - 2)(x + 1)^2$

b.* $g(x) = -x^2(2x + 1)^3$

■ **Answer:**

EXAMPLE 7   Graphing a Polynomial Function

Sketch a graph of the polynomial function $f(x) = 2x^4 - 8x^2$.

Solution:

STEP 1 Determine the y-intercept: $(x = 0)$. $\qquad f(0) = 0$

The y-intercept corresponds to the point $(0, 0)$.

STEP 2 Find the zeros of the polynomial. $\qquad f(x) = 2x^4 - 8x^2$

Factor out the common $2x^2$. $\qquad\qquad = 2x^2(x^2 - 4)$

Factor the quadratic binomial. $\qquad\quad = 2x^2(x - 2)(x + 2)$

Set $f(x) = 0$. $\qquad\qquad\qquad\qquad = 2x^2(x - 2)(x + 2) = 0$

0 is a zero of multiplicity 2. The graph will *touch* the x-axis.

2 is a zero of multiplicity 1. The graph will *cross* the x-axis.

−2 is a zero of multiplicity 1. The graph will *cross* the x-axis.

STEP 3 Determine the end behavior. $\qquad f(x) = 2x^4 - 8x^2$ behaves like $y = 2x^4$.

$y = 2x^4$ is of even degree, and the leading coefficient is positive, so the graph rises without bound as x gets large in both the positive and negative directions.

STEP 4 Sketch the intercepts and end behavior.

STEP 5 Find additional points.

x	-1	$-\frac{1}{2}$	$\frac{1}{2}$	1
f(x)	-6	$-\frac{15}{8}$	$-\frac{15}{8}$	-6

STEP 6 Sketch the graph.

■ Estimate additional points.

■ Connect with a smooth curve.

Note the symmetry about the y-axis. This function is an even function: $f(-x) = f(x)$.

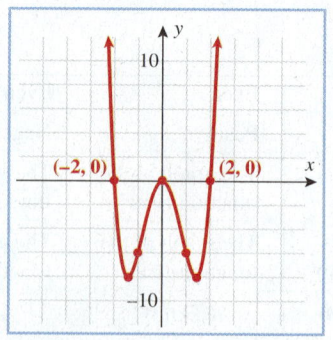

It is important to note that the local minimums occur at $x = \pm\sqrt{2} \approx \pm 1.14$ but at this time can only be illustrated using a graphing utility.

■ **YOUR TURN** Sketch a graph of the polynomial function $f(x) = x^5 - 4x^3$.

SECTION 4.2 SUMMARY

In general, polynomials can be graphed in one of two ways:

- Use graph-shifting techniques with power functions.
- General polynomial function.
  1. Identify intercepts.
  2. Determine each real zero and its multiplicity, and ascertain whether the graph crosses or touches the x-axis there.

3. x-intercepts (real zeros) divide the x-axis into intervals. Test points in the intervals to determine whether the graph is above or below the x-axis.
4. Determine the end behavior by investigating the end behavior of the highest degree monomial.
5. Sketch the graph with a smooth curve.

SECTION 4.2 EXERCISES

▪ SKILLS

In Exercises 1–10, determine which functions are polynomials, and for those that are, state their degree.

1. $f(x) = -3x^2 + 15x - 7$ **2.** $f(x) = 2x^5 - x^2 + 13$ **3.** $g(x) = (x + 2)^3\left(x - \frac{3}{5}\right)^2$ **4.** $g(x) = x^4(x - 1)^2(x + 2.5)^3$

5. $h(x) = \sqrt{x} + 1$ **6.** $h(x) = (x - 1)^{1/2} + 5x$ **7.** $F(x) = x^{1/3} + 7x^2 - 2$ **8.** $F(x) = 3x^2 + 7x - \dfrac{2}{3x}$

9. $G(x) = \dfrac{x + 1}{x^2}$ **10.** $H(x) = \dfrac{x^2 + 1}{2}$

In Exercises 11–18, match the polynomial function with its graph.

11. $f(x) = -3x + 1$ **12.** $f(x) = -3x^2 - x$ **13.** $f(x) = x^2 + x$ **14.** $f(x) = -2x^3 + 4x^2 - 6x$

15. $f(x) = x^3 - x^2$ **16.** $f(x) = 2x^4 - 18x^2$ **17.** $f(x) = -x^4 + 5x^3$ **18.** $f(x) = x^5 - 5x^3 + 4x$

a.

b.

c.

d.

e.

f.

g.

h.

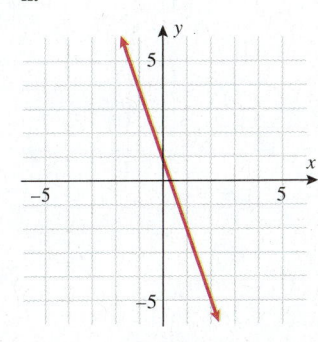

In Exercises 19–26, graph each function by transforming a power function $y = x^n$.

19. $f(x) = -x^5$

20. $f(x) = -x^4$

21. $f(x) = (x - 2)^4$

22. $f(x) = (x + 2)^5$

23. $f(x) = x^5 + 3$

24. $f(x) = -x^4 - 3$

25. $f(x) = 3 - (x + 1)^4$

26. $f(x) = (x - 3)^5 - 2$

In Exercises 27–38, find all the real zeros (and state their multiplicities) of each polynomial function.

27. $f(x) = 2(x - 3)(x + 4)^3$

28. $f(x) = -3(x + 2)^3(x - 1)^2$

29. $f(x) = 4x^2(x - 7)^2(x + 4)$

30. $f(x) = 5x^3(x + 1)^4(x - 6)$

31. $f(x) = 4x^2(x - 1)^2(x^2 + 4)$

32. $f(x) = 4x^2(x^2 - 1)(x^2 + 9)$

33. $f(x) = 8x^3 + 6x^2 - 27x$

34. $f(x) = 2x^4 + 5x^3 - 3x^2$

35. $f(x) = -2.7x^3 - 8.1x^2$

36. $f(x) = 1.2x^6 - 4.6x^4$

37. $f(x) = \frac{1}{3}x^6 + \frac{2}{5}x^4$

38. $f(x) = \frac{2}{7}x^5 - \frac{3}{4}x^4 + \frac{1}{2}x^3$

In Exercises 39–52, find a polynomial (there are many) of minimum degree that has the given zeros.

39. $-3, 0, 1, 2$

40. $-2, 0, 2$

41. $-5, -3, 0, 2, 6$

42. $0, 1, 3, 5, 10$

43. $-\frac{1}{2}, \frac{2}{3}, \frac{3}{4}$

44. $-\frac{3}{4}, -\frac{1}{3}, 0, \frac{1}{2}$

45. $1 - \sqrt{2}, 1 + \sqrt{2}$

46. $1 - \sqrt{3}, 1 + \sqrt{3}$

47. -2 (multiplicity 3), 0 (multiplicity 2)

48. -4 (multiplicity 2), 5 (multiplicity 3)

49. -3 (multiplicity 2), 7 (multiplicity 5)

50. 0 (multiplicity 1), 10 (multiplicity 3)

51. $-\sqrt{3}$ (multiplicity 2), -1 (multiplicity 1), 0 (multiplicity 2), $\sqrt{3}$ (multiplicity 2)

52. $-\sqrt{5}$ (multiplicity 2), 0 (multiplicity 1), 1 (multiplicity 2), $\sqrt{5}$ (multiplicity 2)

In Exercises 53–72, for each polynomial function given: (a) list each real zero and its multiplicity; (b) determine whether the graph touches or crosses at each x-intercept; (c) find the y-intercept and a few points on the graph; (d) determine the end behavior; and (e) sketch the graph.

53. $f(x) = -x^2 - 6x - 9$

54. $f(x) = x^2 + 4x + 4$

55. $f(x) = (x - 2)^3$

56. $f(x) = -(x + 3)^3$

57. $f(x) = x^3 - 9x$

58. $f(x) = -x^3 + 4x^2$

59. $f(x) = -x^3 + x^2 + 2x$

60. $f(x) = x^3 - 6x^2 + 9x$

61. $f(x) = -x^4 - 3x^3$

62. $f(x) = x^5 - x^3$

63. $f(x) = 12x^6 - 36x^5 - 48x^4$

64. $f(x) = 7x^5 - 14x^4 - 21x^3$

65. $f(x) = 2x^5 - 6x^4 - 8x^3$

66. $f(x) = -5x^4 + 10x^3 - 5x^2$

67. $f(x) = x^3 - x^2 - 4x + 4$

68. $f(x) = x^3 - x^2 - x + 1$

69. $f(x) = -(x + 2)^2(x - 1)^2$

70. $f(x) = (x - 2)^3(x + 1)^3$

71. $f(x) = x^2(x - 2)^3(x + 3)^2$

72. $f(x) = -x^3(x - 4)^2(x + 2)^2$

In Exercises 73–76, for each graph given: (a) list each real zero and its smallest possible multiplicity; (b) determine whether the degree of the polynomial is even or odd; (c) determine whether the leading coefficient of the polynomial is positive or negative; (d) find the y-intercept; and (e) write an equation for the polynomial function (assume the least degree possible).

73.

74.

75.

76.

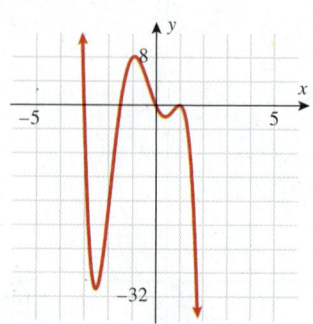

■ **APPLICATIONS**

For Exercises 77 and 78, refer to the following:

The relationship between a company's total revenue R (in millions of dollars) is related to its advertising costs x (in thousands of dollars). The relationship between revenue R and advertising costs x is illustrated in the graph.

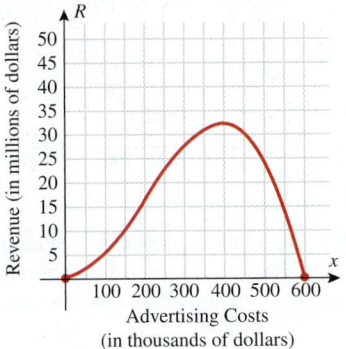

77. **Business.** Analyze the graph of the revenue function.

   a. Determine the intervals on which revenue is increasing and decreasing.

   b. Identify the zeros of the function. Interpret the meaning of zeros for this function.

78. **Business.** Use the graph to identify the maximum revenue for the company and the corresponding advertising costs that produce maximum revenue.

For Exercises 79 and 80, refer to the following:

During a cough, the velocity v (in meters per second) of air in the trachea may be modeled by the function

$$v(r) = -120r^3 + 80r^2$$

where r is the radius of the trachea (in centimeters) during the cough.

79. **Health/Medicine.** Graph the velocity function and estimate the intervals on which the velocity of air in the trachea is increasing and decreasing.

80. **Health/Medicine.** Estimate the radius of the trachea that produces the maximum velocity of air in the trachea. Use this radius to estimate the maximum velocity of air in the trachea.

81. **Weight.** Jennifer has joined a gym to lose weight and feel better. She still likes to cheat a little and will enjoy the occasional bad meal with an ice cream dream dessert and then miss the gym for a couple of days. Given in the table is Jennifer's weight for a period of 8 months. Her weight can be modeled as a polynomial. Plot these data. How many turning points are there? Assuming these are the minimum number of turning points, what is the lowest degree polynomial that can represent Jennifer's weight?

MONTH	WEIGHT
1	169
2	158
3	150
4	161
5	154
6	159
7	148
8	153

82. **Stock Value.** A day trader checks the stock price of Coca-Cola during a 4-hour period (given below). The price of Coca-Cola stock during this 4-hour period can be modeled as a polynomial function. Plot these data. How many turning points are there? Assuming these are the minimum number of turning points, what is the lowest degree polynomial that can represent the Coca-Cola stock price?

PERIOD WATCHING STOCK MARKET	PRICE
1	$53.00
2	$56.00
3	$52.70
4	$51.50

83. **Stock Value.** The price of Tommy Hilfiger stock during a 4-hour period is given in the following table. If a third-degree polynomial models this stock, do you expect the stock to go up or down in the fifth period?

PERIOD WATCHING STOCK MARKET	PRICE
1	$15.10
2	$14.76
3	$15.50
4	$14.85

84. Stock Value. The stock prices for Coca-Cola during a 4-hour period on another day yield the following results. If a third-degree polynomial models this stock, do you expect the stock to go up or down in the fifth period?

PERIOD WATCHING STOCK MARKET	PRICE
1	$52.80
2	$53.00
3	$56.00
4	$52.70

For Exercises 85 and 86, the following graph illustrates the average federal funds rate in the month of January (2000 to 2008).

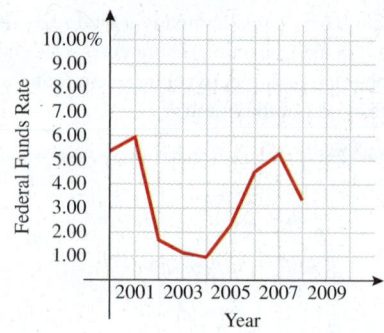

85. Finance. If a polynomial function is used to model the federal funds rate data shown in the graph, determine the degree of the lowest degree polynomial that can be used to model those data.

86. Finance. Should the leading coefficient in the polynomial found in Exercise 85 be positive or negative? Explain.

■ CATCH THE MISTAKE

In Exercises 87–90, explain the mistake that is made.

87. Find a fourth-degree polynomial function with zeros -2, $-1, 3, 4$.

Solution: $$f(x) = (x - 2)(x - 1)(x + 3)(x + 4)$$

This is incorrect. What mistake was made?

88. Determine the end behavior of the polynomial function $f(x) = x(x - 2)^3$.

Solution:

This polynomial has similar end behavior to the graph of $y = x^3$.

End behavior falls to the left and rises to the right.

This is incorrect. What mistake was made?

89. Graph the polynomial function $f(x) = (x - 1)^2(x + 2)^3$.

Solution:

The zeros are -2 and 1, and therefore, the x-intercepts are $(-2, 0)$ and $(1, 0)$.

The y-intercept is $(0, 8)$.

Plotting these points and connecting with a smooth curve yields the graph on the right.

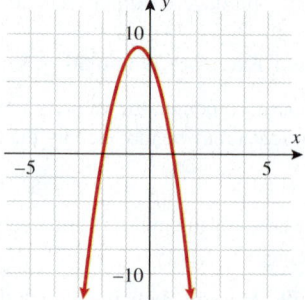

This graph is incorrect. What did we forget to do?

90. Graph the polynomial function $f(x) = (x + 1)^2(x - 1)^2$.

Solution:

The zeros are -1 and 1, so the x-intercepts are $(-1, 0)$ and $(1, 0)$.

The y-intercept is $(0, 1)$.

Plotting these points and connecting with a smooth curve yields the graph on the right.

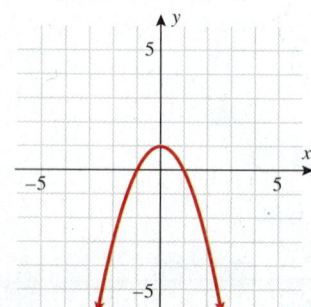

This graph is incorrect. What did we forget to do?

■ CONCEPTUAL

In Exercises 91–94, determine if each statement is true or false.

91. The graph of a polynomial function might not have any y-intercepts.

92. The graph of a polynomial function might not have any x-intercepts.

93. The domain of all polynomial functions is $(-\infty, \infty)$.

94. The range of all polynomial functions is $(-\infty, \infty)$.

95. What is the maximum number of zeros that a polynomial of degree n can have?

96. What is the maximum number of turning points a graph of an nth-degree polynomial can have?

■ CHALLENGE

97. Find a seventh-degree polynomial that has the following graph characteristics: The graph touches the x-axis at $x = -1$, and the graph crosses the x-axis at $x = 3$. Plot this polynomial function.

98. Find a fifth-degree polynomial that has the following graph characteristics: The graph touches the x-axis at $x = 0$ and crosses the x-axis at $x = 4$. Plot the polynomial function.

99. Determine the zeros of the polynomial $f(x) = x^3 + (b - a)x^2 - abx$ for the positive real numbers a and b.

100. Graph the function $f(x) = x^2(x - a)^2(x - b)^2$ for the positive real numbers a, b, where $b > a$.

■ TECHNOLOGY

In Exercises 101 and 102, use a graphing calculator or computer to graph each polynomial. From that graph, estimate the x-intercepts (if any). Set the function equal to zero, and solve for the zeros of the polynomial. Compare the zeros with the x-intercepts.

101. $f(x) = x^4 + 2x^2 + 1$ **102.** $f(x) = 1.1x^3 - 2.4x^2 + 5.2x$

For each polynomial in Exercises 103 and 104, determine the power function that has similar end behavior. Plot this power function and the polynomial. Do they have similar end behavior?

103. $f(x) = -2x^5 - 5x^4 - 3x^3$ **104.** $f(x) = x^4 - 6x^2 + 9$

In Exercises 105 and 106, use a graphing calculator or a computer to graph each polynomial. From the graph, estimate the x-intercepts and state the zeros of the function and their multiplicities.

105. $f(x) = x^4 - 15.9x^3 + 1.31x^2 + 292.905x + 445.7025$

106. $f(x) = -x^5 + 2.2x^4 + 18.49x^3 - 29.878x^2 - 76.5x + 100.8$

In Exercises 107 and 108, use a graphing calculator or a computer to graph each polynomial. From the graph, estimate the coordinates of the relative maximum and minimum points. Round your answers to two decimal places.

107. $f(x) = 2x^4 + 5x^3 - 10x^2 - 15x + 8$

108. $f(x) = 2x^5 - 4x^4 - 12x^3 + 18x^2 + 16x - 7$

SKILLS OBJECTIVES

- Divide polynomials with long division.
- Divide polynomials with synthetic division.

CONCEPTUAL OBJECTIVES

- Extend long division of real numbers to polynomials.
- Understand *when* synthetic division can be used.

$$
\begin{array}{r}
311 \\
21\overline{)6542} \\
-63 \\
\hline
24 \\
-21 \\
\hline
32 \\
-21 \\
\hline
11
\end{array}
$$

To divide polynomials, we rely on the technique we use for dividing real numbers. For example, if you were asked to divide 6542 by 21, the long division method used is illustrated in the margin. This solution can be written two ways: $311\ R11$ or $311 + \dfrac{11}{21}$.

In this example, the **dividend** is 6542, the **divisor** is 21, the **quotient** is 311, and the **remainder** is 11. We employ a similar technique (dividing the leading terms) when dividing polynomials.

Long Division of Polynomials

Let's start with an example whose answer we already know. We know that a quadratic expression can be factored into the product of two linear factors: $x^2 + 4x - 5 = (x + 5)(x - 1)$. Therefore, if we divide both sides of the equation by $(x - 1)$, we get

$$
\frac{x^2 + 4x - 5}{x - 1} = x + 5
$$

We can state this by saying $x^2 + 4x - 5$ divided by $x - 1$ is equal to $x + 5$. Confirm this statement by long division:

$$
x - 1\overline{)x^2 + 4x - 5}
$$

Note that although this is standard division notation, the dividend and the divisor are both polynomials that consist of multiple terms. The *leading* terms of each algebraic expression will guide us.

WORDS	**MATH**
Q: x times what quantity gives x^2?	$x - 1\overline{)x^2 + 4x - 5}$
A: x	$\begin{array}{r} x \\ x - 1\overline{)x^2 + 4x - 5} \end{array}$
Multiply $x(x - 1) = x^2 - x$.	$\begin{array}{r} x \\ x - 1\overline{)x^2 + 4x - 5} \\ x^2 - x \end{array}$
Subtract $(x^2 - x)$ from $x^2 + 4x - 5$. *Note:* $-(x^2 - x) = -x^2 + x$.	$\begin{array}{r} x \\ x - 1\overline{)x^2 + 4x - 5} \\ -x^2 + x \\ \hline 5x - 5 \end{array}$
Bring down the -5.	

WORDS	**MATH**

Q: x times what quantity is $5x$?
A: **5**
Multiply $5(x - 1) = 5x - 5$.

$$
\begin{array}{r}
x + 5 \\
x - 1 \overline{)\,x^2 + 4x - 5} \\
-x^2 + x \\
\hline
5x - 5 \\
\rightarrow 5x - 5
\end{array}
$$

$$
\begin{array}{r}
x + 5 \\
x - 1 \overline{)\,x^2 + 4x - 5} \\
-x^2 + x \\
\hline
5x - 5
\end{array}
$$

Subtract $(5x - 5)$.

Note: $-(5x - 5) = -5x + 5$.

$$
\begin{array}{r}
-5x + 5 \\
\hline
0
\end{array}
$$

As expected, the remainder is 0. By long division we have shown that

$$\boxed{\dfrac{x^2 + 4x - 5}{x - 1} = x + 5}$$

Check: Multiplying the equation by $x - 1$ yields $x^2 + 4x - 5 = (x + 5)(x - 1)$, which we knew to be true.

Technology Tip

A graphing utility can be used to check $(x^2 - 5x + 6)(2x + 1) = 2x^3 - 9x^2 + 7x + 6$ using their graphs.

Notice that the graphs are the same.

▶ **EXAMPLE 1   Dividing Polynomials Using Long Division; Zero Remainder**

Divide $2x^3 - 9x^2 + 7x + 6$ by $2x + 1$.

Solution:

$$
\begin{array}{r}
x^2 - 5x + 6 \\
2x + 1 \overline{)\,2x^3 - 9x^2 + 7x + 6} \\
-(2x^3 + x^2) \\
\hline
-10x^2 + 7x \\
-(-10x^2 - 5x) \\
\hline
12x + 6 \\
-(12x + 6) \\
\hline
0
\end{array}
$$

Multiply: $x^2(2x + 1)$.

Subtract: Bring down the $7x$.

Multiply: $-5x(2x + 1)$.

Subtract: Bring down the 6.

Multiply: $6(2x + 1)$.

Subtract.

Quotient: $\boxed{x^2 - 5x + 6}$

Check: $(2x + 1)(x^2 - 5x + 6) = 2x^3 - 9x^2 + 7x + 6$.

Note: Since the divisor cannot be equal to zero, $2x + 1 \neq 0$, then we say $x \neq -\frac{1}{2}$.

- -

■ **YOUR TURN** Divide $4x^3 + 13x^2 - 2x - 15$ by $4x + 5$.

■ **Answer:** $x^2 + 2x - 3$, remainder 0.

Why are we interested in dividing polynomials? Because it helps us find zeros of polynomials. In Example 1, using long division, we found that

$$2x^3 - 9x^2 + 7x + 6 = (2x + 1)(x^2 - 5x + 6)$$

Factoring the quadratic expression enables us to write the cubic polynomial as a product of three linear factors:

$$2x^3 - 9x^2 + 7x + 6 = (2x + 1)(x^2 - 5x + 6) = (2x + 1)(x - 3)(x - 2)$$

Set the value of the polynomial equal to zero, $(2x + 1)(x - 3)(x - 2) = 0$, and solve for x. The zeros of the polynomial are $-\frac{1}{2}$, 2, and 3. In Example 1 and in the Your Turn, the remainder was 0. Sometimes there is a nonzero remainder (Example 2).

EXAMPLE 2   Dividing Polynomials Using Long Division; Nonzero Remainder

Divide $6x^2 - x - 2$ by $x + 1$.

Solution:

Multiply $6x(x + 1)$.

Subtract and bring down -2.

Multiply $-7(x + 1)$.

Subtract and identify the remainder.

$$
\begin{array}{r}
6x - 7 \\
x + 1 \overline{)6x^2 - x - 2} \\
\underline{-(6x^2 + 6x)} \\
-7x - 2 \\
\underline{-(-7x - 7)} \\
+5
\end{array}
$$

$$\underset{\text{Divisor}}{\underbrace{\frac{\overset{\text{Dividend}}{6x^2 - x - 2}}{x + 1}}} = \underset{}{\underbrace{\overset{\text{Quotient}}{6x - 7}}} + \underset{\text{Divisor}}{\underbrace{\frac{\overset{\text{Remainder}}{5}}{x + 1}}} \qquad \boxed{x \neq -1}$$

Check: Multiply the quotient and remainder by $x + 1$.

The result is the dividend.

$$(6x - 7)(x + 1) + \frac{5}{(x + 1)} \cdot (x + 1)$$

$$= 6x^2 - x - 7 + 5$$

$$= 6x^2 - x - 2 \quad ✓$$

■ **YOUR TURN** Divide $2x^3 + x^2 - 4x - 3$ by $x - 1$.

Classroom Example 4.3.2

a. Divide $-18x^3 + 3x^2 + 3x + 1$ by $3x + 1$.

b.* Let $b \neq 0$. Divide $-18x^3 + 3x^2 + 3x + b$ by $3x + 1$.

Answer:

a. $-6x^2 + 3x + \dfrac{1}{3x + 1}$

b. $-6x^2 + 3x + \dfrac{b}{3x + 1}$

■ **Answer:** $2x^2 + 3x - 1$ $R: -4$ or $2x^2 + 3x - 1 - \dfrac{4}{x - 1}$

In general, when a polynomial is divided by another polynomial, we express the result in the following form:

$$\frac{P(x)}{d(x)} = Q(x) + \frac{r(x)}{d(x)}$$

where $P(x)$ is the **dividend**, $d(x) \neq 0$ is the divisor, $Q(x)$ is the quotient, and $r(x)$ is the remainder. Multiplying this equation by the divisor, $d(x)$, leads us to the division algorithm.

THE DIVISION ALGORITHM

If $P(x)$ and $d(x)$ are polynomials with $d(x) \neq 0$, and if the degree of $P(x)$ is greater than or equal to the degree of $d(x)$, then unique polynomials $Q(x)$ and $r(x)$ exist such that

$$P(x) = d(x) \cdot Q(x) + r(x)$$

If the remainder $r(x) = 0$, then we say that $d(x)$ divides $P(x)$ and that $d(x)$ and $Q(x)$ are factors of $P(x)$.

EXAMPLE 3   **Long Division of Polynomials with "Missing" Terms**

Divide $x^3 - 8$ by $x - 2$.

Solution:

Insert $\mathbf{0x^2 + 0x}$ for placeholders.

Multiply $x^2(x - 2) = x^3 - 2x^2$.

Subtract and bring down $0x$.

Multiply $2x(x - 2) = 2x^2 - 4x$.

Subtract and bring down -8.

Multiply $4(x - 2) = 4x - 8$.

Subtract and get remainder 0.

$$
\begin{array}{r}
x^2 + 2x\ + 4 \\
x - 2\overline{)x^3 + \mathbf{0x^2} + \mathbf{0x} - 8} \\
\underline{-(x^3 - 2x^2)} \\
2x^2 + 0x \\
\underline{-(2x^2 - 4x)} \\
4x - 8 \\
\underline{-(4x - 8)} \\
0
\end{array}
$$

Since the remainder is 0, $x - 2$ is a factor of $x^3 - 8$.

$$\boxed{\frac{x^3 - 8}{x - 2} = x^2 + 2x + 4,\ x \neq 2}$$

Check: $x^3 - 8 = (x^2 + 2x + 4)(x - 2) = x^3 + 2x^2 + 4x - 2x^2 - 4x - 8 = x^3 - 8$ ✓

- -

■ **YOUR TURN** Divide $x^3 - 1$ by $x - 1$.

> **Classroom Example 4.3.3**
> **a.** Divide $x^4 - 81$ by $x + 3$.
> **b.** Divide $x^4 - 81$ by $x^2 + 9$.
> **Answer:**
> **a.** $x^3 - 3x^2 + 9x - 27$
> **b.** $x^2 - 9$

■ **Answer:** $x^2 + x + 1$

EXAMPLE 4   **Long Division of Polynomials**

Divide $3x^4 + 2x^3 + x^2 + 4$ by $x^2 + 1$.

Solution:

Insert $\mathbf{0x}$ as a placeholder in both the divisor and the dividend.

Multiply $3x^2(x^2 + 0x + 1)$.

Subtract and bring down $0x$.

Multiply $2x(x^2 + 0x + 1)$.

Subtract and bring down 4.

Multiply $-2(x^2 - 2x + 1)$.

Subtract and get remainder $-2x + 6$.

$$
\begin{array}{r}
3x^2 + 2x\ - 2 \\
x^2 + \mathbf{0x} + 1\overline{)3x^4 + 2x^3 + x^2 + \mathbf{0x} + 4} \\
\underline{-(3x^4 + 0x^3 + 3x^2)} \\
2x^3 - 2x^2 + 0x \\
\underline{-(2x^3 + 0x^2 + 2x)} \\
-2x^2 - 2x + 4 \\
\underline{-(-2x^2 + 0x - 2)} \\
-2x + 6
\end{array}
$$

$$\boxed{\frac{3x^4 + 2x^3 + x^2 + 4}{x^2 + 1} = 3x^2 + 2x - 2 + \frac{-2x + 6}{x^2 + 1}}$$

> **Classroom Example 4.3.4**
> Divide $x^4 + x^3 + 4x^2 + 3x + 4$ by $x^2 + x + 1$.
> **Answer:**
> $x^3 + 3 + \dfrac{1}{x^2 + x + 1}$

■ **Answer:**
$2x^2 + 6 + \dfrac{11x^2 + 18x + 36}{x^3 - 3x - 4}$

- -

■ **YOUR TURN** Divide $2x^5 + 3x^2 + 12$ by $x^3 - 3x - 4$.

In Examples 1 through 4 the dividends, divisors, and quotients were all polynomials with integer coefficients. In Example 5, however, the resulting quotient has rational (noninteger) coefficients.

EXAMPLE 5 **Long Division of Polynomials Resulting in Quotients with Rational Coefficients**

Divide $8x^4 - 5x^3 + 7x - 2$ by $2x^2 + 1$.

Solution:

Insert $0x^2$ as a placeholder in the dividend and $0x$ as a placeholder in the divisor.

Multiply $4x^2(2x^2 + 0x + 1)$.

Subtract and bring down remaining terms.

Multiply $-\frac{5}{2}x(2x^2 + 0x + 1)$.

Subtract and bring down remaining terms.

Multiply $-2(2x^2 + 0x + 1)$.

Subtract and bring down the remainder $\frac{19}{2}x$.

$$
\begin{array}{r}
4x^2 - \frac{5}{2}x - 2 \\
2x^2 + 0x + 1 \overline{\smash{)}8x^4 - 5x^3 + 0x^2 + 7x - 2} \\
-(8x^4 + 0x^3 + 4x^2) \\
\hline
-5x^3 - 4x^2 + 7x \\
-(-5x^3 + 0x^2 - \frac{5}{2}x) \\
\hline
-4x^2 + \frac{19}{2}x - 2 \\
-(-4x^2 + 0x - 2) \\
\hline
\frac{19}{2}x
\end{array}
$$

$$\frac{8x^4 - 5x^3 + 7x - 2}{2x^2 + 1} = 4x^2 - \frac{5}{2}x - 2 + \frac{\frac{19}{2}x}{2x^2 + 1}$$

■ **YOUR TURN** Divide $10x^4 - 3x^3 + 5x - 4$ by $2x^2 - 1$.

■ **Answer:**

$$5x^2 - \frac{3}{2}x + \frac{5}{2} + \frac{\frac{7}{2}x - \frac{3}{2}}{2x^2 - 1}$$

Synthetic Division of Polynomials

In the special case when the *divisor is a linear factor* of the form $x - a$ or $x + a$, there is another, more efficient way to divide polynomials. This method is called **synthetic division**. It is called synthetic because it is a contrived shorthand way of dividing a polynomial by a linear factor. A detailed step-by-step procedure is given below for synthetic division. Let's divide $x^4 - x^3 - 2x + 2$ by $x + 1$ using synthetic division.

Study Tip

If $(x - a)$ is the divisor, then a is the number used in synthetic division.

STEP 1 Write the division in synthetic form.
- List the coefficients of the dividend. **Remember to use 0 for a placeholder**.
- The divisor is $x + 1$. Since $x + 1 = 0$ $x = -1$ is used.

Coefficients of Dividend

$$-1 \,\big|\, \begin{array}{cccc} 1 & -1 & 0 & -2 & 2 \end{array}$$

STEP 2 *Bring down* the first term (**1**) in the dividend.

$$-1 \,\big|\, \begin{array}{cccc} 1 & -1 & 0 & -2 & 2 \end{array}$$
Bring down the **1**
$$1$$

STEP 3 *Multiply* the **−1** by this leading coefficient (**1**), and place the product up and to the right in the second column.

STEP 4 *Add* the values in the second column.

$$
\begin{array}{r|rrrr}
-1 & 1 & -1 & 0 & -2 & 2 \\
 & & \downarrow -1 \text{ ADD} \\
\hline
 & 1 & -2
\end{array}
$$

STEP 5 Repeat Steps 3 and 4 until all columns are filled.

$$
\begin{array}{r|rrrrr}
-1 & 1 & -1 & 0 & -2 & 2 \\
 & \downarrow & -1\downarrow & 2\downarrow & -2\downarrow & 4\downarrow \\
\hline
 & 1 & -2 & 2 & -4 & 6
\end{array}
$$

STEP 6 Identify the **quotient** by assigning powers of x in descending order, beginning with $x^{n-1} = x^{4-1} = x^3$. The last term is the **remainder**.

$$
\begin{array}{r|rrrrr}
-1 & 1 & -1 & 0 & -2 & 2 \\
 & & -1 & 2 & -2 & 4 \\
\hline
 & 1 & -2 & 2 & -4 & 6
\end{array}
$$

$$\underbrace{1 \quad -2 \quad 2 \quad -4}_{\text{Quotient Coefficients}} \quad \overset{\text{Remainder}}{6}$$

$$x^3 - 2x^2 + 2x - 4$$

Study Tip

Synthetic division can only be used when the divisor is of the form $x - a$ or $x + a$.

We know that the degree of the first term of the quotient is 3 because a fourth-degree polynomial was divided by a first-degree polynomial. Let's compare dividing $x^4 - x^3 - 2x + 2$ by $x + 1$ using both long division and synthetic division.

Long Division

$$
\require{enclose}
\begin{array}{r}
x^3 - 2x^2 + 2x - 4 \\
x + 1 \enclose{longdiv}{x^4 - x^3 + 0x^2 - 2x + 2} \\
\underline{x^4 + x^3} \\
-2x^3 + 0x^2 \\
\underline{-(-2x^3 - 2x^2)} \\
2x^2 - 2x \\
\underline{-(2x^2 + 2x)} \\
-4x + 2 \\
\underline{-(-4x - 4)} \\
+6
\end{array}
$$

Synthetic Division

$$
\begin{array}{r|rrrrr}
-1 & 1 & -1 & 0 & -2 & 2 \\
 & & -1 & 2 & -2 & 4 \\
\hline
 & 1 & -2 & 2 & -4 & 6
\end{array}
$$

$$x^3 - 2x^2 + 2x - 4$$

Both long division and synthetic division yield the same answer.

$$\frac{x^4 - x^3 - 2x + 2}{x + 1} = x^3 - 2x^2 + 2x - 4 + \frac{6}{x + 1}$$

EXAMPLE 6 **Synthetic Division**

Use synthetic division to divide $3x^5 - 2x^3 + x^2 - 7$ by $x + 2$.

Solution:

STEP 1 Write the division in synthetic form.
- List the coefficients of the dividend. Remember to use **0** for a placeholder.
- The divisor of the original problem is $x + 2$. If we set $x + 2 = 0$ we find that $x = -2$, so -2 is the divisor for synthetic division.

$$
\begin{array}{r|rrrrrr}
-2 & 3 & 0 & -2 & 1 & 0 & -7
\end{array}
$$

Classroom Example 4.3.6

Use synthetic division to divide the following:

a. $x^4 - 81$ by $x + 3$

b.* $-18x^3 + 3x^2 + 3x + b$
by $3x + 1$ when $b \neq 0$

Answer:

a. $x^3 - 3x^2 + 9x - 27$

b. $-6x^2 + 3x + \dfrac{b}{3x + 1}$

STEP 2 Perform the synthetic division steps.

$$
\begin{array}{r|rrrrrr}
-2 & 3 & 0 & -2 & 1 & 0 & -7 \\
   &   & -6 & 12 & -20 & 38 & -76 \\
\hline
   & 3 & -6 & 10 & -19 & 38 & -83
\end{array}
$$

STEP 3 Identify the quotient and remainder.

$$
\begin{array}{r|rrrrrr}
-2 & 3 & 0 & -2 & 1 & 0 & -7 \\
   &   & -6 & 12 & -20 & 38 & -76 \\
\hline
   & \underbrace{3 \quad -6 \quad 10 \quad -19 \quad 38}_{3x^4 - 6x^3 + 10x^2 - 19x + 38} & & & & & \boxed{-83}
\end{array}
$$

$$\frac{3x^5 - 2x^3 + x^2 - 7}{x + 2} = 3x^4 - 6x^3 + 10x^2 - 19x + 38 - \frac{83}{x + 2}$$

■ **Answer:** $2x^2 + 2x + 1 + \dfrac{4}{x - 1}$

■ **YOUR TURN** Use synthetic division to divide $2x^3 - x + 3$ by $x - 1$.

SECTION
4.3 SUMMARY

Division of Polynomials

▪ Long division can always be used.
▪ Synthetic division is restricted to when the divisor is of the form $x - a$ or $x + a$.

Expressing Results

▪ $\dfrac{\text{Dividend}}{\text{Divisor}} = \text{quotient} + \dfrac{\text{remainder}}{\text{divisor}}$

▪ Dividend $= $ (quotient)(divisor) $+$ remainder

When Remainder Is Zero

▪ Dividend $= $ (quotient)(divisor)
▪ Quotient and divisor are factors of the dividend.

SECTION
4.3 EXERCISES

▪ **SKILLS**

In Exercises 1–30, divide the polynomials using long division. Use exact values. Express the answer in the form $Q(x) = $?, $r(x) = $?.

1. $(2x^2 + 5x - 3) \div (x + 3)$

2. $(2x^2 + 5x - 3) \div (x - 3)$

3. $(x^2 - 5x + 6) \div (x - 2)$

4. $(2x^2 + 3x + 1) \div (x + 1)$

5. $(3x^2 - 9x - 5) \div (x - 2)$

6. $(x^2 + 4x - 3) \div (x - 1)$

7. $(3x^2 - 13x - 10) \div (x + 5)$

8. $(3x^2 - 13x - 10) \div (x - 5)$

9. $(x^2 - 4) \div (x + 4)$

10. $(x^2 - 9) \div (x - 2)$

11. $(9x^2 - 25) \div (3x - 5)$

12. $(5x^2 - 3) \div (x + 1)$

13. $(4x^2 - 9) \div (2x + 3)$

14. $(8x^3 + 27) \div (2x + 3)$

15. $(11x + 20x^2 + 12x^3 + 2) \div (3x + 2)$

16. $(12x^3 + 2 + 11x + 20x^2) \div (2x + 1)$

17. $(4x^3 - 2x + 7) \div (2x + 1)$

18. $(6x^4 - 2x^2 + 5) \div (-3x + 2)$

19. $(4x^3 - 12x^2 - x + 3) \div \left(x - \frac{1}{2}\right)$

20. $(12x^3 + 1 + 7x + 16x^2) \div \left(x + \frac{1}{3}\right)$

21. $(-2x^5 + 3x^4 - 2x^2) \div (x^3 - 3x^2 + 1)$

22. $(-9x^6 + 7x^4 - 2x^3 + 5) \div (3x^4 - 2x + 1)$

23. $\dfrac{x^4 - 1}{x^2 - 1}$

24. $\dfrac{x^4 - 9}{x^2 + 3}$

25. $\dfrac{40 - 22x + 7x^3 + 6x^4}{6x^2 + x - 2}$

26. $\dfrac{-13x^2 + 4x^4 + 9}{4x^2 - 9}$

27. $\dfrac{-3x^4 + 7x^3 - 2x + 1}{x - 0.6}$

28. $\dfrac{2x^5 - 4x^3 + 3x^2 + 5}{x - 0.9}$

29. $(x^4 + 0.8x^3 - 0.26x^2 - 0.168x + 0.0441) \div (x^2 + 1.4x + 0.49)$

30. $(x^5 + 2.8x^4 + 1.34x^3 - 0.688x^2 - 0.2919x + 0.0882) \div (x^2 - 0.6x + 0.09)$

In Exercises 31–50, divide the polynomial by the linear factor with synthetic division. Indicate the quotient $Q(x)$ and the remainder $r(x)$.

31. $(3x^2 + 7x + 2) \div (x + 2)$

32. $(2x^2 + 7x - 15) \div (x + 5)$

33. $(7x^2 - 3x + 5) \div (x + 1)$

34. $(4x^2 + x + 1) \div (x - 2)$

35. $(3x^2 + 4x - x^4 - 2x^3 - 4) \div (x + 2)$

36. $(3x^2 - 4 + x^3) \div (x - 1)$

37. $(x^4 + 1) \div (x + 1)$

38. $(x^4 + 9) \div (x + 3)$

39. $(x^4 - 16) \div (x + 2)$

40. $(x^4 - 81) \div (x - 3)$

41. $(2x^3 - 5x^2 - x + 1) \div \left(x + \tfrac{1}{2}\right)$

42. $(3x^3 - 8x^2 + 1) \div \left(x + \tfrac{1}{3}\right)$

43. $(2x^4 - 3x^3 + 7x^2 - 4) \div \left(x - \tfrac{2}{3}\right)$

44. $(3x^4 + x^3 + 2x - 3) \div \left(x - \tfrac{3}{4}\right)$

45. $(2x^4 + 9x^3 - 9x^2 - 81x - 81) \div (x + 1.5)$

46. $(5x^3 - x^2 + 6x + 8) \div (x + 0.8)$

47. $\dfrac{x^7 - 8x^4 + 3x^2 + 1}{x - 1}$

48. $\dfrac{x^6 + 4x^5 - 2x^3 + 7}{x + 1}$

49. $(x^6 - 49x^4 - 25x^2 + 1225) \div \left(x - \sqrt{5}\right)$

50. $(x^6 - 4x^4 - 9x^2 + 36) \div \left(x - \sqrt{3}\right)$

In Exercises 51–60, divide the polynomials by either long division or synthetic division.

51. $(6x^2 - 23x + 7) \div (3x - 1)$

52. $(6x^2 + x - 2) \div (2x - 1)$

53. $(x^3 - x^2 - 9x + 9) \div (x - 1)$

54. $(x^3 + 2x^2 - 6x - 12) \div (x + 2)$

55. $(x^5 + 4x^3 + 2x^2 - 1) \div (x - 2)$

56. $(x^4 - x^2 + 3x - 10) \div (x + 5)$

57. $(x^4 - 25) \div (x^2 - 1)$

58. $(x^3 - 8) \div (x^2 - 2)$

59. $(x^7 - 1) \div (x - 1)$

60. $(x^6 - 27) \div (x - 3)$

■ **APPLICATIONS**

61. Geometry. The area of a rectangle is $6x^4 + 4x^3 - x^2 - 2x - 1$ square feet. If the length of the rectangle is $2x^2 - 1$ feet, what is the width of the rectangle?

62. Geometry. If the rectangle in Exercise 61 is the base of a rectangular box with volume $18x^5 + 18x^4 + x^3 - 7x^2 - 5x - 1$ cubic feet, what is the height of the box?

63. Travel. If a car travels a distance of $x^3 + 60x^2 + x + 60$ miles at an average speed of $x + 60$ miles per hour, how long does the trip take?

64. Sports. If a quarterback throws a ball $-x^2 - 5x + 50$ yards in $5 - x$ seconds, how fast is the football traveling?

▪ CATCH THE MISTAKE

In Exercises 65–68, explain the mistake that is made.

65. Divide $x^3 - 4x^2 + x + 6$ by $x^2 + x + 1$.

Solution:

$$
\begin{array}{r}
x - 3 \\
x^2 + x + 1{\overline{\smash{\big)}\,x^3 - 4x^2 + x + 6}} \\
\underline{x^3 + x^2 + x} \\
-3x^2 + 2x + 6 \\
\underline{-3x^2 - 3x - 3} \\
-x + 3
\end{array}
$$

This is incorrect. What mistake was made?

66. Divide $x^4 - 3x^2 + 5x + 2$ by $x - 2$.

Solution:

$$
\begin{array}{r|rrrr}
-2 & 1 & -3 & 5 & 2 \\
& & -2 & 10 & -30 \\
\hline
& 1 & -5 & 15 & \boxed{-28}
\end{array}
$$

$$\underbrace{\quad\quad\quad}_{x^2 - 5x + 15}$$

This is incorrect. What mistake was made?

67. Divide $x^3 + 4x - 12$ by $x - 3$.

Solution:

$$
\begin{array}{r|rrr}
3 & 1 & 4 & -12 \\
& & 3 & 21 \\
\hline
& 1 & 7 & \boxed{9}
\end{array}
$$

$$\underbrace{\quad\quad}_{x + 7}$$

This is incorrect. What mistake was made?

68. Divide $x^3 + 3x^2 - 2x + 1$ by $x^2 + 1$.

Solution:

$$
\begin{array}{r|rrrr}
-1 & 1 & 3 & -2 & 1 \\
& & -1 & -2 & 4 \\
\hline
& 1 & 2 & -4 & \boxed{5}
\end{array}
$$

$$\underbrace{\quad\quad\quad}_{x^2 - 2x - 4}$$

This is incorrect. What mistake was made?

▪ CONCEPTUAL

In Exercises 69–72, determine whether each statement is true or false.

69. A fifth-degree polynomial divided by a third-degree polynomial will yield a quadratic quotient.

70. A third-degree polynomial divided by a linear polynomial will yield a linear quotient.

71. Synthetic division can be used whenever the degree of the dividend is exactly one more than the degree of the divisor.

72. When the remainder is zero, the divisor is a factor of the dividend.

▪ CHALLENGE

73. Is $x + b$ a factor of $x^3 + (2b - a)x^2 + (b^2 - 2ab)x - ab^2$?

74. Is $x + b$ a factor of $x^4 + (b^2 - a^2)x^2 - a^2b^2$?

75. Divide $x^{3n} + x^{2n} - x^n - 1$ by $x^n - 1$.

76. Divide $x^{3n} + 5x^{2n} + 8x^n + 4$ by $x^n + 1$.

▪ TECHNOLOGY

77. Plot $\dfrac{2x^3 - x^2 + 10x - 5}{x^2 + 5}$. What type of function is it? Perform this division using long division, and confirm that the graph corresponds to the quotient.

78. Plot $\dfrac{x^3 - 3x^2 + 4x - 12}{x - 3}$. What type of function is it? Perform this division using synthetic division, and confirm that the graph corresponds to the quotient.

79. Plot $\dfrac{x^4 + 2x^3 - x - 2}{x + 2}$. What type of function is it? Perform this division using synthetic division, and confirm that the graph corresponds to the quotient.

80. Plot $\dfrac{x^5 - 9x^4 + 18x^3 + 2x^2 - 5x - 3}{x^4 - 6x^3 + 2x + 1}$. What type of function is it? Perform this division using long division, and confirm that the graph corresponds to the quotient.

81. Plot $\dfrac{-6x^3 + 7x^2 + 14x - 15}{2x + 3}$. What type of function is it? Perform this division using long division, and confirm that the graph corresponds to the quotient.

82. Plot $\dfrac{-3x^5 - 4x^4 + 29x^3 + 36x^2 - 18x}{3x^2 + 4x - 2}$. What type of function is it? Perform this division using long division, and confirm that the graph corresponds to the quotient.

SKILLS OBJECTIVES

- Apply the remainder theorem to evaluate a polynomial function.
- Apply the factor theorem.
- Use the rational zero (root) theorem to list possible rational zeros.
- Apply Descartes' rule of signs to determine the possible combination of positive and negative real zeros.
- Utilize the upper and lower bound theorems to narrow the search for real zeros.
- Find the real zeros of a polynomial function.
- Factor a polynomial function.
- Employ the intermediate value theorem to approximate a real zero.

CONCEPTUAL OBJECTIVES

- Understand that a polynomial of degree n has at most n real zeros.
- Understand that a real zero can be either rational or irrational and that irrational zeros will not be listed as possible zeros through the rational zero test.
- Realize that rational zeros can be found exactly, whereas irrational zeros must be approximated.

The Remainder Theorem and the Factor Theorem

The zeros of a polynomial function assist us in finding the x-intercepts of the graph of a polynomial function. How do we find the zeros of a polynomial function? For polynomial functions of degree 2, we have the quadratic formula, which allows us to find the two zeros. For polynomial functions whose degree is greater than 2, much more work is required.* In this section we focus our attention on finding the *real* zeros of a polynomial function. Later, in Section 4.5, we expand our discussion to *complex* zeros of polynomial functions.

In this section we start by listing possible rational zeros. As you will see there are sometimes many possibilities. We can then narrow the search using Descartes' rule of signs, which tells us possible combinations of positive and negative real zeros. We can narrow the search even further with the upper and lower bound rules. Once we have tested possible values and determined a zero, we will employ synthetic division to divide the polynomial by the linear factor associated with the zero. We will continue the process until we have factored the polynomial function into a product of either linear factors or irreducible quadratic factors. Last, we will discuss how to find irrational real zeros using the intermediate value theorem.

If we divide the polynomial function $f(x) = x^3 - 2x^2 + x - 3$ by $x - 2$ using synthetic division, we find the remainder is -1.

$$
\begin{array}{r|rrrr}
2 & 1 & -2 & 1 & -3 \\
  &   & 2 & 0 & 2 \\
\hline
  & 1 & 0 & 1 & -1
\end{array}
$$

Notice that if we evaluate the function at $x = 2$, the result is -1. $\qquad f(2) = -1$

This leads us to the *remainder theorem*.

---

*There are complicated formulas for finding the zeros of polynomial functions of degree 3 and 4, but there are no such formulas for degree 5 and higher polynomials (according to the Abel–Ruffini theorem).

WORDS	MATH
Recall the Division Algorithm.	$P(x) = d(x) \cdot Q(x) + r(x)$
Let $d(x) = x - a$ for any real number a. The degree of the remainder is always less than the degree of the divisor: therefore the remainder must be a constant (Call it r, $r(x) = r$).	$P(x) = (x - a) \cdot Q(x) + r(x)$
	$P(x) = (x - a) \cdot Q(x) + r$
Let $x = a$.	$P(a) = \underbrace{(a - a)}_{0} \cdot Q(x) + r$
Simplify.	$\boxed{P(a) = r}$

REMAINDER THEOREM

If a polynomial $P(x)$ is divided by $x - a$, then the remainder is $r = P(a)$.

The remainder theorem tells you that polynomial division can be used to evaluate a polynomial function at a particular point.

EXAMPLE 1 Two Methods for Evaluating Polynomials

Let $P(x) = 4x^5 - 3x^4 + 2x^3 - 7x^2 + 9x - 5$ and evaluate $P(2)$ by

a. Evaluating $P(2)$ directly
b. The remainder theorem and synthetic division

Solution:

a. $P(2) = 4(2)^5 - 3(2)^4 + 2(2)^3 - 7(2)^2 + 9(2) - 5$
$= 4(32) - 3(16) + 2(8) - 7(4) + 9(2) - 5$
$= 128 - 48 + 16 - 28 + 18 - 5$
$= \boxed{81}$

b.
$$
\begin{array}{r|rrrrrr}
2 & 4 & -3 & 2 & -7 & 9 & -5 \\
 & & 8 & 10 & 24 & 34 & 86 \\
\hline
 & 4 & 5 & 12 & 17 & 43 & \boxed{81}
\end{array}
$$

> **Classroom Example 4.4.1**
> Let $P(x) = -x^5 - 3x^3 + 2x - 1$. Evaluate $P(3)$ directly using the remainder theorem and synthetic division.
>
> **Answer:** -319

■ YOUR TURN Let $P(x) = -x^3 + 2x^2 - 5x + 2$ and evaluate $P(-2)$ using the remainder theorem and synthetic division.

Recall that when a polynomial is divided by $x - a$, if the remainder is zero, we say that $x - a$ is a factor of the polynomial. Through the remainder theorem, we now know that the remainder is related to evaluation of the polynomial at the point $x = a$. We are then led to the *factor theorem*.

FACTOR THEOREM

If $P(a) = 0$, then $x - a$ is a factor of $P(x)$. Conversely, if $x - a$ is a factor of $P(x)$, then $P(a) = 0$.

Technology Tip

A graphing utility can be used to evaluate $P(2)$. Enter $P(x) = 4x^5 - 3x^4 + 2x^3 - 7x^2 + 9x - 5$ as Y_1.

To evaluate $P(2)$, press VARS ▶

Y-VARS 1:Function 1:Y_1

(2) ENTER

■ Answer: $P(-2) = 28$

EXAMPLE 2 **Using the Factor Theorem to Factor a Polynomial**

Determine whether $x + 2$ is a factor of $P(x) = x^3 - 2x^2 - 5x + 6$. If so, factor $P(x)$ completely.

Solution:

By the factor theorem, $x + 2$ is a factor of $P(x) = x^3 - 2x^2 - 5x + 6$ if $P(-2) = 0$. By the remainder theorem, if we divide $P(x) = x^3 - 2x^2 - 5x + 6$ by $x + 2$, then the remainder is equal to $P(-2)$.

STEP 1 Divide $P(x) = x^3 - 2x^2 - 5x + 6$
by $x + 2$ using synthetic division.

$$
\begin{array}{r|rrrr}
-2 & 1 & -2 & -5 & 6 \\
   &   & -2 & 8 & -6 \\
\hline
   & 1 & -4 & 3 & \boxed{0} \\
\end{array}
$$
$$x^2 - 4x + 3$$

Since the remainder is zero, $P(-2) = 0$, $\boxed{x + 2 \text{ is a factor}}$ of
$P(x) = x^3 - 2x^2 - 5x + 6$.

STEP 2 Write $P(x)$ as a product.
$$P(x) = (x + 2)(x^2 - 4x + 3)$$

STEP 3 Factor the quadratic polynomial.
$$\boxed{P(x) = (x + 2)(x - 3)(x - 1)}$$

▪ **YOUR TURN** Determine whether $x - 1$ is a factor of $P(x) = x^3 - 4x^2 - 7x + 10$.
If so, factor $P(x)$ completely.

EXAMPLE 3 **Using the Factor Theorem to Factor a Polynomial**

Determine whether $x - 3$ and $x + 2$ are factors of $P(x) = x^4 - 13x^2 + 36$. If so, factor $P(x)$ completely.

Solution:

STEP 1 With synthetic division divide
$P(x) = x^4 - 13x^2 + 36$ by $x - 3$.

$$
\begin{array}{r|rrrrr}
3 & 1 & 0 & -13 & 0 & 36 \\
  &   & 3 & 9 & -12 & -36 \\
\hline
  & 1 & 3 & -4 & -12 & \boxed{0} \\
\end{array}
$$
$$x^3 + 3x^2 - 4x - 12$$

Because the remainder is 0, $\boxed{x - 3 \text{ is a factor}}$, and we can write the polynomial as
$$P(x) = (x - 3)(x^3 + 3x^2 - 4x - 12)$$

STEP 2 With synthetic division divide the
remaining cubic polynomial
$(x^3 + 3x^2 - 4x - 12)$ by $x + 2$.

$$
\begin{array}{r|rrrr}
-2 & 1 & 3 & -4 & -12 \\
   &   & -2 & -2 & 12 \\
\hline
   & 1 & 1 & -6 & \boxed{0} \\
\end{array}
$$
$$x^2 + x - 6$$

Because the remainder is 0, $\boxed{x + 2 \text{ is a factor}}$, and we can now write the polynomial as
$$P(x) = (x - 3)(x + 2)(x^2 + x - 6)$$

STEP 3 Factor the quadratic polynomial: $x^2 + x - 6 = (x + 3)(x - 2)$.

STEP 4 Write $P(x)$ as a product of linear factors: $\boxed{P(x) = (x - 3)(x - 2)(x + 2)(x + 3)}$

▪ **YOUR TURN** Determine whether $x - 3$ and $x + 2$ are factors of
$P(x) = x^4 - x^3 - 7x^2 + x + 6$. If so, factor $P(x)$ completely.

Technology Tip

The three zeros of the function give the three factors $x + 2$, $x - 1$, and $x - 3$. A table of values supports the zeros of the graph.

▪ **Answer:** $(x - 1)$ is a factor;
$P(x) = (x - 5)(x - 1)(x + 2)$

Classroom Example 4.4.2
Determine if $x + 3$ is a factor of these polynomials. If so, factor $P(x)$ completely.
a. $P(x) = x^3 + x^2 - 5x + 3$
b.* $P(x) = x^4 - 5x^3 + 7x^2 - 3x$
Answer:
a. yes, $P(x) = (x + 3)(x - 1)^2$ **b.** no

Technology Tip

The zeros of the function at $x = -3$, $x = -2, x = 2$, and $x = 3$ are shown in the graph. A table of values supports the zeros of the graph.

▪ **Answer:** $(x - 3)$ and $(x + 2)$
are factors;
$P(x) = (x - 3)(x + 2)(x - 1)(x + 1)$

The Search for Real Zeros

In all of the examples thus far, the polynomial function and one or more real zeros (or linear factors) were given. Now, we will not be given any real zeros to start with. Instead, we will develop methods to search for them.

Each real zero corresponds to a linear factor, and each linear factor is of degree 1. Therefore, the largest number of real zeros a polynomial function can have is equal to the degree of the polynomial.

Study Tip

The largest number of zeros a polynomial can have is equal to the degree of the polynomial.

THE NUMBER OF REAL ZEROS

A polynomial function cannot have more real zeros than its degree.

The following functions illustrate that a polynomial function of degree n can have at most n real zeros.

POLYNOMIAL FUNCTION	DEGREE	REAL ZEROS	COMMENTS
$f(x) = x^2 - 9$	2	$x = \pm 3$	**Two** real zeros
$f(x) = x^2 + 4$	2	None	**No** real zeros
$f(x) = x^3 - 1$	3	$x = 1$	**One** real zero
$f(x) = x^3 - x^2 - 6x$	3	$x = -2, 0, 3$	**Three** real zeros

Now that we know the *maximum* number of real zeros a polynomial function can have, let us discuss how to find these zeros.

The Rational Zero Theorem and Descartes' Rule of Signs

When the coefficients of a polynomial are integers, then the *rational zero theorem* (*rational root test*) gives us a list of possible rational zeros. We can then test these possible values to determine whether they really do correspond to actual zeros. *Descartes' rule of signs* tells us the possible combinations of *positive* real zeros and *negative* real zeros. Using Descartes' rule of signs will assist us in narrowing down the large list of possible zeros generated through the rational zero theorem to a (hopefully) shorter list of possible zeros. First, let's look at the rational zero theorem; then we'll turn to Descartes' rule of signs.

THE RATIONAL ZERO THEOREM (RATIONAL ROOT TEST)

If the polynomial function $P(x) = a_n x^n + a_{n-1} x^{n-1} + \cdots + a_2 x^2 + a_1 x + a_0$ has *integer* coefficients, then every rational zero of $P(x)$ has the form:

$$\text{Rational zero} = \frac{\text{integer factors of } a_0}{\text{integer factors of } a_n} = \frac{\text{integer factors of constant term}}{\text{integer factors of leading coefficient}}$$

$$= \pm \frac{\text{positive integer factors of constant term}}{\text{positive integer factors of leading coefficient}}$$

To use this theorem, simply list all combinations of integer factors of both the constant term a_0 and the leading coefficient term a_n and take all appropriate combinations of ratios. This procedure is illustrated in Example 4. Notice that when the leading coefficient is 1, then the possible rational zeros will simply be the possible integer factors of the constant term.

EXAMPLE 4   Using the Rational Zero Theorem

Determine possible rational zeros for the polynomial $P(x) = x^4 - x^3 - 5x^2 - x - 6$ by the rational zero theorem. Test each one to find all rational zeros.

Solution:

STEP 1 List factors of the constant and leading coefficient terms.

$$a_0 = -6 \qquad \pm1, \pm2, \pm3, \pm6$$
$$a_n = 1 \qquad \pm1$$

STEP 2 List possible rational zeros $\dfrac{a_0}{a_n}$.

$$\frac{\pm1}{\pm1}, \frac{\pm2}{\pm1}, \frac{\pm3}{\pm1}, \frac{\pm6}{\pm1} = \pm1, \pm2, \pm3, \pm6$$

There are three ways to test whether any of these are zeros: Substitute these values into the polynomial to see which ones yield zero; use either polynomial division or synthetic division to divide the polynomial by these possible zeros; and look for a zero remainder.

STEP 3 Test possible zeros by looking for zero remainders.

1 is not a zero: $P(1) = (1)^4 - (1)^3 - 5(1)^2 - (1) - 6 = -12$

-1 is not a zero: $P(-1) = (-1)^4 - (-1)^3 - 5(-1)^2 - (-1) - 6 = -8$

We could continue testing with direct substitution, but let us now use synthetic division as an alternative.

2 is not a zero:

$$\begin{array}{r|rrrrr} 2 & 1 & -1 & -5 & -1 & -6 \\ & & 2 & 2 & -6 & -14 \\ \hline & 1 & 1 & -3 & -7 & \boxed{-20} \end{array}$$

-2 is a zero:

$$\begin{array}{r|rrrrr} -2 & 1 & -1 & -5 & -1 & -6 \\ & & -2 & 6 & -2 & 6 \\ \hline & 1 & -3 & 1 & -3 & \boxed{0} \end{array}$$

Since -2 is a zero, then $x + 2$ is a factor of $P(x)$, and the remaining quotient is $x^3 - 3x^2 + x - 3$. Therefore, if there are any other real roots remaining, we can now use the simpler $x^3 - 3x^2 + x - 3$ for the dividend. Also note that the rational zero theorem can be applied to the new dividend and possibly shorten the list of possible rational zeros. In this case, the possible rational zeros of $F(x) = x^3 - 3x^2 + x - 3$ are ±1 and ±3.

3 is a zero:

$$\begin{array}{r|rrrr} 3 & 1 & -3 & 1 & -3 \\ & & 3 & 0 & 3 \\ \hline & 1 & 0 & 1 & \boxed{0} \end{array}$$

We now know that $\boxed{-2}$ and $\boxed{3}$ are confirmed zeros. If we continue testing, we will find that the other possible zeros fail. This is a fourth-degree polynomial, and we have found two rational real zeros. We see in the graph on the right that these two real zeros correspond to the x-intercepts.

Classroom Example 4.4.4
Determine the possible rational zeros for $P(x) = -2x^4 + 16x^3 - 34x^2 + 16x - 32$. Then find all rational zeros.

Answer: Possible rational zeros: $\pm1, \pm2, \pm4, \pm8, \pm16, \pm32, \pm\frac{1}{2}$
Rational zeros of P: only 4

Study Tip

The remainder can be found by evaluating the function or synthetic division. For simple values like $x = \pm1$, it is easier to evaluate the polynomial function. For other values, it is often easier to use synthetic division.

Study Tip

Notice in Step 3 that the polynomial $F(x) = x^3 - 3x^2 + x - 3$ can be factored by grouping:
$F(x) = (x - 3)(x^2 + 1)$.

■ **YOUR TURN** List the possible rational zeros of the polynomial $P(x) = x^4 + 2x^3 - 2x^2 + 2x - 3$, and determine rational real zeros.

■ **Answer:** Possible rational zeros: ±1 and ±3. Rational real zeros: 1 and -3.

Notice in Example 4 that the polynomial function $P(x) = x^4 - x^3 - 5x^2 - x - 6$ had two rational real zeros, $x = -2$ and $x = 3$. This implies that $x + 2$ and $x - 3$ are factors of $P(x)$. Also note in the last step when we divided by the zero $x = 3$, the quotient was $x^2 + 1$. Therefore, we can write the polynomial in factored form as

$$P(x) = \underbrace{(x + 2)}_{\substack{\text{linear} \\ \text{factor}}} \underbrace{(x - 3)}_{\substack{\text{linear} \\ \text{factor}}} \underbrace{(x^2 + 1)}_{\substack{\text{irreducible} \\ \text{quadratic} \\ \text{factor}}}$$

Notice that the first two factors are of degree 1, so we call them **linear factors**. The third expression, $x^2 + 1$, is of degree 2 and cannot be factored in terms of real numbers. We will discuss complex zeros in the next section. For now, we say that a quadratic expression, $ax^2 + bx + c$, is called **irreducible** if it cannot be factored over the real numbers.

■ **Answer:**
$P(x) = (x + 1)(x - 3)(x^2 + 2)$

EXAMPLE 5 Factoring a Polynomial Function

Write the following polynomial function as a product of linear and/or irreducible quadratic factors:

$$P(x) = x^4 - 4x^3 + 4x^2 - 36x - 45.$$

Solution:

Use the rational zero theorem to list possible rational roots. $\qquad x = \pm 1, \pm 3, \pm 5, \pm 9, \pm 15, \pm 45$

Test possible zeros by evaluating the function or by utilizing synthetic division.

$x = 1$ is not a zero.	$P(1) = -80$
$x = -1$ is a zero.	$P(-1) = 0$

Divide $P(x)$ by $x + 1$.

$$
\begin{array}{r|rrrr}
-1 & 1 & -4 & 4 & -36 & -45 \\
   &   & -1 & 5 & -9 & 45 \\
\hline
   & 1 & -5 & 9 & -45 & \boxed{0}
\end{array}
$$

$x = 5$ is a zero.

$$
\begin{array}{r|rrrr}
5 & 1 & -5 & 9 & -45 \\
  &   & 5 & 0 & 45 \\
\hline
  & 1 & 0 & 9 & \boxed{0} \\
  &   &   & x^2 + 9 &
\end{array}
$$

The factor $x^2 + 9$ is irreducible.

Write the polynomial as a product of linear and/or irreducible quadratic factors.

$$\boxed{P(x) = (x - 5)(x + 1)(x^2 + 9)}$$

Notice that the graph of this polynomial function has x-intercepts at $x = -1$ and $x = 5$.

■ **YOUR TURN** Write the following polynomial function as a product of linear and/or irreducible quadratic factors.

$$\boxed{P(x) = x^4 - 2x^3 - x^2 - 4x - 6}$$

The rational zero theorem lists possible zeros. It would be helpful if we could narrow that list. Descartes' rule of signs determines the possible combinations of positive real zeros and negative real zeros through variations of sign. A *variation in sign* is a sign difference seen between consecutive coefficients.

$$\text{Sign Change}$$
$$- \text{ to } +$$
$$P(x) = 2x^6 - 5x^5 - 3x^4 + 2x^3 - x^2 - x - 1$$
$$\text{Sign Change} \qquad \text{Sign Change}$$
$$+ \text{ to } - \qquad + \text{ to } -$$

This polynomial experiences three sign changes or variations in sign.

DESCARTES' RULE OF SIGNS

If the polynomial function $P(x) = a_n x^n + a_{n-1} x^{n-1} + \cdots + a_2 x^2 + a_1 x + a_0$ has real coefficients and $a_0 \neq 0$, then:

- The number of **positive** real zeros of the polynomial is either equal to the number of variations of sign of $P(x)$ or less than that number by an even integer.
- The number of **negative** real zeros of the polynomial is either equal to the number of variations of sign of $P(-x)$ or less than that number by an even integer.

Descartes' rule of signs narrows our search for real zeros because we don't have to test all of the possible rational zeros. For example, if we know there is one positive real zero, then if we find a positive rational zero we no longer need to continue to test possible positive zeros.

EXAMPLE 6 Using Descartes' Rule of Signs

Determine the possible combinations of zeros for $P(x) = x^3 - 2x^2 - 5x + 6$.

Solution:

Determine the number of variations of sign in $P(x)$.

$$\text{Sign Change}$$
$$P(x) = x^3 - 2x^2 - 5x + 6$$
$$\text{Sign Change}$$

$P(x)$ has 2 variations in sign.

Apply Descartes' rule of signs.

$P(x)$ has *either* 2 or 0 **positive** real zeros.

Determine the number of variations of sign in $P(-x)$.

$$(-x)^3 - 2(-x)^2 - 5(-x) + 6$$
$$-x^3 - 2x^2 + 5x + 6$$

$$\text{Sign Change}$$
$$P(x) = -x^3 - 2x^2 + 5x + 6$$

$P(-x)$ has **1** variation in sign.

Apply Descartes' rule of signs.

$P(x)$ must have 1 **negative** real zero.

Since $P(x) = x^3 - 2x^2 - 5x + 6$ is a *third*-degree polynomial, there are at most 3 real zeros. One zero is a negative real number, and there can be either 2 positive real zeros or 0 positive real zeros. Now look back at Example 2 and see that in fact there were 1 negative real zero and 2 positive real zeros.

Classroom Example 4.4.6
Determine the possible combinations of zeros for $P(x) = x^3 - 2x^2 + 3$.

Answer: 1 negative zero and 2 or 0 positive zeros

There are 1 negative real zero and 1 positive real zero.

The number of negative real zeros is 1 and the number of positive real zeros is 1. So, it has 2 complex conjugate zeros.

A table of values supports the zeros of the function.

■ **Answer:** Positive real zeros: 1
Negative real zeros: 3 or 1

EXAMPLE 7 **Using Descartes' Rule of Signs to Find Possible Combinations of Real Zeros**

Determine the possible combinations of real zeros for $P(x) = x^4 - 2x^3 + x^2 + 2x - 2$.

Solution:

$P(x)$ has 3 variations in sign.

$$P(x) = x^4 - 2x^3 + x^2 + 2x - 2$$

Apply Descartes' rule of signs. $P(x)$ has **either** 3 or 1 **positive** real zero.

Find $P(-x)$. $P(-x) = (-x)^4 - 2(-x)^3 + (-x)^2 + 2(-x) - 2$

$$= x^4 + 2x^3 + x^2 - 2x - 2$$

$P(-x)$ has 1 variation in sign. $P(-x) = x^4 + 2x^3 + x^2 - 2x - 2$

Apply Descartes' rule of signs. $P(x)$ has 1 **negative** real zero.

Since $P(x) = x^4 - 2x^3 + x^2 + 2x - 2$ is a *fourth*-degree polynomial, there are at most 4 real zeros.

$P(x)$ has 1 negative real zero and could have 3 or 1 positive real zeros.

The Technology Tip in the margin confirms 1 negative real zero and 1 positive real zero.

■ **YOUR TURN** Determine the possible combinations of zeros for:

$$P(x) = x^4 + 2x^3 + x^2 + 8x - 12$$

Factoring Polynomials

Now let's draw on the tests discussed in this section thus far to help us in finding all real zeros of a polynomial function. Doing so will enable us to factor polynomials.

EXAMPLE 8   Factoring a Polynomial

Write the polynomial $P(x) = x^5 + 2x^4 - x - 2$ as a product of linear and/or irreducible quadratic factors.

Solution:

STEP 1 Determine variations in sign.

$P(x)$ has 1 sign change. $P(x) = x^5 + 2x^4 - x - 2$

$P(-x)$ has 2 sign changes. $P(-x) = -x^5 + 2x^4 + x - 2$

STEP 2 Apply Descartes' rule of signs.   Positive Real Zeros: 1

Negative Real Zeros: 2 or 0

STEP 3 Use the rational zero theorem to
determine the possible rational zeros. $\pm 1, \pm 2$

We know (Step 2) that there is one positive real zero, so test the possible
positive rational zeros first.

STEP 4 Test possible rational zeros.

$$
\begin{array}{r|rrrrrr}
1 & 1 & 2 & 0 & 0 & -1 & -2 \\
  &   & 1 & 3 & 3 & 3 & 2 \\
\hline
  & 1 & 3 & 3 & 3 & 2 & \boxed{0}
\end{array}
$$

1 is a zero:

Now that we have found *the* positive
zero, we can test the other two
possible negative zeros—because
either they both are zeros or neither
is a zero.

$$
\begin{array}{r|rrrrr}
-1 & 1 & 3 & 3 & 3 & 2 \\
   &   & -1 & -2 & -1 & -2 \\
\hline
   & 1 & 2 & 1 & 2 & \boxed{0}
\end{array}
$$

-1 is a zero:

At this point, from Descartes' rule
of signs we know that -2 must also
be a zero, since there are either 2 or 0
negative zeros. Let's confirm this:

$$
\begin{array}{r|rrrr}
-2 & 1 & 2 & 1 & 2 \\
   &   & -2 & 0 & -2 \\
\hline
   & \underbrace{1 \quad 0 \quad 1}_{x^2 + 1} & & & \boxed{0}
\end{array}
$$

-2 is a zero:

STEP 5 Three of the five zeros have been found to be zeros: -1, -2, and 1.

STEP 6 Write the fifth-degree polynomial as a product of 3 linear factors and an irreducible
quadratic factor.

$$\boxed{P(x) = (x - 1)(x + 1)(x + 2)(x^2 + 1)}$$

■ YOUR TURN Write the polynomial $P(x) = x^5 - 2x^4 + x^3 - 2x^2 - 2x + 4$ as a
product of linear and/or irreducible quadratic factors.

■**Answer:**
$P(x) = (x - 2)(x + 1)(x - 1)(x^2 + 2)$

The rational zero theorem gives us possible rational zeros of a polynomial, and
Descartes' rule of signs gives us possible combinations of positive and negative real zeros.
Additional aids that help eliminate possible zeros are the *upper* and *lower bound rules*.
These rules can give you an upper and lower bound on the real zeros of a polynomial
function. If $f(x)$ has a common monomial factor, you should factor it out first, and then
follow the upper and lower bound rules.

Study Tip

If $f(x)$ has a common monomial
factor, it should be factored out first
before applying the bound rules.

UPPER AND LOWER BOUND RULES

Let $f(x)$ be a polynomial with real coefficients and a positive leading coefficient. Suppose $f(x)$ is divided by $x - c$ using synthetic division.

1. If $c > 0$ and each number in the bottom row is either positive or zero, c is an **upper bound** for the real zeros of f.
2. If $c < 0$ and the numbers in the bottom row are alternately positive and negative (zero entries count as either positive or negative), c is a **lower bound** for the real zeros of f.

Classroom Example 4.4.9*
Find the real zeros of $P(x) = 3x^4 - 12x^3 + 17x^2 - 20x + 20$.

Answer: 2 (multiplicity 2)

EXAMPLE 9 **Using Upper and Lower Bounds to Eliminate Possible Zeros**

Find the real zeros of $f(x) = 4x^3 - x^2 + 36x - 9$.

Solution:

STEP 1 The rational zero theorem gives possible rational zeros.

$$\frac{\text{Factors of } 9}{\text{Factors of } 4} = \frac{\pm 1, \pm 3, \pm 9}{\pm 1, \pm 2, \pm 4}$$

$$= \pm 1, \pm\frac{1}{2}, \pm\frac{1}{4}, \pm\frac{3}{4}, \pm\frac{3}{2}, \pm\frac{9}{4}, \pm 3, \pm\frac{9}{2}, \pm 9$$

STEP 2 Apply Descartes' rule of signs:

$f(x)$ has 3 sign variations.   3 or 1 positive real zeros

$f(-x)$ has no sign variations.   no negative real zeros

STEP 3 Try $x = 1$.

$$
\begin{array}{r|rrrr}
1 & 4 & -1 & 36 & -9 \\
  &   & 4 & 3 & 39 \\
\hline
  & 4 & 3 & 39 & 30
\end{array}
$$

$x = 1$ is not a zero, but because the last row contains all positive entries, $x = 1$ is an *upper* bound. Since we know there are no negative real zeros, we restrict our search to between 0 and 1.

STEP 4 Try $x = \frac{1}{4}$.

$$
\begin{array}{r|rrrr}
\frac{1}{4} & 4 & -1 & 36 & -9 \\
            &   & 1 & 0 & 9 \\
\hline
            & 4 & 0 & 36 & 0
\end{array}
$$

$\frac{1}{4}$ is a zero and the quotient $4x^2 + 36$ has all positive coefficients; therefore, $\boxed{\frac{1}{4} \text{ is the only real zero}}$.

Note: If $f(x)$ has a common monomial factor, it should be factored out first before applying the bound rules.

The Intermediate Value Theorem

In our search for zeros, we sometimes encounter irrational zeros, as in, for example, the polynomial

$$f(x) = x^5 - x^4 - 1$$

Descartes' rule of signs tells us there is exactly one real positive zero. However, the rational zero test yields only $x = \pm 1$, neither of which are zeros. So if we know there is a real positive zero and we know it's not rational, it must be irrational. Notice that $f(1) = -1$ and $f(2) = 15$. Since polynomial functions are continuous and the function goes from negative to positive between $x = 1$ and $x = 2$, we expect a zero somewhere in that interval. Generating a graph with a graphing utility, we find that there is a zero around $x = 1.3$.

The *intermediate value theorem* is based on the fact that polynomial functions are continuous.

INTERMEDIATE VALUE THEOREM

Let a and b be real numbers such that $a < b$ and $f(x)$ be a polynomial function. If $f(a)$ and $f(b)$ have opposite signs, then there is at least one real zero between a and b.

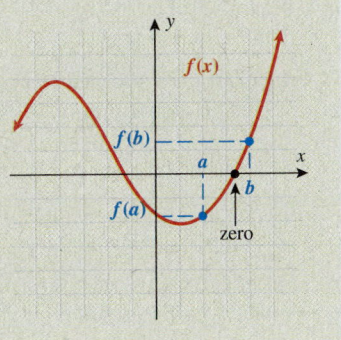

If the intermediate value theorem tells us that there is a real zero in the interval (a, b), how do we approximate that zero? The **bisection method** * is a root-finding algorithm that approximates the solution to the equation $f(x) = 0$. In the bisection method the interval is divided in half, and then the subinterval that contains the zero is selected. This is repeated until the bisection method converges to an approximate root of f.

*In calculus you will learn Newton's method, which is a more efficient approximation technique for finding zeros.

The graph of $f(x) = x^5 - x^4 - 1$ is shown. To find the zero of the function, press

2nd | TRACE | 2 | Zero | 1 | ENTER
2 | ENTER | ENTER

A table of values supports this zero of the function.

Classroom Example 4.4.10

Approximate the real zero(s).
a. $f(x) = 3x^4 + 2x^3 - 10$
b. $g(x) = 16x^4 - 40x - 25$

Answer:
a. -1.55 and 1.21
b. -0.580 and 1.52

EXAMPLE 10 **Approximating Real Zeros of a Polynomial Function**

Approximate the real zero of $f(x) = x^5 - x^4 - 1$.

Note: Descartes' rule of signs tells us that there are no real negative zeros and there is exactly one real positive zero.

Solution:

Find two consecutive integer values for x that have corresponding function values opposite in sign.

x	$f(x)$
1	-1
2	15

Note that a graphing utility would have shown an x-intercept between $x = 1$ and $x = 2$.

Apply the bisection method, with $a = 1$ and $b = 2$.

$$c = \frac{a+b}{2} = \frac{1+2}{2} = \frac{3}{2}$$

Evaluate the function at $x = c$.

$$f(1.5) \approx 1.53$$

Compare the values of f at the endpoints and midpoint.

$$f(1) = -1, f(1.5) \approx 1.53, f(2) = 15$$

Select the subinterval corresponding to the *opposite* signs of f.

$$(1, 1.5)$$

Apply the bisection method again (repeat the algorithm).

$$\frac{1 + 1.5}{2} = 1.25$$

Evaluate the function at $x = 1.25$.

$$f(1.25) \approx -0.38965$$

Compare the values of f at the endpoints and midpoint.

$$f(1) = -1, f(1.25) \approx -0.38965, f(1.5) \approx 1.53$$

Select the subinterval corresponding to the *opposite* signs of f.

$$(1.25, 1.5)$$

Apply the bisection method again (repeat the algorithm).

$$\frac{1.25 + 1.5}{2} = 1.375$$

Evaluate the function at $x = 1.375$.

$$f(1.375) \approx 0.3404$$

Compare the values of f at the endpoints and midpoint.

$$f(1.25) \approx -0.38965, f(1.375) \approx 0.3404, f(1.5) \approx 1.53$$

Select the subinterval corresponding to the *opposite* signs of f.

$$(1.25, 1.375)$$

We can continue this procedure (*applying the bisection method*) to find that the zero is somewhere between $f(1.32) \approx -0.285$ and $f(1.33) \approx 0.0326$.

We find that to three significant digits, $\boxed{1.32}$ is an approximation to the real zero.

Graphing Polynomial Functions

In Section 4.2, we graphed simple polynomial functions that were easily factored. Now that we have procedures for finding real zeros of polynomial functions (rational zero theorem, Descartes' rule of signs, and upper and lower bound rules for rational zeros, and the intermediate value theorem and the bisection method for irrational zeros), let us return to the topic of graphing polynomial functions. Since a real zero of a polynomial function corresponds to an x-intercept of its graph, we now have methods for finding (or estimating) any x-intercepts of the graph of any polynomial function.

EXAMPLE 11   Graphing a Polynomial Function

Graph the function $f(x) = 2x^4 - 2x^3 + 5x^2 + 17x - 22$.

Solution:

STEP 1 Find the y-intercept. $f(0) = -22$

STEP 2 Find any x-intercepts (real zeros).

Apply Descartes' rule of signs.

3 sign changes correspond to
3 or 1 positive real zeros. $f(x) = 2x^4 - 2x^3 + 5x^2 + 17x - 22$

1 sign change corresponds to
1 negative real zero. $f(-x) = 2x^4 + 2x^3 + 5x^2 - 17x - 22$

Apply the rational zero theorem.

Let $a_0 = -22$ and $a_n = 2$. $\dfrac{\text{Factors of } a_0}{\text{Factors of } a_n} = \pm\dfrac{1}{2}, \pm1, \pm2, \pm\dfrac{11}{2}, \pm11, \pm22$

Test the possible zeros.

$x = 1$ is a zero. $f(1) = 0$

There are no other rational zeros.

Apply the upper bound rule.

$$\begin{array}{r|rrrrr} 1 & 2 & -2 & 5 & 17 & -22 \\ & & 2 & 0 & 5 & 22 \\ \hline & 2 & 0 & 5 & 22 & \boxed{0} \end{array}$$

Since $x = 1$ is positive and all of the numbers in the bottom row are positive (or zero), $x = 1$ is an upper bound for the real zeros. We know there is exactly one negative real zero, but none of the possible zeros from the rational zero theorem is a zero. Therefore, the negative real zero is irrational.

Apply the intermediate value theorem and the bisection method.

f is positive at $x = -2$. $f(-2) = 12$

f is negative at $x = -1$. $f(-1) = -30$

Use the bisection method to find the negative
real zero between -2 and -1. $x \approx -1.85$

STEP 3 Determine the end behavior. $y = 2x^4$

Technology Tip

The graph of
$f(x) = 2x^4 - 2x^3 + 5x^2 + 17x - 22$
is shown.

To find the zero of the function, press:

2nd TRACE 2 Zero - 3
ENTER - 1 ENTER ENTER

A table of values supports this zero of
the function and the graph.

Classroom Example 4.4.11
Graph:
a. $f(x) = 3x^5 - 9x^4 - 3x + 9$
b. $g(x) =$
$3x^4 - 12x^3 + 17x^2 - 20x + 20$

Answer:

STEP 4 Find additional points.

x	-2	-1.85	-1	0	1	2
$f(x)$	12	0	-30	-22	0	48
Point	$(-2, 12)$	$(-1.85, 0)$	$(-1, -30)$	$(0, -22)$	$(1, 0)$	$(2, 48)$

STEP 5 Sketch the graph.

In this section we discussed how to find the real zeros of a polynomial function. Once real zeros are known, it is possible to write the polynomial function as a product of linear and/or irreducible quadratic factors.

The Number of Zeros

- A polynomial of degree n has *at most n real zeros.*
- *Descartes' rule of signs* determines the possible combinations of positive and negative real zeros.
- *Upper and lower bounds* help narrow the search for zeros.

How to Find Zeros

- *Rational zero theorem*: List possible rational zeros:

$$\frac{\text{Factors of constant, } a_0}{\text{Factors of leading coefficient, } a_n}$$

- *Irrational zeros:* Approximate zeros by determining when the polynomial function changes sign (intermediate value theorem).

Procedure for Factoring a Polynomial Function

- List possible rational zeros (rational zero theorem).
- List possible combinations of positive and negative real zeros (Descartes' rule of signs).
- Test possible values until a zero is found.*
- Once a real zero is found, repeat testing on the quotient until linear and/or irreducible quadratic factors remain.
- If there is a real zero but all possible rational roots have failed, then approximate the zero using the *intermediate value theorem* and the *bisection method.*

....

*Depending on the form of the quotient, upper and lower bounds may eliminate possible zeros.

SKILLS

In Exercises 1–6, find the following values by using synthetic division. Check by substituting the value into the function.

$$f(x) = 3x^4 - 2x^3 + 7x^2 - 8 \qquad g(x) = 2x^3 + x^2 + 1$$

1. $f(1)$ **2.** $f(-1)$ **3.** $g(1)$ **4.** $g(-1)$ **5.** $f(-2)$ **6.** $g(2)$

In Exercises 7–10, determine whether the number given is a zero of the polynomial.

7. $-7, P(x) = x^3 + 2x^2 - 29x + 42$ **8.** $2, P(x) = x^3 + 2x^2 - 29x + 42$

9. $-3, P(x) = x^3 - x^2 - 8x + 12$ **10.** $1, P(x) = x^3 - x^2 - 8x + 12$

In Exercises 11–20, given a real zero of the polynomial, determine all other real zeros, and write the polynomial in terms of a product of linear and/or irreducible quadratic factors.

	Polynomial	Zero		Polynomial	Zero
11.	$P(x) = x^3 - 13x + 12$	1	**12.**	$P(x) = x^3 + 3x^2 - 10x - 24$	3
13.	$P(x) = 2x^3 + x^2 - 13x + 6$	$\frac{1}{2}$	**14.**	$P(x) = 3x^3 - 14x^2 + 7x + 4$	$-\frac{1}{3}$
15.	$P(x) = x^4 - 2x^3 - 11x^2 - 8x - 60$	$-3, 5$	**16.**	$P(x) = x^4 - x^3 + 7x^2 - 9x - 18$	$-1, 2$
17.	$P(x) = x^4 - 5x^2 + 10x - 6$	$1, -3$	**18.**	$P(x) = x^4 - 4x^3 + x^2 + 6x - 40$	$4, -2$
19.	$P(x) = x^4 + 6x^3 + 13x^2 + 12x + 4$	-2 (multiplicity 2)	**20.**	$P(x) = x^4 + 4x^3 - 2x^2 - 12x + 9$	1 (multiplicity 2)

In Exercises 21–28, use the rational zero theorem to list the *possible* rational zeros.

21. $P(x) = x^4 + 3x^2 - 8x + 4$

22. $P(x) = -x^4 + 2x^3 - 5x + 4$

23. $P(x) = x^5 - 14x^3 + x^2 - 15x + 12$

24. $P(x) = x^5 - x^3 - x^2 + 4x + 9$

25. $P(x) = 2x^6 - 7x^4 + x^3 - 2x + 8$

26. $P(x) = 3x^5 + 2x^4 - 5x^3 + x - 10$

27. $P(x) = 5x^5 + 3x^4 + x^3 - x - 20$

28. $P(x) = 4x^6 - 7x^4 + 4x^3 + x - 21$

In Exercises 29–32, list the possible rational zeros, and test to determine all rational zeros.

29. $P(x) = x^4 + 2x^3 - 9x^2 - 2x + 8$

30. $P(x) = x^4 + 2x^3 - 4x^2 - 2x + 3$

31. $P(x) = 2x^3 - 9x^2 + 10x - 3$

32. $P(x) = 3x^3 - 5x^2 - 26x - 8$

In Exercises 33–44, use Descartes' rule of signs to determine the possible number of positive real zeros and negative real zeros.

33. $P(x) = x^4 - 32$

34. $P(x) = x^4 + 32$

35. $P(x) = x^5 - 1$

36. $P(x) = x^5 + 1$

37. $P(x) = x^5 - 3x^3 - x + 2$

38. $P(x) = x^4 + 2x^2 - 9$

39. $P(x) = 9x^7 + 2x^5 - x^3 - x$

40. $P(x) = 16x^7 - 3x^4 + 2x - 1$

41. $P(x) = x^6 - 16x^4 + 2x^2 + 7$

42. $P(x) = -7x^6 - 5x^4 - x^2 + 2x + 1$

43. $P(x) = -3x^4 + 2x^3 - 4x^2 + x - 11$

44. $P(x) = 2x^4 - 3x^3 + 7x^2 + 3x + 2$

For each polynomial in Exercises 45–62: (a) use Descartes' rule of signs to determine the possible combinations of positive real zeros and negative real zeros; (b) use the rational zero test to determine possible rational zeros; (c) test for rational zeros; and (d) factor as a product of linear and/or irreducible quadratic factors.

45. $P(x) = x^3 + 6x^2 + 11x + 6$

46. $P(x) = x^3 - 6x^2 + 11x - 6$

47. $P(x) = x^3 - 7x^2 - x + 7$

48. $P(x) = x^3 - 5x^2 - 4x + 20$

49. $P(x) = x^4 + 6x^3 + 3x^2 - 10x$

50. $P(x) = x^4 - x^3 - 14x^2 + 24x$

51. $P(x) = x^4 - 7x^3 + 27x^2 - 47x + 26$

52. $P(x) = x^4 - 5x^3 + 5x^2 + 25x - 26$

53. $P(x) = 10x^3 - 7x^2 - 4x + 1$

54. $P(x) = 12x^3 - 13x^2 + 2x - 1$

55. $P(x) = 6x^3 + 17x^2 + x - 10$

56. $P(x) = 6x^3 + x^2 - 5x - 2$

57. $P(x) = x^4 - 2x^3 + 5x^2 - 8x + 4$

58. $P(x) = x^4 + 2x^3 + 10x^2 + 18x + 9$

59. $P(x) = x^6 + 12x^4 + 23x^2 - 36$

60. $P(x) = x^4 - x^2 - 16x^2 + 16$

61. $P(x) = 4x^4 - 20x^3 + 37x^2 - 24x + 5$

62. $P(x) = 4x^4 - 8x^3 + 7x^2 + 30x + 50$

In Exercises 63–66, use the information found in Exercises 47, 51, 55, and 61 to assist in sketching a graph of each polynomial function.

63. Exercise 47.

64. Exercise 51.

65. Exercise 55.

66. Exercise 61.

In Exercises 67–72, use the intermediate value theorem and the bisection method to approximate the real zero in the indicated interval. Approximate to two decimal places.

67. $f(x) = x^4 - 3x^3 + 4$ $[1, 2]$

68. $f(x) = x^5 - 3x^3 + 1$ $[0, 1]$

69. $f(x) = 7x^5 - 2x^2 + 5x - 1$ $[0, 1]$

70. $f(x) = -2x^3 + 3x^2 + 6x - 7$ $[-2, -1]$

71. $f(x) = x^3 - 2x^2 - 8x - 3$ $[-1, 0]$

72. $f(x) = x^4 + 4x^2 - 7x - 13$ $[-2, -1]$

■ APPLICATIONS

73. Profit. A mathematics honorary society wants to sell magazine subscriptions to *Math Weekly*. If there are x hundred subscribers, its monthly revenue and cost are given by:

$$R(x) = 46 - 3x^2 \quad \text{and} \quad C(x) = 20 + 2x$$

a. Determine the profit function. *Hint: P = R − C.*

b. Determine the number of subscribers needed in order to *break even*.

74. Profit. Using the profit equation $P(x) = x^3 - 5x^2 + 3x + 6$, when will the company break even if x represents the units sold?

For Exercises 75 and 76, refer to the following:

The demand function for a product is

$$p(x) = 28 - 0.0002x$$

where p is the unit price (in dollars) of the product and x is the number of units produced and sold. The cost function for the product is

$$C(x) = 20x + 1500$$

where C is the total cost (in dollars) and x is the number of units produced. The total profit obtained by producing and selling x units is

$$P(x) = xp(x) - C(x)$$

75. Business. Find the total profit function when x units are produced and sold. Use Descartes' rule of signs to determine possible combinations of positive zeros for the profit function.

76. Business. Find the break-even point(s) for the product to the nearest unit. Discuss the significance of the break-even point(s) for the product.

77. Health/Medicine. During the course of treatment of an illness the concentration of a dose of a drug (in mcg/mL) in the bloodstream fluctuates according to the model

$$C(t) = 15.4 - 0.05t^2$$

where $t = 0$ is when the drug was administered. Assuming a single dose of the drug is administered, in how many hours (to the nearest hour) after being administered will the drug be eliminated from the bloodstream?

78. Health/Medicine. During the course of treatment of an illness, the concentration of a dose of a drug (in mcg/mL) in the bloodstream fluctuates according to the model

$$C(t) = 60 - 0.75t^2$$

where $t = 0$ is when the drug was administered. Assuming a single dose of the drug is administered, in how many hours (to the nearest hour) after being administered will the drug be eliminated from the bloodstream?

■ CATCH THE MISTAKE

In Exercises 79 and 80, explain the mistake that is made.

79. Use Descartes' rule of signs to determine the possible combinations of zeros of

$$P(x) = 2x^5 + 7x^4 + 9x^3 + 9x^2 + 7x + 2$$

Solution:

No sign changes, so no positive real zeros.

$$P(x) = 2x^5 + 7x^4 + 9x^3 + 9x^2 + 7x + 2$$

Five sign changes, so five negative real zeros.

$$P(-x) = -2x^5 + 7x^4 - 9x^3 + 9x^2 - 7x + 2$$

This is incorrect. What mistake was made?

80. Determine whether $x - 2$ is a factor of
$$P(x) = x^3 - 2x^2 - 5x + 6.$$

Solution:

$$
\begin{array}{r|rrrr}
-2 & 1 & -2 & -5 & 6 \\
   &   & -2 & 8 & -6 \\
\hline
   & 1 & -4 & 3 & 0 \\
\end{array}
$$

Yes, $x - 2$ is a factor of $P(x)$.

This is incorrect. What mistake was made?

▪CONCEPTUAL

In Exercises 81–84 determine whether each statement is true or false.

81. All real zeros of a polynomial correspond to x-intercepts.

82. A polynomial of degree n, $n > 0$, must have at least one zero.

83. A polynomial of degree n, $n > 0$, can be written as a product of n linear factors over real numbers.

84. The number of sign changes in a polynomial is equal to the number of positive real zeros of that polynomial.

▪CHALLENGE

85. Given that $x = a$ is a zero of $P(x) = x^3 - (a + b + c)x^2 + (ab + ac + bc)x - abc$, find the other two zeros, given that a, b, c are real numbers and $a > b > c$.

86. Given that $x = a$ is a zero of $p(x) = x^3 + (-a + b - c)x^2 - (ab + bc - ac)x + abc$, find the other two real zeros, given that a, b, c are real positive numbers.

▪TECHNOLOGY

In Exercises 87 and 88, determine all possible rational zeros of the polynomial. There are many possibilities. Instead of trying them all, use a graphing calculator or software to graph $P(x)$ to help find a zero to test.

87. $P(x) = x^3 - 2x^2 + 16x - 32$

88. $P(x) = x^3 - 3x^2 + 16x - 48$

In Exercises 89 and 90: (a) determine all possible rational zeros of the polynomial, use a graphing calculator or software to graph $P(x)$ to help find the zeros; and (b) factor as a product of linear and/or irreducible quadratic factors.

89. $P(x) = 12x^4 + 25x^3 + 56x^2 - 7x - 30$

90. $P(x) = -3x^3 - x^2 - 7x - 49$

SECTION 4.5
COMPLEX ZEROS: THE FUNDAMENTAL THEOREM OF ALGEBRA

SKILLS OBJECTIVES

▪ Find the complex zeros of a polynomial function.
▪ Use the complex conjugate zeros theorem.
▪ Factor polynomial functions.

CONCEPTUAL OBJECTIVES

▪ Extend the domain of polynomial functions to complex numbers.
▪ Understand how the fundamental theorem of algebra guarantees at least one zero.
▪ Understand why complex zeros occur in conjugate pairs.

Complex Zeros

In Section 4.4, we found the real zeros of a polynomial function. In this section we find the complex zeros of a polynomial function. In this chapter we assume the coefficients of polynomial functions are real numbers. The domain of polynomial functions thus far has been the set of all real numbers. Now, we consider a more general case. In this section, the coefficients of a polynomial function and the domain of a polynomial function are *complex numbers*. Note that the set of real numbers is a subset of the complex numbers. (Choose the imaginary part to be zero.)

It is important to note, however, that when we are discussing *graphs* of polynomial functions, we restrict the domain to the set of real numbers.

Study Tip

The zeros of a polynomial can be complex numbers. Only when the zeros are real numbers do we interpret zeros as x-intercepts.

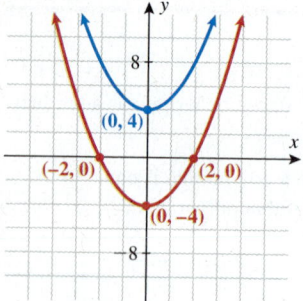

A *zero* of a polynomial $P(x)$ is the *solution* or *root* of the equation $P(x) = 0$. The *zeros of a polynomial can be complex numbers.* However, since the xy-plane represents real numbers, we interpret zeros as x-intercepts only when the zeros are real numbers.

We can illustrate the relationship between real and complex zeros of polynomial functions and their graphs with two similar examples. Let's take the two quadratic functions $f(x) = x^2 - 4$ and $g(x) = x^2 + 4$. The graphs of these two functions are parabolas that open upward with $f(x)$ shifted down four units and $g(x)$ shifted up four units as shown on the left. Setting each function equal to zero and solving for x, we find that the zeros for $f(x)$ are -2 and 2 and the zeros for $g(x)$ are $-2i$ and $2i$. Notice that the x-intercepts for $f(x)$ are $(-2, 0)$ and $(2, 0)$ and $g(x)$ has **no x-intercepts**.

The Fundamental Theorem of Algebra

In Section 4.4, we were able to write a polynomial function as a product of linear and/or irreducible quadratic factors. Now, we consider factors over complex numbers. Therefore, what were irreducible quadratic factors over real numbers will now be a product of two linear factors over the complex numbers.

What are the minimum and maximum number of zeros a polynomial can have? Every polynomial has *at least one zero* (provided the degree is greater than zero). The largest number of zeros a polynomial can have is equal to the degree of the polynomial.

THE FUNDAMENTAL THEOREM OF ALGEBRA

Every polynomial $P(x)$ of degree $n > 0$ has *at least one zero* in the complex number system.

The fundamental theorem of algebra and the factor theorem are used to prove the following n zeros theorem.

n ZEROS THEOREM

Every polynomial $P(x)$ of degree $n > 0$ can be expressed as the product of n linear factors. Hence, $P(x)$ has exactly n zeros, not necessarily distinct.

These two theorems are illustrated with five polynomials below.

a. The **first**-degree polynomial $f(x) = x + 3$ has exactly **one** zero: $x = -3$.

b. The **second**-degree polynomial $f(x) = x^2 + 10x + 25 = (x + 5)(x + 5)$ has exactly **two** zeros: $x = -5$ and $x = -5$. It is customary to write this as a single zero of multiplicity 2 or refer to it as a repeated root.

c. The **third**-degree polynomial $f(x) = x^3 + 16x = x(x^2 + 16) = x(x + 4i)(x - 4i)$ has exactly **three** zeros: $x = 0$, $x = -4i$, and $x = 4i$.

d. The **fourth**-degree polynomial $f(x) = x^4 - 1 = (x^2 - 1)(x^2 + 1)$
$= (x - 1)(x + 1)(x - i)(x + i)$ has exactly **four** zeros:
$x = 1$, $x = -1$, $x = i$, and $x = -i$.

e. The **fifth**-degree polynomial $f(x) = x^5 = x \cdot x \cdot x \cdot x \cdot x$ has exactly **five** zeros:
$x = 0$, which has multiplicity 5.

The fundamental theorem of algebra and the n zeros theorem only tell you that the zeros *exist*—not how to find them. We must rely on techniques discussed in Section 4.4 and additional strategies discussed in this section to determine the zeros.

Complex Conjugate Pairs

Often, at a grocery store or a drugstore, we see signs for special offers—"buy one, get one free." A similar phenomenon occurs for complex zeros of a polynomial function with real coefficients. If we restrict the coefficients of a polynomial to real numbers, complex zeros always come in conjugate pairs. In other words, if a zero of a polynomial function is a complex number, then another zero will always be its complex conjugate. Look at the third-degree polynomial in the above illustration, part (c), where two of the zeros were $-4i$ and $4i$, and in part (d), where two of the zeros were i and $-i$. In general, if we restrict the coefficients of a polynomial to real numbers, complex zeros always come in conjugate pairs.

Study Tip

If we restrict the coefficients of a polynomial to real numbers, complex zeros always come in conjugate pairs.

COMPLEX CONJUGATE ZEROS THEOREM

If a polynomial $P(x)$ has real coefficients, and if $a + bi$ is a zero of $P(x)$, then its complex conjugate $a - bi$ is also a zero of $P(x)$.

EXAMPLE 1 Zeros That Appear as Complex Conjugates

Find the zeros of the polynomial $P(x) = x^2 - 4x + 13$.

Solution:

Set the polynomial equal to zero.

$$P(x) = x^2 - 4x + 13 = 0$$

Use the quadratic formula to solve for x.

$$x = \frac{-(-4) \pm \sqrt{(-4)^2 - 4(1)(13)}}{2(1)}$$

Simplify.

$$\boxed{x = 2 \pm 3i}$$

The zeros are the complex conjugates $2 - 3i$ and $2 + 3i$.

Check: This is a *second*-degree polynomial, so we expect *two* zeros.

Classroom Example 4.5.1
Find the zeros of these polynomials.
a. $P(x) = 3x^2 + 6$
b. $P(x) = 3x^2 + 2x + 1$

Answer:
a. $\pm i\sqrt{2}$

b. $\dfrac{-1 \pm i\sqrt{2}}{3}$

EXAMPLE 2 Finding a Polynomial Given Its Zeros

Find a polynomial of minimum degree that has the zeros: $-2, 1 - i, 1 + i$.

Solution:

Write the factors corresponding to each zero:

-2: $(x + 2)$
$1 + i$: $[x - (1 + i)]$
$1 - i$: $[x - (1 - i)]$

Express the polynomial as the product of the three factors.

$$P(x) = (x + 2)[x - (1 + i)][x - (1 - i)]$$

Regroup inner parentheses.

$$P(x) = (x + 2)[(x - 1) - i][(x - 1) + i]$$

Use the difference of two squares formula $(a - b)(a + b) = a^2 - b^2$ for the product of the latter two factors.

$$P(x) = (x + 2)\underbrace{[(x - 1) - i][(x - 1) + i]}_{(x-1)^2 - i^2}$$

Simplify.

$$P(x) = (x + 2)[\underbrace{(x - 1)^2}_{x^2 - 2x + 1} - \underbrace{i^2}_{+1}]$$

$$P(x) = (x + 2)(x^2 - 2x + 2)$$

$$P(x) = x^3 - 2x^2 + 2x + 2x^2 - 4x + 4$$

$$P(x) = x^3 - 2x + 4$$

■ **YOUR TURN** Find a polynomial of minimum degree that has the zeros: $1, 2 - i, 2 + i$.

■ **Answer:**
$P(x) = x^3 - 5x^2 + 9x - 5$

The graph of
$P(x) = x^4 - x^3 - 5x^2 - x - 6$
is shown.

The real zeros of the function at
$x = -2$ and $x = 3$ give the factors of
$x + 2$ and $x - 3$. Use synthetic
division to find the other factors.

A table of values supports the real
zeros of the function and its factors.

Classroom Example 4.5.3

Factor the polynomial
$P(x) = 2x^4 + 9x^3 + 3x^2 + 36x - 20$
given that $-2i$ is a zero.

Answer: $P(x) =$
$(2x - 1)(x + 5)(x + 2i)(x - 2i)$

■ **Answer:** $P(x) =$
$(x - 2i)(x + 2i)(x - 1)(x - 2)$
Note: The zeros of $P(x)$ are 1, 2, $2i$,
and $-2i$.

EXAMPLE 3   Factoring a Polynomial with Complex Zeros

Factor the polynomial $P(x) = x^4 - x^3 - 5x^2 - x - 6$ given that i is a zero of $P(x)$.

Since $P(x)$ is a *fourth*-degree polynomial we expect *four* zeros. The goal in this problem is to write $P(x)$ as a product of four linear factors: $P(x) = (x - a)(x - b)(x - c)(x - d)$, where a, b, c, and d are complex numbers and represent the zeros of the polynomial.

Solution:

Write known zeros and linear factors.

Since i is a zero, we know that $-i$ is a zero. $x = i$ and $x = -i$

We now know two linear factors of $P(x)$. $(x - i)$ and $(x + i)$

Write $P(x)$ as a product of four factors. $P(x) = (x - i)(x + i)(x - c)(x - d)$

Multiply the two known factors. $(x + i)(x - i) = x^2 - i^2$
$$= x^2 - (-1)$$
$$= x^2 + 1$$

Rewrite the polynomial. $P(x) = (x^2 + 1)(x - c)(x - d)$

Divide both sides of the equation by $x^2 + 1$. $\dfrac{P(x)}{x^2 + 1} = (x - c)(x - d)$

Divide $P(x)$ by $x^2 + 1$ using long division.

$$
\begin{array}{r}
x^2 - x \quad - 6 \\
x^2 + 0x + 1 \overline{)x^4 - x^3 \quad - 5x^2 - x \quad - 6} \\
-(x^4 + 0x^3 + x^2) \\
\hline
-x^3 \quad - 6x^2 - x \\
-(-x^3 + 0x^2 - x) \\
\hline
-6x^2 + 0x - 6 \\
-(-6x^2 + 0x - 6) \\
\hline
0
\end{array}
$$

Since the remainder is 0, $x^2 - x - 6$
is a factor. $P(x) = (x^2 + 1)(x^2 - x - 6)$

Factor the quotient $x^2 - x - 6$. $x^2 - x - 6 = (x - 3)(x + 2)$

Write $P(x)$ as a product of four linear factors. $\boxed{P(x) = (x - i)(x + i)(x - 3)(x + 2)}$

Check: $P(x)$ is a *fourth*-degree polynomial and we found *four* zeros, two of which are complex conjugates.

■ **YOUR TURN** Factor the polynomial $P(x) = x^4 - 3x^3 + 6x^2 - 12x + 8$ given that $x - 2i$ is a factor.

EXAMPLE 4 **Factoring a Polynomial with Complex Zeros**

Factor the polynomial $P(x) = x^4 - 2x^3 + x^2 + 2x - 2$ given that $1 + i$ is a zero of $P(x)$.

Since $P(x)$ is a *fourth*-degree polynomial, we expect *four* zeros. The goal in this problem is to write $P(x)$ as a product of four linear factors: $P(x) = (x - a)(x - b)(x - c)(x - d)$, where a, b, c, and d are complex numbers and represent the zeros of the polynomial.

Solution:

STEP 1 Write known zeros and linear factors.

Since $1 + i$ is a zero, we know that $1 - i$ is a zero.
$$x = 1 + i \text{ and } x = 1 - i$$

We now know two linear factors of $P(x)$.
$$[x - (1 + i)] \text{ and } [x - (1 - i)]$$

STEP 2 Write $P(x)$ as a product of four factors.
$$P(x) = [x - (1 + i)][x - (1 - i)](x - c)(x - d)$$

STEP 3 Multiply the first two terms.
$$[x - (1 + i)][x - (1 - i)]$$

First group the real parts together in each bracket.
$$[(x - 1) - i][(x - 1) + i]$$

Use the special product $(a - b)(a + b) = a^2 - b^2$, where a is $(x - 1)$ and b is i.
$$(x - 1)^2 - i^2$$
$$(x^2 - 2x + 1) - (-1)$$
$$x^2 - 2x + 2$$

STEP 4 Rewrite the polynomial.
$$P(x) = (x^2 - 2x + 2)(x - c)(x - d)$$

STEP 5 Divide both sides of the equation by $x^2 - 2x + 2$, and substitute in the original polynomial $P(x) = x^4 - 2x^3 + x^2 + 2x - 2$.
$$\frac{x^4 - 2x^3 + x^2 + 2x - 2}{x^2 - 2x + 2} = (x - c)(x - d)$$

STEP 6 Divide the left side of the equation using long division.
$$\frac{x^4 - 2x^3 + x^2 + 2x - 2}{x^2 - 2x + 2} = x^2 - 1$$

STEP 7 Factor $x^2 - 1$.
$$(x - 1)(x + 1)$$

STEP 8 Write $P(x)$ as a product of four linear factors.

$$P(x) = [x - (1 + i)][x - (1 - i)][x - 1][x + 1]$$

or $$P(x) = (x - 1 - i)(x - 1 + i)(x - 1)(x + 1)$$

■ **YOUR TURN** Factor the polynomial $P(x) = x^4 - 2x^2 + 16x - 15$ given that $1 + 2i$ is a zero.

Classroom Example 4.5.4
Factor the polynomial $P(x) = 3x^4 - 23x^3 + 67x^2 - 53x - 26$ given that $3 - 2i$ is a zero.

Answer: $P(x) = [x - (3 + 2i)] \cdot [x - (3 - 2i)](3x + 1)(x - 2)$

■ **Answer:** $P(x) = [x - (1 + 2i)] \cdot [x - (1 - 2i)](x - 1)(x + 3)$
Note: The zeros of $P(x)$ are 1, -3, $1 + 2i$, and $1 - 2i$.

Because an n-degree polynomial function has exactly n zeros and since complex zeros always come in conjugate pairs, if the degree of the polynomial is **odd**, there is guaranteed to be **at least one zero that is a real number**. If the degree of the polynomial is even, there is no guarantee that a zero will be real—all the zeros could be complex.

Study Tip

Odd-degree polynomials have at least one real zero.

EXAMPLE 5   Finding Possible Combinations of Real and Complex Zeros

List the possible combinations of real and complex zeros for the given polynomials.

a. $17x^5 + 2x^4 - 3x^3 + x^2 - 5$ **b.** $5x^4 + 2x^3 - x + 2$

Solution:

a.   Since this is a *fifth*-degree polynomial, there are *five* zeros. Because complex zeros come in conjugate pairs, the table describes the possible five zeros.

REAL ZEROS	COMPLEX ZEROS
1	4
3	2
5	0

Applying Descartes' rule of signs, we find that there are 3 or 1 positive real zeros and 2 or 0 negative real zeros.

POSITIVE REAL ZEROS	NEGATIVE REAL ZEROS	COMPLEX ZEROS
1	0	4
3	0	2
1	2	2
3	2	0

b.   Because this is a *fourth*-degree polynomial, there are *four* zeros. Since complex zeros come in conjugate pairs, the table describes the possible four zeros.

REAL ZEROS	COMPLEX ZEROS
0	4
2	2
4	0

Applying Descartes' rule of signs we find that there are 2 or 0 positive real zeros and 2 or 0 negative real zeros.

POSITIVE REAL ZEROS	NEGATIVE REAL ZEROS	COMPLEX ZEROS
0	0	4
2	0	2
0	2	2
2	2	0

■ **YOUR TURN** List the possible combinations of real and complex zeros for:

$$P(x) = x^6 - 7x^5 + 8x^3 - 2x + 1$$

Factoring Polynomials

Now let's draw on the tests discussed in this chapter to help us find all the zeros of a polynomial. Doing so will enable us to write polynomials as a product of linear factors. Before reading Example 5, reread Section 4.4, Example 8.

EXAMPLE 6 Factoring a Polynomial

Factor the polynomial $P(x) = x^5 + 2x^4 - x - 2$.

Solution:

STEP 1 Determine variations in sign.

$P(x)$ has 1 sign change. $P(x) = x^5 + 2x^4 - x - 2$

$P(-x)$ has 2 sign changes. $P(-x) = -x^5 + 2x^4 + x - 2$

STEP 2 Apply Descartes' rule of signs and summarize the results in a table.

POSITIVE REAL ZEROS	NEGATIVE REAL ZEROS	COMPLEX ZEROS
1	2	2
1	0	4

STEP 3 Utilize the rational zero theorem to
determine the possible rational zeros. $\pm 1, \pm 2$

STEP 4 Test possible rational zeros.

1 is a zero:

$$
\begin{array}{r|rrrrrr}
1 & 1 & 2 & 0 & 0 & -1 & -2 \\
  &   & 1 & 3 & 3 & 3 & 2 \\
\hline
  & 1 & 3 & 3 & 3 & 2 & \boxed{0}
\end{array}
$$

-1 is a zero:

$$
\begin{array}{r|rrrrr}
-1 & 1 & 3 & 3 & 3 & 2 \\
   &   & -1 & -2 & -1 & -2 \\
\hline
   & 1 & 2 & 1 & 2 & \boxed{0}
\end{array}
$$

-2 is a zero:

$$
\begin{array}{r|rrrr}
-2 & 1 & 2 & 1 & 2 \\
   &   & -2 & 0 & -2 \\
\hline
   & 1 & 0 & 1 & \boxed{0}
\end{array}
$$

$\underbrace{x^2 + 1}\quad = (x - i)(x + i)$

STEP 5 Write $P(x)$ as a product of linear factors.

$$\boxed{P(x) = (x - 1)(x + 1)(x + 2)(x - i)(x + i)}$$

Classroom Example 4.5.6*

Factor $P(x) = x^5 + 2x^4 + 10x^3 + 20x^2 + 9x + 18$

Answer: $(x + 3i)(x - 3i)(x + i)(x - i)(x + 2)$

SECTION
4.5 SUMMARY

In this section we discussed **complex zeros** of polynomial
functions. A polynomial function, $P(x)$, of degree n with real
coefficients has the following properties:

- $P(x)$ has at least one zero and no more than n zeros.
- If $a + bi$ is a zero, then $a - bi$ is also a zero.
- The polynomial can be written as a product of linear factors,
  not necessarily distinct.

▪ SKILLS

In Exercises 1–8, find all zeros (real and complex). Factor the polynomial as a product of linear factors.

1. $P(x) = x^2 + 4$
2. $P(x) = x^2 + 9$
3. $P(x) = x^2 - 2x + 2$
4. $P(x) = x^2 - 4x + 5$

5. $P(x) = x^4 - 16$
6. $P(x) = x^4 - 81$
7. $P(x) = x^4 - 25$
8. $P(x) = x^4 - 9$

In Exercises 9–16, a polynomial function with real coefficients is described. Find all remaining zeros.

9. Degree: 3   Zeros: $-1, i$
10. Degree: 3   Zeros: $1, -i$

11. Degree: 4   Zeros: $2i, 3 - i$
12. Degree: 4   Zeros: $3i, 2 + i$

13. Degree: 6   Zeros: 2 (multiplicity 2), $1 - 3i, 2 + 5i$
14. Degree: 6   Zeros: -2 (multiplicity 2), $1 - 5i, 2 + 3i$

15. Degree: 6   Zeros: $-i, 1 - i$ (multiplicity 2)
16. Degree: 6   Zeros: $2i, 1 + i$ (multiplicity 2)

In Exercises 17–22, find a polynomial of minimum degree that has the given zeros.

17. $0, 1 - 2i, 1 + 2i$
18. $0, 2 - i, 2 + i$
19. $1, 1 - 5i, 1 + 5i$

20. $2, 4 - i, 4 + i$
21. $1 - i, 1 + i, -3i, 3i$
22. $-i, i, 1 - 2i, 1 + 2i$

In Exercises 23–34, given a zero of the polynomial, determine all other zeros (real and complex) and write the polynomial in terms of a product of linear factors.

Polynomial	**Zero**	**Polynomial**	**Zero**
23. $P(x) = x^4 - 2x^3 - 11x^2 - 8x - 60$	$-2i$	**24.** $P(x) = x^4 - x^3 + 7x^2 - 9x - 18$	$3i$
25. $P(x) = x^4 - 4x^3 + 4x^2 - 4x + 3$	i	**26.** $P(x) = x^4 - x^3 + 2x^2 - 4x - 8$	$-2i$
27. $P(x) = x^4 - 2x^3 + 10x^2 - 18x + 9$	$-3i$	**28.** $P(x) = x^4 - 3x^3 + 21x^2 - 75x - 100$	$5i$
29. $P(x) = x^4 - 9x^2 + 18x - 14$	$1 + i$	**30.** $P(x) = x^4 - 4x^3 + x^2 + 6x - 40$	$1 - 2i$
31. $P(x) = x^4 - 6x^3 + 6x^2 + 24x - 40$	$3 - i$	**32.** $P(x) = x^4 - 4x^3 + 4x^2 + 4x - 5$	$2 + i$
33. $P(x) = x^4 - 9x^3 + 29x^2 - 41x + 20$	$2 - i$	**34.** $P(x) = x^4 - 7x^3 + 14x^2 + 2x - 20$	$3 + i$

In Exercises 35–58, factor each polynomial as a product of linear factors.

35. $P(x) = x^3 - x^2 + 9x - 9$
36. $P(x) = x^3 - 2x^2 + 4x - 8$

37. $P(x) = x^3 - 5x^2 + x - 5$
38. $P(x) = x^3 - 7x^2 + x - 7$

39. $P(x) = x^3 + x^2 + 4x + 4$
40. $P(x) = x^3 + x^2 - 2$

41. $P(x) = x^3 - x^2 - 18$
42. $P(x) = x^4 - 2x^3 - 2x^2 - 2x - 3$

43. $P(x) = x^4 - 2x^3 - 11x^2 - 8x - 60$
44. $P(x) = x^4 - x^3 + 7x^2 - 9x - 18$

45. $P(x) = x^4 - 4x^3 - x^2 - 16x - 20$
46. $P(x) = x^4 - 3x^3 + 11x^2 - 27x + 18$

47. $P(x) = x^4 - 7x^3 + 27x^2 - 47x + 26$
47. $P(x) = x^4 - 5x^3 + 5x^2 + 25x - 26$

49. $P(x) = -x^4 - 3x^3 + x^2 + 13x + 10$
50. $P(x) = -x^4 - x^3 + 12x^2 + 26x + 24$

51. $P(x) = x^4 - 2x^3 + 5x^2 - 8x + 4$

52. $P(x) = x^4 + 2x^3 + 10x^2 + 18x + 9$

53. $P(x) = x^6 + 12x^4 + 23x^2 - 36$

54. $P(x) = x^6 - 2x^5 + 9x^4 - 16x^3 + 24x^2 - 32x + 16$

55. $P(x) = 4x^4 - 20x^3 + 37x^2 - 24x + 5$

56. $P(x) = 4x^4 - 44x^3 + 145x^2 - 114x + 26$

57. $P(x) = 3x^5 - 2x^4 + 9x^3 - 6x^2 - 12x + 8$

58. $P(x) = 2x^5 - 5x^4 + 4x^3 - 26x^2 + 50x - 25$

■ **APPLICATIONS**

In Exercises 59–62, assume the profit model is given by a polynomial function $P(x)$ where x is the number of units sold by the company per year.

59. Profit. If the profit function of a given company has all imaginary zeros and the leading coefficient is positive, would you invest in this company? Explain.

60. Profit. If the profit function of a given company has all imaginary zeros and the leading coefficient is negative, would you invest in this company? Explain.

61. Profit. If the profit function of a company is modeled by a third-degree polynomial with a negative leading coefficient and this polynomial has two complex conjugates as zeros and one positive real zero, would you invest in this company? Explain.

62. Profit. If the profit function of a company is modeled by a third-degree polynomial with a positive leading coefficient and this polynomial has two complex conjugates as zeros and one positive real zero, would you invest in this company? Explain.

For Exercises 63 and 64, refer to the following:

The following graph models the profit P of a company where t is months and $t \geq 0$.

63. Business. If the profit function pictured is a third-degree polynomial, how many real and how many complex zeros does the function have? Discuss the implications of these zeros.

64. Business. If the profit function pictured is a fourth-degree polynomial with a negative leading coefficient, how many real and how many complex zeros does the function have? Discuss the implications of these zeros.

For Exercises 65 and 66, refer to the following:

The following graph models the concentration, C (in μg/mL) of a drug in the bloodstream; and t is time in hours after the drug is administered where $t \geq 0$.

65. Health/Medicine. If the concentration function pictured is a third-degree polynomial, how many real and how many complex zeros does the function have? Discuss the implications of these zeros.

66. Health/Medicine. If the concentration function pictured is a fourth-degree polynomial with a negative leading coefficient, how many real and how many complex zeros does the function have? Discuss the implications of these zeros.

▪CATCH THE MISTAKE

In Exercises 67 and 68, explain the mistake that is made.

67. Given that 1 is a zero of $P(x) = x^3 - 2x^2 + 7x - 6$, find all other zeros.

Solution:

Step 1: $P(x)$ is a third-degree polynomial, so we expect three zeros.

Step 2: Because 1 is a zero, -1 is a zero, so two linear factors are $(x - 1)$ and $(x + 1)$.

Step 3: Write the polynomial as a product of three linear factors.
$$P(x) = (x - 1)(x + 1)(x - c)$$
$$P(x) = (x^2 - 1)(x - c)$$

Step 4: To find the remaining linear factor, we divide $P(x)$ by $x^2 - 1$.
$$\frac{x^3 - 2x^2 + 7x - 6}{x^2 - 1} = x - 2 + \frac{6x - 8}{x^2 - 1}$$

Which has a nonzero remainder? What went wrong?

68. Factor the polynomial $P(x) = 2x^3 + x^2 + 2x + 1$.

Solution:

Step 1: Since $P(x)$ is an odd-degree polynomial, we are guaranteed one real zero (since complex zeros come in conjugate pairs).

Step 2: Apply the rational zero test to develop a list of potential rational zeros.

Possible zeros: ± 1

Step 3: Test possible zeros.

1 is not a zero: $P(x) = 2(1)^3 + (1)^2 + 2(1) + 1$
$$= 6$$

-1 is not a zero: $P(x) = 2(-1)^3 + (-1)^2 + 2(-1) + 1$
$$= -2$$

Note: $-\frac{1}{2}$ is the real zero. Why did we not find it?

▪CONCEPTUAL

In Exercises 69–72, determine whether each statement is true or false.

69. If $x = 1$ is a zero of a polynomial function, then $x = -1$ is also a zero of the polynomial function.

70. All zeros of a polynomial function correspond to x-intercepts.

71. A polynomial function of degree n, $n > 0$ must have at least one zero.

72. A polynomial function of degree n, $n > 0$ can be written as a product of n linear factors.

73. Is it possible for an odd-degree polynomial to have all imaginary complex zeros? Explain.

74. Is it possible for an even-degree polynomial to have all imaginary zeros? Explain.

▪CHALLENGE

In Exercises 75 and 76, assume a and b are nonzero real numbers.

75. Find a polynomial function that has degree 6, and for which bi is a zero of multiplicity 3.

76. Find a polynomial function that has degree 4, and for which $a + bi$ is a zero of multiplicity 2.

▪TECHNOLOGY

For Exercises 77 and 78, determine possible combinations of real and complex zeros. Plot $P(x)$ and identify any real zeros with a graphing calculator or software. Does this agree with your list?

77. $P(x) = x^4 + 13x^2 + 36$

78. $P(x) = x^6 + 2x^4 + 7x^2 - 130x - 288$

For Exercises 79 and 80, find all zeros (real and complex). Factor the polynomial as a product of linear factors.

79. $P(x) = -5x^5 + 3x^4 - 25x^3 + 15x^2 - 20x + 12$

80. $P(x) = x^5 + 2.1x^4 - 5x^3 - 5.592x^2 + 9.792x - 3.456$

SKILLS OBJECTIVES

- Find the domain of a rational function.
- Determine vertical, horizontal, and slant asymptotes of rational functions.
- Graph rational functions.

CONCEPTUAL OBJECTIVES

- Understand arrow notation.
- Interpret the behavior of the graph of a rational function near an asymptote.

Domain of Rational Functions

So far in this chapter we have discussed polynomial functions. We now turn our attention to *rational functions,* which are *ratios* of polynomial functions. Ratios of integers are called *rational numbers.* Similarly, ratios of polynomial functions are called *rational functions.*

DEFINITION   **Rational Function**

A function $f(x)$ is a **rational function** if

$$f(x) = \frac{n(x)}{d(x)} \qquad d(x) \neq 0$$

where the numerator, $n(x)$, and the denominator, $d(x)$, are polynomial functions. The domain of $f(x)$ is the set of all real numbers x such that $d(x) \neq 0$.

Note: If $d(x)$ is a constant, then $f(x)$ is a polynomial function.

The domain of any polynomial function is the set of all real numbers. When we divide two polynomial functions, the result is a *rational function,* and we must exclude any values of x that make the denominator equal to zero.

EXAMPLE 1   Finding the Domain of a Rational Function

Find the domain of the rational function $f(x) = \dfrac{x + 1}{x^2 - x - 6}$. Express the domain in interval notation.

Solution:

Set the denominator equal to zero.	$x^2 - x - 6 = 0$
Factor.	$(x + 2)(x - 3) = 0$
Solve for x.	$x = -2$ or $x = 3$
Eliminate these values from the domain.	$x \neq -2$ or $x \neq 3$
State the domain in interval notation.	$\boxed{(-\infty, -2) \cup (-2, 3) \cup (3, \infty)}$

■ **YOUR TURN** Find the domain of the rational function $f(x) = \dfrac{x - 2}{x^2 - 3x - 4}$. Express the domain in interval notation.

It is important to note that there are not always restrictions on the domain. For example, if the denominator is never equal to zero, the domain is the set of all real numbers.

EXAMPLE 2   When the Domain of a Rational Function Is the Set of All Real Numbers

Find the domain of the rational function $g(x) = \dfrac{3x}{x^2 + 9}$. Express the domain in interval notation.

Solution:

Set the denominator equal to zero.	$x^2 + 9 = 0$
Subtract 9 from both sides.	$x^2 = -9$
Solve for x.	$x = -3i$ or $x = 3i$
There are no *real* solutions; therefore the domain has no restrictions.	$\mathbb{R}$, the set of all real numbers
State the domain in interval notation.	$\boxed{(-\infty, \infty)}$

■ **YOUR TURN** Find the domain of the rational function $g(x) = \dfrac{5x}{x^2 + 4}$. Express the domain in interval notation.

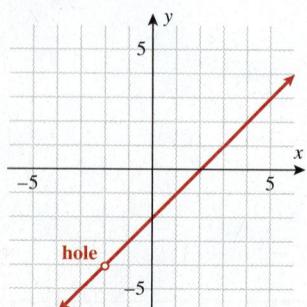

It is important to note that $f(x) = \dfrac{x^2 - 4}{x + 2}$, where $x \neq -2$, and $g(x) = x - 2$ are *not* the same function. Although $f(x)$ can be written in the factored form $f(x) = \dfrac{(x - 2)(x + 2)}{x + 2} = x - 2$, its domain is different. The domain of $g(x)$ is the set of all real numbers, whereas the domain of $f(x)$ is the set of all real numbers such that $x \neq -2$. If we were to plot $f(x)$ and $g(x)$, they would both look like the line $y = x - 2$. However, $f(x)$ would have a hole, or discontinuity, at the point $x = -2$.

Vertical, Horizontal, and Slant Asymptotes

If a function is not defined at a point, then it is still useful to know how the function behaves near that point. Let's start with a simple rational function, the reciprocal function $f(x) = \dfrac{1}{x}$. This function is defined everywhere except at $x = 0$.

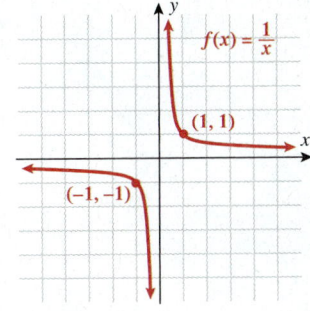

x	$-\dfrac{1}{10}$	$-\dfrac{1}{100}$	$-\dfrac{1}{1000}$	0	$\dfrac{1}{1000}$	$\dfrac{1}{100}$	$\dfrac{1}{10}$
$f(x) = \dfrac{1}{x}$	-10	-100	-1000	undefined	1000	100	10

x approaching 0 from the left

x approaching 0 from the right

x	$f(x) = \dfrac{1}{x}$
-10	$-\dfrac{1}{10}$
-1	-1
1	1
10	$\dfrac{1}{10}$

We cannot let $x = 0$ because that point is not in the domain of the function. We should, however, ask the question, "how does $f(x)$ behave as x *approaches* zero?" Let us take values that get closer and closer to $x = 0$, such as $\frac{1}{10}, \frac{1}{100}, \frac{1}{1000}, \ldots$ (See the table above.) We use an *arrow* to represent the word *approach*, a *positive* superscript to represent from the *right*, and a *negative* superscript to represent from the *left*. A plot of this function can be generated using point-plotting techniques. The following are observations of the graph $f(x) = \dfrac{1}{x}$.

WORDS	**MATH**
As x approaches zero from the *right*, the function $f(x)$ increases without bound.	$x \to 0^+$ $\dfrac{1}{x} \to \infty$
As x approaches zero from the *left*, the function $f(x)$ decreases without bound.	$x \to 0^-$ $\dfrac{1}{x} \to -\infty$
As x approaches infinity (increases without bound), the function $f(x)$ approaches zero from *above*.	$x \to \infty$ $\dfrac{1}{x} \to 0^+$
As x approaches negative infinity (decreases without bound), the function $f(x)$ approaches zero from *below*.	$x \to -\infty$ $\dfrac{1}{x} \to 0^-$

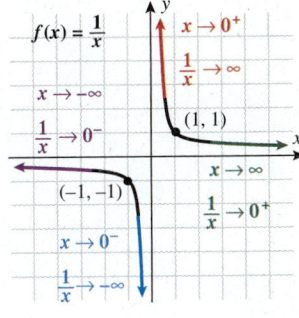

The symbol ∞ does not represent an actual real number. This symbol represents growing without bound.

1. Notice that the function is not defined at $x = 0$. The y-axis, or the vertical line $x = 0$, represents the *vertical asymptote*.
2. Notice that the value of the function is never equal to zero. The x-axis is never touched by the function. The x-axis, or $y = 0$, is a *horizontal asymptote*.

Asymptotes are lines that the graph of a function approaches. Suppose a football team's defense is its own 8 yard line and the team gets an "offsides" penalty that results in loss of "half the distance to the goal." Then the offense would get the ball on the 4 yard line. Suppose the defense gets another penalty on the next play that results in "half the distance to the goal." The offense would then get the ball on the 2 yard line. If the defense received 10 more penalties all resulting in "half the distance to the goal," would the referees *give* the offense a touchdown? No, because although the offense may appear to be snapping the ball from the goal line, technically it has not actually reached the goal line. Asymptotes utilize the same concept.

We will start with *vertical asymptotes*. Although the function $f(x) = \dfrac{1}{x}$ had one vertical

asymptote, in general, rational functions can have *none, one,* or *several* vertical asymptotes. We will first formally define what a vertical asymptote is and then discuss how to find it.

DEFINITION   **Vertical Asymptotes**

The line $x = a$ is a **vertical asymptote** for the graph of a function if $f(x)$ either increases or decreases without bound as x approaches a from either the left or the right.

As $x \to a^+, f(x) \to \infty$ As $x \to a^-, f(x) \to \infty$ As $x \to a^+, f(x) \to -\infty$ As $x \to a^-, f(x) \to -\infty$

Vertical asymptotes assist us in graphing rational functions since they essentially "steer" the function in the vertical direction. How do we locate the vertical asymptotes of a rational function? Set the denominator equal to zero. If the numerator and denominator have no common factors, then any numbers that are excluded from the domain of a rational function locate vertical asymptotes.

A rational function $f(x) = \dfrac{n(x)}{d(x)}$ is said to be in **lowest terms** if the numerator $n(x)$ and

denominator $d(x)$ have no common factors. Let $f(x) = \dfrac{n(x)}{d(x)}$ be a rational function in lowest

terms; then any zeros of the numerator $n(x)$ correspond to *x*-intercepts of the graph of f, and any zeros of the denominator $d(x)$ correspond to vertical asymptotes of the graph of f. If a rational function does have a common factor (is not in lowest terms), then the common factor(s) should be canceled, resulting in an equivalent rational function $R(x)$ in lowest terms. If $(x - a)^p$ is a factor of the numerator and $(x - a)^q$ is a factor of the denominator, then there is a *hole* in the graph at $x = a$ provided $p \geq q$ and $x = a$ is a vertical asymptote if $p < q$.

LOCATING VERTICAL ASYMPTOTES

Let $f(x) = \dfrac{n(x)}{d(x)}$ be a rational function in lowest terms (that is, assume $n(x)$ and $d(x)$ are polynomials with no common factors); then the graph of f has a vertical asymptote at any real zero of the denominator $d(x)$. That is, if $d(a) = 0$, then $x = a$ corresponds to a vertical asymptote on the graph of f.

Note: If f is a rational function that is not in lowest terms, then divide out the common factors, resulting in a rational function R that is in lowest terms. Any common factor $x - a$ of the function f corresponds to a hole in the graph of f at $x = a$ provided the multiplicity of a in the numerator is greater than or equal to the multiplicity of a in the denominator.

Study Tip

The vertical asymptotes of a rational function in *lowest terms* occur at x-values that make the denominator equal to zero.

EXAMPLE 3   Determining Vertical Asymptotes

Locate any vertical asymptotes of the rational function $f(x) = \dfrac{5x + 2}{6x^2 - x - 2}$.

Solution:

Factor the denominator. $f(x) = \dfrac{5x + 2}{(2x + 1)(3x - 2)}$

The numerator and denominator have no common factors.

Set the denominator equal to zero. $2x + 1 = 0$ and $3x - 2 = 0$

Solve for x. $x = -\dfrac{1}{2}$ and $x = \dfrac{2}{3}$

The vertical asymptotes are $\boxed{x = -\tfrac{1}{2}}$ and $\boxed{x = \tfrac{2}{3}}$.

■ YOUR TURN Locate any vertical asymptotes of the following rational function:

$$f(x) = \dfrac{3x - 1}{2x^2 - x - 15}$$

Classroom Example 4.6.3
Locate the vertical asymptotes for these rational functions, if they exist.

a. $f(x) = \dfrac{2 - x}{x^3 + 2x^2 + x}$

b.* $g(x) = \dfrac{2x^2}{3x^3 - 3x^2 + x - 1}$

c. $h(x) = \dfrac{1 - x^2 + x^3}{2x^6 + 3x^4}$

d.* $j(x) = \dfrac{2x}{2x^4 + 15x^2 + 25}$

Answer: a. $x = 0, -1$
　　　　　b. $x = 1$
　　　　　c. $x = 0$
　　　　　d. none

■ Answer: $x = -\tfrac{5}{2}$ and $x = 3$

EXAMPLE 4 **Determining Vertical Asymptotes When the Rational Function Is Not in Lowest Terms**

Locate any vertical asymptotes of the rational function $f(x) = \dfrac{x + 2}{x^3 - 3x^2 - 10x}$.

Solution:

Factor the denominator.

$$x^3 - 3x^2 - 10x = x(x^2 - 3x - 10)$$
$$= x(x - 5)(x + 2)$$

Write the rational function in factored form.

$$f(x) = \dfrac{(x + 2)}{x(x - 5)(x + 2)}$$

Cancel (divide out) the common factor $(x + 2)$.

$$R(x) = \dfrac{1}{x(x - 5)} \quad x \neq -2$$

Find the values when the denominator of R is equal to zero.

$$x = 0 \quad \text{and} \quad x = 5$$

The vertical asymptotes are $\boxed{x = 0}$ and $\boxed{x = 5}$.

Note: $x = -2$ is not in the domain of $f(x)$, even though there is no vertical asymptote there. There is a "hole" in the graph at $x = -2$. Graphing calculators do not always show such "holes."

■ **YOUR TURN** Locate any vertical asymptotes of the following rational function:

$$f(x) = \dfrac{x^2 - 4x}{x^2 - 7x + 12}$$

Classroom Example 4.6.4
Locate the vertical asymptotes for these rational functions, if they exist.

a. $f(x) = \dfrac{2 + x}{x^3 + 4x^2 + 4x}$

b.* $g(x) = \dfrac{2x(x - 1)}{2x^5 + 15x^3 + 25x}$

c.* $h(x) = \dfrac{2x^2 + 7}{2x^4 - 4x^3 + 9x^2 - 14x + 7}$

Answer: a. $x = -2$ and $x = 0$
  b. none
  c. $x = 1$

■ **Answer:** $x = 3$

We now turn our attention to *horizontal asymptotes*. As we have seen, rational functions can have several vertical asymptotes. However, rational functions can have *at most* one horizontal asymptote. Horizontal asymptotes imply that a function approaches a constant value as x becomes large in the positive or negative direction. Another difference between vertical and horizontal asymptotes is that the graph of a function never touches a vertical asymptote but, as you will see in the next box, the graph of a function may cross a horizontal asymptote, just not at the "ends" ($x \to \pm\infty$).

DEFINITION **Horizontal Asymptote**

The line $y = b$ is a **horizontal asymptote** of the graph of a function if $f(x)$ approaches b as x increases or decreases without bound. The following are three examples:

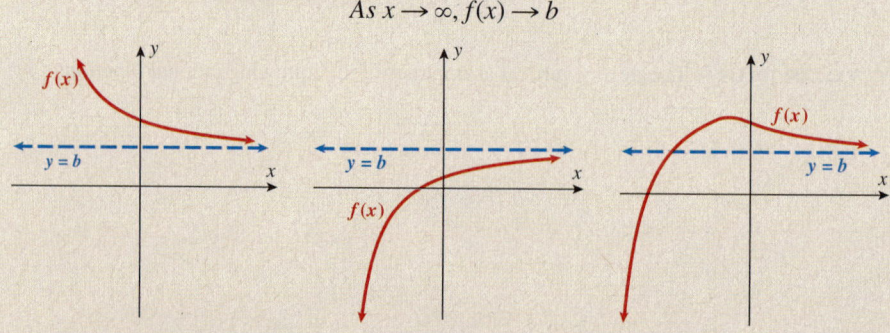

$$\text{As } x \to \infty, f(x) \to b$$

Note: A horizontal asymptote steers a function as x gets large. Therefore, when x is not large, the function may cross the asymptote.

How do we determine whether a horizontal asymptote exists? And, if it does, how do we locate it? We investigate the value of the rational function as $x \to \infty$ or as $x \to -\infty$. One of two things will happen: either the rational function will increase or decrease without bound or the rational function will approach a constant value.

We say that a rational function is **proper** if the degree of the numerator is less than the degree of the denominator. Proper rational functions, like $f(x) = \dfrac{1}{x}$, approach zero as x gets large. Therefore, all proper rational functions have the specific horizontal asymptote, $y = 0$ (see Example 5a).

We say that a rational function is **improper** if the degree of the numerator is greater than or equal to the degree of the denominator. In this case, we can divide the numerator by the denominator and determine how the quotient behaves as x increases without bound.

- If the quotient is a constant (resulting when the degrees of the numerator and denominator are equal), then as $x \to \infty$ or as $x \to -\infty$, the rational function approaches the constant quotient (see Example 5b).
- If the quotient is a polynomial function of degree 1 or higher, then the quotient depends on x and does not approach a constant value as x increases (see Example 5c). In this case, we say that there is no horizontal asymptote.

We find horizontal asymptotes by comparing the degree of the numerator and the degree of the denominator. There are three cases to consider:

1. The degree of the numerator is less than the degree of the denominator.
2. The degree of the numerator is equal to the degree of the denominator.
3. The degree of the numerator is greater than the degree of the denominator.

LOCATING HORIZONTAL ASYMPTOTES

Let f be a rational function given by

$$f(x) = \frac{n(x)}{d(x)} = \frac{a_n x^n + a_{n-1} x^{n-1} + \cdots + a_1 x + a_0}{b_m x^m + b_{m-1} x^{m-1} + \cdots + b_1 x + b_0}$$

where $n(x)$ and $d(x)$ are polynomials.

1. When $n < m$, the x-axis ($y = 0$) is the horizontal asymptote.

2. When $n = m$, the line $y = \dfrac{a_n}{b_m}$ (ratio of leading coefficients) is the horizontal asymptote.

3. When $n > m$, there is no horizontal asymptote.

In other words,

1. When the degree of the numerator is less than the degree of the denominator, then $y = 0$ is the horizontal asymptote.
2. When the degree of the numerator is the same as the degree of the denominator, then the horizontal asymptote is the ratio of the leading coefficients.
3. If the degree of the numerator is greater than the degree of the denominator, then there is no horizontal asymptote.

a. Graph $f(x) = \dfrac{8x + 3}{4x^2 + 1}$.

b. Graph $g(x) = \dfrac{8x^2 + 3}{4x^2 + 1}$.

c. Graph $h(x) = \dfrac{8x^3 + 3}{4x^2 + 1}$.

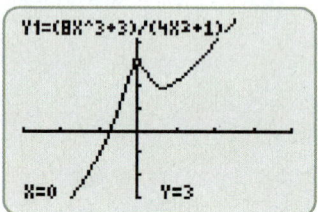

■ **Answer:** $y = -\frac{7}{4}$ is the horizontal
asymptote.

EXAMPLE 5   Finding Horizontal Asymptotes

Determine whether a horizontal asymptote exists for the graph of each of the given rational functions. If it does, locate the horizontal asymptote.

a. $f(x) = \dfrac{8x + 3}{4x^2 + 1}$ **b.** $g(x) = \dfrac{8x^2 + 3}{4x^2 + 1}$ **c.** $h(x) = \dfrac{8x^3 + 3}{4x^2 + 1}$

Solution (a):

The degree of the numerator $8x + 3$ is 1. $n = 1$

The degree of the denominator $4x^2 + 1$ is 2. $m = 2$

The degree of the numerator is less than the
degree of the denominator. $n < m$

The x-axis is the horizontal asymptote for the graph of $f(x)$. $y = 0$

The line $\boxed{y = 0}$ is the horizontal asymptote for the graph of $f(x)$.

Solution (b):

The degree of the numerator $8x^2 + 3$ is 2. $n = 2$

The degree of the denominator $4x^2 + 1$ is 2. $m = 2$

The degree of the numerator is equal to the
degree of the denominator. $n = m$

The ratio of the leading coefficients is the
horizontal asymptote for the graph of $g(x)$. $y = \dfrac{8}{4} = 2$

The line $\boxed{y = 2}$ is the horizontal asymptote for the graph of $g(x)$.

If we divide the numerator by the denominator,
the resulting quotient is the constant 2. $g(x) = \dfrac{8x^2 + 3}{4x^2 + 1} = 2 + \dfrac{1}{4x^2 + 1}$

Solution (c):

The degree of the numerator $8x^3 + 3$ is 3. $n = 3$

The degree of the denominator $4x^2 + 1$ is 2. $m = 2$

The degree of the numerator is greater than
the degree of the denominator. $n > m$

The graph of the rational function $h(x)$ has $\boxed{\text{no horizontal asymptote}}$.

If we divide the numerator by the denominator,
the resulting quotient is a linear function. $h(x) = \dfrac{8x^3 + 3}{4x^2 + 1} = 2x + \dfrac{-2x + 3}{4x^2 + 1}$

■ **YOUR TURN** Find the horizontal asymptote (if one exists) for the graph of the
rational function $f(x) = \dfrac{7x^3 + x - 2}{-4x^3 + 1}$.

Thus far we have discussed linear asymptotes: vertical and horizontal. There are three types of lines: horizontal (slope is zero), vertical (slope is undefined), and slant (nonzero slope). Similarly, there are three types of linear asymptotes: horizontal, vertical, and *slant*.

Recall that when dividing polynomials the degree of the quotient is always the difference between the degree of the numerator and the degree of the denominator. For example, a cubic (third-degree) polynomial divided by a quadratic (second-degree) polynomial results in a linear (first-degree) polynomial. A fifth-degree polynomial divided by a fourth-degree polynomial results in a first-degree (linear) polynomial. When the degree of the numerator is exactly one more than the degree of the denominator, the quotient is linear and represents a *slant asymptote*.

SLANT ASYMPTOTES

Let f be a rational function given by $f(x) = \dfrac{n(x)}{d(x)}$, where $n(x)$ and $d(x)$ are polynomials and the degree of $n(x)$ is *one more than* the degree of $d(x)$. On dividing $n(x)$ by $d(x)$, the rational function can be expressed as

$$f(x) = mx + b + \frac{r(x)}{d(x)}$$

where the degree of the remainder $r(x)$ is less than the degree of $d(x)$ and the line $y = mx + b$ is a **slant asymptote** for the graph of f.

Note that as $x \to -\infty$ or $x \to \infty$, $f(x) \to mx + b$.

EXAMPLE 6   Finding Slant Asymptotes

Determine the slant asymptote of the rational function $f(x) = \dfrac{4x^3 + x^2 + 3}{x^2 - x + 1}$.

Solution:

Divide the numerator by the denominator with long division.

$$
\begin{array}{r}
4x + 5 \\
x^2 - x + 1 \overline{\smash{)}\, 4x^3 + x^2 + 0x + 3} \\
\underline{-(4x^3 - 4x^2 + 4x)} \\
5x^2 - 4x + 3 \\
\underline{-(5x^2 - 5x + 5)} \\
x - 2
\end{array}
$$

Note that as $x \to \pm\infty$ the rational expression approaches 0.

$$f(x) = \mathbf{4x + 5} + \underbrace{\frac{x - 2}{x^2 - x + 1}}_{\substack{\to\, 0 \text{ as} \\ x \to \pm\infty}}$$

The quotient is the slant asymptote.

$$\boxed{y = 4x + 5}$$

■ **YOUR TURN** Find the slant asymptote of the rational function $f(x) = \dfrac{x^2 + 3x + 2}{x - 2}$.

Classroom Example 4.6.5
Find the horizontal asymptote for these rational functions, if it exists.

a. $f(x) = \dfrac{2 + x}{x^3 + 4x^2 + 4x}$

b.* $g(x) = \dfrac{-5x^2(2 + x)}{x^3 + 4x^2 + 4x}$

c.* $h(x) = \dfrac{-5x^2(2 + x)^2}{x^3 + 4x^2 + 4x}$

Answer: a. $y = 0$
b. $y = -5$
c. none

Classroom Example 4.6.6
Find the slant asymptote for these rational functions.

a. $f(x) = \dfrac{3x^3 + 2x^2 + 3x + 4}{x^2 + 1}$

b.* $g(x) = \dfrac{(x + 3)^3}{(x + 2)^2}$

Answer: a. $y = 3x + 2$
b. $y = x + 5$

Technology Tip

The graph of

$$f(x) = \frac{4x^3 + x^2 + 3}{x^2 - x + 1}$$

has a slant asymptote of

$$y = 4x + 5$$

■ **Answer:** $y = x + 5$

Graphing Rational Functions

We can now graph rational functions using asymptotes as graphing aids. The following box summarizes the five-step procedure for graphing rational functions.

Classroom Example 4.6.7

Graph these rational functions.

a. $f(x) = \dfrac{2 - x}{x^3 + 4x^2 + 4x}$

b.* $g(x) = \dfrac{2x}{2x^4 + 15x^2 + 25}$

Answer:

a.

b.

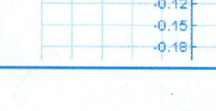

GRAPHING RATIONAL FUNCTIONS

Let f be a rational function given by $f(x) = \dfrac{n(x)}{d(x)}$.

Step 1: Find the domain of the rational function f.

Step 2: Find the **intercept(s)**.
- y-intercept: evaluate $f(0)$.
- x-intercept: solve the equation $n(x) = 0$ for x in the domain of f.

Step 3: Find any **holes**.
- Factor the numerator and denominator.
- Divide out common factors.
- A common factor $x - a$ corresponds to a hole in the graph of f at $x = a$ if the multiplicity of a in the numerator is greater than or equal to the multiplicity of a in the denominator.
- The result is an equivalent rational function $R(x) = \dfrac{p(x)}{q(x)}$ in lowest terms.

Step 4: Find any **asymptotes**.
- Vertical asymptotes: solve $q(x) = 0$.
- Compare the degree of the numerator and the degree of the denominator to determine whether either a horizontal or slant asymptote exists. If one exists, find it.

Step 5: Find **additional points** on the graph of f—particularly near asymptotes.

Step 6: **Sketch** the graph; draw the asymptotes, label the intercept(s) and additional points, and complete the graph with a smooth curve between and beyond the vertical asymptotes.

It is important to note that any real number eliminated from the domain of a rational function corresponds to either a vertical asymptote or a hole on its graph.

EXAMPLE 7   Graphing a Rational Function

Graph the rational function $f(x) = \dfrac{x}{x^2 - 4}$.

Solution:

STEP 1 Find the **domain**.

Set the denominator equal to zero.	$x^2 - 4 = 0$
Solve for x.	$x = \pm 2$
State the domain.	$(-\infty, -2) \cup (-2, 2) \cup (2, \infty)$

STEP 2 Find the **intercepts**.

y-intercept:	$f(0) = \dfrac{0}{-4} = 0$	$y = 0$
x-intercepts:	$f(x) = \dfrac{x}{x^2 - 4} = 0$	$x = 0$

The only intercept is at the point $\boxed{(0, 0)}$.

STEP 3 Find any holes.

$$f(x) = \frac{x}{(x + 2)(x - 2)}$$

There are no common factors, so f is in lowest terms.
Since there are no common factors, there are no holes on the graph of f.

STEP 4 Find any **asymptotes**.

Vertical asymptotes:

$$d(x) = (x + 2)(x - 2) = 0$$

$$\boxed{x = -2} \quad \text{and} \quad \boxed{x = 2}$$

Horizontal asymptote:

Degree of numerator $= 1$
Degree of denominator $= 2$

Degree of numerator $<$ Degree of denominator $\quad \boxed{y = 0}$

STEP 5 Find **additional points** on the graph.

x	-3	-1	1	3
$f(x)$	$-\frac{3}{5}$	$\frac{1}{3}$	$-\frac{1}{3}$	$\frac{3}{5}$

STEP 6 **Sketch** the graph; label the intercepts, asymptotes, and additional points and complete with a smooth curve approaching the asymptotes.

■ **Answer:**

■ **YOUR TURN** Graph the rational function $f(x) = \dfrac{x}{x^2 - 1}$.

EXAMPLE 8 **Graphing a Rational Function with No Horizontal or Slant Asymptotes**

State the asymptotes (if there are any) and graph the rational function $f(x) = \dfrac{x^4 - x^3 - 6x^2}{x^2 - 1}$.

Solution:

STEP 1 Find the domain.

Set the denominator equal to zero. $\qquad x^2 - 1 = 0$

Solve for x. $\qquad x = \pm 1$

State the domain. $\qquad (-\infty, -1) \cup (-1, 1) \cup (1, \infty)$

STEP 2 Find the **intercepts**.

y-intercept: $\qquad f(0) = \dfrac{0}{-1} = 0$

x-intercepts: $\qquad n(x) = x^4 - x^3 - 6x^2 = 0$

Factor. $\qquad x^2(x - 3)(x + 2) = 0$

Solve. $\qquad x = 0, x = 3, \text{ and } x = -2$

The **intercepts** are the points **$(0, 0)$, $(3, 0)$**, and **$(-2, 0)$**.

Technology Tip

The behavior of each function as x approaches ∞ or $-\infty$ can be shown using tables of values.

Graph $f(x) = \dfrac{x^4 - x^3 - 6x^2}{x^2 - 1}$.

```
WINDOW
 Xmin=-4
 Xmax=5
 Xscl=1
 Ymin=-8
 Ymax=8
 Yscl=1
 Xres=1
```

```
Plot1 Plot2 Plot3
\Y1■(X^4-X^3-6X²
)/(X²-1)
\Y2=■
```

The graph of $f(x)$ shows that the vertical asymptotes are at $x = \pm 1$ and there is no horizontal asymptote or slant asymptote.

■ **Answer:** Vertical asymptote: $x = -2$. No horizontal or slant asymptotes.

STEP 3 Find any **holes**.

$$f(x) = \frac{x^2(x - 3)(x + 2)}{(x - 1)(x + 1)}$$

There are no common factors, so f is in lowest terms.

Since there are no common factors, there are no holes on the graph of f.

STEP 4 Find the **asymptotes**.

Vertical asymptote:	$d(x) = x^2 - 1 = 0$
Factor.	$(x + 1)(x - 1) = 0$
Solve.	$x = -1$ and $x = 1$

No horizontal asymptote: degree of $n(x)$ > degree of $d(x)$ [4 > 2]

No slant asymptote: degree of $n(x)$ − degree of $d(x)$ > 1   [4 − 2 = 2 > 1]

The **asymptotes** are $x = -1$ and $x = 1$.

STEP 5 Find **additional points** on the graph.

x	-3	-0.5	0.5	2	4
$f(x)$	6.75	1.75	2.08	-5.33	6.4

STEP 6 Sketch the graph; label the **intercepts** and **asymptotes**, and complete with a smooth curve between and beyond the vertical asymptote.

■ **YOUR TURN** State the asymptotes (if there are any) and graph the rational function
$$f(x) = \frac{x^3 - 2x^2 - 3x}{x + 2}.$$

EXAMPLE 9   **Graphing a Rational Function with a Horizontal Asymptote**

State the asymptotes (if there are any) and graph the rational function

$$f(x) = \frac{4x^3 + 10x^2 - 6x}{8 - x^3}$$

Solution:

STEP 1 Find the **domain**.

Set the denominator equal to zero.	$8 - x^3 = 0$
Solve for x.	$x = 2$
State the domain.	$(-\infty, 2) \cup (2, \infty)$

STEP 2 Find the **intercepts**.

y-intercept:

$$f(0) = \frac{0}{8} = 0$$

x-intercepts:

$$n(x) = 4x^3 + 10x^2 - 6x = 0$$

Factor.

$$2x(2x - 1)(x + 3) = 0$$

Solve.

$$x = 0, x = \frac{1}{2}, \text{ and } x = -3$$

The **intercepts** are the points **(0, 0)**, $\left(\frac{1}{2}, 0\right)$, and **(−3, 0)**.

STEP 3 Find the **holes**.

$$f(x) = \frac{2x(2x - 1)(x + 3)}{(2 - x)(x^2 + 2x + 4)}$$

There are no common factors, so f is in lowest terms (no holes).

STEP 4 Find the **asymptotes**.

Vertical asymptote:

$$d(x) = 8 - x^3 = 0$$

Solve.

$$x = 2$$

Horizontal asymptote:

degree of $n(x)$ = degree of $d(x)$

Use leading coefficients.

$$y = \frac{4}{-1} = -4$$

The **asymptotes** are $x = 2$ and $y = -4$.

STEP 5 Find **additional points** on the graph.

x	-4	-1	$\frac{1}{4}$	1	3
$f(x)$	-1	1.33	-0.10	1.14	-9.47

STEP 6 **Sketch** the graph; label the intercepts and asymptotes and complete with a smooth curve.

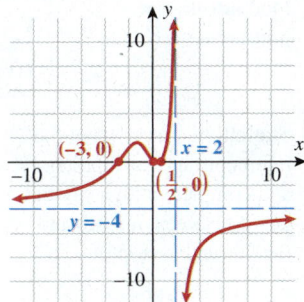

YOUR TURN Graph the rational function $f(x) = \dfrac{2x^2 - 7x + 6}{x^2 - 3x - 4}$. Give equations of the vertical and horizontal asymptotes and state the intercepts.

Technology Tip

The behavior of each function as x approaches ∞ or $-\infty$ can be shown using tables of values.

Graph $f(x) = \dfrac{4x^3 + 10x^2 - 6x}{8 - x^3}$.

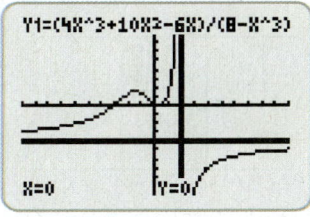

The graph of $f(x)$ shows that the vertical asymptote is at $x = 2$ and the horizontal asymptote is at $y = -4$.

■ **Answer:** Vertical asymptotes: $x = 4, x = -1$
Horizontal asymptote: $y = 2$
Intercepts: $\left(0, -\frac{3}{2}\right), \left(\frac{3}{2}, 0\right), (2, 0)$

Technology Tip

The behavior of each function as x approaches ∞ or $-\infty$ can be shown using tables of values.

Graph $f(x) = \dfrac{x^2 - 3x - 4}{x + 2}$.

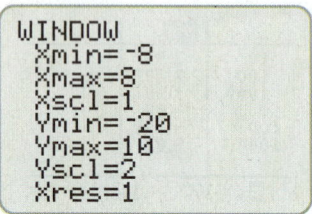

The graph of $f(x)$ shows that the vertical asymptote is at $x = -2$ and the slant asymptote is at $y = x - 5$.

■ **Answer:**
Horizontal asymptote: $x = 3$
Slant asymptote: $y = x + 4$

EXAMPLE 10 Graphing a Rational Function with a Slant Asymptote

Graph the rational function $f(x) = \dfrac{x^2 - 3x - 4}{x + 2}$.

Solution:

STEP 1 Find the **domain**.

Set the denominator equal to zero.	$x + 2 = 0$
Solve for x.	$x = -2$
State the domain.	$(-\infty, -2) \cup (-2, \infty)$

STEP 2 Find the **intercepts**.

y-intercept:	$f(0) = -\dfrac{4}{2} = -2$
x-intercepts:	$n(x) = x^2 - 3x - 4 = 0$
Factor.	$(x + 1)(x - 4) = 0$
Solve.	$x = -1$ and $x = 4$

The **intercepts** are the points $(0, -2)$, $(-1, 0)$, and $(4, 0)$.

STEP 3 Find any **holes**. $f(x) = \dfrac{(x - 4)(x + 1)}{(x + 2)}$

There are no common factors, so f is in lowest terms.
Since there are no common factors, there are no holes on the graph of f.

STEP 4 Find the **asymptotes**.

Vertical asymptote:	$d(x) = x + 2 = 0$
Solve.	$x = -2$
Slant asymptote:	degree of $n(x)$ − degree of $d(x) = 1$
Divide $n(x)$ by $d(x)$.	$f(x) = \dfrac{x^2 - 3x - 4}{x + 2} = x - 5 + \dfrac{6}{x + 2}$
Write the equation of the asymptote.	$y = x - 5$

The **asymptotes** are $x = -2$ and $y = x - 5$.

STEP 5 Find **additional points** on the graph.

x	-6	-5	-3	5	6
$f(x)$	-12.5	-12	-14	0.86	1.75

STEP 6 **Sketch** the graph; label the intercepts and asymptotes, and complete with a smooth curve between and beyond the vertical asymptote.

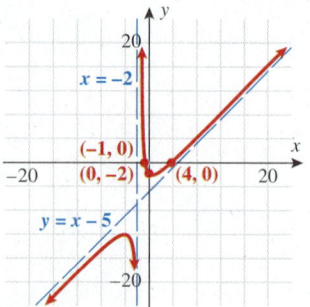

■ **YOUR TURN** For the function $f(x) = \dfrac{x^2 + x - 2}{x - 3}$, state the asymptotes (if any exist) and graph the function.

EXAMPLE 11   Graphing a Rational Function with a Hole in the Graph

Graph the rational function $f(x) = \dfrac{x^2 + x - 6}{x^2 - x - 2}$.

Solution:

STEP 1 Find the **domain**.

Set the denominator equal to zero.	$x^2 - x - 2 = 0$
Solve for x.	$(x - 2)(x + 1) = 0$
	$x = -1$ or $x = 2$
State the domain.	$(-\infty, -1) \cup (-1, 2) \cup (2, \infty)$

STEP 2 Find the **intercepts**.

y-intercept: $f(0) = \dfrac{-6}{-2} = 3 \qquad y = 3$

x-intercepts: $n(x) = x^2 + x - 6 = 0$

$(x + 3)(x - 2) = 0$

$x = -3$ or $x = 2$

The intercepts correspond to the points $\boxed{(0, 3)}$ and $\boxed{(-3, 0)}$. The point $(2, 0)$ appears to be an x-intercept; however, $x = 2$ is not in the domain of the function.

STEP 3 Find any **holes**. $f(x) = \dfrac{(x - 2)(x + 3)}{(x - 2)(x + 1)}$

Since $x - 2$ is a common factor, there is a
hole in the graph of f at $x = 2$.
Dividing out the common factor generates $R(x) = \dfrac{(x + 3)}{(x + 1)}$
an equivalent rational function in lowest terms.

STEP 4 Find the **asymptotes**.

Vertical asymptotes: $x + 1 = 0$

$\boxed{x = -1}$

Horizontal
asymptote:
Degree of numerator of f = Degree of denominator of f = 2
and
Degree of numerator of R = Degree of denominator of R = 1

Since the degree of the numerator
equals the degree of the denominator, $\boxed{y = \dfrac{1}{1} = 1}$
use the leading coefficients.

STEP 5 Find **additional points** on the graph.

x	-4	-2	$-\frac{1}{2}$	1	3
$f(x)$ or $R(x)$	$\frac{1}{3}$	-1	5	2	$\frac{3}{2}$

STEP 6 **Sketch** the graph; label the intercepts,
asymptotes, and additional points and
complete with a smooth curve approaching
asymptotes. Recall the hole at $x = 2$.
Note that $R(2) = \frac{5}{3}$ so the open "hole"
is located at the point $(2, 5/3)$.

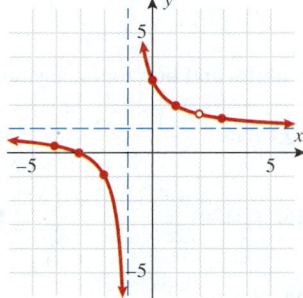

YOUR TURN Graph the rational function $f(x) = \dfrac{x^2 - x - 2}{x^2 + x - 6}$.

Technology Tip

The behavior of each function as x approaches ∞ or $-\infty$ can be shown using tables of values.

Graph $f(x) = \dfrac{x^2 + x - 6}{x^2 - x - 2}$.

The graph of $f(x)$ shows that the vertical asymptote is at $x = -1$ and the horizontal asymptote is at $y = 1$.

Notice that the hole at $x = 2$ is not apparent in the graph. A table of values supports the graph.

Answer:

SECTION
4.6 SUMMARY

In this section, rational functions were discussed.

$$f(x) = \frac{n(x)}{d(x)}$$

- **Domain:** All real numbers except the x-values that make the denominator equal to zero, $d(x) = 0$.
- **Vertical Asymptotes:** Vertical lines, $x = a$, where $d(a) = a$, after all common factors have been divided out. Vertical asymptotes steer the graph and are never touched.
- **Horizontal Asymptotes:** Horizontal lines, $y = b$, that steer the graph as $x \to \pm\infty$.
  1. If degree of the numerator $<$ degree of the denominator, then $y = 0$ is a horizontal asymptote.
  2. If degree of the numerator $=$ degree of the denominator, then $y = c$ is a horizontal asymptote where c is the ratio of the leading coefficients of the numerator and denominator, respectively.
  3. If degree of the numerator $>$ degree of the denominator, then there is no horizontal asymptote.
- **Slant Asymptotes:** Slant lines, $y = mx + b$, that steer the graph as $x \to \pm\infty$.

1. If degree of the numerator $-$ degree of the denominator $= 1$, then there is a slant asymptote.
2. Divide the numerator by the denominator. The quotient corresponds to the equation of the line (slant asymptote).

Procedure for Graphing Rational Functions

1. Find the domain of the function.
2. Find the intercept(s).
   - y-intercept
   - x-intercepts (if any)
3. Find any holes.
   - If $x - a$ is a common factor of the numerator and denominator, then $x = a$ corresponds to a hole in the graph of the rational function if the multiplicity of a in the numerator is greater than or equal to the multiplicity of a in the denominator. The result after the common factor is canceled is an equivalent rational function in lowest terms (no common factor).
4. Find any asymptotes.
   - Vertical asymptotes
   - Horizontal/slant asymptotes
5. Find additional points on the graph.
6. Sketch the graph: draw the asymptotes and label the intercepts and points and connect with a smooth curve.

SECTION
4.6 EXERCISES

▪ SKILLS

In Exercises 1–10, find the domain of each rational function.

1. $f(x) = \dfrac{1}{x + 3}$

2. $f(x) = \dfrac{3}{4 - x}$

3. $f(x) = \dfrac{2x + 1}{(3x + 1)(2x - 1)}$

4. $f(x) = \dfrac{5 - 3x}{(2 - 3x)(x - 7)}$

5. $f(x) = \dfrac{x + 4}{x^2 + x - 12}$

6. $f(x) = \dfrac{x - 1}{x^2 + 2x - 3}$

7. $f(x) = \dfrac{7x}{x^2 + 16}$

8. $f(x) = -\dfrac{2x}{x^2 + 9}$

9. $f(x) = -\dfrac{3(x^2 + x - 2)}{2(x^2 - x - 6)}$

10. $f(x) = \dfrac{5(x^2 - 2x - 3)}{(x^2 - x - 6)}$

In Exercises 11–20, find all vertical asymptotes and horizontal asymptotes (if there are any).

11. $f(x) = \dfrac{1}{x + 2}$

12. $f(x) = \dfrac{1}{5 - x}$

13. $f(x) = \dfrac{7x^3 + 1}{x + 5}$

14. $f(x) = \dfrac{2 - x^3}{2x - 7}$

15. $f(x) = \dfrac{6x^5 - 4x^2 + 5}{6x^2 + 5x - 4}$

16. $f(x) = \dfrac{6x^2 + 3x + 1}{3x^2 - 5x - 2}$

17. $f(x) = \dfrac{\frac{1}{3}x^2 + \frac{1}{3}x - \frac{1}{4}}{x^2 + \frac{1}{9}}$

18. $f(x) = \dfrac{\frac{1}{10}(x^2 - 2x + \frac{3}{10})}{2x - 1}$

19. $f(x) = \dfrac{(0.2x - 3.1)(1.2x + 4.5)}{0.7(x - 0.5)(0.2x + 0.3)}$

20. $f(x) = \dfrac{0.8x^4 - 1}{x^2 - 0.25}$

In Exercises 21–26, find the slant asymptote corresponding to the graph of each rational function.

21. $f(x) = \dfrac{x^2 + 10x + 25}{x + 4}$

22. $f(x) = \dfrac{x^2 + 9x + 20}{x - 3}$

23. $f(x) = \dfrac{2x^2 + 14x + 7}{x - 5}$

24. $f(x) = \dfrac{3x^3 + 4x^2 - 6x + 1}{x^2 - x - 30}$

25. $f(x) = \dfrac{8x^4 + 7x^3 + 2x - 5}{2x^3 - x^2 + 3x - 1}$

26. $f(x) = \dfrac{2x^6 + 1}{x^5 - 1}$

In Exercises 27–32, match the function to the graph.

27. $f(x) = \dfrac{3}{x - 4}$

28. $f(x) = \dfrac{3x}{x - 4}$

29. $f(x) = \dfrac{3x^2}{x^2 - 4}$

30. $f(x) = -\dfrac{3x^2}{x^2 + 4}$

31. $f(x) = \dfrac{3x^2}{4 - x^2}$

32. $f(x) = \dfrac{3x^2}{x + 4}$

a.

b.

c.

d.

e.

f.

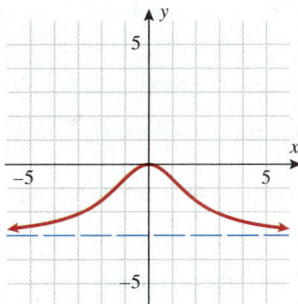

In Exercises 33–58, use the graphing strategy outlined in this section to graph the rational functions.

33. $f(x) = \dfrac{2}{x + 1}$

34. $f(x) = \dfrac{4}{x - 2}$

35. $f(x) = \dfrac{2x}{x - 1}$

36. $f(x) = \dfrac{4x}{x + 2}$

37. $f(x) = \dfrac{x - 1}{x}$

38. $f(x) = \dfrac{2 + x}{x - 1}$

39. $f(x) = \dfrac{2(x^2 - 2x - 3)}{x^2 + 2x}$

40. $f(x) = \dfrac{3(x^2 - 1)}{x^2 - 3x}$

41. $f(x) = \dfrac{x^2}{x + 1}$

42. $f(x) = \dfrac{x^2 - 9}{x + 2}$

43. $f(x) = \dfrac{2x^3 - x^2 - x}{x^2 - 4}$

44. $f(x) = \dfrac{3x^3 + 5x^2 - 2x}{x^2 + 4}$

45. $f(x) = \dfrac{x^2 + 1}{x^2 - 1}$

46. $f(x) = \dfrac{1 - x^2}{x^2 + 1}$

47. $f(x) = \dfrac{7x^2}{(2x + 1)^2}$

48. $f(x) = \dfrac{12x^4}{(3x + 1)^4}$

49. $f(x) = \dfrac{1 - 9x^2}{(1 - 4x^2)^3}$

50. $f(x) = \dfrac{25x^2 - 1}{(16x^2 - 1)^2}$

51. $f(x) = 3x + \dfrac{4}{x}$

52. $f(x) = x - \dfrac{4}{x}$

53. $f(x) = \dfrac{(x - 1)^2}{(x^2 - 1)}$

54. $f(x) = \dfrac{(x + 1)^2}{(x^2 - 1)}$

55. $f(x) = \dfrac{(x - 1)(x^2 - 4)}{(x - 2)(x^2 + 1)}$

56. $f(x) = \dfrac{(x - 1)(x^2 - 9)}{(x - 3)(x^2 + 1)}$

57. $f(x) = \dfrac{3x(x - 1)}{x(x^2 - 4)}$

58. $f(x) = \dfrac{-2x(x - 3)}{x(x^2 + 1)}$

In Exercises 59–62, for each graph of the rational function given determine: (a) all intercepts, (b) all asymptotes, and (c) equation of the rational function.

59.

60.

61.

62.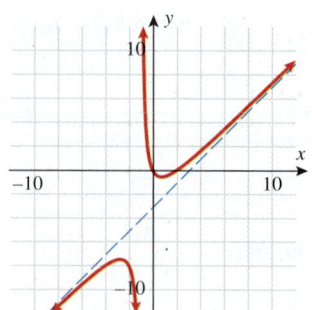

■ **APPLICATIONS**

63. Medicine. The concentration C of a particular drug in a person's bloodstream t minutes after injection is given by

$$C(t) = \frac{2t}{t^2 + 100}$$

a. What is the concentration in the bloodstream after 1 minute?

b. What is the concentration in the bloodstream after 1 hour?

c. What is the concentration in the bloodstream after 5 hours?

d. Find the horizontal asymptote of $C(t)$. What do you expect the concentration to be after several days?

64. Medicine. The concentration C of aspirin in the bloodstream t hours after consumption is given by $C(t) = \dfrac{t}{t^2 + 40}$.

a. What is the concentration in the bloodstream after $\frac{1}{2}$ hour?

b. What is the concentration in the bloodstream after 1 hour?

c. What is the concentration in the bloodstream after 4 hours?

d. Find the horizontal asymptote for $C(t)$. What do you expect the concentration to be after several days?

65. Typing. An administrative assistant is hired after graduating from high school and learns to type on the job. The number of words he can type per minute is given by

$$N(t) = \frac{130t + 260}{t + 5} \qquad t \geq 0$$

where t is the number of months he has been on the job.

a. How many words per minute can he type the day he starts?

b. How many words per minute can he type after 12 months?

c. How many words per minute can he type after 3 years?

d. How many words per minute would you expect him to type if he worked there until he retired?

66. Memorization. A professor teaching a large lecture course tries to learn students' names. The number of names she can remember $N(t)$ increases with each week in the semester t and is given by the rational function:

$$N(t) = \frac{600t}{t + 20}$$

How many students' names does she know by the third week in the semester? How many students' names should she know by the end of the semester (16 weeks)? According to this function, what are the most names she can remember?

67. Food. The amount of food that cats typically eat increases as their weight increases. A rational function that describes this is $F(x) = \dfrac{10x^2}{x^2 + 4}$, where the amount of food $F(x)$ is given in ounces and the weight of the cat x is given in pounds. Calculate the horizontal asymptote. How many ounces of food will most adult cats eat?

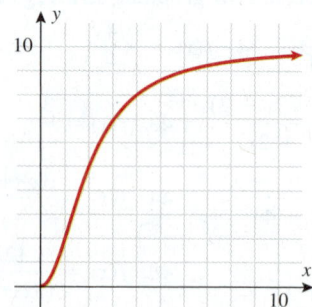

68. Memorization. The *Guinness Book of World Records, 2004* states that Dominic O'Brien (England) memorized on a single sighting a random sequence of 54 separate packs of cards all shuffled together (2808 cards in total) at Simpson's-In-The-Strand, London, England, on May 1, 2002. He memorized the cards in 11 hours 42 minutes, and then recited them in exact sequence in a time of 3 hours 30 minutes. With only a 0.5% margin of error allowed (no more than 14 errors), he broke the record with just 8 errors. If we let x represent the time (hours) it takes to memorize the cards and y represent the number of cards memorized, then a rational function that models this event is given by $y = \dfrac{2800x^2 + x}{x^2 + 2}$. According to this model, how many cards could be memorized in an hour? What is the greatest number of cards that can be memorized?

69. Gardening. A 500-square-foot rectangular garden will be enclosed with fencing. Write a rational function that describes how many linear feet of fence will be needed to enclose the garden as a function of the width of the garden w.

70. Geometry. A rectangular picture has an area of 414 square inches. A border (matting) is used when framing. If the top and bottom borders are each 4 inches and the side borders are 3.5 inches, write a function that represents the area $A(l)$ of the entire frame as a function of the length of the picture l.

For Exercises 71 and 72, refer to the following:

The monthly profit function for a product is given by

$$P(x) = -x^3 + 10x^2$$

where x is the number of units sold measured in thousands and P is profit measured in thousands of dollars. The average profit, which represents the profit per thousand units sold, for this product is given by

$$P(x) = \frac{-x^3 + 10x^2}{x}$$

where x is units sold measured in thousands and P is profit measured in thousands of dollars.

71. Business. Find the number of units that must be sold to produce an average profit of $16,000 per thousand units. Convert the answer to average profit in dollars per unit.

72. Business. Find the number of units that must be sold to produce an average profit of $25,000 per thousand units. Convert the answer to average profit in dollars per unit.

For Exercises 73 and 74, refer to the following:

Some medications, such as Synthroid, are prescribed as a maintenance drug because they are taken regularly for an ongoing condition, such as hypothyroidism. Maintenance drugs function by maintaining a therapeutic drug level in the bloodstream over time. The concentration of a maintenance drug over a 24-hour period is modeled by the function

$$C(t) = \frac{22(t - 1)}{t^2 + 1} + 24$$

where t is time in hours after the dose was administered and C is the concentration of the drug in the bloodstream measured in μg/mL. This medication is designed to maintain a consistent concentration in the bloodstream of approximately 25 μg/mL. *Note:* This drug will become inert; that is, the concentration will drop to 0 μg/mL, during the 25th hour after taking the medication.

73. Health/Medicine. Find the concentration of the drug, to the nearest tenth of μg/mL, in the bloodstream 15 hours after the dose is administered. Is this the only time this concentration of the drug is found in the bloodstream? At what other times is this concentration reached? Round to the nearest hour. Discuss the significance of this answer.

74. Health/Medicine. Find the time, after the first hour and a half, at which the concentration of the drug in the bloodstream has dropped to 25 μg/mL. Find the concentration of the drug 24 hours after taking a dose to the nearest tenth of a μg/mL. Discuss the importance of taking the medication every 24 hours rather than every day.

▪CATCH THE MISTAKE

In Exercises 75–78, explain the mistake that is made.

75. Determine the vertical asymptotes of the function

$$f(x) = \frac{x - 1}{x^2 - 1}.$$

Solution:

Set the denominator equal to zero. $\qquad x^2 - 1 = 0$

Solve for x. $\qquad x = \pm 1$

The vertical asymptotes are $x = -1$ and $x = 1$.

The following is a correct graph of the function. Note that only $x = -1$ is an asymptote. What went wrong?

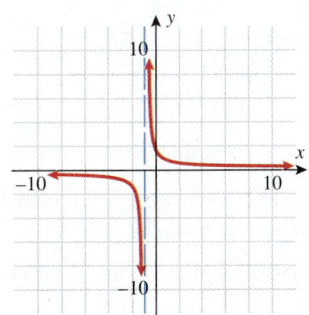

76. Determine the vertical asymptotes of $f(x) = \dfrac{2x}{x^2 + 1}$.

Solution:

Set the denominator equal to zero.　　　$x^2 + 1 = 0$

Solve for x.　　　　　　　　　　　$x = \pm 1$

The vertical asymptotes are $x = -1$ and $x = 1$.

The following is a correct graph of the function. Note that there are no vertical asymptotes. What went wrong?

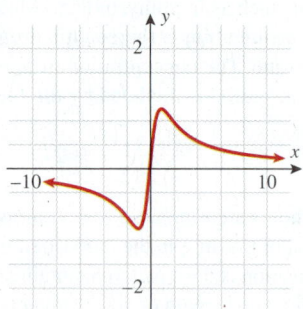

77. Determine whether a horizontal or a slant asymptote exists for the function $f(x) = \dfrac{9 - x^2}{x^2 - 1}$. If one does, find it.

Solution:

Step 1: The degree of the numerator equals the degree of the denominator, so there is a horizontal asymptote.

Step 2: The horizontal asymptote is the ratio of the lead coefficients: $y = \dfrac{9}{1} = 9$.

The horizontal asymptote is $y = 9$.

The following is a correct graph of the function. Note that there is no horizontal asymptote at $y = 9$. What went wrong?

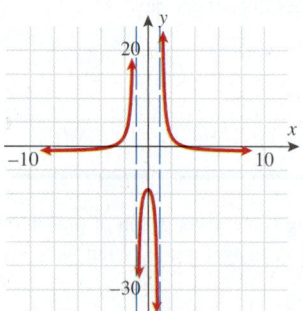

78. Determine whether a horizontal or a slant asymptote exists for the function $f(x) = \dfrac{x^2 + 2x - 1}{3x^3 - 2x^2 - 1}$. If one does, find it.

Solution:

Step 1: The degree of the denominator is exactly one more than the degree of the numerator, so there is a slant asymptote.

Step 2: Divide.

$$
\begin{array}{r}
3x - 8 \\
x^2 + 2x - 1 \overline{)\, 3x^3 - 2x^2 + 0x - 1} \\
\underline{3x^3 + 6x^2 - 3x } \\
-8x^2 + 3x - 1 \\
\underline{8x^2 - 16x + 8} \\
19x - 9
\end{array}
$$

The slant asymptote is $y = 3x - 8$.

The following is the correct graph of the function. Note that $y = 3x - 8$ is not an asymptote. What went wrong?

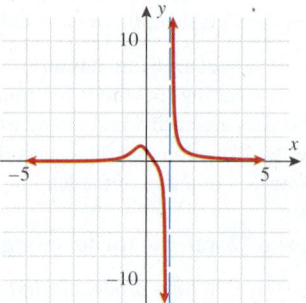

■ CONCEPTUAL

For Exercises 79–82, determine whether each statement is true or false.

79. A rational function can have either a horizontal asymptote or a slant asymptote, but not both.

80. A rational function can have at most one vertical asymptote.

81. A rational function can cross a vertical asymptote.

82. A rational function can cross a horizontal or a slant asymptote.

■ CHALLENGE

83. Determine the asymptotes of the rational function

$$f(x) = \frac{(x - a)(x + b)}{(x - c)(x + d)}.$$

84. Determine the asymptotes of the rational function

$$f(x) = \frac{3x^2 + b^2}{x^2 + a^2}.$$

85. Write a rational function that has vertical asymptotes at $x = -3$ and $x = 1$ and a horizontal asymptote at $y = 4$.

86. Write a rational function that has no vertical asymptotes, approaches the x-axis as a horizontal asymptote, and has an x-intercept of $(3, 0)$.

■ TECHNOLOGY

87. Determine the vertical asymptotes of $f(x) = \dfrac{x - 4}{x^2 - 2x - 8}$.

Graph this function utilizing a graphing utility. Does the graph confirm the asymptotes?

88. Determine the vertical asymptotes of $f(x) = \dfrac{2x + 1}{6x^2 + x - 1}$.

Graph this function utilizing a graphing utility. Does the graph confirm the asymptotes?

89. Find the asymptotes and intercepts of the rational function

$f(x) = \dfrac{1}{3x + 1} - \dfrac{2}{x}$. *Note:* Combine the two expressions

into a single rational expression. Graph this function utilizing a graphing utility. Does the graph confirm what you found?

90. Find the asymptotes and intercepts of the rational function

$f(x) = -\dfrac{1}{x^2 + 1} + \dfrac{1}{x}$. *Note:* Combine the two expressions

into a single rational expression. Graph this function utilizing a graphing utility. Does the graph confirm what you found?

For Exercises 91 and 92: (a) Identify all asymptotes for each function. (b) Plot $f(x)$ and $g(x)$ in the same window. How does the end behavior of the function f differ from g? (c) Plot $g(x)$ and $h(x)$ in the same window. How does the end behavior of g differ from h? (d) Combine the two expressions into a single rational expression for the functions g and h. Does the strategy of finding horizontal and slant asymptotes agree with your findings in (b) and (c)?

91. $f(x) = \dfrac{1}{x - 3}, g(x) = 2 + \dfrac{1}{x - 3}, h(x) = -3 + \dfrac{1}{x - 3}$

92. $f(x) = \dfrac{2x}{x^2 - 1}, g(x) = x + \dfrac{2x}{x^2 - 1}, h(x) = x - 3 + \dfrac{2x}{x^2 - 1}$

Discovering the Connection between the Standard Form of a Quadratic Function and Transformations of the Square Function

In Chapter 3, you saw that if you are familiar with the graphs of a small library of common functions, you can sketch the graphs of many related functions using transformation techniques. These ideas will help you here as you discover the relationship between the standard form of a quadratic function and its graph.

Let G and H be functions with:

$$G(x) = F(x - 1) + 3 \quad \text{and} \quad H(x) = -F(x + 2) - 4$$

where $F(x) = x^2$.

1. For this part, consider the function G.

   a. List the transformation you'd use to sketch the graph of G from the graph of F.

   b. Write an equation for $G(x)$ in the form $G(x) = a(x - h)^2 + k$. This is called the **standard form** of a quadratic function. What are the values of a, h, and k?

   c. The **vertex**, or turning point, of the graph of $F(x) = x^2$ is $(0, 0)$. How can you use the transformations you listed in part (a) to determine the coordinates of the vertex of the graph of G?

   d. The vertical line that passes through the vertex of a parabola is called its **axis of symmetry**. The axis of symmetry of the graph of $F(x) = x^2$ is the y-axis, or the vertical line with equation $x = 0$. How can you determine the axis of symmetry of the graph of G? Write the equation of this line.

   e. Sketch graphs of F and G.

2. Next consider the function H given above.

   a. List the transformations that will produce the graph of H from the graph of F.

   b. Write an equation for $H(x)$ in standard form. What are the values of a, h, and k?

   c. What are the coordinates of the vertex of the graph of H? How do the transformations you listed in part (a) help you determine this?

   d. Determine the equation of the axis of symmetry of the graph of H.

   e. Sketch graphs of F and H.

3. a. What do you know about the graph of a quadratic function just by looking at its equation in standard form, $f(x) = a(x - h)^2 + k$?

   b. Shown below are the graphs of $F(x) = x^2$ and another quadratic function, $y = K(x)$. Write the equation of K in standard form. *Hint:* Think about the transformations.

MODELING OUR WORLD

The following table summarizes average yearly temperature in degrees Fahrenheit (°F) and carbon dioxide emissions in parts per million (ppm) for Mauna Loa, Hawaii.

YEAR	1960	1965	1970	1975	1980	1985	1990	1995	2000	2005
TEMPERATURE	44.45	43.29	43.61	43.35	46.66	45.71	45.53	47.53	45.86	46.23
CO$_2$ EMISSIONS (ppm)	316.9	320.0	325.7	331.1	338.7	345.9	354.2	360.6	369.4	379.7

In the Modeling Our World in Chapter 3, the temperature and carbon emissions were modeled with *linear functions*. Now, let us model these same data using *polynomial functions*.

1. Plot the temperature data with time on the horizontal axis and temperature on the vertical axis. Let $t = 0$ correspond to 1960. Adjust the vertical range of the graph to (43, 48). How many turning points (local maxima and minima) do these data exhibit? What is the lowest degree polynomial function whose graph can pass through these data?

2. Find a *polynomial function* that models the temperature in Mauna Loa.
   a. Find a quadratic function: Let the data from 1995 correspond to the vertex of the graph and apply the 2005 data to determine the function.
   b. Find a quadratic function: Let the data from 2000 correspond to the vertex of the graph and apply the 2005 data to determine the function.
   c. Utilize regression and all data given to find a polynomial function whose degree is found in 1.

3. Predict what the temperature will be in Mauna Loa in 2020.
   a. Use the function found in 2a.
   b. Use the function found in 2b.
   c. Use the function found in 2c.

4. Predict what the temperature will be in Mauna Loa in 2100.
   a. Use the function found in 2a.
   b. Use the function found in 2b.
   c. Use the function found in 2c.

5. Do your models support the claim of "global warming"? Explain. Do these models make similar predictions to the linear models found in Chapter 3?

6. Plot the carbon dioxide emissions data with time on the horizontal axis and carbon dioxide emissions on the vertical axis. Let $t = 0$ correspond to 1960. Adjust the vertical range of the graph to (315, 380).

7. Find a *quadratic function* that models the CO$_2$ emissions (ppm) in Mauna Loa.
   a. Let the data from 1960 correspond to the vertex of the graph and apply the 2005 data to determine the function.
   b. Let the data from 1980 correspond to the vertex of the graph and apply the 2005 data to determine the function.
   c. Utilize regression and all data given.

8. Predict the expected CO_2 levels in Mauna Loa in 2020.

   a. Use the function found in 7a.
   b. Use the function found in 7b.
   c. Use the function found in 7c.

9. Predict the expected CO_2 levels in Mauna Loa in 2100.

   a. Use the function found in 7a.
   b. Use the function found in 7b.
   c. Use the function found in 7c.

10. Do your models support the claim of "global warming"? Explain. Do these models give similar predictions to the linear models found in Chapter 3?

11. Discuss differences in models and predictions found in parts (a), (b), and (c) and also discuss differences in linear and polynomial functions.

SECTION	CONCEPT	KEY IDEAS/FORMULAS
4.1	**Quadratic functions**	
	Graphs of quadratic functions: Parabolas	Parabolas
		Graphing quadratic functions in standard form
		$f(x) = a(x - h)^2 + k$
		■ Vertex: (h, k)
		■ Opens Up: $a > 0$
		■ Opens Down: $a < 0$
		Graphing quadratic functions in general form
		$f(x) = ax^2 + bx + c$, vertex is $(h, k) = \left(-\dfrac{b}{2a}, f\left(-\dfrac{b}{2a}\right)\right)$
	Finding the equation of a parabola	
4.2	**Polynomial functions of higher degree**	
	Identifying polynomial functions	$P(x) = a_n x^n + a_{n-1} x^{n-1} + \cdots + a_2 x^2 + a_1 x + a_0$ is a polynomial of degree n.
	Graphing polynomial functions using transformations of power functions	Shift power functions
		$y = x^n$ behave similar to:
		■ $y = x^2$, when n is even.
		■ $y = x^3$, when n is odd.
	Real zeros of a polynomial function	$P(x) = (x - a)(x - b)^n = 0$
		■ a is a zero of multiplicity 1.
		■ b is a zero of multiplicity n.
	Graphing general polynomial functions	Intercepts; zeros and multiplicities; end behavior
4.3	**Dividing polynomials: Long division and synthetic division**	Use zero placeholders for missing terms.
	Long division of polynomials	Can be used for all polynomial division.
	Synthetic division of polynomials	Can only be used when dividing by $(x \pm a)$.
4.4	**The real zeros of a polynomial function**	$P(x) = a_n x^n + a_{n-1} x^{n-1} + \cdots + a_2 x^2 + a_1 x + a_0$
		If $P(c) = 0$, then c is a zero of $P(x)$.
	The remainder theorem and the factor theorem	If $P(x)$ is divided by $x - a$, then the remainder r is $r = P(a)$.
	The rational zero theorem and Descartes' rule of signs	Possible zeros $= \dfrac{\text{Factors of } a_0}{\text{Factors of } a_n}$
		Number of positive or negative real zeros is related to the number of sign variations in $P(x)$ or $P(-x)$.

SECTION	CONCEPT	KEY IDEAS/FORMULAS
	Factoring polynomials	1. List possible rational zeros (rational zero theorem).
		2. List possible combinations of positive and negative real zeros (Descartes' rule of signs).
		3. Test possible values until a zero is found.
		4. Once a real zero is found, use synthetic division. Then repeat testing on quotient until linear and/or irreducible quadratic factors are reached.
		5. If there is a real zero but all possible rational roots have failed, then approximate the real zero using the intermediate value theorem/bisection method.
	The intermediate value theorem	The intermediate value theorem and the bisection method are used to approximate irrational zeros.
	Graphing polynomial functions	1. Find the intercepts.
		2. Determine end behavior.
		3. Find additional points.
		4. Sketch a smooth curve.
4.5	**Complex zeros: The fundamental theorem of algebra**	$P(x) = a_n x^n + a_{n-1} x^{n-1} + \cdots + a_2 x^2 + a_1 x + a_0$ $P(x) = \underbrace{(x - c_1)(x - c_2) \cdots (x - c_n)}_{n \text{ factors}}$ where the c's represent complex (not necessarily distinct) zeros.
	The fundamental theorem of algebra	▪ $P(x)$ of degree n has at least one zero and at most n zeros.
	Complex zeros	Complex conjugate pairs: ▪ If $a + bi$ is a zero of $P(x)$, then $a - bi$ is also a zero.
	Factoring polynomials	The polynomial can be written as a product of linear factors, not necessarily distinct.
4.6	**Rational functions**	$f(x) = \dfrac{n(x)}{d(x)} \qquad d(x) \neq 0$
	Domain of rational functions	**Domain:** All real numbers except x-values that make the denominator equal to zero; that is, $d(x) = 0$. A rational function $f(x) = \dfrac{n(x)}{d(x)}$ is said to be in *lowest terms* if $n(x)$ and $d(x)$ have no common factors.
	Horizontal, vertical, and slant asymptotes	A rational function that has a common factor $x - a$ in both the numerator and denominator has a hole at $x = a$ in its graph if the multiplicity of a in the numerator is greater than or equal to the multiplicity of a in the denominator. *Vertical Asymptotes* A rational function in lowest terms has a vertical asymptote corresponding to any x-values that make the denominator equal to zero.

Horizontal Asymptotes
- $y = 0$ if degree of $n(x)$ < degree of $d(x)$.
- No horizontal asymptote if degree of $n(x)$ > degree of $d(x)$.
- $y = \dfrac{\text{Leading coefficient of } n(x)}{\text{Leading coefficient of } d(x)}$

  if degree of $n(x)$ = degree of $d(x)$.

Slant Asymptotes

If degree of $n(x)$ − degree of $d(x) = 1$.

Divide $n(x)$ by $d(x)$ and the quotient determines the slant asymptote; that is, $y =$ quotient.

Graphing rational functions

1. Find the domain of the function.
2. Find the intercept(s).
3. Find any holes.
4. Find any asymptotes.
5. Find additional points on the graph.
6. *Sketch the graph:* Draw the asymptotes and label the intercepts and points and connect with a smooth curve.

CHAPTER REVIEW

4.1 Quadratic Functions

Match the quadratic function with its graph.

1. $f(x) = -2(x + 6)^2 + 3$ 2. $f(x) = \frac{1}{4}(x - 4)^2 + 2$

3. $f(x) = x^2 + x - 6$ 4. $f(x) = -3x^2 - 10x + 8$

a.

b.

c.

d.

Graph the quadratic function given in standard form.

5. $f(x) = -(x - 7)^2 + 4$ 6. $f(x) = (x + 3)^2 - 5$

7. $f(x) = -\frac{1}{2}\left(x - \frac{1}{3}\right)^2 + \frac{2}{5}$ 8. $f(x) = 0.6(x - 0.75)^2 + 0.5$

Rewrite the quadratic function in standard form by completing the square.

9. $f(x) = x^2 - 3x - 10$ 10. $f(x) = x^2 - 2x - 24$

11. $f(x) = 4x^2 + 8x - 7$ 12. $f(x) = -\frac{1}{4}x^2 + 2x - 4$

Graph the quadratic function given in general form.

13. $f(x) = x^2 - 3x + 5$

14. $f(x) = -x^2 + 4x + 2$

15. $f(x) = -4x^2 + 2x + 3$

16. $f(x) = -0.75x^2 + 2.5$

Find the vertex of the parabola associated with each quadratic function.

17. $f(x) = 13x^2 - 5x + 12$

18. $f(x) = \frac{2}{5}x^2 - 4x + 3$

19. $f(x) = -0.45x^2 - 0.12x + 3.6$

20. $f(x) = -\frac{3}{4}x^2 + \frac{2}{5}x + 4$

Find the quadratic function that has the given vertex and goes through the given point.

21. vertex: $(-2, 3)$ point: $(1, 4)$

22. vertex: $(4, 7)$ point: $(-3, 1)$

23. vertex: $(2.7, 3.4)$ point: $(3.2, 4.8)$

24. vertex: $\left(-\frac{5}{2}, \frac{7}{4}\right)$ point: $\left(\frac{1}{2}, \frac{3}{5}\right)$

Applications

25. **Profit.** The revenue and the cost of a local business are given below as functions of the number of units x in thousands produced and sold. Use the cost and the revenue to answer the questions that follow.

$$C(x) = \frac{1}{3}x + 2 \quad \text{and} \quad R(x) = -2x^2 + 12x - 12$$

   a. Determine the profit function.

   b. State the break-even points.

   c. Graph the profit function.

   d. What is the range of units to make and sell that will correspond to a profit?

26. **Geometry.** Given the length of a rectangle is $2x - 4$ and the width is $x + 7$, find the area of the rectangle. What dimensions correspond to the largest area?

27. **Geometry.** A triangle has a base of $x + 2$ units and a height of $4 - x$ units. Determine the area of the triangle. What dimensions correspond to the largest area?

28. **Geometry.** A person standing at a ridge in the Grand Canyon throws a penny upward and toward the pit of the canyon. The height of the penny is given by the function:

$$h(t) = -12t^2 + 80t$$

   a. What is the maximum height that the penny will reach?

   b. How many seconds will it take the penny to hit the ground below?

4.2 Polynomial Functions of Higher Degree

Determine which functions are polynomials, and for those, state their degree.

29. $f(x) = x^6 - 2x^5 + 3x^2 + 9x - 42$

30. $f(x) = (3x - 4)^3(x + 6)^2$

31. $f(x) = 3x^4 - x^3 + x^2 + \sqrt[4]{x} + 5$

32. $f(x) = 5x^3 - 2x^2 + \dfrac{4x}{7} - 3$

Match the polynomial function with its graph.

33. $f(x) = 2x - 5$

34. $f(x) = -3x^2 + x - 4$

35. $f(x) = x^4 - 2x^3 + x^2 - 6$

36. $f(x) = x^7 - x^5 + 3x^4 + 3x + 7$

a.

b.

c.

d.

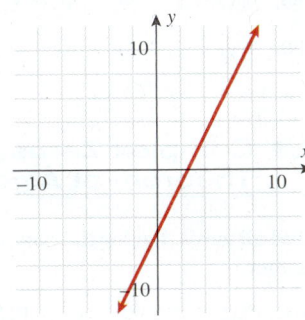

Graph each function by transforming a power function $y = x^n$.

37. $f(x) = -x^7$

38. $f(x) = (x - 3)^3$

39. $f(x) = x^4 - 2$

40. $f(x) = -6 - (x + 7)^5$

Find all the real zeros of each polynomial function, and state their multiplicities.

41. $f(x) = 3(x + 4)^2(x - 6)^5$

42. $f(x) = 7x(2x - 4)^3(x + 5)$

43. $f(x) = x^5 - 13x^3 + 36x$

44. $f(x) = 4.2x^4 - 2.6x^2$

Find a polynomial of minimum degree that has the given zeros.

45. $-3, 0, 4$

46. $2, 4, 6, -8$

47. $-\frac{2}{5}, \frac{3}{4}, 0$

48. $2 - \sqrt{5}, 2 + \sqrt{5}$

49. -2 (multiplicity of 2), 3 (multiplicity of 2)

50. 3 (multiplicity of 2), -1 (multiplicity of 2), 0 (multiplicity of 3)

For each polynomial function given: (a) list each real zero and its multiplicity; (b) determine whether the graph touches or crosses at each x-intercept; (c) find the y-intercept and a few points on the graph; (d) determine the end behavior; and (e) sketch the graph.

51. $f(x) = x^2 - 5x - 14$

52. $f(x) = -(x - 5)^5$

53. $f(x) = 6x^7 + 3x^5 - x^2 + x - 4$

54. $f(x) = -x^4(3x + 6)^3(x - 7)^3$

Applications

55. Salary. Tiffany has started tutoring students x hours per week. The tutoring job corresponds to the following additional income:

$$f(x) = (x - 1)(x - 3)(x - 7)$$

a. Graph the polynomial function.

b. Give any real zeros that occur.

c. How many hours of tutoring are financially beneficial to Tiffany?

56. Profit. The following function is the profit for Walt Disney World, where $P(x)$ represents profit in millions of dollars and x represents the month ($x = 1$ corresponds to January):

$$P(x) = 3(x - 2)^2(x - 5)^2(x - 10)^2 \qquad 1 \le x \le 12$$

Graph the polynomial. When are the peak seasons?

4.3 Dividing Polynomials: Long Division and Synthetic Division

Divide the polynomials with long division. If you choose to use a calculator, do not round off. Keep the exact values instead. Express the answer in the form $Q(x) = ?, \quad r(x) = ?$.

57. $(x^2 + 2x - 6) \div (x - 2)$

58. $(2x^2 - 5x - 1) \div (2x - 3)$

59. $(4x^4 - 16x^3 + x - 9 + 12x^2) \div (2x - 4)$

60. $(6x^2 + 2x^3 - 4x^4 + 2 - x) \div (2x^2 + x - 4)$

Use synthetic division to divide the polynomial by the linear factor. Indicate the quotient $Q(x)$ and the remainder $r(x)$.

61. $(x^4 + 4x^3 + 5x^2 - 2x - 8) \div (x + 2)$

62. $(x^3 - 10x + 3) \div (2 + x)$

63. $(x^6 - 64) \div (x + 8)$

64. $(2x^5 + 4x^4 - 2x^3 + 7x + 5) \div \left(x - \frac{3}{4}\right)$

Divide the polynomials with either long division or synthetic division.

65. $(5x^3 + 8x^2 - 22x + 1) \div (5x^2 - 7x + 3)$

66. $(x^4 + 2x^3 - 5x^2 + 4x + 2) \div (x - 3)$

67. $(x^3 - 4x^2 + 2x - 8) \div (x + 1)$

68. $(x^3 - 5x^2 + 4x - 20) \div (x^2 + 4)$

Applications

69. Geometry. The area of a rectangle is given by the polynomial $6x^4 - 8x^3 - 10x^2 + 12x - 16$. If the width is $2x - 4$, what is the length of the rectangle?

70. Volume. A 10 inch by 15 inch rectangular piece of cardboard is used to make a box. Square pieces x inches on a side are cut out from the corners of the cardboard, and then the sides are folded up. Find the volume of the box.

4.4 The Real Zeros of a Polynomial Function

Find the following values by applying synthetic division. Check by substituting the value into the function.

$$f(x) = 6x^5 + x^4 - 7x^2 + x - 1 \qquad g(x) = x^3 + 2x^2 - 3$$

71. $f(-2)$ **72.** $f(1)$ **73.** $g(1)$ **74.** $g(-1)$

Determine whether the number given is a zero of the polynomial.

75. -3, $P(x) = x^3 - 5x^2 + 4x + 2$

76. 2 and -2, $P(x) = x^4 - 16$

77. 1, $P(x) = 2x^4 - 2x$

78. 4, $P(x) = x^4 - 2x^3 - 8x$

Given a zero of the polynomial, determine all other real zeros, and write the polynomial in terms of a product of linear or irreducible factors.

Polynomial	Zero
79. $P(x) = x^4 - 6x^3 + 32x$	-2
80. $P(x) = x^3 - 7x^2 + 36$	3
81. $P(x) = x^5 - x^4 - 8x^3 + 12x^2$	0
82. $P(x) = x^4 - 32x^2 - 144$	6

Use Descartes' rule of signs to determine the possible number of positive real zeros and negative real zeros.

83. $P(x) = x^4 + 3x^3 - 16$

84. $P(x) = x^5 + 6x^3 - 4x - 2$

85. $P(x) = x^9 - 2x^7 + x^4 - 3x^3 + 2x - 1$

86. $P(x) = 2x^5 - 4x^3 + 2x^2 - 7$

Use the rational zero theorem to list the possible rational zeros.

87. $P(x) = x^3 - 2x^2 + 4x + 6$

88. $P(x) = x^5 - 4x^3 + 2x^2 - 4x - 8$

89. $P(x) = 2x^4 + 2x^3 - 36x^2 - 32x + 64$

90. $P(x) = -4x^5 - 5x^3 + 4x + 2$

List the possible rational zeros, and test to determine all rational zeros.

91. $P(x) = 2x^3 - 5x^2 + 1$

92. $P(x) = 12x^3 + 8x^2 - 13x + 3$

93. $P(x) = x^4 - 5x^3 + 20x - 16$

94. $P(x) = 24x^4 - 4x^3 - 10x^2 + 3x - 2$

For each polynomial: (a) use Descartes' rule of signs to determine the possible combinations of positive real zeros and negative real zeros; (b) use the rational zero test to determine possible rational zeros; (c) use the upper and lower bound rules to eliminate possible zeros; (d) test for rational zeros; (e) factor as a product of linear and/or irreducible quadratic factors; and (f) graph the polynomial function.

95. $P(x) = x^3 + 3x - 5$

96. $P(x) = x^3 + 3x^2 - 6x - 8$

97. $P(x) = x^3 - 9x^2 + 20x - 12$

98. $P(x) = x^4 - x^3 - 7x^2 + x + 6$

99. $P(x) = x^4 - 5x^3 - 10x^2 + 20x + 24$

100. $P(x) = x^5 - 3x^3 - 6x^2 + 8x$

4.5 Complex Zeros: The Fundamental Theorem of Algebra

Find all zeros. Factor the polynomial as a product of linear factors.

101. $P(x) = x^2 + 25$ **102.** $P(x) = x^2 + 16$

103. $P(x) = x^2 - 2x + 5$ **104.** $P(x) = x^2 + 4x + 5$

A polynomial function is described. Find all remaining zeros.

105. Degree: 4   Zeros: $-2i$, $3 + i$

106. Degree: 4   Zeros: $3i$, $2 - i$

107. Degree: 6   Zeros: i, $2 - i$ (multiplicity 2)

108. Degree: 6   Zeros: $2i$, $1 - i$ (multiplicity 2)

Given a zero of the polynomial, determine all other zeros (real and complex) and write the polynomial in terms of a product of linear factors.

Polynomial	Zero
109. $P(x) = x^4 - 3x^3 - 3x^2 - 3x - 4$	i
110. $P(x) = x^4 - 4x^3 + x^2 + 16x - 20$	$2 - i$
111. $P(x) = x^4 - 2x^3 + 11x^2 - 18x + 18$	$-3i$
112. $P(x) = x^4 - 5x^2 + 10x - 6$	$1 + i$

Factor each polynomial as a product of linear factors.

113. $P(x) = x^4 - 81$

114. $P(x) = x^3 - 6x^2 + 12x$

115. $P(x) = x^3 - x^2 + 4x - 4$

116. $P(x) = x^4 - 5x^3 + 12x^2 - 2x - 20$

4.6 Rational Functions

Determine the vertical, horizontal, or slant asymptotes (if they exist) for the following rational functions.

117. $f(x) = \dfrac{7 - x}{x + 2}$

118. $f(x) = \dfrac{2 - x^2}{(x - 1)^3}$

119. $f(x) = \dfrac{4x^2}{x + 1}$

120. $f(x) = \dfrac{3x^2}{x^2 + 9}$

121. $f(x) = \dfrac{2x^2 - 3x + 1}{x^2 + 4}$

122. $f(x) = \dfrac{-2x^2 + 3x + 5}{x + 5}$

Graph the rational functions.

123. $f(x) = -\dfrac{2}{x - 3}$

124. $f(x) = \dfrac{5}{x + 1}$

125. $f(x) = \dfrac{x^2}{x^2 + 4}$

126. $f(x) = \dfrac{x^2 - 36}{x^2 + 25}$

127. $f(x) = \dfrac{x^2 - 49}{x + 7}$

128. $f(x) = \dfrac{2x^2 - 3x - 2}{2x^2 - 5x - 3}$

Technology

Section 4.1

129. On a graphing calculator, plot the quadratic function:

$$f(x) = 0.005x^2 - 4.8x - 59$$

   a. Identify the vertex of this parabola.

   b. Identify the y-intercept.

   c. Identify the x-intercepts (if any).

   d. What is the axis of symmetry?

130. Determine the quadratic function whose vertex is $(2.4, -3.1)$ and passes through the point $(0, 5.54)$.

   a. Write the quadratic function in general form.

   b. Plot this quadratic function with a graphing calculator.

   c. Zoom in on the vertex and y-intercept. Do they agree with the given values?

Section 4.2

In Exercises 131 and 132, use a graphing calculator or a computer to graph each polynomial. From the graph, estimate the x-intercepts and state the zeros of the function and their multiplicities.

131. $f(x) = 5x^3 - 11x^2 - 10.4x + 5.6$

132. $f(x) = -x^3 - 0.9x^2 + 2.16x - 2.16$

Section 4.3

133. Plot $\dfrac{15x^3 - 47x^2 + 38x - 8}{3x^2 - 7x + 2}$. What type of function is it? Perform this division using long division, and confirm that the graph corresponds to the quotient.

134. Plot $\dfrac{-4x^3 + 14x^2 - x - 15}{x - 3}$. What type of function is it? Perform this division using synthetic division, and confirm that the graph corresponds to the quotient.

Section 4.4

In Exercises 135 and 136: (a) From the list of all possible rational zeros of the polynomial, use a graphing calculator or software to graph $P(x)$ to find the rational zeros. (b) Factor as a product of linear and/or irreducible quadratic factors.

135. $P(x) = x^4 - 3x^3 - 12x^2 + 20x + 48$

136. $P(x) = -5x^5 - 18x^4 - 32x^3 - 24x^2 + x + 6$

Section 4.5

Find all zeros (real and complex). Factor the polynomial as a product of linear factors.

137. $P(x) = 2x^3 + x^2 - 2x - 91$

138. $P(x) = -2x^4 + 5x^3 + 37x^2 - 160x + 150$

Section 4.6

In Exercises 139 and 140: (a) graph the function $f(x)$ utilizing a graphing utility to determine if it is a one-to-one function; (b) if it is, find its inverse; and (c) graph both functions in the same viewing window.

139. $f(x) = \dfrac{2x - 3}{x + 1}$

140. $f(x) = \dfrac{4x + 7}{x - 2}$

1. Graph the parabola $y = -(x - 4)^2 + 1$.

2. Write the parabola in standard form $y = -x^2 + 4x - 1$.

3. Find the vertex of the parabola $f(x) = -\frac{1}{2}x^2 + 3x - 4$.

4. Find a quadratic function whose vertex is $(-3, -1)$ and whose graph passes through the point $(-4, 1)$.

5. Find a sixth-degree polynomial function with the given zeros:

   2 of multiplicity 3   1 of multiplicity 2   0 of multiplicity 1

6. For the polynomial function $f(x) = x^4 + 6x^3 - 7x$:

   a. List each real zero and its multiplicity.

   b. Determine whether the graph touches or crosses at each x-intercept.

   c. Find the y-intercept and a few points on the graph.

   d. Determine the end behavior.

   e. Sketch the graph.

7. Divide $-4x^4 + 2x^3 - 7x^2 + 5x - 2$ by $2x^2 - 3x + 1$.

8. Divide $17x^5 - 4x^3 + 2x - 10$ by $x + 2$.

9. Is $x - 3$ a factor of $x^4 + x^3 - 13x^2 - x + 12$?

10. Determine whether -1 is a zero of $P(x) = x^{21} - 2x^{18} + 5x^{12} + 7x^3 + 3x^2 + 2$.

11. Given that $x - 7$ is a factor of $P(x) = x^3 - 6x^2 - 9x + 14$, factor the polynomial in terms of linear factors.

12. Given that $3i$ is a zero of $P(x) = x^4 - 3x^3 + 19x^2 - 27x + 90$, find all other zeros.

13. Can a polynomial have zeros that are not x-intercepts? Explain.

14. Apply Descartes' rule of signs to determine the possible combinations of positive real zeros, negative real zeros, and complex zeros of $P(x) = 3x^5 + 2x^4 - 3x^3 + 2x^2 - x + 1$.

15. From the rational zero test, list all possible rational zeros of $P(x) = 3x^4 - 7x^2 + 3x + 12$.

In Exercises 16–18, determine all zeros of the polynomial function and graph.

16. $P(x) = -x^3 + 4x$

17. $P(x) = 2x^3 - 3x^2 + 8x - 12$

18. $P(x) = x^4 - 6x^3 + 10x^2 - 6x + 9$

19. **Sports.** A football player shows up in August at 300 pounds. After 2 weeks of practice in the hot sun, he is down to 285 pounds. Ten weeks into the season he is up to 315 pounds because of weight training. In the spring he does not work out, and he is back to 300 pounds by the next August. Plot these points on a graph. What degree polynomial could this be?

20. **Profit.** The profit of a company is governed by the polynomial $P(x) = x^3 - 13x^2 + 47x - 35$, where x is the number of units sold in thousands. How many units does the company have to sell to break even?

21. **Interest Rate.** The interest rate for a 30-year fixed mortgage fluctuates with the economy. In 1970, the mortgage interest rate was 8%, and in 1988 it peaked at 13%. In 2002, it dipped down to 4%, and in 2005, it was up to 6%. What is the lowest degree polynomial that can represent this function?

In Exercises 22–25, determine (if any) the:

   a. x- and y-intercepts

   b. vertical asymptotes

   c. horizontal asymptotes

   d. slant asymptotes

   e. graph

22. $f(x) = \dfrac{2x - 9}{x + 3}$

23. $g(x) = \dfrac{x}{x^2 - 4}$

24. $h(x) = \dfrac{3x^3 - 3}{x^2 - 4}$

25. $F(x) = \dfrac{x - 3}{x^2 - 2x - 8}$

26. **Food.** On eating a sugary snack, the average body almost doubles its glucose level. The percentage increase in glucose level y can be approximated by the rational function $y = \dfrac{25x}{x^2 + 50}$, where x represents the number of minutes after eating the snack. Graph the function.

27. a. Use the calculator commands $\boxed{\text{STAT}}$ $\boxed{\text{QuadReg}}$ to model the data using a quadratic function.

   b. Write the quadratic function in standard form and identify the vertex.

   c. Find the x-intercepts.

   d. Plot this quadratic function with a graphing calculator. Do they agree with the given values?

x	-3	2.2	7.5
y	10.01	-9.75	25.76

28. Find the asymptotes and intercepts of the rational function $f(x) = \dfrac{x(2x - 3)}{x^2 - 3x} + 1$. *Note:* Combine the two expressions into a single rational expression. Graph this function utilizing a graphing utility. Does the graph confirm what you found?

1. Simplify $\dfrac{(5x^{-1}y^{-2})^3}{(-10x^2y^2)^2}$ and express in terms of positive exponents.

2. Factor $2xy - 2x + 3y - 3$.

3. Multiply and simplify $\dfrac{4x^2 - 36}{x^2} \cdot \dfrac{x^3 + 6x^2}{x^2 + 9x + 18}$.

4. Solve for x: $|2x - 5| + 3 > 10$.

5. Austin can mow a lawn in 75 minutes. The next week Stacey mows the same lawn in 60 minutes. How long would it take them to mow the lawn working together?

6. Use the discriminant to determine the number and type of roots: $-4x^2 + 3x + 15 = 0$.

7. Solve and check $\sqrt{16 + x^2} = x + 2$.

8. Apply algebraic tests to determine whether the equation's graph is symmetric with respect to the x-axis, y-axis, or origin for $y = |x| - 3$.

9. Write an equation of the line that is parallel to the line $x - 3y = 8$ and has the point $(4, 1)$.

10. Find the x-intercept and y-intercept and sketch the graph for $2y - 6 = 0$.

11. Write the equation of a circle with center $(0, 6)$ and that passes through the point $(1, 5)$.

12. Express the domain of the function $f(x) = \sqrt{6x - 7}$ with interval notation.

13. Determine whether the function $g(x) = \sqrt{x + 10}$ is even, odd, or neither.

14. For the function $y = -(x + 1)^2 + 2$, identify all of the transformations of $y = x^2$.

15. Sketch the graph of $y = \sqrt{x - 1} + 3$ and identify all transformations.

16. Find the composite function $f \circ g$ and state the domain for $f(x) = x^2 - 3$ and $g(x) = \sqrt{x + 2}$.

17. Evaluate $g(f(-1))$ for $f(x) = 7 - 2x^2$ and $g(x) = 2x - 10$.

18. Find the inverse of the function $f(x) = (x - 4)^2 + 2$, where $x \geq 4$.

19. Find a quadratic function that has the vertex $(-2, 3)$ and point $(-1, 4)$.

20. Find all of the real zeros and state their multiplicities of the function $f(x) = -3.7x^4 - 14.8x^3$.

21. Use long division to find the quotient $Q(x)$ and the remainder $r(x)$ of $(-20x^3 - 8x^2 + 7x - 5) \div (-5x + 3)$.

22. Use synthetic division to find the quotient $Q(x)$ and the remainder $r(x)$ of $(2x^3 + 3x^2 - 11x + 6) \div (x - 3)$.

23. List the possible rational zeros, and test to determine all rational zeros for $P(x) = 12x^3 + 29x^2 + 7x - 6$.

24. Given the real zero $x = 5$ of the polynomial $P(x) = 2x^3 - 3x^2 - 32x - 15$, determine all the other zeros and write the polynomial in terms of a product of linear factors.

25. Find all vertical and horizontal asymptotes of $f(x) = \dfrac{3x - 5}{x^2 - 4}$.

26. Graph the function $f(x) = \dfrac{2x^3 - x^2 - x}{x^2 - 1}$.

27. Find the asymptotes and intercepts of the rational function $f(x) = \dfrac{5}{2x - 3} - \dfrac{1}{x}$. *Note:* Combine the two expressions into a single rational expression. Graph this function utilizing a graphing utility. Does the graph confirm what you found?

28. Find the asymptotes and intercepts of the rational function $f(x) = \dfrac{6x}{3x + 1} - \dfrac{6x}{4x - 1}$. *Note:* Combine the two expressions into a single rational expression. Graph this function utilizing a graphing utility. Does the graph confirm what you found?

5

Exponential and Logarithmic Functions

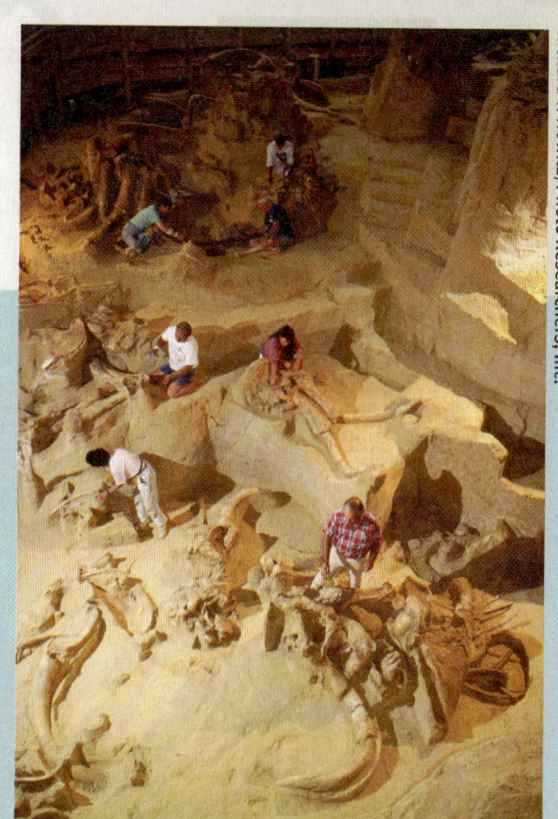

How do archaeologists and anthropologists date fossils? One method is carbon testing. The ratio of carbon 12 (the more stable kind of carbon) to carbon 14 at the moment of death is the same as the ratio in all living things while they are still alive, but the carbon 14 decays and is not replaced. By looking at the ratio of carbon 12 to carbon 14 in the sample and comparing it with the ratio in a living organism, it is possible to determine the age of a formerly living thing fairly precisely.

The amount of carbon in a fossil is a function of how many years the organism has been dead. The amount is modeled by an *exponential function*, and the inverse of the exponential function, a *logarithmic function*, is used to determine the age of the fossil.

IN THIS CHAPTER we will discuss exponential functions and their inverses, logarithmic functions. We will graph these functions and use their properties to solve exponential and logarithmic equations. Last, we will discuss particular exponential and logarithmic models that represent phenomena such as compound interest, world populations, conservation biology models, carbon dating, pH values in chemistry, and the bell curve that is fundamental in statistics for describing how quantities vary in the real world.

EXPONENTIAL AND LOGARITHMIC FUNCTIONS

5.1
Exponential Functions and Their Graphs

- Evaluating Exponential Functions
- Graphs of Exponential Functions
- The Natural Base *e*
- Applications of Exponential Functions

5.2
Logarithmic Functions and Their Graphs

- Evaluating Logarithms
- Common and Natural Logarithms
- Graphs of Logarithmic Functions
- Applications of Logarithms

5.3
Properties of Logarithms

- Properties of Logarithmic Functions
- Change-of-Base Formula

5.4
Exponential and Logarithmic Equations

- Exponential Equations
- Solving Logarithmic Equations

5.5
Exponential and Logarithmic Models

- Exponential Growth Models
- Exponential Decay Models
- Gaussian (Normal) Distribution Models
- Logistic Growth Models
- Logarithmic Models

LEARNING OBJECTIVES

- Graph exponential functions.
- Graph logarithmic functions.
- Apply properties of logarithms.
- Solve exponential and logarithmic equations.
- Use exponential and logarithmic models to represent a variety of real-world phenomena.

SKILLS OBJECTIVES

■ Evaluate exponential functions.
■ Graph exponential functions.
■ Find the domain and range of exponential functions.
■ Utilize the natural base e.
■ Solve compound interest problems.

CONCEPTUAL OBJECTIVES

■ Understand the difference between algebraic and exponential functions.
■ Understand that irrational exponents lead to approximations.

Evaluating Exponential Functions

Most of the functions (polynomial, rational, radical, etc.) we have studied thus far have been **algebraic functions**. Algebraic functions involve basic operations, powers, and roots. In this chapter, we discuss *exponential functions* and *logarithmic functions*, which are called **transcendental functions** because they transcend our ability to define them with a finite number of algebraic expressions. The following table illustrates the difference between algebraic functions and *exponential functions*:

FUNCTION	VARIABLE IS IN THE	CONSTANT IS IN THE	EXAMPLE	EXAMPLE
Algebraic	Base	Exponent	$f(x) = x^2$	$g(x) = x^{1/3}$
Exponential	Exponent	Base	$F(x) = 2^x$	$G(x) = \left(\dfrac{1}{3}\right)^x$

DEFINITION **Exponential Function**

An **exponential function** with **base b** is denoted by

$$f(x) = b^x$$

where b and x are any real numbers such that $b > 0$ and $b \neq 1$.

Note:

■ We eliminate $b = 1$ as a value for the base because it merely yields the constant function $f(x) = 1^x = 1$.
■ We eliminate negative values for b because they would give non–real-number values such as $(-9)^{1/2} = \sqrt{-9} = 3i$.
■ We eliminate $b = 0$ because 0^x corresponds to an undefined value when x is negative.

Sometimes the value of an exponential function for a specific argument can be found by inspection as an *exact* number.

x	-3	-1	0	1	3
$F(x) = 2^x$	$2^{-3} = \dfrac{1}{2^3} = \dfrac{1}{8}$	$2^{-1} = \dfrac{1}{2^1} = \dfrac{1}{2}$	$2^0 = 1$	$2^1 = 2$	$2^3 = 8$

If an exponential function cannot be evaluated exactly, then we find the decimal *approximation* using a calculator. Most calculators have either a base to a power button $\boxed{x^y}$ or a caret $\boxed{\wedge}$ button for working with exponents.

x	-2.7	$-\frac{4}{5}$	$\frac{5}{7}$	2.7
$F(x) = 2^x$	$2^{-2.7} \approx 0.154$	$2^{-4/5} \approx 0.574$	$2^{5/7} \approx 1.641$	$2^{2.7} \approx 6.498$

The domain of exponential functions, $f(x) = b^x$, is the set of all real numbers. It is important to note that in Chapter 0 we discussed exponents of the form b^x, where x is an integer or a rational number. What happens if x is irrational? We can approximate the irrational number with a decimal approximation such as $b^\pi \approx b^{3.14}$ or $b^{\sqrt{2}} \approx b^{1.41}$.

Consider $7^{\sqrt{3}}$, and realize that the irrational number $\sqrt{3}$ is a decimal that never terminates or repeats: $\sqrt{3} = 1.7320508 \ldots$. We can show in advanced mathematics that there is a number $7^{\sqrt{3}}$, and although we cannot write it exactly, we can approximate the number. In fact the closer the exponent is to $\sqrt{3}$, the closer the approximation is to $7^{\sqrt{3}}$.

It is important to note that the properties of exponents (Chapter 0) hold when the exponent is any real number (rational or irrational).

$$7^{1.7} \approx 27.3317$$
$$7^{1.73} \approx 28.9747$$
$$7^{1.732} \approx 29.0877$$
$$\ldots$$
$$7^{\sqrt{3}} \approx 29.0906$$

EXAMPLE 1   Evaluating Exponential Functions

Let $f(x) = 3^x$, $g(x) = \left(\frac{1}{4}\right)^x$, and $h(x) = 10^{x-2}$. Find the following values. If an approximation is required, approximate to four decimal places.

a. $f(2)$ **b.** $f(\pi)$ **c.** $g\left(-\frac{3}{2}\right)$ **d.** $h(2.3)$ **e.** $f(0)$ **f.** $g(0)$

Solution:

a. $f(2) = 3^2 = \boxed{9}$

b. $f(\pi) = 3^\pi \approx \boxed{31.5443}$ *

c. $g\left(-\frac{3}{2}\right) = \left(\frac{1}{4}\right)^{-3/2} = 4^{3/2} = \left(\sqrt{4}\right)^3 = 2^3 = \boxed{8}$

d. $h(2.3) = 10^{2.3-2} = 10^{0.3} \approx \boxed{1.9953}$

e. $f(0) = 3^0 = \boxed{1}$

f. $g(0) = \left(\frac{1}{4}\right)^0 = \boxed{1}$

Notice that parts (a) and (c) were evaluated exactly, whereas parts (b) and (d) required approximation using a calculator.

■ **YOUR TURN** Let $f(x) = 2^x$ and $g(x) = \left(\frac{1}{9}\right)^x$ and $h(x) = 5^{x-2}$. Find the following values. Evaluate exactly when possible, and round to four decimal places when a calculator is needed.

a. $f(4)$ **b.** $f(\pi)$ **c.** $g\left(-\frac{3}{2}\right)$ **d.** $h(2.9)$

Classroom Example 5.1.1
Let $F(x) = 5^x$, $G(x) = \left(\frac{1}{8}\right)^x$ and $H(x) = 3^{1-x}$. Calculate:
a. $F(-1)$ **b.** $H(-2)$
c.* $G\left(-\frac{2}{3}\right)$ **d.** $F(0) - H(1)$

Answer:
a. $\frac{1}{5}$ **b.** 27
c. 4   **d.** 0

■ **Answer: a.** 16   **b.** 8.8250
　　　　c. 27   **d.** 4.2567

*In part (b), the π button on the calculator is selected. If we instead approximate π by 3.14, we get a slightly different approximation for the function value:

$$f(\pi) = 3^\pi \approx 3^{3.14} \approx 31.4891$$

Graphs of Exponential Functions

Let's graph two exponential functions, $y = 2^x$ and $y = 2^{-x} = \left(\frac{1}{2}\right)^x$, by plotting points.

x	$y = 2^x$	(x, y)
-2	$2^{-2} = \dfrac{1}{2^2} = \dfrac{1}{4}$	$\left(-2, \dfrac{1}{4}\right)$
-1	$2^{-1} = \dfrac{1}{2^1} = \dfrac{1}{2}$	$\left(-1, \dfrac{1}{2}\right)$
0	$2^0 = 1$	$(0, 1)$
1	$2^1 = 2$	$(1, 2)$
2	$2^2 = 4$	$(2, 4)$
3	$2^3 = 8$	$(3, 8)$

x	$y = 2^{-x}$	(x, y)
-3	$2^{-(-3)} = 2^3 = 8$	$(-3, 8)$
-2	$2^{-(-2)} = 2^2 = 4$	$(-2, 4)$
-1	$2^{-(-1)} = 2^1 = 2$	$(-1, 2)$
0	$2^0 = 1$	$(0, 1)$
1	$2^{-1} = \dfrac{1}{2^1} = \dfrac{1}{2}$	$\left(1, \dfrac{1}{2}\right)$
2	$2^{-2} = \dfrac{1}{2^2} = \dfrac{1}{4}$	$\left(2, \dfrac{1}{4}\right)$

Notice that both graphs' y-intercepts are $(0, 1)$ (as shown to the left) and neither graph has an x-intercept. The x-axis is a horizontal asymptote for both graphs. The following box summarizes general characteristics of exponential functions.

CHARACTERISTICS OF GRAPHS OF EXPONENTIAL FUNCTIONS

$$f(x) = b^x, \qquad b > 0, \qquad b \neq 1$$

- Domain: $(-\infty, \infty)$
- Range: $(0, \infty)$
- x-intercepts: none
- y-intercept: $(0, 1)$
- Horizontal asymptote: x-axis
- The graph passes through $(1, b)$ and $\left(-1, \dfrac{1}{b}\right)$.
- As x increases, $f(x)$ increases if $b > 1$ and decreases if $0 < b < 1$.
- The function f is one-to-one.

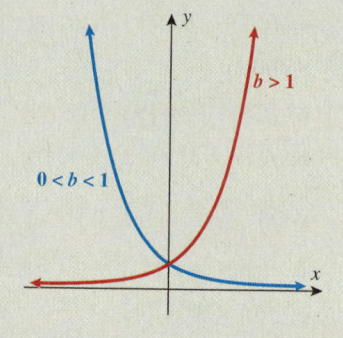

Since exponential functions, $f(x) = b^x$, all go through the point $(0, 1)$ and have the x-axis as a horizontal asymptote, we can find the graph by finding two additional points as outlined in the following procedure.

PROCEDURE FOR GRAPHING $f(x) = b^x$

Step 1: Label the point $(0, 1)$ corresponding to the y-intercept $f(0)$.

Step 2: Find and label two additional points corresponding to $f(-1)$ and $f(1)$.

Step 3: Connect the three points with a *smooth* curve, with the x-axis as the horizontal asymptote.

EXAMPLE 2 Graphing Exponential Functions for $b > 1$

Graph the function $f(x) = 5^x$.

Solution:

STEP 1: Label the y-intercept $(0, 1)$.

STEP 2: Label the point $(1, 5)$.

Label the point $(-1, 0.2)$.

$f(0) = 5^0 = 1$

$f(1) = 5^1 = 5$

$f(-1) = 5^{-1} = \dfrac{1}{5} = 0.2$

STEP 3: Sketch a smooth curve through the three points with the x-axis as a horizontal asymptote.

Domain: $(-\infty, \infty)$

Range: $(0, \infty)$

Technology Tip

The graph of $f(x) = 5^x$ is shown.

■ **YOUR TURN** Graph the function $f(x) = 5^{-x}$.

■ **Answer:**

EXAMPLE 3 Graphing Exponential Functions for $b < 1$

Graph the function $f(x) = \left(\dfrac{2}{5}\right)^x$.

Solution:

STEP 1: Label the y-intercept $(0, 1)$.

STEP 2: Label the point $(-1, 2.5)$.

Label the point $(1, 0.4)$.

$f(0) = \left(\dfrac{2}{5}\right)^0 = 1$

$f(-1) = \left(\dfrac{2}{5}\right)^{-1} = \dfrac{5}{2} = 2.5$

$f(1) = \left(\dfrac{2}{5}\right)^1 = \dfrac{2}{5} = 0.4$

STEP 3: Sketch a smooth curve through the three points with the x-axis as a horizontal asymptote.

Domain: $(-\infty, \infty)$

Range: $(0, \infty)$

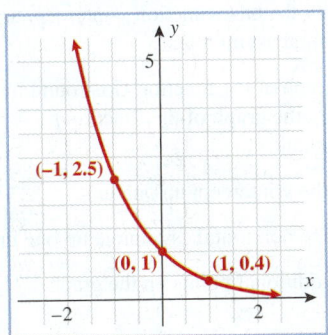

Classroom Example 5.1.3

a. Graph $f(x) = \left(\dfrac{3}{4}\right)^x$

b. Graph $g(x) = \left(\dfrac{4}{3}\right)^x$

Answer:

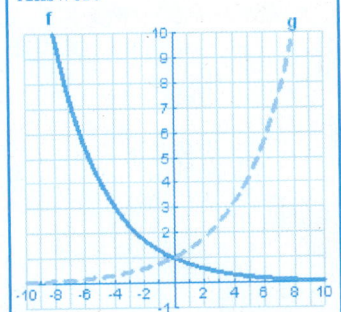

Exponential functions, like all functions, can be graphed by point-plotting. We can also use transformations (horizontal and vertical shifting and reflection; Section 3.3) to graph exponential functions.

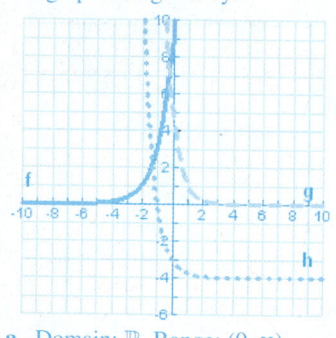
EXAMPLE 4 **Graphing Exponential Functions Using a Horizontal or Vertical Shift**

a. Graph the function $F(x) = 2^{x-1}$. State the domain and range of F.
b. Graph the function $G(x) = 2^x + 1$. State the domain and range of G.

Solution (a):

Identify the base function. $f(x) = 2^x$

Identify the base function y-intercept
and horizontal asymptote. $(0, 1)$ and $y = 0$

The graph of the function F is found by
shifting the graph of the function f
to the right one unit. $F(x) = f(x - 1)$

Shift the y-intercept to the right one unit. $(0, 1)$ shifts to $(1, 1)$

The horizontal asymptote is not altered
by a horizontal shift. $y = 0$

Find additional points on the graph. $F(0) = 2^{0-1} = 2^{-1} = \dfrac{1}{2}$

$y - \text{intercept: } \left(0, \frac{1}{2}\right)$

$F(2) = 2^{2-1} = 2^1 = 2$

Sketch the graph of $F(x) = 2^{x-1}$ with a *smooth* curve.

Domain: $(-\infty, \infty)$
Range: $(0, \infty)$

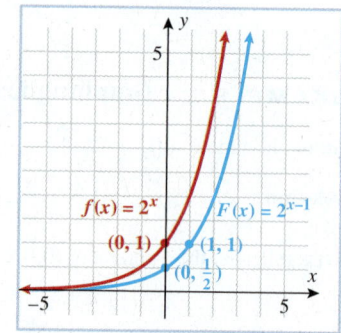

Solution (b):

Identify the base function. $f(x) = 2^x$

Identify the base function y-intercept and
horizontal asymptote. $(0, 1)$ and $y = 0$

The graph of the function G is found by
shifting the graph of the function f
up one unit. $G(x) = f(x) + 1$

Shift the y-intercept up one unit. $(0, 1)$ shifts to $(0, 2)$

Shift the horizontal asymptote up one unit. $y = 0$ shifts to $y = 1$

Find additional points on the graph. $G(1) = 2^1 + 1 = 2 + 1 = 3$

$G(-1) = 2^{-1} + 1 = \dfrac{1}{2} + 1 = \dfrac{3}{2}$

Sketch the graph of $G(x) = 2^x + 1$ with a *smooth* curve.

Domain: $(-\infty, \infty)$
Range: $(1, \infty)$

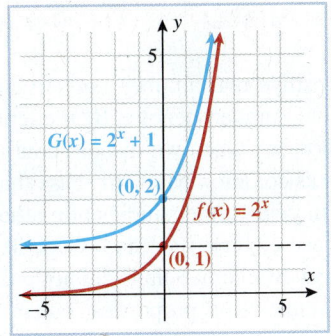

$G(x) = 2^x + 1$
$(0, 2)$
$f(x) = 2^x$
$(0, 1)$

EXAMPLE 5 Graphing Exponential Functions Using Both Horizontal and Vertical Shifts

Graph the function $F(x) = 3^{x+1} - 2$. State the domain and range of F.

Solution:

Identify the base function.

$f(x) = 3^x$

Identify the base function y-intercept and horizontal asymptote.

$(0, 1)$ and $y = 0$

The graph of the function F is found by shifting the graph of the function f to the left one unit and down two units.

$F(x) = f(x + 1) - 2$

Shift the y-intercept to the left one unit and down two units.

$(0, 1)$ shifts to $(-1, -1)$

Shift the horizontal asymptote down two units.

$y = 0$ shifts to $y = -2$

Find additional points on the graph.

$F(0) = 3^{0+1} - 2 = 3 - 2 = 1$
$F(1) = 3^{1+1} - 2 = 9 - 2 = 7$

Sketch the graph of $F(x) = 3^{x+1} - 2$ with a *smooth* curve.

Domain: $(-\infty, \infty)$
Range: $(-2, \infty)$

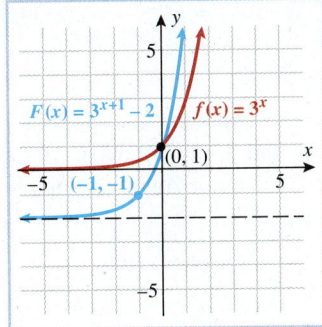

$F(x) = 3^{x+1} - 2$ $f(x) = 3^x$
$(0, 1)$
$(-1, -1)$

Technology Tip

The graph of $f(x) = 3^{x+1} - 2$ is shown.

```
WINDOW
 Xmin=-5
 Xmax=5
 Xscl=1
 Ymin=-3
 Ymax=8
 Yscl=1
 Xres=1
```

```
Plot1 Plot2 Plot3
\Y1■3^(X)
\Y2■3^(X+1)-2
\Y3■-2
```

```
Y2=3^(X+1)-2
X=-1        Y=-1
```

■ **Answer:** Domain: $(-\infty, \infty)$
Range: $(-1, \infty)$

$(0, 7)$
$f(x) = 2^{x+3} - 1$
$(-1, 3)$
$(-2, 1)$
$(-3, 0)$

■ **YOUR TURN** Graph $f(x) = 2^{x+3} - 1$. State the domain and range of f.

The Natural Base *e*

Technology Tip

Use a graphing utility to set up the table for $\left(1 + \dfrac{1}{m}\right)^m$. Enter *x* for *m*.

To find the value of *e*, press the *e* key. This can be done by pressing 2nd ÷ .

Any positive real number can serve as the base for an exponential function. A particular irrational number, denoted by the letter *e*, appears as the base in many applications, as you will soon see when we discuss continuous compounded interest. Although you will see 2 and 10 as common bases, the base that appears most often is *e*, because *e*, as you will come to see in your further studies of mathematics, is the **natural base**. The exponential function with base *e*, $f(x) = e^x$, is called the **exponential function** or the **natural exponential function**. Mathematicians did not pull this irrational number out of a hat. The number *e* has many remarkable properties, but most simply, it comes from evaluating the expression $\left(1 + \dfrac{1}{m}\right)^m$ as *m* increases without bound.

m	$\left(1 + \dfrac{1}{m}\right)^m$
1	2
10	2.59374
100	2.70481
1000	2.71692
10,000	2.71815
100,000	2.71827
1,000,000	**2.71828**

$e \approx 2.71828$

Calculators have an e^x button for approximating the natural exponential function.

Classroom Example 5.1.6

Evaluate the following to three decimal places.

a. $e^{2.3}$ **b.** $e^{-\pi}$ **c.** e^{e^e}

Answer:

a. 9.974

c. 0.043

c. 3,814,279.105

EXAMPLE 6   Evaluating the Natural Exponential Function

Evaluate $f(x) = e^x$ for the given *x*-values. Round your answers to four decimal places.

a. $x = 1$ **b.** $x = -1$ **c.** $x = 1.2$ **d.** $x = -0.47$

Solution:

a. $f(1) = e^1 \approx 2.718281828 \approx \boxed{2.7183}$

b. $f(-1) = e^{-1} \approx 0.367879441 \approx \boxed{0.3679}$

c. $f(1.2) = e^{1.2} \approx 3.320116923 \approx \boxed{3.3201}$

d. $f(-0.47) = e^{-0.47} \approx 0.625002268 \approx \boxed{0.6250}$

Like all exponential functions of the form $f(x) = b^x$, $f(x) = e^x$ and $f(x) = e^{-x}$ have (0, 1) as their *y*-intercept and the *x*-axis as a horizontal asymptote as shown in the figure on the right.

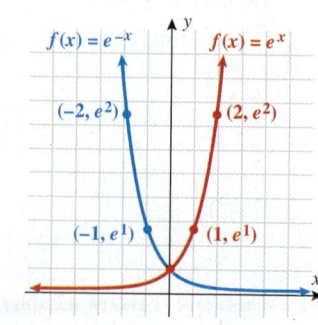

EXAMPLE 7 **Graphing Exponential Functions with Base e**

Graph the function $f(x) = 3 + e^{2x}$.

Solution:

x	$f(x) = 3 + e^{2x}$	(x, y)
-2	3.02	$(-2, 3.02)$
-1	3.14	$(1, 3.14)$
0	4	$(0, 4)$
1	10.39	$(1, 10.39)$
2	57.60	$(2, 57.60)$

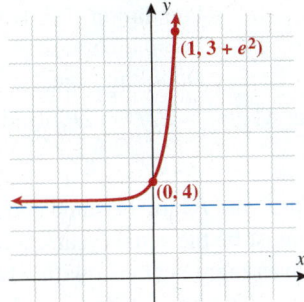

Note: The y-intercept is $(0, 4)$ and the line $y = 3$ is the horizontal asymptote.

■ **YOUR TURN** Graph the function $f(x) = e^{x+1} - 2$.

■ **Answer:**

Applications of Exponential Functions

Exponential functions describe either *growth* or *decay*. Populations and investments are often modeled with exponential growth functions, while the declining value of a used car and the radioactive decay of isotopes are often modeled with exponential decay functions. In Section 5.5, various exponential models will be discussed. In this section we discuss doubling time, half-life, and compound interest.

A successful investment program, growing at about 7.2% per year, will double in size every 10 years. Let's assume that you will retire at the age of 65. There is a saying: *It's not the first time your money doubles, it's the last time that makes such a difference.* As you may already know or as you will soon find, it is important to start investing early.

Suppose Maria invests $5000 at age 25 and David invests $5000 at age 35. Let's calculate how much will accrue from the initial $5000 investment by the time they each retire, assuming their money doubles every 10 years.

AGE	MARIA	DAVID
25	$5,000	
35	$10,000	$5,000
45	$20,000	$10,000
55	$40,000	$20,000
65	**$80,000**	**$40,000**

They each made a one-time investment of $5000. By investing 10 years sooner, Maria made twice what David made.

A measure of growth rate is the *doubling time*, the time it takes for something to double. Often doubling time is used to describe populations.

Classroom Example 5.1.7*

Consider $F(x) = e^{-1-x} - 3$.

a. Find the y-intercept.

b. State the domain and range of F.

c. Graph F.

Answer:

a. $\left(0, \dfrac{1}{e} - 3\right)$

b. Domain: $\mathbb{R}$, Range: $(-3, \infty)$

c.

DOUBLING TIME GROWTH MODEL

The doubling time growth model is given by

$$P = P_0 2^{t/d}$$

where P = Population at time t
P_0 = Population at time $t = 0$
d = Doubling time

Note that when $t = d$, $P = 2P_0$ (population is equal to twice the original).

The units for P and P_0 are the same and can be any quantity (people, dollars, etc.). The units for t and d must be the same (years, weeks, days, hours, seconds, etc.).

In the investment scenario with Maria and David, $P_0 = \$5000$ and $d = 10$ years, so the model used to predict how much money the original $5000 investment yielded is $P = 5000(2)^{t/10}$. Maria retired 40 years after the original investment, $t = 40$, and David retired 30 years after the original investment, $t = 30$.

$$\text{Maria: } P = 5000(2)^{40/10} = 5000(2)^4 = 5000(16) = 80,000$$

$$\text{David: } P = 5000(2)^{30/10} = 5000(2)^3 = 5000(8) = 40,000$$

Classroom Example 5.1.8
The current population of the Island of Doon is 500,000, and it is expected for the population to double in 15 years. Estimate the population in 5 years. Round your answer to the nearest thousandth.

Answer: 630,000

■ Answer: 38 million

EXAMPLE 8   Doubling Time of Populations

In 2004, the population in Kazakhstan, a country in Asia, reached 15 million. It is estimated that the population will double in 30 years. If the population continues to grow at the same rate, what will the population be 20 years from now? Round to the nearest million.

Solution:

Write the doubling model. $P = P_0 2^{t/d}$

Substitute $P_0 = 15$ million,
$d = 30$ years, and $t = 20$ years. $P = 15(2)^{20/30}$

Simplify. $P = 15(2)^{2/3} \approx 23.8110$

In 20 years, there will be approximately $\boxed{24 \text{ million people}}$ in Kazakhstan.

■ YOUR TURN What will the approximate population in Kazakhstan be in 40 years? Round to the nearest million.

We now turn our attention from exponential growth to exponential decay, or negative growth. Suppose you buy a brand-new car from a dealership for $24,000. The value of a car decreases over time according to an exponential decay function. The **half-life** of this particular car, or the time it takes for the car to depreciate 50%, is approximately 3 years. The exponential decay is described by

$$A = A_0 \left(\frac{1}{2}\right)^{t/h}$$

where A_0 is the amount the car is worth (in dollars) when new (that is, when $t = 0$), A is the amount the car is worth (in dollars) after t years, and h is the half-life in years. In our car scenario, $A_0 = 24,000$ and $h = 3$:

$$A = 24,000 \left(\frac{1}{2}\right)^{t/3}$$

How much is the car worth after three years? Six years? Nine years? Twenty-four years?

$$t = 3: \qquad A = 24{,}000\left(\frac{1}{2}\right)^{3/3} = 24{,}000\left(\frac{1}{2}\right) = 12{,}000$$

$$t = 6: \qquad A = 24{,}000\left(\frac{1}{2}\right)^{6/3} = 24{,}000\left(\frac{1}{2}\right)^{2} = 6000$$

$$t = 9: \qquad A = 24{,}000\left(\frac{1}{2}\right)^{9/3} = 24{,}000\left(\frac{1}{2}\right)^{3} = 3000$$

$$t = 24: \qquad A = 24{,}000\left(\frac{1}{2}\right)^{24/3} = 24{,}000\left(\frac{1}{2}\right)^{8} = 93.75 \approx 100$$

The car that was worth \$24,000 new is worth \$12,000 in 3 years, \$6000 in 6 years, \$3000 in 9 years, and about \$100 in the junkyard in 24 years.

EXAMPLE 9 Radioactive Decay

The radioactive isotope of potassium ^{42}K which is used in the diagnosis of brain tumors, has a half-life of 12.36 hours. If 500 milligrams of potassium 42 are taken, how many milligrams will remain after 24 hours? Round to the nearest milligram.

Solution:

Write the half-life formula. $\qquad\qquad\qquad A = A_0\left(\frac{1}{2}\right)^{t/h}$

Substitute $A_0 = 500$ mg, $h = 12.36$ hours, $\qquad A = 500\left(\frac{1}{2}\right)^{24/12.36}$
$t = 24$ hours.

Simplify. $\qquad\qquad\qquad\qquad\qquad A \approx 500(0.2603) \approx 130.15$

After 24 hours, there are approximately $\boxed{130\ \text{milligrams}}$ of potassium 42 left.

■ **Answer:** 0.04 mg (less than 1 mg)

■ **YOUR TURN** How many milligrams of potassium are expected to be left in the body after 1 week?

In Section 1.2, *simple interest* was discussed where the interest I is calculated based on the principal P, the annual interest rate r, and the time t in years, using the formula: $I = Prt$.

If the interest earned in a period is then reinvested at the same rate, future interest is earned on both the principal and the reinvested interest during the next period. Interest paid on both the principal and interest is called *compound interest*.

COMPOUND INTEREST

If a **principal P** is invested at an annual **rate r compounded** n times a year, then the **amount A** in the account at the end of t years is given by

$$A = P\left(1 + \frac{r}{n}\right)^{nt}$$

The annual interest rate r is expressed as a decimal.

The following list shows the typical number of times interest is compounded.

Annually	$n = 1$	Monthly	$n = 12$
Semiannually	$n = 2$	Weekly	$n = 52$
Quarterly	$n = 4$	Daily	$n = 365$

Technology Tip

Enter $3000\left(1 + \dfrac{0.03}{4}\right)^{(4)(7)}$ into the graphing calculator.

```
3000(1+0.03/4)^(
4*7)
         3698.135243
■
```

■ **Answer:** $6312.38

EXAMPLE 10 **Compound Interest**

If $3000 is deposited in an account paying 3% compounded quarterly, how much will you have in the account in 7 years?

Solution:

Write the compound interest formula.

$$A = P\left(1 + \frac{r}{n}\right)^{nt}$$

Substitute $P = 3000$, $r = 0.03$, $n = 4$, and $t = 7$.

$$A = 3000\left(1 + \frac{0.03}{4}\right)^{(4)(7)}$$

Simplify.

$$A = 3000(1.0075)^{28} \approx 3698.14$$

You will have $\boxed{\$3698.14}$ in the account.

■ **YOUR TURN** If $5000 is deposited in an account paying 6% compounded annually, how much will you have in the account in 4 years?

Classroom Example 5.1.10
If $10,000 is deposited in an account paying 4.5% compounding weekly, how much will you have in the account in 3.5 years?

Answer: $11,705.01

Notice in the compound interest formula that as n increases the amount A also increases. In other words, the more times the interest is compounded per year, the more money you make. Ideally, your bank will compound your interest infinitely many times. This is called *compounding continuously.* We will now show the development of the compounding continuously formula, $A = Pe^{rt}$.

WORDS	**MATH**
Write the compound interest formula.	$A = P\left(1 + \dfrac{r}{n}\right)^{nt}$
Note that $\dfrac{r}{n} = \dfrac{1}{n/r}$ and $nt = \left(\dfrac{n}{r}\right)rt$.	$A = P\left(1 + \dfrac{1}{n/r}\right)^{(n/r)rt}$
Let $m = \dfrac{n}{r}$.	$A = P\left(1 + \dfrac{1}{m}\right)^{mrt}$
Use the exponential property: $x^{mrt} = (x^m)^{rt}$.	$A = P\left[\left(1 + \dfrac{1}{m}\right)^{m}\right]^{rt}$

Earlier in this section, we showed that as m increases, $\left(1 + \dfrac{1}{m}\right)^{m}$ approaches e. Therefore, as the number of times the interest is compounded approaches infinity, or as $n \to \infty$, the amount in an account $A = P\left(1 + \dfrac{r}{n}\right)^{nt}$ approaches $A = Pe^{rt}$.

CONTINUOUS COMPOUND INTEREST

If a **principal** P is invested at an annual **rate** r **compounded continuously**, then the **amount** A in the account at the end of t years is given by

$$A = Pe^{rt}$$

The annual interest rate r is expressed as a decimal.

Classroom Example 5.1.11
If $10,000 is deposited in an account paying 4.5% compounding continuously, how much will you have in the account in 3.5 years?

Answer: $11,705.81

It is important to note that for a given interest rate, the highest return you can earn is by compounding continuously.

EXAMPLE 11   Continuously Compounded Interest

If \$3000 is deposited in a savings account paying 3% a year compounded continuously, how much will you have in the account in 7 years?

Solution:

Write the continuous compound interest formula. $A = Pe^{rt}$

Substitute $P = 3000$, $r = 0.03$, and $t = 7$. $A = 3000e^{(0.03)(7)}$

Simplify. $A \approx 3701.034$

There will be $\boxed{\$3701.03}$ in the account in 7 years.

Note: In Example 10, we worked this same problem compounding *quarterly*, and the result was \$3698.14.

If the number of times per year interest is compounded increases, then the total interest earned that year also increases.

■ **YOUR TURN**   If \$5000 is deposited in an account paying 6% compounded continuously, how much will be in the account in 4 years?

Technology Tip

Enter $3000e^{(0.03)(7)}$ into the graphing calculator.

```
3000e^(0.03*7)
          3701.03418
```

Study Tip

If the number of times per year interest is compounded increases, then the total interest earned that year also increases.

■ **Answer:** \$6356.25

SECTION 5.1 SUMMARY

In this section we discussed exponential functions (constant base, variable exponent).

General Exponential Functions: $f(x) = b^x$, $b \neq 1$, and $b > 0$

1. Evaluating exponential functions
   Exact (by inspection): $f(x) = 2^x$ $f(3) = 2^3 = 8$.
   Approximate (with the aid of a calculator): $f(x) = 2^x$
   $f(\sqrt{3}) = 2^{\sqrt{3}} \approx 3.322$
2. Graphs of exponential functions
   Domain: $(-\infty, \infty)$ and range: $(0, \infty)$.
   The point $(0, 1)$ corresponds to the y-intercept.
   The graph passes through the points $(1, b)$ and $\left(-1, \dfrac{1}{b}\right)$.

   The x-axis is a horizontal asymptote.
   The function f is one-to-one.

$$f(x) = b^x, \qquad b > 0, \qquad b \neq 1$$

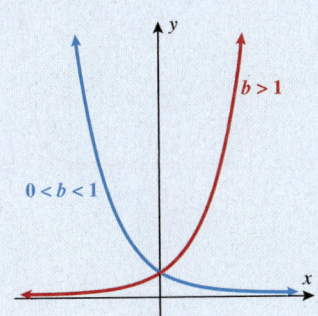

Procedure for Graphing: $f(x) = b^x$

Step 1: Label the point $(0, 1)$ corresponding to the y-intercept $f(0)$.

Step 2: Find and label two additional points corresponding to $f(-1)$ and $f(1)$.

Step 3: Connect the three points with a smooth curve with the x-axis as the horizontal asymptote.

The Natural Exponential Function: $f(x) = e^x$

The irrational number e is called the natural base.

$$e = \left(1 + \frac{1}{m}\right)^m \text{ as } m \to \infty$$

$e \approx 2.71828$

Applications of Exponential Functions (all variables expressed in consistent units)

1. Doubling time: $P = P_0 2^{t/d}$
   d is doubling time.
   P is population at time t.
   P_0 is population at time $t = 0$.

2. Half-life: $A = A_0 \left(\frac{1}{2}\right)^{t/h}$
   h is the half-life.
   A is amount at time t.
   A_0 is amount at time $t = 0$.

3. Compound interest (P = principal, A = amount after t years, r = interest rate)

   Compounded n times a year: $A = P\left(1 + \dfrac{r}{n}\right)^{nt}$

   Compounded continuously: $A = Pe^{rt}$

▪SKILLS

In Exercises 1–10, evaluate *exactly* (without using a calculator).

1. 2^4 **2.** 3^4 **3.** 5^{-2} **4.** 4^{-3} **5.** $8^{2/3}$ **6.** $27^{2/3}$ **7.** $\left(\frac{1}{9}\right)^{-3/2}$ **8.** $\left(\frac{1}{16}\right)^{-3/2}$ **9.** -5^0 **10.** -6^0

In Exercises 11–18, approximate with a calculator. Round your answer to four decimal places.

11. $4^{1.2}$ **12.** $4^{-1.2}$ **13.** $5^{\sqrt{2}}$ **14.** $6^{\sqrt{3}}$

15. e^2 **16.** $e^{1/2}$ **17.** $e^{-\pi}$ **18.** $e^{-\sqrt{2}}$

In Exercises 19–26, for the functions $f(x) = 3^x$, $g(x) = \left(\frac{1}{16}\right)^x$, and $h(x) = 10^{x+1}$, find the function value at the indicated points.

19. $f(3)$ **20.** $h(1)$ **21.** $g(-1)$ **22.** $f(-2)$

23. $g\left(-\frac{1}{2}\right)$ **24.** $g\left(-\frac{3}{2}\right)$ **25.** $f(e)$ **26.** $g(\pi)$

In Exercises 27–32, match the graph with the function.

27. $y = 5^{x-1}$ **28.** $y = 5^{1-x}$ **29.** $y = -5^x$

30. $y = -5^{-x}$ **31.** $y = 1 - 5^{-x}$ **32.** $y = 5^x - 1$

a.

b.

c.

d.

e.

f.

In Exercises 33–48, graph the exponential function using transformations. State the y-intercept, two additional points, the domain, the range, and the horizontal asymptote.

33. $f(x) = 6^x$ 34. $f(x) = 7^x$ 35. $f(x) = 10^{-x}$ 36. $f(x) = 4^{-x}$ 37. $f(x) = e^{-x}$ 38. $f(x) = -e^x$

39. $f(x) = 2^x - 1$ 40. $f(x) = 3^x - 1$ 41. $f(x) = 2 - e^x$ 42. $f(x) = 1 + e^{-x}$ 43. $f(x) = e^{x+1} - 4$ 44. $f(x) = e^{x-1} + 2$

45. $f(x) = 3e^{x/2}$ 46. $f(x) = 2e^{-x}$ 47. $f(x) = 1 + \left(\frac{1}{2}\right)^{x-2}$ 48. $f(x) = 2 - \left(\frac{1}{3}\right)^{x+1}$

▪ APPLICATIONS

49. Population Doubling Time. In 2002, there were 7.1 million people living in London, England. If the population is expected to double by 2090, what is the expected population in London in 2050?

50. Population Doubling Time. In 2004, the population in Morganton, Georgia, was 43,000. The population in Morganton is expected to double by 2010. If the growth rate remains the same, what is the expected population in Morganton in 2020?

51. Investments. Suppose an investor buys land in a rural area for $1500 an acre and sells some of it 5 years later at $3000 an acre and the rest of it 10 years later at $6000. Write a function that models the value of land in that area, assuming the growth rate stays the same. What would the expected cost per acre be 30 years after the initial investment of $1500?

52. Salaries. Twin brothers, Collin and Cameron, get jobs immediately after graduating from college at the age of 22. Collin opts for the higher starting salary, $55,000, and stays with the same company until he retires at 65. His salary doubles every 15 years. Cameron opts for a lower starting salary, $35,000, but moves to a new job every 5 years; he doubles his salary every 10 years until he retires at 65. What is the annual salary of each brother upon retirement?

53. Radioactive Decay. A radioactive isotope of selenium ^{75}Se used in the creation of medical images of the pancreas, has a half-life of 119.77 days. If 200 milligrams are given to a patient, how many milligrams are left after 30 days?

54. Radioactive Decay. The radioactive isotope indium-111 (^{111}In), used as a diagnostic tool for locating tumors associated with prostate cancer, has a half-life of 2.807 days. If 300 milligrams are given to a patient, how many milligrams will be left after a week?

55. Depreciation of Furniture. A couple buys a new bedroom set for $8000 and 10 years later sells it for $4000. If the depreciation continues at the same rate, how much would the bedroom set be worth in 4 more years?

56. Depreciation of a Computer. A student buys a new laptop for $1500 when she arrives as a freshman. A year later, the computer is worth approximately $750. If the depreciation continues at the same rate, how much would she expect to sell her laptop for when she graduates 4 years after she bought it?

57. Compound Interest. If you put $3200 in a savings account that earns 2.5% interest per year compounded quarterly, how much would you expect to have in that account in 3 years?

58. Compound Interest. If you put $10,000 in a savings account that earns 3.5% interest per year compounded annually, how much would you expect to have in that account in 5 years?

59. Compound Interest. How much money should you put in a savings account now that earns 5% a year compounded daily if you want to have $32,000 in 18 years?

60. Compound Interest. How much money should you put in a savings account now that earns 3.0% a year compounded weekly if you want to have $80,000 in 15 years?

61. Compound Interest. If you put $3200 in a savings account that pays 2% a year compounded continuously, how much will you have in the account in 15 years?

62. Compound Interest. If you put $7000 in a money market account that pays 4.3% a year compounded continuously, how much will you have in the account in 10 years?

63. Compound Interest. How much money should you deposit into a money market account that pays 5% a year compounded continuously to have $38,000 in the account in 20 years?

64. Compound Interest. How much money should you deposit into a certificate of deposit that pays 6% a year compounded continuously to have $80,000 in the account in 18 years?

For Exercises 65 and 66, refer to the following:

Exponential functions can be used to model the concentration of a drug in a patient's body. Suppose the concentration of Drug X in a patient's bloodstream is modeled by

$$C(t) = C_0 e^{-rt}$$

where $C(t)$ represents the concentration at time t (in hours), C_0 is the concentration of the drug in the blood immediately after injection, and $r > 0$ is a constant indicating the removal of the drug by the body through metabolism and/or excretion. The rate constant r has units of 1/time (1/hr). It is important to note that this model assumes that the blood concentration of the drug C_0 peaks immediately when the drug is injected.

65. Health/Medicine. After an injection of Drug Y, the concentration of the drug in the bloodstream drops at the rate of 0.020 1/hr. Find the concentration, to the nearest tenth, of the drug 20 hours after receiving an injection with initial concentration of 5.0 mg/L.

66. Health/Medicine. After an injection of Drug Y, the concentration of the drug in the bloodstream drops at the rate of 0.009 1/hr. Find the concentration, to the nearest tenth, of the drug 4 hours after receiving an injection with initial concentration of 4.0 mg/L.

For Exercises 67 and 68, refer to the following:

The demand for a product, in thousands of units, can be expressed by the exponential demand function

$$D(p) = 2300(0.85)^p$$

where p is the price per unit.

67. Economics. Find the demand for the product by completing the following table.

P(PRICE PER UNIT)	D(P)—DEMAND FOR PRODUCT IN UNITS
1.00	
5.00	
10.00	
20.00	
40.00	
60.00	
80.00	
90.00	

68. Economics. Evaluate $D(91)$ and interpret what this means in terms of demand.

■**CATCH THE MISTAKE**

In Exercises 69–72, explain the mistake that is made.

69. Evaluate the expression $4^{-1/2}$.

Solution: $4^{-1/2} = 4^2 = 16$

The correct value is $\frac{1}{2}$. What mistake was made?

70. Evaluate the function for the given x: $f(x) = 4^x$ for $x = \frac{3}{2}$.

Solution: $f\left(\dfrac{3}{2}\right) = 4^{3/2}$

$$= \dfrac{4^3}{4^2} = \dfrac{64}{16} = 4$$

The correct value is 8. What mistake was made?

71. If $2000 is invested in a savings account that earns 2.5% interest compounding continuously, how much will be in the account in one year?

Solution:

Write the compound continuous interest formula. $A = Pe^{rt}$

Substitute $P = 2000$, $r = 2.5$, and $t = 1$. $A = 2000e^{(2.5)(1)}$

Simplify. $A = 24{,}364.99$

This is incorrect. What mistake was made?

72. If $5000 is invested in a savings account that earns 3% interest compounding continuously, how much will be in the account in 6 months?

Solution:

Write the compound continuous interest formula. $A = Pe^{rt}$

Substitute $P = 5000$, $r = 0.03$, and $t = 6$. $A = 5000e^{(0.03)(6)}$

Simplify. $A = 5986.09$

This is incorrect. What mistake was made?

■**CONCEPTUAL**

In Exercises 73–76, determine whether each statement is true or false.

73. The function $f(x) = -e^{-x}$ has the y-intercept $(0,1)$.

74. The function $f(x) = -e^{-x}$ has a horizontal asymptote along the x-axis.

75. The functions $y = 3^{-x}$ and $y = \left(\frac{1}{3}\right)^x$ have the same graphs.

76. $e = 2.718$.

77. Plot $f(x) = 3^x$ and its inverse on the same graph.

78. Plot $f(x) = e^x$ and its inverse on the same graph.

■ **CHALLENGE**

79. Graph $f(x) = e^{|x|}$.

80. Graph $f(x) = e^{-|x|}$.

81. Find the y-intercept and horizontal asymptote of $f(x) = be^{-x+1} - a$.

82. Find the y-intercept and horizontal asymptote of $f(x) = a + be^{x+1}$.

83. Graph $f(x) = b^{|x|}$, $b > 1$, and state the domain.

84. Graph the function $f(x) = \begin{cases} a^x & x < 0 \\ a^{-x} & x \geq 0 \end{cases}$ where $a > 1$.

■ **TECHNOLOGY**

85. Plot the function $y = \left(1 + \dfrac{1}{x}\right)^x$. What is the horizontal asymptote as x increases?

86. Plot the functions $y = 2^x$, $y = e^x$, and $y = 3^x$ in the same viewing screen. Explain why $y = e^x$ lies between the other two graphs.

87. Plot $y_1 = e^x$ and $y_2 = 1 + x + \dfrac{x^2}{2} + \dfrac{x^3}{6} + \dfrac{x^4}{24}$ in the same viewing screen. What do you notice?

88. Plot $y_1 = e^{-x}$ and $y_2 = 1 - x + \dfrac{x^2}{2} - \dfrac{x^3}{6} + \dfrac{x^4}{24}$ in the same viewing screen. What do you notice?

89. Plot the functions $f(x) = \left(1 + \dfrac{1}{x}\right)^x$, $g(x) = \left(1 + \dfrac{2}{x}\right)^x$, and $h(x) = \left(1 + \dfrac{2}{x}\right)^{2x}$ in the same viewing screen. Compare their horizontal asymptotes as x increases. What can you say about the function values of f, g, and h in terms of the powers of e as x increases?

90. Plot the functions $f(x) = \left(1 + \dfrac{1}{x}\right)^x$, $g(x) = \left(1 - \dfrac{1}{x}\right)^x$, and $h(x) = \left(1 - \dfrac{2}{x}\right)^x$ in the same viewing screen. Compare their horizontal asymptotes as x increases. What can you say about the function values of f, g, and h in terms of the powers of e as x increases?

For Exercises 91 and 92, refer to the following:

Newton's Law of Heating and Cooling: Have you ever heated soup in a microwave and, upon taking it out, have it seem to cool considerably in the matter of minutes? Or has your ice-cold soda become tepid in just moments while outside on a hot summer's day? This phenomenon is based on the so-called *Newton's Law of Heating and Cooling*. Eventually, the soup will cool so that its temperature is the same as the temperature of the room in which it is being kept, and the soda will warm until its temperature is the same as the outside temperature.

91. Consider the following data:

Time (in minutes)	1.0	1.5	2.0	2.5	3.0	3.5	4.0	4.5
Temperature of Soup (in degrees Fahrenheit)	203	200	195	188	180	171	160	151

a. Form a scatterplot for this data.

b. Use *ExpReg* to find the best fit exponential function for this data set, and superimpose its graph on the scatterplot. How good is the fit?

c. Use the best fit exponential curve from (b) to answer the following:

   i. What will the predicted temperature of the soup be at 6 minutes?

   ii. What was the temperature of the soup the moment it was taken out of the microwave?

d. Assume the temperature of the house is 72° F. According to Newton's Law of Heating and Cooling, the temperature of the soup should approach 72°. In light of this, comment on the shortcomings of the best fit exponential curve.

92. Consider the following data:

Time (in minutes)	1	2	3	4	5	6	7	8
Temperature of Soda (in degrees Fahrenheit)	45	48	49	53	57	61	68	75

a. Form a scatterplot for this data.

b. Use *ExpReg* to find the best fit exponential function for this data set, and superimpose its graph on the scatterplot. How good is the fit?

c. Use the best fit exponential curve from (b) to answer the following:

   i. What will the predicted temperature of the soda be at 10 minutes?

   ii. What was the temperature of the soda the moment it was taken out of the refrigerator?

d. Assume the temperature of the house is 90° F. According to Newton's Law of Heating and Cooling, the temperature of the soda should approach 90°. In light of this, comment on the shortcomings of the best fit exponential curve.

SKILLS OBJECTIVES

- Convert exponential expressions to logarithmic expressions.
- Convert logarithmic expressions to exponential expressions.
- Evaluate logarithmic expressions exactly by inspection.
- Approximate common and natural logarithms using a calculator.
- Graph logarithmic functions.
- Determine domain restrictions on logarithmic functions.
- Graph functions using a logarithmic scale.

CONCEPTUAL OBJECTIVES

- Interpret logarithmic functions as inverses of exponential functions.
- Understand that logarithmic functions allow very large ranges of numbers in science and engineering applications to be represented on a smaller scale.

Evaluating Logarithms

In Section 5.1, we found that the graph of an exponential function $f(x) = b^x$ passes through the point $(0, 1)$, with the x-axis as a horizontal asymptote. The graph passes both the vertical line test (for a function) and the horizontal line test (for a one-to-one function), and therefore an inverse exists. We will now apply the technique outlined in Section 3.5 to find the inverse of $f(x) = b^x$:

WORDS	**MATH**
Let $y = f(x)$.	$y = b^x$
Interchange x and y.	$x = b^y$
Solve for y.	$y = ?$

We see that y is the exponent that b is raised to in order to obtain x. We call this exponent a **logarithm** (or "log" for short).

WORDS	**MATH**
$x = b^y$ is equivalent to $y = \log_b x$.	$y = \log_b x$
Let $y = f^{-1}(x)$.	$f^{-1}(x) = \log_b x$

DEFINITION   Logarithmic Function

For $x > 0$, $b > 0$, and $b \neq 1$, the **logarithmic function with base b** is denoted $f(x) = \log_b x$, where

$$y = \log_b x \text{ if and only if } x = b^y$$

We read $\log_b x$ as "log base b of x."

Study Tip

$\log_b x = y$ is equivalent to $b^y = x$

This definition says that $x = b^y$ (**exponential form**) and $y = \log_b x$ (**logarithmic form**) are equivalent. One way to remember this relationship is by adding arrows to the logarithmic form:

$$\log_b x = y \iff b^y = x$$

EXAMPLE 1 Changing from Logarithmic Form to Exponential Form

Write each equation in its equivalent exponential form.

a. $\log_2 8 = 3$ **b.** $\log_9 3 = \frac{1}{2}$ **c.** $\log_5 \left(\frac{1}{25}\right) = -2$

Solution:

a. $\log_2 8 = 3$ is equivalent to $\boxed{2^3 = 8}$

b. $\log_9 3 = \frac{1}{2}$ is equivalent to $\boxed{9^{1/2} = 3}$

c. $\log_5 \left(\frac{1}{25}\right) = -2$ is equivalent to $\boxed{5^{-2} = \frac{1}{25}}$

■ **YOUR TURN** Write each equation in its equivalent exponential form.

a. $\log_3 9 = 2$ **b.** $\log_{16} 4 = \frac{1}{2}$ **c.** $\log_2 \left(\frac{1}{8}\right) = -3$

■ **Answer: a.** $3^2 = 9$
   b. $16^{1/2} = 4$
   c. $2^{-3} = \frac{1}{8}$

EXAMPLE 2 Changing from Exponential Form to Logarithmic Form

Write each equation in its equivalent logarithmic form.

a. $16 = 2^4$ **b.** $9 = \sqrt{81}$ **c.** $\frac{1}{9} = 3^{-2}$ **d.** $x^a = z$

Solution:

a. $16 = 2^4$ is equivalent to $\boxed{\log_2 16 = 4}$

b. $9 = \sqrt{81} = 81^{1/2}$ is equivalent to $\boxed{\log_{81} 9 = \frac{1}{2}}$

c. $\frac{1}{9} = 3^{-2}$ is equivalent to $\boxed{\log_3\left(\frac{1}{9}\right) = -2}$

d. $x^a = z$ is equivalent to $\boxed{\log_x z = a}$

■ **YOUR TURN** Write each equation in its equivalent logarithmic form.

a. $81 = 9^2$ **b.** $12 = \sqrt{144}$ **c.** $\frac{1}{49} = 7^{-2}$ **d.** $y^b = w$

■ **Answer: a.** $\log_9 81 = 2$
   b. $\log_{144} 12 = \frac{1}{2}$
   c. $\log_7 \left(\frac{1}{49}\right) = -2$
   d. $\log_y w = b$

Some logarithms can be found exactly, while others must be approximated. Example 3 illustrates how to find the exact value of a logarithm. Example 4 illustrates approximating values of logarithms with a calculator.

EXAMPLE 3   Finding the Exact Value of a Logarithm

Find the exact value of:

a. $\log_3 81$ **b.** $\log_{169} 13$ **c.** $\log_5 \left(\frac{1}{5}\right)$

Solution (a):

The logarithm has some value. Let's call it x.	$\log_3 81 = x$
Change from logarithmic to exponential form.	$3^x = 81$
3 raised to what power is 81?	$3^4 = 81$ $x = 4$
Change from exponential to logarithmic form.	$\boxed{\log_3 81 = 4}$

Solution (b):

The logarithm has some value. Let's call it x.	$\log_{169} 13 = x$
Change from logarithmic to exponential form.	$169^x = 13$
169 raised to what power is 13?	$169^{1/2} = \sqrt{169} = 13$ $x = \dfrac{1}{2}$
Change from exponential to logarithmic form.	$\boxed{\log_{169} 13 = \dfrac{1}{2}}$

Solution (c):

The logarithm has some value. Let's call it x.	$\log_5 \left(\dfrac{1}{5}\right) = x$
Change from logarithmic to exponential form.	$5^x = \dfrac{1}{5}$
5 raised to what power is $\frac{1}{5}$?	$5^{-1} = \dfrac{1}{5}$ $x = -1$
Change from exponential to logarithmic form.	$\boxed{\log_5 \left(\dfrac{1}{5}\right) = -1}$

■**YOUR TURN** Evaluate the given logarithms exactly.

a. $\log_2 \frac{1}{2}$ **b.** $\log_{100} 10$ **c.** $\log_{10} 1000$

Common and Natural Logarithms

Two logarithmic bases that arise frequently are base 10 and base e. The logarithmic function of base 10 is called the **common logarithmic function**. Since it is common, $f(x) = \log_{10} x$ is often expressed as $f(x) = \log x$. Thus, if no explicit base is indicated, base 10 is implied. The logarithmic function of base e is called the **natural logarithmic function**. The natural logarithmic function $f(x) = \log_e x$ is often expressed as $f(x) = \ln x$. Both the LOG and LN buttons appear on scientific and graphing calculators. For the *logarithms* (not the functions) we say "the log" (for base 10) and "the natural log" (for base e).

Earlier in this section, we evaluated logarithms exactly by converting to exponential form and identifying the exponent. For example, to evaluate $\log_{10} 100$, we ask the question, 10 raised to what power is 100? The answer is 2.

Calculators enable us to approximate logarithms. For example, evaluate $\log_{10} 233$. We are unable to evaluate this exactly by asking the question, 10 raised to what power is 233? Since $10^2 < 10^x < 10^3$, we know the answer x must lie between 2 and 3. Instead, we use a calculator to find an approximate value 2.367.

EXAMPLE 4 **Using a Calculator to Evaluate Common and Natural Logarithms**

Use a calculator to evaluate the common and natural logarithms. Round your answers to four decimal places.

a. log 415   **b.** ln 415   **c.** log 1   **d.** ln 1   **e.** log(−2)   **f.** ln(−2)

Solution:

a. $\log(415) \approx 2.618048097 \approx \boxed{2.6180}$ **b.** $\ln(415) \approx 6.02827852 \approx \boxed{6.0283}$

c. $\log(1) = \boxed{0}$ **d.** $\ln(1) = \boxed{0}$

e. $\log(-2)$ $\boxed{\text{undefined}}$ **f.** $\ln(-2)$ $\boxed{\text{undefined}}$

Parts (c) and (d) in Example 4 illustrate that all logarithmic functions pass through the point (1, 0). Parts (e) and (f) in Example 4 illustrate that the domains of logarithmic functions are positive real numbers.

Study Tip

Logarithms can only be evaluated for positive arguments.

Graphs of Logarithmic Functions

The general logarithmic function $y = \log_b x$ is defined as the inverse of the exponential function $y = b^x$. Therefore, when these two functions are plotted on the same graph, they are symmetric about the line $y = x$. Notice the symmetry about the line $y = x$ when $y = b^x$ and $y = \log_b x$ are plotted on the same graph.

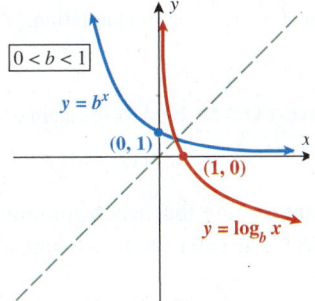

Comparison of Inverse Functions:
$$f(x) = \log_b x \text{ and } f^{-1}(x) = b^x$$

EXPONENTIAL FUNCTION	LOGARITHMIC FUNCTION
$y = b^x$	$y = \log_b x$
y-intercept $(0, 1)$	x-intercept $(1, 0)$
Domain $(-\infty, \infty)$	Domain $(0, \infty)$
Range $(0, \infty)$	Range $(-\infty, \infty)$
Horizontal asymptote: x-axis	Vertical asymptote: y-axis

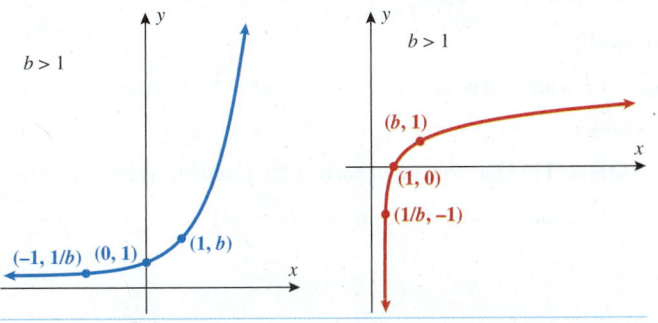

Additionally, the domain of one function is the range of the other, and vice versa. When dealing with logarithmic functions, special attention must be paid to the domain of the function. The domain of $y = \log_b x$ is $(0, \infty)$. In other words, you can only take the log of a positive real number, $x > 0$.

EXAMPLE 5   Finding the Domain of a Shifted Logarithmic Function

Find the domain of each of the given logarithmic functions.

a. $f(x) = \log_b(x - 4)$ **b.** $g(x) = \log_b(5 - 2x)$

Solution (a):

Set the argument greater than zero.	$x - 4 > 0$
Solve the inequality.	$x > 4$
Write the domain in interval notation.	$\boxed{(4, \infty)}$

Solution (b):

Set the argument greater than zero.	$5 - 2x > 0$
Solve the inequality.	$-2x > -5$
	$2x < 5$
	$x < \dfrac{5}{2}$
Write the domain in interval notation.	$\boxed{\left(-\infty, \dfrac{5}{2}\right)}$

■ **Answer: a.** $(-2, \infty)$
        b. $\left(-\infty, \frac{3}{5}\right)$

■ **YOUR TURN** Find the domain of the given logarithmic functions.

a. $f(x) = \log_b(x + 2)$ **b.** $g(x) = \log_b(3 - 5x)$

It is important to note that when finding the domain of a logarithmic function, we set the argument strictly greater than zero and solve.

EXAMPLE 6   Finding the Domain of a Logarithmic Function with a Complicated Argument

Find the domain of each of the given logarithmic functions.
a. $\ln\left(x^2 - 9\right)$ **b.** $\log(|x + 1|)$

Solution (a):

Set the argument greater than zero.	$x^2 - 9 > 0$
Solve the inequality.	$\boxed{(-\infty, -3) \cup (3, \infty)}$

Solution (b):

Set the argument greater than zero.	$|x + 1| > 0$
Solve the inequality.	$x \neq -1$
Write the domain in interval notation.	$\boxed{(-\infty, -1) \cup (-1, \infty)}$

■ **Answer: a.** $(-\infty, -2) \cup (2, \infty)$
        b. $(-\infty, 3) \cup (3, \infty)$

■ **YOUR TURN** Find the domain of each of the given logarithmic functions.

a. $\ln(x^2 - 4)$ **b.** $\log(|x - 3|)$

Recall from Section 3.3 that a technique for graphing general functions is transformation of known functions. For example, to graph $f(x) = (x - 3)^2 + 1$, we start with the known parabola $y = x^2$, whose vertex is at $(0, 0)$, and we shift that graph to the right three units and up one unit. We use the same technique for graphing logarithmic functions. To graph $y = \log_b(x + 2) - 1$, we start with the graph of $y = \log_b(x)$ and shift the graph to the left two units and down one unit.

EXAMPLE 7 **Graphing Logarithmic Functions Using Horizontal and Vertical Shifts**

Graph the functions, and state the domain and range of each.

a. $y = \log_2(x - 3)$ **b.** $\log_2(x) - 3$

Solution:

Identify the base function.

Label key features of $y = \log_2 x$.

x-intercept: $(1, 0)$

Vertical asymptote: $x = 0$

Additional points: $(2, 1)$, $(4, 2)$

$y = \log_2 x$

a. Shift the base function to the *right* three units.

x-intercept: $(4, 0)$

Vertical asymptote: $x = 3$

Additional points: $(5, 1)$, $(7, 2)$

Domain: $(3, \infty)$ Range: $(-\infty, \infty)$

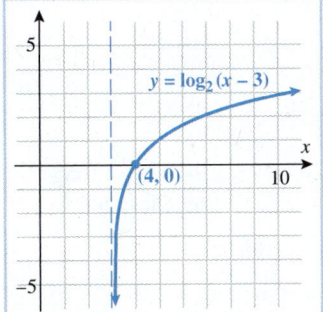

b. Shift the base function *down* three units.

x-intercept: $(1, -3)$

Vertical asymptote: $x = 0$

Additional points: $(2, -2)$, $(4, -1)$

Domain: $(0, \infty)$ Range: $(-\infty, \infty)$

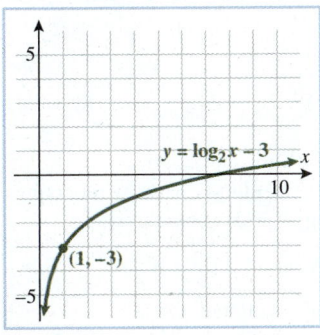

■ **Answer:**
a. Domain: $(0, \infty)$ Range: $(-\infty, \infty)$
b. Domain: $(-3, \infty)$ Range: $(-\infty, \infty)$
c. Domain: $(0, \infty)$ Range: $(-\infty, \infty)$

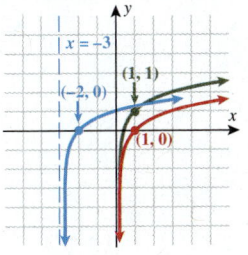

....

■ **YOUR TURN** Graph the functions and state the domain and range of each.

a. $y = \log_3 x$ **b.** $y = \log_3(x + 3)$ **c.** $\log_3(x) + 1$

All of the transformation techniques (shifting, reflection, and compression) discussed in Chapter 3 also apply to logarithmic functions. For example, the graphs of $-\log_2 x$ and $\log_2(-x)$ are found by reflecting the graph of $y = \log_2 x$ about the x-axis and y-axis, respectively.

 EXAMPLE 8 **Graphing Logarithmic Functions Using Transformations**

Graph the function $f(x) = -\log_2(x - 3)$ and state its domain and range.

Solution:

Graph $y = \log_2 x$.

x-intercept: **(1, 0)**

Vertical asymptote: $x = 0$

Additional points: **(2, 1)**, **(4, 2)**

Graph $y = \log_2(x - 3)$ by shifting $y = \log_2 x$ to the *right* three units.

x-intercept: **(4, 0)**

Vertical asymptote: $x = 3$

Additional points: **(5, 1)**, **(7, 2)**

Graph $y = -\log_2(x - 3)$ by reflecting $y = \log_2(x - 3)$ about the x-axis.

x-intercept: **(4, 0)**

Vertical asymptote: $x = 3$

Additional points: **(5, −1)**, **(7, −2)**

Domain: **(3, ∞)**   Range: **(−∞, ∞)**

Applications of Logarithms

Logarithms are used to make a large range of numbers manageable. For example, to create a scale to measure a human's ability to hear, we must have a way to measure the sound intensity of an explosion, even though that intensity can be more than a trillion (10^{12}) times greater than that of a soft whisper. Decibels in engineering and physics, pH in chemistry, and the Richter scale for earthquakes are all applications of logarithmic functions.

The **decibel** is a logarithmic unit used to measure the magnitude of a physical quantity relative to a specified reference level. The *decibel* (dB) is employed in many engineering and science applications. The most common application is the intensity of sound.

DEFINITION   Decibel (Sound)

The **decibel** is defined as
$$D = 10 \log\left(\frac{I}{I_T}\right)$$

where D is the decibel level (dB), I is the intensity of the sound measured in watts per square meter, and I_T is the intensity threshold of the least audible sound a human can hear.

The human average threshold is $I_T = 1 \times 10^{-12}$ W/m².

Notice that when $I = I_T$, then $D = 10 \overbrace{\log 1}^{0} = 0$ dB. People who work professionally with sound, such as acoustics engineers or medical hearing specialists, refer to this threshold level I_T as "0 dB." The following table illustrates typical sounds we hear and their corresponding decibel levels.

SOUND SOURCE	SOUND INTENSITY (W/m²)	DECIBELS (dB)
Threshold of hearing	1.0×10^{-12}	0
Vacuum cleaner	1.0×10^{-4}	80
iPod	1.0×10^{-2}	100
Jet engine	1.0×10^{3}	150

For example, a whisper (approximately 0 dB) from someone standing next to a jet engine (150 dB) might go unheard because when these are added, we get approximately 150 dB (the jet engine).

EXAMPLE 9   Calculating Decibels of Sounds

Suppose you have seats to a concert given by your favorite musical artist. Calculate the approximate decibel level associated with the typical sound intensity, given $I = 1 \times 10^{-2}$ W/m².

Solution:

Write the decibel-scale formula.
$$D = 10 \log\left(\frac{I}{I_T}\right)$$

Substitute $I = 1 \times 10^{-2}$ W/m² and $I_T = 1 \times 10^{-12}$ W/m².
$$D = 10 \log\left(\frac{1 \times 10^{-2}}{1 \times 10^{-12}}\right)$$

Simplify. $\qquad$ $D = 10 \log(10^{10})$

Recall that the implied base for log is 10. $\qquad$ $D = 10 \log_{10}(10^{10})$

Evaluate the right side. $\left[\log_{10}(10^{10}) = 10\right]$ $\qquad$ $D = 10 \cdot 10$

$$D = 100$$

The typical sound level on the front row of a rock concert is $\boxed{100 \text{ dB}}$.

■ **Answer:** 160 dB

▪ **YOUR TURN** Calculate the approximate decibels associated with a sound so loud it will cause instant perforation of the eardrums, $I = 1 \times 10^4 \text{ W/m}^2$.

The Richter scale (earthquakes) is another application of logarithms.

> **DEFINITION**   **Richter Scale**
>
> The magnitude M of an earthquake is measured using the **Richter scale**
>
> $$M = \frac{2}{3} \log\left(\frac{E}{E_0}\right)$$
>
> where: M is the magnitude
> E is the seismic energy released by the earthquake (in joules)
> E_0 is the energy released by a reference earthquake $E_0 = 10^{4.4}$ joules

Technology Tip

Enter the number 1.12×10^{15} using the scientific notation key $\boxed{\text{EXP}}$ or $\boxed{\text{EE}}$.

```
2/3log(1.12ε15/1
0^4.4)
          7.099478682
```

EXAMPLE 10   Calculating the Magnitude of an Earthquake

On October 17, 1989, just moments before game 3 of the World Series between the Oakland A's and the San Francisco Giants was about to start—with 60,000 fans in Candlestick Park—a devastating earthquake erupted. Parts of interstates and bridges collapsed, and President George H. W. Bush declared the area a disaster zone. The earthquake released approximately 1.12×10^{15} joules of energy. Calculate the magnitude of the earthquake using the Richter scale.

Solution:

Write the Richter scale formula. $\qquad$ $M = \frac{2}{3} \log\left(\frac{E}{E_0}\right)$

Substitute $E = 1.12 \times 10^{15}$ and $E_0 = 10^{4.4}$. $\qquad$ $M = \frac{2}{3} \log\left(\frac{1.12 \times 10^{15}}{10^{4.4}}\right)$

Simplify. $\qquad$ $M = \frac{2}{3} \log\left(1.12 \times 10^{10.6}\right)$

Approximate the logarithm using a calculator. $\qquad$ $M \approx \frac{2}{3}(10.65) \approx 7.1$

The 1989 earthquake in California measured $\boxed{7.1}$ on the Richter scale.

■ **Answer:** 5.1

▪ **YOUR TURN** On May 3, 1996, Seattle experienced a moderate earthquake. The energy that the earthquake released was approximately 1.12×10^{12} joules. Calculate the magnitude of the 1996 Seattle earthquake using the Richter scale.

A **logarithmic scale** expresses the logarithm of a physical quantity instead of the quantity itself. In music, the pitch is the perceived fundamental frequency of sound. The note A above middle C on a piano has the pitch associated with a pure tone of 440 hertz (Hz). An octave is the interval between one musical pitch and another with either double or half its frequency. For example, if a note has a frequency of 440 Hz, then the note an octave above it has a frequency of 880 Hz, and the note an octave below it has a frequency of 220 Hz. Therefore, the ratio of two notes an octave apart is 2:1.

The following table lists the frequencies associated with A notes.

NOTE	A_1	A_2	A_3	A_4	A_5	A_6	A_7
Frequency (Hz)	55	110	220	440	880	1760	3520
Octave with respect to A_4	-3	-2	-1	0	$+1$	$+2$	$+3$

We can graph $\dfrac{\text{Frequency of note}}{440 \text{ Hz}}$ on the horizontal axis and the octave (with respect to A_4) on the vertical axis.

If we instead graph the logarithm of this quantity, $\log\left[\dfrac{\text{Frequency of note}}{440 \text{ Hz}}\right]$, we see that using a logarithmic scale expresses octaves linearly (up or down an octave). In other words, an "octave" is a purely logarithmic concept.

When a logarithmic scale is used we typically classify a graph one of two ways:

- Log-log plot (both the horizontal and vertical axes use logarithmic scales)
- Semilog plot (one of the axes uses a logarithmic scale)

The second graph with octaves on the vertical axis and the log of the ratio of frequencies on the horizontal axis is called a semilog plot.

Semilog Plot

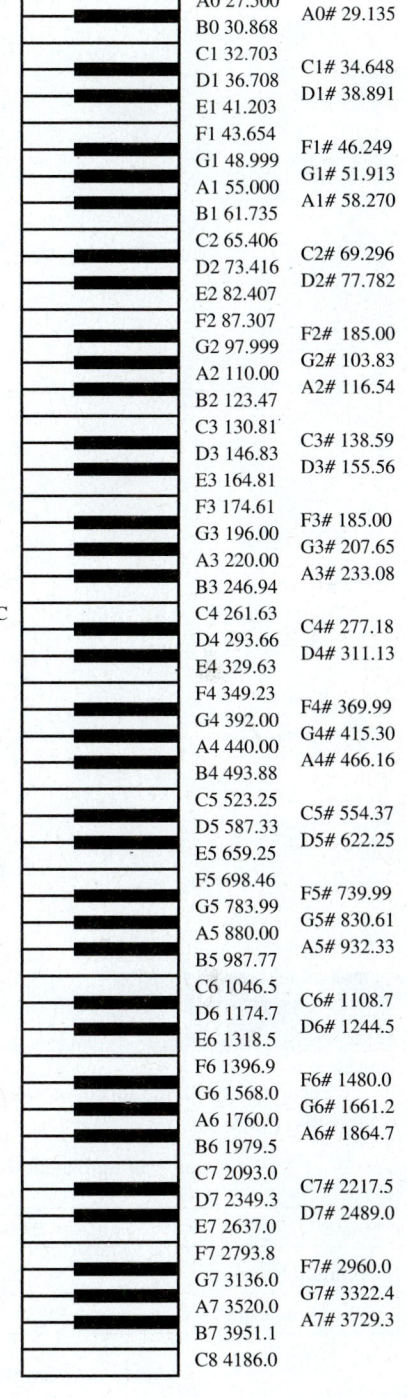

Middle C

EXAMPLE 11   Graphing Using A Logarithmic Scale

Frequency is inversely proportional to the wavelength: In a vacuum $f = \dfrac{c}{\lambda}$, where f is the frequency (in hertz), $c = 3.0 \times 10^8$ m/s is the speed of light in a vacuum, and λ is the wavelength in meters. Graph frequency versus wavelength using a log-log plot.

Solution:

Let wavelength range from microns (10^{-6}) to hundreds of meters (10^2) by powers of 10 along the horizontal axis.

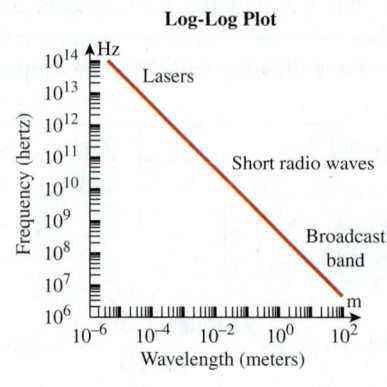

Log-Log Plot

λ	$f = \dfrac{3.0 \times 10^8}{\lambda}$	
10^{-6}	3.0×10^{14}	
10^{-5}	3.0×10^{13}	
10^{-4}	3.0×10^{12}	(TeraHertz: THz)
10^{-3}	3.0×10^{11}	
10^{-2}	3.0×10^{10}	
10^{-1}	3.0×10^{9}	(GigaHertz: GHz)
10^{0}	3.0×10^{8}	
10^{1}	3.0×10^{7}	
10^{2}	3.0×10^{6}	(MegaHertz: MHz)

The logarithmic scales allow us to represent a large range of numbers. In this graph, the x-axis ranges from microns, 10^{-6} meters, to hundreds of meters, and the y-axis ranges from megahertz (MHz), 10^6 hertz, to hundreds of terahertz (THz), 10^{12} hertz.

SECTION 5.2 SUMMARY

In this section, logarithmic functions were defined as inverses of exponential functions.

$$y = \log_b x \text{ is equivalent to } x = b^y$$

NAME	EXPLICIT BASE	IMPLICIT BASE
Common logarithm	$f(x) = \log_{10} x$	$f(x) = \log x$
Natural logarithm	$f(x) = \log_e x$	$f(x) = \ln x$

Evaluating Logarithms

- *Exact:* Convert to exponential form first, then evaluate.
- *Approximate:* Natural and common logarithms with calculators.

Graphs of Logarithmic Functions

EXPONENTIAL FUNCTION	LOGARITHMIC FUNCTION
$y = b^x$	$y = \log_b x$
y-intercept: $(0, 1)$	x-intercept: $(1, 0)$
Domain: $(-\infty, \infty)$	Domain: $(0, \infty)$
Range: $(0, \infty)$	Range: $(-\infty, \infty)$
Horizontal asymptote: x-axis	Vertical asymptote: y-axis

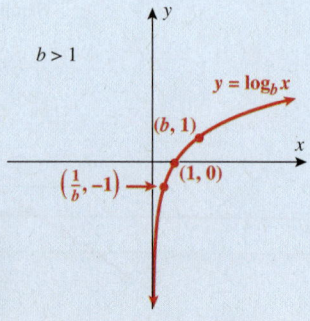

SECTION
5.2 EXERCISES

SKILLS

In Exercises 1–20, write each logarithmic equation in its equivalent exponential form.

1. $\log_5 125 = 3$

2. $\log_3 27 = 3$

3. $\log_{81} 3 = \frac{1}{4}$

4. $\log_{121} 11 = \frac{1}{2}$

5. $\log_2 \left(\frac{1}{32}\right) = -5$

6. $\log_3 \left(\frac{1}{81}\right) = -4$

7. $\log 0.01 = -2$

8. $\log 0.0001 = -4$

9. $\log 10{,}000 = 4$

10. $\log 1000 = 3$

11. $\log_{1/4} (64) = -3$

12. $\log_{1/6} (36) = -2$

13. $-1 = \ln \left(\frac{1}{e}\right)$

14. $1 = \ln e$

15. $\ln 1 = 0$

16. $\log 1 = 0$

17. $\ln 5 = x$

18. $\ln 4 = y$

19. $z = \log_x y$

20. $y = \log_x z$

In Exercises 21–34, write each exponential equation in its equivalent logarithmic form.

21. $8^3 = 512$

22. $2^6 = 64$

23. $0.00001 = 10^{-5}$

24. $100{,}000 = 10^5$

25. $15 = \sqrt{225}$

26. $7 = \sqrt[3]{343}$

27. $\frac{8}{125} = \left(\frac{2}{5}\right)^3$

28. $\frac{8}{27} = \left(\frac{2}{3}\right)^3$

29. $3 = \left(\frac{1}{27}\right)^{-1/3}$

30. $4 = \left(\frac{1}{1024}\right)^{-1/5}$

31. $e^x = 6$

32. $e^{-x} = 4$

33. $x = y^z$

34. $z = y^x$

In Exercises 35–46, evaluate the logarithms exactly (if possible).

35. $\log_2 1$

36. $\log_5 1$

37. $\log_5 3125$

38. $\log_3 729$

39. $\log 10^7$

40. $\log 10^{-2}$

41. $\log_{1/4} 4096$

42. $\log_{1/7} 2401$

43. $\log 0$

44. $\ln 0$

45. $\log(-100)$

46. $\ln(-1)$

In Exercises 47–54, approximate (if possible) the common and natural logarithms using a calculator. Round to two decimal places.

47. $\log 29$

48. $\ln 29$

49. $\ln 380$

50. $\log 380$

51. $\log 0$

52. $\ln 0$

53. $\ln 0.0003$

54. $\log 0.0003$

In Exercises 55–64, state the domain of the logarithmic function in interval notation.

55. $f(x) = \log_2(x + 5)$

56. $f(x) = \log_2(4x - 1)$

57. $f(x) = \log_3(5 - 2x)$

58. $f(x) = \log_3(5 - x)$

59. $f(x) = \ln(7 - 2x)$

60. $f(x) = \ln(3 - x)$

61. $f(x) = \log|x|$

62. $f(x) = \log|x + 1|$

63. $f(x) = \log(x^2 + 1)$

64. $f(x) = \log(1 - x^2)$

In Exercises 65–70, match the graph with the function.

65. $y = \log_5 x$

66. $y = \log_5(-x)$

67. $y = -\log_5(-x)$

68. $y = \log_5(x + 3) - 1$

69. $y = \log_5(1 - x) - 2$

70. $y = -\log_5(3 - x) + 2$

a.

b.

c.

d.

e.

f.

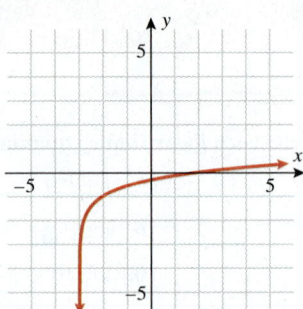

In Exercises 71–82, graph the logarithmic function using transformation techniques. State the domain and range of *f*.

71. $f(x) = \log(x - 1)$ **72.** $f(x) = \log(x + 2)$ **73.** $\ln x + 2$ **74.** $\ln x - 1$

75. $f(x) = \log_3(x + 2) - 1$ **76.** $f(x) = \log_3(x + 1) - 2$ **77.** $f(x) = -\log(x) + 1$ **78.** $f(x) = \log(-x) + 2$

79. $f(x) = \ln(x + 4)$ **80.** $f(x) = \ln(4 - x)$ **81.** $f(x) = \log(2x)$ **82.** $f(x) = 2\ln(-x)$

■ APPLICATIONS

For Exercises 83–86, refer to the following:

$$\text{Decibel: } D = 10\log\left(\frac{I}{I_T}\right) \qquad I_T = 1 \times 10^{-12} \text{ W/m}^2$$

83. Sound. Calculate the decibels associated with *normal conversation* if the intensity is $I = 1 \times 10^{-6}$ W/m².

84. Sound. Calculate the decibels associated with the *onset of pain* if the intensity is $I = 1 \times 10^{1}$ W/m².

85. Sound. Calculate the decibels associated with attending *a football game in a loud college stadium* if the intensity is $I = 1 \times 10^{-0.3}$ W/m².

86. Sound. Calculate the decibels associated with a *doorbell* if the intensity is $I = 1 \times 10^{-4.5}$ W/m².

For Exercises 87–90, refer to the following:

$$\text{Richter Scale: } M = \frac{2}{3}\log\left(\frac{E}{E_0}\right) \qquad E_0 = 10^{4.4} \text{ joules}$$

87. Earthquakes. On Good Friday 1964, one of the most severe North American earthquakes ever recorded struck Alaska. The energy released measured 1.41×10^{17} joules. Calculate the magnitude of the 1964 Alaska earthquake using the Richter scale.

88. Earthquakes. On January 22, 2003, Colima, Mexico, experienced a major earthquake. The energy released measured 6.31×10^{15} joules. Calculate the magnitude of the 2003 Mexican earthquake using the Richter scale.

89. Earthquakes. On December 26, 2003, a major earthquake rocked southeastern Iran. In Bam, 30,000 people were killed, and 85% of buildings were damaged or destroyed. The energy released measured 2×10^{14} joules. Calculate the magnitude of the 2003 Iran earthquake with the Richter scale.

90. Earthquakes. On November 1, 1755, Lisbon was destroyed by an earthquake, which killed 90,000 people and destroyed 85% of the city. It was one of the most destructive earthquakes in history. The energy released measured 8×10^{17} joules. Calculate the magnitude of the 1755 Lisbon earthquake with the Richter scale.

For Exercises 91–96, refer to the following:

The pH of a solution is a measure of the molar concentration of hydrogen ions, H^+ in moles per liter, in the solution, which means that it is a measure of the acidity or basicity of the solution. The letters pH stand for "power of hydrogen," and the numerical value is defined as

$$\text{pH} = -\log_{10}\left[H^+\right]$$

Very acidic corresponds to pH values near 1, neutral corresponds to a pH near 7 (pure water), and very basic corresponds to values near 14. In the next six exercises you will be asked to calculate the pH value of wine, Pepto-Bismol, normal rainwater, bleach, and two fruits. List these six liquids and use your intuition to classify them as neutral, acidic, very acidic, basic, or very basic before you calculate their actual pH values.

91. Chemistry. If wine has an approximate hydrogen ion concentration of 5.01×10^{-4}, calculate its pH value.

92. Chemistry. Pepto-Bismol has a hydrogen ion concentration of about 5.01×10^{-11}. Calculate its pH value.

93. Chemistry. Normal rainwater is slightly acidic and has an approximate hydrogen ion concentration of $10^{-5.6}$. Calculate its pH value. Acid rain and tomato juice have similar approximate hydrogen ion concentrations of 10^{-4}. Calculate the pH value of acid rain and tomato juice.

94. Chemistry. Bleach has an approximate hydrogen ion concentration of 5.0×10^{-13}. Calculate its pH value.

95. Chemistry. An apple has an approximate hydrogen ion concentration of $10^{-3.6}$. Calculate its pH value.

96. Chemistry. An orange has an approximate hydrogen ion concentration of $10^{-4.2}$. Calculate its pH value.

97. Archaeology. Carbon dating is a method used to determine the age of a fossil or other organic remains. The age t in years is related to the mass C (in milligrams) of carbon 14 through a logarithmic equation:

$$t = -\frac{\ln\left(\dfrac{C}{500}\right)}{0.0001216}$$

How old is a fossil that contains 100 milligrams of carbon 14?

98. Archaeology. Repeat Exercise 97, only now the fossil contains 40 milligrams of carbon 14.

99. Broadcasting. Decibels are used to quantify losses associated with atmospheric interference in a communication system. The ratio of the power (watts) received to the power transmitted (watts) is often compared. Often, *watts* are transmitted, but losses due to the atmosphere typically correspond to *milliwatts* being received:

$$dB = 10 \log\left(\frac{\text{Power received}}{\text{Power transmitted}}\right)$$

If 1 W of power is transmitted and 3 mW is received, calculate the power loss in dB.

100. Broadcasting. Repeat Exercise 99 assuming 3 W of power is transmitted and 0.2 mW is received.

For Exercises 101 and 102, refer to the following:

The range of all possible frequencies of electromagnetic radiation is called the electromagnetic spectrum. In a vacuum the frequency of electromagnetic radiation is modeled by

$$f = \frac{c}{\lambda}$$

where c is 3.0×10^8 m/s and λ is wavelength in meters.

101. Physics/Electromagnetic Spectrum. The *radio spectrum* is the portion of the electromagnetic spectrum that corresponds to radio frequencies. The radio spectrum is used for various transmission technologies and is government regulated. Ranges of the radio spectrum are often allocated based on usage; for example, AM radio, cell phones, and television. (Source: http://en.wikipedia.org/wiki/Radio_spectrum)

a. Complete the following table for the various usages of the radio spectrum.

USAGE	WAVELENGTH	FREQUENCY
Super Low Frequency— Communication with Submarines	10,000,000 m	30 Hz
Ultra Low Frequency— Communication within Mines	1,000,000 m	
Very Low Frequency— Avalanche Beacons	100,000 m	
Low Frequency— Navigation, AM Long-wave Broadcasting	10,000 m	
Medium Frequency— AM Broadcasts, Amateur Radio	1000 m	
High Frequency— Shortwave broadcasts, Citizens Band Radio	100 m	
Very High Frequency— FM Radio, Television	10 m	
Ultra High Frequency— Television, Mobile Phones	0.050 m	

b. Graph the frequency within the radio spectrum (in Hertz) as a function of wavelength (in meters).

102. Physics/Electromagnetic Spectrum. The *visible spectrum* is the portion of the electromagnetic spectrum that is visible to the human eye. Typically, the human eye can see wavelengths between 390 and 750 nm (nanometers or 10^{-9} m).

a. Complete the following table for the following colors of the visible spectrum.

COLOR	WAVELENGTH	FREQUENCY
Violet	400 nm	750×10^{12} Hz
Cyan	470 nm	
Green	480 nm	
Yellow	580 nm	
Orange	610 nm	
Red	630 nm	

b. Graph the frequency (in Hertz) of the colors as a function of wavelength (in meters) on a log-log plot.

■ CATCH THE MISTAKE

In Exercises 103–106, explain the mistake that is made.

103. Evaluate the logarithm $\log_2 4$.

Solution:

Set the logarithm equal to x. $\qquad\qquad \log_2 4 = x$

Write the logarithm in exponential form. $\qquad x = 2^4$

Simplify. $\qquad\qquad\qquad\qquad\qquad x = 16$

Answer: $\qquad\qquad\qquad\qquad\qquad \log_2 4 = 16$

This is incorrect. The correct answer is $\log_2 4 = 2$. What went wrong?

104. Evaluate the logarithm $\log_{100} 10$.

Solution:

Set the logarithm equal to x. $\qquad\qquad \log_{100} 10 = x$

Express the equation in exponential form. $\qquad 10^x = 100$

Solve for x. $\qquad\qquad\qquad\qquad\qquad x = 2$

Answer: $\qquad\qquad\qquad\qquad\qquad \log_{100} 10 = 2$

This is incorrect. The correct answer is $\log_{100} 10 = \frac{1}{2}$. What went wrong?

105. State the domain of the logarithmic function $f(x) = \log_2(x + 5)$ in interval notation.

Solution:

The domain of all logarithmic functions is $x > 0$.

Interval notation: $(0, \infty)$

This is incorrect. What went wrong?

106. State the domain of the logarithmic function $f(x) = \ln|x|$ in interval notation.

Solution:

Since the absolute value eliminates all negative numbers, the domain is the set of all real numbers.

Interval notation: $(-\infty, \infty)$

This is incorrect. What went wrong?

■ CONCEPTUAL

In Exercises 107–110, determine whether each statement is true or false.

107. The domain of the standard logarithmic function, $y = \ln x$, is the set of nonnegative real numbers.

108. The horizontal axis is the horizontal asymptote of the graph of $y = \ln x$.

109. The graphs of $y = \log x$ and $y = \ln x$ have the same x-intercept $(1, 0)$.

110. The graphs of $y = \log x$ and $y = \ln x$ have the same vertical asymptote, $x = 0$.

■ CHALLENGE

111. State the domain, range, and x-intercept of the function $f(x) = -\ln(x - a) + b$ for a and b real positive numbers.

112. State the domain, range, and x-intercept of the function $f(x) = \log(a - x) - b$ for a and b real positive numbers.

113. Graph the function $f(x) = \begin{cases} \ln(-x) & x < 0 \\ \ln(x) & x > 0 \end{cases}$.

114. Graph the function $f(x) = \begin{cases} -\ln(-x) & x < 0 \\ -\ln(x) & x > 0 \end{cases}$.

■ **TECHNOLOGY**

115. Use a graphing utility to graph $y = e^x$ and $y = \ln x$ in the same viewing screen. What line are these two graphs symmetric about?

116. Use a graphing utility to graph $y = 10^x$ and $y = \log x$ in the same viewing screen. What line are these two graphs symmetric about?

117. Use a graphing utility to graph $y = \log x$ and $y = \ln x$ in the same viewing screen. What are the two common characteristics?

118. Using a graphing utility, graph $y = \ln|x|$. Is the function defined everywhere?

119. Use a graphing utility to graph $f(x) = \ln(3x)$, $g(x) = \ln 3 + \ln x$, and $h(x) = (\ln 3)(\ln x)$ in the same viewing screen. Determine the domain where two of the functions give the same graph.

120. Use a graphing utility to graph $f(x) = \ln(x^2 - 4)$, $g(x) = \ln(x + 2) + \ln(x - 2)$, and $h(x) = \ln(x + 2)\ln(x - 2)$ in the same viewing screen. Determine the domain where two of the functions give the same graph.

For Exercises 121 and 122, refer to the following:

Experimental data is collected all the time in biology and chemistry labs as scientists seek to understand natural phenomena. In biochemistry, the *Michaelis–Menten* kinetics law describes the rates of enzyme reactions using the relationship between the rate of the reaction and the concentration of the substrate involved.

The following data has been collected, where the velocity v is measured in μmol/min of the enzyme reaction and the substrate level [S] is measured in mol/L.

[S] (IN MOL/L)	v (IN μMOL/MIN)
4.0×10^{-4}	130
2.0×10^{-4}	110
1.0×10^{-4}	89
5.0×10^{-5}	62
4.0×10^{-5}	52
2.5×10^{-5}	38
2.0×10^{-5}	32

121. a. Create a scatterplot of this data by identifying [S] with the x-axis and v with the y-axis.
   b. The graph *seems* to be leveling off. Give an estimate of the maximum value the velocity might achieve. Call this estimate V_{max}.

c. Another constant of importance in describing the relationship between v and [S] is K_m. This is the value of [S] that results in the velocity being half its maximum value. Estimate this value.
   Note: K_m measures the affinity level of a particular enzyme to a particular substrate. The lower the value of K_m, the higher the affinity. The higher the value of K_m, the lower the affinity.

d. The actual equation that governs the relationship between v and [S] is $v = \dfrac{V_{max}}{1 + K_m/[S]}$, which is NOT linear. This is the *simple Michaelis–Menten kinetics equation.*

   i. Use *LnReg* to get a best fit logarithmic curve for this data. Although the relationship between v and [S] is not logarithmic (rather, it is logistic), the best fit logarithmic curve does not grow very quickly and so it serves as a reasonably good fit.
   ii. At what [S] value, approximately, is the velocity 100 μmol/min?

122. The Michaelis–Menten equation can be arranged into various other forms that give a straight line (rather than a logistic curve) when one variable is plotted against another. One such rearrangement is the *double-reciprocal Lineweaver–Burk equation*. This equation plots the data values of the reciprocal of velocity $(1/v)$ versus the reciprocal of the substrate level $(1/[S])$. The equation is as follows:

$$\frac{1}{v} = \frac{K_m}{V_{max}}\frac{1}{[S]} + \frac{1}{V_{max}}$$

Think of y as $\dfrac{1}{v}$ and x as $\dfrac{1}{[S]}$.

a. What is the slope of the "line"? How about its y-intercept?
b. Using the data from Exercise 121, we create two new columns for $1/v$ and $1/[S]$ to obtain the following data set:

[S] (MOL/L)	1/[S]	v (μMOL/MIN)	1/v
4.00E−04	2.50E+03	130	0.00769231
2.00E−04	5.00E+03	110	0.00909091
1.00E−04	1.00E+04	89	0.01123596
5.00E−05	2.00E+04	62	0.01612903
4.00E−05	2.50E+04	53	0.01886792
2.50E−05	4.00E+04	38	0.02631579
2.00E−05	5.00E+04	32	0.03125

Create a scatterplot for the new data, treating x as $1/[S]$ and y as $1/v$.

c. Determine the best fit line and value of r.
d. Use the equation of the best fit line in (c) to calculate V_{max}.
e. Use the above information to determine K_m.

SKILLS OBJECTIVES

- Write a single logarithm as a sum or difference of logarithms.
- Write a logarithmic expression as a single logarithm.
- Evaluate logarithms of a general base (other than base 10 or e).

CONCEPTUAL OBJECTIVES

- Derive the seven basic logarithmic properties.
- Derive the change-of-base formula.

Properties of Logarithmic Functions

Since exponential functions and logarithmic functions are inverses of one another, properties of exponents are related to properties of logarithms. We will start by reviewing properties of exponents, and then proceed to properties of logarithms.

In Chapter 0, properties of exponents were discussed.

PROPERTIES OF EXPONENTS

Let a, b, m, and n be any real numbers and $m > 0$ and $n > 0$, then the following are true.

1. $b^m \cdot b^n = b^{m+n}$ **2.** $b^{-m} = \dfrac{1}{b^m} = \left(\dfrac{1}{b}\right)^m$ **3.** $\dfrac{b^m}{b^n} = b^{m-n}$

4. $(b^m)^n = b^{mn}$ **5.** $(ab)^m = a^m \cdot b^m$ **6.** $b^0 = 1$

7. $b^1 = b$

From these properties of exponents we can develop similar properties for logarithms. We list seven basic properties.

PROPERTIES OF LOGARITHMS

If b, M, and N are positive real numbers, where $b \neq 1$ and p and x are real numbers, then the following are true:

1. $\log_b 1 = 0$ **2.** $\log_b b = 1$

3. $\log_b b^x = x$ **4.** $b^{\log_b x} = x$ $x > 0$

5. $\log_b MN = \log_b M + \log_b N$ *Product rule:* Log of a product is the sum of the logs.

6. $\log_b \left(\dfrac{M}{N}\right) = \log_b M - \log_b N$ *Quotient rule:* Log of a quotient is the difference of the logs.

7. $\log_b M^p = p \log_b M$ *Power rule:* Log of a number raised to an exponent is the exponent times the log of the number.

We will devote this section to proving and illustrating these seven properties.

The first two properties follow directly from the definition of a logarithmic function and properties of exponentials.

Property (1): $\log_b 1 = 0$ since $b^0 = 1$

Property (2): $\log_b b = 1$ since $b^1 = b$

The third and fourth properties follow from the fact that exponential functions and logarithmic functions are inverses of one another. Recall that inverse functions satisfy the relationship that $f^{-1}(f(x)) = x$ for all x in the domain of $f(x)$, and $f(f^{-1}(x)) = x$ for all x in the domain of f^{-1}. Let $f(x) = b^x$ and $f^{-1}(x) = \log_b x$.

Property (3):

Write the inverse identity.	$f^{-1}(f(x)) = x$
Substitute $f^{-1}(x) = \log_b x$.	$\log_b(f(x)) = x$
Substitute $f(x) = b^x$.	$\log_b b^x = x$

Property (4):

Write the inverse identity.	$f(f^{-1}(x)) = x$
Substitute $f(x) = b^x$.	$b^{f^{-1}(x)} = x$
Substitute $f^{-1}(x) = \log_b x$ $x > 0$.	$b^{\log_b x} = x$

The first four properties are summarized for common and natural logarithms.

COMMON AND NATURAL LOGARITHM PROPERTIES

Common Logarithm (base 10)	Natural Logarithm (base e)
1. $\log 1 = 0$	**1.** $\ln 1 = 0$
2. $\log 10 = 1$	**2.** $\ln e = 1$
3. $\log 10^x = x$	**3.** $\ln e^x = x$
4. $10^{\log x} = x$ $x > 0$	**4.** $e^{\ln x} = x$ $x > 0$

EXAMPLE 1   Using Logarithmic Properties

Use properties (1)–(4) to simplify the following expressions.

a. $\log_{10} 10$ **b.** $\ln 1$ **c.** $10^{\log(x+8)}$

d. $e^{\ln(2x+5)}$ **e.** $\log 10^{x^2}$ **f.** $\ln e^{x+3}$

Solution:

a. Use property (2). $\log_{10} 10 = \boxed{1}$

b. Use property (1). $\ln 1 = \boxed{0}$

c. Use property (4). $10^{\log(x+8)} = \boxed{x+8}$ $x > -8$

d. Use property (4). $e^{\ln(2x+5)} = \boxed{2x+5}$ $x > -\frac{5}{2}$

e. Use property (3). $\log 10^{x^2} = \boxed{x^2}$

f. Use property (3). $\ln e^{x+3} = \boxed{x+3}$

Classroom Example 5.3.1
Simplify the expressions.
a. $\ln \sqrt[3]{e^2}$
b. $2^{\log_2(\sqrt{1-x})}$
c.* $\log\left(10^{\log_2 3x}\right)$
d. $\ln(\ln e)$
Answer:
a. $\frac{2}{3}$ **b.** $\sqrt{1-x}$
c. $\log_2 3x, x > 0$ **d.** 0

The fifth through seventh properties follow from the properties of exponents and the definition of logarithms. We will prove the product rule and leave the proofs of the quotient and power rules for the exercises.

$$\text{Property (5):} \quad \log_b MN = \log_b M + \log_b N$$

WORDS	MATH
Assume two logs that have the same base.	Let $u = \log_b M$ and $v = \log_b N$ $M > 0, N > 0$
Change to equivalent exponential forms.	$b^u = M$ and $b^v = N$
Write the log of a product.	$\log_b MN$
Substitute $M = b^u$ and $N = b^v$.	$= \log_b(b^u b^v)$
Use properties of exponents.	$= \log_b(b^{u+v})$
Apply property 3.	$= u + v$
Substitute $u = \log_b M, v = \log_b N$.	$= \log_b M + \log_b N$

$$\boxed{\log_b MN = \log_b M + \log_b N}$$

In other words, the log of a product is the sum of the logs. Let us illustrate this property with a simple example.

$$\overbrace{\log_2 8}^{3} + \overbrace{\log_2 4}^{2} = \overbrace{\log_2 32}^{5}$$

Notice that $\log_2 8 + \log_2 4 \neq \log_2 12$.

Classroom Example 5.3.2
Express as a sum of simpler logarithms.
a. $\log_a\left(x^2 y^3 \sqrt[3]{z^2}\right)$

b.* $\ln\left((ex^2)^2 e^y\right)$

Answer:

a. $2\log_a x + 3\log_a y + \frac{2}{3}\log_a z$

b. $2 + 4\ln x + y$

EXAMPLE 2 **Writing a Logarithmic Expression as a Sum of Logarithms**

Use the logarithmic properties to write the expression $\log_b\left(u^2 \sqrt{v}\right)$ as a sum of simpler logarithms.

Solution:

Convert the radical to exponential form. $\quad \log_b\left(u^2 \sqrt{v}\right) = \log_b(u^2 v^{1/2})$

Use the product property (5). $\quad\quad\quad\quad\quad = \log_b u^2 + \log_b v^{1/2}$

Use the power property (7). $\quad\quad\quad\quad\quad = \boxed{2\log_b u + \frac{1}{2}\log_b v}$

■ **Answer:**
$\log_b\left(x^4 \sqrt[3]{y}\right) = 4\log_b x + \frac{1}{3}\log_b y$

■ **YOUR TURN** Use the logarithmic properties to write the expression $\log_b\left(x^4 \sqrt[3]{y}\right)$ as a sum of simpler logarithms.

EXAMPLE 3 **Writing a Sum of Logarithms as a Single Logarithmic Expression: The Right Way and the Wrong Way**

Use properties of logarithms to write the expression $2\log_b 3 + 4\log_b u$ as a single logarithmic expression.

COMMON MISTAKE

A common mistake is to write the sum of the logs as a log of the sum.

$$\log_b M + \log_b N \neq \log_b(M + N)$$

★ **CORRECT**

Use the power property (7).
$2\log_b 3 + 4\log_b u = \log_b 3^2 + \log_b u^4$
Simplify.

$$\log_b 9 + \log_b u^4$$

Use the product property (5).

$$\boxed{= \log_b(9u^4)}$$

✖ **INCORRECT**

$$\neq \log_b(9 + u^4) \quad \textbf{ERROR}$$

▼ **CAUTION**

$\log_b M + \log_b N = \log_b(MN)$

$\log_b M + \log_b N \neq \log_b(M + N)$

Classroom Example 5.3.3
Write the expressions as single logarithms.

a. $\log_3 x + 2\log_3 y + 8\log_3 \sqrt{z}$

b.* $\ln\sqrt{ex} + 2\ln y^3 + 6\ln\sqrt[4]{e^2 z^2}$

Answer:

a. $\log_3(xy^2 z^4)$

b. $\ln\left(e^4\sqrt{ex}\, y^6 z^4\right)$

▪ **YOUR TURN** Express $2\ln x + 3\ln y$ as a single logarithm.

▪ **Answer:** $\ln(x^2 y^3)$

EXAMPLE 4 **Writing a Logarithmic Expression as a Difference of Logarithms**

Write the expression $\ln\left(\dfrac{x^3}{y^2}\right)$ as a difference of logarithms.

Solution:

Apply the quotient property (6).

$$\ln\left(\frac{x^3}{y^2}\right) = \ln(x^3) - \ln(y^2)$$

Apply the power property (7).

$$= \boxed{3\ln x - 2\ln y}$$

Classroom Example 5.3.4
Write the expressions as differences of logarithms.

a. $\log_5\left(\dfrac{\sqrt[5]{x}}{z^2}\right)$

b.* $\log_6\left(\dfrac{\sqrt[3]{xy^2}}{wz^3}\right)$

Answer:

a. $\frac{1}{5}\log_5 x - 2\log_5 z$

b. $\frac{1}{3}\log_6 x + \frac{2}{3}\log_6 y$
$- \log_6 w - 3\log_6 z$

▪ **Answer:** $4\log a - 5\log b$

▪ **YOUR TURN** Write the expression $\log\left(\dfrac{a^4}{b^5}\right)$ as a difference of logarithms.

Another common mistake is misinterpreting the quotient rule.

$$\log_b M - \log_b N = \log_b\left(\frac{M}{N}\right)$$

$$\log_b M - \log_b N \neq \frac{\log_b M}{\log_b N}$$

Classroom Example 5.3.5
Write the expressions as logarithms of a quotient.

a. $\frac{3}{4}\log_2 x^4 - \log_2 y$

b.* $\log x - \left[\log y - 5\log x^2\right]$

Answer:

a. $\log_2\left(\dfrac{x^3}{y}\right)$ **b.** $\log\left(\dfrac{x^{11}}{y}\right)$

■ **Answer:** $\log\left(\dfrac{a^{1/2}}{b^3}\right)$

EXAMPLE 5 **Writing the Difference of Logarithms as a Logarithm of a Quotient**

Write the expression $\frac{2}{3}\ln x - \frac{1}{2}\ln y$ as a logarithm of a quotient.

COMMON MISTAKE

$$\log_b M - \log_b N \neq \frac{\log_b M}{\log_b N}$$

⭐ **CORRECT**

Use the power property (7).

$$\frac{2}{3}\ln x - \frac{1}{2}\ln y = \ln x^{2/3} - \ln y^{1/2}$$

Use the quotient property (6).

$$\ln\left(\frac{x^{2/3}}{y^{1/2}}\right)$$

❌ **INCORRECT**

$$\frac{\ln x^{2/3}}{\ln y^{1/2}} \quad \textbf{ERROR}$$

■ **YOUR TURN** Write the expression $\frac{1}{2}\log a - 3\log b$ as a single logarithm.

Classroom Example 5.3.6*
Express
$\ln\sqrt[4]{2x} - 2\ln\sqrt[4]{2y^3} + \ln(xy)^2$
as a single logarithm.

Answer:
$\ln\left(\dfrac{x^{9/4}y^{1/2}}{2^{1/4}}\right)$ or equivalently,

$\ln\sqrt[4]{\dfrac{x^9 y^2}{2}}$

■ **Answer:** $\ln\left(\dfrac{x^2 z^3}{3y}\right)$

EXAMPLE 6 **Combining Logarithmic Expressions into a Single Logarithm**

Write the expression $3\log_b x + \log_b(2x + 1) - 2\log_b 4$ as a single logarithm.

Solution:

Use the power property (7) on the first and third terms.

$$= \log_b x^3 + \log_b(2x + 1) - \log_b 4^2$$

Use the product property (5) on the first two terms.

$$= \log_b[x^3(2x + 1)] - \log_b 16$$

Use the quotient property (6).

$$= \log_b\left[\frac{x^3(2x + 1)}{16}\right]$$

■ **YOUR TURN** Write the expression $2\ln x - \ln(3y) + 3\ln z$ as a single logarithm.

Classroom Example 5.3.7
Write
$\log_6\left[\dfrac{x^2 - x}{(x^2 + 1)(x^2 - 4x + 4)}\right]$
as a sum or difference of logarithms.

Answer:
$\log_6 x + \log_6(x - 1)$
$-\log_6(x^2 + 1) - 2\log_6(x - 2)$

EXAMPLE 7 **Expanding a Logarithmic Expression into a Sum or Difference of Logarithms**

Write $\ln\left[\dfrac{x^2 - x - 6}{x^2 + 7x + 6}\right]$ as a sum or difference of logarithms.

Solution:

Factor the numerator and denominator.

$$= \ln\left[\frac{(x - 3)(x + 2)}{(x + 6)(x + 1)}\right]$$

Use the quotient property (6).

$$= \ln[(x - 3)(x + 2)] - \ln[(x + 6)(x + 1)]$$

Use the product property (5).

$$= \ln(x - 3) + \ln(x + 2) - [\ln(x + 6) + \ln(x + 1)]$$

Eliminate brackets.

$$= \ln(x - 3) + \ln(x + 2) - \ln(x + 6) - \ln(x + 1)$$

Change-of-Base Formula

Recall that in the last section we were able to evaluate logarithms two ways: (1) exactly by writing the logarithm in exponential form and identifying the exponent and (2) using a calculator if the logarithms were base 10 or e. How do we evaluate a logarithm of general base if we cannot identify the exponent? We use the *change-of-base formula*.

> **EXAMPLE 8** **Using Properties of Logarithms to Change the Base to Evaluate a General Logarithm**
>
> Evaluate $\log_3 8$. Round the answer to four decimal places.
>
> **Solution:**
>
> | | |
> |---|---|
> | Let $y = \log_3 8$. | $y = \log_3 8$ |
> | Write the logarithm in exponential form. | $3^y = 8$ |
> | Take the log of both sides. | $\log 3^y = \log 8$ |
> | Use the power property (7). | $y \log 3 = \log 8$ |
> | Divide both sides by log 3. | $y = \dfrac{\log 8}{\log 3}$ |
> | Let $y = \log_3 8$. | $\log_3 8 = \dfrac{\log 8}{\log 3}$ |

Example 8 illustrated our ability to use properties of logarithms to change from base 3 to base 10, which our calculators can handle. This leads to the general change-of-base formula.

> **CHANGE-OF-BASE FORMULA**
>
> For any logarithmic bases a and b and any positive number M, the change-of-base formula says that
> $$\log_b M = \frac{\log_a M}{\log_a b}$$
> In the special case when a is either 10 or e, this relationship becomes
>
> | **Common Logarithms** | | **Natural Logarithms** |
> |---|---|---|
> | $\log_b M = \dfrac{\log M}{\log b}$ | or | $\log_b M = \dfrac{\ln M}{\ln b}$ |
>
> It does not matter what base we select (10, e, or any other base), the ratio will be the same.

Proof of Change-of-Base Formula

WORDS	**MATH**
Let y be the logarithm we want to evaluate.	$y = \log_b M$
Write $y = \log_b M$ in exponential form.	$b^y = M$
Let a be any positive real number (where $a \neq 1$).	
Take the log of base a of both sides of the equation.	$\log_a b^y = \log_a M$
Use the power rule on the left side of the equation.	$y \log_a b = \log_a M$
Divide both sides of the equation by $\log_a b$.	$y = \dfrac{\log_a M}{\log_a b}$

Classroom Example 5.3.9
Evaluate:

a. $\log_4 e$

b.* $\log_{(1/2)}\left(\frac{3}{4}\right)$

Answer:

a. $\frac{1}{\ln 4} \approx 0.7213$

b. ≈ 0.4150

EXAMPLE 9   Using the Change-of-Base Formula

Use the change-of-base formula to evaluate $\log_4 17$. Round to four decimal places.

Solution:

We will illustrate this two ways (choosing common and natural logarithms) using a scientific calculator.

Common Logarithms

Use the change-of-base formula with base 10.
$$\log_4 17 = \frac{\log 17}{\log 4}$$

Approximate with a calculator.
$$\approx 2.043731421$$
$$\approx \boxed{2.0437}$$

Natural Logarithms

Use the change-of-base formula with base e.
$$\log_4 17 = \frac{\ln 17}{\ln 4}$$

Approximate with a calculator.
$$\approx 2.043731421$$
$$\approx \boxed{2.0437}$$

■ **Answer:** $\log_7 34 \approx 1.8122$

■ **YOUR TURN** Use the change-of-base formula to approximate $\log_7 34$. Round to four decimal places.

SECTION
5.3 SUMMARY

Properties of Logarithms

If a, b, M, and N are positive real numbers, where $a \neq 1$, $b \neq 1$ and p and x are real numbers, then the following are true:

- ■ Product Property: $\log_b MN = \log_b M + \log_b N$
- ■ Quotient Property: $\log_b\left(\dfrac{M}{N}\right) = \log_b M - \log_b N$
- ■ Power Property: $\log_b M^p = p \log_b M$

GENERAL LOGARITHM	COMMON LOGARITHM	NATURAL LOGARITHM
$\log_b 1 = 0$	$\log 1 = 0$	$\ln 1 = 0$
$\log_b b = 1$	$\log 10 = 1$	$\ln e = 1$
$\log_b b^x = x$	$\log 10^x = x$	$\ln e^x = x$
$b^{\log_b x} = x \qquad x > 0$	$10^{\log x} = x \qquad x > 0$	$e^{\ln x} = x \qquad x > 0$
$\log_b M = \dfrac{\log_a M}{\log_a b}$	$\log_b M = \dfrac{\log M}{\log b}$	$\log_b M = \dfrac{\ln M}{\ln b}$

SECTION
5.3 EXERCISES

■ **SKILLS**

In Exercises 1–18, apply the properties of logarithms to simplify each expression. Do not use a calculator.

1. $\log_9 1$

2. $\log_{69} 1$

3. $\log_{1/2}\left(\frac{1}{2}\right)$

4. $\log_{3.3} 3.3$

5. $\log_{10} 10^8$

6. $\ln e^3$

7. $\log_{10} 0.001$

8. $\log_3 3^7$

9. $\log_2 \sqrt{8}$ **10.** $\log_5 \sqrt[3]{5}$ **11.** $8^{\log_8 5}$ **12.** $2^{\log_2 5}$

13. $e^{\ln(x+5)}$ **14.** $10^{\log(3x^2+2x+1)}$ **15.** $5^{3\log_5 2}$ **16.** $7^{2\log_7 5}$

17. $7^{-2\log_7 3}$ **18.** $e^{-2\ln 10}$

In Exercises 19–32, write each expression as a sum or difference of logarithms.

$$\text{Example: } \log(m^2 n^5) = 2\log m + 5\log n$$

19. $\log_b(x^3 y^5)$ **20.** $\log_b(x^{-3} y^{-5})$ **21.** $\log_b(x^{1/2} y^{1/3})$ **22.** $\log_b(\sqrt{r}\ \sqrt[3]{t})$

23. $\log_b\left(\dfrac{r^{1/3}}{s^{1/2}}\right)$ **24.** $\log_b\left(\dfrac{r^4}{s^2}\right)$ **25.** $\log_b\left(\dfrac{x}{yz}\right)$ **26.** $\log_b\left(\dfrac{xy}{z}\right)$

27. $\log(x^2 \sqrt{x+5})$ **28.** $\log[(x-3)(x+2)]$ **29.** $\ln\left[\dfrac{x^3(x-2)^2}{\sqrt{x^2+5}}\right]$ **30.** $\ln\left[\dfrac{\sqrt{x+3}\ \sqrt[3]{x-4}}{(x+1)^4}\right]$

31. $\log\left[\dfrac{x^2-2x+1}{x^2-9}\right]$ **32.** $\log\left[\dfrac{x^2-x-2}{x^2+3x-4}\right]$

In Exercises 33–44, write each expression as a single logarithm.

$$\text{Example: } 2\log m + 5\log n = \log(m^2 n^5)$$

33. $3\log_b x + 5\log_b y$ **34.** $2\log_b u + 3\log_b v$ **35.** $5\log_b u - 2\log_b v$ **36.** $3\log_b x - \log_b y$

37. $\frac{1}{2}\log_b x + \frac{2}{3}\log_b y$ **38.** $\frac{1}{2}\log_b x - \frac{2}{3}\log_b y$ **39.** $2\log u - 3\log v - 2\log z$ **40.** $3\log u - \log 2v - \log z$

41. $\ln(x+1) + \ln(x-1) - 2\ln(x^2+3)$ **42.** $\ln\sqrt{x-1} + \ln\sqrt{x+1} - 2\ln(x^2-1)$

43. $\frac{1}{2}\ln(x+3) - \frac{1}{3}\ln(x+2) - \ln(x)$ **44.** $\frac{1}{3}\ln(x^2+4) - \frac{1}{2}\ln(x^2-3) - \ln(x-1)$

In Exercises 45–54, evaluate the logarithms using the change-of-base formula. Round to four decimal places.

45. $\log_5 7$ **46.** $\log_4 19$ **47.** $\log_{1/2} 5$ **48.** $\log_5 \frac{1}{2}$

49. $\log_{2.7} 5.2$ **50.** $\log_{7.2} 2.5$ **51.** $\log_\pi 10$ **52.** $\log_\pi 2.7$

53. $\log_{\sqrt{3}} 8$ **54.** $\log_{\sqrt{2}} 9$

■ APPLICATIONS

55. Sound. Sitting in the front row of a rock concert exposes us to a sound pressure (or sound level) of 1×10^{-1} W/m² (or 110 dB), and a normal conversation is typically around 1×10^{-6} W/m² (or 60 dB). How many decibels are you exposed to if a friend is talking in your ear at a rock concert? *Note:* 160 dB causes perforation of the eardrums. *Hint:* Add the sound pressures and convert to dB.

56. Sound. A whisper corresponds to 1×10^{-10} W/m² (or 20 dB), and a normal conversation is typically around 1×10^{-6} W/m² (or 60 dB). How many decibels are you exposed to if one friend is whispering in your ear while the other one is talking at a normal level? *Hint:* Add the sound pressures and convert to dB.

For Exercises 57 and 58, refer to the following:

There are two types of waves associated with an earthquake: *compression* and *shear*. The compression, or longitudinal, waves displace material behind the earthquake's path. Longitudinal waves travel at great speeds and are often called "primary waves" or simply "P" waves. Shear, or transverse, waves displace material at right angles to its path. Transverse waves do not travel as rapidly through the Earth's crust and mantle as do longitudinal waves, and they are called "secondary" or "S" waves.

57. Earthquakes. If a seismologist records the energy of P waves as 4.5×10^{12} joules and the energy of S waves as 7.8×10^8 joules, what is the total energy? What would the combined effect be on the Richter scale?

58. Earthquakes. Repeat Exercise 57 assuming the energy associated with the P waves is 5.2×10^{11} joules and the energy associated with the S waves is 4.1×10^9 joules.

■ **CATCH THE MISTAKE**

In Exercises 59–62, simplify if possible and explain the mistake that is made.

59. $3 \log 5 - \log 25$

Solution:

Apply the quotient property (6). $\dfrac{3 \log 5}{\log 25}$

Write $25 = 5^2$. $\dfrac{3 \log 5}{\log 5^2}$

Apply the power property (7). $\dfrac{3 \log 5}{2 \log 5}$

Simplify. $\dfrac{3}{2}$

This is incorrect. The correct answer is $\log 5$. What mistake was made?

60. $\ln 3 + 2 \ln 4 - 3 \ln 2$

Solution:

Apply the power property (7). $\ln 3 + \ln 4^2 - \ln 2^3$

Simplify. $\ln 3 + \ln 16 - \ln 8$

Apply property (5). $\ln(3 + 16 - 8)$

Simplify. $\ln 11$

This is incorrect. The correct answer is $\ln 64$. What mistake was made?

61. $\log_2 x + \log_3 y - \log_4 z$

Solution:

Apply the product property (5). $\log_6 xy - \log_4 z$

Apply the quotient property (6). $\log_{24} xyz$

This is incorrect. What mistake was made?

62. $2(\log 3 - \log 5)$

Solution:

Apply the quotient property (6). $2\left(\log \dfrac{3}{5}\right)$

Apply the power property (7). $\left(\log \dfrac{3}{5}\right)^2$

Apply a calculator to approximate. ≈ 0.0492

This is incorrect. What mistake was made?

■ **CONCEPTUAL**

In Exercises 63–66, determine whether each statement is true or false.

63. $\log e = \dfrac{1}{\ln 10}$

64. $\ln e = \dfrac{1}{\log 10}$

65. $\ln (xy)^3 = (\ln x + \ln y)^3$

66. $\dfrac{\ln a}{\ln b} = \dfrac{\log a}{\log b}$

■ **CHALLENGE**

67. Prove the quotient rule: $\log_b\left(\dfrac{M}{N}\right) = \log_b M - \log_b N$.

Hint: Let $u = \log_b M$ and $v = \log_b N$. Write both in exponential form and find the quotient $\log_b\left(\dfrac{M}{N}\right)$.

68. Prove the power rule: $\log_b M^p = p \log_b M$.

Hint: Let $u = \log_b M$. Write this log in exponential form, and find $\log_b M^p$.

69. Write in terms of simpler logarithmic forms.

$$\log_b\left(\sqrt{\dfrac{x^2}{y^3 z^{-5}}}\right)^6$$

70. Show that $\log_b\left(\dfrac{1}{x}\right) = -\log_b x$.

▪TECHNOLOGY

71. Use a graphing calculator to plot $y = \ln(2x)$ and $y = \ln 2 + \ln x$. Are they the same graph?

72. Use a graphing calculator to plot $y = \ln(2 + x)$ and $y = \ln 2 + \ln x$. Are they the same graph?

73. Use a graphing calculator to plot $y = \dfrac{\log x}{\log 2}$ and $y = \log x - \log 2$. Are they the same graph?

74. Use a graphing calculator to plot $y = \log\left(\dfrac{x}{2}\right)$ and $y = \log x - \log 2$. Are they the same graph?

75. Use a graphing calculator to plot $y = \ln(x^2)$ and $y = 2 \ln x$. Are they the same graph?

76. Use a graphing calculator to plot $y = (\ln x)^2$ and $y = 2 \ln x$. Are they the same graph?

77. Use a graphing calculator to plot $y = \ln x$ and $y = \dfrac{\log x}{\log e}$. Are they the same graph?

78. Use a graphing calculator to plot $y = \log x$ and $y = \dfrac{\ln x}{\ln 10}$. Are they the same graph?

SECTION 5.4
EXPONENTIAL AND LOGARITHMIC EQUATIONS

SKILLS OBJECTIVES

▪ Solve exponential equations.
▪ Solve logarithmic equations.
▪ Solve application problems using exponential and logarithmic equations.

CONCEPTUAL OBJECTIVE

▪ Understand how exponential and logarithmic equations are solved using properties of one-to-one functions and inverses.

Exponential Equations

In this book you have solved algebraic equations such as $x^2 - 9 = 0$, in which the goal is to solve for x by finding the values of x that make the statement true. Exponential and logarithmic equations have the x buried within an exponent or a logarithm, but the goal is the same. Solve for x.

$$\text{Exponential equation:} \quad e^{2x+1} = 5$$

$$\text{Logarithmic equation:} \quad \log(3x - 1) = 7$$

There are two methods for solving exponential and logarithmic equations that are based on the properties of one-to-one functions and inverses. To solve simple exponential and logarithmic equations, we will use one-to-one properties. To solve more complicated exponential and logarithmic equations, we will use properties of inverses. The following box summarizes the one-to-one and inverse properties that hold true when $b > 0$ and $b \neq 1$.

ONE-TO-ONE PROPERTIES

$b^x = b^y$	if and only if	$x = y$
$\log_b x = \log_b y$	if and only if	$x = y$

INVERSE PROPERTIES

$$b^{\log_b x} = x \qquad x > 0$$

$$\log_b b^x = x$$

The following strategies are outlined for solving simple and complicated exponential equations using the one-to-one and inverse properties.

STRATEGIES FOR SOLVING EXPONENTIAL EQUATIONS

Strategy for Solving Exponential Equations

TYPE OF EQUATION	STRATEGY	EXAMPLE
Simple	1. Rewrite both sides of the equation in terms of the same base.	$2^{x-3} = 32$ $2^{x-3} = 2^5$
	2. Use the one-to-one property to equate the exponents.	$x - 3 = 5$
	3. Solve for the variable.	$\boxed{x = 8}$
Complicated	1. Isolate the exponential expression.	$3e^{2x} - 2 = 7$ $3e^{2x} = 9$ $e^{2x} = 3$
	2. Take the same logarithm* of both sides.	$\ln e^{2x} = \ln 3$
	3. Simplify using the inverse properties.	$2x = \ln 3$
	4. Solve for the variable.	$\boxed{x = \tfrac{1}{2} \ln 3}$

*Take the logarithm with base that is equal to the base of the exponent and use the property $\log_b b^x = x$ or take the natural logarithm and use the property $\ln M^p = p \ln M$.

EXAMPLE 1   Solving a Simple Exponential Equation

Solve the exponential equations using the one-to-one property.

a. $3^x = 81$ **b.** $5^{7-x} = 125$ **c.** $\left(\frac{1}{2}\right)^{4y} = 16$

Solution (a):

Substitute $81 = 3^4$.	$3^x = 3^4$
Use the one-to-one property to identify x.	$\boxed{x = 4}$

Solution (b):

Substitute $125 = 5^3$.	$5^{7-x} = 5^3$
Use the one-to-one property.	$7 - x = 3$
Solve for x.	$\boxed{x = 4}$

Solution (c):

Substitute $\left(\frac{1}{2}\right)^{4y} = \left(\frac{1}{2^{4y}}\right) = 2^{-4y}$.	$2^{-4y} = 16$
Substitute $16 = 2^4$.	$2^{-4y} = 2^4$
Use the one-to-one property to identify y.	$\boxed{y = -1}$

YOUR TURN Solve the following equations.

     a. $2^{x-1} = 8$ **b.** $\left(\frac{1}{3}\right)^y = 27$

Classroom Example 5.4.1
Solve using the one-to-one property.
a. $4^{2-3x} = \frac{1}{64}$
b.* $\left(\frac{1}{3}\right)^{1-y} = \sqrt[3]{81}$

Answer:
a. $x = \frac{5}{3}$ **b.** $y = \frac{7}{3}$

■ **Answer: a.** $x = 4$ **b.** $y = -3$

In Example 1, we were able to rewrite the equation in a form with the same bases so that we could use the one-to-one property. In Example 2, we will not be able to write both sides in a form with the same bases. Instead, we will use properties of inverses.

EXAMPLE 2 Solving a More Complicated Exponential Equation with a Base Other Than 10 or *e*

Solve the exponential equations exactly and then approximate answers to four decimal places.

a. $5^{3x} = 16$ **b.** $4^{3x+2} = 71$

Solution (a):

Take the natural logarithm of both sides of the equation.
$$\ln 5^{3x} = \ln 16$$

Use the power property on the left side of the equation.
$$3x \ln 5 = \ln 16$$

Divide both sides of the equation by 3 ln 5.
$$\boxed{x = \frac{\ln 16}{3 \ln 5}}$$

Use a calculator to approximate x to four decimal places.
$$\boxed{x \approx 0.5742}$$

Solution (b):

Rewrite in logarithmic form.
$$3x + 2 = \log_4 71$$

Subtract 2 from both sides.
$$3x = \log_4 71 - 2$$

Divide both sides by 3.
$$\boxed{x = \frac{\log_4 71 - 2}{3}}$$

Use the change-of-base formula, $\log_4 71 = \dfrac{\ln 71}{\ln 4}$.
$$x = \frac{\dfrac{\ln 71}{\ln 4} - 2}{3}$$

Use a calculator to approximate x to four decimal places.
$$x \approx \frac{3.07487356 - 2}{3} \approx \boxed{0.3583}$$

We could have proceeded in an alternative way by taking either the natural log or the common log of both sides and using the power property (instead of using the change-of-base formula) to evaluate the logarithm with base 4.

Take the natural logarithm of both sides.
$$\ln\left(4^{3x+2}\right) = \ln 71$$

Use the power property (7).
$$(3x + 2)\ln 4 = \ln 71$$

Divide by ln 4.
$$3x + 2 = \frac{\ln 71}{\ln 4}$$

Subtract 2 and divide by 3.
$$\boxed{x = \frac{\dfrac{\ln 71}{\ln 4} - 2}{3}}$$

Use a calculator to approximate x.
$$x \approx \frac{3.07487356 - 2}{3} \approx \boxed{0.3583}$$

▪ **YOUR TURN** Solve the equation $5^{y^2} = 27$ exactly and then approximate the answer to four decimal places.

Technology Tip

```
((ln(71)/ln(4)-2
)/3
          .3582911866
```

▪ **Answer:** $y = \pm\sqrt{\log_5 27}$
$\approx \pm 1.4310$

EXAMPLE 3   Solving a More Complicated Exponential Equation with Base 10 or e

Solve the exponential equation $4e^{x^2} = 64$ exactly and then round the answer to four decimal places.

Solution:

Divide both sides by 4. $e^{x^2} = 16$

Take the natural logarithm (ln) of both sides. $\ln\left(e^{x^2}\right) = \ln 16$

Simplify the left side with the property of inverses. $x^2 = \ln 16$

Solve for x using the square-root method. $\boxed{x = \pm\sqrt{\ln 16}}$

Use a calculator to approximate x to four decimal places. $\boxed{x \approx \pm 1.6651}$

■ **Answer:** $x = \dfrac{\log_{10} 7 + 3}{2}$

≈ 1.9225

■ **YOUR TURN** Solve the equation $10^{2x-3} = 7$ exactly and then round the answer to four decimal places.

Technology Tip

The graph of the function $e^{2x} - 4e^x + 3$ is shown. The x-intercepts are $x \approx 1.10$ and $x = 0$.

■ **Answer:** $x = \log 2 \approx 0.3010$

EXAMPLE 4   Solving an Exponential Equation Quadratic in Form

Solve the equation $e^{2x} - 4e^x + 3 = 0$ exactly and then round the answer to four decimal places.

Solution:

Let $u = e^x$. (*Note:* $u^2 = e^x \cdot e^x = e^{2x}$.) $u^2 - 4u + 3 = 0$

Factor. $(u - 3)(u - 1) = 0$

Solve for u. $u = 3$ or $u = 1$

Substitute $u = e^x$. $e^x = 3$ or $e^x = 1$

Take the natural logarithm (ln) of both sides. $\ln(e^x) = \ln 3$ or $\ln(e^x) = \ln 1$

Simplify with the properties of logarithms. $\boxed{x = \ln 3}$ or $\boxed{x = \ln 1}$

Approximate or evaluate exactly the right sides. $\boxed{x \approx 1.0986}$ or $\boxed{x = 0}$

■ **YOUR TURN** Solve the equation $100^x - 10^x - 2 = 0$ exactly and then round the answer to four decimal places.

Solving Logarithmic Equations

We can solve simple logarithmic equations using the property of one-to-one functions. For more complicated logarithmic equations we can employ properties of logarithms and properties of inverses. **Solutions must be checked to eliminate extraneous solutions.**

Strategy for Solving Logarithmic Equations

TYPE OF EQUATION	STRATEGY	EXAMPLE
Simple	1. Combine logarithms on each side of the equation using properties.	$\log(x-3) + \log x = \log 4$ $\log x(x-3) = \log 4$
	2. Use the one-to-one property to equate the arguments.	$x(x-3) = 4$
	3. Solve for the variable.	$x^2 - 3x - 4 = 0$ $(x-4)(x+1) = 0$ $x = -1, 4$
	4. Check the results and eliminate any extraneous solutions.	Eliminate $x = -1$ because $\log(-1)$ is undefined. $\boxed{x = 4}$
Complicated	1. Combine and isolate the logarithmic expressions.	$\log_5(x+2) - \log_5 x = 2$ $\log_5\left(\dfrac{x+2}{x}\right) = 2$
	2. Rewrite the equation in exponential form.	$\dfrac{x+2}{x} = 5^2$
	3. Solve for the variable.	$x + 2 = 25x$ $24x = 2$ $\boxed{x = \dfrac{1}{12}}$
	4. Check the results and eliminate any extraneous solutions.	$\log_5\left(\dfrac{1}{12}+2\right) - \log_5\left(\dfrac{1}{12}\right)$ $= \log_5\left(\dfrac{25}{12}\right) - \log_5\left(\dfrac{1}{12}\right)$ $= \log_5\left[\dfrac{25/12}{1/12}\right]$ $= \log_5[25] = 2$ ✓

EXAMPLE 5 **Solving a Simple Logarithmic Equation**

Solve the equation $\log_4(2x-3) = \log_4(x) + \log_4(x-2)$.

Solution:

Apply the product property (5) on the right side.
$$\log_4(2x-3) = \log_4[x(x-2)]$$

Apply the property of one-to-one functions.
$$2x - 3 = x(x-2)$$

Distribute and simplify.
$$x^2 - 4x + 3 = 0$$

Factor.
$$(x-3)(x-1) = 0$$

Solve for x.
$$x = 3 \text{ or } x = 1$$

The possible solution $x = 1$ must be eliminated because it is not in the domain of two of the logarithmic functions.

$$x = 1: \log_4(\overbrace{-1}^{\text{undefined}}) \overset{?}{=} \log_4(1) + \log_4(\overbrace{-1}^{\text{undefined}})$$

$$\boxed{x = 3}$$

■ **YOUR TURN** Solve the equation $\ln(x+8) = \ln(x) + \ln(x+3)$.

Classroom Example 5.4.5
Solve the equation
$\log_7(5-2x) = \log_7(x(15-6x))$.

Answer: $x = \frac{1}{3}$ *Note:* $x = \frac{5}{2}$ is an extraneous solution.

Study Tip

Solutions should be checked in the original equation to eliminate extraneous solutions.

■ **Answer:** $x = 2$

EXAMPLE 6 Solving a More Complicated Logarithmic Equation

Solve the equation $\log_3(9x) - \log_3(x - 8) = 4$.

Solution:

Employ the quotient property (6) on the left side.

$$\log_3\left(\frac{9x}{x - 8}\right) = 4$$

Write in exponential form. $\log_b x = y \Rightarrow x = b^y$

$$\frac{9x}{x - 8} = 3^4$$

Simplify the right side.

$$\frac{9x}{x - 8} = 81$$

Multiply the equation by the LCD, $x - 8$.

$$9x = 81(x - 8)$$

Eliminate parentheses.

$$9x = 81x - 648$$

Solve for x.

$$-72x = -648$$

$$\boxed{x = 9}$$

Check: $\log_3[9 \cdot 9] - \log_3[9 - 8] = \log_3[81] - \log_3 1 = 4 - 0 = 4$

■ **YOUR TURN** Solve the equation $\log_2(4x) - \log_2(2) = 2$.

■ **Answer:** $x = 2$

EXAMPLE 7 Solving a Logarithmic Equation with No Solution

Solve the equation $\ln(3 - x^2) = 7$.

Solution:

Write in exponential form.

$$3 - x^2 = e^7$$

Simplify.

$$x^2 = \underbrace{3 - e^7}$$

$3 - e^7$ is negative.

$$x^2 = \text{negative real number}$$

There are no real numbers that when squared yield a negative number. Therefore, there is $\boxed{\text{no real solution}}$.

Technology Tip

Graphs of $y_1 = \ln(3 - x^2)$ and $y_2 = 7$ are shown.

No intersection of the graphs indicates no solution.

Applications

In the chapter opener we saw that archaeologists determine the age of a fossil by how much carbon 14 is present at the time of discovery. The number of grams of carbon 14 based on the radioactive decay of the isotope is given by

$$A = A_0 e^{-0.000124t}$$

where A is the number of grams of carbon 14 at the present time, A_0 is the number of grams of carbon 14 while alive, and t is the number of years since death. Using the inverse properties, we can isolate t.

WORDS	MATH
Divide by A_0.	$\dfrac{A}{A_0} = e^{-0.000124t}$
Take the natural logarithm of both sides.	$\ln\left(\dfrac{A}{A_0}\right) = \ln\left(e^{-0.000124t}\right)$
Simplify the right side utilizing properties of inverses.	$\ln\left(\dfrac{A}{A_0}\right) = -0.000124t$

Solve for t.

$$t = -\frac{1}{0.000124} \ln\left(\frac{A}{A_0}\right)$$

Let's assume that animals have approximately 1000 mg of carbon 14 in their bodies when they are alive. If a fossil has 200 mg of carbon 14, approximately how old is the fossil? Substituting $A = 200$ and $A_0 = 1000$ into our equation for t, we find

$$t = -\frac{1}{0.000124} \ln\left(\frac{1}{5}\right) \approx 12{,}979.338$$

The fossil is approximately 13,000 years old.

EXAMPLE 8   Calculating How Many Years It Will Take for Money to Double

You save $1000 from a summer job and put it in a CD earning 5% compounding continuously. How many years will it take for your money to double? Round to the nearest year.

Solution:

Recall the compound continuous interest formula.	$A = Pe^{rt}$
Substitute $P = 1000$, $A = 2000$, and $r = 0.05$.	$2000 = 1000e^{0.05t}$
Divide by 1000.	$2 = e^{0.05t}$
Take the natural logarithm of both sides.	$\ln 2 = \ln\left(e^{0.05t}\right)$
Simplify with the property $\ln e^x = x$.	$\ln 2 = 0.05t$
Solve for t.	$t = \dfrac{\ln 2}{0.05} \approx 13.8629$

It will take almost 14 years for your money to double.

■ **YOUR TURN** How long will it take $1000 to triple (become $3000) in a savings account earning 10% a year compounding continuously? Round your answer to the nearest year.

■ **Answer:** approximately 11 years

When an investment is compounded continuously, how long will it take for that investment to double?

WORDS	MATH
Write the interest formula for compounding continuously.	$A = Pe^{rt}$
Let $A = 2P$ (investment doubles).	$2P = Pe^{rt}$
Divide both sides of the equation by P.	$2 = e^{rt}$
Take the natural log of both sides of the equation.	$\ln 2 = \ln e^{rt}$
Simplify the right side by applying the property $\ln e^x = x$.	$\ln 2 = rt$
Divide both sides by r.	$t = \dfrac{\ln 2}{r}$
Approximate $\ln 2 \approx 0.7$.	$t \approx \dfrac{0.7}{r}$

Recall that an interest rate such as 4.5% is written in decimal form as $r = 0.045$.

Multiply the numerator and denominator by 100.

$$t \approx \frac{70}{100r}$$

This is the "rule of 70."

If we divide 70 by the interest rate (compounding continuously), we get the approximate time for an investment to double. In Example 8, the interest rate (compounding continuously) is 5%. Dividing 70 by 5 yields 14 years.

SECTION 5.4 SUMMARY

Strategy for Solving Exponential Equations

TYPE OF EQUATION	STRATEGY
Simple	1. Rewrite both sides of the equation in terms of the same base.
	2. Use the one-to-one property to equate the exponents.
	3. Solve for the variable.
Complicated	1. Isolate the exponential expression.
	2. Take the same logarithm* of both sides.
	3. Simplify using the inverse properties.
	4. Solve for the variable.

*Take the logarithm with the base that is equal to the base of the exponent and use the property $\log_b b^x = x$, or take the natural logarithm and use the property $\ln M^p = p \ln M$.

Strategy for Solving Logarithmic Equations

TYPE OF EQUATION	STRATEGY
Simple	1. Combine logarithms on each side of the equation using properties.
	2. Use the one-to-one property to equate the arguments.
	3. Solve for the variable.
	4. Eliminate any extraneous solutions.
Complicated	1. Combine and isolate the logarithmic expressions.
	2. Rewrite the equation in exponential form.
	3. Solve for the variable.
	4. Eliminate any extraneous solutions.

SECTION 5.4 EXERCISES

■ SKILLS

In Exercises 1–18, solve the exponential equations exactly for x.

1. $3^x = 81$

2. $5^x = 125$

3. $7^x = \frac{1}{49}$

4. $4^x = \frac{1}{16}$

5. $2^{x^2} = 16$

6. $169^x = 13$

7. $\left(\frac{2}{3}\right)^{x+1} = \frac{27}{8}$

8. $\left(\frac{3}{5}\right)^{x+1} = \frac{25}{9}$

9. $e^{2x+3} = 1$

10. $10^{x^2-1} = 1$

11. $7^{2x-5} = 7^{3x-4}$

12. $125^x = 5^{2x-3}$

13. $2^{x^2+12} = 2^{7x}$

14. $5^{x^2-3} = 5^{2x}$

15. $9^x = 3^{x^2-4x}$

16. $16^{x-1} = 2^{x^2}$

17. $e^{5x-1} = e^{x^2+3}$

18. $10^{x^2-8} = 100^x$

In Exercises 19–44, solve the exponential equations exactly and then approximate your answers to three decimal places.

19. $10^{2x-3} = 81$

20. $2^{3x+1} = 21$

21. $3^{x+1} = 5$

22. $5^{2x-1} = 35$

23. $27 = 2^{3x-1}$

24. $15 = 7^{3-2x}$

25. $3e^x - 8 = 7$

26. $5e^x + 12 = 27$

27. $9 - 2e^{0.1x} = 1$ **28.** $21 - 4e^{0.1x} = 5$ **29.** $2(3^x) - 11 = 9$ **30.** $3(2^x) + 8 = 35$

31. $e^{3x+4} = 22$ **32.** $e^{x^2} = 73$ **33.** $3e^{2x} = 18$ **34.** $4(10^{3x}) = 20$

35. $e^{2x} + 7e^x - 3 = 0$ **36.** $e^{2x} - 4e^x - 5 = 0$ **37.** $(3^x - 3^{-x})^2 = 0$ **38.** $(3^x - 3^{-x})(3^x + 3^{-x}) = 0$

39. $\dfrac{2}{e^x - 5} = 1$ **40.** $\dfrac{17}{e^x + 4} = 2$ **41.** $\dfrac{20}{6 - e^{2x}} = 4$ **42.** $\dfrac{4}{3 - e^{3x}} = 8$

43. $\dfrac{4}{10^{2x} - 7} = 2$ **44.** $\dfrac{28}{10^x + 3} = 4$

In Exercises 45–62, solve the logarithmic equations exactly.

45. $\log(2x) = 2$ **46.** $\log(5x) = 3$ **47.** $\log_3(2x + 1) = 4$ **48.** $\log_2(3x - 1) = 3$

49. $\log_2(4x - 1) = -3$ **50.** $\log_4(5 - 2x) = -2$ **51.** $\ln x^2 - \ln 9 = 0$ **52.** $\log x^2 + \log x = 3$

53. $\log_5(x - 4) + \log_5 x = 1$ **54.** $\log_2(x - 1) + \log_2(x - 3) = 3$

55. $\log(x - 3) + \log(x + 2) = \log(4x)$ **56.** $\log_2(x + 1) + \log_2(4 - x) = \log_2(6x)$

57. $\log(4 - x) + \log(x + 2) = \log(3 - 2x)$ **58.** $\log(3 - x) + \log(x + 3) = \log(1 - 2x)$

59. $\log_4(4x) - \log_4\left(\dfrac{x}{4}\right) = 3$ **60.** $\log_3(7 - 2x) - \log_3(x + 2) = 2$

61. $\log(2x - 5) - \log(x - 3) = 1$ **62.** $\log_3(10 - x) - \log_3(x + 2) = 1$

In Exercises 63–78, solve the logarithmic equations exactly, if possible; and then approximate your answers to three decimal places.

63. $\ln x^2 = 5$ **64.** $\log 3x = 2$ **65.** $\log(2x + 5) = 2$ **66.** $\ln(4x - 7) = 3$

67. $\ln(x^2 + 1) = 4$ **68.** $\log(x^2 + 4) = 2$ **69.** $\ln(2x + 3) = -2$ **70.** $\log(3x - 5) = -1$

71. $\log(2 - 3x) + \log(3 - 2x) = 1.5$ **72.** $\log_2(3 - x) + \log_2(1 - 2x) = 5$

73. $\ln(x) + \ln(x - 2) = 4$ **74.** $\ln(4x) + \ln(2 + x) = 2$

75. $\log_7(1 - x) - \log_7(x + 2) = \log_7 x$ **76.** $\log_5(x + 1) - \log_5(x - 1) = \log_5(x)$

77. $\ln\sqrt{x + 4} - \ln\sqrt{x - 2} = \ln\sqrt{x + 1}$ **78.** $\log(\sqrt{1 - x}) - \log(\sqrt{x + 2}) = \log x$

■ **APPLICATIONS**

79. Health. After strenuous exercise Sandy's heart rate R (beats per minute) can be modeled by

$$R(t) = 151e^{-0.055t}, \ 0 \le t \le 15$$

where t is the number of minutes that have elapsed after she stops exercising.

  a. Find Sandy's heart rate at the end of exercising (when she stops at time $t = 0$).

  b. Determine how many minutes it takes after Sandy stops exercising for her heart rate to drop to 100 beats per minute. Round to the nearest minute.

  c. Find Sandy's heart rate 15 minutes after she had stopped exercising.

80. Business. A local business purchased a new company van for $45,000. After 2 years the book value of the van is $30,000.

  a. Find an exponential model for the value of the van using $V(t) = V_0 e^{kt}$ where V is the value of the van in dollars and t is time in years.

  b. Approximately how many years will it take for the book value of the van to drop to $20,000?

81. Money. If money is invested in a savings account earning 3.5% interest compounded yearly, how many years will pass until the money triples?

82. Money. If money is invested in a savings account earning 3.5% interest compounded monthly, how many years will pass until the money triples?

83. Money. If $7500 is invested in a savings account earning 5% interest compounded quarterly, how many years will pass until there is $20,000?

84. Money. If $9000 is invested in a savings account earning 6% interest compounded continuously, how many years will pass until there is $15,000?

85. Earthquakes. On September 25, 2003 an earthquake that measured 7.4 on the Richter scale shook Hokkaido, Japan. How much energy (joules) did the earthquake emit?

86. Earthquakes. Again, on that same day (September 25, 2003), a second earthquake that measured 8.3 on the Richter scale shook Hokkaido, Japan. How much energy (joules) did the earthquake emit?

87. Sound. Matt likes to drive around campus in his classic Mustang with the stereo blaring. If his boom stereo has a sound intensity of 120 dB, how many watts per square meter does the stereo emit?

88. Sound. The New York Philharmonic has a sound intensity of 100 dB. How many watts per square meter does the orchestra emit?

89. Anesthesia. When a person has a cavity filled, the dentist typically administers a local anesthetic. After leaving the dentist's office, one's mouth often remains numb for several more hours. If a shot of anesthesia is injected into the bloodstream at the time of the procedure ($t = 0$), and the amount of anesthesia still in the bloodstream t hours after the initial injection is given by $A = A_0 e^{-0.5t}$, in how many hours will only 10% of the original anesthetic still be in the bloodstream?

90. Investments. Money invested in an account that compounds interest continuously at a rate of 3% a year is modeled by $A = A_0 e^{0.03t}$, where A is the amount in the investment after t years and A_0 is the initial investment. How long will it take the initial investment to double?

91. Biology. The U.S. Fish and Wildlife Service is releasing a population of the endangered Mexican gray wolf in a protected area along the New Mexico and Arizona border. They estimate the population of the Mexican gray wolf to be approximated by

$$P(t) = \frac{200}{1 + 24e^{-0.2t}}$$

How many years will it take for the population to reach 100 wolves?

92. Introducing a New Car Model. If the number of new model Honda Accord hybrids purchased in North America is given by $N = \frac{100,000}{1 + 10e^{-2t}}$, where t is the number of weeks after Honda releases the new model, how many weeks will it take after the release until there are 50,000 Honda hybrids from that batch on the road?

93. Earthquakes. A P wave measures 6.2 on the Richter scale, and an S wave measures 3.3 on the Richter scale. What is their combined measure on the Richter scale?

94. Sound. You and a friend get front row seats to a rock concert. The music level is 100 dB, and your normal conversation is 60 dB. If your friend is telling you something during the concert, how many decibels are you subjecting yourself to?

■CATCH THE MISTAKE

In Exercises 95–98, explain the mistake that is made.

95. Solve the equation: $4e^x = 9$.

Solution:

Take the natural log of both sides. $\ln(4e^x) = \ln 9$

Apply the property of inverses. $4x = \ln 9$

Solve for x. $x = \frac{\ln 9}{4} \approx 0.55$

This is incorrect. What mistake was made?

96. Solve the equation: $\log(x) + \log(3) = 1$.

Solution:

Apply the product property (5). $\log(3x) = 1$

Exponentiate (base 10). $10^{\log(3x)} = 1$

Apply the properties of inverses. $3x = 1$

Solve for x. $x = \frac{1}{3}$

This is incorrect. What mistake was made?

97. Solve the equation: $\log(x) + \log(x + 3) = 1$ for x.

Solution:

Apply the product property (5). $\log(x^2 + 3x) = 1$

Exponentiate both sides (base 10). $10^{\log(x^2+3x)} = 10^1$

Apply the property of inverses. $x^2 + 3x = 10$

Factor. $(x + 5)(x - 2) = 0$

Solve for x. $x = -5$ and $x = 2$

This is incorrect. What mistake was made?

98. Solve the equation: $\log x + \log 2 = \log 5$.

Solution:

Combine the logarithms on the left. $\log(x + 2) = \log 5$

Apply the property of one-to-one functions. $x + 2 = 5$

Solve for x. $x = 3$

This is incorrect. What mistake was made?

■ CONCEPTUAL

In Exercises 99–102, determine whether each statement is true or false.

99. The sum of logarithms with the same base is equal to the logarithm of the product.

100. A logarithm squared is equal to two times the logarithm.

101. $e^{\log x} = x$

102. $e^x = -2$ has no solution.

103. Solve for x in terms of b:

$$\frac{1}{3}\log_b(x^3) + \frac{1}{2}\log_b(x^2 - 2x + 1) = 2$$

104. Solve exactly:

$$2\log_b(x) + 2\log_b(1-x) = 4.$$

■ CHALLENGE

105. Solve $y = \dfrac{3000}{1 + 2e^{-0.2t}}$ for t in terms of y.

106. State the range of values of x such that the following identity holds: $e^{\ln(x^2 - a)} = x^2 - a$.

107. A function called the hyperbolic cosine is defined as the average of exponential growth and exponential decay by $f(x) = \dfrac{e^x + e^{-x}}{2}$. Find its inverse.

108. A function called the hyperbolic sine is defined by $f(x) = \dfrac{e^x - e^{-x}}{2}$. Find its inverse.

■ TECHNOLOGY

109. Solve the equation $\ln 3x = \ln(x^2 + 1)$. Using a graphing calculator, plot the graphs $y = \ln(3x)$ and $y = \ln(x^2 + 1)$ in the same viewing rectangle. Zoom in on the point where the graphs intersect. Does this agree with your solution?

110. Solve the equation $10^{x^2} = 0.001^x$. Using a graphing calculator, plot the graphs $y = 10^{x^2}$ and $y = 0.001^x$ in the same viewing rectangle. Does this confirm your solution?

111. Use a graphing utility to help solve $3^x = 5x + 2$.

112. Use a graphing utility to help solve $\log x^2 = \ln(x - 3) + 2$.

113. Use a graphing utility to graph $y = \dfrac{e^x + e^{-x}}{2}$. State the domain. Determine if there are any symmetries and asymptotes.

114. Use a graphing utility to graph $y = \dfrac{e^x + e^{-x}}{e^x - e^{-x}}$. State the domain. Determine if there are any symmetries and asymptotes.

SKILLS OBJECTIVES

- Apply exponential growth and exponential decay models to biological, demographic, and economic phenomena.
- Represent distributions by means of a Gaussian model.
- Use logistic growth models to represent phenomena involving limits to growth.
- Solve problems such as species populations, credit card payoff, and wear off of anesthesia through logarithmic models.

CONCEPTUAL OBJECTIVE

- Recognize exponential growth, exponential decay, Gaussian distributions, logistic growth, and logarithmic models.

The following table summarizes the five primary models that involve exponential and logarithmic functions.

NAME	MODEL	GRAPH	APPLICATIONS
Exponential growth	$f(x) = ce^{kx}$ $k > 0$		World populations, bacteria growth, appreciation, global spread of the HIV virus
Exponential decay	$f(x) = ce^{-kx}$ $k > 0$		Radioactive decay, carbon dating, depreciation
Gaussian (normal) distribution	$f(x) = ce^{-(x-a)^2/k}$		Bell curve (grade distribution), life expectancy, height/weight charts, intensity of a laser beam, IQ tests
Logistic growth	$f(x) = \dfrac{a}{1 + ce^{-kx}}$		Conservation biology, learning curve, spread of virus on an island, carrying capacity
Logarithmic	$f(x) = a + c \log x$ $f(x) = a + c \ln x$		Population of species, anesthesia wearing off, time to pay off credit cards

Exponential Growth Models

Quite often one will hear that something "grows exponentially," meaning that it grows very fast and at increasing speed. In mathematics, the precise meaning of **exponential growth** is a *growth rate of a function that is proportional to its current size*. Let's assume you get a 5% raise every year in a government job. If your starting annual salary out of college is $40,000, then your first raise will be $2000. Fifteen years later your annual salary will be approximately $83,000 and your next 5% raise will be around $4150. The raise is always 5% of the current salary, so the larger the current salary, the larger the raise.

In Section 5.1 we saw that interest that is compounded continuously is modeled by $A = Pe^{rt}$. Here A stands for amount and P stands for principal. There are similar models for populations; these take the form $N(t) = N_0 e^{rt}$, where N_0 represents the number of people at time $t = 0$, r is the growth rate, t is time in years, and N represents the number of people at time t. In general, any model of the form $f(x) = ce^{kx}$, $k > 0$, models exponential growth.

EXAMPLE 1   **World Population Projections**

The world population is the total number of humans on Earth at a given time. In 2000 the world population was 6.1 billion and in 2005 the world population was 6.5 billion. Find the relative growth rate and determine what year the population will reach 9 billion.

Solution:

Assume an exponential growth model.

Let $t = 0$ correspond to 2000.

In 2005, $t = 5$, the population was 6.5 billion.

Solve for r.

$$N(t) = N_0 e^{rt}$$
$$N(0) = N_0 = 6.1$$
$$6.5 = 6.1 e^{5r}$$
$$\frac{6.5}{6.1} = e^{5r}$$
$$\ln\left(\frac{6.5}{6.1}\right) = \ln\left(e^{5r}\right)$$
$$\ln\left(\frac{6.5}{6.1}\right) = 5r$$
$$r \approx 0.012702681$$

The relative growth rate is approximately $\boxed{1.3\%}$ per year.

Assuming the growth rate stays the same, write a population model.

Let $N(t) = 9$.

Solve for t.

$$N(t) = 6.1 e^{0.013t}$$
$$9 = 6.1 e^{0.013t}$$
$$e^{0.013t} = \frac{9}{6.1}$$
$$\ln(e^{0.013t}) = \ln\left(\frac{9}{6.1}\right)$$
$$0.013t = \ln\left(\frac{9}{6.1}\right)$$
$$t \approx 29.91813894$$

In $\boxed{2030}$ the world population will reach 9 billion if the same growth rate is maintained.

■ **YOUR TURN**   The population of North America was 300 million in 1995 and 332 million in 2005. Find the relative growth rate and determine what year the population will reach 1 billion.

■ **Answer:** 1% per year; 2115

Exponential Decay Models

We mentioned radioactive decay briefly in Section 5.1. Radioactive decay is the process in which radioactive elements (atoms) lose energy by emitting radiation in the form of particles. This results in loss of mass. This process is random, but given a large number of atoms, the decay rate is directly proportional to the mass of the radioactive substance. Since the mass is decreasing, we say this represents *exponential decay*, $m = m_0 e^{-rt}$, where m_0 represents the initial mass at time $t = 0$, r is the decay rate, t is time, and m represents the mass at time t. In general, any model of the form $f(x) = ce^{-kx}$, $k > 0$, models **exponential decay**.

Typically, the decay rate r is expressed in terms of the half-life h. Recall (Section 5.1) that half-life is the time it takes for a quantity to decrease by half.

WORDS	MATH
Write the radioactive decay model.	$m = m_0 e^{-rt}$
Divide both sides by m_0.	$\dfrac{m}{m_0} = e^{-rt}$
The remaining mass is half of the initial mass when $t = h$.	$\dfrac{1}{2} = e^{-rh}$
Solve for r.	
Take the natural logarithm of both sides.	$\ln\left(\dfrac{1}{2}\right) = \ln\left(e^{-rh}\right)$
Simplify.	$\underbrace{\ln 1}_{0} - \ln 2 = -rh$
	$rh = \ln 2$
	$\boxed{r = \dfrac{\ln 2}{h}}$

Classroom Example 5.5.2

A radioactive isotope is known to have a half-life of 36 hours.

a. Determine the exponential decay model that represents the mass of this isotope.
b. If 1000 milligrams were taken, how many remain in the body after 48 hours?
c.* How long does it take the original 1000 milligram sample to decay to a mass of 100 milligrams?

Answer:

a. $m = m_0 e^{-0.019t}$, where t is measured in hours and m is measured in milligrams
b. 401.7 mg
c. 121.2 hours

EXAMPLE 2 Radioactive Decay

The radioactive isotope of potassium ^{42}K, which is vital in the diagnosis of brain tumors, has a half-life of 12.36 hours.

a. Determine the exponential decay model that represents the mass of ^{42}K.
b. If 500 milligrams of potassium-42 are taken, how many milligrams will remain after 48 hours?
c. How long will it take for the original 500-milligram sample to decay to a mass of 5 milligrams?

Solution (a):

Write the relationship between rate of decay and half-life.	$r = \dfrac{\ln 2}{h}$
Let $h = 12.36$.	$r \approx 0.056$
Write the exponential decay model for the mass of ^{42}K.	$\boxed{m = m_0 e^{-0.056t}}$

Solution (b):

Let $m_0 = 500$ and $t = 48$.

$$m = 500e^{-(0.056)(48)} \approx 34.00841855$$

There are approximately ⟨ 34 milligrams ⟩ of ^{42}K still in the body after 48 hours.

Note: Had we used the full value of $r = 0.056079868$, the resulting mass would have been $m = 33.8782897$, which is approximately 34 milligrams.

Solution (c):

Write the exponential decay model for the mass of ^{42}K.

$$m = m_0 e^{-0.056t}$$

Let $m = 5$ and $m_0 = 500$.

$$5 = 500e^{-0.056t}$$

Solve for t.

Divide by 500.

$$e^{-0.056t} = \frac{5}{500} = \frac{1}{100}$$

Take the natural logarithm of both sides.

$$\ln\left(e^{-0.056t}\right) = \ln\left(\frac{1}{100}\right)$$

Simplify.

$$-0.056t = \ln\left(\frac{1}{100}\right)$$

Divide by -0.056 and approximate with a calculator.

$$t \approx 82.2352$$

It will take approximately ⟨ 82 hours ⟩ for the original 500-milligram substance to decay to a mass of 5 milligrams.

■ **YOUR TURN** The radioactive element radon-222 has a half-life of 3.8 days.

  a. Determine the exponential decay model that represents the mass of radon-222.

  b. How much of a 64-gram sample of radon-222 will remain after 7 days? Round to the nearest gram.

  c. How long will it take for the original 64-gram sample to decay to a mass of 4 grams? Round to the nearest day.

Gaussian (Normal) Distribution Models

If your instructor plots the grades from the last test, typically you will see a **Gaussian (normal) distribution** of scores, otherwise known as the *bell-shaped curve*. Other examples of phenomena that tend to follow a Gaussian distribution are SAT scores, height distributions of adults, and standardized tests like IQ assessments.

The graph to the right represents a Gaussian distribution of IQ scores. The average score, which for IQ is 100, is the x-value at which the maximum occurs. The typical probability distribution is

$$F(x) = \frac{1}{\sigma\sqrt{2\pi}}\, e^{-(x-\mu)^2/2\sigma^2}$$

where μ is the average or mean value and the variance is σ^2.

Any model of the form $f(x) = ce^{-(x-a)^2/k}$ is classified as a **Gaussian model**.

Number of scores

Intelligence quotient
(Score on Wechsler Adult Intelligence Scale)

EXAMPLE 3   Weight Distributions

Suppose each member of a Little League football team is weighed and the weight distribution follows the Gaussian model $f(x) = 10e^{-(x-100)^2/25}$.

a. Graph the weight distribution.
b. What is the average weight of a member of this team?
c. Approximately how many boys weigh 95 pounds?

Solution:

a.

b. 100 pounds

c. $f(95) = 10e^{-(95-100)^2/25}$

$= 10e^{-25/25}$

$= 10e^{-1}$

≈ 3.6788

Approximately 4 boys weigh 95 pounds.

Logistic Growth Models

Earlier in this section we discussed exponential growth models for populations that experience uninhibited growth. Now we will turn our attention to *logistic growth*, which models population growth when there are factors that impact the ability to grow, such as food and space. For example, if 10 rabbits are dropped off on an uninhabited island, they will reproduce and the population of rabbits on that island will experience rapid growth. The population will continue to increase rapidly until the rabbits start running out of space or food on the island. In other words, under favorable conditions the growth is not restricted, while under less favorable conditions the growth becomes restricted. This type of growth is represented by **logistic growth models**, $f(x) = \dfrac{a}{1 + ce^{-kx}}$. Ultimately, the population of rabbits reaches the island's *carrying capacity, a*.

EXAMPLE 4   Number of Students on a College Campus

In 2011, the University of Central Florida (UCF) was the second largest university in the country. The number of students can be modeled by the function:

$$f(x) = \frac{60,000}{1 + 5e^{-0.12t}},$$ where t is time in years and

$t = 0$ corresponds to 1970.

a. How many students attended UCF in 1990? Round to the nearest thousand.
b. How many students attended UCF in 2010?
c. What is the carrying capacity of the UCF main campus?

Round all answers to the nearest thousand.

Solution (a): Let $t = 20$. $f(x) = \dfrac{60,000}{1 + 5e^{-0.12(20)}} \approx$ 41,000

Solution (b): Let $t = 40$. $f(x) = \dfrac{60,000}{1 + 5e^{-0.12(40)}} \approx$ 58,000

Solution (c): As t increases, the UCF student population approaches 60,000.

Logarithmic Models

Homeowners typically ask the question, "If I increase my payment, how long will it take to pay off my current mortgage?" In general, a loan over t years with an annual interest rate r with n periods per year corresponds to an interest rate per period of $i = \dfrac{r}{n}$. Typically loans are paid in equal payments consisting of the principal P plus total interest divided by the total number of periods over the life of the loan nt. The periodic payment R is given by

$$R = P \, \frac{i}{1 - (1 + i)^{-nt}}$$

We can find the time (in years) it will take to pay off the loan as a function of periodic payment by solving for t.

WORDS	**MATH**
Multiply both sides by $1 - (1 + i)^{-nt}$.	$R[1 - (1 + i)^{-nt}] = Pi$
Eliminate the brackets.	$R - R(1 + i)^{-nt} = Pi$
Subtract R.	$-R(1 + i)^{-nt} = Pi - R$
Divide by $-R$.	$(1 + i)^{-nt} = 1 - \dfrac{Pi}{R}$
Take the natural log of both sides.	$\ln(1 + i)^{-nt} = \ln\left(1 - \dfrac{Pi}{R}\right)$
Use the power property for logarithms.	$-nt \ln(1 + i) = \ln\left(1 - \dfrac{Pi}{R}\right)$
Isolate t.	$t = -\dfrac{\ln(1 - Pi/R)}{n \ln(1 + i)}$
Let $i = \dfrac{r}{n}$.	$\boxed{t = -\dfrac{\ln(1 - Pr/(nR))}{n \ln(1 + r/n)}}$

EXAMPLE 5   Paying Off Credit Cards

James owes \$15,000 on his credit card. The annual interest rate is 13% compounded monthly.

a. Find the time it will take to pay off his credit card if he makes payments of \$200 per month.

b. Find the time it will take to pay off his credit card if he makes payments of \$400 per month.

Let $P = 15{,}000$, $r = 0.13$, and $n = 12$.

$$t = -\frac{\ln\left(1 - \dfrac{15{,}000(0.13)}{12R}\right)}{12 \ln\left(1 + \dfrac{0.13}{12}\right)}$$

Technology Tip

```
Plot1 Plot2 Plot3
\Y1■15000(.13/12
)/(1-(1+0.13/12)
^(-12X)
\Y2■200
\Y3■400
\Y4=■
```

Use the keystrokes 2nd Calc 5:Intersect to find the points of intersection.

Intersection
X=12.946366   Y=200

Intersection
X=4.031658   Y=400

Classroom Example 5.5.5
Ben owes $12,000 on his credit card. The annual interest rate is 9.9% compounded monthly.
a. Find the time it will take to pay off his credit card if he makes payments of $100 per month.
b.* If he tripled his monthly payment, how long would it take?

Answer:
a. approximately 47 years
b. approximately 4 years

Solution (a): Let $R = 200$.

$$t = -\frac{\ln\left(1 - \frac{15{,}000(0.13)}{12(200)}\right)}{12 \ln\left(1 + \frac{0.13}{12}\right)} \approx 13$$

$200 monthly payments will allow James to pay off his credit card in about $\boxed{13 \text{ years}}$.

Solution (b): Let $R = 400$.

$$t = -\frac{\ln\left(1 - \frac{15{,}000(0.13)}{12(400)}\right)}{12 \ln\left(1 + \frac{0.13}{12}\right)} \approx 4$$

$400 monthly payments will allow James to pay off the balance in approximately $\boxed{4 \text{ years}}$. It is important to note that doubling the payment reduced the time to pay off the balance to less than a third.

SECTION 5.5 SUMMARY

In this section we discussed five main types of models that involve exponential and logarithmic functions.

NAME	MODEL	APPLICATIONS
Exponential growth	$f(x) = ce^{kx}, k > 0$	Uninhibited growth (populations/inflation)
Exponential decay	$f(x) = ce^{-kx}, k > 0$	Carbon dating, depreciation
Gaussian (normal) distributions	$f(x) = ce^{-(x-a)^2/k}$	Bell curves (standardized tests, height/weight charts, distribution of power flux of laser beams)
Logistic growth	$f(x) = \dfrac{a}{1 + ce^{-kx}}$	Conservation biology (growth limited by factors like food and space), learning curve
Logarithmic	$f(x) = a + c \log x$ $f(x) = a + c \ln x$ or quotients of logarithmic functions	Time to pay off credit cards, annuity planning

SECTION 5.5 EXERCISES

■ **SKILLS**

In Exercises 1–6, match the function with the graph (a to f) and the model name (i to v).

1. $f(t) = 5e^{2t}$

2. $N(t) = 28e^{-t/2}$

3. $T(x) = 4e^{-(x-80)^2/10}$

4. $P(t) = \dfrac{200}{1 + 5e^{-0.4t}}$

5. $D(x) = 4 + \log(x - 1)$

6. $h(t) = 2 + \ln(t + 3)$

Model Name

i. Logarithmic

ii. Logistic

iii. Gaussian

iv. Exponential growth

v. Exponential decay

Graphs

a.

b.

c.

d.

e.

f.

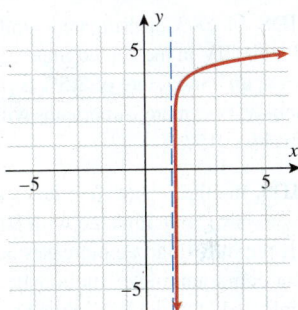

▪ APPLICATIONS

7. Population Growth. The population of the Philippines in 2003 was 80 million. Their population increases 2.36% per year. What was the expected population of the Philippines in 2010? Apply the formula $N = N_0 e^{rt}$, where N represents the number of people.

8. Population Growth. China's urban population is growing at 2.5% a year, compounding continuously. If there were 13.7 million people in Shanghai in 1996, approximately how many people will there be in 2016? Apply the formula $N = N_0 e^{rt}$, where N represents the number of people.

9. Population Growth. Port St. Lucie, Florida, had the United States' fastest growth rate among cities with a population of 100,000 or more between 2003 and 2004. In 2003, the population was 103,800 and increasing at a rate of 12% per year. In what year should the population reach 200,000? (Let $t = 0$ correspond to 2003.) Apply the formula $N = N_0 e^{rt}$, where N represents the number of people.

10. Population Growth. San Francisco's population has been declining since the "dot com" bubble burst. In 2002, the population was 776,000. If the population is declining at a rate of 1.5% per year, in what year will the population be 700,000? (Let $t = 0$ correspond to 2002.) Apply the formula $N = N_0 e^{-rt}$, where N represents the number of people.

11. Cellular Phone Plans. The number of cell phones in China is exploding. In 2007, there were 487.4 million cell phone subscribers and the number was increasing at a rate of 16.5% per year. How many cell phone subscribers were there in 2010 according to this model? Use the formula $N = N_0 e^{rt}$, where N represents the number of cell phone subscribers. Let $t = 0$ correspond to 2007.

12. Bacteria Growth. A colony of bacteria is growing exponentially. Initially 500 bacteria were in the colony. The growth rate is 20% per hour. (a) How many bacteria should be in the colony in 12 hours? (b) How many in one day? Use the formula $N = N_0 e^{rt}$, where N represents the number of bacteria.

13. Real Estate Appreciation. In 2004, the average house in Las Vegas cost $185,000, and real estate prices were increasing at an amazing rate of 30% per year. What was the expected cost of an average house in Las Vegas in 2007? Use the formula $N = N_0 e^{rt}$, where N represents the average cost of a home. Round to the nearest thousand.

14. Real Estate Appreciation. The average cost of a single-family home in California in 2004 was $230,000. In 2005, the average cost was $252,000. If this trend continued, what was the expected cost in 2007? Use the formula $N = N_0 e^{rt}$, where N represents the average cost of a home. Round to the nearest thousand.

15. **Oceanography (Growth of Phytoplankton).** Phytoplankton are microscopic plants that live in the ocean. Phytoplankton grow abundantly in oceans around the world and are the foundation of the marine food chain. One variety of phytoplankton growing in tropical waters is increasing at a rate of 20% per month. If it is estimated that there are 100 million in the water, how many will there be in 6 months? Utilize formula $N = N_0 e^{rt}$, where N represents the population of phytoplankton.

16. **Oceanography (Growth of Phytoplankton).** In Arctic waters there are an estimated 50 million phytoplankton. The growth rate is 12% per month. How many phytoplankton will there be in 3 months? Utilize formula $N = N_0 e^{rt}$, where N represents the population of phytoplankton.

17. **HIV/AIDS.** In 2003, an estimated 1 million people had been infected with HIV in the United States. If the infection rate increases at an annual rate of 2.5% a year compounding continuously, how many Americans will be infected with the HIV virus by 2015?

18. **HIV/AIDS.** In 2003, there were an estimated 25 million people who have been infected with HIV in sub-Saharan Africa. If the infection rate increases at an annual rate of 9% a year compounding continuously, how many Africans will be infected with the HIV virus by 2015?

19. **Anesthesia.** When a person has a cavity filled, the dentist typically gives a local anesthetic. After leaving the dentist's office, one's mouth often is numb for several more hours. If 100 ml of anesthesia is injected into the local tissue at the time of the procedure ($t = 0$), and the amount of anesthesia still in the local tissue t hours after the initial injection is given by $A = 100e^{-0.5t}$, how much is in the local tissue 4 hours later?

20. **Anesthesia.** When a person has a cavity filled, the dentist typically gives a local anesthetic. After leaving the dentist's office, one's mouth often is numb for several more hours. If 100 ml of anesthesia is injected into the local tissue at the time of the procedure ($t = 0$), and the amount of anesthesia still in the local tissue t hours after the initial injection is given by $A = 100e^{-0.5t}$, how much is in the local tissue 12 hours later?

21. **Business.** The sales S (in thousands of units) of a new mp3 player after it has been on the market for t years can be modeled by

$$S(t) = 750(1 - e^{-kt})$$

   a. If 350,000 units of the mp3 player were sold in the first year, find k to four decimal places.
   b. Use the model found in part (a) to estimate the sales of the mp3 player after it has been on the market for 3 years.

22. **Business.** During an economic downturn the annual profits of a company dropped from $850,000 in 2008 to $525,000 in 2010. Assume the exponential model $P(t) = P_0 e^{kt}$ for the annual profit where P is profit in thousands of dollars, and t is time in years.

   a. Find the exponential model for the annual profit.
   b. Assuming the exponential model is applicable in the year 2012, estimate the profit (to the nearest thousand dollars) for the year 2012.

23. **Radioactive Decay.** Carbon-14 has a half-life of 5730 years. How long will it take 5 grams of carbon-14 to be reduced to 2 grams?

24. **Radioactive Decay.** Radium-226 has a half-life of 1600 years. How long will it take 5 grams of radium-226 to be reduced to 2 grams?

25. **Radioactive Decay.** The half-life of uranium-238 is 4.5 billion years. If 98% of uranium-238 remains in a fossil, how old is the fossil?

26. **Radioactive Decay.** A drug has a half-life of 12 hours. If the initial dosage is 5 milligrams, how many milligrams will be in the patient's body in 16 hours?

In Exercises 27–30, use the following formula for Newton's law of cooling.

If you take a hot dinner out of the oven and place it on the kitchen countertop, the dinner cools until it reaches the temperature of the kitchen. Likewise, a glass of ice set on a table in a room eventually melts into a glass of water at that room temperature. The rate at which the hot dinner cools or the ice in the glass melts at any given time is proportional to the difference between its temperature and the temperature of its surroundings (in this case, the room). This is called **Newton's law of cooling** (or warming) and is modeled by

$$T = T_S + (T_0 - T_S)e^{-kt}$$

where T is the temperature of an object at time t, T_s is the temperature of the surrounding medium, T_0 is the temperature of the object at time $t = 0$, t is the time, and k is a constant.

27. **Newton's Law of Cooling.** An apple pie is taken out of the oven with an internal temperature of $325°F$. It is placed on a rack in a room with a temperature of $72°F$. After 10 minutes, the temperature of the pie is $200°F$. What will be the temperature of the pie 30 minutes after coming out of the oven?

28. **Newton's Law of Cooling.** A cold drink is taken out of an ice chest with a temperature of $38°F$ and placed on a picnic table with a surrounding temperature of $75°F$. After 5 minutes the temperature of the drink is $45°F$. What will the temperature of the drink be 20 minutes after it is taken out of the chest?

29. Forensic Science (Time of Death). A body is discovered in a hotel room. At 7:00 A.M. a police detective found the body's temperature to be 85°F. At 8:30 A.M. a medical examiner measures the body's temperature to be 82°F. Assuming the room in which the body was found had a constant temperature of 74°F, how long has the victim been dead? (Normal body temperature is 98.6°F.)

30. Forensic Science (Time of Death). At 4 A.M. a body is found in a park. The police measure the body's temperature to be 90°F. At 5 A.M. the medical examiner arrives and determines the temperature to be 86°F. Assuming the temperature of the park was constant at 60°F, how long has the victim been dead?

31. Depreciation of Automobile. A new Lexus IS250 has a book value of $38,000, and after one year it has a book value of $32,000. What is the car's value in 4 years? Apply the formula $N = N_0 e^{-rt}$, where N represents the value of the car. Round to the nearest hundred.

32. Depreciation of Automobile. A new Hyundai Tiburon has a book value of $22,000, and after 2 years a book value of $14,000. What is the car's value in 4 years? Apply the formula $N = N_0 e^{-rt}$, where N represents the value of the car. Round to the nearest hundred.

33. Automotive. A new model BMW convertible coupe is designed and produced in time to appear in North America in the fall. BMW Corporation has a limited number of new models available. The number of new model BMW convertible coupes purchased in North America is given by $N = \dfrac{100,000}{1 + 10e^{-2t}}$, where t is the number of weeks after the BMW is released.

   a. How many new-model BMW convertible coupes will have been purchased 2 weeks after the new model becomes available?

   b. How many after 30 weeks?

   c. What is the maximum number of new model BMW convertible coupes that will be sold in North America?

34. iPhone. The number of iPhones purchased is given by $N = \dfrac{2,000,000}{1 + 2e^{-4t}}$, where t is the time in weeks after they are made available for purchase.

   a. How many iPhones are purchased within the first 2 weeks?

   b. How many iPhones are purchased within the first month?

35. Spread of a Virus. The number of MRSA (methicillin-resistant *Staphylococcus aureus*) cases has been rising sharply in England and Wales since 1997. In 1997, 2422 cases were reported. The number of cases reported in 2003 was 7684. How many cases will be expected in 2013? (Let $t = 0$ correspond to 1997.) Use the formula $N = N_0 e^{rt}$, where N represents the number of cases reported.

36. Spread of a Virus. Dengue fever, an illness carried by mosquitoes, is occurring in one of the worst outbreaks in decades across Latin America and the Caribbean. In 2004, 300,000 cases were reported, and 630,000 cases in 2007. How many cases might be expected in 2013? (Let $t = 0$ be 2004.) Use the formula $N = N_0 e^{rt}$, where N represents the number of cases.

37. Carrying Capacity. The Virginia Department of Fish and Game stocks a mountain lake with 500 trout. Officials believe the lake can support no more than 10,000 trout. The number of trout is given by $N = \dfrac{10,000}{1 + 19e^{-1.56t}}$, where t is time in years. How many years will it take for the trout population to reach 5000?

38. Carrying Capacity. The World Wildlife Fund has placed 1000 rare pygmy elephants in a conservation area in Borneo. They believe 1600 pygmy elephants can be supported in this environment. The number of elephants is given by $N = \dfrac{1600}{1 + 0.6e^{-0.14t}}$, where t is time in years. How many years will it take the herd to reach 1200 elephants?

39. Lasers. The intensity of a laser beam is given by the ratio of power to area. A particular laser beam has an intensity function given by $I = e^{-r^2}$ mW/cm^2, where r is the radius off the center axis given in cm. Where is the beam brightest (largest intensity)?

40. Lasers. The intensity of a laser beam is given by the ratio of power to area. A particular laser beam has an intensity function given by $I = e^{-r^2}$ mW/cm^2, where r is the radius off the center axis given in cm. What percentage of the on-axis intensity ($r = 0$) corresponds to $r = 2$ cm?

41. Grade Distribution. Suppose the first test in this class has a normal, or bell-shaped, grade distribution of test scores, with an average score of 75. An approximate function that models your class's grades on test 1 is $N(x) = 10e^{-(x-75)^2/25^2}$, where N represents the number of students who received the score x.

   a. Graph this function.

   b. What is the average grade?

   c. Approximately how many students scored a 50?

   d. Approximately how many students scored 100?

42. Grade Distribution. Suppose the final exam in this class has a normal, or bell-shaped, grade distribution of exam scores, with an average score of 80. An approximate function that models your class's grades on the exam is $N(x) = 10e^{-(x-80)^2/16^2}$, where N represents the number of students who received the score x.

   a. Graph this function.

   b. What is the average grade?

   c. Approximately how many students scored a 60?

   d. Approximately how many students scored 100?

43. Time to Pay Off Debt. Diana just graduated from medical school owing $80,000 in student loans. The annual interest rate is 9%.

  a. Approximately how many years will it take to pay off her student loan if she makes a monthly payment of $750?

  b. Approximately how many years will it take to pay off her loan if she makes a monthly payment of $1000?

44. Time to Pay Off Debt. Victor owes $20,000 on his credit card. The annual interest rate is 17%.

  a. Approximately how many years will it take him to pay off this credit card if he makes a monthly payment of $300?

  b. Approximately how many years will it take him to pay off this credit card if he makes a monthly payment of $400?

For Exercises 45 and 46, refer to the following:

A local business borrows $200,000 to purchase property. The loan has an annual interest rate of 8% compounded monthly and a minimum monthly payment of $1467.

45. Time to Pay Off Debt/Business.

  a. Approximately how many years will it take the business to pay off the loans if only the minimum payment is made?

  b. How much interest will the business pay over the life of the loan if only the minimum payment is made?

46. Time to Pay Off Debt/Business.

  a. Approximately how many years will it take the business to pay off the loan if the minimum payment is doubled?

  b. How much interest will the business pay over the life of the loan if the minimum payment is doubled?

  c. How much in interest will the business save by doubling the minimum payment (see Exercise 45, part b)?

■ CATCH THE MISTAKE

In Exercises 47 and 48, explain the mistake that is made.

47. The city of Orlando, Florida, has a population that is growing at 7% a year, compounding continuously. If there were 1.1 million people in greater Orlando in 2006, approximately how many people will there be in 2016? Apply the formula $N = N_0 e^{rt}$, where N represents the number of people.

Solution:

Use the population growth model. $\qquad N = N_0 e^{rt}$

Let $N_0 = 1.1$, $r = 7$, and $t = 10$. $\qquad N = 1.1 e^{(7)(10)}$

Approximate with a calculator. $\qquad 2.8 \times 10^{30}$

This is incorrect. What mistake was made?

48. The city of San Antonio, Texas, has a population that is growing at 5% a year, compounding continuously. If there were 1.3 million people in the greater San Antonio area in 2006, approximately how many people will there be in 2016? Apply the formula $N = N_0 e^{rt}$, where N represents the number of people.

Solution:

Use the population growth model. $\qquad N = N_0 e^{rt}$

Let $N_0 = 1.3$, $r = 5$, and $t = 10$. $\qquad N = 1.3 e^{(5)(10)}$

Approximate with a calculator. $\qquad 6.7 \times 10^{21}$

This is incorrect. What mistake was made?

■ CONCEPTUAL

In Exercises 49–52, determine whether each statement is true or false.

49. When a species gets placed on an endangered species list, the species begins to grow rapidly, and then reaches a carrying capacity. This can be modeled by logistic growth.

50. A professor has 400 students one semester. The number of names (of her students) she can memorize can be modeled by a logarithmic function.

51. The spread of lice at an elementary school can be modeled by exponential growth.

52. If you purchase a laptop computer this year ($t = 0$), then the value of the computer can be modeled with exponential decay.

■ **CHALLENGE**

In Exercises 53 and 54, refer to the logistic model $f(x) = \dfrac{a}{1 + ce^{-kx}}$, where a is the carrying capacity.

53. As c increases, does the model reach the carrying capacity in less time or more time?

54. As k increases, does the model reach the carrying capacity in less time or more time?

■ **TECHNOLOGY**

55. Wing Shan just graduated from dental school owing $80,000 in student loans. The annual interest is 6%. Her time t to pay off the loan is given by

$$t = -\frac{\ln(1 - 80000(0.06)/(nR))}{n \ln(1 + 0.06/n)}$$

where n is the number of payment periods per year and R is the periodic payment.

a. Use a graphing utility to graph

$$t_1 = -\frac{\ln(1 - 80000(0.06)/(12x))}{12 \ln(1 + 0.06/12)} \text{ as } Y_1 \text{ and}$$

$$t_2 = -\frac{\ln(1 - 80000(0.06)/(26x))}{26 \ln(1 + 0.06/26)} \text{ as } Y_2.$$

Explain the difference in the two graphs.

b. Use the $\boxed{\text{TRACE}}$ key to estimate the number of years it will take Wing Shan to pay off her student loan if she can afford a monthly payment of $800.

c. If she can make a biweekly payment of $400, estimate the number of years it will take her to pay off the loan.

d. If she adds $200 more to her monthly or $100 more to her biweekly payment, estimate the number of years it will take her to pay off the loan.

56. Hong has a credit card debt in the amount of $12,000. The annual interest is 18%. His time t to pay off the loan is given by

$$t = -\frac{\ln(1 - 12000(0.18)/(nR))}{n \ln(1 + 0.18/n)}$$

where n is the number of payment periods per year and R is the periodic payment.

a. Use a graphing utility to graph

$$t_1 = -\frac{\ln(1 - 12000(0.18)/(12x))}{12 \ln(1 + 0.18/12)} \text{ as } Y_1 \text{ and}$$

$$t_2 = -\frac{\ln(1 - 12000(0.18)/(26x))}{26 \ln(1 + 0.18/26)} \text{ as } Y_2.$$

Explain the difference in the two graphs.

b. Use the $\boxed{\text{TRACE}}$ key to estimate the number of years it will take Hong to pay off his credit card if he can afford a monthly payment of $300.

c. If he can make a biweekly payment of $150, estimate the number of years it will take him to pay off the credit card.

d. If he adds $100 more to his monthly or $50 more to his biweekly payment, estimate the number of years it will take him to pay off the credit card.

CHAPTER 5 INQUIRY-BASED LEARNING PROJECT

Among other ideas, in Chapters 3 and 4 you studied functions and their inverses. For instance, you worked with this familiar quadratic function: $y = x^2$. In words, this means "squaring x equals y." The equation of its inverse function can be written $x = y^2$; "squaring y equals x." Of course, we call y the "square root of x." In order to write this relationship with y in terms of x, mathematicians devised the symbol for square root, and so we write $y = \sqrt{x}$.

Keep these ideas in mind as you look now at an exponential function and the need to define a new function and new symbol for its inverse.

1. Lef f be the base 10 exponential function, $f(x) = 10^x$.

   a. Graph the exponential function $y = 10^x$ by plotting points.

x	y
−3	
−2	
−1	
0	
1	
2	
3	

   b. Discuss whether or not $f(x) = 10^x$ has an inverse function. How did you determine this?

   c. Using the definition of inverse function, complete the table below for the function $y = f^{-1}(x)$. Then plot the points to make a graph.

x	y

   d. In part (a) $y = 10^x$, so we could say, "10 to the power of x equals y." Write a similar statement about the inverse relationship between x and y in part (d). How would you write this as an equation?

2. If you wanted to obtain the graph of $y = f^{-1}(x)$ using your graphing calculator, you would need to solve for y in the equation you wrote in part (d) above. To this end, we need a new symbol to represent the relationship $x = 10^y$. We will call y the "base 10 logarithm of x" and write $y = \log_{10}(x)$.

   a. The base 10 logarithm is also called the common logarithm. (You will study logarithms of other bases in Chapter 5.) Use the "log" key on your graphing calculator to graph $y = \log_{10}(x)$. How does this graph differ from the one you sketched in part 1(c)?

   b. What are the domain and range of $y = \log_{10}(x)$?

MODELING **OUR WORLD**

The following table summarizes the average yearly temperature in degrees Fahrenheit (°F) and carbon dioxide emissions in parts per million (ppm) for Mauna Loa, Hawaii.

YEAR	1960	1965	1970	1975	1980	1985	1990	1995	2000	2005
TEMPERATURE	44.45	43.29	43.61	43.35	46.66	45.71	45.53	47.53	45.86	46.23
CO_2 EMISSIONS (PPM)	316.9	320.0	325.7	331.1	338.7	345.9	354.2	360.6	369.4	379.7

In the Modeling Our World feature in Chapters 3 and 4, the temperature and carbon emissions were modeled with *linear functions* and *polynomial functions*, respectively. Now, let us model these same data using *exponential* and *logarithmic functions*.

1. Plot the temperature data, with time on the horizontal axis and temperature on the vertical axis. Let $t = 1$ correspond to 1960.

2. Find a *logarithmic function* with base e, $f(t) = A \ln Bt$, that models the temperature in Mauna Loa.

   a. Apply data from 1965 and 2005.
   b. Apply data from 2000 and 2005.
   c. Apply regression and all data given.

3. Predict what the temperature will be in Mauna Loa in 2020.

   a. Use the function found in 2a.
   b. Use the function found in 2b.
   c. Use the function found in 2c.

4. Predict what the temperature will be in Mauna Loa in 2100.

   a. Use the function found in 2a.
   b. Use the function found in 2b.
   c. Use the function found in 2c.

5. Do your models support the claim of "global warming"? Explain. Do these logarithmic models give similar predictions to the linear models found in Chapter 3 and the polynomial models found in Chapter 4?

6. Plot the carbon dioxide emissions data, with time on the horizontal axis and carbon dioxide emissions on the vertical axis. Let $t = 0$ correspond to 1960.

7. Find an *exponential function* with base e, $f(t) = Ae^{bt}$, that models the CO_2 emissions (ppm) in Mauna Loa.

   a. Apply data from 1960 and 2005.
   b. Apply data from 1960 and 2000.
   c. Apply regression and all data given.

8. Predict the expected CO_2 levels in Mauna Loa in 2020.

   a. Use the function found in 7a.
   b. Use the function found in 7b.
   c. Use the function found in 7c.

9. Predict the expected CO_2 levels in Mauna Loa in 2100.
   a. Use the function found in 7a.
   b. Use the function found in 7b.
   c. Use the function found in 7c.

10. Do your models support the claim of "global warming"? Explain. Do these exponential models give similar predictions to the linear models found in Chapter 3 or the polynomial models found in Chapter 4?

11. Comparing the models developed in Chapters 3, 4, and 5, do you believe that global temperatures are best modeled with a linear, polynomial, or logarithmic function?

12. Comparing the models developed in Chapters 3, 4, and 5, do you believe that CO_2 emissions are best modeled by linear, polynomial, or exponential functions?

SECTION	CONCEPT	KEY IDEAS/FORMULAS
5.1	**Exponential functions and their graphs**	
	Evaluating exponential functions	$f(x) = b^x \quad b > 0, b \neq 1$
	Graphs of exponential functions	y-intercept $(0, 1)$ Horizontal asymptote: $y = 0$; the points $(1, b)$ and $(-1, 1/b)$
	The natural base e	$f(x) = e^x$
	Applications of exponential functions	Doubling time: $P = P_0 2^{t/d}$ Compound interest: $A = P\left(1 + \dfrac{r}{n}\right)^{nt}$ Compounded continuously: $A = Pe^{rt}$
5.2	**Logarithmic functions and their graphs**	$y = \log_b x \quad x > 0$ $b > 0, b \neq 1$
	Evaluating logarithms	$y = \log_b x$ and $x = b^y$
	Common and natural logarithms	$y = \log x$ Common (base 10) $y = \ln x$ Natural (base e)
	Graphs of logarithmic functions	x-intercept $(1, 0)$ Vertical asymptote: $x = 0$; the points $(b, 1)$ and $(1/b, -1)$ Domain: $(0, \infty)$
	Applications of logarithms	Decibel scale: $D = 10 \log\left(\dfrac{I}{I_T}\right) \qquad I_T = 1 \times 10^{-12}\ \text{W/m}^2$ Richter scale: $M = \dfrac{2}{3} \log\left(\dfrac{E}{E_0}\right) \qquad E_0 = 10^{4.4}\ \text{joules}$
5.3	**Properties of logarithms**	
	Properties of logarithmic functions	**1.** $\log_b 1 = 0$ **2.** $\log_b b = 1$ **3.** $\log_b b^x = x$ **4.** $b^{\log_b x} = x \quad x > 0$ Product property: **5.** $\log_b MN = \log_b M + \log_b N$ Quotient property: **6.** $\log_b\left(\dfrac{M}{N}\right) = \log_b M - \log_b N$ Power property: **7.** $\log_b M^p = p \log_b M$
	Change-of-base formula	$\log_b M = \dfrac{\log M}{\log b}$ or $\log_b M = \dfrac{\ln M}{\ln b}$
5.4	**Exponential and logarithmic equations**	
	Solving exponential equations	*Simple exponential equations* **1.** Rewrite both sides of the equation in terms of the same base. **2.** Use the one-to-one property to equate the exponents. **3.** Solve for the variable.

		Complicated exponential equations
		1. Isolate the exponential expression.
		2. Take the same logarithm of both sides.
		3. Simplify using the inverse properties.
		4. Solve for the variable.
	Solving logarithmic equations	*Simple logarithmic equations*
		1. Combine logarithms on each side of the equation using properties.
		2. Use the one-to-one property to equate the exponents.
		3. Solve for the variable.
		4. Eliminate any extraneous solutions.
		Complicated logarithmic equations
		1. Combine and isolate the logarithmic expressions.
		2. Rewrite the equation in exponential form.
		3. Solve for the variable.
		4. Eliminate any extraneous solutions.
5.5	**Exponential and logarithmic models**	
	Exponential growth models	$f(x) = ce^{kx} \quad k > 0$
	Exponential decay models	$f(x) = ce^{-kx} \quad k > 0$
	Gaussian (normal) distribution models	$f(x) = ce^{-(x-a)^2/k}$
	Logistic growth models	$f(x) = \dfrac{a}{1 + ce^{-kx}}$
	Logarithmic models	$f(x) = a + c \log x$ $f(x) = a + c \ln x$

CHAPTER REVIEW

5.1 Exponential Functions and Their Graphs

Approximate each number using a calculator and round your answer to two decimal places.

1. $8^{4.7}$ **2.** $\pi^{2/5}$ **3.** $4 \cdot 5^{0.2}$ **4.** $1.2^{1.2}$

Approximate each number using a calculator and round your answer to two decimal places.

5. $e^{3.2}$ **6.** e^{π} **7.** $e^{\sqrt{\pi}}$ **8.** $e^{-2.5\sqrt{3}}$

Evaluate each exponential function for the given values.

9. $f(x) = 2^{4-x}$ $f(-2.2)$
10. $f(x) = -2^{x+4}$ $f(1.3)$
11. $f(x) = \left(\frac{2}{5}\right)^{1-6x}$ $f\left(\frac{1}{2}\right)$
12. $f(x) = \left(\frac{4}{7}\right)^{5x+1}$ $f\left(\frac{1}{5}\right)$

Match the graph with the function.

13. $y = 2^{x-2}$ **14.** $y = -2^{2-x}$
15. $y = 2 + 3^{x+2}$ **16.** $y = -2 - 3^{2-x}$

a.

b.

c.

d.

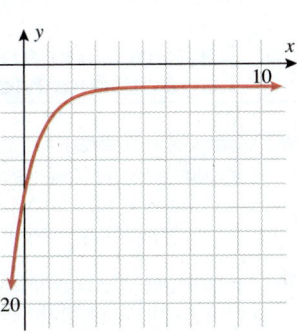

State the y-intercept and the horizontal asymptote and graph the exponential function.

17. $y = -6^{-x}$ **18.** $y = 4 - 3^x$
19. $y = 1 + 10^{-2x}$ **20.** $y = 4^x - 4$

State the y-intercept and horizontal asymptote, and graph the exponential function.

21. $y = e^{-2x}$ **22.** $y = e^{x-1}$
23. $y = 3.2e^{x/3}$ **24.** $y = 2 - e^{1-x}$

Applications

25. Compound Interest. If $4500 is deposited into an account paying 4.5% compounding semiannually, how much will you have in the account in 7 years?

26. Compound Interest. How much money should be put in a savings account now that earns 4.0% a year compounded quarterly if you want $25,000 in 8 years?

27. Compound Interest. If $13,450 is put in a money market account that pays 3.6% a year compounded continuously, how much will be in the account in 15 years?

28. Compound Interest. How much money should be invested today in a money market account that pays 2.5% a year compounded continuously if you desire $15,000 in 10 years?

5.2 Logarithmic Functions and Their Graphs

Write each logarithmic equation in its equivalent exponential form.

29. $\log_4 64 = 3$ **30.** $\log_4 2 = \frac{1}{2}$
31. $\log\left(\frac{1}{100}\right) = -2$ **32.** $\log_{16} 4 = \frac{1}{2}$

Write each exponential equation in its equivalent logarithmic form.

33. $6^3 = 216$ **34.** $10^{-4} = 0.0001$
35. $\frac{4}{169} = \left(\frac{2}{13}\right)^2$ **36.** $\sqrt[3]{512} = 8$

Evaluate the logarithms exactly.

37. $\log_7 1$ **38.** $\log_4 256$
39. $\log_{1/6} 1296$ **40.** $\log 10^{12}$

Approximate the common and natural logarithms utilizing a calculator. Round to two decimal places.

41. $\log 32$ **42.** $\ln 32$
43. $\ln 0.125$ **44.** $\log 0.125$

State the domain of the logarithmic function in interval notation.

45. $f(x) = \log_3(x + 2)$ **46.** $f(x) = \log_2(2 - x)$
47. $f(x) = \log(x^2 + 3)$ **48.** $f(x) = \log(3 - x^2)$

Match the graph with the function.

49. $y = \log_7 x$ **50.** $y = -\log_7(-x)$
51. $y = \log_7(x + 1) - 3$ **52.** $y = -\log_7(1 - x) + 3$

a.

b.

c.

d.

Graph the logarithmic function with transformation techniques.

53. $f(x) = \log_4 (x - 4) + 2$ 54. $f(x) = \log_4 (x + 4) - 3$

55. $f(x) = -\log_4 (x) - 6$ 56. $f(x) = -2 \log_4 (-x) + 4$

Applications

57. **Chemistry.** Calculate the pH value of milk, assuming it has a concentration of hydrogen ions given by $H^+ = 3.16 \times 10^{-7}$.

58. **Chemistry.** Calculate the pH value of Coca-Cola, assuming it has a concentration of hydrogen ions given by $H^+ = 2.0 \times 10^{-3}$.

59. **Sound.** Calculate the decibels associated with a teacher speaking to a medium-sized class if the sound intensity is 1×10^{-7} W/m².

60. **Sound.** Calculate the decibels associated with an alarm clock if the sound intensity is 1×10^{-4} W/m².

5.3 Properties of Logarithms

Use the properties of logarithms to simplify each expression.

61. $\log_{2.5} 2.5$ 62. $\log_2 \sqrt{16}$

63. $2.5^{\log_{2.5} 6}$ 64. $e^{-3 \ln 6}$

Write each expression as a sum or difference of logarithms.

65. $\log_c x^a y^b$ 66. $\log_3 x^2 y^{-3}$

67. $\log_j \left(\dfrac{rs}{t^3} \right)$ 68. $\log x^c \sqrt{x + 5}$

69. $\log \left[\dfrac{a^{1/2}}{b^{3/2} c^{2/5}} \right]$ 70. $\log_7 \left[\dfrac{c^3 d^{1/3}}{e^6} \right]^{1/3}$

Evaluate the logarithms using the change-of-base formula.

71. $\log_8 3$ 72. $\log_5 \frac{1}{2}$

73. $\log_\pi 1.4$ 74. $\log_{\sqrt{3}} 2.5$

5.4 Exponential and Logarithmic Equations

Solve the exponential equations exactly for x.

75. $4^x = \frac{1}{256}$ 76. $3^{x^2} = 81$

77. $e^{3x-4} = 1$ 78. $e^{\sqrt{x}} = e^{4.8}$

79. $\left(\frac{1}{3}\right)^{x+2} = 81$ 80. $100^{x^2 - 3} = 10$

Solve the exponential equation. Round your answer to three decimal places.

81. $e^{2x+3} - 3 = 10$ 82. $2^{2x-1} + 3 = 17$

83. $e^{2x} + 6e^x + 5 = 0$ 84. $4e^{0.1x} = 64$

85. $(2^x - 2^{-x})(2^x + 2^{-x}) = 0$ 86. $5(2^x) = 25$

Solve the logarithmic equations exactly.

87. $\log(3x) = 2$

88. $\log_3(x + 2) = 4$

89. $\log_4 x + \log_4 2x = 8$

90. $\log_6 x + \log_6(2x - 1) = \log_6 3$

Solve the logarithmic equations. Round your answers to three decimal places.

91. $\ln x^2 = 2.2$

92. $\ln(3x - 4) = 7$

93. $\log_3(2 - x) - \log_3(x + 3) = \log_3 x$

94. $4 \log(x + 1) - 2 \log(x + 1) = 1$

5.5 Exponential and Logarithmic Models

95. **Compound Interest.** If Tania needs $30,000 a year from now for a down payment on a new house, how much should she put in a 1-year CD earning 5% a year compounding continuously so that she will have exactly $30,000 a year from now?

96. **Stock Prices.** Jeremy is tracking the stock value of Best Buy (BBY on the NYSE). In 2003, he purchased 100 shares at $28 a share. The stock did not pay dividends because the company reinvested all earnings. In 2005, Jeremy cashed out and sold the stock for $4000. What was the annual rate of return on BBY?

97. **Compound Interest.** Money is invested in a savings account earning 4.2% interest compounded quarterly. How many years will pass until the money doubles?

98. **Compound Interest.** If $9000 is invested in an investment earning 8% interest compounded continuously, how many years will pass until there is $22,500?

99. Population. Nevada has the fastest growing population according to the U.S. Census Bureau. In 2004, the population of Nevada was 2.62 million and increasing at an annual rate of 3.5%. What is the expected population in 2014? (Let $t = 0$ be 2004.) Apply the formula $N = N_0 e^{rt}$, where N is the population.

100. Population. The Hispanic population in the United States is the fastest growing of any ethnic group. In 1996, there were an estimated 28.3 million Hispanics in the United States, and in 2000 there were an estimated 32.5 million. What is the expected population of Hispanics in the United States in 2014? (Let $t = 0$ be 1996.) Apply the formula $N = N_0 e^{rt}$, where N is the population.

101. Bacteria Growth. Bacteria are growing exponentially. Initially, there were 1000 bacteria; after 3 hours there were 2500. How many bacteria should be expected in 6 hours? Apply the formula $N = N_0 e^{rt}$, where N is the number of bacteria.

102. Population. In 2003, the population of Phoenix, Arizona, was 1,388,215. In 2004, the population was 1,418,041. What is the expected population in 2014? (Let $t = 0$ be 2003.) Apply the formula $N = N_0 e^{rt}$, where N is the population.

103. Radioactive Decay. Strontium-90 has a half-life of 28 years. How long will it take for 20 grams of this to decay to 5 grams? Apply the formula $N = N_0 e^{-rt}$, where N is the number of grams.

104. Radioactive Decay. Plutonium-239 has a half-life of 25,000 years. How long will it take for 100 grams to decay to 20 grams? Apply the formula $N = N_0 e^{-rt}$, where N is the number of grams.

105. Wild Life Population. The *Boston Globe* reports that the fish population of the Essex River in Massachusetts is declining. In 2003, it was estimated there were 5600 fish in the river, and in 2004, there were only 2420 fish. How many should there have been in 2010 assuming the same trend? Apply the formula $N = N_0 e^{-rt}$, where N is the number of fish.

106. Car Depreciation. A new Acura TSX costs $28,200. In 2 years the value will be $24,500. What is the expected value in 6 years? Apply the formula $N = N_0 e^{-rt}$, where N is the value of the car.

107. Carrying Capacity. The carrying capacity of a species of beach mice in St. Croix is given by $M = 1000(1 - e^{-0.035t})$ where M is the number of mice and t is time in years ($t = 0$ corresponds to 1998). How many mice were there in 2010 according to this model?

108. Population. The city of Brandon, Florida, had 50,000 residents in 1970, and since the crosstown expressway was built, its population has increased 2.3% per year. If the growth continues at the same rate, how many residents will Brandon have in 2030?

Technology

Section 5.1

109. Use a graphing utility to graph the function $f(x) = \left(1 + \dfrac{\sqrt{2}}{x}\right)^x$. Determine the horizontal asymptote as x increases.

110. Use a graphing utility to graph the functions $y = e^{-x+2}$ and $y = 3^x + 1$ in the same viewing screen. Estimate the coordinates of the point of intersection. Round your answers to three decimal places.

Section 5.2

111. Use a graphing utility to graph the functions $y = \log_{2.4}(3x - 1)$ and $y = \log_{0.8}(x - 1) + 3.5$ in the same viewing screen. Estimate the coordinates of the point of intersection. Round your answers to three decimal places.

112. Use a graphing utility to graph the functions $y = \log_{2.5}(x - 1) + 2$ and $y = 3.5^{x-2}$ in the same viewing screen. Estimate the coordinates of the point(s) of intersection. Round your answers to three decimal places.

Section 5.3

113. Use a graphing utility to graph $f(x) = \log_2\left(\dfrac{x^3}{x^2 - 1}\right)$ and $g(x) = 3\log_2 x - \log_2(x + 1) - \log_2(x - 1)$ in the same viewing screen. Determine the domain where the two functions give the same graph.

114. Use a graphing utility to graph $f(x) = \ln\left(\dfrac{9 - x^2}{x^2 - 1}\right)$ and $g(x) = \ln(3 - x) + \ln(3 + x) - \ln(x + 1) - \ln(x - 1)$ in the same viewing screen. Determine the domain where the two functions give the same graph.

Section 5.4

115. Use a graphing utility to graph $y = \dfrac{e^x - e^{-x}}{e^x + e^{-x}}$. State the domain. Determine if there are any symmetries and asymptotes.

116. Use a graphing utility to graph $y = \dfrac{1}{e^x - e^{-x}}$. State the domain. Determine if there are any symmetries and asymptotes.

Section 5.5

117. A drug with initial dosage of 4 milligrams has a half-life of 18 hours. Let $(0, 4)$ and $(18, 2)$ be two points.

a. Determine the equation of the dosage.

b. Use STAT CALC ExpReg to model the equation of the dosage.

c. Are the equations in (a) and (b) the same?

118. In Exercise 105, let $t = 0$ be 2003 and $(0, 5600)$ and $(1, 2420)$ be the two points.

a. Use STAT CALC ExpReg to model the equation for the fish population.

b. Using the equation found in (a), how many fish should have been expected in 2010?

c. Does the answer in (b) agree with the answer in Exercise 105?

1. Simplify $\log 10^{x^3}$.

2. Use a calculator to evaluate $\log_5 326$ (round to two decimal places).

3. Find the exact value of $\log_{1/3} 81$.

4. Rewrite the expression $\ln\left[\dfrac{e^{5x}}{x(x^4 + 1)}\right]$ in a form with no logarithms of products, quotients, or powers.

In Exercises 5–20, solve for x, exactly if possible. If an approximation is required, round your answer to three decimal places.

5. $e^{x^2-1} = 42$

6. $e^{2x} - 5e^x + 6 = 0$

7. $27e^{0.2x+1} = 300$

8. $3^{2x-1} = 15$

9. $3\ln(x - 4) = 6$

10. $\log(6x + 5) - \log 3 = \log 2 - \log x$

11. $\ln(\ln x) = 1$

12. $\log_2(3x - 1) - \log_2(x - 1) = \log_2(x + 1)$

13. $\log_6 x + \log_6(x - 5) = 2$

14. $\ln(x + 2) - \ln(x - 3) = 2$

15. $\ln x + \ln(x + 3) = 1$

16. $\log_2\left(\dfrac{2x + 3}{x - 1}\right) = 3$

17. $\dfrac{12}{1 + 2e^x} = 6$

18. $\ln x + \ln(x - 3) = 2$

19. State the domain of the function $f(x) = \log\left[\dfrac{x}{x^2 - 1}\right]$.

20. State the range of x values for which the following is true: $10^{\log (4x - a)} = 4x - a$.

In Exercises 21–24, find all intercepts and asymptotes, and graph.

21. $f(x) = 3^{-x} + 1$

22. $f(x) = \left(\tfrac{1}{2}\right)^x - 3$

23. $f(x) = \ln(2x - 3) + 1$

24. $f(x) = \log(1 - x) + 2$

25. **Interest.** If $5000 is invested at a rate of 6% a year, compounded quarterly, what is the amount in the account after 8 years?

26. **Interest.** If $10,000 is invested at a rate of 5%, compounded continuously, what is the amount in the account after 10 years?

27. **Sound.** A lawn mower's sound intensity is approximately 1×10^{-3} W/m². Assuming your threshold of hearing is 1×10^{-12} W/m², calculate the decibels associated with the lawn mower.

28. **Population.** The population in Seattle, Washington, has been increasing at a rate of 5% a year. If the population continues to grow at that rate, and in 2004 there are 800,000 residents, how many residents will there be in 2014? *Hint:* $N = N_0 e^{rt}$.

29. **Earthquake.** An earthquake is considered moderate if it is between 5 and 6 on the Richter scale. What is the energy range in joules for a moderate earthquake?

30. **Radioactive Decay.** The mass $m(t)$ remaining after t hours from a 50-gram sample of a radioactive substance is given by the equation $m(t) = 50e^{-0.0578t}$. After how long will only 30 grams of the substance remain? Round your answer to the nearest hour.

31. **Bacteria Growth.** The number of bacteria in a culture is increasing exponentially. Initially, there were 200 in the culture. After 2 hours there are 500. How many should be expected in 8 hours? Round your answer to the nearest hundred.

32. **Carbon Decay.** Carbon-14 has a half-life of 5730 years. How long will it take for 100 grams to decay to 40 grams?

33. **Spread of a Virus.** The number of people infected by a virus is given by $N = \dfrac{2000}{1 + 3e^{-0.4t}}$, where t is time in days. In how many days will 1000 people be infected?

34. **Oil Consumption.** The world consumption of oil was 76 million barrels per day in 2002. In 2004, the consumption was 83 million barrels per day. How many barrels should be expected to be consumed in 2014?

35. Use a graphing utility to graph $y = \dfrac{e^x - e^{-x}}{2}$. State the domain. Determine if there are any symmetries and asymptotes.

36. Use a graphing utility to help solve the equation $4^{3-x} = 2x - 1$. Round your answer to two decimal places.

1. Simplify $\dfrac{x^{1/3}y^{2/5}}{x^{-1/2}y^{-8/5}}$ and express in terms of positive exponents.

2. Simplify $\dfrac{\dfrac{2}{x-2}+3}{3-\dfrac{2}{x-2}}$.

3. Solve using the quadratic formula: $5x^2 - 4x = 3$.

4. Solve and check: $\sqrt{2x+13} = 2 + \sqrt{x+3}$.

5. Solve and express the solution in interval notation: $\dfrac{x+4}{5} \le -3$.

6. Solve for x: $|x^2 - 4| = 9$.

7. Write an equation of the line that is perpendicular to the line $4x + 3y = 6$ and that passes through the point $(7, 6)$.

8. Using the function $f(x) = 4x - x^2$, evaluate the difference quotient $\dfrac{f(x+h) - f(x)}{h}$, $h \ne 0$.

9. Given the piecewise-defined function
$$f(x) = \begin{cases} 5 & -2 < x \le 0 \\ 2 - \sqrt{x} & 0 < x < 4 \\ x - 3 & x \ge 4 \end{cases}$$
find

   a. $f(4)$ b. $f(0)$ c. $f(1)$ d. $f(-4)$

   e. State the domain and range in interval notation.

   f. Determine the intervals where the function is increasing, decreasing, or constant.

10. Sketch the graph of the function $y = \sqrt{1-x}$ and identify all transformations.

11. Determine whether the function $f(x) = \sqrt{x-4}$ is one-to-one.

12. Write an equation that describes the variation: R is inversely proportional to the square of d and $R = 3.8$ when $d = 0.02$.

13. Find the vertex of the parabola associated with the quadratic function $f(x) = -4x^2 + 8x - 5$.

14. Find a polynomial of minimum degree (there are many) that has the zeros $x = -5$ (multiplicity 2) and $x = 9$ (multiplicity 4).

15. Use synthetic division to find the quotient $Q(x)$ and remainder $r(x)$ of $(3x^2 - 4x^3 - x^4 + 7x - 20) \div (x + 4)$.

16. Given the zero $x = 2 + i$ of the polynomial $P(x) = x^4 - 7x^3 + 13x^2 + x - 20$, determine all the other zeros and write the polynomial as the product of linear factors.

17. Find the vertical and slant asymptotes of $f(x) = \dfrac{x^2 + 7}{x - 3}$.

18. Graph the rational function $f(x) = \dfrac{3x}{x+1}$. Give all asymptotes.

19. Without employing a calculator, give the exact value of $\left(\frac{1}{25}\right)^{-3/2}$.

20. If $5400 is invested at 2.75% compounded monthly, how much is in the account after 4 years?

21. Give the exact value of $\log_3 243$.

22. Write the expression $\frac{1}{2}\ln(x+5) - 2\ln(x+1) - \ln(3x)$ as a single logarithm.

23. Solve the logarithmic equation exactly: $10^{2\log(4x+9)} = 121$.

24. Give an exact solution to the exponential equation $5^{x^2} = 625$.

25. If $8500 is invested at 4% compounded continuously, how many years will pass until there is $12,000?

26. Use a graphing utility to help solve the equation $e^{3-2x} = 2^{x-1}$. Round your answer to two decimal places.

27. Strontium-90 with an initial amount of 6 grams has a half-life of 28 years.

   a. Use [STAT] [CALC] [ExpReg] to model the equation of the amount remaining.

   b. How many grams will remain after 32 years? Round your answer to two decimal places.

6

Trigonometric Functions

© Kathy Yturralde/Alamy

To the ancient Greeks, trigonometry was the study of right triangles. Trigonometric functions (sine, cosine, tangent, cotangent, secant, and cosecant) can be defined as right triangle ratios (ratios of sides of a right triangle). Thousands of years later, we still find applications of right triangle trigonometry today in sports, surveying, navigation, and engineering.

In the eighteenth century, the unit circle approach was formulated. It relies on the correspondence between the values of the cosine and sine functions and the x- and y-coordinates along the unit circle (circle with center at the origin and radius 1).

The sine and cosine functions are used to represent periodic phenomena. Orbits, tide levels, the biological clock in animals and plants, and radio signals are all periodic (repetitive).

Tidal height predictions for Pawleys Island Pier, South Carolina
Plot produced by the Physics Department, Coastal Carolina University
http://kingfish.coastal.edu/physics/

TRIGONOMETRIC FUNCTIONS

6.1	6.2	6.3	6.4	6.5	6.6	6.7	6.8	6.9
Angles, Degrees, and Triangles	Definition 1 of Trigonometric Functions: Right Triangle Ratios	Applications of Right Triangle Trigonometry: Solving Right Triangles	Definition 2 of Trigonometric Functions: Cartesian Plane	Trigonometric Functions of Nonacute Angles	Radian Measure and Applications	Definition 3 of Trigonometric Functions: Unit Circle Approach	Graphs of Sine and Cosine Functions	Graphs of Other Trigonometric Functions

• Angles and Degree Measure Triangles • Special Right Triangles • Similar Triangles	• Trigonometric Functions: Right Triangle Ratios • Reciprocal Identities • Cofunctions • Evaluating Trigonometric Functions Exactly for Special Angle Measures: 30°, 45°, and 60° • Using Calculators to Evaluate (Approximate) Trigonometric Functions	• Accuracy and Significant Digits • Solving a Right Triangle Given the Measure of an Acute Angle and a Side Length • Solving a Right Triangle Given the Length of Two Sides	• Angles in Standard Position • Coterminal Angles • Trigonometric Functions: The Cartesian Plane	• Algebraic Signs of Trigonometric Functions • Ranges of the Trigonometric Functions • Reference Angles and Reference Right Triangles • Evaluating Trigonometric Functions for Nonacute Angles	• The Radian Measure of an Angle • Converting Between Degrees and Radians • Arc Length • Area of a Circular Sector • Linear Speed • Angular Speed • Relationship Between Linear and Angular Speeds	• Trigonometric Functions and the Unit Circle (Circular Functions) • Properties of Circular Functions	• The Graphs of Sinusoidal Functions • Graphing a Shifted Sinusoidal Function: $y = A\sin(Bx + C) + D$ and $y = A\cos(Bx + C) + D$ • Harmonic Motion • Graphing Sums of Functions: Addition of Ordinates	• Graphing the Tangent, Cotangent, Secant, and Cosecant Functions • Translations of Circular Functions

LEARNING OBJECTIVES

■ Understand degree measure.

■ Define the six trigonometric functions as ratios of lengths of the sides of right triangles.

■ Solve right triangles.

■ Define the six trigonometric functions as ratios of x- and y-coordinates and distances in the Cartesian plane.

■ Evaluate trigonometric functions for nonacute angles.

■ Relate degree and radian measures.

■ Draw the unit circle and label the sine and cosine values for special angles (in both degrees and radians).

■ Graph sine and cosine functions (amplitude, period, and translations).

■ Graph tangent, cotangent, secant, and cosecant functions.

SKILLS OBJECTIVES

- Find the complement of an angle.
- Find the supplement of an angle.
- Use the Pythagorean theorem to solve a right triangle.
- Solve 30°-60°-90° and 45°-45°-90° triangles.
- Use similarity to determine the length of a side of a triangle.

CONCEPTUAL OBJECTIVES

- Understand that degrees are a measure of an angle.
- Understand that the Pythagorean theorem applies only to right triangles.
- Understand that the sum of the three angles of any triangle is 180°.

Angles and Degree Measure

The study of trigonometry relies heavily on the concept of angles. Before we define angles, let us review some basic terminology. A **line** is the straight path connecting two points (*A* and *B*) and extending beyond the points in both directions. The portion of the line between the two points (including the points) is called a **line segment**. A **ray** is the portion of the line that starts at one point (*A*) and extends to infinity (beyond *B*). *A* is called the **endpoint of the ray**.

$$A \quad B$$
Line *AB*

$$A \quad B$$
Segment *AB*

$$A \quad B$$
Ray *AB*

In geometry, an **angle** is formed when two rays share the same endpoint. The common endpoint is called the **vertex**.

In trigonometry, we say that an angle is formed when a ray is rotated around its endpoint. The ray in its original position is called the **initial ray** or the **initial side** of an angle. In the Cartesian plane, we usually assume the initial side of an angle is the positive *x*-axis. The ray after it is rotated is called the **terminal ray** or the **terminal side** of an angle. Rotation in a counterclockwise direction corresponds to a **positive angle**, whereas rotation in a clockwise direction corresponds to a **negative angle**.

Study Tip

Positive angle: counterclockwise
Negative angle: clockwise

Lengths, or distances, can be measured in different units: feet, miles, and meters are three common units. In order to compare angles of different sizes, we need a standard unit of measure. One way to measure the size of an angle is with **degree measure**.

We will discuss **degrees** now and in Section 6.6 we will discuss another angle measure called *radians*.

> **DEFINITION**   **Degree Measure of Angles**
>
> An angle formed by one complete counterclockwise rotation has **measure 360 degrees**, denoted 360°.
>
> One complete revolution = 360°

WORDS	MATH
360° represents 1 complete rotation.	$\dfrac{360°}{360°} = 1$
180° represents a $\frac{1}{2}$ rotation.	$\dfrac{180°}{360°} = \dfrac{1}{2}$
90° represents a $\frac{1}{4}$ rotation.	$\dfrac{90°}{360°} = \dfrac{1}{4}$

The Greek letter θ (theta) is the most common name for an angle. Other common names of angles are α (alpha), β (beta), and γ (gamma).

WORDS	MATH
An angle measuring exactly 90° is called a **right angle**.	Right Angle: quarter rotation $\theta = 90°$
A right angle is often represented by the adjacent sides of a rectangle, indicating that the two rays are *perpendicular*.	
An angle measuring exactly 180° is called a **straight angle**.	$\theta = 180°$ Straight Angle: half rotation
An angle measuring greater than 0°, but less than 90°, is called an **acute angle**.	Acute Angle $0° < \theta < 90°$
An angle measuring greater than 90°, but less than 180°, is called an **obtuse angle**.	Obtuse Angle $90° < \theta < 180°$

Study Tip

Greek letters are often used to denote angles in trigonometry.

If the sum of the measures of two positive angles is 90°, the angles are called **complementary**. We say that α is the **complement** of β (and vice versa).

Complementary Angles
$\alpha + \beta = 90°$

If the sum of the measures of two positive angles is 180°, the angles are called **supplementary**. We say that α is the **supplement** of β (and vice versa).

Supplementary Angles
$\alpha + \beta = 180°$

Classroom Example 6.1.1
a. Find the complement of 23.2°.
b. Find the supplement of 163.2°.
c.* Find the complement of $\frac{1}{2}\alpha$, assuming that $0 < \alpha < 180°$.
d.* Find two supplementary angles for which one is 15 degrees less than 4 times the other.

Answer:
a. 66.8°   b. 16.8°
c. $90° - \frac{1}{2}\alpha$ d. 39°, 141°

EXAMPLE 1   **Finding Measures of Complementary and Supplementary Angles**

Find the measure of each angle.

a. Find the complement of 50°.

b. Find the supplement of 110°.

c. Represent the complement of α in terms of α.

d. Find two supplementary angles such that the first angle is twice as large as the second angle.

Solution:

a. The sum of complementary angles is 90°. $\theta + 50° = 90°$

    Solve for θ. $\theta = \boxed{40°}$

b. The sum of supplementary angles is 180°. $\theta + 110° = 180°$

    Solve for θ. $\theta = \boxed{70°}$

c. Let β be the complement of α.

    The sum of complementary angles is 90°. $\alpha + \beta = 90°$

    Solve for β. $\beta = \boxed{90° - \alpha}$

d. The sum of supplementary angles is 180°. $\alpha + \beta = 180°$

    Let $\beta = 2\alpha$. $\alpha + 2\alpha = 180°$

    Solve for α. $3\alpha = 180°$

                                                     $\alpha = 60°$

Substitute $\alpha = 60°$ into $\beta = 2\alpha$. $\beta = 120°$

The angles have measures $\boxed{60°}$ and $\boxed{120°}$.

■**Answer:** The angles have measures 45° and 135°.

■**YOUR TURN** Find two supplementary angles such that the first angle is three times as large as the second angle.

It is important not to confuse an angle with its measure. In Example 1(d), angle α is a rotation and the measure of that rotation is 60°.

Triangles

Trigonometry originated as the study of triangles, with emphasis on calculations involving the lengths of sides and the measures of angles. What do you already know about triangles? You know that triangles are three-sided, closed plane figures. An important property of triangles is that the sum of the measures of the three angles of any triangle is 180°.

ANGLE SUM OF A TRIANGLE

The sum of the measures of the angles of any triangle is 180°.

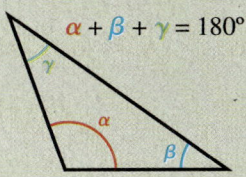

$$\alpha + \beta + \gamma = 180°$$

EXAMPLE 2   Finding an Angle of a Triangle

If two angles of a triangle have measures 32° and 68°, what is the measure of the third angle?

Solution:

The sum of the measures of all three angles is 180°. $32° + 68° + \alpha = 180°$

Solve for α. $\boxed{\alpha = 80°}$

▪ **YOUR TURN** If two angles of a triangle have measures 16° and 96°, what is the measure of the third angle?

Classroom Example 6.1.2*
If $\gamma = 2\beta$ and $\beta = 3\alpha$, find all three angles.

Answer: 18°, 54°, 108°

▪ **Answer:** 68°

Some triangles that may be familiar to you from geometry are equilateral, isosceles, and right. An **equilateral triangle** has three equal sides and three equal angles (60°-60°-60°). An **isosceles triangle** has two equal sides (legs) and two equal angles opposite those legs. The most important triangle that we will discuss in this course is the *right triangle*. A **right triangle** is a triangle in which one of the angles is a right angle (90°). Since one angle is 90°, the other two angles must be complementary (sum to 90°), so that the sum of all three angles is 180°. The longest side of a right triangle, called the **hypotenuse**, is opposite the right angle. The other two sides are called the **legs** of the right triangle.

Study Tip

In this book when we say "equal angles," this implies "equal angle measures." Similarly, when we say an angle is $x°$, this implies that the angle's measure is $x°$.

The *Pythagorean theorem* relates the sides of a right triangle. It says that the sum of the squares of the lengths of the two legs is equal to the square of the length of the hypotenuse. It is important to note that length (a synonym of distance) is always positive.

PYTHAGOREAN THEOREM

In any right triangle, the square of the length of the longest side (hypotenuse) is equal to the sum of the squares of the lengths of the other two sides (legs).

$$a^2 + b^2 = c^2$$

It is important to note that the Pythagorean theorem applies *only* to right triangles.

EXAMPLE 3 Using the Pythagorean Theorem to Find the Side of a Right Triangle

Suppose you have a 10-foot ladder and want to reach a height of 8 feet to clean out the gutters on your house. How far from the base of the house should the base of the ladder be?

Solution:

Label the unknown side x.

Apply the Pythagorean theorem.	$x^2 + 8^2 = 10^2$
Simplify.	$x^2 + 64 = 100$
Solve for x.	$x^2 = 36$
	$x = \pm 6$
Length must be positive.	$\boxed{x = 6}$

The ladder should be $\boxed{6 \text{ feet}}$ from the base of the house along the ground.

Classroom Example 6.1.3
Suppose you have a 13-foot ladder and want to reach a height of 12 feet to clean out the gutters on your house. How far from the base of the house should the base of the ladder be?

Answer: 5 feet

■ **Answer:** 15 feet

■ **YOUR TURN** A steep ramp is being built for skateboarders. The height, horizontal ground distance, and ramp length form a right triangle. If the height is 9 feet and the horizontal ground distance is 12 feet, what is the length of the ramp?

When solving a right triangle exactly, simplification of radicals is often necessary. For example, if a side length of a triangle resulted in $\sqrt{17}$, the radical cannot be simplified any further. However, if a side length of a triangle resulted in $\sqrt{20}$, the radical would be simplified:

$$\sqrt{20} = \sqrt{4 \cdot 5} = \sqrt{4} \cdot \sqrt{5} = \sqrt{2^2} \cdot \sqrt{5} = 2\sqrt{5}$$

EXAMPLE 4 Using the Pythagorean Theorem with Radicals

Use the Pythagorean theorem to solve for the unknown side length in the given right triangle. Express your answer exactly in terms of simplified radicals.

Solution:

Apply the Pythagorean theorem.	$3^2 + x^2 = 7^2$
Simplify known squares.	$9 + x^2 = 49$
Solve for x.	$x^2 = 40$
	$x = \pm\sqrt{40}$
The side length x is a distance that is positive.	$x = \sqrt{40}$
Simplify the radical.	$x = \sqrt{4 \cdot 10} = \underset{2}{\sqrt{4}} \cdot \sqrt{10} = \boxed{2\sqrt{10}}$

■ **YOUR TURN** Use the Pythagorean theorem to solve for the unknown side length in the given right triangle. Express your answer exactly in terms of simplified radicals.

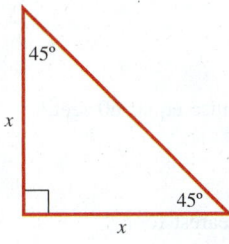

■ **Answer:** $4\sqrt{3}$

Special Right Triangles

Right triangles whose sides are in the ratios of 3-4-5, 5-12-13, and 8-15-17 are examples of right triangles that are special because their side lengths are equal to whole numbers that satisfy the Pythagorean theorem. A Pythagorean triple consists of three positive integers that satisfy the Pythagorean theorem: $3^2 + 4^2 = 5^2$, $5^2 + 12^2 = 13^2$, and $8^2 + 15^2 = 17^2$. There are two other special right triangles that warrant attention: the 30°-60°-90° triangle and the 45°-45°-90° triangle. Although in trigonometry we focus more on the angles than on the side lengths, we are interested in special relationships between the lengths of the sides of these right triangles. We will start with the 45°-45°-90° triangle.

WORDS

A 45°-45°-90° triangle is an isosceles (two legs are equal) right triangle.

MATH

Apply the Pythagorean theorem.	$x^2 + x^2 = \text{hypotenuse}^2$
Simplify the right side of the equation.	$2x^2 = \text{hypotenuse}^2$
Solve for the hypotenuse.	$\text{hypotenuse} = \pm\sqrt{2x^2} = \pm\sqrt{2}|x|$

x and the hypotenuse are lengths and must be positive.

The hypotenuse of a 45°-45°-90° triangle is $\sqrt{2}$ times the length of either leg.

$$\text{hypotenuse} = \sqrt{2}\,x$$

Study Tip

"To solve a triangle" means to find all of the angle measures and side lengths.

If we let $x = 1$, then the triangle will have legs with length equal to 1 and the hypotenuse will have length $\sqrt{2}$. Notice that these lengths satisfy the Pythagorean theorem: $1^2 + 1^2 = \left(\sqrt{2}\right)^2$ or $2 = 2$. Later when we discuss the unit circle approach, we will let the hypotenuse have length 1. The legs will then have lengths $\dfrac{\sqrt{2}}{2}$ and $\dfrac{\sqrt{2}}{2}$:

$$\left(\frac{\sqrt{2}}{2}\right)^2 + \left(\frac{\sqrt{2}}{2}\right)^2 = \frac{1}{2} + \frac{1}{2} = 1^2.$$

 EXAMPLE 5   **Solving a 45°-45°-90° Triangle**

A house has a roof with a 45° pitch (the angle the roof makes with the house). If the house is 60 feet wide, what are the lengths of the sides of the roof that form the attic? Round to the nearest foot.

Solution:

Draw the 45°-45°-90° triangle. Let *x* represent the length of the unknown legs.

Recall that the hypotenuse of a 45°-45°-90° triangle is $\sqrt{2}$ times the length of either leg.

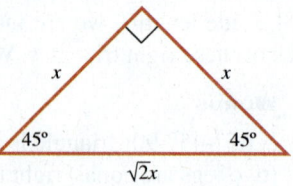

Technology Tip

To calculate $x = \dfrac{60}{\sqrt{2}}$, press

60 ÷ √(2) ENTER .

```
60/√(2)
       42.42640687
```

Let the hypotenuse equal 60 feet. $60 = \sqrt{2}\,x$

Solve for *x*. $x = \dfrac{60}{\sqrt{2}} \approx 42.42641$

Round to the nearest foot. $x \approx \boxed{42 \text{ ft}}$

■ **Answer:** $60\sqrt{2} \approx 85$ feet

■ **YOUR TURN**   A house has a roof with a 45° pitch. If the sides of the roof are 60 feet, how wide is the house? Round to the nearest foot.

Let us now determine the relationship of the sides of a 30°-60°-90° triangle. We start with an equilateral triangle (equal sides and equal angles, 60°).

WORDS	**MATH**

Draw an equilateral triangle with sides $2x$.

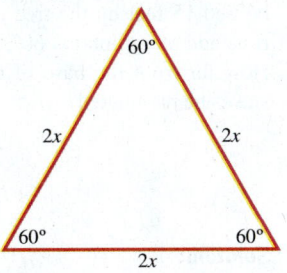

Draw a line segment from one vertex that is perpendicular to the opposite side; this line segment represents the height of the triangle, h, and bisects the base. There are now two identical 30°-60°-90° triangles.

Notice that in each triangle the hypotenuse is twice the shortest leg, which is opposite the 30° angle.

To find the length h, use the Pythagorean theorem.

$$h^2 + x^2 = (2x)^2$$
$$h^2 + x^2 = 4x^2$$

Solve for h.

$$h^2 = 3x^2$$
$$h = \pm\sqrt{3x^2} = \pm\sqrt{3}\,|x|$$

h and x are lengths and must be positive.

$$h = \sqrt{3}x$$

The hypotenuse of a 30°-60°-90° is two times the length of the leg opposite the 30° angle.

The leg opposite the 60° angle is $\sqrt{3}$ times the length of the leg opposite the 30° angle.

If we let $x = 1$, then the triangle will have legs with lengths 1 and $\sqrt{3}$ and hypotenuse of length 2. These satisfy the Pythagorean theorem: $1^2 + \left(\sqrt{3}\right)^2 = 2^2$ or $4 = 4$.

EXAMPLE 6 Solving a 30°-60°-90° Triangle

Before a hurricane, it is wise to stake down trees for additional support during the storm. If the branches allow for the rope to be tied 15 feet up the tree and a desired angle between the rope and the ground is 60°, how much total rope is needed? How far from the base of the tree should each of the two stakes be hammered?

Solution:

Recall the relationship between the sides of a 30°-60°-90° triangle.

In this case, the leg opposite the 60° angle is 15 feet.

Solve for x.

$$\sqrt{3}x = 15$$

$$x = \frac{15}{\sqrt{3}} \approx 8.7$$

Find the length of the hypotenuse.

$$\text{hypotenuse} = 2x = 2\left(\frac{15}{\sqrt{3}}\right) \approx 17$$

The ropes should be staked approximately 8.7 feet from the base of the tree, and approximately $2(17) = 34$ total feet of rope will be needed.

■ Answer: Each rope should be staked approximately 12 feet from the base of the tree. Approximately 46 total feet of rope will be needed.

■ YOUR TURN Rework Example 6 with a height (where the ropes are tied) of 20 feet. How far from the base of the tree should each of the ropes be staked, and how much total rope will be needed?

Similar Triangles

The word *similar* in mathematics means identical in shape, although not necessarily the same size. It is important to note that two triangles can have the exact same shape (same angles) but have different sizes.

DEFINITION   **Similar Triangles**

Similar triangles are triangles with equal corresponding angle measures (equal angles).

The word *congruent* means equal in all corresponding parts; therefore, congruent triangles have all corresponding angles equal and all corresponding sides equal. Two triangles are *congruent* if they have exactly the same shape *and* size. In other words, if one triangle can be picked up and situated on top of another triangle so that the two triangles coincide, they are said to be *congruent*.

DEFINITION   **Congruent Triangles**

Congruent triangles are triangles with equal corresponding angle measures (equal angles) *and* corresponding equal sides.

It is important to note that all congruent triangles are also similar triangles, but not all similar triangles are congruent.

Trigonometry (as you will see in Section 6.2) relies on the properties of similar triangles. Since similar triangles have the same shape (equal corresponding angles), the sides opposite the corresponding angles must be proportional.

Given any 30°-60°-90° triangle (see top figure in the margin), if we let $x = 1$ correspond to a triangle A and $x = 5$ correspond to a triangle B, then we would have the following two triangles.

Notice that the sides opposite the corresponding angles are proportional. This means that we will find that all three ratios of each side of triangle B to its corresponding side of triangle A are equal.

Sides opposite **30°**: $\dfrac{\textbf{Triangle } B}{\textbf{Triangle } A} = \dfrac{5}{1} = 5$

Sides opposite **60°**: $\dfrac{\textbf{Triangle } B}{\textbf{Triangle } A} = \dfrac{5\sqrt{3}}{\sqrt{3}} = 5$

Sides opposite **90°**: $\dfrac{\textbf{Triangle } B}{\textbf{Triangle } A} = \dfrac{10}{2} = 5$

This proportionality property holds for all similar triangles.

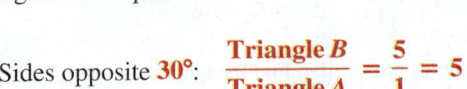

CONDITIONS FOR SIMILAR TRIANGLES

One of the following must be verified in order for two triangles to be similar:

- Corresponding angles must have the same measure.
- Corresponding sides must be proportional (ratios must be equal).

$$\frac{a}{a'} = \frac{b}{b'} = \frac{c}{c'}$$

Let us now use the definition of similar triangles to determine lengths of sides of similar triangles.

▶ **EXAMPLE 7 Finding Lengths of Sides in Similar Triangles**

Given that the two triangles are similar, find the length of each of the unknown sides (b and c).

Solution:

Solve for b.

The corresponding sides are proportional.	$\dfrac{8}{2} = \dfrac{b}{5}$
Multiply by 5.	$b = 4(5)$
Solve for b.	$\boxed{b = 20}$

Solve for c.

The corresponding sides are proportional.	$\dfrac{8}{2} = \dfrac{c}{6}$
Multiply by 6.	$c = 4(6)$
Solve for c.	$\boxed{c = 24}$

Check that the ratios are equal: $\dfrac{8}{2} = \dfrac{20}{5} = \dfrac{24}{6} = 4$

■ **YOUR TURN** Given that the two triangles are similar, find the length of each of the unknown sides (a and b).

■**Answer:** $a = 5$, $b = 8$

Applications Involving Similar Triangles

As you have seen, the common ratios associated with similar triangles are very useful: you can measure the sides of the smaller triangle and then calculate the measures of the sides of the larger triangle by measuring just one side. For example, you can quickly estimate the heights of flagpoles, trees, and any other tall objects by measuring their shadows along the ground because similar triangles are formed. Surveyors rely on the properties of similar triangles to determine distances that are difficult to measure.

EXAMPLE 8   Calculating the Height of a Tree

Billy wants to rent a lift to trim his tall trees. However, he must decide which lift he needs: one that will lift him 25 feet or a more expensive lift that will lift him 50 feet. His wife, Jeanine, hammered a stake into the ground and by measuring found its shadow to be $1\frac{3}{4}$ feet long and the tree's shadow to be 19 feet. If the stake was standing 3 feet above the ground, how tall is the tree? Which lift should Billy rent? (Assume the tree and stake both are perpendicular to the ground.)

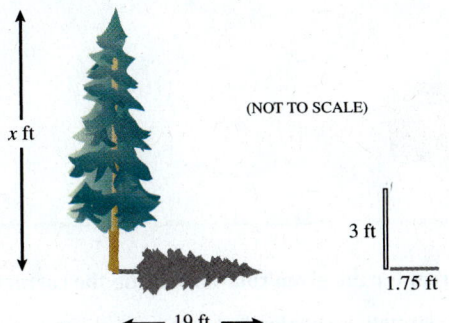

(NOT TO SCALE)

x ft

3 ft

1.75 ft

← 19 ft →

Solution:

Rays of sunlight are straight and parallel to each other. Therefore, the rays make the same angle with the tree that they do with the stake.

Draw and label the two similar triangles.

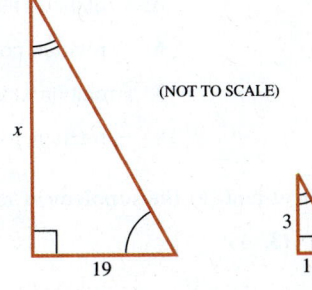

(NOT TO SCALE)

x

19

3

1.75

The ratios of corresponding sides of similar triangles are equal.

$$\frac{x}{3} = \frac{19}{1.75}$$

Solve for x.

$$x = \frac{3(19)}{1.75}$$

Simplify.

$$\boxed{x \approx 32.57}$$

The tree is approximately 33 feet tall. Billy should rent the more expensive lift to be safe.

■ **YOUR TURN**   Billy's neighbor decides to do the same thing. He borrows Jeanine's stake and measures the shadows. If the shadow of his tree is 15 feet and the shadow of the stake (3 feet above the ground) is 1.2 feet, how tall is Billy's neighbor's tree? Round to the nearest foot.

■ **Answer:** approximately 38 feet

SECTION 6.1 SUMMARY

In this section you have practiced working with angles and triangles. One unit of measurement for angles is the degree. An angle measuring exactly 90° is called a right angle. The sum of the three angles of any triangle is always 180°. Triangles that contain a right angle are called right triangles. With the Pythagorean theorem you can solve for one side of a right triangle given the other two sides. The right triangles 30°-60°-90° and 45°-45°-90° are special because of the simple numerical relations of their side lengths.

Similar triangles (the same shape) have corresponding angles with equal measure and the important thing to note is the consistency of their proportions.

SECTION 6.1 EXERCISES

■ SKILLS

In Exercises 1–10, specify the measure of the angle in degrees for the given rotations, using the correct algebraic sign (+ or −).

1. $\frac{1}{2}$ rotation counterclockwise

2. $\frac{1}{4}$ rotation counterclockwise

3. $\frac{1}{3}$ rotation clockwise

4. $\frac{2}{3}$ rotation clockwise

5. $\frac{5}{6}$ rotation counterclockwise

6. $\frac{7}{12}$ rotation counterclockwise

7. $\frac{4}{5}$ rotation clockwise

8. $\frac{5}{9}$ rotation clockwise

9. $\frac{7}{36}$ rotation clockwise

10. $\frac{7}{36}$ rotation counterclockwise

In Exercises 11–16, find (a) the complement and (b) the supplement of the given angles.

11. 18°

12. 39°

13. 42°

14. 57°

15. 89°

16. 75°

In Exercises 17–20, find the measure of each angle.

17.

18.

19. Supplementary angles with measures $8x$ degrees and $4x$ degrees

20. Complementary angles with measures $3x + 15$ degrees and $10x + 10$ degrees

In Exercises 21–26, refer to the triangle in the drawing.

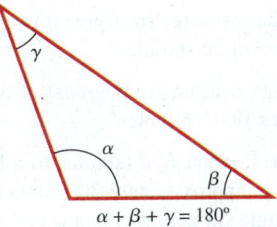

$\alpha + \beta + \gamma = 180°$

21. If $\alpha = 117°$ and $\beta = 33°$, find γ.

22. If $\alpha = 110°$ and $\beta = 45°$, find γ.

23. If $\gamma = \beta$ and $\alpha = 4\beta$, find all three angles.

24. If $\gamma = \beta$ and $\alpha = 3\beta$, find all three angles.

25. If $\gamma = \frac{1}{2}\beta$ and $\beta = 6\alpha$, find all three angles.

26. If $\gamma = \frac{2}{3}\beta$ and $\beta = 3\alpha$, find all three angles.

In Exercises 27–34, refer to the right triangle in the drawing. Express lengths *exactly*.

27. If $a = 4$ and $b = 3$, find c.

28. If $a = 3$ and $b = 3$, find c.

29. If $a = 6$ and $c = 10$, find b.

30. If $b = 7$ and $c = 12$, find a.

31. If $a = 8$ and $b = 5$, find c.

32. If $a = 6$ and $b = 5$, find c.

33. If $a = 24$ and $b = 7$, find c.

34. If $c = 20$ and $a = 10$, find b.

In Exercises 35–38, refer to the 45°-45°-90° triangle.

35. If the two legs have length 10 inches, how long is the hypotenuse?

36. If the two legs have length 8 meters, how long is the hypotenuse?

37. If the hypotenuse has length $2\sqrt{2}$ centimeters, how long are the legs?

38. If the hypotenuse has length $\sqrt{10}$ feet, how long are the legs?

In Exercises 39–44, refer to the 30°-60°-90° triangle.

39. If the shortest leg has length 5 meters, what are the lengths of the other leg and the hypotenuse?

40. If the shortest leg has length 9 feet, what are the lengths of the other leg and the hypotenuse?

41. If the longer leg has length 12 yards, what are the lengths of the other leg and the hypotenuse?

42. If the longer leg has length n units, what are the lengths of the other leg and the hypotenuse?

43. If the hypotenuse has length 10 inches, what are the lengths of the two legs?

44. If the hypotenuse has length 8 centimeters, what are the lengths of the two legs?

In Exercises 45–52, calculate the specified lengths given that the two triangles are similar.

45. $a = 4, c = 6, d = 2, f = ?$

46. $a = 12, b = 9, e = 3, d = ?$

47. $d = 5, e = 2.5, b = 7.5, a = ?$

48. $e = 1.4, f = 2.6, c = 3.9, b = ?$

49. $d = 2.5 \text{ m}, f = 1.1 \text{ m}, a = 26.25 \text{ km}, c = ?$

50. $e = 10 \text{ cm}, f = 14 \text{ cm}, c = 35 \text{ m}, b = ?$

51. If $a = c = \frac{9}{2}$ in. and $d = \frac{1}{3}$ in., $f = ?$

52. If $b = e = 4.2 \text{ cm}$ and $f = 1.1 \text{ cm}, c = ?$

■ **APPLICATIONS**

53. Clock. What is the measure (in degrees) of the angle the minute hand traces in 20 minutes?

54. Clock. What is the measure (in degrees) of the angle the minute hand traces in 25 minutes?

55. London Eye. The London Eye (similar to a bicycle wheel) makes 1 rotation in approximately 30 minutes. What is the measure of the angle (in degrees) that a cart (spoke) will rotate in 12 minutes?

Jason Friend/Photolibrary/Getty Images, Inc.

56. London Eye. The London Eye (similar to a bicycle wheel) makes 1 rotation in approximately 30 minutes. What is the measure of the angle (in degrees) that a cart (spoke) will rotate in 5 minutes?

57. Revolving Restaurant. If a revolving restaurant overlooking a waterfall can rotate 270° in 45 minutes, how long does it take for the restaurant to make a complete revolution?

58. Revolving Restaurant. If a revolving restaurant overlooking a waterfall can rotate 72° in 9 minutes, how long does it take for the restaurant to make a complete revolution?

59. Field Trial. In a Labrador Retriever field trial, dogs are judged by the straight line they take to a fallen bird. They are required to go *through* the water, not along the shore. If the judge wants to calculate how far a dog will travel along a straight path, she walks the two legs of a right triangle as shown in the drawing and uses the Pythagorean theorem. How far would this dog travel (run and swim) if it traveled along the hypotenuse?

60. Field Trial. How far would the dog in Exercise 59 run and swim if it traveled along the hypotenuse? The judge walks 25 feet along the shore and then 100 feet out to the bird.

61. Christmas Lights. A couple want to put up Christmas lights along the roofline of their house. If the front of the house is 100 feet wide and the roof has a 45° pitch, how many linear feet of Christmas lights should the couple buy? Round to the nearest foot.

62. Christmas Lights. Repeat Exercise 61 if the house is 60 feet wide. Round to the nearest foot.

63. Tree Stake. A tree needs to be staked down prior to a storm. If the ropes can be tied on the tree trunk 17 feet above the ground and the staked rope should make a 60° angle with the ground, how far from the base of the tree should each of the ropes be staked?

64. Tree Stake. What is the total amount of rope (feet) that extends to the two stakes supporting the tree in Exercise 63? Round to the nearest foot.

65. Tree Stake. A tree needs to be staked down prior to a storm. If the ropes can be tied on the tree trunk 10 feet above the ground and the staked rope should make a 30° angle with the ground, how far from the base of the tree should the ropes be staked? Round to the nearest foot.

66. Tree Stake. What is the total amount of rope (feet) that extends to the four stakes supporting the tree in Exercise 65? Round to the nearest foot.

67. Party Tent. Steve and Peggy want to rent a 40 ft × 20 ft tent for their backyard to host a barbecue. The base of the tent is supported 7 feet above the ground by poles and then roped stakes are used for support. The ropes make a 60° angle with the ground. How large a footprint in their yard would they need for this tent (and staked ropes)? In other words, what are the dimensions of the rectangle formed by the stakes on the ground? Round to the nearest foot.

68. Party Tent. Ashley's parents are throwing a graduation party and are renting a 40 ft × 80 ft tent for their backyard. The base of the tent is supported 7 feet above the ground by poles and then roped stakes are used for support. The ropes make a 45° angle with the ground. How large a footprint (see Exercise 67) in their yard will they need for this tent (and staked ropes)? Round to the nearest foot.

iStockphoto

69. Height of a Tree. The shadow of a tree measures $14\frac{1}{4}$ feet. At the same time of day, the shadow of a 4-foot pole measures 1.5 feet. How tall is the tree?

70. Height of a Flagpole. The shadow of a flagpole measures 15 feet. At the same time of day, the shadow of a 2-foot stake measures $\frac{3}{4}$ foot. How tall is the flagpole?

71. Height of a Lighthouse. The Cape Hatteras Lighthouse on the Outer Banks of North Carolina is the tallest lighthouse in North America. If a 5-foot woman casts a 1.2-foot shadow and the lighthouse casts a 54-foot shadow, approximately how tall is the Cape Hatteras Lighthouse?

72. Height of a Man. If a 6-foot man casts a 1-foot shadow, how long a shadow will his 4-foot son cast?

For Exercises 73–76, refer to the following:

An archery target consists of a bullseye (at the very center) and concentric circles which form annular rings. The concentric circles have radii 4 inches, 8 inches, and 12 inches, respectively, and a remaining rectangular region that measures 8 inches

vertically and horizontally with respect to the bullseye *beyond* the outermost circle.

Assume an archer releases the bow 50 feet from the target and that the bullseye is a point. Once an archer releases the arrow (assuming it actually hits the target) the score is assigned as follows:

- 50 points: Hits the bullseye (yellow)
- 30 points: Lies in the red ring (misses the bullseye but lies within 4 inches of the bullseye)
- 20 points: Lies in the blue ring (between 4 and 8 inches of the bullseye)
- 10 points: Lies in the black ring (between 8 and 12 inches of the bullseye)
- 5 points: Hits the brown target outside all of the rings

73. Archery. Assume that at the moment prior to releasing the bow, the arrowhead is directly in line with the bullseye. If the force with which the archer releases the bow is such that the arrow falls 3 inches vertically for every 20 feet the arrow travels horizontally, what is her score for this shot?

74. Archery. In Exercise 73, if instead the force with which she releases the bow is such that the arrow rises 3 inches vertically for every 10 feet the arrow travels horizontally, what is her score for this shot?

75. Archery. In Exercise 73, assuming that the arrowhead is located at the origin of a coordinate system when released, what is the distance between its original position and the point at which it hits the target? Express your answer to five significant digits.

76. Archery. In Exercise 73, assuming that the arrowhead is located at the origin of a coordinate system when released, what is the distance between its original position and the point at which it hits the target? Express your answer to five significant digits.

For Exercises 77–80, refer to the following:

A collimator is a device used in radiation treatment that narrows beams or waves, causing the waves to be more aligned in a specific direction. The use of a collimator facilitates the focusing of radiation to treat an affected region of tissue beneath the skin. In the figure, d_s is the distance from the radiation source to the skin, d_t is the distance from the outer layer of skin to the targeted tissue, $2f_s$ is the field size on the skin (diameter of the circular treated skin) and $2f_d$ is the targeted field size at depth d_t (the diameter of the targeted tissue at the specified depth beneath the skin surface).

77. Health/Medicine. Radiation treatment is applied to a field size of 8 centimeters at a depth 2.5 centimeters below the skin surface. If the treatment head is positioned 80 centimeters from the skin, find the targeted field size to the nearest millimeter.

78. Health/Medicine. Radiation treatment is applied to a field size of 4 centimeters lying at a depth of 3.5 centimeters below the skin surface. If the field size on the skin is required to be 3.8 centimeters, find the distance from the skin that the radiation source must be located to the nearest millimeter.

79. Health/Medicine. Radiation treatment is applied to a field size on the skin of 3.75 centimeters to reach an affected region of tissue with field size of 4 centimeters at some depth below the skin. If the treatment head is positioned 60 centimeters from the skin surface, find the desired depth below the skin to the target area to the nearest millimeter.

80. Health/Medicine. Radiation treatment is applied to a field size on the skin of 4.15 centimeters to reach an affected area lying 4.5 centimeters below the skin surface. If the treatment head is positioned 60 centimeters from the skin surface, find the field size of the targeted area to the nearest millimeter.

■CATCH THE MISTAKE

In Exercises 81 and 82, explain the mistake that is made.

81. In a 30°-60°-90° triangle, find the length of the side opposite the 60° angle if the side opposite the 30° angle is 10 inches.

Solution:

The length opposite the 60° angle is twice the length opposite the 30° angle. $2(10) = 20$

The side opposite the 60° angle has length 20 inches.

This is incorrect. What mistake was made?

82. In a 45°-45°-90° triangle, find the length of the hypotenuse if each leg has length 5 cm.

Solution:

Use the Pythagorean theorem. $5^2 + 5^2 = \text{hypotenuse}^2$

Simplify. $50 = \text{hypotenuse}^2$

Solve for the hypotenuse. $\text{hypotenuse} = \pm 5\sqrt{2}$

The hypotenuse has length $\pm 5\sqrt{2}$ cm.

This is incorrect. What mistake was made?

■CONCEPTUAL

In Exercises 83–88, determine whether each statement is true or false.

83. The Pythagorean theorem can be applied to any equilateral triangle.

84. The Pythagorean theorem can be applied to all isosceles triangles.

85. The two angles opposite the legs of a right triangle are complementary.

86. In a 30°-60°-90° triangle, the length of the side opposite the 60° angle is twice the length of the side opposite the 30° angle.

87. If a triangle contains an obtuse angle, the Pythagorean theorem cannot apply.

88. The Pythagorean theorem applies only if a triangle has exactly two acute angles.

■**CHALLENGE**

89. What is the measure (in degrees) of the smaller angle the hour and minute hands make when the time is 12:20?

90. What is the measure (in degrees) of the smaller angle the hour and minute hands make when the time is 9:10?

In Exercises 91–94, use the following figure:

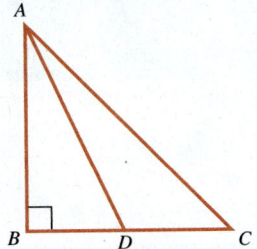

91. If $AB = 3$, $AD = 5$, and $AC = \sqrt{58}$, find DC.

92. If $AB = 4$, $AD = 5$, and $AC = \sqrt{41}$, find DC.

93. If $AC = 30$, $AB = 24$, and $DC = 11$, find AD.

94. If $AB = 60$, $AD = 61$, and $DC = 36$, find AC.

95. Given a square with side length x, draw the two diagonals. The results are four special triangles. Describe these triangles. What are the angle measures?

96. Solve for x in the following triangle.

97. Consider the following diagram and determine y.

98. Consider the following diagram and find y.

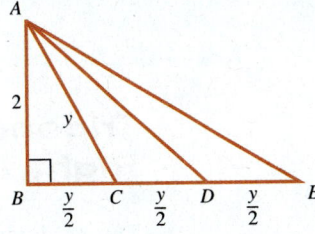

99. Using the diagram in Exercise 98, find BE.

100. Using the diagram in Exercise 98, find AD.

101. Find x.

102. Find x and y.

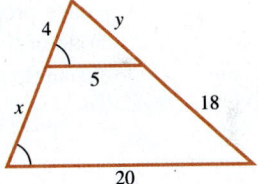

■**TECHNOLOGY**

103. If the shortest leg of a 30°-60°-90° triangle has length 16.68 feet, what are the lengths of the other leg and the hypotenuse? Round answers to two decimal places.

104. If the longer leg of a 30°-60°-90° triangle has length 134.75 cm, what are the lengths of the other leg and the hypotenuse? Round answers to two decimal places.

SKILLS OBJECTIVES

- Calculate trigonometric ratios of general angles.
- Express trigonometric functions in terms of their cofunctions.
- Evaluate trigonometric functions exactly for special angles.
- Evaluate (approximate) trigonometric functions using a calculator.

CONCEPTUAL OBJECTIVES

- Understand that right triangle ratios are based on the properties of similar triangles.
- Learn the trigonometric functions as ratios of sides of a right triangle.
- Understand the difference between evaluating trigonometric functions exactly and using a calculator.

Trigonometric Functions: Right Triangle Ratios

The word **trigonometry** stems from the Greek words *trigonon*, which means triangle, and *metrein*, which means to measure. Trigonometry began as a branch of geometry and was utilized extensively by early Greek mathematicians to determine unknown distances. The major *trigonometric functions*, including *sine*, *cosine*, and *tangent*, were first defined as ratios of sides in a right triangle. This is the way we will define them in this section. Since the two angles, besides the right angle, in a right triangle have to be acute, a second kind of definition was needed to extend the domain of trigonometric functions to nonacute angles in the Cartesian plane (Section 6.4). Starting in the eighteenth century, broader definitions of the trigonometric functions came into use, under which the functions are associated with points along the unit circle (Section 6.7).

In Section 6.1, we learned the fact that triangles with equal corresponding angles also have proportional sides and are called similar triangles. The concept of similar triangles, one of the basic insights in trigonometry, allows us to determine the length of a side of one triangle if we know the length of certain sides of a similar triangle.

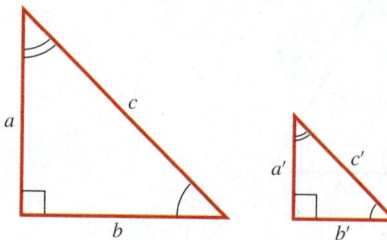

Since the two right triangles above have equal angles, they are similar triangles and the following ratios hold true:

$$\frac{a}{a'} = \frac{b}{b'} = \frac{c}{c'}$$

Separate the common ratio into three equations:

$$\frac{a}{a'} = \frac{b}{b'} \qquad \frac{b}{b'} = \frac{c}{c'} \qquad \frac{a}{a'} = \frac{c}{c'}$$

WORDS	MATH
Start with the first ratio.	$\dfrac{a}{a'} = \dfrac{b}{b'}$
Cross multiply.	$ab' = a'b$
Divide both sides by bb'.	$\dfrac{ab'}{bb'} = \dfrac{a'b}{bb'}$
Simplify.	$\dfrac{a}{b} = \dfrac{a'}{b'}$

Similarly, it can be shown that $\dfrac{b}{c} = \dfrac{b'}{c'}$ and $\dfrac{a}{c} = \dfrac{a'}{c'}$.

Notice that even though the sizes of the triangles are different, since the corresponding angles are equal, the ratio of the lengths of the two legs of the large triangle is equal to the ratio of the lengths of the legs of the small triangle, or $\dfrac{a}{b} = \dfrac{a'}{b'}$. Similarly, the ratios of the lengths of a leg and the hypotenuse of the large triangle and the corresponding leg and hypotenuse of the small triangle are also equal; that is, $\dfrac{b}{c} = \dfrac{b'}{c'}$ and $\dfrac{a}{c} = \dfrac{a'}{c'}$.

For any right triangle, there are six possible ratios of sides that can be calculated for each acute angle θ:

$$\frac{b}{c} \quad \frac{a}{c} \quad \frac{b}{a} \quad \frac{c}{b} \quad \frac{c}{a} \quad \frac{a}{b}$$

These ratios are referred to as **trigonometric ratios** or **trigonometric functions**, since they depend on θ, and each is given a name:

FUNCTION NAME	ABBREVIATION	WORDS	MATH
Sine	sin	The sine of θ	$\sin\theta$
Cosine	cos	The cosine of θ	$\cos\theta$
Tangent	tan	The tangent of θ	$\tan\theta$
Cosecant	csc	The cosecant of θ	$\csc\theta$
Secant	sec	The secant of θ	$\sec\theta$
Cotangent	cot	The cotangent of θ	$\cot\theta$

Sine, cosine, tangent, cotangent, secant, and cosecant are names given to specific ratios of lengths of sides of right triangles.

DEFINITION (1)   **Trigonometric Functions**

Let θ be an acute angle in a right triangle, then

$$\sin\theta = \frac{b}{c} \qquad \cos\theta = \frac{a}{c} \qquad \tan\theta = \frac{b}{a}$$

$$\csc\theta = \frac{c}{b} \qquad \sec\theta = \frac{c}{a} \qquad \cot\theta = \frac{a}{b}$$

The following terminology will be used throughout this text:

- Side c is the **hypotenuse**.
- Side b is the side (leg) **opposite** angle θ.
- Side a is the side (leg) **adjacent** to angle θ.

Also notice that since $\sin\theta = \dfrac{b}{c}$ and $\cos\theta = \dfrac{a}{c}$, then $\tan\theta = \dfrac{\sin\theta}{\cos\theta} = \dfrac{\dfrac{b}{c}}{\dfrac{a}{c}} = \dfrac{b}{a}$.

Using this terminology, we have an alternative definition that is easier to remember.

DEFINITION (1) Trigonometric Functions (Alternate Form)

$$\sin\theta = \frac{\text{opposite}}{\text{hypotenuse}} \qquad \cos\theta = \frac{\text{adjacent}}{\text{hypotenuse}} \qquad \tan\theta = \frac{\text{opposite}}{\text{adjacent}}$$

and their reciprocals:

$$\csc\theta = \frac{1}{\sin\theta} = \frac{\text{hypotenuse}}{\text{opposite}}$$

$$\sec\theta = \frac{1}{\cos\theta} = \frac{\text{hypotenuse}}{\text{adjacent}}$$

$$\cot\theta = \frac{1}{\tan\theta} = \frac{\text{adjacent}}{\text{opposite}}$$

Reciprocal Identities

The three main trigonometric functions should be learned in terms of the ratios.

$$\sin\theta = \frac{\text{opposite}}{\text{hypotenuse}} \qquad \cos\theta = \frac{\text{adjacent}}{\text{hypotenuse}} \qquad \tan\theta = \frac{\text{opposite}}{\text{adjacent}}$$

The remaining three trigonometric functions can be derived from $\sin\theta$, $\cos\theta$, and $\tan\theta$ using the *reciprocal identities*. Recall that the **reciprocal** of x is $\dfrac{1}{x}$ for $x \neq 0$.

RECIPROCAL IDENTITIES

$$\csc\theta = \frac{1}{\sin\theta} \qquad \sec\theta = \frac{1}{\cos\theta} \qquad \cot\theta = \frac{1}{\tan\theta}$$

EXAMPLE 1   Finding Trigonometric Function Values of an Acute Angle θ in a Right Triangle

For the given triangle, calculate $\sin\theta$, $\tan\theta$, and $\csc\theta$.

Solution:

STEP 1 Solve for the hypotenuse.

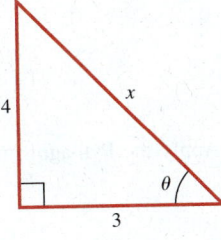

Apply the Pythagorean theorem.

$$4^2 + 3^2 = x^2$$
$$16 + 9 = x^2$$
$$x^2 = 25$$
$$x = \pm 5$$

Lengths of sides can be only positive. $x = 5$

STEP 2 Label the sides of the triangle.
■ with numbers representing lengths.
■ as hypotenuse, or as opposite or adjacent with respect to θ.

STEP 3 Set up the trigonometric functions as ratios.

Sine is opposite over hypotenuse. $\sin\theta = \dfrac{\text{opposite}}{\text{hypotenuse}} = \boxed{\dfrac{4}{5}}$

Tangent is opposite over adjacent. $\tan\theta = \dfrac{\text{opposite}}{\text{adjacent}} = \boxed{\dfrac{4}{3}}$

Cosecant is the reciprocal of sine. $\csc\theta = \dfrac{1}{\sin\theta} = \dfrac{1}{4/5} = \boxed{\dfrac{5}{4}}$

■ **YOUR TURN** For the triangle in Example 1, calculate the values of $\cos\theta$, $\sec\theta$, and $\cot\theta$.

EXAMPLE 2 Finding Trigonometric Function Values of an Acute Angle θ in a Right Triangle

For the given triangle, calculate $\cos\theta$, $\tan\theta$, and $\sec\theta$.

Solution:

STEP 1 Solve for the unknown leg.

Apply the Pythagorean theorem.

$$x^2 + 7^2 = \left(\sqrt{65}\right)^2$$
$$x^2 + 49 = 65$$
$$x^2 = 16$$
$$x = \pm 4$$

Lengths of sides can be only positive.

$$x = 4$$

STEP 2 Label the sides of the triangle.
- with numbers representing lengths.
- as hypotenuse, or as opposite or adjacent with respect to θ.

STEP 3 Set up the trigonometric functions as ratios.

Cosine is adjacent over hypotenuse.

$$\cos\theta = \frac{\text{adjacent}}{\text{hypotenuse}} = \boxed{\frac{4}{\sqrt{65}}}$$

Tangent is opposite over adjacent.

$$\tan\theta = \frac{\text{opposite}}{\text{adjacent}} = \boxed{\frac{7}{4}}$$

Secant is the reciprocal of cosine.

$$\sec\theta = \frac{1}{\cos\theta} = \frac{1}{4/\sqrt{65}} = \boxed{\frac{\sqrt{65}}{4}}$$

Expressions that contain a radical in the denominator like $\dfrac{4}{\sqrt{65}}$ can be rationalized by multiplying both the numerator and the denominator by the radical, $\sqrt{65}$.

$$\frac{4}{\sqrt{65}} \cdot \underbrace{\left(\frac{\sqrt{65}}{\sqrt{65}}\right)}_{1} = \frac{4\sqrt{65}}{65}$$

In this example, the cosine function can now be written as $\boxed{\cos\theta = \dfrac{4\sqrt{65}}{65}}$. In this text, denominators with rational expressions will be rationalized.

YOUR TURN For the triangle in Example 2, calculate $\sin\theta$, $\csc\theta$, and $\cot\theta$.

Cofunctions

Notice the *co* in *co*sine, *co*secant, and *co*tangent functions. These *cofunctions* are based on the relationship of *complementary* angles. Let us look at a right triangle with labeled sides and angles.

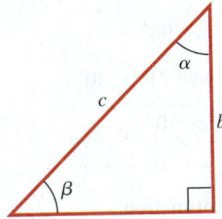

$$\left. \begin{array}{l} \sin\beta = \dfrac{\text{opposite of } \beta}{\text{hypotenuse}} = \dfrac{b}{c} \\[2ex] \cos\alpha = \dfrac{\text{adjacent to } \alpha}{\text{hypotenuse}} = \dfrac{b}{c} \end{array} \right\} \quad \sin\beta = \cos\alpha$$

Recall that the sum of the measures of angles in a triangle is 180°. In a right triangle, one angle is 90°, so the two acute angles are complementary angles (the measures sum to 90°). Therefore in the triangle above, β and α are *complementary* angles. In other words, the sine of an angle is the same as the *co*sine of the *co*mplement of that angle. This is true for all *trigonometric cofunction* pairs.

COFUNCTION THEOREM

A trigonometric function of an angle is always equal to the cofunction of the complement of the angle. If $\alpha + \beta = 90°$, then

$$\sin\beta = \cos\alpha$$
$$\sec\beta = \csc\alpha$$
$$\tan\beta = \cot\alpha$$

COFUNCTION IDENTITIES

$$\sin\theta = \cos(90° - \theta) \qquad \cos\theta = \sin(90° - \theta)$$
$$\tan\theta = \cot(90° - \theta) \qquad \cot\theta = \tan(90° - \theta)$$
$$\sec\theta = \csc(90° - \theta) \qquad \csc\theta = \sec(90° - \theta)$$

EXAMPLE 3 Writing Trigonometric Function Values in Terms of Their Cofunctions

Write each function value in terms of its cofunction.

a. $\sin 30°$ **b.** $\tan x$ **c.** $\csc 40°$

Solution (a):

Cosine is the cofunction of sine. $\sin\theta = \cos(90° - \theta)$

Substitute $\theta = 30°$. $\sin 30° = \cos(90° - 30°)$

Simplify. $\boxed{\sin 30° = \cos 60°}$

■ **Answer: a.** $\sin 45°$
      b. $\sec(90° - y)$

Solution (b):

Cotangent is the cofunction of tangent. $\tan \theta = \cot(90° - \theta)$

Substitute $\theta = x$. $\boxed{\tan x = \cot(90° - x)}$

Solution (c):

Secant is the cofunction of cosecant. $\csc \theta = \sec(90° - \theta)$

Substitute $\theta = 40°$. $\csc 40° = \sec(90° - 40°)$

Simplify. $\boxed{\csc 40° = \sec 50°}$

■ **YOUR TURN** Write each function value in terms of its cofunction.

      a. $\cos 45°$ **b.** $\csc y$

Evaluating Trigonometric Functions Exactly for Special Angle Measures: 30°, 45°, and 60°

Thus far we have defined the trigonometric functions as ratios of sides of right triangles. We have demonstrated how to calculate the value of a trigonometric function for a general angle, but we have not yet discussed trigonometric function values for specific angle measures. We now turn our attention to evaluating trigonometric functions for known angles. We will distinguish between evaluating a trigonometric function *exactly* and *approximating* the value of a trigonometric function *with a calculator*. Throughout this text, instructions will specify which is desired.

There are three special acute angles that are very important in trigonometry: 30°, 45°, and 60°. In Section 6.1, we discussed two important triangles: 30°-60°-90° and 45°-45°-90°. Recall the relationships between the sides of these two right triangles.

We can combine these relationships with the trigonometric ratios developed in this section to evaluate the trigonometric functions for the special angle measures of 30°, 45°, and 60°.

EXAMPLE 4 **Evaluating the Trigonometric Functions Exactly for 30°**

Evaluate the six trigonometric functions for an angle that measures 30°.

Solution:

Label the sides of the 30°-60°-90° triangle with respect to the **30°** angle.

Use the right triangle ratio definitions of sine, cosine, and tangent.

$$\sin 30° = \frac{\text{opposite}}{\text{hypotenuse}} = \frac{x}{2x} = \frac{1}{2}$$

$$\cos 30° = \frac{\text{adjacent}}{\text{hypotenuse}} = \frac{\sqrt{3}x}{2x} = \frac{\sqrt{3}}{2}$$

$$\tan 30° = \frac{\text{opposite}}{\text{adjacent}} = \frac{x}{\sqrt{3}x} = \frac{1}{\sqrt{3}} = \frac{1}{\sqrt{3}} \cdot \frac{\sqrt{3}}{\sqrt{3}} = \frac{\sqrt{3}}{3}$$

Use the reciprocal identities to obtain the values of the cosecant, secant, and cotangent functions.

$$\csc 30° = \frac{1}{\sin 30°} = \frac{1}{\frac{1}{2}} = 2$$

$$\sec 30° = \frac{1}{\cos 30°} = \frac{1}{\frac{\sqrt{3}}{2}} = \frac{2}{\sqrt{3}} = \frac{2}{\sqrt{3}} \cdot \frac{\sqrt{3}}{\sqrt{3}} = \frac{2\sqrt{3}}{3}$$

$$\cot 30° = \frac{1}{\tan 30°} = \frac{1}{\frac{\sqrt{3}}{3}} = \frac{3}{\sqrt{3}} = \frac{3}{\sqrt{3}} \cdot \frac{\sqrt{3}}{\sqrt{3}} = \sqrt{3}$$

The six trigonometric functions evaluated for an angle measuring 30° are

$\sin 30° = \dfrac{1}{2}$	$\cos 30° = \dfrac{\sqrt{3}}{2}$	$\tan 30° = \dfrac{\sqrt{3}}{3}$
$\csc 30° = 2$	$\sec 30° = \dfrac{2\sqrt{3}}{3}$	$\cot 30° = \sqrt{3}$

YOUR TURN Evaluate the six trigonometric functions for an angle that measures 60°.

In comparing our answers in Example 4 and Your Turn, we see that the following cofunction relationships are true:

$$\sin 30° = \cos 60° \qquad \sec 30° = \csc 60° \qquad \tan 30° = \cot 60°$$

$$\sin 60° = \cos 30° \qquad \sec 60° = \csc 30° \qquad \tan 60° = \cot 30°$$

which is expected, since 30° and 60° are complementary angles.

Classroom Example 6.2.4
Assuming a triangle is 30°-60°-90° and the shorter of the two legs has measure $\sqrt{3}$, find the measures of the other two sides.

Answer: 3, $2\sqrt{3}$

Study Tip

$\cos 30° = \dfrac{\sqrt{3}}{2}$ is exact, whereas if we evaluate with a calculator, we get an approximation: $\cos 30° \approx 0.8660$. This decimal approximation from a calculator is a way to "check" your "exact" answer.

■ **Answer:**

$$\sin 60° = \frac{\sqrt{3}}{2} \qquad \cos 60° = \frac{1}{2}$$

$$\tan 60° = \sqrt{3} \qquad \csc 60° = \frac{2\sqrt{3}}{3}$$

$$\sec 60° = 2 \qquad \cot 60° = \frac{\sqrt{3}}{3}$$

EXAMPLE 5   Evaluating the Trigonometric Functions Exactly for 45°

Evaluate the six trigonometric functions for an angle that measures 45°.

Solution:

Label the sides of the 45°-45°-90° right triangle
with respect to one of the **45°** angles.

Use the right triangle ratio definitions of sine, cosine, and tangent.

$$\sin 45° = \frac{\text{opposite}}{\text{hypotenuse}} = \frac{x}{\sqrt{2}x} = \frac{1}{\sqrt{2}} = \frac{1}{\sqrt{2}} \cdot \frac{\sqrt{2}}{\sqrt{2}} = \frac{\sqrt{2}}{2}$$

$$\cos 45° = \frac{\text{adjacent}}{\text{hypotenuse}} = \frac{x}{\sqrt{2}x} = \frac{1}{\sqrt{2}} = \frac{1}{\sqrt{2}} \cdot \frac{\sqrt{2}}{\sqrt{2}} = \frac{\sqrt{2}}{2}$$

$$\tan 45° = \frac{\text{opposite}}{\text{adjacent}} = \frac{x}{x} = 1$$

Use the reciprocal identities to obtain the values of the cosecant, secant, and cotangent functions.

$$\csc 45° = \frac{1}{\sin 45°} = \frac{1}{\frac{\sqrt{2}}{2}} = \frac{2}{\sqrt{2}} = \frac{2}{\sqrt{2}} \cdot \frac{\sqrt{2}}{\sqrt{2}} = \sqrt{2}$$

$$\sec 45° = \frac{1}{\cos 45°} = \frac{1}{\frac{\sqrt{2}}{2}} = \frac{2}{\sqrt{2}} = \frac{2}{\sqrt{2}} \cdot \frac{\sqrt{2}}{\sqrt{2}} = \sqrt{2}$$

$$\cot 45° = \frac{1}{\tan 45°} = \frac{1}{1} = 1$$

The six trigonometric functions evaluated for an angle measuring 45° are

$\sin 45° = \dfrac{\sqrt{2}}{2}$	$\cos 45° = \dfrac{\sqrt{2}}{2}$	$\tan 45° = 1$
$\csc 45° = \sqrt{2}$	$\sec 45° = \sqrt{2}$	$\cot 45° = 1$

We see that the following cofunction relationships are true:

$$\sin 45° = \cos 45° \qquad \sec 45° = \csc 45° \qquad \tan 45° = \cot 45°$$

which is expected, since 45° and 45° are complementary angles.

The trigonometric function values for the three special angle measures (30°, 45°, and 60°) are summarized in the following table.

Study Tip

$\sin 45° = \dfrac{\sqrt{2}}{2}$ is exact, whereas if we evaluate with a calculator, we get an approximation: $\sin 45° \approx 0.7071$. This decimal approximation is a way to "check" your "exact" answer.

Trigonometric Values for Special Angles

θ	$\sin\theta$	$\cos\theta$	$\tan\theta$	$\cot\theta$	$\sec\theta$	$\csc\theta$
30°	$\dfrac{1}{2}$	$\dfrac{\sqrt{3}}{2}$	$\dfrac{\sqrt{3}}{3}$	$\sqrt{3}$	$\dfrac{2\sqrt{3}}{3}$	2
45°	$\dfrac{\sqrt{2}}{2}$	$\dfrac{\sqrt{2}}{2}$	1	1	$\sqrt{2}$	$\sqrt{2}$
60°	$\dfrac{\sqrt{3}}{2}$	$\dfrac{1}{2}$	$\sqrt{3}$	$\dfrac{\sqrt{3}}{3}$	2	$\dfrac{2\sqrt{3}}{3}$

Study Tip

If you memorize the values for sine and cosine for the angles given in the table, then the other trigonometric function values in the table can be found using the quotient and reciprocal identities.

It is important to **learn** the special values in **red** for sine and cosine. All other values in the table can be found through reciprocals or quotients of these two functions. Remember that the tangent function is the ratio of the sine function to the cosine function.

Study Tip

SOHCAHTOA:

- **S**ine is **O**pposite over **H**ypotenuse.
- **C**osine is **A**djacent over **H**ypotenuse.
- **T**angent is **O**pposite over **A**djacent.

$$\sin\theta = \frac{\text{opposite}}{\text{hypotenuse}} \qquad \cos\theta = \frac{\text{adjacent}}{\text{hypotenuse}} \qquad \tan\theta = \frac{\sin\theta}{\cos\theta} = \frac{\dfrac{\text{opposite}}{\text{hypotenuse}}}{\dfrac{\text{adjacent}}{\text{hypotenuse}}} = \frac{\text{opposite}}{\text{adjacent}}$$

Note in the above table that $\sin 30° = \dfrac{1}{2}$ and $\cos 30° = \dfrac{\sqrt{3}}{2}$ so

$$\tan 30° = \frac{\sin 30°}{\cos 30°} = \frac{1/2}{\sqrt{3}/2} = \frac{1}{\sqrt{3}} = \frac{\sqrt{3}}{3}.$$

Technology Tip

```
sin(75)
           .9659258263
tan(67)
           2.355852366
```

```
cos(52)⁻¹
           1.624269245
▪
```

Using Calculators to Evaluate (Approximate) Trigonometric Functions

We now turn our attention to using calculators to evaluate trigonometric functions, which often results in an approximation. Scientific and graphing calculators have buttons for sine (sin), cosine (cos), and tangent (tan) functions. Let us start with what we already know and confirm it with our calculators.

Study Tip

In calculating secant, cosecant, and cotangent function values with a calculator, it is important not to round the number until after using the reciprocal function key 1/x.

EXAMPLE 6   **Evaluating Trigonometric Functions with a Calculator**

Use a calculator to find the values of

a. $\sin 75°$ **b.** $\tan 67°$ **c.** $\sec 52°$ **d.** $\cos 30°$

Round your answers to four decimal places.

Solution:

a. 0.965925826 $\boxed{\approx 0.9659}$ **b.** 2.355852366 $\boxed{\approx 2.3559}$

c. $\dfrac{1}{\cos 52°} \approx \dfrac{1}{0.615661475}$ $\boxed{\approx 1.6247}$ **d.** 0.866025403 $\boxed{\approx 0.8660}$

Note: We know $\cos 30° = \dfrac{\sqrt{3}}{2} \approx 0.8660$.

■ **YOUR TURN**   Use a calculator to find the values of

a. $\cos 22°$ **b.** $\tan 81°$ **c.** $\csc 37°$ **d.** $\sin 45°$

Round your answers to four decimal places.

Classroom Example 6.2.6
Use a calculator to find the values:
a. $\cot 42°$
b. $\csc 16.76°$
Round answers to four decimal places.

Answer: a. 1.1106   **b.** 3.4678

■ **Answer: a.** 0.9272   **b.** 6.3138
c. 1.6616   **d.** 0.7071

When calculating secant, cosecant, and cotangent function values with a calculator, it is important not to round the number until after using the reciprocal function key $1/x$ or x^{-1} in order to be as accurate as possible.

In this section we defined trigonometric functions as ratios of sides of right triangles. This approach is called *right triangle trigonometry*. This is the first of three definitions of trigonometric functions (others will follow in Sections 6.4 and 6.7). We now can find trigonometric functions of an acute angle by taking ratios of the three sides of a right triangle: "adjacent," "opposite," and "hypotenuse." It is important to remember that "adjacent" and "opposite" are defined with respect to one of the acute angles we are considering. We learned that trigonometric functions of an angle are equal to the cofunctions of the complement to the angle. Trigonometric functions can be evaluated exactly for special angles (30°, 45°, and 60°), and calculators can be used to approximate trigonometric function values of any angle.

θ	$\sin\theta$	$\cos\theta$	$\tan\theta$
30°	$\dfrac{1}{2}$	$\dfrac{\sqrt{3}}{2}$	$\dfrac{\sqrt{3}}{3}$
45°	$\dfrac{\sqrt{2}}{2}$	$\dfrac{\sqrt{2}}{2}$	1
60°	$\dfrac{\sqrt{3}}{2}$	$\dfrac{1}{2}$	$\sqrt{3}$

SECTION
6.2 EXERCISES

■ **SKILLS**

In Exercises 1–6, refer to the triangle in the drawing to find the indicated trigonometric function values.

1. $\sin\theta$

2. $\cos\theta$

3. $\csc\theta$

4. $\sec\theta$

5. $\tan\theta$

6. $\cot\theta$

In Exercises 7–12, refer to the triangle in the drawing to find the indicated trigonometric function values. Rationalize any denominators containing radicals that you encounter in the answers.

7. $\cos\theta$

8. $\sin\theta$

9. $\sec\theta$

10. $\csc\theta$

11. $\tan\theta$

12. $\cot\theta$

For Exercises 13–18, refer to the triangle in the drawing to find the indicated trigonometric function values. Rationalize any denominators containing radicals that you encounter in the answers.

13. $\sin\theta$

14. $\cos\theta$

15. $\sec\theta$

16. $\csc\theta$

17. $\cot\theta$

18. $\tan\theta$

In Exercises 19–24, use the cofunction identities to fill in the blanks.

19. $\sin 60° = \cos$ _____

20. $\sin 45° = \cos$ _____

21. $\cos x = \sin$ _____

22. $\cot A = \tan$ _____

23. $\csc 30° = \sec$ _____

24. $\sec B = \csc$ _____

In Exercises 25–32, write the trigonometric function values in terms of its cofunction.

25. $\sin(x + y)$

26. $\sin(60° - x)$

27. $\cos(20° + A)$

28. $\cos(A + B)$

29. $\cot(45° - x)$

30. $\sec(30° - \theta)$

31. $\csc(60° - \theta)$

32. $\tan(40° + \theta)$

In Exercises 33–38, match the trigonometric function values.

$$\text{a. } \frac{1}{2} \quad \text{b. } \frac{\sqrt{3}}{2} \quad \text{c. } \frac{\sqrt{2}}{2}$$

33. $\sin 30°$ **34.** $\sin 60°$ **35.** $\cos 30°$ **36.** $\cos 60°$ **37.** $\sin 45°$ **38.** $\cos 45°$

In Exercises 39–41, use the results in Exercises 33–38 and the trigonometric quotient identity $\tan\theta = \dfrac{\sin\theta}{\cos\theta}$ to calculate the following values.

39. $\tan 30°$ **40.** $\tan 45°$ **41.** $\tan 60°$

In Exercises 42–50, use the results in Exercises 33–41 and the reciprocal identities $\csc\theta = \dfrac{1}{\sin\theta}$, $\sec\theta = \dfrac{1}{\cos\theta}$, and $\cot\theta = \dfrac{1}{\tan\theta}$ to calculate the following values.

42. $\csc 30°$ **43.** $\sec 30°$ **44.** $\cot 30°$ **45.** $\csc 60°$ **46.** $\sec 60°$

47. $\cot 60°$ **48.** $\csc 45°$ **49.** $\sec 45°$ **50.** $\cot 45°$

In Exercises 51–64, use a calculator to evaluate the trigonometric functions for the indicated values. Round your answers to four decimal places.

51. $\sin 37°$ **52.** $\sin 17.8°$ **53.** $\cos 82°$ **54.** $\cos 21.9°$ **55.** $\tan 54°$

56. $\tan 43.2°$ **57.** $\sec 8°$ **58.** $\sec 75°$ **59.** $\csc 89°$ **60.** $\csc 51°$

61. $\cot 55°$ **62.** $\cot 29°$ **63.** $\tan\left(\frac{211}{9}\right)°$ **64.** $\cot\left(\frac{405}{13}\right)°$

■ **APPLICATIONS**

For Exercises 65 and 66, refer to the following:

When traveling through air, a spherical drop of blood with diameter d maintains its spherical shape until hitting a flat surface. The direction of travel of the drop of blood dictates the directionality of the blood splatter on the surface. For this reason, the diameter of the blood drop is equal to the width of the blood splatter on the surface. The angle at which a spherical drop of blood is deposited on a surface, called *angle of impact*, is related to the width w and the length l of the splatter by $\sin\theta = \dfrac{w}{l}$.

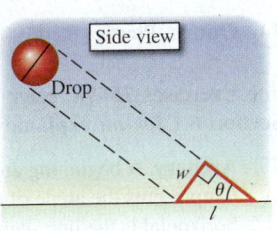

For Exercises 67 and 68, refer to the following:

The monthly profits of PizzaRia are a function of sales, that is, $p(s)$. A financial analysis has determined that the sales s in thousands of dollars of PizzaRia are also related to monthly profits p in thousands of dollars by the relationship

$$\tan\theta = \frac{p}{s} \quad \text{for} \quad 0 \le s \le 55 \quad \text{and} \quad 0 \le p \le 45$$

65. Forensic Science. If a drop of blood found at a crime scene has a width of 6 millimeters and length of 12 millimeters, find the angle θ that represents the directionality.

66. Forensic Science. If a drop of blood found at a crime scene has a width of $\sqrt{2}$ millimeters and length of 2 millimeters, find the angle θ that represents the directionality.

Based on sales and profits, it can be determined that the domain for angle θ is

$$0° \leq \theta \leq 40°$$

The ratio $\tan\theta$ represents the slope of the hypotenuse of the right triangle formed by sales s and profit p (see figure above). The angle θ can be interpreted as a measure of the relative size of s to p. The larger the angle θ is, the greater profit is relative to sales, and conversely, the smaller the angle θ is, the smaller profit is relative to sales.

67. Business. If PizzaRia's monthly sales are $25,000 and monthly profits are $10,000, find

    a. $\tan\theta$ **b.** $\cot\theta$

68. Business. If PizzaRia's monthly sales are s and monthly profits are p:

    a. Determine the hypotenuse in terms of s and p.

    b. Determine a formula for $\cos\theta$ in terms of s and p.

For Exercises 69 and 70, refer to the following:

A man lives in a house that borders a pasture. He decides to go to the grocery store to get some milk. He is trying to decide whether to drive along the roads in his car or take his all-terrain vehicle (ATV) across the pasture. His car is faster than the ATV, but the distance the ATV would travel is less than the distance he would travel in his car.

69. Shortcut. If $\sin\theta = \frac{3}{5}$ and $\cos\theta = \frac{4}{5}$ and he drove his car along the streets, it would be 14 miles round trip. How far would he have to go on his ATV round trip?

70. Shortcut. If $\tan\theta = 1$ and he drove his car along the streets, it would be 200 yards round trip. How far would he have to go on his ATV round trip? Round your answer to the nearest yard.

For Exercises 71–74, refer to the following:

Have you ever noticed that if you put a stick in the water, it looks bent? We know the stick didn't bend. Instead, the light rays bend, which made the image appear to bend. Light rays propagating from one medium (like air) to another medium (like water) experience refraction, or "bending," with respect to the surface. Light bends according to Snell's law, which states

$$n_i \sin(\theta_i) = n_r \sin(\theta_r)$$

where
- n_i is the refractive index of the medium the light is leaving.
- θ_i is the incident angle between the light ray and the normal (perpendicular) to the interface between mediums.
- n_r is the refractive index of the medium the light is entering.
- θ_r is the refractive angle between the light ray and the normal (perpendicular) to the interface between mediums.

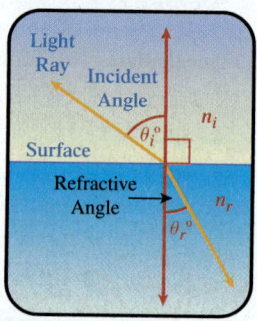

Calculate the index of refraction n_r of the indicated refractive medium given the following assumptions. Round answers to three decimals.

- The incident medium is air.
- Air has an index of refraction of $n_i = 1.00$.
- The incidence angle is $\theta_i = 30°$.

71. Optics. Diamond, $\theta_r = 12°$

72. Optics. Emerald, $\theta_r = 18.5°$

73. Optics. Water, $\theta_r = 22°$

74. Optics. Plastic, $\theta_r = 20°$

For Exercises 75–78, refer back to Exercises 73–76 in Section 6.1 for an explanation of the setting.

75. Archery. A beginning archer stands 50 feet from the target. If θ represents the angle of depression of the bow (where the horizontal is the line segment connecting the original position of the arrowhead, just before release, and the bullseye on the target), and it is known that $\cos\theta = 0.9992$, what would her score be on this shot?

76. Archery. A beginning archer stands 50 feet from the target. If β represents the angle at which the arrow strikes the target, and it is known that $\sin\beta = 0.9995$, what would her score be on this shot?

77. Archery. A beginning archer stands 50 feet from the target. If β represents the angle at which the arrow strikes the target, and it is known that $\tan \beta = 0$, what would her score be on this shot?

78. Archery. A beginning archer stands 50 feet from the target. If β represents the angle at which the arrow strikes the target, and it is known that $\cot \beta = 1$, what would her score be on this shot?

For Exercises 79 and 80, refer to the following:

X-ray crystallography is a method of determining the arrangement of atoms within a crystal. This method has revealed the structure and functioning of many biological molecules including vitamins, drugs, proteins, and nucleic acids (including DNA). The structure of a crystal can be determined experimentally using Bragg's law:

$$n\lambda = 2d \sin \theta$$

where λ is the wavelength of x-ray (measured in angstroms), d is the distance between atomic planes (measured in angstroms), θ is the angle of reflection (in degrees), and n is the order of Bragg reflection (a positive integer).

79. Physics/Life Sciences. A diffractometer was used to make a diffraction pattern for a protein crystal from which it was determined experimentally that x-rays of wavelength 1.54 angstroms produced an angle of reflection of 45° corresponding to a Bragg reflection of order 1. Find the distance between atomic planes for the protein crystal to the nearest hundredth of an angstrom.

80. Physics/Life Sciences. A diffractometer was used to make a diffraction pattern for a salt crystal from which it was determined experimentally that x-rays of wavelength 1.67 angstroms produced an angle of reflection of 71.3° corresponding to a Bragg reflection of order 4. Find the distance between atomic planes for the salt crystal to the nearest hundredth of an angstrom.

▪ CATCH THE MISTAKE

For Exercises 81–84, explain the mistake that is made.

For the triangle in the drawing, calculate the indicated trigonometric function values.

81. Calculate $\sin y$.

Solution:

Formulate sine in terms of trigonometric ratios.
$$\sin y = \frac{\text{opposite}}{\text{hypotenuse}}$$

The opposite side is 4, and the hypotenuse is 5.
$$\sin y = \frac{4}{5}$$

This is incorrect. What mistake was made?

82. Calculate $\tan x$.

Solution:

Formulate tangent in terms of trigonometric ratios.
$$\tan x = \frac{\text{adjacent}}{\text{opposite}}$$

The adjacent side is 3, and the opposite side is 4.
$$\tan x = \frac{3}{4}$$

This is incorrect. What mistake was made?

83. Calculate $\sec x$.

Solution:

Formulate sine in terms of trigonometric ratios.
$$\sin y = \frac{\text{opposite}}{\text{hypotenuse}}$$

The opposite side is 4, and the hypotenuse is 5.
$$\sin x = \frac{4}{5}$$

Write secant as the reciprocal of sine.
$$\sec x = \frac{1}{\sin x}$$

Simplify.
$$\sec x = \frac{1}{\frac{4}{5}} = \frac{5}{4}$$

This is incorrect. What mistake was made?

84. Calculate $\csc y$.

Solution:

Formulate cosine in terms of trigonometric ratios.
$$\cos y = \frac{\text{adjacent}}{\text{hypotenuse}}$$

The adjacent side is 4, and the hypotenuse is 5.
$$\cos y = \frac{4}{5}$$

Write cosecant as the reciprocal of cosine.
$$\csc y = \frac{1}{\cos y}$$

Simplify.
$$\csc y = \frac{1}{\frac{4}{5}} = \frac{5}{4}$$

This is incorrect. What mistake was made?

■ CONCEPTUAL

In Exercises 85–92, use the special triangles (30°-60°-90° and 45°-45°-90°) shown below to help you answer the questions:

In Exercises 85 and 86, determine whether each statement is true or false.

85. $\sin 45° = \cos 45°$

86. $\sin 60° = \cos 30°$

87. Calculate $\sin 30°$ and $\cos 30°$.

88. Calculate $\sin 60°$ and $\cos 60°$.

89. Calculate $\tan 30°$ and $\tan 60°$.

90. Calculate $\sin 45°$, $\cos 45°$, and $\tan 45°$.

91. Calculate $\sec 45°$ and $\csc 45°$.

92. Calculate $\tan 60°$ and $\cot 30°$.

■ CHALLENGE

In Exercises 93–96, use trignometric ratios and the assumption that a is much larger than b.

Thus far, in this text we have discussed trigonometric values only for acute angles, or for $0° < \theta < 90°$. How do we determine these values when θ is approximately 0° or 90°? We will formally consider these cases in Section 6.5, but for now, draw and label a right triangle that has one angle very close to 0°, so that the opposite side is very small compared to the adjacent side. Then the hypotenuse and the adjacent side will be very close to the same length.

93. Approximate $\sin 0°$ without using a calculator.

94. Approximate $\cos 0°$ without using a calculator.

95. Approximate $\cos 90°$ without using a calculator.

96. Approximate $\sin 90°$ without using a calculator.

In Exercises 97–104, use the following diagram. Make certain to rationalize the denominators of all answers involving radicals that you encounter.

97. Calculate BD.

98. Determine y.

99. Determine the exact numerical value for $\sin \beta$.

100. Calculate $\sec(90° - \beta)$.

101. Calculate $\tan^2 \beta + 1$.

102. True or False: $\cos^2 \beta - \sin^2 \beta = 1$.

103. Calculate $\sin^2 \theta + \cos^2 \beta$.

104. Calculate $\csc(90° - \theta) \cdot \sec(90° - \beta)$.

■ TECHNOLOGY

105. Calculate $\sec 70°$ the following two ways:

   a. Find $\cos 70°$ to three decimal places and then divide 1 by that number. Write that number to five decimal places.

   b. In a calculator, enter 70, cos, 1/x, and round the result to five decimal places.

106. Calculate $\csc 40°$ the following two ways:

   a. Find $\sin 40°$, write that down (round to three decimal places), and then divide 1 by that number. Write this last result to five decimal places.

   b. In a calculator, enter 40, sin, 1/x, and round the result to five decimal places.

107. Calculate $\cot 54.9°$ the following two ways:

   a. Find $\tan 54.9°$ to three decimal places and then divide 1 by that number. Write that number to five decimal places.

   b. In a calculator, enter 54.9, tan, 1/x, and round the result to five decimal places.

108. Calculate $\sec 18.6°$ the following two ways:

   a. Find $\cos 18.6°$ to three decimal places and then divide 1 by that number. Write that number to five decimal places.

   b. In a calculator, enter 18.6, cos, 1/x, and round the result to five decimal places.

SKILLS OBJECTIVES

- Identify the number of significant digits to express the lengths of sides and measures of angles when solving right triangles.
- Solve right triangles given the measure of an acute angle and the length of a side.
- Solve right triangles given two side lengths.

CONCEPTUAL OBJECTIVES

- Recognize the importance of significant digits in solving right triangles.
- Understand that the trigonometric inverse keys on a calculator can be used to approximate the measure of an angle given its trigonometric function value.

To *solve a triangle* means to find the measure of the three angles and three sides of the triangle. In this section, we will discuss only right triangles (therefore, we know one angle has a measure of 90°). We will know some information (the measures of two sides or the measures of a side and an acute angle) and we will determine the measures of the unknown sides and angles. However, before we start solving right triangles and determining measures, we must first discuss accuracy and significant digits.

Accuracy and Significant Digits

If we are upgrading our flooring and quickly measure a room as $10\,\text{ft} \times 12\,\text{ft}$ and want to calculate the diagonal length of the room, we use the Pythagorean theorem.

WORDS	**MATH**
Use the Pythagorean theorem.	$10^2 + 12^2 = d^2$
Simplify.	$d^2 = 244$
Use the square root property.	$d = \pm\sqrt{244}$
The length of the diagonal must be positive.	$d = \sqrt{244}$
Approximate the radical with a calculator.	$d \approx 15.62049935$

Would you say that the $10\,\text{ft} \times 12\,\text{ft}$ room has a diagonal of 15.62049935 feet? No, because the known room measurements were given only with an accuracy of 1 foot, and the diagonal above is calculated to eight decimal places. *Your results are no more accurate than the least accurate number in your calculation.* In this example, we round to the nearest foot, and hence we say that the diagonal of the $10\,\text{ft} \times 12\,\text{ft}$ room is about 16 feet.

Significant digits are used to determine the precision of a measurement.

DEFINITION   **Significant Digits**

The number of significant digits in a number is found by counting all of the digits from left to right starting with the first nonzero digit.

NUMBER	SIGNIFICANT DIGITS
0.04	1
0.276	3
0.2076	4
1.23	3
17	2
17.00	4
17.000	5
6.25	3
8000	?

The reason for the question mark next to 8000 is that we don't know. If 8000 is a result from rounding to the nearest thousand, then it has one significant digit. If 8000 is the result of rounding to the nearest ten, then it has three significant digits, and if there are exactly 8000 people surveyed, then 8000 is an exact value and it has four significant digits. In this text we will assume that integers have the greatest number of significant digits. Therefore, 8000 has four significant digits and can be expressed in scientific notation: 8.000×10^3.

In solving right triangles, we first determine which of the given measurements has the *least* number of significant digits and round our answers to the same number of significant digits.

EXAMPLE 1 **Identifying the Least Number of Significant Digits**

Determine the number of significant digits corresponding to the given information in the following triangle: measure of an acute angle and a side length. In solving this right triangle, what number of significant digits should be used to express the remaining measure and side lengths?

Solution:

Determine the significant digits corresponding to $\beta = 45°$.   Two significant digits

Determine the significant digits corresponding to $b = 27.3$ ft.   Three significant digits

In solving this triangle, the remaining side lengths of a and c and the measure of α should be expressed to   two significant digits.

Solving a Right Triangle Given the Measure of an Acute Angle and a Side Length

When solving a right triangle, we already know that one angle has measure 90°. Let us now consider the case when the measure of an acute angle and a side length are given. Since the measure of one of the acute angles is given, the remaining acute angle can be found using the fact that the sum of three angles in a triangle is 180°. Right triangle trigonometry is then used to find a second side length, and then the Pythagorean theorem is used to find the remaining (third) side length.

EXAMPLE 2 **Solving a Right Triangle Given an Angle and a Side**

Solve the right triangle—find a, b, and α.

Solution:

STEP 1 **Determine accuracy.**
Since the given quantities, 15 ft and 56°, are both expressed to two significant digits, we will round all calculated values to two significant digits.

STEP 2 **Solve for α.**
Two acute angles in a right triangle are complementary. $\quad \alpha + 56° = 90°$

Solve for α. $\qquad\boxed{\alpha = 34°}$

STEP 3 **Solve for a.**
Cosine of an angle is equal to the adjacent side over the hypotenuse. $\qquad \cos 56° = \dfrac{a}{15}$

Solve for a. $\qquad a = 15 \cos 56°$

Evaluate the right side of the expression using a calculator. $\qquad a \approx 8.38789$

Round a to two significant digits. $\qquad\boxed{a \approx 8.4 \text{ ft}}$

STEP 4 **Solve for b.**
Notice that there are two ways to solve for b: trigonometric functions or the Pythagorean theorem. Although it is tempting to use the Pythagorean theorem, it is better to use the given information with trigonometric functions than to use a value that has already been rounded, which could make results less accurate.

Sine of an angle is equal to the opposite side over the hypotenuse. $\qquad \sin 56° = \dfrac{b}{15}$

Solve for b. $\qquad b = 15 \sin 56°$

Evaluate the right side of the expression using a calculator. $\qquad b \approx 12.43556$

Round b to two significant digits. $\qquad\boxed{b \approx 12 \text{ ft}}$

STEP 5 **Check.**
Angles and sides are rounded to two significant digits.

Check the trigonometric values of the specific angles by calculating the trigonometric ratios.

$$\sin 34° \overset{?}{=} \frac{8.4}{15} \qquad \cos 34° \overset{?}{=} \frac{12}{15} \qquad \tan 34° \overset{?}{=} \frac{8.4}{12}$$

$$0.5592 \approx 0.56 \qquad 0.8290 \approx 0.80 \qquad 0.6745 \approx 0.70$$

■ **YOUR TURN** Solve the right triangle; find a, b, and θ.

Technology Tip

```
15cos(56)
         8.387893552
```

```
15sin(56)
         12.43556359
```

Classroom Example 6.3.2
Solve the triangle.

Answer:
$a \approx 7.13$, $b \approx 15.71$, $\theta = 63°$

■ **Answer:** $\theta = 53°$, $a = 26$ in., $b = 20$ in.

Solving a Right Triangle Given the Length of Two Sides

When solving a right triangle, we already know that one angle has measure $90°$. Let us now consider the case when the lengths of two sides are given. In this case the third side can be found using the Pythagorean theorem. If we can determine the measure of one of the acute angles, then we can find the measure of the third acute angle using the fact that the sum of the three angle measures in a triangle is $180°$. How do we find the measure of one of the acute angles? Since we know the measure of the side lengths, we can use right triangle ratios to determine the trigonometric function (sine, cosine, or tangent) values and then ask ourselves: What angle corresponds to that value?

Sometimes, we may know the answer exactly. For example, if we determine that $\sin\theta = \frac{1}{2}$ then we know that the acute angle θ is $30°$ because $\sin 30° = \frac{1}{2}$. Other times we may not know the corresponding angle, such as $\sin\theta = 0.9511$. Calculators have three keys ($\sin^{-1}$, $\cos^{-1}$, and $\tan^{-1}$) that help us determine the unknown angle. For example, a calculator can be used to assist us in finding what angle θ corresponds to $\sin\theta = 0.9511$.

$$\sin^{-1}(0.9511) = 72.00806419°$$

At first glance these three keys might appear to yield the reciprocal; however, the -1 superscript corresponds to an inverse function. We will learn more about inverse trigonometric functions in Chapter 7, but for now we will use these three calculator keys to help us solve right triangles.

EXAMPLE 3 **Using a Calculator to Determine an Acute Angle Measure**

Use a calculator to find θ. Round answers to the nearest degree.

a. $\cos\theta = 0.8734$ **b.** $\tan\theta = 2.752$

Solution (a):

Use a calculator to evaluate the inverse cosine function.

$$\cos^{-1}(0.8734) = 29.14382196°$$

Round to the nearest degree.

$$\boxed{\theta \approx 29°}$$

Solution (b):

Use a calculator to evaluate the inverse tangent function.

$$\tan^{-1}(2.752) = 70.03026784°$$

Round to the nearest degree.

$$\boxed{\theta \approx 70°}$$

■ **Answer:** $51°$

■ **YOUR TURN** Use a calculator to find θ, given $\sin\theta = 0.7739$. Round answer to the nearest degree.

EXAMPLE 4 Solving a Right Triangle Given Two Sides

Solve the right triangle—find a, α, and β.

Solution:

STEP 1 Determine accuracy.

The given sides have four significant digits; therefore, round all calculated values to four significant digits.

STEP 2 Solve for α.

Cosine of an angle is equal to the adjacent side over hypotenuse.

$$\cos\alpha = \frac{19.67\,\text{cm}}{37.21\,\text{cm}}$$

Evaluate the right side using a calculator.

$$\cos\alpha \approx 0.528621338$$

Write the angle α in terms of the inverse cosine function.

$$\alpha \approx \cos^{-1}(0.528621338)$$

Use a calculator to evaluate the inverse cosine function.

$$\alpha \approx 58.08764854°$$

Round α to the nearest hundredth of a degree.

$$\boxed{\alpha \approx 58.09°}$$

STEP 3 Solve for β.

The two acute angles in a right triangle are complementary.

$$\alpha + \beta = 90°$$

Substitute $\alpha \approx 58.09°$.

$$58.09° + \beta \approx 90°$$

Solve for β.

$$\boxed{\beta \approx 31.91°}$$

The answer is already rounded to the nearest hundredth of a degree.

STEP 4 Solve for a.

Use the Pythagorean theorem.

$$a^2 + b^2 = c^2$$

Substitute the given values for b and c.

$$a^2 + 19.67^2 = 37.21^2$$

Solve for a.

$$a \approx 31.5859969$$

Round a to four significant digits.

$$\boxed{a \approx 31.59\,\text{cm}}$$

STEP 5 Check.

Check the trigonometric values of the specific angles by calculating the trigonometric ratio.

$$\sin 31.91° \overset{?}{=} \frac{19.67}{37.21} \qquad \sin 58.09° \overset{?}{=} \frac{31.59}{37.21}$$

$$0.5286 \approx 0.5286 \qquad 0.8490 \approx 0.8490$$

■ **YOUR TURN** Solve the right triangle; find a, α, and β.

Technology Tip

To find $\cos^{-1}\left(\dfrac{19.67}{37.21}\right)$, first press

$\boxed{19.67}$ $\boxed{\div}$ $\boxed{37.21}$ $\boxed{\text{ENTER}}$.

Next, press $\boxed{\text{2nd}}$ $\boxed{\text{COS}}$ for $\boxed{\cos^{-1}}$

and $\boxed{\text{2nd}}$ $\boxed{(-)}$ for the answer, $\boxed{\text{ANS}}$

$\boxed{\cos^{-1}}$ $\boxed{\text{ANS}}$ $\boxed{)}$ $\boxed{\text{ENTER}}$.

```
19.67/37.21
        .5286213383
cos⁻¹(Ans)
        58.08764854
■
```

Or enter the entire expression as

$\cos^{-1}\left(\dfrac{19.67}{37.21}\right)$. Press $\boxed{\text{2nd}}$ $\boxed{\text{COS}}$

$\boxed{19.67}$ $\boxed{\div}$ $\boxed{37.21}$ $\boxed{)}$ $\boxed{\text{ENTER}}$.

```
19.67/37.21
        .5286213383
cos⁻¹(Ans)
        58.08764854
cos⁻¹(19.67/37.21
)
        58.08764854
```

Using a scientific calculator, press

$\boxed{\cos^{-1}}$ $\boxed{(}$ $\boxed{19.67}$ $\boxed{\div}$ $\boxed{37.21}$

$\boxed{)}$ $\boxed{\text{ENTER}}$.

Classroom Example 6.3.4
Solve the triangle.

Answer: $a = 9.02$ ft, $\theta = 78.3°$, $\beta = 11.7°$

■ **Answer:** $a = 16.0$ miles, $\alpha = 43.0°$, $\beta = 47.0°$

Study Tip

To find a length in a right triangle, use the sine, cosine, and tangent functions. To find an angle in a right triangle, given the proper ratio, use the inverse sine, inverse cosine, and inverse tangent functions.

Applications

In many applications of solving right triangles, you are given a side and an acute angle and are asked to find one of the other sides. Two common examples involve an observer (or point of reference) located on the horizontal and an object that is either above or below the horizontal. If the object is above the horizontal, then the angle made is called the **angle of elevation**, and if the object is below the horizontal, then the angle made is called the **angle of depression**.

For example, if a race car driver is looking straight ahead (in a horizontal line of sight), then looking up is elevation and looking down is depression.

Angle of elevation

Horizontal

Angle of depression

John Harrelson/Stringer/
Getty Images

If the angle is a physical one (like a skateboard ramp), then the appropriate name is **angle of inclination**.

Angle of Inclination

EXAMPLE 5   Angle of Depression (NASCAR)

In this picture, car 19 is behind the leader car 2. If the angle of depression is 18° from the car 19's driver's eyes to the bottom of the 3 ft high back end of car 2 (side opposite the angle of depression), how far apart are their bumpers? Assume that the horizontal distance from the car 19's driver's eyes to the front of his car is 5 feet.

©Walter G Arce/Icon SMI/Corbis Images

Solution:

Draw an appropriate right triangle and label the known quantities.

Identify the tangent ratio.

$$\tan 18° = \frac{3}{x}$$

Solve for x.

$$x = \frac{3}{\tan 18°}$$

Evaluate the right side.

$$x \approx 9.233$$

Round to the nearest foot.

$$x \approx 9 \text{ ft}$$

Subtract 5 ft (eyes to end of car).

$$9 - 5 = \boxed{4 \text{ ft}}$$

Their bumpers are 4 ft apart.

Suppose NASA wants to talk with the International Space Station (ISS), which is traveling at a speed of 17,700 mph, 400 km (250 miles) above the surface of the Earth. If the antennas at the ground station in Houston have a pointing error of even 1/100 of a degree, that is, 0.01°, the ground station will miss the chance to talk with the astronauts.

EXAMPLE 6   Pointing Error

Assume that the ISS (which is 108 meters long and 73 meters wide) is in a 400-km low earth orbit. If the communications antennas have a 0.01° pointing error, how many meters off will the communications link be?

Solution:

Draw a right triangle that depicts this scenario.

(NOT TO SCALE)

Identify the tangent ratio.

$$\tan 0.01° = \frac{x}{400}$$

Solve for x.

$$x = (400 \text{ km}) \tan 0.01°$$

Evaluate the expression on the right.

$$x \approx 0.06981317 \text{ km}$$

400 km is accurate to three significant digits, so we express the answer to three significant digits.

The pointing error causes the signal to be off by $\boxed{69.8 \text{ meters}}$. Since the ISS is only 108 meters long, it is expected that the signal will be missed by the astronaut crew.

Technology Tip

To calculate $x = \dfrac{3}{\tan 18°}$, press

```
3/tan(18)
        9.233050612
```

Technology Tip

To calculate 400 tan 0.01°, press

```
400tan(0.01)
        .0698131708
```

Classroom Example 6.3.6
Assume that the ISS (which is 108 meters long and 73 meters wide) is in a 400-km low earth orbit.

a. If the communications antennas have a 0.001° pointing error, how many meters off will the communications link be?

b.* Experiment to find the angle pointing error such that the communications link is off by only 1 meter.

Answer: a. 7.0 m
            b. 0.00014°

In navigation, the word **bearing** means the direction a vessel is pointed. **Heading** is the direction the vessel is actually traveling. Heading and bearing are only synonyms when there is no wind on land. Direction is often given as a bearing, which is the measure of an acute angle with respect to the north–south vertical line. "The plane has a bearing N 20° E" means that the plane is pointed 20° to the east of due north.

EXAMPLE 7 Bearing (Navigation)

A jet takes off bearing N 28° E and flies 5 miles and then makes a left (90°) turn and flies 12 miles farther. If the control tower operator wants to locate the plane, what bearing should she use?

Solution:

Draw a picture that represents this scenario.

Identify the tangent ratio.

$$\tan\theta = \frac{12}{5}$$

Use the inverse tangent function to solve for θ.

$$\theta = \tan^{-1}\left(\frac{12}{5}\right) \approx 67.4°$$

Subtract 28° from θ to find the bearing, β.

$$\beta \approx 67.4° - 28° \approx 39.4°$$

Round to the nearest degree.

$$\boxed{\beta \approx \text{N } 39° \text{ W}}$$

SECTION 6.3 SUMMARY

In this section we used right triangle trigonometry to solve right triangles. When either a side length and an acute angle measure are given or two side lengths are given, it is possible to solve the right triangle (find all unknown side and angle measurements). The least accurate number used in your calculations determines the appropriate number of significant digits for your results.

SECTION 6.3 EXERCISES

■ **SKILLS**

In Exercises 1–4, determine the number of significant digits corresponding to each of the given angle measures and side lengths.

1. $\alpha = 37.5°$ **2.** $\beta = 49.76°$ **3.** $a = 0.37$ km   **4.** $b = 0.2$ mi

In Exercises 5–10, use a calculator to find the measure of angle θ. Round answer to the nearest degree.

5. $\sin\theta = 0.7264$ **6.** $\sin\theta = 0.1798$ **7.** $\cos\theta = 0.5674$

8. $\cos\theta = 0.8866$ **9.** $\tan\theta = 8.235$ **10.** $\tan\theta = 3.563$

In Exercises 11–30, refer to the right triangle diagram and the given information to find the indicated measure. Write your answers for angle measures in decimal degrees.

11. $\beta = 35°$, $c = 17$ in.; find a.   **12.** $\beta = 35°$, $c = 17$ in.; find b.

13. $\alpha = 55°$, $c = 22$ ft; find a.   **14.** $\alpha = 55°$, $c = 22$ ft; find b.

15. $\alpha = 20.5°$, $b = 14.7$ miles; find a.   **16.** $\beta = 69.3°$, $a = 0.752$ miles; find b.

17. $\beta = 25°$, $a = 11$ km; find c.   **18.** $\beta = 75°$, $b = 26$ km; find c.

19. $\alpha = 48.25°$, $a = 15.37$ cm; find c.   **20.** $\alpha = 29.80°$, $b = 16.79$ cm; find c.

21. $a = 29$ mm, $c = 38$ mm; find α.   **22.** $a = 89$ mm, $c = 99$ mm; find β.

23. $b = 2.3$ m, $c = 4.9$ m; find α.   **24.** $b = 7.8$ m, $c = 13$ m; find β.

25. $\alpha = 21.283°$, $b = 210.8$ yards; find a.   **26.** $\beta = 27.35°$, $a = 117.0$ yards; find b.

27. $\beta = 15.33°$, $a = 10.2$ km; find c.   **28.** $\beta = 65.5°$, $b = 18.6$ km; find c.

29. $\alpha = 40.4694°$, $a = 12{,}522$ km; find c.   **30.** $\alpha = 28.5472°$, $b = 17{,}986$ km; find c.

In Exercises 31–46, refer to the right triangle diagram and the given information to solve the right triangle. Write your answers for angle measures in decimal degrees.

31. $\alpha = 32°$ and $c = 12$ ft   **32.** $\alpha = 65°$ and $c = 37$ ft

33. $\beta = 72°$ and $c = 9.7$ mm   **34.** $\beta = 45°$ and $c = 7.8$ mm

35. $\alpha = 54.2°$ and $a = 111$ miles   **36.** $\beta = 47.2°$ and $a = 9.75$ miles

37. $\alpha = 28.383°$ and $b = 1734$ ft   **38.** $\alpha = 72.983°$ and $a = 2175$ ft

39. $a = 42.5$ ft and $b = 28.7$ ft   **40.** $a = 19.8$ ft and $c = 48.7$ ft

41. $a = 35{,}236$ km and $c = 42{,}766$ km   **42.** $b = 0.1245$ mm and $c = 0.8763$ mm

43. $\beta = 25.4°$ and $b = 11.6$ in.   **44.** $\beta = 39.21°$ and $b = 6.3$ m

45. $\beta = 16.10°$ and $a = 4.00$ cm   **46.** $\alpha = 50.1°$ and $b = 4$ cm

▪ APPLICATIONS

5 ft

(NOT TO SCALE)

1°

47. Golf. If the flagpole that a golfer aims at on a green measures 5 feet from the ground to the top of the flag and a golfer measures a 1° angle from top to bottom, how far (in horizontal distance) is the golfer from the flag? Round to the nearest foot.

48. Golf. If the flagpole that a golfer aims at on a green measures 5 feet from the ground to the top of the flag and a golfer measures a 3° angle from top to bottom, how far (in horizontal distance) is the golfer from the flag? Round to the nearest foot.

Exercises 49 and 50 illustrate a mid-air refueling scenario that military aircraft often enact. Assume the elevation angle that the hose makes with the plane being fueled is $\theta = 36°$.

49. Mid-Air Refueling. If the hose is 150 feet long, what should be the altitude difference a between the two planes? Round to the nearest foot.

50. Mid-Air Refueling. If the smallest acceptable altitude difference a between the two planes is 100 feet, how long should the hose be? Round to the nearest foot.

Exercises 51–54 are based on the idea of a glide slope (the angle the flight path makes with the ground).

Precision Approach Path Indicator (PAPI) lights are used as a visual approach slope aid for pilots landing aircraft. A typical glide path for commercial jet airliners is 3°. The space shuttle has an outer glide approach of 18°–20°. PAPI lights are typically configured as a row of four lights. All four lights are on, but in different combinations of red or white. If all four lights are white, then the angle of descent is too high; if all four lights are red, then the angle of descent is too low; and if there are two white and two red, then the approach is perfect.

51. Glide Path of a Commercial Jet Airliner. If a commercial jetliner is 5000 feet (about 1 mile) ground distance from the runway, what should the altitude of the plane be to achieve two red and two white PAPI lights? (Assume this corresponds to a 3° glide path.)

52. Glide Path of a Commercial Jet Airliner. If a commercial jetliner is at an altitude of 450 feet when it is 5200 feet from the runway (approximately 1 mile ground distance), what is the glide slope angle? Will the pilot see white lights, red lights, or both?

53. Glide Path of the Space Shuttle *Orbiter*. If the pilot of the space shuttle *Orbiter* is at an altitude of 3000 feet when she is 15,500 ft (approximately 3 miles) from the shuttle landing facility (ground distance), what is her glide slope angle (round to the nearest degree)? Is she too high or too low?

54. Glide Path of the Space Shuttle *Orbiter*. If the same pilot in Exercise 53 raises the nose of the gliding shuttle so that she drops only 500 feet by the time she is 7800 feet from the shuttle landing strip (ground distance), what is her glide angle then (round to the nearest degree)? Is she within the specs to land the shuttle?

In Exercises 55 and 56, refer to the illustration below which shows a search and rescue helicopter with a 30° field of view with a searchlight.

55. Search and Rescue. If the search and rescue helicopter is flying at an altitude of 150 feet above sea level, what is the diameter of the circle illuminated on the surface of the water?

56. Search and Rescue. If the search and rescue helicopter is flying at an altitude of 500 feet above sea level, what is the diameter of the circle illuminated on the surface of the water?

For Exercises 57–60, refer to the following:

Geostationary orbits are useful because they cause a satellite to appear stationary with respect to a fixed point on the rotating Earth. As a result, an antenna (dish TV) can point in a fixed direction and maintain a link with the satellite. The satellite orbits in the direction of the Earth's rotation, at an altitude of approximately 35,000 km.

57. Dish TV. If your dish TV antenna has a pointing error of 0.000278°, how long would the satellite have to be in order to maintain a link? Round your answer to the nearest meter.

58. Dish TV. If your dish TV antenna has a pointing error of 0.000139°, how long would the satellite have to be in order to maintain a link? Round your answer to the nearest meter.

59. Dish TV. If the satellite in a geostationary orbit (at 35,000 km) was only 10 meters long, about how accurate would the pointing of the dish have to be? Give the answer in degrees to two significant digits.

60. Dish TV. If the satellite in a geostationary orbit (at 35,000 km) was only 30 meters long, about how accurate would the pointing of the dish have to be? Give the answer in degrees to two significant digits.

61. Angle of Elevation (Traffic). A person driving in a sedan is driving too close to the back of an 18 wheeler on an interstate highway. He decides to back off until he can see the entire truck (to the top). If the height of the trailer is 15 feet and the sedan driver's angle of elevation (to the top of the trailer from the horizontal line with the bottom of the trailer) is roughly 30°, how far is he sitting from the end of the trailer?

62. Angle of Depression (Opera). The balcony seats at the opera house have an angle of depression of 55° to center stage. If the horizontal (ground) distance to the center of the stage is 50 feet, how far are the patrons in the balcony from the singer at center stage?

63. Angle of Inclination (Skiing). The angle of inclination of a mountain with triple black diamond ski trails is 65°. If a skier at the top of the mountain is at an elevation of 4000 feet, how long is the ski run from the top to the base of the mountain?

64. Bearing (Navigation). If a plane takes off bearing N 33° W and flies 6 miles and then makes a right (90°) turn and flies 10 miles further, what bearing will the traffic controller use to locate the plane?

For Exercises 65 and 66, refer to the following:

With the advent of new technology, tennis racquets can now be constructed to permit a player to serve at speeds in excess of 120 mph (as demonstrated by Andy Roddick and Pete Sampras, to name just two). One of the most effective serves in tennis is a power serve that is hit at top speed directly at the top left corner of the right service court (or top right corner of the left

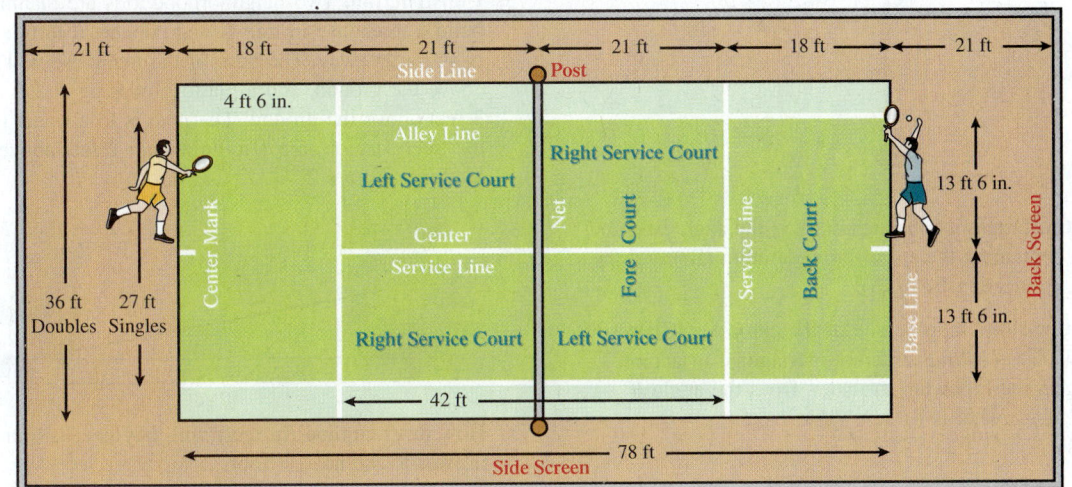

service court). When attempting this serve, a player will toss the ball rather high into the air, bring the racquet back, and then make contact with the ball at the precise moment when the position of the ball in the air coincides with the top of the netted part of the racquet when the player's arm is fully stretched over his or her head.

65. **Tennis.** Assume that the player is serving into the right service court and stands just 2 inches to the right of the center line behind the base line. If, at the moment the racquet strikes the ball, the ball is 9 feet from the ground and the serve actually hits the top left corner of the right service court, determine the angle at which the ball meets the ground in the right service court. Round to the nearest degree.

66. **Tennis.** Assume that the player is serving into the right service court and stands just 2 inches to the right of the center line behind the base line. If the ball hits the top left corner of the right service court at an angle of 8°, at what height above the ground must the ball be struck?

For Exercises 67 and 68, refer to the following:

The structure of molecules is critical to the study of materials science and organic chemistry, and has countless applications to a variety of interesting phenomena. Trigonometry plays a critical role in determining the bonding angles of molecules. For instance, the structure of the $(FeCl_4Br_2)^{-3}$ ion (dibromatetetrachlorideferrate III) is shown in the figure below.

67. **Chemistry.** Determine the angle θ (i.e., the angle between the axis containing the apical bromide atom (Br) and the segment connecting Br to Cl).

68. **Chemistry.** Now, suppose one of the chlorides (Cl) is removed. The resulting structure is triagonal in nature, resulting in the following structure. Does the angle θ change? If so, what is its new value?

For Exercises 69–74, refer back to Exercises 75–78 in Section 6.1 for an explanation of the setting.

69. **Archery.** If the arrow hits the target 13 inches directly below the bullseye, and it has traveled 92 feet from its original position at release to the point of impact with the target, how far was the archer from the target? Express your answer in four significant digits.

70. **Archery.** In Exercise 69, what was the tangent of the angle of depression from the line of sight connecting the arrowhead to the bullseye? Express your answer in four significant digits.

71. **Archery.** Assuming that the angle of depression from the line of sight connecting the arrowhead to the bullseye is 0.7°, how far should the archer stand from the target to ensure she hits the target directly below the bullseye where the red and blue rings meet? Express your answer in two significant digits.

72. **Archery.** In Exercise 71, how far has the arrow traveled? Express your answer in four significant digits.

73. **Archery.** If the archer stands 90 feet from the target and the angle at which the arrow hits the target is 88.4°, what is her score?

74. **Archery.** In Exercise 73, what is the angle of depression of the arrowhead?

75. **Construction.** Two neighborhood kids are planning to build a treehouse in tree 1, and connect it to tree 2 which is 40 yards away. The base of the treehouse will be 20 feet above the ground, and a platform will be nailed into tree 2, 3 feet above the ground. The plan is to connect the base of the treehouse on tree 1 to an anchor 2 feet above the platform on tree 2.

How much zipline (in feet) will they need? Round your answer to the nearest foot.

76. Construction. In Exercise 75, what is the angle of depression β that the zipline makes with tree 1? Express your answer in two significant digits. (*Hint:* Find an expression for $\tan \beta$, and use your calculator to compute $\tan^{-1}$ of this expression.)

77. Construction. A pool that measures 5 feet above the ground is to be placed between the trees directly in the path of the zipline in Exercise 75. Assuming that a rider of the zipline dangles at most 3 feet below the wire anywhere in route, what is the closest the edge of the pool can be placed to tree 2 so that the pool will not impede a rider's trip down the zipline?

78. Construction. If the treehouse is to be built so that its base is now 22 feet above the base of tree 1, where should the anchor on tree 2 (to which the zipline is connected) be placed in order to ensure the same angle of depression found in Exercise 76?

For Exercises 79 and 80, refer to the following:

A canal constructed by a water-users association can be approximated by an isosceles triangle (see the figure below). When the canal was originally constructed, the depth of the canal was 5.0 feet and the angle defining the shape of the canal was 60°.

79. Environmental Science. If the width of the water surface today is 4.0 feet, find the depth of the water running through the canal.

80. Environmental Science. One year later a survey is performed to measure the effects of erosion on the canal. It is determined that when the water depth is 4.0 feet, the width of the water surface is 5.0 feet. Find the angle θ defining the shape of the canal to the nearest degree. Has erosion affected the shape of the canal? Explain.

For Exercises 81 and 82, refer to the following:

After breaking a femur, a patient is placed in traction. The end of a femur of length l is lifted to an elevation forming an angle θ with the horizontal (angle of elevation).

81. Health/Medicine. A femur 18 inches long is placed into traction, forming an angle of 15° with the horizontal. Find the height of elevation at the end of the femur.

82. Health/Medicine. A femur 18 inches long is placed in traction with an elevation of 6.2 inches. What is the angle of elevation of the femur?

■ **CATCH THE MISTAKE**

In Exercises 83 and 84, refer to the right triangle diagram below and explain the mistake that is made:

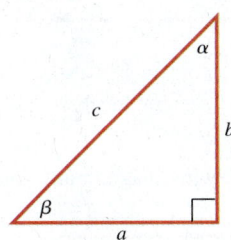

83. If $b = 800$ ft and $a = 10$ ft, find β.

Solution:

Represent tangent as the opposite side over the adjacent side.
$$\tan \beta = \frac{b}{a}$$

Substitute $b = 800$ ft and $a = 10$ ft.
$$\tan \beta = \frac{800}{10}$$

Use a calculator to evaluate β. $\beta = \tan 80° \approx 5.67°$

This is incorrect. What mistake was made?

84. If $\beta = 56°$ and $c = 15$ ft, find b and then find a.

Solution:

Write sine as the opposite side over the hypotenuse.
$$\sin 56° = \frac{b}{15}$$

Solve for b. $b = 15 \sin 56°$

Use a calculator to approximate b. $b \approx 12.4356$

Round the answer to two significant digits. $b \approx 12$ ft

Use the Pythagorean theorem to find a. $a^2 + b^2 = c^2$

Substitute $b = 12$ ft and $c = 15$ ft. $a^2 + 12^2 = 15^2$

Solve for a. $a = 9$

Round the answer to two significant digits. $a = 9.0$ ft

Compare this with the results from Example 2. Why did we get a different value for a here?

■ CONCEPTUAL

In Exercises 85–89, determine whether each statement is true or false.

85. If you are given the measures of two sides of a right triangle, you can solve the right triangle.

86. If you are given the measures of one side and one acute angle of a right triangle, you can solve the right triangle.

87. If you are given the two acute angles of a right triangle, you can solve the right triangle.

88. If you are given the hypotenuse of a right triangle and the angle opposite the hypotenuse, you can solve the right triangle.

89. If both $\sin\theta$ and $\cos\theta$ are known for a given angle θ in a right triangle, you can solve the triangle.

■ CHALLENGE

90. Use the information in the illustration below to determine the height of the mountain.

91. Two friends who are engineers at Kennedy Space Center (KSC) watch the shuttle launch. Carolyn is at the Vehicle Assembly Building (VAB) 3 miles from the launch pad and Jackie is across the Banana River, which is 8 miles from the launch pad. They call each other at liftoff, and after 10 seconds they each estimate the elevation with respect to the ground. Carolyn thinks the elevation is approximately 40°, and Jackie thinks the elevation is approximately 15°. Approximately how high is the shuttle after 10 seconds? (Average their estimates and round to the nearest mile.)

Image Bank/Getty Images, Inc.
Photonica/Getty Images

92. A crew sets out in search of a buried treasure. A diver has a device that informs him where on the ocean floor the treasure is buried, but it has difficulty determining the exact depth that it lies below the surface. The known information

is given in the following diagram. Determine y, the depth at which the treasure lies.

For Exercises 93 and 94, consider the following diagram:

93. Determine x.

94. Determine y.

95. Consider the following diagram and compute $\tan(90° - A)$.

96. Consider the following diagram and compute $\sec A \cdot \csc B$. Express your answer in terms of x.

TECHNOLOGY

97. Use a calculator to find $\sin^{-1}(\sin 40°)$.

98. Use a calculator to find $\cos^{-1}(\cos 17°)$.

99. Use a calculator to find $\cos(\cos^{-1} 0.8)$.

100. Use a calculator to find $\sin(\sin^{-1} 0.3)$.

101. Based on the result from Exercise 97, what would $\sin^{-1}(\sin \theta)$ be for an acute angle θ?

102. Based on the result from Exercise 98, what would $\cos^{-1}(\cos \theta)$ be for an acute angle θ?

103. Compute $\tan^{-1}(\tan 35°)$. What would you expect $\tan^{-1}(\tan \theta)$ to be for an acute angle?

104. Compute $\tan(\tan^{-1} 2)$. What would you expect $\tan(\tan^{-1} x)$ to be for any real number?

105. Use a calculator to find $\sin(\sin^{-1} 0.785)$.

106. Use a calculator to find $\tan(\tan^{-1} 5.321)$.

SECTION 6.4
DEFINITION 2 OF TRIGONOMETRIC FUNCTIONS: CARTESIAN PLANE

SKILLS OBJECTIVES

■ Sketch angles in standard position.
■ Identify coterminal angles.
■ Calculate trigonometric function values for acute angles.
■ Calculate trigonometric function values for nonacute angles.
■ Calculate trigonometric function values for quadrantal angles.

CONCEPTUAL OBJECTIVES

■ Understand that right triangle definitions of trigonometric functions for acute angles are consistent with definitions of trigonometric functions for all angles in the Cartesian plane.
■ Understand why some trigonometric functions are undefined for quadrantal angles.

Angles in Standard Position

In Section 6.1 we introduced angles. A common unit of measure for angles is degrees. We discussed triangles and the fact that the sum of the measures of the three interior angles is always 180°. We also discussed the Pythagorean theorem, which relates the lengths of the three sides of a *right* triangle. Based on what we learned in Section 6.1, the following two angles, which correspond to a right angle or quarter rotation, both have a measurement of 90°.

$\theta = 90°$ $\theta = 90°$

We need a frame of reference. In this section we use the Cartesian plane as our frame of reference by superimposing angles onto the *Cartesian coordinate system*, or *rectangular coordinate system* (Section 2.1).

Study Tip

For a review of the rectangular (Cartesian) coordinate system, see Section 2.1.

Let us now bridge two concepts with which you are already familiar. First, recall that an angle is generated when a ray (the angle's *initial side*) is rotated around an endpoint (which becomes the angle's *vertex*). The ray, in its new position after it is rotated, is called the *terminal side* of the angle. Also recall the Cartesian (rectangular) coordinate system with the *x*-axis, *y*-axis, and origin. Using the *positive x*-axis combined with the origin as a frame of reference, we can graph angles in the Cartesian plane. If the *initial side* of the angle is aligned along the *positive x-axis* and the *vertex* of the angle is positioned at the *origin*, then the angle is said to be in *standard position*.

Study Tip

The initial side of an angle is a ray (initial ray) and the terminal side of an angle is also a ray (terminal ray).

> **DEFINITION   Standard Position**
>
> An angle is said to be in **standard position** if its initial side is along the positive *x*-axis and its vertex is at the origin.

We say that an angle lies in the quadrant in which its terminal side lies. For example, an acute angle ($0° < \theta < 90°$) lies in **quadrant I**, whereas an obtuse angle ($90° < \theta < 180°$) lies in **quadrant II**.

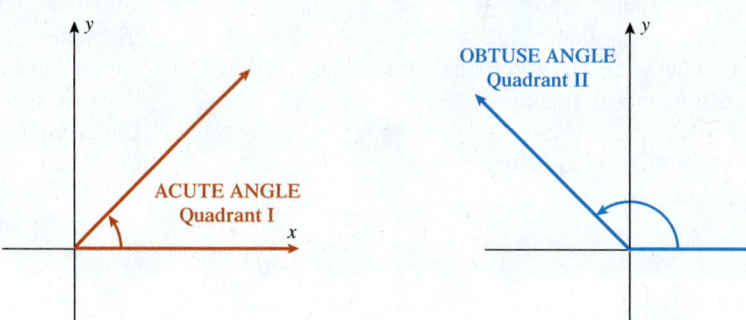

Similarly, angles with measure $180° < \theta < 270°$ lie in **quadrant III** and angles with measure $270° < \theta < 360°$ lie in **quadrant IV**. Angles in standard position with terminal sides along the *x*-axis or *y*-axis (90°, 180°, 270°, 360°, etc.) are called **quadrantal angles**. An abbreviated way to represent an angle θ that lies in quadrant I is $\theta \in$ QI. Similarly, if an angle lies in quadrant II, we say $\theta \in$ QII, and so forth.

Classroom Example 6.4.1
Sketch each angle and state the quadrant or axis on which its terminal side lies.

a. $-270°$

b. $135°$

c.* $-360n°$, where *n* is an integer

Answer:
a. positive *y*-axis

b. QII

c. positive *x*-axis

Recall that rotation in a counterclockwise direction corresponds to a *positive angle*, whereas rotation in a clockwise direction corresponds to a *negative angle*.

EXAMPLE 1 Sketching Angles in Standard Position

Sketch the following angles in standard position, and state the quadrant in (or axis on) which the terminal side lies.

a. $-90°$ **b.** $210°$

Solution (a):

The initial side lies on the positive x-axis.

A negative angle indicates clockwise rotation.

$90°$ is a right angle.

The terminal side lies on the negative y-axis.

Solution (b):

The initial side lies on the positive x-axis.

A positive angle indicates counterclockwise rotation.

$180°$ represents a straight angle, and an additional $30°$ yields a $210°$ angle.

The terminal side lies in quadrant III.

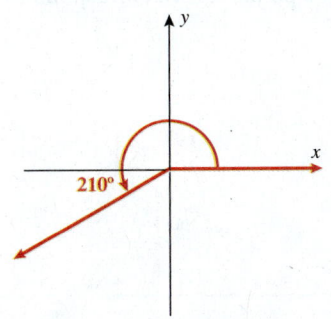

■ **YOUR TURN**   Sketch the following angles in standard position, and state the quadrant in which the terminal side lies.

    a. $-300°$ **b.** $135°$

Common Angles in Standard Position

The common angles for which we determined the exact trigonometric function values in Section 6.2 are $30°$, $45°$, and $60°$. Recall the relationships between the sides of $30°$-$60°$-$90°$ and $45°$-$45°$-$90°$ triangles.

■ **Answer:**

a.

Terminal side lies in quadrant I.

b.

Terminal side lies in quadrant II.

Let us assume the hypotenuse is equal to 1. Then we have the following triangles:

We can position these triangles on the Cartesian plane with one leg along the positive *x*-axis so that we have three angles (30°, 45°, and 60°) in standard position. Remember that we are always assuming that the hypotenuse is equal to 1. Notice that the *x*- and *y*-coordinates of the points shown correspond to the side lengths.

| 30° angle | 45° angle | 60° angle |

If we graph the three angles (30°, 45°, and 60°) on the same Cartesian coordinate system, we get the following in the first quadrant:

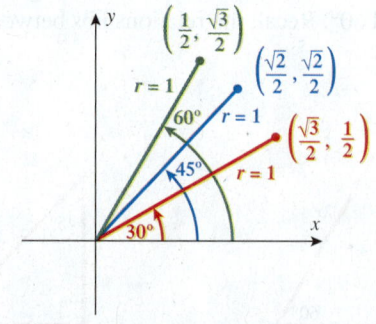

Using symmetry (Section 2.2) and the angles and coordinates in quadrant 1 (QI), we get the following angles and coordinates in QII, QIII, and QIV. Notice that all of these coordinate pairs satisfy the equation of the unit circle (radius equal to 1 and centered at the origin): $x^2 + y^2 = 1$.

Study Tip

In Section 6.7 we will define the values of the sine and cosine functions as the y- and x-coordinates, respectively, along the unit circle.

In Section 6.2 we defined trigonometric functions as ratios of side lengths of right triangles. This definition holds only for acute $(0° < \theta < 90°)$ angles, since the two angles in a right triangle other than the right angle must be acute. We now define trigonometric functions as ratios of x- and y-coordinates and distances in the Cartesian plane, which is consistent with right triangle trigonometry for acute angles. However, this second approach enables us to formulate trigonometric functions for quadrantal angles (whose terminal side lies along an axis) and nonacute angles.

Coterminal Angles

DEFINITION   **Coterminal Angles**

Two angles in standard position with the same terminal side are called **coterminal angles**.

For example, **−40°** and **320°** are coterminal; their terminal rays are identical even though they are formed by rotation in opposite directions. The angles **60°** and **420°** are also coterminal; angles larger than 360° or less than −360° are generated by continuing the rotation beyond a full rotation. Thus, all coterminal angles have the same initial side (positive x-axis) and the same terminal side, just different rotations.

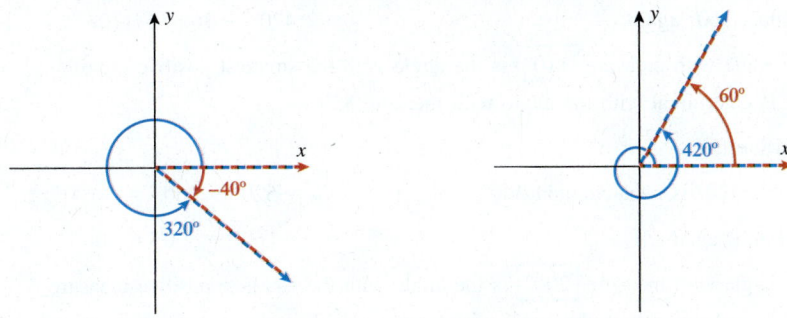

EXAMPLE 2 Recognizing Coterminal Angles

Determine whether the following pairs of angles are coterminal.

a. $\alpha = 120°, \beta = -180°$ **b.** $\alpha = 20°, \beta = 740°$

Solution (a):

The terminal side of α is in QII.

The terminal side of β is along the negative x-axis.

α and β are **not coterminal angles**.

Solution (b):

Since 360° represents one rotation, 720° represents two rotations. Therefore, after 720° of rotation, the angle is again along the positive x-axis. An additional 20° of rotation achieves a 740° angle. Since $\alpha = 20°$ and $\beta = 740°$ have the same terminal side, they are **coterminal angles**.

■ **YOUR TURN** Determine whether the following pairs of angles are coterminal.

    a. $\alpha = 240°, \beta = -120°$ **b.** $\alpha = 20°, \beta = -380°$

To find measures of the smallest positive coterminal angles, if the given angle is positive and greater than 360°, subtract 360° repeatedly until the result is a positive angle less than or equal to 360°. If the given angle is nonpositive, add 360° repeatedly until the result is a positive angle less than or equal to 360°.

EXAMPLE 3 Finding Measures of Coterminal Angles

Determine the angle of the smallest possible positive measure that is coterminal with each of the following angles.

a. 830°   **b.** $-520°$

Solution (a):

Since 830° is positive, subtract 360°. $830° - 360° = 470°$

Subtract 360° again. $470° - 360° = 110°$

The angle with measure $\boxed{110°}$ is the angle with the smallest positive measure that is coterminal with the angle with measure 830°.

Solution (b):

Since $-520°$ is negative, add 360°. $-520° + 360° = -160°$

Add 360° again. $-160° + 360° = 200°$

The angle with measure $\boxed{200°}$ is the angle with the smallest positive measure that is coterminal with the angle with measure $-520°$.

■ **YOUR TURN** Determine the angle of the smallest possible positive measure that is coterminal with each of the following angles.

    a. 900°   **b.** $-430°$

Trigonometric Functions: The Cartesian Plane

To define the trigonometric functions in the Cartesian plane, let us start with an acute angle θ in standard position. Choose any point (x, y) on the terminal side of the angle as long as it is not the vertex (the origin).

A right triangle can be drawn so that the right angle is made when a perpendicular segment connects the point (x, y) to the x-axis. Notice that the side opposite $\boldsymbol{\theta}$ has length $\boldsymbol{y}$ and the other leg of the right triangle has length $\boldsymbol{x}$.

WORDS	**MATH**
The distance r from the origin $(0, 0)$ to the point (x, y) can be found using the distance formula:	$r = \sqrt{(x - 0)^2 + (y - 0)^2}$ $r = \sqrt{x^2 + y^2}$
Since r is a distance, it is always positive.	$r > 0$

Using our first definition of trigonometric functions in terms of right triangle ratios (Section 6.2), we say that $\sin\theta = \dfrac{\text{opposite}}{\text{hypotenuse}}$. From this picture we see that sine can also be defined by the relation $\sin\theta = \dfrac{y}{r}$. Similar reasoning holds for all six trigonometric functions and leads us to the second definition of the trigonometric functions, in terms of ratios of coordinates and distances in the Cartesian plane.

DEFINITION (2)   **Trigonometric Functions**

Let (x, y) be a point other than the origin on the terminal side of an angle θ in standard position. Let r be the distance from the point (x, y) to the origin. Then the six trigonometric functions are defined as

$$\sin\theta = \frac{y}{r} \qquad\qquad \cos\theta = \frac{x}{r} \qquad\qquad \tan\theta = \frac{y}{x} \ \ (x \neq 0)$$

$$\csc\theta = \frac{r}{y} \ \ (y \neq 0) \qquad \sec\theta = \frac{r}{x} \ \ (x \neq 0) \qquad \cot\theta = \frac{x}{y} \ \ (y \neq 0)$$

where $r = \sqrt{x^2 + y^2}$, or $x^2 + y^2 = r^2$. The distance r is positive: $r > 0$.

Study Tip

There is no need to memorize definitions for secant, cosecant, and cotangent functions, since their values can be derived from the reciprocals of the sine, cosine, and tangent function values.

■ **Answer:**

$\sin\theta = \dfrac{7\sqrt{58}}{58}$ $\quad \cos\theta = \dfrac{3\sqrt{58}}{58}$

$\tan\theta = \dfrac{7}{3}$ $\quad \csc\theta = \dfrac{\sqrt{58}}{7}$

$\sec\theta = \dfrac{\sqrt{58}}{3}$ $\quad \cot\theta = \dfrac{3}{7}$

EXAMPLE 4 **Calculating Trigonometric Function Values for Acute Angles**

The terminal side of an angle θ in standard position passes through the point $(2, 5)$. Calculate the values of the six trigonometric functions for angle θ.

Solution:

STEP 1 Draw the angle and label the point **(2, 5)**.

STEP 2 Calculate the distance *r*. $\qquad r = \sqrt{2^2 + 5^2} = \sqrt{29}$

STEP 3 Formulate the trigonometric functions in terms of *x, y,* and *r*.

Let $x = 2, y = 5, r = \sqrt{29}$.

$\sin\theta = \dfrac{y}{r} = \dfrac{5}{\sqrt{29}}$ $\qquad \cos\theta = \dfrac{x}{r} = \dfrac{2}{\sqrt{29}}$ $\qquad \tan\theta = \dfrac{y}{x} = \dfrac{5}{2}$

$\csc\theta = \dfrac{r}{y} = \dfrac{\sqrt{29}}{5}$ $\qquad \sec\theta = \dfrac{r}{x} = \dfrac{\sqrt{29}}{2}$ $\qquad \cot\theta = \dfrac{x}{y} = \dfrac{2}{5}$

STEP 4 Rationalize any denominators containing a radical.

$\sin\theta = \dfrac{5}{\sqrt{29}} \cdot \dfrac{\sqrt{29}}{\sqrt{29}} = \dfrac{5\sqrt{29}}{29}$ $\qquad \cos\theta = \dfrac{2}{\sqrt{29}} \cdot \dfrac{\sqrt{29}}{\sqrt{29}} = \dfrac{2\sqrt{29}}{29}$

STEP 5 Write the values of the six trigonometric functions for θ.

$\sin\theta = \dfrac{5\sqrt{29}}{29}$ $\qquad \cos\theta = \dfrac{2\sqrt{29}}{29}$ $\qquad \tan\theta = \dfrac{5}{2}$

$\csc\theta = \dfrac{\sqrt{29}}{5}$ $\qquad \sec\theta = \dfrac{\sqrt{29}}{2}$ $\qquad \cot\theta = \dfrac{2}{5}$

Note: In Example 4, we could have used the values of the sine, cosine, and tangent functions along with the reciprocal identities to calculate the cosecant, secant, and cotangent function values.

■ **YOUR TURN** The terminal side of an angle θ in standard position passes through the point $(3, 7)$. Calculate the values of the six trigonometric functions for angle θ.

We can now use this second definition of trigonometric functions to find values for nonacute angles (angles with measure greater than or equal to 90°) as well as negative angles.

EXAMPLE 5   Calculating Trigonometric Function Values for Nonacute Angles

The terminal side of an angle θ in standard position passes through the point $(-4, -7)$. Calculate the values of the six trigonometric functions for angle θ.

Solution:

STEP 1   Draw the angle and label the point $(-4, -7)$.

STEP 2   Calculate the distance r. $r = \sqrt{(-4)^2 + (-7)^2} = \sqrt{65}$

STEP 3   Formulate the trigonometric functions in terms of x, y, and r.

Let $x = -4$, $y = -7$, and $r = \sqrt{65}$.

$$\sin\theta = \frac{y}{r} = \frac{-7}{\sqrt{65}} \qquad \cos\theta = \frac{x}{r} = \frac{-4}{\sqrt{65}} \qquad \tan\theta = \frac{y}{x} = \frac{-7}{-4} = \frac{7}{4}$$

$$\csc\theta = \frac{r}{y} = \frac{\sqrt{65}}{-7} \qquad \sec\theta = \frac{r}{x} = \frac{\sqrt{65}}{-4} \qquad \cot\theta = \frac{x}{y} = \frac{-4}{-7} = \frac{4}{7}$$

STEP 4   Rationalize the radical denominators in the sine and cosine functions.

$$\sin\theta = \frac{y}{r} = \frac{-7}{\sqrt{65}} \cdot \frac{\sqrt{65}}{\sqrt{65}} = -\frac{7\sqrt{65}}{65}$$

$$\cos\theta = \frac{x}{r} = \frac{-4}{\sqrt{65}} \cdot \frac{\sqrt{65}}{\sqrt{65}} = -\frac{4\sqrt{65}}{65}$$

STEP 5   Write the values of the six trigonometric functions for θ.

$$\boxed{\sin\theta = -\frac{7\sqrt{65}}{65}} \qquad \boxed{\cos\theta = -\frac{4\sqrt{65}}{65}} \qquad \boxed{\tan\theta = \frac{7}{4}}$$

$$\boxed{\csc\theta = -\frac{\sqrt{65}}{7}} \qquad \boxed{\sec\theta = -\frac{\sqrt{65}}{4}} \qquad \boxed{\cot\theta = \frac{4}{7}}$$

■ **YOUR TURN**   The terminal side of an angle θ in standard position passes through the point $(-3, -5)$. Calculate the values of the six trigonometric functions for angle θ.

Classroom Example 6.4.5*

The terminal side of angle θ passes through the point $\left(-a, -\frac{1}{3}a\right)$, where $a > 0$. Calculate the values of the trigonometric functions.

Answer:

$\sin\theta = -\dfrac{\sqrt{10}}{10}$

$\cot\theta = 3$

$\cos\theta = -\dfrac{3\sqrt{10}}{10}$

$\csc\theta = -\sqrt{10}$

$\tan\theta = \dfrac{1}{3}$

$\sec\theta = -\dfrac{\sqrt{10}}{3}$

■ **Answer:**

$\sin\theta = -\dfrac{5\sqrt{34}}{34} \qquad \cos\theta = -\dfrac{3\sqrt{34}}{34}$

$\tan\theta = \dfrac{5}{3} \qquad \csc\theta = -\dfrac{\sqrt{34}}{5}$

$\sec\theta = -\dfrac{\sqrt{34}}{3} \qquad \cot\theta = \dfrac{3}{5}$

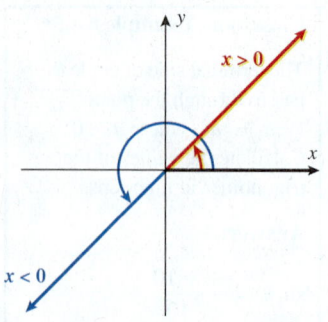

If we say that the terminal side of an angle lies on a line that passes through the origin, we must specify which part of the line represents the terminal ray in order to determine the angle. For example, in the diagram to the left, if we specify $x > 0$, then we know that the terminal side lies in quadrant I. Alternatively, if we specify $x < 0$, then we know that the terminal side lies in quadrant III.

Once we know which part of the line represents the terminal side of the angle, it does not matter which point on the line we use to formulate the trigonometric function values, because corresponding sides of similar triangles are proportional.

Classroom Example 6.4.6*

The terminal side of angle θ lies on the line $y = -mx$, for $m > 0$ and $x > 0$. Calculate the values of the trigonometric functions.

Answer:

$$\sin\theta = -\frac{m\sqrt{m^2+1}}{m^2+1}$$

$$\cos\theta = \frac{\sqrt{m^2+1}}{m^2+1}$$

$$\tan\theta = -m$$

$$\cot\theta = -\frac{1}{m}$$

$$\sec\theta = \sqrt{m^2+1}$$

$$\csc\theta = -\frac{\sqrt{m^2+1}}{m}$$

EXAMPLE 6   **Calculating Trigonometric Function Values for Nonacute Angles**

Calculate the values for the six trigonometric functions of angle θ, given in standard position, if the terminal side of θ lies on the line $y = 3x$, $x \le 0$.

Solution:

STEP 1 Draw the line and label a point on the terminal side.

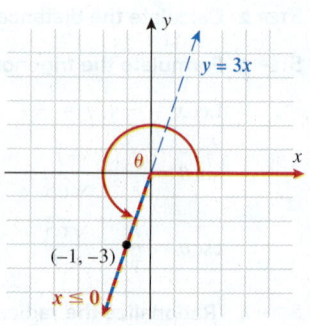

STEP 2 Calculate the distance r.

$$r = \sqrt{(-1)^2 + (-3)^2} = \sqrt{10}$$

STEP 3 Formulate the trigonometric functions in terms of x, y, and r.

Let $x = -1$, $y = -3$, and $r = \sqrt{10}$.

$$\sin\theta = \frac{y}{r} = \frac{-3}{\sqrt{10}} \qquad \cos\theta = \frac{x}{r} = \frac{-1}{\sqrt{10}} \qquad \tan\theta = \frac{y}{x} = \frac{-3}{-1} = 3$$

$$\csc\theta = \frac{r}{y} = \frac{\sqrt{10}}{-3} \qquad \sec\theta = \frac{r}{x} = \frac{\sqrt{10}}{-1} \qquad \cot\theta = \frac{x}{y} = \frac{-1}{-3} = \frac{1}{3}$$

STEP 4 Rationalize the radical denominators in the sine and cosine functions.

$$\sin\theta = \frac{y}{r} = \frac{-3}{\sqrt{10}} \cdot \frac{\sqrt{10}}{\sqrt{10}} = \frac{3\sqrt{10}}{10} \qquad \cos\theta = \frac{x}{r} = \frac{-1}{\sqrt{10}} \cdot \frac{\sqrt{10}}{\sqrt{10}} = -\frac{\sqrt{10}}{10}$$

STEP 5 Write the values of the six trigonometric functions for θ.

$\sin\theta = -\dfrac{3\sqrt{10}}{10}$	$\cos\theta = -\dfrac{\sqrt{10}}{10}$	$\tan\theta = 3$
$\csc\theta = -\dfrac{\sqrt{10}}{3}$	$\sec\theta = -\sqrt{10}$	$\cot\theta = \dfrac{1}{3}$

■ **Answer:**

$$\sin\theta = -\frac{2\sqrt{5}}{5} \qquad \cos\theta = -\frac{\sqrt{5}}{5}$$

$$\tan\theta = 2 \qquad \csc\theta = -\frac{\sqrt{5}}{2}$$

$$\sec\theta = -\sqrt{5} \qquad \cot\theta = \frac{1}{2}$$

■ **YOUR TURN**   Calculate the values for the six trigonometric functions of angle θ, given in standard position, if the terminal side of θ is defined by $y = 2x$, $x \le 0$.

Thus far in Chapter 6 we have avoided evaluating expressions such as $\sin 90°$ because $90°$ is not an acute angle. However, with our second definition of trigonometric functions in the Cartesian plane, we now are able to evaluate the trigonometric functions for quadrantal angles (angles in standard position whose terminal sides coincide with an axis) such as $90°$, $180°$, $270°$, and $360°$. Notice that $90°$ and $270°$ lie along the y-axis and therefore have an x-coordinate value equal to 0. Similarly, $180°$ and $360°$ lie along the x-axis and have a y-coordinate value equal to 0. Some of the trigonometric functions are defined with x- or y-coordinates in the denominator, and since dividing by 0 is undefined in mathematics, not all trigonometric functions are defined for some quadrantal angles.

EXAMPLE 7 Calculating Trigonometric Function Values for Quadrantal Angles

Calculate the values for the six trigonometric functions when $\theta = 90°$.

Solution:

STEP 1 Draw the angle and label a point on the terminal side.

> *Note:* A convenient point on the terminal side is $(0, 1)$.

STEP 2 Calculate the distance r.

$$r = \sqrt{(0)^2 + (1)^2} = \sqrt{1} = 1$$

STEP 3 Formulate the trigonometric functions in terms of x, y, and r.

Let $x = 0$, $y = 1$, and $r = 1$.

$$\sin\theta = \frac{y}{r} = \frac{1}{1} \qquad \cos\theta = \frac{x}{r} = \frac{0}{1} \qquad \tan\theta = \frac{y}{x} = \frac{1}{0}$$

$$\csc\theta = \frac{r}{y} = \frac{1}{1} \qquad \sec\theta = \frac{r}{x} = \frac{1}{0} \qquad \cot\theta = \frac{x}{y} = \frac{0}{1}$$

STEP 4 Write the values of the six trigonometric functions for θ.

$\sin\theta = 1$	$\cos\theta = 0$	$\tan\theta$ is undefined
$\csc\theta = 1$	$\sec\theta$ is undefined	$\cot\theta = 0$

■ **YOUR TURN** Calculate the values for the six trigonometric functions when $\theta = 270°$.

■ **Answer:**
$\sin\theta = -1 \qquad \cos\theta = 0$
$\tan\theta$ is undefined $\quad \csc\theta = -1$
$\sec\theta$ is undefined $\quad \cot\theta = 0$

Technology Tip

```
sin(180)
              0
cos(180)
             -1
tan(180)
              0
```

Use the reciprocal identities to enter $\sec 180°$ as $(\cos 180°)^{-1}$.

```
(cos(180))⁻¹
             -1
```

A TI calculator displays the values of $\csc 180°$ and $\cot 180°$ as

```
(sin(180))⁻¹
(tan(180))⁻¹
■
```

```
ERR:DIVIDE BY 0
1:Quit
2:Goto
```

■ **Answer:**
$\sin\theta = 0 \quad \cos\theta = 1$
$\tan\theta = 0 \quad \csc\theta$ is undefined
$\sec\theta = 1 \quad \cot\theta$ is undefined

EXAMPLE 8 **Calculating Trigonometric Function Values for Quadrantal Angles**

Calculate the values for the six trigonometric functions when $\theta = 180°$.

Solution:

STEP 1 Draw the angle and label a point on the terminal side.

> *Note:* A convenient point on the terminal side is $(-1, 0)$.

STEP 2 Calculate the distance r. $r = \sqrt{(-1)^2 + (0)^2} = \sqrt{1} = 1$

STEP 3 Formulate the trigonometric functions in terms of x, y, and r.

Let $x = -1, y = 0,$ and $r = 1$.

$$\sin\theta = \frac{y}{r} = \frac{0}{1} \qquad \cos\theta = \frac{x}{r} = \frac{-1}{1} \qquad \tan\theta = \frac{y}{x} = \frac{0}{-1}$$

$$\csc\theta = \frac{r}{y} = \frac{1}{0} \qquad \sec\theta = \frac{r}{x} = \frac{1}{-1} \qquad \cot\theta = \frac{x}{y} = \frac{-1}{0}$$

STEP 4 Write the values of the six trigonometric functions for θ.

$\sin\theta = 0$	$\cos\theta = -1$	$\tan\theta = 0$
$\csc\theta$ is undefined	$\sec\theta = -1$	$\cot\theta$ is undefined

■**YOUR TURN** Calculate the values for the six trigonometric functions when $\theta = 360°$.

Study Tip

Note: In Examples 7 and 8, we could have used the definitions of sine, cosine, and tangent along with the reciprocal identities to calculate cosecant, secant, and cotangent. The following table summarizes the trigonometric function values for the quadrantal angles 0°, 90°, 180°, 270°, and 360°.

θ	$\sin\theta$	$\cos\theta$	$\tan\theta$	$\cot\theta$	$\sec\theta$	$\csc\theta$
0°	0	1	0	undefined	1	undefined
90°	1	0	undefined	0	undefined	1
180°	0	-1	0	undefined	-1	undefined
270°	-1	0	undefined	0	undefined	-1
360°	0	1	0	undefined	1	undefined

To confirm these values, evaluate each of these functions with a calculator for the specified angles. Make sure the calculator is set in degree (not radian) mode.

SECTION
6.4 SUMMARY

An angle in the Cartesian plane is in standard position if its initial side lies along the positive x-axis and its vertex is located at the origin. Angles in standard position have terminal sides that lie either in one of the four quadrants or along one of the two axes. The special triangles, 30°-60°-90° and 45°-45°-90°, with hypotenuses having measure 1, were used to develop the coordinates in quadrant I for the special angles, 30°, 45°, and 60°. We then used symmetry to locate similar pairs of coordinates in the other quadrants. All right triangles with one vertex (not the right angle) located at the origin and hypotenuse equal to 1 have

the other nonright angle vertex located along the unit circle ($x^2 + y^2 = 1$). Coterminal angles are angles in standard position that have the same initial side and the same terminal side. Trigonometric functions are defined in the Cartesian plane as ratios of coordinates and distances. Right triangle trigonometric definitions learned in Section 6.2 are consistent with these definitions. We now have the ability to evaluate trigonometric functions for nonacute angles. Trigonometric functions are not always defined for quadrantal angles.

SECTION
6.4 EXERCISES

■ SKILLS

In Exercises 1–22, state in which quadrant or on which axes the angles with given measure in standard position would lie.

1. 89°
2. 91°
3. 145°
4. 175°
5. 310°
6. 355°
7. 270°
8. 180°
9. −540°
10. −450°
11. 210.5°
12. 270.5°
13. 12.34°
14. 100.001°
15. 595°
16. 620°
17. 525°
18. 1085°
19. −905°
20. −640°
21. 1400.0001°
22. −3600°

In Exercises 23–36, sketch each of the following angles with given measure in standard position.

23. 135°
24. 225°
25. −405°
26. −450°
27. −225°
28. −330°
29. 330°
30. −150°
31. 510°
32. −720°
33. 840°
34. −380°
35. −540°
36. 540°

In Exercises 37–42, match the angles (a−f) with the coterminal angles (37–42).

a. 30°   b. −95°   c. 185°   d. −560°   e. 780°   f. 75°

37. −535°
38. −690°
39. 60°
40. 265°
41. −645°
42. 160°

In Exercises 43–52, determine the angle of the smallest possible positive measure that is coterminal with each of the given angles.

43. 412°
44. 379°
45. −92°
46. −187°
47. −390°
48. 945°
49. 510°
50. 1395°
51. 1400.0001°
52. 612.34°

In Exercises 53–72, the terminal side of an angle θ in standard position passes through the indicated point. Calculate the values of the six trigonometric functions for angle θ.

53. (1, 2)
54. (2, 3)
55. (3, 6)
56. (8, 4)
57. $\left(\frac{1}{2}, \frac{2}{5}\right)$
58. $\left(\frac{4}{7}, \frac{2}{3}\right)$
59. (−2, 4)
60. (−1, 3)
61. (−4, −7)
62. (−9, −5)
63. $\left(-\sqrt{2}, \sqrt{3}\right)$
64. $\left(-\sqrt{3}, \sqrt{2}\right)$
65. $\left(-\sqrt{5}, -\sqrt{3}\right)$
66. $\left(-\sqrt{6}, -\sqrt{5}\right)$
67. $\left(-\frac{10}{3}, -\frac{4}{3}\right)$
68. $\left(-\frac{2}{9}, -\frac{1}{3}\right)$
69. (8, −4)
70. (2.1, −4.2)
71. $\left(-\frac{2}{3}, \frac{1}{3}\right)$
72. $\left(-2, \frac{1}{4}\right)$

In Exercises 73–80, calculate the values for the six trigonometric functions of the angle θ given in standard position, if the terminal side of θ lies on each of the indicated lines.

73. $y = 2x,\ x \geq 0$ **74.** $y = 3x,\ x \geq 0$ **75.** $y = \frac{1}{2}x,\ x \geq 0$ **76.** $y = \frac{1}{2}x,\ x \leq 0$

77. $y = -\frac{1}{3}x,\ x \geq 0$ **78.** $y = -\frac{1}{3}x,\ x \leq 0$ **79.** $2x + 3y = 0,\ x \leq 0$ **80.** $2x + 3y = 0,\ x \geq 0$

In Exercises 81–94, calculate (if possible) the values for the six trigonometric functions of the angle θ given in standard position.

81. $\theta = 450°$ **82.** $\theta = 540°$ **83.** $\theta = 630°$ **84.** $\theta = 720°$ **85.** $\theta = -270°$ **86.** $\theta = -180°$

87. $\theta = -90°$ **88.** $\theta = -360°$ **89.** $\theta = -450°$ **90.** $\theta = -540°$ **91.** $\theta = -630°$ **92.** $\theta = -720°$

93. $\theta = 1200°$ **94.** $\theta = 1140°$

▪ APPLICATIONS

95. Clock. What is the measure of the angle swept out by the second hand if it starts on the 3 and continues for 3 minutes and 20 seconds?

96. Clock. What is the measure of the angle swept out by the hour hand if it starts at 3 P.M. on Wednesday and continues until 5 P.M. on Thursday?

For Exercises 97 and 98, refer to the following:

The two-player game of Zim-Zam in which the players swat at a tennis ball tethered to a 4-foot-long string connected atop a 6-foot pole was popular in the mid-80s. One player would be assigned the clockwise direction and the other the counterclockwise direction. A player's goal was to use a sequence of hits to have the ball travel a full 7 revolutions around in the assigned direction. This was complicated by the fact that the person's opponent was attempting to do the same thing in the opposite direction. For each complete revolution the ball made in a given direction, the corresponding player's scoring device would increase by 1; this also had the effect of decreasing the opponent's score by 1. The game was finished once a player reached a score of 7 points.

97. Sports. Suppose the scoring device for the player assigned the clockwise direction goes from 0 to 2 to 1 to 4 to 3 to 7. How many total degrees in the clockwise direction did the ball travel?

98. Sports. If during a game of Zim-Zam the ball traveled 990° counterclockwise in one maneuver, how many points did the player assigned the counterclockwise direction earn?

99. Track. Don and Ron both started running around a circular track, starting at the same point, but Don ran counterclockwise and Ron ran clockwise. The paths they ran swept through angles of 900° and −900°, respectively. Did they end up in the same spot when they finished?

100. Track. Dan and Stan both started running around a circular track, starting at the same point. The paths they ran swept through angles of 3640° and 1890°, respectively. Did they end up in the same spot when they finished?

For Exercises 101 and 102, refer to the following:

A common school locker combination lock is shown. The lock has a dial with forty calibration marks numbered 0 to 39. A combination consists of three of these numbers (e.g., 5-35-20). To open the lock, the following steps are taken:

iStockphoto

- turn the dial clockwise two full turns,
- stop at the first number of the combination,
- turn the dial counterclockwise one full turn,
- continue turning counterclockwise until the second number is reached,
- turn the dial clockwise again until the third number is reached,
- pull the shank and the lock will open.

101. Combination Lock. Given that the initial position of the dial is at zero (shown in the illustration), how many degrees is the dial rotated in total (sum of clockwise and counterclockwise rotations) in opening the lock if the combination is 35-5-20?

102. Combination Lock. Given that the initial position of the dial is at zero (shown in the illustration), how many degrees is the dial rotated in total (sum of clockwise and counterclockwise rotations) in opening the lock if the combination is 20-15-5?

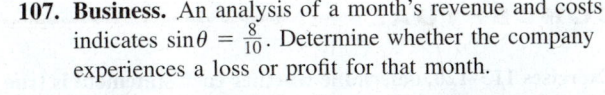

103. Geometry. A right triangle is drawn in QI with one leg on the x-axis and its hypotenuse on the terminal side of angle θ drawn in standard position. If $\sin\theta = \frac{7}{25}$, what is $\tan\theta$?

104. Geometry. A right triangle is drawn in QI with one leg on the x-axis and its hypotenuse on the terminal side of angle θ drawn in standard position. If $\tan\theta = \frac{84}{13}$, what is $\cos\theta$?

For Exercises 105 and 106, refer to the following figure:

105. Forestry. Let θ be the angle of elevation from a point on the ground to the top of a tree. If $\cos\theta = \frac{11}{61}$ and the distance from the point on the ground to the base of the tree is 22 feet, how high is the tree?

106. Forestry. Let θ be the angle of elevation from a point on the ground to the top of a tree. If $\sin\theta = \frac{40}{41}$ and the tree is 20 feet high, how far from the base of the tree is the point on the ground?

For Exercises 107 and 108, refer to the following:

The monthly revenues (measured in thousands of dollars) of PizzaRia are a function of monthly costs (measured in thousands of dollars), that is, $R(c)$. Recall that the angle θ can be interpreted as a measure of the sizes of cost c and revenue R relative to each other, that is,

$$\tan\theta = \frac{\text{revenue}}{\text{costs}}.$$ The larger the angle θ is, the greater revenue is relative to cost, and conversely; the smaller the angle θ is, the smaller revenue is relative to cost.

107. Business. An analysis of a month's revenue and costs indicates $\sin\theta = \frac{8}{10}$. Determine whether the company experiences a loss or profit for that month.

108. Business. An analysis of a month's revenue and costs indicates $\cos\theta = \frac{3}{4}$. Determine whether the company experiences a loss or profit for that month.

For Exercises 109 and 110, refer to the following:

A bunion is a progressive medical disorder of the foot in which the big toe gradually begins to lean toward the second toe. The condition can lead to a misalignment of the metatarsal and phalanges of the big toe when compared to the metatarsal and phalanges of the second toe (see the figure below). The angle θ formed between the metatarsal of the big toe and the metatarsal of the second toe is called the first intermetatarsal angle and is normally less than 9°. In the presence of a bunion, the intermetatarsal angle exceeds 9°, which may be caused by the length of the first metatarsal bone. Exercises 109 and 110 both correspond to the presence of a bunion.

(Source: http://emedicine.medscape.com/article/1235796-media.)

109. Health/Medicine. If $\cos\theta = \dfrac{4.8\,\text{cm}}{5.0\,\text{cm}}$, find the distance between the distal ends of the two metatarsals. Round to the nearest millimeter.

110. Health/Medicine. If $\sec\theta = \dfrac{5.0\,\text{cm}}{4.9\,\text{cm}}$, find the distance between the distal ends of the two metatarsals. Round to the nearest millimeter.

■ CATCH THE MISTAKE

In Exercises 111 and 112, explain the mistake that is made.

111. The terminal side of an angle θ in standard position passes through the point $(1, 2)$. Calculate $\sin\theta$.

Solution:

Label the coordinates.	$x = 1, y = 2$
Calculate r.	$r = 1^2 + 2^2 = 5$
Use the definition of sine.	$\sin\theta = \dfrac{y}{r}$
Substitute $y = 2, r = 5$.	$\sin\theta = \dfrac{2}{5}$

This is incorrect. What mistake was made?

112. Evaluate $\sec 810°$ exactly.

Solution:

Find a coterminal angle that lies between 0° and 360°.
$810° - 360° = 450°$
$450° - 360° = 90°$

810° has the same terminal side as 90°.

Evaluate cosine for 90°. $\cos 90° = 0$

Evaluate secant for 90°. $\sec 90° = \dfrac{1}{\cos 90°} = \dfrac{1}{0} = 0$

This is incorrect. What mistake was made?

▪ CONCEPTUAL

In Exercises 113–120, determine whether each statement is true or false.

113. $\sin 30° + \sin 60° = \sin 90°$

114. $\tan 0° + \tan 90° = \tan 90°$

115. $\cos \theta = \cos(\theta + 360°n)$, where n is an integer.

116. $\sin \theta = \sin(\theta + 360°n)$, where n is an integer.

117. $\cos(90° - \theta) = \sin(\theta - 720°n)$, where n is an integer.

118. $\csc(\theta + 90°n) = \sec(90° - \theta - 90°n)$, where n is an integer.

119. If $\tan \theta = 1$, then $\theta = 45° + 360°n$, for some integer n.

120. If $\cot \theta = -\sqrt{3}$, then $\theta = 150° + 180°n$, for some integer n.

▪ CHALLENGE

121. If the terminal side of angle θ passes through the point $(-3a, 4a)$, find $\cos \theta$.

122. If the terminal side of angle θ passes through the point $(-3a, 4a)$, find $\sin \theta$.

123. If the line $y = mx$ makes an angle θ with the x-axis, which trigonometric function value for θ is equal to m?

124. Find x if $(x, -2)$ is on the terminal side of angle θ and $\csc \theta = -\dfrac{\sqrt{29}}{2}$.

125. Find y if $(-3, y)$ is on the terminal side of angle θ and $\sec \theta = -\dfrac{\sqrt{17}}{3}$.

126. Find the equation of the line with positive slope that passes through the point $(a, 0)$ and makes an acute angle θ with the x-axis. The equation of the line will be in terms of x, a, and a trigonometric function of θ.

127. Find the equation of the line with negative slope that passes through the point $(a, 0)$ and makes an acute angle θ with the x-axis. The equation of the line will be in terms of x, a, and a trigonometric function of θ.

▪ TECHNOLOGY

In Exercises 128–137, use a calculator to evaluate the following expressions. If you get an error, explain why.

128. $\sin 270°$

129. $\cos 270°$

130. $\tan 270°$

131. $\cot 270°$

132. $\sin(-270°)$

133. $\cos(-270°)$

134. $\csc(-270°)$

135. $\sec(-270°)$

136. $\sec 270°$

137. $\csc 270°$

SKILLS OBJECTIVES

- Determine the reference angle of a nonacute angle.
- Evaluate trigonometric functions exactly for common angles.
- Approximate trigonometric functions of nonacute angles.
- Evaluate values of trigonometric functions for quadrantal angles.

CONCEPTUAL OBJECTIVES

- Understand which trigonometric functions are positive or negative in each of the four quadrants.
- Understand that the ranges for the sine and cosine functions are bounded, whereas the ranges for the other four trigonometric functions are unbounded.

In Section 6.4, we defined trigonometric functions in the Cartesian plane as ratios of x, y, and r. We calculated the trigonometric function values given a point on the terminal side of an angle in standard position. We also discussed the special case of a quadrantal angle. In this section, we now evaluate trigonometric functions exactly for common angles. Additionally, we will approximate trigonometric functions for nonacute angles using an acute reference angle and knowledge of algebraic signs of trigonometric functions in particular quadrants.

Algebraic Signs of Trigonometric Functions

In Section 6.4, we defined trigonometric functions as ratios of x, y, and r. Since r is the distance from the origin to the point (x, y) and distance is never negative, r is always taken as the positive solution to $r^2 = x^2 + y^2$, so $r = \sqrt{x^2 + y^2}$.

The x-coordinate is positive in quadrants **I** and **IV** and negative in quadrants **II** and **III**.
The y-coordinate is positive in quadrants **I** and **II** and negative in quadrants **III** and **IV**.
Recall the definition of the six trigonometric functions in the Cartesian plane:

$$\sin\theta = \frac{y}{r} \qquad \cos\theta = \frac{x}{r} \qquad \tan\theta = \frac{y}{x} \quad (x \neq 0)$$

$$\csc\theta = \frac{r}{y} \quad (y \neq 0) \qquad \sec\theta = \frac{r}{x} \quad (x \neq 0) \qquad \cot\theta = \frac{x}{y} \quad (y \neq 0)$$

Therefore, the algebraic sign, $+$ or $-$, of each trigonometric function will depend on which quadrant contains the terminal side of angle θ. Let us look at the three main trigonometric functions: sine, cosine, and tangent. In quadrant I, all three functions are positive since x, y, and r are all positive. However, in quadrant II, only sine is positive since y and r are both positive. In quadrant III, only tangent is positive, and in quadrant IV, only cosine is positive. The phrase "**A**ll **S**tudents **T**ake **C**alculus" helps us remember which of the three main trigonometric functions are positive in each quadrant.

Study Tip

All **S**tudents **T**ake **C**alculus is a phrase that helps us remember which of the three (sine, cosine, tangent) functions are positive in quadrants I, II, III, and IV.

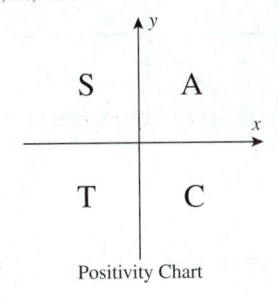

Positivity Chart

PHRASE	QUADRANT	POSITIVE TRIGONOMETRIC FUNCTION
All	I	**A**ll three: sine, cosine, and tangent
Students	II	**S**ine
Take	III	**T**angent
Calculus	IV	**C**osine

The following table indicates the algebraic sign of all six trigonometric functions according to the quadrant in which the terminal side of an angle θ lies. Notice that the reciprocal functions have the same sign.

TERMINAL SIDE OF θ IN QUADRANT	$\sin\theta$	$\cos\theta$	$\tan\theta$	$\cot\theta$	$\sec\theta$	$\csc\theta$
I	+	+	+	+	+	+
II	+	−	−	−	−	+
III	−	−	+	+	−	−
IV	−	+	−	−	+	−

EXAMPLE 1   Evaluating a Trigonometric Function When the Quadrant of the Terminal Side Is Known

If $\cos\theta = -\frac{3}{5}$ and the terminal side of θ lies in quadrant III, find $\sin\theta$.

Solution:

STEP 1 **Draw some angle θ in QIII.**

STEP 2 **Identify known quantities from the information given.**

Recall that $\cos\theta = \dfrac{x}{r}$ $\cos\theta = -\dfrac{3}{5} = \dfrac{-3}{5} = \dfrac{x}{r}$
and $r > 0$.

Identify x and r. $x = -3$ and $r = 5$

STEP 3 **Since x and r are known, find y.**

Substitute $x = -3$ and $r = 5$
into $x^2 + y^2 = r^2$. $(-3)^2 + y^2 = 5^2$

Solve for y. $9 + y^2 = 25$
          $y^2 = 16$
          $y = \pm 4$

STEP 4 **Select the sign of y based on quadrant information.**

Since the terminal side of θ lies in
quadrant III, $y < 0$. $y = -4$

STEP 5 **Find $\sin\theta$.** $\sin\theta = \dfrac{y}{r} = \dfrac{-4}{5}$

$$\sin\theta = -\dfrac{4}{5}$$

■ **YOUR TURN** If $\sin\theta = -\frac{3}{4}$ and the terminal side of θ lies in quadrant III, find $\cos\theta$.

We can also make a table showing the values of the trigonometric functions when the terminal side of θ lies along each of the axes (i.e., when θ is any of the quadrantal angles). We calculated this table for specific angles in Section 6.4, and we will develop it again here using the algebraic signs of the trigonometric functions in the different quadrants.

When the terminal side lies along the x-axis, then $y = 0$. When $y = 0$, notice that $r = \sqrt{x^2 + y^2} = \sqrt{x^2} = |x|$. When the terminal side lies along the positive x-axis, $x > 0$; and when the terminal side lies along the negative x-axis, $x < 0$. Therefore, when the terminal side of the angle lies on the positive x-axis, then $y = 0$, $x > 0$, and $r = x$; and when the terminal side lies along the negative x-axis, then $y = 0$, $x < 0$, and $r = |x|$. A similar argument can be made for the y-axis that results in $r = |y|$.

Terminal Side of θ Lies along the	$\sin\theta$	$\cos\theta$	$\tan\theta$	$\cot\theta$	$\sec\theta$	$\csc\theta$
Positive x-axis ($0°$ or $360°$)	0	1	0	undefined	1	undefined
Positive y-axis ($90°$)	1	0	undefined	0	undefined	1
Negative x-axis ($180°$)	0	-1	0	undefined	-1	undefined
Negative y-axis ($270°$)	-1	0	undefined	0	undefined	-1

EXAMPLE 2 Working with Values of the Trigonometric Functions for Quadrantal Angles

Evaluate the following expressions if possible.

a. $\cos 540° + \sin 270°$ **b.** $\cot 90° + \tan(-90°)$

Solution (a):

The terminal side of an angle with measure $540°$ lies along the negative x-axis.

$$540° - 360° = 180°$$

Evaluate cosine of an angle whose terminal side lies along the negative x-axis.

$$\cos 540° = -1$$

Evaluate sine of an angle whose terminal side lies along the negative y-axis.

$$\sin 270° = -1$$

Sum the cosine and sine terms.

$$\cos 540° + \sin 270° = -1 + (-1)$$

$$\cos 540° + \sin 270° = \boxed{-2}$$

Check: Evaluate this expression with a calculator.

Solution (b):

Evaluate cotangent of an angle whose terminal side lies along the positive y-axis.

$$\cot 90° = 0$$

The terminal side of an angle with measure $-90°$ lies along the negative y-axis.

$$\tan(-90°) = \tan 270°$$

The tangent function is undefined for an angle whose terminal side lies along the negative y-axis.

$$\tan(-90°) \text{ is } \boxed{\text{undefined}}$$

Even though $\cot 90°$ is defined, since $\tan(-90°)$ is undefined, the sum of the two expressions is also undefined.

■ **YOUR TURN** Evaluate the following expressions if possible.

a. $\csc 90° + \sec 180°$ **b.** $\csc(-630°) + \sec(-630°)$

Classroom Example 6.5.2
Evaluate the following, if possible.
a.* $\sin(360°n) + \tan(180°(2n + 1))$, where n is an integer
b. $\cos 1800° + \sec 1800°$
c. $\csc(-270°) + \tan(-270°)$
Answer:
a. 0 **b.** 2 **c.** not possible

■ **Answer: a.** 0 **b.** undefined

Ranges of the Trigonometric Functions

Thus far, we have discussed what the algebraic sign of a trigonometric function value is in a particular quadrant, but we haven't discussed how to find actual values of the trigonometric functions for nonacute angles. We will need to define reference angles and reference right triangles. However, before we proceed, let's get a feel for the ranges (set of values of the functions) we will expect.

Let us start with an angle θ in quadrant I and the sine function defined as the ratio $\sin\theta = \dfrac{y}{r}$.

$$\sin\theta = \frac{y}{r}$$

If we keep the value of r constant, then as the measure of θ increases toward $90°$, y increases. Notice that the value of y approaches the value of r until they are equal when $\theta = 90°$, and y can never be larger than r.

A similar analysis can be conducted in quadrant IV as θ approaches $-90°$ from $0°$ (note that y is negative in quadrant IV). A result that is valid in all four quadrants is $|y| \le r$.

WORDS	**MATH**
Write the absolute value inequality as a double inequality.	$-r \le y \le r$
Divide both sides by r.	$-1 \le \dfrac{y}{r} \le 1$
Let $\sin\theta = \dfrac{y}{r}$.	$-1 \le \sin\theta \le 1$

Similarly, by allowing θ to approach $0°$ and $180°$, we can show that $|x| \le r$, which leads to the range of the cosine function: $-1 \le \cos\theta \le 1$. Sine and cosine values range between

-1 and 1 and since secant and cosecant are reciprocals of the cosine and sine functions, respectively, their ranges are stated as

$$\sec\theta \le -1 \text{ or } \sec\theta \ge 1 \qquad \csc\theta \le -1 \text{ or } \csc\theta \ge 1$$

Since $\tan\theta = \dfrac{y}{x}$ and $\cot\theta = \dfrac{x}{y}$ and since $x < y, x = y,$ and $x > y$ are all possible, the values of the tangent and cotangent functions can be any real numbers (positive, negative, or zero). The following box summarizes the ranges of the trigonometric functions.

RANGES OF THE TRIGONOMETRIC FUNCTIONS

For any angle θ for which the trigonometric functions are defined, the six trigonometric functions have the following ranges:

- $-1 \le \sin\theta \le 1$
- $-1 \le \cos\theta \le 1$
- $\tan\theta$ and $\cot\theta$ can equal any real number.
- $\csc\theta \le -1 \text{ or } \csc\theta \ge 1$
- $\sec\theta \le -1 \text{ or } \sec\theta \ge 1$

EXAMPLE 3   **Determining Whether a Value Is Within the Range of a Trigonometric Function**

Determine whether each statement is possible or not.

a. $\cos\theta = 1.001$ **b.** $\cot\theta = 0$ **c.** $\sec\theta = \dfrac{\sqrt{3}}{2}$

Solution (a): Not possible, because $1.001 > 1$.

Solution (b): Possible, because $\cot 90° = 0$.

Solution (c): Not possible, because $\dfrac{\sqrt{3}}{2} \approx 0.866 < 1$.

■ **YOUR TURN** Determine whether each statement is possible or not.

a. $\sin\theta = -1.1$ **b.** $\tan\theta = 2$ **c.** $\csc\theta = \sqrt{3}$

■ **Answer: a.** not possible
      b. possible
      c. possible

Reference Angles and Reference Right Triangles

Now that we know the trigonometric function ranges and their algebraic signs in each of the four quadrants, we can evaluate the trigonometric functions of nonacute angles. Before we do that, however, we first must discuss *reference angles* and *reference right triangles*.

Every nonquadrantal angle in standard position has a corresponding *reference angle* and *reference right triangle*. We have already calculated the trigonometric function values for quadrantal angles.

DEFINITION   **Reference Angle**

For angle θ, $0° < \theta < 360°$, in standard position whose terminal side lies in one of the four quadrants, there exists a **reference angle** α that is the acute angle with positive measure formed by the terminal side of θ and the *x*-axis.

The reference angle is the positive, acute angle that the terminal side makes with the *x*-axis.

DEFINITION   **Reference Right Triangle**

To form a **reference right triangle** for angle θ, where $0° < \theta < 360°$, drop a perpendicular line from the terminal side of the angle to the *x*-axis. The right triangle now has reference angle α as one of its angles.

EXAMPLE 4 **Finding Reference Angles**

Find the reference angle for each angle given.

a. 210°   **b.** 135°   **c.** 422°

Solution (a):

The terminal side of θ lies in quadrant III.

The reference angle is made by the terminal side and the *x*-axis.

$210° - 180° = \boxed{30°}$

Solution (b):

The terminal side of θ lies in quadrant II.

The reference angle is made by the terminal side and the negative *x*-axis.

$180° - 135° = \boxed{45°}$

Solution (c):

The terminal side of θ lies in quadrant I.

The reference angle is made by the terminal side and the positive *x*-axis.

$422° - 360° = \boxed{62°}$

■ **YOUR TURN** Find the reference angle for each angle given.

a. 160°   **b.** 285°   **c.** 600°

■ **Answer:**
a. 20°   **b.** 75°   **c.** 60°

In Section 6.2, we first defined the trigonometric functions of an acute angle as ratios of lengths of sides of a right triangle. For example, $\sin\theta = \dfrac{\text{opposite}}{\text{hypotenuse}}$. The lengths of the sides of triangles are always positive.

In Section 6.4, we defined the sine function of any angle as $\sin\theta = \dfrac{y}{r}$. For a nonacute angle θ, $\sin\theta = \dfrac{y}{r}$; and for the acute reference angle α, $\sin\alpha = \dfrac{|y|}{r}$. The only difference between these two expressions is the algebraic sign $(+/-)$, since r is always positive and y is positive or negative depending on the quadrant.

Study Tip

To find the trigonometric function values of nonacute angles, first find the trigonometric values of the reference angle and then use the quadrant information to determine the algebraic sign.

Therefore, to calculate the trigonometric function values for a nonacute angle, simply find the trigonometric values for the reference angle and determine the correct algebraic sign according to the quadrant in which the terminal side lies.

Evaluating Trigonometric Functions for Nonacute Angles

Let's look at a specific example before we generalize a procedure for evaluating trigonometric function values for nonacute angles.

Suppose we have the angles in standard position with measure **60°**, **120°**, **240°**, and **300°**. Notice that the reference angle for all of these angles is **60°**.

If we draw reference triangles and let the shortest leg have length **1**, we find that the other leg has length $\sqrt{3}$ and the hypotenuse has length **2**.

Notice that the legs of the triangles have lengths (always positive) 1 and $\sqrt{3}$; however, the coordinates are $(\pm 1, \pm \sqrt{3})$. Therefore, when we calculate the trigonometric functions for any of the angles **60°**, **120°**, **240°**, and **300°**, we can simply calculate the trigonometric functions for the reference angle, **60°**, and determine the algebraic sign (+ or −) for the particular trigonometric function and quadrant.

Study Tip

The value of a trigonometric function of an angle is the same as the trigonometric value of its reference angle, except there may be an algebraic sign (+ or −) difference between the two values.

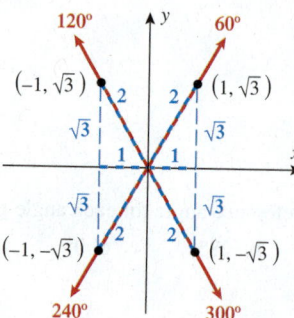

To find the value of $\cos 120°$, we first recognize that the terminal side of an angle with 120° measure lies in quadrant II. We also know that cosine is negative in quadrant II. We then calculate the cosine of the reference angle, **60°**.

$$\cos 60° = \frac{\text{adjacent}}{\text{hypotenuse}} = \frac{1}{2}$$

Since we know $\cos 120°$ is negative because it lies in quadrant II, we know that

$$\cos 120° = -\frac{1}{2}$$

Similarly, we know that $\cos 240° = -\frac{1}{2}$ and $\cos 300° = \frac{1}{2}$.

For any angle whose terminal side lies along one of the axes, we consult the table in this section for the values of the trigonometric functions for quadrantal angles. If the terminal side lies in one of the four quadrants, then the angle is said to be nonquadrantal and the following procedure can be used.

PROCEDURE FOR EVALUATING FUNCTION VALUES FOR ANY NONQUADRANTAL ANGLE θ

Step 1: ▪ If $0° < \theta < 360°$, proceed to Step 2.
  ▪ If $\theta < 0°$, add $360°$ as many times as needed to get a coterminal angle with measure between $0°$ and $360°$.
  ▪ If $\theta > 360°$, subtract $360°$ as many times as needed to get a coterminal angle with measure between $0°$ and $360°$.

Step 2: Find the quadrant in which the terminal side of the angle in Step 1 lies.

Step 3: Find the reference angle α of the coterminal angle found in Step 1.

Step 4: Find the trigonometric function values for the reference angle α.

Step 5: Determine the correct algebraic signs ($+$ or $-$) for the trigonometric function values based on the quadrant identified in Step 2.

Step 6: Combine the trigonometric values found in Step 4 with the algebraic signs in Step 5 to give the trigonometric function values of θ.

It is important to note that the same attention must be given to the difference between evaluating trigonometric functions exactly and approximating with a calculator for non-acute angles as was done with acute angles. We follow the above procedure for all angles except when we get to Step 4. In Step 4, we evaluate exactly if possible ($30°$, $45°$, $60°$) otherwise, we use a calculator to approximate.

EXAMPLE 5   Evaluating the Cosine Function of a Special Angle Exactly

Find the exact value of $\cos 210°$.

Solution:

The terminal side of $\theta = 210°$ lies in quadrant III.

Find the reference angle for $\theta = 210°$.

$$210° - 180° = 30°$$

Find the value of the cosine of the reference angle.

$$\cos 30° = \frac{\sqrt{3}}{2}$$

Determine the algebraic sign for the cosine in quadrant III.

Negative ($-$)

Combine the algebraic sign of the cosine in quadrant III with the value of the cosine of the reference angle.

$$\cos 210° = -\frac{\sqrt{3}}{2}$$

▪ **YOUR TURN** Find the exact value of $\sin 330°$.

Classroom Example 6.5.5
Compute $\cos 225°$ exactly.

Answer: $-\dfrac{\sqrt{2}}{2}$

▪ **Answer:** $-\frac{1}{2}$

Classroom Example 6.5.6
Compute cot 840° exactly.

Answer: $-\dfrac{\sqrt{3}}{3}$

EXAMPLE 6 Evaluating the Tangent Function of a Special Angle Exactly

Find the exact value of tan 495°.

Solution:

Subtract 360° to get a coterminal angle between 0° and 360°.

$$495° - 360° = 135°$$

The terminal side of the angle lies in quadrant II.

Find the reference angle for 135°. $180° - 135° = 45°$

Find the value of the tangent of the reference angle. $\tan 45° = 1$

Determine the algebraic sign for the tangent in quadrant II.   Negative (−)

Combine the algebraic sign of the tangent in quadrant II with the value of the tangent of the reference angle. $\tan 495° = \boxed{-1}$

■ **Answer:** $-\sqrt{3}$

■ **YOUR TURN** Find the exact value of tan 660°.

EXAMPLE 7 Evaluating the Cosecant Function of a Special Angle Exactly

Find the exact value of csc(−210°).

Solution:

Add 360° to get a coterminal angle between 0° and 360°.

$$-210° + 360° = 150°$$

The terminal side of the angle lies in quadrant II.

Classroom Example 6.5.7
Compute sec(−840°) exactly.

Answer: −2

Find the reference angle for 150°.

$$180° - 150° = 30°$$

Find the value of the cosecant of the reference angle.

$$\csc 30° = \frac{1}{\sin 30°} = \frac{1}{\frac{1}{2}} = 2$$

Determine the algebraic sign for the cosecant in quadrant II.

Positive (+)

Combine the algebraic sign of the cosecant in quadrant II with the value of the cosecant of the reference angle.

$\csc(-210°) = \boxed{2}$

■ **YOUR TURN** Find the exact value of $\sec(-330°)$.

■ **Answer:** $\dfrac{2\sqrt{3}}{3}$

EXAMPLE 8   Finding Exact Angle Measures Given Trigonometric Function Values

Find all values of θ where $0° \le \theta \le 360°$, when $\sin\theta = -\dfrac{\sqrt{3}}{2}$.

Classroom Example 6.5.8
Find the values of θ for which $\cos\theta = -\frac{1}{2}$ where $0° \le \theta \le 360°$.

Answer: $120°, 240°$

Solution:

Determine in what quadrants sine is negative.

QIII and QIV

Since the absolute value of $\sin\theta$ is $\dfrac{\sqrt{3}}{2}$, the reference angle is 60°.

$\sin 60° = \dfrac{\sqrt{3}}{2}$

Determine the angles between 180° and 360° in QIII and QIV with reference angle 60°.

Quadrant III:

$180° + 60° = 240°$

Quadrant IV:

$360° - 60° = 300°$

The two angles are $\boxed{240°}$ and $\boxed{300°}$.

■ **YOUR TURN** Find all values of θ where $0° \le \theta \le 360°$, when

$$\cos\theta = -\dfrac{\sqrt{3}}{2}.$$

■ **Answer:** 150° and 210°

We can evaluate trigonometric functions exactly for quadrantal angles and any angle whose reference angle is a special angle (30°, 45°, or 60°). For all other angles, we can approximate the trigonometric function value with a calculator.

In the following examples approximations will be rounded to four decimal places.

Classroom Example 6.5.9
Calculate the following with a calculator.
a. $\cot(-518°)$
b. $\sec 600°$

Answer: a. 2.4751 **b.** -2

EXAMPLE 9   Using a Calculator to Evaluate Trigonometric Values of Angles

Use a calculator to evaluate the following:

a. $\tan 466°$ **b.** $\csc(-313°)$

Solution (a): $\tan 466° \approx \boxed{-3.4874}$

Solution (b): $\sin(-313°) \approx 0.7314$; $\dfrac{1}{\sin(-313°)} \approx \boxed{1.3673}$

Technology Tip

```
tan(466)
        -3.487414444
(sin(-313))⁻¹
         1.367327461
■
```

■ **YOUR TURN** Use a calculator to evaluate the following:

a. $\sec(-118°)$ **b.** $\cot 226°$

■ **Answer: a.** -2.1301
        b. 0.9657

Technology Tip

When $\sin\theta = -0.6293$, use the
$\boxed{\sin^{-1}}$ key and the absolute value to
find the reference angle.

$$\texttt{sin}^{-1}\texttt{(0.6293)}$$
$$\texttt{38.99849667}$$

Classroom Example 6.5.10
Find the value of θ, the smallest
positive angle (rounded to the
nearest degree), if $\cos\theta = 0.716$
and the terminal side of θ lies
in quadrant I.
Answer: 44°

■ **Answer:** 122°

**EXAMPLE 10   Finding Approximate Angle Measures Given
Trigonometric Function Values**

Find the measure of θ (rounded to the nearest degree) if $\sin\theta = -0.6293$ and the terminal
side of θ (in standard position) lies in quadrant III where $0° \leq \theta \leq 360°$.

Solution:

The sine of the reference angle is 0.6293. $\sin\alpha = 0.6293$

Find the reference angle. $\alpha = \sin^{-1}(0.6293) = 38.998°$

Round the reference angle to the
nearest degree. $\alpha = 39°$

Find θ, which lies in quadrant III. $180° + 39° = 219°$

$$\boxed{\theta = 219°}$$

Check with a calculator. $\sin 219° = -0.6293$

■ **YOUR TURN** Find the measure of θ, the smallest positive angle (rounded to the
nearest degree), if $\cos\theta = -0.5299$ and the terminal side of θ (in
standard position) lies in quadrant II.

**SECTION
6.5 SUMMARY**

In this section we identified the algebraic signs of the trigonometric functions in each quadrant. The trigonometric functions
can be evaluated for nonquadrantal angles using reference angles
and knowledge of algebraic signs in each quadrant. We evaluated

the trigonometric functions for quadrantal angles using the
terminal side approach. When reference angles are 30°, 45°, or
60°, nonacute angles can be evaluated exactly; otherwise, we use
a calculator to approximate trigonometric values for any angle.

**SECTION
6.5 EXERCISES**

■ **SKILLS**

In Exercises 1–10, indicate the quadrant in which the terminal side of θ must lie in order for the information to be true.

1. $\cos\theta$ is positive and $\sin\theta$ is negative.

2. $\cos\theta$ is negative and $\sin\theta$ is positive.

3. $\tan\theta$ is negative and $\sin\theta$ is positive.

4. $\tan\theta$ is positive and $\cos\theta$ is negative.

5. $\sec\theta$ and $\csc\theta$ are both positive.

6. $\sec\theta$ and $\csc\theta$ are both negative.

7. $\cot\theta$ and $\cos\theta$ are both positive.

8. $\cot\theta$ and $\sin\theta$ are both negative.

9. $\tan\theta$ is positive and $\sec\theta$ is negative.

10. $\cot\theta$ is negative and $\csc\theta$ is positive.

In Exercises 11–24, find the indicated trigonometric function values.

11. If $\cos\theta = -\dfrac{3}{5}$, and the terminal side of θ lies in quadrant III, find $\sin\theta$.

12. If $\tan\theta = -\dfrac{5}{12}$, and the terminal side of θ lies in quadrant II, find $\cos\theta$.

13. If $\sin\theta = \dfrac{60}{61}$, and the terminal side of θ lies in quadrant II, find $\tan\theta$.

14. If $\cos\theta = \dfrac{40}{41}$, and the terminal side of θ lies in quadrant IV, find $\tan\theta$.

15. If $\tan\theta = \dfrac{84}{13}$, and the terminal side of θ lies in quadrant III, find $\sin\theta$.

16. If $\sin\theta = -\dfrac{7}{25}$, and the terminal side of θ lies in quadrant IV, find $\cos\theta$.

17. If $\sec\theta = -2$, and the terminal side of θ lies in quadrant III, find $\tan\theta$.

18. If $\cot\theta = 1$, and the terminal side of θ lies in quadrant I, find $\sin\theta$.

19. If $\sin\theta = \dfrac{\sqrt{11}}{6}$, and the terminal side of θ lies in quadrant II, find $\tan\theta$.

20. If $\tan\theta = -\dfrac{1}{2}$, and the terminal side of θ lies in quadrant II, find $\cos\theta$.

21. If $\csc\theta = \dfrac{2}{\sqrt{3}}$ and the terminal side of θ lies in quadrant II, find $\cot\theta$.

22. If $\sec\theta = -\dfrac{13}{5}$ and the terminal side of θ lies in quadrant II, find $\csc\theta$.

23. If $\cot\theta = -\sqrt{3}$ and the terminal side of θ lies in quadrant IV, find $\sec\theta$.

24. If $\cot\theta = -\dfrac{13}{84}$ and the terminal side of θ lies in quadrant II, find $\csc\theta$.

In Exercises 25–36, evaluate each expression, if possible:

25. $\cos(-270°) + \sin 450°$

26. $\sin(-270°) + \cos 450°$

27. $\sin 630° + \tan(-540°)$

28. $\cos(-720°) + \tan 720°$

29. $\cos 540° - \sec(-540°)$

30. $\sin(-450°) + \csc 270°$

31. $\csc(-630°) - \cot 630°$

32. $\sec(-540°) + \tan 540°$

33. $\tan 720° + \sec 720°$

34. $\cot 450° - \cos(-450°)$

35. $\csc(3600°) + \sec(-3600°)$

36. $\cot(600°) + \tan(-600°)$

In Exercises 37–46, determine whether each statement is possible or not.

37. $\sin\theta = -0.999$

38. $\cos\theta = 1.0001$

39. $\cos\theta = \dfrac{2\sqrt{6}}{3}$

40. $\sin\theta = \dfrac{\sqrt{2}}{10}$

41. $\tan\theta = 4\sqrt{5}$

42. $\cot\theta = -\dfrac{\sqrt{6}}{7}$

43. $\sec\theta = -\dfrac{4}{\sqrt{7}}$

44. $\csc\theta = \dfrac{\pi}{2}$

45. $\cot\theta = 500$

46. $\sec\theta = 0.9996$

In Exercises 47–58, evaluate the following expressions *exactly*.

47. $\cos 240°$ **48.** $\cos 120°$ **49.** $\sin 300°$ **50.** $\sin 315°$

51. $\tan 210°$ **52.** $\sec 135°$ **53.** $\tan(-315°)$ **54.** $\sec(-330°)$

55. $\csc 330°$ **56.** $\csc(-240°)$ **57.** $\cot(-315°)$ **58.** $\cot(150°)$

In Exercises 59–66, find all possible values of θ, where $0° \leq \theta \leq 360°$.

59. $\cos \theta = \dfrac{\sqrt{3}}{2}$ **60.** $\sin \theta = \dfrac{\sqrt{3}}{2}$ **61.** $\sin \theta = -\dfrac{1}{2}$ **62.** $\cos \theta = -\dfrac{1}{2}$

63. $\cos \theta = 0$ **64.** $\sin \theta = 0$ **65.** $\sin \theta = -1$ **66.** $\cos \theta = -1$

In Exercises 67–76, evaluate the trigonometric expression with a calculator. **Round your result to four decimal places.**

67. $\sin 237°$ **68.** $\cos 317°$ **69.** $\tan(-265°)$ **70.** $\tan 622°$

71. $\sec 421°$ **72.** $\sec(-222°)$ **73.** $\csc(-111°)$ **74.** $\csc 211°$

75. $\cot 159°$ **76.** $\cot(-82°)$

In Exercises 77–90, find the smallest positive measure of θ (rounded to the nearest degree) if the indicated information is true.

77. $\sin \theta = 0.9397$ and the terminal side of θ lies in quadrant II.

78. $\cos \theta = 0.7071$ and the terminal side of θ lies in quadrant IV.

79. $\cos \theta = -0.7986$ and the terminal side of θ lies in quadrant II.

80. $\sin \theta = -0.1746$ and the terminal side of θ lies in quadrant III.

81. $\tan \theta = -0.7813$ and the terminal side of θ lies in quadrant IV.

82. $\cos \theta = -0.3420$ and the terminal side of θ lies in quadrant III.

83. $\tan \theta = -0.8391$ and the terminal side of θ lies in quadrant II.

84. $\tan \theta = 11.4301$ and the terminal side of θ lies in quadrant III.

85. $\sin \theta = -0.3420$ and the terminal side of θ lies in quadrant IV.

86. $\sin \theta = -0.4226$ and the terminal side of θ lies in quadrant III.

87. $\sec \theta = 1.0001$ and the terminal side of θ lies in quadrant I.

88. $\sec \theta = -3.1421$ and the terminal side of θ lies in quadrant II.

89. $\csc \theta = -2.3604$ and the terminal side of θ lies in quadrant IV.

90. $\csc \theta = -1.0001$ and the terminal side of θ lies in quadrant III.

■ APPLICATIONS

In Exercises 91–94, refer to the following:

When light passes from one substance to another, such as from air to water, its path bends. This is called refraction and is what is seen in eyeglass lenses, camera lenses, and gems. The rule governing the change in the path is called *Snell's law*, named after a Dutch astronomer: $n_1 \sin \theta_1 = n_2 \sin \theta_2$, where the n_1 and n_2 are the indices of refraction of the different substances and the θ_1 and θ_2 are the respective angles that light makes with a line perpendicular to the surface at the boundary between substances. The figure shows the path of light rays going from air to water. Assume that the index of refraction in air is 1.

91. If light rays hit the water's surface at an angle of 30° from the perpendicular and are refracted to an angle of 22° from the perpendicular, then what is the refraction index for water?

92. If light rays hit a glass surface at an angle of 30° from the perpendicular and are refracted to an angle of 18° from the perpendicular, then what is the refraction index for that glass?

93. If the refraction index for a diamond is 2.4, then to what angle is light refracted if it enters the diamond at an angle of 30°?

94. If the refraction index for a rhinestone is 1.9, then to what angle is light refracted if it enters the rhinestone at an angle of 30°?

For Exercises 95 and 96, refer to the following:

An orthotic knee brace can be used to treat knee injuries by locking the knee at an angle θ chosen to facilitate healing. The angle θ is measured from the metal bar on the side of the brace on the thigh to the metal bar on the side of the brace on the calf (see the figure). To make working with the brace more convenient, rotate the image such that the thigh aligns with the positive x-axis (see the figure on the right below).

95. Health/Medicine. If $\theta = 165°$, find the measure of the reference angle. What is the physical meaning of the reference angle?

96. Health/Medicine. If $\theta = 160°$, find the measure of the reference angle. What would an angle greater than 180° represent?

For Exercises 97 and 98, refer to the following:

Water covers two-thirds of the Earth's surface and every living thing is dependent on it. For example, the human body is made up of over 70% water. The water molecule is composed of one oxygen atom and two hydrogen atoms and exhibits a bent shape with the oxygen molecule at the center. The angle θ between the O-H bonds is 104.5°.

(*Source*: http://www.wiley.com/college/boyer/0470003790/
reviews/pH/ph_water.htm.)

97. Chemistry. Sketch the water molecule in the xy-coordinate system in a convenient manner for illustrating angles. Find the reference angle. Illustrate both the angle θ and the reference angle on the sketch.

98. Chemistry. Find $\cos(104.5°)$ using the reference angle found in Exercise 97.

■ CATCH THE MISTAKE

In Exercises 99 and 100, explain the mistake that is made.

99. Evaluate the expression $\sec 120°$ exactly.

Solution:

120° lies in quadrant II.
The reference angle is 30°.

Find the cosine of the reference angle.

$$\cos 30° = \frac{\sqrt{3}}{2}$$

Cosine is negative in quadrant II.

$$\cos 120° = -\frac{\sqrt{3}}{2}$$

Secant is the reciprocal of cosine.

$$\sec 120° = -\frac{2}{\sqrt{3}} = -\frac{2\sqrt{3}}{3}$$

This is incorrect. What mistake was made?

100. Find the measure of the smallest positive angle θ (rounded to the nearest degree) if $\cos\theta = -0.2388$ and the terminal side of θ (in standard position) lies in quadrant III.

Solution:

Evaluate with a calculator.

$$\theta = \cos^{-1}(-0.2388) = 103.8157°$$

Approximate to the nearest degree.

$$\theta \approx 104°$$

This is incorrect. What mistake was made?

■**CONCEPTUAL**

In Exercises 101–106, determine whether each statement is true or false.

101. It is possible for all six trigonometric functions of the same angle to have positive values.

102. It is possible for all six trigonometric functions of the same angle to have negative values.

103. The trigonometric function value for any angle with negative measure must be negative.

104. The trigonometric function value for any angle with positive measure must be positive.

105. $\sec^2\theta - 1$ can be negative for some value of θ.

106. $(\sec\theta)(\csc\theta)$ is negative only when the terminal side of θ lies in quadrant II or IV.

■**CHALLENGE**

107. If $\tan\theta = \dfrac{a}{b}$, where a and b are positive, and if θ lies in quadrant III, find $\sin\theta$.

108. If $\tan\theta = -\dfrac{a}{b}$, where a and b are positive, and if θ lies in quadrant II, find $\cos\theta$.

109. If $\csc\theta = -\dfrac{a}{b}$, where a and b are positive, and if θ lies in quadrant IV, find $\cot\theta$.

110. If $\sec\theta = -\dfrac{a}{b}$, where a and b are positive, and if θ lies in quadrant III, find $\tan\theta$.

■**TECHNOLOGY**

111. Use a calculator to evaluate $\sin 80°$ and $\sin 100°$. Now use the calculator to evaluate $\sin^{-1}(0.9848)$. When sine is positive, in which of the quadrants, I or II, does the calculator assume the terminal side of the angle lies?

112. Use a calculator to evaluate $\sin 260°$ and $\sin 280°$. Now use the calculator to evaluate $\sin^{-1}(-0.9848)$. When sine is negative, in which of the quadrants, III or IV, does the calculator assume the terminal side of the angle lies?

113. Use a calculator to evaluate $\cos 75°$ and $\cos(-75°)$. Now use the calculator to evaluate $\cos^{-1}(0.2588)$. When cosine is positive, in which of the quadrants, I or IV, does the calculator assume the terminal side of the angle lies?

114. Use a calculator to evaluate $\cos 105°$ and $\cos 255°$. Now use the calculator to evaluate $\cos^{-1}(-0.2588)$. When cosine is negative, in which of the quadrants, II or III, does the calculator assume the terminal side of the angle lies?

115. Use a calculator to evaluate $\tan 65°$ and $\tan 245°$. Now use the calculator to evaluate $\tan^{-1}(2.1445)$. When tangent is positive, in which of the quadrants, I or III, does the calculator assume the terminal side of the angle lies?

116. Use a calculator to evaluate $\tan 136°$ and $\tan 316°$. Now use the calculator to evaluate $\tan^{-1}(-0.9657)$. When tangent is negative, in which of the quadrants, II or IV, does the calculator assume the terminal side of the angle lies?

SKILLS OBJECTIVES

- Calculate the radian measure of an angle.
- Convert between degrees and radians.
- Calculate the length of an arc along a circle.
- Calculate the area of a circular sector.
- Calculate linear speed.
- Calculate angular speed.

CONCEPTUAL OBJECTIVES

- Understand that degrees and radians are both units for measuring angles.
- Realize that radian measure allows us to write trigonometric functions as functions of real numbers.
- Understand why the angle in the arc length formula must be in radian measure.
- Understand the relationship between angular speed and linear speed.

The Radian Measure of an Angle

In geometry and most everyday applications, angles are measured in degrees. However, *radian measure* is another way to measure angles. Using radian measure allows us to write trigonometric functions as functions not only of angles but also of real numbers in general.

Recall that in Section 6.1 we defined one full rotation as an angle having measure 360°. Now we think of the angle in the context of a circle. A **central angle** is an angle that has its vertex at the center of a circle.

When the intercepted arc's length is equal to the radius, the measure of the central angle is 1 **radian**. From geometry we know that the ratio of the measures of two angles is equal to the ratio of the lengths of the arcs subtended by those angles (along the same circle).

$$\frac{\theta_1}{\theta_2} = \frac{s_1}{s_2}$$

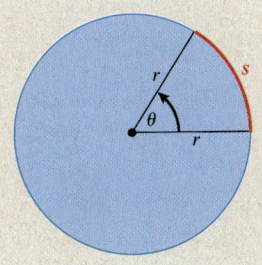

If $\theta_2 = 1$ radian, then the length of the subtended arc is equal to the radius, $s_2 = r$. This leads to $\theta_1 = \dfrac{s_1}{r}$, which is the general definition of *radian measure*.

DEFINITION   **Radian Measure**

If a central angle θ in a circle with radius r intercepts an arc on the circle of length s, then the measure of θ, in **radians**, is given by

$$\theta \,(\text{in radians}) = \frac{s}{r}$$

Note: The formula is valid only if s (arc length) and r (radius) are expressed in the same units.

▼ CAUTION

To correctly calculate radians from the formula $\theta = s/r$, the radius and arc length must be expressed in the same units.

Note that both s and r are measured in units of length. When both are given in the same units, the units cancel, giving the number of radians as a *dimensionless* (unitless) real number. One full rotation corresponds to an arc length equal to the circumference $2\pi r$ of the circle with radius r. We see then that one full rotation is equal to 2π radians.

$$\theta_{\text{full rotation}} = \frac{2\pi r}{r} = 2\pi$$

Classroom Example 6.6.1
Find the measure, in radians, of the central angle θ that intercepts an arc of length 3 yards on a circle of radius 6 yards.
Answer: $\frac{1}{2}$ radian

■ **Answer:** 0.06 rad

EXAMPLE 1   Finding the Radian Measure of an Angle

What is the measure (in radians) of a central angle θ that intercepts an arc of length 4 feet on a circle with radius 10 feet?

Solution:

Write the formula relating radian measure to arc length and radius. $\theta = \dfrac{s}{r}$

Let $s = 4$ ft and $r = 10$ ft. $\theta = \dfrac{4 \text{ ft}}{10 \text{ ft}} = \boxed{0.4 \text{ rad}}$

■ **YOUR TURN** What is the measure (in radians) of a central angle θ that intercepts an arc of length 3 inches on a circle with radius 50 inches?

▶ EXAMPLE 2   Finding the Radian Measure of an Angle

What is the measure (in radians) of a central angle θ that intercepts an arc of length 6 centimeters on a circle with radius 2 meters?

COMMON MISTAKE

A common mistake is to forget to first put the radius and arc length in the same units.

Classroom Example 6.6.2
Find the measure, in radians, of the central angle θ that intercepts an arc of length 3 yards on a circle of radius 6 feet.
Answer: $\frac{3}{2}$ radians

▼ **CAUTION**

Units for arc length and radius must be the same to use $\theta = \dfrac{s}{r}$.

⭐ CORRECT	❌ INCORRECT
Write the formula relating radian measure to arc length and radius.	
θ (in radians) $= \dfrac{s}{r}$	
Substitute $s = 6$ cm and $r = 2$ m into the radian expression.	Substitute $s = 6$ cm and $r = 2$ m into the radian expression.
$\theta = \dfrac{6 \text{ cm}}{2 \text{ m}}$	$\theta = \dfrac{6 \text{ cm}}{2 \text{ m}}$
Convert the radius (2) meters to centimeters: 2 m = 200 cm.	Simplify. $\theta = 3$ rad
$\theta = \dfrac{6 \text{ cm}}{200 \text{ cm}}$	**ERROR:** Did not convert numerator and denominator to the same units.
The units, cm, cancel and the result is a unitless real number.	
$\theta = 0.03$ rad	

■ **Answer:** 0.3 rad

■ **YOUR TURN** What is the measure (in radians) of a central angle θ that intercepts an arc of length 12 mm on a circle with radius 4 cm?

In the above example the units, cm, canceled, therefore correctly giving *radians* as a unitless real number. Because radians are unitless, the word radians (or rad) is often omitted. If an angle measure is given simply as a real number, then radians are implied.

WORDS	MATH
The measure of θ is 4 degrees.	$\theta = 4°$
The measure of θ is 4 radians.	$\theta = 4$

Converting Between Degrees and Radians

In order to convert between degrees and radians, we must first look for a relationship between them. We start by considering one full rotation around the circle. An angle corresponding to one full rotation is said to have measure 360°. Radians are defined as the ratio of the arc length that the angle intercepts on the circle to the radius of the circle. One full rotation corresponds to an arc length equal to the circumference of the circle.

WORDS	MATH
Write the angle measure (in degrees) that corresponds to one full rotation.	$\boldsymbol{\theta = 360°}$
Write the angle measure (in radians) that corresponds to one full rotation.	
Arc length is the circumference of the circle.	$s = 2\pi r$
Substitute $s = 2\pi r$ into θ (in radians) $= \dfrac{s}{r}$.	$\boldsymbol{\theta = \dfrac{2\pi r}{r} = 2\pi}$ **radians**
Equate the measures corresponding to one full rotation.	$\boldsymbol{360° = 2\pi}$ **radians**
Divide by 2.	$180° = \pi$ radians
Divide by 180° or π.	$1 = \dfrac{\pi}{180°}$ or $1 = \dfrac{180°}{\pi}$

This leads us to formulas that convert between degrees and radians. Let θ_d represent an angle measure given in degrees and θ_r represent the corresponding angle measure given in radians.

CONVERTING DEGREES TO RADIANS

To convert degrees to radians, multiply the degree measure by $\dfrac{\pi}{180°}$.

$$\theta_r = \theta_d\left(\dfrac{\pi}{180°}\right)$$

CONVERTING RADIANS TO DEGREES

To convert radians to degrees, multiply the radian measure by $\dfrac{180°}{\pi}$.

$$\theta_d = \theta_r\left(\dfrac{180°}{\pi}\right)$$

Before we begin converting between degrees and radians, let's first get a feel for radians. How many degrees is 1 radian? Use the technique outlined in the box above.

WORDS	MATH
Multiply 1 radian by $\dfrac{180°}{\pi}$.	$1\left(\dfrac{180°}{\pi}\right)$
Approximate π by 3.14.	$1\left(\dfrac{180°}{3.14}\right)$
Use a calculator to evaluate and round to the nearest degree.	$\approx 57°$

$$\boxed{1\,\text{rad} \approx 57°}$$

A radian is much larger than a degree (almost 57 times larger). Let's compare two angles, one measuring 30 radians and the other measuring 30°. Note that $\dfrac{30\ \text{rad}}{2\pi\ \text{rad/rev}} \approx 4.77$ revolutions, whereas $30° = \frac{1}{12}$ revolution.

EXAMPLE 3 Converting Degrees to Radians

Convert 45° to radians.

Solution:

Multiply 45° by $\dfrac{\pi}{180°}$.	$(45°)\left(\dfrac{\pi}{180°}\right) = \dfrac{45°\pi}{180°}$
Simplify.	$= \dfrac{\pi}{4}$ radians

Note: $\dfrac{\pi}{4}$ is the exact value. A calculator can be used to approximate this expression.

Scientific and graphing calculators have a π button (on most scientific calculators it requires using a shift or second command). The decimal approximation rounded to three decimal places is 0.785.

Exact value:	$\dfrac{\pi}{4}$
Approximate value:	0.785

Classroom Example 6.6.3
Convert 135° to radians.

Answer: $\dfrac{3\pi}{4}$

■ **Answer:** $\dfrac{\pi}{3}$ or approximately 1.047

■ **YOUR TURN** Convert 60° to radians.

EXAMPLE 4   Converting Degrees to Radians

Convert $472°$ to radians.

Solution:

Multiply $472°$ by $\dfrac{\pi}{180°}$.

$$472°\left(\dfrac{\pi}{180°}\right)$$

Simplify (factor out the common 4).

$$= \boxed{\dfrac{118}{45}\pi}$$

Approximate with a calculator.

$$\boxed{\approx 8.238}$$

■ **YOUR TURN** Convert $460°$ to radians.

EXAMPLE 5   Converting Radians to Degrees

Convert $\dfrac{2\pi}{3}$ to degrees.

Solution:

Multiply $\dfrac{2\pi}{3}$ by $\dfrac{180°}{\pi}$.

$$\dfrac{2\pi}{3} \cdot \dfrac{180°}{\pi}$$

Simplify.

$$= \boxed{120°}$$

■ **YOUR TURN** Convert $\dfrac{3\pi}{2}$ to degrees.

Since $\pi = 180°$ we know the following special angles

$$\dfrac{\pi}{2} = 90° \qquad \dfrac{\pi}{3} = 60° \qquad \dfrac{\pi}{4} = 45° \qquad \dfrac{\pi}{6} = 30°$$

and we can now draw the unit circle with the special angles in both **degrees** and **radians**.

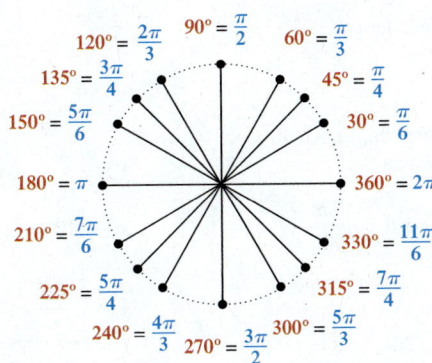

We now look at applications of radian measure that involve calculating *arc lengths*, *areas of circular sectors*, and *angular and linear speeds*. All of these applications are related to the definition of radian measure.

Arc Length

From geometry we know that the length of an arc of a circle is proportional to its central angle. In this section we have learned that for the special case when the arc is equal to the circumference of the circle, the angle measure in radians corresponding to one full rotation is 2π. Let us now assume that we are given the central angle and we want to find the arc length.

WORDS	MATH
Write the definition of radian measure.	$\theta = \dfrac{s}{r}$
Multiply both sides of the equation by r.	$r \cdot \theta = \dfrac{s}{r} \cdot r$
Simplify.	$r\theta = s$

The formula $s = r\theta$ is only true when θ is in radians. To develop a formula when θ is in degrees, we multiply θ by $\dfrac{\pi}{180°}$.

Study Tip

To use the relationship
$$s = r\theta$$
the angle θ must be in radians.

DEFINITION   **Arc Length**

If a central angle θ in a circle with radius r intercepts an arc on the circle of length s, then the **arc length** s is given by

$$s = r\theta_r \qquad\qquad \theta_r \text{ is in radians}$$

or

$$s = r\theta_d\left(\frac{\pi}{180°}\right) \qquad \theta_d \text{ is in degrees}$$

EXAMPLE 6   **Finding Arc Length When the Angle Has Radian Measure**

In a circle with radius 10 centimeters, an arc is intercepted by a central angle with measure $\dfrac{7\pi}{4}$. Find the arc length.

Solution:

Write the formula for arc length when the angle has radian measure.

$$s = r\,\theta_r$$

Substitute $r = 10\,\text{cm}$ and $\theta_r = \dfrac{7\pi}{4}$.

$$s = (10\,\text{cm})\left(\frac{7\pi}{4}\right)$$

Simplify.

$$s = \boxed{\frac{35\pi}{2}}\,\text{cm}$$

Classroom Example 6.6.6
Find the arc length of a sector determined by central angle $\dfrac{11\pi}{6}$ on a circle with radius 24 meters.

Answer: 44π m

■ **Answer:** 5π in.

■ **YOUR TURN** In a circle with radius 15 inches, an arc is intercepted by a central angle with measure $\dfrac{\pi}{3}$. Find the arc length.

EXAMPLE 7 **Finding Arc Length When the Angle Has Degree Measure**

The International Space Station (ISS) is in an approximately circular orbit 400 km above the surface of the Earth. If the ground station tracks the space station when it is within a 45° central angle of this circular orbit above the tracking antenna, how many kilometers does the ISS cover while it is being tracked by the ground station? Assume that the radius of the Earth is 6400 kilometers. Round to the nearest kilometer.

Technology Tip

Use the TI to evaluate the expression for s.

$$s = (6800 \text{ km})(45°)\left(\frac{\pi}{180°}\right)$$

Press 2nd ∧ for π. Type
6800 x 45 x 2nd ∧ ÷
180 ENTER .

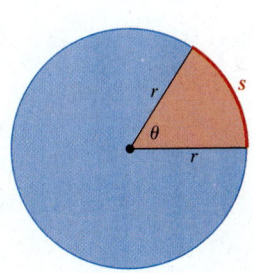

Solution:

Write the formula for arc length when the angle has degree measure.

$$s = r\theta_d\left(\frac{\pi}{180°}\right)$$

Substitute $r = 6400 + 400 = 6800$ km and $\theta_d = 45°$.

$$s = (6800 \text{ km})(45°)\left(\frac{\pi}{180°}\right)$$

Evaluate with a calculator.

$$s \approx 5340.708 \text{ km}$$

Round to the nearest kilometer.

$$s \approx \boxed{5341 \text{ km}}$$

The ISS travels approximately 5341 km during the ground station tracking.

▪ **YOUR TURN** If the ground station in Example 7 could track the ISS within a 60° central angle, how far would the ISS travel during the tracking?

▪ **Answer:** 7121 km

Area of a Circular Sector

A restaurant lists a piece of French Silk Pie as having 400 calories. How does the chef arrive at that number? She calculates the calories of all the ingredients that went into making the entire pie and then divides by the number of slices the pie yields. For example, if an entire pie has 3200 calories and it is sliced into 8 equal pieces, then each piece has 400 calories. Although this example involves volume, the idea is the same with areas of sectors of circles. *Circular sectors* can be thought of as a "piece of a pie."

Recall that arc lengths of a circle are proportional to the central angle (in radians) and the radius. Similarly, a circular sector is a portion of the entire circle. Let A represent the area of the *sector of the circle* and θ_r represent the central angle (in radians) that forms the sector. Then let us consider the entire circle whose area is πr^2 and the angle that represents one full rotation has measure 2π (radians).

WORDS	MATH
Write the ratio of the area of the sector to the area of the entire circle.	$\dfrac{A}{\pi r^2}$
Write the ratio of the central angle θ_r to the measure of one full rotation.	$\dfrac{\theta_r}{2\pi}$
The ratios must be equal (proportionality of sector to circle).	$\dfrac{A}{\pi r^2} = \dfrac{\theta_r}{2\pi}$
Multiply both sides of the equation by πr^2.	$\pi r^2 \cdot \dfrac{A}{\pi r^2} = \dfrac{\theta_r}{2\pi} \cdot \pi r^2$
Simplify.	$A = \dfrac{1}{2}r^2\theta_r$

Study Tip

To use the relationship

$$A = \frac{1}{2}r^2\theta$$

the angle θ must be in radians.

DEFINITION   **Area of a Circular Sector**

The **area of a sector of a circle** with radius r and central angle θ is given by

$$A = \frac{1}{2}r^2\theta_r \qquad\qquad \theta_r \text{ is in radians}$$

or

$$A = \frac{1}{2}r^2\theta_d\left(\frac{\pi}{180°}\right) \qquad \theta_d \text{ is in degrees}$$

Classroom Example 6.6.8
Find the exact area of the sector with diameter 16 ft and central angle $\dfrac{7\pi}{8}$.

Answer: 28π ft^2

EXAMPLE 8   **Finding the Area of a Sector When the Angle Has Radian Measure**

Find the area of the sector associated with a single slice of pizza if the entire pizza has a 14-inch diameter and the pizza is cut into eight equal pieces.

Solution:

The radius is half the diameter.	$r = \dfrac{14}{2} = 7$ in.
Find the angle of each slice if a pizza is cut into eight pieces.	$\theta_r = \dfrac{2\pi}{8} = \dfrac{\pi}{4}$
Write the formula for circular sector area in radians.	$A = \dfrac{1}{2}r^2\theta_r$
Substitute $r = 7$ inches and $\theta_r = \dfrac{\pi}{4}$ into the area equation.	$A = \dfrac{1}{2}(7 \text{ in.})^2\left(\dfrac{\pi}{4}\right)$
Simplify.	$\boxed{A = \dfrac{49\pi}{8} \text{ in.}^2}$
Approximate the area with a calculator (round to the nearest square inch).	$\boxed{A \approx 19 \text{ in.}^2}$

■ **Answer:** 8π in.$^2 \approx 25$ in.^{2}

■ **YOUR TURN** Find the area of a slice of pizza if the entire pizza has a 16-inch diameter. Assume the pizza is cut into eight equal pieces.

EXAMPLE 9 **Finding the Area of a Sector When the Angle Has Degree Measure**

Classroom Example 6.6.9
Find the area of the sector with diameter 1.4 in. and central angle 225°.

Answer: $\dfrac{49\pi}{160}$ in.^{2}

Sprinkler heads come in all different sizes depending on the angle of rotation desired. If a sprinkler head rotates 90° and has enough pressure to keep a constant 25-foot spray, what is the area of the sector of the lawn that gets watered? Round to the nearest square foot.

Solution:

Write the formula for circular sector area in degrees.

$$A = \frac{1}{2}r^2\theta_d\left(\frac{\pi}{180°}\right)$$

Substitute $r = 25\,\text{ft}$ and $\theta_d = 90°$ into the area equation.

$$A = \frac{1}{2}(25\,\text{ft})^2(90°)\left(\frac{\pi}{180°}\right)$$

Simplify.

$$A = \left(\frac{625\pi}{4}\right)\text{ft}^2 \approx 490.87\,\text{ft}^2$$

Round to the nearest square foot.

$$A \approx \boxed{491\,\text{ft}^2}$$

■ Answer: $450\pi\,\text{ft}^2 \approx 1414\,\text{ft}^2$

■ YOUR TURN If a sprinkler head rotates 180° and has enough pressure to keep a constant 30-foot spray, what is the area of the sector of the lawn it can water? Round to the nearest square foot.

Linear Speed

It is important to note that although **velocity** and **speed** are often used as synonyms, speed is how fast you are traveling, whereas velocity is the speed you are traveling *and* the direction you are traveling. In physics the difference between speed and velocity is that velocity has direction and is written as a vector (Chapter 8), and speed is the *magnitude* of the velocity vector which results in a real number. In this chapter, we will only deal with *speed*.

Recall the relationship between distance, rate (assumed to be constant), and time: $d = rt$. Rate is speed, and in words this formula can be rewritten as

$$\text{distance} = \text{speed} \cdot \text{time} \quad \text{or} \quad \text{speed} = \frac{\text{distance}}{\text{time}}$$

It is important to note that we assume speed is constant. If we think of a car driving around a circular track, the distance it travels is the arc length s; and if we let v represent speed and t represent time, we have the formula for speed around a circle (*linear speed*):

$$v = \frac{s}{t}$$

DEFINITION **Linear Speed**

If a point P moves along the circumference of a circle at a constant speed, then the **linear speed** v is given by

$$v = \frac{s}{t}$$

where s is the arc length and t is the time.

EXAMPLE 10   Linear Speed

A car travels at a constant speed around a circular track with circumference equal to 2 miles. If the car records a time of 15 minutes for 9 laps, what is the linear speed of the car in miles per hour?

Solution:

Calculate the distance traveled around the circular track.

$$s = (9 \text{ laps})\left(\frac{2 \text{ miles}}{\text{lap}}\right) = 18 \text{ miles}$$

Substitute $t = 15$ minutes and $s = 18$ miles into $v = \dfrac{s}{t}$.

$$v = \frac{18 \text{ miles}}{15 \text{ min}}$$

Convert the linear speed from miles per minute to miles per hour.

$$v = \left(\frac{18 \text{ miles}}{15 \text{ min}}\right)\left(\frac{60 \text{ min}}{1 \text{ hour}}\right)$$

Simplify.

$$v = \boxed{72 \text{ miles per hour}}$$

■ **YOUR TURN** A car travels at a constant speed around a circular track with circumference equal to 3 miles. If the car records a time of 12 minutes for 7 laps, what is the linear speed of the car in miles per hour?

Angular Speed

To calculate linear speed, we find how fast a position along the circumference of a circle is changing. To calculate **angular speed**, we find how fast the central angle is changing.

> **DEFINITION**   **Angular Speed**
>
> If a point P moves along the circumference of a circle at a constant speed, then the central angle θ that is formed with the terminal side passing through point P also changes over some time t at a constant speed. The **angular speed** ω (omega) is given by
>
> $$\omega = \frac{\theta}{t} \qquad \text{where } \theta \text{ is given in radians}$$

EXAMPLE 11   Angular Speed

A lighthouse in the middle of a channel rotates its light in a circular motion with constant speed. If the beacon of light completes 1 rotation every 10 seconds, what is the angular speed of the beacon in radians per minute?

Solution:

Calculate the angle measure in radians associated with 1 rotation.

$$\theta = 2\pi$$

Substitute $\theta = 2\pi$ and $t = 10$ seconds into $\omega = \dfrac{\theta}{t}$.

$$\omega = \frac{2\pi \,(\text{rad})}{10 \text{ sec}}$$

Convert the angular speed from radians per second to radians per minute.

$$\omega = \frac{2\pi \,(\text{rad})}{10 \text{ sec}} \cdot \frac{60 \text{ sec}}{1 \text{ min}}$$

Simplify.

$$\omega = \boxed{12\pi \text{ rad/min}}$$

■ **Answer:** $\omega = 3\pi$ rad/min

■ **YOUR TURN** If the lighthouse in Example 11 is adjusted so that the beacon rotates 1 time every 40 seconds, what is the angular speed of the beacon in radians per minute?

Relationship Between Linear and Angular Speeds

Angular speed and *linear speed* are related through the *radius*.

WORDS	MATH
Write the definition of radian measure.	$\theta = \dfrac{s}{r}$
Write the definition of arc length (θ in radians).	$s = r\theta$
Divide both sides by t.	$\dfrac{s}{t} = \dfrac{r\theta}{t}$
Rewrite the right side of the equation.	$\dfrac{s}{t} = r\dfrac{\theta}{t}$
Recall the definitions of **linear** and **angular** speeds.	$v = \dfrac{s}{t}$ and $\omega = \dfrac{\theta}{t}$
Substitute $v = \dfrac{s}{t}$ and $\omega = \dfrac{\theta}{t}$ into $\dfrac{s}{t} = r\dfrac{\theta}{t}$.	$v = r\omega$

RELATING LINEAR AND ANGULAR SPEEDS

If a point P moves at a constant speed along the circumference of a circle with radius r, then the **linear speed** v and the **angular speed** ω are related by

$$v = r\omega \qquad \text{or} \qquad \omega = \frac{v}{r}$$

Note: This relationship is true only when θ is given in radians.

Study Tip

This relationship between linear and angular speed assumes the angle is given in radians.

Notice that tires of two different radii with the same angular speed have different linear speeds. The larger tire has the faster linear speed.

EXAMPLE 12 Relating Linear and Angular Speeds

A Ford F-150 truck comes standard with tires that have a diameter of 25.7 inches (17″ rims). If the owner decides to upgrade to tires with a diameter of 28.2 inches (19″ rims) without having the onboard computer updated, how fast will the truck *actually* be traveling when the speedometer reads 75 miles per hour?

Solution:

The computer onboard the F-150 "thinks" the tires are 25.7 inches in diameter and knows the angular speed. Use the programmed tire diameter and speedometer reading to calculate the angular speed. Then use that angular speed and the upgraded tire diameter to get the actual speed (linear speed).

STEP 1 Calculate the angular speed of the tires.

Write the formula for the angular speed.

$$\omega = \frac{v}{r}$$

Substitute $v = 75$ miles per hour and $r = \frac{25.7}{2} = 12.85$ inches into the formula.

$$\omega = \frac{75 \text{ mi/hr}}{12.85 \text{ in.}}$$

1 mile = 5280 feet = 63,360 inches.

$$\omega = \frac{75(63,360) \text{ in./hr}}{12.85 \text{ in.}}$$

Simplify.

$$\omega \approx 369,805 \frac{\text{rad}}{\text{hr}}$$

STEP 2 Calculate the actual linear speed of the truck.

Write the linear speed formula.

$$v = r\omega$$

Substitute $r = \frac{28.2}{2} = 14.1$ inches and $\omega \approx 369,805$ radians per hour.

$$v = (14.1 \text{ in.})\left(369,805 \frac{\text{rad}}{\text{hr}}\right)$$

Simplify.

$$v \approx 5,214,251 \frac{\text{in.}}{\text{hr}}$$

1 mile = 5280 feet = 63,360 inches.

$$v \approx \frac{5,214,251 \frac{\text{in.}}{\text{hr}}}{63,360 \frac{\text{in.}}{\text{mil}}}$$

$$v \approx \boxed{82.296 \frac{\text{mil}}{\text{hr}}}$$

Although the speedometer indicates a speed of 75 miles per hour, the actual speed is approximately $\boxed{82 \text{ miles per hour}}$.

Study Tip

We could have solved Example 12 the following way:

$$\frac{75 \text{ mph}}{25.7} = \frac{x}{28.2}$$

$$x = \left(\frac{28.2}{25.7}\right)(75)$$

$$\approx 82.296 \text{ mph}$$

■ **Answer:** Approximately 62 mph

■ **YOUR TURN** Suppose the owner of the Ford F-150 truck in Example 12 decides to downsize the tires from their original 25.7-inch diameter to a 24.4-inch diameter. If the speedometer indicates a speed of 65 miles per hour, what is the actual speed of the truck?

SECTION
6.6 SUMMARY

In this section a second measure of angles, radians, was introduced. A central angle of a circle has radian measure equal to the ratio of the arc length to the radius of the circle. Radians and degrees are related by the relation $\pi = 180°$. To convert from radians to degrees, multiply the radian measure by $\dfrac{180°}{\pi}$. Similarly, to convert from degrees to radians, multiply the degree measure by $\dfrac{\pi}{180°}$.

Applications of radian measure are arc length, area of a circular sector, and angular speed.

APPLICATION	θ_r IS IN RADIANS	θ_d IS IN DEGREES
Arc length	$s = r\theta_r$	$s = r\theta_d\left(\dfrac{\pi}{180°}\right)$
Area of a circular sector	$A = \dfrac{1}{2}r^2\theta_r$	$A = \dfrac{1}{2}r^2\theta_d\left(\dfrac{\pi}{180°}\right)$
Angular speed	$\omega = \dfrac{\theta_r}{t}$	

Linear speed (speed along the circumference of a circle) v is given by $v = \dfrac{s}{t}$ and is related to angular speed (speed of angle rotation) ω by

$$v = r\omega \qquad \text{or} \qquad \omega = \frac{v}{r}$$

SECTION
6.6 EXERCISES

▪ SKILLS

In Exercises 1–12, find the measure (in radians) of a central angle θ that intercepts an arc of length s on a circle with radius r.

1. $r = 10$ cm, $s = 2$ cm
2. $r = 20$ cm, $s = 2$ cm
3. $r = 22$ in., $s = 4$ in.
4. $r = 6$ in., $s = 1$ in.

5. $r = 100$ cm, $s = 20$ mm
6. $r = 1$ m, $s = 2$ cm
7. $r = \frac{1}{4}$ in., $s = \frac{1}{32}$ in.
8. $r = \frac{3}{4}$ cm, $s = \frac{3}{14}$ cm

9. $r = 2.5$ cm, $s = 5$ mm
10. $r = 1.6$ cm, $s = 0.2$ mm
11. $r = 2\sqrt{3}$ ft, $s = \sqrt{3}$ in.
12. $r = 4\sqrt{5}$ m, $s = 400\sqrt{5}$ cm

In Exercises 13–28, convert from degrees to radians. Leave the answers in terms of π.

13. $30°$
14. $60°$
15. $45°$
16. $90°$
17. $315°$
18. $270°$

19. $75°$
20. $100°$
21. $170°$
22. $340°$
23. $780°$
24. $540°$

25. $-210°$
26. $-320°$
27. $-3600°$
28. $1800°$

In Exercises 29–42, convert from radians to degrees.

29. $\dfrac{\pi}{6}$
30. $\dfrac{\pi}{4}$
31. $\dfrac{3\pi}{4}$
32. $\dfrac{7\pi}{6}$
33. $\dfrac{3\pi}{8}$
34. $\dfrac{11\pi}{9}$
35. $\dfrac{5\pi}{12}$
36. $\dfrac{7\pi}{3}$

37. 9π
38. -6π
39. $\dfrac{19\pi}{20}$
40. $\dfrac{13\pi}{36}$
41. $-\dfrac{7\pi}{15}$
42. $-\dfrac{8\pi}{9}$

In Exercises 43–50, convert from radian measure to degrees. Round your answers to the nearest hundredth of a degree.

43. 4
44. 3
45. 0.85
46. 3.27

47. -2.7989
48. -5.9841
49. $2\sqrt{3}$
50. $5\sqrt{7}$

In Exercises 51–56, convert from degrees to radians. Round your answers to three significant digits.

51. $47°$　　　　**52.** $65°$　　　　**53.** $112°$　　　　**54.** $172°$　　　　**55.** $56.5°$　　　　**56.** $298.7°$

In Exercises 57–66, find the exact length of the arc made by the indicated central angle and radius of each circle.

57. $\theta = 3$, $r = 4\,$mm　　　**58.** $\theta = 4$, $r = 5\,$cm　　　**59.** $\theta = \dfrac{\pi}{12}$, $r = 8\,$ft　　　**60.** $\theta = \dfrac{\pi}{8}$, $r = 6\,$yd

61. $\theta = \frac{1}{2}$, $r = 5\,$in.　　　**62.** $\theta = \frac{3}{4}$, $r = 20\,$m　　　**63.** $\theta = 22°$, $r = 18\,\mu$m　　　**64.** $\theta = 14°$, $r = 15\,\mu$m

65. $\theta = 8°$, $r = 1500\,$km　　　**66.** $\theta = 3°$, $r = 1800\,$km

In Exercises 67–76, find the area of the circular sector given the indicated radius and central angle. Round your answers to three significant digits.

67. $\theta = \dfrac{\pi}{6}$, $r = 7\,$ft　　　**68.** $\theta = \dfrac{\pi}{5}$, $r = 3\,$in.　　　**69.** $\theta = \dfrac{3\pi}{8}$, $r = 2.2\,$km　　　**70.** $\theta = \dfrac{5\pi}{6}$, $r = 13\,$mi

71. $\theta = \frac{5}{2}$, $r = 2\,$yd　　　**72.** $\theta = \sqrt{3}$, $r = 4\,$m　　　**73.** $\theta = 56°$, $r = 4.2\,$cm　　　**74.** $\theta = 27°$, $r = 2.5\,$mm

75. $\theta = 1.2°$, $r = 1.5\,$ft　　　**76.** $\theta = 14°$, $r = 3.0\,$ft

In Exercises 77–80, find the linear speed of a point that moves with constant speed in a circular motion if the point travels along the circle of arc length s in time t.

77. $s = 2\,$m, $t = 5\,$sec　　　　　　　　　**78.** $s = 12\,$ft, $t = 3\,$min

79. $s = 68{,}000\,$km, $t = 250\,$hr　　　　　**80.** $s = 7524\,$mi, $t = 12\,$days

In Exercises 81–84, find the distance traveled (arc length) of a point that moves with constant speed v along a circle in time t.

81. $v = 2.8\,$m/sec, $t = 3.5\,$sec　　　　　　**82.** $v = 6.2\,$km/hr, $t = 4.5\,$hr

83. $v = 4.5\,$mi/hr, $t = 20\,$min　　　　　　　**84.** $v = 5.6\,$ft/sec, $t = 2\,$min

In Exercises 85–88, find the angular speed (radians/second) associated with rotating a central angle θ in time t.

85. $\theta = 25\pi$, $t = 10\,$sec　　　　　　　**86.** $\theta = \dfrac{3\pi}{4}$, $t = \dfrac{1}{6}\,$sec

87. $\theta = 200°$, $t = 5\,$sec　　　　　　　　**88.** $\theta = 60°$, $t = 0.2\,$sec

In Exercises 89–92, find the linear speed of a point traveling at a constant speed along the circumference of a circle with radius r and angular speed ω.

89. $\omega = \dfrac{2\pi\,\text{rad}}{3\,\text{sec}}$, $r = 9\,$in.　　　　　　**90.** $\omega = \dfrac{3\pi\,\text{rad}}{4\,\text{sec}}$, $r = 8\,$cm

91. $\omega = \dfrac{\pi\,\text{rad}}{20\,\text{sec}}$, $r = 5\,$mm　　　　　　**92.** $\omega = \dfrac{5\pi\,\text{rad}}{16\,\text{sec}}$, $r = 24\,$ft

In Exercises 93–96, find the distances a point travels along a circle over a time t, given the angular speed ω and radius r of the circle. Round your answers to three significant digits.

93. $r = 5\,$cm, $\omega = \dfrac{\pi\,\text{rad}}{6\,\text{sec}}$, $t = 10\,$sec　　　　　　**94.** $r = 2\,$mm, $\omega = 6\pi\dfrac{\text{rad}}{\text{sec}}$, $t = 11\,$sec

95. $r = 5.2\,$in., $\omega = \dfrac{\pi\,\text{rad}}{15\,\text{sec}}$, $t = 10\,$min　　　　　**96.** $r = 3.2\,$ft, $\omega = \dfrac{\pi\,\text{rad}}{4\,\text{sec}}$, $t = 3\,$min

▪ APPLICATIONS

97. Low Earth Orbit Satellites. A low earth orbit (LEO) satellite is in an approximately circular orbit 300 km above the surface of the Earth. If the ground station tracks the satellite when it is within a 45° central angle above the tracking antenna (directly overhead), how many kilometers does the satellite cover during the tracking? Assume the radius of the Earth is 6400 km. Round to the nearest kilometers.

98. Low Earth Orbit Satellites. A low earth orbit (LEO) satellite is in an approximately circular orbit 250 km above the surface of the Earth. If the ground station tracks the satellite when it is within a 30° central angle above the tracking antenna (directly overhead), how many kilometers does the satellite cover during the tracking? Assume the radius of the Earth is 6400 km. Round to the nearest kilometers.

99. Big Ben. The famous clock tower in London has a minute hand that is 14 feet long. How far does the tip of the minute hand of Big Ben travel in 25 minutes? Round to two decimal places.

Peter Adams/Digital Vision/Getty Images, Inc.

100. Big Ben. The famous clock tower in London has a minute hand that is 14 feet long. How far does the tip of the minute hand of Big Ben travel in 35 minutes? Round to two decimal places.

101. London Eye. The London Eye has 32 capsules and a diameter of 400 feet. What is the distance you will have traveled once you reach the highest point for the first time?

Jason Friend/PhotolibraryGetty Images, Inc.

102. London Eye. (See Exercise 101.) Assuming every capsule is loaded and unloaded, what is the distance you will have traveled from the time you first get in the capsule to the point when the eye stops for the 6th time (not including the stop made for you)?

103. Gears. In the diagram below the smaller gear has a radius of 5 cm and the larger gear has a radius of 12.1 cm. If the smaller gear rotates 120°, how many degrees has the larger gear rotated? Round your answer to the nearest degree.

12.1 cm   5 cm

104. Gears. In the diagram above the smaller gear has a radius of 3 inches and the larger gear has a radius of 15 inches. If the smaller gear rotates 420°, how many degrees has the larger gear rotated? Round your answer to the nearest degree.

105. Bicycle Low Gear. If a bicycle has 26-inch-diameter wheels (the front chain drive has a radius of 2.2 inches and the back drive has a radius of 3 inches), how far does the bicycle travel every one rotation of the cranks (pedals)? Round to the nearest inch.

106. Bicycle High Gear. If a bicycle has 26-inch-diameter wheels (the front chain drive has a radius of 4 inches and the back drive has a radius of 1 inch), how far does the bicycle travel every one rotation of the cranks (pedals)? Round to the nearest inch.

Photodisc blue/Getty Images, Inc.

107. Odometer. A Ford SUV comes standard with 17″ rims (which corresponds to a tire with 25.7″ diameter). Suppose you decide to later upgrade these tires for 19″ rims (corresponds to a tire with 28.2″ diameter). If you do not get your onboard computer reset for the new tires, the odometer will not be accurate. After your tires have actually driven 1000 miles, how many miles will the odometer report the SUV has been driven? Round to the nearest mile.

108. Odometer. For the same SUV in Exercise 107, after you have driven 50,000 miles, how many miles will the odometer report the SUV has been driven if the computer is not reset to account for the new oversized tires? Round to the nearest mile.

109. Sprinkler Coverage. A sprinkler has a 20-foot spray and it covers an angle of 45°. What is the area that the sprinkler waters? Round to the nearest square foot.

110. Sprinkler Coverage. A sprinkler has a 22-foot spray and it covers an angle of 60°. What is the area that the sprinkler waters? Round to the nearest square foot.

111. Windshield Wiper. A windshield wiper that is 12 inches long (blade and arm) rotates 70°. If the blade is 8 inches long, what is the area cleared by the wiper? Round your answer to the nearest square inch.

112. Windshield Wiper. A windshield wiper that is 11 inches long (blade and arm) rotates 65°. If the blade is 7 inches long, what is the area cleared by the wiper? Round your answer to the nearest square inch.

For Exercises 113 and 114, refer to the following:

Sniffers outside a chemical munitions disposal site monitor the atmosphere surrounding the site to detect any toxic gases. In the event that there is an accidental release of toxic fumes, the data provided by the sniffers make it possible to determine both the distance d that the fumes reach as well as the angle of spread θ that sweep out a circular sector.

113. Environment. If the maximum angle of spread is 105° and the maximum distance at which the toxic fumes were detected was 9 miles from the site, find the area of the circular sector affected by the accidental release.

114. Environment. To protect the public from the fumes, officials must secure the perimeter of this area. Find the perimeter of the circular sector in Exercise 113.

For Exercises 115 and 116, refer to the following:

The structure of human DNA is a *linear* double helix formed of nucleotide base pairs (two nucleotides) that are stacked with spacing of 3.4 angstroms (3.4×10^{-12} m), and each base pair is rotated 36° with respect to an adjacent pair and has 10 base pairs per helical turn. The DNA of a virus or a bacterium, however, is a *circular* double helix (see the figure below) with the structure varying among species.

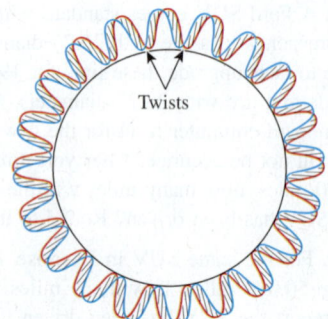

Twists

(*Source*: http://www.biophysics.org/Portals/1/
PDFs/Education/Vologodskii.pdf.)

115. Biology. If the circular DNA of a virus has 10 twists (or turns) per circle and an inner diameter of 4.5 nanometers, find the arc length between consecutive twists of the DNA.

116. Biology. If the circular DNA of a virus has 40 twists (or turns) per circle and an inner diameter of 2.0 nanometers, find the arc length between consecutive twists of the DNA.

117. Tires. A car owner decides to upgrade from tires with a diameter of 24.3 inches to tires with a diameter of 26.1 inches. If she doesn't update the onboard computer, how fast will she actually be traveling when the speedometer reads 65 mph? Round to the nearest mph.

118. Tires. A car owner decides to upgrade from tires with a diameter of 24.8 inches to tires with a diameter of 27.0 inches. If she doesn't update the onboard computer, how fast will she actually be traveling when the speedometer reads 70 mph? Round to the nearest mph.

119. Planets. The Earth rotates every 24 hours (actually 23 hours, 56 minutes, and 4 seconds) and has a diameter of 7926 miles. If you're standing on the equator, how fast are you traveling (how fast is the Earth spinning) in linear speed? Compute this using 24 hours and then with 23 hours, 56 minutes, 4 seconds.

120. Planets. The planet Jupiter rotates every 9.9 hours and has a diameter of 88,846 miles. If you're standing on its equator, how fast are you traveling (linear speed)?

121. Carousel. A boy wants to jump onto a moving carousel that is spinning at the rate of 5 revolutions per minute. If the carousel is 60 feet in diameter, how fast must the boy run, in feet per second, to match the speed of the carousel and jump on? Round to the nearest foot per second.

122. Carousel. A boy wants to jump onto a playground carousel that is spinning at the rate of 30 revolutions per minute. If the carousel is 6 feet in diameter, how fast must the boy run, in feet per second, to match the speed of the carousel and jump on? Round to the nearest foot per second.

123. Music. Some people still have their phonograph collections and play the records on turntables. A phonograph record is a vinyl disc that rotates on the turntable. If a 12-inch-diameter record rotates at $33\frac{1}{3}$ revolutions per minute, what is the angular speed in radians per minute?

Niall McDiarmid/Alamy

124. Music. Some people still have their phonograph collections and play the records on turntables. A phonograph record is a vinyl disc that rotates on the turntable. If a 12-inch-diameter record rotates at $33\frac{1}{3}$ revolutions per minute, what is the linear speed of a point on the outer edge?

125. Bicycle. How fast is a bicyclist traveling in miles per hour if his tires are 27 inches in diameter and his angular speed is 5π radians per second?

126. Bicycle. How fast is a bicyclist traveling in miles per hour if his tires are 22 inches in diameter and his angular speed is 5π radians per second?

127. Electric Motor. If a 2-inch-diameter pulley that is being driven by an electric motor and running at 1600 revolutions per minute is connected by a belt to a 5-inch-diameter pulley to drive a saw, what is the speed of the saw in revolutions per minute?

128. Electric Motor. If a 2.5-inch-diameter pulley that is being driven by an electric motor and running at 1800 revolutions per minute is connected by a belt to a 4-inch-diameter pulley to drive a saw, what is the speed of the saw in revolutions per minute?

For Exercises 129 and 130, refer to the following:

NASA explores artificial gravity as a way to counter the physiologic effects of extended weightlessness for future space exploration. NASA's centrifuge has a 58-foot-diameter arm.

Courtesy NASA

129. NASA. If two humans are on opposite (red and blue) ends of the centrifuge and their linear speed is 200 miles per hour, how fast is the arm rotating?

130. NASA. If two humans are on opposite (red and blue) ends of the centrifuge and they rotate one full rotation every second, what is their linear speed in feet per second?

For Exercises 131 and 132, refer to the following:

To achieve similar weightlessness as NASA's centrifuge, ride the Gravitron at a carnival or fair. The Gravitron has a diameter of 14 meters and in the first 20 seconds it achieves zero gravity and the floor drops.

Patrick Reddy/Americaz4-7/Getty Images, Inc.

131. Gravitron. If the Gravitron rotates 24 times per minute, find the linear speed of the people riding it in meters per second.

132. Gravitron. If the Gravitron rotates 30 times per minute, find the linear speed of the people riding it in kilometers per hour.

133. Biology. A hamster runs around a wheel at a constant rate such that the wheel goes through 1 revolution every 3 seconds. Find the angular speed in radians per minute.

134. Cooking. A chef can turn the hand crank of a pasta maker at a rate of 10π radians every 3 seconds. At this rate, a new batch of spaghetti can be made every 15 seconds. If 40 batches were made, how many revolutions of the hand crank were made?

135. The Price Is Right. On the TV show *The Price Is Right*, the 8-foot-diameter wheel is divided into 20 sectors (corresponding to increments of 5 cents). The three most important amounts on the wheel are the three abutting sectors with amounts $0.15, $1.00, and $0.05; the other amounts are randomly dispersed among the remaining sectors. The beginning of a spin always starts at $1.00. A zealous contestant spins the wheel so that it makes four complete revolutions, landing on $1.00. What is the length of the arc created by this spin?

136. The Price Is Right. If a contestant on *The Price Is Right* lands on the $1.00 space on his first spin, he is granted a second spin. If this spin lands on either the $0.05 or $0.15 space, he wins $5,000; if it lands on the $1.00 space again, he wins $10,000. This time, he tries to control the spin so that the wheel makes exactly one complete revolution, again landing on $1.00. If the arc length of the spin is $\frac{84}{11}\pi$ ft, does the contestant win any money?

CATCH THE MISTAKE

In Exercises 137 and 138, explain the mistake that is made.

137. If the radius of a set of tires on a car is 15 inches and the tires rotate 180° per second, how fast is the car traveling (linear speed) in miles per hour?

Solution:

Write the formula for linear speed.

$$v = r\omega$$

Let $r = 15$ inches and $\omega = 180°$ per second

$$v = (15 \text{ in.})\left(\frac{180°}{\text{sec}}\right)$$

Simplify.

$$v = 2700\,\frac{\text{in.}}{\text{sec}}$$

Let 1 mile = 5280 feet = 63,360 inches and 1 hour = 3600 seconds.

$$v = \left(\frac{2700 \cdot 3600}{63,360}\right)\text{mph}$$

Simplify.

$$v \approx 153.4 \text{ mph}$$

This is incorrect. The correct answer is approximately 2.7 mph. What mistake was made?

138. If a bicycle has tires with radius 10 inches and the tires rotate 90° per $\frac{1}{2}$ second, how fast is the bicycle traveling (linear speed) in miles per hour?

Solution:

Write the formula for linear speed.

$$v = r\omega$$

Let $r = 10$ inches and $\omega = 180°$ per second.

$$v = (10 \text{ in.})\left(\frac{180°}{\text{sec}}\right)$$

Simplify.

$$v = \frac{1800 \text{ in.}}{\text{sec}}$$

Let 1 mile = 5280 feet = 63,360 inches and 1 hour = 3600 seconds.

$$v = \left(\frac{1800 \cdot 3600}{63,360}\right)\text{mph}$$

Simplify.

$$v \approx 102.3 \text{ mph}$$

This is incorrect. The correct answer is approximately 1.8 mph. What mistake was made?

CONCEPTUAL

In Exercises 139–142, determine whether each statement is true or false.

139. If the radius of a circle doubles, then the arc length (associated with a fixed central angle) doubles.

140. If the radius of a circle doubles, then the area of the sector (associated with a fixed central angle) doubles.

141. If the angular speed doubles, then the number of revolutions doubles.

142. If the central angle of a sector doubles, then the area corresponding to the sector is double the area of the original sector.

CHALLENGE

For Exercises 143 and 144, refer to the figure:

The large gear has a radius of 6 cm, the medium gear has a radius of 3 cm, and the small gear has a radius of 1 cm.

143. If the small gear rotates 1 revolution per second, what is the linear speed of a point traveling along the circumference of the large gear?

144. If the small gear rotates 1.5 revolutions per second, what is the linear speed of a point traveling along the circumference of the large gear?

145. A sector has a radius r and central angle θ. If the radius is tripled, but the central angle is reduced by a factor of 3, how does the area of the resulting sector compare to the original one?

146. Doing "ladders" as an exercise is familiar to anyone who has participated on an athletic team. A version of this on a circular track of diameter 400 ft is as follows:

1. Start at point A and run clockwise 180° to point B.
2. Then run from point B counterclockwise 90° to point C.
3. Then run from point C clockwise 45° to point D.
4. Then run from point D counterclockwise 10° to point E. Then stop.

How far has the athlete traveled in one such round of ladders?

147. Find the area of the shaded region below:

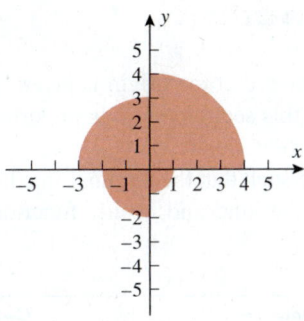

148. Find the perimeter of the shaded region in Exercise 143.

■ **TECHNOLOGY**

In Exercises 149 and 150, find the measure (in degrees, minutes, and nearest seconds) of a central angle θ that intercepts an arc on a circle with radius r and indicated arc length s. With the TI calculator commands $\boxed{\text{ANGLE}}$ and $\boxed{\text{DMS}}$, change to degrees, minutes, and seconds.

149. $r = 78.6\,\text{cm}$, $s = 94.4\,\text{cm}$

150. $r = 14.2\,\text{in.}$, $s = 23.8\,\text{in.}$

SECTION 6.7

DEFINITION 3 OF TRIGONOMETRIC FUNCTIONS: UNIT CIRCLE APPROACH

SKILLS OBJECTIVES

- Draw the unit circle illustrating the special angles and label cosine and sine values.
- Determine the domain and range of circular functions.
- Classify circular functions as even or odd.

CONCEPTUAL OBJECTIVES

- Understand that the definition of trigonometric functions using the unit circle approach is consistent with both of the previous definitions (right triangle trigonometry and trigonometric functions of nonacute angles in the Cartesian plane).
- Relate x-coordinates and y-coordinates of points on the unit circle to the values of cosine and sine functions.
- Visualize the periodic properties of circular functions.

Recall that the first definition of trigonometric functions we developed was in terms of ratios of sides of right triangles (Section 6.2). Then in Section 6.4 we superimposed right triangles on the Cartesian plane, which led to a second definition of trigonometric functions (for any angle) in terms of ratios of x- and y-coordinates of a point and the distance from the origin to that point. In this section we inscribe the right triangles into the unit circle in the Cartesian plane, which will yield a third definition of trigonometric functions. It is important to note that all three definitions are consistent with one another.

Trigonometric Functions and the Unit Circle (Circular Functions)

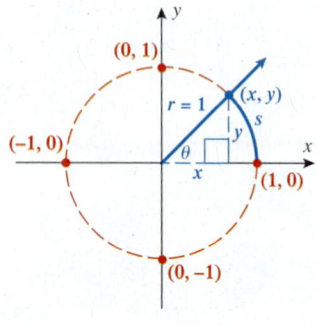

Recall that the equation for the **unit circle** centered at the origin is given by $x^2 + y^2 = 1$. We will use the term *circular function* later in this section, but it is important to note that a circle is not a function (it does not pass the vertical line test).

If we form a central angle θ in the unit circle such that the terminal side lies in quadrant I, we can use the previous two definitions of the sine and cosine functions when $r = 1$ (i.e., in the unit circle).

TRIGONOMETRIC FUNCTION	RIGHT TRIANGLE TRIGONOMETRY	CARTESIAN PLANE
$\sin\theta$	$\dfrac{\text{opposite}}{\text{hypotenuse}} = \dfrac{y}{1} = y$	$\dfrac{y}{r} = \dfrac{y}{1} = y$
$\cos\theta$	$\dfrac{\text{adjacent}}{\text{hypotenuse}} = \dfrac{x}{1} = x$	$\dfrac{x}{r} = \dfrac{x}{1} = x$

Notice that the point (x, y) on the unit circle can be written as $(\cos\theta, \sin\theta)$. If we recall the unit circle coordinate values for special angles (Section 6.4), we can now summarize the exact values for **sine** and **cosine** in the following illustration.

$(x, y) = (\cos\theta, \sin\theta)$, where θ is the central angle whose terminal side intersects the unit circle at (x, y).

The following observations are consistent with properties of trigonometric functions we've studied already:

- $\sin\theta > 0$ in quadrant I and quadrant II
- $\cos\theta > 0$ in quadrant I and quadrant IV
- The unit circle equation $x^2 + y^2 = 1$ leads to the Pythagorean identity
$$\cos^2\theta + \sin^2\theta = 1$$

Circular Functions

Using the unit circle relationship, $(x, y) = (\cos\theta, \sin\theta)$, where θ is the central angle whose terminal side intersects the unit circle at the point (x, y), we can now define the remaining trigonometric functions using this unit circle approach and the quotient and reciprocal identities. Because the trigonometric functions are defined in terms of the unit *circle* the trigonometric functions are often called **circular functions**.

DEFINITION (3)   Trigonometric Functions

Unit Circle Approach

Let (x, y) be any point on the unit circle. If θ is a real number that represents the distance from the point $(1, 0)$ along the circumference to the point (x, y), then

$$\sin\theta = y \qquad\qquad \cos\theta = x \qquad\qquad \tan\theta = \frac{y}{x} \quad x \neq 0$$

$$\csc\theta = \frac{1}{y} \quad y \neq 0 \qquad \sec\theta = \frac{1}{x} \quad x \neq 0 \qquad \cot\theta = \frac{x}{y} \quad y \neq 0$$

Technology Tip

Use TI/scientific calculator to check the values for $\sin\left(\frac{7\pi}{4}\right)$, $\cos\left(\frac{5\pi}{6}\right)$, and $\tan\left(\frac{3\pi}{2}\right)$. Be sure to set the TI in radian mode.

EXAMPLE 1   Finding Exact Trigonometric (Circular) Function Values

Find the exact values for

a. $\sin\left(\frac{7\pi}{4}\right)$ **b.** $\cos\left(\frac{5\pi}{6}\right)$ **c.** $\tan\left(\frac{3\pi}{2}\right)$

Solution (a):

The angle $\frac{7\pi}{4}$ corresponds to the coordinates $\left(\frac{\sqrt{2}}{2}, -\frac{\sqrt{2}}{2}\right)$ on the unit circle.

The value of the sine function is the y-coordinate. $\qquad \boxed{\sin\left(\frac{7\pi}{4}\right) = -\frac{\sqrt{2}}{2}}$

```
sin(7π/4)
        -.7071067812
-√(2)/2
        -.7071067812
```

Solution (b):

The angle $\frac{5\pi}{6}$ corresponds to the coordinate $\left(-\frac{\sqrt{3}}{2}, \frac{1}{2}\right)$ on the unit circle.

The value of the cosine function is the x-coordinate. $\qquad \boxed{\cos\left(\frac{5\pi}{6}\right) = -\frac{\sqrt{3}}{2}}$

```
cos(5π/6)
        -.8660254038
-√(3)/2
        -.8660254038
■
```

Since $\tan\dfrac{3\pi}{2}$ is undefined, TI will display an error message.

Solution (c):

The angle $\frac{3\pi}{2}$ corresponds to the coordinate $(0, -1)$ on the unit circle.

The value of the cosine function is the x-coordinate. $\qquad \cos\left(\frac{3\pi}{2}\right) = 0$

The value of the sine function is the y-coordinate. $\qquad \sin\left(\frac{3\pi}{2}\right) = -1$

```
tan(3π/2)
■
```

```
ERR:DOMAIN
1■Quit
2:Goto
```

Tangent is the ratio of sine to cosine. $\qquad \tan\left(\frac{3\pi}{2}\right) = \dfrac{\sin\left(\dfrac{3\pi}{2}\right)}{\cos\left(\dfrac{3\pi}{2}\right)}$

Let $\cos\left(\frac{3\pi}{2}\right) = 0$ and $\sin\left(\frac{3\pi}{2}\right) = -1$. $\qquad \tan\left(\frac{3\pi}{2}\right) = \dfrac{-1}{0}$

$$\boxed{\tan\left(\frac{3\pi}{2}\right) \text{ is undefined.}}$$

Classroom Example 6.7.1
Compute these values.

a. $\cos\left(\dfrac{4\pi}{3}\right)$ **b.** $\cot\left(\dfrac{3\pi}{2}\right)$

c. $\sec\left(\dfrac{7\pi}{6}\right)$

Answer:
a. $-\frac{1}{2}$ **b.** 0 **c.** $-\dfrac{2\sqrt{3}}{3}$

■ **YOUR TURN** Find the exact values for

a. $\sin\left(\frac{5\pi}{6}\right)$ **b.** $\cos\left(\frac{7\pi}{4}\right)$ **c.** $\tan\left(\frac{2\pi}{3}\right)$

■ **Answer: a.** $\dfrac{1}{2}$ **b.** $\dfrac{\sqrt{2}}{2}$ **c.** $-\sqrt{3}$

EXAMPLE 2 **Solving Equations Involving Circular Functions**

Use the unit circle to find all values of θ, $0 \leq \theta \leq 2\pi$, for which $\sin\theta = -\frac{1}{2}$.

Classroom Example 6.7.2

Solve $\cos\theta = -\dfrac{\sqrt{2}}{2}$ on

$[0, 2\pi]$.

Answer: $\dfrac{3\pi}{4}, \dfrac{5\pi}{4}$

Solution:

The value of sine is the y-coordinate.

Since the value of sine is negative, θ must lie in QIII or QIV.

There are two values for θ that are greater than or equal to zero and less than or equal to 2π that correspond to $\sin\theta = -\frac{1}{2}$.

$$\theta = \frac{7\pi}{6}, \frac{11\pi}{6}$$

■ **Answer:** $\theta = \dfrac{2\pi}{3}, \dfrac{4\pi}{3}$

■ **YOUR TURN** Find all values of θ, $0 \leq \theta \leq 2\pi$, for which $\cos\theta = -\frac{1}{2}$.

Properties of Circular Functions

WORDS	MATH
For a point (x, y) that lies on the unit circle, $x^2 + y^2 = 1$.	$-1 \leq x \leq 1$ and $-1 \leq y \leq 1$
Since $(x, y) = (\cos\theta, \sin\theta)$, the following holds.	$-1 \leq \cos\theta \leq 1$ and $-1 \leq \sin\theta \leq 1$
State the **domain and range of the cosine and sine functions**.	Domain: $(-\infty, \infty)$ Range: $[-1, 1]$
Since $\cot\theta = \dfrac{\cos\theta}{\sin\theta}$ and $\csc\theta = \dfrac{1}{\sin\theta}$, the values for θ that make $\sin\theta = 0$ must be eliminated from the **domain of the cotangent and cosecant functions**.	Domain: $\theta \neq n\pi$, n an integer
Since $\tan\theta = \dfrac{\sin\theta}{\cos\theta}$ and $\sec\theta = \dfrac{1}{\cos\theta}$, the values for θ that make $\cos\theta = 0$ must be eliminated from the **domain of the tangent and secant functions**.	Domain: $\theta \neq \dfrac{(2n + 1)\pi}{2}$, n an integer

The following box summarizes the domains and ranges of the circular functions.

DOMAINS AND RANGES OF THE CIRCULAR FUNCTIONS

For any real number θ and integer n:

FUNCTION	DOMAIN	RANGE
$\sin\theta$	$(-\infty, \infty)$	$[-1, 1]$
$\cos\theta$	$(-\infty, \infty)$	$[-1, 1]$
$\tan\theta$	all real numbers such that $\theta \neq \dfrac{(2n+1)\pi}{2}$	$(-\infty, \infty)$
$\cot\theta$	all real numbers such that $\theta \neq n\pi$	$(-\infty, \infty)$
$\sec\theta$	all real numbers such that $\theta \neq \dfrac{(2n+1)\pi}{2}$	$(-\infty, -1] \cup [1, \infty)$
$\csc\theta$	all real numbers such that $\theta \neq n\pi$	$(-\infty, -1] \cup [1, \infty)$

Recall from algebra that even and odd functions have both an algebraic and a graphical interpretation. **Even functions** are functions for which $f(-x) = f(x)$ and their graphs are symmetric about the y-axis. **Odd functions** are functions for which $f(-x) = -f(x)$ and their graphs are symmetric about the origin.

Classroom Example 6.7.3
Compute:

a. $\sin\left(-\dfrac{5\pi}{3}\right)$

b. $\cos\left(-\dfrac{7\pi}{4}\right)$

Answer: a. $\dfrac{\sqrt{3}}{2}$ **b.** $\dfrac{\sqrt{2}}{2}$

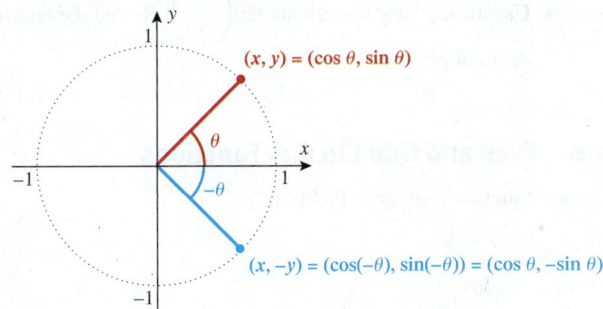

The cosine function is an even function. $\boxed{\cos\theta = \cos(-\theta)}$

The sine function is an odd function. $\boxed{\sin(-\theta) = -\sin\theta}$

Technology Tip

Use a TI/scientific calculator to check the value of $\cos\left(-\dfrac{5\pi}{6}\right)$.

```
cos( -5π/6 )
         -.8660254038
-√(3)/2
         -.8660254038
■
```

EXAMPLE 3   Using Properties of Trigonometric (Circular) Functions

Evaluate $\cos\left(-\dfrac{5\pi}{6}\right)$.

Solution:

The cosine function is an even function. $\cos\left(-\dfrac{5\pi}{6}\right) = \cos\left(\dfrac{5\pi}{6}\right)$

Use the unit circle to evaluate cosine. $\cos\left(\dfrac{5\pi}{6}\right) = -\dfrac{\sqrt{3}}{2}$

$$\cos\left(-\dfrac{5\pi}{6}\right) = \boxed{-\dfrac{\sqrt{3}}{2}}$$

■ **YOUR TURN** Evaluate $\sin\left(-\dfrac{5\pi}{6}\right)$.

■ **Answer:** $-\dfrac{1}{2}$

Study Tip

Set the calculator in radian mode before evaluating circular functions in radians. Alternatively, convert the radian measure to degrees before evaluating the trigonometric function value.

It is important to note that although circular functions can be evaluated exactly for some special angles, a calculator can be used to approximate circular functions for any angle. It is important to set the calculator to radian mode first, since θ is a real number.

 EXAMPLE 4 **Evaluating Trigonometric (Circular) Functions with a Calculator**

Use a calculator to evaluate $\sin\left(\dfrac{7\pi}{12}\right)$. Round the answer to four decimal places.

Technology Tip

Use the TI/scientific calculator to evaluate $\sin\dfrac{7\pi}{12}$. Press 2nd ∧ for π.

```
sin(7π/12)
      .9659258263
```

✪ **CORRECT**	✖ **INCORRECT**
Evaluate with a calculator.	Evaluate with a calculator.
0.965925826	0.031979376 **ERROR**
	(Calculator in degree mode)
Round to four decimal places.	
$\sin\left(\dfrac{7\pi}{12}\right) \approx \boxed{0.9659}$	

Many calculators automatically reset to degree mode after every calculation, so make sure to always check what mode the calculator indicates.

■ **Answer:** -0.7265

■ **YOUR TURN** Use a calculator to evaluate $\tan\left(\dfrac{9\pi}{5}\right)$. Round the answer to four decimal places.

Classroom Example 6.7.4

Compute $\cos\left(\dfrac{8\pi}{11}\right)$ using a calculator.

Answer: -0.6549

EXAMPLE 5 Even and Odd Circular Functions

Show that the secant function is an even function.

Solution:

Show that $\sec(-\theta) = \sec(\theta)$.

Classroom Example 6.7.5

Prove that cosecant is an odd function.

Answer:

$$\csc(-\theta) = \frac{1}{\sin(-\theta)}$$
$$= -\frac{1}{\sin\theta} = -\csc\theta$$

Secant is the reciprocal of cosine. $\qquad \sec(-\theta) = \dfrac{1}{\cos(-\theta)}$

Cosine is an even function, $\cos(-\theta) = \cos(\theta)$. $\qquad \sec(-\theta) = \dfrac{1}{\cos(\theta)}$

Secant is the reciprocal of cosine, $\sec\theta = \dfrac{1}{\cos\theta}$. $\qquad \sec(-\theta) = \dfrac{1}{\cos(\theta)} = \sec\theta$

Since $\sec(-\theta) = \sec(\theta)$, $\boxed{\text{the secant function is an even function}}$.

 SECTION 6.7 SUMMARY

In this section we have defined trigonometric functions as circular functions. Any point (x, y) that lies on the unit circle satisfies the equation $x^2 + y^2 = 1$. The Pythagorean identity $\cos^2\theta + \sin^2\theta = 1$ can also be represented on the unit circle

where $(x, y) = (\cos\theta, \sin\theta)$, and where θ is the central angle whose terminal side intersects the unit circle at the point (x, y). The cosine function is an even function, $\cos(-\theta) = \cos\theta$; and the sine function is an odd function, $\sin(-\theta) = -\sin\theta$.

▪ SKILLS

In Exercises 1–14, find the *exact* values of the indicated trigonometric functions using the unit circle.

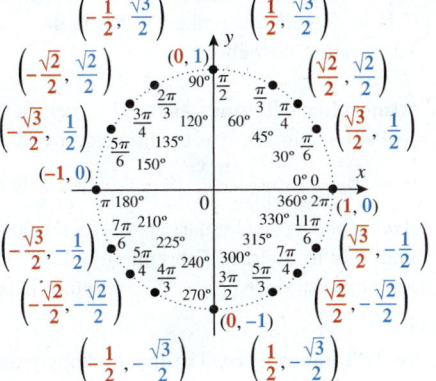

1. $\sin\left(\dfrac{5\pi}{3}\right)$

2. $\cos\left(\dfrac{5\pi}{3}\right)$

3. $\cos\left(\dfrac{7\pi}{6}\right)$

4. $\sin\left(\dfrac{7\pi}{6}\right)$

5. $\sin\left(\dfrac{3\pi}{4}\right)$

6. $\cos\left(\dfrac{3\pi}{4}\right)$

7. $\tan\left(\dfrac{7\pi}{4}\right)$

8. $\cot\left(\dfrac{7\pi}{4}\right)$

9. $\sec 225°$

10. $\csc 300°$

11. $\tan 240°$

12. $\cot 330°$

13. $\csc\dfrac{5\pi}{6}$

14. $\cot 120°$

In Exercises 15–34, use the unit circle and the fact that sine is an odd function and cosine is an even function to find the *exact* values of the indicated functions.

15. $\sin\left(-\dfrac{2\pi}{3}\right)$

16. $\sin\left(-\dfrac{5\pi}{4}\right)$

17. $\sin\left(-\dfrac{\pi}{3}\right)$

18. $\sin\left(-\dfrac{7\pi}{6}\right)$

19. $\cos\left(-\dfrac{3\pi}{4}\right)$

20. $\cos\left(-\dfrac{5\pi}{3}\right)$

21. $\cos\left(-\dfrac{5\pi}{6}\right)$

22. $\cos\left(-\dfrac{7\pi}{4}\right)$

23. $\sin(-225°)$

24. $\sin(-180°)$

25. $\sin(-270°)$

26. $\sin(-60°)$

27. $\cos(-45°)$

28. $\cos(-135°)$

29. $\cos(-90°)$

30. $\cos(-210°)$

31. $\csc(-150°)$

32. $\sec(-315°)$

33. $\tan\left(-\dfrac{11\pi}{6}\right)$

34. $\cot(-330°)$

In Exercises 35–54, use the unit circle to find all of the exact values of θ that make the equation true in the indicated interval.

35. $\cos\theta = \dfrac{\sqrt{3}}{2}, 0 \le \theta \le 2\pi$

36. $\cos\theta = -\dfrac{\sqrt{3}}{2}, 0 \le \theta \le 2\pi$

37. $\sin\theta = -\dfrac{\sqrt{3}}{2}, 0 \le \theta \le 2\pi$

38. $\sin\theta = \dfrac{\sqrt{3}}{2}, 0 \le \theta \le 2\pi$

39. $\sin\theta = 0, 0 \le \theta \le 4\pi$

40. $\sin\theta = -1, 0 \le \theta \le 4\pi$

41. $\cos\theta = -1, 0 \le \theta \le 4\pi$

42. $\cos\theta = 0, 0 \le \theta \le 4\pi$

43. $\tan\theta = -1, 0 \le \theta \le 2\pi$

44. $\cot\theta = 1, 0 \le \theta \le 2\pi$

45. $\sec\theta = -\sqrt{2}, 0 \le \theta \le 2\pi$

46. $\csc\theta = \sqrt{2}, 0 \le \theta \le 2\pi$

47. $\csc\theta$ is undefined, $0 \le \theta \le 2\pi$

48. $\sec\theta$ is undefined, $0 \le \theta \le 2\pi$

49. $\tan\theta$ is undefined, $0 \le \theta \le 2\pi$

50. $\cot\theta$ is undefined, $0 \le \theta \le 2\pi$

51. $\csc\theta = -2, 0 \le \theta \le 2\pi$

52. $\cot\theta = -\sqrt{3}, 0 \le \theta \le 2\pi$

53. $\sec\theta = \dfrac{2\sqrt{3}}{3}, 0 \le \theta \le 2\pi$

54. $\tan\theta = \dfrac{\sqrt{3}}{3}, 0 \le \theta \le 2\pi$

■ **APPLICATIONS**

55. **Atmospheric Temperature.** The average daily temperature in Peoria, Illinois, can be predicted by the formula $T = 50 - 28 \cos\left[\dfrac{2\pi(x - 31)}{365}\right]$, where x is the number of the day in the year (January 1 = 1, February 1 = 32, etc.) and T is in degrees Fahrenheit. What is the expected temperature on February 15?

56. **Atmospheric Temperature.** The average daily temperature in Peoria, Illinois, can be predicted by the formula $T = 50 - 28 \cos\left[\dfrac{2\pi(x - 31)}{365}\right]$, where x is the number of the day in the year (January 1 = 1, February 1 = 32, etc.) and T is in degrees Fahrenheit. What is the expected temperature on August 15? (Assume it is not a leap year.)

57. **Body Temperature.** The human body temperature normally fluctuates during the day. If a person's body temperature can be predicted by the formula $T = 99.1 - 0.5 \sin\left(x + \dfrac{\pi}{12}\right)$, where x is the number of hours since midnight and T is in degrees Fahrenheit, then what is the person's temperature at 6:00 A.M.?

58. **Body Temperature.** The human body temperature normally fluctuates during the day. If a person's body temperature can be predicted by the formula $T = 99.1 - 0.5 \sin\left(x + \dfrac{\pi}{12}\right)$, where x is the number of hours since midnight and T is in degrees Fahrenheit, then what is the person's temperature at 9:00 P.M.?

59. **Tides.** The height of the water in a harbor changes with the tides. If the height of the water at a particular hour during the day can be determined by the formula $h(x) = 5 + 4.8 \sin\left[\dfrac{\pi}{6}(x + 4)\right]$, where x is the number of hours since midnight and h is the height of the tide in feet, then what is the height of the tide at 3:00 P.M.?

60. **Tides.** The height of the water in a harbor changes with the tides. If the height of the water at a particular hour during the day can be determined by the formula $h(x) = 5 + 4.8 \sin\left[\dfrac{\pi}{6}(x + 4)\right]$, where x is the number of hours since midnight and h is the height of the tide in feet, then what is the height of the tide at 5:00 A.M.?

61. **Yo-Yo Dieting.** A woman has been yo-yo dieting for years. Her weight changes throughout the year as she gains and loses weight. Her weight in a particular month can be determined by the formula $w(x) = 145 + 10 \cos\left(\dfrac{\pi}{6}x\right)$,

Bill Brooks/Alamy

where x is the month and w is in pounds. If $x = 1$ corresponds to January, how much does she weigh in June?

62. **Yo-Yo Dieting.** How much does the woman in Exercise 61 weigh in December?

63. **Seasonal Sales.** The average number of guests visiting the Magic Kingdom at Walt Disney World per day is given by $n(x) = 30{,}000 + 20{,}000 \sin\left[\dfrac{\pi}{2}(x + 1)\right]$, where n is the number of guests and x is the month. If January corresponds to $x = 1$, how many people on average are visiting the Magic Kingdom per day in February?

64. **Seasonal Sales.** How many guests are visiting the Magic Kingdom in Exercise 63 in December?

For Exercises 65 and 66, refer to the following:

During the course of treatment of an illness, the concentration of a drug in the bloodstream in micrograms per microliter fluctuates during the dosing period of 8 hours according to the model

$$C(t) = 15.4 - 4.7 \sin\left(\frac{\pi}{4}t + \frac{\pi}{2}\right), \quad 0 \le t \le 8$$

Note: This model does not apply to the first dose of the medication.

65. Health/Medicine. Find the concentration of the drug in the bloodstream at the beginning of a dosing period.

66. Health/Medicine. Find the concentration of the drug in the bloodstream 6 hours after taking a dose of the drug.

In Exercises 67 and 68, refer to the following:

By analyzing available empirical data, it has been determined that the body temperature of a particular species fluctuates during a 24-hour day according to the model

$$T(t) = 36.3 - 1.4 \cos\left[\frac{\pi}{12}(t - 2)\right], \quad 0 \le t \le 24$$

where T represents temperature in degrees Celsius and t represents time in hours measured from 12:00 A.M. (midnight).

67. Biology. Find the approximate body temperature at midnight. Round your answer to the nearest degree.

68. Biology. Find the approximate body temperature at 2:45 P.M. Round your answer to the nearest degree.

▪ CATCH THE MISTAKE

In Exercises 69 and 70, explain the mistake that is made.

69. Use the unit circle to evaluate $\tan\left(\dfrac{5\pi}{6}\right)$ exactly.

Solution:

Tangent is the ratio of sine to cosine.
$$\tan\left(\frac{5\pi}{6}\right) = \frac{\sin\left(\dfrac{5\pi}{6}\right)}{\cos\left(\dfrac{5\pi}{6}\right)}$$

Use the unit circle to identify sine and cosine.
$$\sin\left(\frac{5\pi}{6}\right) = -\frac{\sqrt{3}}{2} \text{ and } \cos\left(\frac{5\pi}{6}\right) = \frac{1}{2}$$

Substitute values for sine and cosine.
$$\tan\left(\frac{5\pi}{6}\right) = \frac{-\dfrac{\sqrt{3}}{2}}{\dfrac{1}{2}}$$

Simplify.
$$\tan\left(\frac{5\pi}{6}\right) = -\sqrt{3}$$

This is incorrect. What mistake was made?

70. Use the unit circle to evaluate $\sec\left(\dfrac{11\pi}{6}\right)$ exactly.

Solution:

Secant is the reciprocal of cosine.
$$\sec\left(\frac{11\pi}{6}\right) = \frac{1}{\cos\left(\dfrac{11\pi}{6}\right)}$$

Use the unit circle to evaluate cosine.
$$\cos\left(\frac{11\pi}{6}\right) = -\frac{1}{2}$$

Substitute the value for cosine.
$$\sec\left(\frac{11\pi}{6}\right) = \frac{1}{-\dfrac{1}{2}}$$

Simplify.
$$\sec\left(\frac{11\pi}{6}\right) = -2$$

This is incorrect. What mistake was made?

▪ CONCEPTUAL

In Exercises 71–78, determine whether each statement is true or false.

71. $\sin(2n\pi + \theta) = \sin\theta$, n an integer.

72. $\cos(2n\pi + \theta) = \cos\theta$, n an integer.

73. $\sin\theta = 1$ when $\theta = \dfrac{(2n+1)\pi}{2}$, n an integer.

74. $\cos\theta = 1$ when $\theta = n\pi$, n an integer.

75. $\tan(\theta + 2n\pi) = \tan\theta$, n an integer.

76. $\tan\theta = 0$ if and only if $\theta = \dfrac{(2n+1)\pi}{2}$, n an integer.

77. Is cosecant an even or an odd function? Justify your answer.

78. Is tangent an even or an odd function? Justify your answer.

▪ CHALLENGE

79. Find all the values of θ, $0 \le \theta \le 2\pi$, for which the equation $\sin\theta = \cos\theta$ is true.

80. Find all the values of θ (θ is any real number) for which the equation $\sin\theta = \cos\theta$ is true.

81. Show that $\csc\left(\dfrac{\pi}{2} + \theta + 2n\pi\right) = \sec\theta$, n an integer.

82. Show that $\sec\left(\dfrac{\pi}{2} - \theta - 2n\pi\right) = \csc\theta$, n an integer.

83. Simplify $\csc\left(2\pi - \dfrac{\pi}{2} - \theta\right) \cdot \sec\left(\theta - \dfrac{\pi}{2}\right) \cdot \sin(-\theta)$.

84. Simplify $\tan\theta \cdot \cot(2\pi - \theta)$.

85. Does there exist an angle $0 \le \theta < 2\pi$ such that $\tan\theta = \cot\theta$?

86. Does there exist an angle $0 \le \theta < 2\pi$ such that $\sec\theta = \csc(-\theta)$?

▪ TECHNOLOGY

87. Use a calculator to approximate $\sin(423°)$. What do you expect $\sin(-423°)$ to be? Verify your answer with a calculator.

88. Use a calculator to approximate $\cos(227°)$. What do you expect $\cos(-227°)$ to be? Verify your answer with a calculator.

89. Use a calculator to approximate $\tan(81°)$. What do you expect $\tan(-81°)$ to be? Verify your answer with a calculator.

90. Use a calculator to approximate $\csc(211°)$. What do you expect $\csc(-211°)$ to be? Verify your answer with a calculator.

For Exercises 91–94, refer to the following:

A graphing calculator can be used to graph the unit circle with parametric equations (these will be covered in more detail in

Section 8.7). For now, set the calculator in parametric and radian modes and let

$$X_1 = \cos T$$
$$Y_1 = \sin T$$

Set the window so that $0 \le T \le 2\pi$, step $= \dfrac{\pi}{15}$, $-2 \le X \le 2$, and $-2 \le Y \le 2$. To approximate the sine or cosine of a T value, use the $\boxed{\text{TRACE}}$ key, type in the T value, and read the corresponding coordinates from the screen.

91. Approximate $\cos\left(\dfrac{\pi}{3}\right)$, take 5 steps of $\dfrac{\pi}{15}$ each, and read the x-coordinate.

92. Approximate $\sin\left(\dfrac{\pi}{3}\right)$, take 5 steps of $\dfrac{\pi}{15}$ each, and read the y-coordinate.

93. Approximate $\sin\left(\dfrac{2\pi}{3}\right)$ to four decimal places.

94. Approximate $\cos\left(\dfrac{5\pi}{4}\right)$ to four decimal places.

GRAPHS OF SINE AND COSINE FUNCTIONS

SKILLS OBJECTIVES

- Determine the amplitude and period of sinusoidal functions.
- Determine the phase shift of a sinusoidal function.
- Graph sinusoidal functions.
- Solve harmonic motion problems.
- Graph sums of functions.

CONCEPTUAL OBJECTIVES

- Understand why the graphs of the sine and cosine functions are called sinusoidal graphs.
- Understand the cyclic nature of periodic functions.
- Visualize harmonic motion as a sinusoidal function.

The Graphs of Sinusoidal Functions

The following are examples of things that repeat in a predictable way (are roughly periodic):

- heartbeat
- tide levels
- time of sunrise
- average outdoor temperature for the time of year

The trigonometric functions are *strictly* periodic. In the unit circle, the value of any of the trigonometric functions is the same for any coterminal angle (same initial and terminal sides no matter how many full rotations the angle makes). For example, if we add (or subtract) multiples of 2π to the angle θ, the values for sine and cosine are unchanged.

$$\sin(\theta + 2n\pi) = \sin\theta \quad \text{or} \quad \cos(\theta + 2n\pi) = \cos\theta \quad (n \text{ is any integer})$$

DEFINITION **Periodic Function**

A function f is called a **periodic function** if there is a positive number p such that

$$f(x + p) = f(x) \quad \text{for all } x \text{ in the domain of } f$$

If p is the smallest such number for which this equation holds, then p is called the **fundamental period**.

You will see in this section and the next that sine, cosine, secant, and cosecant functions have fundamental period 2π, but that tangent and cotangent functions have fundamental period π.

The Graph of $f(x) = \sin x$

Let us start by point-plotting the sine function. We select values for the sine function corresponding to some of the "special angles" we covered in the previous sections of this chapter.

Study Tip

Note that either notation, $y = \sin x$ or $f(x) = \sin x$, can be used.

x	f(x) = sin x	(x, y)
0	$\sin 0 = 0$	$(0, 0)$
$\dfrac{\pi}{4}$	$\sin\left(\dfrac{\pi}{4}\right) = \dfrac{\sqrt{2}}{2}$	$\left(\dfrac{\pi}{4}, \dfrac{\sqrt{2}}{2}\right)$
$\dfrac{\pi}{2}$	$\sin\left(\dfrac{\pi}{2}\right) = 1$	$\left(\dfrac{\pi}{2}, 1\right)$
$\dfrac{3\pi}{4}$	$\sin\left(\dfrac{3\pi}{4}\right) = \dfrac{\sqrt{2}}{2}$	$\left(\dfrac{3\pi}{4}, \dfrac{\sqrt{2}}{2}\right)$
π	$\sin \pi = 0$	$(\pi, 0)$
$\dfrac{5\pi}{4}$	$\sin\left(\dfrac{5\pi}{4}\right) = -\dfrac{\sqrt{2}}{2}$	$\left(\dfrac{5\pi}{4}, -\dfrac{\sqrt{2}}{2}\right)$
$\dfrac{3\pi}{2}$	$\sin\left(\dfrac{3\pi}{2}\right) = -1$	$\left(\dfrac{3\pi}{2}, -1\right)$
$\dfrac{7\pi}{4}$	$\sin\left(\dfrac{7\pi}{4}\right) = -\dfrac{\sqrt{2}}{2}$	$\left(\dfrac{7\pi}{4}, -\dfrac{\sqrt{2}}{2}\right)$
2π	$\sin 2\pi = 0$	$(2\pi, 0)$

By plotting the above coordinates (x, y), we can obtain the graph of one **period**, or **cycle**, of the graph of $y = \sin x$. Note that $\dfrac{\sqrt{2}}{2} \approx 0.7$.

Study Tip

Looking at the graph of $f(x) = \sin x$, we are reminded that $\sin x > 0$ when $0 < x < \pi$ and $\sin x < 0$ when $\pi < x < 2\pi$.

We can extend the graph horizontally in both directions (left and right) since the domain of the sine function is the set of all real numbers.

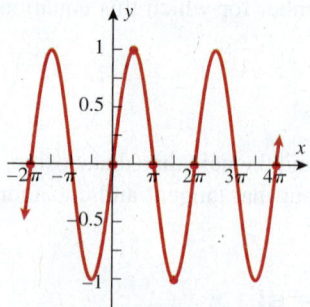

In this section, we are no longer showing angles on the unit circle but are now showing angles as *real numbers* in radians on the x-axis of the *Cartesian* graph. Therefore, we no longer illustrate a "terminal side" to an angle—the physical arcs and angles no longer exist; only their measures exist, as values of the x-coordinate.

If we graph the function $f(x) = \sin x$, the x-intercepts correspond to values of x at which the sine function is equal to zero.

x	$f(x) = \sin x$	(x, y)
0	$\sin 0 = 0$	$(0, 0)$
π	$\sin \pi = 0$	$(\pi, 0)$
2π	$\sin(2\pi) = 0$	$(2\pi, 0)$
3π	$\sin(3\pi) = 0$	$(3\pi, 0)$
4π	$\sin(4\pi) = 0$	$(4\pi, 0)$
$\cdots$	$\cdots$	$\cdots$
$n\pi$	$\sin(n\pi) = 0$	$(n\pi, 0)$

n is an integer.

Notice that the point $(0, 0)$ is both a y-intercept and an x-intercept but all x-intercepts have the form $(n\pi, 0)$. The maximum value of the sine function is 1, and the minimum value of the sine function is -1; these values occur at odd multiples of $\dfrac{\pi}{2}$.

x	$f(x) = \sin x$	(x, y)
$\dfrac{\pi}{2}$	$\sin\left(\dfrac{\pi}{2}\right) = 1$	$\left(\dfrac{\pi}{2}, 1\right)$
$\dfrac{3\pi}{2}$	$\sin\left(\dfrac{3\pi}{2}\right) = -1$	$\left(\dfrac{3\pi}{2}, -1\right)$
$\dfrac{5\pi}{2}$	$\sin\left(\dfrac{5\pi}{2}\right) = 1$	$\left(\dfrac{5\pi}{2}, 1\right)$
$\dfrac{7\pi}{2}$	$\sin\left(\dfrac{7\pi}{2}\right) = -1$	$\left(\dfrac{7\pi}{2}, -1\right)$
$\cdots$	$\cdots$	$\cdots$
$\dfrac{(2n+1)\pi}{2}$	$\sin\left(\dfrac{(2n+1)\pi}{2}\right) = \pm 1$	$\left(\dfrac{(2n+1)\pi}{2}, \pm 1\right)$

n is an integer.

The following box summarizes the sine function:

SINE FUNCTION $f(x) = \sin x$

- Domain: $(-\infty, \infty)$ or $-\infty < x < \infty$
- Range: $[-1, 1]$ or $-1 \le y \le 1$
- The sine function is an odd function:
  - symmetric about the origin
  - $\sin(-x) = -\sin x$
- The sine function is a periodic function with fundamental period 2π.
- The x-intercepts, $0, \pm\pi, \pm 2\pi, \ldots,$ are of the form $n\pi$, where n is an integer.
- The maximum (1) and minimum (-1) values of the sine function correspond to x-values of the form $\dfrac{(2n+1)\pi}{2}$ such as $\pm\dfrac{\pi}{2}, \pm\dfrac{3\pi}{2}, \pm\dfrac{5\pi}{2}, \ldots.$

The Graph of $f(x) = \cos x$

Let us start by point-plotting the cosine function.

x	$f(x) = \cos x$	(x, y)
0	$\cos 0 = 1$	$(0, 1)$
$\dfrac{\pi}{4}$	$\cos\left(\dfrac{\pi}{4}\right) = \dfrac{\sqrt{2}}{2}$	$\left(\dfrac{\pi}{4}, \dfrac{\sqrt{2}}{2}\right)$
$\dfrac{\pi}{2}$	$\cos\left(\dfrac{\pi}{2}\right) = 0$	$\left(\dfrac{\pi}{2}, 0\right)$
$\dfrac{3\pi}{4}$	$\cos\left(\dfrac{3\pi}{4}\right) = -\dfrac{\sqrt{2}}{2}$	$\left(\dfrac{3\pi}{4}, -\dfrac{\sqrt{2}}{2}\right)$
π	$\cos \pi = -1$	$(\pi, -1)$
$\dfrac{5\pi}{4}$	$\cos\left(\dfrac{5\pi}{4}\right) = -\dfrac{\sqrt{2}}{2}$	$\left(\dfrac{5\pi}{4}, -\dfrac{\sqrt{2}}{2}\right)$
$\dfrac{3\pi}{2}$	$\cos\left(\dfrac{3\pi}{2}\right) = 0$	$\left(\dfrac{3\pi}{2}, 0\right)$
$\dfrac{7\pi}{4}$	$\cos\left(\dfrac{7\pi}{4}\right) = \dfrac{\sqrt{2}}{2}$	$\left(\dfrac{7\pi}{4}, \dfrac{\sqrt{2}}{2}\right)$
2π	$\cos(2\pi) = 1$	$(2\pi, 1)$

By plotting the above coordinates (x, y), we can obtain the graph of one period, or cycle, of the graph of $y = \cos x$. Note that $\dfrac{\sqrt{2}}{2} \approx 0.7$.

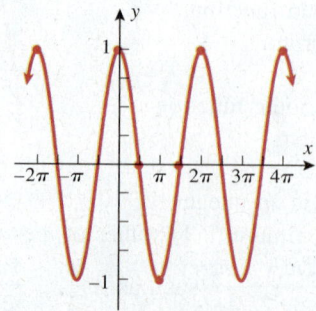

We can extend the graph horizontally in both directions (left and right) since the domain of the cosine function is all real numbers.

If we graph the function $f(x) = \cos x$, the x-intercepts correspond to values of x at which the cosine function is equal to zero.

x	$f(x) = \cos x$	(x, y)	
$\dfrac{\pi}{2}$	$\cos\left(\dfrac{\pi}{2}\right) = 0$	$\left(\dfrac{\pi}{2}, 0\right)$	
$\dfrac{3\pi}{2}$	$\cos\left(\dfrac{3\pi}{2}\right) = 0$	$\left(\dfrac{3\pi}{2}, 0\right)$	
$\dfrac{5\pi}{2}$	$\cos\left(\dfrac{5\pi}{2}\right) = 0$	$\left(\dfrac{5\pi}{2}, 0\right)$	
$\dfrac{7\pi}{2}$	$\cos\left(\dfrac{7\pi}{2}\right) = 0$	$\left(\dfrac{7\pi}{2}, 0\right)$	
$\ldots$	$\ldots$	$\ldots$	
$\dfrac{(2n+1)\pi}{2}$	$\cos\left(\dfrac{(2n+1)\pi}{2}\right) = 0$	$\left(\dfrac{(2n+1)\pi}{2}, 0\right)$	n is an integer.

The point $(0, 1)$ is the y-intercept, and there are several x-intercepts of the form $\left(\dfrac{(2n+1)\pi}{2}, 0\right)$. The maximum value of the cosine function is 1 and the minimum value of the cosine function is -1; these values occur at integer multiples of π; $n\pi$.

x	$f(x) = \cos x$	(x, y)	
0	$\cos 0 = 1$	$(0, 1)$	
π	$\cos \pi = -1$	$(\pi, -1)$	
2π	$\cos 2\pi = 1$	$(2\pi, 1)$	
3π	$\cos 3\pi = -1$	$(3\pi, -1)$	
4π	$\cos 4\pi = 1$	$(4\pi, 1)$	
$\ldots$	$\ldots$	$\ldots$	
$n\pi$	$\cos n\pi = \pm 1$	$(n\pi, \pm 1)$	n is an integer.

The following box summarizes the cosine function:

Technology Tip

Set the window at Xmin at -2π, Xmax at 4π, Xscl at $\dfrac{\pi}{2}$, Ymin at -1, Ymax at 1, and Yscl at 1. Setting Xscl at $\dfrac{\pi}{2}$ will mark the labels on the x-axis in terms of multiples of $\dfrac{\pi}{2}$.

Use ☐ Y= ☐ to enter the function $\cos(X)$.

COSINE FUNCTION $f(x) = \cos x$

- Domain: $(-\infty, \infty)$ or $-\infty < x < \infty$
- Range: $[-1, 1]$ or $-1 \le y \le 1$
- The cosine function is an even function:
  - symmetric about the y-axis
  - $\cos(-x) = \cos x$
- The cosine function is a periodic function with fundamental period 2π.
- The x-intercepts, $\pm\dfrac{\pi}{2}, \pm\dfrac{3\pi}{2}, \pm\dfrac{5\pi}{2}, \ldots$, are odd integer multiples of $\dfrac{\pi}{2}$ that have the form $\dfrac{(2n+1)\pi}{2}$, where n is an integer.
- The maximum (1) and minimum (-1) values of the cosine function correspond to x-values of the form $n\pi$, such as $0, \pm\pi, \pm2\pi, \ldots$.

The Amplitude and Period of Sinusoidal Graphs

In mathematics, the word **sinusoidal** means "resembling the sine function." Let us start by graphing $f(x) = \sin x$ and $f(x) = \cos x$ on the same graph. Notice that they have similar characteristics (domain, range, period, and shape).

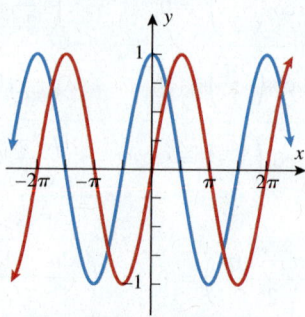

In fact, if we were to shift the cosine graph to the right $\dfrac{\pi}{2}$ units, the two graphs would be identical. For that reason, we refer to any graphs of the form $y = \cos x$ or $y = \sin x$ as **sinusoidal functions**.

We now turn our attention to graphs of the form $y = A\sin(Bx)$ and $y = A\cos(Bx)$, which are graphs like $y = \sin x$ and $y = \cos x$ that have been stretched or compressed vertically and horizontally.

Classroom Example 6.8.1
Plot the functions $y = 2\cos x$, $y = 4\cos x$, $y = \frac{1}{2}\cos x$ on the same set of axes. Discuss the relationship among the graphs.

Answer:

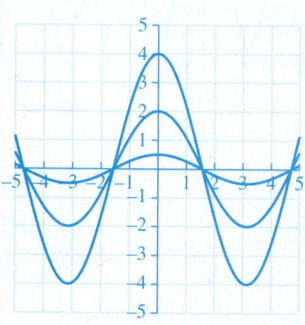

The amplitude changes as the coefficient changes.

EXAMPLE 1   Vertical Stretching and Compressing

Plot the functions $y = 2\sin x$ and $y = \frac{1}{2}\sin x$ on the same graph with $y = \sin x$ on the interval $-4\pi \le x \le 4\pi$.

Solution:

STEP 1 Make a table with the coordinate values of the graphs:

x	0	$\dfrac{\pi}{2}$	π	$\dfrac{3\pi}{2}$	2π
$\sin x$	0	1	0	-1	0
$2\sin x$	0	2	0	-2	0
$\frac{1}{2}\sin x$	0	$\frac{1}{2}$	0	$-\frac{1}{2}$	0

STEP 2 Label the points on the graph and connect with a smooth curve over one period, $0 \le x \le 2\pi$.

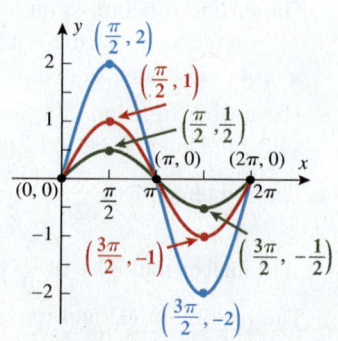

Step 3 Extend the graph in both directions (repeat every 2π).

■ **Answer:**

■ **YOUR TURN** Plot the functions $y = 3\cos x$ and $y = \frac{1}{3}\cos x$ on the same graph with $y = \cos x$ on the interval $-2\pi \le x \le 2\pi$.

Notice in Example 1 and the corresponding Your Turn that:

- $y = 2\sin x$ has the shape and period of $y = \sin x$ but is stretched vertically.
- $y = \frac{1}{2}\sin x$ has the shape and period of $y = \sin x$ but is compressed vertically.
- $y = 3\cos x$ has the shape and period of $y = \cos x$ but is stretched vertically.
- $y = \frac{1}{3}\cos x$ has the shape and period of $y = \cos x$ but is compressed vertically.

In general, functions of the form $y = A\sin x$ and $y = A\cos x$ are stretched vertically when $|A| > 1$ and compressed vertically when $|A| < 1$.

The **amplitude** of a periodic function is half the difference between the maximum value of the function and the minimum value of the function. For the functions $y = \sin x$ and $y = \cos x$, the maximum value is 1 and the minimum value is -1. Therefore, the amplitude of each of these two functions is $|A| = \frac{1}{2}|1 - (-1)| = 1$.

AMPLITUDE OF SINUSOIDAL FUNCTIONS

For sinusoidal functions of the form $y = A\sin(Bx)$ and $y = A\cos(Bx)$, the **amplitude** is $|A|$. When $|A| < 1$, the graph is compressed vertically, and when $|A| > 1$, the graph is stretched vertically.

EXAMPLE 2 Finding the Amplitude of Sinusoidal Functions

State the amplitude of

a. $f(x) = -4\cos x$

b. $g(x) = \frac{1}{5}\sin x$

Solution (a): The amplitude is the magnitude of -4. $A = |-4| = \boxed{4}$

Solution (b): The amplitude is the magnitude of $\frac{1}{5}$. $A = \left|\frac{1}{5}\right| = \boxed{\frac{1}{5}}$

Classroom Example 6.8.2
Determine the amplitude of
$y = -\frac{1}{6}\cos x$.

Answer: $\frac{1}{6}$

EXAMPLE 3 Horizontal Stretching and Compressing

Plot the functions $y = \cos(2x)$ and $y = \cos(\frac{1}{2}x)$ on the same graph with $y = \cos x$ on the interval $-2\pi \leq x \leq 2\pi$.

Solution:

STEP 1 Make a table with the coordinate values of the graphs. It is necessary only to select the points that correspond to x-intercepts, $(y = 0)$, and maximum and minimum points, $(y = \pm 1)$. Usually, the period is divided into four subintervals (which you will see in Examples 5 to 7).

x	0	$\frac{\pi}{4}$	$\frac{\pi}{2}$	$\frac{3\pi}{4}$	π	$\frac{5\pi}{4}$	$\frac{3\pi}{2}$	$\frac{7\pi}{4}$	2π
$\cos x$	1		0		-1		0		1
$\cos(2x)$	1	0	-1	0	1	0	-1	0	1
$\cos(\frac{1}{2}x)$	1				0				-1

STEP 2 Label the points on the graph and connect with a smooth curve.

$y = \cos x$

$y = \cos(2x)$

$y = \cos\left(\frac{1}{2}x\right)$

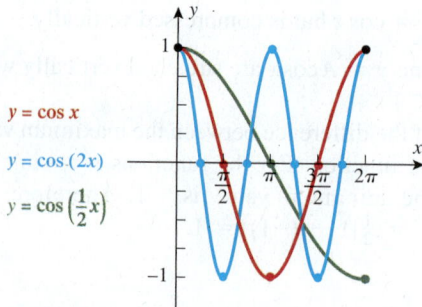

STEP 3 Extend the graph to cover the entire interval: $-2\pi \leq x \leq 2\pi$.

$y = \cos x$

$y = \cos(2x)$

$y = \cos\left(\frac{1}{2}x\right)$

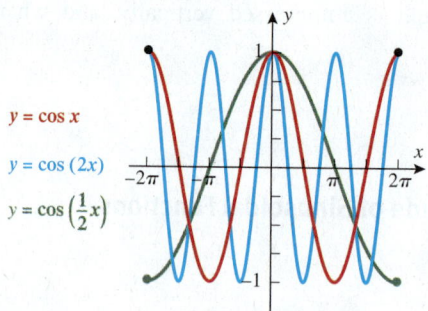

Classroom Example 6.8.3

Plot the functions $y = \cos(3x)$, $y = \cos(6x)$, $y = \cos(\frac{1}{3}x)$ on the same set of axes. Discuss the relationship among the graphs.

Answer:

The coefficient is linked to the period of the function, thereby creating a horizontal compression or expansion.

■ **Answer:**

$y = \sin x$ $y = \sin(2x)$

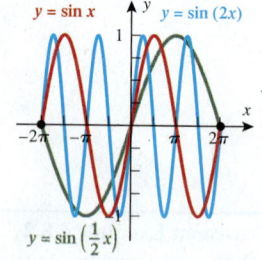

$y = \sin\left(\frac{1}{2}x\right)$

■ **YOUR TURN** Plot the functions $y = \sin(2x)$ and $y = \sin(\frac{1}{2}x)$ on the same graph with $y = \sin x$ on the interval $-2\pi \leq x \leq 2\pi$.

Notice in Example 3 and the corresponding Your Turn that:

■ $y = \cos(2x)$ has the shape and amplitude of $y = \cos x$ but is compressed horizontally.

■ $y = \cos(\frac{1}{2}x)$ has the shape and amplitude of $y = \cos x$ but is stretched horizontally.

■ $y = \sin(2x)$ has the shape and amplitude of $y = \sin x$ but is compressed horizontally.

■ $y = \sin(\frac{1}{2}x)$ has the shape and amplitude of $y = \sin x$ but is stretched horizontally.

In general, functions of the form $y = \sin(Bx)$ and $y = \cos(Bx)$, with $B > 0$, are compressed horizontally when $B > 1$ and stretched horizontally when $0 < B < 1$. We will discuss negative arguments ($B < 0$) later in this section in the context of *reflections*.

The period of the functions $y = \sin x$ and $y = \cos x$ is 2π. To find the period of a function of the form $y = A\sin(Bx)$ or $y = A\cos(Bx)$ set Bx equal to 2π and solve for x.

$$Bx = 2\pi$$
$$x = \frac{2\pi}{B}$$

PERIOD OF SINUSOIDAL FUNCTIONS

For sinusoidal functions of the form $y = A\sin(Bx)$ and $y = A\cos(Bx)$, with $B > 0$, the **period** is $\frac{2\pi}{B}$. When $0 < B < 1$, the graph is stretched horizontally, and when $B > 1$, the graph is compressed horizontally since the period is less than 2π.

EXAMPLE 4   **Finding the Period of a Sinusoidal Function**

State the period of

a. $y = \cos(4x)$

b. $y = \sin(\frac{1}{3}x)$

Solution (a):

Compare $\cos(4x)$ with $\cos(Bx)$ to identify B. $B = 4$

Calculate the period of $\cos(4x)$, using $p = \frac{2\pi}{B}$. $p = \frac{2\pi}{4} = \frac{\pi}{2}$

The period of $\cos(4x)$ is $\boxed{p = \frac{\pi}{2}}$.

Solution (b):

Compare $\sin(\frac{1}{3}x)$ with $\sin(Bx)$ to identify B. $B = \frac{1}{3}$

Calculate the period of $\sin(\frac{1}{3}x)$, using $p = \frac{2\pi}{B}$. $p = \frac{2\pi}{\frac{1}{3}} = 6\pi$

The period of $\sin(\frac{1}{3}x)$ is $\boxed{p = 6\pi}$.

■ **YOUR TURN**   State the period of

a. $y = \sin(3x)$ **b.** $y = \cos(\frac{1}{2}x)$

Classroom Example 6.8.4
Determine the period of the following sinusoidal functions.

a. $f(x) = -\sin(6\pi x)$

b.* $g(x) = \cos((3m - 1)x)$, where m is an integer

c.* $h(x) = \sin(-10x)$

Answer:

a. $\frac{1}{3}$ **b.** $\frac{2\pi}{3m - 1}$ **c.** $\frac{\pi}{5}$

■ **Answer:**

a. $p = \frac{2\pi}{3}$ **b.** $p = 4\pi$

Now that you know the basic graphs of $y = \sin x$ and $y = \cos x$, you can sketch one cycle (period) of these graphs with the following x-values: $0, \dfrac{\pi}{2}, \pi, \dfrac{3\pi}{2}, 2\pi$. For a period of 2π, we used steps of $\dfrac{\pi}{2}$. Therefore, for functions of the form $y = A\sin(Bx)$ or $y = A\cos(Bx)$, when we start at the origin and as long as we include these five basic values (corresponding to four equal intervals) during one period, we are able to sketch the graphs.

STRATEGY FOR SKETCHING GRAPHS OF SINUSOIDAL FUNCTIONS

To graph $y = A\sin(Bx)$ or $y = A\cos(Bx)$ with $B > 0$:

Step 1: Find the amplitude $|A|$ and period $\dfrac{2\pi}{B}$.

Step 2: Divide the period into four subintervals of equal lengths.

Step 3: Make a table and evaluate the function for x-values from Step 2 starting at $x = 0$.

Step 4: Draw the xy-plane (label the y-axis from $-|A|$ to $|A|$) and plot the points found in Step 3.

Step 5: Connect the points with a sinusoidal curve (with amplitude $|A|$).

Step 6: Extend the graph over one or two additional periods in both directions (left and right).

EXAMPLE 5   Graphing Sinusoidal Functions of the Form $y = A\sin(Bx)$

Use the strategy for graphing a sinusoidal function to graph $y = 3\sin(2x)$.

Solution:

STEP 1 Find the amplitude and period for $A = 3$ and $B = 2$.

$$|A| = |3| = 3$$
$$p = \frac{2\pi}{B} = \frac{2\pi}{2} = \pi$$

STEP 2 Divide the period π into four equal steps. $\dfrac{\pi}{4}$

STEP 3 Make a table starting at $x = 0$ to the period $x = \pi$ in steps of $\dfrac{\pi}{4}$.

x	$y = 3\sin 2x$	(x, y)
0	$3[\sin 0] = 3[0] = 0$	$(0, 0)$
$\dfrac{\pi}{4}$	$3\left[\sin\left(\dfrac{\pi}{2}\right)\right] = 3[1] = 3$	$\left(\dfrac{\pi}{4}, 3\right)$
$\dfrac{\pi}{2}$	$3[\sin \pi] = 3[0] = 0$	$\left(\dfrac{\pi}{2}, 0\right)$
$\dfrac{3\pi}{4}$	$3\left[\sin\left(\dfrac{3\pi}{2}\right)\right] = 3[-1] = -3$	$\left(\dfrac{3\pi}{4}, -3\right)$
π	$3[\sin(2\pi)] = 3[0] = 0$	$(\pi, 0)$

STEP 4 Draw the *xy*-plane and label the points in the table.

STEP 5 Connect the points with a sinusoidal curve.

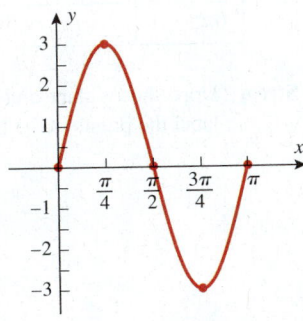

STEP 6 Repeat over several periods (to the left and right).

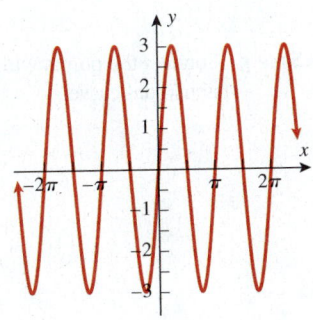

■ **YOUR TURN** Use the strategy for graphing sinusoidal functions to graph $y = 2\sin(3x)$.

EXAMPLE 6 **Graphing Sinusoidal Functions of the Form $y = A\cos(Bx)$**

Use the strategy for graphing a sinusoidal function to graph $y = -2\cos(\frac{1}{3}x)$.

Solution:

STEP 1 Find the amplitude and period for $A = -2$ and $B = \frac{1}{3}$.

$$|A| = |-2| = 2$$

$$p = \frac{2\pi}{B} = \frac{2\pi}{\frac{1}{3}} = 6\pi$$

STEP 2 Divide the period 6π into four equal steps.

$$\frac{6\pi}{4} = \frac{3\pi}{2}$$

Classroom Example 6.8.5
Graph:

a. $y = -\frac{1}{2}\sin\pi x$

b. $y = 3\sin\left(\frac{\pi}{3}x\right)$

Answer:

a.

b.

■ **Answer:**

Technology Tip

Use a TI calculator to check the graph of $y = -2\cos(\frac{1}{3}x)$. Set the window at Xmin at -12π, Xmax at 12π, Xsc1 at $\frac{3\pi}{2}$, Ymin at -2, Ymax at 2, and Ysc1 at 1. Setting Xsc1 at $\frac{3\pi}{2}$ will mark the labels on the *x*-axis in terms of multiples of $\frac{3\pi}{2}$.

Classroom Example 6.8.6
Graph:
$$y = 2\cos\left(\frac{\pi}{4}x\right)$$

Answer:

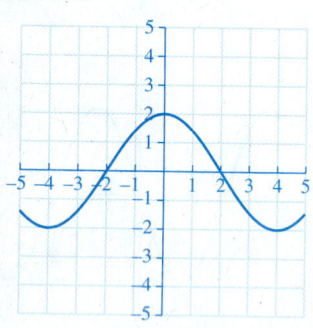

STEP 3 Make a table starting at $x = 0$ and completing one period of 6π in steps of $\frac{3\pi}{2}$.

x	$y = -2\cos(\frac{1}{3}x)$	(x, y)
0	$-2[\cos 0] = -2[1] = -2$	$(0, -2)$
$\frac{3\pi}{2}$	$-2\left[\cos\left(\frac{\pi}{2}\right)\right] = -2[0] = 0$	$\left(\frac{3\pi}{2}, 0\right)$
3π	$-2[\cos \pi] = -2[-1] = 2$	$(3\pi, 2)$
$\frac{9\pi}{2}$	$-2\left[\cos\left(\frac{3\pi}{2}\right)\right] = -2[0] = 0$	$\left(\frac{9\pi}{2}, 0\right)$
6π	$-2[\cos(2\pi)] = -2[1] = -2$	$(6\pi, -2)$

STEP 4 Draw the *xy*-plane and label the points in the table.

STEP 5 Connect the points with a sinusoidal curve.

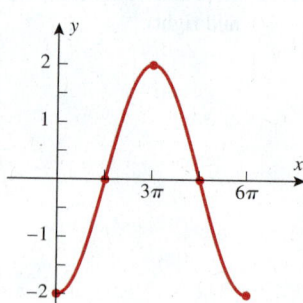

STEP 6 Repeat over several periods (to the left and right).

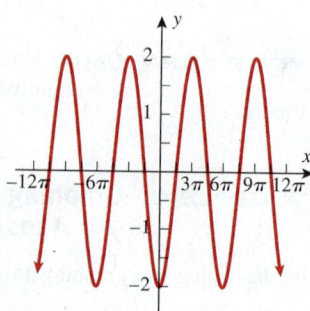

■ YOUR TURN Use the strategy for graphing a sinusoidal function to graph
$$y = -3\cos\left(\frac{1}{2}x\right).$$

■ Answer:

Notice in Example 6 and the corresponding Your Turn that when *A* is negative, the result is a reflection of the original function (sine or cosine) about the *x*-axis.

EXAMPLE 7   Finding an Equation for a Sinusoidal Graph

Find an equation for the graph.

Classroom Example 6.8.7*
Find an equation for the graph.

Answer:

$$y = -3\cos\left(\frac{\pi}{4}x\right)$$

Solution:

This graph represents a cosine function. $y = A\cos(Bx)$

The amplitude is 4 (half the maximum spread). $|A| = 4$

The period $\dfrac{2\pi}{B}$ is equal to 4π. $\dfrac{2\pi}{B} = 4\pi$

Solve for B. $B = \dfrac{1}{2}$

Substitute $A = 4$ and $B = \frac{1}{2}$ into $y = A\cos(Bx)$.

$$y = 4\cos\left(\frac{1}{2}x\right)$$

■ **YOUR TURN** Find an equation for the graph.

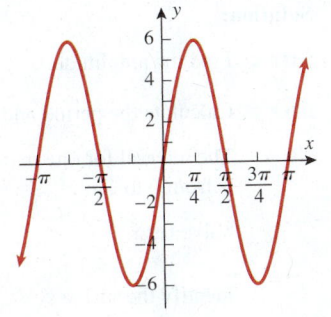

■ **Answer:** $y = 6\sin(2x)$

Graphing a Shifted Sinusoidal Function: $y = A\sin(Bx + C) + D$ and $y = A\cos(Bx + C) + D$

Recall from Section 3.3 that we graph functions using horizontal and vertical translations (shifts) in the following way ($c > 0$):

- To graph $f(x + c)$, shift $f(x)$ to the **left** c units.
- To graph $f(x - c)$, shift $f(x)$ to the **right** c units.
- To graph $f(x) + c$, shift $f(x)$ **up** c units.
- To graph $f(x) - c$, shift $f(x)$ **down** c units.

Classroom Example 6.8.8

Graph $y = \dfrac{1}{2}\sin\left(\dfrac{\pi}{2}x - \dfrac{\pi}{2}\right)$.

Answer:

Study Tip

Rewriting in standard form

$$y = A\sin\left[B\left(x \pm \dfrac{C}{B}\right)\right]$$

makes identifying the phase shift easier.

Study Tip

An alternative method for finding the period and phase shift is to first write the function in standard form.

$$y = 5\cos\left[4\left(x + \dfrac{\pi}{4}\right)\right]$$

$$B = 4$$

$$\text{Period} = \dfrac{2\pi}{B} = \dfrac{2\pi}{4} = \dfrac{\pi}{2}$$

Phase shift $= \dfrac{\pi}{4}$ units to the left.

■ **Answer:**

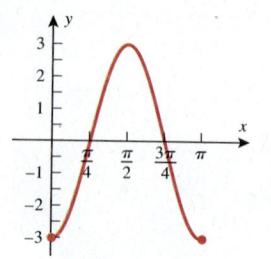

To graph functions of the form $y = A\sin(Bx + C) + D$ and $y = A\cos(Bx + C) + D$, utilize the strategy below.

STRATEGY FOR GRAPHING $y = A\sin(Bx + C) + D$ AND $y = A\cos(Bx + C) + D$

A strategy for graphing $y = A\sin(Bx + C) + D$ is outlined below. The same strategy can be used to graph $y = A\cos(Bx + C) + D$.

Step 1: Find the amplitude $|A|$.

Step 2: Find the period $\dfrac{2\pi}{B}$ and **phase shift** $-\dfrac{C}{B}$.

Step 3: Graph $y = A\sin(Bx + C)$ over one period $\left(\text{from } -\dfrac{C}{B} \text{ to } -\dfrac{C}{B} + \dfrac{2\pi}{B}\right)$.

Step 4: Extend the graph over several periods.

Step 5: Shift the graph of $y = A\sin(Bx + C)$ vertically D units.

Note: If we rewrite the function in **standard form**, we get

$$y = A\sin\left[B\left(x + \dfrac{C}{B}\right)\right] + D$$

which makes it easier to identify the **phase shift**.
If $B < 0$, we can use properties of even and odd functions

$$\sin(-x) = -\sin x \qquad \cos(-x) = \cos x$$

to rewrite the function with $B > 0$.

EXAMPLE 8   Graphing Functions of the Form $y = A\cos(Bx \pm C)$

Graph $y = 5\cos(4x + \pi)$ over one period.

Solution:

STEP 1 Find the amplitude. $|A| = |5| = 5$

STEP 2 Calculate the period and phase shift.

The interval for one period is from **0** to **2π**.

$4x + \pi = 0$ to $4x + \pi = 2\pi$

Solve for x.

$x = -\dfrac{\pi}{4}$ to $x = -\dfrac{\pi}{4} + \dfrac{\pi}{2}$

Identify the **phase shift**.

$-\dfrac{C}{B} = -\dfrac{\pi}{4}$

Identify the **period** $\dfrac{2\pi}{B}$

$\dfrac{2\pi}{B} = \dfrac{2\pi}{4} = \dfrac{\pi}{2}$

STEP 3 Graph.

Draw a cosine function starting at $x = -\dfrac{\pi}{4}$ with period $\dfrac{\pi}{2}$ and amplitude 5.

■ **YOUR TURN** Graph $y = 3\cos(2x - \pi)$ over one period.

EXAMPLE 9 Graphing Sinusoidal Functions

Graph $y = -3 + 2\cos(2x - \pi)$.

Solution:

STEP 1 Find the amplitude.

$|A| = |2| = 2$

STEP 2 Find the phase shift and period.

Set $2x - \pi$ equal to 0 and 2π.

$2x - \pi = 0 \quad$ to $\quad 2x - \pi = 2\pi$

Solve for x.

$x = \dfrac{\pi}{2} \quad$ to $\quad x = \dfrac{\pi}{2} + \pi$

Identify phase shift and period.

$-\dfrac{C}{B} = \dfrac{\pi}{2}$

$\dfrac{2\pi}{B} = \pi$

STEP 3 Graph $y = 2\cos(2x - \pi)$ starting at $x = \dfrac{\pi}{2}$ over one period, π.

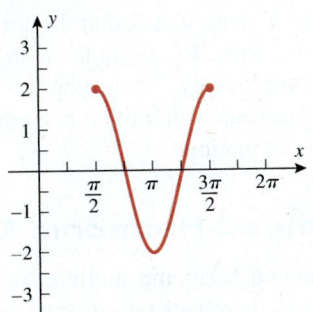

STEP 4 Extend the graph of $y = 2\cos(2x - \pi)$ over several periods.

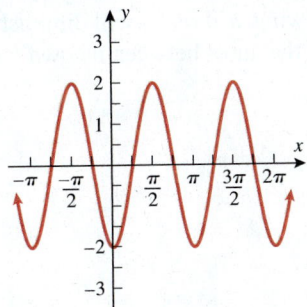

STEP 5 Shift the graph of $y = 2\cos(2x - \pi)$ down 3 units to arrive at the graph of $y = -3 + 2\cos(2x - \pi)$.

■ **YOUR TURN** Graph $y = 2 - 3\sin(2x + \pi)$.

Technology Tip

Use a TI calculator to check the graph of $y = -3 + 2\cos(2x - \pi)$. Set the window at Xmin at $-\pi/2$, Xmax at $3\pi/2$, Xscl at $\pi/4$, Ymin at -5, Ymax at 1, and Yscl at 1. Setting Xscl at $\pi/4$ will mark the labels on the x-axis in terms of multiples of $\pi/4$.

■**Answer:**

Harmonic Motion

One of the most important applications of sinusoidal functions is in describing *harmonic motion*, which we define as the symmetric periodic movement of an object or quantity about a center (equilibrium) position or value. The oscillation of a pendulum is a form of harmonic motion. Other examples are the recoil of a spring balance scale when a weight is placed on the tray and the variation of current or voltage within an AC circuit.

There are three types of harmonic motion: **simple harmonic motion**, **damped harmonic motion**, and **resonance**.

Simple Harmonic Motion

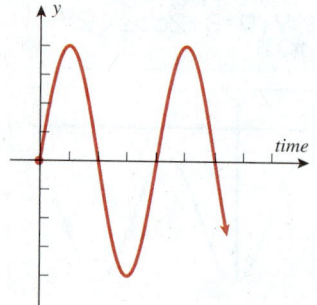

Simple harmonic motion is the kind of *unvarying* periodic motion that would occur in an ideal situation in which no resistive forces, such as friction, cause the amplitude of oscillation to decrease over time: the amplitude stays in exactly the same range in each period as time—the variable on the horizontal axis—increases. It will also occur if energy is being supplied at the correct rate to overcome resistive forces. Simple harmonic motion occurs, for example, in an AC electric circuit when a power source is consistently supplying energy. When you are swinging on a swing and "pumping" energy into the swing to keep it in motion at a constant period and amplitude, you are sustaining simple harmonic motion.

Damped Harmonic Motion

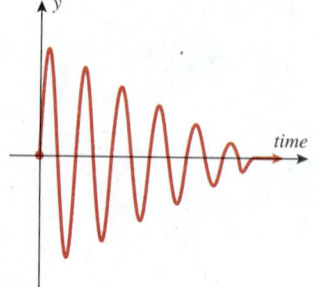

In damped harmonic motion, the amplitude of the periodic motion decreases as time increases. If you are on a moving swing and stop "pumping" new energy into the swing, the swing will continue moving with a constant period, but the amplitude—the height to which the swing will rise—will diminish with each cycle as the swing is slowed down by friction with the air or between its own moving parts.

Resonance

Resonance occurs when the amplitude of periodic motion increases as time increases. It is caused when the energy applied to an oscillating object or system is more than what is needed to oppose friction or other forces and sustain simple harmonic motion. Instead, the applied energy *increases* the amplitude of harmonic motion with each cycle. With resonance, eventually, the amplitude becomes unbounded and the result is disastrous. Bridges have collapsed because of resonance. On the previous page are pictures of the Tacoma Narrows Bridge (near Seattle, Washington) that collapsed due to high winds resulting in resonance. Military soldiers know that when they march across a bridge, they must break cadence to prevent resonance.

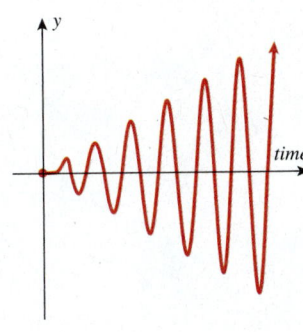

Examples of Harmonic Motion

If we hang a weight from a spring, then while the resulting "system" is at rest we say it is in the equilibrium position.

If we then pull down on the weight and release it, the elasticity in the spring pulls the weight up and causes it to start oscillating up and down.

If we neglect friction and air resistance, we can imagine that the combination of the weight and the spring will oscillate indefinitely; the height of the weight with respect to the equilibrium position can be modeled by a simple sinusoidal function. This is an example of **simple harmonic motion**.

SIMPLE HARMONIC MOTION

The position of a point oscillating around an equilibrium position at time t is modeled by the sinusoidal function

$$y = A \sin(\omega t) \qquad \text{or} \qquad y = A \cos(\omega t)$$

Here $|A|$ is the amplitude and the period is $\dfrac{2\pi}{\omega}$, where $\omega > 0$.

Note: The symbol ω (Greek lowercase Omega) represents the angular frequency.

EXAMPLE 10 Simple Harmonic Motion

Let the height of the seat of a swing be equal to zero when the swing is at rest. Assume that a child starts swinging until she reaches the highest she can swing and keeps her effort constant. The height $h(t)$ of the seat is given by

$$h(t) = 8\sin\left(\frac{\pi}{2}t\right)$$

where t is time in seconds and h is the height in feet. Note that positive h indicates height reached swinging forward and negative h indicates height reached swinging backward. Assume that $t = 0$ is when the child passes through the equilibrium position swinging forward.

a. Graph the height function $h(t)$ for $0 \le t \le 4$.

b. What is the maximum height above the resting level reached by the seat of the swing?

c. What is the period of the swinging child?

Solution (a):

Make a table with integer values of t. $\qquad 0 \le t \le 4$

t (seconds)	$y = h(t) = 8\sin\left(\dfrac{\pi}{2}t\right)$ (feet)	(t, y)
0	$8\sin 0 = 0$	$(0, 0)$
1	$8\sin\left(\dfrac{\pi}{2}\right) = 8$	$(1, 8)$
2	$8\sin \pi = 0$	$(2, 0)$
3	$8\sin\left(\dfrac{3\pi}{2}\right) = -8$	$(3, -8)$
4	$8\sin(2\pi) = 0$	$(4, 0)$

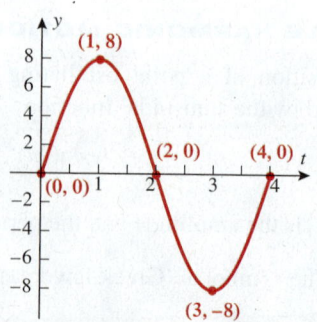

Labeling the time and height on the original diagram, we see that the maximum height is 8 feet and the period is 4 seconds.

$t = 1 \text{ sec}$ $t = 3 \text{ sec}$

8 ft

$t = 0 \text{ sec}$
$= 2 \text{ sec}$
$= 4 \text{ sec}$

Solutions (b) and (c):

$$h(t) = 8 \sin\left(\underset{A}{\underbrace{\frac{\pi}{2}}} t\right) = 8 \sin\left(\underset{\omega}{\underbrace{\frac{2\pi}{4}}} t\right)$$

The maximum height above the equilibrium is the amplitude.

$A = 8 \text{ ft}$

The period is $\dfrac{2\pi}{\omega}$.

$$p = \frac{2\pi}{\frac{2\pi}{4}} = 4 \text{ sec}$$

Damped harmonic motion can be modeled by a sinusoidal function whose amplitude decreases as time increases. If we again hang a weight from a spring so that it is suspended at rest and then pull down on the weight and release, the weight will oscillate about the equilibrium point. This time we will not neglect friction and air resistance: the weight will oscillate closer and closer to the equilibrium point over time until the weight eventually comes to rest at the equilibrium point. This is an example of damped harmonic motion.

The product of any decreasing function and the original periodic function will describe damped oscillatory motion. Here are two examples of functions that describe damped harmonic motion:

$$y = \frac{1}{t} \sin(\omega t) \qquad y = e^{-t} \cos(\omega t)$$

where e^{-t} is a decreasing exponential function (exponential decay).

EXAMPLE 11 **Damped Harmonic Motion**

Assume that the child in Example 10 decides to stop pumping and allows the swing to continue until she eventually comes to rest. Assume that

$$h(t) = \frac{8}{t} \cos\left(\frac{\pi}{2} t\right)$$

where t is time in seconds and h is the height in feet above the resting position. Note that positive h indicates height reached swinging forward and negative h indicates height reached swinging backward, assuming that $t = 1$ is when the child passes through the equilibrium position swinging backward and stops "pumping."

a. Graph the height function $h(t)$ for $1 \le t \le 8$.
b. What is the height above the resting level at 4 seconds? At 8 seconds? After 1 minute?

Technology Tip

To set up a table for $y = \frac{8}{t}\cos\left(\frac{\pi}{2}t\right)$,

enter $Y_1 = \frac{8}{x}\cos\left(\frac{\pi}{2}x\right)$ and select

TBLSET .

Press 2nd TABLE .

Now graph the function in the [0, 10] by [−6, 6] viewing rectangle.

Solution (a):

Make a table with integer values of t. $\qquad$ $1 \le t \le 8$

t (seconds)	$y = h(t) = \frac{8}{t}\cos\left(\frac{\pi}{2}t\right)$ (feet)	(t, y)
1	$\frac{8}{1}\cos\left(\frac{\pi}{2}\right) = 0$	$(1, 0)$
2	$\frac{8}{2}\cos\pi = -4$	$(2, -4)$
3	$\frac{8}{3}\cos\left(\frac{3\pi}{2}\right) = 0$	$(3, 0)$
4	$\frac{8}{4}\cos(2\pi) = 2$	$(4, 2)$
5	$\frac{8}{5}\cos\left(\frac{5\pi}{2}\right) = 0$	$(5, 0)$
6	$\frac{8}{6}\cos(3\pi) = -\frac{4}{3}$	$\left(6, -\frac{4}{3}\right)$
7	$\frac{8}{7}\cos\left(\frac{7\pi}{2}\right) = 0$	$(7, 0)$
8	$\frac{8}{8}\cos(14\pi) = 1$	$(8, 1)$

Solution (b):

The height is 2 feet when t is 4 seconds. $\qquad$ $\frac{8}{4}\cos(2\pi) = 2$

The height is 1 foot when t is 8 seconds. $\qquad$ $\frac{8}{8}\cos(4\pi) = 1$

The height is 0.13 feet when t is 1 minute (60 seconds). $\qquad$ $\frac{8}{60}\cos(30\pi) = 0.1333$

Resonance can be represented by the product of any increasing function and the original sinusoidal function. Here are two examples of functions that result in resonance as time increases:

$$y = t\cos(\omega t) \qquad y = e^t\sin(\omega t)$$

Graphing Sums of Functions: Addition of Ordinates

Since you have the ability to graph sinusoidal functions, let us now consider graphing sums of functions such as

$$y = x - \sin\left(\frac{\pi x}{2}\right) \qquad y = \sin x + \cos x \qquad y = 3\sin x + \cos(2x)$$

The method for graphing these sums is called the **addition of ordinates**, because we add the corresponding y-values (ordinates). The following table illustrates the ordinates (y-values) of the two sinusoidal functions $\sin x$ and $\cos x$; adding the corresponding ordinates leads to the y-values of $y = \sin x + \cos x$.

x	$\sin x$	$\cos x$	$y = \sin x + \cos x$
0	0	1	1
$\dfrac{\pi}{4}$	$\dfrac{\sqrt{2}}{2}$	$\dfrac{\sqrt{2}}{2}$	$\sqrt{2}$
$\dfrac{\pi}{2}$	1	0	1
$\dfrac{3\pi}{4}$	$\dfrac{\sqrt{2}}{2}$	$-\dfrac{\sqrt{2}}{2}$	0
π	0	-1	-1
$\dfrac{5\pi}{4}$	$-\dfrac{\sqrt{2}}{2}$	$-\dfrac{\sqrt{2}}{2}$	$-\sqrt{2}$
$\dfrac{3\pi}{2}$	-1	0	-1
$\dfrac{7\pi}{4}$	$-\dfrac{\sqrt{2}}{2}$	$\dfrac{\sqrt{2}}{2}$	0
2π	0	1	1

Using a graphing utility, we can graph $Y_1 = \sin X$, $Y_2 = \cos X$, and $Y_3 = Y_1 + Y_2$.

EXAMPLE 12 **Graphing Sums of Functions**

Graph $y = x - \sin\left(\dfrac{\pi x}{2}\right)$ on the interval $0 \le x \le 4$.

Solution:

Let $y_1 = x$ and $y_2 = -\sin\left(\dfrac{\pi x}{2}\right)$.

Technology Tip

To display the graphs of $\sin x$, $\cos x$, and $\sin x + \cos x$ in the same $\left[0, \dfrac{9\pi}{4}\right]$ by $[-2, 2]$ viewing window, enter $Y_1 = \sin x$, $Y_2 = \cos x$, and $Y_3 = \sin x + \cos x$.

To graph Y_3 using a thicker line, use the ◀ key to go to the left of Y_3, press ENTER, and select the thicker line.

Technology Tip

To display the graphs of x, $-\sin\dfrac{\pi x}{2}$, and $x - \sin\dfrac{\pi x}{2}$ in the same $[0, 4]$ by $[-1, 5]$ viewing window, enter $Y_1 = x$, $Y_2 = -\sin\dfrac{\pi x}{2}$, and $Y_3 = x - \sin\dfrac{\pi x}{2}$.

To graph Y_3 using a thicker line, use the ◄ key to go to the left of Y_3, press ENTER, and select the thicker line.

State the amplitude and period of the graph of y_2.　　$|A| = |-1| = 1, p = 4$

Make a table of x- and y-values of y_1, y_2, and $y = y_1 + y_2$.

x	$y_1 = x$	$y_2 = -\sin\left(\dfrac{\pi x}{2}\right)$	$y = x + \left[-\sin\left(\dfrac{\pi x}{2}\right)\right]$
0	0	0	0
1	1	−1	0
2	2	0	2
3	3	1	4
4	4	0	4

Graph $y_1 = x$, $y_2 = -\sin\left(\dfrac{\pi x}{2}\right)$, and $y = x - \sin\left(\dfrac{\pi x}{2}\right)$.

Technology Tip

To display the graphs of $3\sin x$, $\cos(2x)$, and $3\sin x + \cos(2x)$ in the same $[0, 2\pi]$ by $[-5, 5]$ viewing window, enter $Y_1 = 3\sin x$, $Y_2 = \cos(2x)$, and $Y_3 = 3\sin x + \cos(2x)$.

EXAMPLE 13　Graphing Sums of Sine and Cosine Functions

Graph $y = 3\sin x + \cos(2x)$ on the interval $0 \le x \le 2\pi$.

Solution:

Let $y_1 = 3\sin x$, and state the amplitude and period of its graph.　　$|A| = 3, p = 2\pi$

Let $y_2 = \cos(2x)$, and state the amplitude and period of its graph.　　$|A| = 1, p = \pi$

Make a table of x- and y-values of y_1, y_2, and $y = y_1 + y_2$.

x	$y_1 = 3\sin x$	$y_2 = \cos(2x)$	$y = 3\sin x + \cos(2x)$
0	0	1	1
$\dfrac{\pi}{4}$	$\dfrac{3\sqrt{2}}{2}$	0	$\dfrac{3\sqrt{2}}{2}$
$\dfrac{\pi}{2}$	3	−1	2
$\dfrac{3\pi}{4}$	$\dfrac{3\sqrt{2}}{2}$	0	$\dfrac{3\sqrt{2}}{2}$
π	0	1	1
$\dfrac{5\pi}{4}$	$-\dfrac{3\sqrt{2}}{2}$	0	$-\dfrac{3\sqrt{2}}{2}$
$\dfrac{3\pi}{2}$	−3	−1	−4
$\dfrac{7\pi}{4}$	$-\dfrac{3\sqrt{2}}{2}$	0	$-\dfrac{3\sqrt{2}}{2}$
2π	0	1	1

Graph $y_1 = 3\sin x$, $y_2 = \cos(2x)$, and
$y = 3\sin x + \cos(2x)$.

EXAMPLE 14   Graphing Sums of Cosine Functions

Graph $y = \cos\left(\dfrac{x}{2}\right) - \cos x$ on the interval $0 \le x \le 4\pi$.

Solution:

Let $y_1 = \cos\left(\dfrac{x}{2}\right)$ and state the amplitude
and period of its graph.

$$|A| = 1, p = 4\pi$$

Let $y_2 = -\cos x$ and state the amplitude
and period of its graph.

$$|A| = |-1| = 1, p = 2\pi$$

Make a table of x- and
y-values of y_1, y_2,
and $y = y_1 + y_2$.

x	$y_1 = \cos\left(\dfrac{x}{2}\right)$	$y_2 = -\cos x$	$y = \cos\left(\dfrac{x}{2}\right) + [-\cos x]$
0	1	-1	0
$\dfrac{\pi}{2}$	$\dfrac{\sqrt{2}}{2}$	0	$\dfrac{\sqrt{2}}{2}$
π	0	1	1
$\dfrac{3\pi}{2}$	$-\dfrac{\sqrt{2}}{2}$	0	$-\dfrac{\sqrt{2}}{2}$
2π	-1	-1	-2
$\dfrac{5\pi}{2}$	$-\dfrac{\sqrt{2}}{2}$	0	$-\dfrac{\sqrt{2}}{2}$
3π	0	1	1
$\dfrac{7\pi}{2}$	$\dfrac{\sqrt{2}}{2}$	0	$\dfrac{\sqrt{2}}{2}$
4π	1	-1	0

Graph $y_1 = \cos\left(\dfrac{x}{2}\right)$, $y_2 = -\cos x$,

and $y = \cos\left(\dfrac{x}{2}\right) - \cos x$.

Technology Tip

To display the graphs of $\cos\left(\dfrac{x}{2}\right)$,

$-\cos x$, and $\cos\left(\dfrac{x}{2}\right) - \cos x$ in the

same $[0, 4\pi]$ by $[-2, 2]$ viewing

window, enter $Y_1 = \cos\left(\dfrac{x}{2}\right)$,

$Y_2 = -\cos x$, and

$Y_3 = \cos\left(\dfrac{x}{2}\right) - \cos x$.

The sine function is an odd function, and its graph is symmetric about the origin. The cosine function is an even function, and its graph is symmetric about the y-axis. Graphs of the form $y = A\sin(Bx)$ and $y = A\cos(Bx)$ have amplitude $|A|$ and period $\dfrac{2\pi}{B}$.

To graph sinusoidal functions, point-plotting can be used. A more efficient way is to first determine the amplitude and period. Divide the period into four equal parts and choose the values of the division points starting at 0 for x. Make a table of those four points and graph them (this is the graph of one period) by labeling the four coordinates and drawing a smooth sinusoidal curve. Extend the graph to the left and right.

To find an equation of a sinusoidal function given its graph, start by first finding the amplitude (half the distance between the maximum and minimum values) so you can find A. Then determine the period so you can find B. Graphs of the form $y = A\sin(Bx + C) + D$ and $y = A\cos(Bx + C) + D$ can be graphed using graph-shifting techniques.

Harmonic motion is one of the primary applications of sinusoidal functions. To graph combinations of trigonometric functions, add the corresponding y-values of the individual functions.

■ SKILLS

In Exercises 1–10, match the function with its graph (a–j).

1. $y = -\sin x$

2. $y = \sin x$

3. $y = \cos x$

4. $y = -\cos x$

5. $y = 2\sin x$

6. $y = 2\cos x$

7. $y = \sin\left(\frac{1}{2}x\right)$

8. $y = \cos\left(\frac{1}{2}x\right)$

9. $y = -2\cos\left(\frac{1}{2}x\right)$

10. $y = -2\sin\left(\frac{1}{2}x\right)$

a.

b.

c.

d.

e.

f.

g.

h.

i.

j.

In Exercises 11–20, state the amplitude and period of each function.

11. $y = \dfrac{3}{2}\cos(3x)$ **12.** $y = \dfrac{2}{3}\sin(4x)$ **13.** $y = -\sin(5x)$ **14.** $y = -\cos(7x)$ **15.** $y = \dfrac{2}{3}\cos\left(\dfrac{3}{2}x\right)$

16. $y = \dfrac{3}{2}\sin\left(\dfrac{2}{3}x\right)$ **17.** $y = -3\cos(\pi x)$ **18.** $y = -2\sin(\pi x)$ **19.** $y = 5\sin\left(\dfrac{\pi}{3}x\right)$ **20.** $y = 4\cos\left(\dfrac{\pi}{4}x\right)$

In Exercises 21–32, graph the given function over one period.

21. $y = 8\cos x$ **22.** $y = 7\sin x$ **23.** $y = \sin(4x)$ **24.** $y = \cos(3x)$

25. $y = -3\cos\left(\dfrac{1}{2}x\right)$ **26.** $y = -2\sin\left(\dfrac{1}{4}x\right)$ **27.** $y = -3\sin(\pi x)$ **28.** $y = -2\cos(\pi x)$

29. $y = 5\cos(2\pi x)$ **30.** $y = 4\sin(2\pi x)$ **31.** $y = -3\sin\left(\dfrac{\pi}{4}x\right)$ **32.** $y = -4\sin\left(\dfrac{\pi}{2}x\right)$

In Exercises 33–40, graph the given function over the interval $[-2p, 2p]$, where p is the period of the function.

33. $y = -4\cos\left(\dfrac{1}{2}x\right)$ **34.** $y = -5\sin\left(\dfrac{1}{2}x\right)$ **35.** $y = -\sin(6x)$ **36.** $y = -\cos(4x)$

37. $y = 3\cos\left(\dfrac{\pi}{4}x\right)$ **38.** $y = 4\sin\left(\dfrac{\pi}{4}x\right)$ **39.** $y = \sin(4\pi x)$ **40.** $y = \cos(6\pi x)$

In Exercises 41–48, find the equation for each graph.

41.

42.

43.

44.

45.

46.

47.

48.

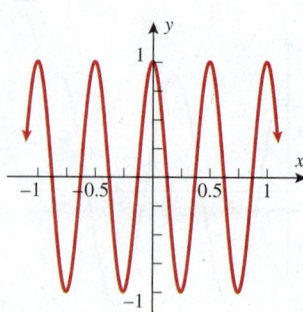

In Exercises 49–60, state the amplitude, period, and phase shift (including direction) of the given function.

49. $y = 2\sin(\pi x - 1)$

50. $y = 4\cos(x + \pi)$

51. $y = -5\cos(3x + 2)$

52. $y = -7\sin(4x - 3)$

53. $y = 6\sin[-\pi(x + 2)]$

54. $y = 3\sin\left[-\dfrac{\pi}{2}(x - 1)\right]$

55. $y = 3\sin(2x + \pi)$

56. $y = -4\cos(2x - \pi)$

57. $y = -\dfrac{1}{4}\cos\left(\dfrac{1}{4}x - \dfrac{\pi}{2}\right)$

58. $y = \dfrac{1}{2}\sin\left(\dfrac{1}{3}x + \pi\right)$

59. $y = 2\cos\left(\dfrac{\pi}{2}(x - 4)\right)$

60. $y = -5\sin(-\pi(x + 1))$

In Exercises 61–66, sketch the graph of the function over the indicated interval.

61. $y = \dfrac{1}{2} + \dfrac{3}{2}\cos[2x + \pi], \left[-\dfrac{3\pi}{2}, \dfrac{3\pi}{2}\right]$

62. $y = \dfrac{1}{3} + \dfrac{2}{3}\sin[2x - \pi], \left[-\dfrac{3\pi}{2}, \dfrac{3\pi}{2}\right]$

63. $y = \dfrac{1}{2} - \dfrac{1}{2}\sin\left[\dfrac{1}{2}x - \dfrac{\pi}{4}\right], \left[-\dfrac{7\pi}{2}, \dfrac{9\pi}{2}\right]$

64. $y = -\dfrac{1}{2} + \dfrac{1}{2}\cos\left[\dfrac{1}{2}x + \dfrac{\pi}{4}\right], \left[-\dfrac{9\pi}{2}, \dfrac{7\pi}{2}\right]$

65. $y = -3 + 4\sin[\pi(x - 2)], [0, 4]$

66. $y = 4 - 3\cos[\pi(x + 1)], [-1, 3]$

In Exercises 67–94, add the ordinates of the individual functions to graph each summed function on the indicated interval.

67. $y = 2x - \cos(\pi x), 0 \le x \le 4$

68. $y = 3x - 2\cos(\pi x), 0 \le x \le 4$

69. $y = \dfrac{1}{3}x + 2\cos(2x), 0 \le x \le 2\pi$

70. $y = \dfrac{1}{4}x + 3\cos\left(\dfrac{x}{2}\right), 0 \le x \le 4\pi$

71. $y = x - \cos\left(\dfrac{3\pi}{2}x\right), 0 \le x \le 6$

72. $y = -2x + 2\sin\left(\dfrac{\pi}{2}x\right), -2 \le x \le 2$

73. $y = \dfrac{1}{4}x - \dfrac{1}{2}\cos[\pi(x - 1)], 2 \le x \le 6$

74. $y = -\dfrac{1}{3}x + \dfrac{1}{3}\sin\left[\dfrac{\pi}{6}(x + 2)\right], -2 \le x \le 10$

75. $y = \sin x - \cos x, 0 \le x \le 2\pi$

76. $y = \cos x - \sin x, 0 \le x \le 2\pi$

77. $y = 3\cos x + \sin x, 0 \le x \le 2\pi$

78. $y = 3\sin x - \cos x, 0 \le x \le 2\pi$

79. $y = 4\cos x - \sin(2x), 0 \le x \le 2\pi$

80. $y = \dfrac{1}{2}\sin x + 2\cos(4x), -\pi \le x \le \pi$

81. $y = 2\sin[\pi(x - 1)] - 2\cos[\pi(x + 1)], -1 \le x \le 2$

82. $y = \sin\left[\dfrac{\pi}{4}(x + 2)\right] + 3\cos\left[\dfrac{3\pi}{3}(x - 1)\right], 1 \le x \le 5$

83. $y = \cos\left(\dfrac{x}{2}\right) + \cos(2x), 0 \le x \le 4\pi$

84. $y = \sin(2x) + \sin(3x), -\pi \le x \le \pi$

85. $y = \sin\left(\dfrac{x}{2}\right) + \sin(2x), 0 \le x \le 4\pi$

86. $y = -\sin\left(\dfrac{\pi}{4}x\right) - 3\sin\left(\dfrac{5\pi}{4}x\right), 0 \le x \le 4$

87. $y = -\dfrac{1}{3}\sin\left(\dfrac{\pi}{6}x\right) + \dfrac{2}{3}\sin\left(\dfrac{5\pi}{6}x\right), 0 \le x \le 3$

88. $y = 8\cos x - 6\cos\left(\dfrac{1}{2}x\right), -2\pi \le x \le 2\pi$

89. $y = -\dfrac{1}{4}\cos\left(\dfrac{\pi}{6}x\right) - \dfrac{1}{2}\cos\left(\dfrac{\pi}{3}x\right), 0 \le x \le 12$

90. $y = 2\cos\left(\dfrac{3}{2}x\right) - \cos\left(\dfrac{1}{2}x\right), -2\pi \le x \le 2\pi$

91. $y = 2\sin\left(\dfrac{x}{2}\right) - \cos(2x), 0 \le x \le 4\pi$

92. $y = 2\cos\left(\dfrac{x}{2}\right) + \sin(2x), 0 \le x \le 4\pi$

93. $y = 2\sin(\pi(x - 1)) + 3\sin\left[2\pi\left(x + \dfrac{1}{2}\right)\right], -2 \le x \le 2$

94. $y = -\dfrac{1}{2}\cos\left(x + \dfrac{\pi}{3}\right) - 2\cos\left(x - \dfrac{\pi}{6}\right), -\pi \le x \le \pi$

▪ APPLICATIONS

For Exercises 95 and 96, refer to the following:

An analysis of demand d for widgets manufactured by WidgetsRUs (measured in thousands of units per week) indicates that demand can be modeled by the graph below, where t is time in months since January 2010 (note that $t = 0$ corresponds to January 2010).

95. Business. Find the amplitude of the graph.

96. Business. Find the period of the graph.

For Exercises 97 and 98, refer to the following:

Researchers have been monitoring oxygen levels (milligrams per liter) in the water of a lake and have found that the oxygen levels fluctuate with an eight-week period. The following tables illustrate data from eight weeks.

97. Environment. Find the amplitude of the oxygen level fluctuations.

Week: t	0 (initial measurement)	1	2	3	4	5	6	7	8
Oxygen levels: mg/L	7	7.7	8	7.7	7	6.3	6	6.3	7

98. Environment. Find the amplitude of the oxygen level fluctuations.

Week: t	0 (initial measurement)	1	2	3	4	5	6	7	8
Oxygen levels: mg/L	7	8.4	9	8.4	7	5.6	5	5.6	7

For Exercises 99–102, refer to the following:

A weight hanging on a spring will oscillate up and down about its equilibrium position after it is pulled down and released.

This is an example of simple harmonic motion. This motion would continue forever if there were not any friction or air resistance. Simple harmonic motion can be described with the function $y = A\cos\left(t\sqrt{\dfrac{k}{m}}\right)$, where $|A|$ is the amplitude, t is the time in seconds, m is the mass of the weight, and k is a constant particular to the spring.

99. Simple Harmonic Motion. If the height of the spring is measured in centimeters and the mass in grams, then what are the amplitude and mass if $y = 4\cos\left(\dfrac{t\sqrt{k}}{2}\right)$?

100. Simple Harmonic Motion. If a spring is measured in centimeters and the mass in grams, then what are the amplitude and mass if $y = 3\cos(3t\sqrt{k})$?

101. Frequency of Oscillations. The frequency of the oscillations in cycles per second is determined by $f = \dfrac{1}{p}$, where p is the period. What is the frequency for the oscillation modeled by $y = 3\cos\left(\dfrac{t}{2}\right)$?

102. Frequency of Oscillations. The frequency of the oscillations f is given by $f = \dfrac{1}{p}$, where p is the period. What is the frequency of oscillation modeled by $y = 3.5\cos(3t)$?

103. Sound Waves. A pure tone created by a vibrating tuning fork shows up as a sine wave on an oscilloscope's screen. A tuning fork vibrating at 256 hertz (Hz) gives the tone middle C and can have the equation $y = 0.005 \sin 2\pi(256t)$, where the amplitude is in centimeters (cm) and the time t in seconds. What are the amplitude and frequency of the wave where the frequency is $\frac{1}{p}$ in cycles per second?

104. Sound Waves. A pure tone created by a vibrating tuning fork shows up as a sine wave on an oscilloscope's screen. A tuning fork vibrating at 288 Hz gives the tone D and can have the equation $y = 0.005 \sin 2\pi(288t)$, where the amplitude is in centimeters (cm) and the time t in seconds. What are the amplitude and frequency of the wave where the frequency is $\frac{1}{p}$ in cycles per second?

105. Sound Waves. If a sound wave is represented by $y = 0.008 \sin(750\pi t)$ cm, what are its amplitude and frequency? See Exercise 103.

106. Sound Waves. If a sound wave is represented by $y = 0.006 \cos(1000\pi t)$ cm, what are its amplitude and frequency? See Exercise 103.

For Exercises 107–110, refer to the following:

When an airplane flies faster than the speed of sound, the sound waves that are formed take on a cone shape, and where the cone hits the ground, a sonic boom is heard. If θ is the angle of the vertex of the cone, then $\sin\dfrac{\theta}{2} = \dfrac{330 \text{ m/sec}}{V} = \dfrac{1}{M}$, where V is the speed of the plane and M is the Mach number.

107. Sonic Booms. What is the speed of the plane if the plane is flying at Mach 2?

108. Sonic Booms. What is the Mach number if the plane is flying at 990 m/sec?

109. Sonic Booms. What is the speed of the plane if the cone angle is 60°?

110. Sonic Booms. What is the speed of the plane if the cone angle is 30°?

For Exercises 111 and 112, refer to the following:

With the advent of summer come fireflies. They are intriguing because they emit a flashing luminescence that beckons their mate to them. It is known that the speed and intensity of the flashing are related to the temperature—the higher the temperature, the quicker and more intense the flashing becomes. If you ever watch a single firefly, you will see that the intensity of the flashing is periodic with time. The intensity of light emitted is measured in *candelas per square meter* (of firefly). To give an idea of this unit of measure, the intensity of a picture on a typical TV screen is about 450 candelas. The measurement for the intensity of the light emitted by a typical firefly at its brightest moment is about 50 candelas. Assume that a typical cycle of this flashing is 4 seconds and that the intensity is essentially zero candelas at the beginning and ending of a cycle.

111. Bioluminescence in Fireflies. Find an equation that describes this flashing. What is the intensity of the flashing at 4 minutes?

112. Bioluminescence in Fireflies. Graph the equation from Exercise 111 for a period of 30 seconds.

■ **CATCH THE MISTAKE**

In Exercises 113 and 114, explain the mistake that is made.

113. Graph the function $y = -2\cos x$.

Solution:

Find the amplitude. $|A| = |-2| = 2$

The graph of $y = -2\cos x$ is similar to the graph of $y = \cos x$ with amplitude 2.

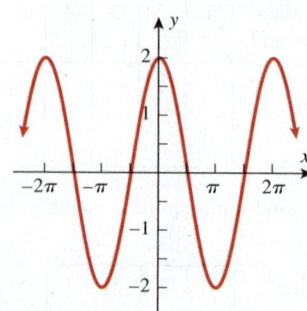

This is incorrect. What mistake was made?

114. Graph the function $y = -\sin(2x)$.

Solution:

Make a table with values.

x	$y = -\sin(2x)$	(x, y)
0	$y = -\sin(0) = 0$	$(0, 0)$
$\dfrac{\pi}{2}$	$y = -\sin(\pi) = 0$	$(0, 0)$
π	$y = -\sin(2\pi) = 0$	$(0, 0)$
$\dfrac{3\pi}{2}$	$y = -\sin(3\pi) = 0$	$(0, 0)$
2π	$y = -\sin(4\pi) = 0$	$(0, 0)$

Graph the function by plotting these points and connecting them with a sinusoidal curve.

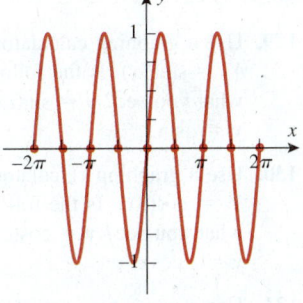

This is incorrect. What mistake was made?

■ **CONCEPTUAL**

In Exercises 115–118, determine whether each statement is true or false. (A and B are positive real numbers.)

115. The graph of $y = -A\cos(Bx)$ is the graph of $y = A\cos Bx$ reflected about the x-axis.

116. The graph of $y = A\sin(-Bx)$ is the graph of $y = A\sin(Bx)$ reflected about the x-axis.

117. The graph of $y = -A\cos(-Bx)$ is the graph of $y = A\cos(Bx)$.

118. The graph of $y = -A\sin(-Bx)$ is the graph of $y = A\sin(Bx)$.

In Exercises 119–122, A and B are positive real numbers.

119. Find the y-intercept of the function $y = A\cos Bx$.

120. Find the y-intercept of the function $y = A\sin Bx$.

121. Find the x-intercepts of the function $y = A\sin Bx$.

122. Find the x-intercepts of the function $y = A\cos Bx$.

■ **CHALLENGE**

123. Find the y-intercept of $y = -A\sin\left(Bx + \dfrac{\pi}{6}\right)$.

124. Find the y-intercept of $y = A\cos(Bx - \pi) + C$.

125. Find the x-intercept(s) of $y = A\sin(Bx) + A$.

126. Find an expression involving C and A that describes the values of C for which the graph of $y = A\cos(Bx) + C$ does not cross the x-axis. (Assume that $A > 0$.)

127. What is the range of $y = 2A\sin(Bx + C) - \dfrac{A}{2}$?

128. Can the y-coordinate of a point on the graph of $y = A\sin(Bx) + 3A\cos\left(\dfrac{B}{2}x\right)$ exceed $4A$? Explain. (Assume that $A > 0$.)

129. Use a graphing calculator to graph $Y_1 = 5\sin x$ and $Y_2 = \sin(5x)$. Is the following statement true based on what you see? $y = \sin(cx)$ has the same graph as $y = c\sin x$.

130. Use a graphing calculator to graph $Y_1 = 3\cos x$ and $Y_2 = \cos(3x)$. Is the following statement true based on what you see? $y = \cos(cx)$ has the same graph as $y = c\cos x$.

131. Use a graphing calculator to graph $Y_1 = \sin x$ and $Y_2 = \cos\left(x - \dfrac{\pi}{2}\right)$. What do you notice?

132. Use a graphing calculator to graph $Y_1 = \cos x$ and $Y_2 = \sin\left(x + \dfrac{\pi}{2}\right)$. What do you notice?

133. Use a graphing calculator to graph $Y_1 = \cos x$ and $Y_2 = \cos(x + c)$, where

a. $c = \dfrac{\pi}{3}$, and explain the relationship between Y_2 and Y_1.

b. $c = -\dfrac{\pi}{3}$, and explain the relationship between Y_2 and Y_1.

134. Use a graphing calculator to graph $Y_1 = \sin x$ and $Y_2 = \sin(x + c)$, where

a. $c = \dfrac{\pi}{3}$, and explain the relationship between Y_2 and Y_1.

b. $c = -\dfrac{\pi}{3}$, and explain the relationship between Y_2 and Y_1.

For Exercises 135 and 136, refer to the following:

Damped oscillatory motion, or *damped oscillation*, occurs when things in oscillatory motion experience friction or resistance. The friction causes the amplitude to decrease as a function of time. Mathematically, we can use a negative exponential function to damp the oscillations in the form of

$$f(t) = e^{-t}\sin t$$

135. Damped Oscillation. Graph the functions $Y_1 = e^{-t}$, $Y_2 = \sin t$, and $Y_3 = e^{-t}\sin t$ in the same viewing window (let t range from 0 to 2π). What happens as t increases?

136. Damped Oscillation. Graph $Y_1 = e^{-t}\sin t$, $Y_2 = e^{-2t}\sin t$, and $Y_3 = e^{-4t}\sin t$ in the same viewing window. What happens to $Y = e^{-kt}\sin t$ as k increases?

137. Use a graphing calculator to graph $Y_1 = \sin x$ and $Y_2 = \sin x + c$, where

a. $c = 1$, and explain the relationship between Y_2 and Y_1.

b. $c = -1$, and explain the relationship between Y_2 and Y_1.

138. Use a graphing calculator to graph $Y_1 = \cos x$ and $Y_2 = \cos x + c$, where

a. $c = \frac{1}{2}$, and explain the relationship between Y_2 and Y_1.

b. $c = -\frac{1}{2}$, and explain the relationship between Y_2 and Y_1.

139. What is the amplitude of the function $y = 3\cos x + 4\sin x$? Use a graphing calculator to graph $Y_1 = 3\cos x$, $Y_2 = 4\sin x$, and $Y_3 = 3\cos x + 4\sin x$ in the same viewing window.

140. What is the amplitude of the function $y = \sqrt{3}\cos x - \sin x$? Use a graphing calculator to graph $Y_1 = \sqrt{3}\cos x$, $Y_2 = \sin x$, and $Y_3 = \sqrt{3}\cos x - \sin x$ in the same viewing window.

SKILLS OBJECTIVES

- Determine the period of tangent, cotangent, secant, and cosecant functions.
- Graph basic tangent, cotangent, secant, and cosecant functions.
- Graph translated tangent, cotangent, secant, and cosecant functions.

CONCEPTUAL OBJECTIVES

- Relate domain restrictions to vertical asymptotes.
- Understand the pattern that vertical asymptotes follow.
- Understand the relationships between the graphs of the cosine and secant functions and the sine and cosecant functions.

Graphing the Tangent, Cotangent, Secant, and Cosecant Functions

Section 6.8 focused on graphing sinusoidal functions (sine and cosine). We now turn our attention to graphing the other circular functions: tangent, cotangent, secant, and cosecant. We know the graphs of the sine and cosine functions, and we can get the graphs of the other circular functions from the sinusoidal functions. Recall the reciprocal and quotient identities:

$$\tan x = \frac{\sin x}{\cos x} \qquad \cot x = \frac{\cos x}{\sin x} \qquad \sec x = \frac{1}{\cos x} \qquad \csc x = \frac{1}{\sin x}$$

Recall that in graphing rational functions, a *vertical asymptote* corresponds to a denominator equal to zero (as long as the numerator and denominator have no common factors). As you will see in this section, tangent and secant functions have graphs with vertical asymptotes at the *x*-values where the cosine function is equal to zero, and cotangent and cosecant functions have graphs with vertical asymptotes at the *x*-values where the sine function is equal to zero.

One important difference between the sinusoidal functions, $y = \sin x$ and $y = \cos x$, and the other four trigonometric functions ($y = \tan x$, $y = \sec x$, $y = \csc x$, and $y = \cot x$) is that the sinusoidal functions have defined amplitudes, whereas the other four trigonometric functions do not (since they are unbounded vertically).

The Tangent Function

Since the tangent function is a quotient that relies on sine and cosine, let us start with a table of values for the quadrantal angles.

x	$\sin x$	$\cos x$	$\tan x = \dfrac{\sin x}{\cos x}$	(x, y) OR ASYMPTOTE
0	0	1	0	$(0, 0)$
$\dfrac{\pi}{2}$	1	0	undefined	vertical asymptote: $x = \dfrac{\pi}{2}$
π	0	-1	0	$(\pi, 0)$
$\dfrac{3\pi}{2}$	-1	0	undefined	vertical asymptote: $x = \dfrac{3\pi}{2}$
2π	0	1	0	$(2\pi, 0)$

Notice that the x-intercepts correspond to integer multiples of π and vertical asymptotes correspond to odd integer multiples of $\dfrac{\pi}{2}$.

We know that the graph of the tangent function is undefined at the odd integer multiples of $\dfrac{\pi}{2}$, so its graph cannot cross the vertical asymptotes. The question is, what happens between the asymptotes? We know the x-intercepts, so let us now make a table for special values of x.

x	$\sin x$	$\cos x$	$y = \tan x = \dfrac{\sin x}{\cos x}$	(x, y)
$\dfrac{\pi}{6}$	$\dfrac{1}{2}$	$\dfrac{\sqrt{3}}{2}$	$\dfrac{1}{\sqrt{3}} = \dfrac{\sqrt{3}}{3} \approx 0.577$	$\left(\dfrac{\pi}{6}, 0.577\right)$
$\dfrac{\pi}{4}$	$\dfrac{\sqrt{2}}{2}$	$\dfrac{\sqrt{2}}{2}$	1	$\left(\dfrac{\pi}{4}, 1\right)$
$\dfrac{\pi}{3}$	$\dfrac{\sqrt{3}}{2}$	$\dfrac{1}{2}$	$\sqrt{3} \approx 1.732$	$\left(\dfrac{\pi}{3}, 1.732\right)$
$\dfrac{2\pi}{3}$	$\dfrac{\sqrt{3}}{2}$	$-\dfrac{1}{2}$	$-\sqrt{3} \approx -1.732$	$\left(\dfrac{2\pi}{3}, -1.732\right)$
$\dfrac{3\pi}{4}$	$\dfrac{\sqrt{2}}{2}$	$-\dfrac{\sqrt{2}}{2}$	-1	$\left(\dfrac{3\pi}{4}, -1\right)$
$\dfrac{5\pi}{6}$	$\dfrac{1}{2}$	$-\dfrac{\sqrt{3}}{2}$	$\dfrac{1}{-\sqrt{3}} = -\dfrac{\sqrt{3}}{3} \approx -0.577$	$\left(\dfrac{5\pi}{6}, -0.577\right)$

What happens to $\tan x$ as x approaches $\dfrac{\pi}{2}$? We know $\tan x$ is undefined at $x = \dfrac{\pi}{2}$ but we must consider x-values both larger and smaller than $\dfrac{\pi}{2} \approx 1.571$.

	approaching from the left $\longrightarrow$			$\dfrac{\pi}{2}$	$\longleftarrow$ approaching from the right		
x	1.5	1.55	1.57	1.571	1.58	1.59	1.65
TAN x	14.1	48.1	1255.8	undefined	-108.6	-52.1	-12.6

$\underbrace{\qquad\qquad\qquad}_{\tan x \text{ gets larger}}$ $\underbrace{\qquad\qquad\qquad}_{\tan x \text{ gets more negative}}$

The arrows on the graph on the right indicate increasing without bound (in the positive and negative directions).

GRAPH OF $y = \tan x$

1. The x-intercepts occur at integer multiples of π. $(n\pi, 0)$

2. Vertical asymptotes occur at odd integer multiples of $\dfrac{\pi}{2}$. $x = \dfrac{(2n+1)\pi}{2}$

3. The domain is the set of all real numbers except odd integer multiples of $\dfrac{\pi}{2}$. $x \neq \dfrac{(2n+1)\pi}{2}$

4. The range is the set of all real numbers. $(-\infty, \infty)$

5. $y = \tan x$ has period π.

6. $y = \tan x$ is an odd function (symmetric about the origin). $\tan(-x) = -\tan x$

7. The graph has no defined amplitude, since the function is unbounded.

Note: n is an integer.

The Cotangent Function

The cotangent function is similar to the tangent function in that it is a quotient involving the sine and cosine functions. The difference is that cotangent has cosine in the numerator and sine in the denominator: $\cot x = \dfrac{\cos x}{\sin x}$. The graph of $y = \tan x$ has x-intercepts corresponding to integer multiples of π and vertical asymptotes corresponding to odd integer multiples of $\dfrac{\pi}{2}$. The graph of the cotangent function is the reverse in that it has x-intercepts corresponding to odd integer multiples of $\dfrac{\pi}{2}$ and vertical asymptotes corresponding to integer multiples of π. This is because the x-intercepts occur when the numerator, $\cos x$, is equal to 0 and the vertical asymptotes occur when the denominator, $\sin x$, is equal to 0.

Study Tip

The graphs of $y = \tan x$ and $y = \cot x$ both have period π, and neither of them has defined amplitude.

GRAPH OF $y = \cot x$

1. The x-intercepts occur at odd integer multiples of $\dfrac{\pi}{2}$. $\left(\dfrac{(2n+1)\pi}{2}, 0\right)$

2. Vertical asymptotes occur at integer multiples of π. $x = n\pi$

3. The domain is the set of all real numbers except integer multiples of π. $x \neq n\pi$

4. The range is the set of all real numbers. $(-\infty, \infty)$

5. $y = \cot x$ has period π.

6. $y = \cot x$ is an odd function (symmetric about the origin). $\cot(-x) = -\cot x$

To graph $y = \cot x$, use the reciprocal property to enter $(\tan(x))^{-1}$.

7. The graph has no defined amplitude, since the function is unbounded.

Note: n is an integer.

The Secant Function

Since $y = \cos x$ has period 2π, the secant function, which is the reciprocal of the cosine function, $\sec x = \dfrac{1}{\cos x}$, also has period 2π. We now illustrate values of the secant function with a table.

x	$\cos x$	$y = \sec x = \dfrac{1}{\cos x}$	(x, y) OR ASYMPTOTE
0	1	1	$(0, 1)$
$\dfrac{\pi}{2}$	0	undefined	vertical asymptote: $x = \dfrac{\pi}{2}$
π	-1	-1	$(\pi, -1)$
$\dfrac{3\pi}{2}$	0	undefined	vertical asymptote: $x = \dfrac{3\pi}{2}$
2π	1	1	$(2\pi, 1)$

Again, we ask the same question: what happens as x approaches the vertical asymptotes? The secant function grows without bound in either the positive or negative direction.

If we graph $y = \cos x$ (the "guide" function) and $y = \sec x$ on the same graph, we notice the following:

- The x-intercepts of $y = \cos x$ correspond to the vertical asymptotes of $y = \sec x$.
- The range of cosine is $[-1, 1]$ and the range of secant is $(-\infty, -1] \cup [1, \infty)$.
- When cosine is positive, secant is positive, and when one is negative, the other is negative.

The cosine function is used as the guide function to graph the secant function.

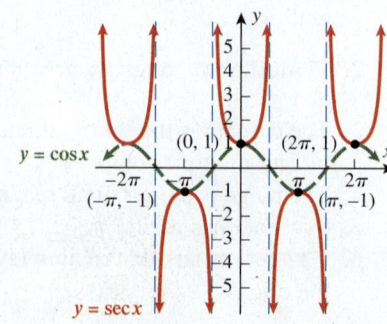

GRAPH OF $y = \sec x$

1. There are no x-intercepts.

   $$\frac{1}{\cos x} \neq 0$$

2. Vertical asymptotes occur at odd integer multiples of $\frac{\pi}{2}$.

   $$x = \frac{(2n + 1)\pi}{2}$$

3. The domain is the set of all real numbers except odd integer multiples of $\frac{\pi}{2}$.

   $$x \neq \frac{(2n + 1)\pi}{2}$$

4. The range is $(-\infty, -1] \cup [1, \infty)$.
5. $y = \sec x$ has period 2π.
6. $y = \sec x$ is an even function (symmetric about the y-axis).

   $$\sec(-x) = \sec x$$

7. The graph has no defined amplitude, since the function is unbounded.

Note: n is an integer.

The Cosecant Function

Since $y = \sin x$ has period 2π, the cosecant function, which is the reciprocal of the sine function, $\csc x = \dfrac{1}{\sin x}$, also has period 2π. We now illustrate values of cosecant with a table.

x	$\sin x$	$y = \csc x = \dfrac{1}{\sin x}$	(x, y) OR ASYMPTOTE
0	0	undefined	vertical asymptote: $x = 0$
$\dfrac{\pi}{2}$	1	1	$\left(\dfrac{\pi}{2}, 1\right)$
π	0	undefined	vertical asymptote: $x = \pi$
$\dfrac{3\pi}{2}$	-1	-1	$\left(\dfrac{3\pi}{2}, -1\right)$
2π	0	undefined	vertical asymptote: $x = 2\pi$

Again, we ask the same question: what happens as x approaches the vertical asymptotes? The cosecant function grows without bound in either the positive or negative direction.

If we graph $y = \sin x$ (the "guide" function) and $y = \csc x$ on the same graph, we notice the following:

- The x-intercepts of $y = \sin x$ correspond to the vertical asymptotes of $y = \csc x$.
- The range of sine is $[-1, 1]$ and the range of cosecant is $(-\infty, -1] \cup [1, \infty)$.
- When sine is positive, cosecant is positive, and when one is negative, the other is negative.

The sine function is used as the guide function to graph the cosecant function.

GRAPH OF $y = \csc x$

1. There are no x-intercepts.

$$\frac{1}{\sin x} \neq 0$$

2. Vertical asymptotes occur at integer multiples of π.

$$x = n\pi$$

3. The domain is the set of all real numbers except integer multiples of π.

$$x \neq n\pi$$

4. The range is $(-\infty, -1] \cup [1, \infty)$.
5. $y = \csc x$ has period 2π.
6. $y = \csc x$ is an odd function (symmetric about the origin).

$$\csc(-x) = -\csc x$$

7. The graph has no defined amplitude, since the function is unbounded.

Note: n is an integer.

Graphing More General Tangent, Cotangent, Secant, and Cosecant Functions

FUNCTION	$y = \sin x$	$y = \cos x$	$y = \tan x$	$y = \cot x$	$y = \sec x$	$y = \csc x$
Graph of One Period						
Domain	$(-\infty, \infty)$	$(-\infty, \infty)$	$x \neq \dfrac{(2n+1)\pi}{2}$	$x \neq n\pi$	$x \neq \dfrac{(2n+1)\pi}{2}$	$x \neq n\pi$
Range	$[-1, 1]$	$[-1, 1]$	$(-\infty, \infty)$	$(-\infty, \infty)$	$(-\infty, -1] \cup [1, \infty)$	$(-\infty, -1] \cup [1, \infty)$
Amplitude	1	1	none	none	none	none
Period	2π	2π	π	π	2π	2π
x-intercepts	$(n\pi, 0)$	$\left(\dfrac{(2n+1)\pi}{2}, 0\right)$	$(n\pi, 0)$	$\left(\dfrac{(2n+1)\pi}{2}, 0\right)$	none	none
Vertical Asymptotes	none	none	$x = \dfrac{(2n+1)\pi}{2}$	$x = n\pi$	$x = \dfrac{(2n+1)\pi}{2}$	$x = n\pi$

Note: n is an integer.

We use these basic functions as the starting point for graphing general tangent, cotangent, secant, and cosecant functions.

GRAPHING TANGENT AND COTANGENT FUNCTIONS

Graphs of $y = A\tan(Bx)$ and $y = A\cot(Bx)$ can be obtained using the following steps (assume $B > 0$):

Step 1: Calculate the period: $\dfrac{\pi}{B}$.

Step 2: Find two neighboring vertical asymptotes.

For $y = A\tan(Bx)$: $Bx = -\dfrac{\pi}{2}$ and $Bx = \dfrac{\pi}{2}$.

For $y = A\cot(Bx)$: $Bx = 0$ and $Bx = \pi$.

Step 3: Find the x-intercept between the two asymptotes.

For $y = A\tan(Bx)$: $Bx = 0$.

For $y = A\cot(Bx)$: $Bx = \dfrac{\pi}{2}$.

Step 4: Draw the vertical asymptotes and label the x-intercept.

Step 5: Divide the interval between the asymptotes into four equal parts. Set up a table with coordinates corresponding to the points in the interval.

Step 6: Connect the points with a smooth curve. Use arrows to indicate the behavior toward the asymptotes.

- If $A > 0$
  - $y = A\tan(Bx)$ increases from left to right.
  - $y = A\cot(Bx)$ decreases from left to right.
- If $A < 0$
  - $y = A\tan(Bx)$ decreases from left to right.
  - $y = A\cot(Bx)$ increases from left to right.

EXAMPLE 1   Graphing $y = A\tan(Bx)$

Graph $y = -3\tan(2x)$ on the interval $-\dfrac{\pi}{2} \le x \le \dfrac{\pi}{2}$.

Solution: $A = -3$, $B = 2$

STEP 1 Calculate the period.

$$\dfrac{\pi}{B} = \dfrac{\pi}{2}$$

STEP 2 Find two vertical asymptotes.

$$Bx = -\dfrac{\pi}{2} \text{ and } Bx = \dfrac{\pi}{2}$$

Substitute $B = 2$ and solve for x.

$$x = -\dfrac{\pi}{4} \text{ and } x = \dfrac{\pi}{4}$$

STEP 3 Find the x-intercept between the asymptotes.

$$Bx = 0$$
$$x = 0$$

STEP 4 Draw the vertical asymptotes $x = -\dfrac{\pi}{4}$ and $x = \dfrac{\pi}{4}$ and label the x-intercept $(0, 0)$.

STEP 5 Divide the period $\dfrac{\pi}{2}$ into four equal parts, in steps of $\dfrac{\pi}{8}$. Set up a table with coordinates corresponding to values of $y = -3\tan(2x)$.

x	$y = -3\tan(2x)$	(x, y)
$-\dfrac{\pi}{4}$	undefined	vertical asymptote, $x = -\dfrac{\pi}{4}$
$-\dfrac{\pi}{8}$	3	$\left(-\dfrac{\pi}{8}, 3\right)$
0	0	$(0, 0)$
$\dfrac{\pi}{8}$	-3	$\left(\dfrac{\pi}{8}, -3\right)$
$\dfrac{\pi}{4}$	undefined	vertical asymptote, $x = \dfrac{\pi}{4}$

STEP 6 Graph the points from the table and connect with a smooth curve. Repeat to the right and left until you reach the interval endpoints. Notice that the distance between the vertical asymptotes is the period length $\frac{\pi}{2}$.

■ **Answer:**

■ **YOUR TURN** Graph $y = \frac{1}{3}\tan\left(\frac{1}{2}x\right)$ on the interval $-\pi \le x \le \pi$.

EXAMPLE 2 Graphing $y = A\cot(Bx)$

Graph $y = 4\cot\left(\frac{1}{2}x\right)$ on the interval $-2\pi \le x \le 2\pi$.

Solution: $A = 4$, $B = \frac{1}{2}$

STEP 1 Calculate the period.

$$\frac{\pi}{B} = 2\pi$$

STEP 2 Find two vertical asymptotes.

$Bx = 0$ and $Bx = \pi$

Substitute $B = \frac{1}{2}$ and solve for x.

$x = 0$ and $x = 2\pi$

STEP 3 Find the x-intercept between the asymptotes.

$Bx = \dfrac{\pi}{2}$

$x = \pi$

STEP 4 Draw the vertical asymptotes $x = 0$ and $x = 2\pi$ and label the x-intercept $(\pi, 0)$.

Technology Tip

To graph $y = 4\cot\left(\frac{1}{2}x\right)$ on the interval $-2\pi \le x \le 2\pi$, enter $y = 4\left(\tan\left(\frac{1}{2}x\right)\right)^{-1}$.

Classroom Example 6.9.2

Graph:

a. $y = \frac{1}{4}\cot\left(\frac{\pi}{4}x\right)$ on $[-8, 8]$

b. $y = -\frac{1}{4}\cot\left(\frac{\pi}{4}x\right)$ on $[-8, 8]$

Answer:

a.

b.

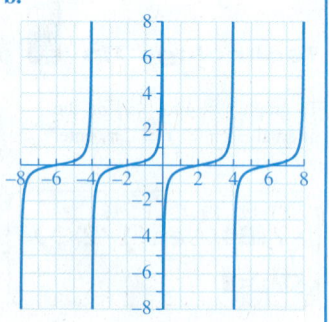

STEP 5 Divide the period 2π into four equal parts, in steps of $\frac{\pi}{2}$. Set up a table with coordinates corresponding to values of $y = 4\cot\left(\frac{1}{2}x\right)$.

x	$y = 4\cot\left(\frac{1}{2}x\right)$	(x, y)
0	undefined	vertical asymptote, $x = 0$
$\dfrac{\pi}{2}$	4	$\left(\dfrac{\pi}{2}, 4\right)$
π	0	$(\pi, 0)$
$\dfrac{3\pi}{2}$	-4	$\left(\dfrac{3\pi}{2}, -4\right)$
2π	undefined	vertical asymptote, $x = 2\pi$

STEP 6 Graph the points from the table and connect with a smooth curve. Repeat to the right and left until you reach the interval endpoints.

■ **Answer:**

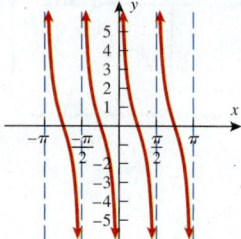

■ **YOUR TURN** Graph $y = 2\cot(2x)$ on the interval $-\pi \le x \le \pi$.

GRAPHING SECANT AND COSECANT FUNCTIONS

Graphs of $y = A\sec(Bx)$ and $y = A\csc(Bx)$ can be obtained using the following steps:

Step 1: Graph the corresponding guide function with a dashed curve.

For $y = A\sec(Bx)$, use $y = A\cos(Bx)$ as a guide.
For $y = A\csc(Bx)$, use $y = A\sin(Bx)$ as a guide.

Step 2: Draw the asymptotes, which correspond to the x-intercepts of the guide function.

Step 3: Draw the U shape between the asymptotes. If the guide function has a positive value between the asymptotes, the U opens upward; and if the guide function has a negative value between the asymptotes, the U opens downward.

EXAMPLE 3 **Graphing $y = A\sec(Bx)$**

Graph $y = 2\sec(\pi x)$ on the interval $-2 \le x \le 2$.

Solution:

STEP 1 Graph the corresponding guide function with a dashed curve.

For $y = 2\sec(\pi x)$, use $y = 2\cos(\pi x)$ as a guide.

STEP 2 Draw the **asymptotes**, which correspond to the x-intercepts of the guide function.

STEP 3 Use the U shape between the asymptotes. If the guide function is positive, the U opens upward, and if the guide function is negative, the U opens downward.

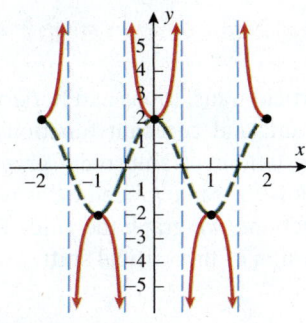

■ **YOUR TURN** Graph $y = -\sec(2\pi x)$ on the interval $-1 \le x \le 1$.

EXAMPLE 4 **Graphing $y = A\csc(Bx)$**

Graph $y = -3\csc(2\pi x)$ on the interval $-1 \le x \le 1$.

Solution:

STEP 1 Graph the corresponding guide function with a dashed curve.

For $y = -3\csc(2\pi x)$, use $y = -3\sin(2\pi x)$ as a guide.

■ **Answer:**

Classroom Example 6.9.3

Graph $y = \frac{1}{2}\sec(2x)$ on $[-\pi, \pi]$.

Answer:

Technology Tip

To graph $y = -3\csc(2\pi x)$ on the interval $-1 \le x \le 1$, enter $y = -3(\sin(2\pi x))^{-1}$.

■ **Answer:**

STEP 2 Draw the **asymptotes**, which correspond to the x-intercepts of the guide function.

STEP 3 Use the U shape between the asymptotes. If the guide function is positive, the U opens upward, and if the guide function is negative, the U opens downward.

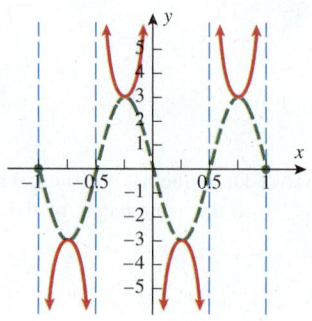

■ **YOUR TURN** Graph $y = \frac{1}{2}\csc(\pi x)$ on the interval $-1 \le x \le 1$.

Translations of Circular Functions

Vertical translations and horizontal translations (phase shifts) of the tangent, cotangent, secant, and cosecant functions are graphed the same way as vertical and horizontal translations of sinusoidal graphs are drawn. For tangent and cotangent functions, we follow the same procedure as we did with sinusoidal functions. For secant and cotangent functions, we graph the guide function first and then translate up or down depending on the sign of the vertical shift.

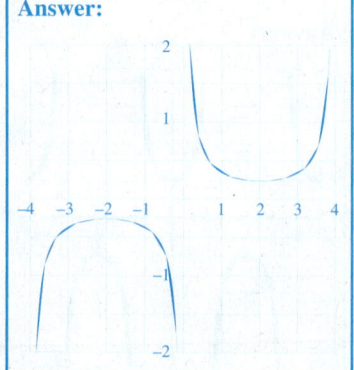
EXAMPLE 5 **Graphing $y = A\tan(Bx + C) + D$**

Graph $y = 1 - \tan\left(x - \frac{\pi}{2}\right)$ on $-\pi \le x \le \pi$. State the domain and range on the interval.

There are two ways to approach graphing this function. Both will be illustrated.

Solution (1):

Plot $y = \tan x$, and then do the following:

- Shift the curve to the right $\frac{\pi}{2}$ units. $y = \tan\left(x - \frac{\pi}{2}\right)$

- Reflect the curve about the x-axis (because of the negative sign). $y = -\tan\left(x - \frac{\pi}{2}\right)$

- Shift the entire graph up one unit. $y = 1 - \tan\left(x - \frac{\pi}{2}\right)$

Solution (2): Graph $y = -\tan\left(x - \dfrac{\pi}{2}\right)$, and then shift the entire graph up one unit, because $D = 1$.

STEP 1 Calculate the period.

$$\dfrac{\pi}{B} = \pi$$

STEP 2 Find two vertical asymptotes.

Solve for x.

$$x - \dfrac{\pi}{2} = -\dfrac{\pi}{2} \text{ and } x - \dfrac{\pi}{2} = \dfrac{\pi}{2}$$
$$x = 0 \qquad \text{and } x = \pi$$

STEP 3 Find the x-intercept between the asymptotes.

$$x - \dfrac{\pi}{2} = 0$$
$$x = \dfrac{\pi}{2}$$

STEP 4 Draw the vertical asymptotes $x = 0$ and $x = \pi$ and label the x-intercept $\left(\dfrac{\pi}{2}, 0\right)$.

STEP 5 Divide the period π into four equal parts, in steps of $\dfrac{\pi}{4}$. Set up a table with coordinates corresponding to values of $y = -\tan\left(x - \dfrac{\pi}{2}\right)$ between the two asymptotes.

x	$y = -\tan\left(x - \dfrac{\pi}{2}\right)$	(x, y)
$x = 0$	undefined	vertical asymptote, $x = 0$
$\dfrac{\pi}{4}$	1	$\left(\dfrac{\pi}{4}, 1\right)$
$\dfrac{\pi}{2}$	0	$\left(\dfrac{\pi}{2}, 0\right)$ (x-intercept)
$\dfrac{3\pi}{4}$	-1	$\left(\dfrac{3\pi}{4}, -1\right)$
$x = \pi$	undefined	vertical asymptote, $x = \pi$

Classroom Example 6.9.5
Graph
$$y = -1 - \cot\left(\dfrac{\pi}{4}x - \dfrac{\pi}{2}\right).$$

Answer:

Technology Tip

To obtain the graph of
$$y = 1 - \tan\left(x - \dfrac{\pi}{2}\right)$$ on the
interval $-\pi \le x \le \pi$, enter
$$y = 1 - \tan\left(x - \dfrac{\pi}{2}\right).$$

STEP 6 Graph the points from the table and connect with a smooth curve. Repeat to the right and left until reaching the interval endpoints.

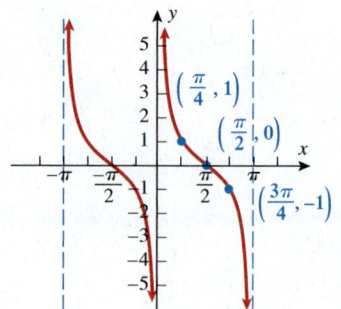

STEP 7 Shift the entire graph up one unit to arrive at the graph of

$$y = 1 - \tan\left(x - \frac{\pi}{2}\right).$$

STEP 8 State the domain and range on the interval.

Domain: $(-\pi, 0) \cup (0, \pi)$
Range: $(-\infty, \infty)$

- **Answer:**

Domain:

$$\left[-\pi, -\frac{\pi}{2}\right) \cup \left(-\frac{\pi}{2}, \frac{\pi}{2}\right) \cup \left(\frac{\pi}{2}, \pi\right]$$

Range: $(-\infty, \infty)$

- **YOUR TURN** Graph $y = -1 + \cot\left(x + \frac{\pi}{2}\right)$ on $-\pi \le x \le \pi$. State the domain and range on the interval.

Classroom Example 6.9.6
Graph

$$y = 2 - \sec\left(\frac{\pi x}{2} - \frac{\pi}{4}\right).$$

Answer:

EXAMPLE 6 Graphing $y = A\csc(Bx + C) + D$

Graph $y = 1 - \csc(2x - \pi)$ on $-\pi \le x \le \pi$. State the domain and range on the interval.

Solution:

Graph $y = -\csc(2x - \pi)$, and shift the entire graph up one unit to arrive at the graph of $y = 1 - \csc(2x - \pi)$.

STEP 1 Draw the guide function,
$y = -\sin(2x - \pi)$.

STEP 2 Draw the **vertical asymptotes** of $y = -\csc(2x - \pi)$ that correspond to the *x*-intercepts of $y = -\sin(2x - \pi)$.

STEP 3 Use the U shape between the asymptotes. If the guide function is positive, the U opens upward, and if the guide function is negative, the U opens downward.

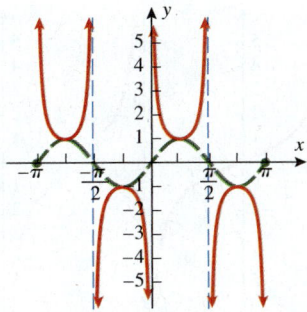

STEP 4 Shift the entire graph up one unit to arrive at the graph of $y = 1 - \csc(2x - \pi)$.

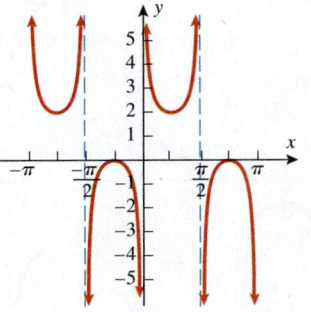

■ **Answer:**
Domain:
$$\left[-1, -\tfrac{1}{2}\right) \cup \left(-\tfrac{1}{2}, \tfrac{1}{2}\right) \cup \left(\tfrac{1}{2}, 1\right]$$
Range: $(-\infty, -3] \cup [-1, \infty)$

STEP 5 State the domain and range on the interval.

Domain: $\left(-\pi, -\dfrac{\pi}{2}\right) \cup \left(-\dfrac{\pi}{2}, 0\right) \cup \left(0, \dfrac{\pi}{2}\right) \cup \left(\dfrac{\pi}{2}, \pi\right)$

Range: $(-\infty, 0] \cup [2, \infty)$

■ **YOUR TURN** Graph $y = -2 + \sec(\pi x - \pi)$ on $-1 \le x \le 1$. State the domain and range on the interval.

SECTION
6.9 SUMMARY

The tangent and cotangent functions have period π, whereas the secant and cosecant functions have period 2π. To graph the tangent and cotangent functions, first identify the vertical asymptotes and *x*-intercepts, and then find values of the function within a period (i.e., between the asymptotes). To find graphs of secant and cosecant functions, first graph their guide functions (cosine and sine, respectively), and then label vertical asymptotes that correspond to *x*-intercepts of the guide function. The graphs of the secant and cosecant functions resemble the letter U opening up or down. The secant and cosecant functions are positive when their guide function is positive and negative when their guide function is negative.

SECTION
6.9 EXERCISES

▪SKILLS

In Exercises 1–8, match the graphs to the functions (a–h).

1. $y = -\tan x$ **2.** $y = -\csc x$ **3.** $y = \sec(2x)$ **4.** $y = \csc(2x)$

5. $y = \cot(\pi x)$ **6.** $y = -\cot(\pi x)$ **7.** $y = 3\sec x$ **8.** $y = 3\csc x$

a.   **b.**   **c.**   **d.**

e.   **f.**   **g.**   **h.**

In Exercises 9–28, graph the functions over the indicated intervals.

9. $y = \tan\left(\frac{1}{2}x\right),\ -2\pi \leq x \leq 2\pi$ **10.** $y = \cot\left(\frac{1}{2}x\right),\ -2\pi \leq x \leq 2\pi$

11. $y = -\cot(2\pi x),\ -1 \leq x \leq 1$ **12.** $y = -\tan(2\pi x),\ -1 \leq x \leq 1$

13. $y = 2\tan(3x),\ -\pi \leq x \leq \pi$ **14.** $y = 2\tan\left(\frac{1}{3}x\right),\ -3\pi \leq x \leq 3\pi$

15. $y = -\frac{1}{4}\cot\left(\frac{x}{2}\right),\ -2\pi \leq x \leq 2\pi$ **16.** $y = -\frac{1}{2}\tan\left(\frac{x}{4}\right),\ -4\pi \leq x \leq 4\pi$

17. $y = -\tan\left(x - \frac{\pi}{2}\right),\ -\pi \leq x \leq \pi$ **18.** $y = \tan\left(x + \frac{\pi}{4}\right),\ -\pi \leq x \leq \pi$

19. $y = 2\tan\left(x + \frac{\pi}{6}\right),\ -\pi \leq x \leq \pi$ **20.** $y = -\frac{1}{2}\tan(x + \pi),\ -\pi \leq x \leq \pi$

21. $y = \cot\left(x - \frac{\pi}{4}\right),\ -\pi \leq x \leq \pi$ **22.** $y = -\cot\left(x + \frac{\pi}{2}\right),\ -\pi \leq x \leq \pi$

23. $y = -\dfrac{1}{2}\cot\left(x + \dfrac{\pi}{3}\right), \ -\pi \le x \le \pi$

24. $y = 3\cot\left(x - \dfrac{\pi}{6}\right), \ -\pi \le x \le \pi$

25. $y = \tan(2x - \pi), \ -2\pi \le x \le 2\pi$

26. $y = \cot(2x - \pi), \ -2\pi \le x \le 2\pi$

27. $y = \cot\left(\dfrac{x}{2} + \dfrac{\pi}{4}\right), \ -\pi \le x \le \pi$

28. $y = \tan\left(\dfrac{x}{3} - \dfrac{\pi}{3}\right), \ -\pi \le x \le \pi$

In Exercises 29–46, graph the functions over the indicated intervals.

29. $y = \sec(\tfrac{1}{2}x), \ -2\pi \le x \le 2\pi$

30. $y = \csc(\tfrac{1}{2}x), \ -2\pi \le x \le 2\pi$

31. $y = -\csc(2\pi x), \ -1 \le x \le 1$

32. $y = -\sec(2\pi x), \ -1 \le x \le 1$

33. $y = \dfrac{1}{3}\sec\left(\dfrac{\pi}{2}x\right), \ -4 \le x \le 4$

34. $y = \dfrac{1}{2}\csc\left(\dfrac{\pi}{3}x\right), \ -6 \le x \le 6$

35. $y = -3\csc\left(\dfrac{x}{3}\right), \ -6\pi \le x \le 0$

36. $y = -4\sec\left(\dfrac{x}{2}\right), \ -4\pi \le x \le 4\pi$

37. $y = 2\sec(3x), \ 0 \le x \le 2\pi$

38. $y = 2\csc(\tfrac{1}{3}x), \ -3\pi \le x \le 3\pi$

39. $y = -3\csc\left(x - \dfrac{\pi}{2}\right),$ over at least one period

40. $y = 5\sec\left(x + \dfrac{\pi}{4}\right),$ over at least one period

41. $y = \tfrac{1}{2}\sec(x - \pi),$ over at least one period

42. $y = -4\csc(x + \pi),$ over at least one period

43. $y = 2\sec(2x - \pi), \ -2\pi \le x \le 2\pi$

44. $y = 2\csc(2x + \pi), \ -2\pi \le x \le 2\pi$

45. $y = -\tfrac{1}{4}\sec(3x + \pi)$

46. $y = -\dfrac{2}{3}\csc\left(4x - \dfrac{\pi}{2}\right), \ -\pi \le x \le \pi$

In Exercises 47–56, graph the functions over at least one period.

47. $y = 3 - 2\sec\left(x - \dfrac{\pi}{2}\right)$

48. $y = -3 + 2\csc\left(x + \dfrac{\pi}{2}\right)$

49. $y = \dfrac{1}{2} + \dfrac{1}{2}\tan\left(x - \dfrac{\pi}{2}\right)$

50. $y = \dfrac{3}{4} - \dfrac{1}{4}\cot\left(x + \dfrac{\pi}{2}\right)$

51. $y = -2 + 3\csc(2x - \pi)$

52. $y = -1 + 4\sec(2x + \pi)$

53. $y = -1 - \sec\left(\dfrac{1}{2}x - \dfrac{\pi}{4}\right)$

54. $y = -2 + \csc\left(\dfrac{1}{2}x + \dfrac{\pi}{4}\right)$

55. $y = -2 - 3\cot\left(2x - \dfrac{\pi}{4}\right), \ -\pi \le x \le \pi$

56. $y = -\dfrac{1}{4} + \dfrac{1}{2}\sec\left(\pi x + \dfrac{\pi}{4}\right), \ -2 \le x \le 2$

In Exercises 57–66, state the domain and range of the functions.

57. $y = \tan\left(\pi x - \dfrac{\pi}{2}\right)$

58. $y = \cot\left(x - \dfrac{\pi}{2}\right)$

59. $y = 2\sec(5x)$

60. $y = -4\sec(3x)$

61. $y = 2 - \csc(\tfrac{1}{2}x - \pi)$

62. $y = 1 - 2\sec(\tfrac{1}{2}x + \pi)$

63. $y = -3\tan\left(\dfrac{\pi}{4}x - \pi\right) + 1$

64. $y = \dfrac{1}{4}\cot\left(2\pi x + \dfrac{\pi}{3}\right) - 3$

65. $y = -2 + \dfrac{1}{2}\sec\left(\pi x + \dfrac{\pi}{2}\right)$

66. $y = \dfrac{1}{2} - \dfrac{1}{3}\csc\left(3x - \dfrac{\pi}{2}\right)$

▪ APPLICATIONS

For Exercises 67 and 68, refer to the following:

When a chalet on the top of a mountain is viewed by a mountain climber from point A on the plain below, the measure of the angle of elevation is α degrees. When the mountain climber moves closer to the mountain, point B, the measure of the angle of elevation is β degrees. If the distance between points A and B is d, then the height of the mountain can be found with $h = \dfrac{d}{\cot\alpha - \cot\beta}$.

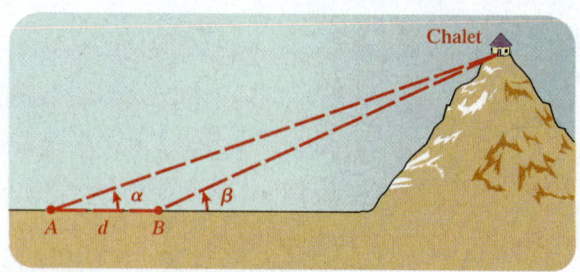

67. Height of a Mountain. What is the height of the mountain if $\alpha = 20°$, $\beta = 25°$, and $d = 2$ miles? Round to the nearest mile.

68. Height of a Mountain. What is the height of the mountain if $\alpha = 15°$, $\beta = 35°$, and $d = 3$ miles?

For Exercises 69 and 70, refer to the following:

The area of a regular polygon that is inscribed by a circle can be determined by $A = nr^2 \tan\dfrac{180°}{n}$, where n is the number of sides of the polygon, and r is the radius of the inscribed circle.

69. Area of a Regular Polygon. Find the area of a regular hexagon if the radius of the inscribed circle is 4 inches.

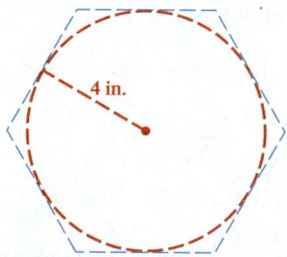

70. Area of a Regular Polygon. Find the area of a regular octagon if the radius of the inscribed circle is 2 feet.

▪ CATCH THE MISTAKE

In Exercises 71 and 72, explain the mistake that is made.

71. Graph $y = 3\csc(2x)$.

Solution:

Graph the guide function, $y = \sin(2x)$.

Draw vertical asymptotes at x-values that correspond to x-intercepts of the guide function.

Draw the cosecant function.

This is incorrect. What mistake was made?

72. Graph $y = \tan(4x)$.

Solution:

Step 1: Calculate the period.

$$\frac{\pi}{B} = \frac{\pi}{4}$$

Step 2: Find two vertical asymptotes.

$$4x = 0 \text{ and } 4x = \pi$$

Solve for x.

$$x = 0 \text{ and } x = \frac{\pi}{4}$$

Step 3: Find the x-intercept between the asymptotes.

$$4x = \frac{\pi}{2}$$

$$x = \frac{\pi}{8}$$

Step 4: Draw the vertical asymptotes $x = 0$ and $x = \frac{\pi}{4}$

and label the x-intercept $\left(\frac{\pi}{8}, 0\right)$.

Step 5: Graph.

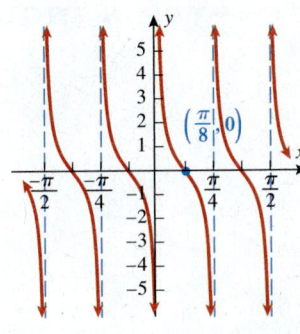

This is incorrect. What mistake was made?

■ **CONCEPTUAL**

In Exercises 73 and 74, determine whether each statement is true or false.

73. $\sec\left(x - \frac{\pi}{2}\right) = \csc x$

74. $\csc\left(x - \frac{\pi}{2}\right) = \sec x$

75. For what values of n do $y = \tan x$ and $y = \tan(x - n\pi)$ have the same graph?

76. For what values of n do $y = \csc x$ and $y = \csc(x - n\pi)$ have the same graph?

77. Solve the equation $\tan(2x - \pi) = 0$ for x in the interval $[-\pi, \pi]$ by graphing.

78. Solve the equation $\csc(2x + \pi) = 0$ for x in the interval $[-\pi, \pi]$ by graphing.

79. Find the x-intercepts of $y = A\tan(Bx + C)$.

80. For what x-values does the graph of $y = -A\sec\left(\frac{\pi}{2}x\right)$ lie above the x-axis? (Assume $A > 0$.)

81. How many solutions are there to the equation $\tan x = x$? Explain.

82. For what values of A do the graphs of $y = A\sin(Bx + C)$ and $y = -2\csc\left(\frac{\pi}{6}x - \pi\right)$ never intersect?

■ **TECHNOLOGY**

83. What is the amplitude of the function $y = \cos x + \sin x$? Use a graphing calculator to graph $Y_1 = \cos x$, $Y_2 = \sin x$, and $Y_3 = \cos x + \sin x$ in the same viewing window.

84. Graph $Y_1 = \cos x + \sin x$ and $Y_2 = \sec x + \csc x$ in the same viewing window. Based on what you see, is $Y_1 = \cos x + \sin x$ the guide function for $Y_2 = \sec x + \csc x$?

85. What is the period of the function $y = \tan x + \cot x$? Use a graphing calculator to graph $Y_1 = \tan x + \cot x$ and $Y_3 = 2\csc(2x)$ in the same viewing window.

86. What is the period of the function $y = \tan\left(2x + \frac{\pi}{2}\right)$? Use a graphing calculator to graph $Y_1 = \tan\left(2x + \frac{\pi}{2}\right)$, $Y_2 = \tan\left(2x - \frac{\pi}{2}\right)$, and $Y_3 = \tan\left(-2x + \frac{\pi}{2}\right)$ in the same viewing window. Describe the relationships of Y_1 and Y_2 and Y_2 and Y_3.

Dr. Parkinson has acquired two 30-foot sections of fence from her neighbor Mr. Wilson. She has decided to build a triangular corral for her animals. She plans to use a barn wall as the third side. (The barn wall is 100 feet long—see the diagram below.) As her contractor, you are expected to maximize the corral area. You have decided to approach this from a trigonometric viewpoint. Hence, you want to find the angle θ that maximizes the area. To do this, follow the steps below. (As a side note: This problem is an example of **optimization** and you will revisit this type of problem in precalculus and/or calculus courses. The outline below is designed to give you an understanding of how to set up and solve this type of problem. Realize that this triangle is an isosceles triangle—two equal side lengths—and that its perpendicular bisector can be used to find the height of the triangle.)

1. Write out the general formula for the area of a triangle.

2. To get an understanding of what happens to the area of the triangle as the angle θ changes, you will calculate the following dimensions using right triangle trigonometry. Do these calculations on a sheet of scratch paper. Since none of the triangles formed below are actually right triangles, you will need to construct a right triangle (using a perpendicular bisector) along with using the sine and cosine relationships to help you identify the base and height. (Use two decimals.)

θ	20°	40°	60°	80°
$b(\theta)$ = base				
$h(\theta)$ = height				
$A(\theta)$ = area				

You don't need to do every example in the chart by hand. However, do as many as you need to see what patterns emerge for calculating each base, height, and area. When you see the pattern, you will hopefully then be able to write a function for each piece of information. You can then use the table in your graphing calculator to list all the answers. However, you are encouraged to do at least two by hand before jumping to the function writing. Also be sure to check that the results you get on your table agree with the numbers you get by hand.

3. Write the base $b(\theta)$ of the triangle as a function of θ. (Show how you arrived at this answer.) Describe what happens to the base values as the θ values increase.

4. Write the height $h(\theta)$ of the triangle as a function of θ. (Show how you arrived at this answer.) Describe what happens to the height values as the θ values increase.

5. Write the area $A(\theta)$ as a function of θ using your results from 3 and 4. Describe what happens to your area values as θ increases.

6. Specify the **domain** for the area within the context of this problem. (**Vocabulary reminder:** The domain for this problem is the set of values of θ that makes sense from a physical standpoint; that is, you wouldn't build a corral using $\theta = -20°$.)

7. Use a graphing utility to graph your area function on its domain.

8. Estimate the maximum area and the θ that this corresponds to.

9. What can you conclude about the shape of the triangle that yields the maximum area in this example?

MODELING **OUR WORLD**

Some would argue that temperatures are oscillatory in nature: that we are not experiencing global warming at all, but instead a natural cycle in Earth's temperature. In the Modeling Our World features in Chapters 3–5, you modeled mean temperatures with linear, polynomial, and logarithmic models—all of which modeled increasing temperatures. Looking at the data another way we can demonstrate—over a short period of time—how it might appear to be oscillatory, which we can model with a sinusoidal curve. Be warned, however, that even the stock market, which increases gradually over time, looks sinusoidal over a short period.

The following table summarizes average yearly temperature in degree Fahrenheit (°F) and carbon dioxide emissions in parts per million (ppm) for the Mauna Loa Observatory in Hawaii.

Year	1960	1665	1970	1975	1980	1985	1990	1995	2000	2005
Temperature	44.45	43.29	43.61	43.35	46.66	45.71	45.53	47.53	45.86	46.23
CO_2 Emissions (ppm)	316.9	320.0	325.7	331.1	338.7	345.9	354.2	360.6	369.4	379.7

1. Find a *sinusoidal function* of the form $f(t) = k + A\sin(Bt)$ that models the temperature in Mauna Loa. Assume that the peak amplitude occurs in 1980 and again in 2010 (at 46.66°). Let $t = 0$ correspond to 1960.

2. Do your models support the claim of global warming? Explain.

3. You have modeled these same data with different types of models—oscillatory and nonoscillatory. Over the short 30-year period from 1980 to 2010 you have a sinusoidal model that can fit the data, but does it prove anything in the long term?

4. Would a resonance-type sinusoidal function of the form $f(t) = k + A(t)\sin(Bt)$ perhaps be a better fit for 1960 to 2005? Develop a function of this form that models the data.

SECTION	CONCEPT	KEY IDEAS/FORMULAS
6.1	**Angles, degrees, and triangles**	

Angles and degree measure

One complete rotation = 360°

Right Angle: quarter rotation
$\theta = 90°$

Acute Angle
$0° < \theta < 90°$

Obtuse Angle
$90° < \theta < 180°$

Complementary Angles
$\alpha + \beta = 90°$

Supplementary Angles
$\alpha + \beta = 180°$

Triangles

$\alpha + \beta + \gamma = 180°$

Right Triangle

Pythagorean Theorem
$a^2 + b^2 = c^2$

Special right triangles

$45° - 45° - 90°$

$30° - 60° - 90°$

Similar triangles

$$\frac{a}{a'} = \frac{b}{b'} = \frac{c}{c'}$$

SECTION	CONCEPT	KEY IDEAS/FORMULAS
6.2	**Definition 1 of trigonometric functions:** **Right triangle ratios**	Definition 1 defines trigonometric functions of acute angles as ratios of sides in a right triangle.

Trigonometric functions:
Right triangle ratios

$$\text{SOH} \quad \sin\theta = \frac{\text{opposite}}{\text{hypotenuse}}$$

$$\text{CAH} \quad \cos\theta = \frac{\text{adjacent}}{\text{hypotenuse}}$$

$$\text{TOA} \quad \tan\theta = \frac{\text{opposite}}{\text{adjacent}}$$

Reciprocal identities

$$\cot\theta = \frac{1}{\tan\theta} \qquad \csc\theta = \frac{1}{\sin\theta} \qquad \sec\theta = \frac{1}{\cos\theta}$$

Cofunctions

If $\alpha + \beta = 90°$, then
$$\sin\alpha = \cos\beta$$
$$\sec\alpha = \csc\beta$$
$$\tan\alpha = \cot\beta$$

Evaluating trigonometric functions exactly for special angle measures: 30°, 45°, and 60°

θ	$\sin\theta$	$\cos\theta$
30°	$\dfrac{1}{2}$	$\dfrac{\sqrt{3}}{2}$
45°	$\dfrac{\sqrt{2}}{2}$	$\dfrac{\sqrt{2}}{2}$
60°	$\dfrac{\sqrt{3}}{2}$	$\dfrac{1}{2}$

The other trigonometric functions can be found for these values using $\tan\theta = \dfrac{\sin\theta}{\cos\theta}$ and reciprocal identities.

Using calculators to evaluate (approximate) trigonometric functions

Make sure the calculator is in degree mode. The sin, cos, and tan buttons can be combined with the reciprocal button, $1/x$, to get csc, sec, and cot.

| 6.3 | **Applications of right triangle trigonometry:** **Solving right triangles** | |

Solving a right triangle given the measure of an acute angle and a side length.

- The third angle measure can be found exactly using $\alpha + \beta + \gamma = 180°$.
- Right triangle trigonometry is used to find the length of the second side.

Solving a right triangle given the lengths of two sides.

- The length of the third side can be found using the Pythagorean theorem.
- The measure of one of the acute angles can be found using right triangle trigonometry.

SECTION	CONCEPT	KEY IDEAS/FORMULAS
6.4	**Definition 2 of trigonometric functions: Cartesian plane**	
	Angles in standard position	An angle is said to be in standard position if its initial side is along the positive x-axis and its vertex is at the origin.

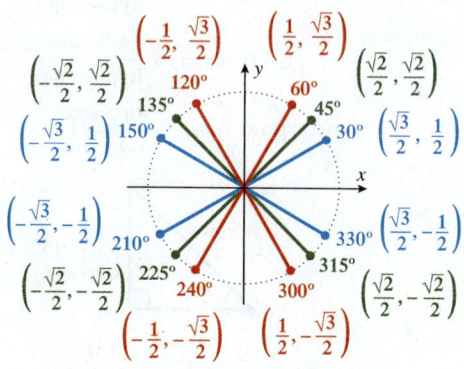

	Coterminal angles	Two angles in standard position with the same terminal side.
	Trigonometric functions: The Cartesian plane	$\sin\theta = \dfrac{y}{r} \qquad \cos\theta = \dfrac{x}{r} \qquad \tan\theta = \dfrac{y}{x}$ $\csc\theta = \dfrac{r}{y} \qquad \sec\theta = \dfrac{r}{x} \qquad \cot\theta = \dfrac{x}{y}$ where $x^2 + y^2 = r^2$. The distance r is positive: $r > 0$.

θ	0°	90°	180°	270°
$\sin\theta$	0	1	0	-1
$\cos\theta$	1	0	-1	0
$\tan\theta$	0	Undefined	0	Undefined
$\cot\theta$	Undefined	0	Undefined	0
$\sec\theta$	1	Undefined	-1	Undefined
$\csc\theta$	Undefined	1	Undefined	-1

Trigonometric function values for quadrantal angles.

6.5	**Trigonometric functions of nonacute angles**	
	Algebraic signs of trigonometric functions	

θ	QI	QII	QIII	QIV
$\sin\theta$	+	+	−	−
$\cos\theta$	+	−	−	+
$\tan\theta$	+	−	+	−

	Ranges of the trigonometric functions	$-1 \le \sin\theta \le 1$ and $-1 \le \cos\theta \le 1$

SECTION	CONCEPT	KEY IDEAS/FORMULAS
	Reference angles and reference right triangles	The reference angle α for angle θ (between $0°$ and $360°$) is given by ■ QI: $\alpha = \theta$ ■ QII: $\alpha = 180° - \theta$ ■ QIII: $\alpha = \theta - 180°$ ■ QIV: $\alpha = 360° - \theta$
	Evaluating trigonometric functions for nonacute angles	

SECTION	CONCEPT	KEY IDEAS/FORMULAS
6.6	**Radian measure and applications**	
	The radian measure of an angle	$\theta\,(\text{in radians}) = \dfrac{s}{r}$ s (arc length) and r (radius) must have the same units.
	Converting between degrees and radians	Degrees to radians: $\theta_r = \theta_d\left(\dfrac{\pi}{180°}\right)$ Radians to degrees: $\theta_d = \theta_r\left(\dfrac{180°}{\pi}\right)$
	Arc length	$s = r\theta_r$ θ_r is in radians. $s = r\theta_d\left(\dfrac{\pi}{180°}\right)$ θ_d is in degrees.
	Area of a circular sector	$A = \dfrac{1}{2}r^2\theta_r$ θ_r is in radians. $A = \dfrac{1}{2}r^2\theta_d\left(\dfrac{\pi}{180°}\right)$ θ_d is in degrees.
	Linear speed	Linear speed v is given by $$v = \dfrac{s}{t}$$ where s is the arc length and t is time.
	Angular speed	Angular speed ω is given by $$\omega = \dfrac{\theta}{t}$$ where θ is given in radians.

SECTION	CONCEPT	KEY IDEAS/FORMULAS
	Relationship between linear and angular speeds	Related through the radius of the circle: $v = r\omega \quad \text{or} \quad \omega = \dfrac{v}{r}$ It is important to note that these formulas hold true only when angular speed is given in *radians* per unit of time.
6.7	**Definition 3 of trigonometric functions: Unit circle approach**	
	Trigonometric functions and the unit circle (circular functions)	 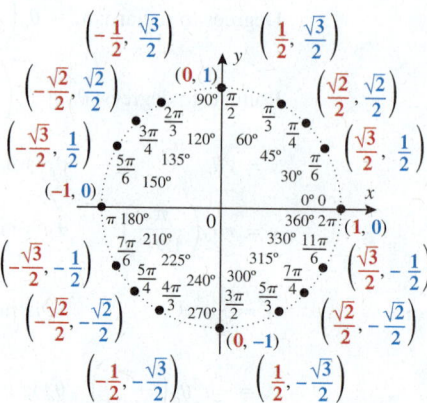
	Properties of circular functions	Cosine is an even function. $\qquad \cos(-\theta) = \cos\theta$ Sine is an odd function. $\qquad \sin(-\theta) = -\sin\theta$

6.8

Graphs of sine and cosine functions

The graph of
$f(x) = \sin x$

Odd function: $f(-x) = -f(x)$

The graph of $f(x) = \cos x$

Even function: $f(-x) = f(x)$

The amplitude and period of sinusoidal graphs

$y = A \sin Bx$ or $y = A \cos(Bx)$, $B > 0$

Amplitude $= |A|$ Period $= \dfrac{2\pi}{B}$

- $|A| > 1$ stretch vertically.
- $|A| < 1$ compress vertically.

- $B > 1$ compress horizontally.
- $B < 1$ stretch horizontally.

Graphing a shifted sinusoidal function:

$y = A \sin(Bx + C) + D$
and
$y = A \cos(Bx + C) + D$

- $y = A \sin(Bx \pm C) = A \sin\left[B\left(x \pm \dfrac{C}{B}\right)\right]$ has period $\dfrac{2\pi}{B}$ and a phase shift

  of $\dfrac{C}{B}$ unit to the left $(+)$ or the right $(-)$.

- $y = A \cos(Bx \pm C) = A \cos\left[B\left(x \pm \dfrac{C}{B}\right)\right]$ has period $\dfrac{2\pi}{B}$ and a phase

  shift of $\dfrac{C}{B}$ unit to the left $(+)$ or the right $(-)$.

- To graph $y = A \sin(Bx + C) + D$ or $y = A \cos(Bx + C) + D$, start with the graph of $y = A \sin(Bx + C)$ or $y = A \cos(Bx + C)$ and shift up or down D units.

Harmonic motion

- Simple
- Damped
- Resonance

Graphing sums of functions:
Addition of ordinates

6.9	**Graphs of other trigonometric functions**	

The tangent function

x-intercepts: $x = n\pi$ n is an integer

Asymptotes: $x = \dfrac{(2n+1)}{2}\pi$

Period: π

Amplitude: none

The cotangent function

x-intercepts: $x = n\pi$ n is an integer

Asymptotes: $x = \dfrac{(2n+1)}{2}\pi$

Period: π

Amplitude: none

The secant function

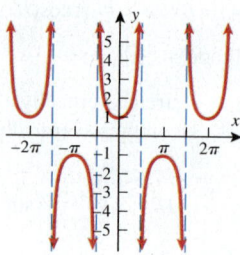

x-intercepts: none n is an integer

Asymptotes: $x = \dfrac{(2n+1)}{2}\pi$

Period: 2π

Amplitude: none

The cosecant function

x-intercepts: none n is an integer

Asymptotes: $x = n\pi$

Period: 2π

Amplitude: none

SECTION	CONCEPT	KEY IDEAS/FORMULAS

Graphing tangent,
cotangent, secant, and
cosecant functions

Translations of
circular functions

$y = A\tan(Bx + C)$ or $y = A\cot(Bx + C)$

To find asymptotes, set $Bx + C$ equal to:

■ $-\dfrac{\pi}{2}$ and $\dfrac{\pi}{2}$ for tangent.

■ 0 and π for cotangent.

To find x-intercepts, set $Bx + C$ equal to:

■ 0 for tangent.

■ $\dfrac{\pi}{2}$ for cotangent.

$y = A\sec(Bx + C)$ or $y = A\csc(Bx + C)$

To graph $y = A\sec(Bx + C)$, use $y = A\cos(Bx + C)$ as the guide.

To graph $y = A\csc(Bx + C)$, use $y = A\sin(Bx + C)$ as the guide.

Intercepts on the guide function correspond to vertical asymptotes of secant or cosecant functions.

6.1 Angles, Degrees, and Triangles

Find (a) the complement and (b) the supplement of the given angles.

1. $28°$ 2. $17°$ 3. $35°$

4. $78°$ 5. $89.01°$ 6. $0.013°$

Refer to the accompanying triangle for the following exercises.

7. If $\alpha = 120°$ and $\beta = 35°$, find γ.

8. If $\alpha = 105°$ and $\beta = 25°$, find γ.

9. If $\gamma = \beta$ and $\alpha = 7\beta$, find the measure of all three angles.

10. If $\gamma = \beta$ and $\alpha = 6\beta$, find the measure of all three angles.

$\alpha + \beta + \gamma = 180°$

Refer to the accompanying right triangle for the following exercises.

11. If $a = 4$ and $c = 12$, find b.

12. If $b = 9$ and $c = 15$, find a.

13. If $a = 7$ and $b = 4$, find c.

14. If $a = 10$ and $b = 8$, find c.

Refer to the accompanying 45°-45°-90° triangle for the following exercises.

15. If the two legs have length 12 yards, how long is the hypotenuse?

16. If the hypotenuse has length $\sqrt{8}$ feet, how long are the legs?

Refer to the accompanying 30°-60°-90° triangle for the following exercises.

17. If the shortest leg has length 3 feet, what are the lengths of the other leg and the hypotenuse?

18. If the hypotenuse has length 12 km, what are the lengths of the two legs?

Calculate the specified lengths given that the two triangles are similar.

19. $A = 10, C = 8,$
    $D = 5, F = ?$

20. $A = 15, B = 12,$
    $E = 4, D = ?$

21. $D = 4.5$ m,
    $F = 8.2$ m,
    $A = 81$ km, $C = ?$

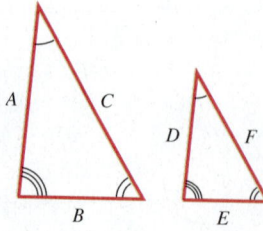

22. $E = 8$ cm, $F = 14$ cm, $C = 84$ m, $B = ?$

23. $B = 2\sqrt{3}$ in., $C = \sqrt{3}$ in., $F = 5\sqrt{3}$ in., $E = ?$

24. $C = 4.2$ ft, $F = 0.2$ ft, $D = 2$ ft, $A = ?$

Applications

25. **Clock.** What is the measure (in degrees) of the angle that the minute hand sweeps in exactly 25 minutes?

26. **Clock.** What is the measure (in degrees) of the angle that the second hand sweeps in exactly 15 seconds?

27. **Height of a Tree.** The shadow of a tree measures 9.6 meters. At the same time of day the shadow of a 4-meter basketball backboard measures 1.2 meters. How tall is the tree?

28. **Height of a Man.** If an NBA center casts a 1-foot 9-inch shadow and his 4-foot son casts a 1-foot shadow, how tall is the NBA center?

6.2 Definition 1 of Trigonometric Functions: Right Triangle Ratios

Use the following triangle to find the indicated trigonometric functions. Rationalize any denominators that you encounter in the answers.

29. $\cos \theta$

30. $\sin \theta$

31. $\sec \theta$

32. $\csc \theta$

33. $\tan \theta$

34. $\cot \theta$

Use the cofunction identities to fill in the blanks.

35. $\sin 30° = \cos$_____

36. $\cos A = \sin$_____

37. $\tan 45° = \cot$_____

38. $\csc 60° = \sec$_____

39. $\sec 30° = \csc$_____

40. $\cot 60° = \tan$_____

Label each trigonometric function value with the corresponding value (a–c).

    a. $\dfrac{\sqrt{3}}{2}$ b. $\dfrac{1}{2}$ c. $\dfrac{\sqrt{2}}{2}$

41. $\sin 30°$ 42. $\cos 30°$ 43. $\cos 60°$

44. $\sin 60°$ 45. $\sin 45°$ 46. $\cos 45°$

Use a calculator to approximate the following trigonometric function values. Round the answers to four decimal places.

47. $\sin 42°$ **48.** $\cos 57°$ **49.** $\cos 17.3°$ **50.** $\tan 25.2°$

51. $\cot 33°$ **52.** $\sec 16.8°$ **53.** $\csc 40.25°$ **54.** $\cot 19.76°$

6.3 Applications of Right Triangle Trigonometry: Solving Right Triangles

The following exercises illustrate a mid-air refueling scenario that U.S. military aircraft often use. Assume the elevation angle that the hose makes with the plane being fueled is $\theta = 30°$.

55. Mid-Air Refueling. If the hose is 150 feet long, what should the altitude difference a be between the two planes?

56. Mid-Air Refueling. If the smallest acceptable altitude difference a between the two planes is 100 feet, how long should the hose be?

6.4 Definition 2 of Trigonometric Functions: Cartesian Plane

In the following exercises, the terminal side of an angle θ in standard position passes through the indicated point. Calculate the values of the six trigonometric functions for angle θ.

57. $(6, -8)$ **58.** $(-24, -7)$ **59.** $(-6, 2)$ **60.** $(-40, 9)$

61. $\left(\sqrt{3}, 1\right)$ **62.** $(-9, -9)$ **63.** $\left(\frac{1}{2}, -\frac{1}{4}\right)$ **64.** $\left(-\frac{3}{4}, \frac{5}{6}\right)$

65. $(-1.2, -2.4)$ **66.** $(0.8, -2.4)$

6.5 Trigonometric Functions of Nonacute Angles

Evaluate the following expressions exactly.

67. $\sin 330°$ **68.** $\cos(-300°)$ **69.** $\tan 150°$

70. $\cot 315°$ **71.** $\sec(-150°)$ **72.** $\csc 210°$

Evaluate the trigonometric expressions with a calculator. Round to four decimal places.

73. $\sin(-14°)$ **74.** $\cos 275°$ **75.** $\tan 46°$ **76.** $\cot 500°$

77. $\sec 111°$ **78.** $\csc 215°$ **79.** $\cot(-57°)$ **80.** $\csc(-59°)$

6.6 Radian Measure and Applications

Convert from degrees to radians. Leave your answers in terms of π.

81. $135°$ **82.** $240°$ **83.** $330°$ **84.** $180°$

85. $216°$ **86.** $108°$ **87.** $1620°$ **88.** $900°$

Convert from radians to degrees.

89. $\dfrac{\pi}{3}$ **90.** $\dfrac{11\pi}{6}$ **91.** $\dfrac{5\pi}{4}$ **92.** $\dfrac{2\pi}{3}$

93. $\dfrac{5\pi}{9}$ **94.** $\dfrac{17\pi}{10}$ **95.** 10π **96.** $\dfrac{31\pi}{2}$

97. A ladybug is clinging to the outer edge of a child's spinning disk. The disk is 4 inches in diameter and is spinning at 60 revolutions per minute. How fast is the ladybug traveling?

98. How fast is a motorcyclist traveling in miles per hour if his tires are 30 inches in diameter and the angular speed of the tire is 10π radians per second?

6.7 Definition 3 of Trigonometric Functions: Unit Circle Approach

Find each trigonometric function value in *exact* form.

99. $\tan\left(\dfrac{5\pi}{6}\right)$ **100.** $\cos\left(\dfrac{5\pi}{6}\right)$ **101.** $\sin\left(\dfrac{11\pi}{6}\right)$

102. $\sec\left(\dfrac{11\pi}{6}\right)$ **103.** $\cot\left(\dfrac{5\pi}{4}\right)$ **104.** $\csc\left(\dfrac{5\pi}{4}\right)$

105. $\sin\left(\dfrac{3\pi}{2}\right)$ **106.** $\cos\left(\dfrac{3\pi}{2}\right)$ **107.** $\cos\pi$

108. $\tan 315°$ **109.** $\cos 60°$ **110.** $\sin 330°$

6.8 Graphs of Sine and Cosine Functions

Refer to the graph of the sinusoidal function to answer the questions.

111. Determine the period of the function.

112. Determine the amplitude of the function.

113. Write an equation for the sinusoidal function.

Refer to the graph of the sinusoidal function to answer the questions.

114. Determine the period of the function.

115. Determine the amplitude of the function.

116. Write an equation for the sinusoidal function.

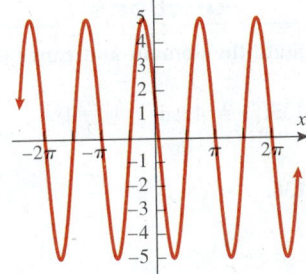

Determine the amplitude and period of each function.

117. $y = -2\cos(2\pi x)$

118. $y = \dfrac{1}{3}\sin\dfrac{\pi}{2}x$

119. $y = \dfrac{1}{5}\sin 3x$

120. $y = -\dfrac{7}{6}\cos 6x$

Graph each function from -2π to 2π.

121. $y = -2\sin\dfrac{x}{2}$

122. $y = 3\sin 3x$

123. $y = \dfrac{1}{2}\cos 2x$

124. $y = -\dfrac{1}{4}\cos\dfrac{x}{2}$

State the amplitude, period, phase shift, and vertical shift of each function.

125. $y = 2 + 3\sin\left(x - \dfrac{\pi}{2}\right)$

126. $y = 3 - \dfrac{1}{2}\sin\left(x + \dfrac{\pi}{4}\right)$

127. $y = -2 - 4\cos 3\left(x + \dfrac{\pi}{4}\right)$

128. $y = -1 + 2\cos 2\left(x - \dfrac{\pi}{3}\right)$

129. $y = -\dfrac{1}{2} + \dfrac{1}{3}\cos\left(\pi x - \dfrac{1}{2}\right)$

130. $y = \dfrac{3}{4} - \dfrac{1}{6}\sin\left(\dfrac{\pi}{6}x + \dfrac{\pi}{3}\right)$

Graph each function from $-\pi$ to π.

131. $y = 3x - \cos 2x$

132. $y = -\dfrac{1}{2}\cos 4x + \dfrac{1}{2}\cos 2x$

133. $y = 2\sin\dfrac{1}{3}x - 3\sin 3x$

134. $y = 5\cos x + 3\sin\dfrac{x}{2}$

6.9 Graphs of Other Trigonometric Functions

State the domain and range of each function.

135. $y = 4\tan\left(x + \dfrac{\pi}{2}\right)$

136. $y = \cot 2\left(x - \dfrac{\pi}{2}\right)$

137. $y = 3\sec(2x)$

138. $y = 1 + 2\csc x$

139. $y = -\dfrac{1}{2} + \dfrac{1}{4}\sec\left(\pi x - \dfrac{2\pi}{3}\right)$

140. $y = 3 - \dfrac{1}{2}\csc(2x - \pi)$

Graph each function on the interval $[-2\pi, 2\pi]$.

141. $y = -\tan\left(x - \dfrac{\pi}{4}\right)$

142. $y = 1 + \cot(2x)$

143. $y = 2 + \sec(x - \pi)$

144. $y = -\csc\left(x + \dfrac{\pi}{4}\right)$

145. $y = \dfrac{1}{2} + 2\csc\left(2x - \dfrac{\pi}{2}\right)$

146. $y = -1 - \dfrac{1}{2}\sec\left(\pi x - \dfrac{3\pi}{4}\right)$

Technology Exercises

Section 6.1

Assume a 30°–60°–90° triangle. Round your answers to two decimal places.

147. If the shorter leg has length 41.32 feet, what are the lengths of the other leg and hypotenuse?

148. If the longer leg has length 87.65 cm, what are the lengths of the other leg and the hypotenuse?

Section 6.2

149. Calculate $\csc 78.4°$ in the following two ways:

    a. Find $\sin 78.4°$ to three decimal places and then divide 1 by that number. Write that number to five decimal places.

    b. In a calculator: 78.4, sin, 1/x, round to five decimal places.

150. Calculate $\cot 34.8°$ in the following two ways:

    a. Find $\tan 34.8°$ to three decimal places and then divide 1 by that number. Write that number to five decimal places.

    b. In a calculator: 34.8, tan, 1/x, round to five decimal places.

151. Use a calculator to find $\tan(\tan^{-1} 2.612)$.

152. Use a calculator to find $\cos(\cos^{-1} 0.125)$.

Use a calculator to evaluate the following expressions. If you get an error, explain why.

153. $\sec 180°$

154. $\csc 180°$

Section 6.5

155. Use a calculator to evaluate $\csc 218°$ and $\csc 322°$. Now use the calculator to evaluate $\csc^{-1}(-1.6243)$. When cosecant is negative, in which of the quadrants, III or IV, does the calculator assume the terminal side of the angle lies?

156. Use a calculator to evaluate $\sec 28°$ and $\sec 332°$. Now use the calculator to evaluate $\sec^{-1}(1.1326)$. When secant is positive, in which of the quadrants, I or IV, does the calculator assume the terminal side of the angle lies?

Section 6.6

Find the measure (in degrees, minutes, and nearest seconds) of a central angle θ that intercepts an arc on a circle with radius r with indicated arc length s. Use the TI calculator comands ANGLE **and** DMS **to change to degrees, minutes, and seconds.**

157. $r = 11.2$ ft, $s = 19.7$ ft

158. $r = 56.9$ cm, $s = 139.2$ cm

Section 6.7

In Exercises 159 and 160, refer to the following:

A graphing calculator can be used to graph the unit circle with parametric equations (these will be covered in more detail in Section 8.7). For now, set the calculator in parametric and radian modes and let

$$X_1 = \cos T$$
$$Y_1 = \sin T$$

Set the window so that $0 \le T \le 2\pi$, step $= \dfrac{\pi}{15}$, $-2 \le X \le 2$, and $-2 \le Y \le 2$. To approximate the sine or cosine of a T value, use the TRACE key, enter the T value, and read the corresponding coordinates from the screen.

159. Use the above steps to approximate $\cos\left(\dfrac{13\pi}{12}\right)$ to four decimal places.

160. Use the above steps to approximate $\sin\left(\dfrac{5\pi}{6}\right)$ to four decimal places.

Section 6.8

161. Use a graphing calculator to graph $Y_1 = \cos x$ and $Y_2 = \cos(x + c)$, where

   a. $c = \dfrac{\pi}{6}$, and explain the relationship between Y_2 and Y_1.

   b. $c = -\dfrac{\pi}{6}$, and explain the relationship between Y_2 and Y_1.

162. Use a graphing calculator to graph $Y_1 = \sin x$ and $Y_2 = \sin x + c$, where

   a. $c = \dfrac{1}{2}$, and explain the relationship between Y_2 and Y_1.

   b. $c = -\dfrac{1}{2}$, and explain the relationship between Y_2 and Y_1.

Section 6.9

163. What is the amplitude of the function $y = 4\cos x - 3\sin x$? Use a graphing calculator to graph $Y_1 = 4\cos x$, $Y_2 = 3\sin x$, and $Y_3 = 4\cos x - 3\sin x$ in the same viewing window.

164. What is the amplitude of the function $y = \sqrt{3}\sin x + \cos x$? Use a graphing calculator to graph $Y_1 = \sqrt{3}\sin x$, $Y_2 = \cos x$, and $Y_3 = \sqrt{3}\sin x + \cos x$ in the same viewing window.

1. A 5-foot girl is standing *in* the Grand Canyon, and she wants to estimate the depth of the canyon. The sun casts her shadow 6 inches along the ground. To measure the shadow cast by the top of the canyon, she walks the length of the shadow. She takes 200 steps and estimates that each step is roughly 3 feet. Approximately how tall is the Grand Canyon?

2. Fill in the values in the table.

θ	sin θ	cos θ	tan θ	cot θ	sec θ	csc θ
30°						
45°						
60°						

3. What is the difference between $\cos \theta = \frac{2}{3}$ and $\cos \theta \approx 0.66$?

4. Fill in the table with exact values for the quadrantal angles and the algebraic signs for the quadrants.

	0°	QI	90°	QII	180°	QIII	270°	QIV	360°
$\sin \theta$									
$\cos \theta$									

5. If $\cot \theta < 0$ and $\sec \theta > 0$, in which quadrant does the terminal side of θ lie?

6. Evaluate $\sin 210°$ exactly.

7. Convert $\dfrac{13\pi}{4}$ to degree measure.

8. Convert 260° to radian measure. Leave the answer in terms of π.

9. What is the area of the sector swept by the second hand of a clock in 25 seconds? Assume the radius of the sector is 3 inches.

10. What is the measure in radians of the smaller angle between the hour and minute hands at 10:10?

11. State the amplitude and period of $y = -5\sin(3x)$.

12. Graph $y = -2\cos(\frac{1}{2}x)$ over $-4\pi \le x \le 4\pi$.

13. Graph $y = \tan\left(\pi x - \dfrac{\pi}{2}\right)$ over two periods.

14. The vertical asymptotes of $y = 2\csc(3x - \pi)$ correspond to the _____ of $y = 2\sin(3x - \pi)$.

15. State the x-intercepts of $y = \tan(2x)$ for all x.

16. State the phase shift and vertical shift for
$$y = -\cot\left(\dfrac{\pi}{3}x - \pi\right).$$

17. State the range of $y = -3\sec\left(2x + \dfrac{\pi}{3}\right) - 1$.

18. State the domain of $y = \tan\left(2x - \dfrac{\pi}{6}\right) + 3$.

19. Graph $y = -2\csc\left(x + \dfrac{\pi}{2}\right)$ over two periods.

20. Find the x-intercept(s) of $y = \dfrac{6}{\sqrt{3}} - 3\sec\left(6x - \dfrac{5\pi}{6}\right)$.

21. True or False: The equation $2\sin\theta = 2.0001$ has no solution.

22. On what x-intervals does the graph of $y = \cos(2x)$ lie below the x-axis?

23. Write the equation of a sine function that has amplitude 4, vertical shift $\frac{1}{2}$ down, phase shift $\frac{3}{2}$ to the left, and period π.

24. Write the equation of a cotangent function that has period π, vertical shift 0.01 up, and no phase shift.

25. Graph $y = \cos 3x - \frac{1}{2}\sin 3x$ for $0 \le x \le \pi$.

26. If the longer leg of a 30°-60°-90° triangle has length 13.95 inches, what are the lengths of the other leg and the hypotenuse? Round your answer to two decimal places.

27. Calculate $\sec(41.9°)$ in the following two ways:

   a. Find $\cos(41.9°)$ to three decimal places and then divide 1 by that number. Write that number to five decimal places.

   b. In a calculator: 41.9, cos, 1/x, round to five decimal places.

1. Simplify $\dfrac{6x^2 - 11x + 5}{30x - 25}$.

2. Perform the operation and write in standard form $a \pm bi$:
$(5 - 7i) - (11 - 9i)$.

3. Solve for x and give any excluded values: $\dfrac{5}{x} - \dfrac{5}{x + 1} = \dfrac{-5}{x^2 + x}$.

4. Solve for the radius r: $A = \pi r^2$.

5. Solve and express the solution in interval notation:
$\dfrac{2t^2}{t - 1} \geq 3t$.

6. Calculate the distance and midpoint between the segment joining the points $\left(-\dfrac{3}{4}, \dfrac{1}{6}\right)$ and $\left(\dfrac{1}{5}, -\dfrac{2}{3}\right)$.

7. Write an equation of a line in slope–intercept form that passes through the two points $(5.6, 6.2)$ and $(3.2, 5.0)$.

8. Use interval notation to express the domain of the function $f(x) = \sqrt{x^2 - 25}$.

9. Find the average rate of change for $f(x) = \dfrac{5}{x}$, from $x = 2$ to $x = 4$.

10. Sketch the graph of the function $y = \dfrac{1}{x - 2} + 1$, and identify all transformations of the function $f(x) = \dfrac{1}{x}$.

11. Evaluate $g(f(-1))$ for $f(x) = \sqrt[3]{x - 7}$ and $g(x) = \dfrac{5}{3 - x}$.

12. Find the inverse of the function $f(x) = \dfrac{5x + 2}{x - 3}$.

13. Find the quadratic function that has the vertex $(0, 7)$ and goes through the point $(2, -1)$.

14. Find all of the real zeros and state the multiplicity of the function $f(x) = \dfrac{1}{7}x^5 + \dfrac{2}{9}x^3$.

15. List the possible rational zeros, and test to determine all rational zeros for $P(x) = 2x^4 + 7x^3 - 18x^2 - 13x + 10$.

16. Factor the polynomial $P(x) = 4x^4 - 4x^3 + 13x^2 + 18x + 5$ as a product of linear factors.

17. Graph the exponential function $f(x) = 5^{-x} - 1$, and state the y-intercept, the domain, the range, and the horizontal asymptote.

18. How much money should be put in a savings account now that earns 5.2% a year compounded daily if you want to have $50,000 in 16 years?

19. Graph the function $f(x) = \ln(x + 1) - 3$ using transformation techniques.

20. Evaluate $\log_{4.7} 8.9$ using the change-of-base formula. Round the answer to three decimal places.

21. Solve the equation $5(10^{2x}) = 37$ for x. Round the answer to three decimal places.

22. **Height of a tree.** The shadow of a tree measures $15\frac{1}{3}$ feet. At the same time of day the shadow of a 6-foot pole measures 2.3 feet. How tall is the tree?

23. The angle θ in standard position has its terminal side defined by the line $3x + 2y = 0$, $x \leq 0$. Calculate the values for the six trigonometric functions of θ.

24. Find the area of a circular sector with radius 3.4 m and central angle $\theta = 35°$. Round the answer to three significant digits.

25. State the amplitude, period, and phase shift of the function $y = -4\sin(2x + \pi)$.

7

Analytic Trigonometry

When you dial a phone number on your iPhone®, how does the smart phone know which key you have pressed? Dual Tone Multi-Frequency (DTMF), also known as touch-tone dialing, was developed by Bell Labs in the 1960s. The Touch-Tone® system also introduced a standardized keypad layout. After testing 18 different layouts, Bell Labs eventually chose the one familiar to us today, with 1 in the upper-left and 0 at the bottom between the star and the pound keys.

The keypad is laid out in a 4 × 3 matrix, with each row representing a low frequency and each column representing a high frequency.

© Jason Brindel Commercial/Alamy

FREQUENCY	1209 Hz	1336 Hz	1477 Hz
697 Hz	1	2	3
770 Hz	4	5	6
852 Hz	7	8	9
941 Hz	*	0	#

When you press the number 8, the phone sends a sinusoidal tone that combines a low-frequency tone of 852 hertz and a high-frequency tone of 1336 hertz. The result can be found using sum-to-product *trigonometric identities*.

IN THIS CHAPTER we will discuss trigonometric identities and use them to simplify trigonometric expressions. We will verify trigonometric identities. Specific identities that we will discuss are sum and difference, double-angle and half-angle, and product-to-sum and sum-to-product. Musical tones and touch-tone keypads are applications of trigonometric identities. We will also define inverse trigonometric functions. Trigonometric identities and inverse trigonometric functions will be used in solving trigonometric equations.

ANALYTIC TRIGONOMETRY

7.1 Basic Trigonometric Identities

- Reciprocal Identities
- Quotient Identities
- Pythagorean Identities

7.2 Verifying Trigonometric Identities

- Trigonometric Identities

7.3 Sum and Difference Identities

- Sum and Difference Identities for the Cosine Function
- Sum and Difference Identities for the Sine Function
- Sum and Difference Identities for the Tangent Function

7.4 Double-Angle Identities

- Applying Double-Angle Identities

7.5 Half-Angle Identities

- Applying Half-Angle Identities

7.6 Product-to-Sum and Sum-to-Product Identities

- Product-to-Sum Identities
- Sum-to-Product Identities

7.7 Inverse Trigonometric Functions

- Inverse Sine Function
- Inverse Cosine Function
- Inverse Tangent Function
- Remaining Inverse Trigonometric Functions
- Finding Exact Values for Expressions Involving Inverse Trigonometric Functions

7.8 Trigonometric Equations

- Solving Trigonometric Equations by Inspection
- Solving Trigonometric Equations Using Algebraic Techniques
- Solving Trigonometric Equations Which Require the Use of Inverse Functions
- Using Trigonometric Identities to Solve Trigonometric Equations

LEARNING OBJECTIVES

- Review basic identities.
- Verify a trigonometric identity.
- Apply the sum and difference identities.
- Apply the double-angle identities.
- Apply half-angle identities.
- Apply the product-to-sum and sum-to-product identities.
- Graph inverse trigonometric functions.
- Solve trigonometric equations.

SKILLS OBJECTIVES

- Apply the reciprocal identities.
- Apply the quotient identities.
- Apply the Pythagorean identities.
- Use the basic identities to simplify expressions.

CONCEPTUAL OBJECTIVES

- Understand that trigonometric reciprocal identities are not always defined.
- Understand that quotient identities are not always defined.

Reciprocal Identities

In this section, we revisit the *reciprocal* and *quotient identities* and develop the *Pythagorean identities*. These identities will allow us to simplify trigonometric expressions. In mathematics, an **identity** is an equation that is true for all values of the variable for which the equation is defined. For example, $x^2 = x \cdot x$ is an identity because the statement is true no matter what values are selected for x. Similarly, $\dfrac{2x}{5x} = \dfrac{2}{5}$ is an identity, but it is important to note that this identity holds for all x except 0 ($x \neq 0$), since the expression on the left is not defined when $x = 0$. In our discussions on identities, it is important to note that the identities hold for all values of the argument, typically chosen to be θ, for which the expressions are defined.

Recall that the **reciprocal** of a nonzero number x is $\dfrac{1}{x}$. For example, the reciprocal of 5 is $\dfrac{1}{5}$ and the reciprocal of $\dfrac{9}{17}$ is $\dfrac{17}{9}$. There is no reciprocal of 0. In mathematics, the expression $\dfrac{1}{0}$ is undefined. Calculators have a reciprocal key typically labeled $1/x$ or x^{-1}. This can be used to find the reciprocal of any number (except 0).

In Section 6.4, the trigonometric functions were defined by the following, remembering that $r > 0$:

$$\sin\theta = \frac{y}{r} \qquad \cos\theta = \frac{x}{r} \qquad \tan\theta = \frac{y}{x} \quad (x \neq 0)$$

$$\csc\theta = \frac{r}{y} \quad (y \neq 0) \qquad \sec\theta = \frac{r}{x} \quad (x \neq 0) \qquad \cot\theta = \frac{x}{y} \quad (y \neq 0)$$

Derivation of the Reciprocal Identities

WORDS	MATH
Write the definition of the cosecant function.	$\csc\theta = \dfrac{r}{y} \quad (y \neq 0)$
The reciprocal identity holds for $y \neq 0$.	$\csc\theta = \dfrac{r}{y} = \dfrac{1}{\frac{y}{r}} \quad (y \neq 0, r > 0)$
Substitute the definition of the sine function, $\sin\theta = \dfrac{y}{r}$.	$\csc\theta = \dfrac{r}{y} = \dfrac{1}{\frac{y}{r}} = \dfrac{1}{\sin\theta}$
Write the condition $y \neq 0$ in terms of $\sin\theta$.	$\sin\theta \neq 0$
State the reciprocal identity.	$\csc\theta = \dfrac{1}{\sin\theta} \quad (\sin\theta \neq 0)$

An equivalent reciprocal identity can be written as $\sin\theta = \dfrac{1}{\csc\theta}$. There is no need to note the restriction that $\csc\theta \neq 0$, since $\csc\theta = \dfrac{r}{y}$ is always greater than or equal to 1.

The following box summarizes the *reciprocal identities* that are true for all values of θ that do not correspond to the trigonometric function in the denominator having a zero value.

RECIPROCAL IDENTITIES

RECIPROCAL IDENTITIES	EQUIVALENT FORMS	DOMAIN RESTRICTIONS	
$\csc\theta = \dfrac{1}{\sin\theta}$	$\sin\theta = \dfrac{1}{\csc\theta}$	$\theta \neq n\pi$	$n =$ integer
$\sec\theta = \dfrac{1}{\cos\theta}$	$\cos\theta = \dfrac{1}{\sec\theta}$	$\theta \neq \dfrac{n\pi}{2}$	$n =$ odd integer
$\cot\theta = \dfrac{1}{\tan\theta}$	$\tan\theta = \dfrac{1}{\cot\theta}$	$\theta \neq \dfrac{n\pi}{2}$	$n =$ integer

Study Tip

$\sin\theta = 0$ when $\theta = n\pi$

$\quad n = 1, 2, 3, \ldots$

$\cos\theta = 0$ when $\theta = \dfrac{n\pi}{2}$

$\quad n = 1, 3, 5, \ldots$

$\tan\theta = \dfrac{\sin\theta}{\cos\theta} = 0$ when $\theta = n\pi$

$\quad n = 1, 2, 3, \ldots$

$\cot\theta = \dfrac{\cos\theta}{\sin\theta} = 0$ when $\theta = \dfrac{n\pi}{2}$

$\quad n = 1, 3, 5, \ldots$

Reciprocals always have the same algebraic sign. Therefore, if sine is positive, then cosecant is positive. If cosine is negative, then secant is negative. In this section we assume that the domain restrictions apply even though not explicitly stated when using the reciprocal identities.

Study Tip

Reciprocals always have the same algebraic sign ($+$ or $-$).

EXAMPLE 1   Using Reciprocal Identities

a. If $\cos\theta = -\dfrac{1}{2}$, find $\sec\theta$.

b. If $\sin\theta = \dfrac{\sqrt{3}}{2}$, find $\csc\theta$.

c. If $\tan\theta = c$, find $\cot\theta$.

Solution (a):

The secant function is the reciprocal of the cosine function. $\sec\theta = \dfrac{1}{\cos\theta}$

Substitute $\cos\theta = -\frac{1}{2}$ into the secant expression. $\sec\theta = \dfrac{1}{-\dfrac{1}{2}}$

Simplify. $\boxed{\sec\theta = -2}$

Solution (b):

The cosecant function is the reciprocal of the sine function. $\csc\theta = \dfrac{1}{\sin\theta}$

Substitute $\sin\theta = \dfrac{\sqrt{3}}{2}$ into the cosecant expression. $\csc\theta = \dfrac{1}{\frac{\sqrt{3}}{2}}$

Simplify. $\csc\theta = \dfrac{2}{\sqrt{3}}$

Rationalize the radical in the denominator. $\csc\theta = \dfrac{2}{\sqrt{3}}\cdot\dfrac{\sqrt{3}}{\sqrt{3}}$

$$\csc\theta = \dfrac{2\sqrt{3}}{3}$$

Solution (c):

The cotangent function is the reciprocal of the tangent function. $\cot\theta = \dfrac{1}{\tan\theta}$

Substitute $\tan\theta = c$ into the cotangent expression. $\cot\theta = \dfrac{1}{c}$

State any restrictions on c. $\boxed{\cot\theta = \dfrac{1}{c} \quad c \neq 0}$

■ **Answer: a.** $\sec\theta = -1$
  b. $\csc\theta = \dfrac{1}{x}$ $(x \neq 0)$

■ **YOUR TURN   a.** If $\cos\theta = -1$, find $\sec\theta$.
       b. If $\sin\theta = x$, find $\csc\theta$.

Quotient Identities

Let us now use the trigonometric definitions to derive the *quotient identities*.

WORDS   **MATH**

Write the definition of the tangent function. $\tan\theta = \dfrac{y}{x}$ $(x \neq 0)$

Multiply the two sides of the equation by $\dfrac{\frac{1}{r}}{\frac{1}{r}}$. $\tan\theta = \dfrac{\frac{y}{r}}{\frac{x}{r}}$ $(x \neq 0, r > 0)$

Substitute $\sin\theta = \dfrac{y}{r}$ and $\cos\theta = \dfrac{x}{r}$. $\tan\theta = \dfrac{\frac{y}{r}}{\frac{x}{r}} = \dfrac{\sin\theta}{\cos\theta}$

Write $x \neq 0$ in terms of θ. $\cos\theta \neq 0$

State the *quotient identity*. $\boxed{\tan\theta = \dfrac{\sin\theta}{\cos\theta} \quad (\cos\theta \neq 0)}$

A similar *quotient identity* can be written as $\cot\theta = \dfrac{\cos\theta}{\sin\theta}$, where $\sin\theta \neq 0$.

The following box summarizes the *quotient identities* that are true for all values of θ that do not correspond to a zero value in the denominator.

QUOTIENT IDENTITIES

QUOTIENT IDENTITIES	DOMAIN RESTRICTIONS
$\tan\theta = \dfrac{\sin\theta}{\cos\theta}$	$\cos\theta \neq 0$ or $\theta \neq \dfrac{n\pi}{2}$ n = odd integer
$\cot\theta = \dfrac{\cos\theta}{\sin\theta}$	$\sin\theta \neq 0$ or $\theta \neq n\pi$ n = integer

Classroom Example 7.1.2

If $\sin\theta = \dfrac{1}{2}$ and

$\cos\theta = -\dfrac{\sqrt{3}}{2}$,

find $\cot\theta$ and $\tan\theta$.

Answer:

$\cot\theta = -\sqrt{3}$ and

$\tan\theta = -\dfrac{\sqrt{3}}{3}$

In this section we assume that the domain restrictions apply even though they are not explicitly stated when using the quotient identities.

EXAMPLE 2   Using the Quotient Identities

If $\sin\theta = \frac{3}{5}$ and $\cos\theta = -\frac{4}{5}$, find $\tan\theta$ and $\cot\theta$.

Solution:

Write the quotient identity involving the tangent and cotangent functions.

$$\tan\theta = \frac{\sin\theta}{\cos\theta} \quad \text{and} \quad \cot\theta = \frac{\cos\theta}{\sin\theta}$$

Substitute $\sin\theta = \frac{3}{5}$ and $\cos\theta = -\frac{4}{5}$.

$$\tan\theta = \frac{\left(\dfrac{3}{5}\right)}{\left(-\dfrac{4}{5}\right)} \quad \text{and} \quad \cot\theta = \frac{\left(-\dfrac{4}{5}\right)}{\left(\dfrac{3}{5}\right)}$$

Simplify.

$$\tan\theta = -\frac{3}{4} \quad \text{and} \quad \cot\theta = -\frac{4}{3}$$

Note: We could have found $\tan\theta$ and then used the reciprocal identity, $\cot\theta = \dfrac{1}{\tan\theta}$, to find the cotangent function value.

■ **YOUR TURN** If $\sin\theta = \frac{4}{5}$ and $\cos\theta = \frac{3}{5}$, find $\tan\theta$ and $\cot\theta$.

Technology Tip

To calculate $\tan\theta = \dfrac{\left(\frac{3}{5}\right)}{\left(-\frac{4}{5}\right)}$ and

$\cot\theta = \dfrac{\left(-\frac{4}{5}\right)}{\left(\frac{3}{5}\right)}$, enter

```
(3/5)/(-4/5)
               -.75
(-4/5)/(3/5)
        -1.333333333
Ans►Frac
               -4/3
```

■ **Answer:** $\tan\theta = \frac{4}{3}$ and $\cot\theta = \frac{3}{4}$

Pythagorean Identities

Before we develop the *Pythagorean identities*, we first need to look at notation.

OPERATION	SHORTHAND NOTATION	WORDS	EXAMPLE
$(\sin\theta)^2$	$\sin^2\theta$	"Sine squared of theta"	$\sin\theta = \dfrac{\sqrt{3}}{2}$, so $\sin^2\theta = \left(\dfrac{\sqrt{3}}{2}\right)^2 = \dfrac{3}{4}$
$(\cos\theta)^3$	$\cos^3\theta$	"Cosine cubed of theta"	$\cos\theta = -\dfrac{1}{2}$, so $\cos^3\theta = \left(-\dfrac{1}{2}\right)^3 = -\dfrac{1}{8}$

Let us now derive our first *Pythagorean identity*. Recall that in the Cartesian plane, the relationship between the coordinates x and y and the distance r from the origin to the point (x, y) is given by $x^2 + y^2 = r^2$.

WORDS	**MATH**
Write the distance relationship in the Cartesian plane.	$x^2 + y^2 = r^2$
Divide the equation by r^2.	$\dfrac{x^2 + y^2}{r^2} = \dfrac{r^2}{r^2}$
Simplify.	$\dfrac{x^2}{r^2} + \dfrac{y^2}{r^2} = 1$
Use the power rule of exponents, $\dfrac{a^2}{b^2} = \left(\dfrac{a}{b}\right)^2$.	$\left(\dfrac{x}{r}\right)^2 + \left(\dfrac{y}{r}\right)^2 = 1$
Substitute the sine and cosine definitions, $\sin\theta = \dfrac{y}{r}$ and $\cos\theta = \dfrac{x}{r}$, into the left side of the equation.	$(\cos\theta)^2 + (\sin\theta)^2 = 1$
Use shorthand notation.	$\boxed{\cos^2\theta + \sin^2\theta = 1}$

Note: $\cos^2\theta + \sin^2\theta = 1$ and $\sin^2\theta + \cos^2\theta = 1$ are equivalent identities.

Two important things to note at this time are:

- The value of θ must remain the same in all expressions for this identity to hold.
- The sum of the squares of the sine and cosine functions of the same angle is always 1.

Let's illustrate the Pythagorean identity with both a special angle and a calculation:

θ	$\sin\theta$	$\cos\theta$	$\sin^2\theta + \cos^2\theta = 1$
$\dfrac{\pi}{6}$	$\dfrac{1}{2}$	$\dfrac{\sqrt{3}}{2}$	$\left(\dfrac{1}{2}\right)^2 + \left(\dfrac{\sqrt{3}}{2}\right)^2 = \dfrac{1}{4} + \dfrac{3}{4} = 1$
$27°$	0.4540	0.8910	$(0.4540)^2 + (0.8910)^2 = 0.999997 \approx 1$

We can derive the second *Pythagorean identity* from the first.

WORDS	**MATH**
Write the first Pythagorean identity.	$\sin^2\theta + \cos^2\theta = 1$
Divide both sides by $\cos^2\theta$.	$\dfrac{\sin^2\theta}{\cos^2\theta} + \dfrac{\cos^2\theta}{\cos^2\theta} = \dfrac{1}{\cos^2\theta}$
Use longhand notation.	$\dfrac{(\sin\theta)^2}{(\cos\theta)^2} + \dfrac{(\cos\theta)^2}{(\cos\theta)^2} = \dfrac{1}{(\cos\theta)^2}$
Use the property of exponents.	$\left(\dfrac{\sin\theta}{\cos\theta}\right)^2 + \left(\dfrac{\cos\theta}{\cos\theta}\right)^2 = \left(\dfrac{1}{\cos\theta}\right)^2$
Use the quotient and reciprocal identities $\tan\theta = \dfrac{\sin\theta}{\cos\theta}$ and $\sec\theta = \dfrac{1}{\cos\theta}$ to write the expressions in terms of the tangent and secant functions.	$(\tan\theta)^2 + 1 = (\sec\theta)^2$
Use shorthand notation.	$\boxed{\tan^2\theta + 1 = \sec^2\theta}$

In this derivation, if we had divided by $\sin^2\theta$ instead of $\cos^2\theta$, we would have arrived at the third *Pythagorean identity*, $1 + \cot^2\theta = \csc^2\theta$.

PYTHAGOREAN IDENTITIES

$$\sin^2\theta + \cos^2\theta = 1 \qquad \tan^2\theta + 1 = \sec^2\theta \qquad 1 + \cot^2\theta = \csc^2\theta$$

EXAMPLE 3   **Using the Identities to Find Trigonometric Function Values**

Find $\cos\theta$ and $\tan\theta$ if $\sin\theta = \frac{4}{5}$ and the terminal side of θ lies in quadrant II.

Solution:

Write the first Pythagorean identity. $\qquad\qquad \sin^2\theta + \cos^2\theta = 1$

Substitute $\sin\theta = \frac{4}{5}$ into the equation. $\qquad \left(\dfrac{4}{5}\right)^2 + \cos^2\theta = 1$

Eliminate the parentheses. $\qquad\qquad\qquad \dfrac{16}{25} + \cos^2\theta = 1$

Subtract $\frac{16}{25}$ from both sides of the equation. $\qquad \cos^2\theta = \dfrac{9}{25}$

Apply the square root property. $\qquad\qquad \cos\theta = \pm\sqrt{\dfrac{9}{25}}$

Simplify. $\qquad\qquad\qquad\qquad\qquad\qquad \cos\theta = \pm\dfrac{3}{5}$

The cosine function is negative in quadrant II. $\qquad \boxed{\cos\theta = -\dfrac{3}{5}}$

Write the quotient identity for the tangent function. $\qquad \tan\theta = \dfrac{\sin\theta}{\cos\theta}$

Substitute $\sin\theta = \frac{4}{5}$ and $\cos\theta = -\frac{3}{5}$. $\qquad \tan\theta = \dfrac{\dfrac{4}{5}}{-\dfrac{3}{5}}$

Simplify. $\qquad\qquad\qquad\qquad\qquad\qquad \boxed{\tan\theta = -\dfrac{4}{3}}$

■ **YOUR TURN**   Find $\sin\theta$ and $\tan\theta$ if $\cos\theta = -\frac{4}{5}$ and the terminal side of θ lies in quadrant III.

■ **Answer:** $\sin\theta = -\frac{3}{5}$ and $\tan\theta = \frac{3}{4}$

Classroom Example 7.1.4
Find $\sin\theta$ and $\cos\theta$ if
$\cot\theta = -\frac{5}{12}$ and the terminal
side of θ is in QIV.

Answer:
$\sin\theta = -\frac{12}{13}, \cos\theta = \frac{5}{13}$

EXAMPLE 4   Using the Identities to Find Trigonometric Function Values

Find $\sin\theta$ and $\cos\theta$ if $\tan\theta = \frac{3}{4}$ and the terminal side of θ lies in quadrant III.

Solution:

Write the second Pythagorean identity. $\tan^2\theta + 1 = \sec^2\theta$

Substitute $\tan\theta = \frac{3}{4}$ into the equation. $\left(\frac{3}{4}\right)^2 + 1 = \sec^2\theta$

Eliminate the parentheses. $\frac{9}{16} + 1 = \sec^2\theta$

Simplify. $\sec^2\theta = \frac{25}{16}$

Apply the square root property. $\sec\theta = \pm\sqrt{\frac{25}{16}} = \pm\frac{5}{4}$

Substitute the reciprocal identity, $\sec\theta = \frac{1}{\cos\theta}$. $\frac{1}{\cos\theta} = \pm\frac{5}{4}$

Solve for $\cos\theta$. $\cos\theta = \pm\frac{4}{5}$

The cosine function is negative in quadrant III. $\boxed{\cos\theta = -\frac{4}{5}}$

Write the first Pythagorean identity. $\sin^2\theta + \cos^2\theta = 1$

Substitute $\cos\theta = -\frac{4}{5}$. $\sin^2\theta + \left(-\frac{4}{5}\right)^2 = 1$

Simplify. $\sin^2\theta = \frac{9}{25}$

Apply the square root property. $\sin\theta = \pm\sqrt{\frac{9}{25}} = \pm\frac{3}{5}$

The sine function is negative in quadrant III. $\boxed{\sin\theta = -\frac{3}{5}}$

Classroom Example 7.1.5
Simplify:
a. $(\sec\theta - 1)(\sec\theta + 1)$
b.* $\dfrac{(\csc\theta - 1)(\csc\theta + 1)}{(1 - \sec\theta)(\sec\theta + 1)}$

Answer: a. $\tan^2\theta$ **b.** $-\cot^4\theta$

EXAMPLE 5   Using Identities to Simplify Expressions

Multiply $(1 - \cos\theta)(1 + \cos\theta)$ and simplify.

Solution:

Multiply using the FOIL method. $(1 - \cos\theta)(1 + \cos\theta) = 1 - \cos\theta + \cos\theta - \cos^2\theta$

Combine like terms. $= 1 - \underline{\cos\theta} + \underline{\cos\theta} - \cos^2\theta$

Rewrite the Pythagorean identity, $= \underbrace{1 - \cos^2\theta}_{\sin^2\theta}$
$\sin^2\theta + \cos^2\theta = 1$ as $\sin^2\theta = 1 - \cos^2\theta$.

$\boxed{(1 - \cos\theta)(1 + \cos\theta) = \sin^2\theta}$

■**Answer:** $\cos^2\theta$

■**YOUR TURN** Multiply $(1 - \sin\theta)(1 + \sin\theta)$ and simplify.

SECTION 7.1 SUMMARY

Reciprocal Identities

$$\csc\theta = \frac{1}{\sin\theta} \qquad \sec\theta = \frac{1}{\cos\theta} \qquad \cot\theta = \frac{1}{\tan\theta}$$

Quotient Identities

$$\tan\theta = \frac{\sin\theta}{\cos\theta} \qquad \cot\theta = \frac{\cos\theta}{\sin\theta}$$

Pythagorean Identities

$$\sin^2\theta + \cos^2\theta = 1 \qquad \tan^2\theta + 1 = \sec^2\theta$$

$$1 + \cot^2\theta = \csc^2\theta$$

These basic trigonometric identities (reciprocal, quotient, and Pythagorean) are used to calculate trigonometric function values given other trigonometric function values and information regarding the quadrant where the terminal side of the angle lies.

SECTION 7.1 EXERCISES

▪ SKILLS

In Exercises 1–12, use a reciprocal identity to find the function value indicated. Rationalize the denominator, if necessary.

1. If $\cos\theta = \dfrac{7}{8}$, find $\sec\theta$.

2. If $\sin\theta = -\dfrac{3}{7}$, find $\csc\theta$.

3. If $\sin\theta = -0.6$, find $\csc\theta$.

4. If $\cos\theta = 0.8$, find $\sec\theta$.

5. If $\tan\theta = -5$, find $\cot\theta$.

6. If $\tan\theta = 0.5$, find $\cot\theta$.

7. If $\csc\theta = \dfrac{\sqrt{5}}{2}$, find $\sin\theta$.

8. If $\sec\theta = \dfrac{\sqrt{11}}{2}$, find $\cos\theta$.

9. If $\cot\theta = -\dfrac{\sqrt{7}}{5}$, find $\tan\theta$.

10. If $\cot\theta = 3.5$, find $\tan\theta$.

11. If $\sec\theta = -4$, find $\cos\theta$.

12. If $\csc\theta = -\sqrt{5}$ find $\sin\theta$.

In Exercises 13–20, use a quotient identity to find the function value indicated. Rationalize the denominator, if necessary.

13. If $\sin\theta = -\dfrac{1}{2}$ and $\cos\theta = \dfrac{\sqrt{3}}{2}$, find $\tan\theta$.

14. If $\sin\theta = -\dfrac{1}{2}$ and $\cos\theta = \dfrac{\sqrt{3}}{2}$, find $\cot\theta$.

15. If $\sin\theta = -0.6$ and $\cos\theta = -0.8$, find $\cot\theta$.

16. If $\sin\theta = -0.6$ and $\cos\theta = -0.8$, find $\tan\theta$.

17. If $\sin\theta = -\dfrac{\sqrt{11}}{6}$ and $\cos\theta = -\dfrac{5}{6}$, find $\tan\theta$.

18. If $\sin\theta = -\dfrac{\sqrt{11}}{6}$ and $\cos\theta = -\dfrac{5}{6}$, find $\cot\theta$.

19. If $\sin\theta = a$ and $\cos\theta = b$, find $\cot\theta$ and state any restrictions on a or b.

20. If $\sin\theta = a$ and $\cos\theta = b$, find $\tan\theta$ and state any restrictions on a or b.

In Exercises 21–26, find the indicated expression.

21. If $\sin\theta = \dfrac{\sqrt{5}}{8}$, find $\sin^2\theta$.

22. If $\cos\theta = 0.1$, find $\cos^3\theta$.

23. If $\csc\theta = -2$, find $\csc^3\theta$.

24. If $\tan\theta = -5$, find $\tan^2\theta$.

25. If $\cot\theta = -\sqrt[3]{5}$, find $\cot^3\theta$.

26. If $\sec\theta = \dfrac{\sqrt{7}}{2}$, find $\sec^4\theta$.

In Exercises 27–46, use a Pythagorean identity to find the function value indicated. Rationalize the denominator, if necessary.

27. If $\sin\theta = -\dfrac{1}{2}$ and the terminal side of θ lies in quadrant III, find $\cos\theta$.

28. If $\sin\theta = -\dfrac{3}{5}$ and the terminal side of θ lies in quadrant III, find $\cos\theta$.

29. If $\cos\theta = \dfrac{2}{5}$ and the terminal side of θ lies in quadrant IV, find $\sin\theta$.

30. If $\cos\theta = \dfrac{2}{7}$ and the terminal side of θ lies in quadrant IV, find $\sin\theta$.

31. If $\tan\theta = 4$ and the terminal side of θ lies in quadrant III, find $\sec\theta$.

32. If $\tan\theta = -5$ and the terminal side of θ lies in quadrant II, find $\sec\theta$.

33. If $\cot\theta = 2$ and the terminal side of θ lies in quadrant III, find $\csc\theta$.

34. If $\cot\theta = -3$ and the terminal side of θ lies in quadrant II, find $\csc\theta$.

35. If $\sin\theta = \dfrac{8}{15}$ and the terminal side of θ lies in quadrant II, find $\tan\theta$.

36. If $\sin\theta = -\dfrac{7}{15}$ and the terminal side of θ lies in quadrant III, find $\tan\theta$.

37. If $\cos\theta = -\dfrac{7}{15}$ and the terminal side of θ lies in quadrant III, find $\csc\theta$.

38. If $\cos\theta = -\dfrac{8}{15}$ and the terminal side of θ lies in quadrant II, find $\csc\theta$.

39. If $\sec\theta = \dfrac{\sqrt{13}}{3}$ and the terminal side of θ lies in quadrant IV, find $\sin\theta$.

40. If $\sec\theta = -\dfrac{\sqrt{61}}{5}$ and the terminal side of θ lies in quadrant II, find $\sin\theta$.

41. If $\sin\theta = -\dfrac{\sqrt{3}}{3}$ and the terminal side of θ lies in quadrant IV, find $\sec\theta$.

42. If $\cos\theta = 0.15$ and the terminal side of θ lies in quadrant IV, find $\csc\theta$.

43. If $\tan\theta = \sqrt{5}$ and the terminal side of θ lies in quadrant I, find $\sec\theta$.

44. If $\cot\theta = -5$ and the terminal side of θ lies in quadrant IV, find $\csc\theta$.

45. If $\csc\theta = -4\sqrt{2}$ and the terminal side of θ lies in quadrant III, find $\cos\theta$.

46. If $\csc\theta = \sqrt{11}$ and the terminal side of θ lies in quadrant I, find $\cos\theta$.

In Exercises 47–56, use identities to find the function values indicated. Rationalize the denominator, if necessary.

47. Find $\sin\theta$ and $\cos\theta$ if $\tan\theta = -\frac{4}{3}$ and the terminal side of θ lies in quadrant II.

48. Find $\sin\theta$ and $\cos\theta$ if $\tan\theta = -\frac{3}{4}$ and the terminal side of θ lies in quadrant II.

49. Find $\sin\theta$ and $\cos\theta$ if $\tan\theta = 2$ and the terminal side of θ lies in quadrant III.

50. Find $\sin\theta$ and $\cos\theta$ if $\tan\theta = 5$ and the terminal side of θ lies in quadrant III.

51. Find $\sin\theta$ and $\cos\theta$ if $\tan\theta = 0.6$ and the terminal side of θ lies in quadrant III.

52. Find $\sin\theta$ and $\cos\theta$ if $\tan\theta = 0.8$ and the terminal side of θ lies in quadrant III.

53. Find $\sin\theta$ and $\cos\theta$ if $\cot\theta = -\frac{5}{2}$ and the terminal side of θ lies in quadrant IV.

54. Find $\sin\theta$ and $\cos\theta$ if $\cot\theta = 0.1$ and the terminal side of θ lies in quadrant III.

55. Find $\sin\theta$ and $\cos\theta$ if $\cot\theta = \frac{1}{4}$ and the terminal side of θ lies in quadrant I.

56. Find $\sin\theta$ and $\cos\theta$ if $\cot\theta = -2.2$ and the terminal side of θ lies in quadrant II.

In Exercises 57–70, perform the indicated operations and simplify your answers, if possible. Leave all answers in terms of $\sin\theta$ and $\cos\theta$.

57. $\sec\theta\cot\theta$

58. $\csc\theta\tan\theta$

59. $\tan^2\theta - \sec^2\theta$

60. $\cot^2\theta - \csc^2\theta$

61. $\csc\theta - \sin\theta$

62. $\sec\theta - \cos\theta$

63. $\dfrac{\sec\theta}{\tan\theta}$

64. $\dfrac{\csc\theta}{\cot\theta}$

65. $(\sin\theta + \cos\theta)^2$

66. $(\sin\theta - \cos\theta)^2$

67. $(1 + \tan\theta)^2$

68. $(1 + \cot\theta)^2$

69. $\dfrac{\cot\theta + \tan\theta}{\sec\theta\csc\theta}$

70. $\dfrac{\sin\theta\cos\theta}{\tan\theta + \cot\theta}$

▪ APPLICATIONS

For Exercises 71 and 72, refer to the following:

The *bifolium* is a curve that can be drawn using either an algebraic equation or an equation involving trigonometric functions. Even though the trigonometric equation uses polar coordinates, it's much easier to solve the trigonometric equation for function values than the algebraic equation. The graph of $r = 8\sin\theta\cos^2\theta$ is shown below. You'll see more on polar coordinates and graphs of polar equations in Chapter 8. The following graph is in polar coordinates.

71. Bifolium. The equation for the bifolium shown on the left is $r = 8\sin\theta\cos^2\theta$. Use a Pythagorean identity to rewrite the equation using only the function $\sin\theta$. Then find r if $\theta = 30°, 60°,$ and $90°$.

72. Bifolium. The equation for the bifolium shown on the left is $r = 8\sin\theta\cos^2\theta$. Use a Pythagorean identity to rewrite the equation using only the function $\sin\theta$. Then find r if $\theta = -30°, -60°,$ and $-90°$.

For Exercises 73 and 74, refer to the following:

The monthly revenues (measured in thousands of dollars) of PizzaRia are a function of cost (measured in thousands of dollars), that is, $R(c)$. Recall the angle θ can be interpreted as a measure of the sizes of c (cost) and R (revenue) relative to each other. The larger the angle θ is, the greater revenue is relative to cost, and conversely, the smaller the angle θ is, the smaller revenue is relative to cost. The ratio of revenue to costs can be approximated by the tangent function, $\tan\theta \approx \dfrac{R}{c}$.

73. Business. An analysis of the revenue and the costs of one month determined that $\cos\theta = \frac{3}{5}$ and $\sin\theta = \frac{4}{5}$. Find $\tan\theta$ and interpret its meaning.

74. Business. An analysis of the revenue and the costs of one month determined that $\cos\theta = \dfrac{2\sqrt{5}}{5}$ and $\sin\theta = \dfrac{\sqrt{5}}{5}$. Find $\tan\theta$ and interpret its meaning.

▪ CATCH THE MISTAKE

In Exercises 75 and 76, explain the mistake that is made.

75. If $\sin\theta = -\frac{1}{3}$ and the terminal side of θ lies in quadrant III, find $\cos\theta$.

Solution:

Write the first Pythagorean identity. $\sin^2\theta + \cos^2\theta = 1$

Substitute $\sin\theta = -\frac{1}{3}$ into the identity. $\left(-\dfrac{1}{3}\right)^2 + \cos^2\theta = 1$

Eliminate parentheses. $\dfrac{1}{9} + \cos^2\theta = 1$

Subtract $\frac{1}{9}$ from both sides. $\cos^2\theta = \dfrac{8}{9}$

Take the square root of both sides. $\cos\theta = \sqrt{\dfrac{8}{9}}$

Simplify. $\cos\theta = \dfrac{2\sqrt{2}}{3}$

This is incorrect. What mistake was made?

76. Find $\sin\theta$ and $\cos\theta$ if $\tan\theta = -\frac{1}{4}$ and the terminal side of θ lies in quadrant II.

Solution:

Write the quotient identity. $\qquad \tan\theta = \dfrac{\sin\theta}{\cos\theta}$

Substitute $\tan\theta = -\frac{1}{4}$ into the identity. $\qquad -\dfrac{1}{4} = \dfrac{\sin\theta}{\cos\theta}$

In quadrant II, sine is positive and cosine is negative. $\qquad \dfrac{1}{-4} = \dfrac{\sin\theta}{\cos\theta}$

Identify the sine and cosine values. $\qquad \sin\theta = 1$ and $\cos\theta = -4$

This is incorrect. What mistake was made?

■ CONCEPTUAL

In Exercises 77 and 78, determine whether each statement is true or false.

77. It is possible for $\sin\theta > 0$ and $\csc\theta < 0$.

78. It is possible for $\cos\theta < 0$ and $\sec\theta > 0$.

79. Find the measure of an angle θ, $0° < \theta \le 180°$, which satisfies the equation $\sin\theta = \csc\theta$.

80. Find the measure of an angle θ, $0° < \theta \le 180°$, which satisfies the equation $\cos\theta = \sec\theta$.

81. Write $\cot\theta$ only in terms of $\sin\theta$.

82. Write $\csc\theta$ only in terms of $\cos\theta$.

■ CHALLENGE

83. Let $x = 8\sin\theta$ in the expression $\sqrt{64 - x^2}$ and simplify.

84. Let $x = 6\cos\theta$ in the expression $\sqrt{36 - x^2}$ and simplify.

85. Let $x = 5\tan\theta$ in the expression $\sqrt{25 + x^2}$ and simplify.

86. Simplify $\dfrac{\cos^2\theta}{\cos^2\theta - 1} \cdot \dfrac{1 - \csc^2\theta}{\tan^2\theta - \sec^2\theta}$.

In Exercises 87 and 88, determine whether each statement is true or false.

87. $\sec^2\theta \ge 1$, for all θ.

88. $1 - \csc^2\theta > 0$, for all θ.

■ TECHNOLOGY

89. Verify the quotient $\tan\theta = \dfrac{\sin\theta}{\cos\theta}$ numerically for $\theta = 22°$. Use a calculator to find

  a. $\sin 22°$ **b.** $\cos 22°$ **c.** $\dfrac{\sin 22°}{\cos 22°}$ **d.** $\tan 22°$

Are the results in (c) and (d) the same?

90. Is $\tan 10° = \dfrac{\sin 160°}{\cos 16°}$? Use a calculator to find

  a. $\sin 160°$ **b.** $\cos 16°$ **c.** $\dfrac{\sin 160°}{\cos 16°}$ **d.** $\tan 10°$

Are the results in (c) and (d) the same?

91. Does $\sin 35° = \sqrt{1 - \cos^2 35°}$? Use a calculator to find:

  a. $\sin 35°$

  b. $1 - \cos 35°$

  c. $\sqrt{1 - \cos^2 35°}$

Which results are the same?

92. Does $\sec 68° = \sqrt{1 + \tan^2 68°}$? Use a calculator to find:

  a. $\sec 68°$

  b. $1 + \tan 68°$

  c. $\sqrt{1 + \tan^2 68°}$

Which results are the same?

SKILLS OBJECTIVES

- Simplify trigonometric expressions using identities.
- Verify trigonometric identities.

CONCEPTUAL OBJECTIVES

- Understand that there is more than one way to verify an identity.
- Understand that identities must hold for all values in the domain of the functions that are related by the identities.

Trigonometric Identities

Basic Identities

In Section 7.1, we discussed the *basic* (fundamental) trigonometric identities: reciprocal, quotient, and Pythagorean. In mathematics, an **identity** is an equation that is true for *all* values of the variable for which the expressions in the equation are defined. If an equation is true for only *some* values of the variable, it is a **conditional equation**. If an equation is true for no values of the variable, it is a **contradiction**.

The following are **identities** (true for all x for which the expressions are defined):

IDENTITY	TRUE FOR THESE VALUES OF x
$x^2 + 3x + 2 = (x + 2)(x + 1)$	All real numbers
$\tan x = \dfrac{\sin x}{\cos x}$	All real numbers except $x = \dfrac{n\pi}{2}$, where n is an odd integer
$\sin^2 x + \cos^2 x = 1$	All real numbers

The following are **conditional equations** (true only for particular values of x):

EQUATION	TRUE FOR ONLY THESE VALUES OF x
$x^2 + 3x + 2 = 0$	$x = -2$ and $x = -1$
$\tan x = 0$	$x = n\pi$, where n is an integer
$\sin^2 x - \cos^2 x = 1$	$x = \dfrac{n\pi}{2}$, where n is an odd integer

The following are **contradictions** (not true for any values of x):

EQUATION	TRUE FOR THESE VALUES OF x
$x + 5 = x + 7$	none
$\sin^2 x + \cos^2 x = 5$	none

The following boxes summarize the identities that were discussed in Section 7.1:

RECIPROCAL IDENTITIES

RECIPROCAL IDENTITIES	EQUIVALENT FORMS	DOMAIN RESTRICTIONS
$\csc x = \dfrac{1}{\sin x}$	$\sin x = \dfrac{1}{\csc x}$	$x \neq n\pi \quad n = $ integer
$\sec x = \dfrac{1}{\cos x}$	$\cos x = \dfrac{1}{\sec x}$	$x \neq \dfrac{n\pi}{2} \quad n = $ odd integer
$\cot x = \dfrac{1}{\tan x}$	$\tan x = \dfrac{1}{\cot x}$	$x \neq \dfrac{n\pi}{2} \quad n = $ integer

Study Tip

Just because an equation is true for *some* values of x does not mean it is an identity.

741

QUOTIENT IDENTITIES

QUOTIENT IDENTITIES	DOMAIN RESTRICTIONS
$\tan x = \dfrac{\sin x}{\cos x}$	$\cos x \neq 0 \quad x \neq \dfrac{n\pi}{2} \quad n = \text{odd integer}$
$\cot x = \dfrac{\cos x}{\sin x}$	$\sin x \neq 0 \quad x \neq n\pi \quad n = \text{integer}$

PYTHAGOREAN IDENTITIES

PYTHAGOREAN IDENTITIES	DOMAIN RESTRICTIONS
$\sin^2 x + \cos^2 x = 1$	
$\tan^2 x + 1 = \sec^2 x$	$\cos x \neq 0 \text{ or } x \neq \dfrac{n\pi}{2} \quad n = \text{odd integer}$
$1 + \cot^2 x = \csc^2 x$	$\sin x \neq 0 \text{ or } x \neq n\pi \quad n = \text{integer}$

In previous chapters, we discussed even and odd functions that have these respective properties:

TYPE OF FUNCTION	ALGEBRAIC IDENTITY	GRAPH
Even	$f(-x) = f(x)$	Symmetry about the y-axis
Odd	$f(-x) = -f(x)$	Symmetry about the origin

We already learned in Chapter 6 that the sine function is an odd function and the cosine function is an even function. Combining this knowledge with the reciprocal and quotient identities, we arrive at the *even-odd identities*, which we can add to our list of basic identities.

EVEN-ODD IDENTITIES

$$\text{Odd} \begin{cases} \sin(-x) = -\sin x \\ \csc(-x) = -\csc x \\ \tan(-x) = -\tan x \\ \cot(-x) = -\cot x \end{cases} \qquad \text{Even} \begin{cases} \cos(-x) = \cos x \\ \sec(-x) = \sec x \end{cases}$$

Simplifying Trigonometric Expressions Using Identities

In Section 7.1, we used the basic trigonometric identities to find values for trigonometric functions, and we simplified trigonometric expressions using the identities. We now will use the basic identities and algebraic manipulation to simplify more complicated trigonometric expressions. In simplifying trigonometric expressions, one approach is to first convert all expressions into sines and cosines and then simplify.

EXAMPLE 1   Simplifying Trigonometric Expressions

Simplify $\tan x \sin x + \cos x$.

Solution:

Write the tangent function in terms of the sine and cosine functions: $\tan x = \dfrac{\sin x}{\cos x}$.

$$\tan x \cdot \sin x + \cos x$$

$$= \left(\dfrac{\sin x}{\cos x}\right)\sin x + \cos x$$

Simplify.

$$= \dfrac{\sin^2 x}{\cos x} + \cos x$$

Write as a fraction with a single quotient by finding a common denominator, $\cos x$.

$$= \dfrac{\sin^2 x + \cos^2 x}{\cos x}$$

Use the Pythagorean identity: $\sin^2 x + \cos^2 x = 1$.

$$= \dfrac{1}{\cos x}$$

Use the reciprocal identity $\sec x = \dfrac{1}{\cos x}$.

$$= \boxed{\sec x}$$

■ **YOUR TURN** Simplify $\cot x \cos x + \sin x$.

> **Classroom Example 7.2.1**
> Simplify $\sec x \cos^3 x - 1$.
>
> **Answer:** $-\sin^2 x$

■ **Answer:** $\csc x$

In Example 1, $\tan x$ and $\sec x$ are not defined for odd integer multiples of $\dfrac{\pi}{2}$. In the Your Turn, $\cot x$ and $\csc x$ are not defined for integer multiples of π. Both the original expression and the simplified form are governed by the same restrictions. There are times when the original expression is subject to more domain restrictions than the simplified form and thus special attention must be given to domain restrictions.

For example, the algebraic expression $\dfrac{x^2 - 1}{x + 1}$ is under the domain restriction $x \neq -1$ because that value for x makes the value of the denominator equal to zero. If we forget to state the domain restrictions, we might simplify the algebraic expression as $\dfrac{x^2 - 1}{x + 1} = \dfrac{(x - 1)(x + 1)}{(x + 1)} = x - 1$ and assume this is true for all values of x. The correct simplification is $\dfrac{x^2 - 1}{x + 1} = x - 1$ for $x \neq -1$. In fact, if we were to graph both the original expression $y = \dfrac{x^2 - 1}{x + 1}$ and the line $y = x - 1$, they would coincide, except that the graph of the original expression would have a "hole" or discontinuity at $x = 1$. In this chapter, it is assumed that the domain of the simplified expression is the same as the domain of the original expression.

EXAMPLE 2 **Simplifying Trigonometric Expressions**

Simplify $\dfrac{1}{\csc^2 x} + \dfrac{1}{\sec^2 x}$.

Solution:

Rewrite the expression in terms of quotients squared.

Use the reciprocal identities:

$\sin x = \dfrac{1}{\csc x}$ and $\cos x = \dfrac{1}{\sec x}$.

Use the Pythagorean identity:
$\sin^2 x + \cos^2 x = 1$.

$$\dfrac{1}{\csc^2 x} + \dfrac{1}{\sec^2 x} = \left(\underbrace{\dfrac{1}{\csc x}}_{\sin x}\right)^2 + \left(\underbrace{\dfrac{1}{\sec x}}_{\cos x}\right)^2$$

$$= \sin^2 x + \cos^2 x$$

$$= \boxed{1}$$

■ **YOUR TURN** Simplify $\dfrac{1}{\cos^2 x} - 1$.

■ **Answer:** $\tan^2 x$

Verifying Identities

We will now use the trigonometric identities to verify, or establish, other trigonometric identities. For example, verify that

$$(\sin x - \cos x)^2 - 1 = -2\sin x \cos x$$

The good news is that we will know we are done when we get there, since we know the desired identity. But how do we get there? How do we verify that the identity is true? Remember that it must be true for *all x*, not just some *x*. Therefore, it is not enough to simply select values for *x* and show it is true for those specific values.

WORDS	MATH
Start with one side of the equation (the more complicated side).	$(\sin x - \cos x)^2 - 1$
Remember that $(a - b)^2 = a^2 - 2ab + b^2$ and expand $(\sin x - \cos x)^2$.	$= \sin^2 x - 2\sin x \cos x + \cos^2 x - 1$
Group the $\sin^2 x$ and $\cos^2 x$ terms and use the Pythagorean identity.	$= -2\sin x \cos x + \underbrace{(\sin^2 x + \cos^2 x)}_{1} - 1$
Simplify.	$= -2\sin x \cos x$

When we arrive at the right side of the equation, then we have succeeded in verifying the identity. In verifying trigonometric identities, there is no one procedure that works for all identities. You must manipulate one side of the equation until it looks like the other side. Here are two suggestions that are generally helpful:

1. Convert all trigonometric expressions to sines and cosines.
2. Write all sums or differences of fractions (quotients) as a single fraction (quotient).

The following suggestions help guide the way to verifying trigonometric identities.

GUIDELINES FOR VERIFYING TRIGONOMETRIC IDENTITIES

- Start with the more complicated side of the equation.
- Combine all sums and differences of fractions (quotients) into a single fraction (quotient).
- Use basic trigonometric identities.
- Use algebraic techniques to manipulate one side of the equation until the other side of the equation is achieved.
- Sometimes it is helpful to convert all trigonometric functions into sines and cosines.

It is important to note that trigonometric identities must be valid for all values of the independent variable (usually x or θ) for which the expressions in the equation are defined (domain of the equation).

EXAMPLE 3   Verifying Trigonometric Identities

Verify the identity $(1 + \sin x)[1 + \sin(-x)] = \cos^2 x$.

Solution:

Start with the more complicated side (left side).

$$(1 + \sin x)[1 + \sin(-x)]$$

The sine function is odd: $\sin(-x) = -\sin x$.

$$= (1 + \sin x)(1 - \sin x)$$

Eliminate the parentheses.

$$= 1 - \sin^2 x$$

Apply the Pythagorean identity, $\sin^2 x + \cos^2 x = 1$.

$$= \cos^2 x$$

Technology Tip

Graphs of $y_1 = \dfrac{\tan x - \cot x}{\tan x + \cot x}$
and $y_2 = \sin^2 x - \cos^2 x$.

EXAMPLE 4   Verifying Trigonometric Identities

Verify the identity $\dfrac{\tan x - \cot x}{\tan x + \cot x} = \sin^2 x - \cos^2 x$.

Solution:

Start with the more complicated side of the equation.

$$\frac{\tan x - \cot x}{\tan x + \cot x}$$

Use the quotient identities to write the tangent and cotangent functions in terms of the sine and cosine functions.

$$= \frac{\dfrac{\sin x}{\cos x} - \dfrac{\cos x}{\sin x}}{\dfrac{\sin x}{\cos x} + \dfrac{\cos x}{\sin x}}$$

Multiply by $\dfrac{\sin x \cos x}{\sin x \cos x}$.

$$= \frac{\left(\dfrac{\sin x}{\cos x} - \dfrac{\cos x}{\sin x} \right)}{\left(\dfrac{\sin x}{\cos x} + \dfrac{\cos x}{\sin x} \right)} \left(\frac{\sin x \cos x}{\sin x \cos x} \right)$$

Simplify.

$$= \frac{\sin^2 x - \cos^2 x}{\sin^2 x + \cos^2 x}$$

Use the Pythagorean identity:
$\sin^2 x + \cos^2 x = 1$.

$$\boxed{= \sin^2 x - \cos^2 x}$$

Classroom Example 7.2.4
Prove that
$(\cos x - 1)(1 + \cos(-x))$
$= -\sin^2 x$.

Answer:
$(\cos x - 1)(1 + \cos(-x))$
$= (\cos x - 1)(\cos x + 1)$
$= \cos^2 x - 1 = -\sin^2 x$

EXAMPLE 5   Determining Whether a Trigonometric Equation Is an Identity

Determine whether $(1 - \cos^2 x)(1 + \cot^2 x) = 0$ is an identity, a conditional equation, or a contradiction.

Solution:

Use the quotient identity to write the cotangent function in terms of the sine and cosine functions.

$$(1 - \cos^2 x)(1 + \cot^2 x) = (1 - \cos^2 x)\left(1 + \frac{\cos^2 x}{\sin^2 x}\right)$$

Combine the expression in the second parentheses so that it is a single quotient.

$$= (1 - \cos^2 x)\left(\frac{\sin^2 x + \cos^2 x}{\sin^2 x}\right)$$

Use the Pythagorean identity.

$$= \underbrace{(1 - \cos^2 x)}_{\sin^2 x}\left(\frac{\overbrace{\sin^2 x + \cos^2 x}^{1}}{\sin^2 x}\right)$$

Eliminate the parentheses.

$$= \frac{\sin^2 x}{\sin^2 x}$$

Simplify.

$$= 1$$

Since $1 \neq 0$, this is not an identity. Therefore, it is either a conditional equation (true for some real values of x) or a contradition (true for no real values of x). Note that for $(1 - \cos^2 x)(1 + \cot^2 x) = 0$ then either $(1 - \cos^2 x) = 0$ or $(1 + \cot^2 x) = 0$

Since $1 + \cot^2 x > 0$, then $1 - \cos^2 x = 0$.
$\cos^2 x = 1$ when $x = n\pi$. Since $\cot x$ is not defined for $x = n\pi$, we say that this equation is a contradiction.

EXAMPLE 6   Verifying Trigonometric Identities

Verify that $\dfrac{\sin(-x)}{\cos(-x)\tan(-x)} = 1.$

Solution:

Start with the left side of the equation.

$$\frac{\sin(-x)}{\cos(-x)\tan(-x)}$$

Use the even-odd identities.

$$= \frac{-\sin x}{-\cos x\tan x}$$

Simplify.

$$= \frac{\sin x}{\cos x\tan x}$$

Use the quotient identity to write the tangent function in terms of the sine and cosine functions.

$$= \frac{\sin x}{\cos x\left(\dfrac{\sin x}{\cos x}\right)}$$

Divide out the cosine term in the denominator.

$$= \frac{\sin x}{\sin x}$$

Simplify.

$$= \boxed{1}$$

EXAMPLE 7   **Verifying a More Complicated Identity**

Verify that $-\dfrac{\cot^2 x}{1 - \csc x} = \dfrac{\sin x + 1}{\sin x}$.

Solution:

Start with the left side of the equation.

$$-\frac{\cot^2 x}{1 - \csc x}$$

Use the Pythagorean identity, $\cot^2 x = \csc^2 x - 1$.

$$= -\frac{(\csc^2 x - 1)}{1 - \csc x}$$

Distribute the negative throughout the numerator.

$$= \frac{1 - \csc^2 x}{1 - \csc x}$$

Factor the numerator (difference of two squares).

$$= \frac{(1 - \csc x)(1 + \csc x)}{1 - \csc x}$$

Divide out the $1 - \csc x$ in both the numerator and denominator.

$$= 1 + \csc x$$

Use the reciprocal identity.

$$= 1 + \frac{1}{\sin x}$$

Combine the two expressions into a single fraction.

$$= \frac{\sin x + 1}{\sin x}$$

Classroom Example 7.2.7
Verify that
$$\sec x \,(\cot(-x) + \tan(-x)) = -\frac{\csc x}{\cos^2 x}.$$

Answer:
$\sec x \,(\cot(-x) + \tan(-x))$
$= -\sec x \,(\cot x + \tan x)$
$= -\dfrac{1}{\cos x}\left(\dfrac{\cos x}{\sin x} + \dfrac{\sin x}{\cos x}\right)$
$= -\dfrac{1}{\cos x}\left(\dfrac{\cos^2 x + \sin^2 x}{\sin x \cos x}\right)$
$= -\dfrac{1}{\cos^2 x} \cdot \dfrac{1}{\sin x} = -\dfrac{\csc x}{\cos^2 x}$

SECTION 7.2 SUMMARY

We combined the basic trigonometric identities—reciprocal, quotient, Pythagorean, and even-odd—with algebraic techniques to simplify trigonometric expressions and verify more complex trigonometric identities. Two steps that we often use in both simplifying trigonometric expressions and verifying trigonometric identities are: (1) writing all of the trigonometric functions in terms of the sine and cosine functions, and (2) combining sums or differences of quotients into a single quotient. When verifying trigonometric identities, we work with the more complicated side (keeping the other side in mind as our goal).

SECTION 7.2 EXERCISES

SKILLS

In Exercises 1–24, simplify each of the trigonometric expressions.

1. $\sin x \csc x$

2. $\tan x \cot x$

3. $\sec(-x) \cot x$

4. $\tan(-x) \cos(-x)$

5. $\csc(-x)\sin x$

6. $\cot(-x)\tan(x)$

7. $\sec(x)\cos(-x) + \tan^2 x$

8. $\sec(-x)\tan(-x)\cos(-x)$

9. $(\sin^2 x)(\cot^2 x + 1)$

10. $(\cos^2 x)(\tan^2 x + 1)$

11. $(\sin x - \cos x)(\sin x + \cos x)$

12. $(\sin x + \cos x)^2$

13. $\dfrac{\csc x}{\cot x}$

14. $\dfrac{\sec x}{\tan x}$

15. $\dfrac{1 - \cot(-x)}{1 + \cot(x)}$

16. $\sec^2(x) - \tan^2(-x)$

17. $\dfrac{1 - \cos^4 x}{1 + \cos^2 x}$

18. $\dfrac{1 - \sin^4 x}{1 + \sin^2 x}$

19. $\dfrac{1 - \cot^4 x}{1 - \cot^2 x}$

20. $\dfrac{1 - \tan^4(-x)}{1 - \tan^2 x}$

21. $1 - \dfrac{\sin^2 x}{1 - \cos x}$

22. $1 - \dfrac{\cos^2 x}{1 + \sin x}$

23. $\dfrac{\tan x - \cot x}{\tan x + \cot x} + 2\cos^2 x$

24. $\dfrac{\tan x - \cot x}{\tan x + \cot x} + \cos^2 x$

In Exercises 25–50, verify each of the trigonometric identities.

25. $(\sin x + \cos x)^2 + (\sin x - \cos x)^2 = 2$

26. $(1 - \sin x)(1 + \sin x) = \cos^2 x$

27. $(\csc x + 1)(\csc x - 1) = \cot^2 x$

28. $(\sec x + 1)(\sec x - 1) = \tan^2 x$

29. $\tan x + \cot x = \csc x \sec x$

30. $\csc x - \sin x = \cot x \cos x$

31. $\dfrac{2 - \sin^2 x}{\cos x} = \sec x + \cos x$

32. $\dfrac{2 - \cos^2 x}{\sin x} = \csc x + \sin x$

33. $[\cos(-x) - 1][1 + \cos x] = -\sin^2 x$

34. $\tan(-x)\cot(x) = -1$

35. $\dfrac{\sec(-x)\cot(x)}{\csc(-x)} = -1$

36. $\csc(-x) - 1 = \dfrac{\cot^2 x}{\csc(-x) + 1}$

37. $\dfrac{1}{\csc^2 x} + \dfrac{1}{\sec^2 x} = 1$

38. $\dfrac{1}{\cot^2 x} - \dfrac{1}{\tan^2 x} = \sec^2 x - \csc^2 x$

39. $\dfrac{1}{1 - \sin x} + \dfrac{1}{1 + \sin x} = 2\sec^2 x$

40. $\dfrac{1}{1 - \cos x} + \dfrac{1}{1 + \cos x} = 2\csc^2 x$

41. $\dfrac{\sin^2 x}{1 - \cos x} = 1 + \cos x$

42. $\dfrac{\cos^2 x}{1 - \sin x} = 1 + \sin x$

43. $\sec x + \tan x = \dfrac{1}{\sec x - \tan x}$

44. $\csc x + \cot x = \dfrac{1}{\csc x - \cot x}$

45. $\dfrac{\csc x - \tan x}{\sec x + \cot x} = \dfrac{\cos x - \sin^2 x}{\sin x + \cos^2 x}$

46. $\dfrac{\sec x + \tan x}{\csc x + 1} = \tan x$

47. $\dfrac{\cos^2 x + 1 + \sin x}{\cos^2 x + 3} = \dfrac{1 + \sin x}{2 + \sin x}$

48. $\dfrac{\sin x + 1 - \cos^2 x}{\cos^2 x} = \dfrac{\sin x}{1 - \sin x}$

49. $\sec x(\tan x + \cot x) = \dfrac{\csc x}{\cos^2 x}$

50. $\tan x(\csc x - \sin x) = \cos x$

In Exercises 51–64, determine whether each equation is an identity, a conditional equation, or a contradiction.

51. $\cos^2 x(\tan x - \sec x)(\tan x + \sec x) = 1$

52. $\cos^2 x(\tan x - \sec x)(\tan x + \sec x) = \sin^2 x - 1$

53. $\dfrac{\csc x \cot x}{\sec x \tan x} = \cot^3 x$

54. $\sin x \cos x = 0$

55. $\sin x + \cos x = \sqrt{2}$

56. $\sin^2 x + \cos^2 x = 1$

57. $\tan^2 x - \sec^2 x = 1$

58. $\sec^2 x - \tan^2 x = 1$

59. $\sin x = \sqrt{1 - \cos^2 x}$

60. $\csc x = \sqrt{1 + \cot^2 x}$

61. $\sqrt{\sin^2 x + \cos^2 x} = 1$

62. $\sqrt{\sin^2 x + \cos^2 x} = \sin x + \cos x$

63. $(\sin x - \cos x)^2 = \sin^2 x - \cos^2 x$

64. $[\sin(-x) - 1][\sin(-x) + 1] = \cos^2 x$

▪ APPLICATIONS

For Exercises 65 and 66, refer to the following:

In calculus, when integrating expressions such as $\sqrt{a^2 - x^2}$, $\sqrt{a^2 + x^2}$, and $\sqrt{x^2 - a^2}$, trigonometric functions are used as "dummy" functions to eliminate the radical. Once the integration is performed, the trigonometric function is "unsubstituted." The trigonometric substitutions shown (and corresponding trigonometric identities) are used to simplify these types of expressions.

When simplifying, it is important to remember that

$$|x| = \begin{cases} x & \text{if } x > 0 \\ -x & \text{if } x < 0 \end{cases}$$

65. Start with the expression $\sqrt{a^2 - x^2}$ and let $x = a\sin\theta$, assuming $-\dfrac{\pi}{2} \le \theta \le \dfrac{\pi}{2}$, and simplify the original expression so that it contains no radicals.

66. Start with the expression $\sqrt{a^2 + x^2}$ and let $x = a\tan\theta$, assuming $-\dfrac{\pi}{2} < \theta < \dfrac{\pi}{2}$, and simplify the original expression so that it contains no radicals.

EXPRESSION	SUBSTITUTION		TRIGONOMETRIC IDENTITY
$\sqrt{a^2 - x^2}$	$x = a\sin\theta$	$-\dfrac{\pi}{2} \le \theta \le \dfrac{\pi}{2}$	$1 - \sin^2\theta = \cos^2\theta$
$\sqrt{a^2 + x^2}$	$x = a\tan\theta$	$-\dfrac{\pi}{2} < \theta < \dfrac{\pi}{2}$	$1 + \tan^2\theta = \sec^2\theta$
$\sqrt{x^2 - a^2}$	$x = a\sec\theta$	$0 \le \theta < \dfrac{\pi}{2}$ or $\pi \le \theta < \dfrac{3\pi}{2}$	$\sec^2\theta - 1 = \tan^2\theta$

■ CATCH THE MISTAKE

In Exercises 67–70, explain the mistake that is made.

67. Verify the identity $\dfrac{\cos x}{1 - \tan x} + \dfrac{\sin x}{1 - \cot x} = \sin x + \cos x.$

Solution:

Start with the left side of the equation.	$\dfrac{\cos x}{1 - \tan x} + \dfrac{\sin x}{1 - \cot x}$
Write the tangent and cotangent functions in terms of sines and cosines.	$= \dfrac{\cos x}{1 - \dfrac{\sin x}{\cos x}} + \dfrac{\sin x}{1 - \dfrac{\cos x}{\sin x}}$
Cancel the common cosine in the first term and sine in the second term.	$= \dfrac{1}{1 - \sin x} + \dfrac{1}{1 - \cos x}$

This is incorrect. What mistake was made?

68. Verify the identity $\dfrac{\cos^3 x \sec x}{1 - \sin x} = 1 + \sin x.$

Solution:

Start with the equation on the left.	$\dfrac{\cos^3 x \sec x}{1 - \sin x}$
Rewrite secant in terms of sine.	$= \dfrac{\cos^3 x \, \dfrac{1}{\sin x}}{1 - \sin x}$
Simplify.	$= \dfrac{\cos^3 x}{1 - \sin^2 x}$
Use the Pythagorean identity.	$= \dfrac{\cos^3 x}{\cos^2 x}$
Simplify.	$= \cos x$

This is incorrect. What mistakes were made?

69. Determine whether the equation is a conditional equation or an identity: $\dfrac{\tan x}{\cot x} = 1.$

Solution:

Start with the left side.	$\dfrac{\tan x}{\cot x}$
Rewrite the tangent and cotangent functions in terms of sines and cosines.	$= \dfrac{\dfrac{\sin x}{\cos x}}{\dfrac{\cos x}{\sin x}}$
Simplify.	$= \dfrac{\sin^2 x}{\cos^2 x} = \tan^2 x$
Let $x = \dfrac{\pi}{4}$. *Note:* $\tan\left(\dfrac{\pi}{4}\right) = 1$.	$= 1$

Since $\dfrac{\tan x}{\cot x} = 1$, this equation is an identity.

This is incorrect. What mistake was made?

70. Determine whether the equation is a conditional equation or an identity: $\left|\sin x\right| - \cos x = 1.$

Solution:

Start with the left side of the equation.	$\left|\sin x\right| - \cos x$
Let $x = \dfrac{n\pi}{2}$, where n is an odd integer.	$\left|\sin\left(\dfrac{n\pi}{2}\right)\right| - \cos\left(\dfrac{n\pi}{2}\right)$
Simplify.	$\left|\pm 1\right| - 0 = 1$

Since $\left|\sin x\right| - \cos x = 0 = 1$, this is an identity.

This is incorrect. What mistake was made?

■ CONCEPTUAL

In Exercises 71 and 72, determine whether each statement is true or false.

71. If an equation is true for some values (but not all values), then it is still an identity.

72. If an equation has an infinite number of solutions, then it is an identity.

73. In which quadrants is the equation $\cos\theta = \sqrt{1 - \sin^2\theta}$ true?

74. In which quadrants is the equation $-\cos\theta = \sqrt{1 - \sin^2\theta}$ true?

75. In which quadrants is the equation $\csc\theta = -\sqrt{1 + \cot^2\theta}$ true?

76. In which quadrants is the equation $\sec\theta = \sqrt{1 + \tan^2\theta}$ true?

▪ CHALLENGE

77. Simplify $(a\sin x + b\cos x)^2 + (b\sin x - a\cos x)^2$.

78. Simplify $\dfrac{1 + \cot^3 x}{1 + \cot x} + \cot x$.

79. Do you think that $\sin(A + B) = \sin A + \sin B$? Why?

80. Do you think that $\cos\left(\frac{1}{2}A\right) = \frac{1}{2}\cos A$? Why?

81. Do you think $\tan(2A) = 2\tan A$? Why?

82. Do you think $\cot(A^2) = (\cot A)^2$? Why?

▪ TECHNOLOGY

In the next section you will learn the sum and difference identities. In Exercises 83–90, we illustrate these identities with graphing calculators.

83. Determine the correct sign ($+$ or $-$) for $\cos(A + B) = \cos A \cos B \underset{?}{\pm} \sin A \sin B$ by graphing $Y_1 = \cos(A + B)$, $Y_2 = \cos A \cos B + \sin A \sin B$, and $Y_3 = \cos A \cos B - \sin A \sin B$ in the same viewing rectangle for several values of A and B.

84. Determine the correct sign ($+$ or $-$) for $\cos(A - B) = \cos A \cos B \underset{?}{\pm} \sin A \sin B$ by graphing $Y_1 = \cos(A - B)$, $Y_2 = \cos A \cos B + \sin A \sin B$, and $Y_3 = \cos A \cos B - \sin A \sin B$ in the same viewing rectangle for several values of A and B.

85. Determine the correct sign ($+$ or $-$) for $\sin(A + B) = \sin A \cos B \underset{?}{\pm} \cos A \sin B$ by graphing $Y_1 = \sin(A + B)$, $Y_2 = \sin A \cos B + \cos A \sin B$, and $Y_3 = \sin A \cos B - \cos A \sin B$ in the same viewing rectangle for several values of A and B.

86. Determine the correct sign ($+$ or $-$) for $\sin(A - B) = \sin A \cos B \underset{?}{\pm} \cos A \sin B$ by graphing $Y_1 = \sin(A - B)$, $Y_2 = \sin A \cos B + \cos A \sin B$, and $Y_3 = \sin A \cos B - \cos A \sin B$ in the same viewing rectangle for several values of A and B.

87. Determine the correct sign ($+$ or $-$) for $\cos^2 A = \dfrac{1 \pm \cos(2A)}{2}$ by graphing $Y_1 = \cos^2 A$, $Y_2 = \dfrac{1 + \cos(2A)}{2}$, and $Y_3 = \dfrac{1 - \cos(2A)}{2}$ in the same viewing rectangle for several values of A.

88. Determine the correct sign ($+$ or $-$) for $\sin^2 A = \dfrac{1 \pm \cos(2A)}{2}$ by graphing $Y_1 = \sin^2 A$, $Y_2 = \dfrac{1 + \cos(2A)}{2}$, and $Y_3 = \dfrac{1 - \cos(2A)}{2}$ in the same viewing rectangle for several values of A.

89. Determine the correct sign ($+$ or $-$) for $\tan\left(x + \dfrac{\pi}{6}\right) = \dfrac{\tan x \underset{?}{\pm} \tan\left(\dfrac{\pi}{6}\right)}{1 \underset{?}{\pm} \tan x \tan\left(\dfrac{\pi}{6}\right)}$ by graphing $Y_1 = \tan\left(x + \dfrac{\pi}{6}\right)$,

$Y_2 = \dfrac{\tan x - \tan\left(\dfrac{\pi}{6}\right)}{1 + \tan x \tan\left(\dfrac{\pi}{6}\right)}$, and $Y_3 = \dfrac{\tan x + \tan\left(\dfrac{\pi}{6}\right)}{1 - \tan x \tan\left(\dfrac{\pi}{6}\right)}$

in the same viewing rectangle. Does $Y_1 = Y_2$ or Y_3?

90. Determine the correct sign ($+$ or $-$) for $\tan\left(x - \dfrac{\pi}{3}\right) = \dfrac{\tan x \underset{?}{\pm} \tan\left(\dfrac{\pi}{3}\right)}{1 \underset{?}{\pm} \tan x \tan\left(\dfrac{\pi}{3}\right)}$ by graphing $Y_1 = \tan\left(x + \dfrac{\pi}{3}\right)$,

$Y_2 = \dfrac{\tan x - \tan\left(\dfrac{\pi}{3}\right)}{1 + \tan x \tan\left(\dfrac{\pi}{3}\right)}$, and $Y_3 = \dfrac{\tan x + \tan\left(\dfrac{\pi}{3}\right)}{1 - \tan x \tan\left(\dfrac{\pi}{3}\right)}$

in the same viewing rectangle. Does $Y_1 = Y_2$ or Y_3?

SKILLS OBJECTIVES

- Find exact values of trigonometric functions of certain rational multiples of π by using the sum and difference identities.
- Develop new identities from the sum and difference identities.

CONCEPTUAL OBJECTIVE

- Understand that a trigonometric function of a sum is not the sum of the trigonometric functions.

In this section, we will consider trigonometric functions with arguments that are sums and differences. In general, $f(A + B) \neq f(A) + f(B)$. First, it is important to note that function notation is not distributive:

$$\cos(A + B) \neq \cos A + \cos B$$

This principle is easy to prove. Let $A = \pi$ and $B = 0$; then

$$\cos(A + B) = \cos(\pi + 0) = \cos \pi = -1$$
$$\cos A + \cos B = \cos \pi + \cos 0 = -1 + 1 = 0$$

Since $-1 \neq 0$, we know that $\cos(A + B) \neq \cos A + \cos B$. In this section, we will derive some new and important identities (sum and difference identities for the cosine, sine, and tangent functions) and revisit cofunction identities. We begin with the familiar distance formula, from which we can derive the sum and difference identities for the cosine function. From there we can derive the sum and difference formulas for the sine and tangent functions.

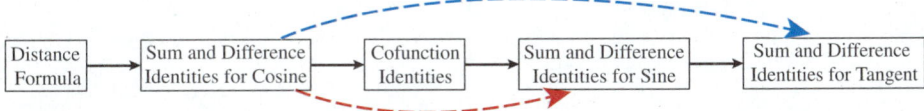

Before we start deriving and working with trigonometric sum and difference identities, let us first discuss why these are important. Sum and difference (and later product-to-sum and sum-to-product) identities are important because they allow calculation in functional (analytic) form and often lead to evaluating expressions *exactly* (as opposed to approximating them with calculators). The identities developed in this chapter are useful in such applications as musical sound, where they allow the determination of the "beat" frequency. In calculus, these identities will simplify the integration and differentiation processes.

Sum and Difference Identities for the Cosine Function

Derivation of the Sum and Difference Identities for the Cosine Function

Recall from Section 6.7 that the unit circle approach gave the relationship between the coordinates along the unit circle and the sine and cosine functions. Specifically, the x-coordinate corresponded to the value of the cosine function and the y-coordinate corresponded to the value of the sine function.

Let us now draw the unit circle with two angles α and β, realizing that the two terminal sides of these angles form a third angle, $\alpha - \beta$.

If we label the points $P_1 = (\cos\alpha, \sin\alpha)$ and $P_2 = (\cos\beta, \sin\beta)$, we can then draw a **segment** connecting points P_1 and P_2.

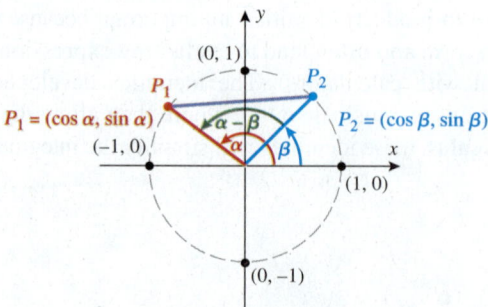

If we rotate the angle clockwise so the central angle $\alpha - \beta$ is in standard position, then the two points where the initial and terminal sides intersect the unit circle are $P_4 = (1, 0)$ and $P_3 = (\cos(\alpha - \beta), \sin(\alpha - \beta))$, respectively.

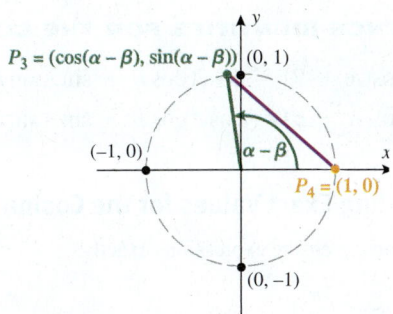

The distance from P_1 to P_2 is equal to the length of the **segment**. Similarly, the distance from P_3 to P_4 is equal to the length of the **segment**. Since the lengths of the **segments** are equal, we say that the distances are equal: $d(P_1, P_2) = d(P_3, P_4)$.

Study Tip

The distance from point $P_1 = (x_1, y_1)$ to $P_2 = (x_2, y_2)$ is given by the distance formula

$$d(P_1, P_2) = \sqrt{(x_2 - x_1)^2 + (y_2 - y_1)^2}.$$

WORDS	MATH
Set the distances (segment lengths) equal.	$d(P_1, P_2) = d(P_3, P_4)$
Apply the distance formula:	$\sqrt{(x_2 - x_1)^2 + (y_2 - y_1)^2} = \sqrt{(x_4 - x_3)^2 + (y_4 - y_3)^2}$

Substitute $P_1 = (x_1, y_1) = (\cos\alpha, \sin\alpha)$ and $P_2 = (x_2, y_2) = (\cos\beta, \sin\beta)$ into the left side of the equation and $P_3 = (x_3, y_3) = (\cos(\alpha - \beta), \sin(\alpha - \beta))$ and $P_4 = (x_4, y_4) = (1, 0)$ into the right side of the equation.

$$\sqrt{[\cos\beta - \cos\alpha]^2 + [\sin\beta - \sin\alpha]^2} = \sqrt{[1 - \cos(\alpha - \beta)]^2 + [0 - \sin(\alpha - \beta)]^2}$$

Square both sides of the equation.	$[\cos\beta - \cos\alpha]^2 + [\sin\beta - \sin\alpha]^2 = [1 - \cos(\alpha - \beta)]^2 + [0 - \sin(\alpha - \beta)]^2$
Eliminate the brackets.	$\cos^2\alpha - 2\cos\alpha\cos\beta + \cos^2\beta + \sin^2\alpha - 2\sin\alpha\sin\beta + \sin^2\beta$ $= 1 - 2\cos(\alpha - \beta) + \cos^2(\alpha - \beta) + \sin^2(\alpha - \beta)$
Regroup terms on each side and use the Pythagorean identity.	$\underbrace{\cos^2\alpha + \sin^2\alpha}_{1} - 2\cos\alpha\cos\beta - 2\sin\alpha\sin\beta + \underbrace{\cos^2\beta + \sin^2\beta}_{1}$ $= 1 - 2\cos(\alpha - \beta) + \underbrace{\cos^2(\alpha - \beta) + \sin^2(\alpha - \beta)}_{1}$
Simplify.	$2 - 2\cos\alpha\cos\beta - 2\sin\alpha\sin\beta = 2 - 2\cos(\alpha - \beta)$
Subtract 2 from both sides.	$-2\cos\alpha\cos\beta - 2\sin\alpha\sin\beta = -2\cos(\alpha - \beta)$
Divide by -2.	$\cos\alpha\cos\beta + \sin\alpha\sin\beta = \cos(\alpha - \beta)$
Write the **difference identity for the cosine function**.	$\boxed{\cos(\alpha - \beta) = \cos\alpha\cos\beta + \sin\alpha\sin\beta}$

We can now derive the sum identity for the cosine function from the difference identity for the cosine function and the properties of even and odd functions.

WORDS	MATH
Replace β with $-\beta$ in the difference identity.	$\cos[\alpha - (-\beta)] = \cos\alpha\cos(-\beta) + \sin\alpha\sin(-\beta)$
Simplify the left side and use properties of even and odd functions on the right side.	$\cos(\alpha + \beta) = \cos\alpha(\cos\beta) + \sin\alpha(-\sin\beta)$
Write the **sum identity for the cosine function**.	$\boxed{\cos(\alpha + \beta) = \cos\alpha\cos\beta - \sin\alpha\sin\beta}$

SUM AND DIFFERENCE IDENTITIES FOR THE COSINE FUNCTION

Sum	$\cos(A + B) = \cos A \cos B - \sin A \sin B$
Difference	$\cos(A - B) = \cos A \cos B + \sin A \sin B$

EXAMPLE 1 **Finding Exact Values for the Cosine Function**

Evaluate each of the following cosine expressions exactly.

a. $\cos 15°$ **b.** $\cos\left(\dfrac{7\pi}{12}\right)$

Solution (a):

Write 15° as a difference of known "special" angles.
$\cos 15° = \cos(45° - 30°)$

Write the difference identity for the cosine function.
$\cos(A - B) = \cos A \cos B + \sin A \sin B$

Substitute $A = 45°$ and $B = 30°$.
$\cos 15° = \cos 45° \cos 30° + \sin 45° \sin 30°$

Evaluate the expressions on the right exactly.
$\cos 15° = \dfrac{\sqrt{2}}{2} \dfrac{\sqrt{3}}{2} + \dfrac{\sqrt{2}}{2} \dfrac{1}{2}$

Simplify.
$\boxed{\cos 15° = \dfrac{\sqrt{6} + \sqrt{2}}{4}}$

Solution (b):

Write $\dfrac{7\pi}{12}$ as a sum of known "special" angles.
$\cos\left(\dfrac{7\pi}{12}\right) = \cos\left(\dfrac{4\pi}{12} + \dfrac{3\pi}{12}\right)$

Simplify.
$\cos\left(\dfrac{7\pi}{12}\right) = \cos\left(\dfrac{\pi}{3} + \dfrac{\pi}{4}\right)$

Write the sum identity for the cosine function.
$\cos(A + B) = \cos A \cos B - \sin A \sin B$

Substitute $A = \dfrac{\pi}{3}$ and $B = \dfrac{\pi}{4}$.
$\cos\left(\dfrac{7\pi}{12}\right) = \cos\left(\dfrac{\pi}{3}\right)\cos\left(\dfrac{\pi}{4}\right) - \sin\left(\dfrac{\pi}{3}\right)\sin\left(\dfrac{\pi}{4}\right)$

Evaluate the expressions on the right exactly.
$\cos\left(\dfrac{7\pi}{12}\right) = \dfrac{1}{2} \dfrac{\sqrt{2}}{2} - \dfrac{\sqrt{3}}{2} \dfrac{\sqrt{2}}{2}$

Simplify.
$\boxed{\cos\left(\dfrac{7\pi}{12}\right) = \dfrac{\sqrt{2} - \sqrt{6}}{4}}$

Note: $\dfrac{7\pi}{12}$ can also be represented as 105°.

■ **YOUR TURN** Use the sum or difference identities for the cosine function to evaluate each cosine expression exactly.

a. $\cos 75°$ **b.** $\cos\left(\dfrac{5\pi}{12}\right)$

Example 1 illustrates an important characteristic of the sum and difference identities: that we can now find the exact trigonometric function value of angles that are multiples of $15°$ (or, equivalently, $\dfrac{\pi}{12}$), since each of these can be written as a sum or difference of angles for which we know the trigonometric function values exactly.

EXAMPLE 2 **Writing a Sum or Difference as a Single Cosine Expression**

Use the sum or the difference identity for the cosine function to write each of the following expressions as a single cosine expression.

a. $\sin(5x)\sin(2x) + \cos(5x)\cos(2x)$ **b.** $\cos x \cos(3x) - \sin x \sin(3x)$

Solution (a):

Because of the positive sign, this will be a cosine of a difference.

Reverse the expression and write the formula.

$$\cos A \cos B + \sin A \sin B = \cos(A - B)$$

Identify A and B.

$$A = 5x \text{ and } B = 2x$$

Substitute $A = 5x$ and $B = 2x$ into the difference identity.

$$\cos(5x)\cos(2x) + \sin(5x)\sin(2x) = \cos(5x - 2x)$$

Simplify.

$$\cos(5x)\cos(2x) + \sin(5x)\sin(2x) = \boxed{\cos(3x)}$$

Notice that if we had selected $A = 2x$ and $B = 5x$ instead, the result would have been $\cos(-3x)$, but since the cosine function is an even function, this would have simplified to $\cos(3x)$.

Solution (b):

Because of the negative sign, this will be a cosine of a sum.

Reverse the expression and write the formula.

$$\cos A \cos B - \sin A \sin B = \cos(A + B)$$

Identify A and B.

$$A = x \text{ and } B = 3x$$

Substitute $A = x$ and $B = 3x$ into the sum identity.

$$\cos x \cos(3x) - \sin x \sin(3x) = \cos(x + 3x)$$

Simplify.

$$\cos x \cos(3x) - \sin x \sin(3x) = \boxed{\cos(4x)}$$

■ YOUR TURN Write $\cos(4x)\cos(7x) + \sin(4x)\sin(7x)$ as a single cosine expression.

Technology Tip

a.

b.

■ **Answer:** $\cos(3x)$

Classroom Example 7.3.2

Write as a single cosine expression.   **Answer:**

a. $\sin(\pi x)\sin\left(\dfrac{x}{2}\right) - \cos\left(\dfrac{x}{2}\right)\cos(\pi x)$ **a.** $-\cos\left[\left(\pi + \tfrac{1}{2}\right)x\right]$

b. $\sin(\pi x)\sin\left(\dfrac{x}{2}\right) + \cos\left(\dfrac{x}{2}\right)\cos(\pi x)$ **b.** $\cos\left[\left(\pi - \tfrac{1}{2}\right)x\right]$

Cofunction Identities

In Section 6.2, we discussed cofunction relationships for acute angles. Recall that a trigonometric function value of an angle is equal to the corresponding cofunction value of its complementary angle. Now we use the sum and difference identities for the cosine function to develop the cofunction identities for any angle, θ.

WORDS	**MATH**
Write the difference identity for the cosine function.	$\cos(A - B) = \cos A \cos B + \sin A \sin B$
Let $A = \dfrac{\pi}{2}$ and $B = \theta$.	$\cos\left(\dfrac{\pi}{2} - \theta\right) = \cos\left(\dfrac{\pi}{2}\right)\cos\theta + \sin\left(\dfrac{\pi}{2}\right)\sin\theta$
Evaluate the known values for the sine and cosine functions.	$\cos\left(\dfrac{\pi}{2} - \theta\right) = 0 \cdot \cos\theta + 1 \cdot \sin\theta$
Simplify.	$\boxed{\cos\left(\dfrac{\pi}{2} - \theta\right) = \sin\theta}$
Similarly, to determine the other corresponding cofunction identity, write the difference identity for the cosine function.	$\cos(A - B) = \cos A \cos B + \sin A \sin B$
Let $A = \dfrac{\pi}{2}$ and $B = \dfrac{\pi}{2} - \theta$.	$\cos\left[\dfrac{\pi}{2} - \left(\dfrac{\pi}{2} - \theta\right)\right] = \cos\left(\dfrac{\pi}{2}\right)\cos\left(\dfrac{\pi}{2} - \theta\right) + \sin\left(\dfrac{\pi}{2}\right)\sin\left(\dfrac{\pi}{2} - \theta\right)$
Evaluate the known values for the sine and cosine functions.	$\cos(\theta) = 0 \cdot \cos\left(\dfrac{\pi}{2} - \theta\right) + 1 \cdot \sin\left(\dfrac{\pi}{2} - \theta\right)$
Simplify.	$\boxed{\cos\theta = \sin\left(\dfrac{\pi}{2} - \theta\right)}$

COFUNCTION IDENTITIES FOR THE SINE AND COSINE FUNCTIONS

$$\cos\left(\frac{\pi}{2} - \theta\right) = \sin\theta \qquad \sin\left(\frac{\pi}{2} - \theta\right) = \cos\theta$$

Sum and Difference Identities for the Sine Function

We can now use the cofunction identities together with the sum and difference identities for the cosine function to develop the sum and difference identities for the sine function.

WORDS	MATH
Start with the cofunction identity.	$\sin\theta = \cos\left(\dfrac{\pi}{2} - \theta\right)$
Let $\theta = A + B$.	$\sin(A + B) = \cos\left[\dfrac{\pi}{2} - (A + B)\right]$
Regroup the terms in the cosine expression.	$\sin(A + B) = \cos\left[\left(\dfrac{\pi}{2} - A\right) - B\right]$
Use the difference identity for the cosine function.	$\sin(A + B) = \cos\left(\dfrac{\pi}{2} - A\right)\cos B + \sin\left(\dfrac{\pi}{2} - A\right)\sin B$
Use the cofunction identities.	$\sin(A + B) = \underbrace{\cos\left(\dfrac{\pi}{2} - A\right)}_{\sin A}\cos B + \underbrace{\sin\left(\dfrac{\pi}{2} - A\right)}_{\cos A}\sin B$
Simplify.	$\boxed{\sin(A + B) = \sin A \cos B + \cos A \sin B}$

Now we can derive the difference identity for the sine function using the sum identity for the sine function and the properties of even and odd functions.

WORDS	MATH
Replace B with $-B$ in the sum identity.	$\sin(A + (-B)) = \sin A \cos(-B) + \cos A \sin(-B)$
Simplify using even and odd identities.	$\boxed{\sin(A - B) = \sin A \cos B - \cos A \sin B}$

SUM AND DIFFERENCE IDENTITIES FOR THE SINE FUNCTION

Sum	$\sin(A + B) = \sin A \cos B + \cos A \sin B$
Difference	$\sin(A - B) = \sin A \cos B - \cos A \sin B$

EXAMPLE 3   Finding Exact Values for the Sine Function

Use the sum or the difference identity for the sine function to evaluate each sine expression exactly.

a. $\sin 75°$ **b.** $\sin\left(\dfrac{5\pi}{12}\right)$

Solution (a):

Write 75° as a sum of known "special" angles.	$\sin 75° = \sin(45° + 30°)$
Write the sum identity for the sine function.	$\sin(A + B) = \sin A \cos B + \cos A \sin B$
Substitute $A = 45°$ and $B = 30°$.	$\sin 75° = \sin 45° \cos 30° + \cos 45° \sin 30°$
Evaluate the expressions on the right exactly.	$\sin 75° = \left(\dfrac{\sqrt{2}}{2}\right)\left(\dfrac{\sqrt{3}}{2}\right) + \left(\dfrac{\sqrt{2}}{2}\right)\left(\dfrac{1}{2}\right)$
Simplify.	$\boxed{\sin 75° = \dfrac{\sqrt{6} + \sqrt{2}}{4}}$

Classroom Example 7.3.3
Compute:

a. $\sin\left(\dfrac{23\pi}{12}\right)$ **b.** $\sin 195°$

Answer:

a. $\dfrac{\sqrt{2} - \sqrt{6}}{4}$

b. $-\dfrac{\sqrt{2} - \sqrt{6}}{4}$

Technology Tip

Use a calculator to check the values of $\sin\left(\dfrac{5\pi}{12}\right)$ and $\dfrac{\sqrt{2}+\sqrt{6}}{4}$. Be sure the calculator is in radian mode.

```
sin(5π/12)
        .9659258263
(√(2)+√(6))/4
        .9659258263
```

Answer: a. $\dfrac{\sqrt{6}-\sqrt{2}}{4}$

b. $\dfrac{\sqrt{6}+\sqrt{2}}{4}$

Solution (b):

Write $\dfrac{5\pi}{12}$ as a sum of known "special" angles.

$$\sin\left(\frac{5\pi}{12}\right)=\sin\left(\frac{2\pi}{12}+\frac{3\pi}{12}\right)$$

Simplify.

$$\sin\left(\frac{5\pi}{12}\right)=\sin\left(\frac{\pi}{6}+\frac{\pi}{4}\right)$$

Write the sum identity for the sine function.

$$\sin(A+B)=\sin A\cos B+\cos A\sin B$$

Substitute $A=\dfrac{\pi}{6}$ and $B=\dfrac{\pi}{4}$.

$$\sin\left(\frac{5\pi}{12}\right)=\sin\left(\frac{\pi}{6}\right)\cos\left(\frac{\pi}{4}\right)+\cos\left(\frac{\pi}{6}\right)\sin\left(\frac{\pi}{4}\right)$$

Evaluate the expressions on the right exactly.

$$\sin\left(\frac{5\pi}{12}\right)=\left(\frac{1}{2}\right)\left(\frac{\sqrt{2}}{2}\right)+\left(\frac{\sqrt{3}}{2}\right)\left(\frac{\sqrt{2}}{2}\right)$$

Simplify.

$$\boxed{\sin\left(\frac{5\pi}{12}\right)=\frac{\sqrt{2}+\sqrt{6}}{4}}$$

Note: $\dfrac{5\pi}{12}$ can also be represented as 75°.

■ **YOUR TURN** Use the sum or the difference identity for the sine function to evaluate each sine expression exactly.

a. $\sin 15°$ **b.** $\sin\left(\dfrac{7\pi}{12}\right)$

We see in Example 3 that the sum and difference identities allow us to calculate exact values for trigonometric functions of angles that are multiples of 15° $\left(\text{or, equivalently, }\dfrac{\pi}{12}\right)$, as we saw with the cosine function.

Technology Tip

Graphs of $y_1 = 3\sin x\cos(3x)+3\cos x\sin(3x)$ and $y_2 = 3\sin(4x)$.

EXAMPLE 4 Writing a Sum or Difference as a Single Sine Expression

Graph $y = 3\sin x\cos(3x)+3\cos x\sin(3x)$.

Solution:

Use the sum identity for the sine function to write the expression as a single sine expression.

Factor out the common 3.

$$y = 3[\sin x\cos(3x)+\cos x\sin(3x)]$$

Write the sum identity for the sine function.

$$\sin A\cos B+\cos A\sin B=\sin(A+B)$$

Identify A and B.

$$A=x \text{ and } B=3x$$

Substitute $A=x$ and $B=3x$ into the sum identity.

$$\sin x\cos(3x)+\cos x\sin(3x)=\sin(x+3x)=\sin(4x)$$

Simplify.

$$y = 3[\underbrace{\sin x\cos(3x)+\cos x\sin(3x)}_{\sin 4x}]$$

Graph $y = 3\sin(4x)$.

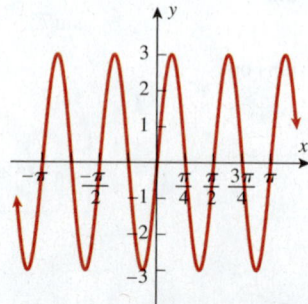

Sum and Difference Identities for the Tangent Function

We now develop the sum and difference identities for the tangent function.

WORDS	**MATH**
Start with the quotient identity.	$\tan x = \dfrac{\sin x}{\cos x}$
Let $x = A + B$.	$\tan(A + B) = \dfrac{\sin(A + B)}{\cos(A + B)}$
Use the sum identities for the sine and cosine functions.	$\tan(A + B) = \dfrac{\sin A \cos B + \cos A \sin B}{\cos A \cos B - \sin A \sin B}$
Multiply the numerator and denominator by $\dfrac{1}{\cos A \cos B}$.	$\tan(A + B) = \dfrac{\dfrac{\sin A \cos B + \cos A \sin B}{\cos A \cos B}}{\dfrac{\cos A \cos B - \sin A \sin B}{\cos A \cos B}} = \dfrac{\dfrac{\sin A \cos B}{\cos A \cos B} + \dfrac{\cos A \sin B}{\cos A \cos B}}{\dfrac{\cos A \cos B}{\cos A \cos B} - \dfrac{\sin A \sin B}{\cos A \cos B}}$
Simplify.	$\tan(A + B) = \dfrac{\left(\dfrac{\sin A}{\cos A}\right) + \left(\dfrac{\sin B}{\cos B}\right)}{1 - \left(\dfrac{\sin A}{\cos A}\right)\left(\dfrac{\sin B}{\cos B}\right)}$
Write the expressions inside the parentheses in terms of the tangent function.	$\boxed{\tan(A + B) = \dfrac{\tan A + \tan B}{1 - \tan A \tan B}}$
Replace B with $-B$.	$\tan[A + (-B)] = \tan(A - B) = \dfrac{\tan A + \tan(-B)}{1 - \tan A \tan(-B)}$
Since the tangent function is an odd function, $\tan(-B) = -\tan B$.	$\boxed{\tan(A - B) = \dfrac{\tan A - \tan B}{1 + \tan A \tan B}}$

SUM AND DIFFERENCE IDENTITIES FOR THE TANGENT FUNCTION

Sum	$\tan(A + B) = \dfrac{\tan A + \tan B}{1 - \tan A \tan B}$
Difference	$\tan(A - B) = \dfrac{\tan A - \tan B}{1 + \tan A \tan B}$

EXAMPLE 5   Finding Exact Values for the Tangent Function

Find the exact value of $\tan(\alpha + \beta)$ if $\sin\alpha = -\frac{1}{3}$ and $\cos\beta = -\frac{1}{4}$ and the terminal side of α lies in quadrant III and the terminal side of β lies in quadrant II.

Solution:

Technology Tip

If $\sin\alpha = -\frac{1}{3}$ and α is in QIII, then $\alpha = \pi + \sin^{-1}(\frac{1}{3})$ (more explanation can be found in Section 7.7). If $\cos\beta = -\frac{1}{4}$ and β is in QII, then $\beta = \pi - \cos^{-1}(\frac{1}{4})$. Now use the graphing calculator to find $\tan(\alpha + \beta)$ by entering

```
tan(π+sin⁻¹(1/3)+
π-cos⁻¹(1/4))
         -1.485426269
```

In Step 4, use the graphing calculator to evaluate both expressions

$$\frac{\frac{\sqrt{2}}{4} - \sqrt{15}}{1 - \left(\frac{\sqrt{2}}{4}\right)\left(-\sqrt{15}\right)} \text{ and}$$

$$\frac{\sqrt{2} - 4\sqrt{15}}{4 + \sqrt{30}} \text{ by entering}$$

```
(√(2)/4-√(15))/(
1-√(2)/4*-√(15))
         -1.485426269
(√(2)-4√(15))/(4
+√(30))
         -1.485426269
■
```

STEP 1 Write the sum identity for the tangent function.

$$\tan(\alpha + \beta) = \frac{\tan\alpha + \tan\beta}{1 - \tan\alpha\,\tan\beta}$$

STEP 2 Find $\tan\alpha$.

The terminal side of α lies in quadrant III.

$$\sin\alpha = \frac{y}{r} = -\frac{1}{3}$$

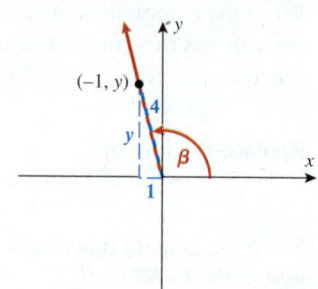

Solve for x. ($x^2 + y^2 = r^2$.)

$$x^2 + 1^2 = 3^2$$
$$x = \pm\sqrt{8}$$

Take the negative sign since we are in quadrant III.

$$x = -2\sqrt{2}$$

Find $\tan\alpha$.

$$\tan\alpha = \frac{y}{x} = \frac{-1}{-2\sqrt{2}} = \frac{1}{2\sqrt{2}}\cdot\frac{\sqrt{2}}{\sqrt{2}} = \frac{\sqrt{2}}{4}$$

STEP 3 Find $\tan\beta$.

The terminal side of β lies in quadrant II.

$$\cos\beta = -\frac{1}{4} = \frac{x}{r}$$

Solve for y. ($x^2 + y^2 = r^2$.)

$$(-1)^2 + y^2 = 4^2$$
$$y = \pm\sqrt{15}$$

Take the positive sign since we are in quadrant II.

$$y = \sqrt{15}$$

Find $\tan\beta$.

$$\tan\beta = \frac{y}{x} = \frac{\sqrt{15}}{-1} = -\sqrt{15}$$

STEP 4 Substitute $\tan\alpha = \dfrac{\sqrt{2}}{4}$ and $\tan\beta = -\sqrt{15}$ into the sum identity for the tangent function.

$$\tan(\alpha + \beta) = \frac{\dfrac{\sqrt{2}}{4} - \sqrt{15}}{1 - \left(\dfrac{\sqrt{2}}{4}\right)(-\sqrt{15})}$$

Multiply the numerator and the denominator by 4.

$$\tan(\alpha + \beta) = \frac{4\left(\dfrac{\sqrt{2}}{4} - \sqrt{15}\right)}{4\left(1 + \dfrac{\sqrt{30}}{4}\right)}$$

$$= \frac{\sqrt{2} - 4\sqrt{15}}{4 + \sqrt{30}}$$

The expression $\boxed{\tan(\alpha + \beta) = \dfrac{\sqrt{2} - 4\sqrt{15}}{4 + \sqrt{30}}}$ can be simplified further if we rationalize the denominator.

It is important to note in Example 5 that right triangles have been superimposed in the Cartesian plane. The coordinate pair (x, y) can have positive or negative values, but the radius r is always positive. When right triangles are superimposed, with one vertex at the point (x, y) and another vertex at the origin, it is important to understand that triangles have positive side lengths.

SECTION 7.3 SUMMARY

In this section we derived the sum and difference identities for the cosine function using the distance formula. We then used these identities to derive the cofunction identities. The cofunction identities and sum and difference identities for the cosine function were used to derive the sum and difference identities for the sine function. We combined the sine and cosine sum and difference identities to determine the tangent sum and difference identities. The sum and difference identities enabled us to evaluate a trigonometric expression exactly for any integer multiple of 15° (i.e., $\pi/12$).

$$\cos(A + B) = \cos A \cos B - \sin A \sin B$$
$$\sin(A + B) = \sin A \cos B + \cos A \sin B$$
$$\tan(A + B) = \frac{\tan A + \tan B}{1 - \tan A \tan B}$$
$$\cos(A - B) = \cos A \cos B + \sin A \sin B$$
$$\sin(A - B) = \sin A \cos B - \cos A \sin B$$
$$\tan(A - B) = \frac{\tan A - \tan B}{1 + \tan A \tan B}$$

SECTION 7.3 EXERCISES

■ SKILLS

In Exercises 1–16, find the exact value for each trigonometric expression.

1. $\sin\left(\dfrac{\pi}{12}\right)$

2. $\cos\left(\dfrac{\pi}{12}\right)$

3. $\cos\left(-\dfrac{5\pi}{12}\right)$

4. $\sin\left(-\dfrac{5\pi}{12}\right)$

5. $\tan\left(-\dfrac{\pi}{12}\right)$

6. $\tan\left(\dfrac{13\pi}{12}\right)$

7. $\sin 105°$

8. $\cos 195°$

9. $\tan(-105°)$

10. $\tan 165°$

11. $\cot\left(\dfrac{\pi}{12}\right)$

12. $\cot\left(-\dfrac{5\pi}{12}\right)$

13. $\sec\left(-\dfrac{11\pi}{12}\right)$

14. $\sec\left(-\dfrac{13\pi}{12}\right)$

15. $\csc(-255°)$

16. $\csc(-15°)$

In Exercises 17–28, write each expression as a single trigonometric function.

17. $\sin(2x)\sin(3x) + \cos(2x)\cos(3x)$

18. $\sin x \sin(2x) - \cos x \cos(2x)$

19. $\sin x \cos(2x) - \cos x \sin(2x)$

20. $\sin(2x)\cos(3x) + \cos(2x)\sin(3x)$

21. $\cos(\pi - x)\sin x + \sin(\pi - x)\cos x$

22. $\sin\left(\dfrac{\pi}{3}x\right)\cos\left(-\dfrac{\pi}{2}x\right) - \cos\left(\dfrac{\pi}{3}x\right)\sin\left(-\dfrac{\pi}{2}x\right)$

23. $(\sin A - \sin B)^2 + (\cos A - \cos B)^2 - 2$

24. $(\sin A + \sin B)^2 + (\cos A + \cos B)^2 - 2$

25. $2 - (\sin A + \cos B)^2 - (\cos A + \sin B)^2$

26. $2 - (\sin A - \cos B)^2 - (\cos A + \sin B)^2$

27. $\dfrac{\tan 49° - \tan 23°}{1 + \tan 49° \tan 23°}$

28. $\dfrac{\tan 49° + \tan 23°}{1 - \tan 49° \tan 23°}$

In Exercises 29–34, find the exact value of the indicated expression using the given information and identities.

29. Find the exact value of $\cos(\alpha + \beta)$ if $\cos\alpha = -\frac{1}{3}$ and $\cos\beta = -\frac{1}{4}$ and the terminal side of α lies in quadrant III and the terminal side of β lies in quadrant II.

30. Find the exact value of $\cos(\alpha - \beta)$ if $\cos\alpha = \frac{1}{3}$ and $\cos\beta = -\frac{1}{4}$ and the terminal side of α lies in quadrant IV and the terminal side of β lies in quadrant II.

31. Find the exact value of $\sin(\alpha - \beta)$ if $\sin\alpha = -\frac{3}{5}$ and $\sin\beta = \frac{1}{5}$ and the terminal side of α lies in quadrant III and the terminal side of β lies in quadrant I.

32. Find the exact value of $\sin(\alpha + \beta)$ if $\sin\alpha = -\frac{3}{5}$ and $\sin\beta = \frac{1}{5}$ and the terminal side of α lies in quadrant III and the terminal side of β lies in quadrant II.

33. Find the exact value of $\tan(\alpha + \beta)$ if $\sin\alpha = -\frac{3}{5}$ and $\cos\beta = -\frac{1}{4}$ and the terminal side of α lies in quadrant III and the terminal side of β lies in quadrant II.

34. Find the exact value of $\tan(\alpha - \beta)$ if $\sin\alpha = -\frac{3}{5}$ and $\cos\beta = -\frac{1}{4}$ and the terminal side of α lies in quadrant III and the terminal side of β lies in quadrant II.

In Exercises 35–52, determine whether each equation is a conditional equation or an identity.

35. $\sin(A + B) + \sin(A - B) = 2\sin A \cos B$

36. $\cos(A + B) + \cos(A - B) = 2\cos A \cos B$

37. $\sin\left(x - \dfrac{\pi}{2}\right) = \cos\left(x + \dfrac{\pi}{2}\right)$

38. $\sin\left(x + \dfrac{\pi}{2}\right) = \cos\left(x + \dfrac{\pi}{2}\right)$

39. $\dfrac{\sqrt{2}}{2}(\sin x + \cos x) = \sin\left(x + \dfrac{\pi}{4}\right)$

40. $\sqrt{3}\cos x + \sin x = 2\cos\left(x + \dfrac{\pi}{3}\right)$

41. $\sin^2 x = \dfrac{1 - \cos(2x)}{2}$

42. $\cos^2 x = \dfrac{1 + \cos(2x)}{2}$

43. $\sin(2x) = 2\sin x \cos x$

44. $\cos(2x) = \cos^2 x - \sin^2 x$

45. $\sin(A + B) = \sin A + \sin B$

46. $\cos(A + B) = \cos A + \cos B$

47. $\tan(\pi + B) = \tan B$

48. $\tan(A - \pi) = \tan A$

49. $\cot(3\pi + x) = \dfrac{1}{\tan x}$

50. $\csc(2x) = 2\sec x \csc x$

51. $\dfrac{1 + \tan x}{1 - \tan x} = \tan\left(x - \dfrac{\pi}{4}\right)$

52. $\cot\left(x + \dfrac{\pi}{4}\right) = \dfrac{1 - \tan x}{1 + \tan x}$

In Exercises 53–62, graph each of the functions by first rewriting it as a sine, cosine, or tangent of a difference or sum.

53. $y = \cos\left(\dfrac{\pi}{3}\right)\sin x + \cos x \sin\left(\dfrac{\pi}{3}\right)$

54. $y = \cos\left(\dfrac{\pi}{3}\right)\sin x - \cos x \sin\left(\dfrac{\pi}{3}\right)$

55. $y = \sin x \sin\left(\dfrac{\pi}{4}\right) + \cos x \cos\left(\dfrac{\pi}{4}\right)$

56. $y = \sin x \sin\left(\dfrac{\pi}{4}\right) - \cos x \cos\left(\dfrac{\pi}{4}\right)$

57. $y = -\sin x \cos(3x) - \cos x \sin(3x)$

58. $y = \sin x \sin(3x) + \cos x \cos(3x)$

59. $y = \dfrac{1 + \tan x}{1 - \tan x}$

60. $y = \dfrac{\sqrt{3} - \tan x}{1 + \sqrt{3}\tan x}$

61. $y = \dfrac{1 + \sqrt{3}\tan x}{\sqrt{3} - \tan x}$

62. $y = \dfrac{1 - \tan x}{1 + \tan x}$

▪ APPLICATIONS

For Exercises 63 and 64, refer to the following:

The difference quotient, $\dfrac{f(x + h) - f(x)}{h}$, is used to approximate the rate of change of the function f and will be used frequently in calculus.

63. Difference Quotient. Show that the difference quotient for $f(x) = \sin x$ is $\cos x\left(\dfrac{\sin h}{h}\right) - \sin x\left(\dfrac{1 - \cos h}{h}\right)$.

64. Difference Quotient. Show that the difference quotient for $f(x) = \cos x$ is $-\sin x\left(\dfrac{\sin h}{h}\right) - \cos x\left(\dfrac{1 - \cos h}{h}\right)$.

For Exercises 65 and 66, use the following:

A nonvertical line makes an angle with the x-axis. In the figure, we see that the line L_1 makes an acute angle θ_1 with the x-axis. Similarly, the line L_2 makes an acute angle θ_2 with the x-axis.

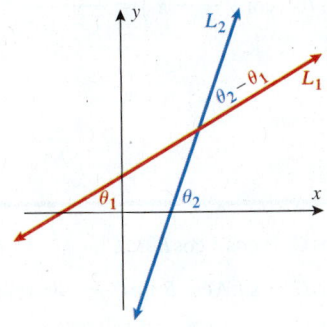

$\tan\theta_1 = $ slope of $L_1 = m_1$

$\tan\theta_2 = $ slope of $L_2 = m_2$

65. Angle Between Two Lines. Show that

$$\tan(\theta_2 - \theta_1) = \dfrac{m_2 - m_1}{1 + m_1 m_2}$$

66. Relating Tangent and Slope. Show that

$$\tan(\theta_1 - \theta_2) = \dfrac{m_1 - m_2}{1 + m_1 m_2}$$

For Exercises 67 and 68, refer to the following:

An electric field E of a wave with constant amplitude A, propagating a distance z, is given by

$$E = A\cos(kz - ct)$$

where k is the propagation wave number, which is related to the wavelength λ by $k = \dfrac{2\pi}{\lambda}$, and where $c = 3.0 \times 10^8$ m/s is the speed of light in a vacuum, and t is time in seconds.

67. Electromagnetic Wave Propagation. Use the cosine difference identity to express the electric field in terms of both sine and cosine functions. When the quotient of the propagation distance z and the wavelength λ are equal to an integer, what do you notice?

68. Electromagnetic Wave Propagation. Use the cosine difference identity to express the electric field in terms of both sine and cosine functions. When $t = 0$, what do you notice?

69. Biology. By analyzing available empirical data, it has been determined that the body temperature of a species fluctuates according to the model

$$T(t) = 38 - 2.5\cos\left[\dfrac{\pi}{6}(t - 3)\right], \qquad 0 \le t \le 24$$

where T represents temperature in degrees Celsius and t represents time (in hours) measured from 12:00 P.M. (noon). Use an identity to express $T(t)$ in terms of the sine function.

70. Health/Medicine. During the course of treatment of an illness, the concentration of a drug (in micrograms per milliliter) in the bloodstream fluctuates during the dosing period of 8 hours according to the model

$$C(t) = 15.4 - 4.7\sin\left(\dfrac{\pi}{4}t + \dfrac{\pi}{2}\right), \qquad 0 \le t \le 8$$

Use an identity to express the concentration $C(t)$ in terms of the cosine function.

Note: This model does not apply to the first dose of the medication as there will be no medication in the bloodstream.

▪ CATCH THE MISTAKE

In Exercises 71 and 72, explain the mistake that is made.

71. Find the exact value of $\tan\left(\dfrac{5\pi}{12}\right)$.

Solution:

Write $\dfrac{5\pi}{12}$ as a sum. $\qquad \tan\left(\dfrac{\pi}{4} + \dfrac{\pi}{6}\right)$

Distribute. $\qquad\qquad \tan\left(\dfrac{\pi}{4}\right) + \tan\left(\dfrac{\pi}{6}\right)$

Evaluate the tangent
function for $\dfrac{\pi}{4}$ and $\dfrac{\pi}{6}$. $\qquad 1 + \dfrac{\sqrt{3}}{3}$

This is incorrect. What mistake was made?

72. Find the exact value of $\tan\left(-\dfrac{7\pi}{6}\right)$.

Solution:

The tangent function is an
even function. $\qquad\qquad\qquad \tan\left(\dfrac{7\pi}{6}\right)$

Write $\dfrac{7\pi}{6}$ as a sum. $\qquad\qquad \tan\left(\pi + \dfrac{\pi}{6}\right)$

Use the tangent sum identity,
$\tan(A + B) = \dfrac{\tan A + \tan B}{1 - \tan A \tan B}$. $\qquad \dfrac{\tan \pi + \tan\left(\dfrac{\pi}{6}\right)}{1 - \tan \pi \, \tan\left(\dfrac{\pi}{6}\right)}$

Evaluate the tangent functions
on the right. $\qquad\qquad\qquad \dfrac{0 + \dfrac{1}{\sqrt{3}}}{1 - 0}$

Simplify. $\qquad\qquad\qquad\qquad \dfrac{\sqrt{3}}{3}$

This is incorrect. What mistake was made?

▪ CONCEPTUAL

In Exercises 73–76, determine whether each statement is true or false.

73. $\cos(15°) = \cos(45°) - \cos(30°)$

74. $\sin\left(\dfrac{\pi}{2}\right) = \sin\left(\dfrac{\pi}{3}\right) + \sin\left(\dfrac{\pi}{6}\right)$

75. $\tan\left(x + \dfrac{\pi}{4}\right) = 1 + \tan x$

76. $\cot\left(\dfrac{\pi}{4} - x\right) = \dfrac{1 + \tan x}{1 - \tan x}$

▪ CHALLENGE

77. Verify that $\sin(A + B + C) = \sin A \cos B \cos C + \cos A \sin B \cos C + \cos A \cos B \sin C - \sin A \sin B \sin C$.

78. Verify that $\cos(A + B + C) = \cos A \cos B \cos C - \sin A \sin B \cos C - \sin A \cos B \sin C - \cos A \sin B \sin C$.

79. Although in general the statement $\sin(A - B) = \sin A - \sin B$ is not true, it is true for some values. Determine some values of A and B which make this statement true.

80. Although in general the statement $\sin(A + B) = \sin A + \sin B$ is not true, it is true for some values. Determine some values of A and B which make this statement true.

■ **TECHNOLOGY**

81. In Exercise 63, you showed that the difference quotient for
$f(x) = \sin x$ is $\cos x \left(\dfrac{\sin h}{h} \right) - \sin x \left(\dfrac{1 - \cos h}{h} \right)$.

Plot $Y_1 = \cos x \left(\dfrac{\sin h}{h} \right) - \sin x \left(\dfrac{1 - \cos h}{h} \right)$ for

a. $h = 1$ **b.** $h = 0.1$ **c.** $h = 0.01$

What function does the difference quotient for $f(x) = \sin x$ resemble when h approaches zero?

82. Show that the difference quotient for $f(x) = \cos x$ is
$-\sin x \left(\dfrac{\sin h}{h} \right) - \cos x \left(\dfrac{1 - \cos h}{h} \right)$.

Plot $Y_1 = -\sin x \left(\dfrac{\sin h}{h} \right) - \cos x \left(\dfrac{1 - \cos h}{h} \right)$ for

a. $h = 1$ **b.** $h = 0.1$ **c.** $h = 0.01$

What function does the difference quotient for $f(x) = \cos x$ resemble when h approaches zero?

83. Show that the difference quotient for $f(x) = \sin(2x)$ is
$\cos(2x) \left[\dfrac{\sin(2h)}{h} \right] - \sin(2x) \left[\dfrac{1 - \cos(2h)}{h} \right]$.

Plot $Y_1 = \cos(2x) \left[\dfrac{\sin(2h)}{h} \right] - \sin(2x) \left[\dfrac{1 - \cos(2h)}{h} \right]$ for

a. $h = 1$ **b.** $h = 0.1$ **c.** $h = 0.01$

What function does the difference quotient for $f(x) = \sin(2x)$ resemble when h approaches zero?

84. Show that the difference quotient for $f(x) = \cos(2x)$ is
$-\sin(2x) \left[\dfrac{\sin(2h)}{h} \right] - \cos(2x) \left[\dfrac{1 - \cos(2h)}{h} \right]$.

Plot $Y_1 = -\sin(2x) \left[\dfrac{\sin(2h)}{h} \right] - \cos(2x) \left[\dfrac{1 - \cos(2h)}{h} \right]$ for

a. $h = 1$ **b.** $h = 0.1$ **c.** $h = 0.01$

What function does the difference quotient for $f(x) = \cos(2x)$ resemble when h approaches zero?

SECTION

7.4 DOUBLE-ANGLE IDENTITIES

SKILLS OBJECTIVES

■ Use the double-angle identities to find exact values of certain trigonometric functions.
■ Use the double-angle identities to help in verifying identities.

CONCEPTUAL OBJECTIVE

■ Understand that the double-angle identities are derived from the sum identities.

Applying Double-Angle Identities

Throughout this text, much attention has been given to distinguishing between evaluating trigonometric functions exactly (for special angles) and approximating values of trigonometric functions with a calculator. In previous chapters we could only evaluate trigonometric functions exactly for reference angles of $30°$, $45°$, and $60°$ or $\dfrac{\pi}{6}, \dfrac{\pi}{4}$, and $\dfrac{\pi}{3}$; note that as of the previous section, we now can include multiples of $\dfrac{\pi}{12}$ among these "special" angles. Now, we can use *double-angle identities* to also evaluate the trigonometric function values for other angles that are even integer multiples of the special angles or to verify other trigonometric identities. One important distinction now is that we will be able to find exact values of many functions using the double-angle identities without needing to know the actual value of the angle.

Derivation of Double-Angle Identities

To derive the *double-angle identities*, we let $A = B$ in the sum identities.

WORDS	MATH
Write the identity for the sine of a sum.	$\sin(A + B) = \sin A \cos B + \cos A \sin B$
Let $B = A$.	$\sin(A + A) = \sin A \cos A + \cos A \sin A$
Simplify.	$\boxed{\sin(2A) = 2\sin A \cos A}$
Write the identity for the cosine of a sum.	$\cos(A + B) = \cos A \cos B - \sin A \sin B$
Let $B = A$.	$\cos(A + A) = \cos A \cos A - \sin A \sin A$
Simplify.	$\boxed{\cos(2A) = \cos^2 A - \sin^2 A}$

We can write the *double-angle identity* for the cosine function two other ways if we use the Pythagorean identity:

1. Write the identity for the cosine function of a double angle.

$$\cos(2A) = \cos^2 A - \sin^2 A$$

Use the Pythagorean identity for the cosine function.

$$\cos(2A) = \underbrace{\cos^2 A}_{1 - \sin^2 A} - \sin^2 A$$

Simplify.

$$\boxed{\cos(2A) = 1 - 2\sin^2 A}$$

2. Write the identity for the cosine function of a double angle.

$$\cos(2A) = \cos^2 A - \sin^2 A$$

Use the Pythagorean identity for the sine function.

$$\cos(2A) = \cos^2 A - \underbrace{\sin^2 A}_{1 - \cos^2 A}$$

Simplify.

$$\boxed{\cos(2A) = 2\cos^2 A - 1}$$

The tangent function can always be written as a quotient, $\tan(2A) = \dfrac{\sin(2A)}{\cos(2A)}$, if $\sin(2A)$ and $\cos(2A)$ are known. Here we write the double-angle identity for the tangent function in terms of only the tangent function.

Write the tangent of a sum identity.	$\tan(A + B) = \dfrac{\tan A + \tan B}{1 - \tan A \tan B}$
Let $B = A$.	$\tan(A + A) = \dfrac{\tan A + \tan A}{1 - \tan A \tan A}$
Simplify.	$\boxed{\tan(2A) = \dfrac{2\tan A}{1 - \tan^2 A}}$

DOUBLE-ANGLE IDENTITIES

SINE	COSINE	TANGENT
$\sin(2A) = 2\sin A \cos A$	$\cos(2A) = \cos^2 A - \sin^2 A$	$\tan(2A) = \dfrac{2\tan A}{1 - \tan^2 A}$
	$\cos(2A) = 1 - 2\sin^2 A$	
	$\cos(2A) = 2\cos^2 A - 1$	

Classroom Example 7.4.1

If $\sin x = \frac{1}{5}$ and $\cos x < 0$, find $\sin(2x)$.

Answer: $-\dfrac{4\sqrt{6}}{25}$

Applying Double-Angle Identities

EXAMPLE 1 **Finding Exact Values of Trigonometric Functions Using Double-Angle Identities**

If $\cos x = \frac{2}{3}$, find $\sin(2x)$ given $\sin x < 0$.

Solution:

Find $\sin x$.

Use the Pythagorean identity.	$\sin^2 x + \cos^2 x = 1$
Substitute $\cos x = \frac{2}{3}$.	$\sin^2 x + \left(\frac{2}{3}\right)^2 = 1$
Solve for $\sin x$, which is negative.	$\sin x = -\sqrt{1 - \frac{4}{9}}$
Simplify.	$\sin x = -\sqrt{\frac{5}{9}} = -\frac{\sqrt{5}}{3}$

Find $\sin 2x$.

Use the double-angle formula for the sine function.	$\sin(2x) = 2\sin x \cos x$
Substitute $\sin x = -\frac{\sqrt{5}}{3}$ and $\cos x = \frac{2}{3}$.	$\sin(2x) = 2\left(-\frac{\sqrt{5}}{3}\right)\left(\frac{2}{3}\right)$
Simplify.	$\boxed{\sin(2x) = -\frac{4\sqrt{5}}{9}}$

■ **YOUR TURN** If $\cos x = -\frac{1}{3}$, find $\sin(2x)$ given $\sin x < 0$.

Technology Tip

If $\cos x = \frac{2}{3}$ and $\sin x < 0$, then x is in QIV and one value for x is $x = 2\pi - \cos^{-1}\left(\frac{2}{3}\right)$ (more explanation can be found in Section 7.7). Now use the graphing calculator to find $\sin(2x)$, entering $\sin\left(2\left[2\pi - \cos^{-1}\left(\frac{2}{3}\right)\right]\right)$, and compare that value to $-\frac{4\sqrt{5}}{9}$.

```
sin(2(2π-cos⁻¹(2/
3)))
          -.99380799
-4√(5)/9
          -.99380799
```

■ **Answer:** $\sin(2x) = \frac{4\sqrt{2}}{9}$

EXAMPLE 2 **Finding Exact Values Using Double-Angle Identities**

If $\sin x = -\frac{4}{5}$ and $\cos x < 0$, find $\sin(2x)$, $\cos(2x)$, and $\tan(2x)$.

Solution:

Solve for $\cos x$.

Use the Pythagorean identity.	$\sin^2 x + \cos^2 x = 1$
Substitute $\sin x = -\frac{4}{5}$.	$\left(-\frac{4}{5}\right)^2 + \cos^2 x = 1$
Simplify.	$\cos^2 x = \frac{9}{25}$
Solve for $\cos x$, which is negative.	$\cos x = -\sqrt{\frac{9}{25}} = -\frac{3}{5}$

Find $\sin(2x)$.

Use the double-angle identity for the sine function.	$\sin(2x) = 2\sin x \cos x$
Substitute $\sin x = -\frac{4}{5}$ and $\cos x = -\frac{3}{5}$.	$\sin(2x) = 2\left(-\frac{4}{5}\right)\left(-\frac{3}{5}\right)$
Simplify.	$\boxed{\sin(2x) = \frac{24}{25}}$

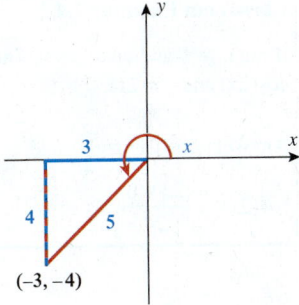

Technology Tip

If $\sin x = -\frac{4}{5}$ and $\cos x < 0$, then x is in QIII and a value for x is $x = \pi + \sin^{-1}\left(\frac{4}{5}\right)$ (more explanation can be found in Section 7.7). Now use the graphing calculator to find $\sin(2x)$, $\cos(2x)$, and $\tan(2x)$.

```
sin(2(π+sin⁻¹(4/5
)))
               .96
Ans▸Frac
             24/25
```

```
cos(2(π+sin⁻¹(4/5
)))
              -.28
Ans▸Frac
             -7/25
```

```
tan(2(π+sin⁻¹(4/5
)))
      -3.428571429
Ans▸Frac
             -24/7
```

■ **Answer:** $\sin(2x) = -\frac{24}{25}$,
$\cos(2x) = -\frac{7}{25}$, $\tan(2x) = \frac{24}{7}$

Classroom Example 7.4.2

If $\sin x = \frac{1}{5}$ and $\cos x > 0$, find $\cos(2x)$ and $\cot(2x)$.

Answer: $\cos(2x) = \dfrac{23}{25}$ and

$\cot(2x) = \dfrac{23\sqrt{6}}{24}$

Find $\cos(2x)$.

Use the double-angle identity for the cosine function. $\cos(2x) = \cos^2 x - \sin^2 x$

Substitute $\sin x = -\frac{4}{5}$ and $\cos x = -\frac{3}{5}$. $\qquad \cos(2x) = \left(-\dfrac{3}{5}\right)^2 - \left(-\dfrac{4}{5}\right)^2$

Simplify. $\qquad \boxed{\cos(2x) = -\dfrac{7}{25}}$

Find $\tan(2x)$.

Use the quotient identity. $\qquad \tan\theta = \dfrac{\sin\theta}{\cos\theta}$

Let $\theta = 2x$. $\qquad \tan(2x) = \dfrac{\sin(2x)}{\cos(2x)}$

Substitute $\sin(2x) = \frac{24}{25}$ and $\cos(2x) = -\frac{7}{25}$. $\qquad \tan(2x) = \dfrac{\frac{24}{25}}{-\frac{7}{25}}$

Simplify. $\qquad \boxed{\tan(2x) = -\dfrac{24}{7}}$

Note: We could also have found $\tan(2x)$ first by finding $\tan x = \dfrac{\sin x}{\cos x}$ and then using the

value for $\tan x$ in the double-angle identity, $\tan(2A) = \dfrac{2\tan A}{1 - \tan^2 A}$.

■ **YOUR TURN** If $\cos x = \frac{3}{5}$ and $\sin x < 0$, find $\sin(2x)$, $\cos(2x)$, and $\tan(2x)$.

EXAMPLE 3 Verifying Trigonometric Identities Using Double-Angle Identities

Verify the identity $(\sin x - \cos x)^2 = 1 - \sin(2x)$.

Solution:

Start with the left side of the equation. $\qquad \boxed{(\sin x - \cos x)^2}$

Expand by squaring. $\qquad = \sin^2 x - 2\sin x\cos x + \cos^2 x$

Group the $\sin^2 x$ and $\cos^2 x$ terms. $\qquad = \sin^2 x + \cos^2 x - 2\sin x\cos x$

Apply the Pythagorean identity. $\qquad = \underbrace{\sin^2 x + \cos^2 x}_{1} - 2\sin x\cos x$

Apply the sine double-angle identity. $\qquad = 1 - \underbrace{2\sin x\cos x}_{\sin(2x)}$

Simplify. $\qquad \boxed{= 1 - \sin(2x)}$

EXAMPLE 4 **Verifying Multiple-Angle Identities, Using Double-Angle Identities**

Verify the identity $\cos(3x) = (1 - 4\sin^2 x)\cos x$.

Solution:

Write the cosine of a sum identity. $\qquad \cos(A + B) = \cos A \cos B - \sin A \sin B$

Let $A = 2x$ and $B = x$. $\qquad \cos(2x + x) = \cos(2x) \cos x - \sin(2x) \sin x$

Apply the double-angle identities. $\qquad \cos(3x) = \underbrace{\cos(2x)}_{1-2\sin^2 x} \cos x - \underbrace{\sin(2x)}_{2\sin x \cos x} \sin x$

Simplify. $\qquad \cos(3x) = \cos x - 2\sin^2 x \cos x - 2\sin^2 x \cos x$

$\cos(3x) = \cos x - 4\sin^2 x \cos x$

Factor out the common cosine term. $\boxed{\cos(3x) = (1 - 4\sin^2 x)\cos x}$

EXAMPLE 5 **Simplifying Trigonometric Expressions Using Double-Angle Identities**

Graph $y = \dfrac{\cot x - \tan x}{\cot x + \tan x}$.

Solution:

Simplify $y = \dfrac{\cot x - \tan x}{\cot x + \tan x}$ first.

Write the cotangent and tangent functions in terms of the sine and cosine functions.

$$y = \frac{\dfrac{\cos x}{\sin x} - \dfrac{\sin x}{\cos x}}{\dfrac{\cos x}{\sin x} + \dfrac{\sin x}{\cos x}}$$

Multiply the numerator and the denominator by $\sin x \cos x$.

$$y = \left(\frac{\dfrac{\cos x}{\sin x} - \dfrac{\sin x}{\cos x}}{\dfrac{\cos x}{\sin x} + \dfrac{\sin x}{\cos x}}\right)\left(\frac{\sin x \cos x}{\sin x \cos x}\right)$$

Simplify.

$$y = \frac{\cos^2 x - \sin^2 x}{\cos^2 x + \sin^2 x}$$

Use the double-angle and Pythagorean identities.

$$y = \frac{\overbrace{\cos^2 x - \sin^2 x}^{\cos(2x)}}{\underbrace{\cos^2 x + \sin^2 x}_{1}}$$

$$\boxed{y = \cos(2x)}$$

Graph $y = \cos(2x)$.

Technology Tip

Graphs of $y = \dfrac{\cot x - \tan x}{\cot x + \tan x}$ and $y = \cos(2x)$.

SECTION
7.4 SUMMARY

In this section we derived the double-angle identities from the sum identities. We then used the double-angle identities to find exact values of trigonometric functions, to verify other trigonometric identities, and to simplify trigonometric expressions. There is no need to memorize the second and third forms of the cosine double-angle identity since they can be derived from the first using the Pythagorean identity.

$$\sin(2A) = 2\sin A\cos A$$

$$\cos(2A) = \cos^2 A - \sin^2 A$$
$$= 1 - 2\sin^2 A$$
$$= 2\cos^2 A - 1$$

$$\tan(2A) = \frac{2\tan A}{1 - \tan^2 A}$$

SECTION
7.4 EXERCISES

■ SKILLS

In Exercises 1–12, use the double-angle identities to answer the following questions.

1. If $\sin x = \dfrac{1}{\sqrt 5}$ and $\cos x < 0$, find $\sin(2x)$.

2. If $\sin x = \dfrac{1}{\sqrt 5}$ and $\cos x < 0$, find $\cos(2x)$.

3. If $\cos x = \dfrac{5}{13}$ and $\sin x < 0$, find $\tan(2x)$.

4. If $\cos x = -\dfrac{5}{13}$ and $\sin x < 0$, find $\tan(2x)$.

5. If $\tan x = \dfrac{12}{5}$ and $\pi < x < \dfrac{3\pi}{2}$, find $\sin(2x)$.

6. If $\tan x = \dfrac{12}{5}$ and $\pi < x < \dfrac{3\pi}{2}$, find $\cos(2x)$.

7. If $\sec x = \sqrt 5$ and $\sin x > 0$, find $\tan(2x)$.

8. If $\sec x = \sqrt 3$ and $\sin x < 0$, find $\tan(2x)$.

9. If $\csc x = -2\sqrt 5$ and $\cos x < 0$, find $\sin(2x)$.

10. If $\csc x = -\sqrt{13}$ and $\cos x > 0$, find $\sin(2x)$.

11. If $\cos x = -\dfrac{12}{13}$ and $\csc x < 0$, find $\cot(2x)$.

12. If $\sin x = \dfrac{12}{13}$ and $\cot x < 0$, find $\csc(2x)$.

In Exercises 13–24, simplify each expression. Evaluate the resulting expression exactly, if possible.

13. $\dfrac{2\tan 15°}{1 - \tan^2 15°}$

14. $\dfrac{2\tan\left(\dfrac{\pi}{8}\right)}{1 - \tan^2\left(\dfrac{\pi}{8}\right)}$

15. $\sin\left(\dfrac{\pi}{8}\right)\cos\left(\dfrac{\pi}{8}\right)$

16. $\sin 15° \cos 15°$

17. $\cos^2(2x) - \sin^2(2x)$

18. $\cos^2(x + 2) - \sin^2(x + 2)$

19. $\dfrac{2\tan\left(\dfrac{5\pi}{12}\right)}{1 - \tan^2\left(\dfrac{5\pi}{12}\right)}$

20. $\dfrac{2\tan\left(\dfrac{x}{2}\right)}{1 - \tan^2\left(\dfrac{x}{2}\right)}$

21. $1 - 2\sin^2\left(\dfrac{7\pi}{12}\right)$

22. $2\sin^2\left(-\dfrac{5\pi}{8}\right) - 1$

23. $2\cos^2\left(-\dfrac{7\pi}{12}\right) - 1$

24. $1 - 2\cos^2\left(-\dfrac{\pi}{8}\right)$

In Exercises 25–40, verify each identity.

25. $\csc(2A) = \frac{1}{2}\csc A \sec A$

26. $\cot(2A) = \frac{1}{2}(\cot A - \tan A)$

27. $(\sin x - \cos x)(\cos x + \sin x) = -\cos(2x)$

28. $(\sin x + \cos x)^2 = 1 + \sin(2x)$

29. $\cos^2 x = \dfrac{1 + \cos(2x)}{2}$

30. $\sin^2 x = \dfrac{1 - \cos(2x)}{2}$

31. $\cos^4 x - \sin^4 x = \cos(2x)$

32. $\cos^4 x + \sin^4 x = 1 - \frac{1}{2}\sin^2(2x)$

33. $8\sin^2 x \cos^2 x = 1 - \cos(4x)$

34. $[\cos(2x) - \sin(2x)][\sin(2x) + \cos(2x)] = \cos(4x)$

35. $-\frac{1}{2}\sec^2 x = -2\sin^2 x \csc^2(2x)$

36. $4\csc(4x) = \dfrac{\sec x \csc x}{\cos(2x)}$

37. $\sin(3x) = \sin x(4\cos^2 x - 1)$

38. $\tan(3x) = \dfrac{\tan x(3 - \tan^2 x)}{(1 - 3\tan^2 x)}$

39. $\frac{1}{2}\sin(4x) = 2\sin x \cos x - 4\sin^3 x \cos x$

40. $\cos(4x) = [\cos(2x) - \sin(2x)][(\cos(2x) + \sin(2x)]$

In Exercises 41–50, graph the functions.

41. $y = \dfrac{\sin(2x)}{1 - \cos(2x)}$

42. $y = \dfrac{2\tan x}{2 - \sec^2 x}$

43. $y = \dfrac{\cot x + \tan x}{\cot x - \tan x}$

44. $y = \frac{1}{2}\tan x \cot x \sec x \csc x$

45. $y = \sin(2x)\cos(2x)$

46. $y = 3\sin(3x)\cos(-3x)$

47. $y = 1 - \dfrac{\tan x \cot x}{\sec x \csc x}$

48. $y = 3 - 2\dfrac{\sec(2x)}{\csc(2x)}$

49. $y = \dfrac{\sin(2x)}{\cos x} - 3\cos(2x)$

50. $y = 2 + \dfrac{\sin(2x)}{\cos x} - 3\cos(2x)$

■ APPLICATIONS

51. Business/Economics. Annual cash flow of a stock fund (measured as a percentage of total assets) has fluctuated in cycles. The highs were roughly $+12\%$ of total assets and lows were roughly -8% of total assets. This cash flow can be modeled by the function

$$C(t) = 12 - 20\sin^2 t$$

Use a double-angle identity to express $C(t)$ in terms of the cosine function.

52. Business. Computer sales are generally subject to seasonal fluctuations. An analysis of the sales of a computer manufacturer during 2008–2010 is approximated by the function

$$s(t) = 0.098\cos^2 t + 0.387 \qquad 1 \le t \le 12$$

where t represents time in quarters ($t = 1$ represents the end of the first quarter of 2008), and $s(t)$ represents computer sales (quarterly revenue) in millions of dollars. Use a double-angle identity to express $s(t)$ in terms of the cosine function.

For Exercises 53 and 54, refer to the following:

An ore-crusher wheel consists of a heavy disk spinning on its axle. The normal (crushing) force F, in pounds, between the wheel and the inclined track is determined by

$$F = W\sin\theta + \frac{1}{2}\psi^2\left[\frac{C}{R}(1 - \cos 2\theta) + \frac{A}{l}\sin 2\theta\right],$$

where W is the weight of the wheel, θ is the angle of the axis, C and A are moments of inertia, R is the radius of the wheel, l is the distance from the wheel to the pin where the axle is attached, and ψ is the speed in rpm that the wheel is spinning. The optimum crushing force occurs when the angle θ is between $45°$ and $90°$.

53. Ore-Crusher Wheel. Find F if the angle is 60°, W is 500 lb, ψ is 200 rpm, $\dfrac{C}{R} = 750$, and $\dfrac{A}{l} = 3.75$.

54. Ore-Crusher Wheel. Find F if the angle is 75°, W is 500 lb, ψ is 200 rpm, $\dfrac{C}{R} = 750$, and $\dfrac{A}{l} = 3.75$.

55. With the information given in the diagram below, compute y.

56. With the information given in the diagram below, compute x.

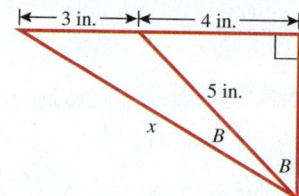

▪ CATCH THE MISTAKE

In Exercises 57 and 58, explain the mistake that is made.

57. If $\cos x = \frac{1}{3}$, find $\sin(2x)$ given $\sin x < 0$.

Solution:

Write the double-angle identity for the sine function. $\sin(2x) = 2\sin x \cos x$

Solve for $\sin x$ using the Pythagorean identity.

$$\sin^2 x + \left(\frac{1}{3}\right)^2 = 1$$

$$\sin x = \frac{2\sqrt{2}}{3}$$

Substitute $\cos x = \frac{1}{3}$ and $\sin x = \dfrac{2\sqrt{2}}{3}$.

$$\sin(2x) = 2\left(\frac{2\sqrt{2}}{3}\right)\left(\frac{1}{3}\right)$$

Simplify.

$$\sin(2x) = \frac{4\sqrt{2}}{9}$$

This is incorrect. What mistake was made?

58. If $\sin x = \frac{1}{3}$, find $\tan(2x)$ given $\cos x < 0$.

Solution:

Use the quotient identity.

$$\tan(2x) = \frac{\sin(2x)}{\cos x}$$

Use the double-angle formula for the sine function.

$$\tan(2x) = \frac{2\sin x \cos x}{\cos x}$$

Cancel the common cosine factors.

$$\tan(2x) = 2\sin x$$

Substitute $\sin x = \frac{1}{3}$.

$$\tan(2x) = \frac{2}{3}$$

This is incorrect. What mistake was made?

▪ CONCEPTUAL

For Exercises 59–67, determine whether each statement is true or false.

59. $\sin(2A) + \sin(2A) = \sin(4A)$

60. $\cos(4A) - \cos(2A) = \cos(2A)$

61. If $\tan x > 0$, then $\tan(2x) > 0$.

62. If $\sin x > 0$, then $\sin(2x) > 0$.

63. If $0 < \cos x < \sin x < 1$, then $\cos(2x) < 0$.

64. $\sin(2x) > 0$ if and only if the terminal side of x lies is quadrant I.

65. $\cos^2(3x) - \sin^2(3x) = \cos(6x)$

66. $1 - 2\sin^2\left(\dfrac{x}{2}\right) = -\cos x$

67. $y = \sin(2x)$ is an even function.

CHALLENGE

68. Express $\tan(4x)$ in terms of functions of $\tan x$.

69. Express $\tan(-4x)$ in terms of functions of $\tan x$.

70. Is the identity $2\csc(2x) = \dfrac{1 + \tan^2 x}{\tan x}$ true for $x = \dfrac{\pi}{2}$? Explain.

71. Is the identity $\tan(2x) = \dfrac{2\tan x}{1 - \tan^2 x}$ true for $x = \dfrac{\pi}{4}$? Explain.

TECHNOLOGY

One cannot *prove* that an equation is an identity using technology, but rather one uses it as a first step to see whether or not the equation *seems* to be an identity.

72. With a graphing calculator, plot $Y_1 = (2x) - \dfrac{(2x)^3}{3!} + \dfrac{(2x)^5}{5!}$ and $Y_2 = \sin(2x)$ for x range $[-1, 1]$. Is Y_1 a good approximation to Y_2?

73. With a graphing calculator, plot $Y_1 = 1 - \dfrac{(2x)^2}{2!} + \dfrac{(2x)^4}{4!}$ and $Y_2 = \cos(2x)$ for x range $[-1, 1]$. Is Y_1 a good approximation to Y_2?

74. With a graphing calculator, determine whether $\dfrac{\tan(4x) - \tan(3x)}{\tan x} = \dfrac{\csc(2x)}{1 - \sec(2x)}$ by plotting each side of the equation and seeing whether the graphs coincide.

75. With a graphing calculator, determine whether $\csc(2x)\sec(2x)[(\cos(2x) - \sin(2x)] = \dfrac{1 - 2\sin^2 x - 2\sin x \cos x}{2\sin x \cos x(\cos^2 x - \sin^2 x)}$ by plotting each side of the equation and seeing whether the graphs coincide.

76. With a graphing calculator, plot $Y_1 = \cos(2x)$, $Y_2 = 2\cos x$, and $Y_3 = 2\cos^2 x - 1$ in the same viewing rectangle $[-2\pi, 2\pi]$ by $[-2, 2]$. Which graphs are the same?

77. With a graphing calculator, plot $Y_1 = \sin(2x)$, $Y_2 = 2\sin x$, and $Y_3 = 2\sin x \cos x$ in the same viewing rectangle $[-2\pi, 2\pi]$ by $[-2, 2]$. Which graphs are the same?

SECTION
7.5 HALF-ANGLE IDENTITIES

SKILLS OBJECTIVES

- Use the half-angle identities to find certain exact values of trigonometric functions.
- Use the half-angle identities to verify other trigonometric identities.

CONCEPTUAL OBJECTIVE

- Understand that the half-angle identities are derived from the double-angle identities.

Applying Half-Angle Identities

We now use the *double-angle identities* from Section 7.4 to develop the *half-angle identities*. Like the double-angle identities, the half-angle identities will allow us to find certain exact values of trigonometric functions and to verify other trigonometric identities. The *half-angle identities* come directly from the double-angle identities. We start by rewriting the second and third forms of the cosine double-angle identity to obtain identities for the square of the sine and cosine functions, $\sin^2$ and $\cos^2$, otherwise known as the power reduction formulas.

WORDS	MATH

Power reduction formula for the sine function

Write the second form of the
cosine double-angle identity.　　　$\cos(2A) = 1 - 2\sin^2 A$

Isolate the $2\sin^2 A$ term on
one side of the equation.　　　$2\sin^2 A = 1 - \cos(2A)$

Divide both sides by 2.　　　$\sin^2 A = \dfrac{1 - \cos(2A)}{2}$

Power reduction formula for the cosine function

Write the third form of the
cosine double-angle identity.　　　$\cos(2A) = 2\cos^2 A - 1$

Isolate the $2\cos^2 A$ term on
one side of the equation.　　　$2\cos^2 A = 1 + \cos^2 A$

Divide both sides by 2.　　　$\cos^2 A = \dfrac{1 + \cos(2A)}{2}$

Power reduction formula for the tangent function

Taking the quotient of these
leads us to another identity.　　　$\tan^2 A = \dfrac{\sin^2 A}{\cos^2 A} = \dfrac{\dfrac{1 - \cos(2A)}{2}}{\dfrac{1 + \cos(2A)}{2}} = \dfrac{1 - \cos(2A)}{1 + \cos(2A)}$

These three identities for the squared functions—really, alternative forms of the double-angle identities—are used in calculus as **power reduction formulas** (identities that allow us to reduce the power of the trigonometric function from 2 to 1):

$$\sin^2 A = \frac{1 - \cos(2A)}{2} \qquad \cos^2 A = \frac{1 + \cos(2A)}{2} \qquad \tan^2 A = \frac{1 - \cos(2A)}{1 + \cos(2A)}$$

Derivation of the Half-Angle Identities

We can now use these forms of the double-angle identities to derive the *half-angle identities*.

WORDS	MATH

Sine half-angle identity

For the **sine half-angle identity**, start with
the double-angle formula involving both
the sine and cosine functions,
$\cos(2x) = 1 - 2\sin^2 x$, and solve for $\sin^2 x$.　　　$\sin^2 x = \dfrac{1 - \cos(2x)}{2}$

Solve for $\sin x$.　　　$\sin x = \pm\sqrt{\dfrac{1 - \cos(2x)}{2}}$

Let $x = \dfrac{A}{2}$.　　　$\sin\left(\dfrac{A}{2}\right) = \pm\sqrt{\dfrac{1 - \cos 2\left(\dfrac{A}{2}\right)}{2}}$

Simplify.　　　$\sin\left(\dfrac{A}{2}\right) = \pm\sqrt{\dfrac{1 - \cos A}{2}}$

Cosine half-angle identity

For the **cosine half-angle identity**, start with the double-angle formula involving only the cosine function, $\cos(2x) = 2\cos^2 x - 1$, and solve for $\cos^2 x$.

$$\cos^2 x = \frac{1 + \cos(2x)}{2}$$

Solve for $\cos x$.

$$\cos x = \pm \sqrt{\frac{1 + \cos(2x)}{2}}$$

Let $x = \dfrac{A}{2}$.

$$\cos\left(\frac{A}{2}\right) = \pm \sqrt{\frac{1 + \cos 2\left(\frac{A}{2}\right)}{2}}$$

Simplify.

$$\boxed{\cos\left(\frac{A}{2}\right) = \pm \sqrt{\frac{1 + \cos A}{2}}}$$

Tangent half-angle identity

For the **tangent half-angle identity**, start with the quotient identity.

$$\tan\left(\frac{A}{2}\right) = \frac{\sin\left(\frac{A}{2}\right)}{\cos\left(\frac{A}{2}\right)}$$

Substitute half-angle identities for the sine and cosine functions.

$$\tan\left(\frac{A}{2}\right) = \frac{\pm\sqrt{\dfrac{1 - \cos A}{2}}}{\pm\sqrt{\dfrac{1 + \cos A}{2}}}$$

Simplify.

$$\boxed{\tan\left(\frac{A}{2}\right) = \pm \sqrt{\frac{1 - \cos A}{1 + \cos A}}}$$

Note: We can also find $\tan\left(\dfrac{A}{2}\right)$ by starting with the identity $\tan^2 x = \dfrac{1 - \cos(2x)}{1 + \cos(2x)}$, solving for $\tan x$, and letting $x = \dfrac{A}{2}$. The tangent function also has two other similar forms for $\tan\dfrac{A}{2}$ (see Exercises 69 and 70).

HALF-ANGLE IDENTITIES

SINE	COSINE	TANGENT
$\sin\left(\dfrac{A}{2}\right) = \pm\sqrt{\dfrac{1 - \cos A}{2}}$	$\cos\left(\dfrac{A}{2}\right) = \pm\sqrt{\dfrac{1 + \cos A}{2}}$	$\tan\left(\dfrac{A}{2}\right) = \pm\sqrt{\dfrac{1 - \cos A}{1 + \cos A}}$
		$\tan\left(\dfrac{A}{2}\right) = \dfrac{\sin A}{1 + \cos A}$
		$\tan\left(\dfrac{A}{2}\right) = \dfrac{1 - \cos A}{\sin A}$

Study Tip

The sign + or − is determined by what quadrant contains $\dfrac{A}{2}$ and what the sign of the particular trigonometric function is in that quadrant.

Applying the Half-Angle Identities

It is important to note that these identities hold for any real number A or any angle with either degree measure or radian measure A as long as both sides of the equation are defined. The sign $(+$ or $-)$ is determined by the sign of the trigonometric function in the quadrant which contains $\frac{A}{2}$.

EXAMPLE 1 Finding Exact Values Using Half-Angle Identities

Use a half-angle identity to find $\cos 15°$.

Solution:

Write $\cos 15°$ in terms of a half-angle.

$$\cos 15° = \cos\left(\frac{30°}{2}\right)$$

Write the half-angle identity for the cosine function.

$$\cos\left(\frac{A}{2}\right) = \pm\sqrt{\frac{1 + \cos A}{2}}$$

Substitute $A = 30°$.

$$\cos\left(\frac{30°}{2}\right) = \pm\sqrt{\frac{1 + \cos 30°}{2}}$$

Simplify.

$$\cos 15° = \pm\sqrt{\frac{1 + \dfrac{\sqrt{3}}{2}}{2}}$$

$15°$ is in quadrant I, where the cosine function is positive.

$$\cos 15° = \sqrt{\frac{2 + \sqrt{3}}{4}} = \boxed{\frac{\sqrt{2 + \sqrt{3}}}{2}}$$

■ **YOUR TURN** Use a half-angle identity to find $\sin(22.5°)$.

> **Classroom Example 7.5.1**
>
> Compute exactly $\sec\left(-\dfrac{5\pi}{8}\right)$.
>
> **Answer:** $-\dfrac{2}{\sqrt{2 - \sqrt{2}}}$

■ **Answer:** $\sin(22.5°) = \dfrac{\sqrt{2 - \sqrt{2}}}{2}$

Look back at Example 1 in Section 7.3 and compare the result we found when evaluating $\cos 15°$ using the cosine of a difference formula, $\dfrac{\sqrt{6} + \sqrt{2}}{4}$, with the result in Example 1 above, $\dfrac{\sqrt{2 + \sqrt{3}}}{2}$. A calculator evaluates both expressions as 0.965925826.

EXAMPLE 2 Finding Exact Values Using Half-Angle Identities

Use a half-angle identity to find $\tan\left(\dfrac{11\pi}{12}\right)$.

Solution:

Write $\tan\left(\dfrac{11\pi}{12}\right)$ in terms of a half-angle.

$$\tan\left(\frac{11\pi}{12}\right) = \tan\left(\frac{\frac{11\pi}{6}}{2}\right)$$

Write the half-angle identity for the tangent function.*

$$\tan\left(\frac{A}{2}\right) = \frac{1 - \cos A}{\sin A}$$

Substitute $A = \dfrac{11\pi}{6}$.

$$\tan\left(\dfrac{\dfrac{11\pi}{6}}{2}\right) = \dfrac{1 - \cos\left(\dfrac{11\pi}{6}\right)}{\sin\left(\dfrac{11\pi}{6}\right)}$$

Simplify.

$$\tan\left(\dfrac{11\pi}{12}\right) = \dfrac{1 - \dfrac{\sqrt{3}}{2}}{-\dfrac{1}{2}}$$

$$\tan\left(\dfrac{11\pi}{12}\right) = \boxed{\sqrt{3} - 2}$$

Use a TI calculator to check the values of $\tan\left(\dfrac{11\pi}{12}\right)$ and $\sqrt{3} - 2$. Be sure the calculator is in radian mode.

```
tan(11π/12)
         -.2679491924
√(3)-2
         -.2679491924
■
```

$\dfrac{11\pi}{12}$ is in quadrant II, where the tangent function is negative. Notice that if we approximate $\tan\left(\dfrac{11\pi}{12}\right)$ with a calculator, we find that $\tan\left(\dfrac{11\pi}{12}\right) \approx -0.2679$ and $\sqrt{3} - 2 \approx -0.2679$.

*This form of the tangent half-angle identity was selected because of mathematical simplicity. If we had selected either of the other forms, we would have obtained an expression that had a square root of a square root or a radical in the denominator (requiring rationalization).

■ **YOUR TURN** Use a half-angle identity to find $\tan\left(\dfrac{\pi}{8}\right)$.

Classroom Example 7.5.2
Compute exactly $\tan 105°$.

Answer: $-2 - \sqrt{3}$

■ **Answer:** $\dfrac{2 - \sqrt{2}}{\sqrt{2}}$ or $\sqrt{2} - 1$

EXAMPLE 3 Finding Exact Values Using Half-Angle Identities

If $\cos x = \dfrac{3}{5}$ and $\dfrac{3\pi}{2} < x < 2\pi$, find $\sin\left(\dfrac{x}{2}\right)$, $\cos\left(\dfrac{x}{2}\right)$, and $\tan\left(\dfrac{x}{2}\right)$.

Solution:

Determine in which quadrant $\dfrac{x}{2}$ lies.

Since $\dfrac{3\pi}{2} < x < 2\pi$, we divide by 2. $\dfrac{3\pi}{4} < \dfrac{x}{2} < \pi$

$\dfrac{x}{2}$ lies in quadrant II; therefore, the sine function is positive and both the cosine and tangent functions are negative.

Write the half-angle identity for the sine function. $\sin\left(\dfrac{x}{2}\right) = \pm\sqrt{\dfrac{1 - \cos x}{2}}$

Substitute $\cos x = \tfrac{3}{5}$. $\sin\left(\dfrac{x}{2}\right) = \pm\sqrt{\dfrac{1 - \dfrac{3}{5}}{2}}$

If $\cos x = \dfrac{3}{5}$ and $\dfrac{3\pi}{2} < x < 2\pi$, then x is in QIV, $\dfrac{x}{2}$ is in QII, and $x = 2\pi - \cos^{-1}\left(\tfrac{3}{5}\right)$ (more explanation can be found in Section 7.7). Now use the graphing calculator to find $\sin\dfrac{x}{2}$, $\cos\dfrac{x}{2}$, and $\tan\dfrac{x}{2}$.

```
sin((2π-cos⁻¹(3/5
))/2)
          .4472135955
√(5)/5
          .4472135955
```

```
cos((2π-cos⁻¹(3/5
))/2)
         -.894427191
-2√(5)/5
         -.894427191
■
```

```
tan((2π-cos⁻¹(3/5
))/2)
         -.5
Ans▶Frac
         -1/2
```

Classroom Example 7.5.3

If $\cot x = \dfrac{24}{7}$ and $\pi < x < \dfrac{3\pi}{2}$, find $\cos\left(\dfrac{x}{2}\right)$.

Answer: $-\dfrac{\sqrt{2}}{10}$

■ **Answer:** $\sin\left(\dfrac{x}{2}\right) = \dfrac{2\sqrt{5}}{5}$,

$\cos\left(\dfrac{x}{2}\right) = -\dfrac{\sqrt{5}}{5}$, $\tan\left(\dfrac{x}{2}\right) = -2$

Simplify.

$$\sin\left(\frac{x}{2}\right) = \pm\sqrt{\frac{1}{5}} = \pm\frac{\sqrt{5}}{5}$$

Since $\dfrac{x}{2}$ lies in quadrant II, choose the positive value for the sine function.

$$\sin\left(\frac{x}{2}\right) = \frac{\sqrt{5}}{5}$$

Write the half-angle identity for the cosine function.

$$\cos\left(\frac{x}{2}\right) = \pm\sqrt{\frac{1 + \cos x}{2}}$$

Substitute $\cos x = \frac{3}{5}$.

$$\cos\left(\frac{x}{2}\right) = \pm\sqrt{\frac{1 + \dfrac{3}{5}}{2}}$$

Simplify.

$$\cos\left(\frac{x}{2}\right) = \pm\sqrt{\frac{4}{5}} = \pm\frac{2\sqrt{5}}{5}$$

Since $\dfrac{x}{2}$ lies in quadrant II, choose the negative value for the cosine function.

$$\cos\left(\frac{x}{2}\right) = -\frac{2\sqrt{5}}{5}$$

Use the quotient identity for tangent.

$$\tan\left(\frac{x}{2}\right) = \frac{\sin\left(\dfrac{x}{2}\right)}{\cos\left(\dfrac{x}{2}\right)}$$

Substitute $\sin\left(\dfrac{x}{2}\right) = \dfrac{\sqrt{5}}{5}$ and $\cos\left(\dfrac{x}{2}\right) = -\dfrac{2\sqrt{5}}{5}$.

$$\tan\left(\frac{x}{2}\right) = \frac{\dfrac{\sqrt{5}}{5}}{-\dfrac{2\sqrt{5}}{5}}$$

Simplify.

$$\tan\left(\frac{x}{2}\right) = -\frac{1}{2}$$

■ **YOUR TURN** If $\cos x = -\dfrac{3}{5}$ and $\pi < x < \dfrac{3\pi}{2}$, find $\sin\left(\dfrac{x}{2}\right)$, $\cos\left(\dfrac{x}{2}\right)$, and $\tan\left(\dfrac{x}{2}\right)$.

Technology Tip

Graphs of $y_1 = \cos^2\left(\dfrac{x}{2}\right)$ and $y_2 = \dfrac{\tan x + \sin x}{2\tan x}$.

```
Plot1 Plot2 Plot3
\Y1◻(cos(X/2))²
\Y2=(tan(X)+sin(
X))/(2tan(X))
```

EXAMPLE 4   Using Half-Angle Identities to Verify Other Identities

Verify the identity $\cos^2\left(\dfrac{x}{2}\right) = \dfrac{\tan x + \sin x}{2\tan x}$.

Solution:

Write the cosine half-angle identity.

$$\cos\left(\frac{x}{2}\right) = \pm\sqrt{\frac{1 + \cos x}{2}}$$

Square both sides of the equation.

$$\cos^2\left(\frac{x}{2}\right) = \frac{1 + \cos x}{2}$$

Multiply the numerator and denominator on the right side by $\tan x$.

$$\cos^2\left(\frac{x}{2}\right) = \left(\frac{1 + \cos x}{2}\right)\left(\frac{\tan x}{\tan x}\right)$$

Simplify.

$$\cos^2\left(\frac{x}{2}\right) = \frac{\tan x + \cos x \tan x}{2\tan x}$$

Note that $\cos x \tan x = \sin x$.

$$\cos^2\left(\frac{x}{2}\right) = \frac{\tan x + \sin x}{2\tan x}$$

An alternative solution is to start with the right-hand side.

Solution (alternative):

Start with the right-hand side.

$$\frac{\tan x + \sin x}{2\tan x}$$

Write this expression as the sum of two expressions.

$$= \frac{\tan x}{2\tan x} + \frac{\sin x}{2\tan x}$$

Simplify.

$$= \frac{1}{2} + \frac{1}{2}\frac{\sin x}{\tan x}$$

Write $\tan x = \frac{\sin x}{\cos x}$.

$$= \frac{1}{2} + \frac{1}{2}\frac{\sin x}{\frac{\sin x}{\cos x}}$$

Simplify.

$$= \frac{1}{2}(1 + \cos x)$$

$$= \cos^2\left(\frac{x}{2}\right)$$

Classroom Example 7.5.4

Verify that

$$\frac{\cot\left(\frac{A}{2}\right)\csc\left(-\frac{A}{2}\right)\tan\left(-\frac{A}{2}\right)}{\cos\left(\frac{A}{2}\right)}$$

$$= 2\csc A.$$

Answer:
Use basic identities to rewrite all expressions in terms of $\sin\left(\frac{A}{2}\right)$ and $\cos\left(\frac{A}{2}\right)$. Then use the double-angle identity.

EXAMPLE 5   Using Half-Angle Identities to Verify Other Trigonometric Identities

Verify the identity $\tan x = \csc(2x) - \cot(2x)$.

Solution:

Write the third half-angle formula for the tangent function.

$$\tan\left(\frac{A}{2}\right) = \frac{1 - \cos A}{\sin A}$$

Write the right side as a difference of two expressions having the same denominator.

$$\tan\left(\frac{A}{2}\right) = \frac{1}{\sin A} - \frac{\cos A}{\sin A}$$

Substitute the reciprocal and quotient identities, respectively, on the right.

$$\tan\left(\frac{A}{2}\right) = \csc A - \cot A$$

Let $A = 2x$.

$$\tan x = \csc(2x) - \cot(2x)$$

Classroom Example 7.5.6

Graph

$$y = -2 + \frac{1}{2}\sqrt{\frac{1 + \cos(5x)}{2}}$$

on $\left[0, \frac{\pi}{10}\right]$.

Answer:
First, reduce this to

$$y = -2 + \frac{1}{2}\cos\left(\frac{5x}{2}\right).$$

Notice in Example 5, we started with the third half-angle identity for the tangent function. In Example 6, we will start with the second half-angle identity for the tangent function. In general, you select the form that appears to lead to the desired expression.

EXAMPLE 6   Using Half-Angle Identities to Simplify Trigonometric Expressions

Graph $y = \frac{\sin(2\pi x)}{1 + \cos(2\pi x)}$ by first simplifying the trigonometric expression to a more recognizable form.

Solution:

Simplify the trigonometric expression using a half-angle identity for the tangent function.

Technology Tip

Graphs of $y_1 = \frac{\sin(2\pi x)}{1 + \cos(2\pi x)}$
and $y_2 = \tan(\pi x)$.

```
Plot1 Plot2 Plot3
\Y1⊟sin(2πX)/(1+
cos(2πX))
\Y2=tan(πX)
```

Write the second half-angle identity for the tangent function.

$$\tan\left(\frac{A}{2}\right) = \frac{\sin A}{1 + \cos A}$$

Let $A = 2\pi x$.

$$\tan(\pi x) = \frac{\sin(2\pi x)}{1 + \cos(2\pi x)}$$

Graph $y = \tan(\pi x)$.

$$y = \tan(\pi x) \qquad y = \frac{\sin(2\pi x)}{1 + \cos(2\pi x)}$$

■ **Answer:**

■ **YOUR TURN** Graph $y = \dfrac{1 - \cos(\pi x)}{\sin(\pi x)}$.

SECTION 7.5 SUMMARY

In this section we used the double-angle identities to derive the half-angle identities. We then used the half-angle identities to find certain exact values of trigonometric functions, verify other trigonometric identities, and simplify trigonometric expressions.

$$\sin\left(\frac{A}{2}\right) = \pm\sqrt{\frac{1 - \cos A}{2}} \qquad \cos\left(\frac{A}{2}\right) = \pm\sqrt{\frac{1 + \cos A}{2}}$$

$$\tan\left(\frac{A}{2}\right) = \pm\sqrt{\frac{1 - \cos A}{1 + \cos A}}$$

We determine the sign, + or −, by first deciding which quadrant contains $\frac{A}{2}$ and then finding the sign of the indicated trigonometric function in that quadrant.

Recall that there are three forms of the tangent half-angle identity. There is no need to memorize the other forms of the tangent half-angle identity, since they can be derived by first using the Pythagorean identity and algebraic manipulation.

SECTION 7.5 EXERCISES

■ **SKILLS**

In Exercises 1–16, use the half-angle identities to find the exact values of the trigonometric expressions.

1. $\sin 15°$

2. $\cos 22.5°$

3. $\cos\left(\dfrac{11\pi}{12}\right)$

4. $\sin\left(\dfrac{\pi}{8}\right)$

5. $\cos 75°$

6. $\sin 75°$

7. $\tan 67.5°$

8. $\tan 202.5°$

9. $\sec\left(-\dfrac{9\pi}{8}\right)$

10. $\csc\left(\dfrac{9\pi}{8}\right)$

11. $\cot\left(\dfrac{13\pi}{8}\right)$

12. $\cot\left(\dfrac{7\pi}{8}\right)$

13. $\sec\left(\dfrac{5\pi}{8}\right)$

14. $\csc\left(-\dfrac{5\pi}{8}\right)$

15. $\cot(-135°)$

16. $\cot 105°$

In Exercises 17–32, use the half-angle identities to find the desired function values.

17. If $\cos x = \dfrac{5}{13}$ and $\sin x < 0$, find $\sin\left(\dfrac{x}{2}\right)$.

18. If $\cos x = -\dfrac{5}{13}$ and $\sin x < 0$, find $\cos\left(\dfrac{x}{2}\right)$.

19. If $\tan x = \dfrac{12}{5}$ and $\pi < x < \dfrac{3\pi}{2}$, find $\sin\left(\dfrac{x}{2}\right)$.

20. If $\tan x = \dfrac{12}{5}$ and $\pi < x < \dfrac{3\pi}{2}$, find $\cos\left(\dfrac{x}{2}\right)$.

21. If $\sec x = \sqrt{5}$ and $\sin x > 0$, find $\tan\left(\dfrac{x}{2}\right)$.

22. If $\sec x = \sqrt{3}$ and $\sin x < 0$, find $\tan\left(\dfrac{x}{2}\right)$.

23. If $\csc x = 3$ and $\cos x < 0$, find $\sin\left(\dfrac{x}{2}\right)$.

24. If $\csc x = -3$ and $\cos x > 0$, find $\cos\left(\dfrac{x}{2}\right)$.

25. If $\cos x = -\dfrac{1}{4}$ and $\csc x < 0$, find $\cot\left(\dfrac{x}{2}\right)$.

26. If $\cos x = \dfrac{1}{4}$ and $\cot x < 0$, find $\csc\left(\dfrac{x}{2}\right)$.

27. If $\cot x = -\dfrac{24}{5}$ and $\dfrac{\pi}{2} < x < \pi$, find $\cos\left(\dfrac{x}{2}\right)$.

28. If $\cot x = -\dfrac{24}{5}$ and $\dfrac{\pi}{2} < x < \pi$, find $\sin\left(\dfrac{x}{2}\right)$.

29. If $\sin x = -0.3$ and $\sec x > 0$, find $\tan\left(\dfrac{x}{2}\right)$.

30. If $\sin x = -0.3$ and $\sec x < 0$, find $\cot\left(\dfrac{x}{2}\right)$.

31. If $\sec x = 2.5$ and $\tan x > 0$, find $\cot\left(\dfrac{x}{2}\right)$.

32. If $\sec x = -3$ and $\cot x < 0$, find $\tan\left(\dfrac{x}{2}\right)$.

In Exercises 33–38, simplify each expression using half-angle identities. Do not evaluate.

33. $\sqrt{\dfrac{1 + \cos\left(\dfrac{5\pi}{6}\right)}{2}}$

34. $\sqrt{\dfrac{1 - \cos\left(\dfrac{\pi}{4}\right)}{2}}$

35. $\dfrac{\sin 150°}{1 + \cos 150°}$

36. $\dfrac{1 - \cos 150°}{\sin 150°}$

37. $\sqrt{\dfrac{1 - \cos\left(\dfrac{5\pi}{4}\right)}{1 + \cos\left(\dfrac{5\pi}{4}\right)}}$

38. $\sqrt{\dfrac{1 - \cos 15°}{1 + \cos 15°}}$

In Exercises 39–50, verify the identities.

39. $\sin^2\left(\dfrac{x}{2}\right) + \cos^2\left(\dfrac{x}{2}\right) = 1$

40. $\cos^2\left(\dfrac{x}{2}\right) - \sin^2\left(\dfrac{x}{2}\right) = \cos x$

41. $\sin(-x) = -2\sin\left(\dfrac{x}{2}\right)\cos\left(\dfrac{x}{2}\right)$

42. $2\cos^2\left(\dfrac{x}{4}\right) = 1 + \cos\left(\dfrac{x}{2}\right)$

43. $\tan^2\left(\dfrac{x}{2}\right) = \dfrac{1 - \cos x}{1 + \cos x}$

44. $\tan^2\left(\dfrac{x}{2}\right) = (\csc x - \cot x)^2$

45. $\tan\left(\dfrac{A}{2}\right) + \cot\left(\dfrac{A}{2}\right) = 2\csc A$

46. $\cot\left(\dfrac{A}{2}\right) - \tan\left(\dfrac{A}{2}\right) = 2\cot A$

47. $\csc^2\left(\dfrac{A}{2}\right) = \dfrac{2(1 + \cos A)}{\sin^2 A}$

48. $\sec^2\left(\dfrac{A}{2}\right) = \dfrac{2(1 - \cos A)}{\sin^2 A}$

49. $\csc\left(\dfrac{A}{2}\right) = \pm|\csc A|\sqrt{2 + 2\cos A}$

50. $\sec\left(\dfrac{A}{2}\right) = \pm|\csc A|\sqrt{2 - 2\cos A}$

In Exercises 51–58, graph the functions.

51. $y = 4\cos^2\left(\dfrac{x}{2}\right)$

52. $y = -6\sin^2\left(\dfrac{x}{2}\right)$

53. $y = \dfrac{1 - \tan^2\left(\dfrac{x}{2}\right)}{1 + \tan^2\left(\dfrac{x}{2}\right)}$

54. $y = 1 - \left[\sin\left(\dfrac{x}{2}\right) + \cos\left(\dfrac{x}{2}\right)\right]^2$

55. $y = 4\sin^2\left(\dfrac{x}{2}\right) - 1$

56. $y = -\dfrac{1}{6}\cos^2\left(\dfrac{x}{2}\right) + 2$

57. $y = \sqrt{\dfrac{1 - \cos(2x)}{1 + \cos(2x)}}$ $0 \le x < \pi$

58. $y = \sqrt{\dfrac{1 + \cos(3x)}{2}} + 3$ $0 \le x \le \dfrac{\pi}{3}$

▪ APPLICATIONS

For Exercises 59 and 60, refer to the following:

Monthly profits can be expressed as a function of sales, that is, $p(s)$. A financial analysis of a company has determined that the sales s in thousands of dollars are also related to monthly profits p in thousands of dollars by the relationship:

$$\tan\theta = \frac{p}{s} \quad \text{for} \quad 0 \le s \le 50, 0 \le p < 40$$

Based on sales and profits, it can be determined that the domain for angle θ is $0 \le \theta \le 38°$.

59. Business. If monthly profits are $3000 and monthly sales are $4000, find $\tan\left(\dfrac{\theta}{2}\right)$.

60. Business. If monthly profits are p and monthly sales are s (where $p < s$), find $\tan\left(\dfrac{\theta}{2}\right)$.

61. Area of an Isosceles Triangle. Consider the triangle below, where the vertex angle measures θ, the equal sides measure a, the height is h, and half the base is b. (In an isosceles triangle, the perpendicular dropped from the vertex angle divides the triangle into two congruent triangles.) The two triangles formed are right triangles.

In the right triangles, $\sin\left(\dfrac{\theta}{2}\right) = \dfrac{b}{a}$ and $\cos\left(\dfrac{\theta}{2}\right) = \dfrac{h}{a}$. Multiply each side of each equation by a to get $b = a\sin\left(\dfrac{\theta}{2}\right)$, $h = a\cos\left(\dfrac{\theta}{2}\right)$. The area of the entire isosceles triangle is $A = \frac{1}{2}(2b)h = bh$. Substitute the values for b and h into the area formula. Show that the area is equivalent to $\left(\dfrac{a^2}{2}\right)\sin\theta$.

62. Area of an Isosceles Triangle. Use the results from Exercise 61 to find the area of an isosceles triangle whose equal sides measure 7 inches and whose base angles each measure 75°.

▪ CATCH THE MISTAKE

In Exercises 63 and 64, explain the mistake that is made.

63. If $\cos x = -\dfrac{1}{3}$, find $\sin\left(\dfrac{x}{2}\right)$ given $\pi < x < \dfrac{3\pi}{2}$.

Solution:

Write the half-angle identity for the sine function.
$$\sin\left(\frac{x}{2}\right) = \pm\sqrt{\frac{1 - \cos x}{2}}$$

Substitute $\cos x = -\frac{1}{3}$.
$$\sin\left(\frac{x}{2}\right) = \pm\sqrt{\frac{1 + \frac{1}{3}}{2}}$$

Simplify.
$$\sin\left(\frac{x}{2}\right) = \pm\sqrt{\frac{4}{3}} = \pm\frac{2}{\sqrt{3}}$$

The sine function is negative.
$$\sin\left(\frac{x}{2}\right) = -\frac{2}{\sqrt{3}}$$

This is incorrect. What mistake was made?

64. If $\cos x = \dfrac{1}{3}$, find $\tan^2\left(\dfrac{x}{2}\right)$.

Solution:

Use the quotient identity.
$$\tan^2\left(\frac{x}{2}\right) = \frac{\sin^2\left(\dfrac{x}{2}\right)}{\cos^2 x}$$

Use the half-angle identity for the sine function.
$$\tan^2\left(\frac{x}{2}\right) = \frac{\dfrac{1 - \cos x}{2}}{\cos^2 x}$$

Simplify.
$$\tan^2\left(\frac{x}{2}\right) = \frac{1}{2}\left(\frac{1}{\cos^2 x} - \frac{\cos x}{\cos^2 x}\right)$$
$$\tan^2\left(\frac{x}{2}\right) = \frac{1}{2}\left(\frac{1}{\cos^2 x} - \frac{1}{\cos x}\right)$$

Substitute $\cos x = \frac{1}{3}$.
$$\tan^2\left(\frac{x}{2}\right) = \frac{1}{2}\left(\frac{1}{\dfrac{1}{9}} - \frac{1}{\dfrac{1}{3}}\right)$$
$$\tan^2\left(\frac{x}{2}\right) = 3$$

This is incorrect. What mistake was made?

▪ CONCEPTUAL

In Exercises 65–68, determine whether each statement is true or false.

65. $\sin\left(\dfrac{A}{2}\right) + \sin\left(\dfrac{A}{2}\right) = \sin A$

66. $\cos\left(\dfrac{A}{2}\right) + \cos\left(\dfrac{A}{2}\right) = \cos A$

67. If $\tan x > 0$, then $\tan\left(\dfrac{x}{2}\right) > 0$.

68. If $\sin x > 0$, then $\sin\left(\dfrac{x}{2}\right) > 0$.

69. Given $\tan\left(\dfrac{A}{2}\right) = \pm\sqrt{\dfrac{1 - \cos A}{1 + \cos A}}$, verify $\tan\left(\dfrac{A}{2}\right) = \dfrac{\sin A}{1 + \cos A}$. Substitute $A = \pi$ into the identity and explain your results.

70. Given $\tan\left(\dfrac{A}{2}\right) = \pm\sqrt{\dfrac{1 - \cos A}{1 + \cos A}}$, verify $\tan\left(\dfrac{A}{2}\right) = \dfrac{1 - \cos A}{\sin A}$. Substitute $A = \pi$ into the identity and explain your results.

▪ CHALLENGE

71. Prove that $\cot\left(\dfrac{A}{4}\right) = \pm\sqrt{\dfrac{1 + \cos\left(\dfrac{A}{2}\right)}{1 - \cos\left(\dfrac{A}{2}\right)}}$.

72. Prove that $\cot\left(-\dfrac{A}{2}\right)\sec\left(\dfrac{A}{2}\right)\csc\left(-\dfrac{A}{2}\right)\tan\left(\dfrac{A}{2}\right) = 2\csc A$.

73. Find the values of x in the interval $[0, 2\pi]$ for which $\tan\left(\dfrac{x}{2}\right) > 0$.

74. Find the values of x in the interval $[0, 2\pi]$ for which $\cot\left(\dfrac{x}{2}\right) \le 0$.

▪ TECHNOLOGY

One cannot *prove* that an equation is an identity using technology, but rather one uses it as a first step to see whether or not the equation *seems* to be an identity.

75. With a graphing calculator, plot $Y_1 = \left(\dfrac{x}{2}\right) - \dfrac{\left(\dfrac{x}{2}\right)^3}{3!} + \dfrac{\left(\dfrac{x}{2}\right)^5}{5!}$ and $Y_2 = \sin\left(\dfrac{x}{2}\right)$ for x range $[-1, 1]$. Is Y_1 a good approximation to Y_2?

76. Using a graphing calculator, plot $Y_1 = 1 - \dfrac{\left(\dfrac{x}{2}\right)^2}{2!} + \dfrac{\left(\dfrac{x}{2}\right)^4}{4!}$ and $Y_2 = \cos\left(\dfrac{x}{2}\right)$ for x range $[-1, 1]$. Is Y_1 a good approximation to Y_2?

77. With a graphing calculator, determine whether $\csc^2\left(\dfrac{x}{2}\right) + \sec^2\left(\dfrac{x}{2}\right) = 4\csc^2 x$ is an identity by plotting each side of the equation and seeing whether the graphs coincide.

78. With a graphing calculator, determine whether $\tan^2\left(\dfrac{x}{2}\right) + \cot^2\left(\dfrac{x}{2}\right) = -2\cot^2 x \sec x$ is an identity by plotting each side of the equation and seeing whether the graphs coincide.

79. With a graphing calculator, plot $Y_1 = \cos\left(\dfrac{x}{2}\right)$, $Y_2 = \dfrac{1}{2}\cos x$, and $Y_3 = \sqrt{\dfrac{1 + \cos x}{2}}$ in the same viewing rectangle $[-\pi, \pi]$ by $[-1, 1]$. Which graphs are the same?

80. With a graphing calculator, plot $Y_1 = \sin\left(\dfrac{x}{2}\right)$, $Y_2 = \dfrac{1}{2}\sin x$, and $Y_3 = \sqrt{\dfrac{1 + \cos x}{2}}$ in the same viewing rectangle $[0, 2\pi]$ by $[-1, 1]$. Which graphs are the same?

SKILLS OBJECTIVES

- Express products of trigonometric functions as sums of trigonometric functions.
- Express sums of trigonometric functions as products of trigonometric functions.

CONCEPTUAL OBJECTIVES

- Understand that the sum and difference identities are used to derive product-to-sum identities.
- Understand that the product-to-sum identities are used to derive the sum-to-product identities.

In calculus, often it is helpful to write products of trigonometric functions as sums of other trigonometric functions, and vice versa. In this section, we discuss the *product-to-sum identities*, which convert products of trigonometric functions to sums of trigonometric functions, and *sum-to-product identities*, which convert sums of trigonometric functions to products of trigonometric functions.

Product-to-Sum Identities

The *product-to-sum identities* are derived from the sum and difference identities.

WORDS	**MATH**
Write the identity for the cosine of a sum.	$\cos A \cos B - \sin A \sin B = \cos(A + B)$
Write the identity for the cosine of a difference.	$\cos A \cos B + \sin A \sin B = \cos(A - B)$
Add the two identities.	$2 \cos A \cos B = \cos(A + B) + \cos(A - B)$
Divide both sides by 2.	$\boxed{\cos A \cos B = \dfrac{1}{2}[\cos(A + B) + \cos(A - B)]}$

$$\cos A \cos B + \sin A \sin B = \cos(A - B)$$

WORDS	**MATH**
Subtract the sum identity from the difference identity.	$-\cos A \cos B + \sin A \sin B = -\cos(A + B)$
	$2 \sin A \sin B = \cos(A - B) - \cos(A + B)$
Divide both sides by 2.	$\boxed{\sin A \sin B = \dfrac{1}{2}[\cos(A - B) - \cos(A + B)]}$
Write the identity for the sine of a sum.	$\sin A \cos B + \cos A \sin B = \sin(A + B)$
Write the identity for the sine of a difference.	$\sin A \cos B - \cos A \sin B = \sin(A - B)$
Add the two identities.	$2 \sin A \cos B = \sin(A + B) + \sin(A - B)$
Divide both sides by 2.	$\boxed{\sin A \cos B = \dfrac{1}{2}[\sin(A + B) + \sin(A - B)]}$

PRODUCT-TO-SUM IDENTITIES

1. $\cos A \cos B = \frac{1}{2}[\cos(A + B) + \cos(A - B)]$

2. $\sin A \sin B = \frac{1}{2}[\cos(A - B) - \cos(A + B)]$

3. $\sin A \cos B = \frac{1}{2}[\sin(A + B) + \sin(A - B)]$

EXAMPLE 1   Illustrating a Product-to-Sum Identity for Specific Values

Show that product-to-sum identity (3) is true when $A = 30°$ and $B = 90°$.

Solution:

Write product-to-sum identity (3).

$$\sin A \cos B = \frac{1}{2}[\sin(A + B) + \sin(A - B)]$$

Let $A = 30°$ and $B = 90°$.

$$\sin 30° \cos 90° = \frac{1}{2}[\sin(30° + 90°) + \sin(30° - 90°)]$$

Simplify.

$$\sin 30° \cos 90° = \frac{1}{2}[\sin 120° + \sin(-60°)]$$

Evaluate the
trigonometric functions.

$$\frac{1}{2} \cdot 0 = \frac{1}{2}\left[\frac{\sqrt{3}}{2} - \frac{\sqrt{3}}{2}\right]$$

Simplify.

$$0 = 0$$

EXAMPLE 2   Convert a Product to a Sum

Convert the product $\cos(4x) \cos(3x)$ to a sum.

Solution:

Write product-to-sum
identity (1).

$$\cos A \cos B = \frac{1}{2}[\cos(A + B) + \cos(A - B)]$$

Let $A = 4x$ and $B = 3x$.

$$\cos(4x) \cos(3x) = \frac{1}{2}[\cos(4x + 3x) + \cos(4x - 3x)]$$

Simplify.

$$\cos(4x) \cos(3x) = \frac{1}{2}[\cos(7x) + \cos(x)]$$

■ **YOUR TURN** Convert the product $\cos(2x) \cos(5x)$ to a sum.

Technology Tip

Graphs of $y_1 = \cos(4x)\cos(3x)$ and
$y_2 = \frac{1}{2}[\cos(7x) + \cos(x)]$.

■ **Answer:** $\frac{1}{2}[\cos(7x) + \cos(3x)]$

Classroom Example 7.6.2

Convert $\sin(\pi x)\cos\left(\dfrac{\pi}{4}x\right)$ to a sum.

Answer:

$\dfrac{1}{2}\left[\sin\left(\dfrac{5\pi}{4}x\right) + \sin\left(\dfrac{3\pi}{4}x\right)\right]$

Classroom Example 7.6.3*

Express $\sin(n\pi x)\sin(m\pi x)$ in terms of cosine. Here, n and m are integers.

Answer:

$\dfrac{1}{2}\{\cos[(n-m)\pi x] - \cos[(n+m)\pi x]\}$

■**Answer:** $\dfrac{1}{2}[\cos x - \cos(3x)]$

EXAMPLE 3 Converting Products to Sums

Express $\sin(2x)\sin(3x)$ in terms of cosines.

Solution:

Write product-to-sum identity (2). $\qquad \sin A \sin B = \dfrac{1}{2}[\cos(A - B) - \cos(A + B)]$

Let $A = 2x$ and $B = 3x$. $\qquad \sin(2x)\sin(3x) = \dfrac{1}{2}[\cos(2x - 3x) - \cos(2x + 3x)]$

Simplify. $\qquad \sin(2x)\sin(3x) = \dfrac{1}{2}[\cos(-x) - \cos(5x)]$

The cosine function is an even function. $\qquad \sin(2x)\sin(3x) = \dfrac{1}{2}[\cos x - \cos(5x)]$

■**YOUR TURN** Express $\sin(x)\sin(2x)$ in terms of cosines.

Sum-to-Product Identities

The *sum-to-product identities* can be obtained from the product-to-sum identities.

WORDS	MATH
Write the identity for the product of the sine and cosine functions.	$\dfrac{1}{2}[\sin(x + y) + \sin(x - y)] = \sin x \cos y$
Let $x + y = A$ and $x - y = B$; then $x = \dfrac{A + B}{2}$ and $y = \dfrac{A - B}{2}$.	
Substitute these values into the identity.	$\dfrac{1}{2}[\sin A + \sin B] = \sin\left(\dfrac{A + B}{2}\right)\cos\left(\dfrac{A - B}{2}\right)$
Multiply by 2.	$\sin A + \sin B = 2\sin\left(\dfrac{A + B}{2}\right)\cos\left(\dfrac{A - B}{2}\right)$

The other three *sum-to-product* identities can be found similarly. All are summarized in the box below.

SUM-TO-PRODUCT IDENTITIES

4. $\sin A + \sin B = 2 \sin\left(\dfrac{A + B}{2}\right) \cos\left(\dfrac{A - B}{2}\right)$

5. $\sin A - \sin B = 2 \sin\left(\dfrac{A - B}{2}\right) \cos\left(\dfrac{A + B}{2}\right)$

6. $\cos A + \cos B = 2 \cos\left(\dfrac{A + B}{2}\right) \cos\left(\dfrac{A - B}{2}\right)$

7. $\cos A - \cos B = -2 \sin\left(\dfrac{A + B}{2}\right) \sin\left(\dfrac{A - B}{2}\right)$

EXAMPLE 4   Illustrating a Sum-to-Product Identity for Specific Values

Show that sum-to-product identity (7) is true when $A = 30°$ and $B = 90°$.

Solution:

Write the sum-to-product identity (7).

$$\cos A - \cos B = -2 \sin\left(\dfrac{A + B}{2}\right) \sin\left(\dfrac{A - B}{2}\right)$$

Let $A = 30°$ and $B = 90°$.

$$\cos 30° - \cos 90° = -2 \sin\left(\dfrac{30° + 90°}{2}\right) \sin\left(\dfrac{30° - 90°}{2}\right)$$

Simplify.

$$\cos 30° - \cos 90° = -2 \sin 60° \sin(-30°)$$

The sine function is an odd function.

$$\cos 30° - \cos 90° = 2 \sin 60° \sin 30°$$

Evaluate the trigonometric functions.

$$\dfrac{\sqrt{3}}{2} - 0 = 2\left(\dfrac{\sqrt{3}}{2}\right)\left(\dfrac{1}{2}\right)$$

Simplify.

$$\dfrac{\sqrt{3}}{2} = \dfrac{\sqrt{3}}{2}$$

EXAMPLE 5   Convert a Sum to a Product

Convert $-9[\sin(2x) - \sin(10x)]$, a trigonometric expression containing a sum, to a product.

Solution:

The expression inside the brackets is in the form of identity (5).

$$\sin A - \sin B = 2 \sin\left(\dfrac{A - B}{2}\right) \cos\left(\dfrac{A + B}{2}\right)$$

Let $A = 2x$ and $B = 10x$.

$$\sin(2x) - \sin(10x) = 2 \sin\left(\dfrac{2x - 10x}{2}\right) \cos\left(\dfrac{2x + 10x}{2}\right)$$

Simplify.

$$\sin(2x) - \sin(10x) = 2 \sin(-4x) \cos(6x)$$

The sine function is an odd function.

$$\sin(2x) - \sin(10x) = -2 \sin(4x) \cos(6x)$$

Multiply both sides by -9.

$$-9[\sin(2x) - \sin(10x)] = 18 \sin(4x) \cos(6x)$$

Classroom Example 7.6.5
Convert the sum,
$6[\cos(\sqrt{5}x) + \cos(3\sqrt{5}x)]$, to a product.

Answer:
$12 \cos(2\sqrt{5}x) \cos(\sqrt{5}x)$

Graphs of
$y_1 = -9[\sin(2x) - \sin(10x)]$ and
$y_2 = 18\sin(4x)\cos(6x)$.

EXAMPLE 6   Simplifying a Trigonometric Expression Using Sum-to-Product Identities

Simplify the expression $\sin\left(\dfrac{x+y}{2}\right)\cos\left(\dfrac{x-y}{2}\right) + \sin\left(\dfrac{x-y}{2}\right)\cos\left(\dfrac{x+y}{2}\right)$.

Solution:

Use identities (4) and (5).

$$\underbrace{\sin\left(\frac{x+y}{2}\right)\cos\left(\frac{x-y}{2}\right)}_{\frac{1}{2}[\sin x + \sin y]} + \underbrace{\sin\left(\frac{x-y}{2}\right)\cos\left(\frac{x+y}{2}\right)}_{\frac{1}{2}[\sin x - \sin y]}$$

$$= \frac{1}{2}\sin x + \frac{1}{2}\sin y + \frac{1}{2}\sin x - \frac{1}{2}\sin y$$

Simplify.

$$= \boxed{\sin x}$$

Applications

In music, a tone is a fixed pitch (frequency) that is given a name. If two notes are sounded simultaneously, then they interfere and produce another tone, often called a "beat." The beat frequency is the difference of the two frequencies. The more rapid the beat, the further apart the two frequencies of the tones are. When musicians tune their instruments, they use a tuning fork to sound a tone and then tune the instrument until the beat is eliminated; then the tuning fork and instrument are in tune with each other. Mathematically, a tone is represented as $A\cos(2\pi ft)$, where A is the amplitude (loudness), f is the frequency in Hz, and t is time in seconds. The following table summarizes common tones and frequencies.

C	D	E	F	G	A	B
262 Hz	294 Hz	330 Hz	349 Hz	392 Hz	440 Hz	494 Hz

EXAMPLE 7 Music

Express the musical tone that is heard when a C and G are simultaneously struck (assume they have the same loudness).

Find the beat frequency, $f_2 - f_1$. Assume uniform loudness, $A = 1$.

Solution:

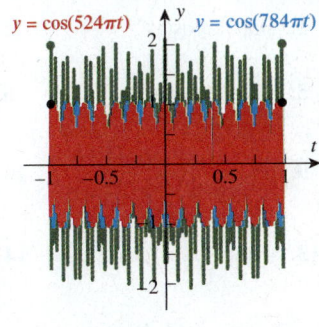

$y = \cos(524\pi t)$ $y = \cos(784\pi t)$

$y = 2\cos(654\pi t)\cos(130\pi t)$

Write the mathematical description of the C tone.	$\cos(2\pi f_1 t)$, $f_1 = 262$ Hz
Write the mathematical description of the G tone.	$\cos(2\pi f_2 t)$, $f_2 = 392$ Hz
Add the two notes.	$\cos(524\pi t) + \cos(784\pi t)$
Use sum-to-product identities.	$\cos(524\pi t) + \cos(784\pi t)$

$$= 2\cos\left(\frac{524\pi t + 784\pi t}{2}\right)\cos\left(\frac{524\pi t - 784\pi t}{2}\right)$$

Simplify.
$$= 2\cos(654\pi t)\cos(-130\pi t)$$
$$= 2\cos(2 \cdot 327 \cdot \pi \cdot t)\cos(130\pi t)$$

Identify the beat frequency. $f_2 - f_1 = 392 - 262 = 130$ Hz

Therefore, the tone of average frequency, 327 Hz, has a beat of 130 Hz (beats/sec).

Notice that an average frequency results, $\cos(2 \cdot \underbrace{327}_{\text{average frequency}} \cdot \pi \cdot t)$, and that frequency is modulated by a beat frequency, 130 Hz.

SECTION 7.6 SUMMARY

In this section, we used the sum and difference identities (Section 7.3) to derive the product-to-sum identities. The product-to-sum identities allowed us to express products as sums.

$$\cos A \cos B = \frac{1}{2}[\cos(A + B) + \cos(A - B)]$$

$$\sin A \sin B = \frac{1}{2}[\cos(A - B) - \cos(A + B)]$$

$$\sin A \cos B = \frac{1}{2}[\sin(A + B) + \sin(A - B)]$$

We then used the product-to-sum identities to derive the sum-to-product identities. The sum-to-product identities allow us to express sums as products.

$$\sin A + \sin B = 2\sin\left(\frac{A + B}{2}\right)\cos\left(\frac{A - B}{2}\right)$$

$$\sin A - \sin B = 2\sin\left(\frac{A - B}{2}\right)\cos\left(\frac{A + B}{2}\right)$$

$$\cos A + \cos B = 2\cos\left(\frac{A + B}{2}\right)\cos\left(\frac{A - B}{2}\right)$$

$$\cos A - \cos B = -2\sin\left(\frac{A + B}{2}\right)\sin\left(\frac{A - B}{2}\right)$$

▪ SKILLS

In Exercises 1–14, write each product as a sum or difference of sines and/or cosines.

1. $\sin(2x)\cos(x)$

2. $\cos(10x)\sin(5x)$

3. $5\sin(4x)\sin(6x)$

4. $-3\sin(2x)\sin(4x)$

5. $4\cos(-x)\cos(2x)$

6. $-8\cos(3x)\cos(5x)$

7. $\sin\left(\dfrac{3x}{2}\right)\sin\left(\dfrac{5x}{2}\right)$

8. $\sin\left(\dfrac{\pi x}{2}\right)\sin\left(\dfrac{5\pi x}{2}\right)$

9. $\cos\left(\dfrac{2x}{3}\right)\cos\left(\dfrac{4x}{3}\right)$

10. $\sin\left(-\dfrac{\pi}{4}x\right)\cos\left(-\dfrac{\pi}{2}x\right)$

11. $-3\cos(0.4x)\cos(1.5x)$

12. $2\sin(2.1x)\sin(3.4x)$

13. $4\sin(-\sqrt{3}x)\cos(3\sqrt{3}x)$

14. $-5\cos\left(-\dfrac{\sqrt{2}}{3}x\right)\sin\left(\dfrac{5\sqrt{2}}{3}x\right)$

In Exercises 15–28, write each expression as a product of sines and/or cosines.

15. $\cos(5x)+\cos(3x)$

16. $\cos(2x)-\cos(4x)$

17. $\sin(3x)-\sin x$

18. $\sin(10x)+\sin(5x)$

19. $\sin\left(\dfrac{x}{2}\right)-\sin\left(\dfrac{5x}{2}\right)$

20. $\cos\left(\dfrac{x}{2}\right)-\cos\left(\dfrac{5x}{2}\right)$

21. $\cos\left(\dfrac{2}{3}x\right)+\cos\left(\dfrac{7}{3}x\right)$

22. $\sin\left(\dfrac{2}{3}x\right)+\sin\left(\dfrac{7}{3}x\right)$

23. $\sin(0.4x)+\sin(0.6x)$

24. $\cos(0.3x)-\cos(0.5x)$

25. $\sin(\sqrt{5}x)-\sin(3\sqrt{5}x)$

26. $\cos(-3\sqrt{7}x)-\cos(2\sqrt{7}x)$

27. $\cos\left(-\dfrac{\pi}{4}x\right)+\cos\left(\dfrac{\pi}{6}x\right)$

28. $\sin\left(\dfrac{3\pi}{4}x\right)+\sin\left(\dfrac{5\pi}{4}x\right)$

In Exercises 29–34, simplify the trigonometric expressions.

29. $\dfrac{\cos(3x)-\cos x}{\sin(3x)+\sin x}$

30. $\dfrac{\sin(4x)+\sin(2x)}{\cos(4x)-\cos(2x)}$

31. $\dfrac{\cos x-\cos(3x)}{\sin(3x)-\sin x}$

32. $\dfrac{\sin(4x)+\sin(2x)}{\cos(4x)+\cos(2x)}$

33. $\dfrac{\cos(5x)+\cos(2x)}{\sin(5x)-\sin(2x)}$

34. $\dfrac{\sin(7x)-\sin(2x)}{\cos(7x)-\cos(2x)}$

In Exercises 35–42, verify the identities.

35. $\dfrac{\sin A+\sin B}{\cos A+\cos B}=\tan\left(\dfrac{A+B}{2}\right)$

36. $\dfrac{\sin A-\sin B}{\cos A+\cos B}=\tan\left(\dfrac{A-B}{2}\right)$

37. $\dfrac{\cos A-\cos B}{\sin A+\sin B}=-\tan\left(\dfrac{A-B}{2}\right)$

38. $\dfrac{\cos A-\cos B}{\sin A-\sin B}=-\tan\left(\dfrac{A+B}{2}\right)$

39. $\dfrac{\sin A+\sin B}{\sin A-\sin B}=\tan\left(\dfrac{A+B}{2}\right)\cot\left(\dfrac{A-B}{2}\right)$

40. $\dfrac{\cos A-\cos B}{\cos A+\cos B}=-\tan\left(\dfrac{A+B}{2}\right)\tan\left(\dfrac{A-B}{2}\right)$

41. $\dfrac{\cos(A+B)+\cos(A-B)}{\sin(A+B)+\sin(A-B)}=\cot A$

42. $\dfrac{\cos(A-B)-\cos(A+B)}{\sin(A+B)+\sin(A-B)}=\tan B$

■ **APPLICATIONS**

43. Business. An analysis of the monthly costs and monthly revenues of a toy store indicates that monthly costs fluctuate (increase and decrease) according to the function

$$C(t) = \sin\left(\frac{\pi}{6}t + \pi\right)$$

and monthly revenues fluctuate (increase and decrease) according to the function

$$R(t) = \sin\left(\frac{\pi}{6}t + \frac{5\pi}{3}\right)$$

Find the function that describes how the monthly profits fluctuate: $P(t) = R(t) - C(t)$. Using identities in this section, express $P(t)$ in terms of a cosine function.

44. Business. An analysis of the monthly costs and monthly revenues of an electronics manufacturer indicates that monthly costs fluctuate (increase and decrease) according to the function

$$C(t) = \cos\left(\frac{\pi}{3}t + \frac{\pi}{3}\right)$$

and monthly revenues fluctuate (increase and decrease) according to the function

$$R(t) = \cos\left(\frac{\pi}{3}t\right)$$

Find the function that describes how the monthly profits fluctuate: $P(t) = R(t) - C(t)$. Using identities in this section, express $P(t)$ in terms of a sine function.

45. Music. Write a mathematical description of a tone that results from simultaneously playing a G and a B. What is the beat frequency? What is the average frequency?

46. Music. Write a mathematical description of a tone that results from simultaneously playing an F and an A. What is the beat frequency? What is the average frequency?

47. Optics. Two optical signals with uniform ($A = 1$) intensities and wavelengths of $1.55 \,\mu$m and $0.63 \,\mu$m are "beat" together. What is the resulting sum if their individual signals are given by $\sin\left(\dfrac{2\pi tc}{1.55 \,\mu\text{m}}\right)$ and $\sin\left(\dfrac{2\pi tc}{0.63 \,\mu\text{m}}\right)$, where $c = 3.0 \times 10^8$ m/s? (*Note:* $1 \,\mu$m $= 10^{-6}$ m.)

48. Optics. The two optical signals in Exercise 47 are beat together. What are the average frequency and the beat frequency?

For Exercises 49 and 50, refer to the following:

Touch-tone keypads have the following simultaneous low and high frequencies:

FREQUENCY	1209 Hz	1336 Hz	1477 Hz
697 Hz	1	2	3
770 Hz	4	5	6
852 Hz	7	8	9
941 Hz	*	0	#

The signal given when a key is pressed is $\sin(2\pi f_1 t) + \sin(2\pi f_2 t)$, where f_1 is the low frequency and f_2 is the high frequency.

49. Touch-Tone Dialing. What is the mathematical function that models the sound of dialing 4?

50. Touch-Tone Dialing. What is the mathematical function that models the sound of dialing 3?

51. Area of a Triangle. A formula for finding the area of a triangle when given the measures of the angles and one side is Area $= \dfrac{a^2 \sin B \sin C}{2 \sin A}$, where a is the side opposite angle A. If the measures of angles B and C are 52.5° and 7.5°, respectively, and if $a = 10$ feet, use the appropriate product-to-sum identity to change the formula so that you can solve for the area of the triangle exactly.

52. Area of a Triangle. If the measures of angles B and C in Exercise 51 are 75° and 45°, respectively, and if $a = 12$ inches, use the appropriate product-to-sum identity to change the formula so that you can solve for the area of the triangle exactly.

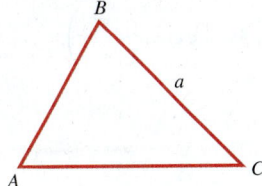

■ **CATCH THE MISTAKE**

In Exercises 53 and 54, explain the mistake that is made.

53. Simplify the expression $(\cos A - \cos B)^2 + (\sin A - \sin B)^2$.

Solution:

Expand by squaring.

$\cos^2 A - 2\cos A \cos B + \cos^2 B + \sin^2 A - 2\sin A \sin B + \sin^2 B$

Group terms.

$\cos^2 A + \sin^2 A - 2\cos A \cos B - 2\sin A \sin B + \cos^2 B + \sin^2 B$

Simplify using the Pythagorean identity.

$\underbrace{\cos^2 A + \sin^2 A}_{1} - 2\cos A \cos B - 2\sin A \sin B + \underbrace{\cos^2 B + \sin^2 B}_{1}$

Factor the common 2. $\quad 2(1 - \cos A \cos B - \sin A \sin B)$

Simplify. $\quad 2(1 - \cos AB - \sin AB)$

This is incorrect. What mistakes were made?

54. Simplify the expression $(\sin A - \sin B)(\cos A + \cos B)$.

Solution:

Multiply the expressions using the distributive property.

$\sin A \cos A + \sin A \cos B - \sin B \cos A - \sin B \cos B$

Cancel the second and third terms.

$\sin A \cos A - \sin B \cos B$

Use the product-to-sum identity.

$\underbrace{\sin A \cos A}_{\frac{1}{2}[\sin(A+A)+\sin(A-A)]} - \underbrace{\sin B \cos B}_{\frac{1}{2}[\sin(B+B)+\sin(B-B)]}$

Simplify. $\quad = \dfrac{1}{2}\sin(2A) - \dfrac{1}{2}\sin(2B)$

This is incorrect. What mistake was made?

■ **CONCEPTUAL**

In Exercises 55–58, determine whether each statement is true or false.

55. $\cos A \cos B = \cos AB$ **56.** $\sin A \sin B = \sin AB$

57. The product of two cosine functions is a sum of two other cosine functions.

58. The product of two sine functions is a difference of two cosine functions.

59. Write $\sin A \sin B \sin C$ as a sum or difference of sines and cosines.

60. Write $\cos A \cos B \cos C$ as a sum or difference of sines and cosines.

■ **CHALLENGE**

61. Prove the addition formula
$\cos(A + B) = \cos A \cos B - \sin A \sin B$
using the identities of this section.

62. Prove the difference formula
$\sin(A - B) = \sin A \cos B - \sin B \cos A$
using the identities of this section.

63. Graph $y = 1 - 3\sin(\pi x)\sin\left(-\dfrac{\pi}{6}x\right)$.

64. Graph $y = 4\sin(2x - 1)\cos(2 - x)$.

65. Graph $y = -\cos\left(\dfrac{2\pi}{3}x\right)\cos\left(\dfrac{5\pi}{6}x\right)$.

66. Graph $y = x - \cos(2x)\sin(3x)$.

■ **TECHNOLOGY**

67. Suggest an identity $4\sin x \cos x \cos(2x) =$ _____ by graphing $Y_1 = 4\sin x \cos x \cos(2x)$ and determining the function based on the graph.

68. Suggest an identity $1 + \tan x \tan(2x) =$ _____ by graphing $Y_1 = 1 + \tan x \tan(2x)$ and determining the function based on the graph.

69. With a graphing calculator, plot $Y_1 = \sin(4x)\sin(2x)$, $Y_2 = \sin(6x)$, and $Y_3 = \frac{1}{2}[\cos(2x) - \cos(6x)]$ in the same viewing rectangle $[0, 2\pi]$ by $[-1, 1]$. Which graphs are the same?

70. With a graphing calculator, plot $Y_1 = \cos(4x)\cos(2x)$, $Y_2 = \cos(6x)$, and $Y_3 = \frac{1}{2}[\cos(6x) + \cos(2x)]$ in the same viewing rectangle $[0, 2\pi]$ by $[-1, 1]$. Which graphs are the same?

SKILLS OBJECTIVES

- Evaluate inverse trigonometric functions for special values.
- Graph inverse trigonometric functions.
- Use inverse identities to evaluate expressions involving inverse trigonometric functions.

CONCEPTUAL OBJECTIVES

- Understand the different notations for inverse trigonometric functions.
- Understand why domain restrictions on trigonometric functions are needed for inverse trigonometric functions to exist.
- Extend properties of inverse functions to develop inverse trigonometric identities.

In Section 3.5, we discussed one-to-one functions and inverse functions. Here we present a summary of that section. A function is one-to-one if it passes the horizontal line test: no two x-values map to the same y-value.

Notice that the sine function does not pass the horizontal line test. However, if we restrict the domain to $-\dfrac{\pi}{2} \le x \le \dfrac{\pi}{2}$, then the restricted function is one-to-one.

Recall that if $y = f(x)$, then $x = f^{-1}(y)$.

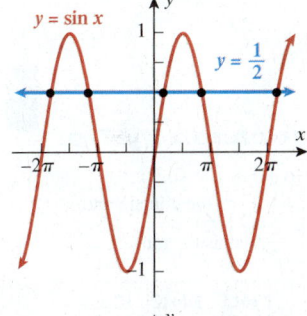

The following are the properties of inverse functions:

1. If f is a one-to-one function, then the inverse function f^{-1} exists.
2. The domain of f^{-1} = the range of f.
   The range of f^{-1} = the domain of f.
3. $f^{-1}(f(x)) = x$ for all x in the domain of f.
   $f(f^{-1}(x)) = x$ for all x in the domain of f^{-1}.
4. The graph of f^{-1} is the reflection of the graph of f about the line $y = x$. If the point (a, b) lies on the graph of a function, then the point (b, a) lies on the graph of its inverse.

Inverse Sine Function

Study Tip

The inverse sine function gives an angle on the right half of the unit circle (QI and QIV).

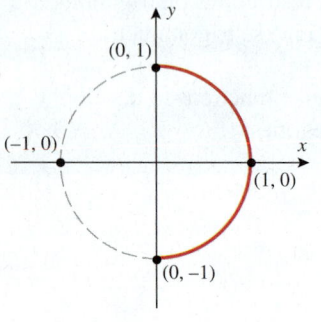

Let us start with the sine function with the restricted domain $\left[-\dfrac{\pi}{2}, \dfrac{\pi}{2}\right]$.

$y = \sin x$ Domain: $\left[-\dfrac{\pi}{2}, \dfrac{\pi}{2}\right]$ Range: $[-1, 1]$

x	y
$-\dfrac{\pi}{2}$	-1
$-\dfrac{\pi}{4}$	$-\dfrac{\sqrt{2}}{2}$
0	0
$\dfrac{\pi}{4}$	$\dfrac{\sqrt{2}}{2}$
$\dfrac{\pi}{2}$	1

Technology Tip

To graph $y = \sin^{-1}x$, use $[-1, 1]$ as the domain and $\left[-\dfrac{\pi}{2}, \dfrac{\pi}{2}\right]$ as the range.

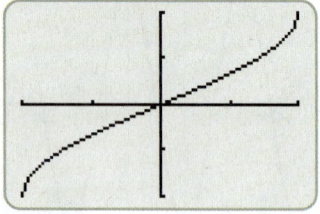

By the properties of inverse functions, the inverse sine function will have a domain of $[-1, 1]$ and a range of $\left[-\dfrac{\pi}{2}, \dfrac{\pi}{2}\right]$. To find the inverse sine function, we interchange the x- and y-values of $y = \sin x$.

$y = \sin^{-1}x$ Domain: $[-1, 1]$ Range: $\left[-\dfrac{\pi}{2}, \dfrac{\pi}{2}\right]$

x	y
-1	$-\dfrac{\pi}{2}$
$-\dfrac{\sqrt{2}}{2}$	$-\dfrac{\pi}{4}$
0	0
$\dfrac{\sqrt{2}}{2}$	$\dfrac{\pi}{4}$
1	$\dfrac{\pi}{2}$

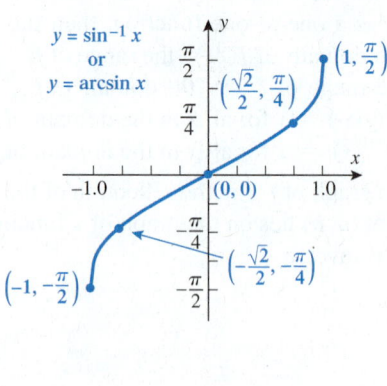

Notice that the inverse sine function, like the sine function, is an odd function (symmetric about the origin).

If the sine of an angle is known, what is the measure of that angle? The inverse sine function determines that angle measure. Another notation for the inverse sine function is arcsin$x$.

Study Tip

Trigonometric functions take angle measures and return real numbers. Inverse trigonometric functions take real numbers and return angle measures.

INVERSE SINE FUNCTION

$$\underbrace{y = \sin^{-1}x \text{ or } y = \arcsin x}_{\text{"}y\text{ is the inverse sine of }x\text{"}} \quad \text{means} \quad \underbrace{x = \sin y}_{\substack{\text{"}y\text{ is the angle measure} \\ \text{whose sine equals }x\text{"}}}$$

where $-1 \le x \le 1$ and $-\dfrac{\pi}{2} \le y \le \dfrac{\pi}{2}$

It is important to note that the -1 superscript indicates an inverse function. Therefore, the inverse sine function should not be interpreted as a reciprocal:

$$\sin^{-1}x \neq \frac{1}{\sin x}$$

EXAMPLE 1   Finding Exact Values of an Inverse Sine Function

Find the exact value of each of the following expressions:

a. $\sin^{-1}\left(\dfrac{\sqrt{3}}{2}\right)$ **b.** $\arcsin\left(-\dfrac{1}{2}\right)$

Solution (a):

Let $\theta = \sin^{-1}\left(\dfrac{\sqrt{3}}{2}\right)$. $\sin\theta = \dfrac{\sqrt{3}}{2}$ when $-\dfrac{\pi}{2} \leq \theta \leq \dfrac{\pi}{2}$

Which value of θ, in the range $-\dfrac{\pi}{2} \leq \theta \leq \dfrac{\pi}{2}$, corresponds to a sine value of $\dfrac{\sqrt{3}}{2}$?

- The range $-\dfrac{\pi}{2} \leq \theta \leq \dfrac{\pi}{2}$ corresponds to quadrants I and IV.
- The sine function is positive in quadrant I.
- We look for a value of θ in quadrant I that has a sine value of $\dfrac{\sqrt{3}}{2}$. $\theta = \dfrac{\pi}{3}$

$\sin\left(\dfrac{\pi}{3}\right) = \dfrac{\sqrt{3}}{2}$ and $\dfrac{\pi}{3}$ is in the interval $\left[-\dfrac{\pi}{2}, \dfrac{\pi}{2}\right]$. $\boxed{\sin^{-1}\left(\dfrac{\sqrt{3}}{2}\right) = \dfrac{\pi}{3}}$

Calculator Confirmation: Since $\dfrac{\pi}{3} = 60°$, if our calculator is set in degree mode, we should find that $\sin^{-1}\left(\dfrac{\sqrt{3}}{2}\right)$ is equal to $60°$.

Solution (b):

Let $\theta = \arcsin\left(-\dfrac{1}{2}\right)$. $\sin\theta = -\dfrac{1}{2}$ when $-\dfrac{\pi}{2} \leq \theta \leq \dfrac{\pi}{2}$

Which value of θ, in the range $-\dfrac{\pi}{2} \leq \theta \leq \dfrac{\pi}{2}$, corresponds to a sine value of $-\dfrac{1}{2}$?

- The range $-\dfrac{\pi}{2} \leq \theta \leq \dfrac{\pi}{2}$ corresponds to quadrants I and IV.
- The sine function is negative in quadrant IV.
- We look for a value of θ in quadrant IV that has a sine value of $-\dfrac{1}{2}$. $\theta = -\dfrac{\pi}{6}$

$\sin\left(-\dfrac{\pi}{6}\right) = -\dfrac{1}{2}$ and $-\dfrac{\pi}{6}$ is in the interval $\left[-\dfrac{\pi}{2}, \dfrac{\pi}{2}\right]$. $\boxed{\arcsin\left(-\dfrac{1}{2}\right) = -\dfrac{\pi}{6}}$

Calculator Confirmation: Since $-\dfrac{\pi}{6} = -30°$, if our calculator is set in degree mode, we should find that $\sin^{-1}\left(-\dfrac{1}{2}\right)$ is equal to $-30°$.

■ **YOUR TURN** Find the exact value of each of the following expressions:

a. $\sin^{-1}\left(-\dfrac{\sqrt{3}}{2}\right)$ **b.** $\arcsin\left(\dfrac{1}{2}\right)$

Classroom Example 7.7.1
Compute:

a. $\sin^{-1}\left(-\dfrac{\sqrt{2}}{2}\right)$

b. $\sin^{-1}\left(\dfrac{\sqrt{2}}{2}\right)$

Answer:

a. $-\dfrac{\pi}{4}$ **b.** $\dfrac{\pi}{4}$

■ **Answer: a.** $-\dfrac{\pi}{3}$ **b.** $\dfrac{\pi}{6}$

In Example 1, it is important to note that in part (a), both 60° and 120° correspond to the sine function equal to $\dfrac{\sqrt{3}}{2}$, which is why the domain restrictions are necessary for inverse functions except for quadrantal angles. There are always two angles (values) from 0 to 360° or 0 to 2π $\left[\text{except for } 90° \left(\dfrac{\pi}{2}\right) \text{ and } 270° \left(\dfrac{3\pi}{2}\right)\right]$ that correspond to the sine function equal to a particular value.

It is important to note that the inverse sine function has a domain $[-1, 1]$. For example, $\sin^{-1}3$ does not exist because 3 is not in the domain of the inverse sine function. Notice that calculator evaluation of $\sin^{-1}3$ says *error*. Calculators can be used to evaluate inverse sine functions when an exact evaluation is not feasible, just as they are for the basic trigonometric functions. For example, $\sin^{-1}0.3 \approx 17.46°$, or 0.305 radians.

We now state the properties relating the sine function and the inverse sine function which follow directly from properties of inverse functions.

SINE–INVERSE SINE IDENTITIES

$$\sin^{-1}(\sin x) = x \qquad \text{for} \qquad -\dfrac{\pi}{2} \le x \le \dfrac{\pi}{2}$$

$$\sin(\sin^{-1}x) = x \qquad \text{for} \qquad -1 \le x \le 1$$

For example, $\sin^{-1}\left[\sin\left(\dfrac{\pi}{12}\right)\right] = \dfrac{\pi}{12}$, since $\dfrac{\pi}{12}$ is in the interval $\left[-\dfrac{\pi}{2}, \dfrac{\pi}{2}\right]$. However, you must be careful not to overlook the domain restriction for which these identities hold, as illustrated in the next example.

 EXAMPLE 2 **Using Inverse Identities to Evaluate Expressions Involving Inverse Sine Functions**

Find the exact value of each of the following trigonometric expressions:

a. $\sin\left[\sin^{-1}\left(\dfrac{\sqrt{2}}{2}\right)\right]$ **b.** $\sin^{-1}\left[\sin\left(\dfrac{3\pi}{4}\right)\right]$

Solution (a):

Write the appropriate identity. $\qquad\qquad \sin(\sin^{-1}x) = x \text{ for } -1 \le x \le 1$

Let $x = \dfrac{\sqrt{2}}{2}$, which is in the interval $[-1, 1]$.

Since the domain restriction is met, the identity can be used. $\qquad\qquad \boxed{\sin\left[\sin^{-1}\left(\dfrac{\sqrt{2}}{2}\right)\right] = \dfrac{\sqrt{2}}{2}}$

Solution (b):

COMMON MISTAKE

Ignoring the domain restrictions on inverse identities.

⭐ CORRECT	❌ INCORRECT
Write the appropriate identity.	$\sin^{-1}(\sin x) = x$ **ERROR**
$\sin^{-1}(\sin x) = x$ for $-\dfrac{\pi}{2} \le x \le \dfrac{\pi}{2}$	
Let $x = \dfrac{3\pi}{4}$, which is *not* in the	Let $x = \dfrac{3\pi}{4}$.
interval $\left[-\dfrac{\pi}{2}, \dfrac{\pi}{2}\right]$.	(Forgot the domain restriction.)
Since the domain restriction is not met, the identity cannot be used. Instead, we look for a value in the domain that corresponds to the same value of sine.	$\sin^{-1}\left[\sin\left(\dfrac{3\pi}{4}\right)\right] = \dfrac{3\pi}{4}$
	INCORRECT

$\left(-\dfrac{\sqrt{2}}{2}, \dfrac{\sqrt{2}}{2}\right)$ $\left(\dfrac{\sqrt{2}}{2}, \dfrac{\sqrt{2}}{2}\right)$

Substitute $\sin\dfrac{3\pi}{4} = \sin\dfrac{\pi}{4}$ into the expression.

$$\sin^{-1}\left[\sin\left(\dfrac{3\pi}{4}\right)\right] = \sin^{-1}\left[\sin\left(\dfrac{\pi}{4}\right)\right]$$

Since $\dfrac{\pi}{4}$ is in the interval $\left[-\dfrac{\pi}{2}, \dfrac{\pi}{2}\right]$, we can use the identity.

$$\sin^{-1}\left[\sin\left(\dfrac{3\pi}{4}\right)\right] = \sin^{-1}\left[\sin\left(\dfrac{\pi}{4}\right)\right] = \boxed{\dfrac{\pi}{4}}$$

■ **YOUR TURN** Find the exact value of each of the following trigonometric expressions:

     a. $\sin\left[\sin^{-1}\left(-\dfrac{1}{2}\right)\right]$ **b.** $\sin^{-1}\left(\sin\dfrac{5\pi}{6}\right)$

Technology Tip

a. Check the answer of $\sin\left[\sin^{-1}\left(\dfrac{\sqrt{2}}{2}\right)\right]$ with a calculator.

```
sin(sin⁻¹(√(2)/2)
)
          .7071067812
√(2)/2
          .7071067812
■
```

b. Check the answer of $\sin^{-1}\left(\sin\dfrac{3\pi}{4}\right)$ with a calculator in radian mode.

```
sin⁻¹(sin(3π/4))
          .7853981634
π/4
          .7853981634
■
```

Classroom Example 7.7.2
Compute:

a. $\sin\left[\sin^{-1}\left(-\dfrac{\sqrt{2}}{2}\right)\right]$

b. $\sin^{-1}(\sin(8\pi))$

c.* $\sin^{-1}(\sin(2n\pi))$, where n is an integer

Answer:
a. $-\dfrac{\sqrt{2}}{2}$ **b.** 0   **c.** 0

■ **Answer: a.** $-\dfrac{1}{2}$ **b.** $\dfrac{\pi}{6}$

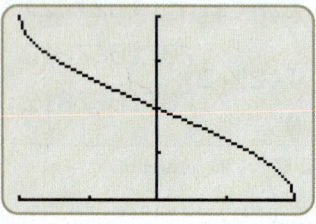
Inverse Cosine Function

The cosine function is not a one-to-one function, so we must restrict the domain in order to develop the inverse cosine function.

$$y = \cos x \qquad \text{Domain: } [0, \pi] \qquad \text{Range: } [-1, 1]$$

x	y
0	1
$\dfrac{\pi}{4}$	$\dfrac{\sqrt{2}}{2}$
$\dfrac{\pi}{2}$	0
$\dfrac{3\pi}{4}$	$-\dfrac{\sqrt{2}}{2}$
π	-1

By the properties of inverses, the inverse cosine function will have a domain of $[-1, 1]$ and a range of $[0, \pi]$. To find the inverse cosine function, we interchange the x- and y-values of $y = \cos x$.

$$y = \cos^{-1}x \qquad \text{Domain: } [-1, 1] \qquad \text{Range: } [0, \pi]$$

x	y
-1	π
$-\dfrac{\sqrt{2}}{2}$	$\dfrac{3\pi}{4}$
0	$\dfrac{\pi}{2}$
$\dfrac{\sqrt{2}}{2}$	$\dfrac{\pi}{4}$
1	0

Study Tip

The inverse cosine function gives an angle on the top half of the unit circle (QI and QII).

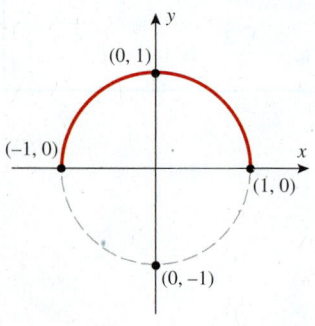

Notice that the inverse cosine function, unlike the cosine function, is not symmetric about the y-axis, or the origin. Although the inverse sine and inverse cosine functions have the same domain, they behave differently. The inverse sine function increases on its domain (from left to right), whereas the inverse cosine function decreases on its domain (from left to right).

If the cosine of an angle is known, what is the measure of that angle? The inverse cosine function determines that angle measure. Another notation for the inverse cosine function is $\arccos x$.

Study Tip

$$\cos^{-1}x \neq \dfrac{1}{\cos x}$$

INVERSE COSINE FUNCTION

$$y = \cos^{-1}x \text{ or } y = \arccos x \qquad \text{means} \qquad x = \cos y$$

"y is the inverse cosine of x"   *"y is the angle measure whose cosine equals x"*

where $-1 \leq x \leq 1$ and $0 \leq y \leq \pi$

EXAMPLE 3   Finding Exact Values of an Inverse Cosine Function

Find the exact value of each of the following expressions:

a. $\cos^{-1}\left(-\dfrac{\sqrt{2}}{2}\right)$ **b.** $\arccos 0$

Solution (a):

Let $\theta = \cos^{-1}\left(-\dfrac{\sqrt{2}}{2}\right)$. $\cos\theta = -\dfrac{\sqrt{2}}{2}$ when $0 \le \theta \le \pi$

Which value of θ, in the range $0 \le \theta \le \pi$, corresponds to a cosine value of $-\dfrac{\sqrt{2}}{2}$?

- ▪ The range $0 \le \theta \le \pi$ corresponds to quadrants I and II.
- ▪ The cosine function is negative in quadrant II.
- ▪ We look for a value of θ in quadrant II that has a cosine value of $-\dfrac{\sqrt{2}}{2}$. $\theta = \dfrac{3\pi}{4}$

$\cos\left(\dfrac{3\pi}{4}\right) = -\dfrac{\sqrt{2}}{2}$ and $\dfrac{3\pi}{4}$ is in the interval $[0, \pi]$. $\boxed{\cos^{-1}\left(-\dfrac{\sqrt{2}}{2}\right) = \dfrac{3\pi}{4}}$

Calculator Confirmation: Since $\dfrac{3\pi}{4} = 135°$, if our calculator is set in degree mode, we should find that $\cos^{-1}\left(-\dfrac{\sqrt{2}}{2}\right)$ is equal to $135°$.

Solution (b):

Let $\theta = \arccos 0$. $\cos\theta = 0$ when $0 \le \theta \le \pi$

Which value of θ, in the range $0 \le \theta \le \pi$, corresponds to a cosine value of 0? $\theta = \dfrac{\pi}{2}$

$\cos\left(\dfrac{\pi}{2}\right) = 0$ and $\dfrac{\pi}{2}$ is in the interval $[0, \pi]$. $\boxed{\arccos 0 = \dfrac{\pi}{2}}$

Calculator Confirmation: Since $\dfrac{\pi}{2} = 90°$, if our calculator is set in degree mode, we should find that $\cos^{-1} 0$ is equal to $90°$.

▪ **YOUR TURN** Find the exact value of each of the following expressions:

a. $\cos^{-1}\left(\dfrac{\sqrt{2}}{2}\right)$ **b.** $\arccos 1$

Technology Tip

a. Check the answer of $\cos^{-1}\left(-\dfrac{\sqrt{2}}{2}\right)$ with a calculator.

```
cos⁻¹(-√(2)/2)
              135
```

b. Check the answer of $\arccos 0$.

```
cos⁻¹(0)
              90
```

Classroom Example 7.7.3
Compute:

a. $\cos^{-1}\left(-\dfrac{\sqrt{3}}{2}\right)$

b. $\cos^{-1}\left(-\dfrac{1}{2}\right)$

Answer:

a. $\dfrac{5\pi}{6}$ **b.** $\dfrac{2\pi}{3}$

▪**Answer: a.** $\dfrac{\pi}{4}$ **b.** 0

We now state the properties relating the cosine function and the inverse cosine function which follow directly from the properties of inverses.

COSINE-INVERSE COSINE IDENTITIES

$$\cos^{-1}(\cos x) = x \quad \text{for} \quad 0 \le x \le \pi$$

$$\cos(\cos^{-1} x) = x \quad \text{for} \quad -1 \le x \le 1$$

As was the case with inverse identities for the sine function, you must be careful not to overlook the domain restrictions governing when each of these identities hold.

EXAMPLE 4 **Using Inverse Identities to Evaluate Expressions Involving Inverse Cosine Functions**

Find the exact value of each of the following trigonometric expressions:

a. $\cos\left[\cos^{-1}\left(-\frac{1}{2}\right)\right]$ **b.** $\cos^{-1}\left[\cos\left(\frac{7\pi}{4}\right)\right]$

Solution (a):

Write the appropriate identity. $\cos(\cos^{-1}x) = x$ for $-1 \le x \le 1$

Let $x = -\frac{1}{2}$, which is in the interval $[-1, 1]$.

Since the domain restriction is met, the identity can be used.

$$\cos\left[\cos^{-1}\left(-\frac{1}{2}\right)\right] = \boxed{-\frac{1}{2}}$$

Solution (b):

Write the appropriate identity. $\cos^{-1}(\cos x) = x$ for $0 \le x \le \pi$

Let $x = \frac{7\pi}{4}$, which is *not* in the interval $[0, \pi]$.

Since the domain restriction is not met, the identity cannot be used.

Instead, we find another angle in the interval that has the same cosine value.

Substitute $\cos\left(\frac{7\pi}{4}\right) = \cos\left(\frac{\pi}{4}\right)$ into the expression.

Since $\frac{\pi}{4}$ is in the interval $[0, \pi]$, we can use the identity.

$$\cos\left(\frac{7\pi}{4}\right) = \cos\left(\frac{\pi}{4}\right)$$

$$\cos^{-1}\left[\cos\left(\frac{7\pi}{4}\right)\right] = \cos^{-1}\left[\cos\left(\frac{\pi}{4}\right)\right]$$

$$= \boxed{\frac{\pi}{4}}$$

■ **Answer: a.** $\frac{1}{2}$ **b.** $\frac{\pi}{6}$

■ **YOUR TURN** Find the exact value of each of the following trigonometric expressions:

a. $\cos\left[\cos^{-1}\left(\frac{1}{2}\right)\right]$ **b.** $\cos^{-1}\left[\cos\left(-\frac{\pi}{6}\right)\right]$

Inverse Tangent Function

Study Tip

The inverse tangent function gives an angle on the right half of the unit circle (QI and QIV).

The tangent function is not a one-to-one function (it fails the horizontal line test). Let us start with the tangent function with a restricted domain:

$$y = \tan x \quad \text{Domain: } \left(-\frac{\pi}{2}, \frac{\pi}{2}\right) \quad \text{Range: } (-\infty, \infty)$$

x	y
$-\dfrac{\pi}{2}$	$-\infty$
$-\dfrac{\pi}{4}$	-1
0	0
$\dfrac{\pi}{4}$	1
$\dfrac{\pi}{2}$	∞

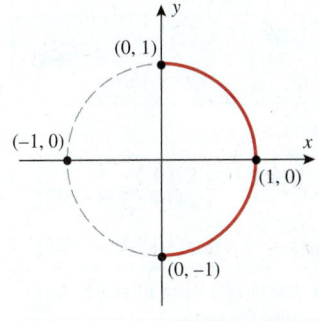

By the properties of inverse functions, the inverse tangent function will have a domain of $(-\infty, \infty)$ and a range of $\left(-\frac{\pi}{2}, \frac{\pi}{2}\right)$. To find the inverse tangent function, interchange x and y values.

$$y = \tan^{-1} x \quad \text{Domain: } (-\infty, \infty) \quad \text{Range: } \left(-\frac{\pi}{2}, \frac{\pi}{2}\right)$$

x	y
$-\infty$	$-\dfrac{\pi}{2}$
-1	$-\dfrac{\pi}{4}$
0	0
1	$\dfrac{\pi}{4}$
∞	$\dfrac{\pi}{2}$

Technology Tip

To graph $y = \tan^{-1} x$, use $(-\infty, \infty)$ as the domain and $\left(-\frac{\pi}{2}, \frac{\pi}{2}\right)$ as the range.

Notice that the inverse tangent function, like the tangent function, is an odd function (it is symmetric about the origin).

The inverse tangent function allows us to answer the question: if the tangent of an angle is known, what is the measure of that angle? Another notation for the inverse tangent function is arctan x.

INVERSE TANGENT FUNCTION

$$\underbrace{y = \tan^{-1} x \text{ or } y = \arctan x}_{\text{"y is the inverse tangent of x"}} \quad \text{means} \quad \underbrace{x = \tan y}_{\substack{\text{"y is the angle measure whose} \\ \text{tangent equals x"}}}$$

$$\text{where } -\frac{\pi}{2} < y < \frac{\pi}{2}$$

EXAMPLE 5 **Finding Exact Values of an Inverse Tangent Function**

Find the exact value of each of the following expressions:

a. $\tan^{-1}(\sqrt{3})$ **b.** $\arctan 0$

Solution (a):

Let $\theta = \tan^{-1}(\sqrt{3})$. $\tan\theta = \sqrt{3}$ when $-\dfrac{\pi}{2} < \theta < \dfrac{\pi}{2}$

Which value of θ, in the range $-\dfrac{\pi}{2} < \theta < \dfrac{\pi}{2}$,

corresponds to a tangent value of $\sqrt{3}$? $\theta = \dfrac{\pi}{3}$

$\tan\left(\dfrac{\pi}{3}\right) = \sqrt{3}$ and $\dfrac{\pi}{3}$ is in the interval $\left(-\dfrac{\pi}{2}, \dfrac{\pi}{2}\right)$. $\boxed{\tan^{-1}(\sqrt{3}) = \dfrac{\pi}{3}}$

Calculator Confirmation: Since $\dfrac{\pi}{3} = 60°$, if our calculator is set in degree mode, we should find that $\tan^{-1}(\sqrt{3})$ is equal to $60°$.

Solution (b):

Let $\theta = \arctan 0$. $\tan\theta = 0$ when $-\dfrac{\pi}{2} < \theta < \dfrac{\pi}{2}$

Which value of θ, in the range $-\dfrac{\pi}{2} < \theta < \dfrac{\pi}{2}$,

corresponds to a tangent value of 0? $\theta = 0$

$\tan 0 = 0$, and 0 is in the interval $\left(-\dfrac{\pi}{2}, \dfrac{\pi}{2}\right)$. $\boxed{\arctan 0 = 0}$

Calculator Confirmation: $\tan^{-1} 0$ is equal to 0.

     We now state the properties relating the tangent function and the inverse tangent function which follow directly from the properties of inverses.

TANGENT-INVERSE TANGENT IDENTITIES

$$\tan^{-1}(\tan x) = x \qquad \text{for} \qquad -\frac{\pi}{2} < x < \frac{\pi}{2}$$

$$\tan(\tan^{-1}x) = x \qquad \text{for} \qquad -\infty < x < \infty$$

EXAMPLE 6 **Using Inverse Identities to Evaluate Expressions Involving Inverse Tangent Functions**

Find the exact value of each of the following trigonometric expressions:

a. $\tan(\tan^{-1} 17)$ **b.** $\tan^{-1}\left[\tan\left(\dfrac{2\pi}{3}\right)\right]$

Solution (a):

Write the appropriate identity. $\tan(\tan^{-1}x) = x$ for $-\infty < x < \infty$

Let $x = 17$, which is in the interval $(-\infty, \infty)$.

Since the domain restriction is met, the identity can be used. $\boxed{\tan(\tan^{-1} 17) = 17}$

Technology Tip

a. Use a calculator to check the answer for $\tan^{-1}(\sqrt{3})$.

```
tan-1(√(3))
                60
■
```

Radian mode:

```
tan-1(√(3))
         1.047197551
π/3
         1.047197551
```

b. Use a calculator to check the answer for $\arctan 0$.

```
tan-1(0)
                 0
■
```

Classroom Example 7.7.5

Compute:

a. $\tan^{-1}\left(-\dfrac{\sqrt{3}}{3}\right)$

b. $\tan^{-1}(-1)$

Answer:

a. $-\dfrac{\pi}{6}$ **b.** $-\dfrac{\pi}{4}$

Technology Tip

a. Use a calculator to check the answer for $\tan(\tan^{-1} 17)$.

```
tan(tan-1(17))
                17
■
```

b. Use a calculator to check the answer for $\tan^{-1}\left[\tan\left(\dfrac{2\pi}{3}\right)\right]$.

```
tan-1(tan(2π/3)
         -1.047197551
-π/3
         -1.047197551
■
```

Solution (b):

Write the appropriate identity. $\tan^{-1}(\tan x) = x$ for $-\dfrac{\pi}{2} < x < \dfrac{\pi}{2}$

Let $x = \dfrac{2\pi}{3}$, which is *not* in the interval $\left(-\dfrac{\pi}{2}, \dfrac{\pi}{2}\right)$.

Since the domain restriction is not met, the identity cannot be used.

Instead, we find another angle
in the interval that has the $\tan\left(\dfrac{2\pi}{3}\right) = \tan\left(-\dfrac{\pi}{3}\right)$
same tangent value.

Substitute $\tan\left(\dfrac{2\pi}{3}\right) = \tan\left(-\dfrac{\pi}{3}\right)$
into the expression. $\tan^{-1}\left[\tan\left(\dfrac{2\pi}{3}\right)\right] = \tan^{-1}\left[\tan\left(-\dfrac{\pi}{3}\right)\right]$

Since $-\dfrac{\pi}{3}$ is in the interval

$\left(-\dfrac{\pi}{2}, \dfrac{\pi}{2}\right)$, $\tan^{-1}\left[\tan\left(-\dfrac{\pi}{3}\right)\right] = -\dfrac{\pi}{3}$ $\tan^{-1}\left[\tan\left(\dfrac{2\pi}{3}\right)\right] = \boxed{-\dfrac{\pi}{3}}$

▪ **YOUR TURN** Find the exact value of $\tan^{-1}\left[\tan\left(\dfrac{7\pi}{6}\right)\right]$.

Classroom Example 7.7.6
Compute:

a. $\tan(\tan^{-1}800)$

b.* $\tan^{-1}[\tan(n\pi)]$, where n is
an interger

Answer: a. 800   **b.** 0

▪ **Answer:** $\dfrac{\pi}{6}$

Remaining Inverse Trigonometric Functions

The remaining three inverse trigonometric functions are defined similarly to the previous ones.

- Inverse cotangent function: $\cot^{-1}x$ or $\text{arccot}\,x$
- Inverse secant function: $\sec^{-1}x$ or $\text{arcsec}\,x$
- Inverse cosecant function: $\csc^{-1}x$ or $\text{arccsc}\,x$

A table summarizing all six of the inverse trigonometric functions is given below:

INVERSE FUNCTION	$y = \sin^{-1}x$	$y = \cos^{-1}x$	$y = \tan^{-1}x$	$y = \cot^{-1}x$	$y = \sec^{-1}x$	$y = \csc^{-1}x$
DOMAIN	$[-1, 1]$	$[-1, 1]$	$(-\infty, \infty)$	$(-\infty, \infty)$	$(-\infty, -1] \cup [1, \infty)$	$(-\infty, -1] \cup [1, \infty)$
RANGE	$\left[-\dfrac{\pi}{2}, \dfrac{\pi}{2}\right]$	$[0, \pi]$	$\left(-\dfrac{\pi}{2}, \dfrac{\pi}{2}\right)$	$(0, \pi)$	$\left[0, \dfrac{\pi}{2}\right) \cup \left(\dfrac{\pi}{2}, \pi\right]$	$\left[-\dfrac{\pi}{2}, 0\right) \cup \left(0, \dfrac{\pi}{2}\right]$
GRAPH						

Technology Tip

$\cot^{-1}(\sqrt{3}) = \tan^{-1}\left(\dfrac{1}{\sqrt{3}}\right)$,

$\csc^{-1}(\sqrt{2}) = \sin^{-1}\left(\dfrac{1}{\sqrt{2}}\right)$, and

$\sec^{-1}(-\sqrt{2}) = \cos^{-1}\left(-\dfrac{1}{\sqrt{2}}\right)$.

```
tan-1(1/√(3))
        .5235987756
π/6
        .5235987756
```

```
sin-1(1/√(2))
        .7853981634
π/4
        .7853981634
```

```
cos-1(-1/√(2))
        2.35619449
3π/4
        2.35619449
■
```

EXAMPLE 7   Finding the Exact Value of Inverse Trigonometric Functions

Find the exact value of the following expressions:

a. $\cot^{-1}(\sqrt{3})$ **b.** $\csc^{-1}(\sqrt{2})$ **c.** $\sec^{-1}(-\sqrt{2})$

Solution (a):

Let $\theta = \cot^{-1}(\sqrt{3})$. $\cot\theta = \sqrt{3}$ when $0 < \theta < \pi$

Which value of θ, in the range $0 < \theta < \pi$,
corresponds to a cotangent value of $\sqrt{3}$? $\theta = \dfrac{\pi}{6}$

$\cot\left(\dfrac{\pi}{6}\right) = \sqrt{3}$ and $\dfrac{\pi}{6}$ is in the interval $\left(-\dfrac{\pi}{2}, \dfrac{\pi}{2}\right)$. $\boxed{\cot^{-1}(\sqrt{3}) = \dfrac{\pi}{6}}$

Solution (b):

Let $\theta = \csc^{-1}(\sqrt{2})$. $\csc\theta = \sqrt{2}$

Which value of θ, in the range
$\left[-\dfrac{\pi}{2}, 0\right) \cup \left(0, \dfrac{\pi}{2}\right]$, corresponds
to a cosecant value of $\sqrt{2}$? $\theta = \dfrac{\pi}{4}$

$\csc\left(\dfrac{\pi}{4}\right) = \sqrt{2}$ and $\dfrac{\pi}{4}$ is in the

interval $\left[-\dfrac{\pi}{2}, 0\right) \cup \left(0, \dfrac{\pi}{2}\right]$. $\boxed{\csc^{-1}(\sqrt{2}) = \dfrac{\pi}{4}}$

Solution (c):

Let $\theta = \sec^{-1}(-\sqrt{2})$. $\sec\theta = -\sqrt{2}$

Which value of θ, in the range
$\left[0, \dfrac{\pi}{2}\right) \cup \left(\dfrac{\pi}{2}, \pi\right]$, corresponds
to a secant value of $-\sqrt{2}$? $\theta = \dfrac{3\pi}{4}$

$\sec\left(\dfrac{3\pi}{4}\right) = -\sqrt{2}$ and $\dfrac{3\pi}{4}$ is in the

interval $\left[0, \dfrac{\pi}{2}\right) \cup \left(\dfrac{\pi}{2}, \pi\right]$. $\boxed{\sec^{-1}(-\sqrt{2}) = \dfrac{3\pi}{4}}$

How do we approximate the inverse secant, inverse cosecant, and inverse cotangent functions with a calculator? Scientific calculators have keys ($\sin^{-1}$, $\cos^{-1}$, and $\tan^{-1}$) for three of the inverse trigonometric functions but not for the other three. Recall that we find the cosecant, secant, and cotangent function values by taking sine, cosine, or tangent, and finding the reciprocal.

$$\csc x = \frac{1}{\sin x} \qquad \sec x = \frac{1}{\cos x} \qquad \cot x = \frac{1}{\tan x}$$

However, *the reciprocal approach cannot be used for inverse functions.* The three inverse trigonometric functions $\csc^{-1}x$, $\sec^{-1}x$, and $\cot^{-1}x$ cannot be found by finding the reciprocal of $\sin^{-1}x$, $\cos^{-1}x$, or $\tan^{-1}x$.

$$\csc^{-1}x \neq \frac{1}{\sin^{-1}x} \qquad \sec^{-1}x \neq \frac{1}{\cos^{-1}x} \qquad \cot^{-1}x \neq \frac{1}{\tan^{-1}x}$$

Instead, we seek the equivalent $\sin^{-1}x$, $\cos^{-1}x$, or $\tan^{-1}x$ values by algebraic means, always remembering to look within the correct domain and range.

Start with the inverse secant function.	$y = \sec^{-1}x$	for	$x \le -1$	or	$x \ge 1$

Write the equivalent secant expression.

$$\sec y = x \qquad \text{for} \qquad 0 \le y < \frac{\pi}{2} \qquad \text{or} \qquad \frac{\pi}{2} < y \le \pi$$

Apply the reciprocal identity: $\sec y = \dfrac{1}{\cos y}$.

$$\frac{1}{\cos y} = x$$

Simplify using algebraic techniques.

$$\cos y = \frac{1}{x}$$

Write the result in terms of the inverse cosine function.

$$y = \cos^{-1}\!\left(\frac{1}{x}\right)$$

Therefore, we have the relationship:

$$\sec^{-1}x = \cos^{-1}\!\left(\frac{1}{x}\right) \qquad \text{for} \qquad x \le -1 \qquad \text{or} \qquad x \ge 1$$

The other relationships will be found in the exercises and are summarized below:

INVERSE SECANT, INVERSE COSECANT, AND INVERSE COTANGENT IDENTITIES

$$\sec^{-1}x = \cos^{-1}\!\left(\frac{1}{x}\right) \qquad \text{for} \qquad x \le -1 \text{ or } x \ge 1$$

$$\csc^{-1}x = \sin^{-1}\!\left(\frac{1}{x}\right) \qquad \text{for} \qquad x \le -1 \text{ or } x \ge 1$$

$$\cot^{-1}x = \begin{cases} \tan^{-1}\!\left(\dfrac{1}{x}\right) & \text{for} \quad x > 0 \\[2mm] \pi + \tan^{-1}\!\left(\dfrac{1}{x}\right) & \text{for} \quad x < 0 \end{cases}$$

Study Tip

$$\sec^{-1}x \ne \frac{1}{\cos^{-1}x}$$

$$\csc^{-1}x \ne \frac{1}{\sin^{-1}x}$$

$$\cot^{-1}x \ne \frac{1}{\tan^{-1}x}$$

EXAMPLE 8 Using Inverse Identities

a. Find the exact value of $\sec^{-1}2$.
b. Use a calculator to find the value of $\cot^{-1}7$.

Solution (a):

Let $\theta = \sec^{-1}2$.

$$\sec\theta = 2 \text{ on } \left[0, \frac{\pi}{2}\right) \cup \left(\frac{\pi}{2}, \pi\right]$$

Substitute the reciprocal identity.

$$\frac{1}{\cos\theta} = 2$$

Solve for $\cos\theta$.

$$\cos\theta = \frac{1}{2}$$

The restricted interval $\left[0, \frac{\pi}{2}\right) \cup \left(\frac{\pi}{2}, \pi\right]$ corresponds to quadrants I and II.

The cosine function is positive in quadrant I.

$$\theta = \frac{\pi}{3}$$

$$\cos\left(\frac{\pi}{3}\right) = \frac{1}{2}$$

$$\boxed{\sec^{-1}2 = \cos^{-1}\!\left(\frac{1}{2}\right) = \frac{\pi}{3}}$$

Technology Tip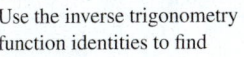

Use the inverse trigonometry function identities to find

a. $\sec^{-1}2 = \cos^{-1}\!\left(\frac{1}{2}\right)$

b. $\cot^{-1}7 = \tan^{-1}\!\left(\frac{1}{7}\right)$

```
cos⁻¹(1/2)
           1.047197551
π/3
           1.047197551
```

```
tan⁻¹(1/7)
           .1418970546
Ans*180/π
           8.130102354
```

Solution (b):

Since we do not know an exact value that would correspond to the cotangent function equal to 7, we proceed using identities and a calculator.

Select the correct identity, given that $x = 7 > 0$. $\cot^{-1}x = \tan^{-1}\left(\dfrac{1}{x}\right)$

Let $x = 7$. $\cot^{-1}7 = \tan^{-1}\left(\dfrac{1}{7}\right)$

Evaluate the right side with a calculator. $\boxed{\cot^{-1}7 \approx 8.13°}$

Finding Exact Values for Expressions Involving Inverse Trigonometric Functions

We will now find exact values of trigonometric expressions that involve inverse trigonometric functions.

EXAMPLE 9 **Finding Exact Values of Trigonometric Expressions Involving Inverse Trigonometric Functions**

Find the exact value of $\cos\left[\sin^{-1}\left(\frac{2}{3}\right)\right]$.

Solution:

STEP 1 Let $\theta = \sin^{-1}\left(\frac{2}{3}\right)$. $\sin\theta = \dfrac{2}{3}$ when $-\dfrac{\pi}{2} \le \theta \le \dfrac{\pi}{2}$

The range $-\dfrac{\pi}{2} \le \theta \le \dfrac{\pi}{2}$ corresponds to quadrants I and IV.

The sine function is positive in quadrant I.

STEP 2 Draw angle θ in quadrant I.

Label the sides known from the sine value. $\sin\theta = \dfrac{2}{3} = \dfrac{\text{opposite}}{\text{hypotenuse}}$

STEP 3 Find the unknown side length a. $a^2 + 2^2 = 3^2$

Solve for a. $a = \pm\sqrt{5}$

Since θ is in quadrant I, a is positive. $a = \sqrt{5}$

STEP 4 Find $\cos\left[\sin^{-1}\left(\frac{2}{3}\right)\right]$.

Substitute $\theta = \sin^{-1}\left(\frac{2}{3}\right)$. $\cos\left[\sin^{-1}\left(\frac{2}{3}\right)\right] = \cos\theta$

Find $\cos\theta$.

$$\cos\theta = \frac{\text{adjacent}}{\text{hypotenuse}} = \frac{\sqrt{5}}{3}$$

$$\boxed{\cos\left[\sin^{-1}\left(\frac{2}{3}\right)\right] = \frac{\sqrt{5}}{3}}$$

■ **YOUR TURN** Find the exact value of $\sin\left[\cos^{-1}\left(\frac{1}{3}\right)\right]$.

■ **Answer:** $\dfrac{2\sqrt{2}}{3}$

▶ **EXAMPLE 10 Finding Exact Values of Trigonometric Expressions Involving Inverse Trigonometric Functions**

Find the exact value of $\tan\left[\cos^{-1}\left(-\frac{7}{12}\right)\right]$.

Solution:

STEP 1 Let $\theta = \cos^{-1}\left(-\frac{7}{12}\right)$. $\cos\theta = -\dfrac{7}{12}$ when $0 \le \theta \le \pi$

The range $0 \le \theta \le \pi$ corresponds to quadrants I and II.

The cosine function is negative in quadrant II.

STEP 2 Draw angle θ in quadrant II.

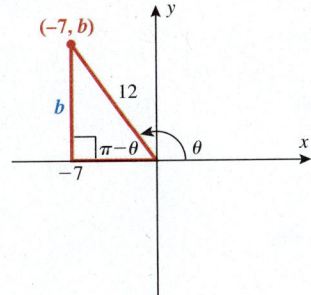

Label the sides known from the cosine value.

$$\cos\theta = -\frac{7}{12} = \frac{\text{adjacent}}{\text{hypotenuse}}$$

STEP 3 Find the length of the unknown side b. $b^2 + (-7)^2 = 12^2$

Solve for b. $b = \pm\sqrt{95}$

Since θ is in quadrant II, b is positive. $b = \sqrt{95}$

STEP 4 Find $\tan\left[\cos^{-1}\left(-\frac{7}{12}\right)\right]$.

Substitute $\theta = \cos^{-1}\left(-\frac{7}{12}\right)$.

$$\tan\left[\cos^{-1}\left(-\frac{7}{12}\right)\right] = \tan\theta$$

Find $\tan\theta$.

$$\tan\theta = \frac{\text{opposite}}{\text{adjacent}} = \frac{\sqrt{95}}{-7}$$

$$\boxed{\tan\left[\cos^{-1}\left(-\frac{7}{12}\right)\right] = -\frac{\sqrt{95}}{7}}$$

■ **Answer:** $-\dfrac{3\sqrt{10}}{20}$

■ **YOUR TURN** Find the exact value of $\tan\left[\sin^{-1}\left(-\frac{3}{7}\right)\right]$.

EXAMPLE 11   Using Identities to Find Exact Values of Trigonometric Expressions Involving Inverse Trigonometric Functions

Find the exact value of $\cos\left[\sin^{-1}\left(\frac{3}{5}\right) + \tan^{-1}1\right]$.

Solution:

Recall the cosine sum identity:

$$\cos(A + B) = \cos A \cos B - \sin A \sin B$$

Let $A = \sin^{-1}\left(\frac{3}{5}\right)$ and $B = \tan^{-1}1$.

$$\cos\left[\sin^{-1}\left(\frac{3}{5}\right) + \tan^{-1}1\right] = \cos\left[\sin^{-1}\left(\frac{3}{5}\right)\right]\cos(\tan^{-1}1) - \sin\left[\sin^{-1}\left(\frac{3}{5}\right)\right]\sin(\tan^{-1}1)$$

From the figure,

$$A = \sin^{-1}\left(\frac{3}{5}\right) \Rightarrow \sin A = \frac{3}{5}$$

we see that

$$\cos\left[\sin^{-1}\left(\frac{3}{5}\right)\right] = \cos A = \frac{4}{5}$$

$$\sin\left[\sin^{-1}\left(\frac{3}{5}\right)\right] = \sin A = \frac{3}{5}$$

From the figure,

$$B = \tan^{-1}(1) \Rightarrow \tan B = 1$$

we see that

$$\cos(\tan^{-1}1) = \cos B = \frac{\sqrt{2}}{2}$$

$$\sin(\tan^{-1}1) = \sin B = \frac{\sqrt{2}}{2}$$

Substitute these values into the cosine sum identity:

$$\cos\left[\sin^{-1}\left(\frac{3}{5}\right) + \tan^{-1}1\right] = \left(\frac{4}{5}\right)\left(\frac{\sqrt{2}}{2}\right) - \left(\frac{3}{5}\right)\left(\frac{\sqrt{2}}{2}\right)$$

Simplify.

$$= \boxed{\frac{\sqrt{2}}{10}}$$

■ **YOUR TURN** Find the exact value of $\sin[\cos^{-1}\left(\frac{3}{5}\right) + \tan^{-1}1]$.

■ **Answer:** $\dfrac{7\sqrt{2}}{10}$

EXAMPLE 12   **Writing Trigonometric Expressions Involving Inverse Trigonometric Functions in Terms of a Single Variable**

Write the expression $\cos(\tan^{-1}u)$ as an equivalent expression in terms of only the variable u.

Solution: Let $\theta = \tan^{-1}u$; therefore, $\tan\theta = u = \dfrac{u}{1}$.

Realize that u can be positive or negative. Since the range of the inverse tangent function is $\left(-\dfrac{\pi}{2}, \dfrac{\pi}{2}\right)$, sketch the angle θ in both quadrants I and IV and draw the corresponding two right triangles. Recalling that the tangent ratio is opposite over adjacent, we label those corresponding sides with u and 1, respectively. Then solving for the hypotenuse using the Pythagorean theorem gives $\sqrt{u^2 + 1}$.

Substitute $\theta = \tan^{-1}u$ into $\cos(\tan^{-1}u)$.

$$\cos(\tan^{-1}u) = \cos\theta$$

Use the right triangle ratio for cosine: adjacent over hypotenuse.

$$= \frac{1}{\sqrt{u^2 + 1}}$$

Rationalize the denominator.

$$\cos(\tan^{-1}u) = \frac{1}{\sqrt{u^2 + 1}} \cdot \frac{\sqrt{u^2 + 1}}{\sqrt{u^2 + 1}}$$

$$= \boxed{\frac{\sqrt{u^2 + 1}}{u^2 + 1}}$$

■ **YOUR TURN** Write the expression $\sin(\tan^{-1}u)$ as an equivalent expression in terms of only the variable u.

■ **Answer:** $\dfrac{u\sqrt{u^2 + 1}}{u^2 + 1}$

SECTION
7.7 SUMMARY

If a trigonometric function value of an angle is known, what is the measure of that angle? Inverse trigonometric functions determine the angle measure. To define the inverse trigonometric relations as functions, we first restrict the trigonometric functions to domains in which they are one-to-one functions. Exact values for inverse trigonometric functions can be found when the function values are those of the special angles. Inverse trigonometric functions also provide a means for evaluating one trigonometric function when we are given the value of another. It is important to note that the -1 as a superscript indicates an inverse function, not a reciprocal.

INVERSE FUNCTION	$y = \sin^{-1}x$	$y = \cos^{-1}x$	$y = \tan^{-1}x$	$y = \cot^{-1}x$	$y = \sec^{-1}x$	$y = \csc^{-1}x$
DOMAIN	$[-1, 1]$	$[-1, 1]$	$(-\infty, \infty)$	$(-\infty, \infty)$	$(-\infty, -1] \cup [1, \infty)$	$(-\infty, -1] \cup [1, \infty)$
RANGE	$\left[-\dfrac{\pi}{2}, \dfrac{\pi}{2}\right]$	$[0, \pi]$	$\left(-\dfrac{\pi}{2}, \dfrac{\pi}{2}\right)$	$(0, \pi)$	$\left[0, \dfrac{\pi}{2}\right) \cup \left(\dfrac{\pi}{2}, \pi\right]$	$\left[-\dfrac{\pi}{2}, 0\right) \cup \left(0, \dfrac{\pi}{2}\right]$
GRAPH						

SECTION
7.7 EXERCISES

▪ SKILLS

In Exercises 1–16, find the exact value of each expression. Give the answer in radians.

1. $\arccos\left(\dfrac{\sqrt{2}}{2}\right)$

2. $\arccos\left(-\dfrac{\sqrt{2}}{2}\right)$

3. $\arcsin\left(-\dfrac{\sqrt{3}}{2}\right)$

4. $\arcsin\left(\dfrac{1}{2}\right)$

5. $\cot^{-1}(-1)$

6. $\tan^{-1}\left(\dfrac{\sqrt{3}}{3}\right)$

7. $\arcsec\left(\dfrac{2\sqrt{3}}{3}\right)$

8. $\arccsc(-1)$

9. $\csc^{-1}2$

10. $\sec^{-1}(-2)$

11. $\arctan(-\sqrt{3})$

12. $\arccot(\sqrt{3})$

13. $\sin^{-1}0$

14. $\tan^{-1}1$

15. $\sec^{-1}(-1)$

16. $\cot^{-1}0$

In Exercises 17–32, find the exact value of each expression. Give the answer in degrees.

17. $\cos^{-1}\left(\dfrac{1}{2}\right)$

18. $\cos^{-1}\left(-\dfrac{\sqrt{3}}{2}\right)$

19. $\sin^{-1}\left(\dfrac{\sqrt{2}}{2}\right)$

20. $\sin^{-1}0$

21. $\cot^{-1}\left(-\dfrac{\sqrt{3}}{3}\right)$

22. $\tan^{-1}(-\sqrt{3})$

23. $\arctan\left(\dfrac{\sqrt{3}}{3}\right)$

24. $\arccot 1$

25. $\arccsc(-2)$

26. $\csc^{-1}\left(-\dfrac{2\sqrt{3}}{3}\right)$

27. $\arcsec(-\sqrt{2})$

28. $\arccsc(-\sqrt{2})$

29. $\sin^{-1}(-1)$

30. $\arctan(-1)$

31. $\arccot 0$

32. $\arcsec(-1)$

In Exercises 33–42, use a calculator to evaluate each expression. Give the answer in degrees and round it to two decimal places.

33. $\cos^{-1}(0.5432)$

34. $\sin^{-1}(0.7821)$

35. $\tan^{-1}(1.895)$

36. $\tan^{-1}(3.2678)$

37. $\sec^{-1}(1.4973)$

38. $\sec^{-1}(2.7864)$

39. $\csc^{-1}(-3.7893)$

40. $\csc^{-1}(-6.1324)$

41. $\cot^{-1}(-4.2319)$

42. $\cot^{-1}(-0.8977)$

In Exercises 43–52, use a calculator to evaluate each expression. Give the answer in radians and round it to two decimal places.

43. $\sin^{-1}(-0.5878)$

44. $\sin^{-1}(0.8660)$

45. $\cos^{-1}(0.1423)$

46. $\tan^{-1}(-0.9279)$

47. $\tan^{-1}(1.3242)$

48. $\cot^{-1}(2.4142)$

49. $\cot^{-1}(-0.5774)$

50. $\sec^{-1}(-1.0422)$

51. $\csc^{-1}(3.2361)$

52. $\csc^{-1}(-2.9238)$

In Exercises 53–76, evaluate each expression exactly, if possible. If not possible, state why.

53. $\sin^{-1}\left[\sin\left(\dfrac{5\pi}{12}\right)\right]$

54. $\sin^{-1}\left[\sin\left(-\dfrac{5\pi}{12}\right)\right]$

55. $\sin[\sin^{-1}(1.03)]$

56. $\sin[\sin^{-1}(1.1)]$

57. $\sin^{-1}\left[\sin\left(-\dfrac{7\pi}{6}\right)\right]$

58. $\sin^{-1}\left[\sin\left(\dfrac{7\pi}{6}\right)\right]$

59. $\cos^{-1}\left[\cos\left(\dfrac{4\pi}{3}\right)\right]$

60. $\cos^{-1}\left[\cos\left(-\dfrac{5\pi}{3}\right)\right]$

61. $\cot\left[\cot^{-1}\left(\sqrt{3}\right)\right]$

62. $\cot^{-1}\left[\cot\left(\dfrac{5\pi}{4}\right)\right]$

63. $\sec^{-1}\left[\sec\left(-\dfrac{\pi}{3}\right)\right]$

64. $\sec\left[\sec^{-1}\left(\dfrac{1}{2}\right)\right]$

65. $\csc\left[\csc^{-1}\left(\dfrac{1}{2}\right)\right]$

66. $\csc^{-1}\left[\csc\left(\dfrac{7\pi}{6}\right)\right]$

67. $\cot(\cot^{-1}0)$

68. $\cot^{-1}\left[\cot\left(-\dfrac{\pi}{4}\right)\right]$

69. $\tan^{-1}\left[\tan\left(-\dfrac{\pi}{4}\right)\right]$

70. $\tan^{-1}\left[\tan\left(\dfrac{\pi}{4}\right)\right]$

71. $\sec(\sec^{-1}0)$

72. $\csc^{-1}[\csc(\pi)]$

73. $\cot^{-1}\left[\cot\left(\dfrac{8\pi}{3}\right)\right]$

74. $\tan^{-1}[\tan(8\pi)]$

75. $\csc^{-1}\left[\csc\left(\dfrac{15\pi}{4}\right)\right]$

76. $\sec^{-1}\left[\sec\left(\dfrac{17\pi}{2}\right)\right]$

In Exercises 77–96, evaluate each expression exactly.

77. $\cos\left[\sin^{-1}\left(\dfrac{3}{4}\right)\right]$

78. $\sin\left[\cos^{-1}\left(\dfrac{2}{3}\right)\right]$

79. $\sin\left[\tan^{-1}\left(\dfrac{12}{5}\right)\right]$

80. $\cos\left[\tan^{-1}\left(\dfrac{7}{24}\right)\right]$

81. $\tan\left[\sin^{-1}\left(\dfrac{3}{5}\right)\right]$

82. $\tan\left[\cos^{-1}\left(\dfrac{2}{5}\right)\right]$

83. $\sec\left[\sin^{-1}\left(\dfrac{\sqrt{2}}{5}\right)\right]$

84. $\sec\left[\cos^{-1}\left(\dfrac{\sqrt{7}}{4}\right)\right]$

85. $\csc\left[\cos^{-1}\left(\dfrac{1}{4}\right)\right]$

86. $\csc\left[\sin^{-1}\left(\dfrac{1}{4}\right)\right]$

87. $\cot\left[\sin^{-1}\left(\dfrac{60}{61}\right)\right]$

88. $\cot\left[\sec^{-1}\left(\dfrac{41}{9}\right)\right]$

89. $\cos\left[\tan^{-1}\left(\dfrac{3}{4}\right) - \sin^{-1}\left(\dfrac{4}{5}\right)\right]$

90. $\cos\left[\tan^{-1}\left(\dfrac{12}{5}\right) + \sin^{-1}\left(\dfrac{3}{5}\right)\right]$

91. $\sin\left[\cos^{-1}\left(\dfrac{5}{13}\right) + \tan^{-1}\left(\dfrac{4}{3}\right)\right]$

92. $\sin\left[\cos^{-1}\left(\dfrac{3}{5}\right) - \tan^{-1}\left(\dfrac{5}{12}\right)\right]$

93. $\sin\left[2\cos^{-1}\left(\dfrac{3}{5}\right)\right]$

94. $\cos\left[2\sin^{-1}\left(\dfrac{3}{5}\right)\right]$

95. $\tan\left[2\sin^{-1}\left(\dfrac{5}{13}\right)\right]$

96. $\tan\left[2\cos^{-1}\left(\dfrac{5}{13}\right)\right]$

For each of the following expressions, write an equivalent expression in terms of only the variable *u*.

97. $\cos(\sin^{-1}u)$

98. $\sin(\cos^{-1}u)$

99. $\tan(\cos^{-1}u)$

100. $\tan(\sin^{-1}u)$

■ **APPLICATIONS**

For Exercises 101 and 102, refer to the following:

Annual sales of a product are generally subject to seasonal fluctuations and are approximated by the function

$$s(t) = 4.3\cos\left(\frac{\pi}{6}t\right) + 56.2 \quad 0 \le t \le 11$$

where t represents time in months ($t = 0$ represents January) and $s(t)$ represents monthly sales of the product in thousands of dollars.

101. Business. Find the month(s) in which monthly sales are $56,200.

102. Business. Find the month(s) in which monthly sales are $51,900.

For Exercises 103 and 104, refer to the following:

Allergy sufferers' symptoms fluctuate with pollen levels. Pollen levels are often reported to the public on a scale of 0–12, which is meant to reflect the levels of pollen in the air. For example, a pollen level between 4.9 and 7.2 indicates that pollen levels will likely cause symptoms for many individuals allergic to the predominant pollen of the season (*Source:* http://www.pollen.com). The pollen levels at a single location were measured and averaged for each month. Over a period of 6 months, the levels fluctuated according to the model

$$p(t) = 5.5 + 1.5\sin\left(\frac{\pi}{6}t\right) \quad 0 \le t \le 6$$

where t is measured in months and $p(t)$ is the pollen level.

103. Biology/Health. In which month(s) was the monthly average pollen level 7.0?

104. Biology/Health. In which month(s) was the monthly average pollen level 6.25?

105. Alternating Current. Alternating electrical current in amperes (A) is modeled by the equation $i = I\sin(2\pi ft)$, where i is the current, I is the maximum current, t is time in seconds, and f is the frequency in hertz (Hz is the number of cycles per second). If the frequency is 5 Hz and maximum current is 115 A, what time t corresponds to a current of 85 A? Find the smallest positive value of t.

106. Alternating Current. If the frequency is 100 Hz and maximum current is 240 A, what time t corresponds to a current of 100 A? Find the smallest positive value of t.

107. Hours of Daylight. The number of hours of daylight in San Diego, California, can be modeled with $H(t) = 12 + 2.4\sin(0.017t - 1.377)$, where t is the day of the year (January 1, $t = 1$, etc.). For what value of t is the number of hours of daylight equal to 14.4? If May 31 is the 151st day of the year, what month and day correspond to that value of t?

108. Hours of Daylight. Repeat Exercise 107. For what value of t is the number of hours of daylight equal to 9.6? What month and day correspond to the value of t? (You may have to count backwards.)

109. Money. A young couple get married and immediately start saving money. They renovate a house and are left with less and less saved money. They have children after 10 years and are in debt until their children are in college. They then save until retirement. A formula that represents the percentage of their annual income that they either save (positive) or are in debt (negative) is given by $P(t) = 12.5\cos(0.157t) + 2.5$, where $t = 0$ corresponds to the year they were married. How many years into their marriage do they first accrue debt?

110. Money. For the couple in Exercise 109, how many years into their marriage are they back to saving 15% of their annual income?

111. Viewing Angle of Painting. A museum patron whose eye level is 5 feet above the floor is studying a painting that is 8 feet in height and mounted on the wall 4 feet above the floor. If the patron is x feet from the wall, use $\tan(\alpha + \beta)$ to express $\tan(\theta)$, where θ is the angle that the patron's eye sweeps from the top to the bottom of the painting.

112. Viewing Angle of Painting. Using the equation for $\tan(\theta)$ in Exercise 111, solve for θ using the inverse tangent. Then find the measure of the angles θ for $x = 10$ and $x = 20$ (to the nearest degree).

113. Earthquake Movement. The horizontal movement of a point that is k kilometers away from an earthquake's fault line can be estimated with

$$M = \frac{f}{2}\left[1 - \frac{2\tan^{-1}\left(\dfrac{k}{d}\right)}{\pi}\right]$$

where M is the movement of the point in meters, f is the total horizontal displacement occurring along the fault line, k is the distance of the point from the fault line, and d is the depth in kilometers of the focal point of the earthquake. If an earthquake produces a displacement f of 2 meters and the depth of the focal point is 4 kilometers, then what is the movement M of a point that is 2 kilometers from the fault line? of a point 10 kilometers from the fault line?

114. Earthquake Movement. Repeat Exercise 113. If an earthquake produces a displacement f of 3 meters and the depth of the focal point is 2.5 kilometers, then what is the movement M of a point that is 5 kilometers from the fault line? of a point 10 kilometers from the fault line?

115. Laser Communication. A laser communication system depends on a narrow beam, and a direct line of sight is necessary for communication links. If a transmitter/receiver for a laser system is placed between two buildings (see the figure) and the other end of the system is located on a low-earth-orbit satellite, then the link is operational only when the satellite and the ground system have a line of sight (when the buildings are not in the way). Find the angle θ that corresponds to the system being operational (i.e., find the maximum value of θ that permits the system to be operational). Express θ in terms of inverse tangent functions and the distance from the shorter building.

Corbis Digital Stock

116. Laser Communication. Repeat Exercise 115, assuming that the ground system is on top of a 20-foot tower.

CATCH THE MISTAKE

In Exercises 117–120, explain the mistake that is made.

117. Evaluate the expression exactly: $\sin^{-1}\left[\sin\left(\dfrac{3\pi}{5}\right)\right]$.

Solution:

Use the identity $\sin^{-1}(\sin x) = x$ on $0 \le x \le \pi$.

Since $\dfrac{3\pi}{5}$ is in the interval

$[0, \pi]$, the identity $\qquad \sin^{-1}\left[\sin\left(\dfrac{3\pi}{5}\right)\right] = \dfrac{3\pi}{5}$
can be used.

This is incorrect. What mistake was made?

118. Evaluate the expression exactly: $\cos^{-1}\left[\cos\left(-\dfrac{\pi}{5}\right)\right]$.

Solution:

Use the identity $\cos^{-1}(\cos x) = x$ on $-\dfrac{\pi}{2} \le x \le \dfrac{\pi}{2}$.

Since $-\dfrac{\pi}{5}$ is in the interval

$\left[-\dfrac{\pi}{2}, \dfrac{\pi}{2}\right]$, the identity

$\cos^{-1}\left[\cos\left(-\dfrac{\pi}{5}\right)\right] = -\dfrac{\pi}{5}$
can be used.

This is incorrect. What mistake was made?

119. Evaluate the expression exactly: $\cot^{-1}(2.5)$.

Solution:

Use the reciprocal identity. $\qquad \cot^{-1}(2.5) = \dfrac{1}{\tan^{-1}(2.5)}$

Evaluate $\tan^{-1}(2.5) = 1.19$. $\qquad \cot^{-1}(2.5) = \dfrac{1}{1.19}$

Simplify. $\qquad \cot^{-1}(2.5) = 0.8403$

This is incorrect. What mistake was made?

120. Evaluate the expression exactly: $\csc^{-1}\left(\dfrac{1}{4}\right)$.

Solution:

Use the reciprocal identity. $\qquad \csc^{-1}\left(\dfrac{1}{4}\right) = \dfrac{1}{\sin^{-1}\left(\dfrac{1}{4}\right)}$

Evaluate $\sin^{-1}\left(\tfrac{1}{4}\right) = 14.478$. $\qquad \csc^{-1}\left(\dfrac{1}{4}\right) = \dfrac{1}{14.478}$

Simplify. $\qquad \csc^{-1}\left(\dfrac{1}{4}\right) = 0.0691$

This is incorrect. What mistake was made?

■ **CONCEPTUAL**

In Exercises 121–124, determine whether each statement is true or false.

121. The inverse secant function is an even function.

122. The inverse cosecant function is an odd function.

123. $\csc^{-1}(\csc\theta) = \theta$, for all θ in the domain of cosecant.

124. $\sin^{-1}(2x) \cdot \csc^{-1}(2x) = 1$, for all x for which both functions are defined.

125. Explain why $\sec^{-1}\left(\frac{1}{2}\right)$ does not exist.

126. Explain why $\csc^{-1}\left(\frac{1}{2}\right)$ does not exist.

■ **CHALLENGE**

127. Find the expression that corresponds to $\sin\left[\cos^{-1}\left(\frac{1}{x}\right)\right]$.

128. Determine the x-values for which

$$\sin^{-1}\left[2\sin\left(\frac{3x}{2}\right)\cos\left(\frac{3x}{2}\right)\right] = 3x$$

129. Evaluate exactly: $\sin(2\sin^{-1} 1)$.

130. Let $f(x) = 2 - 4\sin\left(x - \frac{\pi}{2}\right)$.

    a. State an accepted domain of $f(x)$ so that $f(x)$ is a one-to-one function.

    b. Find $f^{-1}(x)$ and state its domain.

131. Let $f(x) = 3 + \cos\left(x - \frac{\pi}{4}\right)$.

    a. State an accepted domain of $f(x)$ so that $f(x)$ is a one-to-one function.

    b. Find $f^{-1}(x)$ and state its domain.

132. Let $f(x) = 1 - \tan\left(x + \frac{\pi}{3}\right)$.

    a. State an accepted domain of $f(x)$ so that $f(x)$ is a one-to-one function.

    b. Find $f^{-1}(x)$ and state its domain.

133. Let $f(x) = 2 + \frac{1}{4}\cot\left(2x - \frac{\pi}{6}\right)$.

    a. State an accepted domain of $f(x)$ so that $f(x)$ is a one-to-one function.

    b. Find $f^{-1}(x)$ and state its domain.

134. Let $f(x) = -\csc\left(\frac{\pi}{4}x - 1\right)$.

    a. State an accepted domain of $f(x)$ so that $f(x)$ is a one-to-one function.

    b. Find $f^{-1}(x)$ and state its domain.

■ **TECHNOLOGY**

135. Use a graphing calculator to plot $Y_1 = \sin(\sin^{-1}x)$ and $Y_2 = x$ for the domain $-1 \le x \le 1$. If you then increase the domain to $-3 \le x \le 3$, you get a different result. Explain the result.

136. Use a graphing calculator to plot $Y_1 = \cos(\cos^{-1}x)$ and $Y_2 = x$ for the domain $-1 \le x \le 1$. If you then increase the domain to $-3 \le x \le 3$, you get a different result. Explain the result.

137. Use a graphing calculator to plot $Y_1 = \csc^{-1}(\csc x)$ and $Y_2 = x$. Determine the domain for which the following statement is true: $\csc^{-1}(\csc x) = x$. Give the domain in terms of π.

138. Use a graphing calculator to plot $Y_1 = \sec^{-1}(\sec x)$ and $Y_2 = x$. Determine the domain for which the following statement is true: $\sec^{-1}(\sec x) = x$. Give the domain in terms of π.

139. Given $\tan x = \frac{40}{9}$ and $\pi < x < \frac{3\pi}{2}$:

    a. Find $\sin(2x)$ using the double-angle identity.

    b. Use the inverse of tangent to find x in QIII and use a calculator to find $\sin(2x)$. Round to five decimal places.

    c. Are the results in (a) and (b) the same?

140. Given $\sin x = -\frac{1}{\sqrt{10}}$ and $\frac{3\pi}{2} < x < 2\pi$:

    a. Find $\tan(2x)$ using the double-angle identity.

    b. Use the inverse of sine to find x in QIV and find $\tan(2x)$.

    c. Are the results in (a) and (b) the same?

Skills Objectives

- Solve trigonometric equations by inspection.
- Solve trigonometric equations using algebraic techniques.
- Solve trigonometric equations using inverse functions.
- Solve trigonometric equations (involving more than one trigonometric function) using trigonometric identities.

Conceptual Objectives

- Understand that solving trigonometric equations is similar to solving algebraic equations.
- Realize that the goal in solving trigonometric equations is to find the value(s) for the independent variable which make(s) the equation true.

Recall that in solving algebraic equations, the goal is to find the value for the variable which makes the equation true. For example, the linear equation $2x - 5 = 7$ has only one value, $x = 6$, which makes the statement true. A quadratic equation, however, can have two solutions. The equation $x^2 = 9$ has two values, $x = \pm 3$, which make the statement true. With trigonometric equations, the goal is the same: find the value (or values) that makes (or make) the equation true.

Solving Trigonometric Equations by Inspection

To solve the equation

$$\sin x + \cos x = 1 \qquad 0 \leq x < 2\pi$$

we can draw the unit circle and ask the question: "Which points on the unit circle have x- and y-coordinates that sum to 1?"

Inspecting the figure to the left we see that $x = 0$ and $x = \dfrac{\pi}{2}$ correspond to x- and y-coordinates along the unit circle that sum to 1. The solution to $\sin x + \cos x = 1$ on the x interval $[0, 2\pi]$ is $\boxed{x = 0, \dfrac{\pi}{2}}$. This method of solving a trigonometric equation is called **inspection**.

EXAMPLE 1 Solving a Trigonometric Equation by Inspection

Solve each of the following equations over $[0, 2\pi)$:

a. $\sin x = \frac{1}{2}$ **b.** $\cos(2x) = \frac{1}{2}$

Solution (a):

Ask the question, "sine of what angles is $\frac{1}{2}$?"

$$x = \frac{\pi}{6} \text{ or } x = \frac{5\pi}{6}$$

Solution (b):

Ask the question, "cosine of what angles is $\frac{1}{2}$?"

In this case, the angle is equal to $2x$.

$$2x = \frac{\pi}{3} \text{ or } 2x = \frac{5\pi}{3}$$

Solve for x: $x = \frac{\pi}{6}$ or $x = \frac{5\pi}{6}$.

Note: $x = \dfrac{7\pi}{6}$ and $x = \dfrac{11\pi}{6}$ are in the interval $[0, 2\pi)$ and also satisfy the equation $\cos(2x) = \dfrac{1}{2}$.

$$x = \frac{\pi}{6}, \frac{5\pi}{6}, \frac{7\pi}{6}, \text{ and } \frac{11\pi}{6}$$

▪ **YOUR TURN** Solve each of the following equations over $[0, 2\pi)$:

a. $\cos x = \frac{1}{2}$ **b.** $\sin(2x) = \frac{1}{2}$

EXAMPLE 2 Solving a Trigonometric Equation by Inspection

Solve the equation $\sin x = \dfrac{\sqrt{2}}{2}$.

Classroom Example 7.8.2
Find all solutions of
$$\cos x = -\frac{\sqrt{3}}{2}.$$

Answer:
$$\frac{5\pi}{6} + 2n\pi, \frac{7\pi}{6} + 2n\pi,$$
where n is an integer

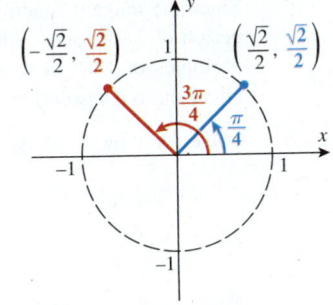

Technology Tip

Use the fact that a solution to the equation $\sin x = \dfrac{\sqrt{2}}{2}$ is the same as a point of intersection of $y = \sin x$ and $y = \dfrac{\sqrt{2}}{2}$ over one period, $[0, 2\pi)$ or $[0, 360°)$.

Solution:

STEP 1 Solve over one period, $[0, 2\pi)$.

Ask the question, "sine of what angles is $\dfrac{\sqrt{2}}{2}$?"

The sine function is positive in quadrants I and II.

DEGREES	$x = 45°$ or $x = 135°$
RADIANS	$x = \dfrac{\pi}{4}$ .or $x = \dfrac{3\pi}{4}$

STEP 2 Solve over all real numbers.

Since the sine function has a period of $360°$ or 2π, adding integer multiples of $360°$ or 2π will give the other infinitely many solutions.

DEGREES	$x = 45° + 360°n$ or $x = 135° + 360°n$
RADIANS	$x = \dfrac{\pi}{4} + 2n\pi$ or $x = \dfrac{3\pi}{4} + 2n\pi$, where n is any integer

Study Tip

Find *all* solutions unless the domain is restricted.

■ **Answer:**

DEGREES	$x = 60° + 360°n$ or $x = 300° + 360°n$
RADIANS	$x = \dfrac{\pi}{3} + 2n\pi$ or $x = \dfrac{5\pi}{3} + 2n\pi$, where n is any integer

■ **YOUR TURN** Solve the equation $\cos x = \frac{1}{2}$.

Notice that the equations in Example 2 and the Your Turn have an infinite number of solutions. Unless the domain is restricted, you must find *all* solutions.

EXAMPLE 3 Solving a Trigonometric Equation by Inspection

Solve the equation $\tan(2x) = -\sqrt{3}$.

Solution:

STEP 1 Solve over one period, $[0, \pi)$.

Ask the question, "tangent of what angles is $-\sqrt{3}$?" Note that the angle in this case is $2x$.

DEGREES	$2x = 120°$
RADIANS	$2x = \dfrac{2\pi}{3}$

The tangent function is negative in quadrants II and IV. Since $[0, \pi)$ includes quadrants I and II, we find only the angle in quadrant II. (The solution corresponding to quadrant IV will be found when we extend the solution over all real numbers.)

Technology Tip

To find the point of intersection, use 2nd TRACE for CALC, move the down arrow to 5: Intersect, type ENTER for the first curve, ENTER for the second curve, 0.8 for guess, and ENTER.

Intersection
X=1.0471976 Y=-1.732051

For the second answer, you need a number close to the answer for the guess. Type 2.5 for guess, and ENTER.

Intersection
X=2.6179939 Y=-1.732051

Classroom Example 7.8.3
Find all solutions of $\cot(3x) = 1$.
Answer:

$\dfrac{\pi}{12} + \dfrac{n\pi}{3}$, where n is an integer

STEP 2 Solve over all x.

Since the tangent function has a period of 180°, or π, adding integer multiples of 180° or π will give all of the other solutions.

Solve for x by dividing by 2.

DEGREES	$2x = 120° + 180°n$
RADIANS	$2x = \dfrac{2\pi}{3} + n\pi$, where n is any integer

DEGREES	$x = 60° + 90°n$
RADIANS	$x = \dfrac{\pi}{3} + \dfrac{n\pi}{2}$, where n is any integer

Note:

- There are infinitely many solutions. If we graph $y = \tan(2x)$ and $y = -\sqrt{3}$, we see that there are infinitely many points of intersection.

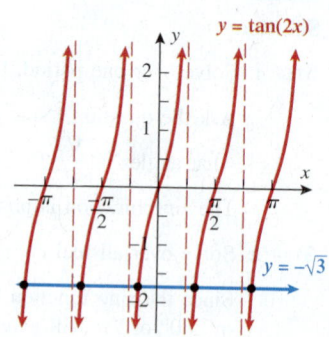

- Had we restricted the domain to $0 \le x < 2\pi$, the solutions (in radians) would be the values given to the right in the table.

Notice that only $n = 0, 1, 2, 3$ yield x-values in the domain $0 \le x < 2\pi$.

n	$x = \dfrac{\pi}{3} + \dfrac{n\pi}{2}$
0	$x = \dfrac{\pi}{3}$
1	$x = \dfrac{5\pi}{6}$
2	$x = \dfrac{4\pi}{3}$
3	$x = \dfrac{11\pi}{6}$

Notice that in Step 2 of Example 2, $2n\pi$ was added to get all of the solutions, whereas in Step 2 of Example 3, we added $n\pi$ to the argument of the tangent function. The reason that we added $2n\pi$ in Example 2 and $n\pi$ in Example 3 is because the sine function has period 2π, whereas the tangent function has period π.

Solving Trigonometric Equations Using Algebraic Techniques

We now will use algebraic techniques to solve trigonometric equations. Let us first start with linear and quadratic equations. For linear equations, we solve for the variable by isolating it. For quadratic equations, we often employ factoring or the quadratic formula. If we can let x represent the trigonometric function and the resulting equation is either linear or quadratic, then we use techniques learned in solving algebraic equations.

TYPE	EQUATION	SUBSTITUTION	ALGEBRAIC EQUATION
Linear trigonometric equation	$4\sin\theta - 2 = -4$	$x = \sin\theta$	$4x - 2 = -4$
Quadratic trigonometric equation	$2\cos^2\theta + \cos\theta - 1 = 0$	$x = \cos\theta$	$2x^2 + x - 1 = 0$

It is not necessary to make the substitution, though it is convenient. Frequently, one can see how to factor a quadratic trigonometric equation without first converting it to an algebraic equation. In Example 4, we will not use the substitution. However, in Example 5, we will illustrate the use of a substitution.

EXAMPLE 4   Solving a Linear Trigonometric Equation

Solve $4\sin\theta - 2 = -4$ on $0 \le \theta < 2\pi$.

Solution:

STEP 1 Solve for $\sin\theta$.

Add 2.

Divide by 4.

$$4\sin\theta - 2 = -4$$
$$4\sin\theta = -2$$
$$\sin\theta = -\frac{1}{2}$$

STEP 2 Find the values of θ on $0 \le \theta < 2\pi$ which satisfy the equation $\sin\theta = -\frac{1}{2}$.

The sine function is negative in quadrants III and IV:

$$\sin\left(\frac{7\pi}{6}\right) = -\frac{1}{2} \text{ and } \sin\left(\frac{11\pi}{6}\right) = -\frac{1}{2}.$$

$$\boxed{\theta = \frac{7\pi}{6}} \text{ or } \boxed{\theta = \frac{11\pi}{6}}$$

■ **YOUR TURN** Solve $2\cos\theta + 1 = 2$ on $0 \le \theta < 2\pi$.

Classroom Example 7.8.4*
Find all solutions of
$2\sin(3x) - 3\sqrt{3} = -4\sqrt{3}$.

Answer:
$\dfrac{4\pi}{9} + \dfrac{2n\pi}{3}, \dfrac{5\pi}{9} + \dfrac{2n\pi}{3}$, where n is an integer

■ **Answer:** $\theta = \dfrac{\pi}{3}$ or $\dfrac{5\pi}{3}$

Technology Tip

EXAMPLE 5   Solving a Quadratic Trigonometric Equation

Solve $2\cos^2\theta + \cos\theta - 1 = 0$ on $0 \le \theta < 2\pi$.

Solution:

STEP 1 Solve for $\cos\theta$.

Let $x = \cos\theta$.

Factor the quadratic equation.

Set each factor equal to 0.

$$2\cos^2\theta + \cos\theta - 1 = 0$$
$$2x^2 + x - 1 = 0$$
$$(2x - 1)(x + 1) = 0$$
$$2x - 1 = 0 \text{ or } x + 1 = 0$$

Study Tip

If Example 5 asked for the solution to the trigonometric equation over all real numbers, then the solutions would be $\theta = \dfrac{\pi}{3} + 2n\pi$,

$\dfrac{5\pi}{3} + 2n\pi, \pi + 2n\pi$.

■ **Answer:** $\theta = \dfrac{\pi}{2}, \dfrac{7\pi}{6}$, or $\dfrac{11\pi}{6}$

Solve each equation for x.
$$x = \frac{1}{2} \text{ or } x = -1$$

Substitute back to θ: $x = \cos\theta$.
$$\cos\theta = \frac{1}{2} \text{ or } \cos\theta = -1$$

STEP 2 Find the values of θ on $0 \leq \theta < 2\pi$ that satisfy the equation $\cos\theta = \frac{1}{2}$.

The cosine function is positive in quadrants I and IV.

$$\cos\left(\frac{\pi}{3}\right) = \frac{1}{2} \text{ and } \cos\left(\frac{5\pi}{3}\right) = \frac{1}{2}.$$

$$\boxed{\theta = \frac{\pi}{3}} \text{ or } \boxed{\theta = \frac{5\pi}{3}}$$

STEP 3 Find the values of θ on $0 \leq \theta < 2\pi$ that satisfy the equation $\cos\theta = -1$.

$$\cos\pi = -1 \qquad\qquad \boxed{\theta = \pi}$$

The solutions to $2\cos^2\theta + \cos\theta - 1 = 0$ on $0 \leq \theta < 2\pi$ are $\theta = \dfrac{\pi}{3}, \theta = \dfrac{5\pi}{3}$, or $\theta = \pi$.

■ **YOUR TURN** Solve $2\sin^2\theta - \sin\theta - 1 = 0$ on $0 \leq \theta < 2\pi$.

Solving Trigonometric Equations Which Require the Use of Inverse Functions

Thus far, we have been able to solve the trigonometric equations exactly. Now we turn our attention to situations that require using a calculator and inverse functions to approximate a solution to a trigonometric equation.

Technology Tip

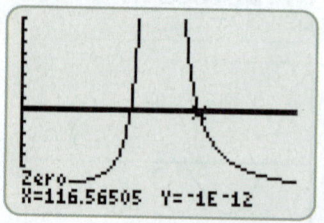

EXAMPLE 6 **Solving a Trigonometric Equation Which Requires the Use of Inverse Functions**

Solve $\tan^2\theta - \tan\theta = 6$ on $0° \leq \theta < 180°$.

Solution:

STEP 1 Solve for $\tan\theta$.

Subtract 6. $\qquad\qquad \tan^2\theta - \tan\theta - 6 = 0$

Factor the quadratic trigonometric expression on the left. $\qquad (\tan\theta - 3)(\tan\theta + 2) = 0$

Set the factors equal to 0. $\qquad \tan\theta - 3 = 0 \text{ or } \tan\theta + 2 = 0$

Solve for $\tan\theta$. $\qquad\qquad \tan\theta = 3 \text{ or } \tan\theta = -2$

STEP 2 Solve $\tan\theta = 3$ on $0° \leq \theta < 180°$.

The tangent function is positive on $0° \leq \theta < 180°$ only in quadrant I.

Write the equivalent inverse notation to $\tan\theta = 3$. $\qquad \theta = \tan^{-1}3$

Use a calculator to evaluate (approximate) θ. $\qquad \boxed{\theta \approx 71.6°}$

STEP 3 Solve $\tan\theta = -2$ on $0° \leq \theta < 180°$.

The tangent function is negative on $0° \leq \theta < 180°$ only in quadrant II.

A calculator gives values of the inverse tangent in quadrants I and IV.

We will call the reference angle in quadrant IV "α."

Write the equivalent inverse notation
to $\tan\alpha = -2$. $\qquad \alpha = \tan^{-1}(-2)$

Use a calculator to evaluate (approximate) α. $\qquad \alpha \approx -63.4°$

To find the value of θ in quadrant II, add $180°$. $\qquad \theta = \alpha + 180°$

$$\boxed{\theta \approx 116.6°}$$

The solutions to $\tan^2\theta - \tan\theta = 6$ on $0° \leq \theta < 180°$ are $\theta = 71.6°$ or $\theta = 116.6°$.

▪ **YOUR TURN** Solve $\tan^2\theta + \tan\theta = 6$ on $0° \leq \theta < 180°$.

▪ **Answer:** $\theta \approx 63.4°$ or $108.4°$

Recall that in solving algebraic quadratic equations, one method (when factoring is not obvious or possible) is to use the Quadratic Formula.

$$ax^2 + bx + c = 0 \text{ has solutions } x = \frac{-b \pm \sqrt{b^2 - 4ac}}{2a} \quad a \neq 0$$

EXAMPLE 7 **Solving a Quadratic Trigonometric Equation Which Requires the Use of the Quadratic Formula and Inverse Functions**

Solve $2\cos^2\theta + 5\cos\theta - 6 = 0$ on $0° \leq \theta < 360°$.

Solution:

STEP 1 Solve for $\cos\theta$. $\qquad 2\cos^2\theta + 5\cos\theta - 6 = 0$

Let $x = \cos\theta$. $\qquad 2x^2 + 5x - 6 = 0$

Use the Quadratic Formula,
$a = 2, b = 5, c = -6$. $\qquad x = \dfrac{-5 \pm \sqrt{5^2 - 4(2)(-6)}}{2(2)}$

Simplify. $\qquad x = \dfrac{-5 \pm \sqrt{73}}{4}$

Use a calculator to approximate
the solution. $\qquad x \approx -3.3860 \text{ or } \quad x \approx 0.8860$

Substitute back to θ: $x = \cos\theta$. $\qquad \cos\theta \approx -3.3860 \text{ or } \cos\theta \approx 0.8860$

STEP 2 Solve $\cos\theta = -3.3860$ on $0° \leq \theta < 360°$.

Recall that the range of the cosine function is $[-1, 1]$; therefore, the cosine function can never equal a number outside that range ($-3.3860 < -1$).

There is $\boxed{\text{no solution}}$ with this equation.

Technology Tip

STEP 3 Solve $\cos\theta = 0.8860$ on $0° \le \theta < 360°$.

The cosine function is positive in quadrants I and IV. Since a calculator gives inverse cosine values only in quadrants I and II, we will have to use a reference angle to get the quadrant IV solution.

Write the equivalent inverse notation for $\cos\theta = 0.8860$.

$$\theta = \cos^{-1}0.8860$$

Use a calculator to evaluate (approximate) the solution.

$$\boxed{\theta \approx 27.6°}$$

To find the second solution (in quadrant IV), subtract the reference angle from 360°.

$$\theta = 360° - 27.6°$$

$$\boxed{\theta \approx 332.4°}$$

The solutions to $2\cos^2\theta + 5\cos\theta - 6 = 0$ on $0° \le \theta < 360°$ are $\theta \approx 27.6°$ or $\theta \approx 332.4°$.

■ **Answer:** $\theta \approx 242.4°$ or 297.6°

■ **YOUR TURN** Solve $2\sin^2\theta - 5\sin\theta - 6 = 0$ on $0° \le \theta < 360°$.

Using Trigonometric Identities to Solve Trigonometric Equations

We now consider trigonometric equations that involve more than one trigonometric function. Trigonometric identities are an important part of solving these types of equations. The trigonometric equation in Example 8 was solved by inspection in the beginning of this section. We solve it again now using trigonometric identities.

EXAMPLE 8 **Using Trigonometric Identities to Solve Trigonometric Equations**

Solve $\sin x + \cos x = 1$ on $0 \le x < 2\pi$.

Solution:

Square both sides.

$$\sin^2 x + 2\sin x\cos x + \cos^2 x = 1$$

Label the Pythagorean identity.

$$\underbrace{\sin^2 x + \cos^2 x}_{1} + 2\sin x\cos x = 1$$

Subtract 1 from both sides.

$$2\sin x\cos x = 0$$

Use the Zero Product Property.

$$\sin x = 0 \quad \text{or} \quad \cos x = 0$$

Solve for x on $0 \le x < 2\pi$.

$$x = 0 \text{ or } x = \pi \text{ or } x = \frac{\pi}{2} \text{ or } x = \frac{3\pi}{2}$$

Because we squared the equation, we have to check for extraneous solutions.

Check $x = 0$.

$$\sin 0 + \cos 0 = 0 + 1 = 1 \ \checkmark$$

Check $x = \pi$.

$$\sin \pi + \cos \pi = 0 - 1 = -1 \ \checkmark$$

Check $x = \dfrac{\pi}{2}$. $\qquad\qquad \sin\left(\dfrac{\pi}{2}\right) + \cos\left(\dfrac{\pi}{2}\right) = 1 + 0 = 1$ ✓

Check $x = \dfrac{3\pi}{2}$. $\qquad\qquad \sin\left(\dfrac{3\pi}{2}\right) + \cos\left(\dfrac{3\pi}{2}\right) = -1 + 0 = -1$ ✓

The solutions to $\sin x + \cos x = 1$ on $0 \le x < 2\pi$ are $\boxed{x = 0}$ or $\boxed{x = \dfrac{\pi}{2}}$.

■ **YOUR TURN** Solve $\sin x - \cos x = 1$ on $0 \le x < 2\pi$.

■ **Answer:** $x = \dfrac{\pi}{2}$ or $x = \pi$

▶ **EXAMPLE 9** **Using Trigonometric Identities to Solve Trigonometric Equations**

Solve $\sin(2x) = \sin x$ on $0 \le x < 2\pi$.

COMMON MISTAKE

Dividing by a trigonometric function (which could be equal to zero).

✪ **CORRECT**

Use the double-angle formula for sine.

$$\underset{2\sin x \cos x}{\underbrace{\sin(2x)}} = \sin x$$

Subtract $\sin x$.

$$2\sin x \cos x - \sin x = 0$$

Factor the common $\sin x$.

$$(\sin x)(2\cos x - 1) = 0$$

Set each factor equal to 0.

$$\sin x = 0 \text{ or } 2\cos x - 1 = 0$$

$$\sin x = 0 \text{ or } \cos x = \dfrac{1}{2}$$

Solve $\sin x = 0$ for x on $0 \le x < 2\pi$.

$$\boxed{x = 0} \quad \text{or} \quad \boxed{x = \pi}$$

Solve $\cos x = \frac{1}{2}$ for x on $0 \le x < 2\pi$.

$$\boxed{x = \dfrac{\pi}{3}} \quad \text{or} \quad \boxed{x = \dfrac{5\pi}{3}}$$

The solutions to $\sin(2x) = \sin x$ are $x = 0, \dfrac{\pi}{3}, \pi$, and $\dfrac{5\pi}{3}$.

✖ **INCORRECT**

$$2\sin x \cos x = \sin x$$

Divide by $\sin x$. **ERROR**

$$2\cos x = 1$$

Solve for $\cos x$.

$$\cos x = \dfrac{1}{2}$$

Missing solutions from $\sin x = 0$.

▼ **CAUTION**

Do not divide equations by trigonometric functions, as they can sometimes equal zero.

Classroom Example 7.8.9*

Solve

$$-\dfrac{1}{4}\csc^2\left(\dfrac{x}{6}\right) + \cos^2\left(\dfrac{x}{6}\right) = 0$$

on $[0, 6\pi]$.

Answer: $\dfrac{3\pi}{2}, \dfrac{9\pi}{2}$

■ **YOUR TURN** Solve $\sin(2x) = \cos x$ on $0 \le x < 2\pi$.

■ **Answer:**

$x = \dfrac{\pi}{2}, \dfrac{3\pi}{2}, \dfrac{\pi}{6}, \text{ or } \dfrac{5\pi}{6}$

EXAMPLE 10 Using Trigonometric Identities to Solve Trigonometric Equations

Solve $\sin x + \csc x = -2$.

Solution:

Use the reciprocal identity.

$$\sin x + \underbrace{\csc x}_{\frac{1}{\sin x}} = -2$$

Add 2.

$$\sin x + 2 + \frac{1}{\sin x} = 0$$

Multiply by $\sin x$. (Note $\sin x \neq 0$.)

$$\sin^2 x + 2\sin x + 1 = 0$$

Factor as a perfect square.

$$(\sin x + 1)^2 = 0$$

Solve for $\sin x$.

$$\sin x = -1$$

Solve for x on one period of the sine function, $[0, 2\pi)$.

$$x = \frac{3\pi}{2}$$

Add integer multiples of 2π to obtain all solutions.

$$\boxed{x = \frac{3\pi}{2} + 2n\pi}$$

EXAMPLE 11 Using Trigonometric Identities and Inverse Functions to Solve Trigonometric Equations

Solve $3\cos^2\theta + \sin\theta = 3$ on $0° \leq \theta < 360°$.

Solution:

Use the Pythagorean identity.

$$3\underbrace{\cos^2\theta}_{1 - \sin^2\theta} + \sin\theta = 3$$

Subtract 3.

$$3(1 - \sin^2\theta) + \sin\theta - 3 = 0$$

Eliminate the parentheses.

$$3 - 3\sin^2\theta + \sin\theta - 3 = 0$$

Simplify.

$$-3\sin^2\theta + \sin\theta = 0$$

Factor the common $\sin\theta$.

$$\sin\theta(1 - 3\sin\theta) = 0$$

Set each factor equal to 0.

$$\sin\theta = 0 \quad \text{or} \quad 1 - 3\sin\theta = 0$$

Solve for $\sin\theta$.

$$\sin\theta = 0 \quad \text{or} \quad \sin\theta = \frac{1}{3}$$

Solve $\sin\theta = 0$ for x on $0° \leq \theta < 360°$.

$$\boxed{\theta = 0°} \text{ or } \boxed{\theta = 180°}$$

Solve $\sin\theta = \frac{1}{3}$ for x on $0° \leq \theta < 360°$.

The sine function is positive in quadrants I and II.

A calculator gives inverse values only in quadrant I.

Write the equivalent inverse notation for $\sin\theta = \frac{1}{3}$.

$$\theta = \sin^{-1}\left(\frac{1}{3}\right)$$

Use a calculator to approximate the quadrant I solution.

$$\boxed{\theta \approx 19.5°}$$

To find the quadrant II solution, subtract the reference angle from 180°.

$$\theta \approx 180° - 19.5°$$

$$\boxed{\theta \approx 160.5°}$$

Applications

EXAMPLE 12   Applications Involving Trigonometric Equations

Light bends (refracts) according to Snell's law, which states

$$n_i \sin(\theta_i) = n_r \sin(\theta_r)$$

where

- n_i is the refractive index of the medium the light is leaving.
- θ_i is the incident angle between the light ray and the normal (perpendicular) to the interface between mediums.
- n_r is the refractive index of the medium the light is entering.
- θ_r is the refractive angle between the light ray and the normal (perpendicular) to the interface between mediums.

Janis Christie/Getty Images, Inc.

Assume that light is going from air into a diamond. Calculate the refractive angle θ_r if the incidence angle is $\theta_i = 32°$ and the index of refraction values for air and diamond are $n_i = 1.00$ and $n_r = 2.417$, respectively.

32°

$n_i = 1.00$ θ_i
Air

Diamond

$n_r = 2.417$

θ_r

Solution:

Write Snell's law.	$n_i \sin(\theta_i) = n_r \sin(\theta_r)$
Substitute $\theta_i = 32°$, $n_i = 1.00$, and $n_r = 2.417$.	$\sin 32° = 2.417 \sin \theta_r$
Isolate $\sin\theta_r$ and simplify.	$\sin\theta_r = \dfrac{\sin 32°}{2.417} \approx 0.21925$
Solve for θ_r using the inverse sine function.	$\theta_r \approx \sin^{-1}(0.21925) \approx 12.665°$
Round to the nearest degree.	$\boxed{\theta_r \approx 13°}$

Musical tones can be represented mathematically with sinusoidal functions. Recall that a tone of 48 Hz (cycles per second) can be represented by the function $A\sin[2\pi(48)t]$, where A is the amplitude (loudness) and 48 Hz is the frequency. If we play two musical tones simultaneously, the combined tone can be found using the sum-to-product identity. For example, if a tone of 48 Hz is simultaneously played with a tone of 56 Hz, the result (assuming uniform amplitude, $A = 1$) is $\sin(96\pi t) + \sin(112\pi t) = 2\sin(104\pi t)\cos(8\pi t)$, whose graph is given on the right.

The $\sin(104\pi t)$ term represents a sound of average frequency 52 Hz. The $2\cos(8\pi t)$ represents a time-varying amplitude and a "beat" frequency that corresponds to 8 beats per second. Notice that in the graph there are 4 beats in $\frac{1}{2}$ second.

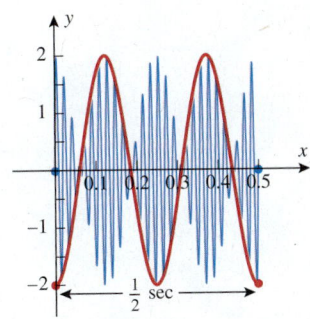

EXAMPLE 13 Tuning Fork

A tuning fork is used to help musicians "tune" their instruments. They simultaneously listen to the vibrating fork and play a tone and adjust their instrument until the beat frequency is eliminated.

If an E (659 Hz) tuning fork is used and the musician hears 4 beats per second, find the frequency of the tone from the instrument if the loudest beat is the third of each group. Assume that each tone has an amplitude of 1.

abzee/iStockphoto

Solution:

Express the tone E mathematically.

$$\sin[2\pi(659)t]$$

Express the instrument's tone mathematically.

$$\sin(2\pi ft)$$

Add the two to get the combined sound.

$$\sin[2\pi(659)t] + \sin(2\pi ft)$$

Use the sum-to-product identity.

$$2\sin\left[2\left(\frac{f+659}{2}\right)\pi t\right]\cos\left[2\left(\frac{f-659}{2}\right)\pi t\right]$$

The beat frequency is 4 Hz.

$$2\sin\left[2\left(\frac{f+659}{2}\right)\pi t\right]\cos[4\pi t]$$

Since the cosine function is an even function, so that $\cos(-x) = \cos x$, we equate the absolute value of the cosine arguments.

$$\left|2\left(\frac{f-659}{2}\right)\pi t\right| = |4\pi t|$$

Solve for f.

$$f = 663 \quad\text{or}\quad f = 655$$

The loudest sound, $|A| = 2$, is heard on the third beat of each four, $t = \frac{3}{4}$ sec.

$$2\sin\left[(f+659)\frac{3}{4}\pi\right]\cos\left[4\pi\frac{3}{4}\right] = 2$$

Simplify.

$$\sin\left[(f+659)\frac{3}{4}\pi\right] = -1$$

The inverse does not exist unless we restrict the domain of the sine function. Therefore, we inspect the two choices, 655 Hz or 663 Hz, and both satisfy the equation. The instrument can be playing either frequency based on the information given.

SECTION
7.8 SUMMARY

In this section we began by solving basic trigonometric equations that contained only one trigonometric function. Some such equations can be solved exactly by inspection, and others can be solved exactly using algebraic techniques similar to those of linear and quadratic equations. Calculators and inverse functions are needed when exact values are not known. It is important to note that calculators give the inverse function in only one of the two relevant quadrants. The other quadrant solutions must be found using reference angles. Trigonometric identities are useful for solving equations that involve more than one trigonometric function. With trigonometric identities we can transform such equations into equations involving only one trigonometric function, and then we can apply algebraic techniques.

SECTION
7.8 EXERCISES

▪ SKILLS

In Exercises 1–20, solve the given trigonometric equation exactly over the indicated interval.

1. $\cos\theta = -\dfrac{\sqrt{2}}{2}, \ 0 \le \theta < 2\pi$

2. $\sin\theta = -\dfrac{\sqrt{2}}{2}, \ 0 \le \theta < 2\pi$

3. $\csc\theta = -2, \ 0 \le \theta < 4\pi$

4. $\sec\theta = -2, \ 0 \le \theta < 4\pi$

5. $\tan\theta = 0,$ all real numbers

6. $\cot\theta = 0,$ all real numbers

7. $\sin(2\theta) = -\dfrac{1}{2}, \ 0 \le \theta < 2\pi$

8. $\cos(2\theta) = \dfrac{\sqrt{3}}{2}, \ 0 \le \theta < 2\pi$

9. $\sin\left(\dfrac{\theta}{2}\right) = -\dfrac{1}{2},$ all real numbers

10. $\cos\left(\dfrac{\theta}{2}\right) = -1,$ all real numbers

11. $\tan(2\theta) = \sqrt{3}, \ -2\pi \le \theta < 2\pi$

12. $\tan(2\theta) = -\sqrt{3},$ all real numbers

13. $\sec\theta = -2, \ -2\pi \le \theta < 0$

14. $\csc\theta = \dfrac{2\sqrt{3}}{3}, \ -\pi \le \theta < \pi$

15. $\cot(4\theta) = -\dfrac{\sqrt{3}}{3},$ all real numbers

16. $\tan(5\theta) = 1,$ all real numbers

17. $\sec(3\theta) = -1, \ -2\pi \le \theta \le 0$

18. $\sec(4\theta) = \sqrt{2}, \ 0 \le \theta \le \pi$

19. $\csc(3\theta) = 1, \ -2\pi \le \theta \le 0$

20. $\csc(6\theta) = -\dfrac{2\sqrt{3}}{3}, \ 0 \le \theta \le \pi$

In Exercises 21–40, solve the given trigonometric equation exactly on $0 \le \theta < 2\pi$.

21. $2\sin(2\theta) = \sqrt{3}$

22. $2\cos\left(\dfrac{\theta}{2}\right) = -\sqrt{2}$

23. $3\tan(2\theta) - \sqrt{3} = 0$

24. $4\tan\left(\dfrac{\theta}{2}\right) - 4 = 0$

25. $2\cos(2\theta) + 1 = 0$

26. $4\csc(2\theta) + 8 = 0$

27. $\sqrt{3}\cot\left(\dfrac{\theta}{2}\right) - 3 = 0$

28. $\sqrt{3}\sec(2\theta) + 2 = 0$

29. $\tan^2\theta - 1 = 0$

30. $\sin^2\theta + 2\sin\theta + 1 = 0$

31. $2\cos^2\theta - \cos\theta = 0$

32. $\tan^2\theta - \sqrt{3}\tan\theta = 0$

33. $\csc^2\theta + 3\csc\theta + 2 = 0$

34. $\cot^2\theta = 1$

35. $\sin^2\theta + 2\sin\theta - 3 = 0$

36. $2\sec^2\theta + \sec\theta - 1 = 0$

37. $\sec^2\theta - 1 = 0$

38. $\csc^2\theta - 1 = 0$

39. $\sec^2(2\theta) - \dfrac{4}{3} = 0$

40. $\csc^2(2\theta) - 4 = 0$

In Exercises 41–60, solve the given trigonometric equation on $0° \le \theta < 360°$ and express the answer in degrees to two decimal places.

41. $\sin(2\theta) = -0.7843$

42. $\cos(2\theta) = 0.5136$

43. $\tan\left(\dfrac{\theta}{2}\right) = -0.2343$

44. $\sec\left(\dfrac{\theta}{2}\right) = 1.4275$

45. $5\cot\theta - 9 = 0$

46. $5\sec\theta + 6 = 0$

47. $4\sin\theta + \sqrt{2} = 0$

48. $3\cos\theta - \sqrt{5} = 0$

49. $4\cos^2\theta + 5\cos\theta - 6 = 0$

50. $6\sin^2\theta - 13\sin\theta - 5 = 0$ **51.** $6\tan^2\theta - \tan\theta - 12 = 0$ **52.** $6\sec^2\theta - 7\sec\theta - 20 = 0$

53. $15\sin^2(2\theta) + \sin(2\theta) - 2 = 0$ **54.** $12\cos^2\left(\dfrac{\theta}{2}\right) - 13\cos\left(\dfrac{\theta}{2}\right) + 3 = 0$ **55.** $\cos^2\theta - 6\cos\theta + 1 = 0$

56. $\sin^2\theta + 3\sin\theta - 3 = 0$ **57.** $2\tan^2\theta - \tan\theta - 7 = 0$ **58.** $3\cot^2\theta + 2\cot\theta - 4 = 0$

59. $\csc^2(3\theta) - 2 = 0$ **60.** $\sec^2\left(\dfrac{\theta}{2}\right) - 2 = 0$

In Exercises 61–88, solve the trigonometric equations exactly on the indicated interval, $0 \le x < 2\pi$.

61. $\sin x = \cos x$ **62.** $\sin x = -\cos x$ **63.** $\sec x + \cos x = -2$ **64.** $\sin x + \csc x = 2$

65. $\sec x - \tan x = \dfrac{\sqrt{3}}{3}$ **66.** $\sec x + \tan x = 1$ **67.** $\csc x + \cot x = \sqrt{3}$ **68.** $\csc x - \cot x = \dfrac{\sqrt{3}}{3}$

69. $2\sin x - \csc x = 0$ **70.** $2\sin x + \csc x = 3$ **71.** $\sin(2x) = 4\cos x$ **72.** $\sin(2x) = \sqrt{3}\sin x$

73. $\sqrt{2}\sin x = \tan x$ **74.** $\cos(2x) = \sin x$ **75.** $\tan(2x) = \cot x$ **76.** $3\cot(2x) = \cot x$

77. $\sqrt{3}\sec x = 4\sin x$ **78.** $\sqrt{3}\tan x = 2\sin x$ **79.** $\sin^2 x - \cos(2x) = -\dfrac{1}{4}$ **80.** $\sin^2 x - 2\sin x = 0$

81. $\cos^2 x + 2\sin x + 2 = 0$ **82.** $2\cos^2 x = \sin x + 1$ **83.** $2\sin^2 x + 3\cos x = 0$ **84.** $4\cos^2 x - 4\sin x = 5$

85. $\cos(2x) + \cos x = 0$ **86.** $2\cot x = \csc x$ **87.** $\dfrac{1}{4}\sec(2x) = \sin(2x)$ **88.** $-\dfrac{1}{4}\csc\left(\dfrac{1}{2}x\right) = \cos\left(\dfrac{1}{2}x\right)$

In Exercises 89–98, solve each trigonometric equation on $0° \le \theta < 360°$. Express solutions in degrees and round to two decimal places.

89. $\cos(2x) + \dfrac{1}{2}\sin x = 0$ **90.** $\sec^2 x = \tan x + 1$ **91.** $6\cos^2 x + \sin x = 5$ **92.** $\sec^2 x = 2\tan x + 4$

93. $\cot^2 x - 3\csc x - 3 = 0$ **94.** $\csc^2 x + \cot x = 7$ **95.** $2\sin^2 x + 2\cos x - 1 = 0$ **96.** $\sec^2 x + \tan x - 2 = 0$

97. $\dfrac{1}{16}\csc^2\left(\dfrac{x}{4}\right) - \cos^2\left(\dfrac{x}{4}\right) = 0$ **98.** $-\dfrac{1}{4}\sec^2\left(\dfrac{x}{8}\right) + \sin^2\left(\dfrac{x}{8}\right) = 0$

■ **APPLICATIONS** ───────────────────

Exercises 99 and 100, refer to the following:

Computer sales are generally subject to seasonal fluctuations. The sales of QualComp computers during 2008–2010 is approximated by the function

$$s(t) = 0.120\sin(0.790t - 2.380) + 0.387 \quad 1 \le t \le 12$$

where t represents time in quarters ($t = 1$ represents the end of the first quarter of 2008), and $s(t)$ represents computer sales (quarterly revenue) in millions of dollars.

99. Business. Find the quarter(s) in which the quarterly sales are $472,000.

100. Business. Find the quarter(s) in which the quarterly sales are $507,000.

For Exercises 101 and 102, refer to the following:

Allergy sufferers' symptoms fluctuate with the concentration of pollen in the air. At one location the pollen concentration, measured in grains per cubic meter, of grasses fluctuates throughout the day according to the function:

$$p(t) = 35 - 26\cos\left(\dfrac{\pi}{12}t - \dfrac{7\pi}{6}\right), \quad 0 \le t \le 24$$

where t is measured in hours and $t = 0$ is 12:00 A.M.

101. Biology/Health. Find the time(s) of day when the grass pollen level is 41 grains per cubic meter. Round to the nearest hour.

102. Biology/Health. Find the time(s) of day when the grass pollen level is 17 grains per cubic meter. Round to the nearest hour.

103. Sales. Monthly sales of soccer balls are approximated by

$$S = 400\sin\left(\dfrac{\pi}{6}x\right) + 2000,$$ where x is the number of the month (January is $x = 1$, etc.). During which month do sales reach 2400?

104. Sales. Monthly sales of soccer balls are approximated by $S = 400\sin\left(\dfrac{\pi}{6}x\right) + 2000$, where x is the number of the month (January is $x = 1$, etc.). During which two months do sales reach 1800?

105. Home Improvement. A rain gutter is constructed from a single strip of sheet metal by bending as shown below, so that the base and sides are the same length. Express the area of the cross section of the rain gutter as a function of the angle θ (note that the expression will also involve x).

106. Home Improvement. A rain gutter is constructed from a single strip of sheet metal by bending as shown above, so that the base and sides are the same length. When the area of the cross section of the rain gutter is expressed as a function of the angle θ, you can then use calculus to determine the value of θ that produces the cross section with the greatest possible area. The angle is found by solving the equation $\cos^2\theta - \sin^2\theta + \cos\theta = 0$. Which angle gives the maximum area?

107. Deer Population. The number of deer on an island is given by $D = 200 + 100\sin\left(\dfrac{\pi}{2}x\right)$, where x is the number of years since 2000. Which is the first year after 2000 that the number of deer reaches 300?

108. Deer Population. The number of deer on an island is given by $D = 200 + 100\sin\left(\dfrac{\pi}{6}x\right)$, where x is the number of years since 2000. Which is the first year after 2000 that the number of deer reaches 150?

109. Optics. Assume that light is going from air into a diamond. Calculate the refractive angle θ_r if the incidence angle is $\theta_i = 75°$ and the index of refraction values for air and diamond are $n_i = 1.00$ and $n_r = 2.417$, respectively. Round to the nearest degree. (See Example 12 for Snell's law.)

110. Optics. Assume that light is going from a diamond into air. Calculate the refractive angle θ_r if the incidence angle is $\theta_i = 15°$ and the index of refraction values for diamond and air are $n_i = 2.417$ and $n_r = 1.00$, respectively. Round to the nearest degree. (See Example 12 for Snell's law.)

111. Air in Lungs. If a person breathes in and out every 3 seconds, the volume of air in the lungs can be modeled by $A = 2\sin\left(\dfrac{\pi}{3}x\right)\cos\left(\dfrac{\pi}{3}x\right) + 3$, where A is in liters of air and x is in seconds. How many seconds into the cycle is the volume of air equal to 4 liters?

112. Air in Lungs. For the function given in Exercise 111, how many seconds into the cycle is the volume of air equal to 2 liters?

For Exercises 113 and 114, refer to the following:

The figure below shows the graph of $y = 2\cos x - \cos(2x)$ between -2π and 2π. The maximum and minimum values of the curve occur at the *turning points* and are found in the solutions of the equation $-2\sin x + 2\sin(2x) = 0$.

113. Finding Turning Points. Solve for the coordinates of the turning points of the curve between 0 and 2π.

114. Finding Turning Points. Solve for the coordinates of the turning points of the curve between -2π and 0.

115. Business. An analysis of a company's costs and revenue shows that annual costs of producing their product as well as annual revenues from the sale of a product are generally subject to seasonal fluctuations and are approximated by the functions

$$C(t) = 2.3 + 0.25\sin\left(\dfrac{\pi}{6}t\right) \qquad 0 \le t \le 11$$

$$R(t) = 2.3 + 0.5\cos\left(\dfrac{\pi}{6}t\right) \qquad 0 \le t \le 11$$

where t represents time in months ($t = 0$ represents January), $C(t)$ represents the monthly costs of producing the product in millions of dollars, and $R(t)$ represents monthly revenue from sales of the product in millions of dollars. Find the month(s) in which the company breaks even. *Hint:* A company breaks even when its profit is zero.

116. Business. An analysis of a company's costs and revenue shows that the annual costs of producing its product as well as annual revenues from the sale of a product are generally subject to seasonal fluctuations and are approximated by the functions

$$C(t) = 25.7 + 0.2\sin\left(\frac{\pi}{6}t\right) \qquad 0 \le t \le 11$$

$$R(t) = 25.7 + 9.6\cos\left(\frac{\pi}{6}t\right) \qquad 0 \le t \le 11$$

where t represents time in months ($t = 0$ represents January), $C(t)$ represents the monthly costs of producing the product in millions of dollars, and $R(t)$ represents monthly revenue from sales of the product in millions of dollars. Find the month(s) in which the company breaks even. *Hint:* A company breaks even when its profit is zero.

For Exercises 117 and 118, refer to the following:

By analyzing available empirical data, it has been determined that the body temperature of a species fluctuates according to the model

$$T(t) = 37.10 + 1.40\sin\left(\frac{\pi}{24}t\right)\cos\left(\frac{\pi}{24}t\right) \qquad 0 \le t \le 24$$

where T represents temperature in degrees Celsius and t represents time (in hours) measured from 12:00 A.M. (midnight).

117. Biology/Health. Find the time(s) of day the body temperature is 37.28 degrees Celsius. Round to the nearest hour.

118. Biology/Health. Find the time(s) of day the body temperature is 36.75 degrees Celsius. Round to the nearest hour.

▪ CATCH THE MISTAKE

In Exercises 119–122, explain the mistake that is made.

119. Solve $\sqrt{2 + \sin\theta} = \sin\theta$ on $0 \le \theta \le 2\pi$.

Solution:

Square both sides.	$2 + \sin\theta = \sin^2\theta$
Gather all terms to one side.	$\sin^2\theta - \sin\theta - 2 = 0$
Factor.	$(\sin\theta - 2)(\sin\theta + 1) = 0$
Set each factor equal to zero.	$\sin\theta - 2 = 0$ or $\sin\theta + 1 = 0$
Solve for $\sin\theta$.	$\sin\theta = 2$ or $\sin\theta = -1$
Solve $\sin\theta = 2$ for θ.	no solution
Solve $\sin\theta = -1$ for θ.	$\theta = \dfrac{3\pi}{2}$

This is incorrect. What mistake was made?

120. Solve $\sqrt{3\sin\theta - 2} = -\sin\theta$ on $0 \le \theta \le 2\pi$.

Solution:

Square both sides.	$3\sin\theta - 2 = \sin^2\theta$
Gather all terms to one side.	$\sin^2\theta - 3\sin\theta + 2 = 0$
Factor.	$(\sin\theta - 2)(\sin\theta - 1) = 0$
Set each factor equal to zero.	$\sin\theta - 2 = 0$ or $\sin\theta - 1 = 0$
Solve for $\sin\theta$.	$\sin\theta = 2$ or $\sin\theta = 1$
Solve $\sin\theta = 2$ for θ.	no solution
Solve $\sin\theta = 1$ for θ.	$\theta = \dfrac{\pi}{2}$

This is incorrect. What mistake was made?

121. Solve $3\sin(2x) = 2\cos x$ on $0° \le \theta \le 180°$.

Solution:

Use the double-angle identity for the sine function.	$3\underset{2\sin x\cos x}{\underline{\sin 2x}} = 2\cos x$
Simplify.	$6\sin x\cos x = 2\cos x$
Divide by $2\cos x$.	$3\sin x = 1$
Divide by 3.	$\sin x = \dfrac{1}{3}$
Write the equivalent inverse notation.	$x = \sin^{-1}\left(\dfrac{1}{3}\right)$
Use a calculator to approximate the solution.	$x \approx 19.47°$, QI solution
The QII solution is:	$x \approx 180° - 19.47° \approx 160.53°$

This is incorrect. What mistake was made?

122. Solve $\sqrt{1 + \sin x} = \cos x$ on $0 \le x \le 2\pi$.

Solution:

Square both sides.	$1 + \sin x = \cos^2 x$
Use the Pythagorean identity.	$1 + \sin x = \underset{1-\sin^2 x}{\underline{\cos^2 x}}$
Simplify.	$\sin^2 x + \sin x = 0$
Factor.	$\sin x(\sin x + 1) = 0$
Set each factor equal to zero.	$\sin x = 0$ or $\sin x + 1 = 0$
Solve for $\sin x$.	$\sin x = 0$ or $\sin x = -1$
Solve for x.	$x = 0, \pi, \dfrac{3\pi}{2}, 2\pi$

This is incorrect. What mistake was made?

■ **CONCEPTUAL**

In Exercises 123–126, determine whether each statement is true or false.

123. Linear trigonometric equations always have one solution on $[0, 2\pi]$.

124. Quadratic trigonometric equations always have two solutions on $[0, 2\pi]$.

125. If a trigonometric equation has all real numbers as its solution, then it is an identity.

126. If a trigonometric equation has an infinite number of solutions, then it is an identity.

■ **CHALLENGE**

127. Solve $16\sin^4\theta - 8\sin^2\theta = -1$ over $0 \le \theta \le 2\pi$.

128. Solve $\left|\cos\left(\theta + \dfrac{\pi}{4}\right)\right| = \dfrac{\sqrt{3}}{2}$ over all real numbers.

129. Solve for the smallest positive x that makes this statement true:
$$\sin\left(x + \frac{\pi}{4}\right) + \sin\left(x - \frac{\pi}{4}\right) = \frac{\sqrt{2}}{2}$$

130. Solve for the smallest positive x that makes this statement true:
$$\cos x\cos 15° + \sin x\sin 15° = 0.7$$

131. Find all real numbers x such that $\dfrac{1 - \cos\left(\dfrac{x}{3}\right)}{1 + \cos\left(\dfrac{x}{3}\right)} + 1 = 0$.

132. Find all real numbers θ such that $\sec^4\left(\dfrac{1}{3}\theta\right) - 1 = 0$.

133. Find all real numbers θ such that $\csc^4\left(\dfrac{\pi}{4}\theta - \pi\right) - 4 = 0$.

134. Find all real numbers x such that
$$2\tan(3x) = \sqrt{3} - \sqrt{3}\tan^2(3x).$$

■ **TECHNOLOGY**

Graphing calculators can be used to find approximate solutions to trigonometric equations. For the equation $f(x) = g(x)$, let $Y_1 = f(x)$ and $Y_2 = g(x)$. The x-values that correspond to points of intersections represent solutions.

135. With a graphing utility, solve the equation $\sin\theta = \cos 2\theta$ on $0 \le \theta \le \pi$.

136. With a graphing utility, solve the equation $\csc\theta = \sec\theta$ on $0 \le \theta \le \dfrac{\pi}{2}$.

137. With a graphing utility, solve the equation $\sin\theta = \sec\theta$ on $0 \le \theta \le \pi$.

138. With a graphing utility, solve the equation $\cos\theta = \csc\theta$ on $0 \le \theta \le \pi$.

139. With a graphing utility, find all of the solutions to the equation $\sin\theta = e^\theta$ for $\theta \ge 0$.

140. With a graphing utility, find all of the solutions to the equation $\cos\theta = e^\theta$ for $\theta \ge 0$.

Find the smallest positive values of x that make the statement true. Give the answer in radians and round to two decimal places.

141. $\sec(3x) + \csc(2x) = 5$

142. $\cot(5x) + \tan(2x) = -3$

143. $e^x - \tan x = 0$

144. $e^x + 2\sin x = 1$

145. $\ln x - \sin x = 0$

146. $\ln x - \cos x = 0$

CHAPTER 7 INQUIRY-BASED LEARNING PROJECT

When it comes to identities, don't always let your intuition be your guide.

1. Suppose a fellow student claims that the equation $\sin(a + b) = \sin a + \sin b$ is an identity. He's wondering if you can help because he's not sure how to verify his claim, but says, "It just seems intuitively so." Can you help this student?

   a. First, let's understand the student's claim. What does it *mean* to say that $\sin(a + b) = \sin a + \sin b$ "is an identity"?

   b. If you decided to try and verify the student's claim, you'd start with one side of his equation and try to manipulate that side until it looks like the other side. But, that may turn out to be a lot of unnecessary work *if*, in fact, the student's claim is false. So, instead, try something else.

   Consider a right triangle with angles a and b. Calculate the values in the chart below, using various values of a and b.

a	b	$a + b$	$\sin(a + b)$	$\sin a$	$\sin b$	$\sin a + \sin b$

   c. What does your data tell you about the student's claim? Explain.

   d. In this example, you discovered that function notation is not distributive. Now, show the student how to write the *sum identity for the sine function*; that is,

   $$\sin(a + b) = \underline{\hspace{4cm}}$$

   (This identity is derived in this chapter.)

2. Word has gotten out that you are really good at helping others understand trigonometric identities. Another of your fellow students asks whether $\sin(2a) = 2\sin a$ is an identity.

   a. How many values of a would you need to check to determine whether the student's equation is an identity? Explain.

   b. Show how to convince the student that his equation is *not* an identity.

   c. Try to discover the *double-angle identity* $\sin(2a) = \underline{\hspace{3cm}}$: For the right triangle below, fill out the chart (exact values) and look for a pattern.

a	$\sin(2a)$	$\sin a$	$\cos a$
30°			
45°			
60°			
90°			

In the Modeling Our World feature in Chapter 6, you modeled mean *temperatures* with sinusoidal models. Now we consider *carbon emissions*, which are greenhouse gases that have been shown to negatively affect the ozone layer. Over the last 50 years, we have increased our global carbon emissions at an alarming rate. Recall that the graph of the inverse tangent function increases rapidly and then levels off at the horizontal asymptote, $y = \dfrac{\pi}{2} \approx 1.57$. To achieve a similar plateau

with carbon emissions, drastic environmental regulations will need to be enacted.

The carbon emissions data over the last 50 years suggest an almost linear climb (similar to the inverse tangent graph from $x = 0$ to $x = 1$). If the world started reducing carbon emissions, they might possibly reach a plateau level.

The following table summarizes average yearly temperature in degrees Fahrenheit (°F) and carbon dioxide emissions in parts per million (ppm) for Mauna Loa, Hawaii.

YEAR	1960	1965	1970	1975	1980	1985	1990	1995	2000	2005
TEMPERATURE	44.45	43.29	43.61	43.35	46.66	45.71	45.53	47.53	45.86	46.23
CO$_2$ EMISSIONS (PPM)	316.9	320.0	325.7	331.1	338.7	345.9	354.2	360.6	369.4	379.7

1. Plot the carbon emissions data with time on the horizontal axis and CO$_2$ emissions (in ppm) on the vertical axis. Let $t = 0$ correspond to 1960.

2. Find an *inverse tangent function* of the form $f(t) = A\tan^{-1}(Bt) + k$ that models the carbon emissions in Mauna Loa.

    a. Use the data from 1960, 1985, and 2005.
    b. Use the data from 1960, 1965, and 1995.

3. According to your model, what are the expected carbon emissions in 2050?

4. Describe the ways by which the world might be able to reach a plateau level of carbon emissions instead of the predicted increased rates.

SECTION	CONCEPT	KEY IDEAS/FORMULAS
7.1	**Basic trigonometric identities**	Reciprocal, quotient, and Pythagorean identities
	Reciprocal identities	$\csc\theta = \dfrac{1}{\sin\theta}$ $\qquad$ $\sec\theta = \dfrac{1}{\cos\theta}$ $\qquad$ $\cot\theta = \dfrac{1}{\tan\theta}$
	Quotient identities	$\tan\theta = \dfrac{\sin\theta}{\cos\theta}$ $\qquad$ $\cot\theta = \dfrac{\cos\theta}{\sin\theta}$
	Pythagorean identities	$\sin^2\theta + \cos^2\theta = 1$ $\qquad$ $\tan^2\theta + 1 = \sec^2\theta$ $\qquad$ $1 + \cot^2\theta = \csc^2\theta$
7.2	**Verifying trigonometric identities**	Identities must hold for *all* values of x (not just some values of x) for which both sides of the equation are defined.
	Trigonometric identities	**Simplifying trigonometric expressions using identities.** Use the reciprocal, quotient, or pythagorean identities to simplify trigonometric expressions.
		Verifying identities
		■ Convert all trigonometric expressions to sines and cosines.
		■ Write all sums or differences of fractions as a single fraction.
7.3	**Sum and difference identities**	$f(A \pm B) \neq f(A) \pm f(B)$ For trigonometric functions, we have the sum and difference identities.
	Sum and difference identities for the cosine function	$\cos(A + B) = \cos A \cos B - \sin A \sin B$ $\cos(A - B) = \cos A \cos B + \sin A \sin B$
	Sum and difference identities for the sine function	$\sin(A + B) = \sin A \cos B + \cos A \sin B$ $\sin(A - B) = \sin A \cos B - \cos A \sin B$
	Sum and difference identities for the tangent function	$\tan(A + B) = \dfrac{\tan A + \tan B}{1 - \tan A \tan B}$ $\tan(A - B) = \dfrac{\tan A - \tan B}{1 + \tan A \tan B}$
7.4	**Double-angle identities**	$\sin(2A) = 2\sin A \cos A$ $\cos(2A) = \cos^2 A - \sin^2 A = 1 - 2\sin^2 A = 2\cos^2 A - 1$
	Applying double-angle identities	$\tan(2A) = \dfrac{2\tan A}{1 - \tan^2 A}$
7.5	**Half-angle identities**	
	Applying half-angle identities	$\sin\left(\dfrac{A}{2}\right) = \pm\sqrt{\dfrac{1 - \cos A}{2}}$
		$\cos\left(\dfrac{A}{2}\right) = \pm\sqrt{\dfrac{1 + \cos A}{2}}$
		$\tan\left(\dfrac{A}{2}\right) = \pm\sqrt{\dfrac{1 - \cos A}{1 + \cos A}} = \dfrac{\sin A}{1 + \cos A} = \dfrac{1 - \cos A}{\sin A}$
7.6	**Product-to-sum and sum-to-product identities**	
	Product-to-sum identities	$\cos A \cos B = \frac{1}{2}\left[\cos(A + B) + \cos(A - B)\right]$
		$\sin A \sin B = \frac{1}{2}\left[\cos(A - B) - \cos(A + B)\right]$
		$\sin A \cos B = \frac{1}{2}\left[\sin(A + B) + \sin(A - B)\right]$

SECTION	CONCEPT	KEY IDEAS/FORMULAS
	Sum-to-product identities	$\sin A + \sin B = 2\sin\left(\dfrac{A+B}{2}\right)\cos\left(\dfrac{A-B}{2}\right)$

$$\sin A - \sin B = 2\sin\left(\frac{A-B}{2}\right)\cos\left(\frac{A+B}{2}\right)$$

$$\cos A + \cos B = 2\cos\left(\frac{A+B}{2}\right)\cos\left(\frac{A-B}{2}\right)$$

$$\cos A - \cos B = -2\sin\left(\frac{A+B}{2}\right)\sin\left(\frac{A-B}{2}\right)$$

7.7 **Inverse trigonometric functions**

$\sin^{-1}x$ or $\arcsin x$ $\qquad$ $\cos^{-1}x$ or $\arccos x$

$\tan^{-1}x$ or $\arctan x$ $\qquad$ $\cot^{-1}x$ or $\text{arccot}\, x$

$\sec^{-1}x$ or $\text{arcsec}\, x$ $\qquad$ $\csc^{-1}x$ or $\text{arccsc}\, x$

Inverse sine function

Definition

$y = \sin^{-1}x$ means $x = \sin y$

$-1 \le x \le 1$ and $-\dfrac{\pi}{2} \le y \le \dfrac{\pi}{2}$

Identities

$\sin^{-1}(\sin x) = x$ for $-\dfrac{\pi}{2} \le x \le \dfrac{\pi}{2}$

$\sin(\sin^{-1}x) = x$ for $-1 \le x \le 1$

Inverse cosine function

Definition

$y = \cos^{-1}x$ means $x = \cos y$

$-1 \le x \le 1$ and $0 \le y \le \pi$

Identities

$\cos^{-1}(\cos x) = x$ for $0 \le x \le \pi$

$\cos(\cos^{-1}x) = x$ for $-1 \le x \le 1$

Inverse tangent function

Definition

$y = \tan^{-1}x$ means $x = \tan y$

$-\infty < x < \infty$ and $-\dfrac{\pi}{2} < y < \dfrac{\pi}{2}$

Identities

$\tan^{-1}(\tan x) = x$ for $-\dfrac{\pi}{2} < x < \dfrac{\pi}{2}$

$\tan(\tan^{-1}x) = x$ for $-\infty < x < \infty$

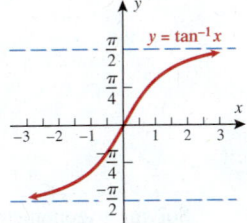

Remaining inverse trigonometric functions

Inverse cotangent function Definition

$y = \cot^{-1}x$ means $x = \cot y$

$-\infty < x < \infty$ and $0 < y < \pi$

Identity

$$\cot^{-1}x = \begin{cases} \tan^{-1}\left(\dfrac{1}{x}\right), & x > 0 \\[2mm] \pi + \tan^{-1}\left(\dfrac{1}{x}\right), & x < 0 \end{cases}$$

Inverse secant function Definition

$y = \sec^{-1} x$ means $x = \sec y$

$x \leq -1$ or $x \geq 1$ and

$0 \leq y < \dfrac{\pi}{2}$ or $\dfrac{\pi}{2} < y \leq \pi$

Identity

$\sec^{-1} x = \cos^{-1}\left(\dfrac{1}{x}\right)$ for $x \leq -1$ or $x \geq 1$

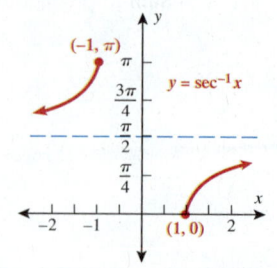

Inverse cosecant function Definition

$y = \csc^{-1} x$ means $x = \csc y$

$x \leq -1$ or $x \geq 1$ and

$-\dfrac{\pi}{2} \leq y < 0$ or $0 < y \leq \dfrac{\pi}{2}$

Identity

$\csc^{-1} x = \sin^{-1}\left(\dfrac{1}{x}\right)$

for $x \leq -1$ or $x \geq 1$

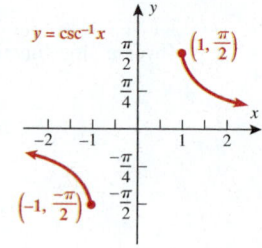

	Finding exact values for expressions involving inverse trigonometric functions	
7.8	**Trigonometric Equations**	Goal: Find the value(s) of the variable that make the equation true.
	Solving trigonometric equations by inspection	Solve: $\sin\theta = \dfrac{\sqrt{2}}{2}$ on $0 \leq \theta \leq 2\pi$. Answer: $\theta = \dfrac{\pi}{4}$ or $\theta = \dfrac{3\pi}{4}$. Solve: $\sin\theta = \dfrac{\sqrt{2}}{2}$ on all real numbers. Answer: $\theta = \begin{cases} \dfrac{\pi}{4} + 2n\pi \\[2mm] \dfrac{3\pi}{4} + 2n\pi \end{cases}$ where n is an integer.
	Solving trigonometric equations using algebraic techniques	Transform trigonometric equations into linear or quadratic algebraic equations by making a substitution such as $x = \sin\theta$. Then use algebraic methods for solving linear and quadratic equations. *Note:* If an expression is squared, always check for extraneous solutions.
	Solving trigonometric equations which require the use of inverse functions	Follow the same procedures outlined by inspection or algebraic methods. Finding the solution requires the use of inverse functions and a calculator. Be careful: calculators only give one solution (the one in the range of the inverse function).
	Using trigonometric identities to solve trigonometric equations	■ Use trigonometric identities to transform an equation with multiple trigonometric functions into an equation with only one trigonometric function. ■ Then use the methods outlined above.

7.1 Basic Trigonometric Identities

Use trigonometric identities to find the indicated value(s).

1. If $\sin\theta = -\dfrac{7}{11}$, find $\csc\theta$.

2. If $\cot\theta = \dfrac{3\sqrt{5}}{5}$, find $\tan\theta$.

3. If $\sin\theta = \dfrac{8}{17}$ and $\cos\theta = -\dfrac{15}{17}$, find $\tan\theta$.

4. If $\sin\theta = -\dfrac{12}{13}$ and $\cos\theta = -\dfrac{5}{13}$, find $\cot\theta$.

5. If $\sin\theta = \dfrac{2\sqrt{3}}{9}$, find $\sin^2\theta$.

6. If $\cos\theta = -0.45$, find $\cos^2\theta$.

7. If $\tan\theta = \sqrt{3}$, find $\tan^3\theta$.

8. If $\sin\theta = \dfrac{11}{61}$ and the terminal side of θ lies in QII, find $\tan\theta$.

9. If $\cos\theta = \dfrac{7}{25}$ and the terminal side of θ lies in QIV, find $\cot\theta$.

10. If $\tan\theta = -1$ and the terminal side of θ lies in QIV, find $\sec\theta$.

11. If $\csc\theta = \dfrac{13}{12}$ and the terminal side of θ lies in QII, find $\cot\theta$.

12. Find $\sin\theta$ and $\cos\theta$ if $\tan\theta = -\dfrac{24}{7}$ and the terminal side of θ lies in QII.

13. Find $\sin\theta$ and $\cos\theta$ if $\cot\theta = \dfrac{\sqrt{3}}{3}$ and the terminal side of θ lies in QIII.

14. Find $\sin\theta$ and $\cos\theta$ if $\cot\theta = \sqrt{3}$ and the terminal side of θ lies in QIII.

Perform the operations and simplify. Write the answers in terms of $\sin\theta$ and $\cos\theta$.

15. $\sec\theta\cot^2\theta$

16. $\sin\theta\cot\theta$

17. $\dfrac{\tan\theta}{\sec\theta}$

18. $(\cos\theta + \sec\theta)^2$

19. $\cot\theta(\sec\theta + \sin\theta)$

20. $\dfrac{\sin^2\theta}{\csc\theta}$

7.2 Verifying Trigonometric Identities

Simplify the following trigonometric expressions.

21. $\tan x(\cot x + \tan x)$

22. $(\sec x + 1)(\sec x - 1)$

23. $\dfrac{\tan^4 x - 1}{\tan^2 x - 1}$

24. $\sec^2 x(\cot^2 x - \cos^2 x)$

25. $\cos x[\cos(-x) - \tan(-x)] - \sin x$

26. $\dfrac{\tan^2 x + 1}{2\sec^2 x}$

27. $\dfrac{\csc^3(-x) + 8}{\csc x - 2}$

28. $\dfrac{\csc^2 x - 1}{\cot x}$

Verify the trigonometric identities.

29. $(\tan x + \cot x)^2 - 2 = \tan^2 x + \cot^2 x$

30. $\csc^2 x - \cot^2 x = 1$

31. $\dfrac{1}{\sin^2 x} - \dfrac{1}{\tan^2 x} = 1$

32. $\dfrac{1}{\csc x + 1} + \dfrac{1}{\csc x - 1} = \dfrac{2\tan x}{\cos x}$

33. $\dfrac{\tan^2 x - 1}{\sec^2 x + 3\tan x + 1} = \dfrac{\tan x - 1}{\tan x + 2}$

34. $\cot x(\sec x - \cos x) = \sin x$

Determine whether each of the following equations is a conditional equation or an identity.

35. $2\tan^2 x + 1 = \dfrac{1 + \sin^2 x}{\cos^2 x}$

36. $\sin x - \cos x = 0$

37. $\cot^2 x - 1 = \tan^2 x$

38. $\cos^2 x(1 + \cot^2 x) = \cot^2 x$

39. $\left(\cot x - \dfrac{1}{\tan x}\right)^2 = 0$

40. $\csc x + \sec x = \dfrac{1}{\sin x + \cos x}$

7.3 Sum and Difference Identities

Find the exact value for each trigonometric expression.

41. $\cos\left(\dfrac{7\pi}{12}\right)$

42. $\sin\left(\dfrac{\pi}{12}\right)$

43. $\tan(-15°)$

44. $\cot 105°$

Write each expression as a single trigonometric function.

45. $\sin(4x)\cos(3x) - \cos(4x)\sin(3x)$

46. $\sin(-x)\sin(-2x) + \cos(-x)\cos(-2x)$

47. $\dfrac{\tan(5x) - \tan(4x)}{1 + \tan(5x)\tan(4x)}$

48. $\dfrac{\tan\left(\dfrac{\pi}{4}\right) + \tan\left(\dfrac{\pi}{3}\right)}{1 - \tan\left(\dfrac{\pi}{4}\right)\tan\left(\dfrac{\pi}{3}\right)}$

Find the exact value of the indicated expression using the given information and identities.

49. Find the exact value of $\tan(\alpha - \beta)$ if $\sin\alpha = -\frac{3}{5}$, $\sin\beta = -\frac{24}{25}$, the terminal side of α lies in quadrant IV, and the terminal side of β lies in quadrant III.

50. Find the exact value of $\cos(\alpha + \beta)$ if $\cos\alpha = -\frac{5}{13}$, $\sin\beta = \frac{7}{25}$, the terminal side of α lies in quadrant II, and the terminal side of β also lies in quadrant II.

51. Find the exact value of $\cos(\alpha - \beta)$ if $\cos\alpha = \frac{9}{41}$, $\cos\beta = \frac{7}{25}$, the terminal side of α lies in quadrant IV, and the terminal side of β lies in quadrant I.

52. Find the exact value of $\sin(\alpha - \beta)$ if $\sin\alpha = -\frac{5}{13}$, $\cos\beta = -\frac{4}{5}$, the terminal side of α lies in quadrant III, and the terminal side of β lies in quadrant II.

Determine whether each of the following equations is a conditional equation or an identity.

53. $2\cos A\cos B = \cos(A + B) + \cos(A - B)$

54. $2\sin A\sin B = \cos(A - B) - \cos(A + B)$

Graph the following functions.

55. $y = \cos\left(\frac{\pi}{2}\right)\cos x - \sin\left(\frac{\pi}{2}\right)\sin x$

56. $y = \sin\left(\frac{2\pi}{3}\right)\cos x + \cos\left(\frac{2\pi}{3}\right)\sin x$

57. $y = \dfrac{2\tan\left(\dfrac{x}{3}\right)}{1 - \tan^2\left(\dfrac{x}{3}\right)}$

58. $y = \dfrac{\tan(\pi x) - \tan x}{1 + \tan(\pi x)\tan x}$

7.4 Double-Angle Identities

Use double-angle identities to answer the following questions.

59. If $\sin x = \frac{3}{5}$ and $\frac{\pi}{2} < x < \pi$, find $\cos(2x)$.

60. If $\cos x = \frac{7}{25}$ and $\frac{3\pi}{2} < x < 2\pi$, find $\sin(2x)$.

61. If $\cot x = -\frac{11}{61}$ and $\frac{3\pi}{2} < x < 2\pi$, find $\tan(2x)$.

62. If $\tan x = -\frac{12}{5}$ and $\frac{\pi}{2} < x < \pi$, find $\cos(2x)$.

63. If $\sec x = \frac{25}{24}$ and $0 < x < \frac{\pi}{2}$, find $\sin(2x)$.

64. If $\csc x = \frac{5}{4}$ and $\frac{\pi}{2} < x < \pi$, find $\tan(2x)$.

Simplify each of the following expressions. Evaluate exactly, if possible.

65. $\cos^2 15° - \sin^2 15°$

66. $\dfrac{2\tan\left(-\dfrac{\pi}{12}\right)}{1 - \tan^2\left(-\dfrac{\pi}{12}\right)}$

67. $6\sin\left(\dfrac{\pi}{12}\right)\cos\left(\dfrac{\pi}{12}\right)$

68. $1 - 2\sin^2\left(\dfrac{\pi}{8}\right)$

Verify the following identities.

69. $\sin^3 A - \cos^3 A = (\sin A - \cos A)\left[1 + \frac{1}{2}\sin(2A)\right]$

70. $2\sin A\cos^3 A - 2\sin^3 A\cos A = \cos(2A)\sin(2A)$

71. $\tan A = \dfrac{\sin(2A)}{1 + \cos(2A)}$

72. $\tan A = \dfrac{1 - \cos(2A)}{\sin(2A)}$

73. Launching a Missile. When launching a missile for a given range, the minimum velocity needed is related to the angle θ of the launch, and the velocity is determined by $V = \dfrac{2\cos(2\theta)}{1 + \cos(2\theta)}$. Show that V is equivalent to $1 - \tan^2\theta$.

74. Launching a Missile. When launching a missile for a given range, the minimum velocity needed is related to the angle θ of the launch, and the velocity is determined by $V = \dfrac{2\cos(2\theta)}{1 + \cos(2\theta)}$. Find the value of V when $\theta = \dfrac{\pi}{6}$.

7.5 Half-Angle Identities

Use half-angle identities to find the exact value of each of the following trigonometric expressions.

75. $\sin(-22.5°)$

76. $\cos 67.5°$

77. $\cot\left(\dfrac{3\pi}{8}\right)$

78. $\csc\left(-\dfrac{7\pi}{8}\right)$

79. $\sec(-165°)$

80. $\tan(-75°)$

Use half-angle identities to find each of the following values.

81. If $\sin x = -\frac{7}{25}$ and $\pi < x < \frac{3\pi}{2}$, find $\sin\left(\dfrac{x}{2}\right)$.

82. If $\cos x = -\frac{4}{5}$ and $\frac{\pi}{2} < x < \pi$, find $\cos\left(\dfrac{x}{2}\right)$.

83. If $\tan x = \frac{40}{9}$ and $\pi < x < \frac{3\pi}{2}$, find $\tan\left(\dfrac{x}{2}\right)$.

84. If $\sec x = \frac{17}{15}$ and $\frac{3\pi}{2} < x < 2\pi$, find $\sin\left(\dfrac{x}{2}\right)$.

Simplify each expression using half-angle identities. Do not evaluate.

85. $\sqrt{\dfrac{1 - \cos\left(\dfrac{\pi}{6}\right)}{2}}$

86. $\sqrt{\dfrac{1 - \cos\left(\dfrac{11\pi}{6}\right)}{1 + \cos\left(\dfrac{11\pi}{6}\right)}}$

Verify each of the following identities.

87. $\left[\sin\left(\dfrac{A}{2}\right) + \cos\left(\dfrac{A}{2}\right)\right]^2 = 1 + \sin A$

88. $\sec^2\left(\dfrac{A}{2}\right) + \tan^2\left(\dfrac{A}{2}\right) = \dfrac{3 - \cos A}{1 + \cos A}$

89. $\csc^2\left(\dfrac{A}{2}\right) + \cot^2\left(\dfrac{A}{2}\right) = \dfrac{3 + \cos A}{1 - \cos A}$

90. $\tan^2\left(\dfrac{A}{2}\right) + 1 = \sec^2\left(\dfrac{A}{2}\right)$

Graph each of the following functions.

91. $y = \sqrt{\dfrac{1 - \cos\left(\dfrac{\pi}{12}x\right)}{2}}$

92. $y = \cos^2\left(\dfrac{x}{2}\right) - \sin^2\left(\dfrac{x}{2}\right)$

93. $y = -\sqrt{\dfrac{1 - \cos x}{1 + \cos x}}$

94. $y = \sqrt{\dfrac{1 + \cos(3x - 1)}{2}}$

7.6 Product-to-Sum and Sum-to-Product Identities

Write each product as a sum or difference of sines and/or cosines.

95. $6\sin(5x)\cos(2x)$

96. $3\sin(4x)\sin(2x)$

Write each expression as a product of sines and/or cosines.

97. $\cos(5x) - \cos(3x)$

98. $\sin\left(\dfrac{5x}{2}\right) + \sin\left(\dfrac{3x}{2}\right)$

99. $\sin\left(\dfrac{4x}{3}\right) - \sin\left(\dfrac{2x}{3}\right)$

100. $\cos(7x) + \cos x$

Simplify each trigonometric expression.

101. $\dfrac{\cos(8x) + \cos(2x)}{\sin(8x) - \sin(2x)}$

102. $\dfrac{\sin(5x) + \sin(3x)}{\cos(5x) + \cos(3x)}$

Verify the identities.

103. $\dfrac{\sin A + \sin B}{\cos A - \cos B} = -\cot\left(\dfrac{A - B}{2}\right)$

104. $\dfrac{\sin A - \sin B}{\cos A - \cos B} = -\cot\left(\dfrac{A + B}{2}\right)$

105. $\csc\left(\dfrac{A - B}{2}\right) = \dfrac{2\sin\left(\dfrac{A + B}{2}\right)}{\cos B - \cos A}$

106. $\sec\left(\dfrac{A + B}{2}\right) = \dfrac{2\sin\left(\dfrac{A - B}{2}\right)}{\sin A - \sin B}$

7.7 Inverse Trigonometric Functions

Find the exact value of each expression. Give the answer in radians.

107. $\arctan 1$

108. $\operatorname{arccsc}(-2)$

109. $\cos^{-1} 0$

110. $\sin^{-1}(-1)$

111. $\sec^{-1}\left(\dfrac{2}{\sqrt{3}}\right)$

112. $\cot^{-1}(-\sqrt{3})$

Find the exact value of each expression. Give the answer in degrees.

113. $\csc^{-1}(-1)$

114. $\arctan(-1)$

115. $\operatorname{arccot}\left(\dfrac{\sqrt{3}}{3}\right)$

116. $\cos^{-1}\left(\dfrac{\sqrt{2}}{2}\right)$

117. $\sin^{-1}\left(-\dfrac{\sqrt{3}}{2}\right)$

118. $\sec^{-1} 1$

Use a calculator to evaluate each expression. Give the answer in degrees and round to two decimal places.

119. $\sin^{-1}(-0.6088)$

120. $\tan^{-1}(1.1918)$

121. $\sec^{-1}(1.0824)$

122. $\cot^{-1}(-3.7321)$

Use a calculator to evaluate each expression. Give the answer in radians and round to two decimal places.

123. $\cos^{-1}(-0.1736)$

124. $\tan^{-1}(0.1584)$

125. $\csc^{-1}(-10.0167)$

126. $\sec^{-1}(-1.1223)$

Evaluate each expression exactly, if possible. If not possible, state why.

127. $\sin^{-1}\left[\sin\left(-\dfrac{\pi}{4}\right)\right]$ **128.** $\cos\left[\cos^{-1}\left(-\dfrac{\sqrt{2}}{2}\right)\right]$

129. $\tan\left[\tan^{-1}(-\sqrt{3})\right]$ **130.** $\cot^{-1}\left[\cot\left(\dfrac{11\pi}{6}\right)\right]$

131. $\csc^{-1}\left[\csc\left(\dfrac{2\pi}{3}\right)\right]$ **132.** $\sec\left[\sec^{-1}\left(-\dfrac{2\sqrt{3}}{3}\right)\right]$

Evaluate each expression exactly.

133. $\sin\left[\cos^{-1}\left(\dfrac{11}{61}\right)\right]$ **134.** $\cos\left[\tan^{-1}\left(\dfrac{40}{9}\right)\right]$

135. $\tan\left[\cot^{-1}\left(\dfrac{6}{7}\right)\right]$ **136.** $\cot\left[\sec^{-1}\left(\dfrac{25}{7}\right)\right]$

137. $\sec\left[\sin^{-1}\left(\dfrac{1}{6}\right)\right]$ **138.** $\csc\left[\cot^{-1}\left(\dfrac{5}{12}\right)\right]$

Applications

139. Average Temperature. If the average temperature in degrees in Chicago, Illinois, can be modeled with the formula $T(m) = 26\sin(0.48m - 1.84) + 47$, where m is the month of the year (January corresponds to $m = 1$, etc.), then during which month is the average temperature $73°F$?

140. Average Temperature. If the average temperature in degrees in Chicago, Illinois, can be modeled with the formula $T(m) = 26\sin(0.48m - 1.84) + 47$, where m is the month of the year (January corresponds to $m = 1$, etc.), then during which month is the average temperature $21°F$?

7.8 Trigonometric Equations

Solve the given trigonometric equation over the indicated interval.

141. $\sin(2\theta) = -\dfrac{\sqrt{3}}{2}, 0 \le \theta \le 2\pi$

142. $\sec\left(\dfrac{\theta}{2}\right) = 2, -2\pi \le \theta \le 2\pi$

143. $\sin\left(\dfrac{\theta}{2}\right) = -\dfrac{\sqrt{2}}{2}, -2\pi \le \theta \le 2\pi$

144. $\csc(2\theta) = 2, 0 \le \theta \le 2\pi$

145. $\tan\left(\dfrac{1}{3}\theta\right) = -1, \ 0 \le \theta \le 6\pi$

146. $\cot(4\theta) = -\sqrt{3}, \ -\pi \le \theta \le \pi$

Solve each trigonometric equation exactly on $0 \le \theta \le 2\pi$.

147. $4\cos(2\theta) + 2 = 0$

148. $\sqrt{3}\tan\left(\dfrac{\theta}{2}\right) - 1 = 0$

149. $2\tan(2\theta) + 2 = 0$

150. $2\sin^2\theta + \sin\theta - 1 = 0$

151. $\tan^2\theta + \tan\theta = 0$

152. $\sec^2\theta - 3\sec\theta + 2 = 0$

Solve the given trigonometric equations on $0° \le \theta \le 360°$ and express the answer in degrees to two decimal places.

153. $\tan(2\theta) = -0.3459$ **154.** $6\sin\theta - 5 = 0$

155. $4\cos^2\theta + 3\cos\theta = 0$ **156.** $12\cos^2\theta - 7\cos\theta + 1 = 0$

157. $\csc^2\theta - 3\csc\theta - 1 = 0$ **158.** $2\cot^2\theta + 5\cot\theta - 4 = 0$

Solve each trigonometric equation exactly on the interval $0 \le \theta \le 2\pi$.

159. $\sec x = 2\sin x$ **160.** $3\tan x + \cot x = 2\sqrt{3}$

161. $\sqrt{3}\tan x - \sec x = 1$ **162.** $2\sin(2x) = \cot x$

163. $\sqrt{3}\tan x = 2\sin x$ **164.** $2\sin x = 3\cot x$

165. $\cos^2 x + \sin x + 1 = 0$ **166.** $2\cos^2 x - \sqrt{3}\cos x = 0$

167. $\cos(2x) + 4\cos x + 3 = 0$ **168.** $\sin(2x) + \sin x = 0$

169. $\tan^2\left(\tfrac{1}{2}x\right) - 1 = 0$ **170.** $\cot^2\left(\tfrac{1}{3}x\right) - 1 = 0$

Solve each trigonometric equation on $0° \le \theta \le 360°$. Give the answers in degrees and round to two decimal places.

171. $\csc^2 x + \cot x = 1$ **172.** $8\cos^2 x + 6\sin x = 9$

173. $\sin^2 x + 2 = 2\cos x$ **174.** $\cos(2x) = 3\sin x - 1$

175. $\cos x - 1 = \cos(2x)$ **176.** $12\cos^2 x + 4\sin x = 11$

Technology Exercises

Section 7.1

177. Is $\cos 73° = \sqrt{1 - \sin^2 73°}$? Use a calculator to find each of the following:

   a. $\cos 73°$

   b. $1 - \sin 73°$

   c. $\sqrt{1 - \sin^2 73°}$

   Which results are the same?

178. Is $\csc 28° = \sqrt{1 + \cot^2 28°}$? Use a calculator to find each of the following.

   a. $\cot 28°$

   b. $1 + \cot 28°$

   c. $\sqrt{1 + \cot^2 28°}$

   Which results are the same?

Section 7.2

179. Determine the correct sign $(+/-)$ for

$$\sin\left(2x + \frac{\pi}{4}\right) = \sin(2x)\cos\left(\frac{\pi}{4}\right) \underset{?}{\pm} \cos(2x)\sin\left(\frac{\pi}{4}\right)$$

by graphing $Y_1 = \sin\left(2x + \frac{\pi}{4}\right)$,

$Y_2 = \sin(2x)\cos\left(\frac{\pi}{4}\right) + \cos(2x)\sin\left(\frac{\pi}{4}\right)$, and

$Y_3 = \sin(2x)\cos\left(\frac{\pi}{4}\right) - \cos(2x)\sin\left(\frac{\pi}{4}\right)$ in the same viewing

rectangle. Does $Y_1 = Y_2$ or Y_3?

180. Determine the correct sign $(+/-)$ for

$$\cos\left(2x - \frac{\pi}{3}\right) = \cos(2x)\cos\left(\frac{\pi}{3}\right) \underset{?}{\pm} \sin(2x)\sin\left(\frac{\pi}{3}\right) \text{ by}$$

graphing $Y_1 = \cos\left(2x - \frac{\pi}{3}\right)$,

$Y_2 = \cos(2x)\cos\left(\frac{\pi}{3}\right) - \sin(2x)\sin\left(\frac{\pi}{3}\right)$, and

$Y_3 = \cos(2x)\cos\left(\frac{\pi}{3}\right) + \sin(2x)\sin\left(\frac{\pi}{3}\right)$ in the same

viewing rectangle. Does $Y_1 = Y_2$ or Y_3?

Section 7.3

Recall that the difference quotient for a function f is given by
$$\frac{f(x + h) + f(x)}{h}.$$

181. Show that the difference quotient for $f(x) = \sin(3x)$ is

$$[\cos(3x)]\left[\frac{\sin(3h)}{h}\right] - [\sin(3x)]\left[\frac{1 - \cos(3h)}{h}\right].$$

Plot $Y_1 = [\cos(3x)]\left[\frac{\sin(3h)}{h}\right] - [\sin(3x)]\left[\frac{1 - \cos(3h)}{h}\right]$ for

a. $h = 1$
b. $h = 0.1$
c. $h = 0.01$

What function does the difference quotient for $f(x) = \sin(3x)$ resemble when h approaches zero?

182. Show that the difference quotient for $f(x) = \cos(3x)$ is

$$[-\sin(3x)]\left[\frac{\sin(3h)}{h}\right] - [\cos(3x)]\left[\frac{1 - \cos(3h)}{h}\right]. \text{ Plot}$$

$Y_1 = [-\sin(3x)]\left[\frac{\sin(3h)}{h}\right] - [\cos(3x)]\left[\frac{1 - \cos(3h)}{h}\right]$ for

a. $h = 1$
b. $h = 0.1$
c. $h = 0.01$

What function does the difference quotient for $f(x) = \cos(3x)$ resemble when h approaches zero?

Section 7.4

183. With a graphing calculator, plot $Y_1 = \tan(2x)$, $Y_2 = 2\tan x$, and $Y_3 = \dfrac{2\tan x}{1 - \tan^2 x}$ in the same viewing rectangle $[-2\pi, 2\pi]$ by $[-10, 10]$. Which graphs are the same?

184. With a graphing calculator, plot $Y_1 = \cos(2x)$, $Y_2 = 2\cos x$, and $Y_3 = 1 - 2\sin^2 x$ in the same viewing rectangle $[-2\pi, 2\pi]$ by $[-2, 2]$. Which graphs are the same?

Section 7.5

185. With a graphing calculator, plot $Y_1 = \cos\left(\dfrac{x}{2}\right)$, $Y_2 = \dfrac{1}{2}\cos x$, and $Y_3 = -\sqrt{\dfrac{1 + \cos x}{2}}$ in the same viewing rectangle $[\pi, 2\pi]$ by $[-1, 1]$. Which graphs are the same?

186. With a graphing calculator, plot $Y_1 = \sin\left(\dfrac{x}{2}\right)$, $Y_2 = \dfrac{1}{2}\sin x$, and $Y_3 = -\sqrt{\dfrac{1 - \cos x}{2}}$ in the same viewing rectangle $[2\pi, 4\pi]$ by $[-1, 1]$. Which graphs are the same?

Section 7.6

187. With a graphing calculator, plot $Y_1 = \sin(5x)\cos(3x)$, $Y_2 = \sin(4x)$, and $Y_3 = \frac{1}{2}[\sin(8x) + \sin(2x)]$ in the same viewing rectangle $[0, 2\pi]$ by $[-1, 1]$. Which graphs are the same?

188. With a graphing calculator, plot $Y_1 = \sin(3x)\cos(5x)$, $Y_2 = \cos(4x)$, and $Y_3 = \frac{1}{2}[\sin(8x) - \sin(2x)]$ in the same viewing rectangle $[0, 2\pi]$ by $[-1, 1]$. Which graphs are the same?

Section 7.7

189. Given $\cos x = -\dfrac{1}{\sqrt{5}}$ and $\dfrac{\pi}{2} < x < \pi$:

a. Find $\cos(2x)$ using the double-angle identity.
b. Use the inverse of cosine to find x in QII and to find $\cos(2x)$.
c. Are the results in (a) and (b) the same?

190. Given $\cos x = \dfrac{5}{12}$ and $\dfrac{3\pi}{2} < x < 2\pi$:

a. Find $\cos\left(\frac{1}{2}x\right)$ using the half-angle identity.
b. Use the inverse of cosine to find x in QIV and to find $\cos\left(\frac{1}{2}x\right)$. Round to five decimal places.
c. Are the results in (a) and (b) the same?

Section 7.8

Find the smallest positive value of x that makes each statement true. Give the answer in radians and round to four decimal places.

191. $\ln x + \sin x = 0$

192. $\ln x + \cos x = 0$

1. For what values of x does the quotient identity $\tan x = \dfrac{\sin x}{\cos x}$ not hold?

2. Is the equation $\sqrt{\sin^2 x + \cos^2 x} = \sin x + \cos x$ a conditional equation or an identity?

3. Evaluate $\sin\left(-\dfrac{\pi}{8}\right)$ exactly.

4. Evaluate $\tan\left(\dfrac{7\pi}{12}\right)$ exactly.

5. If $\cos x = \dfrac{2}{5}$ and $\dfrac{3\pi}{2} < x < 2\pi$, find $\sin\left(\dfrac{x}{2}\right)$.

6. If $\sin x = -\dfrac{1}{5}$ and $\pi < x < \dfrac{3\pi}{2}$, find $\cos(2x)$.

7. Write $\cos(7x)\cos(3x) - \sin(3x)\sin(7x)$ as a cosine or sine of a sum or difference.

8. Write $-\dfrac{2\tan x}{1 - \tan^2 x}$ as a single tangent function.

9. Write $\sqrt{\dfrac{1 + \cos(a + b)}{2}}$ as a single cosine function if $a + b$ is an angle in QII, i.e. $\dfrac{\pi}{2} < a + b < \pi$.

10. Write $2\sin\left(\dfrac{x + 3}{2}\right)\cos\left(\dfrac{x - 3}{2}\right)$ as a sum of two sine functions.

11. Write $10\cos(3 - x) + 10\cos(x + 3)$ as a product of two cosine functions.

12. In the expression $\sqrt{9 - u^2}$, let $u = 3\sin x$. What is the resulting expression?

Solve the trigonometric equations exactly, if possible. Otherwise, use a calculator to approximate solution(s).

13. $2\sin\theta = -\sqrt{3}$ on all real numbers.

14. $2\cos^2\theta + \cos\theta - 1 = 0$ on $0 \le \theta \le 2\pi$

15. $\sin 2\theta = \dfrac{1}{2}\cos\theta$ over $0 \le \theta \le 360°$

16. $\sqrt{\sin x + \cos x} = -1$ over $0 \le \theta \le 2\pi$

17. Determine whether $(1 + \cot x)^2 = \csc^2 x$ is a conditional or an identity.

18. Evaluate $\csc\left(-\dfrac{\pi}{12}\right)$ exactly.

19. If $\sin x = -\dfrac{5}{13}$ and $\pi < x < \dfrac{3\pi}{2}$, find $\cos\left(\dfrac{x}{2}\right)$.

20. If $\cos x = -0.26$ and $\dfrac{\pi}{2} < x < \pi$, find $\sin(2x)$.

21. Express $y = \sqrt{\dfrac{1 + \dfrac{\sqrt{2}}{2}\left[\cos\left(\dfrac{\pi}{3}x\right) + \sin\left(\dfrac{\pi}{3}x\right)\right]}{1 - \dfrac{\sqrt{2}}{2}\left[\cos\left(\dfrac{\pi}{3}x\right) + \sin\left(\dfrac{\pi}{3}x\right)\right]}}$ as a cotangent function.

22. Calculate $\csc\left(\csc^{-1}\sqrt{2}\right)$.

23. Determine an interval on which $f(x) = a + b\csc(\pi x + c)$ is one-to-one, and determine the inverse of $f(x)$ on this interval. Assume that a, b, and c are all positive.

24. Find the range of $y = -\dfrac{\pi}{4} + \arctan(2x - 3)$.

25. Solve $\cos\left(\dfrac{\pi}{4}\theta\right) = -\dfrac{1}{2}$, for all real numbers.

26. Solve $\sqrt{\dfrac{1 - \cos(2\pi x)}{1 + \cos(2\pi x)}} = -\dfrac{1}{\sqrt{3}}$, for all real numbers.

27. Solve $\dfrac{\sqrt{3}}{\csc\left(\dfrac{x}{3}\right)} = \cos\left(\dfrac{x}{3}\right)$, for all real numbers.

28. Show that the difference quotient for $f(x) = \cos\left(\tfrac{1}{2}x\right)$ is

$$-\sin\left(\dfrac{1}{2}x\right)\left[\dfrac{\sin\left(\dfrac{1}{2}h\right)}{h}\right] - \cos\left(\dfrac{1}{2}x\right)\left[\dfrac{1 - \cos\left(\dfrac{1}{2}h\right)}{h}\right].$$

Plot $Y_1 = -\sin\left(\dfrac{1}{2}x\right)\left[\dfrac{\sin\left(\dfrac{1}{2}h\right)}{h}\right] - \cos(2x)\left[\dfrac{1 - \cos\left(\dfrac{1}{2}h\right)}{h}\right]$

for

a. $h = 1$

b. $h = 0.1$

c. $h = 0.01$

What function does the difference quotient for $f(x) = \cos\left(\tfrac{1}{2}x\right)$ resemble when h approaches zero?

29. Given $\tan x = \dfrac{3}{4}$ and $\pi < x < \dfrac{3\pi}{2}$:

a. Find $\sin\left(\tfrac{1}{2}x\right)$ using the half-angle identity.

b. Use the inverse of tangent to find x in QIII and to find $\sin\left(\tfrac{1}{2}x\right)$. Round to five decimal places.

c. Are the results in (a) and (b) the same?

1. Solve by completing the square: $x^2 - 6x = 11$.

2. Write an equation of a line that passes through the point $(-2, 5)$ and is parallel to the y-axis.

3. Write the equation of a circle with center $(-3, -1)$ and passing through the point $(1, 2)$.

4. Determine whether the relation $x^2 - y^2 = 25$ is a function.

5. Determine whether the function $g(x) = \sqrt{2 - x^2}$ is odd or even.

6. For the function $y = 5(x - 4)^2$, identify all of the transformations of $y = x^2$.

7. Find the composite function, $f \circ g$, and state the domain for $f(x) = x^3 - 1$ and $g(x) = \dfrac{1}{x}$.

8. Find the inverse of the function $f(x) = \sqrt[3]{x} - 1$.

9. Find the vertex of the parabola associated with the quadratic function $f(x) = \frac{1}{4}x^2 + \frac{3}{5}x - \frac{6}{25}$.

10. Find a polynomial of minimum degree (there are many) that has the zeros $x = -\sqrt{7}$ (multiplicity 2), $x = 0$ (multiplicity 3), $x = \sqrt{7}$ (multiplicity 2).

11. Use long division to find the quotient $Q(x)$ and the remainder $r(x)$ of $(5x^3 - 4x^2 + 3) \div (x^2 + 1)$.

12. Given the zero $x = 4i$ of the polynomial $P(x) = x^4 + 2x^3 + x^2 + 32x - 240$, determine all the other zeros and write the polynomial in terms of a product of linear factors.

13. Find the vertical and horizontal asymptotes of the function $f(x) = \dfrac{0.7x^2 - 5x + 11}{x^2 - x - 6}$.

14. If \$5,400 is invested at 2.25% compounded continuously, how much is in the account after 4 years?

15. Use interval notation to express the domain of the function $f(x) = \log_4 (x + 3)$.

16. Use properties of logarithms to simplify the expression $\log_\pi 1$.

17. Give an exact solution to the logarithmic equation $\log_5(x + 2) + \log_5(6 - x) = \log_5(3x)$.

18. If money is invested in a saving account earning 4% compounded continuously, how many years will it take for the money to triple?

19. Use a calculator to evaluate $\cos 62°$. Round the answer to four decimal places.

20. **Angle of Inclination (Skiing).** The angle of inclination of a mountain with triple black diamond ski trails is 63°. If a skier at the top of the mountain is at an elevation of 4200 feet, how long is the ski run from the top to the base of the mountain?

21. Convert $-105°$ to radians. Leave the answer in terms of π.

22. Find all of the exact values of θ, when $\tan\theta = 1$ and $0 \le \theta \le 2\pi$.

23. Determine whether the equation $\cos^2 x - \sin^2 x = 1$ is a conditional equation or an identity.

24. Simplify $\dfrac{2\tan\left(-\dfrac{\pi}{8}\right)}{1 - \tan^2\left(-\dfrac{\pi}{8}\right)}$ and evaluate exactly.

25. Evaluate exactly the expression $\tan\left[\sin^{-1}\left(\dfrac{5}{13}\right)\right]$.

26. With a graphing calculator, plot $Y_1 = \sin x \cos(3x)$, $Y_2 = \cos(4x)$, and $Y_3 = \frac{1}{2}[\sin(4x) - \sin(2x)]$ in the same viewing rectangle $[0, 2\pi]$ by $[-1, 1]$. Which graphs are the same?

27. Find the smallest positive value of x that makes the statement true. Give the answer in radians and round to four decimal places.

$$\ln x - \sin(2x) = 0$$

8

Additional Topics in Trigonometry

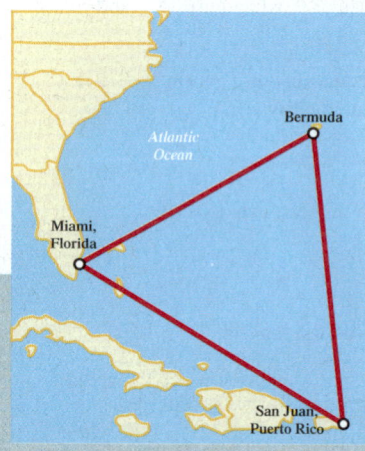

Miami (Florida), Bermuda, Puerto Rico

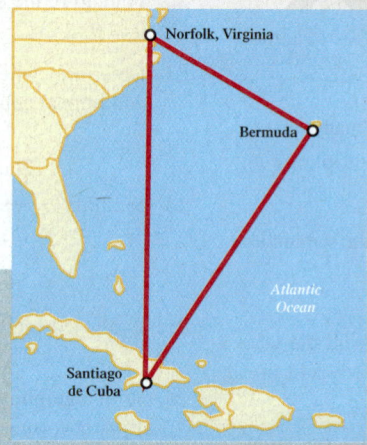

Norfolk (Virginia), Bermuda, Cuba

In recent decades, many people have come to believe that an imaginary area called the "Bermuda Triangle," located off the southeastern Atlantic coast of the United States, has been the site of a high incidence of losses of ships, small boats, and aircraft over the centuries. The U.S. Board of Geographic Names does not recognize the "Bermuda Triangle" as an official name and does not maintain an official file on the area.

Assume for the moment, without judging the merits of the hypothesis, that the "Bermuda Triangle" either has vertices in Miami (Florida), San Juan (Puerto Rico), and Bermuda or it has vertices in Norfolk (Virginia), Bermuda, and Santiago (Cuba). In this chapter you will develop a formula that determines the area of a triangle from its perimeter and side lengths. Which "Bermuda Triangle" has a larger area: Miami–Bermuda–Puerto-Rico or Norfolk–Bermuda–Cuba? You will calculate the answer in this chapter.*

---

*Section 8.3, Exercises 33 and 34.

IN THIS CHAPTER we discuss oblique (nonright) triangles. We use the Law of Sines and the Law of Cosines to solve oblique triangles. Then on the basis of the Law of Sines and the Law of Cosines and with our knowledge of trigonometric identities, we develop formulas for calculating the area of an oblique triangle. We define vectors and use the Law of Cosines and the Law of Sines to determine the resulting velocity and force vectors. We define dot products (the product of two vectors) and see how they are applicable to physical problems such as calculating work. We then use trigonometry to define polar coordinates, and we examine polar equations and their corresponding graphs.

ADDITIONAL TOPICS IN TRIGONOMETRY

8.1 Oblique Triangles and the Law of Sines	8.2 The Law of Cosines	8.3 The Area of a Triangle	8.4 Vectors	8.5 The Dot Product	8.6 Polar (Trigono-metric) Form of Complex Numbers	8.7 Products, Quotients, Powers, and Roots of Complex Numbers; DeMoivre's Theorem	8.8 Polar Equations and Graphs
• Solving Oblique Triangles	• Solving Oblique Triangles	• The Area of a Triangle (SAS Case) • The Area of a Triangle (SSS Case)	• Vectors: Magnitude and Direction • Vector Operations • Horizontal and Vertical Compo-nents of a Vector • Unit Vectors • Resultant Vectors	• Multiplying Two Vectors: The Dot Product • Angle Between Two Vectors • Work	• Complex Numbers in Rectangular Form • Complex Numbers in Polar Form	• Products of Complex Numbers • Quotients of Complex Numbers • Powers of Complex Numbers • Roots of Complex Numbers	• Polar Coordinates • Converting Between Polar and Rectangular Coordinates • Graphs of Polar Equations

LEARNING OBJECTIVES

- Solve oblique triangles using the Law of Sines.
- Solve oblique triangles using the Law of Cosines.
- Find areas of oblique triangles.
- Find the direction and magnitude of a vector; add and subtract vectors.
- Perform scalar multiplication and dot products.
- Express complex numbers in polar form.
- Use DeMoivre's theorem to find a complex number raised to a power.
- Graph polar equations.

Solving Oblique Triangles

Thus far we have discussed only *right* triangles. There are, however, two types of triangles, right and *oblique*. An **oblique triangle** is any triangle that does not have a right angle. An oblique triangle is either an **acute triangle**, having three acute (less than 90°) angles, or an **obtuse triangle**, having one obtuse (between 90° and 180°) angle.

It is customary to label oblique triangles the following way:

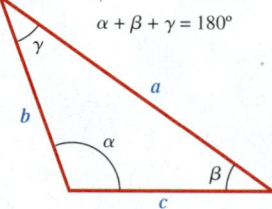

- angle α (alpha) opposite side a.
- angle β (beta) opposite side b.
- angle γ (gamma) opposite side c.

Remember that the sum of the three angles of any triangle must equal 180°. Recall that in Section 6.3 we solved right triangles. In this chapter we solve oblique triangles, which means we find the lengths of all three sides and the measures of all three angles.

Four Cases

To solve an oblique triangle, *we need to know the length of one side* and one of the following three:

- two angles
- one angle and another side
- the other two sides

This requirement leads to the following four possible cases to consider:

REQUIRED INFORMATION TO SOLVE OBLIQUE TRIANGLES

CASE	WHAT'S GIVEN	EXAMPLES/NAMES
Case 1	One side and two angles	AAS: Angle-Angle-Side ASA: Angle-Side-Angle
Case 2	Two sides and the angle opposite one of them	SSA: Side-Side-Angle
Case 3	Two sides and the angle between them	SAS: Side-Angle-Side
Case 4	Three sides	SSS: Side-Side-Side

Notice that there is no AAA case, because two similar triangles can have the same angle measures but different side lengths. That is why at least one side must be known.

In this section, we will derive the Law of Sines, which will enable us to solve Case 1 and Case 2 problems. In the next section, we will derive the Law of Cosines, which will enable us to solve Case 3 and Case 4 problems.

Study Tip

To solve triangles, at least one side must be known.

The Law of Sines

Let us start with two oblique triangles, an acute triangle and an obtuse triangle.

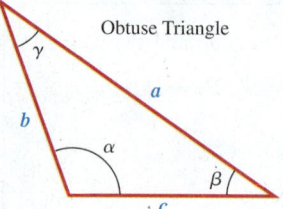

The following discussion applies to both triangles. First, construct an altitude (perpendicular) h from the vertex at angle γ to the side (or its extension) opposite γ.

WORDS	MATH
Formulate sine ratios for an acute triangle.	$\sin\alpha = \dfrac{h}{b}$ and $\sin\beta = \dfrac{h}{a}$
Formulate sine ratios for an obtuse triangle.	$\sin(180° - \alpha) = \dfrac{h}{b}$ and $\sin\beta = \dfrac{h}{a}$
For the obtuse triangle, apply the sine difference identity.	$\begin{aligned} \sin(180° - \alpha) &= \sin 180° \cos\alpha - \cos 180° \sin\alpha \\ &= 0 \cdot \cos\alpha - (-1)\sin\alpha \\ &= \sin\alpha \end{aligned}$
Therefore, in both triangles we find the same equations.	$\sin\alpha = \dfrac{h}{b}$ and $\sin\beta = \dfrac{h}{a}$
Solve for h in both equations.	$h = b\sin\alpha$ and $h = a\sin\beta$
Since h is equal to itself, equate the expressions for h.	$b\sin\alpha = a\sin\beta$
Divide both sides by ab.	$\dfrac{b\sin\alpha}{ab} = \dfrac{a\sin\beta}{ab}$
Divide out common factors.	$\boxed{\dfrac{\sin\alpha}{a} = \dfrac{\sin\beta}{b}}$

In a similar manner, we can extend an altitude (perpendicular) from angle α, and we will find that $\boxed{\dfrac{\sin\gamma}{c} = \dfrac{\sin\beta}{b}}$. Equating these two expressions leads us to the third ratio of the *Law of Sines*: $\boxed{\dfrac{\sin\alpha}{a} = \dfrac{\sin\gamma}{c}}$.

THE LAW OF SINES

For a triangle with sides a, b, and c and opposite angles α, β, and γ, the following is true:

$$\frac{\sin\alpha}{a} = \frac{\sin\beta}{b} = \frac{\sin\gamma}{c}$$

In other words, the ratio of the sine of an angle in a triangle to its opposite side is equal to the ratios of the sines of the other two angles to their opposite sides.

Some things to note before we begin solving oblique triangles are:

- The angles and sides share the same progression of magnitude:
  - The longest side of a triangle is opposite the largest angle.
  - The shortest side of a triangle is opposite the smallest angle.
- Draw the triangle and label the angles and sides.
- If two angles are known, start by determining the third angle.
- Whenever possible, in successive steps always return to given values rather than refer to calculated (approximate) values.

Keeping these pointers in mind will help you determine whether your answers are reasonable.

Case 1: Two Angles and One Side (AAS or ASA)

EXAMPLE 1   Using the Law of Sines to Solve a Triangle (AAS)

Solve the triangle.

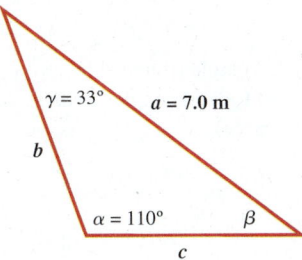

Solution:

This is an AAS (angle-angle-side) case, because two angles and a side are given and the side is opposite one of the angles.

STEP 1 Find β.

The sum of the measures of the angles in a triangle is $180°$.

$$\alpha + \beta + \gamma = 180°$$

Let $\alpha = 110°$ and $\gamma = 33°$.

$$110° + \beta + 33° = 180°$$

Solve for β.

$$\boxed{\beta = 37°}$$

Study Tip

The longest side is opposite the largest angle; the shortest side is opposite the smallest angle.

Study Tip

Always use given values rather than calculated (approximated) values for better accuracy.

Classroom Example 8.1.1
Solve the triangle.

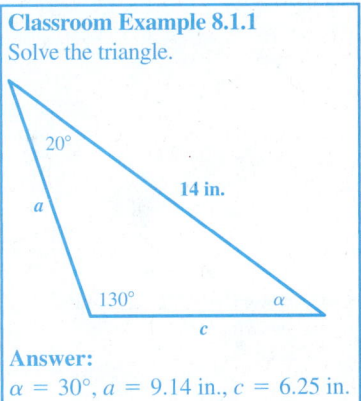

Answer:
$\alpha = 30°$, $a = 9.14$ in., $c = 6.25$ in.

Technology Tip

Step 1: Set the calculator to degree mode by typing MODE.

Step 2: Use the calculator to find
$b = \dfrac{7\sin 37°}{\sin 110°}$.

```
7sin(37)/sin(110
)
         4.483067195
```

Step 3: Use the calculator to find
$c = \dfrac{7\sin 33°}{\sin 110°}$.

```
7sin(33)/sin(110
)
         4.057149286
■
```

Study Tip

Notice in Step 3 that we used *a* which is given, as opposed to *b* which has been calculated (approximated).

STEP 2 Find *b*.

Use the Law of Sines with the known side, *a*.

$$\dfrac{\sin \alpha}{a} = \dfrac{\sin \beta}{b}$$

Isolate *b*.

$$b = \dfrac{a\sin \beta}{\sin \alpha}$$

Let $\alpha = 110°$, $\beta = 37°$, and $a = 7\,\text{m}$.

$$b = \dfrac{7\sin 37°}{\sin 110°}$$

Use a calculator to approximate *b*.

$$b = 4.483067\,\text{m}$$

Round *b* to two significant digits.

$$\boxed{b \approx 4.5\,\text{m}}$$

STEP 3 Find *c*.

Use the Law of Sines with the known side, *a*.

$$\dfrac{\sin \alpha}{a} = \dfrac{\sin \gamma}{c}$$

Isolate *c*.

$$c = \dfrac{a\sin \gamma}{\sin \alpha}$$

Let $\alpha = 110°$, $\gamma = 33°$, and $a = 7\,\text{m}$.

$$c = \dfrac{7\sin 33°}{\sin 110°}$$

Use a calculator to approximate *c*.

$$c = 4.057149\,\text{m}$$

Round *c* to two significant digits.

$$\boxed{c \approx 4.1\,\text{m}}$$

STEP 4 Draw and label the triangle.

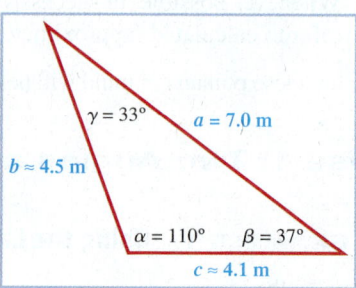

■ **Answer:** $\gamma = 32°$, $a \approx 42\,\text{ft}$, $c \approx 23\,\text{ft}$

■ **YOUR TURN** Solve the triangle (round sides to the nearest foot and angles to the nearest degree).

EXAMPLE 2 **Using the Law of Sines to Solve a Triangle (ASA)**

Solve the triangle.

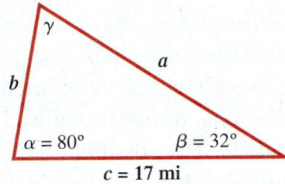

Solution:

This is an ASA (angle-side-angle) case because two angles and a side are given and the side is not opposite one of the angles.

STEP 1 Find γ.

The sum of the measures of the angles in a triangle is $180°$.

$$\alpha + \beta + \gamma = 180°$$

Let $\alpha = 80°$ and $\beta = 32°$.

$$80° + 32° + \gamma = 180°$$

Solve for γ.

$$\boxed{\gamma = 68°}$$

STEP 2 Find b.

Write the Law of Sines to include the known side c.

$$\frac{\sin\beta}{b} = \frac{\sin\gamma}{c}$$

Isolate b.

$$b = \frac{c\sin\beta}{\sin\gamma}$$

Let $\beta = 32°$, $\gamma = 68°$, and $c = 17$ miles.

$$b = \frac{17\sin 32°}{\sin 68°}$$

Use a calculator to approximate b.

$$b = 9.7161177$$

Round b to two significant digits.

$$\boxed{b \approx 9.7 \text{ miles}}$$

STEP 3 Find a.

Write the Law of Sines again incorporating the known side c.

$$\frac{\sin\alpha}{a} = \frac{\sin\gamma}{c}$$

Isolate a.

$$a = \frac{c\sin\alpha}{\sin\gamma}$$

Let $\alpha = 80°$, $\gamma = 68°$, and $c = 17$ miles.

$$a = \frac{17\sin 80°}{\sin 68°}$$

Use a calculator to approximate a.

$$a = 18.056539$$

Round a to two significant digits.

$$\boxed{a \approx 18 \text{ miles}}$$

STEP 4 Draw and label the triangle.

■ YOUR TURN Solve the triangle.

Case 2 (Ambiguous Case): Two Sides and One Angle (SSA)

If we are given two sides and an angle opposite one of the sides, then we call that Case 2, SSA (side-side-angle). This case is called the ambiguous case, because the given information by itself can represent one triangle, two triangles, or no triangle at all. If the angle given is acute, then the possibilities are zero, one, or two triangles. If the angle given is obtuse, then the possibilities are zero or one triangle. The possibilities come from the fact that $\sin \alpha = k$, where $0 < k < 1$, has two solutions for α: one in quadrant I (acute angle) and one in quadrant II (obtuse angle).

In the figure on the left, note that

- $h = b \sin \alpha$ by the definition of the sine ratio; since $0 < \sin \alpha < 1$, then $h < b$.
- a may turn out to be smaller than, equal to, or larger than h.

Given Angle (α) Is Acute

CONDITION	PICTURE	NUMBER OF TRIANGLES
$0 < a < h$ $\sin \beta > 1$	No Triangle	0
$a = h$ $\sin \beta = 1$	Right Triangle	1
$h < a < b$ $0 < \sin \beta < 1$	Acute Triangle / Obtuse Triangle	2
$a \geq b$ $0 < \sin \beta < 1$	Acute Triangle	1

*β is the angle opposite side b.

Given Angle (α) Is Obtuse

CONDITION	PICTURE	NUMBER OF TRIANGLES
$a \leq b$ $\sin \beta \geq 1$	No Triangle	0
$a > b$ $0 < \sin \beta < 1$	One Triangle	1

Study Tip

Notice that if the angle given is obtuse, the side opposite that angle *must* be longer than the other given side (longest side opposite the largest angle).

EXAMPLE 3 Solving the Ambiguous Case (SSA)—One Triangle

Solve the triangle $a = 23\,\text{ft}$, $b = 11\,\text{ft}$, and $\alpha = 122°$.

Solution:

This is the ambiguous case, because two sides and an angle opposite one of those sides are given. Since the given angle α is obtuse and $a > b$, *we expect one triangle.*

STEP 1 Find β.

Use the Law of Sines.

$$\frac{\sin \alpha}{a} = \frac{\sin \beta}{b}$$

Isolate $\sin \beta$.

$$\sin \beta = \frac{b \sin \alpha}{a}$$

Let $a = 23\,\text{ft}$, $b = 11\,\text{ft}$, and $\alpha = 122°$.

$$\sin \beta = \frac{(11\,\text{ft}) \sin 122°}{23\,\text{ft}}$$

Use a calculator to evaluate $\sin \beta$. $\sin \beta = 0.40558822$

Solve for β using the inverse sine function. $\beta = \sin^{-1}(0.40558822)$

Round to the nearest degree. $\boxed{\beta \approx 24°}$

Note: $\beta_1 \approx 24°$ and there is a corresponding $\beta_2 \approx 156°$ but there is no corresponding second triangle because $\alpha + \beta_2 > 180°$.

STEP 2 Find γ.

The measures of angles in a triangle sum to 180°.

$$\alpha + \beta + \gamma = 180°$$

Substitute $\alpha = 122°$ and $\beta \approx 24°$.

$$122° + 24° + \gamma = 180°$$

Solve for γ. $\boxed{\gamma \approx 34°}$

STEP 3 Find c.

Use the Law of Sines.

$$\frac{\sin \alpha}{a} = \frac{\sin \gamma}{c}$$

Isolate c.

$$c = \frac{a \sin \gamma}{\sin \alpha}$$

Substitute $a = 23\,\text{ft}$, $\alpha = 122°$, and $\gamma \approx 34°$.

$$c = \frac{(23\,\text{ft}) \sin 34°}{\sin 122°}$$

Use a calculator to evaluate c. $\boxed{c \approx 15\,\text{ft}}$

Technology Tip

Step 1: Use the calculator to find β.

```
11sin(122)/23
            .4055882199
sin⁻¹(Ans)
           23.92799328
■
```

Step 3: Use the calculator to find c.

```
23sin(34)/sin(12
2)
          15.16592849
```

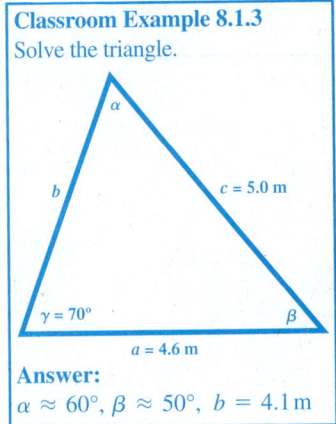

Classroom Example 8.1.3
Solve the triangle.

$c = 5.0\,\text{m}$
b
$\gamma = 70°$
$a = 4.6\,\text{m}$
β
α

Answer:
$\alpha \approx 60°$, $\beta \approx 50°$, $b = 4.1\,\text{m}$

■ **Answer:** $\beta = 32°, \gamma = 15°,$
$b = 35\,\text{mm}$

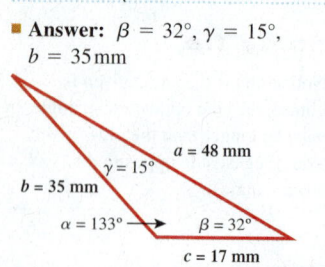

STEP 4 Draw and label the triangle.

■ **YOUR TURN** Solve the triangle $\alpha = 133°, a = 48\,\text{mm}$, and $c = 17\,\text{mm}$.

EXAMPLE 4 Solving the Ambiguous Case (SSA)—Two Triangles

Solve the triangle $a = 8.1\,\text{m}, b = 8.3\,\text{m}$, and $\alpha = 72°$.

Solution:

This is the ambiguous case, because two sides and an angle opposite one of those sides are given. Since the given angle α is acute and $a < b$, *we expect two triangles.*

STEP 1 Find β.

Write the Law of Sines for the given information.	$\dfrac{\sin\alpha}{a} = \dfrac{\sin\beta}{b}$
Isolate $\sin\beta$.	$\sin\beta = \dfrac{b\sin\alpha}{a}$
Let $a = 8.1\,\text{m}, b = 8.3\,\text{m}$, and $\alpha = 72°$.	$\sin\beta = \dfrac{(8.3\,\text{m})\sin72°}{8.1\,\text{m}}$
Use a calculator to evaluate $\sin\beta$.	$\sin\beta \approx 0.974539393$
Solve for β using the inverse sine function. Note that β can be acute or obtuse.	$\beta \approx \sin^{-1}(0.974539393)$
This is the quadrant I solution (β is acute).	$\boxed{\beta_1 \approx 77°}$
The quadrant II (β is obtuse) solution is $\beta_2 = 180 - \beta_1$.	$\boxed{\beta_2 \approx 103°}$

STEP 2 Find γ.

The measures of the angles in a triangle sum to 180°.	$\alpha + \beta + \gamma = 180°$
Solve for Triangle 1. Substitute $\alpha = 72°$ and $\beta_1 \approx 77°$.	$72° + 77° + \gamma_1 = 180°$
Solve for γ_1.	$\boxed{\gamma_1 \approx 31°}$
Solve for Triangle 2. Substitute $\alpha = 72°$ and $\beta_2 \approx 103°$.	$72° + 103° + \gamma_2 = 180°$
Solve for γ_2.	$\boxed{\gamma_2 \approx 5°}$

Classroom Example 8.1.4
Solve the triangle.

Answer:
Two triangles:
$\alpha_1 = 103.2°, \beta_1 = 41.8°,$
$a_1 = 7.3\,\text{in}.$
$\alpha_2 = 6.8°, \beta_2 = 138.2°,$
$a_2 = 0.9\,\text{in}.$

STEP 3 Find c.

Use the Law of Sines.

$$\frac{\sin \alpha}{a} = \frac{\sin \gamma}{c}$$

Isolate c.

$$c = \frac{a \sin \gamma}{\sin \alpha}$$

Solve for Triangle 1.
Substitute $a = 8.1\,\text{m}$, $\alpha = 72°$, and $\gamma_1 \approx 31°$.

$$c_1 \approx \frac{(8.1\,\text{m}) \sin 31°}{\sin 72°}$$

Use a calculator to evaluate c_1.

$$\boxed{c_1 \approx 4.4\,\text{m}}$$

Solve for Triangle 2.
Substitute $a = 8.1\,\text{m}$, $\alpha = 72°$, and $\gamma_2 \approx 5°$.

$$c_2 = \frac{(8.1\,\text{m}) \sin 5°}{\sin 72°}$$

Use a calculator to evaluate c_2.

$$\boxed{c_2 \approx 0.74\,\text{m}}$$

STEP 4 Draw and label the two triangles.

EXAMPLE 5   Solving the Ambiguous Case (SSA)—No Triangle

Solve the triangle $\alpha = 107°$, $a = 6$, and $b = 8$.

Solution:

This is the ambiguous case because two sides and an angle opposite one of those sides are given. Since the given angle α is obtuse and $a < b$, *we know that there is no triangle*. Notice that if we did not realize that there is no triangle and we had proceeded with the Law of Sines, the result would have been the sine function exceeding 1—which is not possible.

Write the Law of Sines.

$$\frac{\sin \alpha}{a} = \frac{\sin \beta}{b}$$

Isolate $\sin \beta$.

$$\sin \beta = \frac{b \sin \alpha}{a}$$

Let $\alpha = 107°$, $a = 6$, and $b = 8$.

$$\sin \beta = \frac{8 \sin 107°}{6}$$

Use a calculator to evaluate $\sin \beta$.

$$\sin \beta \approx 1.28$$

Since the range of the sine function is $[-1, 1]$, there is no angle β such that $\sin \beta \approx 1.28$. Therefore, there is $\boxed{\text{no triangle}}$ with the given measurements.

Applications

Solving oblique triangles is rich with applications in astronomy, surveying, aircraft design, piloting, and many other areas.

EXAMPLE 6 **How Far Over Is the Tower of Pisa Leaning?**

The Tower of Pisa was originally built 56 meters tall. Because of poor soil in the foundation, it started to lean. At a distance 44 meters from the base of the tower, the angle of elevation is 55°. How much is the Tower of Pisa leaning away from the vertical position?

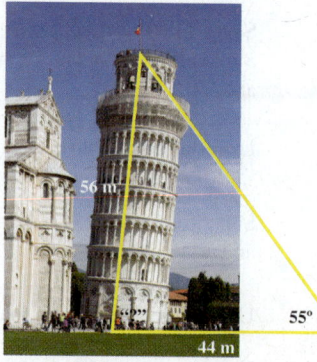

Nathalie Michel/Photodisc/
Getty Images, Inc.

Solution:

$\beta = 55°$, $c = 44$ meters, and $b = 56$ meters is the given information, so this is an SSA problem. There is only one triangle because $b > c$.

STEP 1 Find γ.

Use the Law of Sines.

$$\frac{\sin \beta}{b} = \frac{\sin \gamma}{c}$$

Isolate $\sin \gamma$.

$$\sin \gamma = \frac{c \sin \beta}{b}$$

Let $\beta = 55°$, $c = 44$ meters, and $b = 56$ meters.

$$\sin \gamma = \frac{44 \sin 55°}{56}$$

Evaluate the right side using a calculator.

$$\sin \gamma \approx 0.643619463$$

Solve for γ using the inverse sine function.

$$\gamma \approx \sin^{-1}(0.643619463)$$

Round to two significant digits.

$$\gamma \approx 40°$$

STEP 2 Find α.

The measures of angles in a triangle sum to 180°.

$$\alpha + \beta + \gamma = 180°$$

Let $\beta = 55°$ and $\gamma \approx 40°$.

$$\alpha + 55° + 40° \approx 180°$$

Solve for α.

$$\alpha \approx 85°$$

The Tower of Pisa makes an angle of 85° with the ground. It is leaning at an angle of $\boxed{5°}$.

SECTION 8.1 SUMMARY

In this section, we solved oblique triangles. When given three pieces of information about a triangle, we classify the triangle according to the data (sides and angles). Four cases arise:

- one side and two angles (AAS or ASA)
- two sides and one angle opposite one of the sides (SSA)
- two sides and one angle between sides (SAS)
- three sides (SSS)

The Law of Sines

$$\frac{\sin\alpha}{a} = \frac{\sin\beta}{b} = \frac{\sin\gamma}{c}$$

can be used to solve the first two cases (AAS or ASA, and SSA). It is important to note that the SSA case is called the ambiguous case because any one of three results is possible: no triangle, one triangle, or two triangles.

SECTION 8.1 EXERCISES

■SKILLS

In Exercises 1–6, classify each triangle problem as AAS, ASA, SAS, SSA, or SSS on the basis of the given information.

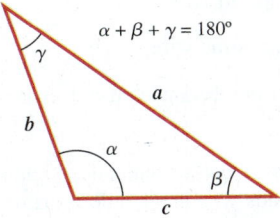

$\alpha + \beta + \gamma = 180°$

1. $c, a,$ and α
2. $c, a,$ and γ
3. $a, b,$ and c
4. $a, b,$ and γ
5. $\alpha, \beta,$ and c
6. $\beta, \gamma,$ and a

In Exercises 7–16, solve the given triangles.

7. $\alpha = 45°, \beta = 60°, a = 10\,\text{m}$
8. $\beta = 75°, \gamma = 60°, b = 25\,\text{in}.$
9. $\alpha = 46°, \gamma = 72°, b = 200\,\text{cm}$
10. $\gamma = 100°, \beta = 40°, a = 16\,\text{ft}$
11. $\alpha = 16.3°, \gamma = 47.6°, c = 211\,\text{yd}$
12. $\beta = 104.2°, \gamma = 33.6°, a = 26\,\text{in}.$
13. $\alpha = 30°, \beta = 30°, c = 12\,\text{m}$
14. $\alpha = 45°, \gamma = 75°, c = 9\,\text{in}.$
15. $\beta = 26°, \gamma = 57°, c = 100\,\text{yd}$
16. $\alpha = 80°, \gamma = 30°, b = 3\,\text{ft}$

In Exercises 17–34, two sides and an angle are given. Determine whether a triangle (or two) exist, and if so, solve the triangle(s).

17. $a = 4, b = 5, \alpha = 16°$
18. $b = 30, c = 20, \beta = 70°$
19. $a = 12, c = 12, \gamma = 40°$
20. $b = 111, a = 80, \alpha = 25°$
21. $a = 21, b = 14, \beta = 100°$
22. $a = 13, b = 26, \alpha = 120°$
23. $\alpha = 30°, b = 18, a = 9$
24. $\alpha = 45°, b = \sqrt{2}, a = 1$
25. $\alpha = 34°, b = 7, a = 10$
26. $\alpha = 71°, b = 5.2, a = 5.2$
27. $\alpha = 21.3°, b = 6.18, a = 6.03$
28. $\alpha = 47.3°, b = 7.3, a = 5.32$
29. $\alpha = 116°, b = 4\sqrt{3}, a = 5\sqrt{2}$
30. $\alpha = 51°, b = 4\sqrt{3}, a = 4\sqrt{5}$
31. $b = 500, c = 330, \gamma = 40°$
32. $b = 16, a = 9, \beta = 137°$
33. $a = \sqrt{2}, b = \sqrt{7}, \beta = 106°$
34. $b = 15.3, c = 27.2, \gamma = 11.6°$

APPLICATIONS

For Exercises 35 and 36, refer to the following:

NASA Kennedy Space Center

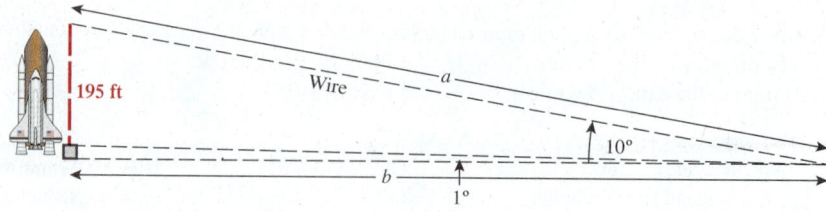

On the launch pad at Kennedy Space Center, there is an escape basket that can hold four astronauts. The basket slides down a wire that is attached 195 feet high, above the base of the launch pad. The angle of inclination measured from where the basket would touch the ground to the base of the launch pad is 1°, and the angle of inclination from that same point to where the wire is attached is 10°.

35. NASA. How long is the wire a?

36. NASA. How far from the launch pad does the basket touch the ground b?

37. Hot-Air Balloon. A hot-air balloon is sighted at the same time by two friends who are 1 mile apart on the same side of the balloon. The angles of elevation from the two friends are 20.5° and 25.5°. How high is the balloon?

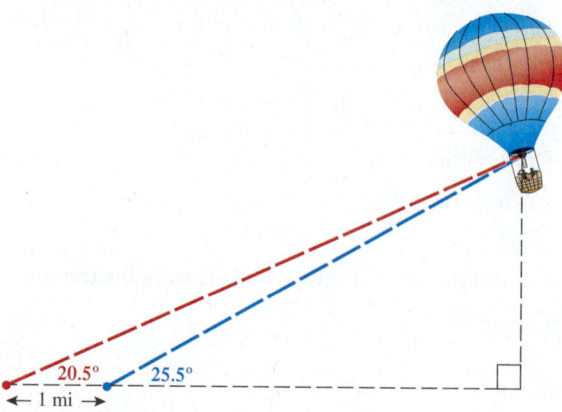

38. Hot-Air Balloon. A hot-air balloon is sighted at the same time by two friends who are 2 miles apart on the same side of the balloon. The angles of elevation from the two friends are 10° and 15°. How high is the balloon?

39. Rocket Tracking. A tracking station has two telescopes that are 1 mile apart. The telescopes can lock onto a rocket after it is launched and record the angles of elevation to the rocket. If the angles of elevation from telescopes A and B are

30° and 80°, respectively, then how far is the rocket from telescope A?

40. Rocket Tracking. Given the data in Exercise 39, how far is the rocket from telescope B?

41. Distance Across River. An engineer wants to construct a bridge across a fast-moving river. Using a straight-line segment between two points that are 100 feet apart along his side of the river, he measures the angles formed when sighting the point on the other side where he wants to have the bridge end. If the angles formed at points A and B are 65° and 15°, respectively, how far is it from point A to the point on the other side of the river? Round to the nearest foot.

42. Distance Across River. Given the data in Exercise 41, how far is it from point B to the point on the other side of the river? Round to the nearest foot.

43. Lifeguard Posts. Two lifeguard chairs, labeled P and Q, are located 400 feet apart. A troubled swimmer is spotted by both lifeguards. If the lifeguard at P reports the swimmer at angle $35°$ (with respect to the line segment connecting P and Q) and the lifeguard at Q reports the swimmer at angle $41°$, how far is the swimmer from P?

44. Rock Climbing. A rock climbing enthusiast is creating a climbing route rated as 5.8 level (i.e., medium difficulty) on the wall at the local rock gym. Given the difficulty of the route, he wants to avoid placing any two holds on the same vertical or horizontal line on the wall. If he places holds at P, Q, and R such that $\angle QPR = 40°$, $QR = 6$ feet, and $QP = 4.5$ feet, how far is the hold at P from the hold at R?

45. Tennis. Two friends playing tennis are both 70 inches tall. After a rather long rally, Player II lobs the ball into Player I's court, enabling him to hit an overhead smash such that the angle between the racquet head (at the point of contact with the ball) and his body is $56°$. The ball travels 20.3 feet to the other side of the court where Player II, with lightning-quick reflexes, volleys the ball 3 inches off the ground at an angle α such that the ball travels directly toward Player I's head. The ball travels 19.4 feet during this return. Find angle α.

46. Tennis. Shocked by the move Player I made in Exercise 45, Player II is forced to quickly deflect the ball straight back directly over the middle of the net (which is 33 inches high) with great speed. Player I reaches behind himself and is able to contact the ball at the same height above the ground with which Player I initially hit it. If the angle between Player I's racquet position at the end of the previous shot (i.e., α from Exercise 45) and its current position at the point of contact of this shot is $130°$, and the angle with which it contacts the ball is $25°$, how far has the ball traveled horizontally as a result of Player II's hit?

47. Archery. (Recall Exercise 73 in Section 6.1 for context.) An expert archer decides to shoot two arrows simultaneously in such a way that the two arrows hit directly above and below the bullseye where the blue and black rings meet (8 inches from the bullseye). If the angle each arrow makes with the target is $84°$, how far did each arrow travel?

48. Archery. An archer fires two arrows simultaneously toward the target. The topmost arrow hits the bullseye dead center. If this arrow traveled 74 feet, and the second arrow hit the target at an angle somewhere between $88.3°$ and $88.7°$, what was the archer's score?

49. Bowling. The 6-8 split is common in bowling. To make this split, a bowler stands dead center and throws the ball hard and straight directly toward the right of the 6 pin. The distance from the ball at the point of release and the 8 pin is 63.2 feet. See the diagram. How far does the ball travel if it hits the 6 pin correctly?

50. Bowling. A bowler is said to get a strike on the "Brooklyn side" of the head pin if he hits the head pin on the opposite side of the pocket. (For a right-handed bowler, the pocket is to the right of the head pin.) There is a small range of angle at which the ball must contact the head pin in order to convert all of the pins. If the measurements are as shown, how far does the ball travel (assuming it is thrown straight with no hook) before it contacts the head pin?

For Exercises 51 and 52, refer to the following:

To quantify the torque (rotational force) of the elbow joint of a human arm (see the figure below), it is necessary to identify angles A, B, and C as well as lengths a, b, and c. Measurements performed on an arm determine that the measure of angle C is $95°$, the measure of angle A is $82°$, and the length of the muscle a is 23 centimeters.

51. Health/Medicine. Find the length of the forearm from the elbow joint to the muscle attachment b.

52. Health/Medicine. Find the length of the upper arm from the muscle attachment to the elbow joint c.

▪CATCH THE MISTAKE

In Exercises 53 and 54, explain the mistake that is made.

53. Solve the triangle $\alpha = 120°, a = 7$, and $b = 9$.

Solution:

Use the Law of Sines to find β.	$\dfrac{\sin\alpha}{a} = \dfrac{\sin\beta}{b}$
Let $\alpha = 120°, a = 7$, and $b = 9$.	$\dfrac{\sin 120°}{7} = \dfrac{\sin\beta}{9}$
Solve for $\sin\beta$.	$\sin\beta = 1.113$
Solve for β.	$\beta = 42°$
Sum the angle measures to 180°.	$120° + 42° + \gamma = 180°$
Solve for γ.	$\gamma = 18°$
Use the Law of Sines to find c.	$\dfrac{\sin\alpha}{a} = \dfrac{\sin\gamma}{c}$
Let $\alpha = 120°, a = 7$, and $\gamma = 18°$.	$\dfrac{\sin 120°}{7} = \dfrac{\sin 18°}{c}$
Solve for c.	$c = 2.5$

$\alpha = 120°, \beta = 42°, \gamma = 18°, a = 7, b = 9$, and $c = 2.5$.

This is incorrect. The longest side is not opposite the longest angle. There is no triangle that makes the original measurements work. What mistake was made?

54. Solve the triangle $\alpha = 40°, a = 7$, and $b = 9$.

Solution:

Use the Law of Sines to find β.	$\dfrac{\sin\alpha}{a} = \dfrac{\sin\beta}{b}$
Let $\alpha = 40°, a = 7$, and $b = 9$.	$\dfrac{\sin 40°}{7} = \dfrac{\sin\beta}{9}$
Solve for $\sin\beta$.	$\sin\beta = 0.826441212$
Solve for β.	$\beta = 56°$
Find γ.	$40° + 56° + \gamma = 180°$
	$\gamma = 84°$
Use the Law of Sines to find c.	$\dfrac{\sin\alpha}{a} = \dfrac{\sin\gamma}{c}$
Let $\alpha = 40°, a = 7$, and $\gamma = 84°$.	$\dfrac{\sin 40°}{7} = \dfrac{\sin 84°}{c}$
Solve for c.	$c = 11$

$\alpha = 40°, \beta = 56°, \gamma = 84°, a = 7, b = 9$, and $c = 11$.

This is incorrect. What mistake was made?

▪CONCEPTUAL

In Exercises 55–60, determine whether each statement is true or false.

55. The Law of Sines applies only to right triangles.

56. If you are given two sides and any angle, there is a unique solution for the triangle.

57. An acute triangle is an oblique triangle.

58. An obtuse triangle is an oblique triangle.

59. If you are given two sides that have the same length, then there can be at most one triangle.

60. If α is obtuse and $\beta = \dfrac{\alpha}{2}$, then the situation is unambiguous.

▪CHALLENGE

61. Mollweide's Identity. For any triangle, the following identity is true. It is often used to check the solution of a triangle since all six pieces of information (three sides and three angles) are involved. Derive the identity using the Law of Sines.

$$(a + b)\sin\left(\tfrac{1}{2}\gamma\right) = c\cos\left[\tfrac{1}{2}(\alpha - \beta)\right]$$

63. Use the Law of Sines to prove that all angles in an equilateral triangle must have the same measure.

62. The Law of Tangents. Use the Law of Sines and trigonometric identities to show that for any triangle, the following is true:

$$\frac{a - b}{a + b} = \frac{\tan\left[\tfrac{1}{2}(\alpha - \beta)\right]}{\tan\left[\tfrac{1}{2}(\alpha + \beta)\right]}$$

64. Suppose that you have a triangle with side lengths a, b, and c, and angles α, β, and γ, respectively, directly across from them. If it is known that $a = \dfrac{1}{\sqrt{2}}b, c = 2, \alpha$ is an acute angle, and $\beta = 2\alpha$, solve the triangle.

■ TECHNOLOGY

For Exercises 65–70, let *A*, *B*, and *C* be the lengths of the three sides with *X*, *Y*, and *Z* as the corresponding angles. Write a program to solve the given triangle with a calculator.

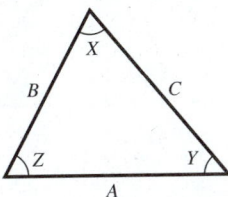

65. $A = 10$, $Y = 40°$, and $Z = 72°$

66. $B = 42.8$, $X = 31.6°$, and $Y = 82.2°$

67. $A = 22$, $B = 17$, and $X = 105°$

68. $B = 16.5$, $C = 9.8$, and $Z = 79.2°$

69. $A = 25.7$, $C = 12.2$, and $X = 65°$

70. $A = 54.6$, $B = 12.9$, and $Y = 23°$

SECTION 8.2 THE LAW OF COSINES

SKILLS OBJECTIVES

■ Solve SAS triangle cases.
■ Solve SSS triangle cases.
■ Solve application problems involving oblique triangles.

CONCEPTUAL OBJECTIVES

■ Understand the derivation of the Law of Cosines.
■ Develop a strategy for which angles (larger or smaller) and which method (the Law of Sines or the Law of Cosines) to select to solve oblique triangles.

Solving Oblique Triangles

In Section 8.1, we learned that to solve oblique triangles means to find all three side lengths and angle measures. At least one side length must be known. We need two additional pieces of information to solve a triangle (combinations of side lengths and/or angles). We found that there are four cases:

■ Case 1: AAS or ASA (two angles and a side are given)
■ Case 2: SSA (two sides and an angle opposite one of the sides are given)
■ Case 3: SAS (two sides and the angle between them are given)
■ Case 4: SSS (three sides are given)

We used the Law of Sines to solve Case 1 and Case 2 triangles. Now, we use the *Law of Cosines* to solve Case 3 and Case 4 triangles.

WORDS	**MATH**

Start with a triangle.

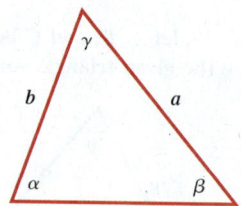

Drop a perpendicular line from γ with height h.

The result is two triangles within the larger triangle.

Write the Pythagorean theorem for both right triangles.

    Triangle 1: $x^2 + h^2 = b^2$

    Triangle 2: $(c - x)^2 + h^2 = a^2$

Solve for h^2.

    Triangle 1: $h^2 = b^2 - x^2$

    Triangle 2: $h^2 = a^2 - (c - x)^2$

Since the segment of length h is shared, set $h^2 = h^2$, for the two triangles. $b^2 - x^2 = a^2 - (c - x)^2$

Multiply out the squared binomial on the right. $b^2 - x^2 = a^2 - (c^2 - 2cx + x^2)$

Eliminate the parentheses. $b^2 - x^2 = a^2 - c^2 + 2cx - x^2$

Add x^2 to both sides. $b^2 = a^2 - c^2 + 2cx$

Isolate a^2. $a^2 = b^2 + c^2 - 2cx$

Notice that $\cos\alpha = \dfrac{x}{b}$. Let $x = b\cos\alpha$. $\boxed{a^2 = b^2 + c^2 - 2bc\cos\alpha}$

Note: If we instead drop the perpendicular line with length h from the angle α or the angle β, we can derive the other two parts of the Law of Cosines:

$$\boxed{b^2 = a^2 + c^2 - 2ac\cos\beta} \quad \text{and} \quad \boxed{c^2 = a^2 + b^2 - 2ab\cos\gamma}$$

THE LAW OF COSINES

For a triangle with sides a, b, and c and opposite angles α, β, and γ, the following are true:

$$a^2 = b^2 + c^2 - 2bc\cos\alpha$$
$$b^2 = a^2 + c^2 - 2ac\cos\beta$$
$$c^2 = a^2 + b^2 - 2ab\cos\gamma$$

It is important to note that the Law of Cosines can be used to find side lengths or angles. As long as three of the four variables in any of the equations are known, the fourth can be calculated.

Notice that in the special case of a right triangle (say, $\gamma = 90°$),

$$c^2 = a^2 + b^2 - 2ab\underbrace{\cos 90°}_{0}$$

one of the components of the Law of Cosines reduces to the Pythagorean theorem:

$$\underbrace{c^2}_{\text{hyp}} = \underbrace{a^2}_{\text{leg}} + \underbrace{b^2}_{\text{leg}}$$

The Pythagorean theorem can thus be regarded as a special case of the Law of Cosines.

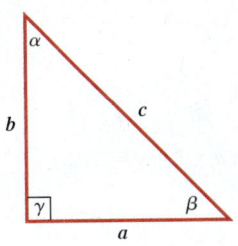

Case 3: Solving Oblique Triangles (SAS)

We now solve SAS triangle problems where the angle between two sides is given. We start by using the Law of Cosines to solve for the side opposite the given angle. We then can apply either the Law of Sines or the Law of Cosines to find the second angle.

Study Tip

The Pythagorean theorem is a special case of the Law of Cosines.

EXAMPLE 1 **Using the Law of Cosines to Solve a Triangle (SAS)**

Solve the triangle $a = 13$, $c = 6.0$, and $\beta = 20°$.

Solution:

Two sides and the angle between them are given (SAS).

Notice that the Law of Sines can't be used, because it requires knowledge of at least one angle and the side opposite that angle.

STEP 1 Find b.

Apply the Law of Cosines that involves β. $b^2 = a^2 + c^2 - 2ac\cos\beta$

Let $a = 13$, $c = 6.0$, and $\beta = 20°$. $b^2 = 13^2 + 6^2 - 2(13)(6)\cos 20°$

Evaluate the right side with a calculator. $b^2 \approx 58.40795$

Solve for b. $b \approx \pm 7.6425$

Round to two significant digits;
b can only be positive. $\boxed{b \approx 7.6}$

STEP 2 Find γ.

Use the Law of Sines. $\dfrac{\sin\gamma}{c} = \dfrac{\sin\beta}{b}$

Isolate $\sin\gamma$. $\sin\gamma = \dfrac{c\sin\beta}{b}$

Let $b = 7.6$, $c = 6.0$, and $\beta = 20°$. $\sin\gamma = \dfrac{6\sin 20°}{7.6}$

Apply the inverse sine function. $\gamma = \sin^{-1}\left(\dfrac{6\sin 20°}{7.6}\right)$

Evaluate the right side with a calculator. $\gamma \approx 15.66521°$

Round to the nearest degree. $\boxed{\gamma \approx 16°}$

Technology Tip

Step 1: Use the calculator to find the value of b.

```
13²+6²-2*13*6*co
s(20)
          58.40795116
√(Ans)
          7.64250948
■
```

Step 2: Use the calculator to find γ.

```
6sin(20)/7.6
          .2700159026
sin⁻¹(Ans)
          15.66521315
sin⁻¹(6sin(20)/7.
6)
          15.66521315
■
```

Classroom Example 8.2.1
Solve the triangle.

Answer:
$\alpha \approx 150°$, $\beta \approx 13°$,
$c \approx 8.7\,\text{cm}$

STEP 3 Find α.

The angle measures must sum to 180°. $\qquad$ $\alpha + 20° + 16° \approx 180°$

Solve for α. $\qquad\qquad\qquad\qquad$ $\boxed{\alpha \approx 144°}$

■ **Answer:** $a = 2.9$, $\gamma = 21°$,
  $\beta = 124°$

■ **YOUR TURN** Solve the triangle $b = 4.2$, $c = 1.8$, and $\alpha = 35°$.

Notice the steps we took in solving an SAS triangle:

1. Find the side opposite the given angle using the Law of Cosines.
2. Solve for the smaller angle using the Law of Sines.
3. Solve for the larger angle using properties of triangles.

Study Tip

Although the Law of Sines is sometimes ambiguous, the Law of Cosines is never ambiguous.

You may be thinking, "Would it matter if we had solved for α before solving for γ?" Yes, it does matter—in this problem you cannot solve for α by the Law of Sines before finding γ. The Law of Sines can be used only on the smaller angle (opposite the shortest side). If we had tried to use the Law of Sines with the obtuse angle α, the inverse sine would have resulted in $\alpha = 36°$. Since the sine function is positive in both quadrants I and II, we would not know whether that angle was $\alpha = 36°$ or its supplementary angle $\alpha = 144°$. Notice that $c < a$; therefore, the angles opposite those sides must have the same relationship, $\gamma < \alpha$. We choose the smaller angle first. Alternatively, if we want to solve for the obtuse angle first, we can use the Law of Cosines to solve for α. If you use the Law of Cosines to find the second angle you can choose either angle. The Law of Cosines can be used for either acute or obtuse angles.

EXAMPLE 2 **Using the Law of Cosines in an Application (SAS)**

Technology Tip

Use a calculator to find the value of c.

```
176²+152²-2*176*
152*cos(117)
          78370.3077
√(Ans)
          279.946973
```

In an AKC (American Kennel Club)-sanctioned field trial, a judge sets up a mark (bird) to which the dog is required to swim (the dogs are judged on how closely they adhere to the straight line to the bird, not the time it takes to retrieve the bird). The judge is trying to calculate how far a swim this mark is for the dogs, so she walks off the two legs across the land and measures the angle as shown in the figure. How far will the dog swim from the starting line to the bird?

152 yd
117°
176 yd
Pond

Solution:

Label the triangle.

$b = 152$ yd
$\gamma = 117°$
c
$a = 176$ yd
Pond

Apply the Law of Cosines. $\qquad\qquad$ $c^2 = a^2 + b^2 - 2ab\cos\gamma$

Let $a = 176$ yd, $b = 152$ yd, and $\gamma = 117°$. $\qquad$ $c^2 = 176^2 + 152^2 - 2(176)(152)\cos 117°$

Simplify. $\qquad\qquad\qquad\qquad$ $c^2 \approx 78370.3077$

Solve for c and round to the nearest yard. $\qquad\qquad$ $\boxed{c \approx 280 \text{ yd}}$

Case 4: Solving Oblique Triangles (SSS)

We now solve oblique triangles when all three side lengths are given (the SSS case). In this case, start by finding the largest angle (opposite the longest side) using the Law of Cosines. Then apply the Law of Sines to find either of the remaining two angles. Last, find the third angle using the sum of the angles in a triangle.

 EXAMPLE 3   Using the Law of Cosines to Solve a Triangle (SSS)

Solve the triangle $a = 8, b = 6$, and $c = 7$.

Solution:

STEP 1 Identify the largest angle, which is α.

Write the component of the Law of Cosines that involves α.

$$a^2 = b^2 + c^2 - 2bc\cos\alpha$$

Let $a = 8, b = 6$, and $c = 7$.

$$8^2 = 6^2 + 7^2 - 2(6)(7)\cos\alpha$$

Simplify and isolate $\cos\alpha$.

$$\cos\alpha = \frac{6^2 + 7^2 - 8^2}{2(6)(7)} = 0.25$$

Write the inverse cosine function.

$$\alpha = \cos^{-1}(0.25)$$

Approximate with a calculator and round to the nearest degree.

$$\boxed{\alpha \approx 76°}$$

STEP 2 Find either of the remaining angles to solve for β.

Write the Law of Sines.

$$\frac{\sin\alpha}{a} = \frac{\sin\beta}{b}$$

Isolate $\sin\beta$.

$$\sin\beta = \frac{b\sin\alpha}{a}$$

Let $a = 8, b = 6$, and $\alpha \approx 76°$.

$$\sin\beta \approx \frac{6\sin 76°}{8}$$

Write the inverse sine function.

$$\beta \approx \sin^{-1}\left(\frac{6\sin 76°}{8}\right)$$

Approximate with a calculator.

$$\boxed{\beta \approx 47°}$$

STEP 3 Find the third angle, γ.

The sum of the angle measures is 180°.

$$76° + 47° + \gamma \approx 180°$$

Solve for γ.

$$\boxed{\gamma \approx 57°}$$

■ **YOUR TURN** Solve the triangle $a = 5, b = 7$, and $c = 8$.

Classroom Example 8.2.3
Solve the triangle.

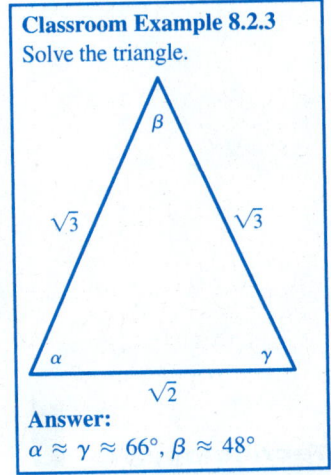

Answer:
$\alpha \approx \gamma \approx 66°, \beta \approx 48°$

Technology Tip

Step 1: Use the calculator to find the value of α.

Step 2: Use the calculator to find β.

■ **Answer:** $\alpha \approx 38.2°, \beta \approx 60.0°,$
$\gamma \approx 81.8°$

In the next example, instead of immediately substituting in values for the Law of Cosines equation, we will solve for the angle in general, and then substitute in values.

EXAMPLE 4   Using the Law of Cosines in an Application (SSS)

Many people believe that an imaginary area called the "Bermuda Triangle" exists on the southeastern Atlantic coast of the United States. Assume that the Bermuda Triangle has vertices in Norfolk (Virginia), Bermuda, and Santiago (Cuba). Find the angles of the Bermuda Triangle given the following distances:

LOCATION	LOCATION	DISTANCE (NAUTICAL MILES)
Norfolk	Bermuda	850
Bermuda	Santiago	810
Norfolk	Santiago	894

Ignore the curvature of the Earth in your calculations.

Solution:

Draw and label the triangle.

VIRGINIA

α $c = 850$ nm

β BERMUDA

$b = 894$ nm

$a = 810$ nm

γ

CUBA

STEP 1 Find β (the largest angle).

Apply the Law of Cosines. $\qquad b^2 = a^2 + c^2 - 2ac\cos\beta$

Isolate $\cos\beta$. $\qquad \cos\beta = \dfrac{a^2 + c^2 - b^2}{2ac}$

Apply the inverse cosine
function to solve for β. $\qquad \beta = \cos^{-1}\!\left(\dfrac{a^2 + c^2 - b^2}{2ac}\right)$

Let $a = 810, b = 894,$ and $c = 850$. $\qquad \beta = \cos^{-1}\!\left[\dfrac{810^2 + 850^2 - 894^2}{2(810)(850)}\right]$

Approximate with a calculator. $\qquad \boxed{\beta \approx 65°}$

STEP 2 Find α.

Apply the Law of Sines. $\qquad \dfrac{\sin\alpha}{a} = \dfrac{\sin\beta}{b}$

Isolate $\sin\alpha$. $\qquad \sin\alpha = \dfrac{a\sin\beta}{b}$

Apply the inverse sine
function to solve for α. $\qquad \alpha = \sin^{-1}\!\left(\dfrac{a\sin\beta}{b}\right)$

Let $a = 810, b = 894,$ and $\beta = 65°$. $\qquad \alpha \approx \sin^{-1}\!\left(\dfrac{810\sin 65°}{894}\right)$

Approximate with a calculator,
and round to the nearest degree. $\qquad \boxed{\alpha \approx 55°}$

STEP 3 Find γ.

The angle measures must sum to 180°. $\qquad 55° + 65° + \gamma \approx 180°$

Solve for γ. $\qquad \boxed{\gamma \approx 60°}$

Technology Tip

Step 1: Use the calculator to find β.

```
cos⁻¹((810²+850²−
894²)/(2*810*850
))
         65.11845415
```

Step 2: Use the calculator to find α.

```
sin⁻¹(810sin(65)/
894)
         55.200215
■
```

Instead of using the approximated value of β to calculate α, you can use ANS to retrieve the actual value of β.

```
cos⁻¹((810²+850²−
894²)/(2*810*850
))
         65.11845415
sin⁻¹(810sin(Ans)
/894)
         55.27959284
```

SECTION
8.2 SUMMARY

We can solve any triangle given three pieces of information, as long as one of the pieces is a side length. Depending on the information given, we either apply the **Law of Sines**

$$\frac{\sin\alpha}{a} = \frac{\sin\beta}{b} = \frac{\sin\gamma}{c}$$

and the angle sum identity, or we apply a combination of the **Law of Cosines**,

$$a^2 = b^2 + c^2 - 2bc\cos\alpha$$
$$b^2 = a^2 + c^2 - 2ac\cos\beta$$
$$c^2 = a^2 + b^2 - 2ab\cos\gamma$$

the Law of Sines, and the angle sum identity. The table below summarizes the strategies for solving oblique triangles:

OBLIQUE TRIANGLE	WHAT'S KNOWN	PROCEDURE FOR SOLVING
AAS or ASA	Two angles and a side	Step 1: Find the remaining angle with $\alpha + \beta + \gamma = 180°$. Step 2: Find the remaining sides with the Law of Sines.
SSA	Two sides and an angle opposite one of the sides	This is the ambiguous case, so there is either no triangle, one triangle, or two triangles. If the given angle is obtuse, then there is either one or no triangle. If the given angle is acute, then there is no triangle, one triangle, or two triangles. Step 1: Apply the Law of Sines to find one of the angles. Step 2: Find the remaining angle with $\alpha + \beta + \gamma = 180°$. Step 3: Find the remaining side with the Law of Sines. If two triangles exist, then the angle found in Step 1 can be either acute or obtuse, and Step 2 and Step 3 must be performed for each triangle.
SAS	Two sides and an angle between the sides	Step 1: Find the third side with the Law of Cosines. Step 2: Find the smaller angle with the Law of Sines. Step 3: Find the remaining angle with $\alpha + \beta + \gamma = 180°$.
SSS	Three sides	Step 1: Find the largest angle with the Law of Cosines. Step 2: Find either remaining angle with the Law of Sines. Step 3: Find the last remaining angle with $\alpha + \beta + \gamma = 180°$.

SECTION
8.2 EXERCISES

■ SKILLS

For each of the triangles in Exercises 1–8, the angle sum identity, $\alpha + \beta + \gamma = 180°$, will be used in solving the triangle. Label the problem as S if only the Law of Sines is needed to solve the triangle. Label the problem as C if the Law of Cosines is needed to solve the triangle.

1. a, b, and c are given.

2. a, b, and γ are given.

3. α, β, and c are given.

4. a, b, and α are given.

5. a, β, and γ are given.

6. α, β, and a are given.

7. b, c, and α are given.

8. b, c, and β are given.

In Exercises 9–42, solve each triangle.

9. $a = 4$, $c = 3$, $\beta = 100°$

10. $a = 6$, $b = 10$, $\gamma = 80°$

11. $b = 7$, $c = 2$, $\alpha = 16°$

12. $b = 5$, $a = 6$, $\gamma = 170°$

13. $b = 5$, $c = 5$, $\alpha = 20°$

14. $a = 4.2$, $b = 7.3$, $\gamma = 25°$

15. $a = 9$, $c = 12$, $\beta = 23°$

16. $b = 6$, $c = 13$, $\alpha = 16°$

17. $a = 4$, $c = 8$, $\beta = 60°$

18. $b = 3$, $c = \sqrt{18}$, $\alpha = 45°$

19. $a = 8, b = 5, c = 6$

20. $a = 6, b = 9, c = 12$

21. $a = 4, b = 4, c = 5$

22. $a = 17, b = 20, c = 33$

23. $a = 8.2, b = 7.1, c = 6.3$

24. $a = 1492, b = 2001, c = 1776$

25. $a = 4, b = 5, c = 10$

26. $a = 1.3, b = 2.7, c = 4.2$

27. $a = 12, b = 5, c = 13$

28. $a = 4, b = 5, c = \sqrt{41}$

29. $\alpha = 40°, \beta = 35°, a = 6$

30. $b = 11.2, a = 19.0, \gamma = 13.3°$

31. $\alpha = 31°, b = 5, a = 12$

32. $a = 11, c = 12, \gamma = 60°$

33. $a = \sqrt{7}, b = \sqrt{8}, c = \sqrt{3}$

34. $\beta = 106°, \gamma = 43°, a = 1$

35. $b = 11, c = 2, \beta = 10°$

36. $\alpha = 25°, a = 6, c = 9$

37. $\beta = 51.62°, a = \frac{48}{5}, c = \frac{71}{5}$

38. $\gamma = 61.03°, a = \frac{22}{9}, c = \frac{22}{9}$

39. $a = 18.21, b = 21.30, c = 24.12$

40. $a = \frac{32}{7}, b = \frac{29}{8}, c = \frac{21}{11}$

41. $\alpha = 71.213°, b = \dfrac{5\sqrt{2}}{7}, c = \dfrac{3\sqrt{2}}{5}$

42. $\beta = 38.21°, a = \sqrt{31}, c = \sqrt{31}$

▪ APPLICATIONS

43. Aviation. A plane flew due north at 500 mph for 3 hours. A second plane, starting at the same point and at the same time, flew southeast at an angle 150° clockwise from due north at 435 mph for 3 hours. At the end of the 3 hours, how far apart were the two planes? Round to the nearest mile.

44. Aviation. A plane flew due north at 400 mph for 4 hours. A second plane, starting at the same point and at the same time, flew southeast at an angle 120° clockwise from due north at 300 mph for 4 hours. At the end of the 4 hours, how far apart were the two planes? Round to the nearest mile.

45. Aviation. A plane flew 30° NW at 350 mph for 2.5 hours. A second plane, starting at the same point and at the same time, flew 35° at an angle clockwise from due north at 550 mph for 2.5 hours. At the end of 2.5 hours, how far apart were the two planes? Round to the nearest mile.

46. Aviation. A plane flew 30° NW at 350 mph for 3 hours. A second plane starts at the same point and takes off at the same time. It is known that after 3 hours, the two planes are 2100 miles apart. Find the original bearing of the second plane, to the nearest hundredth of a degree.

47. Baseball. A baseball diamond is actually a square that is 90 feet on each side. The pitcher's mound is located 60.5 feet from home plate. How far is it from third base to the pitcher's mound?

48. Aircraft Wing. Given the acute angle (14.4°) and two sides (18 feet and 25 feet) of the stealth bomber, what is the unknown length?

Stone/Getty Images, Inc.

49. Sliding Board. A 40-foot slide leaning against the bottom of a building's window makes a 55° angle with the building. The angle formed with the building with the line of sight from the top of the window to the point on the ground where the slide ends is 40°. How tall is the window?

50. Airplane Slide. An airplane door is 6 feet high. If a slide attached to the bottom of the open door is at an angle of 40° with the ground, and the angle formed by the line of sight from where the slide touches the ground to the top of the door is 45°, how long is the slide?

For Exercises 51 and 52, refer to the following:

To quantify the torque (rotational force) of the elbow joint of a human arm (see the figure to the right), it is necessary to identify angles A, B, and C as well as lengths a, b, and c. Measurements performed on an arm determine that the measure of angle C is 105°, the length of the muscle a is 25.5 centimeters, and the length of the forearm from the elbow joint to the muscle attachment b is 1.76 centimeters.

51. Health/Medicine. Find the length of the upper arm from the muscle attachment to the elbow joint c.

52. Health/Medicine. Find the measure of angle B.

For Exercises 53 and 54, refer to the following:

The term "triangulation" is often used to pinpoint a location. A person's cell phone always uses the closest tower. As the cell phone switches towers, authorities can locate the person's location based on known distances from each of the towers. Similarly, firefighters can determine the location of a fire from known distances.

53. Triangulation. Two lead firefighters (Beth and Tim) are 300 yards apart and they estimate each of their distances to the fire as approximately 150 yards (Beth) and 200 yards (Tim). What angle with respect to their line of sight with each other should each lead their team in order to reach the fire?

54. Triangulation. Two cell phone towers are 100 meters apart. When the cell phone switches from tower A to tower B, it is estimated that the phone is 60 meters from tower A and 50 meters from tower B. Authorities looking to locate the owner of the cell phone will head 50 meters from tower B at what angle (with respect to the line of sight between towers A and B)?

55. Law Enforcement. Two members of a SWAT team and the thief they are to apprehend are positioned as follows:

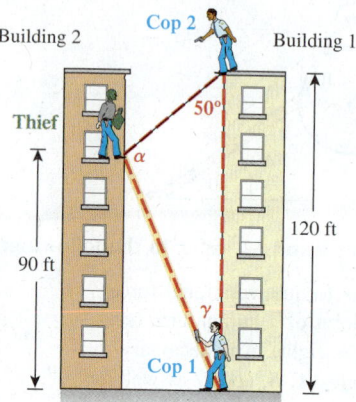

When the signal is given, Cop 2 shoots a zipline across to the window where the thief is spotted (at a 50° angle to Building 1) and Cop 1 shines a very bright light directly at the thief. Find the angle γ at which Cop 1 holds the light to shine it directly at the thief. Round to the nearest hundredth degree.

56. Law Enforcement. In reference to Exercise 55, what angle does the zipline make with respect to Building 2?

57. Rock Climbing. (See Exercise 44 in Section 8.1 for the context.) In order to construct a more advanced route, holds are placed at P and R very close to each other, and a third hold is placed at Q at quite a distance away. Moving from either P or R will require a "dynamo" move! If the person setting the route wants $PQ = 7$ feet, $PR = 1$ foot, and $\angle RPQ = 25°$, find the angle between other holds on the route.

58. Bowling. Consider the following diagram:

From the diagram, we know that the ball is released at an angle of 35° to the line connecting the player to the 10 pin. Find the angle at which the ball hits the head pin, assuming the ball is thrown straight with no hook. Round to the nearest hundredth degree.

59. Bowling. Next, a bowler must convert a 7-8 spare and must hit the 8 pin directly on its right side. If the distance between the bowler and 7 pin is 63 feet, the distance between the bowler and 8 pin is 62.8 feet, and the distance between the 7 and 8 pins is 7 inches, find the angle at which the ball is released (i.e., angle between the segment connecting the bowler to the 7 pin and the segment connecting the bowler to the 8 pin).

60. Surveying. A glaciologist needs to determine the length across a certain crevice on Mendenhall glacier in order to circumvent it with her team. She has the following measurements:

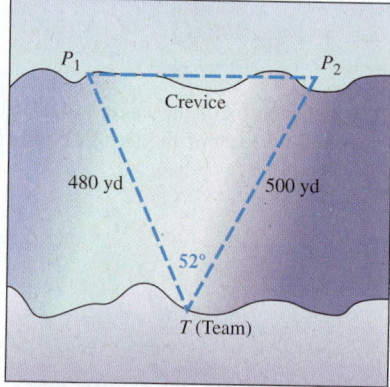

Find the approximate length across the crevice.

61. Surveying. A glaciologist needs to determine the length across a certain crevice on Mendenhall glacier in order to circumvent it with his team. He has the following measurements:

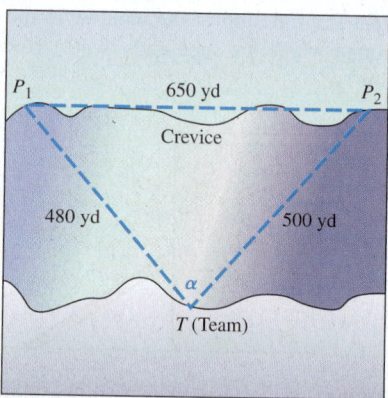

Find the angle α.

62. Home Décor. A table in the shape of a regular octagon is to be constructed to fit on a round oriental rug 16 feet in diameter. Each guest sitting around the perimeter of the table requires 2.5 feet of space. How many people, approximately, can be seated at this table?

▪CATCH THE MISTAKE

In Exercises 63 and 64, explain the mistake that is made.

63. Solve the triangle $b = 3$, $c = 4$, and $\alpha = 30°$.

Solution:

Step 1: Find a.

Apply the Law
of Cosines. $\qquad a^2 = b^2 + c^2 - 2bc\cos\alpha$

Let $b = 3$,
$c = 4$, and
$\alpha = 30°$. $\qquad a^2 = 3^2 + 4^2 - 2(3)(4)\cos 30°$

Solve for a. $\qquad a \approx 2.1$

Step 2: Find γ.

Apply the Law
of Sines. $\qquad \dfrac{\sin\alpha}{a} = \dfrac{\sin\gamma}{c}$

Solve for $\sin\gamma$. $\qquad \sin\gamma = \dfrac{c\sin\alpha}{a}$

Solve for γ. $\qquad \gamma = \sin^{-1}\left(\dfrac{c\sin\alpha}{a}\right)$

Let $a = 2.1$, $c = 4$,
and $\alpha = 30°$. $\qquad \gamma \approx 72°$

Step 3: Find β.

$\alpha + \beta + \gamma = 180° \qquad 30° + \beta + 72° = 180°$

Solve for β. $\qquad \beta \approx 78°$

$a \approx 2.1$, $b = 3$, $c = 4$, $\alpha = 30°$, $\beta \approx 78°$, and $\gamma \approx 72°$

This is incorrect. The longest side is not opposite the
largest angle. What mistake was made?

64. Solve the triangle $a = 6$, $b = 2$, and $c = 5$.

Solution:

Step 1: Find β.

Apply the Law
of Cosines. $\qquad b^2 = a^2 + c^2 - 2ac\cos\beta$

Solve for β. $\qquad \beta = \cos^{-1}\left(\dfrac{a^2 + c^2 - b^2}{2ac}\right)$

Let $a = 6$, $b = 2$,
$c = 5$. $\qquad \beta \approx 18°$

Step 2: Find α.

Apply the Law
of Sines. $\qquad \dfrac{\sin\alpha}{a} = \dfrac{\sin\beta}{b}$

Solve for α. $\qquad \alpha = \sin^{-1}\left(\dfrac{a\sin\beta}{b}\right)$

Let $a = 6$, $b = 2$,
and $\beta = 18°$. $\qquad \alpha \approx 68°$

Step 3: Find γ.

$\alpha + \beta + \gamma = 180°$

$68° + 18° + \gamma = 180°$

$\gamma \approx 94°$

$a = 6$, $b = 2$, $c = 5$, $\alpha \approx 68°$, $\beta \approx 18°$, and $\gamma \approx 94°$

This is incorrect. The longest side is not opposite the
largest angle. What mistake was made?

▪CONCEPTUAL

In Exercises 65–70, determine whether each statement is true or false.

65. Given three sides of a triangle, there is insufficient
information to solve the triangle.

66. Given three angles of a triangle, there is insufficient
information to solve the triangle.

67. The Pythagorean theorem is a special case of the Law of
Cosines.

68. The Law of Cosines is a special case of the Pythagorean
theorem.

69. If an obtuse triangle is isosceles, then knowing the obtuse
angle and a side adjacent to it is sufficient to solve the
triangle.

70. All acute triangles can be solved using the Law of
Cosines.

■ CHALLENGE

71. Show that $\dfrac{\cos\alpha}{a} + \dfrac{\cos\beta}{b} + \dfrac{\cos\gamma}{c} = \dfrac{a^2 + b^2 + c^2}{2abc}$.
Hint: Apply the Law of Cosines.

72. Show that $a = c\cos\beta + b\cos\gamma$. *Hint:* Apply the Law of Cosines.

73. Consider the following diagram. Find $\cos\left(\dfrac{X}{2}\right)$.

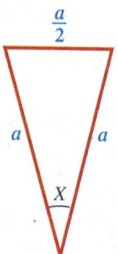

74. Using the diagram in Exercise 73, Find $\tan\left(\dfrac{X}{2}\right)$.

■ TECHNOLOGY

For Exercises 75–80, let A, B, and C be the lengths of the three sides with X, Y, and Z as the corresponding angles. Write a program using a TI calculator to solve the given triangle.

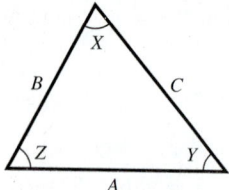

75. $B = 45$, $C = 57$, and $X = 43°$

76. $B = 24.5$, $C = 31.6$, and $X = 81.5°$

77. $A = 29.8$, $B = 37.6$, and $C = 53.2$

78. $A = 100$, $B = 170$, and $C = 250$

79. $A = \sqrt{12}$, $B = \sqrt{21}$, and $Z = 62.8°$

80. $A = 1235$, $B = 987$, and $C = 1456$

SECTION

8.3 THE AREA OF A TRIANGLE

SKILLS OBJECTIVES

- Find the area of a triangle in the SAS case.
- Find the area of a triangle in the SSS case.

CONCEPTUAL OBJECTIVES

- Understand how to derive a formula for the area of a triangle (SAS case) using the Law of Sines.
- Understand how to derive a formula for the area of a triangle (SSS case) using the Law of Cosines.

In Sections 8.1 and 8.2, we applied the Law of Sines and the Law of Cosines to solve oblique triangles, which means to find all of the side lengths and angle measures. Now we use these laws to derive formulas for the area of a triangle (SAS and SSS).

Our starting point for both cases is the standard formula for the area of a triangle:

$$A = \frac{1}{2}bh$$

WORDS	**MATH**
Start with a triangle with base b and height h.	
Construct a rectangle with base b and height h around the triangle.	
The area of the rectangle is base times height.	$A_{\text{rectangle}} = bh$
Triangles **1** and **2** have the same area. Triangles **3** and **4** have the same area. Therefore, the area of the triangle is half the area of the rectangle.	
Area of the triangle with base b and height h.	$A_{\text{triangle}} = \dfrac{1}{2}bh$

The Area of a Triangle (SAS Case)

We now can use the general formula for the area of a triangle and the Law of Sines to develop a formula for the area of a triangle when two sides and the angle between them are given.

WORDS	**MATH**
Start with a triangle, given b, c, and α.	
Write the sine ratio for angle α.	$\sin\alpha = \dfrac{h}{c}$
Solve for h.	$h = c\sin\alpha$
Write the formula for area of a triangle.	$A_{\text{triangle}} = \dfrac{1}{2}bh$
For the SAS case, substitute $h = c\sin\alpha$.	$A_{\text{SAS case}} = \dfrac{1}{2}bc\sin\alpha$

Now we can calculate the area of this triangle with the given information (two sides and the angle between them: b, c, and α). Similarly, it can be shown that the other formulas for SAS triangles are

$$A_{\text{SAS case}} = \frac{1}{2}ab\sin\gamma \qquad \text{and} \qquad A_{\text{SAS case}} = \frac{1}{2}ac\sin\beta$$

AREA OF A TRIANGLE (SAS)

For any triangle where two sides and the angle between them are known, the area for that triangle is given by one of the following formulas (depending on which angle and sides are given):

$$A_{\text{SAS case}} = \frac{1}{2}bc\sin\alpha \qquad \text{when } b, c, \text{ and } \alpha \text{ are known}$$

$$A_{\text{SAS case}} = \frac{1}{2}ab\sin\gamma \qquad \text{when } a, b, \text{ and } \gamma \text{ are known}$$

$$A_{\text{SAS case}} = \frac{1}{2}ac\sin\beta \qquad \text{when } a, c, \text{ and } \beta \text{ are known}$$

In other words, the area of a triangle equals one-half the product of two of its sides and the sine of the angle between them.

Classroom Example 8.3.1
Find the area of a triangle with measurements $b = \sqrt{6}$ cm, $c = 2\sqrt{3}$ cm, and $\alpha = 27.3°$.

Answer: $1.95\,\text{cm}^2$

Classroom Example 8.3.1*
Find the area of an equilateral triangle with side length a.

Answer: $\dfrac{\sqrt{3}}{4}a^2$

▶ **EXAMPLE 1 Finding the Area of a Triangle (SAS Case)**

Find the area of the triangle $a = 7.0$ ft, $b = 9.3$ ft, and $\gamma = 86°$.

Solution:

Apply the area formula where a, b, and γ are given.

$$A = \frac{1}{2}ab\sin\gamma$$

Substitute $a = 7.0$ ft, $b = 9.3$ ft, and $\gamma = 86°$.

$$A = \frac{1}{2}(7.0\text{ ft})(9.3\text{ ft})\sin 86°$$

Approximate with a calculator.

$$A = 32.47071\text{ ft}^2$$

Round to two significant digits.

$$\boxed{A = 32\text{ ft}^2}$$

Technology Tip

Use the calculator to find A.

```
1/2*7*9.3*sin(86
)
       32.47070984
```

■ **Answer:** $6.2\,\text{m}^2$

■ **YOUR TURN** Find the area of the triangle $a = 3.2$ m, $c = 5.1$ m, and $\beta = 49°$.

The Area of a Triangle (SSS Case)

We used the Law of Sines to develop a formula for the area of an SAS triangle. That formula, several trigonometric identities, and the Law of Cosines can be used to develop a formula for the area of an SSS triangle, called **Heron's formula**.

WORDS	**MATH**
Start with any of the formulas for SAS triangles.	$A = \dfrac{1}{2}ab\sin\gamma$
Square both sides.	$A^2 = \dfrac{1}{4}a^2b^2\sin^2\gamma$
Isolate $\sin^2\gamma$.	$\dfrac{4A^2}{a^2b^2} = \sin^2\gamma$
Apply the Pythagorean identity.	$\dfrac{4A^2}{a^2b^2} = 1 - \cos^2\gamma$
Factor the difference of the two squares on the right.	$\dfrac{4A^2}{a^2b^2} = (1 - \cos\gamma)(1 + \cos\gamma)$
Solve the Law of Cosines, $c^2 = a^2 + b^2 - 2ab\cos\gamma$, for $\cos\gamma$.	$\cos\gamma = \dfrac{a^2 + b^2 - c^2}{2ab}$
Substitute $\cos\gamma = \dfrac{a^2 + b^2 - c^2}{2ab}$ into $\dfrac{4A^2}{a^2b^2} = (1 - \cos\gamma)(1 + \cos\gamma)$.	$\dfrac{4A^2}{a^2b^2} = \left[1 - \dfrac{a^2 + b^2 - c^2}{2ab}\right]\left[1 + \dfrac{a^2 + b^2 - c^2}{2ab}\right]$
Combine the expressions in brackets.	$\dfrac{4A^2}{a^2b^2} = \left[\dfrac{2ab - a^2 - b^2 + c^2}{2ab}\right]\left[\dfrac{2ab + a^2 + b^2 - c^2}{2ab}\right]$
Group the terms in the numerators.	$\dfrac{4A^2}{a^2b^2} = \left[\dfrac{-(a^2 - 2ab + b^2) + c^2}{2ab}\right]\left[\dfrac{(a^2 + 2ab + b^2) - c^2}{2ab}\right]$
Write the numerators as the difference of two squares.	$\dfrac{4A^2}{a^2b^2} = \left[\dfrac{c^2 - (a - b)^2}{2ab}\right]\left[\dfrac{(a + b)^2 - c^2}{2ab}\right]$
Factor the numerators: $x^2 - y^2 = (x - y)(x + y)$.	$\dfrac{4A^2}{a^2b^2} = \left[\dfrac{(c - [a - b])(c + [a - b])}{2ab}\right]\left[\dfrac{([a + b] - c)([a + b] + c)}{2ab}\right]$
Simplify.	$\dfrac{4A^2}{a^2b^2} = \left[\dfrac{(c - a + b)(c + a - b)}{2ab}\right]\left[\dfrac{(a + b - c)(a + b + c)}{2ab}\right]$
	$\dfrac{4A^2}{a^2b^2} = \dfrac{(c - a + b)(c + a - b)(a + b - c)(a + b + c)}{4a^2b^2}$
Multiply by $\dfrac{a^2b^2}{4}$.	$A^2 = \dfrac{1}{16}(c - a + b)(c + a - b)(a + b - c)(a + b + c)$
The semiperimeter s is half the perimeter of the triangle.	$2s = a + b + c$
Manipulate each of the four factors:	$c - a + b = a + b + c - 2a = 2s - 2a = 2(s - a)$ $c + a - b = a + b + c - 2b = 2s - 2b = 2(s - b)$ $a + b - c = a + b + c - 2c = 2s - 2c = 2(s - c)$ $a + b + c = 2s$
Substitute in these values for the four factors.	$A^2 = \dfrac{1}{16}\cdot 2(s - a)\cdot 2(s - b)\cdot 2(s - c)\cdot 2s$
Simplify.	$A^2 = s(s - a)(s - b)(s - c)$
Solve for A (area is always positive).	$\boxed{A = \sqrt{s(s - a)(s - b)(s - c)}}$

AREA OF A TRIANGLE (SSS CASE—HERON'S FORMULA)

For any triangle where the lengths of the three sides are known, the area for that triangle is given by the following formula:

$$A_{\text{SSS case}} = \sqrt{s(s-a)(s-b)(s-c)}$$

where a, b, and c are the lengths of the sides of the triangle and s is half the perimeter of the triangle, called the semiperimeter.

$$s = \frac{a+b+c}{2}$$

Technology Tip

Use a calculator to find A.

```
√((10(10-5)(10-6)
(10-9))
          14.14213562
```

Classroom Example 8.3.2
Find the area of a triangle with measurements $a = 2\sqrt{7}$, $b = 4\sqrt{7}$, and $c = 3\sqrt{7}$.

Answer:

$$\frac{\sqrt{6615}}{4} = \frac{21\sqrt{15}}{4} \text{ sq. units}$$

EXAMPLE 2   Finding the Area of a Triangle (SSS Case)

Find the area of the triangle $a = 5$, $b = 6$, and $c = 9$.

Solution:

Find the semiperimeter s.	$s = \dfrac{a+b+c}{2}$
Substitute $a = 5$, $b = 6$, and $c = 9$.	$s = \dfrac{5+6+9}{2}$
Simplify.	$s = 10$
Write the formula for the area of a triangle in the SSS case.	$A = \sqrt{s(s-a)(s-b)(s-c)}$
Substitute $a = 5$, $b = 6$, $c = 9$, and $s = 10$.	$A = \sqrt{10(10-5)(10-6)(10-9)}$
Simplify the radicand.	$A = \sqrt{10 \cdot 5 \cdot 4 \cdot 1}$
Evaluate the radical.	$\boxed{A = 10\sqrt{2} \approx 14 \text{ sq. units}}$

■ **YOUR TURN** Find the area of the triangle $a = 3$, $b = 5$, and $c = 6$.

■ **Answer:** $2\sqrt{14} \approx 7.5$ sq. units

SECTION 8.3 SUMMARY

In this section we derived formulas for calculating the areas of triangles (SAS and SSS cases). The Law of Sines leads to three area formulas for the SAS case depending on which angles and sides are given.

$$A_{\text{SAS}} = \frac{1}{2}bc\sin\alpha$$

$$A_{\text{SAS}} = \frac{1}{2}ab\sin\gamma$$

$$A_{\text{SAS}} = \frac{1}{2}ac\sin\beta$$

The Law of Cosines was instrumental in developing a formula for the area of a triangle (SSS case) when all three sides are given.

$$A_{\text{SSS}} = \sqrt{s(s-a)(s-b)(s-c)}$$

$$\text{where } s = \frac{a+b+c}{2}$$

SECTION
8.3 EXERCISES

SKILLS

In Exercises 1–32, find the area of each triangle described.

1. $a = 8, c = 16, \beta = 60°$

2. $b = 6, c = 4\sqrt{3}, \alpha = 30°$

3. $a = 1, b = \sqrt{2}, \alpha = 45°$

4. $b = 2\sqrt{2}, c = 4, \beta = 45°$

5. $a = 6, b = 8, \gamma = 80°$

6. $b = 9, c = 10, \alpha = 100°$

7. $a = 4, c = 7, \beta = 27°$

8. $a = 6.3, b = 4.8, \gamma = 17°$

9. $b = 100, c = 150, \alpha = 36°$

10. $c = 0.3, a = 0.7, \beta = 145°$

11. $a = \sqrt{5}, b = 5\sqrt{5}, \gamma = 50°$

12. $b = \sqrt{11}, c = \sqrt{11}, \alpha = 21°$

13. $a = 15, b = 15, c = 15$

14. $a = 1, b = 1, c = 1$

15. $a = 7, b = \sqrt{51}, c = 10$

16. $a = 9, b = 40, c = 41$

17. $a = 6, b = 10, c = 9$

18. $a = 40, b = 50, c = 60$

19. $a = 14.3, b = 15.7, c = 20.1$

20. $a = 146.5, b = 146.5, c = 100$

21. $a = 14{,}000, b = 16{,}500, c = 18{,}700$

22. $a = \sqrt{2}, b = \sqrt{3}, c = \sqrt{5}$

23. $a = 80, b = 75, c = 160$

24. $a = 19, b = 23, c = 3$

25. $a = 2\sqrt{3}, b = 5\sqrt{3}, \gamma = 61.23°$

26. $a = \dfrac{7\sqrt{5}}{3}, c = \dfrac{5\sqrt{5}}{2}, \beta = 16.36°$

27. $b = \dfrac{86}{3}, c = \dfrac{51}{4}, \alpha = 73°$

28. $a = 17.61, c = 17.61, \beta = 20.03°$

29. $a = 5.61, b = 6.61, c = 4.31$

30. $a = 7.28, b = 7.28, c = 7.28$

31. $a = \dfrac{63}{8}, b = \dfrac{50}{7}, c = \dfrac{50}{9}$

32. $a = \dfrac{1}{2}, b = \dfrac{1}{3}, c = \dfrac{1}{3}$

APPLICATIONS

33. **Bermuda Triangle.** Calculate the area of the so called Bermuda Triangle, described in Example 4 of Section 8.2, if, as some people define it, its vertices are located in Norfolk, Bermuda, and Santiago.

LOCATION	LOCATION	DISTANCE (NAUTICAL MILES)
Norfolk	Bermuda	850
Bermuda	Santiago	810
Norfolk	Santiago	894

Ignore the curvature of the Earth in your calculations.

34. **Bermuda Triangle.** Calculate the area of the Bermuda Triangle, if, as other people define it, its vertices are located in Miami, Bermuda, and San Juan.

LOCATION	LOCATION	DISTANCE (NAUTICAL MILES)
Miami	Bermuda	898
Bermuda	San Juan	831
Miami	San Juan	890

35. Triangular Tarp. A large triangular tarp is needed to cover a playground when it rains. If the sides of the tarp measure 160 feet, 140 feet, and 175 feet, what is the area of the tarp (to the nearest square foot)?

36. Flower Seed. A triangular garden measures 41 feet × 16 feet × 28 feet. You are going to plant wildflower seed that costs $4 per bag. Each bag of flower seed covers an area of 20 square feet. How much will the seed cost? (Assume you have to buy a whole bag—you can't split one.)

37. Mural. Some students are painting a mural on the side of a building. They have enough paint for a 1000-square-foot-area triangle. If two sides of the triangle measure 60 feet and 120 feet, what angle (to the nearest degree) should the two sides form in order to create a triangle that uses up all the paint?

38. Mural. Some students are painting a mural on the side of a building. They have enough paint for a 500-square-foot-area triangle. If two sides of the triangle measure 40 feet and 60 feet, what angle (to the nearest degree) should the two sides form in order to create a triangle that uses up all the paint?

39. Insect Infestation. Some very destructive beetles have made their way into a forest preserve. The rangers are trying to keep track of their spread and how well preventative measures are working. In a triangular area that is 22.5 miles on one side, 28.1 miles on another, and 38.6 miles on the third, what is the total area the rangers are covering?

40. Real Estate. A real estate agent needs to determine the area of a triangular lot. Two sides of the lot are 150 feet and 60 feet. The angle between the two measured sides is 43°. What is the area of the lot?

For Exercises 41 and 42, refer to the following:

A company is considering purchasing a triangular piece of property (see the figure below) for the construction of a new facility. The purchase is going to be "sight unseen" and the company is using old surveys to approximate area and costs.

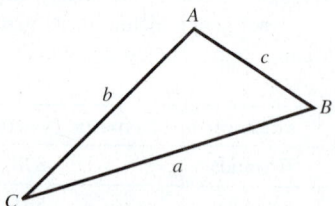

41. Business. If the survey indicates that one side b is 275 feet, a second side a is 310 feet, and the angle between the two sides is 79°,

a. Find the area of the property.

b. If the company wants to offer the seller $2.13 per square foot, what is the total cost of the property?

42. Business. If the survey indicates that one side b is 475 feet, a second side c is 310 feet, and the angle between the two sides is 118°,

a. Find the area of the property.

b. If the company wants to offer the seller $1.97 per square foot, what is the total cost of the property?

43. Parking Lot. A parking lot is to have the shape of a parallelogram that has adjacent sides measuring 200 feet and 260 feet. The angle between the two sides is 65°. What is the area of the parking lot?

44. Parking Lot. A parking lot is to have the shape of a parallelogram that has adjacent sides measuring 250 feet and 300 feet. The angle between the two sides is 55°. What is the area of the parking lot?

45. Regular Hexagon. A regular hexagon has sides measuring 3 feet. What is its area? Recall that the measure of an angle of a regular n-gon is given by the formula

$$\text{angle} = \frac{180(n-2)}{n}.$$

3 ft

5 in.

46. Regular Decagon. A regular decagon has sides measuring 5 inches. What is its area?

47. Geometry. Three spokes Q_1, Q_2, and Q_3 of lengths 8, 6, and 11, respectively, emanate from a common point P with the following angles:

Find the area of the triangles PQ_1Q_2, $Q_1Q_2Q_3$, and PQ_2Q_3. Then, find the area of triangle PQ_1Q_3. Compare the sum of the areas of the smaller three triangles to the area of PQ_1Q_3. Are they equal?

48. Construction. A rectangular great room, 15 ft × 25 ft, has an open beam ceiling. The two parts of the ceiling make angles 50° and 33° with the horizontal, as shown. Find the area of the ceiling.

49. Geometry. A quadrilateral *ABCD* has sides of lengths *AB* = 2, *BC* = 3, *CD* = 4, and *DA* = 5. The angle between *AB* and *BC* is 135°. Find the area of *ABCD*.

50. Geometry. A quadrilateral *ABCD* has sides of lengths *AB* = 5, *BC* = 6, *CD* = 7, and *DA* = 8. The angle between *AB* and *BC* is 135°. Find the area of *ABCD*.

■ **CATCH THE MISTAKE**

In Exercises 51 and 52, explain the mistake that is made.

51. Calculate the area of the triangle $a = 2, b = 6,$ and $c = 7$.

Solution:

Find the semiperimeter. $\qquad s = a + b + c = 2 + 6 + 7 = 15$

Write the formula for the area of the triangle. $\qquad A = \sqrt{s(s - a)(s - b)(s - c)}$

Let $a = 2, b = 6,$ $c = 7,$ and $s = 15.$ $\quad A = \sqrt{15(15 - 2)(15 - 6)(15 - 7)}$

Simplify. $\qquad A = \sqrt{14{,}040} \approx 118$

This is incorrect. What mistake was made?

52. Calculate the area of the triangle $a = 2, b = 6,$ and $c = 5$.

Solution:

Find the semiperimeter. $\qquad s = \dfrac{a + b + c}{2} = \dfrac{2 + 6 + 5}{2} = 6.5$

Write the formula for area of the triangle. $\qquad A = s(s - a)(s - b)(s - c)$

Let $a = 2, b = 6,$ $c = 5,$ and $s = 6.5.$ $\quad A = 6.5(4.5)(0.5)(1.5)$

Simplify. $\qquad A \approx 22$

This is incorrect. What mistake was made?

■ **CONCEPTUAL**

In Exercises 53–58, determine whether each statement is true or false.

53. Heron's formula can be used to find the area of right triangles.

54. Heron's formula can be used to find the area of isosceles triangles.

55. If two triangles have the same side lengths, then they have the same area.

56. If two rhombi (i.e., quadrilateral with four congruent sides) have the same side length, then they have the same area.

57. If two parallelograms have the same area, then the corresponding sides must be the same.

58. The semiperimeter can be less than the length of the largest side.

■CHALLENGE

59. Show that the area for an SAA triangle is given by

$$A = \frac{a^2 \sin \beta \sin \gamma}{2 \sin \alpha}$$

Assume that α, β, and a are given.

60. Show that the area of an isosceles triangle with equal sides of length s is given by

$$A_{\text{isosceles}} = \frac{1}{2} s^2 \sin \theta$$

where θ is the angle between the two equal sides.

61. Find the area of the shaded region.

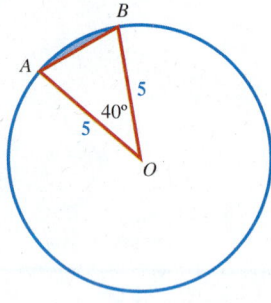

62. Find the area of the shaded region.

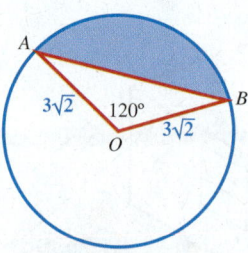

63. Prove that the area of a rhombus that has side length s with an angle of $30°$ is $\frac{1}{2}s^2$.

64. Prove that the area of a parallelogram with side lengths s and $3s$ and an angle of $135°$ is $\frac{3\sqrt{2}}{2}s^2$.

■TECHNOLOGY

For Exercises 65–70, let A, B, and C be the lengths of the three sides with X, Y, and Z as the corresponding angles. Write a program using a TI calculator to calculate the area of the given triangle.

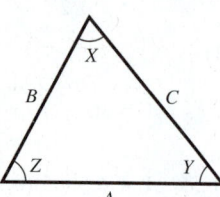

65. $A = 35$, $B = 47$, and $Z = 68°$

66. $A = 1241$, $B = 1472$, and $Z = 56°$

67. $A = 85$, $B = 92$, and $C = 123$

68. $A = \sqrt{167}$, $B = \sqrt{113}$, and $C = \sqrt{203}$

69. $A = 145$, $B = 172$, and $C = 110$

70. $A = 21.8$, $B = 39.6$, and $C = 124.5°$

SKILLS OBJECTIVES

- Find the magnitude and direction of a vector.
- Add and subtract vectors.
- Perform scalar multiplication of a vector.
- Express a vector in terms of its horizontal and vertical components.
- Find unit vectors.
- Find resultant vectors in application problems.

CONCEPTUAL OBJECTIVES

- Understand the difference between scalars and vectors.
- Relate the geometric and algebraic representations of vectors.

Vectors: Magnitude and Direction

What is the difference between velocity and speed? Speed has only *magnitude*, whereas velocity has both *magnitude* and *direction*. We use **scalars**, which are real numbers, to denote magnitudes such as speed and weight. We use **vectors**, which have magnitude *and* direction, to denote quantities such as velocity (speed in a certain direction) and force (weight in a certain direction).

A vector quantity is geometrically denoted by a **directed line segment**, which is a line segment with an arrow representing direction. There are many ways to denote a vector. For example, the vector shown in the margin can be denoted as **u**, $\vec{u}$, or $\overrightarrow{AB}$, where A is the **initial point** and B is the **terminal point**.

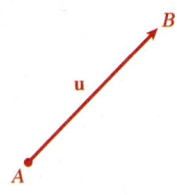

It is customary in books to use the bold letter to represent a vector and when hand-written (as in your class notes and homework) to use the arrow on top to denote a vector.

In this section, we will limit our discussion to vectors in a plane (two-dimensional). It is important to note that geometric representation can be extended to three dimensions and algebraic representation can be extended to any higher dimension, as you will see in the exercises and in later sections.

Geometric Interpretation of Vectors

The *magnitude* of a vector can be denoted one of two ways: $|\mathbf{u}|$ or $\|\mathbf{u}\|$. We will use the former notation.

MAGNITUDE: $|\mathbf{u}|$

The **magnitude** of a vector **u**, denoted $|\mathbf{u}|$, is the length of the directed line segment which is the distance between the initial and terminal points of the vector.

Two vectors have the **same direction** if they are parallel and point in the same direction. Two vectors have **opposite direction** if they are parallel and point in opposite directions.

EQUAL VECTORS: $\mathbf{u} = \mathbf{v}$

Two vectors **u** and **v** are **equal** ($\mathbf{u} = \mathbf{v}$) if and only if they have the same magnitude ($|\mathbf{u}| = |\mathbf{v}|$) and have the same direction.

Study Tip

The magnitude of a vector is the distance between the initial and terminal points of the vector.

Equal Vectors $\mathbf{u} = \mathbf{v}$	Same Magnitude but Opposite Direction $\mathbf{u} = -\mathbf{v}$	Same Magnitude $|\mathbf{u}| = |\mathbf{v}|$	Different Magnitude

It is important to note that vectors do not have to coincide to be equal.

VECTOR ADDITION: U + V

Two vectors, **u** and **v**, can be added together using either of the following approaches:

- The **tail-to-tip** (or head-to-tail) method: Sketch the initial point of one vector at the terminal point of the other vector. The **sum, u + v**, is the **resultant** vector from the tail end of **u** to the tip end of **v**.

[or]

- The parallelogram method: Sketch the initial points of the vectors at the same point. The sum **u + v** is the diagonal of the parallelogram formed by **u** and **v**.

The difference, **u − v**, is the

- Resultant vector from the tip of **v** to the tip of **u**, when the tails of **v** and **u** coincide.

[or]

- The other diagonal formed by the parallelogram method.

Algebraic Interpretation of Vectors

Since vectors that have the same direction and magnitude are equal, any vector can be translated to an equal vector with its initial point located at the origin in the Cartesian plane. Therefore, we will now consider vectors in a rectangular coordinate system.

A vector with its initial point at the origin is called a **position vector**, or a vector in **standard position**. A position vector **u** with its terminal point at the point **(a, b)** is denoted:

$$\mathbf{u} = \langle a, b \rangle$$

where the real numbers a and b are called the **components** of vector **u**.

Notice the subtle difference between coordinate notation and vector notation. The point is denoted with parentheses, (a, b), whereas the vector is denoted with angled brackets, $\langle a, b \rangle$. The notation $\langle a, b \rangle$ denotes a vector whose initial point is $(0, 0)$ and terminal point is (a, b).

The vector with initial point $(3, 4)$ and terminal point $(8, 9)$ is equal to the vector $\langle 5, 5 \rangle$, which has initial point $(0, 0)$ and terminal point $(5, 5)$.

Recall that the geometric definition of the *magnitude* of a vector is the *length* of the vector.

MAGNITUDE: $|\mathbf{u}|$

The **magnitude** (or norm) of a vector, $\mathbf{u} = \langle a, b \rangle$, is

$$|\mathbf{u}| = \sqrt{a^2 + b^2}$$

EXAMPLE 1   Finding the Magnitude of a Vector

Find the magnitude of the vector $\mathbf{u} = \langle 3, -4 \rangle$.

Solution:

Write the formula for magnitude of a vector. $|\mathbf{u}| = \sqrt{a^2 + b^2}$

Let $a = 3$ and $b = -4$. $|\mathbf{u}| = \sqrt{3^2 + (-4)^2}$

Simplify. $|\mathbf{u}| = \boxed{\sqrt{25} = 5}$

Note: If we graph the vector $\mathbf{u} = \langle 3, -4 \rangle$, we see that the distance from the origin to the point $(3, -4)$ is 5 units.

Classroom Example 8.4.1
Find the magnitude of

a. $\vec{u} = \langle 1, -1 \rangle$

b.* $\vec{u} = \left\langle \frac{1}{2}a, -\frac{3}{4}a \right\rangle$

Answer:

a. $\sqrt{2}$ **b.** $\dfrac{a\sqrt{13}}{4}$

■ **YOUR TURN** Find the magnitude of the vector $\mathbf{v} = \langle -1, 5 \rangle$.

■ **Answer:** $\sqrt{26}$

DIRECTION ANGLE OF A VECTOR

The positive angle between the x-axis and a position vector is called the **direction angle**, denoted θ.

$$\tan\theta = \frac{b}{a}, \text{ where } a \neq 0$$

or

$$\theta = \tan^{-1}\left(\frac{b}{a}\right)$$

Classroom Example 8.4.2
Find the direction angle of
a. $\vec{u} = \langle 1, -1 \rangle$
b.* $\vec{u} = \langle \frac{1}{2}a, -\frac{3}{4}a \rangle$,
assuming $a > 0$

Answer:
a. $315°$ b. $303.7°$

■**Answer:** $306.9°$

EXAMPLE 2   Finding the Direction Angle of a Vector

Find the direction angle of the vector $\mathbf{v} = \langle -1, 5 \rangle$.

Solution:

Start with $\tan\theta = \dfrac{b}{a}$ and let $a = -1$ and $b = 5$. $\tan\theta = \dfrac{5}{-1}$

With a calculator, find $\tan^{-1}(-5)$. $\tan^{-1}(-5) = -78.7°$

The calculator gives a **quadrant IV** angle.

The point $(-1, 5)$ lies in quadrant II.

Add $180°$. $\theta = -78.7° + 180° = 101.3°$

$$\boxed{\theta = 101.3°}$$

■**YOUR TURN** Find the direction angle of the vector $\mathbf{u} = \langle 3, -4 \rangle$.

Recall that two vectors are equal if they have the same magnitude and direction. Algebraically, this corresponds to their corresponding components being equal.

EQUAL VECTORS: U = V

The vectors $\mathbf{u} = \langle a, b \rangle$ and $\mathbf{v} = \langle c, d \rangle$ are **equal**; that is, $\mathbf{u} = \mathbf{v}$, if and only if $a = c$ and $b = d$.

Vector Operations

Vector addition is done geometrically with the tail-to-tip rule. Algebraically, vector addition is performed component by component.

VECTOR ADDITION: U + V

If $\mathbf{u} = \langle a, b \rangle$ and $\mathbf{v} = \langle c, d \rangle$, then $\mathbf{u} + \mathbf{v} = \langle a + c, b + d \rangle$.

Classroom Example 8.4.3
Let $\vec{u} = \langle \sqrt{2}, -\sqrt{2} \rangle$,
$\vec{v} = \langle -\sqrt{2}, -\sqrt{2} \rangle$. Find
$\vec{u} + \vec{v}$.

Answer: $\langle 0, -2\sqrt{2} \rangle$

■**Answer:** $\mathbf{u} + \mathbf{v} = \langle -4, -2 \rangle$

EXAMPLE 3   Adding Vectors

Let $\mathbf{u} = \langle 2, -7 \rangle$ and $\mathbf{v} = \langle -3, 4 \rangle$. Find $\mathbf{u} + \mathbf{v}$.

Solution:

Let $\mathbf{u} = \langle 2, -7 \rangle$ and $\mathbf{v} = \langle -3, 4 \rangle$
in the addition formula. $\mathbf{u} + \mathbf{v} = \langle 2 + (-3), -7 + 4 \rangle$

Simplify. $\mathbf{u} + \mathbf{v} = \boxed{\langle -1, -3 \rangle}$

■**YOUR TURN** Let $\mathbf{u} = \langle 1, 2 \rangle$ and $\mathbf{v} = \langle -5, -4 \rangle$. Find $\mathbf{u} + \mathbf{v}$.

We now summarize vector operations. Addition and subtraction of vectors are performed algebraically component by component. Multiplication, however, is not as straightforward. To perform **scalar multiplication** of a vector (to multiply a vector by a real number), we multiply component by component. In Section 8.5 we will study a form of multiplication for two vectors that is defined as long as the vectors have the same number of components; it gives a result known as the *dot product*, and is useful in solving common problems in physics.

SCALAR MULTIPLICATION: $k\mathbf{u}$

If k is a scalar (real number) and $\mathbf{u} = \langle a, b \rangle$, then

$$k\mathbf{u} = k\langle a, b \rangle = \langle ka, kb \rangle$$

Scalar multiplication corresponds to

- Increasing the length of the vector: $|k| > 1$
- Decreasing the length of the vector: $|k| < 1$
- Changing the direction of the vector: $k < 0$

The following box is a summary of vector operations:

VECTOR OPERATIONS

If $\mathbf{u} = \langle a, b \rangle$, $\mathbf{v} = \langle c, d \rangle$, and k is a scalar, then

$$\mathbf{u} + \mathbf{v} = \langle a + c, b + d \rangle$$

$$\mathbf{u} - \mathbf{v} = \langle a - c, b - d \rangle$$

$$k\mathbf{u} = k\langle a, b \rangle = \langle ka, kb \rangle$$

The zero vector, $\mathbf{0} = \langle 0, 0 \rangle$, is a vector in any direction with a magnitude equal to zero. We now can state the algebraic properties (associative, commutative, and distributive) of vectors.

DEFINITION   Algebraic Properties of Vectors

$\mathbf{u} + \mathbf{v} = \mathbf{v} + \mathbf{u}$

$(\mathbf{u} + \mathbf{v}) + \mathbf{w} = \mathbf{u} + (\mathbf{v} + \mathbf{w})$

$(k_1 k_2)\mathbf{u} = k_1(k_2\mathbf{u})$

$k(\mathbf{u} + \mathbf{v}) = k\mathbf{u} + k\mathbf{v}$

$(k_1 + k_2)\mathbf{u} = k_1\mathbf{u} + k_2\mathbf{u}$

$0\mathbf{u} = \mathbf{0} \qquad 1\mathbf{u} = \mathbf{u} \qquad -1\mathbf{u} = -\mathbf{u}$

$\mathbf{u} + (-\mathbf{u}) = \mathbf{0}$

Horizontal and Vertical Components of a Vector

The **horizontal component** a and **vertical component** b of a vector **u** are related to the magnitude of the vector, $|\mathbf{u}|$, through the sine and cosine of the direction angle.

$$\cos\theta = \frac{a}{|\mathbf{u}|} \qquad \sin\theta = \frac{b}{|\mathbf{u}|}$$

HORIZONTAL AND VERTICAL COMPONENTS OF A VECTOR

The horizontal and vertical components of vector **u**, with magnitude $|\mathbf{u}|$ and direction angle θ, are given by

horizontal component: $a = |\mathbf{u}|\cos\theta$

vertical component: $b = |\mathbf{u}|\sin\theta$

The vector **u** can then be written as $\mathbf{u} = \langle a, b \rangle = \langle |\mathbf{u}|\cos\theta, |\mathbf{u}|\sin\theta \rangle$.

Classroom Example 8.4.4*
Find a vector of magnitude $2a$ with direction angle β.

Answer: $\langle 2a\cos\beta, 2a\sin\beta \rangle$

EXAMPLE 4 **Finding the Horizontal and Vertical Components of a Vector**

Find the vector that has a magnitude of 6 and a direction angle of 15°.

Solution:

Write the horizontal and vertical components of a vector **u**. $\quad a = |\mathbf{u}|\cos\theta$ and $b = |\mathbf{u}|\sin\theta$

Let $|\mathbf{u}| = 6$ and $\theta = 15°$. $\quad a = 6\cos15°$ and $b = 6\sin15°$

Evaluate the sine and cosine functions of 15°. $\quad a = 5.8$ and $b = 1.6$

Let $\mathbf{u} = \langle a, b \rangle$. $\quad \mathbf{u} = \boxed{\langle 5.8, 1.6 \rangle}$

■ **Answer:** $\mathbf{u} = \langle 0.78, 2.9 \rangle$

■ **YOUR TURN** Find the vector that has a magnitude of 3 and direction angle of 75°.

Unit Vectors

A **unit vector** is any vector with magnitude equal to 1, or $|\mathbf{u}| = 1$. It is often useful to be able to find a unit vector in the same direction of some vector **v**. A unit vector can be formed from any nonzero vector as follows:

Study Tip

Multiplying a nonzero vector by the reciprocal of its magnitude results in a unit vector.

FINDING A UNIT VECTOR

If **v** is a nonzero vector, then

$$\mathbf{u} = \frac{\mathbf{v}}{|\mathbf{v}|} = \frac{1}{|\mathbf{v}|} \cdot \mathbf{v}$$

is a **unit vector** in the same direction as **v**. In other words, multiplying any nonzero vector by the reciprocal of its magnitude results in a unit vector.

It is important to notice that since the magnitude is always a scalar, then the reciprocal of the magnitude is always a scalar. A scalar times a vector is a vector.

 EXAMPLE 5 Finding a Unit Vector

Find a unit vector in the same direction as $\mathbf{v} = \langle -3, -4 \rangle$.

Solution:

Find the magnitude of the vector
$\mathbf{v} = \langle -3, -4 \rangle$.

$$|\mathbf{v}| = \sqrt{(-3)^2 + (-4)^2}$$

Simplify.

$$|\mathbf{v}| = 5$$

Multiply $\mathbf{v}$ by the reciprocal of its magnitude.

$$\frac{1}{|\mathbf{v}|} \mathbf{v}$$

Let $|\mathbf{v}| = 5$ and $\mathbf{v} = \langle -3, -4 \rangle$.

$$\frac{1}{5} \langle -3, -4 \rangle$$

Simplify.

$$\boxed{\left\langle -\frac{3}{5}, -\frac{4}{5} \right\rangle}$$

Check: The unit vector, $\left\langle -\dfrac{3}{5}, -\dfrac{4}{5} \right\rangle$, should have a magnitude of 1.

$$\sqrt{\left(-\frac{3}{5}\right)^2 + \left(-\frac{4}{5}\right)^2} = \sqrt{\frac{25}{25}} = 1$$

■ **YOUR TURN** Find a unit vector in the same direction as $\mathbf{v} = \langle 5, -12 \rangle$.

Classroom Example 8.4.5
Find a unit vector with the same direction as
a. $\langle 1, -1 \rangle$

b.* $\left\langle -\dfrac{2}{a}, -\dfrac{1}{a^2} \right\rangle$

Answer:

a. $\dfrac{1}{\sqrt{2}} \langle 1, -1 \rangle$

b. $\dfrac{a^2}{\sqrt{4a^2 + 1}} \left\langle -\dfrac{2}{a}, -\dfrac{1}{a^2} \right\rangle$

■ **Answer:** $\left\langle \dfrac{5}{13}, -\dfrac{12}{13} \right\rangle$

Two important unit vectors are the horizontal and vertical unit vectors $\mathbf{i}$ and $\mathbf{j}$. The unit vector $\mathbf{i}$ has an initial point at the origin and terminal point at $(1, 0)$. The unit vector $\mathbf{j}$ has an initial point at the origin and terminal point at $(0, 1)$. We can use these unit vectors to represent vectors algebraically. For example, the vector $\langle 3, -4 \rangle = 3\mathbf{i} - 4\mathbf{j}$.

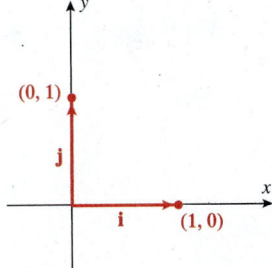

Resultant Vectors

Vectors arise in many applications. **Velocity vectors** and **force vectors** are two that we will discuss. For example, suppose that you are at the beach and "think" that you are swimming straight out at a certain speed (magnitude and direction). This is your **apparent velocity** with respect to the water. After a few minutes you turn around to look at the shore, and you are farther out than you thought and appear to have drifted down the beach. This is because of the current of the water. When the **current velocity** and the apparent velocity are added together, the result is the **actual** or **resultant velocity**. In Chapter 6 (Example 7) the navigation terms *bearing* (the direction a vessel is pointed) and *heading* (the direction the vessel is actually traveling) were used. In the context of vectors, bearing and heading are synonymous with apparent velocity and resultant velocity, respectively.

EXAMPLE 6   Resultant Velocities

A boat's speedometer reads 25 mph (which is relative to the water) and sets a course due east (90° from due north). If the river is moving 10 mph due north, what is the resultant (actual) velocity of the boat?

Solution:

Draw a picture.

Label the horizontal and vertical components of the resultant vector.

$\langle 25, 10 \rangle$

Determine the magnitude of the resultant vector.

$\sqrt{25^2 + 10^2} = 5\sqrt{29} \approx 27 \text{ mph}$

Determine the direction angle.

$\tan\theta = \dfrac{10}{25}$

Solve for θ.

$\theta = \tan^{-1}\left(\dfrac{2}{5}\right) \approx 22°$

The actual velocity of the boat has magnitude $\boxed{27 \text{ mph}}$ and the boat is headed $\boxed{22° \text{ north of east or } 68° \text{ east of north.}}$

In Example 6, the three vectors formed a right triangle. In Example 7, the three vectors form an oblique triangle.

EXAMPLE 7   Resultant Velocities

A speedboat traveling 30 mph has a compass heading of 100° east of north. The current velocity has a magnitude of 15 mph and its heading is 22° east of north. Find the resultant (actual) velocity of the boat.

Solution:

Draw a picture.

Label the supplementary angles to 100°.

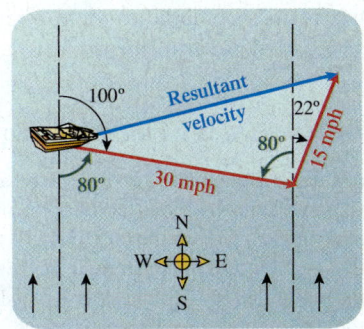

Draw and label the oblique triangle.

The magnitude of the actual (resultant) velocity is b.

The heading of the actual (resultant) velocity is $100° - \alpha$.

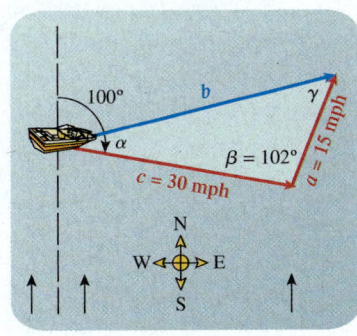

Use the Law of Sines and the Law of Cosines to solve for α and b.

Find b: Apply the Law of Cosines.

$$b^2 = a^2 + c^2 - 2ac\cos\beta$$

Let $a = 15$, $c = 30$, and $\beta = 102°$.

$$b^2 = 15^2 + 30^2 - 2(15)(30)\cos 102°$$

Solve for b.

$$\boxed{b \approx 36 \text{ mph}}$$

Find α: Apply the Law of Sines.

$$\frac{\sin\alpha}{a} = \frac{\sin\beta}{b}$$

Isolate $\sin\alpha$.

$$\sin\alpha = \frac{a}{b}\sin\beta$$

Let $a = 15$, $b = 36$, and $\beta = 102°$.

$$\sin\alpha = \frac{15}{36}\sin 102°$$

Apply the inverse sine function to solve for α.

$$\alpha = \sin^{-1}\left[\frac{15}{36}\sin(102°)\right]$$

Approximate α with a calculator.

$$\alpha \approx 24°$$

Actual heading: $100° - \alpha = 100° - 24° = \boxed{76°}$

The actual velocity vector of the boat has magnitude $\boxed{36 \text{ mph}}$ and the boat is headed $\boxed{76° \text{ east of north}}$.

Two vectors combine to yield a resultant vector. The opposite vector to the resultant vector is called the **equilibrant**.

EXAMPLE 8 Finding an Equilibrant

A skier is being pulled up a slope by a handle lift. Let $\mathbf{F}_1$ represent the vertical force due to gravity and $\mathbf{F}_2$ represent the force of the skier pushing against the side of the mountain, at an angle of 35° to the horizontal. If the weight of the skier is 145 pounds, that is, $|\mathbf{F}_1| = 145$, find the magnitude of the equilibrant force $\mathbf{F}_3$ required to hold the skier in place (that is, to keep the skier from sliding down the mountain). Assume that the side of the mountain is a frictionless surface.

Solution:

The angle between vectors $\mathbf{F}_1$ and $\mathbf{F}_2$ is 35°.

The magnitude of vector $\mathbf{F}_3$ is the force required to hold the skier in place.

Relate the magnitudes (side lengths) to the given angle using the sine ratio.

$$\sin 35° = \frac{|\mathbf{F}_3|}{|\mathbf{F}_1|}$$

Solve for $|\mathbf{F}_3|$.

$$|\mathbf{F}_3| = |\mathbf{F}_1| \sin 35°$$

Let $|\mathbf{F}_1| = 145$.

$$|\mathbf{F}_3| = 145 \sin 35°$$

$$|\mathbf{F}_3| = 83.16858$$

A force of approximately 83 pounds is required to keep the skier from sliding down the hill.

EXAMPLE 9 Resultant Forces

A barge runs aground outside the channel. A single tugboat cannot generate enough force to pull the barge off the sandbar. A second tugboat comes to assist. The following diagram illustrates the force vectors, $\mathbf{F}_1$ and $\mathbf{F}_2$, from the tugboats. What is the resultant force vector of the two tugboats?

Solution:

Using the tail-to-tip rule, we can add these two vectors and form a triangle:

Technology Tip

Use the calculator to find b.

```
15000²+18000²-2*
15000*18000cos(1
60)
          1056434015
√(Ans)
          32502.83088
```

Find b: Apply the Law of Cosines.

Let $a = 15,000$, $c = 18,000$, and $\beta = 160°$.

Solve for b.

$$b^2 = a^2 + c^2 - 2ac\cos\beta$$

$$b^2 = 15,000^2 + 18,000^2 - 2(15,000)(18,000)\cos 160°$$

$$\boxed{b = 32,503 \text{ lb}}$$

Find α: Apply the Law of Sines.

Isolate $\sin\alpha$.

Let $a = 15,000$, $b = 32,503$, and $\beta = 160°$.

Apply the inverse sine function to solve for α.

Approximate α with a calculator.

$$\frac{\sin\alpha}{a} = \frac{\sin\beta}{b}$$

$$\sin\alpha = \frac{a}{b}\sin\beta$$

$$\sin\alpha = \frac{15,000}{32,503}\sin 160°$$

$$\alpha = \sin^{-1}\left[\frac{15,000}{32,503}\sin(160°)\right]$$

$$\boxed{\alpha \approx 9.08°}$$

Use the calculator to find α.

```
15000/32503sin(1
60)
          .1578408808
sin⁻¹(Ans)
          9.08159538
```

The resulting force is $\boxed{32,503 \text{ lb}}$ at an angle of $\boxed{9° \text{ from the tug pulling with a force of 18,000 pounds}}$.

▶ **SECTION**
8.4 SUMMARY

In this section we discussed scalars (real numbers) and vectors. Scalars have only magnitude, whereas vectors have both magnitude and direction.

$$\text{Vector: } \mathbf{u} = \langle a, b \rangle$$

$$\text{Magnitude: } |\mathbf{u}| = \sqrt{a^2 + b^2}$$

$$\text{Direction } (\theta): \tan\theta = \frac{b}{a}$$

We defined vectors both algebraically and geometrically and gave interpretations of magnitude and vector addition in both ways.

Vector addition is performed algebraically component by component.

$$\langle a, b \rangle + \langle c, d \rangle = \langle a + c, b + d \rangle.$$

The trigonometric functions are used to express the horizontal and vertical components of a vector.

$$\text{Horizontal component: } a = |\mathbf{u}|\cos\theta$$

$$\text{Vertical component: } b = |\mathbf{u}|\sin\theta$$

Velocity and force vectors illustrate examples of the Law of Sines and the Law of Cosines.

■**SKILLS**

In Exercises 1–6, find the magnitude of the vector AB.

1. $A = (2, 7)$ and $B = (5, 9)$ **2.** $A = (-2, 3)$ and $B = (3, -4)$ **3.** $A = (4, 1)$ and $B = (-3, 0)$

4. $A = (-1, -1)$ and $B = (2, -5)$ **5.** $A = (0, 7)$ and $B = (-24, 0)$ **6.** $A = (-2, 1)$ and $B = (4, 9)$

In Exercises 7–16, find the magnitude and direction angle of the given vector.

7. $\mathbf{u} = \langle 3, 8 \rangle$ **8.** $\mathbf{u} = \langle 4, 7 \rangle$ **9.** $\mathbf{u} = \langle 5, -1 \rangle$ **10.** $\mathbf{u} = \langle -6, -2 \rangle$

11. $\mathbf{u} = \langle -4, 1 \rangle$ **12.** $\mathbf{u} = \langle -6, 3 \rangle$ **13.** $\mathbf{u} = \langle -8, 0 \rangle$ **14.** $\mathbf{u} = \langle 0, 7 \rangle$

15. $\mathbf{u} = \langle \sqrt{3}, 3 \rangle$ **16.** $\mathbf{u} = \langle -5, -5 \rangle$

In Exercises 17–24, perform the indicated vector operation, given $\mathbf{u} = \langle -4, 3 \rangle$ and $\mathbf{v} = \langle 2, -5 \rangle$.

17. $\mathbf{u} + \mathbf{v}$ **18.** $\mathbf{u} - \mathbf{v}$ **19.** $3\mathbf{u}$ **20.** $-2\mathbf{u}$

21. $2\mathbf{u} + 4\mathbf{v}$ **22.** $5(\mathbf{u} + \mathbf{v})$ **23.** $6(\mathbf{u} - \mathbf{v})$ **24.** $2\mathbf{u} - 3\mathbf{v} + 4\mathbf{u}$

In Exercises 25–34, find the vector, given its magnitude and direction angle.

25. $|\mathbf{u}| = 7, \theta = 25°$ **26.** $|\mathbf{u}| = 5, \theta = 75°$ **27.** $|\mathbf{u}| = 16, \theta = 100°$ **28.** $|\mathbf{u}| = 8, \theta = 200°$

29. $|\mathbf{u}| = 4, \theta = 310°$ **30.** $|\mathbf{u}| = 8, \theta = 225°$ **31.** $|\mathbf{u}| = 9, \theta = 335°$ **32.** $|\mathbf{u}| = 3, \theta = 315°$

33. $|\mathbf{u}| = 2, \theta = 120°$ **34.** $|\mathbf{u}| = 6, \theta = 330°$

In Exercises 35–44, find a unit vector in the direction of the given vector.

35. $\mathbf{v} = \langle -5, -12 \rangle$ **36.** $\mathbf{v} = \langle 3, 4 \rangle$ **37.** $\mathbf{v} = \langle 60, 11 \rangle$ **38.** $\mathbf{v} = \langle -7, 24 \rangle$

39. $\mathbf{v} = \langle 24, -7 \rangle$ **40.** $\mathbf{v} = \langle -10, 24 \rangle$ **41.** $\mathbf{v} = \langle -9, -12 \rangle$ **42.** $\mathbf{v} = \langle 40, -9 \rangle$

43. $\mathbf{v} = \langle \sqrt{2}, 3\sqrt{2} \rangle$ **44.** $\mathbf{v} = \langle -4\sqrt{3}, -2\sqrt{3} \rangle$

In Exercises 45–50, express the vector in terms of unit vectors i and j.

45. $\langle 7, 3 \rangle$ **46.** $\langle -2, 4 \rangle$ **47.** $\langle 5, -3 \rangle$ **48.** $\langle -6, -2 \rangle$ **49.** $\langle -1, 0 \rangle$ **50.** $\langle 0, 2 \rangle$

In Exercises 51–56, perform the indicated vector operation.

51. $(5\mathbf{i} - 2\mathbf{j}) + (-3\mathbf{i} + 2\mathbf{j})$ **52.** $(4\mathbf{i} - 2\mathbf{j}) + (3\mathbf{i} - 5\mathbf{j})$

53. $(-3\mathbf{i} + 3\mathbf{j}) - (2\mathbf{i} - 2\mathbf{j})$ **54.** $(\mathbf{i} - 3\mathbf{j}) - (-2\mathbf{i} + \mathbf{j})$

55. $(5\mathbf{i} + 3\mathbf{j}) + (2\mathbf{i} - 3\mathbf{j})$ **56.** $(-2\mathbf{i} + \mathbf{j}) + (2\mathbf{i} - 4\mathbf{j})$

■**APPLICATIONS**

57. Bullet Speed. A bullet is fired from ground level at a speed of 2200 feet per second at an angle of 30° from the horizontal. Find the magnitude of the horizontal and vertical components of the velocity vector.

58. Weightlifting. A 50-pound weight lies on an inclined bench that makes an angle of 40° with the horizontal. Find the component of the weight directed perpendicular to the bench and also the component of the weight parallel to the inclined bench.

59. Weight of a Boat. A force of 630 pounds is needed to pull a speedboat and its trailer up a ramp that has an incline of 13°. What is the combined weight of the boat and its trailer?

60. Weight of a Boat. A force of 500 pounds is needed to pull a speedboat and its trailer up a ramp that has an incline of 16°. What is the weight of the boat and its trailer?

61. Speed and Direction of a Ship. A ship's captain sets a course due north at 10 mph. The water is moving at 6 mph due west. What is the actual velocity of the ship, and in what direction is it traveling?

62. **Speed and Direction of a Ship.** A ship's captain sets a course due west at 12 mph. The water is moving at 3 mph due north. What is the actual velocity of the ship, and in what direction is it traveling?

63. **Heading and Airspeed.** A plane has a compass heading of 60° east of due north and an airspeed of 300 mph. The wind is blowing at 40 mph with a heading of 30° west of due north. What are the plane's actual heading and airspeed?

64. **Heading and Airspeed.** A plane has a compass heading of 30° east of due north and an airspeed of 400 mph. The wind is blowing at 30 mph with a heading of 60° west of due north. What are the plane's actual heading and airspeed?

65. **Sliding Box.** A box weighing 500 pounds is held in place on an inclined plane that has an angle of 30°. What force is required to hold it in place?

66. **Sliding Box.** A box weighing 500 pounds is held in place on an inclined plane that has an angle of 10°. What force is required to hold it in place?

67. **Baseball.** A baseball player throws a ball with an initial velocity of 80 feet per second at an angle of 40° with the horizontal. What are the vertical and horizontal components of the velocity?

68. **Baseball.** A baseball pitcher throws a ball with an initial velocity of 100 feet per second at an angle of 5° with the horizontal. What are the vertical and horizontal components of the velocity?

For Exercises 69 and 70, refer to the following:

In a post pattern in football, the receiver in motion runs past the quarterback parallel to the line of scrimmage (A), runs 12 yards perpendicular to the line of scrimmage (B), and then cuts toward the goal post (C).

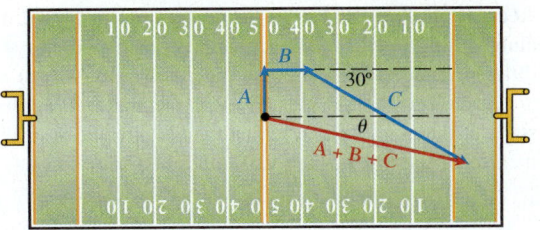

69. **Football.** A receiver runs the post pattern. If the magnitudes of the vectors are $|A| = 4$ yards, $|B| = 12$ yards, and $|C| = 20$ yards, find the magnitude of the resultant vector $\mathbf{A} + \mathbf{B} + \mathbf{C}$.

70. **Football.** A receiver runs the post pattern. If the magnitudes of the vectors are $|A| = 4$ yards, $|B| = 12$ yards, and $|C| = 20$ yards, find the direction angle θ.

71. **Resultant Force.** A force with a magnitude of 100 pounds and another with a magnitude of 400 pounds are acting on an object. The two forces have an angle of 60° between them. What is the direction of the resultant force with respect to the force of 400 pounds?

72. **Resultant Force.** A force with a magnitude of 100 pounds and another with a magnitude of 400 pounds are acting on an object. The two forces have an angle of 60° between them. What is the magnitude of the resultant force?

73. **Resultant Force.** A force of 1000 pounds is acting on an object at an angle of 45° from the horizontal. Another force of 500 pounds is acting at an angle of −40° from the horizontal. What is the magnitude of the resultant force?

74. **Resultant Force.** A force of 1000 pounds is acting on an object at an angle of 45° from the horizontal. Another force of 500 pounds is acting at an angle of −40° from the horizontal. What is the angle of the resultant force?

75. **Resultant Force.** Forces with magnitudes of 200 N and 180 N act on a hook. The angle between these two forces is 45°. Find the direction and magnitude of the resultant of these forces.

76. **Resultant Force.** Forces with magnitudes of 100 N and 50 N act on a hook. The angle between these two forces is 30°. Find the direction and magnitude of the resultant of these forces.

77. **Exercise Equipment.** A tether ball weighing 5 pounds is pulled outward from a pole by a horizontal force $\vec{u}$ until the rope makes a 45° angle with the pole. Determine the resulting tension on the rope and magnitude of $\vec{u}$.

78. **Exercise Equipment.** A tether ball weighing 8 pounds is pulled outward from a pole by a horizontal force $\vec{u}$ until the rope makes a 60° angle with the pole. Determine the resulting tension on the rope and magnitude of $\vec{u}$.

79. Recreation. A freshman wishes to sign up for four different clubs during orientation. Each club is positioned at a different table in the gym, and the clubs of interest to him are positioned at *A*, *B*, *C*, and *D*, as pictured below. He starts at the entrance way *O* and walks directly toward *A*, then to *B*, then to *C*, then to *D*, and then back to *O*.

a. Find the resultant vector of all his movement.

b. How far did he walk during this sign-up adventure?

80. Recreation. A freshman wishes to sign up for three different clubs during orientation. Each club is positioned at a different table in the gym, and the clubs of interest to him are positioned at *A*, *B*, and *C*, as pictured below. He starts at the entrance way *O* at the far end of the gym, walks directly toward *A*, then to *B*, then to *C*, and then exits the gym through the exit *P* at the opposite end.

a. Find the resultant vector of all his movement.

b. How far did he walk during this sign-up adventure?

81. Torque. *Torque* is the tendency for an arm to rotate about a pivot point. If a force $\vec{F}$ is applied at an angle θ to turn an arm of length *L*, as pictured below, then the magnitude of the torque = $L|\vec{F}|\sin\theta$.

Assume that a force of 45 N is applied to a bar 0.2 meters wide on a sewer shut-off valve at an angle 85°. What is the magnitude of the torque?

82. Torque. You walk through a swinging mall door to enter a department store. You exert a force of 40 N applied perpendicular to the door. The door is 0.85 meters wide. Assuming that you pushed the door at its edge and the hinge is the pivot point, find the magnitude of the torque.

83. Torque. You walk through a swinging mall door to enter a department store. You exert a force of 40 N applied at an angle 110° to the door. The door is 0.85 meters wide. Assuming that you pushed the door at its edge and the hinge is the pivot point, find the magnitude of the torque.

84. Torque. Suppose that within the context of Exercises 82 and 83, the magnitude of the torque turned out to be 0 Nm. When can this occur?

85. Resultant Force. A person is walking two dogs fastened to separate leashes that meet in a connective hub, leading to a single leash that she is holding. Dog 1 applies a force 60° NW with a magnitude of 8, and dog 2 applies a force of 45° clockwise of N with a magnitude of 6. Find the magnitude and direction of the force $\vec{w}$ that the walker applies to the leash in order to counterbalance the total force exerted by the dogs.

86. Resultant Force. A person is walking three dogs fastened to separate leashes that meet in a connective hub, leading to a single leash that she is holding. Dog 1 applies a force 60° NW with a magnitude of 8, dog 2 applies a force of 45° clockwise of N with a magnitude of 6, and dog 3 moves directly N with a magnitude of 12. Find the magnitude and direction of the force $\vec{w}$ that the walker applies to the leash in order to counterbalance the total force exerted by the dogs.

For Exercises 87 and 88, refer to the following:

Muscle *A* and muscle *B* are attached to a bone as indicated in the figure below. Muscle *A* exerts a force on the bone at angle α, while muscle *B* exerts a force on the bone at angle β.

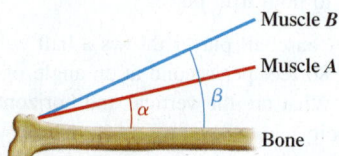

87. Health/Medicine. Assume muscle *A* exerts a force of 900 N on the bone at angle $\alpha = 8°$, while muscle *B* exerts a force of 750 N on the bone at angle $\beta = 33°$. Find the resultant force and the angle of the force due to muscle *A* and muscle *B* on the bone.

88. Health/Medicine. Assume muscle *A* exerts a force of 1000 N on the bone at angle $\alpha = 9°$, while muscle *B* exerts a force of 820 N on the bone at angle $\beta = 38°$. Find the resultant force and the angle of the force due to muscle *A* and muscle *B* on the bone.

▪CATCH THE MISTAKE

In Exercises 89 and 90, explain the mistake that is made.

89. Find the magnitude of the vector $\langle -2, -8 \rangle$.

Solution:

Factor the -1. $-\langle 2, 8 \rangle$

Find the magnitude $|\langle 2, 8 \rangle| = \sqrt{2^2 + 8^2}$
of $\langle 2, 8 \rangle$. $= \sqrt{68} = 2\sqrt{17}$

Write the magnitude
of $\langle -2, -8 \rangle$. $|\langle -2, -8 \rangle| = -2\sqrt{17}$

This is incorrect. What mistake was made?

90. Find the direction angle of the vector $\langle -2, -8 \rangle$.

Solution:

Write the formula for the direction
angle of $\langle a, b \rangle$. $\tan\theta = \dfrac{b}{a}$

Let $a = -2$ and $b = -8$. $\tan\theta = \dfrac{-8}{-2}$

Apply the inverse tangent function. $\theta = \tan^{-1}4$

Evaluate with a calculator. $\theta = 76°$

This is incorrect. What mistake was made?

▪CONCEPTUAL

In Exercises 91–94, determine whether each statment is true or false.

91. The magnitude of the vector **i** is the imaginary number i.

92. The arrow components of equal vectors must coincide.

93. The magnitude of a vector is always greater than or equal to the magnitude of its horizontal component.

94. The magnitude of a vector is always greater than or equal to the magnitude of its vertical component.

95. Would a scalar or a vector represent the following? *The car is driving 72 mph due east (90° with respect to north).*

96. Would a scalar or vector represent the following? *The granite has a mass of 131 kg.*

97. Find the magnitude of the vector $\langle -a, b \rangle$ if $a > 0$ and $b > 0$.

98. Find the direction angle of the vector $\langle -a, b \rangle$ if $a > 0$ and $b > 0$.

▪CHALLENGE

99. Show that if $\vec{u}$ is a unit vector in the direction of $\vec{v}$, then $\vec{v} = |\vec{u}|\,\vec{u}$.

100. Show that if $\vec{u} = a\vec{i} + b\vec{j}$ is a unit vector, then (a, b) lies on the unit circle.

101. A vector $\vec{u}$ is a *linear combination* of $\vec{p}$ and $\vec{q}$ if there exist constants c_1 and c_2 such that $\vec{u} = c_1\vec{p} + c_2\vec{q}$. Show that $\langle -6, 4 \rangle$ is a linear combination of $\langle -8, 4 \rangle$ and $\langle 1, -1 \rangle$.

102. Show that $\left\langle -\frac{2}{9}a, \frac{8}{9}b \right\rangle$ is a linear combination of $\langle a, 3b \rangle$ and $\langle -a, -b \rangle$, for any real constants a and b.

103. Prove that $\vec{u} + 3(2\vec{v} - \vec{u}) = 6\vec{v} - 2\vec{u}$, showing carefully how all relevant properties and definitions enter the proof.

104. Let $\vec{u} = \langle 2a, a \rangle$, $\vec{v} = \langle -a, -2a \rangle$. Compute $\left| \dfrac{2\vec{u}}{|\vec{v}|} - \dfrac{3\vec{v}}{|\vec{u}|} \right|$.

▪TECHNOLOGY

For Exercises 105–110, refer to the following:

Vectors can be represented as column matrices. For example, the vector $\mathbf{u} = \langle 3, -4 \rangle$ can be represented as a 2×1 column matrix $\begin{bmatrix} 3 \\ -4 \end{bmatrix}$. With a TI-83 calculator, vectors can be entered as matrices in two ways, directly or via MATRIX.

Directly:

Matrix:

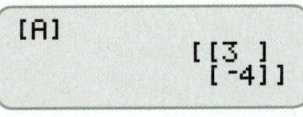

Use a calculator to perform the vector operation given $\mathbf{u} = \langle 8, -5 \rangle$ and $\mathbf{v} = \langle -7, 11 \rangle$.

105. $\mathbf{u} + 3\mathbf{v}$

106. $-9(\mathbf{u} - 2\mathbf{v})$

Use a calculator to find a unit vector in the direction of the given vector.

107. $\mathbf{u} = \langle 10, -24 \rangle$

108. $\mathbf{u} = \langle -9, -40 \rangle$

Use the graphing calculator SUM command to find the magnitude of the given vector. Also, find the direction angle to the nearest degree.

109. $\langle -33, 180 \rangle$

110. $\langle -20, -30\sqrt{5} \rangle$

SKILLS OBJECTIVES

- Find the dot product of two vectors.
- Use the dot product to find the angle between two vectors.
- Determine whether two vectors are parallel or perpendicular.
- Use the dot product to calculate the amount of work associated with a physical problem.

CONCEPTUAL OBJECTIVES

- Understand that the dot products of parallel or perpendicular vectors are respective limiting cases of the dot product of two vectors.

Multiplying Two Vectors: The Dot Product

With two-dimensional vectors there are two types of multiplication defined for vectors: scalar multiplication and the dot product. Scalar multiplication (which we already demonstrated in Section 8.4) is multiplication of a scalar by a vector; the result is a vector. Now we discuss the *dot product* of two vectors. In this case, there are two important things to note: (1) the dot product of two vectors is defined only if the vectors have the same number of components, and (2) if the dot product does exist, then the result is a scalar.

DOT PRODUCT

The **dot product** of two vectors $\mathbf{u} = \langle a, b \rangle$ and $\mathbf{v} = \langle c, d \rangle$ is given by

$$\mathbf{u} \cdot \mathbf{v} = ac + bd$$

$\mathbf{u} \cdot \mathbf{v}$ is pronounced "u dot v."

Classroom Example 8.5.1

a. Find the dot product
$\langle 2, -1 \rangle \cdot \langle -1, 3 \rangle$.

b.* Find the value of x such that
$$\left\langle 2a, -\frac{a}{3} \right\rangle \cdot \langle x, -2 \rangle = 0,$$
assuming $a \neq 0$.

Answer: a. -5 **b.** $-\frac{1}{3}$

Study Tip

The dot product of two vectors is a scalar.

■ **Answer:** -9

EXAMPLE 1   Finding the Dot Product of Two Vectors

Find the dot product $\langle -7, 3 \rangle \cdot \langle 2, 5 \rangle$.

Solution:

Sum the products of the first components and the products of the second components.

$$\langle -7, 3 \rangle \cdot \langle 2, 5 \rangle = (-7)(2) + (3)(5)$$

Simplify.

$$= -14 + 15 = \boxed{1}$$

■ **YOUR TURN** Find the dot product $\langle 6, 1 \rangle \cdot \langle -2, 3 \rangle$.

The following box summarizes the properties of the dot product:

PROPERTIES OF THE DOT PRODUCT

1. $\mathbf{u} \cdot \mathbf{v} = \mathbf{v} \cdot \mathbf{u}$
2. $\mathbf{u} \cdot \mathbf{u} = |\mathbf{u}|^2$
3. $\mathbf{0} \cdot \mathbf{u} = 0$
4. $k(\mathbf{u} \cdot \mathbf{v}) = (k\mathbf{u}) \cdot \mathbf{v} = \mathbf{u} \cdot (k\mathbf{v})$
5. $(\mathbf{u} + \mathbf{v}) \cdot \mathbf{w} = \mathbf{u} \cdot \mathbf{w} + \mathbf{v} \cdot \mathbf{w}$
6. $\mathbf{u} \cdot (\mathbf{v} + \mathbf{w}) = \mathbf{u} \cdot \mathbf{v} + \mathbf{u} \cdot \mathbf{w}$

These properties are verified in the exercises.

Angle Between Two Vectors

We can use these properties to develop an equation that relates the angle between two vectors and the dot product of the vectors.

WORDS	**MATH**
Let $\mathbf{u}$ and $\mathbf{v}$ be two vectors with the same initial point, and let θ be the angle between them.	
The vector $\mathbf{u} - \mathbf{v}$ is opposite angle θ.	
A triangle is formed with side lengths equal to the magnitudes of the three vectors.	

Apply the Law of Cosines.

$$|\mathbf{u} - \mathbf{v}|^2 = |\mathbf{u}|^2 + |\mathbf{v}|^2 - 2|\mathbf{u}||\mathbf{v}|\cos\theta$$

Use properties of the dot product to rewrite the left side of equation.

Property (2): $\qquad |\mathbf{u} - \mathbf{v}|^2 = (\mathbf{u} - \mathbf{v}) \cdot (\mathbf{u} - \mathbf{v})$

Property (6): $\qquad\qquad\quad = \mathbf{u} \cdot (\mathbf{u} - \mathbf{v}) - \mathbf{v} \cdot (\mathbf{u} - \mathbf{v})$

Property (6): $\qquad\qquad\quad = \mathbf{u} \cdot \mathbf{u} - \mathbf{u} \cdot \mathbf{v} - \mathbf{v} \cdot \mathbf{u} + \mathbf{v} \cdot \mathbf{v}$

Property (2): $\qquad\qquad\quad = |\mathbf{u}|^2 - \mathbf{u} \cdot \mathbf{v} - \mathbf{v} \cdot \mathbf{u} + |\mathbf{v}|^2$

Property (1): $\qquad\qquad\quad = |\mathbf{u}|^2 - 2(\mathbf{u} \cdot \mathbf{v}) + |\mathbf{v}|^2$

WORDS	MATH
Substitute this last expression for the left side of the original Law of Cosines equation.	$\|\mathbf{u}\|^2 - 2(\mathbf{u} \cdot \mathbf{v}) + \|\mathbf{v}\|^2 = \|\mathbf{u}\|^2 + \|\mathbf{v}\|^2 - 2\|\mathbf{u}\|\|\mathbf{v}\|\cos\theta$
Simplify.	$-2(\mathbf{u} \cdot \mathbf{v}) = -2\|\mathbf{u}\|\|\mathbf{v}\|\cos\theta$
Isolate $\cos\theta$.	$\boxed{\cos\theta = \dfrac{\mathbf{u} \cdot \mathbf{v}}{\|\mathbf{u}\|\|\mathbf{v}\|}}$

Notice that **u** and **v** have to be nonzero vectors, since we divided by them in the last step.

Study Tip

The angle between two vectors $\theta = \cos^{-1}\left(\dfrac{\mathbf{u} \cdot \mathbf{v}}{\|\mathbf{u}\|\|\mathbf{v}\|}\right)$ is an angle between 0° and 180° (the range of the inverse cosine function).

ANGLE BETWEEN TWO VECTORS

If θ is the angle between two nonzero vectors **u** and **v**, where $0° \le \theta \le 180°$, then

$$\cos\theta = \frac{\mathbf{u} \cdot \mathbf{v}}{\|\mathbf{u}\|\|\mathbf{v}\|}$$

In the Cartesian plane, there are two angles between two vectors, θ and $360° - \theta$. **We assume that θ is the "smaller" angle.**

Technology Tip

Use a TI calculator to find θ, $\cos\theta = \dfrac{-17}{5\sqrt{13}}$.

```
-17/(5*√(13))
       -.9429903336
cos⁻¹(Ans)
       160.5599652
cos⁻¹((-17/(5*√(1
3))
       160.5599652
■
```

Classroom Example 8.5.2
Find the angle between $\langle 2, -1 \rangle$ and $\langle -1, 3 \rangle$.

Answer: 135°

■ Answer: 38°

EXAMPLE 2   Finding the Angle Between Two Vectors

Find the angle between $\langle 2, -3 \rangle$ and $\langle -4, 3 \rangle$.

Solution:

Let $\mathbf{u} = \langle 2, -3 \rangle$ and $\mathbf{v} = \langle -4, 3 \rangle$.

STEP 1 Find $\mathbf{u} \cdot \mathbf{v}$.

$\mathbf{u} \cdot \mathbf{v} = \langle 2, -3 \rangle \cdot \langle -4, 3 \rangle$
$= (2)(-4) + (-3)(3) = -17$

STEP 2 Find $\|\mathbf{u}\|$.

$\|\mathbf{u}\| = \sqrt{\mathbf{u} \cdot \mathbf{u}} = \sqrt{2^2 + (-3)^2} = \sqrt{13}$

STEP 3 Find $\|\mathbf{v}\|$.

$\|\mathbf{v}\| = \sqrt{\mathbf{v} \cdot \mathbf{v}} = \sqrt{(-4)^2 + 3^2} = \sqrt{25} = 5$

STEP 4 Find θ.

$\cos\theta = \dfrac{\mathbf{u} \cdot \mathbf{v}}{\|\mathbf{u}\|\|\mathbf{v}\|} = \dfrac{-17}{5\sqrt{13}}$

Approximate θ with a calculator.

$\theta = \cos^{-1}\left(-\dfrac{17}{5\sqrt{13}}\right) \approx 160.559965°$

$$\boxed{\theta \approx 161°}$$

STEP 5 Draw a picture to confirm the answer.

Draw the vectors $\langle 2, -3 \rangle$ and $\langle -4, 3 \rangle$.
161° appears to be correct.

■ YOUR TURN Find the angle between $\langle 1, 5 \rangle$ and $\langle -2, 4 \rangle$.

When two vectors are **parallel**, the angle between them is 0° or 180°.

When two vectors are **perpendicular (orthogonal)**, the angle between them is 90°.

Note: We did not include 270° because the angle θ between two vectors is taken to be the smaller angle.

When two vectors **u** and **v** are perpendicular, $\theta = 90°$.

$$\cos(90°) = \frac{\mathbf{u} \cdot \mathbf{v}}{|\mathbf{u}||\mathbf{v}|}$$

Substitute $\cos(90°) = 0$.

$$0 = \frac{\mathbf{u} \cdot \mathbf{v}}{|\mathbf{u}||\mathbf{v}|}$$

Therefore, the dot product of **u** and **v** must be zero.

$$\mathbf{u} \cdot \mathbf{v} = \mathbf{0}$$

ORTHOGONAL VECTORS

Two vectors **u** and **v** are **orthogonal** (perpendicular) if and only if their dot product is zero.

$$\mathbf{u} \cdot \mathbf{v} = \mathbf{0}$$

EXAMPLE 3 **Determining Whether Vectors Are Orthogonal**

Determine whether each pair of vectors is orthogonal.

a. $\mathbf{u} = \langle 2, -3 \rangle$ and $\mathbf{v} = \langle 3, 2 \rangle$ **b.** $\mathbf{u} = \langle -7, -3 \rangle$ and $\mathbf{v} = \langle 7, 3 \rangle$

Solution (a):

Find the dot product $\mathbf{u} \cdot \mathbf{v}$. $\mathbf{u} \cdot \mathbf{v} = (2)(3) + (-3)(2)$

Simplify. $\mathbf{u} \cdot \mathbf{v} = 0$

Vectors **u** and **v** are orthogonal, since $\mathbf{u} \cdot \mathbf{v} = 0$.

Solution (b):

Find the dot product $\mathbf{u} \cdot \mathbf{v}$. $\mathbf{u} \cdot \mathbf{v} = (-7)(7) + (-3)(3)$

Simplify. $\mathbf{u} \cdot \mathbf{v} = -58$

Vectors **u** and **v** are not orthogonal, since $\mathbf{u} \cdot \mathbf{v} \neq 0$.

Classroom Example 8.5.3*
Determine the value(s) of a for which $\left\langle 2\sqrt{a}, -\dfrac{a}{3} \right\rangle$ and $\left\langle -\dfrac{1}{4}\sqrt{a}, a \right\rangle$ are perpendicular.

Answer: $0, -\dfrac{3}{2}$

Work

If you had to carry barbells with weights or pillows for 1 mile, which would you choose? You would probably pick the pillows over the barbell and weights, because the pillows are lighter. It requires less work to carry the pillows than it does to carry the weights. If asked to carry either of them 1 mile or 10 miles, you would probably pick 1 mile, because it's a shorter distance and requires less work. **Work** is done when a *force causes an object to move a certain distance.*

The simplest case is when the force is in the same direction as the displacement—for example, a stagecoach (the horses pull with a force in the same direction). In this case the work is defined as the magnitude of the force times the magnitude of the displacement, distance *d*.

$$W = |\mathbf{F}|d$$

Notice that the magnitude of the force is a scalar, the distance *d* is a scalar, and hence the product is a scalar.

If the horses pull with a force of 1000 pounds and they move the stagecoach 100 feet, the work done by the force is

$$W = (1000 \text{ lb})(100 \text{ ft}) = 100{,}000 \text{ ft-lb}$$

In many physical applications, however, the force is not in the same direction as the displacement, and hence vectors (not just their magnitudes) are required.

We often want to know how much of a force is applied in a certain direction. For example, when your car runs out of gasoline and you try to push it, some of the force vector $\mathbf{F}_1$ you generate from pushing translates into the horizontal component $\mathbf{F}_2$; hence, the car moves horizontally.

If we let θ be the angle between the vectors $\mathbf{F}_1$ and $\mathbf{F}_2$, then the horizontal component of $\mathbf{F}_1$ is $\mathbf{F}_2$ where $|F_2| = |F_1| \cos\theta$.

If the woman in the picture pushes at an angle of 25° with a force of 150 pounds, then the horizontal component of the force vector $\mathbf{F}_1$ is

$$(150 \text{ lb})(\cos 25°) \approx \boxed{136 \text{ lb}}$$

WORDS	MATH
To develop a generalized formula when the force exerted and the displacement are not in the same direction, we start with the formula for the angle between two vectors.	$\cos\theta = \dfrac{\mathbf{u}\cdot\mathbf{v}}{|\mathbf{u}||\mathbf{v}|}$

We then isolate the dot product $\mathbf{u}\cdot\mathbf{v}$. $\mathbf{u}\cdot\mathbf{v} = |\mathbf{u}||\mathbf{v}|\cos\theta$

Let $\mathbf{u} = \mathbf{F}$ and $\mathbf{v} = \mathbf{d}$. $W = \mathbf{F}\cdot\mathbf{d} = |\mathbf{F}||\mathbf{d}|\cos\theta = \underbrace{|\mathbf{F}|\cos\theta}_{\substack{\text{magnitude of force} \\ \text{in direction of displacement}}} \cdot \underbrace{|\mathbf{d}|}_{\text{distance}}$

WORK

If an object is moved from point A to point B by a constant force, then the work associated with this displacement is

$$W = \mathbf{F}\cdot\mathbf{d}$$

where $\mathbf{d}$ is the displacement vector and $\mathbf{F}$ is the force vector.

Work is typically expressed in one of two units:

SYSTEM	FORCE	DISTANCE	WORK
U.S. customary	pound	foot	ft-lb
SI	newton	meter	N-m

EXAMPLE 4   Calculating Work

How much work is done when a force (in pounds) $\mathbf{F} = \langle 2, 4 \rangle$ moves an object from $(0, 0)$ to $(5, 9)$ (the distance is in feet)?

Solution:

Find the displacement vector $\mathbf{d}$.	$\mathbf{d} = \langle 5, 9 \rangle$
Apply the work formula, $W = \mathbf{F}\cdot\mathbf{d}$.	$W = \langle 2, 4 \rangle \cdot \langle 5, 9 \rangle$
Calculate the dot product.	$W = (2)(5) + (4)(9)$
Simplify.	$\boxed{W = 46 \text{ ft-lb}}$

▪ **YOUR TURN** How much work is done when a force (in newtons) $\mathbf{F} = \langle 1, 3 \rangle$ moves an object from $(0, 0)$ to $(4, 7)$ (the distance is in meters)?

Classroom Example 8.5.4
Find the work done when a force $\overrightarrow{F} = \left\langle -a, \frac{1}{2}a \right\rangle$ moves an object from $(0, 0)$ to $(a, 3a)$.

Answer: $\frac{1}{2}a^2$

▪ **Answer:** 25 N-m

SECTION
8.5 SUMMARY

In this section we defined the dot product as a form of multiplication of two vectors. A scalar times a vector results in a vector, whereas the dot product of two vectors is a scalar.

$$\langle a, b \rangle \cdot \langle c, d \rangle = ac + bd$$

We developed a formula that determines the angle θ between two vectors **u** and **v**.

$$\cos\theta = \frac{\mathbf{u} \cdot \mathbf{v}}{|\mathbf{u}||\mathbf{v}|}$$

Orthogonal (perpendicular) vectors have an angle of 90° between them, and consequently the dot product of two orthogonal vectors is equal to zero. Work is the result of a force displacing an object. When the force and displacement are in the same direction, the work is equal to the product of the magnitude of the force and the distance (magnitude of the displacement). When the force and displacement are not in the same direction, work is the dot product of the force vector and displacement vector, $W = \mathbf{F} \cdot \mathbf{d}$.

SECTION
8.5 EXERCISES

▪ SKILLS

In Exercises 1–12, find the indicated dot product.

1. $\langle 4, -2 \rangle \cdot \langle 3, 5 \rangle$

2. $\langle 7, 8 \rangle \cdot \langle 2, -1 \rangle$

3. $\langle -5, 6 \rangle \cdot \langle 3, 2 \rangle$

4. $\langle 6, -3 \rangle \cdot \langle 2, 1 \rangle$

5. $\langle -7, -4 \rangle \cdot \langle -2, -7 \rangle$

6. $\langle 5, -2 \rangle \cdot \langle -1, -1 \rangle$

7. $\langle \sqrt{3}, -2 \rangle \cdot \langle 3\sqrt{3}, -1 \rangle$

8. $\langle 4\sqrt{2}, \sqrt{7} \rangle \cdot \langle -\sqrt{2}, -\sqrt{7} \rangle$

9. $\langle 5, a \rangle \cdot \langle -3a, 2 \rangle$

10. $\langle 4x, 3y \rangle \cdot \langle 2y, -5x \rangle$

11. $\langle 0.8, -0.5 \rangle \cdot \langle 2, 6 \rangle$

12. $\langle -18, 3 \rangle \cdot \langle 10, -300 \rangle$

In Exercises 13–24, find the angle (round to the nearest degree) between each pair of vectors.

13. $\langle -4, 3 \rangle$ and $\langle -5, -9 \rangle$

14. $\langle 2, -4 \rangle$ and $\langle 4, -1 \rangle$

15. $\langle -2, -3 \rangle$ and $\langle -3, 4 \rangle$

16. $\langle 6, 5 \rangle$ and $\langle 3, -2 \rangle$

17. $\langle -4, 6 \rangle$ and $\langle -6, 8 \rangle$

18. $\langle 1, 5 \rangle$ and $\langle -3, -2 \rangle$

19. $\langle -2, 2\sqrt{3} \rangle$ and $\langle -\sqrt{3}, 1 \rangle$

20. $\langle -3\sqrt{3}, -3 \rangle$ and $\langle -2\sqrt{3}, 2 \rangle$

21. $\langle -5\sqrt{3}, -5 \rangle$ and $\langle \sqrt{2}, -\sqrt{2} \rangle$

22. $\langle -5, -5\sqrt{3} \rangle$ and $\langle 2, -\sqrt{2} \rangle$

23. $\langle 4, 6 \rangle$ and $\langle -6, -9 \rangle$

24. $\langle 2, 8 \rangle$ and $\langle -12, 3 \rangle$

In Exercises 25–36, determine whether each pair of vectors is orthogonal.

25. $\langle -6, 8 \rangle$ and $\langle -8, 6 \rangle$

26. $\langle 5, -2 \rangle$ and $\langle -5, 2 \rangle$

27. $\langle 6, -4 \rangle$ and $\langle -6, -9 \rangle$

28. $\langle 8, 3 \rangle$ and $\langle -6, 16 \rangle$

29. $\langle 0.8, 4 \rangle$ and $\langle 3, -6 \rangle$

30. $\langle -7, 3 \rangle$ and $\langle \frac{1}{7}, -\frac{1}{3} \rangle$

31. $\langle 5, -0.4 \rangle$ and $\langle 1.6, 20 \rangle$

32. $\langle 12, 9 \rangle$ and $\langle 3, -4 \rangle$

33. $\langle \sqrt{3}, \sqrt{6} \rangle$ and $\langle -\sqrt{2}, 1 \rangle$

34. $\langle \sqrt{7}, -\sqrt{3} \rangle$ and $\langle 3, 7 \rangle$

35. $\langle \frac{4}{3}, \frac{8}{15} \rangle$ and $\langle -\frac{1}{12}, \frac{5}{24} \rangle$

36. $\langle \frac{5}{6}, \frac{6}{7} \rangle$ and $\langle \frac{36}{25}, -\frac{49}{36} \rangle$

▪ APPLICATIONS

37. Lifting Weights. How much work does it take to lift 100 pounds vertically 4 feet?

38. Lifting Weights. How much work does it take to lift 150 pounds vertically 3.5 feet?

39. Raising Wrecks. How much work is done by a crane to lift a 2-ton car to a level of 20 feet?

40. Raising Wrecks. How much work is done by a crane to lift a 2.5-ton car to a level of 25 feet?

41. Work. To slide a crate across the floor, a force of 50 pounds at a 30° angle is needed. How much work is done if the crate is dragged 30 feet?

Direction dragged

42. Work. To slide a crate across the floor, a force of 800 pounds at a 20° angle is needed. How much work is done if the crate is dragged 50 feet?

43. Close a Door. A sliding door is closed by pulling a cord with a constant force of 35 pounds at a constant angle of 45°. The door is moved 6 feet to close it. How much work is done?

44. Close a Door. A sliding door is closed by pulling a cord with a constant force of 45 pounds at a constant angle of 55°. The door is moved 6 feet to close it. How much work is done?

45. Braking Power. A car that weighs 2500 pounds is parked on a hill in San Francisco with a slant of 40° from the horizontal. How much force will keep it from rolling down the hill?

46. Towing Power. A car that weighs 2500 pounds is parked on a hill in San Francisco with a slant of 40° from the horizontal. A tow truck has to remove the car from its parking spot and move it 120 feet up the hill. How much work is required?

47. Towing Power. A semi trailer truck that weighs 40,000 pounds is parked on a hill in San Francisco with a slant of 10° from the horizontal. A tow truck has to remove the truck from its parking spot and move it 100 feet up the hill. How much work is required?

48. Braking Power. A truck that weighs 40,000 pounds is parked on a hill in San Francisco with a slant of 10° from the horizontal. How much force will keep it from rolling down the hill?

49. Business. Suppose that $\vec{u} = \langle 2000, 5000 \rangle$ represents the number of units of battery A and B, respectively, produced by a company and $\vec{v} = \langle 8.40, 6.50 \rangle$ represents the price (in dollars) of a 10-pack of battery A and B, respectively. Compute and interpret $\vec{u} \cdot \vec{v}$.

50. Demographics. Suppose that $\vec{u} = \langle 120, 80 \rangle$ represents the number of males and females in a high school class, and $\vec{v} = \langle 7.2, 5.3 \rangle$ represents the average number of minutes it takes a male and female, respectively, to register. Compute and interpret $\vec{u} \cdot \vec{v}$.

51. Geometry. Use vector methods to show that the diagonals of a rhombus are perpendicular to each other.

52. Geometry. Let $\vec{u}$ be a unit vector, and consider the following diagram:

Compute $\vec{u} \cdot \vec{v}$ and $\vec{u} \cdot \vec{w}$.

53. Geometry. Consider the following diagram:

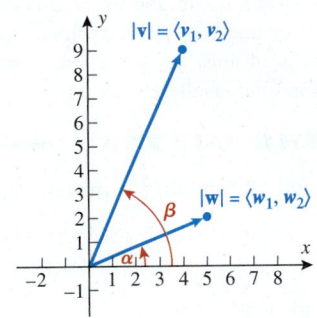

a. Compute $\cos\beta$, $\sin\beta$, $\cos\alpha$, and $\sin\alpha$.

b. Use (a) to show that $\cos(\alpha - \beta) = \dfrac{\vec{v} \cdot \vec{w}}{\sqrt{\vec{v} \cdot \vec{v}}\ \sqrt{\vec{w} \cdot \vec{w}}}$.

54. Geometry. Consider the diagram in Exercise 53.

a. Compute $\cos\beta$, $\sin\beta$, $\cos\alpha$, and $\sin\alpha$.

b. Use (a) to show that $\cos(\alpha + \beta) = \dfrac{\vec{v} \cdot \langle w_1, -w_2 \rangle}{\sqrt{\vec{v} \cdot \vec{v}}\ \sqrt{\vec{w} \cdot \vec{w}}}$.

55. Tennis. A player hits an overhead smash at full arm extension at the top of his racquet, which is 7 feet from the ground. The ball travels 16.3 feet. Consult the following diagram:

a. Determine v_1.

b. Find the angle θ with which the player hits this smash.

56. Tennis. In Exercise 55, use the dot product to determine the angle β with which the ball hits the ground.

57. Optimization. Let $\vec{u} = \langle a, b \rangle$ be a given vector, and suppose that the head of $\vec{n} = \langle n_1, n_2 \rangle$ lies on the circle $x^2 + y^2 = r^2$. Find the vector $\vec{n}$ such that $\vec{u} \cdot \vec{n}$ is as big as possible. Find the actual value of $\vec{u} \cdot \vec{n}$ in this case.

58. Optimization. Let $\vec{u} = \langle a, b \rangle$ be a given vector, and suppose that the head of $\vec{n} = \langle n_1, n_2 \rangle$ lies on the circle $x^2 + y^2 = r^2$. Find the vector $\vec{n}$ such that $\vec{u} \cdot \vec{n}$ is as small as possible. Find the actual value of $\vec{u} \cdot \vec{n}$ in this case.

59. Pursuit Theory. Assume that the head of $\vec{u}$ is restricted so that its tail is at the origin and its head is on the unit circle in quadrant II or quadrant III. A vector $\vec{v}$ has its tail at the origin and its head must lie on the line $y = 2 - x$ in quadrant I. Find the smallest value of $\vec{u} \cdot \vec{v}$.

60. Pursuit Theory. Assume that the head of $\vec{u}$ is restricted so that its tail is at the origin and its head is on the unit circle in quadrant I or quadrant IV. A vector $\vec{v}$ has its tail at the origin and its head must lie on the line $y = 2 - x$ in quadrant I. Find the biggest value of $\vec{u} \cdot \vec{v}$.

▪CATCH THE MISTAKE

In Exercises 61 and 62, explain the mistake that is made.

61. Find the dot product $\langle -3, 2 \rangle \cdot \langle 2, 5 \rangle$.

Solution:

Multiply component by component. $\langle -3, 2 \rangle \cdot \langle 2, 5 \rangle = \langle (-3)(2), (2)(5) \rangle$

Simplify. $\langle -3, 2 \rangle \cdot \langle 2, 5 \rangle = \langle -6, 10 \rangle$

This is incorrect. What mistake was made?

62. Find the dot product $\langle 11, 12 \rangle \cdot \langle -2, 3 \rangle$.

Solution:

Multiply the outer and inner components.

$\langle 11, 12 \rangle \cdot \langle -2, 3 \rangle = (11)(3) + (12)(-2)$

Simplify. $\langle 11, 12 \rangle \cdot \langle -2, 3 \rangle = 9$

This is incorrect. What mistake was made?

▪CONCEPTUAL

In Exercises 63–66, determine whether each statement is true or false.

63. A dot product of two vectors is a vector.

64. A dot product of two vectors is a scalar.

65. Orthogonal vectors have a dot product equal to zero.

66. If the dot product of two nonzero vectors is equal to zero, then the vectors must be perpendicular.

For Exercises 67 and 68, refer to the following to find the dot product:

The dot product of vectors with n components is

$\langle a_1, a_2, \ldots, a_n \rangle \cdot \langle b_1, b_2, \ldots, b_n \rangle = a_1 b_1 + a_2 b_2 + \cdots + a_n b_n.$

67. $\langle 3, 7, -5 \rangle \cdot \langle -2, 4, 1 \rangle$

68. $\langle 1, 0, -2, 3 \rangle \cdot \langle 5, 2, 3, 1 \rangle$

In Exercises 69–72, given $u = \langle a, b \rangle$ and $v = \langle c, d \rangle$, show that the following properties are true.

69. $\mathbf{u} \cdot \mathbf{v} = \mathbf{v} \cdot \mathbf{u}$

70. $\mathbf{u} \cdot \mathbf{u} = |\mathbf{u}|^2$

71. $\mathbf{0} \cdot \mathbf{u} = 0$

72. $k(\mathbf{u} \cdot \mathbf{v}) = (k\mathbf{u}) \cdot \mathbf{v} = \mathbf{u} \cdot (k\mathbf{v})$, k is a scalar

▪CHALLENGE

73. Show that $\vec{u} \cdot (\vec{v} + \vec{w}) = \vec{u} \cdot \vec{v} + \vec{u} \cdot \vec{w}$.

74. Show that $|\vec{u} - \vec{v}|^2 = |\vec{u}|^2 + |\vec{v}|^2 - 2(\vec{u} \cdot \vec{v})$.

75. The *projection of $\vec{v}$ onto $\vec{u}$* is defined by

$\text{proj}_{\vec{u}} \vec{v} = \left(\dfrac{\vec{u} \cdot \vec{v}}{|\vec{u}|^2} \right) \vec{u}$. This vector is depicted below.

Heuristically, this is the "shadow" of $\vec{v}$ on $\vec{u}$.

a. Compute $\text{proj}_{-\vec{u}} 2\vec{u}$.

b. What is $\text{proj}_{-\vec{u}} c\vec{u}$ for any $c > 0$?

Projection

76. a. Compute $\text{proj}_{\vec{u}} 2\vec{u}$.

b. What is $\text{proj}_{\vec{u}} c\vec{u}$ for any $c > 0$?

77. Suppose that you are given a vector $\vec{u}$. For what vectors $\vec{v}$ does $\text{proj}_{\vec{u}} \vec{v} = \vec{0}$?

78. True or False: $\text{proj}_{\vec{u}} (\vec{v} + \vec{w}) = \text{proj}_{\vec{u}} \vec{v} + \text{proj}_{\vec{u}} \vec{w}$.

79. If $\vec{u}$ and $\vec{v}$ are unit vectors, determine the maximum and minimum value of $(-2\vec{u}) \cdot (3\vec{v})$.

80. Assume that the angle between $\vec{u}$ and $\vec{v}$ is $\theta = \dfrac{\pi}{3}$. Show that

$\dfrac{(\vec{u} \cdot \vec{v})\vec{u}}{|\vec{v}|} - \dfrac{(\vec{v} \cdot \vec{u})\vec{v}}{|u|} = \dfrac{|\vec{u}|\vec{u} - |\vec{v}|\vec{v}}{2}.$

■ **TECHNOLOGY** ──────────────────

For Exercises 81 and 82, find the indicated dot product with a calculator.

81. $\langle -11, 34 \rangle \cdot \langle 15, -27 \rangle$

82. $\langle 23, -350 \rangle \cdot \langle 45, 202 \rangle$

83. A rectangle has sides with lengths 18 units and 11 units. Find the angle to one decimal place between the diagonal and the side with length of 18 units. (*Hint:* Set up a rectangular coordinate system, and use vectors $\langle 18, 0 \rangle$ to represent the side of length 18 units and $\langle 18, 11 \rangle$ to represent the diagonal.)

84. The definition of a dot product and the formula to find the angle between two vectors can be extended and applied to vectors with more than two components. A rectangular box has sides with lengths 12 feet, 7 feet, and 9 feet. Find the angle, to the nearest degree, between the diagonal and the side with length 7 feet.

Use the graphing calculator $\boxed{\text{SUM}}$ **command to find the angle (round to the nearest degree) between each pair of vectors.**

85. $\langle -25, 42 \rangle, \langle 10, 35 \rangle$

86. $\langle -12, 9 \rangle, \langle -21, -13 \rangle$

SECTION
8.6
POLAR (TRIGONOMETRIC) FORM OF COMPLEX NUMBERS

SKILLS OBJECTIVES

■ Plot a point in the complex plane.
■ Convert complex numbers from rectangular form to polar form.
■ Convert complex numbers from polar form to rectangular form.

CONCEPTUAL OBJECTIVES

■ Understand that a complex number can be represented either in rectangular or polar form.
■ Relate the horizontal axis in the complex plane to the real component of a complex number.
■ Relate the vertical axis in the complex plane to the imaginary component of a complex number.

Complex Numbers in Rectangular Form

We are already familiar with the **rectangular coordinate system**, where the horizontal axis is called the *x*-axis and the vertical axis is called the *y*-axis. In our study of complex numbers, we refer to the **standard (rectangular) form** as $a + bi$, where a represents the real part and b represents the imaginary part. If we let the horizontal axis be the **real axis** and the vertical axis be the **imaginary axis**, the result is the **complex plane**. The point $a + bi$ is located in the complex plane by finding the coordinates (a, b).

When $b = 0$, the result is a real number, and therefore any numbers along the horizontal axis are real numbers. When $a = 0$, the result is an imaginary number, so any numbers along the vertical axis are imaginary numbers.

The variable z is often used to represent a complex number: $z = x + iy$. Complex numbers are analogous to vectors. Suppose that a vector $\mathbf{z} = \langle x, y \rangle$, whose initial point is the origin and whose terminal point is (x, y); then the magnitude of that vector is $|\mathbf{z}| = \sqrt{x^2 + y^2}$. Similarly, the magnitude, or *modulus*, of a complex number is defined like the magnitude of a position vector in the xy-plane, as the distance from the origin $(0, 0)$ to the point (x, y) in the complex plane.

Technology Tip

To use a TI calculator to find the modulus of a complex number, press | MATH | ▶ | CPX | ▼ |

| 5: ABS (| and enter the complex number.

DEFINITION   **Modulus of a Complex Number**

The **modulus**, or magnitude, of a complex number $z = x + iy$ is the distance from the origin to the point (x, y) in the complex plane given by

$$|z| = \sqrt{x^2 + y^2}$$

Classroom Example 8.6.1*
Compute the modulus of

$z = -a - \frac{3}{5}ai$ (where $a > 0$).

Answer: $\dfrac{a\sqrt{34}}{5}$

Recall that a complex number $z = x + iy$ has a complex conjugate $\bar{z} = x - iy$. The bar above a complex number denotes its conjugate. Notice that

$$z\bar{z} = (x + iy)(x - iy) = x^2 - i^2y^2 = x^2 + y^2$$

and therefore the modulus can also be written as

$$|z| = \sqrt{z\bar{z}}$$

EXAMPLE 1 Finding the Modulus of a Complex Number

Find the modulus of $z = -3 + 2i$.

Technology Tip

Find the modulus of $z = -3 + 2i$.

MATH	▶	CPX	▼	5: *ABS* (
ENTER	(−)	3	+	2	2nd
·	)	ENTER

```
abs(-3+2i)
          3.605551275
√(13)
          3.605551275
```

COMMON MISTAKE

Including the i in the imaginary part.

⭐ CORRECT

Let $x = -3$ and $y = 2$ in
$|z| = \sqrt{x^2 + y^2}$.

$$|-3 + 2i| = \sqrt{(-3)^2 + 2^2}$$

Eliminate the parentheses.

$$|-3 + 2i| = \sqrt{9 + 4}$$

Simplify.

$$\boxed{|z| = |-3 + 2i| = \sqrt{13}}$$

❌ INCORRECT

Let $x = -3$ and $y = 2i$ in
$|z| = \sqrt{x^2 + y^2}$.

$$|-3 + 2i| = \sqrt{(-3)^2 + (2i)^2}$$

ERROR

The i is not included in the formula. Only the imaginary part (the coefficient of i) is used.

■ **YOUR TURN** Find the modulus of $z = 2 - 5i$.

■ **Answer:** $|z| = |2 - 5i| = \sqrt{29}$

Complex Numbers in Polar Form

We say that a complex number $z = x + iy$ is in *rectangular* form because it is located at the point (x, y), which is expressed in rectangular coordinates, in the complex plane. Another convenient way of expressing complex numbers is in *polar* form. Recall from our study of vectors (Section 8.4) that vectors have both magnitude and a direction angle. The same is true of numbers in the complex plane. Let r represent the magnitude, or distance from the origin to the point (x, y), and θ represent the direction angle; then we have the following relationships:

$$r = \sqrt{x^2 + y^2}$$

$$\sin\theta = \frac{y}{r} \qquad \cos\theta = \frac{x}{r} \qquad \text{and} \qquad \tan\theta = \frac{y}{x} \quad (x \neq 0)$$

Isolating x and y in the sinusoidal functions, we find

$$x = r\cos\theta \qquad y = r\sin\theta$$

Using these expressions for x and y, a complex number can be written in *polar* form.

$$z = x + yi = (r\cos\theta) + (r\sin\theta)i = r(\cos\theta + i\sin\theta)$$

POLAR (TRIGONOMETRIC) FORM OF COMPLEX NUMBERS

The following expression is the **polar form** of a complex number:

$$z = r(\cos\theta + i\sin\theta)$$

where r represents the **modulus** (magnitude) of the complex number and θ represents the **argument** of z.

The following is standard notation for modulus and argument:

$$r = \operatorname{mod} z = |z| \quad \text{and} \quad \theta = \operatorname{Arg} z, \qquad 0 \le \theta < 2\pi \quad \text{or} \quad 0° \le \theta < 360°$$

Converting Complex Numbers Between Rectangular and Polar Forms

We can convert back and forth between rectangular and polar (trigonometric) forms of complex numbers using the modulus and trigonometric ratios.

$$r = \sqrt{x^2 + y^2} \qquad \sin\theta = \frac{y}{r} \qquad \cos\theta = \frac{x}{r} \qquad \text{and} \qquad \tan\theta = \frac{y}{x} \quad (x \ne 0)$$

CONVERTING COMPLEX NUMBERS FROM RECTANGULAR FORM TO POLAR FORM

Step 1: Plot the point $z = x + yi$ in the complex plane (note the quadrant).

Step 2: Find r. Use $r = \sqrt{x^2 + y^2}$.

Step 3: Find θ. Apply $\tan\theta = \frac{y}{x}$, $x \ne 0$, where θ is in the quadrant found in Step 1.

Step 4: Write the complex number in polar form: $z = r(\cos\theta + i\sin\theta)$.

Realize that imaginary numbers, $z = 0 + bi$, lie on the imaginary axis. Therefore $\theta = 90°$ if $b > 0$ and $\theta = 270°$ if $b < 0$.

EXAMPLE 2 Converting from Rectangular to Polar Form

Express the complex number $z = \sqrt{3} - i$ in polar form.

Solution:

STEP 1 Plot the point.

The point lies in **quadrant IV**.

STEP 2 Find r.

Let $x = \sqrt{3}$ and $y = -1$
in $r = \sqrt{x^2 + y^2}$. $r = \sqrt{(\sqrt{3})^2 + (-1)^2}$

Eliminate the parentheses. $r = \sqrt{3 + 1}$

Simplify. $\boxed{r = 2}$

STEP 3 Find θ.

Let $x = \sqrt{3}$ and $y = -1$ in $\tan\theta = \dfrac{y}{x}$. $\tan\theta = -\dfrac{1}{\sqrt{3}}$

Solve for θ. $\theta = \tan^{-1}\left(-\dfrac{1}{\sqrt{3}}\right) = -\dfrac{\pi}{6}$

Find the reference angle.   reference angle $= \dfrac{\pi}{6}$

The complex number lies in quadrant IV. $\boxed{\theta = \dfrac{11\pi}{6}}$

STEP 4 Write the complex number in polar form.
$z = r(\cos\theta + i\sin\theta)$ $\boxed{z = 2\left[\cos\left(\dfrac{11\pi}{6}\right) + i\sin\left(\dfrac{11\pi}{6}\right)\right]}$

Note: An alternative form is in degrees: $z = 2(\cos 330° + i\sin 330°)$.

■ **YOUR TURN** Express the complex number $z = 1 - i\sqrt{3}$ in polar form.

Technology Tip

Express the complex number $z = \sqrt{3} - i$ in polar form.

Method I: Use $\tan^{-1}\left(\dfrac{1}{\sqrt{3}}\right)$ to find the reference angle for θ, which is in QIV.

Method II: Use the $\boxed{\text{angle (}}$ feature on the calculator to find θ. You still have to find the actual angle in QIV. Press

$\boxed{\text{MATH}}\boxed{\blacktriangleright}\boxed{\text{CPX}}\boxed{\blacktriangledown}\boxed{\text{4: angle (}}$
$\boxed{\text{ENTER}}\boxed{\text{2nd}}\boxed{x^2}\boxed{3}\boxed{)}\boxed{-}$
$\boxed{\text{2nd}}\boxed{\cdot}\boxed{)}\boxed{\text{ENTER}}$

■ **Answer:**

$z = 2\left[\cos\left(\dfrac{5\pi}{3}\right) + i\sin\left(\dfrac{5\pi}{3}\right)\right]$

or $2(\cos 300° + i\sin 300°)$

You must be very careful in converting from rectangular to polar form. Remember that the inverse tangent function is a one-to-one function and will yield values in quadrants I and IV. If the point lies in quadrant II or III, add 180° to the angle found through the inverse tangent function.

EXAMPLE 3 **Converting from Rectangular to Polar Form**

Technology Tip

Express the complex number $z = -2 + i$ in polar form.

```
abs(-2+i)
       2.236067977
√(5)
       2.236067977
angle(-2+i)
       153.4349488
```

Forgetting to confirm the quadrant, which results in using the reference angle instead of the actual angle.

Express the complex number $z = -2 + i$ in polar form.

✪ CORRECT   **✗ INCORRECT**

Step 1: Plot the point.
            The point lies in **quadrant II**.

Step 2: Find r.
            Let $x = -2$ and $y = 1$ in
            $r = \sqrt{x^2 + y^2}$.

            $r = \sqrt{(-2)^2 + 1^2}$

            Simplify.

            $\boxed{r = \sqrt{5}}$

Step 3: Find θ.

Let $x = -2$ and $y = 1$ in   Evaluate the inverse function with a
$\tan\theta = \dfrac{y}{x}$.   calculator.

$\tan\theta = -\dfrac{1}{2}$ $\theta = \tan^{-1}\left(-\dfrac{1}{2}\right) = -26.565°$

$\theta = \tan^{-1}\left(-\dfrac{1}{2}\right)$ Write the complex number in polar
                                                form.

$= -26.565°$ $z = r(\cos\theta + i\sin\theta)$
                                                $z = \sqrt{5}[\cos(-26.6°) + i\sin(-26.6°)]$
The complex number lies in
quadrant II.   *Note:* $\theta = -26.565°$ lies in quadrant

$\boxed{\theta = -26.6° + 180° = 153.4°}$ IV, whereas the original point lies in
                                                quadrant II. Therefore, we should have
Step 4: Write the complex number in   added 180° to θ in order to arrive at a
            polar form $z = r(\cos\theta + i\sin\theta)$.   point in quadrant II.

$\boxed{z = \sqrt{5}[\cos(153.4°) + i\sin(153.4°)]}$

■ **Answer:**
$z = \sqrt{5}[\cos(116.6°) + i\sin(116.6°)]$

■ **YOUR TURN** Express the complex number $z = -1 + 2i$ in polar form.

To convert from polar to rectangular form, simply evaluate the trigonometric functions.

EXAMPLE 4 Converting from Polar to Rectangular Form

Express $z = 4(\cos 120° + i\sin 120°)$ in rectangular form.

Solution:

Evaluate the trigonometric functions exactly.

$$z = 4\left(\underbrace{\cos 120°}_{-\frac{1}{2}} + i\underbrace{\sin 120°}_{\frac{\sqrt{3}}{2}} \right)$$

Distribute the 4.

$$z = 4\left(-\frac{1}{2}\right) + 4\left(\frac{\sqrt{3}}{2}\right)i$$

Simplify.

$$\boxed{z = -2 + 2\sqrt{3}\,i}$$

▪ **YOUR TURN** Express $z = 2(\cos 210° + i\sin 210°)$ in rectangular form.

Technology Tip

Express $z = 4(\cos 120° + i\sin 120°)$
in rectangular form.

■ **Answer:** $z = -\sqrt{3} - i$

EXAMPLE 5 Using a Calculator to Convert from Polar to Rectangular Form

Express $z = 3(\cos 109° + i\sin 109°)$ in rectangular form. Round to four decimal places.

Solution:

Use a calculator to evaluate the trigonometric functions.

$$z = 3\left(\underbrace{\cos(109°)}_{-0.325568} + i\underbrace{\sin(109°)}_{0.945519} \right)$$

Simplify.

$$\boxed{z \approx -0.9767 + 2.8366\,i}$$

▪ **YOUR TURN** Express $z = 7(\cos 217° + i\sin 217°)$ in rectangular form. Round to four decimal places.

Technology Tip

Express $z = 4(\cos 109° + i\sin 109°)$
in rectangular form.

■ **Answer:** $z \approx -5.5904 - 4.2127i$

SECTION 8.6 SUMMARY

In the complex plane, the horizontal axis is the real axis and the vertical axis is the imaginary axis. We can express complex numbers in either rectangular form, $z = x + iy$, or polar form, $z = r(\cos\theta + i\sin\theta)$. The modulus of a complex number, $z = x + iy$, is given by $|z| = \sqrt{x^2 + y^2}$. To convert from rectangular to polar form, we use the relationships $r = \sqrt{x^2 + y^2}$ and $\tan\theta = \dfrac{y}{x}$, $x \neq 0$ and $0 \leq \theta < 2\pi$. It is important to note in which quadrant the point lies. To convert from polar to rectangular form, simply evaluate the trigonometric functions.

$$x = r\cos\theta \qquad \text{and} \qquad y = r\sin\theta$$

**SECTION
8.6 EXERCISES**

SKILLS

In Exercises 1–8, graph each complex number in the complex plane.

1. $7 + 8i$ | **2.** $3 + 5i$ | **3.** $-2 - 4i$ | **4.** $-3 - 2i$

5. 2 | **6.** 7 | **7.** $-3i$ | **8.** $-5i$

In Exercises 9–22, express each complex number in polar form.

9. $1 - i$ | **10.** $2 + 2i$ | **11.** $1 + \sqrt{3}i$ | **12.** $-3 - \sqrt{3}i$

13. $-4 + 4i$ | **14.** $\sqrt{5} - \sqrt{5}i$ | **15.** $\sqrt{3} - 3i$ | **16.** $-\sqrt{3} + i$

17. $3 + 0i$ | **18.** $-2 + 0i$ | **19.** $-\dfrac{1}{2} - \dfrac{1}{2}i$ | **20.** $\dfrac{1}{6} - \dfrac{1}{6}i$

21. $5.32 - 0i$ | **22.** $-6.48 + 0i$

In Exercises 23–38, use a calculator to express each complex number in polar form.

23. $3 - 7i$ | **24.** $2 + 3i$ | **25.** $-6 + 5i$ | **26.** $-4 - 3i$

27. $-5 + 12i$ | **28.** $24 + 7i$ | **29.** $8 - 6i$ | **30.** $-3 + 4i$

31. $-\dfrac{1}{2} + \dfrac{3}{4}i$ | **32.** $-\dfrac{5}{8} - \dfrac{11}{4}i$ | **33.** $5.1 + 2.3i$ | **34.** $1.8 - 0.9i$

35. $-2\sqrt{3} - \sqrt{5}i$ | **36.** $-\dfrac{4\sqrt{5}}{3} + \dfrac{\sqrt{5}}{2}i$ | **37.** $4.02 - 2.11i$ | **38.** $1.78 - 0.12i$

In Exercises 39–48, express each complex number in rectangular form.

39. $5(\cos 180° + i \sin 180°)$ | **40.** $2(\cos 135° + i \sin 135°)$

41. $2(\cos 315° + i \sin 315°)$ | **42.** $3(\cos 270° + i \sin 270°)$

43. $-4(\cos 60° + i \sin 60°)$ | **44.** $-4(\cos 210° + i \sin 210°)$

45. $\sqrt{3}(\cos 150° + i \sin 150°)$ | **46.** $\sqrt{3}(\cos 330° + i \sin 330°)$

47. $\sqrt{2}\left[\cos\left(\dfrac{\pi}{4}\right) + i \sin\left(\dfrac{\pi}{4}\right)\right]$ | **48.** $2\left[\cos\left(\dfrac{5\pi}{6}\right) + i \sin\left(\dfrac{5\pi}{6}\right)\right]$

In Exercises 49–58, use a calculator to express each complex number in rectangular form.

49. $5(\cos 295° + i \sin 295°)$ | **50.** $4(\cos 35° + i \sin 35°)$

51. $3(\cos 100° + i \sin 100°)$ | **52.** $6(\cos 250° + i \sin 250°)$

53. $-7(\cos 140° + i \sin 140°)$ | **54.** $-5(\cos 320° + i \sin 320°)$

55. $3\left[\cos\left(\dfrac{11\pi}{12}\right) + i \sin\left(\dfrac{11\pi}{12}\right)\right]$ | **56.** $2\left[\cos\left(\dfrac{4\pi}{7}\right) + i \sin\left(\dfrac{4\pi}{7}\right)\right]$

57. $-2\left[\cos\left(\dfrac{3\pi}{5}\right) + i \sin\left(\dfrac{3\pi}{5}\right)\right]$ | **58.** $-4\left[\cos\left(\dfrac{15\pi}{11}\right) + i \sin\left(\dfrac{15\pi}{11}\right)\right]$

■ APPLICATIONS

59. Resultant Force. Force A, at 100 pounds, and force B, at 120 pounds, make an angle of 30° with each other. Represent their respective vectors as complex numbers written in trigonometric form, and solve for the resultant force.

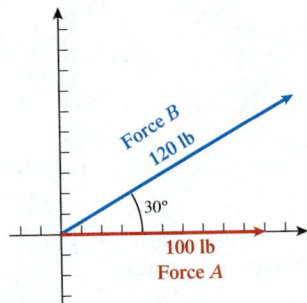

60. Resultant Force. Force A, at 40 pounds, and force B, at 50 pounds, make an angle of 45° with each other. Represent their respective vectors as complex numbers written in trigonometric form, and solve for the resultant force.

61. Resultant Force. Force A, at 80 pounds, and force B, at 150 pounds, make an angle of 30° with each other. Represent their respective vectors as complex numbers written in trigonometric form, and solve for the resultant angle.

62. Resultant Force. Force A, at 20 pounds, and force B, at 60 pounds, make an angle of 60° with each other. Represent their respective vectors as complex numbers written in trigonometric form, and solve for the resultant angle.

63. Resultant Force. Refer to Section 8.4, Exercise 85. Express the vectors $\vec{u}, \vec{v}$ corresponding to the dogs as complex numbers written in trigonometric form, and find the resultant force $\vec{w}$ to keep the system in equilibrium.

64. Resultant Force. Refer to Section 8.4, Exercise 86. Express the vectors $\vec{u}, \vec{v}, \vec{w}$ as complex numbers written in trigonometric form, and find the resultant force $\vec{z}$ to keep the system in equilibrium.

65. Resultant Force. Refer to Section 8.4, Exercise 75. Express the resultant vector $\vec{R}$ as a complex number in trigonometric form.

66. Resultant Force. Refer to Section 8.4, Exercise 76. Express the resultant vector $\vec{R}$ as a complex number in trigonometric form.

■ CATCH THE MISTAKE

In Exercises 67 and 68, explain the mistake that is made.

67. Express $z = -3 - 8i$ in polar form.

Solution:

Find r. $\qquad r = \sqrt{x^2 + y^2} = \sqrt{9 + 64} = \sqrt{73}$

Find θ. $\qquad \tan\theta = \dfrac{8}{3}$

$$\theta = \tan^{-1}\left(\frac{8}{3}\right) = 69.44°$$

Write the complex number in polar form.

$$z = \sqrt{73}(\cos 69.44° + i\sin 69.44°)$$

This is incorrect. What mistake was made?

68. Express $z = -3 + 8i$ in polar form.

Solution:

Find r. $\qquad r = \sqrt{x^2 + y^2} = \sqrt{9 + 64} = \sqrt{73}$

Find θ. $\qquad \tan\theta = -\dfrac{8}{3}$

$$\theta = \tan^{-1}\left(-\frac{8}{3}\right) = -69.44°$$

Write the complex number in polar form.

$$z = \sqrt{73}[\cos(-69.44°) + i\sin(-69.44°)]$$

This is incorrect. What mistake was made?

■ CONCEPTUAL

In Exercises 69–72, determine whether each statement is true or false.

69. In the complex plane, any point that lies along the horizontal axis is a real number.

70. In the complex plane, any point that lies along the vertical axis is an imaginary number.

71. The modulus of z and the modulus of $\bar{z}$ are equal.

72. The argument of z and the argument of $\bar{z}$ are equal.

73. Find the argument of $z = a$, where a is a positive real number.

74. Find the argument of $z = bi$, where b is a positive real number.

75. Find the modulus of $z = bi$, where b is a negative real number.

76. Find the modulus of $z = a$, where a is a negative real number.

In Exercises 77 and 78, use a calculator to express the complex number in polar form.

77. $a - 2ai$, where $a > 0$

78. $-3a - 4ai$, where $a > 0$

▪ CHALLENGE

79. Suppose that a complex number z lies on the circle $x^2 + y^2 = \pi^2$. If $\cos\left(\dfrac{\theta}{2}\right) = \dfrac{1}{2}$ and $\sin\theta < 0$, find the rectangular form of z.

80. Suppose that a complex number z lies on the circle $x^2 + y^2 = 8$. If $\sin\left(\dfrac{\theta}{2}\right) = -\dfrac{\sqrt{3}}{2}$ and $\cos\theta < 0$, find the rectangular form of z.

81. Consider the following diagram:

Find z in trigonometric form.
(*Hint:* Use the Law of Cosines.)

82. Consider the following diagram:

Find z in trigonometric form.
(*Hint:* Use the Law of Cosines.)

▪ TECHNOLOGY

Graphing calculators are able to convert complex numbers from rectangular to polar form using the $\boxed{\text{abs}}$ command to find the modulus and the angle command to find the angle.

83. Find abs$(1 + i)$. Find angle$(1 + i)$. Write $1 + i$ in polar form.

84. Find abs$(1 - i)$. Find angle$(1 - i)$. Write $1 - i$ in polar form.

A second way of using a graphing calculator to convert between rectangular and polar coordinates is with the $\boxed{\text{Pol}}$ and $\boxed{\text{Rec}}$ commands.

85. Find Pol$(2, 1)$. Write $2 + i$ in polar form.

86. Find Rec$(345°)$. Write $3(\cos 45° + i\sin 45°)$ in rectangular form.

Another way of using a graphing calculator to represent complex numbers in rectangular form is to enter the real and imaginary parts as a list of two numbers and use the $\boxed{\text{SUM}}$ command to find the modulus.

87. Write $28 - 21i$ in polar form using the $\boxed{\text{SUM}}$ command to find its modulus, and round the angle to the nearest degree.

88. Write $-\sqrt{21} + 10i$ in polar form using the $\boxed{\text{SUM}}$ command to find its modulus, and round the angle to the nearest degree.

SKILLS OBJECTIVES

- Find the product of two complex numbers.
- Find the quotient of two complex numbers.
- Raise a complex number to an integer power.
- Find the nth root of a complex number.
- Solve a polynomial equation by finding complex roots.

CONCEPTUAL OBJECTIVES

- Derive the identities for products and quotients of complex numbers.
- Relate De Moivre's theorem (the power rule) for complex numbers to the product rule for complex numbers.

In this section we will multiply complex numbers, divide complex numbers, raise complex numbers to powers, and find roots of complex numbers.

Products of Complex Numbers

First, we will derive a formula for the product of two complex numbers.

WORDS	MATH
Start with two complex numbers z_1 and z_2.	$z_1 = r_1(\cos\theta_1 + i\sin\theta_1)$ $z_2 = r_2(\cos\theta_2 + i\sin\theta_2)$
Multiply z_1 and z_2.	$z_1 z_2 = r_1 r_2(\cos\theta_1 + i\sin\theta_1)(\cos\theta_2 + i\sin\theta_2)$
Use the FOIL method to multiply the expressions in parentheses.	$z_1 z_2 = r_1 r_2(\cos\theta_1\cos\theta_2 + i\cos\theta_1\sin\theta_2 + i\sin\theta_1\cos\theta_2 + \underbrace{i^2}_{-1}\sin\theta_1\sin\theta_2)$
Group the real parts and the imaginary parts.	$z_1 z_2 = r_1 r_2[(\cos\theta_1\cos\theta_2 - \sin\theta_1\sin\theta_2) + i(\cos\theta_1\sin\theta_2 + \sin\theta_1\cos\theta_2)]$
Apply the cosine and sine sum identities.	$z_1 z_2 = r_1 r_2\left[\underbrace{(\cos\theta_1\cos\theta_2 - \sin\theta_1\sin\theta_2)}_{\cos(\theta_1 + \theta_2)} + i\underbrace{(\cos\theta_1\sin\theta_2 + \sin\theta_1\cos\theta_2)}_{\sin(\theta_1 + \theta_2)}\right]$
Simplify.	$z_1 z_2 = r_1 r_2[\cos(\theta_1 + \theta_2) + i\sin(\theta_1 + \theta_2)]$

PRODUCT OF TWO COMPLEX NUMBERS

Let $z_1 = r_1(\cos\theta_1 + i\sin\theta_1)$ and $z_2 = r_2(\cos\theta_2 + i\sin\theta_2)$ be two complex numbers. The complex product $z_1 z_2$ is given by

$$z_1 z_2 = r_1 r_2[\cos(\theta_1 + \theta_2) + i\sin(\theta_1 + \theta_2)]$$

In other words, *when multiplying two complex numbers, multiply the magnitudes and add the arguments.*

> **Study Tip**
>
> When two complex numbers are multiplied, the magnitudes are multiplied and the arguments are added.

EXAMPLE 1 Multiplying Complex Numbers

Find the product of $z_1 = 3(\cos 35° + i\sin 35°)$ and $z_2 = 2(\cos 10° + i\sin 10°)$.

Solution:

Set up the product.	$z_1 z_2 = 3(\cos 35° + i\sin 35°) \cdot 2(\cos 10° + i\sin 10°)$
Multiply the magnitudes and add the arguments.	$z_1 z_2 = 3 \cdot 2[\cos(35° + 10°) + i\sin(35° + 10°)]$

> **Classroom Example 8.7.1***
>
> Multiply z_1 and z_2, where m, n are integers and
> $$z_1 = a\left[\cos\left(\frac{n\pi}{6}\right) + i\sin\left(\frac{n\pi}{6}\right)\right],$$
> $$z_2 = 5a\left[\cos\left(\frac{m\pi}{3}\right) + i\sin\left(\frac{m\pi}{3}\right)\right].$$
>
> **Answer:**
> $$5a^2\left[\cos\left(\frac{n + 2m}{6}\pi\right) + i\sin\left(\frac{n + 2m}{6}\pi\right)\right]$$

Simplify.

$$z_1 z_2 = 6(\cos 45° + i \sin 45°)$$

The product is in polar form. To express the product in rectangular form, evaluate the trigonometric functions.

$$z_1 z_2 = 6\left(\frac{\sqrt{2}}{2} + i\frac{\sqrt{2}}{2}\right) = 3\sqrt{2} + 3i\sqrt{2}$$

Product in polar form:

$$z_1 z_2 = 6(\cos 45° + i \sin 45°) = 6\left[\cos\left(\frac{\pi}{4}\right) + i \sin\left(\frac{\pi}{4}\right)\right]$$

Product in rectangular form:

$$z_1 z_2 = 3\sqrt{2} + 3i\sqrt{2}$$

■ **Answer:**

$z_1 z_2 = 10(\cos 120° + i \sin 120°)$
or $z_1 z_2 = -5 + 5i\sqrt{3}$

■ **YOUR TURN** Find the product of $z_1 = 2(\cos 55° + i \sin 55°)$ and $z_2 = 5(\cos 65° + i \sin 65°)$. Express the answer in both polar and rectangular form.

Quotients of Complex Numbers

We now derive a formula for the quotient of two complex numbers.

WORDS	MATH
Start with two complex numbers z_1 and z_2.	$z_1 = r_1(\cos\theta_1 + i\sin\theta_1)$ and $z_2 = r_2(\cos\theta_2 + i\sin\theta_2)$
Divide z_1 by z_2.	$\dfrac{z_1}{z_2} = \dfrac{r_1(\cos\theta_1 + i\sin\theta_1)}{r_2(\cos\theta_2 + i\sin\theta_2)} = \left(\dfrac{r_1}{r_2}\right)\left(\dfrac{\cos\theta_1 + i\sin\theta_1}{\cos\theta_2 + i\sin\theta_2}\right)$
Multiply the numerator and the denominator of the second expression in parentheses by the conjugate of the denominator, $\cos\theta_2 - i\sin\theta_2$.	$\dfrac{z_1}{z_2} = \left(\dfrac{r_1}{r_2}\right)\left(\dfrac{\cos\theta_1 + i\sin\theta_1}{\cos\theta_2 + i\sin\theta_2}\right)\left(\dfrac{\cos\theta_2 - i\sin\theta_2}{\cos\theta_2 - i\sin\theta_2}\right)$
Use the FOIL method to multiply the expressions in parentheses in the last two expressions.	$\dfrac{z_1}{z_2} = \left(\dfrac{r_1}{r_2}\right)\left(\dfrac{\cos\theta_1\cos\theta_2 - i^2\sin\theta_1\sin\theta_2 + i\sin\theta_1\cos\theta_2 - i\sin\theta_2\cos\theta_1}{\cos^2\theta_2 - i^2\sin^2\theta_2}\right)$
Substitute $i^2 = -1$ and group the real parts and the imaginary parts.	$\dfrac{z_1}{z_2} = \left(\dfrac{r_1}{r_2}\right)\left[\dfrac{(\cos\theta_1\cos\theta_2 + \sin\theta_1\sin\theta_2) + i(\sin\theta_1\cos\theta_2 - \sin\theta_2\cos\theta_1)}{\underbrace{\cos^2\theta_2 + \sin^2\theta_2}_{1}}\right]$
Simplify.	$\dfrac{z_1}{z_2} = \left(\dfrac{r_1}{r_2}\right)[(\cos\theta_1\cos\theta_2 + \sin\theta_1\sin\theta_2) + i(\sin\theta_1\cos\theta_2 - \sin\theta_2\cos\theta_1)]$
Use the cosine and sine difference identities.	$\dfrac{z_1}{z_2} = \left(\dfrac{r_1}{r_2}\right)\underbrace{(\cos\theta_1\cos\theta_2 + \sin\theta_1\sin\theta_2)}_{\cos(\theta_1 - \theta_2)} + i\underbrace{(\sin\theta_1\cos\theta_2 - \sin\theta_2\cos\theta_1)}_{\sin(\theta_1 - \theta_2)}$
Simplify.	$\dfrac{z_1}{z_2} = \dfrac{r_1}{r_2}[\cos(\theta_1 - \theta_2) + i\sin(\theta_1 - \theta_2)]$

It is important to notice that the argument difference is the argument of the numerator minus the argument of the denominator.

QUOTIENT OF TWO COMPLEX NUMBERS

Let $z_1 = r_1(\cos\theta_1 + i\sin\theta_1)$ and $z_2 = r_2(\cos\theta_2 + i\sin\theta_2)$ be two complex numbers. The complex quotient $\dfrac{z_1}{z_2}$ is given by

$$\frac{z_1}{z_2} = \frac{r_1}{r_2}[\cos(\theta_1 - \theta_2) + i\sin(\theta_1 - \theta_2)]$$

In other words, *when dividing two complex numbers, divide the magnitudes and subtract the arguments. It is important to note that the argument difference is the argument of the complex number in the numerator minus the argument of the complex number in the denominator.*

Classroom Example 8.7.2*
Divide z_1 and z_2, where m, n are integers and

$$z_1 = a\left[\cos\left(\frac{n\pi}{6}\right) + i\sin\left(\frac{n\pi}{6}\right)\right],$$

$$z_2 = 5a\left[\cos\left(\frac{m\pi}{3}\right) + i\sin\left(\frac{m\pi}{3}\right)\right].$$

Answer:
$$\frac{1}{5}\left[\cos\left(\frac{n - 2m}{6}\pi\right) + i\sin\left(\frac{n - 2m}{6}\pi\right)\right]$$

EXAMPLE 2 Dividing Complex Numbers

Let $z_1 = 6(\cos 125° + i\sin 125°)$ and $z_2 = 3(\cos 65° + i\sin 65°)$. Find $\dfrac{z_1}{z_2}$.

Solution:

Set up the quotient.

$$\frac{z_1}{z_2} = \frac{6(\cos 125° + i\sin 125°)}{3(\cos 65° + i\sin 65°)}$$

Divide the magnitudes and subtract the arguments.

$$\frac{z_1}{z_2} = \frac{6}{3}[\cos(125° - 65°) + i\sin(125° - 65°)]$$

Simplify.

$$\frac{z_1}{z_2} = 2(\cos 60° + i\sin 60°)$$

The quotient is in polar form. To express the quotient in rectangular form, evaluate the trigonometric functions.

$$\frac{z_1}{z_2} = 2\left(\frac{1}{2} + i\frac{\sqrt{3}}{2}\right) = 1 + i\sqrt{3}$$

Polar form:

$$\boxed{\frac{z_1}{z_2} = 2(\cos 60° + i\sin 60°)}$$

Rectangular form:

$$\boxed{\frac{z_1}{z_2} = 1 + i\sqrt{3}}$$

■ **YOUR TURN** Let $z_1 = 10(\cos 275° + i\sin 275°)$ and $z_2 = 5(\cos 65° + i\sin 65°)$, and find $\dfrac{z_1}{z_2}$. Express the answers in both polar and rectangular form.

Technology Tip

Let $z_1 = 6(\cos 125° + i\sin 125°)$ and $z_2 = 3(\cos 65° + i\sin 65°)$; find $\dfrac{z_1}{z_2}$. Be sure to include parentheses for z_1 and z_2.

```
(6(cos(125)+isin
(125)))/(3(cos(6
5)+isin(65)))
      1.0000+1.7321i
√(3)
             1.7321
■
```

■ **Answer:**

$$\frac{z_1}{z_2} = 2(\cos 210° + i\sin 210°)$$

or $\dfrac{z_1}{z_2} = -\sqrt{3} - i$

When multiplying or dividing complex numbers, we have considered only those values of θ such that $0° \leq \theta \leq 360°$. When the value of θ is negative or greater than $360°$, find the coterminal angle in the interval $[0°, 360°]$.

Powers of Complex Numbers

Raising a number to a positive integer power is the same as multiplying that number by itself repeated times.

$$x^3 = x \cdot x \cdot x \qquad (a + b)^2 = (a + b)(a + b)$$

Therefore, raising a complex number to a power that is a positive integer is the same as multiplying the complex number by itself multiple times. Let us illustrate this with the complex number $z = r(\cos\theta + i\sin\theta)$, which we will raise to positive integer powers (n).

WORDS	**MATH**
Take the case **$n = 2$**.	$z^2 = [r(\cos\theta + i\sin\theta)][r(\cos\theta + i\sin\theta)]$
Apply the product rule (multiply the magnitudes and add the arguments).	$z^2 = r^2[\cos(2\theta) + i\sin(2\theta)]$
Take the case **$n = 3$**.	$z^3 = z^2 z = \{r^2[\cos(2\theta) + i\sin(2\theta)]\}[r(\cos\theta + i\sin\theta)]$
Apply the product rule (multiply the magnitudes and add the arguments).	$z^3 = r^3[\cos(3\theta) + i\sin(3\theta)]$
Take the case **$n = 4$**.	$z^4 = z^3 z = \{r^3[\cos(3\theta) + i\sin(3\theta)]\}[r(\cos\theta + i\sin\theta)]$
Apply the product rule (multiply the magnitudes and add the arguments).	$z^4 = r^4[\cos(4\theta) + i\sin(4\theta)]$
The pattern observed for any **n** is:	$z^n = r^n[\cos(n\theta) + i\sin(n\theta)]$

Although we will not prove this generalized representation of a complex number raised to a power, it was proved by Abraham De Moivre, and hence its name.

DE MOIVRE'S THEOREM

If $z = r(\cos\theta + i\sin\theta)$ is a complex number, then

$$z^n = r^n[\cos(n\theta) + i\sin(n\theta)]$$

when n is a positive integer ($n \geq 1$).

In other words, when raising a complex number to a power n, raise the magnitude to the same power n and multiply the argument by n.

Although De Moivre's theorem was proved for all real numbers n, we will only use it for positive integer values of n and their reciprocals (nth roots). This is a very powerful theorem. For example, if asked to find $(\sqrt{3} + i)^{10}$, you have two choices: (1) multiply out the expression algebraically, which we will call the long way or (2) convert to polar coordinates and use De Moivre's theorem, which we will call the short way. We will use De Moivre's theorem.

EXAMPLE 3 **Finding a Power of a Complex Number**

Find $\left(\sqrt{3} + i\right)^{10}$ and express the answer in rectangular form.

Solution:

Technology Tip

Find $\left(\sqrt{3} + i\right)^{10}$ and express the answer in rectangular form.

STEP 1 Convert to polar form. $\left(\sqrt{3} + i\right)^{10} = \left[2(\cos 30° + i\sin 30°)\right]^{10}$

STEP 2 Apply De Moivre's theorem $\left(\sqrt{3} + i\right)^{10} = \left[2(\cos 30° + i\sin 30°)\right]^{10}$
with $n = 10$.
$$= 2^{10}\left[\cos(10 \cdot 30°) + i\sin(10 \cdot 30°)\right]$$

```
((√(3)+i)^10
512-886.8100135i
(2(cos(30)+isin(
30))^10
512-886.8100135i
```

STEP 3 Simplify. $\left(\sqrt{3} + i\right)^{10} = 2^{10}(\cos 300° + i\sin 300°)$

Evaluate 2^{10} and the sine
and cosine functions.
$$= 1024\left(\frac{1}{2} - i\frac{\sqrt{3}}{2}\right)$$

Simplify.
$$\boxed{= 512 - 512i\sqrt{3}}$$

■ **Answer:** $-512 - 512i\sqrt{3}$

■ **YOUR TURN** Find $\left(1 + i\sqrt{3}\right)^{10}$ and express the answer in rectangular form.

Roots of Complex Numbers

De Moivre's theorem is the basis for the *nth root theorem*. Before we proceed, let us motivate it with a problem: Solve $x^3 - 1 = 0$. Recall that a polynomial of degree n has n solutions (roots in the complex number system). So the polynomial $P(x) = x^3 - 1$ is of degree 3 and has three solutions (roots). We can solve it algebraically.

WORDS	**MATH**
List the potential rational roots of the polynomial $P(x) = x^3 - 1$.	$x = \pm 1$

Use synthetic division to test $x = 1$.

$$1 \,\big|\, \begin{array}{rrrr} 1 & 0 & 0 & -1 \\ & 1 & 1 & 1 \\ \hline 1 & 1 & 1 & \boxed{0} \end{array}$$
$$\underbrace{}_{x^2 + x + 1}$$

Since $x = 1$ is a zero, then the polynomial
can be written as a product of the linear $P(x) = (x - 1)(x^2 + x + 1)$
factor $(x - 1)$ and a quadratic factor.

Use the quadratic formula on $x^2 + x + 1 = 0$ $x = \dfrac{-1 \pm \sqrt{1 - 4}}{2} = \dfrac{-1 \pm \sqrt{-3}}{2} = -\dfrac{1}{2} \pm \dfrac{i\sqrt{3}}{2}$
to solve for x.

So the three solutions to the equation $x^3 - 1 = 0$ are $\boxed{x = 1, x = -\dfrac{1}{2} + \dfrac{i\sqrt{3}}{2}, \text{ and } x = -\dfrac{1}{2} - \dfrac{i\sqrt{3}}{2}}$.

An alternative approach to solving $x^3 - 1 = 0$ is to use the *nth root theorem* to find the additional complex cube roots of 1.

Derivation of the *n*th Root Theorem

WORDS	**MATH**
Let z and w be complex numbers such that w is the nth root of z.	$w = z^{1/n}$ or $w = \sqrt[n]{z}$, where n is a positive integer
Raise both sides of the equation to the nth power.	$w^n = z$
Let $z = r(\cos\theta + i\sin\theta)$ and $w = s(\cos\alpha + i\sin\alpha)$.	$[s(\cos\alpha + i\sin\alpha)]^n = r(\cos\theta + i\sin\theta)$
Apply De Moivre's theorem to the left side of the equation.	$s^n[\cos(n\alpha) + i\sin(n\alpha)] = r(\cos\theta + i\sin\theta)$
For these two expressions to be equal, their magnitudes must be equal and their angles must be coterminal.	$s^n = r$ and $n\alpha = \theta + 2k\pi$, where k is any integer
Solve for s and α.	$s = r^{1/n}$ and $\alpha = \dfrac{\theta + 2k\pi}{n}$
Substitute $s = r^{1/n}$ and $\alpha = \dfrac{\theta + 2k\pi}{n}$ into $w = z^{1/n}$.	$z^{1/n} = r^{1/n}\left[\cos\left(\dfrac{\theta + 2k\pi}{n}\right) + i\sin\left(\dfrac{\theta + 2k\pi}{n}\right)\right]$

Notice that when $k = n$, the arguments $\dfrac{\theta}{n} + 2\pi$ and $\dfrac{\theta}{n}$ are coterminal. Therefore, to get distinct roots, let $k = 0, 1, \ldots, n - 1$. If we let z be a given complex number and w be any complex number that satisfies the relationship $z^{1/n} = w$ or $z = w^n$, where $n \geq 2$, then we say that w is a **complex nth root** of z.

NTH ROOT THEOREM

The **nth roots** of the complex number $z = r(\cos\theta + i\sin\theta)$ are given by

$$w_k = r^{1/n}\left[\cos\left(\frac{\theta}{n} + \frac{2k\pi}{n}\right) + i\sin\left(\frac{\theta}{n} + \frac{2k\pi}{n}\right)\right] \qquad \theta \text{ in radians}$$

or

$$w_k = r^{1/n}\left[\cos\left(\frac{\theta}{n} + \frac{k \cdot 360°}{n}\right) + i\sin\left(\frac{\theta}{n} + \frac{k \cdot 360°}{n}\right)\right] \qquad \theta \text{ in degrees}$$

where $k = 0, 1, 2, \ldots, n - 1$.

EXAMPLE 4   **Finding Roots of Complex Numbers**

Find the three distinct cube roots of $-4 - 4i\sqrt{3}$, and plot the roots in the complex plane.

Solution:

STEP 1   Write $-4 - 4i\sqrt{3}$ in polar form. $8(\cos 240° + i\sin 240°)$

STEP 2   Find the three cube roots.

$$w_k = r^{1/n}\left[\cos\left(\frac{\theta}{n} + \frac{k \cdot 360°}{n}\right) + i\sin\left(\frac{\theta}{n} + \frac{k \cdot 360°}{n}\right)\right]$$

$$\theta = 240°, \; r = 8, \; n = 3, \; k = 0, 1, 2$$

$k = 0$:

$$w_0 = 8^{1/3}\left[\cos\left(\frac{240°}{3} + \frac{0 \cdot 360°}{3}\right) + i\sin\left(\frac{240°}{3} + \frac{0 \cdot 360°}{3}\right)\right]$$

Simplify.

$$\boxed{w_0 = 2(\cos 80° + i\sin 80°)}$$

$k = 1$:

$$w_1 = 8^{1/3}\left[\cos\left(\frac{240°}{3} + \frac{1 \cdot 360°}{3}\right) + i\sin\left(\frac{240°}{3} + \frac{1 \cdot 360°}{3}\right)\right]$$

Simplify.

$$\boxed{w_1 = 2(\cos 200° + i\sin 200°)}$$

$k = 2$:

$$w_2 = 8^{1/3}\left[\cos\left(\frac{240°}{3} + \frac{2 \cdot 360°}{3}\right) + i\sin\left(\frac{240°}{3} + \frac{2 \cdot 360°}{3}\right)\right]$$

Simplify.

$$\boxed{w_2 = 2(\cos 320° + i\sin 320°)}$$

STEP 3   Plot the three cube roots in the complex plane.

Notice the following:

- The roots all have a magnitude of 2.

- The roots lie on a circle of radius 2.

- The roots are equally spaced around the circle (120° apart).

···

■ **YOUR TURN**   Find the three distinct cube roots of $4 - 4i\sqrt{3}$, and plot the roots in the complex plane.

Technology Tip

Find the three distinct roots of $-4 - 4i\sqrt{3}$.

Caution: If you use a TI calculator to find $(-4 - 4i\sqrt{3})^{1/3}$, the calculator will return only one root.

To find all three distinct roots, you need to change to polar form and apply the nth root theorem.

·······························

■ **Answer:**
$w_0 = 2(\cos 100° + i\sin 100°)$
$w_1 = 2(\cos 220° + i\sin 220°)$
$w_2 = 2(\cos 340° + i\sin 340°)$

Solving Equations Using Roots of Complex Numbers

Let us return to solving the equation $x^3 - 1 = 0$. As stated, $x = 1$ is the real solution to this cubic equation. However, there are two additional (complex) solutions. Since we are finding the zeros of a third-degree polynomial, we expect three solutions. Furthermore, when complex solutions arise in finding the roots of polynomials, they come in conjugate pairs.

EXAMPLE 5 Solving Equations Using Complex Roots

Find all complex solutions to $x^3 - 1 = 0$.

Solution: $x^3 = 1$

STEP 1 Write 1 in polar form. $1 = 1 + 0i = \cos 0° + i \sin 0°$

STEP 2 Find the three cube roots of 1.

$$w_k = r^{1/n}\left[\cos\left(\frac{\theta}{n} + \frac{k \cdot 360°}{n}\right) + i \sin\left(\frac{\theta}{n} + \frac{k \cdot 360°}{n}\right)\right]$$

$$r = 1, \theta = 0°, n = 3, k = 0, 1, 2$$

$k = 0$: $w_0 = 1^{1/3}\left[\cos\left(\frac{0°}{3} + \frac{0 \cdot 360°}{3}\right) + i \sin\left(\frac{0°}{3} + \frac{0 \cdot 360°}{3}\right)\right]$

Simplify. $w_0 = \cos 0° + i \sin 0°$

$k = 1$: $w_1 = 1^{1/3}\left[\cos\left(\frac{0°}{3} + \frac{1 \cdot 360°}{3}\right) + i \sin\left(\frac{0°}{3} + \frac{1 \cdot 360°}{3}\right)\right]$

Simplify. $w_1 = \cos 120° + i \sin 120°$

$k = 2$: $w_2 = 1^{1/3}\left[\cos\left(\frac{0°}{3} + \frac{2 \cdot 360°}{3}\right) + i \sin\left(\frac{0°}{3} + \frac{2 \cdot 360°}{3}\right)\right]$

Simplify. $w_2 = \cos 240° + i \sin 240°$

STEP 3 Write the roots in rectangular form.

w_0: $w_0 = \underbrace{\cos 0°}_{1} + \underbrace{i \sin 0°}_{0} = 1$

w_1: $w_1 = \underbrace{\cos 120°}_{-\frac{1}{2}} + \underbrace{i \sin 120°}_{\frac{\sqrt{3}}{2}} = -\frac{1}{2} + i\frac{\sqrt{3}}{2}$

w_2: $w_2 = \underbrace{\cos 240°}_{-\frac{1}{2}} + \underbrace{i \sin 240°}_{-\frac{\sqrt{3}}{2}} = -\frac{1}{2} - i\frac{\sqrt{3}}{2}$

STEP 4 Write the solutions to the equation $x^3 - 1 = 0$.

$$\boxed{x = 1} \qquad \boxed{x = -\frac{1}{2} + i\frac{\sqrt{3}}{2}} \qquad \boxed{x = -\frac{1}{2} - i\frac{\sqrt{3}}{2}}$$

Notice that there is one real solution and there are two (nonreal) complex solutions and that the two complex solutions are complex conjugates.

It is always a good idea to check that the solutions indeed satisfy the equation. The equation $x^3 - 1 = 0$ can also be written as $x^3 = 1$, so the check in this case is to cube the three solutions and confirm that the result is 1.

$x = 1$:
$$1^3 = 1 \; \checkmark$$

$x = -\dfrac{1}{2} + \dfrac{i\sqrt{3}}{2}$:
$$\left(-\frac{1}{2} + i\frac{\sqrt{3}}{2}\right)^3 = \left(-\frac{1}{2} + i\frac{\sqrt{3}}{2}\right)^2\left(-\frac{1}{2} + i\frac{\sqrt{3}}{2}\right)$$
$$= \left(-\frac{1}{2} - i\frac{\sqrt{3}}{2}\right)\left(-\frac{1}{2} + i\frac{\sqrt{3}}{2}\right)$$
$$= \frac{1}{4} + \frac{3}{4}$$
$$= 1 \; \checkmark$$

$x = -\dfrac{1}{2} - \dfrac{i\sqrt{3}}{2}$:
$$\left(-\frac{1}{2} - i\frac{\sqrt{3}}{2}\right)^3 = \left(-\frac{1}{2} - i\frac{\sqrt{3}}{2}\right)^2\left(-\frac{1}{2} - i\frac{\sqrt{3}}{2}\right)$$
$$= \left(-\frac{1}{2} + i\frac{\sqrt{3}}{2}\right)\left(-\frac{1}{2} - i\frac{\sqrt{3}}{2}\right)$$
$$= \frac{1}{4} + \frac{3}{4}$$
$$= 1 \; \checkmark$$

SECTION 8.7 SUMMARY

In this section we multiplied and divided complex numbers and, using De Moivre's theorem, raised complex numbers to integer powers and found the nth roots of complex numbers, as follows:

Let $z_1 = r_1(\cos\theta_1 + i\sin\theta_1)$ and $z_2 = r_2(\cos\theta_2 + i\sin\theta_2)$ be two complex numbers.

The **product** $z_1 z_2$ is given by
$$z_1 z_2 = r_1 r_2[\cos(\theta_1 + \theta_2) + i\sin(\theta_1 + \theta_2)]$$

The **quotient** $\dfrac{z_1}{z_2}$ is given by
$$\frac{z_1}{z_2} = \frac{r_1}{r_2}[\cos(\theta_1 - \theta_2) + i\sin(\theta_1 - \theta_2)]$$

Let $z = r(\cos\theta + i\sin\theta)$ be a complex number. Then for a positive integer n,

z raised to a **power** n is given by
$$z^n = r^n[\cos(n\theta) + i\sin(n\theta)]$$

The **nth roots** of z are given by
$$w_k = r^{1/n}\left[\cos\left(\frac{\theta}{n} + \frac{k \cdot 360°}{n}\right) + i\sin\left(\frac{\theta}{n} + \frac{k \cdot 360°}{n}\right)\right]$$ where θ is in degrees or

$$w_k = r^{1/n}\left[\cos\left(\frac{\theta}{n} + \frac{k \cdot 2\pi}{n}\right) + i\sin\left(\frac{\theta}{n} + \frac{k \cdot 2\pi}{n}\right)\right]$$ where θ is in radians

$$k = 0, 1, 2, \ldots, n - 1.$$

■ **SKILLS**

In Exercises 1–10, find the product z_1z_2 and express it in rectangular form.

1. $z_1 = 4(\cos 40° + i\sin 40°)$ and $z_2 = 3(\cos 80° + i\sin 80°)$

2. $z_1 = 2(\cos 100° + i\sin 100°)$ and $z_2 = 5(\cos 50° + i\sin 50°)$

3. $z_1 = 4(\cos 80° + i\sin 80°)$ and $z_2 = 2(\cos 145° + i\sin 145°)$

4. $z_1 = 3(\cos 130° + i\sin 130°)$ and $z_2 = 4(\cos 170° + i\sin 170°)$

5. $z_1 = 2(\cos 10° + i\sin 10°)$ and $z_2 = 4(\cos 80° + i\sin 80°)$

6. $z_1 = 3(\cos 190° + i\sin 190°)$ and $z_2 = 5(\cos 80° + i\sin 80°)$

7. $z_1 = \sqrt{3}\left[\cos\left(\frac{\pi}{12}\right) + i\sin\left(\frac{\pi}{12}\right)\right]$ and $z_2 = \sqrt{27}\left[\cos\left(\frac{\pi}{6}\right) + i\sin\left(\frac{\pi}{6}\right)\right]$

8. $z_1 = \sqrt{5}\left[\cos\left(\frac{\pi}{15}\right) + i\sin\left(\frac{\pi}{15}\right)\right]$ and $z_2 = \sqrt{5}\left[\cos\left(\frac{4\pi}{15}\right) + i\sin\left(\frac{4\pi}{15}\right)\right]$

9. $z_1 = 4\left[\cos\left(\frac{3\pi}{8}\right) + i\sin\left(\frac{3\pi}{8}\right)\right]$ and $z_2 = 3\left[\cos\left(\frac{\pi}{8}\right) + i\sin\left(\frac{\pi}{8}\right)\right]$

10. $z_1 = 6\left[\cos\left(\frac{2\pi}{9}\right) + i\sin\left(\frac{2\pi}{9}\right)\right]$ and $z_2 = 5\left[\cos\left(\frac{\pi}{9}\right) + i\sin\left(\frac{\pi}{9}\right)\right]$

In Exercises 11–20, find the quotient $\frac{z_1}{z_2}$ and express it in rectangular form.

11. $z_1 = 6(\cos 100° + i\sin 100°)$ and $z_2 = 2(\cos 40° + i\sin 40°)$

12. $z_1 = 8(\cos 80° + i\sin 80°)$ and $z_2 = 2(\cos 35° + i\sin 35°)$

13. $z_1 = 10(\cos 200° + i\sin 200°)$ and $z_2 = 5(\cos 65° + i\sin 65°)$

14. $z_1 = 4(\cos 280° + i\sin 280°)$ and $z_2 = 4(\cos 55° + i\sin 55°)$

15. $z_1 = \sqrt{12}(\cos 350° + i\sin 350°)$ and $z_2 = \sqrt{3}(\cos 80° + i\sin 80°)$

16. $z_1 = \sqrt{40}(\cos 110° + i\sin 110°)$ and $z_2 = \sqrt{10}(\cos 20° + i\sin 20°)$

17. $z_1 = 9\left[\cos\left(\frac{5\pi}{12}\right) + i\sin\left(\frac{5\pi}{12}\right)\right]$ and $z_2 = 3\left[\cos\left(\frac{\pi}{12}\right) + i\sin\left(\frac{\pi}{12}\right)\right]$

18. $z_1 = 8\left[\cos\left(\frac{5\pi}{8}\right) + i\sin\left(\frac{5\pi}{8}\right)\right]$ and $z_2 = 4\left[\cos\left(\frac{3\pi}{8}\right) + i\sin\left(\frac{3\pi}{8}\right)\right]$

19. $z_1 = 45\left[\cos\left(\frac{22\pi}{15}\right) + i\sin\left(\frac{22\pi}{15}\right)\right]$ and $z_2 = 9\left[\cos\left(\frac{2\pi}{15}\right) + i\sin\left(\frac{2\pi}{15}\right)\right]$

20. $z_1 = 22\left[\cos\left(\frac{11\pi}{18}\right) + i\sin\left(\frac{11\pi}{18}\right)\right]$ and $z_2 = 11\left[\cos\left(\frac{5\pi}{18}\right) + i\sin\left(\frac{5\pi}{18}\right)\right]$

In Exercises 21–30, find the result of each expression using De Moivre's theorem. Write the answer in rectangular form.

21. $(-1 + i)^5$ **22.** $(1 - i)^4$ **23.** $(-\sqrt{3} + i)^6$ **24.** $(\sqrt{3} - i)^8$

25. $(1 - \sqrt{3}i)^4$ **26.** $(-1 + \sqrt{3}i)^5$ **27.** $(4 - 4i)^8$ **28.** $(-3 + 3i)^{10}$

29. $(4\sqrt{3} + 4i)^7$ **30.** $(-5 + 5\sqrt{3}i)^7$

In Exercises 31–40, find all *n*th roots of *z*. Write the answers in polar form, and plot the roots in the complex plane.

31. $2 - 2i\sqrt{3}$, $n = 2$
32. $2 + 2\sqrt{3}i$, $n = 2$
33. $\sqrt{18} - \sqrt{18}i$, $n = 2$
34. $-\sqrt{2} + \sqrt{2}i$, $n = 2$

35. $4 + 4\sqrt{3}i$, $n = 3$
36. $-\dfrac{27}{2} + \dfrac{27\sqrt{3}}{2}i$, $n = 3$
37. $\sqrt{3} - i$, $n = 3$
38. $4\sqrt{2} + 4\sqrt{2}i$, $n = 3$

39. $8\sqrt{2} - 8\sqrt{2}i$, $n = 4$
40. $-\sqrt{128} + \sqrt{128}i$, $n = 4$

In Exercises 41–56, find all complex solutions to the given equations.

41. $x^4 - 16 = 0$
42. $x^3 - 8 = 0$
43. $x^3 + 8 = 0$
44. $x^3 + 1 = 0$

45. $x^4 + 16 = 0$
46. $x^6 + 1 = 0$
47. $x^6 - 1 = 0$
48. $4x^2 + 1 = 0$

49. $x^2 + i = 0$
50. $x^2 - i = 0$
51. $x^4 - 2i = 0$
52. $x^4 + 2i = 0$

53. $x^5 + 32 = 0$
54. $x^5 - 32 = 0$
55. $x^7 - \pi^{14}i = 0$
56. $x^7 + \pi^{14} = 0$

▪ APPLICATIONS

57. Complex Pentagon. When you graph the five fifth roots of $-\dfrac{\sqrt{2}}{2} - \dfrac{\sqrt{2}}{2}i$ and connect the points, you form a pentagon. Find the roots and draw the pentagon.

58. Complex Square. When you graph the four fourth roots of $16i$ and connect the points, you form a square. Find the roots and draw the square.

59. Hexagon. Compute the six sixth roots of $\dfrac{1}{2} - \dfrac{\sqrt{3}}{2}i$, and form a hexagon by connecting successive roots.

60. Octagon. Compute the eight eighth roots of $2i$, and form an octagon by connecting successive roots.

▪ CATCH THE MISTAKE

In Exercises 61–64, explain the mistake that is made.

61. Let $z_1 = 6(\cos 65° + i\sin 65°)$ and
$z_2 = 3(\cos 125° + i\sin 125°)$. Find $\dfrac{z_1}{z_2}$.

Solution:

Use the quotient formula.

$$\frac{z_1}{z_2} = \frac{r_1}{r_2}[\cos(\theta_1 - \theta_2) + i\sin(\theta_1 - \theta_2)]$$

Substitute values.

$$\frac{z_1}{z_2} = \frac{6}{3}[\cos(125° - 65°) + i\sin(125° - 65°)]$$

Simplify. $\dfrac{z_1}{z_2} = 2(\cos 60° + i\sin 60°)$

Evaluate the trigonometric functions.

$$\frac{z_1}{z_2} = 2\left(\frac{1}{2} + i\frac{\sqrt{3}}{2}\right) = 1 + i\sqrt{3}$$

This is incorrect. What mistake was made?

62. Let $z_1 = 6(\cos 65° + i\sin 65°)$ and
$z_2 = 3(\cos 125° + i\sin 125°)$. Find $z_1 z_2$.

Solution:

Write the product.

$$z_1 z_2 = 6(\cos 65° + i\sin 65°) \cdot 3(\cos 125° + i\sin 125°)$$

Multiply the magnitudes.

$$z_1 z_2 = 18(\cos 65° + i\sin 65°)(\cos 125° + i\sin 125°)$$

Multiply the cosine terms and sine terms (add the arguments).

$$z_1 z_2 = 18[\cos(65° + 125°) + i^2\sin(65° + 125°)]$$

Simplify ($i^2 = -1$).

$$z_1 z_2 = 18(\cos 190° - \sin 190°)$$

This is incorrect. What mistake was made?

63. Find $\left(\sqrt{2} + i\sqrt{2}\right)^6$.

Solution:

Raise each term to the sixth power. $\qquad \left(\sqrt{2}\right)^6 + i^6\left(\sqrt{2}\right)^6$

Simplify. $\qquad 8 + 8i^6$

Let $i^6 = i^4 \cdot i^2 = -1$. $\qquad 8 - 8 = 0$

This is incorrect. What mistake was made?

64. Find all complex solutions to $x^5 - 1 = 0$.

Solution:

Add 1 to both sides. $\qquad x^5 = 1$

Raise both sides to the fifth power. $\qquad x = 1^{1/5}$

Simplify. $\qquad x = 1$

This is incorrect. What mistake was made?

▪ CONCEPTUAL

In Exercises 65–70, determine whether the statement is true or false.

65. The product of two complex numbers is a complex number.

66. The quotient of two complex numbers is a complex number.

67. There are always n distinct real solutions of the equation $x^n - a = 0$, where a is not zero.

68. There are always n distinct complex solutions of the equation $x^n - a = 0$, where a is not zero.

69. There are n distinct complex roots of $\dfrac{1}{a + bi}$, where a and b are positive real numbers.

70. There exists a complex number for which there is no complex square root.

▪ CHALLENGE

In Exercises 71–74, use the following identity:

In calculus you will see an identity called Euler's formula or identity, $e^{i\theta} = \cos\theta + i\sin\theta$. Notice that when $\theta = \pi$, the identity reduces to $e^{i\pi} + 1 = 0$, which is a beautiful identity in that it relates the five fundamental numbers (e, π, 1, i, and 0) and the fundamental operations (multiplication, addition, exponents, and equality) in mathematics.

71. Let $z_1 = r_1(\cos\theta_1 + i\sin\theta_1) = r_1e^{i\theta_1}$ and $z_2 = r_2(\cos\theta_2 + i\sin\theta_2) = r_2e^{i\theta_2}$ be two complex numbers. Use the properties of exponentials to show that $z_1z_2 = r_1r_2[\cos(\theta_1 + \theta_2) + i\sin(\theta_1 + \theta_2)]$.

72. Let $z_1 = r_1(\cos\theta_1 + i\sin\theta_1) = r_1e^{i\theta_1}$ and $z_2 = r_2(\cos\theta_2 + i\sin\theta_2) = r_2e^{i\theta_2}$ be two complex numbers. Use the properties of exponentials to show that $\dfrac{z_1}{z_2} = \dfrac{r_1}{r_2}[\cos(\theta_1 - \theta_2) + i\sin(\theta_1 - \theta_2)]$.

73. Let $z = r(\cos\theta + i\sin\theta) = re^{i\theta}$. Use the properties of exponents to show that $z^n = r^n(\cos n\theta + i\sin n\theta)$.

74. Let $z = r(\cos\theta + i\sin\theta) = re^{i\theta}$. Use the properties of exponents to show that
$$w_k = r^{1/n}\left[\cos\left(\frac{\theta}{n} + \frac{2k\pi}{n}\right) + i\sin\left(\frac{\theta}{n} + \frac{2k\pi}{n}\right)\right].$$

75. Use De Moivre's theorem to prove the identity $\cos 2\theta = \cos^2\theta - \sin^2\theta$.

76. Use De Moivre's theorem to derive an expression for $\sin 3\theta$.

77. Use De Moivre's theorem to derive an expression for $\cos 3\theta$.

78. Calculate $\dfrac{\left(\dfrac{1}{2} + \dfrac{\sqrt{3}}{2}i\right)^{14}}{\left(\dfrac{1}{2} - \dfrac{\sqrt{3}}{2}i\right)^{20}}$.

79. Calculate $(1 - i)^n \cdot (1 + i)^m$, where n and m are positive integers.

80. Calculate $\dfrac{(1 + i)^n}{(1 - i)^m}$, where n and m are positive integers.

▪ TECHNOLOGY

For Exercises 81–86, refer to the following:

According to the nth root theorem, the first of the nth roots of the complex number $z = r(\cos\theta + i\sin\theta)$ is given by

$w_1 = r^{1/n}\left[\cos\left(\dfrac{\theta}{n} + \dfrac{2\pi}{n}\right) + i\sin\left(\dfrac{\theta}{n} + \dfrac{2\pi}{n}\right)\right]$, with θ in radians

or $w_1 = r^{1/n}\left[\cos\left(\dfrac{\theta}{n} + \dfrac{360°}{n}\right) + i\sin\left(\dfrac{\theta}{n} + \dfrac{360°}{n}\right)\right]$,

with θ in degrees.

Using the graphing calculator to plot the n roots of a complex number z, enter $r_1 = r$, $\theta\min = \dfrac{\theta}{n}$, $\theta\max = 2\pi + \dfrac{\theta}{n}$ or $360° + \dfrac{\theta}{n}$, $\theta\text{step} = \dfrac{2\pi}{n}$ or $\dfrac{360°}{n}$, $x\min = -r$, $x\max = r$, $y\min = -r$, $y\max = r$, and $\boxed{\text{MODE}}$ in radians or degrees.

81. Find the fifth roots of $\dfrac{\sqrt{3}}{2} - \dfrac{1}{2}i$, and plot the roots with a calculator.

82. Find the fourth roots of $-\dfrac{\sqrt{2}}{2} + \dfrac{\sqrt{2}}{2}i$, and plot the roots with a calculator.

83. Find the sixth roots of $-\dfrac{1}{2} - \dfrac{\sqrt{3}}{2}i$, and draw the complex hexagon with a calculator.

84. Find the fifth roots of $-4 + 4i$, and draw the complex pentagon with a calculator.

85. Find the cube roots of $\dfrac{27\sqrt{2}}{2} + \dfrac{27\sqrt{2}}{2}i$, and draw the complex triangle with a calculator.

86. Find the fifth roots of $8\sqrt{2}(\sqrt{3} - 1) + 8\sqrt{2}(\sqrt{3} - 1)i$, and draw the complex triangle with a calculator.

SECTION 8.8	POLAR EQUATIONS AND GRAPHS

SKILLS OBJECTIVES

- Plot points in the polar coordinate system.
- Convert between rectangular and polar coordinates.
- Convert equations between polar form and rectangular form.
- Graph polar equations.

CONCEPTUAL OBJECTIVES

- Relate the rectangular coordinate system to the polar coordinate system.
- Classify common shapes that arise from plotting certain types of polar equations.

We have discussed the rectangular and the trigonometric (polar) form of complex numbers in the complex plane. We now turn our attention back to the familiar Cartesian plane, where the horizontal axis represents the x-variable, the vertical axis represents the y-variable, and points in this plane represent pairs of real numbers. It is often convenient to instead represent real-number plots in the *polar coordinate system*.

Polar Coordinates

The **polar coordinate system** is anchored by a point, called the **pole** (taken to be the **origin**), and a ray with a vertex at the pole, called the **polar axis**. The polar axis is normally shown where we expect to find the positive x-axis in Cartesian coordinates.

If you align the pole with the origin on the rectangular graph and the polar axis with the positive x-axis, you can label a point either with rectangular coordinates (x, y) or with an ordered pair (r, θ) in **polar coordinates**.

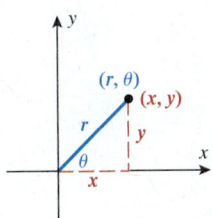

Typically, polar graph paper is used that gives the angles and radii. The graph below gives the angles in radians (the angle also can be given in degrees) and shows the radii from 0 through 5.

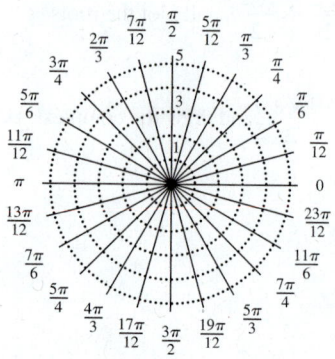

When plotting points in the polar coordinate system, $|r|$ represents the distance from the origin to the point. The following procedure guides us in plotting points in the polar coordinate system.

POINT-PLOTTING POLAR COORDINATES

To plot a point (r, θ):

1. Start on the polar axis and rotate the terminal side of an angle to the value θ.

2. If $r > 0$, the point is r units from the origin in the *same direction* of the terminal side of θ.

3. If $r < 0$, the point is $|r|$ units from the origin in the *opposite direction* of the terminal side of θ.

In polar form it is important to note that (r, θ), the name of the point, is not unique, whereas in rectangular form (x, y) it is unique. For example, $(2, 30°) = (-2, 210°)$.

Classroom Example 8.8.1

Plot the polar points

$$\left(1, \frac{7\pi}{4}\right), \left(\sqrt{2}, 150°\right).$$

Answer:

EXAMPLE 1 Plotting Points in the Polar Coordinate System

Plot the points in a polar coordinate system.

a. $\left(3, \frac{3\pi}{4}\right)$ **b.** $(-2, 60°)$

Solution (a):

Start by placing a pencil along the polar axis.

Rotate the pencil to the angle $\frac{3\pi}{4}$.

Go out (in the direction of the pencil) 3 units.

Solution (b):

Start by placing a pencil along the polar axis.

Rotate the pencil to the angle 60°.

Go out (opposite the direction of
the pencil) 2 units.

■ **YOUR TURN** Plot the points in the polar coordinate system.

a. $\left(-4, \dfrac{3\pi}{2}\right)$ **b.** $(3, 330°)$

■ **Answer:**

Converting Between Polar
and Rectangular Coordinates

The relationships between polar and rectangular coordinates are the familiar ones:

$$\sin\theta = \frac{y}{r}$$

$$\cos\theta = \frac{x}{r}$$

$$\tan\theta = \frac{y}{x} \quad (x \neq 0)$$

$$r^2 = x^2 + y^2$$

**CONVERTING BETWEEN POLAR
AND RECTANGULAR COORDINATES**

FROM	TO	IDENTITIES
Polar (r, θ)	Rectangular (x, y)	$x = r\cos\theta \qquad y = r\sin\theta$
Rectangular (x, y)	Polar (r, θ)	$r = \sqrt{x^2 + y^2} \qquad \tan\theta = \dfrac{y}{x} \quad (x \neq 0)$ Make sure that θ is in the correct quadrant.

Classroom Example 8.8.2*

a. Convert $\left(\dfrac{a\sqrt{34}}{5}, \dfrac{7\pi}{6}\right)$ to rectangular coordinates (assume $a > 0$).

b. Convert $\left(\dfrac{\sqrt{2}}{3}a, -\dfrac{\sqrt{2}}{3}ai\right)$ to polar coordinates.

Answer:

a. $\left(-\dfrac{\sqrt{102}}{10}a, -\dfrac{a\sqrt{34}}{10}\right)$

b. $\left(\dfrac{2}{3}a, \dfrac{7\pi}{4}\right)$

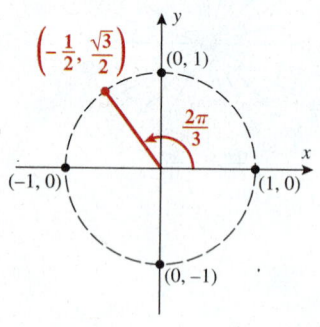

EXAMPLE 2   Converting Between Polar and Rectangular Coordinates

a. Convert $(-1, \sqrt{3})$ to polar coordinates.

b. Convert $(6\sqrt{2}, 135°)$ to rectangular coordinates.

Solution (a): $(-1, \sqrt{3})$ lies in quadrant II.

Identify x and y. $x = -1, \qquad y = \sqrt{3}$

Find r. $r = \sqrt{x^2 + y^2} = \sqrt{(-1)^2 + (\sqrt{3})^2} = \sqrt{4} = 2$

Find θ. $\tan\theta = \dfrac{\sqrt{3}}{-1}$ (θ lies in quadrant II)

Identify θ from the unit circle. $\theta = \dfrac{2\pi}{3}$

Write the point in polar coordinates. $\boxed{\left(2, \dfrac{2\pi}{3}\right)}$

Note: Other polar coordinates like $\left(2, -\dfrac{4\pi}{3}\right)$ and $\left(-2, \dfrac{5\pi}{3}\right)$ also correspond to the point $(-1, \sqrt{3})$.

Solution (b): $(6\sqrt{2}, 135°)$ lies in quadrant II.

Identify r and θ. $r = 6\sqrt{2} \qquad \theta = 135°$

Find x. $x = r\cos\theta = 6\sqrt{2}\cos 135° = 6\sqrt{2}\left(-\dfrac{\sqrt{2}}{2}\right) = -6$

Find y. $y = r\sin\theta = 6\sqrt{2}\sin 135° = 6\sqrt{2}\left(\dfrac{\sqrt{2}}{2}\right) = 6$

Write the point in rectangular coordinates. $\boxed{(-6, 6)}$

Graphs of Polar Equations

Technology Tip

```
abs(-1+√(3)i)
                2.00
angle(-1+√(3)i)
              120.00
Ans/180▸Frac
                 2/3
■
```

The solution is $\left(2, \dfrac{2}{3}\pi\right)$.

We are familiar with equations in rectangular form such as

$$y = 3x + 5 \qquad y = x^2 + 2 \qquad x^2 + y^2 = 9$$
$$\text{(line)} \qquad\quad \text{(parabola)} \qquad \text{(circle)}$$

We now discuss equations in polar form (known as **polar equations**) such as

$$r = 5\theta \qquad r = 2\cos\theta \qquad r = \sin(5\theta)$$

which you will learn to recognize in this section as typical equations whose plots are some general shapes.

Our first example deals with two of the simplest forms of polar equations: when r or θ is constant. The results are a circle centered at the origin and a line that passes through the origin, respectively.

EXAMPLE 3 **Graphing a Polar Equation of the Form $r =$ Constant or $\theta =$ Constant**

Graph the polar equations.

a. $r = 3$ **b.** $\theta = \dfrac{\pi}{4}$

Solution (a): Constant value of r

Approach 1 (polar coordinates): $r = 3$ (θ can take on any value)

Plot points for arbitrary θ and $r = 3$.

Connect the points; a circle with radius 3.

Approach 2 (rectangular coordinates): $r = 3$

Square both sides. $r^2 = 9$

Remember that in rectangular
coordinates $r^2 = x^2 + y^2$. $x^2 + y^2 = 3^2$

This is a circle, centered at
the origin, with radius 3.

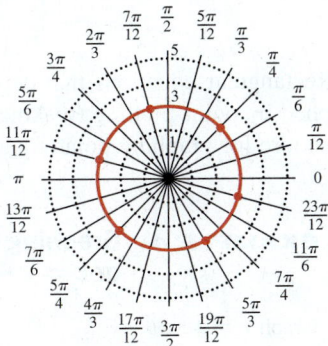

Solution (b): Constant value of θ

Approach 1: $\theta = \dfrac{\pi}{4}$ (r can take on any value, positive or negative)

Plot points for $\theta = \dfrac{\pi}{4}$ at several arbitrary values of r.

Connect the points. The result is a line passing through the origin with

slope $= 1 \left[m = \tan\left(\dfrac{\pi}{4}\right) \right]$.

Technology Tip

a. Graph the polar equation $r = 3$.
To use the TI calculator, set it in
$\boxed{Radian}$ mode and $\boxed{Pol}$ mode.

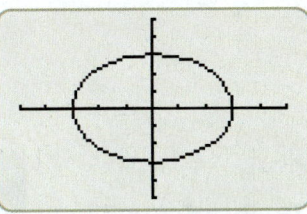

With the window setting, the circle
is displayed as an oval. To display
the circle properly, press

$\boxed{ZOOM}$ $\boxed{\blacktriangledown}$ $\boxed{5:ZSquare}$ $\boxed{ENTER}$.

b. Now graph the polar equation
$\theta = \dfrac{\pi}{4}$.
The TI calculator cannot be used to
graph equations in polar form with
$\theta =$ constant.

Approach 2: $\theta = \dfrac{\pi}{4}$

Take the tangent of both sides. $\tan\theta = \underbrace{\tan\left(\dfrac{\pi}{4}\right)}_{1}$

Use the identity $\tan\theta = \dfrac{y}{x}$. $\dfrac{y}{x} = 1$

Multiply by x. $y = x$

The result is a line passing through the origin with slope $= 1$.

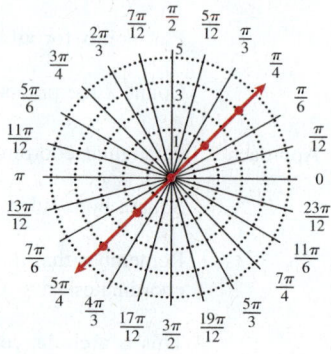

Rectangular equations that depend on varying (not constant) values of x or y can be graphed by point-plotting (making a table and plotting the points). We will use this same procedure for graphing polar equations that depend on varying (not constant) values of r and θ.

  EXAMPLE 4   **Graphing a Polar Equation of the Form $r = c \cdot \cos\theta$ or $r = c \cdot \sin\theta$**

Graph $r = 4\cos\theta$.

Solution:

STEP 1 Make a table and find several key values.

θ	$r = 4\cos\theta$	(r, θ)
0	$4(1) = 4$	$(4, 0)$
$\dfrac{\pi}{4}$	$4\left(\dfrac{\sqrt{2}}{2}\right) \approx 2.8$	$\left(2.8, \dfrac{\pi}{4}\right)$
$\dfrac{\pi}{2}$	$4(0) = 0$	$\left(0, \dfrac{\pi}{2}\right)$
$\dfrac{3\pi}{4}$	$4\left(-\dfrac{\sqrt{2}}{2}\right) \approx -2.8$	$\left(-2.8, \dfrac{3\pi}{4}\right)$
π	$4(-1) = -4$	$(-4, \pi)$
$\dfrac{5\pi}{4}$	$4\left(-\dfrac{\sqrt{2}}{2}\right) \approx -2.8$	$\left(-2.8, \dfrac{5\pi}{4}\right)$
$\dfrac{3\pi}{2}$	$4(0) = 0$	$\left(0, \dfrac{3\pi}{2}\right)$
$\dfrac{7\pi}{4}$	$4\left(\dfrac{\sqrt{2}}{2}\right) \approx 2.8$	$\left(-2.8, \dfrac{7\pi}{4}\right)$
2π	$4(1) = 4$	$(-4, 2\pi)$

STEP 2 Plot the points in polar coordinates.

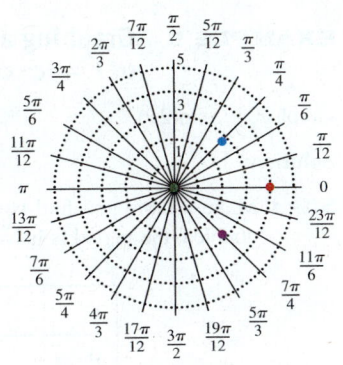

STEP 3 Connect the points with a smooth curve.

Notice that **(4, 0)** and **(−4, π)** correspond to the same point. There is no need to continue with angles beyond π, because the result would be to go around the same circle again.

■ **Answer:**

■ **YOUR TURN** Graph $r = 4\sin\theta$.

Compare the result of Example 4, the graph of $r = 4\cos\theta$, with the result of the Your Turn, the graph of $r = 4\sin\theta$. Notice that they are 90° out of phase (we simply rotate one graph 90° about the pole to get the other graph).

In general, graphs of polar equations of the form $r = a\sin\theta$ and $r = a\cos\theta$ are circles.

WORDS	MATH	
	$r = a\sin\theta$	$r = a\cos\theta$
Apply trigonometric ratios: $\sin\theta = \dfrac{y}{r}$ and $\cos\theta = \dfrac{x}{r}$.	$r = a\dfrac{y}{r}$	$r = a\dfrac{x}{r}$
Multiply the equations by r.	$r^2 = ay$	$r^2 = ax$
Let $r^2 = x^2 + y^2$.	$x^2 + y^2 = ay$	$x^2 + y^2 = ax$
Group x terms together and y terms together.	$x^2 + (y^2 - ay) = 0$	$(x^2 - ax) + y^2 = 0$
Complete the square on the expressions in parentheses.	$x^2 + \left[y^2 - ay + \left(\dfrac{a}{2}\right)^2\right] = \left(\dfrac{a}{2}\right)^2$	$\left[x^2 - ax + \left(\dfrac{a}{2}\right)^2\right] + y^2 = \left(\dfrac{a}{2}\right)^2$
	$x^2 + \left(y - \dfrac{a}{2}\right)^2 = \left(\dfrac{a}{2}\right)^2$	$\left(x - \dfrac{a}{2}\right)^2 + y^2 = \left(\dfrac{a}{2}\right)^2$
Graph of a circle.	Center: $\left(0, \dfrac{a}{2}\right)$ Radius: $\dfrac{a}{2}$	Center: $\left(\dfrac{a}{2}, 0\right)$ Radius: $\dfrac{a}{2}$

Technology Tip

Graph $r = 5\sin(2\theta)$.

EXAMPLE 5 **Graphing a Polar Equation of the Form $r = c \cdot \sin(2\theta)$ or $r = c \cdot \cos(2\theta)$**

Graph $r = 5\sin(2\theta)$.

Solution:

STEP 1 Make a table and find key values. Since the argument of the sine function is doubled, the period is halved. Therefore, instead of steps of $\pi/4$, take steps of $\pi/8$.

θ	$r = 5\sin(2\theta)$	(r, θ)
0	$5(0) = 0$	$(0, 0)$
$\dfrac{\pi}{8}$	$5\left(\dfrac{\sqrt{2}}{2}\right) \approx 3.5$	$\left(3.5, \dfrac{\pi}{8}\right)$
$\dfrac{\pi}{4}$	$5(1) = 5$	$\left(5, \dfrac{\pi}{4}\right)$
$\dfrac{3\pi}{8}$	$5\left(\dfrac{\sqrt{2}}{2}\right) \approx 3.5$	$\left(3.5, \dfrac{3\pi}{8}\right)$
$\dfrac{\pi}{2}$	$5(0) = 0$	$\left(0, \dfrac{\pi}{2}\right)$

STEP 2 Label the polar coordinates.

The values in the table represent what happens in quadrant I. The same pattern repeats in the other three quadrants. The result is a **four-leaved rose**.

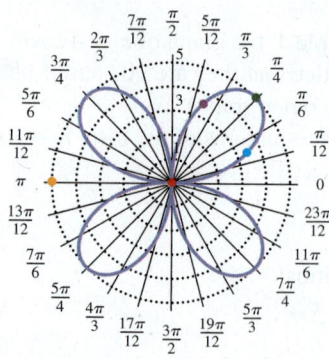

STEP 3 Connect the points with smooth curves.

■ Answer:

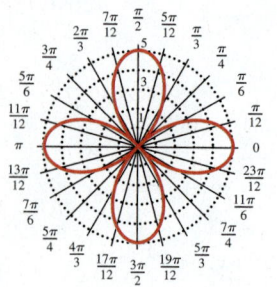

■ YOUR TURN Graph $r = 5\cos(2\theta)$.

Compare the result of Example 5, the graph of $r = 5\sin(2\theta)$, with the result of the Your Turn, the graph of $r = 5\cos(2\theta)$. Notice that they are 45° out of phase (we rotate one graph 45° about the pole to get the other graph).

In general, for $r = a\sin(n\theta)$ or $r = a\cos(n\theta)$, the graph has n leaves (petals) if n is odd and $2n$ leaves (petals) if n is even. As r increases, the leaves (petals) get longer.

The next class of graphs are called **limaçons**, which have equations of the form $r = a \pm b\cos\theta$ or $r = a \pm b\sin\theta$. When $a = b$, the result is a **cardioid** (heart shape).

Study Tip

Graphs of $r = a\sin(n\theta)$ and $r = a\cos(n\theta)$ are roses with n leaves if n is odd and $2n$ leaves if n is even.

EXAMPLE 6 **The Cardioid as a Polar Equation**

Graph $r = 2 + 2\cos\theta$.

Solution:

STEP 1 Make a table and find key values.

This behavior repeats in quadrant III and quadrant IV, because the cosine function has corresponding values in quadrant I and quadrant IV and in quadrant II and quadrant III.

θ	$r = 2 + 2\cos\theta$	(r, θ)
0	$2 + 2(1) = 4$	$(4, 0)$
$\dfrac{\pi}{4}$	$2 + 2\left(\dfrac{\sqrt{2}}{2}\right) \approx 3.4$	$\left(3.4, \dfrac{\pi}{4}\right)$
$\dfrac{\pi}{2}$	$2 + 2(0) = 2$	$\left(2, \dfrac{\pi}{2}\right)$
$\dfrac{3\pi}{4}$	$2 + 2\left(-\dfrac{\sqrt{2}}{2}\right) \approx 0.6$	$\left(0.6, \dfrac{3\pi}{4}\right)$
π	$2 + 2(-1) = 0$	$(0, \pi)$

STEP 2 Plot the points in polar coordinates.

STEP 3 Connect the points with a smooth curve. The curve is a *cardioid*, a term formed from Greek roots meaning "heart-shaped."

EXAMPLE 7 **Graphing a Polar Equation of the Form $r = c \cdot \theta$**

Graph $r = 0.5\theta$.

Solution:

STEP 1 Make a table and find key values.

θ	$r = 0.5\theta$	(r, θ)
0	$0.5(0) = 0$	$(0, 0)$
$\dfrac{\pi}{2}$	$0.5\left(\dfrac{\pi}{2}\right) \approx 0.8$	$\left(0.8, \dfrac{\pi}{2}\right)$
π	$0.5(\pi) \approx 1.6$	$(1.6, \pi)$
$\dfrac{3\pi}{2}$	$0.5\left(\dfrac{3\pi}{2}\right) \approx 2.4$	$\left(2.4, \dfrac{3\pi}{2}\right)$
2π	$0.5(2\pi) \approx 3.1$	$(3.1, 2\pi)$

Classroom Example 8.8.7

Graph $r = 2\theta$, $r = 4\theta$, $r = \frac{1}{2}\theta$, and $r = \frac{1}{4}\theta$. Describe what happens as the coefficient of θ changes.

Answer:

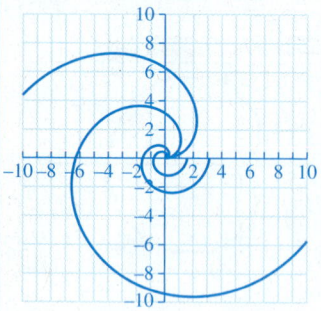

The spiral gets tighter the smaller the coefficient of θ gets.

STEP 2 Plot the points in polar coordinates.

STEP 3 Connect the points with a smooth curve. The curve is a *spiral*.

EXAMPLE 8 **Graphing a Polar Equation of the Form $r^2 = c \cdot \sin(2\theta)$ or $r^2 = c \cdot \cos(2\theta)$**

Graph $r^2 = 4\cos(2\theta)$.

Solution:

STEP 1 Make a table and find key values.

Solving for r yields $r = \pm 2\sqrt{\cos(2\theta)}$. All coordinates $(-r, \theta)$ can be expressed as $(r, \theta + \pi)$. The following table does not have values for $\frac{\pi}{4} < \theta < \frac{3\pi}{4}$, because the corresponding values of $\cos(2\theta)$ are negative and hence r is an imaginary number. The table also does not have values for $\theta > \pi$, because $2\theta > 2\pi$ and the corresponding points are repeated.

θ	$\cos(2\theta)$	$r = \pm 2\sqrt{\cos(2\theta)}$	(r, θ)
0	1	$r = \pm 2$	$(2, 0)$ and $(-2, 0) = (2, \pi)$
$\frac{\pi}{6}$	0.5	$r \approx \pm 1.4$	$\left(1.4, \frac{\pi}{6}\right)$ and $\left(-1.4, \frac{\pi}{6}\right) = \left(1.4, \frac{7\pi}{6}\right)$
$\frac{\pi}{4}$	0	$r = 0$	$\left(0, \frac{\pi}{4}\right)$
$\frac{3\pi}{4}$	0	$r = 0$	$\left(0, \frac{3\pi}{4}\right)$
$\frac{5\pi}{6}$	0.5	$r \approx \pm 1.4$	$\left(1.4, \frac{5\pi}{6}\right)$ and $\left(-1.4, \frac{5\pi}{6}\right) = \left(1.4, \frac{11\pi}{6}\right)$
π	1	$r = \pm 2$	$(2, \pi)$ and $(-2, \pi) = (2, 2\pi)$

STEP 2 Plot the points in polar coordinates.

STEP 3 Connect the points with a smooth curve. The resulting curve is known as a *lemniscate*.

Classroom Example 8.8.8

Graph $r^2 = 6\sin 2\theta$.

Answer:

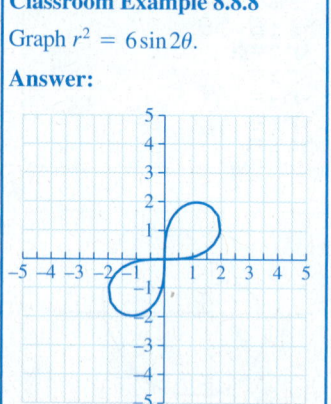

Converting Equations Between Polar and Rectangular Form

It is not always advantageous to plot an equation in the form in which it is given. It is sometimes easier to first convert to rectangular form and then plot. For example, to plot $r = \dfrac{2}{\cos\theta + \sin\theta}$, we could make a table with values. However, as you will see in Example 9, it is much easier to convert this equation to rectangular coordinates.

 EXAMPLE 9   Converting an Equation from Polar Form to Rectangular Form

Graph $r = \dfrac{2}{\cos\theta + \sin\theta}$.

Solution:

Multiply the equation by $\cos\theta + \sin\theta$.	$r(\cos\theta + \sin\theta) = 2$
Eliminate parentheses.	$r\cos\theta + r\sin\theta = 2$
Convert the result to rectangular form.	$\underbrace{r\cos\theta}_{x} + \underbrace{r\sin\theta}_{y} = 2$

Simplify. The result is a straight line. $\boxed{y = -x + 2}$

Graph the line.

■ **YOUR TURN** Graph $r = \dfrac{2}{\cos\theta - \sin\theta}$.

Technology Tip

Graph $r = \dfrac{2}{\cos\theta + \sin\theta}$.

■ **Answer:** $y = x - 2$

 SECTION
8.8 SUMMARY

Graph polar coordinates (r, θ) in the polar coordinate system first by rotating a ray to get the terminal side of the angle. Then if r is positive, go out r units from the origin in the direction of the angle. If r is negative, go out $|r|$ units in the opposite direction of the angle. Conversions between polar and rectangular forms are given by

FROM	TO	IDENTITIES
Polar (r, θ)	Rectangular (x, y)	$x = r\cos\theta \qquad y = r\sin\theta$
Rectangular (x, y)	Polar (r, θ)	$r = \sqrt{x^2 + y^2} \qquad \tan\theta = \dfrac{y}{x}, x \neq 0$ Be careful to note the proper quadrant for θ.

We can graph polar equations by point-plotting. Common shapes that arise are given in the following table. Sine and cosine curves have the same shapes (just rotated). If more than one equation is given, then the top equation corresponds to the actual graph. In this table, a and b are assumed to be positive.

CLASSIFICATION	SPECIAL NAME	POLAR EQUATIONS	GRAPH
Line	Radial line	$\theta = a$	
Circle	Circle centered at the origin	$r = a$	
Circle	Circle that touches the pole and whose center is on the polar axis	$r = a\cos\theta$	
Circle	Circle that touches the pole and whose center is on the line $\theta = \dfrac{\pi}{2}$	$r = a\sin\theta$	
Limaçon	Cardioid	$r = a + a\cos\theta$ $r = a + a\sin\theta$	
Limaçon	Without inner loop $a > b$	$r = -a - b\cos\theta$ $r = a + b\sin\theta$	
Limaçon	With inner loop $a < b$	$r = a + b\sin\theta$ $r = a + b\cos\theta$	
Lemniscate		$r^2 = a^2\cos(2\theta)$ $r^2 = a^2\sin(2\theta)$	
Rose	Three* rose petals	$r = a\sin(3\theta)$ $r = a\cos(3\theta)$	

CLASSIFICATION	SPECIAL NAME	POLAR EQUATIONS	GRAPH
Rose	Four* rose petals	$r = a\sin(2\theta)$ $r = a\cos(2\theta)$	
Spiral		$r = a\theta$	

*In the argument $n\theta$, if n is odd, there are n petals (leaves), and if n is even, there are $2n$ petals (leaves).

SECTION 8.8 EXERCISES

▪ SKILLS

In Exercises 1–10, plot each indicated point in a polar coordinate system.

1. $\left(3, \dfrac{5\pi}{6}\right)$ **2.** $\left(2, \dfrac{5\pi}{4}\right)$ **3.** $\left(4, \dfrac{11\pi}{6}\right)$ **4.** $\left(1, \dfrac{2\pi}{3}\right)$ **5.** $\left(-2, \dfrac{\pi}{6}\right)$ **6.** $\left(-4, \dfrac{7\pi}{4}\right)$

7. $(-4, 270°)$ **8.** $(3, 135°)$ **9.** $(4, 225°)$ **10.** $(-2, 60°)$

In Exercises 11–20, convert each point to exact polar coordinates. Assume that $0 \le \theta < 2\pi$.

11. $\left(2, 2\sqrt{3}\right)$ **12.** $(3, -3)$ **13.** $\left(-1, -\sqrt{3}\right)$ **14.** $\left(6, 6\sqrt{3}\right)$ **15.** $(-4, 4)$ **16.** $\left(0, \sqrt{2}\right)$

17. $(3, 0)$ **18.** $(-7, -7)$ **19.** $\left(-\sqrt{3}, -1\right)$ **20.** $\left(2\sqrt{3}, -2\right)$

In Exercises 21–30, convert each point to exact rectangular coordinates.

21. $\left(4, \dfrac{5\pi}{3}\right)$ **22.** $\left(2, \dfrac{3\pi}{4}\right)$ **23.** $\left(-1, \dfrac{5\pi}{6}\right)$ **24.** $\left(-2, \dfrac{7\pi}{4}\right)$ **25.** $\left(0, \dfrac{11\pi}{6}\right)$ **26.** $(6, 0)$

27. $(2, 240°)$ **28.** $(-3, 150°)$ **29.** $(-1, 135°)$ **30.** $(5, 315°)$

In Exercises 31–34, match the polar graphs with their corresponding equations.

31. $r = 4\cos\theta$ **32.** $r = 2\theta$ **33.** $r = 3 + 3\sin\theta$ **34.** $r = 3\sin(2\theta)$

a.   **b.**   **c.**   **d.**

In Exercises 35–50, graph each equation.

35. $r = 5$

36. $\theta = -\dfrac{\pi}{3}$

37. $r = 2\cos\theta$

38. $r = 3\sin\theta$

39. $r = 4\sin(2\theta)$

40. $r = 5\cos(2\theta)$

41. $r = 3\sin(3\theta)$

42. $r = 4\cos(3\theta)$

43. $r^2 = 9\cos(2\theta)$

44. $r^2 = 16\sin(2\theta)$

45. $r = -2\cos\theta$

46. $r = -3\sin(3\theta)$

47. $r = 4\theta$

48. $r = -2\theta$

49. $r = -3 + 2\cos\theta$

50. $r = 2 + 3\sin\theta$

In Exercises 51–54, convert the equation from polar to rectangular form. Identify the resulting equation as a line, parabola, or circle.

51. $r(\sin\theta + 2\cos\theta) = 1$ **52.** $r(\sin\theta - 3\cos\theta) = 2$ **53.** $r^2\cos^2\theta - 2r\cos\theta + r^2\sin^2\theta = 8$ **54.** $r^2\cos^2\theta - r\sin\theta = -2$

In Exercises 55–60, graph the polar equation.

55. $r = -\frac{1}{3}\theta$

56. $r = \frac{1}{4}\theta$

57. $r = 4\sin(5\theta)$

58. $r = -3\cos(4\theta)$

59. $r = -2 - 3\cos\theta$

60. $r = 4 - 3\sin\theta$

■ APPLICATIONS

61. Halley's Comet. Halley's comet travels an elliptical path that can be modeled with the polar equation $r = \dfrac{0.587(1 + 0.967)}{1 - 0.967\cos\theta}$. Sketch the graph of the path of Halley's comet.

62. Dwarf Planet Pluto. The planet Pluto travels in an elliptical orbit that can be modeled with the polar equation $r = \dfrac{29.62(1 + 0.249)}{1 - 0.249\cos\theta}$. Sketch the graph of Pluto's orbit.

63. Archimedes Spiral. Spirals are seen in nature, as in the swirl of a pine cone; they are also used in machinery to convert motions. An Archimedes spiral has the general equation $r = a\theta$. A more general form for the equation of a spiral is $r = a\theta^{1/n}$, where n is a constant that determines how tightly the spiral is wrapped. Compare the Archimedes spiral $r = \theta$ with the spiral $r = \theta^{1/2}$ by graphing both on the same polar graph.

64. Archimedes Spiral. Spirals are seen in nature, as in the swirl of a pine cone; they are also used in machinery to convert motions. An Archimedes spiral has the general equation $r = a\theta$. A more general form for the equation of a spiral is $r = a\theta^{1/n}$, where n is a constant that determines how tightly the spiral is wrapped. Compare the Archimedes spiral $r = \theta$ with the spiral $r = \theta^{4/3}$ by graphing both on the same polar graph.

For Exercises 65 and 66, refer to the following:

The *lemniscate motion* occurs naturally in the flapping of birds' wings. The bird's vertical lift and wing sweep create the distinctive figure-eight pattern. The patterns vary with the different wing profiles.

65. Flapping Wings of Birds. Compare the following two possible lemniscate patterns by graphing them on the same polar graph: $r^2 = 4\cos(2\theta)$ and $r^2 = \frac{1}{4}\cos(2\theta)$.

66. Flapping Wings of Birds. Compare the following two possible lemniscate patterns by graphing them on the same polar graph: $r^2 = 4\cos(2\theta)$ and $r^2 = 4\cos(2\theta + 2)$.

For Exercises 67 and 68, refer to the following:

Many microphone manufacturers advertise that their microphones' exceptional pickup capabilities isolate the sound source and minimize background noise. These microphones are described as cardioid microphones because of the pattern formed by the range of the pickup.

67. Cardioid Pickup Pattern. Graph the cardioid curve $r = 2 + 2\sin\theta$ to see what the range looks like.

68. Cardioid Pickup Pattern. Graph the cardioid curve $r = -4 - 4\sin\theta$ to see what the range looks like.

For Exercises 69 and 70, refer to the following:

The sword artistry of the Samurai is legendary in Japanese folklore and myth. The elegance with which a samurai could wield a sword rivals the grace exhibited by modern figure skaters. In more modern times, such legends have been rendered digitally in many different video games (e.g., *Onimusha*). In order to make the characters realistically move across the screen, and in particular, wield various sword motions true to the legends, trigonometric functions are extensively used in constructing the underlying graphics module. One famous movement is a figure eight, swept out with two hands on the sword. The actual path of the tip of the blade as the movement progresses in this figure-eight motion depends essentially on the length L of the sword and the speed with which it is swept out. Such a path is modeled using a polar equation of the form

$$r^2(\theta) = L\cos(A\theta) \ \text{ or } \ r^2(\theta) = L\sin(A\theta), \ \ \theta_1 \le \theta \le \theta_2$$

whose graphs are called *lemniscates*.

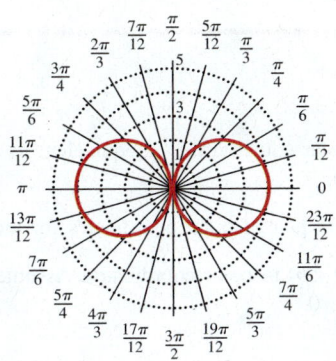

69. Video Games. Graph the following equations:

a. $r^2(\theta) = 5\cos\theta, 0 \le \theta \le 2\pi$

b. $r^2(\theta) = 5\cos(2\theta), 0 \le \theta \le \pi$

c. $r^2(\theta) = 5\cos(4\theta), 0 \le \theta \le \dfrac{\pi}{2}$

What do you notice about all of these graphs? Suppose that the movement of the tip of the sword in a game is governed by these graphs. Describe what happens if you change the domain in (b) and (c) to $0 \le \theta \le 2\pi$.

70. Video Games. Write a polar equation that would describe the motion of a sword 12 units long that makes 8 complete motions in $[0, 2\pi]$.

71. Home Improvement. The owner of a garden maze has decided to replace the square central plot of the maze, which is 60 feet × 60 feet, with a section comprised of a spiral to make the participants literally feel as though they are going in circles as they reach the tall slide in the center of the maze (that leads to the exit). Assume that the center of the maze is at the origin. If the walkway must be 3 feet wide to accommodate a participant, and the bushes on either side of the walkway are 1.5 feet thick,

how many times can the spiral wrap around before the center is reached? Graph the resulting spiral.

72. Home Improvement. Consider the garden maze in Exercise 71, but now suppose that the owner wants the spiral to wrap around exactly 3 times. How wide can the walkway then be throughout the spiral? Graph it.

73. Magnetic Pendulum. A magnetic bob is affixed to an arm of length L, which is fastened to a pivot point. Three magnets of equal strength are positioned on a plane 8 inches from the center; one is placed on the x-axis, one at 120° with respect to the positive x-axis, and the other at 240° with respect to the positive x-axis. The path swept out is a three-petal rose, as shown below:

a. Find the equation of this path.

b. How many times does the path retrace itself on the interval $[0, 100\pi]$?

74. Magnetic Pendulum. In reference to the context of Exercise 73, now position 8 magnets, each 8 units from the origin and at the vertices of a regular octagon, one being on the x-axis. Assume that the path of the pendulum is an eight-petal rose.

a. Find the equation of this path.

b. Graph this equation.

■ CATCH THE MISTAKE

In Exercises 75 and 76, explain the mistake that is made.

75. Convert $(-2, -2)$ to polar coordinates.

Solution:

Label x and y. $\quad x = -2, y = -2$

Find r.

$$r = \sqrt{x^2 + y^2} = \sqrt{4 + 4} = \sqrt{8} = 2\sqrt{2}$$

Find θ. $\qquad \tan\theta = \dfrac{-2}{-2} = 1$

$$\theta = \tan^{-1}(1) = \frac{\pi}{4}$$

Write the point in polar coordinates. $\qquad \left(2\sqrt{2}, \dfrac{\pi}{4}\right)$

This is incorrect. What mistake was made?

76. Convert $\left(-\sqrt{3}, 1\right)$ to polar coordinates.

Solution:

Label x and y. $\quad x = -\sqrt{3}, y = 1$

$$r = \sqrt{x^2 + y^2} = \sqrt{3 + 1} = \sqrt{4} = 2$$

Find θ. $\qquad \tan\theta = \dfrac{1}{-\sqrt{3}} = -\dfrac{1}{\sqrt{3}}$

$$\theta = \tan^{-1}\left(-\frac{1}{\sqrt{3}}\right) = -\frac{\pi}{4}$$

Write the point in polar coordinates. $\qquad \left(2, -\dfrac{\pi}{4}\right)$

This is incorrect. What mistake was made?

■ CONCEPTUAL

In Exercises 77 and 78, determine whether each statement is true or false.

77. All cardioids are limaçons, but not all limaçons are cardioids.

78. All limaçons are cardioids, but not all cardioids are limaçons.

79. Find the polar equation that is equivalent to a vertical line, $x = a$.

80. Find the polar equation that is equivalent to a horizontal line, $y = b$.

81. Give another pair of polar coordinates for the point (a, θ).

82. Convert $(-a, b)$ to polar coordinates. Assume that $a > 0, b > 0$.

■ CHALLENGE

83. Determine the values of θ at which $r = 4\cos\theta$ and $r\cos\theta = 1$ intersect. Graph both equations.

84. Find the Cartesian equation for $r = a\sin\theta + b\cos\theta$, where a and b are positive. Identify the type of graph.

85. Find the Cartesian equation for $r = \dfrac{a\sin(2\theta)}{\cos^3\theta - \sin^3\theta}$.

86. Identify an equation for the following graph:

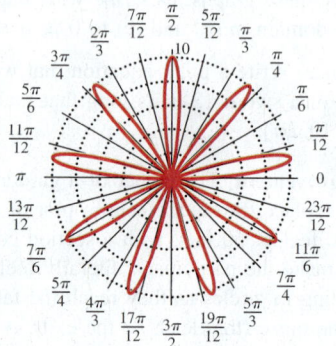

87. Consider the equation $r = 2a\cos(\theta - b)$. Sketch the graph for various values of a and b, and then give a general description of the graph.

88. Consider the equation $r = a\sin(b\theta)$, where $a, b > 0$. Determine the smallest number M for which the graph starts to repeat.

■ TECHNOLOGY

89. Given $r = \cos\left(\dfrac{\theta}{2}\right)$, find the θ-intervals for the inner loop above the x-axis.

90. Given $r = 2\cos\left(\dfrac{3\theta}{2}\right)$, find the θ-intervals for the petal in the first quadrant.

91. Given $r = 1 + 3\cos\theta$, find the θ-intervals for the inner loop.

92. Given $r = 1 + \sin(2\theta)$ and $r = 1 - \cos(2\theta)$, find all points of intersection.

93. Given $r = 2 + \sin(4\theta)$ and $r = 1$, find the angles of all the points of intersection.

94. Given $r = 2 - \cos(3\theta)$ and $r = 1.5$, find the angles of all points of intersection.

CHAPTER 8 INQUIRY-BASED LEARNING PROJECT

When solving triangles using the Law of Sines, it is important to keep in mind the domain and range of the inverse sine function, because it can play a role in determining your answers. Make sure to check your answers for reasonableness, especially when solving an obtuse triangle, as you will see next.

For a triangle ABC, two side lengths and the measure of the angle between them are given below.

$$b = 24 \qquad c = 8 \qquad A = 15°$$

1. How many triangles are possible with these measurements? Explain.

2. Make a careful sketch of a triangle and label the given information. Also label the unknown side a, and the unknown angles opposite sides b and c, as B and C, respectively. What can you say about angle B?

3. Explain why the Law of Cosines is needed for this problem.

4. Find the length of side a. Round to three significant digits.

5. Suppose a fellow student now wants to find the measure of angle B, and decides to use the Law of Sines. Shown below are the steps he wrote out to illustrate his process for computing the measure of B and C. Look at the chart he filled in below.

$$B = \sin^{-1}\left[\frac{24}{16.4}\sin(15)\right] \approx 22.3°$$

$$C = 180° - (15° + 22.3°) = 142.7°$$

Sides		Angles	
a	16.4	A	15°
b	24	B	22.3°
c	8	C	142.7°

What is wrong with the answers given by this student?

6. Because your calculator gives positive values of $\sin^{-1}x$ only between 0 and 90°, the answer this student got for the measure of angle B does not make sense. Show what he needs to do to correct his error.

Many Americans are considering hybrid and electric automobiles as an attractive alternative to traditional automobiles. The following table illustrates the approximate gross vehicle weight of both a large SUV and a small hybrid and the approximate fuel economy rates in miles per gallon:

AUTOMOBILE	WEIGHT	MPG
Ford Expedition	7100 lb	18
Toyota Prius	2800 lb	45

Recall that the amount of work to push an object that weighs F pounds a distance of d feet along a horizontal is $W = \mathbf{F} \cdot \mathbf{d}$.

1. Calculate how much work it would take to move a Ford Expedition 100 feet.

2. Calculate how much work it would take to move a Toyota Prius 100 feet.

3. Compare the values you calculated in Questions 1 and 2. What is the ratio of work to move the Expedition to work required to move the Prius?

4. Compare the result in Question 3 with the ratio of fuel economy (mpg) for these two vehicles. What can you conclude about the relationship between weight of an automobile and fuel economy?

5. Calculate the work required to move both the Ford Expedition and Toyota Prius 100 feet along an incline that makes a 45° angle with the ground (horizontal).

6. Based on your results in Question 5, do you expect the fuel economy ratios to be the same in this inclined scenario compared with the horizontal? In other words, should consumers in Florida (flat) be guided by the same "numbers" as consumers in the Appalachian Mountains (North Carolina)?

SECTION	CONCEPT	KEY IDEAS/FORMULAS
8.1	**Oblique triangles and the Law of Sines**	
	Solving oblique triangles Four cases ■ AAS/ASA ■ SSA ■ SAS ■ SSS	Oblique (Nonright) Triangles Acute Triangle Obtuse Triangle
		The Law of Sines $$\frac{\sin\alpha}{a} = \frac{\sin\beta}{b} = \frac{\sin\gamma}{c}$$ ■ AAS (or ASA) triangles ■ SSA triangles (ambiguous case)
8.2	**The Law of Cosines**	$a^2 = b^2 + c^2 - 2bc\cos\alpha$ $b^2 = a^2 + c^2 - 2ac\cos\beta$ $c^2 = a^2 + b^2 - 2ab\cos\gamma$
	Solving oblique triangles	■ SAS triangles ■ SSS triangles
8.3	**The area of a triangle**	
	The area of a triangle (SAS case)	Use the Law of Sines for the SAS case: $A_{\text{SAS case}} = \frac{1}{2}bc\sin\alpha$ when b, c, and α are known. $A_{\text{SAS case}} = \frac{1}{2}ab\sin\gamma$ when a, b, and γ are known. $A_{\text{SAS case}} = \frac{1}{2}ac\sin\beta$ when a, c, and β are known.
	The area of a triangle (SSS case)	Use Heron's formula for the SSS case: $$A_{\text{SSS case}} = \sqrt{s(s-a)(s-b)(s-c)}$$ where a, b, and c are the lengths of the sides of the triangle and s is half the perimeter of the triangle, called the semiperimeter. $$s = \frac{a+b+c}{2}$$

SECTION	CONCEPT	KEY IDEAS/FORMULAS

8.4 **Vectors**

Vector **u** or $\overrightarrow{AB}$

Vectors: Magnitude and Direction

$\mathbf{u} = \langle a, b \rangle$

	Geometric	*Algebraic*
Magnitude	length of a vector	$|\mathbf{u}| = \sqrt{a^2 + b^2}$
Direction	θ	$\tan\theta = \dfrac{b}{a}$
Addition	tail-to-tip	$\mathbf{u} = \langle a, b \rangle$ and $\mathbf{v} = \langle c, d \rangle$ $\mathbf{u} + \mathbf{v} = \langle a + c, b + d \rangle$

Vector operations

Scalar multiplication: $k \langle a, b \rangle = \langle ka, kb \rangle$

Horizontal and vertical components of a vector

Horizontal component: $a = |\mathbf{u}| \cos\theta$
Vertical component: $b = |\mathbf{u}| \sin\theta$

Unit vectors

$\mathbf{u} = \dfrac{\mathbf{v}}{|\mathbf{v}|}$

Resultant vectors

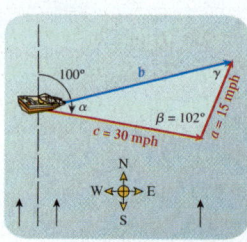

8.5 **The dot product**

- The product of a scalar and a vector is a vector.
- The dot product of two vectors is a scalar.

Multiplying two vectors: the dot product

$\mathbf{u} = \langle a, b \rangle$ and $\mathbf{v} = \langle c, d \rangle$
$\mathbf{u} \cdot \mathbf{v} = ac + bd$

Angle between two vectors

If θ is the angle between two nonzero vectors $\mathbf{u}$ and $\mathbf{v}$, where $0° \leq \theta \leq 180°$, then

$$\cos\theta = \frac{\mathbf{u} \cdot \mathbf{v}}{|\mathbf{u}||\mathbf{v}|}$$

Orthogonal (perpendicular) vectors: $\mathbf{u} \cdot \mathbf{v} = \mathbf{0}$

Work

When force and displacement are in the same direction: $W = |\mathbf{F}||\mathbf{d}|$.

When force and displacement are not in the same direction: $W = \mathbf{F} \cdot \mathbf{d}$.

SECTION	CONCEPT	KEY IDEAS/FORMULAS
8.6	**Polar (trigonometric) form of complex numbers**	
	Complex numbers in rectangular form	The **modulus**, or magnitude, of a complex number $z = x + iy$ is the distance from the origin to the point (x, y) in the complex plane given by $$\|z\| = \sqrt{x^2 + y^2}$$
	Complex numbers in polar form	The **polar form** of a complex number is $$z = r(\cos\theta + i\sin\theta)$$ where r represents the **modulus** (magnitude) of the complex number and θ represents the **argument** of z.
		Converting complex numbers between rectangular and polar forms: **Step 1**: Plot the point $z = x + yi$ in the complex plane (note the quadrant). **Step 2**: Find r. Use $r = \sqrt{x^2 + y^2}$. **Step 3**: Find θ. Use $\tan\theta = \dfrac{y}{x}$, $x \neq 0$, where θ is in the quadrant found in Step 1.
8.7	**Products, quotients, powers, and roots of complex numbers**	
	Products of complex numbers	Let $z_1 = r_1(\cos\theta_1 + i\sin\theta_1)$ and $z_2 = r_2(\cos\theta_2 + i\sin\theta_2)$ be two complex numbers. The product z_1z_2 is given by $$z_1z_2 = r_1r_2[\cos(\theta_1 + \theta_2) + i\sin(\theta_1 + \theta_2)]$$ Multiply the magnitudes and add the arguments.
	Quotients of complex numbers	Let $z_1 = r_1(\cos\theta_1 + i\sin\theta_1)$ and $z_2 = r_2(\cos\theta_2 + i\sin\theta_2)$ be two complex numbers. The quotient $\dfrac{z_1}{z_2}$ is given by $$\frac{z_1}{z_2} = \frac{r_1}{r_2}[\cos(\theta_1 - \theta_2) + i\sin(\theta_1 - \theta_2)]$$ Divide the magnitudes and subtract the arguments.
	Powers of complex numbers	**De Moivre's theorem** If $z = r(\cos\theta + i\sin\theta)$ is a complex number, then $z^n = r^n[\cos(n\theta) + i\sin(n\theta)]$, $n \geq 1$, where n is an integer.
	Roots of complex numbers	The **nth roots** of the complex number $z = r(\cos\theta + i\sin\theta)$ are given by $$w_k = r^{1/n}\left[\cos\left(\frac{\theta}{n} + \frac{k \cdot 360°}{n}\right) + i\sin\left(\frac{\theta}{n} + \frac{k \cdot 360°}{n}\right)\right]$$ θ in degrees, where $k = 0, 1, 2, \ldots, n - 1$.

SECTION	CONCEPT	KEY IDEAS/FORMULAS

8.8 — Polar equations and graphs

Concept	Key Ideas/Formulas
Polar coordinates	To plot a point (r, θ): ■ Start on the polar axis and rotate a ray to form the terminal side of an angle θ. ■ If $r > 0$, the point is r units from the origin in the *same direction* as the terminal side of θ. ■ If $r < 0$, the point is $|r|$ units from the origin in the *opposite direction* of the terminal side of θ.

Concept	Key Ideas/Formulas
Converting between polar and rectangular coordinates	From polar (r, θ) to rectangular (x, y): $$x = r\cos\theta \qquad y = r\sin\theta$$ From rectangular (x, y) to polar (r, θ): $$r = \sqrt{x^2 + y^2} \qquad \tan\theta = \frac{y}{x} \quad x \neq 0$$
Graphs of polar equations	Radial line, circle, spiral, rose petals, lemniscate, and limaçon

8.1 Oblique Triangles and the Law of Sines

Solve the given triangles.

1. $\alpha = 10°, \beta = 20°, a = 4$

2. $\beta = 40°, \gamma = 60°, b = 10$

3. $\alpha = 5°, \beta = 45°, c = 10$

4. $\beta = 60°, \gamma = 70°, a = 20$

5. $\gamma = 11°, \alpha = 11°, c = 11$

6. $\beta = 20°, \gamma = 50°, b = 8$

7. $\alpha = 45°, \gamma = 45°, b = 2$

8. $\alpha = 60°, \beta = 20°, c = 17$

9. $\alpha = 12°, \gamma = 22°, a = 99$

10. $\beta = 102°, \gamma = 27°, a = 24$

Two sides and an angle are given. Determine whether a triangle (or two) exist and, if so, solve the triangle.

11. $a = 7, b = 9, \alpha = 20°$

12. $b = 24, c = 30, \beta = 16°$

13. $a = 10, c = 12, \alpha = 24°$

14. $b = 100, c = 116, \beta = 12°$

15. $a = 40, b = 30, \beta = 150°$

16. $b = 2, c = 3, \gamma = 165°$

17. $a = 4, b = 6, \alpha = 10°$

18. $c = 25, a = 37, \gamma = 4°$

8.2 The Law of Cosines

19. **How Far from Home?** Gary walked 8 miles due north, made a 130° turn to the southeast, and walked for another 6 miles. How far was Gary from home?

20. **How Far from Home?** Mary walked 10 miles due north, made a 55° turn to the northeast, and walked for another 3 miles. How far was Mary from home?

Solve each triangle.

21. $a = 40, b = 60, \gamma = 50°$

22. $b = 15, c = 12, \alpha = 140°$

23. $a = 24, b = 25, c = 30$

24. $a = 6, b = 6, c = 8$

25. $a = \sqrt{11}, b = \sqrt{14}, c = 5$

26. $a = 22, b = 120, c = 122$

27. $b = 7, c = 10, \alpha = 14°$

28. $a = 6, b = 12, \gamma = 80°$

29. $b = 10, c = 4, \alpha = 90°$

30. $a = 4, b = 5, \gamma = 75°$

31. $a = 10, b = 11, c = 12$

32. $a = 22, b = 24, c = 25$

33. $b = 16, c = 18, \alpha = 100°$

34. $a = 25, c = 25, \beta = 9°$

35. $b = 12, c = 40, \alpha = 10°$

36. $a = 26, b = 20, c = 10$

37. $a = 26, b = 40, c = 13$

38. $a = 1, b = 2, c = 3$

39. $a = 6.3, b = 4.2, \alpha = 15°$

40. $b = 5, c = 6, \beta = 35°$

8.3 The Area of a Triangle

Find the area of each triangle described.

41. $b = 16, c = 18, \alpha = 100°$

42. $a = 25, c = 25, \beta = 9°$

43. $a = 10, b = 11, c = 12$

44. $a = 22, b = 24, c = 25$

45. $a = 26, b = 20, c = 10$

46. $a = 24, b = 32, c = 40$

47. $b = 12, c = 40, \alpha = 10°$

48. $a = 21, c = 75, \beta = 60°$

Applications

49. **Area of Inscribed Triangle.** The area of a triangle inscribed in a circle can be found if you know the lengths of the sides of the triangle and the radius of the circle: $A = \dfrac{abc}{4r}$. Find the radius of the circle that circumscribes the triangle if all the sides of the triangle measure 9 inches and the area of the triangle is 35 square inches.

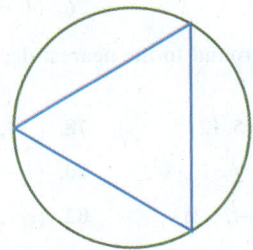

50. **Area of Inscribed Triangle.** The area of a triangle inscribed in a circle can be found if you know the lengths of the sides of the triangle and the radius of the circle: $A = \dfrac{abc}{4r}$. Find the radius of the circle that circumscribes the triangle if the sides of the triangle measure 9, 12, and 15 inches and the area of the triangle is 54 square inches.

8.4 Vectors

Find the magnitude of vector $\overrightarrow{AB}$.

51. $A = (4, -3)$ and $B = (-8, 2)$

52. $A = (-2, 11)$ and $B = (2, 8)$

53. $A = (0, -3)$ and $B = (5, 9)$

54. $A = (3, -11)$ and $B = (9, -3)$

Find the magnitude and direction angle of the given vector.

55. $\mathbf{u} = \langle -10, 24 \rangle$ **56.** $\mathbf{u} = \langle -5, -12 \rangle$

57. $\mathbf{u} = \langle 16, -12 \rangle$ **58.** $\mathbf{u} = \langle 0, 3 \rangle$

Perform the vector operation, given that $\mathbf{u} = \langle 7, -2 \rangle$ and $\mathbf{v} = \langle -4, 5 \rangle$.

59. $2\mathbf{u} + 3\mathbf{v}$ **60.** $\mathbf{u} - \mathbf{v}$

61. $6\mathbf{u} + \mathbf{v}$ **62.** $-3(\mathbf{u} + 2\mathbf{v})$

Find the vector, given its magnitude and direction angle.

63. $|\mathbf{u}| = 10, \theta = 75°$ **64.** $|\mathbf{u}| = 8, \theta = 225°$

65. $|\mathbf{u}| = 12, \theta = 105°$ **66.** $|\mathbf{u}| = 20, \theta = 15°$

Find a unit vector in the direction of the given vector.

67. $\mathbf{v} = \langle \sqrt{6}, -\sqrt{6} \rangle$ **68.** $\mathbf{v} = \langle -11, 60 \rangle$

Perform the indicated vector operation.

69. $(3\mathbf{i} - 4\mathbf{j}) + (2\mathbf{i} + 5\mathbf{j})$ **70.** $(-6\mathbf{i} + \mathbf{j}) - (9\mathbf{i} - \mathbf{j})$

8.5 The Dot Product

Find the indicated dot product.

71. $\langle 6, -3 \rangle \cdot \langle 1, 4 \rangle$ **72.** $\langle -6, 5 \rangle \cdot \langle -4, 2 \rangle$

73. $\langle 3, 3 \rangle \cdot \langle 3, -6 \rangle$ **74.** $\langle -2, -8 \rangle \cdot \langle -1, 1 \rangle$

75. $\langle 0, 8 \rangle \cdot \langle 1, 2 \rangle$ **76.** $\langle 4, -3 \rangle \cdot \langle -1, 0 \rangle$

Find the angle (round to the nearest degree) between each pair of vectors.

77. $\langle 3, 4 \rangle$ and $\langle -5, 12 \rangle$ **78.** $\langle -4, 5 \rangle$ and $\langle 5, -4 \rangle$

79. $\langle 1, \sqrt{2} \rangle$ and $\langle -1, 3\sqrt{2} \rangle$ **80.** $\langle 7, -24 \rangle$ and $\langle -6, 8 \rangle$

81. $\langle 3, 5 \rangle$ and $\langle -4, -4 \rangle$ **82.** $\langle -1, 6 \rangle$ and $\langle 2, -2 \rangle$

Determine whether each pair of vectors is orthogonal.

83. $\langle 8, 3 \rangle$ and $\langle -3, 12 \rangle$ **84.** $\langle -6, 2 \rangle$ and $\langle 4, 12 \rangle$

85. $\langle 5, -6 \rangle$ and $\langle -12, -10 \rangle$ **86.** $\langle 1, 1 \rangle$ and $\langle -4, 4 \rangle$

87. $\langle 0, 4 \rangle$ and $\langle 0, -4 \rangle$ **88.** $\langle -7, 2 \rangle$ and $\langle \frac{1}{7}, -\frac{1}{2} \rangle$

89. $\langle 6z, a - b \rangle$ and $\langle a + b, -6z \rangle$

90. $\langle a - b, -1 \rangle$ and $\langle a + b, a^2 - b^2 \rangle$

8.6 Polar (Trigonometric) Form of Complex Numbers

Graph each complex number in the complex plane.

91. $-6 + 2i$ **92.** $5i$

Express each complex number in polar form.

93. $\sqrt{2} - \sqrt{2}i$ **94.** $\sqrt{3} + i$

95. $-8i$ **96.** $-8 - 8i$

With a calculator, express each complex number in polar form.

97. $-60 + 11i$ **98.** $9 - 40i$

99. $15 + 8i$ **100.** $-10 - 24i$

Express each complex number in rectangular form.

101. $6(\cos 300° + i \sin 300°)$

102. $4(\cos 210° + i \sin 210°)$

103. $\sqrt{2}(\cos 135° + i \sin 135°)$

104. $4(\cos 150° + i \sin 150°)$

With a calculator, express each complex number in rectangular form.

105. $4(\cos 200° + i \sin 200°)$

106. $3(\cos 350° + i \sin 350°)$

8.7 Products, Quotients, Powers, and Roots of Complex Numbers

Find the product $z_1 z_2$.

107. $3(\cos 200° + i \sin 200°)$ and $4(\cos 70° + i \sin 70°)$

108. $3(\cos 20° + i \sin 20°)$ and $4(\cos 220° + i \sin 220°)$

109. $7(\cos 100° + i \sin 100°)$ and $3(\cos 140° + i \sin 140°)$

110. $(\cos 290° + i \sin 290°)$ and $4(\cos 40° + i \sin 40°)$

Find the quotient $\frac{z_1}{z_2}$.

111. $\sqrt{6}(\cos 200° + i \sin 200°)$ and $\sqrt{6}(\cos 50° + i \sin 50°)$

112. $18(\cos 190° + i \sin 190°)$ and $2(\cos 100° + i \sin 100°)$

113. $24(\cos 290° + i \sin 290°)$ and $4(\cos 110° + i \sin 110°)$

114. $\sqrt{200}(\cos 93° + i \sin 93°)$ and $\sqrt{2}(\cos 48° + i \sin 48°)$

Find the result of each expression using De Moivre's theorem. Write the answer in rectangular form.

115. $(3 + 3i)^4$ **116.** $(3 + \sqrt{3}i)^4$

117. $(1 + \sqrt{3}i)^5$ **118.** $(-2 - 2i)^7$

Find all nth roots of z. Write the answers in polar form, and plot the roots in the complex plane.

119. $2 + 2\sqrt{3}i, n = 2$ **120.** $-8 + 8\sqrt{3}i, n = 4$

121. $-256, n = 4$ **122.** $-18i, n = 2$

Find all complex solutions to the given equations.

123. $x^3 + 216 = 0$ **124.** $x^4 - 1 = 0$

125. $x^4 + 1 = 0$ **126.** $x^3 - 125 = 0$

8.8 Polar Equations and Graphs

Convert each point to exact polar coordinates (assuming that $0 \leq \theta < 2\pi$), and then graph the point in the polar coordinate system.

127. $(-2, 2)$

128. $(4, -4\sqrt{3})$

129. $(-5\sqrt{3}, -5)$

130. $(\sqrt{3}, \sqrt{3})$

131. $(0, -2)$

132. $(11, 0)$

Convert each point to exact rectangular coordinates.

133. $\left(-3, \dfrac{5\pi}{3}\right)$

134. $\left(4, \dfrac{5\pi}{4}\right)$

135. $\left(2, \dfrac{\pi}{3}\right)$

136. $\left(6, \dfrac{7\pi}{6}\right)$

137. $\left(1, \dfrac{4\pi}{3}\right)$

138. $\left(-3, \dfrac{7\pi}{4}\right)$

Graph each equation.

139. $r = 4\cos(2\theta)$

140. $r = \sin(3\theta)$

141. $r = -\theta$

142. $r = 4 - 3\sin\theta$

Technology Exercises

Section 8.1

Let A, B, and C be the lengths of the three sides of a triangle with X, Y, and Z as the corresponding angles. Write a program using a TI calculator to solve the given triangle.

143. $A = 31.6$, $C = 23.9$, $X = 42°$

144. $A = 137.2$, $B = 125.1$, $Y = 54°$

Section 8.2

Let A, B, and C be the lengths of the three sides of a triangle with X, Y, and Z as the corresponding angles. Write a program using a TI calculator to solve the given triangle.

145. $A = \sqrt{33}$, $B = \sqrt{29}$, $Z = 41.6°$

146. $A = 3412$, $B = 2178$, $C = 1576$

147. $A = 312$, $B = 267$, $C = 189$

148. $A = 12.7$, $B = 29.9$, $Z = 104.8°$

Section 8.4

With the graphing calculator $\boxed{\text{SUM}}$ command, find the magnitude of the given vector. Also, find the direction angle to the nearest degree.

149. $\langle 25, -60 \rangle$

150. $\langle -70, 10\sqrt{15} \rangle$

Section 8.5

With the graphing calculator $\boxed{\text{SUM}}$ command, find the angle (round to the nearest degree) between each pair of vectors.

151. $\langle 14, 37 \rangle$, $\langle 9, -26 \rangle$

152. $\langle -23, -8 \rangle$, $\langle 18, -32 \rangle$

Section 8.6

Another way of using a graphing calculator to represent complex numbers in rectangular form is to enter the real and imaginary parts as a list of two numbers and use the $\boxed{\text{SUM}}$ command to find the modulus.

153. Write $-\sqrt{23} - 11i$ in polar form using the $\boxed{\text{SUM}}$ command to find its modulus, and round the angle to the nearest degree.

154. Write $11 + \sqrt{23}i$ in polar form using the $\boxed{\text{SUM}}$ command to find its modulus, and round the angle to the nearest degree.

Section 8.7

155. Find the fourth roots of $-8 + 8\sqrt{3}i$, and draw the complex rectangle with the calculator.

156. Find the fourth roots of $8\sqrt{3} + 8i$, and draw the complex rectangle with the calculator.

Section 8.8

157. Given $r = 1 - 2\sin(3\theta)$ find the angles of all points of intersection (where $r = 0$).

158. Given $r = 1 + 2\cos(3\theta)$ find the angles of all points of intersection (where $r = 0$).

Solve the triangles, if possible.

1. $\alpha = 30°, \beta = 40°, b = 10$

2. $\alpha = 47°, \beta = 98°, \gamma = 35°$

3. $a = 7, b = 9, c = 12$

4. $\alpha = 45°, a = 8, b = 10$

5. $a = 1, b = 1, c = 2$

6. $a = \frac{23}{7}, c = \frac{5}{7}, \beta = 61.2°$

7. $\alpha = 110°, \beta = 20°, a = 5$

8. $b = \frac{\sqrt{5}}{2}, c = 3\sqrt{5}, \alpha = 45°$

In Exercises 9 and 10, find the areas of the given triangles.

9. $\gamma = 72°, a = 10, b = 12$

10. $a = 7, b = 10, c = 13$

11. Find the magnitude and direction angle of the vector $\mathbf{u} = \langle -5, 12 \rangle$.

12. Find a unit vector pointing in the same direction as $\mathbf{v} = \langle -3, -4 \rangle$.

13. Perform the indicated operations:

    a. $2\langle -1, 4 \rangle - 3\langle 4, 1 \rangle$

    b. $\langle -7, -1 \rangle \cdot \langle 2, 2 \rangle$

14. In a post pattern in football, the receiver in motion runs past the quarterback parallel to the line of scrimmage (A), runs 12 yards perpendicular to the line of scrimmage (B), and then cuts toward the goal post (C).

A receiver runs the post pattern. If the magnitudes of the vectors are $|A| = 3$ yards, $|B| = 12$ yards, and $|C| = 18$ yards, find the magnitude of the resultant vector $A + B + C$ and the direction angle θ.

For Exercises 15–17, use the complex number
$z = 16(\cos 120° + i\sin 120°)$.

15. Find z^4.

16. Find the four distinct fourth roots of z.

17. Convert the point $(3, 210°)$ to rectangular coordinates.

18. Graph $r = 6\sin(2\theta)$.

19. Graph $r^2 = 9\cos(2\theta)$.

20. Find x such that $\langle x, 1 \rangle$ is perpendicular to $3\vec{i} - 4\vec{j}$.

21. Prove that $\vec{u} \cdot (\vec{v} - \vec{w}) = \vec{u} \cdot \vec{v} - \vec{u} \cdot \vec{w}$.

22. Construct a unit vector in the opposite direction of $\langle 3, 5 \rangle$.

23. Compute $\vec{u} \cdot \vec{v}$ if $|\vec{u}| = 4, |\vec{v}| = 10$, and $\theta = \frac{2\pi}{3}$.

24. Determine whether $\vec{u}$ and $\vec{v}$ are parallel, perpendicular, or neither: $\vec{u} = \langle \sin\theta, \cos\theta \rangle$ $\vec{v} = \langle -\cos\theta, \sin\theta \rangle$.

25. Find the magnitude of $-\vec{i} - \vec{j}$.

26. Determine θ, when a streetlight is formed as follows:

27. Find the lengths of a, b, c, and d in the following diagram:

28. True or False: If $\vec{u} + \vec{v}$ is perpendicular to $\vec{u} - \vec{v}$, then $|\vec{u}| = |\vec{v}|$.

29. Solve $z^4 + 256i = 0$.

30. Convert to a Cartesian equation: $r^2 = \tan\theta$.

31. With the graphing calculator [SUM] command, find the angle (round to the nearest degree) between the pair of vectors: $\langle -8, -11 \rangle$ and $\langle -16, 26 \rangle$.

32. Find the fourth roots of $-8\sqrt{3} - 8i$, and draw the complex rectangle with the calculator.

1. Solve using the square root method: $(2 - x)^2 = -16$.

2. Solve for x, and express the solution in interval notation: $-5x - 3 \leq 12$.

3. Find the slope of the line that passes through the points $\left(\frac{1}{3}, \frac{3}{4}\right)$ and $\left(-\frac{1}{6}, -\frac{1}{4}\right)$.

4. Transform the equation into standard form by completing the square. State the center and radius of the circle: $x^2 + y^2 - 3.2x + 4.4y - 0.44 = 0$.

5. Using the function $f(x) = 5 - x^2$, evaluate the difference quotient $\dfrac{f(x + h) - f(x)}{h}$.

6. Given the piecewise-defined function

$$f(x) = \begin{cases} x^2 & x < 0 \\ 2x - 1 & 0 \leq x < 5 \\ 5 - x & x \geq 5 \end{cases}$$

   find:
   a. $f(0)$ **b.** $f(4)$ **c.** $f(5)$ **d.** $f(-4)$
   e. State the domain and range in interval notation.
   f. Determine the intervals where the function is increasing, decreasing, or constant.

7. Evaluate $f(g(-1))$, $g(x) = \sqrt[3]{x}$, and $f(x) = \dfrac{9}{x + 1}$.

8. Find a quadratic function that has the vertex $(1.5, 2.5)$ and point $(-0.5, -0.5)$.

9. For the function $f(x) = 5x^2(x + 2)^3(x^2 + 7)$, find all of the real zeros and state the multiplicity.

10. List the possible rational zeros, and test to determine all rational zeros for $P(x) = 6x^3 + x^2 - 5x - 2$.

11. Graph the rational function $f(x) = \dfrac{x^2 + 3}{x - 2}$. Give all asymptotes.

12. Graph the exponential function $f(x) = 5^{x-1}$, and state the y-intercept, the domain, the range, and the horizontal asymptote.

13. How much money should be put in a savings account now that earns 5.5% a year compounded continuously, if you want to have $85,000 in 15 years?

14. Graph the function $f(x) = -\log(x) + 2$, using transformation techniques.

15. Use properties of logarithms to simplify the expression $10^{-3 \log 10}$.

16. Solve for x: $\ln \sqrt{6 - 3x} - \frac{1}{2} \ln(x + 2) = \ln(x)$.

17. In a 45°-45°-90° triangle, if the two legs have length 15 feet, how long is the hypotenuse?

18. Find the positive measure of θ (rounded to the nearest degree) when $\tan \theta = 1.4285$ and the terminal side of θ lies in quadrant III.

19. Graph $y = 3\cos(2x)$ over one period.

20. If $\sin \theta = \frac{8}{11}$ and the terminal side of θ lies in quadrant II, find the exact value of $\sec \theta$.

21. Find the exact value of $\cos(\alpha - \beta)$ when $\cos \alpha = \frac{1}{5}$, $\cos \beta = -\frac{3}{5}$, the terminal side of α lies in quadrant IV, and the terminal side of β in quadrant II.

22. Solve the trigonometric equation $\cos(2\theta) = -\frac{1}{2}$ exactly, over the interval $0 \leq \theta < 2\pi$.

23. Solve the triangle with $b = 8$, $a = 13$, and $\gamma = 21°$.

24. Given $\mathbf{u} = \langle 5, -3 \rangle$ and $\mathbf{v} = \langle -2, 6 \rangle$, find $4(\mathbf{v} - \mathbf{u})$.

25. Express the complex number $2\sqrt{2}\left(\cos\dfrac{3\pi}{4} + i\sin\dfrac{3\pi}{4}\right)$ in rectangular form.

26. With the graphing calculator $\boxed{\text{SUM}}$ command, find the magnitude of vector $\langle 40, -96 \rangle$ and the direction angle to the nearest degree.

27. Given $r = 2 - 5\sin(2\theta)$, find the angles of all points of intersection (when $r = 0$). Round the angles to the nearest degree.

9

Systems of Linear Equations and Inequalities

On August 29, 2005, Hurricane Katrina unleashed its fury in Southeast Louisiana. The most significant number of deaths occurred in New Orleans which flooded when its levee system failed.

The Federal Emergency Management Agency (FEMA) coordinated the relief effort, including trucks that hauled generators, water, and tarps. FEMA was faced with an *optimization* problem in mathematics that involved maximizing the number of people aided by the relief subject to the constraints of the weight of the relief items and how much space they would take on the trucks. For example, how many generators, cases of water, and tarps could each truck haul, and how many of each would bring aid to the greatest number of people?

we will solve *systems* of linear equations in two variables and in three variables. We will then perform partial-fraction decomposition, which will be valuable in calculus and is made possible by techniques of solving systems of linear equations. We will discuss linear inequalities in two variables and systems of linear inequalities. Systems of nonlinear equations and inequalities will be discussed in Chapter 8. Finally, we will discuss the linear programming model, which will enable you to address the FEMA relief problem as a result of Florida's multiple hurricanes.

SYSTEMS OF LINEAR EQUATIONS AND INEQUALITIES

9.1 Systems of Linear Equations in Two Variables	**9.2** Systems of Linear Equations in Three Variables	**9.3** Partial Fractions	**9.4** Systems of Linear Inequalities in Two Variables	**9.5** The Linear Programming Model
• Solving Systems of Linear Equations • Three Methods and Three Types of Solutions	• Solving Systems of Linear Equations in Three Variables • Types of Solutions	• Performing Partial-Fraction Decomposition	• Linear Inequalities in Two Variables • Systems of Linear Inequalities in Two Variables	• Solving an Optimization Problem

LEARNING OBJECTIVES

■ Solve systems of linear equations in two variables with the substitution method and the elimination method.

■ Solve systems of linear equations in three variables employing a combination of the elimination and substitution methods.

■ Perform partial-fraction decomposition.

■ Solve a system of linear inequalities by finding the overlapping shaded regions.

■ Use the linear programming model to solve optimization problems subject to constraints.

SKILLS OBJECTIVES

- Solve systems of linear equations in two variables using the substitution method.
- Solve systems of linear equations in two variables using the elimination method.
- Solve systems of linear equations in two variables by graphing.
- Solve applications involving systems of linear equations.

CONCEPTUAL OBJECTIVES

- Understand that a system of linear equations has either one solution, no solution, or infinitely many solutions.
- Visualize two lines that intersect at one point, no points (parallel lines), or infinitely many points (same line).

Solving Systems of Linear Equations

Overview

We learned in Section 2.3 that a linear equation in two variables is given in standard form by

$$Ax + By = C$$

and that the graph of this linear equation is a line, provided that A and B are not both equal to zero. In this section we discuss **systems of linear equations**, which can be thought of as simultaneous equations. To **solve** a system of linear equations means to find the solution that satisfies *both* equations:

$$x + 2y = 6$$
$$3x - y = 11$$

We can interpret the solution to this system of equations both algebraically and graphically.

	ALGEBRAIC	**GRAPHICAL**
Solution	$x = 4$ and $y = 1$	$(4, 1)$
Check	**Equation 1** **Equation 2** $x + 2y = 6$ $3x - y = 11$ $(4) + 2(1) = 6$ ✓ $3(4) - 1 = 11$ ✓	
Interpretation	$x = 4$ and $y = 1$ satisfy both equations.	The point $(4, 1)$ lies on both lines.

This particular example had *one solution*. There are systems of equations that have *no solution* or *infinitely many solutions*. We give these systems special names: **independent**, **inconsistent**, and **dependent**, respectively.

INDEPENDENT SYSTEM	INCONSISTENT SYSTEM	DEPENDENT SYSTEM
One solution	No solution	Infinitely many solutions
Lines have different slopes	Lines are parallel (same slope and different y-intercepts)	Lines coincide (same slope and same y-intercept)

In this section we discuss three methods for solving systems of linear equations: *substitution*, *elimination*, and *graphing*. We use the algebraic methods—substitution and elimination—to find solutions exactly; we then look at a graphical interpretation of the solution (two lines that intersect at one point, parallel lines, or coinciding lines).

We will illustrate each method with the same example given earlier:

$$x + 2y = 6 \qquad \text{Equation (1)}$$

$$3x - y = 11 \qquad \text{Equation (2)}$$

Substitution Method

The following box summarizes the substitution method for solving systems of linear equations in two variables.

SUBSTITUTION METHOD

Step 1: Solve one of the equations for one variable in terms of the other variable.

Equation (2): $y = 3x - 11$

Step 2: Substitute the expression found in Step 1 into the *other* equation. The result is an equation in one variable.

Equation (1): $x + 2(3x - 11) = 6$

Step 3: Solve the equation obtained in Step 2.

$$x + 6x - 22 = 6$$
$$7x = 28$$
$$\boxed{x = 4}$$

Step 4: Back-substitute the value found in Step 3 into the expression found in Step 1.

$$y = 3(4) - 11$$
$$\boxed{y = 1}$$

Step 5: Check that the solution satisfies *both* equations. Substitute (4, 1) into both equations.

Equation (1): $x + 2y = 6$
$$(4) + 2(1) = 6 \checkmark$$

Equation (2): $3x - y = 11$
$$3(4) - 1 = 11 \checkmark$$

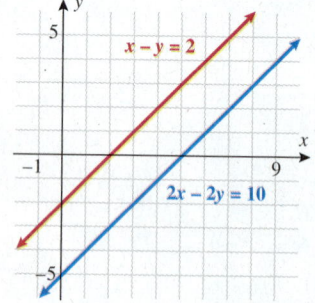

EXAMPLE 1 Determining by Substitution That a System Has One Solution

Use the substitution method to solve the following system of linear equations:

$$x + y = 8 \qquad \text{Equation (1)}$$
$$3x - y = 4 \qquad \text{Equation (2)}$$

Solution:

STEP 1 Solve Equation (2) for y in terms of x. $\qquad y = 3x - 4$

STEP 2 Substitute $y = 3x - 4$ into Equation (1). $\qquad x + (3x - 4) = 8$

STEP 3 Solve for x. $\qquad x + 3x - 4 = 8$
$$4x = 12$$
$$\boxed{x = 3}$$

STEP 4 Back-substitute $x = 3$ into the equation found in Step 1. $\qquad y = 3(3) - 4$
$$\boxed{y = 5}$$

STEP 5 Check that $(3, 5)$ satisfies *both* equations.

$$\text{Equation (1):} \qquad x + y = 8$$
$$3 + 5 = 8$$

$$\text{Equation (2):} \qquad 3x - y = 4$$
$$3(3) - 5 = 4$$

Note: The graphs of the two equations are two lines that intersect at the point $(3, 5)$.

EXAMPLE 2 Determining by Substitution That a System Has No Solution

Use the substitution method to solve the following system of linear equations:

$$x - y = 2 \qquad \text{Equation (1)}$$
$$2x - 2y = 10 \qquad \text{Equation (2)}$$

Solution:

STEP 1 Solve Equation (1) for y in terms of x. $\qquad y = x - 2$

STEP 2 Substitute $y = x - 2$ into Equation (2). $\qquad 2x - 2(x - 2) = 10$

STEP 3 Solve for x. $\qquad 2x - 2x + 4 = 10$
$$4 = 10$$

$4 = 10$ is never true, so this is called an inconsistent system. There is $\boxed{\text{no solution}}$ to this system of linear equations.

Note: The graphs of the two equations are parallel lines.

EXAMPLE 3 **Determining by Substitution That a System Has Infinitely Many Solutions**

Use the substitution method to solve the following system of linear equations:

$$x - y = 2 \qquad \text{Equation (1)}$$
$$-x + y = -2 \qquad \text{Equation (2)}$$

Solution:

STEP 1 Solve Equation (1) for y in terms of x. $\qquad y = x - 2$

STEP 2 Substitute $y = x - 2$ into Equation (2). $\qquad -x + (x - 2) = -2$

STEP 3 Solve for x. $\qquad -x + x - 2 = -2$
$$-2 = -2$$

$-2 = -2$ is always true, so this is called a dependent system. Notice, for instance, that the points $(2, 0)$, $(4, 2)$, and $(7, 5)$ all satisfy both equations. In fact, there are ⟨infinitely many solutions⟩ to this system of linear equations. All solutions are in the form (x, y), where $y = x - 2$ (the graphs of these two equations are the same line). If we let $x = a$, then $y = a - 2$. In other words all of the points $(a, a - 2)$ where a is any real number are solutions to this system of linear equations.

■ **YOUR TURN** Use the substitution method to solve each system of linear equations.

a. $2x + y = 3$
$4x + 2y = 4$

b. $x - y = 2$
$4x - 3y = 10$

c. $x + 2y = 1$
$2x + 4y = 2$

Elimination Method

We now turn our attention to another method, *elimination*, which is often preferred over substitution and will later be used in higher order systems. In a system of two linear equations in two variables, the equations can be combined, resulting in a third equation in one variable, thus *eliminating* one of the variables. The following is an example of when elimination would be preferred because the y terms sum to zero when the two equations are added together:

$$\begin{array}{r} 2x - y = 5 \\ \underline{-x + y = -2} \\ x = 3 \end{array}$$

When you cannot eliminate a variable simply by *adding* the two equations, multiply one equation by a constant that will cause the coefficients of some variable in the two equations to match and be opposite in sign.

The following box summarizes the *elimination method*, also called the *addition method*, for solving systems of linear equations in two variables using the same example given earlier:

$$x + 2y = 6 \qquad \text{Equation (1)}$$
$$3x - y = 11 \qquad \text{Equation (2)}$$

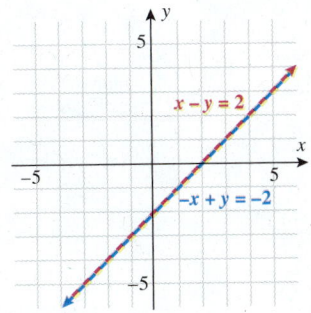

ELIMINATION METHOD

Step 1*: **Multiply** the coefficients of one or both of the equations so that one of the variables will be eliminated when the two equations are added.

Multiply Equation (2) by 2:
$$6x - 2y = 22$$

Step 2: **Eliminate** one of the variables by adding the expression found in Step 1 to the *other* equation. The result is an equation in one variable.

$$\begin{aligned} x + 2y &= 6 \\ \underline{6x - 2y} &= \underline{22} \\ 7x &= 28 \end{aligned}$$

Step 3: **Solve** the equation obtained in Step 2.

$$7x = 28$$
$$\boxed{x = 4}$$

Step 4: **Back-substitute** the value found in Step 3 into either of the two original equations.

$$(4) + 2y = 6$$
$$2y = 2$$
$$\boxed{y = 1}$$

Step 5: **Check** that the solution satisfies *both* equations. Substitute $\boxed{(4, 1)}$ into both equations.

Equation (1):
$$x + 2y = 6$$
$$(4) + 2(1) = 6 ✓$$

Equation (2):
$$3x - y = 11$$
$$3(4) - 1 = 11 ✓$$

*Step 1 is not necessary in cases where corresponding terms already sum to zero.

Study Tip

You can eliminate one variable from the system by addition when (1) the coefficients are equal and (2) the signs are opposite.

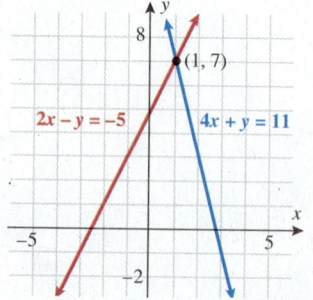

EXAMPLE 4 **Determining by the Elimination Method That a System Has One Solution**

Use the elimination method to solve the following system of linear equations:

$$2x - y = -5 \qquad \text{Equation (1)}$$
$$4x + y = 11 \qquad \text{Equation (2)}$$

Solution:

STEP 1 Not necessary.

STEP 2 Eliminate y by adding Equation (1) to Equation (2).

$$\begin{aligned} 2x - y &= -5 \\ \underline{4x + y} &= \underline{11} \\ 6x &= 6 \end{aligned}$$

STEP 3 Solve for x.

$$\boxed{x = 1}$$

STEP 4 Back-substitute $x = 1$ into Equation (2). Solve for y.

$$4(1) + y = 11$$
$$\boxed{y = 7}$$

STEP 5 Check that $(1, 7)$ satisfies both equations.

Equation (1): $\qquad 2x - y = -5$
$$2(1) - (7) = -5 ✓$$

Equation (2): $\qquad 4x + y = 11$
$$4(1) + (7) = 11 ✓$$

Note: The graphs of the two given equations correspond to two lines that intersect at the point $(1, 7)$.

In Example 4, we eliminated the variable y simply by adding the two equations. Sometimes it is necessary to multiply one (Example 5) or both (Example 6) equations by constants prior to adding.

 EXAMPLE 5 **Applying the Elimination Method When Multiplying One Equation by a Constant Is Necessary**

Use the elimination method to solve the following system of linear equations:

$$-4x + 3y = 23 \qquad \text{Equation (1)}$$
$$12x + 5y = 1 \qquad \text{Equation (2)}$$

Solution:

STEP 1 Multiply Equation (1) by 3.

STEP 2 Eliminate x by adding the modified Equation (1) to Equation (2).

$$-12x + 9y = 69$$

$$\begin{array}{r} -12x + 9y = 69 \\ \underline{12x + 5y = \ 1} \\ 14y = 70 \end{array}$$

STEP 3 Solve for y.

$$\boxed{y = 5}$$

STEP 4 Back-substitute $y = 5$ into Equation (2). Solve for x.

$$12x + 5(5) = 1$$
$$12x + 25 = 1$$
$$12x = -24$$
$$\boxed{x = -2}$$

STEP 5 Check that $(-2, 5)$ satisfies both equations.

Equation (1): $\quad -4(-2) + 3(5) = 23$
$$8 + 15 = 23 \ \checkmark$$

Equation (2): $\quad 12(-2) + 5(5) = 1$
$$-24 + 25 = 1 \ \checkmark$$

Note: The graphs of the two given equations correspond to two lines that intersect at the point $(-2, 5)$.

Study Tip

Be sure to multiply the **entire** equation by the constant.

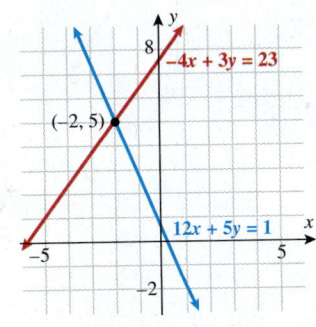

In Example 5, we eliminated x simply by multiplying the first equation by a constant and adding the result to the second equation. In order to eliminate either of the variables in Example 6, we will have to multiply *both* equations by constants prior to adding.

EXAMPLE 6   **Applying the Elimination Method When Multiplying Both Equations by Constants Is Necessary**

Use the elimination method to solve the following system of linear equations:

$$3x + 2y = 1 \qquad \text{Equation (1)}$$
$$5x + 7y = 9 \qquad \text{Equation (2)}$$

Solution:

STEP 1 Multiply Equation (1) by 5 and Equation (2) by -3.

$$15x + 10y = 5$$
$$-15x - 21y = -27$$

STEP 2 Eliminate x by adding the modified Equation (1) to the modified Equation (2).

$$15x + 10y = 5$$
$$\underline{-15x - 21y = -27}$$
$$-11y = -22$$

STEP 3 Solve for y.

$$\boxed{y = 2}$$

STEP 4 Back-substitute $y = 2$ into Equation (1). Solve for x.

$$3x + 2(2) = 1$$
$$3x = -3$$
$$\boxed{x = -1}$$

STEP 5 Check that $(-1, 2)$ satisfies both equations.

Equation (1): $\qquad 3x + 2y = 1$
$\qquad\qquad\qquad 3(-1) + 2(2) = 1 ✓$

Equation (2): $\qquad 5x + 7y = 9$
$\qquad\qquad\qquad 5(-1) + 7(2) = 9 ✓$

Note: The graphs of the two given equations correspond to two lines that intersect at the point $(-1, 2)$.

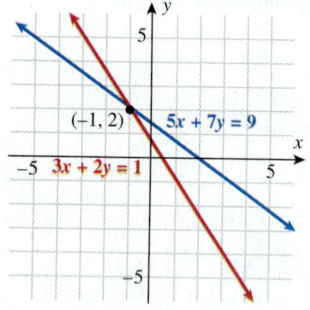

Notice in Example 6 that we could have also eliminated y by multiplying the first equation by 7 and the second equation by -2. Typically, the choice is dictated by which approach will keep the coefficients as simple as possible. In the event that the coefficients contain fractions or decimals, first rewrite the equations in standard form with integer coefficients and then make the decision.

EXAMPLE 7   **Determining by the Elimination Method That a System Has No Solution**

Use the elimination method to solve the following system of linear equations:

$$-x + y = 7 \qquad \text{Equation (1)}$$
$$2x - 2y = 4 \qquad \text{Equation (2)}$$

Solution:

STEP 1 Multiply Equation (1) by 2.

$$-2x + 2y = 14$$

STEP 2 Eliminate y by adding the modified Equation (1) found in Step 1 to Equation (2).

$$-2x + 2y = 14$$
$$\underline{2x - 2y = 4}$$
$$0 = 18$$

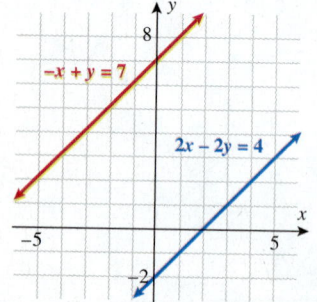

This system is inconsistent since $0 = 18$ is never true. Therefore, there are no values of x and y that satisfy both equations. We say that there is $\boxed{\text{no solution}}$ to this system of linear equations.

Note: The graphs of the two equations are two parallel lines.

EXAMPLE 8 **Determining by the Elimination Method That a System Has Infinitely Many Solutions**

Use the elimination method to solve the following system of linear equations:

$$7x + y = 2 \qquad \text{Equation (1)}$$
$$-14x - 2y = -4 \qquad \text{Equation (2)}$$

Solution:

STEP 1 Multiply Equation (1) by 2.

$$14x + 2y = 4$$

STEP 2 Add the modified Equation (1) found in Step 1 to Equation (2).

$$\begin{array}{r} 14x + 2y = 4 \\ -14x - 2y = -4 \\ \hline 0 = 0 \end{array}$$

This system is dependent since $0 = 0$ is always true. We say that there are $\boxed{\text{infinitely many solutions}}$ to this system of linear equations, and these can be represented by the points $(a, 2 - 7a)$.

Note: The graphs of the two equations are the same line.

▪ **YOUR TURN** Apply the elimination method to solve each system of linear equations.

a. $2x + 3y = 1$
$\ 4x - 3y = -7$

b. $\ x - 5y = 2$
$\ -10x + 50y = -20$

c. $\ x - y = 14$
$\ -x + y = 9$

▪ **Answer: a.** $(-1, 1)$
b. infinitely many
solutions: $\left(a, \dfrac{a-2}{5}\right)$
c. no solution

Graphing Method

A third way to solve a system of linear equations in two variables is to graph the two lines. If the two lines intersect, then the point of intersection is the solution. Graphing is the most labor-intensive method for solving systems of linear equations in two variables. The graphing method is typically not used to solve systems of linear equations when an exact solution is desired. Instead, it is used to interpret or confirm the solution(s) found by the other two methods (substitution and elimination). If you are using a graphing calculator, however, you will get as accurate an answer using the graphing method as you will when applying the other methods.

The following box summarizes the graphing method for solving systems of linear equations in two variables using the same example given earlier:

$$x + 2y = 6 \qquad \text{Equation (1)}$$
$$3x - y = 11 \qquad \text{Equation (2)}$$

Classroom Example 9.1.8
Solve using elimination.
$$-4y - x = -3$$
$$2x + 8y = 6$$

Answer:
(x, y), where $x = -4y + 3$
or $\left(a, \dfrac{3-a}{4}\right)$

GRAPHING METHOD

Step 1*: Write the equations in slope–intercept form.

Equation (1):   Equation (2):

$$y = -\frac{1}{2}x + 3 \qquad y = 3x - 11$$

Step 2: Graph the two lines.

Step 3: Identify the point of intersection.

(4, 1)

Step 4: Check that the solution satisfies *both* equations.

Equation (1):   Equation (2):

$x + \quad 2y = 6$ $3x - y = 11$

$(4) + 2(1) = 6$ ✓ $3(4) - 1 = 11$ ✓

*Step 1 is not necessary when the lines are already in slope–intercept form.

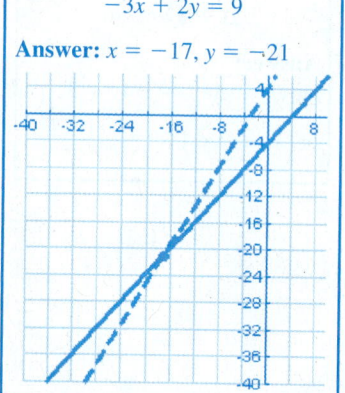
EXAMPLE 9 **Determining by Graphing That a System Has One Solution**

Use graphing to solve the following system of linear equations:

$$x + y = 2 \qquad \text{Equation (1)}$$
$$3x - y = 2 \qquad \text{Equation (2)}$$

Solution:

STEP 1 Write each equation in slope–intercept form.

$y = -x + 2$ Equation (1)

$y = 3x - 2$ Equation (2)

STEP 2 Plot both lines on the same graph.

STEP 3 Identify the point of intersection.

(1, 1)

STEP 4 Check that the point (1, 1) satisfies both equations.

$x + y = 2$

$1 + 1 = 2$ ✓   Equation (1)

$3x - \quad y = 2$

$3(1) - (1) = 2$ ✓   Equation (2)

Note: There is one solution, because the two lines intersect at one point.

EXAMPLE 10   **Determining by Graphing That a System Has No Solution**

Use graphing to solve the following system of linear equations:

$$2x - 3y = 9 \qquad \text{Equation (1)}$$
$$-4x + 6y = 12 \qquad \text{Equation (2)}$$

Solution:

STEP 1 Write each equation in slope–intercept form.

$$y = \frac{2}{3}x - 3 \qquad \text{Equation (1)}$$
$$y = \frac{2}{3}x + 2 \qquad \text{Equation (2)}$$

STEP 2 Plot both lines on the same graph.

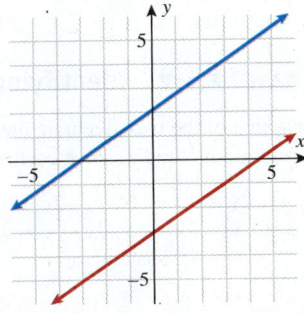

STEP 3 Identify the point of intersection.   None

The two lines are parallel because they have the same slope, but different y-intercepts. For this reason there is ⎸no solution⎸—two parallel lines do not intersect.

EXAMPLE 11   **Determining by Graphing That a System Has Infinitely Many Solutions**

Use graphing to solve the following system of linear equations:

$$3x + 4y = 12 \qquad \text{Equation (1)}$$
$$\frac{3}{4}x + y = 3 \qquad \text{Equation (2)}$$

Solution:

STEP 1 Write each equation in slope–intercept form.

$$y = -\frac{3}{4}x + 3 \quad y = -\frac{3}{4}x + 3$$

STEP 2 Plot both lines on the same graph.

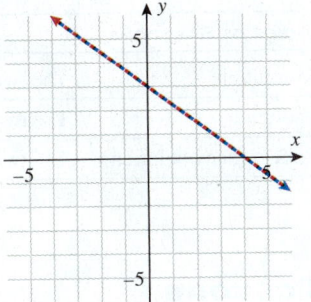

STEP 3 Identify the point of intersection.   Infinitely many points

There are ⎸infinitely many solutions⎸, since the two lines are identical and coincide.

The points that lie along the line are $\left(a, -\frac{3}{4}a + 3 \right)$.

■ **YOUR TURN** Utilize graphing to solve each system of linear equations.

a. $x - 2y = 1$	**b.** $x - 2y = 1$	**c.** $2x + y = 3$
$2x - 4y = 2$	$2x + y = 7$	$2x + y = 7$

Technology Tip

First solve for y in each equation. The graphs of $Y_1 = \frac{1}{3}(2x - 9)$ and $Y_2 = \frac{1}{6}(4x + 12)$ are shown.

The graphs and table show that there is no solution to the system.

Technology Tip

Solve for y in each equation first. The graphs of $Y_1 = \frac{1}{4}(-3x + 12)$ and $Y_2 = -\frac{3}{4}x + 3$ are shown.

Both lines are on the same graph.

■ **Answer:**

a. infinitely many solutions
solutions: $\left(a, \dfrac{a-1}{2} \right)$

b. $(3, 1)$

c. no solution

Three Methods and Three Types of Solutions

Given any system of two linear equations in two variables, any of the three methods (substitution, elimination, or graphing) can be utilized. If you find that it is easy to eliminate a variable by adding multiples of the two equations, then elimination is the preferred choice. If you do not see an obvious elimination, then solve the system by substitution. For exact solutions, choose one of these two algebraic methods. You should typically use graphing to confirm the solution(s) you have found by applying the other two methods or when you are using a graphing utility.

 EXAMPLE 12 Identifying Which Method to Use

State which of the two algebraic methods (elimination or substitution) would be the preferred method to solve each system of linear equations.

a. $x - 2y = 1$ **b.** $x = 2y - 1$ **c.** $7x - 20y = 1$
    $-x + y = 2$ $2x - y = 4$ $5x + 3y = 18$

Solution:

a. Elimination: Because the x variable is eliminated when the two equations are added.

b. Substitution: Because the first equation is easily substituted into the second equation (for x).

c. Either: There is no preferred method, as both elimination and substitution require substantial work.

Regardless of which method is used to solve systems of linear equations in two variables, in general, we can summarize the three types of solutions both algebraically and graphically.

THREE TYPES OF SOLUTIONS TO SYSTEMS OF LINEAR EQUATIONS

NUMBER OF SOLUTIONS	GRAPHICAL INTERPRETATION
One solution	The two lines intersect at one point.
No solution	The two lines are parallel. (*Same* slope/*different* y-intercepts.)
Infinitely many solutions	The two lines coincide. (*Same* slope/*same* y-intercept.)

Applications

Suppose you have two job offers that require sales. One pays a higher base, while the other pays a higher commission. Which job do you take?

EXAMPLE 13   Deciding Which Job to Take

Suppose that upon graduation you are offered a job selling biomolecular devices to laboratories studying DNA. The Beckman-Coulter Company offers you a job selling its DNA sequencer with an annual base salary of $20,000 plus 5% commission on total sales. The MJ Research Corporation offers you a job selling its PCR Machine that makes copies of DNA with an annual base salary of $30,000 plus 3% commission on sales. Determine what the total sales would have to be to make the Beckman-Coulter job the better offer.

Solution:

STEP 1 Identify the question.
When would these two jobs have equal compensations?

STEP 2 Make notes.

| Beckman-Coulter | $20,000 + 5\%$ |
| MJ Research | $30,000 + 3\%$ |

STEP 3 Set up the equations.
Let x = total sales and y = compensation.

Equation (1)   Beckman-Coulter: $y = 20{,}000 + 0.05x$

Equation (2)   MJ Research: $y = 30{,}000 + 0.03x$

STEP 4 Solve the system of equations.
*Substitution method**

Substitute Equation (1)
into Equation (2). $20{,}000 + 0.05x = 30{,}000 + 0.03x$

Solve for x. $0.02x = 10{,}000$
$x = 500{,}000$

If you make $500,000 worth of sales per year, the jobs will yield equal compensations. If you sell less than $500,000, the MJ Research job is the better offer, and more than $500,000, the Beckman-Coulter job is the better offer.

*The elimination method could also have been used.

STEP 5 Check the solution.

Equation (1)   Beckman-Coulter: $y = 20{,}000 + 0.05(500{,}000) = \$45{,}000$

Equation (2)   MJ Research: $y = 30{,}000 + 0.03(500{,}000) = \$45{,}000$

Technology Tip

The graphs of $Y_1 = 20{,}000 + 0.05x$ and $Y_2 = 30{,}000 + 0.03x$ are shown.

The graphs and table support the solution to the system.

Technology Tip

First solve for y in each equation. The graphs of $Y_1 = -x + 5$ and $Y_2 = \frac{1}{7}(-8x + 37)$ are shown.

The graphs and table support the solution to the system.

EXAMPLE 14 Deciding How Many Pounds of Each Meat to Buy at the Deli

The Chi Omega sorority is hosting a party, and the membership committee would like to make sandwiches for the new members. They already have bread and condiments but ask Tara to run to the deli to buy sliced turkey. The membership chair has instructed Tara to buy 5 pounds of the best turkey she can find, and the treasurer has given her $37 and told her to spend it all. When she arrives at Publix supermarket, she has a choice of two types of turkey: Boar's Head ($8 per pound) and the store brand ($7 per pound). How much of each kind should Tara buy to spend the entire $37 on the best quality of turkey?

Solution:

STEP 1 Identify the question.

How much of each type of sliced turkey should Tara buy?

STEP 2 Make notes.

Store brand turkey costs $7/pound.
Boar's Head turkey costs $8/pound.
Tara has $37 to spend on 5 pounds of turkey.

STEP 3 Set up the equations.

Let x = number of pounds of Boar's Head turkey and y = number of pounds of store brand turkey. The system of equations therefore has to include one equation giving a total of 5 pounds and one giving a total of $37.

Equation (1): $x + y = 5$

Equation (2): $8x + 7y = 37$

STEP 4 Solve the system of equations.

Elimination method*

Multiply Equation (1) by -7. $-7x - 7y = -35$

Add this result to Equation (2).

$$\begin{array}{r} -7x - 7y = -35 \\ 8x + 7y = 37 \\ \hline x = 2 \end{array}$$

Substitute $x = 2$ into the original Equation (1). $2 + y = 5$
Solve for y. $y = 3$

> Tara should buy 2 pounds of Boar's Head and 3 pounds of store brand turkey.

*The substitution method could also have been used to solve this system of linear equations.

STEP 5 Check the solution.

Equation (1): $2 + 3 = 5$

Equation (2): $8(2) + 7(3) = 37$

Classroom Example 9.1.14
You have a total of 40 coins of either dimes or nickels amounting to $3.05. How many nickels and dimes do you have?

Answer:
19 nickels and 21 dimes

In this section we discussed two algebraic techniques for solving systems of linear equations in two variables:

- Substitution method
- Elimination method

The algebraic methods are preferred for exact solutions, and the graphing method is typically used to give a visual interpretation and confirmation of the solution. There are three types of solutions to systems of linear equations: one solution, no solution, or infinitely many solutions.

INDEPENDENT SYSTEM	INCONSISTENT SYSTEM	DEPENDENT SYSTEM
One solution	No solution	Infinitely many solutions
Lines have different slopes	Lines are parallel (same slope and different y-intercepts)	Lines coincide (same slope and same y-intercept)

SECTION
9.1 EXERCISES

SKILLS

In Exercises 1–22, solve each system of linear equations by substitution.

1. $x - y = 1$
$x + y = 1$

2. $x - y = 2$
$x + y = -2$

3. $x + y = 7$
$x - y = 9$

4. $x - y = -10$
$x + y = 4$

5. $2x - y = 3$
$x - 3y = 4$

6. $4x + 3y = 3$
$2x + y = 1$

7. $3x + y = 5$
$2x - 5y = -8$

8. $6x - y = -15$
$2x - 4y = -16$

9. $2u + 5v = 7$
$3u - v = 5$

10. $m - 2n = 4$
$3m + 2n = 1$

11. $2x + y = 7$
$-2x - y = 5$

12. $3x - y = 2$
$3x - y = 4$

13. $4r - s = 1$
$8r - 2s = 2$

14. $-3p + q = -4$
$6p - 2q = 8$

15. $5r - 3s = 15$
$-10r + 6s = -30$

16. $-5p - 3q = -1$
$10p + 6q = 2$

17. $2x - 3y = -7$
$3x + 7y = 24$

18. $4x - 5y = -7$
$3x + 8y = 30$

19. $\frac{1}{3}x - \frac{1}{4}y = 0$
$-\frac{2}{3}x + \frac{3}{4}y = 2$

20. $\frac{1}{5}x + \frac{2}{3}y = 10$
$-\frac{1}{2}x - \frac{1}{6}y = -7$

21. $7.2x - 4.1y = 7.0$
$-3.5x + 16.5y = 2.4$

22. $6.3x + 1.5y = 10.5$
$-0.4x + 2.2y = -8.7$

In Exercises 23–42, solve each system of linear equations by elimination.

23. $x - y = 2$
$x + y = 4$

24. $x + y = 2$
$x - y = -2$

25. $x - y = -3$
$x + y = 7$

26. $x - y = -10$
$x + y = 8$

27. $5x + 3y = -3$
$3x - 3y = -21$

28. $-2x + 3y = 1$
$2x - y = 7$

29. $2x - 7y = 4$
$5x + 7y = 3$

30. $3x + 2y = 6$
$-3x + 6y = 18$

31. $2x + 5y = 7$
$3x - 10y = 5$

32. $6x - 2y = 3$
$-3x + 2y = -2$

33. $2x + 5y = 5$
$-4x - 10y = -10$

34. $11x + 3y = 3$
$22x + 6y = 6$

35. $3x - 2y = 12$
$4x + 3y = 16$

36. $5x - 2y = 7$
$3x + 5y = 29$

37. $6x - 3y = -15$
$7x + 2y = -12$

38. $7x - 4y = -1$
$3x - 5y = 16$

39. $0.02x + 0.05y = 1.25$
$-0.06x - 0.15y = -3.75$

40. $-0.5x + 0.3y = 0.8$
$-1.5x + 0.9y = 2.4$

41. $\frac{1}{3}x + \frac{1}{2}y = 1$
$\frac{1}{5}x + \frac{7}{2}y = 2$

42. $\frac{1}{2}x - \frac{1}{3}y = 0$
$\frac{3}{2}x + \frac{1}{2}y = \frac{3}{4}$

In Exercises 43–46, match the systems of equations with the graphs.

43. $3x - y = 1$
$3x + y = 5$

44. $-x + 2y = -1$
$2x + y = 7$

45. $2x + y = 3$
$2x + y = 7$

46. $x - 2y = 1$
$2x - 4y = 2$

a.

b.

c.

d.

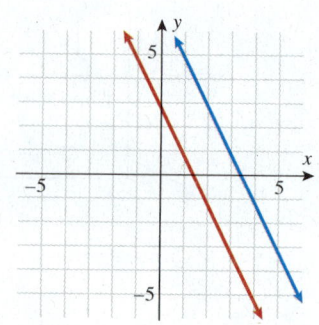

In Exercises 47–54, solve each system of linear equations by graphing.

47. $y = -x$
$y = x$

48. $x - 3y = 0$
$x + 3y = 0$

49. $2x + y = -3$
$x + y = -2$

50. $x - 2y = -1$
$-x - y = -5$

51. $\frac{1}{2}x - \frac{2}{3}y = 4$
$\frac{1}{4}x - y = 6$

52. $\frac{1}{5}x - \frac{5}{2}y = 10$
$\frac{1}{15}x - \frac{5}{6}y = \frac{10}{3}$

53. $1.6x - y = 4.8$
$-0.8x + 0.5y = 1.5$

54. $1.1x - 2.2y = 3.3$
$-3.3x + 6.6y = -6.6$

■ APPLICATIONS

55. Environment. Approximately 2 million dry erase markers are disposed of by teachers each year. Traditional dry erase markers are toxic and nonbiodegradable. EcoSmart World sells AusPens, which are dry erase markers that are recyclable, refillable, and nontoxic. The markers are available in a kit containing x AusPens and x refill ink bottles, each a different color. One refill ink bottle can refill up to 40 AusPens, and one kit is equivalent to y dry erase markers. If one kit is equivalent to 246 traditional dry erase markers, find the number of AusPens that are in each kit.

56. Pharmacy. A pharmacy technician receives an order for 454 grams of a 3% zinc oxide cream. If the pharmacy has 1% and 10% zinc oxide creams in stock, how much of each should be mixed to fill the order?

57. Event Planning. You are in charge of event planning for a major corporation that is having a reception. Your boss has given you a budget of $1000 for gifts and wants you to order Montblanc pens for VIPs and Cross pens for other guests. A Montblanc pen costs $72 and a Cross pen costs $10. You must have 69 pens for guests. How many pens of each type should you order in order to stay within the budget?

58. Health Club Management. A fitness club has a budget of $915 to purchase two different types of dumbbell sets. One set costs $30 each, and the deluxe model dumbbell set costs $45 each. The club wants to purchase 24 new dumbbell sets to use in its gym. How many sets of each type can it purchase?

59. Mixture. In chemistry lab, Stephanie has to make a 37 milliliter solution that is 12% HCl. All that is in the lab is 8% and 15% HCl. How many milliliters of each solution should she use to obtain the desired mix?

60. Mixture. A mechanic has 340 gallons of gasoline and 10 gallons of oil to make gas/oil mixtures. He wants one mixture to be 4% oil and the other mixture to be 2.5% oil. If he wants to use all of the gas and oil, how many gallons of gas and oil are in each of the resulting mixtures?

61. Salary Comparison. Upon graduation with a degree in management of information systems (MIS), you decide to work for a company that buys data from states' departments of motor vehicles and sells to banks and car dealerships customized reports detailing how many cars at each dealership are financed through particular banks. Autocount Corporation offers you a $15,000 base salary and 10% commission on your total annual sales. Polk Corporation offers you a base salary of $30,000 plus a 5%commission on your total annual sales. How many total sales would you have to make per year to make more money at Autocount?

62. Salary Comparison. Two types of residential real estate agents are: those who sell existing houses (resale) and those who sell new homes for developers. Resale of existing homes typically earns 6% commission on every sale, and representing developers in selling new homes typically earns a base salary of $15,000 per year plus an additional 1.5% commission because agents are required to work 5 days a week on site in a new development. How many total dollars would an agent have to sell per year to make more money in resale than in new homes?

63. Gas Mileage. A Honda Accord gets approximately 26 mpg on the highway and 19 mpg in the city. You drove 349.5 miles on a full tank (16 gallons) of gasoline. Approximately how many miles did you drive in the city and how many on the highway?

64. Wireless Plans. AT&T is offering a 600-minute peak plan with free mobile-to-mobile and weekend minutes at $59 per month plus $0.13 per minute for every minute over 600. The next plan up is the 800-minute plan that costs $79 per month. You think you may go over 600 minutes, but are not sure you need 800 minutes. How many minutes would you have to talk for the 800-minute plan to be the better deal?

65. Distance/Rate/Time. A direct flight on Delta Air Lines from Atlanta to Paris is 4000 miles and takes approximately 8 hours going east (Atlanta to Paris) and 10 hours going west (Paris to Atlanta). Although the plane averages the same air speed, there is a headwind while traveling west and a tailwind while traveling east, resulting in different air speeds. What is the average air speed of the plane, and what is the average wind speed?

66. Distance/Rate/Time. A private pilot flies a Cessna 172 on a trip that is 500 miles each way. It takes her approximately 3 hours to get there and 4 hours to return. What is the approximate air speed of the Cessna, and what is the approximate wind speed?

67. Investment Portfolio. Leticia has been tracking two volatile stocks. Stock A over the last year has increased 10%, and stock B has increased 14% (using a simple interest model). She has $10,000 to invest and would like to split it between these two stocks. If the stocks continue to perform at the same rate, how much should she invest in each to result in a balance of $11,260?

68. Investment Portfolio. Toby split his savings into two different investments, one earning 5% and the other earning 7%. He put twice as much in the investment earning the higher rate. In 1 year, he earned $665 in interest. How much money did he invest in each account?

69. Break-Even Analysis. A company produces CD players for a unit cost of $15 per CD player. The company has fixed costs of $120. If each CD player can be sold for $30, how many CD players must be sold to break even? Determine the cost equation first. Next, determine the revenue equation. Use the two equations you have found to determine the break-even point.

70. Managing a Lemonade Stand. An elementary-school-age child wants to have a lemonade stand. She wants to sell each glass of lemonade for $0.25. She has determined that each glass of lemonade costs about $0.10 to make (for lemons and sugar). It costs her $15.00 for materials to make the lemonade stand. How many glasses of lemonade must she sell to break even?

■**CATCH THE MISTAKE**

In Exercises 71–74, explain the mistake that is made.

71. Solve the system of equations by elimination.

$$2x + y = -3$$
$$3x + y = 8$$

Solution:

Multiply Equation (1) by -1. $2x - y = -3$

Add the result to Equation (2). $\dfrac{3x + y = 8}{5x = 5}$

Solve for x. $x = 1$

Substitute $x = 1$ into Equation (2). $3(1) + y = 8$
      $y = 5$

The answer $(1, 5)$ is incorrect. What mistake was made?

72. Solve the system of equations by elimination.

$$4x - y = 12$$
$$4x - y = 24$$

Solution:

Multiply Equation (1) by -1. $-4x + y = -12$

Add the result to Equation (2). $-4x + y = -12$
      $\dfrac{4x - y = 24}{0 = 12}$

Answer: Infinitely many solutions.

This is incorrect. What mistake was made?

73. Solve the system of equations by substitution.

$$x + 3y = -4$$
$$-x + 2y = -6$$

Solution:

Solve Equation (1) for x. $x = -3y - 4$

Substitute $x = -3y - 4$
into Equation (2). $-(-3y - 4) + 2y = -6$

Solve for y. $3y - 4 + 2y = -6$
      $5y = -2$
      $y = -\dfrac{2}{5}$

Substitute $y = -\dfrac{2}{5}$ into Equation (1). $x + 3\left(-\dfrac{2}{5}\right) = -4$

Solve for x. $x = -\dfrac{14}{5}$

The answer $\left(-\dfrac{2}{5}, -\dfrac{14}{5}\right)$ is incorrect. What mistake was made?

74. Solve the system of equations by graphing.

$$2x + 3y = 5$$
$$4x + 6y = 10$$

Solution:

Write both equations in
slope–intercept form. $y = -\dfrac{2}{3}x + \dfrac{5}{3}$

      $y = -\dfrac{2}{3}x + \dfrac{5}{3}$

Since these lines have the same slope, they are parallel lines.
Parallel lines do not intersect, so there is no solution.

This is incorrect. What mistake was made?

■**CONCEPTUAL**

In Exercises 75–78, determine whether each statement is true or false.

75. A system of equations represented by a graph of two lines with the same slope always has no solution.

76. A system of equations represented by a graph of two lines with slopes that are negative reciprocals always has one solution.

77. If two lines do not have exactly one point of intersection, then they must be parallel.

78. The system of equations $Ax - By = 1$ and $-Ax + By = -1$ has no solution.

▪ CHALLENGE

79. The point $(2, -3)$ is a solution to the system of equations

$$Ax + By = -29$$
$$Ax - By = 13$$

Find A and B.

80. If you graph the lines

$$x - 50y = 100$$
$$x - 48y = -98$$

they appear to be parallel lines. However, there is a unique solution. Explain how this might be possible.

81. Energy Drinks. A nutritionist wishes to market a new vitamin-enriched fruit drink and is preparing two versions of it to distribute at a local health club. She has 100 cups of pineapple juice and 4 cups of super vitamin-enriched pomegranate concentrate. One version of the drink is to contain 2% pomegranate and the other version 4% pomegranate. How much of each drink can she create?

82. Easter Eggs. A family is coloring Easter eggs and wants to make two shades of purple, "light purple" and "deep purple." They have 30 tablespoons of deep red solution and 2 tablespoons of blue solution. If "light purple" consists of 2% blue solution and "deep purple" consists of 10% blue solution, how much of each version of purple solution can be created?

▪ TECHNOLOGY

83. Use a graphing utility to graph the two equations $y = -1.25x + 17.5$ and $y = 2.3x - 14.1$. Approximate the solution to this system of linear equations.

84. Use a graphing utility to graph the two equations $y = 14.76x + 19.43$ and $y = 2.76x + 5.22$. Approximate the solution to this system of linear equations.

85. Use a graphing utility to graph the two equations and determine the solution set: $23x + 15y = 7$ and $46x + 30y = 14$.

86. Use a graphing utility to graph the two equations and determine the solution set: $-3x + 7y = 2$ and $6x - 14y = 3$.

87. Use a graphing utility to graph the two equations $\frac{1}{3}x - \frac{5}{12}y = \frac{5}{6}$ and $\frac{3}{7}x + \frac{1}{14}y = \frac{29}{28}$. Approximate the solution to this system of linear equations.

88. Use a graphing utility to graph the two equations $\frac{5}{9}x + \frac{11}{13}y = 2$ and $\frac{3}{4}x + \frac{5}{7}y = \frac{13}{14}$. Approximate the solution to this system of linear equations.

SECTION **9.2**	SYSTEMS OF LINEAR EQUATIONS IN THREE VARIABLES

SKILLS OBJECTIVES

- Solve systems of linear equations in three variables using a combination of the elimination method and the substitution method.
- Solve application problems using systems of linear equations in three variables.

CONCEPTUAL OBJECTIVES

- Understand that a graph of a linear equation in three variables corresponds to a plane.
- Identify three types of solutions: one solution (point), no solution, or infinitely many solutions (a single line in three dimensional space).

Solving Systems of Linear Equations in Three Variables

In Section 9.1, we solved systems of linear equations in two variables. Graphs of linear equations in two variables correspond to lines. Now we turn our attention to linear equations in *three* variables. A **linear equation in three variables** x, y, and z, is given by

$$Ax + By + Cz = D$$

where A, B, C, and D are real numbers that are not all equal to zero. All three variables have degree equal to one, which is why this is called a linear equation in three variables. The graph of any equation in three variables requires a three-dimensional coordinate system.

The *x*-axis, *y*-axis, and *z*-axis are each perpendicular to the other two. For the three-dimensional coordinate system on the right, a point $(x, y, z) = (2, 3, 1)$ is found by starting at the origin, moving two units to the right, three units up, and one unit out toward you.

In two variables, the graph of a linear equation is a line. In three variables, however, the graph of a linear equation is a **plane**. A plane can be thought of as an infinite sheet of paper. When solving systems of linear equations in three variables, we find one of three possibilities: one solution, no solution, and infinitely many solutions.

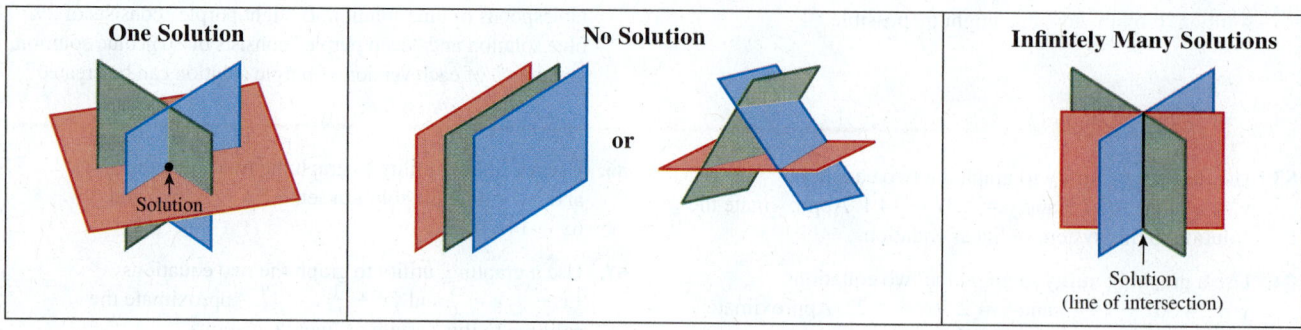

There are many ways to solve systems of linear equations in more than two variables. One method is to combine the elimination and substitution methods, which will be discussed in this section. Other methods involve matrices, which will be discussed in the next chapter. We now outline a procedure for solving systems of linear equations in three variables, which can be extended to solve systems of more than three variables. Solutions are usually given as ordered triples of the form (x, y, z).

SOLVING SYSTEMS OF LINEAR EQUATIONS IN THREE VARIABLES USING ELIMINATION AND SUBSTITUTION

Step 1: Reduce the system of three equations in three variables to two equations in two (of the same) variables by applying elimination.

Step 2: Solve the resulting system of two linear equations in two variables by applying elimination or substitution.

Step 3: Substitute the solutions in Step 2 into *any* of the original equations and solve for the third variable.

Step 4: Check that the solution satisfies *all* three original equations.

EXAMPLE 1 **Solving a System of Linear Equations in Three Variables**

Solve the system:
$$2x + y + 8z = -1 \qquad \text{Equation (1)}$$
$$x - y + z = -2 \qquad \text{Equation (2)}$$
$$3x - 2y - 2z = 2 \qquad \text{Equation (3)}$$

Classroom Example 9.2.1
Solve the system:
$$4x - 3y + 5z = 8$$
$$x + y - 3z = -3$$
$$5x - 3y - 2z = -5$$

Answer:
$x = 1, y = 2, z = 2$

Solution:

Inspecting the three equations, we see that y is easily eliminated when Equations (1) and (2) are added, because the coefficients of y, $+1$ and -1, are equal in magnitude and opposite in sign. We can also eliminate y from Equation (3) by adding Equation (3) to *either* 2 times Equation (1) *or* -2 times Equation (2). Therefore, our plan of attack is to eliminate y from the system of equations, so the result will be two equations in two variables x and z.

STEP 1 Eliminate y in Equation (1) and Equation (2).

Equation (1):	$2x + y + 8z = -1$
Equation (2):	$x - y + z = -2$
Add.	$3x + 9z = -3$

Eliminate y in Equation (2) and Equation (3).

Multiply Equation (2) by -2.	$-2x + 2y - 2z = 4$
Equation (3):	$3x - 2y - 2z = 2$
Add.	$x - 4z = 6$

STEP 2 Solve the system of two linear equations in two variables.
$$3x + 9z = -3$$
$$x - 4z = 6$$

Substitution* method: $x = 4z + 6$	$3(4z + 6) + 9z = -3$
Distribute.	$12z + 18 + 9z = -3$
Combine like terms.	$21z = -21$
Solve for z.	$\boxed{z = -1}$
Substitute $z = -1$ into $x = 4z + 6$.	$x = 4(-1) + 6 = 2$

$\boxed{x = 2}$ and $\boxed{z = -1}$ are the solutions to the system of two equations.

STEP 3 Substitute $x = 2$ and $z = -1$ into any of the three original equations and solve for y.

Substitute $x = 2$ and $z = -1$ into Equation (2).	$2 - y - 1 = -2$
Solve for y.	$\boxed{y = 3}$

STEP 4 Check that $x = 2$, $y = 3$, and $z = -1$ satisfy all three equations.

Equation (1): $2(2) + 3 + 8(-1) = 4 + 3 - 8 = -1$

Equation (2): $2 - 3 - 1 = -2$

Equation (3): $3(2) - 2(3) - 2(-1) = 6 - 6 + 2 = 2$

The solution is $\boxed{x = 2, y = 3, z = -1, \text{ or } (2, 3, -1)}$.

*Elimination method could also be used.

Study Tip

Eliminate the *same* variable.

■ **YOUR TURN** Solve the system:
$$2x - y + 3z = -1$$
$$x + y - z = 0$$
$$3x + 3y - 2z = 1$$

■ **Answer:**
$x = -1, y = 2, z = 1$

In Example 1 and the Your Turn, the variable y was eliminated by adding the first and second equations. In practice, any of the three variables can be eliminated, but typically we select the most convenient variable to eliminate. If a variable is missing from one of the equations (has a coefficient of 0), then we eliminate that variable from the other two equations.

EXAMPLE 2 Solving a System of Linear Equations in Three Variables When One Variable Is Missing

Solve the system:

$$x \qquad + z = 1 \qquad \text{Equation (1)}$$
$$2x + y - z = -3 \qquad \text{Equation (2)}$$
$$x + 2y - z = -1 \qquad \text{Equation (3)}$$

Solution:

Since y is missing from Equation (1), y is the variable to be eliminated in Equation (2) and Equation (3).

STEP 1 Eliminate y.

Multiply Equation (2) by -2. $\qquad\qquad\qquad\qquad -4x - 2y + 2z = 6$
Equation (3): $\qquad\qquad\qquad\qquad\qquad\quadx + 2y - z = -1$
Add. $\qquad\qquad\qquad\qquad\qquad\qquad\qquad\quad -3x + z = 5$

STEP 2 Solve the system of two equations: $\qquad\qquad x + z = 1$
Equation (1) and the resulting equation in Step 1. $\qquad -3x + z = 5$

Multiply the second equation by $\qquad\qquad\qquad\qquad\qquad x + z = 1$
(-1) and add to first equation. $\qquad\qquad\qquad\qquad 3x - z = -5$
$\qquad\qquad\qquad\qquad\qquad\qquad\qquad\qquad\qquad\qquad 4x = -4$

Solve for x. $\qquad\qquad\qquad\qquad\qquad\qquad\qquad\qquad\qquad x = -1$

Substitute $x = -1$ into Equation (1). $\qquad\qquad\qquad -1 + z = 1$

Solve for z. $\qquad\qquad\qquad\qquad\qquad\qquad\qquad\qquad\qquad z = 2$

STEP 3 Substitute $x = -1$ and $z = 2$ into one of the original equations (Equation 2 or Equation 3) and solve for y.

Substitute $x = -1$ and $z = 2$
into $x + 2y - z = -1$. $\qquad\qquad\qquad\qquad\qquad (-1) + 2y - 2 = -1$

Gather like terms. $\qquad\qquad\qquad\qquad\qquad\qquad\qquad\qquad 2y = 2$

Solve for y. $\qquad\qquad\qquad\qquad\qquad\qquad\qquad\qquad\qquad y = 1$

STEP 4 Check that $x = -1$, $y = 1$, and $z = 2$ satisfy all three equations.

Equation (1): $(-1) + 2 = 1$

Equation (2): $2(-1) + (1) - (2) = -3$

Equation (3): $(-1) + 2(1) - (2) = -1$

The solution is $\boxed{x = -1, y = 1, z = 2}$.

Classroom Example 9.2.2
Solve the system:
$$x + y + z = 0$$
$$y + 3x = 0$$
$$y - 2z = 7$$
Answer:
$x = -1, y = 3, z = -2$

■ **Answer:** $x = 1, y = 2, z = -3$

■ **YOUR TURN** Solve the system:

$$x + y + z = 0$$
$$2x + z = -1$$
$$x - y - z = 2$$

Types of Solutions

There are three types of solutions: independent, dependent, and inconsistent. There are three corresponding outcomes: one solution, infinitely many solutions, or no solution. Examples 1 and 2 each had one solution. Examples 3 and 4 illustrate systems with infinitely many solutions and no solution, respectively.

 EXAMPLE 3 **A Dependent System of Linear Equations in Three Variables (Infinitely Many Solutions)**

Solve the system:

$$2x + y - z = 4 \qquad \text{Equation (1)}$$
$$x + y \quad\;\; = 2 \qquad \text{Equation (2)}$$
$$3x + 2y - z = 6 \qquad \text{Equation (3)}$$

Solution:

Since z is missing from Equation (2), z is the variable to be eliminated from Equation (1) and Equation (3).

STEP 1 Eliminate z.

Multiply Equation (1) by (-1).

Equation (3):

Add.

$$\begin{aligned} -2x - y + z &= -4 \\ 3x + 2y - z &= 6 \\ \hline x + y &= 2 \end{aligned}$$

STEP 2 Solve the system of two equations:
Equation (2) and the resulting equation in Step 1.

$$\begin{aligned} x + y &= 2 \\ x + y &= 2 \end{aligned}$$

Multiply the first equation by (-1)
and add to second equation.

$$\begin{aligned} -x - y &= -2 \\ x + y &= 2 \\ \hline 0 &= 0 \end{aligned}$$

This statement is always true; therefore, there are infinitely many solutions . The original system has been reduced to a system of two identical linear equations. Therefore the equations are dependent (share infinitely many solutions). Typically, to define those infinitely many solutions, we let $z = a$, where a stands for any real number, and then find x and y in terms of a. The resulting ordered triple showing the three variables in terms of a is called a **parametric representation** of a line in three dimensions.

STEP 3 Let $z = a$ and find x and y in terms of a.

Solve Equation (2) for y.

Let $y = 2 - x$ and $z = a$ in Equation (1).

Solve for x.

$$y = 2 - x$$
$$2x + (2 - x) - a = 4$$
$$2x + 2 - x - a = 4$$
$$x - a = 2$$
$$\boxed{x = a + 2}$$

Let $x = a + 2$ in Equation (2).

Solve for y.

$$(a + 2) + y = 2$$
$$\boxed{y = -a}$$

The infinitely many solutions are written as $(a + 2, -a, a)$.

STEP 4 Check that $x = a + 2$, $y = -a$, and $z = a$ satisfy all three equations.

Equation (1): $2(a + 2) + (-a) - a = 2a + 4 - a - a = 4$ ✓

Equation (2): $(a + 2) + (-a) = a + 2 - a = 2$ ✓

Equation (3): $3(a + 2) + 2(-a) - a = 3a + 6 - 2a - a = 6$ ✓

■ **YOUR TURN** Solve the system:

$$\begin{aligned} x + y - 2z &= 0 \\ x - z &= -1 \\ x - 2y + z &= -3 \end{aligned}$$

Classroom Example 9.2.4

Solve the system:

$$\begin{aligned} 4x + 3y - z &= 2 \\ x - y + 4z &= 3 \\ 5x + 2y + 3z &= 5 \end{aligned}$$

Answer: $z = a$, $x = \dfrac{(11 - 11a)}{7}$,

$y = \dfrac{(17a - 10)}{7}$

■ **Answer:** $(a - 1, a + 1, a)$

EXAMPLE 4 **An Inconsistent System of Linear Equations in Three Variables (No Solution)**

Solve the system:
$$x + 2y - z = 3 \qquad \text{Equation (1)}$$
$$2x + y + 2z = -1 \qquad \text{Equation (2)}$$
$$-2x - 4y + 2z = 5 \qquad \text{Equation (3)}$$

Solution:

STEP 1 Eliminate x.

Multiply Equation (1) by -2.
Equation (2):
Add.

$$-2x - 4y + 2z = -6$$
$$2x + y + 2z = -1$$
$$\overline{-3y + 4z = -7}$$

Equation (2):
Equation (3):
Add.

$$2x + y + 2z = -1$$
$$-2x - 4y + 2z = 5$$
$$\overline{-3y + 4z = 4}$$

STEP 2 Solve the system of two equations:

$$-3y + 4z = -7$$
$$-3y + 4z = 4$$

Multiply the top equation by (-1)
and add to the second equation.

$$3y - 4z = 7$$
$$-3y + 4z = 4$$
$$\overline{0 = 11}$$

This is a contradiction, or inconsistent statement, and therefore there is no solution .

So far in this section we have discussed only systems of *three* linear equations in *three* variables. What happens if we have a system of *two* linear equations in *three* variables? The two linear equations in three variables will always correspond to two planes in three dimensions. The possibilities are no solution (the two planes are parallel) or infinitely many solutions (the two planes intersect in a line).

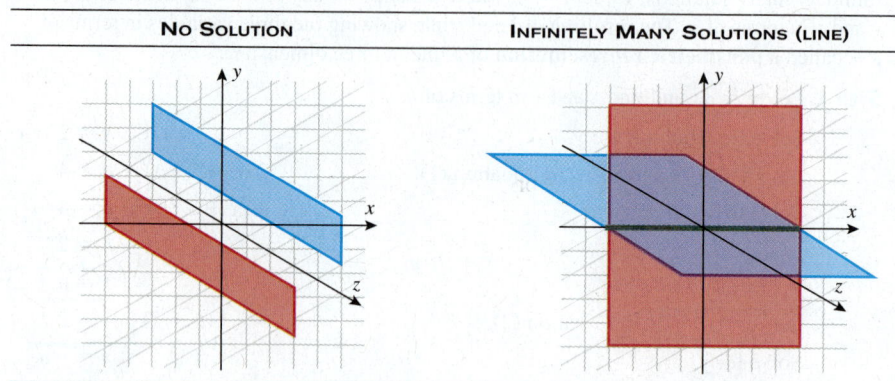

| No Solution | Infinitely Many Solutions (line) |

EXAMPLE 5 **Solving a System of Two Linear Equations in Three Variables**

Solve the system of linear equations:

$x - y + z = 7$ Equation (1)

$x + y + 2z = 2$ Equation (2)

Solution:

Eliminate y by adding the two equations.

$$
\begin{array}{r}
x - y + z = 7 \\
x + y + 2z = 2 \\
\hline
2x \quad\;\; + 3z = 9
\end{array}
$$

Therefore, Equation (1) and Equation (2) are both true if $2x + 3z = 9$. Since we know there is a solution, it must be a line. To define the line of intersection, we again turn to parametric representation.

Let $\boxed{z = a}$, where a is any real number. $2x + 3a = 9$

Solve for x.

$$\boxed{x = \frac{9}{2} - \frac{3}{2}a}$$

Substitute $z = a$ and $x = \frac{9}{2} - \frac{3}{2}a$ into Equation (1).

$$\left(\frac{9}{2} - \frac{3}{2}a\right) - y + a = 7$$

Solve for y.

$$\boxed{y = -\frac{1}{2}a - \frac{5}{2}}$$

The solution is the line in three dimensions given by $\boxed{\left(\frac{9}{2} - \frac{3}{2}a, -\frac{1}{2}a - \frac{5}{2}, a\right)}$, where a is any real number.

Note: Every real number a corresponds to a point on the line of intersection.

a	$\left(\frac{9}{2} - \frac{3}{2}a, -\frac{1}{2}a - \frac{5}{2}, a\right)$
-1	$(6, -2, -1)$
0	$\left(\frac{9}{2}, -\frac{5}{2}, 0\right)$
1	$(3, -3, 1)$

Classroom Example 9.2.5
Solve the system:
$$\begin{aligned} 2x - y &= -1 \\ y + 3z &= 0 \end{aligned}$$

Answer:
$x = \dfrac{-3a - 1}{2}, y = -3a, z = a$
for any real number a.

Classroom Example 9.2.5*
Let $b \neq 0$. Solve the system:
$$\begin{aligned} 2x - by &= -1 \\ by + z &= 0 \end{aligned}$$

Answer:
$x = \dfrac{-a + 1}{2}, y = -\dfrac{a}{b}, z = a$
for any real number a.

Modeling with a System of Three Linear Equations

Suppose you want to model a stock price as a function of time and based on the data you feel a quadratic model would be the best fit. Therefore the model is given by

$$P(t) = at^2 + bt + c$$

where $P(t)$ is the price of the stock at time t. If we have data corresponding to three distinct points $(t, P(t))$, the result is a system of three linear equations in three variables a, b, and c. We can solve the resulting system of linear equations, which determines the coefficients a, b, and c of the quadratic model for stock price.

EXAMPLE 6   Stock Value

The Oracle Corporation's stock (ORCL) over 3 days (Wednesday, October 13, to Friday, October 15, 2004) can be approximately modeled by a quadratic function: $f(t) = at^2 + bt + c$. If Wednesday corresponds to $t = 1$, where t is in days, then the following data points approximately correspond to the stock value.

t	$f(t)$	Days
1	$12.20	Wednesday
2	$12.00	Thursday
3	$12.20	Friday

Determine the function that models this behavior.

Solution:

Substitute the points (1, 12.20), (2, 12.00), and (3, 12.20) into $f(t) = at^2 + bt + c$.

$$a(1)^2 + b(1) + c = 12.20$$
$$a(2)^2 + b(2) + c = 12.00$$
$$a(3)^2 + b(3) + c = 12.20$$

Simplify to a system of three equations in three variables (a, b, and c).

$$a + b + c = 12.20 \quad \text{Equation (1)}$$
$$4a + 2b + c = 12.00 \quad \text{Equation (2)}$$
$$9a + 3b + c = 12.20 \quad \text{Equation (3)}$$

Solve for a, b, and c by applying the technique of this section.

STEP 1 Eliminate c.

Multiply Equation (1) by (−1).	$-a - b - c = -12.20$
Equation (2):	$4a + 2b + c = 12.20$
Add.	$\mathbf{3a + b = -0.20}$
Multiply Equation (1) by −1.	$-a - b - c = -12.20$
Equation (3):	$9a + 3b + c = 12.20$
Add.	$\mathbf{8a + 2b = 0}$

STEP 2 Solve the system of two equations.

$$\mathbf{3a + b = -0.20}$$
$$\mathbf{8a + 2b = 0}$$

Multiply the first equation by −2 and add to the second equation.	$-6a - 2b = 0.40$
	$\underline{ 8a + 2b = 0}$
Add.	$2a = 0.4$
Solve for a.	$a = 0.2$
Substitute $a = 0.2$ into $8a + 2b = 0$.	$8(0.2) + 2b = 0$
Simplify.	$2b = -1.6$
Solve for b.	$b = -0.8$

STEP 3 Substitute $a = 0.2$ and $b = -0.8$ into one of the original three equations.

Substitute $a = 0.2$ and $b = -0.8$ into $a + b + c = 12.20$.	$0.2 - 0.8 + c = 12.20$
Gather like terms.	$-0.6 + c = 12.20$
Solve for c.	$c = 12.80$

Step 4 Check that $a = 0.2$, $b = -0.8$, and $c = 12.80$ satisfy all three equations.

Equation (1): $\quad a + b + c = 0.2 - 0.8 + 12.8 = 12.2$

Equation (2): $\quad 4a + 2b + c = 4(0.2) + 2(-0.8) + 12.80$
$$= 0.8 - 1.6 + 12.8 = 12.00$$

Equation (3): $\quad 9a + 3b + c = 9(0.2) + 3(-0.8) + 12.80$
$$= 1.8 - 2.4 + 12.8 = 12.20$$

The model is given by $\boxed{f(t) = 0.2t^2 - 0.8t + 12.80}$.

SECTION 9.2 SUMMARY

Graphs of linear equations in *two* variables are *lines*, whereas graphs of linear equations in *three* variables are *planes*. Systems of linear equations in three variables have one of three outcomes:

■ One solution (intersection point of the three planes)
■ No solution (no intersection of all three planes)
■ Infinitely many solutions (planes intersect along a line)

When the solution to a system of three linear equations is a line in three dimensions, we use parametric representation to express the solution.

SECTION 9.2 EXERCISES

■ SKILLS

In Exercises 1–30, solve each system of linear equations.

1. $\begin{aligned} x - y + z &= 6 \\ -x + y + z &= 3 \\ -x - y - z &= 0 \end{aligned}$

2. $\begin{aligned} -x - y + z &= -1 \\ -x + y - z &= 3 \\ x - y - z &= 5 \end{aligned}$

3. $\begin{aligned} x + y - z &= 2 \\ -x - y - z &= -3 \\ -x + y - z &= 6 \end{aligned}$

4. $\begin{aligned} x + y + z &= -1 \\ -x + y - z &= 3 \\ -x - y + z &= 8 \end{aligned}$

5. $\begin{aligned} -x + y - z &= -1 \\ x - y - z &= 3 \\ x + y - z &= 9 \end{aligned}$

6. $\begin{aligned} x - y - z &= 2 \\ -x - y + z &= 4 \\ -x + y - z &= 6 \end{aligned}$

7. $\begin{aligned} 2x - 3y + 4z &= -3 \\ -x + y + 2z &= 1 \\ 5x - 2y - 3z &= 7 \end{aligned}$

8. $\begin{aligned} x - 2y + z &= 0 \\ -2x + y - z &= -5 \\ 13x + 7y + 5z &= 6 \end{aligned}$

9. $\begin{aligned} 3y - 4x + 5z &= 2 \\ 2x - 3y - 2z &= -3 \\ 3z + 4y - 2x &= 1 \end{aligned}$

10. $\begin{aligned} 2y + z - x &= 5 \\ 2x + 3z - 2y &= 0 \\ -2z + y - 4x &= 3 \end{aligned}$

11. $\begin{aligned} x - y + z &= -1 \\ y - z &= -1 \\ -x + y + z &= 1 \end{aligned}$

12. $\begin{aligned} -y + z &= 1 \\ x - y + z &= -1 \\ x - y - z &= -1 \end{aligned}$

13. $\begin{aligned} 3x - 2y - 3z &= -1 \\ x - y + z &= -4 \\ 2x + 3y + 5z &= 14 \end{aligned}$

14. $\begin{aligned} 3x - y + z &= 2 \\ x - 2y + 3z &= 1 \\ 2x + y - 3z &= -1 \end{aligned}$

15. $\begin{aligned} -3x - y - z &= 2 \\ x + 2y - 3z &= 4 \\ 2x - y + 4z &= 6 \end{aligned}$

16. $\begin{aligned} 2x - 3y + z &= 1 \\ x + 4y - 2z &= 2 \\ 3x - y + 4z &= -3 \end{aligned}$

17. $\begin{aligned} 3x + 2y + z &= 4 \\ -4x - 3y - z &= -15 \\ x - 2y + 3z &= 12 \end{aligned}$

18. $\begin{aligned} 3x - y + 4z &= 13 \\ -4x - 3y - z &= -15 \\ x - 2y + 3z &= 12 \end{aligned}$

19. $\begin{aligned} -x + 2y + z &= -2 \\ 3x - 2y + z &= 4 \\ 2x - 4y - 2z &= 4 \end{aligned}$

20. $\begin{aligned} 2x - y &= 1 \\ -x + z &= -2 \\ -2x + y &= -1 \end{aligned}$

21. $\begin{aligned} x - z - y &= 10 \\ 2x - 3y + z &= -11 \\ y - x + z &= -10 \end{aligned}$

22. $\begin{aligned} 2x + z + y &= -3 \\ 2y - z + x &= 0 \\ x + y + 2z &= 5 \end{aligned}$

23. $\begin{aligned} 3x_1 + x_2 - x_3 &= 1 \\ x_1 - x_2 + x_3 &= -3 \\ 2x_1 + x_2 + x_3 &= 0 \end{aligned}$

24. $\begin{aligned} 2x_1 + x_2 + x_3 &= -1 \\ x_1 + x_2 - x_3 &= 5 \\ 3x_1 - x_2 - x_3 &= 1 \end{aligned}$

25. $\begin{aligned} 2x + 5y &= 9 \\ x + 2y - z &= 3 \\ -3x - 4y + 7z &= 1 \end{aligned}$

26. $\begin{aligned} x - 2y + 3z &= 1 \\ -2x + 7y - 9z &= 4 \\ x + z &= 9 \end{aligned}$

27. $\begin{aligned} 2x_1 - x_2 + x_3 &= 3 \\ x_1 - x_2 + x_3 &= 2 \\ -2x_1 + 2x_2 - 2x_3 &= -4 \end{aligned}$

28. $\begin{aligned} x_1 - x_2 - 2x_3 &= 0 \\ -2x_1 + 5x_2 + 10x_3 &= -3 \\ 3x_1 + x_2 &= 0 \end{aligned}$

29. $\begin{aligned} 2x + y - z &= 2 \\ x - y - z &= 6 \end{aligned}$

30. $\begin{aligned} 3x + y - z &= 0 \\ x + y + 7z &= 4 \end{aligned}$

▪ APPLICATIONS

31. **Business.** A small company has an assembly line that produces three types of widgets. The basic widget is sold for $10 per unit, the midprice widget for $12 per unit, and the top-of-the-line widget for $15 per unit. The assembly line has a daily capacity of producing 300 widgets that may be sold for a total of $3700. Find the quantity of each type of widget produced on a day when the number of basic widgets and top-of-the-line widgets is the same.

32. **Business.** A small company has an assembly line that produces three types of widgets. The basic widget is sold for $10 per unit, the midprice widget for $12 per unit, and the top-of-the-line widget for $15 per unit. The assembly line has a daily capacity of producing 325 widgets that may be sold for a total of $3825. Find the quantity of each type of widget produced on a day when twice as many basic widgets as top-of-the-line widgets are produced.

33. **Football.** On September 1, 2007, the Appalachian State Mountaineers defeated the University of Michigan Wolverines by a score of 34 to 32. The points came from a total of these types of plays: touchdowns (six points), extra points (one point), and field goals (three points). There were a total of 18 scoring plays. There were four more touchdowns than field goals. How many touchdowns, extra points, and field goals were scored in this football game?

34. **Basketball.** On April 4, 2004, the University of Connecticut Huskies finished the season the same way they started it—as the number one men's basketball team in the NCAA. They defeated the Georgia Tech Yellow Jackets 82–73 in the Final Four championship game. The points came from three types of scoring plays: two-point shots, three-point shots, and one-point free throws. There were seven more two-point shots made than there were one-point free throws completed. The number of successful two-point shots was four more than four times the number of successful three-point shots. How many two-point, three-point, and one-point free throw shots were made in the finals of the 2004 Final Four NCAA tournament?

Exercises 35 and 36 rely on a selection of Subway sandwiches whose nutrition information is given in the table.

Suppose you are going to eat only Subway sandwiches for a week (7 days) for lunch and dinner (total of fourteen meals).

SANDWICH	CALORIES	FAT (GRAMS)
Mediterranean Chicken	350	18
Six Inch Tuna	430	19
Six Inch Roast Beef	290	5
www.subway.com		

35. **Diet.** Your goal is a total of 4840 calories and 190 grams of fat. How many of each sandwich would you eat that week to obtain this goal?

36. **Diet.** Your goal is a total of 4380 calories and 123 grams of fat. How many of each sandwich would you eat that week to obtain this goal?

Exercises 37 and 38 involve vertical motion and the effect of gravity on an object.

Because of gravity, an object that is projected upward will eventually reach a maximum height and then fall to the ground. The equation that determines the height h of a projectile t seconds after it is shot upward is given by

$$h = \frac{1}{2}at^2 + v_0 t + h_0$$

where a is the acceleration due to gravity, h_0 is the initial height of the object at time $t = 0$, and v_0 is the initial velocity of the object at time $t = 0$. Note that a projectile follows the path of a parabola opening down, so $a < 0$.

37. **Vertical Motion.** An object is thrown upward, and the following table depicts the height of the ball t seconds after the projectile is released. Find the initial height, initial velocity, and acceleration due to gravity.

t (SECONDS)	HEIGHT (FEET)
1	36
2	40
3	12

38. **Vertical Motion.** An object is thrown upward, and the following table depicts the height of the ball t seconds after the projectile is released. Find the initial height, initial velocity, and acceleration due to gravity.

t (SECONDS)	HEIGHT (FEET)
1	84
2	136
3	156

39. Data Curve-Fitting. The number of minutes that an average person of age x spends driving a car can be modeled by a quadratic function $y = ax^2 + bx + c$, where $a < 0$ and $18 \le x \le 65$. The following table gives the average number of minutes per day that a person spends driving a car. Determine the quadratic function that models this quantity.

AGE	AVERAGE DAILY MINUTES DRIVING
20	30
40	60
60	40

40. Data Curve-Fitting. The average age when a woman gets married began increasing during the last century. In 1930 the average age was 18.6, in 1950 the average age was 20.2, and in 2002 the average age was 25.3. Find a quadratic function $y = ax^2 + bx + c$, where $a > 0$ and $18 < y < 35$, that models the average age y when a woman gets married as a function of the year x ($x = 0$ corresponds to 1930). What will the average age be in 2020?

41. Money. Tara and Lamar decide to place $20,000 of their savings into investments. They put some in a money market account earning 3% interest, some in a mutual fund that has been averaging 7% a year, and some in a stock that rose 10% last year. If they put $6000 more in the money market than in the mutual fund and the mutual fund and stocks have the same growth in the next year as they did in the previous year, they will earn $1180 in a year. How much money did they put in each of the three investments?

42. Money. Tara talks Lamar into putting less money in the money market and more money in the stock. They place $20,000 of their savings into investments. They put some in a money market account earning 3% interest, some in a mutual fund that has been averaging 7% a year, and some in a stock that rose 10% last year. If they put $6000 more in the stock than in the mutual fund and the mutual fund and stock have the same growth in the next year as they did in the previous year, they will earn $1680 in a year. How much money did they put in each of the three investments?

43. Ski Production. A company produces three types of skis: regular model, trick ski, and slalom ski. They need to fill a customer order of 110 pairs of skis. There are two major production divisions within the company: labor and finishing. Each regular model of skis requires 2 hours of labor and 1 hour of finishing. Each trick ski model requires 3 hours of labor and 2 hours of finishing. Finally, each slalom ski model requires 3 hours of labor and 5 hours of finishing. Suppose the company has only 297 labor hours and 202 finishing hours. How many of each type ski can be made under these restrictions?

44. Automobile Production. An automobile manufacturing company produces three types of automobiles: compact, intermediate, and luxury models. The company has the capability of producing 500 automobiles. Suppose that each compact-model car requires 200 units of steel and 30 units of rubber, each intermediate model requires 300 units of steel and 20 units of rubber, and each luxury model requires 250 units of steel and 45 units of rubber. The number of units of steel available is 128,750, and the number of units of rubber available is 15,625. How many of each type of automobile can be produced with these constraints?

45. Computer versus Man. The *Seattle Times* reported a story on November 18, 2006, about a game of Scrabble played between a human and a computer. The best Scrabble player in the United States was pitted against a computer program designed to play the game. Remarkably, the human beat the computer in the best of two out of three games competition. The total points scored by both computer and the man for all three games was 2591. The difference between the first game's total and second game's total was 62 points. The difference between the first game's total and the third game's total was only 2 points. Determine the total number of points scored by both computer and the man for all three contests.

46. Brain versus Computer. Can the human brain perform more calculations per second than a supercomputer? The calculating speed of the three top supercomputers, IBM's Blue Gene/L, IBM's BGW, and IBM's ASC Purple, has been determined. The speed of IBM's Blue Gene/L is 245 teraflops more than IBM's BGW. The computing speed of IBM's BGW is 22 teraflops more than IBM's ASC Purple. The combined speed of all three top supercomputers is 568 teraflops. Determine the computing speed (in teraflops) of each supercomputer. A **teraflop** is the measure of a computer's speed and can be expressed as one trillion floating-point operations per second. By comparison, it is estimated that the human brain can perform 10 quadrillion calculations per second.

■ **CATCH THE MISTAKE**

In Exercises 47 and 48, explain the mistake that is made.

47. Solve the system of equations.

Equation (1):	$2x - y + z = 2$
Equation (2):	$x - y = 1$
Equation (3):	$x + z = 1$

Solution:

Equation (2):	$x - y = 1$
Equation (3):	$x + z = 1$
Add Equation (2) and Equation (3).	$-y + z = 2$
Multiply Equation (1) by (-1).	$-2x + y - z = -2$
Add.	$-2x = 0$
Solve for x.	$x = 0$
Substitute $x = 0$ into Equation (2).	$0 - y = 1$
Solve for y.	$y = -1$
Substitute $x = 0$ into Equation (3).	$0 + z = 1$
Solve for z.	$z = 1$

The answer is $x = 0$, $y = -1$, and $z = 1$.

This is incorrect. Although $x = 0$, $y = -1$, and $z = 1$ does satisfy the three original equations, it is only one of infinitely many solutions. What mistake was made?

48. Solve the system of equations.

Equation (1):	$x + 3y + 2z = 4$
Equation (2):	$3x + 10y + 9z = 17$
Equation (3):	$2x + 7y + 7z = 17$

Solution:

Multiply Equation (1) by -3.	$-3x - 9y - 6z = -12$
Equation (2):	$3x + 10y + 9z = 17$
Add.	$y + 3z = 5$
Multiply Equation (1) by -2.	$-2x - 6y - 4z = -8$
Equation (3):	$2x + 7y + 7z = 17$
Add.	$y + 3z = 9$

Solve the system of two equations.	$y + 3z = 5$
	$y + 3z = 9$

Infinitely many solutions.

Let $z = a$, then $y = 5 - 3a$.

Substitute $z = a$ and $y = 5 - 3a$ into Equation (1).

$$x + 3y + 2z = 4$$
$$x + 3(5 - 3a) + 2a = 4$$

Eliminate parentheses. $\quad x + 15 - 9a + 2a = 4$

Solve for x. $\quad x = 7a - 11$

The answer is $x = 7a - 11$, $y = 5 - 3a$, and $z = a$.

This is incorrect. There is no solution. What mistake was made?

■ **CONCEPTUAL**

In Exercises 49 and 50, determine whether each statement is true or false.

49. A system of linear equations that has more variables than equations cannot have a unique solution.

50. A system of linear equations that has the same number of equations as variables always has a unique solution.

51. Geometry. The circle given by the equation $x^2 + y^2 + ax + by + c = 0$ passes through the points $(-2, 4)$, $(1, 1)$, and $(-2, -2)$. Find a, b, and c.

52. Geometry. The circle given by the equation $x^2 + y^2 + ax + by + c = 0$ passes through the points $(0, 7)$, $(6, 1)$, and $(5, 4)$. Find a, b, and c.

■ CHALLENGE

53. A fourth-degree polynomial,
$f(x) = ax^4 + bx^3 + cx^2 + dx + e$, with $a < 0$,
can be used to represent the following data on the number
of deaths per year due to lightning strikes. Assume 1999
corresponds to $x = -2$ and 2003 corresponds to $x = 2$.
Use the data to determine a, b, c, d, and e.

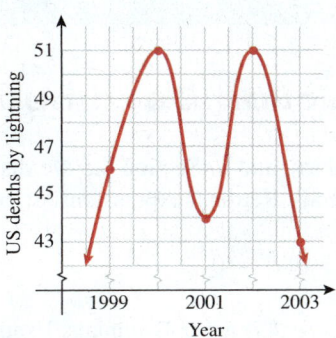

54. A copy machine accepts nickels, dimes, and quarters. After
1 hour, there are 30 coins total and their value is $4.60. If
there are four more quarters than nickels, how many nickels,
quarters, and dimes are in the machine?

In Exercises 55–58, solve the system of linear equations.

55.
$$
\begin{aligned}
2y + z &= 3 \\
4x \quad - z &= -3 \\
7x - 3y - 3z &= 2 \\
x - y - z &= -2
\end{aligned}
$$

56.
$$
\begin{aligned}
-2x - y + 2z &= 3 \\
3x \quad - 4z &= 2 \\
2x + y \quad &= -1 \\
-x + y - z &= -8
\end{aligned}
$$

57.
$$
\begin{aligned}
3x_1 - 2x_2 + x_3 + 2x_4 &= -2 \\
-x_1 + 3x_2 + 4x_3 + 3x_4 &= 4 \\
x_1 + x_2 + x_3 + x_4 &= 0 \\
5x_1 + 3x_2 + x_3 + 2x_4 &= -1
\end{aligned}
$$

58.
$$
\begin{aligned}
5x_1 + 3x_2 + 8x_3 + x_4 &= 1 \\
x_1 + 2x_2 + 5x_3 + 2x_4 &= 3 \\
4x_1 \quad + x_3 - 2x_4 &= -3 \\
x_2 + x_3 + x_4 &= 0
\end{aligned}
$$

■ TECHNOLOGY

In Exercises 59 and 60, employ a graphing calculator to solve the system of linear equations (most graphing calculators have the capability of solving linear systems with the user entering the coefficients).

59.
$$
\begin{aligned}
x - z - y &= 10 \\
2x - 3y + z &= -11 \\
y - x + z &= -10
\end{aligned}
$$

60.
$$
\begin{aligned}
2x + z + y &= -3 \\
2y - z + x &= 0 \\
x + y + 2z &= 5
\end{aligned}
$$

61. Some graphing calculators and graphing utilities have the
ability to graph in three dimensions (3D) as opposed to
the traditional two dimensions (2D). The line must be given
in the form $z = ax + by + c$. Rewrite the system of
equations in Exercise 59 in this form and graph the three
lines in 3D. What is the point of intersection? Compare that
with your answer in Exercise 59.

62. Some graphing calculators and graphing utilities have the
ability to graph in three dimensions (3D) as opposed to the
traditional two dimensions (2D). The line must be given in the
form $z = ax + by + c$. Rewrite the system of equations in
Exercise 60 in this form and graph the three lines in 3D. What is
the point of intersection? Compare that with your answer in
Exercise 60.

**In Exercises 63 and 64, employ a graphing calculator to solve
the system of equations.**

63.
$$
\begin{aligned}
0.2x - 0.7y + 0.8z &= 11.2 \\
-1.2x + 0.3y - 1.5z &= 0 \\
0.8x - 0.1y + 2.1z &= 6.4
\end{aligned}
$$

64.
$$
\begin{aligned}
1.8x - 0.5y + 2.4z &= 1.6 \\
0.3x \quad - 0.6z &= 0.2
\end{aligned}
$$

SKILLS OBJECTIVES

- Decompose rational expressions into sums of partial fractions when the denominators contain: distinct linear factors, repeated linear factors, distinct irreducible quadratic factors, or repeated irreducible quadratic factors.

CONCEPTUAL OBJECTIVE

- Understand the connection between partial-fraction decomposition and systems of linear equations.

Performing Partial-Fraction Decomposition

In Chapter 4 we studied polynomial functions, and in Section 4.5 we discussed ratios of polynomial functions, called rational functions. Rational expressions are of the form:

$$\frac{n(x)}{d(x)} \qquad d(x) \neq 0$$

where the numerator $n(x)$ and the denominator $d(x)$ are polynomials. Examples of rational expressions are

$$\frac{4x - 1}{2x + 3} \qquad \frac{2x + 5}{x^2 - 1} \qquad \frac{3x^4 - 2x + 5}{x^2 + 2x + 4}$$

Suppose we are asked to add two rational expressions: $\dfrac{2}{x + 1} + \dfrac{5}{x - 3}$.

We already possess the skills to accomplish this. We first identify the least common denominator $(x + 1)(x - 3)$ and combine the fractions into a single expression.

$$\frac{2}{x + 1} + \frac{5}{x - 3} = \frac{2(x - 3) + 5(x + 1)}{(x + 1)(x - 3)} = \frac{2x - 6 + 5x + 5}{(x + 1)(x - 3)} = \frac{7x - 1}{x^2 - 2x - 3}$$

How do we do this in reverse? For example, how do we start with $\dfrac{7x - 1}{x^2 - 2x - 3}$ and write this expression as a sum of two simpler expressions?

$$\underbrace{\frac{7x - 1}{x^2 - 2x - 3}}_{} = \overbrace{\underbrace{\frac{2}{x + 1}}_{\text{Partial Fraction}} + \underbrace{\frac{5}{x - 3}}_{\text{Partial Fraction}}}^{\text{Partial-Fraction Decomposition}}$$

Each of the two expressions on the right is called a **partial fraction**. The sum of these fractions is called the **partial-fraction decomposition** of $\dfrac{7x - 1}{x^2 - 2x - 3}$.

Partial-fraction decomposition is an important tool in calculus. Calculus operations such as differentiation and integration are often made simpler if you apply partial fractions. Partial fractions were not discussed until now because partial-fraction decomposition *requires the ability to solve systems of linear equations*. Since partial-fraction decomposition is made possible by the techniques of solving systems of linear equations, we consider partial fractions an important application of systems of linear equations.

As mentioned earlier, a rational expression is the ratio of two polynomial expressions $n(x)/d(x)$, and we assume that $n(x)$ and $d(x)$ are polynomials with no common factors. If the degree of $n(x)$ is less than the degree of $d(x)$, then the rational expression $n(x)/d(x)$ is said to be **proper**. If the degree of $n(x)$ is greater than or equal to the degree of $d(x)$, the rational expression is said to be **improper**. If the rational expression is improper, it should first be divided using long division.

$$\frac{n(x)}{d(x)} = Q(x) + \frac{r(x)}{d(x)}$$

The result is the sum of a quotient $Q(x)$ and a rational expression, which is the ratio of the remainder $r(x)$ and the divisor $d(x)$. The rational expression $r(x)/d(x)$ is proper, and the techniques outlined in this section can be applied to its partial-fraction decomposition.

Partial-fraction decomposition of proper rational expressions always begins with factoring the denominator $d(x)$. The goal is to write $d(x)$ as a product of distinct linear factors, but that may not always be possible. Sometimes $d(x)$ can be factored into a product of linear factors, where one or more are repeated. And sometimes the factored form of $d(x)$ contains irreducible quadratic factors, such as $x^2 + 1$. There are times when the irreducible quadratic factors are repeated, such as $(x^2 + 1)^2$. A procedure is now outlined for partial-fraction decomposition.

PARTIAL-FRACTION DECOMPOSITION

To write a rational expression $\dfrac{n(x)}{d(x)}$ as a sum of partial fractions:

Step 1: Determine whether the rational expression is proper or improper.

- Proper: degree of $n(x) <$ degree of $d(x)$
- Improper: degree of $n(x) \geq$ degree of $d(x)$

Step 2: If proper, proceed to Step 3.

If improper, divide $\dfrac{n(x)}{d(x)}$ using polynomial (long) division and write

the result as $\dfrac{n(x)}{d(x)} = Q(x) + \dfrac{r(x)}{d(x)}$ and proceed to Step 3 with $\dfrac{r(x)}{d(x)}$.

Step 3: Factor $d(x)$. One of four possible cases will arise:

Case 1: Distinct (nonrepeated) *linear* factors: $(ax + b)$

$$d(x) = (3x - 1)(x + 2)$$

Case 2: One or more repeated linear factors: $(ax + b)^m$ $m \geq 2$

$$d(x) = (x + 5)^2(x - 3)$$

Case 3: One or more distinct irreducible ($ax^2 + bx + c = 0$ has no real roots) quadratic factors: $(ax^2 + bx + c)$

$$d(x) = (x^2 + 4)(x + 1)(x - 2)$$

Case 4: One or more repeated irreducible quadratic factors:
$$(ax^2 + bx + c)^m \quad m \geq 2$$
$$d(x) = (x^2 + x + 1)^2(x + 1)(x - 2)$$

Step 4: Decompose the rational expression into a sum of partial fractions according to the procedure outlined in each case in this section.

Step 4 depends on which cases, or types of factors, arise. It is important to note that these four cases are not exclusive and combinations of different types of factors will appear.

Distinct Linear Factors

CASE 1: $d(x)$ HAS ONLY DISTINCT (NONREPEATED) LINEAR FACTORS

If $d(x)$ is a polynomial of degree p, and it can be factored into p linear factors:

$$d(x) = \underbrace{(ax + b)(cx + d)\ldots}_{p \text{ linear factors}}$$

where no two factors are the same, then the partial-fraction decomposition of $\dfrac{n(x)}{d(x)}$ can be written as

$$\frac{n(x)}{d(x)} = \frac{A}{(ax + b)} + \frac{B}{(cx + d)} + \cdots$$

where the numerators, A, B, and so on are constants to be determined.

The goal is to write a proper rational expression as the sum of proper rational expressions. Therefore, if the denominator is a linear factor (degree 1), then the numerator is a constant (degree 0).

EXAMPLE 1 Partial-Fraction Decomposition with Distinct Linear Factors

Find the partial-fraction decomposition of $\dfrac{5x + 13}{x^2 + 4x - 5}$.

Solution:

Factor the denominator.

$$\frac{5x + 13}{(x - 1)(x + 5)}$$

Express as a sum of two partial fractions.

$$\frac{5x + 13}{(x - 1)(x + 5)} = \frac{A}{(x - 1)} + \frac{B}{(x + 5)}$$

Multiply the equation by the LCD $(x - 1)(x + 5)$.

$$5x + 13 = A(x + 5) + B(x - 1)$$

Eliminate the parentheses.

$$5x + 13 = Ax + 5A + Bx - B$$

Group the x's and constants on the right.

$$5x + 13 = (A + B)x + (5A - B)$$

Identify like terms.

$$\color{red}{5x + 13 = (A + B)x + (5A - B)}$$

Equate the **coefficients of x**.

$$\color{red}{5 = A + B}$$

Equate the **constant** terms.

$$\color{blue}{13 = 5A - B}$$

Solve the system of two linear equations using any method to solve for A and B.

$$A = 3, B = 2$$

Substitute $A = 3$, $B = 2$ into partial-fraction decomposition.

$$\frac{5x + 13}{(x - 1)(x + 5)} = \frac{3}{(x - 1)} + \frac{2}{(x + 5)}$$

Check by adding the partial fractions.

$$\frac{3}{(x - 1)} + \frac{2}{(x + 5)} = \frac{3(x + 5) + 2(x - 1)}{(x - 1)(x + 5)} = \frac{5x + 13}{x^2 + 4x - 5}$$

■ **YOUR TURN** Find the partial-fraction decomposition of $\dfrac{4x - 13}{x^2 - 3x - 10}$.

Technology Tip

Use a graphing calculator to check the graph of $Y_1 = \dfrac{5x + 13}{x^2 + 4x - 5}$ and its partial-fraction decomposition $Y_2 = \dfrac{3}{x - 1} + \dfrac{2}{x + 5}$. The graphs and tables of values are shown.

■ **Answer:**

$$\frac{4x - 13}{x^2 - 3x - 10} = \frac{3}{x + 2} + \frac{1}{x - 5}$$

In Example 1, we started with a rational expression that had a numerator of degree 1 and a denominator of degree 2. Partial-fraction decomposition enabled us to write that rational expression as a sum of two rational expressions with degree 0 numerators and degree 1 denominators.

Repeated Linear Factors

CASE 2: $d(x)$ HAS AT LEAST ONE REPEATED LINEAR FACTOR

If $d(x)$ can be factored into a product of linear factors, then the partial-fraction decomposition will proceed as in Case 1, with the exception of a repeated factor $(ax + b)^m$, $m \geq 2$. Any linear factor repeated m times will result in the sum of m partial fractions:

$$\frac{A}{(ax + b)} + \frac{B}{(ax + b)^2} + \frac{C}{(ax + b)^3} + \cdots + \frac{M}{(ax + b)^m}$$

where the numerators, $A, B, C, \ldots, M$ are constants to be determined.

Note that if $d(x)$ is of degree p, the general form of the decomposition will have p partial fractions. If some numerator constants turn out to be zero, then the final decomposition may have fewer than p partial fractions.

EXAMPLE 2   Partial-Fraction Decomposition with a Repeated Linear Factor

Find the partial-fraction decomposition of $\dfrac{-3x^2 + 13x - 12}{x^3 - 4x^2 + 4x}$.

Solution:

Factor the denominator.

$$\frac{-3x^2 + 13x - 12}{x(x - 2)^2}$$

Express as a sum of three partial fractions.

$$\frac{-3x^2 + 13x - 12}{x(x - 2)^2} = \frac{A}{x} + \frac{B}{(x - 2)} + \frac{C}{(x - 2)^2}$$

Multiply the equation by the LCD $x(x - 2)^2$.

$$-3x^2 + 13x - 12 = A(x - 2)^2 + Bx(x - 2) + Cx$$

Eliminate the parentheses.

$$-3x^2 + 13x - 12 = Ax^2 - 4Ax + 4A + Bx^2 - 2Bx + Cx$$

Group like terms on the right.

$$-3x^2 + 13x - 12 = (A + B)x^2 + (-4A - 2B + C)x + 4A$$

Identify like terms on both sides.

$$-3x^2 + 13x - 12 = (A + B)x^2 + (-4A - 2B + C)x + 4A$$

Equate the **coefficients of x^2**.

$$-3 = A + B \qquad (1)$$

Equate the **coefficients of x**.

$$13 = -4A - 2B + C \qquad (2)$$

Equate the **constant** terms.

$$-12 = 4A \qquad (3)$$

Technology Tip

Use a graphing calculator to check the graph of

$$Y_1 = \frac{-3x^2 + 13x - 12}{x^3 - 4x^2 + 4x} \text{ and its}$$

partial-fraction decomposition

$$Y_2 = \frac{-3}{x} + \frac{1}{(x - 2)^2}. \text{ The graphs}$$

and tables of values are shown.

Solve the system of three equations for A, B, and C.

Solve (3) for A.	$A = -3$
Substitute $A = -3$ into (1).	$B = 0$
Substitute $A = -3$ and $B = 0$ into (2).	$C = 1$

Substitute $A = -3$, $B = 0$, $C = 1$ into the partial-fraction decomposition.

$$\frac{-3x^2 + 13x - 12}{x(x - 2)^2} = \frac{-3}{x} + \frac{0}{(x - 2)} + \frac{1}{(x - 2)^2}$$

$$\boxed{\frac{-3x^2 + 13x - 12}{x^3 - 4x^2 + 4x} = \frac{-3}{x} + \frac{1}{(x - 2)^2}}$$

Check by adding the partial fractions.

$$\frac{-3}{x} + \frac{1}{(x - 2)^2} = \frac{-3(x - 2)^2 + 1(x)}{x(x - 2)^2} = \frac{-3x^2 + 13x - 12}{x^3 - 4x^2 + 4x}$$

■ **Answer:**

$\dfrac{x^2 + 1}{x^3 + 2x^2 + x} = \dfrac{1}{x} - \dfrac{2}{(x + 1)^2}$

■ **YOUR TURN** Find the partial-fraction decomposition of $\dfrac{x^2 + 1}{x^3 + 2x^2 + x}$.

EXAMPLE 3 Partial-Fraction Decomposition with Multiple Repeated Linear Factors

Find the partial-fraction decomposition of $\dfrac{2x^3 + 6x^2 + 6x + 9}{x^4 + 6x^3 + 9x^2}$.

Solution:

Factor the denominator.

$$\frac{2x^3 + 6x^2 + 6x + 9}{x^2(x + 3)^2}$$

Express as a sum of four partial fractions.

$$\frac{2x^3 + 6x^2 + 6x + 9}{x^2(x + 3)^2} = \frac{A}{x} + \frac{B}{x^2} + \frac{C}{(x + 3)} + \frac{D}{(x + 3)^2}$$

Multiply the equation by the LCD $x^2 (x + 3)^2$.

$$2x^3 + 6x^2 + 6x + 9 = Ax(x + 3)^2 + B(x + 3)^2 + Cx^2(x + 3) + Dx^2$$

Eliminate the parentheses.

$$2x^3 + 6x^2 + 6x + 9 = Ax^3 + 6Ax^2 + 9Ax + Bx^2 + 6Bx + 9B + Cx^3 + 3Cx^2 + Dx^2$$

Group like terms on the right.

$$2x^3 + 6x^2 + 6x + 9 = (A + C)x^3 + (6A + B + 3C + D)x^2 + (9A + 6B)x + 9B$$

Identify like terms on both sides.

$$\mathbf{2x^3 + 6x^2 + 6x + 9 = (A + C)x^3 + (6A + B + 3C + D)x^2 + (9A + 6B)x + 9B}$$

Equate the **coefficients of x^3**.	$2 = A + C$	(1)
Equate the **coefficients of x^2**.	$6 = 6A + B + 3C + D$	(2)
Equate the **coefficients of x**.	$6 = 9A + 6B$	(3)
Equate the **constant** terms.	$9 = 9B$	(4)

Solve the system of four equations for A, B, C, and D.

Solve Equation (4) for B.	$B = 1$
Substitute $B = 1$ into Equation (3) and solve for A.	$A = 0$
Substitute $A = 0$ into Equation (1) and solve for C.	$C = 2$
Substitute $A = 0$, $B = 1$, and $C = 2$ into Equation (2) and solve for D.	$D = -1$

Substitute $A = 0$, $B = 1$, $C = 2$, $D = -1$ into the partial-fraction decomposition.

$$\frac{2x^3 + 6x^2 + 6x + 9}{x^2(x + 3)^2} = \frac{0}{x} + \frac{1}{x^2} + \frac{2}{(x + 3)} + \frac{-1}{(x + 3)^2}$$

$$\boxed{\frac{2x^3 + 6x^2 + 6x + 9}{x^2(x + 3)^2} = \frac{1}{x^2} + \frac{2}{(x + 3)} - \frac{1}{(x + 3)^2}}$$

Check by adding the partial fractions.

$$\frac{1}{x^2} + \frac{2}{(x + 3)} - \frac{1}{(x + 3)^2} = \frac{(x + 3)^2 + 2x^2(x + 3) - 1(x^2)}{x^2(x + 3)^2}$$

$$= \frac{2x^3 + 6x^2 + 6x + 9}{x^4 + 6x^3 + 9x^2}$$

■ **YOUR TURN** Find the partial-fraction decomposition of $\dfrac{2x^3 + 2x + 1}{x^4 + 2x^3 + x^2}$.

Distinct Irreducible Quadratic Factors

There will be times when a polynomial cannot be factored into a product of linear factors with real coefficients. For example, $x^2 + 4$, $x^2 + x + 1$, and $9x^2 + 3x + 2$ are all examples of *irreducible quadratic* expressions. The general form of an **irreducible quadratic factor** is given by:

$$ax^2 + bx + c \qquad \text{where } ax^2 + bx + c = 0 \text{ has no real roots}$$

CASE 3: d(x) HAS A DISTINCT IRREDUCIBLE QUADRATIC FACTOR

If the factored form of $d(x)$ contains an irreducible quadratic factor $ax^2 + bx + c$, then the partial-fraction decomposition will contain a term of the form:

$$\frac{Ax + B}{ax^2 + bx + c}$$

where A and B are constants to be determined.

Recall that for a proper rational expression, the numerator is less than the denominator. For irreducible quadratic (degree 2) denominators we assume a linear (degree 1) numerator. For example,

$$\underbrace{\frac{7x^2 + 2}{(2x + 1)(x^2 + 1)}}_{} = \underbrace{\frac{A}{(2x + 1)}}_{\substack{\text{Constant numerator} \\ \text{Linear factor}}} + \underbrace{\frac{Bx + C}{(x^2 + 1)}}_{\substack{\text{Linear numerator} \\ \text{Quadratic factor}}}$$

A constant is used in the numerator when the denominator consists of a linear expression and a linear expression is used in the numerator when the denominator consists of a quadratic expression.

Study Tip

The degree of the numerator is always 1 less than the degree of the denominator.

Technology Tip

Use a graphing calculator to check

the graph of $Y_1 = \dfrac{7x^2 + 2}{(2x + 1)(x^2 + 1)}$

and its partial-fraction decomposition

$Y_2 = \dfrac{3}{2x + 1} + \dfrac{2x - 1}{x^2 + 1}$. The graphs

and tables of values are shown.

■ **Answer:** $\dfrac{-2x^2 + x + 6}{(x - 1)(x^2 + 4)}$

$= \dfrac{1}{x - 1} - \dfrac{3x + 2}{x^2 + 4}$

EXAMPLE 4   Partial-Fraction Decomposition with an Irreducible Quadratic Factor

Find the partial-fraction decomposition of $\dfrac{7x^2 + 2}{(2x + 1)(x^2 + 1)}$.

Solution:

The denominator is already in factored form.	$\dfrac{7x^2 + 2}{(2x + 1)(x^2 + 1)}$
Express as a sum of two partial fractions.	$\dfrac{7x^2 + 2}{(2x + 1)(x^2 + 1)} = \dfrac{A}{(2x + 1)} + \dfrac{Bx + C}{(x^2 + 1)}$
Multiply the equation by the LCD $(2x + 1)(x^2 + 1)$.	$7x^2 + 2 = A(x^2 + 1) + (Bx + C)(2x + 1)$
Eliminate the parentheses.	$7x^2 + 2 = Ax^2 + A + 2Bx^2 + Bx + 2Cx + C$
Group like terms on the right.	$7x^2 + 2 = (A + 2B)x^2 + (B + 2C)x + (A + C)$
Identify like terms on both sides.	$7x^2 + 0x + 2 = (A + 2B)x^2 + (B + 2C)x + (A + C)$
Equate the **coefficients of x^2**.	$7 = A + 2B$
Equate the **coefficients of x**.	$0 = B + 2C$
Equate the **constant** terms.	$2 = A + C$
Solve the system of three equations for A, B, and C.	$A = 3, B = 2, C = -1$

Substitute $A = 3, B = 2, C = -1$ into the partial-fraction decomposition.

$$\dfrac{7x^2 + 2}{(2x + 1)(x^2 + 1)} = \dfrac{3}{(2x + 1)} + \dfrac{2x - 1}{(x^2 + 1)}$$

Check by adding the partial fractions.

$$\dfrac{3}{(2x + 1)} + \dfrac{2x - 1}{(x^2 + 1)} = \dfrac{3(x^2 + 1) + (2x - 1)(2x + 1)}{(2x + 1)(x^2 + 1)} = \dfrac{7x^2 + 2}{(2x + 1)(x^2 + 1)}$$

■ **YOUR TURN** Find the partial-fraction decomposition of $\dfrac{-2x^2 + x + 6}{(x - 1)(x^2 + 4)}$.

Classroom Example 9.3.4

Find the partial-fraction decomposition of:

a. $\dfrac{-12x^3 + 13x^2 - 18x + 25}{3x^4 - 3x^3 + 5x^2 - 5x}$

b.* $\dfrac{-x^2 + x + (a - a^2)}{x^3 + ax^2 + a^2x + a^3}$, where $a \neq 0$.

Answer:

a. $\dfrac{-2}{3x^2 + 5} + \dfrac{1}{x - 1} - \dfrac{5}{x}$

b. $\dfrac{1}{x^2 + a^2} - \dfrac{1}{x + a}$

Repeated Irreducible Quadratic Factors

CASE 4: $d(x)$ HAS A REPEATED IRREDUCIBLE QUADRATIC FACTOR

If the factored form of $d(x)$ contains an irreducible quadratic factor $(ax^2 + bx + c)^m$, where $b^2 - 4ac < 0$ and $m \geq 2$, then the partial-fraction decomposition will contain a series of terms of the form:

$$\dfrac{A_1x + B_1}{ax^2 + bx + c} + \dfrac{A_2x + B_2}{(ax^2 + bx + c)^2} + \dfrac{A_3x + B_3}{(ax^2 + bx + c)^3} + \cdots + \dfrac{A_mx + B_m}{(ax^2 + bx + c)^m}$$

where A_i and B_i, $i = 1, 2, \ldots, m$, are constants to be determined.

EXAMPLE 5   Partial-Fraction Decomposition with a Repeated Irreducible Quadratic Factor

Find the partial-fraction decomposition of $\dfrac{x^3 - x^2 + 3x + 2}{(x^2 + 1)^2}$.

Solution:

The denominator is already in factored form.

$$\dfrac{x^3 - x^2 + 3x + 2}{(x^2 + 1)^2}$$

Express as a sum of two partial fractions.

$$\dfrac{x^3 - x^2 + 3x + 2}{(x^2 + 1)^2} = \dfrac{Ax + B}{x^2 + 1} + \dfrac{Cx + D}{(x^2 + 1)^2}$$

Multiply the equation by the LCD $(x^2 + 1)^2$.

$$x^3 - x^2 + 3x + 2 = (Ax + B)(x^2 + 1) + Cx + D$$

Eliminate the parentheses.

$$x^3 - x^2 + 3x + 2 = Ax^3 + Bx^2 + Ax + B + Cx + D$$

Group like terms on the right.

$$x^3 - x^2 + 3x + 2 = Ax^3 + Bx^2 + (A + C)x + (B + D)$$

Identify like terms on both sides.

$$x^3 - x^2 + 3x + 2 = Ax^3 + Bx^2 + (A + C)x + (B + D)$$

Equate the **coefficients of x^3**. $1 = A$ (1)

Equate the **coefficients of x^2**. $-1 = B$ (2)

Equate the **coefficients of x**. $3 = A + C$ (3)

Equate the **constant** terms. $2 = B + D$ (4)

Substitute $A = 1$ into Equation (3) and solve for C. $C = 2$

Substitute $B = -1$ into Equation (4) and solve for D. $D = 3$

Substitute $A = 1, B = -1, C = 2, D = 3$ into the partial-fraction decomposition.

$$\dfrac{x^3 - x^2 + 3x + 2}{(x^2 + 1)^2} = \dfrac{x - 1}{x^2 + 1} + \dfrac{2x + 3}{(x^2 + 1)^2}$$

Check by adding the partial fractions.

$$\dfrac{x - 1}{x^2 + 1} + \dfrac{2x + 3}{(x^2 + 1)^2} = \dfrac{(x - 1)(x^2 + 1) + (2x + 3)}{(x^2 + 1)^2} = \dfrac{x^3 - x^2 + 3x + 2}{(x^2 + 1)^2}$$

■ **YOUR TURN** Find the partial-fraction decomposition of $\dfrac{3x^3 + x^2 + 4x - 1}{(x^2 + 4)^2}$.

■ **Answer:** $\dfrac{3x^3 + x^2 + 4x - 1}{(x^2 + 4)^2}$

$= \dfrac{3x + 1}{x^2 + 4} - \dfrac{8x + 5}{(x^2 + 4)^2}$

Combinations of All Four Cases

As you probably can imagine, there are rational expressions that have combinations of all four cases, which can lead to a system of several equations when solving for the numerator constants.

EXAMPLE 6 Partial-Fraction Decomposition

Find the partial-fraction decomposition of $\dfrac{x^5 + x^4 + 4x^3 - 3x^2 + 4x - 8}{x^2(x^2 + 2)^2}$.

Solution:

The denominator is already in factored form.

$$\dfrac{x^5 + x^4 + 4x^3 - 3x^2 + 4x - 8}{x^2(x^2 + 2)^2}$$

Express as a sum of partial fractions.

There are repeated linear and irreducible quadratic factors.

$$\dfrac{x^5 + x^4 + 4x^3 - 3x^2 + 4x - 8}{x^2(x^2 + 2)^2} = \dfrac{A}{x} + \dfrac{B}{x^2} + \dfrac{Cx + D}{(x^2 + 2)} + \dfrac{Ex + F}{(x^2 + 2)^2}$$

Multiply the equation by the LCD $x^2(x^2 + 2)^2$.

$$x^5 + x^4 + 4x^3 - 3x^2 + 4x - 8$$
$$= Ax(x^2 + 2)^2 + B(x^2 + 2)^2 + (Cx + D)x^2(x^2 + 2) + (Ex + F)x^2$$

Eliminate the parentheses.

$$x^5 + x^4 + 4x^3 - 3x^2 + 4x - 8$$
$$= Ax^5 + 4Ax^3 + 4Ax + Bx^4 + 4Bx^2 + 4B + Cx^5 + 2Cx^3 + Dx^4 + 2Dx^2 + Ex^3 + Fx^2$$

Group like terms on the right.

$$x^5 + x^4 + 4x^3 - 3x^2 + 4x - 8$$
$$= (A + C)x^5 + (B + D)x^4 + (4A + 2C + E)x^3 + (4B + 2D + F)x^2 + 4Ax + 4B$$

Equating the coefficients of like terms leads to six equations.

$$A + C = 1$$
$$B + D = 1$$
$$4A + 2C + E = 4$$
$$4B + 2D + F = -3$$
$$4A = 4$$
$$4B = -8$$

Solve this system of equations.

$$A = 1, \qquad B = -2, \qquad C = 0, \qquad D = 3, \qquad E = 0, \qquad F = -1$$

Substitute $A = 1, B = -2, C = 0, D = 3, E = 0, F = -1$ into the partial-fraction decomposition.

$$\dfrac{x^5 + x^4 + 4x^3 - 3x^2 + 4x - 8}{x^2(x^2 + 2)^2} = \dfrac{1}{x} + \dfrac{-2}{x^2} + \dfrac{0x + 3}{(x^2 + 2)} + \dfrac{0x + -1}{(x^2 + 2)^2}$$

$$\boxed{\dfrac{x^5 + x^4 + 4x^3 - 3x^2 + 4x - 8}{x^2(x^2 + 2)^2} = \dfrac{1}{x} + \dfrac{-2}{x^2} + \dfrac{3}{(x^2 + 2)} - \dfrac{1}{(x^2 + 2)^2}}$$

Check by adding the partial fractions.

A rational expression $\dfrac{n(x)}{d(x)}$ is

■ **Proper:** If the degree of the numerator is less than the degree of the denominator.

■ **Improper:** If the degree of the numerator is equal to or greater than the degree of the denominator.

Partial-Fraction Decomposition of Proper Rational Expressions

1. Distinct (nonrepeated) linear factors

$$\frac{3x - 10}{(x - 5)(x + 4)} = \frac{A}{x - 5} + \frac{B}{x + 4}$$

2. Repeated linear factors

$$\frac{2x + 5}{(x - 3)^2(x + 1)} = \frac{A}{x - 3} + \frac{B}{(x - 3)^2} + \frac{C}{x + 1}$$

3. Distinct irreducible quadratic factors

$$\frac{1 - x}{(x^2 + 1)} = \frac{Ax + B}{x^2 + 1}$$

4. Repeated irreducible quadratic factors

$$\frac{4x^2 - 3x + 2}{(x^2 + 1)^2} = \frac{Ax + B}{x^2 + 1} + \frac{Cx + D}{(x^2 + 1)^2}$$

▪ SKILLS

In Exercises 1–6, match the rational expression (1–6) with the form of the partial-fraction decomposition (a–f).

1. $\dfrac{3x + 2}{x(x^2 - 25)}$ **2.** $\dfrac{3x + 2}{x(x^2 + 25)}$ **3.** $\dfrac{3x + 2}{x^2(x^2 + 25)}$ **4.** $\dfrac{3x + 2}{x^2(x^2 - 25)}$ **5.** $\dfrac{3x + 2}{x(x^2 + 25)^2}$ **6.** $\dfrac{3x + 2}{x^2(x^2 + 25)^2}$

a. $\dfrac{A}{x} + \dfrac{B}{x^2} + \dfrac{Cx + D}{x^2 + 25}$

b. $\dfrac{A}{x} + \dfrac{Bx + C}{x^2 + 25} + \dfrac{Dx + E}{(x^2 + 25)^2}$

c. $\dfrac{A}{x} + \dfrac{Bx + C}{x^2 + 25}$

d. $\dfrac{A}{x} + \dfrac{B}{x + 5} + \dfrac{C}{x - 5}$

e. $\dfrac{A}{x} + \dfrac{B}{x^2} + \dfrac{Cx + D}{x^2 + 25} + \dfrac{Ex + F}{(x^2 + 25)^2}$

f. $\dfrac{A}{x} + \dfrac{B}{x^2} + \dfrac{C}{x + 5} + \dfrac{D}{x - 5}$

In Exercises 7–14, write the form of the partial-fraction decomposition. Do not solve for the constants.

7. $\dfrac{9}{x^2 - x - 20}$

8. $\dfrac{8}{x^2 - 3x - 10}$

9. $\dfrac{2x + 5}{x^3 - 4x^2}$

10. $\dfrac{x^2 + 2x - 1}{x^4 - 9x^2}$

11. $\dfrac{2x^3 - 4x^2 + 7x + 3}{(x^2 + x + 5)}$

12. $\dfrac{2x^3 + 5x^2 + 6}{(x^2 - 3x + 7)}$

13. $\dfrac{3x^3 - x + 9}{(x^2 + 10)^2}$

14. $\dfrac{5x^3 + 2x^2 + 4}{(x^2 + 13)^2}$

In Exercises 15–40, find the partial-fraction decomposition for each rational function.

15. $\dfrac{1}{x(x+1)}$

16. $\dfrac{1}{x(x-1)}$

17. $\dfrac{x}{x(x-1)}$

18. $\dfrac{x}{x(x+1)}$

19. $\dfrac{9x-11}{(x-3)(x+5)}$

20. $\dfrac{8x-13}{(x-2)(x+1)}$

21. $\dfrac{3x+1}{(x-1)^2}$

22. $\dfrac{9y-2}{(y-1)^2}$

23. $\dfrac{4x-3}{x^2+6x+9}$

24. $\dfrac{3x+1}{x^2+4x+4}$

25. $\dfrac{4x^2-32x+72}{(x+1)(x-5)^2}$

26. $\dfrac{4x^2-7x-3}{(x+2)(x-1)^2}$

27. $\dfrac{5x^2+28x-6}{(x+4)(x^2+3)}$

28. $\dfrac{x^2+5x+4}{(x-2)(x^2+2)}$

29. $\dfrac{-2x^2-17x+11}{(x-7)(3x^2-7x+5)}$

30. $\dfrac{14x^2+8x+40}{(x+5)(2x^2-3x+5)}$

31. $\dfrac{x^3}{(x^2+9)^2}$

32. $\dfrac{x^2}{(x^2+9)^2}$

33. $\dfrac{2x^3-3x^2+7x-2}{(x^2+1)^2}$

34. $\dfrac{-x^3+2x^2-3x+15}{(x^2+8)^2}$

35. $\dfrac{3x+1}{x^4-1}$

36. $\dfrac{2-x}{x^4-81}$

37. $\dfrac{5x^2+9x-8}{(x-1)(x^2+2x-1)}$

38. $\dfrac{10x^2-5x+29}{(x-3)(x^2+4x+5)}$

39. $\dfrac{3x}{x^3-1}$

40. $\dfrac{5x+2}{x^3-8}$

■ APPLICATIONS

41. **Optics.** The relationship between the distance of an object to a lens d_o, the distance to the image d_i, and the focal length f of the lens is given by

$$\frac{f(d_i+d_o)}{d_i d_o}=1$$

Use partial-fraction decomposition to write the lens law in terms of sums of fractions. What does each term represent?

42. **Sums.** Find the partial-fraction decomposition of $\dfrac{1}{n(n+1)}$, and apply it to find the sum of

$$\frac{1}{1\cdot 2}+\frac{1}{2\cdot 3}+\frac{1}{3\cdot 4}+\cdots+\frac{1}{999\cdot 1000}.$$

■ CATCH THE MISTAKE

In Exercises 43 and 44, explain the mistake that is made.

43. Find the partial-fraction decomposition of $\dfrac{3x^2+3x+1}{x(x^2+1)}$.

Solution:

Write the partial-fraction decomposition form.
$$\frac{3x^2+3x+1}{x(x^2+1)}=\frac{A}{x}+\frac{B}{x^2+1}$$

Multiply by the LCD $x(x^2+1)$.
$$3x^2+3x+1=A(x^2+1)+Bx$$

Eliminate the parentheses. $3x^2+3x+1=Ax^2+Bx+A$

Matching like terms leads to three equations. $A=3, B=3,$ and $A=1$.

This is incorrect. What mistake was made?

44. Find the partial-fraction decomposition of $\dfrac{3x^4-x-1}{x(x-1)}$.

Solution:

Write the partial-fraction decomposition form.
$$\frac{3x^4-x-1}{x(x-1)}=\frac{A}{x}+\frac{B}{x-1}$$

Multiply by the LCD $x(x-1)$.
$$3x^4-x-1=A(x-1)+Bx$$

Eliminate the parentheses and group like terms.
$$3x^4-x-1=(A+B)x-A$$

Compare like coefficients. $A=1, B=-2$

This is incorrect. What mistake was made?

▪ CONCEPTUAL

In Exercises 45 and 46, determine whether each statement is true or false.

45. Partial-fraction decomposition can only be employed when the degree of the numerator is greater than the degree of the denominator.

46. The degree of the denominator of a reducible rational expression is equal to the number of partial fractions in its decomposition.

▪ CHALLENGE

For Exercises 47–52, find the partial-fraction decomposition.

47. $\dfrac{x^2 + 4x - 8}{x^3 - x^2 - 4x + 4}$

48. $\dfrac{ax + b}{x^2 - c^2}$ a, b, c are real numbers.

49. $\dfrac{2x^3 + x^2 - x - 1}{x^4 + x^3}$

50. $\dfrac{-x^3 + 2x - 2}{x^5 - x^4}$

51. $\dfrac{x^5 + 2}{(x^2 + 1)^3}$

52. $\dfrac{x^2 - 4}{(x^2 + 1)^3}$

▪ TECHNOLOGY

53. Use a graphing utility to graph $y_1 = \dfrac{5x + 4}{x^2 + x - 2}$ and $y_2 = \dfrac{3}{x - 1} + \dfrac{2}{x + 2}$ in the same viewing rectangle. Is y_2 the partial-fraction decomposition of y_1?

54. Use a graphing utility to graph $y_1 = \dfrac{2x^2 + 2x - 5}{x^3 + 5x}$ and $y_2 = \dfrac{3x + 2}{x^2 + 5} - \dfrac{1}{x}$ in the same viewing rectangle. Is y_2 the partial-fraction decomposition of y_1?

55. Use a graphing utility to graph $y_1 = \dfrac{x^9 + 8x - 1}{x^5(x^2 + 1)^3}$ and $y_2 = \dfrac{4}{x} - \dfrac{1}{x^5} + \dfrac{2}{x^2 + 1} - \dfrac{3x + 2}{(x^2 + 1)^2}$ in the same viewing rectangle. Is y_2 the partial-fraction decomposition of y_1?

56. Use a graphing utility to graph $y_1 = \dfrac{x^3 + 2x + 6}{(x + 3)(x^2 - 4)^3}$ and $y_2 = \dfrac{2}{x + 3} + \dfrac{x + 3}{(x^2 - 4)^3}$ in the same viewing rectangle. Is y_2 the partial-fraction decomposition of y_1?

57. Use a graphing utility to graph $y_1 = \dfrac{2x^3 - 8x + 16}{(x - 2)^2(x^2 + 4)}$ and $y_2 = \dfrac{1}{x - 2} + \dfrac{2}{(x - 2)^2} + \dfrac{x + 4}{x^2 + 4}$ in the same viewing rectangle. Is y_2 the partial-fraction decomposition of y_1?

58. Use a graphing utility to graph $y_1 = \dfrac{3x^3 + 14x^2 + 6x + 51}{(x^2 + 3x - 4)(x^2 + 2x + 5)}$ and $y_2 = \dfrac{2}{x - 1} - \dfrac{1}{x + 4} + \dfrac{2x - 3}{x^2 + 2x + 5}$ in the same viewing rectangle. Is y_2 the partial-fraction decomposition of y_1?

SYSTEMS OF LINEAR INEQUALITIES IN TWO VARIABLES

SKILLS OBJECTIVES

- Graph a linear inequality in two variables.
- Graph a system of linear inequalities in two variables.

CONCEPTUAL OBJECTIVES

- Interpret the difference between solid and dashed lines.
- Interpret an overlapped shaded region as a solution.

Linear Inequalities in Two Variables

To graph linear inequalities in two variables, we will bridge together two concepts that we have already learned: *linear inequalities* (Section 1.5) and *lines* (Section 2.3). Recall that in Section 1.5 we discussed linear inequalities in one variable. For example, $3x - 1 < 8$ has a solution $x < 3$, which can be represented graphically on a number line where the red colored area to the left of 3 represents the solution.

Recall in Section 2.3 that $y = 2x + 1$ is an *equation in two variables* whose graph is a line in the xy-plane. We now turn our attention to **linear inequalities in two variables**. For example, if we change the $=$ in $y = 2x + 1$ to $<$ we get $y < 2x + 1$. The solution to this inequality in two variables is the set of all points (x, y) that make this inequality true. Some solutions to this inequality are $(-2, -5), (0, 0), (3, 4), (5, -1), \ldots$

In fact, the entire region *below* the line $y = 2x + 1$ satisfies the inequality $y < 2x + 1$. If we reverse the sign of the inequality to get $y > 2x + 1$, then the entire region *above* the line $y = 2x + 1$ represents the solution to the inequality.

Any line divides the xy-plane into two **half-planes**. For example, the line $y = 2x + 1$ divides the xy-plane into two half-planes represented as $y > 2x + 1$ and $y < 2x + 1$. Recall that with inequalities in one variable we used the notation of parentheses and brackets to denote the type of inequality (strict or nonstrict). We use a similar notation with linear inequalities in two variables. If the inequality is a strict inequality, $<$ or $>$, then the line is *dashed*, and, if the inequality includes the equal sign, $\leq$ or $\geq$, then a *solid* line is used. The following box summarizes the procedure for graphing a linear inequality in two variables.

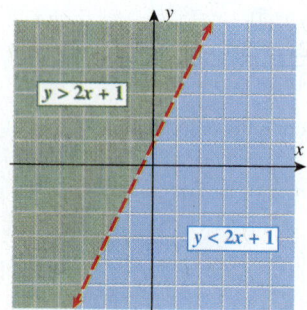

Study Tip

A dashed line means that the points that lie on the line are not included in the solution of the linear inequality.

GRAPHING A LINEAR INEQUALITY IN TWO VARIABLES

Step 1: Change the sign. Change the inequality sign, $<, \leq, \geq,$ or $>$, to an equal sign, $=$.

Step 2: Draw the line that corresponds to the resulting equation in Step 1.
- If the inequality is strict, $<$ or $>$, use a **dashed** line.
- If the inequality is not strict, $\leq$ or $\geq$, use a **solid** line.

Step 3: Test a point.
- Select a point in one half-plane and test to see whether it satisfies the inequality. If it does, then so do all the points in that region (half-plane). If not, then none of the points in that half-plane satisfy the inequality.
- Repeat this step for the other half-plane.

Step 4: Shade the half-plane that satisfies the inequality.

EXAMPLE 1 Graphing a Strict Linear Inequality in Two Variables

Graph the inequality $3x + y < 2$.

Solution:

STEP 1 Change the inequality sign to an equal sign.

$$3x + y = 2$$

STEP 2 Draw the line.
Convert from standard form to slope–intercept form.

$$y = -3x + 2$$

Since the inequality $<$ is a strict inequality, use a **dashed** line.

Technology Tip

The graphing calculator can be used to help in shading the linear inequality $3x + y < 2$. However, it will not show whether the line is solid or dashed. First solve for y, $y < -3x + 2$. Then, enter $y_1 = -3x + 2$. Since $y_1 < -3x + 2$, the region below the dashed line is shaded.

STEP 3 Test points in each half-plane.

Substitute $(3, 0)$ into $3x + y < 2$.

$$3(3) + 0 < 2$$

The point $(3, 0)$ does not satisfy the inequality.

$$9 < 2$$

Substitute $(-2, 0)$ into $3x + y < 2$.

$$3(-2) + 0 < 2$$

The point $(-2, 0)$ does satisfy the inequality.

$$-6 < 2$$

STEP 4 Shade the region containing the point $(-2, 0)$.

■ **Answer:**

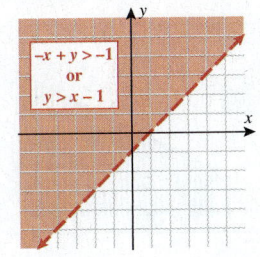

■ **YOUR TURN** Graph the inequality $-x + y > -1$.

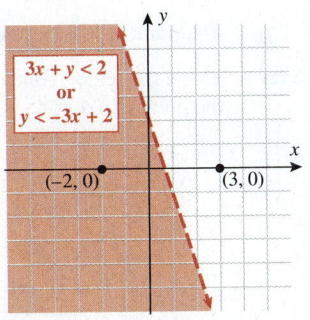

Classroom Example 9.4.1

Graph these inequalities:
a. $y > 3$
b. $-y > -x + 1$

Answer:

a.

b.

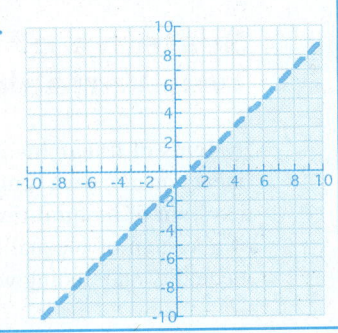

Graph these inequalities:
a. $x \leq 4$ **b.** $-x \leq y - 2$

Answer:
a.

b.

Technology Tip

■ **Answer:**

EXAMPLE 2 **Graphing a Nonstrict Linear Inequality in Two Variables**

Graph the inequality $2x - 3y \geq 6$.

Solution:

STEP 1 Change the inequality sign to an equal sign. $\qquad$ $2x - 3y = 6$

STEP 2 Draw the line.

Convert from standard form to slope–intercept form.
$$y = \frac{2}{3}x - 2$$

Since the inequality $\geq$ is not a strict inequality, use a **solid** line.

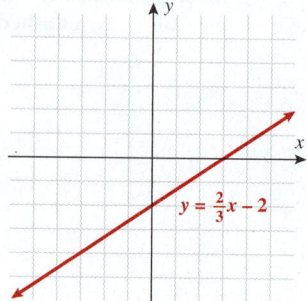

STEP 3 Test points in each half-plane.

Substitute $(5, 0)$ into $2x - 3y \geq 6$. $\qquad$ $2(5) - 3(0) \geq 6$

The point $(5, 0)$ satisfies the inequality. $\qquad$ $10 \geq 6$

Substitute $(0, 0)$ into $2x - 3y \geq 6$. $\qquad$ $2(0) - 3(0) \geq 6$

$$0 \geq 6$$

The point $(0, 0)$ does not satisfy the inequality.

STEP 4 Shade the region containing the point $(5, 0)$.

■ **YOUR TURN** Graph the inequality $x - 2y \leq 6$.

Systems of Linear Inequalities in Two Variables

Systems of linear inequalities are similar to *systems of linear equations.* In systems of linear equations we sought the points that satisfied *all* of the equations. The **solution set of a system of inequalities** contains the points that satisfy *all* of the inequalities. The graph of a system of inequalities can be obtained by simultaneously graphing each individual inequality and finding where the shaded regions intersect (or overlap), if at all.

EXAMPLE 3 **Graphing a System of Two Linear Inequalities**

Graph the system of inequalities:

$$x + y \geq -2$$
$$x + y \leq 2$$

Solution:

STEP 1 Change the inequality signs to equal signs.

$$x + y = -2$$
$$x + y = 2$$

STEP 2 Draw the two lines.

Because the inequality signs are
not strict, use solid lines.

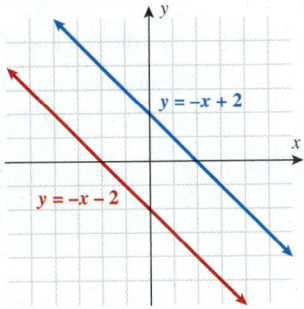

STEP 3 Test points for each inequality.

$x + y \geq -2$

Substitute $(-4, 0)$ into $x + y \geq -2$. $-4 \geq -2$
The point $(-4, 0)$ does not satisfy
the inequality.
Substitute $(0, 0)$ into $x + y \geq -2$. $0 \geq -2$
The point $(0, 0)$ does satisfy the inequality.

$x + y \leq 2$

Substitute $(0, 0)$ into $x + y \leq 2$. $0 \leq 2$
The point $(0, 0)$ does satisfy the inequality.
Substitute $(4, 0)$ into $x + y \leq 2$. $4 \leq 2$
The point $(4, 0)$ does not satisfy the inequality.

STEP 4 For $x + y \geq -2$, shade the region *above* that includes $(0, 0)$.

For $x + y \leq 2$, shade the region *below* that includes $(0, 0)$.

STEP 5 The overlapping region
is the solution.

Notice that the points $(0, 0)$,
$(-1, 1)$, and $(1, -1)$ all lie in the
shaded region and all three satisfy
both inequalities.

Technology Tip

Solve for y in each inequality first.
Enter $y_1 \geq -x - 2$ and $y_2 \leq -x + 2$.

The overlapping region is the
solution.

Solve for y in each inequality first.
Enter $y_1 \leq -x - 2$ and
$y_2 \geq -x + 2$.

There is no overlapping region.
Therefore, there is no solution to the
system of inequalities.

EXAMPLE 4   Graphing a System of Two Linear Inequalities with No Solution

Graph the system of inequalities: $x + y \leq -2$
$\qquad\qquad\qquad\qquad\qquad\qquad\qquad x + y \geq 2$

Solution:

STEP 1   Change the inequality signs to equal signs. $x + y = -2$
$\qquad\qquad\qquad\qquad\qquad\qquad\qquad\qquad\qquad\qquad\qquad\quad x + y = 2$

STEP 2   Draw the two lines.

Because the inequality signs are not
strict, use solid lines.

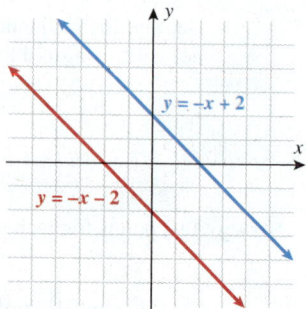

STEP 3   Test points for each inequality.

$x + y \leq -2$

Substitute $(-4, 0)$ into $x + y \leq -2$. $-4 \leq -2$

The point $(-4, 0)$ does satisfy the inequality.

Substitute $(0, 0)$ into $x + y \leq -2$. $0 \leq -2$

The point $(0, 0)$ does not satisfy
the inequality.

$x + y \geq 2$

Substitute $(0, 0)$ into $x + y \geq 2$. $0 \geq 2$

The point $(0, 0)$ does not satisfy
the inequality.

Substitute $(4, 0)$ into $x + y \geq 2$. $4 \geq 2$

The point $(4, 0)$ does satisfy the inequality.

STEP 4   For $x + y \leq -2$, shade the region
below that includes $(-4, 0)$.
For $x + y \leq 2$, shade the region
above that includes $(4, 0)$.

STEP 5   There is no overlapping region.
Therefore, no points satisfy both inequalities.
We say there is   no solution .

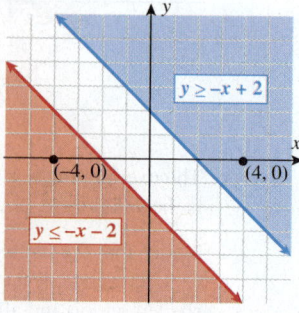

■ **Answer:**
a. No solution.
b.

■ **YOUR TURN**   Graph the solution to the system of inequalities.

a. $y > x + 1$ **b.** $y < x + 1$
$ y < x - 1$ $ y > x - 1$

Thus far we have addressed only systems of two linear inequalities. Systems with more than two inequalities are treated in a similar manner. The solution is the set of all points that satisfy *all* of the inequalities. When there are more than two linear inequalities, the solution may be a **bounded** region. We can algebraically determine where the lines intersect by setting the *y*-values equal to each other.

Technology Tip

Solve for *y* in each inequality first. Enter $y_1 \leq x$, $y_2 \geq -x$, and $y_3 < 3$.

The overlapping region is the solution to the system of inequalities.

 EXAMPLE 5   Graphing a System of Multiple Linear Inequalities

Solve the system of inequalities:
$$y \leq x$$
$$y \geq -x$$
$$y < 3$$

Solution:

STEP 1   Change the inequalities to equal signs.
$$y = x$$
$$y = -x$$
$$y = 3$$

STEP 2   Draw the three lines.

To determine the points of intersection, set the *y*-values equal.

Point where $y = x$ and $y = -x$ intersect:
$$x = -x$$
$$x = 0$$

Substitute $x = 0$ into $y = x$.
$$(0, 0)$$

Point where $y = -x$ and $y = 3$ intersect:
$$-x = 3$$
$$x = -3$$
$$(-3, 3)$$

Point where $y = 3$ and $y = x$ intersect:
$$x = 3$$
$$(3, 3)$$

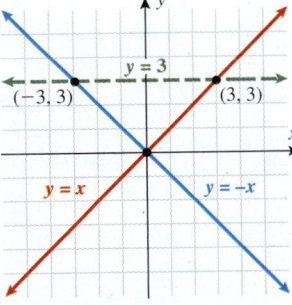

STEP 3   Test points to determine the shaded half-planes corresponding to $y \leq x$, $y \geq -x$, and $y < 3$.

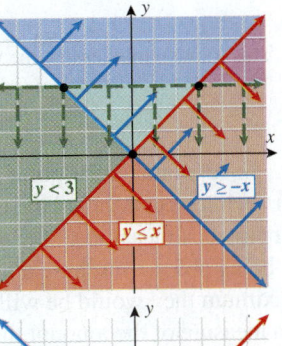

STEP 4   Shade the overlapping region.

Classroom Example 9.4.5*

Graph the systems of inequalities.

a.
$$\begin{cases} y \leq 2x + 5 \\ y \geq 2x + 1 \\ y \leq -2x + 5 \\ y \geq -2x + 1 \end{cases}$$

b.
$$\begin{cases} x < 6, \quad x \geq 1 \\ y \leq 3, \quad y \geq 0 \\ y + \frac{3}{2}x \geq \frac{9}{2} \\ 5y - 3x \geq -3 \end{cases}$$

Answer:

a.

b.

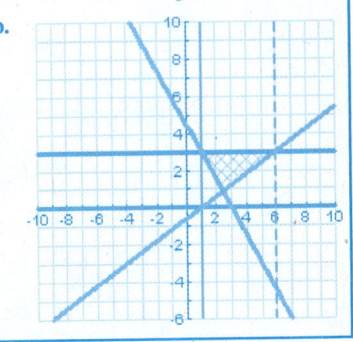

Applications

Systems of linear inequalities arise in many applications, for example, the target zone for heart rate during exercise, normal weight ranges for humans, capacity of a room for an event, and return on investments.

Classroom Example 9.4.6
A couple invites 30 people to a Super Bowl party. The fixed costs are $300, and the variable costs range between $10 and $15 per person. Assuming at least 12 people attend and at most 30 attend, graph the cost of the party as a system of inequalities.

Answer: Let x = number of people who attend the party and y = cost of the party.

$$\begin{cases} x \geq 12 \\ x \leq 30 \\ y \geq 300 + 10x \\ y \leq 300 + 15x \end{cases}$$

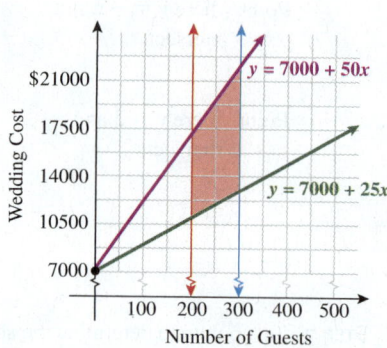

EXAMPLE 6   Cost of a Wedding Reception

A couple has invited 300 guests to their wedding. The fixed costs (such as formal wear, entertainment, flowers, and invitations) are $7000, and the variable costs (party favors, chair covers, food, and drinks) range between $25 and $50 per person, depending on the menu. Assuming at least 200 and at most 300 people attend, graph the cost of the wedding as a system of inequalities.

Solution:

Let x represent the number of people attending the wedding and y represent the cost of the wedding. There are four linear inequalities:

At least 200 guests attend:	$x \geq 200$
No more than 300 guests attend:	$x \leq 300$
Minimum cost of the wedding:	$y \geq 7000 + 25x$
Maximum cost of the wedding:	$y \leq 7000 + 50x$

Graph the system of inequalities.

This solution implies that the wedding cost will be approximately between $12,000 and $22,000.

In economics, the point where the supply and demand curves intersect is called the **equilibrium point**. **Consumer surplus** is a measure of the amount that consumers benefit by being able to purchase a product for a price less than the maximum they would be willing to pay. **Producer surplus** is a measure of the amount that producers benefit by selling at a market price that is higher than the least they would be willing to sell for.

EXAMPLE 7 Consumer Surplus and Producer Surplus

The Tesla Motors Roadster is the first electric car that is able to travel 245 miles on a single charge. The price of a 2011 model is approximately $90,000 (including tax and incentives).

Stefan Falke/Laif/
Redux Pictures

Suppose the supply and demand equations for this electric car are given by

$$P = 90{,}000 - 0.1x \qquad \text{(Demand)}$$
$$P = 10{,}000 + 0.3x \qquad \text{(Supply)}$$

where P is the price in dollars and x is the number of cars produced. Calculate the consumer surplus and the producer surplus for these two equations.

Solution:

Find the equilibrium point.

$$90{,}000 - 0.1x = 10{,}000 + 0.3x$$
$$0.4x = 80{,}000$$
$$x = 200{,}000$$

Let $x = 200{,}000$ in either the supply or demand equation.

$$P = 90{,}000 - 0.1(200{,}000) = 70{,}000$$
$$P = 10{,}000 + 0.3(200{,}000) = 70{,}000$$

According to these models, if the price of a Tesla Motors Roadster is $70,000, then 200,000 cars will be sold and there will be no surplus.

Write the systems of linear inequalities that correspond to consumer surplus and producer surplus.

CONSUMER SURPLUS	PRODUCER SURPLUS
$P \le 90{,}000 - 0.1x$	$P \ge 10{,}000 + 0.3x$
$P \ge 70{,}000$	$P \le 70{,}000$
$x \ge 0$	$x \ge 0$

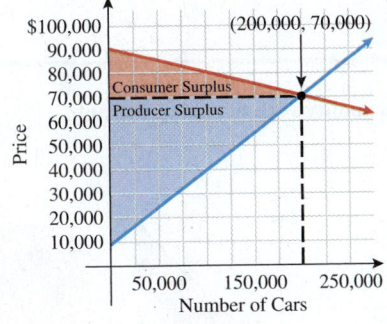

The consumer surplus is the area of the red triangle.

$$A = \frac{1}{2}bh$$
$$= \frac{1}{2}(200{,}000)(20{,}000)$$
$$= 2{,}000{,}000{,}000$$

The consumer surplus is $2B.

The producer surplus is the area of the blue triangle.

$$A = \frac{1}{2}bh$$
$$= \frac{1}{2}(200{,}000)(60{,}000)$$
$$= 6{,}000{,}000{,}000$$

The producer surplus is $6B.

The graphs of the systems of linear inequalities in Examples 6 and 7 are said to be **bounded**, whereas the graphs of the systems of linear inequalities in Examples 3, 4, and 5 are said to be **unbounded**. Any points that correspond to boundary lines intersecting are called **corner points** or **vertices**. In Example 7, the vertices corresponding to the consumer surplus are the points (0, 90,000), (0, 70,000), and (200,000, 70,000), and the vertices corresponding to the producer surplus are the points (0, 70,000), (0, 10,000), and (200,000, 70,000).

SECTION
9.4 SUMMARY

Linear Inequality

1. Change the inequality sign to an equal sign.

2. Draw the line. (Dashed for strict inequalities and solid for nonstrict inequalities.)

3. Test a point. (Select a point in one-half plane and test the inequality. Repeat this step for the other half-plane.)

4. Shade the half-plane that satisfies the linear inequality.

System of Linear Inequalities

▪ Draw the individual linear inequalities.

▪ The overlapping shaded region is the solution.

SECTION
9.4 EXERCISES

■ SKILLS

In Exercises 1–4, match the linear inequality with the correct graph.

1. $y > x$

2. $y \geq x$

3. $y < x$

4. $y \leq x$

a.

b.

c.

d.

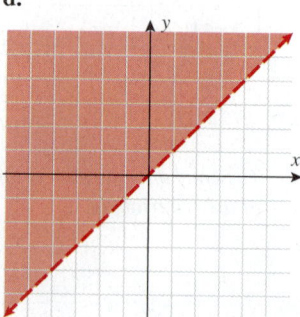

In Exercises 5–20, graph each linear inequality.

5. $y > x - 1$

6. $y \geq -x + 1$

7. $y \leq -x$

8. $y > -x$

9. $y \leq -3x + 2$

10. $y < 2x + 3$

11. $y \leq -2x + 1$

12. $y > 3x - 2$

13. $3x + 4y < 2$

14. $2x + 3y > -6$

15. $5x + 3y < 15$

16. $4x - 5y \leq 20$

17. $4x - 2y \geq 6$

18. $6x - 3y \geq 9$

19. $6x + 4y \leq 12$

20. $5x - 2y \geq 10$

In Exercises 21–48, graph each system of inequalities or indicate that the system has no solution.

21. $y \geq x - 1$
$\quad y \leq x + 1$

22. $y > x + 1$
$\quad y < x - 1$

23. $y > 2x + 1$
$\quad y < 2x - 1$

24. $y \leq 2x - 1$
$\quad y \geq 2x + 1$

25. $y \geq 2x$
$\quad y \leq 2x$

26. $y > 2x$
$\quad y < 2x$

27. $x > -2$
$\quad x < 4$

28. $y < 3$
$\quad y > 0$

29. $x \geq 2$
$\quad y \leq x$

30. $y \leq 3$
$\quad y \geq x$

31. $y > x$
$\quad x < 0$
$\quad y < 4$

32. $y \leq x$
$\quad x \geq 0$
$\quad y \leq 1$

33. $x + y > 2$
$\quad y < 1$
$\quad x > 0$

34. $x + y < 4$
$\quad x > 0$
$\quad y \geq 1$

35. $-x + y > 1$
$\quad y < 3$
$\quad x > 0$

36. $x - y > 2$
$\quad y < 4$
$\quad x \geq 0$

37. $x + 3y > 6$
$\quad y < 1$
$\quad x \geq 1$

38. $x + 2y > 4$
$\quad y < 1$
$\quad x \geq 0$

39. $y \geq x - 1$
$\quad y \leq -x + 3$
$\quad y < x + 2$

40. $y < 4 - x$
$\quad y > x - 4$
$\quad y > -x - 4$

41. $x + y > -4$
$\quad -x + y < 2$
$\quad y \geq -1$
$\quad y \leq 1$

42. $y < x + 2$
$\quad y > x - 2$
$\quad y < -x + 2$
$\quad y > -x - 2$

43. $y < x + 3$
$\quad x + y \geq 1$
$\quad y \geq 1$
$\quad y \leq 3$

44. $y \leq -x + 2$
$\quad y - x \geq -3$
$\quad y \geq -2$
$\quad y \leq 1$

45. $y + x < 2$
$\quad y + x \geq 4$
$\quad y \geq -2$
$\quad y \leq 1$

46. $y - x < 3$
$\quad y + x > 3$
$\quad y \leq -2$
$\quad y \geq -4$

47. $2x - y < 2$
$\quad 2x + y > 2$
$\quad y < 2$

48. $3x - y > 3$
$\quad 3x + y < 3$
$\quad y < -2$

■ APPLICATIONS

49. Area. Find the area enclosed by the system of inequalities.
$\quad y > |x|$
$\quad y < 2$

50. Area. Find the area enclosed by the system of inequalities.
$\quad y < |x|$
$\quad x \geq 0$
$\quad y \geq 0$
$\quad x < 3$

51. Area. Find the area enclosed by the system of linear inequalities (assume $y \geq 0$).
$\quad 5x + y \leq 10$
$\quad x \geq 0$
$\quad x \leq 1$

52. Area. Find the area enclosed by the system of linear inequalities (assume $y \geq 0$).
$\quad -5x + y \leq 0$
$\quad x \geq 1$
$\quad x \leq 2$

53. Hurricanes. After back-to-back-to-back-to-back hurricanes (Charley, Frances, Ivan, and Jeanne) in Florida in the summer of 2004, FEMA sent disaster relief trucks to Florida. Floridians mainly needed drinking water and generators. Each truck could carry no more than 6000 pounds of cargo or 2400 cubic feet of cargo. Each case of bottled water takes up 1 cubic foot of space and weighs 25 pounds. Each generator takes up 20 cubic feet and weighs 150 pounds. Let x represent the number of cases of water and y represent the number of generators, and write a system of linear inequalities that describes the number of generators and cases of water each truck can haul to Florida.

54. Hurricanes. Repeat Exercise 53 with a smaller truck and different supplies. Suppose the smaller trucks that can haul 2000 pounds and 1500 cubic feet of cargo are used to haul plywood and tarps. A case of plywood is 60 cubic feet and weighs 500 pounds. A case of tarps is 10 cubic feet and weighs 50 pounds. Letting x represent the number of cases of plywood and y represent the number of cases of tarps, write a system of linear inequalities that describes the number of cases of tarps and plywood each truck can haul to Florida. Graph the system of linear inequalities.

55. Health. A diet must be designed to provide at least 275 units of calcium, 125 units of iron, and 200 units of Vitamin B. Each ounce of food *A* contains 10 units of calcium, 15 units of iron, and 20 units of vitamin B. Each ounce of food *B* contains 20 units of calcium, 10 units of iron, and 15 units of vitamin B.

   a. Find a system of inequalities to describe the different quantities of food that may be used (let x = the number of ounces of food *A* and y = the number of ounces of food *B*).

   b. Graph the system of inequalities.

   c. Using the graph found in part (b), find two possible solutions (there are infinitely many).

56. Health. A diet must be designed to provide at least 350 units of calcium, 175 units of iron, and 225 units of Vitamin B. Each ounce of food *A* contains 15 units of calcium, 25 units of iron, and 20 units of vitamin B. Each ounce of food *B* contains 25 units of calcium, 10 units of iron, and 10 units of vitamin B.

   a. Find a system of inequalities to describe the different quantities of food that may be used (let x = the number of ounces of food *A* and y = the number of ounces of food *B*).

   b. Graph the system of inequalities.

   c. Using the graph found in part (b), find two possible solutions (there are infinitely many).

57. Business. A manufacturer produces two types of computer mouse: USB wireless mouse and a Bluetooth mouse. Past sales indicate that it is necessary to produce at least twice as many USB wireless mice than Bluetooth mice. To meet demand, the manufacturer must produce at least 1000 computer mice per hour.

   a. Find a system of inequalities describing the production levels of computer mice. Let x be the production level for USB wireless mouse and y be the production level for Bluetooth mouse.

   b. Graph the system of inequalities describing the production levels of computer mice.

   c. Use your graph in part (b) to find two possible solutions.

58. Business. A manufacturer produces two types of mechanical pencil lead: 0.5 millimeter and 0.7 millimeter. Past sales indicate that it is necessary to produce at least 50% more 0.5 millimeter lead than 0.7 millimeter lead. To meet demand, the manufacturer must produce at least 10,000 pieces of pencil lead per hour.

   a. Find a system of inequalities describing the production levels of pencil lead. Let x be the production level for 0.5 millimeter pencil lead and y be the production level for 0.7 millimeter pencil lead.

   b. Graph the system of inequalities describing the production levels of pencil lead.

   c. Use your graph in part (b) to find two possible solutions.

For Exercises 59–62, employ the following supply and demand equations:

$$\text{Demand:} \quad P = 80 - 0.01x$$
$$\text{Supply:} \quad P = 20 + 0.02x$$

59. Consumer Surplus. Write a system of linear inequalities corresponding to the consumer surplus.

60. Producer Surplus. Write a system of linear inequalities corresponding to the producer surplus.

61. Consumer Surplus. Calculate the consumer surplus given the supply and demand equations.

62. Producer Surplus. Calculate the producer surplus given the supply and demand equations.

▪CATCH THE MISTAKE

In Exercises 63 and 64, explain the mistake that is made.

63. Graph the inequality
$y \geq 2x + 1$.

Solution:

Graph the line
$y = 2x + 1$ with
a solid line.

Since the inequality
is $\geq$, shade to the *right*.

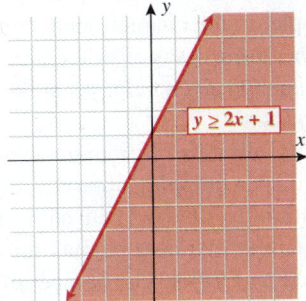

This is incorrect. What
mistake was made?

64. Graph the inequality
$y < 2x + 1$.

Solution:

Graph the line
$y = 2x + 1$ with
a solid line.

Since the inequality
is $<$, shade *below*.

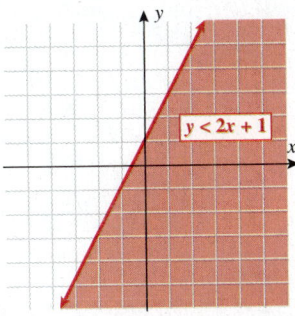

This is incorrect. What
mistake was made?

▪CONCEPTUAL

In Exercises 65–68, determine whether each statement is true or false.

65. A linear inequality always has a solution that is a half-plane.

66. A dashed curve is used for strict inequalities.

67. A solid curve is used for strict inequalities.

68. A system of linear inequalities always has a solution.

▪CHALLENGE

For the system of linear inequalities (Exercises 69 and 70), assume $a, b, c,$ and d are real numbers.

$$x \geq a$$
$$x < b$$
$$y > c$$
$$y \leq d$$

69. Describe the solution when $a < b$ and $c < d$.

70. What will the solution be if $a > b$ and $c > d$, given the system of linear inequalities?

Use the following system of linear inequalities for Exercises 71 and 72:

$$y \leq ax + b$$
$$y \geq -ax + b$$

71. If a and b are positive real numbers, graph the solution.

72. If a and b are negative real numbers, graph the solution.

▪TECHNOLOGY

In Exercises 73 and 74, use a graphing utility to graph the following inequalities.

73. $4x - 2y \geq 6$ (Check with your answer to Exercise 17.)

74. $6x - 3y \geq 9$ (Check with your answer to Exercise 18.)

In Exercises 75 and 76, use a graphing utility to graph each system of inequalities or indicate that the system has no solution.

75. $-0.05x + 0.02y \geq 0.12$
$0.01x + 0.08y \leq 0.08$

76. $y \leq 2x + 3$
$y > -0.5x + 5$

SKILLS OBJECTIVES

- Write an objective function that represents a quantity to be minimized or maximized.
- Utilize inequalities to describe constraints.
- Solve the optimization problem, which combines minimizing or maximizing a function subject to constraints, using linear programming.

CONCEPTUAL OBJECTIVES

- Understand that linear programming is a graphical method that solves optimization problems.

Solving an Optimization Problem

Often we seek to maximize or minimize a function subject to constraints. This process is called **optimization**. For example, in the chapter opener about Hurricane Katrina, FEMA had to determine how many generators, cases of water, and tarps should be in each truck to maximize the number of Louisianians given help, yet at the same time factor in the weight and space constraints on the trucks.

When the function we seek to minimize or maximize is linear and the constraints are given in terms of linear inequalities, a graphing approach to such problems is called **linear programming**. In linear programming, we start with a linear equation, called the **objective function**, that represents the quantity that is to be maximized or minimized, for example, the number of Louisianians aided by FEMA. The number of people aided, however, is subject to constraints represented as linear inequalities, such as how much weight each truck can haul and how much space each truck has for cargo.

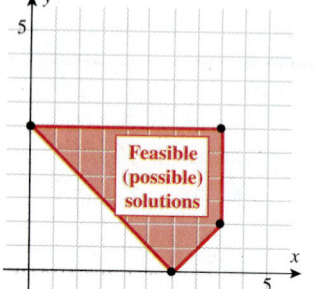

The goal is to minimize or maximize the objective function $z = Ax + By$ subject to *constraints*. In other words, find the points (x, y) that make the value of z the largest (or smallest). The **constraints** are a system of linear inequalities, and the common shaded region represents the **feasible** (possible) **solutions**.

If the constraints form a bounded region, the vertices represent the coordinates (x, y) that correspond to a maximum or minimum value of the objective function $z = Ax + By$. If the region is not bounded, then if an optimal solution exists, it will occur at a vertex. A procedure for solving linear programming problems is outlined below.

SOLVING AN OPTIMIZATION PROBLEM USING LINEAR PROGRAMMING

Step 1: **Write the objective function.** This expression represents the quantity that is to be minimized or maximized.

Step 2: **Write the constraints.** This is a system of linear inequalities.

Step 3: **Graph the constraints.** Graph the system of linear inequalities and shade the common region, which contains the feasible solutions.

Step 4: **Identify the vertices.** The corner points of the shaded region represent possible maximum or minimum values of the objective function.

Step 5: **Evaluate the objective function for each vertex.** For each corner point of the shaded region, substitute the coordinates into the objective function and list the value of the objective function.

Step 6: **Identify the optimal solution.** The largest (maximum) or smallest (minimum) value of the objective function in Step 5 is the optimal solution if the feasible region is bounded.

EXAMPLE 1 **Maximizing an Objective Function**

Find the maximum value of $z = 2x + y$, subject to the constraints:

$$x \geq 1 \qquad x \leq 4 \qquad x + y \leq 5 \qquad y \geq 0$$

Solution:

STEP 1 Write the objective function.

$z = 2x + y$

STEP 2 Write the constraints.

$x \geq 1$
$x \leq 4$
$y \leq -x + 5$
$y \geq 0$

STEP 3 Graph the constraints.

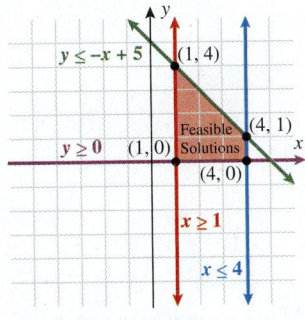

STEP 4 Identify the vertices.

$(1, 4), (4, 1), (1, 0), (4, 0)$

STEP 5 Evaluate the objective function for each vertex.

VERTEX	x	y	OBJECTIVE FUNCTION: $z = 2x + y$
$(1, 4)$	1	4	$2(1) + 4 = \mathbf{6}$
$(4, 1)$	4	1	$2(4) + 1 = \mathbf{9}$
$(1, 0)$	1	0	$2(1) + 0 = \mathbf{2}$
$(4, 0)$	4	0	$2(4) + 0 = \mathbf{8}$

STEP 6 The maximum value of z is **9**, which occurs when $x = 4$ and $y = 1$.

■ **YOUR TURN** Find the maximum value of $z = x + 3y$ subject to the constraints:

$$x \geq 1 \qquad x \leq 3 \qquad y \leq -x + 3 \qquad y \geq 0$$

Study Tip

The bounded region is the region that satisfies *all* of the constraints. Only vertices of the bounded region correspond to possible solutions. Even though $y = -x + 5$ and $y = 0$ intersect at $x = 5$, that point of intersection is outside the shaded region and therefore is *not* one of the vertices.

■ **Answer:** The maximum value of z is **7**, which occurs when $x = 1$ and $y = 2$.

EXAMPLE 2 Minimizing an Objective Function

Find the minimum value of $z = 4x + 5y$, subject to the constraints:

$$x \geq 0 \qquad 2x + y \leq 6 \qquad x + y \leq 5 \qquad y \geq 0$$

Solution:

STEP 1 Write the objective function. $\qquad\qquad z = 4x + 5y$

STEP 2 Write the constraints.

$$x \geq 0$$
$$y \leq -2x + 6$$
$$y \leq -x + 5$$
$$y \geq 0$$

STEP 3 Graph the constraints.

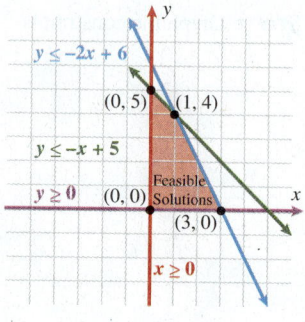

STEP 4 Identify the vertices. $\qquad\qquad (0, 0), (0, 5), (1, 4), (3, 0)$

STEP 5 Evaluate the objective function for each vertex.

VERTEX	x	y	OBJECTIVE FUNCTION: $z = 4x + 5y$
(0, 0)	0	0	$4(0) + 5(0) = \mathbf{0}$
(0, 5)	0	5	$4(0) + 5(5) = \mathbf{25}$
(1, 4)	1	4	$4(1) + 5(4) = \mathbf{24}$
(3, 0)	3	0	$4(3) + 5(0) = \mathbf{12}$

STEP 6 The minimum value of z is **0**, which occurs when $x = 0$ and $y = 0$.

■ **YOUR TURN** Find the minimum value of $z = 2x + 3y$ subject to the constraints:

$$x \geq 1 \qquad 2x + y \leq 8 \qquad x + y \geq 4$$

Classroom Example 9.5.2*
Find the minimum value of $z = -2x + 5y$ subject to the constraints:

$$x \leq 6 \qquad x \geq 1$$
$$y \leq 3 \qquad y \geq 0$$
$$y + \tfrac{3}{2}x \geq \tfrac{9}{2} \qquad 5y - 3x \geq -3$$

Answer:

Vertex	$z = -2x + 5y$
$(1, 3)$	13
$(6, 3)$	3
$\left(\tfrac{17}{7}, \tfrac{6}{7}\right)$	$-\tfrac{4}{7}$

Minimum value is approximately $-\tfrac{4}{7}$.

Study Tip

Maxima or minima of objective functions occur at the vertices of the shaded region corresponding to the constraints.

■ **Answer:** The minimum value of z is **8**, which occurs when $x = 4$ and $y = 0$.

EXAMPLE 3 **Solving an Optimization Problem Using Linear Programming: Unbounded Region**

Find the maximum value and minimum value of $z = 7x + 3y$, subject to the constraints:

$$y \geq 0 \qquad -2x + y \leq 0 \qquad -x + y \geq -4$$

Solution:

STEP 1 Write the objective function.

$z = 7x + 3y$

STEP 2 Write the constraints.

$y \geq 0$
$y \leq 2x$
$y \geq x - 4$

STEP 3 Graph the constraints.

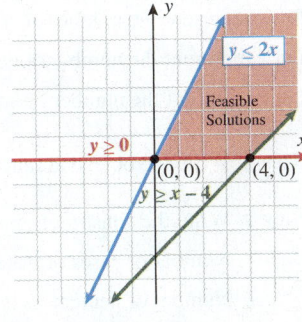

STEP 4 Identify the vertices.

$(0, 0), (4, 0)$

STEP 5 Evaluate the objective function for each vertex.

VERTEX	x	y	OBJECTIVE FUNCTION: $z = 7x + 3y$
$(0, 0)$	0	0	$7(0) + 3(0) = \mathbf{0}$
$(4, 0)$	4	0	$7(4) + 3(0) = \mathbf{28}$

STEP 6 The minimum value of z is **0**, which occurs when $x = 0$ and $y = 0$.

There is no maximum value, because if we select a point in the shaded region, say $(3, 3)$, the objective function at $(3, 3)$ is equal to 30, which is greater than 28.

When the feasible solutions are contained in a bounded region, then a maximum and a minimum exist and are each located at one of the vertices. If the feasible solutions are contained in an unbounded region, then if a maximum or minimum exists, it is located at one of the vertices.

EXAMPLE 4 **Maximizing the Number of People Aided by a Hurricane Relief Effort**

After Hurricane Katrina, FEMA sent disaster relief trucks to Louisiana. Each truck could carry no more than 6000 pounds of cargo. Each case of bottled water weighs 25 pounds and each generator weighs 150 pounds. If each generator helps one household and five cases of water help one household, determine the maximum number of Louisiana households aided by each truck and how many generators and cases of water should be sent in each truck. Due to the number of trucks and the supply of water and generators nationwide, each truck must contain at least 10 times as many cases of water as generators.

Solution:

Let x represent the number of cases of water.

Let y represent the number of generators.

Let z represent the number of households aided per truck.

STEP 1 Write the objective function. $\qquad\qquad z = \dfrac{1}{5}x + y$

STEP 2 Write the constraints.

Number of cases of water is nonnegative. $\qquad x \geq 0$

Number of generators is nonnegative. $\qquad y \geq 0$

At least 10 times as many cases of water as generators. $\qquad \mathbf{x \geq 10y}$

Weight capacity of truck: $\qquad \mathbf{25x + 150y \leq 6000}$

STEP 3 Graph the constraints and determine feasible solutions.

STEP 4 Identify the vertices. $\qquad\qquad$ (0, 0), (150, 15), (240, 0)

STEP 5 Evaluate the objective function for each vertex.

VERTEX	x	y	OBJECTIVE FUNCTION: $z = \frac{1}{5}x + y$
(0, 0)	0	0	**0**
(150, 15)	150	15	**45**
(240, 0)	240	0	**48**

STEP 6 A maximum of **48 households** are aided by each truck. This maximum occurs when there are **240 cases of water** and **0 generators** on each truck.

SECTION
9.5 SUMMARY

In this section the linear programming model was discussed.

1. Write the objective function.
2. Write the constraints.
3. Graph the constraints.

4. Identify the vertices.
5. Evaluate the objective function for each vertex.
6. Identify the optimal solution.

SECTION
9.5 EXERCISES

▪ SKILLS

In Exercises 1–4, find the value of the objective function at each of the vertices. What is the maximum value of the objective function? What is the minimum value of the objective function?

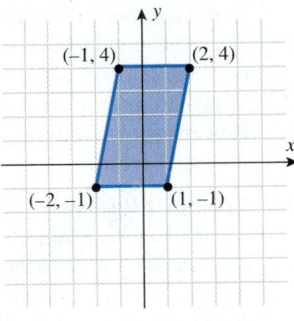

1. Objective function: $z = 2x + 3y$

2. Objective function: $z = 3x + 2y$

3. Objective function: $z = 1.5x + 4.5y$

4. Objective function: $z = \frac{2}{3}x + \frac{3}{5}y$

In Exercises 5–12, minimize or maximize each objective function subject to the constraints.

5. Minimize $z = 7x + 4y$ subject to:
$x \geq 0 \quad y \geq 0 \quad -x + y \leq 4$

6. Maximize $z = 3x + 5y$ subject to:
$x \geq 0 \quad y \geq 0 \quad -x + y \geq 4$

7. Maximize $z = 4x + 3y$ subject to:
$x \geq 0 \quad y \leq -x + 4 \quad y \geq -x$

8. Minimize $z = 4x + 3y$ subject to:
$x \geq 0 \qquad y \geq 0$
$x + y \leq 10 \quad x + y \geq 0$

9. Minimize $z = 2.5x + 3.1y$ subject to:
$x \geq 0 \quad y \geq 0 \quad x \leq 4$
$-x + y \leq 2 \quad x + y \leq 6$

10. Maximize $z = 2.5x - 3.1y$ subject to:
$x \geq 1 \quad y \leq 7 \quad x \leq 3$
$-x + y \geq 2 \quad x + y \geq 6$

11. Maximize $z = \frac{1}{4}x + \frac{2}{5}y$ subject to:
$x + y \geq 5 \quad x + y \leq 7$
$-x + y \leq 5 \quad -x + y \geq 3$

12. Minimize $z = \frac{1}{3}x - \frac{2}{5}y$ subject to:
$x + y \geq 6 \quad x + y \leq 8$
$-x + y \leq 6 \quad -x + y \geq 4$

▪ APPLICATIONS

13. **Hurricanes.** After the 2004 back-to-back-to-back-to-back hurricanes in Florida, a student at Valencia Community College decided to create two T-shirts to sell. One T-shirt says, "I survived Charley on Friday the Thirteenth," and the second says, "I survived Charley, Frances, Ivan, and Jeanne." The Charley T-shirt costs him $7 to make and he sells it for $13. The other T-shirt costs him $5 to make and he sells it for $10. He does not want to invest more than $1000. He estimates that the total demand will not exceed 180 T-shirts. Find the number of each type of T-shirt he should make to yield maximum profit.

14. **Hurricanes.** After Hurricane Charley devastated central Florida unexpectedly, Orlando residents prepared for Hurricane Frances by boarding up windows and filling up their cars with gas. It took 5 hours of standing in line to get plywood, and lines for gas were just as time-consuming. A student at Seminole Community College decides to do a spoof of the "Got Milk" ads and creates two T-shirts: "Got Plywood," with a line of people in a home improvement store, and "Got Gas," with a street lined with cars waiting to pump gasoline. The "Got Plywood" shirts cost $8 to make, and she sells them for $13. The "Got Gas" shirts cost $6 to make, and she sells them for $10. She decides to limit her costs to $1400. She estimates that demand for these T-shirts will not exceed 200 T-shirts. Find the number of each type of T-shirt she should make to yield maximum profit.

15. Computer Business. A computer science major and a business major decide to start a small business that builds and sells a desktop computer and a laptop computer. They buy the parts, assemble them, load the operating system, and sell the computers to other students. The costs for parts, time to assemble the computer, and profit are summarized in the following table:

	DESKTOP	LAPTOP
Cost of Parts	$700	$400
Time to Assemble (hours)	5	3
Profit	$500	$300

They were able to get a small business loan in the amount of $10,000 to cover costs. They plan on making these computers over the summer and selling them the first day of class. They can dedicate at most only 90 hours to assembling these computers. They estimate that the demand for laptops will be at least three times as great as the demand for desktops. How many of each type of computer should they make to maximize profit?

16. Computer Business. Repeat Exercise 15 if the two students are able to get a loan for $30,000 to cover costs and they can dedicate at most 120 hours to assembling the computers.

17. Passenger Ratio. The Eurostar is a high-speed train that travels between London, Brussels, and Paris. There are 30 cars on each departure. Each train car is designated first class or second class. Based on demand for each type of fare, there should always be at least two but no more than four first-class train cars. The management wants to claim that the ratio of first-class to second-class cars never exceeds 1:8. If the profit on each first-class train car is twice as much as the profit of each second-class train car, find the number of each class of train car that will generate a maximum profit.

18. Passenger Ratio. Repeat Exercise 17. This time, assume that there has to be at least one first-class train car and that the profit from each first-class train car is 1.2 times as much as the profit from each second-class train car. The ratio of first class to second class cannot exceed 1:10.

19. Production. A manufacturer of skis produces two models: a regular ski and a slalom ski. A set of regular skis produces a $25 profit, and a set of slalom skis produces a profit of $50. The manufacturer expects a customer demand of at least 200 pairs of regular skis and at least 80 pairs of slalom skis. The maximum number of pairs of skis that can be produced by this company is 400. How many of each model of skis should be produced to maximize profits?

20. Donut Inventory. A well-known donut store makes two popular types of donuts: crème-filled and jelly-filled. The manager knows from past statistics that the number of dozens of donuts sold is at least 10, but no more than 30. To prepare the donuts for frying, the baker needs (on the average) 3 minutes for a dozen crème-filled and 2 minutes for jelly-filled. The baker has at most 2 hours available per day to prepare the donuts. How many dozens of each type should be prepared to maximize the daily profit if there is a $1.20 profit for each dozen crème-filled and $1.80 profit for each dozen jelly-filled donuts?

■CATCH THE MISTAKE

In Exercises 21 and 22, explain the mistake that is made.

21. Maximize the objective function $z = 2x + y$ subject to the following constraints:

$$x \geq 0 \qquad y \geq 0$$
$$-x + y \geq 0 \qquad x + y \leq 2$$

Solution:

Graph the constraints.

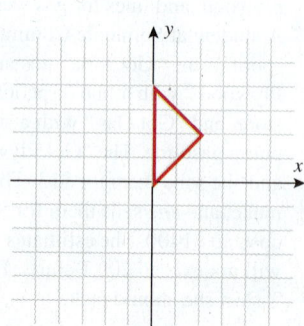

Identify the vertices.   (0, 2), (0, 0), (1, 1)

Comparing the y-coordinates of the vertices, the largest y-value is 2.

The maximum value occurs at (0, 2).

The objective function at that point is equal to **2**.

This is incorrect. The maximum value of the objective function is not 2. What mistake was made?

22. Maximize the objective function $z = 2x + y$ subject to the following constraints:

$$x \geq 0 \qquad y \geq 0$$
$$-x + y \leq 0 \qquad x + y \leq 2$$

Solution:

Graph the constraints.

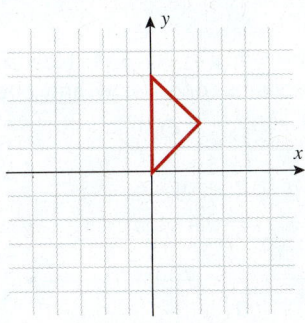

Identify the vertices. (0, 2), (0, 0), (1, 1)

VERTEX	x	y	OBJECTIVE FUNCTION: $z = 2x + y$
(0, 2)	0	2	$2(0) + 2 = $ **2**
(0, 0)	0	0	$2(0) + 0 = $ **0**
(1, 1)	1	1	$2(1) + 1 = $ **3**

The maximum, **3**, is located at the point (1, 1).

This is incorrect. What mistake was made?

■ **CONCEPTUAL**

In Exercises 23 and 24, determine whether each statement is true or false.

23. An objective function always has a maximum or minimum.

24. An objective function subject to constraints that correspond to a bounded region always has a maximum and a minimum.

■ **CHALLENGE**

25. Maximize the objective function $z = 2x + y$ subject to the conditions, where $a > 2$.

$$ax + y \geq -a$$
$$-ax + y \leq a$$
$$ax + y \leq a$$
$$-ax + y \geq -a$$

26. Maximize the objective function $z = x + 2y$ subject to the conditions, where $a > b > 0$.

$$x + y \geq a$$
$$-x + y \leq a$$
$$x + y \leq a + b$$
$$-x + y \geq a - b$$

■ **TECHNOLOGY**

In Exercises 27–32, employ a graphing utility to determine the region of feasible solutions, and apply the intersect feature to find the approximate coordinates of the vertices to solve the linear programming problems.

27. Minimize $z = 2.5x + 3.1y$ subject to:

$$x \geq 0 \qquad y \geq 0 \qquad x \leq 4$$
$$-x + y \leq 2 \qquad x + y \leq 6$$

Compare with your answer to Exercise 9.

28. Maximize $z = 2.5x + 3.1y$ subject to:

$$x \geq 0 \qquad y \geq 0 \qquad x \leq 4$$
$$-x + y \leq 2 \qquad x + y \leq 6$$

Compare with your answer to Exercise 10.

29. Maximize $z = 17x + 14y$ subject to:

$$y \geq 4.5 \qquad -x + y \leq 3.7 \qquad x + y \leq 11.2$$

30. Minimize $z = 1.2x + 1.5y$ subject to:

$$2.3x + y \leq 14.7 \qquad -2.3x + y \leq 14.7$$
$$-5.2x + y \leq 3.7 \qquad -2.3x + y \geq 1.5$$

31. Maximize $z = 4.5x + 1.8y$ subject to:

$$-3.1x + y \leq 11.4 \qquad 1.4x + y \geq 1.5 \qquad 5x + y \leq 15$$

32. Minimize $z = 5.4x - 1.6y$ subject to:

$$y \geq 1.8x - 3.6 \qquad y \geq -2.2x + 1.8$$
$$y \leq 1.8x + 4.2 \qquad y \leq -2.2x + 10.8$$

In Section 9.1, you learned how to solve systems of two linear equations in two variables. That is, you determined the set of all points that satisfy both given equations in a system. In Chapter 1 you solved linear inequalities in one variable. Next, you will put these ideas together as you consider systems of linear inequalities in two variables, and their solutions.

1. The following graph shows the line $y = 2x + 1$. Notice that the line divides the Cartesian plane into two half-planes—one below and one above the line.

The set of all the points in the half-plane below the line are in the shaded region.

a. The set of points *on* the line, together with the points in the shaded region, make up the solution set of the inequality: $y \leq 2x + 1$. To get an idea of what this means, choose several points—a few on the line, a few in the shaded region, and a few in the unshaded region—and evaluate the inequality for each. What do you notice?

b. Write an inequality that has as its solution set the points on the line $y = 2x + 1$ together with all the points above the line.

c. Suppose you wanted to graph the solutions to the strict inequality $y < 2x + 1$. How do you think you could alter the graph shown above to do this? Explain.

2. To graph a linear inequality, first graph the associated line (either solid or dashed), then choose the appropriate half-plane to shade. Remember, ALL of the points in one of the half-planes are in the solution set of the given inequality, and NONE of the points in the other half-plane are in the solution set. To determine which half-plane to shade, one need only check one test point.

a. Graph the linear inequality $3x + 6y < 18$.

b. Graph the linear inequality $3x - 6y \leq 18$.

c. Now consider the system of two linear inequalities:

$$3x + 6y > 18$$
$$3x - 6y \leq 18$$

Graph the two associated lines together at the right.

Shade the region(s) that contain(s) the points in the solution set of the system of inequalities. Use some test points, if needed.

d. How is the region you shaded in part (c) related to the regions you shaded in parts (a) and (b)?

In 2005, hybrid vehicles were introduced into the U.S. market. The demand for hybrids, which are typically powered by a combination of gasoline and electric batteries, was based on popular recognition of petroleum as an increasingly scarce nonrenewable resource, as well as consumers' need to combat rising prices at the gas pumps. In addition to achieving greater fuel economy than conventional internal combustion engine vehicles (ICEVs), their use also results in reduced emissions.

An online "Gas Mileage Impact Calculator," created by the American Council for an Energy-Efficient Economy (www.aceee.org), was used to generate the following tables comparing a conventional sedan (four-door) and an SUV versus their respective hybrid counterparts.

Gas Mileage Impact Calculator

	TOYOTA CAMRY 2.4L 4, AUTO $3.75/GALLON 15,000 MI/YEAR	TOYOTA CAMRY HYBRID 2.4L 4, AUTO $3.75/GALLON 15,000 MI/YEAR
Gas Consumption	611 gallons	449 gallons
Gas Cost	$2289.75	$1681.99
Fuel Economy	25 mpg	33 mpg
EMISSIONS		
Carbon Dioxide (greenhouse gas)	11,601 pounds	8522 pounds
Carbon Monoxide (poisonous gas)	235 pounds	169 pounds
Nitrogen Oxide (lung irritant and smog)	10 pounds	7 pounds
Hydrocarbons (smog)	6 pounds	8 pounds

	TOYOTA HIGHLANDER 3.5L 6, AUTO STK $3.75/GALLON 15,000 MI/YEAR	TOYOTA HIGHLANDER HYBRID 3.3L 6, AUTO AWD $3.75/GALLON 15,000 MI/YEAR
Gas Consumption	740 gallons	576 gallons
Gas Cost	$2773.44	$2158.33
Fuel Economy	20 mpg	26 mpg
EMISSIONS		
Carbon Dioxide (greenhouse gas)	14,052 pounds	10,936 pounds
Carbon Monoxide (poisonous gas)	229 pounds	187 pounds
Nitrogen Oxide (lung irritant and smog)	11 pounds	8 pounds
Hydrocarbons (smog)	7 pounds	16 pounds

The MSRP and mileage comparisons for the 2011 models are given below:

	CAMRY	CAMRY HYBRID	HIGHLANDER	HIGHLANDER HYBRID
MSRP	$19,435	$26,065	$28,035	$34,435
Miles per gallon in city	21	33	18	27
Miles per gallon on highway	31	34	24	25

For the following questions, assume that you drive 15,000 miles per year (all in city) and the price of gasoline is $3.75 per gallon.

1. Write a linear equation that models the total cost of owning and operating each vehicle y as a function of the number of years of ownership x.

   a. Camry
   b. Camry Hybrid
   c. Highlander
   d. Highlander Hybrid

2. Write a linear equation that models the total number of pounds of carbon dioxide each vehicle emits y as a function of the number of years of ownership x.

   a. Camry
   b. Camry Hybrid
   c. Highlander
   d. Highlander Hybrid

3. How many years would you have to own and drive the vehicle for the hybrid to be the better deal?

   a. Camry Hybrid versus Camry
   b. Highlander Hybrid versus Highlander

4. How many years would you have to own and drive the vehicle for the hybrid to emit 50% less carbon dioxide than its conventional counterpart?

   a. Camry Hybrid versus Camry
   b. Highlander Hybrid versus Highlander

CHAPTER 9 REVIEW

SECTION	CONCEPT	KEY IDEAS/FORMULAS
9.1	Systems of linear equations in two variables	$A_1x + B_1y = C_1$ $A_2x + B_2y = C_2$
	Solving systems of linear equations	**Substitution method** Solve for one variable in terms of the other and substitute that expression into the other equation. **Elimination method** Eliminate a variable by adding multiples of the equations. **Graphing method** Graph the two lines. The solution is the point of intersection. Parallel lines have no solution and identical lines have infinitely many solutions.
	Three methods and three types of solutions	One solution, no solution, infinitely many solutions.
9.2	Systems of linear equations in three variables	Planes in three-dimensional coordinate system.
	Solving systems of linear equations in three variables	**Step 1:** Reduce the system to 2 equations and 2 unknowns. **Step 2:** Solve the resulting system from Step 1. **Step 3:** Substitute solutions found in Step 2 into any of the equations to find the third variable. **Step 4:** Check.
	Types of solutions	One solution (point), no solution, infinitely many solutions (line).
9.3	Partial fractions	
	Performing partial-fraction decomposition on a rational expression in proper form	$\dfrac{n(x)}{d(x)}$ Factor $d(x)$ **Distinct linear factors** $\dfrac{n(x)}{d(x)} = \dfrac{A}{(ax+b)} + \dfrac{B}{(cx+d)} + \cdots$ **Repeated linear factors** $\dfrac{n(x)}{d(x)} = \dfrac{A}{(ax+b)} + \dfrac{B}{(ax+b)^2} + \cdots + \dfrac{M}{(ax+b)^m}$ **Distinct irreducible quadratic factors** $\dfrac{n(x)}{d(x)} = \dfrac{Ax+B}{ax^2+bx+c} + \dfrac{Cx+D}{dx^2+ex+f} + \cdots$ **Repeated irreducible quadratic factors** $\dfrac{n(x)}{d(x)} = \dfrac{A_1x+B_1}{ax^2+bx+c} + \dfrac{A_2x+B_2}{(ax^2+bx+c)^2} + \dfrac{A_3x+B_3}{(ax^2+bx+c)^3} + \cdots + \dfrac{A_mx+B_m}{(ax^2+bx+c)^m}$

SECTION	CONCEPT	KEY IDEAS/FORMULAS
9.4	**Systems of linear inequalities in two variables**	
	Linear inequalities in two variables	• $\leq$ or $\geq$ use solid lines • $<$ or $>$ use dashed lines
	Systems of linear inequalities in two variables	Solutions are determined graphically by finding the common shaded regions.
9.5	**The linear programming model**	
	Solving an optimization problem	Finding optimal solutions. Minimizing or maximizing a function subject to constraints (linear inequalities).

9.1 Systems of Linear Equations in Two Variables

Solve each system of linear equations.

1. $r - s = 3$
$r + s = 3$

2. $3x + 4y = 2$
$x - y = 6$

3. $-4x + 2y = 3$
$4x - y = 5$

4. $0.25x - 0.5y = 0.6$
$0.5x + 0.25y = 0.8$

5. $x + y = 3$
$x - y = 1$

6. $3x + y = 4$
$2x + y = 1$

7. $4c - 4d = 3$
$c + d = 4$

8. $5r + 2s = 1$
$r - s = -3$

9. $y = -\frac{1}{2}x$
$y = \frac{1}{2}x + 2$

10. $2x + 4y = -2$
$4x - 2y = 3$

11. $1.3x - 2.4y = 1.6$
$0.7x - 1.2y = 1.4$

12. $\frac{1}{4}x - \frac{3}{4}y = 12$
$\frac{1}{2}y + \frac{1}{4}x = \frac{1}{2}$

13. $5x - 3y = 21$
$-2x + 7y = -20$

14. $6x - 2y = -2$
$4x + 3y = 16$

15. $10x - 7y = -24$
$7x + 4y = 1$

16. $\frac{1}{3}x - \frac{2}{9}y = \frac{2}{9}$
$\frac{4}{5}x + \frac{3}{4}y = -\frac{3}{4}$

Match each system of equations with its graph.

17. $2x - 3y = 4$
$x + 4y = 3$

18. $5x - y = 2$
$5x - y = -2$

19. $x + 2y = -6$
$2x + 4y = -12$

20. $5x + 2y = 3$
$4x - 2y = 6$

a.

b.

c.

d.

Applications

21. Chemistry. In chemistry lab, Alexandra needs to make a 42 milliliter solution that is 15% NaCl. All that is in the lab is 6% and 18% NaCl. How many milliliters of each solution should she use to obtain the desired mix?

22. Gas Mileage. A Nissan Sentra gets approximately 32 mpg on the highway and 18 mpg in the city. Suppose 265 miles were driven on a full tank (12 gallons) of gasoline. Approximately how many miles were driven in the city and how many on the highway?

9.2 Systems of Linear Equations in Three Variables

Solve each system of linear equations.

23. $x + y + z = 1$
$x - y - z = -3$
$-x + y + z = 3$

24. $x - 2y + z = 3$
$2x - y + z = -4$
$3x - 3y - 5z = 2$

25. $x + y + z = 7$
$x - y - z = 17$
$y + z = 5$

26. $x + z = 3$
$-x + y - z = -1$
$x + y + z = 5$

Applications

27. Fitting a Curve to Data. The average number of flights on a commercial plane that a person takes per year can be modeled by a quadratic function $y = ax^2 + bx + c$, where $a < 0$, and x represents age: $16 \leq x \leq 65$. The following table gives the average number of flights per year that a person takes on a commercial airline. Determine the quadratic function that models this quantity.

AGE	NUMBER OF FLIGHTS PER YEAR
16	2
40	6
65	4

28. Investment Portfolio. Danny and Paula decide to invest $20,000 of their savings. They put some in an IRA account earning 4.5% interest, some in a mutual fund that has been averaging 8% a year, and some in a stock that earned 12% last year. If they put $4000 more in the IRA than in the mutual fund and the mutual fund and stock have the same growth in the next year as they did in the previous year, they will earn $1525 in a year. How much money did they put in each of the three investments?

9.3 Partial Fractions

Write the form of each partial-fraction decomposition. Do not solve for the constants.

29. $\dfrac{4}{(x-1)^2(x+3)(x-5)}$

30. $\dfrac{7}{(x-9)(3x+5)^2(x+4)}$

31. $\dfrac{12}{x(4x+5)(2x+1)^2}$

32. $\dfrac{2}{(x+1)(x-5)(x-9)^2}$

33. $\dfrac{3}{x^2+x-12}$

34. $\dfrac{x^2+3x-2}{x^3+6x^2}$

35. $\dfrac{3x^3+4x^2+56x+62}{(x^2+17)^2}$

36. $\dfrac{x^3+7x^2+10}{(x^2+13)^2}$

Find the partial-fraction decomposition for each rational function.

37. $\dfrac{9x+23}{(x-1)(x+7)}$

38. $\dfrac{12x+1}{(3x+2)(2x-1)}$

39. $\dfrac{13x^2+90x-25}{2x^3-50x}$

40. $\dfrac{5x^2+x+24}{x^3+8x}$

41. $\dfrac{2}{x^2+x}$

42. $\dfrac{x}{x(x+3)}$

43. $\dfrac{5x-17}{x^2+4x+4}$

44. $\dfrac{x^3}{(x^2+64)^2}$

9.4 Systems of Linear Inequalities in Two Variables

Graph each linear inequality.

45. $y \geq -2x + 3$

46. $y < x - 4$

47. $2x + 4y > 5$

48. $5x + 2y \leq 4$

49. $y \geq -3x + 2$

50. $y < x - 2$

51. $3x + 8y \leq 16$

52. $4x - 9y \leq 18$

Graph each system of inequalities or indicate that the system has no solution.

53. $y \geq x + 2$
$y \leq x - 2$

54. $y \geq 3x$
$y \leq 3x$

55. $x \leq -2$
$y > x$

56. $x + 3y \geq 6$
$2x - y \leq 8$

57. $3x - 4y \leq 16$
$5x + 3y > 9$

58. $x + y > -4$
$x - y < 3$
$y \geq -2$
$x \leq 8$

9.5 The Linear Programming Model

Minimize or maximize the objective function subject to the constraints.

59. Minimize $z = 2x + y$ subject to:

$$x \geq 0 \qquad y \geq 0 \qquad x + y \leq 3$$

60. Maximize $z = 2x + 3y$ subject to:

$$x \geq 0 \qquad y \geq 0$$
$$-x + y \leq 0 \qquad x \leq 3$$

61. Maximize $z = 2.5x + 3.2y$ subject to:

$$x \geq 0 \qquad y \geq 0$$
$$x + y \leq 8 \qquad -x + y \geq 0$$

62. Minimize $z = 5x + 11y$ subject to:

$$-x + y \leq 10 \qquad x + y \geq 2 \qquad x \leq 2$$

63. Minimize $z = 3x - 5y$ subject to:

$$2x + y \geq 6 \qquad 2x - y \leq 6 \qquad x \geq 0$$

64. Maximize $z = -2x + 7y$ subject to:

$$3x + y \leq 7 \qquad x - 2y \geq 1 \qquad x \geq 0$$

Applications

For Exercises 65 and 66, refer to the following:

An art student decides to hand-paint coasters and sell sets at a flea market. She decides to make two types of coaster sets: an ocean watercolor and black-and-white geometric shapes. The cost, profit, and time it takes her to paint each set are summarized in the following table.

	OCEAN WATERCOLOR	GEOMETRIC SHAPES
Cost	$4	$2
Profit	$15	$8
Hours	3	2

65. Profit. If her costs cannot exceed $100 and she can spend only 90 hours total painting the coasters, determine the number of each type she should make to maximize her profit.

66. Profit. If her costs cannot exceed $300 and she can spend only 90 hours painting, determine the number of each type she should make to maximize her profit.

Technology Exercises

Section 9.1

67. Use a graphing utility to graph the two equations $0.4x + 0.3y = -0.1$ and $0.5x - 0.2y = 1.6$. Find the solution to this system of linear equations.

68. Use a graphing utility to graph the two equations $\frac{1}{2}x + \frac{3}{10}y = \frac{1}{5}$ and $-\frac{5}{3}x + \frac{1}{2}y = \frac{4}{3}$. Find the solution to this system of linear equations.

Section 9.2

In Exercises 69 and 70, employ a graphing calculator to solve the system of equations.

69. $\begin{aligned} 5x - 3y + 15z &= 21 \\ -2x + 0.8y - 4z &= -8 \\ 2.5x - y + 7.5z &= 12 \end{aligned}$

70. $\begin{aligned} 2x - 1.5y + 3z &= 6.5 \\ 0.5x - 0.375y + 0.75z &= 1.5 \end{aligned}$

Section 9.3

71. Apply a graphing utility to graph $y_1 = \dfrac{x^2 + 4}{x^4 - x^2}$ and $y_2 = -\dfrac{4}{x^2} + \dfrac{5/2}{x - 1} - \dfrac{5/2}{x + 1}$ in the same viewing rectangle. Is y_2 the partial-fraction decomposition of y_1?

72. Use a graphing utility to graph
$$y_1 = \frac{x^3 + 6x^2 + 27x + 38}{(x^2 + 8x + 17)(x^2 + 6x + 13)} \text{ and}$$
$$y_2 = \frac{2x + 1}{x^2 + 8x + 17} - \frac{x - 3}{x^2 + 6x + 13} \text{ in the same viewing}$$
rectangle. Is y_2 the partial-fraction decomposition of y_1?

Section 9.4

In Exercises 73 and 74, use a graphing utility to graph each system of inequalities or indicate that the system has no solution.

73. $\begin{aligned} 2x + 5y &\geq -15 \\ y &\leq -\tfrac{2}{3}x - 1 \end{aligned}$

74. $\begin{aligned} y &\leq 0.5x \\ y &> -1.5x + 6 \end{aligned}$

Section 9.5

75. Maximize $z = 6.2x + 1.5y$ subject to:
$$\begin{aligned} 4x - 3y &\leq 5.4 \\ 2x + 4.5y &\leq 6.3 \\ 3x - y &\geq -10.7 \end{aligned}$$

76. Minimize $z = 1.6x - 2.8y$ subject to:
$$\begin{aligned} y &\geq 3.2x - 4.8 & x &\geq -2 \\ y &\leq 3.2x + 4.8 & x &\leq 4 \end{aligned}$$

Solve each system of linear equations.

1. $\begin{aligned} x - 2y &= 1 \\ -x + 3y &= 2 \end{aligned}$

2. $\begin{aligned} 3x + 5y &= -2 \\ 7x + 11y &= -6 \end{aligned}$

3. $\begin{aligned} x - y &= 2 \\ -2x + 2y &= -4 \end{aligned}$

4. $\begin{aligned} 3x - 2y &= 5 \\ 6x - 4y &= 0 \end{aligned}$

5. $\begin{aligned} x + y + z &= -1 \\ 2x + y + z &= 0 \\ -x + y + 2z &= 0 \end{aligned}$

6. $\begin{aligned} 6x + 9y + z &= 5 \\ 2x - 3y + z &= 3 \\ 10x + 12y + 2z &= 9 \end{aligned}$

7. $\begin{aligned} x - 2y + 3z &= 5 \\ -2x + y + 4z &= 21 \\ 3x - 5y + z &= -14 \end{aligned}$

8. $\begin{aligned} x \quad + z &= 1 \\ x + y \quad &= -1 \\ y + z &= 0 \end{aligned}$

Write each rational expression as a sum of partial fractions.

9. $\dfrac{2x + 5}{x^2 + x}$

10. $\dfrac{3x - 13}{(x - 5)^2}$

11. $\dfrac{7x + 5}{(x + 2)^2}$

12. $\dfrac{1}{2x^2 + 5x - 3}$

13. $\dfrac{5x - 3}{x(x^2 - 9)}$

14. $\dfrac{5x - 3}{x(x^2 + 9)}$

Graph the inequalities.

15. $-2x + y < 6$

16. $4x - y \geq 8$

In Exercises 17–20, graph the system of inequalities.

17. $\begin{aligned} x + y &\leq 4 \\ -x + y &\geq -2 \end{aligned}$

18. $\begin{aligned} x + 3y &\leq 6 \\ 2x - y &\leq 4 \end{aligned}$

19. $\begin{aligned} 5x - 2y &\leq 8 \\ 3x + 4y &< 12 \end{aligned}$

20. $\begin{aligned} -x + y &> 6 \\ x + y &\leq 4 \end{aligned}$

21. Minimize the function $z = 5x + 7y$, subject to the constraints:

$$x \geq 0 \qquad y \geq 0 \qquad x + y \leq 3 \qquad -x + y \geq 1$$

22. Find the maximum value of the objective function $z = 3x + 6y$, given the constraints:

$$\begin{aligned} x &\geq 0 & y &\geq 0 \\ x + y &\leq 6 & -x + 2y &\leq 4 \end{aligned}$$

23. Stock Investment. On starting a new job, Cameron gets a $30,000 signing bonus and decides to invest the money. She invests some in a money market account earning 3% and some in two different stocks. The aggressive stock rose 12% last year, and the conservative stock rose 6% last year. She invested $1000 more in the aggressive stock than in the conservative stock, and in 1 year she earned $1890 on the investment. How much did she put in the money market, and how much was invested in each stock?

24. Ranch. A rancher buys a rectangular parcel of property that is 245,000 square feet. She fences the entire border and then adds three internal fences so there are four equal rectangular pastures. If the fence required 3150 linear feet, what are the dimensions of the entire property?

25. Employ a graphing calculator to solve the system of equations.

$$\begin{aligned} 5x - 3y + 3z &= 1 \\ 2x - y + z &= 4 \\ 4x - 2y + 3z &= 9 \end{aligned}$$

26. Minimize $z = x - 1.2y$ subject to:

$$\begin{aligned} 3.2x + y &\leq 15.9 \\ 0.4x - y &\leq 0.3 \\ x &\geq 1.0 \end{aligned}$$

1. Simplify $\dfrac{6x^2 - 11x + 5}{30x - 25}$.

2. Perform the operation and write in standard form: $(5 - 7i) - (11 - 9i)$.

3. Solve for x and give any excluded values:
$$\frac{5}{x} - \frac{5}{x + 1} = -\frac{5}{x^2 + x}.$$

4. Solve for the radius r: $A = \pi r^2$.

5. Solve and express the solution in interval notation:
$$\frac{2t^2}{t - 1} \geq 3t.$$

6. Calculate the distance and midpoint between the line segment joining the points $\left(-\frac{3}{4}, \frac{1}{6}\right)$ and $\left(\frac{1}{5}, -\frac{2}{3}\right)$.

7. Write an equation of a line in slope–intercept form that passes through the two points $(5.6, 6.2)$ and $(3.2, 5.0)$.

8. Use interval notation to express the domain of the function $f(x) = \sqrt{x^2 - 25}$.

9. Find the average rate of change for $f(x) = \dfrac{5}{x}$, from $x = 2$ to $x = 4$.

10. Sketch the graph of the function $y = \dfrac{1}{x - 2} + 1$ and identify all transformations.

11. Evaluate $g(f(-1))$ for $f(x) = \sqrt[3]{x - 7}$ and $g(x) = \dfrac{5}{3 - x}$.

12. Find the inverse of the function $f(x) = \dfrac{5x + 2}{x - 3}$.

13. Find the quadratic function that has the vertex $(0, 7)$ and goes through the point $(2, -1)$.

14. Find all of the real zeros and state the multiplicity of the function $f(x) = 4x^5 + x^3$.

15. List the possible rational zeros and test to determine all rational zeros for $P(x) = 2x^4 + 7x^3 - 18x^2 - 13x + 10$.

16. Factor the polynomial $P(x) = 4x^4 - 4x^3 + 13x^2 + 18x + 5$ as a product of linear factors.

17. Graph the exponential function $f(x) = 5^{-x} - 1$; state the y-intercept, the domain, the range, and the horizontal asymptote.

18. How much money should be put in a savings account now that earns 5.2% a year compounded daily if you want to have $50,000 in 16 years?

19. Graph the function $f(x) = \ln(x + 1) - 3$ using transformation techniques.

20. Evaluate $\log_{4.7} 8.9$ using the change-of-base formula. Round the answer to three decimal places.

21. Solve the equation $5(10^{2x}) = 37$ for x. Round the answers to three decimal places.

22. At a coffee shop, one order of 4 tall coffees and 6 donuts is $14.10. A second order of 5 tall coffees and 4 donuts is $13.11. Find the cost of a tall coffee and the cost of a donut.

23. Solve the system of linear equations.
$$\begin{aligned} x + 6y - z &= -3 \\ 2x - 5y + z &= 9 \\ x + 4y + 2z &= 12 \end{aligned}$$

24. Find the partial-fraction decomposition for the rational expression $\dfrac{-1}{x^2 - 2x + 1}$.

25. Graph the system of linear inequalities.
$$\begin{aligned} y &> -x \\ y &\geq -3 \\ x &\leq 3 \end{aligned}$$

26. Employ a graphing calculator to solve the system of equations.
$$\begin{aligned} 2x - 9y + 8z &= -21.8 \\ 5x + 6y - 11z &= 41.7 \\ 3x - y + 9z &= -11.7 \end{aligned}$$

27. Apply a graphing utility to graph $y_1 = \dfrac{3x^2 + x + 8}{x^3 + 4x}$ and $y_2 = \dfrac{2}{x} + \dfrac{x + 1}{x^2 + 4}$ in the same viewing rectangle. Is y_2 the partial-fraction decomposition of y_1?

10

Matrices

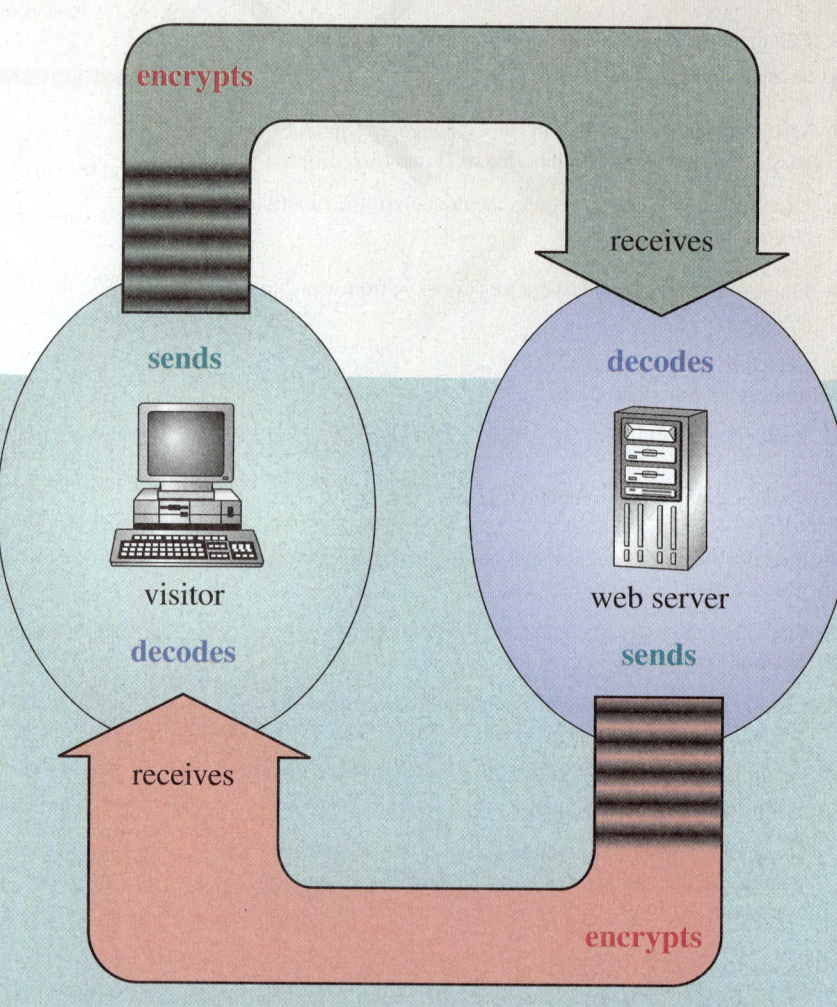

Cryptography is the practice and study of encryption and decryption—encoding data so that it can be decoded only by specific individuals. In other words, it turns a message into gibberish so that only the person who has the deciphering tools can turn that gibberish back into the original message. ATM cards, online shopping sites, and secure military communications all depend on coding and decoding of information. Matrices are used extensively in cryptography. A *matrix* is used as the "key" to encode the data, and then its *inverse matrix* is used as the "key" to decode the data.*

---

*Section 10.3, Exercises 49–54.

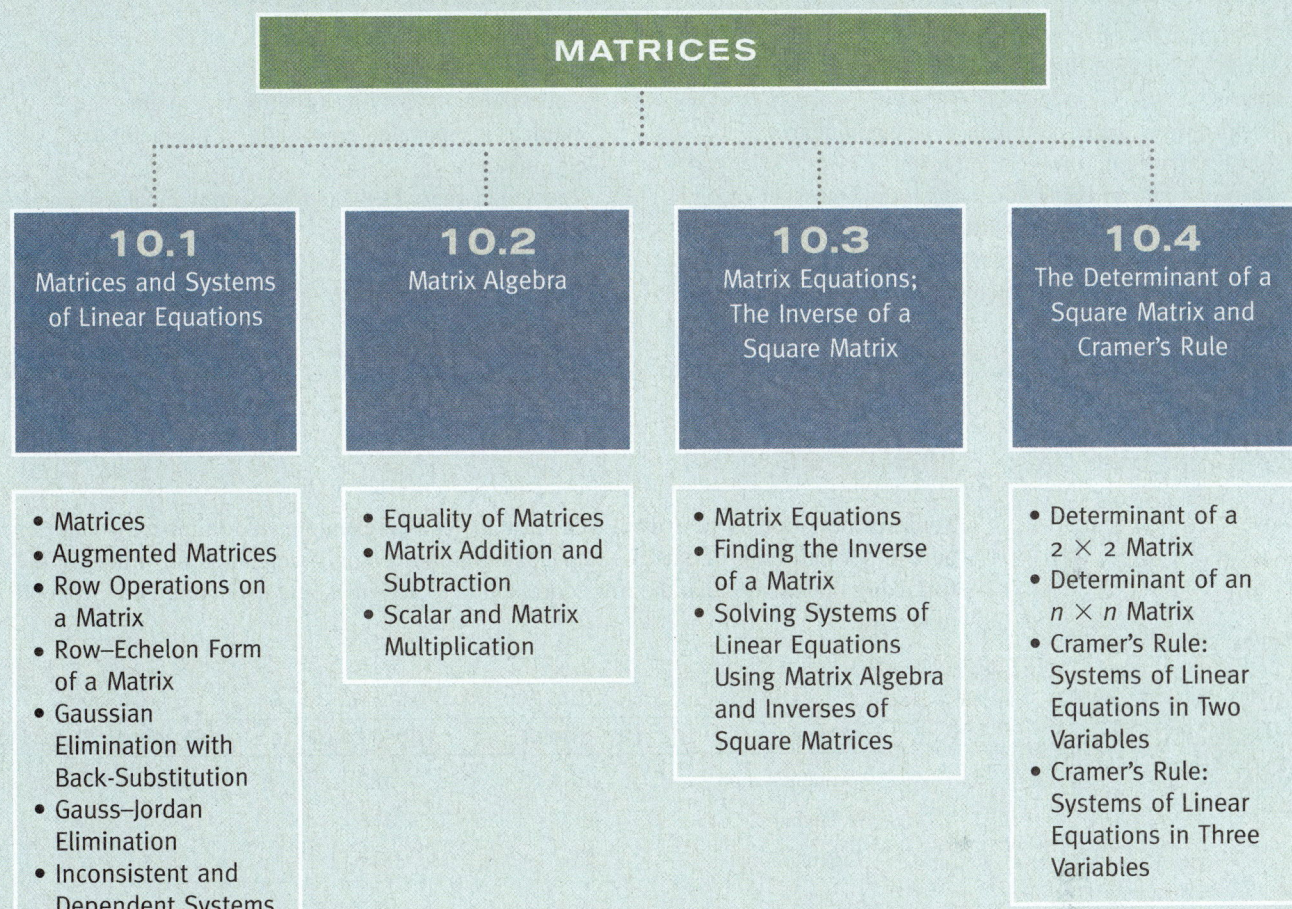

MATRICES

10.1
Matrices and Systems of Linear Equations

- Matrices
- Augmented Matrices
- Row Operations on a Matrix
- Row–Echelon Form of a Matrix
- Gaussian Elimination with Back-Substitution
- Gauss–Jordan Elimination
- Inconsistent and Dependent Systems

10.2
Matrix Algebra

- Equality of Matrices
- Matrix Addition and Subtraction
- Scalar and Matrix Multiplication

10.3
Matrix Equations; The Inverse of a Square Matrix

- Matrix Equations
- Finding the Inverse of a Matrix
- Solving Systems of Linear Equations Using Matrix Algebra and Inverses of Square Matrices

10.4
The Determinant of a Square Matrix and Cramer's Rule

- Determinant of a 2×2 Matrix
- Determinant of an $n \times n$ Matrix
- Cramer's Rule: Systems of Linear Equations in Two Variables
- Cramer's Rule: Systems of Linear Equations in Three Variables

LEARNING OBJECTIVES

- Use Gauss–Jordan elimination to solve systems of linear equations.
- Perform matrix operations: addition, subtraction, and multiplication.
- Utilize matrix algebra and inverses of matrices to solve systems of linear equations.
- Use Cramer's rule (determinants) to solve systems of linear equations.

Matrices

Some information is best displayed in a table. For example, the number of calories burned per half hour of exercise depends upon the person's weight, as illustrated in the following table. Note that the rows correspond to activities and the columns correspond to weight.

ACTIVITY	127–137 LB	160–170 LB	180–200 LB
Walking/4 mph	156	183	204
Volleyball	267	315	348
Jogging/5 mph	276	345	381

Another example is the driving distance in miles from cities in Arizona (columns) to cities outside the state (rows).

CITY	FLAGSTAFF	PHOENIX	TUCSON	YUMA
Albuquerque, NM	325	465	440	650
Las Vegas, NV	250	300	415	295
Los Angeles, CA	470	375	490	285

If we selected only the numbers in each of the preceding tables and placed brackets around them, the result would be a *matrix*.

$$\text{Calories: } \begin{bmatrix} 156 & 183 & 204 \\ 267 & 315 & 348 \\ 276 & 345 & 381 \end{bmatrix} \qquad \text{Miles: } \begin{bmatrix} 325 & 465 & 440 & 650 \\ 250 & 300 & 415 & 295 \\ 470 & 375 & 490 & 285 \end{bmatrix}$$

A *matrix* is a rectangular array of numbers written within brackets.

$$\begin{bmatrix} a_{11} & a_{12} & \cdots & a_{1j} & \cdots & a_{1n} \\ a_{21} & a_{22} & \cdots & a_{2j} & \cdots & a_{2n} \\ \vdots & \vdots & \cdots & \vdots & \cdots & \vdots \\ a_{i1} & a_{i2} & \cdots & a_{ij} & \cdots & a_{in} \\ \vdots & \vdots & \cdots & \vdots & \cdots & \vdots \\ a_{m1} & a_{m2} & \cdots & a_{mj} & \cdots & a_{mn} \end{bmatrix}$$

Each number a_{ij} in the matrix is called an **element** of the matrix. The first subscript i is the **row index,** and the second subscript j is the **column index**. This matrix contains m rows and n columns, and is said to be of **order** $m \times n$.

Study Tip

The order of a matrix is always given as the number of rows by the number of columns.

When the number of rows equals the number of columns (that is, when $m = n$), the matrix is a **square matrix** of order n. In a square matrix, the entries $a_{11}, a_{22}, a_{33}, \ldots, a_{nn}$ are the **main diagonal** entries.

The matrix

$$A_{4\times 3} = \begin{bmatrix} * & * & * \\ * & * & * \\ * & a_{32} & * \\ * & * & * \end{bmatrix}$$

has order (dimensions) 4×3, since there are four rows and three columns. The element a_{32} is in the third row and second column.

EXAMPLE 1 **Finding the Order of a Matrix**

Determine the order of each matrix given.

a. $\begin{bmatrix} 2 & 1 \\ 3 & 0 \end{bmatrix}$

b. $\begin{bmatrix} 1 & -2 & 5 \\ -1 & 3 & 4 \end{bmatrix}$

c. $\begin{bmatrix} -2 & 5 & 4 \\ 1 & -\frac{1}{3} & 0 \\ 3 & 8 & 1 \end{bmatrix}$

d. $\begin{bmatrix} 4 & 9 & -\frac{1}{2} & 3 \end{bmatrix}$

e. $\begin{bmatrix} 3 & -2 \\ 5 & 1 \\ 0 & -\frac{2}{3} \\ 7 & 6 \end{bmatrix}$

Classroom Example 10.1.1
What is the order of these matrices?
a. $[1 \ 1]$
b. $\begin{bmatrix} 3 & 2 & 1 & 4 & 5 \\ 1 & 0 & 1 & 2 & 1 \end{bmatrix}$
c. $[0]$

Answer:
a. 1×2
b. 2×5
c. 1×1

Solution:

a. This matrix has **2** rows and **2** columns, so the order of the matrix is 2×2.

b. This matrix has **2** rows and **3** columns, so the order of the matrix is 2×3.

c. This matrix has **3** rows and **3** columns, so the order of the matrix is 3×3.

d. This matrix has **1** row and **4** columns, so the order of the matrix is 1×4.

e. This matrix has **4** rows and **2** columns, so the order of the matrix is 4×2.

A matrix with only one column is called a **column matrix**, and a matrix that has only one row is called a **row matrix**. Notice that in Example 1 the matrices given in parts (a) and (c) are square matrices and the matrix given in part (d) is a row matrix.

You can use matrices as a shorthand way of writing systems of linear equations. There are two ways we can represent systems of linear equations with matrices: as *augmented matrices* or with *matrix equations*. In this section we will discuss *augmented matrices* and solve systems of linear equations using two methods: *Gaussian elimination with back-substitution* and *Gauss–Jordan elimination*.

Augmented Matrices

A *coefficient matrix* is a matrix whose elements are the coefficients of a system of linear equations. A particular type of matrix that is used in representing a system of linear equations is an **augmented matrix**. It resembles a coefficient matrix with an additional vertical line and column of numbers, hence the name *augmented*. The following table illustrates examples of augmented matrices that represent systems of linear equations.

SYSTEM OF LINEAR EQUATIONS	AUGMENTED MATRIX
$3x + 4y = 1$ $x - 2y = 7$	$\begin{bmatrix} 3 & 4 & \vert & 1 \\ 1 & -2 & \vert & 7 \end{bmatrix}$
$x - y + z = 2$ $2x + 2y - 3z = -3$ $x + y + z = 6$	$\begin{bmatrix} 1 & -1 & 1 & \vert & 2 \\ 2 & 2 & -3 & \vert & -3 \\ 1 & 1 & 1 & \vert & 6 \end{bmatrix}$
$x + y + z = 0$ $3x + 2y - z = 2$	$\begin{bmatrix} 1 & 1 & 1 & \vert & 0 \\ 3 & 2 & -1 & \vert & 2 \end{bmatrix}$

Note the following:

- Each row represents an equation.
- The vertical line represents the equal sign.
- The first column represents the coefficients of the variable x.
- The second column represents the coefficients of the variable y.
- The third column (in the second and third systems) represents the coefficients of the variable z.
- The coefficients of the variables are on the left of the equal sign (vertical line), and the constants are on the right.
- Any variable that does not appear in an equation has an implied coefficient of 0.

EXAMPLE 2 **Writing a System of Linear Equations as an Augmented Matrix**

Write each system of linear equations as an augmented matrix.

a. $2x - y = 5$
$-x + 2y = 3$

b. $3x - 2y + 4z = 5$
$y - 3z = -2$
$7x \quad - z = 1$

c. $x_1 - x_2 + 2x_3 - 3 = 0$
$x_1 + x_2 - 3x_3 + 5 = 0$
$x_1 - x_2 + x_3 - 2 = 0$

Solution:

a.

$$\begin{bmatrix} 2 & -1 & | & 5 \\ -1 & 2 & | & 3 \end{bmatrix}$$

b. Note that all missing terms have a 0 coefficient.

$3x - 2y + 4z = 5$
$0x + y - 3z = -2$
$7x + 0y - z = 1$

$$\begin{bmatrix} 3 & -2 & 4 & | & 5 \\ 0 & 1 & -3 & | & -2 \\ 7 & 0 & -1 & | & 1 \end{bmatrix}$$

c. Write the constants on the right side of the vertical line in the matrix.

$x_1 - x_2 + 2x_3 = 3$
$x_1 + x_2 - 3x_3 = -5$
$x_1 - x_2 + x_3 = 2$

$$\begin{bmatrix} 1 & -1 & 2 & | & 3 \\ 1 & 1 & -3 & | & -5 \\ 1 & -1 & 1 & | & 2 \end{bmatrix}$$

■ **YOUR TURN** Write each system of linear equations as an augmented matrix.

a. $2x + y - 3 = 0$
$x - y = 5$

b. $y - x + z = 7$
$x - y - z = 2$
$z - y = -1$

Classroom Example 10.1.2*

Write these systems as augmented matrices.

a. $x + y + z = 1$
$2 - y = z$
$x + y = z$

b. $a + 3c - b = 1$
$b - a = 2$

Answer:

a. $\begin{bmatrix} 1 & 1 & 1 & | & 1 \\ 0 & -1 & -1 & | & -2 \\ 1 & 1 & -1 & | & 0 \end{bmatrix}$

b. $\begin{bmatrix} 1 & -1 & 3 & | & 1 \\ -1 & 1 & 0 & | & 2 \end{bmatrix}$

■ **Answer:**

a. $\begin{bmatrix} 2 & 1 & | & 3 \\ 1 & -1 & | & 5 \end{bmatrix}$

b. $\begin{bmatrix} -1 & 1 & 1 & | & 7 \\ 1 & -1 & -1 & | & 2 \\ 0 & -1 & 1 & | & -1 \end{bmatrix}$

Row Operations on a Matrix

Row operations on a matrix are used to solve a system of linear equations when the system is written as an augmented matrix. Recall from the elimination method in Chapter 6 that we could interchange equations, multiply an entire equation by a nonzero constant, and add a multiple of one equation to another equation to produce equivalent systems. Because each row in a matrix represents an equation, the operations that produced equivalent systems of equations that were used in the elimination method will also produce equivalent augmented matrices.

Study Tip

Each missing term in the system of linear equations is represented with a zero in the augmented matrix.

ROW OPERATIONS

The following operations on an augmented matrix will yield an equivalent matrix:

1. Interchange any two rows.
2. Multiply a row by a nonzero constant.
3. Add a multiple of one row to another row.

The following symbols describe these row operations:

1. $R_i \leftrightarrow R_j$ Interchange row i with row j.
2. $cR_i \rightarrow R_i$ Multiply row i by the constant c.
3. $cR_i + R_j \rightarrow R_j$ Multiply row i by the constant c and add to row j, writing the results in row j.

EXAMPLE 3   Applying a Row Operation to an Augmented Matrix

For each matrix, perform the given operation.

a. $\begin{bmatrix} 2 & -1 & | & 3 \\ 0 & 2 & | & 1 \end{bmatrix}$ $R_1 \leftrightarrow R_2$ **b.** $\begin{bmatrix} -1 & 0 & 1 & | & -2 \\ 3 & -1 & 2 & | & 3 \\ 0 & 1 & 3 & | & 1 \end{bmatrix}$ $2R_3 \rightarrow R_3$

c. $\begin{bmatrix} 1 & 2 & 0 & 2 & | & 2 \\ 0 & 1 & 2 & 3 & | & 5 \end{bmatrix}$ $R_1 - 2R_2 \rightarrow R_1$

Solution:

a. Interchange the first row with the second row.

$$\begin{bmatrix} 2 & -1 & | & 3 \\ 0 & 2 & | & 1 \end{bmatrix} \quad R_1 \leftrightarrow R_2 \quad \begin{bmatrix} 0 & 2 & | & 1 \\ 2 & -1 & | & 3 \end{bmatrix}$$

b. Multiply the third row by 2.

$$\begin{bmatrix} -1 & 0 & 1 & | & -2 \\ 3 & -1 & 2 & | & 3 \\ 0 & 1 & 3 & | & 1 \end{bmatrix} \quad 2R_3 \rightarrow R_3 \quad \begin{bmatrix} -1 & 0 & 1 & | & -2 \\ 3 & -1 & 2 & | & 3 \\ 0 & 2 & 6 & | & 2 \end{bmatrix}$$

c. From row 1 subtract 2 times row 2, and write the answer in row 1. Note that finding row 1 minus 2 times row 2 is the same as adding row 1 to the product of -2 with row 2.

$$R_1 - 2R_2 \rightarrow R_1 \quad \begin{bmatrix} 1-2(0) & 2-2(1) & 0-2(2) & 2-2(3) & | & 2-2(5) \\ 0 & 1 & 2 & 3 & | & 5 \end{bmatrix}$$

$$\begin{bmatrix} 1 & 0 & -4 & -4 & | & -8 \\ 0 & 1 & 2 & 3 & | & 5 \end{bmatrix}$$

■ **YOUR TURN** Perform the operation $R_1 + 2R_3 \rightarrow R_1$ on the matrix.

$$\begin{bmatrix} 1 & 0 & -2 & | & -3 \\ 0 & 1 & 2 & | & 3 \\ 0 & 0 & 1 & | & 2 \end{bmatrix}$$

Row–Echelon Form of a Matrix

We can solve systems of linear equations using augmented matrices with two procedures: *Gaussian elimination with back-substitution*, which uses row operations to transform a matrix into *row–echelon form*, and *Gauss–Jordan elimination*, which uses row operations to transform a matrix into *reduced row–echelon form*.

Row–Echelon Form

A matrix is in **row–echelon** form if it has all three of the following properties:

1. Any rows consisting entirely of 0s are at the bottom of the matrix.
2. For each row that does not consist entirely of 0s, the first (leftmost) nonzero entry is 1 (called the leading 1).
3. For two successive nonzero rows, the leading 1 in the higher row is farther to the left than the leading 1 in the lower row.

Reduced Row–Echelon Form

If a matrix in row–echelon form has the following additional property, then the matrix is in **reduced row–echelon form**:

4. Every column containing a leading 1 has zeros in every position above and below the leading 1.

EXAMPLE 4 **Determining Whether a Matrix Is in Row–Echelon Form**

Determine whether each matrix is in row–echelon form. If it is in row–echelon form, determine whether it is in reduced row–echelon form.

a. $\begin{bmatrix} 1 & 3 & 2 & | & 3 \\ 0 & 1 & 4 & | & 2 \\ 0 & 0 & 1 & | & -1 \end{bmatrix}$ **b.** $\begin{bmatrix} 1 & 3 & 2 & | & 3 \\ 0 & 1 & 1 & | & 3 \\ 0 & 0 & 0 & | & 0 \end{bmatrix}$ **c.** $\begin{bmatrix} 1 & 0 & 3 & | & 2 \\ 0 & 1 & -1 & | & 5 \end{bmatrix}$

d. $\begin{bmatrix} 1 & 0 & | & 1 \\ 0 & 3 & | & 1 \end{bmatrix}$ **e.** $\begin{bmatrix} 1 & 0 & 0 & | & 3 \\ 0 & 1 & 0 & | & 5 \\ 0 & 0 & 1 & | & 7 \end{bmatrix}$ **f.** $\begin{bmatrix} 1 & 3 & 2 & | & 3 \\ 0 & 0 & 1 & | & 2 \\ 0 & 1 & 0 & | & -3 \end{bmatrix}$

Solution:

The matrices in (a), (b), (c), and (e) are in row–echelon form. The matrix in (d) is not in row–echelon form, by condition 2; the leading nonzero entry is not a 1 in each row. If the "3" were a "1," the matrix would be in row–echelon form. The matrix in (f) is not in row–echelon form, because of condition 3; the leading 1 in row 2 is not to the left of the leading 1 in row 3. The matrices in (c) and (e) are in reduced row–echelon form, because in the columns containing the leading 1s there are zeros in every position (above and below the leading 1).

Classroom Example 10.1.4
Determine if these matrices are in row–echelon form. For those that are, determine if they are also in reduced row–echelon form.

a. $\begin{bmatrix} 1 & 0 & 0 & 0 & | & 1 \\ 0 & 0 & 1 & 0 & | & 0 \\ 0 & 1 & 0 & 0 & | & -1 \end{bmatrix}$

b. $\begin{bmatrix} 1 & 0 & | & 1 \\ 0 & 1 & | & 1 \end{bmatrix}$

Answer: a. neither   **b.** both

Gaussian Elimination with Back-Substitution

Gaussian elimination with back-substitution is a method that uses row operations to transform an augmented matrix into row–echelon form and then uses back-substitution to find the solution to the system of linear equations.

GAUSSIAN ELIMINATION WITH BACK-SUBSTITUTION

Step 1: Write the system of linear equations as an augmented matrix.
Step 2: Use row operations to rewrite the augmented matrix in row–echelon form.
Step 3: Write the system of linear equations that corresponds to the matrix in row–echelon form found in Step 2.
Step 4: Use the system of linear equations found in Step 3 together with back-substitution to find the solution of the system.

Study Tip

For row–echelon form, get 1s along the main diagonal and 0s below these 1s.

The order in which we perform row operations is important. You should move from left to right. Here is an example of Step 2 in the procedure.

$$\begin{bmatrix} 1 & * & * & | & * \\ * & * & * & | & * \\ * & * & * & | & * \end{bmatrix} \rightarrow \begin{bmatrix} 1 & * & * & | & * \\ 0 & * & * & | & * \\ 0 & * & * & | & * \end{bmatrix} \rightarrow \begin{bmatrix} 1 & * & * & | & * \\ 0 & 1 & * & | & * \\ 0 & * & * & | & * \end{bmatrix} \rightarrow \begin{bmatrix} 1 & * & * & | & * \\ 0 & 1 & * & | & * \\ 0 & 0 & * & | & * \end{bmatrix} \rightarrow \begin{bmatrix} 1 & * & * & | & * \\ 0 & 1 & * & | & * \\ 0 & 0 & 1 & | & * \end{bmatrix}$$

Matrices are not typically used for systems of linear equations in two variables because the methods from Chapter 6 (substitution and elimination) are more efficient. Example 5 illustrates this procedure with a simple system of linear equations in two variables.

EXAMPLE 5   **Using Gaussian Elimination with Back-Substitution to Solve a System of Linear Equations in Two Variables**

Apply Gaussian elimination with back-substitution to solve the system of linear equations.

$$2x + y = -8$$
$$x + 3y = 6$$

Solution:

STEP 1   Write the system of linear equations as an augmented matrix. $\begin{bmatrix} 2 & 1 & | & -8 \\ 1 & 3 & | & 6 \end{bmatrix}$

STEP 2   Use row operations to rewrite the matrix in row–echelon form.

Get a 1 in the top left. Interchange rows 1 and 2.

$$\begin{bmatrix} 2 & 1 & | & -8 \\ 1 & 3 & | & 6 \end{bmatrix} \quad R_1 \leftrightarrow R_2 \quad \begin{bmatrix} 1 & 3 & | & 6 \\ 2 & 1 & | & -8 \end{bmatrix}$$

Get a 0 below the leading 1 in row 1.

$$\begin{bmatrix} 1 & 3 & | & 6 \\ 2 & 1 & | & -8 \end{bmatrix} \quad R_2 - 2R_1 \rightarrow R_2 \quad \begin{bmatrix} 1 & 3 & | & 6 \\ 0 & -5 & | & -20 \end{bmatrix}$$

Get a leading 1 in row 2. Make the "-5" a "1" by dividing by -5. Dividing by -5 is the same as multiplying by its reciprocal $-\frac{1}{5}$.

$$\begin{bmatrix} 1 & 3 & | & 6 \\ 0 & -5 & | & -20 \end{bmatrix} \quad -\tfrac{1}{5}R_2 \rightarrow R_2 \quad \begin{bmatrix} 1 & 3 & | & 6 \\ 0 & 1 & | & 4 \end{bmatrix}$$

The resulting matrix is in row–echelon form.

STEP 3   Write the system of linear equations corresponding to the row–echelon form of the matrix resulting in Step 2.

$$\begin{bmatrix} 1 & 3 & | & 6 \\ 0 & 1 & | & 4 \end{bmatrix} \quad \rightarrow \quad \begin{aligned} x + 3y &= 6 \\ y &= 4 \end{aligned}$$

STEP 4   Use back-substitution to find the solution to the system.

Let $y = 4$ in the first equation $x + 3y = 6$. $x + 3(4) = 6$

Solve for x. $x = -6$

The solution to the system of linear equations is $\boxed{x = -6, y = 4}$.

EXAMPLE 6   **Using Gaussian Elimination with Back-Substitution to Solve a System of Linear Equations in Three Variables**

Use Gaussian elimination with back-substitution to solve the system of linear equations.

$$2x + y + 8z = -1$$
$$x - y + z = -2$$
$$3x - 2y - 2z = 2$$

Solution:

STEP 1   Write the system of linear equations as an augmented matrix. $\begin{bmatrix} 2 & 1 & 8 & | & -1 \\ 1 & -1 & 1 & | & -2 \\ 3 & -2 & -2 & | & 2 \end{bmatrix}$

STEP 2   Use row operations to rewrite the matrix in row–echelon form.

Get a 1 in the top left.
Interchange rows 1 and 2.

$$R_1 \leftrightarrow R_2 \quad \begin{bmatrix} 1 & -1 & 1 & | & -2 \\ 2 & 1 & 8 & | & -1 \\ 3 & -2 & -2 & | & 2 \end{bmatrix}$$

Get 0s below the leading 1
in row 1.

$$R_2 - 2R_1 \rightarrow R_2 \quad \begin{bmatrix} 1 & -1 & 1 & | & -2 \\ 0 & 3 & 6 & | & 3 \\ 3 & -2 & -2 & | & 2 \end{bmatrix}$$

$$R_3 - 3R_1 \rightarrow R_3 \quad \begin{bmatrix} 1 & -1 & 1 & | & -2 \\ 0 & 3 & 6 & | & 3 \\ 0 & 1 & -5 & | & 8 \end{bmatrix}$$

Get a leading 1 in row 2. Make the
"3" a "1" by dividing by 3.

$$\tfrac{1}{3}R_2 \rightarrow R_2 \quad \begin{bmatrix} 1 & -1 & 1 & | & -2 \\ 0 & 1 & 2 & | & 1 \\ 0 & 1 & -5 & | & 8 \end{bmatrix}$$

Get a zero below the leading
1 in row 2.

$$R_3 - R_2 \rightarrow R_3 \quad \begin{bmatrix} 1 & -1 & 1 & | & -2 \\ 0 & 1 & 2 & | & 1 \\ 0 & 0 & -7 & | & 7 \end{bmatrix}$$

Get a leading 1 in row 3. Make the
"−7" a "1" by dividing by −7.

$$-\tfrac{1}{7}R_3 \rightarrow R_3 \quad \begin{bmatrix} 1 & -1 & 1 & | & -2 \\ 0 & 1 & 2 & | & 1 \\ 0 & 0 & 1 & | & -1 \end{bmatrix}$$

STEP 3   Write the system of linear equations corresponding
to the row–echelon form of the matrix resulting in Step 2.

$$\begin{aligned} x - y + z &= -2 \\ y + 2z &= 1 \\ z &= -1 \end{aligned}$$

STEP 4   Use back-substitution to find the solution to the system.

Let $\boxed{z = -1}$ in the second equation $y + 2z = 1$.

$$y + 2(-1) = 1$$

Solve for y.

$$\boxed{y = 3}$$

Let $y = 3$ and $z = -1$ in the first equation
$x - y + z = -2$.

$$x - (3) + (-1) = -2$$

Solve for x.

$$\boxed{x = 2}$$

The solution to the system of linear equations is $\boxed{x = 2, y = 3, \text{ and } z = -1}$.

■ YOUR TURN   Use Gaussian elimination with back-substitution to solve the system
of linear equations.

$$\begin{aligned} x + y - z &= 0 \\ 2x + y + z &= 1 \\ 2x - y + 3z &= -1 \end{aligned}$$

■ Answer: $x = -1$, $y = 2$, and $z = 1$

Gauss–Jordan Elimination

In Gaussian elimination with back-substitution, we used row operations to rewrite the matrix in an equivalent row–echelon form. If we continue using row operations until the matrix is in *reduced* row–echelon form, this eliminates the need for back-substitution, and we call this process *Gauss–Jordan elimination*.

GAUSS–JORDAN ELIMINATION

Step 1: Write the system of linear equations as an augmented matrix.
Step 2: Use row operations to rewrite the augmented matrix in *reduced* row–echelon form.
Step 3: Write the system of linear equations that corresponds to the matrix in reduced row–echelon form found in Step 2. The result is the solution to the system.

Study Tip

For reduced row–echelon form, get 1s along the main diagonal and 0s above and below these 1s.

The order in which we perform row operations is important. You should move from left to right. Think of this process as climbing *down* a set of stairs first and then back up the stairs second. On the way *down* the stairs always use operations with rows *above* where you currently are, and on the way back *up* the stairs always use rows *below* where you currently are.

Down the stairs:

$$
\begin{bmatrix} 1 & * & * & | & * \\ * & * & * & | & * \\ * & * & * & | & * \end{bmatrix}
\rightarrow
\begin{bmatrix} 1 & * & * & | & * \\ 0 & * & * & | & * \\ 0 & * & * & | & * \end{bmatrix}
\rightarrow
\begin{bmatrix} 1 & * & * & | & * \\ 0 & 1 & * & | & * \\ 0 & * & * & | & * \end{bmatrix}
\rightarrow
\begin{bmatrix} 1 & * & * & | & * \\ 0 & 1 & * & | & * \\ 0 & 0 & * & | & * \end{bmatrix}
\rightarrow
\begin{bmatrix} 1 & * & * & | & * \\ 0 & 1 & * & | & * \\ 0 & 0 & 1 & | & * \end{bmatrix}
$$

Up the stairs:

$$
\begin{bmatrix} 1 & * & * & | & * \\ 0 & 1 & * & | & * \\ 0 & 0 & 1 & | & * \end{bmatrix}
\rightarrow
\begin{bmatrix} 1 & * & * & | & * \\ 0 & 1 & 0 & | & * \\ 0 & 0 & 1 & | & * \end{bmatrix}
\rightarrow
\begin{bmatrix} 1 & * & 0 & | & * \\ 0 & 1 & 0 & | & * \\ 0 & 0 & 1 & | & * \end{bmatrix}
\rightarrow
\begin{bmatrix} 1 & 0 & 0 & | & * \\ 0 & 1 & 0 & | & * \\ 0 & 0 & 1 & | & * \end{bmatrix}
$$

EXAMPLE 7   Using Gauss–Jordan Elimination to Solve a System of Linear Equations in Three Variables

Apply Gauss–Jordan elimination to solve the system of linear equations.

$$
\begin{aligned}
x - y + 2z &= -1 \\
3x + 2y - 6z &= 1 \\
2x + 3y + 4z &= 8
\end{aligned}
$$

Solution:

STEP 1 Write the system as an augmented matrix.

$$\begin{bmatrix} 1 & -1 & 2 & | & -1 \\ 3 & 2 & -6 & | & 1 \\ 2 & 3 & 4 & | & 8 \end{bmatrix}$$

STEP 2 Utilize row operations to rewrite the matrix in reduced row–echelon form.

There is already a 1 in the first row/first column.

Get 0s below the leading 1 in row 1.

$$\begin{matrix} R_2 - 3R_1 \rightarrow R_2 \\ R_3 - 2R_1 \rightarrow R_3 \end{matrix} \quad \begin{bmatrix} 1 & -1 & 2 & | & -1 \\ 0 & 5 & -12 & | & 4 \\ 0 & 5 & 0 & | & 10 \end{bmatrix}$$

Get a 1 in row 2/column 2.

$$R_2 \leftrightarrow R_3 \quad \begin{bmatrix} 1 & -1 & 2 & | & -1 \\ 0 & 5 & 0 & | & 10 \\ 0 & 5 & -12 & | & 4 \end{bmatrix}$$

$$\tfrac{1}{5}R_2 \rightarrow R_2 \quad \begin{bmatrix} 1 & -1 & 2 & | & -1 \\ 0 & 1 & 0 & | & 2 \\ 0 & 5 & -12 & | & 4 \end{bmatrix}$$

Get a 0 in row 3/column 2.

$$R_3 - 5R_2 \rightarrow R_3 \quad \begin{bmatrix} 1 & -1 & 2 & | & -1 \\ 0 & 1 & 0 & | & 2 \\ 0 & 0 & -12 & | & -6 \end{bmatrix}$$

Get a 1 in row 3/column 3.

$$-\tfrac{1}{12}R_3 \rightarrow R_3 \quad \begin{bmatrix} 1 & -1 & 2 & | & -1 \\ 0 & 1 & 0 & | & 2 \\ 0 & 0 & 1 & | & \tfrac{1}{2} \end{bmatrix}$$

Now, go back up the stairs.

Get 0s above the 1 in row 3/column 3.

$$R_1 - 2R_3 \rightarrow R_1 \quad \begin{bmatrix} 1 & -1 & 0 & | & -2 \\ 0 & 1 & 0 & | & 2 \\ 0 & 0 & 1 & | & \tfrac{1}{2} \end{bmatrix}$$

Get a 0 in row 1/column 2.

$$R_1 + R_2 \rightarrow R_1 \quad \begin{bmatrix} 1 & 0 & 0 & | & 0 \\ 0 & 1 & 0 & | & 2 \\ 0 & 0 & 1 & | & \tfrac{1}{2} \end{bmatrix}$$

STEP 3 Identify the solution. $\quad \boxed{x = 0, y = 2, z = \tfrac{1}{2}}$

> **Classroom Example 10.1.7***
> Solve using Gauss–Jordan elimination.
>
> $-\tfrac{1}{4}x + \tfrac{1}{2}y - \tfrac{1}{2}z = -2$
> $\tfrac{1}{2}x + \tfrac{1}{3}y - \tfrac{1}{4}z = 2$
> $\tfrac{1}{2}x - \tfrac{1}{2}y + \tfrac{1}{4}z = 1$
>
> **Answer:**
> $x = 4, y = 6, z = 8$

■ **YOUR TURN** Use an augmented matrix and Gaussian elimination to solve the system of equations.

$$\begin{aligned} x + y - z &= -2 \\ 3x + y - z &= -4 \\ 2x - 2y + 3z &= 3 \end{aligned}$$

■ **Answer:** $x = -1, y = 2,$ and $z = 3$

Classroom Example 10.1.8
Solve using Gauss–Jordan elimination.

$$\begin{aligned} x + y + z - w &= 5 \\ 2x + y - z + w &= 3 \\ x - 2y + 3z + w &= 18 \\ -x - y + z + 2w &= 8 \end{aligned}$$

Answer:
$x = 2, y = 1, z = 5, w = 3$

EXAMPLE 8   Solving a System of Linear Equations in Four Variables

Solve the system of equations with Gauss–Jordan elimination.

$$\begin{aligned} x_1 + x_2 - x_3 + 3x_4 &= 3 \\ 3x_2 - 2x_4 &= 4 \\ 2x_1 - 3x_3 &= -1 \\ 4x_4 + 2x_1 &= -6 \end{aligned}$$

Solution:

STEP 1 Write the system as an augmented matrix.

$$\left[\begin{array}{cccc|c} 1 & 1 & -1 & 3 & 3 \\ 0 & 3 & 0 & -2 & 4 \\ 2 & 0 & -3 & 0 & -1 \\ 2 & 0 & 0 & 4 & -6 \end{array}\right]$$

STEP 2 Use row operations to rewrite the matrix in reduced row–echelon form.

There is already a 1 in the first row/first column.

Get 0s below the 1 in row 1/column 1.

$\begin{aligned} R_3 - 2R_1 &\rightarrow R_3 \\ R_4 - 2R_1 &\rightarrow R_4 \end{aligned}$
$$\left[\begin{array}{cccc|c} 1 & 1 & -1 & 3 & 3 \\ 0 & 3 & 0 & -2 & 4 \\ 0 & -2 & -1 & -6 & -7 \\ 0 & -2 & 2 & -2 & -12 \end{array}\right]$$

Get a 1 in row 2/column 2.

$R_2 \leftrightarrow R_4$
$$\left[\begin{array}{cccc|c} 1 & 1 & -1 & 3 & 3 \\ 0 & -2 & 2 & -2 & -12 \\ 0 & -2 & -1 & -6 & -7 \\ 0 & 3 & 0 & -2 & 4 \end{array}\right]$$

$-\frac{1}{2}R_2 \leftrightarrow R_2$
$$\left[\begin{array}{cccc|c} 1 & 1 & -1 & 3 & 3 \\ 0 & 1 & -1 & 1 & 6 \\ 0 & -2 & -1 & -6 & -7 \\ 0 & 3 & 0 & -2 & 4 \end{array}\right]$$

Get 0s below the 1 in row 2/column 2.

$\begin{aligned} R_3 + 2R_2 &\rightarrow R_3 \\ R_4 - 3R_2 &\rightarrow R_4 \end{aligned}$
$$\left[\begin{array}{cccc|c} 1 & 1 & -1 & 3 & 3 \\ 0 & 1 & -1 & 1 & 6 \\ 0 & 0 & -3 & -4 & 5 \\ 0 & 0 & 3 & -5 & -14 \end{array}\right]$$

Get a 1 in row 3/column 3.

$-\frac{1}{3}R_3 \rightarrow R_3$
$$\left[\begin{array}{cccc|c} 1 & 1 & -1 & 3 & 3 \\ 0 & 1 & -1 & 1 & 6 \\ 0 & 0 & 1 & \frac{4}{3} & -\frac{5}{3} \\ 0 & 0 & 3 & -5 & -14 \end{array}\right]$$

Get a 0 in row 4/column 3.

$R_4 - 3R_3 \rightarrow R_4$
$$\left[\begin{array}{cccc|c} 1 & 1 & -1 & 3 & 3 \\ 0 & 1 & -1 & 1 & 6 \\ 0 & 0 & 1 & \frac{4}{3} & -\frac{5}{3} \\ 0 & 0 & 0 & -9 & -9 \end{array}\right]$$

Get a 1 in row 4/column 4.

$-\frac{1}{9}R_4 \rightarrow R_4$
$$\left[\begin{array}{cccc|c} 1 & 1 & -1 & 3 & 3 \\ 0 & 1 & -1 & 1 & 6 \\ 0 & 0 & 1 & \frac{4}{3} & -\frac{5}{3} \\ 0 & 0 & 0 & 1 & 1 \end{array}\right]$$

Now go back up the stairs.

Get 0s above the 1 in
row 4/column 4.

$$R_3 - \tfrac{4}{3}R_4 \rightarrow R_3$$
$$R_2 - R_4 \rightarrow R_2$$
$$R_1 - 3R_4 \rightarrow R_1$$

$$\begin{bmatrix} 1 & 1 & -1 & 0 & | & 0 \\ 0 & 1 & -1 & 0 & | & 5 \\ 0 & 0 & 1 & 0 & | & -3 \\ 0 & 0 & 0 & 1 & | & 1 \end{bmatrix}$$

Get 0s above the 1 in
row 3/column 3.

$$R_2 + R_3 \rightarrow R_2$$
$$R_1 + R_3 \rightarrow R_1$$

$$\begin{bmatrix} 1 & 1 & 0 & 0 & | & -3 \\ 0 & 1 & 0 & 0 & | & 2 \\ 0 & 0 & 1 & 0 & | & -3 \\ 0 & 0 & 0 & 1 & | & 1 \end{bmatrix}$$

Get a 0 in row 1/column 2.

$$R_1 - R_2 \rightarrow R_1$$

$$\begin{bmatrix} 1 & 0 & 0 & 0 & | & -5 \\ 0 & 1 & 0 & 0 & | & 2 \\ 0 & 0 & 1 & 0 & | & -3 \\ 0 & 0 & 0 & 1 & | & 1 \end{bmatrix}$$

STEP 3 Identify the solution.

$$x_1 = -5, x_2 = 2, x_3 = -3, x_4 = 1$$

Inconsistent and Dependent Systems

Recall that systems of linear equations can be independent, inconsistent, or dependent systems and therefore have *one solution*, *no solution*, or *infinitely many solutions*. All of the systems we have solved so far in this section have been independent systems (unique solution). When solving a system of linear equations using Gaussian elimination or Gauss–Jordan elimination, the following will indicate the three possible types of solutions.

SYSTEM	TYPE OF SOLUTION	MATRIX DURING GAUSS–JORDAN ELIMINATION	EXAMPLE	
Independent	One (unique) solution	Diagonal entries are all 1s and 0s elsewhere.	$\begin{bmatrix} 1 & 0 & 0 & \vert & 1 \\ 0 & 1 & 0 & \vert & -3 \\ 0 & 0 & 1 & \vert & 2 \end{bmatrix}$	or $\begin{aligned} x &= 1 \\ y &= -3 \\ z &= 2 \end{aligned}$
Inconsistent	No solution	One row will have only zero entries for coefficients and a nonzero entry for the constant.	$\begin{bmatrix} 1 & 0 & 0 & \vert & 1 \\ 0 & 1 & 0 & \vert & -3 \\ 0 & 0 & 0 & \vert & 2 \end{bmatrix}$	or $\begin{aligned} x &= 1 \\ y &= -3 \\ 0 &= 2 \end{aligned}$
Dependent	Infinitely many solutions	One row will be entirely 0s when the number of equations equals the number of variables.	$\begin{bmatrix} 1 & 0 & -2 & \vert & 1 \\ 0 & 1 & 1 & \vert & -3 \\ 0 & 0 & 0 & \vert & 0 \end{bmatrix}$	or $\begin{aligned} x - 2z &= 1 \\ y + z &= -3 \\ 0 &= 0 \end{aligned}$

EXAMPLE 9   Solving an Inconsistent System: No Solution

Solve the system of equations.

$$\begin{aligned} x + 2y - z &= 3 \\ 2x + y + 2z &= -1 \\ -2x - 4y + 2z &= 5 \end{aligned}$$

Solution:

STEP 1 Write the system of equations as an augmented matrix.

$$\begin{bmatrix} 1 & 2 & -1 & | & 3 \\ 2 & 1 & 2 & | & -1 \\ -2 & -4 & 2 & | & 5 \end{bmatrix}$$

STEP 2 Apply row operations to rewrite the matrix in row–echelon form.

Get 0s below the 1 in column 1.

$$\begin{matrix} R_2 - 2R_1 \to R_2 \\ R_3 + 2R_1 \to R_3 \end{matrix} \begin{bmatrix} 1 & 2 & -1 & | & 3 \\ 0 & -3 & 4 & | & -7 \\ 0 & 0 & 0 & | & 11 \end{bmatrix}$$

There is no need to continue because $0x + 0y + 0z = 11$ or $0 = 11$
row 3 is a contradiction.

Since this is inconsistent, there is *no solution* to this system of equations.

EXAMPLE 10   Solving a Dependent System: Infinitely Many Solutions

Solve the system of equations.

$$\begin{aligned} x \quad + z &= 3 \\ 2x + y + 4z &= 8 \\ 3x + y + 5z &= 11 \end{aligned}$$

Solution:

STEP 1 Write the system of equations as an augmented matrix.

$$\begin{bmatrix} 1 & 0 & 1 & | & 3 \\ 2 & 1 & 4 & | & 8 \\ 3 & 1 & 5 & | & 11 \end{bmatrix}$$

STEP 2 Use row operations to rewrite the matrix in reduced row–echelon form.

Get the 0s below the 1 in column 1.

$$\begin{matrix} R_2 - 2R_1 \to R_2 \\ R_3 - 3R_1 \to R_3 \end{matrix} \begin{bmatrix} 1 & 0 & 1 & | & 3 \\ 0 & 1 & 2 & | & 2 \\ 0 & 1 & 2 & | & 2 \end{bmatrix}$$

Get a 0 in row 3/column 2.

$$R_3 - R_2 \to R_3 \begin{bmatrix} 1 & 0 & 1 & | & 3 \\ 0 & 1 & 2 & | & 2 \\ 0 & 0 & 0 & | & 0 \end{bmatrix}$$

This matrix is in reduced row–echelon form. This matrix corresponds to a dependent system of linear equations and has infinitely many solutions.

STEP 3 Write the augmented matrix as a system of linear equations.

$$\begin{aligned} x + z &= 3 \\ y + 2z &= 2 \end{aligned}$$

Let $z = a$, where a is any real number, and substitute this into the two equations.

We find that $x = 3 - a$ and $y = 2 - 2a$. The general solution is $\boxed{x = 3 - a, y = 2 - 2a, z = a}$ for a any real number. Note that $(2, 0, 1)$ and $(3, 2, 0)$ are particular solutions when $a = 1$ and $a = 0$, respectively.

Study Tip

In a system with three variables, say x, y, and z, we typically let $z = a$ (where a is called a parameter) and then solve for x and y in terms of a.

A common mistake that is made is to identify a unique solution as no solution when one of the variables is equal to zero. For example, what is the difference between the following two matrices?

$$\left[\begin{array}{ccc|c} 1 & 0 & 2 & 1 \\ 0 & 1 & 3 & 2 \\ 0 & 0 & 3 & 0 \end{array}\right] \quad \text{and} \quad \left[\begin{array}{ccc|c} 1 & 0 & 2 & 1 \\ 0 & 1 & 3 & 2 \\ 0 & 0 & 0 & 3 \end{array}\right]$$

The first matrix has a *unique solution,* whereas the second matrix has *no solution.* The third row of the first matrix corresponds to the equation $3z = 0$, which implies that $z = 0$. The third row of the second matrix corresponds to the equation $0x + 0y + 0z = 3$ or $0 = 3$, which is inconsistent, and therefore the system has no solution.

EXAMPLE 11   Solving a Dependent System: Infinitely Many Solutions

Solve the system of linear equations.

$$2x + y + z = 8$$
$$x + y - z = -3$$

Solution:

STEP 1 Write the system of equations as an augmented matrix. $\left[\begin{array}{ccc|c} 2 & 1 & 1 & 8 \\ 1 & 1 & -1 & -3 \end{array}\right]$

STEP 2 Use row operations to rewrite the matrix in reduced row–echelon form.

Get a 1 in row 1/column 1. $R_1 \leftrightarrow R_2$ $\left[\begin{array}{ccc|c} 1 & 1 & -1 & -3 \\ 2 & 1 & 1 & 8 \end{array}\right]$

Get a 0 in row 2/column 1. $R_2 - 2R_1 \rightarrow R_2$ $\left[\begin{array}{ccc|c} 1 & 1 & -1 & -3 \\ 0 & -1 & 3 & 14 \end{array}\right]$

Get a 1 in row 2/column 1. $-R_2 \rightarrow R_2$ $\left[\begin{array}{ccc|c} 1 & 1 & -1 & -3 \\ 0 & 1 & -3 & -14 \end{array}\right]$

Get a 0 in row 1/column 2. $R_1 - R_2 \rightarrow R_1$ $\left[\begin{array}{ccc|c} 1 & 0 & 2 & 11 \\ 0 & 1 & -3 & -14 \end{array}\right]$

This matrix is in reduced row–echelon form.

STEP 3 Identify the solution. $x + 2z = 11$
$y - 3z = -14$

Let $z = a$, where a is any real number. Substituting $z = a$ into these two equations gives the infinitely many solutions $\boxed{x = 11 - 2a,\ y = 3a - 14,\ z = a}$.

....

■ **YOUR TURN** Solve the system of equations using an augmented matrix.

$$x + y + z = 0$$
$$3x + 2y - z = 2$$

Applications

Remember Jared who lost all that weight eating at Subway and is still keeping it off 10 years later? He ate Subway sandwiches for lunch and dinner for one year and lost 235 pounds! The following table gives nutritional information for Subway's 6-inch sandwiches.

Classroom Example 10.1.11

Solve these systems.

a. $x - y - z = -1$
$\qquad y + z = x$

b.* $ax - 3az = 1$
$\qquad y - az = a$,
where $a \neq 0$

Answer:

a. no solution

b. $x = 3t + \dfrac{1}{a}$
$\quad y = at + a$
$\quad z = t$,

where t is a real number.

■ **Answer:** $x = 3a + 2, y = -4a - 2,$ $z = a$, where a is any real number.

SANDWICH	CALORIES	FAT (g)	CARBOHYDRATES (g)	PROTEIN (g)
Veggie Delight	350	18	17	36
Oven-roasted chicken breast	430	19	46	20
Ham (Black Forest without cheese)	290	5	45	19

EXAMPLE 12 Subway Diet

Suppose you are going to eat only Subway 6-inch sandwiches for a week (seven days) for both lunch and dinner (total of 14 meals). If your goal is to eat 388 grams of protein and 4900 calories in those 14 sandwiches, how many of each sandwich should you eat that week?

Solution:

STEP 1 Determine the system of linear equations.

Let three variables represent the number of each type of sandwich you eat in a week.

x = number of Veggie Delight sandwiches
y = number of chicken breast sandwiches
z = number of ham sandwiches

The total number of sandwiches eaten is 14. $\qquad x + y + z = 14$

The total number of calories consumed is 4900. $\qquad 350x + 430y + 290z = 4900$

The total number of grams of protein consumed is 388. $\qquad 36x + 20y + 19z = 388$

Write an augmented matrix representing this system of linear equations.

$$\begin{bmatrix} 1 & 1 & 1 & | & 14 \\ 350 & 430 & 290 & | & 4900 \\ 36 & 20 & 19 & | & 388 \end{bmatrix}$$

STEP 2 Utilize row operations to rewrite the matrix in reduced row–echelon form.

$$\begin{matrix} R_2 - 350R_1 \to R_2 \\ R_3 - 36R_1 \to R_3 \end{matrix} \quad \begin{bmatrix} 1 & 1 & 1 & | & 14 \\ 0 & 80 & -60 & | & 0 \\ 0 & -16 & -17 & | & -116 \end{bmatrix}$$

$$\tfrac{1}{80}R_2 \to R_2 \quad \begin{bmatrix} 1 & 1 & 1 & | & 14 \\ 0 & 1 & -\tfrac{3}{4} & | & 0 \\ 0 & -16 & -17 & | & -116 \end{bmatrix}$$

$$R_3 + 16R_2 \to R_3 \quad \begin{bmatrix} 1 & 1 & 1 & | & 14 \\ 0 & 1 & -\tfrac{3}{4} & | & 0 \\ 0 & 0 & -29 & | & -116 \end{bmatrix}$$

$$-\tfrac{1}{29}R_3 \to R_3 \quad \begin{bmatrix} 1 & 1 & 1 & | & 14 \\ 0 & 1 & -\tfrac{3}{4} & | & 0 \\ 0 & 0 & 1 & | & 4 \end{bmatrix}$$

$$\begin{matrix} R_2 + \tfrac{3}{4}R_3 \to R_2 \\ R_1 - R_3 \to R_1 \end{matrix} \quad \begin{bmatrix} 1 & 1 & 0 & | & 10 \\ 0 & 1 & 0 & | & 3 \\ 0 & 0 & 1 & | & 4 \end{bmatrix}$$

$$R_1 - R_2 \to R_1 \quad \begin{bmatrix} 1 & 0 & 0 & | & 7 \\ 0 & 1 & 0 & | & 3 \\ 0 & 0 & 1 & | & 4 \end{bmatrix}$$

STEP 3 Identify the solution. $\boxed{x = 7, \ y = 3, \ z = 4}$

You should eat $\boxed{\text{7 Veggie Delights, 3 oven-roasted chicken breast, and 4 ham sandwiches}}$.

Many times in the real world we see a relationship that looks like a particular function such as a quadratic function and we know particular data points, but we do not know the function. We start with the general function, fit the curve to particular data points, and solve a system of linear equations to determine the specific function parameters.

EXAMPLE 13   Fitting a Curve to Data

The amount of money awarded in medical malpractice suits is rising. This can be modeled with a quadratic function $y = at^2 + bt + c$, where $t > 0$ and $a > 0$. Determine a quadratic function that passes through the three points shown on the graph.

Technology Tip

Enter the matrix into the TI or graphing calculator. Press [2nd] [MATRIX]. Use [▶] [EDIT] [ENTER].

Now enter the size by typing the number of rows first and the number of columns second. Enter the elements of the matrix one row at a time by pressing the number and [ENTER] each time.

When done, press [2nd] [QUIT]. To show the matrix A, press [2nd] [MATRIX] [ENTER] [ENTER].

To find the reduced row–echelon form of the original matrix directly use [rref] (matrix) command. Press [2nd] [MATRIX] [▶] [MATH] [▼] [B:rref] [ENTER] [2nd] [MATRIX] [ENTER] [)] [MATH] [1:Frac] [ENTER].

The solution to the system is $a = \frac{203}{90}, b = -\frac{151}{30}, c = \frac{2302}{9}$.

Solution:

Let 1991 correspond to $t = 0$ and y represent the number of dollars awarded for malpractice suits. The following data are reflected in the illustration above:

YEAR	t	y (THOUSANDS OF DOLLARS)	(t, y)
1992	1	253	$(1, 253)$
1996	5	287	$(5, 287)$
2001	10	431	$(10, 431)$

Substitute the three points, $(1, 253)$, $(5, 287)$, and $(10, 431)$ into the general quadratic equation: $y = at^2 + bt + c$.

POINT	$y = at^2 + bt + c$	SYSTEM OF EQUATIONS
$(1, 253)$	$253 = a(1)^2 + b(1) + c$	$a + b + c = 253$
$(5, 287)$	$287 = a(5)^2 + b(5) + c$	$25a + 5b + c = 287$
$(10, 431)$	$431 = a(10)^2 + b(10) + c$	$100a + 10b + c = 431$

STEP 1 Write this system of linear equations as an augmented matrix.

$$\begin{bmatrix} 1 & 1 & 1 & | & 253 \\ 25 & 5 & 1 & | & 287 \\ 100 & 10 & 1 & | & 431 \end{bmatrix}$$

STEP 2 Apply row operations to rewrite the matrix in reduced row–echelon form.

$$R_2 - 25R_1 \rightarrow R_2 \quad \begin{bmatrix} 1 & 1 & 1 & | & 253 \\ 0 & -20 & -24 & | & -6{,}038 \\ 100 & 10 & 1 & | & 431 \end{bmatrix}$$

$$R_3 - 100R_1 \rightarrow R_3 \quad \begin{bmatrix} 1 & 1 & 1 & | & 253 \\ 0 & -20 & -24 & | & -6{,}038 \\ 0 & -90 & -99 & | & -24{,}869 \end{bmatrix}$$

$$-\tfrac{1}{20}R_2 \rightarrow R_2 \quad \begin{bmatrix} 1 & 1 & 1 & | & 253 \\ 0 & 1 & \frac{6}{5} & | & \frac{3{,}019}{10} \\ 0 & -90 & -99 & | & -24{,}869 \end{bmatrix}$$

$$R_3 + 90R_2 \rightarrow R_3 \quad \begin{bmatrix} 1 & 1 & 1 & 253 \\ 0 & 1 & \frac{6}{5} & \frac{3,019}{10} \\ 0 & 0 & 9 & 2,302 \end{bmatrix}$$

$$\frac{1}{9}R_3 \rightarrow R_3 \quad \begin{bmatrix} 1 & 1 & 1 & 253 \\ 0 & 1 & \frac{6}{5} & \frac{3,019}{10} \\ 0 & 0 & 1 & \frac{2,302}{9} \end{bmatrix}$$

$$R_2 - \frac{6}{5}R_3 \rightarrow R_2 \quad \begin{bmatrix} 1 & 1 & 1 & 253 \\ 0 & 1 & 0 & -\frac{151}{30} \\ 0 & 0 & 1 & \frac{2,302}{9} \end{bmatrix}$$

$$R_1 - R_3 \rightarrow R_1 \quad \begin{bmatrix} 1 & 1 & 0 & -\frac{25}{9} \\ 0 & 1 & 0 & -\frac{151}{30} \\ 0 & 0 & 1 & \frac{2,302}{9} \end{bmatrix}$$

$$R_1 - R_2 \rightarrow R_1 \quad \begin{bmatrix} 1 & 0 & 0 & \frac{203}{90} \\ 0 & 1 & 0 & -\frac{151}{30} \\ 0 & 0 & 1 & \frac{2,302}{9} \end{bmatrix}$$

STEP 3 Identify the solution. $\qquad a = \dfrac{203}{90} \quad b = -\dfrac{151}{30} \quad c = \dfrac{2,302}{9}$

Substituting $a = 203/90$, $b = -151/30$, $c = 2302/9$ into $y = at^2 + bt + c$, we find that the thousands of dollars spent on malpractice suits as a function of year is given by

$$y = \frac{203}{90}t^2 - \frac{151}{30}t + \frac{2302}{9} \qquad \text{1991 is } t = 0$$

Notice that all three points lie on this curve.

SECTION
10.1 SUMMARY

In this section, we used augmented matrices to represent a system of linear equations.

$$\begin{aligned} a_1x + b_1y + c_1z &= d_1 \\ a_2x + b_2y + c_2z &= d_2 \\ a_3x + b_3y + c_3z &= d_3 \end{aligned} \iff \begin{bmatrix} a_1 & b_1 & c_1 & d_1 \\ a_2 & b_2 & c_2 & d_2 \\ a_3 & b_3 & c_3 & d_3 \end{bmatrix}$$

Any missing terms correspond to a 0 in the matrix. A matrix is in **row–echelon** form if it has all three of the following properties:

1. Any rows consisting entirely of 0s are at the bottom of the matrix.
2. For each row that does not consist entirely of 0s, the first (leftmost) nonzero entry is 1 (called the leading 1).
3. For two successive nonzero rows, the leading 1 in the higher row is farther to the left than the leading 1 in the lower row.

If a matrix in row–echelon form has the following additional property, then the matrix is in **reduced row–echelon form**:

4. Every column containing a leading 1 has zeros in every position above and below the leading 1.

The two methods used for solving systems of linear equations represented as augmented matrices are Gaussian elimination with back-substitution and Gauss–Jordan elimination. In both cases, we represent the system of linear equations as an augmented matrix and then use row operations to rewrite in row–echelon form. With Gaussian elimination we then stop and perform back-substitution to solve the system, and with Gauss–Jordan elimination we continue with row operations until the matrix is in reduced row–echelon form and then identify the solution to the system.

▪ **SKILLS**

In Exercises 1–6, determine the order of each matrix.

1. $\begin{bmatrix} -1 & 3 & 4 \\ 2 & 7 & 9 \end{bmatrix}$
2. $\begin{bmatrix} 0 & 1 \\ 3 & 9 \\ 7 & 8 \end{bmatrix}$
3. $\begin{bmatrix} 1 & 2 & 3 & 4 \end{bmatrix}$
4. $\begin{bmatrix} 3 \\ 7 \\ -1 \\ 10 \end{bmatrix}$
5. $[0]$
6. $\begin{bmatrix} -1 & 3 & 6 & 8 \\ 2 & 9 & 7 & 3 \\ 5 & 4 & -2 & -10 \\ 6 & 3 & 1 & 5 \end{bmatrix}$

In Exercises 7–14, write the augmented matrix for each system of linear equations.

7. $\begin{aligned} 3x - 2y &= 7 \\ -4x + 6y &= -3 \end{aligned}$

8. $\begin{aligned} -x + y &= 2 \\ x - y &= -4 \end{aligned}$

9. $\begin{aligned} 2x - 3y + 4z &= -3 \\ -x + y + 2z &= 1 \\ 5x - 2y - 3z &= 7 \end{aligned}$

10. $\begin{aligned} x - 2y + z &= 0 \\ -2x + y - z &= -5 \\ 13x + 7y + 5z &= 6 \end{aligned}$

11. $\begin{aligned} x + y &= 3 \\ x - z &= 2 \\ y + z &= 5 \end{aligned}$

12. $\begin{aligned} x - y &= -4 \\ y + z &= 3 \end{aligned}$

13. $\begin{aligned} 3y - 4x + 5z - 2 &= 0 \\ 2x - 3y - 2z &= -3 \\ 3z + 4y - 2x - 1 &= 0 \end{aligned}$

14. $\begin{aligned} 2y + z - x - 3 &= 2 \\ 2x + 3z - 2y &= 0 \\ -2z + y - 4x - 3 &= 0 \end{aligned}$

In Exercises 15–20, write the system of linear equations represented by the augmented matrix. Utilize the variables x, y, and z.

15. $\left[\begin{array}{cc|c} -3 & 7 & 2 \\ 1 & 5 & 8 \end{array}\right]$

16. $\left[\begin{array}{ccc|c} -1 & 2 & 4 & 4 \\ 7 & 9 & 3 & -3 \\ 4 & 6 & -5 & 8 \end{array}\right]$

17. $\left[\begin{array}{ccc|c} -1 & 0 & 0 & 4 \\ 7 & 9 & 3 & -3 \\ 4 & 6 & -5 & 8 \end{array}\right]$

18. $\left[\begin{array}{ccc|c} 2 & 3 & -4 & 6 \\ 7 & -1 & 5 & 9 \end{array}\right]$

19. $\left[\begin{array}{cc|c} 1 & 0 & a \\ 0 & 1 & b \end{array}\right]$

20. $\left[\begin{array}{ccc|c} 3 & 0 & 5 & 1 \\ 0 & -4 & 7 & -3 \\ 2 & -1 & 0 & 8 \end{array}\right]$

In Exercises 21–30, indicate whether each matrix is in row–echelon form. If it is, determine whether it is in reduced row–echelon form.

21. $\left[\begin{array}{cc|c} 1 & 0 & 3 \\ 1 & 1 & 2 \end{array}\right]$

22. $\left[\begin{array}{cc|c} 0 & 1 & 3 \\ 1 & 0 & 2 \end{array}\right]$

23. $\left[\begin{array}{cc|c} 1 & 0 & -1 & -3 \\ 0 & 1 & 3 & 14 \end{array}\right]$

24. $\left[\begin{array}{ccc|c} 1 & 0 & 0 & -3 \\ 0 & 1 & 3 & 14 \end{array}\right]$

25. $\left[\begin{array}{ccc|c} 1 & 0 & 1 & 3 \\ 0 & 0 & 0 & 0 \\ 0 & 1 & 2 & 2 \end{array}\right]$

26. $\left[\begin{array}{ccc|c} 1 & 0 & 1 & 3 \\ 0 & 1 & 2 & 2 \\ 0 & 0 & 0 & 0 \end{array}\right]$

27. $\left[\begin{array}{ccc|c} 1 & 0 & 0 & 3 \\ 0 & 1 & 0 & 2 \\ 0 & 0 & 1 & 5 \end{array}\right]$

28. $\left[\begin{array}{ccc|c} -1 & 0 & 0 & 3 \\ 0 & -1 & 0 & 2 \\ 0 & 0 & -1 & 5 \end{array}\right]$

29. $\left[\begin{array}{cccc|c} 1 & 0 & 0 & 1 & 3 \\ 0 & 1 & 0 & 3 & 2 \\ 0 & 0 & 1 & 0 & 5 \\ 0 & 0 & 0 & 1 & 0 \end{array}\right]$

30. $\left[\begin{array}{cccc|c} 1 & 0 & 0 & 1 & 3 \\ 0 & 1 & 0 & 3 & 2 \\ 0 & 0 & 1 & 0 & 5 \\ 0 & 0 & 0 & 0 & 0 \end{array}\right]$

In Exercises 31–40, perform the indicated row operations on each augmented matrix.

31. $\left[\begin{array}{cc|c} 1 & -2 & -3 \\ 2 & 3 & -1 \end{array}\right]$ $\quad R_2 - 2R_1 \rightarrow R_2$

32. $\left[\begin{array}{cc|c} 2 & -3 & -4 \\ 1 & 2 & 5 \end{array}\right]$ $\quad R_1 \leftrightarrow R_2$

33. $\left[\begin{array}{ccc|c} 1 & -2 & -1 & 3 \\ 2 & 1 & -3 & 6 \\ 3 & -2 & 5 & -8 \end{array}\right]$ $\quad R_2 - 2R_1 \rightarrow R_2$

34. $\left[\begin{array}{ccc|c} 1 & -2 & 1 & 3 \\ 0 & 1 & -2 & 6 \\ -3 & 0 & -1 & -5 \end{array}\right]$ $\quad R_3 + 3R_1 \rightarrow R_3$

35. $\left[\begin{array}{cccc|c} 1 & -2 & 5 & -1 & 2 \\ 0 & 3 & 0 & -1 & -2 \\ 0 & -2 & 1 & -2 & 5 \\ 0 & 0 & 1 & -1 & -6 \end{array}\right]$ $\quad R_3 + R_2 \rightarrow R_2$

36. $\left[\begin{array}{cccc|c} 1 & 0 & 5 & -10 & 15 \\ 0 & 1 & 2 & -3 & 4 \\ 0 & 2 & -3 & 0 & -1 \\ 0 & 0 & 1 & -1 & -3 \end{array}\right]$ $\quad R_2 - \frac{1}{2}R_3 \rightarrow R_3$

37. $\begin{bmatrix} 1 & 0 & 5 & -10 & | & -5 \\ 0 & 1 & 2 & -3 & | & -2 \\ 0 & 2 & -3 & 0 & | & -1 \\ 0 & -3 & 2 & -1 & | & -3 \end{bmatrix}$ $\quad \begin{matrix} R_3 - 2R_2 \rightarrow R_3 \\ R_4 + 3R_2 \rightarrow R_4 \end{matrix}$

38. $\begin{bmatrix} 1 & 0 & 4 & 0 & | & 1 \\ 0 & 1 & 2 & 0 & | & -2 \\ 0 & 0 & 1 & 0 & | & 0 \\ 0 & 0 & 0 & 1 & | & -3 \end{bmatrix}$ $\quad \begin{matrix} R_2 - 2R_3 \rightarrow R_2 \\ R_1 - 4R_3 \rightarrow R_1 \end{matrix}$

39. $\begin{bmatrix} 1 & 0 & 4 & 8 & | & 3 \\ 0 & 1 & 2 & -3 & | & -2 \\ 0 & 0 & 1 & 6 & | & 3 \\ 0 & 0 & 0 & 1 & | & -3 \end{bmatrix}$ $\quad \begin{matrix} R_3 - 6R_4 \rightarrow R_3 \\ R_2 + 3R_4 \rightarrow R_2 \\ R_1 - 8R_4 \rightarrow R_1 \end{matrix}$

40. $\begin{bmatrix} 1 & 0 & -1 & 5 & | & 2 \\ 0 & 1 & 2 & 3 & | & -5 \\ 0 & 0 & 1 & -2 & | & 2 \\ 0 & 0 & 0 & 1 & | & 1 \end{bmatrix}$ $\quad \begin{matrix} R_3 + 2R_4 \rightarrow R_3 \\ R_2 - 3R_4 \rightarrow R_2 \\ R_1 - 5R_4 \rightarrow R_1 \end{matrix}$

In Exercises 41–50, use row operations to transform each matrix to reduced row–echelon form.

41. $\begin{bmatrix} 1 & 2 & | & 4 \\ 2 & 3 & | & 2 \end{bmatrix}$

42. $\begin{bmatrix} 1 & -1 & | & 3 \\ -3 & 2 & | & 2 \end{bmatrix}$

43. $\begin{bmatrix} 1 & -1 & 1 & | & -1 \\ 0 & 1 & -1 & | & -1 \\ -1 & 1 & 1 & | & 1 \end{bmatrix}$

44. $\begin{bmatrix} 0 & -1 & 1 & | & 1 \\ 1 & -1 & 1 & | & -1 \\ 1 & -1 & -1 & | & -1 \end{bmatrix}$

45. $\begin{bmatrix} 3 & -2 & -3 & | & -1 \\ 1 & -1 & 1 & | & -4 \\ 2 & 3 & 5 & | & 14 \end{bmatrix}$

46. $\begin{bmatrix} 3 & -1 & 1 & | & 2 \\ 1 & -2 & 3 & | & 1 \\ 2 & 1 & -3 & | & -1 \end{bmatrix}$

47. $\begin{bmatrix} 2 & 1 & -6 & | & 4 \\ 1 & -2 & 2 & | & -3 \end{bmatrix}$

48. $\begin{bmatrix} -3 & -1 & 2 & | & -1 \\ -1 & -2 & 1 & | & -3 \end{bmatrix}$

49. $\begin{bmatrix} -1 & 2 & 1 & | & -2 \\ 3 & -2 & 1 & | & 4 \\ 2 & -4 & -2 & | & 4 \end{bmatrix}$

50. $\begin{bmatrix} 2 & -1 & 0 & | & 1 \\ -1 & 0 & 1 & | & -2 \\ -2 & 1 & 0 & | & -1 \end{bmatrix}$

In Exercises 51–70, solve the system of linear equations using Gaussian elimination with back-substitution.

51. $2x + 3y = 1$
$\quad x + y = -2$

52. $3x + 2y = 11$
$\quad x - y = 12$

53. $-x + 2y = 3$
$\quad 2x - 4y = -6$

54. $3x - y = -1$
$\quad 2y + 6x = 2$

55. $\frac{2}{3}x + \frac{1}{3}y = \frac{8}{9}$
$\quad \frac{1}{2}x + \frac{1}{4}y = \frac{3}{4}$

56. $0.4x - 0.5y = 2.08$
$\quad -0.3x + 0.7y = 1.88$

57. $x - z - y = 10$
$\quad 2x - 3y + z = -11$
$\quad y - x + z = -10$

58. $2x + z + y = -3$
$\quad 2y - z + x = 0$
$\quad x + y + 2z = 5$

59. $3x_1 + x_2 - x_3 = 1$
$\quad x_1 - x_2 + x_3 = -3$
$\quad 2x_1 + x_2 + x_3 = 0$

60. $2x_1 + x_2 + x_3 = -1$
$\quad x_1 + x_2 - x_3 = 5$
$\quad 3x_1 - x_2 - x_3 = 1$

61. $\quad 2x + 5y = 9$
$\quad x + 2y - z = 3$
$\quad -3x - 4y + 7z = 1$

62. $x - 2y + 3z = 1$
$\quad -2x + 7y - 9z = 4$
$\quad x + z = 9$

63. $2x_1 - x_2 + x_3 = 3$
$\quad x_1 - x_2 + x_3 = 2$
$\quad -2x_1 + 2x_2 - 2x_3 = -4$

64. $x_1 - x_2 - 2x_3 = 0$
$\quad -2x_1 + 5x_2 + 10x_3 = -3$
$\quad 3x_1 + x_2 = 0$

65. $2x + y - z = 2$
$\quad x - y - z = 6$

66. $3x + y - z = 0$
$\quad x + y + 7z = 4$

67. $\quad 2y + z = 3$
$\quad 4x - z = -3$
$\quad 7x - 3y - 3z = 2$
$\quad x - y - z = -2$

68. $-2x - y + 2z = 3$
$\quad 3x - 4z = 2$
$\quad 2x + y = -1$
$\quad -x + y - z = -8$

69. $3x_1 - 2x_2 + x_3 + 2x_4 = -2$
$\quad -x_1 + 3x_2 + 4x_3 + 3x_4 = 4$
$\quad x_1 + x_2 + x_3 + x_4 = 0$
$\quad 5x_1 + 3x_2 + x_3 + 2x_4 = -1$

70. $5x_1 + 3x_2 + 8x_3 + x_4 = 1$
$\quad x_1 + 2x_2 + 5x_3 + 2x_4 = 3$
$\quad 4x_1 + x_3 - 2x_4 = -3$
$\quad x_2 + x_3 + x_4 = 0$

In Exercises 71–86, solve the system of linear equations using Gauss–Jordan elimination.

71.
$$\begin{aligned} x + 3y &= -5 \\ -2x - y &= 0 \end{aligned}$$

72.
$$\begin{aligned} 5x - 4y &= 31 \\ 3x + 7y &= -19 \end{aligned}$$

73.
$$\begin{aligned} x + y &= 4 \\ -3x - 3y &= 10 \end{aligned}$$

74.
$$\begin{aligned} 3x - 4y &= 12 \\ -6x + 8y &= -24 \end{aligned}$$

75.
$$\begin{aligned} x - 2y + 3z &= 5 \\ 3x + 6y - 4z &= -12 \\ -x - 4y + 6z &= 16 \end{aligned}$$

76.
$$\begin{aligned} x + 2y - z &= 6 \\ 2x - y + 3z &= -13 \\ 3x - 2y + 3z &= -16 \end{aligned}$$

77.
$$\begin{aligned} x + y + z &= 3 \\ x - z &= 1 \\ y - z &= -4 \end{aligned}$$

78.
$$\begin{aligned} x - 2y + 4z &= 2 \\ 2x - 3y - 2z &= -3 \\ \tfrac{1}{2}x + \tfrac{1}{4}y + z &= -2 \end{aligned}$$

79.
$$\begin{aligned} x + 2y + z &= 3 \\ 2x - y + 3z &= 7 \\ 3x + y + 4z &= 5 \end{aligned}$$

80.
$$\begin{aligned} x + 2y + z &= 3 \\ 2x - y + 3z &= 7 \\ 3x + y + 4z &= 10 \end{aligned}$$

81.
$$\begin{aligned} 3x - y + z &= 8 \\ x + y - 2z &= 4 \end{aligned}$$

82.
$$\begin{aligned} x - 2y + 3z &= 10 \\ -3x + z &= 9 \end{aligned}$$

83.
$$\begin{aligned} 4x - 2y + 5z &= 20 \\ x + 3y - 2z &= 6 \end{aligned}$$

84.
$$\begin{aligned} y + z &= 4 \\ x + y &= 8 \end{aligned}$$

85.
$$\begin{aligned} x - y - z - w &= 1 \\ 2x + y + z + 2w &= 3 \\ x - 2y - 2z - 3w &= 0 \\ 3x - 4y + z + 5w &= -3 \end{aligned}$$

86.
$$\begin{aligned} x - 3y + 3z - 2w &= 4 \\ x + 2y - z &= -3 \\ x + 3z + 2w &= 3 \\ y + z + 5w &= 6 \end{aligned}$$

■ APPLICATIONS

87. Football. In Super Bowl XXXVIII, the New England Patriots defeated the Carolina Panthers 32–29. The points came from a total of four types of plays: touchdowns (6 points), extra points (1 point), two-point conversions (2 points), and field goals (3 points). There were a total of 16 scoring plays. There were four times as many touchdowns as field goals, and there were five times as many extra points as 2-point conversions. How many touchdowns, extra points, 2-point conversions, and field goals were scored in Super Bowl XXXVIII?

88. Basketball. In the 2004 Summer Olympics in Athens, Greece, the U.S. men's basketball team, consisting of NBA superstars, was defeated by the Puerto Rican team 92–73. The points came from three types of scoring plays: 2-point shots, 3-point shots, and 1-point free throws. There were six more 2-point shots made than there were 1-point free throws. The number of successful 2-point shots was three less than four times the number of successful 3-point shots. How many 2-point and 3-point shots, and 1-point free throws were made in that Olympic competition?

Exercises 89 and 90 rely on a selection of Subway sandwiches whose nutrition information is given in the following table. Suppose you are going to eat only Subway sandwiches for a week (seven days) for lunch and dinner (a total of 14 meals).

Sandwich	Calories	Fat (g)	Carbohydrates (g)	Protein (g)
Mediterranean chicken	350	18	17	36
6" tuna	430	19	46	20
6" roast beef	290	5	45	19
Turkey–bacon wrap	430	27	20	34

www.subway.com

89. Diet. Your goal is a low-fat diet consisting of 526 grams of carbohydrates, 168 grams of fat, and 332 grams of protein. How many of each sandwich would you eat that week to obtain this goal?

90. Diet. Your goal is a low-carb diet consisting of 5180 calories, 335 grams of carbohydrates, and 263 grams of fat. How many of each sandwich would you eat that week to obtain this goal?

Exercises 91 and 92 involve vertical motion and the effect of gravity on an object.

Because of gravity, an object that is projected upward will eventually reach a maximum height and then fall to the ground. The equation that relates the height h of a projectile t seconds after it is shot upward is given by

$$h = \frac{1}{2}at^2 + v_0 t + h_0$$

where a is the acceleration due to gravity, h_0 is the initial height of the object at time $t = 0$, and v_0 is the initial velocity of the object at time $t = 0$. Note that a projectile follows the path of a parabola opening down, so $a < 0$.

91. Vertical Motion. An object is thrown upward, and the following table depicts the height of the ball t seconds after the projectile is released. Find the initial height, initial velocity, and acceleration due to gravity.

t (SECONDS)	HEIGHT (FEET)
1	34
2	36
3	6

92. Vertical Motion. An object is thrown upward, and the following table depicts the height of the ball t seconds after the projectile is released. Find the initial height, initial velocity, and acceleration due to gravity.

t (SECONDS)	HEIGHT (FEET)
1	54
2	66
3	46

93. Data Curve-Fitting. The average number of minutes that a person spends driving a car can be modeled by a quadratic function $y = ax^2 + bx + c$, where $a < 0$ and $15 < x < 65$. The following table gives the average number of minutes a day that a person spends driving a car. Determine the quadratic function that models this quantity.

AGE	AVERAGE DAILY MINUTES DRIVING
16	25
40	64
65	40

94. Data Curve-Fitting. The average age when a woman gets married has been increasing during the last century. In 1920 the average age was 18.4, in 1960 the average age was 20.3, and in 2002 the average age was 25.30. Find a quadratic function $y = ax^2 + bx + c$, where $a > 0$ and $18 < x < 35$, that models the average age y when a woman gets married as a function of the year x ($x = 0$ corresponds to 1920). What will the average age be in 2020?

95. Chemistry/Pharmacy. A pharmacy receives an order for 100 milliliters of 5% hydrogen peroxide solution. The pharmacy has a 1.5% and a 30% solution on hand. A technician will mix the 1.5% and 30% solutions to make the 5% solution. How much of the 1.5% and 30% solutions, respectively, will be needed to fill this order? Round to the nearest ml.

96. Chemistry/Pharmacy. A pharmacy receives an order for 60 grams of a 0.7% hydrocortisone cream. The pharmacy has 1% and 0.5% hydrocortisone creams as well as a Eucerin cream for use as a base (0% hydrocortisone). The technician must use twice as much 0.5% hydrocortisone cream than the Eucerin base. How much of the 1% and 0.5% hydrocortisone creams and Eucerin cream are needed to fill this order?

97. Business. A small company has an assembly line that produces three types of widgets. The basic widget is sold for $12 per unit, the midprice widget for $15 per unit, and the top-of-the-line widget for $18 per unit. The assembly line has a daily capacity of producing 375 widgets that may be sold for a total of $5250. Find the quantity of each type of widget produced on a day when twice as many basic widgets as midprice widgets are produced.

98. Business. A small company has an assembly line that produces three types of widgets. The basic widget is sold for $10 per unit, the midprice widget for $12 per unit, and the top-of-the-line widget for $15 per unit. The assembly line has a daily capacity of producing 350 widgets that may be sold for a total of $4600. Find the quantity of each type of widget produced on a day when twice as many top-of-the-line widgets as basic widgets are produced.

99. Money. Gary and Ginger decide to place $10,000 of their savings into investments. They put some in a money market account earning 3% interest, some in a mutual fund that has been averaging 7% a year, and some in a stock that rose 10% last year. If they put $3000 more in the money market than in the mutual fund and the mutual fund and stocks have the same growth in the next year as they did in the previous year, they will earn $540 in a year. How much money did they put in each of the three investments?

100. Money. Ginger talks Gary into putting less money in the money market and more money in the stock. They place $10,000 of their savings into investments. They put some in a money market account earning 3% interest, some in a mutual fund that has been averaging 7% a year, and some in a stock that rose 10% last year. If they put $3000 more in the stock than in the mutual fund and the mutual fund and stock have the same growth in the next year as they did in the previous year, they will earn $840 in a year. How much money did they put in each of the three investments?

101. Manufacturing. A company produces three products x, y, and z. Each item of product x requires 20 units of steel, 2 units of plastic, and 1 unit of glass. Each item of product y requires 25 units of steel, 5 units of plastic, and no units of glass. Each item of product z requires 150 units of steel, 10 units of plastic, and 0.5 units of glass. The available amounts of steel, plastic, and glass are 2400, 310, and 28, respectively. How many items of each type can the company produce and utilize all the available raw materials?

102. Geometry. Find the values of a, b, and c such that the graph of the quadratic function $y = ax^2 + bx + c$ passes through the points $(1, 5)$, $(-2, -10)$, and $(0, 4)$.

103. Ticket Sales. One hundred students decide to buy tickets to a football game. There are three types of tickets: general admission, reserved, and end zone. Each general admission ticket costs $20, each reserved ticket costs $40, and each end zone ticket costs $15. The students spend a total of $2375 for all the tickets. There are five more reserved tickets than general admission tickets, and 20 more end zone tickets than general admission tickets. How many of each type of ticket were purchased by the students?

104. Exercise and Nutrition. Ann would like to exercise one hour per day to burn calories and lose weight. She would like to engage in three activities: walking, step-up exercise, and weight training. She knows she can burn 85 calories walking at a certain pace in 15 minutes, 45 calories doing the step-up exercise in 10 minutes, and 137 calories by weight training for 20 minutes. (a) Determine the number of calories per minute she can burn doing each activity. (b) Suppose she has time to exercise for only one hour (60 minutes). She sets a goal of burning 358 calories in one hour and would like to weight train twice as long as walking. How many minutes must she engage in each exercise to burn the required number of calories in one hour?

105. Geometry. The circle given by the equation $x^2 + y^2 + ax + by + c = 0$ passes through the point $(4, 4)$, $(-3, -1)$, and $(1, -3)$. Find a, b, and c.

106. Geometry. The circle given by the equation $x^2 + y^2 + ax + by + c = 0$ passes through the point $(0, 7)$, $(6, 1)$, and $(5, 4)$. Find a, b, and c.

■CATCH THE MISTAKE

In Exercises 107–110, explain the mistake that is made.

107. Solve the system of equations using the augmented matrices.

$$\begin{aligned} y - x + z &= 2 \\ x - 2z + y &= -3 \\ x + y + z &= 6 \end{aligned}$$

Solution:

Step 1: Write as an augmented matrix.
$$\begin{bmatrix} 1 & -1 & 1 & | & 2 \\ 1 & -2 & 1 & | & -3 \\ 1 & 1 & 1 & | & 6 \end{bmatrix}$$

Step 2: Reduce the matrix using Gaussian elimination.
$$\begin{bmatrix} 1 & -1 & 1 & | & 2 \\ 0 & 1 & 0 & | & 5 \\ 0 & 0 & 0 & | & -6 \end{bmatrix}$$

Step 3: Identify the solution. Row 3 is inconsistent, so there is no solution.

This is incorrect. The correct answer is $x = 1$, $y = 2$, $z = 3$. What mistake was made?

108. Perform the indicated row operations on the matrix.
$$\begin{bmatrix} 1 & -1 & 1 & | & 2 \\ 2 & -3 & 1 & | & 4 \\ 3 & 1 & 2 & | & -6 \end{bmatrix}$$

a. $R_2 - 2R_1 \rightarrow R_2$
b. $R_3 - 3R_1 \rightarrow R_3$

Solution:

a. $\begin{bmatrix} 1 & -1 & 1 & | & 2 \\ 0 & -3 & 1 & | & 4 \\ 3 & 1 & 2 & | & -6 \end{bmatrix}$

b. $\begin{bmatrix} 1 & -1 & 1 & | & 2 \\ 2 & -3 & 1 & | & 4 \\ 0 & 1 & 2 & | & -6 \end{bmatrix}$

This is incorrect. What mistake was made?

109. Solve the system of equations using an augmented matrix.

$$\begin{aligned} 3x - 2y + z &= -1 \\ x + y - z &= 3 \\ 2x - y + 3z &= 0 \end{aligned}$$

Solution:

Step 1: Write the system as an augmented matrix.
$$\begin{bmatrix} 3 & -2 & 1 & | & -1 \\ 1 & 1 & -1 & | & 3 \\ 2 & -1 & 3 & | & 0 \end{bmatrix}$$

Step 2: Reduce the matrix using Gaussian elimination.
$$\begin{bmatrix} 1 & 0 & 0 & | & 1 \\ 0 & 1 & 0 & | & 2 \\ 0 & 0 & 1 & | & 0 \end{bmatrix}$$

Step 3: Identify the answer. Row 3 is inconsistent $1 = 0$, therefore there is no solution.

This is incorrect. What mistake was made?

110. Solve the system of equations using an augmented matrix.

$$\begin{aligned} x + 3y + 2z &= 4 \\ 3x + 10y + 9z &= 17 \\ 2x + 7y + 7z &= 17 \end{aligned}$$

Solution:

Step 1: Write the system as an augmented matrix.
$$\begin{bmatrix} 1 & 3 & 2 & | & 4 \\ 3 & 10 & 9 & | & 17 \\ 2 & 7 & 7 & | & 17 \end{bmatrix}$$

Step 2: Reduce the matrix using Gaussian elimination.
$$\begin{bmatrix} 1 & 0 & -7 & | & -11 \\ 0 & 1 & 3 & | & 5 \\ 0 & 0 & 0 & | & 4 \end{bmatrix}$$

Step 3: Identify the answer:
Infinitely many solutions.
$$\begin{aligned} x &= 7t - 11 \\ y &= -3t + 5 \\ z &= t \end{aligned}$$

This is incorrect. What mistake was made?

■ CONCEPTUAL

Determine whether each of the following statements is true or false.

111. A nonsquare matrix cannot have a unique solution.

112. The procedure for Gaussian elimination can be used only for square matrices.

113. A square matrix that has a unique solution has a reduced matrix with 1s along the diagonal and 0s above and below the 1s.

114. A square matrix with an all-zero row has infinitely many solutions.

■ CHALLENGE

115. A fourth-degree polynomial $f(x) = ax^4 + bx^3 + cx^2 + dx + e$, with $a < 0$, can be used to represent the following data on the number of deaths per year due to lightning strikes (assume 1999 corresponds to $x = 0$).

Use the data to determine a, b, c, d, and e.

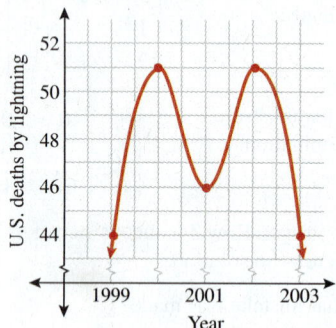

116. A copy machine accepts nickels, dimes, and quarters. After 1 hour, there are 30 coins total, and their value is $4.60. How many nickels, quarters, and dimes are in the machine?

■ TECHNOLOGY

117. In Exercise 57, you were asked to solve this system of equations using an augmented matrix.

$$x - z - y = 10$$
$$2x - 3y + z = -11$$
$$y - x + z = -10$$

A graphing calculator or graphing utility can be used to solve systems of linear equations by entering the coefficients of the matrix. Solve this system and confirm your answer with the calculator's answer.

118. In Exercise 58, you were asked to solve this system of equations using an augmented matrix.

$$2x + z + y = -3$$
$$2y - z + x = 0$$
$$x + y + 2z = 5$$

A graphing calculator or graphing utility can be used to solve systems of linear equations by entering the coefficients of the matrix. Solve this system and confirm your answer with the calculator's answer.

In Exercises 119 and 120, you are asked to model a set of three points with a quadratic function $y = ax^2 + bx + c$ and determine the quadratic function.

a. Set up a system of equations, use a graphing utility or graphing calculator to solve the system by entering the coefficients of the augmented matrix.

b. Use the graphing calculator commands STAT QuadReg to model the data using a quadratic function. Round your answers to two decimal places.

119. $(-6, -8)$, $(2, 7)$, $(7, 1)$

120. $(-9, 20)$, $(2, -18)$, $(11, 16)$

SKILLS OBJECTIVES

- Use equality of matrices.
- Add and subtract matrices.
- Perform scalar multiplication.
- Multiply two matrices.

CONCEPTUAL OBJECTIVES

- Understand what is meant by equal matrices.
- Understand why multiplication of some matrices is undefined.
- Realize that matrix multiplication is *not* commutative.

Equality of Matrices

In Section 10.1, we defined a matrix with m rows and n columns to have order $m \times n$:

$$A = \begin{bmatrix} a_{11} & a_{12} & \cdots & a_{1n} \\ a_{21} & a_{22} & \cdots & a_{2n} \\ \vdots & \vdots & \cdots & \vdots \\ a_{m1} & a_{m2} & \cdots & a_{mn} \end{bmatrix}$$

Capital letters are used to represent (or name) a matrix, and lowercase letters are used to represent the entries (elements) of the matrix. The subscripts are used to denote the location (row/column) of each entry. The order of a matrix is often written as a subscript of the matrix name: $A_{m \times n}$. Other words like "size" and "dimension" are used as synonyms of "order." Matrices are a convenient way to represent data.

Average Marriage Age on the Rise

A survey conducted by the Census Bureau found that the average marriage age in the United States is rising.

	1960	2009
Men	22.8	28.4
Women	20.3	26.5

These data can be represented by the 2×2 matrix $\begin{bmatrix} 22.8 & 28.4 \\ 20.3 & 26.5 \end{bmatrix}$, with two rows (gender) and two columns (year).

U.S. Homeowner Bailout

According to a Gallup poll conducted March 24–27, 2008 of adult Americans were asked the question, "Do you favor or oppose the Federal Government taking steps to help prevent people from losing their homes because they can't pay their mortgages?"

The results were along party lines:

	REPUBLICAN (%)	DEMOCRAT (%)
Favor	40	70
Oppose	58	27
No Opinion	2	3

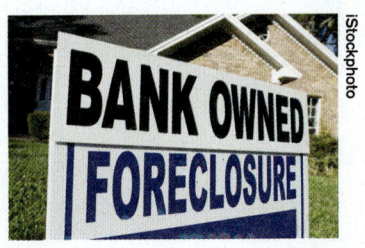

These data can be represented with the 3×2 matrix $\begin{bmatrix} 40 & 70 \\ 58 & 27 \\ 2 & 3 \end{bmatrix}$, with three rows (opinions on bailout) and two columns (registered political party).

There is an entire field of study called **matrix algebra** that treats matrices similarly to functions and variables in traditional algebra. This section serves as an introduction to matrix algebra. We will discuss how to

- determine whether two matrices are equal,
- add and subtract matrices,
- multiply a matrix by a scalar,
- multiply two matrices.

It is important to pay special attention to the *order* of a matrix because it determines whether certain operations are defined.

Two matrices are equal if and only if they have the same order, $m \times n$, and all of their corresponding elements are equal.

DEFINITION **Equality of Matrices**

Two matrices A and B are **equal**, written as $A = B$ if and only if *both* of the following are true:

- A and B have the same order $m \times n$,
- Every pair of corresponding elements is equal: $a_{ij} = b_{ij}$ for all $i = 1, 2, \ldots, m$ and $j = 1, 2, \ldots, n$.

**Classroom Example
10.2.1–10.2.2**
a. Solve for x and y.
$\begin{bmatrix} 1 & x^2 \\ 2y & 0 \end{bmatrix} = \begin{bmatrix} y & 25 \\ 2 & 0 \end{bmatrix}$
b.* Solve for x, y, and z.
$\begin{bmatrix} x+y & 0 & 0 \\ 0 & y+z & 0 \\ 0 & 0 & x+z \end{bmatrix}$
$= \begin{bmatrix} -1 & 0 & 0 \\ 0 & 0 & 0 \\ 0 & 0 & 1 \end{bmatrix}$

Answer:
a. $x = \pm 5, y = 1$
b. $x = 0, y = -1, z = 1$

EXAMPLE 1 **Equality of Matrices**

Referring to the definition of equality of matrices, find the indicated elements.

$$\begin{bmatrix} a_{11} & a_{12} & a_{13} \\ a_{21} & a_{22} & a_{23} \\ a_{31} & a_{32} & a_{33} \end{bmatrix} = \begin{bmatrix} 2 & -7 & 1 \\ 0 & 5 & -3 \\ -1 & 8 & 9 \end{bmatrix}$$

Find the main diagonal elements: a_{11}, a_{22}, and a_{33}.

Solution:

Since the matrices are equal, their corresponding elements are equal.
$a_{11} = 2$ $a_{22} = 5$ $a_{33} = 9$

EXAMPLE 2 **Equality of Matrices**

Given that $\begin{bmatrix} 2x & y-1 \\ 1 & 2 \end{bmatrix} = \begin{bmatrix} 6 & 7 \\ 1 & 2 \end{bmatrix}$, solve for x and y.

Solution:

Equate corresponding elements. $2x = 6$ $y - 1 = 7$

Solve for x and y. $x = 3$ $y = 8$

Matrix Addition and Subtraction

Two matrices A and B can be added or subtracted only if they have the *same order*. Suppose A and B are both of order $m \times n$; then the *sum* $A + B$ is found by adding corresponding elements, or taking $a_{ij} + b_{ij}$. The *difference* $A - B$ is found by subtracting the elements in B from the corresponding elements in A, or finding $a_{ij} - b_{ij}$.

> **DEFINITION**   **Matrix Addition and Matrix Subtraction**
>
> If A is an $m \times n$ matrix and B is an $m \times n$ matrix, then their **sum** $A + B$ is an $m \times n$ matrix whose elements are given by
>
> $$a_{ij} + b_{ij}$$
>
> and their **difference** $A - B$ is an $m \times n$ matrix whose elements are given by
>
> $$a_{ij} - b_{ij}$$

EXAMPLE 3   **Adding and Subtracting Matrices**

Given that $A = \begin{bmatrix} -1 & 3 & 4 \\ -5 & 2 & 0 \end{bmatrix}$ and $B = \begin{bmatrix} 2 & 1 & -3 \\ 0 & -5 & 4 \end{bmatrix}$, find the following:

a. $A + B$ **b.** $A - B$

Solution:

Since $A_{2 \times 3}$ and $B_{2 \times 3}$ have the same order, they can be added or subtracted.

a. Write the sum.

$$A + B = \begin{bmatrix} -1 & 3 & 4 \\ -5 & 2 & 0 \end{bmatrix} + \begin{bmatrix} 2 & 1 & -3 \\ 0 & -5 & 4 \end{bmatrix}$$

Add the corresponding elements.

$$= \begin{bmatrix} -1+2 & 3+1 & 4+(-3) \\ -5+0 & 2+(-5) & 0+4 \end{bmatrix}$$

Simplify.

$$= \begin{bmatrix} 1 & 4 & 1 \\ -5 & -3 & 4 \end{bmatrix}$$

b. Write the difference.

$$A - B = \begin{bmatrix} -1 & 3 & 4 \\ -5 & 2 & 0 \end{bmatrix} - \begin{bmatrix} 2 & 1 & -3 \\ 0 & -5 & 4 \end{bmatrix}$$

Subtract the corresponding elements.

$$= \begin{bmatrix} -1-2 & 3-1 & 4-(-3) \\ -5-0 & 2-(-5) & 0-4 \end{bmatrix}$$

Simplify.

$$= \begin{bmatrix} -3 & 2 & 7 \\ -5 & 7 & -4 \end{bmatrix}$$

■ **YOUR TURN** Perform the indicated matrix operations, if possible.

$$A = \begin{bmatrix} -4 & 0 \\ 1 & 2 \end{bmatrix} \quad B = \begin{bmatrix} 2 & 3 \\ -4 & 0 \end{bmatrix} \quad C = [2 \quad 9 \quad 5 \quad -1] \quad D = \begin{bmatrix} 0 \\ -3 \\ 4 \\ 2 \end{bmatrix}$$

     a. $B - A$ **b.** $C + D$ **c.** $A + B$ **d.** $A + D$

Technology Tip

Enter the matrices as A and B.

```
[A]
        [[-1  3  4]
         [-5  2  0]]
[B]
        [[2  1  -3]
         [0 -5  4 ]]
```

Now enter $[A] + [B]$ and $[A] - [B]$.

```
[A]+[B]
        [[1   4   1]
         [-5  -3  4]]
[A]-[B]
        [[-3  2   7]
         [-5  7  -4]]
```

■ **Answer:**

a. $\begin{bmatrix} 6 & 3 \\ -5 & -2 \end{bmatrix}$ **b.** not defined

c. $\begin{bmatrix} -2 & 3 \\ -3 & 2 \end{bmatrix}$ **d.** not defined

Study Tip

Only matrices of the same order can be added or subtracted.

Classroom Example 10.2.3*
Explain why if A and B are any 3×2 matrices, then $A + B = B + A$.

Answer: Since addition in $\mathbb{R}$ is commutative, we can apply this property component-wise to show $A + B = B + A$.

Classroom Example 10.2.4
Let a and b be real numbers and define the matrices.

$A = \begin{bmatrix} a & -b \\ -2b & -a \\ 0 & a \end{bmatrix}$

$B = \begin{bmatrix} -a & b \\ 0 & b + a \\ 2b & 0 \end{bmatrix}$

$C = \begin{bmatrix} 0 & 0 \\ 0 & -a \\ 0 & -b \end{bmatrix}$

Perform the operations.
a. $-\frac{1}{2}A$
b.* $-B - 2(A + 2C)$
c.* $A - B - C$

Answer:

a. $\begin{bmatrix} -\dfrac{a}{2} & \dfrac{b}{2} \\ b & \dfrac{a}{2} \\ 0 & -\dfrac{a}{2} \end{bmatrix}$

b. $\begin{bmatrix} -a & b \\ 4b & 5a - b \\ -2b & 4b - 2a \end{bmatrix}$

c. $\begin{bmatrix} 2a & -2b \\ -2b & -a - b \\ -2b & a + b \end{bmatrix}$

It is important to note that only matrices of the same order can be added or subtracted. For example, if $A = \begin{bmatrix} -1 & 3 & 4 \\ -5 & 2 & 0 \end{bmatrix}$ and $B = \begin{bmatrix} 5 & -3 \\ 12 & 1 \end{bmatrix}$, the sum and difference of these matrices is undefined because $A_{2 \times 3}$ and $B_{2 \times 2}$ do not have the same order.

A matrix whose elements are all equal to 0 is called a **zero matrix**. The following are examples of zero matrices:

2×2 square zero matrix $\qquad \begin{bmatrix} 0 & 0 \\ 0 & 0 \end{bmatrix}$

3×2 zero matrix $\qquad \begin{bmatrix} 0 & 0 \\ 0 & 0 \\ 0 & 0 \end{bmatrix}$

1×4 zero matrix $\qquad \begin{bmatrix} 0 & 0 & 0 & 0 \end{bmatrix}$

If A, an $m \times n$ matrix, is added to the $m \times n$ zero matrix, the result is A.

$$A + 0 = A$$

For example,

$$\begin{bmatrix} 1 & -3 \\ 2 & 5 \end{bmatrix} + \begin{bmatrix} 0 & 0 \\ 0 & 0 \end{bmatrix} = \begin{bmatrix} 1 & -3 \\ 2 & 5 \end{bmatrix}$$

Because of this result, an $m \times n$ zero matrix is called the **additive identity** for $m \times n$ matrices. Similarly, for any matrix A, there exists an **additive inverse** $-A$ such that every element of $-A$ is the negative of every element of A.

For example, $A = \begin{bmatrix} 1 & -3 \\ 2 & 5 \end{bmatrix}$ and $-A = \begin{bmatrix} -1 & 3 \\ -2 & -5 \end{bmatrix}$, and adding these two matrices results in a zero matrix: $A + (-A) = 0$.

The same properties that hold for adding real numbers also hold for adding matrices, provided that addition of matrices is defined.

PROPERTIES OF MATRIX ADDITION

If A, B, and C are all $m \times n$ matrices and 0 is the $m \times n$ zero matrix, then the following are true:

Commutative property:	$A + B = B + A$
Associative property:	$(A + B) + C = A + (B + C)$
Additive identity property:	$A + 0 = A$
Additive inverse property:	$A + (-A) = 0$

Scalar and Matrix Multiplication

There are two types of multiplication involving matrices: *scalar multiplication* and *matrix multiplication*. A **scalar** is any real number. *Scalar multiplication* is the multiplication of a matrix by a scalar, or real number, and is defined for all matrices. *Matrix multiplication* is the multiplication of two matrices and is defined only for certain pairs of matrices, depending on the order of each matrix.

Scalar Multiplication

To multiply a matrix A by a scalar k, multiply every element in A by k.

$$3 \begin{bmatrix} -1 & 0 & 4 \\ 7 & 5 & -2 \end{bmatrix} = \begin{bmatrix} 3(-1) & 3(0) & 3(4) \\ 3(7) & 3(5) & 3(-2) \end{bmatrix} = \begin{bmatrix} -3 & 0 & 12 \\ 21 & 15 & -6 \end{bmatrix}$$

Here, the scalar is $k = 3$.

DEFINITION   Scalar Multiplication

If A is an $m \times n$ matrix and k is any real number, then their product kA is an $m \times n$ matrix whose elements are given by

$$ka_{ij}$$

In other words, every element a_{ij} that is in the ith row and jth column of A is multiplied by k.

In general, uppercase letters are used to denote a matrix, and lowercase letters are used to denote scalars. Notice that the elements of each matrix are also represented with lowercase letters, since they are real numbers.

EXAMPLE 4   Multiplying a Matrix by a Scalar

Given that $A = \begin{bmatrix} -1 & 2 \\ -3 & 4 \end{bmatrix}$ and $B = \begin{bmatrix} 0 & 1 \\ -2 & 3 \end{bmatrix}$, perform the following operations.

a. $2A$ **b.** $-3B$ **c.** $2A - 3B$

Solution (a):

Write the scalar multiplication.

$$2A = 2\begin{bmatrix} -1 & 2 \\ -3 & 4 \end{bmatrix}$$

Multiply all elements of A by 2.

$$2A = \begin{bmatrix} 2(-1) & 2(2) \\ 2(-3) & 2(4) \end{bmatrix}$$

Simplify.

$$2A = \begin{bmatrix} -2 & 4 \\ -6 & 8 \end{bmatrix}$$

Solution (b):

Write the scalar multiplication.

$$-3B = -3\begin{bmatrix} 0 & 1 \\ -2 & 3 \end{bmatrix}$$

Multiply all elements of B by -3.

$$-3B = \begin{bmatrix} -3(0) & -3(1) \\ -3(-2) & -3(3) \end{bmatrix}$$

Simplify.

$$-3B = \begin{bmatrix} 0 & -3 \\ 6 & -9 \end{bmatrix}$$

Solution (c):

Add the results of parts (a) and (b). $2A - 3B = 2A + (-3B)$

$$2A - 3B = \begin{bmatrix} -2 & 4 \\ -6 & 8 \end{bmatrix} + \begin{bmatrix} 0 & -3 \\ 6 & -9 \end{bmatrix}$$

Add the corresponding elements.

$$2A - 3B = \begin{bmatrix} -2+0 & 4+(-3) \\ -6+6 & 8+(-9) \end{bmatrix}$$

Simplify.

$$2A - 3B = \begin{bmatrix} -2 & 1 \\ 0 & -1 \end{bmatrix}$$

■ **YOUR TURN** For the matrices A and B given in Example 4, find $-5A + 2B$.

Technology Tip

Enter matrices as A and B.

```
[A]
           [[-1 2]
            [-3 4]]
[B]
           [[0  1]
            [-2 3]]
```

Now enter $2A$, $-3B$, $2A - 3B$.

```
2[A]
           [[-2 4]
            [-6 8]]
-3[B]
           [[0 -3]
            [6 -9]]
■
```

```
2[A]-3[B]
           [[-2 1 ]
            [0  -1]]
■
```

■ **Answer:**

$$-5A + 2B = \begin{bmatrix} 5 & -8 \\ 11 & -14 \end{bmatrix}$$

Matrix Multiplication

Scalar multiplication is straightforward in that it is defined for all matrices and is performed by multiplying every element in the matrix by the scalar. Addition of matrices is also an element-by-element operation. *Matrix multiplication*, on the other hand, is not as straightforward in that we *do not multiply the corresponding elements* and it is not defined for all matrices. Matrices are multiplied using a row-by-column method.

Before we even try to find the product AB of two matrices A and B, we first have to determine whether the product is defined. For the product AB to exist, **the number of columns in the first matrix A must equal the number of rows in the second matrix B**. In other words, if the matrix $A_{m \times n}$ has m rows and n columns and the matrix $B_{n \times p}$ has n rows and p columns, then the product $(AB)_{m \times p}$ is defined and has m rows and p columns.

Matrix: A B AB
Order: $m \times n$ $n \times p$ $m \times p$

Equal

Order of AB

EXAMPLE 5   **Determining Whether the Product of Two Matrices Is Defined**

Given the matrices

$$A = \begin{bmatrix} 1 & -2 & 0 \\ 5 & -1 & 3 \end{bmatrix} \quad B = \begin{bmatrix} 2 & 3 \\ 0 & 7 \\ 4 & 9 \end{bmatrix} \quad C = \begin{bmatrix} 6 & -1 \\ 5 & 2 \end{bmatrix} \quad D = \begin{bmatrix} -3 & -2 \end{bmatrix}$$

state whether each of the following products exists. If the product exists, state the order of the product matrix.

a. AB **b.** AC **c.** BC **d.** CD **e.** DC

Solution:

Label the order of each matrix: $A_{2 \times 3}$, $B_{3 \times 2}$, $C_{2 \times 2}$, and $D_{1 \times 2}$.

a. AB is defined, because A has 3 columns and B has 3 rows. $A_{2 \times 3} B_{3 \times 2}$
AB is order $\boxed{2 \times 2}$. $(AB)_{2 \times 2}$

b. AC is $\boxed{\text{not defined}}$, because A has 3 columns and C has 2 rows.

c. BC is defined, because B has 2 columns and C has 2 rows. $B_{3 \times 2} C_{2 \times 2}$
BC is order $\boxed{3 \times 2}$. $(BC)_{3 \times 2}$

d. CD is $\boxed{\text{not defined}}$, because C has 2 columns and D has 1 row.

e. DC is defined, because D has 2 columns and C has 2 rows. $D_{1 \times 2} C_{2 \times 2}$
DC is order $\boxed{1 \times 2}$. $(DC)_{1 \times 2}$

Notice that in part (d) we found that CD is not defined, but in part (e) we found that DC is defined. **Matrix multiplication is not commutative**. Therefore, the order in which matrices are multiplied is important in determining whether the product is defined or undefined. For the product of two matrices to exist, the number of *columns* in the *first* matrix A must equal the number of *rows* in the *second* matrix B.

■ **YOUR TURN**   For the matrices given in Example 5, state whether the following products exist. If the product exists, state the order of the product matrix.

a. DA **b.** CB **c.** BA

Now that we can determine whether a product of two matrices is defined and if so, what the order of the resulting product is, let us turn our attention to how to multiply two matrices.

DEFINITION | **Matrix Multiplication**

If A is an $m \times n$ matrix and B is an $n \times p$ matrix, then their product AB is an $m \times p$ matrix whose elements are given by

$$(ab)_{ij} = a_{i1}b_{1j} + a_{i2}b_{2j} + \cdots + a_{in}b_{nj}$$

In other words, the element $(ab)_{ij}$, which is in the ith row and jth column of AB, is the sum of the products of the corresponding elements in the ith row of A and the jth column of B.

$$(ab)_{ij} = a_{i1}b_{1j} + a_{i2}b_{2j} + a_{i3}b_{3j}$$

EXAMPLE 6 **Multiplication of Two 2 × 2 Matrices**

Given $A = \begin{bmatrix} 1 & 2 \\ 3 & 4 \end{bmatrix}$ and $B = \begin{bmatrix} 5 & 6 \\ 7 & 8 \end{bmatrix}$, find AB.

COMMON MISTAKE

Do not multiply element by element.

✪ **CORRECT**

Write the product of the two matrices A and B.

$$AB = \begin{bmatrix} 1 & 2 \\ 3 & 4 \end{bmatrix}\begin{bmatrix} 5 & 6 \\ 7 & 8 \end{bmatrix}$$

Perform the row-by-column multiplication.

$$AB = \begin{bmatrix} (1)(5) + (2)(7) & (1)(6) + (2)(8) \\ (3)(5) + (4)(7) & (3)(6) + (4)(8) \end{bmatrix}$$

Simplify.

$$AB = \begin{bmatrix} 19 & 22 \\ 43 & 50 \end{bmatrix}$$

✖ **INCORRECT**

Multiply the corresponding elements.

ERROR

$$AB \neq \begin{bmatrix} (1)(5) & (2)(6) \\ (3)(7) & (4)(8) \end{bmatrix}$$

▪ **YOUR TURN** For matrices A and B given in Example 6, find BA.

Classroom Example 10.2.6–10.2.7

Let a and b be real numbers and define the matrices.

$$A = \begin{bmatrix} -a & b \\ -b & a \end{bmatrix} \quad \begin{aligned} B &= -A \\ C &= 2A + B \end{aligned}$$

$$D = \begin{bmatrix} 0 & 1 & 0 \\ 1 & 1 & 0 \end{bmatrix} \quad E = \begin{bmatrix} 1 & 0 \\ 1 & 1 \\ 0 & -1 \end{bmatrix}$$

Compute the following:
a.* AB **b.*** CB **c.** $(-2D)E$
d.* EC **e.** $(-B)(4D)$

Answer:

a. $\begin{bmatrix} b^2 - a^2 & 0 \\ 0 & b^2 - a^2 \end{bmatrix}$

b. $\begin{bmatrix} b^2 - a^2 & 0 \\ 0 & b^2 - a^2 \end{bmatrix}$

c. $\begin{bmatrix} -2 & -2 \\ -4 & -2 \end{bmatrix}$

d. $\begin{bmatrix} -a & b \\ -a - b & b + a \\ b & -a \end{bmatrix}$

e. $\begin{bmatrix} 4b & 4(b - a) & 0 \\ 4a & 4(a - b) & 0 \end{bmatrix}$

Technology Tip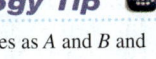

Enter the matrices as A and B and calculate AB.

```
[A]
          [[1 2]
           [3 4]]
[B]
          [[5 6]
           [7 8]]
■
```

```
[A][B]
          [[19 22]
           [43 50]]
```

▪ **Answer:**

$$BA = \begin{bmatrix} 5 & 6 \\ 7 & 8 \end{bmatrix}\begin{bmatrix} 1 & 2 \\ 3 & 4 \end{bmatrix}$$

$$= \begin{bmatrix} 23 & 34 \\ 31 & 46 \end{bmatrix}$$

Compare the products obtained in Example 6 and the preceding Your Turn. Note that $AB \neq BA$. Therefore, there is **no commutative property for matrix multiplication**.

Technology Tip

Enter the matrices as A and B and calculate AB.

```
[A]
      [[-1 2 -3]
       [-2 0 4 ]]
[B]
      [[2  0 ]
       [1  3 ]
       [-1 -2]]
■
```

```
[A][B]
      [[3   12]
       [-8  -8]]
```

■ **Answer:** $AB = \begin{bmatrix} 0 & -5 \\ -1 & -7 \end{bmatrix}$

EXAMPLE 7   Multiplying Matrices

For $A = \begin{bmatrix} -1 & 2 & -3 \\ -2 & 0 & 4 \end{bmatrix}$ and $B = \begin{bmatrix} 2 & 0 \\ 1 & 3 \\ -1 & -2 \end{bmatrix}$, find AB.

Solution:

Since A is order 2×3 and B is order 3×2, the product AB is defined and has order 2×2.

$$A_{2\times3}B_{3\times2} = (AB)_{2\times2}$$

Write the product of the two matrices.

$$AB = \begin{bmatrix} -1 & 2 & -3 \\ -2 & 0 & 4 \end{bmatrix}\begin{bmatrix} 2 & 0 \\ 1 & 3 \\ -1 & -2 \end{bmatrix}$$

Perform the row-by-column multiplication.

$$AB = \begin{bmatrix} (-1)(2) + (2)(1) + (-3)(-1) & (-1)(0) + (2)(3) + (-3)(-2) \\ (-2)(2) + (0)(1) + (4)(-1) & (-2)(0) + (0)(3) + (4)(-2) \end{bmatrix}$$

Simplify.

$$AB = \begin{bmatrix} 3 & 12 \\ -8 & -8 \end{bmatrix}$$

■ **YOUR TURN** For $A = \begin{bmatrix} 1 & 0 & 2 \\ -3 & -1 & 4 \end{bmatrix}$ and $B = \begin{bmatrix} 0 & -1 \\ 1 & 2 \\ 0 & -2 \end{bmatrix}$, find AB.

EXAMPLE 8   Multiplying Matrices

For $A = \begin{bmatrix} 1 & 0 & 3 \\ -2 & 5 & -1 \end{bmatrix}$ and $B = \begin{bmatrix} -2 & 0 & 1 \\ -3 & -1 & 4 \\ 0 & 2 & 5 \end{bmatrix}$, find AB.

Solution:

Since A is order 2×3 and B is order 3×3, the product AB is defined and has order 2×3.

$$A_{2\times3}B_{3\times3} = (AB)_{2\times3}$$

Write the product of the two matrices.

$$AB = \begin{bmatrix} 1 & 0 & 3 \\ -2 & 5 & -1 \end{bmatrix}\begin{bmatrix} -2 & 0 & 1 \\ -3 & -1 & 4 \\ 0 & 2 & 5 \end{bmatrix}$$

Perform the row-by-column multiplication.

$$AB = \begin{bmatrix} (1)(-2) + (0)(-3) + (3)(0) & (1)(0) + (0)(-1) + (3)(2) & (1)(1) + (0)(4) + (3)(5) \\ (-2)(-2) + (5)(-3) + (-1)(0) & (-2)(0) + (5)(-1) + (-1)(2) & (-2)(1) + (5)(4) + (-1)(5) \end{bmatrix}$$

Simplify.

$$AB = \begin{bmatrix} -2 & 6 & 16 \\ -11 & -7 & 13 \end{bmatrix}$$

■ **YOUR TURN** Given $A = \begin{bmatrix} 1 \\ 2 \\ 3 \end{bmatrix}$ and $B = [4 \quad 5]$:

a. Find AB, if it exists.   b. Find BA, if it exists.

■ **Answer: a.** $AB = \begin{bmatrix} 4 & 5 \\ 8 & 10 \\ 12 & 15 \end{bmatrix}$

   b. does not exist

Although we have shown repeatedly that there is no commutative property for matrices, matrices do have an associative property of multiplication, as well as a distributive property of multiplication similar to real numbers.

PROPERTIES OF MATRIX MULTIPLICATION

If A, B, and C are all matrices for which AB, AC, and BC are all defined, then the following properties are true:

Associative property: $A(BC) = (AB)C$

Distributive property: $A(B + C) = AB + AC$ or $(A + B)C = AC + BC$

EXAMPLE 9   Application of Matrix Multiplication

The following table gives fuel and electric requirements per mile associated with gasoline and electric automobiles.

	NUMBER OF GALLONS/MILE	NUMBER OF kW-hr/MILE
Gas Car	0.05	0
Hybrid Car	0.02	0.1
Electric Car	0	0.25

The following table gives an average cost for gasoline and electricity.

Cost per gallon of gasoline	$3.00
Cost per kW-hr of electricity	$0.05

a. Let matrix A represent the gasoline and electricity consumption and matrix B represent the costs of gasoline and electricity.

b. Find AB and describe what the elements of the product matrix represent.

c. Assume you drive 12,000 miles per year. What are the yearly costs associated with driving the three types of cars?

Solution (a):

A has order 3×2.

$$A = \begin{bmatrix} 0.05 & 0 \\ 0.02 & 0.1 \\ 0 & 0.25 \end{bmatrix}$$

B has order 2×1.

$$B = \begin{bmatrix} \$3.00 \\ \$0.05 \end{bmatrix}$$

Solution (b):

Find the order of the product matrix AB.

$$A_{3\times2}B_{2\times1} = (AB)_{3\times1}$$

$$AB = \begin{bmatrix} 0.05 & 0 \\ 0.02 & 0.1 \\ 0 & 0.25 \end{bmatrix} \begin{bmatrix} \$3.00 \\ \$0.05 \end{bmatrix}$$

Calculate AB.

$$= \begin{bmatrix} (0.05)(\$3.00) + (0)(\$0.05) \\ (0.02)(\$3.00) + (0.1)(\$0.05) \\ (0)(\$3.00) + (0.25)(\$0.05) \end{bmatrix}$$

$$AB = \begin{bmatrix} \$0.15 \\ \$0.065 \\ \$0.0125 \end{bmatrix}$$

Technology Tip

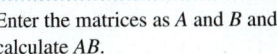

Enter the matrices as A and B and calculate AB.

```
[A]
        [[.05 0  ]
         [.02 .1 ]
         [0   .25]]
[B]
          [[3   ]
           [.05]]
```

```
[A][B]
          [[.15  ]
           [.065 ]
           [.0125]]
```

Interpret the product matrix.

$$AB = \begin{bmatrix} \text{Cost per mile to drive the gas car} \\ \text{Cost per mile to drive the hybrid car} \\ \text{Cost per mile to drive the electric car} \end{bmatrix}$$

Solution (c):

Find 12,000*AB*.

$$12,000 \begin{bmatrix} \$0.15 \\ \$0.065 \\ \$0.0125 \end{bmatrix} = \begin{bmatrix} \$1800 \\ \$780 \\ \$150 \end{bmatrix}$$

GAS/ELECTRIC COSTS PER YEAR ($)	
Gas Car	1800
Hybrid Car	780
Electric Car	150

SECTION
10.2 SUMMARY

Matrices can be used to represent data. Operations such as equality, addition, subtraction, and scalar multiplication are performed element by element. Two matrices can be added or subtracted only if they have the same order. Matrix multiplication, however, requires that the number of columns in the first matrix is equal to the number of rows in the second matrix and is performed using a row by column procedure.

Matrix Multiplication Is Not Commutative: $AB \neq BA$

OPERATION	ORDER REQUIREMENT
Equality	Same: $A_{m \times n} = B_{m \times n}$
Addition	Same: $A_{m \times n} + B_{m \times n}$
Subtraction	Same: $A_{m \times n} - B_{m \times n}$
Scalar multiplication	None: $kA_{m \times n}$
Matrix multiplication	$A_{m \times n} B_{n \times p} = (AB)_{m \times p}$

SECTION
10.2 EXERCISES

■ SKILLS

In Exercises 1–10, state the order of each matrix.

1. $\begin{bmatrix} -1 & 2 & 4 \\ 7 & -3 & 9 \end{bmatrix}$

2. $\begin{bmatrix} 3 & 5 \\ 2 & 6 \\ -1 & -4 \end{bmatrix}$

3. $\begin{bmatrix} -4 & 5 \\ 0 & 1 \end{bmatrix}$

4. $\begin{bmatrix} -4 & 5 & 3 & 7 \end{bmatrix}$

5. $\begin{bmatrix} -3 & 4 & 1 \\ 10 & 8 & 0 \\ -2 & 5 & 7 \end{bmatrix}$

6. $\begin{bmatrix} 1 \\ 2 \\ 3 \\ 4 \end{bmatrix}$

7. $\begin{bmatrix} -2 \end{bmatrix}$

8. $\begin{bmatrix} -1 & 3 & 6 & 9 \\ 2 & 5 & -7 & 8 \end{bmatrix}$

9. $\begin{bmatrix} -3 & 6 & 0 & 5 \\ 4 & -9 & 2 & 7 \\ 1 & 8 & 3 & 6 \\ 5 & 0 & -4 & 11 \end{bmatrix}$

10. $\begin{bmatrix} 3 & 7 & -1 & 5 & 8 \end{bmatrix}$

In Exercises 11–16, solve for the indicated variable.

11. $\begin{bmatrix} 2 & x \\ y & 3 \end{bmatrix} = \begin{bmatrix} 2 & -5 \\ 1 & 3 \end{bmatrix}$

12. $\begin{bmatrix} -3 & 17 \\ x & y \end{bmatrix} = \begin{bmatrix} -3 & 17 \\ 10 & 12 \end{bmatrix}$

13. $\begin{bmatrix} x+y & 3 \\ x-y & 9 \end{bmatrix} = \begin{bmatrix} -5 & z \\ -1 & 9 \end{bmatrix}$

14. $\begin{bmatrix} x & -4 \\ y & 7 \end{bmatrix} = \begin{bmatrix} 2+y & -4 \\ 5 & 7 \end{bmatrix}$

15. $\begin{bmatrix} 3 & 4 \\ 0 & 12 \end{bmatrix} = \begin{bmatrix} x-y & 4 \\ 0 & 2y+x \end{bmatrix}$

16. $\begin{bmatrix} 9 & 2b+1 \\ -5 & 16 \end{bmatrix} = \begin{bmatrix} a^2 & 9 \\ 2a+1 & b^2 \end{bmatrix}$

In Exercises 17–30, perform the indicated operations for each expression, if possible.

$$A = \begin{bmatrix} -1 & 3 & 0 \\ 2 & 4 & 1 \end{bmatrix} \quad B = \begin{bmatrix} 0 & 2 & 1 \\ 3 & -2 & 4 \end{bmatrix} \quad C = \begin{bmatrix} 0 & 1 \\ 2 & -1 \\ 3 & 1 \end{bmatrix} \quad D = \begin{bmatrix} 2 & -3 \\ 0 & 1 \\ 4 & -2 \end{bmatrix}$$

17. $A + B$

18. $C + D$

19. $C - D$

20. $A - B$

21. $B + C$

22. $A + D$

23. $D - B$

24. $C - A$

25. $2A$

26. $4D$

27. $-5C$

28. $-2B$

29. $2A + 3B$

30. $2B - 3A$

In Exercises 31–50, perform the indicated operations for each expression, if possible.

$$A = \begin{bmatrix} 1 & 2 & -1 \\ 0 & 3 & 1 \\ 5 & 0 & -2 \end{bmatrix} \quad B = \begin{bmatrix} 2 & 0 & -3 \end{bmatrix} \quad C = \begin{bmatrix} -1 & 7 & 2 \\ 3 & 0 & 1 \end{bmatrix} \quad D = \begin{bmatrix} 3 & 0 \\ 1 & -1 \\ 2 & 5 \end{bmatrix}$$

$$E = \begin{bmatrix} -1 & 0 & 1 \\ 2 & 1 & 4 \\ -3 & 1 & 5 \end{bmatrix} \quad F = \begin{bmatrix} 1 \\ 0 \\ -1 \end{bmatrix} \quad G = \begin{bmatrix} 1 & 2 \\ 3 & 4 \end{bmatrix}$$

31. CD

32. BF

33. DC

34. $(A + E)D$

35. DG

36. $2A + 3E$

37. GD

38. $ED + C$

39. $-4BD$

40. $-3ED$

41. $B(A + E)$

42. $GC + 5C$

43. $FB + 5A$

44. A^2

45. $G^2 + 5G$

46. $C \cdot (2E)$

47. $(2E) \cdot F$

48. $CA + 5C$

49. DF

50. AE

▪ APPLICATIONS

51. Smoking. On January 6 and 10, 2000, the Harris Poll conducted a survey of adult smokers in the United States. When asked, "Have you ever tried to quit smoking?" 70% said yes and 30% said no. Write a 2×1 matrix—call it A—that represents those smokers. When asked what consequences smoking would have on their lives, 89% believed it would increase their chance of getting lung cancer and 84% believed smoking would shorten their lives. Write a 2×1 matrix—call it B—that represents those smokers. If there are 46 million adult smokers in the United States:

a. What does $46A$ tell us?
b. What does $46B$ tell us?

52. Women in Science. According to the study of science and engineering indicators by the National Science Foundation (www.nsf.gov), the number of female graduate students in science and engineering disciplines has increased over the last 30 years. In 1981, 24% of mathematics graduate students were female and 23% of graduate students in computer

science were female. In 1991, 32% of mathematics graduate students and 21% of computer science graduate students were female. In 2001, 38% of mathematics graduate students and 30% of computer science graduate students were female. Write three 2×1 matrices representing the percentage of female graduate students.

$$A = \begin{bmatrix} \% \text{ female–math–1981} \\ \% \text{ female–C.S.–1981} \end{bmatrix}$$

$$B = \begin{bmatrix} \% \text{ female–math–1991} \\ \% \text{ female–C.S.–1991} \end{bmatrix}$$

$$C = \begin{bmatrix} \% \text{ female–math–2001} \\ \% \text{ female–C.S.–2001} \end{bmatrix}$$

What does $C - B$ tell us? What does $B - A$ tell us? What can you conclude about the number of women pursuing mathematics and computer science graduate degrees?

Note: C.S. = computer science.

53. Registered Voters. According to the U.S. Census Bureau (www.census.gov), in the 2000 national election, 58.9% of the men over the age of 18 were registered voters, but only 41.4% voted; and 62.8% of women over 18 were registered voters, but only 43% actually voted. Write a 2 × 2 matrix with the following data:

$$A = \begin{bmatrix} \text{Percentage of registered} & \text{Percentage of registered} \\ \text{male voters} & \text{female voters} \\ \text{Percent of males} & \text{Percent of females} \\ \text{that voted} & \text{that voted} \end{bmatrix}$$

If we let B be a 2 × 1 matrix representing the total population of males and females over the age of 18 in the United States, or $B = \begin{bmatrix} 100 \text{ M} \\ 110 \text{ M} \end{bmatrix}$, what does AB tell us?

54. Job Application. A company has two rubrics for scoring job applicants based on weighting education, experience, and the interview differently.

Matrix A

	Rubric 1	Rubric 2
Education	0.5	0.6
Experience	0.3	0.1
Interview	0.2	0.3

Applicants receive a score from 1 to 10 in each category (education, experience, and interview). Two applicants are shown in matrix B:

Matrix B

	Education	Experience	Interview
Applicant 1	8	7	5
Applicant 2	6	8	8

What is the order of BA? What does each entry in BA tell us?

55. Taxes. The IRS allows an individual to deduct business expenses in the following way: $0.45 per mile driven, 50% of entertainment costs, and 100% of actual expenses. Represent these deductions as a row matrix A. In 2006, Jamie had the following business expenses: $2700 in entertainment, $15,200 actual expenses, and he drove 7523 miles. Represent Jamie's expenses as a column matrix B. Multiply these matrices to find the total amount of business expenses Jamie can claim on his 2006 federal tax form: AB.

56. Tips on Service. Marilyn decides to go to the Safety Harbor Spa for a day of pampering. She is treated to a Hot Stone Massage ($85), a manicure and pedicure ($75), and a haircut and style ($100). Represent the costs of the individual services as a row matrix A. She decides to tip her masseur 25%, her nail tech 20%, and her hair stylist 15%. Represent the tipping percentages as a column matrix B. Multiply these matrices to find the total amount in tips AB she needs to add to her final bill.

Use the following nutritional chart for Exercises 57 and 58:

SANDWICH	CALORIES	FAT (g)	CARBOHYDRATES (g)	PROTEIN (g)
6" Veggie Delight	230	3	44	9
6" Tuna	430	19	46	20
6" Roast beef	290	5	45	19
6" Chicken	330	5	47	24

57. Nutrition. Utilize the table to write a 4 × 4 matrix A. Find $2A$. What would $2A$ represent? Find $0.5A$. What would $0.5A$ represent?

58. Nutrition. Don decides to film a documentary similar to *Supersize Me*, but instead of eating only at McDonalds, he decides to eat only at Subway. He is hoping to lose weight and name the film *Minimize Me*. He is going to eat only three meals a day from the four options listed in the preceding table. Each week he will consume the following sandwiches: 7 veggie, 5 tuna, 8 roast beef, and 1 chicken. Let A be the matrix (found in Exercise 57) that represents the nutrition values and let B be a row matrix that represents the number of each type of sandwich consumed in a week. Find BA. What does BA represent? Find $\frac{1}{7}BA$. What does $\frac{1}{7}BA$ represent?

Use the following tables for Exercises 59 and 60:

The following table gives fuel and electric requirements per mile associated with gasoline and electric automobiles.

	NUMBER OF GALLONS/MILE	NUMBER OF kW-hr/MILE
SUV full size	0.06	0
Hybrid car	0.02	0.1
Electric car	0	0.3

The following table gives an average cost for gasoline and electricity.

Cost per gallon of gasoline	$3.80
Cost per kW-hr of electricity	$0.05

59. Environment. Let matrix A represent the gasoline and electricity consumption and matrix B represent the costs of gasoline and electricity. Find AB and describe what the elements of the product matrix represent.

60. Environment. Assume you drive 12,000 miles per year. What are the yearly costs associated with driving the three types of cars in Exercise 59?

For Exercises 61 and 62, refer to the following:

The results of a nutritional analysis of one serving of three foods A, B, and C were

$$X = \begin{matrix} \text{Carbohydrates (g)} & \text{Protein (g)} & \text{Fat (g)} \\ \begin{bmatrix} 5 & 0 & 2 \\ 5 & 6 & 5 \\ 8 & 4 & 4 \end{bmatrix} & \begin{matrix} A \\ B \\ C \end{matrix} \end{matrix}$$

It is possible to find the nutritional content of a meal consisting of a combination of the foods A, B, and C by multiplying the matrix X by a second matrix $N = \begin{bmatrix} r \\ s \\ t \end{bmatrix}$, that is, XN, where r is the number of servings of food A, s is the number of servings of food B, and t is the number of servings of food C.

61. Health/Nutrition. Find the matrix N that represents a meal consisting of two servings of food A and one serving of food B. Find the nutritional content of that meal.

62. Health/Nutrition. Find the matrix N that represents a meal consisting of one serving of food A and two servings of food C. Find the nutritional content of that meal.

For Exercises 63 and 64, refer to the following:

Cell phone companies charge users based on the number of minutes talked, the number of text messages sent, and the number of megabytes of data used. The costs for three cell phone providers are given in the following table:

	MINUTES	TEXT MESSAGES	MEGABYTE OF DATA
C_1	$0.04	$0.05	$0.15
C_2	$0.06	$0.05	$0.18
C_3	$0.07	$0.07	$0.13

It is possible to find the cost to a cell phone user for each of the three providers by creating a matrix X whose rows are the rows of data in the table and multiplying the matrix X by a second matrix

$N = \begin{bmatrix} m \\ t \\ d \end{bmatrix}$, that is, XN, where m is the number of minutes talked, t is the number of text messages sent, and d is the megabytes of data used.

63. Telecommunications/Business. A local business is looking at providing an employee a cell phone for business use. Find the matrix N that represents expected normal cell phone usage of 200 minutes, 25 text messages, and no data usage. Find and interpret XN. Which is the better cell phone provider for this employee?

64. Telecommunications/Business. A local business is looking at providing an employee a cell phone for business use. Find the matrix N that represents expected normal cell phone usage of 125 minutes, 125 text messages, and 320 megabytes of data usage. Find and interpret XN. Which is the better cell phone provider for this employee?

■**CATCH THE MISTAKE**

In Exercises 65 and 66, explain the mistake that is made.

65. Multiply $\begin{bmatrix} 3 & 2 \\ 1 & 4 \end{bmatrix}\begin{bmatrix} -1 & 3 \\ -2 & 5 \end{bmatrix}$.

Solution:

Multiply corresponding elements.

$$\begin{bmatrix} 3 & 2 \\ 1 & 4 \end{bmatrix}\begin{bmatrix} -1 & 3 \\ -2 & 5 \end{bmatrix} = \begin{bmatrix} (3)(-1) & (2)(3) \\ (1)(-2) & (4)(5) \end{bmatrix}$$

Simplify.

$$\begin{bmatrix} 3 & 2 \\ 1 & 4 \end{bmatrix}\begin{bmatrix} -1 & 3 \\ -2 & 5 \end{bmatrix} = \begin{bmatrix} -3 & 6 \\ -2 & 20 \end{bmatrix}$$

This is incorrect. What mistake was made?

66. Multiply $\begin{bmatrix} 3 & 2 \\ 1 & 4 \end{bmatrix}\begin{bmatrix} -1 & 3 \\ -2 & 5 \end{bmatrix}$.

Solution:

Multiply using column-by-row method.

$$\begin{bmatrix} 3 & 2 \\ 1 & 4 \end{bmatrix}\begin{bmatrix} -1 & 3 \\ -2 & 5 \end{bmatrix} = \begin{bmatrix} (3)(-1) + (1)(3) & (2)(-1) + (4)(3) \\ (3)(-2) + (1)(5) & (2)(-2) + (4)(5) \end{bmatrix}$$

Simplify. $\begin{bmatrix} 3 & 2 \\ 1 & 4 \end{bmatrix}\begin{bmatrix} -1 & 3 \\ -2 & 5 \end{bmatrix} = \begin{bmatrix} 0 & 10 \\ -1 & 16 \end{bmatrix}$

This is incorrect. What mistake was made?

■ **CONCEPTUAL**

In Exercises 67–70, determine whether the statements are true or false.

67. If $A = \begin{bmatrix} a_{11} & a_{12} \\ a_{21} & a_{22} \end{bmatrix}$ and $B = \begin{bmatrix} b_{11} & b_{12} \\ b_{21} & b_{22} \end{bmatrix}$, then

$$AB = \begin{bmatrix} a_{11}b_{11} & a_{12}b_{12} \\ a_{21}b_{21} & a_{22}b_{22} \end{bmatrix}.$$

68. If AB is defined, then $AB = BA$.

69. AB is defined only if the number of columns in A equals the number of rows in B.

70. $A + B$ is defined only if A and B have the same order.

71. For $A = \begin{bmatrix} a_{11} & a_{12} \\ a_{21} & a_{22} \end{bmatrix}$, find A^2.

72. In order for $A^2_{m \times n}$ to be defined, what condition (with respect to m and n) must be met?

■ **CHALLENGE**

73. For $A = \begin{bmatrix} 1 & 1 \\ 1 & 1 \end{bmatrix}$ find $A, A^2, A^3, \ldots$. What is A^n?

74. For $A = \begin{bmatrix} 1 & 0 \\ 0 & 1 \end{bmatrix}$ find $A, A^2, A^3, \ldots$. What is A^n?

75. If $A_{m \times n}B_{n \times p}$ is defined, explain why $(A_{m \times n}B_{n \times p})^2$ is not defined for $m \neq p$.

76. If $A_{m \times n} = B_{m \times n}$ and $C_{n \times m}$, explain why $AC \neq CB$, if $m \neq n$.

■ **TECHNOLOGY**

In Exercises 77–82, apply a graphing utility to perform the indicated matrix operations, if possible.

$$A = \begin{bmatrix} 1 & 7 & 9 & 2 \\ -3 & -6 & 15 & 11 \\ 0 & 3 & 2 & 5 \\ 9 & 8 & -4 & 1 \end{bmatrix} \qquad B = \begin{bmatrix} 7 & 9 \\ 8 & 6 \\ -4 & -2 \\ 3 & 1 \end{bmatrix}$$

77. AB 78. BA

79. BB 80. AA

$$A = \begin{bmatrix} 2 & 1 & 1 \\ -3 & 0 & 2 \\ 4 & -6 & 0 \end{bmatrix}$$

81. A^2 82. A^5

SKILLS OBJECTIVES

- Write a system of linear equations as a matrix equation.
- Find the inverse of a square matrix.
- Solve systems of linear equations using inverse matrices.

CONCEPTUAL OBJECTIVES

- Visualize a system of linear equations as a matrix equation.
- Understand that only a square matrix can have an inverse.
- Realize that not every square matrix has an inverse.

Matrix Equations

Matrix equations are another way of writing systems of linear equations.

WORDS	**MATH**
Start with a matrix equation.	$\begin{bmatrix} 2 & -3 \\ 1 & 5 \end{bmatrix} \begin{bmatrix} x \\ y \end{bmatrix} = \begin{bmatrix} -7 \\ 9 \end{bmatrix}$
Multiply the two matrices on the left.	$\begin{bmatrix} 2x - 3y \\ x + 5y \end{bmatrix} = \begin{bmatrix} -7 \\ 9 \end{bmatrix}$
Apply equality of two matrices.	$2x - 3y = -7$ $x + 5y = 9$

Let A be a matrix with m rows and n columns, which represents the coefficients in the system; let X represent the variables in the system; and let B represent the constants in the system. Then, a system of linear equations can be written as $AX = B$.

System of Linear Equations	A	X	B	Matrix Equation: $AX = B$
$3x + 4y = 1$ $x - 2y = 7$	$\begin{bmatrix} 3 & 4 \\ 1 & -2 \end{bmatrix}$	$\begin{bmatrix} x \\ y \end{bmatrix}$	$\begin{bmatrix} 1 \\ 7 \end{bmatrix}$	$\begin{bmatrix} 3 & 4 \\ 1 & -2 \end{bmatrix} \begin{bmatrix} x \\ y \end{bmatrix} = \begin{bmatrix} 1 \\ 7 \end{bmatrix}$
$x - y + z = 2$ $2x + 2y - 3z = -3$ $x + y + z = 6$	$\begin{bmatrix} 1 & -1 & 1 \\ 2 & 2 & -3 \\ 1 & 1 & 1 \end{bmatrix}$	$\begin{bmatrix} x \\ y \\ z \end{bmatrix}$	$\begin{bmatrix} 2 \\ -3 \\ 6 \end{bmatrix}$	$\begin{bmatrix} 1 & -1 & 1 \\ 2 & 2 & -3 \\ 1 & 1 & 1 \end{bmatrix} \begin{bmatrix} x \\ y \\ z \end{bmatrix} = \begin{bmatrix} 2 \\ -3 \\ 6 \end{bmatrix}$
$x + y + z = 0$ $3x + 2y - z = 2$	$\begin{bmatrix} 1 & 1 & 1 \\ 3 & 2 & -1 \end{bmatrix}$	$\begin{bmatrix} x \\ y \\ z \end{bmatrix}$	$\begin{bmatrix} 0 \\ 2 \end{bmatrix}$	$\begin{bmatrix} 1 & 1 & 1 \\ 3 & 2 & -1 \end{bmatrix} \begin{bmatrix} x \\ y \\ z \end{bmatrix} = \begin{bmatrix} 0 \\ 2 \end{bmatrix}$

EXAMPLE 1 **Writing a System of Linear Equations as a Matrix Equation**

Write each system of linear equations as a matrix equation.

a. $2x - y = 5$
$-x + 2y = 3$

b. $3x - 2y + 4z = 5$
$y - 3z = -2$
$7x - z = 1$

c. $x_1 - x_2 + 2x_3 - 3 = 0$
$x_1 + x_2 - 3x_3 + 5 = 0$
$x_1 - x_2 + x_3 - 2 = 0$

Solution:

a.

$$\begin{bmatrix} 2 & -1 \\ -1 & 2 \end{bmatrix} \begin{bmatrix} x \\ y \end{bmatrix} = \begin{bmatrix} 5 \\ 3 \end{bmatrix}$$

b. Note that all missing terms have 0 coefficients.

$3x - 2y + 4z = 5$
$0x + y - 3z = -2$
$7x + 0y - z = 1$

$$\begin{bmatrix} 3 & -2 & 4 \\ 0 & 1 & -3 \\ 7 & 0 & -1 \end{bmatrix} \begin{bmatrix} x \\ y \\ z \end{bmatrix} = \begin{bmatrix} 5 \\ -2 \\ 1 \end{bmatrix}$$

c. Write the constants on the right side of the equal sign.

$x_1 - x_2 + 2x_3 = 3$
$x_1 + x_2 - 3x_3 = -5$
$x_1 - x_2 + x_3 = 2$

$$\begin{bmatrix} 1 & -1 & 2 \\ 1 & 1 & -3 \\ 1 & -1 & 1 \end{bmatrix} \begin{bmatrix} x_1 \\ x_2 \\ x_3 \end{bmatrix} = \begin{bmatrix} 3 \\ -5 \\ 2 \end{bmatrix}$$

■ **YOUR TURN** Write each system of linear equations as a matrix equation.

a. $2x + y - 3 = 0$
$x - y = 5$

b. $y - x + z = 7$
$x - y - z = 2$
$z - y = -1$

Finding the Inverse of a Matrix

Before we discuss solving systems of linear equations in the form $AX = B$, let us first recall how we solve $ax = b$, where a and b are real numbers (not matrices).

WORDS	**MATH**
Write the linear equation in one variable.	$ax = b$
Multiply both sides by a^{-1} (same as dividing by a), provided $a \neq 0$.	$a^{-1}ax = a^{-1}b$
Simplify.	$\underset{1}{\underline{a^{-1}a}}x = a^{-1}b$
	$x = a^{-1}b$

Recall that a^{-1}, or $\dfrac{1}{a}$, is the *multiplicative inverse* (Chapter 0) of a because $a^{-1}a = 1$.
And we call 1 the *multiplicative identity*, because any number multiplied by 1 is itself.
Before we solve matrix equations, we need to define the *multiplicative identity matrix* and the *multiplicative inverse matrix*.

A square matrix of order n with 1s along the **main diagonal** (a_{ii}) and 0s for all other elements is called the **multiplicative identity matrix I_n**.

$$I_2 = \begin{bmatrix} 1 & 0 \\ 0 & 1 \end{bmatrix} \qquad I_3 = \begin{bmatrix} 1 & 0 & 0 \\ 0 & 1 & 0 \\ 0 & 0 & 1 \end{bmatrix} \qquad I_4 = \begin{bmatrix} 1 & 0 & 0 & 0 \\ 0 & 1 & 0 & 0 \\ 0 & 0 & 1 & 0 \\ 0 & 0 & 0 & 1 \end{bmatrix}$$

Since a real number multiplied by 1 is itself ($a \cdot 1 = a$), we expect that a matrix multiplied by the appropriate identity matrix should result in itself:

$$A_{m \times n} I_n = A_{m \times n} \qquad \text{and} \qquad I_m A_{m \times n} = A_{m \times n}$$

EXAMPLE 2   **Multiplying a Matrix by the Multiplicative Identity Matrix I_n**

For $A = \begin{bmatrix} -2 & 4 & 1 \\ 3 & 7 & -1 \end{bmatrix}$, find $I_2 A$.

Solution:

Write the two matrices. $A = \begin{bmatrix} -2 & 4 & 1 \\ 3 & 7 & -1 \end{bmatrix}$ $I_2 = \begin{bmatrix} 1 & 0 \\ 0 & 1 \end{bmatrix}$

Find the product $I_2 A$. $I_2 A = \begin{bmatrix} 1 & 0 \\ 0 & 1 \end{bmatrix}\begin{bmatrix} -2 & 4 & 1 \\ 3 & 7 & -1 \end{bmatrix}$

$$I_2 A = \begin{bmatrix} (1)(-2) + (0)(3) & (1)(4) + (0)(7) & (1)(1) + (0)(-1) \\ (0)(-2) + (1)(3) & (0)(4) + (1)(7) & (0)(1) + (1)(-1) \end{bmatrix}$$

$$I_2 A = \begin{bmatrix} -2 & 4 & 1 \\ 3 & 7 & -1 \end{bmatrix} = A$$

■ **YOUR TURN** For A in Example 2, find AI_3.

■**Answer:**

$AI_3 = \begin{bmatrix} -2 & 4 & 1 \\ 3 & 7 & -1 \end{bmatrix} = A$

The identity matrix I_n will assist us in developing the concept of an *inverse of a square matrix*.

DEFINITION   **Inverse of a Square Matrix**

Let A be a square $n \times n$ matrix. If there exists a square $n \times n$ matrix A^{-1} such that

$$AA^{-1} = I_n \qquad \text{and} \qquad A^{-1}A = I_n$$

then A^{-1}, stated as "A inverse," is the **inverse** of A.

Study Tip

- Only a *square* matrix can have an inverse.
- Not all square matrices have inverses.

It is important to note that only a square matrix can have an inverse. Even then, not all square matrices have inverses.

Enter the matrices as A, and A^{-1} as B.

```
[A]
        [[1 3]
         [2 5]]
[B]
        [[-5 3 ]
         [2  -1]]
■
```

Enter AA^{-1} as AB, and $A^{-1}A$ as BA.

```
[A][B]
        [[1 0]
         [0 1]]
[B][A]
        [[1 0]
         [0 1]]
```

■ **Answer:** $AA^{-1} = A^{-1}A = I_2$

Classroom Example 10.3.3*

Let $a_1, a_2, \ldots, a_6$ be nonzero real numbers. Show that:

$$\begin{bmatrix} a_1 & 0 & 0 & 0 & 0 & 0 \\ 0 & a_2 & 0 & 0 & 0 & 0 \\ 0 & 0 & a_3 & 0 & 0 & 0 \\ 0 & 0 & 0 & a_4 & 0 & 0 \\ 0 & 0 & 0 & 0 & a_5 & 0 \\ 0 & 0 & 0 & 0 & 0 & a_6 \end{bmatrix}.$$

$$\begin{bmatrix} \dfrac{1}{a_1} & 0 & 0 & 0 & 0 & 0 \\ 0 & \dfrac{1}{a_2} & 0 & 0 & 0 & 0 \\ 0 & 0 & \dfrac{1}{a_3} & 0 & 0 & 0 \\ 0 & 0 & 0 & \dfrac{1}{a_4} & 0 & 0 \\ 0 & 0 & 0 & 0 & \dfrac{1}{a_5} & 0 \\ 0 & 0 & 0 & 0 & 0 & \dfrac{1}{a_6} \end{bmatrix} = I_6$$

Answer: Just multiply the matrices as usual.

EXAMPLE 3 Multiplying a Matrix by Its Inverse

Verify that the inverse of $A = \begin{bmatrix} 1 & 3 \\ 2 & 5 \end{bmatrix}$ is $A^{-1} = \begin{bmatrix} -5 & 3 \\ 2 & -1 \end{bmatrix}$.

Solution:

Show that $AA^{-1} = I_2$ and $A^{-1}A = I_2$.

Find the product AA^{-1}.

$$AA^{-1} = \begin{bmatrix} 1 & 3 \\ 2 & 5 \end{bmatrix}\begin{bmatrix} -5 & 3 \\ 2 & -1 \end{bmatrix}$$

$$= \begin{bmatrix} (1)(-5) + (3)(2) & (1)(3) + (3)(-1) \\ (2)(-5) + (5)(2) & (2)(3) + (5)(-1) \end{bmatrix}$$

$$= \begin{bmatrix} 1 & 0 \\ 0 & 1 \end{bmatrix} = I_2$$

Find the product $A^{-1}A$.

$$A^{-1}A = \begin{bmatrix} -5 & 3 \\ 2 & -1 \end{bmatrix}\begin{bmatrix} 1 & 3 \\ 2 & 5 \end{bmatrix}$$

$$= \begin{bmatrix} (-5)(1) + (3)(2) & (-5)(3) + (3)(5) \\ (2)(1) + (-1)(2) & (2)(3) + (-1)(5) \end{bmatrix}$$

$$= \begin{bmatrix} 1 & 0 \\ 0 & 1 \end{bmatrix} = I_2$$

■ **YOUR TURN** Verify that the inverse of $A = \begin{bmatrix} 1 & 4 \\ 2 & 9 \end{bmatrix}$ is $A^{-1} = \begin{bmatrix} 9 & -4 \\ -2 & 1 \end{bmatrix}$.

Now that we can show that two matrices are inverses of one another, let us describe the process for finding an inverse, if it exists. If an inverse A^{-1} exists, then the matrix A is said to be **nonsingular**. If the inverse does not exist, then the matrix A is said to be **singular**.

Let $A = \begin{bmatrix} 1 & -1 \\ 2 & -3 \end{bmatrix}$ and the inverse be $A^{-1} = \begin{bmatrix} w & x \\ y & z \end{bmatrix}$, where $w, x, y,$ and z are variables to be determined. A matrix and its inverse must satisfy the identity $AA^{-1} = I_2$.

WORDS	MATH
The product of a matrix and its inverse is the identity matrix.	$\begin{bmatrix} 1 & -1 \\ 2 & -3 \end{bmatrix}\begin{bmatrix} w & x \\ y & z \end{bmatrix} = \begin{bmatrix} 1 & 0 \\ 0 & 1 \end{bmatrix}$
Multiply the two matrices on the left.	$\begin{bmatrix} w - y & x - z \\ 2w - 3y & 2x - 3z \end{bmatrix} = \begin{bmatrix} 1 & 0 \\ 0 & 1 \end{bmatrix}$
Equate corresponding matrix elements.	$\begin{aligned} w - y &= 1 \\ 2w - 3y &= 0 \end{aligned}$ and $\begin{aligned} x - z &= 0 \\ 2x - 3z &= 1 \end{aligned}$

Notice that there are two systems of equations, both of which can be solved by several methods (elimination, substitution, or augmented matrices). We will find that $w = 3$, $x = -1$, $y = 2$, and $z = -1$. Therefore, we know the inverse is $A^{-1} = \begin{bmatrix} 3 & -1 \\ 2 & -1 \end{bmatrix}$. But, instead, let us use augmented matrices in order to develop the general procedure.

Write the two systems of equations as two augmented matrices:

$$\begin{matrix} w & y \\ \begin{bmatrix} 1 & -1 & | & 1 \\ 2 & -3 & | & 0 \end{bmatrix} \end{matrix} \qquad \begin{matrix} x & z \\ \begin{bmatrix} 1 & -1 & | & 0 \\ 2 & -3 & | & 1 \end{bmatrix} \end{matrix}$$

Since the left side is the same for each augmented matrix, we can combine these two matrices into one matrix, thereby simultaneously solving both systems of equations:

$$\begin{bmatrix} 1 & -1 & | & 1 & 0 \\ 2 & -3 & | & 0 & 1 \end{bmatrix}$$

Notice that the right side of the vertical line is the identity matrix I_2.

Using Gauss–Jordan elimination, transform the matrix on the left to the identity matrix.

$$\begin{bmatrix} 1 & -1 & | & 1 & 0 \\ 2 & -3 & | & 0 & 1 \end{bmatrix}$$

$$R_2 - 2R_1 \rightarrow R_2 \quad \begin{bmatrix} 1 & -1 & | & 1 & 0 \\ 0 & -1 & | & -2 & 1 \end{bmatrix}$$

$$-R_2 \rightarrow R_2 \quad \begin{bmatrix} 1 & -1 & | & 1 & 0 \\ 0 & 1 & | & 2 & -1 \end{bmatrix}$$

$$R_1 + R_2 \rightarrow R_1 \quad \begin{bmatrix} 1 & 0 & | & 3 & -1 \\ 0 & 1 & | & 2 & -1 \end{bmatrix}$$

The matrix on the right of the vertical line is the inverse $A^{-1} = \begin{bmatrix} 3 & -1 \\ 2 & -1 \end{bmatrix}$.

FINDING THE INVERSE OF A SQUARE MATRIX

To find the inverse of an $n \times n$ matrix A:

Step 1: Form the augmented matrix $[A \mid I_n]$.
Step 2: Use row operations to transform this entire matrix to $[I_n \mid A^{-1}]$. This is done by applying Gauss–Jordan elimination to reduce A to the identity matrix I_n (which is in reduced row–echelon form). If this is not possible, then A is a singular matrix and no inverse exists.
Step 3: Verify the result by showing that $AA^{-1} = I_n$ and $A^{-1}A = I_n$.

EXAMPLE 4   Finding the Inverse of a 2 × 2 Matrix

Find the inverse of $A = \begin{bmatrix} 1 & 2 \\ 3 & 5 \end{bmatrix}$.

Solution:

STEP 1 Form the matrix $[A \mid I_2]$.

$$\begin{bmatrix} 1 & 2 & | & 1 & 0 \\ 3 & 5 & | & 0 & 1 \end{bmatrix}$$

STEP 2 Use row operations to transform A into I_2.

$$R_2 - 3R_1 \rightarrow R_2 \quad \begin{bmatrix} 1 & 2 & | & 1 & 0 \\ 0 & -1 & | & -3 & 1 \end{bmatrix}$$

$$-R_2 \rightarrow R_2 \quad \begin{bmatrix} 1 & 2 & | & 1 & 0 \\ 0 & 1 & | & 3 & -1 \end{bmatrix}$$

$$R_1 - 2R_2 \rightarrow R_1 \quad \begin{bmatrix} 1 & 0 & | & -5 & 2 \\ 0 & 1 & | & 3 & -1 \end{bmatrix}$$

Identify the inverse.

$$A^{-1} = \begin{bmatrix} -5 & 2 \\ 3 & -1 \end{bmatrix}$$

Technology Tip

A graphing calculator can be used to find the inverse of A. Enter the matrix A.

```
[A]
              [[1 2]
               [3 5]]
■
```

To find A^{-1}, press $\boxed{2^{nd}}$ $\boxed{MATRIX}$ $\boxed{1:[A]}$ $\boxed{ENTER}$ $\boxed{x^{-1}}$ $\boxed{ENTER}$.

```
[A]
              [[1 2]
               [3 5]]
[A]⁻¹
              [[-5 2 ]
               [3  -1]]
```

STEP 3 Check.

$$AA^{-1} = \begin{bmatrix} 1 & 2 \\ 3 & 5 \end{bmatrix}\begin{bmatrix} -5 & 2 \\ 3 & -1 \end{bmatrix} = \begin{bmatrix} 1 & 0 \\ 0 & 1 \end{bmatrix} = I_2$$

$$A^{-1}A = \begin{bmatrix} -5 & 2 \\ 3 & -1 \end{bmatrix}\begin{bmatrix} 1 & 2 \\ 3 & 5 \end{bmatrix} = \begin{bmatrix} 1 & 0 \\ 0 & 1 \end{bmatrix} = I_2$$

■ **Answer:** $A^{-1} = \begin{bmatrix} 8 & -3 \\ -5 & 2 \end{bmatrix}$

■ **YOUR TURN** Find the inverse of $A = \begin{bmatrix} 2 & 3 \\ 5 & 8 \end{bmatrix}$.

Classroom Example 10.3.4–10.3.5

Find the inverse of the matrices, if it exists.

a. $\begin{bmatrix} 1 & 1 \\ 1 & -1 \end{bmatrix}$

b.* $\begin{bmatrix} a & 1 \\ 0 & a \end{bmatrix}$, where a is a nonzero real number.

c.* $\begin{bmatrix} 2a & a \\ 1 & \frac{1}{2} \end{bmatrix}$, where a is a nonzero real number.

Answer:

a. $-\dfrac{1}{2}\begin{bmatrix} -1 & -1 \\ -1 & 1 \end{bmatrix}$

b. $\dfrac{1}{a^2}\begin{bmatrix} a & -1 \\ 0 & a \end{bmatrix}$

c. no inverse

This procedure for finding an inverse of a square matrix is used for all square matrices of order n. For the special case of a 2×2 matrix, there is a formula (that will be derived in Exercises 65 and 66) for finding the inverse.

Let $A = \begin{bmatrix} a & b \\ c & d \end{bmatrix}$ represent any 2×2 matrix; then the inverse matrix is given by

$$A^{-1} = \frac{1}{ad - bc}\begin{bmatrix} d & -b \\ -c & a \end{bmatrix} \qquad ad - bc \neq 0$$

The denominator $ad - bc$ is called the *determinant* and will be discussed in Section 10.4.

We found the inverse of $A = \begin{bmatrix} 1 & 2 \\ 3 & 5 \end{bmatrix}$ in Example 4. Let us now find the inverse using this formula.

WORDS	**MATH**
Write the formula for A^{-1}.	$A^{-1} = \dfrac{1}{ad - bc}\begin{bmatrix} d & -b \\ -c & a \end{bmatrix}$
Substitute $a = 1$, $b = 2$, $c = 3$, $d = 5$ into the formula.	$A^{-1} = \dfrac{1}{(1)(5) - (2)(3)}\begin{bmatrix} 5 & -2 \\ -3 & 1 \end{bmatrix}$
Simplify.	$A^{-1} = (-1)\begin{bmatrix} 5 & -2 \\ -3 & 1 \end{bmatrix}$
	$A^{-1} = \begin{bmatrix} -5 & 2 \\ 3 & -1 \end{bmatrix}$

The result is the same as what we found in Example 4.

EXAMPLE 5 Finding That No Inverse Exists: Singular Matrix

Find the inverse of $A = \begin{bmatrix} 1 & -5 \\ -1 & 5 \end{bmatrix}$.

Solution:

STEP 1 Form the matrix $[A \mid I_2]$.

$$\begin{bmatrix} 1 & -5 & | & 1 & 0 \\ -1 & 5 & | & 0 & 1 \end{bmatrix}$$

STEP 2 Apply row operations to transform A into I_2. $\qquad R_2 + R_1 \to R_2 \qquad \begin{bmatrix} 1 & -5 & | & 1 & 0 \\ 0 & 0 & | & 1 & 1 \end{bmatrix}$

We cannot convert the left-hand side of the augmented matrix to I_2 because of the all-zero row on the left-hand side. Therefore, $\boxed{A \text{ is not invertible}}$.

EXAMPLE 6 Finding the Inverse of a 3 × 3 Matrix

Find the inverse of $A = \begin{bmatrix} 1 & 2 & -1 \\ 0 & 1 & -1 \\ -1 & 0 & -2 \end{bmatrix}$.

Solution:

STEP 1 Form the matrix $[A \mid I_3]$.

$$\begin{bmatrix} 1 & 2 & -1 & 1 & 0 & 0 \\ 0 & 1 & -1 & 0 & 1 & 0 \\ -1 & 0 & -2 & 0 & 0 & 1 \end{bmatrix}$$

STEP 2 Apply row operations to transform A into I_3.

$$R_3 + R_1 \rightarrow R_3 \quad \begin{bmatrix} 1 & 2 & -1 & 1 & 0 & 0 \\ 0 & 1 & -1 & 0 & 1 & 0 \\ 0 & 2 & -3 & 1 & 0 & 1 \end{bmatrix}$$

$$R_3 - 2R_2 \rightarrow R_3 \quad \begin{bmatrix} 1 & 2 & -1 & 1 & 0 & 0 \\ 0 & 1 & -1 & 0 & 1 & 0 \\ 0 & 0 & -1 & 1 & -2 & 1 \end{bmatrix}$$

$$-R_3 \rightarrow R_3 \quad \begin{bmatrix} 1 & 2 & -1 & 1 & 0 & 0 \\ 0 & 1 & -1 & 0 & 1 & 0 \\ 0 & 0 & 1 & -1 & 2 & -1 \end{bmatrix}$$

$$\begin{matrix} R_2 + R_3 \rightarrow R_2 \\ R_1 + R_3 \rightarrow R_1 \end{matrix} \quad \begin{bmatrix} 1 & 2 & 0 & 0 & 2 & -1 \\ 0 & 1 & 0 & -1 & 3 & -1 \\ 0 & 0 & 1 & -1 & 2 & -1 \end{bmatrix}$$

$$R_1 - 2R_2 \rightarrow R_1 \quad \begin{bmatrix} 1 & 0 & 0 & 2 & -4 & 1 \\ 0 & 1 & 0 & -1 & 3 & -1 \\ 0 & 0 & 1 & -1 & 2 & -1 \end{bmatrix}$$

Identify the inverse.

$$A^{-1} = \begin{bmatrix} 2 & -4 & 1 \\ -1 & 3 & -1 \\ -1 & 2 & -1 \end{bmatrix}$$

STEP 3 Check. $AA^{-1} = \begin{bmatrix} 1 & 2 & -1 \\ 0 & 1 & -1 \\ -1 & 0 & -2 \end{bmatrix} \begin{bmatrix} 2 & -4 & 1 \\ -1 & 3 & -1 \\ -1 & 2 & -1 \end{bmatrix} = \begin{bmatrix} 1 & 0 & 0 \\ 0 & 1 & 0 \\ 0 & 0 & 1 \end{bmatrix} = I_3$

$A^{-1}A = \begin{bmatrix} 2 & -4 & 1 \\ -1 & 3 & -1 \\ -1 & 2 & -1 \end{bmatrix} \begin{bmatrix} 1 & 2 & -1 \\ 0 & 1 & -1 \\ -1 & 0 & -2 \end{bmatrix} = \begin{bmatrix} 1 & 0 & 0 \\ 0 & 1 & 0 \\ 0 & 0 & 1 \end{bmatrix} = I_3$

■ **YOUR TURN** Find the inverse of $A = \begin{bmatrix} 1 & 1 & 0 \\ -1 & 0 & 1 \\ 2 & 0 & -1 \end{bmatrix}$.

Technology Tip

A graphing calculator can be used to find the inverse of A. Enter the matrix A.

```
[A]
      [[1    2  -1]
       [0    1  -1]
       [-1  0  -2]]
```

To find A^{-1}, press 2nd MATRIX 1:[A] ENTER x^{-1} ENTER.

```
[A]-1
      [[2   -4   1 ]
       [-1   3  -1 ]
       [-1   2  -1 ]]
```

Classroom Example 10.3.6
Find the inverse of the matrix

$\begin{bmatrix} a & b & 1 \\ 0 & a & b \\ 0 & 0 & a \end{bmatrix}$, where $a, b \neq 0$.

Answer:

$\begin{bmatrix} \dfrac{1}{a} & -\dfrac{b}{a^2} & -\dfrac{1}{a^2} + \dfrac{b^2}{a^3} \\ 0 & \dfrac{1}{a} & -\dfrac{b}{a^2} \\ 0 & 0 & \dfrac{1}{a} \end{bmatrix}$

■ **Answer:** $A^{-1} = \begin{bmatrix} 0 & 1 & 1 \\ 1 & -1 & -1 \\ 0 & 2 & 1 \end{bmatrix}$

Solving Systems of Linear Equations Using Matrix Algebra and Inverses of Square Matrices

We can solve systems of linear equations using matrix algebra. We will use a system of three equations and three variables to demonstrate the procedure. However, it can be extended to any square system.

Linear System of Equations	**Matrix Form of the System**

$$\begin{aligned} a_1x + b_1y + c_1z &= d_1 \\ a_2x + b_2y + c_2z &= d_2 \\ a_3x + b_3y + c_3z &= d_3 \end{aligned} \qquad \underbrace{\begin{bmatrix} a_1 & b_1 & c_1 \\ a_2 & b_2 & c_2 \\ a_3 & b_3 & c_3 \end{bmatrix}}_{A} \underbrace{\begin{bmatrix} x \\ y \\ z \end{bmatrix}}_{X} = \underbrace{\begin{bmatrix} d_1 \\ d_2 \\ d_3 \end{bmatrix}}_{B}$$

Recall that a system of linear equations has a unique solution, no solution, or infinitely many solutions. If a system of n equations in n variables has a unique solution, it can be found using the following procedure.

WORDS	**MATH**
Write the system of linear equations as a matrix equation.	$A_{n \times n} X_{n \times 1} = B_{n \times 1}$
Multiply both sides of the equation by A^{-1}.	$A^{-1}AX = A^{-1}B$
A matrix times its inverse is the identity matrix.	$I_n X = A^{-1}B$
A matrix times the identity matrix is equal to itself.	$X = A^{-1}B$

Notice the order in which the right side is multiplied, $X_{n \times 1} = A_{n \times n}^{-1} B_{n \times 1}$, and remember that matrix multiplication is not commutative.

SOLVING A SYSTEM OF LINEAR EQUATIONS USING MATRIX ALGEBRA: UNIQUE SOLUTION

If a system of linear equations is represented by $AX = B$, where A is a nonsingular square matrix, then the system has a unique solution given by

$$X = A^{-1}B$$

EXAMPLE 7 **Solving a System of Linear Equations Using Matrix Algebra**

Solve the system of equations using matrix algebra.

$$\begin{aligned} x + y + z &= 2 \\ x \phantom{{}+ y} + z &= 1 \\ x - y - z &= -4 \end{aligned}$$

Solution:

Write the system in matrix form. $AX = B$

$$A = \begin{bmatrix} 1 & 1 & 1 \\ 1 & 0 & 1 \\ 1 & -1 & -1 \end{bmatrix} \quad X = \begin{bmatrix} x \\ y \\ z \end{bmatrix} \quad B = \begin{bmatrix} 2 \\ 1 \\ -4 \end{bmatrix}$$

Find the inverse of A.

Form the matrix $[A \mid I_3]$.

$$\begin{bmatrix} 1 & 1 & 1 & 1 & 0 & 0 \\ 1 & 0 & 1 & 0 & 1 & 0 \\ 1 & -1 & -1 & 0 & 0 & 1 \end{bmatrix}$$

$\begin{matrix} R_2 - R_1 \to R_2 \\ R_3 - R_1 \to R_3 \end{matrix}$ $\begin{bmatrix} 1 & 1 & 1 & 1 & 0 & 0 \\ 0 & -1 & 0 & -1 & 1 & 0 \\ 0 & -2 & -2 & -1 & 0 & 1 \end{bmatrix}$

$-R_2 \to R_2$ $\begin{bmatrix} 1 & 1 & 1 & 1 & 0 & 0 \\ 0 & 1 & 0 & 1 & -1 & 0 \\ 0 & -2 & -2 & -1 & 0 & 1 \end{bmatrix}$

$R_3 + 2R_2 \to R_3$ $\begin{bmatrix} 1 & 1 & 1 & 1 & 0 & 0 \\ 0 & 1 & 0 & 1 & -1 & 0 \\ 0 & 0 & -2 & 1 & -2 & 1 \end{bmatrix}$

$-\frac{1}{2}R_3 \to R_3$ $\begin{bmatrix} 1 & 1 & 1 & 1 & 0 & 0 \\ 0 & 1 & 0 & 1 & -1 & 0 \\ 0 & 0 & 1 & -\frac{1}{2} & 1 & -\frac{1}{2} \end{bmatrix}$

$R_1 - R_3 \to R_1$ $\begin{bmatrix} 1 & 1 & 0 & \frac{3}{2} & -1 & \frac{1}{2} \\ 0 & 1 & 0 & 1 & -1 & 0 \\ 0 & 0 & 1 & -\frac{1}{2} & 1 & -\frac{1}{2} \end{bmatrix}$

$R_1 - R_2 \to R_1$ $\begin{bmatrix} 1 & 0 & 0 & \frac{1}{2} & 0 & \frac{1}{2} \\ 0 & 1 & 0 & 1 & -1 & 0 \\ 0 & 0 & 1 & -\frac{1}{2} & 1 & -\frac{1}{2} \end{bmatrix}$

Identify the inverse.

$$A^{-1} = \begin{bmatrix} \frac{1}{2} & 0 & \frac{1}{2} \\ 1 & -1 & 0 \\ -\frac{1}{2} & 1 & -\frac{1}{2} \end{bmatrix}$$

The solution to the system is
$X = A^{-1}B$.

$$X = A^{-1}B = \begin{bmatrix} \frac{1}{2} & 0 & \frac{1}{2} \\ 1 & -1 & 0 \\ -\frac{1}{2} & 1 & -\frac{1}{2} \end{bmatrix} \begin{bmatrix} 2 \\ 1 \\ -4 \end{bmatrix}$$

Simplify.

$$X = \begin{bmatrix} x \\ y \\ z \end{bmatrix} = \begin{bmatrix} -1 \\ 1 \\ 2 \end{bmatrix} \quad \boxed{x = -1, y = 1, z = 2}$$

> **Classroom Example 10.3.7**
> Solve these systems using matrix algebra.
>
> **a.** $-2y = 2 - x$
> $x = 3 - 4y$
> **b.** $x + y + z = 0$
> $y + 3x = 0$
> $y - 2z - 7 = 0$
> **c.** $-\frac{1}{4}x + \frac{1}{2}y - \frac{1}{2}z = -2$
> $\frac{1}{2}x + \frac{1}{3}y - \frac{1}{4}z = 2$
> $\frac{1}{2}x - \frac{1}{2}y + \frac{1}{4}z = 1$
> **d.** $ax - 2ay = 1$
> $x + ay = 0$, where $a \neq 0, -2$
>
> **Answer:**
> **a.** $x = \frac{7}{3}, y = \frac{1}{6}$
> **b.** $x = -1, y = 3, z = -2$
> **c.** $x = 4, y = 6, z = 8$
> **d.** $x = \dfrac{1}{a + 2}, y = \dfrac{-1}{a(a + 2)}$

■ **YOUR TURN** Solve the system of equations using matrix algebra.

$$x + y - z = 3$$
$$y + z = 1$$
$$2x + 3y + z = 5$$

Technology Tip

Use TI to find the inverse of A and X.
Enter the matrices A and B.

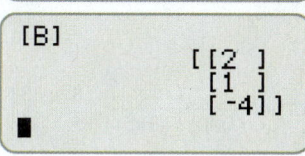

Now use the graphing calculator to find the inverse of A, A^{-1}.
$\boxed{\text{2nd}} \boxed{MATRIX} \boxed{1:[A]} \boxed{\text{ENTER}}$
$\boxed{x^{-1}} \boxed{\text{ENTER}}$.

To show elements using fractions, press $\boxed{\text{2nd}} \boxed{MATRIX} \boxed{1:[A]} \boxed{\text{ENTER}}$
$\boxed{x^{-1}} \boxed{\text{MATH}} \boxed{1: Frac} \boxed{\text{ENTER}}$
$\boxed{\text{ENTER}}$.

To enter $A^{-1}B$, press $\boxed{\text{2nd}} \boxed{ANS}$
$\boxed{\text{2nd}} \boxed{MATRIX} \boxed{2:B}$
$\boxed{\text{ENTER}} \boxed{\text{ENTER}}$.

```
Ans [B]
             [ [ -1]
               [1 ]
               [2 ]]
■
```

The solution to the system is
$x = -1, y = 1$, and $z = 2$.

■**Answer:** $x = 0, y = 2, z = -1$

Cryptography Applications

Cryptography is the practice of hiding information, or secret communication. Let's assume you want to send your ATM PIN code over the Internet, but you don't want hackers to be able to retrieve it. You can represent the PIN code in a matrix and then multiply that PIN matrix by a "key" matrix so that it is encrypted. If the person you send it to has the "inverse key" matrix, he can multiply the encrypted matrix he receives by the inverse key matrix and the result will be the original PIN matrix. Although PIN numbers are typically four digits, we will assume two digits to illustrate the process.

Study Tip

$$K = \begin{bmatrix} 2 & 3 \\ 5 & 8 \end{bmatrix}$$

$$K^{-1} = \frac{1}{(2)(8) - (3)(5)} \begin{bmatrix} 8 & -3 \\ -5 & 2 \end{bmatrix}$$

$$= \begin{bmatrix} 8 & -3 \\ -5 & 2 \end{bmatrix}$$

WORDS	**MATH**
Suppose the two-digit ATM PIN is 13.	$P = [1 \quad 3]_{1 \times 2}$
Apply any 2×2 nonsingular matrix as the "key" (encryption) matrix.	$K = \begin{bmatrix} 2 & 3 \\ 5 & 8 \end{bmatrix}$
Multiply the PIN and encryption matrices.	$PK = [1 \quad 3]_{1 \times 2}\begin{bmatrix} 2 & 3 \\ 5 & 8 \end{bmatrix}_{2 \times 2}$
	$= [1(2) + 3(5) \quad 1(3) + 3(8)]$
	$= [17 \quad 27]$
The receiver of the encrypted matrix sees only $[17 \quad 27]_{1 \times 2}$.	
The decoding "key" is the inverse matrix K^{-1}.	$K^{-1} = \begin{bmatrix} 8 & -3 \\ -5 & 2 \end{bmatrix}$

Any receiver who has the decoding key can multiply the received encrypted matrix by the decoding "key" matrix. The result is the original transmitted PIN number.

$$[17 \quad 27]_{1 \times 2}\begin{bmatrix} 8 & -3 \\ -5 & 2 \end{bmatrix} = [17(8) + 27(-5) \quad 17(-3) + 27(2)] = [1 \quad 3]$$

SECTION
10.3 SUMMARY

Systems of linear equations can be solved using matrix equations.

SYSTEM OF LINEAR EQUATIONS	A	X	B	MATRIX EQUATION: $AX = B$
$\begin{aligned} x - y + z &= 2 \\ 2x + 2y - 3z &= -3 \\ x + y + z &= 6 \end{aligned}$	$\begin{bmatrix} 1 & -1 & 1 \\ 2 & 2 & -3 \\ 1 & 1 & 1 \end{bmatrix}$	$\begin{bmatrix} x \\ y \\ z \end{bmatrix}$	$\begin{bmatrix} 2 \\ -3 \\ 6 \end{bmatrix}$	$\begin{bmatrix} 1 & -1 & 1 \\ 2 & 2 & -3 \\ 1 & 1 & 1 \end{bmatrix}\begin{bmatrix} x \\ y \\ z \end{bmatrix} = \begin{bmatrix} 2 \\ -3 \\ 6 \end{bmatrix}$

If this system of linear equations has a unique solution, then it is represented by

$$X = A^{-1}B$$

A^{-1} is the inverse of A, that is, $AA^{-1} = A^{-1}A = I$, and is found by

$$[A_{n \times n} | I_n] \rightarrow [I_n | A^{-1}_{n \times n}]$$

SKILLS

In Exercises 1–8, write the system of linear equations as a matrix equation. (Do not solve the system.)

1. $-2x + 5y = 10$
   $7x - 2y = -4$

2. $4x - 8y = 10$
   $3x + 5y = 15$

3. $x - 2y = 8$
   $-3x + y = 6$

4. $7x - 2y = 28$
   $3x + 7y = 42$

5. $3x + 5y - z = 2$
   $x \quad\quad + 2z = 17$
   $-x + y - z = 4$

6. $x - y + z = 12$
   $2x + y - 3z = 6$
   $-3x + 2y + z = 18$

7. $3x \quad\quad + z = 10$
   $y - 2z = 4$
   $x + 2y \quad\quad = 6$

8. $x + y - 2z + w = 11$
   $2x - y + 3z \quad\quad = 17$
   $-x + 2y - 3z + 4w = 12$
   $y + 4z + 6w = 19$

In Exercises 9–18, determine whether B is the multiplicative inverse of A using $AA^{-1} = I$.

9. $A = \begin{bmatrix} 8 & -11 \\ -5 & 7 \end{bmatrix}$ $B = \begin{bmatrix} 7 & 11 \\ 5 & 8 \end{bmatrix}$

10. $A = \begin{bmatrix} 7 & -9 \\ -3 & 4 \end{bmatrix}$ $B = \begin{bmatrix} 4 & 9 \\ 3 & 7 \end{bmatrix}$

11. $A = \begin{bmatrix} 3 & 1 \\ 1 & -2 \end{bmatrix}$ $B = \begin{bmatrix} \frac{2}{7} & \frac{1}{7} \\ \frac{1}{7} & -\frac{3}{7} \end{bmatrix}$

12. $A = \begin{bmatrix} 2 & 3 \\ 1 & -1 \end{bmatrix}$ $B = \begin{bmatrix} \frac{1}{5} & \frac{3}{5} \\ \frac{1}{5} & -\frac{2}{5} \end{bmatrix}$

13. $A = \begin{bmatrix} 1 & 2 \\ 3 & 4 \end{bmatrix}$ $B = \begin{bmatrix} 4 & -2 \\ -3 & 1 \end{bmatrix}$

14. $A = \begin{bmatrix} 1 & 2 \\ 3 & 4 \end{bmatrix}$ $B = \begin{bmatrix} 1 & \frac{1}{2} \\ \frac{1}{3} & \frac{1}{4} \end{bmatrix}$

15. $A = \begin{bmatrix} 1 & -1 & 1 \\ 1 & 0 & -1 \\ 0 & 1 & -1 \end{bmatrix}$ $B = \begin{bmatrix} 1 & 0 & 1 \\ 1 & -1 & 2 \\ 1 & -1 & 1 \end{bmatrix}$

16. $A = \begin{bmatrix} -1 & 0 & -1 \\ -1 & 1 & -2 \\ -1 & 1 & -1 \end{bmatrix}$ $B = \begin{bmatrix} -1 & 1 & -1 \\ -1 & 0 & 1 \\ 0 & -1 & 1 \end{bmatrix}$

17. $A = \begin{bmatrix} 2 & 0 & 1 \\ 0 & 3 & 1 \\ 0 & 2 & -1 \end{bmatrix}$ $B = \begin{bmatrix} 0 & 2 & 1 \\ 0 & 3 & 0 \\ 2 & 0 & 2 \end{bmatrix}$

18. $A = \begin{bmatrix} 1 & 0 & 0 \\ 0 & 2 & 0 \\ 0 & 0 & 3 \end{bmatrix}$ $B = \begin{bmatrix} 1 & 0 & 0 \\ 0 & \frac{1}{2} & 0 \\ 0 & 0 & \frac{1}{3} \end{bmatrix}$

In Exercises 19–32, find A^{-1}, if possible.

19. $A = \begin{bmatrix} 2 & 1 \\ -1 & 0 \end{bmatrix}$

20. $A = \begin{bmatrix} 3 & 1 \\ 2 & 1 \end{bmatrix}$

21. $A = \begin{bmatrix} \frac{1}{3} & 2 \\ 5 & \frac{3}{4} \end{bmatrix}$

22. $A = \begin{bmatrix} \frac{1}{4} & 2 \\ \frac{1}{3} & \frac{2}{3} \end{bmatrix}$

23. $A = \begin{bmatrix} 1.3 & 2.4 \\ 5.3 & 1.7 \end{bmatrix}$

24. $A = \begin{bmatrix} -2.3 & 1.1 \\ 4.6 & -3.2 \end{bmatrix}$

25. $A = \begin{bmatrix} 1 & 1 & 1 \\ 1 & -1 & -1 \\ -1 & 1 & -1 \end{bmatrix}$

26. $A = \begin{bmatrix} 1 & -1 & 1 \\ 1 & 1 & 1 \\ -1 & 2 & -3 \end{bmatrix}$

27. $A = \begin{bmatrix} 1 & 0 & 1 \\ 0 & 1 & 1 \\ 1 & -1 & 0 \end{bmatrix}$

28. $A = \begin{bmatrix} 1 & 2 & -3 \\ 1 & -1 & -1 \\ 1 & 0 & -4 \end{bmatrix}$

29. $A = \begin{bmatrix} 2 & 4 & 1 \\ 1 & 1 & -1 \\ 1 & 1 & 0 \end{bmatrix}$

30. $A = \begin{bmatrix} 1 & 0 & 1 \\ 1 & 1 & -1 \\ 2 & 1 & -1 \end{bmatrix}$

31. $A = \begin{bmatrix} 1 & 1 & -1 \\ 1 & -1 & 1 \\ 2 & -1 & -1 \end{bmatrix}$

32. $A = \begin{bmatrix} 1 & -1 & -1 \\ 1 & 1 & -3 \\ 3 & -5 & 1 \end{bmatrix}$

In Exercises 33–46, apply matrix algebra to solve the system of linear equations.

33. $2x - y = 5$
   $x + y = 1$

34. $2x - 3y = 12$
   $x + y = 1$

35. $4x - 9y = -1$
   $7x - 3y = \frac{5}{2}$

36. $7x - 3y = 1$
   $4x - 5y = -\frac{7}{5}$

37. $\frac{3}{4}x - \frac{2}{3}y = 5$
   $-\frac{1}{2}x - \frac{5}{3}y = 3$

38. $\frac{2}{5}x + \frac{3}{7}y = 1$
   $-\frac{1}{2}x - \frac{1}{3}y = \frac{1}{6}$

39. $x + y + z = 1$
   $x - y - z = -1$
   $-x + y - z = -1$

40. $x - y + z = 0$
   $x + y + z = 2$
   $-x + 2y - 3z = 1$

41. $x \quad\;\; + z = 3$
$\quad\;\; y + z = 1$
$\quad\; x - y \quad\;\; = 2$

42. $x + 2y - 3z = 1$
$\quad x - y - z = 3$
$\quad x \quad\quad - 4z = 0$

43. $2x + 4y + z = -5$
$\quad x + y - z = 7$
$\quad x + y \quad\; = 0$

44. $x \quad\;\; + z = \;\; 3$
$\quad x + y - z = -3$
$\quad 2x + y - z = -5$

45. $x + y - z = \;\; 4$
$\quad x - y + z = \;\; 2$
$\quad 2x - y - z = -3$

46. $x - y - z = 0$
$\quad x + y - 3z = 2$
$\quad 3x - 5y + z = 4$

■ APPLICATIONS

47. NCAA. University of Florida apparel sales associated with the Final Four Basketball Tournament and the BCS Championship Game in Football are represented with the following matrix:

$$\begin{array}{cc} & \textbf{Sweatshirts} \quad \textbf{T-shirts} \end{array}$$
$$\begin{array}{c} \textbf{Final Four} \\ \textbf{BCS} \end{array} \begin{bmatrix} 20{,}000 & 100{,}000 \\ 100{,}000 & 50{,}000 \end{bmatrix} = A$$

The revenue generated by the sales of these T-shirts and sweatshirts is given by the following matrix:

$$\begin{array}{c} \textbf{Final Four} \\ \textbf{BCS} \end{array} \begin{bmatrix} \$3{,}000{,}000 \\ \$6{,}000{,}000 \end{bmatrix} = B$$

a. Find $A^{-1}B$.

b. What does $A^{-1}B$ represent?

48. NASCAR. Tony Stewart (NASCAR driver) often drives in two races in the same weekend. The Saturday race is called the Nationwide Series, and the Sunday race is called the Sprint Cup. The numbers of Tony Stewart hat and jacket sales associated with the Daytona 500 weekend of races are represented with the following matrix:

$$\begin{array}{cc} & \textbf{Hats} \quad\quad \textbf{Jackets} \end{array}$$
$$\begin{array}{c} \textbf{Nationwide} \\ \textbf{Sprint Cup} \end{array} \begin{bmatrix} 30{,}000 & 2{,}000 \\ 50{,}000 & 5{,}000 \end{bmatrix} = A$$

The revenue generated by the sales of these hats and jackets is given by the following matrix:

$$\begin{array}{c} \textbf{Nationwide} \\ \textbf{Sprint Cup} \end{array} \begin{bmatrix} \$750{,}000 \\ \$1{,}375{,}000 \end{bmatrix} = B$$

a. Find $A^{-1}B$.

b. What does $A^{-1}B$ represent?

For Exercises 49–54, apply the following decoding scheme:

1	A	10	J	19	S
2	B	11	K	20	T
3	C	12	L	21	U
4	D	13	M	22	V
5	E	14	N	23	W
6	F	15	O	24	X
7	G	16	P	25	Y
8	H	17	Q	26	Z
9	I	18	R		

The encoding matrix is $\begin{bmatrix} 1 & 1 & 0 \\ -1 & 0 & 1 \\ 2 & 0 & -1 \end{bmatrix}$. The encrypted matrices are given below. For each of the following, determine the three-letter word that is originally transmitted. *Hint:* All four words are parts of the body.

49. Cryptography. $[55 \quad 10 \quad -22]$

50. Cryptography. $[31 \quad 8 \quad -7]$

51. Cryptography. $[21 \quad 12 \quad -2]$

52. Cryptography. $[9 \quad 1 \quad 5]$

53. Cryptography. $[-10 \quad 5 \quad 20]$

54. Cryptography. $[40 \quad 5 \quad -17]$

For Exercises 55 and 56, refer to the following:

The results of a nutritional analysis of one serving of three foods A, B, and C were:

$$\begin{array}{c} \quad\quad \text{Carbohydrates (g)} \;\; \text{Protein (g)} \;\; \text{Fat (g)} \end{array}$$
$$Y = \begin{bmatrix} 8 & 4 & 6 \\ 6 & 10 & 5 \\ 10 & 4 & 8 \end{bmatrix} \begin{array}{c} A \\ B \\ C \end{array}$$

The nutritional content of a meal consisting of a combination of the foods A, B, and C is the product of the matrix Y and a second

matrix $N = \begin{bmatrix} r \\ s \\ t \end{bmatrix}$, that is, YN, where r is the number of servings of

food A, s is the number of servings of food B, and t is the number of servings of food C.

55. Health/Nutrition. Use the inverse matrix technique to find the number of servings of foods A, B, and C necessary to create a meal of 18 grams of carbohydrates, 21 grams of protein, and 22 grams of fat.

56. Health/Nutrition. Use the inverse matrix technique to find the number of servings of foods A, B, and C necessary to create a meal of 14 grams of carbohydrates, 25 grams of protein, and 16 grams of fat.

For Exercises 57 and 58, refer to the following:

Cell phone companies charge users based on the number of minutes talked, the number of text messages sent, and the number of megabytes of data used. The costs for three cell phone providers are given in the table:

	MINUTES	TEXT MESSAGES	MEGABYTES OF DATA
C_1	$0.03	$0.06	$0.15
C_2	$0.04	$0.05	$0.18
C_3	$0.05	$0.07	$0.13

The cost to a cell phone user for each of the three providers is the product of the matrix X whose rows are the rows of data in the table and the matrix $N = \begin{bmatrix} m \\ t \\ d \end{bmatrix}$ where m is the number of minutes talked, t is the number of text messages sent, and d is the megabytes of data used.

57. Telecommunications/Business. A local business is looking at providing an employee a cell phone for business use. The business solicits estimates for their normal monthly usage from three cell phone providers. Company 1 estimates the cost to be $49.50, Company 2 estimates the cost to be $52.00, and Company 3 estimates the cost to be $58.50. Use the inverse matrix technique to find the normal monthly usage for the employee.

58. Telecommunications/Business. A local business is looking at providing an employee a cell phone for business use. The business solicits estimates for their normal monthly usage from three cell phone providers. Company 1 estimates the cost to be $82.50, Company 2 estimates the cost to be $85.00, and Company 3 estimates the cost to be $92.50. Use the inverse matrix technique to find the normal monthly usage for the employee.

■ **CATCH THE MISTAKE**

In Exercises 59 and 60, explain the mistake that is made.

59. Find the inverse of $A = \begin{bmatrix} 1 & 0 & 1 \\ -1 & 0 & -1 \\ 1 & 2 & 0 \end{bmatrix}$.

Solution:

Write the matrix $[A \mid I_3]$.
$$\begin{bmatrix} 1 & 0 & 1 & | & 1 & 0 & 0 \\ -1 & 0 & -1 & | & 0 & 1 & 0 \\ 1 & 2 & 0 & | & 0 & 0 & 1 \end{bmatrix}$$

Use Gaussian elimination to reduce A.

$\begin{matrix} R_2 + R_1 \to R_2 \\ R_3 - R_1 \to R_3 \end{matrix}$ $\begin{bmatrix} 1 & 0 & 1 & | & 1 & 0 & 0 \\ 0 & 0 & 0 & | & 1 & 1 & 0 \\ 0 & 2 & -1 & | & -1 & 0 & 1 \end{bmatrix}$

$R_2 \leftrightarrow R_3$ $\begin{bmatrix} 1 & 0 & 1 & | & 1 & 0 & 0 \\ 0 & 2 & -1 & | & -1 & 0 & 1 \\ 0 & 0 & 0 & | & 1 & 1 & 0 \end{bmatrix}$

$\frac{1}{2}R_2 \to R_2$ $\begin{bmatrix} 1 & 0 & 1 & | & 1 & 0 & 0 \\ 0 & 1 & -\frac{1}{2} & | & -\frac{1}{2} & 0 & \frac{1}{2} \\ 0 & 0 & 0 & | & 1 & 1 & 0 \end{bmatrix}$

$A^{-1} = \begin{bmatrix} 1 & 0 & 0 \\ -\frac{1}{2} & 0 & \frac{1}{2} \\ 1 & 1 & 0 \end{bmatrix}$ is incorrect because $AA^{-1} \neq I_3$.

What mistake was made?

60. Find the inverse of A given that $A = \begin{bmatrix} 2 & 5 \\ 3 & 10 \end{bmatrix}$.

Solution:
$$A^{-1} = \frac{1}{A} \qquad A^{-1} = \frac{1}{\begin{bmatrix} 2 & 5 \\ 3 & 10 \end{bmatrix}}$$

Simplify.
$$A^{-1} = \begin{bmatrix} \frac{1}{2} & \frac{1}{5} \\ \frac{1}{3} & \frac{1}{10} \end{bmatrix}$$

This is incorrect. What mistake was made?

■ **CONCEPTUAL**

In Exercises 61 and 62, determine whether each statement is true or false.

61. If $A = \begin{bmatrix} a_{11} & a_{12} \\ a_{21} & a_{22} \end{bmatrix}$, then $A^{-1} = \begin{bmatrix} \dfrac{1}{a_{11}} & \dfrac{1}{a_{12}} \\ \dfrac{1}{a_{21}} & \dfrac{1}{a_{22}} \end{bmatrix}$.

62. All square matrices have inverses.

63. For what values of x does the inverse of A not exist, given
$A = \begin{bmatrix} x & 6 \\ 3 & 2 \end{bmatrix}$?

64. Let $A = \begin{bmatrix} a & 0 & 0 \\ 0 & b & 0 \\ 0 & 0 & c \end{bmatrix}$. Find A^{-1}.

■ **CHALLENGE**

65. Verify that $A^{-1} = \dfrac{1}{ad - bc} \begin{bmatrix} d & -b \\ -c & a \end{bmatrix}$ is the inverse of
$A = \begin{bmatrix} a & b \\ c & d \end{bmatrix}$, provided $ad - bc \neq 0$.

66. Let $A = \begin{bmatrix} a & b \\ c & d \end{bmatrix}$ and form the matrix $[A \mid I_2]$. Apply row

operations to transform into $[I_2 \mid A^{-1}]$, where

$A^{-1} = \dfrac{1}{ad - bc} \begin{bmatrix} d & -b \\ -c & a \end{bmatrix}$ such that $ad - bc \neq 0$.

67. Why does the square matrix $A = \begin{bmatrix} 2 & 3 \\ 4 & 6 \end{bmatrix}$ not have an inverse?

68. Why does the square matrix $A = \begin{bmatrix} 1 & 2 & -1 \\ 2 & 4 & -2 \\ 0 & 1 & 3 \end{bmatrix}$ not have an inverse?

■ **TECHNOLOGY**

In Exercises 69 and 70, apply a graphing utility to perform the indicated matrix operations.

$$A = \begin{bmatrix} 1 & 7 & 9 & 2 \\ -3 & -6 & 15 & 11 \\ 0 & 3 & 2 & 5 \\ 9 & 8 & -4 & 1 \end{bmatrix}$$

69. Find A^{-1}. **70.** Find AA^{-1}.

In Exercises 71–74, apply a graphing utility and matrix algebra to solve the system of linear equations.

71. $2.7x - 3.1y = 9.82$
$1.5x + 2.7y = -1.62$

72. $3.7x - 2.5y = 31.77$
$-5.1x + 1.3y = -39.07$

73. $5.1x + 7.3y + 1.2z = 12.51$
$2.3x - 1.5y + 4.5z = 53.96$
$-8.1x + 5.4y - 9.4z = -130.35$

74. $12.4x - 5.8y + 2.7z = -60.92$
$-3.9x + 1.9y - 0.6z = 18.73$
$6.4x - 4.3y + 8.5z = -62.79$

SKILLS OBJECTIVES

- Find the determinant of a 2 × 2 matrix.
- Find the determinant of a $n \times n$ matrix.
- Use Cramer's rule to solve a square system of linear equations.

CONCEPTUAL OBJECTIVES

- Derive Cramer's rule.
- Understand that if a determinant of a matrix is equal to zero, then that matrix does not have an inverse.
- Understand that Cramer's rule can be used to find only a unique solution.

In Section 10.1, we discussed Gauss–Jordan elimination as a way to solve systems of linear equations using augmented matrices. Then in Section 10.3, we employed matrix algebra and inverses to solve systems of linear equations that are square (same number of equations as variables). In this section we will describe another method, called Cramer's rule, for solving systems of linear equations. Cramer's rule is applicable only to square systems. *Determinants* of square matrices play a vital role in Cramer's rule and indicate whether a matrix has an inverse.

Determinant of a 2 × 2 Matrix

Every square matrix A has a number associated with it called its *determinant*, denoted $\det(A)$ or $|A|$.

DEFINITION   **Determinant of a 2 × 2 Matrix**

The **determinant** of the 2 × 2 matrix $A = \begin{bmatrix} a & b \\ c & d \end{bmatrix}$ is given by

$$\det(A) = |A| = \begin{vmatrix} a & b \\ c & d \end{vmatrix} = ad - bc.$$

Although the symbol for determinant, $|\;|$, looks like absolute value bars, the determinant can be any real number (positive, negative, or zero). The determinant of a 2 × 2 matrix is found by finding the product of the main diagonal entries (top left to bottom right) and subtracting the product of the entries along the other diagonal (bottom left to top right).

$$\begin{vmatrix} a & b \\ c & d \end{vmatrix} = ad - bc$$

Study Tip

The determinant of a 2 × 2 matrix is found by finding the product of the main diagonal entries and subtracting the product of the other diagonal entries.

Technology Tip

A graphing calculator can be used to find the determinant of each matrix.

Enter the matrix $\begin{bmatrix} 2 & -5 \\ -1 & 3 \end{bmatrix}$ as A.

Press 2nd MATRIX . Use ► to highlight MATH 1:*det(*
ENTER 2nd MATRIX ENTER)
ENTER .

```
[A]
      [[2  -5]
       [-1  3]]
det([A])
             1
■
```

■ **Answer:** -1

EXAMPLE 1   Finding the Determinant of a 2 × 2 Matrix

Find the determinant of each matrix.

a. $\begin{bmatrix} 2 & -5 \\ -1 & 3 \end{bmatrix}$ **b.** $\begin{bmatrix} 0.5 & 0.2 \\ -3.0 & -4.2 \end{bmatrix}$ **c.** $\begin{bmatrix} \frac{2}{3} & 1 \\ 2 & 3 \end{bmatrix}$

Solution:

a. $\begin{vmatrix} 2 & -5 \\ -1 & 3 \end{vmatrix} = (2)(3) - (-1)(-5) = 6 - 5 = \boxed{1}$

b. $\begin{vmatrix} 0.5 & 0.2 \\ -3 & -4.2 \end{vmatrix} = (0.5)(-4.2) - (-3)(0.2) = -2.1 + 0.6 = \boxed{-1.5}$

c. $\begin{vmatrix} \frac{2}{3} & 1 \\ 2 & 3 \end{vmatrix} = \left(\frac{2}{3}\right)(3) - (2)(1) = 2 - 2 = \boxed{0}$

In Example 1, we see that determinants are real numbers that can be positive, negative, or zero. Although evaluating determinants of 2 × 2 matrices is a simple process, one **common mistake** is reversing the difference: $\begin{vmatrix} a & b \\ c & d \end{vmatrix} \neq bc - ad$.

■ **YOUR TURN** Evaluate the determinant $\begin{vmatrix} -2 & 1 \\ -3 & 2 \end{vmatrix}$.

Classroom Example 10.4.1
Compute the following determinants.

a. $\begin{vmatrix} \pi & 0 \\ -3 & \dfrac{e}{\pi} \end{vmatrix}$

b.* $\begin{vmatrix} 2a & -a \\ -a & a \end{vmatrix}$
where a is a real number.

Answer: a. e **b.** a^2

Determinant of an *n* × *n* Matrix

In order to define the *determinant* of a 3 × 3 or a general $n \times n$ (where $n \geq 3$) matrix, we first define *minors* and *cofactors* of a square matrix.

DEFINITION   Minor and Cofactor

Let A be a square matrix of order $n \times n$. Then

■ The **minor** M_{ij} of the element a_{ij} is the determinant of the $(n-1) \times (n-1)$ matrix obtained when the ith row and jth column of A are deleted.

■ The **cofactor** C_{ij} of the element a_{ij} is given by $C_{ij} = (-1)^{i+j} M_{ij}$.

$A = \begin{bmatrix} 1 & -3 & 2 \\ 4 & -1 & 0 \\ 5 & -2 & 3 \end{bmatrix}$

ELEMENT, a_{ij}	MINOR, M_{ij}	COFACTOR, C_{ij}
$a_{11} = 1$	For M_{11}, delete the first row and first column: $\begin{bmatrix} 1 & -3 & 2 \\ 4 & -1 & 0 \\ 5 & -2 & 3 \end{bmatrix}$ $M_{11} = \begin{vmatrix} -1 & 0 \\ -2 & 3 \end{vmatrix} = -3 - 0 = -3$	$C_{11} = (-1)^{1+1} M_{11}$ $= (1)(-3)$ $= -3$
$a_{32} = -2$	For M_{32}, delete the third row and second column: $\begin{bmatrix} 1 & -3 & 2 \\ 4 & -1 & 0 \\ 5 & -2 & 3 \end{bmatrix}$ $M_{32} = \begin{vmatrix} 1 & 2 \\ 4 & 0 \end{vmatrix} = 0 - 8 = -8$	$C_{32} = (-1)^{3+2} M_{32}$ $= (-1)(-8)$ $= 8$

Notice that the cofactor is simply the minor multiplied by either 1 or -1, depending on whether $i + j$ is even or odd. Therefore, we can make the following sign pattern for 3×3 and 4×4 matrices and obtain the cofactor by multiplying the minor with the appropriate sign ($+1$ or -1).

$$\begin{bmatrix} + & - & + \\ - & + & - \\ + & - & + \end{bmatrix} \qquad \begin{bmatrix} + & - & + & - \\ - & + & - & + \\ + & - & + & - \\ - & + & - & + \end{bmatrix}$$

DEFINITION   **Determinant of an $n \times n$ Matrix**

Let A be an $n \times n$ matrix. Then the **determinant** of A is found by summing the elements in *any* row of A (or column of A), multiplied by each element's respective cofactor.

If A is a 3×3, the determinant can be given by $\det(A) = a_{11}C_{11} + a_{12}C_{12} + a_{13}C_{13}$; this is called **expanding the determinant by the first row**. It is important to note that any row or column can be used. Typically, the row or column with the most zeros is selected because it makes the arithmetic simpler.

Combining the definitions of minors, cofactors, and determinants, we now give a general definition for the determinant of a 3×3 matrix.

Row 1 expansion:
$$\begin{vmatrix} a_1 & b_1 & c_1 \\ a_2 & b_2 & c_2 \\ a_3 & b_3 & c_3 \end{vmatrix} = a_1 \begin{vmatrix} b_2 & c_2 \\ b_3 & c_3 \end{vmatrix} - b_1 \begin{vmatrix} a_2 & c_2 \\ a_3 & c_3 \end{vmatrix} + c_1 \begin{vmatrix} a_2 & b_2 \\ a_3 & b_3 \end{vmatrix}$$

Column 1 expansion:
$$\begin{vmatrix} a_1 & b_1 & c_1 \\ a_2 & b_2 & c_2 \\ a_3 & b_3 & c_3 \end{vmatrix} = a_1 \begin{vmatrix} b_2 & c_2 \\ b_3 & c_3 \end{vmatrix} - a_2 \begin{vmatrix} b_1 & c_1 \\ b_3 & c_3 \end{vmatrix} + a_3 \begin{vmatrix} b_1 & c_1 \\ b_2 & c_2 \end{vmatrix}$$

Whichever row or column is expanded, an alternating sign scheme is used (see sign arrays above). Notice that in either of the expansions above, the 2×2 determinant obtained is found by crossing out the row and column containing the element that is multiplying the determinant.

Study Tip

The determinant by the third column is also 102. It does not matter on which row or column the expansion occurs.

■ **Answer: 156**

Technology Tip

Enter the matrix as A and find the determinant.

Press 2nd MATRIX. Use ▶
to highlight MATH 1:*det(* 2nd
MATRIX ENTER) ENTER.

```
[A]
        [[-1 2 0]
         [4 7 1]
         [5 3 0]]
det([A])
                13
■
```

■ **Answer: 20**

EXAMPLE 2   Finding the Determinant of a 3 × 3 Matrix

For the given matrix, expand the determinant by the *first row*.

$$\begin{bmatrix} 2 & 1 & 3 \\ -1 & 5 & -2 \\ -3 & 7 & 4 \end{bmatrix}$$

Solution:

Expand the determinant by the **first** row. Remember the alternating **sign**.

$$\begin{vmatrix} 2 & 1 & 3 \\ -1 & 5 & -2 \\ -3 & 7 & 4 \end{vmatrix} = +\,2\begin{vmatrix} 5 & -2 \\ 7 & 4 \end{vmatrix} - 1\begin{vmatrix} -1 & -2 \\ -3 & 4 \end{vmatrix} + 3\begin{vmatrix} -1 & 5 \\ -3 & 7 \end{vmatrix}$$

Evaluate the resulting 2 × 2 determinants.

$$= 2[(5)(4) - (7)(-2)] - 1[(-1)(4) - (-3)(-2)] + 3[(-1)(7) - (-3)(5)]$$

$$= 2[20 + 14] - [-4 - 6] + 3[-7 + 15]$$

Simplify.

$$= 2(34) - (-10) + 3(8)$$

$$= 68 + 10 + 24$$

$$= \boxed{102}$$

■ **YOUR TURN** For the given matrix, expand the determinant by the first row.

$$\begin{bmatrix} 1 & 3 & -2 \\ 2 & 5 & 4 \\ 7 & -1 & 6 \end{bmatrix}$$

Determinants can be expanded by any row *or* column. Typically, the row or column with the most zeros is selected.

EXAMPLE 3   Finding the Determinant of a 3 × 3 Matrix

Find the determinant of the matrix $\begin{vmatrix} -1 & 2 & 0 \\ 4 & 7 & 1 \\ 5 & 3 & 0 \end{vmatrix}$.

Solution:

Since there are two 0s in the third column, expand the determinant by the third column. Recall the sign array.

$$\begin{bmatrix} + & - & + \\ - & + & - \\ + & - & + \end{bmatrix}$$

$$\begin{vmatrix} -1 & 2 & 0 \\ 4 & 7 & 1 \\ 5 & 3 & 0 \end{vmatrix} = +\,0\begin{vmatrix} 4 & 7 \\ 5 & 3 \end{vmatrix} - 1\begin{vmatrix} -1 & 2 \\ 5 & 3 \end{vmatrix} + 0\begin{vmatrix} -1 & 2 \\ 4 & 7 \end{vmatrix}$$

There is no need to calculate the two determinants that are multiplied by 0s, since 0 times any real number is zero.

$$\begin{vmatrix} -1 & 2 & 0 \\ 4 & 7 & 1 \\ 5 & 3 & 0 \end{vmatrix} = 0 - 1\underbrace{\begin{vmatrix} -1 & 2 \\ 5 & 3 \end{vmatrix}}_{-3-10} + 0$$

Simplify.

$$= -1(-13) = \boxed{13}$$

■ **YOUR TURN** Evaluate the determinant $\begin{vmatrix} 1 & -2 & 1 \\ -1 & 0 & 3 \\ -4 & 0 & 2 \end{vmatrix}$.

EXAMPLE 4   Finding the Determinant of a 4 × 4 Matrix

Find the determinant of the matrix $\begin{vmatrix} 1 & -2 & 3 & 4 \\ -4 & 0 & -1 & 0 \\ -3 & 9 & 6 & 5 \\ -5 & 7 & 2 & 1 \end{vmatrix}$.

Solution:

Since there are two 0s in the second row, expand the determinant by the second row. Recall the sign array for a 4 × 4 matrix.

$$\begin{bmatrix} + & - & + & - \\ - & + & - & + \\ + & - & + & - \\ - & + & - & + \end{bmatrix}$$

$$\begin{bmatrix} 1 & -2 & 3 & 4 \\ -4 & 0 & -1 & 0 \\ -3 & 9 & 6 & 5 \\ -5 & 7 & 2 & 1 \end{bmatrix} = -\,-(-4)\begin{vmatrix} -2 & 3 & 4 \\ 9 & 6 & 5 \\ 7 & 2 & 1 \end{vmatrix} + 0 - (-1)\begin{vmatrix} 1 & -2 & 4 \\ -3 & 9 & 5 \\ -5 & 7 & 1 \end{vmatrix} + 0$$

Evaluate the two 3 × 3 determinants.

$$\begin{vmatrix} -2 & 3 & 4 \\ 9 & 6 & 5 \\ 7 & 2 & 1 \end{vmatrix} = -2\begin{vmatrix} 6 & 5 \\ 2 & 1 \end{vmatrix} - 3\begin{vmatrix} 9 & 5 \\ 7 & 1 \end{vmatrix} + 4\begin{vmatrix} 9 & 6 \\ 7 & 2 \end{vmatrix}$$

$$= -2(6 - 10) - 3(9 - 35) + 4(18 - 42)$$
$$= -2(-4) - 3(-26) + 4(-24)$$
$$= 8 + 78 - 96$$
$$= -10$$

$$\begin{vmatrix} 1 & -2 & 4 \\ -3 & 9 & 5 \\ -5 & 7 & 1 \end{vmatrix} = 1\begin{vmatrix} 9 & 5 \\ 7 & 1 \end{vmatrix} - (-2)\begin{vmatrix} -3 & 5 \\ -5 & 1 \end{vmatrix} + 4\begin{vmatrix} -3 & 9 \\ -5 & 7 \end{vmatrix}$$

$$= 1(9 - 35) + 2(-3 + 25) + 4(-21 + 45)$$
$$= -26 + 2(22) + 4(24)$$
$$= -26 + 44 + 96$$
$$= 114$$

$$\begin{bmatrix} 1 & -2 & 3 & 4 \\ -4 & 0 & -1 & 0 \\ -3 & 9 & 6 & 5 \\ -5 & 7 & 2 & 1 \end{bmatrix} = 4\underbrace{\begin{vmatrix} -2 & 3 & 4 \\ 9 & 6 & 5 \\ 7 & 2 & 1 \end{vmatrix}}_{-10} + \underbrace{\begin{vmatrix} 1 & -2 & 4 \\ -3 & 9 & 5 \\ -5 & 7 & 1 \end{vmatrix}}_{114} = 4(-10) + 114 = \boxed{74}$$

Cramer's Rule: Systems of Linear Equations in Two Variables

Let's now apply determinants of 2×2 matrices to solve systems of linear equations in two variables. We begin by solving the general system of two linear equations in two variables:

$$(1) \quad a_1 x + b_1 y = c_1$$

$$(2) \quad a_2 x + b_2 y = c_2$$

Solve for x using elimination (eliminate y).

Multiply (1) by b_2.
$$b_2 a_1 x + b_2 b_1 y = b_2 c_1$$

Multiply (2) by $-b_1$.
$$-b_1 a_2 x - b_1 b_2 y = -b_1 c_2$$

Add the two new equations to eliminate y.
$$(a_1 b_2 - a_2 b_1)x = (b_2 c_1 - b_1 c_2)$$

Divide both sides by $(a_1 b_2 - a_2 b_1)$.
$$x = \frac{(b_2 c_1 - b_1 c_2)}{(a_1 b_2 - a_2 b_1)}$$

Write both the numerator and the denominator as determinants.
$$x = \frac{\begin{vmatrix} c_1 & b_1 \\ c_2 & b_2 \end{vmatrix}}{\begin{vmatrix} a_1 & b_1 \\ a_2 & b_2 \end{vmatrix}}$$

Solve for y using elimination (eliminate x).

Multiply (1) by $-a_2$.
$$-a_2 a_1 x - a_2 b_1 y = -a_2 c_1$$

Multiply (2) by a_1.
$$a_1 a_2 x + a_1 b_2 y = a_1 c_2$$

Add the two new equations to eliminate x.
$$(a_1 b_2 - a_2 b_1)y = (a_1 c_2 - a_2 c_1)$$

Divide both sides by $(a_1 b_2 - a_2 b_1)$.
$$y = \frac{(a_1 c_2 - a_2 c_1)}{(a_1 b_2 - a_2 b_1)}$$

Write both the numerator and the denominator as determinants.
$$y = \frac{\begin{vmatrix} a_1 & c_1 \\ a_2 & c_2 \end{vmatrix}}{\begin{vmatrix} a_1 & b_1 \\ a_2 & b_2 \end{vmatrix}}$$

Notice that the solutions for x and y involve three determinants. If we let

$$D = \begin{vmatrix} a_1 & b_1 \\ a_2 & b_2 \end{vmatrix} \qquad D_x = \begin{vmatrix} c_1 & b_1 \\ c_2 & b_2 \end{vmatrix} \qquad D_y = \begin{vmatrix} a_1 & c_1 \\ a_2 & c_2 \end{vmatrix}$$

then

$$x = \frac{D_x}{D} \text{ and } y = \frac{D_y}{D}.$$

Notice that the matrix D is the determinant of the coefficient matrix of the system and cannot equal zero ($D \neq 0$) or there will be no unique solution. These formulas for solving a system of two linear equations in two variables are known as *Cramer's rule*.

CRAMER'S RULE FOR SOLVING SYSTEMS OF TWO LINEAR EQUATIONS IN TWO VARIABLES

For the system of linear equations

$$a_1 x + b_1 y = c_1$$
$$a_2 x + b_2 y = c_2$$

let

$$D = \begin{vmatrix} a_1 & b_1 \\ a_2 & b_2 \end{vmatrix} \qquad D_x = \begin{vmatrix} c_1 & b_1 \\ c_2 & b_2 \end{vmatrix} \qquad D_y = \begin{vmatrix} a_1 & c_1 \\ a_2 & c_2 \end{vmatrix}$$

If $D \neq 0$, then the solution to the system of linear equations is

$$x = \frac{D_x}{D}, \qquad y = \frac{D_y}{D}$$

If $D = 0$, then the system of linear equations has either no solution or infinitely many solutions.

Study Tip

Cramer's rule is only applicable to square systems of linear equations.

Notice that the determinants D_x and D_y are similar to the determinant D. A three-step procedure is outlined for setting up the three determinants for a system of two linear equations in two variables.

$$a_1 x + b_1 y = \mathbf{c_1}$$
$$a_2 x + b_2 y = \mathbf{c_2}$$

Step 1: Set up D.

Apply the coefficients of x and y.

$$D = \begin{vmatrix} a_1 & b_1 \\ a_2 & b_2 \end{vmatrix}$$

Step 2: Set up D_x.

Start with D and replace the coefficients of x (column 1) with the constants on the right side of the equal sign.

$$D_x = \begin{vmatrix} \mathbf{c_1} & b_1 \\ \mathbf{c_2} & b_2 \end{vmatrix}$$

Step 3: Set up D_y.

Start with D and replace the coefficients of y (column 2) with the constants on the right side of the equal sign.

$$D_y = \begin{vmatrix} a_1 & \mathbf{c_1} \\ a_2 & \mathbf{c_2} \end{vmatrix}$$

EXAMPLE 5   Using Cramer's Rule to Solve a System of Two Linear Equations

Apply Cramer's rule to solve the system.

$$x + 3y = 1$$
$$2x + y = -3$$

Technology Tip

A graphing calculator can be used to solve the system using Cramer's rule. Enter the matrix A for the determinant D_x, B for D_y, C for D.

[A]
$$\begin{bmatrix} 1 & 3 \\ -3 & 1 \end{bmatrix}$$

[B]
$$\begin{bmatrix} 1 & 1 \\ 2 & -3 \end{bmatrix}$$

[C]
$$\begin{bmatrix} 1 & 3 \\ 2 & 1 \end{bmatrix}$$

To solve for x and y, enter D_x/D as A/C for x and D_y/D as B/C for y.

```
det([A])/det([C]
)
               -2
det([B])/det([C]
)
                1
```

■ **Answer:** $x = 5, y = -6$

Solution:

Set up the three determinants.

$$D = \begin{vmatrix} 1 & 3 \\ 2 & 1 \end{vmatrix}$$

$$D_x = \begin{vmatrix} 1 & 3 \\ -3 & 1 \end{vmatrix}$$

$$D_y = \begin{vmatrix} 1 & 1 \\ 2 & -3 \end{vmatrix}$$

Evaluate the determinants.

$$D = 1 - 6 = -5$$
$$D_x = 1 - (-9) = 10$$
$$D_y = -3 - 2 = -5$$

Solve for x and y.

$$x = \frac{D_x}{D} = \frac{10}{-5} = -2$$

$$y = \frac{D_y}{D} = \frac{-5}{-5} = 1$$

$$\boxed{x = -2, y = 1}$$

■ **YOUR TURN** Apply Cramer's rule to solve the system.

$$5x + 4y = 1$$
$$-3x - 2y = -3$$

Classroom Example 10.4.5
Solve the systems using Cramer's rule.

a. $-2y + x = 2$
$\quad x + 4y = 3$

b. $3x - y = 2$
$\quad 2y = 6x - 1$

c.* $ax - 2ay = 1$
$\quad ay = -x$

where $a \neq 0, -2$

Answer:

a. $x = \frac{7}{3}, y = \frac{1}{6}$

b. no solution

c. $x = \dfrac{1}{a + 2}$

$\quad y = -\dfrac{1}{a(a + 2)}$

Recall from Section 6.1 that systems of two equations and two variables led to one of three possible outcomes: a unique solution, no solution, and infinitely many solutions. When $D = 0$, Cramer's rule does not apply and the system is either inconsistent (has no solution) or contains dependent equations (has infinitely many solutions).

Cramer's Rule: Systems of Linear Equations in Three Variables

Cramer's rule can also be used to solve higher order systems of linear equations. The following box summarizes Cramer's rule for solving a system of three equations in three variables.

CRAMER'S RULE: SOLUTION FOR SYSTEMS OF THREE EQUATIONS IN THREE VARIABLES

The system of linear equations

$$a_1x + b_1y + c_1z = d_1$$
$$a_2x + b_2y + c_2z = d_2$$
$$a_3x + b_3y + c_3z = d_3$$

has the solution

$$x = \frac{D_x}{D} \qquad y = \frac{D_y}{D} \qquad z = \frac{D_z}{D} \qquad D \neq 0$$

where the determinants are given as follows:

WORDS	MATH
Display the coefficients of x, y, and z.	$D = \begin{vmatrix} a_1 & b_1 & c_1 \\ a_2 & b_2 & c_2 \\ a_3 & b_3 & c_3 \end{vmatrix}$
Replace the coefficients of x (column 1) in D with the constants on the right side of the equal sign.	$D_x = \begin{vmatrix} d_1 & b_1 & c_1 \\ d_2 & b_2 & c_2 \\ d_3 & b_3 & c_3 \end{vmatrix}$
Replace the coefficients of y (column 2) in D with the constants on the right side of the equal sign.	$D_y = \begin{vmatrix} a_1 & d_1 & c_1 \\ a_2 & d_2 & c_2 \\ a_3 & d_3 & c_3 \end{vmatrix}$
Replace the coefficients of z (column 3) in D with the constants on the right side of the equal sign.	$D_z = \begin{vmatrix} a_1 & b_1 & d_1 \\ a_2 & b_2 & d_2 \\ a_3 & b_3 & d_3 \end{vmatrix}$

EXAMPLE 6   Using Cramer's Rule to Solve a System of Three Linear Equations

Use Cramer's rule to solve the system.

$$3x - 2y + 3z = -3$$
$$5x + 3y + 8z = -2$$
$$x + y + 3z = 1$$

Solution:

Set up the four determinants.

D contains the coefficients of x, y, and z.
$$D = \begin{vmatrix} 3 & -2 & 3 \\ 5 & 3 & 8 \\ 1 & 1 & 3 \end{vmatrix}$$

Replace a column with constants on the right side of the equation.

$$D_x = \begin{vmatrix} -3 & -2 & 3 \\ -2 & 3 & 8 \\ 1 & 1 & 3 \end{vmatrix} \qquad D_y = \begin{vmatrix} 3 & -3 & 3 \\ 5 & -2 & 8 \\ 1 & 1 & 3 \end{vmatrix} \qquad D_z = \begin{vmatrix} 3 & -2 & -3 \\ 5 & 3 & -2 \\ 1 & 1 & 1 \end{vmatrix}$$

Evaluate the determinants.

$$D = 3(9 - 8) - (-2)(15 - 8) + 3(5 - 3) = 23$$
$$D_x = -3(9 - 8) - (-2)(-6 - 8) + 3(-2 - 3) = -46$$
$$D_y = 3(-6 - 8) - (-3)(15 - 8) + 3(5 + 2) = 0$$
$$D_z = 3(3 + 2) - (-2)(5 + 2) - 3(5 - 3) = 23$$

Solve for x, y, and z.

$$x = \frac{D_x}{D} = \frac{-46}{23} = -2 \qquad y = \frac{D_y}{D} = \frac{0}{23} = 0 \qquad z = \frac{D_z}{D} = \frac{23}{23} = 1$$

$$\boxed{x = -2, y = 0, z = 1}$$

YOUR TURN Use Cramer's rule to solve the system.

$$2x + 3y + z = -1$$
$$x - y - z = 0$$
$$-3x - 2y + 3z = 10$$

▪ **Answer:** $x = 1, y = -2, z = 3$

As was the case in two equations, when $D = 0$, Cramer's rule does not apply and the system of three equations is either inconsistent (no solution) or contains dependent equations (infinitely many solutions).

All of the coefficients and constants in the systems we have solved have been nice (rational). One of the advantages to Cramer's rule is the ability to solve with ease systems of linear equations with irrational coefficients.

EXAMPLE 7   Using Cramer's Rule to Solve a System of Linear Equations with Irrational Coefficients and Constants

Apply Cramer's rule to solve the system.

$$5x - ey + z = -\sqrt{2}$$
$$\sqrt{3}x + 2y - \pi z = 1$$
$$4x - 7y + \sqrt{5}z = 0$$

Solution:

Set up the four determinants.

D contains the coefficients of x, y, and z.
$$D = \begin{vmatrix} 5 & -e & 1 \\ \sqrt{3} & 2 & -\pi \\ 4 & -7 & \sqrt{5} \end{vmatrix}$$

Replace a column with constants on the right side of the equation.

$$D_x = \begin{vmatrix} -\sqrt{2} & -e & 1 \\ 1 & 2 & -\pi \\ 0 & -7 & \sqrt{5} \end{vmatrix} \qquad D_y = \begin{vmatrix} 5 & -\sqrt{2} & 1 \\ \sqrt{3} & 1 & -\pi \\ 4 & 0 & \sqrt{5} \end{vmatrix} \qquad D_z = \begin{vmatrix} 5 & -e & -\sqrt{2} \\ \sqrt{3} & 2 & 1 \\ 4 & -7 & 0 \end{vmatrix}$$

Evaluate the determinants (along the third row because of the zero).

$$D_x = 0 - (-7)\begin{vmatrix} -\sqrt{2} & 1 \\ 1 & -\pi \end{vmatrix} + \sqrt{5}\begin{vmatrix} -\sqrt{2} & -e \\ 1 & 2 \end{vmatrix}$$
$$= 7\left[(-\sqrt{2})(-\pi) - (1)(1)\right] + \sqrt{5}\left[(-\sqrt{2})(2) - (1)(-e)\right]$$
$$= 7\left[\pi\sqrt{2} - 1\right] + \sqrt{5}\left[-2\sqrt{2} + e\right]$$
$$= 7\pi\sqrt{2} - 7 - 2\sqrt{10} + e\sqrt{5}$$

$$D_y = 4\begin{vmatrix} -\sqrt{2} & 1 \\ 1 & -\pi \end{vmatrix} - 0 + \sqrt{5}\begin{vmatrix} 5 & -\sqrt{2} \\ \sqrt{3} & 1 \end{vmatrix}$$
$$= 4\left[(-\sqrt{2})(-\pi) - (1)(1)\right] + \sqrt{5}\left[(5)(1) - (\sqrt{3})(-\sqrt{2})\right]$$
$$= 4\left[\pi\sqrt{2} - 1\right] + \sqrt{5}\left[5 + \sqrt{6}\right]$$
$$= 4\pi\sqrt{2} - 4 + 5\sqrt{5} + \sqrt{30}$$

$$D_z = 4\begin{vmatrix} -e & -\sqrt{2} \\ 2 & 1 \end{vmatrix} - (-7)\begin{vmatrix} 5 & -\sqrt{2} \\ \sqrt{3} & 1 \end{vmatrix} + 0$$
$$= 4\left[(-e)(1) - (2)(-\sqrt{2})\right] + 7\left[(5)(1) - (\sqrt{3})(-\sqrt{2})\right]$$
$$= 4\left[-e + 2\sqrt{2}\right] + 7\left[5 + \sqrt{6}\right]$$
$$= -4e + 8\sqrt{2} + 35 + 7\sqrt{6}$$

$$D = 4\begin{vmatrix} -e & 1 \\ 2 & -\pi \end{vmatrix} - (-7)\begin{vmatrix} 5 & 1 \\ \sqrt{3} & -\pi \end{vmatrix} + \sqrt{5}\begin{vmatrix} 5 & -e \\ \sqrt{3} & 2 \end{vmatrix}$$
$$= 4[(-e)(-\pi) - (2)(1)] + 7\left[(5)(-\pi) - (\sqrt{3})(1)\right] + \sqrt{5}\left[(5)(2) - (\sqrt{3})(-e)\right]$$
$$= 4[e\pi - 2] + 7\left[-5\pi - \sqrt{3}\right] + \sqrt{5}\left[10 + e\sqrt{3}\right]$$
$$= 4e\pi - 8 - 35\pi - 7\sqrt{3} + 10\sqrt{5} + e\sqrt{15}$$

Solve for x, y, and z.

$$x = \frac{D_x}{D} = \frac{7\pi\sqrt{2} - 7 - 2\sqrt{10} + e\sqrt{5}}{4e\pi - 8 - 35\pi - 7\sqrt{3} + 10\sqrt{5} + e\sqrt{15}}$$

$$y = \frac{D_y}{D} = \frac{4\pi\sqrt{2} - 4 + 5\sqrt{5} + \sqrt{30}}{4e\pi - 8 - 35\pi - 7\sqrt{3} + 10\sqrt{5} + e\sqrt{15}}$$

$$z = \frac{D_z}{D} = \frac{-4e + 8\sqrt{2} + 35 + 7\sqrt{6}}{4e\pi - 8 - 35\pi - 7\sqrt{3} + 10\sqrt{5} + e\sqrt{15}}$$

SECTION 10.4   SUMMARY

In this section **determinants** were discussed for square matrices.

ORDER	DETERMINANT	ARRAY
2×2	$\det(A) = \lvert A \rvert = \begin{vmatrix} a & b \\ c & d \end{vmatrix} = ad - bc$	
3×3	$\begin{vmatrix} a_1 & b_1 & c_1 \\ a_2 & b_2 & c_2 \\ a_3 & b_3 & c_3 \end{vmatrix} = a_1 \begin{vmatrix} b_2 & c_2 \\ b_3 & c_3 \end{vmatrix} - b_1 \begin{vmatrix} a_2 & c_2 \\ a_3 & c_3 \end{vmatrix} + c_1 \begin{vmatrix} a_2 & b_2 \\ a_3 & b_3 \end{vmatrix}$ Expansion by first row (any row or column can be used)	$\begin{bmatrix} + & - & + \\ - & + & - \\ + & - & + \end{bmatrix}$

Cramer's rule was developed for 2×2 and 3×3 matrices, but it can be extended to general $n \times n$ matrices. When the coefficient determinant is equal to zero ($D = 0$), then the system is either inconsistent (and has no solution) or represents dependent equations (and has infinitely many solutions) and Cramer's rule does not apply.

SYSTEM	ORDER	SOLUTION	DETERMINANTS
$a_1x + b_1y = c_1$ $a_2x + b_2y = c_2$	2×2	$x = \dfrac{D_x}{D} \qquad y = \dfrac{D_y}{D}$	$D = \begin{vmatrix} a_1 & b_1 \\ a_2 & b_2 \end{vmatrix} \neq 0$ $D_x = \begin{vmatrix} c_1 & b_1 \\ c_2 & b_2 \end{vmatrix}$ $D_y = \begin{vmatrix} a_1 & c_1 \\ a_2 & c_2 \end{vmatrix}$
$a_1x + b_1y + c_1z = d_1$ $a_2x + b_2y + c_2z = d_2$ $a_3x + b_3y + c_3z = d_3$	3×3	$x = \dfrac{D_x}{D} \qquad y = \dfrac{D_y}{D} \qquad z = \dfrac{D_z}{D}$	$D = \begin{vmatrix} a_1 & b_1 & c_1 \\ a_2 & b_2 & c_2 \\ a_3 & b_3 & c_3 \end{vmatrix} \neq 0$ $D_x = \begin{vmatrix} d_1 & b_1 & c_1 \\ d_2 & b_2 & c_2 \\ d_3 & b_3 & c_3 \end{vmatrix}$ $D_y = \begin{vmatrix} a_1 & d_1 & c_1 \\ a_2 & d_2 & c_2 \\ a_3 & d_3 & c_3 \end{vmatrix}$ $D_z = \begin{vmatrix} a_1 & b_1 & d_1 \\ a_2 & b_2 & d_2 \\ a_3 & b_3 & d_3 \end{vmatrix}$

SKILLS

In Exercises 1–10, evaluate each 2×2 determinant.

1. $\begin{vmatrix} 1 & 2 \\ 3 & 4 \end{vmatrix}$

2. $\begin{vmatrix} 1 & -2 \\ -3 & -4 \end{vmatrix}$

3. $\begin{vmatrix} 7 & 9 \\ -5 & -2 \end{vmatrix}$

4. $\begin{vmatrix} -3 & -11 \\ 7 & 15 \end{vmatrix}$

5. $\begin{vmatrix} 0 & 7 \\ 4 & -1 \end{vmatrix}$

6. $\begin{vmatrix} 0 & 0 \\ 1 & 0 \end{vmatrix}$

7. $\begin{vmatrix} -1.2 & 2.4 \\ -0.5 & 1.5 \end{vmatrix}$

8. $\begin{vmatrix} -1.0 & 1.4 \\ 1.5 & -2.8 \end{vmatrix}$

9. $\begin{vmatrix} \frac{3}{4} & \frac{1}{3} \\ 2 & \frac{8}{9} \end{vmatrix}$

10. $\begin{vmatrix} -\frac{1}{2} & \frac{1}{4} \\ \frac{2}{3} & -\frac{8}{9} \end{vmatrix}$

In Exercises 11–32, use Cramer's rule to solve each system of equations, if possible.

11. $x + y = -1$
$x - y = 11$

12. $x + y = -1$
$x - y = -9$

13. $3x + 2y = -4$
$-2x + y = 5$

14. $5x + 3y = 1$
$4x - 7y = -18$

15. $3x - 2y = -1$
$5x + 4y = -31$

16. $x - 4y = -7$
$3x + 8y = 19$

17. $7x - 3y = -29$
$5x + 2y = 0$

18. $6x - 2y = 24$
$4x + 7y = 41$

19. $3x + 5y = 16$
$y - x = 0$

20. $-2x - 3y = 15$
$7y + 4x = -33$

21. $3x - 5y = 7$
$-6x + 10y = -21$

22. $3x - 5y = 7$
$6x - 10y = 14$

23. $2x - 3y = 4$
$-10x + 15y = -20$

24. $2x - 3y = 2$
$10x - 15y = 20$

25. $3x + \frac{1}{2}y = 1$
$4x + \frac{1}{3}y = \frac{5}{3}$

26. $\frac{3}{2}x + \frac{9}{4}y = \frac{9}{8}$
$\frac{1}{3}x + \frac{1}{4}y = \frac{1}{12}$

27. $0.3x - 0.5y = -0.6$
$0.2x + y = 2.4$

28. $0.5x - 0.4y = -3.6$
$10x + 3.6y = -14$

29. $y = 17x + 7$
$y = -15x + 7$

30. $9x = -45 - 2y$
$4x = -3y - 20$

31. $\dfrac{2}{x} - \dfrac{3}{y} = 2$

$\dfrac{5}{x} - \dfrac{6}{y} = 7$

32. $\dfrac{2}{x} - \dfrac{3}{y} = -12$

$\dfrac{3}{x} + \dfrac{1}{2y} = 7$

In Exercises 33–46, evaluate each 3×3 determinant.

33. $\begin{vmatrix} 3 & 1 & 0 \\ 2 & 0 & -1 \\ -4 & 1 & 0 \end{vmatrix}$

34. $\begin{vmatrix} 1 & 1 & 0 \\ 0 & 2 & -1 \\ 0 & -3 & 5 \end{vmatrix}$

35. $\begin{vmatrix} 2 & 1 & -5 \\ 3 & 0 & -1 \\ 4 & 0 & 7 \end{vmatrix}$

36. $\begin{vmatrix} 2 & 1 & -5 \\ 3 & -7 & 0 \\ 4 & -6 & 0 \end{vmatrix}$

37. $\begin{vmatrix} 1 & 1 & -5 \\ 3 & -7 & -4 \\ 4 & -6 & 9 \end{vmatrix}$

38. $\begin{vmatrix} -3 & 2 & -5 \\ 1 & 8 & 2 \\ 4 & -6 & 9 \end{vmatrix}$

39. $\begin{vmatrix} 1 & 3 & 4 \\ 2 & -1 & 1 \\ 3 & -2 & 1 \end{vmatrix}$

40. $\begin{vmatrix} -7 & 2 & 5 \\ \frac{7}{8} & 3 & 4 \\ -1 & 4 & 6 \end{vmatrix}$

41. $\begin{vmatrix} -3 & 1 & 5 \\ 2 & 0 & 6 \\ 4 & 7 & -9 \end{vmatrix}$

42. $\begin{vmatrix} 1 & -1 & 5 \\ 3 & -3 & 6 \\ 4 & 9 & 0 \end{vmatrix}$

43. $\begin{vmatrix} -2 & 1 & -7 \\ 4 & -2 & 14 \\ 0 & 1 & 8 \end{vmatrix}$

44. $\begin{vmatrix} 5 & -2 & -1 \\ 4 & -9 & -3 \\ 2 & 8 & -6 \end{vmatrix}$

45. $\begin{vmatrix} \frac{3}{4} & -1 & 0 \\ 0 & \frac{1}{5} & -12 \\ 8 & 0 & -2 \end{vmatrix}$

46. $\begin{vmatrix} 0.2 & 0 & 3 \\ 5 & -1.4 & 2 \\ -1 & 0 & -3 \end{vmatrix}$

In Exercises 47–60, apply Cramer's rule to solve each system of equations, if possible.

47. $\begin{aligned} x + y - z &= 0 \\ x - y + z &= 4 \\ x + y + z &= 10 \end{aligned}$

48. $\begin{aligned} -x + y + z &= -4 \\ x + y - z &= 0 \\ x + y + z &= 2 \end{aligned}$

49. $\begin{aligned} 3x + 8y + 2z &= 28 \\ -2x + 5y + 3z &= 34 \\ 4x + 9y + 2z &= 29 \end{aligned}$

50. $\begin{aligned} 7x + 2y - z &= -1 \\ 6x + 5y + z &= 16 \\ -5x - 4y + 3z &= -5 \end{aligned}$

51. $\begin{aligned} 3x \quad\quad + 5z &= 11 \\ 4y + 3z &= -9 \\ 2x - y \quad\quad &= 7 \end{aligned}$

52. $\begin{aligned} 3x \quad\quad - 2z &= 7 \\ 4x \quad\quad + z &= 24 \\ 6x - 2y \quad\quad &= 10 \end{aligned}$

53. $\begin{aligned} x + y - z &= 5 \\ x - y + z &= -1 \\ -2x - 2y + 2z &= -10 \end{aligned}$

54. $\begin{aligned} x + y - z &= 3 \\ x - y + z &= -2 \\ -2x - 2y + 2z &= -6 \end{aligned}$

55. $\begin{aligned} x + y + z &= 9 \\ x - y + z &= 3 \\ -x + y - z &= 5 \end{aligned}$

56. $\begin{aligned} x + y + z &= 6 \\ x - y - z &= 0 \\ -x + y + z &= 7 \end{aligned}$

57. $\begin{aligned} x + 2y + 3z &= 11 \\ -2x + 3y + 5z &= 29 \\ 4x - y + 8z &= 19 \end{aligned}$

58. $\begin{aligned} 8x - 2y + 5z &= 36 \\ 3x + y - z &= 17 \\ 2x - 6y + 4z &= -2 \end{aligned}$

59. $\begin{aligned} x - 4y + 7z &= 49 \\ -3x + 2y - z &= -17 \\ 5x + 8y - 2z &= -24 \end{aligned}$

60. $\begin{aligned} \tfrac{1}{2}x - 2y + 7z &= 25 \\ x + \tfrac{1}{4}y - 4z &= -2 \\ -4x + 5y \quad\quad &= -56 \end{aligned}$

▪ APPLICATIONS

For Exercises 61–64, the area of a triangle with vertices (x_1, y_1), (x_2, y_2), and (x_3, y_3) is given by

$$\text{Area} = \pm\frac{1}{2}\begin{vmatrix} x_1 & y_1 & 1 \\ x_2 & y_2 & 1 \\ x_3 & y_3 & 1 \end{vmatrix}$$

where the sign is chosen so that the area is positive.

61. Geometry. Apply determinants to find the area of a triangle with vertices $(3, 2)$, $(5, 2)$, and $(3, -4)$. Check your answer by plotting these vertices in a Cartesian plane and using the area of a right triangle.

62. Geometry. Apply determinants to find the area of a triangle with vertices $(2, 3)$, $(7, 3)$, and $(7, 7)$. Check your answer by plotting these vertices in a Cartesian plane and using the area of a right triangle.

63. Geometry. Apply determinants to find the area of a triangle with vertices $(1, 2)$, $(3, 4)$, and $(-2, 5)$.

64. Geometry. Apply determinants to find the area of a triangle with vertices $(-1, -2)$, $(3, 4)$, and $(2, 1)$.

65. Geometry. An equation of a line that passes through two points (x_1, y_1) and (x_2, y_2) can be expressed as a determinant:

$$\begin{vmatrix} x & y & 1 \\ x_1 & y_1 & 1 \\ x_2 & y_2 & 1 \end{vmatrix} = 0$$

Apply the determinant to write an equation of the line passing through the points $(1, 2)$ and $(2, 4)$. Expand the determinant and express the equation of the line in slope–intercept form.

66. Geometry. If three points (x_1, y_1), (x_2, y_2), and (x_3, y_3) are collinear (lie on the same line), then the following determinant must be satisfied:

$$\begin{vmatrix} x_1 & y_1 & 1 \\ x_2 & y_2 & 1 \\ x_3 & y_3 & 1 \end{vmatrix} = 0$$

Determine whether $(0, 5)$, $(2, 0)$, and $(1, 2)$ are collinear.

67. Electricity: Circuit Theory. The following equations come from circuit theory. Find the currents I_1, I_2, and I_3.

$$I_1 = I_2 + I_3$$
$$16 = 4I_1 + 2I_3$$
$$24 = 4I_1 + 4I_2$$

68. Electricity: Circuit Theory. The following equations come from circuit theory. Find the currents I_1, I_2, and I_3.

$$I_1 = I_2 + I_3$$
$$24 = 6I_1 + 3I_3$$
$$36 = 6I_1 + 6I_2$$

▪CATCH THE MISTAKE

In Exercises 69–72, explain the mistake that is made.

69. Evaluate the determinant $\begin{vmatrix} 2 & 1 & 3 \\ -3 & 0 & 2 \\ 1 & 4 & -1 \end{vmatrix}$.

Solution:

Expand the 3×3 determinant in terms of the 2×2 determinants.

$$\begin{vmatrix} 2 & 1 & 3 \\ -3 & 0 & 2 \\ 1 & 4 & -1 \end{vmatrix} = 2\begin{vmatrix} 0 & 2 \\ 4 & -1 \end{vmatrix} + 1\begin{vmatrix} -3 & 2 \\ 1 & -1 \end{vmatrix} + 3\begin{vmatrix} -3 & 0 \\ 1 & 4 \end{vmatrix}$$

Expand the 2×2 determinants.
$$= 2(0 - 8) + 1(3 - 2) + 3(-12 - 0)$$

Simplify.
$$= -16 + 1 - 36 = -51$$

This is incorrect. What mistake was made?

70. Evaluate the determinant $\begin{vmatrix} 2 & 1 & 3 \\ -3 & 0 & 2 \\ 1 & 4 & -1 \end{vmatrix}$.

Solution:

Expand the 3×3 determinant in terms of the 2×2 determinants.

$$\begin{vmatrix} 2 & 1 & 3 \\ -3 & 0 & 2 \\ 1 & 4 & -1 \end{vmatrix} = 2\begin{vmatrix} 0 & 2 \\ 4 & -1 \end{vmatrix} - 1\begin{vmatrix} -3 & 2 \\ 1 & -1 \end{vmatrix} + 3\begin{vmatrix} -3 & 2 \\ 1 & -1 \end{vmatrix}$$

Expand the 2×2 determinants.
$$= 2(0 - 8) - 1(3 - 2) + 3(3 - 2)$$

Simplify.
$$= -16 - 1 + 3 = -14$$

This is incorrect. What mistake was made?

71. Solve the system of linear equations.
$$\begin{aligned} 2x + 3y &= 6 \\ -x - y &= -3 \end{aligned}$$

Solution:

Set up the determinants.

$$D = \begin{vmatrix} 2 & 3 \\ -1 & -1 \end{vmatrix}, D_x = \begin{vmatrix} 2 & 6 \\ -1 & -3 \end{vmatrix}, \text{ and } D_y = \begin{vmatrix} 6 & 3 \\ -3 & -1 \end{vmatrix}$$

Evaluate the determinants.
$$D = 1, D_x = 0, \text{ and } D_y = 3$$

Solve for x and y.
$$x = \frac{D_x}{D} = \frac{0}{1} = 0 \text{ and } y = \frac{D_y}{D} = \frac{3}{1} = 3$$

$x = 0, y = 3$ is incorrect. What mistake was made?

72. Solve the system of linear equations.
$$\begin{aligned} 4x - 6y &= 0 \\ 4x + 6y &= 4 \end{aligned}$$

Solution:

Set up the determinants.

$$D = \begin{vmatrix} 4 & -6 \\ 4 & 6 \end{vmatrix}, D_x = \begin{vmatrix} 0 & -6 \\ 4 & 6 \end{vmatrix}, \text{ and } D_y = \begin{vmatrix} 4 & 0 \\ 4 & 4 \end{vmatrix}$$

Evaluate the determinants.
$$D = 48, D_x = 24, \text{ and } D_y = 16$$

Solve for x and y.
$$x = \frac{D}{D_x} = \frac{48}{24} = 2 \text{ and } y = \frac{D_y}{D} = \frac{48}{16} = 3$$

$x = 2, y = 3$ is incorrect. What mistake was made?

▪CONCEPTUAL

In Exercises 73–76, determine whether each statement is true or false.

73. The value of a determinant changes sign if any two rows are interchanged.

74. If all the entries in any column are equal to zero, the value of the determinant is 0.

75. $\begin{vmatrix} 2 & 6 & 4 \\ 0 & 2 & 8 \\ 4 & 0 & 10 \end{vmatrix} = 2\begin{vmatrix} 1 & 3 & 2 \\ 0 & 1 & 4 \\ 2 & 0 & 5 \end{vmatrix}$

76. $\begin{vmatrix} 3 & 1 & 2 \\ 0 & 2 & 8 \\ 3 & 1 & 2 \end{vmatrix} = 0$

77. Calculate the determinant $\begin{vmatrix} a & 0 & 0 \\ 0 & b & 0 \\ 0 & 0 & c \end{vmatrix}$.

78. Calculate the determinant $\begin{vmatrix} a_1 & b_1 & c_1 \\ 0 & b_2 & c_2 \\ 0 & 0 & c_3 \end{vmatrix}$.

■ CHALLENGE

79. Evaluate the determinant:

$$\begin{vmatrix} 1 & -2 & -1 & 3 \\ 4 & 0 & 1 & 2 \\ 0 & 3 & 2 & 4 \\ 1 & -3 & 5 & -4 \end{vmatrix}$$

80. For the system of equations

$$3x + 2y = 5$$
$$ax - 4y = 1$$

find a that guarantees no unique solution.

81. Show that

$$\begin{vmatrix} a_1 & b_1 & c_1 \\ a_2 & b_2 & c_2 \\ a_3 & b_3 & c_3 \end{vmatrix} = a_1b_2c_3 + b_1c_2a_3 + c_1a_2b_3 - a_3b_2c_1 - b_3c_2a_1 - b_1a_2c_3$$

by expanding down the second column.

82. Show that

$$\begin{vmatrix} a_1 & b_1 & c_1 \\ a_2 & b_2 & c_2 \\ a_3 & b_3 & c_3 \end{vmatrix} = a_1b_2c_3 + b_1c_2a_3 + c_1a_2b_3 - a_3b_2c_1 - b_3c_2a_1 - c_3b_1a_2$$

by expanding across the third row.

■ TECHNOLOGY

In Exercises 83–86, use a graphing utility to evaluate the determinants.

83. $\begin{vmatrix} 1 & 1 & -5 \\ 3 & -7 & -4 \\ 4 & -6 & 9 \end{vmatrix}$ Compare with your answer to Exercise 37.

84. $\begin{vmatrix} -3 & 2 & -5 \\ 1 & 8 & 2 \\ 4 & -6 & 9 \end{vmatrix}$ Compare with your answer to Exercise 38.

85. $\begin{vmatrix} -3 & 2 & -1 & 3 \\ 4 & 1 & 5 & 2 \\ 17 & 2 & 2 & 8 \\ 13 & -4 & 10 & -11 \end{vmatrix}$

86. $\begin{vmatrix} -3 & 21 & 19 & 3 \\ 4 & 1 & 16 & 2 \\ 17 & 31 & 2 & 5 \\ 13 & -4 & 10 & 2 \end{vmatrix}$

In Exercises 87 and 88, apply Cramer's rule to solve each system of equations and a graphing utility to evaluate the determinants.

87. $3.1x + 1.6y - 4.8z = -33.76$
$5.2x - 3.4y + 0.5z = -36.68$
$0.5x - 6.4y + 11.4z = 25.96$

88. $-9.2x + 2.7y + 5.1z = -89.2$
$4.3x - 6.9y - 7.6z = 38.89$
$2.8x - 3.9y - 3.5z = 34.08$

Recall from Chapter 0 the following properties of real number multiplication:

1. The multiplicative identity property, which states that $a \cdot 1 = a = 1 \cdot a$, for any real number a. In other words, the product of a real number and 1 (the multiplicative identity) is that number.

2. The multiplicative inverse property, which states that $a \cdot \dfrac{1}{a} = 1$, for all nonzero real numbers a. That is, the product of a nonzero real number and its multiplicative inverse (reciprocal) is 1, the multiplicative identity.

Next you will discover how these ideas carry over to matrix multiplication.

1. Let I be the 2 × 2 matrix with 1s on the main diagonal (upper left to lower right) and 0s for the other elements.

$$I = \begin{bmatrix} 1 & 0 \\ 0 & 1 \end{bmatrix}$$

  a. Multiply the matrix I by $A = \begin{bmatrix} 1 & 3 \\ -2 & 4 \end{bmatrix}$ in either order; that is, find the two products IA and AI. What do you notice?

  b. Let $B = \begin{bmatrix} 1 \\ -2 \end{bmatrix}$. What do you notice about the product IB?

  c. Let $C = \begin{bmatrix} 1 & 3 & -5 \\ -2 & 4 & 6 \end{bmatrix}$. What do you notice about the product IC?

  d. Why do you suppose you were not asked to consider the products BI and CI?

2. The examples in part (1) suggest a new definition. The matrix I is the **2 × 2 multiplicative identity matrix**, denoted I_2. When I_2 is multiplied by any matrix A (of the appropriate size), the product is A.

  a. What do you think is the 3 × 3 multiplicative identity matrix, I_3?

  b. If your answer in part 2(a) is correct, what should be the product of CI_3, where $C = \begin{bmatrix} 1 & 3 & -5 \\ -2 & 4 & 6 \end{bmatrix}$? Carry out this multiplication to check.

  c. Try to find a matrix D so that $DI_3 = D$ and $I_3D = D$.

3. Again consider the matrix $A = \begin{bmatrix} 1 & 3 \\ -2 & 4 \end{bmatrix}$ and let $E = \begin{bmatrix} 2/5 & -3/10 \\ 1/5 & 1/10 \end{bmatrix}$.

  a. Find the products AE and EA. What do you notice?

  b. In the example in part (a), E is the multiplicative inverse of A, denoted $E = A^{-1}$. Suppose A is *any* 2 × 2 matrix. Try to write a definition of the multiplicative inverse of A.

  c. Just as not all real numbers have a multiplicative inverse (namely, 0), not all matrices have multiplicative inverses. In order for a multiplicative inverse to exist, the matrix *must be a square matrix*. Can you explain why that is so?

While only a square matrix can have a multiplicative inverse, not all square matrices have inverses. In Section 10.3 you learned how to find the multiplicative inverse of a square matrix, if it exists.

In 2008, Honda introduced the Civic GX NGV—the NGV standing for *Natural Gas Vehicle.* Honda also sells a refueling station for your home that runs off the same natural gas that you use to power a stove or water heater.

© Karen Bieler/AFP/Getty Images

The typical range per tank is approximately 170 miles. In addition, to the home fueling option, there are over 1300 natural gas fueling stations in the United States, primarily in California, as of 2008.

Compared to the Civic Hybrid (gas/electric) Sedan which gets an estimated 40 mpg in the city and 45 mpg on the highway, the Civic NGV gets an estimated gasoline equivalent of 24 mpg in the city and 36 mpg on the highway.

The MSRP and mileage comparisons for the 2008 Honda Civic models are given below, where Gas represents the traditional sedan, Hybrid represents the electric/gasoline model, and NGV represents the Civic that runs on natural gas.

	GAS	HYBRID	NGV
MSRP	$15,010	$22,600	$24,590
MILES PER GALLON IN CITY	21	40	24
MILES PER GALLON ON HIGHWAY	31	45	36

For the following questions, assume that you drive 15,000 miles per year—12,000 in the city and 3000 on the highway—and the price of gasoline is $4 per gallon while the cost of natural gas is $2 per gallon.

1. Determine an initial purchase price matrix *A*, which has order 3 × 1. Let the first row represent the traditional gasoline model, the second row represent the hybrid model, and the third row represent the natural gas model.

2. Determine a mileage matrix *B*, which has order 3 × 2. Let the first column represent mpg in the city and the second column represent the mpg on the highway. Also let the first through third rows represent the gas, hybrid, and natural gas models, respectively. Let

$$C = \begin{bmatrix} 12{,}000 \\ 3000 \end{bmatrix}$$

What does *BC* represent?

3. Assume the gas and hybrid models are filled with regular gasoline at $4 per gallon and the natural gas vehicle is filled with natural gas at $2 per gallon. What does the matrix D represent?

$$D = \begin{bmatrix} 4 \cdot (BC)_{11} \\ 4 \cdot (BC)_{21} \\ 2 \cdot (BC)_{31} \end{bmatrix}$$

4. What does the matrix $A + D$ represent?

5. Let n represent the number of years you own the car (assuming you bought it new). What does $A + nD$ represent?

SECTION	CONCEPT	KEY IDEAS/FORMULAS
10.1	**Matrices and systems of linear equations**	

Matrices

$$\begin{array}{ccccc} & \text{Column 1} & \text{Column 2} & \cdots & \text{Column } j & \cdots & \text{Column } n \end{array}$$

$$\begin{array}{c} \text{Row 1} \\ \text{Row 2} \\ \vdots \\ \text{Row } i \\ \vdots \\ \text{Row } m \end{array} \begin{bmatrix} a_{11} & a_{12} & \cdots & a_{1j} & \cdots & a_{1n} \\ a_{21} & a_{22} & \cdots & a_{2j} & \cdots & a_{2n} \\ \vdots & \vdots & \cdots & \vdots & \cdots & \vdots \\ a_{i1} & a_{i2} & \cdots & a_{ij} & \cdots & a_{in} \\ \vdots & \vdots & \cdots & \vdots & \cdots & \vdots \\ a_{m1} & a_{m2} & \cdots & a_{mj} & \cdots & a_{mn} \end{bmatrix} \quad \text{Order: } m \times n$$

Augmented matrices

$$\begin{aligned} a_1x + b_1y + c_1z &= d_1 \\ a_2x + b_2y + c_2z &= d_2 \implies \\ a_3x + b_3y + c_3z &= d_3 \end{aligned} \quad \begin{bmatrix} a_1 & b_1 & c_1 & | & d_1 \\ a_2 & b_2 & c_2 & | & d_2 \\ a_3 & b_3 & c_3 & | & d_3 \end{bmatrix}$$

Row operations on a matrix

1. $R_i \leftrightarrow R_j$ Interchange row i with row j.
2. $cR_i \rightarrow R_i$ Multiply row i by the constant c.
3. $cR_i + R_j \rightarrow R_j$ Multiply row i by the constant c and add to row j, writing the results in row j.

Row–echelon form of a matrix

A matrix is in **row–echelon form** if it has all three of the following properties:

1. Any rows consisting entirely of 0s are at the bottom of the matrix.
2. For each row that does not consist entirely of 0s, the first (leftmost) nonzero entry is 1 (called the leading 1).
3. For two successive nonzero rows, the leading 1 in the higher row is farther to the left than the leading 1 in the lower row.

If a matrix in row–echelon form has the following additional property, then the matrix is in **reduced row–echelon form**:

4. Every column containing a leading 1 has zeros in every position above and below the leading 1.

Gaussian elimination with back-substitution

Step 1: Write the system of equations as an augmented matrix.
Step 2: Apply row operations to transform the matrix into row–echelon form.
Step 3: Apply back-substitution to identify the solution.

Gauss–Jordan elimination

Step 1: Write the system of equations as an augmented matrix.
Step 2: Apply row operations to transform the matrix into *reduced* row–echelon form.
Step 3: Identify the solution.

Inconsistent and dependent systems

No solution or infinitely many solutions

10.2 Matrix algebra

Equality of matrices	The orders must be the same: $A_{m \times n}$ and $B_{m \times n}$.
Matrix addition and subtraction	The orders must be the same: $A_{m \times n}$ and $B_{m \times n}$.
Scalar and matrix multiplication	**Scalar Multiplication** Perform operation element by element. Each element is multiplied by the scalar. **Matrix Multiplication** • The orders must satisfy the relationship: $A_{m \times n}$ and $B_{n \times p}$, resulting in $(AB)_{m \times p}$. • Perform multiplication row by column. • Matrix multiplication is not commutative: $AB \neq BA$.

10.3 Matrix equations; the inverse of a square matrix

Matrix equations	Linear system: $AX = B$.
Finding the inverse of a matrix	Only square matrices, $n \times n$, can have inverses. $A^{-1}A = I_n$ **Step 1:** Form the matrix $[A \mid I_n]$. **Step 2:** Use row operations to transform this matrix to $[I_n \mid A^{-1}]$. Note: Not every square matrix has an inverse.
Solving systems of linear equations using matrix algebra and inverses of square matrices	$AX = B$ **Step 1:** Find A^{-1}. **Step 2:** $X = A^{-1}B$.

10.4 The determinant of a square matrix and Cramer's rule

	Cramer's rule can be used to solve only a system of linear equations with a unique solution.
Determinant of a 2×2 matrix	$\begin{vmatrix} a & b \\ c & d \end{vmatrix} = ad - bc$
Determinant of an $n \times n$ matrix	Let A be a square matrix of order $n \times n$; then • The **minor** M_{ij} of the element a_{ij} is the determinant of the $(n-1) \times (n-1)$ matrix obtained when the ith row and jth column of A are deleted. • The **cofactor** C_{ij} of the element a_{ij} is given by $C_{ij} = (-1)^{i+j}M_{ij}$. $\begin{bmatrix} 1 & -3 & 2 \\ 4 & -1 & 0 \\ 5 & -2 & 3 \end{bmatrix}$ $M_{11} = \begin{vmatrix} -1 & 0 \\ -2 & 3 \end{vmatrix} = -3 - 0 = -3$ $C_{11} = (-1)^{1+1}M_{11} = (1)(-3) = -3$ Sign Array of a 3×3 matrix: $\begin{bmatrix} + & - & + \\ - & + & - \\ + & - & + \end{bmatrix}$

If A is a 3×3 matrix, the determinant can be given by $\det(A) = a_{11}C_{11} + a_{12}C_{12} + a_{13}C_{13}$. This is called **expanding the determinant by the first row**. (Note that any row or column can be used.)

$$\begin{vmatrix} a_1 & b_1 & c_1 \\ a_2 & b_2 & c_2 \\ a_3 & b_3 & c_3 \end{vmatrix} = a_1 \begin{vmatrix} b_2 & c_2 \\ b_3 & c_3 \end{vmatrix} - b_1 \begin{vmatrix} a_2 & c_2 \\ a_3 & c_3 \end{vmatrix} + c_1 \begin{vmatrix} a_2 & b_2 \\ a_3 & b_3 \end{vmatrix}$$

Cramer's rule: Systems of linear equations in two variables

The system

$$a_1 x + b_1 y = c_1$$
$$a_2 x + b_2 y = c_2$$

has the solution

$$x = \frac{D_x}{D} \qquad y = \frac{D_y}{D} \quad \text{if } D \neq 0$$

where

$$D = \begin{vmatrix} a_1 & b_1 \\ a_2 & b_2 \end{vmatrix} \qquad D_x = \begin{vmatrix} c_1 & b_1 \\ c_2 & b_2 \end{vmatrix} \qquad D_y = \begin{vmatrix} a_1 & c_1 \\ a_2 & c_2 \end{vmatrix}$$

Cramer's rule: Systems of linear equations in three variables

The system

$$a_1 x + b_1 y + c_1 z = d_1$$
$$a_2 x + b_2 y + c_2 z = d_2$$
$$a_3 x + b_3 y + c_3 z = d_3$$

has the solution

$$x = \frac{D_x}{D} \qquad y = \frac{D_y}{D} \qquad z = \frac{D_z}{D} \quad \text{if } D \neq 0$$

where

$$D = \begin{vmatrix} a_1 & b_1 & c_1 \\ a_2 & b_2 & c_2 \\ a_3 & b_3 & c_3 \end{vmatrix} \qquad D_x = \begin{vmatrix} d_1 & b_1 & c_1 \\ d_2 & b_2 & c_2 \\ d_3 & b_3 & c_3 \end{vmatrix}$$

$$D_y = \begin{vmatrix} a_1 & d_1 & c_1 \\ a_2 & d_2 & c_2 \\ a_3 & d_3 & c_3 \end{vmatrix} \qquad D_z = \begin{vmatrix} a_1 & b_1 & d_1 \\ a_2 & b_2 & d_2 \\ a_3 & b_3 & d_3 \end{vmatrix}$$

CHAPTER REVIEW

10.1 Matrices and Systems of Linear Equations

Write the augmented matrix for each system of linear equations.

1. $5x + 7y = 2$
$3x - 4y = -2$

2. $2.3x - 4.5y = 6.8$
$-0.4x + 2.1y = -9.1$

3. $2x - z = 3$
$y - 3z = -2$
$x + 4z = -3$

4. $2y - x + 3z = 1$
$4z - 2y + 3x = -2$
$x - y - 4z = 0$

Indicate whether each matrix is in row–echelon form. If it is, state whether it is in *reduced* row–echelon form.

5. $\begin{bmatrix} 1 & 1 & | & 0 \\ 0 & 1 & | & 2 \end{bmatrix}$

6. $\begin{bmatrix} 1 & 2 & | & 0 \\ 0 & 0 & | & 1 \end{bmatrix}$

7. $\begin{bmatrix} 2 & 0 & 1 & | & 1 \\ 0 & -2 & 0 & | & 2 \\ 0 & 0 & 2 & | & 3 \end{bmatrix}$

8. $\begin{bmatrix} 1 & 0 & 1 & 0 & | & 2 \\ 0 & 0 & 1 & 1 & | & -3 \\ 0 & 1 & 0 & 0 & | & 2 \\ 0 & 0 & 0 & 1 & | & 1 \end{bmatrix}$

Perform the indicated row operations on each matrix.

9. $\begin{bmatrix} 1 & -2 & | & 1 \\ 0 & -2 & | & 2 \end{bmatrix}$ $\frac{-1}{2}R_2 \to R_2$

10. $\begin{bmatrix} 1 & 4 & | & 1 \\ 2 & -2 & | & 3 \end{bmatrix}$ $R_2 - 2R_1 \to R_2$

11. $\begin{bmatrix} 1 & -2 & 0 & | & 1 \\ 0 & -2 & 3 & | & -2 \\ 0 & 1 & -4 & | & 8 \end{bmatrix}$ $R_2 + R_1 \to R_1$

12. $\begin{bmatrix} 1 & 1 & 1 & 6 & | & 0 \\ 0 & 2 & -2 & 3 & | & -2 \\ 0 & 0 & 1 & -2 & | & 4 \\ 0 & -1 & 3 & -3 & | & 3 \end{bmatrix}$ $\begin{array}{c} -2R_1 + R_2 \to R_1 \\ R_4 + R_3 \to R_4 \end{array}$

Apply row operations to transform each matrix to reduced row–echelon form.

13. $\begin{bmatrix} 1 & 3 & | & 0 \\ 3 & 4 & | & 1 \end{bmatrix}$

14. $\begin{bmatrix} 1 & 2 & -1 & | & 0 \\ 0 & 1 & -1 & | & -1 \\ -2 & 0 & 1 & | & -2 \end{bmatrix}$

15. $\begin{bmatrix} 4 & 1 & -2 & | & 0 \\ 1 & 0 & -1 & | & 0 \\ -2 & 1 & 1 & | & 12 \end{bmatrix}$

16. $\begin{bmatrix} 2 & 3 & 2 & | & 1 \\ 0 & -1 & 1 & | & -2 \\ 1 & 1 & -1 & | & 6 \end{bmatrix}$

Solve the system of linear equations using augmented matrices.

17. $3x - 2y = 2$
$-2x + 4y = 1$

18. $2x - 7y = 22$
$x + 5y = -23$

19. $5x - y = 9$
$x + 4y = 6$

20. $8x + 7y = 10$
$-3x + 5y = 42$

21. $x - 2y + z = 3$
$2x - y + z = -4$
$3x - 3y - 5z = 2$

22. $3x - y + 4z = 18$
$5x + 2y - z = -20$
$x + 7y - 6z = -38$

23. $x - 4y + 10z = -61$
$3x - 5y + 8z = -52$
$-5x + y - 2z = 8$

24. $4x - 2y + 5z = 17$
$x + 6y - 3z = -\frac{17}{2}$
$-2x + 5y + z = 2$

25. $3x + y + z = -4$
$x - 2y + z = -6$

26. $2x - y + 3z = 6$
$3x + 2y - z = 12$

Applications

27. Fitting a Curve to Data. The average number of flights on a commercial plane that a person takes a year can be modeled by a quadratic function $y = ax^2 + bx + c$, where $a < 0$ and x represents age: $16 < x < 65$. The following table gives the average number of flights per year that a person takes on a commercial airline. Determine the quadratic function that models this quantity.

AGE	NUMBER OF FLIGHTS PER YEAR
16	2
40	6
65	4

28. Investment Portfolio. Danny and Paula decide to invest $20,000 of their savings in investments. They put some in an IRA account earning 4.5% interest, some in a mutual fund that has been averaging 8% a year, and some in a stock that earned 12% last year. If they put $3000 more in the mutual fund than in the IRA and the mutual fund and stock have the same growth in the next year as they did in the previous year, they will earn $1877.50 in a year. How much money did they put in each of the three investments?

10.2 Matrix Algebra

Calculate the given expression, if possible.

$A = \begin{bmatrix} 2 & -3 \\ 0 & 1 \end{bmatrix}$ $B = \begin{bmatrix} 1 & 5 & -1 \\ 3 & 7 & 2 \end{bmatrix}$ $C = \begin{bmatrix} 5 & 0 & 1 \\ 2 & -1 & 4 \\ 0 & 3 & 6 \end{bmatrix}$

$D = \begin{bmatrix} 5 & 2 \\ 9 & 7 \end{bmatrix}$ $E = \begin{bmatrix} 2 & 0 & 3 \\ 4 & 1 & -1 \end{bmatrix}$

29. $A + C$

30. $B + A$

31. $B + E$

32. $A + D$

33. $2A + D$

34. $3E + B$

35. $2D - 3A$

36. $3B - 4E$

37. $5A - 2D$

38. $5B - 4E$

39. AB

40. BC

41. DA

42. AD

43. $BC + E$

44. DB

45. EC

46. CE

10.3 Matrix Equations; The Inverse of a Square Matrix

Determine whether B is the multiplicative inverse of A using $AA^{-1} = I$.

47. $A = \begin{bmatrix} 6 & 4 \\ 4 & 2 \end{bmatrix}$ $\quad B = \begin{bmatrix} -0.5 & 1 \\ 1 & -1.5 \end{bmatrix}$

48. $A = \begin{bmatrix} 1 & -2 \\ 2 & -4 \end{bmatrix}$ $\quad B = \begin{bmatrix} 1 & 2 \\ 2 & -2 \end{bmatrix}$

49. $A = \begin{bmatrix} 1 & -2 & 6 \\ 2 & 3 & -2 \\ 0 & -1 & 1 \end{bmatrix}$ $\quad B = \begin{bmatrix} -\frac{1}{7} & \frac{4}{7} & 2 \\ \frac{2}{7} & -\frac{1}{7} & -2 \\ \frac{2}{7} & -\frac{1}{7} & -1 \end{bmatrix}$

50. $A = \begin{bmatrix} 0 & 7 & 6 \\ 1 & 0 & -4 \\ -2 & 1 & 0 \end{bmatrix}$ $\quad B = \begin{bmatrix} 1 & 1 & 1 \\ -2 & -2 & -2 \\ 2 & 0 & 6 \end{bmatrix}$

Find A^{-1}, if it exists.

51. $A = \begin{bmatrix} 1 & 2 \\ -3 & 4 \end{bmatrix}$

52. $A = \begin{bmatrix} -2 & 7 \\ -4 & 6 \end{bmatrix}$

53. $A = \begin{bmatrix} 0 & 1 \\ -2 & 0 \end{bmatrix}$

54. $A = \begin{bmatrix} 3 & -1 \\ -2 & 2 \end{bmatrix}$

55. $A = \begin{bmatrix} 1 & 3 & -2 \\ 2 & 1 & -1 \\ 0 & 1 & -3 \end{bmatrix}$

56. $A = \begin{bmatrix} 0 & 1 & 0 \\ 4 & 1 & 2 \\ -3 & -2 & 1 \end{bmatrix}$

57. $A = \begin{bmatrix} -1 & 1 & 0 \\ -2 & 1 & 2 \\ 1 & 2 & 4 \end{bmatrix}$

58. $A = \begin{bmatrix} -4 & 4 & 3 \\ 1 & 2 & 2 \\ 3 & -1 & 6 \end{bmatrix}$

Solve the system of linear equations using matrix algebra.

59. $3x - y = 11$
$\quad 5x + 2y = 33$

60. $6x + 4y = 15$
$\quad -3x - 2y = -1$

61. $\frac{5}{8}x - \frac{2}{3}y = -3$
$\quad \frac{3}{4}x + \frac{5}{6}y = 16$

62. $x + y - z = 0$
$\quad 2x - y + 3z = 18$
$\quad 3x - 2y + z = 17$

63. $3x - 2y + 4z = 11$
$\quad 6x + 3y - 2z = 6$
$\quad x - y + 7z = 20$

64. $2x + 6y - 4z = 11$
$\quad -x - 3y + 2z = -\frac{11}{2}$
$\quad 4x + 5y + 6z = 20$

10.4 The Determinant of a Square Matrix and Cramer's Rule

Evaluate each 2×2 determinant.

65. $\begin{vmatrix} 2 & 4 \\ 3 & 2 \end{vmatrix}$

66. $\begin{vmatrix} -2 & -4 \\ -3 & 2 \end{vmatrix}$

67. $\begin{vmatrix} 2.4 & -2.3 \\ 3.6 & -1.2 \end{vmatrix}$

68. $\begin{vmatrix} -\frac{1}{4} & 4 \\ \frac{3}{4} & -4 \end{vmatrix}$

Employ Cramer's rule to solve each system of equations, if possible.

69. $x - y = 2$
$\quad x + y = 4$

70. $3x - y = -17$
$\quad -x + 5y = 43$

71. $2x + 4y = 12$
$\quad x - 2y = 6$

72. $-x + y = 4$
$\quad 2x - 6y = -5$

73. $-3x = 40 - 2y$
$\quad 2x = 25 + y$

74. $3x = 20 + 4y$
$\quad y - x = -6$

Evaluate each 3×3 determinant.

75. $\begin{vmatrix} 1 & 2 & 2 \\ 0 & 1 & 3 \\ 2 & -1 & 0 \end{vmatrix}$

76. $\begin{vmatrix} 0 & -2 & 1 \\ 0 & -3 & 7 \\ 1 & -10 & -3 \end{vmatrix}$

77. $\begin{vmatrix} a & 0 & -b \\ -a & b & c \\ 0 & 0 & -d \end{vmatrix}$

78. $\begin{vmatrix} -2 & -4 & 6 \\ 2 & 0 & 3 \\ -1 & 2 & \frac{3}{4} \end{vmatrix}$

Employ Cramer's rule to solve each system of equations, if possible.

79. $x + y - 2z = -2$
$\quad 2x - y + z = 3$
$\quad x + y + z = 4$

80. $-x - y + z = 3$
$\quad x + 2y - 2z = 8$
$\quad 2x + y + 4z = -4$

81. $3x + 4z = -1$
$\quad x + y + 2z = -3$
$\quad y - 4z = -9$

82. $x + y + z = 0$
$\quad -x - 3y + 5z = -2$
$\quad 2x + y - 3z = -4$

Applications

83. Apply determinants to find the area of a triangle with vertices $(2, 4)$, $(4, 4)$, and $(-4, 3)$.

84. If three points (x_1, y_1), (x_2, y_2), and (x_3, y_3) are collinear (lie on the same line), then the following determinant must be satisfied:

$$\begin{vmatrix} x_1 & y_1 & 1 \\ x_2 & y_2 & 1 \\ x_3 & y_3 & 1 \end{vmatrix} = 0$$

Determine whether $(0, -3)$, $(3, 0)$, and $(1, 6)$ are collinear.

Technology

Section 10.1

In Exercises 85 and 86, refer to the following:

You are asked to model a set of three points with a quadratic function $y = ax^2 + bx + c$ and determine the quadratic function.

a. Set up a system of equations; use a graphing utility or graphing calculator to solve the system by entering the coefficients of the augmented matrix.

b. Use the graphing calculator commands $\boxed{\text{STAT QuadReg}}$ to model the data using a quadratic function. Round your answers to two decimal places.

85. $(-10, 12.5), (3, -2.8), (9, 8.5)$

86. $(-4, 10), (2.5, -9.5), (13.5, 12.6)$

Section 10.2

Apply a graphing utility to perform the indicated matrix operations, if possible.

$$A = \begin{bmatrix} -6 & 0 & 4 \\ 1 & 3 & 5 \\ 2 & -1 & 0 \end{bmatrix} \quad B = \begin{bmatrix} 5 & 1 \\ 0 & 2 \\ -8 & 4 \end{bmatrix} \quad C = \begin{bmatrix} 4 & -3 & 0 \\ 1 & 2 & 5 \end{bmatrix}$$

87. ABC

88. CAB

Section 10.3

Apply a graphing utility and matrix algebra to solve the system of linear equations.

89.
$$\begin{aligned} 6.1x - 14.2y &= 75.495 \\ -2.3x + 7.2y &= -36.495 \end{aligned}$$

90.
$$\begin{aligned} 7.2x + 3.2y - 1.7z &= 5.53 \\ -1.3x + 4.1y + 2.8z &= -23.949 \end{aligned}$$

Section 10.4

Apply Cramer's rule to solve each system of equations and a graphing utility to evaluate the determinants.

91.
$$\begin{aligned} 4.5x - 8.7y &= -72.33 \\ -1.4x + 5.3y &= 31.32 \end{aligned}$$

92.
$$\begin{aligned} 1.4x + 3.6y + 7.5z &= 42.08 \\ 2.1x - 5.7y - 4.2z &= 5.37 \\ 1.8x - 2.8y - 6.2z &= -9.86 \end{aligned}$$

Write each of the following systems of linear equations as an augmented matrix.

1. $x - 2y = 1$
$-x + 3y = 2$

2. $3x + 5y = -2$
$7x + 11y = -6$

3. $6x + 9y + z = 5$
$2x - 3y + z = 3$
$10x + 12y + 2z = 9$

4. $3x + 2y - 10z = 2$
$x + y - z = 5$

5. Perform the following row operations:

$$\begin{bmatrix} 1 & 3 & 5 \\ 2 & 7 & -1 \\ -3 & -2 & 0 \end{bmatrix} \begin{matrix} R_2 - 2R_1 \rightarrow R_2 \\ R_3 + 3R_1 \rightarrow R_3 \end{matrix}$$

6. Rewrite the following matrix in reduced row–echelon form:

$$\begin{bmatrix} 2 & -1 & 1 & | & 3 \\ 1 & 1 & -1 & | & 0 \\ 3 & 2 & -2 & | & 1 \end{bmatrix}$$

In Exercises 7 and 8, solve the systems of linear equations using augmented matrices.

7. $6x + 9y + z = 5$
$2x - 3y + z = 3$
$10x + 12y + 2z = 9$

8. $3x + 2y - 10z = 2$
$x + y - z = 5$

9. Multiply the matrices, if possible.

$$\begin{bmatrix} 1 & -2 & 5 \\ 0 & -1 & 3 \end{bmatrix} \begin{bmatrix} 0 & 4 \\ 3 & -5 \\ -1 & 1 \end{bmatrix}$$

10. Add the matrices, if possible.

$$\begin{bmatrix} 1 & -2 & 5 \\ 0 & -1 & 3 \end{bmatrix} + \begin{bmatrix} 0 & 4 \\ 3 & -5 \\ -1 & 1 \end{bmatrix}$$

11. Find the inverse of $\begin{bmatrix} 4 & 3 \\ 5 & -1 \end{bmatrix}$, if it exists.

12. Find the inverse of $\begin{bmatrix} 1 & -3 & 2 \\ 4 & 2 & 0 \\ -1 & 2 & 5 \end{bmatrix}$, if it exists.

13. Find the inverse of $\begin{bmatrix} 3 & 1 & 0 \\ 5 & 2 & -1 \end{bmatrix}$, if it exists.

14. Solve the system of linear equations with matrix algebra (inverses).

$$x + 3y = -1$$
$$-2x - 5y = 4$$

15. Solve the system of linear equations with matrix algebra (inverses).

$$3x - y + 4z = 18$$
$$x + 2y + 3z = 20$$
$$-4x + 6y - z = 11$$

Calculate the determinant.

16. $\begin{vmatrix} 7 & -5 \\ 2 & -1 \end{vmatrix}$

17. $\begin{vmatrix} 1 & -2 & -1 \\ 3 & -5 & 2 \\ 4 & -1 & 0 \end{vmatrix}$

In Exercises 18 and 19, solve the system of linear equations using Cramer's rule.

18. $x - 2y = 1$
$-x + 3y = 2$

19. $3x + 5y - 2z = -6$
$7x + 11y + 3z = 2$
$x - y + z = 4$

20. A company has two rubrics for scoring job applicants based on weighting education, experience, and the interview differently.

Matrix A:

	Rubric 1	Rubric 2
Education	0.4	0.6
Experience	0.5	0.1
Interview	0.1	0.3

Applicants receive a score from 1 to 10 in each category (education, experience, and interview). Two applicants are shown in the matrix B:

Matrix B:

	Education	Experience	Interview
Applicant 1	4	7	3
Applicant 2	6	5	4

What is the order of BA? What does each entry in BA tell us?

21. A college student inherits $15,000 from his favorite aunt. He decides to invest it with a diversified scheme. He divides the money into a money market account paying 2% annual simple interest, a conservative stock that rose 4% last year, and an aggressive stock that rose 22% last year. He places $1000 more in the conservative stock than in the money market and twice as much in the aggressive stock than in the money market. If the stocks perform the same way this year, he will make $1790 in one year, assuming a simple interest model. How much did he put in each of the three investments?

22. You are asked to model a set of three points with a quadratic function $y = ax^2 + bx + c$ and determine the quadratic function.

a. Set up a system of equations; use a graphing utility or graphing calculator to solve the system by entering the coefficients of the augmented matrix.

b. Use the graphing calculator commands $\boxed{\text{STAT}}$ $\boxed{\text{QuadReg}}$ to model the data using a quadratic function.

$(-3, 6), (1, 12), (5, 7)$

23. Apply a graphing utility and matrix algebra to solve the system of linear equations.

$$5.6x - 2.7y = 87.28$$
$$-4.2x + 8.4y = -106.26$$

24. Apply Cramer's rule to solve the system of equations and a graphing utility to evaluate the determinants.

$$1.5x + 2.6y = 18.34$$
$$-2.3x + 1.5y = 28.94$$

1. Solve by completing the square: $x^2 - 6x = 11$.

2. Write an equation of a line that passes through the point $(-2, 5)$ and is parallel to the y-axis.

3. Write the equation of a circle with center $(-3, -1)$ and passing through the point $(1, 2)$.

4. Determine whether the relation $x^2 - y^2 = 25$ is a function.

5. Determine whether the function $g(x) = \sqrt{2 - x^2}$ is odd or even.

6. For the function $y = 5(x - 4)^2$, identify all of the transformations of $y = x^2$.

7. Find the composite function $f \circ g$, and state the domain, for $f(x) = x^3 - 1$ and $g(x) = \dfrac{1}{x}$.

8. Find the inverse of the function $f(x) = \sqrt[3]{x} - 1$.

9. Find the vertex of the parabola associated with the quadratic function $f(x) = \frac{1}{4}x^2 + \frac{3}{5}x - \frac{6}{25}$.

10. Find a polynomial of minimum degree (there are many) that has the zeros $x = -\sqrt{7}$ (multiplicity 2), $x = 0$ (multiplicity 3), $x = \sqrt{7}$ (multiplicity 2).

11. Use long division to find the quotient $Q(x)$ and the remainder $r(x)$ of $(5x^3 - 4x^2 + 3) \div (x^2 + 1)$.

12. Given the zero $x = 4i$ of the polynomial $P(x) = x^4 + 2x^3 + x^2 + 32x - 240$, determine all the other zeros and write the polynomial in terms of a product of linear factors.

13. Find the vertical and horizontal asymptotes of the function $f(x) = \dfrac{0.7x^2 - 5x + 11}{x^2 - x - 6}$.

14. If \$5400 is invested at 2.25% compounded continuously, how much is in the account after 4 years?

15. Use interval notation to express the domain of the function $f(x) = \log_4 (x + 3)$.

16. Use the properties of logarithms to simplify the expression $\log_\pi 1$.

17. Give an exact solution to the logarithmic equation $\log_5 (x + 2) + \log_5 (6 - x) = \log_5 (3x)$.

18. If money is invested in a savings account earning 4% compounded continuously, how many years will pass until the money triples?

19. Solve the following system of linear equations:
$$2x + 3y = 6$$
$$x = -1.5y - 5.5$$

20. At the student union, one group of students bought 6 deluxe burgers, 5 orders of fries, and 5 sodas for \$24.34. A second group of students ordered 8 deluxe burgers, 4 orders of fries, and 6 sodas for \$28.42. A third group of students ordered 3 deluxe burgers, 2 orders of fries, and 4 sodas for \$13.51. Determine the price of each food item.

21. Maximize the objective function $z = 5x + 7y$, subject to the constraints:
$$2x + 5y \le 10 \qquad x \ge 0$$
$$5x + 2y \le 10 \qquad y \ge 0$$

22. Solve the system with Gauss–Jordan elimination.
$$x - 2y + 3z = 11$$
$$4x + 5y - z = -8$$
$$3x + y - 2z = 1$$

23. Given
$$A = \begin{bmatrix} 3 & 4 & -7 \\ 0 & 1 & 5 \end{bmatrix} \quad B = \begin{bmatrix} 8 & -2 & 6 \\ 9 & 0 & -1 \end{bmatrix} \quad C = \begin{bmatrix} 9 & 0 \\ 1 & 2 \end{bmatrix}$$
find $2A + CB$.

24. Write the matrix equation, find the inverse of the coefficient matrix, and solve the system using matrix algebra.
$$2x + 5y = -1$$
$$-x + 4y = 7$$

25. Apply Cramer's rule to solve the system of equations.
$$7x + 5y = 1$$
$$-x + 4y = -1$$

26. Use a graphing calculator to graph the given polynomial. From the graph, estimate x-intercepts and the coordinates of the relative maximum and minimum points. Round your answers to two decimal places.
$$f(x) = x^3 + 5x^2 - 12x - 13$$

27. Given
$$A = \begin{bmatrix} 0 & -1 & 6 \\ 4 & 3 & 1 \end{bmatrix} \quad B = \begin{bmatrix} -7 & -4 \\ 1 & 0 \\ 6 & 2 \end{bmatrix}$$
find AB and $(AB)^{-1}$.

Analytic Geometry and Systems of Nonlinear Equations and Inequalities

© F1 online digitale Bildagentur GmbH/Alamy

A satellite dish is a *paraboloid*.

Paul Souders/The Image Bank/Getty Images

Some buildings have a *hyperbolic* shape.

In this chapter we will study the three types of conic sections (conics): the parabola, the ellipse, and the hyperbola. We have already studied the circle (Section 2.4), which is a special form of the ellipse. We see these shapes all around us: satellite dishes (parabolas), cooling towers (hyperbolas), and planetary orbits (ellipses).

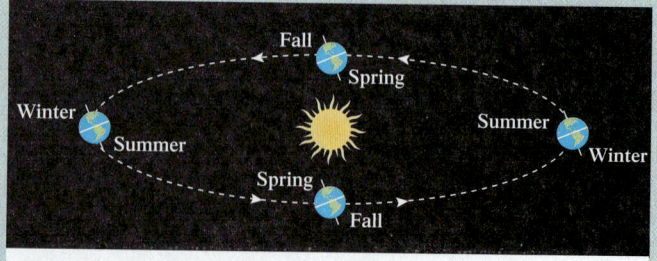

The Earth's orbit around the Sun is an *ellipse*.

▶ **IN THIS CHAPTER** we define the three conic sections: the parabola, the ellipse, and the hyperbola. Algebraic equations and the graphs of these conics are discussed. We solve systems of nonlinear equations and inequalities involving parabolas, ellipses, and hyperbolas. We then will rotate the axes and identify the graph of a general second-degree equation as one of the three conics. We will discuss the equations of the conics first in rectangular coordinates and then in polar coordinates. Last, we will look at parametric equations, which give orientation along a plane curve.

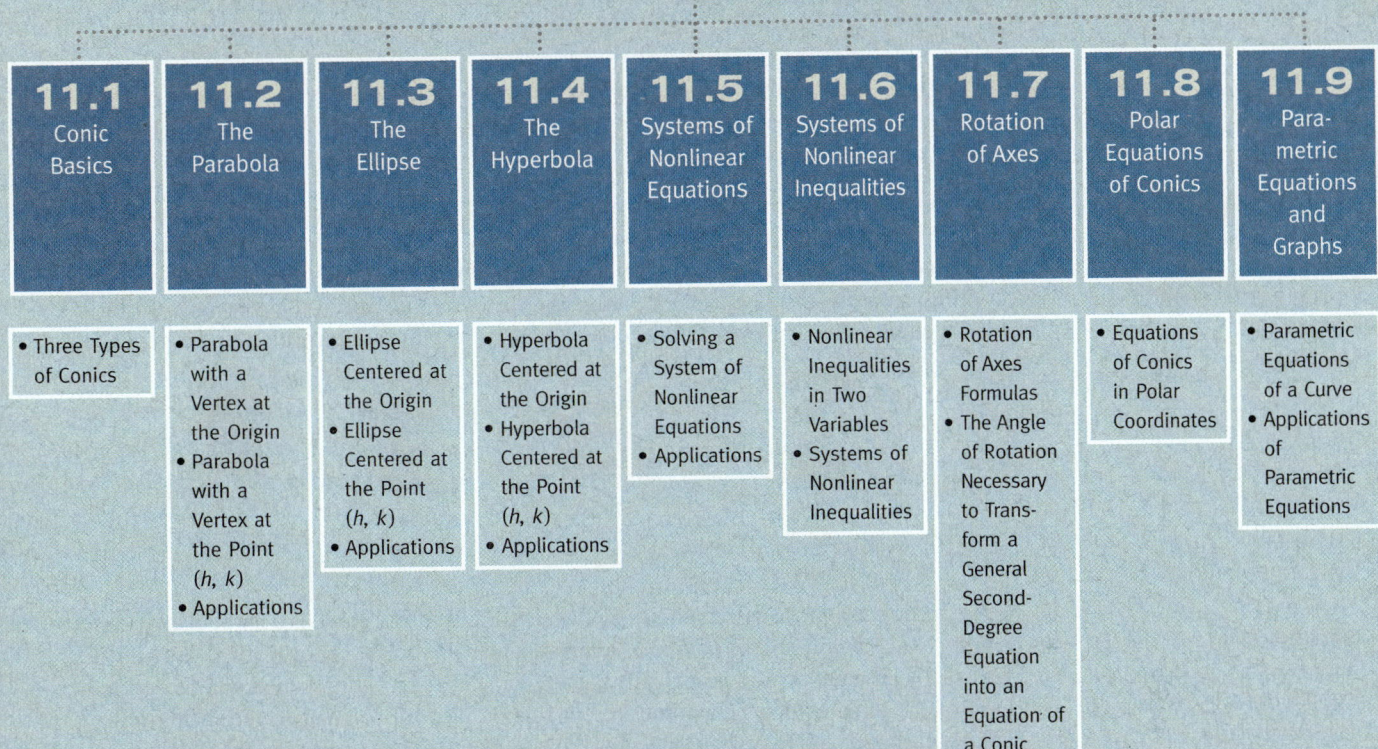

ANALYTIC GEOMETRY AND SYSTEMS OF NONLINEAR EQUATIONS AND INEQUALITIES

11.1 Conic Basics	11.2 The Parabola	11.3 The Ellipse	11.4 The Hyperbola	11.5 Systems of Nonlinear Equations	11.6 Systems of Nonlinear Inequalities	11.7 Rotation of Axes	11.8 Polar Equations of Conics	11.9 Parametric Equations and Graphs
• Three Types of Conics	• Parabola with a Vertex at the Origin • Parabola with a Vertex at the Point (h, k) • Applications	• Ellipse Centered at the Origin • Ellipse Centered at the Point (h, k) • Applications	• Hyperbola Centered at the Origin • Hyperbola Centered at the Point (h, k) • Applications	• Solving a System of Nonlinear Equations • Applications	• Nonlinear Inequalities in Two Variables • Systems of Nonlinear Inequalities	• Rotation of Axes Formulas • The Angle of Rotation Necessary to Transform a General Second-Degree Equation into an Equation of a Conic	• Equations of Conics in Polar Coordinates	• Parametric Equations of a Curve • Applications of Parametric Equations

LEARNING OBJECTIVES

- ■ Determine if a general second-degree equation in two variables corresponds to a parabola, ellipse, or hyperbola.
- ■ Graph a parabola in rectangular coordinates.
- ■ Graph an ellipse in rectangular coordinates.
- ■ Graph a hyperbola in rectangular coordinates.
- ■ Solve systems of nonlinear equations and interpret some graphically in terms of conics.
- ■ Solve systems of nonlinear inequalities and interpret some graphically in terms of conics.
- ■ Graph general second-degree polynomial functions by rotation of axes.
- ■ Graph parabolas, ellipses, and hyperbolas in polar coordinates.
- ■ Use parametric equations to describe orientation along a plane curve.

SKILLS OBJECTIVES

- Learn the name of each conic section.
- Define conics.
- Recognize the algebraic equation associated with each conic.

CONCEPTUAL OBJECTIVES

- Understand each conic as an intersection of a plane and a cone.
- Understand how the three equations of the conic sections are related to the general form of a second-degree equation in two variables.

Three Types of Conics

Names of Conics

The word *conic* is derived from the word *cone*. Let's start with a (right circular) **double cone** (see the figure on the left). **Conic sections** are curves that result from the intersection of a plane and a double cone. **Conics** is an abbreviation for conic sections. Although there appear to be four conic sections: a *circle*, an *ellipse*, a *parabola*, and a *hyperbola*, a circle is a special ellipse. We say there are three conic sections: an **ellipse**, a **parabola**, and a **hyperbola**.

Circle

Ellipse

Parabola

Hyperbola

In Section 2.4 we discussed circles, and we will show that a circle is a particular type of an ellipse. Now we will discuss parabolas, ellipses, and hyperbolas. There are two ways in which we usually describe conics: graphically and algebraically. An entire section will be devoted to each of the three conics, but in the present section we will summarize the definitions of a parabola, ellipse, and hyperbola and show how to identify the equations of these conics.

Definitions

You already know that a circle consists of all points equidistant (at a distance equal to the radius) from a point (the center). Ellipses, parabolas, and hyperbolas have similar definitions in that they all have a constant distance (or a sum or difference of distances) to some reference point(s).

A **parabola** is the set of all points that are **equidistant from both a line and a point**. An **ellipse** is the set of all points, the **sum of whose distances to two fixed points is constant**. A **hyperbola** is the set of all points, the **difference of whose distances to two fixed points is a nonzero constant**.

Parabola

$d_1 + d_2 = d_3 + d_4$
Ellipse

$|d_2 - d_1| = |d_4 - d_3|$
Hyperbola

The **general form of a second-degree equation in two variables**, x and y, is given by

$$Ax^2 + Bxy + Cy^2 + Dx + Ey + F = 0$$

If we let $A = 1$, $B = 0$, $C = 1$, $D = 0$, $E = 0$, and $F = -r^2$, this general equation reduces to the equation of a circle centered at the origin: $x^2 + y^2 = r^2$. In fact, all three conics (parabolas, ellipses, and hyperbolas) are special cases of the general second-degree equation.

Recall from Section 1.3 (Quadratic Equations) that the discriminant, $b^2 - 4ac$, determines what types of solutions result from solving a second-degree equation in one variable. If the discriminant is positive, the solutions are two distinct real roots. If the discriminant is zero, the solution is a real repeated root. If the discriminant is negative, the solutions are two complex conjugate roots.

The concept of discriminant has been generalized to conic sections. The discriminant determines the *shape* of the conic section.

CONIC	DISCRIMINANT
Ellipse	$B^2 - 4AC < 0$
Parabola	$B^2 - 4AC = 0$
Hyperbola	$B^2 - 4AC > 0$

Study Tip

All circles are ellipses since $B^2 - 4AC < 0$.

Using the discriminant to identify the shape of the conic will not work for degenerate cases (when the polynomial factors). For example,

$$2x^2 - xy - y^2 = 0$$

At first glance, one may think this is a hyperbola because $B^2 - 4AC > 0$, but this is a degenerative case.

$$(2x + y)(x - y) = 0$$
$$2x + y = 0 \qquad \text{or} \qquad x - y = 0$$
$$y = -2x \qquad \text{or} \qquad y = x$$

The graph is two intersecting lines.

We now identify conics from the general form of a second-degree equation in two variables.

EXAMPLE 1 **Determining the Type of Conic**

Determine what type of conic corresponds to each of the following equations:

a. $\dfrac{x^2}{a^2} + \dfrac{y^2}{b^2} = 1$ **b.** $y = x^2$ **c.** $\dfrac{x^2}{a^2} - \dfrac{y^2}{b^2} = 1$

Solution:

Write the general form of the second-degree equation:

$$Ax^2 + Bxy + Cy^2 + Dx + Ey + F = 0$$

a. Identify A, B, C, D, E, and F. $A = \dfrac{1}{a^2}, B = 0, C = \dfrac{1}{b^2}, D = 0, E = 0, F = -1$

Calculate the discriminant. $B^2 - 4AC = -\dfrac{4}{a^2 b^2} < 0$

Since the discriminant is negative, the equation $\dfrac{x^2}{a^2} + \dfrac{y^2}{b^2} = 1$ is that of an $\boxed{\text{ellipse}}$.

Notice that if $a = b = r$, then this equation of an ellipse reduces to the general equation of a circle, $x^2 + y^2 = r^2$, centered at the origin, with radius r.

Classroom Example 11.1.1
Determine what type of conic section corresponds to each of the equations.

a. $1 - \dfrac{(x + 3)^2}{\sqrt{2}} = 3y^2$

b.* $2x(x + 5) - y = 1 + y^2$

c.* $(3x + y)^2 - 2y^2 = x$

d. $3y - x^2 = 2x(4 - x)$

Answer:
a. ellipse
b. hyperbola
c. hyperbola
d. parabola

b. Identify $A, B, C, D, E,$ and F. $\qquad A = 1, B = 0, C = 0, D = 0, E = -1, F = 0$

Calculate the discriminant. $\qquad B^2 - 4AC = 0$

Since the discriminant is zero, the equation $y = x^2$ is a $\boxed{\textbf{parabola}}$.

c. Identify $A, B, C, D, E,$ and F. $\qquad A = \dfrac{1}{a^2}, B = 0, C = -\dfrac{1}{b^2}, D = 0, E = 0, F = -1$

Calculate the discriminant. $\qquad B^2 - 4AC = \dfrac{4}{a^2 b^2} > 0$

Since the discriminant is positive, the equation $\dfrac{x^2}{a^2} - \dfrac{y^2}{b^2} = 1$ is a $\boxed{\textbf{hyperbola}}$.

■ **Answer: a.** ellipse
       b. hyperbola
       c. parabola

■ **YOUR TURN** Determine what type of conic corresponds to each of the following equations:

    a. $2x^2 + y^2 = 4$ **b.** $2x^2 = y^2 + 4$ **c.** $2y^2 = x$

In the next three sections, we will discuss the standard forms of equations and the graphs of parabolas, ellipses, and hyperbolas.

SECTION 11.1 SUMMARY

In this section we defined the three conic sections and determined their general equations with respect to the general form of a second-degree equation in two variables:

$$Ax^2 + Bxy + Cy^2 + Dx + Ey + F = 0$$

The following table summarizes the three conics: ellipse, parabola, and hyperbola.

CONIC	GEOMETRIC DEFINITION: THE SET OF ALL POINTS	DISCRIMINANT
Ellipse	the sum of whose distances to two fixed points is constant	Negative: $B^2 - 4AC < 0$
Parabola	equidistant to both a line and a point	Zero: $B^2 - 4AC = 0$
Hyperbola	the difference of whose distances to two fixed points is a constant	Positive: $B^2 - 4AC > 0$

It is important to note that a circle is a special type of ellipse.

SECTION 11.1 EXERCISES

■ **SKILLS**

In Exercises 1–12, identify the conic section as a parabola, ellipse, circle, or hyperbola.

1. $x^2 + xy - y^2 + 2x = -3$ **2.** $x^2 + xy + y^2 + 2x = -3$ **3.** $2x^2 + 2y^2 = 10$ **4.** $x^2 - 4x + y^2 + 2y = 4$

5. $2x^2 - y^2 = 4$ **6.** $2y^2 - x^2 = 16$ **7.** $5x^2 + 20y^2 = 25$ **8.** $4x^2 + 8y^2 = 30$

9. $x^2 - y = 1$ **10.** $y^2 - x = 2$ **11.** $x^2 + y^2 = 10$ **12.** $x^2 + y^2 = 100$

SKILLS OBJECTIVES

- Graph a parabola given the focus, directrix, and vertex.
- Find the equation of a parabola whose vertex is at the origin.
- Find the equation of a parabola whose vertex is at the point (h, k).
- Solve applied problems that involve parabolas.

CONCEPTUAL OBJECTIVES

- Derive the general equation of a parabola.
- Identify, draw, and use the focus, directrix, and axis of symmetry.

Parabola with a Vertex at the Origin

Recall from Section 4.1 that the graphs of quadratic functions such as

$$f(x) = a(x - h)^2 + k \quad \text{or} \quad y = ax^2 + bx + c$$

were *parabolas* that opened either upward or downward. We now expand our discussion to *parabolas* that open to the **right** or **left**. We did not discuss these types of parabolas before because they are not functions (they fail the vertical line test).

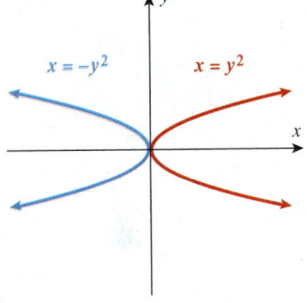

DEFINITION | **Parabola**

A **parabola** is the set of all points in a plane that are equidistant from a fixed line, the **directrix**, and a fixed point not on the line, the **focus**. The line through the focus and perpendicular to the directrix is the **axis of symmetry**. The **vertex** of the parabola is located at the midpoint between the directrix and the focus along the axis of symmetry.

Here p is the distance along the axis of symmetry from the directrix to the vertex and from the vertex to the focus.

Let's consider a parabola with the vertex at the origin and the focus on the positive x-axis. Let the distance from the vertex to the focus be p. Therefore, the focus is located at the point $(p, 0)$. Since the distance from the vertex to the focus is p, the distance from the vertex to the directrix must also be p. Since the axis of symmetry is the positive axis, the directrix must be perpendicular to the positive axis. Therefore, the directrix is given by $x = -p$. Any point, (x, y), must have the same distance to the focus, $(p, 0)$, as it does to the directrix, $(-p, y)$.

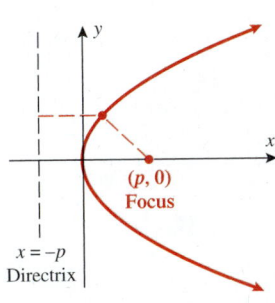

Derivation of the Equation of a Parabola

WORDS	MATH
Calculate the distance from (x, y) to $(p, 0)$ with the distance formula.	$\sqrt{(x - p)^2 + y^2}$
Calculate the distance from (x, y) to $(-p, y)$ with the distance formula.	$\sqrt{(x - (-p))^2 + 0^2}$
Set the two distances equal to one another.	$\sqrt{(x - p)^2 + y^2} = \sqrt{(x + p)^2}$
Recall that $\sqrt{x^2} = |x|$.	$\sqrt{(x - p)^2 + y^2} = |x + p|$
Square both sides of the equation.	$(x - p)^2 + y^2 = (x + p)^2$
Square the binomials inside the parentheses.	$x^2 - 2px + p^2 + y^2 = x^2 + 2px + p^2$
Simplify.	$\boxed{y^2 = 4px}$

The equation $y^2 = 4px$ represents a parabola opening right with the vertex at the origin. The following box summarizes parabolas that have a vertex at the origin and a focus along either the x-axis or the y-axis:

EQUATION OF A PARABOLA WITH VERTEX AT THE ORIGIN

The standard (conic) form of the equation of a **parabola** with vertex at the origin is given by

EQUATION	$y^2 = 4px$	$x^2 = 4py$
VERTEX	$(0, 0)$	$(0, 0)$
FOCUS	$(p, 0)$	$(0, p)$
DIRECTRIX	$x = -p$	$y = -p$
AXIS OF SYMMETRY	x-axis	y-axis
$p > 0$	opens to the right	opens upward
$p < 0$	opens to the left	opens downward
GRAPH ($p > 0$)		

EXAMPLE 1 Finding the Focus and Directrix of a Parabola Whose Vertex Is Located at the Origin

Find the focus and directrix of a parabola whose equation is $y^2 = 8x$.

Solution:

Compare this parabola with the general equation of a parabola.

$$y^2 = \mathbf{4px}$$
$$y^2 = \mathbf{8x}$$

Let $y^2 = 8x$.

$$4px = 8x$$

Solve for p.

$$4p = 8$$
$$p = 2$$

The focus of a parabola of the form $y^2 = 4px$ is $(p, 0)$.

Focus: $(2, 0)$

The directrix of a parabola of the form $y^2 = 4px$ is $x = -p$.

Directrix: $x = -2$

■ **YOUR TURN** Find the focus and directrix of a parabola whose equation is $y^2 = 16x$.

Classroom Example 11.2.1
Find the focus and directrix of a parabola whose equation is $3x^2 = 16y$.

Answer:

Focus: $\left(0, \frac{4}{3}\right)$

Directrix: $y = -\frac{4}{3}$

■ **Answer:** The focus is $(4, 0)$ and the directrix is $x = -4$.

Study Tip

The focus and directrix define a parabola, but do not appear on its graph.

Graphing a Parabola with a Vertex at the Origin

We can draw an approximate sketch of a parabola whose vertex is at the origin with three pieces of information:

1. the vertex is located at $(0, 0)$
2. the direction in which the parabola opens
3. approximately how wide or narrow to draw the parabolic curve.

Classroom Example 11.2.1*
If the focus of a parabola with vertex at the origin and which opens down is $(0, -4)$, what is the equation of the parabola?

Answer: $x^2 = -16y$

The direction toward which the parabola opens is found from the equation. An equation of the form $y^2 = 4px$ opens either left or right. It opens right if $p > 0$ and opens left if $p < 0$. An equation of the form $x^2 = 4py$ opens either up or down. It opens up if $p > 0$ and opens down if $p < 0$.

How narrow or wide should we draw the parabolic curve? If we select a few points that satisfy the equation, we can use those as graphing aids.

The line segment that passes through the focus $(2, 0)$, is parallel to the directrix $x = -2$, and whose endpoints are on the parabola is called the **latus rectum**. The latus rectum in this case has length 8. The latus rectum is a graphing aid that assists us in finding the width of a parabola.

In general, the points on a parabola of the form $y^2 = 4px$ that lie above and below the focus $(p, 0)$ satisfy the equation $y^2 = 4p^2$ and are located at $(p, -2p)$ and $(p, 2p)$. The latus rectum will have length $4|p|$. Similarly, a parabola of the form $x^2 = 4py$ will have a horizontal latus rectum of length $4|p|$. We will use the latus rectum as a graphing aid to determine the parabola's width.

To graph $x^2 = -12y$ with a graphing calculator, solve for y first. That is, $y = -\frac{1}{12}x^2$.

EXAMPLE 2 Graphing a Parabola Whose Vertex Is at the Origin, Using the Focus, Directrix, and Latus Rectum as Graphing Aids

Determine the focus, directrix, and length of the latus rectum of the parabola $x^2 = -12y$. Employ these to assist in graphing the parabola.

Solution:

Compare this parabola with the general equation of a parabola. $x^2 = 4py$ $x^2 = -12y$

Solve for p.

$$4p = -12$$
$$p = -3$$

A parabola of the form $x^2 = 4py$ has focus $(0, p)$, directrix $y = -p$, and a latus rectum of length $4|p|$. For this parabola, $p = -3$; therefore, the focus is $(0, -3)$, the directrix is $y = 3$, and the length of the latus rectum is 12.

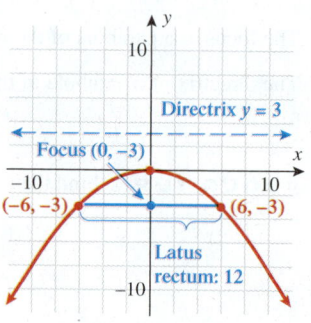

■ **YOUR TURN** Find the focus, directrix, and length of the latus rectum of the parabola $y^2 = -8x$, and use these to graph the parabola.

■ **Answer:**
The focus is $(-2, 0)$.
The directrix is $x = 2$.
The length of the latus rectum is 8.

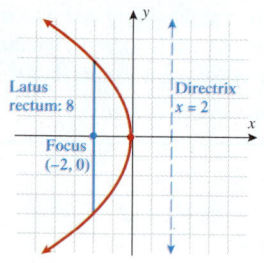

Finding the Equation of a Parabola with a Vertex at the Origin

Thus far we have started with the equation of a parabola and then determined its focus and directrix. Let's now reverse the process. For example, if we know the focus and directrix of a parabola, how do we find the equation of the parabola? If we are given the focus and directrix, then we can find the vertex, which is the midpoint between the focus and the directrix. If the vertex is at the origin, then we know the general equation of the parabola that corresponds with the focus.

EXAMPLE 3 Finding the Equation of a Parabola Given the Focus and Directrix When the Vertex Is at the Origin

Find the equation of a parabola whose focus is at the point $\left(0, \frac{1}{2}\right)$ and whose directrix is $y = -\frac{1}{2}$. Graph the equation.

Solution:

The midpoint of the segment joining the focus and the directrix along the axis of symmetry is the vertex.

Calculate the midpoint between $\left(0, \frac{1}{2}\right)$ and $\left(0, -\frac{1}{2}\right)$.

$$\text{Vertex} = \left(\frac{0 + 0}{2}, \frac{\frac{1}{2} - \frac{1}{2}}{2}\right) = (0, 0).$$

A parabola with vertex at $(0, 0)$, focus at $(0, p)$, and directrix $y = -p$ corresponds to the equation $x^2 = 4py$.

Identify p given that the focus is $\mathbf{(0, p)} = \left(\mathbf{0, \frac{1}{2}}\right)$.

$$p = \frac{1}{2}$$

Substitute $p = \frac{1}{2}$ into the standard equation of a parabola with vertex at the origin $x^2 = 4py$.

$$x^2 = 2y$$

Now that the equation is known, a few points can be selected, and the parabola can be point-plotted. Alternatively, the length of the latus rectum can be calculated to sketch the approximate width of the parabola.

To graph $x^2 = 2y$, first calculate the latus rectum.

$$4|p| = 4\left(\frac{1}{2}\right) = 2$$

Label the focus, directrix, and latus rectum, and draw a parabolic curve whose vertex is at the origin that intersects with the latus rectum's endpoints.

■ **YOUR TURN** Find the equation of a parabola whose focus is at the point $(-5, 0)$ and whose directrix is $x = 5$.

■ **Answer:** $y^2 = -20x$

Before we proceed to parabolas with general vertices, let's first make a few observations: the larger the latus rectum, the wider the parabola. An alternative approach for graphing the parabola is to plot a few points that satisfy the equation of the parabola, which is the approach in most textbooks.

Parabola with a Vertex at the Point (h, k)

Recall (Section 2.4) that the graph of $x^2 + y^2 = r^2$ is a circle with radius r centered at the origin, whereas the graph of $(x - h)^2 + (y - k)^2 = r^2$ is a circle with radius r centered at the point (h, k). In other words, the center is shifted from the origin to the point (h, k). This same translation (shift) can be used to describe parabolas whose vertex is at the point (h, k).

Study Tip

When $(h, k) = (0, 0)$, the vertex of the parabola is located at the origin.

EQUATION OF A PARABOLA WITH VERTEX AT THE POINT (h, k)

The standard (conic) form of the equation of a parabola with vertex at the point (h, k) is given by

EQUATION	$(y - k)^2 = 4p(x - h)$	$(x - h)^2 = 4p(y - k)$
VERTEX	(h, k)	(h, k)
FOCUS	$(p + h, k)$	$(h, p + k)$
DIRECTRIX	$x = -p + h$	$y = -p + k$
AXIS OF SYMMETRY	$y = k$	$x = h$
$p > 0$	opens to the right	opens upward
$p < 0$	opens to the left	opens downward

In order to find the vertex of a parabola given a general second-degree equation, first complete the square (Section 1.3) in order to identify (h, k). Then determine whether the parabola opens up, down, left, or right. Identify points that lie on the graph of the parabola. Intercepts are often the easiest points to find, since one of the variables is set equal to zero.

Technology Tip

Use a graphing calculator to check the graph of $y^2 - 6y - 2x + 8 = 0$. Use $(y - 3)^2 = 2x + 1$ to solve for y first. That is, $y_1 = 3 + \sqrt{2x + 1}$ or $y_2 = 3 - \sqrt{2x + 1}$.

Study Tip

It is often easier to find the intercepts by converting to the general form of the equation.

■ **Answer:**
Vertex: $\left(-\frac{9}{5}, -2\right)$
x-intercept: $x = -1$
y-intercepts: $y = -5$ and $y = 1$

EXAMPLE 4 Graphing a Parabola with Vertex (h, k)

Graph the parabola given by the equation $y^2 - 6y - 2x + 8 = 0$.

Solution:

Transform this equation into the form $(y - k)^2 = 4p(x - h)$, since this equation is of degree two in y and degree one in x. We know this parabola opens either to the left or right.

Complete the square on y: $\qquad\qquad\qquad\qquad\qquad y^2 - 6y - 2x + 8 = 0$

Isolate the y terms. $\qquad\qquad\qquad\qquad\qquad\qquad y^2 - 6y = 2x - 8$

Add 9 to both sides to complete the square. $\qquad\quad y^2 - 6y + 9 = 2x - 8 + 9$

Write the left side as a perfect square. $\qquad\qquad\quad (y - 3)^2 = 2x + 1$

Factor out a 2 on the right side. $\qquad\qquad\qquad\quad (y - 3)^2 = 2\left(x + \frac{1}{2}\right)$

Compare with $(y - k)^2 = 4p(x - h)$ and identify (h, k) and p.

$$(h, k) = \left(-\frac{1}{2}, 3\right)$$

$$4p = 2 \Rightarrow p = \frac{1}{2}$$

The vertex is at the point $\left(-\frac{1}{2}, 3\right)$, and since $p = \frac{1}{2}$ is positive, the parabola opens to the right. Since the parabola's vertex lies in quadrant II and it opens to the right, we know there are two y-intercepts and one x-intercept. Apply the general equation $y^2 - 6y - 2x + 8 = 0$ to find the intercepts.

Find the y-intercepts (set $x = 0$). $\qquad\qquad\qquad\qquad\quad y^2 - 6y + 8 = 0$

Factor. $\qquad\qquad\qquad\qquad\qquad\qquad\qquad\qquad (y - 2)(y - 4) = 0$

Solve for y. $\qquad\qquad\qquad\qquad\qquad\qquad\qquad\quad y = 2 \text{ or } y = 4$

Find the x-intercept (set $y = 0$). $\qquad\qquad\qquad\qquad\quad -2x + 8 = 0$

Solve for x. $\qquad\qquad\qquad\qquad\qquad\qquad\qquad\qquad x = 4$

Label the following points and connect them with a smooth curve:

Vertex: $\qquad\quad \left(-\frac{1}{2}, 3\right)$

y-intercepts: $\quad (0, 2)$ and $(0, 4)$

x-intercept: $\quad\; (4, 0)$

■ **YOUR TURN** For the equation $y^2 + 4y - 5x - 5 = 0$, identify the vertex and the intercepts, and graph.

EXAMPLE 5   **Graphing a Parabola with Vertex (*h*, *k*)**

Graph the parabola given by the equation $x^2 - 2x - 8y - 7 = 0$.

Solution:

Transform this equation into the form $(x - h)^2 = 4p(y - k)$, since this equation is degree two in x and degree one in y. We know this parabola opens either upward or downward.

Complete the square on x: $\qquad\qquad\qquad\qquad\quad x^2 - 2x - 8y - 7 = 0$

    Isolate the x terms. $\qquad\qquad\qquad\qquad\qquad\quad x^2 - 2x = 8y + 7$

    Add 1 to both sides to complete the square. $\qquad x^2 - 2x + 1 = 8y + 7 + 1$

    Write the left side as a perfect square. $\qquad\qquad (x - 1)^2 = 8y + 8$

    Factor out the 8 on the right side. $\qquad\qquad\quad (x - 1)^2 = 8(y + 1)$

Compare with $(x - h)^2 = 4p(y - k)$ and identify (h, k) and p.

$$(h, k) = (1, -1)$$

$$4p = 8 \Rightarrow p = 2$$

The vertex is at the point $(1, -1)$, and since $p = 2$ is positive, the parabola opens upward. Since the parabola's vertex lies in quadrant IV and it opens upward, we know there are two x-intercepts and one y-intercept. Use the general equation $x^2 - 2x - 8y - 7 = 0$ to find the intercepts.

Find the y-intercept (set $x = 0$). $\qquad\qquad\qquad\qquad -8y - 7 = 0$

    Solve for y. $\qquad\qquad\qquad\qquad\qquad\qquad\qquad\qquad y = -\dfrac{7}{8}$

Find the x-intercepts (set $y = 0$). $\qquad\qquad\qquad\quad x^2 - 2x - 7 = 0$

    Solve for x. $\qquad x = \dfrac{2 \pm \sqrt{4 + 28}}{2} = \dfrac{2 \pm \sqrt{32}}{2} = \dfrac{2 \pm 4\sqrt{2}}{2} = 1 \pm 2\sqrt{2}$

Label the following points and connect with a smooth curve:

    Vertex: $\qquad\quad (1, -1)$

    y-intercept: $\quad \left(0, -\dfrac{7}{8}\right)$

    x-intercepts: $\;\; (1 - 2\sqrt{2}, 0)$ and
$\qquad\qquad\qquad\;\; (1 + 2\sqrt{2}, 0)$

■ **YOUR TURN**   For the equation $x^2 + 2x + 8y - 7 = 0$, identify the vertex and the intercepts, and graph.

Technology Tip

Use a graphing calculator to check the graph of $x^2 - 2x - 8y - 7 = 0$.

Solve for y first: $y = \frac{1}{8}(x^2 - 2x - 7)$

■ **Answer:**
Vertex: $(-1, 1)$
x-intercepts: $x = -1 \pm 2\sqrt{2}$
y-intercept: $y = \dfrac{7}{8}$

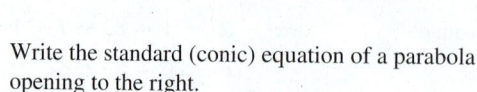

EXAMPLE 6   Finding the Equation of a Parabola with Vertex (h, k)

Find the equation of a parabola whose vertex is located at the point $(2, -3)$ and whose focus is located at the point $(5, -3)$.

Solution:

Draw a Cartesian plane and label the vertex and focus. The vertex and focus share the same axis of symmetry $y = -3$, and indicate a parabola opening to the right.

Technology Tip

Use a graphing calculator to check the graph of
$y^2 + 6y - 12x + 33 = 0$. Use
$y^2 + 6y - (12x - 33) = 0$
to solve for y first. That is,
$y_1 = -3 + 2\sqrt{3x - 6}$ or
$y_2 = -3 - 2\sqrt{3x - 6}$.

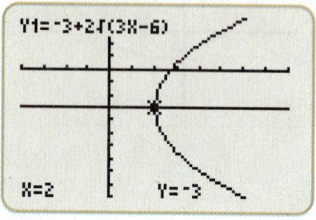

Write the standard (conic) equation of a parabola opening to the right.

$$(y - k)^2 = 4p(x - h) \qquad p > 0$$

Substitute the vertex $(h, k) = (2, -3)$, into the standard equation.

$$[y - (-3)]^2 = 4p(x - 2)$$

Find p.

The general form of the vertex is (h, k) and the focus is ($h + p$, k).

For this parabola, the vertex is (2, -3) and the focus is (5, -3).

Find p by taking the difference of the x-coordinates. $p = 3$

Substitute $p = 3$ into $[y - (-3)]^2 = 4p(x - 2)$. $(y + 3)^2 = 4(3)(x - 2)$

Eliminate the parentheses. $y^2 + 6y + 9 = 12x - 24$

Simplify. $\boxed{y^2 + 6y - 12x + 33 = 0}$

■ **Answer:** $y^2 + 6y + 8x - 7 = 0$

■ **YOUR TURN** Find the equation of the parabola whose vertex is located at $(2, -3)$ and whose focus is located at $(0, -3)$.

Applications

If we start with a parabola in the xy-plane and rotate it around its axis of symmetry, the result will be a three-dimensional paraboloid. Solar cookers illustrate the physical property that the rays of light coming into a parabola should be reflected to the focus. A flashlight reverses this process in that its light source at the focus illuminates a parabolic reflector to direct the beam outward.

Satellite dish

A satellite dish is in the shape of a paraboloid. Functioning as an antenna, the parabolic dish collects all of the incoming signals and reflects them to a single point, the focal point, which is where the receiver is located. In Examples 7 and 8, and in the Applications Exercises, the intention is not to find the three-dimensional equation of the paraboloid, but rather the equation of the plane parabola that's rotated to generate the paraboloid.

EXAMPLE 7   Finding the Location of the Receiver in a Satellite Dish

A satellite dish is 24 feet in diameter at its opening and 4 feet deep in its center. Where should the receiver be placed?

(−12, 4)   (12, 4)   24 ft   4 ft   y   x

Solution:

Draw a parabola with a vertex at the origin representing the center cross section of the satellite dish.

Write the standard equation of a parabola opening upward with vertex at $(0, 0)$.

$$x^2 = 4py$$

The point $(12, 4)$ lies on the parabola, so substitute $(12, 4)$ into $x^2 = 4py$.

$$(12)^2 = 4p(4)$$

Simplify.

$$144 = 16p$$

Solve for p.

$$p = 9$$

Substitute $p = 9$ into the focus $(0, p)$.

focus: $(0, 9)$

The receiver should be placed 9 feet from the vertex of the dish.

Parabolic antennas work for sound in addition to light. Have you ever wondered how the sound of the quarterback calling audible plays is heard by the sideline crew? The crew holds a parabolic system with a microphone at the focus. All of the sound in the direction of the parabolic system is reflected toward the focus, where the microphone amplifies and records the sound.

EXAMPLE 8   Finding the Equation of a Parabolic Sound Dish

If the parabolic sound dish the sideline crew is holding has a 2-foot diameter at the opening and the microphone is located 6 inches from the vertex, find the equation that governs the center cross section of the parabolic sound dish.

Solution:

Write the standard equation of a parabola opening to the right with the vertex at the origin $(0, 0)$.

$$x = 4py^2$$

The focus is located 6 inches $\left(\frac{1}{2} \text{ foot}\right)$ from the vertex.

$$(p, 0) = \left(\frac{1}{2}, 0\right)$$

Solve for p.

$$p = \frac{1}{2}$$

Let $p = \frac{1}{2}$ in $x = 4py^2$.

$$x = 4\left(\frac{1}{2}\right)y^2$$

Simplify.

$$x = 2y^2$$

In this section we discussed parabolas whose vertex is at the origin.

EQUATION	$y^2 = 4px$	$x^2 = 4py$
VERTEX	$(0, 0)$	$(0, 0)$
FOCUS	$(p, 0)$	$(0, p)$
DIRECTRIX	$x = -p$	$y = -p$
AXIS OF SYMMETRY	x-axis	y-axis
$p > 0$	opens to the right	opens upward
$p < 0$	opens to the left	opens downward
GRAPH		

For parabolas whose vertex is at the point, (h, k):

EQUATION	$(y - k)^2 = 4p(x - h)$	$(x - h)^2 = 4p(y - k)$
VERTEX	(h, k)	(h, k)
FOCUS	$(p + h, k)$	$(h, p + k)$
DIRECTRIX	$x = -p + h$	$y = -p + k$
AXIS OF SYMMETRY	$y = k$	$x = h$
$p > 0$	opens to the right	opens upward
$p < 0$	opens to the left	opens downward

SECTION
11.2 EXERCISES

■ SKILLS

In Exercises 1–4, match the parabola to the equation.

1. $y^2 = 4x$

2. $y^2 = -4x$

3. $x^2 = -4y$

4. $x^2 = 4y$

a.

b.

c.

d.

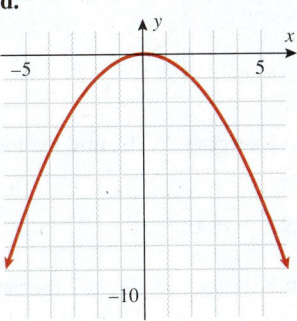

In Exercises 5–8, match the parabola to the equation.

5. $(y - 1)^2 = 4(x - 1)$

6. $(y + 1)^2 = -4(x - 1)$

7. $(x + 1)^2 = -4(y + 1)$

8. $(x - 1)^2 = 4(y - 1)$

a.

b.

c.

d.

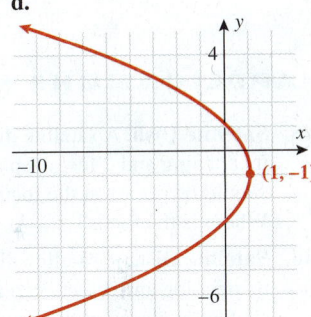

In Exercises 9–20, find an equation for the parabola described.

9. Vertex at $(0, 0)$; focus at $(0, 3)$

10. Vertex at $(0, 0)$; focus at $(2, 0)$

11. Vertex at $(0, 0)$; focus at $(-5, 0)$

12. Vertex at $(0, 0)$; focus at $(0, -4)$

13. Vertex at $(3, 5)$; focus at $(3, 7)$

14. Vertex at $(3, 5)$; focus at $(7, 5)$

15. Vertex at $(2, 4)$; focus at $(0, 4)$

16. Vertex at $(2, 4)$; focus at $(2, -1)$

17. Focus at $(2, 4)$; directrix at $y = -2$

18. Focus at $(2, -2)$; directrix at $y = 4$

19. Focus at $(3, -1)$; directrix at $x = 1$

20. Focus at $(-1, 5)$; directrix at $x = 5$

In Exercises 21–24, write an equation for each parabola.

21.

22.

23.

24.

In Exercises 25–32, find the focus, vertex, directrix, and length of latus rectum and graph the parabola.

25. $x^2 = 8y$

26. $x^2 = -12y$

27. $y^2 = -2x$

28. $y^2 = 6x$

29. $x^2 = 16y$

30. $x^2 = -8y$

31. $y^2 = 4x$

32. $y^2 = -16x$

In Exercises 33–44, find the vertex and graph the parabola.

33. $(y - 2)^2 = 4(x + 3)$

34. $(y + 2)^2 = -4(x - 1)$

35. $(x - 3)^2 = -8(y + 1)$

36. $(x + 3)^2 = -8(y - 2)$

37. $(x + 5)^2 = -2y$

38. $y^2 = -16(x + 1)$

39. $y^2 - 4y - 2x + 4 = 0$

40. $x^2 - 6x + 2y + 9 = 0$

41. $y^2 + 2y - 8x - 23 = 0$

42. $x^2 - 6x - 4y + 10 = 0$

43. $x^2 - x + y - 1 = 0$

44. $y^2 + y - x + 1 = 0$

▪ APPLICATIONS

45. Satellite Dish. A satellite dish measures 8 feet across its opening and 2 feet deep at its center. The receiver should be placed at the focus of the parabolic dish. Where is the focus?

46. Satellite Dish. A satellite dish measures 30 feet across its opening and 5 feet deep at its center. The receiver should be placed at the focus of the parabolic dish. Where is the focus?

47. Eyeglass Lens. Eyeglass lenses can be thought of as very wide parabolic curves. If the focus occurs 2 centimeters from the center of the lens and the lens at its opening is 5 centimeters, find an equation that governs the shape of the center cross section of the lens.

48. Optical Lens. A parabolic lens focuses light onto a focal point 3 centimeters from the vertex of the lens. How wide is the lens 0.5 centimeter from the vertex?

Exercises 49 and 50 are examples of solar cookers. Parabolic shapes are often used to generate intense heat by collecting sun rays and focusing all of them at a focal point.

49. Solar Cooker. The parabolic cooker MS-ST10 is delivered as a kit, handily packed in a single carton, with complete assembly instructions and even the necessary tools.

Solar cooker, Ubuntu Village, Johannesburg, South Africa

Thanks to the reflector diameter of 1 meter, it develops an immense power: 1 liter of water boils in significantly less than half an hour. If the rays are focused 40 centimeters from the vertex, find the equation for the parabolic cooker.

50. Le Four Solaire at Font-Romeur "Mirrors of the Solar Furnace." There is a reflector in the Pyrenees Mountains that is eight stories high. It cost $2 million and took 10 years to build. Made of 9000 mirrors arranged in a parabolic formation, it can reach 6000°F just from the Sun hitting it!

Solar furnace, Odellio, France

Mark Antman/The Image Works

If the diameter of the parabolic mirror is 100 meters and the sunlight is focused 25 meters from the vertex, find the equation for the parabolic dish.

51. Sailing under a Bridge. A bridge with a parabolic shape has an opening 80 feet wide at the base (where the bridge meets the water), and the height in the center of the bridge is 20 feet. A sailboat whose mast reaches 17 feet above the water is traveling under the bridge 10 feet from the center of the bridge. Will it clear the bridge without scraping its mast? Justify your answer.

52. Driving under a Bridge. A bridge with a parabolic shape reaches a height of 25 feet in the center of the road, and the width of the bridge opening at ground level is 20 feet combined (both lanes). If an RV is 10 feet tall and 8 feet wide, it won't make it under the bridge if it hugs the center line. Will it clear the bridge if it straddles the center line? Justify your answer.

53. Parabolic Telescope. The Arecibo radio telescope in Puerto Rico has an enormous reflecting surface, or radio mirror. The huge "dish" is 1000 feet in diameter and 167 feet deep and covers an area of about 20 acres. Using these dimensions, determine the focal length of the telescope. Find the equation for the dish portion of the telescope.

54. Suspension Bridge. If one parabolic segment of a suspension bridge is 300 feet and if the cables at the vertex are suspended 10 feet above the bridge, whereas the height of the cables 150 feet from the vertex reaches 60 feet, find the equation of the parabolic path of the suspension cables.

55. Business. The profit, in thousands of dollars, for a product is $P(x) = -x^2 + 60x - 500$ where x is the production level in hundreds of units. Find the production level that maximizes the profit. Find the maximum profit.

56. Business. The profit, in thousands of dollars, for a product is $P(x) = -x^2 + 80x - 1200$ where x is the production level in hundreds of units. Find the production level that maximizes the profit. Find the maximum profit.

■ CATCH THE MISTAKE

In Exercises 57 and 58, explain the mistake that is made.

57. Find an equation for a parabola whose vertex is at the origin and whose focus is at the point (3, 0).

Solution:

Write the general equation for a parabola whose vertex is at the origin. $\qquad x^2 = 4py$

The focus of this parabola is $(p, 0) = (3, 0)$. $\quad p = 3$

Substitute $p = 3$ into $x^2 = 4py$. $\qquad x^2 = 12y$

This is incorrect. What mistake was made?

58. Find an equation for a parabola whose vertex is at the point (3, 2) and whose focus is located at (5, 2).

Solution:

Write the equation associated with a parabola whose vertex is (3, 2). $\quad (x - h)^2 = 4p(y - k)$

Substitute (3, 2) into $(x - h)^2 = 4p(y - k)$. $\qquad (x - 3)^2 = 4p(y - 2)$

The focus is located at (5, 2); therefore, $p = 5$.

Substitute $p = 5$ into $(x - 3)^2 = 4p(y - 2)$. $\qquad (x - 3)^2 = 20(y - 2)$

This is incorrect. What mistake(s) was made?

■ CONCEPTUAL

In Exercises 59–62, determine whether each statement is true or false.

59. The vertex lies on the graph of a parabola.

60. The focus lies on the graph of a parabola.

61. The directrix lies on the graph of a parabola.

62. The endpoints of the latus rectum lie on the graph of a parabola.

■ CHALLENGE

63. Derive the standard equation of a parabola with its vertex at the origin, opening upward $x^2 = 4py$. [Calculate the distance d_1 from any point on the parabola (x, y) to the focus $(0, p)$. Calculate the distance d_2 from any point on the parabola (x, y) to the directrix $(-p, y)$. Set $d_1 = d_2$.]

64. Derive the standard equation of a parabola opening right, $y^2 = 4px$. [Calculate the distance d_1 from any point on the parabola (x, y) to the focus $(p, 0)$. Calculate the distance d_2 from any point on the parabola (x, y) to the directrix $(x, -p)$. Set $d_1 = d_2$.]

■ TECHNOLOGY

65. With a graphing utility, plot the parabola $x^2 - x + y - 1 = 0$. Compare with the sketch you drew for Exercise 43.

66. With a graphing utility, plot the parabola $y^2 + y - x + 1 = 0$. Compare with the sketch you drew for Exercise 44.

67. In your mind, picture the parabola given by $(y + 3.5)^2 = 10(x - 2.5)$. Where is the vertex? Which way does this parabola open? Now plot the parabola with a graphing utility.

68. In your mind, picture the parabola given by $(x + 1.4)^2 = -5(y + 1.7)$. Where is the vertex? Which way does this parabola open? Now plot the parabola with a graphing utility.

69. In your mind, picture the parabola given by $(y - 1.5)^2 = -8(x - 1.8)$. Where is the vertex? Which way does this parabola open? Now plot the parabola with a graphing utility.

70. In your mind, picture the parabola given by $(x + 2.4)^2 = 6(y - 3.2)$. Where is the vertex? Which way does this parabola open? Now plot the parabola with a graphing utility.

71. Given is the parabola $y^2 - 4.5y - 4x - 8.9375 = 0$.
   a. Solve the equation for y and use a graphing utility to plot the parabola.
   b. Transform the equation into the form $(y - k)^2 = 4p(x - h)$. Find the vertex. Which way does the parabola open?
   c. Do (a) and (b) agree with each other?

72. Given is the parabola $2x^2 + 6.4x + y - 2.08 = 0$.
   a. Solve the equation for y and use a graphing utility to plot the parabola.
   b. Transform the equation into the form $(x - h)^2 = 4p(y - k)$. Find the vertex. Which way does the parabola open?
   c. Do (a) and (b) agree with each other?

SKILLS OBJECTIVES

- Graph an ellipse given the center, major axis, and minor axis.
- Find the equation of an ellipse centered at the origin.
- Find the equation of an ellipse centered at the point (h, k).
- Solve applied problems that involve ellipses.

CONCEPTUAL OBJECTIVES

- Derive the general equation of an ellipse.
- Understand the meaning of major and minor axes and foci.
- Understand the properties of an ellipse that result in a circle.
- Interpret eccentricity in terms of the shape of the ellipse.

Ellipse Centered at the Origin

Definition of an Ellipse

If we were to take a piece of string, tie loops at both ends, and tack the ends down so that the string had lots of slack, we would have the picture on the left. If we then took a pencil and pulled the string taut and traced our way around for one full rotation, the result would be an ellipse. See the second figure on the left.

| **DEFINITION** | **Ellipse** |

An **ellipse** is the set of all points in a plane, the sum of whose distances from two fixed points is constant. These two fixed points are called **foci** (plural of focus). A line segment through the foci called the **major axis** intersects the ellipse at the **vertices**. The midpoint of the line segment joining the vertices is called the **center**. The line segment that intersects the center and joins two points on the ellipse and is perpendicular to the major axis is called the **minor axis**.

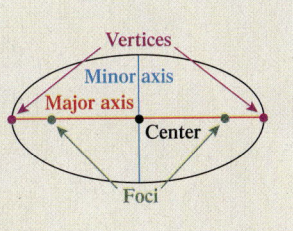

Let's start with an ellipse whose center is located at the origin. Using graph-shifting techniques, we can later extend the characteristics of an ellipse centered at a point other than the origin. Ellipses can vary from the shape of circles to something quite elongated, either horizontally or vertically, that resembles the shape of a racetrack. We say that the ellipse has either greater or lesser *eccentricity*; as we will see, there is a simple mathematical definition of *eccentricity*. It can be shown that the standard equation of an ellipse with its center at the origin is given by one of two forms, depending on whether the orientation of the major axis of the ellipse is horizontal or vertical. For $a > b > 0$, if the major axis is horizontal, then the equation is given by $\dfrac{x^2}{a^2} + \dfrac{y^2}{b^2} = 1$, and if the major axis is vertical, then the equation is given by $\dfrac{x^2}{b^2} + \dfrac{y^2}{a^2} = 1$.

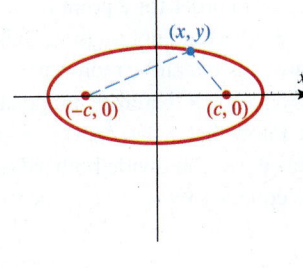

Let's consider an ellipse with center at the origin and the foci on the x-axis. Let the distance from the center to the focus be c. Therefore, the foci are located at the points $(-c, 0)$ and $(c, 0)$. The line segment containing the foci is called the major axis, and it lies along the x-axis. The sum of the two distances from the foci to any point (x, y) must be constant.

Derivation of the Equation of an Ellipse

WORDS	MATH
Calculate the distance from (x, y) to $(-c, 0)$ by applying the distance formula.	$\sqrt{[x - (-c)]^2 + y^2}$
Calculate the distance from (x, y) to $(c, 0)$ by applying the distance formula.	$\sqrt{(x - c)^2 + y^2}$
The sum of these two distances is equal to a constant ($2a$ for convenience).	$\sqrt{[x - (-c)]^2 + y^2} + \sqrt{(x - c)^2 + y^2} = 2a$
Isolate one radical.	$\sqrt{[x - (-c)]^2 + y^2} = 2a - \sqrt{(x - c)^2 + y^2}$
Square both sides of the equation.	$(x + c)^2 + y^2 = 4a^2 - 4a\sqrt{(x - c)^2 + y^2} + (x - c)^2 + y^2$
Expand the squares.	$x^2 + 2cx + c^2 + y^2 = 4a^2 - 4a\sqrt{(x - c)^2 + y^2}$ $+ x^2 - 2cx + c^2 + y^2$
Simplify.	$4cx - 4a^2 = -4a\sqrt{(x - c)^2 + y^2}$
Divide both sides of the equation by -4.	$a^2 - cx = a\sqrt{(x - c)^2 + y^2}$
Square both sides of the equation.	$(a^2 - cx)^2 = a^2[(x - c)^2 + y^2]$
Expand the squares.	$a^4 - 2a^2cx + c^2x^2 = a^2[x^2 - 2cx + c^2 + y^2]$
Distribute the a^2 term.	$a^4 - 2a^2cx + c^2x^2 = a^2x^2 - 2a^2cx + a^2c^2 + a^2y^2$
Group the x and y terms together, respectively, on one side and constants on the other side.	$c^2x^2 - a^2x^2 - a^2y^2 = a^2c^2 - a^4$
Factor out the common factors.	$(c^2 - a^2)x^2 - a^2y^2 = a^2(c^2 - a^2)$
Multiply both sides of the equation by -1.	$(a^2 - c^2)x^2 + a^2y^2 = a^2(a^2 - c^2)$
We can make the argument that $a > c$ in order for a point to be on the ellipse (and not on the x-axis). Thus, since a and c represent distances and therefore are positive, we know that $a^2 > c^2$, or $a^2 - c^2 > 0$. Hence, we can divide both sides of the equation by $a^2 - c^2$, since $a^2 - c^2 \neq 0$.	$x^2 + \dfrac{a^2y^2}{(a^2 - c^2)} = a^2$
Let $b^2 = a^2 - c^2$.	$x^2 + \dfrac{a^2y^2}{b^2} = a^2$
Divide both sides of the equation by a^2.	$\boxed{\dfrac{x^2}{a^2} + \dfrac{y^2}{b^2} = 1}$

The equation $\dfrac{x^2}{a^2} + \dfrac{y^2}{b^2} = 1$ represents an ellipse with center at the origin with the foci along the x-axis, since $a > b$. The following box summarizes ellipses that have their center at the origin and foci along either the x-axis or y-axis:

EQUATION OF AN ELLIPSE WITH CENTER AT THE ORIGIN

The **standard form of the equation of an ellipse** with its center at the origin is given by

ORIENTATION OF MAJOR AXIS	Horizontal (along the x-axis)	Vertical (along the y-axis)
EQUATION	$\dfrac{x^2}{a^2} + \dfrac{y^2}{b^2} = 1 \quad a > b > 0$	$\dfrac{x^2}{b^2} + \dfrac{y^2}{a^2} = 1 \quad a > b > 0$
FOCI	$(-c, 0) \quad (c, 0)$ where $c^2 = a^2 - b^2$	$(0, -c) \quad (0, c)$ where $c^2 = a^2 - b^2$
VERTICES	$(-a, 0) \quad (a, 0)$	$(0, -a) \quad (0, a)$
OTHER INTERCEPTS	$(0, b) \quad (0, -b)$	$(b, 0) \quad (-b, 0)$
GRAPH	*(graph of horizontal ellipse with points $(0, b)$, $(-c, 0)$, $(c, 0)$, $(0, 0)$, $(-a, 0)$, $(a, 0)$, $(0, -b)$)*	*(graph of vertical ellipse with points $(0, a)$, $(0, c)$, $(-b, 0)$, $(0, 0)$, $(b, 0)$, $(0, -c)$, $(0, -a)$)*

In both cases, the value of c, the distance along the major axis from the center to the focus, is given by $c^2 = a^2 - b^2$. The length of the major axis is $2a$ and the length of the minor axis is $2b$.

Notice that when $a = b$ the equation $\dfrac{x^2}{a^2} + \dfrac{y^2}{b^2} = 1$ simplifies to $\dfrac{x^2}{a^2} + \dfrac{y^2}{a^2} = 1$ or $x^2 + y^2 = a^2$, which corresponds to a circle. The vertices correspond to intercepts when an ellipse is centered at the origin. One of the first things we notice about an ellipse is its *eccentricity*. The **eccentricity**, denoted e, is given by $e = \dfrac{c}{a}$, where $0 \le e < 1$. The circle is a limiting form of an ellipse, $c = 0$. In other words, if the eccentricity is close to 0, then the ellipse resembles a circle, whereas if the eccentricity is close to 1, then the ellipse is quite elongated, or eccentric.

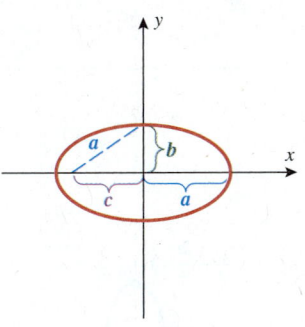

Graphing an Ellipse with Center at the Origin

The equation of an ellipse in standard form can be used to graph an ellipse. Although an ellipse is defined in terms of the foci, the foci are not part of the graph. It is important to note that if the divisor of the term with x^2 is larger than the divisor of the term with y^2, then the ellipse is elongated horizontally.

Use a graphing calculator to check the graph of $\dfrac{x^2}{25} + \dfrac{y^2}{9} = 1$.

Solve for y first. That is,

$$y_1 = 3\sqrt{1 - \dfrac{x^2}{25}} \text{ and}$$

$$y_2 = -3\sqrt{1 - \dfrac{x^2}{25}}.$$

Study Tip

If the divisor of x^2 is larger than the divisor of y^2, then the major axis is horizontal along the x-axis, as in Example 1. If the divisor of y^2 is larger than the divisor of x^2, then the major axis is vertical along the y-axis, as you will see in Example 2.

■ **Answer:**

a.

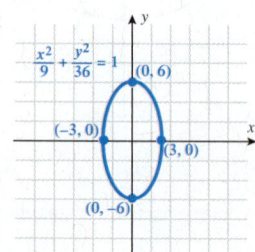

b.

EXAMPLE 1   Graphing an Ellipse with a Horizontal Major Axis

Graph the ellipse given by $\dfrac{x^2}{25} + \dfrac{y^2}{9} = 1$.

Solution:

Since $25 > 9$, the major axis is horizontal. $\qquad a^2 = 25$ and $b^2 = 9$

Solve for a and b. $\qquad\qquad\qquad\qquad\qquad a = 5$ and $b = 3$

Identify the vertices: $(-a, 0)$ and $(a, 0)$. $\qquad (-5, 0)$ and $(5, 0)$

Identify the endpoints (y-intercepts) on the minor axis: $(0, -b)$ and $(0, b)$. $\qquad (0, -3)$ and $(0, 3)$

Graph by labeling the points $(-5, 0)$, $(5, 0)$, $(0, -3)$, and $(0, 3)$ and connecting them with a smooth curve.

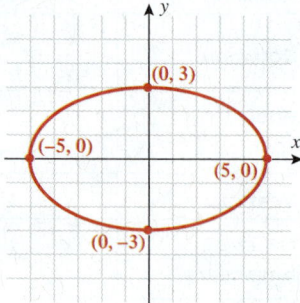

If the divisor of x^2 is larger than the divisor of y^2, then the major axis is horizontal along the x-axis, as in Example 1. If the divisor of y^2 is larger than the divisor of x^2, then the major axis is vertical along the y-axis, as you will see in Example 2.

EXAMPLE 2   Graphing an Ellipse with a Vertical Major Axis

Graph the ellipse given by $16x^2 + y^2 = 16$.

Solution:

Write the equation in standard form by dividing by 16. $\qquad \dfrac{x^2}{1} + \dfrac{y^2}{16} = 1$

Since $16 > 1$, this ellipse is elongated vertically. $\qquad a^2 = 16$ and $b^2 = 1$

Solve for a and b. $\qquad\qquad\qquad\qquad\qquad a = 4$ and $b = 1$

Identify the vertices: $(0, -a)$ and $(0, a)$. $\qquad (0, -4)$ and $(0, 4)$

Identify the x-intercepts on the minor axis: $(-b, 0)$ and $(b, 0)$. $\qquad (-1, 0)$ and $(1, 0)$

Graph by labeling the points $(0, -4)$, $(0, 4)$, $(-1, 0)$, and $(1, 0)$ and connecting them with a smooth curve.

> **Classroom Example 11.3.2***
> Determine the range of a values for which the ellipse $\dfrac{x^2}{2} + \dfrac{y^2}{a^2} = 1$ has a vertical major axis.
> **Answer:** $(-\infty, -\sqrt{2}) \cup (\sqrt{2}, \infty)$

■ **YOUR TURN** Graph the ellipses:

$\qquad$ **a.** $\dfrac{x^2}{9} + \dfrac{y^2}{4} = 1$ $\qquad$ **b.** $\dfrac{x^2}{9} + \dfrac{y^2}{36} = 1$

Finding the Equation of an Ellipse with Center at the Origin

What if we know the vertices and the foci of an ellipse and want to find the equation to which it corresponds? The axis on which the foci and vertices are located is the major axis. Therefore, we will have the standard equation of an ellipse, and a will be known (from the vertices). Since c is known from the foci, we can use the relation $c^2 = a^2 - b^2$ to determine the unknown b.

EXAMPLE 3   Finding the Equation of an Ellipse Centered at the Origin

Find the standard form of the equation of an ellipse with foci at $(-3, 0)$ and $(3, 0)$ and vertices $(-4, 0)$ and $(4, 0)$.

Solution:

The major axis lies along the x-axis, since it contains the foci and vertices.

Write the corresponding general equation of an ellipse. $\dfrac{x^2}{a^2} + \dfrac{y^2}{b^2} = 1$

Identify a from the vertices:

   Match vertices $(-4, 0) = (-a, 0)$ and $(4, 0) = (a, 0)$. $a = 4$

Identify c from the foci:

   Match foci $(-3, 0) = (-c, 0)$ and $(3, 0) = (c, 0)$. $c = 3$

Substitute $a = 4$ and $c = 3$ into $b^2 = a^2 - c^2$. $b^2 = 4^2 - 3^2$

Simplify. $b^2 = 7$

Substitute $a^2 = 16$ and $b^2 = 7$ into $\dfrac{x^2}{a^2} + \dfrac{y^2}{b^2} = 1$. $\dfrac{x^2}{16} + \dfrac{y^2}{7} = 1$

The equation of the ellipse is $\boxed{\dfrac{x^2}{16} + \dfrac{y^2}{7} = 1}$.

YOUR TURN Find the standard form of the equation of an ellipse with vertices at $(0, -6)$ and $(0, 6)$ and foci $(0, -5)$ and $(0, 5)$.

Classroom Example 11.3.3
Find the equation of the ellipse with foci $\left(0, \frac{1}{2}\right)$ and $\left(0, -\frac{1}{2}\right)$ and vertices $(0, -2)$ and $(0, 2)$.

Answer:

$\dfrac{4}{15}x^2 + \dfrac{y^2}{4} = 1$

Answer: $\dfrac{x^2}{11} + \dfrac{y^2}{36} = 1$

Ellipse Centered at the Point (h, k)

We can use graph-shifting techniques to graph ellipses that are centered at a point other than the origin. For example, to graph $\dfrac{(x - h)^2}{a^2} + \dfrac{(y - k)^2}{b^2} = 1$ (assuming h and k are positive constants), start with the graph of $\dfrac{x^2}{a^2} + \dfrac{y^2}{b^2} = 1$ and shift to the right h units and up k units. The center, the vertices, the foci, and the major and minor axes all shift. In other words, the two ellipses are identical in shape and size, except that the ellipse $\dfrac{(x - h)^2}{a^2} + \dfrac{(y - k)^2}{b^2} = 1$ is centered at the point (h, k).

The following table summarizes the characteristics of ellipses centered at a point other than the origin:

EQUATION OF AN ELLIPSE WITH CENTER AT THE POINT (*h, k*)

The **standard form of the equation of an ellipse** with its center at the point (*h, k*) is given by

ORIENTATION OF MAJOR AXIS	Horizontal (parallel to the *x*-axis)	Vertical (parallel to the *y*-axis)
EQUATION	$\dfrac{(x-h)^2}{a^2} + \dfrac{(y-k)^2}{b^2} = 1$	$\dfrac{(x-h)^2}{b^2} + \dfrac{(y-k)^2}{a^2} = 1$
GRAPH	(graph of horizontal ellipse with points $(h-a,k)$, $(h+a,k)$, $(h,k+b)$, $(h,k-b)$, $(h-c,k)$, $(h+c,k)$, center (h,k))	(graph of vertical ellipse with points $(h,k+a)$, $(h,k-a)$, $(h-b,k)$, $(h+b,k)$, $(h,k+c)$, $(h,k-c)$, center (h,k))
FOCI	$(h-c, k)$ $(h+c, k)$	$(h, k-c)$ $(h, k+c)$
VERTICES	$(h-a, k)$ $(h+a, k)$	$(h, k-a)$ $(h, k+a)$

In both cases, $a > b > 0$, $c^2 = a^2 - b^2$, the length of the major axis is $2a$, and the length of the minor axis is $2b$.

Technology Tip

Use a graphing calculator to check the graph of $\dfrac{(x-2)^2}{9} + \dfrac{(y+1)^2}{16} = 1$.
Solve for *y* first. That is,

$$y_1 = -1 + 4\sqrt{1 - \frac{(x-2)^2}{9}} \text{ or}$$

$$y_2 = -1 - 4\sqrt{1 - \frac{(x-2)^2}{9}}.$$

■ **Answer:**

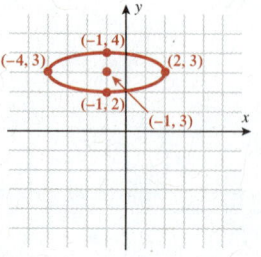

EXAMPLE 4   Graphing an Ellipse with Center (*h, k*) Given the Equation in Standard Form

Graph the ellipse given by $\dfrac{(x-2)^2}{9} + \dfrac{(y+1)^2}{16} = 1$.

Solution:

Write the equation in the form

$$\frac{(x-h)^2}{b^2} + \frac{(y-k)^2}{a^2} = 1. \qquad \frac{(x-2)^2}{3^2} + \frac{[y-(-1)]^2}{4^2} = 1$$

Identify *a*, *b*, and the center (*h, k*). $a = 4$, $b = 3$, and $(h, k) = (2, -1)$

Draw a graph and label the center: $(2, -1)$.

Since $a = 4$, the vertices are up 4 and down 4 units from the center: $(2, -5)$ and $(2, 3)$.

Since $b = 3$, the endpoints of the minor axis are to the left and right 3 units: $(-1, -1)$ and $(5, -1)$.

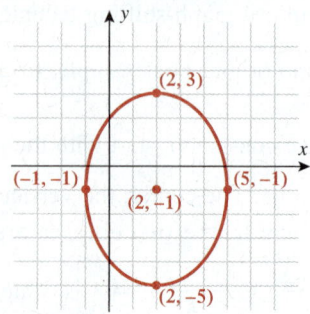

■ **YOUR TURN**   Graph the ellipse given by $\dfrac{(x+1)^2}{9} + \dfrac{(y-3)^2}{1} = 1$.

Recall that when we are given the equation of a circle in general form, we first complete the square in order to express the equation in standard form, which allows the center and radius to be identified. We use that same approach when the equation of an ellipse is given in a general form.

EXAMPLE 5   Graphing an Ellipse with Center (h, k) Given an Equation in General Form

Graph the ellipse given by $4x^2 + 24x + 25y^2 - 50y - 39 = 0$.

Solution:

Transform the general equation into standard form.

Group x terms together and y terms together and add 39 to both sides.

$$(4x^2 + 24x) + (25y^2 - 50y) = 39$$

Factor out the 4 common to the x terms and the 25 common to the y terms.

$$4(x^2 + 6x) + 25(y^2 - 2y) = 39$$

Complete the square on x and y.

$$4(x^2 + 6x + \mathbf{9}) + 25(y^2 - 2y + \mathbf{1}) = 39 + 4(\mathbf{9}) + 25(\mathbf{1})$$

Simplify.

$$4(x + 3)^2 + 25(y - 1)^2 = 100$$

Divide by 100.

$$\frac{(x + 3)^2}{25} + \frac{(y - 1)^2}{4} = 1$$

Since $25 > 4$, this is an ellipse with a horizontal major axis.

Now that the equation of the ellipse is in standard form, compare to $\frac{(x - h)^2}{a^2} + \frac{(y - k)^2}{b^2} = 1$ and identify a, b, h, k.

$$a = 5, b = 2, \text{ and } (h, k) = (-3, 1)$$

Since $a = 5$, the vertices are 5 units to left and right of the center.

$$(-8, 1) \text{ and } (2, 1)$$

Since $b = 2$, the endpoints of the minor axis are up and down 2 units from the center.

$$(-3, -1) \text{ and } (-3, 3)$$

Graph.

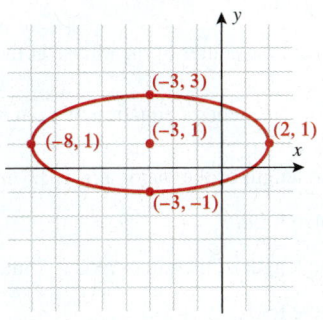

YOUR TURN Write the equation $4x^2 + 32x + y^2 - 2y + 61 = 0$ in standard form. Identify the center, vertices, and endpoints of the minor axis, and graph.

Technology Tip

Use a graphing calculator to check the graph of $4x^2 + 24x + 25y^2 - 50y - 39 = 0$.

Use $\frac{(x + 3)^2}{25} + \frac{(y - 1)^2}{4} = 1$ to solve for y first. That is,

$$y_1 = 1 + 2\sqrt{1 - \frac{(x + 3)^2}{25}} \text{ or}$$

$$y_2 = 1 - 2\sqrt{1 - \frac{(x + 3)^2}{25}}.$$

■ **Answer:**

$$\frac{(x + 4)^2}{1} + \frac{(y - 1)^2}{4} = 1$$

Center: $(-4, 1)$

Vertices: $(-4, -1)$ and $(-4, 3)$

Endpoints of minor axis: $(-5, 1)$ and $(-3, 1)$

Applications

There are many examples of ellipses all around us. On Earth we have racetracks, and in our solar system, the planets travel in elliptical orbits with the Sun as a focus. Satellites are in elliptical orbits around the Earth. Most communications satellites are in a *geosynchronous* (GEO) orbit—they orbit the Earth once each day. In order to stay over the same spot on Earth, a *geostationary* satellite has to be directly above the equator; it circles the Earth in exactly the time it takes the Earth to turn once on its axis, and its orbit has to follow the path of the equator as the Earth rotates. Otherwise, from the Earth the satellite would appear to move in a north–south line every day.

If we start with an ellipse in the *xy*-plane and rotate it around its major axis, the result is a three-dimensional ellipsoid.

A football and a blimp are two examples of ellipsoids. The ellipsoidal shape allows for a more aerodynamic path.

iStockphoto

Joseph Sohm/Visions of America/
Age Fotostock America, Inc.

Technology Tip

Use a graphing calculator to check the graph of $\dfrac{x^2}{5.5^2} + \dfrac{y^2}{3.5^2} = 1$. Solve for *y* first.

That is, $y_1 = 3.5\sqrt{1 - \dfrac{x^2}{5.5^2}}$ or $y_2 = -3.5\sqrt{1 - \dfrac{x^2}{5.5^2}}$.

EXAMPLE 6   An Official NFL Football

A longitudinal section (that includes the two vertices and the center) of an official Wilson NFL football is an ellipse. The longitudinal section is approximately 11 inches long and 7 inches wide. Write an equation governing the elliptical longitudinal section.

Solution:

Locate the center of the ellipse at the origin and orient the football horizontally.

Write the general equation of a circle centered at the origin.	$\dfrac{x^2}{a^2} + \dfrac{y^2}{b^2} = 1$
The length of the major axis is 11 inches.	$2a = 11$
Solve for *a*.	$a = 5.5$
The length of the minor axis is 7 inches.	$2b = 7$
Solve for *b*.	$b = 3.5$
Substitute $a = 5.5$ and $b = 3.5$ into $\dfrac{x^2}{a^2} + \dfrac{y^2}{b^2} = 1$.	$\boxed{\dfrac{x^2}{5.5^2} + \dfrac{y^2}{3.5^2} = 1}$

In this section we first analyzed ellipses that are centered at the origin.

ORIENTATION OF MAJOR AXIS	Horizontal along the x-axis	Vertical along the y-axis
EQUATION	$\dfrac{x^2}{a^2} + \dfrac{y^2}{b^2} = 1 \quad a > b > 0$	$\dfrac{x^2}{b^2} + \dfrac{y^2}{a^2} = 1 \quad a > b > 0$
FOCI*	$(-c, 0) \quad (c, 0)$	$(0, -c) \quad (0, c)$
VERTICES	$(-a, 0) \quad (a, 0)$	$(0, -a) \quad (0, a)$
OTHER INTERCEPTS	$(0, -b) \quad (0, b)$	$(-b, 0) \quad (b, 0)$
GRAPH		

*$c^2 = a^2 - b^2$

For ellipses centered at the origin, we can graph an ellipse by finding all four intercepts.
For ellipses centered at the point (h, k), the major and minor axes and endpoints of the ellipse all shift accordingly.

ORIENTATION OF MAJOR AXIS	Horizontal (parallel to the x-axis)	Vertical (parallel to the y-axis)
EQUATION	$\dfrac{(x - h)^2}{a^2} + \dfrac{(y - k)^2}{b^2} = 1$	$\dfrac{(x - h)^2}{b^2} + \dfrac{(y - k)^2}{a^2} = 1$
GRAPH		
FOCI	$(h - c, k) \quad (h + c, k)$	$(h, k - c) \quad (h, k + c)$
VERTICES	$(h - a, k) \quad (h + a, k)$	$(h, k - a) \quad (h, k + a)$

When $a = b$, the ellipse is a circle.

SECTION
11.3 EXERCISES

■ SKILLS

In Exercises 1–4, match the equation to the ellipse.

1. $\dfrac{x^2}{36} + \dfrac{y^2}{16} = 1$

2. $\dfrac{x^2}{16} + \dfrac{y^2}{36} = 1$

3. $\dfrac{x^2}{8} + \dfrac{y^2}{72} = 1$

4. $4x^2 + y^2 = 1$

a.

b.

c.

d.

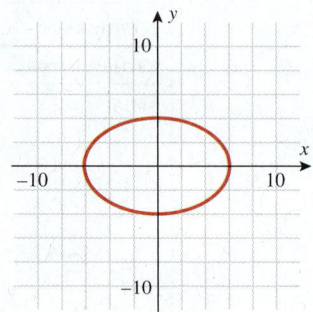

In Exercises 5–16, graph each ellipse. Label the center and vertices.

5. $\dfrac{x^2}{25} + \dfrac{y^2}{16} = 1$

6. $\dfrac{x^2}{49} + \dfrac{y^2}{9} = 1$

7. $\dfrac{x^2}{16} + \dfrac{y^2}{64} = 1$

8. $\dfrac{x^2}{25} + \dfrac{y^2}{144} = 1$

9. $\dfrac{x^2}{100} + y^2 = 1$

10. $9x^2 + 4y^2 = 36$

11. $\dfrac{4}{9}x^2 + 81y^2 = 1$

12. $\dfrac{4}{25}x^2 + \dfrac{100}{9}y^2 = 1$

13. $4x^2 + y^2 = 16$

14. $x^2 + y^2 = 81$

15. $8x^2 + 16y^2 = 32$

16. $10x^2 + 25y^2 = 50$

In Exercises 17–24, find the standard form of the equation of an ellipse with the given characteristics.

17. Foci: $(-4, 0)$ and $(4, 0)$ Vertices: $(-6, 0)$ and $(6, 0)$

18. Foci: $(-1, 0)$ and $(1, 0)$ Vertices: $(-3, 0)$ and $(3, 0)$

19. Foci: $(0, -3)$ and $(0, 3)$ Vertices: $(0, -4)$ and $(0, 4)$

20. Foci: $(0, -1)$ and $(0, 1)$ Vertices: $(0, -2)$ and $(0, 2)$

21. Major axis vertical with length of 8, minor axis length of 4, and centered at $(0, 0)$

22. Major axis horizontal with length of 10, minor axis length of 2, and centered at $(0, 0)$

23. Vertices $(0, -7)$ and $(0, 7)$ and endpoints of minor axis $(-3, 0)$ and $(3, 0)$

24. Vertices $(-9, 0)$ and $(9, 0)$ and endpoints of minor axis $(0, -4)$ and $(0, 4)$

In Exercises 25–28, match each equation with the ellipse.

25. $\dfrac{(x-3)^2}{4} + \dfrac{(y+2)^2}{25} = 1$

26. $\dfrac{(x+3)^2}{4} + \dfrac{(y-2)^2}{25} = 1$

27. $\dfrac{(x-3)^2}{25} + \dfrac{(y+2)^2}{4} = 1$

28. $\dfrac{(x+3)^2}{25} + \dfrac{(y-2)^2}{4} = 1$

a.

b.

c.

d.

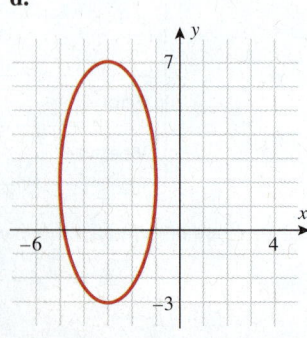

In Exercises 29–38, graph each ellipse. Label the center and vertices.

29. $\dfrac{(x-1)^2}{16} + \dfrac{(y-2)^2}{4} = 1$

30. $\dfrac{(x+1)^2}{36} + \dfrac{(y+2)^2}{9} = 1$

31. $10(x+3)^2 + (y-4)^2 = 80$

32. $3(x+3)^2 + 12(y-4)^2 = 36$

33. $x^2 + 4y^2 - 24y + 32 = 0$

34. $25x^2 + 2y^2 - 4y - 48 = 0$

35. $x^2 - 2x + 2y^2 - 4y - 5 = 0$

36. $9x^2 - 18x + 4y^2 - 27 = 0$

37. $5x^2 + 20x + y^2 + 6y - 21 = 0$

38. $9x^2 + 36x + y^2 + 2y + 36 = 0$

In Exercises 39–46, find the standard form of the equation of an ellipse with the given characteristics.

39. Foci: $(-2, 5)$ and $(6, 5)$ Vertices: $(-3, 5)$ and $(7, 5)$

40. Foci: $(2, -2)$ and $(4, -2)$ Vertices: $(0, -2)$ and $(6, -2)$

41. Foci: $(4, -7)$ and $(4, -1)$ Vertices: $(4, -8)$ and $(4, 0)$

42. Foci: $(2, -6)$ and $(2, -4)$ Vertices: $(2, -7)$ and $(2, -3)$

43. Major axis vertical with length of 8, minor axis length of 4, and centered at $(3, 2)$

44. Major axis horizontal with length of 10, minor axis length of 2, and centered at $(-4, 3)$

45. Vertices $(-1, -9)$ and $(-1, 1)$ and endpoints of minor axis $(-4, -4)$ and $(2, -4)$

46. Vertices $(-2, 3)$ and $(6, 3)$ and endpoints of minor axis $(2, 1)$ and $(2, 5)$

▪ APPLICATIONS

47. Carnival Ride. The Zipper, a favorite carnival ride, maintains an elliptical shape with a major axis of 150 feet and a minor axis of 30 feet. Assuming it is centered at the origin, find an equation for the ellipse.

Zipper

48. Carnival Ride. A Ferris wheel traces an elliptical path with both a major and minor axis of 180 feet. Assuming it is centered at the origin, find an equation for the ellipse (circle).

Ferris wheel, Barcelona, Spain

For Exercises 49 and 50, refer to the following information:

A high school wants to build a football field surrounded by an elliptical track. A regulation football field must be 120 yards long and 30 yards wide.

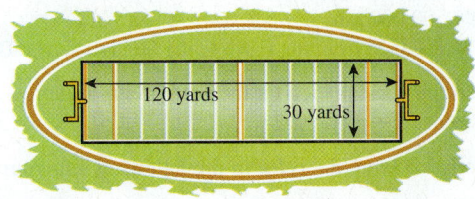

49. Sports Field. Suppose the elliptical track is centered at the origin and has a horizontal major axis of length 150 yards and a minor axis length of 40 yards.

 a. Write an equation for the ellipse.
 b. Find the width of the track at the end of the field. Will the track completely enclose the football field?

50. Sports Field. Suppose the elliptical track is centered at the origin and has a horizontal major axis of length 150 yards. How long should the minor axis be in order to enclose the field?

For Exercises 51 and 52, refer to orbits in our solar system:

The planets have elliptical orbits with the Sun as one of the foci. Pluto (orange), the planet furthest from the Sun, has a very elongated, or flattened, elliptical orbit, whereas the Earth (royal blue) has an almost circular orbit. Because of Pluto's flattened path, it is not always the planet furthest from the Sun.

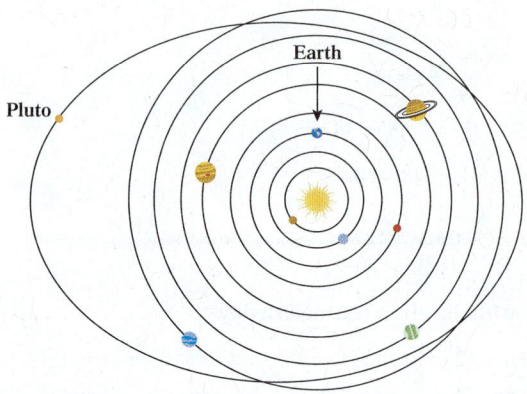

51. Planetary Orbits. The orbit of the dwarf planet Pluto has approximately the following characteristics (assume the Sun is the focus):

- The length of the major axis $2a$ is approximately 11,827,000,000 km.
- The distance from the dwarf planet to the Sun is 4,447,000,000 km.

Determine the equation for Pluto's elliptical orbit around the Sun.

52. Planetary Orbits. The Earth's orbit has approximately the following characteristics (assume the Sun is the focus):

- The length of the major axis $2a$ is approximately 299,700,000 km.
- The distance from the Earth to the Sun is 147,100,000 km.

Determine the equation for the Earth's elliptical orbit around the Sun.

For Exercises 53 and 54, refer to the following information:

Asteroids orbit the Sun in elliptical patterns and often cross paths with the Earth's orbit, making life a little tense now and again. A few asteroids have orbits that cross the Earth's orbit—called "Apollo asteroids" or "Earth-crossing asteroids." In recent years, asteroids have passed within 100,000 kilometers of the Earth!

53. Asteroids. Asteroid 433, or Eros, is the second largest near-Earth asteroid. The semimajor axis is 150 million kilometers and the eccentricity is 0.223, where eccentricity is defined as $e = \sqrt{1 - \dfrac{b^2}{a^2}}$, where a is the semimajor axis or $2a$ is the major axis, and b is the semiminor axis or $2b$ is the minor axis. Find the equation of Eros's orbit. Round to the nearest million kilometers.

54. Asteroids. The asteroid Toutatis is the largest near-Earth asteroid. The semimajor axis is 350 million kilometers and the eccentricity is 0.634, where eccentricity is defined as $e = \sqrt{1 - \dfrac{b^2}{a^2}}$, where a is the semimajor axis or $2a$ is the major axis, and b is the semimajor axis or $2b$ is the minor axis. On September 29, 2004, it missed Earth by 961,000 miles. Find the equation of Toutatis's orbit.

55. Halley's Comet. The eccentricity of Halley's Comet is approximately 0.967. If a comet had $e = 1$, what would its orbit appear to be from Earth?

56. Halley's Comet. The length of the semimajor axis is 17.8 AU (astronomical units) and the eccentricity is the approximately 0.967. Find the equation of Halley's Comet. (Assume 1 AU = 150 million km.)

For Exercises 57 and 58 refer to the following:

An elliptical trainer is an exercise machine that can be used to simulate stair climbing, walking, or running. The stride length is the length of a step on the trainer (forward foot to rear foot). The minimum step-up height is the height of a pedal at its lowest point, while the maximum step-up height is the height of a pedal at its highest point.

57. Health/Exercise. An elliptical trainer has a stride length of 16 inches. The maximum step-up height is 12.5 inches, while the minimum step-up height is 2.5 inches.
   a. Find the equation of the ellipse traced by the pedals assuming the origin lies at the pedal axle (center of the ellipse is at the origin).
   b. Use the approximation to the perimeter of an ellipse $p = \pi \sqrt{2(a^2 + b^2)}$ to find the distance, to the nearest inch, traveled in one complete step (revolution of a pedal).
   c. How many steps, to the nearest step, are necessary to travel a distance of one mile?

58. Health/Exercise. An elliptical trainer has a stride length of 18 inches. The maximum step-up height is 13.5 inches, while the minimum step-up height is 3.5 inches. Find the equation of ellipse traced by the pedals assuming the origin lies at the pedal axle.
   a. Find the equation of the ellipse traced by the pedals assuming the origin lies at the pedal axle (center of the ellipse is at the origin).
   b. Use the approximation to the perimeter of an ellipse $p = \pi \sqrt{2(a^2 + b^2)}$ to find the distance, to the nearest inch, traveled in one complete step (revolution of a pedal).
   c. How many steps, to the nearest step, are necessary to travel a distance of one mile?

■ CATCH THE MISTAKE

In Exercises 59 and 60, explain the mistake that is made.

59. Graph the ellipse given by $\dfrac{x^2}{6} + \dfrac{y^2}{4} = 1$.

Solution:

Write the standard form of the equation of an ellipse.

$$\dfrac{x^2}{a^2} + \dfrac{y^2}{b^2} = 1$$

Identify a and b.

$$a = 6, b = 4$$

Label the vertices and the endpoints of the minor axis, $(-6, 0)$, $(6, 0)$, $(0, -4)$, $(0, 4)$, and connect with an elliptical curve.

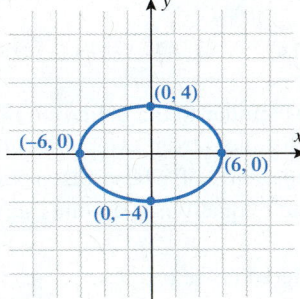

This is incorrect. What mistake was made?

60. Determine the foci of the ellipse $\dfrac{x^2}{16} + \dfrac{y^2}{9} = 1$.

Solution:

Write the general equation of a horizontal ellipse.

$$\dfrac{x^2}{a^2} + \dfrac{y^2}{b^2} = 1$$

Identify a and b.

$$a = 4, b = 3$$

Substitute $a = 4, b = 3$ into $c^2 = a^2 + b^2$.

$$c^2 = 4^2 + 3^2$$

Solve for c.

$$c = 5$$

Foci are located at $(-5, 0)$ and $(5, 0)$.

The points $(-5, 0)$ and $(5, 0)$ are located outside of the ellipse.

This is incorrect. What mistake was made?

■ CONCEPTUAL

In Exercises 61–64, determine whether each statement is true or false.

61. If you know the vertices of an ellipse, you can determine the equation for the ellipse.

62. If you know the foci and the endpoints of the minor axis, you can determine the equation for the ellipse.

63. Ellipses centered at the origin have symmetry with respect to the x-axis, y-axis, and the origin.

64. All ellipses are circles, but not all circles are ellipses.

■ CHALLENGE

65. The eccentricity of an ellipse is defined as $e = \dfrac{c}{a}$. Compare the eccentricity of the orbit of Pluto to that of the Earth (refer to Exercises 51 and 52).

66. The eccentricity of an ellipse is defined as $e = \dfrac{c}{a}$. Since $a > c \geq 0$, then $0 \leq e < 1$. Describe the shape of an ellipse when

a. e is close to zero   **b.** e is close to one   **c.** $e = 0.5$

■ TECHNOLOGY

67. Graph the following three ellipses: $x^2 + y^2 = 1$, $x^2 + 5y^2 = 1$, and $x^2 + 10y^2 = 1$. What can be said to happen to the ellipse $x^2 + cy^2 = 1$ as c increases?

68. Graph the following three ellipses: $x^2 + y^2 = 1$, $5x^2 + y^2 = 1$, and $10x^2 + y^2 = 1$. What can be said to happen to the ellipse $cx^2 + y^2 = 1$ as c increases?

69. Graph the following three ellipses: $x^2 + y^2 = 1$, $5x^2 + 5y^2 = 1$, and $10x^2 + 10y^2 = 1$. What can be said to happen to the ellipse $cx^2 + cy^2 = 1$ as c increases?

70. Graph the equation $\dfrac{x^2}{9} - \dfrac{y^2}{16} = 1$. Notice what a difference the sign makes. Is this an ellipse?

71. Graph the following three ellipses: $x^2 + y^2 = 1$, $0.5x^2 + y^2 = 1$, and $0.05x^2 + y^2 = 1$. What can be said to happen to ellipse $cx^2 + y^2 = 1$ as c decreases?

72. Graph the following three ellipses: $x^2 + y^2 = 1$, $x^2 + 0.5y^2 = 1$, and $x^2 + 0.05y^2 = 1$. What can be said to happen to ellipse $x^2 + cy^2 = 1$ as c decreases?

SKILLS OBJECTIVES

- Find a hyperbola's foci and vertices.
- Find the equation of a hyperbola centered at the origin.
- Graph a hyperbola using asymptotes as graphing aids.
- Find the equation of a hyperbola centered at the point (h, k).
- Solve applied problems that involve hyperbolas.

CONCEPTUAL OBJECTIVES

- Derive the general equation of a hyperbola.
- Identify, apply, and graph the transverse axis, vertices, and foci.
- Use asymptotes to determine the shape of a hyperbola.

Hyperbola Centered at the Origin

The definition of a hyperbola is similar to the definition of an ellipse. An ellipse is the set of all points the *sum* of whose distances from two points (the foci) is constant. A *hyperbola* is the set of all points, the *difference* of whose distances from two points (the foci) is constant. What distinguishes their equations is a minus sign.

Ellipse centered at the origin: $\dfrac{x^2}{a^2} + \dfrac{y^2}{b^2} = 1$

Hyperbola centered at the origin: $\dfrac{x^2}{a^2} - \dfrac{y^2}{b^2} = 1$

DEFINITION   **Hyperbola**

A **hyperbola** is the set of all points in a plane, the difference of whose distances from two fixed points is a positive constant. These two fixed points are called **foci**. The hyperbola has two separate curves called **branches**. The two points where the hyperbola intersects the line joining the foci are called **vertices**. The line segment joining the vertices is called the **transverse axis of the hyperbola**. The midpoint of the transverse axis is called the **center**.

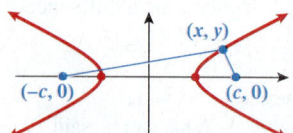

Let's consider a hyperbola with the center at the origin and the foci on the x-axis. Let the distance from the center to the focus be c. Therefore, the foci are located at the points $(-c, 0)$ and $(c, 0)$. The difference of the two distances from the foci to any point (x, y) must be constant. We then can follow a similar analysis as done with an ellipse.

Derivation of the Equation of a Hyperbola

WORDS	MATH
The difference of these two distances is equal to a constant (2*a* for convenience).	$\sqrt{[x-(-c)]^2 + y^2} - \sqrt{(x-c)^2 + y^2} = \pm 2a$
Following the same procedure that we did with an ellipse leads to:	$(c^2 - a^2)x^2 - a^2y^2 = a^2(c^2 - a^2)$
We can make the argument that $c > a$ in order for a point to be on the hyperbola (and not on the *x*-axis). Therefore, since *a* and *c* represent distances and therefore are positive, we know that $c^2 > a^2$, or $c^2 - a^2 > 0$. Hence, we can divide both sides of the equation by $c^2 - a^2$, since $c^2 - a^2 \neq 0$.	$x^2 - \dfrac{a^2y^2}{(c^2 - a^2)} = a^2$
Let $b^2 = c^2 - a^2$.	$x^2 - \dfrac{a^2y^2}{b^2} = a^2$
Divide both sides of the equation by a^2.	$\dfrac{x^2}{a^2} - \dfrac{y^2}{b^2} = 1$

The equation $\dfrac{x^2}{a^2} - \dfrac{y^2}{b^2} = 1$ represents a hyperbola with center at the origin, with the foci along the *x*-axis. The following box summarizes hyperbolas that have their center at the origin and foci along either the *x*-axis or *y*-axis:

EQUATION OF A HYPERBOLA WITH CENTER AT THE ORIGIN

The **standard form of the equation of a hyperbola** with its center at the origin is given by

ORIENTATION OF TRANSVERSE AXIS	Horizontal (along the *x*-axis)	Vertical (along the *y*-axis)
EQUATION	$\dfrac{x^2}{a^2} - \dfrac{y^2}{b^2} = 1$	$\dfrac{y^2}{a^2} - \dfrac{x^2}{b^2} = 1$
FOCI	$(-c, 0) \quad (c, 0)$ where $c^2 = a^2 + b^2$	$(0, -c) \quad (0, c)$ where $c^2 = a^2 + b^2$
ASYMPTOTES	$y = \dfrac{b}{a}x \quad y = -\dfrac{b}{a}x$	$y = \dfrac{a}{b}x \quad y = -\dfrac{a}{b}x$
VERTICES	$(-a, 0) \quad (a, 0)$	$(0, -a) \quad (0, a)$
TRANSVERSE AXIS	Horizontal length 2*a*	Vertical length 2*a*
GRAPH		

Note that for $\dfrac{x^2}{a^2} - \dfrac{y^2}{b^2} = 1$, if $x = 0$, then $-\dfrac{y^2}{b^2} = 1$, which yields an imaginary number for y.

However, when $y = 0$, $\dfrac{x^2}{a^2} = 1$, and therefore $x = \pm a$. The vertices for this hyperbola are

$(-a, 0)$ and $(a, 0)$.

Technology Tip

Use a graphing calculator to check the graph of $\dfrac{x^2}{9} - \dfrac{y^2}{4} = 1$. Solve for

y first. That is, $y_1 = 2\sqrt{\dfrac{x^2}{9} - 1}$ or

$y_2 = -2\sqrt{\dfrac{x^2}{9} - 1}$.

■ **Answer:** Vertices: $(0, -4)$ and $(0, 4)$
Foci: $(0, -6)$ and $(0, 6)$

Technology Tip

Use a graphing calculator to check the graph of $\dfrac{y^2}{16} - \dfrac{x^2}{9} = 1$. Solve for

y first. That is, $y_1 = 4\sqrt{1 + \dfrac{x^2}{9}}$ or

$y_2 = -4\sqrt{1 - \dfrac{x^2}{9}}$.

EXAMPLE 1 Finding the Foci and Vertices of a Hyperbola Given the Equation

Find the foci and vertices of the hyperbola given by $\dfrac{x^2}{9} - \dfrac{y^2}{4} = 1$.

Solution:

Compare to the standard equation of a hyperbola, $\dfrac{x^2}{a^2} - \dfrac{y^2}{b^2} = 1$.	$a^2 = 9, b^2 = 4$
Solve for a and b.	$a = 3, b = 2$
Substitute $a = 3$ into the vertices, $(-a, 0)$ and $(a, 0)$.	$(-3, 0)$ and $(3, 0)$
Substitute $a = 3, b = 2$ into $c^2 = a^2 + b^2$.	$c^2 = 3^2 + 2^2$
Solve for c.	$c^2 = 13$
	$c = \sqrt{13}$
Substitute $c = \sqrt{13}$ into the foci, $(-c, 0)$ and $(c, 0)$.	$\left(-\sqrt{13}, 0\right)$ and $\left(\sqrt{13}, 0\right)$

The vertices are $\boxed{(-3, 0)}$ and $\boxed{(3, 0)}$, and the foci are $\boxed{\left(-\sqrt{13}, 0\right)}$ and $\boxed{\left(\sqrt{13}, 0\right)}$.

■ **YOUR TURN** Find the vertices and foci of the hyperbola $\dfrac{y^2}{16} - \dfrac{x^2}{20} = 1$.

EXAMPLE 2 Finding the Equation of a Hyperbola Given Foci and Vertices

Find the equation of a hyperbola whose vertices are located at $(0, -4)$ and $(0, 4)$ and whose foci are located at $(0, -5)$ and $(0, 5)$.

Solution:

The center is located at the midpoint of the segment joining the vertices.	$\left(\dfrac{0 + 0}{2}, \dfrac{-4 + 4}{2}\right) = (0, 0)$
Since the foci and vertices are located on the y-axis, the standard equation is given by:	$\dfrac{y^2}{a^2} - \dfrac{x^2}{b^2} = 1$
The vertices $(0, \pm a)$ and the foci $(0, \pm c)$ can be used to identify a and c.	$a = 4, c = 5$
Substitute $a = 4, c = 5$ into $b^2 = c^2 - a^2$.	$b^2 = 5^2 - 4^2$

Solve for b.

$$b^2 = 25 - 16 = 9$$

$$b = 3$$

Substitute $a = 4$ and $b = 3$ into $\dfrac{y^2}{a^2} - \dfrac{x^2}{b^2} = 1$.

$$\boxed{\dfrac{y^2}{16} - \dfrac{x^2}{9} = 1}$$

■ **YOUR TURN** Find the equation of a hyperbola whose vertices are located at $(-2, 0)$ and $(2, 0)$ and whose foci are located at $(-4, 0)$ and $(4, 0)$.

■ **Answer:** $\dfrac{x^2}{4} - \dfrac{y^2}{12} = 1$

Graphing a Hyperbola Centered at the Origin

To graph a hyperbola, we use the vertices and asymptotes. The asymptotes are found by the equations $y = \pm \dfrac{b}{a} x$ or $y = \pm \dfrac{a}{b} x$, depending on whether the transverse axis is horizontal or vertical. An easy way to draw these graphing aids is to first draw the rectangular box that passes through the vertices and the points $(0, \pm b)$ or $(\pm b, 0)$. The **conjugate axis** is perpendicular to the transverse axis and has length $2b$. The asymptotes pass through the center of the hyperbola and the corners of the rectangular box.

$$\dfrac{x^2}{a^2} - \dfrac{y^2}{b^2} = 1$$

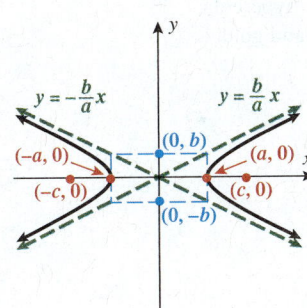

Classroom Example 11.4.2
Find the equation of a hyperbola whose vertices are $(\pm 2, 0)$ and whose foci are $(\pm 5, 0)$.

Answer: $\dfrac{x^2}{4} - \dfrac{y^2}{21} = 1$

$$\dfrac{y^2}{a^2} - \dfrac{x^2}{b^2} = 1$$

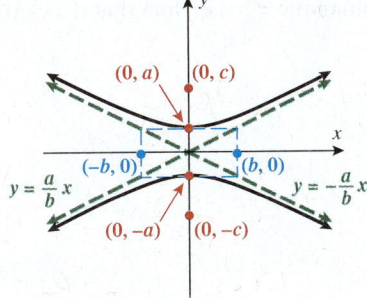

Classroom Example 11.4.2*
What are the foci of the hyperbola whose equation is $a^2 x^2 - 9a^2 y^2 = 1$, where $a > 0$?

Answer: $\left(\dfrac{\pm\sqrt{10}}{3a}, 0 \right)$

Technology Tip

Use a graphing calculator to check the graph of $\dfrac{x^2}{4} - \dfrac{y^2}{9} = 1$. Solve for y first. That is, $y_1 = 3\sqrt{\dfrac{x^2}{4} - 1}$ or $y_2 = -3\sqrt{\dfrac{x^2}{4} - 1}$.

To add the two asymptotes, enter $y_3 = \frac{3}{2}x$ or $y_4 = -\frac{3}{2}x$.

EXAMPLE 3 Graphing a Hyperbola Centered at the Origin with a Horizontal Transverse Axis

Graph the hyperbola given by $\dfrac{x^2}{4} - \dfrac{y^2}{9} = 1$.

Solution:

Compare $\dfrac{x^2}{2^2} - \dfrac{y^2}{3^2} = 1$ to the general equation $\dfrac{x^2}{a^2} - \dfrac{y^2}{b^2} = 1$.

Identify a and b. $a = 2$ and $b = 3$

The transverse axis of this hyperbola lies on the x-axis.

Label the vertices $(-a, 0) = \mathbf{(-2, 0)}$ and $(a, 0) = \mathbf{(2, 0)}$ and the points $(0, -b) = (0, -3)$ and $(0, b) = (0, 3)$. Draw the rectangular box that passes through those points. Draw the **asymptotes** that pass through the center and the corners of the rectangle.

Draw the two **branches** of the hyperbola, each passing through a vertex and guided by the asymptotes.

In Example 3, if we let $y = 0$, then $\dfrac{x^2}{4} = 1$ or $x = \pm 2$. Thus the vertices are $(-2, 0)$ and $(2, 0)$, and the transverse axis lies along the x-axis. Note that if $x = 0$, $y = \pm 3i$.

EXAMPLE 4 **Graphing a Hyperbola Centered at the Origin with a Vertical Transverse Axis**

Graph the hyperbola given by $\dfrac{y^2}{16} - \dfrac{x^2}{4} = 1$.

Solution:

Compare $\dfrac{y^2}{4^2} - \dfrac{x^2}{2^2} = 1$ to the general equation $\dfrac{y^2}{a^2} - \dfrac{x^2}{b^2} = 1$.

Identify a and b. $a = 4$ and $b = 2$

The transverse axis of this hyperbola lies along the y-axis.

Label the vertices $(0, -a) = \mathbf{(0, -4)}$ and $(0, a) = \mathbf{(0, 4)}$, and the points $(-b, 0) = (-2, 0)$ and $(b, 0) = (2, 0)$. Draw the rectangular box that passes through those points. Draw the **asymptotes** that pass through the center and the corners of the rectangle.

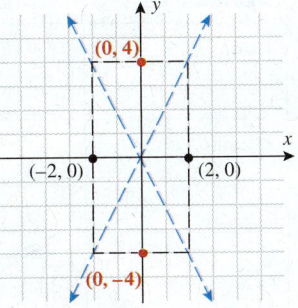

Draw the two **branches** of the hyperbola, each passing through a vertex and guided by the asymptotes.

Technology Tip

Use a graphing calculator to check the graph of $\dfrac{y^2}{16} - \dfrac{x^2}{4} = 1$. Solve for y first. That is, $y_1 = 4\sqrt{1 + \dfrac{x^2}{4}}$ or $y_2 = -4\sqrt{1 - \dfrac{x^2}{4}}$.

To add the two asymptotes, enter $y_3 = 2x$ or $y_4 = -2x$.

■ **Answer:**

a.

b.

■ **YOUR TURN** Graph the hyperbolas:

    a. $\dfrac{y^2}{1} - \dfrac{x^2}{4} = 1$ **b.** $\dfrac{x^2}{4} - \dfrac{y^2}{1} = 1$

Hyperbola Centered at the Point (h, k)

We can use graph-shifting techniques to graph hyperbolas that are centered at a point other than the origin—say, (h, k). For example, to graph $\dfrac{(x - h)^2}{a^2} - \dfrac{(y - k)^2}{b^2} = 1$, start with the graph of $\dfrac{x^2}{a^2} - \dfrac{y^2}{b^2} = 1$ and shift to the right h units and up k units. The center, the vertices, the foci, the transverse and conjugate axes, and the asymptotes all shift. The following table summarizes the characteristics of hyperbolas centered at a point other than the origin:

EQUATION OF A HYPERBOLA WITH CENTER AT THE POINT (h, k)

The **standard form of the equation of a hyperbola** with its center at the point (h, k) is given by

ORIENTATION OF TRANSVERSE AXIS	Horizontal (parallel to the x-axis)	Vertical (parallel to the y-axis)
EQUATION	$\dfrac{(x-h)^2}{a^2} - \dfrac{(y-k)^2}{b^2} = 1$	$\dfrac{(y-k)^2}{a^2} - \dfrac{(x-h)^2}{b^2} = 1$
VERTICES	$(h-a, k) \quad (h+a, k)$	$(h, k-a) \quad (h, k+a)$
FOCI	$(h-c, k) \quad (h+c, k)$ where $c^2 = a^2 + b^2$	$(h, k-c) \quad (h, k+c)$ where $c^2 = a^2 + b^2$
GRAPH		

Horizontal graph: $y = \dfrac{b}{a}(x-h) + k$, $y = -\dfrac{b}{a}(x-h) + k$, with points $(h-a, k)$, $(h+a, k)$, $(h, k+b)$, $(h, k-b)$, center (h, k).

Vertical graph: $y = \dfrac{a}{b}(x-h) + k$, $y = -\dfrac{a}{b}(x-h) + k$, with points $(h-b, k)$, $(h+b, k)$, $(h, k+a)$, $(h, k-a)$, center (h, k).

Technology Tip

Use a graphing calculator to check the graph of $\dfrac{(y-2)^2}{16} - \dfrac{(x-1)^2}{9} = 1$.

Solve for y first. That is,

$$y_1 = 2 + 4\sqrt{1 + \dfrac{(x-1)^2}{9}} \quad \text{or}$$

$$y_2 = 2 - 4\sqrt{1 + \dfrac{(x-1)^2}{9}}.$$

To add the two asymptotes, enter

$$y_3 = \frac{4}{3}(x-1) + 2 \quad \text{or}$$

$$y_4 = -\frac{4}{3}(x-1) + 2.$$

EXAMPLE 5 Graphing a Hyperbola with Center Not at the Origin

Graph the hyperbola $\dfrac{(y-2)^2}{16} - \dfrac{(x-1)^2}{9} = 1$.

Solution:

Compare $\dfrac{(y-2)^2}{4^2} - \dfrac{(x-1)^2}{3^2} = 1$ to the general equation $\dfrac{(y-k)^2}{a^2} - \dfrac{(x-h)^2}{b^2} = 1$.

Identify a, b, and (h, k). $\qquad a = 4$, $b = 3$, and $(h, k) = (1, 2)$

The transverse axis of this hyperbola lies along $x = 2$, which is parallel to the y-axis.

Label the vertices $(h, k-a) = \mathbf{(1, -2)}$ and $(h, k+a) = \mathbf{(1, 6)}$ and the points $(h-b, k) = (-2, 2)$ and $(h+b, k) = (4, 2)$. Draw the rectangular box that passes through those points. Draw the **asymptotes** that pass through the center $(h, k) = (1, 2)$ and the corners of the rectangle. Draw the two **branches** of the hyperbola, each passing through a vertex and guided by the asymptotes.

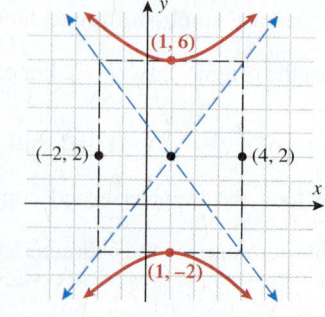

EXAMPLE 6   Transforming an Equation of a Hyperbola to Standard Form

Graph the hyperbola $9x^2 - 16y^2 - 18x + 32y - 151 = 0$.

Solution:

Complete the square on the x

$$9(x^2 - 2x) - 16(y^2 - 2y) = 151$$

terms and y terms, respectively.

$$9(x^2 - 2x + 1) - 16(y^2 - 2y + 1) = 151 + 9 - 16$$

$$9(x - 1)^2 - 16(y - 1)^2 = 144$$

$$\frac{(x - 1)^2}{16} - \frac{(y - 1)^2}{9} = 1$$

Compare $\dfrac{(x - 1)^2}{16} - \dfrac{(y - 1)^2}{9} = 1$ to the general form $\dfrac{(x - h)^2}{a^2} - \dfrac{(y - k)^2}{b^2} = 1$.

Identify a, b, and (h, k). $a = 4$, $b = 3$, and $(h, k) = (1, 1)$

The transverse axis of this hyperbola lies along $y = 1$.

Label the vertices $(h - a, k) = (\mathbf{-3, 1})$ and
$(h + a, k) = (\mathbf{5, 1})$ and the points
$(h, k - b) = (1, -2)$ and $(h, k + b) = (1, 4)$.
Draw the rectangular box that passes through these
points. Draw the **asymptotes** that pass through the
center $(1, 1)$ and the corners of the box. Draw the
two **branches** of the hyperbola, each passing through
a vertex and guided by the asymptotes.

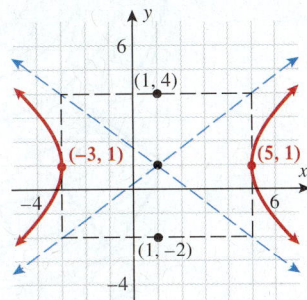

Applications

Nautical navigation is assisted by hyperbolas. For example, suppose that two radio stations
on a coast are emitting simultaneous signals. If a boat is at sea, it will be slightly closer to
one station than the other station, which results in a small time difference between the
received signals from the two stations. Recall that a hyperbola is the set of all points whose
differences of the distances from two points (the foci—or the radio stations) is constant.
Therefore, if the boat follows the path associated with a constant time difference, that path
will be hyperbolic.

The synchronized signals would intersect one another in associated hyperbolas. Each
time difference corresponds to a different path. The radio stations are the foci of the
hyperbolas. This principle forms the basis of a hyperbolic radio navigation system known
as *loran* (**LO**ng-**RA**nge **N**avigation).

There are navigational charts that correspond to different time differences. A ship
selects the hyperbolic path that will take it to the desired port, and the loran chart lists the
corresponding time difference.

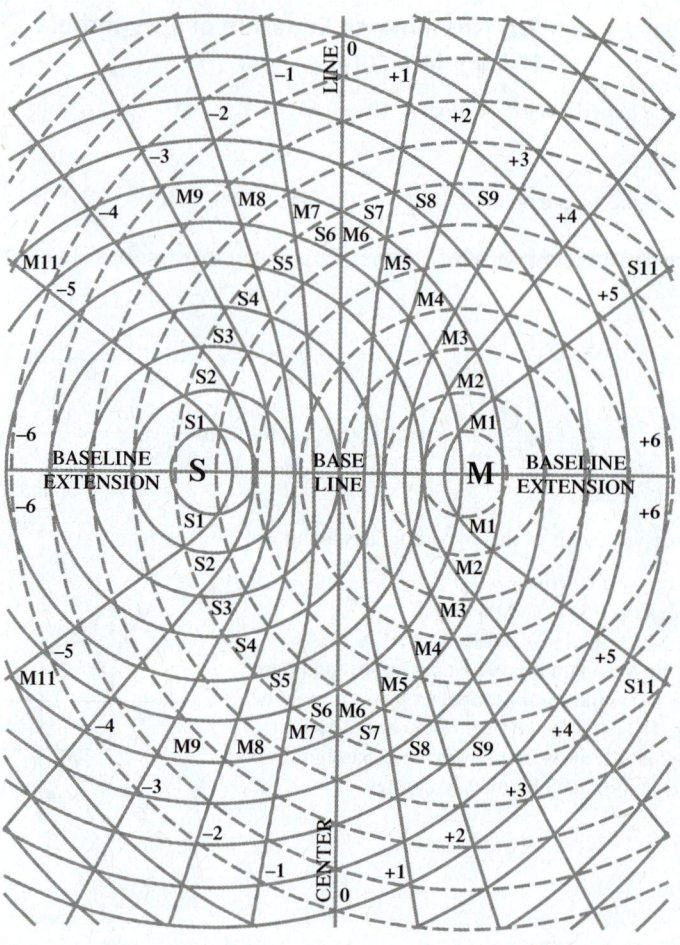

EXAMPLE 7   Nautical Navigation Using Loran

Two loran stations are located 200 miles apart along a coast. If a ship records a time difference of 0.00043 second and continues on the hyperbolic path corresponding to that difference, where does it reach shore?

Solution:

Draw the xy-plane and the two stations corresponding to the foci at $(-100, 0)$ and $(100, 0)$. Draw the ship somewhere in quadrant I.

The hyperbola corresponds to a path where the difference of the distances between the ship and the respective stations remains constant. The constant is $2a$, where $(a, 0)$ is a vertex. Find that difference by using $d = rt$. Assume that the speed of the radio signal is 186,000 miles per second.

Substitute $r = 186,000$ miles/second and $t = 0.00043$ second into $d = rt$.

$$d = (186,000 \text{ miles/second})(0.00043 \text{ second}) \approx 80 \text{ miles}$$

Set the constant equal to $2a$. $2a = 80$

Find a vertex $(a, 0)$. $(40, 0)$

| The ship reaches shore between the two stations, 60 miles from station B and 140 miles from station A. |

In this section we discussed hyperbolas centered at the origin.

EQUATION	$\dfrac{x^2}{a^2} - \dfrac{y^2}{b^2} = 1$	$\dfrac{y^2}{a^2} - \dfrac{x^2}{b^2} = 1$
TRANSVERSE AXIS	Horizontal (x-axis), length $2a$	Vertical (y-axis), length $2a$
CONJUGATE AXIS	Vertical (y-axis), length $2b$	Horizontal (x-axis), length $2b$
VERTICES	$(-a, 0)$ and $(a, 0)$	$(0, -a)$ and $(0, a)$
FOCI	$(-c, 0)$ and $(c, 0)$ where $c^2 = a^2 + b^2$	$(0, -c)$ and $(0, c)$ where $c^2 = a^2 + b^2$
ASYMPTOTE	$y = \dfrac{b}{a}x$ and $y = -\dfrac{b}{a}x$	$y = \dfrac{a}{b}x$ and $y = -\dfrac{a}{b}x$
GRAPH		

For a hyperbola centered at (h, k), the vertices, foci, and asymptotes all shift accordingly.

ORIENTATION OF TRANSVERSE AXIS	Horizontal (parallel to the x-axis)	Vertical (parallel to the y-axis)
EQUATION	$\dfrac{(x - h)^2}{a^2} - \dfrac{(y - k)^2}{b^2} = 1$	$\dfrac{(y - k)^2}{a^2} - \dfrac{(x - h)^2}{b^2} = 1$
VERTICES	$(h - a, k) \quad (h + a, k)$	$(h, k - a) \quad (h, k + a)$
FOCI	$(h - c, k) \quad (h + c, k)$ where $c^2 = a^2 + b^2$	$(h, k - c) \quad (h, k + c)$ where $c^2 = a^2 + b^2$
GRAPH		

SECTION
11.4 EXERCISES

■ **SKILLS**

In Exercises 1–4, match each equation with the corresponding hyperbola.

1. $\dfrac{x^2}{36} - \dfrac{y^2}{16} = 1$

2. $\dfrac{y^2}{36} - \dfrac{x^2}{16} = 1$

3. $\dfrac{x^2}{8} - \dfrac{y^2}{72} = 1$

4. $4y^2 - x^2 = 1$

a.

b.

c.

d.

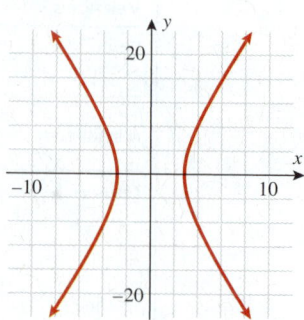

In Exercises 5–16, graph each hyperbola.

5. $\dfrac{x^2}{25} - \dfrac{y^2}{16} = 1$

6. $\dfrac{x^2}{49} - \dfrac{y^2}{9} = 1$

7. $\dfrac{y^2}{16} - \dfrac{x^2}{64} = 1$

8. $\dfrac{y^2}{144} - \dfrac{x^2}{25} = 1$

9. $\dfrac{x^2}{100} - y^2 = 1$

10. $9y^2 - 4x^2 = 36$

11. $\dfrac{4y^2}{9} - 81x^2 = 1$

12. $\dfrac{4}{25}x^2 - \dfrac{100}{9}y^2 = 1$

13. $4x^2 - y^2 = 16$

14. $y^2 - x^2 = 81$

15. $8y^2 - 16x^2 = 32$

16. $10x^2 - 25y^2 = 50$

In Exercises 17–24, find the standard form of an equation of the hyperbola with the given characteristics.

17. Vertices: $(-4, 0)$ and $(4, 0)$ Foci: $(-6, 0)$ and $(6, 0)$

18. Vertices: $(-1, 0)$ and $(1, 0)$ Foci: $(-3, 0)$ and $(3, 0)$

19. Vertices: $(0, -3)$ and $(0, 3)$ Foci: $(0, -4)$ and $(0, 4)$

20. Vertices: $(0, -1)$ and $(0, 1)$ Foci: $(0, -2)$ and $(0, 2)$

21. Center: $(0, 0)$; transverse: x-axis; asymptotes: $y = x$ and $y = -x$

22. Center: $(0, 0)$; transverse: y-axis; asymptotes: $y = x$ and $y = -x$

23. Center: $(0, 0)$; transverse axis: y-axis; asymptotes: $y = 2x$ and $y = -2x$

24. Center: $(0, 0)$; transverse axis: x-axis; asymptotes: $y = 2x$ and $y = -2x$

In Exercises 25–28, match each equation with the hyperbola.

25. $\dfrac{(x - 3)^2}{4} - \dfrac{(y + 2)^2}{25} = 1$

26. $\dfrac{(x + 3)^2}{4} - \dfrac{(y - 2)^2}{25} = 1$

27. $\dfrac{(y - 3)^2}{25} - \dfrac{(x + 2)^2}{4} = 1$

28. $\dfrac{(y + 3)^2}{25} - \dfrac{(x - 2)^2}{4} = 1$

a.

b.

c.

d.

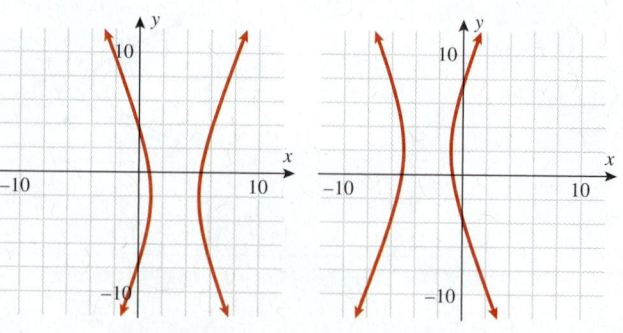

In Exercises 29–38, graph each hyperbola.

29. $\dfrac{(x-1)^2}{16} - \dfrac{(y-2)^2}{4} = 1$

30. $\dfrac{(y+1)^2}{36} - \dfrac{(x+2)^2}{9} = 1$

31. $10(y+3)^2 - (x-4)^2 = 80$

32. $3(x+3)^2 - 12(y-4)^2 = 36$

33. $x^2 - 4x - 4y^2 = 0$

34. $-9x^2 + y^2 + 2y - 8 = 0$

35. $-9x^2 - 18x + 4y^2 - 8y - 41 = 0$

36. $25x^2 - 50x - 4y^2 - 8y - 79 = 0$

37. $x^2 - 6x - 4y^2 - 16y - 8 = 0$

38. $-4x^2 - 16x + y^2 - 2y - 19 = 0$

In Exercises 39–42, find the standard form of the equation of a hyperbola with the given characteristics.

39. Vertices: $(-2, 5)$ and $(6, 5)$ Foci: $(-3, 5)$ and $(7, 5)$

40. Vertices: $(1, -2)$ and $(3, -2)$ Foci: $(0, -2)$ and $(4, -2)$

41. Vertices: $(4, -7)$ and $(4, -1)$ Foci: $(4, -8)$ and $(4, 0)$

42. Vertices: $(2, -6)$ and $(2, -4)$ Foci: $(2, -7)$ and $(2, -3)$

■ APPLICATIONS

43. **Ship Navigation.** Two loran stations are located 150 miles apart along a coast. If a ship records a time difference of 0.0005 second and continues on the hyperbolic path corresponding to that difference, where will it reach shore?

44. **Ship Navigation.** Two loran stations are located 300 miles apart along a coast. If a ship records a time difference of 0.0007 second and continues on the hyperbolic path corresponding to that difference, where will it reach shore? Round to the nearest mile.

45. **Ship Navigation.** If the captain of the ship in Exercise 43 wants to reach shore between the stations and 30 miles from one of them, what time difference should he look for?

46. **Ship Navigation.** If the captain of the ship in Exercise 44 wants to reach shore between the stations and 50 miles from one of them, what time difference should he look for?

47. **Light.** If the light from a lamp casts a hyperbolic pattern on the wall due to its lampshade, calculate the equation of the hyperbola if the distance between the vertices is 2 feet and the foci are half a foot from the vertices.

48. **Special Ops.** A military special ops team is calibrating its recording devices used for passive ascertaining of enemy location. They place two recording stations, alpha and bravo, 3000 feet apart (alpha is due east of bravo). The team detonates small explosives 300 feet west of alpha and records the time it takes each station to register an explosion. The team also sets up a second set of explosives directly north of the alpha station. How many feet north of alpha should the team set off the explosives if it wants to record the same times as on the first explosion?

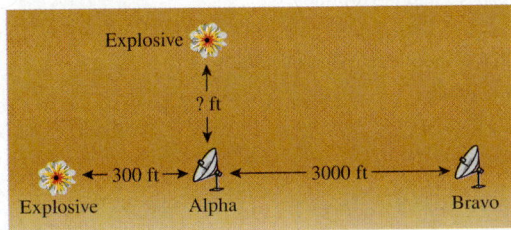

For Exercises 49 and 50 refer to the following:

Nuclear cooling towers are typically built in the shape of a hyperboloid. The cross section of a cooling tower forms a hyperbola. The cooling water pictured in 450 feet tall and modeled by the equation

$$\dfrac{x^2}{8100} - \dfrac{y^2}{16900} = 1$$

49. **Engineering/Design.** Find the diameter of the top of the cooling tower to the nearest foot.

50. **Engineering/Design.** Find the diameter of the base of the tower to the nearest foot.

▪ CATCH THE MISTAKE

In Exercises 51 and 52, explain the mistake that is made.

51. Graph the hyperbola $\dfrac{y^2}{4} - \dfrac{x^2}{9} = 1$.

Solution:

Compare the equation to the standard form and solve for a and b.　　$a = 2, b = 3$

Label the vertices $(-a, 0)$ and $(a, 0)$.　$(-2, 0)$ and $(2, 0)$

Label the points $(0, -b)$ and $(0, b)$.　$(0, -3)$ and $(0, 3)$

Draw the rectangle connecting these four points, and align the asymptotes so that they pass through the center and the corner of the boxes. Then draw the hyperbola using the vertices and asymptotes.

This is incorrect. What mistake was made?

52. Graph the hyperbola $\dfrac{x^2}{1} - \dfrac{y^2}{4} = 1$.

Solution:

Compare the equation to the general form and solve for a and b.　　$a = 2, b = 1$

Label the vertices $(-a, 0)$ and $(a, 0)$.　$(-2, 0)$ and $(2, 0)$

Label the points $(0, -b)$ and $(0, b)$.　$(0, -1)$ and $(0, 1)$

Draw the rectangle connecting these four points, and align the asymptotes so that they pass through the center and the corner of the boxes. Then draw the hyperbola using the vertices and asymptotes.

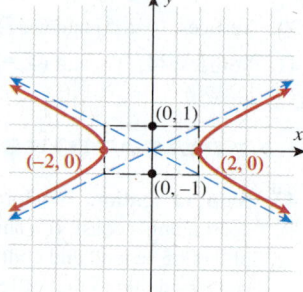

This is incorrect. What mistake was made?

▪ CONCEPTUAL

In Exercises 53–56, determine whether each statement is true or false.

53. If you know the vertices of a hyperbola, you can determine the equation for the hyperbola.

54. If you know the foci and vertices, you can determine the equation for the hyperbola.

55. Hyperbolas centered at the origin have symmetry with respect to the x-axis, y-axis, and the origin.

56. The center and foci are part of the graph of a hyperbola.

▪ CHALLENGE

57. Find the general equation of a hyperbola whose asymptotes are perpendicular.

58. Find the general equation of a hyperbola whose vertices are $(3, -2)$ and $(-1, -2)$ and whose asymptotes are the lines $y = 2x - 4$ and $y = -2x$.

▪ TECHNOLOGY

59. Graph the following three hyperbolas: $x^2 - y^2 = 1$, $x^2 - 5y^2 = 1$, and $x^2 - 10y^2 = 1$. What can be said to happen to the hyperbola $x^2 - cy^2 = 1$ as c increases?

60. Graph the following three hyperbolas: $x^2 - y^2 = 1$, $5x^2 - y^2 = 1$, and $10x^2 - y^2 = 1$. What can be said to happen to the hyperbola $cx^2 - y^2 = 1$ as c increases?

61. Graph the following three hyperbolas: $x^2 - y^2 = 1$, $0.5x^2 - y^2 = 1$, and $0.05x^2 - y^2 = 1$. What can be said to happen to the hyperbola $cx^2 - y^2 = 1$ as c decreases?

62. Graph the following three hyperbolas: $x^2 - y^2 = 1$, $x^2 - 0.5y^2 = 1$, and $x^2 - 0.05y^2 = 1$. What can be said to happen to the hyperbola $x^2 - cy^2 = 1$ as c decreases?

Solving a System of Nonlinear Equations

In Chapters 9 and 10, we discussed solving systems of *linear* equations. In Chapter 9, we applied elimination and substitution to solve systems of linear equations in two variables, and in Chapter 10, we employed matrices to solve systems of linear equations. Recall that a system of linear equations in two variables has one of three types of solutions:

One solution	Two lines that intersect at one point	
No solution	Two parallel lines (never intersect)	
Infinitely many solutions	Two lines that coincide (same line)	

Notice that systems of *linear* equations in two variables always corresponded to *lines*. Now we turn our attention to systems of *nonlinear* equations in two variables. If any of the equations in a system of equations is nonlinear, then the system is a nonlinear system. The following are systems of nonlinear equations:

$$\begin{cases} y = x^2 + 1 \ (\text{Parabola}) \\ y = 2x + 2 \ (\text{Line}) \end{cases} \quad \begin{cases} x^2 + y^2 = 25 \ (\text{Circle}) \\ y = x \ (\text{Line}) \end{cases} \quad \begin{cases} \dfrac{x^2}{9} + \dfrac{y^2}{4} = 1 \ (\text{Ellipse}) \\ \dfrac{y^2}{16} - \dfrac{x^2}{25} = 1 \ (\text{Hyperbola}) \end{cases}$$

To find the solution to these systems, we ask the question, "at what point(s)—if any—do the graphs of these equations intersect?" Since some nonlinear equations represent conics, this is a convenient time to discuss systems of nonlinear equations.

How many points of intersection do a line and a parabola have? The answer depends on which line and which parabola. As we see in the following graphs, the answer can be one, two, or none.

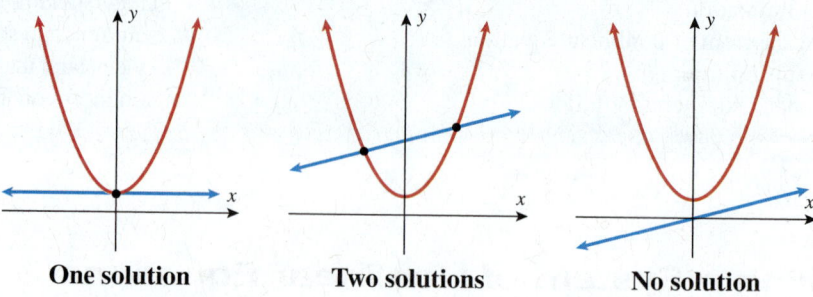

One solution   **Two solutions**   **No solution**

How many points of intersection do a parabola and an ellipse have? One, two, three, four, or no points of intersection correspond to one solution, two solutions, three solutions, four solutions, or no solution, respectively.

One solution   **Two solutions**   **Three solutions**   **Four solutions**   **No solution**

How many points of intersection do a parabola and a hyperbola have? The answer depends on which parabola and which hyperbola. As we see in the following graphs, the answer can be one, two, three, four, or none.

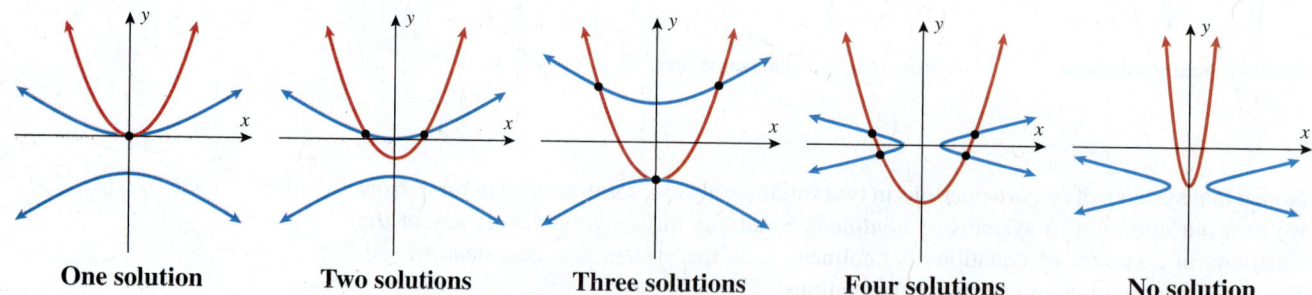

One solution   **Two solutions**   **Three solutions**   **Four solutions**   **No solution**

Using Elimination to Solve Systems of Nonlinear Equations

The first three examples in this section use elimination to solve systems of two nonlinear equations. In linear systems, we can eliminate either variable. In nonlinear systems, the variable to eliminate is the one that is raised to the same power in both equations.

EXAMPLE 1 Solving a System of Two Nonlinear Equations by Elimination: One Solution

Solve the system of equations, and graph the corresponding line and parabola to verify the answer.

$$\text{Equation (1):} \quad 2x - y = 3$$

$$\text{Equation (2):} \quad x^2 - y = 2$$

Solution:

Equation (1):

Multiply Equation (2) by -1.

Add.

Gather all terms to one side.

Factor.

Solve for x.

Substitute $x = 1$ into original Equation (1).

Solve for y.

$$2x - y = 3$$
$$-x^2 + y = -2$$
$$\overline{2x - x^2 = 1}$$
$$x^2 - 2x + 1 = 0$$
$$(x - 1)^2 = 0$$
$$x = 1$$
$$2(1) - y = 3$$
$$y = -1$$

The solution is $x = 1$, $y = -1$, or $(1, -1)$.

Graph the line $y = 2x - 3$ and the parabola $y = x^2 - 2$ and confirm that the point of intersection is $(1, -1)$.

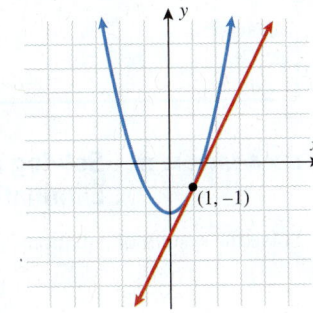

$(1, -1)$

Classroom Example 11.5.1
Solve the following systems:
a. $2y - x = 0$
   $y^2 + 1 = x$
b.* $(x - 1)^2 + \dfrac{(y + 2)^2}{4} = 1$
   $4 + y^2 = 0$

Answer:
a. $x = 2$, $y = 1$
b. no solution

EXAMPLE 2 Solving a System of Two Nonlinear Equations with Elimination: More Than One Solution

Solve the system of equations, and graph the corresponding parabola and circle to verify the answer.

$$\text{Equation (1):} \quad -x^2 + y = -7$$

$$\text{Equation (2):} \quad x^2 + y^2 = 9$$

Solution:

Equation (1):

Equation (2):

Add.

$$-x^2 + y = -7$$
$$x^2 + y^2 = 9$$
$$\overline{y^2 + y = 2}$$

Note: The circle appears elliptical because the *x*- and *y*-axis are not of equal scale. The zoom square feature can give the appearance of a circle.

Gather all terms to one side.	$y^2 + y - 2 = 0$
Factor.	$(y + 2)(y - 1) = 0$
Solve for *y*.	$y = -2$ or $y = 1$
Substitute $y = -2$ into Equation (2).	$x^2 + (-2)^2 = 9$
Solve for *x*.	$x = \pm\sqrt{5}$
Substitute $y = 1$ into Equation (2).	$x^2 + (1)^2 = 9$
Solve for *x*.	$x = \pm\sqrt{8} = \pm2\sqrt{2}$

There are four solutions: $\boxed{(-\sqrt{5}, -2), (\sqrt{5}, -2), (-2\sqrt{2}, 1), \text{ and } (2\sqrt{2}, 1)}$.

Graph the parabola $y = x^2 - 7$ and the circle $x^2 + y^2 = 9$ and confirm the four points of intersection.

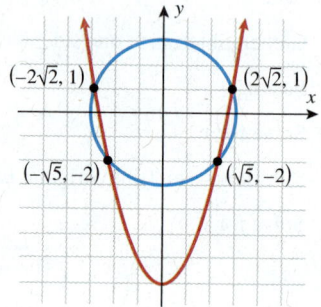

> **Classroom Example 11.5.2**
> Solve the following systems:
> **a.** $y - x^2 = a^2$
> $\quad y + x^2 = a^2$, where $a \neq 0$
> **b.*** $(x - 1)^2 + y^2 = 1$
> $\quad (x - 1)^2 + (y + 1)^2 = 1$
>
> **Answer:**
> **a.** $(0, a^2)$ **b.** $\left(1 \pm \dfrac{\sqrt{3}}{2}, \dfrac{-1}{2}\right)$

EXAMPLE 3 Solving a System of Two Nonlinear Equations with Elimination: No Solution

Solve the system of equations, and graph the corresponding parabolas to verify the answer.

$$\text{Equation (1):} \qquad x^2 + y = 3$$
$$\text{Equation (2):} \qquad -x^2 + y = 5$$

Technology Tip

Use a graphing calculator to solve the system of equations. Solve for *y* in each equation first; that is, $y_1 = -x^2 + 3$ and $y_2 = x^2 + 5$.

Note: The graphs do not intersect each other. There is no solution to the system.

Solution:

Equation (1):	$x^2 + y = 3$
Equation (2):	$-x^2 + y = 5$
Add.	$2y = 8$
Solve for *y*.	$y = 4$
Substitute $y = 4$ into Equation (1).	$x^2 + 4 = 3$
Simplify.	$x^2 = -1$

$x^2 = -1$ has no real solution.

There is $\boxed{\text{no solution}}$ to this system of nonlinear equations.

Graph the parabola $x^2 + y = 3$ and the parabola $y = x^2 + 5$ and confirm there are no points of intersection.

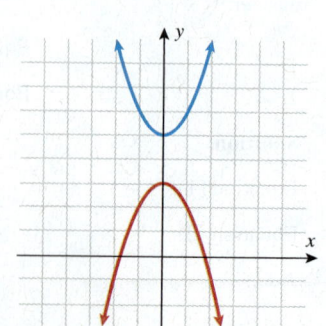

EXAMPLE 4   **Solving a System of Nonlinear Equations with Elimination**

Solve the system of nonlinear equations with elimination.

Equation (1): $\dfrac{x^2}{4} + y^2 = 1$

Equation (2): $x^2 - y^2 = 1$

Solution:

Add Equations (1) and (2) to eliminate y^2.

$$\dfrac{x^2}{4} + y^2 = 1$$
$$\underline{x^2 - y^2 = 1}$$
$$\dfrac{5}{4}x^2 = 2$$

Solve for x.

$$x^2 = \dfrac{8}{5}$$

$$x = \pm\sqrt{\dfrac{8}{5}}$$

Let $x = \pm\sqrt{\dfrac{8}{5}}$ in Equation (2).

$$\left(\pm\sqrt{\dfrac{8}{5}}\right)^2 - y^2 = 1$$

Solve for y.

$$y^2 = \dfrac{8}{5} - 1 = \dfrac{3}{5}$$

$$y = \pm\sqrt{\dfrac{3}{5}}$$

There are four solutions:

$$\left(-\sqrt{\dfrac{8}{5}}, -\sqrt{\dfrac{3}{5}}\right), \left(-\sqrt{\dfrac{8}{5}}, \sqrt{\dfrac{3}{5}}\right), \left(\sqrt{\dfrac{8}{5}}, -\sqrt{\dfrac{3}{5}}\right), \text{ and } \left(\sqrt{\dfrac{8}{5}}, \sqrt{\dfrac{3}{5}}\right)$$

A calculator can be used to approximate these solutions:

$$\sqrt{\dfrac{8}{5}} \approx 1.26$$

$$\sqrt{\dfrac{3}{5}} \approx 0.77$$

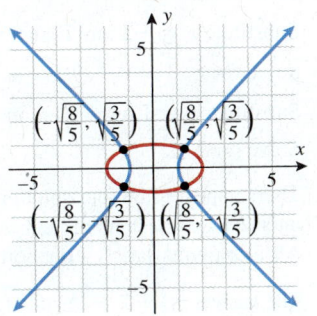

■ **Answer:**
a. $(-1, 2)$ and $(2, 5)$

b. no solution

c. no solution

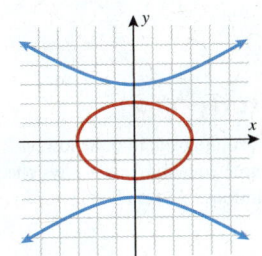

■ **YOUR TURN**   Solve the following systems of nonlinear equations:

a. $-x + y = 3$
$\quad x^2 - y = -1$

b. $x^2 + y = 2$
$\quad -x + y = 3$

c. $\dfrac{x^2}{9} + \dfrac{y^2}{4} = 1$
$\quad \dfrac{y^2}{9} - \dfrac{x^2}{16} = 1$

Using Substitution to Solve Systems of Nonlinear Equations

Elimination is based on the idea of eliminating one of the variables and solving the remaining equation in one variable. This is not always possible with nonlinear systems. For example, a system consisting of a circle and a line

$$x^2 + y^2 = 5$$
$$-x + y = 1$$

cannot be solved with elimination, because both variables are raised to different powers in each equation. We now turn to the substitution method. It is important to always check solutions, because extraneous solutions are possible.

Technology Tip

Use a graphing calculator to solve the system of equations. Solve for *y* in each equation first; that is,

$y_1 = \sqrt{5 - x^2}, y_2 = -\sqrt{5 - x^2}$,

and $y_3 = x + 1$.

Note: The graphs support the solutions to the system.

EXAMPLE 5 Solving a System of Nonlinear Equations with Substitution

Solve the system of equations, and graph the corresponding circle and line to verify the answer.

$$\text{Equation (1):} \quad x^2 + y^2 = 5$$
$$\text{Equation (2):} \quad -x + y = 1$$

Solution:

Rewrite Equation (2) with *y* isolated.

Equation (1):	$x^2 + y^2 = 5$
Equation (2):	$y = x + 1$
Substitute Equation (2), $y = x + 1$, into Equation (1).	$x^2 + (x + 1)^2 = 5$
Eliminate the parentheses.	$x^2 + x^2 + 2x + 1 = 5$
Gather like terms.	$2x^2 + 2x - 4 = 0$
Divide by 2.	$x^2 + x - 2 = 0$
Factor.	$(x + 2)(x - 1) = 0$
Solve for *x*.	$x = -2 \text{ or } x = 1$
Substitute $x = -2$ into Equation (1).	$(-2)^2 + y^2 = 5$
Solve for *y*.	$y = -1 \text{ or } y = 1$
Substitute $x = 1$ into Equation (1).	$(1)^2 + y^2 = 5$
Solve for *y*.	$y = -2 \text{ or } y = 2$

There appear to be four solutions: $(-2, -1), (-2, 1), (1, -2),$ and $(1, 2)$, but a line can intersect a circle in no more than two points. Therefore, at least two solutions are *extraneous*. All four points satisfy Equation (1), but only $(-2, -1)$ and $(1, 2)$ also satisfy Equation (2).

The answer is $\boxed{(-2, -1) \text{ and } (1, 2)}$.

Graph the circle $x^2 + y^2 = 5$ and the line $y = x + 1$ and confirm the two points of intersection.

Note: After solving for *x*, had we substituted back into the linear Equation (2) instead of Equation (1), extraneous solutions would not have appeared. In general, **substitute back into the lowest degree equation, and always check solutions.**

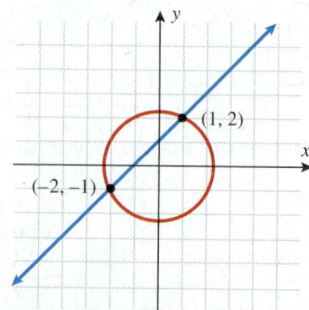

■ **Answer:** (2, 3) and (3, 2)

■ **YOUR TURN** Solve the system of equations $x^2 + y^2 = 13$ and $x + y = 5$.

In Example 6, the equation $xy = 2$ can also be shown to be a rotated hyperbola (Sec. 11.7). For now, we can express this equation in terms of a reciprocal function $y = \dfrac{2}{x}$, which we discussed in Section 3.2.

EXAMPLE 6 Solving a System of Nonlinear Equations with Substitution

Solve the system of equations.

$$\text{Equation (1):} \quad x^2 + y^2 = 5$$

$$\text{Equation (2):} \quad xy = 2$$

Solution:

Since Equation (2) tells us that $xy = 2$, we know that neither x nor y can be zero.

Solve Equation (2) for y.

$$y = \frac{2}{x}$$

Substitute $y = \dfrac{2}{x}$ into Equation (1).

$$x^2 + \left(\frac{2}{x}\right)^2 = 5$$

Eliminate the parentheses.

$$x^2 + \frac{4}{x^2} = 5$$

Multiply by x^2.

$$x^4 + 4 = 5x^2$$

Collect the terms to one side.

$$x^4 - 5x^2 + 4 = 0$$

Factor.

$$(x^2 - 4)(x^2 - 1) = 0$$

Solve for x.

$$x = \pm 2 \text{ or } x = \pm 1$$

Substitute $x = -2$ into Equation (2), $xy = 2$, and solve for y.

$$y = -1$$

Substitute $x = 2$ into Equation (2), $xy = 2$, and solve for y.

$$y = 1$$

Substitute $x = -1$ into Equation (2), $xy = 2$, and solve for y.

$$y = -2$$

Substitute $x = 1$ into Equation (2), $xy = 2$, and solve for y.

$$y = 2$$

Check to see that there are four solutions: $\boxed{(-2, -1), (-1, -2), (2, 1), \text{ and } (1, 2)}$.

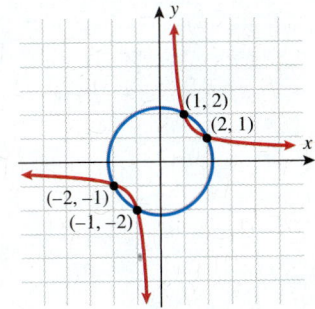

Note: It is important to check the solutions either algebraically or graphically (see the graph).

■ **YOUR TURN** Solve the system of equations $x^2 + y^2 = 2$ and $xy = 1$.

■ **Answer:** $(-1, -1)$ and $(1, 1)$

Applications

Technology Tip

Use a graphing calculator to solve the system of equations. Solve for y in each equation first; that is,

$$y_1 = \frac{400{,}000}{x} \text{ and } y_2 = 40 + \frac{237{,}600}{x - 40}.$$

Note: The graphs support the solutions to the system.

EXAMPLE 7   Calculating How Much Fence to Buy

A couple buy a rectangular piece of property advertised as 10 acres (approximately 400,000 square feet). They want two fences to divide the land into an internal grazing area and a surrounding riding path. If they want the riding path to be 20 feet wide, one fence will enclose the property and one internal fence will sit 20 feet inside the outer fence. If the internal grazing field is 237,600 square feet, how many linear feet of fencing should they buy?

Solution:

Use the five-step procedure for solving word problems from Section 1.2, and use two variables.

STEP 1   Identify the question.

How many linear feet of fence should they buy? Or, what is the sum of the perimeters of the two fences?

STEP 2   Make notes or draw a sketch.

STEP 3   Set up the equations.

$x = $ length of property	$x - 40 = $ length of internal field
$y = $ width of property	$y - 40 = $ width of internal field

$$\text{Equation (1):} \qquad xy = 400{,}000$$

$$\text{Equation (2):} \qquad (x - 40)(y - 40) = 237{,}600$$

STEP 4   Solve the system of equations.

Substitution Method

Since Equation (1) tells us that $xy = 400{,}000$, we know that neither x nor y can be zero.

Solve Equation (1) for y.
$$y = \frac{400{,}000}{x}$$

Substitute $y = \dfrac{400{,}000}{x}$ into Equation (2).
$$(x - 40)\left(\frac{400{,}000}{x} - 40\right) = 237{,}600$$

Eliminate the parentheses.
$$400{,}000 - 40x - \frac{16{,}000{,}000}{x} + 1600 = 237{,}600$$

Multiply by the LCD, x.
$$400{,}000x - 40x^2 - 16{,}000{,}000 + 1600x = 237{,}600x$$

Collect like terms on one side.
$$40x^2 - 164{,}000x + 16{,}000{,}000 = 0$$

Divide by 40.
$$x^2 - 4100x + 400{,}000 = 0$$

Factor.
$$(x - 4000)(x - 100) = 0$$

Solve for x.
$$x = 4000 \text{ or } x = 100$$

Substitute $x = 4000$ into the original Equation (1). $\qquad$ $4000y = 400{,}000$

Solve for y. $\qquad$ $y = 100$

Substitute $x = 100$ into the original Equation (1). $\qquad$ $100y = 400{,}000$

Solve for y. $\qquad$ $y = 4000$

The two solutions yield the same dimensions: 4000×100. The inner field has the dimensions 3960×60. Therefore, the perimeter of both fences is

$$2(4000) + 2(100) + 2(3960) + 2(60) = 8000 + 200 + 7920 + 120 = 16{,}240$$

The couple should buy 16,240 linear feet of fencing.

STEP 5 **Check the solution.**

The point $(4000, 100)$ satisfies both Equation (1) and Equation (2).

It is important to note that some nonlinear equations are not conic sections (exponential, logarithmic, and higher degree polynomial equations). These systems of linear equations are typically solved by the substitution method (see the exercises).

SECTION 11.5 SUMMARY

In this section, systems of two equations were discussed when at least one of the equations is nonlinear (for example, conics). The substitution method and elimination method from Section 6.1 can *sometimes* be applied to nonlinear systems. When graphing the two equations, the points of intersection are the solutions of the system. Systems of nonlinear equations can have more than one solution. Also, extraneous solutions can appear, so it is important to always check solutions.

SECTION 11.5 EXERCISES

▪ SKILLS

In Exercises 1–12, solve the system of equations by applying the elimination method.

1. $x^2 - y = -2$
$-x + y = 4$

2. $x^2 + y = 2$
$2x + y = -1$

3. $x^2 + y = 1$
$2x + y = 2$

4. $x^2 - y = 2$
$-2x + y = -3$

5. $x^2 + y = -5$
$-x + y = 3$

6. $x^2 - y = -7$
$x + y = -2$

7. $x^2 + y^2 = 1$
$x^2 - y = -1$

8. $x^2 + y^2 = 1$
$x^2 + y = -1$

9. $x^2 + y^2 = 3$
$4x^2 + y = 0$

10. $x^2 + y^2 = 6$
$-7x^2 + y = 0$

11. $x^2 + y^2 = -6$
$-2x^2 + y = 7$

12. $x^2 + y^2 = 5$
$3x^2 + y = 9$

In Exercises 13–24, solve the system of equations by applying the substitution method.

13. $x + y = 2$
$x^2 + y^2 = 2$

14. $x - y = -2$
$x^2 + y^2 = 2$

15. $xy = 4$
$x^2 + y^2 = 10$

16. $xy = -3$
$x^2 + y^2 = 12$

17. $y = x^2 - 3$
$y = -4x + 9$

18. $y = -x^2 + 5$
$y = 3x - 4$

19. $x^2 + xy - y^2 = 5$
$x - y = -1$

20. $x^2 + xy + y^2 = 13$
$x + y = -1$

21. $2x - y = 3$
$x^2 + y^2 - 2x + 6y = -9$

22. $x^2 + y^2 - 2x - 4y = 0$
$-2x + y = -3$

23. $4x^2 + 12xy + 9y^2 = 25$
$-2x + y = 1$

24. $-4xy + 4y^2 = 8$
$3x + y = 8$

In Exercises 25–34, solve the system of equations by applying any method.

25. $x^3 - y^3 = 63$
$x - y = 3$

26. $x^3 + y^3 = -26$
$x + y = -2$

27. $4x^2 - 3xy = -5$
$-x^2 + 3xy = 8$

28. $2x^2 + 5xy = 2$
$x^2 - xy = 1$

29. $\log_x(2y) = 3$
$\log_x(y) = 2$

30. $\log_x(y) = 1$
$\log_x(2y) = \frac{1}{2}$

31. $\dfrac{1}{x^3} + \dfrac{1}{y^2} = 17$
$\dfrac{1}{x^3} - \dfrac{1}{y^2} = -1$

32. $\dfrac{2}{x^2} + \dfrac{3}{y^2} = \dfrac{5}{6}$
$\dfrac{4}{x^2} - \dfrac{9}{y^2} = 0$

33. $2x^2 + 4y^4 = -2$
$6x^2 + 3y^4 = -1$

34. $x^2 + y^2 = -2$
$x^2 + y^2 = -1$

In Exercises 35–38, graph each equation and find the point(s) of intersection.

35. The parabola $y = x^2 - 6x + 11$ and the line $y = -x + 7$

36. The circle $x^2 + y^2 - 4x - 2y + 5 = 0$ and the line $-x + 3y = 6$

37. The ellipse $9x^2 - 18x + 4y^2 + 8y - 23 = 0$ and the line $-3x + 2y = 1$.

38. The parabola $y = -x^2 + 2x$ and the circle $x^2 + 6x + y^2 - 4y + 12 = 0$.

■ APPLICATIONS

39. Numbers. The sum of two numbers is 10, and the difference of their squares is 40. Find the numbers.

40. Numbers. The difference of two numbers is 3, and the difference of their squares is 51. Find the numbers.

41. Numbers. The product of two numbers is equal to the reciprocal of the difference of their reciprocals. The product of the two numbers is 72. Find the numbers.

42. Numbers. The ratio of the sum of two numbers to the difference of the two numbers is 9. The product of the two numbers is 80. Find the numbers.

43. Geometry. A rectangle has a perimeter of 36 centimeters and an area of 80 square centimeters. Find the dimensions of the rectangle.

44. Geometry. Two concentric circles have perimeters that add up to 16π and areas that add up to 34π. Find the radii of the two circles.

45. Horse Paddock. An equestrian buys a 5-acre rectangular parcel (approximately 200,000 square feet) and is going to fence in the entire property and then divide the parcel into two halves with a fence. If 2200 linear feet of fencing is required, what are the dimensions of the parcel?

46. Dog Run. A family moves into a new home and decides to fence in the yard to give its dog room to roam. If the area that will be fenced in is rectangular and has an area of 11,250 square feet, and the length is twice as much as the width, how many linear feet of fence should the family buy?

47. Footrace. Your college algebra professor and Jeremy Wariner (2004 Olympic Gold Medalist in the men's 400 meter) decided to race. The race was 400 meters and Jeremy gave your professor a 1-minute head start, and still crossed the finish line 1 minute 40 seconds before your professor. If Jeremy ran five times faster than your professor, what was each person's average speed?

48. Footrace. You decided to race Jeremy Wariner for 800 meters. At that distance, Jeremy runs approximately twice as fast as you. He gave you a 1-minute head start and crossed the finish line 20 seconds before you. What were each of your average speeds?

CATCH THE MISTAKE

In Exercises 49 and 50, explain the mistake that is made.

49. Solve the system of equations:
$$x^2 + y^2 = 4$$
$$x + y = 2$$

Solution:

Multiply the second equation by (-1) and add to the first equation.	$x^2 - x = 2$
Subtract 2.	$x^2 - x - 2 = 0$
Factor.	$(x + 1)(x - 2) = 0$
Solve for x.	$x = -1$ and $x = 2$
Substitute $x = -1$ and $x = 2$ into $x + y = 2$.	$-1 + y = 2$ and $2 + y = 2$
Solve for y.	$y = 3$ and $y = 0$

The answer is $(-1, 3)$ and $(2, 0)$.

This is incorrect. What mistake was made?

50. Solve the system of equations:
$$x^2 + y^2 = 5$$
$$2x - y = 0$$

Solution:

Solve the second equation for y.	$y = 2x$
Substitute $y = 2x$ into the first equation.	$x^2 + (2x)^2 = 5$
Eliminate the parentheses.	$x^2 + 4x^2 = 5$
Gather like terms.	$5x^2 = 5$
Solve for x.	$x = -1$ and $x = 1$
Substitute $x = -1$ into the first equation.	$(-1)^2 + y^2 = 5$
Solve for y.	$y = -2$ and $y = 2$
Substitute $x = 1$ into the first equation.	$(1)^2 + y^2 = 5$
Solve for y.	$y = -2$ and $y = 2$

The answers are $(-1, -2)$, $(-1, 2)$, $(1, -2)$, and $(1, 2)$.

This is incorrect. What mistake was made?

CONCEPTUAL

In Exercises 51–54, determine whether each statement is true or false.

51. A system of equations representing a line and a parabola can intersect in at most three points.

52. A system of equations representing a line and a cubic function can intersect in at most three places.

53. The elimination method can always be used to solve systems of two nonlinear equations.

54. The substitution method always works for solving systems of nonlinear equations.

CHALLENGE

55. A circle and a line have at most two points of intersection. A circle and a parabola have at most four points of intersection. What is the greatest number of points of intersection that a circle and an nth-degree polynomial can have?

56. A line and a parabola have at most two points of intersection. A line and a cubic function have at most three points of intersection. What is the greatest number of points of intersection that a line and an nth-degree polynomial can have?

57. Find a system of equations representing a line and a parabola that has only one real solution.

58. Find a system of equations representing a circle and a parabola that has only one real solution.

TECHNOLOGY

In Exercises 59–64, use a graphing utility to solve the following systems of equations.

59. $y = e^x$
$y = \ln x$

60. $y = 10^x$
$y = \log x$

61. $2x^3 + 4y^2 = 3$
$xy^3 = 7$

62. $3x^4 - 2xy + 5y^2 = 19$
$x^4 y = 5$

63. $5x^3 + 2y^2 = 40$
$x^3 y = 5$

64. $4x^4 + 2xy + 3y^2 = 60$
$x^4 y = 8 - 3x^4$

SKILLS OBJECTIVES

- Graph a nonlinear inequality in two variables.
- Graph a system of nonlinear inequalities in two variables.

CONCEPTUAL OBJECTIVES

- Understand that a nonlinear inequality in two variables may be represented by either a bounded or an unbounded region.
- Interpret an overlapping shaded region as a solution.

Classroom Example 11.6.1
Graph the following inequalities:
a. $(x - 1)^2 + (y + 3)^2 > 4$
b.* $x^2 + 4y^2 > 4$
c.* $2y - 3(x + 1)^2 < 0$

Answer:
a.

b.

c.

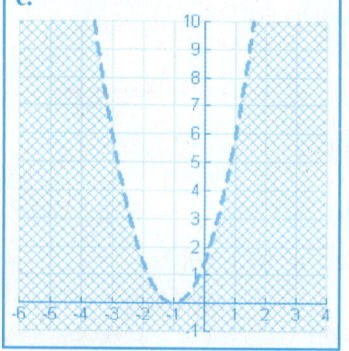

Nonlinear Inequalities in Two Variables

Linear inequalities are expressed in the form $Ax + By \leq C$. Specific expressions can involve either of the strict or either of the nonstrict inequalities. Examples of **nonlinear inequalities in two variables** are

$$9x^2 + 16y^2 \geq 1 \qquad x^2 + y^2 > 1 \qquad y \leq -x^2 + 3 \qquad \text{and} \qquad \frac{x^2}{20} - \frac{y^2}{81} < 1$$

We follow the same procedure as we did with linear inequalities. We change the inequality to an equal sign, graph the resulting nonlinear equation, test points from the two regions, and shade the region that makes the inequality true. For strict inequalities, $<$ or $>$, we use dashed curves, and for nonstrict inequalities, $\leq$ or $\geq$, we use solid curves.

EXAMPLE 1 **Graphing a Strict Nonlinear Inequality in Two Variables**

Graph the inequality $x^2 + y^2 > 1$.

Solution:

STEP 1 Change the inequality sign to an equal sign. $\qquad x^2 + y^2 = 1$

The equation is the equation of a circle.

STEP 2 Draw the graph of the circle.

The center is $(0, 0)$ and the radius is 1.

Since the inequality $>$ is a strict inequality, draw the circle as a **dashed** curve.

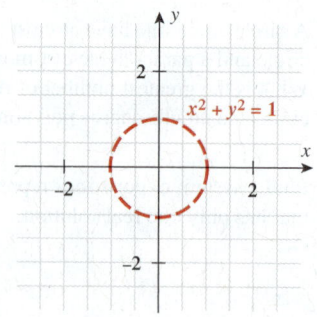

STEP 3 Test points in each region (outside the circle and inside the circle).

Substitute $(2, 0)$ into $x^2 + y^2 > 1$. $\qquad 4 \geq 1$

The point $(2, 0)$ satisfies the inequality.

Substitute $(0, 0)$ into $x^2 + y^2 > 1$. $\qquad 0 \geq 1$

The point $(0, 0)$ does not satisfy the inequality.

STEP 4 Shade the region containing the point (2, 0).

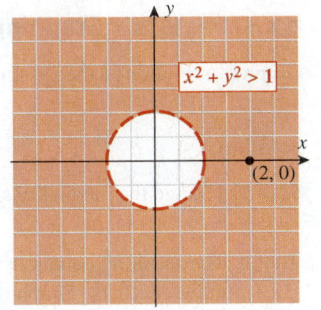

EXAMPLE 2 Graphing a Nonstrict Nonlinear Inequality in Two Variables

Graph the inequality $y \le -x^2 + 3$.

Solution:

STEP 1 Change the inequality sign to an equal sign.

The equation is that of a parabola.

$$y = -x^2 + 3$$

STEP 2 Graph the parabola.

Reflect the base function, $f(x) = x^2$, about the x-axis and shift up 3 units. Since the inequality $\le$ is a nonstrict inequality, draw the parabola as a **solid** curve.

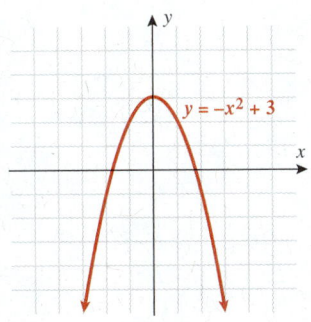

STEP 3 Test points in each region (inside the parabola and outside the parabola).

Substitute (3, 0) into $y \le -x^2 + 3$. $0 \le -6$

The point (3, 0) does not satisfy the inequality.

Substitute (0, 0) into $y \le -x^2 + 3$. $0 \le 3$

The point (0, 0) does satisfy the inequality.

STEP 4 Shade the region containing the point (0, 0).

■ **YOUR TURN** Graph the following inequalities:

    a. $x^2 + y^2 \le 9$ **b.** $y > -x^2 + 2$

Technology Tip

Use a graphing calculator to graph the inequality $y \le -x^2 + 3$. Enter $y_1 = -x^2 + 3$. For $\le$, use the arrow key to move the cursor to the left of Y_1 and type ENTER until you see ◣.

Note: The parabola should be drawn solid.

■ **Answer:**

  a.

  b.

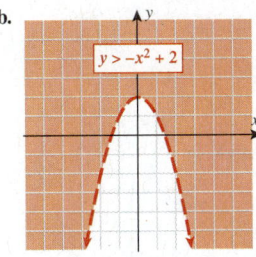

Systems of Nonlinear Inequalities

To solve a system of inequalities, first graph the inequalities and shade the region containing the points that satisfy each inequality. The overlap of all the shaded regions is the solution.

EXAMPLE 3 Graphing a System of Inequalities

Graph the solution to the system of inequalities: $y \geq x^2 - 1$
$y < x + 1$

Solution:

STEP 1 Change the inequality signs to equal signs.

$$y = x^2 - 1$$
$$y = x + 1$$

STEP 2 The resulting equations represent a parabola (to be drawn solid) and a line (to be drawn dashed). Graph the two equations.

To determine the points of intersection, set the y-values equal to each other.

$$x^2 - 1 = x + 1$$

Write the quadratic equation in standard form.

$$x^2 - x - 2 = 0$$

Factor.

$$(x - 2)(x + 1) = 0$$

Solve for x.

$$x = 2 \text{ or } x = -1$$

Substitute $x = 2$ into $y = x + 1$.

$(2, 3)$

Substitute $x = -1$ into $y = x + 1$.

$(-1, 0)$

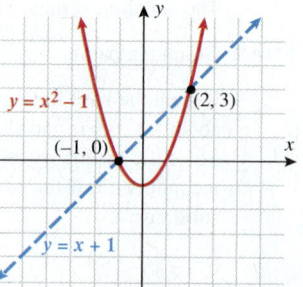

STEP 3 Test points and shade the regions.

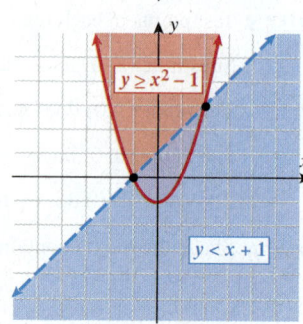

STEP 4 Shade the common region.

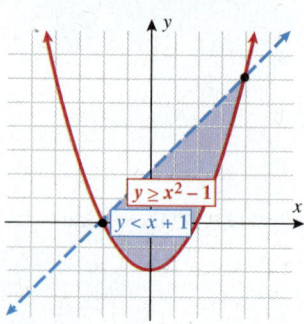

■ **YOUR TURN** Graph the solution to the system of inequalities: $x^2 + y^2 < 9$
$y > 0$

Technology Tip

Use a graphing calculator to graph the solution to the system of inequalities $y \geq x^2 - 1$ and $y < x + 1$.

First enter $y_2 = x^2 - 1$. For $\geq$, use the arrow key to move the cursor to the left of Y_1 and type ENTER until you see ◤. Next enter $y_2 = x + 1$.

For $<$, use the arrow key to move the cursor to the left of Y_1 and type ENTER until you see ◣.

Note: The parabola should be drawn solid, and the line should be drawn dashed.

■ **Answer:**

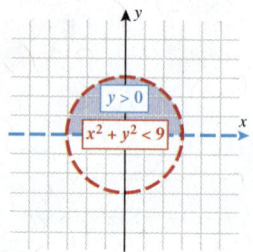

EXAMPLE 4 Solving a System of Nonlinear Inequalities

Solve the system of inequalities: $x^2 + y^2 < 2$
$y \geq x^2$

Solution:

STEP 1 Change the inequality signs to equal signs.

$$x^2 + y^2 = 2$$
$$y = x^2$$

STEP 2 The resulting equations correspond to a circle (to be drawn dashed) and a parabola (to be drawn solid). Graph the two inequalities.

To determine the points of intersection, solve the system of equations by substitution.

$$x^2 + (\underbrace{x^2}_{y})^2 = 2$$

$$x^4 + x^2 - 2 = 0$$

Factor.

$$(x^2 + 2)(x^2 - 1) = 0$$

Solve for x.

$$\underbrace{x^2 = -2}_{\text{no solution}} \text{ or } \underbrace{x^2 = 1}_{x = \pm 1}$$

The points of intersection are $(-1, 1)$ and $(1, 1)$.

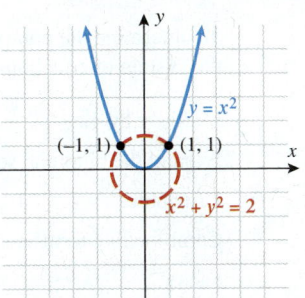

STEP 3 Test points and shade the region.

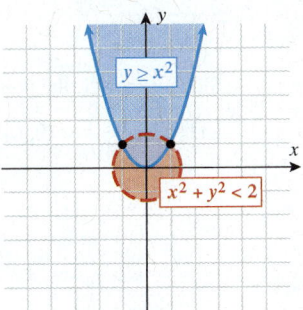

STEP 4 Identify the common region as the solution.

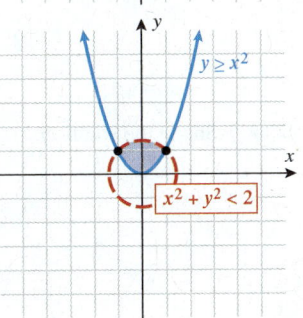

■ **YOUR TURN** Solve the system of inequalities: $x^2 + y^2 < 2$
$y < x^2$

■ **Answer:**

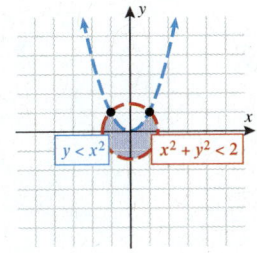

It is important to note that any inequality based on an equation whose graph is not a line is considered a nonlinear inequality.

EXAMPLE 5   Solving a System of Nonlinear Inequalities

Solve the system of inequalities: $(x - 1)^2 + \dfrac{y^2}{4} < 1$

$$y \geq \sqrt{x}$$

Solution:

STEP 1 Change the inequality signs to equal signs.

$$(x - 1)^2 + \dfrac{y^2}{4} = 1$$
$$y = \sqrt{x}$$

STEP 2 The resulting equations correspond to an ellipse (to be drawn dashed) and the square-root function (to be drawn solid). Graph the two inequalities.

To determine the points of intersection, solve the system of equations by substitution.

$$(x - 1)^2 + \dfrac{\left(\sqrt{x}\right)^2}{4} = 1$$

Multiply by 4.

$$4(x - 1)^2 + x = 4$$

Expand the binomial squared.

$$4(x^2 - 2x + 1) + x = 4$$

Distribute.

$$4x^2 - 8x + 4 + x = 4$$

Combine like terms and gather terms to one side.

$$4x^2 - 7x = 0$$

Factor.

$$x(4x - 7) = 0$$

Solve for x.

$$x = 0 \text{ and } x = \dfrac{7}{4}$$

Since $y = \sqrt{x}$, the points of intersection are $(0, 0)$ and $\left(\dfrac{7}{4}, \sqrt{\dfrac{7}{4}}\right)$ or $(0, 0)$ and $\left(\dfrac{7}{4}, \dfrac{\sqrt{7}}{2}\right)$

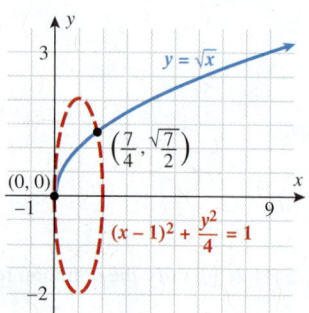

STEP 3 Shade the solution.

SECTION
11.6 SUMMARY

In this section we discussed nonlinear inequalities in two variables. Sometimes these result in bounded regions (e.g., $x^2 + y^2 \le 1$), and sometimes these result in unbounded regions (e.g., $x^2 + y^2 > 1$).

When solving systems of inequalities, we first graph each of the inequalities separately and then look for the intersection (overlap) of all shaded regions.

SECTION
11.6 EXERCISES

■ SKILLS

In Exercises 1–12, match the nonlinear inequality with the correct graph.

1. $x^2 + y^2 < 25$

2. $x^2 + y^2 \le 9$

3. $\dfrac{x^2}{9} + \dfrac{y^2}{16} \ge 1$

4. $\dfrac{x^2}{4} + \dfrac{y^2}{9} > 1$

5. $y \ge x^2 - 3$

6. $x^2 \ge 16y$

7. $x \ge y^2 - 4$

8. $\dfrac{x^2}{9} + \dfrac{y^2}{25} \ge 1$

9. $9x^2 + 9y^2 < 36$

10. $(x - 2)^2 + (y + 3)^2 \le 9$

11. $\dfrac{x^2}{4} - \dfrac{y^2}{9} \ge 1$

12. $\dfrac{y^2}{16} - \dfrac{x^2}{9} < 1$

a.

b.

c.

d.

e.

f.

g.

h.

i.

j.

k.

l.

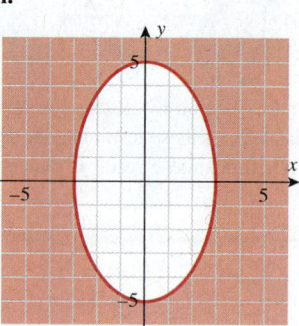

In Exercises 13–30, graph the nonlinear inequality.

13. $y \le x^2 - 2$

14. $y \ge -x^2 + 3$

15. $x^2 + y^2 > 4$

16. $x^2 + y^2 < 16$

17. $x^2 + y^2 - 2x + 4y + 4 \ge 0$

18. $x^2 + y^2 + 2x - 2y - 2 \le 0$

19. $3x^2 + 4y^2 \le 12$

20. $\dfrac{(x-2)^2}{9} + \dfrac{(y+1)^2}{25} > 1$

21. $9x^2 + 16y^2 - 18x + 96y + 9 > 0$

22. $\dfrac{(x-2)^2}{4} - \dfrac{(y+3)^2}{1} \ge 1$

23. $9x^2 - 4y^2 \ge 26$

24. $\dfrac{(y+1)^2}{9} - \dfrac{(x+2)^2}{16} < 1$

25. $36x^2 - 9y^2 \ge 324$

26. $25x^2 - 36y^2 + 200x + 144y - 644 \ge 0$

27. $y \ge e^x$

28. $y \le \ln x$

29. $y < -x^3$

30. $y > -x^4$

In Exercises 31–50, graph each system of inequalities or indicate that the system has no solution.

31. $y < x + 1$
$y \le x^2$

32. $y < x^2 + 4x$
$y \le 3 - x$

33. $y \ge 2 + x$
$y \le 4 - x^2$

34. $y \ge (x-2)^2$
$y \le 4 - x$

35. $y \le -(x+2)^2$
$y > -5 + x$

36. $y \ge (x-1)^2 + 2$
$y \le 10 - x$

37. $-x^2 + y > -1$
$x^2 + y < 1$

38. $x < -y^2 + 1$
$x > y^2 - 1$

39. $y \ge x^2$
$x \ge y^2$

40. $y < x^2$
$x > y^2$

41. $x^2 + y^2 < 36$
$2x + y > 3$

42. $x^2 + y^2 < 36$
$y > 6$

43. $x^2 + y^2 < 25$
$y \ge 6 + x$

44. $(x-1)^2 + (y+2)^2 \le 36$
$y \ge x - 3$

45. $x^2 + y^2 \le 9$
$y \ge 1 + x^2$

46. $x^2 + y^2 \ge 16$
$x^2 + (y-3)^2 \le 9$

47. $x^2 - y^2 < 4$
$y > 1 - x^2$

48. $\dfrac{x^2}{4} - \dfrac{y^2}{9} \le 1$
$y \ge x - 5$

49. $y < e^x$
$y > \ln x \qquad x > 0$

50. $y < 10^x$
$y > \log x \qquad x > 0$

■ **APPLICATIONS**

51. Find the area enclosed by the system of inequalities:
$x^2 + y^2 < 9$
$x > 0$

52. Find the area enclosed by the system of inequalities:
$x^2 + y^2 \le 5$
$x \le 0$
$y \ge 0$

■ CATCH THE MISTAKE

In Exercises 53 and 54, explain the mistake that is made.

53. Graph the system of inequalities: $\quad x^2 + y^2 < 1$
$\qquad\qquad\qquad\qquad\qquad\qquad\qquad x^2 + y^2 > 4$

Solution:

Draw the circles
$x^2 + y^2 = 1$ and
$x^2 + y^2 = 4$

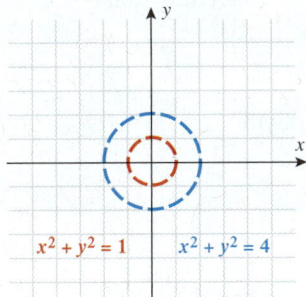

Shade outside
$x^2 + y^2 = 1$ and
inside $x^2 + y^2 = 4$

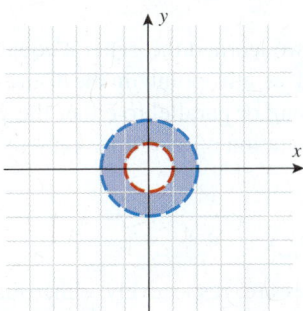

This is incorrect. What mistake was made?

54. Graph the system of inequalities: $\quad x > -y^2 + 1$
$\qquad\qquad\qquad\qquad\qquad\qquad\qquad x < \quad y^2 - 1$

Solution:

Draw the parabolas
$x = -y^2 + 1$ and
$x = y^2 - 1$.
Shade the region
between the curves.

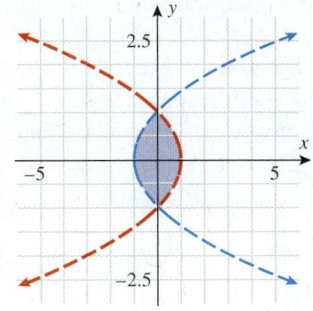

This is incorrect. What mistake was made?

■ CONCEPTUAL

In Exercises 55 and 56, determine whether each statement is true or false.

55. A nonlinear inequality always represents a bounded region.

56. A system of inequalities always has a solution.

■ CHALLENGE

57. For the system of nonlinear inequalities $\dfrac{x^2 + y^2 \geq a^2}{x^2 + y^2 \leq b^2}$, what restriction must be placed on the values of a and b for this system to have a solution? Assume that a and b are real numbers.

58. Can $x^2 + y^2 < -1$ ever have a real solution? What types of numbers would x and/or y have to be to satisfy this inequality?

■ TECHNOLOGY

In Exercises 59–66, use a graphing utility to graph the inequalities.

59. $x^2 + y^2 - 2x + 4y + 4 \geq 0$

60. $x^2 + y^2 + 2x - 2y - 2 \leq 0$

61. $y \geq e^x$

62. $y \leq \ln x$

63. $y < e^x$
$\quad y > \ln x \qquad x > 0$

64. $y < 10^x$
$\quad y > \log x \qquad x > 0$

65. $x^2 - 4y^2 + 5x - 6y + 18 \geq 0$

66. $x^2 - 2xy + 4y^2 + 10x - 25 \leq 0$

SKILLS OBJECTIVES

- Transform general second-degree equations into recognizable equations of conics by analyzing rotation of axes.
- Determine the angle of rotation that will transform a general second-degree equation into a familiar equation of a conic section.
- Graph a rotated conic.

CONCEPTUAL OBJECTIVE

- Understand how the equation of a conic section is altered by rotation of axes.

Rotation of Axes Formulas

In Sections 11.1 through 11.4 we learned the general equations of parabolas, ellipses, and hyperbolas that were centered at any point in the Cartesian plane and whose vertices and foci were aligned either along or parallel to either the x-axis or the y-axis. We learned, for example, that the general equation of an ellipse centered at the origin is given by

$$\frac{x^2}{a^2} + \frac{y^2}{b^2} = 1$$

where the major and minor axes are, respectively, either the x- or the y-axis depending on whether a is greater than or less than b. Now let us look at an equation of a conic section whose graph is *not* aligned with the x- or y-axis: the equation $5x^2 - 8xy + 5y^2 - 9 = 0$.

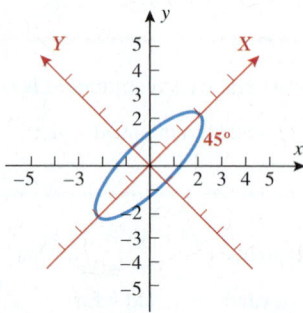

This graph can be thought of as an ellipse that started with the major axis along the x-axis and the minor axis along the y-axis and then was rotated counterclockwise 45°. A new XY-coordinate system can be introduced such that this system has the same origin but the XY-coordinate system is rotated by a certain amount from the standard xy-coordinate system. In this example, the major axis of the ellipse lies along the new X-axis and the minor axis lies along the new Y-axis. We will see that we can write the equation of this ellipse as

$$\frac{X^2}{9} - \frac{Y^2}{1} = 1$$

We will now develop the *rotation of axes formulas*, which allow us to transform the generalized second-degree equation in xy, that is, $Ax^2 + Bxy + Cy^2 + Dx + Ey + F = 0$, into an equation in XY of a conic that is familiar to us.

Let the new **XY**-coordinate system be displaced from the *xy*-coordinate system by rotation through an angle **θ**. Let **P** represent some point a distance *r* from the origin.

We can represent the point **P** as the point (*x, y*) or the point (**X, Y**).

We define the angle α as the angle *r* makes with the *X*-axis and $\alpha + \theta$ as the angle *r* makes with the *x*-axis.

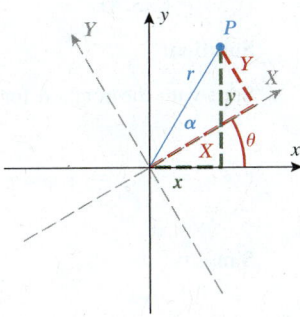

We can represent the point **P** in polar coordinates using the following relationships:

$$x = r\cos(\alpha + \theta)$$
$$y = r\sin(\alpha + \theta)$$
$$X = r\cos\alpha$$
$$Y = r\sin\alpha$$

Words	**Math**
Start with the *x*-term and write the cosine identity for a sum.	$x = r\cos(\alpha + \theta)$ $\quad = r(\cos\alpha\cos\theta - \sin\alpha\sin\theta)$
Eliminate the parentheses and group *r* with the α-terms.	$x = (r\cos\alpha)\cos\theta - (r\sin\alpha)\sin\theta$
Substitute according to the relationships $X = r\cos\alpha$ and $Y = r\sin\alpha$.	$\boxed{x = X\cos\theta - Y\sin\theta}$
Start with the *y*-term and write the sine identity for a sum.	$y = r\sin(\alpha + \theta)$ $\quad = r(\sin\alpha\cos\theta + \cos\alpha\sin\theta)$
Eliminate the parentheses and group *r* with the α-terms.	$y = (r\sin\alpha)\cos\theta + (r\cos\alpha)\sin\theta$
Substitute according to the relationships $X = r\cos\alpha$ and $Y = r\sin\alpha$.	$\boxed{y = Y\cos\theta + X\sin\theta}$

By treating the highlighted equations for *x* and *y* as a system of linear equations in *X* and *Y*, we can then solve for *X* and *Y* in terms of *x* and *y*. The results are summarized in the following box:

ROTATION OF AXES FORMULAS

Suppose that the x- and y-axes in the rectangular coordinate plane are rotated through an acute angle θ to produce the X- and Y-axes. Then, the coordinates (x, y) and (X, Y) are related according to the following equations:

$$x = X\cos\theta - Y\sin\theta \qquad\qquad X = x\cos\theta + y\sin\theta$$
$$y = X\sin\theta + Y\cos\theta \quad \text{or} \quad Y = -x\sin\theta + y\cos\theta$$

Classroom Example 11.7.1*

If the xy-coordinate axes are rotated 45°, find the XY-coordinates of $(x, y) = (-a, -a)$, where $a > 0$.

Answer: $\left(-a\sqrt{2}, 0\right)$

EXAMPLE 1 Rotating the Axes

If the xy-coordinate axes are rotated 60°, find the XY-coordinates of the point $(x, y) = (-3, 4)$.

Solution:

Start with the rotation formulas.

$$X = x\cos\theta + y\sin\theta$$
$$Y = -x\sin\theta + y\cos\theta$$

Let $x = -3, y = 4,$ and $\theta = 60°.$

$$X = -3\cos(60°) + 4\sin(60°)$$
$$Y = -(-3)\sin(60°) + 4\cos(60°)$$

Simplify.

$$X = -3\underbrace{\cos(60°)}_{\frac{1}{2}} + 4\underbrace{\sin(60°)}_{\frac{\sqrt{3}}{2}}$$

$$Y = 3\underbrace{\sin(60°)}_{\frac{\sqrt{3}}{2}} + 4\underbrace{\cos(60°)}_{\frac{1}{2}}$$

$$X = -\frac{3}{2} + 2\sqrt{3}$$

$$Y = \frac{3\sqrt{3}}{2} + 2$$

The XY-coordinates are $\boxed{\left(-\dfrac{3}{2} + 2\sqrt{3}, \dfrac{3\sqrt{3}}{2} + 2\right)}.$

■ **Answer:**
$$\left(\frac{3\sqrt{3}}{2} - 2, -\frac{3}{2} - 2\sqrt{3}\right)$$

■ **YOUR TURN:** If the xy-coordinate axes are rotated 30°, find the XY-coordinates of the point $(x, y) = (3, -4)$.

EXAMPLE 2 Rotating an Ellipse

Show that the graph of the equation $5x^2 - 8xy + 5y^2 - 9 = 0$ is an ellipse aligning with coordinate axes that are rotated by 45°.

Solution:

Start with the rotation formulas.

$$x = X\cos\theta - Y\sin\theta$$
$$y = X\sin\theta + Y\cos\theta$$

Let $\theta = 45°.$

$$x = X\underbrace{\cos(45°)}_{\frac{\sqrt{2}}{2}} - Y\underbrace{\sin(45°)}_{\frac{\sqrt{2}}{2}}$$

$$y = X\underbrace{\sin(45°)}_{\frac{\sqrt{2}}{2}} + Y\underbrace{\cos(45°)}_{\frac{\sqrt{2}}{2}}$$

Simplify.

$$x = \frac{\sqrt{2}}{2}(X - Y)$$

$$y = \frac{\sqrt{2}}{2}(X + Y)$$

Substitute $x = \frac{\sqrt{2}}{2}(X - Y)$ and $y = \frac{\sqrt{2}}{2}(X + Y)$ into $5x^2 - 8xy + 5y^2 - 9 = 0$.

$$5\left[\frac{\sqrt{2}}{2}(X - Y)\right]^2 - 8\left[\frac{\sqrt{2}}{2}(X - Y)\right]\left[\frac{\sqrt{2}}{2}(X + Y)\right] + 5\left[\frac{\sqrt{2}}{2}(X + Y)\right]^2 - 9 = 0$$

Simplify.

$$\frac{5}{2}(X^2 - 2XY + Y^2) - 4(X^2 - Y^2) + \frac{5}{2}(X^2 + 2XY + Y^2) - 9 = 0$$

$$\frac{5}{2}X^2 - 5XY + \frac{5}{2}Y^2 - 4X^2 + 4Y^2 + \frac{5}{2}X^2 + 5XY + \frac{5}{2}Y^2 = 9$$

Combine like terms.

$$X^2 + 9Y^2 = 9$$

Divide by 9.

$$\boxed{\frac{X^2}{9} + \frac{Y^2}{1} = 1}$$

This (as discussed earlier) is an ellipse whose major axis is along the X-axis.

The vertices are at the points $(X, Y) = (\pm 3, 0)$.

The Angle of Rotation Necessary to Transform a General Second-Degree Equation into an Equation of a Conic

In Section 11.1 we stated that the general second-degree equation

$$Ax^2 + Bxy + Cy^2 + Dx + Ey + F = 0$$

corresponds to a graph of a conic. Which type of conic it is depends on the value of the discriminant, $B^2 - 4AC$. In Sections 11.2–11.4 we discussed graphs of parabolas, ellipses, and hyperbolas with vertices along either the axes or lines parallel (or perpendicular) to the axes. In all cases the value of B was taken to be zero. When the value of B is nonzero, the result is a conic with vertices along the new XY-axes (or, respectively, parallel and perpendicular to them), which are the original xy-axes rotated through an angle θ. If given θ, we can determine the rotation equations as illustrated in Example 2, but how do we find the angle θ that represents the *angle of rotation*?

To find the angle of rotation, let us start with a general second-degree polynomial equation:

$$Ax^2 + Bxy + Cy^2 + Dx + Ey + F = 0$$

We want to transform this equation into an equation in X and Y that does not contain an XY-term. Suppose we rotate our coordinates by an angle θ and then use the rotation equations

$$x = X\cos\theta - Y\sin\theta \qquad y = X\sin\theta + Y\cos\theta$$

in the general second-degree polynomial equation; then the result is

$$A(X\cos\theta - Y\sin\theta)^2 + B(X\cos\theta - Y\sin\theta)(X\sin\theta + Y\cos\theta)$$
$$+ C(X\cos\theta + Y\sin\theta)^2 + D(X\cos\theta - Y\sin\theta) + E(X\sin\theta + Y\cos\theta) + F = 0$$

If we expand these expressions and collect like terms, the result is an equation of the form

$$aX^2 + bXY + cY^2 + dX + eY + f = 0$$

where

$$a = A\cos^2\theta + B\sin\theta\cos\theta + C\sin^2\theta$$

$$b = B(\cos^2\theta - \sin^2\theta) + 2(C - A)\sin\theta\cos\theta$$

$$c = A\sin^2\theta - B\sin\theta\cos\theta + C\cos^2\theta$$

$$d = D\cos\theta + E\sin\theta$$

$$e = -D\sin\theta + E\cos\theta$$

$$f = F$$

WORDS	**MATH**
We do not want this new equation to have an XY-term, so we set $b = 0$.	$B(\cos^2\theta - \sin^2\theta) + 2(C - A)\sin\theta\cos\theta = 0$
We can use the double-angle formulas to simplify.	$B\underbrace{(\cos^2\theta - \sin^2\theta)}_{\cos 2\theta} + (C - A)\underbrace{2\sin\theta\cos\theta}_{\sin 2\theta} = 0$
Subtract the sine term.	$B\cos 2\theta = (A - C)\sin 2\theta$
Divide by $B\sin 2\theta$.	$\dfrac{B\cos 2\theta}{B\sin 2\theta} = \dfrac{(A - C)\sin 2\theta}{B\sin 2\theta}$
Simplify.	$\boxed{\cot 2\theta = \dfrac{A - C}{B}}$

ANGLE OF ROTATION FORMULA

To transform the equation of a conic

$$Ax^2 + Bxy + Cy^2 + Dx + Ey + F = 0$$

into an equation in X and Y without an XY-term, rotate the xy-axes by an acute angle θ that satisfies the equation

$$\cot 2\theta = \frac{A - C}{B}$$

Notice that the trigonometric equation $\cot 2\theta = \dfrac{A - C}{B}$ can be solved exactly for some values of θ (Example 3) and will have to be approximated with a calculator for other values of θ (Example 4).

EXAMPLE 3 Determining the Angle of Rotation I: The Value of the Cotangent Function Is That of a Known (Special) Angle

Determine the angle of rotation necessary to transform the following equation into an equation in X and Y with no XY-term.

$$3x^2 + 2\sqrt{3}xy + y^2 + 2x - 2\sqrt{3}y = 0$$

Solution:

Identify the A, B, and C parameters in the equation.

$$\underset{A}{\underline{3x^2}} + \underset{B}{\underline{2\sqrt{3}}}\,xy + \underset{C}{\underline{1y^2}} + 2x - 2\sqrt{3}y = 0$$

Write the rotation formula.

$$\cot 2\theta = \frac{A - C}{B}$$

Let $A = 3$, $B = 2\sqrt{3}$, and $C = 1$.

$$\cot 2\theta = \frac{3 - 1}{2\sqrt{3}}$$

Simplify.

$$\cot 2\theta = \frac{1}{\sqrt{3}}$$

Apply the reciprocal identity.

$$\tan 2\theta = \sqrt{3}$$

From our knowledge of trigonometric exact values, we know that $2\theta = 60°$ or $\boxed{\theta = 30°}$.

EXAMPLE 4 Determining the Angle of Rotation II: The Argument of the Cotangent Function Needs to Be Approximated with a Calculator

Determine the angle of rotation necessary to transform the following equation into an equation in X and Y with no XY-term. Round to the nearest tenth of a degree.

$$4x^2 + 2xy - 6y^2 - 5x + y - 2 = 0$$

Solution:

Identify the A, B, and C parameters in the equation.

$$\underset{A}{\underline{4x^2}} + \underset{B}{\underline{2xy}} \underset{C}{\underline{- 6y^2}} - 5x + y - 2 = 0$$

Write the rotation formula.

$$\cot 2\theta = \frac{A - C}{B}$$

Let $A = 4$, $B = 2$, and $C = -6$.

$$\cot 2\theta = \frac{4 - (-6)}{2}$$

Simplify.

$$\cot 2\theta = 5$$

Apply the reciprocal identity.

$$\tan 2\theta = \frac{1}{5} = 0.2$$

Write the result as an inverse tangent function.

$$2\theta = \tan^{-1}0.2$$

With a calculator evaluate the right side of the equation.

$$2\theta \approx 11.31°$$

Solve for θ and round to the nearest tenth of a degree.

$$\boxed{\theta = 5.7°}$$

Classroom Example 11.7.3*
Assume $a > 0$. Determine the angle of rotation needed to transform $ax^2 + a\sqrt{5}xy + \frac{1}{2}ay^2 + \frac{1}{3}ax - ay = 0$ into an XY-equation with no XY-term.

Answer:

$$\theta = \frac{1}{2}\cot^{-1}\left(\frac{1}{2\sqrt{5}}\right)$$

Technology Tip

To find the angle of rotation, use

$$\theta = \frac{1}{2}\tan^{-1}\left(\frac{B}{A - C}\right)$$ and set the

calculator to degree mode.
Substitute $A = 1$, $B = 2$, and
$C = 1$.

The calculator displays an error
message, which means that
$\cos 2\theta = 0$. Therefore, $2\theta = 90°$
or $\theta = 45°$.

To graph the equation $x^2 + 2xy + y^2 - \sqrt{2}x - 3\sqrt{2}y + 6 = 0$
with a TI-83 or TI-83 Plus calculator,
you need to solve for y using the
quadratic formula.

$$y^2 + y(2x - 3\sqrt{2}) +$$
$$(x^2 - \sqrt{2}x + 6) = 0$$

$a = 1$, $b = 2x - 3\sqrt{2}$,

$c = x^2 - \sqrt{2}x + 6$

$$y = \frac{-b \pm \sqrt{b^2 - 4ac}}{2a}$$

$$y = \frac{-2x + 3\sqrt{2} \pm \sqrt{-8\sqrt{2}x - 6}}{2}$$

Now enter

$$y1 = \frac{-2x + 3\sqrt{2} + \sqrt{-8\sqrt{2}x - 6}}{2}$$

and

$$y2 = \frac{-2x + 3\sqrt{2} - \sqrt{-8\sqrt{2}x - 6}}{2}.$$

To graph the X- and Y-axes, enter
$y3 = \tan(45)x$ and

$$y4 = -\frac{1}{\tan(45)}x.$$

Special attention must be given when evaluating the inverse tangent function on a calculator, as the result is always in quadrant I or IV. If 2θ turns out to be negative, then $180°$ must be added so that 2θ is in quadrant II (as opposed to quadrant IV). Then θ will be an acute angle lying in quadrant I.

Recall that we stated (without proof) in Section 11.1 that we can identify a general equation of the form

$$Ax^2 + Bxy + Cy^2 + Dx + Ey + F = 0$$

as that of a particular conic depending on the discriminant.

Parabola	$B^2 - 4AC = 0$
Ellipse	$B^2 - 4AC < 0$
Hyperbola	$B^2 - 4AC > 0$

EXAMPLE 5   Graphing a Rotated Conic

For the equation $x^2 + 2xy + y^2 - \sqrt{2}x - 3\sqrt{2}y + 6 = 0$:

a. Determine which conic the equation represents.

b. Find the rotation angle required to eliminate the XY-term in the new coordinate system.

c. Transform the equation in x and y into an equation in X and Y.

d. Graph the resulting conic.

Solution (a):

Identify A, B, and C.

$$\underset{A}{1x^2} + \underset{B}{2xy} + \underset{C}{1y^2} - \sqrt{2}x - 3\sqrt{2}y + 6 = 0$$

$$A = 1, B = 2, C = 1$$

Compute the discriminant.

$$B^2 - 4AC = 2^2 - 4(1)(1) = 0$$

Since the discriminant equals zero, the equation represents a **parabola**.

Solution (b):

Write the rotation formula.

$$\cot 2\theta = \frac{A - C}{B}$$

Let $A = 1$, $B = 2$, and $C = 1$.

$$\cot 2\theta = \frac{1 - 1}{2}$$

Simplify.

$$\cot 2\theta = 0$$

Write the cotangent function in terms of the sine and cosine functions.

$$\frac{\cos 2\theta}{\sin 2\theta} = 0$$

The numerator must equal zero.

$$\cos 2\theta = 0$$

From our knowledge of trigonometric exact values, we know that $2\theta = 90°$ or $\boxed{\theta = 45°}$.

Solution (c):

Start with the equation
$x^2 + 2xy + y^2 - \sqrt{2}x - 3\sqrt{2}y + 6 = 0$,
and use the rotation formulas
with $\theta = 45°$.

$$x = X\cos(45°) - Y\sin(45°) = \frac{\sqrt{2}}{2}(X - Y)$$

$$y = X\sin(45°) + Y\cos(45°) = \frac{\sqrt{2}}{2}(X + Y)$$

Find x^2, xy, and y^2.

$$x^2 = \left[\frac{\sqrt{2}}{2}(X - Y)\right]^2 = \frac{1}{2}(X^2 - 2XY + Y^2)$$

$$xy = \left[\frac{\sqrt{2}}{2}(X - Y)\right]\left[\frac{\sqrt{2}}{2}(X + Y)\right] = \frac{1}{2}(X^2 - Y^2)$$

$$y^2 = \left[\frac{\sqrt{2}}{2}(X + Y)\right]^2 = \frac{1}{2}(X^2 + 2XY + Y^2)$$

Substitute the values
for x, y, x^2, xy,
and y^2 into the
original equation.

$$x^2 + 2xy + y^2 - \sqrt{2}x - 3\sqrt{2}y + 6 = 0$$

$$\frac{1}{2}(X^2 - 2XY + Y^2) + 2\frac{1}{2}(X^2 - Y^2)$$

$$+ \frac{1}{2}(X^2 + 2XY + Y^2) - \sqrt{2}\left[\frac{\sqrt{2}}{2}(X - Y)\right]$$

$$- 3\sqrt{2}\left[\frac{\sqrt{2}}{2}(X + Y)\right] + 6 = 0$$

Eliminate the parentheses
and combine like terms.

$$2X^2 - 4X - 2Y + 6 = 0$$

Divide by 2.

$$X^2 - 2X - Y + 3 = 0$$

Add Y.

$$Y = (X^2 - 2X) + 3$$

Complete the square on X.

$$Y = (X - 1)^2 + 2$$

Solution (d):

This is a parabola opening upward in the
XY-coordinate system shifted to the
right 1 unit and up 2 units.

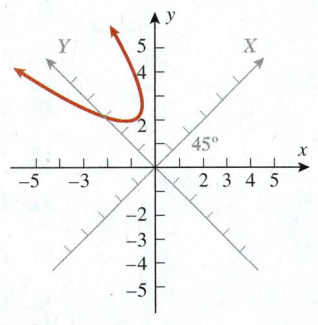

SECTION
11.7 SUMMARY

In this section we found that the graph of the general second-degree equation

$$Ax^2 + Bxy + Cy^2 + Dx + Ey + F = 0$$

can represent conics in a system of rotated axes.

The following are the rotation formulas relating the xy-coordinate system to a rotated coordinate system with axes X and Y

$$x = X\cos\theta - Y\sin\theta$$

$$y = X\sin\theta + Y\cos\theta$$

where the rotation angle θ is found from the equation

$$\cot 2\theta = \frac{A - C}{B}.$$

■ SKILLS

In Exercises 1–8, the coordinates of a point in the xy-coordinate system are given. Assuming that the XY-axes are found by rotating the xy-axes by an angle θ, find the corresponding coordinates for the point in the XY-system.

1. $(2, 4), \theta = 45°$ **2.** $(5, 1), \theta = 60°$ **3.** $(-3, 2), \theta = 30°$ **4.** $(-4, 6), \theta = 45°$

5. $(-1, -3), \theta = 60°$ **6.** $(4, -4), \theta = 45°$ **7.** $(0, 3), \theta = 60°$ **8.** $(-2, 0), \theta = 30°$

In Exercises 9–24, (a) identify the type of conic from the discriminant, (b) transform the equation in x and y into an equation in X and Y (without an XY-term) by rotating the x- and y-axes by an angle θ to arrive at the new X- and Y-axes, and (c) graph the resulting equation (showing both sets of axes).

9. $xy - 1 = 0, \theta = 45°$ **10.** $xy - 4 = 0, \theta = 45°$

11. $x^2 + 2xy + y^2 + \sqrt{2}x - \sqrt{2}y - 1 = 0, \theta = 45°$ **12.** $2x^2 - 4xy + 2y^2 - \sqrt{2}x + 1 = 0, \theta = 45°$

13. $y^2 - \sqrt{3}xy + 3 = 0, \theta = 30°$ **14.** $x^2 - \sqrt{3}xy - 3 = 0, \theta = 60°$

15. $7x^2 - 2\sqrt{3}xy + 5y^2 - 8 = 0, \theta = 60°$ **16.** $4x^2 + \sqrt{3}xy + 3y^2 - 45 = 0, \theta = 30°$

17. $3x^2 + 2\sqrt{3}xy + y^2 + 2x - 2\sqrt{3}y - 2 = 0, \theta = 30°$ **18.** $x^2 + 2\sqrt{3}xy + 3y^2 - 2\sqrt{3}x + 2y - 4 = 0, \theta = 60°$

19. $7x^2 + 4\sqrt{3}xy + 3y^2 - 9 = 0, \theta = \dfrac{\pi}{6}$ **20.** $37x^2 + 42\sqrt{3}xy + 79y^2 - 400 = 0, \theta = \dfrac{\pi}{3}$

21. $7x^2 - 10\sqrt{3}xy - 3y^2 + 24 = 0, \theta = \dfrac{\pi}{3}$ **22.** $9x^2 + 14\sqrt{3}xy - 5y^2 + 48 = 0, \theta = \dfrac{\pi}{6}$

23. $x^2 - 2xy + y^2 - \sqrt{2}x - \sqrt{2}y - 8 = 0, \theta = \dfrac{\pi}{4}$ **24.** $x^2 + 2xy + y^2 + 3\sqrt{2}x + \sqrt{2}y = 0, \theta = \dfrac{\pi}{4}$

In Exercises 25–38, determine the angle of rotation necessary to transform the equation in x and y into an equation in X and Y with no XY-term.

25. $x^2 + 4xy + y^2 - 4 = 0$ **26.** $3x^2 + 5xy + 3y^2 - 2 = 0$

27. $2x^2 + \sqrt{3}xy + 3y^2 - 1 = 0$ **28.** $4x^2 + \sqrt{3}xy + 3y^2 - 1 = 0$

29. $2x^2 + \sqrt{3}xy + y^2 - 5 = 0$ **30.** $2\sqrt{3}x^2 + xy + 3\sqrt{3}y^2 + 1 = 0$

31. $\sqrt{2}x^2 + xy + \sqrt{2}y^2 - 1 = 0$ **32.** $x^2 + 10xy + y^2 + 2 = 0$

33. $12\sqrt{3}x^2 + 4xy + 8\sqrt{3}y^2 - 1 = 0$ **34.** $4x^2 + 2xy + 2y^2 - 7 = 0$

35. $5x^2 + 6xy + 4y^2 - 1 = 0$ **36.** $x^2 + 2xy + 12y^2 + 3 = 0$

37. $3x^2 + 10xy + 5y^2 - 1 = 0$ **38.** $10x^2 + 3xy + 2y^2 + 3 = 0$

In Exercises 39–48, graph the second-degree equation. (*Hint:* Transform the equation into an equation that contains no xy-term.)

39. $21x^2 + 10\sqrt{3}xy + 31y^2 - 144 = 0$ **40.** $5x^2 + 6xy + 5y^2 - 8 = 0$

41. $8x^2 - 20xy + 8y^2 + 18 = 0$ **42.** $3y^2 - 26\sqrt{3}xy - 23x^2 - 144 = 0$

43. $3x^2 + 2\sqrt{3}xy + y^2 + 2x - 2\sqrt{3}y - 12 = 0$ **44.** $3x^2 - 2\sqrt{3}xy + y^2 - 2x - 2\sqrt{3}y - 4 = 0$

45. $37x^2 - 42\sqrt{3}xy + 79y^2 - 400 = 0$ **46.** $71x^2 - 58\sqrt{3}xy + 13y^2 + 400 = 0$

47. $x^2 + 2xy + y^2 + 5\sqrt{2}x + 3\sqrt{2}y = 0$ **48.** $7x^2 - 4\sqrt{3}xy + 3y^2 - 9 = 0$

▪ CONCEPTUAL

In Exercises 49–52, determine whether each statement is true or false.

49. The graph of the equation $x^2 + kxy + 9y^2 = 5$, where k is any positive constant less than 6, is an ellipse.

50. The graph of the equation $x^2 + kxy + 9y^2 = 5$, where k is any constant greater than 6, is a parabola.

51. The reciprocal function is a rotated hyperbola.

52. The equation $\sqrt{x} + \sqrt{y} = 3$ can be transformed into the equation $X^2 + Y^2 = 9$.

▪ CHALLENGE

53. Determine the equation in X and Y that corresponds to $\dfrac{x^2}{a^2} + \dfrac{y^2}{b^2} = 1$ when the axes are rotated through

a. 90°   **b.** 180°

54. Determine the equation in X and Y that corresponds to $\dfrac{x^2}{a^2} - \dfrac{y^2}{b^2} = 1$ when the axes are rotated through

a. 90°   **b.** 180°

55. Identify the conic section with equation $y^2 + ax^2 = x$ for various values of a.

56. Identify the conic section with equation $x^2 - ay^2 = y$ for various values of a.

▪ TECHNOLOGY

For Exercises 57–62, refer to the following:

To use a TI-83 or TI-83 Plus (function-driven software or graphing utility) to graph a general second-degree equation, you need to solve for y. Let us consider a general second-degree equation $Ax^2 + Bxy + Cy^2 + Dx + Ey + F = 0$.

Group y^2 terms together, y terms together, and the remaining terms together.

$$Ax^2 + \underline{Bxy} + \underline{Cy^2} + Dx + \underline{\underline{Ey}} + F = 0$$

$$Cy^2 + (Bxy + Ey) + (Ax^2 + Dx + F) = 0$$

Factor out the common y in the first set of parentheses.

$$Cy^2 + y(Bx + E) + (Ax^2 + Dx + F) = 0$$

Now this is a quadratic equation in y: $ay^2 + by + c = 0$.

Use the quadratic formula to solve for y.

$$Cy^2 + y(Bx + E) + (Ax^2 + Dx + F) = 0$$

$$a = C, b = Bx + E, c = Ax^2 + Dx + F$$

$$y = \frac{-b \pm \sqrt{b^2 - 4ac}}{2a} \qquad y = \frac{-(Bx + E) \pm \sqrt{(Bx + E)^2 - 4(C)(Ax^2 + Dx + F)}}{2(C)}$$

$$y = \frac{-(Bx + E) \pm \sqrt{B^2x^2 + 2BEx + E^2 - 4ACx^2 - 4CDx - 4CF}}{2C}$$

$$y = \frac{-(Bx + E) \pm \sqrt{(B^2 - 4AC)x^2 + (2BE - 4CD)x + (E^2 - 4CF)}}{2C}$$

Case I: $B^2 - 4AC = 0 \rightarrow$ The second-degree equation $Ax^2 + Bxy + Cy^2 + Dx + Ey + F = 0$ is a parabola.

$$y = \frac{-(Bx + E) \pm \sqrt{(2BE - 4CD)x + (E^2 - 4CF)}}{2C}$$

Case II: $B^2 - 4AC < 0 \rightarrow$ The second-degree equation $Ax^2 + Bxy + Cy^2 + Dx + Ey + F = 0$ is an ellipse.

$$y = \frac{-(Bx + E) \pm \sqrt{(B^2 - 4AC)x^2 + (2BE - 4CD)x + (E^2 - 4CF)}}{2C}$$

Case III: $B^2 - 4AC > 0 \rightarrow$ The second-degree equation $Ax^2 + Bxy + Cy^2 + Dx + Ey + F = 0$ is a hyperbola.

$$y = \frac{-(Bx + E) \pm \sqrt{(B^2 - 4AC)x^2 + (2BE - 4CD)x + (E^2 - 4CF)}}{2C}$$

57. Use a graphing utility to explore the second-degree equation $3x^2 + 2\sqrt{3}xy + y^2 + Dx + Ey + F = 0$ for the following values of D, E, and F:

a. $D = 1, E = 3, F = 2$

b. $D = -1, E = -3, F = 2$

Show the angle of rotation to the nearest degree. Explain the differences.

58. Use a graphing utility to explore the second-degree equation $x^2 + 3xy + 3y^2 + Dx + Ey + F = 0$ for the following values of D, E, and F:

a. $D = 2, E = 6, F = -1$

b. $D = 6, E = 2, F = -1$

Show the angle of rotation to the nearest degree. Explain the differences.

59. Use a graphing utility to explore the second-degree equation $2x^2 + 3xy + y^2 + Dx + Ey + F = 0$ for the following values of D, E, and F:

a. $D = 2, E = 1, F = -2$

b. $D = 2, E = 1, F = 2$

Show the angle of rotation to the nearest degree. Explain the differences.

60. Use a graphing utility to explore the second-degree equation $2\sqrt{3}x^2 + xy + \sqrt{3}y^2 + Dx + Ey + F = 0$ for the following values of D, E, and F:

a. $D = 2, E = 1, F = -1$

b. $D = 2, E = 6, F = -1$

Show the angle of rotation to the nearest degree. Explain the differences.

61. Use a graphing utility to explore the second-degree equation $Ax^2 + Bxy + Cy^2 + 2x + y - 1 = 0$ for the following values of A, B, and C:

a. $A = 4, B = -4, C = 1$

b. $A = 4, B = 4, C = -1$

c. $A = 1, B = -4, C = 4$

Show the angle of rotation to the nearest degree. Explain the differences.

62. Use a graphing utility to explore the second-degree equation $Ax^2 + Bxy + Cy^2 + 3x + 5y - 2 = 0$ for the following values of A, B, and C:

a. $A = 1, B = -4, C = 4$

b. $A = 1, B = 4, D = -4$

Show the angle of rotation to the nearest degree. Explain the differences.

SKILLS OBJECTIVES

- Define conics in terms of eccentricity.
- Express equations of conics in polar form.
- Graph the polar equations of conics.

CONCEPTUAL OBJECTIVE

- Define all conics in terms of a focus and a directrix.

Equations of Conics in Polar Coordinates

In Section 11.1 we discussed parabolas, ellipses, and hyperbolas in terms of geometric definitions. Then in Sections 11.2–11.4 we examined the rectangular equations of these conics. The equations for the conics when their centers are at the origin are simpler than when they are not (when conics are shifted). In Section 8.8 we discussed polar coordinates and graphing of polar equations. In this section we develop a more unified definition of the three conics in terms of a single focus and a directrix. You will see in this section that if the *focus* is located at the origin, then equations of conics are simpler when written in polar coordinates.

Alternative Definition of Conics

Recall that when we work with rectangular coordinates we define a parabola (Sections 11.1 and 11.2) in terms of a fixed point (focus) and a line (directrix), whereas we define an ellipse and hyperbola (Sections 11.1, 11.3, and 11.4) in terms of two fixed points (the foci). However, it is possible to define all three conics in terms of a single focus and a directrix.

The following alternative representation of conics depends on a parameter called *eccentricity*.

ALTERNATIVE DESCRIPTION OF CONICS

Let D be a fixed line (the **directrix**), F be a fixed point (a **focus**) not on D, and e be a fixed positive number (**eccentricity**). The set of all points P such that the ratio of the distance from P to F to the distance from P to D equals the constant e defines a conic section.

$$\frac{d(P, F)}{d(P, D)} = e$$

- If $e = 1$, the conic is a **parabola**.
- If $e < 1$, the conic is an **ellipse**.
- If $e > 1$, the conic is a **hyperbola**.

When $e = 1$, the result is a parabola, described by the same definition we used previously in Section 11.1. When $e \neq 1$, the result is either an ellipse or a hyperbola. The major axis of an ellipse passes through the focus and is perpendicular to the directrix. The transverse axis of a hyperbola also passes through the focus and is perpendicular to the directrix. If we let c represent the distance from the focus to the center and a represent the distance from the vertex to the center, then eccentricity is given by

$$e = \frac{c}{a}$$

In polar coordinates, if we locate the focus of a conic at the pole and the directrix is either perpendicular or parallel to the polar axis, then we have four possible scenarios:

- The directrix is *perpendicular* to the polar axis and p units to the *right* of the pole.
- The directrix is *perpendicular* to the polar axis and p units to the *left* of the pole.
- The directrix is *parallel* to the polar axis and p units *above* the pole.
- The directrix is *parallel* to the polar axis and p units *below* the pole.

Let us take the case in which the directrix is perpendicular to the polar axis and p units to the right of the pole.

In polar coordinates (r, θ), we see that the distance from the focus to a point P is equal to r, that is, $d(P, F) = r$, and the distance from P to the closest point on the directrix is $d(P, D) = p - r\cos\theta$.

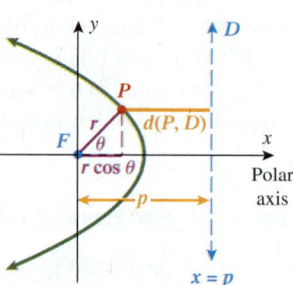

WORDS	MATH
Substitute $d(P, F) = r$ and $d(P, D) = p - r\cos\theta$ into the formula for eccentricity, $\dfrac{d(P, F)}{d(P, D)} = e$.	$\dfrac{r}{p - r\cos\theta} = e$
Multiply the result by $p - r\cos\theta$.	$r = e(p - r\cos\theta)$
Eliminate the parentheses.	$r = ep - er\cos\theta$
Add $er\cos\theta$.	$r + er\cos\theta = ep$
Factor out the common r.	$r(1 + e\cos\theta) = ep$
Divide by $1 + e\cos\theta$.	$\boxed{r = \dfrac{ep}{1 + e\cos\theta}}$

We need not derive the other three cases here, but note that if the directrix is perpendicular to the polar axis and p units to the *left* of the pole, the resulting polar equation is

$$r = \frac{ep}{1 - e\cos\theta}$$

If the directrix is parallel to the polar axis, the directrix is either above ($y = p$) or below ($y = -p$) the polar axis and we get the sine function instead of the cosine function, as summarized in the following box:

POLAR EQUATIONS OF CONICS

The following polar equations represent conics with one focus at the origin and with eccentricity e. It is assumed that the positive x-axis represents the polar axis.

EQUATION	DESCRIPTION
$r = \dfrac{ep}{1 + e\cos\theta}$	The directrix is *vertical* and p units to the *right* of the pole.
$r = \dfrac{ep}{1 - e\cos\theta}$	The directrix is *vertical* and p units to the *left* of the pole.
$r = \dfrac{ep}{1 + e\sin\theta}$	The directrix is *horizontal* and p units *above* the pole.
$r = \dfrac{ep}{1 - e\sin\theta}$	The directrix is *horizontal* and p units *below* the pole.

ECCENTRICITY	THE CONIC IS A _____	THE ____ IS PERPENDICULAR TO THE DIRECTRIX
$e = 1$	Parabola	Axis of symmetry
$e < 1$	Ellipse	Major axis
$e > 1$	Hyperbola	Transverse axis

EXAMPLE 1 Finding the Polar Equation of a Conic

Find a polar equation for a parabola that has its focus at the origin and whose directrix is the line $y = 3$.

Solution:

The directrix is horizontal and above the pole.

$$r = \frac{ep}{1 + e\sin\theta}$$

A parabola has eccentricity $e = 1$ and $p = 3$.

$$r = \frac{3}{1 + \sin\theta}$$

■ **YOUR TURN** Find a polar equation for a parabola that has its focus at the origin and whose directrix is the line $x = -3$.

Classroom Example 11.8.1
Find the polar equation for a parabola with focus at the origin and whose directrix is $y = -a$ (where $a > 0$).

Answer: $r = \dfrac{a}{1 - \sin\theta}$

■ **Answer:** $r = \dfrac{3}{1 - 6\cos\theta}$

EXAMPLE 2   Identifying a Conic from Its Equation

Identify the type of conic represented by the equation $\dfrac{10}{3 + 2\cos\theta}$.

Solution:

To identify the type of conic, we need to rewrite the equation in the form:

$$r = \frac{ep}{1 \pm e\cos\theta}$$

Divide the numerator and denominator by 3.

$$\dfrac{\dfrac{10}{3}}{\left(1 + \dfrac{2}{3}\cos\theta\right)}$$

Identify e in the denominator.

$$= \dfrac{\dfrac{10}{3}}{\left(1 + \underset{e}{\dfrac{2}{3}}\cos\theta\right)}$$

The numerator is equal to ep.

$$= \dfrac{\overset{p}{5}\cdot\overset{e}{\dfrac{2}{3}}}{\left(1 + \underset{e}{\dfrac{2}{3}}\cos\theta\right)}$$

Since $e = \dfrac{2}{3} < 1$, the conic is an $\boxed{\text{ellipse}}$. The directrix is $x = 5$, so the major axis is along the x-axis (perpendicular to the directrix).

■ **YOUR TURN** Identify the type of conic represented by the equation

$$r = \frac{10}{2 - 10\sin\theta}$$

In Example 2 we found that the polar equation $r = \dfrac{10}{3 + 2\cos\theta}$ is an ellipse with major axis along the x-axis. We will graph this ellipse in Example 3.

EXAMPLE 3   Graphing a Conic from Its Equation

The graph of the polar equation $r = \dfrac{10}{3 + 2\cos\theta}$ is an ellipse.

a. Find the vertices.
b. Find the center of the ellipse.
c. Find the lengths of the major and minor axes.
d. Graph the ellipse.

Solution (a):

From Example 2 we see that $e = \frac{2}{3}$, which corresponds to an ellipse, and $x = 5$ is the directrix.

The major axis is perpendicular to the directrix. Therefore, the major axis lies along the polar axis. To find the vertices (which lie along the major axis), let $\theta = 0$ and $\theta = \pi$.

$\theta = 0$:
$$r = \frac{10}{3 + 2\cos\theta} = \frac{10}{5} = 2$$

$\theta = \pi$:
$$r = \frac{10}{3 + 2\cos\pi} = \frac{10}{1} = 10$$

The vertices are the points $\boxed{V_1 = (2, 0)}$ and $\boxed{V_2 = (10, \pi)}$.

Solution (b):

The vertices in rectangular coordinates are $V_1 = (2, 0)$ and $V_2 = (-10, 0)$.

The midpoint (in rectangular coordinates) between the two vertices is the point $(-4, 0)$, which corresponds to the point $\boxed{(4, \pi)}$ in polar coordinates.

Solution (c):

The length of the major axis, $2a$, is the distance between the vertices.

$$\boxed{2a = 12}$$

The length $a = 6$ corresponds to the distance from the center to a vertex.

Apply the formula $e = \dfrac{c}{a}$ with $a = 6$ and $e = \frac{2}{3}$ to find c.

$$c = ae = 6\frac{2}{3} = 4$$

Let $a = 6$ and $c = 4$ in $b^2 = a^2 - c^2$.

$$b^2 = 6^2 - 4^2 = 20$$

Solve for b.

$$b = 2\sqrt{5}$$

The length of the minor axis is $\boxed{2b = 4\sqrt{5}}$.

Solution (d):

Graph the ellipse.

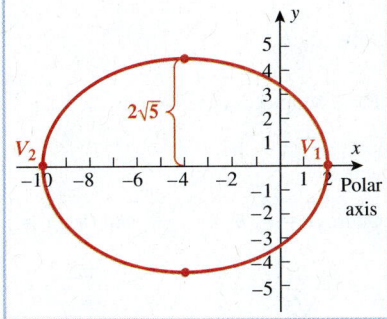

Classroom Example 11.8.3*

The graph of $r = \dfrac{a}{\sqrt{5} - \cos\theta}$ is an ellipse.

a. Find the vertices.
b. Find the center.

Answer:

a. $\left(\dfrac{a}{\sqrt{5} - 1}, 0\right), \left(\dfrac{a}{\sqrt{5} + 1}, \pi\right)$

b. $\left[\dfrac{1}{2}\left(\dfrac{a}{\sqrt{5} - 1} + \dfrac{a}{\sqrt{5} + 1}\right), 0\right]$.

Technology Tip

Use $\boxed{Y =}$ to enter the polar equation $r = \dfrac{10}{3 + 2\cos\theta}$.

$\boxed{r1 =} \boxed{10} \boxed{\div} \boxed{(} \boxed{3} \boxed{+} \boxed{2}$
$\boxed{\cos} \boxed{X, T, \theta, n} \boxed{)} \boxed{)}$

Use the $\boxed{TRACE}$ key to trace the vertices of the ellipse.

Technology Tip

Use $\boxed{Y=}$ to enter the polar

equation $r = \dfrac{2}{2 + 3\sin\theta}$.

$\boxed{\text{r1}} = \boxed{2}\,\boxed{\div}\,\boxed{(}\,\boxed{2}\,\boxed{+}\,\boxed{3}$
$\boxed{\sin}\,\boxed{X, T, \theta, n}\,\boxed{)}\,\boxed{)}$

Use the $\boxed{\text{TRACE}}$ key to trace the vertices of the hyperbola.

Classroom Example 11.8.4*
Identify the type of conic, assuming $a > 0$:

$$r = \dfrac{a\sqrt{7}}{1 + \sqrt{7}\sin\theta}$$

Answer: hyperbola

EXAMPLE 4   Identifying and Graphing a Conic from Its Equation

Identify and graph the conic defined by the equation $r = \dfrac{2}{2 + 3\sin\theta}$.

Solution:

Rewrite the equation in the form $r = \dfrac{ep}{1 + e\sin\theta}$.

$$r = \dfrac{2}{2 + 3\sin\theta} = \dfrac{\overset{p}{\left(\dfrac{2}{3}\right)}\overset{e}{\left(\dfrac{3}{2}\right)}}{1 + \underset{e}{\left(\dfrac{3}{2}\right)}\sin\theta}$$

The conic is a *hyperbola* since $e = \dfrac{3}{2} > 1$.

The directrix is horizontal and $\dfrac{2}{3}$ unit above the pole (origin).

To find the vertices, let $\theta = \dfrac{\pi}{2}$ and $\theta = \dfrac{3\pi}{2}$.

$\theta = \dfrac{\pi}{2}$: $r = \dfrac{2}{2 + 3\sin\left(\dfrac{\pi}{2}\right)} = \dfrac{2}{5}$

$\theta = \dfrac{3\pi}{2}$: $r = \dfrac{2}{2 + 3\sin\left(\dfrac{3\pi}{2}\right)} = \dfrac{2}{-1} = -2$

The vertices in polar coordinates are $\left(\dfrac{2}{5}, \dfrac{\pi}{2}\right)$ and $\left(-2, \dfrac{3\pi}{2}\right)$.

The vertices in rectangular coordinates are $V_1 = \left(0, \dfrac{2}{5}\right)$ and $V_2 = (0, 2)$.

The center is the midpoint between the vertices: $\left(0, \dfrac{6}{5}\right)$.

The distance from the center to a focus is $c = \dfrac{6}{5}$.

Apply the formula $e = \dfrac{c}{a}$ with $c = \dfrac{6}{5}$ and $e = \dfrac{3}{2}$ to find a.

$$a = \dfrac{c}{e} = \dfrac{\dfrac{6}{5}}{\dfrac{3}{2}} = \dfrac{4}{5}$$

Let $a = \dfrac{4}{5}$ and $c = \dfrac{6}{5}$ in $b^2 = c^2 - a^2$.

$$b^2 = \left(\dfrac{6}{5}\right)^2 - \left(\dfrac{4}{5}\right)^2 = \dfrac{20}{25}$$

Solve for b.

$$b = \dfrac{2\sqrt{5}}{5}$$

The asymptotes are given by

$$y = \pm\dfrac{a}{b}(x - h) + k,$$

where $a = \dfrac{4}{5}$, $b = \dfrac{2\sqrt{5}}{5}$, and $(h, k) = \left(0, \dfrac{6}{5}\right)$.

$$y = \pm\dfrac{2}{\sqrt{5}}x + \dfrac{6}{5}$$

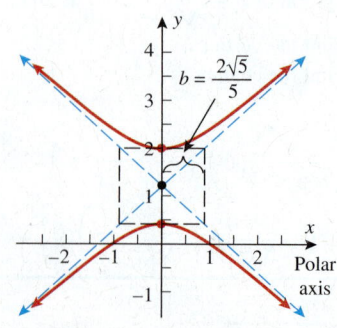

It is important to note that although we relate specific points (vertices, foci, etc.) to rectangular coordinates, another approach to finding a rough sketch is to simply point-plot the equation in polar coordinates.

EXAMPLE 5 Graphing a Conic by Point-Plotting in Polar Coordinates

Sketch a graph of the conic $r = \dfrac{4}{1 - \sin\theta}$.

Solution:

STEP 1 The conic is a parabola because the equation is in the form

$$r = \frac{(4)(1)}{1 - (1)\sin\theta}$$

Make a table with key values for θ and r.

θ	$r = \dfrac{4}{1 - \sin\theta}$	(r, θ)
0	$r = \dfrac{4}{1 - \sin 0} = \dfrac{4}{1} = 4$	$(4, 0)$
$\dfrac{\pi}{2}$	$r = \dfrac{4}{1 - \sin\dfrac{\pi}{2}} = \dfrac{4}{1 - 1} = \dfrac{4}{0}$	undefined
π	$r = \dfrac{4}{1 - \sin\pi} = \dfrac{4}{1} = 4$	$(4, \pi)$
$\dfrac{3\pi}{2}$	$r = \dfrac{4}{1 - \sin\dfrac{3\pi}{2}} = \dfrac{4}{1 - (-1)} = \dfrac{4}{2} = 2$	$\left(2, \dfrac{3\pi}{2}\right)$
2π	$r = \dfrac{4}{1 - \sin 2\pi} = \dfrac{4}{1} = 4$	$(4, 2\pi)$

STEP 2 Plot the points on a polar graph and connect them with a smooth parabolic curve.

Classroom Example 11.8.5

Graph $r = \dfrac{1}{2 - 3\cos\theta}$.

Answer:

SECTION
11.8 SUMMARY

In this section we found that we could graph polar equations of conics by identifying a single focus and the directrix. There are four possible equations in terms of eccentricity e:

EQUATION	DESCRIPTION
$r = \dfrac{ep}{1 + e\cos\theta}$	The directrix is *vertical* and p units to the *right* of the pole.
$r = \dfrac{ep}{1 - e\cos\theta}$	The directrix is *vertical* and p units to the *left* of the pole.
$r = \dfrac{ep}{1 + e\sin\theta}$	The directrix is *horizontal* and p units *above* the pole.
$r = \dfrac{ep}{1 - e\sin\theta}$	The directrix is *horizontal* and p units *below* the pole.

SECTION
11.8 EXERCISES

▪ SKILLS

In Exercises 1–14, find the polar equation that represents the conic described (assume that a focus is at the origin).

	Conic	Eccentricity	Directrix		Conic	Eccentricity	Directrix
1.	Ellipse	$e = \frac{1}{2}$	$y = -5$	**2.**	Ellipse	$e = \frac{1}{3}$	$y = 3$
3.	Hyperbola	$e = 2$	$y = 4$	**4.**	Hyperbola	$e = 3$	$y = -2$
5.	Parabola	$e = 1$	$x = 1$	**6.**	Parabola	$e = 1$	$x = -1$
7.	Ellipse	$e = \frac{3}{4}$	$x = 2$	**8.**	Ellipse	$e = \frac{2}{3}$	$x = -4$
9.	Hyperbola	$e = \frac{4}{3}$	$x = -3$	**10.**	Hyperbola	$e = \frac{3}{2}$	$x = 5$
11.	Parabola	$e = 1$	$y = -3$	**12.**	Parabola	$e = 1$	$y = 4$
13.	Ellipse	$e = \frac{3}{5}$	$y = 6$	**14.**	Hyperbola	$e = \frac{8}{5}$	$y = 5$

In Exercises 15–26, identify the conic (parabola, ellipse, or hyperbola) that each polar equation represents.

15. $r = \dfrac{4}{1 + \cos\theta}$

16. $r = \dfrac{3}{2 - 3\sin\theta}$

17. $r = \dfrac{2}{3 + 2\sin\theta}$

18. $r = \dfrac{3}{2 - 2\cos\theta}$

19. $r = \dfrac{2}{4 + 8\cos\theta}$

20. $r = \dfrac{1}{4 - \cos\theta}$

21. $r = \dfrac{7}{3 + \cos\theta}$

22. $r = \dfrac{4}{5 + 6\sin\theta}$

23. $r = \dfrac{40}{5 + 5\sin\theta}$

24. $r = \dfrac{5}{5 - 4\sin\theta}$

25. $r = \dfrac{1}{1 - 6\cos\theta}$

26. $r = \dfrac{5}{3 - 3\sin\theta}$

In Exercises 27–40, for the given polar equations: (a) identify the conic as either a parabola, ellipse, or hyperbola; (b) find the eccentricity and vertex (or vertices); and (c) graph.

27. $r = \dfrac{2}{1 + \sin\theta}$

28. $r = \dfrac{4}{1 - \cos\theta}$

29. $r = \dfrac{4}{1 - 2\sin\theta}$

30. $r = \dfrac{3}{3 + 8\cos\theta}$

31. $r = \dfrac{2}{2 + \sin\theta}$

32. $r = \dfrac{1}{3 - \sin\theta}$

33. $r = \dfrac{1}{2 - 2\sin\theta}$

34. $r = \dfrac{1}{1 - 2\sin\theta}$

35. $r = \dfrac{4}{3 + \cos\theta}$

36. $r = \dfrac{2}{5 + 4\sin\theta}$

37. $r = \dfrac{6}{2 + 3\sin\theta}$

38. $r = \dfrac{6}{1 + \cos\theta}$

39. $r = \dfrac{2}{5 + 5\cos\theta}$

40. $r = \dfrac{10}{6 - 3\cos\theta}$

■ **APPLICATIONS**

For Exercises 41 and 42, refer to the following:

Planets travel in elliptical orbits around a single focus, the Sun. Pluto (orange), the dwarf planet furthest from the Sun, has a pronounced elliptical orbit, whereas Earth (royal blue) has an almost circular orbit. The polar equation of a planet's orbit can be expressed as

$$r = \frac{a(1 - e^2)}{(1 - e\cos\theta)}$$

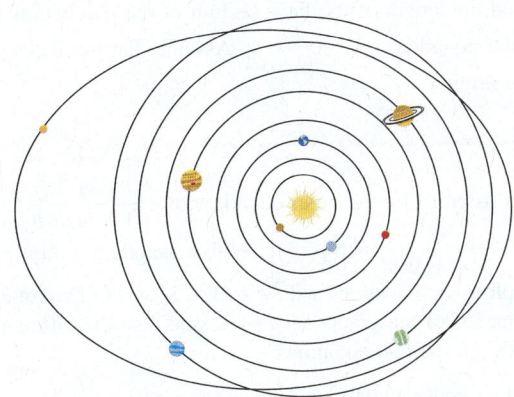

where e is the eccentricity and $2a$ is the length of the major axis. It can also be shown that the perihelion distance (minimum distance from the Sun to a planet) and the aphelion distance (maximum distance from the Sun to the planet) can be represented by $r = a(1 - e)$ and $r = a(1 + e)$, respectively.

41. Planetary Orbits. Pluto's orbit is summarized in the picture below. Find the eccentricity of Pluto's orbit. Find the polar equation that governs Pluto's orbit.

Pluto's Orbit

Perihelion
4,447,000,000 km

Aphelion
7,380,000,000 km

Pluto has a very elliptical orbit, which is highly inclined with respect to the plane of the ecliptic.

The Sun and Pluto are not to scale in this drawing.

42. Planetary Orbits. Earth's orbit is summarized in the picture below. Find the eccentricity of Earth's orbit. Find the polar equation that governs Earth's orbit.

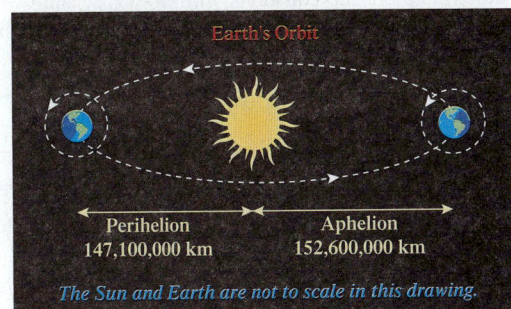

Earth's Orbit

Perihelion
147,100,000 km

Aphelion
152,600,000 km

The Sun and Earth are not to scale in this drawing.

For Exercises 43 and 44, refer to the following:

Asteroids, meteors, and comets all orbit the Sun in elliptical patterns and often cross paths with Earth's orbit, making life a little tense now and again. Asteroids are large rocks (bodies under 1000 km across), meteors range from sand particles to rocks, and comets are masses of debris. A few asteroids have orbits that cross the Earth's orbits—called Apollos or Earth-crossing asteroids. In recent years, asteroids have passed within 100,000 km of Earth!

43. Asteroids. The asteroid 433 or Eros is the second largest near-Earth asteroid. The semimajor axis of its orbit is 150 million km and the eccentricity is 0.223. Find the polar equation of Eros's orbit.

44. Asteroids. The asteroid Toutatis is the largest near-Earth asteroid. The semimajor axis of its orbit is 350 million km and the eccentricity is 0.634. On September 29, 2004, it missed Earth by 961,000 miles. Find the polar equation of Toutatis's orbit.

■ CONCEPTUAL

45. When $0 < e < 1$, the conic is an ellipse. Does the conic become more elongated or elliptical as e approaches 1 or as e approaches 0?

46. Show that $r = \dfrac{ep}{1 - e\sin\theta}$ is the polar equation of a conic with a horizontal directrix that is p units *below* the pole.

47. Convert from rectangular to polar coordinates to show that the equation of a hyperbola, $\dfrac{x^2}{a^2} - \dfrac{y^2}{b^2} = 1$, in polar form is $r^2 = -\dfrac{b^2}{1 - e^2\cos^2\theta}$.

48. Convert from rectangular to polar coordinates to show that the equation of an ellipse, $\dfrac{x^2}{a^2} + \dfrac{y^2}{b^2} = 1$, in polar form is $r^2 = \dfrac{b^2}{1 - e^2\cos^2\theta}$.

■ CHALLENGE

49. Find the major diameter of the ellipse with polar equation $r = \dfrac{ep}{1 + e\cos\theta}$ in terms of e and p.

50. Find the minor diameter of the ellipse with polar equation $r = \dfrac{ep}{1 + e\cos\theta}$ in terms of e and p.

51. Find the center of the ellipse with polar equation $r = \dfrac{ep}{1 + e\cos\theta}$ in terms of e and p.

52. Find the length of the latus rectum of the parabola with polar equation $r = \dfrac{p}{1 + \cos\theta}$. Assume that the focus is at the origin.

■ TECHNOLOGY

53. Let us consider the polar equations $r = \dfrac{ep}{1 + e\cos\theta}$ and $r = \dfrac{ep}{1 - e\cos\theta}$ with eccentricity $e = 1$. With a graphing utility, explore the equations with $p = 1, 2,$ and 6. Describe the behavior of the graphs as $p \to \infty$ and also the difference between the two equations.

54. Let us consider the polar equations $r = \dfrac{ep}{1 + e\sin\theta}$ and $r = \dfrac{ep}{1 - e\sin\theta}$ with eccentricity $e = 1$. With a graphing utility, explore the equations with $p = 1, 2,$ and 6. Describe the behavior of the graphs as $p \to \infty$ and also the difference between the two equations.

55. Let us consider the polar equations $r = \dfrac{ep}{1 + e\cos\theta}$ and $r = \dfrac{ep}{1 - e\cos\theta}$ with $p = 1$. With a graphing utility, explore the equations with $e = 1.5, 3,$ and 6. Describe the behavior of the graphs as $e \to \infty$ and also the difference between the two equations.

56. Let us consider the polar equations $r = \dfrac{ep}{1 + e\sin\theta}$ and $r = \dfrac{ep}{1 - e\sin\theta}$ with $p = 1$. With a graphing utility, explore the equations with $e = 1.5, 3,$ and 6. Describe the behavior of the graphs as $e \to \infty$ and also the difference between the two equations.

57. Let us consider the polar equations $r = \dfrac{ep}{1 + e\cos\theta}$ and $r = \dfrac{ep}{1 - e\cos\theta}$ with $p = 1$. With a graphing utility, explore the equations with $e = 0.001, 0.5, 0.9,$ and 0.99. Describe the behavior of the graphs as $e \to 1$ and also the difference between the two equations. Be sure to set the window parameters properly.

58. Let us consider the polar equations $r = \dfrac{ep}{1 + e\sin\theta}$ and $r = \dfrac{ep}{1 - e\sin\theta}$ with $p = 1$. With a graphing utility, explore the equations with $e = 0.001, 0.5, 0.9,$ and 0.99. Describe the behavior of the graphs as $e \to 1$ and also the difference between the two equations. Be sure to set the window parameters properly.

59. Let us consider the polar equation $r = \dfrac{5}{5 + 2\sin\theta}$. Explain why the graphing utility gives the following graphs with the specified window parameters:

a. $[-2, 2]$ by $[-2, 2]$ with θ step $= \dfrac{\pi}{2}$

b. $[-2, 2]$ by $[-2, 2]$ with θ step $= \dfrac{\pi}{3}$

60. Let us consider the polar equation $r = \dfrac{2}{1 + \cos\theta}$. Explain why a graphing utility gives the following graphs with the specified window parameters:

a. $[-2, 2]$ by $[-4, 4]$ with θ step $= \dfrac{\pi}{2}$

b. $[-2, 2]$ by $[-4, 4]$ with θ step $= \dfrac{\pi}{3}$

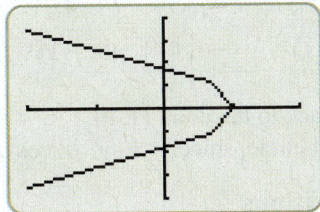

61. Let us consider the polar equation $r = \dfrac{6}{1 + 3\sin\theta}$. Explain why a graphing utility gives the following graphs with the specified window parameters:

a. $[-8, 8]$ by $[-2, 4]$ with θ step $= \dfrac{\pi}{2}$

b. $[-4, 8]$ by $[-2, 6]$ with θ step $= 0.4\pi$

62. Let us consider the polar equation $r = \dfrac{2}{1 - \sin\theta}$. Explain why a graphing utility gives the following graphs with the specified window parameters:

a. $[-4, 4]$ by $[-2, 4]$ with θ step $= \dfrac{\pi}{3}$

b. $[-4, 4]$ by $[-2, 6]$ with θ step $= 0.8\pi$

SKILLS OBJECTIVES

■ Graph parametric equations.
■ Find an equation (in rectangular form) that corresponds to a graph defined parametrically.
■ Find parametric equations for a graph that is defined by an equation in rectangular form.

CONCEPTUAL OBJECTIVES

■ Understand that the results of increasing the value of the parameter reveal the orientation of a curve or the direction of motion along it.
■ Use time as a parameter in parametric equations such as cycloids and projectile motion.

Parametric Equations of a Curve

Thus far we have talked about graphs in planes. For example, the equation $x^2 + y^2 = 1$ when graphed in a plane is the unit circle. Similarly, the function $f(x) = \sin x$ when graphed in a plane is a sinusoidal curve. Now, we consider the **path along a curve**. For example, if a car is being driven on a circular racetrack, we want to see the movement along the circle. We can determine where (position) along the circle the car is at some time t using *parametric equations*. Before we define *parametric equations* in general, let us start with a simple example.

Let $x = \cos t$ and $y = \sin t$ and $t \geq 0$. We then can make a table of some corresponding values.

t SECONDS	$x = \cos t$	$y = \sin t$	(x, y)
0	$x = \cos 0 = 1$	$y = \sin 0 = 0$	$(1, 0)$
$\dfrac{\pi}{2}$	$x = \cos\left(\dfrac{\pi}{2}\right) = 0$	$y = \sin\left(\dfrac{\pi}{2}\right) = 1$	$(0, 1)$
π	$x = \cos \pi = -1$	$y = \sin \pi = 0$	$(-1, 0)$
$\dfrac{3\pi}{2}$	$x = \cos\left(\dfrac{3\pi}{2}\right) = 0$	$y = \sin\left(\dfrac{3\pi}{2}\right) = -1$	$(0, -1)$
2π	$x = \cos(2\pi) = 1$	$y = \sin(2\pi) = 0$	$(1, 0)$

If we plot these points and note the correspondence to time (by converting all numbers to decimals), we will be tracing a *path* counterclockwise along the unit circle.

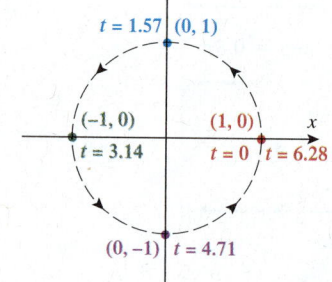

TIME (SECONDS)	$t = 0$	$t = 1.57$	$t = 3.14$	$t = 4.71$
POSITION	$(1, 0)$	$(0, 1)$	$(-1, 0)$	$(0, -1)$

Notice that at time $t = 6.28$ seconds we are back to the point $(1, 0)$.

We can see that the path represents the unit circle, since $x^2 + y^2 = \cos^2 t + \sin^2 t = 1$.

DEFINITION   Parametric Equations

Let $x = f(t)$ and $y = g(t)$ be functions defined for t on some interval. The set of points $(x, y) = (f(t), g(t))$ represents a **plane curve**. The equations

$$x = f(t) \qquad \text{and} \qquad y = g(t)$$

are called **parametric equations** of the curve. The variable t is called the **parameter**.

Parametric equations are useful for showing movement along a curve. We insert arrows in the graph to show **direction**, or **orientation**, along the curve as t increases.

EXAMPLE 1 Graphing a Curve Defined by Parametric Equations

Graph the curve defined by the parametric equations

$$x = t^2 \qquad y = (t - 1) \qquad t \text{ in } [-2, 2]$$

Indicate the orientation with arrows.

Solution:

STEP 1 Make a table and find values for t, x, and y.

t	$x = t^2$	$y = (t - 1)$	(x, y)
$t = -2$	$x = (-2)^2 = 4$	$y = (-2 - 1) = -3$	$(4, -3)$
$t = -1$	$x = (-1)^2 = 1$	$y = (-1 - 1) = -2$	$(1, -2)$
$t = 0$	$x = 0^2 = 0$	$y = (0 - 1) = -1$	$(0, -1)$
$t = 1$	$x = 1^2 = 1$	$y = (1 - 1) = 0$	$(1, 0)$
$t = 2$	$x = 2^2 = 4$	$y = (2 - 1) = 1$	$(4, 1)$

STEP 2 Plot the points in the xy-plane.

STEP 3 Connect the points with a smooth curve and use arrows to indicate direction.

The shape of the graph appears to be a parabola. The parametric equations are $x = t^2$ and $y = (t - 1)$. If we solve the second equation for t, getting $t = y + 1$, and substitute this expression into $x = t^2$, the result is $x = (y + 1)^2$. The graph of $x = (y + 1)^2$ is a parabola with vertex at the point $(0, -1)$ and opening to the right.

■ **YOUR TURN** Graph the curve defined by the parametric equations

$$x = t + 1 \qquad y = t^2 \qquad t \text{ in } [-2, 2]$$

Indicate the orientation with arrows.

Classroom Example 11.9.1
Graph $x = 3t^2$, $y = 2 - t$,
$-2 \le t \le 2$.

Answer:

■ **Answer:**

Sometimes it is easier to show the rectangular equivalent of the curve and eliminate the parameter.

 EXAMPLE 2 Graphing a Curve Defined by Parametric Equations by First Finding an Equivalent Rectangular Equation

Graph the curve defined by the parametric equations

$$x = 4\cos t \qquad y = 3\sin t \qquad t \text{ is any real number}$$

Indicate the orientation with arrows.

Solution:

One approach is to point-plot as in Example 1. A second approach is to find the equivalent rectangular equation that represents the curve.

We apply the Pythagorean identity.	$\sin^2 t + \cos^2 t = 1$
Find $\sin^2 t$ from the parametric equation for y.	$y = 3\sin t$
Square both sides.	$y^2 = 9\sin^2 t$
Divide by 9.	$\sin^2 t = \dfrac{y^2}{9}$
Similarly, find $\cos^2 t$.	$x = 4\cos t$
Square both sides.	$x^2 = 16\cos^2 t$
Divide by 16.	$\cos^2 t = \dfrac{x^2}{16}$
Substitute $\sin^2 t = \dfrac{y^2}{9}$ and $\cos^2 t = \dfrac{x^2}{16}$ into $\sin^2 t + \cos^2 t = 1$.	$\dfrac{y^2}{9} + \dfrac{x^2}{16} = 1$

The curve is an ellipse centered at the origin and elongated horizontally.

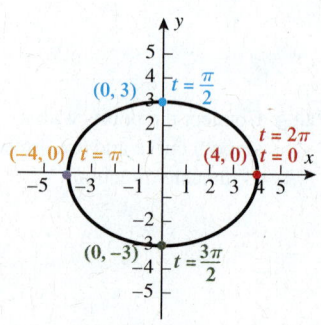

The orientation is counterclockwise. For example, when $t = 0$, the position is $(4, 0)$, when $t = \dfrac{\pi}{2}$, the position is $(0, 3)$, and when $t = \pi$, the position is $(-4, 0)$.

Study Tip

For open curves the orientation can be determined from two values of t. However, for closed curves three points should be chosen to ensure clockwise or counterclockwise orientation.

Applications of Parametric Equations

Parametric equations can be used to describe motion in many applications. Two that we will discuss are the *cycloid* and a *projectile*. Suppose that you paint a red **X** on a bicycle tire. As the bicycle moves in a straight line, if you watch the motion of the red **X**, you will see that it follows the path of a **cycloid**.

The parametric equations that define a cycloid are

$$x = a(t - \sin t) \quad \text{and} \quad y = a(1 - \cos t)$$

where t is any real number.

EXAMPLE 3   Graphing a Cycloid

Graph the cycloid given by $x = 2(t - \sin t)$ and $y = 2(1 - \cos t)$ for t in $[0, 4\pi]$.

Solution:

STEP 1   Make a table and find key values for t, x, and y.

t	$x = 2(t - \sin t)$	$y = 2(1 - \cos t)$	(x, y)
$t = 0$	$x = 2(0 - 0) = 0$	$y = 2(1 - 1) = 0$	$(0, 0)$
$t = \pi$	$x = 2(\pi - 0) = 2\pi$	$y = 2[1 - (-1)] = 4$	$(2\pi, 4)$
$t = 2\pi$	$x = 2(2\pi - 0) = 4\pi$	$y = 2(1 - 1) = 0$	$(4\pi, 0)$
$t = 3\pi$	$x = 2(3\pi - 0) = 6\pi$	$y = 2[1 - (-1)] = 4$	$(6\pi, 4)$
$t = 4\pi$	$x = 2(4\pi - 0) = 8\pi$	$y = 2(1 - 1) = 0$	$(8\pi, 0)$

STEP 2   Plot points in a plane and connect them with a smooth curve.

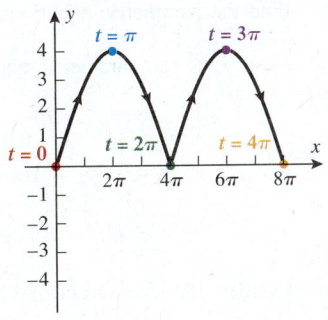

Classroom Example 11.9.3

a. Graph $x = 5(t - \sin t)$,
$y = 5(1 - \cos t)$,
$0 \le t \le 4\pi$.

b.* Describe what happens to the graph of $x = a(t - \sin t)$,
$y = a(1 - \cos t)$, $0 \le t$
as a increases.

Answer:

a.

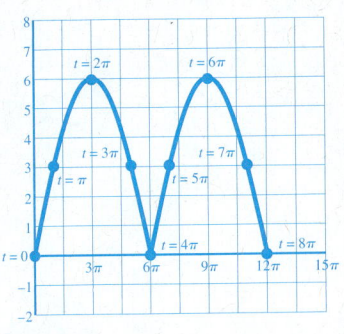

b. The maxima of respective "humps" increases, while the x-intercepts are a function of a, but the general periodic nature remain the same.

Classroom Example 11.9.3*

Write the parametric equations that would generate the following cycloid:

Answer:

$$x = 3\left[\frac{t}{2} - \sin\left(\frac{t}{2}\right)\right],$$

$$y = 3\left[1 - \cos\left(\frac{t}{2}\right)\right],$$

$$0 \le t \le 8\pi$$

Another example of parametric equations describing real-world phenomena is *projectile motion*. The accompanying photo of a golfer hitting a golf ball illustrates an example of a projectile.

Joshua Dalsimer/©Corbis; iStockphoto (golf ball)

Let v_0 be the initial velocity of an object, θ be the initial angle of inclination with the horizontal, and h be the initial height above the ground. Then the parametric equations describing the **projectile motion** (which will be developed in calculus) are

$$x = (v_0\cos\theta)t \quad \text{and} \quad y = -\tfrac{1}{2}gt^2 + (v_0\sin\theta)t + h$$

where t is the time and g is the constant acceleration due to gravity (9.8 meters per square second or 32 feet per square second).

EXAMPLE 4   Graphing Projectile Motion

Suppose a golfer hits his golf ball with an initial velocity of 160 feet per second at an angle of 30° with the ground. How far is his drive, assuming the length of the drive is from the tee to where the ball first hits the ground? Graph the curve representing the path of the golf ball. Assume that he hits the ball straight off the tee and down the fairway.

Solution:

STEP 1   Find the parametric equations that describe the golf ball that the golfer drove.

First, write the parametric equations for projectile motion.

$$x = (v_0\cos\theta)t \quad \text{and} \quad y = -\tfrac{1}{2}gt^2 + (v_0\sin\theta)t + h$$

Let $g = 32\,\text{ft/sec}^2$, $v_0 = 160\,\text{ft/sec}$, $h = 0$, and $\theta = 30°$.

$$x = (160 \cdot \cos 30°)t \quad \text{and} \quad y = -16t^2 + (160 \cdot \sin 30°)t$$

Evaluate the sine and cosine functions and simplify.

$$x = 80\sqrt{3}\,t \quad \text{and} \quad y = -16t^2 + 80t$$

STEP 2 Graph the projectile motion.

t	$x = 80\sqrt{3}\,t$	$y = -16t^2 + 80t$	(x, y)
$t = 0$	$x = 80\sqrt{3}(0) = 0$	$y = -16(0)^2 + 80(0) = 0$	**(0, 0)**
$t = 1$	$x = 80\sqrt{3}(1) \approx 139$	$y = -16(1)^2 + 80(1) = 64$	**(139, 64)**
$t = 2$	$x = 80\sqrt{3}(2) \approx 277$	$y = -16(2)^2 + 80(2) = 96$	**(277, 96)**
$t = 3$	$x = 80\sqrt{3}(3) \approx 416$	$y = -16(3)^2 + 80(3) = 96$	**(416, 96)**
$t = 4$	$x = 80\sqrt{3}(4) \approx 554$	$y = -16(4)^2 + 80(4) = 64$	**(554, 64)**
$t = 5$	$x = 80\sqrt{3}(5) \approx 693$	$y = -16(5)^2 + 80(5) = 0$	**(693, 0)**

Technology Tip

Graph $x = 80\sqrt{3}\,t$ and $y = -16t^2 + 80t$.

We can see that we selected our time increments well (the last point, **(693, 0)**, corresponds to the ball hitting the ground 693 feet from the tee).

STEP 3 Identify the horizontal distance from the tee to where the ball first hits the ground.

Algebraically, we can determine the distance of the tee shot by setting the height y equal to zero.

$$y = -16t^2 + 80t = 0$$

Factor (divide) the common, $-16t$.

$$-16t(t - 5) = 0$$

Solve for t.

$$t = 0 \text{ or } t = 5$$

The ball hits the ground after 5 seconds.

Let $t = 5$ in the horizontal distance, $x = 80\sqrt{3}\,t$.

$$x = 80\sqrt{3}(5) \approx 693$$

> The ball hits the ground 693 feet from the tee.

With parametric equations, we can also determine when the ball lands (5 seconds).

Parametric equations are a way of describing the path an object takes along a curve in the xy-plane. Parametric equations have equivalent rectangular equations. Typically, the method of graphing a set of parametric equations is to eliminate t and graph the corresponding rectangular equation. Once the curve is found, orientation along the curve can be determined by finding points corresponding to different t-values. Two important applications are cycloids and projectiles, whose paths we can trace using parametric equations.

SECTION
11.9 EXERCISES

■ SKILLS

In Exercises 1–30, graph the curve defined by the parametric equations.

1. $x = t + 1$, $y = \sqrt{t}$, $t \geq 0$

2. $x = 3t$, $y = t^2 - 1$, t in $[0, 4]$

3. $x = -3t$, $y = t^2 + 1$, t in $[0, 4]$

4. $x = t^2 - 1$, $y = t^2 + 1$, t in $[-3, 3]$

5. $x = t^2$, $y = t^3$, t in $[-2, 2]$

6. $x = t^3 + 1$, $y = t^3 - 1$, t in $[-2, 2]$

7. $x = \sqrt{t}$, $y = t$, t in $[0, 10]$

8. $x = t$, $y = \sqrt{t^2 + 1}$, t in $[0, 10]$

9. $x = (t + 1)^2$, $y = (t + 2)^3$, t in $[0, 1]$

10. $x = (t - 1)^3$, $y = (t - 2)^2$, t in $[0, 4]$

11. $x = e^t$, $y = e^{-t}$, $-\ln 3 \leq t \leq \ln 3$

12. $x = e^{-2t}$, $y = e^{2t} + 4$, $-\ln 2 \leq t \leq \ln 3$

13. $x = 2t^4 - 1$, $y = t^8 + 1$, $0 \leq t \leq 4$

14. $x = 3t^6 - 1$, $y = 2t^3$, $-1 \leq t \leq 1$

15. $x = t(t - 2)^3$, $y = t(t - 2)^3$, $0 \leq t \leq 4$

16. $x = -t\sqrt[3]{t}$, $y = -5t^8 - 2$, $-3 \leq t \leq 3$

17. $x = 3\sin t$, $y = 2\cos t$, t in $[0, 2\pi]$

18. $x = \cos(2t)$, $y = \sin t$, t in $[0, 2\pi]$

19. $x = \sin t + 1$, $y = \cos t - 2$, t in $[0, 2\pi]$

20. $x = \tan t$, $y = 1$, t in $\left[-\dfrac{\pi}{4}, \dfrac{\pi}{4}\right]$

21. $x = 1$, $y = \sin t$, t in $[-2\pi, 2\pi]$

22. $x = \sin t$, $y = 2$, t in $[0, 2\pi]$

23. $x = \sin^2 t$, $y = \cos^2 t$, t in $[0, 2\pi]$

24. $x = 2\sin^2 t$, $y = 2\cos^2 t$, t in $[0, 2\pi]$

25. $x = 2\sin(3t)$, $y = 3\cos(2t)$, t in $[0, 2\pi]$

26. $x = 4\cos(2t)$, $y = t$, t in $[0, 2\pi]$

27. $x = \cos\left(\dfrac{t}{2}\right) - 1$, $y = \sin\left(\dfrac{t}{2}\right) + 1$, $-2\pi \leq t \leq 2\pi$

28. $x = \sin\left(\dfrac{t}{3}\right) + 3$, $y = \cos\left(\dfrac{t}{3}\right) - 1$, $0 \leq t \leq 6\pi$

29. $x = 2\sin\left(t + \dfrac{\pi}{4}\right)$, $y = -2\cos\left(t + \dfrac{\pi}{4}\right)$, $-\dfrac{\pi}{4} \leq t \leq \dfrac{7\pi}{4}$

30. $x = -3\cos^2(3t)$, $y = 2\cos(3t)$, $-\dfrac{\pi}{3} \leq t \leq \dfrac{\pi}{3}$

In Exercises 31–40, the given parametric equations define a plane curve. Find an equation in rectangular form that also corresponds to the plane curve.

31. $x = \dfrac{1}{t}$, $y = t^2$

32. $x = t^2 - 1$, $y = t^2 + 1$

33. $x = t^3 + 1$, $y = t^3 - 1$

34. $x = 3t$, $y = t^2 - 1$

35. $x = t$, $y = \sqrt{t^2 + 1}$

36. $x = \sin^2 t$, $y = \cos^2 t$

37. $x = 2\sin^2 t$, $y = 2\cos^2 t$

38. $x = \sec^2 t$, $y = \tan^2 t$

39. $x = 4(t^2 + 1)$, $y = 1 - t^2$

40. $x = \sqrt{t - 1}$, $y = \sqrt{t}$

■ APPLICATIONS

For Exercises 41–50, recall that the flight of a projectile can be modeled with the parametric equations

$$x = (v_0\cos\theta)t \qquad y = -16t^2 + (v_0\sin\theta)t + h$$

where t is in seconds, v_0 is the initial velocity, θ is the angle with the horizontal, and x and y are in feet.

41. **Flight of a Projectile.** A projectile is launched from the ground at a speed of 400 ft/sec at an angle of 45° with the horizontal. After how many seconds does the projectile hit the ground?

42. **Flight of a Projectile.** A projectile is launched from the ground at a speed of 400 ft/sec at an angle of 45° with the horizontal. How far does the projectile travel (what is the horizontal distance), and what is its maximum altitude?

43. **Flight of a Baseball.** A baseball is hit at an initial speed of 105 mph and an angle of 20° at a height of 3 feet above the ground. If home plate is 420 feet from the back fence, which is 15 feet tall, will the baseball clear the back fence for a home run?

44. **Flight of a Baseball.** A baseball is hit at an initial speed of 105 mph and an angle of 20° at a height of 3 feet above the ground. If there is no back fence or other obstruction, how far does the baseball travel (horizontal distance), and what is its maximum height?

45. **Bullet Fired.** A gun is fired from the ground at an angle of 60°, and the bullet has an initial speed of 700 ft/sec. How high does the bullet go? What is the horizontal (ground) distance between the point where the gun is fired and the point where the bullet hits the ground?

46. **Bullet Fired.** A gun is fired from the ground at an angle of 60°, and the bullet has an initial speed of 2000 ft/sec. How high does the bullet go? What is the horizontal (ground) distance between the point where the gun is fired and the point where the bullet hits the ground?

47. **Missile Fired.** A missile is fired from a ship at an angle of 30°, an initial height of 20 feet above the water's surface, and a speed of 4000 ft/sec. How long will it be before the missile hits the water?

48. **Missile Fired.** A missile is fired from a ship at an angle of 40°, an initial height of 20 feet above the water's surface, and a speed of 5000 ft/sec. Will the missile be able to hit a target that is 2 miles away?

49. **Path of a Projectile.** A projectile is launched at a speed of 100 ft/sec at an angle of 35° with the horizontal. Plot the path of the projectile on a graph. Assume that $h = 0$.

50. **Path of a Projectile.** A projectile is launched at a speed of 150 ft/sec at an angle of 55° with the horizontal. Plot the path of the projectile on a graph. Assume that $h = 0$.

For Exercises 51 and 52, refer to the following:

Modern amusement park rides are often designed to push the envelope in terms of speed, angle, and ultimately *g*s, and usually take the form of gargantuan roller coasters or skyscraping towers. However, even just a couple of decades ago, such creations were depicted only in fantasy-type drawings, with their creators never truly believing their construction would become a reality. Nevertheless, thrill rides still capable of nauseating any would-be rider were still able to be constructed; one example is the *Calypso*. This ride is a not-too-distant cousin of the more well-known *Scrambler*. It consists of four rotating arms (instead of three like the Scrambler), and on each of these arms, four cars (equally spaced around the circumference of a circular frame) are attached. Once in motion, the main piston to which the four arms are connected rotates clockwise, while each of the four arms themselves rotates counterclockwise. The combined motion appears as a blur to any onlooker from the crowd, but the motion of a single rider is much less chaotic. In fact, a single rider's path can be modeled by the following graph:

The equation of this graph is defined parametrically by

$$x(t) = A\cos t + B\cos(-3t)$$
$$y(t) = A\sin t + B\sin(-3t) \quad 0 \le t \le 2\pi$$

51. **Amusement Rides.** What is the location of the rider at $t = 0$, $t = \dfrac{\pi}{2}$, $t = \pi$, $t = \dfrac{3\pi}{2}$, and $t = 2\pi$?

52. **Amusement Rides.** Suppose that the ride conductor was rather sinister and speeded up the ride to twice the speed. How would you modify the parametric equations to model such a change? Now vary the values of A and B. What do you think these parameters are modeling in this problem?

▪ CATCH THE MISTAKE

In Exercises 53 and 54, explain the mistake that is made.

53. Find the rectangular equation that corresponds to the plane curve defined by the parametric equations $x = t + 1$ and $y = \sqrt{t}$. Describe the plane curve.

Solution:

Square $y = \sqrt{t}$. $\qquad\qquad\qquad y^2 = t$

Substitute $t = y^2$ into $x = t + 1$. $\quad x = y^2 + 1$

The graph of $x = y^2 + 1$ is a parabola opening to the right with vertex at $(1, 0)$.

This is incorrect. What mistake was made?

54. Find the rectangular equation that corresponds to the plane curve defined by the parametric equations $x = \sqrt{t}$ and $y = t - 1$. Describe the plane curve.

Solution:

Square $x = \sqrt{t}$. $\qquad\qquad\qquad x^2 = t$

Substitute $t = x^2$ into $y = t - 1$. $\quad y = x^2 - 1$

The graph of $y = x^2 - 1$ is a parabola opening up with vertex at $(0, -1)$.

This is incorrect. What mistake was made?

▪ CONCEPTUAL

In Exercises 55 and 56, determine whether each statement is true or false.

55. Curves given by equations in rectangular form have orientation.

56. Curves given by parametric equations have orientation.

57. Determine what type of curve the parametric equations $x = \sqrt{t}$ and $y = \sqrt{1 - t}$ define.

58. Determine what type of curve the parametric equations $x = \ln t$ and $y = t$ define.

▪ CHALLENGE

59. Prove that $x = a\tan t$, $y = b\sec t$, $0 \le t \le 2\pi$, $t \ne \dfrac{\pi}{2}, \dfrac{3\pi}{2}$

are parametric equations for a hyperbola. Assume that a and b are nonzero constants.

60. Prove that $x = a\csc\left(\dfrac{t}{2}\right)$, $y = b\cot\left(\dfrac{t}{2}\right)$, $0 \le t \le 4\pi$,

$t \ne \pi, 3\pi$ are parametric equations for a hyperbola. Assume that a and b are nonzero constants.

61. Consider the parametric curve $x = a\sin^2 t - b\cos^2 t$,

$y = b\cos^2 t + a\sin^2 t$, $0 \le t \le \dfrac{\pi}{2}$.

Assume that a and b are nonzero constants. Find the Cartesian equation for this curve.

62. Consider the parametric curve $x = a\sin t + a\cos t$, $y = a\cos t - a\sin t$, $0 \le t \le 2\pi$. Assume that a is not zero. Find the Cartesian equation for this curve.

63. Consider the parametric curve $x = e^{at}$, $y = be^t$, $t > 0$. Assume that a is a positive integer and b is a positive real number. Determine the Cartesian equation.

64. Consider the parametric curve $x = a\ln t$, $y = \ln(bt)$, $t > 0$. Assume that b is a positive integer and a is a positive real number. Determine the Cartesian equation.

▪ TECHNOLOGY

65. Consider the parametric equations: $x = a\sin t - \sin(at)$ and $y = a\cos t + \cos(at)$. With a graphing utility, explore the graphs for $a = 2, 3$, and 4.

66. Consider the parametric equations: $x = a\cos t - b\cos(at)$ and $y = a\sin t + \sin(at)$. With a graphing utility, explore the graphs for $a = 3$ and $b = 1$, $a = 4$ and $b = 2$, and $a = 6$ and $b = 2$. Find the t-interval that gives one cycle of the curve.

67. Consider the parametric equations: $x = \cos(at)$ and $y = \sin(bt)$. With a graphing utility, explore the graphs for $a = 2$ and $b = 4$, $a = 4$ and $b = 2$, $a = 1$ and $b = 3$, and $a = 3$ and $b = 1$. Find the t-interval that gives one cycle of the curve.

68. Consider the parametric equations: $x = a\sin(at) - \sin t$ and $y = a\cos(at) - \cos t$. With a graphing utility, explore the graphs for $a = 2$ and 3. Describe the t-interval for each case.

69. Consider the parametric equations $x = a\cos(at) - \sin t$ and $y = a\sin(at) - \cos t$. With a graphing utility, explore the graphs for $a = 2$ and 3. Describe the t-interval for each case.

70. Consider the parametric equations $x = a\sin(at) - \cos t$ and $y = a\cos(at) - \sin t$. With a graphing utility, explore the graphs for $a = 2$ and 3. Describe the t-interval for each case.

"And the Rockets Red Glare . . ."

Scientists at Vandenberg Air Force Base are interested in tracing the path of some newly designed rockets. They will launch two rockets at 100 feet per second. One will depart at 45°, the other at 60°. From vector analysis and gravity, you determine the following coordinates (x, y) as a function of time t where y stands for height in feet above the ground and x stands for lateral distance traveled.

45°	60°
$x = 100\cos(45°)t$	$x = 100\cos(60°)t$
$y = -16t^2 + 100\sin(45°)t$	$y = -16t^2 + 100\sin(60°)t$

1. For each angle (45°, 60°), fill in the chart (round to one decimal place). You can do this by hand (very slowly) or use the table capabilities of a calculator or similar device.

t	0	0.5	1.0	1.5	2.0	2.5	3.0	3.5	4.0	4.5	5.0
X_{45}											
Y_{45}											
X_{60}											
Y_{60}											

2. What are a few things worth noting when looking at your table values? Think about the big picture and the fact that you are dealing with projectiles.

Show your work for the following:

3. Which rocket traveled higher and by how much? Recall that the y variable is the height. (List the heights of each rocket.)

4. When each rocket first hits the ground, which one has traveled farther laterally and by how much? (List distances of each.)

5. Which rocket was in the air longer and by how much? (List the times of each.)

6. For each rocket, write t in terms of x. Then substitute this value into the y equation. This is called "eliminating the parameter" and puts y as a function of x. Simplify completely. Use exact values. Reduce the fractions to their lowest terms.

MODELING OUR WORLD

In the Modeling Our World features in Chapters 3–5, you used the average yearly temperature in degrees Fahrenheit (°F) and carbon dioxide emissions in parts per million (ppm) collected by NOAA in Mauna Loa, Hawaii, to develop linear (Chapter 3) and nonlinear (Chapters 4 and 5) models. In the following exercises you will determine when these different models actually predict the same temperatures and carbon emissions. It is important to realize that not only can different models be used to predict trends, but also the choice of data those models are fitted to also affects the models and hence the predicted values.

YEAR	1960	1965	1970	1975	1980	1985	1990	1995	2000	2005
TEMPERATURE	44.45	43.29	43.61	43.35	46.66	45.71	45.53	47.53	45.86	46.23
CO_2 EMISSIONS (PPM)	316.9	320.0	325.7	331.1	338.7	345.9	354.2	360.6	369.4	379.7

1. Solve the system of nonlinear equations governing mean temperature that was found by using two data points:

   Equation (1): Use the linear model developed in Modeling Our World, Chapter 3, Problem 2a.

   Equation (2): Use the quadratic model found in Modeling Our World, Chapter 4, Problem 2a.

2. For what year do the models used in Problem 1 agree? Compare the value given by the models that year to the actual data for the year.

3. Solve the system of nonlinear equations governing mean temperature that was found by applying regression (all data points):

   Equation (1): Use the linear model developed in Modeling Our World, Chapter 3, Problem 2c.

   Equation (2): Use the quadratic model found in Modeling Our World, Chapter 4, Problem 2c.

4. For what year do the models used in Problem 3 agree? Compare the value given by the models that year to the actual data for the year.

5. Solve the system of nonlinear equations governing carbon dioxide emissions that was found by using two data points:

   Equation (1): Use the linear model developed in Modeling Our World, Chapter 3, Problem 7a.

   Equation (2): Use the quadratic model found in Modeling Our World, Chapter 4, Problem 7a.

6. For what year do the models used in Problem 5 agree? Compare the value given by the models that year to the actual data for the year.

7. Solve the system of nonlinear equations governing carbon emissions that was found by applying regression (all data points):

   Equation (1): Use the linear model developed in Modeling Our World, Chapter 3, Problem 7c.

   Equation (2): Use the quadratic model found in Modeling Our World, Chapter 4, Problem 7c.

8. For what year do the models used in Problem 7 agree? Compare the value given by the models that year to the actual data for the year.

SECTION	CONCEPT	KEY IDEAS/FORMULAS

11.1   **Conic basics**

Three types of conics

Names of conics: Parabola, Ellipse, and Hyperbola

Definitions:
Parabola: Distance to a reference point (focus) and a reference line (directrix) is constant.
Ellipse: Sum of the distances between the point and two reference points (foci) is constant.
Hyperbola: Difference of the distances between the point and two reference points (foci) is constant.

11.2   **The parabola**

A parabola with a vertex at the origin.

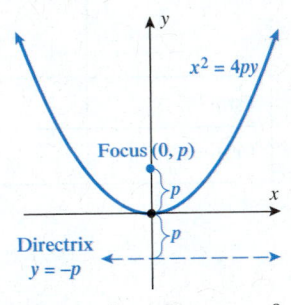

Up: $p > 0$ Down: $p < 0$

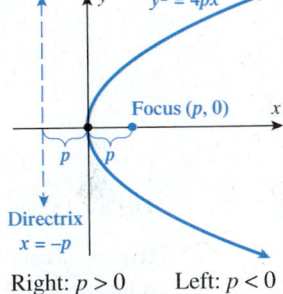

Right: $p > 0$ Left: $p < 0$

A parabola with a vertex at the point (h, k).

EQUATION	$(y - k)^2 = 4p(x - h)$	$(x - h)^2 = 4p(y - k)$
VERTEX	(h, k)	(h, k)
FOCUS	$(p + h, k)$	$(h, p + k)$
DIRECTRIX	$x = -p + h$	$y = -p + k$
AXIS OF SYMMETRY	$y = k$	$x = h$
$p > 0$	opens to the right	opens upward
$p < 0$	opens to the left	opens downward

11.3   **The ellipse**

Ellipse centered at the origin

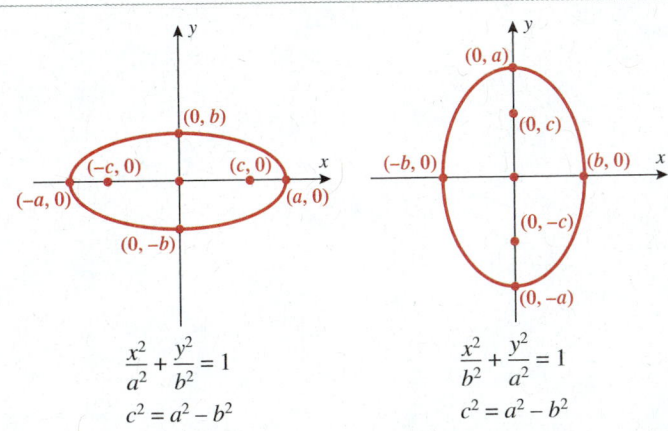

$$\frac{x^2}{a^2} + \frac{y^2}{b^2} = 1$$
$$c^2 = a^2 - b^2$$

$$\frac{x^2}{b^2} + \frac{y^2}{a^2} = 1$$
$$c^2 = a^2 - b^2$$

Ellipse centered at the point (h, k)

ORIENTATION OF MAJOR AXIS	Horizontal (parallel to the x-axis)	Vertical (parallel to the y-axis)
EQUATION	$\dfrac{(x - h)^2}{a^2} + \dfrac{(y - k)^2}{b^2} = 1$	$\dfrac{(x - h)^2}{b^2} + \dfrac{(y - k)^2}{a^2} = 1$
GRAPH		
FOCI	$(h - c, k)$ $(h + c, k)$	$(h, k - c)$ $(h, k + c)$
VERTICES	$(h - a, k)$ $(h + a, k)$	$(h, k - a)$ $(h, k + a)$

11.4 The hyperbola

Hyperbola centered at the origin.

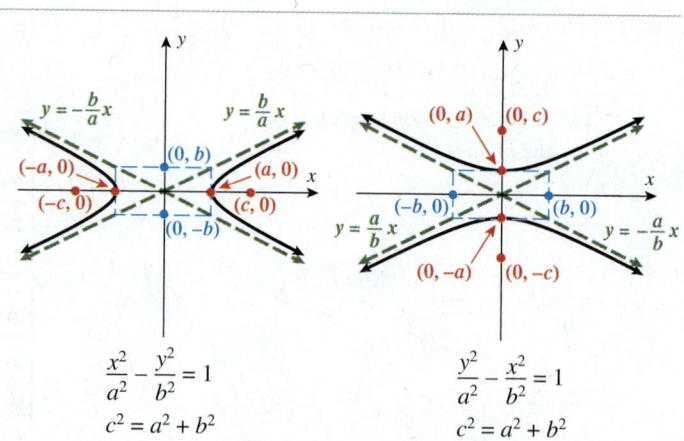

$$\frac{x^2}{a^2} - \frac{y^2}{b^2} = 1$$
$$c^2 = a^2 + b^2$$

$$\frac{y^2}{a^2} - \frac{x^2}{b^2} = 1$$
$$c^2 = a^2 + b^2$$

	Hyperbola centered at the point (h, k)			

ORIENTATION OF TRANSVERSE AXIS	Horizontal (parallel to the x-axis)	Vertical (parallel to the y-axis)
EQUATION	$\dfrac{(x-h)^2}{a^2} - \dfrac{(y-k)^2}{b^2} = 1$	$\dfrac{(y-k)^2}{a^2} - \dfrac{(x-h)^2}{b^2} = 1$
VERTICES	$(h-a, k)$ $(h+a, k)$	$(h, k-a)$ $(h, k+a)$
FOCI	$(h-c, k)$ $(h+c, k)$ where $c^2 = a^2 + b^2$	$(h, k-c)$ $(h, k+c)$ where $c^2 = a^2 + b^2$
GRAPH		

11.5	Systems of nonlinear equations	There is no procedure guaranteed to solve nonlinear equations.
		Using elimination to solve systems of nonlinear equations. Eliminate a variable by either adding one equation to or subtracting one equation from the other. **Using substitution to solve systems of nonlinear equations.** Solve for one variable in terms of the other and substitute into the second equation.
11.6	Systems of nonlinear inequalities	Solutions are determined graphically by finding the common shaded regions. ■ $\leq$ or $\geq$ use solid curves ■ $<$ or $>$ use dashed curves
	Nonlinear inequalities in two variables	Step 1: Rewrite the inequality as an equation. Step 2: Graph the equation. Step 3: Test points. Step 4: Shade.
	Systems of nonlinear inequalities	Graph the individual inequalities and the solution in the common (overlapping) shaded region.
11.7	**Rotation of axes**	
	Rotation of axes formulas	$x = X\cos\theta - Y\sin\theta$ $x = X\sin\theta + Y\cos\theta$
	The angle of rotation necessary to transform a general second-degree equation into an equation of a conic	$\cot 2\theta = \dfrac{A - C}{B}$

SECTION	CONCEPT	KEY IDEAS/FORMULAS
11.8	**Polar equations of conics**	All three conics (parabolas, ellipses, and hyperbolas) are defined in terms of a single focus and a directrix.
	Equations of conics in polar coordinates	The directrix is *vertical* and p units to the *right* of the pole. $$r = \frac{ep}{1 + e\cos\theta}$$ The directrix is *vertical* and p units to the *left* of the pole. $$r = \frac{ep}{1 - e\cos\theta}$$ The directrix is *horizontal* and p units *above* the pole. $$r = \frac{ep}{1 + e\sin\theta}$$ The directrix is *horizontal* and p units *below* the pole. $$r = \frac{ep}{1 - e\sin\theta}$$
11.9	**Parametric equations and graphs**	
	Parametric equations of a curve	Parametric equations: $x = f(t)$ and $y = g(t)$ Plane curve: $(x, y) = (f(t), g(t))$
	Applications of parametric equations	**Cycloid:** $x = a(t - \sin t)$ $y = a(1 - \cos t)$ **Projectile motion:** $x = (v_0\cos\theta)t$ $y = -\frac{1}{2}gt^2 + (v_0\sin\theta)t + h$

11.1 Conic Basics

Determine whether each statement is true or false.

1. The focus is a point on the graph of the parabola.

2. The graph of $y^2 = 8x$ is a parabola that opens upward.

3. $\dfrac{x^2}{9} - \dfrac{y^2}{1} = 1$ is the graph of a hyperbola that has a horizontal transverse axis.

4. $\dfrac{(x+1)^2}{9} + \dfrac{(y-3)^2}{16} = 1$ is a graph of an ellipse whose center is $(1, 3)$.

11.2 The Parabola

Find an equation for the parabola described.

5. Vertex at $(0, 0)$; Focus at $(3, 0)$

6. Vertex at $(0, 0)$; Focus at $(0, 2)$

7. Vertex at $(0, 0)$; Directrix at $x = 5$

8. Vertex at $(0, 0)$; Directrix at $y = 4$

9. Vertex at $(2, 3)$; Focus at $(2, 5)$

10. Vertex at $(-1, -2)$; Focus at $(1, -2)$

11. Focus at $(1, 5)$; Directrix at $y = 7$

12. Focus at $(2, 2)$; Directrix at $x = 0$

Find the focus, vertex, directrix, and length of the latus rectum, and graph the parabola.

13. $x^2 = -12y$

14. $x^2 = 8y$

15. $y^2 = x$

16. $y^2 = -6x$

17. $(y + 2)^2 = 4(x - 2)$

18. $(y - 2)^2 = -4(x + 1)$

19. $(x + 3)^2 = -8(y - 1)$

20. $(x - 3)^2 = -8(y + 2)$

21. $x^2 + 5x + 2y + 25 = 0$

22. $y^2 + 2y - 16x + 1 = 0$

Applications

23. **Satellite Dish.** A satellite dish measures 10 feet across its opening and 2 feet deep at its center. The receiver should be placed at the focus of the parabolic dish. Where should the receiver be placed?

24. **Clearance Under a Bridge.** A bridge with a parabolic shape reaches a height of 40 feet in the center of the road, and the width of the bridge opening at ground level is 30 feet combined (both lanes). If an RV is 14 feet tall and 8 feet wide, will it make it through the tunnel?

11.3 The Ellipse

Graph each ellipse.

25. $\dfrac{x^2}{9} + \dfrac{y^2}{64} = 1$

26. $\dfrac{x^2}{81} + \dfrac{y^2}{49} = 1$

27. $25x^2 + y^2 = 25$

28. $4x^2 + 8y^2 = 64$

Find the standard form of an equation of the ellipse with the given characteristics.

29. Foci: $(-3, 0)$ and $(3, 0)$ Vertices: $(-5, 0)$ and $(5, 0)$

30. Foci: $(0, -2)$ and $(0, 2)$ Vertices: $(0, -3)$ and $(0, 3)$

31. Major axis vertical with length of 16, minor axis length of 6 and centered at $(0, 0)$.

32. Major axis horizontal with length of 30, minor axis length of 20 and centered at $(0, 0)$.

Graph each ellipse.

33. $\dfrac{(x - 7)^2}{100} + \dfrac{(y + 5)^2}{36} = 1$

34. $20(x + 3)^2 + (y - 4)^2 = 120$

35. $4x^2 - 16x + 12y^2 + 72y + 123 = 0$

36. $4x^2 - 8x + 9y^2 - 72y + 147 = 0$

Find the standard form of an equation of the ellipse with the given characteristics.

37. Foci: $(-1, 3)$ and $(7, 3)$ Vertices: $(-2, 3)$ and $(8, 3)$

38. Foci: $(1, -3)$ and $(1, -1)$ Vertices: $(1, -4)$ and $(1, 0)$

Applications

39. **Planetary Orbits.** Jupiter's orbit is summarized in the picture. Utilize the fact that the Sun is a focus to determine an equation for Jupiter's elliptical orbit around the Sun. Round to the nearest hundred thousand kilometers.

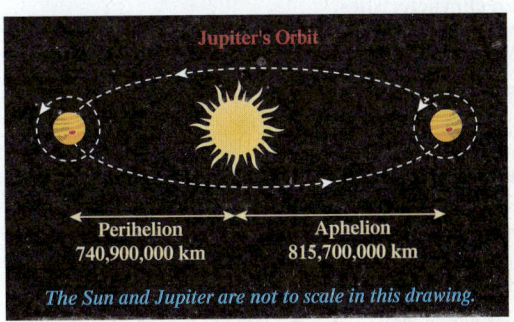

Jupiter's Orbit

Perihelion
740,900,000 km

Aphelion
815,700,000 km

The Sun and Jupiter are not to scale in this drawing.

40. Planetary Orbits. Mars's orbit is summarized in the picture that follows. Utilize the fact that the Sun is a focus to determine an equation for Mars's elliptical orbit around the Sun. Round to the nearest million kilometers.

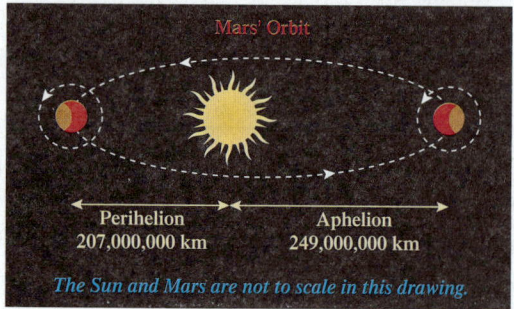

Mars' Orbit

Perihelion
207,000,000 km

Aphelion
249,000,000 km

The Sun and Mars are not to scale in this drawing.

11.4   The Hyperbola

Graph each hyperbola.

41. $\dfrac{x^2}{9} - \dfrac{y^2}{64} = 1$

42. $\dfrac{x^2}{81} - \dfrac{y^2}{49} = 1$

43. $x^2 - 25y^2 = 25$

44. $8y^2 - 4x^2 = 64$

Find the standard form of an equation of the hyperbola with the given characteristics.

45. Vertices: $(-3, 0)$ and $(3, 0)$ Foci: $(-5, 0)$ and $(5, 0)$

46. Vertices: $(0, -1)$ and $(0, 1)$ Foci: $(0, -3)$ and $(0, 3)$

47. Center: $(0, 0)$; Transverse: y-axis; Asymptotes: $y = 3x$ and $y = -3x$

48. Center: $(0, 0)$; Transverse axis: y-axis; Asymptotes: $y = \frac{1}{2}x$ and $y = -\frac{1}{2}x$

Graph each hyperbola.

49. $\dfrac{(y - 1)^2}{36} - \dfrac{(x - 2)^2}{9} = 1$

50. $3(x + 3)^2 - 12(y - 4)^2 = 72$

51. $8x^2 - 32x - 10y^2 - 60y - 138 = 0$

52. $2x^2 + 12x - 8y^2 + 16y + 6 = 0$

Find the standard form of an equation of the hyperbola with the given characteristics.

53. Vertices: $(0, 3)$ and $(8, 3)$ Foci: $(-1, 3)$ and $(9, 3)$

54. Vertices: $(4, -2)$ and $(4, 0)$ Foci: $(4, -3)$ and $(4, 1)$

Applications

55. Ship Navigation. Two loran stations are located 220 miles apart along a coast. If a ship records a time difference of 0.00048 second and continues on the hyperbolic path corresponding to that difference, where would it reach shore? Assume that the speed of radio signals is 186,000 miles per second.

56. Ship Navigation. Two loran stations are located 400 miles apart along a coast. If a ship records a time difference of 0.0008 second and continues on the hyperbolic path corresponding to that difference, where would it reach shore?

11.5   Systems of Nonlinear Equations

Solve the system of equations with the elimination method.

57. $\begin{aligned} x^2 + y &= -3 \\ x - y &= 5 \end{aligned}$

58. $\begin{aligned} x^2 + y^2 &= 4 \\ x^2 + y &= 2 \end{aligned}$

59. $\begin{aligned} x^2 + y^2 &= 5 \\ 2x^2 - y &= 0 \end{aligned}$

60. $\begin{aligned} x^2 + y^2 &= 16 \\ 6x^2 + y^2 &= 16 \end{aligned}$

Solve the system of equations with the substitution method.

61. $\begin{aligned} x + y &= 3 \\ x^2 + y^2 &= 4 \end{aligned}$

62. $\begin{aligned} xy &= 4 \\ x^2 + y^2 &= 16 \end{aligned}$

63. $\begin{aligned} x^2 + xy + y^2 &= -12 \\ x - y &= 2 \end{aligned}$

64. $\begin{aligned} 3x + y &= 3 \\ x - y^2 &= -9 \end{aligned}$

Solve the system of equations by applying any method.

65. $\begin{aligned} x^3 - y^3 &= -19 \\ x - y &= -1 \end{aligned}$

66. $\begin{aligned} 2x^2 + 4xy &= 9 \\ x^2 - 2xy &= 0 \end{aligned}$

67. $\begin{aligned} \dfrac{2}{x^2} + \dfrac{1}{y^2} &= 15 \\ \dfrac{1}{x^2} - \dfrac{1}{y^2} &= -3 \end{aligned}$

68. $\begin{aligned} x^2 + y^2 &= 2 \\ x^2 + y^2 &= 4 \end{aligned}$

11.6   Systems of Nonlinear Inequalities

Graph the nonlinear inequality.

69. $y \geq x^2 + 3$

70. $x^2 + y^2 > 16$

71. $y \leq e^x$

72. $y < -x^3 + 2$

73. $y \geq \ln(x - 1)$

74. $9x^2 + 4y^2 \leq 36$

Solve each system of inequalities and shade the region on a graph, or indicate that the system has no solution.

75. $\begin{aligned} y &\geq x^2 - 2 \\ y &\leq -x^2 + 2 \end{aligned}$

76. $\begin{aligned} x^2 + y^2 &\leq 4 \\ y &\leq x \end{aligned}$

77. $\begin{aligned} y &\geq (x + 1)^2 - 2 \\ y &\leq 10 - x \end{aligned}$

78. $\begin{aligned} 3x^2 + 3y^2 &\leq 27 \\ y &\geq x - 1 \end{aligned}$

79. $\begin{aligned} 4y^2 - 9x^2 &\leq 36 \\ y &\geq x + 1 \end{aligned}$

80. $\begin{aligned} 9x^2 + 16y^2 &\leq 144 \\ y &\geq 1 - x^2 \end{aligned}$

11.7   Rotation of Axes

The coordinates of a point in the xy-coordinate system are given. Assuming the X- and Y-axes are found by rotating the x- and y-axes by an angle θ, find the corresponding coordinates for the point in the XY-system.

81. $(-3, 2)$, $\theta = 60°$

82. $(4, -3)$, $\theta = 45°$

Transform the equation of the conic into an equation in X and Y (without an XY-term) by rotating the x- and y-axes through an angle θ. Then graph the resulting equation.

83. $2x^2 + 4\sqrt{3}xy - 2y^2 - 16 = 0$, $\theta = 30°$

84. $25x^2 + 14xy + 25y^2 - 288 = 0$, $\theta = \dfrac{\pi}{4}$

Determine the angle of rotation necessary to transform the equation in x and y into an equation in X and Y with no XY-term.

85. $4x^2 + 2\sqrt{3}xy + 6y^2 - 9 = 0$

86. $4x^2 + 5xy + 4y^2 - 11 = 0$

Graph the second-degree equation.

87. $x^2 + 2xy + y^2 + \sqrt{2}x - \sqrt{2}y + 8 = 0$

88. $76x^2 + 48\sqrt{3}xy + 28y^2 - 100 = 0$

11.8 Polar Equations of Conics

Find the polar equation that represents the conic described.

89. An ellipse with eccentricity $e = \frac{3}{7}$ and directrix $y = -7$

90. A parabola with directrix $x = 2$

Identify the conic (parabola, ellipse, or hyperbola) that each polar equation represents.

91. $r = \dfrac{6}{4 - 5\cos\theta}$ **92.** $r = \dfrac{2}{5 + 3\sin\theta}$

For the given polar equations, find the eccentricity and vertex (or vertices), and graph the curve.

93. $r = \dfrac{4}{2 + \cos\theta}$ **94.** $r = \dfrac{6}{1 - \sin\theta}$

11.9 Parametric Equations and Graphs

Graph the curve defined by the parametric equations.

95. $x = \sin t, y = 4\cos t$ for t in $[-\pi, \pi]$

96. $x = 5\sin^2 t, y = 2\cos^2 t$ for t in $[-\pi, \pi]$

97. $x = 4 - t^2, y = t^2$ for t in $[-3, 3]$

98. $x = t + 3, y = 4$ for t in $[-4, 4]$

The given parametric equations define a plane curve. Find an equation in rectangular form that also corresponds to the plane curve.

99. $x = 4 - t^2, y = t$ **100.** $x = 5\sin^2 t, y = 2\cos^2 t$

101. $x = 2\tan^2 t, y = 4\sec^2 t$ **102.** $x = 3t^2 + 4, y = 3t^2 - 5$

Technology Exercises

Section 11.2

103. In your mind, picture the parabola given by $(x - 0.6)^2 = -4(y + 1.2)$. Where is the vertex? Which way does this parabola open? Now plot the parabola with a graphing utility.

104. In your mind, picture the parabola given by $(y - 0.2)^2 = 3(x - 2.8)$. Where is the vertex? Which way does this parabola open? Now plot the parabola with a graphing utility.

105. Given is the parabola $y^2 + 2.8y + 3x - 6.85 = 0$.
  a. Solve the equation for y, and use a graphing utility to plot the parabola.
  b. Transform the equation into the form $(y - k)^2 = 4p(x - h)$. Find the vertex. Which way does the parabola open?
  c. Do (a) and (b) agree with each other?

106. Given is the parabola $x^2 - 10.2x - y + 24.8 = 0$.
  a. Solve the equation for y, and use a graphing utility to plot the parabola.
  b. Transform the equation into the form $(x - h)^2 = 4p(y - k)$. Find the vertex. Which way does the parabola open?
  c. Do (a) and (b) agree with each other?

Section 11.3

107. Graph the following three ellipses: $4x^2 + y^2 = 1$, $4(2x)^2 + y^2 = 1$, and $4(3x)^2 + y^2 = 1$. What can be said to happen to ellipse $4(cx)^2 + y^2 = 1$ as c increases?

108. Graph the following three ellipses: $x^2 + 4y^2 = 1$, $x^2 + 4(2y)^2 = 1$, and $x^2 + 4(3y)^2 = 1$. What can be said to happen to ellipse $x^2 + 4(cy)^2 = 1$ as c increases?

Section 11.4

109. Graph the following three hyperbolas: $4x^2 - y^2 = 1$, $4(2x)^2 - y^2 = 1$, and $4(3x)^2 - y^2 = 1$. What can be said to happen to hyperbola $4(cx)^2 - y^2 = 1$ as c increases?

110. Graph the following three hyperbolas: $x^2 - 4y^2 = 1$, $x^2 - 4(2y)^2 = 1$, and $x^2 - 4(3y)^2 = 1$. What can be said to happen to hyperbola $x^2 - 4(cy)^2 = 1$ as c increases?

Section 11.5

With a graphing utility, solve the following systems of equations.

111. $7.5x^2 + 1.5y^2 = 12.25$
$x^2y = 1$

112. $4x^2 + 2xy + 3y^2 = 12$
$x^3y = 3 - 3x^3$

Section 11.6

With a graphing utility, graph the following systems of nonlinear inequalities.

113. $y \geq 10^x - 1$
$y \leq 1 - x^2$

114. $x^2 + 4y^2 \leq 36$
$y \geq e^x$

Section 11.7

115. With a graphing utility, explore the second-degree equation $Ax^2 + Bxy + Cy^2 + 10x - 8y - 5 = 0$ for the following values of A, B, and C:

a. $A = 2, B = -3, C = 5$
b. $A = 2, B = 3, C = -5$

Show the angle of rotation to one decimal place. Explain the differences.

116. With a graphing utility, explore the second-degree equation $Ax^2 + Bxy + Cy^2 + 2x - y = 0$ for the following values of A, B, and C:

a. $A = 1, B = -2, C = -1$
b. $A = 1, B = 2, D = 1$

Show the angle of rotation to the nearest degree. Explain the differences.

Section 11.8

117. Let us consider the polar equation $r = \dfrac{8}{4 + 5\sin\theta}$. Explain why a graphing utility gives the following graph with the specified window parameters:

$$[-6, 6] \text{ by } [-3, 9] \text{ with } \theta \text{ step} = \frac{\pi}{4}$$

118. Let us consider the polar equation $r = \dfrac{9}{3 - 2\sin\theta}$. Explain why a graphing utility gives the following graph with the specified window parameters:

$$[-6, 6] \text{ by } [-3, 9] \text{ with } \theta \text{ step} = \frac{\pi}{2}$$

Section 11.9

119. Consider the parametric equations $x = a\cos at + b\sin bt$ and $y = a\sin at + b\cos bt$. Use a graphing utility to explore the graphs for $(a, b) = (2, 3)$ and $(a, b) = (3, 2)$. Describe the t-interval for each case.

120. Consider the parametric equations $x = a\sin at - b\cos bt$ and $y = a\cos at - b\sin bt$. Use a graphing utility to explore the graphs for $(a, b) = (1, 2)$ and $(a, b) = (2, 1)$. Describe the t-interval for each case.

Match the equation to the graph.

1. $x = 16y^2$

2. $y = 16x^2$

3. $x^2 + 16y^2 = 1$

4. $x^2 - 16y^2 = 1$

5. $16x^2 + y^2 = 1$

6. $16y^2 - x^2 = 1$

a.

b.

c.

d.

e.

f.

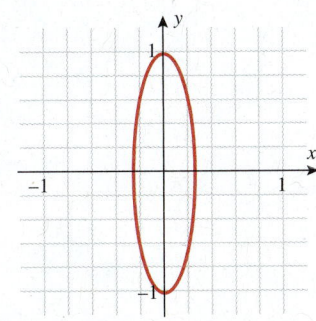

Find the equation of the conic with the given characteristics.

7. Parabola   vertex: $(0, 0)$ focus: $(-4, 0)$

8. Parabola   vertex: $(0, 0)$ directrix: $y = 2$

9. Parabola   vertex: $(-1, 5)$ focus: $(-1, 2)$

10. Parabola   vertex: $(2, -3)$ directrix: $x = 0$

11. Ellipse   center: $(0, 0)$
vertices: $(0, -4), (0, 4)$
foci: $(0, -3), (0, 3)$

12. Ellipse   center: $(0, 0)$
vertices: $(-3, 0), (3, 0)$
foci: $(-1, 0), (1, 0)$

13. Ellipse   vertices: $(2, -6), (2, 6)$
foci: $(2, -4), (2, 4)$

14. Ellipse   vertices: $(-7, -3), (-4, -3)$
foci: $(-6, -3), (-5, -3)$

15. Hyperbola   vertices: $(-1, 0)$ and $(1, 0)$
asymptotes: $y = -2x$ and $y = 2x$

16. Hyperbola   vertices: $(0, -1)$ and $(0, 1)$
asymptotes: $y = -\frac{1}{3}x$ and $y = \frac{1}{3}x$

17. Hyperbola   foci: $(2, -6), (2, 6)$
vertices: $(2, -4), (2, 4)$

18. Hyperbola   foci: $(-7, -3), (-4, -3)$
vertices: $(-6, -3), (-5, -3)$

Graph the following equations.

19. $9x^2 + 18x - 4y^2 + 16y - 43 = 0$

20. $4x^2 - 8x + y^2 + 10y + 28 = 0$

21. $y^2 + 4y - 16x + 20 = 0$

22. $x^2 - 4x + y + 1 = 0$

23. Eyeglass Lens. Eyeglass lenses can be thought of as very wide parabolic curves. If the focus occurs 1.5 centimeters from the center of the lens, and the lens at its opening is 4 centimeters across, find an equation that governs the shape of the lens.

24. Planetary Orbits. The planet Uranus's orbit is described in the following picture with the Sun as a focus of the elliptical orbit. Write an equation for the orbit.

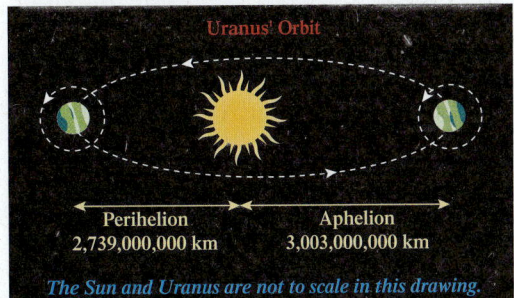

Uranus' Orbit

Perihelion 2,739,000,000 km   Aphelion 3,003,000,000 km

The Sun and Uranus are not to scale in this drawing.

Graph the following nonlinear inequalities.

25. $y < x^3 + 1$

26. $y^2 \geq 16x$

Graph the following systems of nonlinear inequalities.

27. $y \leq 4 - x^2$
$16x^2 + 25y^2 \leq 400$

28. $y \leq e^{-x}$
$y \geq x^2 - 4$

29. Identify the conic represented by the equation

$$r = \frac{12}{3 + 2\sin\theta}.$$ State the eccentricity.

30. Use rotation of axes to transform the equation in x and y into an equation in X and Y that has no XY-term: $6\sqrt{3}x^2 + 6xy + 4\sqrt{3}y^2 = 21\sqrt{3}$. State the rotation angle.

31. A golf ball is hit with an initial speed of 120 feet per second at an angle of $45°$ with the ground. How long will the ball stay in the air? How far will the ball travel (horizontal distance) before it hits the ground?

32. Describe (classify) the plane curve defined by the parametric equations $x = \sqrt{1-t}$ and $y = \sqrt{t}$ for t in $[0, 1]$.

33. Use a graphing utility to graph the following nonlinear inequality:

$$x^2 + 4xy - 9y^2 - 6x + 8y + 28 \le 0$$

34. Use a graphing utility to solve the following systems of equations:

$$0.1225x^2 + 0.0289y^2 = 1$$
$$y^3 = 11x$$

Round your answers to three decimal places.

35. Given is the parabola $x^2 + 4.2x - y + 5.61 = 0$.
  a. Solve the equation for y and use a graphing utility to plot the parabola.
  b. Transform the equation into the form $(x - h)^2 = 4p(y - k)$. Find the vertex. Which way does the parabola open?
  c. Do (a) and (b) agree with each other?

36. With a graphing utility, explore the second-degree equation $Ax^2 + Bxy + Cy^2 + 10x - 8y - 5 = 0$ for the following values of A, B, and C:
  a. $A = 2, B = -\sqrt{3}, C = 1$
  b. $A = 2, B = \sqrt{3}, C = -1$

Show the angle of rotation to one decimal place. Explain the differences.

PRACTICE TEST

1. Solve for x: $(x + 2)^2 - (x + 2) - 20 = 0$.

2. Find an equation of a circle centered at $(5, 1)$ and passing through the point $(6, -2)$.

3. Evaluate the difference quotient $\dfrac{f(x + h) - f(x)}{h}$ for the function $f(x) = 8 - 7x$.

4. Write an equation that describes the following variation: I is directly proportional to both P and t, and $I = 90$ when $P = 1500$ and $t = 2$.

5. Find the quadratic function that has vertex $(7, 7)$ and goes through the point $(10, 10)$.

6. **Compound Interest.** How much money should you put in a savings account now that earns 4.7% interest a year compounded weekly if you want to have $65,000 in 17 years?

7. Solve the logarithmic equation exactly: $\log x^2 - \log 16 = 0$.

8. In a 30°-60°-90° triangle, if the shortest leg has length 8 inches, what are the lengths of the other leg and the hypotenuse?

9. Use a calculator to evaluate $\cot(-27°)$. Round your answer to four decimal places.

10. **Sound Waves.** If a sound wave is represented by $y = 0.007 \sin 850\pi t$ cm, what are its amplitude and frequency?

11. For the trigonometric expression $\tan\theta(\csc\theta + \cos\theta)$, perform the operations and simplify. Write the answer in terms of $\sin\theta$ and $\cos\theta$.

12. Find the exact value of $\cos\left(-\dfrac{11\pi}{12}\right)$.

13. Solve the trigonometric equation $4\cos^2 x + 4\cos 2x + 1 = 0$ exactly over the interval $0 \le \theta \le 2\pi$.

14. **Airplane Speed.** A plane flew due north at 450 mph for 2 hours. A second plane, starting at the same point and at the same time, flew southeast at an angle of 135° clockwise from due north at 375 mph for 2 hours. At the end of 2 hours, how far apart were the two planes? Round to the nearest mile.

15. Find the vector with magnitude $|\mathbf{u}| = 15$ and direction angle $\theta = 110°$.

16. Given $z_1 = 5[\cos(15°) + i\sin(15°)]$ and $z_2 = 2[\cos(75°) + i\sin(75°)]$, find the product $z_1 z_2$ and express it in rectangular form.

17. At a food court, 3 medium sodas and 2 soft pretzels cost $6.77. A second order of 5 medium sodas and 4 soft pretzels costs $12.25. Find the cost of a soda and the cost of a soft pretzel.

18. Find the partial fraction decomposition for the rational expression $\dfrac{3x + 5}{(x - 3)(x^2 + 5)}$.

19. Graph the system of inequalities or indicate that the system has no solution.
$$y \ge 3x - 2$$
$$y \le 3x + 2$$

20. Solve the system using Gauss–Jordan elimination.
$$\begin{aligned} x - 2y + z &= 7 \\ -3x + y + 2z &= -11 \end{aligned}$$

21. Given $A = \begin{bmatrix} 3 & 4 & -7 \\ 0 & 1 & 5 \end{bmatrix}$, $B = \begin{bmatrix} 8 & -2 & 6 \\ 9 & 0 & -1 \end{bmatrix}$, and $C = \begin{bmatrix} 9 & 0 \\ 1 & 2 \end{bmatrix}$, find $2B - 3A$.

22. Use Cramer's rule to solve the system of equations.
$$\begin{aligned} 25x + 40y &= -12 \\ 75x - 105y &= 69 \end{aligned}$$

23. Find the standard form of the equation of an ellipse with foci $(6, 2)$ and $(6, -6)$ and vertices $(6, 3)$ and $(6, -7)$.

24. Find the standard form of the equation of a hyperbola with vertices $(5, -2)$ and $(5, 0)$ and foci $(5, -3)$ and $(5, 1)$.

25. Solve the system of equations.
$$\begin{aligned} x + y &= 6 \\ x^2 + y^2 &= 20 \end{aligned}$$

26. Use a graphing utility to graph the following equation:
$$x^2 - 3xy + 10y^2 - 1 = 0$$

27. Use a graphing utility to graph the following system of nonlinear inequalities:
$$y \ge e^{-0.3x} - 3.5$$
$$y \le 4 - x^2$$

Sequences, Series, and Probability

A BASIC STRATEGY FOR BLACKJACK

← Dealer's Up Card →

Your Hand		2	3	4	5	6	7	8	9	10	A
I	17+	S	S	S	S	S	S	S	S	S	S
	16	S	S	S	S	S	H	H	H	H	H
	15	S	S	S	S	S	H	H	H	H	H
	14	S	S	S	S	S	H	H	H	H	H
	13	S	S	S	S	S	H	H	H	H	H
	12	H	H	S	S	S	H	H	H	H	H
II	11	D	D	D	D	D	D	D	D	D	H
	10	D	D	D	D	D	D	D	D	D	H
	9	H	D	D	D	D	H	H	H	H	H
	5 - 8	H	H	H	H	H	H	H	H	H	H
III	A, 8 - 10	S	S	S	S	S	S	S	S	S	S
	A, 7	S	D	D	D	D	S	S	H	H	H
	A, 6	H	D	D	D	D	H	H	H	H	H
	A, 5	H	H	D	D	D	H	H	H	H	H
	A, 4	H	H	D	D	D	H	H	H	H	H
	A, 3	H	H	H	D	D	H	H	H	H	H
	A, 2	H	H	H	D	D	H	H	H	H	H
IV	A, A; 8, 8	SP	SP	SP	SP	SP	SP	SP	SP	SP	SP
	10, 10	S	S	S	S	S	S	S	S	S	S
	9, 9	SP	SP	SP	SP	SP	S	SP	SP	S	S
	7, 7	SP	SP	SP	SP	SP	SP	H	H	H	H
	6, 6	H	SP	SP	SP	SP	H	H	H	H	H
	5, 5	D	D	D	D	D	D	D	D	H	H
	4, 4	H	H	H	H	H	H	H	H	H	H
	3, 3	H	H	SP	SP	SP	SP	H	H	H	H
	2, 2	H	H	SP	SP	SP	SP	H	H	H	H

HIT **STAND** **DOUBLE DOWN** **SPLIT**

When surrender is allowed, surrender 9, 7 or 10, 6 vs 9, 10, A; 9, 6 or 10, 5 vs 10
When doubling down after splitting is allowed, split: 2's, 3's, 7's vs 2-7; 4's vs 5 or 6; 6's vs 2-6

Have you ever been to a casino and played blackjack? It is the only game in the casino that you can win based on the law of large numbers. In the early 1990s a group of math and science majors from the Massachusetts Institute of Technology (MIT) devised a foolproof scheme to win at blackjack. A professor at MIT developed a basic strategy outlined in the figure on the right that is based on the probability of combinations of particular cards being dealt, given certain cards already showing. To play blackjack (also called 21), each person is dealt two cards with the option of taking additional cards. The goal is to get a combination of cards that is worth 21 points (or less) without going over (called a bust). You have to avoid going over 21 or staying too far below 21. All face cards (jacks, queens, and kings) are worth 10 points, and an ace in blackjack is worth either 1 or 11 points. The students used the professor's strategy along with a card-counting technique to place higher bets when there were more high-value cards left in the deck.

It is reported that in 1992 the team won $4 million from Las Vegas casinos. The casinos caught on, and the students were all banned within 2 years. The 2008 movie *21* was based on this event.

SEQUENCES, SERIES, AND PROBABILITY

12.1 Sequences and Series	**12.2** Arithmetic Sequences and Series	**12.3** Geometric Sequences and Series	**12.4** Mathematical Induction	**12.5** The Binomial Theorem	**12.6** Counting, Permutations, and Combinations	**12.7** Probability
• Sequences • Factorial Notation • Recursion Formulas • Sums and Series	• Arithmetic Sequences • The General (*n*th) Term of an Arithmetic Sequence • The Sum of an Arithmetic Sequence	• Geometric Sequences • The General (*n*th) Term of a Geometric Sequence • Geometric Series	• Mathematical Induction	• Binomial Coefficients • Binomial Expansion • Pascal's Triangle • Finding a Particular Term of a Binomial Expansion	• The Fundamental Counting Principle • Permutations • Combinations • Permutations with Repetition	• Sample Space • Probability of an Event • Probability of an Event Not Occurring • Mutually Exclusive Events • Independent Events

LEARNING OBJECTIVES

- Find the general *n*th term of a sequence or series.
- Evaluate a finite arithmetic series.
- Determine if an infinite geometric series converges or diverges.
- Prove a mathematical statement using induction.
- Use the binomial theorem to expand a binomial raised to a positive integer power.
- Understand the difference between permutations and combinations.
- Calculate the probability of an event.

Sequences

The word *sequence* means an order in which one thing follows another in succession. In mathematics, it means the same thing. For example, if we write $x, 2x^2, 3x^3, 4x^4, 5x^5, ?$, what would the next term in the *sequence* be, the one where the question mark now stands? The answer is $6x^6$.

DEFINITION   **A Sequence**

A **sequence** is a function whose domain is a set of positive integers. The function values, or **terms**, of the sequence are written as

$$a_1, a_2, a_3, \ldots, a_n, \ldots$$

Rather than using function notation, sequences are usually written with subscript (or index) notation, $a_{subscript}$.

A **finite sequence** has the domain $\{1, 2, 3, \ldots, n\}$ for some positive integer n. An **infinite sequence** has the domain of all positive integers $\{1, 2, 3, \ldots\}$. There are times when it is convenient to start the indexing at 0 instead of 1:

$$a_0, a_1, a_2, a_3, \ldots, a_n, \ldots$$

Sometimes a pattern in the sequence can be obtained and the sequence can be written using a *general term*. In the previous example, $x, 2x^2, 3x^3, 4x^4, 5x^5, 6x^6, \ldots$, each term has the same exponent and coefficient. We can write this sequence as $a_n = nx^n$, $n = 1, 2, 3, 4, 3, 6, \ldots$, where a_n is called the **general term**.

EXAMPLE 1   Finding the Sequence, Given the General Term

Find the first four ($n = 1, 2, 3, 4$) terms of the sequences, given the general term.

a. $a_n = 2n - 1$

b. $b_n = \dfrac{(-1)^n}{n + 1}$

Solution (a): $a_n = 2n - 1$

Find the first term, $n = 1$. $a_1 = 2(\mathbf{1}) - 1 = 1$

Find the second term, $n = 2$. $a_2 = 2(\mathbf{2}) - 1 = 3$

Find the third term, $n = 3$. $a_3 = 2(\mathbf{3}) - 1 = 5$

Find the fourth term, $n = 4$. $a_4 = 2(\mathbf{4}) - 1 = 7$

The first four terms of the sequence are $\boxed{1, 3, 5, 7}$.

Solution (b): $b_n = \dfrac{(-1)^n}{n + 1}$

Find the first term, $n = 1$. $b_1 = \dfrac{(-1)^{\mathbf{1}}}{\mathbf{1} + 1} = -\dfrac{1}{2}$

Find the second term, $n = 2$. $b_2 = \dfrac{(-1)^{\mathbf{2}}}{\mathbf{2} + 1} = \dfrac{1}{3}$

Find the third term, $n = 3$. $b_3 = \dfrac{(-1)^{\mathbf{3}}}{\mathbf{3} + 1} = -\dfrac{1}{4}$

Find the fourth term, $n = 4$. $b_4 = \dfrac{(-1)^{\mathbf{4}}}{\mathbf{4} + 1} = \dfrac{1}{5}$

The first four terms of the sequence are $\boxed{-\dfrac{1}{2}, \dfrac{1}{3}, -\dfrac{1}{4}, \dfrac{1}{5}}$.

■ **YOUR TURN** Find the first four terms of the sequence $a_n = \dfrac{(-1)^n}{n^2}$.

EXAMPLE 2   Finding the General Term, Given Several Terms of the Sequence

Find the general term of the sequence, given the first five terms.

a. $1, \dfrac{1}{4}, \dfrac{1}{9}, \dfrac{1}{16}, \dfrac{1}{25}, \cdots$ **b.** $-1, 4, -9, 16, -25, \cdots$

Solution (a):

Write 1 as $\dfrac{1}{1}$. $\dfrac{1}{1}, \dfrac{1}{4}, \dfrac{1}{9}, \dfrac{1}{16}, \dfrac{1}{25}, \cdots$

Notice that each denominator is an integer squared. $\dfrac{1}{1^2}, \dfrac{1}{2^2}, \dfrac{1}{3^2}, \dfrac{1}{4^2}, \dfrac{1}{5^2}, \cdots$

Identify the general term. $\boxed{a_n = \dfrac{1}{n^2} \qquad n = 1, 2, 3, 4, 5, \cdots}$

Solution (b):

Notice that each term includes an integer squared. $-1^2, 2^2, -3^2, 4^2, -5^2, \cdots$

Identify the general term. $\boxed{b_n = (-1)^n n^2 \qquad n = 1, 2, 3, 4, 5, \cdots}$

■ **YOUR TURN** Find the general term of the sequence, given the first five terms.

a. $-\dfrac{1}{2}, \dfrac{1}{4}, -\dfrac{1}{6}, \dfrac{1}{8}, -\dfrac{1}{10}, \cdots$ **b.** $\dfrac{1}{2}, \dfrac{1}{4}, \dfrac{1}{8}, \dfrac{1}{16}, \dfrac{1}{32}, \cdots$

Classroom Example 12.1.1
Find the first four terms of the sequences.

a. $a_n = -1 - 2n^2$

b.* $b_n = \dfrac{(-1)^n n}{1 - 3(-1)^n}$

Answer:

a. $-3, -9, -19, -33$

b. $-\dfrac{1}{4}, -1, -\dfrac{3}{4}, -2$

■ **Answer:** $-1, \dfrac{1}{4}, -\dfrac{1}{9}, \dfrac{1}{16}$

Classroom Example 12.1.2
Find the general term of the sequences.

a. $\dfrac{1}{2}, -\dfrac{1}{4}, \dfrac{1}{8}, -\dfrac{1}{16}, \cdots$

b.* $3, 0, 5, -2, 7, -4, \cdots$

c. $\sqrt{2}, \sqrt[4]{2}, \sqrt[8]{2}, \cdots$

Answer:

a. $(-1)^{n+1}\left(\dfrac{1}{2}\right)^n$

b. $2 - (-1)^n n$

c. $2^{1/2^n}$

Study Tip

$(-1)^n$ or $(-1)^{n+1}$ is a way to represent an alternating sequence.

■ **Answer:**

a. $a_n = \dfrac{(-1)^n}{2n}$ **b.** $a_n = \dfrac{1}{2^n}$

Parts (b) in both Example 1 and Example 2 are called **alternating** sequences because the terms alternate signs (positive and negative). If the odd terms, $a_1, a_3, a_5, \ldots$ are negative and the even terms, $a_2, a_4, a_6, \ldots$, are positive, we include $(-1)^n$ in the general term. If the opposite is true, and the odd terms are positive and the even terms are negative, we include $(-1)^{n+1}$ in the general term.

Factorial Notation

Many important sequences that arise in mathematics involve terms that are defined with products of consecutive positive integers. The products are expressed in *factorial notation*.

DEFINITION Factorial

If n is a positive integer, then $n!$ (stated as "n factorial") is the product of all positive integers from n down to 1.

$$n! = n(n-1)(n-2)\cdots 3 \cdot 2 \cdot 1 \qquad n \geq 2$$

and $0! = 1$ and $1! = 1$.

The values of $n!$ for the first six nonnegative integers are

$$0! = 1$$
$$1! = 1$$
$$2! = 2 \cdot 1 = 2$$
$$3! = 3 \cdot 2 \cdot 1 = 6$$
$$4! = 4 \cdot 3 \cdot 2 \cdot 1 = 24$$
$$5! = 5 \cdot 4 \cdot 3 \cdot 2 \cdot 1 = 120$$

Notice that $4! = 4 \cdot 3 \cdot 2 \cdot 1 = 4 \cdot 3!$. In general, we can apply the formula $n! = n[(n-1)!]$. Often the brackets are not used, and the notation $n! = n(n-1)!$ implies calculating the factorial $(n-1)!$ and then multiplying that quantity by n. For example, to find $6!$, we employ the relationship $n! = n(n-1)!$ and set $n = 6$:

$$6! = 6 \cdot 5! = 6 \cdot 120 = 720$$

EXAMPLE 3 Finding the Terms of a Sequence Involving Factorials

Find the first four terms of the sequence, given the general term $a_n = \dfrac{x^n}{n!}$.

Solution:

Find the first term, $n = 1$. $\qquad a_1 = \dfrac{x^1}{1!} = x$

Find the second term, $n = 2$. $\qquad a_2 = \dfrac{x^2}{2!} = \dfrac{x^2}{2 \cdot 1} = \dfrac{x^2}{2}$

Find the third term, $n = 3$. $\qquad a_3 = \dfrac{x^3}{3!} = \dfrac{x^3}{3 \cdot 2 \cdot 1} = \dfrac{x^3}{6}$

Find the fourth term, $n = 4$. $\qquad a_4 = \dfrac{x^4}{4!} = \dfrac{x^4}{4 \cdot 3 \cdot 2 \cdot 1} = \dfrac{x^4}{24}$

The first four terms of the sequence are $\boxed{x, \dfrac{x^2}{2}, \dfrac{x^3}{6}, \dfrac{x^4}{24}}$.

Technology Tip

Find $0!, 1!, 2!, 3!, 4!,$ and $5!$.

Scientific calculators:

Press	Display
0 ! =	1
1 ! =	1
2 ! =	2
3 ! =	6
4 ! =	24
5 ! =	120

Graphing calculators:

Press	Display
0 MATH PRB 4:! ENTER	1
1 MATH PRB 4:! ENTER	1
2 MATH PRB 4:! ENTER	2
3 MATH PRB 4:! ENTER	6
4 MATH PRB 4:! ENTER	24
5 MATH PRB 4:! ENTER	120

```
MATH NUM CPX PRB
1:rand
2:nPr
3:nCr
4:!
5:randInt(
6:randNorm(
7:randBin(
```

```
0!
              1
1!
              1
2!
              2
```

```
3!
              6
4!
             24
5!
            120
```

Classroom Example 12.1.3

Find the first three terms of the sequences.

a. $a_n = \dfrac{(-1)^n x^{2n}}{(2n)!}$ **b.*** $b_n = \dfrac{n!}{(-n)^n}$

Answer:

a. $-\dfrac{x^2}{2}, \dfrac{x^4}{24}, -\dfrac{x^6}{720}$ **b.** $-1, \dfrac{1}{2}, -\dfrac{2}{9}$

EXAMPLE 4   Evaluating Expressions with Factorials

Evaluate each factorial expression.

a. $\dfrac{6!}{2! \cdot 3!}$ **b.** $\dfrac{(n+1)!}{(n-1)!}$

Solution (a):

Expand each factorial in the numerator and denominator.

$$\frac{6!}{2! \cdot 3!} = \frac{6 \cdot 5 \cdot 4 \cdot \cancel{3} \cdot \cancel{2} \cdot \cancel{1}}{2 \cdot 1 \cdot \cancel{3} \cdot \cancel{2} \cdot \cancel{1}}$$

Cancel the $3 \cdot 2 \cdot 1$ in both the numerator and denominator.

$$= \frac{6 \cdot 5 \cdot 4}{2 \cdot 1}$$

Simplify.

$$= \frac{6 \cdot 5 \cdot 2}{1} = 60$$

$$\boxed{\frac{6!}{2! \cdot 3!} = 60}$$

Solution (b):

Expand each factorial in the numerator and denominator.

$$\frac{(n+1)!}{(n-1)!} = \frac{(n+1)(n)(n-1)(n-2)\cdots 3 \cdot 2 \cdot 1}{(n-1)(n-2)\cdots 3 \cdot 2 \cdot 1}$$

Cancel the $(n-1)(n-2)\cdots 3 \cdot 2 \cdot 1$ in the numerator and denominator.

$$\frac{(n+1)!}{(n-1)!} = (n+1)(n)$$

Alternatively,

$$\frac{(n+1)!}{(n-1)!} = \frac{(n+1)(n)\cancel{(n-1)!}}{\cancel{(n-1)!}}$$

$$\boxed{\frac{(n+1)!}{(n-1)!} = n^2 + n}$$

COMMON MISTAKE

In Example 4 we found $\dfrac{6!}{2! \cdot 3!} = 60$. It is important to note that $2! \cdot 3! \neq 6!$.

■ **YOUR TURN** Evaluate the factorial expressions.

a. $\dfrac{3! \cdot 4!}{2! \cdot 6!}$.   **b.** $\dfrac{(n+2)!}{n!}$

Recursion Formulas

Another way to define a sequence is **recursively**, or using a **recursion formula**. The first few terms are listed, and the recursion formula determines the remaining terms based on previous terms. For example, the famous Fibonacci sequence is 1, 1, 2, 3, 5, 8, 13, 21, 34, 55, 89,…. Each term in the Fibonacci sequence is found by adding the previous two terms.

$$1 + 1 = 2 \qquad 1 + 2 = 3 \qquad 2 + 3 = 5$$
$$3 + 5 = 8 \qquad 5 + 8 = 13 \qquad 8 + 13 = 21$$
$$13 + 21 = 34 \qquad 21 + 34 = 55 \qquad 34 + 55 = 89$$

We can define the Fibonacci sequence using a general term:

$$a_1 = 1, \, a_2 = 1, \text{ and } a_n = a_{n-2} + a_{n-1} \qquad n \geq 3$$

The Fibonacci sequence is found in places we least expect them (for example, pineapples, broccoli, and flowers). The number of petals in a flower is a Fibonacci number. For example, a wild rose has 5 petals, lilies and irises have 3 petals, and daisies have 34, 55, or even 89 petals. The number of spirals in an Italian broccoli is a Fibonacci number (13).

EXAMPLE 5 **Using a Recursion Formula to Find a Sequence**

Find the first four terms of the sequence: $a_1 = 2$ and $a_n = 2a_{n-1} - 1$, $n \geq 2$.

Solution:

Write the first term, $n = 1$.	$a_1 = \mathbf{2}$
Find the second term, $n = 2$.	$a_2 = 2a_1 - 1 = 2(\mathbf{2}) - 1 = \mathbf{3}$
Find the third term, $n = 3$.	$a_3 = 2a_2 - 1 = 2(\mathbf{3}) - 1 = \mathbf{5}$
Find the fourth term, $n = 4$.	$a_4 = 2a_3 - 1 = 2(\mathbf{5}) - 1 = 9$

The first four terms of the sequence are $\boxed{2, 3, 5, 9}$.

■ **Answer:** $1, \frac{1}{2}, \frac{1}{12}, \frac{1}{288}$

■ **YOUR TURN** Find the first four terms of the sequence:

$$a_1 = 1 \quad \text{and} \quad a_n = \frac{a_{n-1}}{n!} \quad n \geq 2$$

Classroom Example 12.1.5
Find the first four terms of the sequences.

a. $a_1 = 1, a_n = \sqrt{a_{n-1}} + 1, n \geq 2$
b. $a_1 = -3, a_n = (-1)^n a_{n-1}, n \geq 2$
c.* $a_1 = 2, a_n = n^{-n} a_{n-1}, n \geq 2$

Answer:

a. $1, 2, 1 + \sqrt{2}, 1 + \sqrt{1 + \sqrt{2}}$
b. $-3, -3, 3, 3$
c. $2, \frac{1}{2}, \frac{1}{54}, \frac{1}{13824}$

Sums and Series

When we add the terms in a sequence, the result is a *series*.

DEFINITION **Series**

Given the infinite sequence $a_1, a_2, a_3, \ldots, a_n, \ldots$ the sum of all of the terms in the infinite sequence is called an **infinite series** and is denoted by

$$a_1 + a_2 + a_3 + \cdots + a_n + \cdots$$

and the sum of only the first n terms is called a **finite series**, or **nth partial sum**, and is denoted by

$$S_n = a_1 + a_2 + a_3 + \cdots + a_n$$

The capital Greek letter Σ (sigma) corresponds to the capital S in our alphabet. Therefore, we use Σ as a shorthand way to represent a sum (series). For example, the sum of the first five terms of the sequence $1, 4, 9, 16, 25, \ldots, n^2, \ldots$ can be represented using **sigma (or summation) notation**:

$$\sum_{n=1}^{5} n^2 = (1)^2 + (2)^2 + (3)^2 + (4)^2 + (5)^2$$

$$= 1 + 4 + 9 + 16 + 25$$

This is read "the sum as n goes from 1 to 5 of n^2." The letter n is called the **index of summation**, and often other letters are used instead of n. It is important to note that the sum can start at other numbers besides 1.

If we wanted the sum of all of the terms in the sequence, we would represent that infinite series using summation notation as

$$\sum_{n=1}^{\infty} n^2 = 1 + 4 + 9 + 16 + 25 + \cdots$$

EXAMPLE 6 **Writing a Series Using Sigma Notation**

Write the following series using sigma notation.

a. $1 + 1 + \frac{1}{2} + \frac{1}{6} + \frac{1}{24} + \frac{1}{120}$ **b.** $8 + 27 + 64 + 125 + \cdots$

Solution (a):

Write 1 as $\frac{1}{1}$.

$$\frac{1}{1} + \frac{1}{1} + \frac{1}{2} + \frac{1}{6} + \frac{1}{24} + \frac{1}{120}$$

Notice that we can write the denominators using factorials.

$$= \frac{1}{1} + \frac{1}{1} + \frac{1}{2!} + \frac{1}{3!} + \frac{1}{4!} + \frac{1}{5!}$$

Recall that $0! = 1$ and $1! = 1$.

$$= \frac{1}{0!} + \frac{1}{1!} + \frac{1}{2!} + \frac{1}{3!} + \frac{1}{4!} + \frac{1}{5!}$$

Identify the general term.

$$a_n = \frac{1}{n!} \qquad n = 0, 1, 2, 3, 4, 5$$

Write the finite series using sigma notation.

$$\sum_{n=0}^{5} \frac{1}{n!}$$

Solution (b):

Write the infinite series as a sum of terms cubed.

$$8 + 27 + 64 + 125 + \cdots$$
$$= 2^3 + 3^3 + 4^3 + 5^3 + \cdots$$

Identify the general term of the series.

$$a_n = n^3 \qquad n \geq 2$$

Write the infinite series using sigma notation.

$$\sum_{n=2}^{\infty} n^3$$

■ **YOUR TURN** Write the following series using sigma notation.

a. $1 - \frac{1}{2} + \frac{1}{6} - \frac{1}{24} + \frac{1}{120} - \cdots$

b. $4 + 8 + 16 + 32 + 64 + \cdots$

■ **Answer:**

a. $\displaystyle\sum_{n=1}^{\infty} \frac{(-1)^{n+1}}{n!}$ **b.** $\displaystyle\sum_{n=2}^{\infty} 2^n$

Now that we are comfortable with sigma (summation) notation, let's turn our attention to evaluating a series (calculating the sum). You can always evaluate a finite series. However, you cannot always evaluate an infinite series.

EXAMPLE 7 **Evaluating a Finite Series**

Evaluate the series $\displaystyle\sum_{i=0}^{4} (2i + 1)$.

Solution:

Write out the partial sum.

$$\overset{(i=1)}{\underset{(i=0)}{\downarrow}} \qquad \overset{(i=3)}{\underset{(i=2)}{\downarrow}}$$
$$\sum_{i=0}^{4} (2i + 1) = 1 + 3 + 5 + 7 + 9$$
$$\underset{(i=0)}{\uparrow} \quad \underset{(i=2)}{\uparrow} \quad \underset{(i=4)}{\uparrow}$$

$$= 25$$

Simplify.

$$\sum_{i=0}^{4} (2i + 1) = 25$$

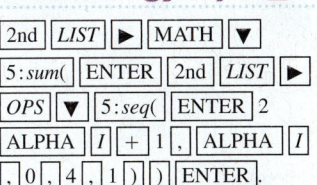
■ **Answer:** -3

■ **YOUR TURN** Evaluate the series $\displaystyle\sum_{n=1}^{5} (-1)^n n$.

Study Tip

The sum of a finite series always exists. The sum of an infinite series may or may not exist.

Infinite series may or may not have a finite sum. For example, if we keep adding $1 + 1 + 1 + 1 + \ldots$, then there is no single real number that the series sums to because the sum continues to grow without bound. However, if we add $0.9 + 0.09 + 0.009 + 0.0009 + \ldots$, this sum is $0.9999\ldots = 0.\overline{9}$, which is a rational number, and it can be proven that $0.\overline{9} = 1$.

Technology Tip

a. [2nd] [LIST] [▶] [MATH] [▼]
[5:sum(] [ENTER] [2nd] [LIST] [▶]
[OPS] [▼] [5:seq(] [ENTER] [3] [÷]
[10] [^] [ALPHA] [N] [,] [ALPHA]
[N] [,] [1] [,] [10] [,] [1] [)] [)] [ENTER].

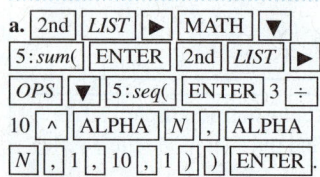

b. [2nd] [LIST] [▶] [MATH] [▼]
[5:sum(] [ENTER] [2nd] [LIST] [▶]
[OPS] [▼] [5:seq(] [ALPHA] [N]
[x^2] [,] [^] [ALPHA] [N] [,] [1] [,]
[100] [,] [1] [)] [)] [ENTER].

■ **Answer: a.** Series diverges.
  b. Series converges to $\frac{2}{3}$.

EXAMPLE 8 Evaluating an Infinite Series, If Possible

Evaluate the following infinite series, if possible.

a. $\displaystyle\sum_{n=1}^{\infty} \frac{3}{10^n}$ **b.** $\displaystyle\sum_{n=1}^{\infty} n^2$

Solution (a):

Expand the series.

$$\sum_{n=1}^{\infty} \frac{3}{10^n} = \frac{3}{10} + \frac{3}{100} + \frac{3}{1000} + \frac{3}{10,000} + \cdots$$

Write in decimal form.

$$\sum_{n=1}^{\infty} \frac{3}{10^n} = 0.3 + 0.03 + 0.003 + 0.0003 + \cdots$$

Calculate the sum.

$$\sum_{n=1}^{\infty} \frac{3}{10^n} = 0.333333\overline{3} = \frac{1}{3}$$

$$\boxed{\sum_{n=1}^{\infty} \frac{3}{10^n} = \frac{1}{3}}$$

Solution (b):

Expand the series.

$$\sum_{n=1}^{\infty} n^2 = 1 + 4 + 9 + 16 + 25 + 36 + \cdots$$

This sum is infinite since it continues to grow without any bound.

In part (a) we say that the series **converges** to $\frac{1}{3}$, and in part (b) we say that the series **diverges**.

■ **YOUR TURN** Evaluate the following infinite series, if possible.

a. $\displaystyle\sum_{n=1}^{\infty} 2n$ **b.** $\displaystyle\sum_{n=1}^{\infty} 6\left(\frac{1}{10}\right)^n$

Applications

The annual sales at Home Depot from 2000 to 2002 can be approximated by the model $a_n = 45.7 + 9.5n - 1.6n^2$, where a_n is the yearly sales in billions of dollars and $n = 0, 1, 2$.

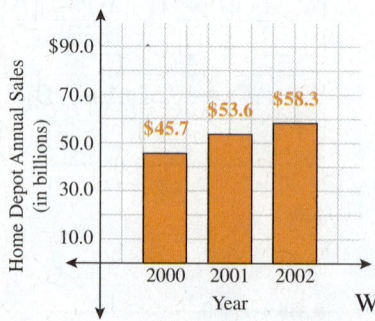

YEAR	n	$a_n = 45.7 + 9.5n - 1.6n^2$	TOTAL SALES IN BILLIONS
2000	0	$a_0 = 45.7 + 9.5(0) - 1.6(0)^2$	$45.7
2001	1	$a_1 = 45.7 + 9.5(1) - 1.6(1)^2$	$53.6
2002	2	$a_2 = 45.7 + 9.5(2) - 1.6(2)^2$	$58.3

What does the finite series $\frac{1}{3}\sum_{n=0}^{2} a_n$ tell us? It tells us the average yearly sales over 3 years.

SECTION
12.1 SUMMARY

In this section we discussed finite and infinite sequences and series. When the terms of a sequence are added together, the result is a series.

Finite sequence: $a_1, a_2, a_3, \ldots, a_n$

Infinite sequence: $a_1, a_2, a_3, \ldots, a_n, \ldots$

Finite series: $a_1 + a_2 + a_3 + \cdots + a_n$

Infinite series: $a_1 + a_2 + a_3 + \cdots + a_n + \cdots$

Factorial notation was also introduced:

$$n! = n \cdot (n - 1) \cdot \cdots \cdot 3 \cdot 2 \cdot 1 \quad n \geq 2$$

and $0! = 1$ and $1! = 1$.

The sum of a finite series is always finite.

The sum of an infinite series is either:

- convergent or
- divergent

Sigma notation is used to express a series.

Finite series: $\displaystyle\sum_{i=1}^{n} a_i = a_1 + a_2 + \cdots + a_n$

Infinite series: $\displaystyle\sum_{n=1}^{\infty} a_n = a_1 + a_2 + a_3 + \cdots$

SECTION
12.1 EXERCISES

■ SKILLS

In Exercises 1–12, write the first four terms of the sequence. Assume n starts at 1.

1. $a_n = n$

2. $a_n = n^2$

3. $a_n = 2n - 1$

4. $a_n = x^n$

5. $a_n = \dfrac{n}{(n + 1)}$

6. $a_n = \dfrac{(n + 1)}{n}$

7. $a_n = \dfrac{2^n}{n!}$

8. $a_n = \dfrac{n!}{(n + 1)!}$

9. $a_n = (-1)^n x^{n+1}$

10. $a_n = (-1)^{n+1} n^2$

11. $a_n = \dfrac{(-1)^n}{(n + 1)(n + 2)}$

12. $a_n = \dfrac{(n - 1)^2}{(n + 1)^2}$

In Exercises 13–20, find the indicated term of the sequence.

13. $a_n = \left(\dfrac{1}{2}\right)^n \quad a_9 = ?$

14. $a_n = \dfrac{n}{(n + 1)^2} \quad a_{15} = ?$

15. $a_n = \dfrac{(-1)^n n!}{(n + 2)!} \quad a_{19} = ?$

16. $a_n = \dfrac{(-1)^{n+1}(n - 1)(n + 2)}{n} \quad a_{13} = ?$

17. $a_n = \left(1 + \dfrac{1}{n}\right)^2 \quad a_{100} = ?$

18. $a_n = 1 - \dfrac{1}{n^2} \quad a_{10} = ?$

19. $a_n = \log 10^n \quad a_{23} = ?$

20. $a_n = e^{\ln n} \quad a_{49} = ?$

In Exercises 21–28, write an expression for the nth term of the given sequence.

21. $2, 4, 6, 8, 10, \ldots$

22. $3, 6, 9, 12, 15, \ldots$

23. $\dfrac{1}{2 \cdot 1}, \dfrac{1}{3 \cdot 2}, \dfrac{1}{4 \cdot 3}, \dfrac{1}{5 \cdot 4}, \dfrac{1}{6 \cdot 5}, \ldots$

24. $\dfrac{1}{2}, \dfrac{1}{4}, \dfrac{1}{8}, \dfrac{1}{16}, \dfrac{1}{32}, \ldots$

25. $-\dfrac{2}{3}, \dfrac{4}{9}, -\dfrac{8}{27}, \dfrac{16}{81}, \ldots$

26. $\dfrac{1}{2}, \dfrac{3}{4}, \dfrac{9}{8}, \dfrac{27}{16}, \dfrac{81}{32}, \ldots$

27. $1, -1, 1, -1, 1, \ldots$

28. $\dfrac{1}{3}, -\dfrac{2}{4}, \dfrac{3}{5}, -\dfrac{4}{6}, \dfrac{5}{7}, \ldots$

In Exercises 29–40, simplify the ratio of factorials.

29. $\dfrac{9!}{7!}$ **30.** $\dfrac{4!}{6!}$ **31.** $\dfrac{29!}{27!}$ **32.** $\dfrac{32!}{30!}$ **33.** $\dfrac{75!}{77!}$ **34.** $\dfrac{100!}{103!}$

35. $\dfrac{97!}{93!}$ **36.** $\dfrac{101!}{98!}$ **37.** $\dfrac{(n-1)!}{(n+1)!}$ **38.** $\dfrac{(n+2)!}{n!}$ **39.** $\dfrac{(2n+3)!}{(2n+1)!}$ **40.** $\dfrac{(2n+2)!}{(2n-1)!}$

In Exercises 41–50, write the first four terms of the sequence defined by the recursion formula. Assume the sequence begins at 1.

41. $a_1 = 7$ $a_n = a_{n-1} + 3$ **42.** $a_1 = 2$ $a_n = a_{n-1} + 1$

43. $a_1 = 1$ $a_n = n \cdot a_{n-1}$ **44.** $a_1 = 2$ $a_n = (n+1) \cdot a_{n-1}$

45. $a_1 = 100$ $a_n = \dfrac{a_{n-1}}{n!}$ **46.** $a_1 = 20$ $a_n = \dfrac{a_{n-1}}{n^2}$

47. $a_1 = 1, a_2 = 2$ $a_n = a_{n-1} \cdot a_{n-2}$ **48.** $a_1 = 1, a_2 = 2$ $a_n = \dfrac{a_{n-2}}{a_{n-1}}$

49. $a_1 = 1, a_2 = -1$ $a_n = (-1)^n\left[a_{n-1}^2 + a_{n-2}^2\right]$ **50.** $a_1 = 1, a_2 = -1$ $a_n = (n-1)a_{n-1} + (n-2)a_{n-2}$

In Exercises 51–64, evaluate the finite series.

51. $\displaystyle\sum_{n=1}^{5} 2$ **52.** $\displaystyle\sum_{n=1}^{5} 7$ **53.** $\displaystyle\sum_{n=0}^{4} n^2$ **54.** $\displaystyle\sum_{n=1}^{4} \dfrac{1}{n}$ **55.** $\displaystyle\sum_{n=1}^{6} (2n-1)$ **56.** $\displaystyle\sum_{n=1}^{6} (n+1)$

57. $\displaystyle\sum_{n=0}^{4} 1^n$ **58.** $\displaystyle\sum_{n=0}^{4} 2^n$ **59.** $\displaystyle\sum_{n=0}^{3} (-x)^n$ **60.** $\displaystyle\sum_{n=0}^{3} (-x)^{n+1}$ **61.** $\displaystyle\sum_{k=0}^{5} \dfrac{2^k}{k!}$ **62.** $\displaystyle\sum_{k=0}^{5} \dfrac{(-1)^k}{k!}$

63. $\displaystyle\sum_{k=0}^{4} \dfrac{x^k}{k!}$ **64.** $\displaystyle\sum_{k=0}^{4} \dfrac{(-1)^k x^k}{k!}$

In Exercises 65–68, evaluate the infinite series, if possible.

65. $\displaystyle\sum_{j=0}^{\infty} 2 \cdot (0.1)^j$ **66.** $\displaystyle\sum_{j=0}^{\infty} 5 \cdot \left(\dfrac{1}{10}\right)^j$ **67.** $\displaystyle\sum_{j=0}^{\infty} n^j$ $n \geq 1$ **68.** $\displaystyle\sum_{j=0}^{\infty} 1^j$

In Exercises 69–76, apply sigma notation to write the sum.

69. $1 - \dfrac{1}{2} + \dfrac{1}{4} - \dfrac{1}{8} + \cdots + \dfrac{1}{64}$ **70.** $1 + \dfrac{1}{2} + \dfrac{1}{4} + \dfrac{1}{8} + \cdots + \dfrac{1}{64} + \cdots$

71. $1 - 2 + 3 - 4 + 5 - 6 + \cdots$ **72.** $1 + 2 + 3 + 4 + 5 + \cdots + 21 + 22 + 23$

73. $\dfrac{2 \cdot 1}{1} + \dfrac{3 \cdot 2 \cdot 1}{1} + \dfrac{4 \cdot 3 \cdot 2 \cdot 1}{2 \cdot 1} + \dfrac{5 \cdot 4 \cdot 3 \cdot 2 \cdot 1}{3 \cdot 2 \cdot 1} + \dfrac{6 \cdot 5 \cdot 4 \cdot 3 \cdot 2 \cdot 1}{4 \cdot 3 \cdot 2 \cdot 1}$

74. $1 + \dfrac{2}{1} + \dfrac{2^2}{2 \cdot 1} + \dfrac{2^3}{3 \cdot 2 \cdot 1} + \dfrac{2^4}{4 \cdot 3 \cdot 2 \cdot 1} + \cdots$

75. $1 - x + \dfrac{x^2}{2} - \dfrac{x^3}{6} + \dfrac{x^4}{24} - \dfrac{x^5}{120} + \cdots$

76. $x + x^2 + \dfrac{x^3}{2} + \dfrac{x^4}{6} + \dfrac{x^5}{24} + \dfrac{x^6}{120}$

APPLICATIONS

77. Money. Upon graduation Jessica receives a commission from the U.S. Navy to become an officer and a $20,000 signing bonus for selecting aviation. She puts the entire bonus in an account that earns 6% interest compounded monthly. The balance in the account after n months is

$$A_n = 20,000 \left(1 + \frac{0.06}{12}\right)^n \qquad n = 1, 2, 3, \ldots$$

Her commitment to the Navy is 6 years. Calculate A_{72}. What does A_{72} represent?

78. Money. Dylan sells his car in his freshman year and puts $7000 in an account that earns 5% interest compounded quarterly. The balance in the account after n quarters is

$$A_n = 7000 \left(1 + \frac{0.05}{4}\right)^n \qquad n = 1, 2, 3, \ldots$$

Calculate A_{12}. What does A_{12} represent?

79. Salary. An attorney is trying to calculate the costs associated with going into private practice. If she hires a paralegal to assist her, she will have to pay the paralegal $20 per hour. To be competitive with most firms, she will have to give her paralegal a $2 per hour raise per year. Find a general term of a sequence a_n, which represents the hourly salary of a paralegal with n years of experience. What will be the paralegal's salary with 20 years of experience?

80. NFL Salaries. A player in the NFL typically has a career that lasts 3 years. The practice squad makes the league minimum of $275,000 (2004) in the first year, with a $75,000 raise per year. Write the general term of a sequence a_n that represents the salary of an NFL player making the league minimum during his entire career. Assuming $n = 1$ corresponds to the first year, what does $\sum_{n=1}^{3} a_n$ represent?

81. Salary. Upon graduation Sheldon decides to go to work for a local police department. His starting salary is $30,000 per year, and he expects to get a 3% raise per year. Write the recursion formula for a sequence that represents his salary n years on the job. Assume $n = 0$ represents his first year making $30,000.

82. *Escherichia coli.* A single cell of bacteria reproduces through a process called binary fission. *Escherichia coli* cells divide into two every 20 minutes. Suppose the same rate of division is maintained for 12 hours after the original cell enters the body. How many *E. coli* bacteria cells would be in the body 12 hours later? Suppose there is an infinite nutrient source so that the *E. coli* bacteria maintain the same rate of division for 48 hours after the original cell enters the body. How many *E. coli* bacteria cells would be in the body 48 hours later?

83. AIDS/HIV. A typical person has 500 to 1500 T cells per drop of blood in the body. HIV destroys the T cell count at a rate of 50–100 cells per drop of blood per year, depending on how aggressive it is in the body. Generally, the onset of AIDS occurs once the body's T cell count drops below 200. Write a sequence that represents the total number of T cells in a person infected with HIV. Assume that before infection the person has a 1000 T cell count ($a_0 = 1000$) and the rate at which the infection spreads corresponds to a loss of 75 T cells per drop of blood per year. How much time will elapse until this person has full-blown AIDS?

84. Company Sales. Lowe's reported total sales from 2003 through 2004 in the billions. The sequence $a_n = 3.8 + 1.6n$ represents the total sales in billions of dollars. Assuming $n = 3$ corresponds to 2003, what were the reported sales in 2003 and 2004? What does $\frac{1}{2} \cdot \sum_{n=3}^{4} a_n$ represent?

85. Cost of Eating Out. A college student tries to save money by bringing a bag lunch instead of eating out. He will be able to save $100 per month. He puts the money into his savings account, which draws 1.2% interest and is compounded monthly. The balance in his account after n months of bagging his lunch is

$$A_n = 100,000[(1.001)^n - 1] \qquad n = 1, 2, \ldots$$

Calculate the first four terms of this sequence. Calculate the amount after 3 years (36 months).

86. Cost of Acrylic Nails. A college student tries to save money by growing her own nails out and not spending $50 per month on acrylic fills. She will be able to save $50 per month. She puts the money into her savings account, which draws 1.2% interest and is compounded monthly. The balance in her account after n months of natural nails is

$$A_n = 50,000[(1.001)^n - 1] \qquad n = 1, 2, \ldots$$

Calculate the first four terms of this sequence. Calculate the amount after 4 years (48 months).

87. Math and Engineering. The formula $\sum_{n=0}^{\infty} \frac{x^n}{n!}$ can be used to approximate the function $y = e^x$. Compute the first five terms of this formula to approximate e^x. Apply the result to find e^2 and compare this result with the calculator value of e^2.

88. Home Prices. If the inflation rate is 3.5% per year and the average price of a home is $195,000, the average price of a home after n years is given by $A_n = 195,000(1.035)^n$. Find the average price of the home after 6 years.

89. Approximating Functions. Polynomials can be used to approximate transcendental functions such as $\ln(x)$ and e^x, which are found in advanced mathematics and engineering.

For example, $\sum_{n=0}^{\infty}(-1)^n\dfrac{(x-1)^{n+1}}{n+1}$ can be used to approximate $\ln(x)$, where x is close to 1. Use the first five terms of the series to approximate $\ln(x)$. Next, find $\ln(1.1)$ and compare with the value given by your calculator.

90. Future Value of an Annuity. The future value of an ordinary annuity is given by the formula $FV = PMT[((1+i)^n - 1)/i]$, where PMT = amount paid into the account at the end of each period, i = interest rate per period, and n = number of compounding periods. If you invest \$5000 at the end of each year for 5 years, you will have an accumulated value of FV as given in the above equation at the end of the nth year. Determine how much is in the account at the end of each year for the next 5 years if $i = 0.06$.

■ CATCH THE MISTAKE

In Exercises 91–94, explain the mistake that is made.

91. Simplify the ratio of factorials: $\dfrac{(3!)(5!)}{6!}$.

Solution:

Express 6! in factored form. $\qquad \dfrac{(3!)(5!)}{(3!)(2!)}$

Cancel the 3! in the numerator and denominator. $\qquad \dfrac{(5!)}{(2!)}$

Write out the factorials. $\qquad \dfrac{5\cdot 4\cdot 3\cdot 2\cdot 1}{2\cdot 1}$

Simplify. $\qquad 5\cdot 4\cdot 3 = 60$

$\dfrac{(3!)(5!)}{(3!)(2!)} \neq 60$. What mistake was made?

92. Simplify the factorial expression: $\dfrac{2n(2n-2)!}{(2n+2)!}$.

Solution:

Express factorials in factored form.

$$\frac{2n(2n-2)(2n-4)(2n-6)\cdots}{(2n+2)(2n)(2n-2)(2n-4)(2n-6)\cdots}$$

Cancel common terms.

$$\frac{1}{2n+2}$$

This is incorrect. What mistake was made?

93. Find the first four terms of the sequence defined by $a_n = (-1)^{n+1}n^2$.

Solution:

Find the $n = 1$ term. $\qquad a_1 = -1$

Find the $n = 2$ term. $\qquad a_2 = 4$

Find the $n = 3$ term. $\qquad a_3 = -9$

Find the $n = 4$ term. $\qquad a_4 = 16$

The sequence $-1, 4, -9, 16, \ldots$, is incorrect. What mistake was made?

94. Evaluate the series $\sum_{k=0}^{3}(-1)^k k^2$.

Solution:

Write out the sum. $\qquad \sum_{k=0}^{3}(-1)^k k^2 = -1 + 4 - 9$

Simplify the sum. $\qquad \sum_{k=6}^{3}(-1)^k k^2 = -6$

This is incorrect. What mistake was made?

■ CONCEPTUAL

In Exercises 95–98, determine whether each statement is true or false.

95. $\sum_{k=0}^{n}cx^k = c\sum_{k=0}^{n}x^k$

96. $\sum_{i=1}^{n}a_i + b_i = \sum_{i=1}^{n}a_i + \sum_{i=1}^{n}b_i$

97. $\sum_{k=1}^{n}a_k b_k = \sum_{k=1}^{n}a_k \cdot \sum_{k=1}^{n}b_k$

98. $(a!)(b!) = (ab)!$

▪ CHALLENGE

99. Write the first four terms of the sequence defined by the recursion formula:
$$a_1 = C \qquad a_n = a_{n-1} + D \qquad D \neq 0$$

100. Write the first four terms of the sequence defined by the recursion formula:
$$a_1 = C \qquad a_n = Da_{n-1} \qquad D \neq 0$$

101. Fibonacci Sequence. An explicit formula for the nth term of the Fibonacci sequence is:
$$F_n = \frac{\left(1 + \sqrt{5}\right)^n - \left(1 - \sqrt{5}\right)^n}{2^n \sqrt{5}}$$

Apply algebra (not your calculator) to find the first two terms of this sequence and verify that these are indeed the first two terms of the Fibonacci sequence.

102. Let $a_n = \sqrt{a_{n-1}}$ for $n \geq 2$ and $a_1 = 7$. Find the first five terms of this sequence and make a generalization for the nth term.

▪ TECHNOLOGY

103. The sequence defined by $a_n = \left(1 + \dfrac{1}{n}\right)^n$ approaches the number e as n gets large. Use a graphing calculator to find a_{100}, a_{1000}, $a_{10,000}$, and keep increasing n until the terms in the sequence approach 2.7183.

104. The Fibonacci sequence is defined by $a_1 = 1$, $a_2 = 1$, and $a_n = a_{n-2} + a_{n-1}$ for $n \geq 3$. The ratio $\dfrac{a_{n+1}}{a_n}$ is an approximation of the golden ratio. The ratio approaches a constant ϕ (phi) as n gets large. Find the golden ratio using a graphing utility.

105. Use a graphing calculator "SUM" to sum $\displaystyle\sum_{k=0}^{5} \frac{2^k}{k!}$. Compare it with your answer to Exercise 61.

106. Use a graphing calculator "SUM" to sum $\displaystyle\sum_{k=0}^{5} \frac{(-1)^k}{k!}$. Compare it with your answer to Exercise 62.

SECTION 12.2 ARITHMETIC SEQUENCES AND SERIES

SKILLS OBJECTIVES

- Recognize an arithmetic sequence.
- Find the general nth term of an arithmetic sequence.
- Evaluate a finite arithmetic series.
- Use arithmetic sequences and series to model real-world problems.

CONCEPTUAL OBJECTIVE

- Understand the difference between an arithmetic sequence and an arithmetic series.

Arithmetic Sequences

The word *arithmetic* (with emphasis on the third syllable) often implies adding or subtracting of numbers. *Arithmetic sequences* are sequences whose terms are found by adding a constant to each previous term. The sequence 1, 3, 5, 7, 9, . . . is arithmetic because each successive term is found by adding 2 to the previous term.

DEFINITION   Arithmetic Sequences

A sequence is **arithmetic** if each term in the sequence is found by adding a real number d to the previous term, so that $a_{n+1} = a_n + d$. Because $a_{n+1} - a_n = d$, the number d is called the **common difference**.

EXAMPLE 1   Identifying the Common Difference in Arithmetic Sequences

Determine whether each sequence is arithmetic. If so, find the common difference for each of the arithmetic sequences.

a. $5, 9, 13, 17, \ldots$ **b.** $18, 9, 0, -9, \ldots$ **c.** $\frac{1}{2}, \frac{5}{4}, 2, \frac{11}{4}, \ldots$

Solution (a):

Label the terms. $a_1 = 5, a_2 = 9, a_3 = 13, a_4 = 17, \ldots$

Find the difference $d = a_{n+1} - a_n$. $d = a_2 - a_1 = 9 - 5 = \boxed{4}$

Check that the difference for the $d = a_3 - a_2 = 13 - 9 = 4$
next two terms is also 4.
    $d = a_4 - a_3 = 17 - 13 = 4$

There is a common difference of $\boxed{4}$. Therefore, this sequence is arithmetic, and each successive term is found by adding 4 to the previous term.

Solution (b):

Label the terms. $a_1 = 18, a_2 = 9, a_3 = 0, a_4 = -9, \ldots$

Find the difference $d = a_{n+1} - a_n$. $d = a_2 - a_1 = 9 - 18 = \boxed{-9}$

Check that the difference for $d = a_3 - a_2 = 0 - 9 = -9$
the next two terms is also -9.
    $d = a_4 - a_3 = -9 - 0 = -9$

There is a common difference of $\boxed{-9}$. Therefore, this sequence is arithmetic, and each successive term is found by subtracting 9 from (that is, adding -9 to) the previous term.

Solution (c):

Label the terms. $a_1 = \dfrac{1}{2}, a_2 = \dfrac{5}{4}, a_3 = 2, a_4 = \dfrac{11}{4}, \ldots$

Find the difference $d = a_{n+1} - a_n$. $d = a_2 - a_1 = \dfrac{5}{4} - \dfrac{1}{2} = \boxed{\dfrac{3}{4}}$

Check that the difference
for the next two terms is also $\frac{3}{4}$. $d = a_3 - a_2 = 2 - \dfrac{5}{4} = \dfrac{3}{4}$

    $d = a_4 - a_3 = \dfrac{11}{4} - 2 = \dfrac{3}{4}$

There is a common difference of $\boxed{\dfrac{3}{4}}$. Therefore, this sequence is arithmetic, and each successive term is found by adding $\frac{3}{4}$ to the previous term.

■ **YOUR TURN** Find the common difference for each of the arithmetic sequences.

a. $7, 2, -3, -8, \ldots$ **b.** $1, \frac{5}{3}, \frac{7}{3}, 3, \ldots$

The General (*n*th) Term of an Arithmetic Sequence

To find a formula for the general, or *n*th, term of an arithmetic sequence, write out the first several terms and look for a pattern.

First term, $n = 1$. a_1

Second term, $n = 2$. $a_2 = a_1 + d$

Third term, $n = 3$. $a_3 = a_2 + d = (a_1 + d) + d = a_1 + 2d$

Fourth term, $n = 4$. $a_4 = a_3 + d = (a_1 + 2d) + d = a_1 + 3d$

In general, the *n*th term is given by $a_n = a_1 + (n - 1)d$.

THE *n*TH TERM OF AN ARITHMETIC SEQUENCE

The ***n*th term** of an arithmetic sequence with common difference d is given by

$$a_n = a_1 + (n - 1)d \qquad \text{for } n \geq 1$$

EXAMPLE 2   Finding the *n*th Term of an Arithmetic Sequence

Find the 13th term of the arithmetic sequence 2, 5, 8, 11, ...

Solution:

Identify the common difference. $d = 5 - 2 = 3$

Identify the first ($n = 1$) term. $a_1 = 2$

Substitute $a_1 = 2$ and $d = 3$ into $a_n = a_1 + (n - 1)d$. $a_n = 2 + 3(n - 1)$

Substitute $n = 13$ into $a_n = 2 + 3(n - 1)$. $a_{13} = 2 + 3(13 - 1) = \boxed{38}$

■ **YOUR TURN** Find the 10th term of the arithmetic sequence 3, 10, 17, 24, ...

EXAMPLE 3   Finding the Arithmetic Sequence

The 4th term of an arithmetic sequence is 16, and the 21st term is 67. Find a_1 and d and construct the sequence.

Solution:

Write the 4th and 21st terms. $a_4 = 16$ and $a_{21} = 67$

Adding d 17 times to a_4 results in a_{21}. $a_{21} = a_4 + 17d$

Substitute $a_4 = 16$ and $a_{21} = 67$. $67 = 16 + 17d$

Solve for d. $\boxed{d = 3}$

Substitute $d = 3$ into $a_n = a_1 + (n - 1)d$. $a_n = a_1 + 3(n - 1)$

Let $a_4 = 16$. $16 = a_1 + 3(4 - 1)$

Solve for a_1. $\boxed{a_1 = 7}$

The arithmetic sequence that starts at 7 and has a common difference of 3 is $\boxed{7, 10, 13, 16, \ldots}$.

■ **YOUR TURN** Construct the arithmetic sequence whose 7th term is 26 and whose 13th term is 50.

The Sum of an Arithmetic Sequence

What is the sum of the first 100 counting numbers

$$1 + 2 + 3 + 4 + \cdots + 99 + 100 = ?$$

If we write this sum twice (one in ascending order and one in descending order) and add, we get 100 pairs of 101.

$$
\begin{array}{l}
1 + 2 + 3 + 4 + \cdots + 99 + 100 \\
100 + 99 + 98 + 97 + \cdots + 2 + 1 \\
\hline
101 + 101 + 101 + 101 + \cdots + 101 + 101 = 100\,(101)
\end{array}
$$

Since we added twice the sum, we divide by 2.

$$1 + 2 + 3 + 4 + \cdots + 99 + 100 = \frac{(101)(100)}{2} = 5050$$

Now, let us develop the sum of a general arithmetic series.

The sum of the first n terms of an arithmetic sequence is called the **nth partial sum**, or **finite arithmetic series**, and is denoted by S_n. An arithmetic sequence can be found by starting at the first term and adding the common difference to each successive term, and so the nth partial sum, or finite series, can be found the same way, but terminating the sum at the nth term:

$$S_n = a_1 + a_2 + a_3 + a_4 + \cdots$$
$$S_n = a_1 + (a_1 + d) + (a_1 + 2d) + (a_1 + 3d) + \cdots + (a_n)$$

Similarly, we can start with the nth term and find terms going backward by subtracting the common difference until we arrive at the first term:

$$S_n = a_n + a_{n-1} + a_{n-2} + a_{n-3} + \cdots$$
$$S_n = a_n + (a_n - d) + (a_n - 2d) + (a_n - 3d) + \cdots + (a_1)$$

Add these two representations of the nth partial sum. Notice that the d terms are eliminated:

$$
\begin{array}{l}
S_n = a_1 + (a_1 + d) + (a_1 + 2d) + (a_1 + 3d) + \cdots + (a_n) \\
S_n = a_n + (a_n - d) + (a_n - 2d) + (a_n - 3d) + \cdots + (a_1) \\
\hline
2S_n = (a_1 + a_n) + (a_1 + a_n) + (a_1 + a_n) + \cdots + (a_1 + a_n) \\
\underbrace{}_{n(a_1 + a_n)}
\end{array}
$$

$$2S_n = n(a_1 + a_n) \qquad \text{or} \qquad S_n = \frac{n}{2}(a_1 + a_n)$$

Let $a_n = a_1 + (n - 1)d$.

$$S_n = \frac{n}{2}\,[a_1 + a_1 + (n - 1)d]$$

$$= \frac{n}{2}\,[2a_1 + (n - 1)d] = na_1 + \frac{n(n - 1)d}{2}$$

Classroom Example 12.2.4

Compute the sums.

a. $\displaystyle\sum_{k=1}^{45} 2k$ b. $\displaystyle\sum_{k=1}^{20} (3k - 2)$

Answer:
a. 2070 b. 590

Study Tip

S_n can also be written as

$S_n = \dfrac{n}{2}[2a_1 + (n - 1)d].$

DEFINITION Evaluating a Finite Arithmetic Series

The sum of the first n terms of an arithmetic sequence, called a **finite arithmetic series**, is given by the formula

$$S_n = \frac{n}{2}[2a_1 + (n - 1)d] \qquad \text{or} \qquad S_n = \frac{n}{2}(a_1 + a_n) \quad n \geq 2$$

EXAMPLE 4   Evaluating a Finite Arithmetic Series

Evaluate the finite arithmetic series $\sum_{k=1}^{100} k$.

Solution:

Expand the arithmetic series.

$$\sum_{k=1}^{100} k = 1 + 2 + 3 + \cdots + 99 + 100$$

This is the sum of an arithmetic sequence of numbers with a common difference of 1.

Identify the parameters of the arithmetic sequence.

$$a_1 = 1, a_n = 100, \text{ and } n = 100$$

Substitute these values into $S_n = \dfrac{n}{2}(a_1 + a_n)$.

$$S_{100} = \frac{100}{2}(1 + 100)$$

Simplify.

$$\boxed{S_{100} = 5050}$$

The sum of the first 100 natural numbers is 5050.

■ YOUR TURN Evaluate the following finite arithmetic series.

$$\text{a. } \sum_{k=1}^{30} k \qquad \text{b. } \sum_{k=1}^{20} (2k + 1)$$

Technology Tip

To find the sum of the series $\sum_{k=1}^{100} k$, press

[2nd] [LIST] [▶] [MATH] [▼]
[5:sum(] [ENTER] [2nd] [LIST]
[▶] [OPS] [▼] [5:seq(] [ENTER]
[ALPHA] [K] [,] [ALPHA] [K] [,]
[1] [,] [100] [,] [1] [)] [)] [ENTER].

```
sum(seq(K,K,1,10
0,1))
             5050
```

■ **Answer: a.** 465   **b.** 440

EXAMPLE 5   Finding the *n*th Partial Sum of an Arithmetic Sequence

Find the sum of the first 20 terms of the arithmetic sequence 3, 8, 13, 18, 23, ...

Solution:

Recall the partial-sum formula.

$$S_n = \frac{n}{2}(a_1 + a_n)$$

Find the 20th partial sum of this arithmetic sequence.

$$S_{20} = \frac{20}{2}(a_1 + a_{20})$$

Recall that the general *n*th term of an arithmetic sequence is given by:

$$a_n = a_1 + (n - 1)d$$

Note that the first term of the arithmetic sequence is 3.

$$a_1 = 3$$

This is an arithmetic sequence with a common difference of 5.

$$d = 5$$

Substitute $a_1 = 3$ and $d = 5$ into $a_n = a_1 + (n - 1)d$.

$$a_n = 3 + (n - 1)5$$

Substitute $n = 20$ to find the 20th term.

$$a_{20} = 3 + (20 - 1)5 = 98$$

Substitute $a_1 = 3$ and $a_{20} = 98$ into the partial sum.

$$\boxed{S_{20} = 10(3 + 98) = 1010}$$

The sum of the first 20 terms of this arithmetic sequence is 1010.

■ YOUR TURN Find the sum of the first 25 terms of the arithmetic sequence 2, 6, 10, 14, 18, ...

Technology Tip

To find the sum of the series $\sum_{n=1}^{20} 3 + (n - 1)5$, press

[2nd] [LIST] [▶] [MATH] [▼]
[5:sum(] [ENTER] [2nd] [LIST]
[▶] [OPS] [▼] [5:seq(] [ENTER]
[3] [+] [(] [ALPHA] [N] [−] [1] [)]
[5] [,] [ALPHA] [N] [,] [1] [,] [20] [,] [1]
[)] [)] [ENTER].

```
sum(seq(3+(N-1)5
,N,1,20,1))
             1010
■
```

■ **Answer:** 1250

Applications

EXAMPLE 6 **Marching Band Formation**

Suppose a band has 18 members in the first row, 22 members in the second row, and 26 members in the third row and continues with that pattern for a total of nine rows. How many marchers are there all together?

Brian Bahr/Getty Images, Inc.

Solution:

The number of members in each row forms an arithmetic sequence with a common difference of 4, and the first row has 18 members.

$$a_1 = 18 \quad d = 4$$

Calculate the nth term of the sequence $a_n = a_1 + (n - 1)d$.

$$a_n = 18 + (n - 1)4$$

Find the 9th term $n = 9$.

$$a_9 = 18 + (9 - 1)4 = 50$$

Calculate the sum $S_n = \frac{n}{2}(a_1 + a_n)$ of the nine rows.

$$S_9 = \frac{9}{2}(a_1 + a_9)$$

$$= \frac{9}{2}(18 + 50)$$

$$= \frac{9}{2}(68)$$

$$= \boxed{306}$$

There are 306 members in the marching band.

■ **Answer:** 328

■ **YOUR TURN** Suppose a bed of tulips is arranged in a garden so that there are 20 tulips in the first row, 26 tulips in the second row, and 32 tulips in the third row and the rows continue with that pattern for a total of 8 rows. How many tulips are there all together?

SECTION
12.2 SUMMARY

In this section, arithmetic sequences were defined as sequences of which each successive term is found by adding the same constant d to the previous term. Formulas were developed for the general, or nth, term of an arithmetic sequence, and for the nth partial sum of an arithmetic sequence, also called a finite arithmetic series.

$$a_n = a_1 + (n - 1)d \qquad n \geq 1$$

$$S_n = \frac{n}{2}(a_1 + a_n) = \frac{n}{2}[2a_1 + (n - 1)d]$$

SECTION
12.2 EXERCISES

▪ SKILLS

In Exercises 1–10, determine whether the sequence is arithmetic. If it is, find the common difference.

1. $2, 5, 8, 11, 14, \ldots$

2. $9, 6, 3, 0, -3, -6, \ldots$

3. $1^2 + 2^2 + 3^2 + \cdots$

4. $1! + 2! + 3! + \cdots$

5. $3.33, 3.30, 3.27, 3.24, \ldots$

6. $0.7, 1.2, 1.7, 2.2, \ldots$

7. $4, \frac{14}{3}, \frac{16}{3}, 6, \ldots$

8. $2, \frac{7}{3}, \frac{8}{3}, 3, \ldots$

9. $10^1, 10^2, 10^3, 10^4, \ldots$

10. $120, 60, 30, 15, \ldots$

In Exercises 11–20, find the first four terms of the sequence. Determine whether the sequence is arithmetic, and if so, find the common difference.

11. $a_n = -2n + 5$

12. $a_n = 3n - 10$

13. $a_n = n^2$

14. $a_n = \frac{n^2}{n!}$

15. $a_n = 5n - 3$

16. $a_n = -4n + 5$

17. $a_n = 10(n - 1)$

18. $a_n = 8n - 4$

19. $a_n = (-1)^n n$

20. $a_n = (-1)^{n+1} 2n$

In Exercises 21–28, find the general, or nth, term of the arithmetic sequence given the first term and the common difference.

21. $a_1 = 11 \qquad d = 5$

22. $a_1 = 5 \qquad d = 11$

23. $a_1 = -4 \qquad d = 2$

24. $a_1 = 2 \qquad d = -4$

25. $a_1 = 0 \qquad d = \frac{2}{3}$

26. $a_1 = -1 \qquad d = -\frac{3}{4}$

27. $a_1 = 0 \qquad d = e$

28. $a_1 = 1.1 \qquad d = -0.3$

In Exercises 29–34, find the specified term for each arithmetic sequence.

29. The 10th term of the sequence $7, 20, 33, 46, \ldots$

30. The 19th term of the sequence $7, 1, -5, -11, \ldots$

31. The 100th term of the sequence $9, 2, -5, -12, \ldots$

32. The 90th term of the sequence $13, 19, 25, 31, \ldots$

33. The 21st term of the sequence $\frac{1}{3}, \frac{7}{12}, \frac{5}{6}, \frac{13}{12}, \ldots$

34. The 33rd term of the sequence $\frac{1}{5}, \frac{8}{15}, \frac{13}{15}, \frac{6}{5}, \ldots$

In Exercises 35–40, for each arithmetic sequence described, find a_1 and d and construct the sequence by stating the general, or nth, term.

35. The 5th term is 44 and the 17th term is 152.

36. The 9th term is -19 and the 21st term is -55.

37. The 7th term is -1 and the 17th term is -41.

38. The 8th term is 47 and the 21st term is 112.

39. The 4th term is 3 and the 22nd term is 15.

40. The 11th term is -3 and the 31st term is -13.

In Exercises 41–52, find the sum.

41. $\displaystyle\sum_{k=1}^{23} 2k$
42. $\displaystyle\sum_{k=0}^{20} 5k$
43. $\displaystyle\sum_{n=1}^{30} (-2n + 5)$
44. $\displaystyle\sum_{n=0}^{17} (3n - 10)$
45. $\displaystyle\sum_{j=3}^{14} 0.5j$
46. $\displaystyle\sum_{j=1}^{33} \frac{j}{4}$

47. $2 + 7 + 12 + 17 + \cdots + 62$

48. $1 - 3 - 7 - \cdots - 75$

49. $4 + 7 + 10 + \cdots + 151$

50. $2 + 0 - 2 - \cdots - 56$

51. $\frac{1}{6} - \frac{1}{6} - \frac{1}{2} - \cdots - \frac{13}{2}$

52. $\frac{11}{12} + \frac{7}{6} + \frac{17}{12} + \cdots + \frac{14}{3}$

In Exercises 53–58, find the partial sum of the arithmetic series.

53. The first 18 terms of $1 + 5 + 9 + 13 + \cdots$.

54. The first 21 terms of $2 + 5 + 8 + 11 + \cdots$.

55. The first 43 terms of $1 + \frac{1}{2} + 0 - \frac{1}{2} - \cdots$.

56. The first 37 terms of $3 + \frac{3}{2} + 0 - \frac{3}{2} - \cdots$.

57. The first 18 terms of $-9 + 1 + 11 + 21 + 31 + \cdots$.

58. The first 21 terms of $-2 + 8 + 18 + 28 + \cdots$.

▪ APPLICATIONS

59. Comparing Salaries. Colin and Camden are twin brothers graduating with B.S. degrees in biology. Colin takes a job at the San Diego Zoo making $28,000 for his first year with a $1500 raise per year every year after that. Camden accepts a job at Florida Fish and Wildlife making $25,000 with a guaranteed $2000 raise per year. How much will each of the brothers have made in a total of 10 years?

60. Comparing Salaries. On graduating with a Ph.D. in optical sciences, Jasmine and Megan choose different career paths. Jasmine accepts a faculty position at the University of Arizona making $80,000 with a guaranteed $2000 raise every year. Megan takes a job with the Boeing Corporation making $90,000 with a guaranteed $5000 raise each year. Calculate how many total dollars each woman will have made after 15 years.

61. Theater Seating. You walk into the premiere of Brad Pitt's new movie, and the theater is packed, with almost every seat filled. You want to estimate the number of people in the theater. You quickly count to find that there are 22 seats in the front row, and there are 25 rows in the theater. Each row appears to have 1 more seat than the row in front of it. How many seats are in that theater?

62. Field of Tulips. Every spring the Skagit County Tulip Festival plants more than 100,000 bulbs. In honor of the Tri-Delta sorority that has sent 120 sisters from the University of Washington to volunteer for the festival, Skagit County has planted tulips in the shape of $\Delta\Delta\Delta$. In each of the triangles there are 20 rows of tulips, each row having one less than the row before. How many tulips are planted in each delta if there is 1 tulip in the first row?

63. World's Largest Champagne Fountain. From December 28 to 30, 1999, Luuk Broos, director of Maison Luuk-Chalet Fontain, constructed a 56-story champagne fountain at the Steigenberger Kurhaus Hotel, Scheveningen, Netherlands. The fountain consisted of 30,856 champagne glasses. Assuming there was one glass at the top and the number of glasses in each row forms an arithmetic sequence, how many were on the bottom row (story)? How many glasses less did each successive row (story) have? Assume each story is one row.

64. Stacking of Logs. If 25 logs are laid side by side on the ground, and 24 logs are placed on top of those, and 23 logs are placed on the 3rd row, and the pattern continues until there is a single log on the 25th row, how many logs are in the stack?

65. Falling Object. When a skydiver jumps out of an airplane, she falls approximately 16 feet in the 1st second, 48 feet during the 2nd second, 80 feet during the 3rd second, 112 feet during the 4th second, and 144 feet during the 5th second, and this pattern continues. If she deploys her parachute after 10 seconds have elapsed, how far will she have fallen during those 10 seconds?

66. Falling Object. If a penny is dropped out of a plane, it falls approximately 4.9 meters during the 1st second, 14.7 meters during the 2nd second, 24.5 meters during the 3rd second, and 34.3 meters during the 4th second. Assuming this pattern continues, how many meters will the penny have fallen after 10 seconds?

67. Grocery Store. A grocer has a triangular display of oranges in a window. There are 20 oranges in the bottom row, and the number of oranges decreases by one in each row above this row. How many oranges are in the display?

68. Salary. Suppose your salary is $45,000 and you receive a $1500 raise for each year you work for 35 years.
 a. How much will you earn during the 35th year?
 b. What is the total amount you earned over your 35-year career?

69. Theater Seating. At a theater, seats are arranged in a triangular pattern of rows with each succeeding row having one more seat than the previous row. You count the number of seats in the fourth row and determine that there are 26 seats.
 a. How many seats are in the first row?
 b. Now, suppose there are 30 rows of seats. How many total seats are there in the theater?

70. Mathematics. Find the exact sum of

$$\frac{1}{e} + \frac{3}{e} + \frac{5}{e} + \cdots + \frac{23}{e}$$

■ CATCH THE MISTAKE

In Exercises 71–74, explain the mistake that is made.

71. Find the general or nth term of the arithmetic sequence
3, 4, 5, 6, 7, ...

Solution:

The common difference of this
sequence is 1. $d = 1$

The first term is 3. $a_1 = 3$

The general term is $a_n = a_1 + nd$. $a_n = 3 + n$

This is incorrect. What mistake was made?

72. Find the general or nth term of the arithmetic sequence
10, 8, 6, ...

Solution:

The common difference of this
sequence is 2. $d = 2$

The first term is 10. $a_1 = 10$

The general term is
$a_n = a_1 + (n - 1)d$. $a_n = 10 + 2(n - 1)$

This is incorrect. What mistake was made?

73. Find the sum $\sum_{k=0}^{10} 2n + 1$.

Solution:

The sum is given by $S_n = \dfrac{n}{2}(a_1 + a_n)$, where $n = 10$.

Identify the 1st and 10th terms. $a_1 = 1$ $a_{10} = 21$

Substitute $a_1 = 1$, $a_{10} = 21$,
and $n = 10$ into
$S_n = \dfrac{n}{2}(a_1 + a_n)$. $S_{10} = \dfrac{10}{2}(1 + 21) = 110$

This is incorrect. What mistake was made?

74. Find the sum $3 + 9 + 15 + 21 + 27 + 33 + \cdots + 87$.

Solution:

This is an arithmetic sequence
with common difference of 6. $d = 6$

The general term is given by
$a_n = a_1 + (n - 1)d$. $a_n = 3 + (n - 1)6$

87 is the 15th term of the
series. $a_{15} = 3 + (15 - 1)6 = 87$

The sum of the series is
$S_n = \dfrac{n}{2}(a_n - a_1)$. $S_{15} = \dfrac{15}{2}(87 - 3) = 630$

This is incorrect. What mistake was made?

■ CONCEPTUAL

In Exercises 75–78, determine whether each statement is true or false.

75. An arithmetic sequence and a finite arithmetic series are the same.

76. The sum of all infinite and finite arithmetic series can always be found.

77. An alternating sequence cannot be an arithmetic sequence.

78. The common difference of an arithmetic sequence is always positive.

■ CHALLENGE

79. Find the sum $a + (a + b) + (a + 2b) + \cdots + (a + nb)$.

80. Find the sum $\sum_{k=-29}^{30} \ln e^k$.

81. The wave number, v (reciprocal of wave length) of certain light waves in the spectrum of light emitted by hydrogen is given by $v = R\left(\dfrac{1}{k^2} - \dfrac{1}{n^2}\right)$, $n > k$, where $R = 109{,}678$. A series of lines is given by holding k constant and varying the value of n. Suppose $k = 2$ and $n = 3, 4, 5, \ldots$. Find the limit of the wave number of the series.

82. In a certain arithmetic sequence $a_1 = -4$ and $d = 6$. If $S_n = 570$, find the value of n.

■ TECHNOLOGY

83. Use a graphing calculator "SUM" to sum the natural numbers from 1 to 100.

84. Use a graphing calculator to sum the even natural numbers from 1 to 100.

85. Use a graphing calculator to sum the odd natural numbers from 1 to 100.

86. Use a graphing calculator to sum $\sum_{n=1}^{30} (-2n + 5)$. Compare it with your answer to Exercise 43.

87. Use a graphing calculator to sum $\sum_{n=1}^{100} [-59 + 5(n - 1)]$.

88. Use a graphing calculator to sum $\sum_{n=1}^{200} \left[-18 + \frac{4}{5}(n - 1)\right]$.

SECTION 12.3 GEOMETRIC SEQUENCES AND SERIES

SKILLS OBJECTIVES

- Recognize a geometric sequence.
- Find the general, nth term of a geometric sequence.
- Evaluate a finite geometric series.
- Evaluate an infinite geometric series, if the sum exists.
- Use geometric sequences and series to model real-world problems.

CONCEPTUAL OBJECTIVES

- Understand the difference between a geometric sequence and a geometric series.
- Distinguish between an arithmetic sequence and a geometric sequence.
- Understand why it is not possible to evaluate all infinite geometric series.

Geometric Sequences

In Section 12.2, we discussed *arithmetic* sequences, where successive terms had a *common difference*. In other words, each term was found by adding the same constant to the previous term. In this section we discuss *geometric* sequences, where successive terms have a *common ratio*. In other words, each term is found by multiplying the previous term by the same constant. The sequence $4, 12, 36, 108, \ldots$ is geometric because each successive term is found by multiplying the previous term by 3.

DEFINITION   **Geometric Sequences**

A sequence is **geometric** if each term in the sequence is found by multiplying the previous term by a number r, so that $a_{n+1} = r \cdot a_n$. Because $\dfrac{a_{n+1}}{a_n} = r$, the number r is called the **common ratio**.

EXAMPLE 1   **Identifying the Common Ratio in Geometric Sequences**

Find the common ratio for each of the geometric sequences.

a. $5, 20, 80, 320, \ldots$ **b.** $1, -\dfrac{1}{2}, \dfrac{1}{4}, -\dfrac{1}{8}, \ldots$ **c.** $\$5000, \$5500, \$6050, \$6655, \ldots$

Solution (a):

Label the terms. $a_1 = 5, a_2 = 20, a_3 = 80, a_4 = 320, \ldots$

Find the ratio $r = \dfrac{a_{n+1}}{a_n}$. $r = \dfrac{a_2}{a_1} = \dfrac{20}{5} = 4$

$r = \dfrac{a_3}{a_2} = \dfrac{80}{20} = 4$

$r = \dfrac{a_4}{a_3} = \dfrac{320}{80} = 4$

The common ratio is 4.

Solution (b):

Label the terms.

$$a_1 = 1, a_2 = -\frac{1}{2}, a_3 = \frac{1}{4}, a_4 = -\frac{1}{8}, \ldots$$

Find the ratio $r = \dfrac{a_{n+1}}{a_n}$.

$$r = \frac{a_2}{a_1} = \frac{-1/2}{1} = -\frac{1}{2}$$

$$r = \frac{a_3}{a_2} = \frac{1/4}{-1/2} = -\frac{1}{2}$$

$$r = \frac{a_4}{a_3} = \frac{-1/8}{1/4} = -\frac{1}{2}$$

The common ratio is $-\frac{1}{2}$.

Solution (c):

Label the terms.

$$a_1 = \$5000, a_2 = \$5500, a_3 = \$6050, a_4 = \$6655, \ldots$$

Find the ratio $r = \dfrac{a_{n+1}}{a_n}$.

$$r = \frac{a_2}{a_1} = \frac{\$5500}{\$5000} = 1.1$$

$$r = \frac{a_3}{a_2} = \frac{\$6050}{\$5500} = 1.1$$

$$r = \frac{a_4}{a_3} = \frac{\$6655}{\$6050} = 1.1$$

The common ratio is 1.1.

■ **YOUR TURN** Find the common ratio of each geometric sequence.

a. $1, -3, 9, -27, \ldots$ **b.** $320, 80, 20, 5, \ldots$

■ **Answer: a.** -3 **b.** $\frac{1}{4}$ or 0.25

The General (*n*th) Term of a Geometric Sequence

To find a formula for the general, or *n*th, term of a geometric sequence, write out the first several terms and look for a pattern.

First term, $n = 1$. a_1

Second term, $n = 2$. $a_2 = a_1 \cdot r$

Third term, $n = 3$. $a_3 = a_2 \cdot r = (a_1 \cdot r) \cdot r = a_1 \cdot r^2$

Fourth term, $n = 4$. $a_4 = a_3 \cdot r = (a_1 \cdot r^2) \cdot r = a_1 \cdot r^3$

In general, the *n*th term is given by $a_n = a_1 \cdot r^{n-1}$.

THE *n*TH TERM OF A GEOMETRIC SEQUENCE

The ***n*th term** of a geometric sequence with common ratio r is given by

$$a_n = a_1 \cdot r^{n-1} \text{ for } n \geq 1$$

Use $\boxed{seq}$ to find the nth term of the sequence by setting the initial index value equal to the final index value. To find the 7th term of the geometric sequence $a_n = 2 \cdot 5^{n-1}$, press

$\boxed{2nd}$ $\boxed{LIST}$ $\boxed{\blacktriangleright}$ $\boxed{OPS}$ $\boxed{\blacktriangledown}$ $\boxed{5:seq(}$
$\boxed{ENTER}$ 2 $\boxed{\times}$ 5 $\boxed{^}$ $\boxed{(}$ $\boxed{ALPHA}$
$\boxed{N}$ $\boxed{-}$ 1 $\boxed{)}$ $\boxed{,}$ $\boxed{ALPHA}$ $\boxed{N}$ $\boxed{,}$
7 $\boxed{,}$ 7 $\boxed{,}$ 1 $\boxed{)}$ $\boxed{ENTER}$.

```
seq(2*5^(N-1),N,
7,7,1)
          {31250}
■
```

■ **Answer:** 49,152

EXAMPLE 2   Finding the nth Term of a Geometric Sequence

Find the 7th term of the sequence 2, 10, 50, 250, ...

Solution:

Identify the common ratio. $r = \dfrac{10}{2} = \dfrac{50}{10} = \dfrac{250}{50} = 5$

Identify the first ($n = 1$) term. $a_1 = 2$

Substitute $a_1 = 2$ and $r = 5$ into $a_n = a_1 \cdot r^{n-1}$. $a_n = 2 \cdot 5^{n-1}$

Substitute $n = 7$ into $a_n = 2 \cdot 5^{n-1}$. $a_7 = 2 \cdot 5^{7-1} = 2 \cdot 5^6 = 31{,}250$

> The 7th term of the geometric sequence is 31,250.

■ **YOUR TURN** Find the 8th term of the sequence 3, 12, 48, 192, ...

Classroom Example 12.3.2
Find the nth term of the geometric sequences.
a. $\frac{6}{5}, \frac{18}{25}, \frac{54}{125}, \cdots$
b. $-\frac{2}{3}, \frac{4}{9}, -\frac{8}{27}, \frac{16}{81}, \cdots$
Answer:
a. $2\left(\frac{3}{5}\right)^n$ b. $\left(-\frac{2}{3}\right)^n$.

■ **Answer:** 81, 27, 9, 3, 1, . . .

EXAMPLE 3   Finding the Geometric Sequence

Find the geometric sequence whose 5th term is 0.01 and whose common ratio is 0.1.

Solution:

Label the common ratio and 5th term. $a_5 = 0.01$ and $r = 0.1$

Substitute $a_5 = 0.01$, $n = 5$, and $r = 0.1$ into $a_n = a_1 \cdot r^{n-1}$. $0.01 = a_1 \cdot (0.1)^{5-1}$

Solve for a_1. $a_1 = \dfrac{0.01}{(0.1)^4} = \dfrac{0.01}{0.0001} = 100$

> The geometric sequence that starts at 100 and has a common ratio of 0.1 is 100, 10, 1, 0.1, 0.01, ...

■ **YOUR TURN** Find the geometric sequence whose 4th term is 3 and whose common ratio is $\frac{1}{3}$.

Classroom Example 12.3.3*
Find the geometric sequence whose 9th term is $\frac{1}{243}$ and whose common ratio is $-\frac{1}{9}$.
Answer:
$a_n = 3^{11} \cdot \left(-\dfrac{1}{3^2}\right)^{n-1}$
$\quad = (-1)^{n-1}\dfrac{1}{3^{2n-13}}$

Geometric Series

The sum of the terms of a geometric sequence is called a **geometric series**.

$$a_1 + a_1 \cdot r + a_1 \cdot r^2 + a_1 \cdot r^3 + \cdots$$

If we only sum the first n terms of a geometric sequence, the result is a **finite geometric series** given by

$$S_n = a_1 + a_1 \cdot r + a_1 \cdot r^2 + a_1 \cdot r^3 + \cdots + a_1 \cdot r^{n-1}$$

To develop a formula for the nth partial sum, we multiply the above equation by r:

$$r \cdot S_n = a_1 \cdot r + a_1 \cdot r^2 + a_1 \cdot r^3 + \cdots + a_1 \cdot r^{n-1} + a_1 \cdot r^n$$

Subtracting the **second** equation from the **first** equation, we find that all of the terms on the right side drop out except the *first* term in the **first** equation and the *last* term in the **second** equation:

$$S_n = a_1 + a_1 \cdot r + a_1 \cdot r^2 + \cdots + a_1 r^{n-1}$$
$$-rS_n = \quad\quad - a_1 \cdot r - a_1 \cdot r^2 - \cdots - a_1 r^{n-1} - a_1 r^n$$
$$\overline{}$$
$$S_n - rS_n = a_1 \quad\quad\quad\quad\quad\quad\quad\quad\quad\quad\quad\quad -a_1 r^n$$

Factor the S_n out of the left side and the a_1 out of the right side:

$$S_n(1 - r) = a_1(1 - r^n)$$

Divide both sides by $(1 - r)$, assuming $r \neq 1$. The result is a general formula for the sum of a finite geometric series:

$$S_n = a_1 \frac{(1 - r^n)}{(1 - r)} \quad\quad r \neq 1$$

EVALUATING A FINITE GEOMETRIC SERIES

The sum of the first n terms of a geometric sequence, called a **finite geometric series**, is given by the formula

$$S_n = a_1 \frac{(1 - r^n)}{(1 - r)} \quad\quad r \neq 1$$

It is important to note that a finite geometric series can also be written in sigma (summation) notation:

$$S_n = \sum_{k=1}^{n} a_1 \cdot r^{k-1} = a_1 + a_1 \cdot r + a_1 \cdot r^2 + a_1 \cdot r^3 + \cdots + a_1 \cdot r^{n-1}$$

Study Tip

The underscript $k = 1$ applies only when the summation starts at the a_1 term. It is important to note which term is the starting term.

Technology Tip

a. To find the sum of the series

$\sum_{k=1}^{13} 3 \cdot (0.4)^{k-1}$, press

| 2nd | LIST | ▶ | MATH | ▼ |

| 5:sum(| ENTER | 2nd | LIST | ▶ |

| OPS | ▼ | 5:seq(| ENTER | 3 | × |

| (| 0.4 |) | ^ | (| ALPHA | K | − |

| 1 |) | , | ALPHA | K | , | 1 | , | 13 | , |

| 1 |) |) | ENTER |.

```
sum(seq(3*(0.4)^
(K-1),K,1,13,1))
      4.999966446
```

b. To find the sum of the first nine terms of the series $1 + 2 + 4 + 8 + 16 + 32 + 64 + \cdots$, press

| 2nd | LIST | ▶ | MATH | ▼ |

| 5:sum(| ENTER | 2nd | LIST | ▶ |

| OPS | ▼ | 5:seq(| ENTER | 2 | ^ |

| (| ALPHA | K | − | 1 |) | , |

| ALPHA | K | , | 1 | , | 9 | , | 1 |) |) |

| ENTER |.

```
sum(seq(2^(K-1),
K,1,9,1))
             511
```

EXAMPLE 4   Evaluating a Finite Geometric Series

Evaluate the finite geometric series.

a. $\displaystyle\sum_{k=1}^{13} 3 \cdot (0.4)^{k-1}$

b. The first nine terms of the series $1 + 2 + 4 + 8 + 16 + 32 + 64 + \cdots$

Solution (a):

Identify a_1, n, and r. $\qquad a_1 = 3, n = 13,$ and $r = 0.4$

Substitute $a_1 = 3, n = 13,$ and $r = 0.4$

into $s_n = a_1 \dfrac{(1 - r^n)}{(1 - r)}$. $\qquad S_{13} = 3 \dfrac{\left(1 - 0.4^{13}\right)}{(1 - 0.4)}$

Simplify. $\qquad\qquad\qquad \boxed{S_{13} \approx 4.99997}$

Solution (b):

Identify the first term and common ratio. $\qquad a_1 = 1$ and $r = 2$

Substitute $a_1 = 1$ and $r = 2$ into $S_n = a_1 \dfrac{(1 - r^n)}{(1 - r)}$. $\quad S_n = \dfrac{(1 - 2^n)}{(1 - 2)}$

To sum the first nine terms, let $n = 9$. $\qquad S_9 = \dfrac{\left(1 - 2^9\right)}{(1 - 2)}$

Simplify. $\qquad\qquad\qquad \boxed{S_9 = 511}$

Classroom Example 12.3.4

Compute the sums.

a. $\displaystyle\sum_{n=1}^{30} 2\left(\dfrac{3}{5}\right)^n$

b. $\displaystyle\sum_{n=1}^{100} \left(-\dfrac{2}{3}\right)^n$

Answers:

a. $3\left[1 - \left(\dfrac{3}{5}\right)^{30}\right]$

b. $-\dfrac{2}{5}\left[1 - \left(-\dfrac{2}{3}\right)^{100}\right]$

The sum of an infinite geometric sequence is called an **infinite geometric series**. Some infinite geometric series converge (yield a finite sum), and some diverge (do not have a finite sum). For example,

$$\frac{1}{2} + \frac{1}{4} + \frac{1}{8} + \frac{1}{16} + \frac{1}{32} + \cdots + \frac{1}{2^n} + \cdots = 1 \text{ (converges)}$$

$$2 + 4 + 8 + 16 + 32 + \cdots + 2^n + \cdots \text{ (diverges)}$$

For infinite geometric series that converge, the partial sum S_n approaches a single number as n gets large. The formula used to evaluate a finite geometric series

$$S_n = a_1 \frac{(1 - r^n)}{(1 - r)}$$

can be extended to an infinite geometric series for certain values of r. If $|r| < 1$, then when r is raised to a power, it continues to get smaller, approaching 0. For those values of r, the infinite geometric series converges to a finite sum.

$$\text{Let } n \to \infty; \text{ then } a_1 \frac{(1 - r^n)}{(1 - r)} \to a_1 \frac{(1 - 0)}{(1 - r)} = a_1 \frac{1}{1 - r}, \text{ if } |r| < 1.$$

EVALUATING AN INFINITE GEOMETRIC SERIES

The **sum of an infinite geometric series** is given by the formula

$$\sum_{n=0}^{\infty} a_1 \cdot r^n = a_1 \frac{1}{(1-r)} \qquad |r| < 1$$

EXAMPLE 5 **Determining Whether the Sum of an Infinite Series Exists**

Determine whether the sum exists for each of the geometric series.

a. $3 + 15 + 75 + 375 + \cdots$ **b.** $8 + 4 + 2 + 1 + \frac{1}{2} + \frac{1}{4} + \frac{1}{8} + \cdots$

Solution (a):

Identify the common ratio. $r = 5$

Since 5 is greater than 1, the $\boxed{\text{sum does not exist}}$. $r = 5 > 1$

Solution (b):

Identify the common ratio. $r = \dfrac{1}{2}$

Since $\frac{1}{2}$ is less than 1, the $\boxed{\text{sum exists}}$. $r = \dfrac{1}{2} < 1$

■ **YOUR TURN** Determine whether the sum exists for the following geometric series.

a. $81, 9, 1, \frac{1}{9}, \ldots$ **b.** $1, 5, 25, 125, \ldots$

Do you expect $\frac{1}{4} + \frac{1}{12} + \frac{1}{36} + \frac{1}{64} + \cdots$ and $\frac{1}{4} - \frac{1}{12} + \frac{1}{36} - \frac{1}{64} + \cdots$ to sum to the same number? The answer is no, because the second series is an alternating series and terms are both added and subtracted. Hence, we would expect the second series to sum to a smaller number than the first series sums to.

EXAMPLE 6 **Evaluating an Infinite Geometric Series**

Evaluate each infinite geometric series.

a. $1 + \frac{1}{3} + \frac{1}{9} + \frac{1}{27} + \cdots$ **b.** $1 - \frac{1}{3} + \frac{1}{9} - \frac{1}{27} + \cdots$

Solution (a):

Identify the first term and the common ratio. $a_1 = 1 \qquad r = \dfrac{1}{3}$

Since $|r| = \left|\frac{1}{3}\right| < 1$, the sum of the series exists.

Substitute $a_1 = 1$ and $r = \frac{1}{3}$ into $\sum_{n=0}^{\infty} a_1 \cdot r^n = a_1 \dfrac{1}{(1-r)}$. $\dfrac{1}{1 - 1/3}$

Simplify. $= \dfrac{1}{2/3} = \dfrac{3}{2}$

$$\boxed{1 + \frac{1}{3} + \frac{1}{9} + \frac{1}{27} + \cdots = \frac{3}{2}}$$

Solution (b):

Identify the first term and the common ratio. $a_1 = 1 \qquad r = -\frac{1}{3}$

Since $|r| = |-\frac{1}{3}| < 1$, the sum of the series exists.

Substitute $a_1 = 1$ and $r = -\frac{1}{3}$ into $\sum_{n=0}^{\infty} a_1 \cdot r^n = a_1 \frac{1}{(1-r)}$.

Simplify. $= \frac{1}{1-(-1/3)} = \frac{1}{1+(1/3)} = \frac{1}{4/3} = \frac{3}{4}$

$$\boxed{1 - \frac{1}{3} + \frac{1}{9} - \frac{1}{27} + \cdots = \frac{3}{4}}$$

Notice that the alternating series summed to $\frac{3}{4}$, whereas the positive series summed to $\frac{3}{2}$.

■ **Answer: a.** $\frac{3}{8}$ **b.** $\frac{3}{16}$

■ **YOUR TURN** Find the sum of each infinite geometric series.

a. $\frac{1}{4} + \frac{1}{12} + \frac{1}{36} + \frac{1}{108} + \cdots$ **b.** $\frac{1}{4} - \frac{1}{12} + \frac{1}{36} - \frac{1}{108} + \cdots$

It is important to note the restriction on the common ratio r. The absolute value of the common ratio has to be strictly less than 1 for an infinite geometric series to converge. Otherwise the infinite geometric series diverges.

EXAMPLE 7   Evaluating an Infinite Geometric Series

Evaluate the infinite geometric series, if possible.

a. $\displaystyle\sum_{n=0}^{\infty} 2\left(-\frac{1}{4}\right)^n$ **b.** $\displaystyle\sum_{n=1}^{\infty} 3 \cdot (2)^{n-1}$

Solution (a):

Identify a_1 and r.

$$\sum_{n=0}^{\infty} 2\left(-\frac{1}{4}\right)^n = \underset{a_1}{2} - \frac{1}{2} + \frac{1}{8} - \frac{1}{32} + \frac{1}{128} - \cdots$$

with $r = -\frac{1}{4}$ throughout.

Since $|r| = |-\frac{1}{4}| = \frac{1}{4} < 1$, the infinite geometric series converges.

$$\sum_{n=0}^{\infty} a_1 \cdot r^n = \frac{a_1}{(1-r)}$$

Let $a_1 = 2$ and $r = -\frac{1}{4}$.

$$= \frac{2}{[1-(-1/4)]}$$

Simplify.

$$= \frac{2}{1 + 1/4} = \frac{2}{5/4} = \frac{8}{5}$$

This infinite geometric series converges.

$$\boxed{\sum_{n=0}^{\infty} 2\left(-\frac{1}{4}\right)^n = \frac{8}{5}}$$

Solution (b):

Identify a_1 and r.

$$\sum_{n=1}^{\infty} 3 \cdot (2)^{n-1} = \underset{a_1}{3} + 6 + 12 + 24 + 48 + \cdots$$

with $r = 2$.

$$\boxed{\text{Since } r = 2 > 1, \text{ this infinite geometric series diverges.}}$$

Applications

Suppose you are given a job offer with a guaranteed percentage raise per year. What will your annual salary be 10 years from now? That answer can be obtained using a geometric sequence. Suppose you want to make voluntary contributions to a retirement account directly debited from your paycheck every month. Suppose the account earns a fixed percentage rate: How much will you have in 30 years if you deposit $50 a month? What is the difference in the total you will have in 30 years if you deposit $100 a month instead? These important questions about your personal finances can be answered using geometric sequences and series.

EXAMPLE 8   Future Salary: Geometric Sequence

Suppose you are offered a job as an event planner for the PGA Tour. The starting salary is $45,000, and employees are given a 5% raise per year. What will your annual salary be during the 10th year with the PGA Tour?

Solution:

Every year the salary is 5% more than the previous year.

Label the year 1 salary.	$a_1 = 45,000$
Calculate the year 2 salary.	$a_2 = 1.05 \cdot a_1$
Calculate the year 3 salary.	$a_3 = 1.05 \cdot a_2$
	$= 1.05(1.05 \cdot a_1) = (1.05)^2 a_1$
Calculate the year 4 salary.	$a_4 = 1.05 \cdot a_3$
	$= 1.05(1.05)^2 a_1 = (1.05)^3 a_1$
Identify the year n salary.	$a_n = 1.05^{n-1} a_1$
Substitute $n = 10$ and $a_1 = 45,000$.	$a_{10} = (1.05)^9 \cdot 45,000$
Simplify.	$a_{10} \approx 69,809.77$

During your 10th year with the company your salary will be $69,809.77.

■ **YOUR TURN** Suppose you are offered a job with AT&T at $37,000 per year with a guaranteed raise of 4% after every year. What will your annual salary be after 15 years with the company?

Classroom Example 12.3.8
Suppose the starting salary of an assistant professor is $55,000 and she receives a 4% raise per year. What is her annual salary during the 6th year at the university?

Answer:
$55,000(1.04)^5 = $66,915.91$

■ **Answer:** $64,072.03

EXAMPLE 9   Savings Growth: Geometric Series

Karen has maintained acrylic nails by paying for them with money earned from a part-time job. After hearing a lecture from her economics professor on the importance of investing early in life, she decides to remove the acrylic nails, which cost $50 per month, and do her own manicures. She has that $50 automatically debited from her checking account on the first of every month and put into a money market account that receives 3% interest compounded monthly. What will the balance be in the money market account exactly 2 years from the day of her initial $50 deposit?

Technology Tip

Use a calculator to find
$$S_{24} = 50(1.0025)\frac{(1 - 1.0025^{24})}{(1 - 1.0025)}.$$

Scientific calculators:

Press	Display
50 [x] 1.0025 [x]	1238.23
[(] 1 [−] 1.0025	
[x^y] 24 [)] [÷] [(]	
1 [−] 1.0025 [)] [=]	

Graphing calculators:

50 [x] 1.0025 [x] [(] 1 [−] 1.0025

[^] 24 [)] [÷] [(] 1 [−] 1.0025 [)]

[ENTER]

```
50*1.0025*(1-1.0
025^24)/(1-1.002
5)
        1238.228737
■
```

■ **Answer:** $5105.85

Solution:

Recall the compound interest formula.

$$A = P\left(1 + \frac{r}{n}\right)^{nt}$$

Substitute $r = 0.03$ and $n = 12$ into the compound interest formula.

$$A = P\left(1 + \frac{0.03}{12}\right)^{12t}$$

$$= P(1.0025)^{12t}$$

Let $t = \frac{n}{12}$, where n is the number of months of the investment:

$$A_n = P(1.0025)^n$$

The first deposit of $50 will gain interest for 24 months. $\quad A_{24} = 50(1.0025)^{24}$

The second deposit of $50 will gain interest for 23 months. $\quad A_{23} = 50(1.0025)^{23}$

The third deposit of $50 will gain interest for 22 months. $\quad A_{22} = 50(1.0025)^{22}$

The last deposit of $50 will gain interest for 1 month. $\quad A_1 = 50(1.0025)^1$

Sum the amounts accrued from the 24 deposits.

$$A_1 + A_2 + \cdots + A_{24} = 50(1.0025) + 50(1.0025)^2 + 50(1.0025)^3 + \cdots + 50(1.0025)^{24}$$

Identify the first term and common ratio. $\quad a_1 = 50(1.0025)$ and $r = 1.0025$

Sum the first n terms of a geometric series.

$$S_n = a_1\frac{(1 - r^n)}{(1 - r)}$$

Substitute $n = 24$, $a_1 = 50(1.0025)$, and $r = 1.0025$.

$$S_{24} = 50(1.0025)\frac{(1 - 1.0025^{24})}{(1 - 1.0025)}$$

Simplify. $\quad S_{24} \approx 1238.23$

Karen will have $1238.23 saved in her money market account in 2 years.

■ **YOUR TURN** Repeat Example 9 with Karen putting $100 (instead of $50) in the same money market. Assume she does this for 4 years (instead of 2 years).

▶ **SECTION**
12.3 SUMMARY

In this section, we discussed geometric sequences, in which each successive term is found by multiplying the previous term by a constant, so that $a_{n+1} = r \cdot a_n$. That constant, r, is called the common ratio. The nth term of a geometric sequence is given by $a_n = a_1 r^{n-1}$, $n \geq 1$ or $a_{n+1} = a_1 r^n$, $n \geq 0$. The sum of the terms of a geometric sequence is called a geometric series. Finite geometric series converge to a number. Infinite geometric series converge to a number if the absolute value of the common ratio is less than 1. If the absolute value of the common ratio is greater

than or equal to 1, the infinite geometric series diverges and the sum does not exist. Many real-world applications involve geometric sequences and series, such as growth of salaries and annuities through percentage increases.

Finite Geometric Series: $\quad \displaystyle\sum_{i=0}^{n} a_1 r^i = a_1\frac{(1 - r^n)}{(1 - r)}$

Infinite Geometric Series: $\quad \displaystyle\sum_{i=0}^{\infty} a_1 r^i = a_1\frac{1}{(1 - r)} \quad |r| < 1$

SECTION 12.3 EXERCISES

▪ SKILLS

In Exercises 1–8, determine whether the sequence is geometric. If it is, find the common ratio.

1. $1, 3, 9, 27, \ldots$

2. $2, 4, 8, 16, \ldots$

3. $1, 4, 9, 16, 25, \ldots$

4. $1, \frac{1}{4}, \frac{1}{9}, \frac{1}{16}, \ldots$

5. $8, 4, 2, 1, \ldots$

6. $8, -4, 2, -1, \ldots$

7. $800, 1360, 2312, 3930.4, \ldots$

8. $7, 15.4, 33.88, 74.536, \ldots$

In Exercises 9–16, write the first five terms of the geometric series.

9. $a_1 = 6 \quad r = 3$

10. $a_1 = 17 \quad r = 2$

11. $a_1 = 1 \quad r = -4$

12. $a_1 = -3 \quad r = -2$

13. $a_1 = 10{,}000 \quad r = 1.06$

14. $a_1 = 10{,}000 \quad r = 0.8$

15. $a_1 = \frac{2}{3} \quad r = \frac{1}{2}$

16. $a_1 = \frac{1}{10} \quad r = -\frac{1}{5}$

In Exercises 17–24, write the formula for the nth term of the geometric series.

17. $a_1 = 5 \quad r = 2$

18. $a_1 = 12 \quad r = 3$

19. $a_1 = 1 \quad r = -3$

20. $a_1 = -4 \quad r = -2$

21. $a_1 = 1000 \quad r = 1.07$

22. $a_1 = 1000 \quad r = 0.5$

23. $a_1 = \frac{16}{3} \quad r = -\frac{1}{4}$

24. $a_1 = \frac{1}{200} \quad r = 5$

In Exercises 25–30, find the indicated term of the geometric sequence.

25. 7th term of the sequence $-2, 4, -8, 16, \ldots$

26. 10th term of the sequence $1, -5, 25, -225, \ldots$

27. 13th term of the sequence $\frac{1}{3}, \frac{2}{3}, \frac{4}{3}, \frac{8}{3}, \ldots$

28. 9th term of the sequence $100, 20, 4, 0.8, \ldots$

29. 15th term of the sequence $1000, 50, 2.5, 0.125, \ldots$

30. 8th term of the sequence $1000, -800, 640, -512, \ldots$

In Exercises 31–40, find the sum of the finite geometric series.

31. $\dfrac{1}{3} + \dfrac{2}{3} + \dfrac{2^2}{3} + \cdots + \dfrac{2^{12}}{3}$

32. $1 + \dfrac{1}{3} + \dfrac{1}{3^2} + \dfrac{1}{3^3} + \cdots + \dfrac{1}{3^{10}}$

33. $2 + 6 + 18 + 54 + \cdots + 2(3^9)$

34. $1 + 4 + 16 + 64 + \cdots + 4^9$

35. $\displaystyle\sum_{n=0}^{10} 2(0.1)^n$

36. $\displaystyle\sum_{n=0}^{11} 3(0.2)^n$

37. $\displaystyle\sum_{n=1}^{8} 2(3)^{n-1}$

38. $\displaystyle\sum_{n=1}^{9} \frac{2}{3}(5)^{n-1}$

39. $\displaystyle\sum_{k=0}^{13} 2^k$

40. $\displaystyle\sum_{k=0}^{13} \left(\frac{1}{2}\right)^k$

In Exercises 41–54, find the sum of the infinite geometric series, if possible.

41. $\displaystyle\sum_{n=0}^{\infty} \left(\frac{1}{2}\right)^n$

42. $\displaystyle\sum_{n=1}^{\infty} \left(\frac{1}{3}\right)^n$

43. $\displaystyle\sum_{n=1}^{\infty} \left(-\frac{1}{3}\right)^n$

44. $\displaystyle\sum_{n=0}^{\infty} \left(-\frac{1}{2}\right)^n$

45. $\displaystyle\sum_{n=0}^{\infty} 1^n$

46. $\displaystyle\sum_{n=0}^{\infty} 1.01^n$

47. $\displaystyle\sum_{n=0}^{\infty} -9\left(\frac{1}{3}\right)^n$

48. $\displaystyle\sum_{n=0}^{\infty} -8\left(-\frac{1}{2}\right)^n$

49. $\displaystyle\sum_{n=0}^{\infty} 10{,}000(0.05)^n$

50. $\displaystyle\sum_{n=0}^{\infty} 200(0.04)^n$

51. $\displaystyle\sum_{n=1}^{\infty} 0.4^n$

52. $0.3 + 0.03 + 0.003 + 0.0003 + \cdots$

53. $\displaystyle\sum_{n=0}^{\infty} 0.99^n$

54. $\displaystyle\sum_{n=0}^{\infty} \left(\frac{5}{4}\right)^n$

▪ APPLICATIONS

55. Salary. Jeremy is offered a government job with the Department of Commerce. He is hired on the "GS" scale at a base rate of \$34,000 with 2.5% increases in his salary per year. Calculate what his salary will be after he has been with the Department of Commerce for 12 years.

56. Salary. Alison is offered a job with a small start-up company that wants to promote loyalty to the company with incentives for employees to stay with the company. The company offers her a starting salary of \$22,000 with a guaranteed 15% raise per year. What will her salary be after she has been with the company for 10 years?

57. Depreciation. Brittany, a graduating senior in high school, receives a laptop computer as a graduation gift from her Aunt Jeanine so that she can use it when she gets to the University of Alabama. If the laptop costs $2000 new and depreciates 50% per year, write a formula for the value of the laptop n years after it was purchased. How much will the laptop be worth when Brittany graduates from college (assuming she will graduate in 4 years)? How much will it be worth when she finishes graduate school? Assume graduate school is another 3 years.

58. Depreciation. Derek is deciding between a new Honda Accord and the BMW 325 series. The BMW costs $35,000 and the Honda costs $25,000. If the BMW depreciates at 20% per year and the Honda depreciates at 10% per year, find formulas for the value of each car n years after it is purchased. Which car is worth more in 10 years?

59. Bungee Jumping. A bungee jumper rebounds 70% of the height jumped. Assuming the bungee jump is made with a cord that stretches to 100 feet, how far will the bungee jumper travel upward on the fifth rebound?

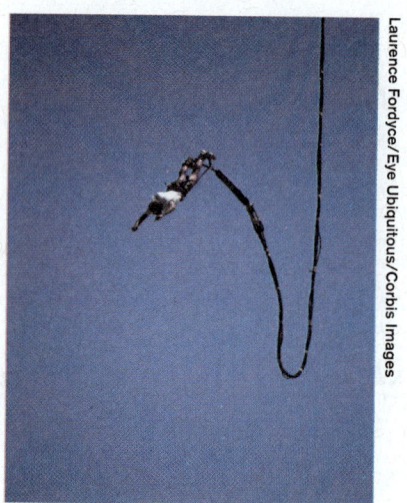

Laurence Fordyce/Eye Ubiquitous/Corbis Images

60. Bungee Jumping. A bungee jumper rebounds 65% of the height jumped. Assuming the bungee cord stretches 200 feet, how far will the bungee jumper travel upward on the eighth rebound?

61. Population Growth. One of the fastest-growing universities in the country is the University of Central Florida. The student populations each year starting in 2000 were 36,000, 37,800, 39,690, 41,675, ... If this rate continued, how many students were at UCF in 2010?

62. Website Hits. The website for Matchbox 20 (www.matchboxtwenty.com) has noticed that every week the number of hits to its website increases 5%. If there were 20,000 hits this week, how many will there be exactly 52 weeks from now?

63. Rich Man's Promise. A rich man promises that he will give you $1000 on January 1, and every day after that, he will pay you 90% of what he paid you the day before. How many days will it take before you are making less than $1? How much will the rich man pay out for the entire month of January? Round answers to the nearest dollar.

64. Poor Man's Clever Deal. A poor man promises to work for you for $0.01 the first day, $0.02 on the second day, $0.04 on the third day; his salary will continue to double each day. If he started on January 1, how much would he be paid to work on January 31? How much total would he make during the month? Round answers to the nearest dollar.

65. Investing Lunch. A newlywed couple decides to stop going out to lunch every day and instead brings their lunch. They estimate it will save them $100 per month. They invest that $100 on the first of every month into an account that is compounded monthly and pays 5% interest. How much will be in the account at the end of 3 years?

66. Pizza as an Investment. A college freshman decides to stop ordering late-night pizzas (for both health and cost reasons). He realizes that he has been spending $50 a week on pizzas. Instead, he deposits $50 into an account that compounds weekly and pays 4% interest. How much money will be in the account in 52 weeks?

67. Tax-Deferred Annuity. Dr. Schober contributes $500 from her paycheck (weekly) to a tax-deferred investment account. Assuming the investment earns 6% and is compounded weekly, how much will be in the account in 26 weeks? 52 weeks?

68. Saving for a House. If a new graduate decides she wants to save for a house and she is able to put $300 every month into an account that earns 5% compounded monthly, how much will she have in the account in 5 years?

69. House Values. In 2008, you buy a house for $195,000. The value of the house appreciates 6.5% per year, on the average. How much is the house worth after 15 years?

70. The Bouncing Ball Problem. A ball is dropped from a height of 9 feet. Assume that on each bounce, the ball rebounds to one-third of its previous height. Find the total distance that the ball travels.

71. Probability. A fair coin is tossed repeatedly. The probability that the first head occurs on the nth toss is given by the function $p(n) = \left(\frac{1}{2}\right)^n$, where $n \geq 1$. Show that

$$\sum_{n=1}^{\infty} \left(\frac{1}{2}\right)^n = 1.0$$

72. Salary. Suppose you work for a supervisor who gives you two different options to choose from for your monthly pay. Option 1: The company pays you $0.01 for the first day of work, $0.02 the second day, $0.04 for the third day, $0.08 for the fourth day, and so on for 30 days. Option 2: You can receive a check right now for $10 million. Which pay option is better? How much better is it?

▪ CATCH THE MISTAKE

In Exercises 73–76, explain the mistake that is made.

73. Find the nth term of the geometric sequence:

$$-1, \tfrac{1}{3}, -\tfrac{1}{9}, \tfrac{1}{27}, \ldots .$$

Solution:

Identify the first term and common ratio.

$$a_1 = -1 \text{ and } r = \frac{1}{3}$$

Substitute $a_1 = -1$ and $r = \frac{1}{3}$ into $a_n = a_1 \cdot r^{n-1}$.

$$a_n = (-1) \cdot \left(\frac{1}{3}\right)^{n-1}$$

Simplify.

$$a_n = \frac{-1}{3^{n-1}}$$

This is incorrect. What mistake was made?

74. Find the sum of the first n terms of the finite geometric series:

$$2, 4, 8, 16, \ldots .$$

Solution:

Write the sum in sigma notation.

$$\sum_{k=1}^{n} (2)^k$$

Identify the first term and common ratio.

$$a_1 = 1 \text{ and } r = 2$$

Substitute $a_1 = 1$ and $r = 2$ into $S_n = a_1 \dfrac{(1 - r^n)}{(1 - r)}$.

$$S_n = 1\frac{(1 - 2^n)}{(1 - 2)}$$

Simplify.

$$S_n = 2^n - 1$$

This is incorrect. What mistake was made?

75. Find the sum of the finite geometric series $\displaystyle\sum_{n=1}^{8} 4(-3)^n$.

Solution:

Identify the first term and common ratio.

$$a_1 = 4 \text{ and } r = -3$$

Substitute $a_1 = 4$ and $r = -3$ into $S_n = a_1 \dfrac{(1 - r^n)}{(1 - r)}$.

$$S_n = 4\frac{[1 - (-3)^n]}{[1 - (-3)]}$$

$$= 4\frac{[1 - (-3)^n]}{4}$$

Simplify.

$$S_n = [1 - (-3)^n]$$

Substitute $n = 8$.

$$S_8 = [1 - (-3)^8] = -6{,}560$$

This is incorrect. What mistake was made?

76. Find the sum of the infinite geometric series $\displaystyle\sum_{n=1}^{\infty} 2 \cdot 3^{n-1}$.

Solution:

Identify the first term and common ratio.

$$a_1 = 2 \text{ and } r = 3$$

Substitute $a_1 = 2$ and $r = 3$ into $S_\infty = a_1 \dfrac{1}{(1 - r)}$.

$$S_\infty = 2\frac{1}{(1 - 3)}$$

Simplify.

$$S_\infty = -1$$

This is incorrect. The series does not sum to -1. What mistake was made?

▪ CONCEPTUAL

In Exercises 77–80, determine whether each statement is true or false.

77. An alternating sequence cannot be a geometric sequence.

78. All finite and infinite geometric series can always be evaluated.

79. The common ratio of a geometric sequence can be positive or negative.

80. An infinite geometric series can be evaluated if the common ratio is less than or equal to 1.

▪ CHALLENGE

81. State the conditions for the sum

$$a + a \cdot b + a \cdot b^2 + \cdots + a \cdot b^n + \cdots$$

to exist. Assuming those conditions are met, find the sum.

82. Find the sum of $\sum_{k=0}^{20} \log 10^{2^k}$.

83. Represent the repeating decimal $0.474747\ldots$ as a fraction (ratio of two integers).

84. Suppose the sum of an infinite geometric series is

$$S = \frac{2}{1 - x}, \text{ where } x \text{ is a variable.}$$

a. Write out the first five terms of the series.

b. For what values of x will the series converge?

TECHNOLOGY

85. Sum the series: $\sum_{k=1}^{50} (-2)^{k-1}$. Apply a graphing utility to confirm your answer.

86. Does the sum of the infinite series $\sum_{n=0}^{\infty} \left(\frac{1}{3}\right)^n$ exist? Use a graphing calculator to find it.

87. Apply a graphing utility to plot $y_1 = 1 + x + x^2 + x^3 + x^4$ and $y_2 = \dfrac{1}{1 - x}$, and let the range of x be $[-0.5, 0.5]$. Based on what you see, what do you expect the geometric series $\sum_{n=0}^{\infty} x^n$ to sum to?

88. Apply a graphing utility to plot $y_1 = 1 - x + x^2 - x^3 + x^4$ and $y_2 = \dfrac{1}{1 + x}$, and let x range from $[-0.5, 0.5]$. Based on what you see, what do you expect the geometric series $\sum_{n=0}^{\infty} (-1)^n x^n$ to sum to?

89. Apply a graphing utility to plot $y_1 = 1 + 2x + 4x^2 + 8x^3 + 16x^4$ and $y_2 = \dfrac{1}{1 - 2x}$, and let x range from $[-0.3, 0.3]$. Based on what you see, what do you expect the geometric series $\sum_{n=0}^{\infty} (2x)^n$ to sum to?

SECTION
12.4 MATHEMATICAL INDUCTION

SKILLS OBJECTIVE

■ Prove mathematical statements using mathematical induction.

CONCEPTUAL OBJECTIVE

■ Understand that just because there appears to be a pattern, the pattern is not necessarily true for all values.

Mathematical Induction

n	$n^2 - n + 41$	PRIME?
1	41	Yes
2	43	Yes
3	47	Yes
4	53	Yes
5	61	Yes

Is the expression $n^2 - n + 41$ *always* a prime number if n is a natural number? Your instinct may lead you to try a few values for n.

It appears that the statement might be true for all natural numbers. However, what about when $n = 41$?

$$n^2 - n + 41 = (41)^2 - 41 + 41 = 41^2$$

We find that when $n = 41$, $n^2 - n + 41$ is not prime. The moral of the story is that just because a pattern seems to exist for *some* values, the pattern is not necessarily true for *all* values. We must look for a way to show whether a statement is true for all values. In this section we talk about *mathematical induction*, which is a way to show a statement is true for all values.

Mathematics is based on logic and proof (not assumptions or belief). One of the most famous mathematical statements was Fermat's last theorem. Pierre de Fermat (1601–1665) conjectured that there are no positive integer values for x, y, and z such that $x^n + y^n = z^n$, if $n \geq 3$. Although mathematicians *believed* that this theorem was true, no one was able to *prove* it until 350 years after the assumption was made. Professor Andrew Wiles at Princeton University received a $50,000 prize for successfully proving Fermat's last theorem in 1994.

Mathematical induction is a technique used in college algebra and even in very advanced mathematics to prove many kinds of mathematical statements. In this section you will use it to prove statements like "if $x > 1$, then $x^n > 1$ for all natural numbers n."

The principle of mathematical induction can be illustrated by a row of standing dominos, as in the picture. We make two assumptions:

1. The first domino is knocked down.
2. If a domino is knocked down, then the domino immediately following it will also be knocked down.

If both of these assumptions are true, then it is also true that all of the dominos will fall.

PRINCIPLE OF MATHEMATICAL INDUCTION

Let S_n be a statement involving the positive integer n. To prove that S_n is true for all positive integers, the following steps are required.

Step 1: Show that S_1 is true.
Step 2: Assume S_k is true and show that S_{k+1} is true (k = positive integer).

Combining Steps 1 and 2 proves the statement is true for all positive integers.

EXAMPLE 1   Using Mathematical Induction

Apply the principle of mathematical induction to prove this statement:

If $x > 1$, then $x^n > 1$ for all natural numbers n.

Solution:

STEP 1 Show the statement is true for $n = 1$. $x^1 > 1$ because $x > 1$

STEP 2 Assume the statement is true for $n = k$. $x^k > 1$

Show the statement is true for $k + 1$.

Multiply both sides by x. $x^k \cdot x > 1 \cdot x$

(Since $x > 1$, this step does not reverse the inequality sign.)

Simplify. $x^{k+1} > x$

Recall that $x > 1$. $x^{k+1} > x > 1$

Therefore, we have shown that $x^{k+1} > 1$.

This completes the induction proof. Thus, the following statement is true.

"If $x > 1$, then $x^n > 1$ for **all** natural numbers n."

EXAMPLE 2   Using Mathematical Induction

Use mathematical induction to prove that $n^2 + n$ is divisible by 2 for all natural numbers n.

Solution:

STEP 1 Show the statement we are testing is true for $n = 1$. $1^2 + 1 = 2$

2 is divisible by 2. $\dfrac{2}{2} = 1$

Classroom Example 12.4.1
Let $a, b, c \neq 0$ and define

$$A = \begin{bmatrix} a & 0 & 0 \\ 0 & b & 0 \\ 0 & 0 & c \end{bmatrix}.$$

Prove $A^n = \begin{bmatrix} a^n & 0 & 0 \\ 0 & b^n & 0 \\ 0 & 0 & c^n \end{bmatrix}$

for every $n \geq 2$.

Answer:
Proof:

$$A^2 = \begin{bmatrix} a & 0 & 0 \\ 0 & b & 0 \\ 0 & 0 & c \end{bmatrix} \cdot \begin{bmatrix} a & 0 & 0 \\ 0 & b & 0 \\ 0 & 0 & c \end{bmatrix}$$

$$= \begin{bmatrix} a^2 & 0 & 0 \\ 0 & b^2 & 0 \\ 0 & 0 & c^2 \end{bmatrix}$$

Assume $A^n = \begin{bmatrix} a^n & 0 & 0 \\ 0 & b^n & 0 \\ 0 & 0 & c^n \end{bmatrix}$.

Observe that
$A^{n+1} = A^n \cdot A$

$$= \begin{bmatrix} a^{n+1} & 0 & 0 \\ 0 & b^{n+1} & 0 \\ 0 & 0 & c^{n+1} \end{bmatrix}$$

Classroom Example 12.4.2*
Prove that

$$\sum_{k=1}^{n} \ln\left(\frac{k}{k+1}\right) = -\ln(k+1)$$

for every $n \geq 1$.

Answer:
Note that

$$\ln\left(\frac{k}{k+1}\right) = \ln k - \ln(k+1).$$

Use this to simplify the given sum.

STEP 2 Assume the statement is true for $n = k$. $\qquad \dfrac{k^2 + k}{2} =$ an integer

Show it is true for $k + 1$ where $k \geq 1$. $\qquad \dfrac{(k + 1)^2 + (k + 1)}{2} \stackrel{?}{=}$ an integer

$$\dfrac{k^2 + 2k + 1 + k + 1}{2} \stackrel{?}{=} \text{an integer}$$

Regroup terms. $\qquad \dfrac{(k^2 + k) + 2(k + 1)}{2} \stackrel{?}{=}$ an integer

$$\dfrac{(k^2 + k)}{2} + \dfrac{2(k + 1)}{2} \stackrel{?}{=} \text{an integer}$$

We assumed $\dfrac{k^2 + k}{2} =$ an integer. $\qquad$ an integer $+ (k + 1) \stackrel{?}{=}$ an integer

Since k is a natural number, $\qquad$ an integer $+$ an integer $=$ an integer

This completes the induction proof. The following statement is true:

"$n^2 + n$ is divisible by 2 for all natural numbers n."

Mathematical induction is often used to prove formulas for partial sums.

EXAMPLE 3 **Proving a Partial-Sum Formula with Mathematical Induction**

Apply mathematical induction to prove the following partial-sum formula:

$$1 + 2 + 3 + \cdots + n = \dfrac{n(n + 1)}{2} \quad \text{for all positive integers } n$$

Solution:

STEP 1 Show the formula is true for $n = 1$. $\qquad 1 = \dfrac{1(1 + 1)}{2} = \dfrac{2}{2} = 1$

STEP 2 Assume the formula is true for $n = k$. $\qquad 1 + 2 + 3 + \cdots + k = \dfrac{k(k + 1)}{2}$

Show it is true for $n = k + 1$. $\qquad 1 + 2 + 3 + \cdots + k + (k + 1) \stackrel{?}{=} \dfrac{(k + 1)(k + 2)}{2}$

$$\underbrace{1 + 2 + 3 + \cdots + k}_{\frac{k(k + 1)}{2}} + (k + 1) \stackrel{?}{=} \dfrac{(k + 1)(k + 2)}{2}$$

$$\dfrac{k(k + 1)}{2} + (k + 1) \stackrel{?}{=} \dfrac{(k + 1)(k + 2)}{2}$$

$$\dfrac{k(k + 1) + 2(k + 1)}{2} \stackrel{?}{=} \dfrac{(k + 1)(k + 2)}{2}$$

$$\dfrac{k^2 + 3k + 2}{2} \stackrel{?}{=} \dfrac{(k + 1)(k + 2)}{2}$$

$$\dfrac{(k + 1)(k + 2)}{2} = \dfrac{(k + 1)(k + 2)}{2}$$

This completes the induction proof. The following statement is true:

"$1 + 2 + 3 + \cdots + n = \dfrac{n(n + 1)}{2}$ for all positive integers n."

SECTION
12.4 SUMMARY

Just because we believe something is true does not mean that it is. In mathematics we rely on proof. In this section we discussed *mathematical induction*, a process of proving mathematical statements. The two-step procedure for mathematical induction is to (1) show the statement is true for $n = 1$, then (2) assume the statement is true for $n = k$ and show the statement must be true for $n = k + 1$. The combination of Steps 1 and 2 proves the statement.

SECTION
12.4 EXERCISES

■ SKILLS

In Exercises 1–24, prove the statements using mathematical induction for all positive integers n.

1. $n^2 \leq n^3$

2. If $0 < x < 1$, then $0 < x^n < 1$.

3. $2n \leq 2^n$

4. $5^n < 5^{n+1}$

5. $n! > 2^n$ $n \geq 4$ (Show it is true for $n = 4$, instead of $n = 1$.)

6. $(1 + c)^n \geq nc$ $c > 1$

7. $n(n + 1)(n - 1)$ is divisible by 3.

8. $n^3 - n$ is divisible by 3.

9. $n^2 + 3n$ is divisible by 2.

10. $n(n + 1)(n + 2)$ is divisible by 6.

11. $2 + 4 + 6 + 8 + \cdots + 2n = n(n + 1)$

12. $1 + 3 + 5 + 7 + \cdots + (2n - 1) = n^2$

13. $1 + 3 + 3^2 + 3^3 + \cdots + 3^n = \dfrac{3^{n+1} - 1}{2}$

14. $2 + 4 + 8 + \cdots + 2^n = 2^{n+1} - 2$

15. $1^2 + 2^2 + 3^2 + \cdots + n^2 = \dfrac{n(n + 1)(2n + 1)}{6}$

16. $1^3 + 2^3 + 3^3 + \cdots + n^3 = \dfrac{n^2(n + 1)^2}{4}$

17. $\dfrac{1}{1 \cdot 2} + \dfrac{1}{2 \cdot 3} + \dfrac{1}{3 \cdot 4} + \cdots + \dfrac{1}{n(n + 1)} = \dfrac{n}{n + 1}$

18. $\dfrac{1}{2 \cdot 3} + \dfrac{1}{3 \cdot 4} + \cdots + \dfrac{1}{(n + 1)(n + 2)} = \dfrac{n}{2(n + 2)}$

19. $(1 \cdot 2) + (2 \cdot 3) + (3 \cdot 4) + \cdots + n(n + 1) = \dfrac{n(n + 1)(n + 2)}{3}$

20. $(1 \cdot 3) + (2 \cdot 4) + (3 \cdot 5) + \cdots + n(n + 2) = \dfrac{n(n + 1)(2n + 7)}{6}$

21. $1 + x + x^2 + x^3 + \cdots + x^{n-1} = \dfrac{1 - x^n}{1 - x}$ $x \neq 1$

22. $\dfrac{1}{2} + \dfrac{1}{4} + \dfrac{1}{8} + \cdots + \dfrac{1}{2^n} = 1 - \dfrac{1}{2^n}$

23. The sum of an arithmetic sequence: $a_1 + (a_1 + d) + (a_1 + 2d) + \cdots + [a_1 + (n - 1)d] = \dfrac{n}{2}[2a_1 + (n - 1)d]$.

24. The sum of a geometric sequence: $a_1 + a_1 r + a_1 r^2 + \cdots + a_1 r^{n-1} = a_1\left(\dfrac{1 - r^n}{1 - r}\right)$.

■ **APPLICATIONS**

The Tower of Hanoi. This is a game with three pegs and n disks (largest on the bottom and smallest on the top). The goal is to move this entire tower of disks to another peg (in the same order). The challenge is that you may move only one disk at a time, and at no time can a larger disk be resting on a smaller disk. You may want to first go online to www.mazeworks.com/hanoi/index/htm and play the game.

Tower of Hanoi

25. What is the smallest number of moves needed if there are three disks?

26. What is the smallest number of moves needed if there are four disks?

27. What is the smallest number of moves needed if there are five disks?

28. What is the smallest number of moves needed if there are n disks? Prove it by mathematical induction.

29. **Telephone Infrastructure.** Suppose there are n cities that are to be connected with telephone wires. Apply mathematical induction to prove that the number of telephone wires required to connect the n cities is given by $\dfrac{n(n-1)}{2}$. Assume each city has to connect directly with any other city.

30. **Geometry.** Prove, with mathematical induction, that the sum of the interior angles of a regular polygon of n sides is given by the formula: $(n-2)(180)$ for $n \geq 3$. *Hint:* Divide a polygon into triangles. For example, a four-sided polygon can be divided into two triangles. A five-sided polygon can be divided into three triangles. A six-sided polygon can be divided into four triangles, and so on.

■ **CONCEPTUAL**

In Exercises 31 and 32, determine whether each statement is true or false.

31. Assume S_k is true. If it can be shown that S_{k+1} is true, then S_n is true for all n, where n is any positive integer.

32. Assume S_1 is true. If it can be shown that S_2 and S_3 are true, then S_n is true for all n, where n is any positive integer.

■ **CHALLENGE**

33. Apply mathematical induction to prove:
$$\sum_{k=1}^{n} k^4 = \frac{n(n+1)(2n+1)(3n^2+3n-1)}{30}$$

34. Apply mathematical induction to prove:
$$\sum_{k=1}^{n} k^5 = \frac{n^2(n+1)^2(2n^2+2n-1)}{12}$$

35. Apply mathematical induction to prove:
$$\left(1+\frac{1}{1}\right)\left(1+\frac{1}{2}\right)\left(1+\frac{1}{3}\right)\cdots\left(1+\frac{1}{n}\right) = n+1$$

36. Apply mathematical induction to prove that $x+y$ is a factor of $x^{2n}-y^{2n}$.

37. Apply mathematical induction to prove:
$$\ln(c_1 \cdot c_2 \cdot c_3 \cdots c_n) = \ln c_1 + \ln c_2 + \cdots + \ln c_n$$

■ **TECHNOLOGY**

38. Use a graphing calculator to sum the series $(1 \cdot 2) + (2 \cdot 3) + (3 \cdot 4) + \cdots + n(n+1)$ on the left side, and evaluate the expression $\dfrac{n(n+1)(n+2)}{3}$ on the right side for $n = 200$. Do they agree with each other? Do your answers confirm the proof for Exercise 19?

39. Use a graphing calculator to sum the series $\dfrac{1}{2} + \dfrac{1}{4} + \dfrac{1}{8} + \cdots + \dfrac{1}{2^n}$ on the left side, and evaluate the expression $1 - \dfrac{1}{2^n}$ on the right side for $n = 8$. Do they agree with each other? Do your answers confirm the proof for Exercise 22?

SKILLS OBJECTIVES

■ Evaluate a binomial coefficient with the binomial theorem.
■ Evaluate a binomial coefficient with Pascal's triangle.
■ Expand a binomial raised to a positive integer power.
■ Find a particular term of a binomial expansion.

CONCEPTUAL OBJECTIVE

■ Recognize patterns in binomial expansions.

Binomial Coefficients

A **binomial** is a polynomial that has two terms. The following are all examples of binomials:

$$x^2 + 2y \qquad a + 3b \qquad 4x^2 + 9$$

In this section we will develop a formula for raising a binomial to a power n, where n is a positive integer.

$$(x^2 + 2y)^6 \qquad (a + 3b)^4 \qquad (4x^2 + 9)^5$$

To begin, let's start by writing out the expansions of $(a + b)^n$ for several values of n.

$$(a + b)^0 = 1$$
$$(a + b)^1 = a + b$$
$$(a + b)^2 = a^2 + 2ab + b^2$$
$$(a + b)^3 = a^3 + 3a^2b + 3ab^2 + b^3$$
$$(a + b)^4 = a^4 + 4a^3b + 6a^2b^2 + 4ab^3 + b^4$$
$$(a + b)^5 = a^5 + 5a^4b + 10a^3b^2 + 10a^2b^3 + 5ab^4 + b^5$$

There are several *patterns* that all of the **binomial expansions** have.

1. The number of terms in each resulting polynomial is always *one more* than the power of the binomial n. Thus, there are $n + 1$ terms in each expansion.

$$n = 3: \qquad (a + b)^3 = \underbrace{a^3 + 3a^2b + 3ab^2 + b^3}_{\text{four terms}}$$

2. Each expansion has symmetry. For example, a and b can be interchanged and you will arrive at the same expansion. Furthermore, the powers of ***a* decrease** by 1 in each successive term, and the powers of ***b* increase** by 1 in each successive term.

$$(a + b)^3 = a^3b^0 + 3a^2b^1 + 3a^1b^2 + a^0b^3$$

3. The sum of the powers of each term in the expansion is n.

$$n = 3: \qquad (a + b)^3 = \overset{3+0=3}{a^3b^0} + \overset{2+1=3}{3a^2b^1} + \overset{1+2=3}{3a^1b^2} + \overset{0+3=3}{a^0b^3}$$

4. The coefficients **increase** and **decrease** in a symmetric manner.

$$(a + b)^5 = \mathbf{1}a^5 + \mathbf{5}a^4b + \mathbf{10}a^3b^2 + \mathbf{10}a^2b^3 + \mathbf{5}ab^4 + \mathbf{1}b^5$$

Using these patterns, we can develop a generalized formula for $(a + b)^n$.

$$(a + b)^n = \square a^n + \square a^{n-1}b + \square a^{n-2}b^2 + \cdots + \square a^2b^{n-2} + \square ab^{n-1} + \square b^n$$

We know that there are $n + 1$ terms in the expansion. We also know that the sum of the powers of each term must equal n. The powers increase and decrease by 1 in each successive term, and if we interchanged a and b, the result would be the same expansion. The question that remains is, what coefficients go in the blanks?

We know that the coefficients must increase and then decrease in a symmetric order (similar to walking up and then down a hill). It turns out that the *binomial coefficients* are represented by a symbol that we will now define.

Classroom Example 12.5.1
Evaluate:

a. $\dbinom{99}{97}$

b. $\dbinom{1002}{1002}$

c.* $\dbinom{2k}{2k-1}$ for $k \geq 1$

Answer:
a. 4851   **b.** 1   **c.** $2k$

DEFINITION   **Binomial Coefficients**

For nonnegative integers n and k, where $n \geq k$, the symbol $\dbinom{n}{k}$ is called the **binomial coefficient** and is defined by

$$\binom{n}{k} = \frac{n!}{(n-k)!k!} \qquad \binom{n}{k} \text{ is read "}n \text{ choose } k\text{."}$$

You will see in the following sections that "n choose k" comes from combinations.

Technology Tip

Press	Display
6 $\boxed{nCr}$ 4 $\boxed{=}$	15
5 $\boxed{nCr}$ 5 $\boxed{=}$	1
4 $\boxed{nCr}$ 0 $\boxed{=}$	1
10 $\boxed{nCr}$ 9 $\boxed{=}$	10

```
6 nCr 4
              15
5 nCr 5
               1
4 nCr 0
               1
```

```
10 nCr 9
              10
■
```

EXAMPLE 1   **Evaluating a Binomial Coefficient**

Evaluate the following binomial coefficients.

a. $\dbinom{6}{4}$ **b.** $\dbinom{5}{5}$ **c.** $\dbinom{4}{0}$ **d.** $\dbinom{10}{9}$

Solution:

Select the top number as n and the bottom number as k and substitute into the binomial coefficient formula $\dbinom{n}{k} = \dfrac{n!}{(n-k)!k!}$.

a. $\dbinom{6}{4} = \dfrac{6!}{(6-4)!4!} = \dfrac{6!}{2!4!} = \dfrac{6 \cdot 5 \cdot 4 \cdot 3 \cdot 2 \cdot 1}{(2 \cdot 1)(4 \cdot 3 \cdot 2 \cdot 1)} = \dfrac{6 \cdot 5}{2} = \boxed{15}$

b. $\dbinom{5}{5} = \dfrac{5!}{(5-5)!5!} = \dfrac{5!}{0!5!} = \dfrac{1}{0!} = \dfrac{1}{1} = \boxed{1}$

c. $\dbinom{4}{0} = \dfrac{4!}{(4-0)!0!} = \dfrac{4!}{4!0!} = \dfrac{1}{0!} = \boxed{1}$

d. $\dbinom{10}{9} = \dfrac{10!}{(10-9)!9!} = \dfrac{10!}{1!9!} = \dfrac{10 \cdot 9!}{9!} = \boxed{10}$

■ **Answer: a.** 84   **b.** 28

■ **YOUR TURN** Evaluate the following binomial coefficients.

a. $\dbinom{9}{6}$ **b.** $\dbinom{8}{6}$

Parts (b) and (c) of Example 1 lead to the general formulas:

$$\binom{n}{n} = 1 \qquad \text{and} \qquad \binom{n}{0} = 1$$

Binomial Expansion

Let's return to the question of the binomial expansion and how to determine the coefficients:

$$(a + b)^n = \square a^n + \square a^{n-1}b + \square a^{n-2}b^2 + \cdots + \square a^2 b^{n-2} + \square ab^{n-1} + \square b^n$$

The symbol $\binom{n}{k}$ is called a binomial coefficient because the coefficients in the blanks in the binomial expansion are equivalent to this symbol.

THE BINOMIAL THEOREM

Let a and b be real numbers; then for any positive integer n,

$$(a + b)^n = \binom{n}{0}a^n + \binom{n}{1}a^{n-1}b + \binom{n}{2}a^{n-2}b^2 + \cdots + \binom{n}{n-2}a^2 b^{n-2} + \binom{n}{n-1}ab^{n-1} + \binom{n}{n}b^n$$

or in sigma (summation) notation as

$$(a + b)^n = \sum_{k=0}^{n}\binom{n}{k}a^{n-k}b^k$$

EXAMPLE 2 **Applying the Binomial Theorem**

Expand $(x + 2)^3$ with the binomial theorem.

Solution:

Substitute $a = x, b = 2, n = 3$ into the equation of the binomial theorem.

$$(x + 2)^3 = \sum_{k=0}^{3}\binom{3}{k}x^{3-k}2^k$$

Expand the summation.

$$= \binom{3}{0}x^3 + \binom{3}{1}x^2 \cdot 2 + \binom{3}{2}x \cdot 2^2 + \binom{3}{3}2^3$$

Find the binomial coefficients.

$$= x^3 + 3x^2 \cdot 2 + 3x \cdot 2^2 + 2^3$$

Simplify.

$$= \boxed{x^3 + 6x^2 + 12x + 8}$$

■ **YOUR TURN** Expand $(x + 5)^4$ with the binomial theorem.

EXAMPLE 3 **Applying the Binomial Theorem**

Expand $(2x - 3)^4$ with the binomial theorem.

Solution:

Substitute $a = 2x, b = -3, n = 4$ into the equation of the binomial theorem.

$$(2x - 3)^4 = \sum_{k=0}^{4}\binom{4}{k}(2x)^{4-k}(-3)^k$$

Expand the summation.

$$= \binom{4}{0}(2x)^4 + \binom{4}{1}(2x)^3(-3) + \binom{4}{2}(2x)^2(-3)^2 + \binom{4}{3}(2x)(-3)^3 + \binom{4}{4}(-3)^4$$

Find the binomial coefficients.

$$= (2x)^4 + 4(2x)^3(-3) + 6(2x)^2(-3)^2 + 4(2x)(-3)^3 + (-3)^4$$

Simplify.

$$= \boxed{16x^4 - 96x^3 + 216x^2 - 216x + 81}$$

■ **YOUR TURN** Expand $(3x - 2)^4$ with the binomial theorem.

Classroom Example 12.5.2–12.5.3

Expand these using the binomial theorem.

a. $(2x + 5)^4$

b. $(2x - 5)^4$

Answer:

a. $16x^4 + 160x^3 + 600x^2 + 1000x + 625$

b. $16x^4 - 160x^3 + 600x^2 - 1000x + 625$

■ **Answer:**
$x^4 + 20x^3 + 150x^2 + 500x + 625$

■ **Answer:**
$81x^4 - 216x^3 + 216x^2 - 96x + 16$

Pascal's Triangle

1

1   1

1   2   1

1   3   3   1

1   4   6   4   1

1   5   10   10   5   1

Instead of writing out the binomial theorem and calculating the binomial coefficients using factorials every time you want to do a binomial expansion, we now present an alternative, more convenient way of remembering the binomial coefficients, called **Pascal's triangle**.

Notice that the first and last number in every row is 1. Each of the other numbers is found by adding the two numbers directly above it. For example,

$$3 = 2 + 1 \qquad 4 = 1 + 3 \qquad 10 = 6 + 4$$

Let's arrange values of $\binom{n}{k}$ in a triangular pattern.

$$\binom{0}{0}$$

$$\binom{1}{0} \quad \binom{1}{1}$$

$$\binom{2}{0} \quad \binom{2}{1} \quad \binom{2}{2}$$

$$\binom{3}{0} \quad \binom{3}{1} \quad \binom{3}{2} \quad \binom{3}{3}$$

$$\binom{4}{0} \quad \binom{4}{1} \quad \binom{4}{2} \quad \binom{4}{3} \quad \binom{4}{4}$$

$$\binom{5}{0} \quad \binom{5}{1} \quad \binom{5}{2} \quad \binom{5}{3} \quad \binom{5}{4} \quad \binom{5}{5}$$

$$\binom{6}{0} \quad \binom{6}{1} \quad \binom{6}{2} \quad \binom{6}{3} \quad \binom{6}{4} \quad \binom{6}{5} \quad \binom{6}{6}$$

It turns out that these numbers in Pascal's triangle are exactly the coefficients in a binomial expansion.

$$1$$

$$1a + 1b$$

$$1a^2 + 2ab + 1b^2$$

$$1a^3 + 3a^2b + 3ab^2 + 1b^3$$

$$1a^4 + 4a^3b + 6a^2b^2 + 4ab^3 + 1b^4$$

$$1a^5 + 5a^4b + 10a^3b^2 + 10a^2b^3 + 5ab^4 + 1b^5$$

The top row is called the *zero row* because it corresponds to the binomial raised to the zero power, $n = 0$. Since each row in Pascal's triangle starts and ends with a 1 and all other values are found by adding the two numbers directly above it, we can now easily calculate the sixth row.

$$(a + b)^5 = 1a^5 + 5a^4b + 10a^3b^2 + 10a^2b^3 + 5ab^4 + 1b^5$$

$$(a + b)^6 = 1a^6 + 6a^5b + 15a^4b^2 + 20a^3b^3 + 15a^2b^4 + 6ab^5 + 1b^6$$

Study Tip

Since the top row of Pascal's triangle is called the zero row, the fifth row is the row with 6 coefficients. The *n*th row is the row with $n + 1$ coefficients.

EXAMPLE 4 **Applying Pascal's Triangle in a Binomial Expansion**

Use Pascal's triangle to determine the binomial expansion of $(x + 2)^5$.

Solution:

Write the binomial expansion with blanks for coefficients.

$$(x + 2)^5 = \square x^5 + \square x^4 \cdot 2 + \square x^3 \cdot 2^2 + \square x^2 \cdot 2^3 + \square x \cdot 2^4 + \square 2^5$$

Write the binomial coefficients in the *fifth* row of Pascal's triangle.

1, 5, 10, 10, 5, 1

Substitute these coefficients into the blanks of the binomial expansion.

$$(x + 2)^5 = \mathbf{1}x^5 + \mathbf{5}x^4 \cdot 2 + \mathbf{10}x^3 \cdot 2^2 + \mathbf{10}x^2 \cdot 2^3 + \mathbf{5}x \cdot 2^4 + \mathbf{1} \cdot 2^5$$

Simplify. $\boxed{(x + 2)^5 = x^5 + 10x^4 + 40x^3 + 80x^2 + 80x + 32}$

■ **YOUR TURN** Apply Pascal's triangle to determine the binomial expansion of $(x + 3)^4$.

■ **Answer:**
$x^4 + 12x^3 + 54x^2 + 108x + 81$

EXAMPLE 5 **Applying Pascal's Triangle in a Binomial Expansion**

Use Pascal's triangle to determine the binomial expansion of $(2x + 5)^4$.

Solution:

Write the binomial expansion with blanks for coefficients.

$$(2x + 5)^4 = \square(2x)^4 + \square(2x)^3 \cdot 5 + \square(2x)^2 \cdot 5^2 + \square(2x) \cdot 5^3 + \square 5^4$$

Write the binomial coefficients in the *fourth* row of Pascal's triangle.

1, 4, 6, 4, 1

Substitute these coefficients into the blanks of the binomial expansion.

$$(2x + 5)^4 = \mathbf{1}(2x)^4 + \mathbf{4}(2x)^3 \cdot 5 + \mathbf{6}(2x)^2 \cdot 5^2 + \mathbf{4}(2x) \cdot 5^3 + \mathbf{1} \cdot 5^4$$

Simplify. $\boxed{(2x + 5)^4 = 16x^4 + 160x^3 + 600x^2 + 1000x + 625}$

■ **YOUR TURN** Use Pascal's triangle to determine the binomial expansion of:

a. $(3x + 2)^3$ **b.** $(3x - 2)^5$

■ **Answer:**
a. $27x^3 + 54x^2 + 36x + 8$
b. $243x^5 - 810x^4 + 1080x^3 - 720x^2 + 240x - 32$

Finding a Particular Term of a Binomial Expansion

What if we don't want to find the entire expansion, but instead want just a single term? For example, what is the fourth term of $(a + b)^5$?

WORDS	MATH
Recall the sigma notation.	$(a + b)^n = \sum_{k=0}^{n} \binom{n}{k} a^{n-k} b^k$
Let $n = 5$.	$(a + b)^5 = \sum_{k=0}^{5} \binom{5}{k} a^{5-k} b^k$
Expand.	$(a + b)^5 = \binom{5}{0}a^5 + \binom{5}{1}a^4b + \binom{5}{2}a^3b^2 + \underbrace{\binom{5}{3}a^2b^3}_{\text{fourth term}} + \binom{5}{4}ab^4 + \binom{5}{5}b^5$
Simplify the fourth term.	$10a^2b^3$

> **Classroom Example 12.5.6**
> Determine the 6th term of the binomial expansion of $(5x - 4)^{11}$.
>
> **Answer:** $\binom{11}{5}(5x)^6 (-4)^5$

> **Classroom Example 12.5.6***
> Determine the 10th term of the binomial expansion of $[(1 - 2x) + 3y]^{20}$.
>
> **Answer:** $\binom{20}{9}(1 - 2x)^{11} (3y)^9$

FINDING A PARTICULAR TERM OF A BINOMIAL EXPANSION

The $(r + 1)$ term of the expansion $(a + b)^n$ is $\binom{n}{r}a^{n-r}b^r$.

EXAMPLE 6 Finding a Particular Term of a Binomial Expansion

Find the 5th term of the binomial expansion of $(2x - 7)^6$.

Solution:

Recall that the $r + 1$ term of $(a + b)^n$ is $\binom{n}{r}a^{n-r}b^r$.	
For the 5th term, let $r = 4$.	$\binom{n}{4}a^{n-4}b^4$
For this expansion, let $a = 2x, b = -7, n = 6$.	$\binom{6}{4}(2x)^{6-4}(-7)^4$
Note that $\binom{6}{4} = 15$.	$15(2x)^2(-7)^4$
Simplify.	$\boxed{144{,}060x^2}$

Technology Tip

6 MATH ▶ PRB ▼ 3:nCr
ENTER 4 × 2 x² × ((−)
7) ^ 4 ENTER

```
6 nCr 4*2²*(-7)^
4
            144060
```

■ **Answer:** $1080x^3$

■ **YOUR TURN** What is the third term of the binomial expansion of $(3x - 2)^5$?

In this section we developed a formula for raising a binomial expression to an integer power, $n \geq 0$. The patterns that surfaced were that the expansion displays symmetry between the two terms; that is, every expansion has $n + 1$ terms, the powers sum to n, and the coefficients, called binomial coefficients, are ratios of factorials:

$$(a + b)^n = \sum_{k=0}^{n} \binom{n}{k} a^{n-k} b^k$$

$$\binom{n}{k} = \frac{n!}{(n - k)!\, k!}$$

Also, Pascal's triangle, a shortcut method for evaluating the binomial coefficients, was discussed. The patterns in the triangle are that every row begins and ends with 1 and all other numbers are found by adding the two numbers above the entry.

```
          1
        1   1
      1   2   1
    1   3   3   1
  1   4   6   4   1
1   5  10  10   5   1
```

Last, a formula was given for finding a particular term of a binomial expansion; the $(r + 1)$ term of $(a + b)^n$ is $\binom{n}{r} a^{n-r} b^r$.

SECTION
12.5 EXERCISES

▪ SKILLS

In Exercises 1–10, evaluate the binomial coefficients.

1. $\dbinom{7}{3}$
2. $\dbinom{8}{2}$
3. $\dbinom{10}{8}$
4. $\dbinom{23}{21}$

5. $\dbinom{17}{0}$
6. $\dbinom{100}{0}$
7. $\dbinom{99}{99}$
8. $\dbinom{52}{52}$

9. $\dbinom{48}{45}$
10. $\dbinom{29}{26}$

In Exercises 11–32, expand the expression using the binomial theorem.

11. $(x + 2)^4$
12. $(x + 3)^5$
13. $(y - 3)^5$
14. $(y - 4)^4$

15. $(x + y)^5$
16. $(x - y)^6$
17. $(x + 3y)^3$
18. $(2x - y)^3$

19. $(5x - 2)^3$
20. $(a - 7b)^3$
21. $\left(\dfrac{1}{x} + 5y\right)^4$
22. $\left(2x + \dfrac{3}{y}\right)^4$

23. $(x^2 + y^2)^4$
24. $(r^3 - s^3)^3$
25. $(ax + by)^5$
26. $(ax - by)^5$

27. $\left(\sqrt{x} + 2\right)^6$
28. $\left(3 + \sqrt{y}\right)^4$
29. $(a^{3/4} + b^{1/4})^4$
30. $(x^{2/3} + y^{1/3})^3$

31. $\left(x^{1/4} + 2\sqrt{y}\right)^4$
32. $\left(\sqrt{x} - 3y^{1/4}\right)^8$

In Exercises 33–36, expand the expression using Pascal's triangle.

33. $(r - s)^4$
34. $(x^2 + y^2)^7$
35. $(ax + by)^6$
36. $(x + 3y)^4$

In Exercises 37–44, find the coefficient C of the term in the binomial expansion.

Binomial	Term	Binomial	Term	Binomial	Term	Binomial	Term
37. $(x + 2)^{10}$	Cx^6	38. $(3 + y)^9$	Cy^5	39. $(y - 3)^8$	Cy^4	40. $(x - 1)^{12}$	Cx^5
41. $(2x + 3y)^7$	Cx^3y^4	42. $(3x - 5y)^9$	Cx^2y^7	43. $(x^2 + y)^8$	Cx^8y^4	44. $(r - s^2)^{10}$	Cr^6s^8

▪ APPLICATIONS

In later sections, you will learn the "*n* choose *k*" notation for combinations.

45. **Lottery.** In a state lottery in which six numbers are drawn from a possible 40 numbers, the number of possible six-number combinations is equal to $\dbinom{40}{6}$. How many possible combinations are there?

46. **Lottery.** In a state lottery in which six numbers are drawn from a possible 60 numbers, the number of possible six-number combinations is equal to $\dbinom{60}{6}$. How many possible combinations are there?

47. **Poker.** With a deck of 52 cards, 5 cards are dealt in a game of poker. There are a total of $\dbinom{52}{5}$ different 5-card poker hands that can be dealt. How many possible hands are there?

48. **Canasta.** In the card game canasta, two decks of cards including the jokers are used and 11 cards are dealt to each person. There are a total of $\dbinom{108}{11}$ different 11-card canasta hands that can be dealt. How many possible hands are there?

■**CATCH THE MISTAKE**

In Exercises 49 and 50, explain the mistake that is made.

49. Evaluate the expression $\begin{pmatrix} 7 \\ 5 \end{pmatrix}$.

Solution:

Write out the binomial coefficient in terms of factorials.

$$\begin{pmatrix} 7 \\ 5 \end{pmatrix} = \frac{7!}{5!}$$

Write out the factorials.

$$\begin{pmatrix} 7 \\ 5 \end{pmatrix} = \frac{7!}{5!} = \frac{7 \cdot 6 \cdot 5 \cdot 4 \cdot 3 \cdot 2 \cdot 1}{5 \cdot 4 \cdot 3 \cdot 2 \cdot 1}$$

Simplify.

$$\begin{pmatrix} 7 \\ 5 \end{pmatrix} = \frac{7!}{5!} = \frac{7 \cdot 6}{1} = 42$$

This is incorrect. What mistake was made?

50. Expand $(x + 2y)^4$.

Solution:

Write out with blanks.

$$(x + 2y)^4 = \square x^4 + \square x^3 y + \square x^2 y^2 + \square xy^3 + \square y^4$$

Write out the terms from the fifth row of Pascal's triangle.

$$1, 4, 6, 4, 1$$

Substitute these coefficients into the binomial expansion.

$$(x + 2y)^4 = x^4 + 4x^3 y + 6x^2 y^2 + 4xy^3 + y^4$$

This is incorrect. What mistake was made?

■**CONCEPTUAL**

In Exercises 51–54, determine whether each statement is true or false.

51. The binomial expansion of $(x + y)^{10}$ has 10 terms.

52. The binomial expansion of $(x^2 + y^2)^{15}$ has 16 terms.

53. $\begin{pmatrix} n \\ n \end{pmatrix} = 1$

54. $\begin{pmatrix} n \\ -n \end{pmatrix} = -1$

■**CHALLENGE**

55. Show that $\begin{pmatrix} n \\ k \end{pmatrix} = \begin{pmatrix} n \\ n - k \end{pmatrix}$, if $0 \le k \le n$.

56. Show that if n is a positive integer, then:

$$\begin{pmatrix} n \\ 0 \end{pmatrix} + \begin{pmatrix} n \\ 1 \end{pmatrix} + \begin{pmatrix} n \\ 2 \end{pmatrix} + \cdots + \begin{pmatrix} n \\ n \end{pmatrix} = 2^n$$

Hint: Let $2^n = (1 + 1)^n$ and use the binomial theorem to expand.

■**TECHNOLOGY**

57. With a graphing utility, plot $y_1 = 1 - 3x + 3x^2 - x^3$, $y_2 = -1 + 3x - 3x^2 + x^3$, and $y_3 = (1 - x)^3$ in the same viewing screen. What is the binomial expansion of $(1 - x)^3$?

58. With a graphing utility, plot $y_1 = (x + 3)^4$, $y_2 = x^4 + 4x^3 + 6x^2 + 4x + 1$, and $y_3 = x^4 + 12x^3 + 54x^2 + 108x + 81$. What is the binomial expansion of $(x + 3)^4$?

59. With a graphing utility, plot $y_1 = 1 - 3x$, $y_2 = 1 - 3x + 3x^2$, $y_3 = 1 - 3x + 3x^2 - x^3$, and $y_4 = (1 - x)^3$ for $-1 < x < 1$. What do you notice happening each time an additional term is added? Now, let $1 < x < 2$. Does the same thing happen?

60. With a graphing utility, plot $y_1 = 1 - \dfrac{3}{x}$, $y_2 = 1 - \dfrac{3}{x} + \dfrac{3}{x^2}$, $y_3 = 1 - \dfrac{3}{x} + \dfrac{3}{x^2} - \dfrac{1}{x^3}$, and $y_4 = \left(1 - \dfrac{1}{x}\right)^3$ for $1 < x < 2$. What do you notice happening each time an additional term is added? Now, let $0 < x < 1$. Does the same thing happen?

61. With a graphing utility, plot $y_1 = 1 + \dfrac{3}{x}$, $y_2 = 1 + \dfrac{3}{x} + \dfrac{3}{x^2}$, $y_3 = 1 + \dfrac{3}{x} + \dfrac{3}{x^2} - \dfrac{1}{x^3}$, and $y_4 = \left(1 + \dfrac{1}{x}\right)^3$ for $1 < x < 2$. What do you notice happening each time an additional term is added? Now, let $0 < x < 1$. Does the same thing happen?

62. With a graphing utility, plot $y_1 = 1 + \dfrac{x}{1!}$, $y_2 = 1 + \dfrac{x}{1!} + \dfrac{x^2}{2!}$, $y_3 = 1 + \dfrac{x}{1!} + \dfrac{x^2}{2!} - \dfrac{x^3}{3!}$, and $y_4 = e^x$ for $-1 < x < 1$. What do you notice happening each time an additional term is added? Now, let $1 < x < 2$. Does the same thing happen?

SKILLS OBJECTIVES

- Apply the fundamental counting principle to solve counting problems.
- Apply permutations to solve counting problems.
- Apply combinations to solve counting problems.

CONCEPTUAL OBJECTIVE

- Understand the difference between permutations and combinations.

The Fundamental Counting Principle

You are traveling through Europe for the summer and decide the best packing option is to select separates that can be mixed and matched. You pack one pair of shorts and one pair of khaki pants. You pack a pair of Teva sport sandals and a pair of sneakers. You have three shirts (red, blue, and white). How many different outfits can be worn using only the clothes mentioned above?

The answer is 12. There are two options for bottoms (pants or shorts), three options for shirts, and two options for shoes. The product of these is

$$2 \cdot 3 \cdot 2 = 12$$

The general formula for counting possibilities is given by the *fundamental counting principle*.

FUNDAMENTAL COUNTING PRINCIPLE

Let E_1 and E_2 be two independent events. The first event E_1 can occur in m_1 ways. The second event E_2 can occur in m_2 ways. The number of ways that the combination of the two events can occur is $m_1 \cdot m_2$.

In other words, *the number of ways in which successive things can occur is found by multiplying the number of ways each thing can occur.*

Study Tip

The fundamental counting principle can be extended to more than two events.

EXAMPLE 1 Possible Meals Served at a Restaurant

A restaurant is rented for a retirement party. The owner offers an appetizer, an entrée, and a dessert for a set price. The following are the choices that people attending the party may choose from. How many possible dinners could be served that night?

Appetizers:	calamari, stuffed mushrooms, or caesar salad
Entrées:	tortellini alfredo, shrimp scampi, eggplant parmesan, or chicken marsala
Desserts:	tiramisu or flan

Solution:

There are three possible appetizers, four possible entrées, and two possible desserts.

Write the product of possible options. $3 \cdot 4 \cdot 2 = 24$

There are 24 possible dinners for the retirement party.

■ **YOUR TURN** In Example 1, the restaurant will lower the cost per person for the retirement party if the number of appetizers and entrées is reduced. Suppose the appetizers are reduced to either soup or salad and the entrées are reduced to either tortellini or eggplant parmesan. How many possible dinners could be served at the party?

EXAMPLE 2 Telephone Numbers (When to Require 10-Digit Dialing)

In many towns in the United States, residents can call one another using a 7-digit dialing system. In some large cities, 10-digit dialing is required because two or more area codes coexist. Determine how many telephone numbers can be allocated in a 7-digit dialing area.

Solution:

With 7-digit telephone numbers, the first number cannot be a 0 or a 1, but each of the following six numbers can be 0, 1, 2, 3, 4, 5, 6, 7, 8, or 9.

First number: 2, 3, 4, 5, 6, 7, 8, or 9.	8 possible digits
Second number: 0, 1, 2, 3, 4, 5, 6, 7, 8, or 9.	10 possible digits
Third number:	10 possible digits
Fourth number:	10 possible digits
Fifth number:	10 possible digits
Sixth number:	10 possible digits
Seventh number:	10 possible digits
Counting principle:	$8 \cdot 10 \cdot 10 \cdot 10 \cdot 10 \cdot 10 \cdot 10$
Possible telephone numbers:	8,000,000

Eight million 7-digit telephone numbers can be allocated within one area code.

■ **YOUR TURN** If the first digit of an area code cannot be 0 or 1, but the second and third numbers of an area code can be 0, 1, 2, 3, 4, 5, 6, 7, 8, or 9, how many 10-digit telephone numbers can be allocated in the United States?

The fundamental counting principle applies when an event can occur more than once. We now introduce two other concepts, *permutations* and *combinations*, which allow individual events to occur only once. For example, in Example 2, the allowable telephone numbers can include the same number in two or more digit places, as in 555-1212. However, in many state lottery games, once a number is selected, it cannot be used again.

An important distinction between a *permutation* and a *combination* is that in a *permutation* order matters, but in a combination order does not matter. For example, the Florida winning lotto numbers one week could be 2–3–5–11–19–27. This would be a *combination* because the order in which they are drawn does not matter. However, if you were betting on a trifecta at the Kentucky Derby, to win you must not only select the first, second, and third place horses, you must select them in the order in which they finished. This would be a *permutation*.

Permutations

DEFINITION   **Permutation**

A **permutation** is an *ordered* arrangement of distinct objects without repetition.

EXAMPLE 3   Finding the Number of Permutations of *n* Objects

How many permutations are possible for the letters A, B, C, and D?

Solution:

ABCD	ABDC	ACBD	ACDB	ADCB	ADBC
BACD	BADC	BCAD	BCDA	BDCA	BDAC
CABD	CADB	CBAD	CBDA	CDAB	CDBA
DABC	DACB	DBCA	DBAC	DCAB	DCBA

There are 24 (or 4!) possible permutations of the letters A, B, C, and D.

Classroom Example 12.6.3
How many permutations of the symbols *AEIOUZ* are possible?

Answer: 6!

Notice that in the first row of permutations in Example 3, A was selected for the first space. That left one of the remaining three letters to fill the second space. Once that was selected there remained two letters to choose from for the third space, and then the last space was filled with the unselected letter. In general, there are *n*! ways to order *n* objects.

NUMBER OF PERMUTATIONS OF *n* OBJECTS

The number of permutations of *n* objects is

$$n! = n \cdot (n-1) \cdot (n-2) \cdots 2 \cdot 1$$

Study Tip

In a permutation of objects, order matters. That is, the same objects arranged in a different order are considered to be a distinct permutation.

EXAMPLE 4   Running Order of Dogs

In an American Kennel Club (AKC) sponsored field trial, the dogs compete in random order. If there are nine dogs competing in the trials, how many possible running orders are there?

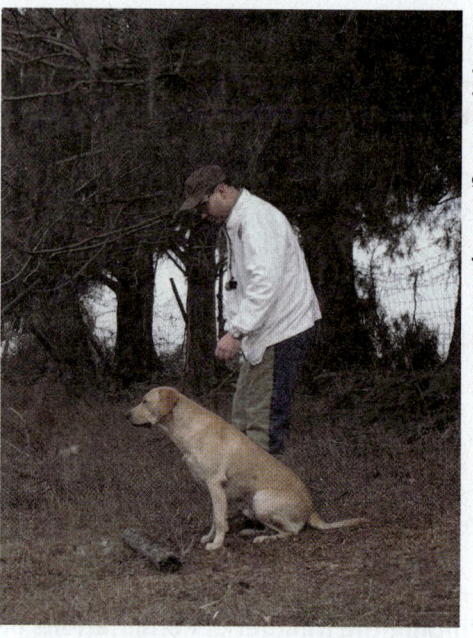

Courtesy Cynthia Young, University of Central Florida

Solution:

There are nine dogs that will run. $n = 9$

The number of possible running orders is $n!$ $n! = 9! = \boxed{362,880}$

There are $\boxed{362,880}$ different possible running orders of nine dogs.

■ **Answer:** 120

■ **YOUR TURN**   Five contestants in the Miss America pageant reach the live television interview round. In how many possible orders can the contestants compete in the interview round?

In Examples 3 and 4, we were interested in all possible permutations. Sometimes, we are interested in only some permutations. For instance, 20 horses usually run in the Kentucky Derby. If you bet on a trifecta, you must pick the top three places in the correct order to win. Therefore, we do not consider all possible permutations of 20 horses finishing (places 1–20). Instead, we only consider the possible permutations of first, second, and third place finishes of the 20 horses. We would call this *a permutation of 20 objects taken 3 at a time*. In general, this ordering is called *a permutation of n objects taken r at a time*.

If 20 horses are entered in the Kentucky Derby, there are 20 possible first-place finishers. We consider the permutations with one horse in first place. That leaves 19 possible horses for second place, and then 18 possible horses for third place. Therefore, there are $20 \cdot 19 \cdot 18 = 6840$ possible winning scenarios for the trifecta. This can also be represented as $\dfrac{20!}{(20 - 3)!} = \dfrac{20!}{17!}$.

NUMBER OF PERMUTATIONS OF *n* OBJECTS TAKEN *r* AT A TIME

The number of permutations of *n* objects taken *r* at a time is

$$_nP_r = \frac{n!}{(n-r)!} = n(n-1)(n-2)\cdots(n-r+1)$$

EXAMPLE 5 Starting Lineup for a Volleyball Team

The starters for a six-woman volleyball team have to be listed in a particular order (1–6). If there are 13 women on the team, how many possible starting lineups are there?

Solution:

Identify the total number of players. $n = 13$

Identify the total number of starters in the lineup. $r = 6$

Substitute $n = 13$ and $r = 6$ into $_nP_r = \dfrac{n!}{(n-r)!}$.

$$_{13}P_6 = \frac{13!}{(13-6)!} = \frac{13!}{7!} = 13\cdot12\cdot11\cdot10\cdot9\cdot8 = \boxed{1{,}235{,}520}$$

There are $\boxed{1{,}235{,}520}$ possible combinations.

■ **YOUR TURN** A softball team has 12 players, 10 of whom will be in the starting lineup (batters 1–10). How many possible starting lineups are there for this team?

Technology Tip

Use a calculator to find $_{13}P_6$.

Scientific calculators:

Press	Display
13 $\boxed{nPr}$ 6 $\boxed{=}$	1235520

Graphing calculators:

13 $\boxed{MATH}$ $\boxed{\blacktriangleright}$ $\boxed{PRB}$ $\boxed{\blacktriangledown}$ $\boxed{2:nPr}$
6 $\boxed{ENTER}$

```
13 nPr 6
            1235520
■
```

■ **Answer:** 239,500,800

Combinations

The difference between a permutation and a combination is that a permutation has an order associated with it, whereas a *combination* does not have an order associated with it.

DEFINITION Combination

A **combination** is an arrangement, without specific order, of distinct objects without repetition.

The six winning Florida lotto numbers and the NCAA men's Final Four basketball tournament are examples of combinations (six numbers and four teams) without regard to order. The number of combinations of *n* objects taken *r* at a time is equal to the binomial coefficient $\dbinom{n}{r}$.

Classroom Example 12.6.5(1)
The president, vice president, treasurer, and secretary are chosen from a pool of 11 candidates. How many different administrations exist?

Answer:
$$_{11}P_4 = \frac{11!}{7!}$$

Classroom Example 12.6.5(2)
On "So You Think You Can Dance" eight contestants take their turns on the dance floor. How many ways can they be ordered 1–8?

Answer: 8!

NUMBER OF COMBINATIONS OF *n* OBJECTS TAKEN *r* AT A TIME

The number of combinations of n objects taken r at a time is

$$_nC_r = \binom{n}{r} = \frac{n!}{(n-r)!r!}$$

Classroom Example 12.6.6(1)
Towels are on sale "Buy 4 get 1 free." There are 36 different towels. How many ways can a customer buy 5 different towels?

Answer: $_{36}C_5$

Compare the number of permutations $_nP_r = \dfrac{n!}{(n-r)!}$ and the number of combinations $_nC_r = \dfrac{n!}{(n-r)!r!}$. It makes sense that the number of combinations is less than the number of permutations. The denominator is larger because there are no separate orders associated with a combination.

EXAMPLE 6 Possible Combinations to the Lottery

If there are a possible 59 numbers and the lottery officials draw 6 numbers, how many possible combinations are there?

Solution:

Technology Tip

Use a calculator to find $_{59}C_6$.

Scientific calculators:

Press	Display
59 $\boxed{nCr}$ 6 $\boxed{=}$	45057474

Graphing calculators:

59 $\boxed{\text{MATH}}$ $\boxed{\blacktriangleright}$ $\boxed{PRB}$ $\boxed{\blacktriangledown}$ $\boxed{3:nCr}$
6 $\boxed{\text{ENTER}}$.

```
59 nCr 6
          45057474
```

Identify how many numbers are in the drawing. $n = 59$

Identify how many numbers are chosen. $r = 6$

Substitute $n = 59$ and $r = 6$ into $_nC_r = \dfrac{n!}{(n-r)!r!}$.

$$_{59}C_6 = \frac{59!}{(59-6)!6!}$$

Simplify.

$$= \frac{59 \cdot 58 \cdot 57 \cdot 56 \cdot 55 \cdot 54 \cdot (53)!}{53! \cdot 6!}$$

$$= \frac{59 \cdot 58 \cdot 57 \cdot 56 \cdot 55 \cdot 54}{6 \cdot 5 \cdot 4 \cdot 3 \cdot 2}$$

$$= \boxed{45{,}057{,}474}$$

There are $\boxed{45{,}057{,}474}$ possible combinations.

■ **Answer:** 13,983,816

■ **YOUR TURN** What are the possible combinations for a lottery with 49 possible numbers and 6 drawn numbers?

Classroom Example 12.6.6(2)
A virus detection program randomly samples 12 words from an e-mail to determine if it is spam. If a typical e-mail contains 200 words, how many different ways can the program test the e-mail?

Answer: $_{200}C_{12}$

Permutations with Repetition

Permutations and combinations are arrangements of distinct (nonrepeated) objects. A permutation in which some of the objects are repeated is called a **permutation with repetition** or a **nondistinguishable permutation**. For example, if a sack has three red marbles, two blue marbles, and one white marble, how many possible permutations would there be when drawing six marbles, one at a time?

This is a different problem from writing the numbers 1 through 6 on pieces of paper, putting them in a hat, and drawing them out. The reason the problems are different is that the two blue balls are indistinguishable and the three red balls are also indistinguishable. The possible permutations for drawing numbers out of the hat are 6!, whereas the possible permutations for drawing balls out of the sack are given by

$$\frac{6!}{3! \cdot 2! \cdot 1!}$$

NUMBER OF DISTINGUISHABLE PERMUTATIONS

If a set of n objects has n_1 of one kind of object, n_2 of another kind of object, n_3 of a third kind of object, and so on for k different types of objects so that $n = n_1 + n_2 + \cdots + n_k$, then the number of **distinguishable permutations** of the n objects is

$$\frac{n!}{n_1! \cdot n_2! \cdot n_3! \cdots n_k!}$$

In our sack of marbles, there were six marbles $n = 6$. Specifically, there were three red marbles ($n_1 = 3$), two blue marbles ($n_2 = 2$), and one white marble ($n_3 = 1$). Notice that $n = n_1 + n_2 + n_3$ and that the number of distinguishable permutations is equal to

$$\frac{6!}{3! \cdot 2! \cdot 1!} = 60$$

EXAMPLE 7   Peg Game at Cracker Barrel

The peg game on the tables at Cracker Barrel is a triangle with 15 holes drilled in it, in which pegs are placed. There are 5 red pegs, 5 white pegs, 3 blue pegs, and 2 yellow pegs. If all 15 pegs are in the holes, how many different ways can the pegs be aligned?

Andy Washnik

Solution:

There are four different colors of pegs (red, white, blue, and yellow).

5 red pegs:	$n_1 = 5$
5 white pegs:	$n_2 = 5$
3 blue pegs:	$n_3 = 3$
2 yellow pegs:	$n_4 = 2$

There are 15 pegs total: $n = 15$

Substitute $n = 15$, $n_1 = 5$, $n_2 = 5$, $n_3 = 3$, and $n_4 = 2$ into $\dfrac{n!}{n_1! \cdot n_2! \cdot n_3! \cdots n_k!}$. $\dfrac{15!}{5! \cdot 5! \cdot 3! \cdot 2!}$

Simplify. $= \boxed{7{,}567{,}560}$

There are a possible $\boxed{7{,}567{,}560}$ ways to insert the 15 colored pegs at the Cracker Barrel.

■ **YOUR TURN** Suppose a similar game to the peg game at Cracker Barrel is set up with only 10 holes in a triangle. With 5 red pegs, 2 white pegs, and 3 blue pegs, how many different permutations can fill that board?

Technology Tip

Use a calculator to calculate $\dfrac{15!}{5! \cdot 5! \cdot 3! \cdot 2!}$

Scientific calculators:

Press	Display
15 ! ÷ ((5 ! ×	7567560
5 ! × 3 ! ×	
2 !)	

Graphing calculators:

15 MATH ▶ PRB ▼ 4: !
ENTER ÷ (5 MATH
▶ PRB ▼ 4: ! 5 MATH
▶ PRB ▼ 4: ! 3 MATH
▶ PRB ▼ 4: ! 2 MATH
▶ PRB ▼ 4: ! ENTER)
ENTER

```
15!/(5!5!3!2!)
          7567560
■
```

■ **Answer:** 2520

In this section we discussed the fundamental counting principle, permutations, and combinations.

The Fundamental Counting Principle Is Applicable When

- Objects can be repeated.
- The objects can occur in any order.
- The first event E_1 can occur m_1 ways, and the second event E_2 can occur m_2 ways: the number of ways successive events can occur is $m_1 m_2$ ways.

Permutations

- Objects *cannot* be repeated.
- Order *matters*.
- Number of permutations of n objects: $n!$.
- Number of permutations of n objects taken r at a time:

$$_nP_r = \frac{n!}{(n-r)!}$$

Combinations

- Objects *cannot* be repeated.
- Order *does not* matter.
- Number of combinations of n objects taken r at a time:

$$_nC_r = \frac{n!}{(n-r)!\,r!}$$

Nondistinguishable Permutations

- Some objects are repeated because they are not distinguishable.
- For n objects with k different types of objects:

$$\frac{n!}{n_1!n_2!n_3!\cdots n_k!}$$

▪ SKILLS

In Exercises 1–8, use the formula for $_nP_r$ to evaluate each expression.

1. $_6P_4$ **2.** $_7P_3$ **3.** $_9P_5$ **4.** $_9P_4$ **5.** $_8P_8$ **6.** $_6P_6$ **7.** $_{13}P_3$ **8.** $_{20}P_3$

In Exercises 9–18, use the formula for $_nC_r$ to evaluate each expression.

9. $_{10}C_5$ **10.** $_9C_4$ **11.** $_{50}C_6$ **12.** $_{50}C_{10}$ **13.** $_7C_7$ **14.** $_8C_8$

15. $_{30}C_4$ **16.** $_{13}C_5$ **17.** $_{45}C_8$ **18.** $_{30}C_4$

▪ APPLICATIONS

19. Computers. At the www.dell.com website, a customer can "build" a system. If there are four monitors to choose from, three different computers, and two different keyboards, how many possible system configurations are there?

20. Houses. In a "new home" community, a person can select from one of four models, five paint colors, three tile selections, and two landscaping options. How many different houses (interior and exterior) are there to choose from?

21. Wedding Invitations. An engaged couple is ordering wedding invitations. The wedding invitations come in white or ivory. The writing can be printed, embossed, or engraved. The envelopes can come with liners or without. How many possible designs of wedding invitations are there to choose from?

22. Dinner. Siblings are planning their father's 65th birthday dinner and have to select one of four main courses (baked chicken, grilled mahi-mahi, beef Wellington, or lasagna), one of two starches (rosemary potatoes or rice), one of three vegetables (green beans, carrots, or zucchini), and one of five appetizers (soup, salad, pot stickers, artichoke dip, or calamari). How many possible dinner combinations are there?

23. PIN Number. Most banks require a 4-digit ATM PIN code for each customer's bank card. How many possible four-digit PIN codes are there to choose from?

24. Password. All e-mail accounts require passwords. If a four-character password is required that can contain letters (but no numbers), how many possible passwords can there be? (Assume letters are not case sensitive.)

25. **Leadership.** There are 15 professors in a department and there are four leadership positions (chair, assistant chair, undergraduate coordinator, and graduate coordinator). How many possible leadership teams are there?

26. **Fraternity Elections.** A fraternity is having elections. There are three men running for president, two men running for vice-president, four men running for secretary, and one man running for treasurer. How many possible outcomes do the elections have?

27. **Multiple-Choice Tests.** There are 20 questions on a multiple-choice exam, and each question has four possible answers (A, B, C, and D). Assuming no answers are left blank, how many different ways can you answer the questions on the exam?

28. **Multiple-Choice Tests.** There are 25 questions on a multiple-choice exam, and each question has five possible answers (A, B, C, D, and E). Assuming no answers are left blank, how many different ways can you answer the questions on the exam?

29. **Zip Codes.** In the United States a 5-digit zip code is used to route mail. How many possible 5-digit zip codes are possible? (All numbers can be used.) If 0s were eliminated from the first and last digits, how many possible zip codes would there be?

30. **License Plates.** In a particular state there are six characters in a license plate: three letters followed by three numbers. If 0s and 1s are eliminated from possible numbers and Os and Is are eliminated from possible letters, how many different license plates can be made?

31. **Class Seating.** If there are 30 students in a class and there are exactly 30 seats, how many possible seating charts can be made, assuming all 30 students are present?

32. **Season Tickets.** Four friends buy four season tickets to the Green Bay Packers. To be fair, they change the seating arrangement every game. How many different seating arrangements are there for the four friends?

33. **Combination "Permutation" Lock.** A combination lock on most lockers will open when the correct choice of three numbers (1 to 40) is selected and entered in the correct order. Therefore, a combination lock should really be called a permutation lock. How many possible permutations are there, assuming no numbers can be repeated?

34. **Safe.** A safe will open when the correct choice of three numbers (1 to 50) is selected and entered in the correct order. How many possible permutations are there, assuming no numbers can be repeated?

35. **Raffle.** A fundraiser raffle is held to benefit cystic fibrosis research, and 1000 raffle tickets are sold. There are three prizes raffled off. First prize is a round-trip ticket on Delta Air Lines, second prize is a round of golf for four people at a local golf course, and third prize is a $50 gift certificate to Chili's. How many possible winning scenarios are there if all 1000 tickets are sold to different people?

36. **Ironman Triathlon.** If 100 people compete in an ironman triathlon, how many possible placings are there (first, second, and third place)?

37. **Lotto.** If a state lottery picks from 53 numbers and 6 numbers are selected, how many possible 6-number combinations are there?

38. **Lotto.** If a state lottery picks from 53 numbers and 5 numbers are selected, how many possible 5-number combinations are there?

39. **Cards.** In a deck of 52 cards, how many different 5-card hands can be dealt?

40. **Cards.** In a deck of 52 cards, how many different 7-card hands can be dealt?

41. **Blackjack.** In a single-deck blackjack game (52 cards), how many different 2-card combinations are there?

42. **Blackjack.** In a single deck, how many two-card combinations are there that equal 21: ace (worth 11) and a 10 or face card—jack, queen, or king?

43. **March Madness.** Every spring, the NCAA men's basketball tournament starts with 64 teams. After two rounds, it is down to the Sweet Sixteen, and after two more rounds, it is reduced to the Final Four. Once 64 teams are selected (but not yet put in brackets), how many possible scenarios are there for the Sweet Sixteen?

44. **March Madness.** Every spring, the NCAA men's basketball tournament starts with 64 teams. After two rounds, it is down to the Sweet Sixteen, and after two more rounds, it is reduced to the Final Four. Once the 64 teams are identified (but not yet put in brackets), how many possible scenarios are there for the Final Four?

45. **NFL Playoffs.** There are 32 teams in the National Football League (16 AFC and 16 NFC). How many possible combinations are there for the Superbowl? (Assume one team from the AFC plays one team from the NFC in the Superbowl.)

46. **NFL Playoffs.** After the regular season in the National Football League, 12 teams make the playoffs (6 from the AFC and 6 from the NFC). How many possible combinations are there for the Superbowl once the 6 teams in each conference are identified?

47. *Survivor.* On the television show *Survivor*, one person is voted off the island every week. When it is down to six contestants, how many possible voting combinations remain, if no one will vote themselves off the island? Assume that the order (who votes for whom) makes a difference. How many total possible voting outcomes are there?

48. *American Idol.* On the television show *American Idol*, a young rising star is eliminated from the competition every week. The first week, each of the 12 contestants sings one song. How many possible ways could the contestants be ordered 1–12? How many possible ways could 6 men and 6 women be ordered to alternate female and male contestants?

49. *Dancing with the Stars.* In the popular TV show *Dancing with the Stars*, 12 entertainers (6 men and 6 women) compete in a dancing contest. The first night, the show decides to select 3 men and 3 women. How many ways can this be done?

50. *Dancing with the Stars.* See Exercise 49. How many ways can six male celebrities line up for a picture alongside six female celebrities?

■ **CATCH THE MISTAKE**

In Exercises 51 and 52, explain the mistake that is made.

51. In a lottery that picks from 30 numbers, how many five-number combinations are there?

Solution:

Let $n = 30$ and $r = 5$.

Calculate $_nP_r = \dfrac{n!}{(n-r)!}$. $_{30}P_5 = \dfrac{30!}{25!}$

Simplify. $_{30}P_5 = 17,100,720$

This is incorrect. What mistake was made?

52. A homeowners association has 12 members on the board of directors. How many ways can the board elect a president, vice-president, secretary, and treasurer?

Solution:

Let $n = 12$ and $r = 4$.

Calculate $_nC_r = \dfrac{n!}{(n-r)! \cdot r!}$. $_{12}C_4 = \dfrac{12!}{8! \cdot 4!}$

Simplify. $_{12}C_4 = 495$

This is incorrect. What mistake was made?

■ **CONCEPTUAL**

In Exercises 53–56, determine whether each statement is true or false.

53. The number of permutations of n objects is always greater than the number of combinations of n objects if the objects are distinct.

54. The number of permutations of n objects is always greater than the number of combinations of n objects even when the objects are indistinguishable.

55. The number of four-letter permutations of the letters A, B, C, and D is equal to the number of four-letter permutations of ABBA.

56. The number of possible answers to a true/false question is a permutation problem.

■ **CHALLENGE**

57. What is the relationship between $_nC_r$ and $_nC_{r+1}$?

58. What is the relationship between $_nP_r$ and $_nP_{r-1}$?

59. Simplify the expression $_nC_r \cdot r!$.

60. What is the relationship between $_nC_r$ and $_nC_{n-r}$?

■ **TECHNOLOGY**

61. Employ a graphing utility with a $_nP_r$ feature and compare it with answers to Exercises 1–8.

62. Employ a graphing utility with a $_nC_r$ feature and compare it with answers to Exercises 9–18.

63. Use a graphing calculator to evaluate:

a. $_{10}P_4$

b. $4!(_{10}C_4)$

c. Are answers in (a) and (b) the same?

d. Why?

64. Use a graphing calculator to evaluate:

a. $_{12}P_5$

b. $5!(_{12}C_5)$

c. Are answers in (a) and (b) the same?

d. Why?

SKILLS OBJECTIVES

- Determine the sample space of an outcome.
- Find the probability of an event.
- Find the probability that an event will not occur.
- Find the probability of mutually exclusive events.
- Find the probability of independent events.

CONCEPTUAL OBJECTIVES

- Understand that the mathematics of probability gives us a good sense of how likely it is that a certain event will happen.
- Understand the difference between the probability of:
  - event 1 *and* event 2
  - event 1 *or* event 2

Sample Space

You are sitting at a blackjack table at Caesar's Palace, and the dealer is showing a 7. You have a 9 and a 7; should you hit? Will it rain today? What will the lotto numbers be this week? Will the coin toss at the Superbowl result in a head or a tail? Will Derek Jeter get a hit at his next trip to the plate? These are all questions where *probability* is used to guide us.

Anything that happens for which the result is uncertain is called an **experiment**. Each trial of an experiment is called an **outcome**. All of the possible outcomes of an experiment constitute the **sample space**. The term **event** is used to describe the kind of possible outcomes. For example, a coin toss is an experiment. Every outcome is either heads or tails. The sample space of a single toss is {heads, tails}.

The result of one experiment has no certain outcome. However, if the experiment is performed many times, the results will produce regular patterns. For example, if you toss a fair coin, you don't know whether it will come up heads or tails. You can toss a coin 10 times and get 10 heads. However, if you made 1,000,000 tosses, you would get about 500,000 heads and 500,000 tails. Therefore, since we assume a head is equally likely as a tail and there are only two possible events (heads or tails), we assign a *probability of a head* equal to $\frac{1}{2}$ and a *probability of a tail* equal to $\frac{1}{2}$.

EXAMPLE 1 Finding the Sample Space

Find the sample space for each of the following outcomes.

a. Tossing a coin once   **b.** Tossing a coin twice   **c.** Tossing a coin three times

Solution (a):

Tossing a coin one time will result in one of two events: heads (H) or tails (T).

The sample space S is written as $\boxed{S = \{H, T\}}$.

Solution (b):

Tossing a coin twice can result in one of four possible outcomes.

The sample space consists of all possible outcomes.

$$\boxed{S = \{HH, HT, TH, TT\}}$$

Note that TH and HT are two different outcomes.

Solution (c):

There are eight possible outcomes when a coin is tossed three times.

$$\boxed{S = \{HHH, HHT, HTH, HTT, TTT, TTH, THT, THH\}}$$

■ **YOUR TURN** Find the sample space associated with having three children (B boys or G girls).

Study Tip

If the coin is tossed n times, there are 2^n possible outcomes.

■ **Answer:**
$S = \{GGG, GGB, GBG, GBB, BBB, BBG, BGB, BGG\}$

Probability of an Event

To calculate the probability of an event, start by counting the number of outcomes in the event and the number of outcomes in the sample space. The ratio is equal to the probability if all outcomes are equally likely.

> **DEFINITION**   **Probability of an Event**
>
> If an event E has $n(E)$ equally likely outcomes and its sample space S has $n(S)$ equally likely outcomes, then the **probability of event** E, denoted $P(E)$, is
>
> $$P(E) = \frac{n(E)}{n(S)} = \frac{\text{Number of outcomes in event } E}{\text{Number of outcomes in sample space } S}$$

Since the number of outcomes in an event must be less than or equal to the number of outcomes in the sample space, the probability of an event must be a number between 0 and 1 or equal to 0 or 1; to be precise, $0 \leq P(E) \leq 1$. If $P(E) = 0$, then the event can never happen, and if $P(E) = 1$, the event is certain to happen.

EXAMPLE 2   Finding the Probability of Two Girls

If two children are born, what is the probability that they are both girls?

Solution:

The event is both children being girls.	$E = \{GG\}$
The sample space is all four possible outcomes.	$S = \{BB, BG, GB, GG\}$
The number of outcomes in the event is 1.	$n(E) = 1$
The number of events in the sample space is 4.	$n(S) = 4$
Compute the probability using $P(E) = \dfrac{n(E)}{n(S)}$.	$P(E) = \boxed{\dfrac{1}{4}}$

> The probability that both children are girls is $\frac{1}{4}$, or 0.25.

EXAMPLE 3   Finding the Probability of Drawing a Face Card

Find the probability of drawing a face card (jack, queen, or king) out of a 52-card deck.

Solution:

There are 12 face cards in a deck.	$n(E) = 12$
There are 52 cards in a deck.	$n(S) = 52$
Compute the probability using $P(E) = \dfrac{n(E)}{n(S)}$.	$P(E) = \dfrac{12}{52} = \boxed{\dfrac{3}{13}}$

> The probability of drawing a face card out of a 52-card deck is $\frac{3}{13}$ or ≈ 0.23.

■ **YOUR TURN** **a.** Find the probability that an ace is drawn from a deck of 52 cards.
   b. Find the probability that a jack, queen, king, or ace is drawn from a deck of 52 cards.

Classroom Example 12.7.3
a. Find the probability of drawing a red even-numbered card from a 52-card deck.
b.* Find the probability of not getting an odd-numbered club or diamond.
c. Find the probability of getting a red card or ace of clubs.

Answer:

a. $\frac{10}{52}$ **b.** $\frac{42}{52}$ **c.** $\frac{27}{52}$

■ **Answer: a.** $\frac{4}{52} = \frac{1}{13}$ **b.** $\frac{16}{52} = \frac{4}{13}$

EXAMPLE 4   **Finding the Probability of Rolling a 7 or an 11**

If you bet on the "pass line" at a craps table and the person's first roll is a 7 or an 11, using a pair of dice, then you win. Find the probability of winning a pass line bet on the first roll.

Solution:

The fundamental counting principle tells us that there
will be $6 \cdot 6 = 36$ possible rolls of the pair of dice. $n(S) = 36$

Draw a table listing possible sums of the two dice.

DICE VALUE	1	2	3	4	5	6
1	2	3	4	5	6	7
2	3	4	5	6	7	8
3	4	5	6	7	8	9
4	5	6	7	8	9	10
5	6	7	8	9	10	11
6	7	8	9	10	11	12

Of the 36 rolls, there are 8 rolls that will produce a 7 or an 11. $n(E) = 8$

Compute the probability using $P(E) = \dfrac{n(E)}{n(S)}$. $P(E) = \dfrac{8}{36} = \boxed{\dfrac{2}{9}}$

> The probability of winning a pass line bet is $\frac{2}{9}$ or ≈ 0.22 (22%).

■ **YOUR TURN**   If a 2, 3, or 12 is rolled on the first roll, then the pass line bet loses.
Find the probability of losing a pass line bet on the first roll.

Pass
Line

■ **Answer:** $\frac{4}{36} = \frac{1}{12} \approx 8.3\%$

Probability of an Event Not Occurring

The sum of the probabilities of all possible outcomes is 1. For example, when a die is rolled, if the outcomes are equally likely, then the probabilities are all $\frac{1}{6}$:

$$P(1) = \frac{1}{6},\ P(2) = \frac{1}{6},\ P(3) = \frac{1}{6},\ P(4) = \frac{1}{6},\ P(5) = \frac{1}{6},\ P(6) = \frac{1}{6}$$

The sum of these six probabilities is 1.

Since the sum of the probabilities of all possible outcomes sums to 1, we can find the probability that an event won't occur by subtracting the probability that the event will occur from 1.

$$P(E) + P(\text{not } E) = 1 \quad \text{or} \quad P(\text{not } E) = 1 - P(E)$$

PROBABILITY OF AN EVENT NOT OCCURRING

The **probability** that an event E will **not occur** is equal to 1 minus the probability that E will occur.

$$P(\text{not } E) = 1 - P(E)$$

The **complement of an event** E is the collection of all outcomes in the sample space that are not in E. The complement of E is denoted E' or $\overline{E}$.

EXAMPLE 5   Finding the Probability of Not Winning the Lottery

Find the probability of not winning the lottery if six numbers are selected from 1 to 49.

Solution:

Calculate the number of possible six-number combinations.

$$_{49}C_6 = \frac{49!}{43! \cdot 6!} = 13{,}983{,}816$$

Calculate the probability of winning.

$$P(\text{winning}) = \frac{1}{13{,}983{,}816}$$

Calculate the probability of not winning.

$$P(\text{not winning}) = 1 - P(\text{winning})$$

$$P(\text{not winning}) = 1 - \frac{1}{13{,}983{,}816}$$

$$P(\text{not winning}) = \frac{13{,}983{,}816}{13{,}983{,}816} - \frac{1}{13{,}983{,}816}$$

$$\approx \boxed{0.999999928}$$

The probability of *not* winning the lottery is very close to 1.

■ **YOUR TURN** Find the probability of not winning the lottery if 6 numbers are selected from 1 to 39.

Mutually Exclusive Events

Recall the definition of union and intersection in Section 1.5. The probability of one event E_1 or a second event E_2 occurring is given by the probability of the union of the two events.

$$P(E_1 \cup E_2)$$

If there is any overlap between the two events, we must be careful not to count those twice. For example, what is the probability of drawing *either* a face card *or* a spade out of a deck of 52 cards? We must be careful not to count twice any face cards that are spades.

PROBABILITY OF THE UNION OF TWO EVENTS

If E_1 and E_2 are two events in the same sample space, the probability of *either E_1 or E_2* occurring is given by

$$P(E_1 \cup E_2) = P(E_1) + P(E_2) - P(E_1 \cap E_2)$$

If E_1 and E_2 are disjoint, $P(E_1 \cap E_2) = 0$, then E_1 and E_2 are **mutually exclusive**. In that case, the probability of *either E_1 or E_2* occurring is given by

$$P(E_1 \cup E_2) = P(E_1) + P(E_2)$$

EXAMPLE 6   Finding the Probability of Drawing a Face Card or a Spade

Find the probability of drawing either a face card or a spade out of a deck of 52 cards.

Solution:

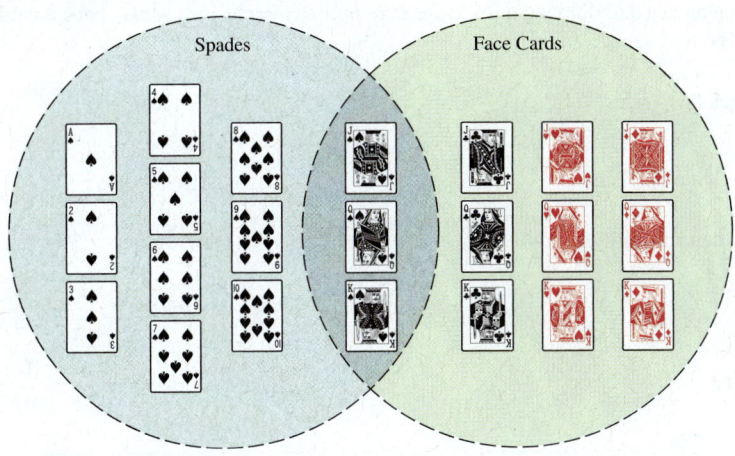

The deck has 12 face cards.

$$P(\text{face card}) = \frac{12}{52}$$

The deck has 13 spades.

$$P(\text{spade}) = \frac{13}{52}$$

The deck has 3 face cards that are spades.

$$P(\text{face card and spade}) = \frac{3}{52}$$

Apply the probability formula.

$$P(E_1 \cup E_2) = P(E_1) + P(E_2) - P(E_1 \cap E_2)$$

$$P(\text{face card or a spade}) = \frac{12}{52} + \frac{13}{52} - \frac{3}{52}$$

Simplify.

$$P(\text{face card or a spade}) = \frac{22}{52} = \boxed{\frac{11}{26}}$$

The probability of either a face card or a spade being drawn is $\frac{11}{26} \approx 0.42$.

▪ **YOUR TURN**   Find the probability of drawing either a heart or an ace out of a deck of 52 cards.

▪ **Answer:** $\frac{4}{13}$

■ **Answer:** $\frac{7}{9}$ or ≈ 0.78

Classroom Example 12.7.7*
Consider the set of integers between 1 and 999, inclusive. Find the probability of choosing, at random, either a number ending in 2 or a number ending in an odd number.

Answer: $\frac{599}{999}$

EXAMPLE 7   Finding the Probability of Mutually Exclusive Events

Find the probability of drawing either an ace or a joker in a 54-card deck (a deck with two jokers).

Solution:

Drawing an ace and drawing a joker are two mutually exclusive events, since a card cannot be both an ace and a joker.

The deck has 4 aces. $P(\text{ace}) = \dfrac{4}{54}$

The deck has 2 jokers. $P(\text{joker}) = \dfrac{2}{54}$

Apply the probability formula. $P(E_1 \cup E_2) = P(E_1) + P(E_2)$

                                         $P(\text{ace or a joker}) = \dfrac{4}{54} + \dfrac{2}{54}$

Simplify. $P(\text{ace or a joker}) = \dfrac{6}{54} = \dfrac{1}{9} \approx 0.11\overline{1}$

> The probability of drawing either an ace or a joker is $\frac{1}{9} \approx 0.11$.

■ **YOUR TURN** If there are 10 women mathematicians, 8 men mathematicians, 6 women engineers, and 12 men engineers, what is the probability that a selected person is *either* a woman *or* an engineer?

Independent Events

Suppose you have two children. The sex of the second child is not affected by the sex of the first child. For example, if your first child is a boy, then it is no less likely that the second child will be a boy. We say that two events are **independent** if the occurrence of either of them has no effect on the probability of the other occurring.

PROBABILITIES OF INDEPENDENT EVENTS

If E_1 and E_2 are independent events, then the probability of both occurring is the product of the individual probabilities:

$$P(E_1 \cap E_2) = P(E_1) \cdot P(E_2)$$

Scientists used to believe that one gene controlled human eye color. Each parent gives one chromosome (either blue, green, or brown), and the result is a child with an eye-color gene composed of combinations, with brown being dominant over blue and blue being dominant over green. The genetic basis for eye color is actually far more complex, but we use this simpler model in the next example.

EXAMPLE 8   Probabilities of Blue-Eyed Children of Brown-Eyed Parents

If two brown-eyed parents have a blue-eyed child, then the parents must each have one blue- and one brown-eye gene. In order to have blue eyes, the child must get the blue-eye genes from both parents. What is the probability that two brown-eyed parents can have three children, all with blue eyes?

Solution:

The sample space for childrens' eye-color genes from these parents is

$$S = \{\text{Blue/Blue, Blue/Brown, Brown/Brown, Brown/Blue}\}$$

The only way for a child to have blue eyes is if that child inherits two blue-eyed genes.

$$P(\text{blue-eyed child}) = \frac{1}{4}$$

The eye color of each child is independent of that of the other. The probability of having three blue-eyed children is the product of the three individual probabilities.

$$P(\text{all three children with blue eyes}) = \left(\frac{1}{4}\right) \cdot \left(\frac{1}{4}\right) \cdot \left(\frac{1}{4}\right)$$

$$= \frac{1}{64}$$

The probability of two brown-eyed parents having three blue-eyed children is $\boxed{\frac{1}{64} \approx 0.016}$.

■ **YOUR TURN** Find the probability of the brown-eyed parents in Example 8 having three brown-eyed children.

Classroom Example 12.7.8
An urn contains 6 balls (2 red, 4 blue). You choose a ball and record its color and put the ball back into the urn. You do this twice. Find the probability that:
a. you choose 2 blue balls.
b. neither ball is red.

Answer:
a. $\frac{16}{36}$ **b.** $\frac{16}{36}$

■ **Answer:** $\frac{27}{64} \approx 0.422$

SECTION 12.7 SUMMARY

In this section we discussed the probability, or likelihood, of an event. It is found by dividing the total number of possible equally likely outcomes in the event by all of the possible outcomes in the sample space.

$$P(E) = \frac{n(E)}{n(S)} = \frac{\text{Number of outcomes in event } E}{\text{Number of outcomes in sample space } S}$$

Probability is a number between 0 and 1 or equal to 0 or 1:

$$0 \leq P(E) \leq 1$$

The probability of an event not occurring is 1 minus the probability of the event occurring:

$$P(\text{not } E) = 1 - P(E)$$

The probability of one event *or* another event occurring is found by adding the individual probabilities of each event and subtracting the probability of both:

"or"
$$P(E_1 \cup E_2) = P(E_1) + P(E_2) - P(E_1 \cap E_2)$$

If two events are mutually exclusive, they have no outcomes in common:

$$P(E_1 \cup E_2) = P(E_1) + P(E_2)$$

The probability of two events occurring is the product of the individual probabilities, provided the two events are independent or do not affect one another:

"and"
$$P(E_1 \cap E_2) = P(E_1) \cdot P(E_2)$$

▪SKILLS

In Exercises 1–6, find the sample space for each experiment.

1. The sum of two dice rolled simultaneously.

2. A coin tossed three times in a row.

3. The sex (boy or girl) of four children born to the same parents.

4. Tossing a coin and rolling a die.

5. Two balls selected from a container that has 3 red balls, 2 blue balls, and 1 white ball.

6. The grade (freshman, sophomore, or junior) of two high school students who work at a local restaurant.

Heads or Tails. In Exercises 7–10, find the probability for the experiment of tossing a coin three times.

7. Getting all heads.

8. Getting exactly one heads.

9. Getting at least one heads.

10. Getting more than one heads.

Tossing a Die. In Exercises 11–16, find the probability for the experiment of tossing two dice.

11. The sum is 3.

12. The sum is odd.

13. The sum is even.

14. The sum is prime.

15. The sum is more than 7.

16. The sum is less than 7.

Drawing a Card. In Exercises 17–20, find the probability for the experiment of drawing a single card from a deck of 52 cards.

17. Drawing a non-face card.

18. Drawing a black card.

19. Drawing a 2, 4, 6, or 8.

20. Drawing a 3, 5, 7, 9, or ace.

In Exercises 21–26, let $P(E_1) = \frac{1}{4}$ and $P(E_2) = \frac{1}{2}$ and find the probability of the event.

21. Probability of E_1 not occurring.

22. Probability of E_2 not occurring.

23. Probability of either E_1 or E_2 occurring if E_1 and E_2 are mutually exclusive.

24. Probability of either E_1 or E_2 occurring if E_1 and E_2 are not mutually exclusive and $P(E_1 \cap E_2) = \frac{1}{8}$.

25. Probability of both E_1 and E_2 occurring if E_1 and E_2 are mutually exclusive.

26. Probability of both E_1 and E_2 occurring if E_1 and E_2 are independent.

▪APPLICATIONS

27. Cards. A deck of 52 cards is dealt.

   a. How many possible combinations of four-card hands are there?

   b. What is the probability of having all spades?

   c. What is the probability of having four of a kind?

28. Blackjack. A deck of 52 cards is dealt for blackjack.

   a. How many possible combinations of two-card hands are there?

   b. What is the probability of having 21 points (ace with a 10 or face card)?

29. Cards. With a 52-card deck, what is the probability of drawing a 7 or an 8?

30. Cards. With a 52-card deck, what is the probability of drawing a red 7 or a black 8?

31. Cards. By drawing twice, what is the probability of drawing a 7 and then an 8?

32. Cards. By drawing twice, what is the probability of drawing a red 7 and then a black 8?

33. Children. What is the probability of having five daughters in a row and no sons?

34. Children. What is the probability of having four sons in a row and no daughters?

35. Children. What is the probability that of five children at least one is a boy? *Note:* P(at least one boy) $= 1 - P$(no boys).

36. Children. What is the probability that of six children at least one is a girl? *Note:* P(at least one girl) $= 1 - P$(no girls).

37. **Roulette.** In roulette, there are 38 numbered slots (1–36, 0, and 00). Eighteen are red, 18 are black, and the 0 and 00 are green. What is the probability of having 4 reds in a row?

38. **Roulette.** What is the probability of having 2 greens in a row on a roulette table?

39. **Item Defectiveness.** For a particular brand of DVD players, 10% of the ones on the market are defective. If a company has ordered 8 DVD players, what is the probability that none of the 8 DVD players is defective?

40. **Item Defectiveness.** For a particular brand of generators 20% of the ones on the market are defective. If a company buys 10 generators, what is the probability that none of the 10 generators is defective?

41. **Number Generator.** A random-number generator (computer program that selects numbers in no particular order) is used to select two numbers between 1 and 10. What is the probability that both numbers are even?

42. **Number Generator.** A random-number generator is used to select two numbers between 1 and 15. What is the probability that both numbers are odd?

In Exercises 43 and 44, assume each deal is from a complete (shuffled) single deck of cards.

43. **Blackjack.** What is the probability of being dealt a blackjack (any ace and any face card) with a single deck?

44. **Blackjack.** What is the probability of being dealt two blackjacks in a row with a single deck?

45. **Sports.** With the salary cap in the NFL, it is said that on "any given Sunday" any team could beat any other team. If we assume every week a team has a 50% chance of winning, what is the probability that a team will go 16–0?

46. **Sports.** With the salary cap in the NFL, it is said that on "any given Sunday" any team could beat any other team. If we assume every week a team has a 50% chance of winning, what is the probability that a team will have at least 1 win?

47. **Genetics.** Suppose both parents have the brown/blue pair of eye-color genes, and each parent contributes one gene to the child. Suppose the brown eye-color gene is dominant so that if the child has at least one brown gene, the color will dominate and the eyes will be brown.

   a. List the possible outcomes (sample space). Assume each outcome is equally likely.

   b. What is the probability that the child will have the blue/blue pair of genes?

   c. What is the probability that the child will have brown eyes?

48. **Genetics.** Refer to Exercise 47. In this exercise, the father has a brown/brown pair of eye-color genes, while the mother has a brown/blue pair of eye-color genes.

   a. List the possible outcomes (sample space). Assume each outcome is equally likely.

   b. What is the probability that the child will have a blue/blue pair of genes?

   c. What is the probability that the child will have brown eyes?

49. **Playing Cards.** How many 5-card hands can be drawn from a 52-card deck (no jokers)?

50. **Playing Cards.**

   a. How many ways can you select 2 aces and 3 other cards (non-aces) from a standard deck of 52 cards (no jokers)?

   b. What is the probability that you draw a 5-card hand with 2 aces and 3 non-aces?

51. **Playing Cards.** Find the probability of drawing 5 clubs from a standard deck of 52 cards.

52. **Poker.** Find the probability of getting 2 fives and 3 kings when drawing 5 cards from a standard deck of 52 cards.

■ CATCH THE MISTAKE

In Exercises 53 and 54, explain the mistake that is made.

53. Calculate the probability of drawing a 2 or a spade from a deck of 52 cards.

   Solution:

   The probability of drawing a 2 from a deck of 52 cards is $\frac{4}{52}$.

   The probability of drawing a spade from a deck of 52 cards is $\frac{13}{52}$.

   The probability of drawing a 2 or a spade is $\frac{4}{52} + \frac{13}{52} = \frac{17}{52}$.

   This is incorrect. What mistake was made?

54. Calculate the probability of having two boys and one girl.

   Solution:

   The probability of having a boy is $\frac{1}{2}$.

   The probability of having a girl is $\frac{1}{2}$.

   These are independent, so the probability of having two boys and a girl is $\left(\frac{1}{2}\right)\left(\frac{1}{2}\right)\left(\frac{1}{2}\right) = \frac{1}{8}$.

   This is incorrect. What mistake was made?

▪ CONCEPTUAL

In Exercises 55–58, determine whether each statement is true or false.

55. If $P(E_1) = 0.5$ and $P(E_2) = 0.4$, then $P(E_3)$ must equal 0.1 if there are three possible events and they are all mutually exclusive.

56. If two events are mutually exclusive, then they cannot be independent.

57. If two events are independent, then they are not mutually exclusive.

58. The probability of having five sons and no daughters is 1 minus the probability of having five daughters and no sons.

▪ CHALLENGE

59. If two people are selected at random, what is the probability that they have the same birthday? Assume 365 days per year.

60. If 30 people are selected at random, what is the probability that at least two of them will have the same birthday?

61. If one die is weighted so that 3 and 4 are the only numbers that the die will roll, and the other die is fair, what is the probability of rolling two dice that sum to 2, 5, or 6?

62. If one die is weighted so that 3 and 4 are the only numbers that the dice will roll, and 3 comes up twice as often as 4, what is the probability of rolling a 3?

▪ TECHNOLOGY

63. Use a random-number generator on a graphing utility to select two numbers between 1 and 10. Run this generator 50 times. How many times (out of 50 trials) were both of the two numbers even? Compare with your answer from Exercise 41.

64. Use a random-number generator on a graphing utility to select two numbers between 1 and 15. Run this 50 times. How many times (out of 50 trials) were both of the two numbers odd? Compare with your answer from Exercise 42.

In Exercises 65 and 66, when a die is rolled once, the probability of getting a 2 is $\frac{1}{6}$ and the probability of not getting a 2 is $\frac{5}{6}$. If a die is rolled n times, the probability of getting a 2 exactly k times can be found by using the binomial theorem:

$$_nC_k\left(\frac{1}{6}\right)^k\left(\frac{5}{6}\right)^{n-k}$$

65. If a die is rolled 10 times, find the probability of getting a 2 exactly two times. Round your answer to four decimal places.

66. If a die is rolled 8 times, find the probability of getting a 2 at most two times. Round your answer to four decimal places.

Pascal's triangle, which you first learned about in Section 12.5, is useful for expanding binomials. As you shall see next, it can also help you compute probabilities of outcomes and events in a certain type of multistage experiment.

1. Suppose a math teacher gives her students a pop quiz, including 2 true/false questions. Since Cameron didn't study, he decides to randomly guess for the true/false portion of the quiz. Each of his answers will either be right (R) or wrong (W). There are four possible outcomes for his answers: RR, RW, WR, and WW. These outcomes may be organized according to the number of correct answers given in each possible outcome, as shown below.

RANDOMLY GUESSING ON 2 TRUE/FALSE QUIZ QUESTIONS			
Number of correct answers given	2	1	0
Outcomes	RR	RW, WR	WW
Number of ways	1	2	1

a. Cameron's math teacher will give several more pop quizzes this term, with various numbers of true/false questions on each quiz. Complete the following tables for randomly guessing on true/false quizzes with 1 question, 3 questions, and 4 questions.

RANDOMLY GUESSING ON 1 TRUE/FALSE QUIZ QUESTION		
Number of correct answers given	1	0
Outcomes		
Number of ways		

RANDOMLY GUESSING ON 3 TRUE/FALSE QUIZ QUESTIONS	
Number of correct answers given	
Outcomes	
Number of ways	

RANDOMLY GUESSING ON 4 TRUE/FALSE QUIZ QUESTIONS	
Number of correct answers given	
Outcomes	
Number of ways	

```
              1
           1     1
        1     2     1
     1     3     3     1
  1     4     6     4     1
```

b. Notice that the "Number of ways" row of each table in part (a) are the rows of Pascal's triangle, shown above. Notice that the first and last entry of each row is a 1. Describe how to find the other entries.

c. Extend Pascal's triangle for four more rows, to the row that begins 1   8 .... How can you interpret each of the entries in the 1   8 ... row, in the context of randomly guessing on a true/false quiz?

d. What is the sum of the entries in each of the rows of Pascal's triangle? Try to find a formula for the sum of the entries in the nth row. What do these numbers represent, in terms of randomly guessing on a true/false quiz?

2. Use Pascal's triangle to find the probability of each event E given in parts (a) and (b) below.

a. E is the event of guessing 4 answers correctly on a 6-question true/false quiz.

b. E is the event of guessing *at least* 5 answers correctly on an 8-question true/false quiz.

3. Randomly guessing on a true/false quiz is an example of a multistage experiment in which there are *two equally likely outcomes* at each stage. Pascal's triangle may also be used to solve similar probability problems, as with flipping a coin or births of boys and girls.

Show how to use Pascal's triangle to find the probability of each of the following events.

a. No boys in a family with 4 children.

b. At least one boy in a family with 6 children.

c. All 10 heads when flipping a coin 10 times.

MODELING OUR WORLD

In the Modeling Our World feature in Chapter 2, you found the equations of lines that correspond to the seven wedges and the stabilization triangle. In 2005, the world was producing 7 billion tons of carbon emissions per year. In 2055, this number is projected to double, with the worldwide production of carbon emissions equaling 14 billion tons per year.

The Stabilization Triangle

1. Determine the equation of the line for the projected path: increase of 7 gigatons of carbon (GtC) over 50 years (2005 to 2055). Calculate the slope of the line.

2. What is the increase (per year) in the rate of carbon emissions per year based on the projected path model?

3. Develop a model in terms of a finite series that yields the total additional billions of tons of carbon emitted over the 50-year period (2005–2055) for the projected path over the flat path.

4. Calculate the total additional billions of tons of carbon of the projected path over the flat path (i.e., sum the series in (3)).

5. Discuss possible ways to provide the reduction between the projected path and the flat path based on the proposals given by Pacala and Socolow (professors at Princeton).*

*S. Pacala and R. Socolow, "Stabilization Wedges: Solving the Climate Problem for the Next 50 Years with Current Technologies," *Science*, Volume 305 (2004).

SECTION	CONCEPT	KEY IDEAS/FORMULAS
12.1	**Sequences and series**	
	Sequences	$a_1, a_2, a_3, \ldots, a_n, \ldots$ a_n is the general term.
	Factorial notation	$n! = n(n-1)(n-2) \cdots 3 \cdot 2 \cdot 1 \quad n \geq 2$ $0! = 1 \quad \text{and} \quad 1! = 1$
	Recursion formulas	When a_n is defined by previous terms a_{n-1}.
	Sums and series	Sigma Notation: $$\sum_{n=1}^{\infty} a_n = a_1 + a_2 + a_3 + \cdots + a_n + \cdots$$ Finite series, or nth partial sum, S_n. $S_n = a_1 + a_2 + a_3 + \cdots + a_n$
12.2	**Arithmetic sequences and series**	
	Arithmetic sequences	$a_{n+1} = a_n + d$ or $a_{n+1} - a_n = d$ d is called the **common difference**.
	The general (nth) term of an arithmetic sequence	$a_n = a_1 + (n-1)d$ for $n \geq 1$
	The sum of an arithmetic sequence	$S_n = \dfrac{n}{2}(a_1 + a_n)$
12.3	**Geometric sequences and series**	
	Geometric sequences	$a_{n+1} = r \cdot a_n$ or $\dfrac{a_{n+1}}{a_n} = r$ r is called the **common ratio**.
	The general (nth) term of a geometric sequence	$a_n = a_1 \cdot r^{n-1}$ for $n \geq 1$
	Geometric series	Finite series: $S_n = a_1 \dfrac{(1-r^n)}{(1-r)} \quad r \neq 1$ Infinite series: $\displaystyle\sum_{n=0}^{\infty} a_1 r^n = a_1 \dfrac{1}{(1-r)} \quad |r| < 1$
12.4	**Mathematical induction**	
	Mathematical induction	Prove that S_n is true for all positive integers. Step 1: Show that S_1 is true. Step 2: Assume S_n is true for S_k and show it is true for S_{k+1} (k = integer).
12.5	**The Binomial theorem**	
	Binomial coefficients	$\dbinom{n}{k} = \dfrac{n!}{(n-k)!k!}$ $\dbinom{n}{n} = 1$ and $\dbinom{n}{0} = 1$
	Binomial expansion	$(a+b)^n = \displaystyle\sum_{k=0}^{n} \dbinom{n}{k} a^{n-k} b^k$

SECTION	CONCEPT	KEY IDEAS/FORMULAS
	Pascal's triangle	Shortcut way of remembering binomial coefficients. Each term is found by adding the two numbers above it.
	Finding a particular term of a binomial expansion	The $(r + 1)$ term of the expansion $(a + b)^n$ is $\binom{n}{r} a^{n-r} b^r$.
12.6	**Counting, permutations, and combinations**	
	The fundamental counting principle	The number of ways in which successive independent things can occur is found by multiplying the number of ways each thing can occur.
	Permutations	The number of permutations of n objects is $$n! = n \cdot (n - 1) \cdot (n - 2) \cdots 2 \cdot 1$$ The number of permutations of n objects taken r at a time is $$_nP_r = \frac{n!}{(n - r)!} = n(n - 1)(n - 2) \cdots (n - r + 1)$$
	Combinations	The number of combinations of n objects taken r at a time $$_nC_r = \frac{n!}{(n - r)!r!}$$
	Permutations with repetition	The number of distinguishable permutations of n objects is $$\frac{n!}{n_1! \cdot n_2! \cdot n_3! \cdots n_k!}$$
12.7	**Probability**	
	Sample space	All of the possible outcomes of an experiment.
	Probability of an event	The probability of event E, denoted $P(E)$, is $$P(E) = \frac{n(E)}{n(S)} = \frac{\text{number of outcomes in event } E}{\text{number of outcomes in sample space } S}$$
	Probability of an event *not* occurring	$P(\text{not } E) = 1 - P(E)$
	Mutually exclusive events	The probability of *either* E_1 *or* E_2 occurring is $$P(E_1 \cup E_2) = P(E_1) + P(E_2) - P(E_1 \cap E_2)$$ If E_1 and E_2 are mutually exclusive, $P(E_1 \cap E_2) = 0$.
	Independent events	If E_1 and E_2 are independent events, then the probability of both occurring is the product of the individual probabilities: $$P(E_1 \cap E_2) = P(E_1) \cdot P(E_2)$$

12.1 Sequences and Series

Write the first four terms of the sequence. Assume n starts at 1.

1. $a_n = n^3$

2. $a_n = \dfrac{n!}{n}$

3. $a_n = 3n + 2$

4. $a_n = (-1)^n x^{n+2}$

Find the indicated term of the sequence.

5. $a_n = \left(\dfrac{2}{3}\right)^n \qquad a_5 = ?$

6. $a_n = \dfrac{n^2}{3^n} \qquad a_8 = ?$

7. $a_n = \dfrac{(-1)^n (n-1)!}{n(n+1)!} \qquad a_{15} = ?$

8. $a_n = 1 + \dfrac{1}{n} \qquad a_{10} = ?$

Write an expression for the nth term of the given sequence.

9. $3, -6, 9, -12, \ldots$

10. $1, \frac{1}{2}, 3, \frac{1}{4}, 5, \frac{1}{6}, 7, \frac{1}{8}, \ldots$

11. $-1, 1, -1, 1, \ldots$

12. $1, 10, 10^2, 10^3, \ldots$

Simplify the ratio of factorials.

13. $\dfrac{8!}{6!}$

14. $\dfrac{20!}{23!}$

15. $\dfrac{n(n-1)!}{(n+1)!}$

16. $\dfrac{(n-2)!}{n!}$

Write the first four terms of the sequence defined by the recursion formula.

17. $a_1 = 5 \qquad a_n = a_{n-1} - 2$

18. $a_1 = 1 \qquad a_n = n^2 \cdot a_{n-1}$

19. $a_1 = 1, a_2 = 2 \qquad a_n = (a_{n-1})^2 \cdot (a_{n-2})$

20. $a_1 = 1, a_2 = 2 \qquad a_n = \dfrac{a_{n-2}}{(a_{n-1})^2}$

Evaluate the finite series.

21. $\displaystyle\sum_{n=1}^{5} 3$

22. $\displaystyle\sum_{n=1}^{4} \dfrac{1}{n^2}$

23. $\displaystyle\sum_{n=1}^{6} (3n + 1)$

24. $\displaystyle\sum_{k=0}^{5} \dfrac{2^{k+1}}{k!}$

Use sigma (summation) notation to write the sum.

25. $-1 + \frac{1}{2} - \frac{1}{4} + \frac{1}{8} + \cdots - \frac{1}{64}$

26. $2 + 4 + 6 + 8 + 10 + \cdots + 20$

27. $1 + x + \dfrac{x^2}{2} + \dfrac{x^3}{6} + \dfrac{x^4}{24} + \cdots$

28. $x - x^2 + \dfrac{x^3}{2} - \dfrac{x^4}{6} + \dfrac{x^5}{24} - \dfrac{x^6}{120} + \cdots$

Applications

29. Marines Investment. With the prospect of continued fighting in Iraq, in December 2004, the Marine Corps offered bonuses of as much as \$30,000—in some cases, tax-free—to persuade enlisted personnel with combat experience and training to reenlist. Suppose a Marine put her entire \$30,000 reenlistment bonus in an account that earns 4% interest compounded monthly. The balance in the account after n months is

$$A_n = 30{,}000\left(1 + \dfrac{0.04}{12}\right)^n \qquad n = 1, 2, 3, \ldots$$

Her commitment with the Marines is 5 years. Calculate A_{60}. What does A_{60} represent?

30. Sports. The NFL minimum salary for a rookie is \$180,000. Suppose a rookie comes into the league making the minimum and gets a \$30,000 raise every year he plays. Write the general term a_n of a sequence that represents the salary of an NFL player making the league minimum during his entire career. Assuming $n = 1$ corresponds to the first year, what does $\displaystyle\sum_{n=1}^{4} a_n$ represent?

12.2 Arithmetic Sequences and Series

Determine whether the sequence is arithmetic. If it is, find the common difference.

31. $7, 5, 3, 1, -1, \ldots$

32. $1^3 + 2^3 + 3^3 + \cdots$

33. $1, \frac{3}{2}, 2, \frac{5}{2}, \ldots$

34. $a_n = -n + 3$

35. $a_n = \dfrac{(n+1)!}{n!}$

36. $a_n = 5(n-1)$

Find the general, or nth, term of the arithmetic sequence given the first term and the common difference.

37. $a_1 = -4 \qquad d = 5$

38. $a_1 = 5 \qquad d = 6$

39. $a_1 = 1 \qquad d = -\frac{2}{3}$

40. $a_1 = 0.001 \qquad d = 0.01$

For each arithmetic sequence described below, find a_1 and d and construct the sequence by stating the general, or nth, term.

41. The 5th term is 13 and the 17th term is 37.

42. The 7th term is -14 and the 10th term is -23.

43. The 8th term is 52 and the 21st term is 130.

44. The 11th term is -30 and the 21st term is -80.

Find the sum.

45. $\displaystyle\sum_{k=1}^{20} 3k$

46. $\displaystyle\sum_{n=1}^{15} n + 5$

47. $2 + 8 + 14 + 20 + \cdots + 68$

48. $\frac{1}{4} - \frac{1}{4} - \frac{3}{4} - \cdots - \frac{31}{4}$

Applications

49. Salary. Upon graduating with M.B.A.s, Bob and Tania opt for different career paths. Bob accepts a job with the U.S. Department of Transportation making $45,000 with a guaranteed $2000 raise every year. Tania takes a job with Templeton Corporation making $38,000 with a guaranteed $4000 raise every year. Calculate how many total dollars both Bob and Tania will have each made after 15 years.

50. Gravity. When a skydiver jumps out of an airplane, she falls approximately 16 feet in the 1st second, 48 feet during the 2nd second, 80 feet during the 3rd second, 112 feet during the 4th second, 144 feet during the 5th second, and this pattern continues. If she deploys her parachute after 5 seconds have elapsed, how far will she have fallen during those 5 seconds?

12.3 Geometric Sequences and Series

Determine whether the sequence is geometric. If it is, find the common ratio.

51. $2, -4, 8, -16, \ldots$

52. $1, \dfrac{1}{2^2}, \dfrac{1}{3^2}, \dfrac{1}{4^2}, \ldots$

53. $20, 10, 5, \dfrac{5}{2}, \ldots$

54. $\dfrac{1}{100}, \dfrac{1}{10}, 1, 10, \ldots$

Write the first five terms of the geometric series.

55. $a_1 = 3 \qquad r = 2$

56. $a_1 = 10 \qquad r = \dfrac{1}{4}$

57. $a_1 = 100 \qquad r = -4$

58. $a_1 = -60 \qquad r = -\dfrac{1}{2}$

Write the formula for the nth term of the geometric series.

59. $a_1 = 7 \qquad r = 2$

60. $a_1 = 12 \qquad r = \dfrac{1}{3}$

61. $a_1 = 1 \qquad r = -2$

62. $a_1 = \dfrac{32}{5} \qquad r = -\dfrac{1}{4}$

Find the indicated term of the geometric sequence.

63. 25th term of the sequence $2, 4, 8, 16, \ldots$

64. 10th term of the sequence $\dfrac{1}{2}, 1, 2, 4, \ldots$

65. 12th term of the sequence $100, -20, 4, -0.8, \ldots$

66. 11th term of the sequence $1000, -500, 250, -125, \ldots$

Evaluate the geometric series, if possible.

67. $\dfrac{1}{2} + \dfrac{3}{2} + \dfrac{3^2}{2} + \cdots + \dfrac{3^8}{2}$

68. $1 + \dfrac{1}{2} + \dfrac{1}{2^2} + \dfrac{1}{2^3} + \cdots + \dfrac{1}{2^{10}}$

69. $\displaystyle\sum_{n=1}^{8} 5(3)^{n-1}$

70. $\displaystyle\sum_{n=1}^{7} \dfrac{2}{3}(5)^n$

71. $\displaystyle\sum_{n=0}^{\infty} \left(\dfrac{2}{3}\right)^n$

72. $\displaystyle\sum_{n=1}^{\infty} \left(-\dfrac{1}{5}\right)^{n+1}$

Applications

73. Salary. Murad is fluent in four languages and is offered a job with the U.S. government as a translator. He is hired on the "GS" scale at a base rate of $48,000 with 2% increases in his salary per year. Calculate what his salary will be *after* he has been with the U.S. government for 12 years.

74. Boat Depreciation. Upon graduating from Auburn University, Philip and Steve get jobs at Disney Ride and Show Engineering and decide to buy a ski boat together. If the boat costs $15,000 new, and depreciates 20% per year, write a formula for the value of the boat n years after it was purchased. How much will the boat be worth when Philip and Steve have been working at Disney for 3 years?

12.4 Mathematical Induction

Prove the statements using mathematical induction for all positive integers n.

75. $3n \le 3^n$

76. $4^n < 4^{n+1}$

77. $2 + 7 + 12 + 17 + \cdots + (5n - 3) = \dfrac{n}{2}(5n - 1)$

78. $2n^2 > (n + 1)^2 \qquad n \ge 3$

12.5 The Binomial Theorem

Evaluate the binomial coefficients.

79. $\dbinom{11}{8}$

80. $\dbinom{10}{0}$

81. $\dbinom{22}{22}$

82. $\dbinom{47}{45}$

Expand the expression using the binomial theorem.

83. $(x - 5)^4$

84. $(x + y)^5$

85. $(2x - 5)^3$

86. $(x^2 + y^3)^4$

87. $\left(\sqrt{x} + 1\right)^5$

88. $(x^{2/3} + y^{1/3})^6$

Expand the expression using Pascal's triangle.

89. $(r - s)^5$

90. $(ax + by)^4$

Find the coefficient C of the term in the binomial expansion.

Binomial	Term
91. $(x - 2)^8$	Cx^6
92. $(3 + y)^7$	Cy^4
93. $(2x + 5y)^6$	Cx^2y^4
94. $(r^2 - s)^8$	Cr^8s^4

Applications

95. Lottery. In a state lottery in which 6 numbers are drawn from a possible 53 numbers, the number of possible 6-number combinations is equal to $\binom{53}{6}$. How many possible combinations are there?

96. Canasta. In the card game canasta, two decks of cards including the jokers are used, and 13 cards are dealt to each person. A total of $\binom{108}{13}$ different 13-card canasta hands can be dealt. How many possible hands are there?

12.6 Counting, Permutations, and Combinations

Use the formula for $_nP_r$ to evaluate each expression.

97. $_7P_4$ **98.** $_9P_9$ **99.** $_{12}P_5$ **100.** $_{10}P_1$

Use the formula for $_nC_r$ to evaluate each expression.

101. $_{12}C_7$ **102.** $_{40}C_5$ **103.** $_9C_9$ **104.** $_{53}C_6$

Applications

105. Car Options. A new Honda Accord comes in three models (LX, VX, and EX). Each of those models comes with either a cloth or a leather interior, and the exterior comes in either silver, white, black, red, or blue. How many different cars (models, interior seat upholstery, and exterior color) are there to choose from?

106. E-mail Passwords. All e-mail accounts require passwords. If a six-character password is required that can contain letters (but no numbers), how many possible passwords can there be if letters can be repeated? (Assume no letters are case-sensitive.)

107. Team Arrangements. There are 10 candidates for the board of directors, and there are four leadership positions (president, vice president, secretary, and treasurer). How many possible leadership teams are there?

108. License Plates. In a particular state, there are six characters in a license plate consisting of letters and numbers. If 0s and 1s are eliminated from possible numbers and Os and Is are eliminated from possible letters, how many different license plates can be made?

109. Seating Arrangements. Five friends buy five season tickets to the Philadelphia Eagles. To be fair, they change the seating arrangement every game. How many different seating arrangements are there for the five friends? How many seasons would they have to buy tickets in order to sit in all of the combinations (each season has eight home games)?

110. Safe. A safe will open when the correct choice of three numbers (1 to 60) is selected in a specific order. How many possible permutations are there?

111. Raffle. A fundraiser raffle is held to benefit the Make a Wish Foundation, and 100 raffle tickets are sold. Four prizes are raffled off. First prize is a round-trip ticket on American Airlines, second prize is a round of golf for four people at a Links golf course, third prize is a $100 gift certificate to the Outback Steakhouse, and fourth prize is a half-hour massage. How many possible winning scenarios are there if all 100 tickets are sold to different people?

112. Sports. There are 117 Division 1-A football teams in the United States. At the end of the regular season is the Bowl Championship Series, and the top two teams play each other in the championship game. Assuming that any two Division 1-A teams can advance to the championship, how many possible matchups are there for the championship game?

113. Cards. In a deck of 52 cards, how many different 6-card hands can be dealt?

114. Blackjack. In a game of two-deck blackjack (104 cards), how many 2-card combinations are there that equal 21, that is, ace and a 10 or face card—jack, queen, or king?

12.7 Probability

115. Coin Tossing. For the experiment of tossing a coin four times, what is the probability of getting all heads?

116. Dice. For an experiment of tossing two dice, what is the probability that the sum of the dice is odd?

117. Dice. For an experiment of tossing two dice, what is the probability of not rolling a combined 7?

118. Cards. For a deck of 52 cards, what is the probability of drawing a diamond?

For Exercises 119–122, let $P(E_1) = \frac{1}{3}$ and $P(E_2) = \frac{1}{2}$ and find the probability of the event.

119. Probability. Find the probability of an event E_1 not occurring.

120. Probability. Find the probability of either E_1 or E_2 occurring if E_1 and E_2 are mutually exclusive.

121. Probability. Find the probability of either E_1 or E_2 occurring if E_1 and E_2 are not mutually exclusive and $P(E_1 \cap E_2) = \frac{1}{4}$.

122. Probability. Find the probability of both E_1 and E_2 occurring if E_1 and E_2 are independent.

123. Cards. With a 52-card deck, what is the probability of drawing an ace or a 2?

124. Cards. By drawing twice, what is the probability of drawing an ace and then a 2? (Assume that after the first card is drawn it is not put back into the deck.)

125. Children. What is the probability that in a family of five children at least one is a girl?

126. Sports. With the salary cap in the NFL, it is said that on "any given Sunday" any team could beat any other team. If we assume every week a team has a 50% chance of winning, what is the probability that a team will go 11–1?

Technology

Section 12.1

127. Use a graphing calculator "SUM" to find the sum of the series $\sum_{n=1}^{6} \frac{1}{n^2}$.

128. Use a graphing calculator "SUM" to find the sum of the infinite series $\sum_{n=1}^{\infty} \frac{1}{n}$, if possible.

Section 12.2

129. Use a graphing calculator to sum $\sum_{n=1}^{75} \left[\frac{3}{2} + \frac{6}{7}(n-1)\right]$.

130. Use a graphing calculator to sum $\sum_{n=1}^{264} \left[-19 + \frac{1}{3}(n-1)\right]$.

Section 12.3

131. Apply a graphing utility to plot
$y_1 = 1 - 2x + 4x^2 - 8x^3 + 16x^4$ and $y_2 = \frac{1}{1 + 2x}$, and let x range from $[-0.3, 0.3]$. Based on what you see, what do you expect the geometric series $\sum_{n=0}^{\infty} (-1)^n (2x)^n$ to sum to in this range of x values?

132. Does the sum of the infinite series $\sum_{n=0}^{\infty} \left(\frac{e}{\pi}\right)^n$ exist? Use a graphing calculator to find it and round to four decimal places.

Section 12.4

133. Use a graphing calculator to sum the series
$2 + 7 + 12 + 17 + \cdots + (5n - 3)$ on the left side, and evaluate the expression $\frac{n}{2}(5n - 1)$ on the right side for $n = 200$. Do they agree with each other? Do your answers confirm the proof for Exercise 77?

134. Use a graphing calculator to plot the graphs of $y_1 = 2x^2$ and $y_2 = (x + 1)^2$ in the $[100, 1000]$ by $[10,000, 2,500,000]$ viewing rectangle. Do your answers confirm the proof for Exercise 78?

Section 12.5

135. With a graphing utility, plot $y_1 = 1 + 8x$, $y_2 = 1 + 8x + 24x^2$, $y_3 = 1 + 8x + 24x^2 + 32x^3$,
$y_4 = 1 + 8x + 24x^2 + 32x^3 + 16x^4$, and $y_5 = (1 + 2x)^4$ for $-0.1 < x < 0.1$. What do you notice happening each time an additional term is added? Now, let $0.1 < x < 1$. Does the same thing happen?

136. With a graphing utility, plot $y_1 = 1 - 8x$, $y_2 = 1 - 8x + 24x^2$, $y_3 = 1 - 8x + 24x^2 - 32x^3$,
$y_4 = 1 - 8x + 24x^2 - 32x^3 + 16x^4$, and $y_5 = (1 - 2x)^4$ for $-0.1 < x < 0.1$. What do you notice happening each time an additional term is added? Now, let $0.1 < x < 1$. Does the same thing happen?

Section 12.6

137. Use a graphing calculator to evaluate:

a. $\frac{_{16}P_7}{7!}$

b. $_{16}C_7$

c. Are answers in (a) and (b) the same?

d. Why?

138. Use a graphing calculator to evaluate:

a. $\frac{_{52}P_6}{6!}$

b. $_{52}C_6$

c. Are answers in (a) and (b) the same?

d. Why?

Section 12.7

In Exercises 139 and 140, when two dice are rolled, the probability of getting a sum of 9 is $\frac{1}{9}$ and the probability of not getting a sum of 9 is $\frac{8}{9}$. If two dice are rolled n times, the probability of getting a sum of 9 exactly k times can be found by using the binomial theorem $_nC_k \left(\frac{1}{9}\right)^k \left(\frac{8}{9}\right)^{n-k}$.

139. If two dice are rolled 10 times, find the probability of getting a sum of 9 exactly three times. Round your answer to four decimal places.

140. If a die is rolled 8 times, find the probability of getting a sum of 9 at least two times. Round your answer to four decimal places.

For Exercises 1–5, use the sequence $1, x, x^2, x^3, \ldots$

1. Write the nth term of the sequence.

2. Classify this sequence as arithmetic, geometric, or neither.

3. Find the nth partial sum of the series S_n.

4. Assuming this sequence is infinite, write the series using sigma notation.

5. Assuming this sequence is infinite, what condition would have to be satisfied in order for the sum to exist?

6. Find the following sum: $\frac{1}{3} + \frac{1}{9} + \frac{1}{27} + \frac{1}{81} + \cdots$.

7. Find the following sum: $\sum_{n=1}^{10} 3 \cdot \left(\frac{1}{4}\right)^n$.

8. Find the following sum: $\sum_{k=1}^{50} (2k + 1)$.

9. Write the series using sigma notation, then find its sum: $2 + 7 + 12 + 17 + \cdots + 497$.

10. Use mathematical induction to prove that $2 + 4 + 6 + \cdots + 2n = n^2 + n$.

11. Evaluate $\frac{7!}{2!}$.

12. Find the third term of $(2x + y)^5$.

In Exercises 13–16, evaluate the expressions.

13. $\binom{15}{12}$

14. $\binom{k}{k}$

15. $_{14}P_3$

16. $_{200}C_3$

17. Expand the expression $\left(x^2 + \frac{1}{x}\right)^5$.

18. Use the binomial theorem to expand the binomial $(3x - 2)^4$.

19. Explain why there are always more permutations than combinations.

20. What is the probability of not winning a trifecta (selecting the first-, second-, and third-place finishers) in a horse race with 15 horses?

For Exercises 21–23, refer to a roulette wheel with 18 red, 18 black, and 2 green slots.

21. **Roulette.** What is the probability of the ball landing in a red slot?

22. **Roulette.** What is the probability of the ball landing in a red slot 5 times in a row?

23. **Roulette.** If the four previous rolls landed on red, what is the probability that the next roll will land on red?

24. **Marbles.** If there are four red marbles, three blue marbles, two green marbles, and one black marble in a sack, find the probability of pulling out the following order: black, blue, red, red, green.

25. **Cards.** What is the probability of drawing an ace or a diamond from a deck of 52 cards?

26. **Human Anatomy.** Vasopressin is a relatively simple protein that is found in the human liver. It consists of eight amino acids that must be joined together in one particular order for the effective functioning of the protein.
    a. How many different arrangements of the eight amino acids are possible?
    b. What is the probability of randomly selecting one of these arrangements and obtaining the correct arrangement to make vasopressin?

27. Find the constant term in the expression $\left(x^3 + \frac{1}{x^3}\right)^{20}$.

28. Use a graphing calculator to sum $\sum_{n=1}^{125}\left[-\frac{11}{4} + \frac{5}{6}(n - 1)\right]$.

1. Simplify $\dfrac{10x + 1}{3x - 2} + \dfrac{5 + 4x}{2 - 3x}$.

2. The length of a rectangle is 5 less than twice the width and the perimeter is 38 inches. What are the dimensions of the rectangle?

3. Solve using the quadratic formula: $2x^2 - 5x + 11 = 0$.

4. Solve and express the solution in interval notation: $|x - 5| > 3$.

5. Write an equation of the line with slope undefined and x-intercept $(-8, 0)$.

6. Find the x-intercept and y-intercept and slope of the line $3x - 5y = 15$.

7. Using the function $f(x) = x^2 - 3x + 2$, evaluate the difference quotient $\dfrac{f(x + h) - f(x)}{h}$.

8. Find the composite function $f \circ g$, and state the domain for
$$f(x) = 4 - \frac{1}{x} \text{ and } g(x) = \frac{1}{2x + 1}.$$

9. Find the vertex of the parabola $f(x) = -0.04x^2 + 1.2x - 3$.

10. Factor the polynomial $P(x) = x^4 + 8x^2 - 9$ as a product of linear factors.

11. Find the vertical and horizontal asymptotes of the function:
$$f(x) = \frac{5x - 7}{3 - x}$$

12. How much money should be put in a savings account now that earns 4.7% a year compounded weekly, if you want to have $65,000 in 17 years?

13. Evaluate $\log_2 6$ using the change-of-base formula. Round your answer to three decimal places.

14. Solve $\ln(5x - 6) = 2$. Round your answer to three decimal places.

15. Solve the system of linear equations.
$$8x - 5y = 15$$
$$y = \tfrac{8}{5}x + 10$$

16. Solve the system of linear equations.
$$2x - y + z = 1$$
$$x - y + 4z = 3$$

17. Maximize the objective function $z = 4x + 5y$, subject to the constraints $x + y \le 5$, $x \ge 1$, $y \ge 2$.

18. Solve the system using Gauss–Jordan elimination.
$$x + 5y - 2z = 3$$
$$3x + y + 2z = -3$$
$$2x - 4y + 4z = 10$$

19. Given
$$A = \begin{bmatrix} 3 & 4 & -7 \\ 0 & 1 & 5 \end{bmatrix} \quad B = \begin{bmatrix} 8 & -2 & 6 \\ 9 & 0 & -1 \end{bmatrix} \quad C = \begin{bmatrix} 9 & 0 \\ 1 & 2 \end{bmatrix}$$
find $C(A + B)$.

20. Calculate the determinant.
$$\begin{vmatrix} 2 & 5 & -1 \\ 1 & 4 & 0 \\ -2 & 1 & 3 \end{vmatrix}$$

21. Find the equation of a parabola with vertex $(3, 5)$ and directrix $x = 7$.

22. Graph $x^2 + y^2 < 4$.

23. Find the sum of the finite series $\sum_{n=1}^{4} \dfrac{2^{n-1}}{n!}$.

24. Classify the sequence as arithmetic, geometric, or neither:
$$5, 15, 45, 135, \ldots$$

25. There are 10 true/false questions on a quiz. Assuming no answers are left blank, how many different ways can you answer the questions on the quiz?

26. Use a graphing calculator to sum $\sum_{n=1}^{164} \left[\tfrac{1}{4} + \tfrac{3}{2}(n - 1) \right]$.

27. Find the constant term in the expression $\left(x^3 - \dfrac{1}{x^3} \right)^{14}$.

ANSWERS TO ALL EXERCISES*

*Answers that require a proof, graph, or otherwise lengthy solution are not included. See *Instructor's Solutions Manual*.

1295

CHAPTER 0

Section 0.1

1. rational
2. rational
3. irrational
4. irrational
5. rational
6. rational
7. irrational
8. irrational
9. **a.** 7.347 **b.** 7.347
10. **a.** 9.255 **b.** 9.254
11. **a.** 2.995 **b.** 2.994
12. **a.** 6.995 **b.** 6.995
13. **a.** 0.234 **b.** 0.234
14. **a.** 1.327 **b.** 1.327
15. **a.** 5.238 **b.** 5.238
16. **a.** 2.118 **b.** 2.118
17. 4
18. 40
19. 26
20. 13
21. -130
22. -96
23. 17
24. -1
25. 3
26. 0
27. $x + y - z$
28. $-a + b + c$
29. $-3x - y$
30. $-4a + 2b$
31. $\frac{3}{5}$
32. -1
33. 68
34. -2
35. $-9x - y$
36. $-\dfrac{x}{2}$
37. -10
38. $-3x + 38$
39. 5
40. $-15x - 23y + 2$
41. $\frac{19}{12}$
42. $\frac{3}{10}$
43. $\frac{1}{2}$
44. $\frac{13}{6}$
45. $\frac{23}{12}$
46. $\frac{8}{9}$
47. $\frac{1}{27}$
48. $\frac{13}{14}$
49. $\dfrac{x}{3}$
50. $\dfrac{y}{6}$
51. $\dfrac{x}{21}$
52. $\dfrac{y}{30}$
53. $\dfrac{61y}{60}$
54. $\dfrac{3x}{20}$
55. $\frac{11}{30}$
56. $\frac{17}{60}$
57. $\frac{4}{3}$
58. $\frac{3}{5}$
59. $\frac{3}{35}$
60. $\frac{8}{7}$
61. $\dfrac{12b}{a}$
62. $\dfrac{9a}{b}$
63. $\frac{3}{4}$
64. $\frac{28}{47}$
65. $\dfrac{4xy}{3}$
66. $4m$
67. $\dfrac{8x}{y}$
68. $\frac{329}{18}$
69. $\frac{2}{3}$
70. 30
71. $\frac{3}{25}$
72. 2
73. $9,176,366,000,000
74. 303,818,000
75. $30,203
76. $30,203.46
77. Rounded incorrectly: 13.27

78. Added incorrectly in very first step. Get an LCD. Should be $\dfrac{2(3)+1}{9} = \dfrac{7}{9}$.
79. The -2 did not distribute to the second term.
80. The -1 did not distribute to the second term.
81. false
82. true
83. true
84. false
85. $x \neq 0$
86. $x \neq 0$
87. $-7x + 12y - 75$
88. $14x + 10y - 19$
89. irrational
90. rational
91. rational
92. irrational

Section 0.2

1. 256
2. 125
3. -243
4. 16
5. -25
6. -49
7. -16
8. -45
9. 1
10. -8
11. 0.1
12. $\frac{1}{a}$
13. $\frac{1}{64}$
14. $\frac{1}{81}$
15. -150
16. -32
17. 5
18. 10
19. -54
20. $\frac{3}{8}$
21. x^5
22. y^8
23. $\dfrac{1}{x}$
24. $\dfrac{1}{y^4}$
25. x^6
26. y^6
27. $64a^3$
28. $64x^6$
29. $-8t^3$
30. $81b^4$
31. $75x^5y^5$
32. $16x^4y^7$
33. $\dfrac{y^2}{x^2}$
34. x^7y^7
35. $-\dfrac{1}{2xy}$
36. $\dfrac{3}{4x^3y^8}$
37. $\dfrac{16}{b^4}$
38. $\dfrac{9}{c^2}$
39. $\dfrac{a^4}{81b^6}$
40. $\dfrac{x^{12}}{6561y^8}$
41. $\dfrac{1}{a^6b^2}$
42. $\dfrac{y^6}{x^8}$
43. x^8y^2
44. $x^{18}y^5$
45. $\dfrac{x^{14}}{4y}$
46. $\dfrac{64y^7x^2}{z^3}$
47. $\dfrac{y^{10}}{-32x^{22}}$
48. $64x^{17}$

49. $-\dfrac{y^{24}}{a^{12}x^3}$ **50.** $-\dfrac{y^{30}}{b^{45}x^{15}}$ **51.** 2^{26}

52. 3^{31} **53.** 2.76×10^7 **54.** 1.44×10^{11}

55. 9.3×10^7 **56.** 1.2345×10^9

57. 5.67×10^{-8} **58.** 8.28×10^{-6}

59. 1.23×10^{-7} **60.** 5.0×10^{-9}

61. $47,000,000$ **62.** $390,000$

63. $23,000$ **64.** 0.0078

65. 0.000041 **66.** 0.000000092

67. a. 2.08×10^9 ft   **b.** Yes-almost 16 times

68. a. 3.95×10^5 miles   **b.** Yes

69. 2.0×10^8 miles   **70.** 1.42×10^8 miles

71. 0.00000155 meter

72. 0.000000693 meter

73. Should be adding exponents here: y^5

74. Forgot to apply the power to "2".

75. Computed $(y^3)^2$ incorrectly. Should be y^6.

76. Should be subtracting the powers, not
dividing them.

77. false   **78.** true   **79.** false

80. false   **81.** a^{mnk} **82.** a^{-mnk}

83. -16 **84.** 16 **85.** 156

86. $\frac{35}{6}$ **87.** 5.6 acres/person

88. 8.09 acres/person   **89.** 0.2

90. 1000

91. same answer using the calculator

92. same answer using the calculator

93. 5.11×10^{14} **94.** 6.25×10^{23}

Section 0.3

1. $-7x^4-2x^3+5x^2+16$ Degree 4

2. In standard form. Degree 3

3. $-6x^3+4x+3$ Degree 3

4. $5x^5+8x^4-7x^3-x^2+10$ Degree 5

5. 15   Degree 0

6. In standard form. Degree 0

7. $y-2$ Degree 1

8. In standard form. Degree 1

9. $-x^2+5x+5$ **10.** $5x^2+x-7$

11. $2x^2-x-18$ **12.** $x^3-15x^2+9x-10$

13. $4x^4-10x^2-1$ **14.** $9x^2-6x-7$

15. $2z^2+2z-3$

16. $25y^3-21y^2+16y-2$

17. $-11y^3-16y^2+16y-4$

18. $x^2-5xy+7y^2$ **19.** $-4x+12y$

20. $-6a^2+8a+3$ **21.** x^2-x+2

22. $3x^3-3x^2-5x+5$ **23.** $-4t^2+6t-1$

24. $-5z^3-4z^2-4z-1$ **25.** $35x^2y^3$

26. $24z^4$ **27.** $2x^5-2x^4+2x^3$

28. $4z^4-4z^3-8z^2$ **29.** $10x^4-2x^3-10x^2$

30. $-2z^3-z^2+5z$ **31.** $2x^5+2x^4-4x^3$

32. $3x^5-3x^4+6x^3$ **33.** $2a^3b^2+4a^2b^3-6ab^4$

34. $b^3c^4d^2+bc^4d^5-b^3c^3d^6$ **35.** $6x^2-5x-4$

36. $12z^2+17z-7$ **37.** x^2-4

38. y^2-25 **39.** $4x^2-9$ **40.** $25y^2-1$

41. $-4x^2+4x-1$ **42.** $16b^2-25y^2$

43. $4x^4-9$ **44.** $16x^2y^2-81$

45. $2y^4-9y^3+5y^2+7y$

46. $t^4-6t^3-5t^2+24t+4$

47. x^3-x^2+x+3 **48.** x^3+27

49. t^2-4t+4 **50.** t^2-6t+9

51. z^2+4z+4 **52.** z^2+6z+9

53. $x^2+2xy+y^2-6x-6y+9$

54. $4x^4+12x^2y+9y^2$

55. $25x^2-20x+4$ **56.** x^3+2x^2+2x+1

57. $6y^3+5y^2-4y$ **58.** $p^4-p^3-2p^2$

59. x^4-1 **60.** t^4-50t^2+625

61. $ab^2+2b^3-9a^3-18a^2b$ **62.** x^3-8y^3

63. $2x^2-3y^2-xy+3xz+8yz-5z^2$

64. $-5b^4+17b^3+3b^2-b+2$

65. $P=11x-100$ **66.** $P(x)=25x-75$

67. $P=-x^2+200x-7500$

68. $P=-\frac{1}{2}x^2+200x-8000$

69. $V=(15-2x)(8-2x)x$
          $=4x^3-46x^2+120x$

70. $V(x)=2(x-4)^2=2x^2-16x+32$

71. a. $P = (2\pi x + 4x + 10)$ feet

b. $A = (\pi x^2 + 4x^2 + 10x)$ feet^2

72. a. $\frac{8}{3}\pi r^3$ units^3 **b.** $7\pi r^2$ units^2

73. $F = \dfrac{3k}{100}$

74. a. $S(t) = -16t^2 + 96t + 192$

b. 320 feet; no

75. Forgot to distribute the negative sign through the entire second polynomial.
$2x^5 - 5 - 3x + x^2 - 1 = 3x^2 - 3x - 6$

76. Forgot the middle (inner and outer) terms.
$$(2 + x)(2 + x) = 4 + 2x + 2x + x^2$$
$$= x^2 + 4x + 4$$

77. true   **78.** false   **79.** false

80. false   **81.** $m + n$ **82.** m

83. $2401x^4 - 1568x^2y^4 + 256y^8$

84. $81x^4 - 450x^2y^4 + 625y^8$ **85.** $x^3 - a^3$

86. $x^3 + a^3$

87. $(2x + 3)(x - 4)$ and $2x^2 - 5x - 12$

88. $(x + 5)^2$ and $x^2 + 10x + 25$

Section 0.4

1. $5(x + 5)$ **2.** $x(x + 2)$ **3.** $2(2t^2 - 1)$

4. $4z(4z - 5)$ **5.** $2x(x - 5)(x + 5)$

6. $4xy(x - 2y + 4xy)$ **7.** $3x(x^2 - 3x + 4)$

8. $7x(2x^3 - x + 3)$ **9.** $x(x - 8)(x + 5)$

10. $-9y(y - 5)$ **11.** $2xy(2xy^2 + 3)$

12. $3z(z^2 - 2z + 6)$ **13.** $(x - 3)(x + 3)$

14. $(x - 5)(x + 5)$ **15.** $(2x - 3)(2x + 3)$

16. $(1 - x)(1 + x)\left(1 + x^2\right)$

17. $2(x - 7)(x + 7)$ **18.** $9(4 - 3y)(4 + 3y)$

19. $(15x - 13y)(15x + 13y)$

20. $(11y - 7x)(11y + 7x)$ **21.** $(x + 4)^2$

22. $(y - 5)^2$ **23.** $(x^2 - 2)^2$ **24.** $(1 - 3y)^2$

25. $(2x + 3y)^2$ **26.** $(x - 3y)^2$

27. $(x - 3)^2$ **28.** $(5x - 2y)^2$

29. $(x^2 + 1)^2$ **30.** $(x^3 - 3)^2$ **31.** $(p + q)^2$

32. $(p - q)^2$ **33.** $(t + 3)(t^2 - 3t + 9)$

34. $(z + 4)(z^2 - 4z + 16)$

35. $(y - 4)(y^2 + 4y + 16)$

36. $(x - 1)(x^2 + x + 1)$

37. $(2 - x)(4 + 2x + x^2)$

38. $(3 - y)(9 + 3y + y^2)$

39. $(y + 5)(y^2 - 5y + 25)$

40. $x(4 - x)(16 + 4x + x^2)$

41. $(3 + x)(9 - 3x + x^2)$

42. $(6x - y)(36x^2 + 6xy + y^2)$

43. $(x - 5)(x - 1)$ **44.** $(t + 1)(t - 6)$

45. $(y - 3)(y + 1)$ **46.** $(y - 5)(y + 2)$

47. $(2y + 1)(y - 3)$ **48.** $2(z - 3)(z + 1)$

49. $(3t + 1)(t + 2)$ **50.** $2(2x + 3)(x - 2)$

51. $(-3t + 2)(2t + 1)$ **52.** $(-2x + 1)(3x + 10)$

53. $(x^2 + 2)(x - 3)$ **54.** $(x^3 - 3)(x^2 + 5)$

55. $(a^3 - 8)(a + 2)$
$$= (a + 2)(a - 2)(a^2 + 2a + 4)$$

56. $(x - 1)(x^2 + x + 1)(x - 3)$

57. $(3y - 5r)(x + 2s)$ **58.** $(2x + 1)(3x - 5)$

59. $(4x - y)(5x + 2y)$

60. $(x - 1)(x^2 + x + 1)(3x - a)(3x + a)$

61. $(x - 2y)(x + 2y)$ **62.** $(a + 3)(a + 2)$

63. $(3a + 7)(a - 2)$ **64.** $(x + 1)(a + b)$

65. prime   **66.** prime   **67.** prime

68. $\left[\frac{1}{2} - b\right]\left[\frac{1}{2} + b\right]\left[\frac{1}{4} + b^2\right]$

69. $2(3x + 2)(x + 1)$ **70.** prime

71. $(3x - y)(2x + 5y)$ **72.** $15x(1 + y)$

73. $9(2s - t)(2s + t)$ **74.** $3x(x - 6)(x + 6)$

75. $(ab - 5c)(ab + 5c)$

76. $2(x + 3)(x^2 - 3x + 9)$

77. $(x - 2)(4x + 5)$ **78.** $-(x - 5)^2$

79. $x(3x + 1)(x - 2)$ **80.** $y(2y - 1)(y + 2)$

81. $x(x-3)(x+3)$ **82.** $w(w-5)(w+5)$

83. $(x-1)(y-1)$ **84.** $(1+b)(a+b)$

85. $(x^2+3)(x^2+2)$

86. $(x-2)(x^2+2x+4)(x+1)(x^2-x+1)$

87. $(x-6)(x+4)$ **88.** $(5x+3)^2$

89. $x(x+5)(x^2-5x+25)$

90. $(x-1)(x+1)(x^2+1)$

91. $(x-3)(x+3)(x^2+9)$

92. $(2x-5)(5x-3)$ **93.** $2(3x+4)$

94. $x(x+4)(x+3)$

95. $(x+2)(2x-15)$ **96.** $(3x-4)(x+3)$

97. $-2(8t-1)(t+5)$ **98.** $x(10-x)$

99. Last step.
$(x-1)(x^2-9)=(x-1)(x-3)(x+3)$

100. A 2 should have been factored out of both
binomials, resulting in a 4.
$(2x-4)(2x+10)=2(x-2)2(x+5)=4(x-2)(x+5)$

101. false   **102.** false   **103.** true

104. false   **105.** $(a^n-b^n)(a^n+b^n)$

106. $5, 13, -5, -13$

107. $8x^3+1$ and $(2x+1)(4x^2-2x+1)$

108. $27x^3-1$ and $(3x-1)(9x^2+3x+1)$

Section 0.5

1. $x\neq 0$ **2.** $x\neq 0$ **3.** $x\neq 1$

4. $y\neq 1$ **5.** $x\neq -1$ **6.** $x\neq 3$

7. $p\neq \pm 1$ **8.** $t\neq -3,3$

9. no restrictions   **10.** no restrictions

11. $\dfrac{(x-9)}{2(x+9)}; x\neq -9,-3$

12. $\dfrac{(y-8)}{2(y+8)}; y\neq -8,-7,0$

13. $\dfrac{x-3}{2}; x\neq -1$ **14.** $\dfrac{2x+1}{3}; x\neq 3$

15. $\dfrac{2(3y+1)}{9y}; y\neq 0,\frac{1}{2}$ **16.** $\dfrac{7(2y+1)}{10y}; y\neq 0,\frac{1}{3}$

17. $\dfrac{y+1}{5}; y\neq \frac{1}{5}$ **18.** $\dfrac{2t-1}{4}; t\neq -2$

19. $\dfrac{3x+7}{4}; x\neq 4$ **20.** $\dfrac{2t+5}{3}; t\neq 7$

21. $x+2; x\neq 2$ **22.** $t(t+1); t\neq 1$

23. $1; x\neq -7$ **24.** $1; y\neq \frac{9}{2}$

25. $\dfrac{x^2+9}{2x+9}; x\neq -\frac{9}{2}$ **26.** $\dfrac{x^2+4}{2(x+2)}; x\neq -2$

27. $\dfrac{x+3}{x-5}; x\neq -2,5$ **28.** $\dfrac{x+15}{x+4}; x\neq -4$

29. $\dfrac{3x+1}{x+5}; x\neq -5,\frac{1}{2}$

30. $\dfrac{3x+1}{x+3}; x\neq -3,\frac{2}{5}$

31. $\dfrac{3x+5}{x+1}; x\neq -1,2$ **32.** $\dfrac{3x+4}{x-2}; x\neq -\frac{5}{4},2$

33. $\dfrac{2(5x+6)}{5x-6}; x\neq 0,\frac{6}{5}$

34. $\dfrac{8(x-2)}{x-5}; x\neq 0,\pm 5$

35. $\frac{2}{3}; x\neq 0,\pm 1$ **36.** $\frac{1}{2}; x\neq 0,\pm 1$

37. $3(x+2); x\neq -5,0,2$ **38.** $4x; x\neq -3,0,8$

39. $\dfrac{t-3}{3(t+2)}; t\neq -2,3$

40. $\dfrac{y-5}{3(y+8)}; y\neq -8,-3,5$

41. $\dfrac{3t(t^2+4)}{(t-3)(t+2)}; t\neq -2,3$

42. $\dfrac{a(a+3)}{14(a-3)}; a\neq \pm 3$

43. $\dfrac{3y(y-2)}{y-3}; y\neq -2,3$

44. $\dfrac{4(t+3)}{t(t+2)}; t\neq 0,\pm 2$

45. $\dfrac{(2x+3)(x-5)}{2x(x+5)^2}; x\neq 0,\pm 5$

46. $\dfrac{5t-1}{4(4t-3)}; t\neq 0,\pm\frac{3}{4}$

47. $\dfrac{(2x-7)(2x+7)}{(3x-5)(4x+3)}; x\neq -\frac{3}{4},\frac{7}{2},\pm\frac{5}{3}$

48. $\dfrac{x+2}{8x(x+9)}$; $x \neq 0, \frac{2}{3}, \pm 9$

49. $\dfrac{x}{4}$; $x \neq 0$ **50.** $\dfrac{x}{2}$; $x \neq 0$

51. $\dfrac{x+2}{2}$; $x \neq \pm 2$ **52.** $\dfrac{1}{5(x-6)}$; $x \neq \pm 6$

53. $\dfrac{x+1}{5}$; $x \neq \pm 1$ **54.** $\dfrac{3x+4}{2}$; $x \neq \pm \frac{4}{3}$

55. $\dfrac{1}{2(1-p)}$; $p \neq 2, \pm 1$

56. $\dfrac{1}{3(x+4)}$; $x \neq \pm 4$ **57.** $\dfrac{6-n}{n-3}$; $n \neq -6, \pm 3$

58. $\dfrac{2(7-y)}{y-5}$; $y \neq -7, \pm 5$

59. $\dfrac{2t(t-3)}{5}$; $t \neq -1, 2$

60. $\dfrac{(x+6)(x+2)}{20}$; $x \neq 0, \pm 2$

61. $\dfrac{5w^2}{w+1}$; $w \neq 0, \pm 1$ **62.** $\dfrac{4}{y}$; $y \neq 0, 3$

63. $\dfrac{(x-3)(x-4)}{(x-2)(x-9)}$; $x \neq -7, -5, 2, 4, 9$

64. $\dfrac{y}{3}$; $y \neq -\frac{1}{2}, 0, 3, 5$

65. $\dfrac{x}{5x+2}$; $x \neq -\frac{5}{3}, -\frac{1}{4}, 0, \pm \frac{2}{5}$

66. $\dfrac{(x+5)(x+3)}{(x+7)(2x+3)}$; $x \neq -7, -5, -\frac{3}{2}, \frac{1}{2}, 9$

67. $\dfrac{13}{5x}$; $x \neq 0$ **68.** $-\dfrac{16}{7x}$; $x \neq 0$

69. $\dfrac{5p^2 - 7p + 3}{(p-2)(p+1)}$; $p \neq -1, 2$

70. $\dfrac{-(5x+1)(x+8)}{(x+9)(x-2)}$; $x \neq -9, 2$

71. $\dfrac{4}{5x-1}$; $x \neq \frac{1}{5}$ **72.** $\dfrac{12}{2x-1}$; $x \neq \frac{1}{2}$

73. $\dfrac{3y^3 - 5y^2 - y + 1}{y^2 - 1}$; $y \neq \pm 1$

74. $\dfrac{1}{x-1}$; $x \neq 1$ **75.** $\dfrac{x^2 + 4x - 6}{x^2 - 4}$; $x \neq \pm 2$

76. $\dfrac{x^2 - 3}{(2-x)(x+2)}$; $x \neq \pm 2$

77. $\dfrac{x+4}{x+2}$; $x \neq \pm 2$

78. $\dfrac{9y - 17}{(y-3)(y+2)}$; $y \neq -2, 3$

79. $\dfrac{12 + 7b}{a^2 - b^2}$; $a \neq \pm b$ **80.** $\dfrac{y+4}{y(y+2)}$; $y \neq 0, \pm 2$

81. $\dfrac{7x - 20}{x-3}$; $x \neq 3$

82. $\dfrac{y^2 - 18y - 30}{(5y+6)(y-2)}$; $y \neq -\frac{6}{5}, 2$

83. $\dfrac{1-x}{x-2}$; $x \neq 0, 2$ **84.** $\dfrac{3 - 5y}{4y - 2}$; $y \neq 0, \frac{1}{2}$

85. $\dfrac{x}{3x - 1}$; $x \neq 0, \pm \frac{1}{3}$ **86.** $\dfrac{x+2}{9x - 5}$; $x \neq 0, \frac{5}{9}$

87. $\dfrac{x+1}{x-1}$; $x \neq 0, \pm 1$ **88.** $-y$; $y \neq -7, 0$

89. $\dfrac{x(x+1)}{(x-1)(x+2)}$; $x \neq -2, \pm 1$

90. $-\dfrac{6}{5}$; $x \neq \pm 1$ **91.** $A = \dfrac{\pi(1+i)^5}{(1+i)^5 - 1}$

92. \$948.10 **93.** $\dfrac{R_1 R_2}{R_2 + R_1}$ **94.** $f = \dfrac{pq}{q+p}$

95. Initially, $\dfrac{x^2 + 2x + 1}{x + 1}$ has the restriction $x \neq -1$.

$\dfrac{x^2 + 2x + 1}{x+1} = x + 1$

96. Cannot cancel the 1's. Must factor first.

$\dfrac{\cancel{x+1}}{(\cancel{x+1})(x+1)} = \dfrac{1}{x+1}$

97. false **98.** true **99.** false

100. false **101.** $\dfrac{(x+a)(x+d)}{(x+b)(x+c)}$; $x \neq -b, -c, -d$

102. $a^n + b^n$, $a \neq b$

103. yes

104. yes

105. a. $\dfrac{(x-1)(x+2)}{(x-2)(x+1)};\ x\neq -2,-1,2$

b.

c. agree as long as $x\neq -1,\pm 2$

106. a. $\dfrac{(x-4)(x+1)}{(x-3)(x+3)};\ x\neq -3,3,4$

b.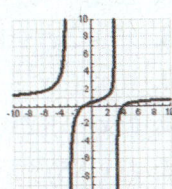

c. agree as long as $x\neq 4,\pm 3$

Section 0.6

1. 10   **2.** 11   **3.** −12

4. not real   **5.** −6   **6.** −5

7. 7   **8.** 3   **9.** 1

10. −1   **11.** 0   **12.** 0

13. not real   **14.** −1   **15.** −3

16. −4   **17.** 4   **18.** 16

19. −2   **20.** −3   **21.** −1

22. 1   **23.** 27   **24.** 9

25. $-4\sqrt{2}$ **26.** $-4\sqrt{5}$ **27.** $8\sqrt{5}$

28. $3\sqrt{7}$ **29.** $2\sqrt{6}$ **30.** $60\sqrt{2}$

31. $2\sqrt[3]{6}$ **32.** $2\sqrt[4]{2}$ **33.** $\sqrt{21}$

34. $\sqrt{10}$ **35.** $40|x|$ **36.** $96y^2$

37. $2|x|\sqrt{y}$ **38.** $4|x|\sqrt{xy}$ **39.** $-3x^2y^2\sqrt[3]{3y^2}$

40. $-2x^2y\sqrt[5]{y^3}$ **41.** $\dfrac{\sqrt{3}}{3}$ **42.** $\dfrac{\sqrt{10}}{5}$

43. $\dfrac{2\sqrt{11}}{33}$ **44.** $\dfrac{5\sqrt{2}}{6}$ **45.** $\dfrac{3+3\sqrt{5}}{-4}$

46. $-1+\sqrt{3}$ **47.** $-3-2\sqrt{2}$ **48.** $\dfrac{7-3\sqrt{5}}{2}$

49. $-3(\sqrt{2}+\sqrt{3})$ **50.** $\dfrac{5(\sqrt{5}-\sqrt{2})}{3}$

51. $\dfrac{2(3\sqrt{2}-2\sqrt{3})}{3}$ **52.** $\dfrac{7(3\sqrt{2}-2\sqrt{3})}{6}$

53. $\dfrac{5\sqrt{5}-2}{11}$ **54.** $3(3\sqrt{2}-4)$

55. $\dfrac{(\sqrt{7}+\sqrt{3})(\sqrt{2}+\sqrt{5})}{-3}$

56. $\dfrac{\sqrt{xy}+y}{x-y}$ **57.** x^3y^4 **58.** y^{11}

59. $\dfrac{1}{y^2}$ **60.** $y^{1/2}$ **61.** $x^{7/6}y^2$

62. $\dfrac{1}{y^{15}x^{64}}$ **63.** $\dfrac{x^{8/3}}{2}$ **64.** $\dfrac{x^{2/3}}{2}$

65. $x^{1/3}(x-2)(x+1)$ **66.** $4x^{1/4}(x+2)$

67. $7x^{3/7}(1-2x^{3/7}+3x)$ **68.** $7x^{-1/3}(1+10x^{4/3})$

69. 9 sec   **70.** 11 sec

71. 9.54 astronomical units   **72.** $2\pi\sqrt{2}$ sec

73. Forgot to square the 4.

74. Should have multiplied by the conjugate $5+\sqrt{11}$.

75. false   **76.** false   **77.** false

78. false   **79.** a^{mnk} **80.** a

81. $\frac{1}{2}$ **82.** 1   **83.** $\dfrac{a-2\sqrt{ab}+b}{(a-b)^2}$

84. $\dfrac{2a+b-2\sqrt{a(a+b)}}{b}$ **85.** 3.317

86. 1.913

87. a. $10\sqrt{2}-8\sqrt{3}$ **b.** 0.2857291632   **c.** yes

88. a. $\dfrac{3\sqrt{6}+4\sqrt{5}}{13}$ **b.** 1.25328778   **c.** yes

Section 0.7

1. $4i$ 2. $10i$ 3. $2i\sqrt{5}$ 4. $2\sqrt{6}\cdot i$

5. -4 6. -3 7. $8i$ 8. $3i\sqrt{3}$ 9. $3-10i$

10. $4-11i$ 11. $-10-12i$ 12. 12

13. $2-9i$ 14. $10-2i$ 15. $10-14i$

16. $-5+5i$ 17. $2-2i$ 18. $-3+2i$

19. $-1+2i$ 20. $-1+4i$ 21. $12-6i$

22. $28-24i$ 23. $96-60i$ 24. $-48-12i$

25. $-48+27i$ 26. $15-30i$ 27. $-102+30i$

28. $-96-36i$ 29. $5-i$ 30. $3+11i$

31. $13+41i$ 32. $-37-6i$ 33. $87+33i$

34. $-29-2i$ 35. $-6+48i$ 36. 25

37. $\frac{56}{9}-\frac{11}{18}i$ 38. $-\frac{3}{4}+\frac{1}{24}i$ 39. $37+49i$

40. $-5+12i$ 41. $4-7i;\ 65$ 42. $2-5i;\ 29$

43. $2+3i;\ 13$ 44. $5+3i;\ 34$ 45. $6-4i;\ 52$

46. $-2-7i;\ 53$ 47. $-2+6i;\ 40$

48. $-3+9i;\ 90$ 49. $-2i$ 50. $-3i$

51. $\frac{3}{10}+\frac{1}{10}i$ 52. $\frac{7}{25}+\frac{1}{25}i$ 53. $\frac{3}{13}-\frac{2}{13}i$

54. $\frac{4}{25}+\frac{3}{25}i$ 55. $\frac{14}{53}-\frac{4}{53}i$ 56. $\frac{8}{37}-\frac{48}{37}i$

57. $-i$ 58. $\frac{4}{5}-\frac{3}{5}i$ 59. $-\frac{9}{34}+\frac{19}{34}i$

60. $\frac{1}{2}+\frac{1}{2}i$ 61. $\frac{18}{53}-\frac{43}{53}i$ 62. $\frac{17}{30}+\frac{19}{30}i$

63. $\frac{66}{85}+\frac{43}{85}i$ 64. $\frac{115}{169}-\frac{62}{169}i$ 65. $-i$

66. $-i$ 67. 1 68. -1

69. $21-20i$ 70. $-16-30i$ 71. $-5+12i$

72. $-65-72i$ 73. $18+26i$ 74. $2+11i$

75. $-2-2i$ 76. $-44-117i$

77. $8-2i$ ohms   78. $\frac{116}{615}+\frac{22}{615}i$

79. Should have multiplied by the conjugate $4+i$ (not $4-i$).
$$\frac{2}{4-i}\cdot\frac{(4+i)}{(4+i)}=\frac{8+2i}{16-i^2}=\frac{8+2i}{17}=\frac{8}{17}+\frac{2}{17}i$$

80. $10-7i-12i^2=10-7i+12=22-7i$

81. true   82. true   83. true

84. false   85. $(x+i)^2(x-i)^2$

86. $(x+3i)^2(x-3i)^2$ 87. $41-38i$

88. $-352-936i$ 89. $\frac{2}{125}+\frac{11}{125}i$

90. $\frac{7}{625}-\frac{24}{625}i$

Review Exercises

1. a. 5.22 b. 5.21 2. a. 7.36 b. 7.36

3. 4 4. -65 5. -2 6. $9x-5y$ 7. $-\dfrac{x}{12}$

8. $\dfrac{11y}{30}$ 9. 9 10. $\dfrac{a}{2b}$ 11. $-8z^3$

12. $-64z^6$ 13. $\dfrac{9}{2x^2}$ 14. $\dfrac{xy^3}{16}$

15. 2.15×10^{-6} 16. $7{,}200{,}000{,}000$

17. $14z^2+3z-2$ 18. $26y^2-9y+9$

19. $45x^2-10x-15$ 20. $-2x^2+11x-4$

21. $15x^2y^2-20xy^3$

22. $2st^3-2s^2t^2+4s^2t^3$

23. $x^2+2x-63$ 24. $6x^2-x-2$

25. $4x^2-12x+9$ 26. $25x^2-49$

27. x^4+2x^2+1 28. x^4-2x^2+1

29. $2xy^2(7x-5y)$ 30. $10x^2(3x^2-2x+1)$

31. $(x+5)(2x-1)$ 32. $(3x+1)(2x-7)$

33. $(4x+5)(4x-5)$ 34. $(3x-5)^2$

35. $(x+5)(x^2-5x+25)$

36. $(1-2x)(1+2x+4x^2)$

37. $2x(x+5)(x-3)$ 38. $x(3x-1)(2x-1)$

39. $(x^2-2)(x+1)$ 40. $(x^2+3)(2x-1)$

41. $x\neq\pm3$ 42. x is any real number

43. $x+2;\ x\neq2$ 44. $1;\ x\neq5$

45. $\dfrac{t+3}{t+1};\ t\neq-1,2$ 46. $z-1;\ z\neq-1,0$

47. $\dfrac{(x+5)(x+2)}{(x+3)^2};x\neq-3,1,2$ 48. $\dfrac{(x-2)(x+2)}{x(x+3)};x\neq-3,-2,-1,0$

49. $\dfrac{2}{(x+1)(x+3)};\ x\neq-1,-3$ 50. $\dfrac{x^2+2x+2}{x(x+1)(x+2)};x\neq0,-1,-2$

51. $\dfrac{10x-25}{20x-59};\ x\neq3,\frac{59}{20}$ 52. $\dfrac{x+2}{3x^2-1};\ x\neq0,\pm\sqrt{\frac{1}{3}}$

53. $2\sqrt{5}$ 54. $4\sqrt{5}$ 55. $-5xy\sqrt[3]{x^2y}$

56. $2|xy|\sqrt[4]{2y}$ 57. $26\sqrt{5}$ 58. $-4\sqrt{3x}$

59. $-3-\sqrt{5}$ **60.** $12-x+\sqrt{x}$

61. $2+\sqrt{3}$ **62.** $\dfrac{3+\sqrt{x}}{9-x}$ **63.** $\dfrac{9x^{2/3}}{16}$

64. $4x^{13/6}$ **65.** $5^{\frac{1}{6}}$ **66.** $\dfrac{y^3}{x^8}$

67. $13i$ **68.** $4i\sqrt{2}$ **69.** $-i$
70. i **71.** $8-6i$ **72.** $-6+4i$
73. $14+4i$ **74.** $5+10i$
75. 12 **76.** $-29+11i$
77. $-33+56i$ **78.** $48-14i$ **79.** $\frac{2}{5}+\frac{1}{5}i$

80. $\frac{3}{10}-\frac{1}{10}i$ **81.** $\dfrac{38}{41}-\dfrac{27}{41}i$ **82.** $\frac{28}{13}-\frac{3}{13}i$

83. $-\frac{10}{3}i$ **84.** $-\frac{7}{2}i$ **85.** rational

86. irrational   **87.** 1.945×10^{-6}
88. 1.5625×10^{3}

89. The solid curve represents the graph of $y=(2x+3)^3$ and $y=8x^3+36x^2+54x+27$.

90. The solid curve represents the graph of both $y=(x-3)^2$ and $y=x^2-6x+9$.

91. All three graphs are different.

92. The solid curve represents the graph of $y=x^2-8x+16$ and $y=(x-4)^2$.

93. **a.** $\dfrac{x(x-4)}{(x+2)(x-2)}$ **b.**

c. $x\neq0,\pm2$

94. **a.** $\dfrac{x(x-3)}{x^2+9}$ **b.**

c. $x\neq0$

95. **a.** $2\sqrt{5}+2\sqrt{2}$ **b.** 7.30056308 **c.** yes
96. **a.** $2\sqrt{6}-\sqrt{13}$ **b.** 1.29342821 **c.** yes
97. $2868-6100i$ **98.** $\frac{7}{2500}+\frac{6}{625}i$

99. $\frac{7}{40000}-\frac{3}{5000}i$

Practice Test

1. 4 **2.** $3x^2\sqrt[3]{2}$ **3.** -97 **4.** -2

5. $2i|x|\sqrt{3}$ **6.** i **7.** $\dfrac{y^5z^{1/2}}{x^{7/2}}$ **8.** $4\sqrt{x}$

9. $-7\sqrt{2}$ **10.** $18+26\sqrt{3}$
11. $2y^2-12y+20$ **12.** $10x^2-x-21$
13. $(x-4)(x+4)$ **14.** $3(x+3)(x+2)$
15. $(2x+3y)^2$ **16.** $(x-1)^2(x+1)^2$
17. $(2x+1)(x-1)$ **18.** $(2y-1)(3y+1)$
19. $t(t+1)(2t-3)$ **20.** $x(2x+1)(x-3)$
21. $(x+4y)(x-3y)$
22. $\left(x-\sqrt{3}\right)\left(x+\sqrt{3}\right)\left(x^2+5\right)$
23. $3(3+x)\left(9-3x+x^2\right)$
24. $x(3-x)\left(9+3x+x^2\right)$
25. $\dfrac{5x-2}{x(x-1)};\ x\neq0,1$
26. $\dfrac{5x^2+21x+8}{(x-5)(x+5)(x-2)};\ x\neq2,\pm5$
27. $\dfrac{1}{(x+1)(x-1)};\ x\neq\pm1$

28. $\dfrac{(2x-3)(x-4)}{(x-15)}$; $x \neq -4, -\frac{3}{2}, 15$

29. $-\dfrac{1}{x+3}$; $x \neq \frac{5}{2}, \pm 3$　　**30.** $\dfrac{-7t}{t-1}$; $t \neq -\frac{1}{3}, 1$

31. $-8-26i$　**32.** $-\frac{3}{17}-\frac{46}{17}i$　**33.** $\dfrac{2-27\sqrt{3}}{59}$

34. 1.55×10^{-5}　**35.** $-\dfrac{1}{x(x+1)}$; $x \neq 0, \pm 1$

36. a. $\dfrac{x}{x-5}$　**b.**　　　　**c.** $x \neq 0, \pm 5$

37. 2.330

CHAPTER 1

Section 1.1

1. $x = 7$　　**2.** $t = 8$　　**3.** $n = 15$

4. $y = 9$　　**5.** $x = -8$　　**6.** $t = 10$

7. $n = 15$　　**8.** $p = 18$　　**9.** $x = 4$

10. $p = 1$　　**11.** $m = 2$　　**12.** $x = \frac{1}{2}$

13. $t = \frac{7}{5}$　　**14.** $x = -1$　　**15.** $x = -10$

16. $x = 5$　　**17.** $n = 2$　　**18.** $c = 4$

19. $x = 12$　　**20.** $y = -\frac{1}{2}$　　**21.** $t = -\frac{15}{2}$

22. $n = -\frac{10}{7}$　　**23.** $x = -1$　　**24.** $y = -4$

25. $p = -\frac{9}{2}$　　**26.** $z = 3$　　**27.** $x = \frac{1}{4}$

28. $x = \frac{3}{2}$　　**29.** $x = -\frac{3}{2}$　　**30.** $x = -6$

31. $a = -8$　　**32.** $y = -2$　　**33.** $x = -15$

34. $m = 2$　　**35.** $c = -\frac{35}{13}$　　**36.** $y = 2$

37. $m = \frac{60}{11}$　　**38.** $z = 72$　　**39.** $x = 36$

40. $a = 198$　　**41.** $p = 8$　　**42.** $x = 5$

43. $y = -2$　　**44.** $m = 1$　　**45.** $p = 2$

46. $c = -1$　　**47.** no solution

48. $x = \frac{33}{2}$　　**49.** $y \neq 0, y = \frac{3}{10}$

50. $x \neq 0, x = -\frac{1}{3}$　　**51.** $x \neq 0, x = \frac{1}{2}$

52. $t \neq 0, t = -\frac{1}{4}$　　**53.** $a \neq 0, a = \frac{1}{6}$

54. $x \neq 0, x = \frac{3}{4}$　　**55.** $x \neq 2$, no solution

56. $n \neq 5$, no solution　**57.** $p \neq 1$, no solution

58. $t \neq -2$, no solution　**59.** $x \neq -2, x = -10$

60. $y \neq \frac{1}{2}, y = -9$　　**61.** $n \neq -1, 0$, no solution

62. $x \neq 0, 1$, no solution

63. $a \neq 0, -3$, no solution

64. $c \neq 0, 2$, no solution　　**65.** $n \neq 1, n = \frac{53}{11}$

66. $m \neq 0, 2$, no solution

67. $x \neq -\frac{1}{5}, \frac{1}{2}, x = -3$　**68.** $n \neq \frac{1}{4}, \frac{5}{2}, n = -\frac{13}{2}$

69. $t \neq 1$, no solution　**70.** $x \neq 2$, no solution

71. $C = \frac{5}{9}F - \frac{160}{9}$　　**72.** $W = \dfrac{P-2L}{2}$

73. 84 min　　**74.** 600 miles　　**75.** 17 min

76. 145 subscriptions

77. a. $C = 15,000 + 2500x$　**b.** 2200 days

78. a. $R = 5000 + 0.75x$　**b.** 125,000 minutes

79. 24 ml of the 125 mg/5 ml suspension of amoxicillin

80. 30 ml of the 100 mg/5 ml suspension of carbamazepine

81. $\lambda \neq 0$　　**82.** $d_0 = \dfrac{d_i f}{d_i - f}$

83. Should have subtracted $4x$ and added 7 to both sides; $x = 5$.

84. Forgot to distribute the negative sign through the parentheses.

85. Cannot cross multiply- must multiply by LCD first; $p = \frac{6}{5}$.

86. Should have eliminated $x = 0$, $x = 1$ from the domain first.

87. false　　**88.** false　　**89.** true

90. false　　**91.** $x = \dfrac{c-b}{a}$　**92.** $x = \dfrac{a-b}{c}$

93. $x = \dfrac{ba}{c}$, $x \neq \pm a$　**94.** $y \geq 0$, $y \neq -a, a, 1$

95. no solution　　**96.** $t = -1$

97. $x = \dfrac{by}{a-y-cy}$, $x \neq 0, -\dfrac{b}{c+1}$　**98.** $a = 1$

99. $x = 2$　　**100.** $x = 3$　**101.** all real numbers

102. all real numbers   **103.** no solution
104. no solution   **105.** $x = 5426$ **106.** $x = 7.95$

Section 1.2

1. $242.17   **2.** 30% markdown

3. $13.76   **4.** $8000

5. $147,058.82 ($22,058.82 saved)

6. $63.20   **7.** 12 miles

8. 60 miles   **9.** 9 hours of sleep

10. Breakfast: 300; Lunch: 400; Dinner: 600

11. 270 units   **12.** 146   **13.** 24

14. 2   **15.** 8, 10   **16.** 1, 2, 3   **17.** 20 inches

18. width = 3 feet, length = 7 feet

19. width = 30 yards, length = 100 yards

20. width = 3 inches, length = 11 inches

21. $r_1 = 3$ feet, $r_2 = 6$ feet   **22.** 1

23. 300 feet   **24.** 352 feet

25. 5.25 feet   **26.** 46.8 inches

27. $20,000 at 4%, $100,000 at 7%

28. $6000 at 10%, $7000 at 14%

29. $3000 at 10%, $5500 at 2%, $5500 at 40%

30. Money market: 1.5%, Stock market: 4.5%

31. 6 trees, 27 shrubs

32. 1.9 lbs of turkey, 1.3 lbs of cheese

33. 70 ml of 5% HCl, 30 ml of 15% HCl

34. 3 gallons   **35.** $3\frac{1}{3}$ gallons

36. On campus (42.5%) grant: $921,818
    Off campus (26%) grant: $248,182

37. No caramels, 1.25 lb of gummy bears

38. 0.75 lbs Jamaican Blue Mountain, 1.25 lbs
    of regular coffee beans

39. 9 minutes   **40.** 2.4 seconds

41. $3.07 per gallon   **42.** $498

43. 233 ml   **44.** 33 ml   **45.** 2.3 mph

46. 30 mph   **47.** walker: 4 mph, jogger: 6 mph

48. S route = 2700 miles
    N route = 3000 miles

49. bicyclist: 6 minutes, walker: 18 minutes

50. 75 minutes   **51.** 22.5 hours

52. 1.5 hours   **53.** 2.4 hours   **54.** 12 minutes

55. 330 hertz, 396 hertz   **56.** 264, 330

57. 77.5, 92.5

58. 85   **59.** 2 field goals, 6 touchdowns

60. 45 yard line   **61.** 3.5 feet from the center

62. 5 feet from the center

63. Fulcrum is 0.4 units from Maria and
    0.6 units from Max

64. Fulcrum is 0.56 units from Maria and
    0.44 units from Max/Martin.

65. 7.5 cm in front of lens

66. 2.67 cm behind lens

67. Object distance = 6 cm
    image distance = 3 cm

68. 24 cm   **69.** $\dfrac{P - 2l}{2} = w$ **70.** $\dfrac{P - 2w}{2} = l$

71. $\dfrac{2A}{b} = h$ **72.** $\dfrac{C}{2\pi} = r$ **73.** $\dfrac{A}{l} = w$

74. $\dfrac{d}{r} = t$ **75.** $\dfrac{V}{lw} = h$ **76.** $\dfrac{V}{\pi r^2} = h$

77. Janine's average speed is 58 mph, Tricia's
    average speed is 70 mph

78. Rick's average speed is 66 mph, Mike's
    average speed is 74 mph

79. $191,983.35   **80.** $144,397

81. Option B: better for 5 or few plays/month
    Option A: better for 6 or more plays/month

82. Option A: better for 285 min or less/month
    Option B: better for more than 285
    min/month

Section 1.3

1. $x = 3$ or $x = 2$ **2.** $v = -6$ or $v = -1$

3. $p = 5$ or $p = 3$ **4.** $u = 6$ or $u = -4$

5. $x = -4$ or $x = 3$ **6.** $x = \frac{3}{2}$ or $x = 4$

7. $x = -\frac{1}{4}$ **8.** $x = \frac{2}{3}$ or $x = -4$

9. $y = \frac{1}{3}$ **10.** $x = \frac{1}{2}$ **11.** $y = 0$ or $y = 2$

12. $A = 0$ or $A = -4$ **13.** $p = \frac{2}{3}$

14. $u = \frac{5}{2}$ **15.** $x = -3$ or $x = 3$

16. $v = \frac{5}{4}$ or $v = -\frac{5}{4}$ **17.** $x = -6$ or $x = 2$

18. $t = -4$ or $t = 4$ **19.** $p = -5$ or $p = 5$

20. $y = -3$ or $y = 3$ **21.** $x = -2$ or $x = 2$

22. $v = -2$ or $v = 2$ **23.** $p = \pm 2\sqrt{2}$

24. $y = \pm 6\sqrt{2}$ **25.** $x = \pm 3i$ **26.** $v = \pm 4i$

27. $x = -3, 9$ **28.** $x = -4, 6$ **29.** $x = \dfrac{-3 \pm 2i}{2}$

30. $x = \frac{1}{4} \pm i$ **31.** $x = \dfrac{2 \pm 3\sqrt{3}}{5}$

32. $x = \dfrac{-8 \pm 2\sqrt{3}}{3}$ **33.** $x = -2, 4$

34. $x = 1 \pm 3i$ **35.** $x^2 + 6x + \underline{9}$

36. $x^2 - 8x + \underline{16}$ **37.** $x^2 - 12x + \underline{36}$

38. $x^2 + 20x + \underline{100}$ **39.** $x^2 - \frac{1}{2}x + \underline{\frac{1}{16}}$

40. $x^2 - \frac{1}{3}x + \underline{\frac{1}{36}}$ **41.** $x^2 + \frac{2}{5}x + \underline{\frac{1}{25}}$

42. $x^2 + \frac{4}{5}x + \underline{\frac{4}{25}}$ **43.** $x^2 - 2.4 + \underline{1.44}$

44. $x^2 + 1.6x + \underline{0.64}$ **45.** $x = -3, 1$

46. $y = -4 \pm 3\sqrt{2}$ **47.** $t = 1, 5$

48. $x = -7, -3$ **49.** $y = 1, 3$

50. $x = 3, 4$ **51.** $p = \dfrac{-4 \pm \sqrt{10}}{2}$

52. $x = 1 \pm \dfrac{\sqrt{2}}{2} i$ **53.** $x = \frac{1}{2}, 3$

54. $x = \dfrac{5 \pm \sqrt{145}}{6}$ **55.** $x = \dfrac{4 \pm 3\sqrt{2}}{2}$

56. $t = -1 \pm \dfrac{\sqrt{6}}{2} i$ **57.** $t = \dfrac{-3 \pm \sqrt{13}}{2}$

58. $t = -1 \pm \sqrt{2}$ **59.** $s = \dfrac{-1 \pm \sqrt{3}\, i}{2}$

60. $s = -2, -\frac{1}{2}$ **61.** $x = \dfrac{3 \pm \sqrt{57}}{6}$

62. $x = \dfrac{1 \pm \sqrt{29}}{4}$ **63.** $x = 1 \pm 4i$

64. $m = -\dfrac{7}{8} \pm \dfrac{\sqrt{79}}{8} i$ **65.** $x = \dfrac{-7 \pm \sqrt{109}}{10}$

66. $x = -\dfrac{5}{6} \pm \dfrac{\sqrt{107}}{6} i$ **67.** $x = \dfrac{-4 \pm \sqrt{34}}{3}$

68. $x = \dfrac{4 \pm 2\sqrt{7}}{3}$ **69.** 1 real solution

70. 1 real solution   **71.** 2 real solutions

72. 2 real solutions   **73.** 2 complex solutions

74. 2 complex solutions

75. $v = -2, 10$ **76.** $v = 4 \pm 2i$

77. $t = -6, 1$ **78.** $t = -3, -2$

79. $x = -7, 1$ **80.** $x = -3 \pm 4i$

81. $p = 4 \pm 2\sqrt{3}$ **82.** $u = 3 \pm 4i$

83. $w = \dfrac{-1 \pm \sqrt{167}\, i}{8}$ **84.** $w = -\frac{7}{4}, \frac{3}{2}$

85. $p = \dfrac{9 \pm \sqrt{69}}{6}$ **86.** $p = \dfrac{9 \pm \sqrt{93}}{6}$

87. $t = \dfrac{10 \pm \sqrt{130}}{10}$ **88.** $x = \dfrac{-10 \pm 2\sqrt{70}}{15}$

89. $x = 3$ or $x = 4$ **90.** $x = -5$ or $x = 2$

91. $x = -\frac{3}{4}$ or $x = 2$ **92.** $y = -\frac{5}{2}$ or $y = 1$

93. $x = -0.3, 0.4$ **94.** $y = 0.2, 0.3$

95. $t = 8$ (August 2003) and $t = 12$ (Dec. 2003)

96. $t = 1$ (Nov. 2003) and $t = 5$ (March 2004)

97. 31,000 units   **98.** 35,000 units

99. \$1 per bottle   **100.** \$3 per bottle

101. 3 days   **102.** 6 days

103. a. 55.25 square inches

 b. $4x^2 + 30x + 55.25$

 c. $4x^2 + 30x$ represents the increase in useable area of the paper.

 d. $x \approx 0.2$ inches

104. a. 58.5 square inches   **b.** $4x^2 + 31x + 58.5$

 c. $4x^2 + 31x$ represents the increase in useable area of the paper.

 d. $x \approx 0.2$ inches

105. 20 inches   **106.** 37 inches   **107.** 17, 18

108. 11, 13   **109.** Length: 15 ft, width: 9 ft

110. Length: 9 m, width: $\frac{7}{2}$ m

111. Base: 6, height: 20   **112.** 10 yards

113. Impact with ground in 2.5 seconds

114. 2.3 seconds **115.** 21.2 feet **116.** 127 feet
117. 5 ft × 5 ft **118.** 4 ft × 8 ft **119.** 2.3 feet
120. 1 foot **121.** 10 days **122.** 2.4 hours
123. The problem is factored incorrectly. The correction would be $t = -1, 6$.
124. When taking the square root of both sides, forgot the $\pm$ on the right side. The correction would be $y = -1, 4$.
125. When taking the square root of both sides, the i is missing from the right side. The correction would be $a = \pm\frac{3}{4}i$.
126. In completing the square we should add 2 (not 1) to the right side. Solution is $1 \pm \sqrt{\frac{5}{2}}$.
127. false **128.** true
129. true **130.** true
131. $x^2 - 2ax + a^2 = 0$ **132.** $x^2 + b^2 = 0$
133. $x^2 - 7x + 10 = 0$ **134.** $x^2 + 3x = 0$
135. $t = \pm\sqrt{\dfrac{2s}{g}}$ **136.** $r = -1 \pm \sqrt{\dfrac{A}{P}}$
137. $c = \pm\sqrt{a^2 + b^2}$ **138.** $I = \dfrac{E \pm \sqrt{E^2 - 4RP}}{2R}$
139. $x = 0, \pm 2$ **140.** $x = 0, \frac{1}{2}$
141. $x = -1, \pm 2$ **142.** $x = \pm 1, -2$
143. $\dfrac{-b}{2a} + \dfrac{\sqrt{b^2 - 4ac}}{2a} - \dfrac{b}{2a} - \dfrac{\sqrt{b^2 - 4ac}}{2a}$

$= \dfrac{-2b}{2a} = \dfrac{-b}{a}$

144. $\dfrac{\left(-b + \sqrt{b^2 - 4ac}\right)\left(-b - \sqrt{b^2 - 4ac}\right)}{2a \cdot 2a}$

$= \dfrac{b^2 - (b^2 - 4ac)}{4a^2} = \dfrac{4ac}{4a^2} = \dfrac{c}{a}$

145. $x^2 - 6x + 4 = 0$ **146.** $x^2 - 4x + 5 = 0$
147. 250 mph **148.** 2 mph
149. $ax^2 - bx + c = 0$ **150.** $cx^2 + bx + a = 0$
151. Small jet: 300 mph, 757: 400 mph
152. Small boat: 30 mph, large boat: 40 mph

153. $x = -1, 2$

154. $x = 1 \pm i$

155. a. $x = -2, 4$ **b.** $b = -3: x = 1 \pm \sqrt{2}\,i$

$b = -1, \ x = 1$ $b = 0, \ x = 0, 2$

$b = 5, \ x = 1 \pm \sqrt{6}$

156. a. $x = 2, -4$ **b.** $b = -3: x = -1 \pm \sqrt{2}\,i$

$b = -1: \ x = -1$ $b = 0: \ x = 0, -2$

$b = 5: \ x = -1 \pm \sqrt{6}$

Section 1.4

1. $t = 9$ **2.** $t = 8$ **3.** $p = 8$

4. $p = -100$ **5.** no solution **6.** no solution

7. $x = 5$ **8.** $x = 9$ **9.** $y = -\frac{1}{2}$

10. $x = 13$ **11.** $x = 4$ **12.** $x = 7$

13. $y = 0, 25$ **14.** $y = 0, 16$ **15.** $s = 3, 6$

16. $s = -1$ **17.** $x = -3, -1$

18. $x = 2$ **19.** $x = 0$ **20.** $x = 2$

21. $x = 1$ and $x = 5$ **22.** $x = -2$ and $x = 2$

23. $x = 7$ **24.** $x = 8$ **25.** $x = -3$ and $x = -\frac{15}{4}$

26. $x = 3$ **27.** $x = \frac{5}{2}$ **28.** $x = 3$

29. no solution   **30.** $x = 4$ **31.** $x = 1$

32. $x = -2$ **33.** $x = 4$ and $x = -8$

34. $x = 1$ and $x = 5$ **35.** $x = 1, 5$

36. $x = -1$ **37.** $x = 7$ **38.** $x = 11$

39. $x = 4$ **40.** $x = 1$ **41.** $x = 0, x = -8$

42. $x = 0, x = 16$ **43.** $x = \pm 1, x = \pm \sqrt{2}$

44. $x = \pm 2$ **45.** $x = \dfrac{\pm \sqrt{6}\, i}{2}, x = \pm \sqrt{2}\, i$

46. $x = \pm 1, x = \pm i, \ x = \pm 2, x = \pm 2i$

47. $x = -\frac{5}{2}, \ x = -1$ **48.** $x = \pm 1$

49. $t = \frac{5}{4}, \ t = 3$ **50.** $y = -\frac{1}{2}, \ y = 5$

51. $x = \pm 1, \pm i, \ x = \pm \frac{1}{2}, \pm \frac{1}{2}i$

52. $u = \frac{2}{3}, \ u = -\frac{1}{4}$ **53.** $y = -\frac{3}{4}, \ y = 1$

54. $a = -5, \ a = -\frac{1}{2}$ **55.** $z = 1$

56. $x = \frac{1}{16}$ **57.** $x = 5$

58. $x = 6$ **59.** $x = -9$ or $x = 7$

60. $x = -20$ or $x = 34$ **61.** $t = -27, \ t = 8$

62. $t = \frac{1}{27}, \ t = -\frac{1}{8}$ **63.** $x = -\frac{4}{3}, x = 0$

64. $x = -\frac{3}{2}$ **65.** $x = \frac{3}{8}, x = \frac{2}{3}$

66. $x = -\frac{7}{4}, \ x = 0$ **67.** $u = \pm 8, \ u = \pm 1$

68. $u = \pm 8i, \ u = \pm i$ **69.** $t = \sqrt{3}$

70. $u = i$ **71.** $x = 0, -3, 4$

72. $y = 0, 4, \frac{3}{2}$ **73.** $p = 0, \pm \frac{3}{2}$

74. $x = 0, \pm \frac{2}{5}$ **75.** $u = 0, \pm 2, \pm 2i$

76. $t = 0, \pm 3, \pm 3i$ **77.** $x = \pm 3, 5$

78. $p = \pm 2, \frac{3}{2}$ **79.** $y = -2, 5, 7$

80. $v = -8, -3, 5$

81. $x = 0, 3$ ($x = -1$ does not check)

82. $u = -5, 0, 4$ **83.** $t = \pm 5$

84. $x = \pm \frac{3}{2}$ **85.** $y = 2, 3$

86. $p = -\frac{3}{4}, 2$ **87.** January and September

88. Never (the demand for the product is never 4M units)

89. 162 cm   **90.** 90 kg   **91.** $a = 71$ years old

92. 79.8 years old

93. $t = 4\sqrt{3} \approx 7$ months (Oct. 2004)

94. In the year 2255   **95.** 132 feet

96. 144 feet   **97.** 25 cm

98. 10 inches   **99.** 80% of the speed of light

100. about 98.6% of the speed of light

101. no solution   **102.** $x = 2$

103. Forgot about the substitution $u = x^{1/3}$. The correct answer is $x = -64, 125$.

104. $x = \pm i$ **105.** true   **106.** false

107. false   **108.** false   **109.** $[0, \infty)$

110. $(-\infty, 0]$ **111.** $x = 0, \ x = -\frac{2}{3}, \ x = -1, \ x = \frac{1}{3}$

112. $x = 16, \ x = 1, \ x = 0$ **113.** $x = -2$

114. $x = 4$

115. $x = \frac{313}{64} \cong 4.891$ **116.** $x \cong 0.62$

 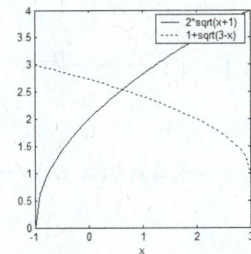

117. no solution   **118.** $x \cong 22.2$

119. $x = 81$ $y1 = x^{1/2}$, $y2 = -4x^{1/4} + 21$

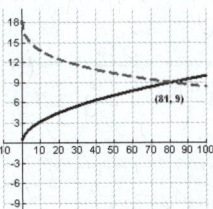

120. $x = -\frac{3}{5}$, $x = \frac{1}{2}$ $y1 = x^{-1}$, $y2 = 3x^{-2} - 10$

121. no real solution $y1 = x^{-2}$, $y2 = 3x^{-1} - 10$

Section 1.5

1. $[3, \infty)$

... 0 1 2 3 4 5 6 ...

2. $(-\infty, -2)$

... −4 −3 −2 −1 0 1 2 ...

3. $(-\infty, -5]$

... −7 −6 −5 −4 −3 −2 −1 0 ...

4. $(-7, \infty)$

... −7 −6 −5 −4 −3 −2 −1 0 1 ...

5. $[-2, 3)$

... −3 −2 −1 0 1 2 3 4 ...

6. $[-4, -1]$

... −5 −4 −3 −2 −1 0 ...

7. $(-3, 5]$

... −3 −2 −1 0 1 2 3 4 5 ...

8. $(0, 6)$

... −2 −1 0 1 2 3 4 5 6 7 8 ...

9. $[0, 0]$

... −3 −2 −1 0 1 2 3 ...

10. $[-7]$

... −9 −8 −7 −6 −5 −4 ...

11. $[4, 6]$

... 3 4 5 6 7 ...

12. $(-3, 2]$

... −3 −2 −1 0 1 2 ...

13. $[-8, -6]$

... −9 −8 −7 −6 −5 ...

14. $(-\infty, 2)$

... 0 1 2 3 4 ...

15. $\varnothing$

... −3 −2 −1 0 1 2 3 ...

16. $\varnothing$ ...−3−2 −1 0 1 2 3 ...

17. $\{x \mid 0 \le x < 2\}$ **18.** $\{x \mid 0 < x \le 3\}$

19. $\{x \mid -7 < x < -2\}$ **20.** $\{x \mid -3 \le x \le 2\}$

21. $\{x \mid x \le 6\}$ **22.** $\{x \mid x > 5\}$

23. $\{x \mid -\infty < x < \infty\}$ **24.** $\{x \mid 4 \le x \le 4\}$

25. $-3 < x \le 7$ $(-3,7]$

26. $-\frac{1}{2} \le x < \frac{7}{8}$ $\left[-\frac{1}{2}, \frac{7}{8}\right)$

27. $3 \le x < 5$ $[3,5)$ **28.** $4 < x \le 8$ $(4,8]$

29. $-2 \le x$ $[-2,\infty)$ **30.** $x < -3$ $(-\infty,-3)$

31. $-\infty < x < 8$ $(-\infty,8)$

32. $-2 \le x < \infty$ $[-2,\infty)$

33. $(-5,3)$ ...−5−4 −3 −2 −1 0 1 2 3 ...

34. $[-5,7)$ . ..−5−4 −3 −2 −1 0 ...7 8 9 ...

35. $[-6,5)$ ...−7−6 −5 ... 0 1 2 3 4 5...

36. $[-6,1)$ ...−7−6 −5 −4 −3−2 −1 0 1 2 ...

37. $[-1,1]$ ...−2−1 0 1 2 3 ...

38. $(-\infty,-5)$ ... −7 −6 −5 −3 −2 −1 0...

39. $[1,4)$ ... 0 1 2 3 4 ...

40. $[-3,\infty)$ ...−3−2 −1 0 1 2 3 4 ...

41. $[-1,2)$ ...−2−1 0 1 2 3...

42. $[-2,5)$ ...−3−2 −1 0 1 2 3 4 5...

43. $(-\infty,4) \cup (4,\infty)$ ...2 3 4 5 6 7 ...

44. $(-\infty,\infty)$ ...−3−2 −1 0 1 2 3 ...

45. $(-\infty,-3] \cup [3,\infty)$ ...−3−2 −1 0 1 2 3 ...

46. $(-2,1]$ ...−3−2 −1 0 1 2 3 4 ...

47. $(-3,2]$ ...−4−3 −2 −1 0 1 2 3 4 ...

48. $(-\infty,\infty)$ ...−3−2 −1 0 1 2 3 ...

49. $\varnothing$ ...−3−2 −1 0 1 2 3 ...

50. $\varnothing$ ...−3−2 −1 0 1 2 3 ...

51. $(-\infty,2) \cup [3,5)$ **52.** $(-\infty,-5) \cup [0,2]$

53. $(-\infty,-4] \cup (2,5]$ **54.** $[-12,-5) \cup (-2,\infty)$

55. $[-4,-2] \cup (3,7]$ **56.** $(-\infty,-2) \cup (2,\infty)$

57. $(-6,-3] \cup [0,4)$ **58.** $(-\infty,-5] \cup [-1,\infty)$

59. $(-\infty,10)$ **60.** $(5,\infty)$ **61.** $(-\infty,2]$

62. $[-5,\infty)$ **63.** $(-\infty,-2]$ **64.** $(-3,\infty)$

65. $[-2,\infty)$ **66.** $(-\infty,7)$ **67.** $(-\infty,-0.5)$

68. $(-\infty,3)$ **69.** $(-3,\infty)$

70. $[22,\infty)$ **71.** $(-\infty,6)$ **72.** $\left(-\frac{7}{3},\infty\right)$

73. $(-\infty,-8]$ **74.** $[-52,\infty)$ **75.** $(-\infty,-7]$

76. $(-8,\infty)$ **77.** $(-\infty,1)$ **78.** $(-\infty,13)$

79. $(-5,2)$ **80.** $(-5,6)$ **81.** $[-6,2)$

82. $(-2,3]$ **83.** $[-8,4)$ **84.** $[-3,-1]$

85. $(-6,6)$ **86.** $(12,20)$ **87.** $\left[\frac{1}{2},\frac{5}{4}\right]$

88. $\left[\frac{6}{5},6\right)$ **89.** $[-4.5,0.5]$

90. $(-2,3)$ **91.** $128 \le w \le 164$

92. $114 \le w \le 150$ **93.** More than 50 dresses

94. \$4000 to \$18,400 **95.** 285,700 units

96. 142,900 units

97. between 33% and 71% intensities

98. between 35% and 75% intensities

99. Least: 878 minutes, most: 1013 minutes

100. Least: 1080 minutes, most: 1215 minutes

101. 92 **102.** between 49% and 99%

103. $\$21,537.69 < \text{invoice price} < \$24,346.96$

104. $\$32,768.46 < \text{invoice price} < \$37,042.61$

105. $0.9\,r_T \le r_R \le 1.1\,r_T$ **106.** $S > 2.2\,N$

107. $0.85L \le B \le 0.95L$

108. $0.95h_t \le h_m \le 1.05h_t$

109. 4 times **110.** 5 times

111. $4386.25 \le T \le 15,698.75$

112. $15,698.75 \le T \le 39,148.75$

113. Mixed up parenthesis and brackets $[-1,4)$ and the graph is not correct.

114. Performed union instead of intersection. $(3,4)$

115. Forgot to flip the sign when dividing by -3. Answer should be $[2,\infty)$.

116. $x \ge -2$ corresponds to $[-2,\infty)$

117. true **118.** false **119.** a and b

120. c and d **121.** a and b **122.** c and d

123. c **124.** a and b **125.** $(-\infty,0]$

126. $(0,\infty)$ **127.** no solution

128. all real numbers

129. a. $x > 0.83582$ (rounded) **b.**

c. yes

130. a. $x < 1.36957$ (rounded) **b.**

c. yes

131. a. $(-2,5)$ **b.**

c. yes

132. a. $(-3,2]$

b. $y1 = x - 2,\ y2 = 3x + 4,\ y3 = 2x + 6$

c. agree

133. a. $(-\infty,\infty)$

b.

c. yes

134. a. $\left(\frac{24}{7},\infty\right)$

b. $y1 = \frac{1}{2}x - 3,\ y2 = -\frac{2}{3}x + 1$

c. agree

Section 1.6

1. $(-\infty, -2] \cup [5, \infty)$ **2.** $(-3, 1)$

3. $[-1, 6]$ **4.** $(-\infty, -4) \cup (10, \infty)$

5. $(-3, -1)$ **6.** $(-\infty, -3] \cup [5, \infty)$

7. $\left[-1, \frac{3}{2}\right]$ **8.** $(-\infty, -2] \cup \left[\frac{1}{3}, \infty\right)$

9. $\left(\frac{1}{3}, \frac{1}{2}\right)$ **10.** $\left(-\frac{1}{4}, \frac{10}{3}\right)$

11. $\left(-\infty, -\frac{1}{2}\right] \cup [3, \infty)$ **12.** $[-6, -2]$

13. $\left(-\infty, -1-\sqrt{5}\right] \cup \left[-1+\sqrt{5}, \infty\right)$

14. $\left[\dfrac{-3-\sqrt{13}}{2}, \dfrac{-3+\sqrt{13}}{2}\right]$

15. $\left(2-\sqrt{10}, 2+\sqrt{10}\right)$

16. $\left(-\infty, 1-\sqrt{6}\right) \cup \left(1+\sqrt{6}, \infty\right)$

17. $(-\infty, 0] \cup [3, \infty)$ **18.** $[-4, 0]$ **19.** $[0, 2]$

20. $(-\infty, -3] \cup [0, \infty)$ **21.** $(-\infty, -3) \cup (3, \infty)$

22. $(-\infty, -4] \cup [4, \infty)$ **23.** $(-9, 9)$ **24.** $[-7, 7]$

25. $(-\infty, \infty)$ **26.** $(-\infty, \infty)$

27. no real solution **28.** no real solution

29. $(0, \infty)$ **30.** $(-\infty, 0)$ **31.** $(-\infty, -3) \cup (0, \infty)$

32. $(-\infty, 0] \cup (2, \infty)$ **33.** $(-\infty, -3] \cup (4, \infty)$

34. $\left(\frac{5}{2}, 6\right)$ **35.** $(-\infty, -2) \cup [-1, 2)$

36. $[-5, -2) \cup (2, \infty)$ **37.** $(-5, 3] \cup (5, \infty)$

38. $(-3, 1] \cup (3, \infty)$ **39.** $\left(-\frac{3}{2}, 1\right)$

40. $(-\infty, -3] \cup [6, \infty)$ **41.** $(-\infty, -5] \cup (-2, 0]$

42. $(-\infty, 0] \cup (4, \infty)$ **43.** $(-2, 2)$

44. $(-\infty, -10) \cup [-4, 5] \cup (10, \infty)$

45. no real solution **46.** $x = 0$

47. $(-\infty, \infty)$ **48.** $(-\infty, \infty)$

49. $[-3, 3) \cup (3, \infty)$ **50.** $(-\infty, -1) \cup (-1, -1]$

51. $(-3, -1] \cup (3, \infty)$ **52.** $(-\infty, -2) \cup [0, 2)$

53. $(-\infty, -4) \cup (2, 5]$ **54.** $[-3, 1) \cup (5, \infty)$

55. $(-6, -4) \cup (4, 8)$

56. $\left(-\infty, -2\sqrt{3}\right] \cup (-3, 3) \cup \left[2\sqrt{3}, \infty\right)$

57. $(-\infty, -2) \cup (2, \infty)$ **58.** $\left(-1, \frac{3}{2}\right)$

59. Between 30 and 100 orders

60. Less than 40 or more than 90 orders

61. For years 3-5, the car is worth more than you owe. In the first 3 years you owe more than the car is worth.

62. For between 2 and 4 years, the car is worth more than you owe. In the first 2 years and after the 4th year you owe more than the car is worth.

63. 75 seconds **64.** 37.5 seconds

65. Between 20 and 30 feet

66. From Sept 2003 – Nov 2003 the stock value was at least $36.

67. A price increase less than $1 per bottle or greater than $20 per bottle

68. A price increase between $2 per bottle and $19 per bottle

69. $3342 to $3461 per acre

70. $990 to $1010 per acre

71. Cannot divide by x. $(-\infty, 0) \cup (3, \infty)$

72. Cannot take square root. $(-5, 5)$

73. Should have considered $x = -2$ a critical point.

74. Cannot cross-multiply. $(-3, 0)$

75. false **76.** false

77. infinitely many solutions or no real solution

78. infinitely many solutions or no real solution

79. $(-\infty, \infty)$ **80.** $(-\infty, -b) \cup (-b, b)$

81. $(-\infty, \infty)$ **82.** no solution

83. $(-\infty, \infty)$ **84.** $(-6.65, 0.4)$

85. $(-0.8960, 1.6233)$ **86.** $(-6, 6.33)$

87. $(-2, 0)$ **88.** $(-\infty, -3] \cup [2.5, \infty)$

89. $\left(\frac{5}{3}, 5\right)$ **90.** $(-\infty, 1) \cup (4, \infty)$

Section 1.7

1. $x = -3$ or $x = 3$ **2.** $x = -2$ or $x = 2$
3. no solution   **4.** no solution
5. $t = -5$ or $t = -1$ **6.** $t = 1$ or $t = 5$
7. $p = 10$ or $p = 4$ **8.** $p = -4$ or $p = -10$
9. $y = 5$ or $y = 3$ **10.** $y = 13$ or $y = -9$
11. $x = -3$ or $x = 3$ **12.** $x = -10$ or $x = 10$
13. $x = -8$ or $x = 1$ **14.** $x = 6$ or $x = -1$
15. $t = 4$ or $t = 2$ **16.** $t = 0$ or $t = -1$
17. $x = 8$ or $x = -1$ **18.** $y = -2$ or $y = 6$
19. $y = 0$ or $y = \frac{2}{3}$ **20.** $x = 3$ or $x = 7$
21. $x = \frac{80}{21}$ or $x = \frac{2}{3}$ **22.** $x = -\frac{61}{52}$ or $x = -\frac{1}{4}$
23. $x = -\frac{23}{14}$ or $x = \frac{47}{14}$
24. $x = -\frac{13}{8}$ or $x = -\frac{11}{8}$
25. $x = 13$ or $x = -3$ **26.** $x = 8$ or $x = -14$
27. $x = -4$ or $x = 8$ **28.** $x = -2$ or $x = 4$
29. $x = 0$ or $x = 4$ **30.** $x = -1$ or $x = 7$
31. $p = 7$ or $p = -13$ **32.** $p = 6$ or $p = 2$
33. $y = 9$ or $y = -5$ **34.** $y = -5$ or $y = -13$
35. $x = \pm\sqrt{5}$ or $x = \pm\sqrt{3}$
36. $x = \pm\sqrt{10}$ or $x = \pm 2$ **37.** $x = \pm 2$
38. $x = \pm 2i$ or $x = \pm\sqrt{6}$ **39.** $(-7, 7)$
40. $(-9, 9)$ **41.** $(-\infty, -5] \cup [5, \infty)$
42. $(-\infty, -2] \cup [2, \infty)$ **43.** $(-10, 4)$ **44.** $[-6, 2]$
45. $(-\infty, 2) \cup (6, \infty)$ **46.** $(-2, 4)$ **47.** $[3, 5]$
48. $(-2, 4)$ **49.** $(-\infty, \infty)$ **50.** no real solution
51. $(-4, 1)$ **52.** $(-\infty, \frac{4}{3}) \cup (2, \infty)$
53. $(-\infty, 2] \cup [5, \infty)$ **54.** $\left[1, \frac{7}{5}\right]$ **55.** $(-\infty, \infty)$
56. $(-\infty, 1] \cup \left[\frac{5}{3}, \infty\right)$ **57.** $(-\infty, -\frac{3}{2}] \cup [\frac{3}{2}, \infty)$
58. $[0, 2]$ **59.** $[-6, 4]$ **60.** $(-\infty, -2) \cup (4, \infty)$

61. $(-\infty, \infty)$ **62.** $[-9, 5]$ **63.** $(-\infty, -3) \cup (3, \infty)$
64. $(-4, 2)$ **65.** $\left(\frac{1}{4}, \frac{3}{4}\right)$
66. $(-\infty, 0] \cup \left[\frac{4}{3}, \infty\right)$ **67.** $(-2.769, -1.385)$
68. $\left(-\infty, -\frac{6}{55}\right) \cup \left(\frac{16}{11}, \infty\right)$ **69.** $[-3, 3]$
70. $(-\infty, -5] \cup [5, \infty)$ **71.** $|x - 2| < 7$
72. $|x + 2| > 3$ **73.** $\left|x - \frac{3}{2}\right| \geq 1/2$
74. $\left|x - \frac{11}{3}\right| \leq \frac{5}{3}$ **75.** $|x - a| \leq 2$
76. $|x + 3| \geq a$ **77.** $|T - 83| \leq 15$
78. $|x - 97.8| \leq 1.2$ **79.** Win: $d < 4$, tie: $d = 4$
80. $|f - f_c| \leq 15$
81. When the number of units sold was between 25 and 75 units
82. When the number of units sold was between 35 and 65 units
83. The mistake was that $x - 3 = -7$ was not considered, solution $x = -4$.
84. $-7 < x - 3 < 7$ is the appropriate inequality. The answer is $(-4, 10)$.
85. Did not reverse the inequality sign when divided by a negative. The answer is $[2, 3]$.
86. Absolute value can never yield a negative number, so no solution.
87. true   **88.** true   **89.** false
90. $x \geq 7$ **91.** $(a - b, a + b)$
92. $(-\infty, a - b) \cup (a + b, \infty)$ **93.** $(-\infty, \infty)$
94. no solution   **95.** $x = a - b$, $x = a + b$
96. no solution   **97.** no solution
98. $\left(-\infty, -\frac{2}{3}\right) \cup (3, \infty)$
99. $x \geq 7$, yes   **100.** Do not coincide, yes

101. $\left(-\infty, -\frac{2}{3}\right) \cup (3, \infty)$, yes

102. $(-\infty, -4.31) \cup (-0.38, \infty)$　**103.** $\left(-\frac{1}{2}, \infty\right)$

104. $(-\infty, -2) \cup \left(-\frac{2}{3}, \infty\right)$

Review Exercises

1. $x = \frac{16}{7}$　　**2.** $d = -1$　　**3.** $p = -\frac{8}{25}$

4. $x = 9$　　**5.** $x = 27$　　**6.** $c = \frac{7}{8}$

7. $y = -\frac{17}{5}$　　**8.** $x = -\frac{15}{22}$　　**9.** $b = \frac{6}{7}$

10. $g = \frac{7}{12}$　　**11.** $x = -\frac{6}{17}$　　**12.** $b = -1$

13. $x = 2,\ x \neq 0$

14. $x = -\frac{2}{3} \pm \frac{\sqrt{23}}{3} i$　　**15.** $t = -\frac{34}{5}$

16. $x = \frac{11}{13}$　　**17.** $x = -\frac{1}{2}$　　**18.** $m = 10$

19. $x = \frac{29}{17}$　　**20.** $x = -\frac{93}{2}$　　**21.** $x = 8 - 7y$

22. $-(x+1)$　　**23.** 96 miles

24. $B = 400,\ L = 500,\ D = 600,\ S = 125$

25. 144　　　**26.** 9, 11, 13, 15

27. Width: 3 inches, length: 7 inches

28. 20 inches

29. \$5000 @ 20%, \$20,000 @ 8%

30. Mutual Fund: 2%, Stock: 8%

31. 60 ml of 5%, 90 ml of 10%

32. 8 ounces of 8%　　**33.** At least 91

34. \$31,250　　**35.** $b = -3, 7$　　**36.** $x = -6, 9$

37. $x = 0, 8$　　**38.** $y = \frac{5}{3}$ or $y = -\frac{1}{2}$

39. $q = \pm 13$　　**40.** $c = \pm 6i$　　**41.** $x = 2 \pm 4i$

42. $d = -9, -5$　　　**43.** $x = -2, 6$

44. $x = -1, \frac{7}{2}$　　**45.** $x = \dfrac{1 \pm \sqrt{33}}{2}$

46. $m = 3, 5$　　**47.** $t = -1, \frac{7}{3}$

48. $x = \dfrac{-5 \pm i\sqrt{87}}{8}$　　**49.** $f = \dfrac{1 \pm \sqrt{337}}{48}$

50. $x = -3 \pm \sqrt{15}$　　**51.** $q = \dfrac{3 \pm \sqrt{69}}{10}$

52. $x = 7 \pm 2i\sqrt{3}$　　**53.** $x = -1, \frac{5}{2}$

54. $g = \dfrac{-3 \pm \sqrt{21}}{2}$　　**55.** $x = -3, \frac{2}{7}$

56. $b = \pm \dfrac{\sqrt{10}}{2}$　　**57.** $r = \sqrt{\dfrac{S}{\pi h}}$

58. $r = \sqrt[3]{\dfrac{3V}{\pi h}}$　　**59.** $v = \dfrac{h + 16t^2}{t} = \dfrac{h}{t} + 16t$

60. $h = \dfrac{A - 2\pi r^2}{2\pi r}$　　**61.** $h = 1$ ft, $b = 4$ ft

62. approximately 5.6 seconds　　　**63.** $x = 6$

64. no solution　　**65.** $x = 125$

66. $x = 2, 5$　　**67.** no solution

68. $x = 10$　　**69.** $\dfrac{36 \pm \sqrt{1664}}{8}, x \cong -0.6$

70. no solution　　**71.** $\dfrac{4 \pm \sqrt{376}}{4}, x \cong 5.85$

72. $x \cong 65.2$　　**73.** $\dfrac{-1 \pm \sqrt{13}}{2}, x \cong -2.303$

74. $x = 13.9$　　**75.** $x = 2,\ x = 3$

76. $x = \pm\sqrt{3}$　　**77.** $x = \frac{5}{4},\ x = \frac{3}{4}$

78. $x = 4 \pm \sqrt{5}$ **79.** $y = \frac{1}{4}, y = 1$

80. $p = -\frac{1}{6}, p = \frac{1}{2}$ **81.** $x = -\frac{125}{8}, x = 1$

82. $x = \frac{125}{8}, x = -1$ **83.** $x = -\frac{1}{8}, x = -1$

84. $y = 1$ **85.** $x = \pm 3i, x = \pm 2$

86. $x = 1, x = \frac{1}{9}$ **87.** $x = 0, -8, 4$

88. $t = 0, \pm\frac{5}{3}$ **89.** $p = \pm 2, 3$

90. $x = \pm i, \frac{9}{4}$ **91.** $p = -\frac{1}{2}, \frac{5}{2}, 3$

92. $t = \pm 3, \pm\sqrt{19}$ **93.** $y = \pm 9$

94. $x = \frac{1}{9}, 4$ **95.** $(-\infty, -4]$ **96.** $(-1, 7]$

97. $[2, 6]$ **98.** $(-1, \infty)$ **99.** $x > -6$

100. $x \le 0$ **101.** $-3 \le x \le 7$

102. $-5 < x \le 2$ **103.** $x \ge -4, [-4, \infty)$

104. $-4 \le x \le x, [-4, 4]$

105.

...0 1 2 3 4 5 6...

$, (4, \infty)$

106.

...0 1 2 3 4 ...

$, (-\infty, 2]$

107.

...7 8 9 10 11 12 ...

$, [8, 12]$

108.

...−3−2 −1 0 1 2 3 ...

$, \varnothing$

109. $(-\infty, \frac{5}{3})$,

...0 $\frac{1}{3}$ $\frac{2}{3}$ 1 $\frac{4}{3}$ $\frac{5}{3}$ ...

110. $(-\infty, -\frac{1}{3}]$,

...−1 $-\frac{2}{3}$ $-\frac{1}{3}$ 0 $\frac{1}{3}$...

111. $(-\frac{3}{2}, \infty)$,

... -2 $-\frac{3}{2}$ -1 $-\frac{1}{2}$ 0...

112. $[15, \infty)$,

... 13 14 15 16 ...

113. $(4, 9]$,

...3 4 5 6 7 8 9 10...

114. $\left[-\frac{23}{4}, -\frac{1}{4}\right]$,

...-6 $-\frac{23}{4}$ $-\frac{22}{4}$ ... $-\frac{1}{4}$ 0...

115. $\left[3, \frac{7}{2}\right]$,

... $\frac{5}{2}$ 3 $\frac{7}{2}$ 4 ...

116. $\left(-\frac{14}{5}, \infty\right)$,

... -3 $-\frac{14}{5}$ $-\frac{13}{5}$ $-\frac{12}{5}$ ... 0...

117. $x \ge 74$ **118.** greater than 34 suits

119. $[-6, 6]$ **120.** $\left(-\frac{4}{3}, \frac{5}{2}\right)$

121. $(-\infty, 0] \cup [4, \infty)$ **122.** $[-7, -2]$

123. $(-\infty, 0) \cup (7, \infty)$ **124.** no solution

125. $\left(-\infty, -\frac{3}{4}\right) \cup (4, \infty)$

126. $(-\infty, 1] \cup [2, \infty)$ **127.** $(0, 3)$

128. $(-\infty, 1) \cup (4, \infty)$ **129.** $(-\infty, -6] \cup [9, \infty)$

130. $[-7, 7) \cup (7, \infty)$ **131.** $(-\infty, 2) \cup (4, 5]$

132. $(-\infty, -7] \cup (-3, 1)$ **133.** $(3, \infty)$

134. $(-\infty, -1) \cup (0, 6)$ **135.** no solution

136. $x = -7$ or $x = 3$

137. $x = 1.7$ and $x = 0.9667$

138. $x = \pm\sqrt{3}, x = \pm 3$ **139.** $(-4, 4)$

140. $(3, 9)$ **141.** $(-\infty, -11) \cup (3, \infty)$

142. $[3, 11]$ **143.** $(-\infty, -3) \cup (3, \infty)$

144. $\left(-\infty, -\frac{31}{14}\right] \cup \left[-\frac{25}{14}, \infty\right)$ **145.** $(-\infty, \infty)$

146. $\left[-\frac{3}{2}, \frac{5}{2}\right]$ **147.** $75 \le T \le 95$ or $|T - 85| \le 10$

148. $|B - 0.08| \le 0.007$ **149.** $x = 2,510$

150. $x = 24.2$

151. a. $x = -5, 1$

b. $b = -5 : x = -2 \pm i$ $b = 0 : x = 0, -4$

$b = 7 : x = -2 \pm \sqrt{11}$ $b = 12 : x = -6, 2$

152. a. $x = -1, 5$

b. $b = -5 : x = 2 \pm i$ $b = 0 : x = 0, 4$

     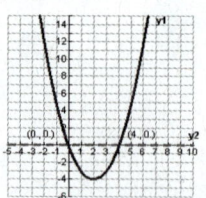

$b = 7 : x = 2 \pm \sqrt{11}$ $b = 12 : x = 6, -2$

153. $x = \left(-1 + \sqrt{7}\right)^4 \approx 7.34$ $y1 = 2x^{1/4}$

$y2 = -x^{1/2} + 6$

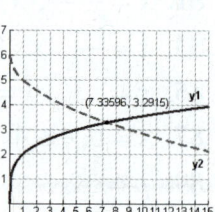

154. $x = \frac{1}{16}$ $y1 = 2x^{-1/2}, \ y2 = x^{-1/4} + 6$

155. a. $(-\infty, 15.4)$

b. $y1 = -0.61x + 7.62, \ y2 = 0.24x - 5.47$

c. agree

156. a. $(9.6, \infty)$

b. $y1 = -0.5x + 7, \ y2 = 0.75x - 5$

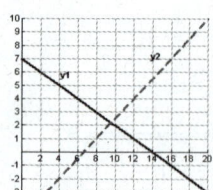

c. agree

157. $(-\infty, -5) \cup (5.25, \infty)$ **158.** $\left(-\frac{3}{5}, \frac{3}{2}\right)$

159. $\left(\frac{7}{5},\frac{7}{2}\right)$ **160.** $\left(-\infty,\frac{5}{3}\right)\cup\left(\frac{15}{2},\infty\right)$

161. $(-2.19,-0.9)\cup(0.9,2.19)$

162. $(-\infty,-0.75)\cup(0.97,5.78)\cup(7.5,\infty)$

Practice Test

1. $-3=p$ **2.** $z=4$ **3.** $t=-4,7$

4. $x=-\frac{3}{8},2$ **5.** $x=-\frac{1}{2},\frac{8}{3}$ **6.** $x=\frac{11}{2}$

7. $y=-8$ **8.** $x=\pm3,\pm2i$

9. $x=4$ **10.** $x=-8,\frac{1}{8}$ **11.** $y=1$

12. $x=\frac{5}{3},-\frac{1}{3},2$ **13.** $x=0,2,6$

14. $C=\frac{5}{9}\left(F-32\right)$ **15.** $L=\dfrac{P-2W}{2}$

16. $(-\infty,5)$ **17.** $(-\infty,17]$ **18.** $[-2,7)$

---

19. $\left(-\frac{32}{5},-6\right]$ **20.** $\left[0,\frac{3}{2}\right]$

21. $(-\infty,-1]\cup\left[\frac{4}{3},\infty\right)$ **22.** $(-\infty,2)\cup(3,\infty)$

23. $\left(-\frac{1}{2},3\right]$ **24.** $[-4,-3)\cup(3,\infty)$

25. 1000 feet

26. $\$9450\le$ commission $\le\$11{,}550$

27. $627\le$ minutes ≤722

28. Movie: 20 inches $\times8.6$ inches,
bars: 20 inches $\times3.2$ inches each

29. $x=7.95$ **30.** $[-1.768,1.768]$

Cumulative Test

1. -15 **2.** $\dfrac{x^9}{64b^{12}}$ **3.** $\dfrac{x^{12}}{y^3}$

4. $-6x^4-x^3+x-14$ **5.** $x^4+2x^3-15x^2$

6. $3x(x-5)(x+4)$ **7.** $2(a+10)(a^2-10a+100)$

8. $-\dfrac{1}{5(x-1)}$ **9.** $\dfrac{x^2+22x}{x^2-4}$ **10.** $x=-5,0,6$

11. $x=56$ **12.** $6+3i$ **13.** $x=-4$

14. no solution   **15.** 11.25 hours

16. $y=\pm6i$ **17.** $x=-6\pm2i$

18. $x=\dfrac{-1\pm i\sqrt{35}}{2}$ **19.** $x=4$

20. $x=-\frac{3}{2},x=-\frac{1}{2}$ **21.** $[-3,4)$

22. $\dfrac{9\pm\sqrt{95}\,i}{8}$ **23.** $(-\infty,-3)\cup[-2,3)$

24. $\left(-\infty,\frac{1}{2}\right]\cup\left[\frac{11}{10},\infty\right)$ **25.** $x=-1,x=-\frac{17}{3}$

26. $x = -2$, $x = \frac{3}{2}$

$y1 = x^6 + \frac{37}{8}x^3$,

$y2 = 27$

27. $\left(-1, \frac{1}{2}\right)$

$y1 = \left|\dfrac{3x}{x-2}\right|$, $y2 = 1$

CHAPTER 2

Section 2.1

1. $(4, 2)$ **2.** $(-2, 3)$ **3.** $(-3, 0)$

4. $(-4, -2)$ **5.** $(0, -3)$ **6.** $(3, -1)$

7. A: quadrant II
B: quadrant I
C: quadrant III
D: quadrant IV
E: on negative y-axis
F: on positive x-axis

8. A: quadrant II
B: quadrant I
C: quadrant III
D: quadrant IV
E: on positive y-axis
F: on negative x-axis

9.

10.

The line being described is $x = -3$.

The line being described is $y = 2$.

11. $d = 4$, $(3, 3)$ **12.** $d = 8$, $(-2, 0)$

13. $d = 4\sqrt{2}$, $(1, 2)$ **14.** $d = 4\sqrt{2}$, $(-1, 1)$

15. $d = 3\sqrt{10}$, $\left(-\frac{17}{2}, \frac{7}{2}\right)$

16. $d = 3\sqrt{10}$, $\left(\frac{5}{2}, \frac{27}{2}\right)$

17. $d = 5$, $\left(-5, \frac{1}{2}\right)$ **18.** $d = 13$, $\left(-\frac{13}{2}, -1\right)$

19. $d = 4\sqrt{2}$, $(-4, -6)$

20. $d = 2\sqrt{5}$, $(-2, -6)$

21. $d = 5$, $\left(\frac{3}{2}, \frac{11}{6}\right)$ **22.** $d = \frac{17}{5}$, $\left(1, \frac{5}{6}\right)$

23. $d = \dfrac{\sqrt{4049}}{60}$, $\left(-\frac{5}{24}, \frac{1}{15}\right)$

24. $d = \dfrac{\sqrt{54{,}961}}{90}$, $\left(\frac{19}{20}, -\frac{10}{9}\right)$

25. $d = 3.9$, $(0.3, 3.95)$

26. $d = 8.627$, $(1.25, 1.05)$

27. $d = 44.64$, $(1.05, -1.2)$

28. $d = 3.111$, $(2.2, 3.3)$

29. $d = 4\sqrt{2}$, $\left(\sqrt{3}, 3\sqrt{2}\right)$

30. $d = 2\sqrt{23}$, $\left(\sqrt{5}, -2\sqrt{3}\right)$

31. $d = \sqrt{10 + 2\sqrt{2} + 4\sqrt{3}}$, $\left(\dfrac{1 - \sqrt{2}}{2}, \dfrac{-2 + \sqrt{3}}{2}\right)$

32. $d = \sqrt{49 - 4\sqrt{5} - 16\sqrt{3}}$, $\left(\dfrac{1 + 2\sqrt{5}}{2}, 2 + \sqrt{3}\right)$

33. The perimeter of the triangle rounded to two decimal places is 21.84.

34. The perimeter of the triangle rounded to two decimal places is 16.71.

35. right triangle   **36.** isosceles

37. isosceles   **38.** right triangle, isosceles

39. 128.06 miles   **40.** 64.03 miles from Tampa

41. 268 miles   **42.** 50.29 yards

43. (2003, 330); $330 million   **44.** $42

45.

46.

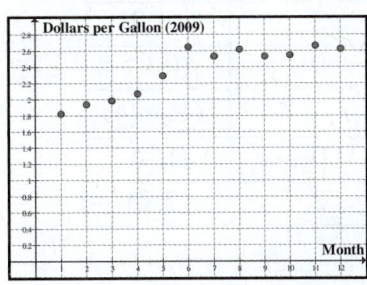

47. Substituted the values incorrectly. The correct answer is $d = \sqrt{58}$.

48. The value of x_1 is incorrect. The quantity $(3 - (-2))^2$ should be $(3 + 2)^2$.

49. Substituted the values incorrectly. It should have been $\left(\dfrac{-3 + 7}{2}, \dfrac{4 + 9}{2} \right) = \left(2, \frac{13}{2} \right)$.

50. The midpoint formula is incorrect. It should be $\left(\dfrac{x_1 + x_2}{2}, \dfrac{y_1 + y_2}{2} \right)$. So, $M = (-2, -3)$.

51. true   **52.** true
53. true   **54.** false

55. $d = \sqrt{2}\,|a - b|, \left(\dfrac{a + b}{2}, \dfrac{b + a}{2} \right)$

56. $d = 2\sqrt{a^2 + b^2}$, $(0, 0)$

57.

$$d = \sqrt{\left(x_1 - \dfrac{x_1 + x_2}{2} \right)^2 + \left(y_1 - \dfrac{y_1 + y_2}{2} \right)^2}$$

$$= \sqrt{\left(\dfrac{2x_1 - x_1 - x_2}{2} \right)^2 + \left(\dfrac{2y_1 - y_1 - y_2}{2} \right)^2}$$

$$= \sqrt{\left(\dfrac{x_1 - x_2}{2} \right)^2 + \left(\dfrac{y_1 - y_2}{2} \right)^2}$$

$$= \dfrac{1}{2} \sqrt{\left(x_1 - x_2 \right)^2 + \left(y_1 - y_2 \right)^2}$$

Using (x_2, y_2) with the midpoint yields the same result.

58. D_1: $(0, 0)$ to $(a + b, c)$; D_2: (a, c) to $(b, 0)$

Midpoint of D_1: $\left(\dfrac{a + b}{2}, \dfrac{c}{2} \right)$, same as D_2 midpoint.

59. $P_1 = (a, b)$, $P_2 = (c, d)$,
$d_1 =$ distance from P_1 to P_2, and
$d_2 =$ distance from P_2 to P_1.

$d_1 = \sqrt{(a - c)^2 + (b - d)^2}$

$d_2 = \sqrt{(c - a)^2 + (d - b)^2}$

since $(c - a)^2 = \left(-(a - c) \right)^2 = (a - c)^2$,

$(d - b)^2 = \left(-(b - d) \right)^2 = (b - d)^2 \therefore d_1 = d_2$,

so does not matter what point is labeled "first."

60. $d_1 =$ distance between $(-1, -1)$ and $(0, 0)$
$d_2 =$ distance between $(0, 0)$ and $(2, 2)$
$d_3 =$ distance between $(2, 2)$ and $(-1, -1)$.

$d_1 = \sqrt{(-1 - 0)^2 + (-1 - 0)^2} = \sqrt{2}$

$d_2 = \sqrt{(0 - 2)^2 + (0 - 2)^2} = 2\sqrt{2}$

$d_3 = \sqrt{(-1 - 2)^2 + (-1 - 2)^2} = 3\sqrt{2}$

$d_1 + d_2 = d_3$

61. $d \cong 6.357$ **62.** $d \cong 12.065$
$(0.7, 5.15)$ $(0.15, 0.1)$

63. $d \cong 3.111$ **64.** $d \cong 12.241$
$(2.2, 3.3)$ $(0.5, 1.35)$

  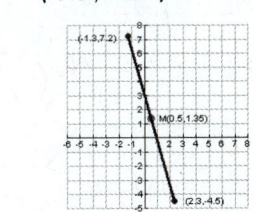

Section 2.2

1. a. no   **b.** yes
2. a. yes   **b.** no
3. a. yes   **b.** no
4. a. no   **b.** yes
5. a. yes   **b.** no
6. a. no   **b.** yes
7. a. yes   **b.** no
8. a. no   **b.** yes

9.

x	$y = 2 + x$	(x, y)
-2	0	$(-2, 0)$
0	2	$(0, 2)$
1	3	$(1, 3)$

10.

x	$y = 3x - 1$	(x, y)
-1	-4	$(-1, -4)$
0	-1	$(0, -1)$
2	5	$(2, 5)$

11.

x	$y = x^2 - x$	(x, y)
-1	2	$(-1, 2)$
0	0	$(0, 0)$
$\frac{1}{2}$	$-\frac{1}{4}$	$\left(\frac{1}{2}, -\frac{1}{4}\right)$
1	0	$(1, 0)$
2	2	$(2, 2)$

12.

x	$y = 1 - 2x - x^2$	(x, y)
-3	-2	$(-3, -2)$
-2	1	$(-2, 1)$
-1	2	$(-1, 2)$
0	1	$(0, 1)$
1	-2	$(1, -2)$

13.

x	$y = \sqrt{x - 1}$	(x, y)
1	0	$(1, 0)$
2	1	$(2, 1)$
5	2	$(5, 2)$
10	3	$(10, 3)$

14.

x	$y = -\sqrt{x} + 2$	(x, y)
-2	0	$(-2, 0)$
-1	-1	$(-1, -1)$
2	-2	$(2, -2)$
7	-3	$(7, -3)$

15.

16.

17.

18.

19.

20.

21.

22.

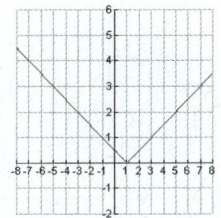

23. x-intercept: $(3, 0)$, y-intercept: $(0, -6)$

24. x-intercept: $(\frac{5}{2}, 0)$, y-intercept: $(0, 5)$

25. x-intercept: $(\pm 3, 0)$, y-intercept: $(0, -9)$

26. x-intercept: $(\pm \frac{1}{2}, 0)$, y-intercept: $(0, -1)$

27. x-intercept: $(4, 0)$, no y-intercept

28. x-intercept: $(8, 0)$, y-intercept: $(0, -2)$

29. no x-intercept, y-intercept: $(0, \frac{1}{4})$

30. x-intercept: $(4, 0)$, $(-3, 0)$, no y-intercept

31. x-intercept: $(\pm 2, 0)$, y-intercept: $(0, \pm 4)$

32. x-intercept: $(\pm 3, 0)$, no y-intercept

33. d   **34.** e   **35.** a   **36.** c   **37.** b   **38.** d

39. $(-1, -3)$ **40.** $(2, 4)$ **41.** $(-7, 10)$

42. $(1, 1)$ **43.** $(3, 2)$, $(-3, 2)$, $(-3, -2)$

44. $(-1, -7)$, $(1, 7)$, $(1, -7)$

45. x-axis   **46.** x-axis   **47.** origin

48. none   **49.** x-axis   **50.** x-axis

51. x-axis, y-axis, origin

52. x-axis, y-axis, origin

53. y-axis   **54.** x-axis   **55.** y-axis

56. y-axis   **57.** origin   **58.** origin

59.   **60.**

61.

62.

63.

64.

65.

66.

67.

68.

69.

70.

71.

72.

73.

74.

75.

76.

77.

Break even units: 2000
$2000 < x < 4000$

78.

$1000 < x < 3000$

79. $x \geq 1$ or $[1, \infty)$; the demand model is defined when at least 1000 units per day are demanded.

80. $x \geq 4$ or $[4, \infty)$; the demand model is defined when at least 4000 units per day are demanded.

81. The equation is not linear – you need more than two points to plot the graph.

82. Note that $-(-x)^2 \neq x^2$. Rather, $-(-x)^2 \neq -x^2$. As such, the graph IS symmetric with respect to the y-axis.

83. To test for symmetry about the y-axis, one should replace x by $-x$, NOT y by $-y$. Doing so here yields the equation $-x = |y|$, which is not equivalent to the original equation $x = |y|$.

84. To test for symmetry about the x-axis, replace y by $-y$, NOT x by $-x$. The conclusion should be that the graph is symmetric about the y-axis. The correct graph is

85. false   **86.** true   **87.** true

88. false   **89.** origin

90. x-intercept: $(a+b, 0)$, $(a-b, 0)$, y-intercept: $\left(0, a^2 - b^2\right)$

91. y-axis   **92.** origin

93. x-axis, y-axis, origin

94. x-axis, y-axis, origin

95. *x*-axis, *y*-axis, origin

96. *x*-axis

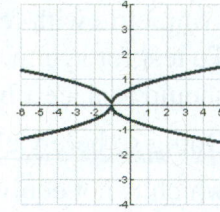

Section 2.3

1. $m = 3$ **2.** $m = 4$ **3.** $m = -2$

4. $m = 2$ **5.** $m = -\frac{19}{10}$ **6.** $m = -\frac{9}{13}$

7. $m \approx 2.379$ **8.** $m = 1.25$ **9.** $m = -3$

10. $m = -\frac{16}{25}$

11. *x*-intercept: $(0.5, 0)$, *y*-intercept: $(0, -1)$, $m = 2$, rising

12. *x*-intercept: $(-2, 0)$, *y*-intercept: $(0, 3)$, $m = \frac{3}{2}$, rising

13. *x*-intercept: $(1, 0)$, *y*-intercept: $(0, 1)$, $m = -1$, falling

14. *x*-intercept: $(0, 0)$, *y*-intercept: $(0, 0)$, $m = -1$, falling

15. *x*-intercept: none, *y*-intercept: $(0, 1)$, $m = 0$, horizontal

16. *x*-intercept: $(-4, 0)$, *y*-intercept: none, slope undefined, vertical

17. *x*-intercept: $\left(\frac{3}{2}, 0\right)$, *y*-intercept: $(0, -3)$

18. *x*-intercept: $\left(\frac{2}{3}, 0\right)$, *y*-intercept: $\left(\frac{2}{3}, 0\right)$

19. *x*-intercept: $(4, 0)$, *y*-intercept: $(0, 2)$

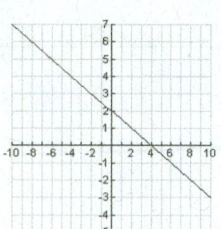

20. *x*-intercept: $(3, 0)$, *y*-intercept: $(0, -1)$

21. *x*-intercept: $(2, 0)$, *y*-intercept: $\left(0, -\frac{4}{3}\right)$

22. *x*-intercept: $(1, 0)$, *y*-intercept: $(0, -1)$

23. *x*-intercept: $(-2, 0)$, *y*-intercept: $(0, -2)$

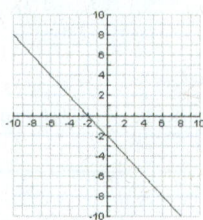

24. *x*-intercept: $\left(\frac{1}{4}, 0\right)$, *y*-intercept: $\left(0, -\frac{1}{3}\right)$

25. *x*-intercept: $(-1, 0)$, *y*-intercept: none

26. *x*-intercept: none, *y*-intercept: $(0, -3)$

27. x-intercept: none
y-intercept: $(0, 1.5)$

28. x-intercept: $(-7.5, 0)$
y-intercept: none

29. x-intercept: $\left(-\frac{7}{2}, 0\right)$
y-intercept: none

30. x-intercept: none
y-intercept: $\left(0, \frac{5}{3}\right)$

31. $y = \frac{2}{5}x - 2$ $m = \frac{2}{5}$ y-intercept: $(0, -2)$

32. $y = \frac{3}{4}x - 3$ $m = \frac{3}{4}$ y-intercept: $(0, -3)$

33. $y = -\frac{1}{3}x + 2$ $m = -\frac{1}{3}$ y-intercept: $(0, 2)$

34. $y = -\frac{1}{2}x + 4$ $m = -\frac{1}{2}$ y-intercept: $(0, 4)$

35. $y = 4x - 3$ $m = 4$ y-intercept: $(0, -3)$

36. $y = x - 5$ $m = 1$ y-intercept: $(0, -5)$

37. $y = -2x + 4$ $m = -2$ y-intercept: $(0, 4)$

38. $y = \frac{1}{4}x - \frac{1}{2}$ $m = \frac{1}{4}$ y-intercept: $\left(0, -\frac{1}{2}\right)$

39. $y = \frac{2}{3}x - 2$ $m = \frac{2}{3}$ y-intercept: $(0, -2)$

40. $y = -4x + 3$ $m = -4$ y-intercept: $(0, 3)$

41. $y = -\frac{3}{4}x + 6$ $m = -\frac{3}{4}$ y-intercept: $(0, 6)$

42. $y = -\frac{5}{8}x + 5$ $m = -\frac{5}{8}$ y-intercept: $(0, 5)$

43. $y = 2x + 3$

44. $y = -2x + 1$

45. $y = -\frac{1}{3}x$

46. $y = \frac{1}{2}x - 3$

47. $y = 2$

48. $y = -1.5$

49. $x = \frac{3}{2}$

50. $x = -3.5$

51. $y = 5x + 2$

52. $y = 2x - 3$

53. $y = -3x - 4$

54. $y = -x - 1$

55. $y = \frac{3}{4}x - \frac{7}{4}$

56. $y = -\frac{1}{7}x + \frac{16}{7}$

57. $y = 4$

58. $y = -3$

59. $x = -1$

60. $x = 4$

61. $y = \frac{3}{5}x + \frac{1}{5}$

62. $y = \frac{4}{9}x - \frac{11}{9}$

63. $y = -5x - 16$

64. $y = \frac{1}{2}x - \frac{11}{2}$

65. $y = \frac{1}{6}x - \frac{121}{3}$

66. $y = 2x + 28$

67. $y = -3x + 1$

68. $y = -\frac{3}{2}x$

69. $y = \frac{3}{2}x$

70. $y = \frac{1}{3}x - \frac{5}{18}$

71. $x = 3$

72. $x = -5$

73. $y = 7$

74. $y = -1$

75. $y = \frac{6}{5}x + 6$

76. $x = 0$

77. $x = -6$

78. $x = -9$

79. $x = \frac{2}{5}$

80. $x = \frac{1}{3}$

81. $y = x - 1$

82. $y = x + 1$

83. $y = -2x + 3$

84. $y = -4x + 2$

85. $y = -\frac{1}{2}x + 1$

86. $y = -\frac{1}{3}x + 2$

87. $y = 2x + 7$

88. $y = -x + 4$

89. $y = \frac{3}{2}x$

90. $y = -x + 6$

91. $y = 5$

92. $x = 3$

93. $y = 2$

94. $x = -1$

95. $y = \frac{3}{2}x - 4$

96. $y = -\frac{9}{4}x + \frac{25}{4}$

97. $y = \frac{5}{4}x + \frac{3}{2}$

98. $y = \frac{7}{3}x + \frac{1}{5}$

99. $y = \frac{3}{7}x + \frac{5}{2}$

100. $y = -\frac{2}{9}x - \frac{3}{2}$

101. 32-hour job will cost \$2,000

102. $C = 50 + 0.39x$ **103.** \$375

104. monthly loan payment: \$400, fill-up: \$40

105. 347 units   **106.** 873 units

107. $F = \frac{9}{5}C + 32$, $-40°$ C $= -40°$ F

108. $C = -\frac{1}{500}x + 15$, $10°$ C

109. $\frac{1}{50}$ inches per year

110. 0.03 inches per year

111. 0.06 ounces per year, 6 pounds 12.4 oz

112. −0.01 mile speed per year

113. The y-intercept represents the flat monthly fee of $35.

114. x-intercept: (6,0), represents the age at which the car is worth $0, y-intercept: (0, 11100), represents the initial value of the car when it is brand new.

115. −0.35 in./yr, 2.75 inches

116. 0.75°F/yr, 46.75°F

117. 2.4 plastic bags per year (in billions), 404 billion

118. 22 dollars per year, 832 dollars

119. a. (1, 31.93) (2, 51.18) (5,111.83)
b. $m = 25.59$. This means that when you buy one bottle of Hoisin it costs $31.93 per bottle.
c. $m = 31.93$. This means that when you buy two bottles of Hoisin it costs $25.59 per bottle.
d. $m = 22.366$. This means that when you buy five bottles of Hoisin it costs $22.37 per bottle.

120. a. (1, 18.27) (2, 22.77) (5, 35.93)
b. $m = 18.27$. This means that when you buy one bottle of Plum sauce it costs $18.27 per bottle.
c. $m = 11.385$. This means that when you buy two bottles of Plum sauce it costs $11.39 per bottle.
d. $m = 7.186$. This means that when you buy five bottles of Plum sauce it costs $7.19 per bottle.

121. The computations used to calculate the x- and y-intercepts should be reversed. So, the x-intercept is (3, 0) and the y-intercept is (0,−2).

122. The denominator of the slope should be $4−(−2)$, resulting in $m = -\frac{1}{3}$.

123. The denominator and numerator in the slope computation should be switched, resulting in the slope being undefined.

124. These two are listed incorrectly:
a. horizontal   **b.** vertical

125. true   **126.** false

127. false   **128.** true

129. Any vertical line is perpendicular to a line with slope 0.

130. Any vertical line is parallel to a line with no slope.

131. $y = -\frac{A}{B}x + 1$ **132.** $y = -\frac{A}{B}x + (2A-1)$

133. $y = \frac{B}{A}x + (2B-1)$

134. Case 1: $A \ne 0$ and $B \ne 0$: $y = \frac{B}{A}x + 1$;
Case 2: $A = 0$ and $B \ne 0$: $x = 0$;
Case 3: $A \ne 0$ and $B = 0$: $y = 1$

135. Let $y_1 = mx + b_1$ and $y_2 = mx + b_2$, assuming that $b_1 \ne b_2$. At a point of intersection of these two lines, $y_1 = y_2$. This is equivalent to $mx + b_1 = mx + b_2$, which implies $b_1 = b_2$, which contradicts our assumption. Hence, there are no points of intersection.

136. $x = \dfrac{b_2 - b_1}{m_1 - m_2}$

137. perpendicular   **138.** parallel

139. perpendicular   **140.** neither

141. neither   **142.** perpendicular

Section 2.4

1. $(x-1)^2 + (y-2)^2 = 9$

2. $(x-3)^2 + (y-4)^2 = 25$

3. $(x+3)^2 + (y+4)^2 = 100$

4. $(x+1)^2 + (y+2)^2 = 16$

5. $(x-5)^2 + (y-7)^2 = 81$

6. $(x-2)^2 + (y-8)^2 = 36$

7. $(x+11)^2 + (y-12)^2 = 169$

8. $(x-6)^2 + (y+7)^2 = 64$

9. $x^2 + y^2 = 4$ 10. $x^2 + y^2 = 9$

11. $x^2 + (y-2)^2 = 9$ 12. $(x-3)^2 + y^2 = 4$

13. $x^2 + y^2 = 2$ 14. $(x+1)^2 + (y-2)^2 = 7$

15. $(x-5)^2 + (y+3)^2 = 12$

16. $(x+4)^2 + (y+1)^2 = 45$

17. $\left(x-\frac{2}{3}\right)^2 + \left(y+\frac{3}{5}\right)^2 = \frac{1}{16}$

18. $\left(x+\frac{1}{3}\right)^2 + \left(y+\frac{2}{7}\right)^2 = \frac{4}{25}$

19. $(x-1.3)^2 + (y-2.7)^2 = 10.24$

20. $(x+3.1)^2 + (y-4.2)^2 = 30.25$

21. $C(1,3)$, $r=5$ 22. $C(-1,-3)$, $r=\sqrt{11}$

23. $C(2,-5)$, $r=7$ 24. $C(-3,7)$, $r=9$

25. $C(4,9)$, $r=2\sqrt{5}$ 26. $C(-1,-2)$, $r=2\sqrt{2}$

27. $C\left(\frac{2}{5},\frac{1}{7}\right)$, $r=\frac{2}{3}$ 28. $C\left(\frac{1}{2},\frac{1}{3}\right)$, $r=\frac{3}{5}$

29. $C(1.5,-2.7)$, $r=1.3$

30. $C(-3.1,7.4)$, $r=7.5$

31. $C(0,0)$, $r=5\sqrt{2}$ 32. $C(0,0)$, $r=2\sqrt{2}$

33. $C(-2,-3)$, $r=4$ 34. $C(-1,-5)$, $r=3$

35. $C(-3,-4)$, $r=10$ 36. $C(-1,-2)$, $r=\sqrt{14}$

37. $C(5,7)$, $r=9$ 38. $C(2,8)$, $r=6$

39. $C(0,1)$, $r=4$ 40. $C(-1,0)$, $r=3$

41. $C(1,3)$, $r=3$ 42. $C(4,3)$, $r=2$

43. $C(5,-3)$, $r=2\sqrt{3}$

44. $C(-4,-1)$, $r=3\sqrt{5}$

45. $C(3,2)$, $r=2\sqrt{3}$ 46. $C(1,5)$, $r=2\sqrt{6}$

47. $C\left(\frac{1}{2},-\frac{1}{2}\right)$, $r=\frac{1}{2}$ 48. $C\left(\frac{1}{4},\frac{3}{4}\right)$, $r=\frac{1}{2}$

49. $C(1.3,2.7)$, $r=3.2$

50. $C(3.1,4.2)$, $r=5.5$

51. $(x+1)^2 + (y+2)^2 = 8$

52. $(x-4)^2 + (y-9)^2 = 20$

53. $(x+2)^2 + (y-3)^2 = 41$

54. $(x-1)^2 + (y-1)^2 = 117$

55. $(x+2)^2 + (y+5)^2 = 25$

56. $(x+3)^2 + (y+4)^2 = 20$

57. no $\left(\sqrt{95^2 + 33^2} \cong 100.568\right)$

58. yes $\left(\sqrt{87^2 + 45^2} \cong 97.949\right)$

59. $x^2 + y^2 = 2500$

60. $(x-25)^2 + (y-25)^2 = 625$, $C(25,25)$, $r=25$

61. $x^2 + y^2 = 2{,}250{,}000$

62. $x^2 + y^2 = 9\times 10^6$

63. $x^2 + y^2 = 40{,}000$

64. $x^2 + y^2 = 4$

65. a.

     Tower 1: $(x-2.5)^2 + (y-2.5)^2 = 3.5^2$

     Tower 2: $(x-2.5)^2 + (y-7.5)^2 = 3.5^2$

     Tower 3: $(x-7.5)^2 + (y-2.5)^2 = 3.5^2$

     Tower 4: $(x-7.5)^2 + (y-7.5)^2 = 3.5^2$

b. This placement of cell phone towers will provide cell phone coverage for the entire 10 mile by 10 mile square.

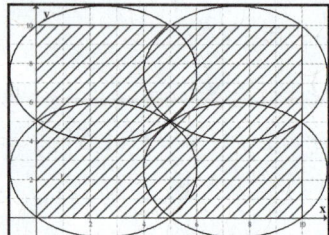

66. a.

Tower 1: $(x-3)^2+(y-3)^2=3.5^2$

Tower 2: $(x-3)^2+(y-7)^2=3.5^2$

Tower 3: $(x-7)^2+(y-3)^2=3.5^2$

Tower 4: $(x-7)^2+(y-7)^2=3.5^2$

b. This placement of cell phone towers will not provide cell phone coverage for the entire 10 mile by 10 mile square.

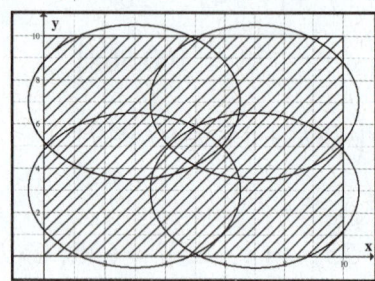

67. The center should be $(4,-3)$.

68. The radius should be $\sqrt{2}$.

69. The standard form of the equation of a circle requires that the right-side be non-negative.
Since the radius would have to be $\sqrt{-16}$, which is not a real number; the result cannot be a circle.

70. The right-side, upon completing the square, should be $3+9+4=16$. So, the radius would be 4 rather than $2\sqrt{3}$.

74. true   **75.** single point $(-5,3)$

71. true   **72.** true   **73.** true

76. no graph   **77.** $(x-3)^2+(y+2)^2=20$

78. $(x-1)^2+(y+2)^2=8$

79. $4c=a^2+b^2$ **80.** $4c>a^2+b^2$

81. $C(a,0),\ r=10$ **82.** $C(0,-b),\ r=7$

83. no graph (because no solution)

84. single point $(1,-3)$ **85.** single point $(-5,3)$

86. no graph (because no solution)

87. a. $(x-5.5)^2+(y+1.5)^2=39.69$,

$(5.5,\ -1.5),\ r=6.3$

b. $y=-1.5\pm\sqrt{39.69-(x-5.5)^2}$

c.

d. The graphs in (a) and (c) are the same.

88. a. $(x+0.6)^2+(y-1.6)^2=0.81$,

$(-0.6,1.6),\ r=0.9$

b. $y=1.6\pm\sqrt{0.81-(x+0.6)^2}$

c.

d. The graphs in (a) and (c) are the same.

Section 2.5*

1. Negative linear association because the data closely cluster around what is reasonably described as a linear with negative slope.

2. There is no identifiable direction of association, and the data is certainly not linear and there is not even a recognizable nonlinear curve that describes a pattern to which the data conform.

3. Although the data seem to be comprised of two *linear* segments, the overall data set cannot be described as having a positive or negative direction of association. Moreover, the pattern of the data is not linear, per se; rather, it is nonlinear and conforms to an identifiable curve (an upside down V called the *absolute value* function).

4. Positive nonlinear association because the data are rising from left to right and cluster relatively closely around what is reasonably described as a nonlinear curve.

5. **B** because the association is positive, thereby eliminating choices A and C. And, since the data are closely clustered around a linear curve, the bigger of the two correlation coefficients, 0.80 and 0.20, is more appropriate.

6. **A** because the association is negative, thereby eliminating choices B and D. And, the since the data are closely clustered around a linear curve, the correlation coefficient closer to -1 is more appropriate.

7. **C** because the association is negative, thereby eliminating choices B and D. And, the data are more loosely clustered around a linear curve than are those pictured in #6. So, the correlation coefficient is the negative choice closer to 0.

8. **D** because there is no real identifiable association, so that the correlation coefficient closest to 0 is the most appropriate.

9. **a.**

b. The data seem to be nearly perfectly aligned to a line with negative slope. So, it is reasonable to guess that the correlation coefficient is very close to -1.

c. The equation of the best fit line is $y = -3x + 5$ with a correlation coefficient of $r = -1$.

d. There is a perfect negative linear association between x and y.

10. **a.**

b. The data seem to be nearly perfectly aligned to a line with positive slope. So, it is reasonable to guess that the correlation coefficient is very close to -1.

c. The equation of the best fit line is $y = 2x$ with a correlation coefficient of $r = 1$.

d. There is a perfect positive linear relationship between x and y.

11. a.

b. The data tends to fall from left to right, so that the correlation coefficient should be negative. Also, the data do not seem to stray too far from a linear curve, so the r value should be reasonably close to -1, but not equal to it. A reasonable guess would be around -0.90.

c. The equation of the best fit line is approximately $y = -0.5844x - 3.801$ with a correlation coefficient of about $r = 0.9833$.

d. There is a strong (but not perfect) negative linear relationship between x and y.

12. a.

b. The data tends to rise from left to right, so that the correlation coefficient should be positive. Also, the data do not seem to stray too far from a linear curve, so the r value should be reasonably close to 1, but not equal to it. A reasonable guess would be around 0.90.

c. The equation of the best fit line is approximately $y = 15.717x - 0.2945$ with a correlation coefficient of about $r = 0.9569$.

d. There is a strong (but not perfect) positive linear relationship between x and y.

13. a.

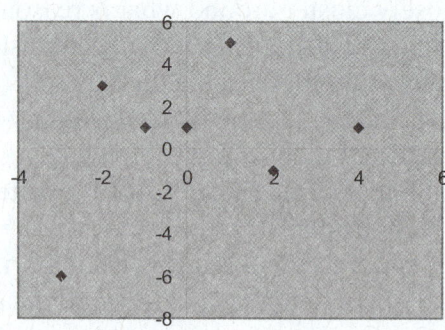

b. The data seems to rise from left to right, but it is difficult to be certain about this relationship since the data stray considerably away from an identifiable line. As such, it is reasonable to guess that r is a rather small value close to 0, say around 0.30.

c. The equation of the best fit line is approximately $y = 0.5x + 0.5$ with a correlation coefficient of about 0.349.

d. There is a very loose (bordering on unidentifiable) positive linear relationship between x and y.

14. a.

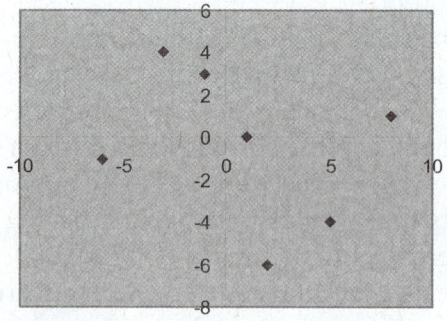

b. It is very difficult to discern if there is a rising or falling trend in the data as you move from left to right, and it does not appear to be reasonably described by a line As such, it is reasonable to guess that r is a rather small value close to 0, say around 0.25 or less.

c. The equation of the best fit line is approximately $y = -0.2256x - 0.2352$ with a correlation coefficient of about 0.297.

d. The relationship between x and y is not really discernible.

15. a.

The equation of the best fit line is about $y = 0.7553x - 0.4392$ with a correlation coefficient of about $r = 0.868$.

b. The values $x = 0$ and $x = -6$ are within the range of the data set, so that using the best fit line for predictive purposes is reasonable. This is not the case for the values $x = 12$ and $x = -15$. The predicted value of y when $x = 0$ is approximately -0.4392, and the predicted value of y when $x = -6$ is -4.971.

c. Solve the equation $2 = 0.7553x - 0.4392$ for x to obtain: $2.4392 = 0.7553x$ so that $x = 3.229$. So, using the best fit line, you would expect to get a y-value of 2 when x is approximately 3.229.

16. a.

The equation of the best fit line is about $y = 2.9832x - 13.508$ with a correlation coefficient of about $r = 0.909$.

b. None of the given x-values are within reasonable enough range of the data set to ensure that the best fit line should be used for predictive purposes for them.

c. Solve the equation $2 = 2.9832x - 13.508$ for x to obtain: $15.508 = 2.9832x$ so that $x = 5.198$. So, using the best fit line, you would expect to get a y-value of 2 when x is approximately 5.198.

17. a.

The equation of the best fit line is about $y = -1.2631x - 11.979$ with a correlation coefficient of about $r = -0.980$.

b. The values $x = -15$, -6, and 0 are within the range of the data set, so that using the best fit line for predictive purposes is reasonable. This is not the case for the value $x = 12$. The predicted value of y when $x = -15$ is approximately 6.9675, the predicted value of y when $x = -6$ is about -4.4004, and the predicted value of y when $x = 0$ is -11.979.

c. Solve the equation $2 = -1.2631x - 11.979$ for x to obtain: $13.979 = -1.2631x$ so that $x = -11.067$. So, using the best fit line, you would expect to get a y-value of 2 when x is approximately -11.067.

18. a.

The equation of the best fit line is about $y = -0.4077x + 2.9457$ with a correlation coefficient of about $r = 0.715$.

b. All of the given x-values are within reasonable range of the data set to be able to obtain a reasonable predicted y-value for each of them. The predicted value of y when $x = -15$ is about 9.0612, the predicted value of y when $x = -6$ is about 5.3919, the predicted value of y when $x = 0$ is 2.9457, and the predicted value of y when $x = 12$ is about -1.9467.

c. Solve the equation $2 = -0.4077x + 2.9457$ for x to obtain: $4.9457 = -0.4077x$ so that $x = -12.1307$. So, using the best fit line, you would expect to get a y-value of 2 when x is approximately -12.1307.

19. a. The scatterplot for the entire data set is:

The equation of the best fit line is $y = -3x + 5$ with a correlation coefficient of $r = -1$.

b. The scatterplot for the data set obtained by removing the starred data point $(5, -10)$ is:

The equation of the best fit line of this modified data set is $y = -3x + 5$ with a correlation coefficient of $r = -1$.

c. Removing the data point did not result in the slightest change in either the equation of the best fit line or the correlation coefficient. This is reasonable since the relationship between x and y in the original data set is perfectly linear, so that all of the points lie ON the same line. As such, removing one of them has no effect on the line itself.

20. a. The scatterplot for the entire data set is:

The equation of the best fit line is $y = -0.5844x - 3.801$ with a correlation coefficient of about $r = -0.983$.

b. The scatterplot for the data set obtained by removing the starred data point $(20, -16)$ is:

The equation of the best fit line of this modified data set is $y = -0.5597x - 3.7284$ with a correlation coefficient of $r = -0.971$.

c. Removing the data point did change both the best fit line and the correlation coefficient, but only very slightly.

21. a. The scatterplot for the entire data set is:

The equation of the best fit line is $y = -3.4776x + 4.6076$ with a correlation coefficient of about $r = -0.993$.

b. The scatterplot for the data set obtained by removing the starred data points $(3, -4)$ and $(6, -16)$ is:

The equation of the best fit line of this modified data set is $y = -3.6534x + 4.2614$ with a correlation coefficient of $r = -0.995$.

c. Removing the data point did change both the best fit line and the correlation coefficient, but only very slightly.

22. a. The scatterplot for the entire data set is:

The equation of the best fit line is $y = 3.5357x - 6$ with a correlation coefficient of about $r = 0.908$.

b. The scatterplot for the data set obtained by removing the starred data point $(7, 25)$ is:

The equation of the best fit line of this modified data set is $y = 2x$ with a correlation coefficient of $r = 1$.

c. Removing the data point resulted in significant changes in both the best fit line and the correlation coefficient.

23. a.

b. The correlation coefficient is approximately $r = -0.980$. This is identical to the r-value from Problem 17. This makes sense because simply interchanging the x and y-values does not change how the points cluster together in the xy-plane.

c. The equation of the best fit line for the paired data (y, x) is $x = -0.7607y - 9.4957$.

d. It is not reasonable to use the best fit line in (c) to find the predicted value of x when $y = 23$ because this value falls outside the range of the given data. However, it is okay to use the best fit line to find the predicted values of x when $y = 2$ or $y = -16$. Indeed, the predicted value of x when $y = 2$ is about -11.0171, and the predicted value of x when $y = -16$ is about 2.6755.

24. a.

b. The correlation coefficient is approximately 0.909. This is identical to the r-value from Problem 16. This makes sense because simply interchanging the x

and y-values does not change how the points cluster together in the xy-plane.

c. The equation of the best fit line for the paired data (y, x) is $x = 0.2773y + 5.0654$.

d. The only y-value for which it is reasonable to use the best fit line in (c) for predictive purposes is $y = 2$ because the other two values lie outside the range of values present in the data set. The predicted value of x when $y = 2$ is about $x = 5.62$.

25. First, note that the scatterplot is given by

The paired data all lie identically on the vertical line $x = 3$. As such, you might think that the square of the correlation coefficient would be 1 and the best fit line is, in fact, $x = 3$. However, since there is absolutely no *variation* in the x-values for this data set, it turns out that in the formula for the correlation coefficient

$$r = \frac{n\sum xy - \left(\sum x\right)\left(\sum y\right)}{\sqrt{n\sum x^2 - \left(\sum x\right)^2} \cdot \sqrt{n\sum y^2 - \left(\sum y\right)^2}}$$

the quantity $\sqrt{n\sum x^2 - \left(\sum x\right)^2}$ turns out to be zero. (Check this on Excel for this data set!) As such, there is no meaningful r-value for this data set.

Also, the best fit line is definitely the vertical line $x = 3$, but the technology cannot provide it because its slope is undefined.

26. First, note that the scatterplot is given by

The paired data all lie identically on the horizontal line $y = -2$. As such, you might think that the square of the correlation coefficient would be 1 and the best fit line is, in fact, $y = -2$. However, since there is absolutely no *variation* in the y-values for this data set, it turns out that in the formula for the correlation coefficient

$$r = \frac{n\sum xy - \left(\sum x\right)\left(\sum y\right)}{\sqrt{n\sum x^2 - \left(\sum x\right)^2} \cdot \sqrt{n\sum y^2 - \left(\sum y\right)^2}}$$

the quantity $\sqrt{n\sum y^2 - \left(\sum y\right)^2}$ turns out to be zero. (Check this on Excel for this data set!) As such, there is no meaningful r-value for this data set.

However, this time, the technology is able to produce the equation of the best fit line (unlike in Problem #25) because the slope is defined – it is simply 0.

27. The y-intercept 1.257 is mistakenly interpreted as the slope. The correct interpretation is that for every unit increase in x, the y-value increases by about 5.175.

28. Since the slope of the best fit line will be negative, the correlation coefficient r should be negative as well. As such, r should equal -0.9913, not 0.9913.

29. a. Here is a table listing all of the correlation coefficients between each of the events and the total points:

Event	r
100 m	−0.714
Long Jump	0.768
Shot Put	0.621
High Jump	0.627
400 m	−0.704
1500 m	−0.289
110 m hurdle	−0.653
Discus	0.505
Pole Vault	0.283
Javeline	0.421

Long jump has the strongest relationship to the total points.

b. The correlation coefficient between *long jump* and *total events* is $r = 0.768$.

c. The equation of the best fit line between the two events in (b) is $y = 838.70x + 1957.77$.

d. Evaluate the equation in (c) at $x = 40$ to get the total points are about 35,506.

30. a. Here is a table listing all of the correlation coefficients between each of the events and the total points:

Event	r
100 m	−0.714
Long Jump	0.768
Shot Put	0.621
High Jump	0.627
400 m	−0.704
1500 m	−0.289
110 m hurdle	−0.653
Discus	0.505
Pole Vault	0.283
Javeline	0.421

The 100 m event has the second strongest relationship to the total points.

b. The correlation coefficient between 100m and total points is $r = -0.714$

c. The equation of the best fit line between the two events in (b) is $y = -1151.19x + 20617.61$.

d. Since $r = -0.714$ indicates a reasonably strong relationship between x100m and total points, we can make a reasonably accurate prediction. However, it will not be completely accurate.

e. Evaluating the equation of the best fit line in (c) at $x = 40$ yields the total points of $-25,429.99$, which is not realistic at all.

31. a. The following is a scatterplot illustrating the relationship between *left thumb length* and *total both scores*.

b. $r = -0.105$

c. The correlation coefficient ($r = -0.105$) indicates a weak relationship between left thumb length and total both scores.

d. The equation of the best fit line for these two events is $y = -7.62x + 127.73$.

e. No, given the weak correlation coefficient between total both scores and left thumb length, one could not use the best fit line to produce accurate predictions.

32. a. A scatterplot illustrating the relationship between *right thumb length* and *total both scores* is shown below.

b. The correlation coefficient between *right thumb length* and *total both scores* is $r = -0.095$.

c. The correlation coefficient ($r = 0.095$) indicates a weak relationship between left thumb length and total both scores.

d. The equation of the best fit line that describes the relationship between *right thumb length* and *total both scores* is $y = -7.91x + 129.21$.

e. No, given the weak correlation coefficient between total both scores and right thumb length, one could not use the best fit line to produce accurate predictions.

33. a. A scatterplot illustrating the relationship between *% residents immunized* and *% residents with influenza* is shown below.

b. The correlation coefficient between *% residents immunized* and *% residents with influenza* is $r = -0.812$

c. Based on the correlation coefficient ($r = -0.812$), we would believe that there i a strong relationship between % residents immunized and % residents with influenz

d. The equation of the best fit line that describes the relationship between *% residents immunized* and *% residents with influenza* is $y = -0.27x + 32.20$.

e. Since $r = -0.812$ indicates a strong relationship between % residents immunized and % residents with influenza, we can make a reasonably accurate prediction. However it will not be completely accurate.

34. a. The nursing home number of the outlier is 21.

  b. The scatterplot illustrating the relationship between *% residents immunized* and *% residents with influenza* WITH THE OUTLIER REMOVED is shown below.

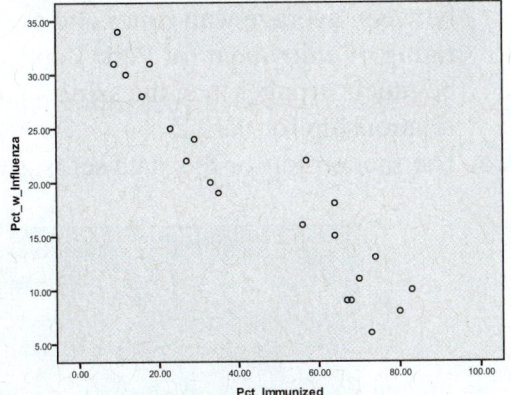

  c. The revised correlation coefficient between *% residents immunized* and *% residents with influenza* is $r = -0.938$

  d. By removing the outlier, the correlation coefficient goes from -0.812 to -0.938 and as such, the strength of the relationship between *% residents immunized* and *% residents with influenza* is increased.

  e. The revised equation of the best fit line that describes the relationship between *% residents immunized* and *% residents with influenza* is $y = -0.31x + 33.54$.

35. a. A scatterplot illustrating the relationship between *average wait times* and *average rating of enjoyment* is shown below.

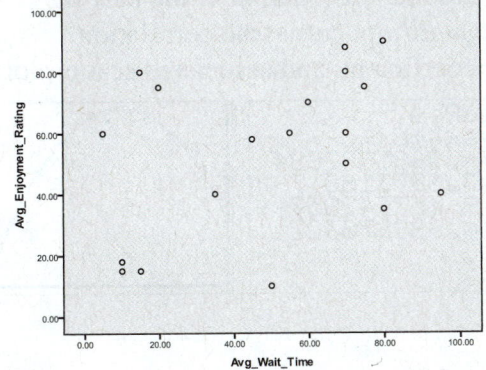

  b. The correlation coefficient between *average wait times* and *average rating of enjoyment* is $r = 0.348$.

  c. The correlation coefficient ($r = 0.348$) indicates a somewhat weak relationship between average wait times and average rating of enjoyment.

  d. The equation of the best fit line that describes the relationship between *average wait times* and *average rating of enjoyment* is $y = 0.31x + 37.83$.

  e. No, given the somewhat weak correlation coefficient between average wait times and average rating of enjoyment, one could not use the best fit line to produce accurate predictions.

36. a. A scatterplot illustrating the relationship between *average wait times* and *average rating of enjoyment* for *Park 1* is shown below.

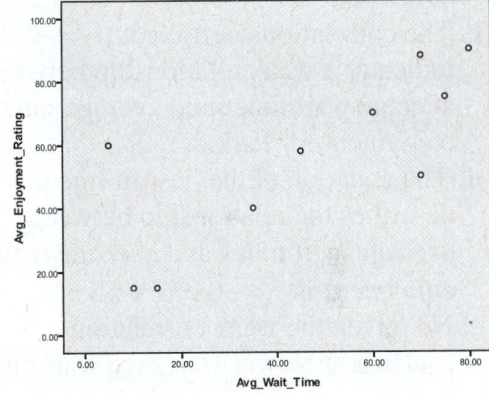

  b. The correlation coefficient between *average wait times* and *average rating of enjoyment* is $r = 0.767$.

  c. The correlation coefficient ($r = 0.767$) indicates a strong relationship between average wait times and average rating of enjoyment for Park 1.

  d. The equation of the best fit line that describes the relationship between *average wait times* and *average rating of enjoyment* is $y = 0.71x + 22.91$.

  e. Yes, given the strong correlation coefficient between average wait times and average rating of enjoyment for

Park 1, one could use the best fit line to produce accurate predictions.

37. a. A scatterplot illustrating the relationship between *average wait times* and *average rating of enjoyment* for *Park 2* is shown below.

b. The correlation coefficient between *average wait times* and *average rating of enjoyment* is $r = -0.064$.

c. The correlation coefficient ($r = -0.064$) indicates a weak relationship between average wait times and average rating of enjoyment for Park 2.

d. The equation of the best fit line that describes the relationship between *average wait times* and *average rating of enjoyment* is $y = -0.05x + 52.53$.

e. No, given the weak correlation coefficient between average wait times and average rating of enjoyment for Park 2, one could not use the best fit line to produce accurate predictions.

38. Consider the combined scatterplot with best fit lines for Park 1 and Park 2 below.

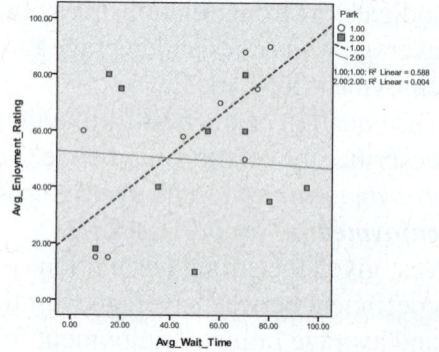

As can be seen from the scatterplot and from reviewing the correlation coefficients, the relationship between average wait times and average rating of enjoyment for Park 1 in Florida ($r = 0.767$) is quite different from this relationship for Park 2 in California ($r = -0.064$). Specifically relationship between average wait times and average rating of enjoyment for Park 1 appears to be much stronger than the same relationship for Park 2.

39. a. The scatterplot for this data set is given by

b. The equation of the best fit *line* is $y = -1.9867x + 27.211$ with a correlation coefficient of $r = -0.671$. This line does not seem to accurately describe the data because some of the points rise as you move left to right, while others fall as you move left to right; a line cannot capture both types of behavior simultaneously. Also, r being negative has no meaning here.

c. The best fit is provided by QuadReg. The associated equation of the best fit *quadratic* curve, the correlation coefficient, and associated scatterplot are:

40. a. The scatterplot for this data set is given by

b. The equation of the best fit *line* is
$y = -0.3733x + 1.1647$ with a correlation coefficient of $r = -0.923$. This line seems to provide a reasonably good fit for this data, although not perfect.

c. The best fit is provided by ExpReg. The associated equation of the best fit *exponential* curve, the correlation coefficient, and associated scatterplot are:

41. a. The scatterplot for this data set is given by

b. The equation of the best fit *line* is
$y = 0.3537x + 0.5593$ with a correlation coefficient of $r = 0.971$. This line seems to provide a very good fit for this data, although not perfect.

c. The best fit is provided by LnReg. The associated equation of the best fit *logarithmic* curve, the correlation coefficient, and associated scatterplot are:

42. a. The scatterplot for this data set is given by z

b. The equation of the best fit *line* is
$y = 2.9237x - 5.2956$ with a correlation coefficient of $r = 0.206$. This line seems to provide a terrible fit for the data.

c. The best fit is provided by CubicReg. The associated equation of the best fit *cubic polynomial* curve, the correlation coefficient, and associated scatterplot are:

Review Exercises

1. quadrant II
2. quadrant I
3. quadrant III
4. quadrant IV

5. $d = 3\sqrt{5}$ **6.** $d = 3$ **7.** $d = \sqrt{205}$

8. $d = 2.418$ **9.** $\left(\frac{5}{2}, 6\right)$ **10.** $\left(\frac{3}{2}, \frac{13}{2}\right)$

11. $(3.85, 5.3)$ **12.** $(0, 3)$ **13.** $d \approx 52.20$ units

14. $\left(\frac{5}{2}, 5\right)$

15. x-intercepts: $(\pm 2, 0)$, y-intercepts: $(0, \pm 1)$

16. no x-intercepts, y-intercepts: $(0, 2)$

17. x-intercepts: $(\pm 3, 0)$, no y-intercepts

18. x-intercepts: $(-3, 0)$, $(4, 0)$

    y-intercepts: $(0, 1)$

19. y-axis   **20.** y-axis
21. origin   **22.** x-axis
23.   **24.**

25.   **26.**

27.   **28.**

29.   **30.**

31. $y = -3x + 6$ $m = -3$ y-intercept: $(0, 6)$

32. $y = -\frac{3}{4}x + \frac{9}{4}$ $m = -\frac{3}{4}$ y-intercept: $(0, \frac{9}{4})$

33. $y = -\frac{3}{2}x - \frac{1}{2}$ $m = -\frac{3}{2}$ y-intercept: $(0, -\frac{1}{2})$

34. $y = -\frac{8}{3}x - \frac{1}{2}$ $m = -\frac{8}{3}$ y-intercept: $(0, -\frac{1}{2})$

35. x-intercepts: $\left(\frac{5}{4}, 0\right)$, y-intercepts: $(0, -5)$,

    $m = 4$

36. x-intercepts: $(-4, 0)$, y-intercepts: $(0, -3)$,

    $m = -\frac{3}{4}$

37. x-intercepts: $(4, 0)$, y-intercepts: $(0, 4)$,

    $m = -1$

38. x-intercepts: $(-4,0)$, no y-intercepts, slope undefined

39. no x-intercepts, y-intercepts: $(0,2)$, $m=0$

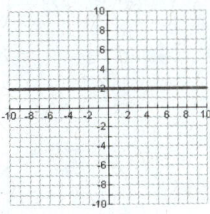

40. x-intercepts: $(-6,0)$, y-intercepts: $(0,-6)$, $m=-1$

41. $y=4x-3$

42. $y=4$

43. $x=-3$

44. $y=-\frac{2}{3}x+\frac{3}{4}$

45. $y=-2x-2$

46. $y=\frac{3}{4}x+\frac{29}{2}$

47. $y=6$

48. $x=2$

49. $y=\frac{5}{6}x+\frac{4}{3}$

50. $y=-x+3$

51. $y=-2x-1$

52. $y=-\frac{1}{3}x-1$

53. $y=\frac{2}{3}x+\frac{1}{3}$

54. $y=-\frac{3}{5}x+9$

55. $y=-\frac{3}{4}x+\frac{31}{16}$

56. Case 1: $B\neq 0$: $y=-\frac{A}{B}x+\left(b-1+\frac{A}{B}(a+2)\right)$

  Case 2: $B=0$: $x=a+2$

57. $y=1.2x+100$ (x is pretest score, y is posttest score)

58. $C=250+38t$, \$307

59. $(x+2)^2+(y-3)^2=36$

60. $(x+6)^2+(y+8)^2=54$

61. $\left(x-\frac{3}{4}\right)^2+\left(y-\frac{5}{2}\right)^2=\frac{4}{25}$

62. $(x-1.2)^2+(y+2.4)^2=12.96$

63. $C(-2,-3)$, $r=9$ **64.** $C(4,-2)$, $r=4\sqrt{2}$

65. $C\left(-\frac{3}{4},\frac{1}{2}\right)$, $r=\frac{2}{3}$ **66.** $C(-2,1)$, $r=\sqrt{5}$

67. not a circle   **68.** $C(1,0)$, $r=\sqrt{\frac{10}{3}}$

69. $C\left(\frac{1}{3},-\frac{2}{3}\right)$, $r=3$

70. $C(-1.6,3.3)$, $r=\sqrt{15.85}\cong 3.98$

71. $(x-2)^2+(y-7)^2=2$

72. $\left(x-\frac{3}{2}\right)^2+(y-2)^2=\frac{85}{4}=21.25$

73. right triangle   **74.** neither

75. x-axis, y-axis, origin   **76.** x-axis, y-axis, origin

77. perpendicular   **78.** parallel

  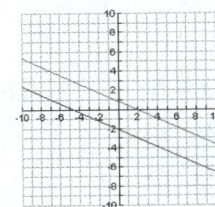

79. $y = -\frac{2}{3} \pm \sqrt{9 - \left(x - \frac{1}{3}\right)^2}$

80. $y = 3.3 \pm \sqrt{15.85 - \left(x + 1.6\right)^2}$

Practice Test

1. $d = \sqrt{82}$ **2.** $(1,2)$ **3.** $d = \sqrt{29}$, $\left(\frac{1}{2}, 5\right)$

4. 77 miles **5.** $y = 1, 9$

6. $(-3, -4)$ **7.** x-axis

8. x-intercepts: $(\pm 3, 0)$, no y-intercepts

9.   **10.**

11. x-intercepts: $(6,0)$, y-intercepts: $(0,-2)$

12. $y = \frac{2}{3}x - 2$ **13.** $y = \frac{8}{3}x - 8$

14. $y = 4x + 3$

15. $y = x + 5$ **16.** $y = 4x + 3$

17. $y = -2x + 3$ **18.** $y = -2x + 6$

19. $y = 2x + 2$ **20.** $y = -x + 1$

21. $(x - 6)^2 + (y + 7)^2 = 64$

22. $C(5, -3)$, $r = 2\sqrt{3}$

23. $(x - 4)^2 + (y - 9)^2 = 20$ **24.** $x^2 + y^2 = 8649$

25. both   **26.** x-axis, y-axis, origin

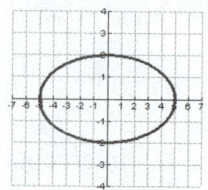

Cumulative Test

1. $\frac{1}{2}$ **2.** $25x^{13/4}$ **3.** $x^4 - 32x^2 + 256$

4. $(2x - 3y)(4x^2 + 6xy + 9y^2)$ **5.** $\frac{5 - x}{5 + x}$

6. $x = \pm 2, 5$ **7.** $12 + 30i$ **8.** $x = -\frac{11}{10}$

9. $x = 1$ **10.** CD: \$9500, stock: \$7500

11. $x = \pm 3$ **12.** $x = -1 \pm \sqrt{\frac{10}{3}}$

13. discriminant is negative; two complex (conjugate) roots

14. $r = \pm \sqrt{p^2 + q^2}$ **15.** $x = 2$ **16.** $x = -\frac{7}{4}, -1$

17. $(0, 12)$ **18.** $(-\infty, -4] \cup [5, \infty)$ **19.** $(-2, 6)$

20. $x = -\frac{9}{2}$, $x = 7$ **21.** origin   **22.** $y = \frac{4}{5}x - 3$

23. $x = 5$ **24.** $y = \frac{7}{3}x + \frac{4}{3}$

25. $C(-5, -3)$, $r = \sqrt{30}$ **26.** $d \approx 5.8$, $(-0.7, 3.8)$

27. neither

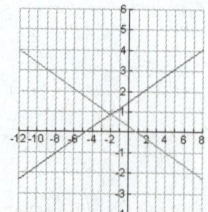

CHAPTER 3

Section 3.1

1. function
2. function
3. not a function
4. not a function
5. function
6. not a function
7. not a function
8. not a function
9. not a function
10. function
11. function
12. function
13. not a function
14. not a function
15. not a function
16. function
17. function
18. function
19. not a function
20. not a function
21. function
22. function
23. not a function
24. function
25. a. 5 **b.** 1 **c.** −3 **26. a.** 1 **b.** −5 **c.** 0
27. a. 3 **b.** 2 **c.** 5 **28. a.** 0 **b.** 4 **c.** −5
29. a. −5 **b.** −5 **c.** −5 **30. a.** −2 **b.** −6 **c.** −4
31. a. 2 **b.** −8 **c.** −5 **32. a.** DNE **b.** 0 **c.** 3
33. 1
34. −1.5 and 3
35. −3 and 1 **36.** −7 **37.** $[-4, 4]$
38. $[-4, 0) \cup [4]$ **39.** 6
40. −3 **41.** −7 **42.** −4
43. 6 **44.** 3 **45.** −1
46. −7 **47.** −33 **48.** 14
49. $-\frac{7}{6}$ **50.** $-\frac{4}{3}$ **51.** $\frac{2}{3}$
52. −1 **53.** 4 **54.** $-4t$
55. $8 - x - a$ **56.** $x^2 + 2bx + 2x + 2b - 3$
57. 2 **58.** $-(2t + h)$ **59.** 1
60. $2x + h + 2$ **61.** 2 **62.** $2 - h$
63. 1 **64.** $h - 4$ **65.** $(-\infty, \infty)$
66. $(-\infty, \infty)$ **67.** $(-\infty, \infty)$ **68.** $(-\infty, \infty)$
69. $(-\infty, 5) \cup (5, \infty)$ **70.** $(-\infty, -3) \cup (-3, \infty)$
71. $(-\infty, -2) \cup (-2, 2) \cup (2, \infty)$
72. $(-\infty, -1) \cup (-1, 1) \cup (1, \infty)$
73. $(-\infty, \infty)$ **74.** $(-\infty, \infty)$ **75.** $(-\infty, 7]$
76. $[7, \infty)$ **77.** $\left[-\frac{5}{2}, \infty\right)$ **78.** $\left(-\infty, \frac{5}{2}\right]$
79. $(-\infty, -2] \cup [2, \infty)$ **80.** $(-\infty, -5] \cup [5, \infty)$
81. $(3, \infty)$ **82.** $(-\infty, 5)$ **83.** $(-\infty, \infty)$

84. $(-\infty, \infty)$ **85.** $(-\infty, -4) \cup (-4, \infty)$
86. $(-\infty, -3) \cup (-3, 3) \cup (3, \infty)$
87. $\left(-\infty, \frac{3}{2}\right)$ **88.** $(-5, 5)$
89. $(-\infty, -2) \cup (3, \infty)$ **90.** $(-\infty, \infty)$
91. $(-\infty, -4] \cup [4, \infty)$ **92.** $(-\infty, \infty)$
93. $\left(-\infty, \frac{3}{2}\right)$ **94.** $(-\infty, -3) \cup (-3, 3) \cup (3, \infty)$
95. $(-\infty, \infty)$ **96.** $(-\infty, \infty)$ **97.** $x = -2, 4$
98. $x = \frac{17}{10}$ **99.** $x = -1, 5, 6$
100. $x = -3, -6$ **101.** $y = 45x$, $(75, \infty)$
102. $y = 35 + 0.10x$, $[0, \infty)$
103. $T(6) = 64.8°F$, $T(12) = 90°F$
104. $256\,\text{ft}$, $[0, \infty)$
105. $P(10) \cong \$641.66$, $P(100) \cong \$634.50$
106. lowest price: $10, 4000 available cards, highest price: $642.38, 1 available card
107. $V(x) = x(10 - 2x)^2$, $(0, 5)$
108. $V(h) = 100\pi h$, $4700\,\text{gallons}$
109. $E(4) \cong 84$ Yen

$E(7) \cong 84$ Yen

$E(8) \cong 83$ Yen

110. a. The number of Japanese Yen to the US Dollar exchange rate increased by approximately 1 Japanese Yen to US Dollar from Week 2 to Week 3.
   b. The number of Japanese Yen to the US Dollar exchange rate decrease by approximately 2 Japanese Yen to US Dollar from Week 6 to Week 7.
111. 229 people
112. 213 people
113. a. $A(x) = x(x - 3.375) = x^2 - 3.375x$
   b. $A(4.5) = 5.0625$. The area of the window is approximately 5 square inches.
   c. $A(8.5) = 43.5625$. This is not possible because the window would be larger than the entire envelope (32 square inches).

114. a. $A(x) = x(x - 2.875) = x^2 - 2.875x$

b. $A(5.25) = 12.46875$. The area of the window is approximately 12.5 square inches.

c. $A(10) = 71.25$. This is not possible because the window would be larger than the entire envelope (57 square inches).

115. Yes, for every input (year), there corresponds a unique output (federal funds rate).

116. (2000, 5.45), (2002, 1.73), (2004, 1.00), (2006, 4.50), (2008, 3.50)

117. (1989, 4000), (1993, 6000), (1997, 6000), (2001, 8000), (2005, 11000)

118. Yes, for every input there corresponds a unique output.

119. a. $F(50) = 0$ **b.** $g(50) = 1000$
c. $H(50) = 2000$

120. $F(100) + g(100) + G(100)$ represents the total amount (in millions of metric tons) of carbon emitted in 2000 by natural gas, coal, and petroleum.

121. Should apply the <u>vertical</u> line test to determine if the relationship describes a function, which it is a function.

122. $H(3) - H(-1) \neq H(3) + H(1)$, in general.

123. $f(x + 1) \neq f(x) + f(1)$, in general.

124. $3 - t > 0$ should be $3 - t \geq 0$. The statement directly preceding the computation should be, "What can $3 - t$ be?" The domain should be $(-\infty, 3]$.

125. $G(-1 + h) \neq G(-1) + G(h)$, in general.

126. $f(1) = -1$ means the point $(1, -1)$ must satisfy the function. So,
$-1 = |1 - A| - 1 \Rightarrow 0 = |1 - A| \Rightarrow 1 = A$

127. false   **128.** false   **129.** false
130. true   **131.** $A = 2$ **132.** $b = 3$
133. $C = -5$, $D = -2$

134. answers may vary, $g(x) = \dfrac{4}{x - 5}$

135. $(-\infty, -a) \cup (-a, a) \cup (a, \infty)$

136. $(-\infty, -a] \cup [a, \infty)$

137.

Warmest: noon, 90° F The values of T outside the interval [6,18] are too small to be considered temperatures in Florida.

138.

Airborne: 8 seconds, $h = 256$ ft The firecracker will land after 8 seconds.

139.

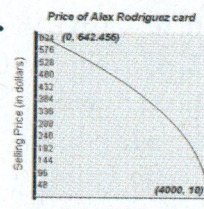

Lowest: is $10, highest: $642.38, agree

140. $4\pi(6r + 9)$ mm

141.

shift graph of $f(x)$ two units to the right

142.

shift graph of $f(x)$ two units to the left

Section 3.2

1. neither   **2.** neither   **3.** even
4. even   **5.** odd   **6.** odd

7. neither **8.** neither **9.** odd
10. odd **11.** even **12.** even
13. even **14.** even **15.** neither
16. neither **17.** neither **18.** neither
19. neither **20.** even **21.** neither
22. odd **23.** neither **24.** even

25. a. $(-\infty,\infty)$ **b.** $[-1,\infty)$ **c.** increasing: $(-1,\infty)$, decreasing: $(-3,-2)$, constant: $(-\infty,-3)\cup(-2,-1)$ **d.** 0 **e.** -1 **f.** 2

26. a. $[-4,\infty)$ **b.** $(-\infty,3]$ **c.** increasing: $(1,2)$, decreasing: $(-3,0)\cup(2,\infty)$, constant: $[-4,-3)\cup(0,1)$ **d.** -1 **e.** approximately 1.8 **f.** 1

27. a. $[-7,2]$ **b.** $[-5,4]$ **c.** increasing: $(-4,0)$, decreasing: $(-7,-4)\cup(0,2)$, constant: nowhere **d.** 4 **e.** 1 **f.** -5

28. a. $(-\infty,\infty)$ **b.** $(-\infty,\infty)$ **c.** increasing: $(-\infty,-3)\cup(3,\infty)$, decreasing: $(-3,3)$, constant: nowhere **d.** 0 **e.** 3.5 **f.** approximately -3.3

29. a. $(-\infty,\infty)$ **b.** $(-\infty,\infty)$ **c.** increasing: $(-\infty,-3)\cup(4,\infty)$, decreasing: nowhere, constant: $(-3,4)$ **d.** 2 **e.** 2 **f.** 2

30. a. $(-\infty,\infty)$ **b.** $(-\infty,\infty)$ **c.** increasing: nowhere, decreasing: $(-\infty,\infty)$, constant: nowhere **d.** 2 **e.** 1 **f.** -1

31. a. $(-\infty,\infty)$ **b.** $[-4,\infty)$ **c.** increasing: $(0,\infty)$, decreasing: $(-\infty,0)$, constant: nowhere **d.** -4 **e.** 0 **f.** 0

32. a. $(-\infty,\infty)$ **b.** $[0,\infty)$ **c.** increasing: $(3,\infty)$, decreasing: $(-\infty,-3)$, constant: $(-3,3)$ **d.** 0 **e.** 0 **f.** 0

33. a. $(-\infty,0)\cup(0,\infty)$ **b.** $(-\infty,0)\cup(0,\infty)$ **c.** increasing: $(-\infty,0)\cup(0,\infty)$, decreasing: nowhere, constant: nowhere **d.** undefined **e.** 3 **f.** -3

34. a. $(-\infty,4)\cup(4,\infty)$ **b.** $(-\infty,\infty)$ **c.** increasing: $(-\infty,0)\cup(4,\infty)$, decreasing: $(0,4)$, constant: nowhere **d.** 4 **e.** approximately 3.5 **f.** approximately 2.5

35. a. $(-\infty,0)\cup(0,\infty)$ **b.** $(-\infty,5)\cup[7]$ **c.** increasing: $(-\infty,0)$, decreasing: $(5,\infty)$, constant: $(0,5)$ **d.** undefined **e.** 3 **f.** 7

36. a. $(-8,0)\cup(0,4]$ **b.** $(-4,3]$ **c.** increasing: $(-8,-5)\cup(0,4)$, decreasing: $(-5,0)$, constant: nowhere **d.** undefined **e.** -0.8 **f.** 0

37. $2x+h-1$ **38.** $2x+h+2$
39. $2x+h+3$ **40.** $-2x-h+5$
41. $2x+h-3$ **42.** $2x+h-2$
43. $-6x-3h+5$ **44.** $-8x-4h+2$
45. 13 **46.** $-\frac{1}{3}$ **47.** 1
48. 2 **49.** -2 **50.** -4
51. -1 **52.** 1

53.

domain: $(-\infty,\infty)$
range: $(-\infty,2]$
increasing: $(-\infty,2)$
decreasing: nowhere
constant: $(2,\infty)$

54.

domain: $(-\infty,\infty)$
range: $\{-1\}\cup(1,\infty)$
increasing: nowhere
decreasing: $(-\infty,-1)$
constant: $(-1,\infty)$

open hole $(-1,1)$, closed hole $(-1,-1)$

55.

domain: $(-\infty,\infty)$
range: $[0,\infty)$
increasing: $(0,\infty)$
decreasing: $(-1,0)$
constant: $(-\infty,-1)$

56.

domain: $(-\infty, \infty)$

range: $[0, \infty)$

increasing: $(0, 2)$

decreasing: $(-\infty, 0)$

constant: $(2, \infty)$

57.

domain: $(-\infty, \infty)$

range: $(-\infty, \infty)$

increasing: $(-\infty, \infty)$

decreasing: nowhere
constant: nowhere

58.

domain: $(-\infty, \infty)$

range: $[0, \infty)$

increasing: $(0, \infty)$

decreasing: $(-\infty, 0)$

constant: nowhere

59.

domain: $(-\infty, \infty)$

range: $[1, \infty)$

increasing: $(1, \infty)$

decreasing: $(-\infty, 1)$

constant: nowhere

60.

domain: $(-\infty, \infty)$

range: $(-\infty, \infty)$

increasing:
$(-\infty, -1) \cup (0, \infty)$

decreasing: $(-1, 0)$

constant: nowhere

61.

domain:
$(-\infty, 2) \cup (2, \infty)$

range: $(1, \infty)$

increasing: $(2, \infty)$

decreasing: $(-\infty, 2)$

62.

domain:
$(-\infty, 2) \cup (-2, \infty)$

range: $(1, \infty)$

increasing: $(-2, \infty)$

decreasing: $(-\infty, -2)$

63.

domain: $(-\infty, \infty)$

range: $[-1, 3]$

increasing: $(-1, 3)$

decreasing: nowhere
constant:
$(-\infty, -1) \cup (3, \infty)$

64.

domain:
$(-\infty, -1) \cup (-1, 3) \cup (3, \infty)$

range: $[-1, 3]$

increasing: $(-1, 3)$

decreasing: nowhere
constant:

open holes $(-1, -1)$, $(3, 3)$ $(-\infty, -1) \cup (3, \infty)$

65.

domain: $(-\infty, \infty)$

range: $[1, 4]$

increasing: $(1, 2)$

decreasing: nowhere
constant:
$(-\infty, 1) \cup (2, \infty)$

66.

domain:
$(-\infty, 1) \cup (1, 2) \cup (2, \infty)$

range: $[1, 4]$

increasing: $(1, 2)$

decreasing: nowhere

open holes $(1, 1)$, $(2, 4)$ constant: $(-\infty, 1) \cup (2, \infty)$

67.

open holes $(-2,1), (-2,-1), (1,2)$

closed hole $(1,0)$

domain:
$$(-\infty,-2)\cup(-2,\infty)$$
range: $(-\infty,\infty)$

increasing: $(-2,1)$

decreasing:
$$(-\infty,-2)\cup(1,\infty)$$
constant: nowhere

68.

open holes $(-2,-1), (1,2), (1,0)$

closed hole $(-2,1)$

domain: $(-\infty,1)\cup(1,\infty)$

range: $(-\infty,\infty)$

increasing: $(-2,1)$

decreasing:
$$(-\infty,-2)\cup(1,\infty)$$
constant: nowhere

69.

domain: $(-\infty,\infty)$

range: $[0,\infty)$

increasing: $(0,\infty)$

decreasing: nowhere

constant: $(-\infty,0)$

70.

open hole $(1,1)$

domain: $(-\infty,1)\cup(1,\infty)$

range: $[1,\infty)$

increasing: $(1,\infty)$

decreasing: nowhere

constant: $(-\infty,1)$

71.

closed hole $(0,0)$

domain: $(-\infty,\infty)$

range: $(-\infty,\infty)$

increasing: nowhere

decreasing:
$$(-\infty,0)\cup(0,\infty)$$
constant: nowhere

72.

closed hole $(0,0)$

domain: $(-\infty,\infty)$

range: $(-\infty,\infty)$

increasing:
$$(-\infty,0)\cup(0,\infty)$$
decreasing: nowhere

constant: nowhere

73.

open holes $(-1,-1), (1,1), (1,-1)$

graph of $-\sqrt[3]{x}$ on $(-\infty,-1)$,

closed hole $(-1,1)$

domain:
$$(-\infty,1)\cup(1,\infty)$$
range:
$$(-\infty,-1)\cup(-1,\infty)$$
increasing: $(-1,1)$

decreasing:
$$(-\infty,-1)\cup(1,\infty)$$
constant: nowhere

74.

graph of $-\sqrt[3]{x}$ should appear
on the interval $(-\infty,-1)$

domain: $(-\infty,1)\cup(1,\infty)$

range: $(-\infty,1)\cup(1,\infty)$

increasing: $(1,\infty)$

decreasing: $(-\infty,-1)$

constant: nowhere

75.

open holes $(-2,2), (2,2)$

closed holes $(-2,1), (2,4)$

domain: $(-\infty,\infty)$

range: $(-\infty,2)\cup[4,\infty)$

increasing:
$$(-\infty,-2)\cup(0,2)\cup(2,\infty)$$
decreasing: $(-2,0)$

constant: nowhere

76.

open holes $(-1,1), (1,1)$

domain:
$$(-\infty,-1)\cup(-1,1)\cup(1,\infty)$$
range: $[1,\infty)$

increasing: $(1,\infty)$

decreasing: $(-\infty,-1)$

constant: $(-1,1)$

77.

open hole (1, 1)

domain:
$$(-\infty,1)\cup(1,\infty)$$
range: $(-\infty,1)\cup(1,\infty)$
increasing:
$$(-\infty,1)\cup(1,\infty)$$
decreasing: nowhere
constant: nowhere

78.

open hole (−1, −1), closed hole (−1,1)

domain: $(-\infty,\infty)$
range: $(-1,\infty)$
increasing: $(-1,\infty)$
decreasing: $(-\infty,-1)$
constant: nowhere

79. Profit is increasing from October through December and decreasing from January through October.

80. Cost is increasing from January through August and decreasing from August through December.

81. $C(x)=\begin{cases}10x, & 0\le x\le 50\\ 9x, & 50<x\le100\\ 8x, & x>100\end{cases}$

82. $C(x)=\begin{cases}176.12x, & 0\le x\le 50\\ 159.73x, & 50<x\le100\end{cases}$

83. $C(x)=\begin{cases}250x, & 0\le x\le 10\\ 175x+750, & x>10\end{cases}$

84. $C(x)=\begin{cases}0.39x, & 0\le x\le 10\\ 0.12x+2.7, & x>10\end{cases}$

85. $C(x)=\begin{cases}1000+35x, & 0\le x\le 100\\ 2000+25x, & x>100\end{cases}$

86. $C(x)=1400+25x$

87. $R(x)=\begin{cases}50,000+3x, & 0\le x\le 100,000\\ 4x-50,000, & x>100,000\end{cases}$

88. $R(x)=\begin{cases}35,000+3x, & 0\le x\le 100,000\\ -165,000+5x, & x>100,000\end{cases}$

89. $P(x)=65x-800$ **90.** $P(x)=5x-30$

91. $f(x)=0.80+0.17[\![x]\!], \; x\ge0$

92. $f(x)=1.13+0.17[\![x]\!], \; x\ge0$

93. $f(t)=3(-1)^{[\![t]\!]}, \; t\ge0$

94. $f(x)=(-1)^{\left(1+\left[\!\left[\frac{x}{100}\right]\!\right]\right)}, \; x\ge0$

95. a. 20 million tons per year
  b. 110 million tons per year
96. a. 140 million tons per year
  b. 80 million tons per year
97. 0 ft/sec   **98.** −16 ft/sec
99. Demand for the product is increasing at an approximate rate of 236 units over the first quarter.
100. Demand for the product is increasing at an approximate rate of 250 units over the fourth quarter.
101. Should exclude the origin since $x=0$ is not in the domain. The range should be $(0,\infty)$. The graph is also incorrect. It should contain an open circle at (0,0).
102. The open and closed holes at $x=1$ should be switched, and then the range should be changed to $[-1,\infty)$.
103. The portion of $C(x)$ for $x>30$ should be:
$15+\underset{\substack{\text{Number miles}\\\text{beyond first 30}}}{x-30}$
104. The portion of $C(x)$ for $x>10,000$ should be: $0.02(10,000)+0.04(x-10,000)$
105. true   **106.** true   **107.** false
108. true   **109.** yes, if $a=2b$
110. no   **111.** odd   **112.** even
113. odd   **114.** graph of tan x

115. domain: $\mathbb{R}$
range:
    set of integers

116. domain: $\mathbb{R}$
range:
    set of integers

31.

32.

Section 3.3

1. l   **2.** j   **3.** a   **4.** d   **5.** b   **6.** k
7. i   **8.** h   **9.** c   **10.** e   **11.** g   **12.** f

13. $y = |x| + 3$ **14.** $y = |x + 4|$

15. $y = |-x| = |x|$ **16.** $y = -|x|$

17. $y = 3|x|$ **18.** $y = \frac{1}{3}|x|$

19. $y = x^3 - 4$ **20.** $y = (x-3)^3$

21. $y = (x+1)^3 + 3$ **22.** $y = -x^3$

23. $y = (-x)^3$ **24.** $y = x^3$

33.

34.

35.

36.

25.

26.

37.

38.

27.

28.

39.

40.

29.

30.

41.

42.

43.

44.

45.

46.

47.

48.

49.

50.

51.

52.

53.

54.

55.

56.

57.

58.

59.

60.

61.

62.

63.

64.

65.

66.

67.

68.

69.

70.

71.

72.

73.

74.

75. $f(x) = (x-3)^2 + 2$

76. $f(x) = (x+1)^2 - 3$

77. $f(x) = -(x+1)^2 + 1$

78. $f(x) = -(x-3)^2 + 2$

79. $f(x) = 2(x-2)^2 - 5$

80. $f(x) = 3(x-1)^2 + 2$

81. $S(x) = 10x$ and $S(x) = 10x + 50$

82. $P(x-10)$ **83.** $T(x) = 0.33(x - 6500)$

84. $A(x+3) = \sqrt{x+3} + 2$

85. a. $BSA(w) = \sqrt{\dfrac{9w}{200}}$;

b. $BSA(w-3) = \sqrt{\dfrac{9(w-3)}{200}}$

86. a. $BSA(w) = \sqrt{\dfrac{w}{20}}$;

b. $BSA(w+5) = \sqrt{\dfrac{w+5}{20}}$

87. (b) is wrong – shift right 3 units.

88. (c) is wrong – reflect over x-axis.

89. (b) should be deleted since $|3-x| = |x-3|$.

The correct sequence of steps would be:
$(a) \rightarrow (c)^* \rightarrow (d)$, where

$(c)^*$: Shift to the right 3.

90. (b) is wrong and (d) is misplaced. The correct sequence of steps would be:
$(a) \rightarrow (d) \rightarrow (*) \rightarrow (c)$,

where $(*)$ = reflect over x-axis.

91. true   **92.** false   **93.** true

94. true   **95.** $(a+3, b+2)$

96. $(-a, -b+1)$

97. Any part of the graph of $y = f(x)$ that is below the x-axis is reflected above it for the graph of $y = |f(x)|$.

a.   **b.**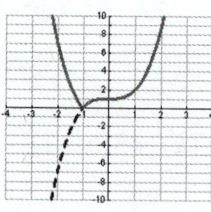

98. The relationship is described by:
$$f(|x|) = \begin{cases} f(x), & x \geq 0 \\ f(-x), & x < 0 \end{cases}$$

a.   **b.**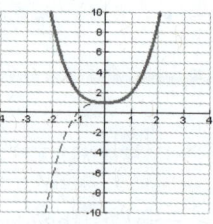

99. If $a > 1$, then the graph is a horizontal compression. If $0 < a < 1$, then the graph is a horizontal expansion.

a.   **b.**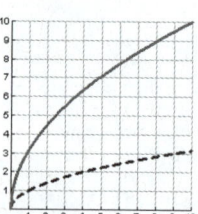

100. If $0 < a < 1$, then the graph is a vertical compression. If $a > 1$, then the graph is a vertical expansion.

a.   **b.**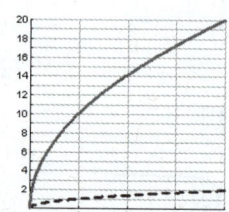

101. Each horizontal line in the graph of $y = [\![x]\!]$ is stretched by a factor of 2. Also, there is a vertical shift up of one unit.

102. Any portion of the graph of $y = [\![x]\!]$ that is below the x-axis is reflected above it. Also, there is a vertical shift up of one unit and a vertical compression by a factor of ½.

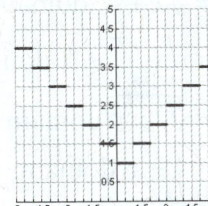

Section 3.4

1.

$$
\left.\begin{array}{l}
f(x) + g(x) = x + 2 \\
f(x) - g(x) = 3x \\
f(x) \cdot g(x) = -2x^2 + x + 1
\end{array}\right\} \text{domain: } (-\infty, \infty)
$$

$$\frac{f(x)}{g(x)} = \frac{2x+1}{1-x} \text{ domain: } (-\infty, 1) \cup (1, \infty)$$

2.

$$
\left.\begin{array}{l}
f(x) + g(x) = 5x - 2 \\
f(x) - g(x) = x + 6 \\
f(x) \cdot g(x) = 6x^2 - 8x - 8
\end{array}\right\} \text{domain: } (-\infty, \infty)
$$

$$\frac{f(x)}{g(x)} = \frac{3x+2}{2x-4} \text{ domain: } (-\infty, 2) \cup (2, \infty)$$

3.

$$
\left.\begin{array}{l}
f(x) + g(x) = 3x^2 - x - 4 \\
f(x) - g(x) = x^2 - x + 4 \\
f(x) \cdot g(x) = 2x^4 - x^3 - 8x^2 + 4x
\end{array}\right\} \text{domain:} (-\infty, \infty)
$$

$$\frac{f(x)}{g(x)} = \frac{2x^2 - x}{x^2 - 4} \text{ domain: } (-\infty, -2) \cup (-2, 2) \cup (2, \infty)$$

4.

$$
\left.\begin{array}{l}
f(x) + g(x) = x^2 + 3x - 23 \\
f(x) - g(x) = -x^2 + 3x + 27 \\
f(x) \cdot g(x) = 3x^3 + 2x^2 - 75x - 50
\end{array}\right\} \text{domain:} (-\infty, \infty)
$$

$$\frac{f(x)}{g(x)} = \frac{3x+2}{x^2 - 25} \text{ domain: } (-\infty, -5) \cup (-5, 5) \cup (5, \infty)$$

5.

$$
\left.\begin{array}{l}
f(x) + g(x) = \dfrac{1+x^2}{x} \\[2mm]
f(x) - g(x) = \dfrac{1-x^2}{x} \\[2mm]
f(x) \cdot g(x) = 1 \\[2mm]
\dfrac{f(x)}{g(x)} = \dfrac{1}{x^2}
\end{array}\right\} \text{domain: } (-\infty, 0) \cup (0, \infty)
$$

6.

$$
\left.\begin{array}{l}
f(x) + g(x) = \dfrac{7x^2 + 5x + 22}{(x-4)(3x+2)} \\[3mm]
f(x) - g(x) = \dfrac{5x^2 + 21x - 10}{(x-4)(3x+2)} \\[3mm]
f(x) \cdot g(x) = \dfrac{2x+3}{3x+2} \\[3mm]
\dfrac{f(x)}{g(x)} = \dfrac{(2x+3)(3x+2)}{(x-4)^2}
\end{array}\right\}
$$

$$\text{domain: } \left(-\infty, -\tfrac{2}{3}\right) \cup \left(-\tfrac{2}{3}, 4\right) \cup (4, \infty)$$

7.

$$
\left.\begin{array}{l}
f(x) + g(x) = 3\sqrt{x} \\
f(x) - g(x) = -\sqrt{x} \\
f(x) \cdot g(x) = 2x
\end{array}\right\} \text{domain: } [0, \infty)
$$

$$\frac{f(x)}{g(x)} = \frac{1}{2} \text{ domain: } (0, \infty)$$

8.

$$f(x)+g(x)=\sqrt{x-1}+2x^2$$
$$f(x)-g(x)=\sqrt{x-1}-2x^2$$
$$f(x)\cdot g(x)=2x^2\sqrt{x-1}$$
$$\frac{f(x)}{g(x)}=\frac{\sqrt{x-1}}{2x^2}$$
domain: $[1,\infty)$

9.

$$f(x)+g(x)=\sqrt{4-x}+\sqrt{x+3}$$
$$f(x)-g(x)=\sqrt{4-x}-\sqrt{x+3}$$
$$f(x)\cdot g(x)=\sqrt{4-x}\cdot\sqrt{x+3}$$
domain: $[-3,4]$

$$\frac{f(x)}{g(x)}=\frac{\sqrt{4-x}\sqrt{x+3}}{x+3}\quad\text{domain: }(-3,4]$$

10.

$$f(x)+g(x)=\sqrt{1-2x}+\tfrac{1}{x}$$
$$f(x)-g(x)=\sqrt{1-2x}-\tfrac{1}{x}$$
$$f(x)\cdot g(x)=\sqrt{1-2x}\cdot\tfrac{1}{x}$$
domain: $\left(-\infty,0\right)\cup\left(0,\tfrac{1}{2}\right]$
$$\frac{f(x)}{g(x)}=x\sqrt{1-2x}$$

11. $(f\circ g)(x)=2x^2-5$ domain: $(-\infty,\infty)$

$(g\circ f)(x)=4x^2+4x-2$ domain: $(-\infty,\infty)$

12. $(f\circ g)(x)=x^2-4x+3$ domain: $(-\infty,\infty)$

$(g\circ f)(x)=-x^2+3$ domain: $(-\infty,\infty)$

13. $(f\circ g)(x)=\dfrac{1}{x+1}$

domain: $(-\infty,-1)\cup(-1,\infty)$

$(g\circ f)(x)=\dfrac{1}{x-1}+2$ domain:

$(-\infty,1)\cup(1,\infty)$

14. $(f\circ g)(x)=\dfrac{2}{x-1}$

domain: $(-\infty,1)\cup(1,3)\cup(3,\infty)$

$(g\circ f)(x)=\dfrac{2x-4}{x-3}$ domain: $(-\infty,3)\cup(3,\infty)$

15. $(f\circ g)(x)=\dfrac{1}{|x-1|}$ domain: $(-\infty,1)\cup(1,\infty)$

$(g\circ f)(x)=\dfrac{1}{|x|-1}$

domain: $(-\infty,-1)\cup(-1,1)\cup(1,\infty)$

16. $(f\circ g)(x)=\left|\dfrac{1-x}{x}\right|$ domain: $(-\infty,0)\cup(0,\infty)$

$(g\circ f)(x)=\dfrac{1}{|x-1|}$

domain: $(-\infty,0)\cup(0,1)\cup(1,\infty)$

17. $(f\circ g)(x)=\sqrt{x+4}$ domain: $[-4,\infty)$

$(g\circ f)(x)=\sqrt{x-1}+5$ domain: $[1,\infty)$

18. $(f\circ g)(x)=\sqrt{-x^2}$ domain: $[0]$

$(g\circ f)(x)=4-x$ domain: $(-\infty,2]$

19. $(f\circ g)(x)=x$ domain: $(-\infty,\infty)$

$(g\circ f)(x)=x$ domain: $(-\infty,\infty)$

20. $(f\circ g)(x)=\sqrt[3]{x^{2/3}\left(x^{2/3}+2\right)}$ domain: $(-\infty,\infty)$

$(g\circ f)(x)=x^4-2x^2+2$ domain: $(-\infty,\infty)$

21. 15   **22.** 113   **23.** 13

24. 33   **25.** $26\sqrt{3}$ **26.** 70

27. $\frac{110}{3}$ **28.** 14   **29.** 11

30. 10   **31.** $3\sqrt{2}$ **32.** 5

33. undefined   **34.** 3   **35.** undefined

36. 4   **37.** 13   **38.** $3\sqrt{2}$

39. $f(g(1))=\tfrac{1}{3}$ $g(f(2))=2$

40. $f(g(1))=2$ $g(f(2))=-\tfrac{1}{3}$

41. $f(g(1))=$ undefined $g(f(2))=$ undefined

42. $f(g(1))=1$ $g(f(2))=2$

43. $f(g(1))=\tfrac{1}{3}$ $g(f(2))=4$

44. $f(g(1))=1$ $g(f(2))=2$

45. $f(g(1))=\sqrt{5}$ $g(f(2))=6$

46. $f(g(1))=\sqrt[3]{-\tfrac{7}{2}}$ $g(f(2))=-\tfrac{1}{4}$

47. $f(g(1)) =$ undefined $\quad g(f(2)) =$ undefined

48. $f(g(1)) = -3 \quad g(f(2)) =$ undefined

49. $f(g(1)) = \sqrt[3]{3} \quad g(f(2)) = 4$

50. $f(g(1)) =$ undefined $\quad g(f(2)) =$ undefined

51. $f(g(x)) = 2\left(\dfrac{x-1}{2}\right) + 1 = x - 1 + 1 = x$

$g(f(x)) = \dfrac{(2x+1)-1}{2} = \dfrac{2x}{2} = x$

52. $f(g(x)) = \dfrac{(3x+2)-2}{3} = \dfrac{3x}{3} = x$

$g(f(x)) = 3\left(\dfrac{x-2}{3}\right) + 2 = x - 2 + 2 = x$

53. $f(g(x)) = \sqrt{(x^2+1)-1} = \sqrt{x^2} = \underset{\text{Since } x \geq 1}{|x| = x}$

$g(f(x)) = \left(\sqrt{x-1}\right)^2 + 1 = (x-1) + 1 = x$

54. $f(g(x)) = 2 - \left(\sqrt{2-x}\right)^2 = 2 - (2-x)$

$\qquad = 2 - 2 + x = x$

$g(f(x)) = \sqrt{2 - \left(2 - x^2\right)} = \sqrt{2 - 2 + x^2}$

$\qquad = \sqrt{x^2} = x$

55. $f(g(x)) = \dfrac{1}{\frac{1}{x}} = x \qquad g(f(x)) = \dfrac{1}{\frac{1}{x}} = x$

56. $f(g(x)) = \left[5 - \left(5 - x^3\right)\right]^{1/3} = \left[5 - 5 + x^3\right]^{1/3}$

$\qquad = \left[x^3\right]^{1/3} = x$

$g(f(x)) = 5 - \left[(5-x)^{1/3}\right]^3 = 5 - (5-x)$

$\qquad = 5 - 5 + x = x$

57. $f(g(x)) = 4\left(\dfrac{\sqrt{x+9}}{2}\right)^2 - 9$

$\qquad = 4\left(\dfrac{x+9}{4}\right) - 9 = x$

$g(f(x)) = \dfrac{\sqrt{(4x^2-9)+9}}{2}$

$\qquad = \dfrac{\sqrt{4x^2}}{2} = \dfrac{2x}{2} = x$

58. $f(g(x)) = \sqrt[3]{8\left(\dfrac{x^3+1}{8}\right) - 1} = \sqrt[3]{x^3} = x$

$g(f(x)) = \dfrac{\left(\sqrt[3]{8x-1}\right)^3 + 1}{8} = \dfrac{8x-1+1}{8} = x$

59. $f(g(x)) = \dfrac{1}{\frac{x+1}{x} - 1} = \dfrac{1}{\frac{x+1-x}{x}} = \dfrac{1}{\frac{1}{x}} = x$

$g(f(x)) = \dfrac{\frac{1}{x-1} + 1}{\frac{1}{x-1}} = \dfrac{\frac{1+x-1}{x-1}}{\frac{1}{x-1}} = \dfrac{\frac{x}{x-1}}{\frac{1}{x-1}} = x$

60. $f(g(x)) = g(f(x)) = \sqrt{25 - \left(\sqrt{25-x^2}\right)^2}$

$\qquad = \sqrt{25 - \left(25 - x^2\right)} = \sqrt{x^2} = x$ since $x \geq 0$

61. $f(x) = 2x^2 + 5x \quad g(x) = 3x - 1$

62. $f(x) = \frac{1}{x} \quad g(x) = x^2 + 1$ or

$\qquad f(x) = \frac{1}{x+1} \quad g(x) = x^2$

63. $f(x) = \frac{2}{|x|} \quad g(x) = x - 3$

64. $f(x) = \sqrt{x} \quad g(x) = 1 - x^2$

65. $f(x) = \dfrac{3}{\sqrt{x} - 2} \quad g(x) = x + 1$

66. $f(x) = \dfrac{x}{3x+2} \quad g(x) = \sqrt{x}$

67. $F(C(K)) = \frac{9}{5}(K - 273.15) + 32$

68. $32°\text{F} = 273.15K, \quad 212°\text{F} = 373.15K$

69. a. $A(x) = \left(\frac{x}{4}\right)^2$ **b.** $A(100) = 625 \text{ ft}^2$

  c. $A(200) = 2500 \text{ ft}^2$

70. a. $A(x) = \dfrac{x^2}{4\pi}$ **b.** $A(100) = \frac{2500}{\pi}$

  c. $A(200) = \frac{10,000}{\pi}$

71. a. $C(p) = 62,000 - 20p$

  b. $R(p) = 600,000 - 200p$

  c. $P(p) = 538,000 - 180p$

72. a. $C(p) = 230,000 - 20p$

  b. $R(p) = 40,000,000 - 4000p$

  c. $P(p) = 39,770,000 - 3980p$

73. a. $C\big(n(t)\big) = -10t^2 + 500t + 1375$

    b. $C\big(n(16)\big) = 6815$

    The cost of production on a day when the assembly line was running for 16 hours is $6,815,000.

74. a. $C\big(n(t)\big) = -32t^2 + 800t + 2375$

    b. $C\big(n(24)\big) = 3143$

    The cost of production on a day when the assembly line was running for 24 hours is $8,615,000.

75. a. $A\big(r(t)\big) = \pi\big(10t - 0.2t^2\big)^2$

    b. 11,385 square miles

76. a. $A\big(r(t)\big) = \pi\big(8t - 0.1t^2\big)^2$

    b. 4418 square miles

77. $A(t) = \pi\Big[150\sqrt{t}\,\Big]^2 = 22{,}500\pi t \ \text{ft}^2$

78. $h = \frac{t}{4}$ **79.** $d(h) = \sqrt{h^2 + 4}$

80. $R(p) = p - (172{,}000 + 0.6p)$

81. Must exclude -2 from the domain.

82. Must exclude -2 from the domain.

83. $(f \circ g)(x) = f(g(x))$, not $f(x) \cdot g(x)$

84. Domain is $[3, \infty)$.

85. Function notation, not multiplication.

86. Didn't distribute " $-$ " to all parts of $g(x)$.

87. false   **88.** false   **89.** true

90. false   **91.** $(g \circ f)(x) = \dfrac{1}{x}$ domain: $x \neq 0, a$

92. $(g \circ f)(x) = \dfrac{1}{x(ax + b)}$ domain: $x \neq 0, -\frac{b}{a}, c$

93. $(g \circ f)(x) = x$ domain: $[-a, \infty)$

94. $(g \circ f)(x) = x^{ab}$ domain: $(0, \infty)$

95.

domain: $[-7, 9]$

96.

domain: $(-\infty, 3)$

97.

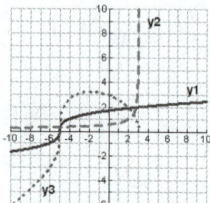

domain: $(-\infty, 3) \cup (-3, -1] \cup [4, 6) \cup (6, \infty)$

98.

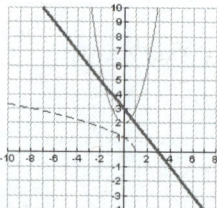

Section 3.5

1. function, not one-to-one

2. function, one-to-one

3. function, one-to-one

4. function, not one-to-one

5. function, one-to-one

6. function, one-to-one

7. not a function

8. function, not one-to-one

9. function, not one-to-one

10. function, one-to-one

11. function, not one-to-one

12. function, not one-to-one

13. function, one-to-one

14. function, one-to-one

15. function, not one-to-one

16. function, one-to-one
17. not one-to-one function
18. not one-to-one function
19. one-to-one function
20. one-to-one function
21. not one-to-one function
22. one-to-one function
23. one-to-one function
24. one-to-one function

25.

26.

27.

28.

29.

30.

31.

32.

33.

34.

35.

36.

37.

38.

39.

40.

41.

42.
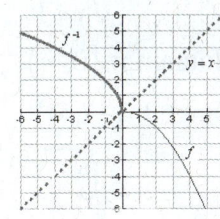

43. $f^{-1}(x) = x + 1$

domain f: $(-\infty, \infty)$ domain f^{-1}: $(-\infty, \infty)$

range f: $(-\infty, \infty)$ range f^{-1}: $(-\infty, \infty)$

44. $f^{-1}(x) = \frac{1}{7} x$

domain f: $(-\infty, \infty)$ domain f^{-1}: $(-\infty, \infty)$

range f: $(-\infty, \infty)$ range f^{-1}: $(-\infty, \infty)$

45. $f^{-1}(x) = -\frac{1}{3} x + \frac{2}{3}$

domain f: $(-\infty, \infty)$ domain f^{-1}: $(-\infty, \infty)$

range f: $(-\infty, \infty)$ range f^{-1}: $(-\infty, \infty)$

46. $f^{-1}(x) = \frac{1}{2}(x-3)$

domain f: $(-\infty, \infty)$ domain f^{-1}: $(-\infty, \infty)$

range f: $(-\infty, \infty)$ range f^{-1}: $(-\infty, \infty)$

47. $f^{-1}(x) = \sqrt[3]{x-1}$

domain f: $(-\infty, \infty)$ domain f^{-1}: $(-\infty, \infty)$

range f: $(-\infty, \infty)$ range f^{-1}: $(-\infty, \infty)$

48. $f^{-1}(x) = \sqrt[3]{x+1}$

domain f: $(-\infty, \infty)$ domain f^{-1}: $(-\infty, \infty)$

range f: $(-\infty, \infty)$ range f^{-1}: $(-\infty, \infty)$

49. $f^{-1}(x) = x^2 + 3$

domain f: $[3, \infty)$ domain f^{-1}: $[0, \infty)$

range f: $[0, \infty)$ range f^{-1}: $[3, \infty)$

50. $f^{-1}(x) = 3 - x^2$

domain f: $(-\infty, 3]$ domain f^{-1}: $[0, \infty)$

range f: $[0, \infty)$ range f^{-1}: $(-\infty, 3]$

51. $f^{-1}(x) = \sqrt{x+1}$

domain f: $[0, \infty)$ domain f^{-1}: $[-1, \infty)$

range f: $[-1, \infty)$ range f^{-1}: $[0, \infty)$

52. $f^{-1}(x) = \sqrt{\dfrac{x-1}{2}}$

domain f: $[0, \infty)$ domain f^{-1}: $[1, \infty)$

range f: $[1, \infty)$ range f^{-1}: $[0, \infty)$

53. $f^{-1}(x) = -2 + \sqrt{x+3}$

domain f: $[-2, \infty)$ domain f^{-1}: $[-3, \infty)$

range f: $[-3, \infty)$ range f^{-1}: $[-2, \infty)$

54. $f^{-1}(x) = 3 + \sqrt{x+2}$

domain f: $[3, \infty)$ domain f^{-1}: $[-2, \infty)$

range f: $[-2, \infty)$ range f^{-1}: $[3, \infty)$

55. $f^{-1}(x) = \frac{2}{x}$

domain f: $(-\infty, 0) \cup (0, \infty)$

range f: $(-\infty, 0) \cup (0, \infty)$

domain f^{-1}: $(-\infty, 0) \cup (0, \infty)$

range f^{-1}: $(-\infty, 0) \cup (0, \infty)$

56. $f^{-1}(x) = -\frac{3}{x}$

domain f: $(-\infty, 0) \cup (0, \infty)$

range f: $(-\infty, 0) \cup (0, \infty)$

domain f^{-1}: $(-\infty, 0) \cup (0, \infty)$

range f^{-1}: $(-\infty, 0) \cup (0, \infty)$

57. $f^{-1}(x) = \frac{3x-2}{x} = 3 - \frac{2}{x}$

domain f: $(-\infty, 3) \cup (3, \infty)$

range f: $(-\infty, 0) \cup (0, \infty)$

domain f^{-1}: $(-\infty, 0) \cup (0, \infty)$

range f^{-1}: $(-\infty, 3) \cup (3, \infty)$

58. $f^{-1}(x) = \frac{7-2x}{x}$

domain f: $(-\infty, -2) \cup (-2, \infty)$

range f: $(-\infty, 0) \cup (0, \infty)$

domain f^{-1}: $(-\infty, 0) \cup (0, \infty)$

range f^{-1}: $(-\infty, -2) \cup (-2, \infty)$

59. $f^{-1}(x) = \frac{5x-1}{x+7}$

domain f: $(-\infty, 5) \cup (5, \infty)$

range f: $(-\infty, -7) \cup (-7, \infty)$

domain f^{-1}: $(-\infty, -7) \cup (-7, \infty)$

range f^{-1}: $(-\infty, 5) \cup (5, \infty)$

60. $f^{-1}(x) = \frac{5-7x}{x-2}$

domain f: $(-\infty, -7) \cup (-7, \infty)$

range f: $(-\infty, 2) \cup (2, \infty)$

domain f^{-1}: $(-\infty, 2) \cup (2, \infty)$

range f^{-1}: $(-\infty, -7) \cup (-7, \infty)$

61. not one-to-one

62. one-to-one

$$G^{-1}(x) = \begin{cases} \frac{1}{x}, & x < 0 \\ x^2, & x \geq 0 \end{cases}$$

63. one-to-one

64. not one-to-one

$$f^{-1}(x) = \begin{cases} x & x \leq -1 \\ \sqrt[3]{x} & -1 < x < 1 \\ x & x \geq 1 \end{cases}$$

65. $f^{-1}(x) = \frac{5}{9}(x - 32)$ The inverse function represents the conversion from degrees Fahrenheit to degrees Celsius.

66. $C^{-1}(x) = \frac{9}{5}x + 32$ The inverse function represents the conversion from degrees Celsius to degrees Fahrenheit.

67. $C(x) = \begin{cases} 250x, & 0 \leq x \leq 10 \\ 175x + 750, & x > 10 \end{cases}$

$C^{-1}(x) = \begin{cases} \frac{x}{250}, & 0 \leq x \leq 2500 \\ \frac{x-750}{175}, & x > 2500 \end{cases}$

68. $C(x) = \begin{cases} 0.39x, & 0 \leq x \leq 10 \\ 3.9 + 0.12(x - 10), & x > 10 \end{cases}$

$C^{-1}(x) = \begin{cases} \frac{x}{0.39}, & 0 \leq x \leq 3.9 \\ \frac{x-2.7}{0.12}, & x > 3.9 \end{cases}$

69. $E(x) = 7.5x$, $E^{-1}(x) = \frac{x}{7.5}$, $x \geq 0$ The inverse function tells you how many hours you need to work to attain a certain take home pay.

70. $E(x) = \begin{cases} 8x, & 0 \leq x \leq 40 \\ 12x - 160, & x > 40 \end{cases}$,

$E^{-1}(x) = \begin{cases} \frac{x}{8}, & 0 \leq x \leq 320 \\ \frac{x+160}{12}, & x > 320 \end{cases}$ The inverse function tells you how many hours you need to work to attain a certain take home pay.

71. domain: $[0, 24]$ range: $[97.5528, 101.70]$

72. $t(T) = 24 + \sqrt[3]{\frac{10,000(T - 101.70)}{3}}$

73. domain: $[97.5528, 101.70]$ range: $[0, 24]$

74. 5:00 am

75. Not a function since the graph does not pass the vertical line test.

76. To determine the points on the graph of the inverse of f, switch the order of the x and y in the ordered pairs rather than multiplying by -1.

77. Must restrict the domain to a portion on which f is one-to-one, say $x \geq 0$. Then, the calculation will be valid.

78. domain f^{-1} = range $f = [0, \infty)$, not $[2, \infty)$.

79. false   **80.** false   **81.** false

82. true   **83.** $(b, 0)$ **84.** $(0, a)$

85. $f(x) = \sqrt{1 - x^2}$, $0 \leq x \leq 1, 0 \leq y \leq 1$,

$f^{-1}(x) = \sqrt{1 - x^2}$, $0 \leq x \leq 1, 0 \leq y \leq 1$, domain and range of both are $[0, 1]$

86. $f(x) = f^{-1}(x)$, $x \neq 0$ **87.** $m \neq 0$

88. $f^{-1}(x) = \frac{x-b}{m}$

89. not one-to-one   **90.** one-to-one

91. not one-to-one   **92.** one-to-one

93. no

94. yes

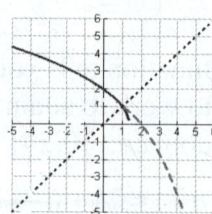

To be inverses, restrict the domain of the parabola to $[0, \infty)$.

95. yes

96. no

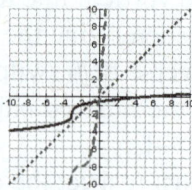

Section 3.6

1. $y = kx$

2. $s = kt$

3. $V = kx^3$

4. $A = kx^2$

5. $z = km$

6. $h = k\sqrt{t}$

7. $f = \dfrac{k}{\lambda}$

8. $P = \dfrac{k}{r^2}$

9. $F = \dfrac{kw}{L}$

10. $V = \dfrac{kT}{P}$

11. $v = kgt$

12. $S = ktd$

13. $R = \dfrac{k}{PT}$

14. $y = \dfrac{k}{xz}$

15. $y = k\sqrt{x}$

16. $y = \dfrac{k}{t^3}$

17. $d = rt$

18. $F = ma$

19. $V = lwh$

20. $A = \dfrac{1}{2}bh$

21. $A = \pi r^2$

22. $V = \dfrac{4}{3}\pi r^3$

23. $V = \dfrac{\pi}{16}hr^2$

24. $W = \dfrac{16}{25}RI^2$

25. $V = \dfrac{400,000}{P}$

26. $I = \dfrac{10,752}{d^2}$

27. $F = \dfrac{2\pi}{\lambda L}$

28. $y = \dfrac{6.4}{xz}$

29. $t = \dfrac{19.2}{s}$

30. $W = \dfrac{7.2}{d^2}$

31. $R = \dfrac{4.9}{I^2}$

32. $y = \dfrac{4.8}{x\sqrt{z}}$

33. $R = \dfrac{0.01L}{A}$

34. $F = \dfrac{12.8m}{d}$

35. $F = \dfrac{0.025m_1 m_2}{d^2}$

36. $w = \dfrac{1.25\sqrt{g}}{t^2}$

37. $W = 7.5H$

38. Orange County: $T = 0.065P$
Seminole County: $T = 0.07P$

39. 1292 mph **40.** 1444 mph

41. $F = 1.618H$ **42.** $S_1 = 1.6S_2$

43. 24 cm **44.** 54N **45.** \$37.50

46. 8800 units **47.** 20,000

48. 75,000 **49.** 600 W/m^2 **50.** 9400 W/m^2

51. Bank of America: 1.5%
Navy Federal Credit Union: 3%

52. 5% **53.** $\frac{11}{12}$ or 0.92 atm

54. $\frac{4}{3}$ or 1.33 atm

55. Should be y is <u>inversely</u> proportional to x.

56. y varies directly with the <u>square</u> of x (x^2), NOT the square root ($\sqrt{x}$).

57. true **58.** false **59.** b **60.** a

61. $\sigma_{pl}^2 = 1.23C_n^2 k^{7/6} L^{11/6}$

62. $\sigma_{sp}^2 = 0.399C_n^2 k^{7/6} L^{11/6}$

63. a. $y = 2.93x + 201.72$

b. 120.07, $y = 120.074x^{0.259}$

c. When the oil price is \$72.70 per barrel in September 2006, the predicted stock index obtained from the least squares regression line is 415, and the value from the equation of direct variation is 364. In this case, the least squares regression line provides a closer approximation to the actual value, 417.

64. a. $y = -0.04x + 6.08$ **b.** 11.53, $y = \dfrac{11.53}{x^{0.27}}$

c. When oil price is \$72.70 per barrel in September 2006, the predicted 5-year maturity rate obtained from the least squares regression line is 3.25, and the equation of inverse variation is 3.61. The equation of the inverse variation provides a closer approximation to the actual value, 5.02.

65. a. $y = -141.73x + 2,419.35$

b. $3,217.69, \quad y = \dfrac{3217.69}{x^{0.41}}$ **c.** When the

5-year maturity rate is 5.02% in September 2006, the predicted number of housing units obtained from the least squares regression line is 1708, and the equation of inverse variation is 1661. The equation of the least squares regression line provides a closer approximation to the actual value, 1861.

66. a. $y = 0.15x + 22.60$ **b.** $0.32, \quad y = 0.32x^{0.91}$

c. There are 1861 housing units in September 2006. The predicted utilities stock index obtained from the least squares regression line is 307, and the equation of direct variation is 304. The equation of the least squares regression line provides a closer approximation to the actual value, 417.

67. a. $y = 0.218x + 0.898$

b. About $2.427 per gallon. Yes. **c.** $3.083

68. a. $0.346, \quad y = 1.163x^{0.346}$

b. About $2.283 per gallon. No. **c.** $2.583

Review Exercises

1. yes **2.** no **3.** yes **4.** yes **5.** no
6. no **7.** yes **8.** yes **9.** no **10.** yes
11. a. 2 **b.** 4 **c.** $x = -3, 4$
12. a. 0 **b.** -4 **c.** $x = -2, 3.2$
13. a. 0 **b.** -2 **c.** $x = -5, 2$
14. a. 7 **b.** -1.5 **c.** never
15. 5 **16.** 29 **17.** -665 **18.** $-\frac{3}{4}$
19. -2 **20.** $5 + 4h$ **21.** 4
22. $2t + h + 4$ **23.** $(-\infty, \infty)$ **24.** $(-\infty, \infty)$
25. $(-\infty, -4) \cup (-4, \infty)$ **26.** $(-\infty, \infty)$
27. $[4, \infty)$ **28.** $(3, \infty)$ **29.** $D = 18$

30. Answers may vary, $f(x) = \dfrac{24}{(x+3)(x-2)}$

31. neither **32.** odd **33.** odd

34. even **35.** neither **36.** neither
37. odd **38.** even
39. a. $[-4, 7]$ **b.** $[-2, 4]$ **c.** increasing: $(3, 7)$, decreasing: $(0, 3)$, constant: $(-4, 0)$
40. a. $(-\infty, -3) \cup (3, \infty)$ **b.** $(-\infty, -2) \cup (3, \infty)$
c. increasing: nowhere, decreasing: $(-\infty, -3) \cup (3, \infty)$, constant: nowhere
41. -2 **42.** 2

43.

domain: $(-\infty, \infty)$

range: $(0, \infty)$

open hole $(0,0)$, closed hole $(0,2)$

44.

domain: $(-\infty, \infty)$

range: $[-3, \infty)$

open holes $(0,4)$, $(1,5)$
closed holes $(1,4)$, $(0,-3)$

45.

domain: $(-\infty, \infty)$

range: $[-1, \infty)$

open hole $(1,3)$
closed hole $(1,-1)$

46.

domain: $(-\infty, 0) \cup (0, \infty)$

range: $(-\infty, -2] \cup (0, \infty)$

open holes $(0,0)$, $(1,1)$
closed hole $(1,-2)$

47. $C(x) = \begin{cases} 25, & x \le 2, \\ 25 + 10.50(x-2), & x > 2 \end{cases}$

48. $E(x) = \begin{cases} 30x, & x \ge 40 \\ 1200 + 45(x-40), & x > 40 \end{cases}$

49.

50.

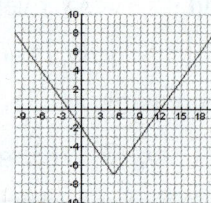

61. $y = \sqrt{x-2} + 3$ domain: $[2, \infty)$

62. $y = \sqrt{-x}$ domain: $(-\infty, 0]$

63. $y = 5\sqrt{x} - 6$ domain: $[0, \infty)$

64. $y = \frac{1}{2}\sqrt{x} + 3$ domain: $[0, \infty)$

65. $y = (x+2)^2 - 12$ **66.** $y = 2\left(x + \frac{3}{2}\right)^2 - \frac{19}{2}$

51.

52.

53.

54.

67.

$$\left.\begin{array}{l} g(x) + h(x) = -2x - 7 \\ g(x) - h(x) = -4x - 1 \\ g(x) \cdot h(x) = -3x^2 + 5x + 12 \end{array}\right\}$$

domain: $(-\infty, \infty)$

$\dfrac{g(x)}{h(x)} = \dfrac{-3x-4}{x-3}$ domain: $(-\infty, 3) \cup (3, \infty)$

68.

$$\left.\begin{array}{l} g(x) + h(x) = x^2 + 2x + 9 \\ g(x) - h(x) = -x^2 + 2x - 3 \\ g(x) \cdot h(x) = 2x^3 + 3x^2 + 12x + 18 \\ \dfrac{g(x)}{h(x)} = \dfrac{2x+3}{x^2+6} \end{array}\right\}$$

domain: $(-\infty, \infty)$

55.

56.

57.

58.

69.

$$\left.\begin{array}{l} g(x) + h(x) = \frac{1}{x^2} + \sqrt{x} \\ g(x) - h(x) = \frac{1}{x^2} - \sqrt{x} \\ g(x) \cdot h(x) = \frac{1}{x^{3/2}} \\ \dfrac{g(x)}{h(x)} = \dfrac{1}{x^{5/2}} \end{array}\right\}$$

domain: $(0, \infty)$

59. $y = \sqrt{x+3}$ domain: $[-3, \infty)$

60. $y = \sqrt{x} - 4$ domain: $[0, \infty)$

70.

$$g(x)+h(x)=\frac{7x+1}{2(x-2)}$$

$$g(x)-h(x)=\frac{-5x+5}{2(x-2)}$$

$$g(x)\cdot h(x)=\frac{(x+3)\cdot(3x-1)}{2(x-2)^2}$$

domain: $(-\infty,2)\cup(2,\infty)$

$$\frac{g(x)}{h(x)}=\frac{x+3}{2(3x-1)}$$ domain: $\left(-\infty,\tfrac{1}{3}\right)\cup\left(\tfrac{1}{3},2\right)\cup(2,\infty)$

71.

$$g(x)+h(x)=\sqrt{x-4}+\sqrt{2x+1}$$
$$g(x)-h(x)=\sqrt{x-4}-\sqrt{2x+1}$$ domain:$[4,\infty)$
$$g(x)\cdot h(x)=\sqrt{x-4}\cdot\sqrt{2x+1}$$

$$\frac{g(x)}{h(x)}=\frac{\sqrt{x-4}}{\sqrt{2x+1}}$$ domain: $[4,\infty)$

72.

$$g(x)+h(x)=x^2+x-2$$
$$g(x)-h(x)=x^2-x-6$$ domain:$(-\infty,\infty)$
$$g(x)\cdot h(x)=(x^2-4)\cdot(x+2)$$

$$\frac{g(x)}{h(x)}=x-2$$ domain: $(-\infty,-2)\cup(-2,\infty)$

73. $(f\circ g)(x)=6x-1$ domain: $(-\infty,\infty)$

$(g\circ f)(x)=6x-7$ domain: $(-\infty,\infty)$

74. $(f\circ g)(x)=x^3+9x^2+29x+3$

domain: $(-\infty,\infty)$

$(g\circ f)(x)=x^3+2x+2$ domain: $(-\infty,\infty)$

75. $(f\circ g)(x)=\dfrac{8-2x}{13-3x}$

domain: $\left(-\infty,4\right)\cup\left(4,\tfrac{13}{3}\right)\cup\left(\tfrac{13}{3},\infty\right)$

$(g\circ f)(x)=\dfrac{x+3}{4x+10}$

domain: $\left(-\infty,-3\right)\cup\left(-3,-\tfrac{5}{2}\right)\cup\left(-\tfrac{5}{2},\infty\right)$

76. $(f\circ g)(x)=\sqrt{2x+7}$ domain: $\left[-\tfrac{7}{2},\infty\right)$

$(g\circ f)(x)=\sqrt{\sqrt{2x^2-5}+6}$

domain: $\left(-\infty,-\sqrt{\tfrac{5}{2}}\right)\cup\left(\sqrt{\tfrac{5}{2}},\infty\right)$

77. $(f\circ g)(x)=\sqrt{x^2-9}$

domain: $(-\infty,-3]\cup[3,\infty)$

$(g\circ f)(x)=x-9$ domain: $[5,\infty)$

78. $(f\circ g)(x)=\sqrt{x^2-4}$

domain: $(-\infty,-2)\cup(2,\infty)$

$(g\circ f)(x)=\dfrac{x}{1-4x}$

domain: $(-\infty,0)\cup\left(0,\tfrac{1}{4}\right)\cup\left(\tfrac{1}{4},\infty\right)$

79. $f(g(3))=857$, $g(f(-1))=51$

80. $f(g(3))=$ undefined, $g(f(-1))=10$

81. $f(g(3))=\tfrac{17}{31}$, $g(f(-1))=1$

82. $f(g(3))=\tfrac{1}{7}$, $g(f(-1))=-\tfrac{3}{4}$

83. $f(g(3))=12$, $g(f(-1))=2$

84. $f(g(3))=$ undefined, $g(f(-1))=\tfrac{1}{7}$

85. $f(x)=3x^2+4x+7$, $g(x)=x-2$

86. $f(x)=\dfrac{x}{1-x}$, $g(x)=\sqrt[3]{x}$

87. $f(x)=\dfrac{1}{\sqrt{x}}$, $g(x)=x^2+7$

88. $f(x)=\sqrt{x}$, $g(x)=|3x+4|$

89. $A(t)=625\pi(t+2)$ in^2

90. $w^2-18w+42=0$

91. yes **92.** no **93.** no

94. no **95.** yes **96.** yes

97. yes **98.** no **99.** yes

100. no

101. **102.**

103.

104.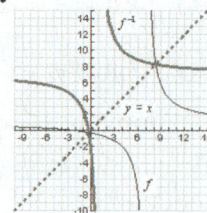

105. $f^{-1}(x) = \frac{1}{2}(x-1) = \frac{x-1}{2}$

domain f: $(-\infty,\infty)$ domain f^{-1}: $(-\infty,\infty)$

range f: $(-\infty,\infty)$ range f^{-1}: $(-\infty,\infty)$

106. $f^{-1}(x) = \sqrt[5]{x-2}$

domain f: $(-\infty,\infty)$ domain f^{-1}: $(-\infty,\infty)$

range f: $(-\infty,\infty)$ range f^{-1}: $(-\infty,\infty)$

107. $f^{-1}(x) = x^2 - 4$

domain f: $[-4,\infty)$ domain f^{-1}: $[0,\infty)$

range f: $[0,\infty)$ range f^{-1}: $[-4,\infty)$

108. $f^{-1}(x) = -4 + \sqrt{x-3}$

domain f: $[-4,\infty)$ domain f^{-1}: $[3,\infty)$

range f: $[3,\infty)$ range f^{-1}: $[-4,\infty)$

109. $f^{-1}(x) = \frac{6-3x}{x-1}$

domain f: $(-\infty,-3)\cup(-3,\infty)$

range f: $(-\infty,1)\cup(1,\infty)$

domain f^{-1}: $(-\infty,1)\cup(1,\infty)$

range f^{-1}: $(-\infty,-3)\cup(-3,\infty)$

110. $f^{-1}(x) = 5 + \left(\frac{1}{2}(x+8)\right)^3$

domain f: $(-\infty,\infty)$ domain f^{-1}: $(-\infty,\infty)$

range f: $(-\infty,\infty)$ range f^{-1}: $(-\infty,\infty)$

111. $S(x) = 22,000 + 0.08x$, $S^{-1}(x) = \frac{x-22,000}{0.08}$,

sales required to earn a desired income

112. $V(s) = 3s^2$, $s \geq 0$, $V^{-1}(s) = \sqrt{\frac{1}{3}s}$, length s

of a side of a base required to get a desired
volume

113. $C = 2\pi r$ **114.** $V = lwh$ **115.** $A = \pi r^2$

116. $F = \dfrac{\pi}{50\lambda L}$ **117.** $W = 8.5H$

118. County A: $T(P) = 0.07P$

County B: $T(P) = 0.08P$

119. domain: $(-\infty,-1)\cup(3,\infty)$

120. domain:

$(-\infty,-3)\cup(-3,3)\cup(3,\infty)$

121. a. $(-\infty,2)\cup(2,\infty)$ **b.** $\{-1,0,1\}\cup(2,\infty)$

c. increasing: $(2,\infty)$, decreasing:

$(-\infty,-1)$, constant: $(-1,0)\cup(0,1)\cup(1,2)$

122. a. $(-\infty,2)\cup(2,\infty)$ **b.** $[0,3)\cup(4,\infty)$

c. increasing: $(-1,0)\cup(1,2)\cup(2,\infty)$,

decreasing: $(-2,-1)\cup(0,1)$,

constant: nowhere

123. The graph of f can be obtained by shifting the graph of g two units to the left. That is, $f(x) = g(x+2)$.

124. The graph of f can be obtained by shifting the graph of g one unit to the right and then reflecting it over the x-axis. That is, $f(x) = -g(x-1)$.

125. domain: $[-1.5, 4)$ **126.** domain:

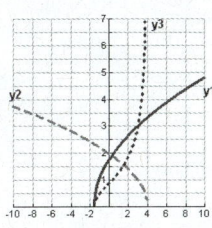

$(-\infty, -2] \cup [2, \infty)$

127. yes   **128.** yes

129. a. $y = 64.03 + 127.06$ **b.** no **c.** $767.36
130. a. 0.464, $y = 215.35 x^{0.464}$ **b.** yes
 c. $626.45

Practice Test

1. b   **2.** a   **3.** c   **4.** −21

5. $\dfrac{\sqrt{x-2}}{x^2+11}$ domain: $[2, \infty)$

6. $\dfrac{x^2+11}{\sqrt{x-2}}$ domain: $(2, \infty)$

7. $x+9$ domain: $(2, \infty)$ **8.** 49

9. 4   **10.** even   **11.** neither **12.** odd

13.   domain: $[3, \infty)$

range: $(-\infty, 2]$

14.   domain: $(-\infty, \infty)$

range: $(-\infty, 0]$

15.   domain: $(-\infty, -1) \cup (-1, \infty)$

range: $[1, \infty)$

16. a. 5 **b.** 2 **c.** 7 **d.** $x = -1, 1, 5.5$ **e.** $x = 7$
17. a. −2 **b.** 4 **c.** −3 **d.** $x = -3, 2$
18. a. −3 **b.** never **c.** −1 **d.** 2
19. $6x + 3h - 4$ **20.** −7 **21.** −32 **22.** $\frac{1}{4}$
23. $f^{-1}(x) = x^2 + 5$

domain f: $[5, \infty)$ domain f^{-1}: $[0, \infty)$

range f: $[0, \infty)$ range f^{-1}: $[5, \infty)$

24. $f^{-1}(x) = \sqrt{x-5}$

domain f: $[0, \infty)$ domain f^{-1}: $[5, \infty)$

range f: $[5, \infty)$ range f^{-1}: $[0, \infty)$

25. $f^{-1}(x) = \dfrac{5x-1}{x+2}$

domain f: $(-\infty, 5) \cup (5, \infty)$

range f: $(-\infty, -2) \cup (-2, \infty)$

domain f^{-1}: $(-\infty, -2) \cup (-2, \infty)$

range f^{-1}: $(-\infty, 5) \cup (5, \infty)$

26. $f^{-1}(x) = \begin{cases} -x, & x \geq 0 \\ -\sqrt{-x}, & x < 0 \end{cases}$

27. $[0, \infty)$ **28.** $(5, -2)$

29. $S(x) = 0.42x$ where x is the original price of a suit

30. $F(K) = \frac{9}{5}K - 459.67$

31. quadrant III, "quarter of unit circle"

32. 2.25 hours

33. $C(x) = \begin{cases} 15, & 0 \leq x \leq 30 \\ x - 15, & x > 30 \end{cases}$

34. $y = \frac{8}{25}x^2$ **35.** $F = \dfrac{30m}{P}$

36. a. $[-4, -2) \cup (-2, 4]$ **b.** $[0, 5]$ **c.** increasing: $(-1, 1) \cup (3, 4)$, decreasing: $(-2, -1) \cup (1, 3)$, constant: $(-4, -2)$

37. yes

Cumulative Test

1. $\dfrac{3 + \sqrt{5}}{2}$ **2.** $(5x + 3)(2x - 7)$

3. $x(x - 2), \ x \neq -2$

4. $x = 30$ **5.** $145 + 0i$ **6.** $x = \frac{1}{6}$

7. 40%   **8.** $x = \frac{4}{3}, -\frac{3}{2}$ **9.** $x = 1 \pm \frac{\sqrt{35}}{5}$

10. $x = -29$ **11.** $x = \pm 2, \pm i\sqrt{3}$

12. $[-1, 5)$ **13.** $(0, 5)$ **14.** $[0.4375, 1.25]$

15. $d = 11.03, (1.25, 2.45)$ **16.** $m = 2.375$

17. $y = -3$ **18.** $C(-6, 9), r = 11$

19. $(x + 2)^2 + (y + 1)^2 = 20$

20. yes, $\sqrt{85^2 + 23^2} \approx 88.06 < 100$

21. $(-\infty, 1) \cup (1, \infty)$ **22.** 30   **23.** 46

24. $f^{-1}(x) = \sqrt{x - 3}$ **25.** $r = \dfrac{135}{t}$

26. a. $[-1, 1) \cup (1, 3]$ **b.** $[0, 1]$ **c.** increasing: $(-1, 0) \cup (1, 2)$, decreasing: $(0, 1) \cup (2, 3)$, constant: nowhere

open hole at $x = 1$

27. $h(x) = x - 2$

CHAPTER 4

Section 4.1

1. b **2.** d **3.** a **4.** c **5.** b **6.** d
7. c **8.** a

9.

10.

11.

12.

13.

14.

15.

16.

17.

18.

19.

20.

21.

22.

23. $f(x) = (x+3)^2 - 12$

24. $f(x) = (x+4)^2 - 14$

25. $f(x) = -(x+5)^2 + 28$

26. $f(x) = -(x+6)^2 + 42$

27. $f(x) = 2(x+2)^2 - 10$

28. $f(x) = 3\left(x - \frac{3}{2}\right)^2 + \frac{17}{4}$

29. $f(x) = -4(x-2)^2 + 9$

30. $f(x) = -5(x-10)^2 + 464$

31. $f(x) = \frac{1}{2}(x-4)^2 - 5$

32. $f(x) = -\frac{1}{3}(x-9)^2 + 31$

33.

34.

35.

36.

37.

38.

39.

40.

41. $\left(\frac{1}{33}, \frac{494}{33}\right)$ **42.** $\left(-\frac{2}{17}, -\frac{55}{17}\right)$

43. $\left(7, -\frac{39}{2}\right)$ **44.** $\left(\frac{3}{5}, \frac{103}{25}\right)$

45. $(-75, 12.95)$ **46.** $(-25, -32.75)$

47. $\left(\frac{15}{28}, \frac{829}{392}\right)$ **48.** $\left(-\frac{7}{3}, \frac{8}{9}\right)$

49. $y = -2(x+1)^2 + 4$ **50.** $y = (x-2)^2 - 3$

51. $y = -5(x-2)^2 + 5$ **52.** $y = -\frac{1}{3}(x-1)^2 + 3$

53. $y = \frac{5}{9}(x+1)^2 - 3$ **54.** $y = \frac{4}{3}(x-0)^2 - 2$

55. $y = 12\left(x - \frac{1}{2}\right)^2 - \frac{3}{4}$ **56.** $y = -\frac{24}{25}\left(x + \frac{5}{6}\right)^2 + \frac{2}{3}$

57. $y = \frac{5}{4}(x-2.5)^2 - 3.5$

58. $y = -0.3(x-1.8)^2 + 2.7$

59. a. 350,000 units   **b.** \$12,262,500

60. a. 20,000 units   **b.** \$140,000

61. He is gaining weight during Jan of 2010 and losing weight from Feb 2010 to Jun 2011.

62. 89 kg

63. a. 120 feet   **b.** 50 yards

64. a. 150 feet   **b.** 40 yards

65. 2,083,333 sq. ft   **66.** 5,625,000 sq. ft.

67. a. 1 second, 116 feet   **b.** 3.69 seconds

68. 75 seconds

69. a. 26,000 feet   **b.** 8944 feet

70. a. 19.53125 feet   **b.** 78.125 feet

71. a. 100 boards   **b.** \$24,000

72. a. $y(x) = -0.0125(x-50)^2 + 30$ **b.** 10 mpg

73. 15 to 16 or 64 to 65 units to break even.

74. \$600

75. a. $f(t) = \frac{28}{27}t^2 + 16$ **b.** 219 million

76. a. $y = -\frac{13}{64}x^2 + 49$ **b.** 19.75%

77. a. $y = -0.01(t-225)^2 + 400$ **b.** 425 minutes

78. a. $p(x) = -\frac{1}{4}x + \frac{75}{2}$ **b.** $R(x) = -\frac{1}{4}x^2 + \frac{75}{2}x$

   c. 75 cars   **d.** \$18.75

79. Step 2 is wrong: Vertex is $(-3, -1)$ Step 4 is wrong: The x-intercepts are $(-2, 0), (-4, 0)$. Should graph $y = (x+3)^2 - 1$.

80. $b = -6$, not 6

81. Step 2 is wrong: $\left(-x^2 + 2x\right) = -\left(x^2 - 2x\right)$

$$f(9) = a(9-2)^2 - 3 = 0$$

82. Step 3 is wrong: $49a - 3 = 0$

$$a = \frac{3}{49}$$

So, $f(x) = \frac{3}{49}(x-2)^2 - 3$.

83. true   **84.** false

85. false   **86.** true

87. $f(x) = a\left(x + \frac{b}{2a}\right)^2 + \frac{4ac - b^2}{4a}$

88. x-intercepts: $\left(h + \sqrt{-\frac{k}{a}}, 0\right), \left(h - \sqrt{-\frac{k}{a}}, 0\right)$

   y-intercept: $\left(0, ah^2 + k\right)$

89. a. 62,500 sq ft   **b.** 79,577 sq ft

90. $195, $76,050

91. a. $(1425, 4038.25)$ **b.** $(0, -23)$

   c. $(4.04, 0), (2845.96, 0)$

   d. $x = 1425$

92. a. $y = 9.2(x + 0.5)^2 + 1.7$ **b.**

   c. yes

93. a. $y = -2x^2 + 12.8x + 4.32$

   b. $y = -2(x - 3.2)^2 + 24.8$, $(3.2, 24.8)$

   c. yes

94. a. $f(x) = 0.44x^2 + 2.92x - 12.10$

   b. $f(x) = 0.44(x + 3.32)^2 - 16.95$, $(-3.32, -16.95)$

   c. yes

95. a.

   b. The equation of the best fit parabola is $y = -1.0589x^2 + 12.268x + 4.3042$ with $r^2 = 0.9814$. This is shown below on the scatterplot:

   c. (i) Using the equation from part (b), the initial height from which the pumpkin was thrown is about 4.3 feet, and the initial velocity is about 12.3 feet per second.

   (ii) The maximum height occurs at the vertex and is approximately 39.8 feet; this occurs about 5.85 seconds into the flight.

   (iii) The pumpkin lands approximately 12 seconds after it is launched.

96. a.

b. The equation of the best fit parabola is $y = -0.2315x^2 + 3.8292x + 4.2375$ with $r^2 = 0.9945$. This is shown below on the scatterplot:

c. (i) Using the equation from part (b), the initial height from which the pumpkin was thrown is about 4.2 feet, and the initial velocity is about 3.83 feet per second.

(ii) The maximum height occurs at the vertex and is approximately 20.07 feet; this occurs about 8.3 seconds into the flight.

(iii) The pumpkin lands approximately 17.5 seconds after it is launched.

Section 4.2

1. polynomial; degree 2 **2.** polynomial; degree 5
3. polynomial; degree 5 **4.** polynomial; degree 9
5. not a polynomial   **6.** not a polynomial
7. not a polynomial   **8.** not a polynomial
9. not a polynomial   **10.** polynomial; degree 2
11. h **12.** g **13.** b **14.** f **15.** e **16.** d
17. c **18.** a

19.   **20.**

21.   **22.**

23.   **24.**

25.   **26.**

27. 3 (multiplicity 1), -4 (multiplicity 3)
28. -2 (multiplicity 3), 1 (multiplicity 2)
29. 0 (multiplicity 2), 7 (multiplicity 2), -4 (multiplicity 1)
30. 0 (multiplicity 3), -1 (multiplicity 4), 6 (multiplicity 1)

31. 0 (multiplicity 2), 1 (multiplicity 2)

32. 0 (multiplicity 2), -1(multiplicity 1), 1 (multiplicity 1)

33. 0 (multiplicity 1), $\frac{3}{2}$ (multiplicity 1), $-\frac{9}{4}$ (multiplicity 1)

34. 0 (multiplicity 2), $\frac{1}{2}$ (multiplicity 1), -3 (multiplicity 1)

35. 0 (multiplicity 2), -3 (multiplicity 1)

36. 0 (multiplicity 1), 1.957 (multiplicity 1), -1.957 (multiplicity 1)

37. 0 (multiplicity 4)　　**38.** 0 (multiplicity 3)

39. $P(x) = x(x+3)(x-1)(x-2)$

40. $P(x) = x(x+2)(x-2)$

41. $P(x) = x(x+5)(x+3)(x-2)(x-6)$

42. $P(x) = x(x-1)(x-3)(x-5)(x-10)$

43. $P(x) = (2x+1)(3x-2)(4x-3)$

44. $P(x) = x(4x+3)(3x+1)(2x-1)$

45. $P(x) = x^2 - 2x - 1$　**46.** $P(x) = x^2 - 2x - 2$

47. $P(x) = x^2(x+2)^3$　**48.** $P(x) = (x+4)^2(x-5)^3$

49. $P(x) = (x+3)^2(x-7)^5$　**50.** $P(x) = x(x-10)^3$

51. $P(x) = x^2(x+1)(x+\sqrt{3})^2(x-\sqrt{3})^2$

52. $P(x) = x(x-1)^2(x+\sqrt{5})^2(x-\sqrt{5})^2$

53. $f(x) = -(x+3)^2$ **a.** -3 (multiplicity 2) **b.** touches at -3 **c.** $(0,-9)$ **d.** falls left and right, without bound **e.**

54. $f(x) = (x+2)^2$ **a.** -2 (multiplicity 2) **b.** touches at -2 **c.** $(0,4)$ **d.** rises left and right, without bound **e.**

55. $f(x) = (x-2)^3$ **a.** 2 (multiplicity 3) **b.** crosses at 2 **c.** $(0,-8)$ **d.** falls left, rises right **e.**

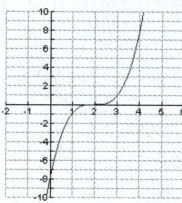

56. $f(x) = -(x+3)^3$ **a.** -3 (multiplicity 3) **b.** crosses at -3 **c.** $(0,-27)$ **d.** falls right, rises left **e.**

57. $f(x) = x(x-3)(x+3)$ **a.** 0, 3, -3 (multiplicity 1) **b.** crosses at each zero **c.** $(0,0)$ **d.** falls left, rises right **e.**

58. $f(x) = -x^2(x-4)$ **a.** 0 (multiplicity 2), 4 (multiplicity 1) **b.** crosses at 4, touches at 0 **c.** $(0,0)$ **d.** falls right, rises left **e.**

59. $f(x) = -x(x-2)(x+1)$ **a.** 0, 2, -1 (multiplicity 1) **b.** crosses at each zero **c.** $(0,0)$ **d.** falls right, rises left **e.**

60. $f(x) = x(x-3)^2$ **a.** 0 (multiplicity 1), 3 (multiplicity 2) **b.** crosses at 0, touches at 3 **c.** (0,0) **d.** falls left, rises right **e.**

61. $f(x) = -x^3(x+3)$ **a.** 0 (multiplicity 3), −3 (multiplicity 1) **b.** crosses at both 0 and −3 **c.** (0,0) **d.** falls left and right, without bound **e.**

62. $f(x) = x^3(x-1)(x+1)$ **a.** 0 (multiplicity 3), 1 (multiplicity 1), −1 (multiplicity 1) **b.** crosses at each of 0, 1, and −1 **c.** (0,0) **d.** falls left, rises right **e.**

63. $f(x) = 12x^4(x-4)(x+1)$ **a.** 0 (multiplicity 4), 4 (multiplicity 1), −1 (multiplicity 1) **b.** touches at 0 and crosses at 4 and −1 **c.** (0,0) **d.** rises left and right, without bound **e.**

64. $f(x) = 7x^3(x-3)(x+1)$ **a.** 0 (multiplicity 3), 3 (multiplicity 1), −1 (multiplicity 1) **b.** crosses at each of 0, 3, and −1 **c.** (0,0) **d.** falls left, rises right **e.**

65. $f(x) = 2x^3(x-4)(x+1)$ **a.** 0 (multiplicity 3), 4 (multiplicity 1), −1 (multiplicity 1) **b.** crosses at each zero **c.** (0,0) **d.** falls left, rises right **e.**

66. $f(x) = -5x^2(x-1)^2$ **a.** 0 (multiplicity 2), 1 (multiplicity 2) **b.** touches at each zero **c.** (0,0) **d.** falls right and left, without bound **e.**

67. $f(x) = (x-2)(x+2)(x-1)$ **a.** 1, 2, −2 (multiplicity 1) **b.** crosses at each zero **c.** (0,4) **d.** falls left, rises right **e.**

68. $f(x) = (x-1)^2(x+1)$ **a.** −1 (multiplicity 1), 1 (multiplicity 2) **b.** crosses at −1, touches at 1 **c.** (0,1) **d.** falls left, rises right **e.**

69. $f(x) = -(x+2)^2(x-1)^2$ **a.** −2 (multiplicity 2), 1 (multiplicity 2) **b.** touches at both −2 and 1 **c.** (0,−4) **d.** falls left and right, without bound **e.**

70. $f(x) = (x-2)^3(x+1)^3$ **a.** −1 (multiplicity 3), 2 (multiplicity 3) **b.** crosses at both −1 and 2 **c.** $(0,-8)$ **d.** rises right and left, without bound **e.**

71. $f(x) = x^2(x-2)^3(x+3)^2$ **a.** 0 (multiplicity 2), 2 (multiplicity 3), −3 (multiplicity 2) **b.** touches at both 0 and −3, and crosses at 2 **c.** $(0,0)$ **d.** falls left, and rises right **e.**

72. $f(x) = -x^3(x-4)^2(x+2)^2$ **a.** 0 (multiplicity 3), 4 (multiplicity 2), −2 (multiplicity 2) **b.** touches at 4 and −2, and crosses at 0 **c.** $(0,0)$ **d.** falls right, rises left **e.**

73. **a.** −3 (multiplicity 1), −1 (multiplicity 2), 2 (multiplicity 1) **b.** even **c.** negative **d.** $(0, 6)$ **e.** $f(x) = -(x+1)^2(x-2)(x+3)$

74. **a.** −2 (multiplicity 1), 2 (multiplicity 2), 0 (multiplicity 1) **b.** even **c.** positive **d.** $(0,0)$ **e.** $f(x) = x(x+2)(x-2)^2$

75. **a.** 0 (multiplicity 2), −2 (multiplicity 2), $\frac{3}{2}$ (multiplicity 1) **b.** odd **c.** positive **d.** $(0,0)$ **e.** $f(x) = x^2(2x-3)(x+2)^2$

76. **a.** −3 (multiplicity 1), 0 (multiplicity 1), $-\frac{3}{2}$ (multiplicity 1), 1 (multiplicity 2) **b.** odd **c.** negative **d.** $(0,0)$ **e.** $f(x) = -x(2x+3)(x+2)(x-1)^2$

77. **a.** Revenue for the company is increasing when advertising costs are less than

$400,000. Revenue for the company is decreasing when advertising costs are between $400,000 and $600,000.

b. The zeros of the revenue function occur when $0 and $600,000 are spent on advertising. When either $0 or $600,000 is spent on advertising the company's revenue is $0.

78. The company's maximum revenue is $32,000,000 when $400,000 is spent on advertising.

79. The velocity of air in the trachea is increasing when the radius of the trachea is between 0 and 0.45 cm and decreasing when the radius of the trachea is between 0.45 cm and 0.65 cm.

80. $r \cong 0.45 \text{ cm}$; $v \approx 5.265$ meters per second.

81. sixth-degree polynomial

82. second degree polynomial

83. down **84.** up **85.** 4th degree **86.** positive

87. If h is a zero of a polynomial, then $(x-h)$ is a factor of it. So, in this case the function would be: $P(x) = (x+2)(x+1)(x-3)(x-4)$

88. It should be similar to $y = x^4$, which rises to the left and right without bound.

89. The zeros are correct. But, it is a fifth degree polynomial. The graph should touch at 1 (since even multiplicity) and cross at −2. The graph should look like:

90. Should touch, not cross, at 1 and −1 since both have even multiplicity. The graph should look like:

91. false **92.** true **93.** true

94. false **95.** n **96.** $n - 1$

97.

$f(x) = (x+1)^2(x-3)^5$, $g(x) = (x+1)^4(x-3)^3$

$h(x) = (x+1)^6(x-3)$

98. $P_1(x) = (x-0)^4(x-4)$

$P_2(x) = (x-0)^2(x-4)^3$

99. $0, a, -b$ **100.** $0, a, b$

101. no x-intercepts   **102.** 0

103. $y = -2x^5$, yes   **104.** $y = x^4$, yes

105. x-intercepts: $(-2.25, 0)$, $(6.2, 0)$, $(14.2, 0)$
zeros: -2.25 (multiplicity 2), 6.2 (multiplicity 1), 14.2 (multiplicity 1)

106. x-intercepts: $(-3.2, 0)$, $(-2.5, 0)$, $(1.2, 0)$, $(2.5, 0)$, $(4.2, 0)$ zeros: -3.2, -2.5, 1.2, 2.5, 4.2 (all have multiplicity 1)

107.

$(-2.56, -17.12)$,
$(-0.58, 12.59)$,
$(1.27, -11.73)$

108.

$(-1.61, 15.47)$,
$(-0.35, -9.95)$,
$(1.21, 14.07)$,
$(2.36, -4.38)$.

Section 4.3

1. $Q(x) = 2x - 1, \quad r(x) = 0$

2. $Q(x) = 2x + 11, \quad r(x) = \frac{30}{x-3}$

3. $Q(x) = x - 3, \quad r(x) = 0$

4. $Q(x) = 2x + 1, \quad r(x) = 0$

5. $Q(x) = 3x - 3, \quad r(x) = -11$

6. $Q(x) = x + 5, \quad r(x) = 2$

7. $Q(x) = 3x - 28, \quad r(x) = 130$

8. $Q(x) = 3x + 2, \quad r(x) = 0$

9. $Q(x) = x - 4, \quad r(x) = 12$

10. $Q(x) = x + 2, \quad r(x) = -5$

11. $Q(x) = 3x + 5, \quad r(x) = 0$

12. $Q(x) = 5x - 5, \quad r(x) = 2$

13. $Q(x) = 2x - 3, \quad r(x) = 0$

14. $Q(x) = 4x^2 - 6x + 9, \quad r(x) = 0$

15. $Q(x) = 4x^2 + 4x + 1, \quad r(x) = 0$

16. $Q(x) = 6x^2 + 7x + 2, \quad r(x) = 0$

17. $Q(x) = 2x^2 - x - \frac{1}{2}, \quad r(x) = \frac{15}{2}$

18. $Q(x) = -2x^3 - \frac{4}{3}x^2 - \frac{2}{9}x - \frac{4}{27}, \quad r(x) = \frac{143}{27}$

19. $Q(x) = 4x^2 - 10x - 6, \quad r(x) = 0$

20. $Q(x) = 12x^2 + 12x + 3, \quad r(x) = 0$

21. $Q(x) = -2x^2 - 3x - 9,$
    $r(x) = -27x^2 + 3x + 9$

22. $Q(x) = -3x^2 + \frac{7}{3},$
    $r(x) = -8x^3 + 3x^2 + \frac{14}{3}x + \frac{8}{3}$

23. $Q(x) = x^2 + 1, \quad r(x) = 0$

24. $Q(x) = x^2 - 3, \quad r(x) = 0$

25. $Q(x) = x^2 + x + \frac{1}{6}, \quad r(x) = -\frac{121}{6}x + \frac{121}{3}$

26. $Q(x) = x^2 - 1, \quad r(x) = 0$

27. $Q(x) = -3x^3 + 5.2x^2 + 3.12x - 0.128,$
    $r(x) = 0.9232$

28.
$Q(x) = 2x^4 + 1.8x^3 - 2.38x^2 + 0.858x + 0.7722,$
    $r(x) = \frac{5.69498}{x-0.9}$

29. $Q(x) = x^2 - 0.6x + 0.09, \quad r(x) = 0$

30. $Q(x) = x^3 + 3.4x^2 + 3.29x + 0.98, \quad r(x) = 0$

31. $Q(x) = 3x + 1, \quad r(x) = 0$

32. $Q(x) = 2x - 3, \quad r(x) = 0$

33. $Q(x) = 7x - 10, \quad r(x) = 15$

34. $Q(x) = 4x + 9, \quad r(x) = 19$

35. $Q(x) = -x^3 + 3x - 2, \quad r(x) = 0$

36. $Q(x) = x^2 + 4x + 4, \quad r(x) = 0$

37. $Q(x) = x^3 - x^2 + x - 1, \quad r(x) = 2$

38. $Q(x) = x^3 - 3x^2 + 9x - 27, \quad r(x) = 90$

39. $Q(x) = x^3 - 2x^2 + 4x - 8, \quad r(x) = 0$

40. $Q(x) = x^3 + 3x^2 + 9x + 27, \quad r(x) = 0$

41. $Q(x) = 2x^2 - 6x + 2, \quad r(x) = 0$

42. $Q(x) = 3x^2 - 9x + 3, \quad r(x) = 0$

43. $Q(x) = 2x^3 - \frac{5}{3}x^2 + \frac{53}{9}x + \frac{106}{27}, \quad r(x) = -\frac{112}{81}$

44. $Q(x) = 3x^3 + \frac{13}{4}x^2 + \frac{39}{16}x + \frac{245}{64}, \quad r(x) = -\frac{33}{256}$

45. $Q(x) = 2x^3 + 6x^2 - 18x - 54, \quad r(x) = 0$

46. $Q(x) = 5x^2 - 5x + 10, \quad r(x) = 0$

47. $Q(x) = x^6 + x^5 + x^4 - 7x^3 - 7x^2 - 4x - 4,$
    $r(x) = -3$

48. $Q(x) = x^5 + 3x^4 - 3x^3 + x^2 - x + 1, \quad r(x) = 6$

49.
$Q(x) = x^5 + \sqrt{5}x^4 - 44x^3 - 44\sqrt{5}x^2 - 245x - 245\sqrt{5},$
    $r(x) = 0$

50. $Q(x) = x^5 + \sqrt{3}x^4 - x^3 - \sqrt{3}x^2 - 12x - 12\sqrt{3},$
    $r(x) = 0$

51. $Q(x) = 2x - 7, \quad r(x) = 0$

52. $Q(x) = 3x + 2, \quad r(x) = 0$

53. $Q(x) = x^2 - 9, \quad r(x) = 0$

54. $Q(x) = x^2 - 6, \quad r(x) = 0$

55. $Q(x) = x^4 + 2x^3 + 8x^2 + 18x + 36, \quad r(x) = 71$

56. $Q(x) = x^3 - 5x^2 + 24x - 117, \quad r(x) = 575$

57. $Q(x) = x^2 + 1, \quad r(x) = -24$

58. $Q(x) = x, \quad r(x) = 2x - 8$

59. $Q(x) = x^6 + x^5 + x^4 + x^3 + x^2 + x + 1,$
$r(x) = 0$

60. $Q(x) = x^5 + 3x^4 + 9x^3 + 27x^2 + 81x + 243,$
$r(x) = 702$

61. $3x^2 + 2x + 1$ feet   **62.** $3x + 1$ feet

63. $x^2 + 1$ hours   **64.** $x + 10$ yards per second

65. Should have subtracted each term in the long division rather than adding them.

66. The zero of the divisor is used in synthetic division. So, 2 should replace -2 as the divisor.

67. Forgot the "0" placeholder.

68. Cannot use synthetic division with a quadratic divisor. Use long division instead.

69. true   **70.** false   **71.** false

72. true   **73.** yes   **74.** no

75. $Q(x) = x^{2n} + 2x^n + 1,\ r(x) = 0$

76. $x^{3n} + 5x^{2n} + 8x^n + 4 = (x^n + 1)(x^n + 2)^2$

77. $2x - 1$; linear   **78.** $y = x^2 + 4$ with a hole at 3

79. $x^3 - 1$; cubic   **80.** $x - 3$; linear

81. $-6x^2 + 6x - 10$;
quadratic function

82. third-degree
polynomial

Section 4.4

1. $f(1) = 0$ **2.** $f(-1) = 4$ **3.** $g(1) = 4$

4. $g(-1) = 0$ **5.** $f(-2) = 84$ **6.** $g(2) = 21$

7. yes **8.** yes **9.** yes **10.** no

11. $-4, 1, 3$; $P(x) = (x - 1)(x + 4)(x - 3)$

12. $-4, -2, 3$; $P(x) = (x - 3)(x + 4)(x + 2)$

13. $-3, \frac{1}{2}, 2$; $P(x) = (2x - 1)(x + 3)(x - 2)$

14. $-\frac{1}{3}, 1, 4$; $P(x) = (3x + 1)(x - 4)(x - 1)$

15. $-3, 5$; $P(x) = (x^2 + 4)(x - 5)(x + 3)$

16. $-1, 2$; $P(x) = (x^2 + 9)(x - 2)(x + 1)$

17. $-3, 1$; $P(x) = (x - 1)(x + 3)(x^2 - 2x + 2)$

18. $-2, 4$; $P(x) = (x^2 - 2x + 5)(x - 4)(x + 2)$

19. $-2, -1$ (both multiplicity 2);
$P(x) = (x + 2)^2(x + 1)^2$

20. $-3, 1$ (both multiplicity 2);
$P(x) = (x - 1)^2(x + 3)^2$

21. $\pm 1, \pm 2, \pm 4$ **22.** $\pm 1, \pm 2, \pm 4$

23. $\pm 1, \pm 2, \pm 3, \pm 4, \pm 6, \pm 12$

24. $\pm 1, \pm 3, \pm 9$ **25.** $\pm \frac{1}{2}, \pm 1, \pm 2, \pm 4, \pm 8$

26. $\pm 1, \pm 2, \pm 5, \pm 10, \pm \frac{1}{3}, \pm \frac{2}{3}, \pm \frac{5}{3}, \pm \frac{10}{3}$

27. $\pm 1, \pm 2, \pm 4, \pm 5, \pm 10, \pm 20, \pm \frac{1}{5}, \pm \frac{2}{5}, \pm \frac{4}{5}$

28. $\pm 1, \pm 3, \pm 7, \pm 21, \pm \frac{1}{2}, \pm \frac{1}{4}, \pm \frac{3}{2}, \pm \frac{7}{2}, \pm \frac{21}{2},$
$\pm \frac{3}{4}, \pm \frac{7}{4}, \pm \frac{21}{4}$

29. $\pm 1, \pm 2, \pm 4, \pm 8$; rational zeros: $-4, -1, 2, 1$

30. $\pm 1, \pm 3$; rational zeros: $-3, -1$

31. $\pm 1, \pm 3, \pm \frac{1}{2}, \pm \frac{3}{2}$; rational zeros: $\frac{1}{2}, 1, 3$

32. $\pm 1, \pm 2, \pm 4, \pm 8, \pm \frac{1}{3}, \pm \frac{2}{3}, \pm \frac{4}{3}, \pm \frac{8}{3}$; rational zeros: $-2, -\frac{1}{3}, 4$

33.

Positive Real Zeros	Negative Real Zeros
1	1

34.

Positive Real Zeros	Negative Real Zeros
0	0

35.

Positive Real Zeros	Negative Real Zeros
1	0

36.

Positive Real Zeros	Negative Real Zeros
0	1

37.

Positive Real Zeros	Negative Real Zeros
2	1
0	1

38.

Positive Real Zeros	Negative Real Zeros
1	1

39.

Positive Real Zeros	Negative Real Zeros
1	1

40.

Positive Real Zeros	Negative Real Zeros
3	0
1	0

41.

Positive Real Zeros	Negative Real Zeros
2	2
0	2
2	0
0	0

42.

Positive Real Zeros	Negative Real Zeros
1	1

43.

Positive Real Zeros	Negative Real Zeros
4	0
2	0
0	0

44.

Positive Real Zeros	Negative Real Zeros
2	2
2	0
0	2
0	0

45. a. Number of sign variations for $P(x)$: 0

Number of sign variations for $P(-x)$: 3

Positive Real Zeros	Negative Real Zeros
0	3
0	1

b. possible rational zeros: $\pm 1, \pm 2, \pm 3, \pm 6$

c. rational zeros: $-1, -2, -3$

d. $P(x) = (x+1)(x+2)(x+3)$

46. a. Number of sign variations for $P(x)$: 3

Number of sign variations for $P(-x)$: 0

Positive Real Zeros	Negative Real Zeros
3	0
1	0

b. possible rational zeros: $\pm 1, \pm 2, \pm 3, \pm 6$

c. rational zeros: $1, 2, 3$

d. $P(x) = (x-1)(x-2)(x-3)$

47. a. Number of sign variations for $P(x)$: 2

Number of sign variations for $P(-x)$: 1

Positive Real Zeros	Negative Real Zeros
2	1
0	1

b. possible rational zeros: $\pm 1, \pm 7$

c. rational zeros: $-1, 1, 7$

d. $P(x) = (x+1)(x-1)(x-7)$

48. a. Number of sign variations for $P(x)$: 2

Number of sign variations for $P(-x)$: 1

Positive Real Zeros	Negative Real Zeros
2	1
0	1

b. possible rational zeros:
$\pm 1, \pm 2, \pm 4, \pm 5, \pm 10, \pm 20$

c. rational zeros: $-2, 2, 5$

d. $P(x) = (x+2)(x-2)(x-5)$

49. a. Number of sign variations for $P(x)$: 1

Number of sign variations for $P(-x)$: 2

Positive Real Zeros	Negative Real Zeros
1	2
1	0

b. possible rational zeros: $\pm 1, \pm 2, \pm 5, \pm 10$

c. rational zeros: $0, 1, -2, -5$

d. $P(x) = x(x-1)(x+2)(x+5)$

50. a. Number of sign variations for $P(x)$: 2

Number of sign variations for $P(-x)$: 1

Positive Real Zeros	Negative Real Zeros
2	1
0	1

b. possible rational zeros:
$\pm 1, \pm 2, \pm 3, \pm 4, \pm 6, \pm 8, \pm 12, \pm 24$

c. rational zeros: $0, -4, 2, 3$

d. $P(x) = x(x+4)(x-2)(x-3)$

51. a. Number of sign variations for $P(x)$: 4

Number of sign variations for $P(-x)$: 0

Positive Real Zeros	Negative Real Zeros
4	0
2	0
0	0

b. possible rational zeros: $\pm 1, \pm 2, \pm 13, \pm 26$

c. rational zeros: $1, 2$

d. $P(x) = (x-1)(x-2)\left(x^2 - 4x + 13\right)$

52. a. Number of sign variations for $P(x)$: 3

Number of sign variations for $P(-x)$: 1

Positive Real Zeros	Negative Real Zeros
3	1
1	1

b. possible rational zeros: $\pm 1, \pm 2, \pm 13, \pm 26$

c. rational zeros: $-2, 1$

d. $P(x) = (x-1)(x+2)\left(x^2 - 6x + 13\right)$

53. a. Number of sign variations for $P(x)$: 2

Number of sign variations for $P(-x)$: 1

Positive Real Zeros	Negative Real Zeros
2	1
0	1

b. possible rational zeros: $\pm 1, \pm \frac{1}{2}, \pm \frac{1}{5}, \pm \frac{1}{10}$

c. rational zeros: $-1, -\frac{1}{2}, \frac{1}{5}$

d. $P(x) = (x-1)(2x+1)(5x-1)$

54. a. Number of sign variations for $P(x)$: 3

Number of sign variations for $P(-x)$: 0

Positive Real Zeros	Negative Real Zeros
3	0
1	0

b. possible rational zeros:
$\pm 1, \pm \frac{1}{2}, \pm \frac{1}{3}, \pm \frac{1}{4}, \pm \frac{1}{6}, \pm \frac{1}{12}$

c. rational zeros: 1

d. $P(x) = (x-1)\left(12x^2 - x + 1\right)$

55. a. Number of sign variations for $P(x)$: 1

Number of sign variations for $P(-x)$: 2

Positive Real Zeros	Negative Real Zeros
1	2
1	0

b. possible rational zeros:
$\pm 1, \pm 2, \pm 5, \pm 10, \pm \frac{1}{2}, \pm \frac{1}{3}, \pm \frac{1}{6}, \pm \frac{2}{3},$
$\pm \frac{5}{2}, \pm \frac{5}{3}, \pm \frac{5}{6}, \pm \frac{10}{3}$

c. rational zeros: $-1, -\frac{5}{2}, \frac{2}{3}$

d. $P(x) = 6(x+1)\left(x+\frac{5}{2}\right)\left(x-\frac{2}{3}\right)$

56. a. Number of sign variations for $P(x)$: 1

Number of sign variations for $P(-x)$: 2

Positive Real Zeros	Negative Real Zeros
1	2
1	0

b. possible rational zeros:
$\pm 1, \pm 2, \pm \frac{1}{2}, \pm \frac{1}{3}, \pm \frac{1}{6}, \pm \frac{2}{3}$

c. rational zeros: $1, -\frac{1}{2}, -\frac{2}{3}$

d. $P(x) = 6(x-1)\left(x+\frac{2}{3}\right)\left(x+\frac{1}{2}\right)$

57. a. Number of sign variations for $P(x)$: 4

Number of sign variations for $P(-x)$: 0

Positive Real Zeros	Negative Real Zeros
4	0
2	0
0	0

b. possible rational zeros: $\pm 1, \pm 2, \pm 4$

c. rational zeros: 1

d. $P(x) = (x-1)^2(x^2 + 4)$

58. a. Number of sign variations for $P(x)$: 0

Number of sign variations for $P(-x)$: 4

Positive Real Zeros	Negative Real Zeros
0	4
0	2
0	0

b. possible rational zeros: $\pm 1, \pm 3, \pm 9$

c. rational zeros: -1

d. $P(x) = (x+1)^2(x^2+9)$

59. a. Number of sign variations for $P(x)$: 1

Number of sign variations for $P(-x)$: 1

Positive Real Zeros	Negative Real Zeros
1	1

b. possible rational zeros:

$\pm 1, \pm 2, \pm 3, \pm 4, \pm 6, \pm 9, \pm 12, \pm 18, \pm 36$

c. rational zeros: $-1, 1$

d. $P(x) = (x+1)(x-1)(x^2+9)(x^2+4)$

60. a. Number of sign variations for $P(x)$: 2

Number of sign variations for $P(-x)$: 2

Positive Real Zeros	Negative Real Zeros
2	2
2	0
0	2
0	0

b. possible rational zeros:

$\pm 1, \pm 2, \pm 4, \pm 8, \pm 16$

c. rational zeros: 1

d.

$P(x) = (x-1)(x+3.35026)(x+1.07838)(x-4.2864)$

61. a. Number of sign variations for $P(x)$: 4

Number of sign variations for $P(-x)$: 0

Positive Real Zeros	Negative Real Zeros
4	0
2	0
0	0

b. possible rational zeros:

$\pm 1, \pm 5, \pm \frac{1}{2}, \pm \frac{1}{4}, \pm \frac{5}{2}, \pm \frac{5}{4}$

c. rational zeros: $\frac{1}{2}$

d. $P(x) = 4(x-\frac{1}{2})^2(x^2-4x+5)$

62. a. Number of sign variations for $P(x)$: 2

Number of sign variations for $P(-x)$: 2

Positive Real Zeros	Negative Real Zeros
2	2
0	2
0	0
2	0

b. possible rational zeros:

$\pm 1, \pm 2, \pm 5, \pm 10, \pm 25, \pm 50,$

$\pm \frac{1}{2}, \pm \frac{5}{2}, \pm \frac{25}{2}, \pm \frac{1}{4}, \pm \frac{5}{4}, \pm \frac{25}{4}$

c. rational zeros: none

d. factorization not possible

63.

64.

65.

66.

67. $x = 1.34$

68. $x = 0.74$

69. $x = 0.22$

70. $x = -1.64$

71. $x = -0.43$

72. $x = -1.05$

73. a. $P(x) = -3x^2 - 2x + 26, \quad x \geq 0$

b. 263 subscribers

74. 2 or approximately 3.79 units sold

75. $P(x) = -0.0002x^2 + 8x - 1500$; 0 or 2 positive real zeros.

76. 188 and 39,812 units: when fewer than 188 units or more than 39,812 units are produced and sold, profit is negative–money is lost. When the number of units being produced and sold is between 188 and 39,812 a profit is being made on the product.

77. 18 hours

78. 9 hours

79. It is true that one can get 5 negative zeros here, but there may be just 1 or 3.

Positive Real Zeros	Negative Real Zeros
0	5
0	3
0	1

80. Use 2, not -2.   **81.** true   **82.** false

83. false   **84.** false   **85.** b, c

86. $-b, c$

87. possible rational zeros:
$\pm 1, \pm 2, \pm 4, \pm 8, \pm 16, \pm 32$
zeros: 2

88. possible rational zeros:
$\pm 1, \pm 2, \pm 3, \pm 4, \pm 6, \pm 8,$
$\pm 12, \pm 16, \pm 24, \pm 48$
zeros: 3

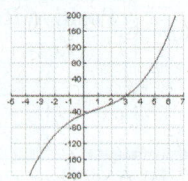

89. a. $-\frac{3}{4}, \frac{2}{3}$

b. $P(x) = (3x - 2)(4x + 3)\left(x^2 + 2x + 5\right)$

90. a. $-\frac{7}{3}$ **b.** $P(x) = -(3x + 7)\left(x^2 - 2x + 7\right)$

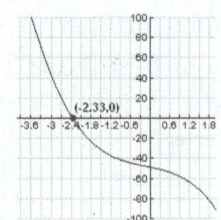

Section 4.5

1. $x = \pm 2i$; $P(x) = (x + 2i)(x - 2i)$

2. $x = \pm 3i$; $P(x) = (x + 3i)(x - 3i)$

3. $x = 1 \pm i$; $P(x) = (x - (1 - i))(x - (1 + i))$

4. $x = 2 \pm i$; $P(x) = (x - (2 - i))(x - (2 + i))$

5. $x = \pm 2, \pm 2i$;
$P(x) = (x - 2)(x + 2)(x - 2i)(x + 2i)$

6. $x = \pm 3, \pm 3i$;
$P(x) = (x - 3)(x + 3)(x - 3i)(x + 3i)$

7. $x = \pm\sqrt{5}, \pm\sqrt{5}\,i$;
$P(x) = \left(x - \sqrt{5}\right)\left(x + \sqrt{5}\right)\left(x - \sqrt{5}\,i\right)\left(x + \sqrt{5}\,i\right)$

8. $x = \pm\sqrt{3},\ \pm\sqrt{3}\,i$;

$P(x) = \left(x - \sqrt{3}\right)\left(x + \sqrt{3}\right)\left(x - \sqrt{3}\,i\right)\left(x + \sqrt{3}\,i\right)$

9. $-i$ **10.** i **11.** $-2i,\ 3+i$ **12.** $-3i,\ 2-i$

13. $1+3i,\ 2-5i$ **14.** $1+5i,\ 2-3i$

15. $i,\ 1+i$ **16.** $-2i,\ 1-i$

17. $P(x) = x^3 - 2x^2 + 5x$

18. $P(x) = x^3 - 4x^2 + 5x$

19. $P(x) = x^3 - 3x^2 + 28x - 26$

20. $P(x) = x^3 - 10x^2 + 33x - 34$

21. $P(x) = x^4 - 2x^3 + 11x^2 - 18x + 18$

22. $P(x) = x^4 - 2x^3 + 46x^2 - 2x + 5$

23. $\pm 2i,\ -3,\ 5$;

$P(x) = (x - 2i)(x + 2i)(x - 5)(x + 3)$

24. $\pm 3i,\ -1,\ 2$;

$P(x) = (x - 3i)(x + 3i)(x - 2)(x + 1)$

25. $\pm i,\ 1,\ 3$; $P(x) = (x - i)(x + i)(x - 3)(x - 1)$

26. $\pm 2i,\ -1,\ 2$;

$P(x) = (x - 2i)(x + 2i)(x - 2)(x + 1)$

27. $\pm 3i,\ 1$ (multiplicity 2) ;

$P(x) = (x - 3i)(x + 3i)(x - 1)^2$

28. $\pm 5i,\ -1,\ 4$;

$P(x) = (x - 5i)(x + 5i)(x - 4)(x + 1)$

29. $1 \pm i,\ -1 \pm 2\sqrt{2}$;

$P(x) = (x - (1 + i))(x - (1 - i)) \cdot$

$\qquad (x - (-1 - 2\sqrt{2}))(x - (-1 + 2\sqrt{2}))$

30. $1 \pm 2i,\ -2,\ 4$;

$P(x) = (x - (1 + 2i))(x - (1 - 2i))(x - 4)(x + 2)$

31. $3 \pm i,\ \pm 2$;

$P(x) = (x - (3 + i))(x - (3 - i))(x - 2)(x + 2)$

32. $2 \pm i,\ \pm 1$;

$P(x) = (x - (2 + i))(x - (2 - i))(x - 1)(x + 1)$

33. $2 \pm i,\ 1, 4$;

$\ P(x) = (x - (2 + i))(x - (2 - i))(x - 1)(x - 4)$

34. $3 \pm i,\ -1,\ 2$;

$P(x) = (x - (3 + i))(x - (3 - i))(x - 2)(x + 1)$

35. $P(x) = (x + 3i)(x - 3i)(x - 1)$

36. $P(x) = (x + 2i)(x - 2i)(x - 2)$

37. $P(x) = (x + i)(x - i)(x - 5)$

38. $P(x) = (x + i)(x - i)(x - 7)$

39. $P(x) = (x + 2i)(x - 2i)(x + 1)$

40. $P(x) = (x - 1)(x - (-1 + i))(x - (-1 - i))$

41. $P(x) = (x - 3)(x - (-1 + \sqrt{5}\,i))(x - (-1 - \sqrt{5}\,i))$

42. $P(x) = (x + 1)(x - 3)(x + i)(x - i)$

43. $P(x) = (x + 3)(x - 5)(x + 2i)(x - 2i)$

44. $P(x) = (x + 1)(x - 2)(x + 3i)(x - 3i)$

45. $P(x) = (x + 1)(x - 5)(x + 2i)(x - 2i)$

46. $P(x) = (x - 1)(x - 2)(x + 3i)(x - 3i)$

47. $P(x) = (x - 1)(x - 2)(x - (2 - 3i))(x - (2 + 3i))$

48. $P(x) = (x - 1)(x + 2)(x - (3 - 2i))(x - (3 + 2i))$

49. $P(x) = -(x + 1)(x - 2)(x - (2 - i))(x - (2 + i))$

50. $P(x) = -(x + 3)(x - 4)(x - (-1 - i))(x - (-1 + i))$

51. $P(x) = (x - 1)^2 (x + 2i)(x - 2i)$

52. $P(x) = (x + 1)^2 (x + 3i)(x - 3i)$

53. $P(x) = (x - 1)(x + 1)(x - 2i)(x + 2i) \cdot$

$\qquad\qquad\qquad\qquad\qquad (x - 3i)(x + 3i)$

54. $P(x) = (x - 1)^2 (x - 2i)^2 (x + 2i)^2$

55. $P(x) = (2x - 1)^2 (x - (2 - i))(x - (2 + i))$

56. $P(x) = 4\left(x - \tfrac{1}{2}\right)^2 (x - (5 - i))(x - (5 + i))$

57. $P(x) = (x - 1)(x + 1)(3x - 2)(x - 2i)(x + 2i)$

58. $P(x) = (x - 1)^2 (2x - 5)(x + 1 - 2i)(x + 1 + 2i)$

59. Yes. In such case, $P(x)$ is always above the x-axis since the leading coefficient is positive, indicating that the end behavior should resemble that of $y = x^{2n}$, for some positive integer n. So, profit is always positive and increasing.

60. No. In such case, $P(x)$ is always below the x-axis since the leading coefficient is negative. So, never have a positive profit.

61. No. In such case, it crosses the x-axis and looks like $y = -x^3$. So, profit is decreasing.

62. Yes. In such case, it crosses the x-axis and looks like $y = x^3$. So, profit is increasing.

63. Since the profit function is a third-degree polynomial we know that the function has three zeros and at most two turning points. Looking at the graph we can see there is one real zero where $t \le 0$. There are no real zeros when $t > 0$, therefore the other two zeros must be complex conjugates. Therefore, the company always has a profit greater than approximately 5.1 million dollars and, in fact, the profit will increase towards infinity as t increases.

64. Since the profit function is a fourth-degree polynomial with a negative leading we know the function has four zeros and at most three turning points. The end behavior is towards negative infinity because of the negative leading coefficient of an even degree polynomial and there will be two real zeros; one where $t \le 0$ and one where $t \ge 6$. The remaining two zeros are a complex conjugate pair. Therefore, the company will have profits of greater than approximately 5.1 million dollars during the first six months. Sometime later ($t \ge 6$) the company's profit will be zero. Then the company will start losing money and, in fact, the profit will decrease towards negative infinity as time increases.

65. Since the concentration function is a third degree polynomial we know the function has three zeros and at most two turning points. Looking at the graph we can see there will be one real zero where $t \ge 8$. The remaining zeros are a pair of complex conjugates. Therefore, the concentration of the drug in the blood stream will decrease to zero as the hours go by. Note that the concentration will not approach negative infinity since concentration is a non-negative quantity.

66. Since the concentration function is a fourth degree polynomial with a positive leading coefficient we know the function has four zeros. The negative leading coefficient indicates negative end behavior (opening down). Since the function opens down there

is a real zero for $t \le 0$ and there will be a real zero for $t \ge 8$. Note that the concentration will not approach negative infinity since concentration is a non-negative quantity.

67. Step 2 is an error. In general, the additive inverse of a real root need not be a root. This is being confused with the fact that complex roots occur in conjugate pairs.

68. Possible rational roots include $\pm \frac{1}{2}$ from the Rational Zero theorem.

69. false   **70.** false

71. true   **72.** true

73. No. Complex zeros occur in conjugate pairs. So, the collection of complex solutions contribute an even number of zeros, thereby requiring there to be at least one real zero.

74. Yes. For example, $P(x) = x^2 + 4$ has zeros $\pm 2i$, both of which are imaginary.

75. $P(x) = x^6 + 3b^2 x^4 + 3b^4 x^2 + b^6$

76. $P(x) = x^2 - 2ax + (a^2 + b^2)$

77. all roots are complex

Real Zeros	Complex Zeros
0	4
2	2
4	0

78. two real roots

Real Zeros	Complex Zeros
0	6
2	4
4	2
6	0

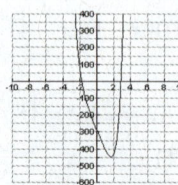

79. $\frac{3}{5}, \pm i, \pm 2i$;

    $P(x) = -5(x - 0.6)(x - 2i)(x + 2i)(x - i)(x + i)$

80. $-2.4, 1.5, 0.6 \pm 0.2i$;

    $P(x) = (x - 1.5)(x + 2.4)^2 \cdot$

        $(x - (0.6 + 0.2i))(x - (0.6 - 0.2i))$

Section 4.6

1. $(-\infty, -3) \cup (-3, \infty)$ **2.** $(-\infty, 4) \cup (4, \infty)$

3. $\left(-\infty, -\frac{1}{3}\right) \cup \left(-\frac{1}{3}, \frac{1}{2}\right) \cup \left(\frac{1}{2}, \infty\right)$

4. $\left(-\infty, \frac{2}{3}\right) \cup \left(\frac{2}{3}, 7\right) \cup (7, \infty)$

5. $(-\infty, -4) \cup (-4, 3) \cup (3, \infty)$

6. $(-\infty, -3) \cup (-3, 1) \cup (1, \infty)$

7. $(-\infty, \infty)$ **8.** $(-\infty, \infty)$

9. $(-\infty, -2) \cup (-2, 3) \cup (3, \infty)$

10. $(-\infty, -2) \cup (-2, 3) \cup (3, \infty)$

11. HA: $y = 0$ VA: $x = -2$

12. HA: $y = 0$ VA: $x = 5$

13. HA: none VA: $x = -5$

14. HA: none VA: $x = \frac{7}{2}$

15. HA: none VA: $x = \frac{1}{2}$, $x = -\frac{4}{3}$

16. HA: $y = 2$ VA: $x = 2$, $x = -\frac{1}{3}$

17. HA: $y = \frac{1}{3}$ VA: none

18. HA: none VA: $x = \frac{1}{2}$

19. HA: $y = \frac{12}{7}$ VA: $x = 0.5$, $x = -1.5$

20. HA: none VA: $x = 0.5$, $x = -0.5$

21. $y = x + 6$ **22.** $y = x + 12$

23. $y = 2x + 24$ **24.** $y = 3x + 7$

25. $y = 4x + \frac{11}{2}$ **26.** $y = 2x$

27. b   **28.** d   **29.** a   **30.** f   **31.** e   **32.** c

33.   **34.**

35.   **36.**

37.   **38.**

39.   **40.**

41.   **42.**

43.   **44.**

45.

46.

47.

48.

49.

50.

51.

52.

53.

54.

55.

56.

57.

58.

59. a. x-intercept: $(2,0)$; y-intercept: $(0,0.5)$
   b. HA: $y=0$ VA: $x=-1$, $x=4$
   c. $f(x) = \dfrac{x-2}{(x+1)(x-4)}$

60. a. x-intercept: $(-0.5,0)$, $(3,0)$; y-intercept: $(0,0.5)$ **b.** HA: $y=2$ VA: $x=-4$, $x=3$
   c. $f(x) = \dfrac{(2x+1)(x-3)}{(x+3)(x-2)}$

61. a. x-intercept: $(0,0)$; y-intercept: $(0,0)$
   b. HA: $y=-3$ VA: $x=-4$, $x=4$
   c. $f(x) = \dfrac{-3x^2}{(x+4)(x-4)}$

62. a. x-intercept: $(0,0)$, $(2,0)$; y-intercept: $(0,0)$ **b.** HA: none   VA: $x=-1$
   Slant: $y=x-3$
   c. $f(x) = \dfrac{x(x-2)}{x+1}$

63. a. $C(1) \cong 0.0198$ **b.** $C(60) \cong 0.0324$
   c. $C(300) \cong 0$
   d. $y=0$; after several days, $C(t) \cong 0$

64. a. $C\left(\frac{1}{2}\right) \cong 0.0124$ **b.** $C(1) \cong 0.0243$
   c. $C(4) \cong 0.0714$
   d. $y=0$; after several days, $C(t) \cong 0$

65. a. $N(0) = 52$ wpm   **b.** $N(12) \cong 107$ wpm
   c. $N(36) \cong 120$ wpm **d.** $y=130$; 130 wpm

66. $N(3) \cong 78$, $N(16) \cong 267$ 600 names

67. $y=10$, 10 ounces of food

68. $y(1) \cong 934$ 2800 cards

69. $\dfrac{2w^2 + 1000}{w}$ **70.** $\dfrac{(8l+414)(7+l)}{l}$

71. 2000 or 8000 units; average profit of $16 per unit.

72. 5000 units; average profit of $25 per unit.

73. The concentration of the drug in the bloodstream 15 hours after taking the dose is approximately 25.4 μg/mL. There are two times, 1 hour and 15 hours, after taking the medication at which the concentration of the drug in the bloodstream is approximately 25.4 μg/mL. The first time, approximately 1 hour, occurs as the concentration of the drug is increasing to a level high enough that the body will be able to maintain a concentration of approximately 25 μg/mL throughout the day. The second time, approximately 15 hours, occurs many hours later in the day as the concentration of the medication in the bloodstream drops.

74. The concentration of the drug in the bloodstream is 25 μg/mL approximately 21 hours after taking the medication. After 24 hours the concentration of the medication in the bloodstream has dropped to 24.9 μg/mL. As the drug becomes inert during the 25th hour this concentration will drop to 0 μg/mL. Thus it is important to take the next dose 24 hours after the previous dose so that as the previous dose becomes inert the new dose has time to build up the concentration of the drug in the bloodstream. At the end of the 25th hour the previous dose will no longer be in the patients system but the new dose will provide a concentration of 24 μg/mL.

75. $f(x) = \dfrac{x-1}{x^2-1} = \dfrac{\cancel{x-1}}{(\cancel{x-1})(x+1)} = \dfrac{1}{x+1}$ with a hole at $x = 1$. So, $x = 1$ is not a vertical asymptote.

76. $x^2 + 1 = 0$ has no real solution. So, there is no vertical asymptote.

77. In Step 2, the ratio of the leading coefficients should be $\frac{-1}{1}$. So, the horizontal asymptote is $y = -1$.

78. "Degree of numerator = Degree of denominator – 1" is not the criterion for the existence of an oblique asymptote. In this case, there is a horizontal asymptote, namely $y = 0$, but no oblique asymptote.

79. true   **80.** false

81. false   **82.** true

83. HA: $y = 1$ VA: $x = c$, $x = -d$

84. HA: $y = 3$ VA: none

85. Two possibilities: $y = \dfrac{4x^2}{(x+3)(x-1)}$ and

$$y = \dfrac{4x^5}{(x+3)^3(x-1)^2}$$

86. $f(x) = \dfrac{x-3}{x^2+1}$

87. VA: $x = -2$, yes   **88.** VA: $x = \frac{1}{3}$, yes

89. HA: $y = 0$ VA: $x = 0$, $x = -\frac{1}{3}$

Intercepts: $\left(-\frac{2}{5}, 0\right)$, yes

90. HA: $y = 0$ VA: $x = 0$

Intercepts: none, yes

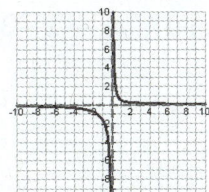

91. a. *f*: HA: $y = 0$ VA: $x = 3$
   g: HA: $y = 2$ VA: $x = 3$
   h: HA: $y = -3$ VA: $x = 3$

b. graphs of *f* and *g*: as
   $x \to \pm\infty$, $f(x) \to 0$ and $g(x) \to 2$

c. graphs of *g* and *h* below:
   as $x \to \pm\infty$, $g(x) \to 2$ and $h(x) \to -3$

d. $g(x) = \dfrac{2x - 5}{x - 3}$, $h(x) = \dfrac{-3x + 10}{x - 3}$

   Yes, if the degree of the numerator is the
   same as the degree of the denominator,
   then the horizontal asymptote is the ratio
   of the leading coefficients for both *g* and *h*.

92. a. *f*: HA: $y = 0$ VA: $x = \pm 1$
   g: HA: none VA: $x = \pm 1$ Slant: $y = x$
   h: HA: none VA: $x = \pm 1$ Slant: $y = x - 3$

b. graphs of *f* and *g*: as
   $x \to \pm\infty$, $f(x) \to 0$ and $g(x) \to x$

c. graphs of *g* and *h* below:
   as $x \to \pm\infty$, $g(x) \to x$ and $h(x) \to x - 3$

d. $g(x) = \dfrac{x^3 + x}{x^2 - 1}$, $h(x) = \dfrac{x^3 - 3x^2 + x + 3}{x^2 - 1}$.

Yes, if the degree of the numerator is the
exactly one more than the degree of the
denominator, then the quotient is the
slant asymptote.

Review Exercises

1. b   **2.** c   **3.** a   **4.** d

5.

6.

7.

8.

9. $f(x) = \left(x - \tfrac{3}{2}\right)^2 - \tfrac{49}{4}$

10. $f(x) = (x - 1)^2 - 25$

11. $f(x) = 4(x + 1)^2 - 11$

12. $f(x) = -\tfrac{1}{4}(x - 4)^2$

13.

14.

15.

16.

17. $\left(\tfrac{5}{26}, \tfrac{599}{52}\right)$ **18.** $(5, -7)$ **19.** $\left(-\tfrac{2}{15}, \tfrac{451}{125}\right)$

20. $\left(\tfrac{4}{15}, \tfrac{304}{75}\right)$ **21.** $y = \tfrac{1}{9}(x + 2)^2 + 3$

22. $y = -\tfrac{6}{49}(x - 4)^2 + 7$

23. $y = 5.6(x - 2.7)^2 + 3.4$

24. $y = -\frac{23}{180}\left(x + \frac{5}{2}\right)^2 + \frac{7}{4}$

25. a. $P(x) = -2x^2 + \frac{35}{3}x - 14$

　　b. $x \cong 4.1442433,\ 1.68909$

　　c. 　**d.** $(1.6891, 4.144)$ or 1689 to 4144

26. $A(x) = 2x^2 + 10x - 28$

27. $A(x) = -\frac{1}{2}(x-1)^2 + \frac{9}{2}$, maximum at $x = 1$
base: 3 units, height: 3 units

28. a. approximately 133.33 units
　　b. 6.7 seconds

29. yes, 6　　**30.** yes, 5　　**31.** no

32. yes, 3　　**33.** d　　**34.** b

35. a　　**36.** c

37. 　　**38.**

39. 　　**40.**

41. 6 (multiplicity 5), -4 (multiplicity 2)

42. 0 (multiplicity 1), 2 (multiplicity 3), -5 (multiplicity 1)

43. $0, -2, 2, 3, -3$, all multiplicity 1

44. 0 (multiplicity 2), 0.786795, -0.786795 (multiplicity 1)

45. $f(x) = x(x+3)(x-4)$

46. $f(x) = (x-2)(x-4)(x-6)(x+8)$

47. $f(x) = x(5x+2)(4x-3)$

48. $f(x) = (x - (2 - \sqrt{5})(x - (2 + \sqrt{5}))$

49. $f(x) = x^4 - 2x^3 - 11x^2 + 12x + 36$

50. $f(x) = (x-3)^2 x^3 (x+1)^2$

51. $f(x) = (x-7)(x+2)$　**a.** $-2, 7$ (both multiplicity 1)　**b.** crosses at $-2, 7$
　　c. $(0, -14)$　**d.** rises right and left
　　e.

52. $f(x) = -(x-5)^5$　**a.** 5 (multiplicity 5)
　　b. crosses at 5　**c.** $(0, 3125)$　**d.** falls right and rises left　**e.**

53. $f(x) = 6x^7 + 3x^5 - x^2 + x - 4$　**a.** $(0.8748, 0)$ with multiplicity 1　**b.** crosses at its only real zero　**c.** $(0, -4)$　**d.** falls left and rises right　**e.**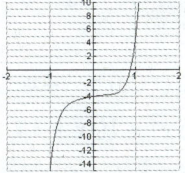

54. $f(x) = -x^4 (3x+6)^3 (x-7)^3$　**a.** 0 (multiplicity 4), -2 (multiplicity 3), 7 (multiplicity 3)　**b.** touches at 0, crosses at $-2, 7$　**c.** $(0, 0)$　**d.** falls right and left, without bound
　　e.

55. a.

b. 1, 3, 7 (all with multiplicity 1)
c. between 1 and 3 hours, and more than 7 hours is financially beneficial

56. peak seasons between 2 & 5, and 7 & 9

57. $Q(x) = x + 4, \ r(x) = 2$

58. $Q(x) = x - 1, \ r(x) = -4$

59. $Q(x) = 2x^3 - 4x^2 - 2x - \frac{7}{2}, \ r(x) = -23$

60. $Q(x) = -2x^2 + 2x - 2, \ r(x) = 9x - 6$

61. $Q(x) = x^3 + 2x^2 + x - 4, \ r(x) = 0$

62. $Q(x) = x^2 - 2x - 6, \ r(x) = 15$

63.
$Q(x) = x^5 - 8x^4 + 64x^3 - 512x^2 + 4096x - 32{,}768,$
$\quad r(x) = 262{,}080$

64. $Q(x) = 2x^4 + \frac{11}{2}x^3 + \frac{17}{8}x^2 + \frac{51}{32}x + \frac{1049}{128},$
$\quad r(x) = \frac{5707}{512}$

65. $Q(x) = x + 3, \ r(x) = -4x - 8$

66. $Q(x) = x^3 + 5x^2 + 10x + 34, \ r(x) = 104$

67. $Q(x) = x^2 - 5x + 7, \ r(x) = -15$

68. $Q(x) = x - 5, \ r(x) = 0$

69. $3x^3 + 2x^2 - x + 4$ feet

70. $V(x) = x(10 - 2x)(15 - 2x)$

71. $f(-2) = -207$ **72.** $f(1) = 0$

73. $g(1) = 0$ **74.** $g(-1) = -2$

75. no   **76.** yes   **77.** yes   **78.** no

79. $P(x) = x(x + 2)(x - 4)^2$

80. $P(x) = (x - 3)(x - 6)(x + 2)$

81. $P(x) = x^2(x + 3)(x - 2)^2$

82. $P(x) = (x - 6)(x + 6)(x - 2i)(x + 2i)$

83.

Positive Real Zeros	Negative Real Zeros
1	1

84.

Positive Real Zeros	Negative Real Zeros
1	2
1	0

85.

Positive Real Zeros	Negative Real Zeros
5	2
5	0
3	2
3	0
1	2
1	0

86.

Positive Real Zeros	Negative Real Zeros
3	2
1	2
3	0
1	0

87. possible rational zeros: $\pm 1, \pm 2, \pm 3, \pm 6$

88. possible rational zeros: $\pm 1, \pm 2, \pm 4, \pm 8$

89. possible rational zeros:
$\pm 1, \pm 2, \pm 4, \pm 8, \pm 16, \pm 32, \pm 64, \pm \frac{1}{2}$

90. possible rational zeros: $\pm 1, \pm 2, \pm \frac{1}{2}, \pm \frac{1}{4}$

91. possible rational zeros: $\pm 1, \pm \frac{1}{2}$; zeros: $\frac{1}{2}$

92. possible rational zeros:
$\pm 1, \pm \frac{1}{2}, \pm \frac{1}{3}, \pm \frac{1}{4}, \pm \frac{1}{6}, \pm \frac{1}{12}, \pm 3, \pm \frac{3}{2}, \pm \frac{3}{4}$;
zeros: $-\frac{3}{2}, \frac{1}{3}, \frac{1}{2}$

93. possible rational zeros:
$\pm 1, \pm 2, \pm 4, \pm 8, \pm 16$; zeros: $1, 2, 4, -2$

94. possible rational zeros:
$\pm 1, \pm \frac{1}{2}, \pm \frac{1}{3}, \pm \frac{1}{4}, \pm \frac{1}{6}, \pm \frac{1}{8}, \pm \frac{1}{12}, \pm \frac{1}{24}, \pm 2, \pm \frac{2}{3}$;
zeros: none

95. a.

Positive Real Zeros	Negative Real Zeros
1	0

b. ±1, ±5

c. −1 is a lower bound, 5 is an upper bound

d. none   **e.** not possible

f.

96. a.

Positive Real Zeros	Negative Real Zeros
1	0
1	2

b. ±1, ±2, ±4, ±8

c. −8 is a lower bound, 4 is upper bound

d. −4, 1, 2   **e.** $P(x) = (x-2)(x+4)(x+1)$

f.

97. a.

Positive Real Zeros	Negative Real Zeros
3	0
1	0

b. ±1, ±2, ±3, ±4, ±6, ±12

c. −4 is a lower bound, 12 is upper bound

d. 1, 2, 6   **e.** $P(x) = (x-1)(x-6)(x-2)$

f.

98. a.

Positive Real Zeros	Negative Real Zeros
2	2
2	0
0	2
0	0

b. ±1, ±2, ±3, ±6

c. −3 is a lower bound, 6 is upper bound

d. −2, −1, 1, 3

e. $P(x) = (x-1)(x+2)(x-3)(x+1)$

f.

99. a.

Positive Real Zeros	Negative Real Zeros
0	0
0	2
2	2
2	0

b. ±1, ±2, ±3, ±4, ±6, ±8, ±12, ±24

c. −4 is a lower bound, 8 is upper bound

d. −2, −1, 1, 6

e. $P(x) = (x-2)(x+1)(x+2)(x-6)$

f.

100. a.

Positive Real Zeros	Negative Real Zeros
2	2
2	0
0	2
0	0

b. ±1, ±2, ±4, ±8

c. −2 is a lower bound

d. 0, 1, 2

e.

$$P(x) = x(x-1)(x-2)(x-(\frac{-3+\sqrt{7}\,i}{2}))(x-(\frac{-3-\sqrt{7}\,i}{2}))$$

f.

101. $P(x) = (x - 5i)(x + 5i)$

102. $P(x) = (x - 4i)(x + 4i)$

103. $P(x) = (x - (1 - 2i))(x - (1 + 2i))$

104. $P(x) = (x - (-2 - i))(x - (-2 + i))$

105. $2i, 3 - i$ **106.** $-3i, 2 + i$

107. $-i, 2 + I$ **108.** $-2i, 1 + i$

109. $-i, 4, -1; P(x) = (x - i)(x + i)(x - 4)(x + 1)$

110. $2 + i, \pm 2$;
$P(x) = (x - (2 + i))(x - (2 - i))(x - 2)(x + 2)$

111. $3i, 1 \pm i$;
$P(x) = (x - 3i)(x + 3i)(x - (1 + i))(x - (1 - i))$

112. $1 - i, -3, 1$;
$P(x) = (x - (1 + i))(x - (1 - i))(x - 1)(x + 3)$

113. $P(x) = (x - 3)(x + 3)(x - 3i)(x + 3i)$

114. $P(x) = x(x - (3 + \sqrt{3}\,i))(x - (3 - \sqrt{3}\,i))$

115. $P(x) = (x - 2i)(x + 2i)(x - 1)$

116. $P(x) = (x + 1)(x - 2)(x - (2 + \sqrt{6}\,i))(x - (2 - \sqrt{6}\,i))$

117. HA: $y = -1$ VA: $x = -2$

118. HA: $y = 0$ VA: $x = 1$

119. HA: none   VA: $x = -1$ Slant: $y = 4x - 4$

120. HA: $y = 3$ VA: none

121. HA: $y = 2$ VA: none

122. HA: none   VA: $x = -5$
     Slant: $y = -2x + 13$

123.   **124.**

125.   **126.**

127.   **128.**

129. a. $(480, -1211)$ **b.** $(0, -59)$ **c.** $(-12.14, 0)$,
     $(972.14, 0)$ **d.** $x = 480$

130. a. $f(x) = 1.5(x - 2.4)^2 - 3.1$
     b.   **c.** yes

131. x-intercepts: $(-1, 0)$, $(0.4, 0)$, $(2.8, 0)$;
     zeros: $-1, 0.4, 2.8$, each with multiplicity 1

132. x-intercepts: $(-2.27, 0)$;
     zeros: -2.27 (multiplicity 1)

133. linear function   **134.** quadratic function

135. a. -2 (multiplicity 2), 3, 4
     b. $P(x) = (x + 2)^2(x - 3)(x - 4)$

136. a. -1 (multiplicity 2) and $\frac{2}{5}$

b. $P(x)=-(x+1)^2(5x-2)(x^2+2x+3)$

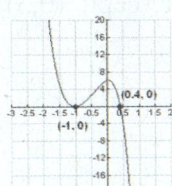

137. $\frac{7}{2}$, $-2\pm 3i$;

$P(x)=(2x-7)(x+2-3i)(x+2+3i)$

138. -5, $\frac{3}{2}$, $3\pm i$;

$P(x)=-(2x-3)(x+5)(x-3-i)(x-3+i)$

139. a. yes, one-to-one **b.** $f^{-1}(x)=\dfrac{x+3}{2-x}$

c.

140. a. yes, one-to-one **b.** $f^{-1}(x)=\dfrac{2x+7}{x-4}$

c.

Practice Test

1.

2. $y=-(x-2)^2+3$

3. $(3,\frac{1}{2})$ **4.** $y=2(x+3)^2-1$

5. $f(x)=x(x-2)^3(x-1)^2$

6. a. $x=\dfrac{-7\pm\sqrt{21}}{2}$ **b.** crosses at all four zeros

c. $(0,0)$ **d.** rises left and right, without

bound **e.**

7. $Q(x)=-2x^2-2x-\frac{11}{2}$, $r(x)=-\frac{19}{2}x+\frac{7}{2}$

8. $Q(x)=17x^4-34x^3+64x^2-128x+258$,

$r(x)=-526$

9. yes   **10.** yes

11. $P(x)=(x-7)(x+2)(x-1)$

12. $-3i,-2,5$

13. yes, complex zero

14.

Positive Real Zeros	Negative Real Zeros	Imaginary Zeros
4	1	0
2	1	2
0	1	4

15. possible rational zeros:

$\pm 1,\ \pm 2,\ \pm 3,\ \pm 4, \pm 6,\ \pm 12,\ \pm\frac{1}{3},\ \pm\frac{2}{3},\ \pm\frac{4}{3}$

16. $0,\pm 2$ **17.** $\frac{3}{2}$, $\pm 2i$

18. 3 (multiplicity 2)   **19.** degree 3

20. 1, 5, or 7 units   **21.** degree 3

22. a. x-intercept: $(\frac{9}{2},0)$, y-intercept: $(0,-3)$

b. $x=-3$ **c.** $y=2$ **d.** none

e.

23. a. x-intercept: $(0,0)$, y-intercept: $(0,0)$

b. $x=\pm 2$ **c.** $y=0$ **d.** none

e.

24. a. x-intercept: $(1,0)$, y-intercept: $(0, \frac{3}{4})$
    b. $x = \pm 2$ **c.** none   **d.** $y = 3x$
    e.

25. a. x-intercept: $(3,0)$, y-intercept: $(0, \frac{3}{8})$
    b. $x = -2$, $x = 4$ **c.** $y = 0$ **d.** none
    e.

26.

27. a. $y = x^2 - 3x - 7.99$
    b. $y = (x-1.5)^2 - 10.24$ **c.** $(-1.7, 0)$ and
    $(4.7, 0)$ **d.** yes

28. x-intercept: $(2,0)$, y-intercept: none,
    HA: $y = 3$,   VA: $x = 3$

Cumulative Test

1. $\dfrac{5}{4x^7 y^{10}}$ **2.** $(2x+3)(y-1)$ **3.** $4x - 12$

4. $(-\infty, -1) \cup (6, \infty)$ **5.** $33\frac{1}{3}$ min.

6. two distinct real solutions   **7.** $x = 3$
8. y-axis   **9.** $y = \frac{1}{3}x - \frac{1}{3}$
10. x-intercept: none, y-intercept: $(0,3)$

11. $x^2 + (y-6)^2 = 2$ **12.** $\left[\frac{7}{6}, \infty\right)$
13. neither
14. reflect over the x-axis, shift left 1 unit, move
    up 2 units
15. right 1 unit and   **16.** $f(g(x)) = x - 1$,
    then up 3 units $x \ge -2$

17. $g(f(-1)) = 0$ **18.** $f^{-1}(x) = 4 + \sqrt{x-2}$, $x \ge 2$
19. $f(x) = (x+2)^2 + 3$
20. 0 (multiplicity 3), -4 (multiplicity 1)
21. $Q(x) = 4x^2 + 4x + 1$, $r(x) = -8$
22. $Q(x) = 2x^2 + 9x + 16$, $r(x) = 54$
23. possible rational zeros:
    $\pm 1, \pm 2, \pm 3, \pm 4, \pm 6, \pm\frac{1}{2}, \pm\frac{1}{3},$
    $\pm\frac{1}{4}, \pm\frac{1}{6}, \pm\frac{1}{12}, \pm\frac{2}{3}, \pm\frac{3}{2}, \pm\frac{3}{4}, \pm\frac{1}{4}$
    zeros: $-2, -\frac{3}{4}, \frac{1}{3}$
24. $-3, -\frac{1}{2}, 5$; $P(x) = (x-5)(2x+1)(x+3)$
25. HA: $y = 0$ VA: $x = \pm 2$ **26.**

27. $f(x) = \dfrac{3(x+1)}{x(2x-3)}$ x-intercept: $(-1,0)$
    HA: $y = 0$

    VA: $x = 0$, $x = \frac{3}{2}$
    yes

28. $f(x) = \dfrac{6x(x-2)}{(3x+1)(4x-1)}$ x-intercept: (2,0),

(0,0)   y-intercept: (0,0)

HA: $y = 0.5$

VA: $x = -\frac{1}{3},\ x = \frac{1}{4}$

yes

CHAPTER 5

Section 5.1

1. 16   **2.** 81   **3.** $\frac{1}{25}$ **4.** $\frac{1}{64}$ **5.** 4   **6.** 9
7. 27   **8.** 64   **9.** −1   **10.** −1   **11.** 5.2780
12. 0.1895   **13.** 9.7385   **14.** 22.2740
15. 7.3891   **16.** 1.6487   **17.** 0.0432
18. 0.2431   **19.** 27   **20.** 100
21. 16   **22.** $\frac{1}{9}$ **23.** 4   **24.** 64   **25.** 19.81
26. 0.0002   **27.** f   **28.** c   **29.** e   **30.** d
31. b   **32.** a

33. y-intercept: (0,1)   HA: $y = 0$

domain: $(-\infty,\infty)$ range: $(0,\infty)$

other points: $\left(-1,\frac{1}{6}\right)$, (1,6)

34. y-intercept: (0,1)   HA: $y = 0$

domain: $(-\infty,\infty)$ range: $(0,\infty)$

other points: $\left(-1,\frac{1}{7}\right)$, (1,7)

35. y-intercept: (0,1)   HA: $y = 0$

domain: $(-\infty,\infty)$ range: $(0,\infty)$

other points: (1, 0.1), (−1, 10)

36. y-intercept: (0, 1)   HA: $y = 0$

domain: $(-\infty,\infty)$ range: $(0,\infty)$

other points: $\left(1,\frac{1}{4}\right)$, (−1, 4)

37. y-intercept: (0,1)   HA: $y = 0$

domain: $(-\infty,\infty)$ range: $(0,\infty)$

other points: $\left(1,\frac{1}{e}\right)$, (−1, e)

38. y-intercept: (0,−1)   HA: $y = 0$

domain: $(-\infty,\infty)$ range: $(0,\infty)$

other points: $\left(-1,-\frac{1}{e}\right)$, (1, −e)

39. y-intercept: (0,0)   HA: $y = -1$

domain: $(-\infty,\infty)$ range: $(-1,\infty)$

other points: (2,3), (1,1)

40. y-intercept: $(0,0)$ HA: $y = -1$
domain: $(-\infty, \infty)$ range: $(-1, \infty)$
other points: $(2,8), (1,2)$

41. y-intercept: $(0,1)$ HA: $y = 2$
domain: $(-\infty, \infty)$ range: $(-\infty, 2)$
other points: $(1, 2-e), \left(-1, 2-\frac{1}{e}\right)$

42. y-intercept: $(0,1)$ HA: $y = 1$
domain: $(-\infty, \infty)$ range: $(1, \infty)$
other points: $(-1, 1+e), \left(1, 1+\frac{1}{e}\right)$

43. y-intercept: $(0, e-4)$ HA: $y = -4$
domain: $(-\infty, \infty)$ range: $(-4, \infty)$
other points: $(-1, -3), (1, e^2 - 4)$

44. y-intercept: $\left(0, 2+\frac{1}{e}\right)$ HA: $y = 2$
domain: $(-\infty, \infty)$ range: $(2, \infty)$
other points: $(1, 3), (2, e+2)$

45. y-intercept: $(0,3)$ HA: $y = 0$
domain: $(-\infty, \infty)$ range: $(0, \infty)$
other points: $(2, 3e), (1, 3\sqrt{e})$

46. y-intercept: $(0,2)$ HA: $y = 0$
domain: $(-\infty, \infty)$ range: $(0, \infty)$
other points: $(0, 2), (-1, 2e)$

47. y-intercept: $(0,5)$ HA: $y = 1$
domain: $(-\infty, \infty)$ range: $(1, \infty)$
other points: $(0, 5), (2, 2)$

48. y-intercept: $\left(0, \frac{5}{3}\right)$ HA: $y = 2$
domain: $(-\infty, \infty)$ range: $(-\infty, 2)$
other points: $\left(0, \frac{5}{3}\right), (-1, 1)$

49. 10.4 million

50. expected population in 2020 is approximately 273,033

51. $P(30) = 1500\left(2^{30/5}\right) \cong 96,000$

52. Colin: \$401,158 Cameron: \$689,441

53. 168 mg   **54.** 53 mg   **55.** \$3031

56. \$94   **57.** \$3,448.42 **58.** \$11,876.86

59. $13,011.03   **60.** $51,016.87

61. $4319.55   **62.** $10,760.80

63. $13,979.42   **64.** $27,167.64

65. 3.4 mg/L

66. 3.9 mg/L

67.

p (price per unit)	$D(p)$—approximate demand for product in units
1.00	1,955,000
5.00	1,020,500
10.00	452,810
20.00	89,147
40.00	3455
60.00	134
80.00	5
90.00	1

68. $D(91) \approx 0.9$. When the cost is $91 per unit, the demand for these units is less than a single unit (no demand at all).

69. The mistake is that $4^{-\frac{1}{2}} \neq 4^2$. Rather,

$$4^{-\frac{1}{2}} = \frac{1}{4^{\frac{1}{2}}} = \frac{1}{2}.$$

70. The mistake is that $4^{\frac{3}{2}} \neq \frac{4^3}{4^2}$. Rather,

$$4^{\frac{3}{2}} = \left(\sqrt{4}\right)^3 = 2^3 = 8.$$

71. $r = 0.025$ rather than 2.5

72. Time is measured in years, so 6 months should be converted to $\frac{1}{2}$. Using $t = \frac{1}{2}$ in the formula yields $A = 5075.57$.

73. false   **74.** true

75. true   **76.** false

77.   **78.**

79.   **80.**

81. y-intercept: $(0, be - a)$ HA: $y = -a$

82. y-intercept: $(0, a + be)$ HA: $y = a$

83. Domain: $(-\infty, \infty)$

84.   **85.**

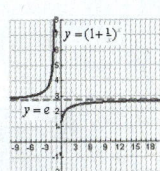

86. Since $2 < e < 3$, we have $2^x < e^x < 3^x$

87. close on the interval $(-3, 3)$

88. close on the interval $(-3, 3)$

89. as x increases,
$$f(x) \to e, \ g(x) \to e^2, h(x) \to e^4$$

90. as x increases,
$$f(x) \to e, \ g(x) \to e^{-1}, h(x) \to e^{-2}$$

91. a.

b. The best fit exponential curve is
$y = 228.34 \, (0.9173)^x$ with $r^2 = 0.9628$.
This best fit curve is shown below on the scatterplot. The fit is very good, as evidenced by the fact that the square of the correlation coefficient is very close to 1.

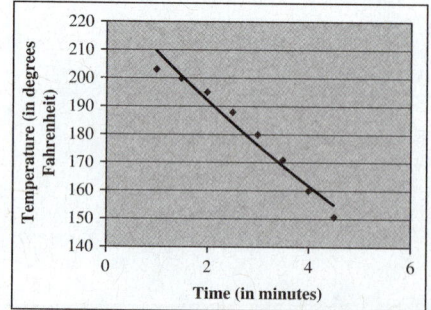

c. (i) Compute the y-value when $x = 6$ to obtain about 136 degrees Fahrenheit.
(ii) The temperature of the soup the moment it was taken out of the microwave is the y-value at $x = 0$, namely about 228 degrees Fahrenheit.

d. The shortcoming of this model for large values of x is that the curve approaches the x-axis, not 72 degrees. As such, it is no longer useful for describing the temperature beyond the x-value at which the temperature is 72 degrees.

92. a.

b. The best fit exponential curve is
$y = 40.646e^{0.072x}$ with $r^2 = 0.976$. This best fit curve is shown below on the scatterplot. The fit is very good, as evidenced by the fact that the square of the correlation coefficient is very close to 1.

c. Use the best fit exponential curve from (b) to answer the following:
(i) Compute the y-value when $x = 10$ to obtain about 84 degrees Fahrenheit.

(ii) The temperature of the soda the moment it was taken out of the refrigerator is the y-value at $x = 0$, namely about 41 degrees Fahrenheit.

d. The shortcoming of this model for large values of x is that the curve continues to grow indefinitely in the y-direction rather than stopping once it hits 90 degrees. As such, it is no longer useful for describing the temperature beyond the x-value at which the temperature is 90 degrees.

Section 5.2

1. $5^3 = 125$ **2.** $3^3 = 27$ **3.** $81^{\frac{1}{4}} = 3$

4. $121^{\frac{1}{2}} = 11$ **5.** $2^{-5} = \frac{1}{32}$ **6.** $3^{-4} = \frac{1}{81}$

7. $10^{-2} = 0.01$ **8.** $10^{-4} = 0.0001$

9. $10^4 = 10,000$ **10.** $10^3 = 1000$

11. $\left(\frac{1}{4}\right)^{-3} = 64$ **12.** $\left(\frac{1}{6}\right)^{-2} = 36$

13. $e^{-1} = \frac{1}{e}$ **14.** $e^1 = e$ **15.** $e^0 = 1$

16. $10^0 = 1$ **17.** $e^x = 5$ **18.** $e^y = 4$

19. $x^z = y$ **20.** $x^y = z$ **21.** $\log_8(512) = 3$

22. $\log_2(64) = 6$ **23.** $\log(0.00001) = -5$

24. $\log(100,000) = 5$ **25.** $\log_{225}(15) = \frac{1}{2}$

26. $\log_{343}(7) = \frac{1}{3}$ **27.** $\log_{2/5}\left(\frac{8}{125}\right) = 3$

28. $\log_{2/3}\left(\frac{8}{27}\right) = 3$ **29.** $\log_{1/27}(3) = -\frac{1}{3}$

30. $\log_{1/1024}(4) = -\frac{1}{5}$ **31.** $\ln 6 = x$

32. $\ln 4 = -x$ **33.** $\log_y x = z$

34. $\log_y z = x$ **35.** 0 **36.** 0

37. 5 **38.** 6 **39.** 7

40. -2 **41.** -6 **42.** -4

43. undefined   **44.** undefined

45. undefined   **46.** undefined

47. 1.46 **48.** 3.37 **49.** 5.94

50. 2.58 **51.** undefined

52. undefined   **53.** -8.11 **54.** -3.52

55. $(-5, \infty)$ **56.** $\left(\frac{1}{4}, \infty\right)$ **57.** $\left(-\infty, \frac{5}{2}\right)$

58. $(-\infty, 5)$ **59.** $\left(-\infty, \frac{7}{2}\right)$ **60.** $(-\infty, 3)$

61. $(-\infty, 0) \cup (0, \infty)$ **62.** $(-\infty, -1) \cup (-1, \infty)$

63. $\mathbb{R}$ **64.** $(-1, 1)$ **65.** b   **66.** e   **67.** c

68. f   **69.** d   **70.** a

71. domain: $(1, \infty)$ **72.** domain: $(-2, \infty)$

range: $(-\infty, \infty)$ range: $(-\infty, \infty)$

73. domain: $(0, \infty)$ **74.** domain: $(0, \infty)$

range: $(-\infty, \infty)$ range: $(-\infty, \infty)$

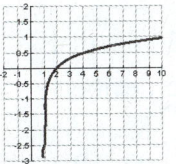

75. domain: $(-2, \infty)$ **76.** domain: $(-1, \infty)$

range: $(-\infty, \infty)$ range: $(-\infty, \infty)$

77. domain: $(0, \infty)$ **78.** domain: $(-\infty, 0)$

range: $(-\infty, \infty)$ range: $(-\infty, \infty)$

79. domain: $(-4, \infty)$ **80.** domain: $(-\infty, 4)$

range: $(-\infty, \infty)$ range: $(-\infty, \infty)$

81. domain: $(0, \infty)$

range: $(-\infty, \infty)$

82. domain: $(-\infty, 0)$

range: $(-\infty, \infty)$

b.

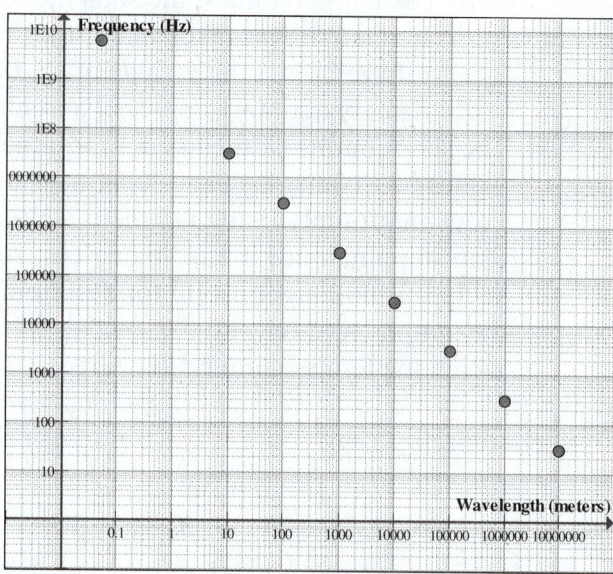

83. 60 decibels   **84.** 130 decibels

85. 117 decibels   **86.** 75 decibels

87. 8.5   **88.** 7.6   **89.** 6.6

90. 9.0   **91.** 3.3   **92.** 10.3

93. Normal Rainwater: 5.6

    Acid rain/tomato juice: 4

94. 12.3   **95.** 3.6   **96.** 4.2

97. 13,236 years   **98.** 20,771 years

99. 25 dB loss   **100.** −42 dB

101. a.

Usage	Wavelength	Frequency
Super Low Frequency—Communication with Submarines	10,000,000 m	30 Hz
Ultra Low Frequency—Communication within Mines	1,000,000 m	300 Hz
Very Low Frequency—Avalanche Beacons	100,000 m	3000 Hz
Low Frequency—Navigation, AM Longwave Broadcasting	10,000 m	30,000 Hz
Medium Frequency—AM Bradcasts, Amatuer Radio	1,000 m	300,000 Hz
High Frequency—Shortwave broadcasts, Citizens Band Radio	100 m	3,000,000 Hz
Very High Frequency—FM Radio, Television	10 m	30,000,000 Hz
Ultra High Frequency—Television, Mobile Phones	0.050 m	6,000,000,000 Hz

102. a.

Color	Wavelength	Frequency
Violet	400 nm	750×10^{12} Hz
Cyan	470 nm	638×10^{12} Hz
Green	480 nm	625×10^{12} Hz
Yellow	580 nm	517×10^{12} Hz
Orange	610 nm	491×10^{12} Hz
Red	630 nm	476×10^{12} Hz

b.

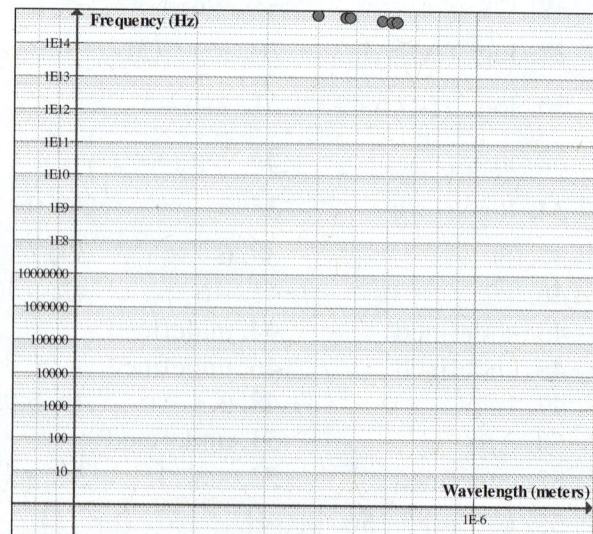

103. $\log_2 4 = x$ is equivalent to $2^x = 4$ (not $x = 2^4$).

104. $\log_{100} 10 = x$ is equivalent to $100^x = 10$ (not $10^x = 100$).

105. The domain is the set of all real numbers such that $x + 5 > 0$, which is written as $(-5, \infty)$.

106. Must exclude $x = 0$ from the domain since $\ln 0$ is not defined.

107. false   **108.** false

109. true   **110.** true

111. domain: (a, ∞) range: $(-\infty, \infty)$

x-intercept: $\left(a + e^b, 0\right)$

112. domain: $(-\infty, a)$ range: $(-\infty, \infty)$

x-intercept: $\left(a - 10^b, 0\right)$

113.   **114.**

115. $y = x$ **116.** $y = x$

117. x-intercept: $(1, 0)$ **118.** except at 0

VA: $x = 0$

119. $(0, \infty)$ **120.** $(2, \infty)$

121. a.

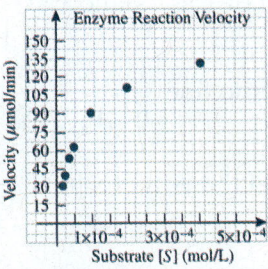

b. A reasonable estimate for $V_{\max}$ is about 156 µmol/min.

c. K_m is the value of $[S]$ that results in the velocity being half of its maximum value, which by (b) is about 156. So, we need the value of $[S]$ that corresponds to $v = 78$. From the graph, this is very difficult to ascertain because of the very small units. We can simply say that it occurs between 0.0001 and 0.0002. A more accurate estimate can be obtained if a best fit curve is known.

d. (i) $v = 33.70 \ln([S]) + 395.80$ with $r^2 = 0.9984$. It is shown on the scatterplot below.

(ii) Using the equation, we must solve the following equation for $[S]$:

$$100 = 33.70 \ln([S]) + 395.80$$
$$-295.80 = 33.70 \ln[S]$$
$$-8.77745 = \ln[S]$$
$$e^{-8.77745} = S$$
$$S \approx 0.000154171$$

122. a. Thinking of y as $\dfrac{1}{v}$ and x as $\dfrac{1}{[S]}$, the

slope of the line $\dfrac{1}{v}=\dfrac{K_m}{V_{max}}\dfrac{1}{[S]}+\dfrac{1}{V_{max}}$ is

$\dfrac{K_m}{V_{max}}$ and the y-intercept is $\dfrac{1}{V_{max}}$.

b.

Lineweaver-Burk plot

c. The equation of the best fit line is $y = 0.0000005x + 0.0064$ with $r^2 = 0.9998$. It is shown on the scatterplot below.

Lineweaver-Burk plot

d. Using the information in (a), we solve the following equation:

$$\frac{1}{V_{max}} = 0.0064$$

$$V_{max} = 156.25$$

This value is very close to what we obtained in the previous problem.

e. Using the information in (a) and (d), we solve the following equation:

$$\frac{K_m}{V_{max}} = \frac{K_m}{156.25} = 0.0000005$$

$$K_m = 0.0000078$$

Section 5.3

1. 0   **2.** 0   **3.** 1   **4.** 1   **5.** 8   **6.** 3

7. −3   **8.** 7   **9.** $\frac{3}{2}$ **10.** $\frac{1}{3}$ **11.** 5   **12.** 5

13. $x+5$ **14.** $3x^2+2x+1$ **15.** 8

16. 25   **17.** $\frac{1}{9}$ **18.** $\frac{1}{100}$ **19.** $3\log_b(x)+5\log_b(y)$

20. $-3\log_b(x)-5\log_b(y)$

21. $\frac{1}{2}\log_b(x)+\frac{1}{3}\log_b(y)$

22. $\frac{1}{2}\log_b(r)+\frac{1}{2}\log_b(r)=\log_b(r)$

23. $\frac{1}{3}\log_b(r)-\frac{1}{2}\log_b(s)$

24. $4\log_b(r)-2\log_b(s)$

25. $\log_b(x)-\log_b(y)-\log_b(z)$

26. $\log_b(x)+\log_b(y)-\log_b(z)$

27. $2\log x+\frac{1}{2}\log(x+5)$

28. $\log(x-3)+\log(x+2)$

29. $3\ln(x)+2\ln(x-2)-\frac{1}{2}\ln\left(x^2+5\right)$

30. $\frac{1}{2}\ln(x+3)+\frac{1}{3}\ln(x-4)-4\ln(x+1)$

31. $2\log(x-1)-\log(x-3)-\log(x+3)$

32. $\log(x-2)+\log(x+1)-\log(x-1)$
$$-\log(x+4)$$

33. $\log_b\left(x^3y^5\right)$ **34.** $\log_b\left(u^2v^3\right)$

35. $\log_b\left(\dfrac{u^5}{v^2}\right)$ **36.** $\log_b\left(\dfrac{x^3}{y}\right)$

37. $\log_b\left(x^{\frac{1}{2}}y^{\frac{2}{3}}\right)$ **38.** $\log_b\left(\dfrac{x^{\frac{1}{2}}}{y^{\frac{2}{3}}}\right)$

39. $\log\left(\dfrac{u^2}{v^3z^2}\right)$ **40.** $\log\left(\dfrac{u^3}{2vz}\right)$

41. $\ln\left(\dfrac{x^2-1}{(x^2+3)^2}\right)$ **42.** $\ln\left(\dfrac{\sqrt{(x-1)(x+1)}}{(x^2-1)^2}\right)$

43. $\ln\left(\dfrac{(x+3)^{\frac{1}{2}}}{x(x+2)^{\frac{1}{3}}}\right)$ **44.** $\ln\left(\dfrac{(x^2+4)^{\frac{1}{3}}}{(x-1)(x^2-3)^{\frac{1}{2}}}\right)$

45. 1.2091   **46.** 2.1240   **47.** −2.3219

48. −0.4307   **49.** 1.6599   **50.** 0.4642

51. 2.0115   **52.** 0.8677   **53.** 3.7856

54. 6.3400   **55.** 110 decibels

56. 60 decibels   **57.** 5.5   **58.** 4.9

59. $3\log 5 - \log 5^2 = 3\log 5 - 2\log 5 = \log 5$

60. Step 3 is wrong. Apply the product and quotient properties:

$\ln 3 + \ln 16 - \ln 8 = \ln(3 \cdot 16) - \ln 8$

$= \ln\left(\dfrac{48}{8}\right) = \ln 6$

61. Cannot apply the product and quotient properties to logarithms with different bases. So, you cannot reduce the given expression further without using the change of base formula.

62. Applied the power property incorrectly. Should be $2\log\left(\frac{3}{5}\right) = \log\left(\frac{3}{5}\right)^2$, not $\left(\log\frac{3}{5}\right)^2$.

63. true   **64.** true

65. false   **66.** true

67. *Proof*: Let $u = \log_b M$, $v = \log_b N$. Then, $b^u = M$, $b^v = N$. Observe that

$\log_b\left(\frac{M}{N}\right) = \log_b\left(\frac{b^u}{b^v}\right)$

$= \log_b\left(b^{u-v}\right) = u - v = \log_b M - \log_b N$

68. *Proof*: Let $u = \log_b M$. Then, $b^u = M$. Observe that

$\log_b\left(M^p\right) = \log_b\left(b^u\right)^p = \log_b\left(b^{u \cdot p}\right)$

$= u \cdot p = (\log_b M) \cdot p = p\log_b M$

69. $6\log_b x - 9\log_b y + 15\log_b z$

70. $\log_b\left(\frac{1}{x}\right) = \log_b\left(x^{-1}\right) = -\log_b x$

71. yes   **72.** no

73. no   **74.** yes

75. no   **76.** no

77. yes   **78.** yes

Section 5.4

1. $x = 4$ **2.** $x = 3$ **3.** $x = -2$

4. $x = -2$ **5.** $x = \pm 2$ **6.** $x = \frac{1}{2}$

7. $x = -4$ **8.** $x = -3$ **9.** $x = -\frac{3}{2}$

10. $x = \pm 1$ **11.** $x = -1$ **12.** $x = -3$

13. $x = 3, 4$ **14.** $x = -1, 3$ **15.** $x = 0, 6$

16. $x = 2$ **17.** $x = 1, 4$ **18.** $x = 0, 10$

19. $x = \dfrac{3 + \log(81)}{2} \approx 2.454$

20. $x = \dfrac{-1 + \log_2(21)}{3} \approx 1.131$

21. $x = \log_3(5) - 1 \approx 0.465$

22. $x = \dfrac{\log_5(35) + 1}{2} \approx 1.605$

23. $x = \dfrac{\log_2(27) + 1}{3} \approx 1.918$

24. $x = \dfrac{3 - \log_7(15)}{2} \approx 0.804$

25. $x = \ln 5 \approx 1.609$ **26.** $x = \ln 3 \approx 1.100$

27. $x = 10\ln 4 \approx 13.863$

28. $x = 10\ln 4 \approx 13.863$

29. $x = \log_3(10) \approx 2.096$

30. $x = \log_2(9) \approx 3.170$

31. $x = \dfrac{\ln(22) - 4}{3} \approx -0.303$

32. $x = \pm\sqrt{\ln(73)} \approx \pm 2.071$

33. $x = \dfrac{\ln 6}{2} \approx 0.896$

34. $x = \dfrac{\log 5}{3} \approx 0.233$

35. $x = \ln\left(\dfrac{-7+\sqrt{61}}{2}\right) \approx -0.904$

36. $x = \ln 5 \approx 1.609$

37. $x = 0$ **38.** $x = 0$

39. $x = \ln 7 \approx 1.946$

40. $x = \ln\left(\dfrac{9}{2}\right) \approx 1.504$

41. $x = 0$ **42.** $x = \dfrac{\ln\left(\frac{5}{2}\right)}{3} \approx 0.305$

43. $x = \dfrac{\log_{10}(9)}{2} \approx 0.477$

44. $x = \log_{10}(4) \approx 0.602$

45. $x = 50$ **46.** $x = 200$ **47.** $x = 40$

48. $x = 3$ **49.** $x = \frac{9}{32}$ **50.** $x = \frac{79}{32}$

51. $x = \pm 3$ **52.** $x = 10$ **53.** $x = 5$

54. $x = 5$ **55.** $x = 6$ **56.** $x = 1$

57. $x = -1$ **58.** $x = -2$

59. no solution   **60.** $x = -1$

61. $x = \frac{25}{8}$ **62.** $x = 1$

63. $x = \pm\sqrt{e^5} \approx \pm 12.182$

64. $x = \frac{100}{3}$ **65.** $x = 47.5$ **66.** $x \approx 6.771$

67. $x = \pm\sqrt{e^4 - 1} \approx \pm 7.321$ **68.** $x \approx \pm 9.798$

69. $x = \dfrac{1}{2}(-3 + e^{-2}) \approx -1.432$ **70.** $x = 1.7$

71. $x \approx -1.25$ **72.** $x \approx -2.441$

73. $x = \dfrac{2 + \sqrt{4 + 4e^4}}{2} \approx 8.456$ **74.** $x \cong 0.6875$

75. $x = \dfrac{-3 + \sqrt{13}}{2} \approx 0.303$ **76.** $x \approx 2.414$

77. $x = 1 + \sqrt{7} \approx 3.646$ **78.** $x \approx 0.4656$

79. a. 151 beats per minute
   b. 7 minutes
   c. 66 beats per minute

80. a. $V(t) = 45{,}000e^{-0.2027t}$
   b. four years

81. 31.9 years   **82.** 31.62 years

83. 19.74 years   **84.** 8.514 years

85. 3.16×10^{15} joules   **86.** 7.08×10^{16} joules

87. $1\,{}^W\!/_{m^2}$ **88.** $10^{-2}\,{}^W\!/_{m^2}$

89. 4.61 hours   **90.** 23.105 years

91. 15.89 years   **92.** $\frac{\ln 10}{2}$ weeks

93. 6.2   **94.** 10^4 watts

95. $\ln(4e^x) \neq 4x$. Should first divide both sides by 4, then take the natural log:
$$4e^x = 9$$
$$e^x = \tfrac{9}{4}$$
$$\ln(e^x) = \ln(\tfrac{9}{4})$$
$$x = \ln(\tfrac{9}{4})$$

96. Step 2 should be:
$$10^{\log(3x)} = 10$$
$$\log(3x) = 1$$
$$3x = 10$$
$$x = \tfrac{10}{3}$$

97. $x = -5$ is not a solution since $\log(-5)$ is not defined.

98. $\log x + \log 2 \neq \log(x+2)$, in general. The computation should be:
$$\log(2x) = \log 5$$
$$2x = 5$$
$$x = \tfrac{5}{2}$$

99. true   **100.** false

101. false   **102.** true

103. $x = \dfrac{1 + \sqrt{1 + 4b^2}}{2}$

104. $x = \dfrac{1 + \sqrt{1 - 4b^2}}{2},\ \dfrac{1 - \sqrt{1 - 4b^2}}{2}$

105. $t = -5\ln\left(\frac{3000-y}{2y}\right)$ **106.** $(-\infty, -a) \cup (a, \infty)$

107. $f^{-1}(x) = \ln\left(x + \sqrt{x^2 - 1}\right); x \geq 1$

108. $f^{-1}(x) = \ln\left(x + \sqrt{x^2 + 1}\right)$

109. $x = \frac{3 \pm \sqrt{5}}{2}$

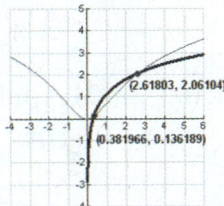

110. $x = 0, -3$

111.   **112.**

113. domain: $(-\infty, \infty)$ y-axis symmetry

114. domain: $(-\infty, 0) \cup (0, \infty)$ origin symmetry
HA: $y = 1, y = -1$

Section 5.5

1. c (iv)   **2.** d (v)   **3.** a (iii)
4. b (ii)   **5.** f (i)   **6.** e (i)
7. 94 million **8.** 23 million **9.** 5.5 years, 2008
10. 6.871 years, 2008
11. 799.6 subscribers **12.** 60,756 bacteria
13. $455,000   **14.** $302,000
15. 332 million   **16.** 72 million
17. 1.45 million   **18.** 96.44 million
19. 13.53 ml   **20.** 0.25 ml

21. a. $k = -\ln\left(\frac{8}{15}\right) \cong 0.6286$

b. 636,000 mp3 players
22. a. $P(t) = 850e^{-0.2409t}$ **b.** $324,000
23. 7575 years   **24.** 2115 years
25. 131,158,556 years old

26. 1.984 mg **27.** $105°F$ **28.** $59°F$
29. 3.8 hours before 7 am   **30.** 2.76 hours
31. $19,100   **32.** $8900
33. a. 84,520 **b.** 100,000 **c.** 100,000
34. a. 1,998,659 **b.** 2,000,000
35. 11,439,406 cases **36.** 2,778,300 cases
37. 1.89 years   **38.** 4.20 years
39. $r = 0$ **40.** 1.83%
41. a.   **b.** 75 **c.** 4 **d.** 4

42. a.   **b.** 80 **c.** 2 **d.** 2

43. a. 18 years **b.** 10 years
44. a. 17.12 years **b.** 7.30 years
45. a. 30 years **b.** \$328,120
46. a. 7.6 years **b.** \$67,581 **c.** \$260,539
47. $r = 0.07$, not 7   **48.** $r = 0.05$, not 5
49. true   **50.** false
51. false   **52.** true
53. less time   **54.** more time
55. a. For the same periodic payment, it will take Wing Shan fewer years to pay off the loan if she can afford to pay biweekly.

b. 11.58 years **c.** 10.33 years **d.** 8.54 years, 7.69 years, respectively
56. a. For the same periodic payment, it will take Hong fewer years to pay off the loan if he can afford to pay biweekly.

b. 5.12 years **c.** 4.50 years **d.** 3.35 years, 2.99 years, respectively

Review Exercises

1. 17,559.94   **2.** 1.58   **3.** 5.52
4. 1.24   **5.** 24.53   **6.** 23.14
7. 5.89   **8.** 0.01   **9.** 73.52
10. −39.40   **11.** 6.25   **12.** $\frac{16}{49}$
13. b   **14.** a   **15.** c   **16.** d
17. y-intercept: $(0, -1)$
     HA: $y = 0$

18. y-intercept: $(0, 3)$ **19.** y-intercept: $(0, 2)$
     HA: $y = 4$ HA: $y = 1$

20. y-intercept: $(0, -3)$ **21.** y-intercept: $(0, 1)$
     HA: $y = -4$ HA: $y = 0$

22. y-intercept: $(0, e^{-1})$ **23.** y-intercept: $(0, 3.2)$
     HA: $y = 0$ HA: $y = 0$

24. y-intercept: $(0, 2 - e)$
     HA: $y = 2$

25. \$6144.68   **26.** \$18,182.60

27. \$23,080.29   **28.** \$11,682.01

29. $4^3 = 64$ **30.** $4^{\frac{1}{2}} = 2$ **31.** $10^{-2} = \frac{1}{100}$

32. $16^{\frac{1}{2}} = 4$ **33.** $\log_6 216 = 3$

34. $\log_{10} 0.0001 = -4$ **35.** $\log_{\frac{2}{13}}\left(\frac{4}{169}\right) = 2$

36. $\log_{512} 8 = \frac{1}{3}$ **37.** 0 **38.** 4

39. −4   **40.** 12   **41.** 1.51
42. 3.47   **43.** −2.08   **44.** −0.90

45. $(-2, \infty)$ **46.** $(-\infty, 2)$ **47.** $(-\infty, \infty)$

48. $\left(-\sqrt{3}, \sqrt{3}\right)$ **49.** b **50.** a **51.** d **52.** c

53.

54.

55.

56.

57. 6.5 **58.** 2.7 **59.** 50 decibels

60. 80 decibels **61.** 1 **62.** 2

63. 6 **64.** $\frac{1}{216}$

65. $a\log_c(x) + b\log_c(y)$

66. $2\log_3(x) - 3\log_3(y)$

67. $\log_j(r) + \log_j(s) - 3\log_j(t)$

68. $c\log x + \frac{1}{2}\log(x+5)$

69. $\frac{1}{2}\log(a) - \frac{3}{2}\log(b) - \frac{2}{5}\log(c)$

70. $\frac{1}{3}\left[3\log_7(c) + \frac{1}{3}\log_7(d) - 6\log_7(e)\right]$

71. 0.5283 **72.** −0.4307 **73.** 0.2939

74. 1.6681 **75.** $x = -4$ **76.** $x = \pm 2$

77. $x = \frac{4}{3}$ **78.** $x = 23.04$ **79.** $x = -6$

80. $x = \pm\sqrt{\frac{7}{2}}$ **81.** $x \approx -0.218$

82. $x \cong 2.404$ **83.** no solution

84. $x \approx 27.726$ **85.** $x = 0$

86. $x \cong 2.3219$ **87.** $x = \frac{100}{3}$

88. $x = 79$ **89.** $x = 128\sqrt{2}$

90. $x = \frac{3}{2}$ **91.** $x \approx \pm 3.004$

92. $x \cong 366.88$ **93.** $x \approx 0.449$

94. $x \approx 2.162$ **95.** $28,536.88

96. 17.8% **97.** 16.6 years

98. 11.453 years **99.** 3.72 million

100. 52.75 million **101.** 6250 bacteria

102. 1,753,918 **103.** 56 years

104. 58,048.2 years **105.** 16 fish

106. $18,492.69 **107.** 343 mice

108. population in 2030 is about 195,745

109. HA: $y = e^{\sqrt{2}} \approx 4.11$ **110.** (0.785, 3.369)

111. (2.376, 2.071)

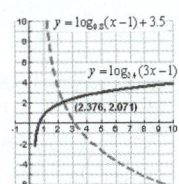

112. (28.09, 2.753), (1.227, 0.379)

113. $(0, \infty)$ **114.** (1, 3)

115. domain: $(-\infty, \infty)$ symmetric origin

HA: $y = -1$ (as $x \to -\infty$), $y = 1$ (as $x \to \infty$)

116. domain: $(-\infty, 0) \cup (0, \infty)$ symmetric origin

HA: $y = 0$ VA: $x = 0$

117. a. $N = 4e^{-0.038508t} \approx 4(0.9622)^t$
   b. $N = 4(0.9622)^t$ **c.** yes
118. a. $N = 5600e^{-0.8390t} \approx 5600(0.4321428571)^t$
   b. $N(7) \approx 15.76 = 16$ fish **c.** yes

Practice Test

1. x^3 **2.** 3.60   **3.** −4
4. $5x - \ln x - \ln(x^4 + 1)$
5. $x = \pm\sqrt{1 + \ln 42} \approx \pm 2.177$
6. $x = \ln 3, \ x = \ln 2$
7. $x = \dfrac{-1 + \ln\left(\frac{300}{27}\right)}{0.2} \cong 7.04$
8. $x = \frac{1}{2}\left(1 + \log_3 15\right) \approx 1.732$
9. $x = 4 + e^2 \approx 11.389$
10. $x = \frac{2}{3}$ **11.** $x = e^e \approx 15.154$
12. $x = 3$ **13.** $x = 9$
14. $x = \dfrac{3e^2 + 2}{e^2 - 1} \approx 3.783$
15. $x = \dfrac{-3 \pm \sqrt{9 + 4e}}{2} \approx 0.729$
16. $x = \frac{11}{6}$ **17.** $x = \ln\left(\frac{1}{2}\right) = -\ln(2) \approx -0.693$
18. $x = \dfrac{3 \pm \sqrt{9 - 4(1)(-e^2)}}{2} \approx 4.605$
19. $(-1, 0) \cup (1, \infty)$ **20.** $x > \frac{a}{4}$
21. x-intercept: none
   y-intercept: $(0, 2)$
   HA: $y = 1$

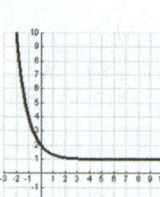

22. x-intercept: $(-1.5850, 0)$
   y-intercept: $(0, -2)$
   HA: $y = -3$

23. x-intercept: $\left(\dfrac{3 + \frac{1}{e}}{2}, 0\right)$
   y-intercept: none
   VA: $x = \frac{3}{2}$

24. x-intercept: $\left(\frac{99}{100}, 0\right)$
   y-intercept: $(0, 2)$
   VA: $x = 1$

25. \$8051.62   **26.** \$16,487
27. 90 decibels   **28.** 1,079,887
29. $7.9 \times 10^{11} < E < 2.5 \times 10^{13}$ joules
30. 9 hours   **31.** 7800 bacteria
32. 7574.65 years   **33.** 3 days
34. 108.11 million barrels
35. domain: $(-\infty, \infty)$ **36.** $x = 2.22$
   symmetric origin

Cumulative Test

1. $x^{5/6} y^2$ **2.** $\dfrac{3x - 4}{3x - 8}$ **3.** $x = \dfrac{2 \pm \sqrt{19}}{5}$
4. $x = -2, 6$ **5.** $(-\infty, -19]$ **6.** $x = \pm\sqrt{13}$
7. $y = \frac{3}{4}x + \frac{3}{4}$ **8.** $4 - 2x - h$
9. a. 1   **b.** 5   **c.** 1   **d.** undefined
   e. domain: $(-2, \infty)$ range: $(0, \infty)$
   f. increasing: $(4, \infty)$, decreasing: $(0, 4)$,
   constant: $(-2, 0)$
10. Shift the graph of $y = \sqrt{x}$ left 1 unit, and
   then reflect over y-axis.

11. yes   **12.** $R = \frac{0.00152}{d^2}$ **13.** $(1, -1)$

14. $P(x) = C(x+5)^2(x-9)^4$

15. $Q(x) = -x^3 + 3x - 5, \ r(x) = 0$

16. $P(x) = (x-(2+i))(x-(2-i))(x-4)(x+1)$

17. HA: none   VA: $x = 3$ Slant: $y = x + 3$

18.   **19.** 125

20. \$6027.90   **21.** 5

22. $\ln\left(\dfrac{(x+5)^{1/2}}{(x+1)^2(3x)} \right)$ **23.** $x = 0.5$

24. $x = \pm 2$ **25.** 8.62 years   **26.** $x \cong 1.21$

27. a. $N = 6e^{-0.247553t} \approx 6(0.9755486421)^t$

    b. 2.72 grams

CHAPTER 6

Section 6.1

1. $180°$ **2.** $90°$

3. $-120°$ **4.** $-240°$

5. $300°$ **6.** $210°$

7. $-288°$ **8.** $-200°$

9. $-70°$ **10.** $70°$

11. a. $72°$ **b.** $162°$ **12. a.** $51°$ **b.** $141°$

13. a. $48°$ **b.** $138°$ **14. a.** $83°$ **b.** $123°$

15. a. $1°$ **b.** $91°$ **16. a.** $15°$ **b.** $105°$

17. $54°/36°$ **18.** $150°/30°$

19. $120°/60°$ **20.** $60°/30°$

21. $\gamma = 30°$ **22.** $\gamma = 25°$

23. $\alpha = 120°, \ \beta = 30°, \ \gamma = 30°$

24. $\alpha = 108°, \ \beta = 36°, \ \gamma = 36°$

25. $\alpha = 18°, \ \beta = 108°, \ \gamma = 54°$

26. $\alpha = 30°, \ \beta = 90°, \ \gamma = 60°$

27. $c = 5$ **28.** $c = 3\sqrt{2}$

29. $b = 8$ **30.** $a = \sqrt{95}$

31. $c = \sqrt{89}$ **32.** $c = \sqrt{61}$

33. $c = 25$ **34.** $b = 10\sqrt{3}$

35. $10\sqrt{2} \approx 14.14$ inches   **36.** 11.31 meters

37. 2 cm   **38.** $\sqrt{5}$ feet

39. Other leg: $5\sqrt{3} \approx 8.66$ meters;
     Hypotenuse: 10 meters

40. Other leg: $9\sqrt{3} \approx 15.59$ feet;
     Hypotenuse: 18 feet

41. Other leg: $4\sqrt{3} \approx 6.93$ yards;
     Hypotenuse: $8\sqrt{3} \approx 13.9$ yards

42. Other leg: $\dfrac{n\sqrt{3}}{3}$ units;

     Hypotenuse: $\dfrac{2n\sqrt{3}}{3}$ units

43. Short leg: 5 inches;
     Long leg: $5\sqrt{3} \approx 8.66$ inches

44. Short leg: 4 cm;
     Long leg: $4\sqrt{3} \approx 6.93$ cm

45. $f = 3$ **46.** $d = 4$

47. $a = 15$ **48.** $b = 2.1$

49. $c = 11.55$ km   **50.** $b = 25$ m

51. $f = \frac{1}{3}$ in.   **52.** $c = 1.1$ cm

53. $120°$ **54.** $150°$

55. $144°$ **56.** $60°$

57. 60 minutes or 1 hour   **58.** 45 minutes

59. $\sqrt{7300} \approx 85$ feet

60. $\sqrt{10{,}625} \approx 103$ feet

61. 241 feet   **62.** 145 feet

63. 9.8 feet   **64.** 39.3 feet

65. 17 feet   **66.** 80 feet

67. 48 ft × 28 ft   **68.** 54 ft × 94 ft

69. 38 feet   **70.** 40 feet

71. 225 feet   **72.** 8 inches

73. 20 points   **74.** 5 points

75. 50.004 feet   **76.** 50.016 feet

77. 7.8 cm   **78.** 66.5 cm

79. 4 cm   **80.** 4.5 cm

81. The length opposite the 60° angle is $\sqrt{3}$ times (not twice) the length opposite 30° angle.

82. The length of the hypotenuse is positive. The length must be $5\sqrt{2}$ cm.

83. false   **84.** false

85. true   **86.** false

87. true   **88.** true

89. 110°   **90.** 145°

91. $DC = 3$ **92.** $DC = 2$

93. 25   **94.** 76.2

95. The triangles are isosceles right triangles 45°–45°–90°.

96. $\frac{10}{3}$ **97.** 8

98. $\frac{4\sqrt{3}}{3}$ **99.** $2\sqrt{3}$

100. $\frac{2\sqrt{21}}{3}$ **101.** $x = 2$

102. $x = 12$, $y = 6$

103. 28.89 feet, 33.36 feet

104. Shorter leg: 77.80 cm; Hypotenuse: 155.60 cm

Section 6.2

1. $\frac{4}{5}$ **2.** $\frac{3}{5}$

3. $\frac{5}{4}$ **4.** $\frac{5}{3}$

5. $\frac{4}{3}$ **6.** $\frac{3}{4}$

7. $\frac{\sqrt{5}}{5}$ **8.** $\frac{2\sqrt{5}}{5}$

9. $\sqrt{5}$ **10.** $\frac{\sqrt{5}}{2}$

11. 2   **12.** $\frac{1}{2}$

13. $\frac{2\sqrt{10}}{7}$ **14.** $\frac{3}{7}$

15. $\frac{7}{3}$ **16.** $\frac{7\sqrt{10}}{20}$

17. $\frac{3\sqrt{10}}{20}$ **18.** $\frac{2\sqrt{10}}{3}$

19. 30°   **20.** 45°

21. $90° - x$ **22.** $90° - A$

23. 60°   **24.** $90° - B$

25. $\cos(90° - x - y)$ **26.** $\cos(30° + x)$

27. $\sin(70° - A)$ **28.** $\sin(90° - A - B)$

29. $\tan(45° + x)$ **30.** $\csc(60° + \theta)$

31. $\sec(30° + \theta)$ **32.** $\cot(50° - \theta)$

33. a   **34.** b

35. b   **36.** a

37. c   **38.** c

39. $\frac{\sqrt{3}}{3}$ **40.** 1

41. $\sqrt{3}$ **42.** 2

43. $\frac{2\sqrt{3}}{3}$ **44.** $\sqrt{3}$

45. $\frac{2\sqrt{3}}{3}$ **46.** 2

47. $\frac{\sqrt{3}}{3}$ **48.** $\sqrt{2}$

49. $\sqrt{2}$ **50.** 1

51. 0.6018   **52.** 0.3057

53. 0.1392   **54.** 0.9278

55. 1.3764   **56.** 0.9391

57. 1.0098   **58.** 3.8637

59. 1.0002   **60.** 1.2868

61. 0.7002   **62.** 1.8040

63. 0.4337   **64.** 1.6542

65. 30°   **66.** 45°

67. a. $\frac{2}{5}$ b. $\frac{1}{2}$

68. a. $\sqrt{s^2 + p^2}$;   b. $\dfrac{s}{\sqrt{s^2 + p^2}}$

69. 10 miles   **70.** 141 yards

71. 2.405   **72.** 1.576

73. 1.335   **74.** 1.462

75. 0 points   **76.** 0 points

77. 50 points (a bullseye)   **78.** 0 points

79. 1.09 angstroms

80. 3.53 angstroms

81. The opposite of angle y is 3 (not 4).

82. $\tan x = \dfrac{\text{opp}}{\text{adj}}$, not $\dfrac{\text{adj}}{\text{opp}}$

83. Secant is the reciprocal of cosine (not sine).

84. Cosecant is the reciprocal of sine (not cosine).

85. true   **86.** true

87. $\sin 30° = \dfrac{1}{2}$ and $\cos 30° = \dfrac{\sqrt{3}}{2}$

88. $\sin 60° = \dfrac{\sqrt{3}}{2}$ and $\cos 60° = \dfrac{1}{2}$

89. $\tan 30° = \dfrac{\sqrt{3}}{3}$ and $\tan 60° = \sqrt{3}$

90. $\sin 45° = \cos 45° = \dfrac{\sqrt{2}}{2}$ and $\tan 45° = 1$ 1

91. $\sec 45° = \sqrt{2}$ and $\csc 45° = \sqrt{2}$

92. $\cot 30° = \tan 60° = \sqrt{3}$

93. 0   **94.** approximately 1

95. 0   **96.** 1

97. $\dfrac{3\sqrt{5}}{5}$ **98.** $\dfrac{\sqrt{65}}{5}$

99. $\dfrac{2\sqrt{13}}{13}$ **100.** $\dfrac{\sqrt{13}}{2}$

101. $\frac{13}{9}$ **102.** false

103. $\frac{97}{65}$ **104.** $\sqrt{5} + \dfrac{\sqrt{13}}{2}$

105. a. 2.92398   **b.** 2.92380; (b) is more accurate.

106. a. 1.55521   **b.** 1.55572; (b) is more accurate.

107. a. 0.70274   **b.** 0.70281; (b) is more accurate.

108. a. 1.05511   **b.** 1.05511

Section 6.3

1. three   **2.** four
3. two   **4.** one
5. 47°   **6.** 10°
7. 55°   **8.** 28°
9. 83°   **10.** 74°
11. $a \approx 14$ in.   **12.** $b \approx 10$ in.
13. $a \approx 18$ ft   **14.** $b \approx 13$ ft
15. $a \approx 5.50$ miles   **16.** $b \approx 1.99$ miles
17. $c \approx 12$ km   **18.** $c \approx 27$ km

19. $c \approx 20.60$ cm   **20.** $c \approx 19.35$ cm
21. $\alpha \approx 50°$ **22.** $\beta \approx 26°$
23. $\alpha \approx 62°$ **24.** $\beta \approx 37°$
25. $a \approx 82.12$ yards   **26.** $b \approx 60.52$ yards
27. $c \approx 10.6$ km   **28.** $c \approx 20.4$ km
29. $c \approx 19,293$ km   **30.** $c \approx 20,475$ km
31. $\beta \approx 58°$, $a \approx 6.4$ feet, $b \approx 10$ feet
32. $\beta \approx 25°$, $a \approx 34$ feet, $b \approx 16$ feet
33. $\alpha \approx 18°$, $a \approx 3.0$ mm, $b \approx 9.2$ mm
34. $\alpha \approx 45°$, $a \approx 5.5$ mm, $b \approx 5.5$ mm
35. $\beta \approx 35.8°$, $b \approx 80.1$ miles, $c \approx 137$ miles
36. $\alpha \approx 42.8°$, $b \approx 10.5$ miles, $c \approx 14.4$ miles
37. $\beta \approx 61.62°$, $a \approx 936.9$ feet, $c \approx 1971$ feet
38. $\beta \approx 17.017°$, $b \approx 665.7$ feet, $c \approx 2275$ feet
39. $\alpha \approx 56.0°$, $\beta \approx 34.0°$, $c \approx 51.3$ feet
40. $\alpha \approx 24.0°$, $\beta \approx 66.0°$, $b \approx 44.5$ feet
41. $\alpha \approx 55.480°$, $\beta \approx 34.520°$, $b \approx 24{,}235$ km
42. $\alpha \approx 81.83°$, $\beta \approx 8.17°$, $a \approx 0.8674$ mm
43. $c = 27.0$ in., $a = 24.4$ in., $\alpha = 64.6°$
44. $c = 10.0$ m, $a = 7.7$ m, $\alpha = 50.79°$
45. $c = 4.16$ cm, $b = 1.15$ cm, $\alpha = 73.90°$
46. $c = 6.2$ dm, $a = 4.8$ dm, $\alpha \approx 39.90°$
47. 286 feet   **48.** 95 feet
49. $a = 88$ feet   **50.** 170 feet
51. 260 ft (262 rounded to two significant digits)
52. 5° (pilot only sees white lights)
53. 11° (she is too low)
54. 19° (within safety specifications to land)
55. 80 feet   **56.** 270 feet
57. 170 meters   **58.** 85 meters
59. 0.000016°   **60.** 0.000049°
61. 26 feet   **62.** 87 feet
63. 4400 feet   **64.** N 26° E
65. 8°   **66.** 8.65 feet
67. 136.7°   **68.** 138.30°
69. 91.99 feet   **70.** 0.0118
71. 27 feet   **72.** 27 feet
73. 0 points (just misses the entire target)
74. 1.6°   **75.** 121 feet
76. 82.87°   **77.** above 24 feet

78. above 4 feet
79. $d \approx 3.5$ feet
80. Yes, the angle has increased by
    4 degrees.
81. 4.7 inches   **82.** 20°
83. $\tan^{-1}$ should have been used (not tan).
84. Calculation of b is correct, but should not be
    used because of round-off error.
85. true   **86.** true
87. false   **88.** false
89. false   **90.** 4000 feet
91. 2 miles   **92.** 300 feet
93. 0.5   **94.** $\sqrt{17}$
95. $\frac{2}{7}$ **96.** $\dfrac{\sqrt{5}}{4x}$
97. 40°   **98.** 17°
99. 0.8   **100.** 0.3
101. θ **102.** θ
103. 35°; θ **104.** 2; y
105. 0.785   **106.** 5.321

Section 6.4

1. QI   **2.** QII
3. QII   **4.** QII
5. QIV   **6.** QIV
7. negative y-axis **8.** negative x-axis
9. x-axis   **10.** negative y-axis
11. QIII   **12.** QIV
13. QI   **14.** QII
15. QIII   **16.** QIII
17. QII   **18.** QI
19. QII   **20.** QI
21. QIV   **22.** positive x-axis
23.

24.

25.

26.

27.

28.

29.

30.

31.

32.

33.

34.

35.

36.

37. c	**38.** a	**39.** e	**40.** b
41. f	**42.** d	**43.** 52°	**44.** 19°
45. 268°	**46.** 173°		
47. 330°	**48.** 225°		
49. 150°	**50.** 315°		
51. 320.0001°	**52.** 252.34°		

	$\sin\theta$	$\cos\theta$	$\tan\theta$	$\cot\theta$	$\sec\theta$	$\csc\theta$
53.	$\dfrac{2\sqrt5}{5}$	$\dfrac{\sqrt5}{5}$	2	$\dfrac12$	$\sqrt5$	$\dfrac{\sqrt5}{2}$
54.	$\dfrac{3\sqrt{13}}{13}$	$\dfrac{2\sqrt{13}}{13}$	$\dfrac32$	$\dfrac23$	$\dfrac{\sqrt{13}}{2}$	$\dfrac{\sqrt{13}}{3}$
55.	$\dfrac{2\sqrt5}{5}$	$\dfrac{\sqrt5}{5}$	2	$\dfrac12$	$\sqrt5$	$\dfrac{\sqrt5}{2}$
56.	$\dfrac{\sqrt5}{5}$	$\dfrac{2\sqrt5}{5}$	$\dfrac12$	2	$\dfrac{\sqrt5}{2}$	$\sqrt5$
57.	$\dfrac{4\sqrt{41}}{41}$	$\dfrac{5\sqrt{41}}{41}$	$\dfrac45$	$\dfrac54$	$\dfrac{\sqrt{41}}{5}$	$\dfrac{\sqrt{41}}{4}$
58.	$\dfrac{7\sqrt{85}}{85}$	$\dfrac{6\sqrt{85}}{85}$	$\dfrac76$	$\dfrac67$	$\dfrac{\sqrt{85}}{6}$	$\dfrac{\sqrt{85}}{7}$
59.	$\dfrac{2\sqrt5}{5}$	$-\dfrac{\sqrt5}{5}$	-2	$-\dfrac12$	$-\sqrt5$	$\dfrac{\sqrt5}{2}$
60.	$\dfrac{3\sqrt{10}}{10}$	$-\dfrac{\sqrt{10}}{10}$	-3	$-\dfrac13$	$-\sqrt{10}$	$\dfrac{\sqrt{10}}{3}$
61.	$-\dfrac{7\sqrt{65}}{65}$	$\dfrac{4\sqrt{65}}{65}$	$-\dfrac74$	$-\dfrac47$	$\dfrac{\sqrt{65}}{4}$	$-\dfrac{\sqrt{65}}{7}$
62.	$-\dfrac{5\sqrt{106}}{106}$	$-\dfrac{9\sqrt{106}}{106}$	$\dfrac59$	$\dfrac95$	$-\dfrac{\sqrt{106}}{9}$	$-\dfrac{\sqrt{106}}{5}$
63.	$\dfrac{\sqrt{15}}{5}$	$\dfrac{\sqrt{10}}{5}$	$-\dfrac{\sqrt6}{2}$	$-\dfrac{\sqrt6}{3}$	$-\dfrac{\sqrt{10}}{2}$	$\dfrac{\sqrt{15}}{3}$
64.	$\dfrac{\sqrt{10}}{5}$	$\dfrac{\sqrt{15}}{5}$	$-\dfrac{\sqrt6}{3}$	$-\dfrac{\sqrt6}{2}$	$-\dfrac{\sqrt{15}}{3}$	$\dfrac{\sqrt{10}}{2}$
65.	$-\dfrac{\sqrt6}{4}$	$\dfrac{\sqrt{10}}{4}$	$-\dfrac{\sqrt{15}}{5}$	$-\dfrac{\sqrt{15}}{3}$	$\dfrac{2\sqrt{10}}{5}$	$-\dfrac{2\sqrt6}{3}$
66.	$-\dfrac{\sqrt{55}}{11}$	$\dfrac{\sqrt{66}}{11}$	$-\dfrac{\sqrt{30}}{6}$	$-\dfrac{\sqrt{30}}{5}$	$\dfrac{\sqrt{66}}{6}$	$-\dfrac{\sqrt{55}}{5}$
67.	$-\dfrac{2\sqrt{29}}{29}$	$-\dfrac{5\sqrt{29}}{29}$	$\dfrac25$	$\dfrac52$	$-\dfrac{\sqrt{29}}{5}$	$-\dfrac{\sqrt{29}}{2}$
68.	$-\dfrac{3\sqrt{13}}{13}$	$-\dfrac{2\sqrt{13}}{13}$	$\dfrac32$	$\dfrac23$	$-\dfrac{\sqrt{13}}{2}$	$-\dfrac{\sqrt{13}}{3}$
69.	$-\dfrac{\sqrt5}{5}$	$\dfrac{2\sqrt5}{5}$	$-\dfrac12$	-2	$\dfrac{\sqrt5}{2}$	$-\sqrt5$
70.	0.4472	-0.8944	$-\dfrac12$	2	≈ 1.1180	≈ 2.2361
71.	$\dfrac{\sqrt5}{5}$	$-\dfrac{2\sqrt5}{5}$	$-\dfrac12$	-2	$-\dfrac{\sqrt5}{2}$	$\sqrt5$
72.	$\dfrac{\sqrt{65}}{65}$	$-\dfrac{8\sqrt{65}}{65}$	$-\dfrac18$	-8	$-\dfrac{\sqrt{65}}{8}$	$2\sqrt{65}$
73.	$\dfrac{2\sqrt5}{5}$	$\dfrac{\sqrt5}{5}$	2	$\dfrac12$	$\sqrt5$	$\dfrac{\sqrt5}{2}$
74.	$\dfrac{3\sqrt{10}}{10}$	$\dfrac{\sqrt{10}}{10}$	3	$\dfrac13$	$\sqrt{10}$	$\dfrac{\sqrt{10}}{3}$
75.	$\dfrac{\sqrt5}{5}$	$\dfrac{2\sqrt5}{5}$	$\dfrac12$	2	$\dfrac{\sqrt5}{2}$	$\sqrt5$

	$\sin\theta$	$\cos\theta$	$\tan\theta$	$\cot\theta$	$\sec\theta$	$\csc\theta$
76.	$-\dfrac{\sqrt5}{5}$	$-\dfrac{2\sqrt5}{5}$	$\dfrac12$	2	$-\dfrac{\sqrt5}{2}$	$-\sqrt5$
77.	$-\dfrac{\sqrt{10}}{10}$	$\dfrac{3\sqrt{10}}{10}$	$-\dfrac13$	-3	$\dfrac{\sqrt{10}}{3}$	$-\sqrt{10}$
78.	$\dfrac{\sqrt{10}}{10}$	$-\dfrac{3\sqrt{10}}{10}$	$-\dfrac13$	-3	$-\dfrac{\sqrt{10}}{3}$	$\sqrt{10}$
79.	$\dfrac{2\sqrt{13}}{13}$	$\dfrac{3\sqrt{13}}{13}$	$-\dfrac23$	$-\dfrac32$	$\dfrac{\sqrt{13}}{3}$	$\dfrac{\sqrt{13}}{2}$
80.	$\dfrac{2\sqrt{13}}{13}$	$\dfrac{3\sqrt{13}}{13}$	$-\dfrac23$	$-\dfrac32$	$\dfrac{\sqrt{13}}{3}$	$-\dfrac{\sqrt{13}}{2}$
81.	1	0	Undefined	0	Undefined	1
82.	0	-1	0	Undefined	-1	Undefined
83.	-1	0	Undefined	0	Undefined	-1
84.	0	1	0	Undefined	1	Undefined
85.	1	0	Undefined	0	Undefined	1
86.	0	-1	0	Undefined	-1	Undefined
87.	-1	0	Undefined	0	Undefined	-1
88.	0	1	0	Undefined	1	Undefined
89.	-1	0	Undefined	0	Undefined	-1
90.	0	-1	0	Undefined	-1	Undefined
91.	1	0	Undefined	0	Undefined	1
92.	0	1	0	Undefined	1	Undefined
93.	$-\dfrac{\sqrt3}{2}$	$-\dfrac12$	$\sqrt3$	$\dfrac{\sqrt3}{3}$	-2	$\dfrac{2\sqrt3}{3}$
94.	$\dfrac{\sqrt3}{2}$	$\dfrac12$	$\sqrt3$	$\dfrac{\sqrt3}{3}$	2	$\dfrac{2\sqrt3}{3}$

95. $-1200°$ **96.** $-780°$

97. $-3240°$ **98.** 2.75 points

99. Don and Ron end up in the same place, 180° from where they started.

100. Dan and Stan do not end at the same spot. Dan ends 40° counterclockwise from where he started, while Stan ends 90° counterclockwise from where he started.

101. $1440°$ **102.** $1575°$

103. $\tan\theta = \frac{7}{24}$ **104.** $\cos\theta = \frac{13}{85}$

105. 120 feet   **106.** 4.5 feet

107. profit   **108.** loss

109. 1.4 cm   **110.** 1.0 cm

111. $r = \sqrt5$ (not 5)   **112.** undefined

113. false   **114.** false

115. true   **116.** true
117. true   **118.** true
119. false   **120.** true
121. $-\frac{3}{5}$ **122.** $\frac{4}{5}$
123. $m = \tan\theta$ **124.** ± 5
125. $y = 2\sqrt{2}$ **126.** $y = (\tan\theta)(x - a)$
127. $y = -(\tan\theta)(x - a)$
128. $\sin 270° = -1$
129. $\cos 270° = 0$ **130.** undefined
131. 0   **132.** 1
133. 0   **134.** 1
135. Does not exist, because we are dividing by 0.
136. Does not exist, because we are dividing by 0.
137. -1

Section 6.5

1. QIV   **2.** QII
3. QII   **4.** QIII
5. QI   **6.** QIII
7. QI   **8.** QIV
9. QIII   **10.** QII
11. $\sin\theta = -\frac{4}{5}$ **12.** $\cos\theta = -\frac{12}{13}$
13. $\tan\theta = -\frac{60}{11}$ **14.** $\tan\theta = -\frac{9}{40}$
15. $\sin\theta = -\frac{84}{85}$ **16.** $\cos\theta = \frac{24}{25}$
17. $\tan\theta = \sqrt{3}$ **18.** $\sin\theta = \frac{\sqrt{2}}{2}$
19. $\tan\theta = -\frac{\sqrt{11}}{5}$ **20.** $\cos\theta = -\frac{2\sqrt{5}}{5}$
21. $-\frac{\sqrt{3}}{3}$ **22.** $\frac{13}{12}$
23. $\frac{2\sqrt{3}}{3}$ **24.** $\frac{85}{84}$
25. 1   **26.** 1
27. -1 **28.** 1
29. 0   **30.** -2
31. 1   **32.** -1
33. 1   **34.** 0
35. not possible   **36.** $-\frac{2\sqrt{3}}{3}$
37. possible   **38.** not possible

39. not possible   **40.** possible
41. possible   **42.** possible
43. possible   **44.** possible
45. possible $\theta \approx 0.1146°$ **46.** not possible
47. $-\frac{1}{2}$ **48.** $-\frac{1}{2}$
49. $-\frac{\sqrt{3}}{2}$ **50.** $-\frac{\sqrt{2}}{2}$
51. $\frac{\sqrt{3}}{3}$ **52.** $-\sqrt{2}$
53. 1   **54.** $\frac{2\sqrt{3}}{3}$
55. -2 **56.** $\frac{2\sqrt{3}}{3}$
57. 1   **58.** $-\sqrt{3}$
59. 30° and 330°   **60.** 60° and 120°
61. 210° and 330°   **62.** 120° and 240°
63. 90° and 270°   **64.** 0°, 180°, and 360°
65. 270°   **66.** 180°
67. -0.8387 **68.** 0.7314
69. -11.4301 **70.** 7.1154
71. 2.0627   **72.** -1.3456
73. -1.0711 **74.** -1.9416
75. -2.6051 **76.** -0.1405
77. 110°   **78.** 315°
79. 143°   **80.** 190°
81. 322°   **82.** 250°
83. 140°   **84.** 265°
85. 340°   **86.** 205°
87. 1°   **88.** 109°
89. 335°   **90.** 269°
91. 1.3   **92.** 1.6
93. 12°   **94.** 15°
95. Lower leg is bent at knee at an angle of 15°.
96. Lower leg is bent at knee at an angle of 20°.
97. 75.5°   **98.** -0.2504
99. The reference angle is made with the terminal side and the negative x-axis (not y-axis).
100. 256°
101. true   **102.** false
103. false   **104.** false
105. false   **106.** true

107. $-\dfrac{a}{\sqrt{a^2+b^2}}$

108. $-\dfrac{b}{\sqrt{a^2+b^2}}$

109. $-\dfrac{\sqrt{a^2-b^2}}{b}$

110. $\dfrac{\sqrt{a^2-b^2}}{b}$

111. QI **112.** QIV
113. QI **114.** QII
115. QI **116.** QIV

Section 6.6

1. $\frac{1}{5}$ or 0.2 **2.** $\frac{1}{10}$ or 0.1

3. $\frac{2}{11}$ or ≈ 0.18 **4.** $\frac{1}{6}$ or ≈ 0.17

5. $\frac{1}{50}$ or 0.02 **6.** $\frac{1}{50}$ or 0.02

7. $\frac{1}{8}$ or 0.125 **8.** $\frac{4}{14}$ or ≈ 0.29

9. $\frac{1}{5}$ or 0.2 **10.** $\frac{1}{80}$ or 0.0125

11. $\frac{1}{24}$ or 0.042 **12.** 1

13. $\dfrac{\pi}{6}$ **14.** $\dfrac{\pi}{3}$

15. $\dfrac{\pi}{4}$ **16.** $\dfrac{\pi}{2}$

17. $\dfrac{7\pi}{4}$ **18.** $\dfrac{3\pi}{2}$

19. $\dfrac{5\pi}{12}$ **20.** $\dfrac{5\pi}{9}$

21. $\dfrac{17\pi}{18}$ **22.** $\dfrac{17\pi}{9}$

23. $\dfrac{13\pi}{3}$ **24.** 3π

25. $-\dfrac{7\pi}{6}$ **26.** $-\dfrac{16\pi}{9}$

27. -20π **28.** 10π
29. 30° **30.** 45°
31. 135° **32.** 210°
33. 67.5° **34.** 220°
35. 75° **36.** 420°
37. 1620° **38.** −1080°
39. 171° **40.** 65°
41. −84° **42.** −160°
43. 229.18° **44.** $\approx 171.89°$
45. 48.70° **46.** $\approx 187.36°$
47. −160.37° **48.** $\approx -342.86°$

49. 198.48° **50.** $\approx 757.95°$
51. 0.820 **52.** ≈ 1.13
53. 1.95 **54.** ≈ 3.00
55. 0.986 **56.** ≈ 5.21
57. 12 mm **58.** 20 cm
59. $\frac{2\pi}{3}$ ft **60.** $\frac{3\pi}{4}$ yd
61. $\frac{5}{2}$ in. **62.** 15 m
63. $\frac{11\pi}{5}$ μm **64.** $\frac{7\pi}{6}$ μm
65. $\frac{200\pi}{3}$ km **66.** 30π km
67. 12.8 sq ft **68.** ≈ 2.83 sq in.
69. 2.85 sq km **70.** ≈ 221 sq mi.
71. 5 sq yd **72.** ≈ 13.856 sq m
73. 8.62 sq cm **74.** ≈ 1.47 sq mm
75. 0.0236 sq ft **76.** ≈ 1.10 sq ft
77. $\frac{2}{5}$ m/sec **78.** 4 ft/min
79. 272 km/hr **80.** 627 mi./day
81. 9.8 meters **82.** 27.9 km
83. 1.5 miles **84.** 672 feet

85. $\dfrac{5\pi}{2}\dfrac{\text{rad}}{\text{sec}}$ **86.** $\dfrac{9\pi}{2}\dfrac{\text{rad}}{\text{sec}}$

87. $\dfrac{2\pi}{9}\dfrac{\text{rad}}{\text{sec}}$ **88.** $\dfrac{5\pi}{3}\dfrac{\text{rad}}{\text{sec}}$

89. 6π inches per second

90. 6π cm per second

91. $\dfrac{\pi}{4}$ mm per second

92. $\dfrac{15\pi}{2}$ feet per second

93. 26.2 cm **94.** 415 mm
95. 653 in. (or 54.5 ft) **96.** 452 ft
97. 5262 km **98.** 3482 km
99. 37 ft **100.** 51.31 ft
101. $200\pi \approx 628$ ft **102.** $75\pi \approx 236$ ft
103. 50° **104.** 84°
105. 60 inches (5 feet) **106.** 327 in.
107. 911 miles **108.** $\approx 45{,}567$ miles
109. 157 sq ft **110.** ≈ 253 sq ft
111. 78 sq in. **112.** 60 sq in.
113. 74 sq mi. **114.** 34 mi.
115. 1.4 nm **116.** 0.2 nm

117. 70 mph   **118.** 76.21 mph

119. 1037.51 mph; 1040.35 mph

120. 28,193.73 mph

121. 16 ft/sec   **122.** 9.42 ft/sec

123. $66\frac{2}{3}\pi$ rad/min

124. 1256.64 in/min

125. 12.05 mph   **126.** 9.82 mph

127. 640 rev/min

128. 1125 rev/min

129. 10.11 rad/sec or 1.6 rotations per second

130. 182 ft/sec

131. 17.59 m/s   **132.** 79.17 km/hr

133. 40π rad/min   **134.** 1000 revolutions

135. 32π ft   **136.** wins $5000

137. Angular speed needs to be in radians (not degrees) per second.

138. Angular speed needs to be in radians (not degrees) per second.

139. true   **140.** false

141. true   **142.** true

143. $2\pi \approx 6.3$ cm/sec   **144.** $3\pi \approx 9.4$ cm/sec

145. tripled   **146.** $\frac{3250}{9}\pi$ ft

147. $\frac{15}{2}\pi$ square units   **148.** $\left(6+5\pi\right)$ units

149. 68°45′44″   **150.** 96°1′51.436″

Section 6.7

1. $-\dfrac{\sqrt{3}}{2}$ **2.** $\dfrac{1}{2}$

3. $-\dfrac{\sqrt{3}}{2}$ **4.** $-\dfrac{1}{2}$

5. $\dfrac{\sqrt{2}}{2}$ **6.** $-\dfrac{\sqrt{2}}{2}$

7. -1 **8.** -1

9. $-\sqrt{2}$ **10.** $-\dfrac{2\sqrt{3}}{3}$

11. $\sqrt{3}$ **12.** $-\sqrt{3}$

13. 2   **14.** $-\dfrac{\sqrt{3}}{3}$

15. $-\dfrac{\sqrt{3}}{2}$ **16.** $\dfrac{\sqrt{2}}{2}$

17. $-\dfrac{\sqrt{3}}{2}$ **18.** $\dfrac{1}{2}$

19. $-\dfrac{\sqrt{2}}{2}$ **20.** $\dfrac{1}{2}$

21. $-\dfrac{\sqrt{3}}{2}$ **22.** $\dfrac{\sqrt{2}}{2}$

23. $\dfrac{\sqrt{2}}{2}$ **24.** 0

25. 1   **26.** $-\dfrac{\sqrt{3}}{2}$

27. $\dfrac{\sqrt{2}}{2}$ **28.** $-\dfrac{\sqrt{2}}{2}$

29. 0   **30.** $-\dfrac{\sqrt{3}}{2}$

31. -2 **32.** $\sqrt{2}$

33. $\dfrac{\sqrt{3}}{3}$ **34.** $\sqrt{3}$

35. $\dfrac{\pi}{6}, \dfrac{11\pi}{6}$ **36.** $\dfrac{5\pi}{6}, \dfrac{7\pi}{6}$

37. $\dfrac{4\pi}{3}, \dfrac{5\pi}{3}$ **38.** $\dfrac{\pi}{3}, \dfrac{2\pi}{3}$

39. $0, \pi, 2\pi, 3\pi, 4\pi$ **40.** $\dfrac{3\pi}{2}, \dfrac{7\pi}{2}$

41. $\pi, 3\pi$ **42.** $\dfrac{\pi}{2}, \dfrac{3\pi}{2}, \dfrac{5\pi}{2}, \dfrac{7\pi}{2}$

43. $\dfrac{3\pi}{4}, \dfrac{7\pi}{4}$ **44.** $\dfrac{\pi}{4}, \dfrac{5\pi}{4}$

45. $\dfrac{3\pi}{4}, \dfrac{5\pi}{4}$ **46.** $\dfrac{\pi}{4}, \dfrac{3\pi}{4}$

47. $0, \pi, 2\pi$ **48.** $\dfrac{\pi}{2}, \dfrac{3\pi}{2}$

49. $\dfrac{\pi}{2}, \dfrac{3\pi}{2}$ **50.** $0, \pi, 2\pi$

51. $\dfrac{7\pi}{6}, \dfrac{11\pi}{6}$ **52.** $\dfrac{5\pi}{6}, \dfrac{11\pi}{6}$

53. $\dfrac{\pi}{6}, \dfrac{11\pi}{6}$ **54.** $\dfrac{\pi}{6}, \dfrac{7\pi}{6}$

55. 22.0°F **56.** 77.2°F

57. 99.05°F **58.** 98.9°F

59. 2.6 feet **60.** 0.2 feet

61. 135 lb **62.** 155 lb

63. 10,000 guests **64.** 50,000 guests

65. 10.7 mcg/mL **66.** 15.4 mcg/mL

67. 35°C **68.** 38°C

69. Used the *x*-coordinate for sine and the *y*-coordinate for cosine; should have done the opposite.

70. Should have used $\cos\left(\dfrac{11\pi}{6}\right) = \dfrac{\sqrt{3}}{2}$.

71. true **72.** true

73. false **74.** false

75. true **76.** false

77. odd **78.** odd **79.** $\dfrac{\pi}{4}, \dfrac{5\pi}{4}$

80. $\dfrac{\pi}{4} + n\pi$, where *n* is an integer.

81. $\sec\theta$ **82.** $\csc\theta$

83. $\sec\theta$ **84.** −1

85. Yes, $\theta = \dfrac{\pi}{4}, \dfrac{3\pi}{4}, \dfrac{5\pi}{4}, \dfrac{7\pi}{4}$

86. Yes, $\theta = \dfrac{3\pi}{4}, \dfrac{7\pi}{4}$

87. $\sin(423°) \approx 0.891$ and $\sin(-423°) \approx -0.891$

88. $\cos(227°) \approx -0.682$ and $\cos(-227°)$
$$= \cos(227°) \approx -0.682$$

89. $6.314, -6.314$

90. $-1.942, 1.942$ **91.** 0.5 **92.** 0.866

93. 0.8660 **94.** −0.707

Section 6.8

1. c **2.** d **3.** a **4.** j

5. h **6.** i **7.** b **8.** f

9. e **10.** g **11.** $A = \dfrac{3}{2}, p = \dfrac{2\pi}{3}$

12. $A = \dfrac{2}{3}, p = \dfrac{\pi}{2}$ **13.** $A = 1, p = \dfrac{2\pi}{5}$

14. $A = 1, p = \dfrac{2\pi}{7}$ **15.** $A = \dfrac{2}{3}, p = \dfrac{4\pi}{3}$

16. $A = \dfrac{3}{2}, p = 3\pi$ **17.** $A = 3, p = 2$

18. $A = 2, p = 2$ **19.** $A = 5, p = 6$

20. $A = 4, p = 8$

21.

22.

23.

24.

25.

26.

27.

28.

29.

30.

31.

32.

33.

34.

35.

36.

37.

38.

39.

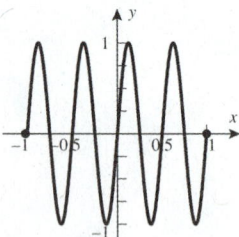

40.

41. $y = -\sin 2x$ **42.** $y = -2\cos x$

43. $y = \cos \pi x$ **44.** $y = \sin \pi x$

45. $y = -2\sin\left(\dfrac{\pi}{2}x\right)$ **46.** $y = -3\cos\left(\dfrac{\pi}{2}x\right)$

47. $y = \sin 8\pi x$ **48.** $y = \cos 4\pi x$

49. Amplitude: 2; period: 2;

phase shift: $\dfrac{1}{\pi}$ (right)

50. Amplitude: 4; period: 2π;

phase shift: $-\pi$ (left)

51. Amplitude: 5; period: $\dfrac{2\pi}{3}$;

phase shift: $-\frac{2}{3}$ (left)

52. Amplitude: 7; period: $\dfrac{\pi}{2}$;

phase shift: $\frac{3}{4}$ (right)

53. Amplitude: 6; period: 2;
phase shift: -2 (left)

54. Amplitude: 3; period: 4;
phase shift 1 (right)

55. Amplitude: 3; period: π;

phase shift: $-\dfrac{\pi}{2}$ (left)

56. Amplitude: 4; period: π;

phase shift: $\frac{\pi}{2}$ (right)

57. Amplitude: $\frac{1}{4}$; period: 8π;

phase shift: 2π (right)

58. Amplitude: $\frac{1}{2}$; period: 6π;

phase shift: -3π (left)

59. Amplitude: 2; period: 4;
phase shift: 4 (right)

60. Amplitude: 5; period: 2;
phase shift: -1 (left)

61.

62.

63.

64.

65.

66.

67.

68.

69.

70.

71.

72.

73.

74.

75.

76.

77.

78.

79.

80.

81.

82.

83.

84.

85.

86.

87.

88.

89.

90.

91.

92.

93.

94.

95. 3500 widgets   **96.** 12 months

97. Amplitude: 1 mg/L; period 8 weeks

98. Amplitude: 2 mg/L; period 8 weeks

99. Amplitude: 4 cm; Mass: 4 g

100. Amplitude: 3 cm; Mass: 1/9 g

101. $\dfrac{1}{4\pi}$ cycles per second

102. $\dfrac{3}{2\pi}$ cycles per second

103. Amplitude: 0.005 cm; Frequency: 256 hertz

104. Amplitude: 0.005 cm; Frequency: 375 hertz

105. Amplitude: 0.008 cm; Frequency: 375 hertz

106. Amplitude: 0.006 cm; Frequency: 500 hertz

107. 660 m/sec   **108.** $M = 3$

109. 660 m/sec   **110.** 1275 m/sec

111. $y = 25 + 25\sin\left[\dfrac{2\pi}{4}(t-1)\right]$ or

$y = 25 - 25\cos\left(\dfrac{\pi t}{2}\right)$

112.

113. Forgot to reflect about the x-axis

114. Forgot to reflect about the x-axis

115. true   **116.** true

117. false   **118.** true

119. $(0, A)$ **120.** $(0, 0)$

121. $x = \dfrac{n\pi}{B}$, where n is an integer.

122. $x = \dfrac{(2n+1)\pi}{2B}$, where n is an integer.

123. $\left(0, -\dfrac{A}{2}\right)$ **124.** $(0, -A + C)$

125. $\left(\dfrac{\pi(3+4n)}{2B}, 0\right)$, where n an integer.

126. If $|C| > A$, then there are no x-intercepts.

127. $\left[-\dfrac{5A}{2}, \dfrac{3A}{2}\right]$ **128.** no

129. no

130. no

131. coincide

132. coincide

133. a. Y_2 is Y_1 shifted to the left $\dfrac{\pi}{3}$.

    b. Y_2 is Y_1 shifted to the right $\dfrac{\pi}{3}$.

134. a. Y_2 is Y_1 shifted to the left $\dfrac{\pi}{3}$.

    b. Y_2 is Y_1 shifted to the right $\dfrac{\pi}{3}$.

135. As t increases, the amplitude goes to zero for Y_3.

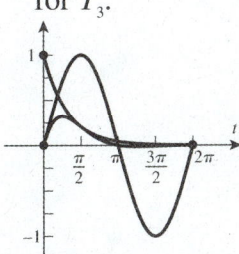

136. As the value of k increases, the effect of the damping increases, causing the graph to approach the t-axis more quickly.

137. a. Y_2 is Y_1 shifted upward by 1 unit.

    b. Y_2 is Y_2 shifted downward by 1 unit.

138. a. Y_2 is Y_1 shifted upwards by $\frac{1}{2}$ unit.

    b. Y_2 is Y_1 shifted downwards by $\frac{1}{2}$ unit.

139. 5   **140.** 2

Section 6.9

1. b   **2.** f   **3.** h   **4.** a

5. c   **6.** e   **7.** d   **8.** g

9.

10.

11.

12.

13.

14.

15.

16.

17.

18.

19.

20.

21.

22.

23.

24.

25.

26.

27.

28.

29.

30.

31.

32.

33.

34.

35.

36.

37.

38.

39.

40.

41.

42.

43.

44.

45.

46.

47.

48.

49.

50.

51.

52.

53.

54.

55.

56.

57. Domain: all real numbers x, such that $x \neq n$, where n is an integer
Range: all real numbers

58. Domain: all real numbers x, such that
$x \neq \dfrac{2n+1}{2}\pi$, where n is an integer
Range: all real numbers

59. Domain: all real numbers x, such that
$x \neq \dfrac{2n+1}{10}\pi$, where n is an integer
Range: $(-\infty, -2] \cup [2, \infty)$

60. Domain: all real numbers x, such that
$x \neq \dfrac{2n+1}{6}\pi$, where n is an integer
Range: $(-\infty, -4] \cup [4, \infty)$

61. Domain: all real numbers x, such that
$x \neq 2n\pi$, where n is an integer
Range: $(-\infty, 1] \cup [3, \infty)$

62. Domain: all real numbers x, such that
$x \neq (2n-1)\pi$, where n is an integer
Range: $(-\infty, -1] \cup [3, \infty)$

63. Domain: $\{x : x \neq 4n+6,\ n$ an integer$\}$
Range: all real numbers.

64. Domain: All real numbers x, such that
$x \neq \dfrac{3n-1}{6}$, where n is an integer
Range: all real numbers

65. Domain: $\{x : x \neq n,\ n$ an integer$\}$
Range: $\left(-\infty, -\frac{5}{2}\right] \cup \left[-\frac{3}{2}, \infty\right)$

66. Domain: all real numbers x, such that
$x \neq \dfrac{(2n+1)\pi}{6}$, where n is an integer
Range: $\left(-\infty, \frac{1}{6}\right] \cup \left[\frac{5}{6}, \infty\right)$

67. 3 miles   **68.** 1 mile

69. 55.4 sq in.   **70.** 13.3 sq ft

71. Forgot the amplitude of 3.

72. The quantities used in Steps 2 and 3 are incorrect

73. true   **74.** false   **75.** $n =$ integer

76. Equality holds whenever n is an even integer.

77. $x = -\pi, -\dfrac{\pi}{2}, 0, \dfrac{\pi}{2}, \pi$ **78.** no solution

79. $\left(\dfrac{n\pi - C}{B}, 0\right)$, where n an integer

80. $2 + 4n < x < 3 + 4n$, where n is an integer

81. Infinitely many. The tangent function is an increasing, periodic function and $y = x$ is increasing.

82. If $|A| < 2$, then the graph of
$y = A\sin(Bx + C)$ never intersects this graph.

83. $A = \sqrt{2}$

84. y_1 is not the guide function of y_2

85. π

86. The graphs of y_1 and y_2 coincide, while the graph of y_3 is the reflection of the graph of y_2 (and hence of y_1) over the x-axis.

Review Exercises

1. a. 62° **b.** 152° **2. a.** 73° **b.** 163°
3. a. 55° **b.** 145° **4. a.** 12° **b.** 102°
5. a. 0.99° **b.** 90.99°
6. a. 89.987° **b.** 179.987°
7. $\gamma = 25°$ **8.** $\gamma = 50°$
9. $\alpha = 140°,\ \beta = 20°,\ \gamma = 20°$
10. $\alpha = 135°,\ \beta = 22.5°,\ \gamma = 22.5°$
11. $b = \sqrt{128} = 8\sqrt{2}$ **12.** $a = 12$
13. $c = \sqrt{65}$ **14.** $c = 2\sqrt{41}$
15. $12\sqrt{2}$ yards **16.** 2 feet
17. Leg: $3\sqrt{3}$ feet; Hypotenuse: 6 feet
18. Shorter leg: 6 km; Longer leg: $6\sqrt{3}$ km
19. $F = 4$ **20.** $D = 5$
21. $C = 147.6$ km **22.** $B = 0.048$ m
23. $10\sqrt{3}$ in. **24.** 42 feet
25. 150° **26.** 90°
27. 32 m **28.** 7 feet
29. $\dfrac{2\sqrt{13}}{13}$ **30.** $\dfrac{3\sqrt{13}}{13}$
31. $\dfrac{\sqrt{13}}{2}$ **32.** $\dfrac{\sqrt{13}}{3}$
33. $\frac{3}{2}$ **34.** $\frac{2}{3}$
35. $\cos 60°$ **36.** $\sin(90° - A)$
37. $\cot 45°$ **38.** $\csc(30°)$
39. $\csc 60°$ **40.** $\tan(30°)$
41. b **42.** a
43. b **44.** a
45. c **46.** c
47. 0.6691 **48.** 0.5446
49. 0.9548 **50.** 0.4706
51. 1.5399 **52.** 1.0446
53. 1.5477 **54.** 2.7837
55. 75 feet **56.** 200 feet

	$\sin\theta$	$\cos\theta$	$\tan\theta$	$\cot\theta$	$\sec\theta$	$\csc\theta$
57.	$-\dfrac{4}{5}$	$\dfrac{3}{5}$	$-\dfrac{4}{3}$	$-\dfrac{3}{4}$	$\dfrac{5}{3}$	$-\dfrac{5}{4}$
58.	$-\dfrac{7}{25}$	$-\dfrac{24}{25}$	$\dfrac{7}{24}$	$\dfrac{24}{7}$	$-\dfrac{25}{24}$	$-\dfrac{25}{7}$
59.	$\dfrac{\sqrt{10}}{10}$	$-\dfrac{3\sqrt{10}}{10}$	$-\dfrac{1}{3}$	-3	$-\dfrac{\sqrt{10}}{3}$	$\sqrt{10}$
60.	$\dfrac{9}{41}$	$-\dfrac{40}{41}$	$-\dfrac{9}{40}$	$-\dfrac{40}{9}$	$-\dfrac{41}{40}$	$\dfrac{41}{9}$
61.	$\dfrac{1}{2}$	$\dfrac{\sqrt{3}}{2}$	$\dfrac{\sqrt{3}}{3}$	$\sqrt{3}$	$\dfrac{2\sqrt{3}}{3}$	2
62.	$-\dfrac{\sqrt{2}}{2}$	$-\dfrac{\sqrt{2}}{2}$	1	1	$-\sqrt{2}$	$-\sqrt{2}$
63.	$-\dfrac{\sqrt{5}}{5}$	$\dfrac{2\sqrt{5}}{5}$	$-\dfrac{1}{2}$	-2	$\dfrac{\sqrt{5}}{2}$	$-\sqrt{5}$
64.	$\dfrac{10\sqrt{181}}{181}$	$-\dfrac{9\sqrt{181}}{181}$	$-\dfrac{10}{9}$	$-\dfrac{9}{10}$	$-\dfrac{\sqrt{181}}{9}$	$\dfrac{\sqrt{181}}{10}$
65.	$-\dfrac{\sqrt{7.2}}{3}$	$-\dfrac{\sqrt{7.2}}{6}$	2	$\dfrac{1}{2}$	$-\dfrac{\sqrt{7.2}}{1.2}$	$-\dfrac{\sqrt{7.2}}{2.4}$
66.	$-\dfrac{3\sqrt{10}}{10}$	$\dfrac{\sqrt{10}}{10}$	-3	$-\dfrac{1}{3}$	$\sqrt{10}$	$-\dfrac{\sqrt{10}}{3}$

67. $-\frac{1}{2}$ **68.** $\dfrac{1}{2}$
69. $-\dfrac{\sqrt{3}}{3}$ **70.** -1
71. $-\dfrac{2\sqrt{3}}{3}$ **72.** -2
73. -0.2419 **74.** 0.0872
75. 1.0355 **76.** -1.1918
77. -2.7904 **78.** -1.7434
79. -0.6494 **80.** -1.1666
81. $\dfrac{3\pi}{4}$ **82.** $\dfrac{4\pi}{3}$
83. $\dfrac{11\pi}{6}$ **84.** π
85. $\dfrac{6\pi}{5}$ **86.** $\dfrac{3\pi}{5}$
87. 9π **88.** 5π
89. 60° **90.** 330°

91. 225° **92.** 120°
93. 100° **94.** 306°
95. 1800° **96.** 2790°
97. 240π in./min $\approx$ 754 in./min **98.** 27 mph

99. $-\dfrac{\sqrt{3}}{3}$ **100.** $-\dfrac{\sqrt{3}}{2}$

101. $-\dfrac{1}{2}$ **102.** $\dfrac{2\sqrt{3}}{3}$

103. 1 **104.** $-\sqrt{2}$
105. -1 **106.** 0
107. -1 **108.** -1
109. $\dfrac{1}{2}$ **110.** $-\dfrac{1}{2}$
111. 2π **112.** 4
113. $y = 4\cos x$ **114.** π
115. 5 **116.** $y = -5\sin 2x$

117. Amplitude: 2; Period: 1
118. Amplitude: $\frac{1}{3}$; Period: 4

119. Amplitude: $\frac{1}{5}$; Period: $\dfrac{2\pi}{3}$

120. Amplitude: $\frac{7}{6}$; Period: $\dfrac{\pi}{3}$

121.

122.

123.

124.

	AMPLITUDE	PERIOD	PHASE SHIFT	VERTICAL SHIFT
125.	3	2π	$\dfrac{\pi}{2}$ (right)	+2
126.	$\dfrac{1}{2}$	2π	$-\dfrac{\pi}{4}$ (left)	+3
127.	4	$\dfrac{2\pi}{3}$	$-\dfrac{\pi}{4}$ (left)	-2
128.	2	π	$\dfrac{\pi}{3}$ (right)	-1
129.	$\dfrac{1}{3}$	2	$\dfrac{1}{2\pi}$ (right)	$\dfrac{1}{2}$ (down)
130.	$\dfrac{1}{6}$	12	2 (right)	$\dfrac{3}{4}$ up

131.

132.

133.

134.

135. Domain: all real numbers such that $x \neq n\pi$, where n is an integer;
Range: all real numbers

136. Domain: all real numbers such that
$x \neq \dfrac{n\pi}{2}$, where n is an integer;
Range: all real numbers

137. Domain: all real numbers such that
$x \neq \dfrac{2n+1}{4}\pi$, where n is an integer;
Range: $(-\infty, -3] \cup [3, \infty)$

138. Domain: all real numbers such that $x \neq n\pi$, where n is an integer;
Range: $(-\infty, -1] \cup [3, \infty)$

139. Domain: $\left\{ x : x \neq \dfrac{6n+7}{6},\ n \text{ an integer} \right\}$
Range: $\left(-\infty, -\dfrac{3}{4} \right] \cup \left[-\dfrac{1}{4}, \infty \right)$.

140. Domain: $\left\{ x : x \neq \dfrac{n\pi}{2},\ n \text{ an integer} \right\}$
Range: $\left(-\infty, \dfrac{5}{2} \right] \cup \left[\dfrac{7}{2}, \infty \right)$

141.

142.

143.

144.

145.

146.

147. 71.57 feet, 82.64 feet   **148.** 101.20 cm
149. a. 1.02041   **b.** 1.02085, (b) is more
      accurate.
150. a. 1.43885   **b.** 1.43885
151. 2.612   **152.** 0.125
153. −1
154. Error since you are dividing by zero
155. QIV   **156.** QI
157. 100°46′45″   **158.** 140°10′6″
159. −0.9659   **160.** 0.5000

161. a. Y_2 is Y_1 shifted to the left by $\dfrac{\pi}{6}$ unit.

  b. Y_2 is Y_1 shifted to the right $\dfrac{\pi}{6}$ unit.

162. a. Y_2 is Y_1 shifted up by $\dfrac{1}{2}$ unit.

  b. Y_2 is Y_1 shifted down by $\dfrac{1}{2}$ unit.

163. 5   **164.** 2

Practice Test

1. 6000 feet
2.

θ	$\sin\theta$	$\cos\theta$	$\tan\theta$	$\cot\theta$	$\sec\theta$	$\csc\theta$
30°	$\dfrac{1}{2}$	$\dfrac{\sqrt{3}}{2}$	$\dfrac{\sqrt{3}}{3}$	$\sqrt{3}$	$\dfrac{2\sqrt{3}}{3}$	2
45°	$\dfrac{\sqrt{2}}{2}$	$\dfrac{\sqrt{2}}{2}$	1	1	$\sqrt{2}$	$\sqrt{2}$
60°	$\dfrac{\sqrt{3}}{2}$	$\dfrac{1}{2}$	$\sqrt{3}$	$\dfrac{\sqrt{3}}{3}$	2	$\dfrac{2\sqrt{3}}{3}$

3. The first is an exact value of the cosine
   function, the second is an approximation.
4.

	0°	QI	90°	QII	180°	QIII	270°	QIV	360°
$\sin\theta$	0	+	1	+	0	−	−1	−	0
$\cos\theta$	1	+	0	−	−1	−	0	+	1

5. QIV   **6.** $-\dfrac{1}{2}$ [QIII]   **7.** 585°

8. $\dfrac{13\pi}{9}$ **9.** $\dfrac{15\pi}{4}$ sq in.   **10.** $\dfrac{23\pi}{36}$

11. Amplitude: 5; Period: $\dfrac{2\pi}{3}$

12.

13.

14. The vertical asymptotes of
   $y = 2\csc(3x - \pi)$ correspond to the
   x-intercepts of $y = 2\sin(3x - \pi)$.

15. $x = \dfrac{n\pi}{2}$ or $\left(\dfrac{n\pi}{2}, 0\right)$, where n is an integer.

16. Phase shift: −3; Vertical shift: none
17. $(-\infty, -4] \cup [2, \infty)$
18. Domain is the set of all real numbers for
   which $x \neq \dfrac{(3n + 2)\pi}{6}$, where n is an integer.

19.

20. $x = \left(\dfrac{(2n+1)\pi}{6}, 0 \right), \left(\dfrac{(3n+4)\pi}{9}, 0 \right)$, where n
is an integer.

21. true

22. Graph is below the x-axis precisely when $\dfrac{\pi}{4} + n\pi < x < \dfrac{3\pi}{4} + n\pi$, where n is an integer.

23. $y = 4\sin\left[2\left(x + \tfrac{3}{2} \right) \right] - \tfrac{1}{2}$

24. $y = \cot x + 0.01$

25.

26. 16.10 in.

27. a. 1.34409

 b. 1.34352, (b) is more accurate.

Cumulative Test

1. $\dfrac{x-1}{5}$ **2.** $-6 + 2i$

3. $x \neq -1, 0$; no solution **4.** $r = \sqrt{\dfrac{A}{\pi}}$

5. $(-\infty, 0] \cup (1, 3]$

6. $d \approx 1.264 \quad M = \left(-\dfrac{11}{40}, -\dfrac{1}{4} \right)$

7. $y = 0.5x + 3.4$ **8.** $(-\infty, -5] \cup [5, \infty)$

9. $-\dfrac{5}{8}$

10.

11. 1 **12.** $f^{-1}(x) = \dfrac{3x+2}{x-5}$

13. $f(x) = -2x^2 + 7$

14. The only real zero is 0, with multiplicity 3, the other factor is an irreducible quadratic factor.

15. Factors of $a_0 = 10$; $\pm 1, \pm 2, \pm 5, \pm 10$;

 Factors of $a_n = 2$; $\pm 1, \pm 2$

 Possible rational zeros:

 $\pm 1, \pm 2, \pm 5, \pm 10, \pm \tfrac{1}{2}, \pm \tfrac{5}{2}$

 Testing the zeros:

 $P(-5) = 0, P(-1) = 0, P\left(\tfrac{1}{2} \right) = 0, P(2) = 0$

16. $(2x+1)^2(x^2 - 2x + 5)$

17. y-intercept: $(0, 0)$; domain: $(-\infty, \infty)$; range: $(-1, \infty)$; horizontal asymptote: $y = -1$

18. $21,760.19

19.

20. 1.413

21. 0.435 **22.** $T = 40$ feet

23. $\sin\theta = \dfrac{3\sqrt{13}}{13}, \cos\theta = -\dfrac{2\sqrt{13}}{13}, \tan\theta = -\dfrac{3}{2},$

 $\cot\theta = -\tfrac{2}{3}, \sec\theta = -\dfrac{\sqrt{13}}{2}, \csc\theta = \dfrac{\sqrt{13}}{3}$

24. 3.53 sq m

25. Amplitude: 4; Period: π;

 Phase shift: $-\dfrac{\pi}{2}$ (left)

CHAPTER 7

Section 7.1

1. $\dfrac{8}{7}$

2. $-\dfrac{7}{3}$

3. $-\dfrac{5}{3}$

4. $\dfrac{5}{4}$

5. $-\dfrac{1}{5}$

6. 2

7. $\dfrac{2\sqrt{5}}{5}$

8. $\dfrac{2\sqrt{11}}{11}$

9. $-\dfrac{5\sqrt{7}}{7}$

10. $\dfrac{2}{7}$

11. $-\dfrac{1}{4}$

12. $-\dfrac{\sqrt{5}}{5}$

13. $-\dfrac{\sqrt{3}}{3}$

14. $-\sqrt{3}$

15. $\dfrac{4}{3}$

16. $\dfrac{3}{4}$

17. $\dfrac{\sqrt{11}}{5}$

18. $\dfrac{5\sqrt{11}}{11}$

19. $\dfrac{b}{a}$ $(a \neq 0)$

20. $\dfrac{a}{b}$ $(b \neq 0)$

21. $\dfrac{5}{64}$

22. 0.001

23. -8

24. 25

25. -5

26. $\dfrac{49}{16}$

27. $-\dfrac{\sqrt{3}}{2}$

28. $-\dfrac{4}{5}$

29. $-\dfrac{\sqrt{21}}{5}$

30. $-\dfrac{3\sqrt{5}}{7}$

31. $-\sqrt{17}$

32. $-\sqrt{26}$

33. $-\sqrt{5}$

34. $\sqrt{10}$

35. $-\dfrac{8\sqrt{161}}{161}$

36. $\dfrac{7\sqrt{11}}{44}$

37. $-\dfrac{15\sqrt{11}}{44}$

38. $\dfrac{15\sqrt{161}}{161}$

39. $-\dfrac{2\sqrt{13}}{13}$

40. $\dfrac{6\sqrt{61}}{61}$

41. $\dfrac{\sqrt{6}}{2}$

42. $-\sqrt{0.9775}$

43. $\sqrt{6}$

44. $-\dfrac{\sqrt{26}}{26}$

45. $-\dfrac{\sqrt{62}}{8}$

46. $\dfrac{\sqrt{110}}{11}$

47. $\cos\theta = -\dfrac{3}{5}$, $\sin\theta = \dfrac{4}{5}$

48. $\cos\theta = -\dfrac{4}{5}$, $\sin\theta = \dfrac{3}{5}$

49. $\cos\theta = -\dfrac{\sqrt{5}}{5}$, $\sin\theta = -\dfrac{2\sqrt{5}}{5}$

50. $\cos\theta = -\dfrac{\sqrt{26}}{26}$, $\sin\theta = -\dfrac{5\sqrt{26}}{26}$

51. $\cos\theta = -\dfrac{5\sqrt{34}}{34}$, $\sin\theta = -\dfrac{3\sqrt{34}}{34}$

52. $\cos\theta = -\dfrac{5\sqrt{41}}{41}$, $\sin\theta = -\dfrac{4\sqrt{41}}{41}$

53. $\sin\theta = -\dfrac{2\sqrt{29}}{29}$, $\cos\theta = \dfrac{5\sqrt{29}}{29}$

54. $\cos\theta = -\dfrac{0.1\sqrt{1.01}}{1.01}$, $\sin\theta = -\dfrac{\sqrt{1.01}}{1.01}$

55. $\sin\theta = \dfrac{4\sqrt{17}}{17}$, $\cos\theta = \dfrac{\sqrt{17}}{17}$

56. $\cos\theta = -\dfrac{2.2\sqrt{5.84}}{5.84}$, $\sin\theta = \dfrac{\sqrt{5.84}}{5.84}$

57. $\dfrac{1}{\sin\theta}$

58. $\dfrac{1}{\cos\theta}$

59. -1

60. -1

61. $\dfrac{\cos^2\theta}{\sin\theta}$

62. $\dfrac{\sin^2\theta}{\cos\theta}$

63. $\dfrac{1}{\sin\theta}$

64. $\dfrac{1}{\cos\theta}$

65. $1 + 2\sin\theta\cos\theta$

66. $1 - 2\sin\theta\cos\theta$

67. $\dfrac{1 + 2\sin\theta\cos\theta}{\cos^2\theta}$

68. $\dfrac{1 + 2\sin\theta\cos\theta}{\sin^2\theta}$

69. 1

70. $\sin^2\theta\cos^2\theta$

71. $r = 8\sin\theta(1 - \sin^2\theta) = 8\sin\theta - 8\sin^3\theta$

$\theta = 30°, \quad r = 3$

$\theta = 60°, \quad r = \sqrt{3}$

$\theta = 90°, \quad r = 0$

72. $r = 8\sin\theta\cos^2\theta = 8\sin\theta - 8\sin^3\theta$

$\theta = -30°, \quad r = -3$

$\theta = -60°, \quad r = -\sqrt{3}$

$\theta = -90°, \quad r = 0$

73. Each dollar spent on costs produces $1.33 in revenue.

74. Each dollar spent on costs produces $0.50 in revenue.

75. Cosine is negative in quadrant III.

76. $\sin\theta = \dfrac{\sqrt{17}}{17}, \cos\theta = -\dfrac{4\sqrt{17}}{17}$

77. false **78.** false

79. 90° **80.** 180°

81. $\cot\theta = \pm\dfrac{\sqrt{1 - \sin^2\theta}}{\sin\theta}$

82. $\csc\theta = \pm\dfrac{1}{\sqrt{1 - \cos^2\theta}}$

83. $8|\cos\theta|$ **84.** $6|\sin\theta|$

85. $5|\sec\theta|$ **86.** $-\cot^4\theta$

87. true **88.** false

89. yes **a.** 0.3746 **b.** 0.9272 **c.** 0.4040
d. 0.4040 (c) and (d) are the same.

90. no **a.** 0.3420 **b.** 0.9613 **c.** 0.3558
d. 0.1763 (c) and (d) are not the same.

91. (a) and (c)

92. (a) and (c) are same.

Section 7.2

1. 1 **2.** 1

3. $\csc x$ **4.** $-\sin x$

5. -1 **6.** -1

7. $\sec^2 x$ **8.** $-\tan x$

9. 1 **10.** 1

11. $\sin^2 x - \cos^2 x$ **12.** $1 + 2\sin x\cos x$

13. $\sec x$ **14.** $\csc x$

15. 1 **16.** 1 **17.** $\sin^2 x$ **18.** $\cos^2 x$

19. $\csc^2 x$ **20.** $\sec^2 x$ **21.** $-\cos x$

22. $\sin x$ **23.** 1

24. – 50. See Instructor's Solution Manual

51. conditional **52.** identity

53. identity **54.** conditional

55. conditional **56.** identity

57. conditional **58.** identity

59. conditional **60.** conditional

61. identity **62.** conditional

63. conditional **64.** conditional

65. $|a|\cos\theta$ **66.** $|a|\sec\theta$

67. The $\cos x$ and $\sin x$ terms do not cancel. The numerators become $\cos^2 x$ and $\sin^2 x$, respectively.

68. There are two errors. $\sec x = \dfrac{1}{\cos x}$, not $\dfrac{1}{\sin x}$, and $\dfrac{\sin x}{1 - \sin x} = \dfrac{1}{\sin x - \sin^2 x}$, not $\dfrac{1}{1 - \sin^2 x}$.

69. This is a conditional equation. Just because the equation is true for $\dfrac{\pi}{4}$ does not mean it is true for all x.

70. This is a conditional equation.

71. false **72.** false

73. QI, QIV **74.** QII, QIII

75. QIII and QIV **76.** QI and QIV

77. $a^2 + b^2$ **78.** $\csc^2 x$

79. No, let $A = 30°$ and $B = 60°$.

80. No, take $A = \dfrac{\pi}{2}$. **81.** No, take $A = \dfrac{\pi}{4}$.

82. No, they are not the same.

83. $\cos(A + B) = \cos A\cos B - \sin A\sin B$

84. $\cos(A - B) = \cos A\cos B + \sin A\sin B$

85. $\sin(A + B) = \sin A\cos B + \cos A\sin B$

86. $\sin(A - B) = \sin A\cos B - \cos A\sin B$

87. $\cos^2 A = \dfrac{1 + \cos(2A)}{2}$

88. $\sin^2 A = \dfrac{1 - \cos(2A)}{2}$

89. $Y_1 = Y_3$ **90.** $Y_1 = Y_2$

Section 7.3

1. $\dfrac{\sqrt{6}-\sqrt{2}}{4}$

2. $\dfrac{\sqrt{2}+\sqrt{6}}{4}$

3. $\dfrac{\sqrt{6}-\sqrt{2}}{4}$

4. $\dfrac{-\sqrt{2}-\sqrt{6}}{4}$

5. $-2+\sqrt{3}$

6. $2-\sqrt{3}$

7. $\dfrac{\sqrt{2}+\sqrt{6}}{4}$

8. $\dfrac{-\sqrt{2}-\sqrt{6}}{4}$

9. $2+\sqrt{3}$

10. $-2+\sqrt{3}$

11. $2+\sqrt{3}$

12. $-2+\sqrt{3}$

13. $\sqrt{2}-\sqrt{6}$

14. $\sqrt{2}-\sqrt{6}$

15. $\dfrac{4}{\sqrt{2}(1+\sqrt{3})}=\sqrt{6}-\sqrt{2}$

16. $-(\sqrt{2}+\sqrt{6})$

17. $\cos x$

18. $-\cos 3x$

19. $-\sin x$

20. $\sin 5x$

21. 0

22. $\sin\left(\frac{5\pi}{6}x\right)$

23. $-2\cos(A-B)$

24. $2\cos(A-B)$

25. $-2\sin(A+B)$

26. $2\sin(A-B)$

27. $\tan(26°)$

28. $\tan(72°)$

29. $\dfrac{1+2\sqrt{30}}{12}$

30. $\dfrac{-(1+2\sqrt{30})}{12}$

31. $\dfrac{-6\sqrt{6}+4}{25}$

32. $\dfrac{6\sqrt{6}-4}{25}$

33. $\dfrac{3-4\sqrt{15}}{4+3\sqrt{15}}=\dfrac{192-25\sqrt{15}}{-119}$

34. $\dfrac{192+25\sqrt{15}}{-119}$

35. identity

36. identity

37. conditional

38. conditional

39. identity

40. conditional

41. identity

42. identity

43. identity

44. identity

45. conditional

46. conditional

47. identity

48. identity

49. identity

50. conditional

51. conditional

52. identity

53. $y=\sin\left(x+\dfrac{\pi}{3}\right)$

54. $y=\sin\left(x-\dfrac{\pi}{3}\right)$

55. $y=\cos\left(x-\dfrac{\pi}{4}\right)$

56. $y=-\cos\left(x+\dfrac{\pi}{4}\right)$

57. $y=-\sin(4x)$

58. $y = \cos 2x$

59. $y = \tan\left(x + \dfrac{\pi}{4}\right)$

60. $y = \tan\left(\dfrac{\pi}{3} - x\right)$

61. $y = \tan\left(\dfrac{\pi}{6} + x\right)$

62. – 66. See Instructor's Solution Manual

67. $\cos(kz - ct) = \cos(kz)\cos(ct) +$

$\sin(kz)\sin(ct)$; when $\dfrac{z}{\lambda}$ = integer, then

$kz = 2\pi n$ and the $\sin kz$ term goes to zero.

68. $E(0) = A\cos(kz)$

69. $T(t) = 38 - 2.5\sin\left(\dfrac{\pi}{6}t\right)$

70. $C(t) = 15.4 - 4.7\cos\left(\dfrac{\pi}{4}t\right)$

71. Tangent of a sum is not the sum of the tangents. Needed to use the tangent of a sum identity.

72. Tangent is an odd function, not an even function, so $\tan(-\theta) = -\tan\theta$, other calculations are same.

73. false   **74.** false   **75.** false

76. – 78. See Instructor's Solution Manual

79. $A = n\pi$ and $B = m\pi$, where n and m are integers, or A and B, where $B = A \pm 2n\pi$.

80. $A = n\pi$ and $B = m\pi$, where n and m are integers, or A and B, where $B = A + 2n\pi$.

81. a.

b.

c.

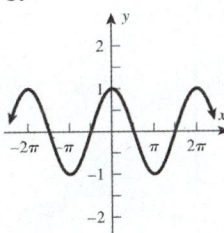

The difference quotients of $y = \sin x$ better approximate $y = \cos x$ as h goes to zero.

82. a.

b.

c.

The difference quotients of
$y = \cos x$ better approximate
$y = -\sin x$ as h goes to zero.

83. a.

b.

c.

The graphs better approximate $y = 2\cos 2x$
as h goes to zero.

84. a.

b.

c.

The graphs better approximate $y = -2\sin 2x$
as h goes to zero.

Section 7.4

1. $-\dfrac{4}{5}$ **2.** $\dfrac{3}{5}$

3. $\dfrac{120}{119}$ **4.** $-\dfrac{120}{119}$

5. $\dfrac{120}{169}$ **6.** $-\dfrac{119}{169}$

7. $-\dfrac{4}{3}$ **8.** $2\sqrt{2}$

9. $\dfrac{\sqrt{19}}{10}$ **10.** $-\dfrac{4\sqrt{3}}{13}$

11. $\dfrac{119}{120}$ **12.** $-\dfrac{169}{120}$

13. $\tan(30°) = \dfrac{\sqrt{3}}{3}$ **14.** $\tan\left(\dfrac{\pi}{4}\right) = 1$

15. $\frac{1}{2}\sin\left(\frac{\pi}{4}\right) = \frac{\sqrt{2}}{4}$ **16.** $\frac{1}{4}$

17. $\cos(4x)$ **18.** $\cos(2x+4)$

19. $-\frac{\sqrt{3}}{3}$ **20.** $\tan x$

21. $-\frac{\sqrt{3}}{2}$ **22.** $\frac{\sqrt{2}}{2}$

23. $-\frac{\sqrt{3}}{2}$

24. – 40. See Instructor's Solution Manual

41. $y = \cot x$

42. $y = \tan(2x)$

43. $y = \sec(2x)$

44. $y = \csc(2x)$

45. $y = \frac{1}{2}\sin(4x)$

46. $y = \frac{3}{2}\sin(6x)$

47. $y = 1 - \frac{1}{2}\sin(2x)$

48. $y = 2 - 3\tan(2x)$

49. $y = 2\sin x - 3\cos(2x)$

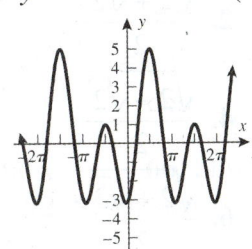

77. identity   **78.** not an identity

79. Y_1 and Y_3 **80.** Y_1 and Y_3

Section 7.6

1. $\frac{1}{2}[\sin(3x)+\sin x]$

2. $\frac{1}{2}[\sin(15x)-\sin(5x)]$

3. $\frac{5}{2}[\cos(2x)-\cos(10x)]$

4. $-\frac{3}{2}[\cos(2x)-\cos(6x)]$

5. $2[\cos x+\cos(3x)]$

6. $-4[\cos(8x)+\cos(2x)]$

7. $\frac{1}{2}[\cos x-\cos(4x)]$

8. $\frac{1}{2}[\cos(2\pi x)-\cos(3\pi x)]$

9. $\frac{1}{2}\left[\cos\left(\frac{2x}{3}\right)+\cos(2x)\right]$

10. $\frac{1}{2}\left[\sin\left(\frac{\pi}{4}x\right)-\sin\left(\frac{3\pi}{4}x\right)\right]$

11. $-\frac{3}{2}[\cos(1.9x)+\cos(1.1x)]$

12. $\cos(1.3x)-\cos(5.5x)$

13. $2[\sin(2\sqrt{3}x)-\sin(4\sqrt{3}x)]$

14. $-\frac{5}{2}\left[\sin\left(\frac{4\sqrt{2}}{3}x\right)-\sin\left(2\sqrt{2}x\right)\right]$

15. $2\cos(4x)\cos x$ **16.** $2\sin(3x)\sin x$

17. $2\sin x\cos(2x)$

18. $2\sin\left(\frac{15}{2}x\right)\cos\left(\frac{5}{2}x\right)$

19. $-2\sin x\cos\left(\frac{3}{2}x\right)$ **20.** $2\sin\left(\frac{3x}{2}\right)\sin x$

21. $2\cos\left(\frac{3}{2}x\right)\cos\left(\frac{5}{6}x\right)$

22. $2\sin\left(\frac{3}{2}x\right)\cos\left(\frac{5}{6}x\right)$

23. $2\sin(0.5x)\cos(0.1x)$

24. $2\sin(0.4x)\sin(0.1x)$

25. $-2\sin(\sqrt{5}x)\cos(2\sqrt{5}x)$

26. $-2\sin\left(\frac{\sqrt{7}x}{2}\right)\sin\left(\frac{5\sqrt{7}x}{2}\right)$

27. $2\cos\left(\frac{\pi}{24}x\right)\cos\left(\frac{5\pi}{24}x\right)$

28. $2\sin(\pi x)\cos\left(\frac{\pi}{4}x\right)$

29. $-\tan x$ **30.** $-\cot x$

31. $\tan(2x)$ **32.** $\tan(3x)$

33. $\cot\left(\frac{3x}{2}\right)$

34. – 42. See Instructor's Solution Manual

43. $R(t)=\sqrt{3}\cos\left(\frac{\pi}{6}t+\frac{4\pi}{3}\right)$

44. $\text{Profit}=\sin\left(\frac{\pi}{3}t+\frac{\pi}{6}\right)$

45. Average frequency: 443 Hz; Beat frequency: 102 Hz

46. Average frequency: 394.5 Hz; Beat frequency: 91 Hz

47. $2\sin\left[\frac{2\pi ct}{2}\left(\frac{1}{1.55}+\frac{1}{0.63}\right)10^6\right]$

$\cos\left[\frac{2\pi ct}{2}\left(\frac{1}{1.55}-\frac{1}{0.63}\right)10^6\right]$

48. Average frequency: $c\left[\frac{1}{1.55}-\frac{1}{0.63}\right]10^6$ Hz;

Beat frequency: $\frac{c}{2}\left[\frac{1}{1.55}+\frac{1}{0.63}\right]10^6$ Hz

49. $2\sin\left[\frac{2\pi(1979)t}{2}\right]\cos\left[\frac{2\pi(439)t}{2}\right]=$ $2\sin(1979\pi t)\cos(439\pi t)$

50. $\sin(2\pi(697)t)+\sin(2\pi(1477)t)=$ $2\sin(2174\pi t)\cos(780\pi t)$

51. $\frac{25(\sqrt{6}-\sqrt{3})}{3}\approx 5.98$ ft^2

52. $(36+12\sqrt{3})\approx 56.78$ sq in.

53. $\cos A\cos B\neq\cos(AB)$ and $\sin A\sin B\neq\sin(AB)$. Should have used the product-to-sum identity.

54. $\sin A \cos B - \sin B \cos A \neq 0.$ Should have used the product-to-sum identity.

55. false   **56.** false

57. true   **58.** true

59. Answers will vary. Here is one approach.

$$\underbrace{\sin A \sin B}_{\frac{1}{2}[\cos(A-B)-\cos(A+B)]} \sin C$$

$$= \frac{1}{2}\left[\underbrace{\cos(A-B)\sin C}_{\frac{1}{2}[\sin(A-B+C)+\sin(C-A+B)]} - \underbrace{\cos(A+B)\sin C}_{\frac{1}{2}[\sin(A-B+C)+\sin(C-A+B)]} \right]$$

$$\sin A \sin B \sin C =$$

$$\frac{1}{4}[\sin(A-B+C)+\sin(C-A+B)]$$

$$-\sin(A+B+C)-\sin(C-A-B)]$$

60. – 62. See Instructor's Solution Manual

63. $y = 1 - \dfrac{3}{2}\cos\left(\dfrac{7\pi}{6}x\right) + \dfrac{3}{2}\cos\left(\dfrac{5\pi}{6}x\right)$

64. $y = 2[\sin(x+1)+\sin(3x-3)]$

65. $y = -\dfrac{1}{2}\cos\left(\dfrac{3\pi}{2}x\right) - \dfrac{1}{2}\cos\left(\dfrac{\pi}{6}x\right)$

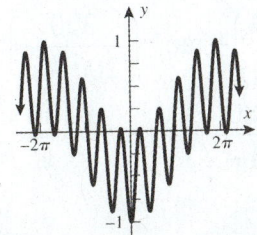

66. $y = x - \dfrac{1}{2}[\sin(5x)+\sin x]$

67. $y = \sin(4x)$

68. $y = \sec(2x)$ **69.** Y_1 and Y_3

70. Y_1 and Y_3

Section 7.7

1. $\dfrac{\pi}{4}$ **2.** $\dfrac{3\pi}{4}$

3. $-\dfrac{\pi}{3}$ **4.** $\dfrac{\pi}{6}$

5. $\dfrac{3\pi}{4}$ **6.** $\dfrac{\pi}{6}$

7. $\dfrac{\pi}{6}$ **8.** $-\dfrac{\pi}{2}$

9. $\dfrac{\pi}{6}$ **10.** $\dfrac{2\pi}{3}$

11. $-\dfrac{\pi}{3}$ **12.** $\dfrac{\pi}{6}$

13. 0 **14.** $\dfrac{\pi}{4}$

15. π **16.** $\dfrac{\pi}{2}$

17. 60°
18. 150°
19. 45°
20. 0°
21. 120°
22. −60°
23. 30°
24. 45°
25. −30°
26. −60°
27. 135°
28. −45°
29. −90°
30. −45°
31. 90°
32. 180°
33. 57.10°
34. 51.45°
35. 62.18°
36. 72.99°
37. 48.10°
38. 68.97°
39. −15.30°
40. −9.39°
41. 166.70°
42. 131.91°
43. −0.63
44. 1.05
45. 1.43
46. −0.75
47. 0.92
48. 0.39
49. 2.09
50. 2.86
51. 0.31
52. −0.35
53. $\dfrac{5\pi}{12}$
54. $-\dfrac{5\pi}{12}$
55. not possible
56. not possible
57. $\dfrac{\pi}{6}$
58. $-\dfrac{\pi}{6}$
59. $\dfrac{2\pi}{3}$
60. $\dfrac{\pi}{3}$
61. $\sqrt{3}$
62. $\dfrac{\pi}{4}$
63. $\dfrac{\pi}{3}$
64. not possible
65. not possible
66. $-\dfrac{\pi}{6}$
67. 0
68. $\dfrac{3\pi}{4}$
69. $-\dfrac{\pi}{4}$
70. $\dfrac{\pi}{4}$
71. not possible
72. not possible
73. $\dfrac{2\pi}{3}$
74. 0
75. $-\dfrac{\pi}{4}$
76. not possible
77. $\dfrac{\sqrt{7}}{4}$
78. $\dfrac{\sqrt{5}}{3}$

79. $\dfrac{12}{13}$
80. $\dfrac{24}{25}$
81. $\dfrac{3}{4}$
82. $\dfrac{\sqrt{21}}{2}$
83. $\dfrac{5\sqrt{23}}{23}$
84. $\dfrac{4\sqrt{7}}{7}$
85. $\dfrac{4\sqrt{15}}{15}$
86. 4
87. $\dfrac{11}{60}$
88. $\dfrac{9}{40}$
89. $\dfrac{24}{25}$
90. $-\dfrac{16}{65}$
91. $\dfrac{56}{65}$
92. $\dfrac{63}{65}$
93. $\dfrac{24}{25}$
94. $\dfrac{7}{25}$
95. $\dfrac{120}{119}$
96. $-\dfrac{120}{119}$
97. $\sqrt{1-u^2}$
98. $\sqrt{1-u^2}$
99. $\dfrac{\sqrt{1-u^2}}{u}$
100. $\dfrac{u\sqrt{1-u^2}}{1-u^2}$
101. April or October
102. July
103. In month 3
104. Months 1 and 5
105. $t \approx 0.026476$ sec or 26 ms
106. $t \approx 0.000684$ sec ≈ 0.68 ms
107. 173.4; June 22–23
108. −11.4; Dec. 19
109. $11.3 \approx 11$ years
110. 40 years
111. $\tan\theta = \dfrac{\dfrac{7}{x}+\dfrac{1}{x}}{1-\dfrac{7}{x}\cdot\dfrac{1}{x}} = \dfrac{8x}{x^2-7}$

112. $x-10$: $\theta = 41°$; $x = 20$: $\theta = 22°$

113. 0.70 m; 0.24 m
114. 0.44 m; 0.23 m

115. $\theta = \pi - \tan^{-1}\left(\dfrac{150}{x}\right) - \tan^{-1}\left(\dfrac{300}{200-x}\right)$

116. $\theta = \pi - \tan^{-1}\left(\dfrac{280}{200-x}\right) - \tan^{-1}\left(\dfrac{130}{x}\right)$

117. The wrong interval for the identity was used. The correct domain is $\left[-\dfrac{\pi}{2}, \dfrac{\pi}{2}\right]$.

118. The wrong interval for the indentity was used. The correct domain is $[0, \pi]$.

119. $\cot^{-1} x \neq \dfrac{1}{\tan^{-1} x}$

120. $\csc^{-1} x \neq \dfrac{1}{\sin^{-1} x}$

121. false　　　　**122.** true

123. false　　　　**124.** false

125. $\frac{1}{2}$ is not in the domain of the inverse secant function.

126. $\frac{1}{2}$ is not in the domain of the inverse cosecant function.

127. $\dfrac{\sqrt{x^2 - 1}}{x}$　　　**128.** $-\dfrac{\pi}{6} \leq x \leq \dfrac{\pi}{6}$

129. 0

130. a. f is 1-1 on $[0, \pi]$.

  b. $f^{-1}(x) = \dfrac{\pi}{2} + \sin^{-1}\left(\dfrac{2 - x}{4}\right)$ with domain $[-2, 6]$.

131. a. $\left(\dfrac{\pi}{4}, \dfrac{5\pi}{4}\right)$

  b. $f^{-1}(x) = \cos^{-1}(x - 3) + \dfrac{\pi}{4}$

  Domain: [2, 4]

132. a. f is 1-1 on $\left(-\dfrac{\pi}{2}, \dfrac{\pi}{2}\right)$

  b. $f^{-1}(x) = -\dfrac{\pi}{3} + \tan^{-1}(1 - x)$ with domain $\mathbb{R}$

133. a. $\left(\dfrac{\pi}{12}, \dfrac{7\pi}{12}\right)$

  b. $f^{-1}(x) = \dfrac{\pi}{12} + \dfrac{1}{2}\operatorname{arccot}(4x - 8)$

  Domain: All real numbers.

134. a. f is 1-1 on $\left[-\dfrac{\pi}{2}, 0\right) \cup \left(0, \dfrac{\pi}{2}\right]$

  b. $f^{-1}(x) = \dfrac{4}{\pi}[1 + \csc^{-1}(-x)]$ with domain $(-\infty, -1] \cup [1, \infty)$.

135. The identity $\sin(\sin^{-1} x) = x$ only holds for $-1 \leq x \leq 1$.

136. The identity $\cos(\cos^{-1} x) = x$ only holds for $-1 \leq x \leq 1$.

137. $\left[-\dfrac{\pi}{2}, 0\right) \cup \left(0, \dfrac{\pi}{2}\right]$

138. $0 \leq x \leq \pi$

139. a. $\dfrac{720}{1681}$　　**b.** 0.42832　**c.** yes

140. a. $-\dfrac{3}{4}$　　**b.** −0.7500　**c.** yes

Section 7.8

1. $\dfrac{3\pi}{4}, \dfrac{5\pi}{4}$　　　　**2.** $\dfrac{5\pi}{4}, \dfrac{7\pi}{4}$

3. $\dfrac{7\pi}{6}, \dfrac{11\pi}{6}, \dfrac{19\pi}{6}, \dfrac{23\pi}{6}$

4. $\dfrac{2\pi}{3}, \dfrac{4\pi}{3}, \dfrac{8\pi}{3}, \dfrac{10\pi}{3}$

5. $n\pi$, where n is any integer

6. $\dfrac{(2n+1)\pi}{2}$, where n is an integer

7. $\dfrac{7\pi}{12}, \dfrac{11\pi}{12}, \dfrac{19\pi}{12}, \dfrac{23\pi}{12}$

8. $\dfrac{\pi}{12}, \dfrac{11\pi}{12}, \dfrac{13\pi}{12}, \dfrac{23\pi}{12}$

9. $\dfrac{7\pi}{3} + 4n\pi$ or $\dfrac{11\pi}{3} + 4n\pi$, where n is any integer

10. $2(2n+1)\pi$, where n is an integer

11. $-\dfrac{4\pi}{3}, -\dfrac{5\pi}{6}, -\dfrac{11\pi}{6}, -\dfrac{\pi}{3}, \dfrac{\pi}{6}, \dfrac{2\pi}{3}, \dfrac{7\pi}{6}, \dfrac{5\pi}{3}$

12. $\dfrac{\pi(2 + 3n)}{6}$, where n is an integer

13. $-\dfrac{4\pi}{3}, -\dfrac{2\pi}{3}$　　　**14.** $\dfrac{\pi}{3}, \dfrac{2\pi}{3}$

15. $\dfrac{\pi}{6}+\dfrac{n\pi}{4}$, where n is any integer

16. $\dfrac{\pi}{20}+\dfrac{n\pi}{5}$, where n is an integer

17. $-\pi,\ -\dfrac{5\pi}{3},\ -\dfrac{\pi}{3}$

18. $\dfrac{\pi}{16},\ \dfrac{7\pi}{16},\ \dfrac{9\pi}{16},\ \dfrac{15\pi}{16}$

19. $-\dfrac{\pi}{2},\ -\dfrac{7\pi}{6},\ -\dfrac{11\pi}{6}$ **20.** $\dfrac{2\pi}{9},\ \dfrac{5\pi}{18}$

21. $\dfrac{\pi}{6},\ \dfrac{\pi}{3},\ \dfrac{7\pi}{6},\ \dfrac{4\pi}{3}$ **22.** $\dfrac{3\pi}{2}$

23. $\dfrac{\pi}{12},\ \dfrac{7\pi}{12},\ \dfrac{13\pi}{12},\ \dfrac{19\pi}{12}$ **24.** $\dfrac{\pi}{2}$

25. $\dfrac{\pi}{3},\ \dfrac{2\pi}{3},\ \dfrac{4\pi}{3},\ \dfrac{5\pi}{3}$

26. $\dfrac{7\pi}{12},\ \dfrac{11\pi}{12},\ \dfrac{19\pi}{12},\ \dfrac{23\pi}{12}$

27. $\dfrac{2\pi}{3}$

28. $\dfrac{5\pi}{12},\ \dfrac{7\pi}{12},\ \dfrac{17\pi}{12},\ \dfrac{19\pi}{12}$

29. $\dfrac{\pi}{4},\ \dfrac{3\pi}{4},\ \dfrac{5\pi}{4},\ \dfrac{7\pi}{4}$

30. $\dfrac{3\pi}{2}$

31. $\dfrac{\pi}{2},\ \dfrac{3\pi}{2},\ \dfrac{\pi}{3},\ \dfrac{5\pi}{3}$

32. $0,\ \pi,\ 2\pi,\ \dfrac{\pi}{3},\ \dfrac{4\pi}{3}$

33. $\dfrac{3\pi}{2},\ \dfrac{7\pi}{6},\ \dfrac{11\pi}{6}$

34. $\dfrac{\pi}{4},\ \dfrac{3\pi}{4},\ \dfrac{5\pi}{4},\ \dfrac{7\pi}{4}$

35. $\dfrac{\pi}{2}$ **36.** π

37. $0,\ \pi$ **38.** $\dfrac{\pi}{2},\ \dfrac{3\pi}{2}$

39. $\dfrac{\pi}{12},\ \dfrac{5\pi}{12},\ \dfrac{7\pi}{12},\ \dfrac{11\pi}{12},\ \dfrac{13\pi}{12},\ \dfrac{17\pi}{12},\ \dfrac{19\pi}{12},\ \dfrac{23\pi}{12}$

40. $\dfrac{\pi}{12},\ \dfrac{5\pi}{12},\ \dfrac{7\pi}{12},\ \dfrac{11\pi}{12},\ \dfrac{13\pi}{12},\ \dfrac{17\pi}{12},\ \dfrac{19\pi}{12},\ \dfrac{23\pi}{12}$

41. 115.83°, 154.17°, 295.83°, 334.17°

42. 29.55°, 150.45°, 209.55°, 330.45°

43. 333.63°   **44.** 91.06°

45. 29.05°, 209.05°

46. 146.44°, 213.56°

47. 200.70°, 339.30°

48. 41.81°, 318.19°

49. 41.41°, 318.59°

50. 199.47°, 340.53°

51. 56.31°, 126.87°, 236.31°, 306.87°

52. 66.42°, 138.59°, 221.41°, 293.58°

53. 9.74°, 80.26°, 101.79°, 168.21°, 189.74°, 260.26°, 281.79°, 348.21°

54. 82.82°, 141.06°

55. 80.12°, 279.88°

56. 52.306°, 127.69°

57. 64.93°, 121.41°, 244.93°, 301.41°

58. 49.03°, 146.92°, 229.03°, 326.92°

59. 15°, 45°, 75°, 105°, 135°, 165°, 195°, 225°, 255°, 285°, 315°, 345°

60. 90°, 270°

61. $\dfrac{\pi}{4},\ \dfrac{5\pi}{4}$ **62.** $\dfrac{3\pi}{4},\ \dfrac{7\pi}{4}$

63. π **64.** $\dfrac{\pi}{2}$

65. $\dfrac{\pi}{6}$ **66.** $0,\ 2\pi$

67. $\dfrac{\pi}{3}$ **68.** $\dfrac{5\pi}{3}$

69. $\dfrac{\pi}{4},\ \dfrac{3\pi}{4},\ \dfrac{5\pi}{4},\ \dfrac{7\pi}{4}$ **70.** $\dfrac{\pi}{6},\ \dfrac{5\pi}{6},\ \dfrac{\pi}{2}$

71. $\dfrac{\pi}{2}, \dfrac{3\pi}{2}$

72. $0, \dfrac{\pi}{6}, \pi, \dfrac{11\pi}{6}, 2\pi$

73. $0, \dfrac{\pi}{4}, \pi, \dfrac{7\pi}{4}$ **74.** $\dfrac{\pi}{6}, \dfrac{5\pi}{6}, \dfrac{3\pi}{2}$

75. $\dfrac{\pi}{6}, \dfrac{5\pi}{6}, \dfrac{7\pi}{6}, \dfrac{11\pi}{6}, \dfrac{\pi}{2}, \dfrac{3\pi}{2}$

76. $\dfrac{\pi}{6}, \dfrac{5\pi}{6}, \dfrac{7\pi}{6}, \dfrac{11\pi}{6}$

77. $\dfrac{\pi}{6}, \dfrac{\pi}{3}, \dfrac{7\pi}{6}, \dfrac{4\pi}{3}$

78. $0, \dfrac{\pi}{6}, \pi, \dfrac{11\pi}{6}, 2\pi$

79. $\dfrac{\pi}{6}, \dfrac{5\pi}{6}, \dfrac{7\pi}{6}, \dfrac{11\pi}{6}$ **80.** $0, \pi, 2\pi$

81. $\dfrac{3\pi}{2}$ **82.** $\dfrac{\pi}{6}, \dfrac{5\pi}{6}, \dfrac{3\pi}{2}$

83. $\dfrac{2\pi}{3}, \dfrac{4\pi}{3}$ **84.** $\dfrac{7\pi}{6}, \dfrac{11\pi}{6}$

85. $\dfrac{\pi}{3}, \dfrac{5\pi}{3}, \pi$ **86.** $\dfrac{\pi}{3}, \dfrac{5\pi}{3}$

87. $\dfrac{\pi}{24}, \dfrac{5\pi}{24}, \dfrac{13\pi}{24}, \dfrac{17\pi}{24}, \dfrac{25\pi}{24}, \dfrac{29\pi}{24},$ $\dfrac{37\pi}{24}, \dfrac{41\pi}{24}$

88. $\dfrac{7\pi}{6}, \dfrac{11\pi}{6}$

89. $57.47°, 122.53°, 323.62°, 216.38°$

90. $0°, 45°, 180°, 225°, 360°$

91. $30°, 150°, 199.47°, 340.53°$

92. $71.57°, 135°, 251.57°, 315°$

93. $14.48°, 165.52°, 270°$

94. $26.57°, 161.57°, 206.57°, 341.57°$

95. $111.47°, 248.53°$

96. $31.72°, 121.72°, 211.72°, 301.72°$

97. $\dfrac{\pi}{3}, \dfrac{5\pi}{3}$ **98.** 2π

99. Fourth quarter of 2008, second quarter of 2009, and fourth quarter of 2010

100. First quarter of 2009

101. around 9 pm   **102.** around 5 pm

103. March

104. July and November

105. $A = \frac{1}{2}h(b_1 + b_2)$
$= \frac{1}{2}(x\sin\theta)(x + (x\cos\theta + x + x\cos\theta))$
$= x^2 \sin\theta(1 + \cos\theta)$

106. $\dfrac{\pi}{3}, \pi, \dfrac{5\pi}{3}$

107. 2001   **108.** 2007

109. $24°$ **110.** $39°$

111. $\dfrac{3}{4}$ sec   **112.** $\dfrac{9}{4}$ sec

113. $(0, 1),\ \left(\dfrac{\pi}{3}, \dfrac{3}{2}\right),\ (\pi, -3),\ \left(\dfrac{5\pi}{3}, \dfrac{3}{2}\right)$

114. $(0, 1),\ \left(-\dfrac{\pi}{3}, \dfrac{3}{2}\right),\ \left(-\dfrac{5\pi}{3}, \dfrac{3}{2}\right),$
$(-\pi, -3),\ (-2\pi, 1)$

115. March and September

116. April and October

117. Around 1 am and 11 am

118. Around 2 pm and 10 pm

119. Extraneous solution. Forgot to check.

120. Extraneous solution. Forgot to check.

121. Can't divide by cos x. Must factor.

122. Forgot to check for extraneous solutions. Indeed, $x = \pi$ is not a solution to the equation.

123. false   **124.** false

125. true   **126.** false

127. $\dfrac{\pi}{6}, \dfrac{5\pi}{6}, \dfrac{7\pi}{6}, \dfrac{11\pi}{6}$

128. $-\dfrac{\pi}{12} + n\pi, \dfrac{7\pi}{12} + 2n\pi$, where n is an integer

129. $\dfrac{\pi}{6}$ or $30°$ **130.** $60.57°$

131. no solution

132. $3n\pi$, where n is an integer

133. $5 + 2n$, where n is any integer

134. $\dfrac{(1 + 3n)\pi}{18}$, where n is an integer

135. $x = \dfrac{\pi}{6} \approx 0.524$ and $x = \dfrac{5\pi}{6} \approx 2.618$

136. $x = \dfrac{\pi}{4} \approx 0.785$

137. no solution

138. no solution

139. no solution

140. $\theta = 0$

141.

142.

143. $x \approx 1.3$

144. There is no positive solution.

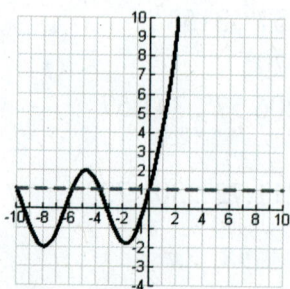

145. 2.21911
146. 1.303

Review Exercises

1. $-\frac{11}{7}$

2. $\frac{\sqrt{5}}{3}$

3. $-\frac{8}{15}$

4. $\frac{5}{12}$

5. $\frac{4}{27}$

6. 0.2025

7. $3\sqrt{3}$

8. $-\frac{11}{60}$

9. $-\frac{7}{24}$

10. $\sqrt{2}$

11. $-\frac{5}{12}$

12. $\sin\theta = \frac{24}{25}, \cos\theta = -\frac{7}{25}$

13. $\sin\theta = -\frac{\sqrt{3}}{2}, \cos\theta = -\frac{1}{2}$

14. $\sin\theta = -\frac{1}{2}, \cos\theta = -\frac{\sqrt{3}}{2}$ **15.** $\frac{\cos\theta}{\sin^2\theta}$

16. $\cos\theta$

17. $\sin\theta$

18. $\cos^2\theta + 2 + \frac{1}{\cos^2\theta}$

19. $\frac{1}{\sin\theta} + \cos\theta$

20. $\sin^3\theta$

21. $\sec^2 x$

22. $\tan^2 x$

23. $\sec^2 x$

24. $\cot^2 x$

25. $\cos^2 x$

26. $\frac{1}{2}$

27. $-(4 + 2\csc x + \csc^2 x)$

28. – 34. See Instructor's Solution Manual

35. identity

36. conditional

37. conditional

38. identity

39. identity

40. conditional

41. $\frac{\sqrt{2} - \sqrt{6}}{4}$

42. $\frac{\sqrt{6} - \sqrt{2}}{4}$

43. $\frac{\sqrt{3} - 3}{3 + \sqrt{3}} = \sqrt{3} - 2$

44. $-2 + \sqrt{3}$

45. $\sin x$

46. $\cos x$

47. $\tan x$

48. $\tan\left(\frac{7\pi}{12}\right)$

49. $\frac{117}{44}$

50. $\frac{36}{325}$

51. $-\frac{897}{1025}$

52. $\frac{56}{65}$

53. identity

54. identity

55. $y = \cos\left(x + \frac{\pi}{2}\right)$

56. $y = \sin\left(x + \frac{2\pi}{3}\right)$

57. $y = \tan\left(\frac{2x}{3}\right)$

58. $y = \tan((\pi - 1)x)$

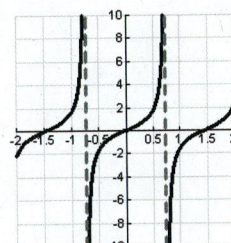

59. $\frac{7}{25}$

60. $-\frac{326}{625}$

61. $\frac{671}{1800}$

62. $-\frac{119}{169}$

63. $\frac{336}{625}$

64. $\frac{24}{7}$

65. $\frac{\sqrt{3}}{2}$

66. $-\frac{\sqrt{3}}{3}$

67. $\frac{3}{2}$

68. – 74. See Instructor's Solution Manual

75. $-\dfrac{\sqrt{2-\sqrt{2}}}{2}$

76. $\dfrac{\sqrt{2-\sqrt{2}}}{2}$

77. $\sqrt{2}-1$

78. $\dfrac{2}{\sqrt{2-\sqrt{2}}}$

79. $-\dfrac{2}{\sqrt{2+\sqrt{3}}}$

80. $-\sqrt{\dfrac{2+\sqrt{3}}{2-\sqrt{3}}}$

81. $\dfrac{7\sqrt{2}}{10}$

82. $\dfrac{\sqrt{10}}{10}$

83. $-\dfrac{5}{4}$

84. $\dfrac{\sqrt{17}}{17}$

85. $\sin\left(\dfrac{\pi}{12}\right)$

86. – 90. See Instructor's Solution Manual

91. $y=\left|\sin\left(\dfrac{\pi}{24}x\right)\right|$

92. $y=\cos x$

93. $y=-\tan\left(\dfrac{x}{2}\right)$

94. $y=\cos\left(\dfrac{3x-1}{2}\right)$

95. $3[\sin(7x)+\sin(3x)]$

96. $\dfrac{3}{2}\big[\cos 2x-\cos 6x\big]$

97. $-2\sin(4x)\sin x$

98. $2\sin(2x)\cos\left(\dfrac{x}{2}\right)$

99. $2\sin\left(\dfrac{x}{3}\right)\cos x$

100. $2\cos(4x)\cos(3x)$ **101.** $\cot(3x)$

102. – 106. See Instructor's Solution Manual

107. $\dfrac{\pi}{4}$

108. $-\dfrac{\pi}{6}$

109. $\dfrac{\pi}{2}$

110. $-\dfrac{\pi}{2}$

111. $\dfrac{\pi}{6}$

112. $\dfrac{5\pi}{6}$

113. $-90°$ **114.** $-45°$

115. $60°$ **116.** $45°$

117. $-60°$ **118.** $0°$

119. $-37.50°$ **120.** $50.00°$

121. $22.50°$ **122.** $165.00°$

123. 1.75 **124.** 0.16

125. -0.10 **126.** 2.67

127. $-\dfrac{\pi}{4}$

128. $-\dfrac{\sqrt{2}}{2}$

129. $-\sqrt{3}$

130. $\dfrac{5\pi}{6}$

131. $\dfrac{\pi}{3}$

132. $-\dfrac{2\sqrt{3}}{3}$

133. $\dfrac{60}{61}$

134. $\dfrac{9}{41}$

135. $\dfrac{7}{6}$ **136.** $\dfrac{7}{24}$

137. $\dfrac{6\sqrt{35}}{35}$ **138.** $\dfrac{13}{12}$

139. July   **140.** January

141. $\dfrac{2\pi}{3}, \dfrac{5\pi}{6}, \dfrac{5\pi}{3}, \dfrac{11\pi}{6}$ **142.** $\pm\dfrac{2\pi}{3}$

143. $-\dfrac{3\pi}{2}, -\dfrac{\pi}{2}$

144. $\dfrac{\pi}{12}, \dfrac{5\pi}{12}, \dfrac{13\pi}{12}, \dfrac{17\pi}{12}$

145. $\dfrac{9\pi}{4}, \dfrac{21\pi}{4}$

146. $-\dfrac{7\pi}{24}, -\dfrac{\pi}{24}, \dfrac{5\pi}{24}, \dfrac{11\pi}{24}$

147. $\dfrac{\pi}{3}, \dfrac{2\pi}{3}, \dfrac{4\pi}{3}, \dfrac{5\pi}{3}$

148. $\dfrac{\pi}{3}$

149. $\dfrac{3\pi}{8}, \dfrac{7\pi}{8}, \dfrac{11\pi}{8}, \dfrac{15\pi}{8}$

150. $\dfrac{\pi}{6}, \dfrac{5\pi}{6}, \dfrac{3\pi}{2}$

151. $0, \pi, \dfrac{3\pi}{4}, \dfrac{7\pi}{4}$

152. $0, \dfrac{\pi}{3}, \dfrac{5\pi}{3}, 2\pi$

153. $80.46°, 170.46°, 260.46°, 350.46°$

154. $56.44°, 123.56°$

155. $90°, 138.59°, 221.41°, 270°$

156. $70.53°, 289.47°$

157. $17.62°, 162.38°$

158. $57.48°, 162.32°, 237.48°, 342.32°$

159. $\dfrac{\pi}{4}, \dfrac{5\pi}{4}$ **160.** $\dfrac{\pi}{6}, \dfrac{7\pi}{6}$

161. $\dfrac{\pi}{3}, \pi$

162. $\dfrac{\pi}{2}, \dfrac{3\pi}{2}, \dfrac{\pi}{6}, \dfrac{5\pi}{6}, \dfrac{7\pi}{6}, \dfrac{11\pi}{6}$

163. $0, \dfrac{\pi}{6}, \pi, \dfrac{11\pi}{6}$ **164.** $\dfrac{\pi}{3}, \dfrac{5\pi}{3}$

165. $\dfrac{3\pi}{2}$

166. $\dfrac{\pi}{2}, \dfrac{3\pi}{2}, \dfrac{\pi}{6}, \dfrac{11\pi}{6}$ **167.** π

168. $0, \pi, 2\pi, \dfrac{2\pi}{3}, \dfrac{4\pi}{3}$

169. $\dfrac{\pi}{2}, \dfrac{3\pi}{2}$ **170.** $\dfrac{3\pi}{4}$

171. $90°, 135°, 270°, 315°$

172. $14.48°, 30°, 150°, 165.52°$

173. $0°$ **174.** $30°, 150°$

175. $60°, 90°, 270°, 300°$

176. $30°, 150°, 189.59°, 350.41°$

177. (a) and (c)

178. None of the values are same.

179. $Y_1 = Y_2$ **180.** $Y_1 = Y_3$

181. $3\cos(3x)$ **182.** $-3\sin 3x$

183. Y_1 and Y_3 **184.** Y_1 and Y_3

185. Y_1 and Y_3 **186.** Y_1 and Y_3

187. Y_1 and Y_3 **188.** Y_1 and Y_3

189. a. $-\dfrac{3}{5}$ **b.** -0.6 **c.** yes

190. a. $-\sqrt{\dfrac{17}{24}}$ **b.** 0.8416 **c.** yes

191. 0.5787 **192.** $22.80°$

Practice Test

1. $x = \dfrac{(2n+1)}{2}\pi$, where n is an integer

2. conditional   **3.** $-\dfrac{\sqrt{2-\sqrt{2}}}{2}$

4. $-2-\sqrt{3}$ **5.** $\sqrt{\dfrac{3}{10}} = \dfrac{\sqrt{30}}{10}$

6. $\dfrac{23}{25}$ **7.** $\cos(10x)$

8. $-\tan(2x)$ **9.** $\cos\left(\dfrac{a+b}{2}\right)$

10. $\sin x + \sin 3$

11. $20 \cos x \cos 3$

12. $3|\cos x|$

13. $\begin{cases} \dfrac{4\pi}{3} + 2n\pi \\ \dfrac{5\pi}{3} + 2n\pi \end{cases}$, where n is an integer

14. $\pi, \dfrac{\pi}{3}, \dfrac{5\pi}{3}$

15. $14.48°, 90°, 165.52°, 270°$

16. no solution

17. conditional

18. $\dfrac{4}{\sqrt{2}\left(1 - \sqrt{3}\right)}$

19. $-\dfrac{\sqrt{26}}{26}$

20. -0.502

21. $\cot\left(\dfrac{\pi}{6}x + \dfrac{\pi}{8}\right)$

22. $\sqrt{2}$

23. One-to-one on the interval

$\left[c - \dfrac{1}{2}, c\right) \cup \left(c, c + \dfrac{1}{2}\right].$

$f^{-1}(x) = \dfrac{1}{\pi}\csc^{-1}\left(\dfrac{x-a}{b}\right) - \dfrac{c}{\pi}.$

24. $\left(-\dfrac{3\pi}{4}, \dfrac{\pi}{4}\right)$

25. $\dfrac{8}{3} + 8n, \dfrac{16}{3} + 8n$, where n is an integer

26. $\dfrac{5 + 12n}{6}, \dfrac{11 + 12n}{6}$ where n is an integer.

27. $\dfrac{\pi}{2} + 6n\pi, \dfrac{7\pi}{2} + 6n\pi$, where n is an integer

28. a.

b.

c.

This sequence of graphs better approximates the graph of $y = -\frac{1}{2}\sin(\frac{1}{2}x)$ as $h \to 0$.

29. a. $\dfrac{3\sqrt{10}}{10}$ **b.** 0.94868 **c.** yes

Cumulative Test

1. $x = 3 \pm 2\sqrt{5}$ **2.** -2

3. $(x+3)^2 + (y+1)^2 = 25$

4. not a function **5.** even

6. First translate the graph of $y = x^2$ to the right 4 units. Then, vertically stretch it by a factor of 5.

7. $f \circ g = \dfrac{1}{x^3} - 1$; domain is $x \neq 0$

8. $f^{-1}(x) = (x+1)^3$ **9.** $\left(-\dfrac{6}{5}, -\dfrac{3}{5}\right)$

10. $P(x) = cx^3\left(x + \sqrt{7}\right)^2\left(x - \sqrt{7}\right)^2$, for any constant c.

11. $Q(x) = 5x - 4, r(x) = -5x + 7$

12. $(x^2 + 16)(x - 3)(x + 5)$

13. HA: $y = 0.7$; VA: $x = -2, x = 3$

14. $5,908.54$ **15.** $(-3, \infty)$ **16.** 0

17. 4 **18.** 27.47 years

19. 0.4695 **20.** $4,713.77$ ft

21. $-\dfrac{7\pi}{12}$ **22.** $\dfrac{\pi}{4}, \dfrac{5\pi}{4}$

23. conditional   **24.** -1 **25.** $\dfrac{5}{12}$

26. The graphs of y_1 and y_3 are the same.

27. 1.3994

CHAPTER 8

Section 8.1

1. SSA   **2.** SSA   **3.** SSS

4. SAS   **5.** ASA   **6.** ASA

7. $\gamma = 75°$, $b = 12.2$ m, $c = 13.7$ m

8. $\alpha = 45°$, $a = 18$ in., $c = 22$ in.

9. $\beta = 62°$, $a = 163$ cm, $c = 215$ cm

10. $\alpha = 40°$, $b = 16$ ft, $c = 25$ ft

11. $\beta = 116.1°$, $a = 80.2$ yd, $b = 256.6$ yd

12. $\alpha = 42.2°$, $b = 37.5$ in., $c = 21.4$ in.

13. $\gamma = 120°$, $a = 7$ m, $b = 7$ m

14. $\beta = 60°$, $a = 7$ in., $b = 8$ in.

15. $\alpha = 97°$, $a = 118$ yd, $b = 52$ yd

16. $\beta = 70°$, $a = 3$ ft, $c = 2$ ft.

17. $\beta_1 = 20°$, $\gamma_1 = 144°$, $c_1 = 9$;

    $\beta_2 = 160°$, $\gamma_2 = 4°$, $c_2 = 1$

18. $\gamma = 39°$, $\alpha = 71°$, $a = 30$

19. $\alpha = 40°$, $\beta = 100°$, $b = 18$

20. $\beta_1 = 36°$, $\gamma_1 = 119°$, $c_1 = 166$

    $\beta_2 = 144°$, $\gamma_2 = 11°$, $c_2 = 36$

21. no triangle   **22.** no triangle

23. $\beta = 90°$, $\gamma = 60°$, $c = 16$

24. $\beta = 90°$, $\gamma = 45°$, $c = 1$

25. $\beta = 23°$, $\gamma = 123°$, $c = 15$

26. $\beta = 71°$, $\gamma = 38°$, $c = 3.4$

27. $\beta_1 = 21.9°$, $\gamma_1 = 136.8°$, $c_1 = 11.36$

    $\beta_2 = 158.1°$, $\gamma_1 = 0.6°$, $c_1 = 0.17$

28. no triangle   **29.** $\beta = 62°$, $\gamma = 2°$, $c = 0.3$

30. $\beta = 37°$, $\gamma = 92°$, $c = 11.50$

31. $\beta_1 = 77°$, $\alpha_1 = 63°$, $a = 457$

    $\beta_2 = 103°$, $\alpha_2 = 37°$, $a = 309$

32. $\alpha = 23°$, $\gamma = 20°$, $c = 8$

33. $\alpha = 31°$, $\gamma = 43°$, $c = 2$

34. $\beta = 6.5°$, $\gamma = 161.9°$, $a = 42.0$

35. 1246 ft   **36.** 1106 ft   **37.** 1.7 mi

38. 1.0 mi   **39.** 1.3 mi   **40.** 0.7 mi

41. 26 ft   **42.** 92 ft   **43.** 270 ft

44. 8.7 ft   **45.** 63.83°   **46.** 35 ft

47. 76.5 ft

48. The bullseye counts for 50 points, the overall score is either 55 points or 60 points.

49. 60.14 ft   **50.** 60.797 ft

51. 1.2 cm   **52.** $c = 23$ cm

53. No solution, because $b > a$ and $\alpha < 90°$.

54. $\beta = 124°$

55. false   **56.** false   **57.** true

58. true   **59.** true   **60.** false

61. – 70. See Instructor's Solution Manual

Section 8.2

1. C   **2.** C   **3.** S   **4.** S

5. S   **6.** S   **7.** C   **8.** S

9. $b = 5$, $\alpha = 47°$, $\gamma = 33°$

10. $c = 11$, $\alpha = 33°$, $\beta = 67°$

11. $a = 5$, $\beta = 158°$, $\gamma = 6°$

12. $c = 11$, $\alpha = 5°$, $\beta = 5°$

13. $a = 2$, $\beta = 80°$, $\gamma = 80°$

14. $c = 3.9$, $\alpha = 26.9°$, $\beta = 128.1°$

15. $b = 5$, $\alpha = 43°$, $\gamma = 114°$

16. $a = 7$, $\beta = 13°$, $\gamma = 151°$

17. $b = 7$, $\alpha = 30°$, $\gamma = 90°$

18. $a = 3$, $\beta = 45°$, $\gamma = 90°$

19. $\alpha = 93°$, $\beta = 39°$, $\gamma = 48°$

20. $\alpha = 29°$, $\beta = 47°$, $\gamma = 104°$

21. $\alpha = 51.3°$, $\beta = 51.3°$, $\gamma = 77.4°$

22. $\alpha = 25°$, $\beta = 29°$, $\gamma = 126°$

23. $\alpha = 75°$, $\beta = 57°$, $\gamma = 48°$

24. $\alpha = 46°$, $\beta = 75°$, $\gamma = 59°$

25. no triangle   **26.** no triangle

27. $\alpha = 67°$, $\beta = 23°$, $\gamma = 90°$

28. $\alpha = 39°$, $\beta = 51°$, $\gamma = 90°$

29. $\gamma = 105°$, $b = 5$, $c = 9$

30. $\alpha = 149.1°$, $\beta = 17.6°$, $c = 8.5$

31. $\beta = 12°$, $\gamma = 137°$, $c = 16$

32. $\alpha = 53°$, $\gamma = 67°$, $b = 13$

33. $\alpha = 66°$, $\beta = 77°$, $\gamma = 37°$

34. $\alpha = 31°$, $b = 2$, $c = 1$

35. $\gamma = 2°$, $\alpha = 168°$, $a = 13$

36. $\beta_1 = 116°$, $\gamma_1 = 39°$, $b_1 = 13$
$\beta_2 = 14°$, $\gamma_2 = 141°$, $b_2 = 3$

37. $b = 11.16$, $\alpha = 42.40°$, $\gamma = 85.91°$

38. $c = 2.48$, $\alpha = 59.39°$, $\beta = 59.58°$

39. $\alpha = 46.76°$, $\beta = 58.45°$, $\gamma = 74.79°$

40. $\alpha = 72.71°$, $\beta = 49.21°$, $\gamma = 58.08°$

41. $\alpha = 1.09$, $\beta = 61.327°$, $\gamma = 47.460°$

42. $\alpha = 70.88°$, $\gamma = 70.88°$, $b = 3.645$

43. 2710 mi **44.** 2433 mi **45.** 1280 mi

46. 69.72° clockwise of north **47.** 63.7 ft

48. 8.8 ft **49.** 16 ft **50.** 49 ft

51. 0.8 mi **52.** 0.9 mi **53.** 26°

54. 27° **55.** 21.67° **56.** 50°

57. $\angle QRP = 151.03°$, $\angle RQP = 3.97°$

58. 76.85° **59.** 0.5° **60.** 430 yards

61. about 83° **62.** 16 guests

63. Should have used the smaller angle β in Step 2.

64. Should have used the smaller angle α in Step 1.

65. false **66.** true **67.** true

68. false **69.** true

70. – 72. See Instructor's Solution Manual

73. $\sqrt{\dfrac{1 - \cos\left(2\cos^{-1}\left(\frac{1}{4}\right)\right)}{2}}$

74. $\sqrt{\dfrac{1 + \cos\left(2\cos^{-1}\left(\frac{1}{4}\right)\right)}{1 - \cos\left(2\cos^{-1}\left(\frac{1}{4}\right)\right)}}$

75. – 80. See Instructor's Solution Manual

Section 8.3

1. 55.4 **2.** 10.4 **3.** 0.5 **4.** 4

5. 23.6 **6.** 44.3 **7.** 6.4 **8.** 4.42

9. 4408.4 **10.** 0.06 **11.** 9.6 **12.** 2.0

13. 97.4 **14.** 0.4 **15.** 25.0 **16.** 180

17. 26.7 **18.** 992.2 **19.** 111.64

20. 6885.17 **21.** 111,632,076 **22.** 1.2

23. no triangle **24.** no triangle

25. 13.15 **26.** 4.11 **27.** 174.76

28. 53.11 **29.** 11.98 **30.** 22.95

31. 19.21 **32.** 0.55 **33.** 312,297 nm^2

34. 328,885 nm^2 **35.** 10,591 ft^2

36. \$32 **37.** 16° or 164° **38.** 25°

39. 312.4 mi^2 **40.** 3068 sq ft

41. a. 41,842 sq ft **b.** \$89,123

42. a. 65,007 sq ft **b.** \$128,064

43. 47,128 sq ft **44.** 61,436 sq ft

45. 23.38 ft^2 **46.** 192.36 sq in

47. Area of $PQ_1Q_2 = 12$

Area of $PQ_2Q_3 = 22.55$

Area of $Q_1Q_2Q_3 = 15.21$

Area of $PQ_1Q_3 = 49.76$

The sum of the areas of the smaller three triangles does not equal the area of the outer triangle.

48. 47.29 sq ft **49.** 10.86 **50.** 38.504 sq units

51. Semiperimeter is *half* the perimeter.

52. $\sqrt{s(s-a)(s-b)(s-c)}$

53. true **54.** true **55.** true

56. true **57.** false

58. – 60. See Instructor's Solution Manual

61. 0.69 **62.** 11.06 sq. units

63. – 70. See Instructor's Solution Manual

Section 8.4

1. $\sqrt{13}$ **2.** $\sqrt{74}$ **3.** $5\sqrt{2}$ **4.** 5

5. 25 **6.** 10 **7.** $\sqrt{73}$; $\theta = 69.4°$

8. $\sqrt{65}$; $\theta = 60.3°$

9. $\sqrt{26}$; $\theta = 348.7°$

10. $2\sqrt{10}$; $\theta = 198.4°$

11. $\sqrt{17}$; $\theta = 166.0°$

12. $3\sqrt{5}$; $\theta = 153.4°$

13. 8; $\theta = 180°$ **14.** 7; $\theta = 90°$

15. $2\sqrt{3}$; $\theta = 60°$ **16.** $5\sqrt{2}$; $\theta = 225°$

17. $\langle -2, -2 \rangle$ **18.** $\langle -6, 8 \rangle$

19. $\langle -12, 9 \rangle$ **20.** $\langle 8, -6 \rangle$

21. $\langle 0, -14 \rangle$ **22.** $\langle -10, -10 \rangle$

23. $\langle -36, 48 \rangle$ **24.** $\langle -30, 33 \rangle$

25. $\langle 6.3, 3.0 \rangle$ **26.** $\langle 1.3, 4.8 \rangle$

27. $\langle -2.8, 15.8 \rangle$ **28.** $\langle -7.5, -2.7 \rangle$

29. $\langle 2.6, -3.1 \rangle$ **30.** $\langle -5.7, -5.7 \rangle$

31. $\langle 8.2, -3.8 \rangle$ **32.** $\langle 2.1, -2.1 \rangle$

33. $\langle -1, 1.7 \rangle$ **34.** $\langle 5.2, -3 \rangle$

35. $\langle -\frac{5}{13}, -\frac{12}{13} \rangle$ **36.** $\langle \frac{3}{5}, \frac{4}{5} \rangle$

37. $\langle \frac{60}{11}, \frac{11}{61} \rangle$ **38.** $\langle \frac{-7}{25}, \frac{24}{25} \rangle$

39. $\langle \frac{24}{25}, -\frac{7}{25} \rangle$ **40.** $\langle \frac{-5}{13}, \frac{12}{13} \rangle$

41. $\langle -\frac{3}{5}, -\frac{4}{5} \rangle$ **42.** $\langle \frac{40}{41}, \frac{-9}{41} \rangle$

43. $\langle \frac{\sqrt{10}}{10}, \frac{3\sqrt{10}}{10} \rangle$ **44.** $\langle \frac{-2\sqrt{5}}{5}, \frac{-\sqrt{5}}{5} \rangle$

45. $7\mathbf{i} + 3\mathbf{j}$ **46.** $-2\mathbf{i} + 4\mathbf{j}$ **47.** $5\mathbf{i} - 3\mathbf{j}$

48. $-6\mathbf{i} - 2\mathbf{j}$ **49.** $-\mathbf{i} + 0\mathbf{j}$ **50.** $0\mathbf{i} + 2\mathbf{j}$

51. $2\mathbf{i} + 0\mathbf{j}$ **52.** $7\mathbf{i} - 7\mathbf{j}$ **53.** $-5\mathbf{i} + 5\mathbf{j}$

54. $3\mathbf{i} - 4\mathbf{j}$ **55.** $7\mathbf{i} + 0\mathbf{j}$ **56.** $0\mathbf{i} - 3\mathbf{j}$

57. Vertical: 1100 ft/sec
Horizontal: 1905 ft/sec

58. Perpendicular = 38.30 lb
Parallel = 32.14 lb

59. 2801 lb **60.** 1814 lb

61. 11.7 mph; 31° west of due north

62. 12.4 mph; 76° west of due north

63. 52.41° east of north; 303 mph

64. 25.71° east of north; 401 mph

65. 250 lb **66.** 87 lb

67. Vertical: 51.4 ft/sec
Horizontal: 61.3 ft/sec

68. Vertical: 8.7 ft/sec
Horizontal: 99.6 ft/sec

69. 29.93 yd **70.** −11.57° **71.** 10.9°

72. $c = 458$ lb **73.** 1156 lb **74.** 25.5°

75. Magnitude: 351.16; Angle: 23.75° from 180° N force

76. Magnitude: 145.47; Angle: 20.10° from 50° N force.

77. Tension: $5\sqrt{2}$; $|\vec{u}| = 5$

78. Tension: $\dfrac{16\sqrt{3}}{3}$; $|\vec{u}| = \dfrac{8\sqrt{3}}{3}$

79. a. $\vec{0}$ **b.** $10 + 3\sqrt{2} + \sqrt{10}$ units.

80. a. $\overrightarrow{OP}$ **b.** $19 + \sqrt{61} + \sqrt{164}$ units

81. 8.97 Nm **82.** 34 Nm

83. 31.95 Nm **84.** 0°

85. Magnitude: 8.67; Direction: 18.05° counterclockwise of S

86. Magnitude: 20.42; Direction 7.56° counterclockwise of S

87. 19° **88.** 22°

89. Magnitude is never negative. Should not have factored out the negative but instead squared it in finding the magnitude.

90. Inverse tangent gives an angle whose terminal side is in QI. As such, we must add 180° to obtain the QIII solution.

91. false **92.** false **93.** true

94. true **95.** vector **96.** scalar

97. $\sqrt{a^2 + b^2}$

98. – 100. See Instructor's Solution Manual

101. $\langle -6, 4 \rangle = \frac{1}{2}\langle -8, 4 \rangle - 2\langle 1, -1 \rangle$

102. – 104. See Instructor's Solution Manual

105. Directly

106. Enter the following to compute the given quantity: $-9([[8][-5]] - 2[[-7][11]])$. The output is $[[-198][243]]$.

107.

108. $[[-9/41][-40/41]]$ **109.** 183, 100.4°

110. 70, −73.4°

Section 8.5

1. 2 **2.** 6 **3.** −3 **4.** 9

5. 42 **6.** −3 **7.** 11 **8.** −15

9. $-13a$ **10.** $-7xy$ **11.** −1.4 **12.** −1080

13. 98° **14.** 49° **15.** 109° **16.** 73°

17. 3° **18.** 135° **19.** 30° **20.** 60°

21. 105° 22. 85° 23. 180° 24. 90°
25. no 26. no 27. yes 28. yes
29. no 30. no 31. yes 32. yes
33. yes 34. no 35. yes 36. no
37. 400 ft-lb 38. 525 ft-lb
39. 80,000 ft-lb 40. 125,000 ft-lb
41. 1299 ft-lb 42. 37,588 ft-lb
43. 148 ft-lb 44. 155 ft-lb
45. 1607 lb 46. 229,813 ft-lb
47. 694,593 ft-lb 48. 6946 lb
49. $\vec{u} \cdot \vec{v} = 49{,}300$, and it represents the total
cost of buying the prescribed number of
10-packs of both types of battery.
50. – 54. See Instructor's Solution Manual
55. a. $v_1 = -14.72$ b. 64.57° 56. 25.43°
57. $\vec{n} = \langle r, 0 \rangle$; $\vec{u} \cdot \vec{n} = r|\vec{u}|$
58. $\vec{n} = \langle -r, 0 \rangle$; $\vec{u} \cdot \vec{n} = -|\vec{u}|r$
59. –2 60. 2
61. The dot product of two vectors is a scalar
(not a vector). Should have summed the
products of components.
62. The dot product is the sum of the products
of corresponding components
(first) × (first) + (second) × (second).
63. false 64. true 65. true
66. true 67. 17
68. – 74. See Instructor's Solution Manual
75. a. $-2\vec{u}$ b. $-c\vec{u}$ 76. a. $2\vec{u}$ b. $c\vec{u}$
77. Any vector that is perpendicular to $\vec{u}$.
78. true
79. –6 is minimum and 6 is maximum.
80. See Instructor's Solution Manual
81. –1083 82. –69,665 83. 31.43°
84. 65.0° 85. 47°

Section 8.6

The answers to Exercises 1 to 8 are all plotted
on the same graph, below.

9. $\sqrt{2}\left[\cos\left(\dfrac{7\pi}{4}\right) + i\sin\left(\dfrac{7\pi}{4}\right)\right] =$

$\sqrt{2}(\cos 315° + i\sin 315°)$

10. $2\sqrt{2}\left[\cos\left(\dfrac{\pi}{4}\right) + i\sin\left(\dfrac{\pi}{4}\right)\right] =$

$2\sqrt{2}(\cos 45° + i\sin 45°)$

11. $2\left[\cos\left(\dfrac{\pi}{3}\right) + i\sin\left(\dfrac{\pi}{3}\right)\right] =$

$2(\cos 60° + i\sin 60°)$

12. $2\sqrt{3}\left[\cos\left(\dfrac{7\pi}{6}\right) + i\sin\left(\dfrac{7\pi}{6}\right)\right] =$

$2\sqrt{3}(\cos 210° + i\sin 210°)$

13. $4\sqrt{2}\left[\cos\left(\dfrac{3\pi}{4}\right) + i\sin\left(\dfrac{3\pi}{4}\right)\right] =$

$4\sqrt{2}(\cos 135° + i\sin 135°)$

14. $\sqrt{10}\left[\cos\left(\dfrac{7\pi}{4}\right) + i\sin\left(\dfrac{7\pi}{4}\right)\right] =$

$\sqrt{10}(\cos 315° + i\sin 315°)$

15. $2\sqrt{3}\left[\cos\left(\dfrac{5\pi}{3}\right) + i\sin\left(\dfrac{5\pi}{3}\right)\right] =$

$2\sqrt{3}(\cos 300° + i\sin 300°)$

16. $2\left[\cos\left(\dfrac{5\pi}{6}\right) + i\sin\left(\dfrac{5\pi}{6}\right)\right] =$

$2(\cos 150° + i\sin 150°)$

17. $3(\cos 0 + i\sin 0) = 3(\cos 0° + i\sin 0°)$
18. $2(\cos \pi + i\sin \pi) = 2(\cos 180° + i\sin 180°)$

19. $\dfrac{\sqrt{2}}{2}\left[\cos\left(\dfrac{5\pi}{4}\right) + i\sin\left(\dfrac{5\pi}{4}\right)\right] =$

$\dfrac{\sqrt{2}}{2}(\cos 225° + i\sin 225°)$

20. $\dfrac{\sqrt{2}}{6}\left[\cos\left(\dfrac{7\pi}{4}\right) + i\sin\left(\dfrac{7\pi}{4}\right)\right] =$

$\dfrac{\sqrt{2}}{6}(\cos 315° + i\sin 315°)$

21. $5.32(\cos 0 + i\sin 0)$

22. $6.48(\cos\pi + i\sin\pi) =$
$6.48(\cos 180° + i\sin 180°)$

23. $\sqrt{58}(\cos 293.2° + i\sin 293.2°)$

24. $\sqrt{13}(\cos 56.3° + i\sin 56.3°)$

25. $\sqrt{61}(\cos 140.2° + i\sin 140.2°)$

26. $5(\cos 216.9° + i\sin 216.9°)$

27. $13(\cos 112.6° + i\sin 112.6°)$

28. $25(\cos 16.3° + i\sin 16.3°)$

29. $10(\cos 323.1° + i\sin 323.1°)$

30. $5(\cos 126.9° + i\sin 126.9°)$

31. $\dfrac{\sqrt{13}}{4}(\cos 123.7° + i\sin 123.7°)$

32. $\dfrac{\sqrt{509}}{8}(\cos 257.20° + i\sin 257.20°)$

33. $5.59(\cos 24.27° + i\sin 24.27°)$

34. $2.01(\cos 333.43° + i\sin 333.43°)$

35. $\sqrt{17}(\cos 212.84° + i\sin 212.84°)$

36. $\dfrac{\sqrt{85}}{2}(\cos 159.44° + i\sin 159.44°)$

37. $4.54(\cos 332.31° + i\sin 332.31°)$

38. $1.78(\cos 356.14° + i\sin 356.14°)$

39. -5 **40.** $-\sqrt{2} + \sqrt{2}i$

41. $\sqrt{2} - \sqrt{2}i$ **42.** $-3i$

43. $-2 - 2\sqrt{3}i$ **44.** $2\sqrt{3} + 2i$

45. $-\dfrac{3}{2} + \dfrac{\sqrt{3}}{2}i$ **46.** $\dfrac{3}{2} - \dfrac{\sqrt{3}}{2}i$

47. $1 + i$ **48.** $-\sqrt{3} + i$

49. $2.1131 - 4.5315i$

50. $3.2766 + 2.2943i$

51. $-0.5209 + 2.9544i$

52. $-2.0521 - 5.6382i$

53. $5.3623 - 4.4995i$

54. $-3.8302 + 3.2139i$

55. $-2.8978 + 0.7765i$

56. $-0.4450 + 1.9499i$

57. $0.6180 - 1.9021i$

58. $1.6617 + 3.6385i$

59. $100(\cos 0° + i\sin 0°) +$
$120(\cos 30° + i\sin 30°)$; 212.6 pounds

60. $40(\cos 0° + i\sin 0°) +$
$50(\cos 45° + i\sin 45°)$; 83.2 pounds

61. $80(\cos 0° + i\sin 0°) +$
$150(\cos 30° + i\sin 30°)$; 19.7°

62. $20(\cos 0° + i\sin 0°) +$
$60(\cos 60° + i\sin 60°)$; 46.1°

63. $\vec{u} = 8(\cos 150° + i\sin 150°)$;
$\vec{v} = 6(\cos 45° + i\sin 45°)$;
$\vec{w} = |\vec{w}|(\cos\theta + i\sin\theta)$, where
$\vec{w} = \sqrt{(4\sqrt{3} - 3\sqrt{2})^2 + (-4 - 3\sqrt{2})^2}$;
$\theta = \tan^{-1}\left(\dfrac{-4 - 3\sqrt{2}}{4\sqrt{3} - 3\sqrt{2}}\right)$

64. $\vec{u} = 8(\cos 150° + i\sin 150°)$;
$\vec{v} = 12(\cos 90° + i\sin 90°)$;
$\vec{w} = 6(\cos 45° + i\sin 45°)$;
$c = 8$; $\vec{z} = |\vec{z}|(\cos\theta + i\sin\theta)$, where
$|\vec{z}| = \sqrt{(4\sqrt{3} - 3\sqrt{2})^2 + (-16 - 3\sqrt{2})^2}$;
$\theta = \tan^{-1}\left(\dfrac{-16 - 3\sqrt{2}}{4\sqrt{3} - 3\sqrt{2}}\right)$;
$\vec{z} = 20.42(\cos 82.44° - i\sin 82.44°)$

65. $\vec{R} = |\vec{R}|(\cos\theta + i\sin\theta)$, where
$|\vec{R}| = \sqrt{(100\sqrt{2} + 180)^2 + (100\sqrt{2})^2}$;
$\theta = \tan^{-1}\left(\dfrac{100\sqrt{2}}{100\sqrt{2} + 180}\right)$;
$|\vec{R}| = 351.16(\cos(23.75°) + i\sin(23.75°))$

66. $\vec{R} = |\vec{R}|(\cos\theta + i\sin\theta)$, where
$|\vec{R}| = \sqrt{(50\sqrt{3} + 50)^2 + (50)^2}$;
$\theta = \tan^{-1}\left(\dfrac{50}{50\sqrt{3} + 50}\right)$;
$|\vec{R}| = 145.47(\cos 20.10° + i\sin 20.10°)$

67. The point is in QIII (not QI).

68. The point is in QII (not QIV).

69. true **70.** true

71. true **72.** false

73. $0°$ **74.** $90°$ **75.** $|b|$ **76.** $|a|$

77. $a\sqrt{5}(\cos 296.6° + i\sin 296.6°)$

78. $5a(\cos 233.1° + i\sin(233.1°))$

79. $\pi\left(\dfrac{1}{2} - \dfrac{\sqrt{3}}{2}i\right)$ **80.** $-\sqrt{2} + i\sqrt{6}$

81. $z = 8.79(\cos 28° + i\sin 28°)$

82. $z = 4.99(\cos 248.13° + i\sin 248.13°)$

83. $1.414, 45°, 1.414(\cos 45° + i\sin 45°)$

84. $1.414, 315°, 1.414(\cos 315° + i\sin 315°)$

85. $2.236(\cos 26.6° + i\sin 26.6°)$

86. $\dfrac{3\sqrt{2}}{2} + \dfrac{3\sqrt{2}}{2}i$ **87.** $35,323°$

88. $11, 114.6°, 11(\cos 114.6° + i\sin 114.6°)$

Section 8.7

1. $-6 + 6\sqrt{3}i$ **2.** $-5\sqrt{3} + 5i$

3. $-4\sqrt{2} - 4\sqrt{2}i$ **4.** $6 - 6\sqrt{3}i$

5. $0 + 8i$ **6.** $0 - 15i$

7. $\dfrac{9\sqrt{2}}{2} + \dfrac{9\sqrt{2}}{2}i$ **8.** $\dfrac{5}{2} + \dfrac{5\sqrt{3}}{2}i$

9. $0 + 12i$ **10.** $15 + 15\sqrt{3}i$

11. $\dfrac{3}{2} + \dfrac{3\sqrt{3}}{2}i$ **12.** $2\sqrt{2} + 2\sqrt{2}i$

13. $-\sqrt{2} + \sqrt{2}i$ **14.** $-\dfrac{\sqrt{2}}{2} - \dfrac{\sqrt{2}}{2}i$

15. $0 - 2i$ **16.** $0 + 2i$

17. $\dfrac{3}{2} + \dfrac{3\sqrt{3}}{2}i$ **18.** $\sqrt{2} + \sqrt{2}i$

19. $-\dfrac{5}{2} - \dfrac{5\sqrt{3}}{2}i$ **20.** $1 + \sqrt{3}i$

21. $4 - 4i$ **22.** $-4 + 0i$

23. $-64 + 0i$ **24.** $-128 + 128\sqrt{3}i$

25. $-8 + 8\sqrt{3}i$ **26.** $-16 - 16\sqrt{3}i$

27. $1,048,576 + 0i$ **28.** $0 - 1,889,568i$

29. $-1,048,576\sqrt{3} - 1,048,576i$

30. $-5,000,000 + 5,000,000\sqrt{3}i$

31. $2(\cos 150° + i\sin 150°)$ and $2(\cos 330° + i\sin 330°)$

32. $2(\cos 30° + i\sin 30°)$ and $2(\cos 210° + i\sin 210°)$

33. $\sqrt{6}(\cos 157.5° + i\sin 157.5°),$ $\sqrt{6}(\cos 337.5° + i\sin 337.5°)$

34. $\sqrt{2}(\cos 67.5° + i\sin 67.5°)$ and $\sqrt{2}(\cos 247.5° + i\sin 247.5°)$

35. $2(\cos 20° + i\sin 20°),$ $2(\cos 140° + i\sin 140°),$ $2(\cos 260° + i\sin 260°)$

36. $3(\cos 40° + i\sin 40°),$
$3(\cos 160° + i\sin 160°),$
$3(\cos 280° + i\sin 280°)$

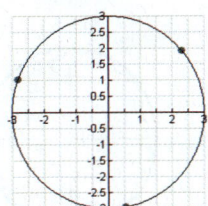

37. $\sqrt[3]{2}(\cos 110° + i\sin 110°),$
$\sqrt[3]{2}(\cos 230° + i\sin 230°),$
$\sqrt[3]{2}(\cos 350° + i\sin 350°)$

38. $2(\cos 15° + i\sin 15°),$
$2(\cos 135° + i\sin 135°),$
$2(\cos 255° + i\sin 255°)$

39. $2(\cos 78.75° + i\sin 78.75°),$
$2(\cos 168.75° + i\sin 168.75°),$
$2(\cos 258.75° + i\sin 258.75°),$
$2(\cos 348.75° + i\sin 348.75°)$

40. $2(\cos 33.75° + i\sin 33.75°),$
$2(\cos 123.75° + i\sin 123.75°),$
$2(\cos 213.75° + i\sin 213.75°),$
$2(\cos 303.75° + i\sin 303.75°)$

41. $x = \pm 2, \ \ x = \pm 2i$

42. $2, \ -1 + \sqrt{3}i, \ -1 - \sqrt{3}i$

43. $-2, \ 1 - \sqrt{3}i, \ 1 + \sqrt{3}i$

44. $-1, \ \dfrac{1}{2} - \dfrac{\sqrt{3}}{2}i, \ \dfrac{1}{2} + \dfrac{\sqrt{3}}{2}i$

45. $\sqrt{2} - \sqrt{2}i, \ \sqrt{2} + \sqrt{2}i, \ -\sqrt{2} - \sqrt{2}i,$
$-\sqrt{2} + \sqrt{2}i$

46. $i, \ -i, \ \dfrac{\sqrt{3}}{2} + \dfrac{1}{2}i, \ \dfrac{\sqrt{3}}{2} - \dfrac{1}{2}i, \ -\dfrac{\sqrt{3}}{2} + \dfrac{1}{2}i,$
$-\dfrac{\sqrt{3}}{2} - \dfrac{1}{2}i$

47. $1, \ -1, \ \dfrac{1}{2} + \dfrac{\sqrt{3}}{2}i, \ \dfrac{1}{2} - \dfrac{\sqrt{3}}{2}i, \ -\dfrac{1}{2} + \dfrac{\sqrt{3}}{2}i,$
$-\dfrac{1}{2} - \dfrac{\sqrt{3}}{2}i$

48. $-\dfrac{1}{2}i, \ \dfrac{1}{2}i$

49. $-\dfrac{\sqrt{2}}{2} + \dfrac{\sqrt{2}}{2}i, \ \dfrac{\sqrt{2}}{2} - \dfrac{\sqrt{2}}{2}i$

50. $\dfrac{\sqrt{2}}{2} + \dfrac{\sqrt{2}}{2}i, \ -\dfrac{\sqrt{2}}{2} - \dfrac{\sqrt{2}}{2}i$

51. $\sqrt[4]{2}\left[\cos\left(\dfrac{\pi}{8} + \dfrac{\pi k}{2}\right) + i\sin\left(\dfrac{\pi}{8} + \dfrac{\pi k}{2}\right)\right],$
$k = 0, 1, 2, 3$

52. $\sqrt[4]{2}\left[\cos\left(\dfrac{3\pi}{8} + \dfrac{\pi k}{2}\right) + i\sin\left(\dfrac{3\pi}{8} + \dfrac{\pi k}{2}\right)\right],$
$k = 0, 1, 2, 3$

53. $2\left[\cos\left(\dfrac{\pi}{5}+\dfrac{2\pi k}{5}\right)+i\sin\left(\dfrac{\pi}{5}+\dfrac{2\pi k}{5}\right)\right]$,

$k = 0, 1, 2, 3, 4$

54. $2\left[\cos\left(\dfrac{2\pi k}{5}\right)+i\sin\left(\dfrac{2\pi k}{5}\right)\right]$,

$k = 0, 1, 2, 3, 4$

55. $\pi^2\left[\cos\left(\dfrac{\pi}{14}+\dfrac{2\pi k}{7}\right)+i\sin\left(\dfrac{\pi}{14}+\dfrac{2\pi k}{7}\right)\right]$,

$k = 0, 1, 2, 3, 4, 5, 6$

56. $\pi^2\left[\cos\left(\dfrac{3\pi}{14}+\dfrac{2\pi k}{7}\right)+i\sin\left(\dfrac{3\pi}{2}+\dfrac{2\pi k}{7}\right)\right]$,

$k = 0, 1, 2, 3, 4, 5, 6$

57. $(\cos 45° + i\sin 45°)$,

$(\cos 117° + i\sin 117°)$, $(\cos 189° + i\sin 189°)$,

$(\cos 261° + i\sin 261°)$, $(\cos 333° + i\sin 333°)$

58. $2(\cos 22.5° + i\sin 22.5°)$,

$2(\cos 112.5° + i\sin 112.5°)$,

$2(\cos 202.5° + i\sin 202.5°)$,

$2(\cos 292.5° + i\sin 292.5°)$

59. $\left[\cos\left(5\dfrac{\pi}{18}+\dfrac{\pi k}{3}\right)+i\sin\left(5\dfrac{\pi}{18}+\dfrac{\pi k}{3}\right)\right]$,

$k = 0, 1, 2, 3, 4, 5$

60. $\sqrt[8]{2}=\left[\cos\left(\dfrac{\pi}{16}+\dfrac{\pi k}{4}\right)+i\sin\left(\dfrac{\pi}{16}+\dfrac{\pi k}{4}\right)\right]$,

$k = 0, 1, 2, 3, 4, 5, 6, 7$

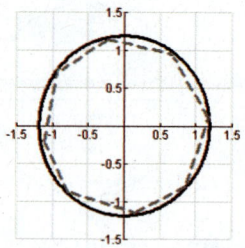

61. Reversed the order of angles being subtracted.

62. Should use the product formula, namely $z_1 z_2 = r_1 r_2[\cos(\theta_1 + \theta_2)+i\sin(\theta_1 + \theta_2)]$, to multiply.

63. Use De Moivre's theorem. In general, $(a + b)^6 \neq a^6 + b^6$.

64. Only found one of the five solutions.

65. true **66.** true

67. false **68.** true

69. true

70. – 78. See Instructor's Solution Manual

79. $2^{\frac{n+m}{2}} e^{\frac{\pi}{4}(m-n)i}$ **80.** $2^{\frac{n-m}{2}} e^{\frac{\pi}{4}(m+n)i}$

81. $\cos 66° + i\sin 66°$, $\cos 138° + i\sin 138°$,

$\cos 210° + i\sin 210°$, $\cos 282° + i\sin 282°$,

$\cos 354° + i\sin 354°$

82. $\cos\left(\dfrac{3\pi}{16}+\dfrac{k\pi}{2}\right)+i\sin\left(\dfrac{3\pi}{16}+\dfrac{k\pi}{2}\right)$,

$k = 0, 1, 2, 3$

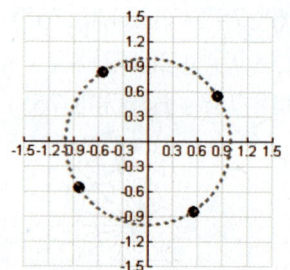

83. $\cos 40° + i\sin 40°$, $\cos 100° + i\sin 100°$,

$\cos 160° + i\sin 160°$, $\cos 220° + i\sin 220°$,

$\cos 280° + i\sin 280°$, $\cos 340° + i\sin 340°$

84. $\sqrt{2}\left[\cos\left(\dfrac{3\pi}{20}+\dfrac{2k\pi}{5}\right)+i\sin\left(\dfrac{3\pi}{20}+\dfrac{2k\pi}{5}\right)\right]$,

$k = 0, 1, 2, 3, 4$

85. $3(\cos 15° + \sin 15°)$, $3(\cos 135° + \sin 135°)$,

$3(\cos 255° + \sin 255°)$

86. $(11.712)^{\frac{1}{5}}\left[\cos\left(\dfrac{\pi}{20}+\dfrac{2k\pi}{5}\right)+\right.$

$\left. i\sin\left(\dfrac{\pi}{20}+\dfrac{2k\pi}{5}\right)\right]$, $k = 0, 1, 2, 3, 4$

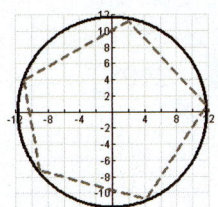

Section 8.8

The answers to Exercises 1 to 10 are all plotted on the same graph, below.

11. $\left(4, \dfrac{\pi}{3}\right)$ **12.** $\left(3\sqrt{2}, \dfrac{7\pi}{4}\right)$ **13.** $\left(2, \dfrac{4\pi}{3}\right)$

14. $\left(12, \dfrac{\pi}{3}\right)$ **15.** $\left(4\sqrt{2}, \dfrac{3\pi}{4}\right)$ **16.** $\left(\sqrt{2}, \dfrac{\pi}{2}\right)$

17. $(3, 0)$ **18.** $\left(7\sqrt{2}, \dfrac{5\pi}{4}\right)$ **19.** $\left(2, \dfrac{7\pi}{6}\right)$

20. $\left(4, \dfrac{11\pi}{6}\right)$ **21.** $(2, -2\sqrt{3})$ **22.** $(-\sqrt{2}, \sqrt{2})$

23. $\left(\dfrac{\sqrt{3}}{2}, -\dfrac{1}{2}\right)$ **24.** $(-\sqrt{2}, \sqrt{2})$ **25.** $(0, 0)$

26. $(6, 0)$ **27.** $(-1, -\sqrt{3})$ **28.** $\left(\dfrac{3\sqrt{3}}{2}, -\dfrac{3}{2}\right)$

29. $\left(\dfrac{\sqrt{2}}{2}, -\dfrac{\sqrt{2}}{2}\right)$ **30.** $\left(\dfrac{5\sqrt{2}}{2}, -\dfrac{5\sqrt{2}}{2}\right)$

31. d   **32.** b   **33.** a   **34.** c

35.

36.

37.

38.

39.

40.

41.

42.

43.

44.

45.

46.

47.

48.

49.

50.

51. Line: $y = -2x + 1$

52. Line: $y = 3x + 2$

53. Circle: $(x - 1)^2 + y^2 = 9$

54. Parabola: $y = x^2 + 2$

55.

56.

57.

58.

59.

60.

61.

62.

63.

Archimedes Spiral

64.

65.

Multiplier 4
Multiplier $\frac{1}{4}$

66.

67.

68.

69. For (a)–(c) all three graphs generate the same set of points, as seen below:

Note that all three graphs are figure eights. Extending the domain in (b) results in twice as fast movement, while doing so in (c) results in movement that is four times as fast.

70. $r^2 = 12\cos(8\theta)$ **71.** 6 times

72.

73. a. $r = 8\cos 3\theta$
   b. 50 times, because the period is 2π.
74. a. $r = 8\cos 4\theta$
   b.

75. The point is in QIII; the angle found was the reference angle (needed to add π).
76. The point is in QII; the angle found was the reference angle (needed to add π).

77. true   **78.** false   **79.** $r = \dfrac{a}{\cos\theta}$

80. $r = \dfrac{b}{\sin\theta}$ **81.** $(-a, \theta \pm 180°)$

82. $\left(\sqrt{a^2 + b^2},\, 180° + \tan^{-1}\left(-\dfrac{b}{a}\right)\right)$

83. The graphs intersect when
$$\theta = \frac{\pi}{3}, \frac{2\pi}{3}, \frac{4\pi}{3}, \frac{5\pi}{3}.$$

84. $\left(x - \dfrac{b}{2}\right)^2 + \left(y - \dfrac{a}{2}\right)^2 = \dfrac{a^2 + b^2}{4}$. This graph represents a circle.

85. $x^3 - y^3 - 2axy = 0$

86. $r = 10\sin(9\theta),\ 0 \le \theta \le 2\pi$

87. It is a circle of radius a and center $(a\cos b, a\sin b)$.

88. $\dfrac{2\pi}{b}$

89. The inner loop is generated beginning with $\theta = \frac{\pi}{2}$ and ending with $\frac{3\pi}{2}$.

90. The patel is generated beginning with $\theta = 0$ and ending with $\frac{\pi}{3}$.

91. The very tip of the inner loop begins with $\theta = \frac{\pi}{2}$, then it crosses the origin (the first time) at $\theta = \cos^{-1}\left(-\frac{1}{3}\right)$, winds around, and eventually ends with $\theta = \frac{3\pi}{2}$.

92. $\dfrac{3\pi}{8}, \dfrac{7\pi}{8}, \dfrac{11\pi}{8}, \dfrac{15\pi}{8}$

93. 67.5°, 157.5°, 247.5°, 337.5°

94. $\dfrac{\pi}{9} + \dfrac{2n\pi}{3},\ \dfrac{5\pi}{9} + \dfrac{2n\pi}{3}$, n an integer

Review Exercises

1. $\gamma = 150°,\ c = 12,\ b = 8$

2. $\alpha = 80°,\ a = 15,\ c = 13$

3. $\gamma = 130°,\ a = 1,\ b = 9$

4. $\alpha = 50°,\ b = 23,\ c = 25$

5. $\beta = 158°,\ a = 11,\ b = 22$

6. $\alpha = 110°,\ a = 22,\ c = 18$

7. $\beta = 90°,\ a = \sqrt{2},\ c = \sqrt{2}$

8. $\gamma = 100°,\ a = 15,\ b = 6$

9. $\beta = 146°,\ b = 266,\ c = 178$

10. $\alpha = 51°,\ b = 30,\ c = 14$

11. $\beta = 26°,\ \gamma = 134°,\ c = 15$ or $\beta = 154°,\ \gamma = 6°,\ c = 2$

12. $\alpha = 144°,\ \gamma = 20°,\ a = 51$ or $\alpha = 4°,\ \gamma = 160°,\ a = 6$

13. $\beta = 127°,\ \gamma = 29°,\ b = 20$ or $\beta = 5°,\ \gamma = 151°,\ b = 2$

14. $\alpha = 154°,\ \gamma = 14°,\ a = 211$ or $\alpha = 2°,\ \gamma = 166°,\ a = 17$

15. no triangle   **16.** $\alpha = 5°,\ \beta = 10°,\ a = 1$

17. $\beta = 15°,\ \gamma = 155°,\ c = 10$ or $\beta = 165°,\ \gamma = 5°,\ c = 2$

18. $\alpha = 6°,\ \beta = 170°,\ b = 62$ or $\alpha = 174°,\ \beta = 2°,\ b = 13$

19. 12.2 mi   **20.** 12 mi

21. $\alpha = 42°,\ \beta = 88°,\ c = 46$

22. $\beta = 22°,\ \gamma = 18°,\ a = 25$

23. $\alpha = 51°,\ \beta = 54°,\ \gamma = 75°$

24. $\alpha = 48°,\ \beta = 48°,\ \gamma = 84°$

25. $\alpha = 42°,\ \beta = 48°,\ \gamma = 90°$

26. $\alpha = 10°,\ \beta = 80°,\ \gamma = 90°$

27. $\beta = 28°,\ \gamma = 138°,\ a = 4$

28. $\alpha = 28°,\ \beta = 72°,\ c = 12$

29. $\beta = 68°,\ \gamma = 22°,\ a = 11$

30. $\alpha = 44°,\ \beta = 61°,\ c = 6$

31. $\alpha = 51°,\ \beta = 59°,\ \gamma = 70°$

32. $\alpha = 53°,\ \beta = 61°,\ \gamma = 66°$

33. $\beta = 37°,\ \gamma = 43°,\ a = 26$

34. $\alpha = 85.5°,\ \gamma = 85.5°,\ b = 4$

35. $\beta = 4°,\ \gamma = 166°,\ a = 28$

36. $\alpha = 116°,\ \beta = 44°,\ \gamma = 20°$

37. no triangle   **38.** no triangle

39. $\beta = 10°,\ \gamma = 155°,\ c = 10.3$

40. $\alpha = 102°,\ \gamma = 43°,\ a = 9$

41. 141.8   **42.** 48.89   **43.** 51.5

44. 240.6   **45.** 89.8   **46.** 384

47. 41.7   **48.** 682.0   **49.** 5.2 in.

50. 7.5 in.   **51.** 13   **52.** 5

53. 13   **54.** 10   **55.** 26; 112.6°

56. 13; 247.4°   **57.** 20; 323.1°   **58.** 3; 90°

59. $\langle 2, 11 \rangle$ **60.** $\langle 11, -7 \rangle$ **61.** $\langle 38, -7 \rangle$

62. $\langle 3, -24 \rangle$ **63.** $\langle 2.6, 9.7 \rangle$

64. $\langle -5.7, -5.7 \rangle$ **65.** $\langle -3.1, 11.6 \rangle$

66. $\langle 19.3, 5.2 \rangle$ **67.** $\left\langle \dfrac{\sqrt{2}}{2}, -\dfrac{\sqrt{2}}{2} \right\rangle$

68. $\left\langle \dfrac{-11}{61}, \dfrac{60}{61} \right\rangle$ **69.** $(5\mathbf{i} + \mathbf{j})$ **70.** $(-15\mathbf{i} + 2\mathbf{j})$

71. -6 **72.** 34   **73.** -9 **74.** -6

1465

75. 16   **76.** −4   **77.** 59°   **78.** 167°

79. 49°   **80.** 159°   **81.** 166°   **82.** 144°

83. no   **84.** yes   **85.** yes   **86.** yes

87. no   **88.** no   **89.** no   **90.** yes

91.

92.

93. $2(\cos 315° + i \sin 315°)$

94. $2(\cos 30° + i \sin 30°)$

95. $8(\cos 270° + i \sin 270°)$

96. $8\sqrt{2}(\cos 225° + i \sin 225°)$

97. $61(\cos 169.6° + i \sin 169.6°)$

98. $41(\cos 282.7° + i \sin 282.7°)$

99. $17(\cos 28.1° + i \sin 28.1°)$

100. $26(\cos 247.4° + i \sin 247.4°)$

101. $3 - 3\sqrt{3}i$ **102.** $-2\sqrt{3} - 2i$

103. $-1 + i$ **104.** $-2\sqrt{3} + 2i$

105. $-3.7588 - 1.3681i$ **106.** $2.9544 - 0.5209i$

107. $-12i$ **108.** $-6 - 6\sqrt{3}i$

109. $-\dfrac{21}{2} - \dfrac{21\sqrt{3}}{2}i$ **110.** $2\sqrt{3} - 2i$

111. $-\dfrac{\sqrt{3}}{2} + \dfrac{1}{2}i$ **112.** $9i$

113. -6 **114.** $5\sqrt{2} + 5\sqrt{2}i$

115. -324 **116.** $-72 + 72\sqrt{3}i$

117. $16 - 16\sqrt{3}i$ **118.** $-1024 + 1024i$

119. $2(\cos 30° + i \sin 30°)$,
     $2(\cos 210° + i \sin 210°)$

120. $2(\cos 30° + i \sin 30°)$,
     $2(\cos 120° + i \sin 120°)$,
     $2(\cos 210° + i \sin 210°)$,
     $2(\cos 300° + i \sin 300°)$

121. $4(\cos 45° + i \sin 45°)$,
     $4(\cos 135° + i \sin 135°)$,
     $4(\cos 225° + i \sin 225°)$,
     $4(\cos 315° + i \sin 315°)$

122. $3\sqrt{2}(\cos 135° + i\sin 135°),$
$3\sqrt{2}(\cos 315° + i\sin 315°)$

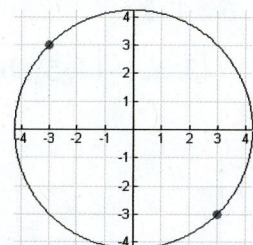

123. $-6,\ 3 + 3\sqrt{3}i,\ 3 - 3\sqrt{3}i$

124. $\pm 1,\ \pm i$

125. $\dfrac{\sqrt{2}}{2} + \dfrac{\sqrt{2}}{2}i,\ \dfrac{\sqrt{2}}{2} - \dfrac{\sqrt{2}}{2}i,\ -\dfrac{\sqrt{2}}{2} + \dfrac{\sqrt{2}}{2}i,$
$-\dfrac{\sqrt{2}}{2} - \dfrac{\sqrt{2}}{2}i$

126. $5,\ -\dfrac{5}{2} + \dfrac{5\sqrt{3}}{2}i,\ -\dfrac{5}{2} - \dfrac{5\sqrt{3}}{2}i$

All the points from Exercises 127–132 are graphed on the single graph below.

127. $\left(2\sqrt{2}, \dfrac{3\pi}{4}\right)$

128. $\left(8, \dfrac{5\pi}{3}\right)$

129. $\left(10, \dfrac{7\pi}{6}\right)$

130. $\left(\sqrt{6}, \dfrac{\pi}{4}\right)$

131. $\left(2, \dfrac{3\pi}{2}\right)$

132. $(11, 0)$

133. $\left(-\dfrac{3}{2}, \dfrac{3\sqrt{3}}{2}\right)$ **134.** $(-2\sqrt{2}, -2\sqrt{2})$

135. $(1, \sqrt{3})$ **136.** $(-3\sqrt{3}, -3)$

137. $\left(-\dfrac{1}{2}, -\dfrac{\sqrt{3}}{2}\right)$ **138.** $\left(-\dfrac{3\sqrt{2}}{2}, \dfrac{3\sqrt{2}}{2}\right)$

139.

140.

141.

142.

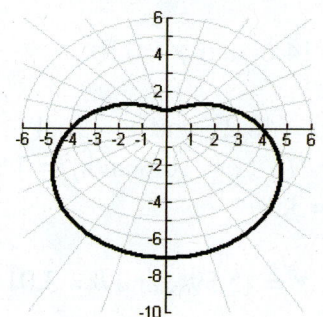

143. $Z = 30.4°$, $Y = 107.6°$, $B = 45.0$

144. $X = 47.5°$, $Z = 78.5°$, $C = 166.2$

145. $C = 4.0$, $X = 74.1°$, $Y = 64.3°$

146. $X = 130.6°$, $Y = 29.3°$, $Z = 20.7°$

147. $84.5°$, $58.4°$, $37.1°$

149. 65, $293°$ **150.** 80, 331 **151.** $140°$

152. $100°$ **153.** $12{,}246°$ **154.** 12, $24°$

155. $2(\cos 30° + \sin 30°)$, $2(\cos 120° + \sin 120°)$, $2(\cos 210° + \sin 210°)$, $2(\cos 300° + \sin 300°)$

156. $2\left[\cos\left(\dfrac{30°}{4} + \dfrac{360°k}{4}\right) + i\sin\left(\dfrac{30°}{4} + \dfrac{360°k}{4}\right)\right]$, $k = 0, 1, 2, 3$

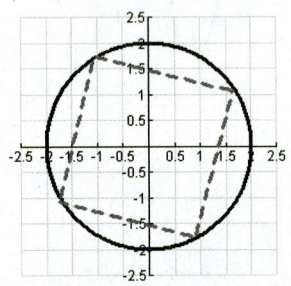

157. $10°$, $50°$, $130°$, $170°$, $250°$, $290°$

158. $20°$, $100°$, $140°$, $220°$, $260°$, $340°$

Practice Test

1. $a = 7.8$, $c = 14.6$, and $\gamma = 110°$

2. not possible

3. $\alpha = 35.4°$, $\beta = 48.2°$, and $\gamma = 96.4°$

4. $\beta = 62°$, $\gamma = 73°$, $c = 10.8$ and $\beta = 118°$, $\gamma = 17°$, $c = 3.3$

5. no triangle

6. $\alpha = 73.21°$, $\gamma = 45.59°$, and $b = 3.01$

7. $b = 1.82$, $c = 4.08$, $\gamma = 50°$

8. $\beta = 7.61°$, $\gamma = 127.39°$, $a = 5.97$

9. 57   **10.** 34.64

11. magnitude = 13, $\theta = 112.6°$

12. $\left\langle \dfrac{-3}{5}, \dfrac{-4}{5} \right\rangle$ **13. a.** $\langle -14, 5 \rangle$; **b.** -16

14. 28.23 yards, $-12.27°$

15. $32{,}768\left[-1 + i\sqrt{3}\right]$

16. $\sqrt{3} + i$, $-1 + \sqrt{3}i$, $-\sqrt{3} - i$, $1 - \sqrt{3}i$

17. $\left(\dfrac{-3\sqrt{3}}{2}, \dfrac{3}{2}\right)$

18.

19.

20. – 22. See Instructor's Solution Manual

23. -20 **24.** perpendicular

25. $\sqrt{2}$ **26.** $49.46°$

27. $a = 5.49$, $b = 4.48$, $c = 4.23$, $d = 1.16$

28. true

29. $4\left[\cos\left(\dfrac{3\pi}{8} + \dfrac{\pi k}{2}\right) + i\sin\left(\dfrac{3\pi}{8} + \dfrac{\pi k}{2}\right)\right]$, $k = 0, 1, 2, 3$

30. $x^2 + y^2 = \dfrac{y}{x}$ **31.** $112°$

32. $2\left[\cos\left(\dfrac{210°}{4}+\dfrac{360°k}{4}\right)+\right.$

$\left.i\sin\left(\dfrac{210°}{4}+\dfrac{360°k}{4}\right)\right],\ k=0,1,2,3$

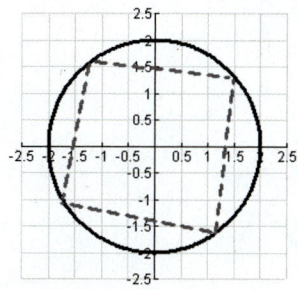

Cumulative Test

1. $x=2\pm4i$ **2.** $[-3,\infty)$ **3.** 2

4. 2.8   **5.** $-2x+h$

6. a. -1;   **b.** 7;   **c.** 0;   **d.** 16;   **e.** The domain and range are both all reals;   **f.** increasing on (0, 5) and decreasing on $(-\infty,0)\cup(5,\infty)$. Never constant.

7. undefined   **8.** $y=-0.75(x-1.5)^2+2.5$

9. $x=-2$, multiplicity 3; $x=0$, multiplicity 2

10. $\pm1,\pm2,\pm\frac{1}{2},\pm\frac{1}{3},\pm\frac{2}{3},\pm\frac{1}{6}$;

     $(x-1)(2x+1)(3x+2)$

11. VA: $x=2$; slant asymptote: $y=x+2$

12. y-intercept: $\left(0,\dfrac{1}{5}\right)$

    Domain: all reals

    Range: $(0,\infty)$

    Horizontal asymptote: $y=0$

13. $37,250

14.

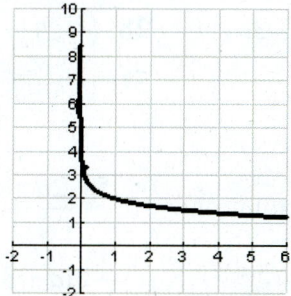

15. 0.001   **16.** 1

17. $15\sqrt{2}$ ft ≈ 21.21 ft   **18.** $0.96005+\pi$

19.

20. $-\dfrac{11\sqrt{57}}{57}$ **21.** $-\dfrac{3+8\sqrt{6}}{25}$

22. 60°, 120°, 240°, 300°

23. $c=6$, $\alpha=132°$, $\beta=27°$

24. $\langle-28,36\rangle$ **25.** $-2+2i$

26. $-67°$ **27.** 12°, 78°, 192°, 248°

CHAPTER 9

Section 9.1

1. $(1,0)$ **2.** $(0,-2)$ **3.** $(8,-1)$

4. $(-3,7)$ **5.** $(1,-1)$ **6.** $(0,1)$

7. $(1,2)$ **8.** $(-2,3)$

9. $u=\frac{32}{17}$, $v=\frac{11}{17}$ **10.** $m=\frac{5}{4}$, $n=-\frac{11}{8}$

11. no solution   **12.** no solution

13. infinitely many solutions: $(a,4a-1)$

14. infinitely many solutions: $(a,3a-4)$

15. infinitely many solutions: $\left(a,\dfrac{5a-15}{3}\right)$

16. infinitely many solutions: $\left(a,\dfrac{-5a+1}{3}\right)$

17. $(1,3)$ **18.** $(2,3)$ **19.** $(6,8)$

20. $(10, 12)$ **21.** $(1.2, 0.04)$ **22.** $(2.5, -3.5)$

23. $(3, 1)$ **24.** $(0, 2)$ **25.** $(2, 5)$

26. $(-1, 9)$ **27.** $(-3, 4)$ **28.** $\left(\frac{11}{2}, 4\right)$

29. $\left(1, -\frac{2}{7}\right)$ **30.** $(0, 3)$ **31.** $\left(\frac{19}{7}, \frac{11}{35}\right)$

32. $\left(\frac{1}{3}, -\frac{1}{2}\right)$

33. infinitely many solutions: $\left(a, \dfrac{5-2a}{5}\right)$

34. infinitely many solutions: $\left(a, \dfrac{-11a+3}{3}\right)$

35. $(4, 0)$ **36.** $(3, 4)$ **37.** $(-2, 1)$ **38.** $(-3, -5)$

39. infinitely many solutions: $\left(a, \dfrac{1.25-0.02a}{0.05}\right)$

40. infinitely many solutions: $\left(a, \dfrac{0.5a+0.8}{0.3}\right)$

41. $\left(\frac{75}{32}, \frac{7}{16}\right)$ **42.** $\left(\frac{1}{3}, \frac{1}{2}\right)$

43. c **44.** b **45.** d **46.** a

47. $(0, 0)$

48. $(0, 0)$

49. $(-1, -1)$

50. $(3, 2)$

51. $(0, -6)$

52. infinitely many solutions

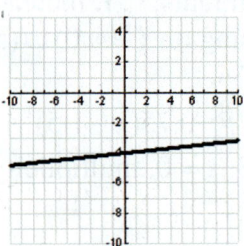

53. no solution **54.** no solution

55. 6 AusPens per kit

56. 353 g of 1% zinc oxide cream and 101 g of 10% zinc oxide cream.

57. 5 Montblanc pens, 64 Cross pens

58. 11 standard dumbbell sets
13 deluxe dumbbell sets

59. 15.86 ml of 8% HCl, 21.14 ml of 15% HCl

60. 2.5% mixture: $\frac{800}{3}$ gallons

4% mixture: $\frac{250}{3}$ gallons

61. $300,000 of sales

62. \$333,333.33 worth of homes

63. 169 highway miles, 180.5 city miles

64. 753.85 minutes

65. plane speed: 450 mph, wind speed: 50 mph

66. plane speed: 145.84 mph
wind speed: 20.85 mph

67. 10% stock: \$3500, 14% stock: \$6500

68. 5% stock: \$3500, 7% stock: \$7000

69. 8 CD players   **70.** 100 glasses of lemonade

71. Every term in the first equation in not multiplied by -1 correctly. The equation should be $-2x - y = 3$, and the resulting solution should be $x = 11$, $y = -25$.

72. $0 = 12$ is an inconsistent statement. Hence, there is no solution to the system.

73. Did not distribute -1 correctly. In Step 3, the calculation should be $-(-3y - 4) = 3y + 4$.

74. Actually, the lines are coincident, so that there are infinitely many solutions to the system.

75. false   **76.** true

77. false   **78.** false

79. $A = -4$, $B = 7$

80. The slopes are very close together, but are not equal. It may be difficult to distinguish on a graphing calculator unless just the right window is chosen.

81. 2% drink: 8 cups, 4% drink: 96 cups

82. light purple: 15 tablespoons
deep purple: 17 tablespoons

83. (8.9, 6.4)   **84.** $(-1.2, 2.0)$

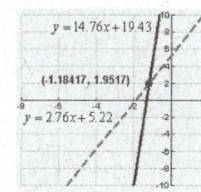

85. infinitely many solutions: $\left(a, \dfrac{7 - 23a}{5}\right)$

86. no solution   **87.** (2.426, -0.059)

88. $(-2.703, 4.139)$

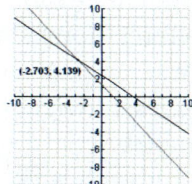

Section 9.2

1. $x = -\frac{3}{2}$, $y = -3$, $z = \frac{9}{2}$

2. $x = -1$, $y = -2$, $z = -4$

3. $x = -2$, $y = \frac{9}{2}$, $z = \frac{1}{2}$

4. $x = -\frac{11}{2}$, $y = 1$, $z = \frac{7}{2}$

5. $x = 5$, $y = 3$, $z = -1$

6. $x = -5$, $y = -3$, $z = -4$

7. $x = \frac{90}{31}$, $y = \frac{103}{31}$, $z = \frac{9}{31}$

8. $x = \frac{79}{9}$, $y = -\frac{34}{9}$, $z = -\frac{49}{3}$

9. $x = -\frac{13}{4}$, $y = \frac{1}{2}$, $z = -\frac{5}{2}$

10. $x = -\frac{29}{23}$, $y = \frac{25}{23}$, $z = \frac{36}{23}$

11. $x = -2$, $y = -1$, $z = 0$

12. $x = -2$, $y = -1$, $z = 0$

13. $x = 2$, $y = 5$, $z = -1$

14. $x = 1$, $y = 3$, $z = 2$ **15.** no solution

16. $x = \frac{2}{3}$, $y = -\frac{1}{3}$, $z = -\frac{4}{3}$ **17.** no solution

18. no solution

19. $x = 1 - a$, $y = -(a + \frac{1}{2})$, $z = a$

20. $x = \frac{1+a}{2}$, $y = a$, $z = \frac{a-3}{2}$

21. $x = 41 + 4a$, $y = 31 + 3a$, $z = a$

22. $x = -5$, $y = 4$, $z = 3$

23. $x_1 = -\frac{1}{2}$, $x_2 = \frac{7}{4}$, $x_3 = -\frac{3}{4}$

24. $x_1 = 0$, $x_2 = 2$, $x_3 = -3$

25. no solution   **26.** no solution

27. $x_1 = 1$, $x_2 = -1 + a$, $x_3 = a$

28. $x_1 = -1,\ x_2 = -3,\ x_3 = -2$

29. $x = \frac{2}{3}a + \frac{8}{3},\ y = -\frac{1}{3}a - \frac{10}{3},\ z = a$

30. $x = a,\ y = \frac{1}{4}(1 - 11a),\ z = \frac{1}{4}(a + 2)$

31. 100 basic widgets, 100 mid-price widgets, and 100 top-of-the-line widgets produced.

32. 150 basic widgets, 100 mid-price widgets, and 75 top-of-the-line widgets produced.

33. 8 touchdowns, 6 extra points, 4 field goals

34. 44 two-point shots, 10 three-point shots, 37 one-point free throws

35. 6 Mediterranean chicken sandwiches
3 Six-inch tuna sandwiches
5 Six-inch roast beef sandwiches

36. 3 Mediterranean chicken sandwiches
1 Six-inch tuna sandwiches
10 Six-inch roast beef sandwiches

37. $h_0 = 0,\ v_0 = 52,\ a = -32$

38. $h_0 = 0,\ v_0 = 100,\ a = -32$

39. $y = -0.0625x^2 + 5.25x - 50$

40. $y = 0.000251x^2 + 0.07498x + 18.6$

41. money market: \$10,000,
mutual fund: \$4000, stock: \$6000

42. money market: \$2000, mutual fund: \$6000, stock: \$12,000

43. 33 regular model skis, 72 trick skis, 5 slalom skis

44. 125 compact cars, 200 intermediate cars, 175 luxury models

45. game 1: 885 points, game 2: 823 points, game 3: 883 points

46. Blue Gene/L: 360 teraflops
BGW: 115 teraflops
ASC Purple: 93 teraflops

47. Equation (2) and Equation (3) must be added correctly – should be $2x - y + z = 2$.

Also, should begin by eliminating one variable from Equation (1).

48. From the system
$$\begin{cases} y + 3z = 5 \\ y + 3z = 9 \end{cases}$$
one should (upon subtracting the equations) conclude that $0 = -4$. So, the system is inconsistent, and hence, has no solution.

49. true   **50.** false

51. $a = 4,\ b = -2,\ c = -4$

52. $a = 1,\ b = -1,\ c = -42$

53. $a = -\frac{55}{24},\ b = -\frac{1}{4},\ c = \frac{223}{24},\ d = \frac{1}{4},\ e = 44$

54. 10 nickels, 6 dimes, 14 quarters

55. no solution

56. $x = 2,\ y = -5,\ z = 1$

57. $x_1 = -2,\ x_2 = 1,\ x_3 = -4,\ x_4 = 5$

58. $x_1 = 0,\ x_2 = -3,\ x_3 = 1,\ x_4 = 2$

59. $x = 41 + 4a,\ y = 31 + 3a,\ z = a$

60. $x = -5,\ y = 4,\ z = 3$

61. same as answer in Ex. 59 **62.** no solution

63. $\left(-\frac{80}{7}, -\frac{80}{7}, \frac{48}{7}\right)$ **64.** $x = a,\ y = \frac{24 - 30a}{5},\ z = \frac{3a - 2}{6}$

Section 9.3

1. d **2.** c **3.** a **4.** f **5.** b **6.** e

7. $\dfrac{A}{x - 5} + \dfrac{B}{x + 4}$ **8.** $\dfrac{A}{x - 5} + \dfrac{B}{x + 2}$

9. $\dfrac{A}{x - 4} + \dfrac{B}{x} + \dfrac{C}{x^2}$

10. $\dfrac{A}{x} + \dfrac{B}{x^2} + \dfrac{C}{x - 3} + \dfrac{D}{x + 3}$

11. $2x - 6 + \dfrac{3x + 33}{x^2 + x + 5}$

12. $2x + 11 + \dfrac{19x - 71}{x^2 - 3x + 7}$

13. $\dfrac{Ax+B}{x^2+10}+\dfrac{Cx+D}{\left(x^2+10\right)^2}$

14. $\dfrac{Ax+B}{x^2+13}+\dfrac{Cx+D}{\left(x^2+13\right)^2}$

15. $\dfrac{1}{x}-\dfrac{1}{x+1}$

16. $\dfrac{1}{x-1}-\dfrac{1}{x}$

17. $\dfrac{1}{x-1}$

18. $\dfrac{1}{x+1}$

19. $\dfrac{2}{x-3}+\dfrac{7}{x+5}$

20. $\dfrac{1}{x-2}+\dfrac{7}{x+1}$

21. $\dfrac{3}{x-1}+\dfrac{4}{\left(x-1\right)^2}$

22. $\dfrac{9}{y-1}+\dfrac{7}{\left(y-1\right)^2}$

23. $\dfrac{4}{x+3}-\dfrac{15}{\left(x+3\right)^2}$

24. $\dfrac{3}{x+2}-\dfrac{5}{\left(x+2\right)^2}$

25. $\dfrac{3}{x+1}+\dfrac{1}{x-5}+\dfrac{2}{\left(x-5\right)^2}$

26. $\dfrac{3}{x+2}+\dfrac{1}{x-1}-\dfrac{2}{\left(x-1\right)^2}$

27. $\dfrac{-2}{x+4}+\dfrac{7x}{x^2+3}$

28. $\dfrac{3}{x-2}+\dfrac{-2x+1}{x^2+2}$

29. $\dfrac{-2}{x-7}+\dfrac{4x-3}{3x^2-7x+5}$

30. $\dfrac{5}{x+5}+\dfrac{4x+3}{2x^2-3x+5}$

31. $\dfrac{x}{x^2+9}-\dfrac{9x}{\left(x^2+9\right)^2}$

32. $\dfrac{1}{x^2+9}-\dfrac{9}{\left(x^2+9\right)^2}$

33. $\dfrac{2x-3}{x^2+1}+\dfrac{5x+1}{\left(x^2+1\right)^2}$

34. $\dfrac{-x+2}{x^2+8}+\dfrac{5x-1}{\left(x^2+8\right)^2}$

35. $\dfrac{1}{x-1}+\dfrac{1}{2\left(x+1\right)}+\dfrac{-3x-1}{2\left(x^2+1\right)}$

36. $\dfrac{-\frac{1}{108}}{x-3}+\dfrac{-\frac{5}{108}}{x+3}+\dfrac{\frac{1}{18}x-\frac{1}{9}}{x^2+9}$

37. $\dfrac{3}{x-1}+\dfrac{2x+5}{x^2+2x-1}$

38. $\dfrac{4}{x-3}+\dfrac{6x-3}{x^2+4x+5}$

39. $\dfrac{1}{x-1}+\dfrac{1-x}{x^2+x+1}$

40. $\dfrac{1}{x-2}+\dfrac{-x+1}{x^2+2x+4}$

41. $\dfrac{1}{d_0}+\dfrac{1}{d_i}=\dfrac{1}{f}$

42. $\frac{999}{1000}$

43. The form of the decomposition is incorrect. It should be $\dfrac{A}{x}+\dfrac{Bx+C}{x^2+1}$. Once this correction is made, the correct decomposition is $\dfrac{1}{x}+\dfrac{2x+3}{x^2+1}$.

44. Should long divide first since the degree of the numerator is larger than the degree of the denominator.

45. false **46.** true

47. $\dfrac{1}{x-1}-\dfrac{1}{x+2}+\dfrac{1}{x-2}$

48. Case 1 $c\neq0$: $\dfrac{1}{2c}\left[\dfrac{ac+b}{x-c}+\dfrac{ac-b}{x+c}\right]$

Case 2 $c=0$: $\dfrac{a}{x}+\dfrac{b}{x^2}$

49. $\dfrac{1}{x}+\dfrac{1}{x+1}-\dfrac{1}{x^3}$

50. $\dfrac{1}{x}+\dfrac{2}{x^4}-\dfrac{1}{x-1}$

51. $\dfrac{x}{x^2+1}-\dfrac{2x}{\left(x^2+1\right)^2}+\dfrac{x+2}{\left(x^2+1\right)^3}$

52. $\dfrac{1}{\left(x^2+1\right)^2}-\dfrac{5}{\left(x^2+1\right)^3}$

53. yes

54. yes

55. no

56. no

57. yes

58. no

Section 9.4

1. d   **2.** c   **3.** b   **4.** a

5.

6.

7.

8.

9.

10.

11.

12.

13.

14.

15.

16.

17.

18.

19.

20.

21.

22.

23. no solution

24.

25.

26. no solution

27.

28.

29.

30.

31.

32.

33.

34.

35.

36.

37.

38.

39.

40.

41.

42.

43.

44.

45. no solution

46.

47.

48.

49. 4 units^2

50. $\frac{9}{2}$ units^2

51. 7.5 units^2

52. 7.5 units^2

53. $\begin{cases} x \geq 0,\ y \geq 0 \\ x + 20y \leq 2400 \\ 25x + 150y \leq 6000 \end{cases}$

54. $\begin{cases} x \geq 0,\ y \geq 0 \\ 60x + 10y \leq 1500 \\ 500x + 50y \leq 2000 \end{cases}$

55. a.
$$275 \leq 10x + 20y$$
$$125 \leq 15x + 10y$$
$$200 \leq 20x + 15y$$
$$x \geq 0, \ y \geq 0$$

b.

c. Two possible diet combinations are 2 ounces of food A and 14 ounces of food B or 10 ounces of food A and 10 ounces of food B.

56. a.
$$350 \leq 15x + 25y$$
$$175 \leq 25x + 10y$$
$$225 \leq 20x + 10y$$

b.

c. Two possible diet combinations are 4 ounces of food A and 15 ounces of food B or 6 ounces of food A and 12 ounces of food B.

57. a.
$$x \geq 2y$$
$$x + y \geq 1000$$
$$x \geq 0, \ y \geq 0$$

b.

c. Two possible solutions would be for the manufacturer to produce 700 USB wireless mice and 300 Bluetooth mice or 800 USB wireless mice and 300 Bluetooth mice.

58. a.
$$x \geq 1.5y$$
$$x + y \geq 10000$$

b.

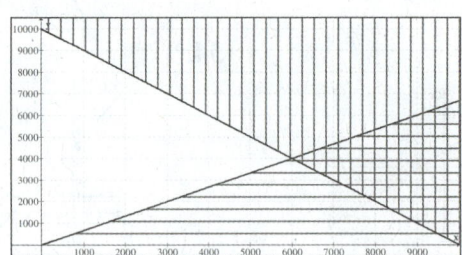

c. Two possible solutions would be for the manufacturer to produce 7000 pieces of 0.5 mm pencil lead and 4000 pieces of 0.7mm pencil lead or 8000 pieces of 0.5 mm pencil lead and 3000 pieces of 0.7 mm pencil lead.

59. $\begin{cases} P \leq 80 - 0.01x \\ P \geq 60 \\ x \geq 0 \end{cases}$ **60.** $\begin{cases} P \geq 20 + 0.02x \\ P \leq 60 \\ x \geq 0 \end{cases}$

61. 20,000   **62.** 40,000

63. The shading should be <u>above</u> the line.

64. The shading should not include the actual line (the line should be dashed).

65. true   **66.** true   **67.** false   **68.** false

69. shaded rectangle   **70.** no solution

71.

72.

73.

74.

75.

76.

Section 9.5

1. $f(x,y) = z = 2x + 3y$ $f(-1,4) = 10$
$f(2,4) = 16$ (MAX) $f(-2,-1) = -7$ (MIN)
$f(1,-1) = -1$

2. $f(x,y) = z = 3x + 2y$ $f(-1,4) = 5$
$f(2,4) = 14$ (MAX) $f(-2,-1) = -8$ (MIN)
$f(1,-1) = 0$

3. $f(x,y) = z = 1.5x + 4.5y$ $f(-1,4) = 16.5$
$f(2,4) = 21$ (MAX) $f(-2,-1) = -7.5$ (MIN)
$f(1,-1) = -3$

4. $f(x,y) = z = \frac{2}{3}x + \frac{3}{5}y$ $f(-1,4) = \frac{26}{15}$
$f(2,4) = \frac{56}{15}$ (MAX) $f(-2,-1) = -\frac{29}{15}$ (MIN)
$f(1,-1) = \frac{1}{15}$

5. minimize at $f(0,0) = 0$

6. no maximum   **7.** no maximum

8. minimize at $f(0,0) = 0$

9. minimize at $f(0,0) = 0$

10. maximize at $f(2,4) = -7.4$

11. maximize at $f(1,6) = \frac{53}{20} = 2.65$

12. minimize at $f(1,7) = -\frac{37}{15}$

13. Francis T-shirts: 130
Charley T-shirts: 50 (profit $950)

14. both types of T-shirts: 100 (profit $900)

15. laptops: 25, desktops: 0 (profit $7500)

16. laptops:30, desktops: 6 (profit $12,000)

17. first class cars: 3, second class cars: 27

18. first class cars: 2, second class cars: 28

19. 200 of each type of ski

20. crème-filled donuts: 0, jelly-filled donuts: 30

21. Should compare the values of the objective function at the vertices rather than comparing the y-values of the vertices.

22. Since $-x + y \le 0$ is a constraint, you would need to shade on AND below that line. The region should be:

23. false   **24.** true

25. maximum at $(0,a)$ and is a

26. maximum at $\left(\frac{b}{2}, a + \frac{b}{2}\right)$ and is $2a + \frac{3b}{2}$

27. minimum at (0, 0) and is 0

28. maximum is 28.6

29. maximum at (6.7, 4.5) and is 176.9

30. minimum at $(-0.76, -0.24)$ and is -1.27

31. maximum at $\left(\frac{4}{9}, \frac{115}{9}\right)$ and is approximately 25

32. minimum occurs at (–0.6, 3.12) as is approximately –8.232

Review Exercises

1. $(3,0)$

2. $\left(\frac{26}{7}, -\frac{16}{7}\right)$ **3.** $\left(\frac{13}{4}, 8\right)$

4. $(1.76, -0.32)$ **5.** $(2,1)$ **6.** $(3, -5)$

7. $\left(\frac{19}{8}, \frac{13}{8}\right)$ **8.** $\left(-\frac{5}{7}, \frac{16}{7}\right)$

9. $(-2,1)$

10. $(0.4,-0.7)$

11. $\left(12, \frac{35}{6}\right)$

12. $(20.4,-9.2)$

13. $(3,-2)$　　**14.** $(1,4)$　　**15.** $(-1,2)$

16. $(0,-1)$　　**17.** c　**18.** b　**19.** d　**20.** a

21. 6% NaCl: 10.5 ml, 18% NaCl: 31.5 ml

22. highway miles: 112, city miles: 153

23. $x=-1$, $y=-a+2$, $z=a$

24. $x=-3.524$, $y=-3.476$, $z=-0.429$

25. no solution　　**26.** $x=a$, $y=2$, $z=3-a$

27. $y=-0.0050x^2+0.4486x-3.8884$

28. IRA: \$9000, mutual fund: \$5000, stock: \$6000

29. $\dfrac{A}{x-1}+\dfrac{B}{(x-1)^2}+\dfrac{C}{x+3}+\dfrac{D}{x-5}$

30. $\dfrac{A}{x-9}+\dfrac{B}{3x+5}+\dfrac{C}{(3x+5)^2}+\dfrac{D}{x+4}$

31. $\dfrac{A}{x}+\dfrac{B}{4x+5}+\dfrac{C}{2x+1}+\dfrac{D}{(2x+1)^2}$

32. $\dfrac{A}{x+1}+\dfrac{B}{x-5}+\dfrac{C}{x-9}+\dfrac{D}{(x-9)^2}$

33. $\dfrac{A}{x-3}+\dfrac{B}{x+4}$　　**34.** $\dfrac{A}{x}+\dfrac{B}{x^2}+\dfrac{C}{x+6}$

35. $\dfrac{Ax+B}{x^2+17}+\dfrac{Cx+D}{(x^2+17)^2}$

36. $\dfrac{Ax+B}{x^2+13}+\dfrac{Cx+D}{(x^2+13)^2}$

37. $\dfrac{4}{x-1}+\dfrac{5}{x+7}$　　**38.** $\dfrac{3}{3x+2}+\dfrac{2}{2x-1}$

39. $\dfrac{1}{2x}+\dfrac{15}{2(x-5)}-\dfrac{3}{2(x+5)}$

40. $\dfrac{3}{x}+\dfrac{2x+1}{x^2+8}$　　**41.** $\dfrac{-2}{x+1}+\dfrac{2}{x}$

42. $\dfrac{1}{x+3}$　　**43.** $\dfrac{5}{x+2}-\dfrac{27}{(x+2)^2}$

44. $\dfrac{x}{x^2+64}-\dfrac{64x}{(x^2+64)^2}$

45.

46.

47.

48.

49.

50.

51.

52.

53. no solution

54. $y=3x$

55.

56.

57.

58.

59. minimum value of z is 0, occurs at $(0, 0)$

60. maximum value of z is 15, occurs at $(3, 3)$

61. maximum value of z is 25.6, occurs at $(0, 8)$

62. minimum value of z is 10, occurs at $(2, 0)$

63. minimum value of z is -30, occurs at $(0, 6)$

64. maximum value of z is $-\frac{2}{7}$, occurs at $\left(\frac{15}{7}, \frac{4}{7}\right)$

65. ocean watercolor: 10
geometric shape: 30 (profit $390)

66. ocean watercolor: 30
geometric shape: 0 (profit $450)

67. $(2, -3)$

68. $\left(-\frac{2}{5}, \frac{4}{3}\right)$

69. $(3.6, 3, 0.8)$

70. $\left(0.75a - 1.5b - 13, a, b\right)$

71. yes

72. no

73.

74.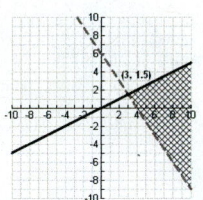

75. maximum is 12.06, occurs at $(1.8, 0.6)$

76. minimum is -42.88, occurs at $(4, 17.6)$

Practice Test

1. $(7, 3)$ **2.** $(-4, 2)$

3. $x = a,\ y = a - 2$ **4.** no solution

5. $x = 1,\ y = -5,\ z = 3$

6. $x = -\frac{1}{3}a + \frac{7}{6},\ y = \frac{1}{9}(a - 2),\ z = a$

7. $x = -1,\ y = 3,\ z = 4$

8. $x = 0,\ y = -1,\ z = 1$

9. $\dfrac{5}{x} - \dfrac{3}{x + 1}$ **10.** $\dfrac{3}{x - 5} + \dfrac{2}{(x - 5)^2}$

11. $\dfrac{7}{x + 2} - \dfrac{9}{(x + 2)^2}$ **12.** $\dfrac{\frac{2}{7}}{2x - 1} - \dfrac{\frac{1}{7}}{x + 3}$

13. $\dfrac{1}{3x} + \dfrac{2}{3(x - 3)} - \dfrac{1}{x + 3}$

14. $\dfrac{-\frac{1}{3}}{x} + \dfrac{\frac{1}{3}x + 5}{x^2 + 9}$

15.

16.

17.

18.

19.

20.

21. minimum value of z is 7, occurs at $(0, 1)$

22. maximum value of z is 28, occurs at $\left(\frac{8}{3}, \frac{10}{3}\right)$

23. money market: \$14,000, aggressive stock: \$8500, conservative stock: \$7500

24. $175 \times 350\,\text{ft}$ or $218.75 \times 280\,\text{ft}$

25. $(11, 19, 1)$

26. minimum value is -14.24, occurs at $(1, 12.7)$

Cumulative Test

1. $\dfrac{x-1}{5}$, $x \neq \frac{5}{6}$ **2.** $-6 + 2i$ **3.** no solution

4. $r = \sqrt{\dfrac{A}{\pi}}$ **5.** $(-\infty, 0] \cup (1, 3]$

6. $d \approx 1.263$, $\left(-\frac{11}{40}, -\frac{1}{4}\right)$ **7.** $y = 0.5x + 3.4$

8. $(-\infty, -5] \cup [5, \infty)$ **9.** $-\frac{5}{8}$

10. Shift the graph of $y = \frac{1}{x}$ right 2 units and then up 1 unit.

11. 1   **12.** $f^{-1}(x) = \dfrac{3x+2}{x-5}$

13. $f(x) = -2x^2 + 7$ **14.** 0 (multiplicity 3)

15. possible rational zeros:
$$\pm 1, \pm 2, \pm 5, \pm 10, \pm \tfrac{1}{2}, \pm \tfrac{5}{2}$$
rational zeros: $-5, -1, \frac{1}{2}, 2$

16. $P(x) = 4\left(x + \tfrac{1}{2}\right)^2 (x - (1 + 2i))(x - (1 - 2i))$

17. domain: $(-\infty, \infty)$

range: $(-1, \infty)$

y-intercept: $(0, 0)$

HA: $y = -1$

18. \$21,760.19

19. Shift the graph of $y = \ln x$ left 1 unit, and then down 3 units.

20. 1.413   **21.** 0.435

22. tall coffee: \$1.59, donut: \$1.29

23. $x = 2$, $y = 0$, $z = 5$ **24.** $-\dfrac{1}{(x-1)^2}$

25.

26. $(2.5, 1.2, -2)$

27. yes

CHAPTER 10

Section 10.1

1. 2×3 **2.** 3×2 **3.** 1×4

4. 4×1 **5.** 1×1 **6.** 4×4

7. $\begin{bmatrix} 3 & -2 & | & 7 \\ -4 & 6 & | & -3 \end{bmatrix}$ **8.** $\begin{bmatrix} -1 & 1 & | & 2 \\ 1 & -1 & | & -4 \end{bmatrix}$

9. $\begin{bmatrix} 2 & -3 & 4 & | & -3 \\ -1 & 1 & 2 & | & 1 \\ 5 & -2 & -3 & | & 7 \end{bmatrix}$ **10.** $\begin{bmatrix} 1 & -2 & 1 & | & 0 \\ -2 & 1 & -1 & | & -5 \\ 13 & 7 & 5 & | & 6 \end{bmatrix}$

11. $\begin{bmatrix} 1 & 1 & 0 & | & 3 \\ 1 & 0 & -1 & | & 2 \\ 0 & 1 & 1 & | & 5 \end{bmatrix}$ **12.** $\begin{bmatrix} 1 & -1 & 0 & | & -4 \\ 0 & 1 & 1 & | & 3 \end{bmatrix}$

13. $\begin{bmatrix} -4 & 3 & 5 & | & 2 \\ 2 & -3 & -2 & | & -3 \\ -2 & 4 & 3 & | & 1 \end{bmatrix}$ **14.** $\begin{bmatrix} -1 & 2 & 1 & | & 5 \\ 2 & -2 & 3 & | & 0 \\ -4 & 1 & -2 & | & 3 \end{bmatrix}$

15. $\begin{cases} -3x + 7y = 2 \\ x + 5y = 8 \end{cases}$ **16.** $\begin{cases} -x + 2y + 4z = 4 \\ 7x + 9y + 3z = -3 \\ 4x + 6y - 5z = 8 \end{cases}$

17. $\begin{cases} -x = 4 \\ 7x + 9y + 3z = -3 \\ 4x + 6y - 5z = 8 \end{cases}$ **18.** $\begin{cases} 2x + 3y - 4z = 6 \\ 7x - y + 5z = 9 \end{cases}$

19. $\begin{cases} x = a \\ y = b \end{cases}$ **20.** $\begin{cases} 3x + 5z = 1 \\ -4y + 7z = -3 \\ 2x - y = 8 \end{cases}$

21. not reduced form **22.** not reduced form
23. reduced form **24.** reduced form
25. not reduced form **26.** reduced form
27. reduced form **28.** not reduced form
29. row-echelon form **30.** reduced form

31. $\begin{bmatrix} 1 & -2 & | & -3 \\ 0 & 7 & | & 5 \end{bmatrix}$ **32.** $\begin{bmatrix} 1 & 2 & | & 5 \\ 2 & -3 & | & -4 \end{bmatrix}$

33. $\begin{bmatrix} 1 & -2 & -1 & | & 3 \\ 0 & 5 & -1 & | & 0 \\ 3 & -2 & 5 & | & 8 \end{bmatrix}$ **34.** $\begin{bmatrix} 1 & -2 & 1 & | & 3 \\ 0 & 1 & -2 & | & 6 \\ 0 & -6 & 2 & | & 4 \end{bmatrix}$

35. $\begin{bmatrix} 1 & -2 & 5 & -1 & | & 2 \\ 0 & 1 & 1 & -3 & | & 3 \\ 0 & -2 & 1 & -2 & | & 5 \\ 0 & 0 & 1 & -1 & | & -6 \end{bmatrix}$

36. $\begin{bmatrix} 1 & 0 & 5 & -10 & | & 15 \\ 0 & 1 & 2 & -3 & | & 4 \\ 0 & 0 & \frac{7}{2} & -3 & | & \frac{9}{2} \\ 0 & 0 & 1 & -1 & | & -3 \end{bmatrix}$

37. $\begin{bmatrix} 1 & 0 & 5 & -10 & | & -5 \\ 0 & 1 & 2 & -3 & | & -2 \\ 0 & 0 & -7 & 6 & | & 3 \\ 0 & 0 & 8 & -10 & | & -9 \end{bmatrix}$

38. $\begin{bmatrix} 1 & 0 & 0 & 0 & | & 1 \\ 0 & 1 & 0 & 0 & | & -2 \\ 0 & 0 & 1 & 0 & | & 0 \\ 0 & 0 & 0 & 1 & | & -3 \end{bmatrix}$ **39.** $\begin{bmatrix} 1 & 0 & 4 & 0 & | & 27 \\ 0 & 1 & 2 & 0 & | & -11 \\ 0 & 0 & 1 & 0 & | & 21 \\ 0 & 0 & 0 & 1 & | & -3 \end{bmatrix}$

40. $\begin{bmatrix} 1 & 0 & -1 & 0 & | & -3 \\ 0 & 1 & 2 & 0 & | & -8 \\ 0 & 0 & 1 & 0 & | & 4 \\ 0 & 0 & 0 & 1 & | & 1 \end{bmatrix}$ **41.** $\begin{bmatrix} 1 & 0 & | & -8 \\ 0 & 1 & | & 6 \end{bmatrix}$

42. $\begin{bmatrix} 1 & 0 & | & -8 \\ 0 & 1 & | & -11 \end{bmatrix}$ **43.** $\begin{bmatrix} 1 & 0 & 0 & | & -2 \\ 0 & 1 & 0 & | & -1 \\ 0 & 0 & 1 & | & 0 \end{bmatrix}$

44. $\begin{bmatrix} 1 & 0 & 0 & | & 0 \\ 0 & 1 & 0 & | & 1 \\ 0 & 0 & 1 & | & 0 \end{bmatrix}$ **45.** $\begin{bmatrix} 1 & 0 & 0 & | & 2 \\ 0 & 1 & 0 & | & 5 \\ 0 & 0 & 1 & | & -1 \end{bmatrix}$

46. $\begin{bmatrix} 1 & 0 & 0 & | & 1 \\ 0 & 1 & 0 & | & 3 \\ 0 & 0 & 1 & | & 2 \end{bmatrix}$ **47.** $\begin{bmatrix} 1 & 0 & -2 & | & 1 \\ 0 & 1 & -2 & | & 2 \end{bmatrix}$

48. $\begin{bmatrix} 1 & 0 & -\frac{3}{5} & -\frac{1}{5} \\ 0 & 1 & -\frac{1}{5} & \frac{8}{5} \end{bmatrix}$ **49.** $\begin{bmatrix} 1 & 0 & 1 & 1 \\ 0 & 1 & 1 & -\frac{1}{2} \\ 0 & 0 & 0 & 0 \end{bmatrix}$

50. $\begin{bmatrix} 1 & 0 & -1 & 2 \\ 0 & 1 & -2 & 3 \\ 0 & 0 & 0 & 0 \end{bmatrix}$ **51.** $x = -7, \ y = 5$

52. $x = 7, \ y = -5$

53. $x - 2y = -3$ or $x = 2a - 3, \ y = a$

54. $x = 0, \ y = 1$ **55.** no solution

56. $x = 18.425, \ y = 10.58$

57. $x = 4a + 41, \ y = 3a + 31, \ z = a$

58. $x = -5, \ y = 4, \ z = 3$

59. $x_1 = -\frac{1}{2}, \ x_2 = \frac{7}{4}, \ x_3 = -\frac{3}{4}$

60. $x = 0, \ y = 2, \ z = -3$

61. no solution **62.** no solution

63. $x_1 = 1, \ x_2 = a - 1, \ x_3 = a$

64. $x_1 = -1, \ x_2 = 3, \ x_3 = -2$

65. $x = \frac{2}{3}a + \frac{8}{3}, \ y = -\frac{1}{3}a - \frac{10}{3}, z = a$

66. $x = 4a - 2, \ y = -11a + 6, \ z = a$

67. no solution **68.** $x = 3, \ y = -7, \ z = 1$

69. $x_1 = -2, \ x_2 = 1, \ x_3 = -4, \ x_4 = 5$

70. $x_1 = 0, \ x_2 = -3, \ x_3 = 1, \ x_4 = 2$

71. $(1, -2)$ **72.** $(3, -4)$ **73.** no solution

74. $x = a, \ y = \frac{3}{4}a - 3$ **75.** $(-2, 1, 3)$

76. $(-1, 2, -3)$ **77.** $(3, -2, 2)$

78. $\left(-4, -2, \frac{1}{2}\right)$ **79.** no solution

80. $x = a, \ y = \frac{2-a}{7}, \ z = \frac{17-5a}{7}$

81. $x = \frac{a}{4} + 3, \ y = \frac{7a}{4} + 1, z = a$

82. $x = a, \ y = \frac{30a+51}{6}, \ z = 3a + 9$

83. $x = \frac{72-11a}{14}, \ y = \frac{13a+4}{14}, \ z = a$

84. $x = a, \ y = 4 - a, \ z = a - 4$

85. $x = 1, y = 2, z = -3, w = 1$

86. $w = 2, x = -22, y = -7, z = 3$

87. 8 touchdowns, 5 extra points, 1 two-point conversion, and 2 field goals

88. 45 two-point shots, 12 three-point shots, 39 one-point free throws

89. 2 chicken, 2 Tuna, 8 roast beef, 2 turkey bacon

90. 7 chicken, 1 Tuna, 2 roast beef, 4 turkey bacon

91. initial height: 0 ft, initial velocity: 50 ft/sec acceleration: -32 ft/sec^2

92. initial height: 10 ft, initial velocity: 60 ft/sec acceleration: -32 ft/sec^2

93. $y = -0.053x^2 + 4.58x - 34.76$

94. $y = 0.0008725x^2 + 0.0126x + 18.4$, 26.60 years

95. about 88 ml of the 1.5% solution and 12 ml of the 30% solution

96. 33 grams of the 1% hydrocortisone cream, 18 grams of the 0.5% hydrocortisone cream, and 9 grams of the Eucerin cream.

97. 200 basic widgets, 100 mid-price widgets, and 75 top-of-the-line widgets produced

98. 100 basic widgets, 50 mid-price widgets, and 200 top-of-the-line widgets produced

99. money market: $5500, mutual fund: $2500, stock: $2000

100. money market: $1000, mutual fund: $3000, stock: $6000

101. 25 units product x, 40 units product y, 6 units product z

102. $y = -\frac{7}{6}x^2 + \frac{13}{6}x + 4$

103. general admission: 25, reserved: 30, end zone: 45

104. 15 minutes walking, 15 minutes step-up exercise, 30 minutes weight training

105. $a = -\frac{22}{17}, b = -\frac{44}{17}, c = -\frac{280}{17}$

106. $a = 1, b = -1, c = -42$

107. Need to line up a single variable in a given column before forming the augmented matrix. The correct matrix is

$\begin{bmatrix} -1 & 1 & 1 & 2 \\ 1 & 1 & -2 & -3 \\ 1 & 1 & 1 & 6 \end{bmatrix}$, after reducing, $\begin{bmatrix} 1 & 0 & 0 & 2 \\ 0 & 1 & 0 & 1 \\ 0 & 0 & 1 & 3 \end{bmatrix}$.

108. a. Need to subtract all entries of $2R_1$ from the respective entries of R_2. So, the second row should be $0 \quad -1 \quad -1 \mid 0$.

b. Same error as in part **a**. The third row should be $0 \quad 4 \quad -1 \mid -12$.

109. Row 3 is not inconsistent. It implies $z = 0$.

110. Row 3 implies that the system is inconsistent since it requires $0z = 4$.

111. false   **112.** false

113. true   **114.** true

115. $f(x) = -\frac{11}{6}x^4 + \frac{44}{3}x^3 - \frac{223}{6}x^2 + \frac{94}{3}x + 44$

116. possible combinations of nickels, dimes, quarters: (1, 18, 11), (4, 14, 12), (7, 10, 13), (10, 6, 14), (13, 2, 15)

117.
```
rref([A])
[[1 0 -4 41]
 [0 1 -3 31]
 [0 0 0 0 ]]
```

118.
```
rref([A])
[[1 0 0 -5]
 [0 1 0 4 ]
 [0 0 1 3 ]]
```

119. a.
```
rref([A])
[[1 0 0 -.23653...
 [0 1 0 .928846...
 [0 0 1 6.08846...
```
$y = -0.24x^2 + 0.93x + 6.09$

b.
```
QuadReg
y=ax²+bx+c
a=-.2365384615
b=.9288461538
c=6.088461538
```
$y = -0.24x^2 + 0.93x + 6.09$

120. a.
```
rref([A])
[[1 0 0 .361616...
 [0 1 0 -.92323...
 [0 0 1 -17.6 ...
```
$y = 0.36x^2 - 0.92x - 17.6$

b.
```
QuadReg
y=ax²+bx+c
a=.3616161616
b=-.9232323232
c=-17.6
```
$y = 0.36x^2 - 0.92x - 17.6$

Section 10.2

1. 2×3 **2.** 3×2 **3.** 2×2

4. 1×4 **5.** 3×3 **6.** 4×1

7. 1×1 **8.** 2×4 **9.** 4×4

10. 1×5 **11.** $x = -5, y = 1$

12. $x = 10, y = 12$ **13.** $x = -3, \; y = -2, \; z = 3$

14. $x = 7, \; y = 5$ **15.** $x = 6, \; y = 3$

16. $a = -3, b = 4$ **17.** $\begin{bmatrix} -1 & 5 & 1 \\ 5 & 2 & 5 \end{bmatrix}$

18. $\begin{bmatrix} 2 & -2 \\ 2 & 0 \\ 7 & -1 \end{bmatrix}$ **19.** $\begin{bmatrix} -2 & 4 \\ 2 & -2 \\ -1 & 3 \end{bmatrix}$ **20.** $\begin{bmatrix} -1 & 1 & -1 \\ -1 & 6 & -3 \end{bmatrix}$

21. not defined   **22.** not defined

23. not defined   **24.** not defined

25. $\begin{bmatrix} -2 & 6 & 0 \\ 4 & 8 & 2 \end{bmatrix}$ **26.** $\begin{bmatrix} 8 & -12 \\ 0 & 4 \\ 16 & -8 \end{bmatrix}$

27. $\begin{bmatrix} 0 & -5 \\ -10 & 5 \\ -15 & -5 \end{bmatrix}$ **28.** $\begin{bmatrix} 0 & -4 & -2 \\ -6 & 4 & -8 \end{bmatrix}$

29. $\begin{bmatrix} -2 & 12 & 3 \\ 13 & 2 & 14 \end{bmatrix}$ **30.** $\begin{bmatrix} 3 & -5 & 2 \\ 0 & -16 & 5 \end{bmatrix}$

31. $\begin{bmatrix} 8 & 3 \\ 11 & 5 \end{bmatrix}$ **32.** $\begin{bmatrix} 5 \end{bmatrix}$

33. $\begin{bmatrix} -3 & 21 & 6 \\ -4 & 7 & 1 \\ 13 & 14 & 9 \end{bmatrix}$ **34.** $\begin{bmatrix} 2 & -2 \\ 20 & 21 \\ 13 & 14 \end{bmatrix}$

35. $\begin{bmatrix} 3 & 6 \\ -2 & -2 \\ 17 & 24 \end{bmatrix}$ **36.** $\begin{bmatrix} -1 & 4 & 1 \\ 6 & 9 & 14 \\ 1 & 3 & 11 \end{bmatrix}$

37. not defined   **38.** not defined

39. $\begin{bmatrix} 0 & 60 \end{bmatrix}$ **40.** $\begin{bmatrix} 3 & -15 \\ -45 & -57 \\ -6 & -72 \end{bmatrix}$

41. $\begin{bmatrix} -6 & 1 & -9 \end{bmatrix}$ **42.** $\begin{bmatrix} 0 & 42 & 14 \\ 24 & 21 & 15 \end{bmatrix}$

43. $\begin{bmatrix} 7 & 10 & -8 \\ 0 & 15 & 5 \\ 23 & 0 & -7 \end{bmatrix}$ **44.** $\begin{bmatrix} -4 & 8 & 3 \\ 5 & 9 & 1 \\ -5 & 10 & -1 \end{bmatrix}$

45. $\begin{bmatrix} 12 & 20 \\ 30 & 42 \end{bmatrix}$ **46.** $\begin{bmatrix} 18 & 18 & 74 \\ -12 & 2 & 16 \end{bmatrix}$

47. $\begin{bmatrix} -4 \\ -4 \\ -16 \end{bmatrix}$ **48.** $\begin{bmatrix} 4 & 54 & 14 \\ 23 & 6 & 0 \end{bmatrix}$

49. not defined   **50.** $\begin{bmatrix} 6 & 1 & 4 \\ 3 & 4 & 17 \\ 1 & -2 & -5 \end{bmatrix}$

51. $A = \begin{bmatrix} 0.70 \\ 0.30 \end{bmatrix}$, $B = \begin{bmatrix} 0.89 \\ 0.84 \end{bmatrix}$

a. $46A = \begin{bmatrix} 32.3 \\ 13.8 \end{bmatrix}$, out of 46 million people,

32.2 million said that they had tried to quit smoking, while 13.8 million said that they had not.

b. $46B = \begin{bmatrix} 40.94 \\ 38.64 \end{bmatrix}$, out of 46 million people,

40.94 million believed that smoking would increase the chance of getting lung cancer, and that 38.64 million believed that smoking would shorten their lives.

52. $A = \begin{bmatrix} 0.24 \\ 0.23 \end{bmatrix}$ (1981), $B = \begin{bmatrix} 0.32 \\ 0.21 \end{bmatrix}$ (1991),

$C = \begin{bmatrix} 0.38 \\ 0.30 \end{bmatrix}$ (2001): $C - B = \begin{bmatrix} 0.06 \\ 0.09 \end{bmatrix}$

percent increase from 1991 to 2001 of female graduate students in mathematics and computer science. 6% increase of female graduate students in mathematics, 9% increase of female graduate students in computer science

$B - A = \begin{bmatrix} 0.08 \\ -0.02 \end{bmatrix}$ the percent increase from

1981 to 1991 of female graduate students in mathematics and computer science. 8% increase of female graduate students in mathematics, 2% decrease of female graduate students in computer science

53. $A = \begin{bmatrix} 0.589 & 0.628 \\ 0.414 & 0.430 \end{bmatrix}$, $B = \begin{bmatrix} 100M \\ 110M \end{bmatrix}$

$AB = \begin{bmatrix} 127.98M \\ 88.7M \end{bmatrix}$ 127.98 million registered

voters, of those 88.7 million actually vote.

54. 2×2, the respective applicant's total score according to each of the two rubrics

55. $A = \begin{bmatrix} 0.45 & 0.50 & 1.00 \end{bmatrix}$

$B = \begin{bmatrix} 7523 \\ 2700 \\ 15200 \end{bmatrix}$ $AB = \begin{bmatrix} 19{,}935.35 \end{bmatrix}$

56. $A = \begin{bmatrix} 85 & 75 & 100 \end{bmatrix}$, $B = \begin{bmatrix} 0.25 \\ 0.20 \\ 0.15 \end{bmatrix}$

$AB = \begin{bmatrix} 51.25 \end{bmatrix}$

57. $A = \begin{bmatrix} 230 & 3 & 44 & 9 \\ 430 & 19 & 46 & 20 \\ 290 & 5 & 45 & 19 \\ 330 & 5 & 47 & 24 \end{bmatrix}$

$2A = \begin{bmatrix} 460 & 6 & 88 & 18 \\ 860 & 38 & 92 & 40 \\ 580 & 10 & 90 & 38 \\ 660 & 10 & 94 & 48 \end{bmatrix}$, nutritional

information corresponding to 2 sandwiches

$0.5A = \begin{bmatrix} 115 & 1.5 & 22 & 4.5 \\ 215 & 9.5 & 23 & 10 \\ 145 & 2.5 & 22.5 & 9.5 \\ 165 & 2.5 & 23.5 & 12 \end{bmatrix}$, nutritional

information corresponding to one-half of a sandwich

58. $BA = \begin{bmatrix} 6,410 & 161 & 945 & 339 \end{bmatrix}$, nutritional information for food consumed during one week

$\frac{1}{7}BA = \begin{bmatrix} 915.7 & 23 & 135 & 48.4 \end{bmatrix}$,

corresponds to one day's information

59. $AB = \begin{bmatrix} \$0.228 \\ \$0.081 \\ \$0.015 \end{bmatrix}$, total cost per mile to run

each type of automobile

60. SUV: \$2736, Hybrid: \$972,
Electric Car: \$180

61. $N = \begin{bmatrix} 2 \\ 1 \\ 0 \end{bmatrix}$ $XN = \begin{bmatrix} 10 \\ 16 \\ 20 \end{bmatrix}$

The nutritional content of the meal is 10 g of carbohydrates, 16 g of protein, and 20 g of fat.

62. $N = \begin{bmatrix} 1 \\ 0 \\ 2 \end{bmatrix}$ $XN = \begin{bmatrix} 9 \\ 15 \\ 16 \end{bmatrix}$

The nutritional content of the meal is 9 g of carbohydrates, 15 g of protein, and 16 g of fat.

63. $N = \begin{bmatrix} 200 \\ 25 \\ 0 \end{bmatrix}$ $XN = \begin{bmatrix} 9.25 \\ 13.25 \\ 15.75 \end{bmatrix}$

Company 1 would charge \$9.25, Company 2 would charge \$13.25, and Company 3 would charge \$15.75 respectively for 200 minutes of talking and 25 text messages. The better cell phone provider for this employee would be Company 1.

64. $N = \begin{bmatrix} 125 \\ 125 \\ 320 \end{bmatrix}$ $XN = \begin{bmatrix} 59.25 \\ 71.35 \\ 59.10 \end{bmatrix}$

Company 1 would charge \$59.25, Company 2 would charge \$71.35, and Company 3 would charge \$59.10 for 125 minutes of

talking, 125 text messages, and 320 megabytes of data usage. The better cell phone provider for this employee would be Company 3.

65. Not multiplying correctly. It should be:

$\begin{bmatrix} 3 & 2 \\ 1 & 4 \end{bmatrix} \cdot \begin{bmatrix} -1 & 3 \\ -2 & 5 \end{bmatrix} = \begin{bmatrix} -7 & 19 \\ -9 & 23 \end{bmatrix}$

66. Not multiplying correctly. It should be:

$\begin{bmatrix} 3(-1)+2(-2) & 3(3)+2(5) \\ 1(-1)+4(-2) & 1(3)+4(5) \end{bmatrix} = \begin{bmatrix} -7 & 19 \\ -9 & 23 \end{bmatrix}$

67. false   **68.** false

69. true   **70.** true

71. $\begin{bmatrix} a_{11}^2 + a_{12}a_{21} & a_{11}a_{12} + a_{12}a_{22} \\ a_{21}a_{11} + a_{22}a_{21} & a_{22}^2 + a_{21}a_{12} \end{bmatrix}$

72. $m = n$

73. $A = \begin{bmatrix} 1 & 1 \\ 1 & 1 \end{bmatrix}$, $A^2 = \begin{bmatrix} 2 & 2 \\ 2 & 2 \end{bmatrix}$, $A^n = 2^{n-1}A, n \geq 1$

74. $A = \begin{bmatrix} 1 & 0 \\ 0 & 1 \end{bmatrix}$, $A^2 = A, \cdots A^n = A, \ n \geq 1$

75. must have $m = p$ **76.** have different orders

77. $\begin{bmatrix} 33 & 35 \\ -96 & -82 \\ 31 & 19 \\ 146 & 138 \end{bmatrix}$ **78.** not defined

79. not defined   **80.**

$\begin{bmatrix} -2 & 8 & 124 & 126 \\ 114 & 148 & -131 & 14 \\ 36 & 28 & 29 & 48 \\ -6 & 11 & 189 & 87 \end{bmatrix}$

81. $\begin{bmatrix} 5 & -4 & 4 \\ 2 & -15 & -3 \\ 26 & 4 & -8 \end{bmatrix}$ **82.** $\begin{bmatrix} 74 & 121 & 233 \\ -503 & -560 & 312 \\ 1072 & -1006 & -462 \end{bmatrix}$

Section 10.3

1. $\begin{bmatrix} -2 & 5 \\ 7 & -2 \end{bmatrix} \begin{bmatrix} x \\ y \end{bmatrix} = \begin{bmatrix} 10 \\ -4 \end{bmatrix}$

2. $\begin{bmatrix} 4 & -8 \\ 3 & 5 \end{bmatrix} \begin{bmatrix} x \\ y \end{bmatrix} = \begin{bmatrix} 10 \\ 15 \end{bmatrix}$

3. $\begin{bmatrix} 1 & -2 \\ -3 & 1 \end{bmatrix} \begin{bmatrix} x \\ y \end{bmatrix} = \begin{bmatrix} 8 \\ 6 \end{bmatrix}$

4. $\begin{bmatrix} 7 & -2 \\ 3 & 7 \end{bmatrix} \begin{bmatrix} x \\ y \end{bmatrix} = \begin{bmatrix} 28 \\ 42 \end{bmatrix}$

5. $\begin{bmatrix} 3 & 5 & -1 \\ 1 & 0 & 2 \\ -1 & 1 & -1 \end{bmatrix} \begin{bmatrix} x \\ y \\ z \end{bmatrix} = \begin{bmatrix} 2 \\ 17 \\ 4 \end{bmatrix}$

6. $\begin{bmatrix} 1 & -1 & 1 \\ 2 & 1 & -3 \\ -3 & 2 & 1 \end{bmatrix} \begin{bmatrix} x \\ y \\ z \end{bmatrix} = \begin{bmatrix} 12 \\ 6 \\ 18 \end{bmatrix}$

7. $\begin{bmatrix} 3 & 0 & 1 \\ 0 & 1 & -2 \\ 1 & 2 & 0 \end{bmatrix} \begin{bmatrix} x \\ y \\ z \end{bmatrix} = \begin{bmatrix} 10 \\ 4 \\ 6 \end{bmatrix}$

8. $\begin{bmatrix} 1 & 1 & -2 & 1 \\ 2 & -1 & 3 & 0 \\ -1 & 2 & -3 & 4 \\ 0 & 1 & 4 & 6 \end{bmatrix} \begin{bmatrix} x \\ y \\ z \\ w \end{bmatrix} = \begin{bmatrix} 11 \\ 17 \\ 12 \\ 19 \end{bmatrix}$

9. yes 10. yes 11. yes 12. yes 13. no
14. no 15. yes 16. yes 17. no 18. yes

19. $\begin{bmatrix} 0 & -1 \\ 1 & 2 \end{bmatrix}$ 20. $\begin{bmatrix} 1 & -1 \\ -2 & 3 \end{bmatrix}$ 21. $\begin{bmatrix} -\frac{1}{13} & \frac{8}{39} \\ \frac{20}{39} & -\frac{4}{117} \end{bmatrix}$

22. $\begin{bmatrix} -\frac{4}{3} & 4 \\ \frac{2}{3} & -\frac{1}{2} \end{bmatrix}$ 23. $\begin{bmatrix} -0.1618 & 0.2284 \\ 0.5043 & 0.1237 \end{bmatrix}$

24. $\begin{bmatrix} 1.391 & -0.4783 \\ -2 & -1 \end{bmatrix}$ 25. $\begin{bmatrix} \frac{1}{2} & \frac{1}{2} & 0 \\ \frac{1}{2} & 0 & \frac{1}{2} \\ 0 & -\frac{1}{2} & -\frac{1}{2} \end{bmatrix}$

26. $\begin{bmatrix} \frac{5}{4} & \frac{1}{4} & \frac{1}{2} \\ -\frac{1}{2} & \frac{1}{2} & 0 \\ -\frac{3}{4} & \frac{1}{4} & -\frac{1}{2} \end{bmatrix}$ 27. A^{-1} does not exist.

28. $\begin{bmatrix} \frac{4}{7} & \frac{8}{7} & -\frac{5}{7} \\ \frac{3}{7} & -\frac{1}{7} & -\frac{2}{7} \\ \frac{1}{7} & \frac{2}{7} & -\frac{3}{7} \end{bmatrix}$ 29. $\begin{bmatrix} -\frac{1}{2} & -\frac{1}{2} & \frac{5}{2} \\ \frac{1}{2} & \frac{1}{2} & -\frac{3}{2} \\ 0 & -1 & 1 \end{bmatrix}$

30. $\begin{bmatrix} 0 & -1 & 1 \\ 1 & 3 & -2 \\ 1 & 1 & -1 \end{bmatrix}$ 31. $\begin{bmatrix} \frac{1}{2} & \frac{1}{2} & 0 \\ \frac{3}{4} & \frac{1}{4} & -\frac{1}{2} \\ \frac{1}{4} & \frac{3}{4} & -\frac{1}{2} \end{bmatrix}$

32. $\begin{bmatrix} -\frac{7}{2} & \frac{3}{2} & 1 \\ -\frac{5}{2} & 1 & \frac{1}{2} \\ -2 & \frac{1}{2} & \frac{1}{2} \end{bmatrix}$ 33. $x = 2, y = -1$

34. $x = 3, y = -2$ 35. $x = \frac{1}{2}, y = \frac{1}{3}$
36. $x = \frac{2}{5}, y = \frac{3}{5}$ 37. $x = 4, y = -3$
38. $x = -5, y = 7$ 39. $x = 0, y = 0, z = 1$
40. $x = 1, y = 1, z = 0$
41. A^{-1} does not exist
42. $x = 4, y = 0, z = 1$
43. $x = -1, y = 1, z = -7$
44. $x = -2, y = 4, z = 5$
45. $x = 3, y = 5, z = 4$
46. $x = 7, y = 4, z = 3$

47. a. $\begin{bmatrix} \$50 \\ \$20 \end{bmatrix}$ b. sweatshirt: $50, t-shirt: $20

48. a. $\begin{bmatrix} 20 \\ 75 \end{bmatrix}$ b. 20 hats and 75 jackets sold

49. JAW 50. HIP 51. LEG
52. ARM 53. EYE 54. EAR

55. $X = \begin{bmatrix} 8 & 4 & 6 \\ 6 & 10 & 5 \\ 10 & 4 & 8 \end{bmatrix}^{-1} \begin{bmatrix} 18 \\ 21 \\ 22 \end{bmatrix} = \begin{bmatrix} 1 \\ 1 \\ 1 \end{bmatrix}$

The combination of 1 serving each of food A, B, and C will create a meal of 18 g carbohydrates, 21 g of protein, and 22 g of fat.

56. $X = \begin{bmatrix} 8 & 4 & 6 \\ 6 & 10 & 5 \\ 10 & 4 & 8 \end{bmatrix}^{-1} \begin{bmatrix} 14 \\ 25 \\ 16 \end{bmatrix} = \begin{bmatrix} 0 \\ 2 \\ 1 \end{bmatrix}$

The combination of 2 servings of food B and 1 serving of food C will create a meal of 14 g carbohydrates, 25 g of protein, and 16 g of fat.

57. $X = \begin{bmatrix} 0.03 & 0.06 & 0.15 \\ 0.04 & 0.05 & 0.18 \\ 0.05 & 0.07 & 0.13 \end{bmatrix}^{-1} \begin{bmatrix} 49.50 \\ 52.00 \\ 58.50 \end{bmatrix} = \begin{bmatrix} 350 \\ 400 \\ 100 \end{bmatrix}$

The employee's normal monthly usage is 350 minutes talking, 400 text messages, and 100 megabytes of data usage.

58. $X = \begin{bmatrix} 0.03 & 0.06 & 0.15 \\ 0.04 & 0.05 & 0.18 \\ 0.05 & 0.07 & 0.13 \end{bmatrix}^{-1} \begin{bmatrix} 82.50 \\ 85.00 \\ 92.50 \end{bmatrix} = \begin{bmatrix} 350 \\ 700 \\ 200 \end{bmatrix}$

The employee's normal monthly usage is 350 minutes talking, 700 text messages, and 200 megabytes of data usage.

59. A is not invertible because the identity matrix was not reached.

60. This is not how the inverse is defined.

Rather, $A^{-1} = \frac{1}{2(10)-5(3)} \begin{bmatrix} 10 & -5 \\ -3 & 2 \end{bmatrix} = \begin{bmatrix} 2 & -1 \\ -\frac{3}{5} & \frac{2}{5} \end{bmatrix}$.

61. false   **62.** false

63. $x = 9$ **64.** $\begin{bmatrix} \frac{1}{a} & 0 & 0 \\ 0 & \frac{1}{b} & 0 \\ 0 & 0 & \frac{1}{c} \end{bmatrix}$

65. $A \cdot A^{-1} = \begin{bmatrix} a & b \\ c & d \end{bmatrix} \cdot \left(\frac{1}{ad-bc} \begin{bmatrix} d & -b \\ -c & a \end{bmatrix} \right)$

$= \frac{1}{ad-bc} \left(\begin{bmatrix} a & b \\ c & d \end{bmatrix} \begin{bmatrix} d & -b \\ -c & a \end{bmatrix} \right) = \frac{1}{ad-bc} \begin{bmatrix} ad-bc & 0 \\ 0 & ad-bc \end{bmatrix}$

$= \begin{bmatrix} \frac{ad-bc}{ad-bc} & 0 \\ 0 & \frac{ad-bc}{ad-bc} \end{bmatrix} = \begin{bmatrix} 1 & 0 \\ 0 & 1 \end{bmatrix} = I$

66. $A^{-1} = \begin{bmatrix} \frac{d}{da-bc} & -\frac{b}{da-bc} \\ -\frac{c}{da-bc} & \frac{a}{da-bc} \end{bmatrix} = \frac{1}{da-bc} \begin{bmatrix} d & -b \\ -c & a \end{bmatrix}$

67. $ad - bc = 0$

68. unable to transform into the identity matrix

69. $\begin{bmatrix} -\frac{115}{6008} & \frac{431}{6008} & -\frac{1067}{6008} & \frac{103}{751} \\ \frac{411}{6008} & \frac{-391}{6008} & \frac{731}{6008} & -\frac{22}{751} \\ \frac{57}{751} & \frac{28}{751} & -\frac{85}{751} & \frac{3}{751} \\ -\frac{429}{6008} & \frac{145}{6008} & \frac{1035}{6008} & \frac{12}{751} \end{bmatrix}$

70. $\begin{bmatrix} 1 & 0 & 0 & 0 \\ 0 & 1 & 0 & 0 \\ 0 & 0 & 1 & 0 \\ 0 & 0 & 0 & 1 \end{bmatrix} = I$

71. $x = 1.8, \ y = -1.6$ **72.** $x = 7.1, y = -2.2$

73. $(3.7, -2.4, 9.3)$ **74.** $(-1.6, 5.5, -3.4)$

Section 10.4

1. -2 **2.** -10 **3.** 31

4. 32   **5.** -28 **6.** 0

7. -0.6 **8.** 0.7   **9.** 0

10. $\frac{5}{18}$ **11.** $x = 5, \ y = -6$

12. $x = -5, \ y = 4$ **13.** $x = -2, \ y = 1$

14. $x = -1, \ y = 2$ **15.** $x = -3, \ y = -4$

16. $x = 1, \ y = 2$ **17.** $x = -2, \ y = 5$

18. $x = 5, \ y = 3$ **19.** $x = 2, \ y = 2$

20. $x = -3, \ y = -3$

21. $D = 0$, inconsistent or dependent system

22. $D = 0$, inconsistent or dependent system

23. $D = 0$, inconsistent or dependent system

24. $D = 0$, inconsistent or dependent system

25. $x = \frac{1}{2}, \ y = -1$ **26.** $x = -\frac{1}{4}, \ y = \frac{2}{3}$

27. $x = 1.5, \ y = 2.1$ **28.** $x \cong -3.2, \ y \cong 5$

29. $x = 0, \ y = 7$ **30.** $x = -5, \ y = 0$

31. $x = \frac{1}{3}, \ y = \frac{3}{4}$ **32.** $x = \frac{2}{3}, \ y = \frac{1}{5}$

33. 7   **34.** 7   **35.** -25

36. -50 **37.** -180 **38.** -64

39. 0   **40.** 0   **41.** 238

42. -225 **43.** 0   **44.** 304

45. 95.7   **46.** -3.36

47. $x = 2, \ y = 3, \ z = 5$ **48.** $x = 3, \ y = -2, \ z = 1$

49. $x = -2, \ y = 3, \ z = 5$ **50.** $x = -1, \ y = 4, \ z = 2$

51. $x = 2, \ y = -3, \ z = 1$ **52.** $x = 5, \ y = 10, \ z = 4$

53. $D = 0$, inconsistent or dependent system

54. $D = 0$, inconsistent or dependent system

55. $D = 0$, inconsistent or dependent system

56. $D = 0$, inconsistent or dependent system

57. $x = -3, \ y = 1, \ z = 4$

58. $x = -\frac{23}{2}, \ y = 2, \ z = \frac{204}{11}$

59. $x = 2$, $y = -3$, $z = 5$

60. $x = 4$, $y = -8$, $z = 1$

61. 6 units^2 **62.** 10 units^2 **63.** 6 units^2

64. 3 units^2 **65.** $y = 2x$ **66.** not collinear

67. $I_1 = \frac{7}{2}$, $I_2 = \frac{5}{2}$, $I_3 = 1$

68. $I_1 = \frac{7}{2}$, $I_2 = \frac{5}{2}$, $I_3 = 1$

69. The second determinant should be subtracted; that is, it should be $-1\begin{vmatrix} -3 & 2 \\ 1 & -1 \end{vmatrix}$.

70. The third determinant should be $3\begin{vmatrix} -3 & 0 \\ 1 & 4 \end{vmatrix}$.

71. In D_x and D_y, the column $\begin{bmatrix} 6 \\ -3 \end{bmatrix}$ should replace the column corresponding to the variable that is being solved for in each case. Precisely, D_x should be $\begin{vmatrix} 6 & 3 \\ -3 & -1 \end{vmatrix}$ and D_y should be $\begin{vmatrix} 2 & 6 \\ -1 & -3 \end{vmatrix}$.

72. $x = \dfrac{D_x}{D}$, not $\dfrac{D}{D_x}$.   **73.** true   **74.** true

75. false   **76.** true   **77.** abc

78. $a_1 b_2 c_3$ **79.** -419 **80.** $a = -6$

81. $-b_1 \begin{vmatrix} a_2 & c_2 \\ a_3 & c_3 \end{vmatrix} + b_2 \begin{vmatrix} a_1 & c_1 \\ a_3 & c_3 \end{vmatrix} - b_3 \begin{vmatrix} a_1 & c_1 \\ a_2 & c_2 \end{vmatrix}$

$= -b_1 \left[(a_2)(c_3) - (a_3)(c_2)\right] + b_2 \left[(a_1)(c_3) - (a_3)(c_1)\right] - b_3 \left[(a_1)(c_2) - (a_2)(c_1)\right]$

$= -a_2 b_1 c_3 + a_3 b_1 c_2 + a_1 b_2 c_3 - a_3 b_2 c_1 - a_1 b_3 c_2 + a_2 b_3 c_1$

82. $a_3 \begin{vmatrix} b_1 & c_1 \\ b_2 & c_2 \end{vmatrix} - b_3 \begin{vmatrix} a_1 & c_1 \\ a_2 & c_2 \end{vmatrix} + c_3 \begin{vmatrix} a_1 & b_1 \\ a_2 & b_2 \end{vmatrix}$

$= a_3 \left[(b_1)(c_2) - (b_2)(c_1)\right] - b_3 \left[(a_1)(c_2) - (a_2)(c_1)\right] + c_3 \left[(a_1)(b_2) - (a_2)(b_1)\right]$

$= a_3 b_1 c_2 - a_3 b_2 c_1 - a_1 b_3 c_2 + a_2 b_3 c_1 + a_1 b_2 c_3 - a_2 b_1 c_3$

$= a_1 b_2 c_3 + a_2 b_3 c_1 + a_3 b_1 c_2 - a_1 b_3 c_2 - a_2 b_1 c_3 - a_3 b_2 c_1$

83. -180 **84.** -64 **85.** -1019

86. -2287 **87.** $x = -6.4$, $y = 1.5$, $z = 3.4$

88. $x = 12.5$, $y = -8.2$, $z = 9.4$

Review Exercises

1. $\begin{bmatrix} 5 & 7 & | & 2 \\ 3 & -4 & | & -2 \end{bmatrix}$ **2.** $\begin{bmatrix} 2.3 & -4.5 & | & 6.8 \\ -0.4 & 2.1 & | & -9.1 \end{bmatrix}$

3. $\begin{bmatrix} 2 & 0 & -1 & | & 3 \\ 0 & 1 & -3 & | & -2 \\ 1 & 0 & 4 & | & -3 \end{bmatrix}$ **4.** $\begin{bmatrix} -1 & 2 & 3 & | & 1 \\ 3 & -2 & 4 & | & -2 \\ 1 & -1 & -4 & | & 0 \end{bmatrix}$

5. no   **6.** yes   **7.** no   **8.** no

9. $\begin{bmatrix} 1 & -2 & | & 1 \\ 0 & 1 & | & -1 \end{bmatrix}$ **10.** $\begin{bmatrix} 1 & 4 & | & 1 \\ 0 & -10 & | & 1 \end{bmatrix}$

11. $\begin{bmatrix} 1 & -4 & 3 & | & -1 \\ 0 & -2 & 3 & | & -2 \\ 0 & 1 & -4 & | & 8 \end{bmatrix}$

12. $\begin{bmatrix} -2 & 0 & -4 & -9 & | & -2 \\ 0 & 2 & 2 & 3 & | & -2 \\ 0 & 0 & 1 & -2 & | & 4 \\ 0 & -1 & 4 & -5 & | & 7 \end{bmatrix}$

13. $\begin{bmatrix} 1 & 0 & | & \frac{3}{5} \\ 0 & 1 & | & -\frac{1}{5} \end{bmatrix}$ **14.** $\begin{bmatrix} 1 & 0 & 0 & | & \frac{4}{5} \\ 0 & 1 & 0 & | & -\frac{3}{5} \\ 0 & 0 & 1 & | & -\frac{2}{5} \end{bmatrix}$

15. $\begin{bmatrix} 1 & 0 & 0 & | & -4 \\ 0 & 1 & 0 & | & 8 \\ 0 & 0 & 1 & | & -4 \end{bmatrix}$ **16.** $\begin{bmatrix} 1 & 0 & 0 & | & 4 \\ 0 & 1 & 0 & | & -\frac{3}{5} \\ 0 & 0 & 1 & | & -\frac{13}{5} \end{bmatrix}$

17. $x = \frac{5}{4}$, $y = \frac{7}{8}$ **18.** $x = -3$, $y = -4$

19. $x = 2$, $y = 1$ **20.** $x = -4$, $y = 6$

21. $x = -\frac{74}{21}$, $y = -\frac{73}{21}$, $z = -\frac{3}{7}$

22. $x = -3$, $y = 1$, $z = 7$

23. $x = 1$, $y = 3$, $z = -5$

24. $x = \frac{1}{2}$, $y = 0$, $z = 3$

25. $x = -\frac{3}{7}a - 2$, $y = \frac{2}{7}a + 2$, $z = a$

26. $x = -\frac{5}{11}a + \frac{42}{11}$, $y = a$, $z = \frac{7}{11}a - \frac{6}{11}$

27. $y = -0.005x^2 + 0.45x - 3.89$

28. IRA: \$3500, mutual fund: \$6500, stock: \$10,000

29. not defined

30. not defined

31. $\begin{bmatrix} 3 & 5 & 2 \\ 7 & 8 & 1 \end{bmatrix}$

32. $\begin{bmatrix} 7 & -1 \\ 9 & 8 \end{bmatrix}$

33. $\begin{bmatrix} 9 & -4 \\ 9 & 9 \end{bmatrix}$

34. $\begin{bmatrix} 7 & 5 & 8 \\ 15 & 10 & -1 \end{bmatrix}$

35. $\begin{bmatrix} 4 & 13 \\ 18 & 11 \end{bmatrix}$

36. $\begin{bmatrix} -5 & 15 & -15 \\ -7 & 17 & 10 \end{bmatrix}$

37. $\begin{bmatrix} 0 & -19 \\ -18 & -9 \end{bmatrix}$

38. $\begin{bmatrix} -3 & 25 & -17 \\ -1 & 31 & 14 \end{bmatrix}$

39. $\begin{bmatrix} -7 & -11 & -8 \\ 3 & 7 & 2 \end{bmatrix}$

40. $\begin{bmatrix} 15 & -8 & 15 \\ 29 & -1 & 43 \end{bmatrix}$

41. $\begin{bmatrix} 10 & -13 \\ 18 & -20 \end{bmatrix}$

42. $\begin{bmatrix} -17 & -17 \\ 9 & 7 \end{bmatrix}$

43. $\begin{bmatrix} 17 & -8 & 18 \\ 33 & 0 & 42 \end{bmatrix}$

44. $\begin{bmatrix} 11 & 39 & -1 \\ 30 & 94 & 5 \end{bmatrix}$

45. $\begin{bmatrix} 10 & 9 & 20 \\ 22 & -4 & 2 \end{bmatrix}$

46. not defined

47. yes   **48.** no   **49.** yes

50. no

51. $\begin{bmatrix} \frac{2}{5} & -\frac{1}{5} \\ \frac{3}{10} & \frac{1}{10} \end{bmatrix}$

52. $\begin{bmatrix} \frac{3}{8} & -\frac{7}{16} \\ \frac{1}{4} & -\frac{1}{8} \end{bmatrix}$

53. $\begin{bmatrix} 0 & -\frac{1}{2} \\ 1 & 0 \end{bmatrix}$

54. $\begin{bmatrix} \frac{1}{2} & \frac{1}{4} \\ \frac{1}{2} & \frac{3}{4} \end{bmatrix}$

55. $\begin{bmatrix} -\frac{1}{6} & \frac{7}{12} & -\frac{1}{12} \\ \frac{1}{2} & -\frac{1}{4} & -\frac{1}{4} \\ \frac{1}{6} & -\frac{1}{12} & -\frac{5}{12} \end{bmatrix}$

56. $\begin{bmatrix} -\frac{1}{2} & \frac{1}{10} & -\frac{1}{5} \\ 1 & 0 & 0 \\ \frac{1}{2} & \frac{3}{10} & \frac{2}{5} \end{bmatrix}$

57. $\begin{bmatrix} 0 & -\frac{2}{5} & \frac{1}{5} \\ 1 & -\frac{2}{5} & \frac{1}{5} \\ -\frac{1}{2} & \frac{3}{10} & \frac{1}{10} \end{bmatrix}$

58. $\begin{bmatrix} -\frac{2}{11} & \frac{27}{77} & -\frac{2}{77} \\ 0 & \frac{3}{7} & -\frac{1}{7} \\ \frac{1}{11} & -\frac{8}{77} & \frac{12}{77} \end{bmatrix}$

59. $x = 5, y = 4$

60. no solution   **61.** $x = 8, y = 12$

62. $x = \frac{112}{13}, y = -\frac{97}{13}, z = 3$

63. $x = 1, y = 2, z = 3$

64. $x = a, y = \frac{1}{2}a - \frac{1}{7}, z = \frac{5}{4}a - \frac{83}{28}$

65. -8 **66.** -16 **67.** 5.4

68. -2 **69.** $x = 3, y = 1$

70. $x = -3, y = 8$ **71.** $x = 6, y = 0$

72. $x = -\frac{19}{4}, y = -\frac{3}{4}$ **73.** $x = 90, y = 155$

74. $x = 4, y = -2$ **75.** 11 **76.** -11

77. $-abd$ **78.** 54 **79.** $x = 1, y = 1, z = 2$

80. $x = -14, y = 13.6, z = 2.6$

81. $x = -\frac{15}{7}, y = -\frac{25}{7}, z = \frac{19}{14}$

82. $x = -\frac{5}{2}, y = \frac{17}{8}, z = \frac{3}{8}$ **83.** 1 unit^2

84. not collinear

85. a. rref([A])
```
[[1  0  0   .161066…
 [0  1  0  -.04946…
 [0  0  1  -4.1012…
```
$y = 0.16x^2 - 0.05x - 4.10$

b. QuadReg
```
y=ax²+bx+c
a=.1610661269
b=-.0494601889
c=-4.101214575
```
$y = 0.16x^2 - 0.05x - 4.10$

86. a. QuadReg
```
y=ax²+bx+c
a=.2862337662
b=-2.570649351
c=-4.862337662
```
$y = 0.29x^2 - 2.57x - 4.86$

b. QuadReg
```
y=ax²+bx+c
a=.3616161616
b=-.9232323232
c=-17.6
```
$y = 0.29x^2 - 2.57x - 4.86$

87. $\begin{bmatrix} -238 & 206 & 50 \\ -113 & 159 & 135 \\ 40 & -30 & 0 \end{bmatrix}$

88. $\begin{bmatrix} -143 & -41 \\ -82 & 64 \end{bmatrix}$

89. $x = 2.25, y = -4.35$

90. $x = 4.15, y = -6.26, z = 2.54$

91. $x = -9.5, y = 3.4$

92. $x = 8.5, y = -1.2, z = 4.6$

Practice Test

1. $\begin{bmatrix} 1 & -2 & | & 1 \\ -2 & 3 & | & 2 \end{bmatrix}$

2. $\begin{bmatrix} 3 & 5 & | & -2 \\ 7 & 11 & | & -6 \end{bmatrix}$

3. $\begin{bmatrix} 6 & 9 & 1 & | & 5 \\ 2 & -3 & 1 & | & 3 \\ 10 & 12 & 2 & | & 9 \end{bmatrix}$ **4.** $\begin{bmatrix} 3 & 2 & -10 & | & 2 \\ 1 & 1 & -1 & | & 5 \end{bmatrix}$

5. $\begin{bmatrix} 1 & 3 & 5 \\ 0 & 1 & -11 \\ 0 & 7 & 15 \end{bmatrix}$ **6.** $\begin{bmatrix} 1 & 0 & 0 & | & 1 \\ 0 & 1 & -1 & | & -1 \\ 0 & 0 & 0 & | & 0 \end{bmatrix}$

7. $x = -\frac{1}{3}a + \frac{7}{6}$, $y = \frac{1}{9}a - \frac{2}{9}$, $z = a$

8. $x = 8t - 8$, $y = -7t + 13$, $z = t$

9. $\begin{bmatrix} -11 & 19 \\ -6 & 8 \end{bmatrix}$ **10.** not possible

11. $\begin{bmatrix} \frac{1}{19} & \frac{3}{19} \\ \frac{5}{19} & -\frac{4}{19} \end{bmatrix}$ **12.** $\begin{bmatrix} \frac{1}{9} & \frac{19}{90} & -\frac{2}{45} \\ -\frac{2}{9} & \frac{7}{90} & \frac{4}{45} \\ \frac{1}{9} & \frac{1}{90} & \frac{7}{45} \end{bmatrix}$

13. no inverse   **14.** $x = -7$, $y = 2$

15. $x = -3, y = 1, z = 7$ **16.** 3

17. -31 **18.** $x = 7$, $y = 3$

19. $x = 1, y = -1, z = 2$

20. 2×2; gives the respective applicant's total score according to each of the two rubrics

21. money market: \$3500, conservative stock: \$4500, aggressive stock: \$7000

22. a. $\begin{cases} 9a - 3b + c = 6 \\ a + b + c = 12 \\ 25a + 5b + c = 7 \end{cases}$

   b. $y = -0.34375x^2 + 0.8125x + 11.53125$

23. $(12.5, -6.4)$ **24.** $x = -5.8$, $y = 10.4$

Cumulative Test

1. $x = 3 \pm 2\sqrt{5}$ **2.** $x = -2$

3. $(x + 3)^2 + (y + 1)^2 = 25$ **4.** not a function

5. even

6. Translate to the right 4 units and vertically compress by a factor of 5.

7. $f(g(x)) = \left(\frac{1}{x}\right)^3 - 1 = \frac{1}{x^3} - 1$

   domain: $(-\infty, 0) \cup (0, \infty)$

8. $f^{-1}(x) = (x + 1)^3$ **9.** $\left(-\frac{6}{5}, -\frac{3}{5}\right)$

10. $p(x) = C\left(x + \sqrt{7}\right)^2 (x - 0)^3 \left(x - \sqrt{7}\right)^2$

11. $Q(x) = 5x - 4$, $r(x) = -5x + 7$

12. $P(x) = (x + 4i)(x - 4i)(x + 5)(x - 3)$

13. VA: $x = -2$, $x = 3$ HA: $y = 0.7$

14. \$5,967.92 **15.** $(-3, \infty)$ **16.** 0

17. $x = 4$ **18.** 27.47 years **19.** no solution

20. burger: \$2.07, fries: \$1.19, soda: \$1.29

21. maximum $z\left(\frac{10}{7}, \frac{10}{7}\right) = \frac{120}{7}$

22. $x = 2, y = -3, z = 1$

23. $\begin{bmatrix} 78 & -10 & 40 \\ 26 & 0 & 14 \end{bmatrix}$ **24.** $x = -3$, $y = 1$

25. $x = \frac{3}{11}$, $y = -\frac{2}{11}$

26. x-intercepts: $(-6.53, 0)$, $(-0.84, 0)$, $(2.37, 0)$ relative maximum $(-4.27, 51.55)$,

27. $\begin{bmatrix} 35 & 12 \\ -19 & -14 \end{bmatrix}$, $\begin{bmatrix} \frac{7}{131} & \frac{6}{131} \\ -\frac{19}{262} & -\frac{35}{262} \end{bmatrix}$

CHAPTER 11

Section 11.1

1. Hyperbola   **2.** Ellipse
3. Circle   **4.** Circle
5. Hyperbola   **6.** Hyperbola
7. Ellipse   **8.** Ellipse
9. Parabola   **10.** Parabola
11. Circle   **12.** Circle

Section 11.2

1. c   **2.** a
3. d   **4.** b
5. c   **6.** d
7. a   **8.** b
9. $x^2 = 12y$ **10.** $y^2 = 8x$
11. $y^2 = -20x$ **12.** $x^2 = -16y$
13. $(x - 3)^2 = 8(y - 5)$ **14.** $(y - 5)^2 = 16(x - 3)$
15. $(y - 4)^2 = -8(x - 2)$
16. $(x - 2)^2 = -20(y - 4)$
17. $(x - 2)^2 = 12(y - 1)$

18. $(x-2)^2 = -12(y-1)$

19. $(y+1)^2 = 4(x-2)$

20. $(y-5)^2 = -12(x-2)$

21. $(y-2)^2 = 8(x+1)$

22. $(x-2)^2 = 8(y+1)$

23. $(x-2)^2 = -8(y+1)$

24. $(y-2)^2 = -8(x+1)$

25. Vertex: $(0,0)$
Focus: $(0,2)$
Directrix: $y = -2$
Length of latus rectum: 8

26. Vertex: $(0,0)$
Focus: $(0,-3)$
Directrix: $y = 3$
Length of latus rectum: 12

27. Vertex: $(0,0)$
Focus: $(-\frac{1}{2}, 0)$
Directrix: $x = \frac{1}{2}$
Length of latus rectum: 2

28. Vertex: $(0,0)$
Focus: $(\frac{3}{2}, 0)$
Directrix: $x = -\frac{3}{2}$
Length of latus rectum: 6

29. Vertex: $(0,0)$
Focus: $(0,4)$
Directrix: $y = -4$
Length of latus rectum: 16

30. Vertex: $(0,0)$
Focus: $(0,-2)$
Directrix: $y = 2$
Length of latus rectum: 8

31. Vertex: $(0,0)$
Focus: $(1,0)$
Directrix: $x = -1$
Length of latus rectum: 4

32. Vertex: $(0, 0)$
Focus: $(-4, 0)$
Directrix: $x = 4$
Length of latus rectum: 16

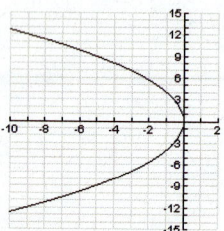

33. Vertex: $(-3, 2)$
Focus: $(-2, 2)$
Directrix: $x = -4$
Length of latus rectum: 4

34. Vertex: $(1, -2)$
Focus: $(0, -2)$
Directrix: $x = 2$
Length of latus rectum: 4

35. Vertex: $(3, -1)$
Focus: $(3, -3)$
Directrix: $y = 1$
Length of latus rectum: 8

36. Vertex: $(-3, 2)$
Focus: $(-3, 0)$
Directrix: $y = 4$
Length of latus rectum: 8

37. Vertex: $(-5, 0)$
Focus: $(-5, -\frac{1}{2})$
Directrix: $y = \frac{1}{2}$
Length of latus rectum: 2

38. Vertex: $(-1, 0)$
Focus: $(-5, 0)$
Directrix: $x = 3$
Length of latus rectum: 16

39. Vertex: $(0, 2)$
Focus: $(\frac{1}{2}, 2)$
Directrix: $x = -\frac{1}{2}$
Length of latus rectum: 2

40. Vertex: $(3, 0)$

Focus: $(3, -\frac{1}{2})$

Directrix: $y = \frac{1}{2}$

Length of latus rectum: 2

41. Vertex: $(-3, -1)$

Focus: $(-1, -1)$

Directrix: $x = -5$

Length of latus rectum: 8

42. Vertex: $(3, \frac{1}{4})$

Focus: $(3, \frac{5}{4})$

Directrix: $y = -\frac{3}{4}$

Length of latus rectum: 4

43. Vertex: $(\frac{1}{2}, \frac{5}{4})$

Focus: $(\frac{1}{2}, 1)$

Directrix: $y = \frac{3}{2}$

Length of latus rectum: 1

44. Vertex: $(\frac{3}{4}, -\frac{1}{2})$

Focus: $(1, -\frac{1}{2})$

Directrix: $x = \frac{1}{2}$

Length of latus rectum: 1

45. The focus will be at $(0, 2)$ so the receiver should be placed 2 feet from the vertex.

46. The focus will be at $(0, 5)$ so the receiver should be placed 5 feet from the vertex.

47. $x = \frac{1}{8}y^2$, $-2.5 \leq y \leq 2.5$

48. The lens is $2\sqrt{6}$ cm wide, and is 0.5 cm from the vertex.

49. $x^2 = 160y$, $-50 \leq x \leq 50$

50. $x^2 = 100y$

51. Yes. The opening height is 18.75 feet, and the mast is only 17 feet.

52. Yes. The opening height is 21 feet, and the RV straddle is only 10 feet.

53. Focal length = 374.25 ft

Equation: $x^2 = 1497y$ (measurement in feet)

54. $(y - 10) = 0.0022x^2$

55. Maximum profit is achieved when 3,000 units are produced.

56. Maximum profit is achieved when 4,000 units are produced.

57. The correction that needs to be made is that the formula $y^2 = 4px$ should be used.

58. The correction that needs to be made is that the formula $(y-2)^2 = 4p(x-3)$ should be used.

59. true   **60.** false   **61.** false

62. – 64. See Instructor's Solution Manual

65.

66.

67. The vertex is located at $(2.5, -3.5)$. The parabola opens to the right.

68. The vertex is located at $(-1.4, -1.7)$. The parabola opens downwards.

69. The vertex is located at $(1.8, 1.5)$. The parabola opens to the left.

70. The vertex is at $(-2.4, 3.2)$. The parabola open upwards.

71. $y = 2.25 \pm \sqrt{4x + 14}$

b. Vertex at $(-3.5, 2.25)$, opens to the right.
c. Yes, (a) and (b) agree with each other.

72. a. $y = 7.2 - 2(x + 1.6)^2$

b. Vertex at $(-1.6, 7.2)$, opens downwards.
c. Yes, (a) and (b) agree with each other.

Section 11.3

1. d   **2.** b   **3.** a   **4.** c

5.

6.

7.

8.

9.

10.

11.

12.

13.

14.

15.

16.

17. $\dfrac{x^2}{36} + \dfrac{y^2}{20} = 1$ **18.** $\dfrac{x^2}{9} + \dfrac{y^2}{8} = 1$

19. $\dfrac{x^2}{7} + \dfrac{y^2}{16} = 1$ **20.** $\dfrac{x^2}{3} + \dfrac{y^2}{4} = 1$

21. $\dfrac{x^2}{4} + \dfrac{y^2}{16} = 1$ **22.** $\dfrac{x^2}{25} + y^2 = 1$

23. $\dfrac{x^2}{9} + \dfrac{y^2}{49} = 1$ **24.** $\dfrac{x^2}{81} + \dfrac{y^2}{16} = 1$

25. c **26.** d **27.** b **28.** a

29.

30.

31.

32.

33.

34.

35.

36.

Center (1, 0)

37.

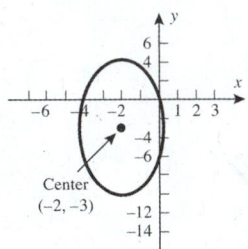

Center (–2, –3)

38.

Center (–2, –1)

39. $\dfrac{(x-2)^2}{25}+\dfrac{(y-5)^2}{9}=1$

40. $\dfrac{(x-3)^2}{9}+\dfrac{(y+2)^2}{8}=1$

41. $\dfrac{(x-4)^2}{7}+\dfrac{(y+4)^2}{16}=1$

42. $\dfrac{(x-2)^2}{3}+\dfrac{(y+5)^2}{4}=1$

43. $\dfrac{(x-3)^2}{4}+\dfrac{(y-2)^2}{16}=1$

44. $\dfrac{(x+4)^2}{25}+(y-3)^2=1$

45. $\dfrac{(x+1)^2}{9}+\dfrac{(y+4)^2}{25}=1$

46. $\dfrac{(x-2)^2}{16}+\dfrac{(y-3)^2}{4}=1$

47. $\dfrac{x^2}{225}+\dfrac{y^2}{5625}=1$

48. $x^2+y^2=8100$

49. a. $\dfrac{x^2}{5625}+\dfrac{y^2}{400}=1$ **b.** Let $x=60$, then $y=12$. The field extends 15 feet in that direction so the track will not encompass the field.

50. The minor axis should be at least 100 yards wide in order to enclose the field.

51. $\dfrac{x^2}{5.914{,}000{,}000^2}+\dfrac{y^2}{5{,}729{,}000{,}000^2}=1$

52. $\dfrac{(x-(2.75\times10^6))^2}{2.246\times10^{16}}+\dfrac{(y-0)^2}{2.245\times10^{16}}=1$

53. $\dfrac{x^2}{150{,}000{,}000^2}+\dfrac{y^2}{146{,}000{,}000^2}=1$

54. $\dfrac{x^2}{1.225\times10^{17}}+\dfrac{y^2}{7.326\times10^{16}}=1$

55. straight line

56. $\dfrac{x^2}{316.84}+\dfrac{y^2}{20.5664}=1$ (in AU units)

57. a. $\dfrac{x^2}{64}+\dfrac{y^2}{25}=1$ **b.** $p\approx42$ inches **c.** about 1509

58. a. $\dfrac{x^2}{81}+\dfrac{y^2}{25}=1$ **b.** 46 inches **c.** 1378

59. $a=6$ and $b=4$ is incorrect. In the formula a and b are being squared; therefore $a=\sqrt6$ and $b=2$.

60. $c^2=a^2-b^2$ in place of $c^2=a^2+b^2$.

61. false **62.** true

63. true **64.** false

65. Pluto: $e\approx0.25$ Earth: $e\approx0.02$

66. a. Nearly circular **b.** Very narrow in the y-direction **c.** $b=\sqrt3\,c$

67. Ellipse becomes more elongated.

68. Ellipse becomes more elongated.

```
............    10x^2+y^2=1
_____     5x^2+y^2=1
-----------     x^2+y^2=1
```

69. Circle gets smaller as c increases.

70. This is the graph of a hyperbola, not an ellipse.

71. As c decreases, the major axis along the x-axis increases.

72. As c decreases, the major axis along the y-axis increases.

Section 11.4

1. b   **2.** a   **3.** d   **4.** c

5.

6.

7.

8.

9.

10.

11.

12.

13.

14.

15.

16.

17. $\dfrac{x^2}{16} - \dfrac{y^2}{20} = 1$ **18.** $x^2 - \dfrac{y^2}{8} = 1$

19. $\dfrac{y^2}{9} - \dfrac{x^2}{7} = 1$ **20.** $y^2 - \dfrac{x^2}{3} = 1$

21. $x^2 - y^2 = a^2$ **22.** $y^2 - x^2 = a^2$

23. $\dfrac{y^2}{4} - x^2 = b^2$ **24.** $x^2 - \dfrac{y^2}{4} = a^2$

25. c   **26.** d   **27.** b   **28.** a

29.

30.

31.

32.

33.

34.

35.

36.

37.

38.

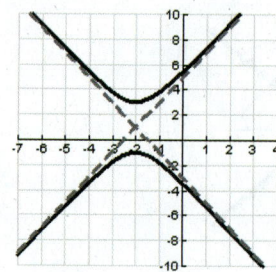

39. $\dfrac{(x-2)^2}{16} - \dfrac{(y-5)^2}{9} = 1$

40. $(x-2)^2 - \dfrac{(y+2)^2}{3} = 1$

41. $\dfrac{(y+4)^2}{9} - \dfrac{(x-4)^2}{7} = 1$

42. $(y+5)^2 - \dfrac{(x-2)^2}{3} = 1$

43. The ship will come ashore between the two stations (28.5 miles from one and 121.5 miles from the other).

44. The ship will come ashore between the two stations 84.9 miles from one and 215.1 miles from the other.

45. 0.000484 seconds   **46.** 0.00107 seconds

47. $y^2 - \frac{4}{5}x^2 = 1$

48. The explosive should be placed at $(1500, 300)$.

49. 275 feet   **50.** 453 feet

51. The transverse axis is vertical. The points are $(\pm 3, 0)$. The vertices are $(0, \pm 2)$.

52. The vertices are $(1, 0)$, $(-1, 0)$ and the points are $(0, 2)$, $(0, -2)$.

53. false   **54.** true

55. true   **56.** false

57. $x^2 - y^2 = a^2$ or $y^2 - x^2 = a^2$; note that $a = b$.

58. $\dfrac{(x-1)^2}{4} - \dfrac{(y+2)^2}{16} = 1$

59.

60.

.... Innermost curve

- - - - - - - Middle Curve

———— Outermost Curve

61. As c decreases, the vertices of the hyperbolas located at $\left(\pm\dfrac{1}{c}, 0\right)$ are moving away from the origin.

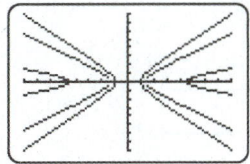

62. As c decreases, the vertices of the hyperbola remain at $(\pm 1, 0)$, but open outward more narrowly.

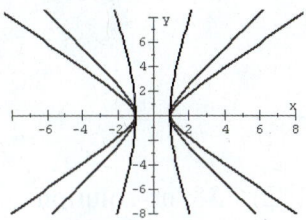

Section 11.5

1. $(2, 6), (-1, 3)$ **2.** $(3, -7), (-1, 1)$

3. $(1, 0)$ **4.** $(1, -1)$

5. no solution   **6.** no solution

7. $(0, 1)$ **8.** $(0, -1)$

9. $(-0.63, -1.61), (0.63, -1.61)$

10. $(0.583, 2.379), (-0.583, 2.379)$

11. no solution

12. $(1.572, 1.591), (-1.572, 1.591),$
$(1.849, -1.257), (-1.849, -1.257)$

13. $(1, 1)$ **14.** $(-1, 1)$

15. $\left(-\sqrt{2}, -2\sqrt{2}\right), \left(2\sqrt{2}, \sqrt{2}\right),$
$\left(\sqrt{2}, 2\sqrt{2}\right), \left(-2\sqrt{2}, -\sqrt{2}\right)$

16. $\left(-3.346, 0.897\right), \left(3.346, -0.897\right),$
$\left(-0.896, 3.348\right), \left(0.896, -3.348\right)$

17. $(-6, 33), (2, 1)$

18. $(1.854, 1.562), (-4.854, -18.562)$

19. $(3, 4)\ (-2, -1)$

20. $(3, -4), (-4, 3)$

21. $(0, -3), \left(\frac{2}{5}, -\frac{11}{5}\right)$

22. $(3, 3), \left(\frac{7}{5}, -\frac{1}{5}\right)$ **23.** $(-1, -1), \left(\frac{1}{4}, \frac{3}{2}\right)$

24. $(1, -1), \left(\frac{1}{6}, \frac{3}{2}\right)$ **25.** $(-1, -4), (4, 1)$

26. $(-3, 1), (1, -3)$ **27.** $(1, 3), (-1, -3)$

28. $(1, 0), (-1, 0)$ **29.** $(2, 4)$

30. $\left(\frac{1}{4}, \frac{1}{4}\right)$ **31.** $\left(\frac{1}{2}, \frac{1}{3}\right), \left(\frac{1}{2}, -\frac{1}{3}\right)$

32. $(2, 3), (2, -3), (-2, 3), (-2, -3)$

33. no solution   **34.** no solution

35.

36. no solution

37. $(-1, -1)$ and $(1, 2)$ **38.** no solution

39. 3 and 7 **40.** 7 and 10

41. 8 and 9; -8 and -9

42. 8 and 10; -8 and -10

43. $8\,\text{cm} \times 10\,\text{cm}$

44. The radii of the two concentric circles should be 3 units and 5 units.

45. $400\,\text{ft} \times 500\,\text{ft}$ or $\frac{1000}{3}\,\text{ft} \times 600\,\text{ft}$

46. $75\,\text{ft} \times 150\,\text{ft}$

47. Professor: 2 meters per second
Jeremy: 10 meters per second

48. You: 300 meters per minute
Jeremy: 600 meters per minute

49. Can't use elimination. Should have used substitution.

50. $(1, 2), (-1, -2)$

51. false **52.** true

53. false **54.** false

55. $2n$ **56.** n

57. $\left.\begin{array}{l} y = x^2 + 1 \\ y = 1 \end{array}\right\}$ other answers possible

58. other answer possible

59. no solution

60. no solution

61. $(-1.57, -1.64)$

62. no solution

The first curve is right above.

63. $(1.067, 4.119), (1.986, 0.638),$
$(-1.017, -4.757)$

64. $(1.035, 3.966), (1.899, -2.385), (-1.011,$
$4.664), (-1.759, -2.165).$

Section 11.6

1. b　　**2.** i　　**3.** j　　**4.** g
5. h　　**6.** f　　**7.** c　　**8.** l
9. d　　**10.** a　　**11.** k　　**12.** e

13.

14.

15.

16.

17.

18.

19.

20.

21.

22.

23.

24.

25.

26.

27.

28.

29.

30.

31.

32.

33.

34.

35.

36.

37.

38.

39.

40.

41.

42.

43.

44.

45.

46.

47.

48.

49.

50.

51. $\dfrac{9\pi}{2}$ **52.** $\dfrac{5\pi}{4}$

53. No solution (should have shaded inside the inner circle and outside the outer circle, resulting in no overlap).

54. The shading is wrong.

55. false   **56.** false   **57.** $b \geq a$

58. No solution for real numbers.

59.

60.

61.

62.

63.

64.

65.

66.

Section 11.7

1. $\left(3\sqrt{2}, \sqrt{2}\right)$ **2.** $\left(\dfrac{5+\sqrt{3}}{2}, \dfrac{1-5\sqrt{3}}{2}\right)$

3. $\left(-\dfrac{3\sqrt{3}}{2}+1, \dfrac{3}{2}+\sqrt{3}\right)$ **4.** $\left(\sqrt{2}, 5\sqrt{2}\right)$

5. $\left(-\dfrac{1}{2}-\dfrac{3\sqrt{3}}{2}, \dfrac{\sqrt{3}}{2}-\dfrac{3}{2}\right)$ **6.** $\left(0, -4\sqrt{2}\right)$

7. $\left(\dfrac{3\sqrt{3}}{2}, \dfrac{3}{2}\right)$ **8.** $\left(-\sqrt{3}, 1\right)$

9. hyperbola; $\dfrac{X^2}{2}-\dfrac{Y^2}{2}=1$

10. hyperbola; $\dfrac{X^2}{8}-\dfrac{Y^2}{8}=1$

11. parabola; $2X^2 - 2Y - 1 = 0$

12. parabola; $4Y^2 + Y - X + 1 = 0$

13. hyperbola; $\dfrac{X^2}{6} - \dfrac{Y^2}{2} = 1$

14. hyperbola; $\dfrac{Y^2}{2} - \dfrac{X^2}{6} = 1$

15. ellipse; $\dfrac{X^2}{2} + \dfrac{Y^2}{1} = 1$

16. ellipse; $\dfrac{X^2}{10} + \dfrac{Y^2}{18} = 1$

17. parabola; $2X^2 - 2Y - 1 = 0$

18. parabola; $X^2 + Y - 1 = 0$

19. ellipse; $\dfrac{X^2}{1} + \dfrac{Y^2}{9} = 1$

20. $\dfrac{X^2}{4} + \dfrac{Y^2}{25} = 1$

21. hyperbola; $\dfrac{X^2}{3} - \dfrac{Y^2}{2} = 1$

22. hyperbola; $\dfrac{Y^2}{4} - \dfrac{X^2}{3} = 1$

23. parabola; $Y^2 - X - 4 = 0$

24. parabola; $X^2 + 2X - Y = 0$

25. 45°   **26.** 45°
27. 60°   **28.** 30°
29. 30°   **30.** 75°
31. 45°   **32.** 45°
33. 15°   **34.** 22.5°
35. ≈ 40.3°   **36.** ≈ 84.8°
37. ≈ 50.7°   **38.** ≈ 10.3°

39. $\dfrac{X^2}{4} + \dfrac{Y^2}{9} = 1$;
rotation of 60°

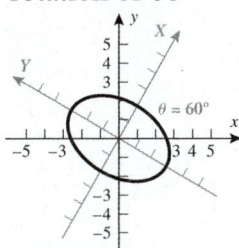

40. $\dfrac{X^2}{1} + \dfrac{Y^2}{4} = 1$;
rotation of 45°

41. $\dfrac{X^2}{9} - \dfrac{Y^2}{1} = 1$;
rotation of 45°

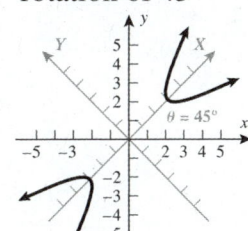

42. $\dfrac{Y^2}{9} - \dfrac{X^2}{4} = 1$;
rotation of 30°

43. $X^2 - Y - 3 = 0$;
rotation of 30°

44. $Y^2 - X - 1 = 0$;
rotation of 60°

45. $\dfrac{X^2}{25} + \dfrac{Y^2}{4} = 1$;
rotation of 30°

46. $\dfrac{X^2}{25} - \dfrac{Y^2}{4} = 1$;
rotation of 60°

47. $X^2 + 4X - Y = 0$;
rotation of 45°

48. $\dfrac{X^2}{9} + \dfrac{Y^2}{1} = 1$;
rotation of 60°

49. true   **50.** false
51. true   **52.** false

53. a. For 90°, the new equation is $\dfrac{x^2}{b^2} + \dfrac{y^2}{a^2} = 1$.

    b. For 180°, the original equation results.

54. a. For 90°, the new equation is $\dfrac{x^2}{b^2} - \dfrac{y^2}{a^2} = 1$.

    b. For 180°, the original equation results.

55. $a < 0$ hyperbola; $a = 0$ parabola; $a > 0$,
$a \ne 1$ ellipse; $a = 1$ circle.

56. $a > 0$ hyperbola; $a = 0$ parabola; $a < 0$,
$a \ne -1$ ellipse; $a = -1$ circle.

57. Amount of rotation is 30°.

    a.

    b.

58. a.

b.

59. The amount of rotation is about 36°.

a.

b.

60. a.

b.

61. a. The amount of rotation is about 63°. It is a parabola.

b. The amount of rotation is about 19°. It is a hyperbola.

c. The amount of rotation is about 26.5°.

62. a.

b.

Section 11.8

1. $r = \dfrac{5}{2 - \sin\theta}$ **2.** $r = \dfrac{3}{3 + \sin\theta}$

3. $r = \dfrac{8}{1 + 2\sin\theta}$ **4.** $r = \dfrac{6}{1 - 3\sin\theta}$

5. $r = \dfrac{1}{1 + \cos\theta}$ **6.** $r = \dfrac{1}{1 - \cos\theta}$

7. $r = \dfrac{6}{4 + 3\cos\theta}$ **8.** $r = \dfrac{8}{3 - 2\cos\theta}$

9. $r = \dfrac{12}{3 - 4\cos\theta}$ **10.** $r = \dfrac{15}{2 + 3\cos\theta}$

11. $r = \dfrac{3}{1 - \sin\theta}$ **12.** $r = \dfrac{4}{1 + \sin\theta}$

13. $r = \dfrac{18}{5 + 3\sin\theta}$ **14.** $r = \dfrac{40}{5 + 8\sin\theta}$

15. parabola   **16.** hyperbola
17. ellipse   **18.** parabola
19. hyperbola   **20.** ellipse
21. ellipse   **22.** hyperbola
23. parabola   **24.** ellipse
25. hyperbola   **26.** parabola

27. parabola; $e = 1$

28. parabola; $e = 1$

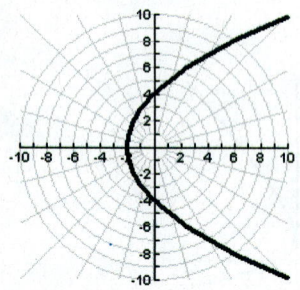

29. hyperbola; $e = 2$

30. hyperbola; $e = \frac{8}{3} > 1$

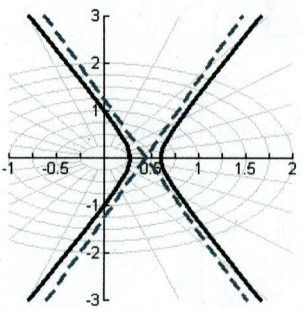

31. ellipse; $e = \frac{1}{2}$

32. ellipse; $e = \frac{1}{3} < 1$

33. parabola; $e = 1$

34. hyperbola; $e = 2$

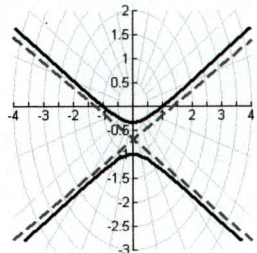

35. ellipse; $e = \frac{1}{3}$

36. ellipse; $e = \frac{4}{5} < 1$

37. hyperbola; $e = \frac{3}{2}$

38. parabola; $e = 1$

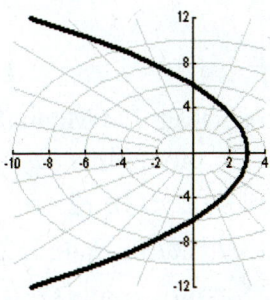

39. parabola; $e = 1$

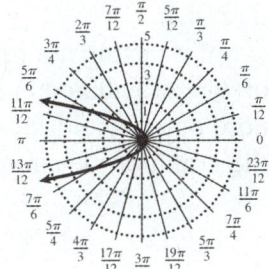

40. ellipse; $e = \frac{1}{2} < 1$

41. $e = 0.248$; $a = 5{,}913{,}500{,}000$ km;

$$r = \frac{5{,}913{,}500{,}000(1 - 0.248^2)}{1 - 0.248\cos\theta}$$

42. $e \approx 0.0184$; $a \approx 149{,}850{,}000$

$$r = \frac{149{,}850{,}000(1 - 0.0184^2)}{1 - 0.0184\cos\theta}$$

43. $r = \dfrac{75{,}000{,}000(1-0.223^2)}{1-0.223\cos\theta}$

44. $r = \dfrac{350{,}000{,}000(1-0.634^2)}{1-0.634\cos\theta}$

45. Conic becomes more elliptic as $e \to 1$ and more circular as $e \to 0$.

46. – 48. See Instructor's Solution Manual

49. $\dfrac{2ep}{1-e^2}$ **50.** $\dfrac{2ep\sqrt{1-e^2}}{1-e^2}$

51. $\left(\dfrac{\dfrac{ep}{1-e} - \dfrac{ep}{1+e}}{2}, \pi \right)$

52. – 60. See Instructor's Solution Manual

61. a. With θ step $= \pi/3$, plot points $(2, 0)$, $(14.93, \pi/3)$, $(14.93, 2\pi/3)$, $(2, \pi)$, $(1.07, 4\pi/3)$, $(1.07, 5\pi/3)$, and $(2, 2\pi)$.

   b. With θ step $= 0.8\pi$, plot points $(2, 0)$, $(4.85, 0.8\pi)$, $(1.03, 1.6\pi)$, $(40.86, 2.4\pi)$, $(1.26, 3.2\pi)$, and $(2, 4\pi)$.

62. The calculator is sampling using a more refined partition for $\theta = \pi/3$ than it is for $\theta = 0.8\pi$. Using more points yields a better approximation to the graph of the curve.

Section 11.9

1.

2.

3.

4.

5.

6.

7.

8.

9.

10.

11.

12.

13.

14.

15.

16.

17.

18.

19.

20.

21.

22.

23. Arrow in different directions, depending on *t*.

24.

25.

26.

27.

28.

29.

30.

31. $y = \dfrac{1}{x^2}$ **32.** $y = x + 2$

33. $y = x - 2$ **34.** $y = \dfrac{1}{9}x^2 - 1$

35. $y = \sqrt{x^2 + 1}$ **36.** $x + y = 1$

37. $x + y = 2$ **38.** $y = x - 1$

39. $x + 4y = 8$ **40.** $y^2 - x^2 = 1$

41. 17.7 seconds

42. Height: 5000 ft; Horizontal distance: 1250 ft

43. yes

44. Height: 46.3 ft; Horizontal distance: 484.5 ft

45. Height: 5742 ft; Horizontal distance: 13,261 ft

46. Height: 46,875 ft; Horizontal distance: 108,253 ft

47. 125 seconds

48. It can definitely hit a target 2 miles away.

49.

50.

51.

t	x	y
0	$A + B$	0
$\dfrac{\pi}{2}$	0	$A + B$
π	$-A - B$	0
$\dfrac{3\pi}{2}$	0	$-A - B$
2π	$A + B$	0

52. In general, multiply the arguments of the trigonometric functions by coefficients larger than 1 in order to increase speed. Note that $A + B$ represents the length of the arm when fully extended.

53. The original domain must be $t \geq 0$; therefore, only the part of the parabola where $y \geq 0$ is part of the plane curve.

54. The original domain must be $t \geq 0$; therefore, only the portion of the parabola where $x \geq 0$ is part of the plane curve.

55. false   **56.** true   **57.** Quarter circle in QI

58. – 60. See Instructor's Solution Manual

61. $\dfrac{x + y}{2a} + \dfrac{y - x}{2b} = 1$ **62.** $x^2 + y^2 = 2a^2$

63. $y = b\sqrt[a]{x}$ **64.** $y = \ln b + \dfrac{x}{a}$

65.

66.

Graph for $a = 3$, $b = 1$

Graph for $a = 4$, $b = 2$

Graph for $a = 6$, $b = 2$

67.

68.

Graph for $a = 2$

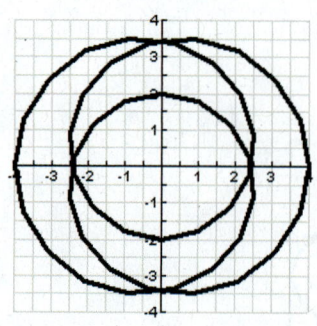

Graph for $a = 3$

69. $a = 2, 0 \le t < 2\pi$

$a = 3, 0 \le t < 2\pi$

70. $a = 2, 0 \le t \le 2\pi$

$a = 3, 0 \le t \le 2\pi$

Review Exercises

1. false **2.** false
3. true **4.** false
5. $y^2 = 12x$ **6.** $x^2 = 8y$
7. $y^2 = -20x$ **8.** $x^2 = -16y$
9. $(x-2)^2 = 8(y-3)$ **10.** $(y+2)^2 = 8(x+1)$
11. $(x-1)^2 = -4(y-6)$
12. $(y-2)^2 = 4(x-1)$
13. Focus: $(0, -3)$
   Directrix: $y = 3$
   Vertex: $(0, 0)$
   Length of latus rectum: 12

14. Focus: $(0, 2)$
   Directrix: $y = -2$
   Vertex: $(0, 0)$
   Length of latus rectum: 8

15. Focus: $\left(\frac{1}{4}, 0\right)$
   Directrix: $x = -\frac{1}{4}$
   Vertex: $(0, 0)$
   Length of latus rectum: 1

16. Focus: $\left(-\frac{3}{2}, 0\right)$
   Directrix: $x = \frac{3}{2}$
   Vertex: $(0, 0)$
   Length of latus rectum: 6

17. Focus: $(3, -2)$
   Directrix: $x = 1$
   Vertex: $(2, -2)$
   Length of latus rectum: 4

18. Focus: $(-2, 2)$
   Directrix: $x = 0$
   Vertex: $(-1, 2)$
   Length of latus rectum: 4

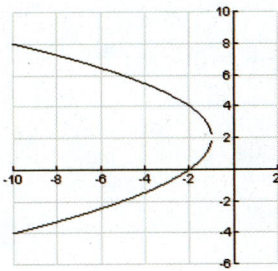

19. Focus: $(-3, -1)$

Directrix: $y = 3$

Vertex: $(-3, 1)$

Length of latus rectum: 8

20. Focus: $(3, -4)$

Directrix: $y = 0$

Vertex: $(3, -2)$

Length of latus rectum: 8

21. Focus: $\left(-\frac{5}{2}, -\frac{79}{8}\right)$

Directrix: $y = -\frac{71}{8}$

Vertex: $\left(-\frac{5}{2}, -\frac{75}{8}\right)$

Length of latus rectum: 2

22. Focus: $(4, -1)$

Directrix: $x = -4$

Vertex: $(0, -1)$

Length of latus rectum: 16

23. The receiver should be placed $\frac{25}{8} = 3.125$ feet from the vertex.

24. The RV will pass under the bridge without a problem.

25.

26.

27.

28.

29. $\dfrac{x^2}{25} + \dfrac{y^2}{16} = 1$ **30.** $\dfrac{x^2}{5} + \dfrac{y^2}{9} = 1$

31. $\dfrac{x^2}{9} + \dfrac{y^2}{64} = 1$ **32.** $\dfrac{x^2}{225} + \dfrac{y^2}{100} = 1$

33.

34.

35.

36. Left point (0.5, 4), center (1, 4), right point (1.5, 4)

37. $\dfrac{(x-3)^2}{25} + \dfrac{(y-3)^2}{9} = 1$

38. $\dfrac{(x-1)^2}{3} + \dfrac{(y+2)^2}{4} = 1$

39. $\dfrac{x^2}{778{,}300{,}000^2} + \dfrac{y^2}{777{,}400{,}000^2} = 1$

40. $\dfrac{(x-(2.1\times10^7))^2}{5.1984\times10^{16}} + \dfrac{(y-0)^2}{5.154\times10^{16}} = 1$

41.

42.

43.

44.

45. $\dfrac{x^2}{9} - \dfrac{y^2}{16} = 1$ **46.** $y^2 - \dfrac{x^2}{8} = 1$

47. $\dfrac{y^2}{9} - x^2 = 1$ (Other answers are possible.)

48. $y^2 - \dfrac{x^2}{4} = a^2$

49.

50.

51.

52.

53. $\dfrac{(x-4)^2}{16} - \dfrac{(y-3)^2}{9} = 1$

54. $\dfrac{(y+1)^2}{1} - \dfrac{(x-4)^2}{3} = 1$

55. The ship will get to shore between the two stations: 65.36 miles from one station and 154.64 miles from the other.

56. The ship will come ashore between the two stations: 125.6 miles from one station and 274.4 miles from the other.

57. $(1, -4)$ and $(-2, -7)$

58. $\left(\pm\sqrt{3}, -1\right), (0, 2)$

59. $(-1, 2)$ and $(1, 2)$ **60.** $(0, 4)$ and $(0, -4)$

61. no solution

62. $(3.864, 1.035), (-3.864, -1.035),$
$(1.035, 3.865), (-1.035, -3.865)$

63. no solution

64. $(0, 3)$ and $\left(-\dfrac{10}{3}, \dfrac{19}{9}\right)$

65. $(-3, -2)$ and $(2, 3)$

66. $\left(\dfrac{3}{2}, \dfrac{3}{4}\right)$ and $\left(-\dfrac{3}{2}, -\dfrac{3}{4}\right)$

67. $\left(-\dfrac{1}{2}, -\dfrac{1}{\sqrt{7}}\right), \left(-\dfrac{1}{2}, \dfrac{1}{\sqrt{7}}\right), \left(\dfrac{1}{2}, -\dfrac{1}{\sqrt{7}}\right),$
$\left(\dfrac{1}{2}, \dfrac{1}{\sqrt{7}}\right)$

68. no solution

69.

70.

71.

72.

73.

74.

75.

76.

77.

78.

79.

80.

81. $\left(-\dfrac{3}{2}+\sqrt{3}, \dfrac{-3\sqrt{3}}{2}+1\right)$

82. $\left(\dfrac{\sqrt{2}}{2}, -\dfrac{7\sqrt{2}}{2}\right)$

83. $\dfrac{X^2}{4} - \dfrac{Y^2}{4} = 1$

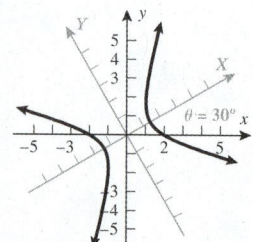

84. $\dfrac{7X^2}{288} - \dfrac{7Y^2}{288} = 1;\ \theta = 45°$

85. $60°$ **86.** $45°$

87. $X^2 + 4 = Y$

88. $\dfrac{X^2}{1} + \dfrac{Y^2}{25} = 1$

89. $r = \dfrac{21}{7 - 3\sin\theta}$ **90.** $r = \dfrac{2}{1 + \cos\theta}$

91. hyperbola   **92.** ellipse

93. $e = \frac{1}{2}$; vertices $\left(\frac{4}{3}, 0\right), (4, \pi)$ or in rectangular form $\left(\frac{4}{3}, 0\right), (-4, 0)$

94. $e = 1$; the vertex in rectangular form $(0, -3)$

95.

96.

97.

98.

99. $x = 4 - y^2$ **100.** $2x + 5y = 10$

101. $y = 2x + 4$ **102.** $y = x - 9$

103. The vertex is located at $(0.6, -1.2)$. The parabola opens down.

104. The vertex is located at $(2.8, 0.2)$. The parabola open to the right.

105. a. $y = -1.4 \pm \sqrt{8.81 - 3x}$

b. Vertex at $(2.94, -1.4)$, opens to the left.

c. Yes, (a) and (b) agree with each other.

106. a. $y = x^2 - 10.2x + 24.8$

b. Vertex at $(5.1, -1.21)$. The parabola opens up.

c. Yes. (a) and (b) agree with each other.

107. As c increases, the minor axis along the x-axis decreases.

108. As c increases the minor axis along the y-axis decreases.

109. As c increases, the vertices of the hyperbolas located at $\left(\pm\dfrac{1}{2c}, 0\right)$ are moving toward the origin.

110. As c increases, the vertices of the hyperbola remain at $(\pm 1, 0)$, but the graphs open more narrowly.

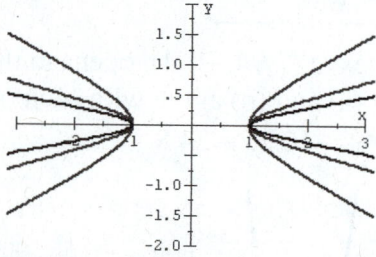

111. $(0.635, 2.480)$, $(-0.635, 2.480)$, $(-1.245, 0.645)$, $(1.245, 0.645)$

112. $(0.876, 1.458)$ and $(1.350, -1.781)$

113.

114.

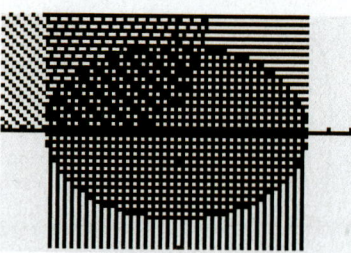

115. a. The amount of rotation is about $22.5°$. It is an ellipse.

b. The amount of rotation is about $11.6°$. It is a hyperbola.

116. a. The amount of rotation is about 1.2 rad. It is a hyperbola.

b. The amount of rotation is about 0.8 rad. It is a parabola.

117. With θ step $= \dfrac{\pi}{4}$, points $(2, 0)$,

$$\left(1.06, \frac{\pi}{4}\right), \left(0.89, \frac{\pi}{2}\right), \left(1.06, \frac{3\pi}{4}\right), (2, \pi),$$

$$\left(17.22, \frac{5\pi}{4}\right), \left(-8, \frac{3\pi}{2}\right), \left(17.22, \frac{7\pi}{4}\right),$$

and $(2, 2\pi)$ are plotted.

118. The calculator is sampling using too wide a step size, and is jumping over parts of the graph. Using more points yields a better approximation to the graph of the curve.

119. $a = 2, b = 3, t$ in $[0, 2\pi]$

$a = 3, b = 2, t$ in $[0, 2\pi]$

120. a. $a = 1, b = 2, t$ in $[0, 2\pi]$

b. $a = 2, b = 1, t$ in $[0, 2\pi]$

Practice Test

1. c **2.** b **3.** d **4.** e

5. f **6.** a **7.** $y^2 = -16x$

8. $x^2 = -8y$ **9.** $(x+1)^2 = -12(y-5)$

10. $(y+3)^2 = 8(x-2)$

11. $\dfrac{x^2}{7} + \dfrac{y^2}{16} = 1$ **12.** $\dfrac{x^2}{9} + \dfrac{y^2}{8} = 1$

13. $\dfrac{(x-2)^2}{20} + \dfrac{y^2}{36} = 1$

14. $\dfrac{(x+5.5)^2}{2.25} + \dfrac{(y+3)^2}{2} = 1$

15. $x^2 - \dfrac{y^2}{4} = 1$ **16.** $y^2 - \dfrac{x^2}{9} = 1$

17. $\dfrac{y^2}{16} - \dfrac{(x-2)^2}{20} = 1$

18. $\dfrac{(x+5.5)^2}{0.25} - \dfrac{(y+3)^2}{2} = 1$

19.

20.

21.

22.

23. $x^2 = 6y; -2 \le x \le 2$

24. $\dfrac{(x - (1.32 \times 10^8))^2}{8.24 \times 10^{18}} + \dfrac{y^2}{8.225 \times 10^{18}} = 1$

25.

26.

27.

28.

29. Ellipse: $e = \frac{2}{3}$

30. This is an ellipse with equation

$$\frac{X^2}{3} + \frac{Y^2}{7} = 1$$

31. 5.3 seconds, 450 ft

32. Quarter circle in QI

33.

34.

35. The vertex is located at $(-2.1, 1.2)$. The parabola opens upward.

36. a. The amount of rotation is about 1.047 rad. It is an ellipse.

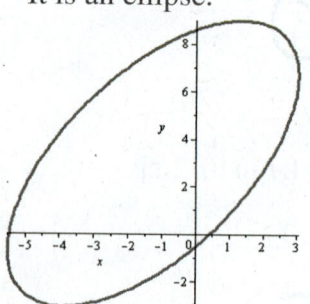

b. The amount of rotation is about 0.26 rad. It is a hyperbola.

Cumulative Test

1. $-6, 3$ **2.** $(x-5)^2 + (y-1)^2 = 10$

3. -7 **4.** $I = \dfrac{3}{100} Pt$

5. $f(x) = \frac{1}{3}(x-7)^2 + 7$ **6.** $\$29,246.16$

7. $-4, 4$ **8.** Other leg: $8\sqrt{3}$ in.; Hypotenuse: 16 in.

9. -1.9626

10. Amplitude: 0.007; Frequency: $\frac{1}{425}$

11. $\dfrac{1}{\cos\theta} + \sin\theta$ **12.** $-\dfrac{\sqrt{2+\sqrt{3}}}{2}$

13. $\dfrac{\pi}{3}, \dfrac{2\pi}{3}, \dfrac{4\pi}{3}, \dfrac{5\pi}{3}$

14. The two planes are approximately 1525 miles apart after two hours.

15. $\langle -5.1, 14.1 \rangle$ **16.** $10i$

17. Soda: $\$1.29$; Soft pretzel: $\$1.45$

18. $\dfrac{3x+5}{(x-3)(x^2+5)} = \dfrac{1}{x-3} + \dfrac{-x}{x^2+5}$

19.

20. $x = a-2;\ y = a;\ z = a+9$

21. $\begin{bmatrix} 7 & -16 & 33 \\ 18 & -3 & -17 \end{bmatrix}$

22. $x = \dfrac{4}{15};\ y = -\dfrac{7}{15}$

23. $\dfrac{(x-6)^2}{9} + \dfrac{(y+2)^2}{25} = 1$

24. $(y+1)^2 - \dfrac{(x-5)^2}{3} = 1$

25. $(2, 4)\ (4, 2)$

26.

27.

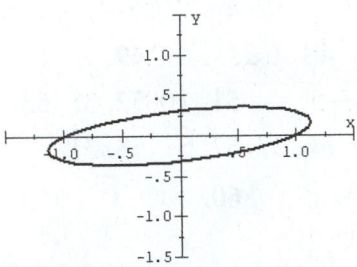

CHAPTER 12

Section 12.1

1. $1, 2, 3, 4$ **2.** $1, 4, 9, 16$ **3.** $1, 3, 5, 7$

4. x, x^2, x^3, x^4 **5.** $\frac{1}{2}, \frac{2}{3}, \frac{3}{4}, \frac{4}{5}$ **6.** $2, \frac{3}{2}, \frac{4}{3}, \frac{5}{4}$

7. $2, 2, \frac{4}{3}, \frac{2}{3}$ **8.** $\frac{1}{2}, \frac{1}{3}, \frac{1}{4}, \frac{1}{5}$ **9.** $-x^2, x^3, -x^4, x^5$

10. $1, -4, 9, -16$ **11.** $-\frac{1}{6}, \frac{1}{12}, -\frac{1}{20}, \frac{1}{30}$

12. $0, \frac{1}{9}, \frac{1}{4}, \frac{9}{25}$ **13.** $\frac{1}{512}$ **14.** $\frac{15}{256}$ **15.** $-\frac{1}{420}$

16. $\frac{180}{13}$ **17.** $\frac{10201}{10000} = 1.0201$ **18.** $\frac{99}{100}$

19. 23 **20.** 49 **21.** $a_n = 2n$ **22.** $a_n = 3n$

23. $a_n = \dfrac{1}{(n+1)n}$ **24.** $a_n = \left(\dfrac{1}{2}\right)^n$

25. $a_n = \dfrac{(-1)^n 2^n}{3^n}$ **26.** $a_n = \dfrac{3^{n-1}}{2^n}$

27. $a_n = (-1)^{n+1}$ **28.** $a_n = (-1)^{n+1} \dfrac{n}{n+2}$

29. 72 **30.** $\frac{1}{30}$ **31.** 812 **32.** 992 **33.** $\frac{1}{5852}$

34. $\frac{1}{1,061,106}$ **35.** $83,156,160$ **36.** $999,900$

37. $\dfrac{1}{n(n+1)}$ **38.** $(n+2)(n+1)$

39. $(2n+3)(2n+2)$ **40.** $(2n+2)(2n+1)(2n)$

41. $7,10,13,16$ **42.** $2,3,4,5$

43. $1,2,6,24$ **44.** $2,6,24,120$

45. $100, 50, \frac{25}{3}, \frac{25}{72}$ **46.** $20, 5, \frac{5}{9}, \frac{5}{144}$

47. $1,2,2,4$ **48.** $1, 2, \frac{1}{2}, 4$ **49.** $1,-1,-2,5$

50. $1, -1, -1, -5$ **51.** 10 **52.** 35 **53.** 30

54. $\frac{25}{12}$ **55.** 36 **56.** 27 **57.** 5 **58.** 31

59. $1-x+x^2-x^3$ **60.** $-x+x^2-x^3+x^4$

61. $\frac{109}{15}$ **62.** $\frac{11}{30}$ **63.** $1+x+\dfrac{x^2}{2}+\dfrac{x^3}{6}+\dfrac{x^4}{24}$

64. $\dfrac{1}{0!}-\dfrac{(1)x}{1!}-\dfrac{(2)x^2}{2!}-\dfrac{(3)x^3}{3!}-\dfrac{(4)x^4}{4!}=$

    $1-x-x^2-\dfrac{x^3}{2}-\dfrac{x^4}{6}$

65. $\frac{20}{9}$ **66.** $\frac{50}{9}$

67. not possible   **68.** not possible

69. $\displaystyle\sum_{n=0}^{6}\dfrac{(-1)^n}{2^n}$ **70.** $\displaystyle\sum_{n=0}^{\infty}\dfrac{1}{2^n}$

71. $\displaystyle\sum_{n=1}^{\infty}(-1)^{n-1}n$ **72.** $\displaystyle\sum_{n=1}^{23}n$

73. $\displaystyle\sum_{n=1}^{6}\dfrac{(n+1)!}{(n-1)!}=\sum_{n=1}^{5}n(n+1)$ **74.** $\displaystyle\sum_{n=1}^{\infty}\dfrac{2^{n-1}}{(n-1)!}$

75. $\displaystyle\sum_{n=1}^{\infty}(-1)^{n-1}\dfrac{x^{n-1}}{(n-1)!}=\sum_{n=0}^{\infty}(-1)^n\dfrac{x^n}{n!}$

76. $\displaystyle\sum_{n=1}^{6}\dfrac{x^n}{(n-1)!}$

77. $\$28,640.89$; total balance in account after 6 years (or 72 months)

78. $\$8125.28$; total balance in account after 3 years (or 36 months)

79. $s_n = 20 + 2n$; a paralegal with 20 years experience would make $\$60$ per hour

80. $a_n = 275,000 + 75,000(n-1)$;

    $\displaystyle\sum_{n=1}^{3}a_n =$ salary for a 3 year career

81. $a_n = 1.03a_{n-1}$; $a_0 = 30,000$

82. $2^{37} = 1.374 \times 10^{11}$; $2^{145} \approx 4.46 \times 10^{43}$

83. $a_{n+1} = 1000 - 75n$; approximately 10.7 years

84. $a_3 = 8.6$ billion in sales,

    $a_4 = 10.2$ billion in sales, represents the average sales for the years 2003 and 2004

85. $A_1 = 100$, $A_2 = 200.10$, $A_3 = 300.30$, $A_4 = 400.60$, $A_{36} = 3663.72$

86. $A_1 = 50$, $A_2 = 100.05$, $A_3 = 150.15$, $A_4 = 200.30$, $A_{48} = 2457.27$

87. 7; 7.38906 **88.** $\$239,705$

89. 0.095310; 0.095310

90. $\$5000$; $\$10,300$; $\$15,918$; $\$21,873.08$; $\$28,185.46$

91. The mistake is that $6! \neq 3!2!$, but rather $6! = 6\cdot 5\cdot 4\cdot 3\cdot 2\cdot 1$.

92. The mistake is that
$(2n-2)! \neq (2n-2)(2n-4)(2n-6)\ldots\ldots(2)$
The terms should be consecutive and decrease by 1 (not 2). Therefore, it should be
$(2n-2)! = (2n-2)(2n-3)(2n-4)\ldots\ldots(1)$
The same error is made when computing $(2n+2)!$

93. $(-1)^{n+1} = \begin{cases} 1, & n=1,3,5,\ldots \\ -1, & n=2,4,6,\ldots \end{cases}$

So, the terms should all be the opposite sign.

94. Same error as 93.

95. true   **96.** true

97. false   **98.** false

99. $C, C+D, C+2D, C+3D$

100. $C, D(C), D^2C, D^3C$ **101.** 1 and 1

102. $7, 7^{1/2}, 7^{1/4}, 7^{1/8}, 7^{1/16}$; $a_n = 7^{\frac{1}{2^{n-1}}}$

103. ≈ 2.705; ≈ 2.717; ≈ 2.718;

104. 1.618036 **105.** $\frac{109}{15}$ **106.** $\frac{11}{30}$

Section 12.2

1. arithmetic, $d=3$ **2.** arithmetic, $d=-3$

3. not arithmetic   **4.** not arithmetic

5. arithmetic, $d=-0.03$ **6.** arithmetic, $d=0.5$

7. arithmetic, $d=\frac{2}{3}$ **8.** arithmetic, $d=\frac{1}{3}$

9. not arithmetic   **10.** not arithmetic

11. $3,1,-1,-3$; arithmetic; $d=-2$

12. $-7,-4,-1,2$; arithmetic; $d=3$

13. $1,4,9,16$; not arithmetic

14. $1,2,\frac{3}{2},\frac{2}{3}$; not arithmetic

15. $2,7,12,17$; arithmetic; $d=5$

16. $1,-3,-7,-11$; arithmetic; $d=-4$

17. $0,10,20,30$; arithmetic; $d=10$

18. $4,12,20,28$; arithmetic; $d=8$

19. $-1,2,-3,4$; not arithmetic

20. $2,-4,6,-8$; not arithmetic

21. $a_n=11+(n-1)5=5n+6$

22. $a_n=5+(n-1)11=11n-6$

23. $a_n=-4+(n-1)(2)=-6+2n$

24. $a_n=2+(n-1)(-4)=-4n+6$

25. $a_n=0+(n-1)\frac{2}{3}=\frac{2}{3}n-\frac{2}{3}$

26. $a_n=-1+(n-1)\left(-\frac{3}{4}\right)=-\frac{3}{4}n-\frac{1}{4}$

27. $a_n=0+(n-1)e=en-e$

28. $a_n=1.1+(n-1)(-0.3)=-0.3n+1.4$

29. 124   **30.** -101 **31.** -684 **32.** 547

33. $\frac{16}{3}$ **34.** $\frac{163}{15}$

35. $a_5=44$, $a_{17}=152$; $a_n=8+(n-1)9=9n-1$

36. $a_9=-19$, $a_{21}=-55$;

$\quad a_n=5+(n-1)(-3)=-3n+8$

37. $a_7=-1$, $a_{17}=-41$;

$\quad a_n=23+(n-1)(-4)=-4n+27$

38. $a_8=47$, $a_{21}=112$;

$\quad a_n=12+(n-1)(5)=5n+7$

39. $a_4=3$, $a_{22}=15$; $a_n=1+(n-1)\frac{2}{3}=\frac{2}{3}n+\frac{1}{3}$

40. $a_{11}=-3$, $a_{31}=-13$;

$\quad a_n=2+(n-1)\left(-\frac{1}{2}\right)=-\frac{1}{2}n+\frac{5}{2}$

41. 552   **42.** 1050   **43.** -780

44. 282   **45.** 51   **46.** 140.25

47. 416   **48.** -740 **49.** 3875

50. -810 **51.** $\frac{21}{2}\left[\frac{1}{6}-\frac{13}{2}\right]=\frac{21}{2}\left(\frac{1-39}{6}\right)=-\frac{133}{2}$

52. $\frac{134}{3}$ **53.** 630   **54.** 672

55. $S_{43}=\frac{43}{4}(5-43)=-\frac{817}{2}$

56. -888 **57.** 1368   **58.** 2058

59. Colin: \$347,500; Camden: \$340,000

60. Jasmine: \$1,410,000; Megan: \$1,875,000

61. 850 seats   **62.** 210 tulips

63. 1101 glasses on the bottom row, each row had 20 fewer glasses than the one before

64. 325 logs in the pile   **65.** 1600 feet

66. 490 meters   **67.** 210 oranges

68. a. \$96,000 **b.** \$2,467,500

69. a. 23 seats in the first row **b.** 1125 seats

70. $S_{12}=132+\frac{12}{e}$

71. $a_n=a_1+(n-1)d$, not a_1+nd

72. Here, $d=-2$ since consecutive terms decrease by 2.

73. There are 11 terms, not 10. So, $n=11$, and thus, $S_{11}=\dfrac{11}{2}(1+21)=121$.

74. $S_n=\dfrac{n}{2}(a_n+a_1)$, not $S_n=\dfrac{n}{2}(a_n-a_1)$

75. false   **76.** false

77. true   **78.** false

79. $\dfrac{(n+1)(2a+nb)}{2}$ **80.** 30 **81.** 27,420

82. 15   **83.** 5050   **84.** 2550

85. 2500   **86.** Same answer

87. 18,850   **88.** 12,320

Section 12.3

1. yes, $r=3$ **2.** yes, $r=2$ **3.** no

4. no   **5.** yes, $r=\frac{1}{2}$ **6.** yes, $r=-\frac{1}{2}$

7. yes, $r=1.7$ **8.** yes, $r=2.2$

9. 6,18,54,162,486   **10.** 17,34,68,136,272

11. $1,-4,16,-64,256$ **12.** $-3,6,-12,24,-48$

13. 10,000, 10,600, 11,236,
11,910.16, 12,624.77

14. 10000, 8000, 6400, 5120, 4096

15. $\frac{2}{3},\frac{1}{3},\frac{1}{6},\frac{1}{12},\frac{1}{24}$ **16.** $\frac{1}{10},-\frac{1}{50},\frac{1}{250},-\frac{1}{1250},\frac{1}{6250}$

17. $a_n = 5(2)^{n-1}$ **18.** $a_n = 12(3)^{n-1}$

19. $a_n = 1(-3)^{n-1}$ **20.** $a_n = -4(-2)^{n-1}$

21. $a_n = 1000(1.07)^{n-1}$

22. $a_n = 1000(0.5)^{n-1}$ **23.** $a_n = \frac{16}{3}\left(-\frac{1}{4}\right)^{n-1}$

24. $a_n = \frac{1}{200}(5)^{n-1}$ **25.** $a_7 = -128$

26. $a_{10} = -1,953,125$ **27.** $a_{13} = \frac{4096}{3}$

28. $a_9 = 2.56\times10^{-4}$ **29.** $a_{15} = 6.10\times10^{-16}$

30. $a_8 = -209.7152$ **31.** $\frac{8191}{3}$ **32.** 1.5

33. 59,048

34. 349,525 **35.** $2.\overline{2}$

36. 3.75

37. 6560

38. 325,520.6667

39. 16,383 **40.** 2.0

41. 2   **42.** $\frac{1}{2}$ **43.** $-\frac{1}{4}$ **44.** $\frac{2}{3}$

45. not possible, diverges
46. not possible, diverges
47. $-\frac{27}{2}$ **48.** $-\frac{16}{3}$ **49.** 10,526

50. $S_\infty = \dfrac{200}{1-0.04} = \frac{625}{3}$ **51.** $\frac{2}{3}$

52. $\frac{1}{3}$ **53.** 100   **54.** diverges, infinite sum
55. $44,610.95   **56.** $77,393.28
57. $a_n = 2000(0.5)^n$; $a_4 = 125, a_7 = 16$

58. BMW: $a_n = 35,000(0.8)^{n-1}$

Honda: $a_n = 25,000(0.9)^{n-1}$; Honda

59. 17 feet   **60.** 9.80 feet
61. 58,640 students   **62.** 240,815 hits
63. 67 days; $9618
64. $10,737,418.24; $21,474,836.47
65. $3877.64   **66.** $2651.71
67. 26 weeks: $13,196.88
  52 weeks: $26,811.75
68. $20,422.66   **69.** $501,509

70. 18 feet   **71.** $\dfrac{\frac{1}{2}}{1-\frac{1}{2}} = 1$

72. Option 1; $1,063,741,823 more
73. Should be $r = -\frac{1}{3}$.

74. Should be $a_1 = 2$, so that you have $\displaystyle\sum_{k=0}^{n}2\cdot2^k$.

75. Should use $r = -3$ all the way
  through the calculation. Also, $a_1 - 12$ (not 4).

76. Formula for S_∞ only applies if $|r| < 1$.

77. false   **78.** false
79. true   **80.** false

81. $|b| < 1,\ \dfrac{a}{1-b}$ **82.** 2,097,151 **83.** $\frac{47}{99}$

84. a. $2 + 2x + 2x^2 + 2x^3 + 2x^4$ **b.** $|x| < 1$
85. $-37,529,996,894,754$

86. yes, $S_\infty = \dfrac{1}{1-\frac{1}{3}} = \frac{3}{2}$

87. $\displaystyle\sum_{n=0}^{\infty} x^n = \dfrac{1}{1-x}$ for $|x| \le 1$

88. $\dfrac{1}{1+x}$ **89.** $\dfrac{1}{1-2x}$ for $|x| < \dfrac{1}{2}$

Section 12.4

1. – 24. See Instructor's Solution Manual
25. 7 steps   **26.** 15 steps   **27.** 31 steps
28. $2^n - 1$ steps
29. – 30. See Instructor's Solution Manual
31. false   **32.** false
33. – 37. See Instructor's Solution Manual
38. 2,706,800, yes   **39.** $\frac{255}{256}$, yes

Section 12.5

1. 35　**2.** 28　**3.** 45　**4.** 253　**5.** 1　**6.** 1　**7.** 1

8. 1　**9.** 17,296　**10.** 3654

11. $x^4 + 8x^3 + 24x^2 + 32x + 16$

12. $x^5 + 15x^4 + 90x^3 + 270x^2 + 405x + 243$

13. $y^5 - 15y^4 + 90y^3 - 270y^2 + 405y - 243$

14. $y^4 - 16y^3 + 96y^2 - 256y + 256$

15. $x^5 + 5x^4y + 10x^3y^2 + 10x^2y^3 + 5xy^4 + y^5$

16. $x^6 - 6x^5y + 15x^4y^2 - 20x^3y^3$
$\quad + 15x^2y^4 - 6xy^5 + y^6$

17. $x^3 + 9x^2y + 27xy^2 + 27y^3$

18. $8x^3 - 12x^2y + 6xy^2 - y^3$

19. $125x^3 - 150x^2 + 60x - 8$

20. $a^3 - 21a^2b + 147ab^2 - 343b^3$

21. $\dfrac{1}{x^4} + \dfrac{20y}{x^3} + \dfrac{150y^2}{x^2} + \dfrac{500y^3}{x} + 625y^4$

22. $16x^4 + \dfrac{96x^3}{y} + \dfrac{216x^2}{y^2} + \dfrac{216x}{y^3} + \dfrac{811}{y^4}$

23. $x^8 + 4x^6y^2 + 6x^4y^4 + 4x^2y^6 + y^8$

24. $r^9 - 3r^6s^3 + 3r^3s^6 - s^9$

25. $a^5x^5 + 5a^4bx^4y + 10a^3b^2x^3y^2 + 10a^2b^2x^2y^3$
$\quad + 5ab^4xy^4 + b^5y^5$

26. $a^5x^5 - 5a^4bx^4y + 10a^3b^2x^3y^2 - 10a^2b^2x^2y^3$
$\quad + 5ab^4xy^4 - b^5y^5$

27. $x^3 + 12x^{5/2} + 60x^2 + 160x^{3/2} + 240x$
$\quad + 192x^{1/2} + 64$

28. $81 + 108y^{1/2} + 54y + 12y^{3/2} + y^2$

29. $a^3 + 4a^{9/4}b^{1/4} + 6a^{3/2}b^{1/2} + 4a^{3/4}b^{3/4} + b$

30. $x^2 + 3x^{4/3}y^{1/3} + 3x^{2/3}y^{2/3} + y$

31. $x + 8x^{3/4}y^{1/2} + 24x^{1/2}y + 32x^{1/4}y^{3/2} + 16y^2$

32. $x^4 - 24x^{7/2}y^{1/4} + 252x^3y^{1/2} - 1512x^{5/2}y^{3/4}$
$\quad + 5670x^2y - 13{,}608x^{3/2}y^{5/4} + 20{,}412xy^{3/2}$
$\quad - 17{,}496x^{1/2}y^{7/4} + 6561y^2$

33. $r^4 - 4r^3s + 6r^2s^2 - 4rs^3 + s^4$

34. $x^{14} + 7x^{12}y^2 + 21x^{10}y^4 + 35x^8y^6 + 35x^6y^8$
$\quad + 21x^4y^{10} + 7x^2y^{12} + y^{14}$

35. $a^6x^6 + 6a^5bx^5y + 15a^4b^2x^4y^2 + 20a^3b^3x^3y^3$
$\quad + 15a^2b^4x^2y^4 + 6ab^5xy^5 + b^6y^6$

36. $x^4 + 12x^3y + 54x^2y^2 + 108xy^3 + 81y^4$

37. 3360　**38.** 10,206　**39.** 5670

40. -792　**41.** 22,680　**42.** $-25{,}312{,}500$

43. 70　**44.** 210　**45.** 3,838,380

46. 50,063,860　　　**47.** 2,598,960

48. 344,985,116,800,000

49. $\begin{pmatrix} 7 \\ 5 \end{pmatrix} \neq \dfrac{7!}{5!}$, but rather $\dfrac{7!}{5!2!} = \dfrac{7 \cdot 6 \cdot \cancel{5!}}{\cancel{5!}(2 \cdot 1)} = 21$.

50. Should replace powers of y by powers of $2y$. More precisely, the expansion should be $(x + 2y)^4$
$\quad = x^4 + 4x^3(2y) + 6x^2(2y)^2 + 4x(2y)^3 + (2y)^4$
$\quad = x^4 + 8x^3y + 24x^2y^2 + 32xy^3 + 16y^4$

51. false　　　　　**52.** true

53. true　　　　　**54.** false

55. – 56. See Instructor's Solution Manual

57. $1 - 3x + 3x^2 - x^3$

58. $x^4 + 12x^3 + 54x^2 + 108x + 81$

59. graphs of the respective functions get closer to the graph of $y_4 = (1-x)^3$ when $1 < x < 2$, when $x > 1$, no longer true

60. graphs of the respective functions get closer to the graph of $y_4 = (1 - \frac{1}{x})^3$ when $1 < x < 2$, when $0 < x < 1$, no longer true

61. graph of the curve better approximation to the graph of $y = (1 + \frac{1}{x})^3$, for $1 < x < 2$; no, does not get closer to this graph if $0 < x < 1$

62. graph of the curve better approximation to the graph of $y = e^x$, for $-1 < x < 1$; no, does not get closer to this graph if $1 < x < 2$

Section 12.6

1. 360   **2.** 210   **3.** 15,120

4. 3024   **5.** 40,320   **6.** 720

7. 1716   **8.** 6840   **9.** 252

10. 126   **11.** 15,890,700

12. 10,272,278,170   **13.** 1

14. 1   **15.** 27,405   **16.** 1287

17. 215,553,195   **18.** 27,405

19. 24   **20.** 120   **21.** 12

22. 120   **23.** 10,000   **24.** 456,976

25. 32,760   **26.** 24

27. Each of 20 questions has 4 answer choices. So, there are $4^{20} \approx 1.1 \times 10^{12}$ possible ways to answer the questions on the exam.

28. Each of 25 questions has 5 answer choices. So, there are $5^{25} (= 2,980,232,239 \times 10^8)$ different ways to complete the exam.

29. 100,000; 81,000   **30.** 7,077,888

31. 2.65×10^{32} **32.** 24   **33.** 59,280

34. 117,600   **35.** 997,002,000

36. 970,200   **37.** 22,957,480

38. 2,869,685   **39.** 2,598,960

40. 133,784,560   **41.** 1326   **42.** 64

43. 4.9×10^{14} **44.** 635,376   **45.** 256

46. 36   **47.** 15,625   **48.** 518,400

49. 400   **50.** 518,400

51. The combination formula $_nC_r$ should be used instead.

52. Order is important here (since president, vice president, secretary, and treasurer are all different positions with different possibilities). So, should use permutation formula instead.

53. true   **54.** false

55. false   **56.** true

57. $\dfrac{_nC_r}{_nC_{r+1}} = \dfrac{r+1}{n-r}$ **58.** $\dfrac{_nP_r}{_nP_{r-1}} = (n-r+1)$

59. $C(n,r) \cdot r! = \dfrac{n!}{(n-r)!}$

60. $C(n,r) = C(n, n-r)$

61. answers the same **62.** answers the same

63. **a.** 5,040 **b.** 5,040 **c.** yes **d.** $_nP_r = r! \, _nC_r$

64. **a.** 95,040 **b.** 95,040 **c.** yes **d.** $_nP_r = r! \, _nC_r$

Section 12.7

1. $\{2, 3, \ldots, 12\}$

2. $\{HHH, HHT, HTH, HTT, THH, THT, TTH, TTT\}$

3. $\{BBBB, BBBG, BBGB, BBGG, BGBB, BGBG, BGGB, BGGG, GBBB, GBBG, GBGB, GBGG, GGBB, GGBG, GGGB, GGGG\}$

4. $\{H1, H2, \ldots, H6, T1, T2, \ldots, T6\}$

5. $\{RR, RB, RW, BB, BR, BW, WR, WB\}$

6. $\{FF, FS, FJ, SF, SS, SJ, JF, JS, JJ\}$

7. $\frac{1}{8}$ **8.** $\frac{3}{8}$ **9.** $\frac{7}{8}$ **10.** $\frac{1}{2}$ **11.** $\frac{1}{18}$ **12.** $\frac{1}{2}$

13. $\frac{1}{2}$ **14.** $\frac{1}{2}$ **15.** $\frac{5}{12}$ **16.** $\frac{5}{12}$ **17.** $\frac{10}{13}$ **18.** $\frac{1}{2}$

19. $\frac{4}{13}$ **20.** $\frac{5}{13}$ **21.** $\frac{3}{4}$ **22.** $\frac{1}{2}$ **23.** $\frac{3}{4}$ **24.** $\frac{5}{8}$

25. 0   **26.** $\frac{1}{8}$

27. **a.** 270,725 **b.** $\frac{715}{270,725} \cong 0.26\%$

    c. $\frac{13}{270,725} \approx 0.005\%$

28. a. 1326 **b.** $\frac{48}{1326} \approx 0.037$

29. $\frac{8}{52} = \frac{2}{13} \approx 15.4\%$ **30.** $\frac{4}{52} = \frac{1}{13} \approx 7.7\%$

31. $\frac{4}{663} \approx 0.6\%$ **32.** $\frac{1}{663} \approx 0.2\%$

33. $\frac{1}{32} \approx 3.1\%$ **34.** $\frac{1}{16} = 6.25\%$

35. $\frac{31}{32} \approx 96.9\%$ **36.** $\frac{63}{64} \approx 98.4\%$

37. $\left(\frac{18}{38}\right)^4 \approx 5.03\%$ **38.** $\left(\frac{2}{38}\right)^2 \approx 0.28\%$

39. 20% **40.** 80%

41. 25% **42.** $\frac{64}{225} \approx 28.4\%$

43. $\frac{48}{1326} \approx 3.6\%$ **44.** 0.13%

45. 0.001526% **46.** 99.99847%

47. a. {(Brown, Blue), (Brown, Brown), (Blue, Brown), (Blue, Blue)} **b.** ¼ **c.** ¾

48. a. {(Brown, Brown), (Blue, Brown)} **b.** 0 **c.** 1

49. 2,598,960 **50. a.** 103,776 **b.** 4%

51. $\frac{1287}{2,598,960}$ **52.** 0.00001

53. The events aren't mutually exclusive. So, probability $= \frac{4}{52} + \frac{13}{52} - \underbrace{\frac{1}{52}}_{2 \text{ of spades}} = \frac{16}{52} = \frac{4}{13}$

54. Using #49 for the total number of 5-card hands, we obtain
$$\frac{\binom{4}{2}\binom{4}{3}}{2,598,960} = \frac{\frac{4!}{2!2!} \cdot \frac{4!}{3!1!}}{2,598,960} = 0.00001$$

55. true **56.** false

57. false **58.** false

59. 0.0027 **60.** 0.999924733

61. $\frac{1}{3} \approx 0.333$ **62.** $\frac{2}{3} \approx 0.667$

63. – 64. use random number generator and compare

65. 0.2907 **66.** 0.8652

Review Exercises

1. 1, 8, 27, 64 **2.** 1, 1, 2, 6 **3.** 5, 8, 11, 14

4. $-x^3, x^4, -x^5, x^6$ **5.** $a_5 = \frac{32}{243} \approx 0.13$

6. $a_8 = \frac{64}{6561}$ **7.** $a_{15} = -\frac{1}{3600}$ **8.** $a_{10} = \frac{11}{10}$

9. $a_n = (-1)^{n+1} 3n$ **10.** $a_n = \begin{cases} n, & n \text{ odd} \\ \dfrac{1}{2^{n/2}}, & n \text{ even} \end{cases}$

11. $a_n = (-1)^n$ **12.** $a_n = 10^{n-1}$

13. 56 **14.** $\frac{1}{10,626}$ **15.** $\frac{1}{n+1}$

16. $\frac{1}{n(n-1)}$ **17.** $5, 3, 1, -1$ **18.** $1, 4, 16, 256$

19. $1, 2, 4, 32$ **20.** $1, 2, \frac{1}{4}, 32$ **21.** 15

22. $\frac{205}{144}$ **23.** 69 **24.** $\frac{218}{15}$

25. $\sum_{n=1}^{7} \frac{(-1)^n}{2^{n-1}}$ **26.** $\sum_{n=1}^{10} 2n$ **27.** $\sum_{n=0}^{\infty} \frac{x^n}{n!}$

28. $\sum_{n=0}^{\infty} (-1)^{n+1} \frac{x^n}{n!}$

29. $36,639.90; amount in the account after 5 years

30. total salary for 4 years is $900,000

31. arithmetic, $d = -2$ **32.** not arithmetic

33. arithmetic, $d = \frac{1}{2}$ **34.** arithmetic, $d = -1$

35. arithmetic, $d = 1$ **36.** arithmetic, $d = 5$

37. $a_n = -4 + (n-1)(5) = 5n - 9$

38. $a_n = 5 + (n-1)(6) = 6n - 1$

39. $a_n = 1 + (n-1)\left(-\frac{2}{3}\right) = -\frac{2}{3}n + \frac{5}{3}$

40. $a_n = 0.001 + (n-1)(0.01) = 0.01n - 0.009$

41. $a_1 = 5$, $d = 2$, $a_n = 5 + (n-1)(2) = 2n + 3$

42. $a_1 = 4$, $d = -3$, $a_n = 4 + (n-1)(-3) = -3n + 7$

43. $a_1 = 10$, $d = 6$, $a_n = 10 + (n-1)6 = 6n + 4$

44. $a_1 = 20$, $d = -5$, $a_n = 20 + (n-1)(-5) = -5n + 25$

45. 630 **46.** 125 **47.** 420 **48.** −63.75

49. Bob: $885,000 Tania: $990,000

50. 400 feet **51.** geometric, $r = -2$

52. not geometric **53.** geometric, $r = \frac{1}{2}$

54. geometric, $r = 10$ **55.** $3, 6, 12, 24, 48$

56. $10, \frac{10}{4}, \frac{10}{16}, \frac{10}{64}, \frac{10}{256}$

57. $100, -400, 1600, -6400, 25,600$

58. $-60, 30, -15, \frac{15}{2}, -\frac{15}{4}$

59. $a_n = a_1 r^{n-1} = 7 \cdot 2^{n-1}$ **60.** $a_n = 12\left(\frac{1}{3}\right)^{n-1}$

61. $a_n = (-2)^{n-1}$ **62.** $a_n = \frac{32}{5}\left(-\frac{1}{4}\right)^{n-1}$

63. $a_{25} = 33,554,432$ **64.** $a_n = \frac{1}{2}(2)^{n-1}$

65. $a_{12} = -2.048 \times 10^{-6}$ **66.** $a_{11} = \frac{1000}{1024}$

67. 4920.50 **68.** $\frac{4094}{2048}$ **69.** $16,400$

70. $\frac{195,310}{3}$ **71.** 3 **72.** $\frac{1}{30}$ **73.** $\$60,875.61$

74. $\$9600$

75. – 78. See Instructor's Solution Manual

79. 165 **80.** 1 **81.** 1 **82.** 1081

83. $x^4 - 20x^3 + 150x^2 - 500x + 625$

84. $x^5 + 5x^4 y + 10x^3 y^2 + 10x^2 y^3 + 5xy^4 + y^5$

85. $8x^3 - 60x^2 + 150x - 125$

86. $x^8 + 4x^6 y^3 + 6x^4 y^6 + 4x^2 y^9 + y^{12}$

87. $x^{5/2} + 5x^2 + 10x^{3/2} + 10x + 5x^{1/2} + 1$

88. $x^4 + 6x^{10/3} y^{1/3} + 15x^{8/3} y^{2/3} + 20x^2 y + 15x^{4/3} y^{4/3}$

    $+ 6x^{2/3} y^{5/3} + y^2$

89. $r^5 - 5r^4 s + 10r^3 s^2 - 10r^2 s^3 + 5rs^4 - s^5$

90. $a^4 x^4 + 4a^3 b x^3 y + 6a^2 b^2 x^2 y^2$

    $+ 4ab^3 xy^3 + b^4 y^4$

91. 112 **92.** 945 **93.** $37,500$

94. 70 **95.** $22,957,480$

96. $20,592,957,740,000,000$

97. 840 **98.** $362,880$ **99.** $95,040$

100. 10 **101.** 792 **102.** $658,008$

103. 1 **104.** $22,957,480$ **105.** 30

106. 26^6 **107.** 5040 **108.** 32^6

109. 120 seating arrangements, 15 years

110. $205,320$ **111.** $94,109,400$ **112.** 6786

113. $20,358,520$ **114.** 256

115. $\frac{1}{16} \approx 6.25\%$ **116.** $\frac{1}{2}$

117. $\frac{30}{36} \approx 83.3\%$ **118.** $\frac{1}{4}$

119. $\frac{2}{3} \approx 66.7\%$ **120.** $\frac{5}{6}$

121. $\frac{7}{12} \approx 58.3\%$ **122.** $\frac{1}{6}$

123. $\frac{2}{13} \approx 15.4\%$ **124.** $\frac{16}{2652}$

125. $\frac{31}{32} \approx 96.88\%$ **126.** $\frac{1}{4096}$ **127.** $\frac{5369}{3600}$

128. infinite   **129.** $\frac{34,875}{14}$ **130.** 6556

131. $\dfrac{1}{1+2x}$

132. yes, 7.4215 **133.** $99,900$, yes

134. yes

135. graphs become better approximations of the graph of $y = (1+2x)^4$ for $-0.1 < x < 0.1$; does not get closer for $0.1 < x < 1$

136. graphs become better approximations of the graph of $y = (1-2x)^4$ for $-0.1 < x < 0.1$; does not get closer to this graph for $0.1 < x < 1$

137. a. $11,440$ **b.** $11,440$ **c.** yes   **d.** $\frac{{}_n P_r}{r!} = {}_n C_r$

138. a. $20,358,520$ **b.** $20,358,520$ **c.** yes

    d. $\frac{{}_n P_r}{r!} = {}_n C_r$

139. 0.0722 **140.** 0.7795

Practice Test

1. x^{n-1} **2.** geometric   **3.** $S_n = \dfrac{1 - x^n}{1 - x}$

4. $\displaystyle\sum_{n=0}^{\infty} x^n$ **5.** $|x| < 1$ **6.** $\frac{1}{2}$

7. $1 - \left(\frac{1}{4}\right)^{10} \approx 1$ **8.** 2600 **9.** $24,950$

10. See Instructor's Solution Manual

11. 2520 **12.** $80x^3 y^2$ **13.** 455

14. 1 **15.** 2184 **16.** $1,313,400$

17. $x^{10} + 5x^7 + 10x^4 + 10x + \frac{5}{x^2} + \frac{1}{x^5}$

18. $81x^4 - 216x^3 + 216x^2 - 96x + 16$

19. more permutations than combinations since order is taken into account when determining the number of permutations

20. $1 - \frac{2730}{15!} \approx 1$ **21.** $\frac{18}{38} \approx 0.47$ **22.** $\left(\frac{18}{38}\right)^5 \approx 2.38\%$

23. $\frac{18}{38} \cong 0.47$ **24.** $\frac{1}{420}$ **25.** $\frac{4}{13} \approx 0.308$

26. a. $40,320$ **b.** $\frac{1}{8!} \approx 0.00002$

27. $184,756$ **28.** $\frac{73,375}{12}$

Cumulative Test

1. 2 **2.** 8 inches by 11 inches

3. $\dfrac{5 \pm 3\sqrt{7}\,i}{4}$ **4.** $(-\infty, 2) \cup (8, \infty)$

5. $x = -8$

6. x-intercept: $(5,0)$, y-intercept: $(0,-3)$

7. $2x + h - 3$

8. $f(g(x)) = -2x + 3$

domain: $\left(-\infty, -\frac{1}{2}\right) \cup \left(-\frac{1}{2}, \infty\right)$

9. $(15,6)$

10. $P(x) = (x + 3i)(x - 3i)(x - 1)(x + 1)$

11. VA: $x = 3$ HA: $y = -5$

12. $\$29,246.21$ **13.** 2.585

14. 2.678 **15.** no solution

16. $x = 7a - 5$, $y = a$, $z = 11 - 13a$

17. $z(1,4) = 24$

18. no solution **19.** $\begin{bmatrix} 99 & 18 & -9 \\ 29 & 4 & 7 \end{bmatrix}$

20. 0 **21.** $-16(x - 3) = (y - 5)^2$

22.

23. 3 **24.** geometric **25.** $2^{10} = 1024$

26. $20,090$ **27.** -3432

Applications Index

Subject Index